Excretory/ osmoregulatory system	Nervous system	Some distinctive features
absent	absent	body organized around pores and cavities through which a current of water flows; choanocytes, provided with flagella, create current and also capture microscopic food particles; supporting skeleton of spicules or fibers secreted by amoebocytes
absent	networks of neurons, some nervelike concentrations, but no brain	body in the form of a polyp (often colonial) or medusa; many have both polyp and medusa stage in the life cycle; secreted mesoglea (particularly abundant in medusae) between epidermis and gastrodermis; carnivorous, capturing food with the aid of nematocysts, especially those on tentacles; colonies may consist of two or more kinds of members.
flame-bulb protonephridia, in many	brainlike concentration of ganglia, or nerve ring, near anterior end; one or more pairs of prominent longitudinal nerves	complex reproductive system (nearly always hermaphroditic); body usually at least slightly flattened; many parasitic species
flame-cell protonephridia, many of which are closely associated with blood vessels	brainlike concentration of ganglia near anterior end; a pair of prominent longitudinal nerves, and often additional unpaired nerves	proboscis–rhynchocoel complex used for capture of animal prey (tip of everted proboscis may have stylet for making a wound into which venom may penetrate)
renette cells or derivatives of these	nerve ring around pharynx, several longitudinal nerves	tough cuticle; cilia absent (except for modified cilia in some sense organs); body wall with only longitudinal muscles; many species parasitic in animals, plants
flame-bulb protonephridia	brain, several longitudinal nerve cords	ciliated corona used in locomotion and for directing microscopic food particles to mouth
solenocytic protonephridia or metanephridia, often joined to coelomoducts	brain; ventral nerve cord, consisting of a pair of trunks, segmentally ganglionated	setae secreted in epidermal pockets of most segments (except in leeches, which lack setae); many diverse lifestyles
renal organs derived from coelomoducts	typically several pairs of prominent ganglia, these joined by cords; cerebral ganglia, often with other ganglia, form brain (brain highly developed in class Cephalopoda; nervous system not markedly ganglionated in classes Polyplacophora and Aplacophora)	usually with a calcareous protective shell (1, 2, or 8 pieces); body organized into head, foot, mantle, visceral mass; ctenidia
probably absent, although certain structures in some freshwater species slightly resemble funnel organs	ganglion between mouth and anus, nerves extending to all parts of body	colonial (few exceptions), often with two or more kinds of member zooids, some specialized for functions other than feeding; feeding zooids with tentacles organized into a lophophore, the cilia on the tentacles driving food particles to the mouth
various, but usually either coelomoducts or Malpighian tubules (the latter opening into the posteriormost part of the midgut)	well developed brain; ventral nerve cord, consisting of a pair of nerve trunks, segmentally ganglionated	secreted cuticle, containing chitin, forms exoskeleton; segments organized into tagmata (head, thorax, abdomen, etc.); jointed appendages functioning as mouthparts, legs, etc.
none in the conventional sense	nerve ring (no definite brain), main nerves extending radially from this	water-vascular system (derived from part of embryonic coelom); tube feet; skeletal structures (spines, ossicles, etc.) secreted by cells derived from mesoderm, and thus constituting an endoskeleton
none in the conventional sense	brain; nerve cord, when definite, dorsal and tubular	solitary or colonial; pharynx, used for processing water from which microscopic food particles are collected, with perforations that usually open into an atrial cavity rather than directly to the outside; notochord, when present, restricted to tail region of tadpolelike larva (or adult that resembles a tadpole)

INVERTEBRATES

INVERTEBRATES

EUGENE N. KOZLOFF
University of Washington

SAUNDERS COLLEGE PUBLISHING
Philadelphia New York Chicago
San Francisco Montreal Toronto
London Sydney Tokyo

Cover photo: *Anthopleura xanthogrammica* (giant green anemone). © Richard Herrman.

ISBN: 0-03-046204-5
Library of Congress Catalog Card Number: 89-84701
Printed in the United States of America

8901 041 987654321

PREFACE

When I began writing this book, I had in mind a small volume in which the important features of invertebrate animals would be explained briefly and simply. Soon, however, I realized that I could not be satisfied with a superficial treatment of the many distinctive phyla, some of which are exceedingly diverse. Furthermore, perusal of some of the shorter textbooks that have been published in the last decade persuaded me that it would be better to describe and illustrate the invertebrates with considerable thoroughness and let teachers decide—on the basis of their geographic location, the needs of their students, and other criteria—which portions of my book will be most useful in their courses.

It is not likely that all of the material presented here can be covered in one quarter or one semester of an academic year, and even in more extensive courses it may be necessary to omit some topics. But I have tried to provide a well-balanced book that will serve as a useful reference after the course is over, and one that will prepare students for future encounters with specialized treatises and research papers.

Some books take up the phyla in order of increasing complexity or possible phylogenetic sequence; others are organized around systems or functions. Each approach has its advantages, but I have elected to follow the phylum-by-phylum arrangement because it minimizes the chance that smaller groups will be overlooked or accorded only perfunctory treatment. There are no "minor" phyla; each group, even if it has only a few species, helps us to understand every other group, and each has its unique role to play in the biosphere.

The vocabulary of biology, especially the portion concerned with structure, is necessarily large. While some students may wish that they could understand diversity without learning terminology, they cannot. Nevertheless, certain terms are more widely used than others and I have tried to stress those that are indispensable, just as I have emphasized the cardinal features of the phyla and their subdivisions.

In preparing illustrations for this book, I have avoided diagrams that are so simple as to be misleading. Every biological function is controlled by processes going on in cells and tissues. In explaining structure and function, it is essential to give "credit" to specific muscle layers, tracts of cilia, glands, and other features upon which the operation of a particular organ, or of the body as a whole, depends.

Many of the illustrations taken from books and periodicals have been reproduced without modification or redrawn with only slight changes. Some of the sources are in the public domain; others are used with permission. I thank the authors and publishers of all of these works. I am also indebted to several colleagues and to Carolina Biological Supply Company for the use of photographs and drawings.

Most of my own photographs of living invertebrates were taken in the Puget Sound region of Washington.

Many of my associates very kindly allowed me to use material they had collected for their own work. For the photography I did in San Francisco Bay and at other localities in California, I received much help from my friend Linda Price. She provided transportation, hunted for specimens, and made available a convenient workplace. For an opportunity to photograph a few shells, preserved specimens, and fossils, I am indebted to the California Academy of Sciences and to four of its curators: Peter Rodda, Welton Lee, Barry Roth, and Dustin Chivers.

The majority of the line drawings prepared for this book were made by Mary Keeler and Carol Noyes. Their intelligent and skillful collaboration over a period of many months is warmly appreciated. Bonnie Hungate, Carol Jerome, and Robin Lefberg drew some of the illustrations, and their efforts are also acknowledged with gratitude. Beatrice Vogel, Suki Patten, and Ann Lesher were of great help in working out preliminary sketches.

I owe much to many well-informed colleagues who read portions of the manuscript or who gave advice when I asked for it: Paul Illg, University of Washington; John Edwards, University of Washington; George Shinn, Northeast Missouri State University; Charles Lambert, California State University at Fullerton; Arthur Martin, University of Washington; Ronald Shimek, Monroe, Washington; Chris Reed, Dartmouth College; W. T. Edmondson, University of Washington; Michael Hadfield, Kewalo Marine Laboratory; Scott Smiley, University of Alaska; Stephen Stricker, University of New Mexico; Richard Cloney, University of Washington; Robert Rausch, University of Washington; Dennis Willows, University of Washington; Richard Strathmann, University of Washington; Claudia Mills, Friday Harbor Laboratories; Norman McLean, California State University at San Diego; Sandra Carlson, University of California at Davis; Kraig Derstler, University of California at Davis; Catherine McFadden, University of California at Davis; Esther Leise, Georgia State University; Richard Palmer, University of Alberta; Roger Seapy, California State University at Fullerton; Jennifer Purcell, Horn Point Laboratory; Alan Kohn, University of Washington; Kenneth King, Rosenstiel School of Marine and Atmospheric Science; Beatrice Vogel, Seattle, Washington; Meredith Jones, Smithsonian Institution; and Gail Stratton, Bradley University. Although I attempted to patch up errors or misconceptions they recognized, they are not responsible for any problems that remain. I apologize for these in advance and ask users of the textbook to tell me of any difficulties they may encounter.

I thank also the following reviewers of the various versions of the manuscript, whose suggestions were always appreciated if not always followed: Christopher Bayne, Oregon State University; John Farley, Dalhousie University; Ju-shey Ho, California State University at Long Beach; Robert Loveland, Rutgers University; Charles McMillan, Alabama A&M University; James Morin, UCLA; Mark Rausher, Duke University; Edward Ruppert, Clemson University; Barbara Welsh, University of Connecticut at Storrs; and Mary Wicksten, Texas A&M University.

I am grateful to LaRena Christenson for promptly and expertly typing much of the manuscript. Her patience and desire to be of service are sincerely appreciated. My daughter, Rae, was also an efficient typist, and my wife, Anne, carefully read all proofs. Linda Price recorded index entries in her computer, which miraculously alphabetized them on command. Many colleagues not mentioned previously assisted me in one way or another. Estelle Johnson, Kathryn Hahn, Beverly Paul, Joyce Smith, Sally Dickman, and Laura Norris did many favors for me in the office at Friday Harbor Laboratories, and Kathy Carr, Doris Jones, Pamela Mofjeld, and Diane Wilkinson helped enormously in connection with library materials.

It is a pleasure to acknowledge the highly professional and enthusiastic support of the College Department of Harcourt Brace Jovanovich. Karl Yambert applied his exceptional editorial skill and good judgment to improving the manuscript, and Candy Young supervised the arranging and labeling of illustrations and obtained permissions to use figures published in other works. Lynne Bush saw that the many parts and stages of the book remained on schedule. The typesetting was so accurate that there were relatively few errors, and Debra Lilly conscientiously caught nearly all of them before the proofs reached us. Cathy Reynolds designed the attractive format of the book, as well as the cover. Cindy Simpson kept the whole system functioning smoothly, critically reviewing all materials before they were sent to me and after they were returned. She and Mary Tanner made sure that I got a lot of mail and that I attended to it right away. Finally, Cathleen Petree, Biology Editor, deserves credit for not abandoning hope. She still can't believe I finished the writing.

Eugene N. Kozloff
Friday Harbor, Washington

Brief Contents

Contents

A FEW NOTES THAT MAY BE HELPFUL

When important terms are introduced, they are printed in **boldface** type. Furthermore, terms such as **coelom, gastrula,** and **stomodeum,** which appear in many contexts, are defined in the Glossary. For definitions of terms whose application is limited to a single chapter or single group of invertebrates, the Index should be consulted.

The outlines of classification provided at the beginning of most chapters will give the reader some idea of the diversity of each assemblage of invertebrates. The reader will also be able to see which categories will be dealt with in each chapter and to establish the position of a particular group with respect to groups of higher, lower, or equal rank. Chapter 1 offers suggestions on how to approach classification with understanding.

The lists of references at the ends of chapters, or at the ends of sections within chapters, include books and research papers that will provide more information about certain groups of invertebrates. Some of the references were selected because they have extensive bibliographies. The major multivolume treatises listed in Chapter 1 are not, as a rule, cited again. Most of them, however, have comprehensive accounts of nearly all phyla.

Credit for drawings that were lent, and for photographs other than those taken by the author, is given in the legends. Sources of illustrations taken from books and periodicals are also cited, because many of these works are important sources of additional information. An author's name that is preceded by *after* indicates that the illustration has been redrawn and perhaps modified.

1

GENERALITIES

Invertebrate means ''nonvertebrate,'' and refers to animals that do not have a bony or cartilaginous skeleton. The vertebrates, from hagfishes and lampreys to mammals, are remarkably diversified, but they are united by several basic features and are believed to derive from one ancestral stock. All of them are therefore placed in a single major division of the animal kingdom, called a **phylum.** The situation is different in the case of invertebrates. They belong to many phyla, some of which show no close relationship to any others. Furthermore, when multicellular invertebrates and the one-celled organisms called *protozoans* are added together, they outnumber vertebrates by more than 100 to 1.

This chapter introduces many basic terms used in describing the structure, function, and development of invertebrates. Eventually, you should have a working knowledge of all of the terms. (Some of the terms do not apply to the first few groups of animals that are covered in succeeding chapters; however, such terms are discussed again in connection with the phyla to which they apply.) Look this chapter over carefully so that you will know where to find general definitions and comparisons of terms when you need them.

PHYLOGENY AND CLASSIFICATION

One of the goals of biology is to understand **phylogeny,** the ancestry of living and extinct organisms, and to develop schemes of classification based on evolutionary relationships. In this endeavor, we must depend to a considerable extent on the study of fossils, as well as on attributes of living organisms. Unfortunately, the fossil record for very early forms of life is poor, partly because conditions in the shallow seas that covered the earth for a long time after life evolved were not suitable for preservation of plants and animals. An insufficiency of mineral salts and the rapid recycling by decomposers are just two reasons why organisms that evolved early did not fossilize well. Another important consideration is that for many millions of years, most organisms in the seas were microbes, simple plants, and the herbivores that fed on them. Without carnivores, there was little selective pressure favoring the evolution of animals with protective hard parts. Once exoskeletons and shells appeared, the fossil record became richer.

In spite of the fact that we do not yet know the evolutionary connections between many of the phyla, or even between major subdivisions within some phyla, we should try to classify animals, plants, and microorganisms in ways that reflect their phylogeny. The problems of classification are generally more difficult in large phyla than in small ones. For instance, the phylum Phoronida, with about 20 species and little diversity, is much easier to deal with than the phylum Mollusca, which has more than 100,000 species and enormous diversity.

The outline below shows the main categories in the sequence of classification of one animal, a planarian flatworm, in descending order:

Phylum: Platyhelminthes
Class: Turbellaria
Order (Latin name usually ending in -ida): Tricladida
Family (Latin name always ending in -idae): Planariidae
Genus (conventionally written in italics and capitalized): *Planaria*
Species (conventionally written in italics, but not capitalized): *torva*

This scheme is adequate for most purposes. In the case of large, greatly diversified groups, such as arthropods and molluscs, intermediate categories are essential. The more important of these are subclass and suborder, but there are others.

Is it important to memorize the ranks given to the various subdivisions of a phylum? The answer is generally no. The status accorded to a particular assemblage may vary from one system of classification to another. In dealing with the diversity of animals, most students and instructors will be satisfied to know which groups are subsidiary to others. One should appreciate why a crayfish is an arthropod and why it belongs to certain successively lower categories: Crustacea, Malacostraca, and Eucarida. Relatively few zoologists will need to concern themselves with labeling these categories as class, subclass, order, or whatever.

Grades of Construction

In characterizing an animal phylum, there are many features to consider. What structures are present and how do they develop? What are their functions? Are there distinctive physiological, biochemical, or behavioral attributes? What do fossils tell us about the history of the group? The answer to the first of these questions, concerned with morphology and embryology, depends to a large extent on the **grade of construction** the phylum has attained.

The organisms called **protozoans,** which have almost certainly evolved along several separate lines, are placed together because they consist of single cells. Multicellular animals, including vertebrates and all invertebrates other than protozoans, are called **metazoans.** Sponges are generally considered to be the simplest metazoans, even though many of them are large and nearly all of them have an intricate arrangement of pores and passageways through which water is circulated. They have several kinds of cells, but these are not organized into definite **tissues.**

The cnidarians—hydras, jellyfishes, sea anemones, and their relatives—are rather simple, too, but they do have tissues that exhibit a variety of structural and physiological specializations. Cnidarians have not, however, evolved clearly beyond the tissue grade of construction. In other words, they do not often have what may be called true organs—structures that consist of several different kinds of tissues. Another important point is that their tissues are basically epithelial and originate from only two embryonic germ layers, called **endoderm** and **ectoderm.** Thus cnidarians are said to be **diploblastic** (from *diplo-* meaning "double" and *blast-* meaning "germ").

In the flatworms, which make up the phylum Platyhelminthes, there are definitely true **organs.** These generally include a muscular pharynx for swallowing or sucking in food, and structures used in copulation. Certain of the organs characteristic of flatworms consist not only of epithelial tissue and muscle, but also of connective tissue and glands. The various tissues of flatworms, moreover, originate not only from ectoderm and endoderm but also from a third embryonic germ layer, **mesoderm.** Flatworms and higher phyla are therefore said to be at the organ grade of construction and to be **triploblastic.** From flatworms onward, then, the advances we see are largely embellishments on the organ grade–triploblastic plan. The formation of tissues derived from mesoderm and the appearance of organs were truly major events in the evolution of the animal kingdom.

Embryology

Embryonic development tells us much about probable relationships of animals, and it supports evidence obtained in studies of adult forms. Even within a much-diversified phylum, development is likely to be conservative and to proceed along similar lines. For example, the presence of gill slits, or intimations of these, in embryos of mammals, birds, reptiles, and amphibians links these vertebrates with fishes, in which embryonic gill slits persist to adulthood. Likewise, a sea star and a sea cucumber, which superficially do not resemble

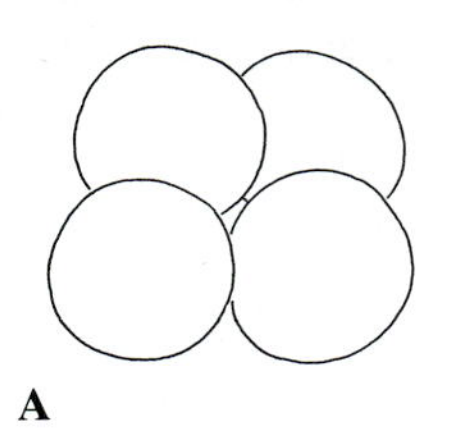

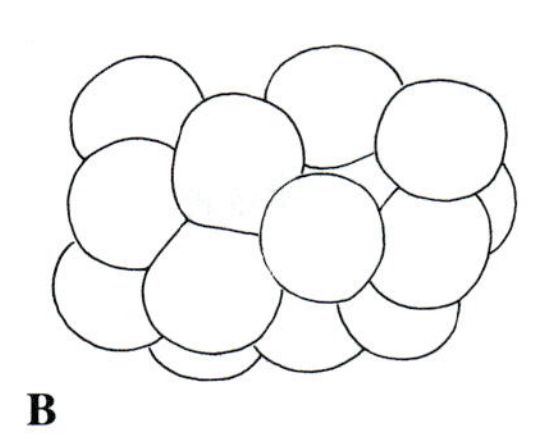

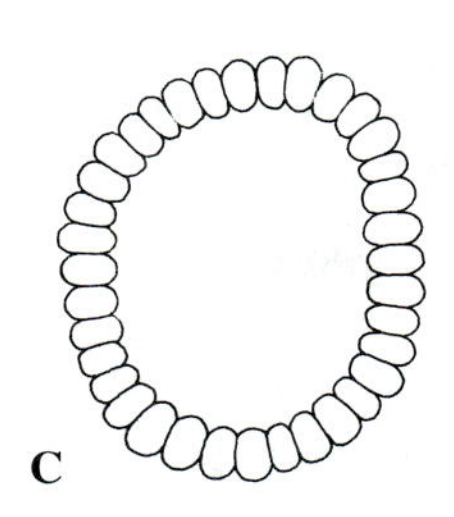

FIGURE 1.1 Cleavage in a cnidarian, *Eucheilota* (class Hydrozoa). The early divisions (A, B) are somewhat irregular, but the blastula (C), seen here in optical section, is a typical one. (After Werner, Helgoländer Wissenschaftliche Meeresuntersuchungen, *18*, 1968.)

one another and which belong to separate classes, are shown to be closely related when their early developments are compared and when structures present in adults are traced back to their embryonic origins.

CLEAVAGE: RADIAL AND SPIRAL

The term **cleavage** refers to the first few divisions of a fertilized egg. In some invertebrates, particularly certain cnidarians, cleavage is somewhat irregular, and for a time the embryo may seem to be a chaotic jumble of cells (Figure 1.1). As a rule, however, cleavage follows a definite pattern, and this is usually either radial or spiral.

In **radial cleavage** (Figure 1.2), characteristic of invertebrates as diverse as cnidarians, ctenophores (comb jellies), phoronids, and sea urchins, the second division of the fertilized egg takes place along axes that are at right angles to the axis of the first division. By the third or fourth division, the cleavage is in a different plane: the axes of the mitotic spindles are parallel to the top-to-bottom axis of the fertilized egg. Thus two tiers of cells are produced, and each cell of the upper tier lies directly above a cell of the lower tier. The next few divisions may conform to the same general pattern, whether they take place in the plane of the first two divisions or at right angles to this plane.

In **spiral cleavage** (Figure 1.3), the first two divisions typically proceed as they do in radial cleavage. In the third division, however, the axes of the spindles are oblique to the top-to-bottom axis of the fertilized egg. After this division is completed, the cells of the upper quartet lie over the furrows that separate the cells of the lower quartet. Subsequent divisions follow the pattern established in the third division, so the cells appear to be arranged spirally. Spiral cleavage is typical of flatworms, annelids, molluscs, and several other phyla.

The cells produced by early divisions of a fertilized egg are called **blastomeres.** As a rule, the third cleavage marks the point at which unequal division becomes apparent. The smaller cells of the young embryo are called **micromeres,** and the larger cells are called **macromeres.**

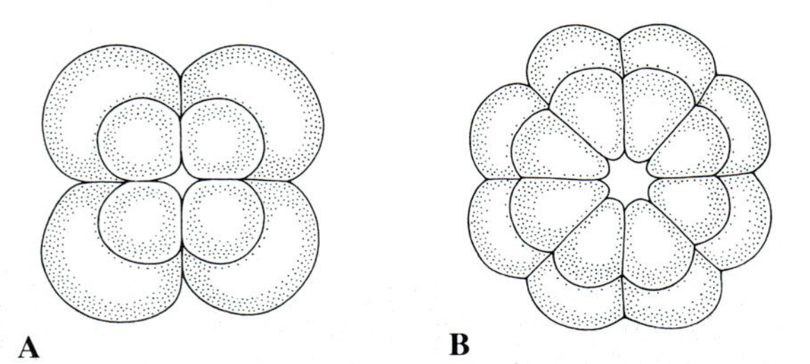

FIGURE 1.2 Radial cleavage, such as is characteristic of echinoderms. A. Eight-cell stage (third cleavage completed). B. Sixteen-cell stage (fourth cleavage completed).

INDETERMINATE AND DETERMINATE DEVELOPMENT

In some animals, the first two or four blastomeres have equal potentials; for example, if the four blastomeres resulting from the first two divisions of the egg

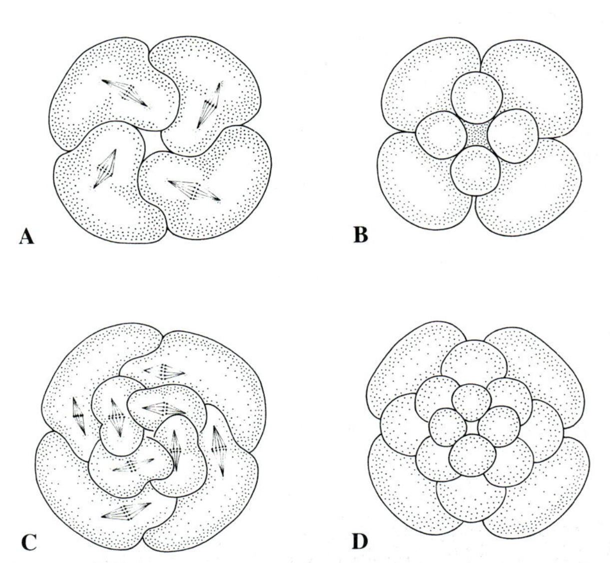

FIGURE 1.3 Spiral cleavage, such as is characteristic of annelids and molluscs. A. Third cleavage in progress. B. Eight-cell stage (third cleavage completed). Note that the four smaller cells (micromeres) lie on the furrows separating the larger cells (macromeres). C. Fourth cleavage in progress. D. Sixteen-cell stage (fourth cleavage completed).

of a sea urchin are separated, each may develop as if it were a fertilized egg. It has everything it needs to produce a normal embryo and then a complete sea urchin. In other words, the first four blastomeres are substantially alike, and the development of the sea urchin at this point is said to be **indeterminate.** However, through subsequent divisions the constituents of the early blastomeres become distributed less and less equally.

In contrast, the isolated blastomeres of a four-celled embryo of an annelid or mollusc will not develop into complete embryos. The embryo derived from each cell will correspond to just a portion of a normal embryo. As early as the four-cell stage, then, development is strictly **determinate.**

You may already have deduced that animals with spiral cleavage are likely to have determinate development, and that those with radial cleavage are likely to have indeterminate development. In general, that is correct, but there are exceptions. In ctenophores, for instance, early cleavage is radial, but development is determinate in the sense that a blastomere isolated at the two-cell or four-cell stage produces only a partial embryo. Even if it fills in the missing parts later, its development is not the same as that of a blastomere that can directly produce a whole embryo.

THE BLASTULA

Many animals, including most vertebrates, go through a stage of embryonic development called the **blastula.** This is arrived at after the fertilized egg has undergone at least several cleavages. A blastula (Figure 1.1, C; Figure 1.4, A and C) consists of a layer of cells that enclose a fluid-filled cavity called the **blastocoel.** Sometimes this cavity is small in proportion to the cells that surround it.

GASTRULATION

Gastrulation is the process by which endoderm is segregated from ectoderm. There are several ways in which this is achieved. In animals whose eggs do not contain a great deal of yolk, gastrulation is usually accomplished by delamination or invagination. **Delamination** (Figure 1.4, A and B) is the usual method for cnidarians, including jellyfishes and sea anemones, and also for some other invertebrates. As the cells that form the wall of the blastula divide, some of the products begin to occupy the blastocoel. They, too, multiply, and eventually all or most of the free space is filled up. The cells of the inner mass constitute endoderm; those forming the surrounding jacket constitute ectoderm. This particular type of gastrula is called a **stereogas-**

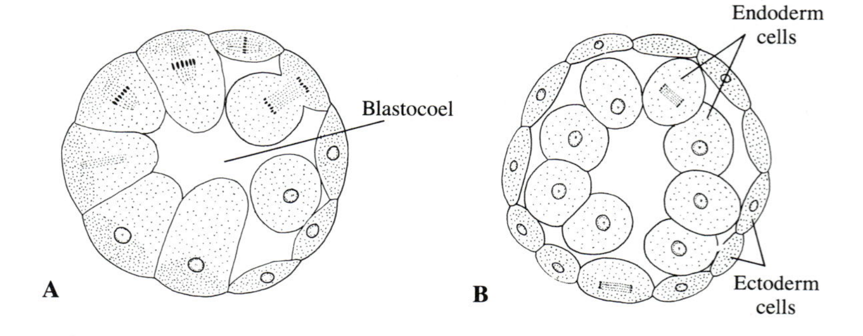

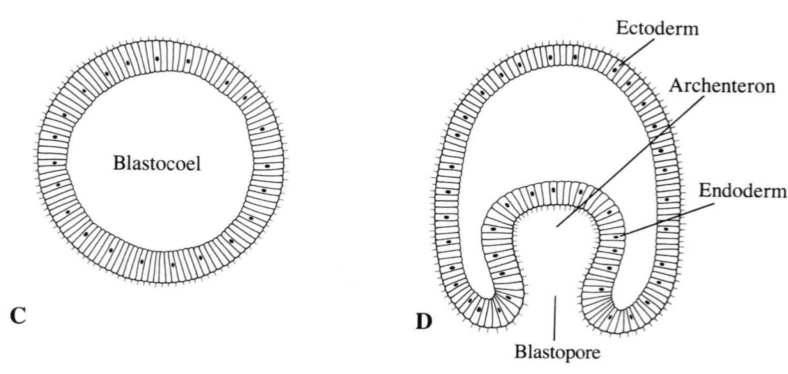

Figure 1.4 A, B. Gastrulation by delamination, in a medusa of the order Trachylina (class Hydrozoa). As the cells that form the wall of the blastula divide, some of the products become endoderm cells. (After Metschnikoff, Zeitschrift für Wissenschaftliche Zoologie, *36*.) C, D. Gastrulation by invagination, such as is characteristic of echinoderms. A portion of the wall of the blastula (C) turns inward, forming a gastrula (D).

trula (*stereos* is Greek for "solid"). Eventually a gut cavity appears in the endoderm and a mouth opens up.

Gastrulation by **invagination** (Figure 1.4, C and D) is characteristic of many invertebrates, but the classical examples are seen in echinoderms—sea urchins, sand dollars, sea stars, and their relatives. The cells in certain portions of the blastula multiply more rapidly than others. As a result, part of the blastula wall is pushed into the blastocoel as a short, blind tube. This is the primitive gut, or **archenteron.** The opening into the gut cavity is called the **blastopore.** The fate of the blastopore varies from phylum to phylum. In echinoderms, it typically becomes the larval anus; the mouth appears later. Animals with this kind of development are called **deuterostomes,** because the mouth is a secondary opening. Molluscs, annelids, and some other phyla are said to be **protostomes,** because the mouth is either derived directly from the blastopore or develops in the same region. In protostomes, then, the anus, if there is one, is the secondary opening.

A third method of gastrulation, characteristic of some animals that have moderately yolky eggs, is **epiboly** (Figure 1.5). The smaller blastomeres at the upper pole of the blastula, whose blastocoel is likely to be reduced if not lacking altogether, multiply more rapidly than the larger and yolkier cells at the lower pole. The smaller cells eventually cover the larger cells, which can now be called endoderm cells, for they form the first traces of the gut. The gut cavity, however, may not appear right away.

In insects and other animals whose eggs are large and contain much yolk, development—including gastrulation—is greatly modified. The yolk interferes, at least for a time, with total cleavage, but eventually a disk of cells is segregated on the upper pole, or a jacket of cells forms around a central yolky mass.

THE FORMATION OF MESODERM

It has already been pointed out that the tissues of flatworms and most other invertebrates derive not only from embryonic ectoderm and endoderm but also from a third germ layer called mesoderm. Mesoderm may originate in more than one way, and its appearance during development may be closely linked with the formation of a type of body cavity called the **coelom.** This is a fluid-filled space, or series of spaces, that lies between the gut and body wall. A true coelom, such as is found in the vertebrates and most of the higher phyla of invertebrates, is lined by an **epithelium** of mesodermal origin.

Some invertebrates are said to be **enterocoelous** because their coelom develops from **evaginations** ("outpocketings") of the archenteron (Figure 1.6). The walls of the evaginations, which soon become separated from the archenteron, consist of mesoderm cells. Thus coelom formation coincides with the origin of mesoderm. Classical examples of enterocoely are found in the echinoderms, but the phenomenon is also characteristic of chaetognaths and some other invertebrate phyla, as well as of vertebrates. It does not necessarily account for all of the early mesoderm, however. In echinoderms, for instance, some mesodermal cells may bud into the blastocoel from where the archenteric invagination is just becoming recognizable, or comparable cells may be produced from the blind end of the archenteron after this tube is fairly well developed. This mesoderm originates before the enterocoelic pockets appear.

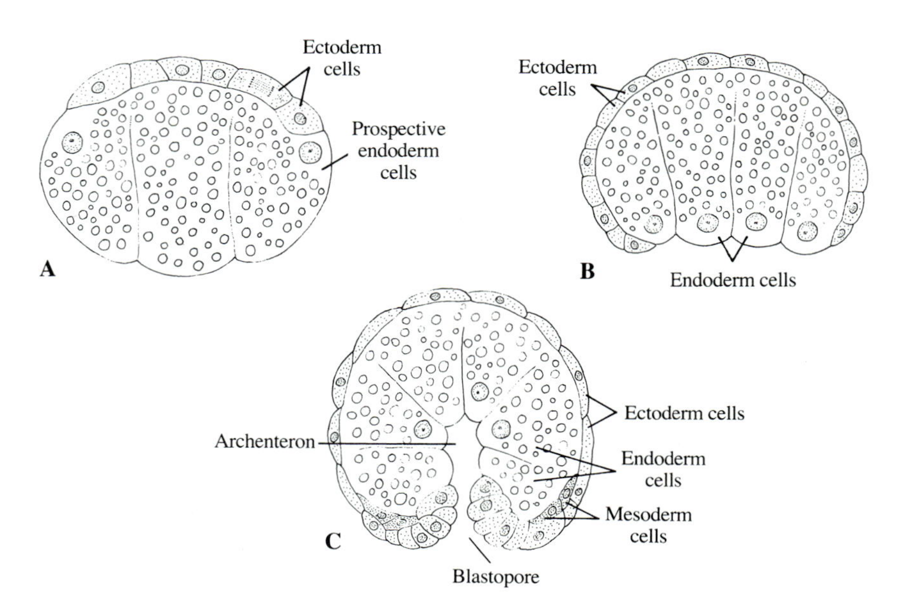

FIGURE 1.5 Gastrulation by epiboly, in the gastropod mollusc *Crepidula*. A, B. Ectodermal cells grow down to enclose the yolky prospective endoderm cells. C. Gastrula, already with a few mesoderm cells derived from a cell that was located at the lip of the blastopore. (After Conklin, Journal of Morphology, *13*.)

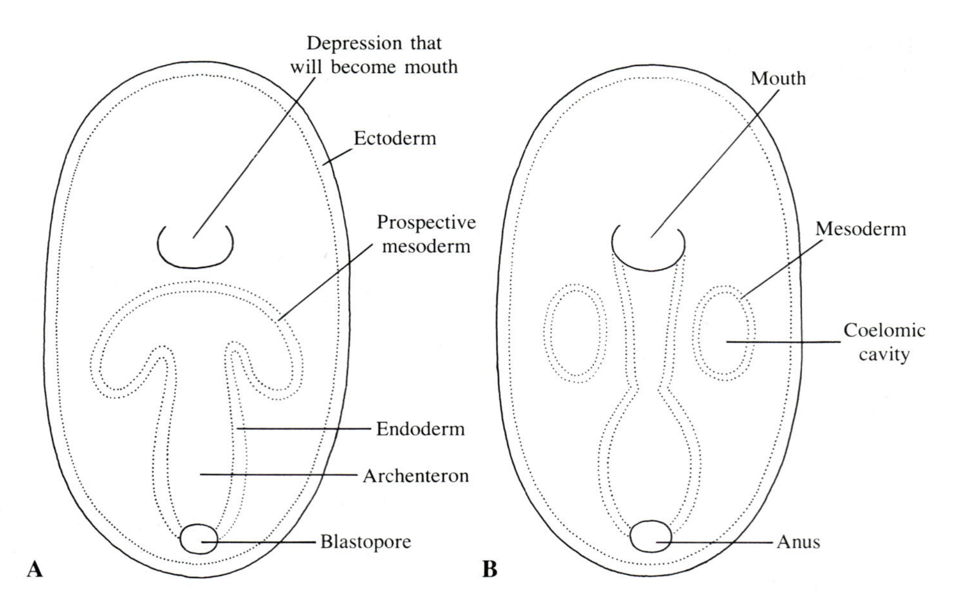

FIGURE 1.6 A, B. Formation of mesoderm and coelomic cavities by evagination of the archenteron.

A different method of mesoderm formation is observed in several other phyla, some of which have a coelom (annelids and molluscs, for instance) and some of which do not. The mesoderm can be traced back to a particular cell, called the 4d cell, at the lip of the blastopore (Figure 1.7, A). The progeny of this cell sink into the blastocoel and multiply, filling up much of the space between the endoderm and ectoderm. In annelids, molluscs, and related groups, spaces appear within the mesoderm (Figure 1.7, B), and these are the rudiments of the coelom. Because the spaces appear in solid masses of mesoderm, animals showing this kind of coelom formation are said to be **schizocoelous** (*schizo-* refers to "splitting").

There are some less clear-cut styles of mesoderm formation. Arthropods, which almost always have large, yolky eggs, are among the invertebrates in which the origin of mesoderm does not conform clearly to the patterns discussed above. In insects, for example, after the plate of cells that will form the definitive embryo becomes differentiated from the rest of the cells that enclose an inner yolky mass, both endoderm and meso-

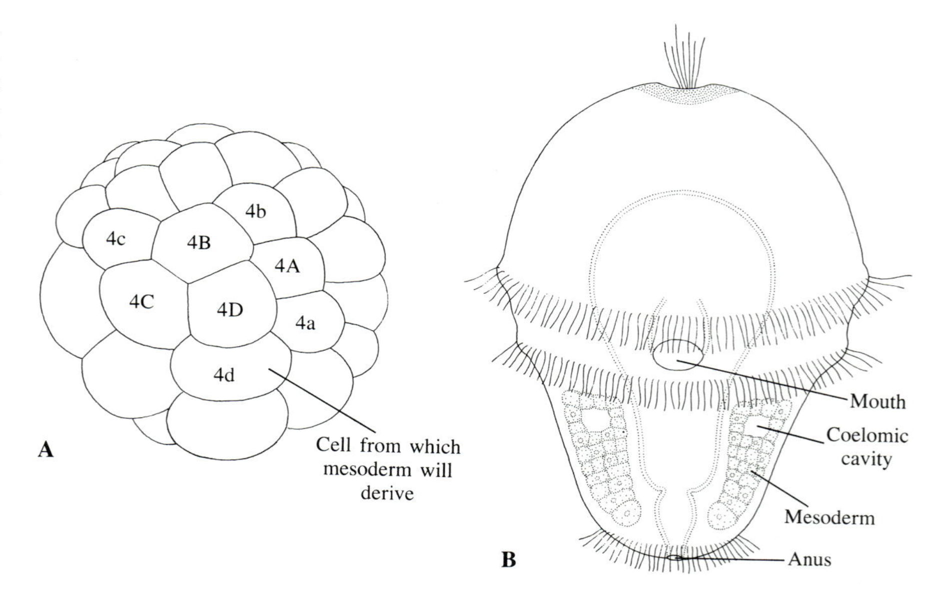

FIGURE 1.7 An embryo of the annelid *Arenicola* (class Polychaeta), showing the 4d cell from which mesoderm will derive. (After Child, Archiv für Entwicklungsmechanik, *9.)* B. Trochophore larva of the archiannelid *Polygordius,* showing the mesoderm derived from the 4d cell and the appearance of the first coelomic cavities within it.

derm bud from this plate. After the mass of mesoderm becomes substantial, coelomic spaces appear in it.

Not all methods of mesoderm and coelom formation are covered by the examples just given. There are, moreover, body cavities of types other than the coelom. These are discussed next.

BODY CAVITIES: COELOM, PSEUDOCOEL, AND HEMOCOEL

In flatworms and a few other lower triploblastic phyla, the space between the gut and epidermis is almost fully occupied by tissue (Figure 1.8, A). Certain components of the reproductive system and some other ducts are hollow, but there is not an extensive, fluid-filled body cavity. These animals are therefore said to be **acoelomate** (''lacking a cavity'').

Most of the triploblastic phyla do have a body cavity of one type or another. This may serve as a hydrostatic skeleton against which muscles operate to effect locomotion, burrowing, or protrusion of certain organs. A body cavity frees some internal organs from the influence of body movements, and it may also function in distribution of nutrients and oxygen. Among the structures that lie in the body cavities of invertebrates are the gonads and organs concerned with the excretion of nitrogenous wastes and the maintenance of a proper water–salt balance. These structures usually have ducts that lead to the outside or to the lumen (cavity) of the gut.

One type of body cavity, the coelom (Figure 1.8, C), develops within a sac of mesoderm or within a solid mass of mesoderm, as explained in the preceding section. The coelom enlarges until it replaces the blastocoel, and it has an epithelial lining called the **peritoneum.** A true coelom is characteristic of vertebrates, as well as of annelids, molluscs, echinoderms, and most of the other higher phyla of invertebrates, although the extent to which it is developed varies considerably.

Another type of body cavity is the **pseudocoel** (or pseudocoelom) (Figure 1.8, B), characteristic of nematodes, rotifers, acanthocephalans, and several other phyla. It lacks a peritoneal lining, and is conventionally regarded as a blastocoel that has persisted because muscle and other tissues derived from mesoderm have not completely filled up the space between the gut and epidermis. This explanation may apply to the pseudocoel of at least some phyla, but it needs to be re-evaluated.

In molluscs, arthropods, urochordates, and a few other groups of invertebrates, a system of blood-filled

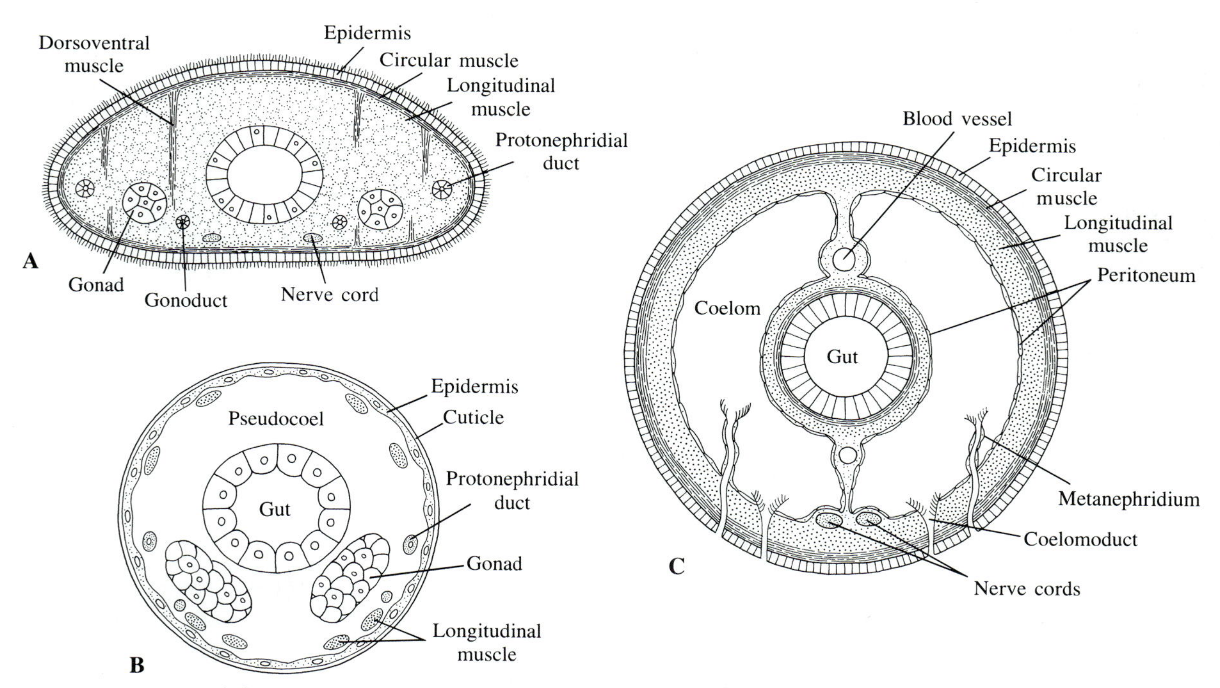

FIGURE 1.8 Three main body plans of triploblastic invertebrates. A. Acoelomate type, such as is characteristic of flatworms. B. Pseudocoelomate type. C. Coelomate type, such as is characteristic of annelids.

spaces forms what is called a **hemocoel.** Collectively, the spaces may make up a considerable part of the volume of the body. In animals that have a hemocoel, the circulatory system is open; that is, the arteries that lead away from the heart soon end and allow blood to enter the spaces. The tissues are therefore bathed by blood, rather than pervaded by small vessels functionally comparable to the capillaries of vertebrates. Depending on the phylum, blood may be returned to the heart by veins, or it may enter the heart directly through openings called ostia.

In general, animals characterized by a hemocoel also have a true coelom, but the coelom is restricted to only a few structures. In molluscs, for instance, the principal coelomic derivatives are the gonads and gonoducts, the renal organs that function in excretion, and the pericardial cavity that surrounds the heart. The hemocoel, on the other hand, is extensive, and may be as important in locomotion, burrowing, and other movements as the true coelom is in annelids. For example, the rapid changes in the shape of the foot of a clam as the animal burrows in mud or sand are due to displacements of hemocoelic fluid brought about by muscular activity.

SYMMETRY

Relatively few animals are perfectly symmetrical. Their body plans are adaptations to many different life styles, so deviations from the major patterns are to be expected.

Vertebrates and most groups of invertebrates exhibit **bilateral symmetry** (Figure 1.9, A). They have an anterior end, posterior end, ventral (lower) surface, dorsal (upper) surface, right side, and left side. Cutting through the animal longitudinally along the midline produces two halves that are mirror images of one another, although some internal or external structures may not be distributed quite equally. As a rule, bilateral symmetry is accompanied by **cephalization**—the presence of a head end, with nerve cells concentrated into a large ganglion, or brain, and also with well-developed sense organs.

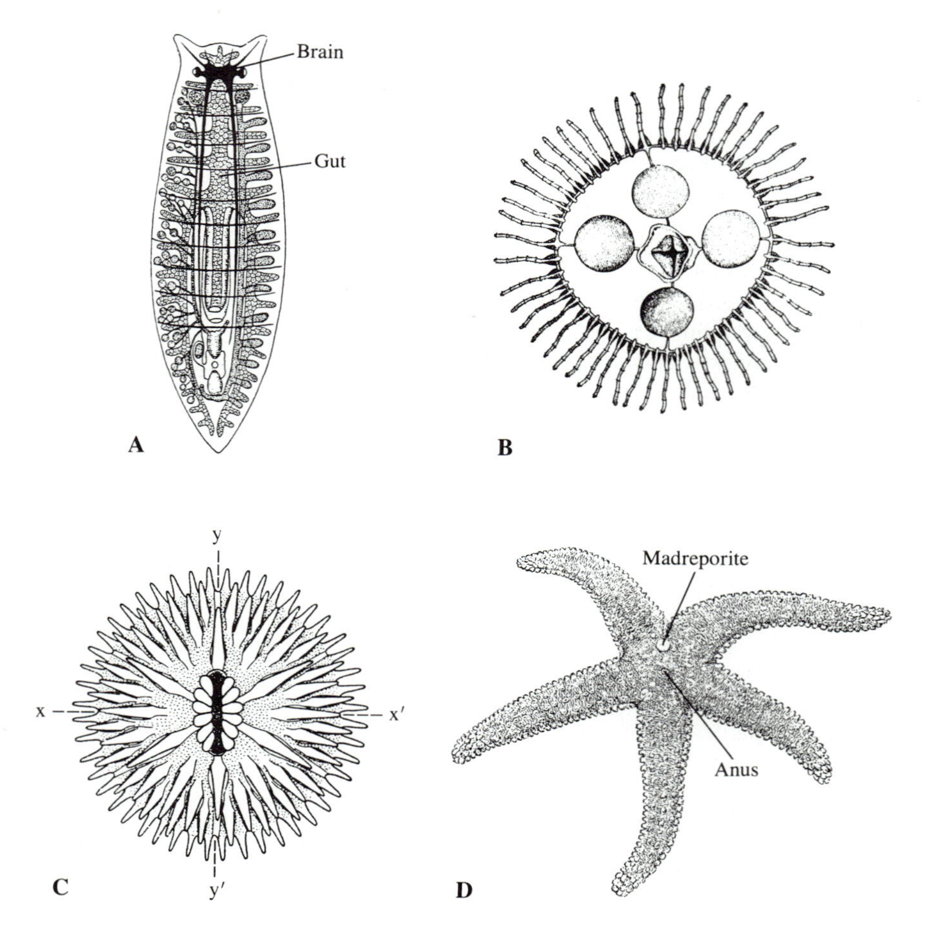

FIGURE 1.9 Types of symmetry. A. Bilateral symmetry, with cephalization, in a flatworm. B. Radial symmetry in a jellyfish. C. Biradial symmetry in a sea anemone. The two halves produced by cuts along the line x–x′ will be different from the two produced by cuts along the line y–y′. D. Secondary radial symmetry in a sea star. Of the structures visible externally, only one (the madreporite) disturbs the radial symmetry; internally, however, several structures are similarly placed.

A different type of symmetry, called **radial symmetry** (Figure 1.9, B) is typical of most jellyfishes and some other members of the phylum Cnidaria. Any cut going from top to bottom through the center of the animal will produce identical equal halves. Many jellyfishes can, in fact, be cut like a pie into perfectly equal quarters, eighths, or even smaller pieces.

There is a modification of radial symmetry, called **biradial symmetry** (Figure 1.9, D) in which the two equal halves produced by a cut along one axis are different from the equal halves produced by a cut made at right angles to it. This type of symmetry is characteristic of most sea anemones and coral polyps, as well as of ctenophores and some jellyfishes.

Both radial and biradial symmetry may be marred by eccentric placement of certain internal structures, or by a tendency toward bilaterality. This situation is especially common in the phylum Echinodermata, which includes sea stars, sea urchins, sand dollars, and sea cucumbers. These animals undeniably have radial symmetry to a greater or lesser extent, but it is imperfect (Figure 1.9, D). In sand dollars and sea cucumbers, in fact, the body shows a rather distinctly bilateral organization in certain respects, even though most features of a basically radial symmetry are conserved. In echinoderms, moreover, the apparent radial symmetry is secondarily derived from a primary bilateral symmetry. This is evident in the fossil record, as well as in the fact that the free-swimming larvae characteristic of many species are bilaterally symmetrical. A larva undergoes a drastic metamorphosis before the radial symmetry characteristic of the adult is attained.

Among the animals that do not fit any of the plans of symmetry described above are sponges. They generally form sprawling encrustations or lumpy growths without symmetry. Even those sponges that look like vases or goblets are not really symmetrical when their systems of canals and chambers are examined on a microscopic level. Many protozoans, even those that have a constant body form, are asymmetrical, too.

SEGMENTATION

In some animals, notably the annelids and arthropods, the body is divided into a series of units called **segments** (or *metameres*). The extent to which successive segments resemble one another varies. It is generally assumed that when most of the segments are similar externally and internally, this represents the ancestral condition. When, on the other hand, every segment is highly specialized and markedly different from every other segment, this is considered to be a derived feature.

The structure of an earthworm provides a good introduction to segmentation in annelids (Figure 1.10, A). Externally, nearly all segments are much alike in terms of their general appearance as well as in the arrangement of the bristles used in locomotion. The genital pores and certain epidermal glands, however, are present only on a few of the segments. Internally, the paired organs concerned with excretion are found in most of the segments, and the ganglia of the ventral nerve cord and branches of the major longitudinal blood vessels are

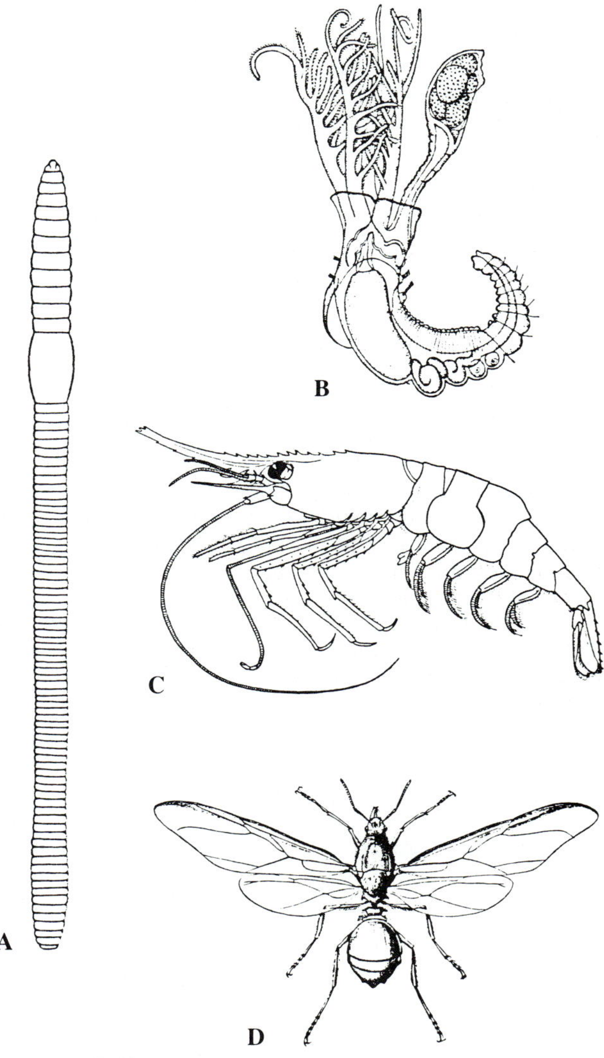

FIGURE 1.10 Segmentation and tagmatization. A. An earthworm; most of the segments are similar externally. B. A serpulid polychaete worm, with segments specialized to the extent that the body is organized into regions. C. A shrimp, showing tagmata: a cephalothorax in which the segments are obscure, and a distinctly segmented abdomen. D. An insect, with three tagmata: head, thorax, and abdomen. (The three pairs of legs and two pairs of wings are on the thorax.)

also arranged according to a segmental plan. The gut runs uninterrupted through all segments, but the coelom is divided by transverse partitions, so that the body cavity of one segment is almost completely separate from that of the preceding and succeeding segments. It is true that the reproductive organs of an earthworm are concentrated in relatively few segments, yet there are many annelids in which the reproductive structures are repeated serially in almost all segments.

Various other annelids are somewhat similar to earthworms in that most of their body segments resemble one another. Not all annelids are locked into this pattern, however. There are many in which almost every segment is distinctive in one way or another. There are also many in which most of the partitions between segments have disappeared, so that the coelom is nearly continuous. Another tendency seen in the phylum Annelida is the organization of groups of segments into two or more rather definite body regions, with the segments belonging to one region being markedly different from those of other regions (Figure 1.10, B).

It is in the Arthropoda that the grouping of segments into body divisions reaches its highest level (Figure 1.10, C and D). In insects, for example, several segments are united into a head, three form the thorax (with legs and wings), and several more form the abdomen. Each of these divisions, decisively different from the others, is called a **tagma**.

The term **somite** is often used interchangeably with segment. It is best, however, to restrict its use to an incipient segment that is recognizable during development. Not all embryonic somites necessarily become segments. Some of them may lose their identity. Moreover, what looks like one segment in the adult may have been formed by two or more embryonic somites that have fused. These exceptions are mentioned so that you will use the terms *segment* and *embryonic somite* with precision.

Juvenile or Larva?

If an immature animal is similar to the adult in most respects, it is best to call it a **juvenile**. The term **larva** is generally reserved for a stage that is so distinctly different from the adult that it must undergo a drastic metamorphosis before it reaches maturity. Thus the caterpillar of a butterfly, the maggot of a fly, and the tadpole of a frog are larvae, but young spiders or young nematodes are juveniles.

It must be admitted that there are cases in which it is difficult to decide whether *juvenile* or *larva* is the more appropriate term. A newly hatched octopus, for instance, resembles the adult in many ways, but because it is small and may lead a free-swimming existence for several weeks, it is often called a larva. The situation is complicated by the fact that some zoologists use the term *larva* for all obviously immature stages; a few even apply it to fertilized eggs. In this book, however, *larva* and *juvenile* will be used according to the definitions given above.

Another term, **nymph,** should also be mentioned here. It is applied only to the young of certain insects, such as grasshoppers and dragonflies. When these insects hatch, they have legs and the rudiments of wings, but they are very different from the adults in other respects. The young may, in fact, be aquatic, whereas the adults are terrestrial. They attain maturity gradually through a series of stages that look progressively more like the adult.

Homology

Homologous structures show similarities in basic design and embryonic origin. The arm of a human, the flipper of a seal, and the wing of a bird are unquestionably homologous, even though they are specialized for different purposes. The wing of an insect and the wing of a bird, on the other hand, are **analogous structures,** for while they serve the same general function, they are structurally dissimilar and originate in completely different ways. Zoologists stress homology between structures, when there is good reason to believe that it exists, for it is a clue to understanding relationships between groups of animals. Many terms used in zoology imply homology, and they should be used with precision. If they are applied carelessly to unrelated structures, one may be misled to believe that these structures are homologous when in fact they are not. Among the terms that are commonly misused is *metanephridium,* properly applied only to a particular type of excretory organ that originates as an invagination of ectoderm, and not to similar structures that derive from mesoderm.

Ancestral or Derived?

As you read about various groups of invertebrates, you will note that certain representatives are said to show ancestral characteristics, and that others are said to show derived characteristics. An **ancestral** (or **primitive**) **feature** is presumed to have been typical of a group for a long time, perhaps ever since the group became distinct. A **derived** (or **advanced**) **feature** is a marked departure from any that might have been characteristic of ancestral types. In crustaceans, for instance, the presence of numerous pairs of similar limbs used for swimming and for collecting food is considered to be an ancestral trait; the presence of only a few pairs of limbs, each with very specific functions, is considered to be derived.

Related to the terms just explained are the terms *specialized* and *generalized,* and *higher* and *lower*. A **specialized structure** is one that adapts an animal to life in a certain situation or under certain conditions. A specialized animal, then, is one whose structure, reproduction, development, or physiological attributes—or all of these—fit it for life under rather restricted circumstances. A tapeworm that lives in the intestine of a vertebrate is extremely specialized. It has no gut and depends for its nutrition on amino acids, sugars, and other organic materials that the digestive processes of its host make available to it. A barnacle that is attached to a rock and that uses its legs to comb microscopic food from sea water is specialized when compared to some other crustaceans. Certain parasitic relatives of barnacles, moreover, are so specialized that it is almost impossible to tell that they are crustaceans. As adults, they do not even have appendages. Their free-living larval stages, however, tie them unmistakably to barnacles and to other crustaceans as well.

Generalized is not exactly the opposite of specialized. It is applied to animals that exhibit most of the basic features of the group to which they belong. Planarians, for instance, are often used for introducing students to the phylum Platyhelminthes. They are by no means the simplest flatworms, but they do have most of the structures that one can expect to see in the phylum as a whole.

Lower and *higher* are terms that may be applied to entire phyla as well as to subordinate taxa. Thus, for example, annelids are considered to be higher than flatworms because they are segmented and typically have a complete gut, a coelom, and a circulatory system. Yet some annelids have more derived features than others, so they are called higher annelids. There are always problems connected with applying such labels, however. A group of organisms that shows certain derived characteristics may also have some ancestral features. Furthermore, a particular organ system may be more complex in a lower phylum than in a higher phylum. The reproductive system of most flatworms, for instance, is more complex than that of many annelids.

"Major" and "Minor" Phyla

The term *major phylum* is widely used in referring to an assemblage that has thousands or many thousands of species and that is greatly diversified and successful in a variety of habitats. A *minor phylum,* on the other hand, is one that has relatively few species and that may be restricted to a single habitat, such as the intestine of vertebrate animals or marine sediment. From the standpoint of a zoologist interested in evolutionary relationships and in the diversity of the animal kingdom as a whole, all groups are important. Some so-called minor phyla are, moreover, extremely successful in the habitats they occupy, and in these habitats they may be of considerable ecological importance. It is perhaps best, therefore, not to overdo the use of these terms. *Large phylum* and *small phylum* are, in general, more suitable.

Symbiosis

The term **symbiosis** (''living together'') is applied to intimate nutritional or behavioral relationships between different species. Some symbiotic associations benefit both members of the partnership (the *symbionts*), and may even be essential for the survival of both. Others are beneficial only to one partner. There are many diverse kinds of permanent and temporary relationships between species, and relatively few of them have been investigated thoroughly. A brief description of three major categories of symbiosis will, however, make it easier to understand some of the cases that are discussed in succeeding chapters.

MUTUALISM

In **mutualism,** both members of the partnership receive benefit. A classical case is that involving termites and certain species of bacteria and flagellate protozoans that live in the termite hindgut. Termites chew and swallow wood particles, but they cannot digest the cellulose. Digestion is accomplished by the microorganisms, which use some of the soluble products of the breakdown and make the rest available to their hosts. The termites thus need the bacteria and flagellates to digest the wood, while the bacteria and flagellates depend on the termites for a supply of wood particles to digest.

Another well-known type of mutualism is seen in corals and other invertebrates that harbor photosynthetic dinoflagellates or unicellular green algae in their cells. Some of the organic compounds produced by the photosynthetic symbionts diffuse out of them and are utilized by the host cells.

PARASITISM

In a different category of symbiosis, called **parasitism**, one member of the association derives benefit at the other's expense. A tapeworm that lives in the intestine of a vertebrate is an example of a parasite. It has no gut of its own and it depends on soluble organic nutrients produced by the host's digestive processes. The nutrition of the host may therefore be impaired.

Many parasites, such as nematodes and flukes, do have a digestive tract. They suck in blood or other tissues from the organs in which they reside. There are also intracellular parasites. Among these are the protozoans that cause malaria.

Some external parasites are transitory. They do not feed continuously, and they may not remain long on one host before they drop off. Certain insects, ticks, and crustaceans are parasites of this general type.

COMMENSALISM

Commensalism, defined broadly, covers symbiotic associations in which one species obtains benefit from living with another, but in which the host seems not to be helped or harmed. Many invertebrates that live in tubes or burrows have other animals associated with them. The commensals require the protection of the tube or burrow and they may profit from other conditions that are maintained by the host, such as a current of water that brings in suitable food. In other words, they ''eat at the same table,'' the literal meaning of *commensal*. Other good examples of commensals are organisms that feed on bacteria or dead epithelial cells on the body surface or in the gut of an animal. One must always be aware of the possibility, however, that some relationships long considered to be commensal are really cases of mutualism or parasitism. Indeed, true commensalism may be rather rare.

Fossils and the Geological Time Scale

Although this book is concerned almost entirely with invertebrates that are living now, it occasionally refers to fossils. These relics provide the only direct evidence for the existence of many groups of animals and plants that are extinct. The structure of these fossils—to the extent that we understand it—has considerable bearing on how we interpret the evolutionary relationships of living organisms. A zoologist or botanist should therefore have some knowledge of the time periods during which extinct groups were prevalent.

Geologists and paleontologists use a time scale in which four or five eras in the history of the earth are subdivided into a number of periods and epochs. No one scale is universally accepted, so the layout shown in Figure 1.11 may not agree exactly with those you find in other books. It will, however, enable you to correlate time with some important events in geological history and with the appearance of certain forms of life.

Comprehensive References

Extensive lists of specialized research papers are not necessary in a text whose purpose is to present the more important facts concerning the functional organization and classification of invertebrate animals. In general, the references provided in subsequent chapters have been selected because they deal in considerable detail with at least one aspect of the biology of a particular assemblage of organisms. Many of these references contain bibliographies that will be useful to those who need additional information.

The references given in this chapter are primarily comprehensive works that cover at least several invertebrate phyla. Some of the works are not likely to be available except in large libraries. The comments provided in connection with certain of the references will perhaps be helpful to prospective users. It is important to appreciate that in almost any multivolume work, some parts will be more nearly up-to-date than others. The usefulness of a book cannot always be judged by its date. What one user thinks is old or superfluous may be essential to the studies of another person.

MAJOR WORKS DEALING WITH ALL OR MOST INVERTEBRATE PHYLA

Bronn, H. G. (editor), 1866– . Klassen und Ordnungen des Tierreichs (Thier-Reichs). Gustav Fischer Verlag, Jena.
In many volumes and still incomplete, this work deals with morphology, classification, and other aspects of the biology of invertebrates. Many parts are now of little use except for documenting nineteenth-century concepts of taxa. Some of the more recent sections, however, provide the best comprehensive accounts that are available for certain groups.

Grassé, P.-P. (editor), 1948– . Traité de Zoologie. Masson, Paris.
Most groups of invertebrates have already been covered in this series of many volumes. It will be nearly complete when the parts on ciliates, cnidarians, ctenophores, and crustaceans are finished.

Hyman, L. H., 1940–67. The Invertebrates. McGraw-Hill Book Co., New York.
In six volumes, this work admirably covers nearly all invertebrates other than annelids, arthropods, and tunicates; the section on molluscs is incomplete.

Kaestner, A., 1967–70. Invertebrate Zoology. Translated and adapted from the German by H. W. and L. R. Levi. Wiley-Interscience Publishers, New York. (Reprinted by Robert E. Krieger Publishing Co., Huntington, New York.)
Volume 1, covering most invertebrates other than arthropods, is not especially good; volumes 2 (arthropods in general, chelicerates, myriapods, and ''near-arthropods'') and 3 (crustaceans) are detailed and useful.

Kükenthal, W., and Krumbach, T. (editors), 1923– . Handbuch der Zoologie. W. de Gruyter, Berlin.
Long the standard encyclopedic work, this has to a considerable extent been superseded by Grassé: Traité de Zoologie (see above).

Moore, R. C. (editor), 1952– . Treatise on Invertebrate Paleontology. Geological Society of America and University of Kansas Press, Lawrence, Kansas.

Era	Period	Epoch	Millions of Years Ago	Important Geological and Biological Events
Cenozoic	Quaternary	Recent		Abnormally high provincialism and few epicontinental seas
			0.011	
		Pleistocene		Great Ice Age (glacial and interglacial periods)
			2.5	
	Tertiary	Pliocene		Renewed uplift of Rocky Mountains; Grand Canyon, Cascades begin to form
			7.0	
		Miocene		
			26	Antarctica freezes
		Oligocene		
			38	Major uplift of Himalayan Mountains as India collides with Asia
		Eocene		
			54	
		Paleocene		
			65	Marine and terrestrial mass extinction
Mesozoic	Cretaceous			
			136	Major uplift of Rocky Mountains
	Jurassic			
			190	
	Triassic			Break-up of Pangea begins
			225	Marine mass extinction
Paleozoic	Permian			Completion of uplift of Appalachian Mountains: supercontinent Pangea assembled; glaciation
			280	
	Carboniferous			Extensive coal swamps
			345	
	Devonian			
			395	
	Silurian			
			435	Glaciation
	Ordovician			
			500	Second marine diversification
	Cambrian			
			570	Major marine diversification
Proterozoic ("Pre-Cambrian")				Glaciation; widespread continental rifting
			1000	
			2000	Free oxygen appears in atmosphere
			3500	Oldest known fossils
			3700	Oldest known terrestrial rocks
			4600	Oldest known lunar rocks (? origin of solar system)

Temporal Distribution of Life Forms

- Bacteria and Blue-green algae
- Simpler eukaryotic organisms: protozoa, algae, fungi
- ? (probably older than fossil record shows) Mosses, liverworts
- Ferns and fern-allies
- Gymnosperms
- Angiosperms
- Marine invertebrates
- Terrestrial invertebrates
- Fishes (including ostracoderms)
- Amphibians
- Reptiles
- Birds, mammals
- Modern humans

FIGURE 1.11 Geological time-table, showing some of the major events in the history of the earth and the temporal distribution of life forms.

Although concerned primarily with fossils, this work, now nearly complete, deals in considerable detail with morphology and classification of living invertebrates. It is important because it brings much evidence from fossils to bear on classification.

Parker, S. P. (editor), 1982. Synopsis and Classification of Living Organisms. 2 volumes. McGraw-Hill Book Company, New York.

TEXTS AND BOOKS ON COMPARATIVE MORPHOLOGY

Barnes, R. D., 1987. Invertebrate Zoology. 5th edition. Saunders College Publishing, Philadelphia.

Barnes, R. S. K., Calow, P., and Olive, P. J. W., 1988. The Invertebrates: A New Synthesis. Blackwell Scientific Publications, Oxford, London, and Palo Alto.

Barrington, E. J. W., 1979. Invertebrate Structure and Function. 2nd edition. John Wiley & Sons, New York.

Bayer, F. M., and Owre, H. B., 1968. The Free-Living Lower Invertebrates. Macmillan, New York.

Beklemishev, V. N., 1969. Principles of Comparative Anatomy of Invertebrates. 2 volumes. Translated from the Russian by J. M. MacLennan; edited by Z. Kabata. Oliver and Boyd, Edinburgh; University of Chicago Press, Chicago.

Clark, R. B., 1964. Dynamics in Metazoan Evolution: The Origin of the Coelom and Segmentation. Clarendon Press, Oxford.

Dougherty, E. C. (editor), 1973. The Lower Metazoa: Comparative Biology and Phylogeny. University of California Press, Berkeley, Los Angeles, and London.

Fretter, V., and Graham, A., 1976. A Functional Anatomy of Invertebrates. Academic Press, London and New York.

Gardiner, M. S., 1972. The Biology of Invertebrates. McGraw-Hill Book Company, New York.

Laverack, M. S., and Dando, J., 1987. Lecture Notes on Invertebrate Zoology. Blackwell Scientific Publications, Palo Alto, California.

Leake, L. D., 1975. Comparative Histology. An Introduction to the Microscopic Structure of Animals. Academic Press, London and New York.

Lutz, P. E., 1985. Invertebrate Zoology. Addison Wesley Publishing Co., Reading, Massachusetts.

Marshall, A. J., and Williams, W. D. (editors), 1972. A Text-Book of Zoology, volume 1. 7th edition. Macmillan, London.

Meglitsch, P. A., 1972. Invertebrate Zoology. 2nd edition. Oxford University Press, London and New York.

Pearse, V., Pearse, J., Buchsbaum, M., and Buchsbaum, R., 1987. Living Invertebrates. Blackwell Scientific Publications, Palo Alto, California.

Pechenik, J. A., 1985. Biology of the Invertebrates. Prindle, Weber, and Schmidt, Boston.

Russell-Hunter, W. D., 1979. A Life of Invertebrates. Macmillan, New York; Collier-Macmillan, London.

LABORATORY MANUALS

Beck, E., and Braithwaite, L. F., 1968. Invertebrate Zoology Laboratory Workbook. Burgess Publishing Company, Minneapolis.

Dales, R. P. (editor), 1981. Practical Invertebrate Zoology. 2nd edition. Blackwell Scientific Publishers, Oxford, London, and Boston.

Freeman, W. H., and Bracegirdle, B., 1971. An Atlas of Invertebrate Structure. Heinemann Educational Books, London.

Pierce, S. K., and Maugel, T. K., 1987. Illustrated Invertebrate Anatomy. Oxford University Press, Oxford and New York.

Sherman, I. W., and Sherman, V. G., 1976. The Invertebrates. Function and Form. 2nd edition. Macmillan, New York.

Wallace, R. L., Taylor, W. K., and Litton, J. R., 1989. Beck and Braithwaite's Invertebrate Zoology. A Lab Manual. New York, Macmillan.

REPRODUCTION AND DEVELOPMENT

Adiyodi, K. G., and Adiyodi, R. G. (editors), 1983– . Reproductive Biology of Invertebrates. John Wiley & Sons, New York.
Three volumes have been published.

Chia, F.-S., and Rice, M. E. (editors), 1978. Settlement and Metamorphosis of Marine Invertebrate Larvae. Elsevier, New York.

Giese, A. D., and Pearse, J. S., 1973– . Reproduction of Marine Invertebrates. Academic Press, New York (volumes 1–5), and Blackwell Scientific Publishers, Palo Alto, and Boxwood Press, Pacific Grove, California (volume 9). The set should be completed soon. Nearly all aspects of reproduction and development, from formation of gametes to larval metamorphosis, are covered on a phylum-by-phylum basis.

Kume, M., and Dan, K., 1968. Invertebrate Embryology. Translated by J. C. Dan. NOLIT, Belgrade.
(Out of print, but available in photocopy form from Clearinghouse for Federal Scientific and Technical Information, Springfield, Virginia.)

Mileikovsky, S. A., 1971. Types of larval development in marine bottom invertebrates, their distribution and ecological significance: a re-evaluation. Marine Biology, *10:* 193–213.

Reverberi, G. (editor), 1971. Experimental Embryology of Marine and Fresh-Water Invertebrates. North Holland Publishing Company, Amsterdam and London; American Elsevier Publishing Company, New York.

LOCOMOTION, PHYSIOLOGY, AND BEHAVIOR

Alexander, R. McN., 1979. The Invertebrates. Cambridge University Press, Cambridge.

Bullock, T. H., and Horridge, G. A., 1965. Structure and Function of the Nervous Systems of Invertebrates. 2 volumes. W. H. Freeman and Co., San Francisco. The standard reference work on invertebrate nervous systems.

Corning, W. C., Dyal, J. A., and Willows, A. O. D., 1973–75. Invertebrate Learning. 3 volumes. Plenum Press, New York and London.

Florkin, M., and Scheer, B. T. (editors), 1967–79. Chemical Zoology. 11 volumes. Academic Press, Inc., New York.
A collection of reviews on biochemistry and physiology, covering most groups of invertebrates and vertebrates.

Some perspective is provided by brief treatments on functional morphology. The quality is uneven, and certain of the chapters seem to have suffered from poor translation.
Grant, P. T., and Mackie, A. M., 1974. Chemoreception in Marine Organisms. Academic Press, Inc., London and New York.
Gray, J., 1968. Animal Locomotion. Weidenfeld and Nicholson, London.
Highnam, K. C., and Hill, L., 1977. The Comparative Endocrinology of the Invertebrates. 2nd edition. University Park Press, Baltimore.
Jennings, J. B., 1972. Feeding, Digestion, and Assimilation in Animals. 2nd edition. Macmillan, London; St. Martin's Press, New York.
Jorgenson, C. B., 1966. Biology of Suspension Feeding. Pergamon Press, Oxford and New York.
Laverack, M. S., 1968. On the receptors of marine invertebrates. Oceanography and Marine Biology, An Annual Review, *6:*249–324.
Mill, P. J., 1972. Respiration in the Invertebrates. Macmillan, London.
Potts, W. T. W., 1964. Osmotic and Ionic Regulation in Animals. Pergamon Press, Oxford and New York.
Prosser, C. L. (editor), 1973. Comparative Animal Physiology. 3rd edition. W. B. Saunders Co., Philadelphia.
Tombes, A. S., 1970. An Introduction to Invertebrate Endocrinology. Academic Press, Inc., New York.
Trueman, E. R., 1975. The Locomotion of Soft-Bodied Animals. Edward Arnold, London.
Vernberg, W. B., and Vernberg, F. J., 1972. Environmental Physiology of Marine Animals. Springer-Verlag, New York.
Vogel, S., 1981. Life in Moving Fluids: The Physical Biology of Flow. W. Grant Press, Boston.
Wainwright, S. A., Biggs, W. D., Currey, J. D., and Gosline, J. M., 1976. Mechanical Design in Organisms. Edward Arnold, London.
Wells, M. J., 1965. Learning by marine invertebrates. Advances in Marine Biology, *1:*1–62.

ECOLOGY

Carefoot, T., 1977. Pacific Seashores. A Guide to Intertidal Ecology. J. J. Douglas, Ltd., North Vancouver, British Columbia; University of Washington Press, Seattle and London.
Eltringham, S. K., 1971. Life in Mud and Sand. Crane, Russak and Company, New York; English Universities Press, London.
Green, J., 1968. The Biology of Estuarine Animals. University of Washington Press, Seattle and London.
Levinton, J. S., 1982. Marine Ecology. Prentice-Hall, Englewood Cliffs, New Jersey.
Newell, R. C., 1979. Biology of Intertidal Animals. 3rd edition. Marine Ecological Surveys Ltd., Faversham, Kent, England.
Schaller, F., 1968. Soil Animals. University of Michigan Press, Ann Arbor.
Stephenson, T. A., and Stephenson, A., 1972. Life Between Tidemarks on Rocky Shores. W. H. Freeman, San Francisco.
Swedmark, B., 1964. The interstitial fauna of marine sand. Biological Reviews, *39:*1–42.

PLANKTON

Hardy, A., 1965. The Open Sea: Its Natural History. Houghton Mifflin Company, Boston.
Newell, B. E., and Newell, R. C., 1963. Marine Plankton. Hutchinson Educational Ltd., London.

PARASITOLOGY

Cheng, T. C., 1986. General Parasitology. 2nd edition. Academic Press, Orlando, Florida.
Noble, E. R., and Noble, G. A., 1971. Parasitology. The Biology of Animal Parasites. 3rd edition. Lea and Febiger, Philadelphia.
Olsen, O. W., 1974. Animal Parasites. Their Life Cycles and Ecology. 3rd edition. University Park Press, Baltimore.
Schmidt, G. D., and Roberts, L. S., 1989. Foundations of Parasitology. 4th edition. Times Mirror/Mosby College Publishing, Saint Louis.

PALEONTOLOGY

Boardman, R. S., Cheetham, A. H., and Rowell, J., 1987. Fossil Invertebrates. Blackwell Scientific Publications, Palo Alto, California.
Moore, R. C. (editor), 1952– . Treatise on Invertebrate Paleontology. Geological Society of America and University of Kansas Press, Lawrence, Kansas.
Tasch, P., 1973. Paleobiology of the Invertebrates. John Wiley & Sons, New York and London.

EVOLUTION AND PHYLOGENY

Hadzi, J., 1963. The Evolution of the Metazoa. Pergamon Press, Oxford.
Hanson, E. D., 1977. The Origin and Early Evolution of Animals. Wesleyan University Press, Middletown, Connecticut.
Jagersten, G., 1972. Evolution of the Metazoan Life Cycle. A Comprehensive Theory. Academic Press, Inc., London and New York.
Morris, S. C., George, J. D., Gibson, R., and Platt, H. M. (editors), 1985. The Origins and Relationships of Lower Invertebrates. (Systematics Association Special Volume no. 28.) Clarendon Press, Oxford.
Salvini-Plawen, L. von, 1978. On the origin and evolution of the lower Metazoa. Zeitschrift für Zoologische Systematik und Evolutionsforschung, *16:*40–87.

GUIDES FOR IDENTIFICATION

Fresh Water

Edmondson, W. T., Ward, H. B., and Whipple, G. C. (editors), 1959. Freshwater Biology. 2nd edition. John Wiley and Sons, New York.

Pennak, R. W., 1978. Freshwater Invertebrates of the United States. 2nd edition. John Wiley and Sons, New York.

Marine

North America

Brusca, R. C., 1980. Common Intertidal Invertebrates of the Gulf of California. University of Arizona Press, Tucson.

Gosner, K. L., 1971. Guide to Identification of Marine and Estuarine Invertebrates. Interscience Publishers, New York.
Covers much of the Atlantic coast of North America.

Gosner, K. L., 1978. A Field Guide to the Atlantic Seashore from the Bay of Fundy to Cape Hatteras. Houghton Mifflin Company, Boston.

Kozloff, E. N., 1983. Seashore Life of the Northern Pacific Coast. University of Washington Press, Seattle.

Kozloff, E. N., 1987. Marine Invertebrates of the Pacific Northwest. University of Washington Press, Seattle.

Morris, R. H., Abbott, D. P., and Haderlie, E. C. (editors), 1980. Intertidal Invertebrates of California. Stanford University Press, Stanford, California.

Ruppert, E. E., and Fox, R. S., 1988. Seashore Animals of the Southeast. University of South Carolina Press, Columbia.

Smith, R. I., and Carleton, J. T. (editors), 1975. Light's Manual: Intertidal Invertebrates of the Central California Coast. University of California Press, Berkeley, Los Angeles, and London.

Caribbean Region

Colin, P. C., 1978. Caribbean Reef Invertebrates and Plants. T. H. F. Publishers, Neptune City, Florida.

Kaplan, E. H., 1982. A Field Guide to Coral Reefs of the Caribbean and Florida Including Bermuda and the Bahamas. Houghton Mifflin Company, Boston.

Sefton, N., and Webster, S. K., 1986. A Field Guide to Caribbean Reef Invertebrates. Sea Challengers, Monterey, California; E. J. Brill, Leiden.

Voss, R. H., 1976. Seashore Life of Florida and the Caribbean. E. A. Seeman, Miami.

Zeiller, W., 1974. Tropical Marine Invertebrates of Southern Florida and the Bahama Islands. Wiley-Interscience, New York.

European Atlantic Region

Barrett, J. H., and Yonge, C. M., 1958. Collins Pocket Guide to the Seashore. Collins, London.

Campbell, A. C., 1976. The Hamlyn Guide to the Seashore and Shallow Seas of Britain and Europe. Hamlyn Publishing Group, London.

Eales, N. B., 1967. The Littoral Fauna of the British Isles. 4th edition. Cambridge University Press.

Yonge, C. M., 1966. The Seashore. 2nd edition. Collins, London.

Mediterranean Region

Luther, W., and Fiedler, K., 1976. A Field Guide to the Mediterranean Seashore. Collins, London.

Riedl, R. (editor), 1970. Fauna und Flora der Mittelmeeres. Paul Parey, Hamburg.

South African Region

Branch, G., and Branch, M., 1981. The Living Shores of Southern Africa. C. Struik Publishers, Cape Town.

Day, J. H., 1969. A Guide to Marine Life on South African Shores. 2nd edition. A. A. Balkema, Cape Town.

Indo-Pacific, Australian, and Zelandic Region

Dakin, W. J., 1960. Australian Seashores. Angus and Robertson, Sydney.

Hobson, C., and Chave, E. G., 1972. Hawaiian Reef Animals. Hawaii University Press, Honolulu.

Morton, J., and Miller, M., 1968. The New Zealand Seashore. Collins, Auckland.

GUIDES TO PERIODICAL LITERATURE

Biological Abstracts. 1926– . BioSciences Information Service, Philadelphia.
Reasonably up-to-date, providing (whenever possible) a brief summary of the contents of each article to which a reference is given. Annual indexes according to author and subject.

Current Contents: Life Sciences, 1957– ; Agriculture, Biology and Environmental Sciences, 1969– . Institute for Scientific Information, Philadelphia.
Published weekly. Reproduces tables of contents of hundreds of journals about the same time as the journals themselves appear. Indexing, according to author and subject, is provided only for individual issues.

Marine Science Contents Tables. 1964– . Food and Agriculture Organization of the United Nations, Rome.
Published monthly, reproducing tables of contents of current issues of journals that deal with marine biology and oceanography.

Zoological Record. 1864– . Zoological Society of London.
The most useful single source of the literature on zoology. An attempt is made to list substantially all publications on an annual basis. Issued in parts, each part dealing with a single phylum, group of phyla, or other major assemblage (Mollusca, "Vermes," Crustacea, etc.). Marvelously cross-indexed, so that contributions on ecology, physiology, reproduction, systematics, and other topics are easily located. The coverage is nearly complete, but generally the parts for a particular year do not begin to appear until two or three years later.

2

PROTOZOANS

- Phylum Sarcomastigophora
  - Subphylum Mastigophora
    - Class Phytomastigophorea
      - Order Dinoflagellida
      - Order Euglenida
      - Order Volvocida
    - Class Zoomastigophorea
      - Order Choanoflagellida
      - Order Kinetoplastida
      - Order Diplomonadida
      - Order Trichomonadida
      - Order Hypermastigida
  - Subphylum Sarcodina
    - Superclass Rhizopodea
      - Class Lobosea
        - Subclass Gymnamoebia
        - Subclass Testacealobosia
      - Class Filosea
      - Class Granuloreticulosea
    - Superclass Actinopodea
      - Class Polycystinea
      - Class Phaeodarea
      - Class Acantharea
      - Class Heliozoea
- Phylum Apicomplexa
  - Class Gregarinea
  - Class Coccidea
    - Order Coccidiida
    - Order Haemosporida
- Phylum Microspora
- Phylum Myxozoa
  - Class Myxosporea
  - Class Actinosporea
- Phylum Ciliophora
  - Class Kinetofragminophorea
    - Subclass Gymnostomatia
    - Subclass Vestibuliferia
    - Subclass Hypostomatia
    - Subclass Suctoria
  - Class Oligohymenophorea
    - Subclass Hymenostomatia
    - Subclass Peritrichia
  - Class Polyhymenophorea
    - Order Heterotrichida
    - Order Oligotrichida
    - Order Hypotrichida
- Phylum Opalinata

INTRODUCTION

Most biologists concerned with the diversity of living things recognize five kingdoms of organisms. The kingdom **Monera** consists of bacteria and blue-green algae, which are said to be **prokaryotic** (''prenuclear'') because the nuclear material in their cells is not enclosed by a membrane. Monerans also lack mitochondria, and their chlorophyll, when they have any, is not localized in plastids. If flagella are present, as they are in many bacteria, they are of a simple type found only in this group of microbes.

The other four kingdoms—**Protoctista, Fungi, Plantae,** and **Animalia**—consist of **eukaryotic** (''true nuclear'') organisms. Their cells have true nuclei, and they almost always have mitochondria. Chlorophyll, if present, is packaged in chloroplasts, and the flagella or cilia of eukaryotes conform, with few exceptions, to a structural plan that includes two central microtubules and nine pairs of peripheral microtubules.

The name Protoctista has been suggested as a replacement for *Protista,* long applied to one-celled organisms in general. The definition of protists has been broadened to include all algae, even large seaweeds, so the new name is a desirable change. The kingdom Fungi includes molds, mushrooms, smuts, and yeasts, and the scope of the kingdom Plantae extends from mosses and liverworts to flowering plants. Animalia covers all multicellular animals, from sponges to vertebrates.

The five-kingdom arrangement of living things is, for the most part, satisfactory. Monera, Fungi, Plantae, and Animalia are logical groupings. Protoctista, however, is a catch-all. It has enormous diversity, and some of the organisms in it are perhaps closer to certain fungi, lower plants, or lower animals than they are to one another. Nevertheless, our concept of Protoctista is sound in that many organisms placed in this kingdom are probably very similar to some of the early forms of eukaryotic life. It is now generally believed that eukaryotic cells, with nuclei and mitochondria, and sometimes with chloroplasts and flagella, originated through symbiotic associations of prokaryotic cells. There are, in fact, some unicellular prokaryotes in which symbiotic prokaryotes function in much the same way as mitochondria, chloroplasts, and flagella. Mitosis and meiosis also undoubtedly became established in protoctists before multicellular organisms appeared.

This chapter is concerned only with the protoctists that are called *protozoans,* of which there are about 65,000 described species. Protozoans belong to several diverse phyla, and it is therefore difficult to define them unequivocally. To say that they are one-celled and motile, and that their nutrition is more nearly like that of animals than that of plants, covers many of the protozoans. But there are protozoans that form colonies of cells, some that have no means of locomotion, and some that are **photosynthetic**. Photosynthetic protozoans, however, have flagella, at least during certain stages of the life cycle, and this separates them from one-celled photosynthetic eukaryotes that are not motile and that are conventionally classified as algae.

Protozoans will here be dealt with as one-celled organisms, because the cells of colonial types usually show little or no division of labor. Nevertheless, unicellularity must not be equated with structural or physiological simplicity. Most protozoans are more complex and more versatile than any cells of multicellular organisms, and what some of them do with the resources of a single cell is remarkable.

CHARACTERISTICS OF PROTOZOAN CELLS

Among the many different kinds of protozoans, we find all of the basic components of plant and animal cells. Besides one or more nuclei, there are the organelles called mitochondria, chloroplasts, and flagella or cilia. Golgi material and an endoplasmic reticulum are also present, and there are often centrioles. Lysosomes, which are derived from Golgi material and which contain hydrolytic enzymes, are found in protozoans that ingest their food.

Almost unique to protozoans are **contractile vacuoles,** concerned with regulation of the water and salt balance of the cell. If the salt content of the cell is higher than that of the surrounding medium, water enters the cell by diffusion more rapidly than it leaves. The function of the contractile vacuole is to collect the excess water and eliminate it. A contractile vacuole may be formed by the coalescence of several smaller vacuoles, or excess water may be fed to it from a complex of fine tubules that pervade the cytoplasm. After it fills, it discharges the water through a pore in the cell membrane. Contractile vacuoles are almost always characteristic of freshwater protozoans, but they also occur in many marine species.

The devices that protozoans use for feeding are varied. Pseudopodia, suctorial tentacles, and mouthlike apertures for localized ingestion of food are the major types of structures involved; these are discussed in some detail in connection with particular groups.

FLAGELLA AND CILIA

Next to nuclei, the most nearly universal protozoan organelles readily visible with the light microscope are **flagella** and **cilia,** which are used in locomotion. The ultrastructure of flagella and cilia is similar, and some biologists apply the term *undulipodium* to both. It is more conventional, however, to use the term *flagellum* when a cell has only one or a few of them, when they are relatively long, and when their activity is undulatory, either in one plane or in a spiral (Figure 2.1). *Cilium* is the preferred term when a cell has many of them, when they are relatively short, and when they

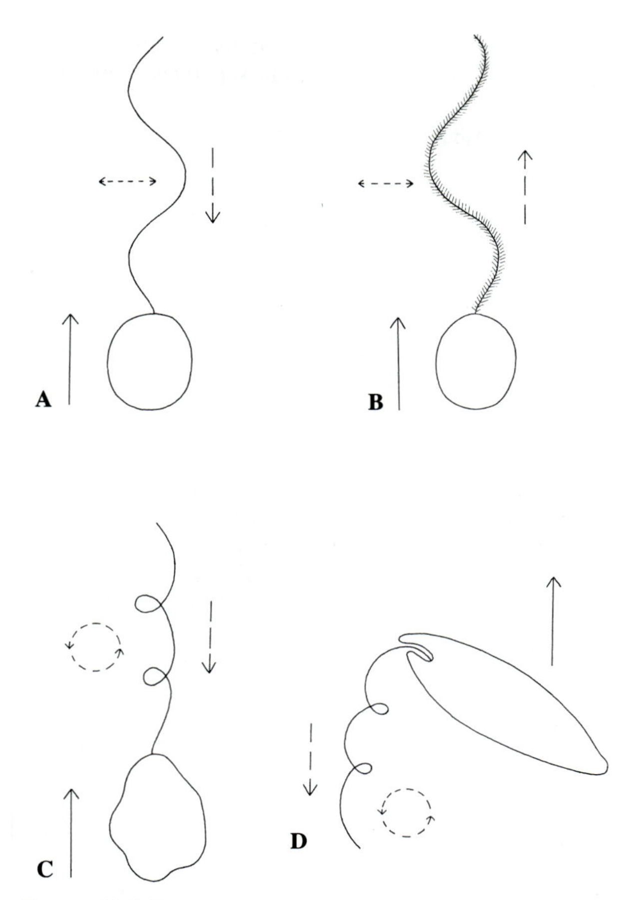

FIGURE 2.1 Some ways in which flagella bring about movement in protozoans. The larger arrow with a dashed stem indicates the direction of the flagellar wave, and the smaller arrow with dashed stem indicates the plane in which the flagellum beats (back and forth or in a circular pattern); the arrow with a solid stem indicates the direction of movement of the organism. A. *Strigomonas*. B. *Ochromonas* (its flagellum is of the tinsel type, with delicate outgrowths called mastigonemes). C. *Mastigamoeba*. D. *Euglena*. (After Sleigh, in Bittar, Cell Biology in Medicine.)

make rapid effective strokes that alternate with slower recovery strokes (Figure 2.2). To some extent, the rule just stated for distinguishing flagella from cilia is followed in connection with vibratile organelles of multicellular animals, although there are exceptions. For instance, even though a cell has only one vibratile organelle, it is usually called a cilium if it is short and beats like a typical cilium of a protozoan.

In some protozoans, one or more flagella or cilia are continuously present; in others they appear only at certain stages in the life cycle or under certain environmental conditions. Regardless of how many are characteristic of a particular species, each has its own **kinetosome,** or basal body, usually located close to the cell surface (Figure 2.3). Electron microscopy reveals that the flagellum or cilium proper, which is a slender cytoplasmic outgrowth, has nine pairs of peripheral microtubules and one pair of microtubules at the center. Transverse sections of a kinetosome show nine trios of peripheral microtubules, but no microtubules in the center. In other words, when the flagellum or cilium reaches the kinetosome—there is commonly an axial granule at this point—the central microtubules drop out, and a third microtubule is added to each of the peripheral pairs. Extending away from the kinetosome is a cross-banded fiber called the **kinetodesma,** the flagellar or ciliary rootlet. In specimens that have been stained for light microscopy, the kinetosome sometimes shows up as a tiny dot, and if the kinetodesma is relatively large, it may also be visible.

NUTRITION

The remarkable diversity of protozoans is evident in what they eat and how they get it. Most of these organisms are **heterotrophic,** which means they require organic foods they cannot make for themselves. Some take up soluble nutrients or macromolecules by **pinocytosis,** accomplished by filling membrane-bound vesicles at the surface of the cell. The vesicles are then

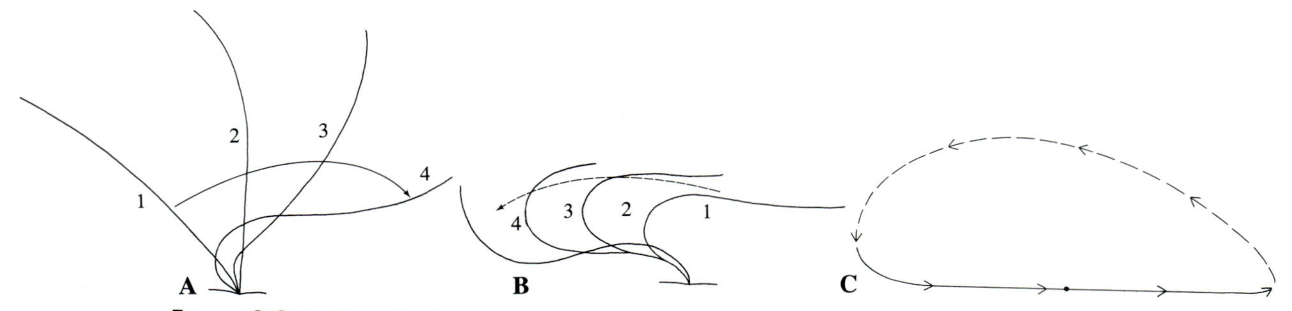

FIGURE 2.2 A. The effective, motion-producing stroke of a cilium of a ciliate. B. The recovery stroke. C. The path of the *tip* of the cilium during the effective stroke (solid line) and recovery stroke (dashed line). While the cilium is making the recovery stroke, its tip traces a more curved path than it does during the effective stroke.

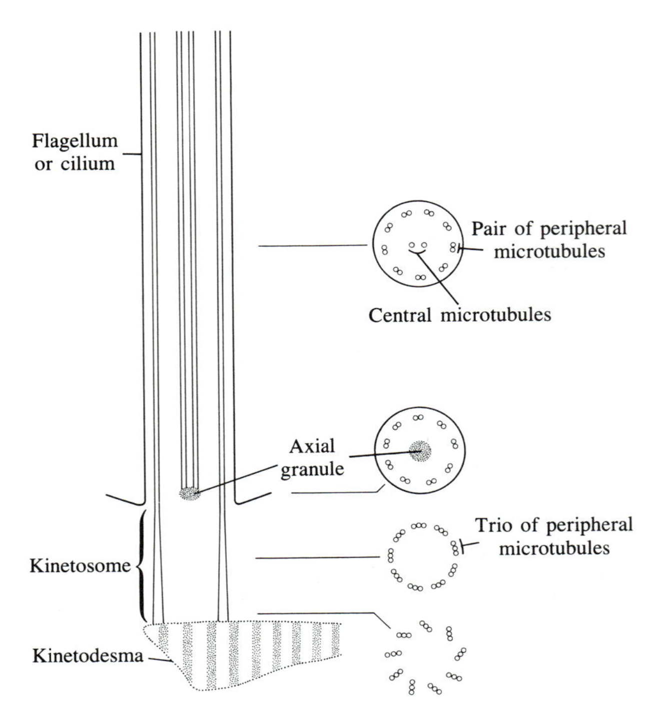

FIGURE 2.3 Structure of a flagellum (or cilium) and its kinetosome and kinetodesma, based on electron micrographs. At the left, a longitudinal section; at the right, transverse sections taken at four levels.

pinched off and sink into the cytoplasm, carrying the nutrients with them. The formation of food vacuoles in protozoans that eat other protozoans, bacteria, or the cells of disintegrating organisms is somewhat similar. A food vacuole develops, for instance, when an amoeba surrounds its prey with cytoplasmic extensions called pseudopodia, or when a ciliate such as *Paramecium* concentrates bacteria in a food-gathering depression called the **buccal cavity.** Membrane-bound vacuoles formed in the deepest part of the buccal cavity are pinched off and circulate through the cytoplasm.

As a rule, the earlier phases of enzymatic digestion, which take place when lysosomes fuse with a food vacuole, proceed under acid conditions. Later phases are completed under alkaline conditions. In time, small subsidiary vesicles are pinched off around the food vacuole, and it is apparently by way of these that the products of digestion are turned over to the cytoplasm. Undigested residues are usually eliminated through the cell membrane, sometimes at a particular place specialized for this purpose.

All of the **autotrophic** protozoans belong to the general group called flagellates. Having chlorophyll, they are photosynthetic and make the organic compounds they need from carbon dioxide, water, and mineral salts. Not all of them are completely autotrophic, however. Some, such as *Euglena,* require at least one vitamin that they are unable to synthesize for themselves.

RESPIRATION AND EXCRETION

Because protozoans are small, the amount of surface area in proportion to volume is adequate for inward diffusion of oxygen and outward diffusion of carbon dioxide. Some protozoans live in habitats where there is little or no free oxygen and they must obtain energy from metabolic processes that are carried out anaerobically. In this respect they are similar to many bacteria and also to brewer's yeast, whose fermentative conversion of sugar to ethyl alcohol and carbon dioxide is its way of getting energy when free oxygen is not available.

Small cell size favors rapid elimination not only of carbon dioxide but also of other soluble metabolic wastes, such as ammonia. Many species, nevertheless, store up wastes in the form of crystals that have such low solubility that they are nontoxic.

REPRODUCTION

One good reason for studying protozoans, as well as other Protoctista, is to appreciate the variety of methods, sexual and asexual, by which they reproduce. Most of the diverse groups are defined partly, if not primarily, by their life cycles, often characterized by regular alternation of sexual and asexual processes.

ASEXUAL REPRODUCTION

A common form of asexual reproduction is **binary fission,** in which nuclear division is followed by division of the cytoplasm, so that two more or less equal individuals are produced. The process is achieved rather simply in some protozoans. In others, especially ciliates, fission involves extensive reorganization in both prospective daughters.

Multiple fission, or **merogony** (also called *schizogony*), is a type of asexual reproduction in which a few to many nuclear divisions are followed by division of the cytoplasm, resulting in the formation of numerous progeny. Among the organisms that reproduce by merogony are those that cause malaria.

SEXUAL REPRODUCTION

Sexual processes, essential to recombination of genes, occur in most of the major assemblages of protozoans. Sexual union sometimes takes place between two individuals that look like all other members of the species. Many protozoans, however, produce distinct **gametes** at a particular stage in the life cycle. The two kinds of gametes may be nearly or completely similar in appearance, or they may be of different size and structure and therefore comparable to eggs and sperm.

In the protozoans called ciliates, sexual union of a type called conjugation typically continues only as long as required for reciprocal transfer of gametic nuclei. It should be emphasized that sexual processes do not always lead directly to an increase in numbers, but they do maintain genetic diversity.

Most protozoans that have sexual reproduction are haploid except for the zygote stage, which, having been produced by fusion of two cells, is diploid. Meiosis—the process of nuclear division by which the chromosome complement is reduced from diploid to haploid—normally takes place in the first two divisions of the zygote. Ciliates are radically different in this respect. They are like metazoan animals in being diploid. Their meiotic divisions are accomplished during the formation of nuclei that function as gametes, and diploidy is restored when the gametic nuclei unite.

CLASSIFICATION

The characterization of major groups of protozoans is based primarily on their structure, the ways in which they obtain food, their means of locomotion, and their reproduction. It is important to appreciate, however, that a particular group may show wide variation in one or more of these aspects of its biology. For instance, among the many kinds of flagellates that subsist heterotrophically on organic matter, some are nearly identical to certain types that have chloroplasts and make their own food by photosynthesis. There are also flagellates that lose their flagella and assume the appearance and style of locomotion typical of some amoebae.

Because we do not know how closely or how distantly most of the major groups and their subdivisions are related, any comprehensive system of classification of protozoans and other protoctists is necessarily tentative. A practical scheme for protozoans was developed by a committee of the Society of Protozoologists (Levine *et al.*, 1980). This is adhered to, with some modifications, in a guide to the protozoans published by the Society (Lee, Hutner, and Bovee, editors, 1985). The scheme is to some extent artificial, for the reasons just given, but it is the best we have at the present time. In the following sections of this chapter, whose objective is to provide a broad view of the protozoans, all of the phyla and subphyla recognized in the Society of Protozoologists' classification are mentioned but some of the classes and other subsidiary taxa are omitted.

GENERAL REFERENCES ON PROTOZOANS

Aikawa, M., and Sterling, C. H., 1974. Intracellular Parasitic Protozoa. Academic Press, New York and London.

Anderson, O. R., 1988. Comparative Protozoology: Ecology, Physiology, Life History. Springer-Verlag, New York and Berlin.

Baker, J. R., 1969. Parasitic Protozoa, Hutchinson University Library, London.

Chen, T. T. (editor), 1967–1970. Research in Protozoology. 4 volumes. Pergamon Press, London.

Dodge, J. D., 1973. The Fine Structure of Algal Cells. Academic Press, London and New York.

Farmer, J. N., 1980. The Protozoa: Introduction to Protozoology. C. V. Mosby, St. Louis.

Fenchel, T., 1986. Ecology of Protozoa: The Biology of Free-living Phagotrophic Protists. Science Tech Publishers, Madison, Wisconsin.

Grell, K. G., 1973. Protozoology. Springer-Verlag, New York, Heidelberg, Berlin.

Kudo, R. R., 1966. Protozoology. 5th edition. Charles C Thomas, Springfield, Illinois.

Lee, J. J., Hutner, S. H., and Bovee, E. C. (editors), 1985. An Illustrated Guide to the Protozoa. Society of Protozoologists, Lawrence, Kansas.

Levandowsky, M., and Hutner, S. H. (editors), 1979–1981. Biochemistry and Physiology of Protozoa. 2nd edition. 3 volumes. Academic Press, New York and London.

Levine, N. D., 1973. Protozoan Parasites of Domestic Animals and of Man. 2nd edition. Burgess Publishing Company, Minneapolis.

Levine, N. D. *et al.* (The Committee on Systematics and Evolution of the Society of Protozoologists), 1980. A newly revised classification of Protozoa. Journal of Protozoology, *27:* 37–58.

Mackinnon, D. L., and Hawes, R. S. J., 1961. An Introduction to the Study of Protozoa. Clarendon Press, Oxford.

O'Day, D. H., and Horgen, P. A. (editors), 1981. Sexual Interactions in Eukaryotic Microbes. Academic Press, New York and London.

Puytorac, P. de, Grain, J., and Mignot, J.-P., 1987. Précis de Protistologie. Éditions N. Boubée, Paris.

Ragan, M. A., and Chapman, O. J. (editors), 1978. A Biochemical Phylogeny of the Protists. Academic Press, New York and London.

Sleigh, M. A., 1973. The Biology of Protozoa. Edward Arnold, London.

Sleigh, M. A. (editor), 1974. Cilia and Flagella. Academic Press, London and New York.

Vickerman, K., and Cox, F. E. G., 1967. The Protozoa. John Murray, London.

Westphal, A., 1976. Protozoa. Blackie, Glasgow and London; International Ideas, Inc., Philadelphia.

PHYLUM SARCOMASTIGOPHORA

The Sarcomastigophora includes nearly all protozoans that have either flagella or pseudopodia, or both. It is an enormously diversified assemblage, and some biologists prefer to divide it into several phyla, one for each distinctive and perhaps well-unified group.

SUBPHYLUM MASTIGOPHORA

Mastigophora means "whip-bearers" and is synonymous with an older name, Flagellata. The organisms in this group are hereafter referred to as *flagellates*.

Most of them have flagella at some phase of the life cycle, if not permanently; the few that lack flagella exhibit other characteristics, such as the pattern of nuclear division, that indicates a close relationship with flagellates. The number of flagella usually ranges from one to several, and when there are more than one, they generally originate close together. The only exceptions to this statement are certain flagellates that inhabit the guts of termites and roaches; they have many flagella distributed over much of the body surface. These remarkable and highly evolved protozoans are dealt with later on.

While most flagella are cylindrical, some are ribbonlike. In various groups of flagellates, moreover, there are so-called tinsel flagella, characterized by extremely slender outgrowths called **mastigonemes** (Figure 2.1, B).

Nutrition is varied. Some flagellates are autotrophic; they are like green plants in having chlorophyll and in being able to synthesize the organic nutrients they need from inorganic compounds. Others are strictly heterotrophic; they must ingest or absorb organic material. A number of flagellates are said to be **mixotrophic,** because they have the capacity for both autotrophic and heterotrophic nutrition.

CLASS PHYTOMASTIGOPHOREA

The Phytomastigophorea includes all flagellates that are photosynthetic, and some that lack chlorophyll but that are believed to be derived from photosynthetic types. Only a few orders of this large and varied assemblage are described here.

Order Dinoflagellida

Dinoflagellates (Figures 2.4, 2.5) are found in the sea and in freshwater lakes and ponds. They typically have two flagella, and these are arranged in a distinctive way: one, which is usually ribbonlike, lies in the **annulus,** or girdle, a groove that encircles the body; the other lies in the **sulcus,** a groove that extends backward. This arrangement of flagella gives dinoflagellates a characteristically spiral swimming motion. The nucleus is interesting in that its chromosomes are recognizable during interphase, as well as during stages of mitosis. Some genera, such as *Ceratium* (Figure 2.4, B), are said to be armored, because they have a cellulose wall that consists of two or more plates; others, including *Gymnodinium* (Figure 2.4, A), lack a cell wall.

The plastids of photosynthetic dinoflagellates contain chlorophylls *a* and *c,* rather than chlorophylls *a* and *b,* characteristic of euglenid and volvocid flagellates (dealt with in the next two sections) and also of green algae and higher plants. The chlorophylls are partly masked by a brown pigment called **peridinin,** so the overall color of photosynthetic types is yellow-brown or olive. Many dinoflagellates lack chlorophyll, and these either ingest their food, which consists of other unicellular organisms, or absorb soluble organic matter.

Certain of the marine dinoflagellates, especially species of *Gonyaulax* and *Gymnodinium,* produce potent toxins responsible for mass mortality of other marine organisms. "Red tides," in which the color of the water may actually become reddish because of a high concentration of particular species of dinoflagellates, are dangerous to sea life. So-called paralytic shellfish poisoning, or PSP, which may be fatal to humans, is due to consumption of mussels or clams that have accumulated a neurotoxin present in the dinoflagellates they have eaten. The paralytic effect of the toxin is similar to that of curare, which certain South American Indians extract from plants and use on their arrows and darts. "Blooms" of dinoflagellates, including dangerous types, usually occur when the water has a good supply of vitamin B_{12}, which the dinoflagellates are unable to synthesize themselves. The vitamin is probably produced mostly by bacteria in terrestrial and freshwater habitats, then washed into the sea.

Some photosynthetic dinoflagellates live as intracellular or intercellular symbionts in corals, sea anemones, and other invertebrates. The name given to these mutualistic symbionts (Figure 2.5, A) is **zooxanthellae;** it alludes to their yellow-brown color and to the fact that they live in animals. During the symbiotic stage, they may not resemble dinoflagellates, especially if they lack flagella, but their dispersal stages are typical. Some facts about the nutritional relationships between zooxanthellae and their coral and sea anemone hosts are given in Chapter 4.

The order Dinoflagellida includes some strange parasites. For example, *Piscinoodinium limneticum* (Figure 2.4, D) lives on the skin of freshwater fishes. The parasitic phase has rootlike processes that penetrate the skin. After the organism has grown to full size, it falls off and divides repeatedly to form hundreds of **swarmers,** which are unmistakably dinoflagellates. Species of several related genera live on marine fishes, chaetognaths, and other hosts. *Haplozoon* (Figure 2.5, B) is a bizarre genus found in the gut of marine polychaete annelids. It forms colonies in which several to many cells are arranged in one or more rows behind a cell that is attached to the gut epithelium by a protrusible **stylet.** Like *Piscinoodinium* and its relatives, *Haplozoon* has been observed to produce dinoflagellate swarmers.

Noctiluca (Figure 2.5, C), though far from a typical dinoflagellate, merits discussion because it is one of the more common organisms of marine plankton. It is large—sometimes up to 1.5 mm across—and its cytoplasm is concentrated in threads that branch and interconnect (anastomose). On one side of the body is a slit-

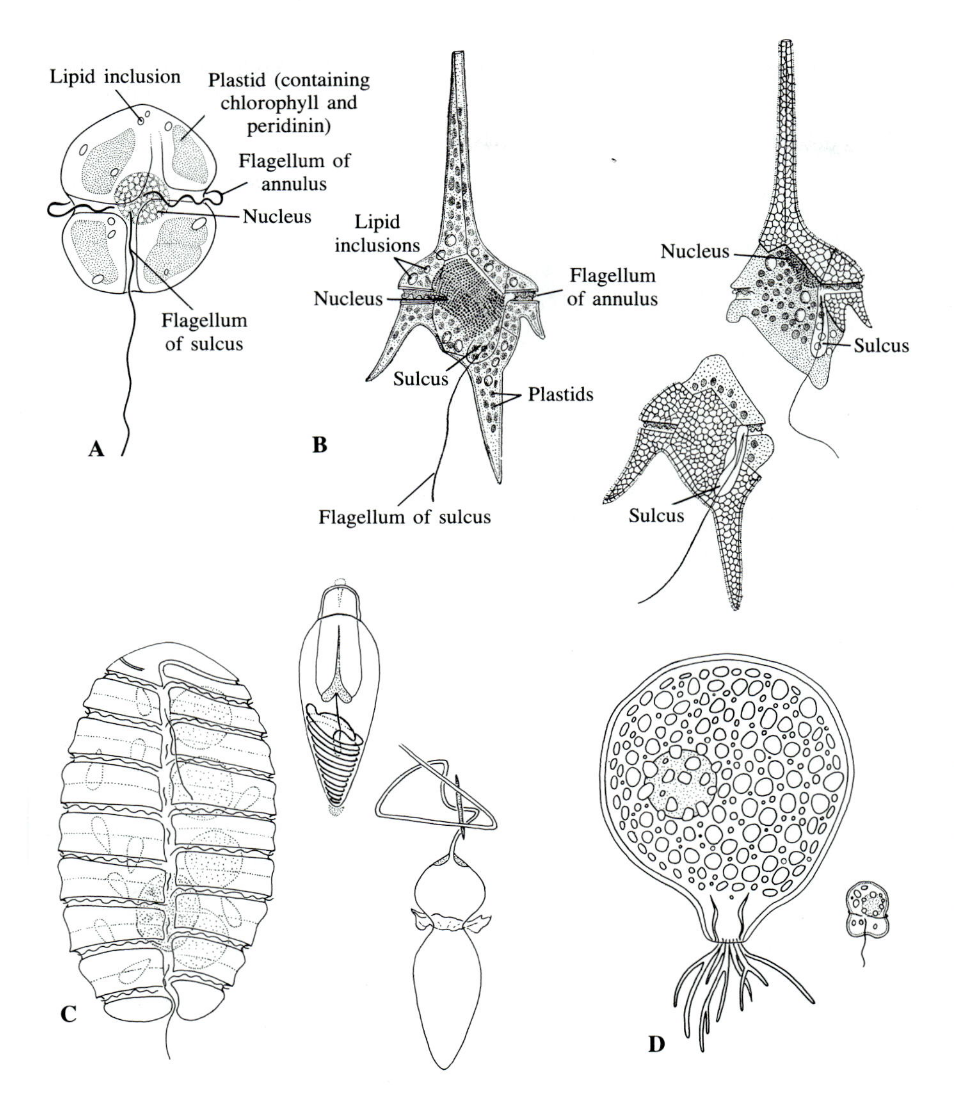

FIGURE 2.4 Order Dinoflagellida. A. *Gymnodinium.* (After Ballentine, Journal of the Marine Biological Association of the United Kingdom, *35.*) B. *Ceratium hirundinella,* at the beginning of binary fission and after the separation of the two progeny, which still do not have the full complement of plates. (Dogiel, General Protistology.) C. *Polykrikos schwartzi,* which has up to four nuclei and up to eight sets of flagella, as well as structures that function in much the same way as nematocysts of cnidarians. One of these is shown with its thread still inverted, the other after it has been "fired." (After Chatton, in Grassé, Traité de Zoologie, *1.*) D. *Piscinoodinium limneticum,* a parasite of freshwater fishes. At the left, a large trophic individual containing considerable starch and characterized by rootlike processes that penetrate the skin of the host; at the right, one of the many swarmers produced by repeated divisions of a trophic individual after it has fallen off the host. (After Jacobs, Transactions of the American Microscopical Society, *65.*)

like opening that serves for ingestion of food, mostly diatoms and small invertebrate larvae. Originating close to the mouth is a single flagellum and a thick tentacle that moves around slowly and causes the organism to rotate. The tentacle, being sticky, is also effective in capture of food. *Noctiluca* reproduces by binary fission, which is the usual mode of asexual reproduction in dinoflagellates, but it sometimes divides up into many swarmers, each with a single flagellum; these swarmers function as gametes and unite to form zygotes.

Noctiluca is able to maintain its position close to the surface because its cell sap, between the cytoplasmic threads, is less dense than seawater, due to retention of certain salts of ammonium, which are relatively light. The cell sap is very acid, having a pH of about 3. The name *Noctiluca* ("night light") was given to the organism because some of the bioluminescence observed in the sea was attributed to it. In certain areas, however, other dinoflagellates, such as those of the genus *Gonyaulax,* produce more light than *Noctiluca* does, and some strains of *Noctiluca* seem not to be bioluminescent at all.

Among the more highly specialized marine dinoflagellates are those characterized by structures that resemble **nematocysts,** the stinging capsules found in the phylum Cnidaria. A common genus with this feature is *Polykrikos* (Figure 2.4, C), which is interesting for another reason, too: it is a multiple dinoflagellate, for it usually has two or four nuclei and four or eight sets of flagella. There are many other remarkable dinoflagellates, including some that have a complicated light-sensitive organelle, complete with a glassy lens.

A

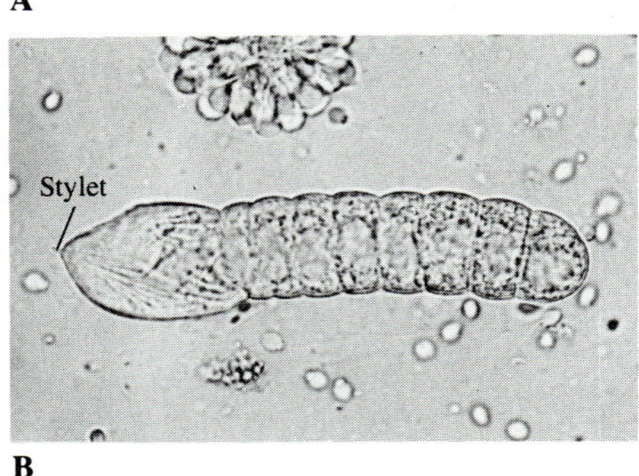

B

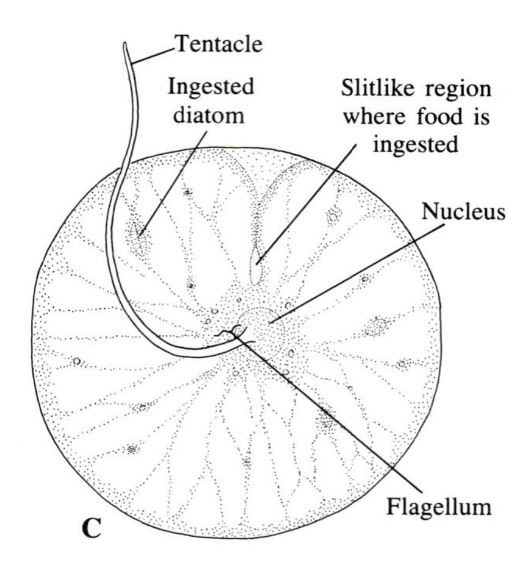

FIGURE 2.5 Order Dinoflagellida. A. Symbiotic zooxanthellae in the gastrodermal tissue of a tentacle of a sea anemone, *Anthopleura elegantissima;* photomicrograph. B. *Haplozoon axiothellae,* from the gut of *Axiothella rubrocincta,* a polychaete annelid; photomicrograph. The stylet enables the organism to maintain contact with an epithelial cell. C. *Noctiluca scintillans.*

Order Euglenida

In euglenids, the two or more flagella originate within a deep depression called the **reservoir.** These organisms are typically photosynthetic and have chlorophylls *a* and *b*. They accumulate excess carbohydrate in the form of a polysaccharide called **paramylon.** Granules of this substance form around **pyrenoid bodies,** which may be within the chloroplasts or external to them.

Species of *Euglena* (Figure 2.6, A), found mostly in fresh or brackish water, illustrate the characteristics of this group. The body is commonly 40 or 50 μm long and elongated. The shape is not constant, however, because the locomotion of *Euglena* is effected by changes in shape as well as by the action of the flagella. Beneath the cell membrane are strips of proteinaceous material arranged in a spiral pattern; these make up much of the **pellicle,** the somewhat firm but nevertheless flexible outer layer of the cell. The two flagella are decidedly unequal. The longer one, important in locomotion, extends for a considerable distance outside the reservoir; the shorter one is restricted to the inner half of the reservoir and is difficult to distinguish, especially when its tip seems to be touching the longer flagellum.

The numerous chloroplasts are oval or elongated, depending on the species. In some members of the genus, including the one shown, the chloroplasts contain the pyrenoid bodies; in others, the pyrenoid bodies, of which there may be only two, are outside the chloroplasts. The paramylon granules are often so numerous that they obscure the nucleus, usually located near the middle of the body or slightly behind the middle. Next to the deepest part of the reservoir is a contractile vacuole. Soon after it discharges its fluid into the reservoir, it is replaced by a new one, formed by coalescence of several small vacuoles that appear in the cytoplasm.

To one side of the reservoir is a mass of bright red granules. This is the **stigma,** or eyespot, which consists largely of carotenoid pigments. The actual photoreceptor is a little swelling near the base of the longer flagellum. The stigma probably serves only as a shield, essential if the photoreceptor is to detect light coming from certain directions and not from others. *Euglena* usually orients itself so that its photoreceptor is exposed to light, but it moves away from intense illumination; a moderate amount of light suits it best.

In the light, *Euglena* is photosynthetic, using carbon dioxide and water as raw materials for synthesis of carbohydrate, and getting nitrogen and other elements from mineral salts. It is not a perfect autotroph, however, for it must be supplied with vitamin B_{12}, which is

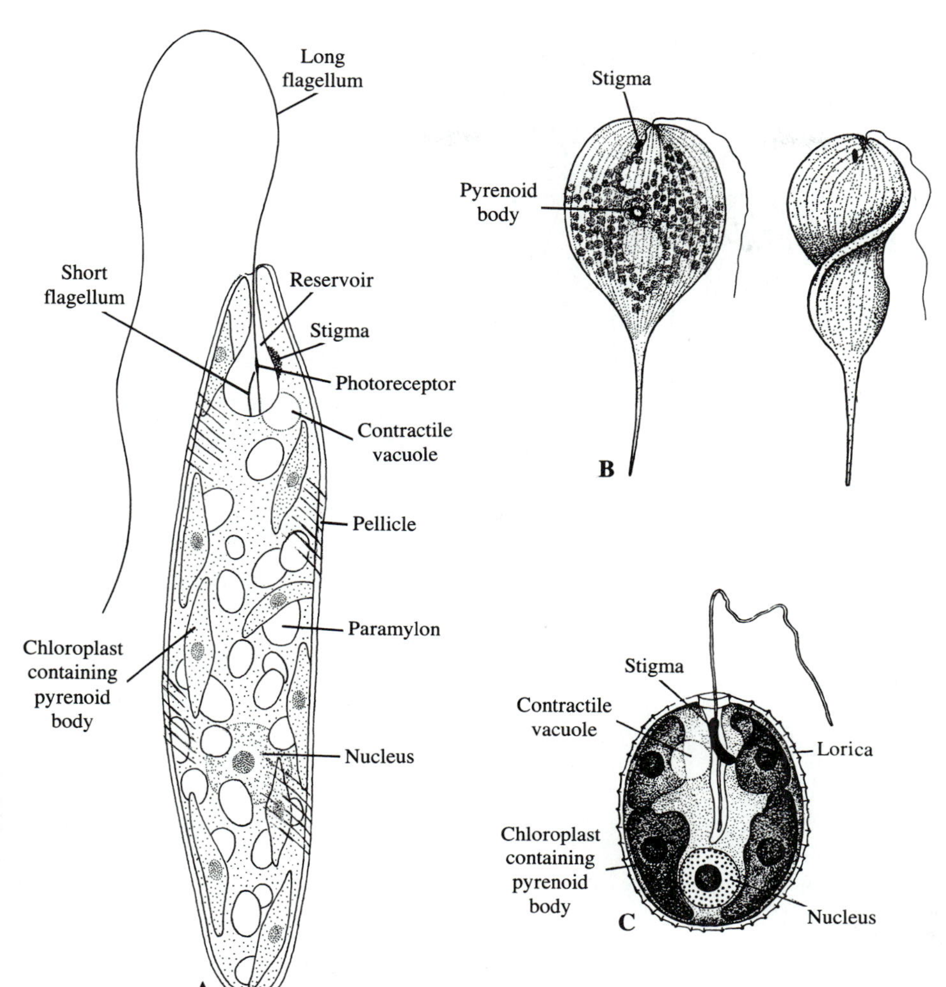

FIGURE 2.6 Order Euglenida. A. *Euglena gracilis*. (Only a few of the pellicular strips are indicated.) B. *Phacus*. (Allegre and Jahn, Transactions of the American Microscopical Society, *62*.) C. *Trachelomonas*. (Kühn, Morphologie der Tiere in Bildern, *1*.)

synthesized by bacteria and other microorganisms. When deprived of light, *Euglena* can subsist on dissolved organic substances.

Reproduction is by equal binary fission. Mitosis of the nucleus is followed by lengthwise division of the body. This requires formation of a second set of flagella within the reservoir of the parent flagellate, so that one long and one short flagellum end up in the reservoir of each daughter *Euglena*.

In some euglenids, both flagella are long, and in certain genera there are several flagella. The pellicle is not always as flexible as it is in *Euglena;* in *Phacus* (Figure 2.6, B), for instance, the body cannot change shape to any appreciable extent. *Trachelomonas* (Figure 2.6, C) and some other genera secrete a rigid proteinaceous covering, called a **lorica.**

A few genera lack chloroplasts and depend on organic nutrients. *Astasia* is especially similar to *Euglena,* and almost certainly has evolved from this or a closely related genus. Species of *Euglena* will lose their chlorophyll if they are cultivated in a rich organic medium, kept in the dark too long, or treated with certain drugs, such as streptomycin. Being unable to form chloroplasts, they become permanently heterotrophic.

Order Volvocida

Volvocid flagellates—botanists often place these in the Volvocales, an order of green algae—usually have two or four equal flagella and a proportionately large, cup-shaped chloroplast. They have chlorophylls *a* and *b* and convert excess carbohydrate into starch. The pyrenoid bodies, around which the starch accumulates, are within the chloroplast. When a reddish stigma, or eyespot, is present, this is also inside the chloroplast. Like

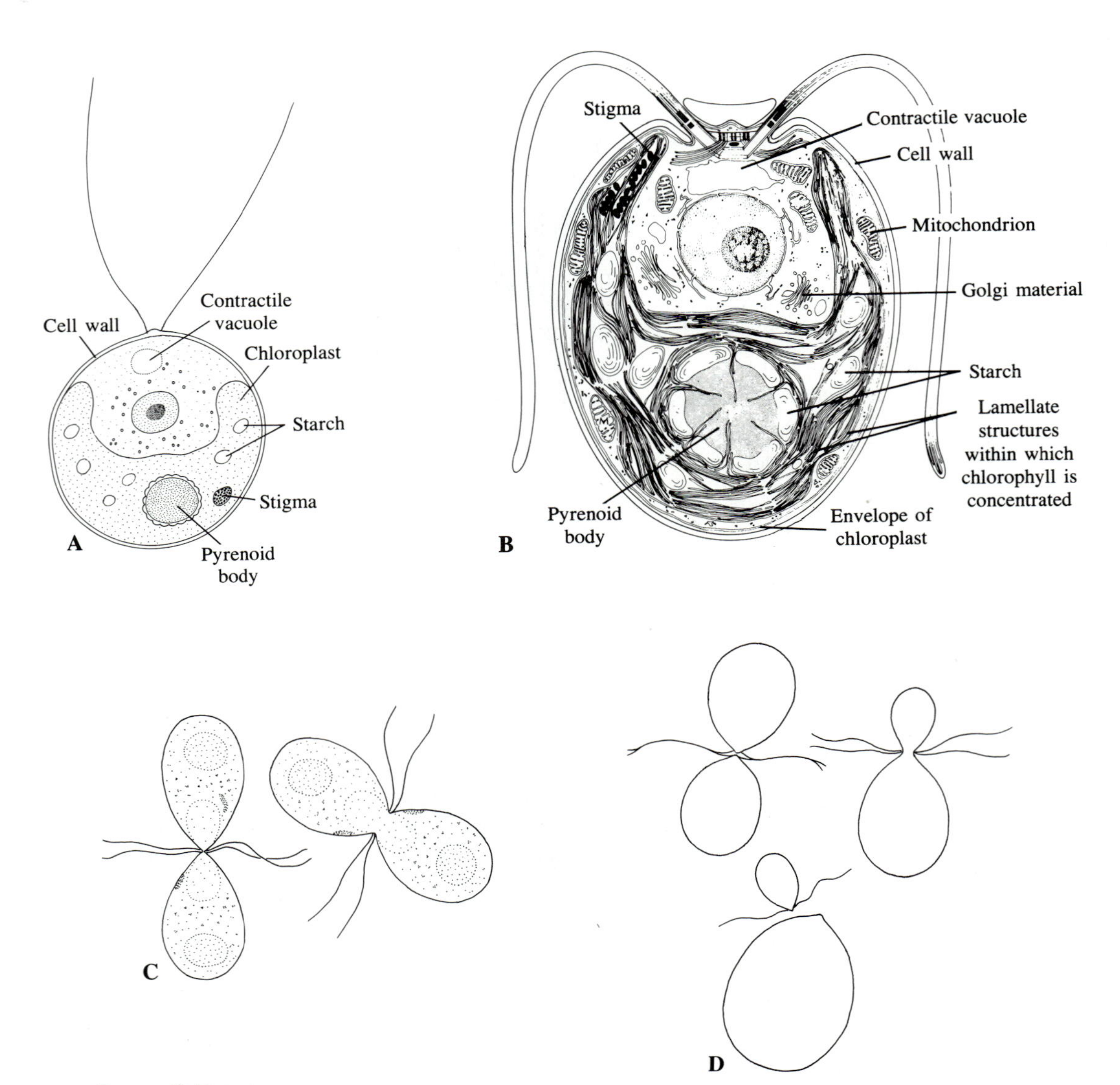

FIGURE 2.7 Order Volvocida. A. *Chlamydomonas*. B. *Chlamydomonas;* diagram based on electron micrographs. The components of the complicated kinetosomal region are not labelled. (Sager, Cytoplasmic Genes and Organelles.) C. *Dunaliella salina;* sexual union of equal gametes. (After Lerche, Archiv für Protistenkunde, *88*.) D. *Chlamydomonas;* sexual union in three species, one with equal gametes, the others with unequal gametes. (After Skuja, Svensk Botanisk Tidskrift, *43*.)

the stigma of *Euglena,* it is believed to shield the actual photoreceptor in such a way that this is stimulated by light coming from certain directions but not others. A cell wall, consisting of a carbohydrate closely related to cellulose, is present in certain genera, including *Chlamydomonas* (Figure 2.7, A and B), which is typical of the group.

Binary fission is the usual mode of asexual reproduction. It is sometimes repeated within the cell wall of the parent flagellate, so that several progeny are released at the same time. In sexual union, the two flagellates that fuse may be identical in size and form **(isogamous)** or they may be dissimilar **(anisogamous)** (Figure 2.7, C and D).

Some volvocids form colonies that consist of several to many cells embedded in a gelatinous matrix. The cells may be interconnected by cytoplasmic bridges. The genus *Volvox* (Figures 2.8, 2.9, 2.10), whose col-

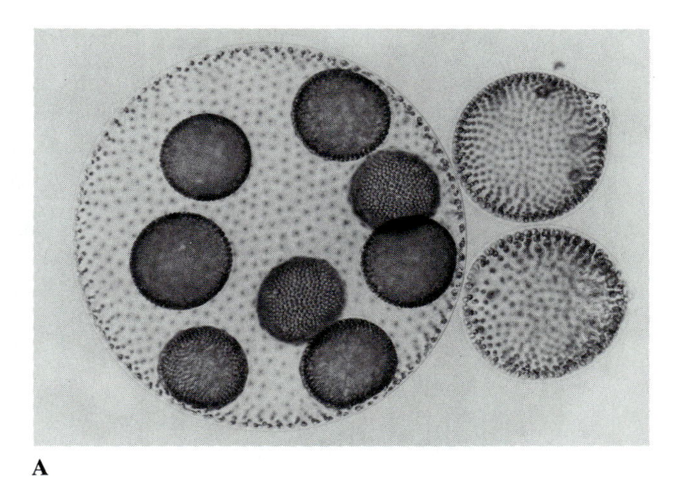

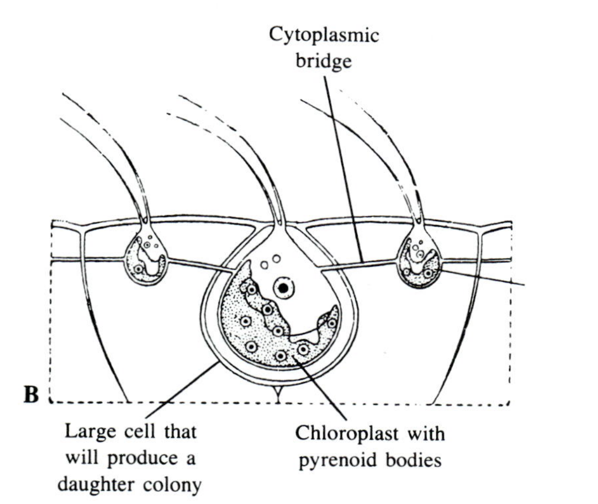

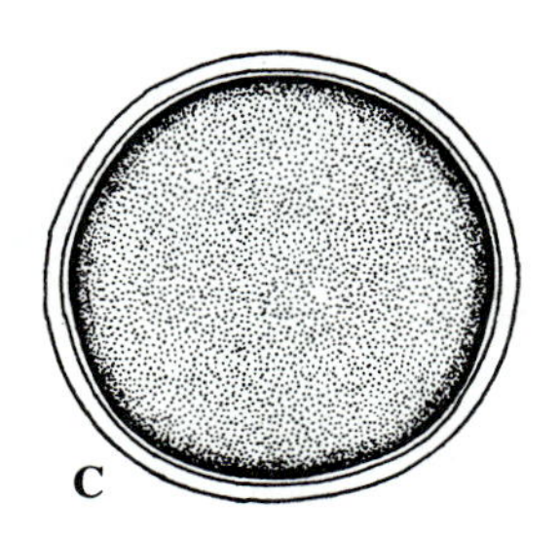

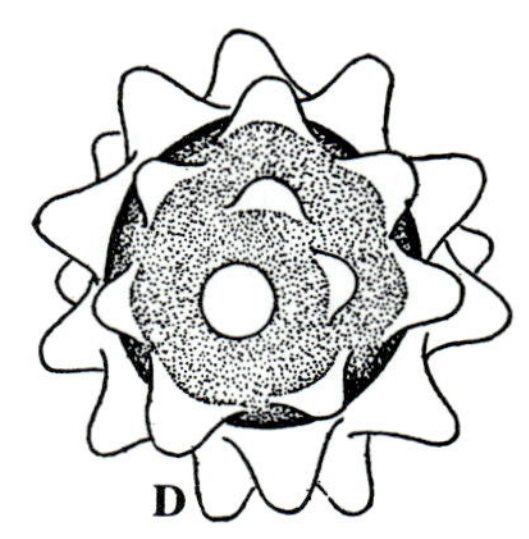

FIGURE 2.8 *Volvox aureus,* three colonies, the largest of which has asexually-produced daughter colonies in its central cavity; photomicrograph. (Courtesy of Carolina Biological Supply Co., Inc.) B. *Volvox aureus,* a portion of the wall of a colony. C. *Volvox aureus,* zygote. D. *Volvox globator,* zygote. (C and D, Smith, Transactions of the American Microscopical Society, *63*.)

onies are like hollow balls, is a classic example. New colonies of *Volvox* are produced asexually (Figures 2.8, A; 2.9, A; 2.10) by division of certain cells that have lost their flagella and that grow to a large size. At first, the cells of the daughter colonies are oriented in such a way that the flagella face inward, but the colonies open up and turn inside-out (evert), so the flagella now face outward (Figure 2.10, D and E). Eventually the young colonies escape through a break in the wall of the parent colony. No further cell division takes place; increase in colony size is achieved by secretion of matrix material.

In sexual reproduction of *Volvox,* the male and female gametes, produced by separate colonies, are dissimilar. The female gametes (eggs) develop by enlargement of certain cells in the wall of a colony (Figure 2.9, B), whereas the small male gametes (sperm) are produced by division of relatively small cells (Figure 2.9, C and D). The bundles of sperm escape and swim about, and those that reach female colonies stick to them. The bundles break up and the individual sperm must work their way through the colony matrix to reach the eggs. The zygotes secrete cysts around themselves (Figure 2.8, C and D) and they may remain dormant for some time before they divide to form new colonies.

Most volvocids, including *Chlamydomonas* and the colonial types, are freshwater organisms. Some, however, are marine. Among them is *Dunaliella,* which is often abundant in brine evaporation ponds from which salt is extracted for commercial purposes.

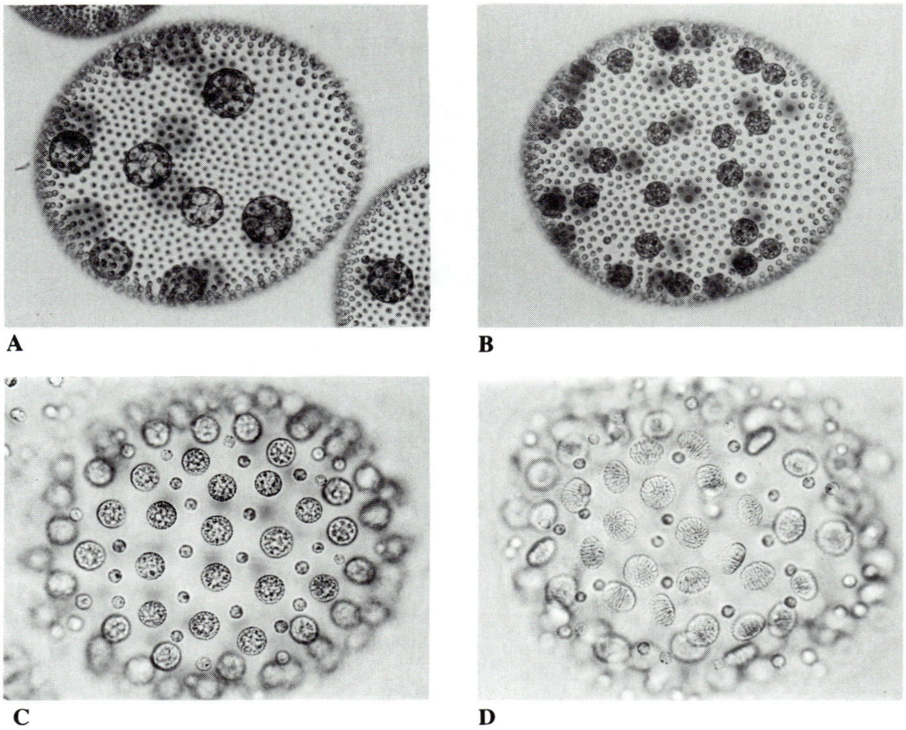

A

B

C

D

FIGURE 2.9 Order Volvocida. *Volvox carteri* form *nagariensis*. A. An asexual colony containing enlarged cells that have sunk into the central cavity; these will divide to produce daughter colonies. B. A sexual colony containing female gametes (eggs). C. A sexual colony with enlarged cells destined to produce male gametes (sperm); these cells, still in the wall of the colony, alternate almost regularly with vegetative cells. D. A sexual colony containing packets of male gametes. (Photomicrographs courtesy of Richard Starr.)

CLASS ZOOMASTIGOPHOREA

This large class should be viewed as a category of convenience rather than as a phylogenetically unified assemblage. The one feature characteristic of all of its members is the absence of chloroplasts. Remember, however, that some representatives of the class Phytomastigophorea also lack chloroplasts, although they show affinities with photosynthetic flagellates in other ways. The Zoomastigophorea is an ecologically important group. Besides many free-living types, it includes nearly all of the symbiotic flagellates. Only five of the orders are considered here.

Order Choanoflagellida

Choanoflagellates (Figure 2.11) are interesting in that they resemble the **choanocytes,** or collar cells, that are characteristic of sponges and that have been reported from echinoderms and some other invertebrates. The collar, which surrounds the basal portion of the single long flagellum, consists of a ring of **microvilli.** It is essentially a sieve, because although the microvilli are close together and may be joined by cross-bridges, there are spaces between them. As bacteria and other small particles touch the collar, they are incorporated into food vacuoles.

Choanoflagellates occur in fresh water as well as in the sea. Some form colonies, and these may be free-floating or attached by a stalk. In the free colonies of one of the marine genera, *Proterospongia,* the cells are embedded in a gelatinous matrix (Figure 2.11, C). *Proterospongia,* as its name suggests, has been thought of as the type of choanoflagellate that could, with further differentiation of cells, have given rise to sponges. In certain genera, the body is enclosed within a lorica; this may be strictly organic, or it may have overlapping siliceous rods that support an organic membrane (Figure 2.11, B).

Order Kinetoplastida

With few exceptions, the kinetoplastid flagellates are symbiotic organisms—parasites—that live in various invertebrates and vertebrates, and also in certain plants that have milky sap. They are characterized by one or two flagella and by a **kinetoplast,** which is the DNA-rich portion of a large mitochondrion. The standard methods for staining nuclei bring out the kinetoplast, but not the mitochondrion as a whole, so the spatial relationship of the two structures was not appreciated until a few years ago, when it was demonstrated by electron microscopy.

Good examples of the kinetoplastid flagellates are the trypanosomes. These live and reproduce in the

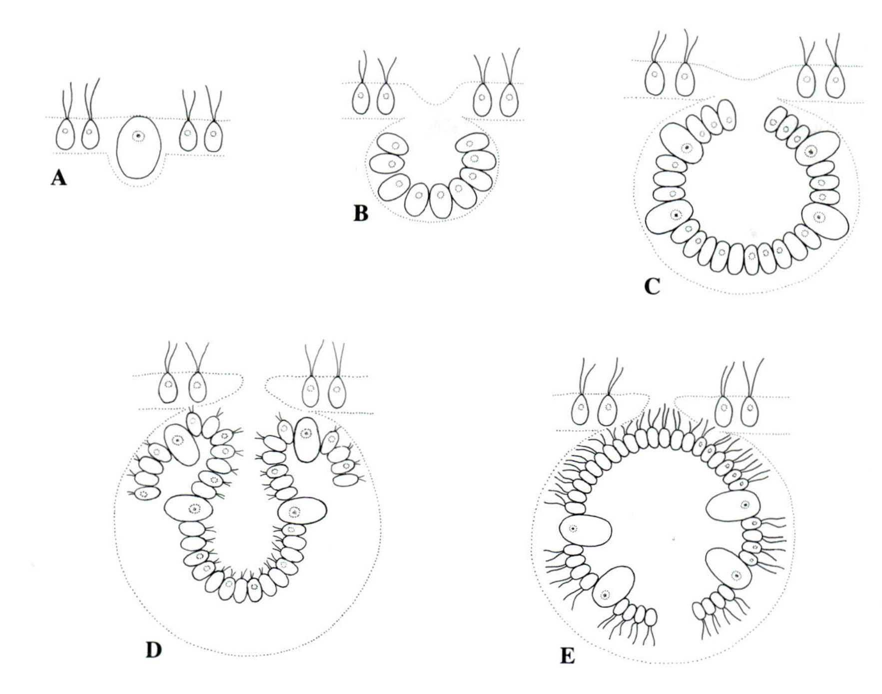

FIGURE 2.10 Order Volvocida. Stages in the development of a daughter colony, produced asexually, in *Volvox;* diagrammatic. A. An enlarged, nonflagellated cell in the wall of the parent colony. B. Cells produced by early cleavages. C. Nearly spherical stage; large cells destined to produce another generation of daughter colonies already differentiated. D. Smaller cells in wall of young colony now have short flagella; the colony is beginning to turn inside-out. E. Eversion of colony nearly complete.

blood, lymph, intercellular fluids, and sometimes within the cells of vertebrates. Nearly all of them, however, also have an invertebrate host, which serves as a vector. In general, trypanosomes of reptiles, birds, and mammals are transmitted by bloodsucking insects or ticks, whereas those of fishes and amphibians are transmitted by leeches.

When examined with the light microscope, a typical blood-dwelling trypanosome (Figure 2.12, A) shows a nucleus, a single flagellum that adheres to the body along a thin sheet of cytoplasm called the **undulating membrane,** and the kinetoplast. The latter lies close to the kinetosome from which the flagellum originates. During division (Figure 2.12, B), one daughter gets the original flagellum and a new flagellum grows out in the other daughter. Electron microscopy reveals the large

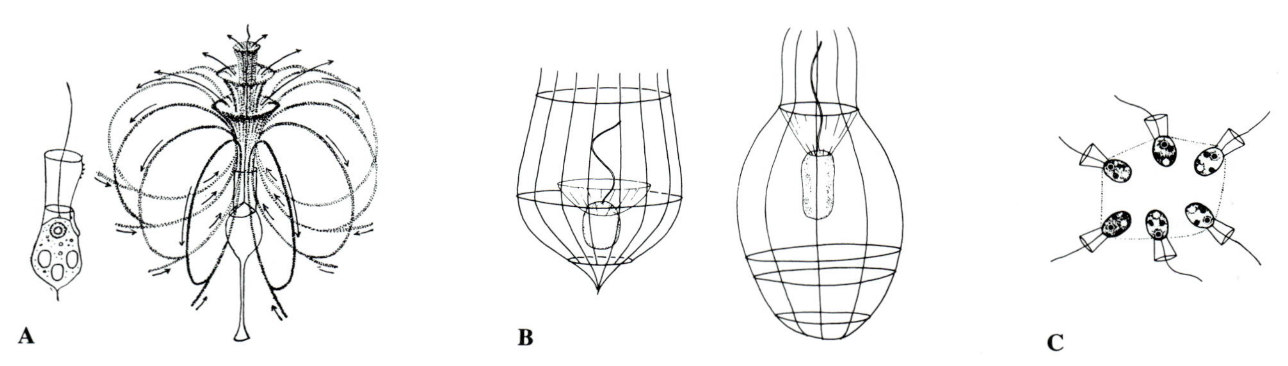

FIGURE 2.11 Order Choanoflagellida. A. *Codosiga,* with its flagellum producing elaborate currents (right) for trapping food particles on the collar (left). (Dogiel, General Protistology.) B. *Campanoeca,* which secretes a lorica consisting of an organic membrane and overlapping siliceous rods. (The membrane, not shown, is supported much like the fabric of a lampshade.) (Throndsen, Sarsia, *56.*) C. *Proterospongia haeckelii,* its cells embedded in a secreted jelly. (Lackey, Transactions of the American Microscopical Society, *78.*)

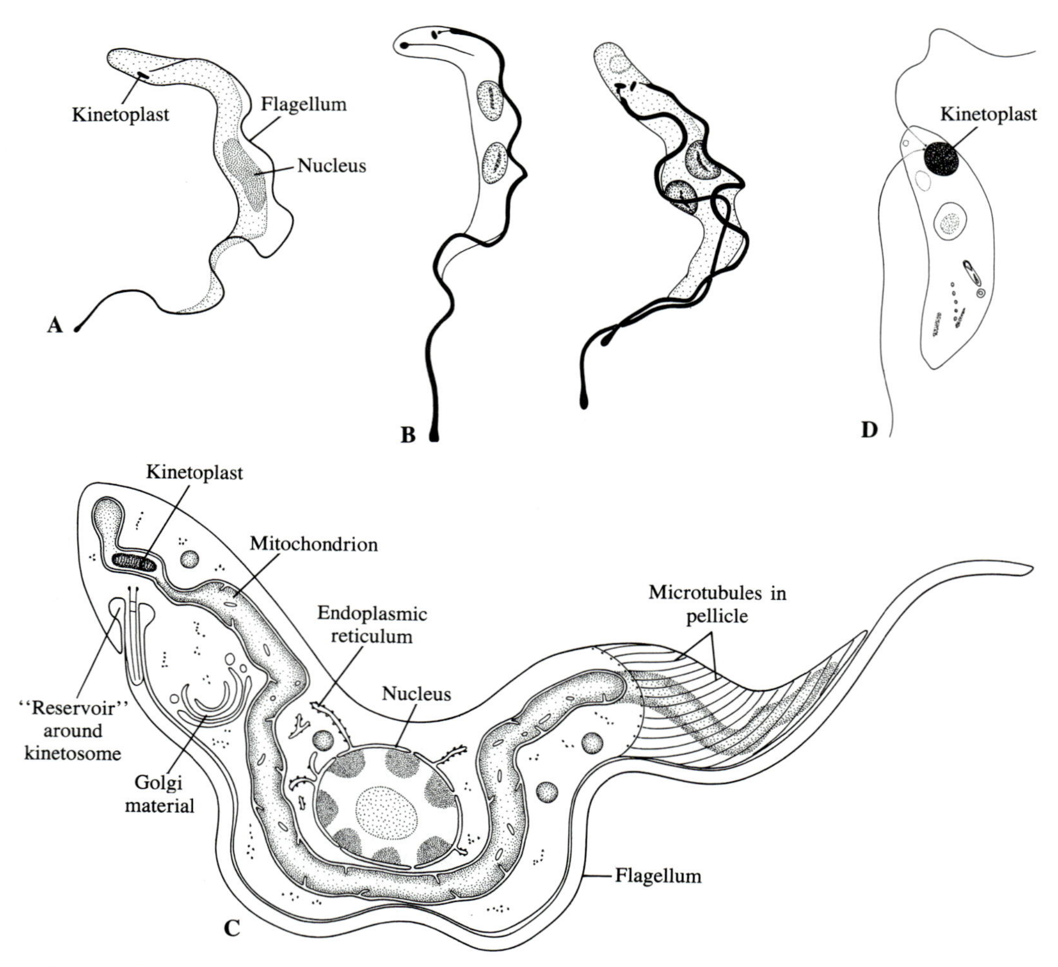

FIGURE 2.12 Order Kinetoplastida. A. *Trypanosoma brucei,* stained to show structures that can be seen with the light microscope. B. Specimens in division, impregnated with silver, which makes the flagellum appear thicker. Note that the original flagellum is retained by one daughter, and a new flagellum is formed in the other daughter. (After Kirby, Journal of Morphology, *75*.) C. *Trypanosoma congolense;* diagram showing details seen in electron micrographs. (After Vickerman, Journal of Protozoology, *16*.) D. *Bodo;* diagrammatic.

mitochondrion of which the kinetoplast is a part, as well as other details, such as Golgi material, microtubules, and various vesicles (Figure 2.12, C).

During their life cycles, most trypanosomes go through stages different from the blood-inhabiting phase. In the invertebrate host, the flagellum may adhere to the body for only a short distance, or not at all, although the infective stage does typically have an undulating membrane. An ovoid stage in which the flagellum is rudimentary may occur, and in a few species a comparable stage multiplies within cells of the vertebrate host. To make things more interesting, many insects, whether or not they suck blood, have flagellates that are related to trypanosomes, although they rarely or never reach the stage with a fully developed undulating membrane. It is likely that trypanosomes and similar kinetoplastid flagellates that parasitize vertebrates are derived from flagellates that were originally associated with insects and other invertebrates.

Humans are affected by trypanosomes of two rather different types. One type is represented by *Trypanosoma gambiense* and *T. rhodesiense,* which cause sleeping sickness in parts of Africa. They are transmitted by tsetse flies. After a tsetse fly draws blood containing trypanosomes, the flagellates multiply first in the midgut, then move to the salivary glands, where they multiply some more. It is through salivary secretions, injected when the fly punctures the skin, that the trypanosomes reach the bloodstream of a new host. *Trypanosoma gambiense* parasitizes a wide variety of mammals, and tsetse flies that suck blood from these can transmit the disease to humans; *T. rhodesiense* is primarily a parasite of humans. Both of these trypanosomes are morphologically indistinguishable from *T. brucei,* which is found in many wild animals and which can be transmitted to some domestic animals, but not to humans.

The other type of trypanosome found in humans is represented by *T. cruzi,* which causes Chagas' disease in tropical America. It is transmitted by bloodsucking bugs of the family Reduviidae, and the infective stage is reached in the hindgut. Infection of humans is therefore achieved through the wet fecal material the bugs deposit while they are feeding, which takes place mostly at night. When the puncture made by a bug is rubbed, some of the flagellates in the fecal deposit nearby may reach the wound, or they may be brought into contact with a mucous membrane, especially that of an eye, that they can penetrate. This species is one of the trypanosomes that invades cells, usually of muscle, nerve, or glands, and multiplies in these cells. The intracellular phase is much smaller than the blood phase, and it lacks an emergent flagellum, although the basal portion of the flagellum is present.

A trypanosome parasite of horses and donkeys, *T. equiperdum,* becomes abundant in the fluid associated with swellings on the skin and sex organs. It is transmitted during sexual intercourse and has no known insect host. There are also a few mammalian trypanosomes that are transmitted by biting flies, which are purely mechanical vectors; the flagellates do not multiply in the flies.

Members of the genus *Leishmania*—important parasites of mammals, including humans—normally do not have an emergent flagellum while in the intracellular phase in their mammalian hosts, but they revert to the flagellated stage in the gut of sandflies and related insects. The flagellum is free, however, rather than adherent as it is in trypanosomes.

The leishmanias of humans are all very similar in structure, though the diseases they cause are different. One kind causes a skin disease called oriental sore; it is especially prevalent in North Africa and the Middle East. Another mostly affects the mucous membranes of the nose and mouth. Still another is responsible for kala-azar, a disease in which the parasites multiply in the spleen, bone marrow, and other visceral organs, as well as in the skin and mucous membranes.

The genus *Phytomonas,* living in the milky sap of milkweeds and various other plants, is characterized by a form similar to that of the insect-inhabiting stage of leishmanias. In the sapsucking insects that transmit *Phytomonas,* multiplication may take place in a nonflagellated stage as well as in a stage that resembles the one found in plant hosts.

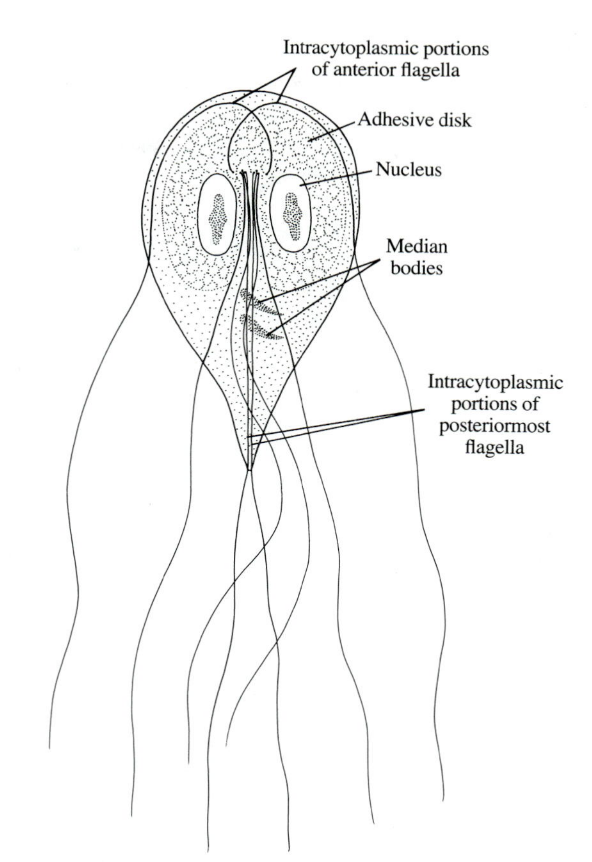

FIGURE 2.13 Order Diplomonadida. *Giardia;* somewhat diagrammatic.

Cryptobia and *Trypanoplasma* slightly resemble trypanosomes, but they have two flagella. One adheres to the body for much of its length; the other is free. Cryptobias are found in the seminal receptacles of pulmonate gastropod molluscs and are believed to be transmitted venereally; the life cycle of trypanoplasmas alternates between the blood of fishes and the gut of leeches, which serve as vectors.

Some common free-living kinetoplastid flagellates are those belonging to the genus *Bodo* (Figure 2.12, D). They are often found in stagnant freshwater habitats, where they feed on bacteria and also take up dissolved organic nutrients. Like cryptobias and trypanoplasmas, they have two flagella, one of which is free, the other partly adherent.

Order Diplomonadida

The diplomonadids are literally double flagellates, and are unusual in being bilaterally symmetrical. *Giardia* (Figure 2.13), a genus represented in the digestive

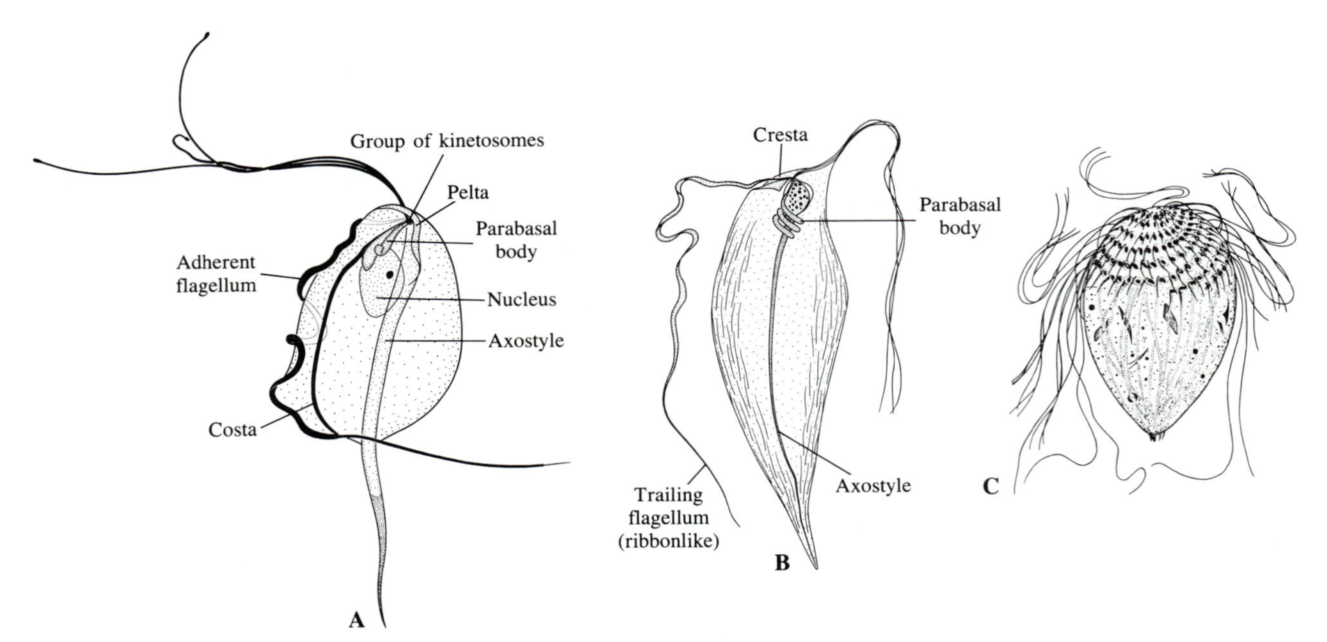

FIGURE 2.14 Order Trichomonadida. A. *Trichomonas batrachorum,* from the large intestine of various frogs and toads. The drawing is based primarily on specimens impregnated with silver, but shows some details that have been worked out with the aid of electron microscopy. (After Honigberg, Journal of Parasitology, *39.*) B. *Devescovina,* from the hindgut of a termite. (After Kirby, University of California Publications in Zoology, *45.*) C. *Metacoronympha,* a ''multiple devescovinid,'' from the hindgut of a termite. (Kirby, Proceedings of the California Academy of Sciences, series 4, *22.*)

tract of many vertebrates, especially mammals, illustrates the idea. Note that there is a nucleus on each side of the midline. The eight flagella are arranged in four pairs, and they run through the cytoplasm for varying distances before they become free. The symmetry is disturbed only by a pair of obliquely placed bodies (''median bodies'') near the middle of the body. The significance of these is not known. In most species of *Giardia,* one of the broader surfaces (arbitrarily called the ''ventral'' surface) is slightly concave and is applied to the epithelium of the host's gut. *Giardia intestinalis,* which inhabits the small intestine of humans, can cause severe diarrhea. It is also thought to interfere with absorption of fats and therefore of fat-soluble vitamins. Giardias voided with feces are usually encysted, and it is through the encysted stage that these flagellates are transmitted.

Order Trichomonadida

The trichomonadids are structurally complex flagellates. Most of their organelles can be seen in specimens that have been properly stained or impregnated with silver, but the composition and exact relationships of these structures can be explained only by electron microscopy. There are characteristically three to five free flagella arising from a cluster of kinetosomes, and another flagellum that adheres to the body for part or all of its length. In most genera, such as *Trichomonas* (Figure 2.14, A), the adherent portion of this flagellum borders an undulating membrane. Other organelles, besides the nucleus, are the axostyle, pelta, costa, and parabasal body. The **axostyle,** which runs like a rod through the body and usually protrudes from the posterior end, consists of a bundle of microtubules. The **pelta** is an extension of the axostyle, and is likewise composed of microtubules. The **costa** is thought to be a large ciliary rootlet arising from the basal body of the adherent flagellum, and the **parabasal body** is interpreted as a concentration of Golgi material linked with one or two rootlets of the other flagella.

Trichomonadids are entirely symbiotic. They inhabit the digestive tracts of many vertebrates and some invertebrates, especially insects. Three species are known to occur in man: *Trichomonas tenax* inhabits the mouth; *T. vaginalis* lives in the vagina, urethra, prostate gland, and seminal vesicles; and *Pentatrichomonas hominis* lives in the intestine. They all feed to a large extent on bacteria.

An easily available and thoroughly typical trichomonadid for laboratory study is *Trichomonas augusta,* almost routinely found in the large intestine of frogs and toads. *Trichomonas termopsidis* is a mutualistic

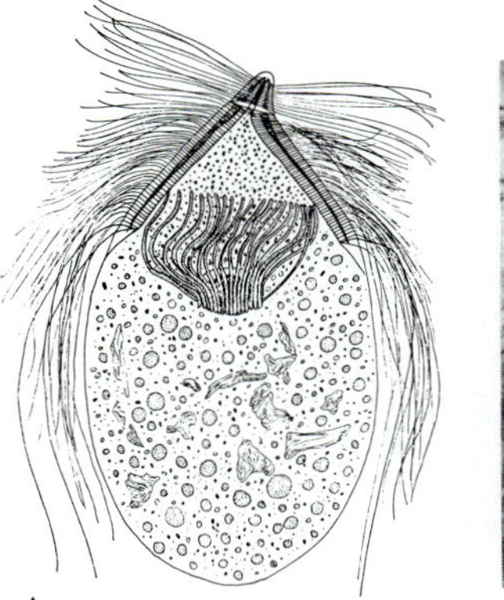

A

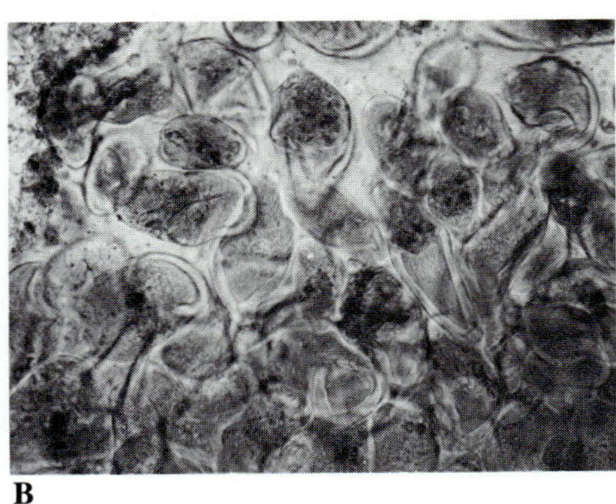

B

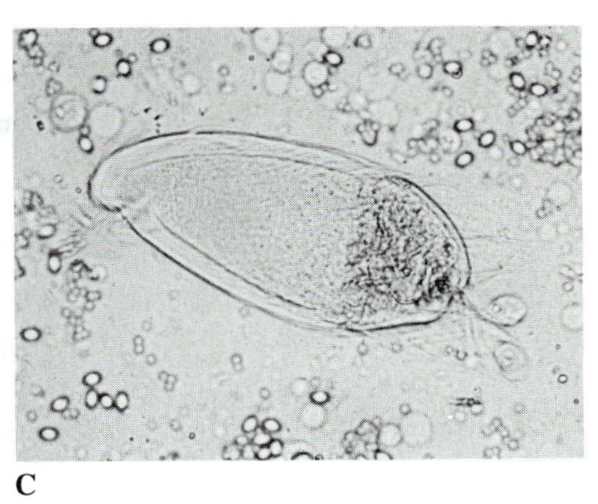

C

FIGURE 2.15 Order Hypermastigida. A. *Trichonympha corbula,* from the hindgut of a termite; optical section of a stained specimen, showing the complex parabasal body. (Kirby, University of California Publications in Zoology, *49*.) B. A mass of *Trichonympha campanula* in a smear of the contents of the hindgut of a termite. C. *Trichonympha campanula;* a living specimen, showing wood particles concentrated at the broader end.

symbiont in the hindgut of many termites. These insects cannot themselves digest the wood they consume, so it is up to *T. termopsidis* and other flagellates, especially hypermastigids (next section), to ingest the particles of wood and convert the cellulose to soluble carbon compounds that can be assimilated or used for energy. Bacteria assist in the process, and they must do all of the work in certain termites that do not have wood-digesting flagellates. In any case, termites are absolutely dependent on certain of the organisms living in the hindgut.

There are many trichomonadids in which the trailing flagellum is largely or completely free of the body, so there is no undulating membrane. This is the case in the flagellates called devescovinids (Figure 2.14, B) which are also among the wood-digesting symbionts found in certain termites. The trailing flagellum of devescovinids is sometimes ribbonlike for much of its length. There is an axostyle, a parabasal body, and a structure called the **cresta,** which is perhaps an elaborate counterpart of the costa of *Trichomonas* and its close allies. Some other flagellates found in termites are apparently "multiple" devescovinids (Figure 2.14, C). They have a number of nuclei, each accompanied by a flagellar–axostylar complex similar to that of devescovinids. Although these flagellates may be nearly covered by flagella, they must not be confused with the hypermastigids, described in the following section.

Order Hypermastigida

The hypermastigids—the most complex of all flagellates—are almost strictly limited to the guts of termites and wood-eating roaches. The majority of them digest cellulose, thus providing themselves and their hosts with soluble carbon compounds. *Trichonympha* (Figure 2.15), found in many termites, illustrates the structure of a hypermastigid. Note that although there is only one nucleus, the body is almost completely covered by flagella, except at the broader end, where food particles are ingested. In *Trichonympha* the flagella are arranged in nearly straight rows; in some genera the rows follow a spiral course, and in others all of the flagella are concentrated in a circle near one end of the body. The parabasal body of hypermastigids may be elaborate, and the cytoplasm sometimes contains filaments that are perhaps homologues of the axostyles of trichomonadids.

SUBPHYLUM SARCODINA

The organisms placed in the Sarcodina are characterized by cytoplasmic extensions called **pseudopodia,** used for locomotion or capture of food, or both. There are three main types of pseudopodia: **lobopodia,** which are relatively broad, **filopodia,** which are slender threads, and **axopodia,** which are slender and supported by axial bundles of microtubules, so that they stick out stiffly like needles. In many of the organisms that have filopodia, the threads are joined in such a way that they form complex networks called **reticulopodia.**

All pseudopodia exhibit protoplasmic streaming, and they are labile to at least some extent. Their usefulness to the organism depends on their ability to change. Even stiff axopodia may be withdrawn and reformed.

SUPERCLASS RHIZOPODEA

Rhizopods are characterized by lobopodia or filopodia. The group includes the organisms called amoebae, as well as foraminiferans and slime molds. Slime molds are not dealt with in this book.

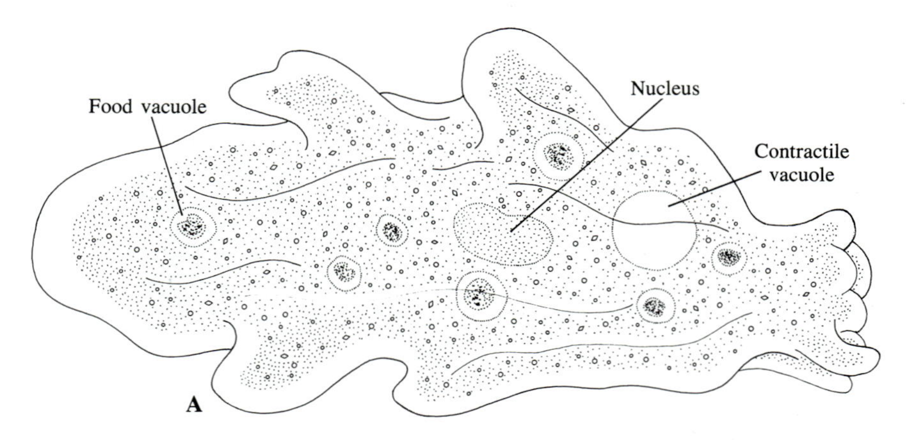

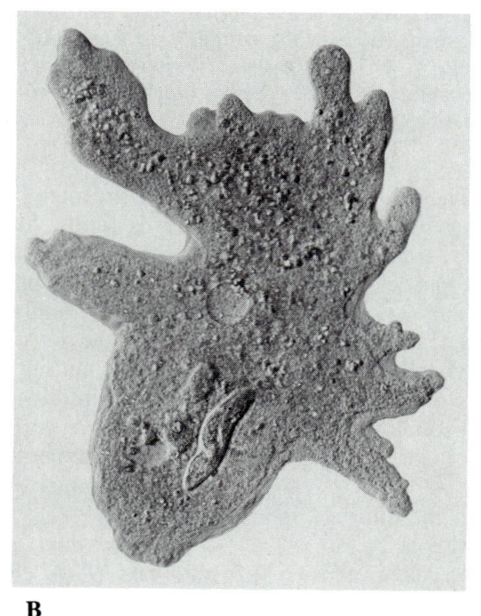

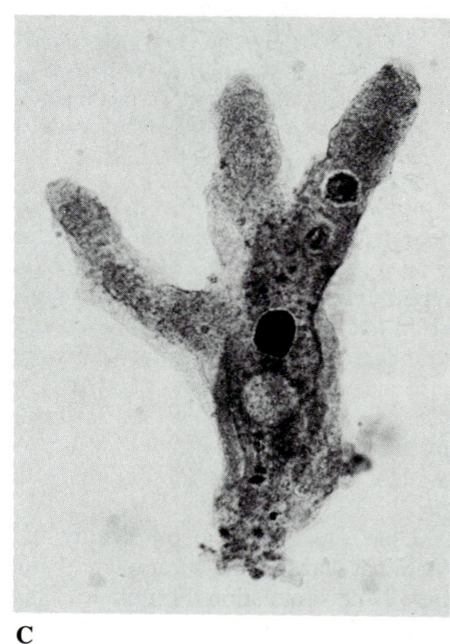

FIGURE 2.16 Subclass Gymnamoebia. A. *Amoeba proteus;* somewhat diagrammatic. Only a few of the crystals and other inclusions of the cytoplasm are shown. B. *Amoeba;* photomicrograph of a living specimen. The contractile vacuole is near the center; note the still-intact flagellate inside a food vacuole. C. *Amoeba;* photomicrograph of a specimen stained to show the nucleus. (B and C, Courtesy of Carolina Biological Supply Co., Inc.)

CLASS LOBOSEA

The Lobosea includes the amoebae that have either relatively broad pseudopodia or slender pseudopodia that originate from a broad lobe. There are two subclasses, one consisting of "naked" types, the other consisting of amoebae that secrete a "house."

SUBCLASS GYMNAMOEBIA

Amoeba proteus (Figure 2.16), a freshwater species to which most students of biology are introduced sooner or later, is a good example of a gymnamoeba ("naked amoeba"). It is relatively large—up to 0.5 mm long—and freely produces pseudopodia. Looking at an amoeba with a compound microscope, one sees that the relatively fluid inner cytoplasm, called **endoplasm,** contains food vacuoles, lipid droplets, and various kinds of granules and crystals; the crystals are metabolic wastes. Collectively, the inclusions nearly obscure the large nucleus. A thin outer layer of more viscous cytoplasm, called **ectoplasm,** is almost free of inclusions.

Formation of a pseudopodium involves not only active streaming of the cytoplasm, but also localized alterations in the viscosity of the ectoplasm. Essentially what happens at the tip of a pseudopodium, or where a pseudopodium is starting to develop, is that the ectoplasm becomes more fluid; in other words, it is converted into endoplasm that tends to flow out. Behind the advancing tip of the pseudopodium, endoplasm at the surface is changed back into ectoplasm, so that over most of the length of the pseudopodium a cylinder of ectoplasm surrounds a core of endoplasm streaming towards the tip.

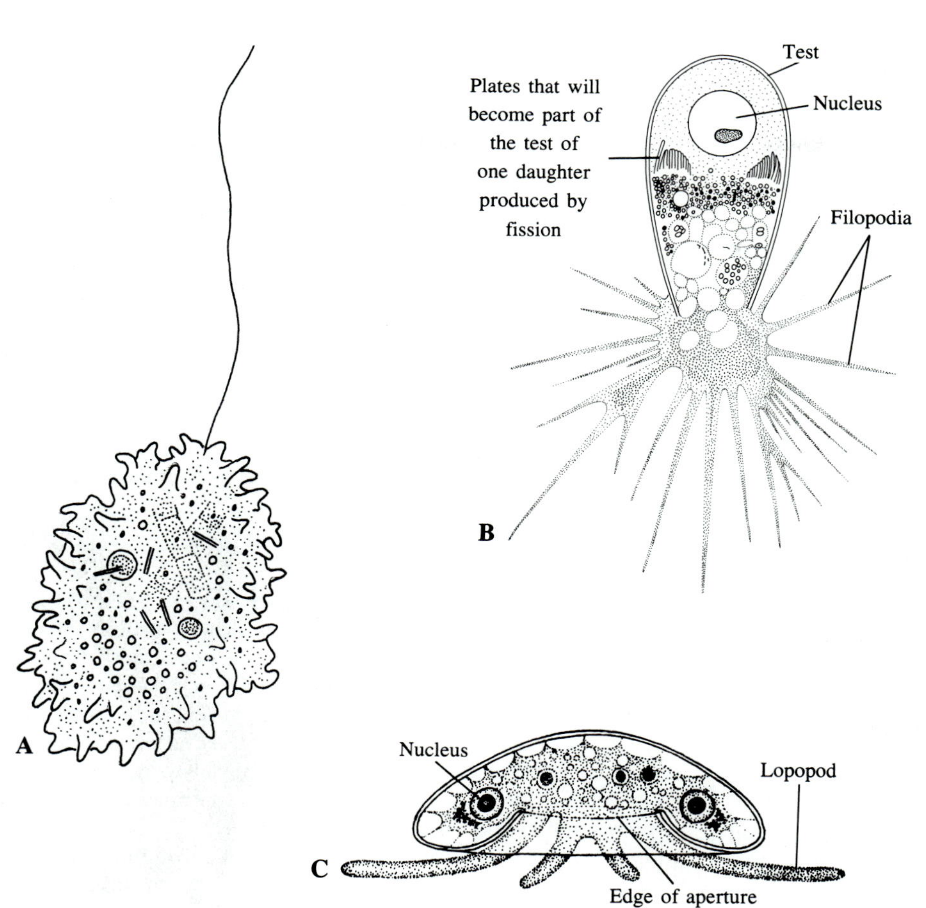

FIGURE 2.17 A. Subclass Gymnamoebia. *Mastigina vitrea,* an amoeboid organism that has a flagellum. B. Subclass Testacealobosia. *Euglypha.* C. Subclass Testacealobosia. *Arcella.* (Kühn, Morphologie der Tiere in Bildern, *2.*)

What causes the endoplasm within an amoeba to flow? This question cannot yet be answered conclusively, but there is some evidence favoring the idea that, as endoplasm converts to ectoplasm, there is a shortening of protein chains. If this is indeed true, the formation of ectoplasm in advancing pseudopodia would pull the amoeba forward and also maintain enough pressure on the endoplasm to keep it flowing.

When pseudopodia are used to capture flagellates and other small food organisms, they surround the prey and fuse, incorporating the prey into a food vacuole where it can be digested.

A contractile vacuole at the edge of the more fluid inner cytoplasm alternately fills with fluid, discharges the fluid through a temporary pore in the cell membrane, then fills up again. This vacuole is concerned with **osmoregulation,** removing water that diffuses into the amoeba because its protoplasm has a higher content of salts than the medium in which it lives. Freshwater amoebae rather routinely have a contractile vacuole, but marine species and symbiotic types that live in the guts of other animals usually do not.

So far as is known, *Amoeba proteus* reproduces only by fission.

Not many amoebae are quite like *Amoeba proteus.* Some extend only one broad pseudopodium at a time; others have a lumpy appearance because they produce many wartlike pseudopodia. A few, such as *Chaos carolinense,* may be more than 3 mm in diameter and have hundreds of nuclei and contractile vacuoles. Numerous species develop flagella under particular environmental conditions or at certain stages in the life cycle (Figure 2.17, A), but as a rule the flagellated stage is temporary. Some amoebae can encyst, and encystment may be a prelude to multiplication by multiple fission, so that several to many individuals emerge from the cyst.

Besides the genera found in fresh water, soil, and marine habitats, there are some that live in the guts of various invertebrates and vertebrates. Most of these are commensals, for they feed only on bacteria. A few, however, destroy tissue and are therefore viewed as parasites. Of the species in man, *Entamoeba histolytica* is the one most often associated with the symptoms of amoebic dysentery, which range from diarrhea to ulcers of the large intestine and even abscesses of the liver and other organs. *Entamoeba gingivalis* inhabits the tartar and debris on teeth and gums. It is probably not pathogenic, although it is often present in mouths that are unhealthy.

The Strange Case of *Pelomyxa*

Living in the black anaerobic mud at the bottom of ponds and ditches is a multinucleate amoeboid organism called *Pelomyxa palustris*. It sometimes reaches a length of 5 mm, and because of its large size and multinucleate condition, it has been confused with species of *Chaos,* referred to above. Unlike *Chaos,* however, it produces only one broad pseudopodium at a time. Its food is also different; instead of ingesting small motile organisms the way *Chaos* does, it consumes one-celled algae, filamentous algae, and fragments of decaying plant material. The food is taken up at the posterior end. Unlike other freshwater amoebae, *Pelomyxa* lacks a contractile vacuole.

The most remarkable features of *Pelomyxa* were not discovered until the organism was studied with the electron microscope. The nuclei, though enclosed by membranes, divide without going through the usual stages of chromosome formation and mitosis. Furthermore, the cytoplasm does not have mitochondria, an endoplasmic reticulum, or Golgi bodies. Symbiotic bacteria of two types perhaps perform the functions of mitochondria. If they are destroyed by treatment with antibiotics, lactic acid and other intermediate products of cellular respiration accumulate and the amoeba dies.

Scattered over its body surface, *Pelomyxa* has stubby structures that are interpreted as flagella. They have the "9 pairs + 2" pattern of microtubules found in typical flagella, but they do not undulate. The organism has been found in an encysted stage.

Pelomyxa is truly enigmatic and its placement here in Gymnamoebia is provisional. It has been suggested that this is not an amoeba at all but a relic of some otherwise extinct group of heterotrophs that flourished before aerobic respiration evolved. Some biologists go so far as to give it a phylum of its own.

SUBCLASS TESTACEALOBOSIA

Most amoebae that secrete a **test** are found in fresh water, especially in situations where decay of plant material makes the water slightly acid. They are generally abundant in sphagnum bogs, in mats of moss at the edges of ponds, and in the sediment that accumulates on the floor of aquaria.

The test of some genera is glassy and consists primarily of organic material, although this may be colored by salts of metals. Certain types secrete siliceous scales and incorporate these into the test. There are also genera in which the test is a collection of foreign particles, such as sand grains and diatom frustules, that have been cemented together. Testate amoebae extend their pseudopodia through the opening of the test. The pseudopodia may be numerous and threadlike (Figure 2.17, B), in which case they are called filopodia, or there may be just a few rather stout lobopodia (Figure 2.17, C).

During binary fission, the test of the parent usually goes with one daughter and a new test develops around the other daughter. Some species survive drying by plugging the opening of the test with secreted material, into which debris may be incorporated. Encystment also occurs; the cyst is produced between the amoeba and its test.

In a marine genus, *Trichosphaerium,* the test is a fibrous structure and has calcareous spicules part of the time. It is also different from the test of other genera in that it has several openings.

CLASS FILOSEA

Members of the class Filosea generally have filopodia that branch, but the branches do not usually anastomose to form networks. In this respect, these organisms differ from foraminiferans and most other members of the next class, Granuloreticulosea. Another point of distinction is that filoseans, as far as is known, have no flagellated stage in their life history. In certain genera, such as *Gromia,* there is an organic test that resembles that of some foraminiferans.

Gromia oviformis (Figure 2.18) is a common marine species that lives on rocks, seaweeds, worm tubes, and similar habitats. Its brownish, egg-shaped or spherical test may reach a diameter of 5 mm. At one end of it is an opening through which the filopodia emerge. Asexual reproduction has not been observed, but there is sexual reproduction. This is initiated when two individuals come together and produce nonflagellated gametes. The gametes unite within a chamber formed by the fusion of the tests of the parents. The zygotes secrete tests of their own and then escape by way of pores that develop in the wall of the chamber.

CLASS GRANULORETICULOSEA

The members of this class, all marine, typically have reticulopodia—filopodia united in such a way that they form complex networks. Of the several orders, the Foraminiferida is the largest and most important. It is the only one discussed here.

Foraminiferans secrete a test, and this is generally calcareous and subdivided into several to many chambers. The chambers are added sequentially, the newest and largest one being formed around protoplasm that protrudes when the organism has outgrown the existing chambers.

In some genera the chambers are arranged in a spiral (Figures 2.19, A; 2.20, A, B, and C); in others, they

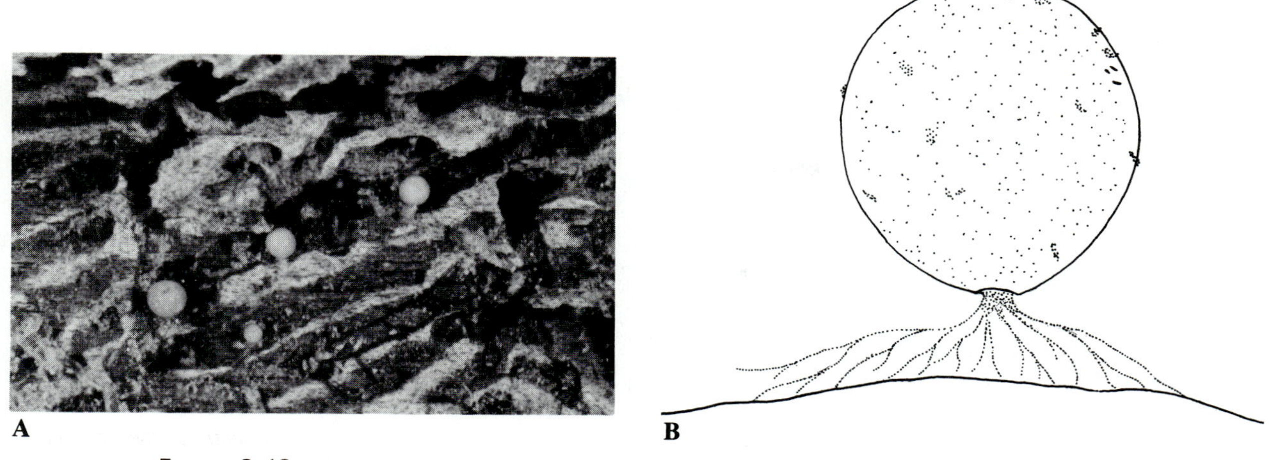

FIGURE 2.18 Class Filosea. *Gromia oviformis*. A. Tests of four specimens on a piece of wood riddled by *Limnoria,* a marine isopod. B. Drawing of a specimen showing its pseudopodia in contact with the substratum.

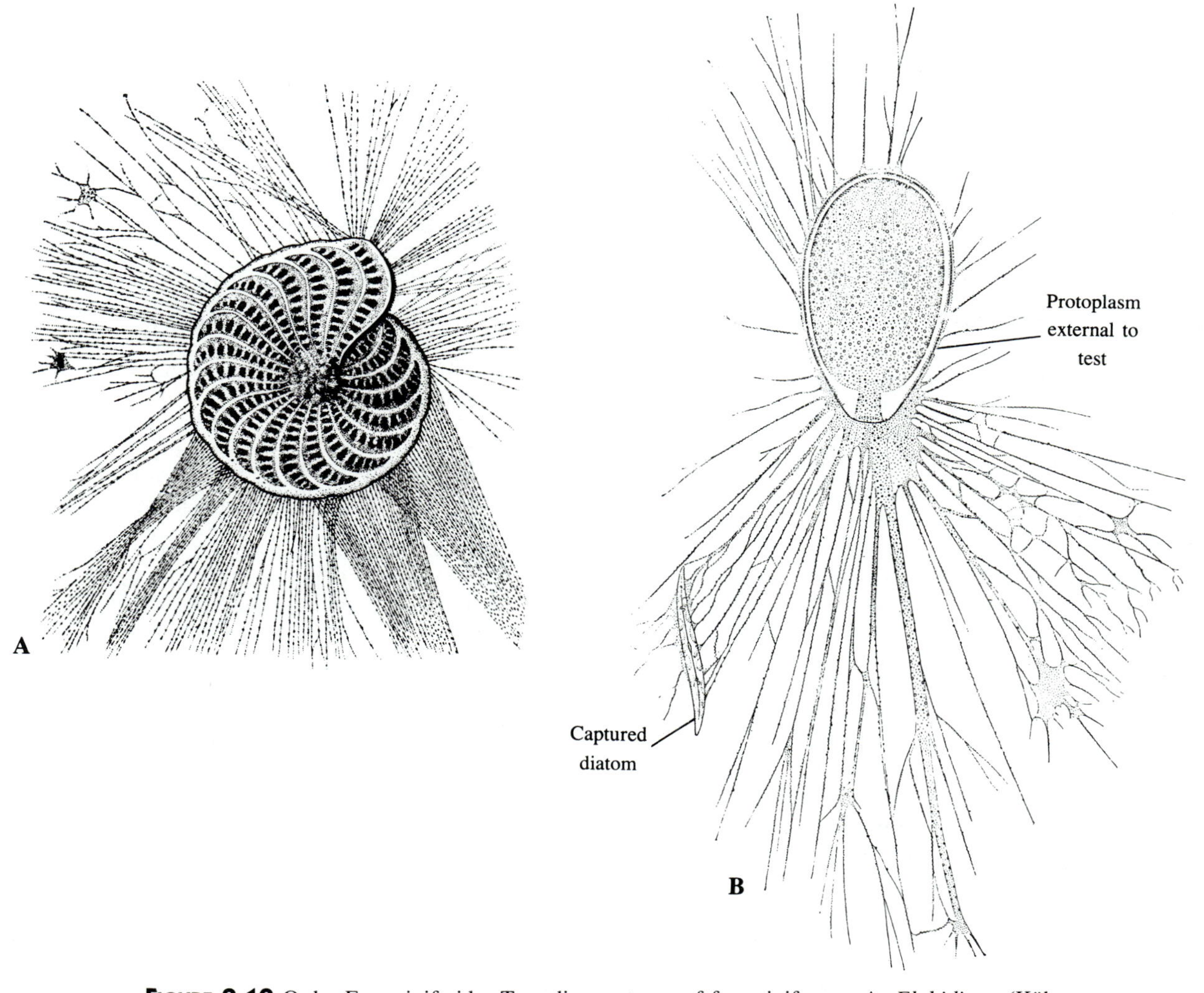

FIGURE 2.19 Order Foraminiferida. Two diverse types of foraminiferans. A. *Elphidium*. (Kühn, Morphologie der Tiere in Bildern, *1*.) B. *Allogromia ovoidea*. (Lankester, A Treatise on Zoology, *1*.)

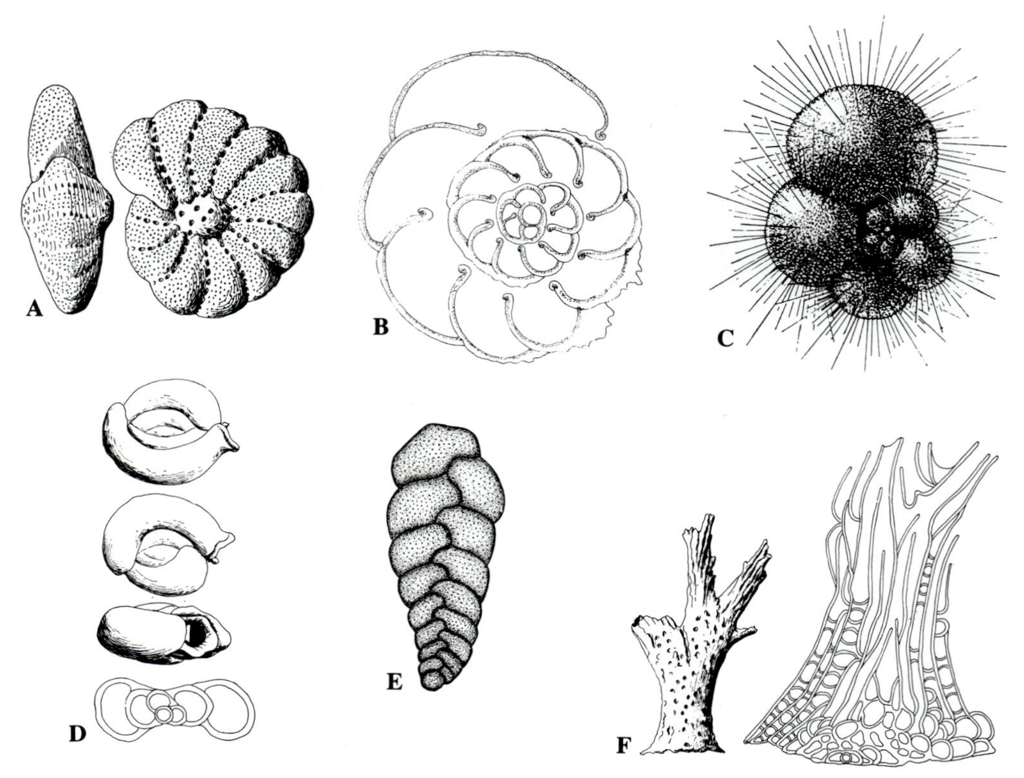

FIGURE 2.20 Order Foraminiferida. Tests of a variety of foraminiferans. A. *Elphidiononion*. (Hofker, Studies on the Fauna of Curaçao, *35*.) B. *Planulina*, sectioned to show the arrangement of the chambers. (Hofker, Studies on the Fauna of Curaçao, *31*.) C. *Globigerina*, a planktonic genus. (Bé, in Fiches d'Identification du Zooplancton, no. 108.) D. *Bolivina*. (Cushman, Proceedings of the Boston Society of Natural History, *34*.) E. *Massilina*. (Hofker, Studies on the Fauna of Curaçao, *49*.) F. *Miniacina*. (Hofker, Studies on the Fauna of Curaçao, *31*.)

are laid down along a straight axis, alternately on opposite sides of the original chamber (Figure 2.20, D and E), with each new chamber enclosing its predecessor, or in an almost chaotic way (Figure 2.20, F). There is an opening (foramen) where one chamber joins another, so the protoplasmic mass is continuous. The number of nuclei increases as a foraminiferan grows. The test is often perforated, so that the cytoplasm emerges from the many small openings as well as from the main aperture.

Not all foraminiferans have tests of the general type described above. In the genus *Allogromia,* for instance, the test is organic and has only one aperture (Figure 2.19, B). There are also foraminiferans in which the test, sometimes a tube, consists to a large extent of foreign material, such as sand grains and the spicules that form the skeletons of sponges. A variation on the plan of the multichambered test is the formation of a complex system of tubes that lie in a calcareous deposit outside the test proper.

The food of foraminiferans consists mostly of microorganisms, such as bacteria and diatoms. Some species, however, are known to use their reticulopodia to trap and digest small invertebrates that may be up to 0.5 mm long.

Both asexual and sexual reproduction occur in this group. In asexual reproduction, multiple fission of the protoplasmic mass produces many small amoeboid individuals that escape from the parent before or after they develop small tests of their own. The parent protoplasm may be entirely consumed by this process, or much of it may survive to reproduce asexually again.

In sexual reproduction it is usual for two individuals to pair and for each to produce numerous gametes; these may have flagella or they may be amoeboid (Figure 2.21). A sexual process known as **plastogamy** occurs in some species: two individuals come together, their several nuclei unite, and the cytoplasm divides, each zygote getting some of it. In *Allogromia,* which has amoeboid gametes, fertilization is **autogamous,** that is, the gametes that unite are produced by the same individual.

Many foraminiferans have two distinct phases in their life history. These can be distinguished on the ba-

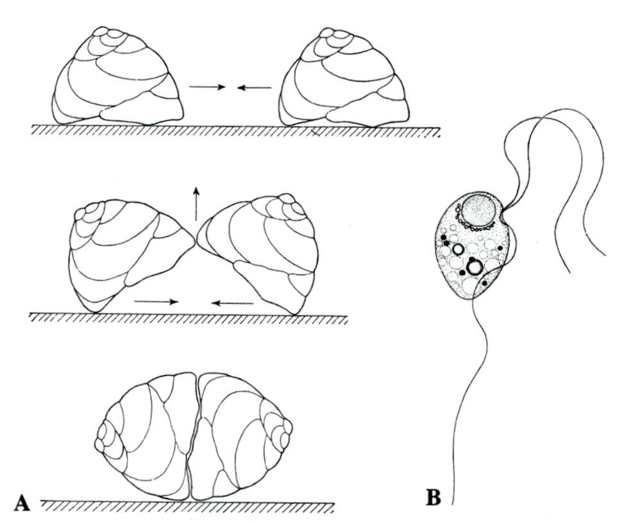

FIGURE 2.21 Order Foraminiferida. *Discorbis mediterranensis*. A. Pairing of sexual individuals. B. A flagellated gamete. (Le Calvez, Archives de Zoologie Expérimentale et Générale, *87*.)

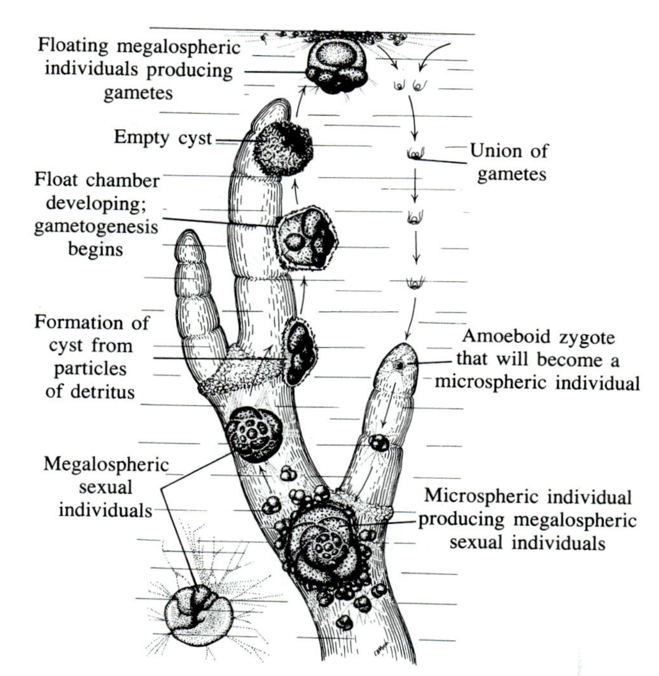

FIGURE 2.22 Order Foraminiferida. Life history of *Tretomphalus bulloides*, a foraminiferan with pelagic and attached stages. (Myers, Stanford University Publications, Biological Science, *9*.)

sis of the size of the initial chamber in proportion to that of other chambers. In *Tretomphalus* (Figure 2.22), for instance, an individual of the **megalospheric** type, with a large initial chamber, produces flagellated gametes. The zygotes develop into **microspheric** individuals, characterized by a small initial chamber. These reproduce asexually, giving rise to many megalospheric individuals. *Tretomphalus* differs from most other foraminiferans in that its megalospheric phase, after a period of encystment, becomes pelagic.

Alternation of phases is not always regular; the megalospheric phase may produce more of its own kind for several generations. In certain genera that have amoeboid gametes, it is the microspheric phase that reproduces sexually and the megalospheric phase that reproduces asexually.

Although some foraminiferans are only 20 μm in diameter, most species are in the size range of 0.1 to 1 mm. Certain types, however, reach diameters of several millimeters. The fossil record reveals foraminiferans whose tests were 14 cm across. Many of the larger extinct species belonged to a group called nummulites, whose tests were so flat as to be coin-shaped. Chalk deposits, such as those in the White Cliffs of Dover, in England, consist mostly of calcareous tests of foraminiferans, and some limestones also contain abundant remains of these protozoans. Paleontologists employed by oil companies pay close attention to foraminiferans and other microfossils found in core samples, because many of them are reliable indicators of the age of ancient deposits. Geologists, biologists, and others interested in climatic changes during the earth's history depend to some extent on foraminiferan fossils because species with small, porous tests are characteristic of colder waters, whereas those with larger and less porous tests are characteristic of warmer waters.

Most foraminiferans are found at the surface of marine sediments and in the coatings formed by mixtures of small organisms on rocks, shells, seaweeds, and other firm substrata. A few, such as *Tretomphalus*, mentioned above, are temporarily pelagic, and certain genera, especially *Globigerina* (Figure 2.20, C) and its relatives, are permanently pelagic. When they die, their tests sink to the bottom; a type of mud rich in these tests is called *Globigerina* ooze.

SUPERCLASS ACTINOPODEA

The members of this group differ from other Sarcodina in having axopodia. They may also, however, have filopodia and reticulopodia. Actinopodea is a large assemblage that includes the three classes of marine organisms called radiolarians, as well as the class Heliozoea, found mostly in fresh water.

RADIOLARIANS

Radiolarians are strictly marine, and nearly all are planktonic. They are usually abundant in the open ocean, but are not likely to be found in quiet waters of

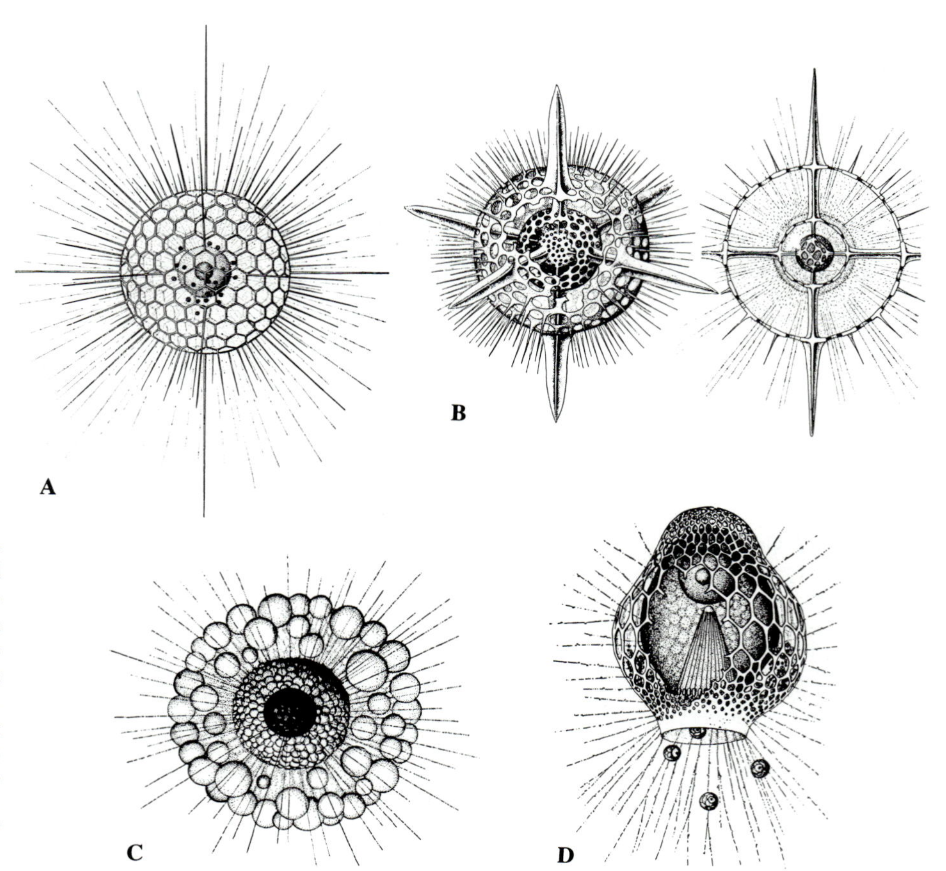

FIGURE 2.23 Class Polycystinea. A. *Heliosphaera actinota,* showing axopodia radiating outward from the central capsule. B. *Hexacontium asteracanthion,* with three latticelike spheres, one inside the other; the innermost sphere is within the nucleus. C. *Thalassicolla nucleata,* showing axopodia radiating outward from the central capsule. There is no skeleton. D. *Cyrtocalpis urceolus.* (A–D, Kühn, Morphologie der Tiere in Bildern, *1.*)

bays or estuaries. The hard skeletons of these protozoans, consisting of silica or strontium sulfate, are important components of the fossil record. Rich deposits of them have been exploited in the manufacture of abrasives.

There are three classes of radiolarians. Two of them, Polycystinea and Phaeodaria, are characterized by a **central capsule** that separates the endoplasm from the ectoplasm. In the third class, Acantharea, what looks like a central capsule is within the ectoplasm.

A substantial portion of the outer ectoplasm is almost always differentiated as a "gelatinous layer," or **calymma.** When this is especially prominent, the organism as a whole may reach a relatively large size for a protozoan. Some unicellular radiolarians reach a diameter of more than 2 cm, and there are colonial types that form masses more than 20 cm long. The character of the gelatinous layer varies considerably. It may be relatively homogeneous or it may be frothy due to the presence of many vacuoles. There are axopodia as well as filopodia and reticulopodia.

The food of radiolarians consists mostly of diatoms, other one-celled algae, and protozoans, but some species can trap small animals such as copepods. Symbiotic dinoflagellates (zooxanthellae) may be present in the cytoplasm.

Binary fission is probably the most widespread mode of reproduction. Multiple fission can also occur, however. This may lead to production of numerous biflagellated cells. These are thought to represent an asexual dispersal stage, but there exists a possibility that they may sometimes function as gametes. Relatively little is known about sexuality in radiolarians. (Certain flagellated cells once thought to be gametes turned out to be motile stages of symbiotic dinoflagellates.)

CLASS POLYCYSTINEA

In the Polycystinea there is a true central capsule. It separates the endoplasm, which contains one or more nuclei, from the ectoplasm. A siliceous skeleton is usually present, but is lacking in some genera, such as *Thalassicolla* (Figure 2.23, C).

In one large assemblage within the class, the central capsule is typically round and has many evenly distributed pores (Figure 2.23, A). The skeleton in this group

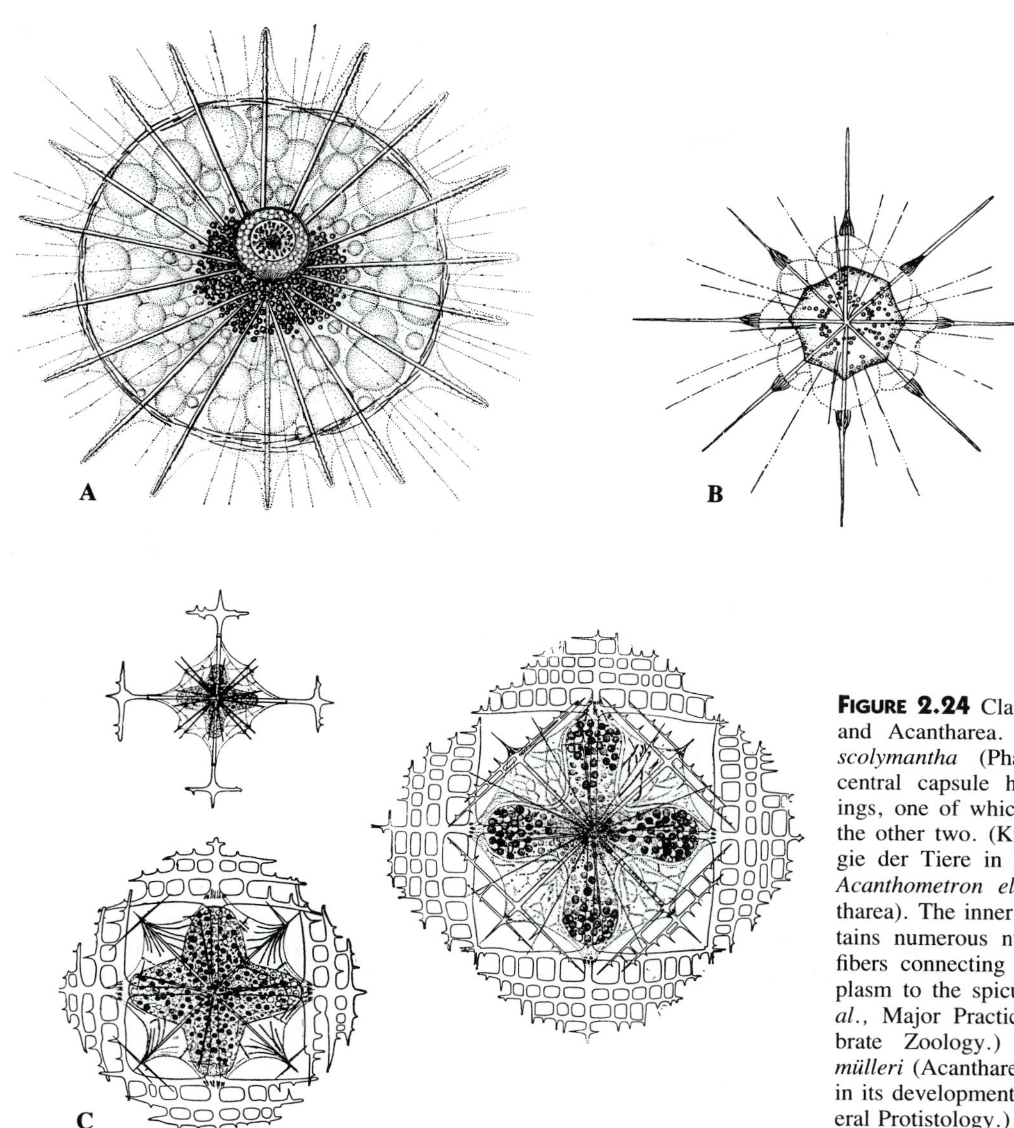

FIGURE 2.24 Classes Phaeodarea and Acantharea. A. *Aulacantha scolymantha* (Phaeodarea). The central capsule has three openings, one of which is larger than the other two. (Kühn, Morphologie der Tiere in Bildern, *1*.) B. *Acanthometron elasticum* (Acantharea). The inner cytoplasm contains numerous nuclei. Note the fibers connecting the outer cytoplasm to the spicules. (Ivanov *et al.*, Major Practicum in Invertebrate Zoology.) C. *Lithoptera mülleri* (Acantharea), three stages in its development. (Dogiel, General Protistology.)

may be a sphere in which the silica is organized into a latticework and projecting spines, or it may consist of separate spicules. Certain of the genera are remarkable in having two or three latticework spheres, one inside the other (Figure 2.23, B). There are also some genera that are viewed as colonial organisms because they have a number of central capsules, each with a nucleus and each surrounded by radiating spicules or by a latticework.

In another assemblage, the central capsule is usually ovoid, and the numerous pores are concentrated at one end (Figure 2.23, D). In the genera forming this group, the skeleton, if present, has the shape of a vase, beehive, or cone in which the silica forms a latticework.

CLASS PHAEODAREA

In the Phaeodarea, there is a thick central capsule that usually has one large opening **(astropyle)** and two smaller openings **(parapyles)** (Figure 2.24, A). The openings are covered by nipplelike structures prolonged as tubes which run for some distance into the ectoplasm. Within the ectoplasm, especially around the larger opening of the central capsule, is a deposit of brownish material, some of it particulate, called the **phaeodium.** This material is believed to be a concentration of waste products.

The skeleton of Phaeodarea, unless absent or consisting of an external agglomeration of foreign material, is siliceous. It is generally organized into needlelike spicules, hollow spines (often much-branched), or a complicated shell that has hollow ducts running through it; the shell may be bilaterally symmetrical.

CLASS ACANTHAREA

A distinctive feature of the Acantharea is a skeleton that consists largely of strontium sulfate. It is generally organized into 10 or 16 spicules that run the full diameter of the organism, or into 20 or 32 radial spicules that converge at the center (Figure 2.24, B and C). The spicules are usually fused together where they cross or converge. The spicules may have side branches, and in certain genera the branches unite to form a latticed internal sphere. Sometimes there are two spheres, one inside the other.

The endoplasm contains one to many nuclei, depending on the species and the age of the organism. There is a structure that resembles the central capsule of the Polycystinea and Phaeodarea, but it is entirely within the ectoplasmic layer, not between the endoplasm and ectoplasm. There are usually *zooxanthellae* (symbiotic dinoflagellates) in the ectoplasm that is enclosed by the capsule. The portion external to the capsule sends out filopodia, reticulopodia, and axopodia; the latter stand out stiffly, supported by bundles of microtubules. The microtubules originate near the center of the organism and project through the capsule, just as the skeletal spicules do.

The outer, gelatinous layer of the ectoplasm is interesting in that it has fibers attached to the spicules. It has been proposed that these fibers are contractile and are used to increase the buoyancy of the organism by dilating the gelatinous layer.

CLASS HELIOZOEA

Heliozoans, which are planktonic or benthic organisms found mostly in fresh water, have slender, needlelike axopodia. The best-known genera, and the ones most likely to be seen in the course of laboratory study, are *Actinophrys* (Figure 2.25, A) and *Actinosphaerium* (Figure 2.25, B and C). Both have clear vacuoles in the peripheral protoplasm, and some of these vacuoles periodically discharge the fluid they collect; they are almost certainly osmoregulatory and are comparable to contractile vacuoles. *Actinophrys* has a single nucleus, but *Actinosphaerium* has a number of nuclei. In both types, feeding follows contact with food organisms, such as ciliates or rotifers, by one or more axopodia. The axopodia are labile and their bundles of microtubules may break down to permit changes of shape. Several axopodia may operate collectively to engulf the prey. The prey is gradually shifted to a food vacuole in the more fluid inner protoplasm of the main mass, where digestion takes place. *Actinophrys* and *Actinosphaerium* may float freely, or they may become attached to a firm substratum by means of their axopodia.

Some heliozoans become united by protoplasmic bridges into networklike colonies. Others secrete a gelatinous envelope, which may contain siliceous spicules produced by the organism itself, or it may contain foreign material, such as the siliceous walls of diatoms. In one beautiful genus, *Clathrulina,* the main part of the body, supported by a slender stalk, is enclosed by a perforated organic capsule.

Asexual reproduction is by fission, formation of buds, or multiple fission. The last-named process may take place within a cyst. In some genera the products of multiple fission temporarily have flagella. Sexual union of an odd type occurs in *Actinophrys*. Encystment is followed by binary fission. The two daughter cells now divide twice, but their divisions are unequal, so that each produces one gamete and three small, nonfunctional cells. The gametes unite to form a zygote, which secretes a cyst of its own. Eventually the cysts break down and the zygote emerges. Note that in this sexual process, called autogamy, both gametes come from the same parent. A similar phenomenon occurs in *Actinosphaerium,* but in this genus it is more complicated because the parent individual is multinucleate.

REFERENCES ON SARCOMASTIGOPHORA

MASTIGOPHORA

Buetow, D. C., 1968. The biology of *Euglena*. 2 volumes. Academic Press, New York and London.

Cachon, J., 1964. Contribution à l'étude des péridiniens parasites. Cytologie, cycles évolutifs. Annales des Sciences Naturelles, Zoologie, series 12, *6*:1–158.

Cleveland, L. R., 1956. Brief accounts of the sexual cycles of the flagellates of *Cryptocercus*. Journal of Protozoology, *3*:161–180.

Cox, E. R. (editor), 1980. Phytoflagellates. Elsevier/North Holland, New York and Amsterdam.

Dodge, J. D., 1986. Atlas of Dinoflagellates. Blackwell Scientific Publications, Palo Alto, California.

Ettl, H., 1976. Die gattung *Chlamydomonas* (*Chlamydomonas* und die nächstverwandten Gattungen 2). Nova Hedwigia, Beiheft 49.

Hoare, C. A., 1972. The Trypanosomes of Mammals: A Zoological Monograph. Blackwell, Oxford.

Leadbeater, B. S. C., 1972. Fine-structural observations on some marine choanoflagellates from the coast of Norway. Journal of the Marine Biological Association of the United Kingdom, *52*:67–79.

Leedale, G. F., 1967. Euglenoid Flagellates. Prentice-Hall, Englewood Cliffs, New Jersey.

Lumsden, W. H. R., and Evans, D. A. (editors), 1976–1979. Biology of the Kinetoplastida. 2 volumes. Academic Press, London and New York.

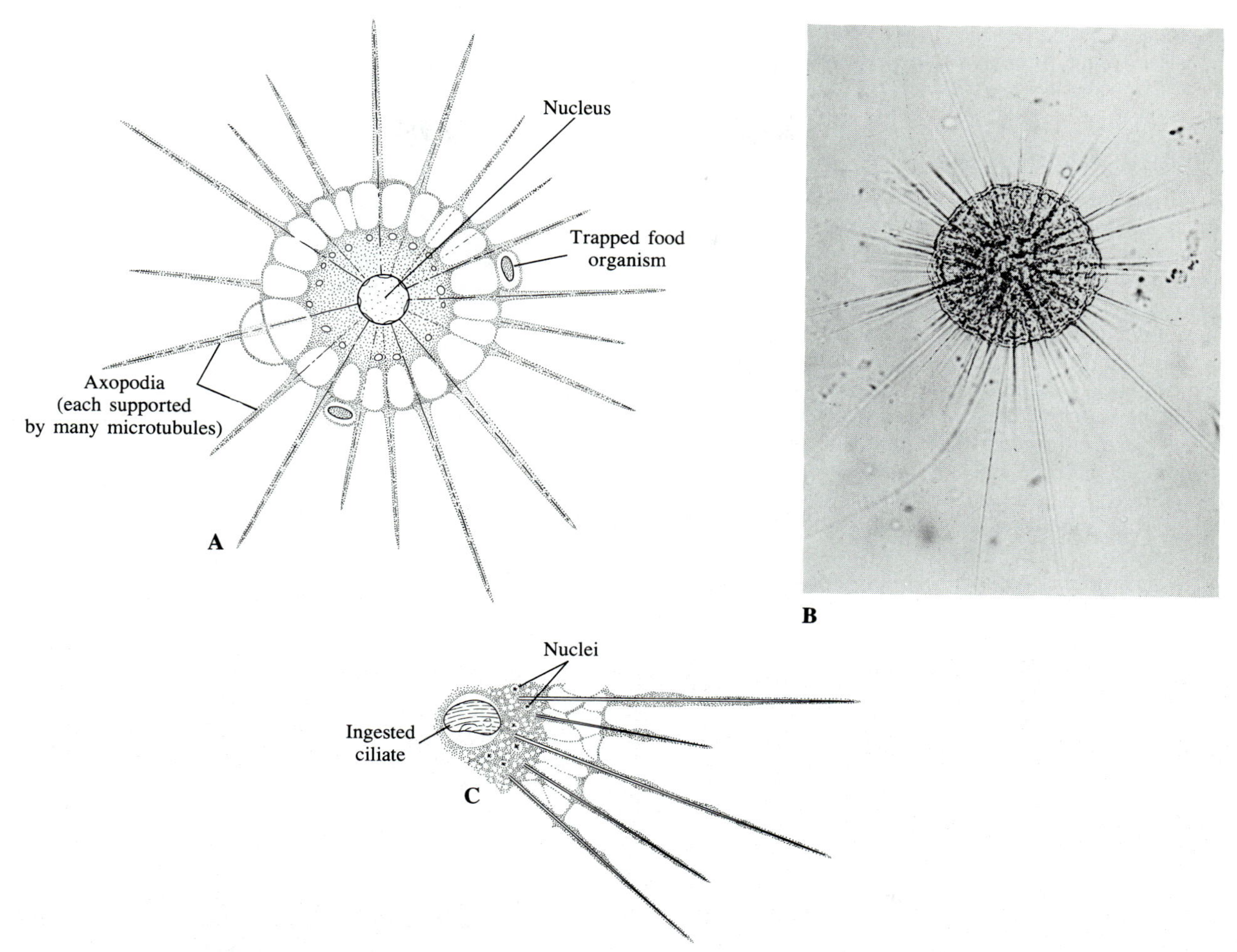

FIGURE 2.25 Class Heliozoea. A. *Actinophrys*. B. *Actinosphaerium;* photomicrograph of a living specimen. (Courtesy of Carolina Biological Supply Co., Inc.) C. *Actinosphaerium,* peripheral portion.

Sargeant, W. A. S., 1974. Fossil and Living Dinoflagellates. Academic Press, London and New York.

Taylor, D. L., and Seliger, H. H. (editors), 1979. Toxic Dinoflagellate Blooms. Elsevier/North Holland, New York and Amsterdam.

Taylor, F. J. R., 1987. The Biology of Dinoflagellates. Blackwell Scientific Publications, Palo Alto, California.

Vickerman, K., 1969. The fine structure of *Trypanosoma congolense* in its bloodstream phase. Journal of Protozoology, *16*: 54–69.

SARCODINA

Anderson, O. R., 1983. Radiolaria. Springer-Verlag, New York and Heidelberg.

Boltovskoy, E., and Wright, R., 1976. Recent Foraminifera. W. Junk, The Hague.

Cushman, J. A., 1948. Foraminifera. Their Classification and Economic Use. 4th edition. Harvard University Press, Cambridge.

Grospietsch, T., 1958. Wechseltierchen (Rhizopoden). Kosmos, Stuttgart.

Haynes, J. R., 1981. Foraminifera. John Wiley & Sons, New York.

Hedley, R. H., and Adams, C. G. 1974–1976. Foraminifera. 2 volumes. Academic Press, London and New York.

Jeon, K. W. (editor), 1973. The Biology of Amoeba. Academic Press, New York and London.

Jepps, M. W., 1967. The Protozoa, Sarcodina. Oliver and Boyd, Edinburgh and London.

Murray, J. W., 1971. An Atlas of British Recent Foraminifera. Heinemann, London.

Murray, J. W., 1979. British Nearshore Foraminifera. Synopses of the British Fauna, no. 16. Linnean Society of London and Academic Press, London and New York.

Ogden, C. G., and Hedley, R. H., 1980. An Atlas of Freshwater Testate Amoebae. Oxford University Press, London and New York.

PHYLUM APICOMPLEXA

Gregarines, coccidians, hemosporidians, and a few other groups of parasitic protozoans are thought to be related because their infective stages have what is called an **apical complex** (Figure 2.26). This is elaborate, but also very small, so that it cannot be recognized or interpreted without the help of an electron microscope. Moreover, it disappears after the infective stage establishes its association with a host cell. It is, nevertheless, what the infective stage uses to penetrate the host cell, either for becoming partly embedded in the cell or for becoming a completely intracellular parasite, at least temporarily. The apical complex generally includes a polar ring and some longitudinally oriented peripheral microtubules, as well as a few electron-opaque bodies called **rhoptries** and one or more pores at the surface. The numerous elongate bodies **(micronemes)** that fill up much of the inside of the apical complex appear to have fine ducts that join the rhoptries, and it is thought that this part of the system is concerned with secreting material that helps the infective body penetrate a host cell.

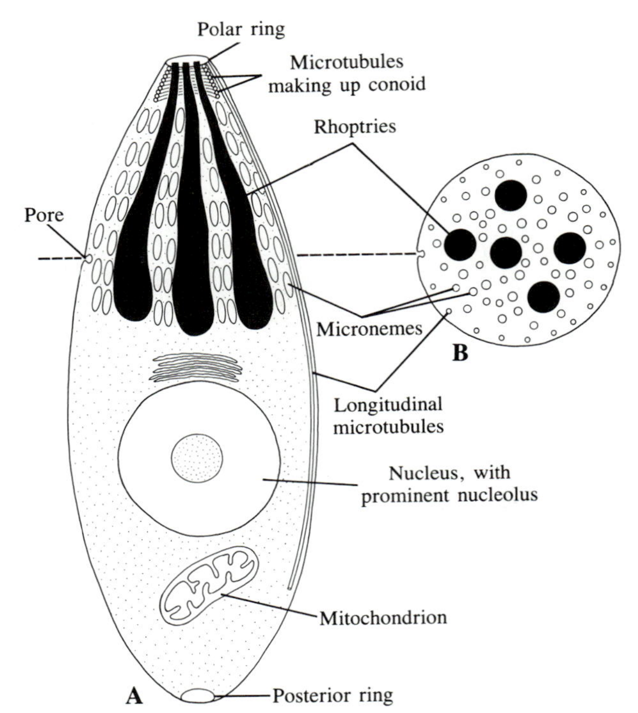

FIGURE 2.26 An infective body of an apicomplexan, generalized; diagrams based on electron micrographs. A. Longitudinal section. B. Transverse section at the level indicated by the dashed line.

CLASS GREGARINEA

Gregarines are most often found in insects, crustaceans, annelids, and ascidians, but a few occur in other invertebrates. Those that live in the guts of insects and crustaceans are characteristically divided by a transverse septum into an anterior **protomerite** and posterior **deutomerite** (Figure 2.27, A–C); the deutomerite contains the nucleus. Growing out of the protomerite, at least in a young septate gregarine of this type, is a structure called the **epimerite.** This is embedded in the gut epithelium and is sometimes rather complicated (Figure 2.27, C). Gregarines that parasitize hosts other than insects and crustaceans are not divided (Figure 2.27, D; Figure 2.29, A), and most of them lack an epimerite.

Gregarines usually have a thick pellicle. This consists of the cell membrane and fibrous elements directly under it. The fibers presumably give the pellicle strength and elasticity. In some species, electron microscopy reveals what are probably contractile filaments beneath the pellicle. Most of the cytoplasm is very fluid, but the outer layer of it, next to the pellicle, is rather firm. In gregarines that are divided into protomerite and deutomerite, and in some one-piece gregarines, locomotion is by a peculiar sort of gliding. The pellicle is folded into closely spaced longitudinal ridges and it seems likely that subtle undulations of these ridges result in forward progression. Secretion of mucus or some other material has also been seriously considered as the cause of the gliding movement. Many of the gregarines without an epimerite can expand and contract in such a way that the more fluid inner cytoplasm is squeezed from one part of the body to another. Movements of this type are often peristaltic, and almost certainly depend upon contractility of myofilaments in the firm outer layer of cytoplasm.

The pattern of sexual reproduction is basically the same in all true gregarines. Two mature individuals, called **gamonts,** secrete a common cyst **(gamontocyst)** around themselves, and each individual produces many gametes. The gametes pair off and fuse to form zygotes. The nucleus of a zygote divides three times, and then the cytoplasm is apportioned equally to the eight nuclei. The resulting slender cells, conventionally called **sporozoites,**[1] represent the infective stage. They are enclosed within a **zygocyst** secreted by the zygote. Meiosis takes place right after the zygote is formed, so

[1]The term *sporozoite,* although deeply entrenched, is somewhat unsatisfactory because it alludes to the term *spore,* which has many different meanings in microbiology, botany, and zoology.

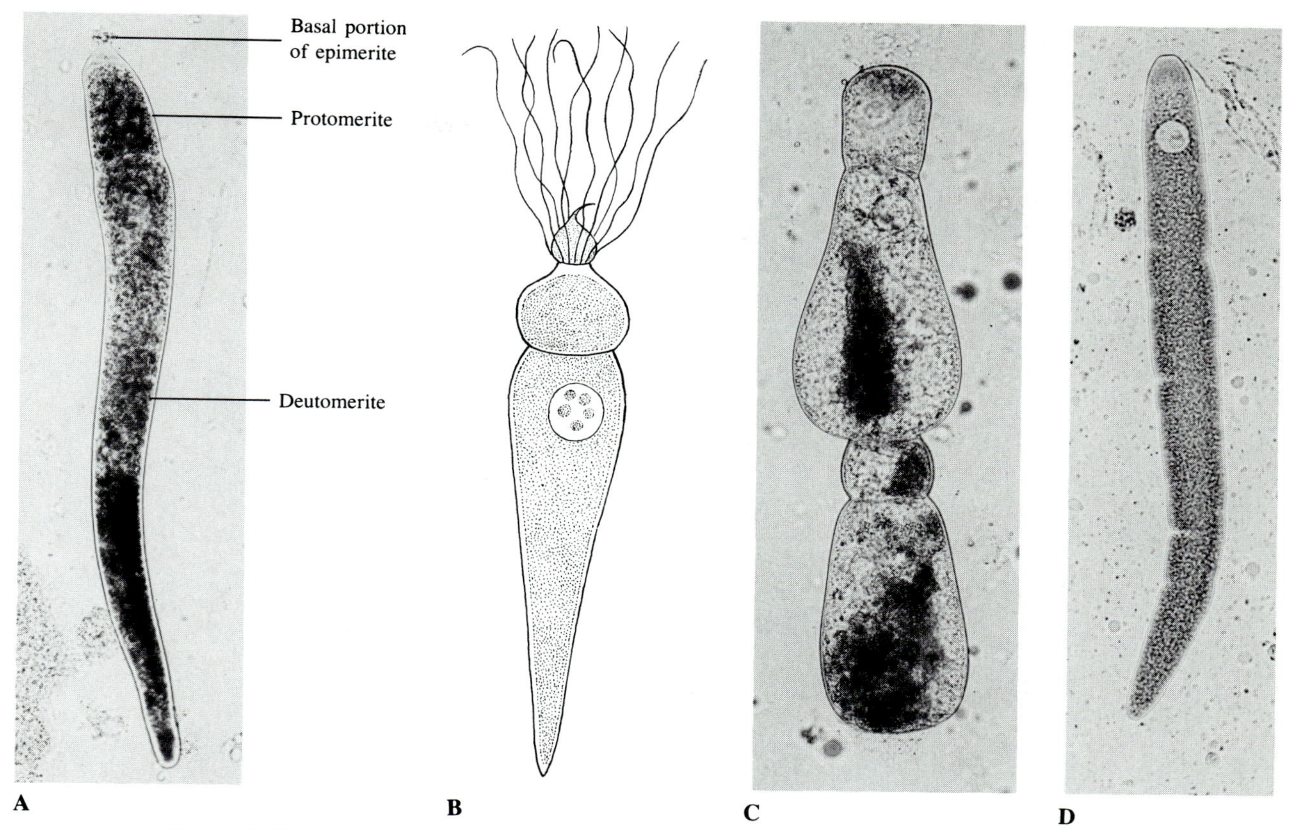

FIGURE 2.27 Class Gregarinea. A. *Pyxinia crystalligera,* a septate type from the midgut of the hide beetle, *Dermestes vulpinus;* photomicrograph. The slender terminal portion of the epimerite, not quite as long as the protomerite, has been broken off, but the crownlike basal portion is still present. B. *Pogonites capitatus,* which has a complex epimerite. (Léger, Tablettes Zoologiques, *3.*) C. *Gregarina polymorpha,* a septate type from the the midgut of the mealworm beetle, *Tenebrio molitor;* photomicrograph. The two specimens are in syzygy. This species loses its epimerite early. D. *Lecudina,* an aseptate gregarine from the gut of *Lumbrineris,* a polychaete annelid.

the sporozoites and the gregarines into which they develop are haploid. Thus the zygote is the only diploid stage in the life cycle.

To illustrate a little of the diversity of this group of protozoans, the life history of three genera will be described briefly. The first two, *Pyxinia* and *Gregarina,* are divided into protomerite and deutomerite, have an epimerite (at least when young), and move by gliding. The third genus, *Apolocystis,* is not divided, has no epimerite, and moves by actively changing its shape.

Pyxinia (Figure 2.27, A) is found in the midgut of various beetles. The epimerite is prominent in a growing individual but is shed as the organism approaches maturity. After two individuals pair, they secrete a gamontocyst around themselves, and in the process of doing this each becomes hemispherical (Figure 2.28, A). The partition between protomerite and deutomerite of both partners disappears. Repeated divisions of the nuclei are followed by cytoplasmic division, so many small gametes are produced (B, C). Because the original covering of each gregarine has broken down, the gametes derived from one individual are free to unite with those from the other (D). Each zygote (E, F) secretes a zygocyst around itself, and its diploid nucleus divides three times to form eight haploid nuclei. The cytoplasm then divides, so that there are now eight sporozoites within the zygocyst. Although the pairing of two mature gregarines and the formation of a gamontocyst around them takes place while they are still in the gut, the production of gametes, the fusion of gametes, and the divisions of the zygotes to form sporozoites take place after the gamontocyst has been voided

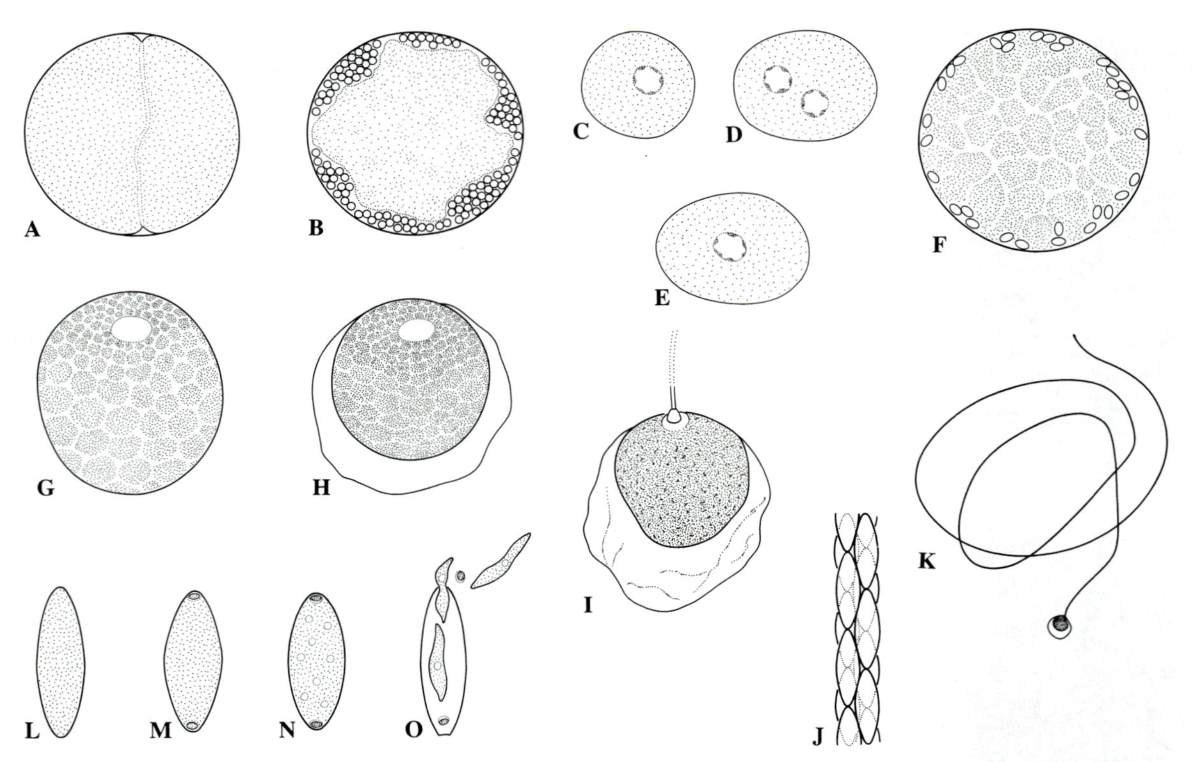

FIGURE 2.28 Class Gregarinea. Life history of *Pyxinia crystalligera*. A. Two gamonts after pairing and secreting a cyst. B. Numerous gametes produced by the gamonts, which are no longer distinct. C. A gamete. D. Gametes fused, but their nuclei still separate. E. Zygote. F. Zygotes secreting zygocysts. Each zygote undergoes two meiotic divisions and one mitotic division, producing eight haploid sporozoites. G. A clear area developing on one side of the mass of crystals and paraglycogen granules that encloses the mature zygocysts, which fill the interior. H. Shrinkage of the mass of crystals and paraglycogen granules. I. Emergence of the thread of zygocysts from the center of the clear area. J. Detail of a portion of a thread of zygocysts. K. Thread of zygocysts completely extruded. L–O. Stages in the germination of a zygocyst. After a zygocyst has swelled due to absorption of water (N), the nuclei of the sporozoites become distinct. In O, only three of the eight sporozoites are shown; the others have moved away after escaping from the zygocyst.

with feces. The inner mass of the gamontocyst shrinks (G, H), causing the zygocysts to be pressed against the membrane until they puncture it. A long thread of zygocysts emerges from the neat hole in the gamontocyst membrane (I–K). The thread emerges slowly and steadily. It is brittle and soon breaks up, so that the zygocysts are scattered by air currents or movements of animals. If eaten by a beetle, a zygocyst swells (L–N) and soon a plug at either end pops out. The sporozoites then crawl out of the tiny opening (L–O). They are motile to some extent, and after they become attached to the epithelium of the midgut they grow rapidly and become differentiated into epimerite, protomerite, and deutomerite.

The genus *Gregarina,* found in many insects (including the common mealworm cultured for bird and lizard food), follows a slightly different pattern. Prospective gamete-forming individuals become associated while they are young (Figure 2.27, C); this precocious association is called **syzygy.** The protomerite of one individual adheres to the deutomerite of the other. (By this time, the epimerites have disappeared.) After maturity, the two gregarines secrete a cyst around themselves. The rest of the reproductive cycle is essentially similar to that described for *Pyxinia,* although details of cyst rupture and zygocyst germination are different.

Apolocystis (Figure 2.29) is found in the seminal vesicles of the common earthworms of the genus *Lumbricus*. After mature individuals have paired, they secrete a cyst around themselves. Gamete formation, union of gametes, and division of the zygotes within zygocyst envelopes follow the pattern described for *Pyxinia* and *Gregarina*. The cysts accumulate in the seminal vesicles, and these probably cannot get out and

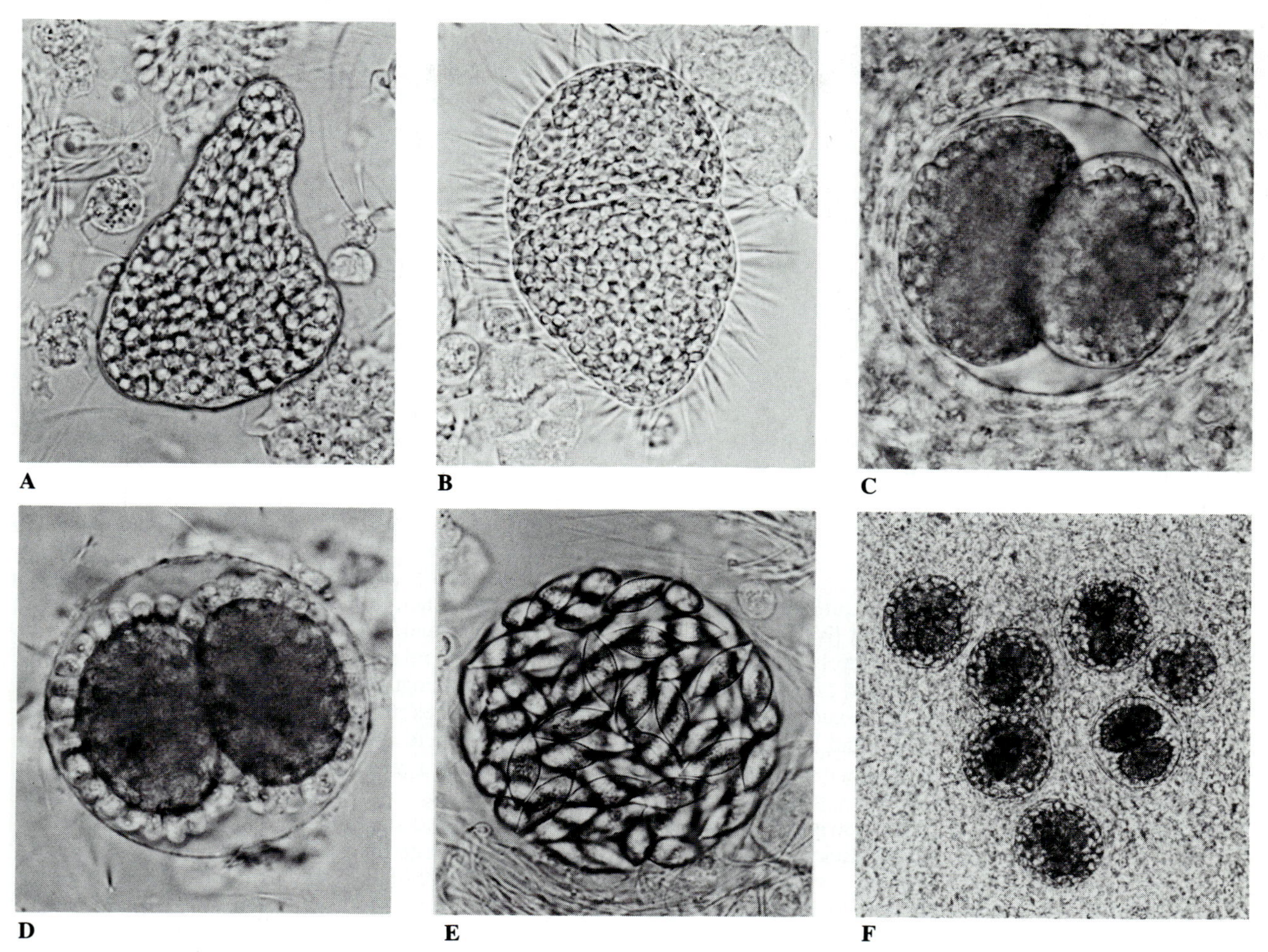

FIGURE 2.29 Class Gregarinea. Stages in the life history of *Apolocystis minuta*, an aseptate gregarine found in the seminal vesicles of the earthworm *Lumbricus terrestris* and related species. A. A mature gamont, which moves in much the same way as an amoeba. The large granules in the cytoplasm consist of paraglycogen. B. Gamonts paired and ready to encyst; both still have earthworm sperm attached to them. C. A pair of gamonts encysted. D. Numerous gametes produced by both gamonts. E. Zygocysts filling up the cyst secreted by the gamonts. F. A group of cysts. One, at the right, contains gametes; the others already have zygocysts.

release their zygocysts until the earthworm dies and disintegrates. When zygocysts are eaten by a prospective host, the sporozoites inside them are released and somehow get into a seminal vesicle. They eventually penetrate clusters of developing sperm and grow to maturity. The sperm of parasitized clusters do not develop normally.

Apolocystis is not the only genus of gregarines found in *Lumbricus*. One often sees trophic individuals and reproductive stages of some of its relatives, especially *Monocystis*, and two or more species may parasitize the same earthworm.

The types that have been mentioned are only a few of many that are called true gregarines, assigned to the order Eugregarinida. The orders Archigregarinida and Neogregarinida include organisms that are much like true gregarines in the way they reproduce sexually, but that also reproduce asexually by multiple fission (merogony). Some of them structurally resemble true gregarines; others do not. They live in a wide variety of hosts: archigregarines live mostly in polychaete annelids, sipunculans, hemichordates, and ascidians; neogregarines parasitize insects. Archigregarines, of which *Meroselenidium* (Figure 2.30) is an example, are especially distinctive because the gamonts exhibit active coiling and undulating movements that resemble those

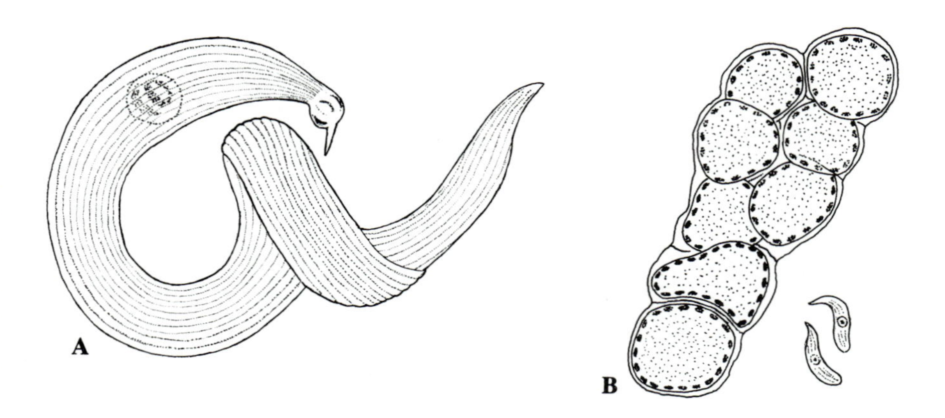

FIGURE 2.30 Class Gregarinea. *Meroselenidium keilini,* an archigregarine, from the gut of *Potamilla reniformis,* a polychaete annelid. A. A large trophic individual, showing the longitudinal bands. B. A late stage of merogony and two merozoites. (After Mackinnon and Ray, Parasitology, *25.*)

of nematode worms. Part of the cytoplasm is differentiated into several longitudinal bands; the contractile properties almost certainly reside in these.

CLASS COCCIDEA

Members of the class Coccidea, unlike gregarines, are strictly intracellular parasites that live in the gut epithelium and other tissues of both vertebrates and invertebrates. In most of them, a period of asexual reproduction by merogony is followed by sexual reproduction, in which distinctly different male and female gametes are formed. Division of the zygote is meiotic and leads to production of infective sporozoites, just as in gregarines. If the sporozoites get into a susceptible host, they develop into a new generation of individuals that reproduce asexually.

The class is here divided into two orders, Coccidiida and Haemosporida. These are placed into a single order, Eucoccidiida, in the Society of Protozoologists' classification. Two other small orders in that scheme—Agamococcidiida and Protococcidiida—are not discussed here: neither of them exhibits merogony, and placing them in the Coccidea may not be justified.

Order Coccidiida

This order includes a number of species that are of importance in veterinary medicine, and one species, *Isospora hominis,* is a parasite of humans. The life cycle of *Eimeria tenella* (Figure 2.31), which inhabits the intestinal epithelium of chicks, is typical of the group. After a sporozoite of *E. tenella* enters an epithelial cell, it grows rapidly, becoming what is called a **meront** (A), because it will undergo merogony. The parasitized cell and its nucleus also enlarge. In a few days the nucleus of the meront divides repeatedly, forming many daughter nuclei, and then a little cytoplasm is apportioned to each of them. The uninucleate bodies produced by merogony are called **merozoites.** They are released as the host cell breaks down and may then enter previously unparasitized cells, grow, and repeat the process of merogony, producing a second generation of merozoites. The epithelium may be severely damaged by the parasitism.

Although merogony may continue, some merozoites of the second generation develop into individuals called **microgamonts** and **macrogamonts,** which give rise to gametes (C). A microgamont divides to produce many flagellated **microgametes** (sperm); a macrogamont gradually matures into a single **macrogamete** (egg). Each macrogamete develops a rather thick wall around itself, except in a restricted area at one end, where the sperm may enter. Following fertilization, the envelope is completed, and it may now be considered a homologue of the gregarine zygocyst, though it is usually called an **oocyst.** This falls out of the diseased tissue and leaves the host along with fecal material. The nucleus of the zygote divides twice, forming four nuclei, and the cytoplasm is now distributed evenly to these. Each of the units secretes a thin envelope around itself and divides again, resulting in eight infective sporozoites, in four pairs, within the oocyst. When an oocyst is eaten by another chick, its outer envelope and the envelopes enclosing pairs of sporozoites are broken down by digestive enzymes, and probably also by abrasion. The sporozoites are thus freed to enter cells of the intestinal epithelium and start a new infection.

Toxoplasma gondii, an intracellular parasite of many mammals (including humans) and certain birds, is a coccidian whose life cycle has stages in addition to those just described for *Eimeria.* In the intestinal epithelium of the domestic cat and other felines, there is asexual multiplication by merogony, and finally sexual multiplication, just as in *Eimeria.* After an oocyst has left the body of the cat with fecal material, eight sporozoites—in two groups of four each, as is characteristic of the genus *Isospora,* rather than the four groups of two each characteristic of *Eimeria*—develop from the zygote. When the oocyst is eaten by another cat, the

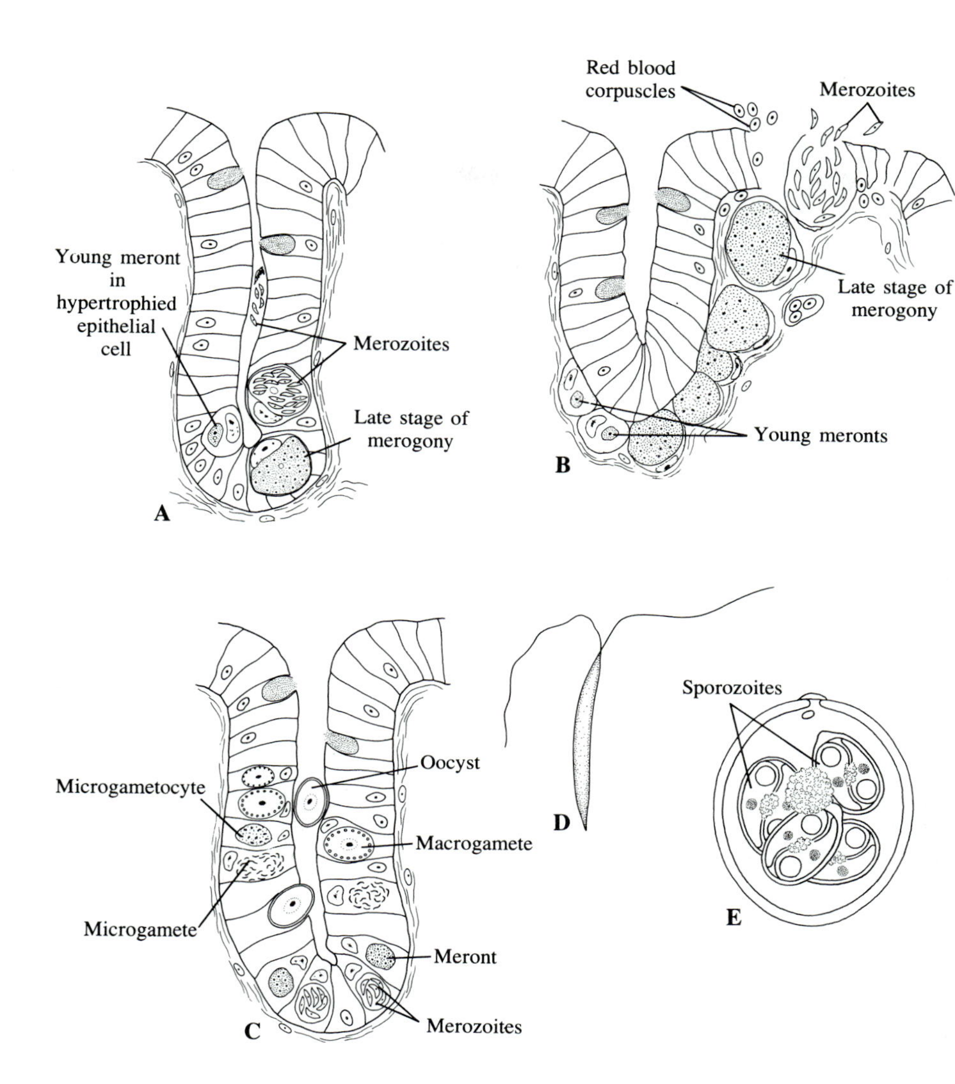

FIGURE 2.31 Class Coccidea. Life history of the coccidian *Eimeria tenella,* a parasite of the intestine of chickens. A. Merogony leading to production of first generation of merozoites. B. Merogony leading to production of second generation of merozoites. C. Production of microgametes and macrogametes, and escape of young oocysts; some merogony leading to production of a third generation of merozoites. (A–C, after Tyzzer, American Journal of Hygiene, *10.*) D. Microgamete. (This is only a few μm long.) E. An oocyst after formation of its eight sporozoites. The outer wall is a two-layered structure.

cycle of merogony and sexual reproduction may be repeated. The parasites may, however, get into other tissues, where they multiply rapidly by binary fission. Eight to 16 parasites may be formed within a single cell before this cell breaks down and liberates them. Additional cells become infected, and the process of multiplication is repeated. As the disease becomes chronic, multiplication slows down, but now the parasites produced by fission accumulate in large numbers within cystlike structures. When these rupture, the parasites are released and may multiply again.

If oocysts expelled with a cat's feces are accidentally eaten by humans, sheep, cows, pigs, rodents, or other mammals, there is no typical coccidian life cycle in the intestinal epithelium; the disease moves directly to the stages in which there is repeated binary fission within muscle, lymphatic organs, the brain, and other tissues.

The symptoms of an acute infection in humans vary. Sometimes they resemble those of influenza: headache, fever, and muscle pain are often noted. The disease in adults is rarely fatal or dangerous. A pregnant woman, however, can confer the disease on an unborn child, and this may result in the serious disability or death of the child. Pregnant women should definitely avoid close contact with cats.

A further complication is that a mammal in whose tissues binary fission has taken place may serve as food for some other animal. The parasites released from a cyst are resistant to digestion, and if they penetrate the gut epithelium of a carnivore, they may start new infections. Thus humans can get toxoplasmosis by eating raw meat, cats and dogs can get it from raw meat or from rodents, carnivorous birds can get it from rodents, and so on.

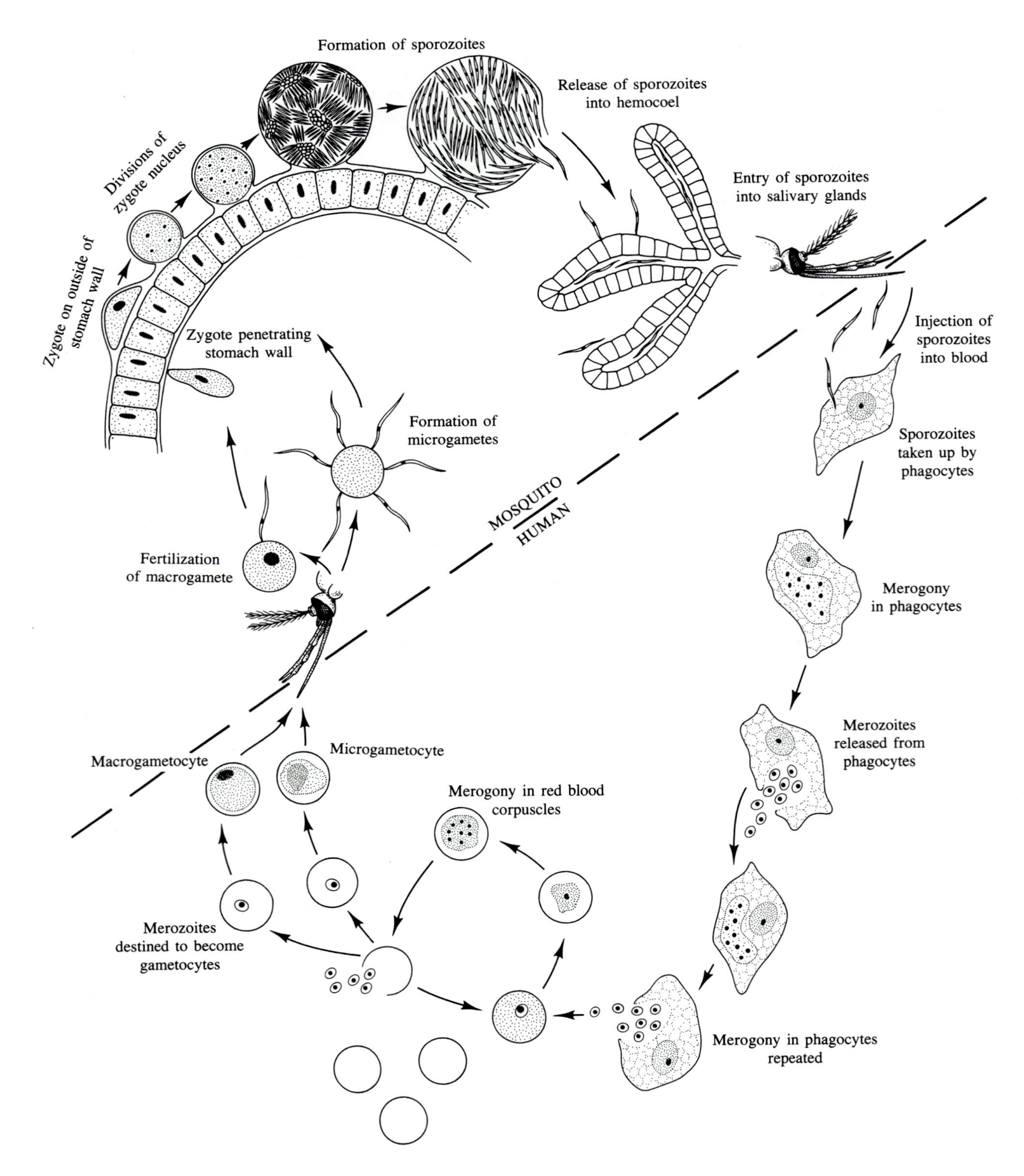

FIGURE 2.32 Class Coccidea. Life history of *Plasmodium vivax,* one of the malarial parasites of humans. Explanation in text.

In summary, *Toxoplasma* is one of the most versatile of parasites. It lives in many species of mammals and birds, and is capable not only of completing a life cycle similar to that of *Eimeria* and *Isospora,* but also of undergoing intracellular binary fission.

Order Haemosporida

Hemosporidans (or haemosporidans) are apparently close to coccidians, although completion of their life cycle requires two hosts. The merogonic phase takes place in the blood or other tissues of a reptile, bird, or mammal; sexual reproduction takes place in an invertebrate (usually an arthropod) that sucks the blood of the vertebrate host and thus acts as a transmitting agent. Malarial infections in humans are caused by hemosporidans of the genus *Plasmodium,* transmitted by certain mosquitoes. *Plasmodium vivax* (Figure 2.32) is the most prevalent species infecting humans.

A malarial infection begins when a female *Anopheles* mosquito in which sporozoites have developed pierces the skin and injects a salivary secretion that prevents blood-clotting. (Male *Anopheles* do not feed on blood, so they are not involved in the life cycle.) The sporozoites entering the blood are taken up by phagocytic cells, including those closely associated with capillaries in the liver. Within the phagocytes, the parasites enlarge and undergo merogony to produce a number of merozoites. These break out of their host cells and are taken up by other phagocytes, so the process of growth and merogony is repeated. Eventually, some merozoites enter red blood corpuscles, where merogony again produces merozoites. Merogony in red blood corpuscles, as well as in phagocytic cells, may continue indefinitely. The chills and fever associated with a malarial infection are due to the synchronous release of toxic substances from red blood corpuscles at the time these disintegrate and release the merozoites that have developed within them. In *P. vivax,* chills and fever are noted about every 48 hours.

Eventually some merozoites that have entered red blood corpuscles become microgamonts and macrogamonts instead of meronts. Both types of gamonts may live for several weeks. If the right kind of female *Anopheles* draws blood from an infected person, it may ingest a few of them. The nucleus of a microgamont divides several times, and each of the products of this nuclear division moves to the periphery, acquires some cytoplasm, and breaks away as a slender, motile microgamete. A macrogamont simply matures into a macrogamete, and is fertilized. The zygote works its way to the outside of the thin stomach wall, where it grows. Its nucleus undergoes several divisions, forming centers where subsequent divisions will lead to the production of numerous infective sporozoites. The delicate envelope around the mass of sporozoites ruptures, and the sporozoites are released in the hemocoel of the mosquito. They migrate to the salivary glands and become concentrated in these organs, ready to enter the bloodstream when the mosquito pierces the skin of another person.

Note that the life cycle of *Plasmodium* is basically similar to that of *Eimeria* in that repeated merogony is followed by formation of gametes. The zygote, however, produces hundreds of sporozoites, instead of only eight, and it neither leaves its insect host nor develops a thick zygocyst.

REFERENCES ON APICOMPLEXA

Garnham, P. C. C., 1966. Malarial Parasites and Other Haemosporidia. Blackwell Scientific Publications, Oxford.

Hammond, D. M., and Long, P. L. (editors), 1972. The Coccidia: *Eimeria, Isospora, Toxoplasma,* and Related Genera. University Park Press, Baltimore.

Kreier, J. P. (editor), 1980. Malaria. 3 volumes. Academic Press, New York and London.

PHYLUM MICROSPORA

Microsporans (or microsporidians in older accounts) are intracellular parasites of various invertebrates and vertebrates, especially insects and fishes. They have been found in mammals, however, and at least one species causes encephalitis-like symptoms in humans. It is unlikely that the assemblage is monophyletic.

A new host is probably infected by way of the gut, at least in most cases. The infective cell must get into a host cell of the type in which it can develop further. There it grows and divides by successive binary fissions or multiple fission (Figure 2.33, A). Sooner or later individual cells differentiate into cysts (Figure 2.33, B).

The cyst (''spore'') that contains an infective cell is small—only a few micrometers long—but it is generally fairly complex. Besides the infective cell, a cyst contains, among other things, a **hollow thread.** When a cyst germinates, the thread turns inside-out and its lumen serves as a conduit for the escape of the infective cell.

Nosema bombycis and *N. apis* are economically important parasites of the caterpillars of silkworm moths and honeybees, respectively. The former is famous for another reason, too: when Louis Pasteur was called upon to help the French silkworm industry, troubled by a disease called *pébrine,* his researches proved that the malady was caused by a specific microorganism. His findings fit with the results of his other studies, as well as with the work of other scientists of the nineteenth century, and led to the acceptance of the idea that many diseases or biological processes, such as fermentation, are caused by bacteria, yeasts, fungi, and other microorganisms.

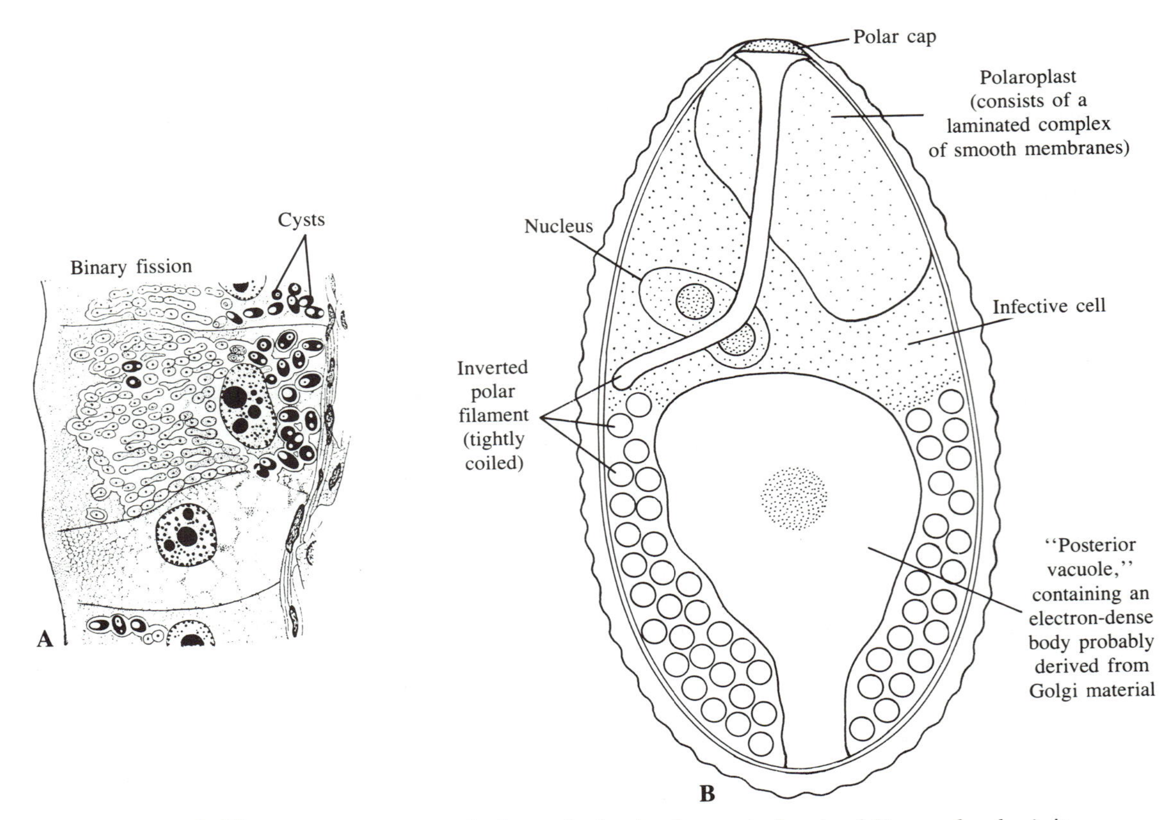

FIGURE 2.33 Phylum Microspora. A. Stages in the development of cysts of *Nosema bombycis* in the gut of a silkworm. The products of repeated binary fission, which takes place almost entirely along one plane, differentiate into cysts, each containing an infective body. (Ivanov *et al.*, Major Practicum in Invertebrate Zoology.) B. Cyst of *Plistophora hyphessobryconis*, a parasite of certain freshwater fishes; diagram based on electron micrographs. (Lom and Vavra, Wiadmosci Parazytologiczne, 7.)

REFERENCES ON MICROSPORA

Canning, E. V., Lom, J., and Dykova, I., 1986. The Microsporidia of Vertebrates. Academic Press, New York and London.

Issi, I. V., and Shulman, S. S., 1968. The systematic position of Microsporidia. Acta Protozoologica, *6:* 121–135.

Sprague, V., 1977. Classification and phylogeny of the Microsporidia. In Bulla, L. A., Jr., and Cheng, T. C. (editors), Comparative Pathobiology, volume 2, Systematics of the Microsporidia, pages 1–30. Plenum Press, New York.

Vivier, E., 1975. The Microsporidia of the Protozoa. Protistologica, *11:* 345–361.

PHYLUM MYXOZOA

In myxozoans, cysts derive from multinucleate masses of protoplasm. The cyst wall encloses not only one or more infective bodies ("sporoplasms") but also one or more **polar capsules.** These resemble nematocysts of jellyfishes and other cnidarians in that they contain hollow threads that are extruded when they turn inside-out under appropriate stimulation. The two classes recognized in the Society of Protozoologists' classification may not be closely related. Perhaps both groups should be elevated to the rank of phylum.

CLASS MYXOSPOREA

Myxosporeans (or myxosporidians, in older systems of classification) are parasites mostly of fishes, but some occur in amphibians and reptiles. They live in a variety of tissues and organs, including the gall bladder, urinary bladder, kidneys, gills, connective tissue, and muscle. Certain species form large masses that become visible externally as lumps.

The wall of a myxosporean cyst (Figure 2.34) is glassy and consists of two equal valves. Within it is an infective cell and one to several (usually two) polar cap-

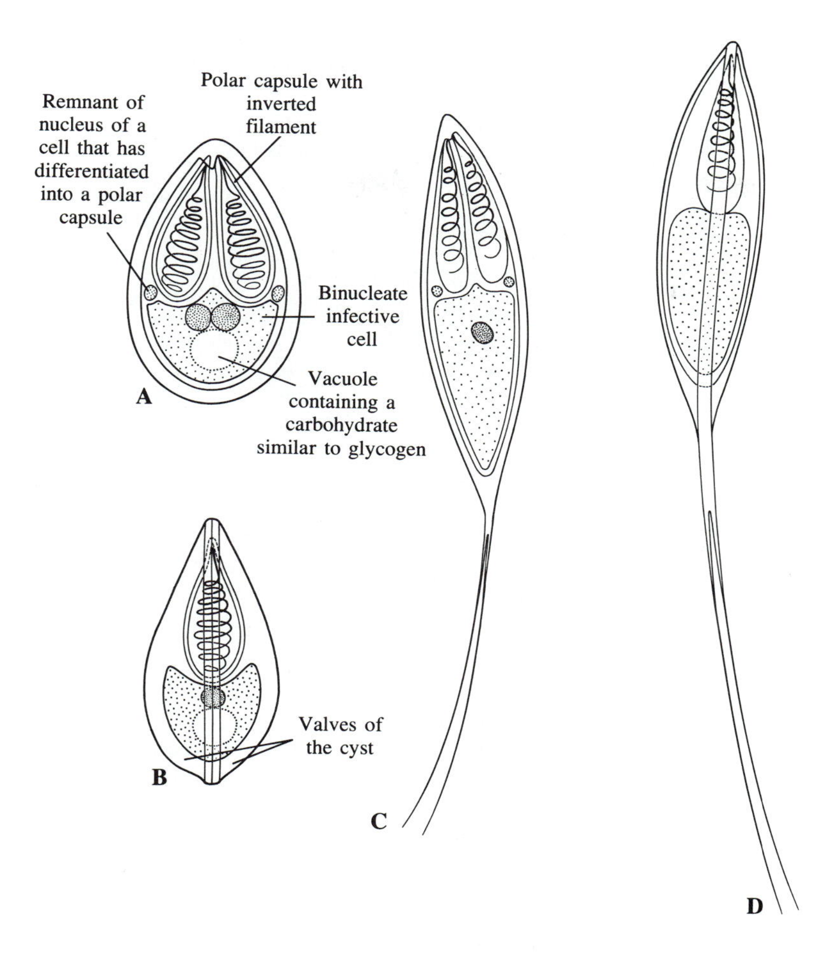

FIGURE 2.34 Class Myxosporea. A, B. Cyst, as seen from different aspects. C, D. Cyst of *Henneguya alekseevi*. (A–D, after Shulman, Myxosporidia of the Fauna of the USSR.)

sules. The threads of the polar capsules are extruded when a cyst enters the gut of a suitable host. (Extrusion can often be induced in the laboratory by placing cysts in a strong solution of hydrogen peroxide.) The discharged threads probably serve to anchor the cyst to the epithelium of the gut, so that when the infective cell is liberated it will be close to tissue it can penetrate. It is not certain, however, that infection necessarily begins in the gut in all species. It is important to understand that the infective cell does not escape by way of the filament, as happens in microsporans.

During its growth phase, a myxosporean forms a plasmodium, a mass that contains many nuclei. Some of the nuclei are enclosed by cells, others are not. When and how the cellular and noncellular lines become separated is not known. In *Sphaeromyxa sabrazesi* (Figure 2.35), a cell of one type becomes enveloped by a cell of another type, then undergoes division to produce two clusters of five cells. Each cluster differentiates into a cyst, two cells forming the valves, two forming polar capsules, and one becoming the infective cell. Later, the infective cell becomes binucleate. The enveloping cells do not contribute to the formation of cysts; they probably nourish the cyst-forming cells. Cells of another type found in the plasmodium are phagocytic; they attack and digest some of the cyst-forming and enveloping cells after these have become associated.

Results of a recent study on the DNA content of plasmodial nuclei suggest that those of the cyst-forming, enveloping, and phagocytic cells are diploid, whereas the nuclei not enclosed by cells are polyploid. Furthermore, the nuclei in the cells that differentiate into cysts appear to be haploid. Chromosomes have not been detected, however, during nuclear divisions, and how meiosis takes place remains a mystery.

Cysts of some myxosporeans, especially those localized in muscle and connective tissue, escape from the host when the flesh is damaged to the extent that it ruptures. Species living in the gall bladder or urinary bladder are thought to leave the body with feces or urine. Some myxosporeans living in deep tissues probably

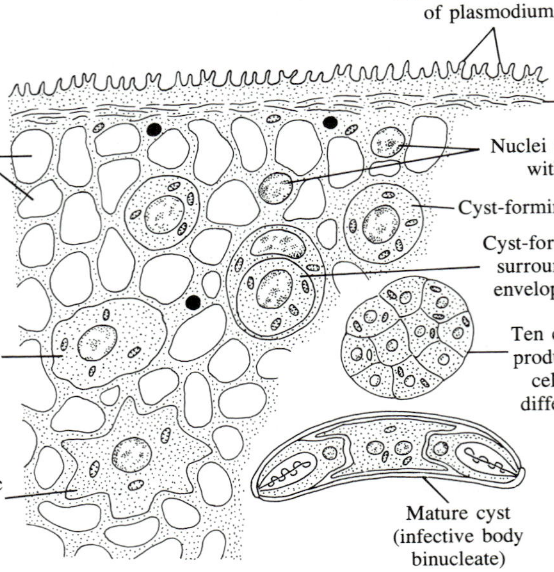

FIGURE 2.35 Class Myxosporea. Stages in the life history of *Sphaeromyxa sabrazesi;* diagrammatic, based on electron micrographs. A cyst-forming cell becomes surrounded by an enveloping cell and produces two clusters of five cells. Each cluster differentiates into a cyst. The infective body, derived from one of the five cells, then becomes binucleate. (After Grassé and Lavette, Annales des Sciences Naturelles, Zoologie, series 12, *20*.)

cannot release cysts except when the host dies and disintegrates, or is eaten and digested.

CLASS ACTINOSPOREA

Actinosporeans live in the gut of oligochaete annelids and in the coelom of sipunculans. Their cysts, which contain one or more infective cells, have three valves (Figure 2.36, A). These are extended into spinelike processes (sometimes forked), and each one has a polar capsule.

The life cycle briefly explained here is that of *Neoactinomyxon eiseniellae* (Figure 2.36, B), a parasite of a freshwater oligochaete. The earliest known stage is a binucleate cell in the epithelium of the intestine (1). When this cell divides, one of the products is destined to be the source of gametes; the other cell gives rise to the four cells that form an external envelope. The first division of the primordial germ cell establishes two lines of cells that may arbitrarily be designated A and B cells (2), for they are of separate sexes. Both divide mitotically three times, so eight A cells and eight B cells are now enclosed within the envelope (3). (The A cells complete the three divisions more quickly than the B cells, however.) The A cells, then the B cells, undergo the first meiotic division, each casting off a polar body (4). Now the cells are at the secondary gametocyte stage. The A and B cells pair and fuse, but before their nuclei unite, they undergo the second meiotic division, and each casts off another polar body (5). After the nuclei fuse, the zygote (6) divides twice, producing three small cells and one large cell (7, 8). Progeny of the small cells differentiate into the polar capsules and valves of the cysts; the large cell produces the infective cells.

Recently, it was reported that a species of *Triactinomyxon* found in a freshwater oligochaete annelid of the genus *Tubifex* is a phase in the life cycle of *Myxosoma cerebralis,* a myxosporean that causes "whirling disease" in young salmonid fishes. This needs verification by rigorous experimental tests. If it is true, it is most remarkable, because the switch from diploidy to haploidy would presumably occur twice, and also because processes of nuclear division are different in myxosporeans and actinosporeans. Hundreds of species of marine and freshwater fishes are parasitized by myxosporeans, but actinosporeans are known only from freshwater oligochaetes and marine sipunculans. It does not, therefore, seem likely that all myxosporeans have an actinosporean phase.

REFERENCES ON MYXOZOA

Grassé, P.-P., and Lavette, A., 1978. La Myxosporidie *Sphaeromyxa sabrazesi* et le nouvel embranchement de Myxozoaires (Myxozoa). Recherches sur l'état pluricellulaire primitif et considérations phylogénétiques. Annales des Sciences Naturelles, Zoologie, series 12, *20*:193–285.

Marquès, A., 1987. La sexualité chez les Actinomyxidies: étude chez *Neoactinomyxon eiseniellae* (Ormières et Frézil, 1969), Actinosporea, Noble, 1980; Myxozoa, Grassé, 1970. Annales des Sciences Naturelles, Zoologie, series 13, *8*:81–101.

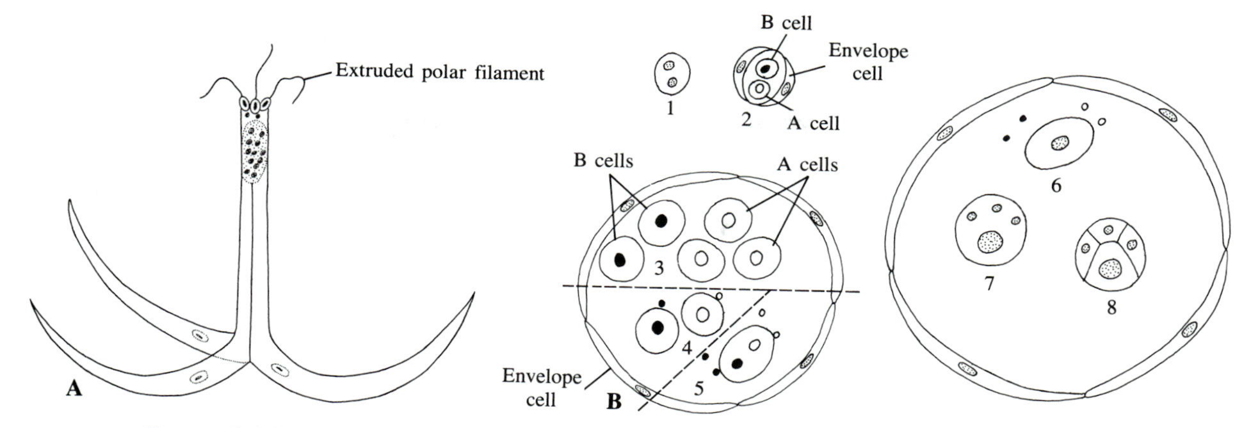

FIGURE 2.36 Class Actinosporea. A. Cyst of *Triactinomyxon ignotum,* a parasite of *Tubifex tubifex,* a freshwater oligochaete annelid. B. Stages in the life history of *Neoactinomyxon eiseniellae;* diagrammatic. The binucleate cell (earliest stage known) (1) divides. One of the cells produces the envelope cells (four at maturity), the other produces primordial germ cells designated A and B types (2). A and B cells each produce 8 cells (3) (only a few are shown), which undergo the first meiotic division (4), then unite and undergo the second meiotic division (5), after which the zygote is formed (6). At the end of the second division of the zygote, the large cell that will produce infective cells can be distinguished from the three whose progeny will contribute the valves and polar capsule of the cyst. (After Marquès, Annales des Sciences Naturelles, Zoologie, series 13, *8*.)

Shulman, S. S., 1959. [Protozoa of freshwater fishes of the USSR. In Russian.] Opredeliteli po Faune SSSR, *80*:7–151.

Uspenskaya, A. V., 1982. New data on the life cycle and biology of Myxosporidia. Archiv für Protiskenkunde, *126*:309–338.

Wolf, K., and Markiw, M. E., 1984. Biology contravenes taxonomy in the Myxozoa: new discoveries show alternation of invertebrate and vertebrate hosts. Science, *225*:1449–1452.

PHYLUM CILIOPHORA

The ciliates are so well unified and so different from other protozoans that they certainly deserve the rank of phylum. Although they typically have cilia at all times, in some the cilia are present only at a certain stage in the life cycle. A few highly specialized types lack cilia altogether, but they meet other criteria for membership in the phylum.

In ciliates, there are generally hundreds or thousands of cilia, arranged in long rows, short rows, tufts, or tonguelike or spinelike aggregates. *Paramecium* (Figure 2.37) provides a good example of the organization of the ciliary system and other structures, though other, decidedly different plans are also described below in connection with the major subgroups of the phylum.

The body of a *Paramecium*—about 150 to 200 μm long, depending on the species—is approximately cigar-shaped and is covered with **somatic cilia** arranged in longitudinal rows, which come together in sutures (Figure 2.38, A). The action of the somatic cilia reminds one of the waves that a gentle breeze produces in a wheatfield. In one part of the body, most of the cilia are just beginning their effective stroke, while in another part they are completing their effective strokes or making their recovery strokes. This synchronized pattern of ciliary activity, called **metachrony,** indicates that the beating of the cilia is coordinated.

A shallow groove on one side of the body leads to an obvious depression. The outer part of this is entered by rows of somatic cilia. The inner part, the buccal cavity, has four specialized groups of buccal cilia that do not belong to the somatic system. One group consists of a membranelike row of cilia that partly encircles the opening to the buccal cavity; the other three groups consist of four rows each and run down the wall of the buccal cavity toward the cytostome, where food is incorporated into a food vacuole. This vacuole, with which lysosomes fuse, is moved around by cytoplasmic streaming, and its contents are gradually digested by enzymes and assimilated. Any residues that remain are ejected through the **cytoproct,** a pore near the posterior end of the body.

The pellicle that covers the body of *Paramecium* (Figure 2.37, B) and other ciliates consists of three membranes, made visible by an electron microscope.

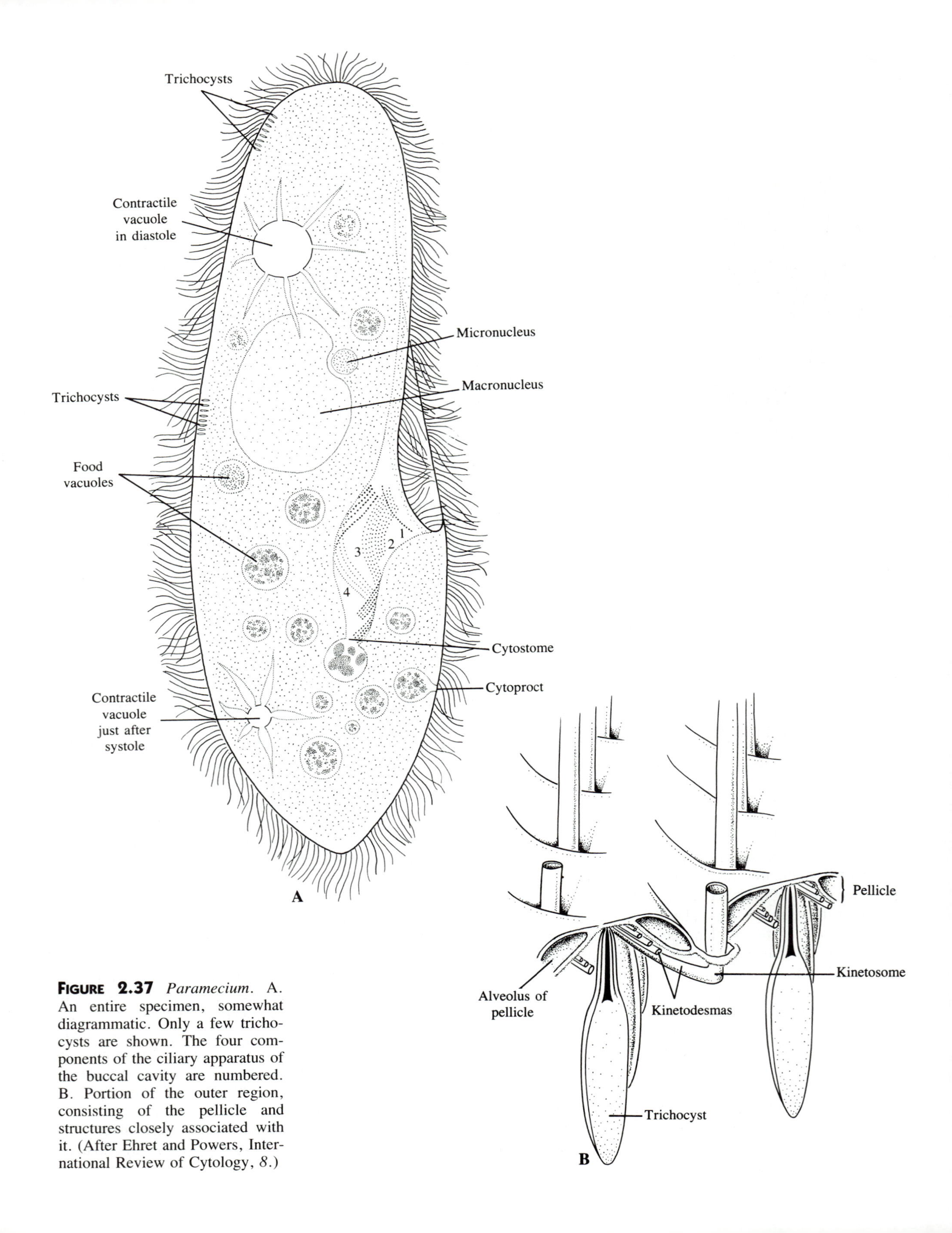

FIGURE 2.37 *Paramecium*. A. An entire specimen, somewhat diagrammatic. Only a few trichocysts are shown. The four components of the ciliary apparatus of the buccal cavity are numbered. B. Portion of the outer region, consisting of the pellicle and structures closely associated with it. (After Ehret and Powers, International Review of Cytology, *8*.)

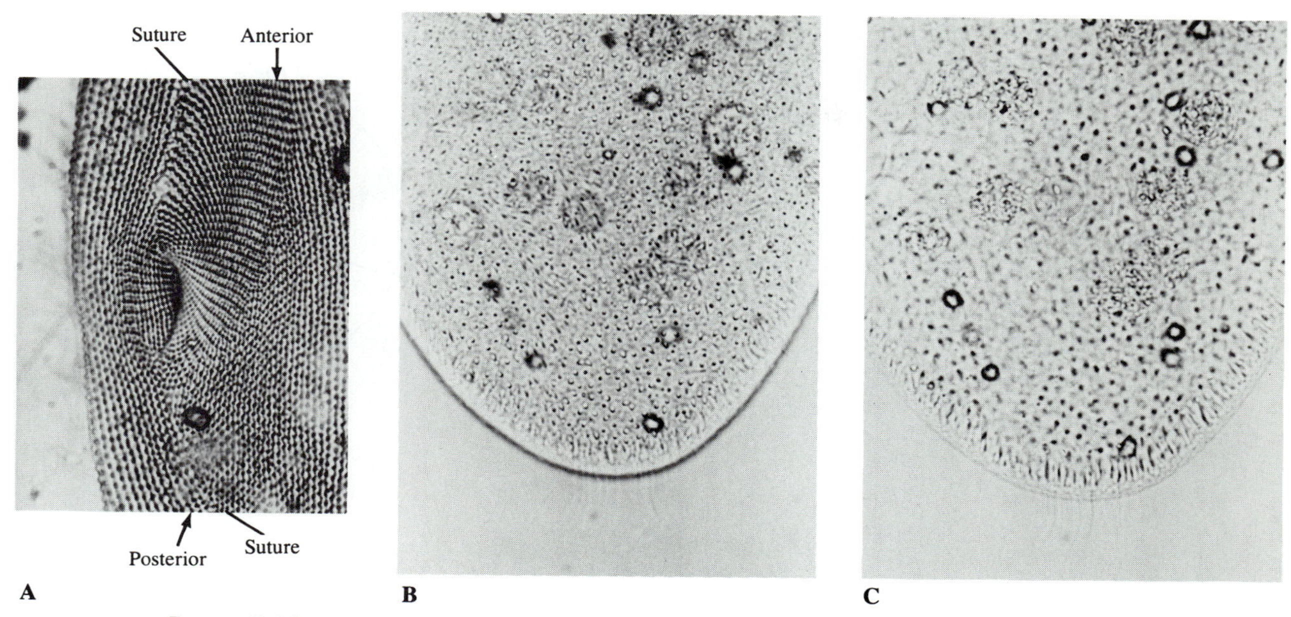

FIGURE 2.38 *Paramecium.* A. Kinetosomes of ciliary rows in the vicinity of entrance to the buccal cavity; photomicrograph of a specimen impregnated with silver. B. Trichocysts in surface view of a living specimen. C. Trichocysts in optical section of a living specimen.

The outermost membrane is continuous and extends to the tips of the cilia. The other two membranes form a system of "cushions," called **alveoli.** The pellicle is flexible but strong; its strength and form-giving qualities are perhaps in large part related to the fact that it is composed of alveoli. The kinetosomes of the cilia reach deeper than the alveoli. The kinetodesmas (rootlets) of successive cilia of a particular row overlap one another, so that they form bundles. There are various other fibers linked with the kinetosomes, too.

Contractile vacuoles—most species of *Paramecium* have two of them—are concerned primarily with osmoregulation. By eliminating excess water that enters the cell by diffusion, they keep the salt concentration of the protoplasm stable. A contractile vacuole is fed by canals that collect fluid from a spongy network of tubules in the cytoplasm (Figure 2.39, A). The canals may swell as fluid accumulates in them, but then the fluid moves into the **ampullae** (dilations) of the canals and finally into the contractile vacuole itself. The accumulated fluid is discharged periodically through a pore in the pellicle. Nitrogenous wastes probably diffuse away all over the surface of the body but perhaps some are also carried away with the water that leaves the contractile vacuole.

Nearly all ciliates, no matter what their habitat, have at least one contractile vacuole. Some have ten or twelve or more, in which case the vacuoles are likely to be in two rows, one on each side of the body.

Many ciliates, including *Paramecium,* have structures called **trichocysts** (Figures 2.37, B; 2.38, B and C). These are perpendicular to the pellicle, and when stimulated by an irritant of some sort, such as dilute methylene blue or acetic acid, they discharge long threads from their tips. Their function remains uncertain, although it is possible that they are useful in defense against certain kinds of predators. However, when *Paramecium* is attacked by *Didinium,* another ciliate of about the same size, the trichocysts seem not to do much good.

With few exceptions, ciliates have two types of nuclei. These differ in size and appearance as well as in their biological roles. The **micronucleus,** during asexual reproduction by binary fission, divides mitotically; it can also divide meiotically, and it does so when a ciliate is preparing for sexual union of haploid nuclei. The **macronucleus,** generally much larger than the micronucleus and commonly polyploid, does not undergo meiosis and does not participate in formation of gametic nuclei. It divides without a mitotic spindle.

There may be two or more micronuclei—this is the case in some species of *Paramecium*—and in many ciliates there is more than one macronucleus. The patterns of nuclear behavior will therefore vary, especially during sexual processes.

During binary fission (Figure 2.40), micronuclei divide by mitosis; macronuclei divide simply by elongating and separating into two parts, often "spilling" a

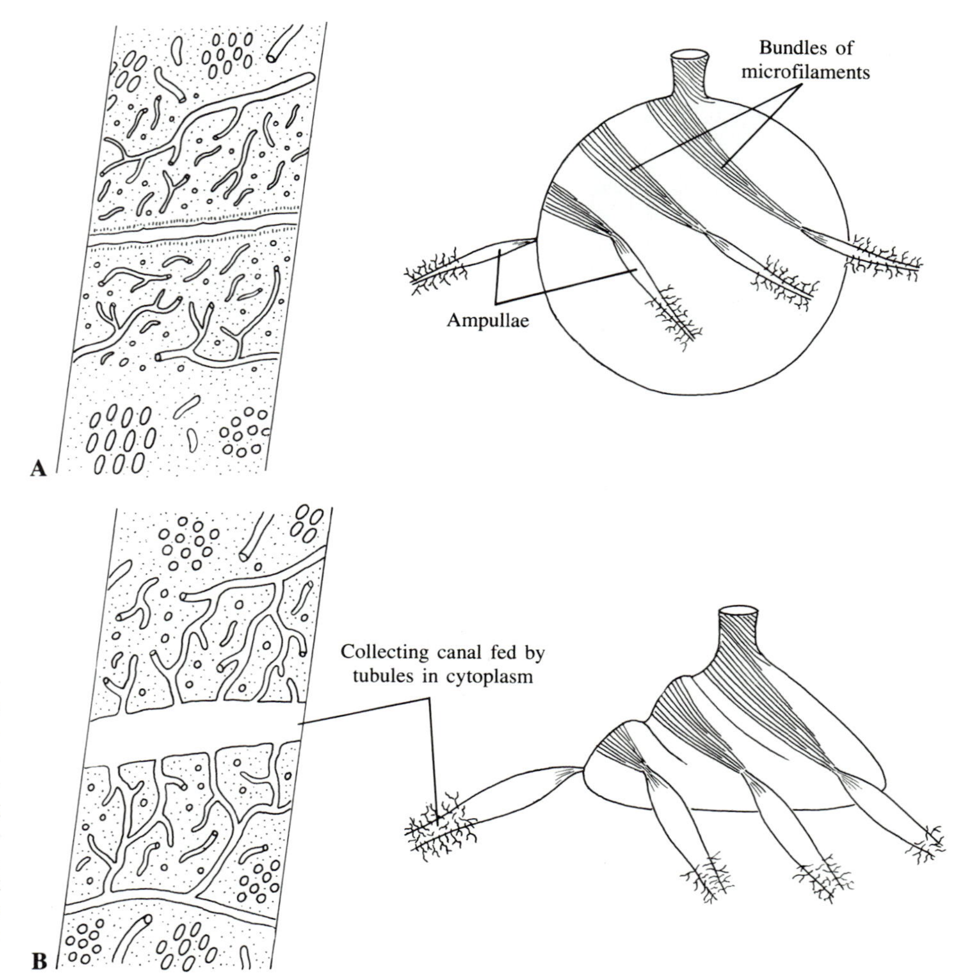

FIGURE 2.39 *Paramecium,* contractile vacuole. A. Diastole. The reservoir is filled, but the ampullae and collecting canal are narrowed; note also that the microtubules that feed the collecting canals (enlarged section) are temporarily disconnected. B. Systole. The reservoir is almost empty, but the ampullae and collecting canals are filled. (After Schneider, Journal of Protozoology, 7.)

little chromatin. By the time fission has been completed, the anterior and posterior daughters are essentially like the parent. Nevertheless, during the later stages of the process of division, both daughters must undergo extensive reorganization, so that each will have a buccal cavity, one or more contractile vacuoles, and a complete ciliary system.

The usual mode of sexual union in ciliates is **conjugation.** It has no exact counterpart in any other group of organisms. The following account describes the events that take place during conjugation in *Tetrahymena* and a number of other ciliates that have one micronucleus and one macronucleus. (It does not apply perfectly to any species of *Paramecium,* but the discrepancies are only in details, not in the main theme.) The essential features of the process are illustrated diagrammatically in Figure 2.41. Note that after two ciliates come together and become joined, usually by their oral surfaces (A, B), their macronuclei are gradually

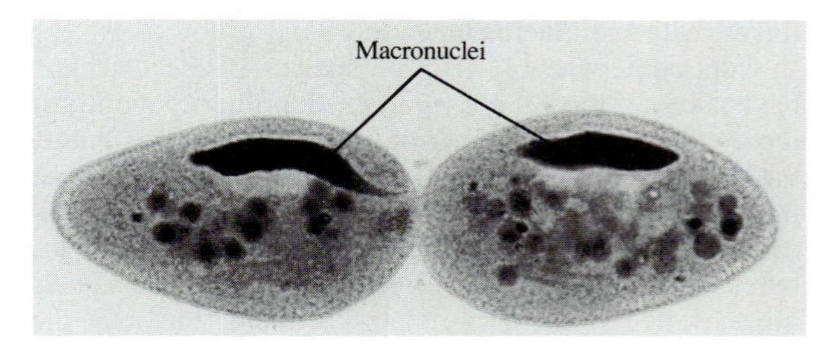

FIGURE 2.40 *Paramecium,* binary fission, late stage; photomicrograph of a stained specimen. The micronuclei are not distinct, because of the food vacuoles. (Courtesy of Carolina Biological Supply Co., Inc.)

resorbed. The micronucleus of each partner is diploid, but it undergoes two meiotic divisions, so four haploid micronuclei are produced (C, D). Three of these disappear (E), but the remaining one divides again, this time

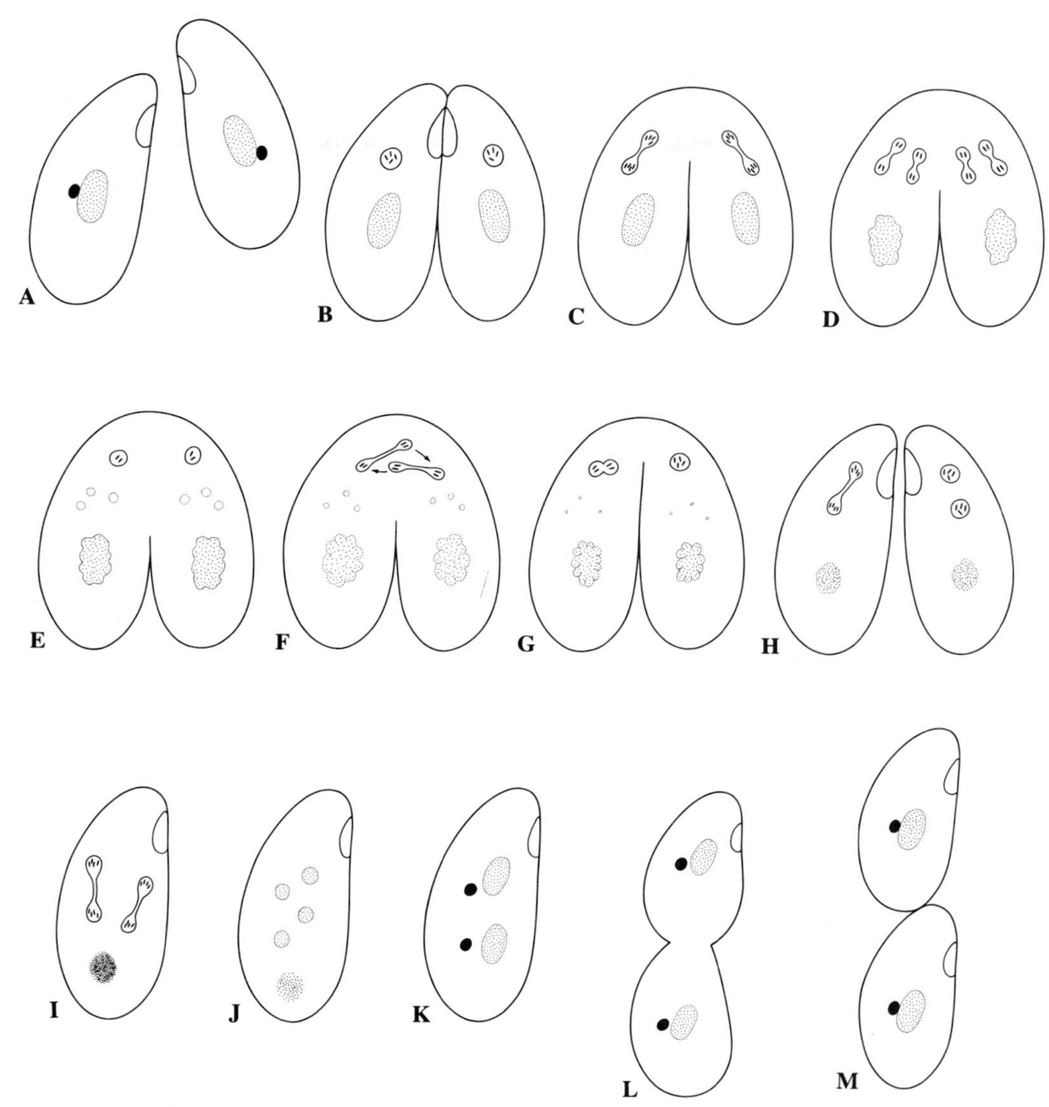

FIGURE 2.41 Conjugation in a ciliate, generalized. A. Prospective partners. B. Partners joined by their oral regions. C. First meiotic division of the micronuclei. D. Second meiotic division; macronuclei beginning to degenerate. E. One haploid micronucleus in each partner surviving, the other three degenerating. F. Haploid micronuclei of partners dividing; arrows indicate that one product of the division in the left partner will unite with one product in the right partner, and *vice versa*. G. Union of micronuclei in progress in left partner; union completed in right partner, and diploidy restored. H. Partners separating; first division of diploid micronuclei. I. Second division of micronuclei; old macronucleus scarcely recognizable. J. Four micronuclei. K. Two micronuclei developing into macronuclei, the other two remaining as micronuclei. L. Cytoplasmic division in progress. M. Cytoplasmic division completed.

mitotically (F). The two micronuclei in each ciliate now function as nuclei of gametes. One of them migrates into the partner ciliate and unites with a micronucleus that has remained stationary; diploidy is thus restored (G). The partners separate and the zygote nucleus of each divides mitotically twice (H, I), forming four micronuclei (J). Two of these become macronuclei, and two persist as micronuclei (K). Fission (L, M) results in two ciliates with the usual nuclear complement of one micronucleus and one macronucleus.

The nuclear events taking place during conjugation indicate that micronuclei are genetically complete, much like the nuclei of germ cells of metazoans. When a micronucleus transforms into a macronucleus, some genetic components are probably lost. The transformation cannot be reversed. The macronucleus of some ciliates is incapable even of dividing during binary fission; it disappears, and the macronuclei of the daughter cells are derived from micronuclei.

The process of conjugation slightly resembles copulation by metazoans, for the partners usually separate and go their own ways after mating. One other important feature of conjugation of ciliates must be stressed: the micronucleus, before conjugation, is diploid, and meiosis takes place during the production of the migratory and stationary micronuclei that will function as gametes. In this respect ciliates resemble metazoans more than they do nearly all other protozoans, in which the zygote is the only diploid stage in the life cycle.

Symbiotic bacteria occur in the cytoplasm or nucleus of many ciliates. The significance of these remains largely unexplored. In *Paramecium aurelia,* however, there is a much-studied situation in which certain bacterialike bodies, called **kappa particles,** multiply in the cytoplasm of individuals that have a particular dominant gene. Ciliates with kappa particles have an adverse effect on those that lack them; thus there are "killer" stocks and "sensitive" stocks of *P. aurelia*. The symbiotic association of kappa particles with *P. aurelia* could conceivably help a "killer" stock to survive when the population of *P. aurelia* is threatened by a shortage of food, because it would eliminate competition from "sensitive" stocks. This is only a conjecture, however.

Some ciliates, especially freshwater types, harbor unicellular green algae, including species of *Chlorella*. Collectively, these symbionts, which occur also in invertebrates of other phyla, are referred to as **zoochlorellae.** *Paramecium bursaria* typically has them. If they are caused, by one method or another, to disappear, *P. bursaria* will multiply more or less normally, provided that it has an ample supply of bacteria to eat. Under conditions of nutritional stress, however, a clone devoid of zoochlorellae will not do as well as one in which they are present.

In *Paramecium,* as well as in other ciliates whose conjugation has been studied, there are mating groups that consist of two or more mating types. Normally, conjugation takes place only between individuals belonging to different mating types within the same mating group. Thus, although members of one mating group are morphologically similar to those of another mating group, they are genetically isolated, and some protozoologists and geneticists view them as distinct species.

Two variations on the theme of conjugation occur in ciliates. **Autogamy** is a process that takes place in a single individual. After meiosis has taken place, the micronuclei comparable to the migratory and stationary micronuclei of a conjugating individual unite to form a zygote nucleus, and this is followed by mitosis and differentiation and distribution of macro- and micronuclei, just as if conjugation had taken place. **Cytogamy** is somewhat similar; it is really just autogamy that takes place in each of the two individuals that have become joined as if they were going to conjugate.

CLASSIFICATION

In the past 30 years, classification of ciliates has been changed substantially. As revisions have been suggested, unfamiliar and relatively cumbersome names have been introduced for some of the higher taxa, including the three classes into which the phylum is divided in the Society of Protozoologists' scheme. An effort is made here to explain where each of the more important and distinctive groups of ciliates fits into the new system, but without an elaborate breakdown into orders, suborders, and so on.

Classification of ciliates is based primarily on the arrangement of cilia and aggregates of cilia, and on the organization of the feeding apparatus. It must be borne in mind that ciliates exhibit great diversity, and that the degree of kinship between many of the major groups is not clear. There are, moreover, bizarre types that have no feeding apparatus, and some that do not even have cilia. These organisms usually show at least a few features that enable protozoologists to link them with more nearly typical ciliates.

CLASS KINETOFRAGMINOPHOREA

This class is almost a hodgepodge and is therefore difficult to characterize. When a cytostome is present, however, this is at the surface, or within a depression that is not a true buccal cavity, because any cilia or aggregates of cilia within it are essentially part of the somatic ciliature. The assemblage consists mostly of ciliates that have long been known as gymnostomes and trichostomes, as well as the suctorians and some other groups.

SUBCLASS GYMNOSTOMATIA

The name Gymnostomatia ("naked mouth") refers to the fact that the cytostome, usually at the anterior tip of the body or close to it, is flush with the body surface. Most members of the group are carnivores or scavengers. Some representative genera are *Prorodon* (Figure 2.42, A), *Enchelyodon* (Figure 2.42, B), and *Coleps* (Figure 2.42, C).

One ciliate assigned to the Gymnostomatia deserves special mention because it does something no other member of the phylum Ciliophora has been observed to do. It forms dense aggregations (Figure 2.42, D) that become elevated into stalked aerial structures, up to nearly 0.5 mm high, which resemble sorocarps, the spore-producing bodies of certain slime molds. The secreted stalks, consisting of polysaccharides and proteins, eventually become thin and brittle, but the globular mass at the top of each one is filled with encysted ciliates. When the cysts are placed in water, swimming ciliates emerge.

This remarkable organism, named *Sorogena,* was discovered only a few years ago. It feeds voraciously on ciliates of the genus *Colpoda,* which belong to the subclass Vestibuliferia (next section). In culture, a scarcity of *Colpoda* is what triggers the aggregation, secretion of a stalk, and encystment. The process will occur at the side of a dish and even on the bacterial film that coats the surface of the water. In nature, however, it takes place mostly on dead stems of plants, seed pods, and similar material.

SUBCLASS VESTIBULIFERIA

This group includes most of the ciliates that have, in the past, been called trichostomes, plus a few others. The cytostome, generally at or close to the anterior end, lies in a depression. This may look like a buccal cavity, but it is not a true buccal cavity because, as explained in the definition of the Kinetofragminophorea, the cilia within it are part of the somatic system. *Colpoda* (Figure 2.43, A) is a good example. One or more species of this genus almost always appear in bacteria-rich infusions of soil.

The entodiniomorphids (Figure 2.43, B–D) are placed here tentatively. They are limited to the digestive tract of mammals, especially ungulates. Their cilia, usually in aggregates that resemble the membranelles characteristic of the class Polyhymenophorea (discussed below), are in circles or tufts near the anterior or posterior end, or both. The body is characteristically firm and often has lobes or spinelike outgrowths. The outer cytoplasm (ectoplasm) may have one or more skeletal rods or plates; these consist of polygonal units of a substance related to cellulose, and are also impregnated by another carbohydrate similar to glycogen. Whether or not there are skeletal elements in the ectoplasm, the endoplasm is clearly set apart from the latter. The cytostome delivers ingested food to the endoplasm, and the food vacuoles are limited to this part of the ciliate. In many entodiniomorphids a distinct canal leads from the endoplasm to the pore ("anus") through which undigested residues may be eliminated. Entodiniomorphids are definitely among the more complex members of the phylum Ciliophora.

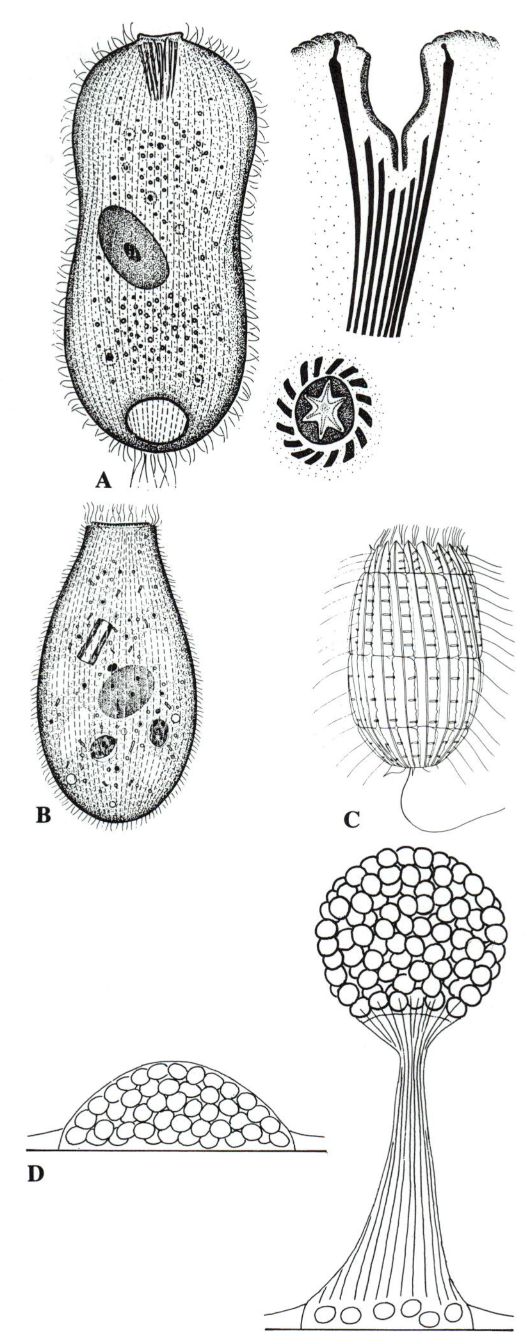

FIGURE 2.42 Subclass Gymnostomatia. A. *Prorodon teres.* The drawings at the right of the entire specimen show the rodlike supports of the buccal region in longitudinal and transverse sections. B. *Enchelyodon vacuolatus.* (A and B, Dragesco, Travaux de la Station Biologique de Roscoff, *12.*) C. *Coleps spiralis* (Borror, Transactions of the American Microscopical Society, *87.*) D. *Sorogena stoianovitchae,* aerial phase. While still in a thin film of water, the ciliates aggregate, then secrete a stalk and encyst. (After Olive, Science, *202.*)

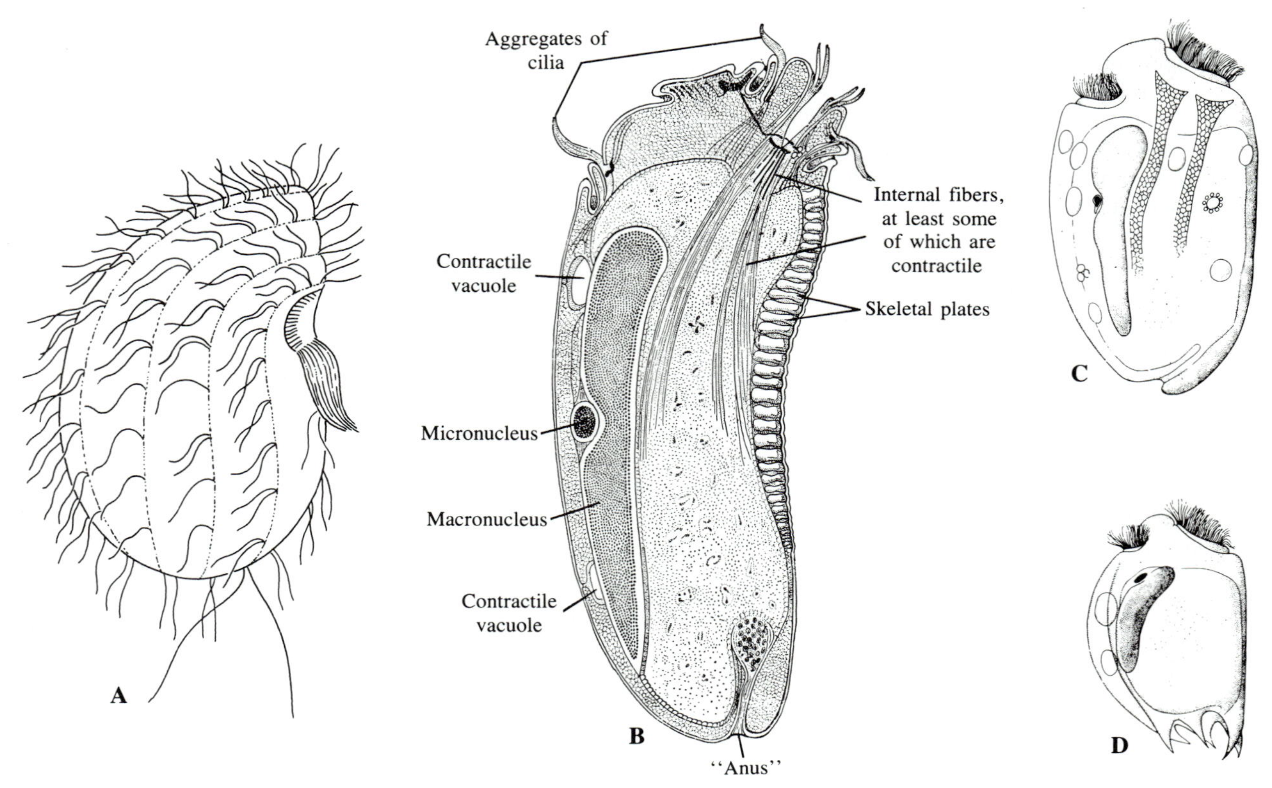

FIGURE 2.43 Subclass Vestibuliferia. A. *Colpoda steinii,* a "trichostome." (Burt, Transactions of the American Microscopical Society, *59*.) Entodiniomorphids: B. *Epidinium ecaudatum*. (Dogiel, General Protistology.) C. *Polyplastron multivesiculatum*. D. *Diplodinium dentatum*. (C and D, Noirot-Timotheé, Annales des Sciences Naturelles, Zoologie, series 12, *2*.)

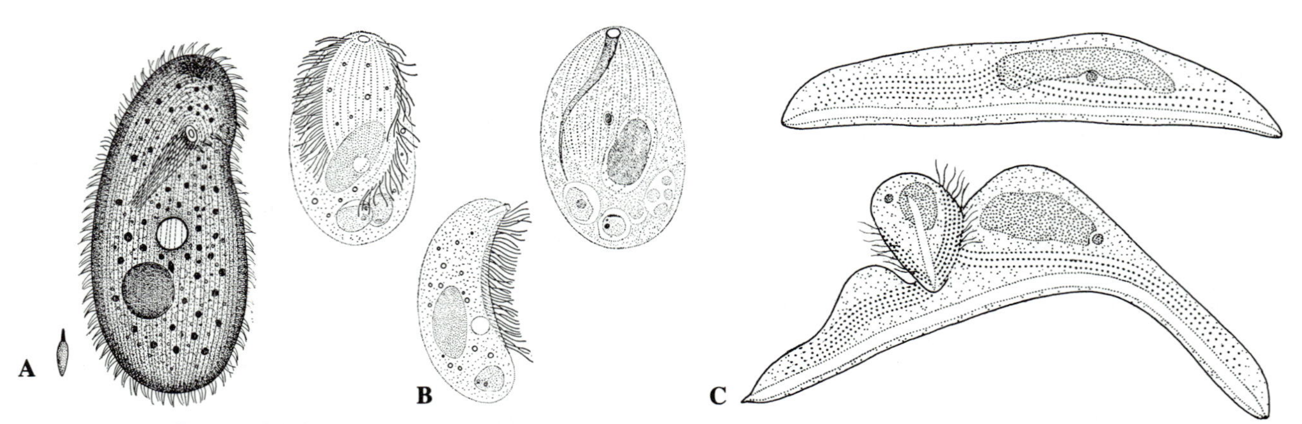

FIGURE 2.44 Subclass Hypostomatia. A. *Nassula georgiana* and one of its trichocysts. (Dragesco, Annales de la Faculté des Sciences, Université Fédérale du Cameroun, no. 9.) B. *Hypocomella katharinae,* a parasite of the ctenidia of the chiton *Katharina tunicata*. The stained specimen (upper right) shows the internal canal. (Kozloff, Journal of Protozoology, *8*.) C. *Sphenophrya dosiniae,* which lives on the ctenidia of certain marine bivalve molluscs. The adult has rows of kinetosomes on one side, but no cilia; cilia develop, however, on the portions of the rows that become incorporated into the bud. (Chatton and Lwoff, Archives de Zoologie Expérimentale et Générale, *86*.)

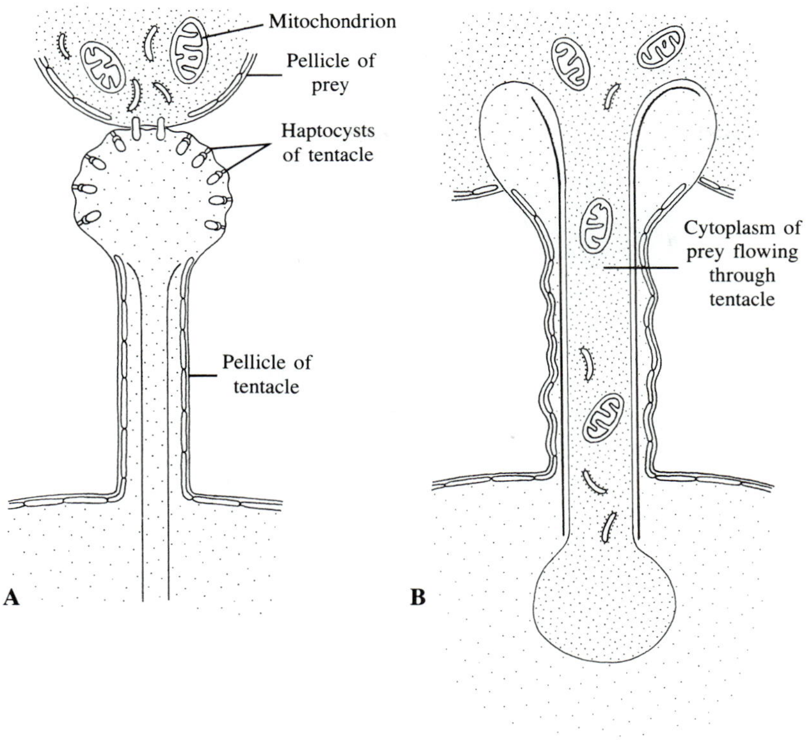

FIGURE 2.45 Subclass Suctoria. Diagrammatic representation of a suctorian tentacle. A. Contact of the tentacle with a prey ciliate. B. Flow of prey cytoplasm through the tentacle. Some details, including the microtubules that run longitudinally in the portion of the tentacle that functions as a tube, are omitted. (After Bardele, in Symposia of the Society for Experimental Biology, Cambridge, no. 28; the figures are based mostly on work of Bardele and Grell on *Acineta tuberosa*.)

SUBCLASS HYPOSTOMATIA

The cytostome in ciliates of this large assemblage is on what may be called, arbitrarily, the ventral surface; in any case, it is decidedly not apical. It is sometimes within a depression more or less comparable to that of certain members of the Vestibuliferia. The somatic ciliature is often reduced, and in flattened types it may be lacking altogether on the surface opposite that in contact with the substratum. A representative genus is *Nassula* (Figure 2.44, A).

Some ciliates called ancistrocomids (Figure 2.44, B), usually characterized by partial loss of ciliation and by a tube for sucking contents out of epithelial cells of various invertebrates, probably belong here. They were long thought to be thigmotrichs that had lost the buccal cavity and adopted a suctorial method of obtaining food. The structure of the feeding tube, however, is so strikingly similar to that of gymnostomes that the two groups are probably closely related.

The remarkable *Sphenophrya* (Figure 2.44, C) and its allies, which live on the ctenidia (gills) of bivalve molluscs, may be highly specialized ancistrocomids. With one exception, they lack cilia, but they have rows of kinetosomes. *Sphenophrya,* by unequal binary fission, produces a little bud, and the kinetosomes of this sprout cilia. The bud is a dispersal stage. If it finds a suitable place to settle, in the same clam or a different one, it loses its cilia and becomes an adult that could not be recognized as a ciliate except for the fact that it has a micronucleus and macronucleus and a few rows of kinetosomes. How *Sphenophrya* and its relatives feed needs to be studied. In any case, they do not have a tube like that of ancistrocomids.

SUBCLASS SUCTORIA

As adults, suctorians have no cilia. These organisms occur in both marine and freshwater habitats and generally attach to something firm, such as the exoskeleton of a crustacean, the perisarc of a hydroid, or an alga. There are, however, some unattached types that just float around. Suctorians have rather stiff cytoplasmic **tentacles** that pierce ciliates, algal spores, or other organisms and conduct protoplasm of the prey into their own bodies. Electron microscopy reveals that each tentacle contains one or more structures called **haptocysts,** or ''missilelike bodies'' (Figure 2.45). These appear to facilitate tight contact between tentacle and prey and to initiate lysis of the cell membrane. Following these events, protoplasm of the prey flows into the suctorian.

One of the common marine suctorians is *Ephelota* (Figure 2.46, A), found on marine hydroids, bryozoans, crustaceans, and seaweeds. Its bulbous body, with many sharp-tipped tentacles, is at the end of a

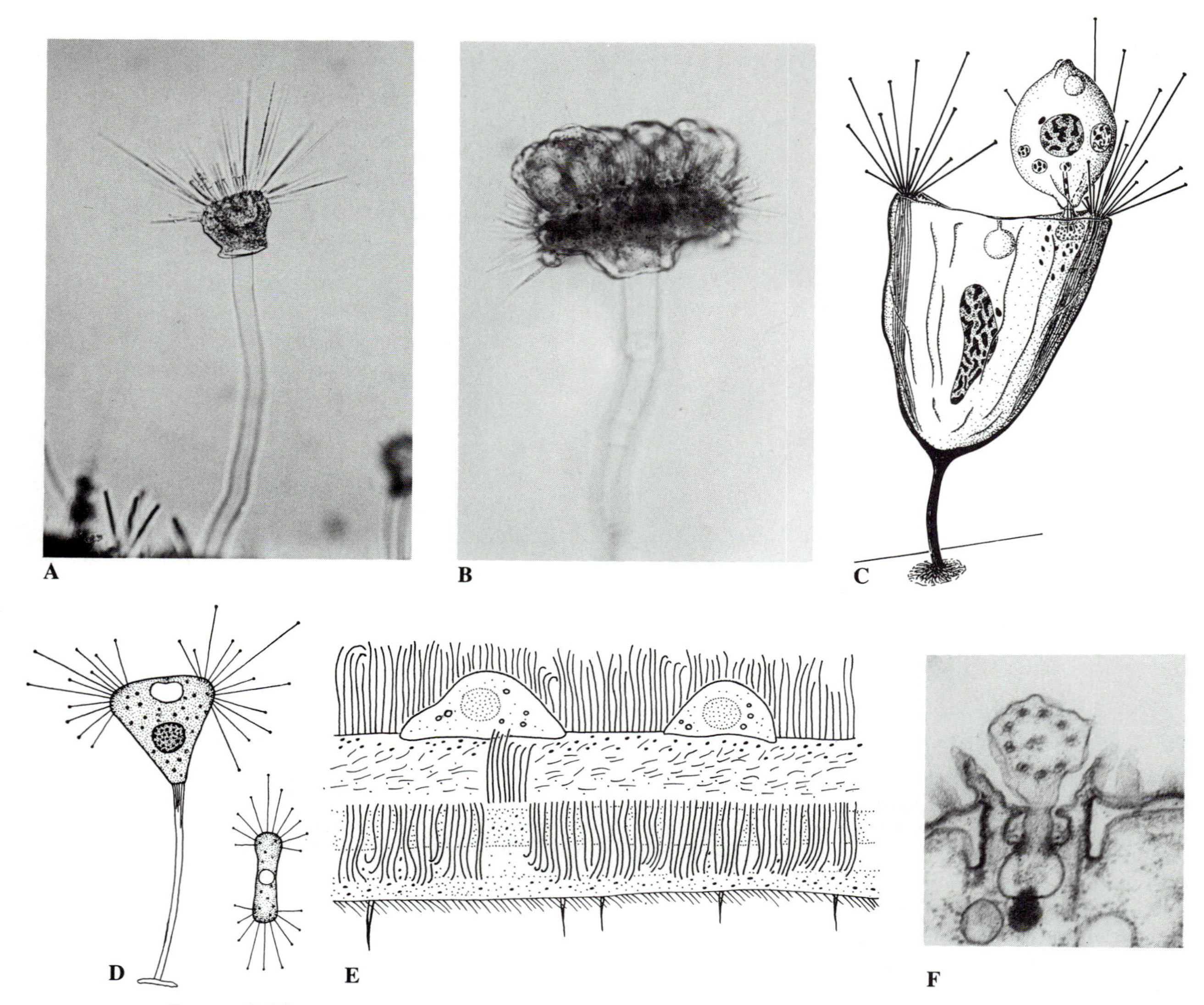

FIGURE 2.46 Subclass Suctoria. A, B. *Ephelota gemmipara,* a common marine suctorian. The specimen in B is producing ciliated buds. C. *Acineta tuberosa* parasitized by another suctorian, *Pseudogemma pachystyla;* the single tentacle of the latter has no haptocysts. (Batisse, Protistologica, *4*.) D. *Tokophrya lemnarum,* a widely distributed freshwater species. (Nozawa, Annotationes Zoologicae Japonenses, *18*.) E, F. *Phalacrocleptes verruciformis,* a suctorian that apparently never has cilia. It lives on the peristomial radioles of *Schizobranchia insignis,* a sabellid polychaete. The many tentacles distributed over the body surface are too short to be seen clearly with the light microscope. Each, however, has a haptocyst that may establish contact with a cilium of the host, as shown in the electron micrograph (F). The pellicle is raised up into projections, and perhaps these absorb some organic nutrients from sea water.

stalk of secreted material. Eventually it produces, by nuclear proliferation and complex morphogenetic changes, a number of ciliated buds. These swim off in search of a suitable place to settle. They lose their cilia, develop tentacles, and secrete a stalk. There are a number of other genera, such as *Tokophyra* and *Acineta,* that are similar to *Ephelota;* in some of them the body is partly enclosed by a cup-shaped lorica.

Pseudogemma (Figure 2.46, B) is unusual in that it parasitizes another suctorian, *Acineta*. *Phalacrocleptes* (Figure 2.46, D) is also a noteworthy type that lives on the peristomial radioles of certain sabellid polychaetes.

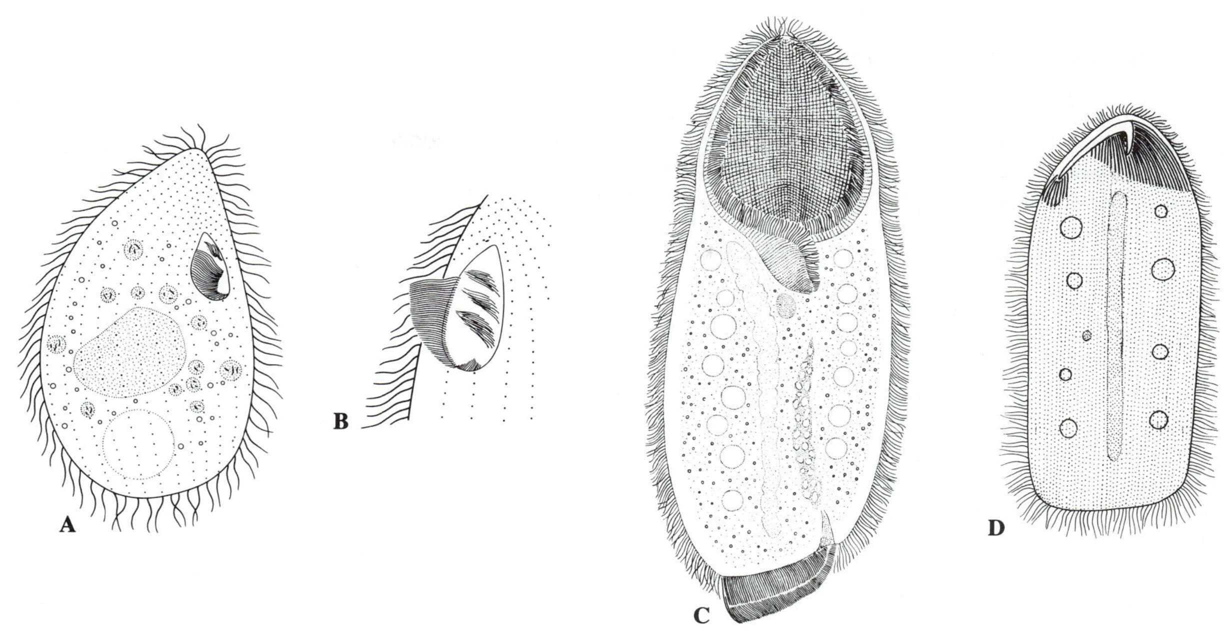

FIGURE 2.47 Subclass Hymenostomatia. A. *Tetrahymena limacis,* a facultative parasite of various terrestrial pulmonate snails (free-living species, such as *T. pyriformis,* are similar). B. Detail of the buccal region of *Tetrahymena,* showing the hyaline ciliary membrane on the right side of the buccal cavity (observer's left) and the three membranelles within the cavity. C. *Epicharocotyle kyburzi,* from the gut of an oligochaete annelid. The buccal cavity is at the posterior end; the complex ''sucker'' in the anterior part of the body is for attachment. (Kozloff, Journal of Protozoology, *7.*) D. *Metaradiophrya,* from the gut of an oligochaete annelid. There is no buccal cavity; at the anterior end, there are skeletal ''ribs'' underlying many of the ciliary rows, and a hooklike structure whose tooth projects. (After Beers, Journal of the Elisha Mitchell Scientific Society, *54.*)

It has no cilia of its own at any stage. Its short tentacles, of which there are many, attach to the cilia of the host, in which case continuity of cytoplasm in the tentacles and cilia is established. *Phalacrocleptes* seems quite different from any ciliate, yet it undergoes binary fission and conjugation according to classical ciliate patterns. One genus, *Endosphaera,* is an internal parasite of other ciliates.

CLASS OLIGOHYMENOPHOREA

Most members of the Oligohymenophorea have a buccal cavity; the specialized ciliature associated with this is distinct from the somatic ciliature. The group includes the many kinds of ciliates called hymenostomes, and some apparently close allies, the peritrichs.

SUBCLASS HYMENOSTOMATIA

In typical hymenostomes, the well-defined buccal cavity is located in the anterior third of the body. Its right side (as the ciliate is viewed with its buccal surface down) is bordered by a membrane consisting of a row of closely spaced, fused cilia. Within the buccal cavity are aggregates of cilia that form tonguelike **membranelles** or small tufts called **peniculi.** None of the buccal ciliature belongs to the system of somatic cilia, which is usually extensive and laid out in distinct rows.

Paramecium, already described in some detail, belongs to this subclass. Its buccal cavity lies at the bottom of a deep depression traversed by rows of somatic cilia, so the membrane and membranelles are difficult to see. *Tetrahymena* (Figure 2.47, A and B) and allied genera, however, are more satisfactory for observing the type of buccal ciliature that is characteristic of hymenostomes.

Tetrahymenas are common in fresh water, especially in slimy coatings on vegetation and in dead or dying organisms. They feed on bacteria and fragments of disintegrating cells. Certain species are important in biochemical and physiological research, for they can be grown in bacteria-free media in which all ingredients—sugars, amino acids, and vitamins—are known. A few tetrahymenas are benign parasites of the digestive gland, kidney, and other tissues of land molluscs. They may reach high concentrations within their hosts, but seem to feed largely on fluids rather than on cells. They can be cultivated apart from their hosts, and they can survive and multiply in soil or water, taking up parasitism again when the opportunity presents itself. Some free-living tetrahymenas will destroy small animals or embryos if they can get into them through a wound. These must be regarded as predators rather than as parasites.

In a *Tetrahymena* the buccal cavity is flush with the surface of the body. One side of it is bordered by the typical membranelike complex of adherent cilia, and deeper in the cavity are three tonguelike membranelles. Bacteria and other particles collected by the buccal cilia become incorporated into food vacuoles that are pinched off at the cytostome. Like *Paramecium, Tetrahymena* has a contractile vacuole and a cytoproct for eliminating undigested residues. It has no trichocysts, however.

A few parasitic ciliates without a buccal cavity belong to this group. In one genus found in freshwater pulmonate snails, the arrangement of ciliary rows is essentially identical to that of *Tetrahymena*. It is likely that the buccal cavity has disappeared as these ciliates have become specialized for feeding on soluble nutrients.

In certain hymenostomes called thigmotrichs, which are commensals of various invertebrates, especially molluscs and annelids, the buccal cavity is displaced to the posterior end. Running toward it and into it is an extensive membrane of fused cilia, and this is paralleled by two rows of adherent cilia that form a brushlike complex. The body is generally flattened. The name *thigmotrich* refers to the fact that the cilia of an extensive area on the lower surface are used for clinging to epithelia. (Thigmotactism is by no means limited to this group of ciliates, however.) In some genera that inhabit the gut of oligochaete annelids and freshwater snails, there is a complicated sucker (Figure 2.47, C). Most thigmotrichs feed on bacteria and sloughed-off epithelial cells.

An allied group, in which there is no buccal cavity, is represented by genera such as *Anoplophrya* and *Metaradiophyra* (Figure 2.47, D), both found in the gut of earthworms of the genus *Lumbricus*. Presumably they subsist on dissolved organic nutrients. Although they have no buccal cavity, they are almost certainly closely related to certain thigmotrichs that do.

SUBCLASS PERITRICHIA

In peritrichs, there is a broad buccal field. A membrane consisting of fused cilia runs counterclockwise around the edge of the buccal field and then turns inward, spiralling down into the buccal cavity toward the cytostome. Running alongside the membrane are two other rows of cilia that collectively form a brushlike complex. These also spiral into the buccal cavity, where they may be augmented by additional rows of cilia (Figure 2.48, B). The somatic ciliature is reduced to one or a few circles, and may be absent altogether except in a motile dispersal stage.

Vorticella (Figure 2.48, A–C) is the best-known genus, and is represented in both fresh and salt water. It is a sessile peritrich, attached by a long stalk to a stick, aquatic plant, crustacean, hydroid, or some other firm object. The cytoplasm of the stalk has contractile filaments, and when *Vorticella* is touched or otherwise stimulated it pulls itself closer to the substratum, the stalk becoming coiled in the process. There are also contractile filaments in the outer cytoplasm of the main part of the body. An attached *Vorticella* has no cilia other than those concerned with feeding, but when it divides by binary fission, it produces a free-swimming larva that has a circle of cilia at the end opposite that on which the buccal cavity is located. This end, which is directed forward when the larva swims, also has some kinetosomes that collectively form a structure called the **scopula.** It is from the scopula that the stalk grows out after the larva settles. The circle of cilia soon disappears.

Vorticella feeds largely on bacteria, and the way it captures these has been neatly analyzed with the help of motion pictures (Figure 2.48, C). The two inner rows of buccal cilia, which form a brushlike complex, produce currents of water that move past the organism. The cilia of the outer membrane intercept the particles that touch them. Particles that happen to land between the inner rows and outer row are effectively trapped, too. The combined activity of all three rows moves the particles toward the depression where the cytostome is located. The cilia also produce a rejection current that carries unacceptable particles away from the buccal field.

There are several genera that are similar to *Vorticella*. All are stalked, and some of them form branching colonies. In another group of peritrichs the stalk is absent but there is permanently a circle of somatic cilia and a complex attachment disk reinforced with hard, toothlike elements. Peritrichs of this type are mobile and can glide around on the skins of many aquatic invertebrates, and of fishes and amphibians as well. A few even live on the wet, slimy surfaces of land snails and slugs. *Trichodina* (Figure 2.48, D) is an especially common genus, found on the body surfaces of many marine and freshwater organisms, and in the urinary bladder of some amphibians.

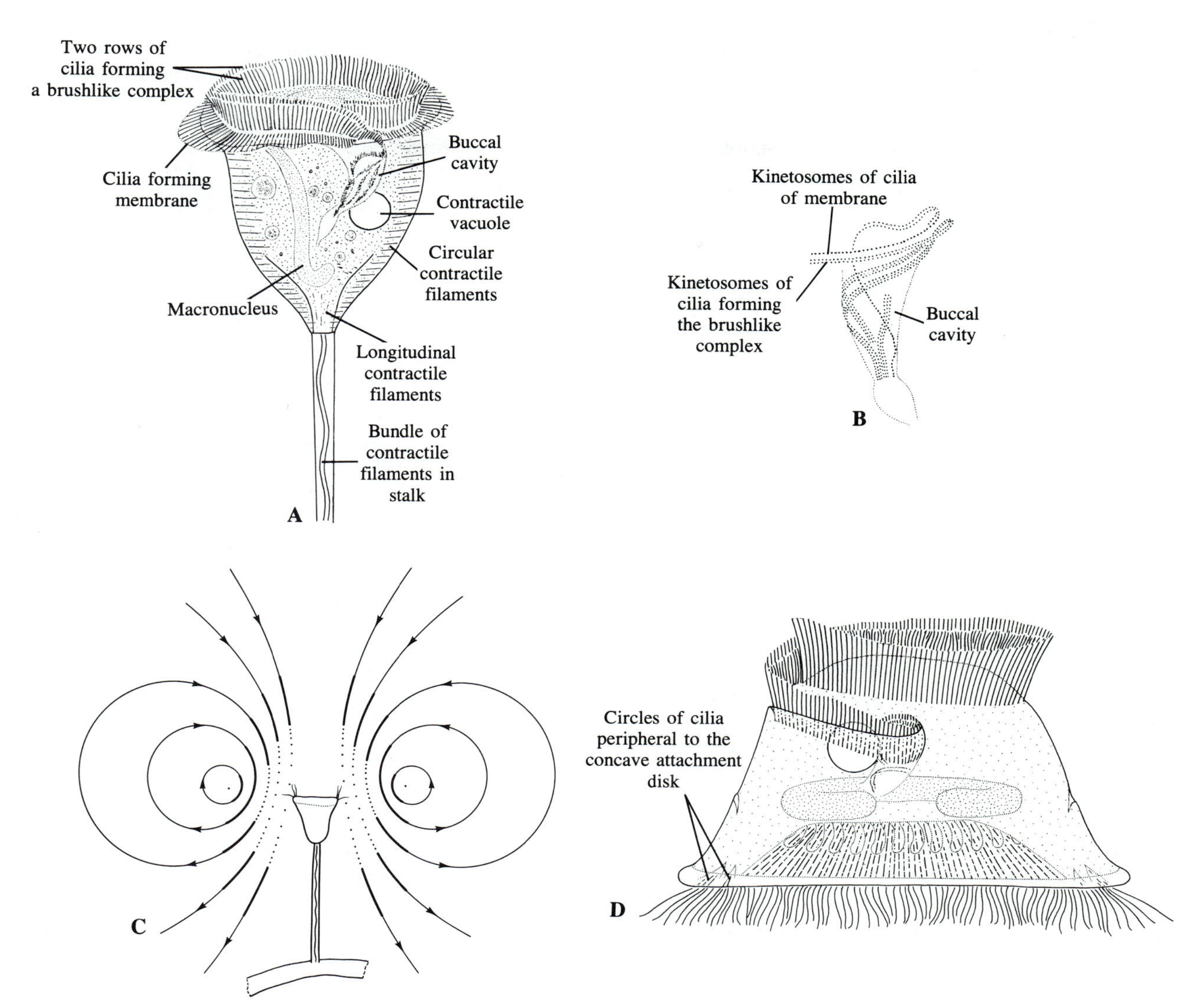

FIGURE 2.48 Subclass Peritrichia. A. *Vorticella*. B. Arrangement of kinetosomes of ciliary rows entering the buccal cavity of *Vorticella*. (After Lom, Archiv für Protistenkunde, *107*.) C. Feeding currents set up by *Vorticella*. (After Sleigh and Barlow, Transactions of the American Microscopical Society, *95*.) D. *Trichodina urinicola,* a mobile peritrich that inhabits the urinary bladder of newts. As in *Vorticella,* there are three rows of cilia running around the oral end of the animal and into the buccal cavity; those of the outermost rows form a hyaline membrane. Similar species are found on the skin of many aquatic vertebrates and invertebrates. (After Lom, Journal of Protozoology, *5*.)

CLASS POLYHYMENOPHOREA

Polyhymenophorea is essentially synonymous with Spirotricha, long applied to ciliates characterized by a series of membranelles that follows a clockwise, spiral path into the buccal cavity. The individual membranelles are similar to those of hymenostomes, but they are more numerous, and the series usually begins well outside the buccal cavity. The members of this group inhabit both fresh and salt water, and most of them are free-living. A few, however, are commensal with various invertebrates. Three of the four orders are mentioned briefly.

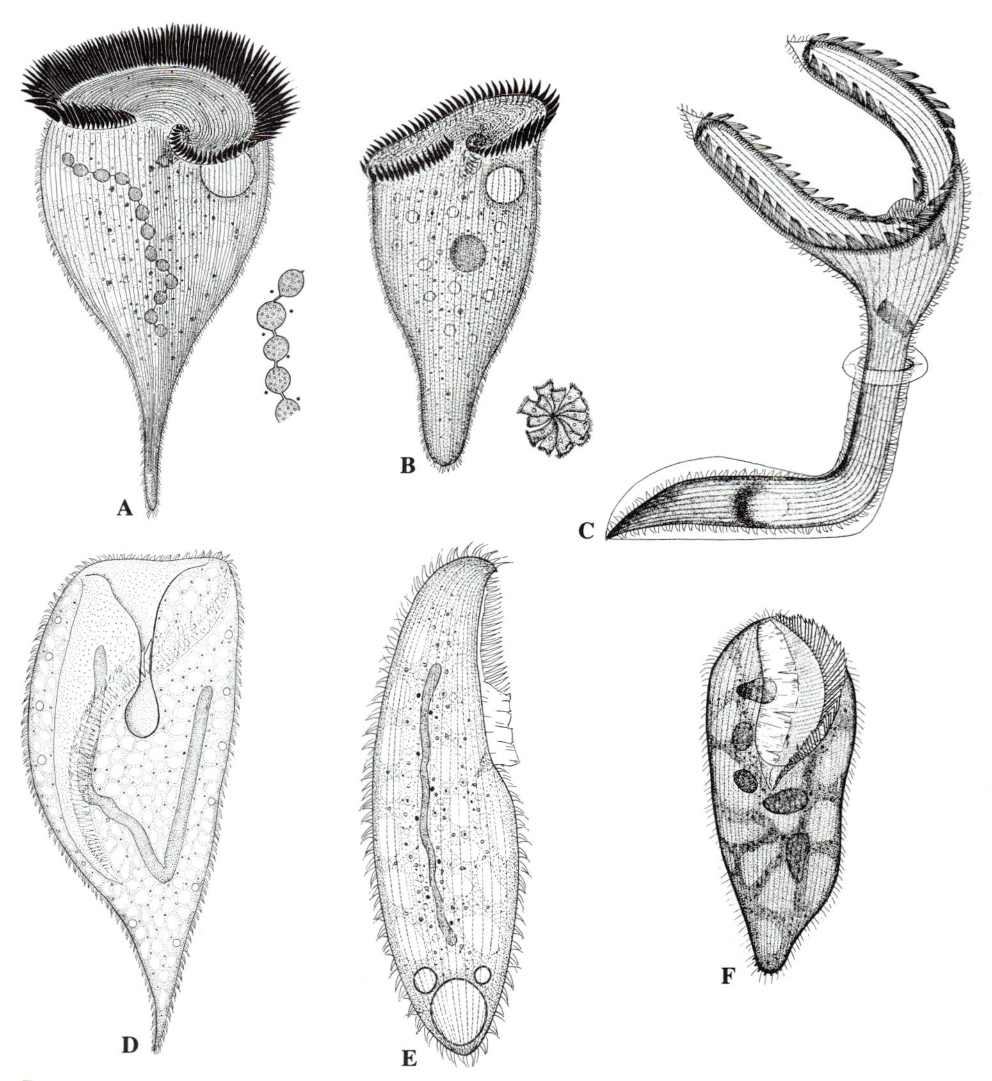

FIGURE 2.49 Order Heterotrichida. A. *Stentor caudatus*. The figure at the right shows several of the many micronuclei and a portion of the macronucleus. B. *Stentor multimicronucleatus*. (A and B, Dragesco, Ciliés Libres du Cameroun, Annales de la Faculté des Sciences, Université Fédérale du Cameroun, 1970.) C. *Folliculina aculeata*. (Dewey, Biological Bulletin, *77*.) D. *Bursaria caudata*. (Dragesco, Annales de la Faculté des Sciences, Université Fédérale du Cameroun, no. 9.) E. *Blepharisma japonicum*. (Dragesco, Ciliés Libres du Cameroun.) F. *Condylostoma kahli*. (Dragesco, Travaux de la Station Biologique de Roscoff, *12*.)

Order Heterotrichida

Heterotrichs not only have an extensive zone of membranelles, but also a somatic ciliature laid out in rows. *Stentor* (Figure 2.49, A and B) is a famous example of the group, and has been much used in research on regeneration in ciliates. *Folliculina* (Figure 2.49, C) is similar, but lives within a secreted lorica that is attached to shells and other objects. *Bursaria* (Figure 2.49, D), *Blepharisma* (Figure 2.49, E), and *Condylostoma* (Figure 2.49, F) are other common genera. The last two are unusual among heterotrichs in that they have a hyaline ciliary membrane alongside the buccal cavity, as well as membranelles. The macronucleus is often ''beaded,'' especially in large species, whose length may exceed 2 mm.

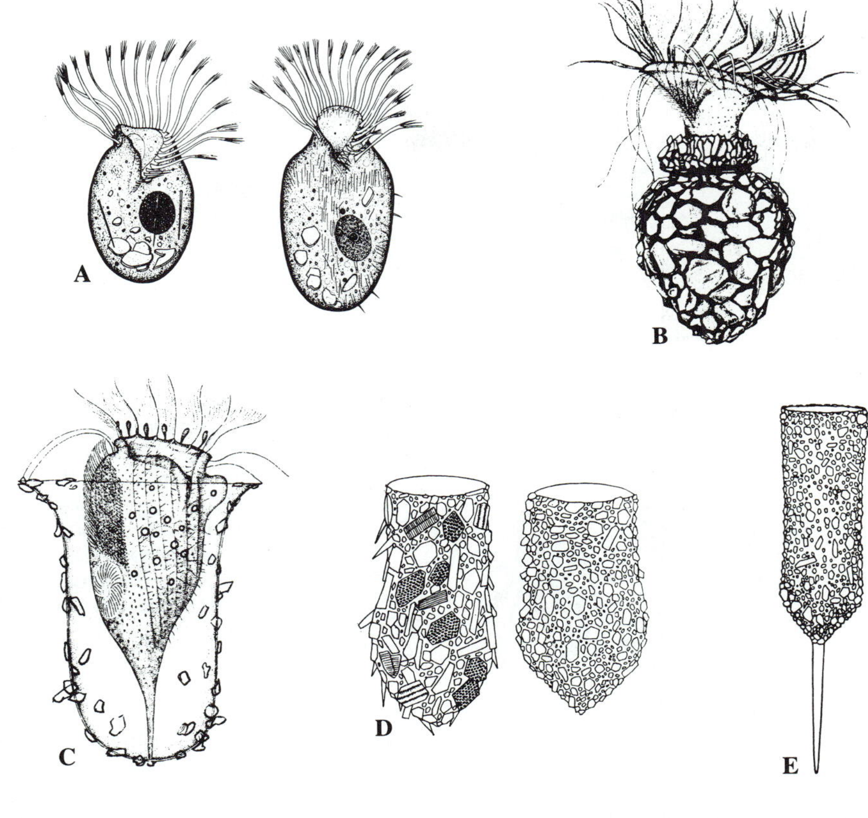

Figure 2.50 Order Oligotrichida. A. *Strombidium arenicola* (left) and *S. faurei* (right). (Dragesco, Travaux de la Station Biologique de Roscoff, *12*.) B. *Stenosemella ventricosa*. C. *Tintinnopsis bütschlii*. (B and C, Ushakov *et al.*, Atlas of Invertebrates of the Far-Eastern Seas of the USSR.) D. Loricas of *Tintinnopsis tubulosa*. (Hada, Journal of the Faculty of Science, Hokkaido University, Zoology, *5*.) E. Lorica of *Tintinnopsis platensis* (Hada, Journal of the Faculty of Science, Hokkaido University, Zoology, *6*.)

Order Oligotrichida

Oligotrichs are more difficult than heterotrichs to define. In general, however, they have reduced or specialized somatic ciliature. In *Strombidium* (Figure 2.50, A), for example, the few somatic cilia are in the form of long bristles. *Tintinnopsis* and its allies, collectively called tintinnids (Figure 2.50, B–E), are planktonic and secrete a lorica. This may be gelatinous and thin, or it may be composed largely of foreign particles cemented together. In some genera of tintinnids, aggregates of cilia called **tentaculoids** alternate with the membranelles.

Order Hypotrichida

Hypotrichs, exemplified by *Euplotes* (Figure 2.51, A) and *Oxytricha* (Figure 2.51, B), are characterized by short, bristlelike cilia on the upper surface, and by some nearly spinelike aggregates of cilia, called **cirri,** that are mostly on the lower surface and at the posterior end. In addition to the series of membranelles that leads to the buccal cavity, there may be one or two rows of cilia forming membranelike structures within the buccal cavity.

References on Ciliophora

Beale, G. H., 1954. The Genetics of *Paramecium aurelia*. Monographs of Experimental Biology, vol. 2. Cambridge University Press, London.

Borror, A. C., 1973. Marine Flora and Fauna of the Northeastern United States. Protozoa: Ciliophora. National Oceanic and Atmospheric Administration Technical Report, National Marine Fisheries Service Circular 378.

Corliss, J. O., 1979. The Ciliated Protozoa. Characterization, Classification, and Guide to the Literature. 2nd edition, Pergamon Press, Oxford and New York.

Dragesco, J., 1960. Les ciliés mesopsammiques littoraux (systématique, morphologie, écologie). Travaux de la Station Biologique de Roscoff, *12*:1–356.

Dragesco, J., and Dragesco-Kerneis, A., 1985. Ciliés Libres de l'Afrique Intertropicale. Editions de l'ORSTOM, Bondy, France.

Elliott, A. M. (editor), 1973. Biology of *Tetrahymena*. Dowden, Hutchinson, and Ross, Stroudsberg, Pennsylvania.

Giese, A. C., 1973. *Blepharisma:* The Biology of a Light-Sensitive Protozoan. Stanford University Press, Stanford, California.

Jones, A. R., 1974. The Ciliates. Hutchinson University Library, London; St. Martin's Press, New York.

Jurand, A., and Selman, C. C., 1969. The Anatomy of *Paramecium aurelia*. Macmillan, London.

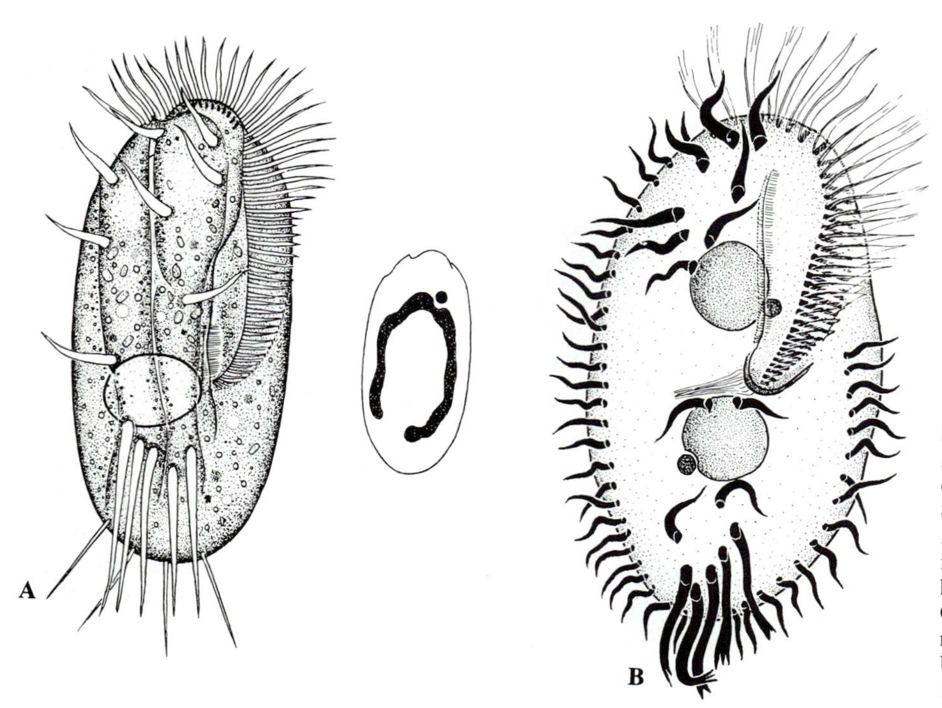

FIGURE 2.51 Order Hypotrichida. A. *Euplotes aberrans* and its nuclei. (Dragesco, Travaux de la Station Biologique de Roscoff, *12*.) B. *Oxytricha elliptica,* after impregnation with silver, which has blackened the cirri (Dragesco, Ciliés Libres du Cameroun, Annales de la Faculté des Sciences, Université Fédérale du Cameroun, 1970.)

Nanney, D. L., 1980. Experimental Ciliatology. An Introduction to Genetic and Developmental Analysis in Ciliates. John Wiley & Sons, Somerset, New Jersey.

Tuffrau, M., 1987. Proposition d'une classification nouvelle de l'ordre Hypotrichida (Protozoa, Ciliophora). Annales des Sciences Naturelles, Zoologie, series 13, *8:*111–117.

PHYLUM OPALINATA

Most opalinids live in the rectum of frogs and toads, but a few species have been reported from salamanders, aquatic reptiles, and fishes. For a long time these organisms were thought to be primitive ciliates, but from about 1950 onward it became the fashion to view them as flagellates. They really do not fit well in either group. Although the Society of Protozoologists' classification places them among the flagellates, they are so unusual that it seems best to give them the rank of phylum.

Opalinids, whether flattened (as most of them are) or cylindrical, are covered with cilia arranged in lengthwise rows. In this respect they resemble many ciliates. Between successive rows the cell membrane is raised into several high ridges. Electron microscopy reveals that within each of these ridges are numerous fibers, one above the other (Figure 2.52, F). In the principal genus, *Opalina* (Figure 2.52, C and D), there are many nuclei, all of which are alike; in *Zelleriella* (Figure 2.52, A and B) and a few other genera there are only two nuclei. Contractile vacuoles are absent, and there is no buccal cavity or cytostome. Pinocytic vesicles have been seen, however, at the bases of the longitudinal folds that lie between the rows of cilia. The folds themselves would presumably increase the body surface and thus favor absorption of soluble materials.

When an opalinid divides, the plane of fission may be parallel to the ciliary rows, thus longitudinal; or it may be transverse with respect to the ciliary rows, as it is in most ciliates. At the anterior end of the body is a crescentic area called the **falx** (Figure 2.52, D and E). This is a morphogenetically active zone. When division of an opalinid is longitudinal, the falx is bisected. New rows of cilia that originate along the falx are added between the existing rows, and the normal complement is eventually restored. Division of the nuclei is definitely mitotic, involving a moderate number of chromosomes and showing distinct spindles.

Sexual reproduction is known to occur, but it is only part of an interesting sequence of events. *Opalina ranarum* is the most intensively studied species. It lives in a European frog, and during most of the year, while its host is living on land, it reproduces occasionally by fission. In late winter, as the frog's breeding season approaches, *O. ranarum* divides repeatedly to produce small individuals with only a few nuclei. These small opalinids encyst and the cysts leave the body with

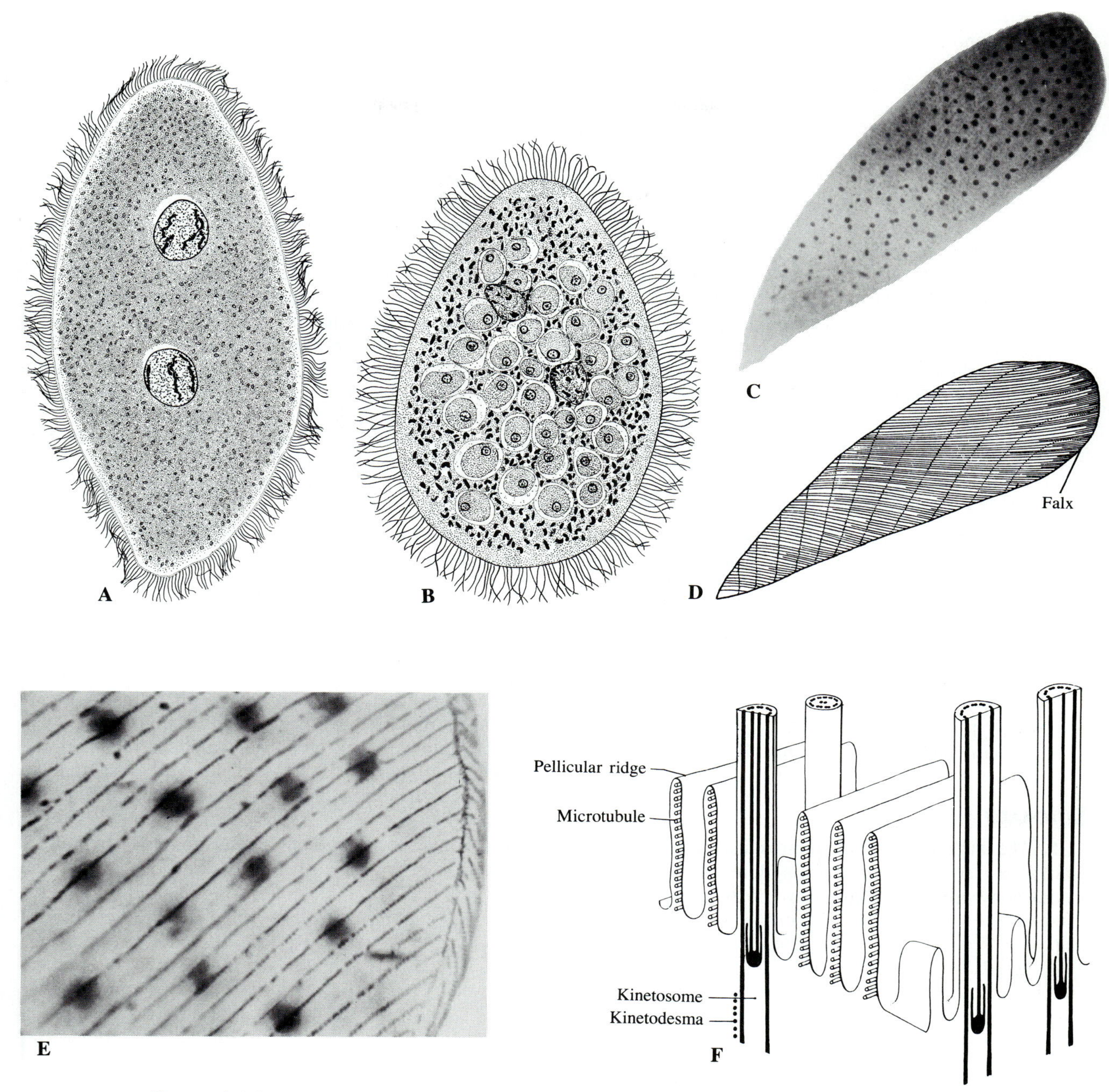

FIGURE 2.52 A. *Zelleriella elliptica* (arrangement of ciliary rows not shown). (Chen, Journal of Morphology, *83*.) B. A species of *Zelleriella* parasitized by an *Entamoeba*. (After Stabler and Chen, Biological Bulletin, *70*.) C. *Opalina virguloidea;* photomicrograph of a stained specimen, showing numerous nuclei. D. *Opalina virguloidea,* diagram of the arrangement of ciliary rows. All of the rows on the observer's side are shown, but only a few of the rows on the opposite side are indicated. Note that many of the rows originating at the falx are short. E. *Opalina virguloidea,* photomicrograph of a portion of the anterior end, showing the ciliary rows originating at the falx. F. *Opalina ranarum,* section through the pellicle and ciliary rows. (Noirot-Timotheé, Annales des Sciences Naturelles, Zoologie, Series 12, *1*.)

feces. After the cysts are eaten by tadpoles, the opalinids emerge and their nuclei divide meiotically, producing haploid nuclei that function as gametes. Pairing and complete union of the small individuals take place, and the nuclei also pair and fuse. Thus nuclear diploidy is restored.

What happens next is not certain. It has been reported that opalinids formed by union of two small individuals also encyst and pass out of the tadpole's gut. If eaten by another tadpole, such a zygote would presumably be the founder of a population of opalinids that will persist into the frog's adult life. Another account explains that the zygote merely divides repeatedly, giving rise to the larger opalinids found in the frog after metamorphosis. In any case, the production of small encysted individuals during the breeding season of the frog is linked to changes in the frog's concentrations of sex hormones and of the gonad-stimulating hormones produced by the pituitary gland.

References on Opalinata

El Mofty, M. M., and Smyth, J. D., 1964. Endocrine control of encystation in *Opalina ranarum* parasitic in *Rana temporaria*. Experimental Parasitology, *15*:185–199.

Noirot-Timothée, C., 1959. Récherches sur l'ultrastructure d'*Opalina ranarum*. Annales des Sciences Naturelles, Zoologie, series 12, *1*:265–281.

Wessenberg, H., 1961. Studies on the life cycle and morphogenesis of *Opalina*. University of California Publications in Zoology, *61*:315–370.

Wessenberg, H., 1966. Observations on cortical ultrastructure in *Opalina*. Journal de Microscopie, *5*:471–492.

Summary

1. The organisms called protozoans belong to several diverse phyla within the kingdom Protoctista. Although they are one-celled, many of them are structurally complex. Most are heterotrophs that require organic food, but some are photosynthetic autotrophs. Locomotion is primarily by the action of flagella, cilia, or pseudopodia.

2. Asexual reproduction is by binary fission or multiple fission. Sexual reproduction occurs in nearly all of the major groups, but the pattern varies; some protozoans produce gametes that are distinctly different from feeding individuals, others do not. Sexual reproduction often alternates, moreover, with asexual reproduction. In one group, the phylum Ciliophora, transfer of gametic nuclei is accomplished by a process called conjugation, in which two individuals typically become joined temporarily. Each member of the pair produces two gamete nuclei. One of the nuclei remains stationary; the other migrates into the partner and unites with its stationary nucleus. The partners then separate.

3. Free-living protozoans are found in marine and freshwater habitats, and also in soil, moss, and similar substrata that are thoroughly wet part of the time. Many species produce cysts that enable them to withstand desiccation. The cysts may enclose vegetative individuals that have become dormant, or they may be formed at a particular stage of the reproductive cycle.

4. Symbiotic protozoans occur as parasites, mutualists, and commensals of nearly all groups of invertebrates and vertebrates, and a few occur in tissues of plants. Certain phyla of protozoans, or major subdivisions of phyla, are exclusively symbiotic. The life cycle of some parasitic types, such as those that cause malaria in humans, requires two hosts for its completion.

5. Classification is based on structure, locomotion, methods of obtaining food, and patterns of reproduction. The schemes in general use are to some extent tentative and are being improved as our knowledge of relationships between and within major groups increases.

3

Phylum Porifera

Sponges

- Phylum Porifera
  - Class Calcarea
  - Class Demospongiae
  - Class Sclerospongiae
  - Class Hexactinellida

INTRODUCTION

Sponges are multicellular, but their structure is very different from that of any other metazoans. They are the only animals whose bodies are organized around a system of pores, passageways, and chambers through which water flows continually. The pores, which give the phylum its name (''pore-bearing''), are scattered all over the surface. They are microscopic, but there are thousands of them. In certain simple sponges, water that enters through the pores goes directly to a central cavity and then leaves by way of an opening called the **osculum.** In most types, however, the water circulates through passageways and chambers of varying complexity before it reaches a cavity that has an osculum. It is important to understand that an osculum is in no way comparable to a mouth; it is just an exhalant opening.

While all cells of a sponge independently take up oxygen and give off carbon dioxide and nitrogenous wastes, the circulation of water and much of the trapping of microscopic particles of food, such as bacteria, are accomplished by cells called **choanocytes,** or collar cells. In the simplest sponges, choanocytes line the central cavity; in others, they line certain of the chambers through which water passes on its way to a central cavity. The distinctive features of choanocytes are discussed in the next section.

Sponges that have a tubular, urn-shaped, or vaselike form (Figure 3.1, A and D; Figure 3.7, A; Color Plate 1) may seem to be radially symmetrical, but their pores and passageways do not fit a perfectly symmetrical pattern. Most sponges are sprawling encrustations or lumpy or branching growths in which there is no discernible plan of symmetry (Figure 3.1, B and C; Figure 3.4; Figure 3.17; Color Plate 1). Although each species has a characteristic form and texture, growth may continue indefinitely. So long as new cells are added and new pores, passageways, and chambers are formed, it is inevitable that there will be almost continual reorganization.

The cells that cover the outside of a sponge, and that line the internal passageways and chambers, form layers that resemble epithelia. They do not, however, secrete a basal lamina comparable to that which underlies epithelial tissues of most metazoan animals, and there are relatively few intercellular junctions. For these reasons, cell layers of sponges are not regarded as true tissues.

In spite of their simplicity, sponges are successful and diversified. More than 5000 living species have been described, and there are many others in the fossil record. Nearly all of the living species are marine, but two families of the largest class, Demospongiae, are found in fresh water.

TYPES OF CELLS AND OTHER STRUCTURAL COMPONENTS OF SPONGES

CHOANOCYTES

Choanocytes (Figure 3.2; Figure 3.10, C) are distinctive because they have a flagellum whose basal portion is encircled by a delicate collar that consists of numerous cytoplasmic extensions called **microvilli.** When united by cross-connections, as has been reported for some sponges, the microvilli form a meshwork (Figure 3.2, C). As water circulates through and around the collar, bacteria and other small food particles are trapped and incorporated into food vacuoles, which are moved to the main body of the cell. The microvilli thus function in much the same way as slender pseudopodia. Collars can sometimes be seen with a good light microscope—they show up especially well in choanocytes of some sponges of the class Calcarea—but the details of their organization can be understood only with the aid of electron microscopy. Cells with a structure similar to that of choanocytes have been observed in various other invertebrates, including echinoderms, but it is in the sponges and the protozoans called choanoflagellates that they are an especially characteristic feature (Figure 3.3).

PINACOCYTES AND POROCYTES

In most sponges, the outer surface of the body, as well as any internal passageways or chambers not lined by choanocytes, are covered by cells called **pinacocytes** (Figures 3.4, 3.11). The pinacocytes at the base secrete material that affixes the sponge firmly to the substratum. Pinacocytes are generally flattened, but they can change shape to the extent that they collectively modify the shape of the sponge as a whole, or at least a portion of it. When a sponge is touched or exposed to certain chemicals or to silt, it may slowly contract slightly. It is thought that pinacocytes, or other cells that have been stimulated, influence their neighbors to react also. Thus, in spite of the fact that there are few intercellular junctional complexes between sponge cells and no cells comparable to neurons of other metazoans, there is definitely cell-to-cell conduction.

Porocytes (Figure 3.4), which resemble doughnuts because they are pierced by an opening, are specialized pinacocytes. They occur in only certain species, but when they are present they are found among the other pinacocytes that form the surface covering. It is through the openings in the porocytes that water enters the sponge. These cells regulate water intake by redistributing their cytoplasm and thus making their apertures

FIGURE 3.1 Some diverse growth forms of sponges. (All four species belong to the class Demospongiae.) A. *Stylissa stipitata*. B. A species that branches extensively. C. *Halichondria panicea*. D. A large Caribbean species, which forms urn-shaped growths. (Photograph by Thomas Suchanek.) (See also Color Plate 1.)

larger or smaller. In sponges that do not have porocytes, water enters through spaces between pinacocytes.

THE MESOHYL

Between the cell layers of a sponge there is a somewhat gelatinous matrix called the **mesohyl.** This usually contains considerable **collagen,** a structural protein that is widespread in the animal kingdom. The collagen generally appears in threads too fine to be seen with a light microscope, but in bath sponges much of it is concentrated into coarse but supple fibers (Figure 3.16). Collagen of this type, called **spongin,** is what makes the cleaned skeleton of a bath sponge strong, squeezable, and absorbent.

The living components of the mesohyl are mostly **amoebocytes,** so called because they resemble some kinds of amoebae in their form and locomotion (Figures

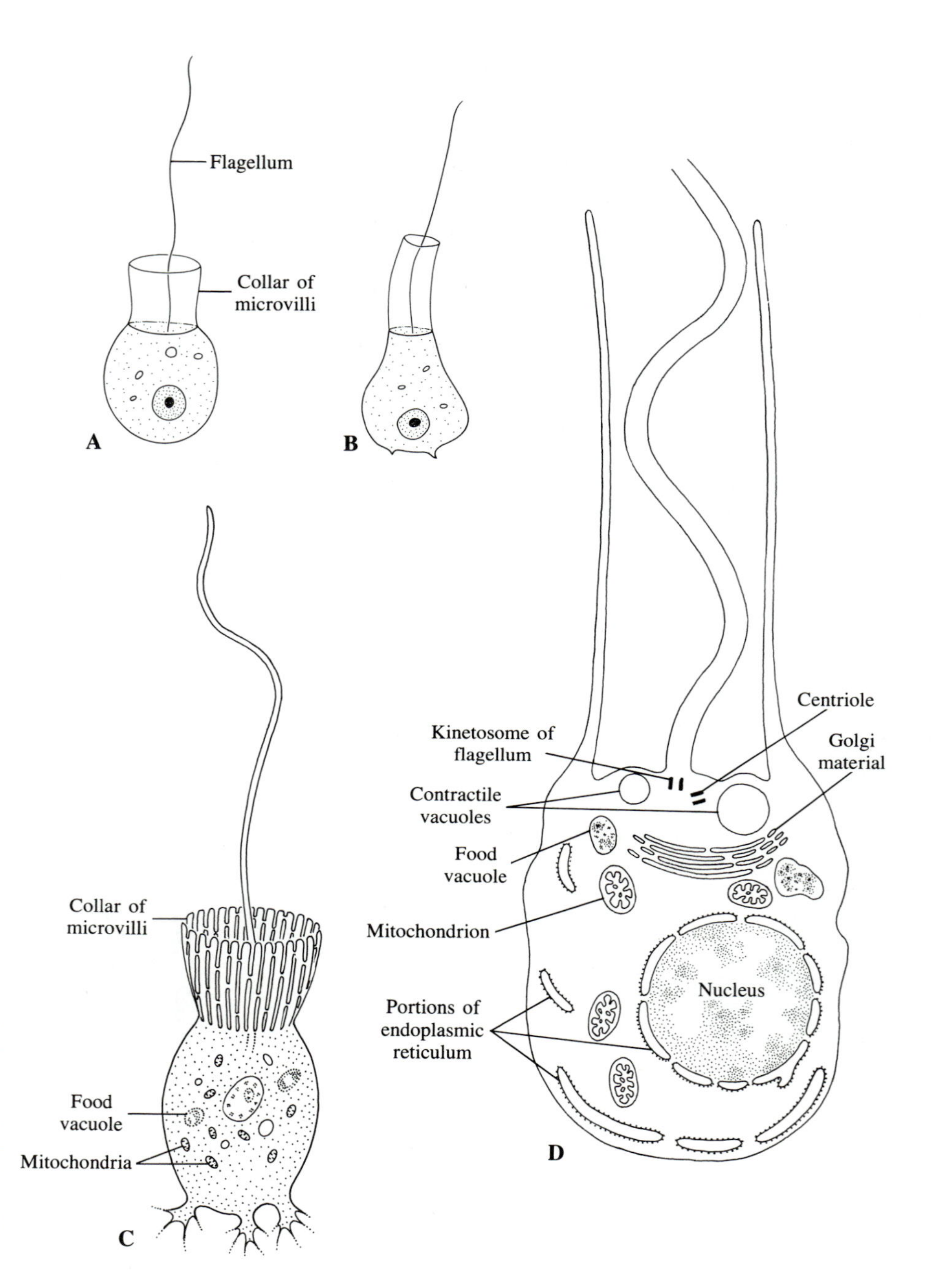

FIGURE 3.2 A, B. Choanocytes of calcareous sponges, as seen with the light microscope. C. Diagram of a choanocyte, as reconstructed from electron micrographs. D. Section of a choanocyte of *Spongilla,* based on electron micrographs. (After Brill, Zeitschrift für Zellforschung und Mikroskopische Anatomie, *144*.)

3.4, 3.11). Many of these are specialized for certain functions, such as storage and distribution of food, secretion of spicules (discussed below), and production of collagen; some of the collagen-secreting cells are anchored by rootlike strands of cytoplasm. The numerous undifferentiated amoebocytes, called **archeocytes,** can transform into any of the specialized types. Furthermore, choanocytes and pinacocytes may sink into the mesohyl and become amoeboid. It should be kept in mind, then, that sponges are labile organisms in which certain cells can change their appearance and function.

In the mesohyl of some parts of a sponge, especially around oscula, are contractile cells called **myocytes.** Electron microscopy reveals that they contain myofilaments of the type found in muscle cells of other animals. Although myocytes may be grouped to form sphincterlike complexes, they remain separate from one another.

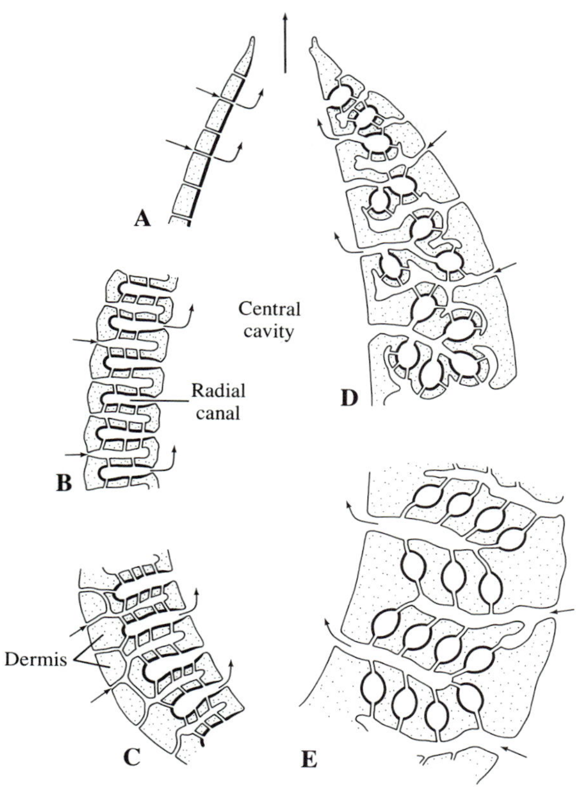

FIGURE 3.3 General types of sponge structure. In each type, the location of choanocytes is indicated by a thick black line. Arrows indicate the direction of water flow. A. Asconoid. B, C. Syconoid. D, E. Leuconoid.

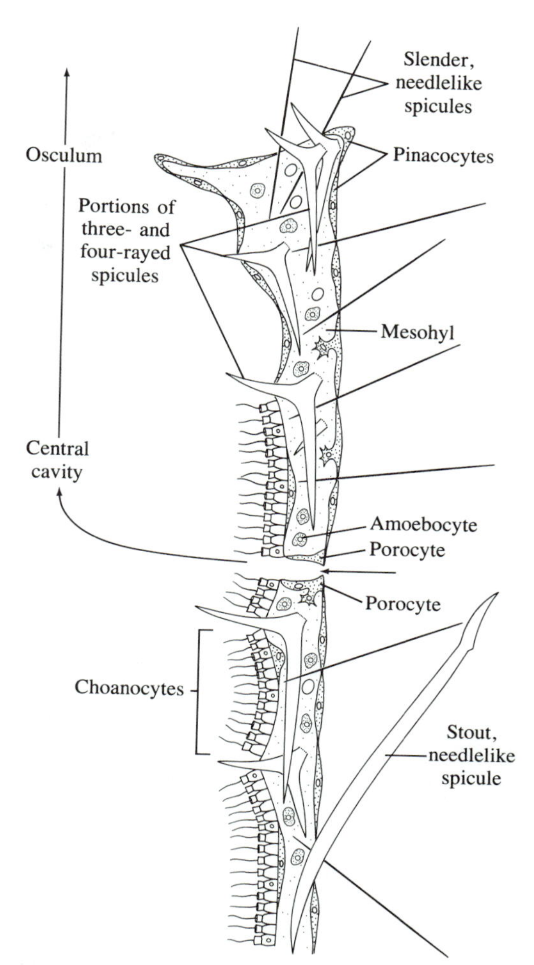

FIGURE 3.4 Portion of the body wall of *Leucosolenia complicata*. (After Jones, Journal of the Royal Microscopical Society, series 3, *85*.)

SPICULES

Spicules consist of either calcium carbonate or silica (silicon dioxide). The way in which certain cells produce spicules is described in the next section, in connection with calcareous sponges in which spicule formation has been most carefully studied. Although spicules frequently protrude externally, forming a fuzzy or even prickly protective covering, they originate in the mesohyl. They are held together by collagen fibers and are sometimes enclosed within tracts of collagen. In the hexactinellid sponges, many of the spicules are fused together to form a rigid latticework.

ASCONOID SPONGES

The simplest sponges are of the **asconoid** type (Figure 3.3, A). Their structure is illustrated by *Leucosolenia,* which forms mats of whitish tubes that are usually about 2 or 3 mm in diameter (Figure 3.5). The tubes branch and may reunite, but each larger tube has an osculum at its free end.

In sections through the body wall of *Leucosolenia* (Figure 3.4), it can be seen that porocytes are scattered among the pinacocytes that make up the outer covering. The apertures of the porocytes are only about 2 μm in diameter, and they can be made even smaller, so they effectively block the movement of particles larger than bacteria. The water and suspended material that move through them go directly into the central cavity, sometimes called the **spongocoel.** This is lined by choanocytes, whose flagella provide the force that brings water into the sponge and drives it out of the osculum. The diameter of the osculum is controlled by myocytes.

As was explained previously, capture of food is to a large extent the responsibility of the choanocytes. In some sponges these cells initiate and perhaps even complete digestion of the particles they collect. In others

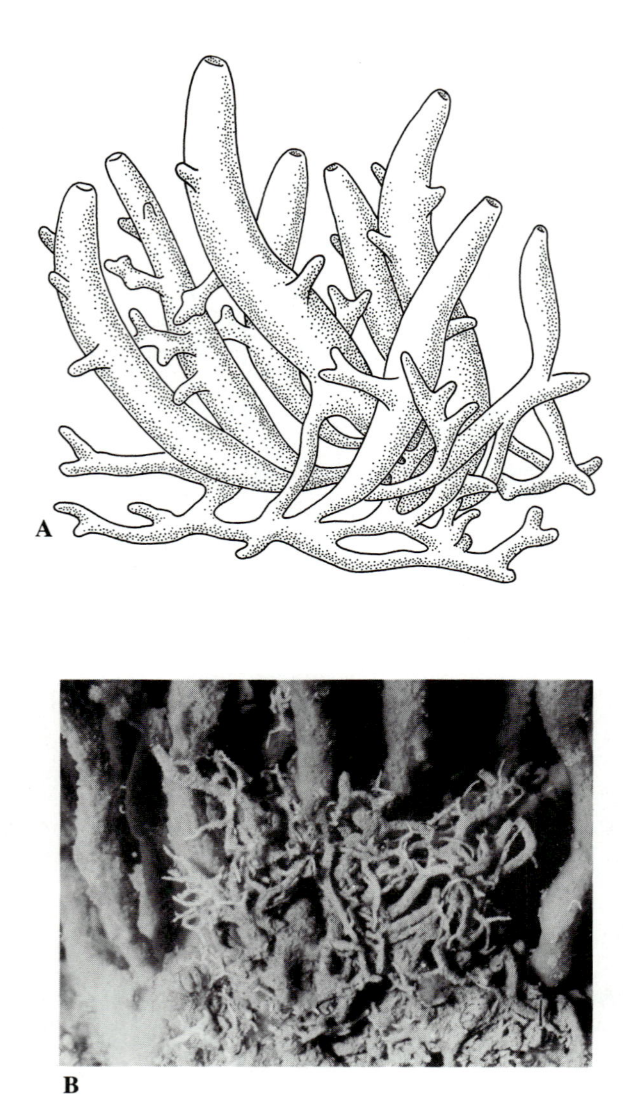

FIGURE 3.5 *Leucosolenia*. A. Drawing of a colony. Only the larger tubes have oscula. B. Photograph of a colony growing on *Microciona* (class Demospongiae).

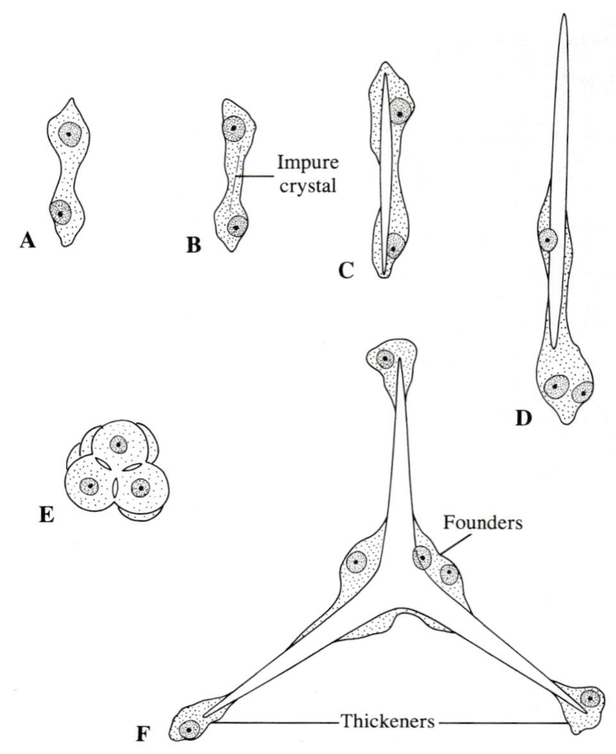

FIGURE 3.6 Spicule formation in calcareous sponges. A–D. Formation of a monaxon spicule. E, F. Formation of a three-rayed spicule. (After Woodland, Quarterly Journal of Microscopical Science, *49*.)

they only capture food and pass it to amoebocytes, which take care of digestion as well as storage and distribution of useful products.

The fuzzy appearance of *Leucosolenia* under low magnification is due to the needlelike spicules that stick out between the pinacocytes. Some of these spicules are extremely fine; others are stout and have enlarged and slightly curved tips. There are also spicules with three or four rays, but they are largely confined to the mesohyl, although one ray of each four-rayed spicule typically protrudes into the central cavity (Figure 3.4).

The spicules of *Leucosolenia* and other calcareous sponges consist largely of calcium carbonate, but they contain slight amounts of magnesium, strontium, and other elements. They are secreted extracellularly by amoebocytes called **scleroblasts.** These are frequently, if not generally, derived from pinacocytes. A scleroblast preparing to produce a spicule becomes binucleate and lays down an impure crystal of calcium carbonate outside the cell (Figure 3.6, A and B). Then successive layers of calcium carbonate are deposited (Figure 3.6, C and D). In time, the cell divides. One daughter, called the **founder,** controls the shape and size of the spicule; the other daughter, called the **thickener,** is concerned with adding more calcium carbonate.

In the formation of three-rayed spicules (Figure 3.6, E and F), a founder–thickener team is responsible for each ray. Four-rayed spicules are essentially three-rayed spicules to which a fourth ray is added. The addition is not made by a typical founder–thickener pair but by a pinacocyte that wanders into the mesohyl and joins the enterprise. It becomes binucleate only after work on the first three rays is well under way, and it does not divide into two cells.

A

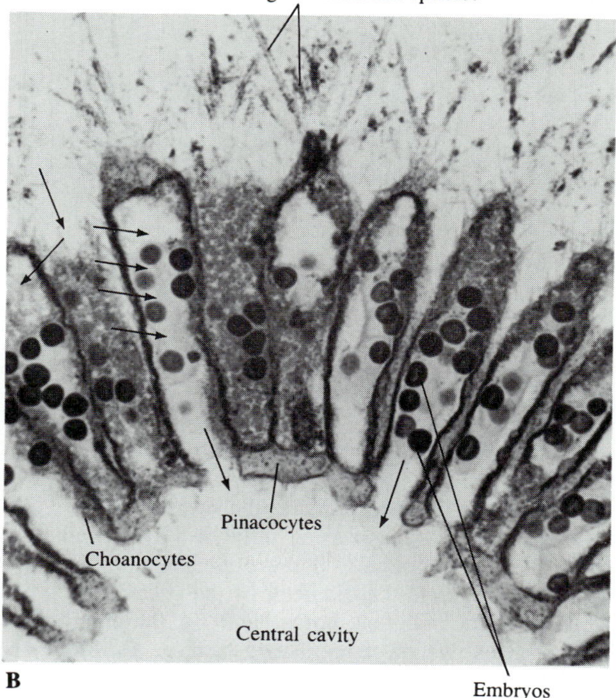

B

FIGURE 3.7 Calcareous sponges of the syconoid type. A. *Scypha,* living. B. *Sycon,* portion of a transverse section. The individual pinacocytes and choanocytes cannot be distinguished, but their locations are marked.

SYCONOID SPONGES

Syconoid sponges, represented by *Sycon, Scypha,* and related genera (Figure 3.7), are more complex than those of the asconoid type. The body wall is pushed out into hundreds of fingerlike projections (Figure 3.3, B; Figure 3.7, B). The radial canal in each projection is lined by choanocytes and leads to the central cavity. The combined internal surface of the radial canals far exceeds that of the central cavity, so that a syconoid sponge has many more choanocytes in proportion to its size than an asconoid type does. This increases its efficiency in circulating water and in extracting microscopic food particles.

In all syconoid sponges, as in asconoid sponges, there are calcareous spicules, many of which protrude from the body wall. Especially prominent spicules are located around the osculum. In addition, there are three- and four-rayed spicules. The three-rayed spicules form the skeleton that supports the radial canals; the four-rayed spicules are concentrated beneath the pinacocytes that line the central cavity, and one ray of each projects into the central cavity.

In simpler syconoid sponges of the type just described, the body wall is thin and the pores that admit water and minute particles of food into the radial canals are controlled by porocytes. In more complex syconoids (Figure 3.3, C), however, water moves into the radial canals by a less direct route. The spaces between the projections that close the radial canals are partly blocked by a secondary thickening of the body wall, sometimes called the **dermis.** The secondary layer, consisting of pinacocytes and mesohyl, is interrupted by small openings that permit the water to move inward and reach the pores that lead to the radial canals.

LEUCONOID SPONGES

The majority of sponges are more elaborate than either asconoid or syconoid types. The choanocytes line small chambers located between the outer surface and the main cavities that open to the outside by oscula (Figure 3.3, D and E; Figure 3.8). Such sponges are said to be **leuconoid.** Some of them, such as *Leuconia,* have calcareous spicules similar to those of asconoid and syconoid sponges, but most have siliceous spicules. These, unlike calcareous spicules, are secreted intracellularly. They come in a wide variety of shapes and sizes (Figure 3.9). The ones called **megascleres** are, in general, larger than those called **microscleres,** but the shapes of the spicules assigned to these major categories are also different. There may be more than one type of spicule from each category in a particular species (Figure 3.9, C).

In almost any marine habitat where there is a diversified fauna of sponges, siliceous leuconoid types predominate. They generally form rather thin encrustations (Color Plate 1), lumpy growths (Figure 3.1, C; Color Plate 1), or masses with fingerlike projections (Figure 3.1, B; Color Plate 1). Some, however, are vase-shaped

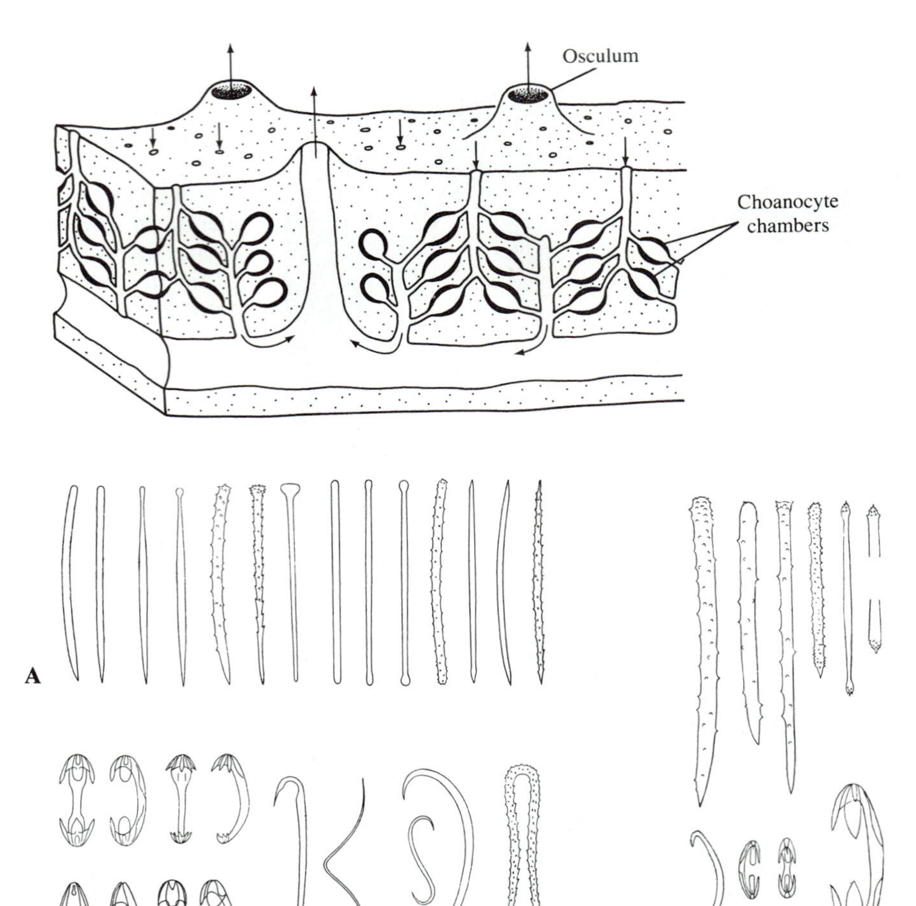

FIGURE 3.8 Diagram of a portion of an encrusting leuconoid sponge.

FIGURE 3.9 Siliceous spicules found in sponges of the class Demospongiae. A. Various types of megascleres. B. Various types of microscleres. C. The megascleres and microscleres found in one species, *Myxilla incrustans*. (Koltun, Keys to the Fauna of the Far-Eastern Seas of the USSR, *61*.)

or urn-shaped (Figure 3.1, A and D; Color Plate 1). They are often bright yellow, orange, red, violet, or green. Their oscula may be obvious and raised up on volcanolike eminences, or they may be small and obscured by spicules. The pores by which water enters the system of canals are strictly microscopic and cannot be seen except in thin sections.

SPONGE PHYSIOLOGY

The functioning of a sponge depends to a large extent on its ability to maintain and control the current of water flowing through it. The current brings in food and oxygen and takes away carbon dioxide and nitrogenous wastes. In a sponge that is releasing gametes or larvae, the current also carries these out of the body.

So far as is known, the beat of the flagella of neighboring choanocytes is neither synchronized nor coordinated, but water is nevertheless moved efficiently from the numerous inhalant pores to the one or more oscula. The amount of water passing through a sponge may be controlled by porocytes, if these are present, by myocytes that regulate the diameter of the oscula and probably of some inhalant pores, and by the number of choanocytes whose flagella are active. In *Suberites massa,* a leuconoid type, and perhaps in other sponges, there is an amoeboid cell on the exhalant side of each choanocyte-lined chamber (Figure 3.10, A and B). This cell is remarkable in that it is pierced by canals that run all the way through it, so it is much like a sieve. It is believed that when the amoebocyte plugs the exhalant passageway, water flows more slowly out of the chamber. Presumably, this gives the choanocytes a better chance to trap particles of food moving past them. The fact that the long flagella of some of the choanocytes actually enter the ducts and sometimes contact pseudopodia of the amoeboid cell suggest that this cell may physically slow down the beat of the flagella. The

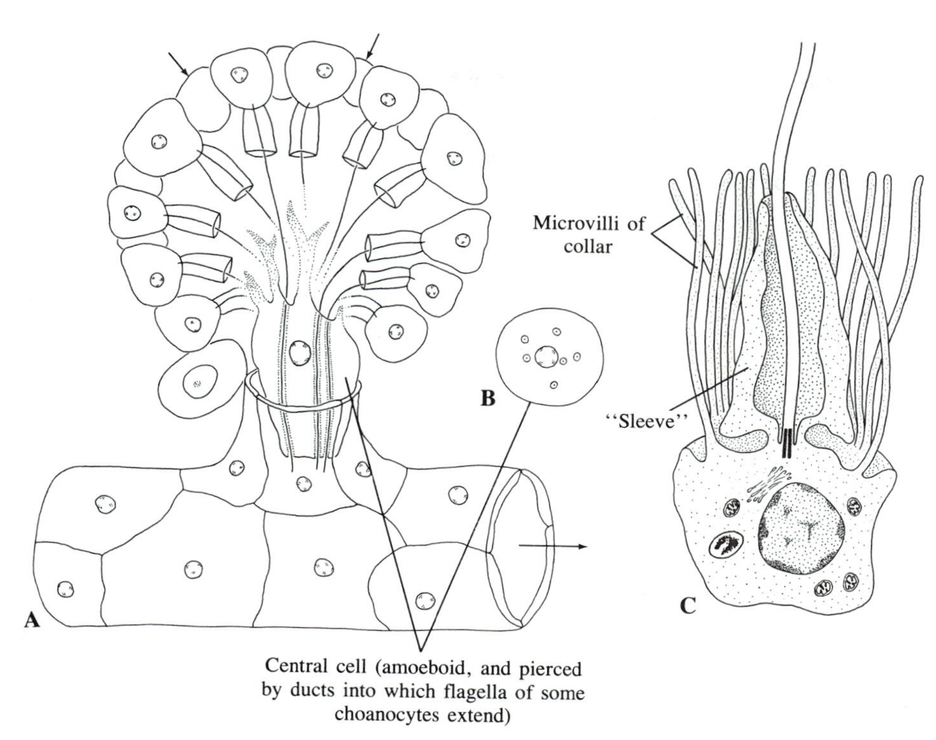

FIGURE 3.10 A. Choanocyte chamber, amoeboid central cell, and exhalant canal of *Suberites*. B. Section through a central cell, showing flagella of choanocytes in ducts running through the cell. C. Diagram of a section of a choanocyte of *Suberites,* based on electron micrographs. (After Connes, Diaz, and Paris, Comptes rendus de l'Académie des Sciences, *273*.)

choanocytes of *Suberites,* by the way, are unusual in that they have a sleeve of cytoplasm between the flagellum and the collar formed by microvilli (Figure 3.10, C).

In a complex sponge, the total cross-sectional area of the choanocyte chambers is much greater than that of the oscula or even that of the thousands of inhalant pores. Thus water moves more slowly through the choanocyte chambers than it does through other passageways. The current is always strongest at the oscula.

In some sponges, the organism passes its own weight in water every five seconds. Flagellar activity accounts for much of the flow, needed not only for feeding but for respiration and removal of wastes. Sponges also take advantage of currents that impinge on them. The faster the current, in fact, the faster the rate of flow through the sponge. To use a current most effectively, and to keep water moving from the inhalant pores to the oscula, a sponge opens the pores on the side that faces "upstream" and closes down the others.

Another way in which an external current can affect the flow through a sponge is by creating negative pressure as it moves past the oscula. It thus draws water out of the oscula. In this respect, the current operates in much the same way as a breeze that draws air out of a chimney or that ventilates the underground passages inhabited by prairie dogs. The form of some species of sponges varies according to the strength of the prevailing current. In the case of *Halichondria panicea,* for instance, specimens from relatively calm water have their oscula raised higher than do those swept by strong currents. This is a good adaptation, because in calm water the current close to the substratum is generally slower, and therefore less effective in creating negative pressure, than the current farther away.

Many sponges probably subsist to a large extent on bacteria, which are generally in the size range of 1 to 3 μm, and on various other small organisms, such as dinoflagellates. Studies on certain tropical sponges have shown, however, that most of the organic matter they filter out of the water consists of particles even smaller than bacteria. Although choanocytes are responsible for trapping much of a sponge's food, pinacocytes and amoebocytes also take up food. Amoebocytes are perhaps the only cells that can ingest dinoflagellates and other algal cells that are substantially larger than bacteria. It has been pointed out earlier that food can be transferred from one cell to another—from a choanocyte to an amoebocyte, for instance. Not all food is taken from the water current; bacteria and other particles that somehow get into the mesohyl are also ingested. Figure 3.11 shows, diagrammatically, how bacteria entering *Ephydatia,* a freshwater sponge, are ingested and passed from cell to cell.

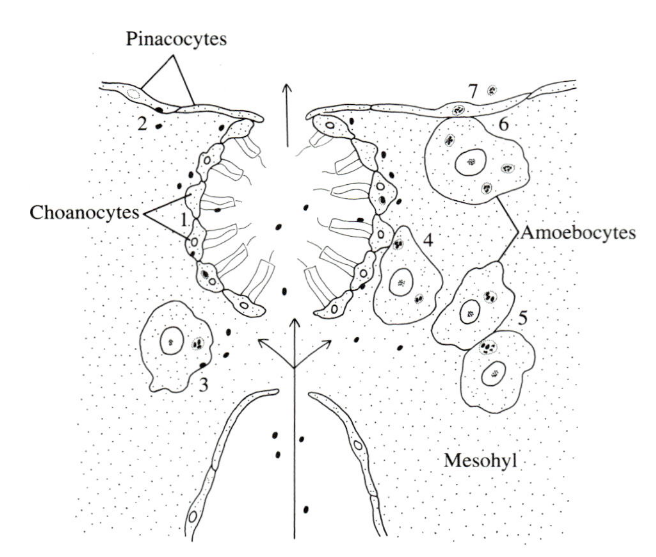

Figure 3.11 Capture of bacteria by choanocytes (1), a pinacocyte (2), and an amoebocyte of *Ephydatia fluviatilis* (3). Transfer of food vacuoles from a choanocyte to an amoebocyte (4), from one amoebocyte to another (5), and from an amoebocyte to a pinacocyte (6) is also shown, as well as elimination of material by a pinacocyte (7). (Weissenfels, Zoomorphologie, *85*.)

Sexual Reproduction and Development

Most sponges are hermaphroditic, but a particular specimen usually does not produce sperm and oocytes at the same time. In older literature, gametes were generally reported to originate from amoebocytes, and this interpretation probably does apply to some sponges. However, studies on gametogenesis in certain members of the class Calcarea have shown that sperm and oocytes derive from choanocytes that lose their collars and flagella and sink into the mesohyl. The same thing happens in at least some siliceous sponges, including those of the freshwater genus *Spongilla*.

A cell destined to form male gametes undergoes two meiotic divisions, and each of the four cells produced differentiates into a sperm with a flagellar tail. Oocytes usually do not undergo meiosis until they have grown considerably and have been penetrated by sperm. During later phases of their enlargement, they actually consume certain cells, called nurse cells, that become associated with them. Fertilization is accomplished with the help of choanocytes, which trap sperm that have entered the canal system. A choanocyte that has captured a sperm becomes amoeboid and carries the sperm to a receptive oocyte, or gives the sperm to an amoebocyte that acts as a carrier. Both patterns of sperm transfer may occur in different species of the same genus.

Cleavage of a fertilized egg takes place in the mesohyl; the swimming larva that is eventually produced enters a canal or the central cavity and is carried out with the exhalant current. The kind of larva varies from group to group, but there are two main types: the amphiblastula and the parenchymula.

The **amphiblastula** is seldom found outside of the class Calcarea. The way in which it develops and becomes transformed into a young sponge is decidedly unlike any other embryological sequence in multicellular animals. The process is illustrated here by an account of the reproduction of *Sycon raphanus*.

The sperm of *Sycon* is transmitted to the egg by a choanocyte that has become amoeboid (Figure 3.12, A–C). Following meiosis and the casting off of polar bodies, cleavage begins (Figure 3.12, D–F). At the 16-cell stage (Figure 3.12, G), the embryo is almost discoid, with eight of its cells being appreciably larger than the other eight. As development continues, the smaller cells divide more rapidly than the larger cells. In time, the embryo becomes ball-shaped, but it has an opening among the larger cells (Figure 3.12, H). The opening serves as a mouth, permitting the embryo to feed on special nutritive cells in the mesohyl. At this stage the embryo is called a **stomoblastula** (''blastula with a mouth'').

The smaller cells develop flagella on the side that faces the inner cavity of the stomoblastula, and the opening closes (Figure 3.13, A). Then a break develops among the larger cells and the embryo literally turns itself inside-out, so the flagella are now external (Figure 3.13, B). (The process of inversion is somewhat similar to that which takes place in a young colony of *Volvox*—see Chapter 2—but this must not be construed to mean that there is a close relationship between *Volvox* and sponges.)

With further development, the embryo reaches the amphiblastula stage (Figure 3.13, C and D), when it leaves the mesohyl and is swept out of the sponge by way of the osculum. It swims with its flagellated side forward. The next important event, which takes place about the time the amphiblastula prepares to settle, is a process comparable to gastrulation. The flagellated cells invaginate, so that they are enclosed by the nonflagellated cells, which become pinacocytes. The larva settles on the side that has the opening (Figure 3.13, E), and the pinacocytes that border the opening establish a tight contact with the substratum. These pinacocytes have been observed to form pseudopodia.

The opening formed by the invagination of the flagellated cells soon closes. At this stage the young sponge is called an **olynthus.** Some of its flagellated cells transform into choanocytes; others lose their flagella and become amoebocytes. Most of the cells on the outside persist as pinacocytes, but some sink inward and become amoebocytes. Thus the wandering cells of

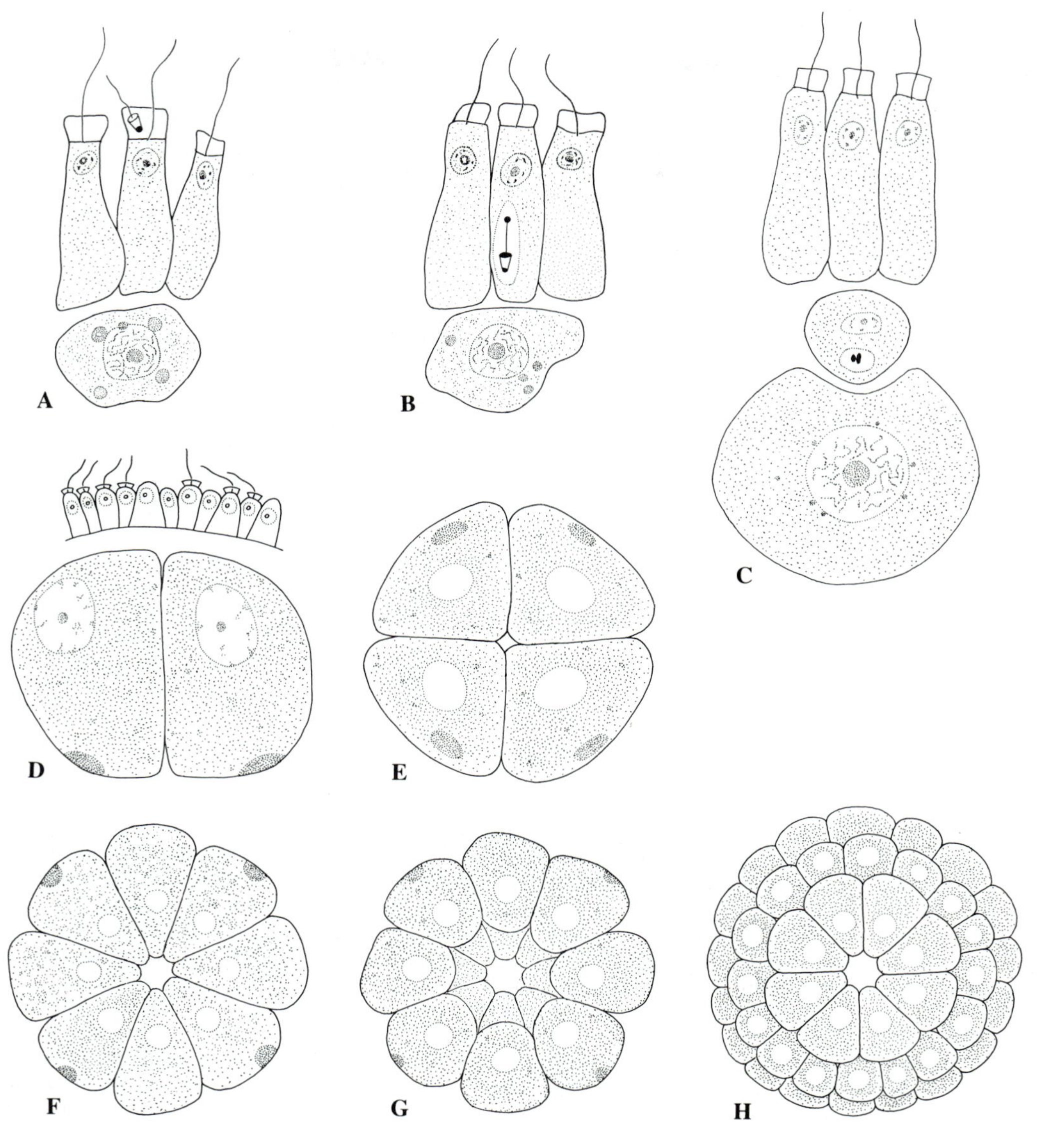

FIGURE 3.12 A, B. Capture of sperm by a choanocyte of *Sycon raphanus* (class Calcarea.) C. Choanocyte transformed into an amoebocyte and carrying the sperm to an enlarged oocyte. D. Embryo at the 2-cell stage. E. Four-cell stage. F. Eight-cell stage. G. Sixteen-cell stage. H. Stomoblastula, much like a blastula but with an opening. (After Duboscq and Tuzet, Archives de Zoologie Expérimentale et Générale, *79*.)

the mesohyl are derived from both the flagellated and the nonflagellated layers of the recently settled sponge. For a while the sponge resembles the asconoid type, although it still has no osculum. By the time an osculum appears, on the side away from the substratum, side pockets of the central cavity have formed and the syconoid level of complexity is attained.

Among the flagellated cells of the amphiblastula (Figure 3.13, C and D) are four nonflagellated cells that are believed to be photoreceptors. They contain rings of granules derived from cytoplasmic constituents that can be recognized at early stages of embryonic development (Figure 3.12, E–G). The four presumed photoreceptor cells are cast off as the amphiblastula prepares to settle. At this stage the larva, previously phototropic, becomes negatively phototropic.

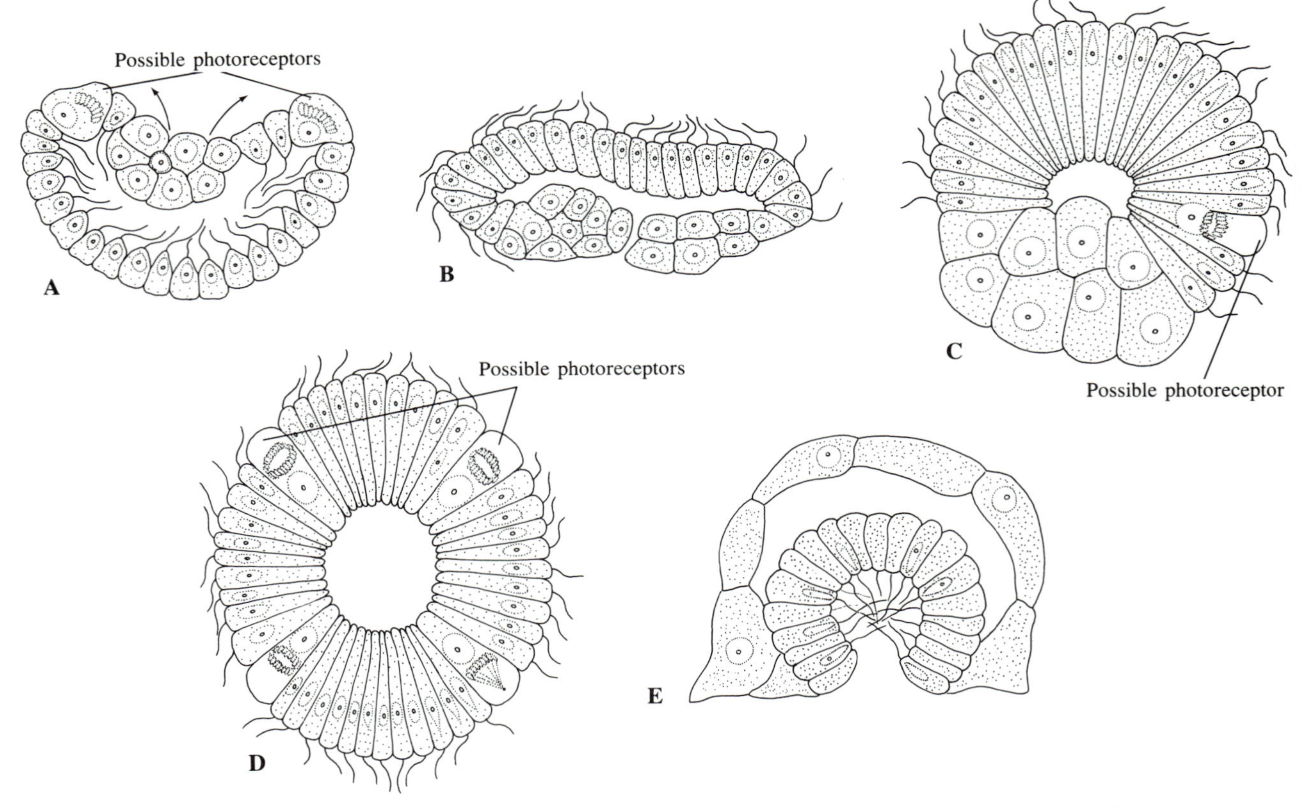

FIGURE 3.13 A. Section through a stomoblastula of *Sycon* after its opening has closed. The smaller cells now have flagella. One of the four cells thought to be photoreceptors is shown. B. The embryo after it has turned inside-out. C. Amphiblastula. D. Amphiblastula, transverse section at the level of the four photoreceptor cells. (After Duboscq and Tuzet, Archives de Zoologie Expérimentale et Générale, *79*.) E. A young *Sycon* soon after settling. The flagellated cells have invaginated. The opening on the lower side will close, and an osculum will open on the upper side. (After Hammen, Archiv für Biontologie, *2*.)

The **parenchymula** (or parenchymella) larva occurs in both calcareous and siliceous sponges. It is typically solid, although it may be preceded by a hollow-ball stage. For instance, in *Clathrina,* a genus of leuconoid sponges of the class Calcarea, the 16-cell stage is a blastula in which all of the cells have flagella on the side that faces the inner cavity. A little later on, one or a few nonflagellated cells differentiate, and soon the embryo turns inside-out. The flagella are now external (Figure 3.14, A). The free-swimming larva thus produced slightly resembles an amphiblastula, although nearly all of its cells have flagella. Some of these cells lose their flagella and sink into the central cavity; products of division of the few nonflagellated cells also do this. The parenchymula stage is reached when the cavity has been filled up. The larva continues to swim for a time, then settles by the end where the nonflagellated cells are located. As it flattens out over the substratum, it undergoes extensive reorganization (Figure 3.14, B). The flagellated cells, which are precursors of choanocytes, migrate inward, and many of the cells that had previously filled the inside move outward to form the pinacocyte layer. When the definitive central cavity appears within the young sponge, it is lined by future porocytes (Figure 3.14, C). Soon, however, the porocytes move outward to their functional positions at the surface, and the choanocytes take their places around the cavity (Figure 3.14, D). When the definitive osculum opens up, the young sponge is essentially of the asconoid type. With further changes, it reaches the leuconoid level of complexity.

In most members of the class Demospongiae, all of which are leuconoid, the embryo is solid from the beginning. Repeated cleavages of the egg result in the formation of a parenchymula in which a layer of small, externally flagellated cells surrounds a mass of larger

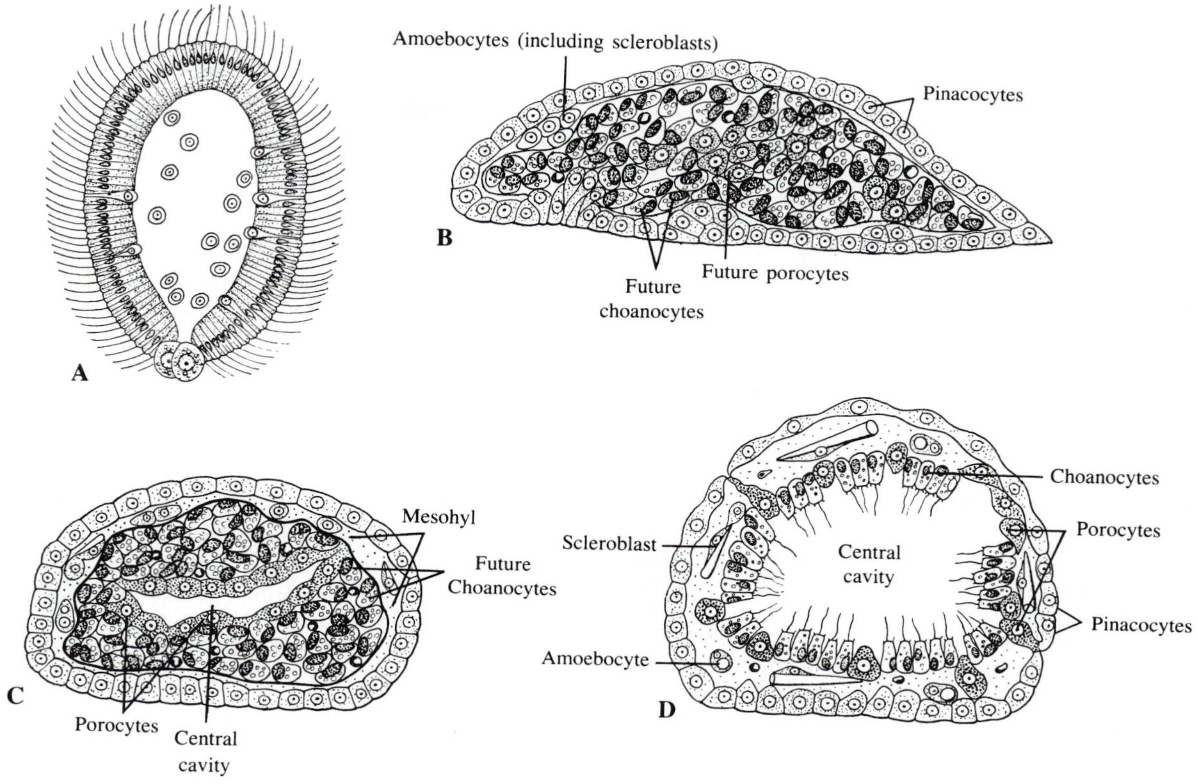

FIGURE 3.14 Some stages in the development of *Clathrina,* a leuconoid sponge of the class Calcarea. A. Free-swimming larva. (It will be called a parenchymula after the inner cavity fills with cells.) B. Parenchymula after its flagellated cells have migrated inward and its pinacocytes, previously in the interior, have migrated outward. C. The central cavity has appeared and is lined by future porocytes. D. The porocytes have moved to their functional positions and the choanocytes now line the central cavity. Spicules are being produced by some of the scleroblasts. (Brien, in Florkin and Scheer, Chemical Zoology, *2,* after Minchin, in Lankester, A Treatise on Zoology.)

cells. There is mesohyl between the larger cells, and some of the amoebocytes are already active in spicule formation. In certain demosponges, the cells of the flagellated layer are phagocytosed when the larva settles, so that all constituents of the adult sponge originate from the inner mass of larger cells. In others, at least many of the outer cells survive and migrate inward, giving rise to choanocytes.

ASEXUAL REPRODUCTION AND GEMMULE FORMATION

Asexual reproduction in sponges is usually accomplished by **budding,** in which aggregations of cells (mostly amoebocytes) differentiate into small sponges that are released from the surface of the colony or liberated through oscula. Pieces that break off when a sponge is damaged may form complete colonies if they happen to settle on a suitable substratum. Commercially important bath sponges have been propagated by cutting off pieces and affixing these to firm objects, such as concrete blocks, on the ocean floor.

A peculiar type of bud called the **gemmule** (Figure 3.15) is found in many species of the freshwater family Spongillidae. The production of gemmules takes place in the mesohyl and is achieved by the cooperative effort of a variety of amoebocytes. Those of one type break down to provide food for the ones that will form the germinal mass in the interior of the gemmule. Other amoebocytes lay down the outer covering of the gemmule. When finished, the covering consists of a hard coat of collagen reinforced by spicules.

Gemmules set free by a dying sponge survive extreme cold and desiccation. Thus they serve as overwintering bodies in situations where spongillids are unable to maintain themselves—for lack of microbial food or

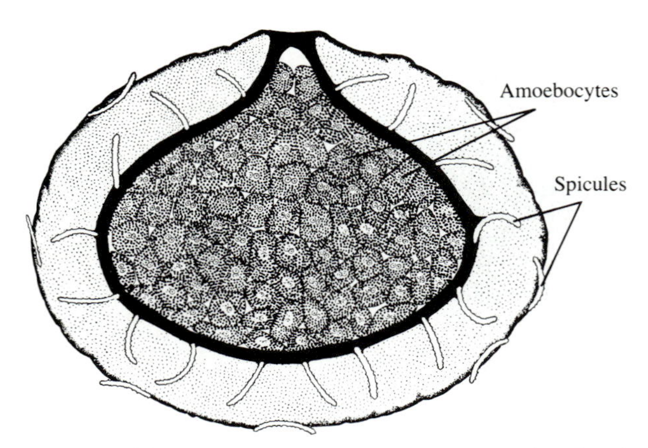

FIGURE 3.15 Gemmule of *Spongilla*. (Leveaux, Annales de la Societé Royale Zoologique de Belgique, *70*.)

other reasons—when the water is cold. They are also oversummering bodies for species that live in ponds that dry up. When conditions are favorable for germination of a gemmule, an opening, called the **micropyle,** appears at a predetermined place in the outer covering. The amoebocytes escape and differentiate in the various kinds of cells needed to build a complete young sponge.

REGENERATION AND RECONSTITUTION

If a sponge is injured, the cells in the damaged portion will, under favorable circumstances, become reorganized so that healing is complete. In certain species, the capacity for regeneration is so great that even if a sponge is squeezed through a fine-meshed cloth, the individual dissociated cells will regroup and form one or more new sponges. This process is referred to as **reconstitution.** Archeocytes are believed to play an important role in the reaggregation process, and calcium and magnesium ions, as well as at least one organic molecule present on cell surfaces, are required.

Experiments have shown that some sponges will not accept a graft unless the piece has come from the same specimen. Others, however, will accept a graft so long as it is from the same species. Moreover, when gemmules of *Spongilla* germinate, two or more young colonies may fuse to form a single mass, and a similar phenomenon has been observed in certain marine species whose larvae have settled close together. In such cases, genetic differences between the young colonies seem not to interfere with the process of integration.

CLASSIFICATION

The classification of sponges has traditionally been based on their general organization and the nature of their skeletal components. In recent years, much use has been made of patterns of reproduction and development. Electron microscopy also reveals details helpful to taxonomists, and comparative biochemistry provides some interesting facts about sterols, free amino acids, and other compounds in various groups of sponges.

CLASS CALCAREA

Sponges of this class are marine and mostly restricted to shallow water. The spicules, which consist of calcium carbonate, include three- and four-rayed types as well as needlelike types. Collagen is present but does not form conspicuous fibers such as are characteristic of some members of the class Demospongiae. The structure of calcareous sponges may be asconoid (as in *Leucosolenia*), syconoid (as in *Scypha* and *Sycon*), or leuconoid (as in *Leuconia* and *Leucandra*).

A few genera referred to this class secrete substantial deposits of calcite in the form of plates or anastomosing strands. They also have spicules. *Murrayona* and *Petrobiona,* found in grottoes in the Mediterranean Sea, are examples of this type of construction, which apparently was characteristic of some fossil sponges called pharetronids. One genus, *Neocoelia,* has no spicules, only an aragonite skeleton that divides the sponge into compartments.

CLASS DEMOSPONGIAE

About 95% of the known species of sponges belong to this class. The name given to the group is a tolerated misspelling of Desmospongiae, which means "sponges held together by bonds." It alludes to the collagenous fibers that form the skeletons of bath sponges (Figure 3.16). These large sponges, found in the Mediterranean Sea, Gulf of Mexico, Caribbean region, and a few other places, do not have spicules. They are harvested by divers, and after their living components have died and been washed out, the skeletons are bleached and marketed. Although the collagenous skeletons are strong, they are also resilient, so bath sponges are soft as well as absorbent.

Demosponges have a leuconoid structure, and, as a rule, their skeletons consist of siliceous spicules held together by collagen. The collagen fibers may be too fine to see except with an electron microscope, but in some species they are visible with the light microscope. Several genera outside the family Spongiidae, which includes the bath sponges, lack spicules, and at least two genera have neither spicules nor collagen.

The spicules of demosponges are of many forms. Those classified as microscleres are, on the whole, smaller than those called megascleres, but the variety of forms belonging to the two categories is also different. Microscleres are not always present, but when they

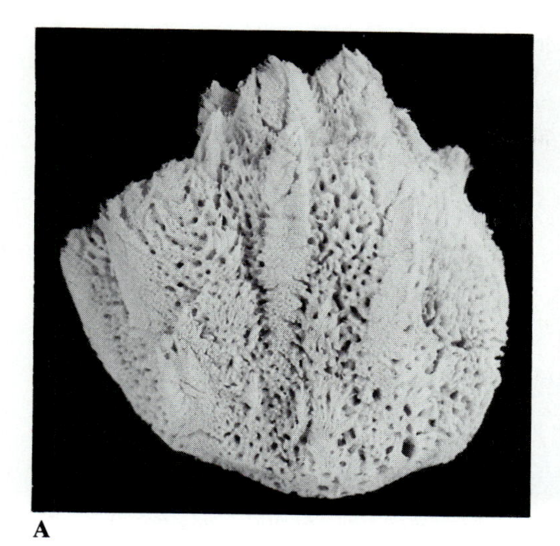
A

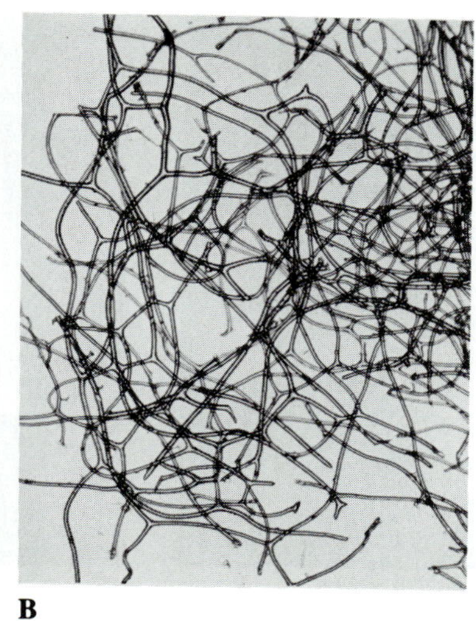
B

FIGURE 3.16 Bath sponges. A. Dried spongin skeleton. B. Photomicrograph of spongin fibers.

do occur, many of them underlie and support the sheets of pinacocytes that line internal passageways. Demosponges have no six-rayed spicules, so they are easily distinguished from members of the class Hexactinellida.

All freshwater sponges belong to the Demospongiae. The more widespread of the two families is the Spongillidae, which has about 120 species in temperate and tropical regions. Spongillids are usually attached to wood and stones. Some of them have zoochlorellae—unicellular green algae—living and multiplying in their amoebocytes. The species that harbor these symbionts are usually bright or dark green (Color Plate 1). Results of experiments with *Spongilla lacustris* show that exclusion of light, which makes it impossible for the zoochlorellae to be photosynthetic, reduces growth of the sponge as much as 40%, even though the sponge has continued to feed normally on microorganisms.

The gemmules produced by spongillids may be viewed as reproductive bodies, but they also enable these freshwater sponges to survive periods of drought and extreme cold.

Freshwater sponges are the only members of the phylum that have contractile vacuoles. These vacuoles have been found in pinacocytes, porocytes, choanocytes, and various kinds of amoebocytes. In *Ephydatia fluviatilis,* in fact, they are present in all types of cells, and perhaps this is to be expected in other species.

Many marine demosponges are involved in close associations with other organisms. Blue-green algae, zooxanthellae (dinoflagellates), bacteria, and microbes of uncertain systematic position have been found in various species. Some of the bacterial symbionts are localized within amoebocytes, but others, as well as the blue-green algae and zooxanthellae, are extracellular in the mesohyl. The blue-green algae inhabiting certain marine sponges, such as *Verongia,* make up about one-third of the weight of the sponge mass, and they are known to release glycerol and other organic compounds that can be utilized by their hosts.

The shells of some live scallops are commonly encrusted by sponges (Figure 3.17, B). The sponges do not harm the scallops but in fact provide protection because their spicules or their taste, or both, discourage octopuses and other predators. There are, nevertheless, animals that subsist mainly on sponges. Among them are certain sea stars, fishes, and sea slugs of the suborder Doridacea.

Species of *Cliona* (Figure 3.17, C and D), which bore into coral skeletons and the shells of molluscs and barnacles, show some interesting specializations. Soon after a young *Cliona* begins to grow, its amoebocytes attack the calcareous substratum, secreting a substance that dissolves it, to establish anchorage. Chips that have been undercut are then ingested by the amoebocytes. Somehow the chips enter the main cavities of the sponge and are carried out with the exhalant current. The oscula of *Cliona,* like the inhalant pores through which water and food enter, are located on the small, yellow patches visible at the surface of the shell or coral. Much of the sponge is beneath the surface, however, and the shell or coral skeleton may be honeycombed to the extent that it is weakened.

Certain hermit crabs typically inhabit the firm, rather dense masses formed by species of *Suberites* (Figure 3.17, A). A hermit crab first seeks out an empty snail shell, but this becomes overgrown by a *Suberites*. After the shell has dissolved or been worn away, the hermit crab occupies and maintains the cavity that remains. A hermit crab only 3 or 4 cm long is sometimes found in a *Suberites* that is 10 cm in diameter.

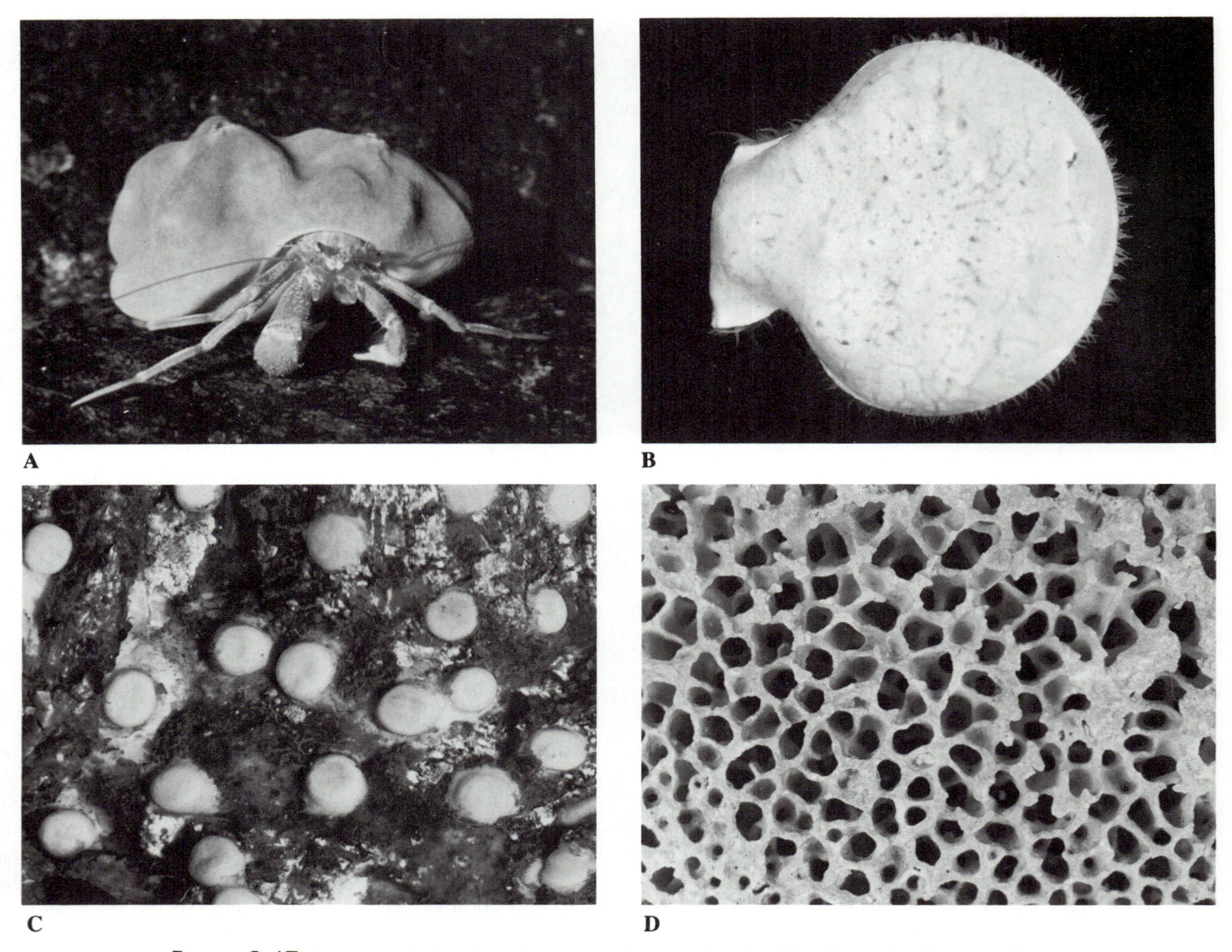

FIGURE 3.17 Sponges of the class Demospongiae associated with other animals. A. *Suberites ficus* covering a gastropod shell inhabited by a hermit crab. B. *Myxilla incrustans* coating one valve of a scallop. C. *Cliona celata* in the calcareous shell of a large barnacle. Most of the sponge is beneath the surface of the shell. D. Portion of a valve of the rock scallop, *Hinnites,* showing the damage caused by *Cliona*.

CLASS SCLEROSPONGIAE

In the Mesozoic era, a considerable amount of reef-building in the seas was carried on by organisms that paleontologists call **stromatoporoids.** Certain of the hills in central Europe, for instance, have substantial deposits of these calcareous fossils, which were for a long time thought to be corals of some sort. It now appears that they are the remains of sponges related to some living species found in the West Indies, South Pacific, and elsewhere. Most of these were discovered rather recently, and a new class was established for them.

Sclerosponges deposit a basal skeleton of calcite or aragonite. This calcareous material, which may have pits, upright projections, or various other elaborations, some of which are found in fossil stromatoporoids, is beneath the mass of living cells and is therefore an exoskeleton. The material is independent of the siliceous spicules secreted in the living portion, which is organized according to the leuconoid plan. In certain species, however, as the base becomes thicker and the living portion is pushed upward, spicules and collagen fibers are left behind and are incorporated into the calcareous matrix.

It has been argued that sclerosponges should be placed in the Demospongiae. This view is based on a possibly valid assumption that living demosponges derive from now-extinct groups that secreted basal exoskeletons comparable to those of sclerosponges.

Class Hexactinellida

Hexactinellid sponges are strictly marine, and most of them are found at depths greater than 50 m. Their firm, upright growths are usually large, and some species reach a height of more than 1 m. The form of these sponges ranges from that of a cup, vase, or simple tube to a mass of branching tubes. The basic plan of construction is somewhat comparable to that of syconoid sponges because the chambers comparable to choanocyte chambers are fairly close to the spacious central cavities.

Hexactinellids differ so markedly from other sponges that some experts think they should be placed in a separate phylum. The name Symplasma has, in fact, been proposed for the phylum, because the species whose structure has been studied by modern methods consists to a large extent of **syncytia:** multinucleate sheets and networks (Figure 3.18, C). Single cells functionally comparable to archeocytes and other kinds of amoebocytes are present, but they have limited mobility and are linked closely with the syncytia. There is little mesohyl, and myocytes have not been seen. The absence of myocytes is correlated with the inability of hexactinellids to contract.

An extremely interesting feature of hexactinellids is the fact that the protoplasmic units that function as choanocytes lack nuclei. They are, moreover, separate from the syncytial complexes, although they are embedded in the latter, and one may be connected to another by slender protoplasmic bridges. Because they lack nuclei, it is more appropriate to call them **collar bodies** than collar cells or choanocytes. Collar bodies are thought to derive from archeocytes that penetrate the syncytium and then bud off masses that develop microvillar collars and flagella. The absence of nuclei in the collar bodies suggests that they have limited life spans.

The secondary network within the chambers into which the flagella and collars project consists of slender extensions of the syncytium in which the collar bodies are located. It has been proposed that the network supports the edges of the collars and, by occupying much of the space between collars, prevents backflow.

The spicules of hexactinellids are siliceous, and there are both megascleres and microscleres. Megascleres typically have three axes, and therefore a total of six points, a configuration that gives the class its name. Many of the megascleres are united by cross-bridges, so a stiff, stable latticework is formed (Figure 3.18, A, B, and D). There are also, however, abundant free spicules (Figure 3.18, E).

The attractive skeletons of some hexactinellids, such as *Euplectella aspergillus* (Figure 3.18, A and B), are prized as curios. A species of *Euplectella* found in Japanese waters often has a pair of red shrimps living in its central cavity. The male and female enter the sponge when they are young, and after they reach a certain size their escape is prevented by the grate of spicules that covers the exhalant opening. Cleaned skeletons, with the dried shrimps intact, have been given to newlyweds as symbols of undying love and devotion. Various other hexactinellids harbor crustaceans, small fishes, or other associates.

Fossil Sponges and Speculation on the Origins of Sponges

Fossil sponges are found in deposits that date to the Cambrian period, the earliest part of the Paleozoic era. In general, however, the only sponges that fossilized well were those that secreted a calcareous base (such as stromatoporoids) and those whose spicules were fused into a latticework (such as hexactinellids). Some of the ancient deposits also contain the calcareous skeletons of spongelike animals called archaeocyathans, which became extinct soon afterward. The greatest diversity of true sponges is thought to have occurred during the Cretaceous period, which ended the Mesozoic era. It was during this period, which lasted for about 25 million years, that the last dinosaurs flourished and finally became extinct and that birds and mammals started to become the dominant terrestrial vertebrates.

The fact that sponges do not have tissues or organs suggests that they are not far removed from some group of colonial protozoans. Choanoflagellates, to which choanocytes are so similar, are almost certainly the ancestors of sponges. One can imagine that after colonies of choanoflagellates became larger, more cohesive, and sedentary, cells with flagella and collars could have become organized around a central cavity, whereas other cells assumed functions comparable to those of pinacocytes and amoebocytes of modern sponges. Coincidentally, or later, pores could have opened up in the wall, allowing water to move into the central cavity from the sides. Eventually, more complex arrangements of choanocytes and passageways characteristic of syconoid and leuconoid sponges could have evolved. The stages through which some sponges pass during their development suggest that this is what may have happened.

The adoption of a sessile habit may not have been one of the earliest steps in the evolution of sponges. It is possible that small organisms with a central cavity

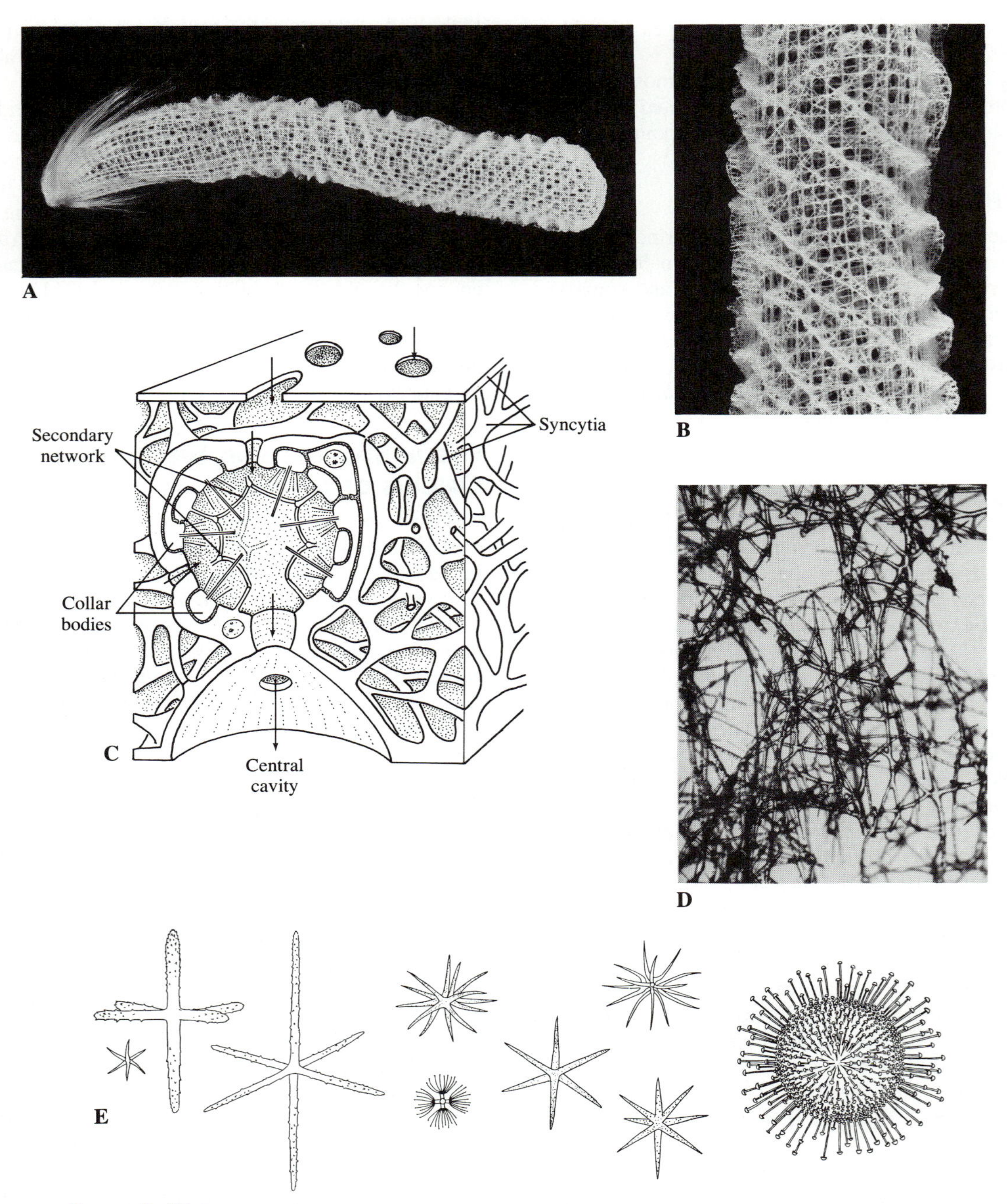

Figure 3.18 Sponges of the class Hexactinellida. A. Dried skeleton of *Euplectella aspergillus*, "Venus' flower basket," from the Philippine region. The long spicules of the basal portion help to anchor the sponge. B. Closer view of the latticework formed by the cross-bridged spicules of *Euplectella*. C. Portion of the body wall of an idealized hexactinellid sponge (spicules omitted), showing syncytial sheets and networks, and one collar-body chamber. (After Mackie, Lawn, and Pavans de Ceccaty, Philosophical Transactions of the Royal Society of London, B, *301*.) D. Portion of the latticework formed by cross-bridged spicules of *Aphrocallistes vastus;* photomicrograph. E. Free spicules of *Aulosaccus solaster*. (Koltun, Keys to the Fauna of the Far-Eastern Seas of the USSR, *94*.)

lined by choanocytes could have been creepers or swimmers long before any of them became sedentary.

The origin of hexactinellids would seem to be more difficult to explain that that of other sponges. They are largely syncytial, have nonnucleated collar bodies embedded in the syncytia, and lack pinacocytes. It is nevertheless probable that choanoflagellates are in their ancestry, too.

Are sponges ancestral to other metazoan phyla? Probably not. They are an ancient group, but—in lacking true tissues and true symmetry and in being organized around a system of pores and chambers through which water flows—they are unlike any other metazoan animals. Choanoflagellates, however, may very well have been involved in the evolution of the stocks that gave rise to all metazoans. This possibility is discussed in Chapter 26.

SUMMARY

1. Sponges are aquatic animals, and most of the approximately 5000 species are marine. Except for a few that are anchored in soft substrata, they are permanently attached to rocks, shells, wood, vegetation, and other firm surfaces.

2. Sponges do not have a digestive tract, nervous system, or any specialized structures for respiration and excretion. Capture and digestion of bacteria and other microscopic food organisms, absorption of oxygen, and elimination of metabolic wastes are accomplished by individual cells. There is, however, some cooperation between cells. For example, ingested food and nutritive reserves may be transferred from one cell to another.

3. Most of the cells of a sponge, other than those that cover the outside of the body, are organized around passageways through which a current of water flows. The current, maintained by the activity of flagella on cells called choanocytes, enters a sponge through thousands of microscopic pores and leaves through one or more major openings. Choanocytes are distinctive in that each one has a collar of microvilli around the base of the flagellum. Microorganisms that come into contact with the collar are trapped and incorporated into food vacuoles.

4. Other important cell types are pinacocytes, amoebocytes, and myocytes. Pinacocytes, which are usually flattened, cover the outside and also line passageways that are not lined by choanocytes. Amoebocytes are mostly in the mesohyl, a semifluid matrix that occupies the space between the cellular layers formed by choanocytes and pinacocytes. Certain of them store and distribute nutritive materials; others secrete collagen and the calcareous or siliceous spicules that form the skeletons of sponges. Myocytes are contractile cells that form sphincterlike complexes around exhalant openings, and thus partly control the rate of flow through a sponge.

5. The various kinds of cells do not form true tissues. Even the epitheliumlike layers that consist of choanocytes or pinacocytes do not have a secreted basal lamina, and there are fewer intercellular junctions than are typical of epithelia of other metazoans.

6. Most sponges are hermaphroditic, but produce eggs and sperm at different times. Gametes are derived from choanocytes or amoebocytes. Sperm are released with the exhalant current, but eggs are usually fertilized in the mesohyl and undergo early stages of embryonic development in this matrix. There is generally a free-swimming larval stage.

7. Asexual reproduction by budding or fragmentation occurs in some species. Freshwater sponges produce resistant bodies called gemmules. When a gemmule germinates, the amoebocytes within it are transformed into choanocytes, pinacocytes, and other cell types, and these become organized into a new sponge.

8. Hexactinellid sponges differ from those of other groups in that they are primarily syncytial. The protoplasmic units that function as choanocytes are embedded in the syncytia and lack nuclei. Hexactinellids are also distinctive in having six-rayed siliceous spicules that are united to form a rigid latticework.

9. In spite of their relative simplicity, sponges are efficient in collecting microscopic food from the water that they pass through their bodies. In some habitats, they form a considerable part of the biomass.

10. The similarity of choanocytes to choanoflagellates suggests that sponges have evolved from these protozoans. Although sponges appeared early in the history of animal life on earth, it is not likely that they are in the line of evolution of other existing metazoan phyla.

REFERENCES ON PORIFERA

Bergquist, P. R., 1978. Sponges. University of California Press, Berkeley and Los Angeles; Hutchinson University Press, London.

Brauer, E. B., 1975. Osmoregulation in the fresh water sponge, *Spongilla lacustris*. Journal of Experimental Zoology, *192*:181–192.

Brill, B., 1973. Untersuchungen zur Ultrastruktur der Choanocyte von *Ephydatia fluviatilis* L. Zeitschrift für Zellforschung und Mikroskopische Anatomie, *144*:231–245.

Burton, M., 1963. A Revision of the Classification of Calcareous Sponges. British Museum (Natural History), London.

Elvin, D. W., 1976. Seasonal growth and reproduction of an intertidal sponge, *Haliclona permollis* (Bowerbank). Biological Bulletin, *151*:108–125.

Frost, T. M., and Williamson, C. E., 1980. *In situ* determination of the effect of symbiotic algae on the growth of the freshwater sponge *Spongilla lacustris*. Ecology, *61*:1361–1370.

Fry, W. G. (editor), 1970. The Biology of the Porifera. Symposia of the Zoological Society of London, 25. Academic Press, London and New York.

Harrison, F. W., and Cowden, R. R. (editors), 1976. Aspects of Sponge Biology. Academic Press, New York and London.

Harrison, F. W., and Cowden, R. R., 1983. Dormancy release and development from gemmules of the fresh-water sponge, *Spongilla lacustris* L.: a supravital study with acridine orange. Transactions of the American Microscopical Society, *102*:309–318.

Hartman, W. D., and Goreau, T. F., 1975. A Pacific tabulate sponge, living representative of a new order of sclerosponges. Postilla, *167*:1–14.

Hartman, W. D., Wendt, J. W., and Wiedenmayer, F., 1980. Living and Fossil Sponges (Notes for a Short Course). Comparative Sedimentology Laboratory, University of Miami, Miami Beach, Florida.

Höhr, D., 1977. Differenzierungsvorgänge in der keimenden Gemmula von *Ephydatia fluviatilis*. Wilhelm Roux's Archives of Developmental Biology, *182*:329–346.

Jones, E. C., 1962. Is there a nervous system in sponges? Biological Reviews, *37*:1–50.

Koehl, M. A. R., 1982. Mechanical design of spicule-reinforced connective tissue: stiffness. Journal of Experimental Biology, *98*:239–267.

Langenbruch, P.-F., and Scalera-Liaci, L., 1986. Body structure of marine sponges. IV. Aquiferous system and choanocyte chambers in *Haliclona elegans*. Zoomorphology, *106*:205–211.

Langenbruch, P.-F., and Weissenfels, N., 1987. Canal systems and choanocyte chambers in freshwater sponges (Porifera, Spongillidae). Zoomorphology, *107*:11–16.

Lawn, I. D., Mackie, G. O., and Silver, G., 1981. Conduction system in a sponge. Science, *211*:1169–1171.

Lemche, H., and Tendal, O. S., 1977. An interpretation of the sex cells and the early development of sponges, with a note on the terms acrocoel and spongocoel. Zeitschrift für Zoologische Systematik und Evolutionsforschung, *15*:241–252.

Levi, C., and Boury-Esnault, N., 1979. Biologie des Spongiares (Sponge Biology). Colloques Internationaux du Centre National de la Recherche Scientifique, no. 291. Editions du Centre National de la Recherche Scientifique, Paris.

Mackie, G. O., and Singla, C. L., 1983. Studies on hexactinellid sponges. I. Histology of *Rhabdocalyptus dawsoni* (Lambe, 1873). Philosophical Transactions of the Royal Society of London, B, *301*:365–400.

Mackie, G. O., Lawn, I. D., and Pavans de Ceccaty, M., 1983. Studies on hexactinellid sponges. II. Excitability, conduction and coordination of responses in *Rhabdocalyptus dawsoni* (Lambe, 1873). Philosophical Transactions of the Royal Society of London, B, *301*:401–418.

Pavans de Ceccaty, M., 1986. Cytoskeletal organization and tissue patterns of epithelia in the sponge *Ephydatia mülleri*. Journal of Morphology, *189*:45–65.

Reiswig, H. M., 1971. Particle feeding in natural populations of three marine demosponges. Biological Bulletin, *141*:568–591.

Reiswig, H. M., 1974. Water transport, respiration and energetics of three tropical marine sponges. Journal of Experimental Marine Biology and Ecology, *14*:231–249.

Reiswig, H. M., 1975. The aquiferous systems of three marine Demospongiae. Journal of Morphology, *145*:493–502.

Reiswig, H. M., 1975. Bacteria as food for temperate-water marine sponges. Canadian Journal of Zoology, *53*:582–589.

Reiswig, H. M., and Mackie, G. O., 1983. Studies on hexactinellid sponges. III. The taxonomic status of Hexactinellida within the Porifera. Philosophical Transactions of the Royal Society of London, B, *301*:419–428.

Rützler, K., and Rieger, C., 1973. Sponge burrowing: fine structure of *Cliona lampa* penetrating calcareous substrata. Marine Biology, *21*:144–162.

Simpson, T. L., 1980. Reproductive processes in sponges: a critical evaluation of current data and views. International Journal of Invertebrate Reproduction, *2*:251–269.

Simpson, T. L., 1984. The Cell Biology of Sponges. Springer-Verlag, New York.

Simpson, T. L., and Gilbert, J. J., 1973. Gemmulation, gemmule hatching, and sexual reproduction in fresh-water sponges. 1. The life cycle of *Spongilla lacustris* and *Tubella pennsylvanica*. Transactions of the American Microscopical Society, *92*:422–433.

Stearn, C. W., 1975. The stromatoporoid animal. Lethaia, *8*:89–100.

Vogel, S., 1978. Evidence for one-way valves in the water-flow system of sponges. Journal of Experimental Biology, *76*:137–148.

Vogel, S., 1978. Organisms that capture currents. Scientific American, *239*:129–139.

Willenz, P., 1980. Kinetic and morphological aspects of particle ingestion by the freshwater sponge *Ephydatia fluviatilis*. *In* Smith, D. C., and Tiffon, Y. (editors), Nutrition in the Lower Metazoa, pp. 163–178. Pergamon Press, Oxford.

4

Phylum Cnidaria

Hydroids, Jellyfishes, Sea Anemones, and Their Relatives

- Phylum Cnidaria
  - Class Hydrozoa
    - Order Hydroida
      - Suborder Thecata (Leptomedusae)
      - Suborder Athecata (Anthomedusae)
      - Suborder Limnomedusae
    - Order Milleporina
    - Order Stylasterina
    - Order Trachylina
      - Suborder Trachymedusae
      - Suborder Narcomedusae
    - Order Siphonophora
    - Order Actinulida
  - Class Scyphozoa
    - Order Semaeostomeae
    - Order Coronatae
    - Order Rhizostomeae
    - Order Stauromedusae
  - Class Cubozoa
  - Class Anthozoa
    - Subclass Zoantharia (Hexacorallia)
      - Order Actiniaria
      - Order Corallimorpharia
      - Order Ptychodactiaria
      - Order Zoanthidea
      - Order Scleractinia (Madreporaria)
    - Subclass Ceriantipatharia
      - Order Ceriantharia
      - Order Antipatharia
    - Subclass Alcyonaria (Octocorallia)
      - Order Alcyonacea
        - Stoloniferans
        - Telestaceans
        - Organ-pipe coral
        - Soft corals
        - Gorgonians
      - Order Pennatulacea
      - Order Helioporacea (Coenothecalia)

INTRODUCTION

The cnidarians constitute a large phylum of aquatic animals. The majority of them are marine but a few occur in fresh water. About 10,000 living species have been described, and skeletons of many others—mostly corals—are found in the fossil record. The name Coelenterata is sometimes used for this phylum, or for the phyla Cnidaria and Ctenophora combined. In the past, however, it had a broader meaning, for it also included sponges.

Although cnidarians are comparatively simple animals, they have a definite shape and symmetry, their cells are organized into tissues, and they have a gut and nervous system. The gut is basically a sac (Figure 4.1, A) but it may branch out into radial canals (Figure 4.1, B) or be subdivided by vertical partitions. The mouth serves both for ingesting food and for eliminating undigested residues. Sometimes the gut has pores to the outside, at the tips of the tentacles or elsewhere. These pores permit fluid to leave the gut, but they are not true anal openings. The gut of cnidarians is often called a **gastrovascular cavity,** a term that indicates that it functions as a kind of circulatory system, distributing oxygen as well as products of digestion and assisting in the elimination of carbon dioxide and nitrogenous wastes.

GENERAL FEATURES OF CNIDARIANS

ORIENTATION AND SYMMETRY

Because cnidarians do not have a head, the surface on which the mouth is located is referred to as the **oral surface;** the term **aboral surface** is applied to the opposite side (Figure 4.1). These animals are therefore organized around an oral–aboral axis rather than around an anterior–posterior axis. Their symmetry, in general, is radial, or nearly so. Perfect radial symmetry can be seen in many jellyfishes. Biradial symmetry is a variation that characterizes certain groups, such as sea anemones.

TISSUES

The tissues of cnidarians are epithelia or modified epithelia. A layer of cells called the **epidermis** covers the outside of the body, and a similar layer forms the **gastrodermis** that lines the gut (Figure 4.1). When cilia are present on these epithelia, each cell has only one cilium. In nearly all other phyla that are more advanced than cnidarians, ciliated cells have several to many cilia.

Between the epidermis and gastrodermis is a secreted layer called the **mesoglea** (''jelly in the middle''). The organic components of this are mostly mucopolysaccharides and collagen. Gonads and muscles may appear to be part of the mesoglea, but they too are epidermal or gastrodermal derivatives. The muscles of cnidarians are contractile basal portions of certain epithelial cells. Such cells are called *myoepithelial cells,* or *epitheliomuscular cells*. The contractile bases are often concentrated into bundles, and those capable of fast action may have cross-striations.

MESOGLEA

Mesoglea is a key feature of the phylum Cnidaria. Its nature, and the amount of it in proportion to the cellular components of the body, vary greatly. Most polyps, for instance, have relatively little mesoglea between the epidermis and gastrodermis (Figure 4.1, A), but this layer nevertheless provides some support for the cellular layers. In a jellyfish, on the other hand,

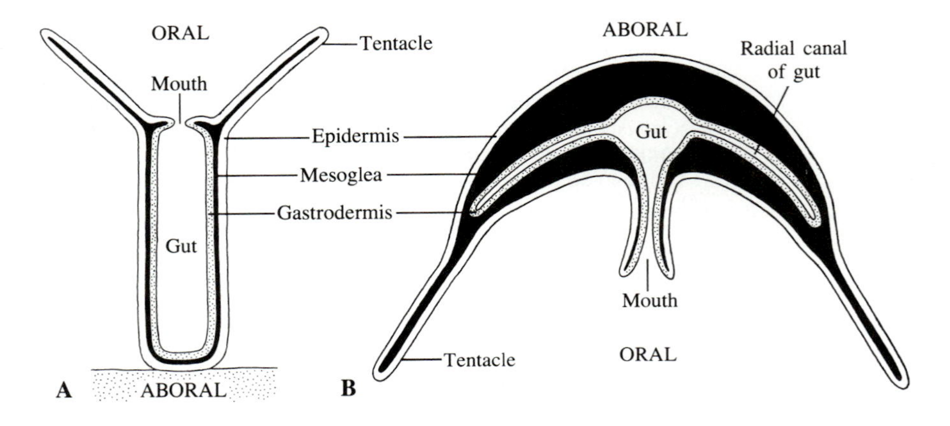

FIGURE 4.1 Sections of the two main body forms in cnidarians, showing tissue layers and mesoglea. A. Polyp (attached to substratum). B. Medusa (free-swimming).

mesoglea makes up most of the body (Figure 4.1, B), giving the animal form and buoyancy and also serving as a skeleton that plays an important role in locomotion. A jellyfish swims by jet propulsion, its rhythmic contractions forcing water from the space beneath its bell-shaped or saucer-shaped body. With each contraction, the muscles temporarily deform the mesoglea, as if it were made of soft rubber. When the muscles relax, the mesoglea springs back to its original shape, and the jellyfish is ready for another contraction. In some cnidarians, including sea anemones, the mesoglea may resemble a tissue, for it has cells and conspicuous fibrous components in it. The cells, however, originate in the epidermal and gastrodermal layers.

COLONY FORMATION AND POLYMORPHISM

By budding repeatedly, many cnidarians form colonies in which the members remain connected and share a communal gut and nervous system. The colonial habit is often accompanied by **polymorphism,** that is, the specialization of certain members for feeding and of the rest for reproduction or other functions. Polymorphism provides for division of labor and thus compensates for the absence of complex organs in a single individual.

A different kind of polymorphism is shown by species that have alternating phases in the life cycle; one phase is attached, the other free-swimming. This situation is often called *alternation of generations* or *metagenesis,* but it is not comparable to the alternation of haploid and diploid phases in the life cycles of plants. The term **serial polymorphism** is more suitable.

To understand polymorphism and cnidarian life cycles, it is essential to understand the two principal body forms characteristic of the phylum (Figure 4.1). Both forms, called **polyp** and **medusa,** may occur in the same species. The differences between them, as summarized in Table 4.1, are not absolute; there are deviations from the rules in certain groups.

NEMATOCYSTS

Cnidarians, as a group, are carnivores. They owe much of their success in capturing prey (and repelling predators) to microscopic structures called **nematocysts.** A nematocyst shoots out a hollow thread that engages the prey or predator, either by penetrating it, entangling it, or sticking to it. The epidermis of the tentacles almost always has large numbers of nematocysts, and these structures may also be plentiful elsewhere, even in the gastrodermis.

A nematocyst (Figure 4.2, A and B; Figure 4.3, A) is a firm capsule consisting of material chemically related to collagen, and it is secreted within a single cell called a **nematocyte** (or *cnidocyte*) (Figure 4.3, B). The wall of the capsule is invaginated at one point to form a hollow thread, which is not only coiled but folded. A small lid, or **operculum,** may cover the pore that leads into the narrow cavity within the thread (Figure 4.3, B; Figure 4.4, A). When the nematocyst "fires," the thread rapidly turns inside-out. At the same time it uncoils and unfolds (Figure 4.4, A, B, and D), generating considerable force. A fully discharged (everted) thread is several to many times as long as the capsule.

A nematocyst and the nematocyte that secretes it form a functional complex. The nematocyst will not

TABLE 4.1 POLYPS AND MEDUSAE CONTRASTED

Polyp	Medusa
Attached to substratum by aboral surface (but may be mobile to some extent)	Usually free-swimming and with oral surface usually facing downwards (in some cnidarians, however, the medusa is not released, and it may not look much like a medusa)
Shape cylindrical, with tentacles close to the mouth, or bordering a flattened area around the mouth	Generally saucer-shaped or bell-shaped, with tentacles at or near the margins
With a thin layer of mesoglea, or at least without much jelly in the mesoglea	With a thick layer of mesoglea
Often with a hard skeleton	Without a hard skeleton
Reproducing sexually or asexually, sometimes both ways	Always reproducing sexually, sometimes also asexually
Solitary or colonial (colonies may include two or more kinds of polyps)	Usually solitary

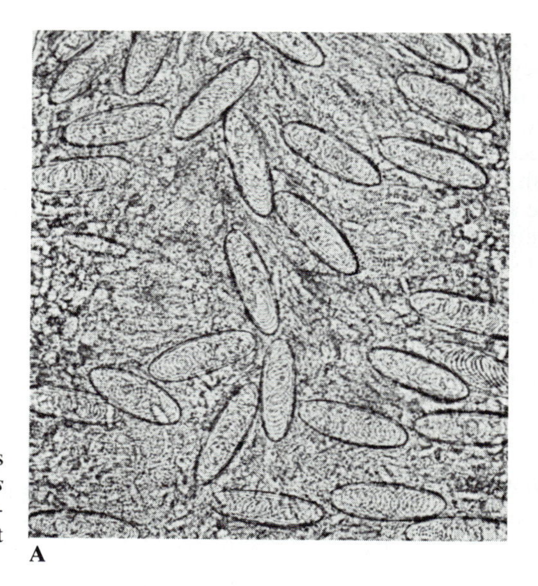

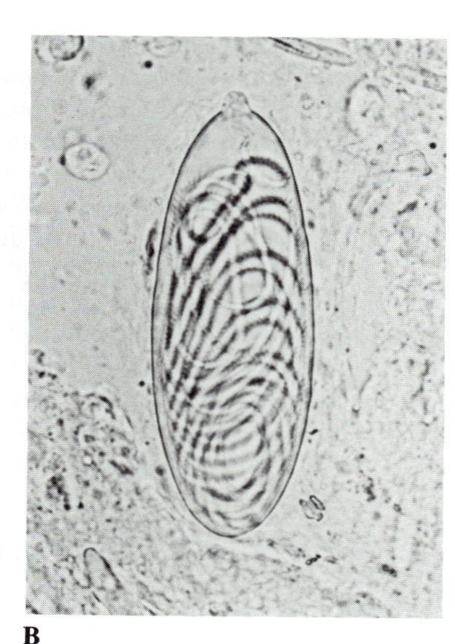

FIGURE 4.2 A, B. Nematocysts from the tentacles of *Corynactis* (class Anthozoa, order Corallimorpharia); photomicrographs at different magnifications.

normally discharge its thread unless the nematocyte has been stimulated. Appropriate stimulation may be chemical, in the form of a substance produced by a prey or predator, or it may be tactile, such as when the prey or predator touches a nematocyte. Many nematocytes have a stiff bristle—a modified cilium—that serves as a trigger (Figure 4.3, C). Discharge of the thread is generally believed to depend largely on osmotic uptake of water by a nematocyte whose permeability has been increased by chemical or tactile stimulation. If the filaments found in some nematocytes are contractile, as some zoologists think they are, these could conceivably put pressure on the capsule.

The nematocyte–nematocyst complex, responding to stimulation, operates as an independent effector; that is, it will function without input from the nervous system.

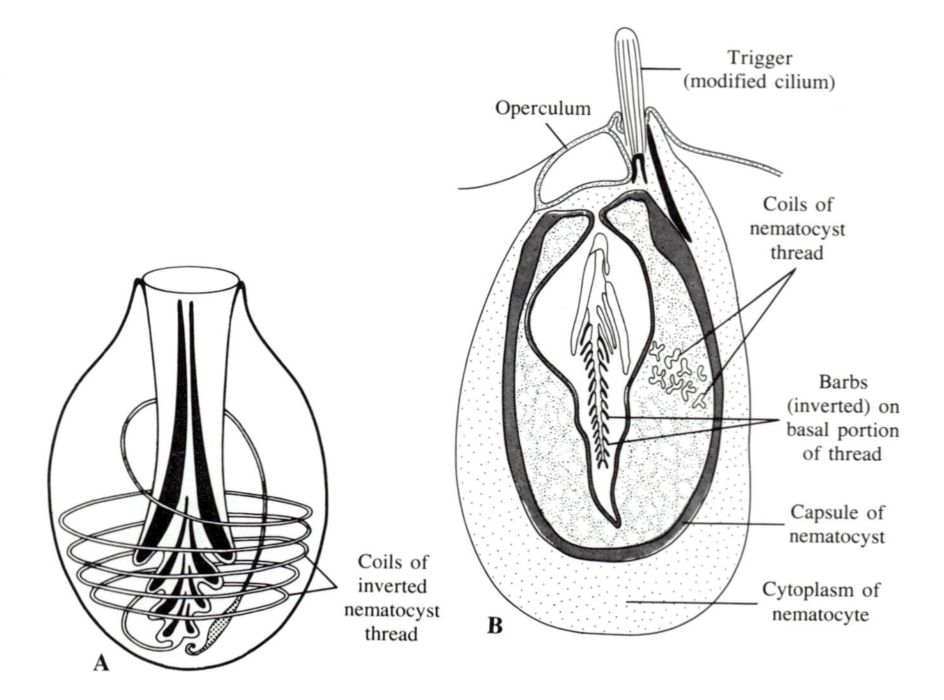

FIGURE 4.3 A. Diagram of a nematocyst (penetrant type) of a hydra. (Campbell, Transactions of the American Microscopical Society, *96*.) B. Lengthwise section of a nematocyte and the nematocyst (penetrant type) that it has secreted; diagram based on electron micrographs. (After Fretter and Graham, A Functional Anatomy of Invertebrates.)

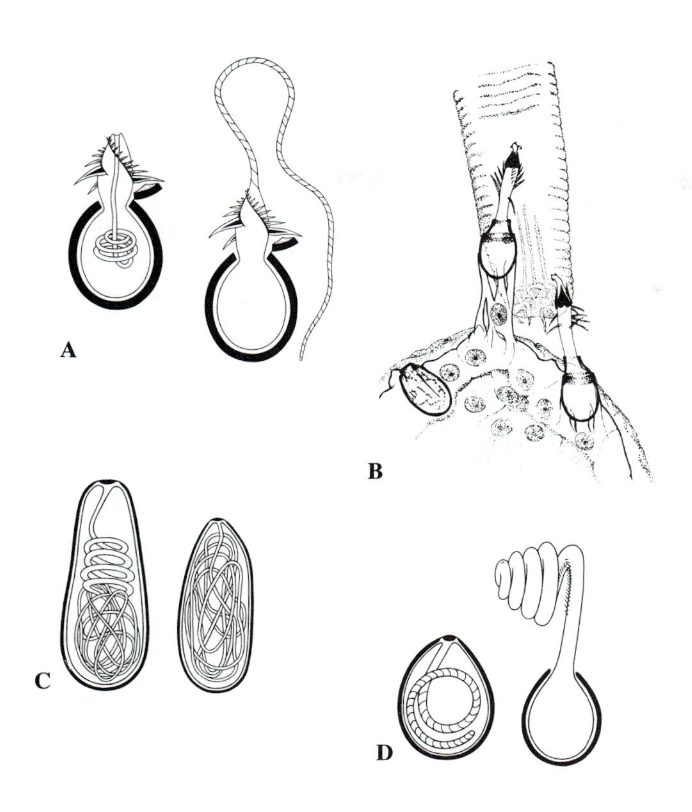

Figure 4.4 A. Eversion of the thread of a penetrant nematocyst. (Mergner, Natur und Museum, *94*.) B. Penetrants of *Microhydrula pontica,* embedded in a nematode (Bouillon and Deroux, Cahiers de Biologie Marine, *8*.) C. Two intact glutinant nematocysts. D. A volvent nematocyst, before and after eversion of its thread. (C and D, Mergner, Natur und Museum, *94*.)

There is evidence that the nervous system does exert some control over nematocytes, but perhaps mostly in an inhibitory way. This is suggested, for instance, by the behavior of a sea anemone, *Calliactis parasitica,* which lives on snail shells, particularly those occupied by hermit crabs. When the anemone comes in contact with the right kind of shell, it attaches itself with the help of nematocyst threads fired from its tentacles. Once fixed by its base to the shell, however, it does not use its nematocysts when it is offered another shell of the same type. It seems likely that information traveling from the base to the tentacles suppresses further discharge of nematocysts.

A nematocyte does not necessarily originate where it will eventually reside. In many cnidarians, nematocytes migrate among the epithelial cells until they reach the tentacles or other parts of the body where they are needed. Sometimes they are pushed passively by slow mass migrations of other cells.

When structural details are considered, there are many kinds of nematocysts. In terms of their function, however, they fall into three general types. **Penetrants** (Figure 4.4, A and B), whose threads are armed with spines and are open at the tip, are used for piercing and delivering a mixture of toxins. **Glutinants** (Figure 4.4, C) also have open-tipped threads. Most of them discharge a sticky substance that enables the tentacles of the animal to get a firm hold on the substratum, but some are believed to exude defensive toxins. The threads of **volvents** (Figure 4.4, D), which become tangled up with the prey, are closed at the tip.

Nematocysts may serve in ways other than those mentioned. In ceriantharians, which are relatives of sea anemones, discharged nematocysts are used in the construction of a tube. Certain sea slugs and a few other animals that eat cnidarians concentrate nematocysts in their own bodies. In at least some cases, these stored-up nematocysts are probably used against predators.

The toxins of cnidarian nematocysts consist largely of proteins of moderately high molecular weight. An extract of the contents of nematocysts is not likely to be homogeneous, however. There are usually two or more important fractions, with high toxicity residing in only certain of them. While the more commonly observed effects of potent toxins on humans are mild skin irritations, some cnidarians are extremely dangerous. Their stings may lead to severe lesions, general or local paralysis, and even death. In fact, the most virulent animal known is a jellyfish, *Chironex fleckeri*.

FEEDING AND DIGESTION

Feeding almost always involves the tentacles, except in a few cnidarians in which these are absent. The tentacles usually have extensive concentrations of nematocysts, and in some jellyfishes the lobes of the mouth area are richly supplied with them. An individual nematocyst has little effect on the prey, but by firing hundreds of them, the animal may immobilize a crustacean, fish, or other organism. In most cnidarians, the tentacles bend to deliver the prey to the mouth. Some species move small organisms toward the mouth by action of cilia.

Digestion in cnidarians is partly extracellular and partly intracellular. In extracellular digestion, certain cells of the gastrodermis secrete protein-splitting enzymes that break the prey down to particles small enough to be taken into food vacuoles in gastrodermal cells, where digestion is completed intracellularly. Some foodstuffs, however, may be totally digested extracellularly.

NERVOUS SYSTEM AND SENSORY STRUCTURES

The nervous system of cnidarians is generally called a **nerve net.** While the neural pattern often does resemble a network, in many cases the neurons are organized into condensed, nervelike tracts.

Neurons can be either nonpolarized or polarized. A **nonpolarized neuron** may transmit an impulse in either direction. Both of its two terminals, or all of several terminals, are capable of secreting a transmitter substance such as acetylcholine, which initiates an impulse in the succeeding neuron. Electron micrographs of the synapses of such neurons show secretory vesicles in all of the terminals. In a **polarized neuron,** only one of the two terminals, or only certain of several terminals, are capable of secreting the transmitter substance, and the impulse moves only in one direction. An interesting feature of the nervous system of some cnidarians is that synapses do not necessarily involve two distinct terminals; a terminal of one neuron may be applied to almost any part of the next neuron.

Neurons with just two terminals are more likely than those with several terminals to be organized into distinct rings or other condensed tracts. Such tracts are concerned primarily with rapid transmission over long distances. Whether concentrated or not, neurons of cnidarians tend to be closely associated with epidermal and gastrodermal cells (Figure 4.7, B). They may weave between the bases of these cells, and they are frequently insinuated into them. Neurons probably do not often run freely through the mesoglea, but the epidermal and gastrodermal components of the nervous system are joined in the region of the mouth, where the two epithelial layers meet.

Epithelial cells themselves may transmit impulses, a process called **neuroid transmission.** In at least some jellyfishes and siphonophores, for instance, impulses travel through cells of the epidermis. The phenomenon has been reported for various phyla. (In sponges, which have no nervous system, all impulse transmission is believed to be of this type.)

In cnidarians, as in higher animals, the function of the nervous system is to process information supplied by sensory receptors and send out appropriate signals to the effectors, such as contractile cells. Thus the animal moves, captures prey, or avoids being eaten. Sensory cells, which may properly be considered to be part of the nervous system, are mostly in the epidermis. They include touch receptors, chemoreceptors, photoreceptors, and receptors concerned with maintaining proper orientation. The sensory cells that have these functions are not necessarily recognizable as such, although touch receptors often have small bristles, and concentrations of photoreceptors generally are disclosed by the shields of dark pigment (''eyespots'') that accompany them. Cnidarians are among the several invertebrate phyla whose photoreceptors, like those of vertebrates, are characterized by modified cilia.

Structures concerned with detecting changes in orientation are found in many jellyfishes. The most common type, called a **statocyst,** is a marginal vesicle lined by receptor cells. The cavity of the vesicle is partly filled by a cell that has secreted a crystalline, calcareous body, the **statolith.** As the animal bobs about, the statolith weighs more heavily on certain receptor cells than on others, and impulses generated by the receptors lead to a righting reflex or some other adjustment. A different type of structure with the same general function is an open pit that has several to many statolith-producing cells attached to the sensory epithelium. More complex devices are discussed below in connection with jellyfishes of the class Scyphozoa.

BIOLUMINESCENCE

Many invertebrates, as well as some fishes, dinoflagellates, and bacteria, emit light. This phenomenon, called **bioluminescence,** is much more common in marine environments than in terrestrial or freshwater habitats. Because it is a characteristic of numerous cnidarians—and also of ctenophores, dealt with in the next chapter—this is a good place to discuss it.

Bioluminescence results from the interaction of a substance called **luciferin** with an enzyme called **luciferase,** in the presence of appropriate cofactors such as free calcium or hydrogen ions. The terms *luciferin* and *luciferase* are applied to a variety of biochemical compounds; in other words, the luciferin and luciferase of one species or group of species are not likely to be chemically identical with those of other organisms. Bioluminescence has probably evolved many times, so it is to be expected that the exact properties of the light-emitting substance and the enzyme that acts upon it will vary considerably.

Most bioluminescent animals, including cnidarians and ctenophores, produce light within cells, usually of the epidermis or gastrodermis. Sometimes, however, the glow does not come from the animal itself but from a secretion, discharged as slime or in the form of a cloud. Light may also be produced by bacteria that are localized in the skin or lining of the gut. Symbiotic associations of this type are found in some fishes, squids, and pelagic tunicates called thaliaceans.

What functions does bioluminescence serve? One is communication, such as is involved in courtship and mating, between individuals of the same species. This is especially well documented for fireflies, but in marine animals it has been reported for fishes, a polychaete annelid, and an ostracode. Another function of bioluminescence is advertisement. For instance, a fungus that glows in the dark will be recognized by certain insects seeking a place to lay their eggs. Although the fungus will be eaten by the larvae that hatch, it is destined to decay anyway, and the visits of insects will help disseminate its spores. Luminescent bacteria on fecal pellets or dead organisms might attract scavengers, thus reintroducing the bacteria into the digestive tracts of their usual hosts.

Many marine animals, especially deep-water fishes, use light to attract prey, which may be other fishes, crustaceans, or polychaete annelids. A few fishes emit flashes bright enough to illuminate the crustaceans and fishes they prey on. There is some evidence that flashes of light produced by certain animals, after they have made contact with a prey organism, may momentarily stun the prey and assist in its capture.

Probably the most common function of bioluminescence is the evasion of predators. Production of light may simply startle or frighten a predator, or blind it temporarily. It could also inform a predator that the animal emitting light is unpalatable or able to retaliate. Sea pens are believed to avoid certain predators in this way. Furthermore, a light signal produced by a vulnerable prey organism may mimic the signal of less-vulnerable or less-palatable prey.

Some bioluminescent marine animals protect themselves by shedding a piece of the body when they are in the clutches of a predator or are otherwise provoked. A brittle star, for example, may shed one of its rays, and a polynoid polychaete may part with one of its scalelike structures that cover its dorsal surface. Such decoys continue to glow and may distract the predator long enough to allow the animal to escape. A cloud of luminescent material, or a luminous slime trail, may also function as decoys.

This brief discussion has not covered all possible functions of bioluminescence. The list of references at the end of this section of the chapter includes four comprehensive reviews of the subject (Hastings, 1983; Johnson, 1988; Morin, 1983; Young, 1983).

SKELETONS

Some cnidarians have hard skeletons, but both the gut and mesoglea may also function as skeletal elements. The gut is filled with water, which is incompressible, so if the mouth and any other openings to the outside are closed, the contraction of certain muscle cells and the relaxation of others will result in changes in body shape without any noticeable change in volume. As previously explained, the jellylike mesoglea of medusae, which consists mostly of water, is a resilient skeleton that makes possible the pulsating swimming movements of these animals. After a muscular contraction deforms the bell and drives out a jet of water from the underside, the bell springs back to its original shape.

Hard skeletons are of several types. One kind consists of a sclerotized protein, usually transparent or translucent, secreted by glands in the epidermis. The polyps, and the branches from which polyps arise, are thus partly or largely covered. Sclerotized protein may also be deposited in the axial rod that supports a fleshy colony, such as that of a sea pen. As a rule, however, a skeleton of this type is reinforced by calcification. In corals, there is generally an external calcareous skeleton, though in certain groups the skeleton consists of calcareous spicules laid down in the mesoglea.

SEXUAL REPRODUCTION AND DEVELOPMENT

Gametes of cnidarians originate from certain undifferentiated epithelial cells. Usually, simple gonads are recognizable in reproductive individuals. In the class Hydrozoa, the gonads come from epidermal tissue, although they may sink into the mesoglea; in Scyphozoa and Anthozoa they are derived from the gastrodermis. There are no specialized ducts for carrying out gametes; these are simply discharged through breaks in the epidermis or gastrodermis. Gametes shed into the gut leave the body by way of the mouth or through pores located at the tips of the tentacles or elsewhere.

Sexes are almost always separate. Fertilization may take place while the eggs are still in the ovary or in the gut, or after they have been released. In some species, the eggs exude substances that attract sperm of the same species. Many cnidarians brood their young, at least for a time.

Patterns of reproduction and life cycles in the Cnidaria show great variation. Embryonic development is

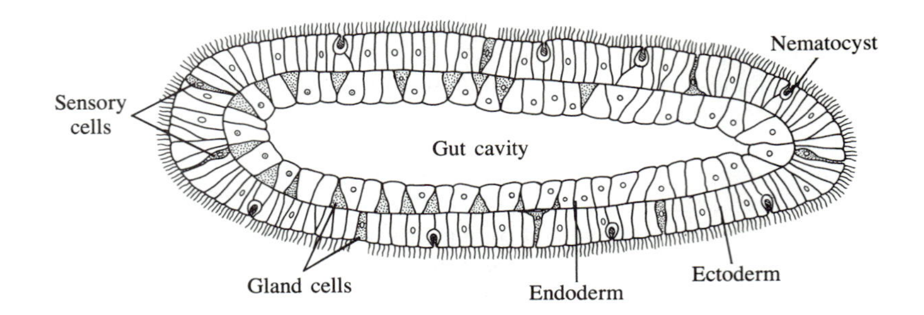

FIGURE 4.5 A planula larva. The gut cavity within the mass of endoderm cells appears earlier in some planulae than in others.

by no means the same in all groups, but it usually adheres to a plan that is typical of the phylum. Eggs with relatively little yolk are likely to show more nearly regular and more nearly equal cleavage than eggs with considerable yolk. As a rule, however, even if the first two cleavages produce a symmetrical group of four cells, the more numerous cells of succeeding stages often appear to be jumbled together. The berrylike early embryo, or **morula,** may transform gradually into a solid **gastrula** without going through a **blastula** stage; the cells on the outside form ectoderm and those inside become the endoderm. When there is a blastula, characterized by a spacious central cavity **(blastocoel),** the cavity is soon partly or completely filled by endoderm cells. The endoderm cells may originate from one portion of the wall of the blastula or from all parts of it. Sometimes a more or less continuous layer of endoderm cells is produced by division of the cells that make up the wall of the blastula. There are also cases in which endoderm is formed by invagination, but when this happens the **blastopore** does not usually remain open.

Whether or not there is a definite blastula stage, the gastrula is generally nearly or completely solid, belonging to a type called a **stereogastrula.** With little further differentiation it becomes a **planula** (Figure 4.5), a kind of larva characteristic of the phylum. The planula typically swims or glides with the aid of cilia on its epidermis, and usually it has some nerve cells, sensory cells, and nematocysts. One end of it consistently behaves as the anterior end. The planula of the cup coral *Balanophyllia* is reported to have epidermal cells that resemble sponge choanocytes; these collect food by phagocytosis. Perhaps the capacity for feeding in this way is widespread among planulae.

In most cnidarians, the planula develops into a polyp. It settles on its anterior end and a mouth opens up on the opposite side. The endodermal cells then become organized around a gut cavity, which may already be evident to some extent during the planula stage. In some members of the phylum, the planula becomes a small polyp before it settles. Such a polyp, complete with mouth and tentacles, is called an **actinula.** It may drift in the plankton or swim actively by means of cilia. In cnidarians that have no polyp stage, the planula differentiates into a medusa, generally by way of an actinula stage.

HYDRAS AS EXAMPLES OF CNIDARIANS

Among the more intensively studied cnidarians are *Hydra* and *Chlorohydra,* which belong to the class Hydrozoa. They occur in fresh water in many parts of the world, and some of them are easily cultivated. In certain respects, hydras are atypical members of their phylum and class, yet much of what we know about cells and tissues of cnidarians has been learned from them.

GENERAL STRUCTURE AND HISTOLOGY

A hydra (Figure 4.6) is a slender, solitary polyp, attached by a basal disk to an aquatic plant, stick, or stone. The height of a well-fed specimen is generally about 3 to 8 mm, but a few species are larger. The mouth is in the center of a small cone, the **hypostome,** which is set off from the column by a circle of six to eight tentacles. The tentacles are hollow, and a branch of the gut extends up into each one.

The histology of a hydra (Figure 4.7) illustrates some important aspects of cnidarian structure. The epidermis that covers the outside of the body and the gastrodermis that lines the gut meet at the mouth, but elsewhere these tissues are separated by a thin layer of mesoglea. Both epidermis and gastrodermis consist to a large extent of myoepithelial cells whose bases are contractile. When relaxed, the base of a myoepithelial cell is elongated along one axis; when it contracts, it becomes shorter. The property of contractility resides in myofilaments.

In the myoepithelial cells of the epidermis, the myofilaments are mostly oriented parallel to the oral–aboral axis of the column of the animal; in the gastrodermal counterparts, they are mostly at right angles to those in the epidermal cells (Figure 4.7, B). This arrangement

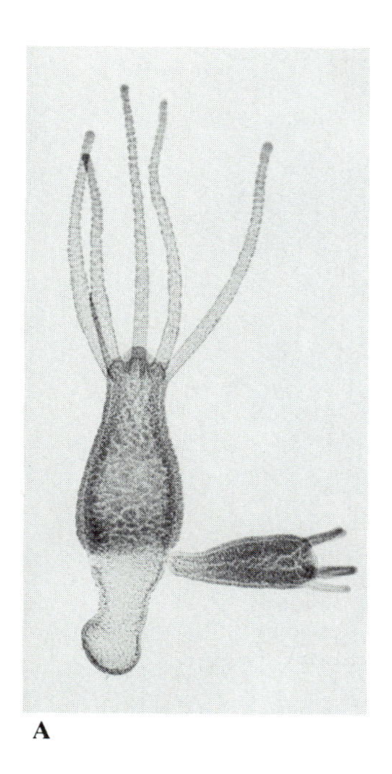

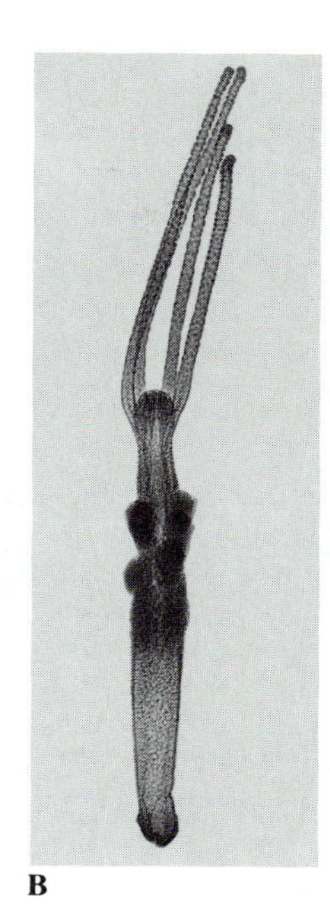

FIGURE 4.6 Hydra; photomicrographs. A. Specimen with a bud. B. Specimen with spermaries in which sperm develop. C. Specimen producing an egg. (Courtesy of Carolina Biological Supply Co., Inc.)

makes the myoepithelial cells of the epidermis antagonistic to those of the gastrodermis. When the mouth of a hydra is closed, so that the volume of water in the gut remains constant, contraction of myoepithelial cells of the epidermis, coupled with relaxation of comparable cells in the gastrodermis, results in shortening and thickening of the column. If the contractile cells of the epidermis relax and those in the gastrodermis contract, the column becomes long and slender. The tentacles, whose histology is similar, operate in the same general way. Bending of the column, or of a tentacle, occurs when the epidermal cells on only one side of the cylinder contract.

Although what has just been said about myoepithelial cells is specifically concerned with hydras, it expresses an important principle of cnidarian structure: contractile elements belong to the epidermis or gastrodermis, even if the parts of the cells in which the myofilaments are concentrated are internal to the rest of the epithelial cells. In other words, there is no true muscle tissue separate from epidermis and gastrodermis.

The myoepithelial cells of the gastrodermis of hydras have functions other than contraction. They can take in small particles of food and digest them intracellularly. For this reason they are called **nutritive-myoepithelial cells.** In addition, they are generally equipped with cilia whose activity helps distribute materials in the gut cavity. The gastrodermis also has secretory cells that produce enzymes for extracellular digestion in the gut cavity. These enzymes reduce the meaty portions of small crustaceans to particles that can be taken up by the phagocytic activity of the nutritive-myoepithelial cells, within which the digestive process is completed. Much of the breakdown of proteins, and perhaps all fat digestion, is intracellular. How the epidermal cells are nourished needs to be studied. Whatever the mechanism, it must involve transfer of nutrients from the gastrodermis to the epidermis.

Myoepithelial cells that secrete mucus are found in both the epidermis and gastrodermis. They are especially abundant in the region of the mouth, where mucus assists in swallowing food, and in the epidermis of the basal disk, where the secretion functions as an adhesive substance.

Scattered among the epidermal cells are undifferentiated interstitial cells. These are often in groups and they are of two sizes. The larger ones can differentiate into epidermal myoepithelial cells or move to the gastrodermis and become nematocytes or another of the cell types of that layer. They are also the source of gametes in sexual individuals. The smaller interstitial cells differentiate into nerve cells.

NEMATOCYTES AND NEMATOCYSTS

Most of the nematocytes derived from interstitial cells end up on the tentacles, where they become concentrated among myoepithelial cells of the epidermis.

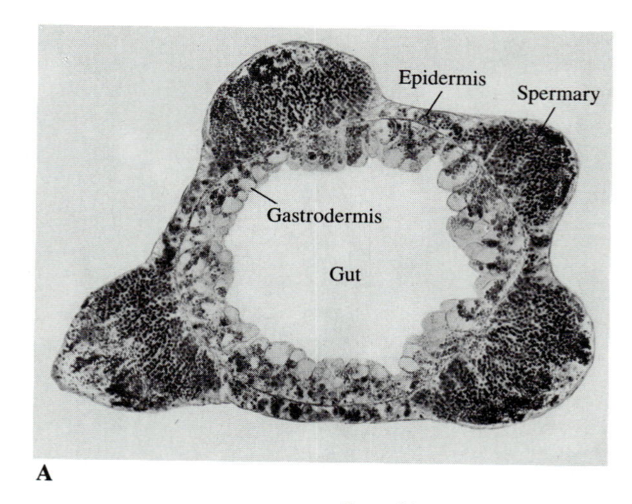

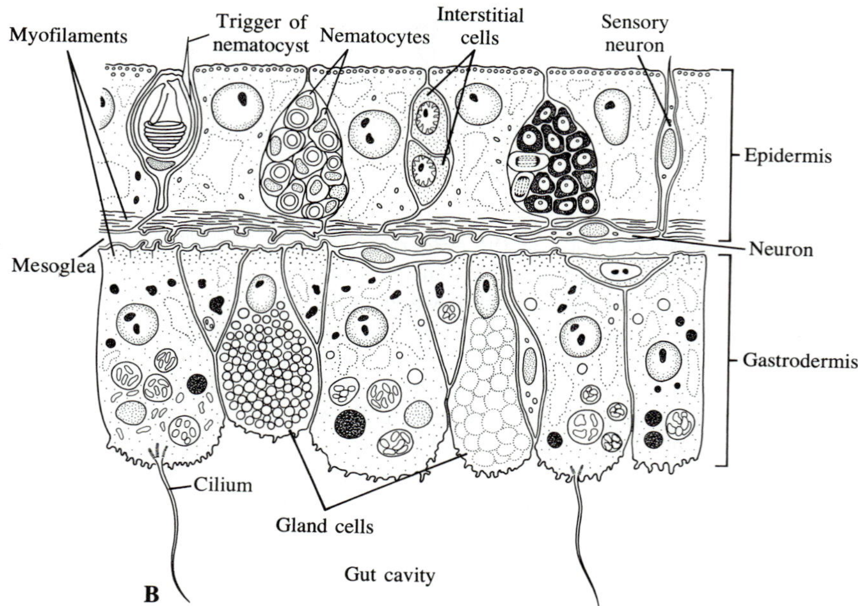

Figure 4.7 Histology of hydra. A. Transverse section through the column of a specimen with spermaries; photomicrograph. (Courtesy of Carolina Biological Supply Co., Inc.) B. Diagrammatic representation of a portion of the column, based on electron micrographs of a longitudinal section. (After a drawing lent by Pierre Tardent.)

The wartlike bumps on the tentacles mark where there are large numbers of nematocytes, their capsules ready to fire.

Hydras have all three of the functional types of nematocysts. Volvents (Figure 4.4, D) are used for entangling prey; their threads lack barbs and are closed at the tip. The threads of penetrants (Figure 4.4, A and B), open at the tip and with prominent basal barbs, are used for injecting a toxin that paralyzes small crustaceans. The threads of glutinants (Figure 4.4, C), also open at the tip, may be smooth or have small bristles. The bristly ones are thought to discharge toxins and to function mostly in defense; the smooth ones are sticky and are used to grip the substratum when a hydra is moving with the help of its tentacles.

NERVOUS SYSTEM

The nervous system of a hydra is probably as simple as any found in cnidarians. Of the neurons that constitute the nerve net, the majority are associated with the epidermis (Figure 4.7, B); some, however, are closely linked with the gastrodermis. Hydras respond to touch, to the presence of crustaceans that they eat, and to light. At least some of the tactile receptors in the epidermis can be recognized because they have little bristles. Processes of the neurons of the nerve net are joined synaptically both with sensory cells and with contractile cells, nematocytes, and other effectors.

By treating a hydra intermittently with colchicine, a substance that interferes with mitosis, it is possible to eliminate the smaller interstitial cells that differentiate into nerve cells. Some hydras treated in this manner end up without a nervous system. They are relatively nonreactive and must be force-fed to keep them alive. Some survive indefinitely and provide a unique system for studying the role of nerves in development and behavior.

SYMBIOTIC ALGAL CELLS

The green hydra *(Chlorohydra viridissima)* has unicellular green algae living in its gastrodermal cells. These algae, called **zoochlorellae,** fix carbon photosynthetically, and about 85% of the organic material they produce diffuses out of them into the tissue of the hydra. (Similar free-living algae also release soluble organic compounds, but the amounts are small, less than 5% of the total produced.) The principal product exported by the zoochlorellae in the green hydra is maltose. If the hydra is deprived of its zoochlorellae by treatment with streptomycin or by other techniques, it will grow and multiply normally as long as it is kept supplied with small crustaceans. If starved, however, it dies much more rapidly than a hydra that has a normal complement of zoochlorellae.

REPRODUCTION AND REGENERATION

When they are well nourished, hydras readily reproduce asexually. A bud starts out as a small protuberance, usually in the lower half of the column (Figure 4.6, A). The gut of the parent extends up into the bud, and after a mouth opens up and tentacles develop, the young hydra becomes detached.

If a fine thread is run lengthwise through a hydra and knotted at the basal end, the animal can sometimes be turned inside-out by carefully pulling on the thread. This is one of the experiments performed by Abraham Trembley in 1744, about the time biologists were recognizing that hydras were animals and not plants. An inside-out hydra will reorganize itself into a normal animal; its original gastrodermal cells move inward and the epidermal cells migrate outward.

Complete animals will regenerate from small pieces of the column, or sometimes even from strictly epidermal or gastrodermal components. Interstitial cells, being undifferentiated to start with, play an important role in regeneration, but differentiated epidermal and gastrodermal cells are versatile, too. In regeneration, the oral end of a hydra exerts dominance over the aboral end. This phenomenon, called **polarity,** is demonstrated by cutting several consecutive pieces out of the column. Each piece develops a mouth and tentacles on the side that was originally closer to the oral end, and produces a basal disk on the side that was closer to the aboral end.

Sexual reproduction is most likely to take place in the autumn. As a rule, a particular hydra produces either eggs or sperm, but some species are hermaphroditic. The so-called **spermaries** or "testes" in which sperm develop are usually located near the middle of the column (Figure 4.6, B; Figure 4.7, A); the **ovaries** (each produces a single egg) are in the lower half (Figure 4.6, C). These simple gonads, if they may be called that, originate from large interstitial cells in the epidermis. Sperm are released from an opening at the tip of each spermary. An egg becomes exposed to sperm when a portion of the epidermis that covers it ruptures. During its early development, the embryo becomes covered by a hard coating and soon falls away from the parent to remain dormant until spring. The covering then softens and a small hydra emerges.

Note that development of the embryo proceeds directly to the form characteristic of the adult. There is no swimming or creeping planula such as is typical of most cnidarians. The hydras are among the few members of the phylum adapted for life in fresh water, and perhaps the planula stage disappeared during the course of that evolutionary adaptation.

GENERAL REFERENCES ON CNIDARIA

Hand, C., 1959. On the origin and phylogeny of the coelenterates. Systematic Zoology, *8:*191–202.

Hastings, J. W., 1983. Chemistry and control of luminescence in marine organisms. Bulletin of Marine Science, *33:*818–828.

Johnson, F. H., 1988. Luminescence, Narcosis, and Life in the Deep Sea. Vantage Press, New York.

Lenhoff, H. M., and Loomis, W. F. (editors), 1961. The Biology of *Hydra* and of Some Other Coelenterates. University of Miami Press, Coral Gables, Florida.

Mackie, G. W. (editor), 1976. Coelenterate Ecology and Behavior. Plenum Press, New York and London.

Miller, R. L., and Wyttenbach, C. R. (editors), 1974. The developmental biology of the Cnidaria. American Zoologist, *14:*540–866.

Morin, J. G., 1983. Coastal bioluminescence: patterns and functions. Bulletin of Marine Science, *33:*787–817.

Muscatine, L., and Lenhoff, H. M. (editors), 1974. Coelenterate Biology. Academic Press, New York and London.

Rees, W. J. (editor), 1966. The Cnidaria and Their Evolution. Symposia of the Zoological Society of London, 16. Academic Press, London and New York.

Russell, F. S., 1953, 1970. The Medusae of the British Isles. 2 volumes. Cambridge University Press, Cambridge.

Tardent, P., 1978. Coelenterata, Cnidaria. *In* Seidel, F. (editor), Morphogenese der Tiere, 1, A-1:69–415. Gustav Fischer Verlag, Jena.

Taylor, D. L., 1973. The cellular interaction of algal–invertebrate symbiosis. Advances in Marine Biology, *11*:1–56.

Tokioka, T., and Nishimura, S. (editors), 1973. The Proceedings of the Second International Symposium on Cnidaria. Seto Marine Biological Laboratory, Japan.

Widersten, B., 1968. On the morphology and development in some cnidarian larvae. Zoologiska Bidrag från Uppsala, *37*:139–182.

Young, R. E., 1983. Oceanic bioluminescence: an overview of general functions. Bulletin of Marine Science, *33*:829–847.

CLASSIFICATION OF CNIDARIA

The Cnidaria have conventionally been divided into three classes: Hydrozoa, Scyphozoa, and Anthozoa. Some specialists have argued, however, that the Cubomedusae (or Cubozoa), formerly assigned to the Scyphozoa, deserve the ranking of class. Their view is accepted in this book.

The sequence in which the following classes, orders, and suborders are presented is based to a large extent on pedagogical convenience. Certain examples within the phylum are more likely than others to be studied in formal courses. Thus, while thecate hydrozoans and sea anemones are believed to be more advanced than certain other representatives of the classes to which they belong, they are used to introduce the classes.

CLASS HYDROZOA

Hydrozoans, of which there are about 3000 living species, are difficult to characterize in a few words. Most representatives of the class have both a polyp and a medusa stage in the life history, but some have only one stage or the other. When the life cycle includes both stages, the medusa may remain attached to the polyp, or even be embedded in it. It is often so extensively modified that it does not look like a medusa. In certain subdivisions of the Hydrozoa, the colonies formed by polyps, or by mixtures of polyps and medusae, exhibit a high degree of polymorphism; the members may be of three or more distinctly different types.

As a rule, free-swimming medusae and individual feeding polyps, called **gastrozooids,** have perfect or nearly perfect radial symmetry. The mesoglea, which does not often contain fibrous material, usually forms a thick layer in medusae and a thin layer in polyps. If a firm skeleton is present, this is almost always strictly external. It generally consists of hardened proteins, but a calcareous skeleton is characteristic of certain groups. Sex cells are of epidermal origin, although they may mature in deep folds that become closely associated with branches of the gut. Nematocysts are rarely present in the gastrodermis.

The medusa of a hydrozoan typically has a **velum** (Figure 4.13, A and B; Figure 4.20). This is a delicate flange that borders the inner margin of the edge of the bell. Because the velum reduces the diameter of the aperture through which water is squeezed out when the animal contracts, it strengthens the force of the jet and increases the efficiency of jet propulsion. In the Scyphozoa, the only other class with a medusa stage, a true velum is lacking.

Obelia as a Representative Hydrozoan

The best way to approach the diversity of hydrozoans is to begin with a type that is somewhat generalized in a morphological sense and that exhibits a free medusa stage as well as an attached polyp stage. *Obelia* (Figure 4.8) fulfills these qualifications. It is a marine genus, abundant on floating docks, pilings, boat bottoms, rocks, and seaweeds. Colonies of some species form bushy growths that are attached at just one point; others spread over the substratum by ramifying stolons and send up many short branches. *Obelia* emits light when it is touched. In a well-established colony, there will be hundreds or thousands of gastrozooids, the polyps concerned with feeding. The part of a gastrozooid that bears the mouth and tentacles is called a **hydranth.** This term is a convenient one to use when dealing with colonial hydroids because it is often hard to define the exact limits of each member polyp; a hydranth, at least, has some individuality.

Each stolon, branch, or polyp has two main components: the coenosarc and perisarc (Figure 4.8, A and B; Figure 4.9, A). The **coenosarc** consists of gastrodermis, epidermis, and mesoglea. The **perisarc** is a transparent, nonliving covering—a hardened protein—secreted by the epidermis. At the base of a hydranth, the perisarc becomes flared out into a cuplike structure called the **hydrotheca.** Each hydranth has about twenty tentacles that surround an elevation on which the mouth is located. When the hydranth is expanded, the tentacles and mouth are extended beyond the edge of the hydrotheca, but they can be withdrawn almost completely into the cup when the hydranth contracts.

If a small crustacean or other suitable food organism touches one or more tentacles, some nematocysts will discharge their threads and the prey will be immobilized. The tentacles bend to bring the food to the mouth, which can be distended. Once the prey is in the gut cavity of the gastrozooid, digestion begins. The gut cavities of all or most polyps of the colony are continuous, and cilia on the gastrodermal cells serve to circulate fluid, so that nutritive materials are distributed to

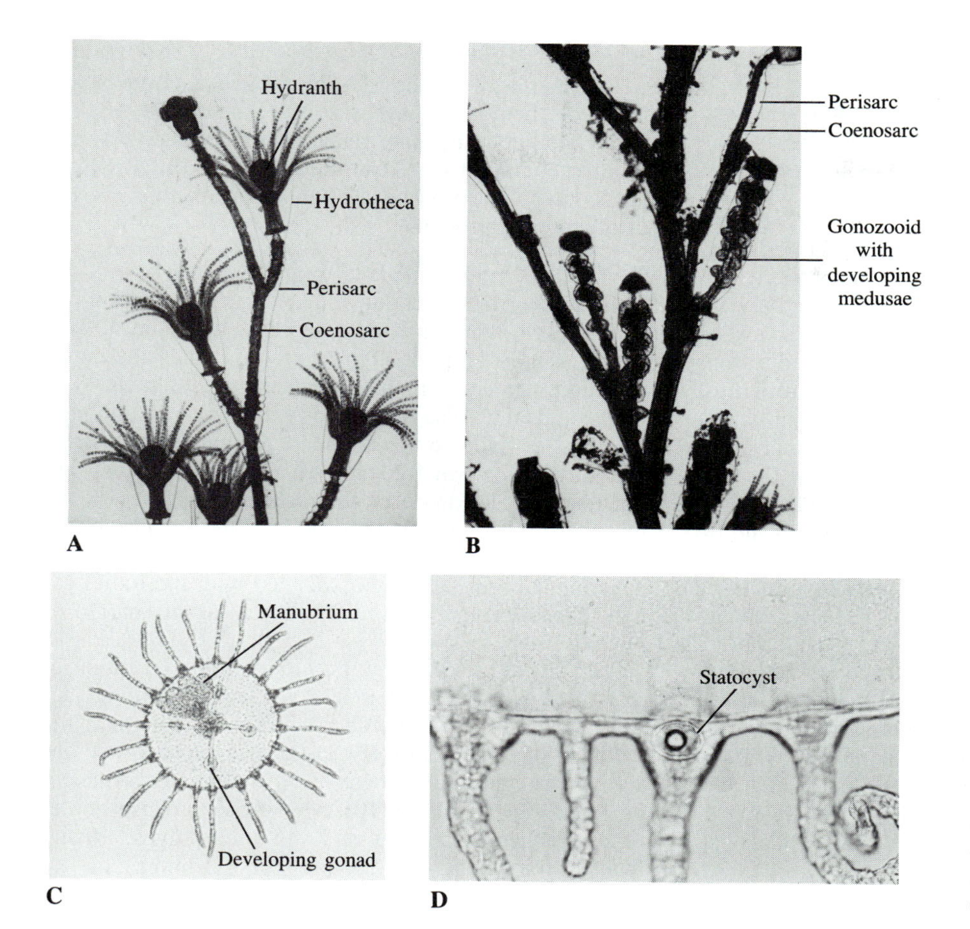

FIGURE 4.8 Class Hydrozoa, order Hydroida, suborder Thecata. *Obelia;* photomicrographs. A. Portion of a colony, showing several hydranths. B. Portion of a colony, showing several gonozooids. C. Recently released medusa. D. Statocyst of a medusa.

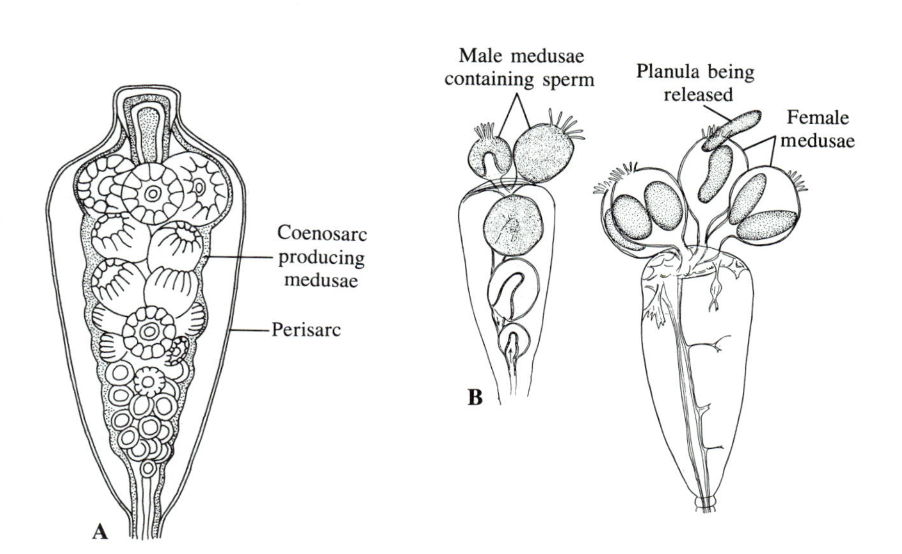

FIGURE 4.9 A gonozooid of *Obelia,* showing developing medusae. B. Gonozooids of *Gonothyraea,* showing extruded medusae. (Berrill, Journal of Morphology, *87*.)

differentiating polyps that have not yet begun to feed, as well as to polyps that produce medusae.

A reproductive polyp is called a **gonozooid** (or gonangium) (Figure 4.8, B; Figure 4.9, A). It consists of a clublike mass of tissue, the **blastostyle.** Because the blastostyle is continuous with the coenosarc of the stem from which it arises, it is provided with both gastrodermis and epidermis. Surrounding it is a thin perisarcal covering, or **gonotheca.** Buds on the blastostyle develop into medusae, whose oral surfaces face outward (Figure 4.9, A). By the time the connection between the blastostyle and medusa is severed, the medusa has sixteen tentacles, a prominent extension **(manubrium)** on which the mouth is located, and four radial canals that join a continuous circular canal near the margin of the bell. There is no velum, and in this respect the medusa of *Obelia* is not typical of the class Hydrozoa.

After it escapes from a pore that opens up at the tip of the gonotheca, the medusa starts to feed on small crustaceans and other organisms. It grows to a diameter of about 3 mm and develops additional tentacles (Figure 4.8, C). Simple statocysts are located at the bases of some of the tentacles (Figure 4.8, D). Four compact gonads appear. These hang from the underside of the bell, and although they originate from the epidermis, they are closely associated with the radial canals. Food brought to the mouth by the tentacles is partly digested in the manubrium. The products of preliminary digestion are then distributed by way of the radial canals and circular canal.

The medusae of *Obelia* are of separate sexes. Sperm released by males fertilize the eggs while these are still within the gonads of females. Cleavage and development into planula larvae also take place in the gonad. After a planula is released, it settles on its anterior end, and in time a mouth and tentacles appear at the opposite end. Buds that form at the base of the first polyp become stolons or branches from which new members of the colony arise.

Classification of Hydrozoa

In an assemblage as amply diversified as the Hydrozoa, classification is necessarily complicated. It is also not universally agreed upon. Some of the major subdivisions generally recognized may be polyphyletic. As the different orders and suborders are discussed here, life histories as well as morphology are emphasized. The beginner may at first be frustrated by the variability of the life histories, but many of them do conform to a single basic plan, even if it is obscured to some extent by morphological specialization or simplification of certain stages. Even the most aberrant cycles will not be difficult to understand, however, if one appreciates which stages are comparable to those in the life history of *Obelia*.

Order Hydroida

All members of the Hydroida have a polyp stage. Most of them also have medusae, but these may be much modified and may remain within the polyps that produce them. In that case they are called **medusoids.** If the polyp stage has a secreted perisarc, this consists of a hardened protein.

Suborder Thecata (Leptomedusae)

In thecate hydroids, of which *Obelia* is a good example, each hydranth is enclosed by a hydrotheca. This cup is usually large enough to accommodate the entire hydranth, tentacles included, when it withdraws. The gonozooids are also covered by a perisarc, although openings develop so that the medusae can escape, or so that medusoids can function sexually even if they are not released. Medusae of this group, when free-swimming, usually have shallow bells. The gonads of the medusae originate in folds of the epidermis, but they generally become closely associated with the radial canals. Statocysts are commonly present at the margin of the bell.

The life cycle of many thecate hydroids is similar to that of *Obelia* in that it includes a free-swimming medusa stage. In some members of Thecata, in fact, the medusa stage is relatively large and more often observed than the polyp stage. This is true, for instance, of *Aequorea* (Figure 4.10, A), whose medusa sometimes reaches a diameter of 7 cm. The polyps, arising from a creeping stolon, are only about 3 mm high.

Aequorea, like many other jellyfishes, is strikingly luminescent. The light-emitting substance in this particular genus is a luciferin called aequorin. It is produced within epidermal cells, especially those close to the edge of the bell (Figure 4.10, B). When fragments of the epidermis are rubbed off, they continue to glow for a few seconds. Luminescence of aequorin requires free calcium ions, and the sensitivity of aequorin to calcium is so high that purified preparations of it are widely used in research on physiology of muscle and nerve.

There are varying degrees of reduction of medusae among the Thecata. In *Gonothyraea,* the medusae that emerge from the tip of a gonozooid remain connected to the column of tissue that produced them (Figure 4.9, B). They do not expand into full-fledged medusae but nevertheless function sexually. The fertilized eggs develop into planulae within the females. Eventually they swim away, settle, and transform into young polyps. The polyp stage of *Gonothyraea* does not differ substantially from that of *Obelia*.

Orthopyxis (Figure 4.10, C) is an interesting genus whose short polyps, arising from branching stolons, are similar to those of *Obelia*. The medusae originate in proportionately large gonozooids and are released one at a time. They have neither mouths nor manubria,

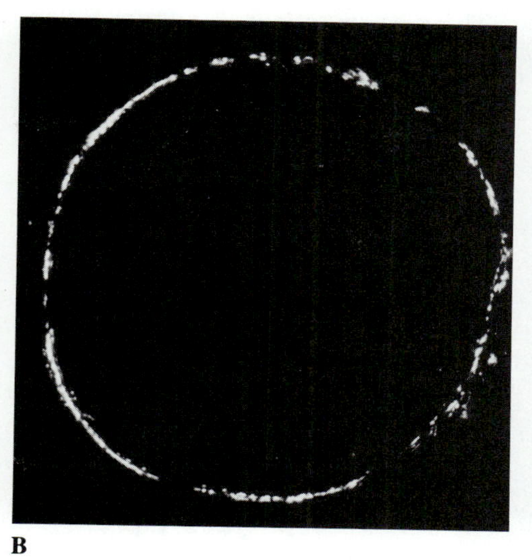

A B C

FIGURE 4.10 Order Hydroida, suborder Thecata. A. *Aequorea,* medusa stage. (Photograph by H. F. Dietrich.) B. An *Aequorea* stimulated to luminesce in the dark. The most intense light is produced near the margin of the bell. C. *Orthopyxis,* an extensive colony growing on a seaweed. The gonozooids are in the lower half of the picture.

so they are unable to feed. They spawn immediately, then die.

Drastic reduction of medusae to medusoids is seen in *Plumularia* (Figure 4.11, A and B) and *Aglaophenia* (Figure 4.11, C and D). Both of these hydroids produce upright growths that branch in a featherlike pattern, and both have small, nonfeeding polyps specialized to function as batteries of nematocysts. They are called **nematozooids,** or guard polyps, and they outnumber the feeding polyps. The medusoids of *Plumularia* and *Aglaophenia* lack tentacles and remain embedded in the gonozooids, and the female medusoids release planulae. In *Aglaophenia,* the gonozooids are protected by a basketlike structure called the **corbula.** This consists of a number of "leaves," which are elaborations of the stem and which bear abundant nematozooids.

Suborder Athecata (Anthomedusae)

In athecate hydroids, a perisarc is usually present, but it either stops short of the hydranths or encloses only their lower portions. The gonozooids, at least during the later phases of medusa formation, have no perisarc at all. Most members of this suborder form branching colonies, but some are solitary polyps. Free-swimming medusae liberated by athecate hydroids, unlike those produced by thecate hydroids, are generally taller than wide. They commonly have eyespots but lack statocysts. The gonads are borne on the surface of the manubrium. In some types the medusae reproduce asexually, budding young medusae from the manubrium or from near the base of a tentacle. In the case of certain athecate hydroids, the medusae are so modified that their resemblance to typical free-swimming medusae is minimal. Like the reduced medusae of many thecate hydroids, they are referred to as medusoids.

A few examples of Athecata show how diverse this group is. The hydroid stage of the genus *Sarsia* (Figure 4.12, A) has elongated hydranths that bear scattered tentacles tipped with knobs in which nematocysts are concentrated. The buds formed below the tentacles develop into medusae characterized by a tall bell, four tentacles, and a long manubrium (Figure 4.12, B). The medusae produce eggs or sperm on the manubrium, but in certain species this is also the site of asexual proliferation of more medusae.

Several other genera, including *Rathkea* (Figure 4.13, A) and *Teissiera* (Figure 4.13, B), also bud medusae from the manubrium. *Teissiera* is unusual because its two tentacles have slender outgrowths that bear concentrations of nematocysts at their tips.

Among the numerous genera whose medusae do not produce asexual buds are *Stomotoca* (Figure 4.13, C) and *Polyorchis* (Figure 4.13, D). The latter has an especially beautiful medusa, with many tentacles, four pinnately branched radial canals, and gonads that hang like sausages from the basal portion of the manubrium. It is usually found in beds of eelgrass and tends to stay close to the bottom.

In some genera, there is no free-swimming medusa stage; the sexual medusoids are retained within the gonozooids that are part of the hydroid colony. *Hydractinia*

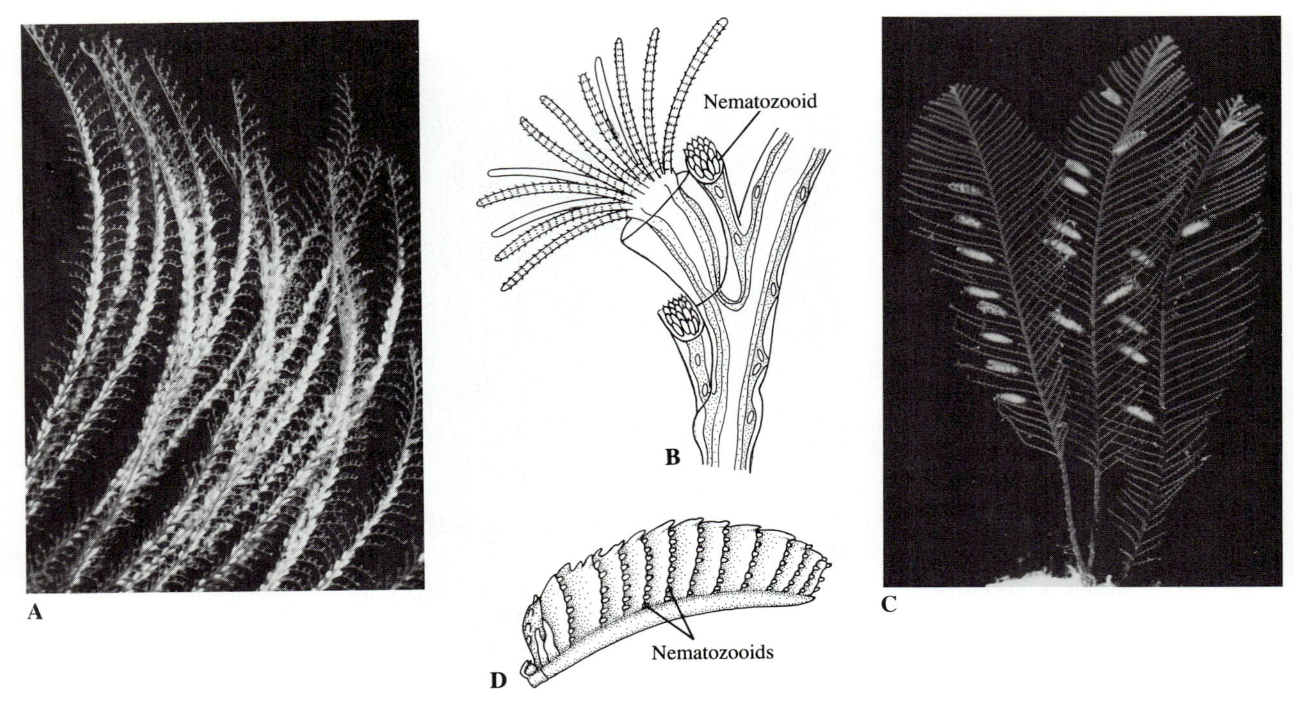

FIGURE 4.11 Order Hydroida, suborder Thecata. A. *Plumularia*. The pinnately branched stems, some with gonozooids, arise from a creeping stolon. B. *Plumularia,* a hydranth and two nematozooids. C. *Aglaophenia*. D. A corbula of *Aglaophenia*. The gonozooids are enclosed by the leaflike elaborations of the branch. (After Fraser, Hydroids of the Pacific Coast of Canada and the United States.)

FIGURE 4.12 Order Hydroida, suborder Athecata. A. *Sarsia,* hydroid stage, with buds that will develop into medusae. B. *Sarsia,* medusa stage. (Photograph by Claudia Mills.)

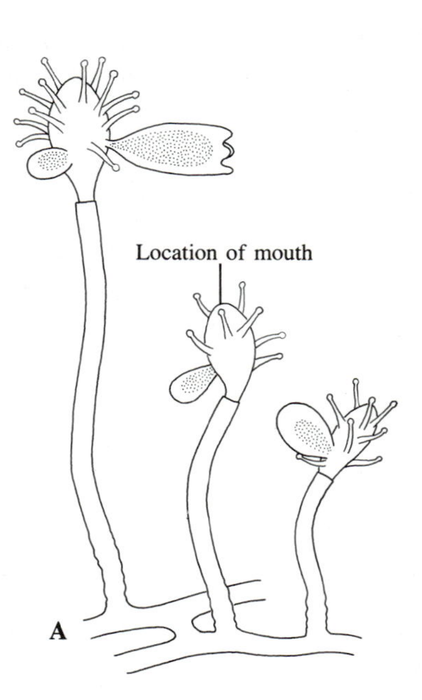

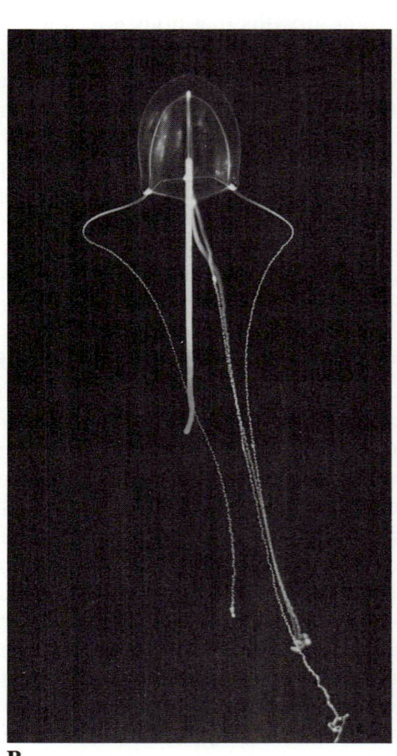

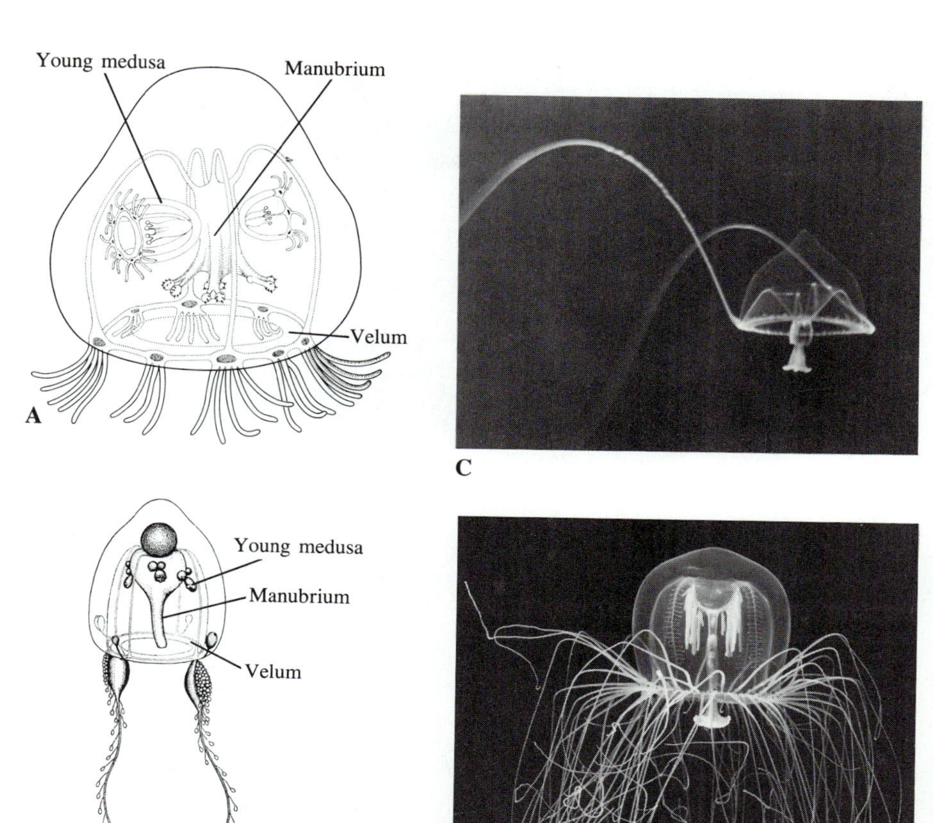

FIGURE 4.13 Order Hydroida, suborder Athecata. A. *Rathkea octopunctata,* mature medusa producing young medusae on its manubrium. (After Naumov, Keys to the Fauna of the USSR, *70*.) B. *Teissiera medusifera*. This species also produces medusae on the manubrium. Nematocysts are especially concentrated at the tips of slender branches of the tentacles. (Bouillon, Cahiers de Biologie Marine, *19*.) C. *Stomotoca atra,* which has only two tentacles. D. *Polyorchis penicillatus*. The gonads hang from the proximal portion of the manubrium. (C and D, photographs by Claudia Mills.)

(Figure 4.14), most commonly found on snail shells inhabited by hermit crabs, fits into this category and is interesting in several other respects, too. The coenosarcal tubes in a young colony unite in such a way that a communal epidermis encloses branching and anastomosing tubes of gastrodermis. The lower epidermis coats the shell with a continuous sheet of brownish perisarcal material, some of which is raised up into spines. In addition to the gastrozooids, which have knobbed tentacles, and the gonozooids, there are sometimes two other types of polyps: dactylozooids and spiral zooids. The **dactylozooids,** when present, may be scattered throughout the colony or may be concentrated at the edges, especially if other organisms have taken up residence nearby. Dactylozooids have been observed to touch snails that are coming too close, and in at least one species, adjacent colonies maintain their separation and sovereignty by ''fighting'' with their dactylozooids. They are slender structures, without mouths, but their tips are richly supplied with nematocysts. They are analogous to the nematozooids, or guard polyps, of *Plumularia, Aglaophenia,* and related thecate genera. The **spiral zooids** have small, mouthlike openings and rudimentary tentacles. They are found only around the aperture of the shell the hermit crab occupies. Their role in the life of the colony needs study.

The significance of the relationship between *Hydractinia* and its hosts is not fully understood. It is known, however, that some hermit crabs prefer shells on which colonies of the hydroid have been established. It is conceivable that the hydroid provides camouflage and that its nematocysts discourage certain predators. *Hydractinia* feeds on a wide variety of small organisms, including nematodes and copepods (Figure 4.14, B and C).

Tubularia is another genus in which medusoids are retained by the gonozooids within which they develop. The large polyps (Figure 4.15, A; Color Plate 2) have two circles of tentacles, one around the basal part of the

Figure 4.14 Order Hydroida, suborder Athecata. *Hydractinia.* A. A dense colony on a snail shell occupied by a hermit crab. B. Gastrozooids capturing a nematode. C. Gastrozooid consuming a copepod. (B and C, Christensen, Ophelia, *4.*)

hydranth, the other around the mouth. From just above the basal tentacles, there arise numerous branched structures that end in slight swellings. Each swelling is a gonozooid occupied by a much-reduced medusoid (Figure 4.15, B and C). All of the medusoids on a particular polyp are either male or female. Fertilization of the eggs and development through the planula stage to small actinulae (Figure 4.15, D) take place in the female medusoids. The actinulae migrate out of the medusoids, settle, and grow to maturity (Figure 4.15, E).

The fact that some structurally similar hydroids produce free medusae, whereas others do not, suggests that life cycles in hydrozoans have been, in an evolutionary sense, highly labile. It is likely that many of the diverse cases of reduction of medusae to medusoids have evolved independently.

In the family Velellidae, the polyp stage is free-floating. The members of this group were for a long time placed in the order Siphonophora, not only because they have a float but because they may appear at first glance to be colonies rather than single polyps. More recently, they were assigned to their own order, Chondrophora, but this isolates them from other athecate hydroids to which they are probably related.

The genus most commonly encountered in temperate waters is *Velella,* the "purple sailor" or "by-the-wind sailor" (Figure 4.16). Its float consists of proteinaceous perisarcal material organized into concentric air-filled tubes; some of the tubes have openings to the exterior. A part of the float is raised to form a sail. Aborally, the float and sail are covered not only by the epidermis that secretes them and that remains in contact with them, but also by a layer of mesoglea pervaded by gastrodermal tubes and then by another layer of epidermis at the surface. The texture of the float resembles that of cartilage, but after a *Velella* has been cast up on the shore, the float dries to the consistency of a soft fish scale.

Although the sail of *Velella* is thin, its triangular shape makes it less likely to collapse in a strong wind than if it were rectangular. When viewed from above, moreover, the sail has a slightly S-shaped contour, giving the sail a concave and a convex portion on both sides. Since a concave surface provides more resistance to wind than a convex surface, the sail of *Velella* is

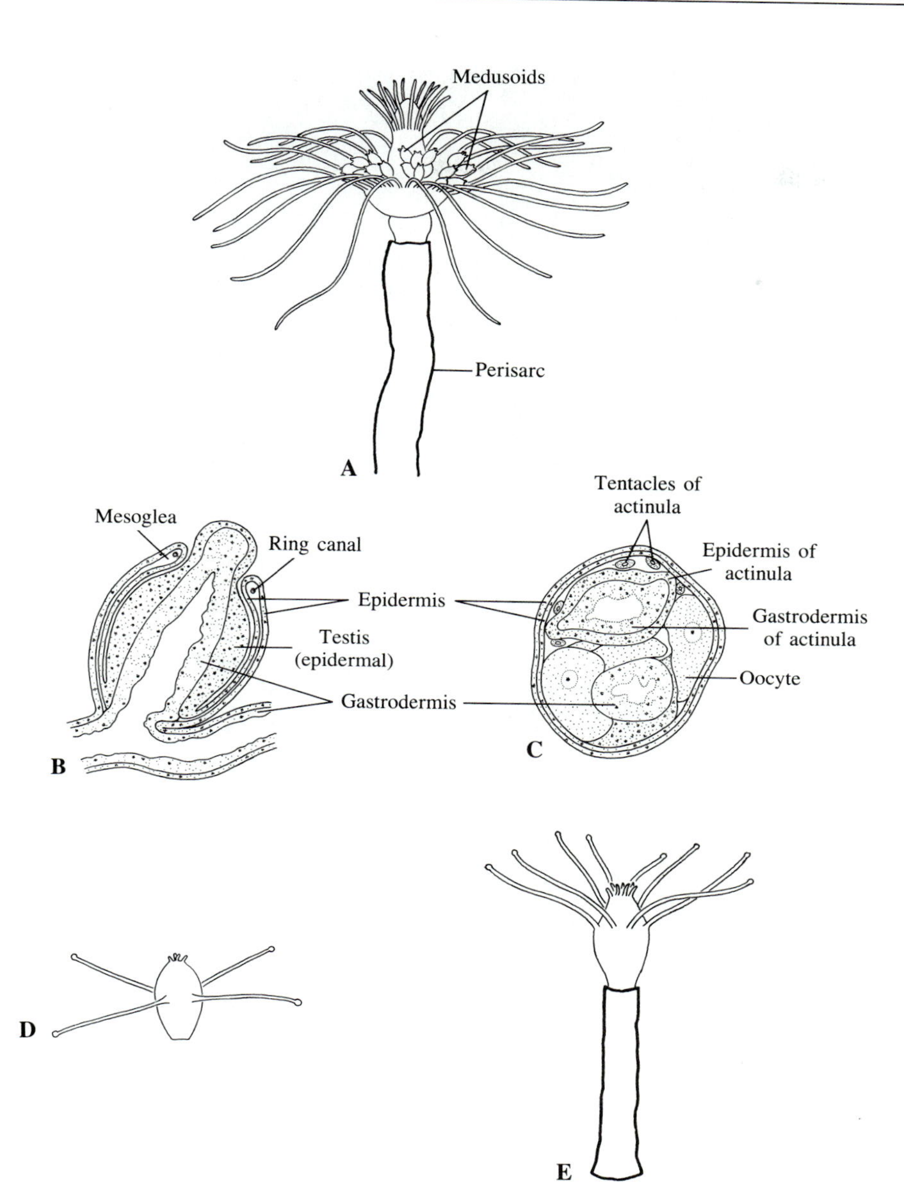

FIGURE 4.15 Order Hydroida, suborder Athecata. *Tubularia*. A. Hydranth of a mature polyp. B. Male medusoid, sectioned parallel to the oral–aboral axis. C. Female medusoid, sectioned perpendicular to the oral–aboral axis. D. Actinula. E. Young polyp. (See also Color Plate 1.)

equally responsive to wind coming from either direction.

The sail is set diagonally on its elliptical base. In about half the individuals, it is canted in one direction, and in the other half it is canted in the opposite direction. This explains why some specimens veer to the right and others to the left when driven by the wind. The deviation from the wind direction may be as much as 60°. This may be of significance in dispersing the species. In the Northern Hemisphere, winds tend to follow a clockwise pattern, whereas in the Southern Hemisphere they are mostly counterclockwise. Specimens of *Velella* with the sail canted in one direction will spin toward the outside of the gyre, whereas those with the sail canted in the opposite direction will spin toward the center of the gyre. The fact that *Velella* veers further off the course of the wind in dirty water than in clean water may help it avoid being trapped in slicks of dust and other matter that generally move in the same direction as the surface water being blown by the wind.

The mouth of *Velella* is centrally located beneath the float. It opens into a cavity from which a complex system of gastrodermal tubes spreads through the soft parts of the animal. Near the edges of the underside of the body, a fringe of tentacles is richly supplied with nematocysts. Between the tentacles and the mouth are

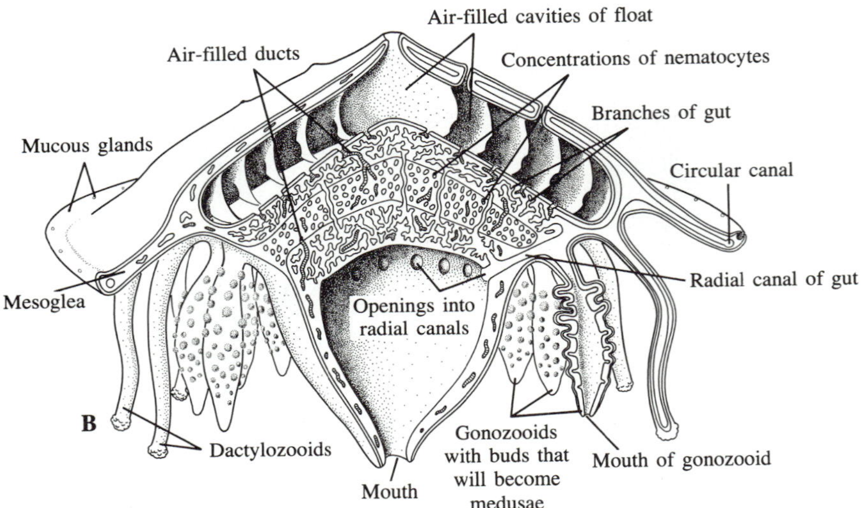

FIGURE 4.16 Order Hydroida, suborder Athecata. *Velella*. A. Mature specimen. B. Young specimen, sectioned to show details of internal anatomy; somewhat diagrammatic. (After Delage and Hérouard, Traité de Zoologie Concrète.)

club-shaped bodies that bud off medusae and have small openings that serve as secondary mouths. No wonder, then, that *Velella* was long viewed as a colony, in which a central gastrozooid appeared to be surrounded by gonozooids capable of feeding (''gastrogonozooids''), and, farther peripherally, by dactylozooids.

The morphology, histology, and development of *Velella* strongly suggest that it is closely related to *Tubularia*. The gut of *Tubularia,* for instance, breaks up into gastrodermal tubes much like those of *Velella*. The marginal tentacles and the medusa-producing structures of *Velella* are probably homologous, respectively, to the lower whorl of tentacles and the gonozooids of *Tubularia*. Thus *Velella* can be compared with an upside-down *Tubularia* in which the perisarc forms an expansive float.

The small medusae of *Velella* have no tentacles when they are released, but they later develop two, then two more. They are golden brown because of the symbiotic dinoflagellates (zooxanthellae) in their tissues. The medusae have a brief free-living existence, during which they release their gametes. The zygotes develop into pelagic planulae and then into individuals in which the mouth, the rudiment of the float, and buds that will become marginal tentacles are apparent.

Hydras, discussed previously, are also generally assigned to the suborder Athecata. The simplicity of their polyps and the absence of medusa and planula stages are believed to be adaptations for life in fresh water.

A

B

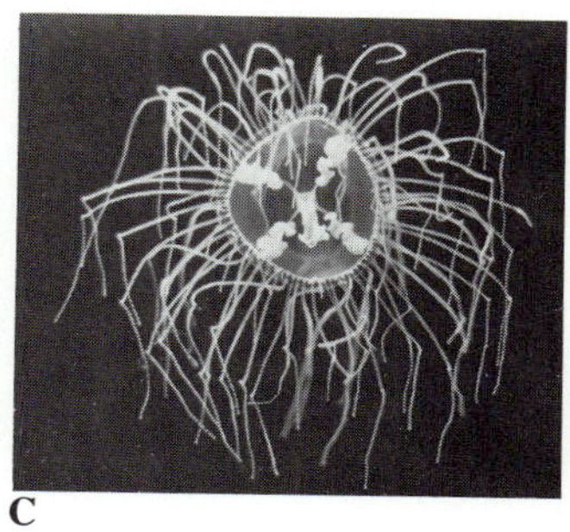

C

FIGURE 4.17 Order Hydroida, suborder Limnomedusae. A. Gastrozooids of *Proboscidactyla;* photomicrograph. The colony is growing at the lip of the tube of *Schizobranchia,* a sabellid polychaete. B. Gonozooids of *Proboscidactyla,* budding off medusae; photomicrograph. C. *Gonionemus vertens,* oral view. (Photograph by Claudia Mills.)

Suborder Limnomedusae

There are relatively few hydrozoans in the Limnomedusae, and they probably do not constitute a monophyletic assemblage. Almost all of the genera have been assigned, at one time or another, to other orders or suborders.

In general, the polyp stage of Limnomedusae has no more than two tentacles, and sometimes only one or none. The polyp may also be so small and inconspicuous as to be rarely observed. The name Limnomedusae indicates that there are freshwater representatives. The classical freshwater example is *Craspedacusta sowerbii,* whose medusa was first discovered in tanks of a giant tropical waterlily at the Royal Botanical Garden near London. It has since been found in water-filled quarries, slow rivers, and similar habitats in many parts of the world. Its polyps are extremely small, have no tentacles, and usually bud to form a small group of connected individuals.

Marine examples of the Limnomedusae include *Gonionemus* (Figure 4.17, C). Each of the many tentacles of the medusa has a sticky "knee" used for attachment to eelgrass or algae. The polyp phase is minute and has only a few tentacles, usually four. *Proboscidactyla* (Figure 4.17, A and B) forms dense colonies at the lips of the leathery tubes of certain polychaete annelid worms that belong to the family Sabellidae. The colonies consist of two-tentacled gastrozooids, fingerlike dactylozooids, and gonozooids that bud off medusae with four tentacles and four radial canals. After being released, the medusae add tentacles and the radial canals bifurcate repeatedly, so that eventually there may be over 50 tentacles and an approximately equal number of canal branches. If the sabellid is removed from a tube on which *Proboscidactyla* is growing, the colony will regress. It can be maintained for a time in a quasinormal state if it is periodically brushed with mucus from a live worm of the right species. The hydroid is obviously dependent on its host, but exactly what it gets from the association remains obscure. If it actually feeds on mucus, as some have suggested, this would be, in the light of the carnivorous food habits of cnidarians, an eccentricity.

Order Milleporina

Two orders of hydrozoans, Milleporina and Stylasterina, secrete calcareous skeletons, so they are often referred to as hydrocorals. The interior of a colony is honeycombed by a network of tubes that consist of both epidermis and gastrodermis. The tubes are therefore comparable to those forming the coenosarc of *Obelia* and other hydroids. The skeleton is essentially external, for it is outside the epidermis of the coenosarcal tubes. There is, however, another sheet of epidermis at the surface of the colony.

The Milleporina are represented by several species of a single genus, *Millepora* (Figure 4.18). They are characteristically yellow-brown because of symbiotic zooxanthellae in their tissues. The dactylozooids are branched, and there are several of them in a regular or irregular circle around each gastrozooid. Like the gastrozooids, they can be completely withdrawn into the pits from which they emerge. The dactylozooids are thought to function mainly in defense, but they may assist in the capture of prey and they are probably also sensory.

Medusae develop from the tips of the dactylozooids or in special depressions at the surface of the colony. Although they are set free, they are not quite complete. They lack tentacles, a mouth, and radial canals, but in the short time they live they fulfill their role as producers of gametes.

Species of *Millepora* tend to overgrow other sessile organisms on reefs. They are often called fire corals,

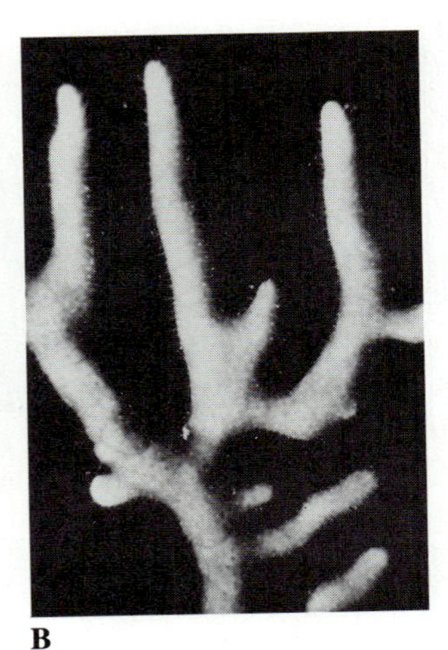

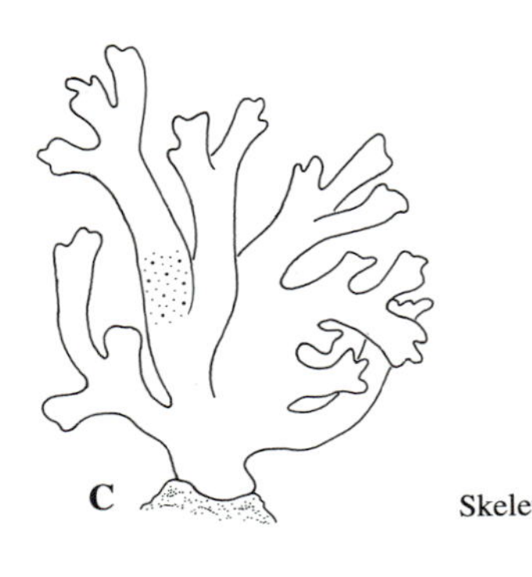

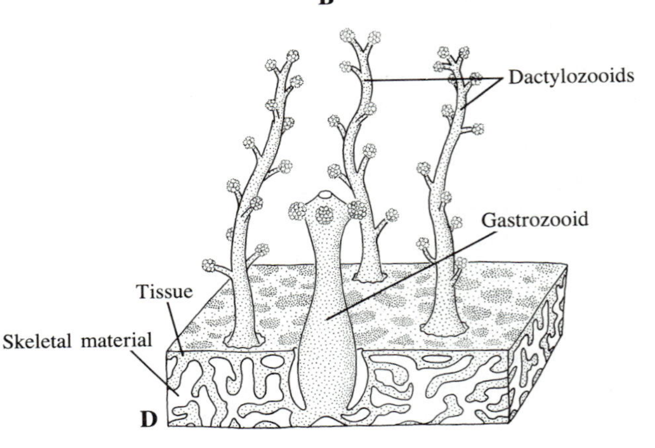

FIGURE 4.18 Order Milleporina. A. *Millepora complanata,* a substantial colony (arrows) in a tropical reef. (Photograph by Raymond Highsmith.) B. *Millepora alcicornis,* small portion of a colony, showing extended dactylozooids. (Photograph by Thomas Suchanek.) C. *Millepora,* a young colony, showing pores through which gastrozooids and dactylozooids emerge. D. Diagram of a small portion of a colony of *Millepora,* showing a gastrozooid, three dactylozooids, and skeletal material deposited between tissue connections. The dactylozooids are arranged in circles around the gastrozooids. Not shown are little pits, called ampullae, from which modified medusae, lacking tentacles, are released.

because their large nematocysts have a potent toxic effect on human skin.

Order Stylasterina

Stylasterines, like milleporines, are colonial and have a calcareous skeleton. Their dactylozooids, however, do not branch, and the circles they form are directly adjacent to the gastrozooids. The bottom of the pit occupied by a gastrozooid has a sharp skeletal projection, or a cluster of projections; these support the basal portion of the polyp. Somewhat similar skeletal structures support the bases of the dactylozooids.

Allopora (Figure 4.19), found in colder waters, is a representative genus. It generally forms thin encrustations or lumpy growths attached to rocks. Some stylasterines, including a few that contribute to coral reefs, make rather large colonies.

Instead of free medusae, stylasterines produce medusoids that originate in the ampullae that lie close to the surface. Fertilization takes place in the ampullae, perhaps while the eggs are still within the female medusoids. The planulae migrate through the tubes of the communal gut until they reach gastrozooids, and then they emerge through the mouths of these. If a planula finds a suitable place to settle, it flattens out and differentiates into a gastrozooid, thereby founding a new colony.

Order Trachylina

The trachylines do not have a genuine polyp stage in the life history. The planula is succeeded by an actinula, and this is either reorganized into a medusa or buds off more actinulae that become medusae. Most hydrozoans have two or more kinds of nematocysts, but a particular species of trachyline will have only one type. A brief comparison of the two suborders should provide an adequate overview of the group as a whole.

In the suborder Trachymedusae, the tentacles originate at the margin of the bell. Most of what looks like a long manubrium is really a downward extension of the bell proper and contains the stomach and proximal portions of the radial canals.

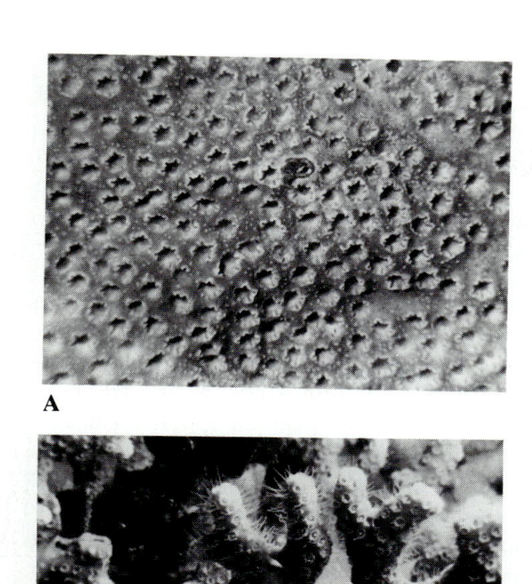

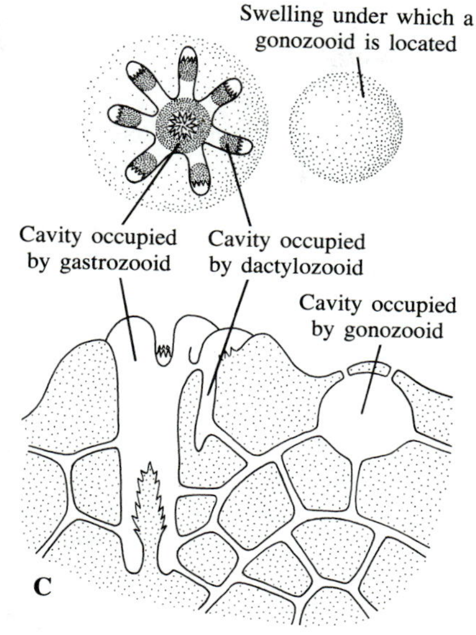

FIGURE 4.19 Order Stylasterina. *Allopora*. A. Portion of a colony forming a thin crust on a rock. The dactylozooids emerge from the "notches" at the edge of each pit where a gastrozooid is situated. (All polyps are retracted, however.) B. Portion of a colony that produces upright branches. Many of the dactylozooids are extended. (Photograph by Thomas Suchanek.) C. Diagram of a portion of a colony, as seen in surface view (above) and in section (below). The sexual medusoids that develop in the ampullae are not released, but are nevertheless comparable to medusae. (After Naumov, Keys to the Fauna of the USSR, *70*.)

Aglantha (Figure 4.20, A) is a representative trachymedusan. It has a tall bell and for this reason could be mistaken for a medusa of the sort produced by many athecate hydroids. There are eight radial canals and eight gonads hanging down into the space enclosed by the bell. *Aglantha* is an unusual jellyfish because it does not move by regular pulsations. For a time it just drifts, but when the bell does contract, the animal "leaps" several centimeters. The planula of *Aglantha* becomes a ciliated actinula that swims in the plankton for a short

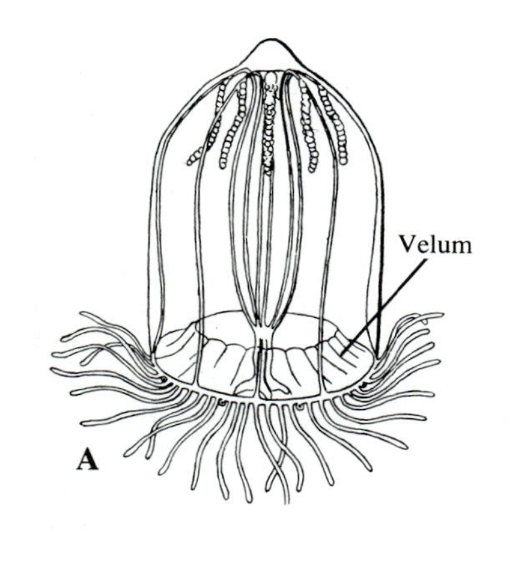

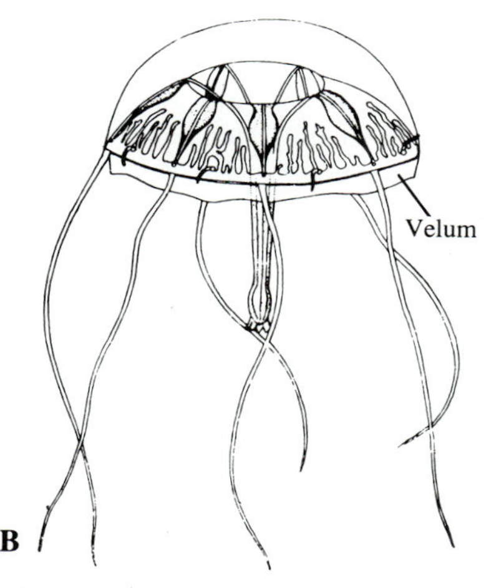

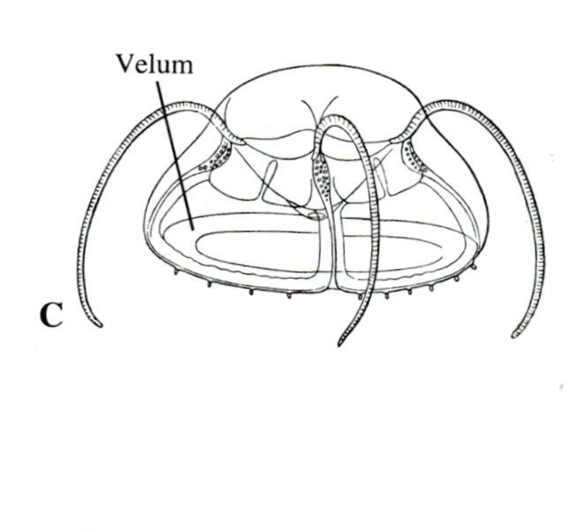

FIGURE 4.20 Order Trachylina. A. *Aglantha digitale* (suborder Trachymedusae). (Kramp, Dana Reports, no. 72.) B. *Geryonia proboscidialis* (suborder Trachymedusae). (Uchida, Japanese Journal of Zoology, *2*.) C. *Aegina citrea* (suborder Narcomedusae). (Kramp, Dana Reports, no. 72.)

time before it becomes transformed into a young medusa.

Geryonia (Figure 4.20, B) has a shallower bell than *Aglantha,* and its gonads are associated with the radial canals.

In the suborder Narcomedusae, the tentacles originate well above the margin of the bell, because the margin continues to grow after the tentacles begin to form. There is no manubrium, and the mouth opens into a large stomach from which a number of broad pockets radiate. Radial canals are lacking. The gonads are adjacent to the floor of the stomach pockets, but they are of epidermal origin, as is typical of hydrozoans in general. In most Narcomedusae the actinulae remain on the underside of the bell of the parent, or settle on another medusa, sometimes of a different species. In certain members of this group the actinula buds off replicas. The actinulae or daughter actinulae eventually become medusae. *Aegina* (Figure 4.20, C) and *Solmissus* are good examples of the Narcomedusae.

Order Siphonophora

Siphonophores form elaborate colonies that swim or drift. Their gastrozooids, sometimes called siphons, usually have a branched tentacle, richly supplied with nematocysts, arising near the base. In at least one genus, *Nanomia,* the nematocyst batteries are accompanied by hooklike spicules. Those are secreted by cells in the mesoglea, and consist mostly of calcium phosphate. After the nematocysts fire in response to contact with prey, the hooks protrude and help maintain a tight hold on the prey. The gonophores, which produce gametes, are reduced medusae. Both of these types of zooids are invariably present.

Two other types of zooids—nectophores and dactylozooids—are characteristic of certain genera. **Nectophores,** which are essentially medusae, propel the colony by pulsating. Some siphonophores have just one nectophore; others have two or several. The dactylozooids are generally fingerlike and have an unbranched tentacle originating at the base. They assist the feeding zooids in procurement of food and are probably also defensive weapons. The organisms consumed include copepods, crustacean and molluscan larvae, chaetognaths, and fish larvae. Many siphonophores capture a wide variety of prey; others tend to be selective.

At the top of the colony of some genera, there is a gas-filled float called the **pneumatophore.** The gas consists mostly of nitrogen. The pneumatophore should probably not be viewed as a true zooid, for it develops from part of the planula, rather than as a bud produced by the planula.

There are three suborders of siphonophores, and the distinctions between them are based largely on the composition of the colonies. In the Physonectae, to which *Physophora* (Figure 4.21, A), *Agalma* (Figure 4.21, B), *Nanomia,* and *Halistemma* belong, there is a small pneumatophore subtended by several nectophores. The latter arise from the main stem that hangs down from the pneumatophore. Below the nectophores, the stem proliferates into gastrozooids, dactylozooids, and gonophores. These three types of zooids together form several units called **cormidia.**

Each cluster of gonophores is partly covered by a leaflike outgrowth called a **bract.** After an egg is fertilized, it develops into a free-swimming planula. A portion of this invaginates to form a pocket that becomes the gas-filled chamber of the pneumatophore. The rest of it starts to bud off other kinds of zooids characteristic of the mature colony. In time, the stem becomes greatly elongated. In some genera, formation of the colony does not proceed in quite the same fashion; the pneumatophore does not differentiate until at least one of the other zooids has been produced.

The suborder Calycophorae, in which there is no pneumatophore, is represented by *Muggiaea* (Figure 4.22). The top member of the colony is an angular, asymmetrical nectophore. The stem of the colony runs up into the mesoglea of the nectophore as a narrow canal, and at the tip of this is a drop of oil that provides buoyancy. Each of the cormidia into which the rest of the stem is organized consists of a gastrozooid, a gonophore, and a bract.

Some of the cormidia eventually break away from the main stem and swim around on their own. The impetus for the movement is provided by the pulsating bell of the gonophore. An independent cormidium of this type, which serves to disperse gametes, is referred to as a **eudoxid** (Figure 4.22, C). The gonophores of calycophorans do not mature except on free eudoxids.

The muscles in the main stem and cormidia of *Muggiaea* are efficiently coordinated by nervelike bundles of neurons. With an elegant swirl of the stem, the colony sets its complicated net, consisting of the branched tentacles of the feeding zooids. As the colony sinks, the net captures small organisms, especially copepods and other crustaceans, in the plankton.

The suborder Cystonectae includes *Physalia,* the infamous Portuguese man-of-war (Figure 4.23). This remarkable drifting animal lacks nectophores, but the length of its crested pneumatophore may exceed 15 cm. The pneumatophore, moreover, floats above the surface. (In other groups, it provides buoyancy, but does not really float.) The pneumatophore is usually bright blue but is often tinged with orange or pink. The shape of the pneumatophore can be changed by the action of muscles in its wall. *Physalia* is primarily a warm-water organism, but currents sometimes carry it up to the Atlantic coasts of North America and Europe.

Beneath the pneumatophore are several groups of

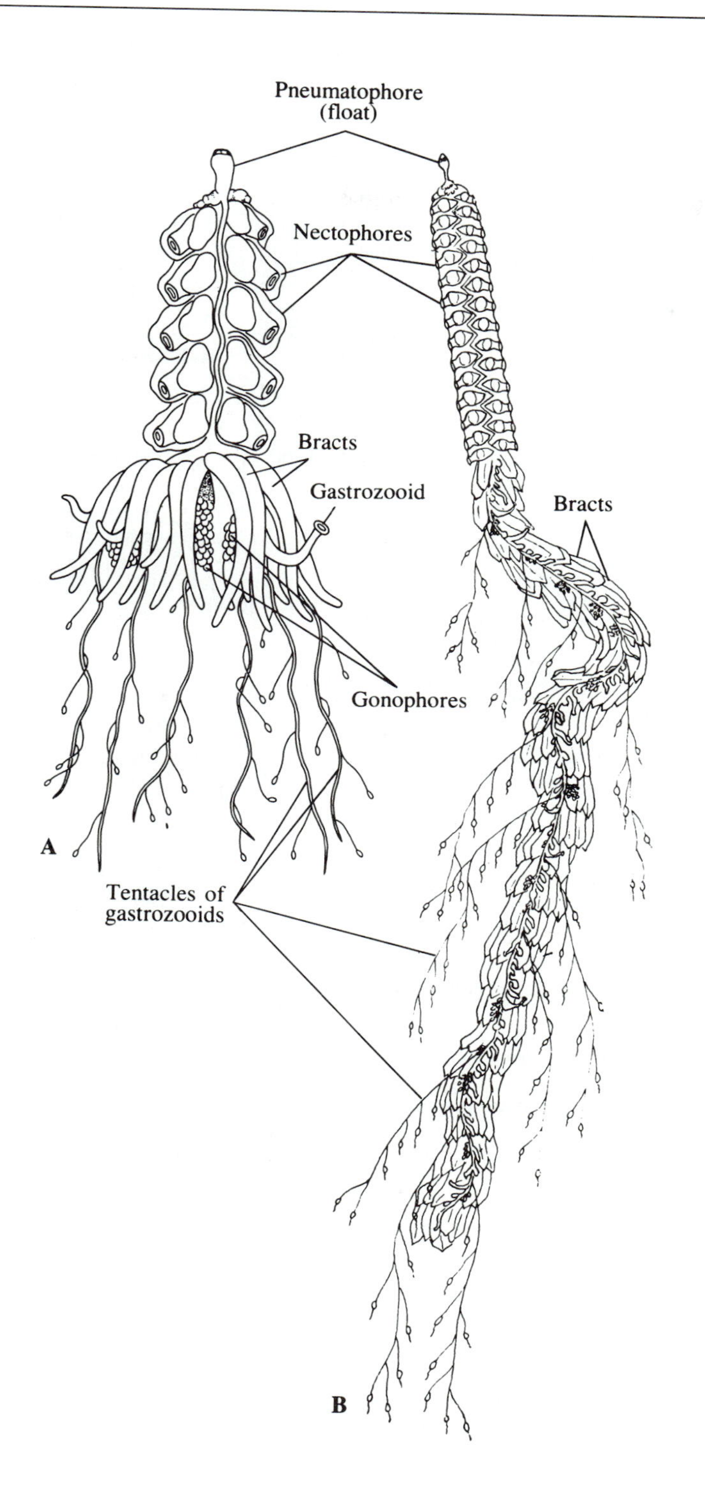

FIGURE 4.21 Order Siphonophora, suborder Physonectae. A. *Physophora hydrostatica*. (After Lankester, A Treatise on Zoology.) B. *Agalma*. (Künne, Natur und Volk, *71*.)

gastrozooids (these lack tentacles), as well as dactylozooids and gonophores. There is no main stem such as is typical of the suborders discussed previously. The tentacles of the dactylozooids, when fully extended, may reach a length of several meters. The toxin in the nematocysts of these tentacles is extremely potent, and *Physalia* is one of the more dangerous animals likely to be contacted by bathers and fishermen. Its food consists

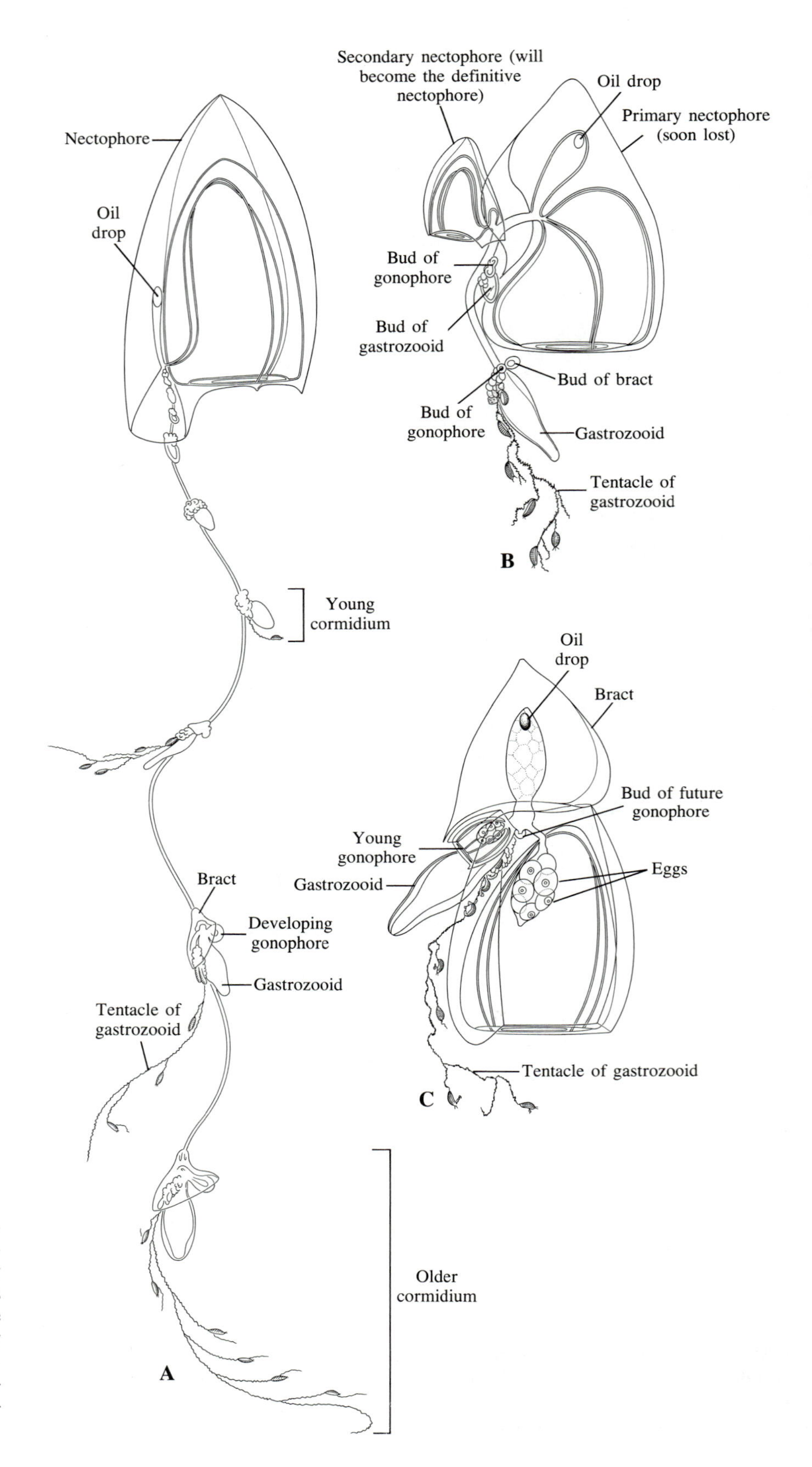

FIGURE 4.22 Order Siphonophora, suborder Calycophorae. *Muggiaea kochii*. A. Colony with several cormidia in various stages of development. (In a mature colony, there may be many cormidia.) B. Young colony. C. Eudoxid. (After Chun, Verhandlungen der Deutschen Zoologischen Gesellschaft, *7*.)

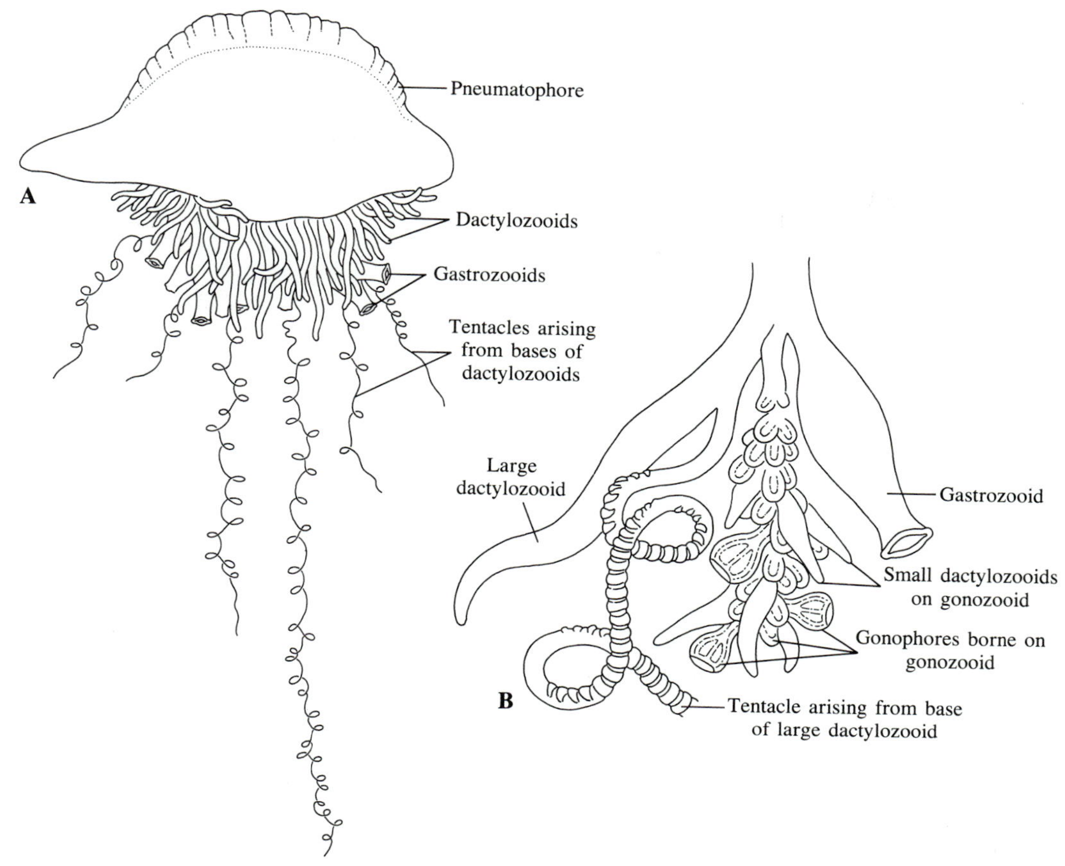

FIGURE 4.23 Order Siphonophora, suborder Cystonectae. *Physalia*. A. Entire colony, floating on the surface. B. Cormidium.

of fishes, which it paralyzes and brings up to the sucking mouths of the gastrozooids.

The gonophores originate in clusters on branching stems. Groups of them become detached and are propelled by gonophores modified to serve as swimming bells.

Order Actinulida

At first glance, actinulids resemble freshwater hydras. They are ciliated externally, however, and can swim. Also unusual is the fact that their tentacles, arranged in two circles, are close to the aboral end, not the oral end. Alternating with the tentacles of the circle that is nearer the mouth are stalked statocysts similar to those found in some genera of the order Trachylina. The height of these animals does not exceed 2 mm.

The genus *Halammohydra* is represented by several species found between sand grains in ocean beaches. The maximum number of tentacles in each circle is sixteen, but usually the number is smaller. *Halammohydra vermiformis* (Figure 4.24, A) has four in one circle and three or four in the other. There is a prominent epidermal nerve ring at the level of the statocysts and the tentacles that alternate with them, and a cuplike attachment device at the aboral end (Figure 4.24, B). Sexes are separate, as is usual in cnidarians, and eggs, still unfertilized, escape through the mouth. Development to the adult form is gradual and essentially direct; cilia appear early and persist.

The single species of *Otohydra* has up to sixteen tentacles. It lacks the cuplike aboral attachment device of *Halammohydra,* and there is no distinct nerve ring, either. This actinulid, found subtidally on the Atlantic coast of France, is hermaphroditic. It releases actinulae, already ciliated, through the epidermis.

The nerve ring of *Halammohydra,* and the statocysts of both genera, are features found in many hydrozoan medusae, but not in polyps. The ciliation of the body is

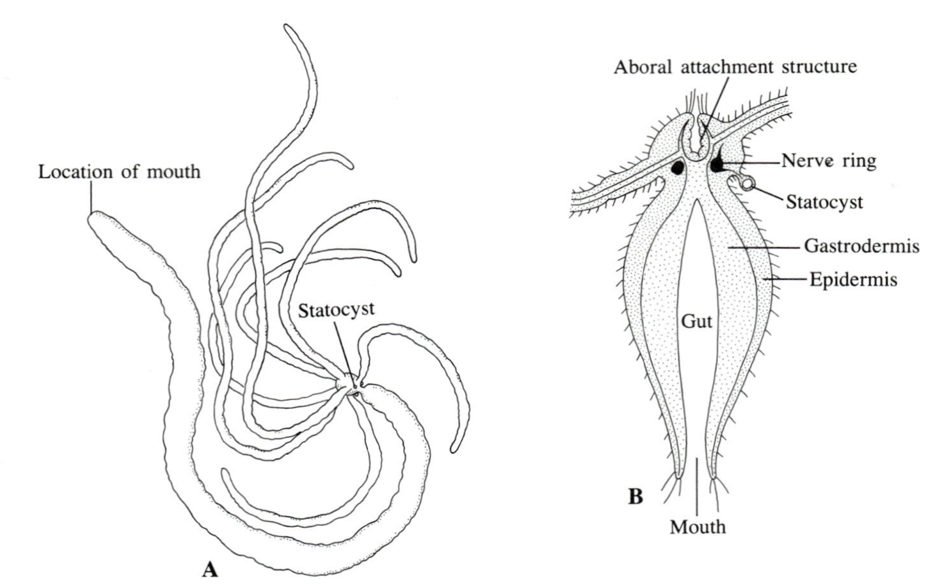

Figure 4.24 Order Actinulida. A. *Halammohydra vermiformis.* B. Diagram of a lengthwise section of *Halammohydra,* showing a portion of a tentacle from each of the two circles, as well as other structures. (After Swedmark and Teissier, in Rees (editor), The Cnidaria and Their Evolution.)

characteristic of planulae and some actinulae, but it is not typical of either polyps or medusae. Thus actinulids have an interesting mixture of traits.

References on Hydrozoa

Biggs, D. C., 1977. Field studies on fishing, feeding and digestion in siphonophores. Marine Behaviour and Physiology, *4:*261–274.

Bouillon, J., Boero, F., Cigogna, F., and Cornelius, P. F. S. (editors), 1987. Modern Trends in the Systematics, Ecology, and Evolution of Hydroids and Hydromedusae. Clarendon Press, Oxford.

Burnett, A. (editor), 1973. Biology of *Hydra.* Academic Press, New York and London.

Christensen, H. E., 1967. Ecology of *Hydractinia echinata.* I. Feeding biology. Ophelia, *4:*245–275.

Fritchman, H. K., 1974. The planula of the stylasterine hydrocoral *Allopora petrograpta* Fisher: Its structure, metamorphosis and development of the primary cyclosystem. Proceedings of the Second International Coral Reef Symposium, *2:*245–258.

Gierer, A., 1974. Hydra as a model for the development of biological form. Scientific American, *231*(6):44–54.

Kirkpatrick, P. A., and Pugh, P. R., 1984. Siphonophores and Velellids. Synopses of the British Fauna, no. 29. E. J. Brill/W. Backhuys, London and Leiden.

Kramp, P. L., 1961. Synopsis of the medusae of the world. Journal of the Marine Biological Association of the United Kingdom, *40:*1–469.

Kruijf, H. A. M. de, 1975. General morphology and behavior of gastrozoids and dactylozoids in two species of *Millepora* (Milleporina, Coelenterata). Marine Behaviour and Physiology, *3:*181–192.

Mackie, G. O., 1965. Analysis of locomotion in a siphonophore colony. Proceedings of the Royal Society of London, B, 159:366–391.

Mackie, G. O., and Marx, R. M. 1988. Phosphatic spicules in the nematocyst batteries of *Nanomia cara* (Hydrozoa, Siphonophora). Zoomorphology, *108:*85–91.

Mackie, G. O., and Passano, L. M., 1968. Epithelial conduction in hydromedusae. Journal of Comparative Physiology, *52:*600–621.

Müller, W., 1964. Experimentelle Untersuchungen über Stockentwicklung, Polypendifferenzierung und Sexualchimären bei *Hydractinia echinata.* Wilhelm Roux' Archiv für Entwicklungsmechanik, *155:*181–268.

Pardy, R. L., and White, B. N., 1977. Metabolic relationships between green hydra and its symbiotic algae. Biological Bulletin, *153:*228–236.

Russell, F. S., 1953. The Medusae of the British Isles. Volume 1. Cambridge University Press, Cambridge.

Singla, C. L., 1975. Statocysts of Hydromedusae. Cell and Tissue Research, *158:*391–407.

Westfall, J. A., 1970. The nematocyte complex in a hydromedusan, *Gonionemus vertens.* Zeitschrift für Zellforschung, *110:*457–470.

CLASS SCYPHOZOA

Jellyfishes that lack a true velum and whose gonads originate from the gastrodermis rather than from the epidermis belong to the Scyphozoa. Although this group has only about 200 species, it is so diversified that it is difficult to make hard and fast rules by which a scyphozoan medusa can be recognized. In most species found swimming near the surface in temperate and colder seas, the four corners of the mouth are drawn

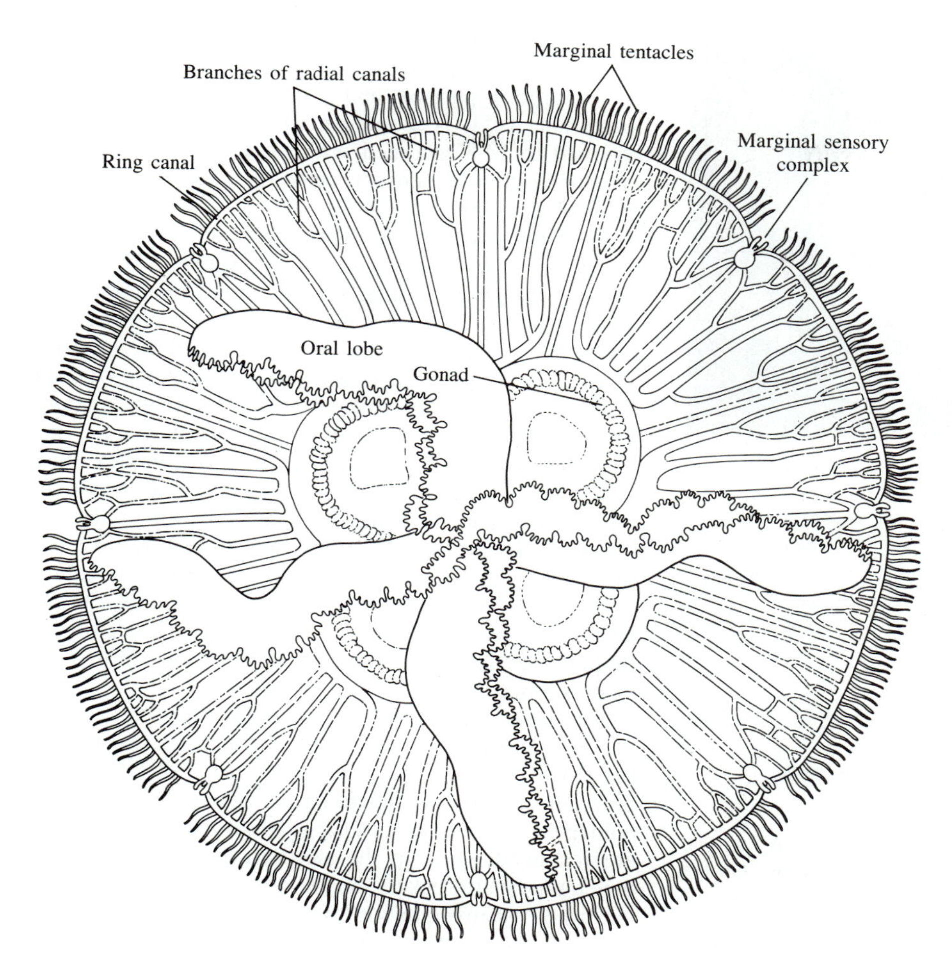

FIGURE 4.25 Class Scyphozoa, order Semaeostomeae. *Aurelia*, oral view; diagrammatic.

out into lobes that may trail for some distance behind the pulsating bell. Scyphozoans of this type belong to the order Semaeostomeae. One of them, *Aurelia,* is widely distributed and illustrates the main features of the class as a whole.

Aurelia as an Example of the Scyphozoa

The medusa of *Aurelia* (Figures 4.25, 4.26) is of moderate size for a scyphozoan—generally between 10 and 20 cm in diameter. The margin of the rather shallow bell is subdivided into eight lobes and is fringed by hundreds of short tentacles. The sensory complexes between the lobes are described later. Most of the animal consists of firm mesoglea. As in hydrozoan medusae, the mesoglea constitutes an elastic skeleton against which muscles operate. The muscles, of course, are contractile bases of myoepithelial cells. The more prominent ones involved in locomotion form a ring near the edge of the bell. After a contraction of the muscles narrows the diameter of the bell and forces water out from the underside, the bell springs back as if it were made of rubber. There is no true velum in scyphozoans, but some of these animals, including *Aurelia,* have a superficially similar structure called a **velarium.** This does not originate in quite the same way as a velum. It is, moreover, pervaded by branches of the gut, which is never the case in the velum of hydrozoan medusae.

The manubrium of *Aurelia* is short. From the corners of the nearly square mouth arise long processes called **oral lobes,** well supplied with nematocysts. The mouth opens almost directly into a stomach that has four main pouches, each with a number of filaments in which nematocysts are concentrated. From the pouches, as well as from between them, the stomach sends out radial canals, most of which branch repeatedly. The ultimate branches join the ring canal that lies close to the margin of the bell. *Aurelia* feeds primarily on smaller organisms, especially copepods and other planktonic crustaceans, which it traps in mucus on both the oral and aboral sides of the bell. Food is moved by ciliary activ-

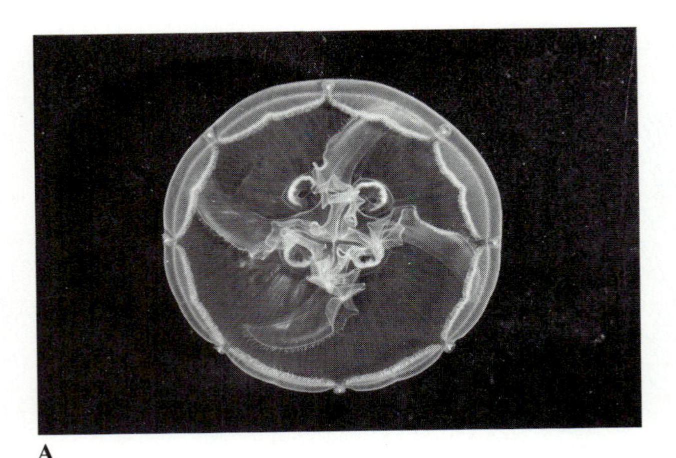

A

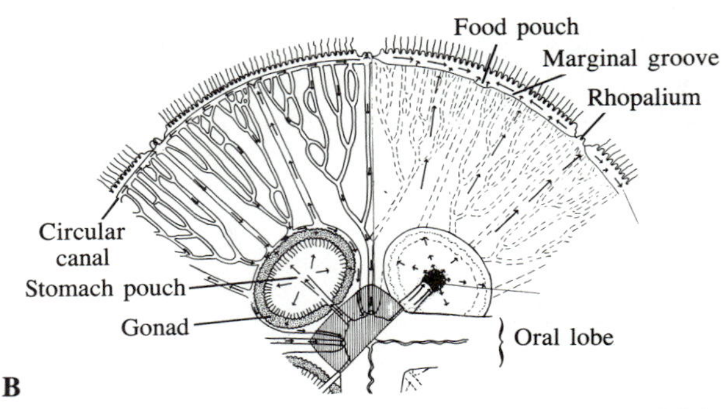

B

FIGURE 4.26 Order Semaeostomeae. *Aurelia*. A. Living specimen, oral view. B. Part of the disk, oral view. On the right, arrows indicate ciliary currents on the underside of the bell. These currents bring food to eight places, called food pouches, where it is picked up by the oral lobes. On the left, arrows show the principal currents within the gut. (Southward, Journal of the Marine Biological Association of the United Kingdom, *34*.)

ity to eight **food pouches** near the edge of the bell (Figure 4.26, B), and is then picked up by the oral lobes. Ciliary tracts on the lobes carry the mixture of food and mucus to the mouth. By reversing the direction of the ciliary beat on its oral lobes, *Aurelia* can reject unsuitable material. Once inside the gut cavity, food is circulated through the various canals. The ciliary method of feeding used by *Aurelia* is not typical of scyphozoans in general. As a rule these jellyfishes capture larger prey, using their nematocyst-rich tentacles and oral lobes.

The sensory complexes situated in the notches between marginal lobes consist of several distinct elements (Figure 4.27, B). First, there is the **rhopalium,** a fingerlike outgrowth that is entered by a branch of the ring canal (Figure 4.27, C). Gastrodermal cells at the tip of this secrete crystals that consist mostly of calcium sulfate. On the upper side of the rhopalium is an eyespot with dark pigment and photoreceptor cells. On the lower side of it is a more complicated structure also believed to be concerned with photoreception; it consists of epidermal sensory cells enclosed by a cup of pigmented gastrodermal cells.

On both sides of the rhopalium, and just below it, are two fleshy outgrowths called **lappets.** These are joined by a structure that forms a hood over much of the rhopalium. From near the sides of the proximal portion of the hood there arise two small auxiliary lappets. These lie, therefore, above the larger lappets. Patches of epidermal sensory tissue, other than those already mentioned in connection with photoreceptors, are located in various structures of the sensory complex. Just proximal to the hood, for instance, there is a prominent sensory pit, and there are concentrations of epidermal sensory cells on both the upper and lower sides of the rhopalium. Much of the sensory complex is certainly concerned with equilibrium and with initiation of impulses that lead to pulsations of the bell. If all of the rhopalia are removed, the animal will not continue to swim spontaneously.

The sexes of *Aurelia* are separate. The four gonads are conspicuous horseshoe-shaped structures that border the stomach pouches. Unlike the gonads of Hydrozoa, they derive from the gastrodermis rather than from the epidermis. Sperm are discharged into the sea, but eggs are fertilized while in the stomach and undergo early development in pockets on the oral lobes. The gastrula is formed by invagination of the blastula, and the blastopore forms the future mouth. The gastrula becomes a ciliated planula, which settles by its aboral end. Within two or three days it becomes a type of polyp called a **scyphistoma** (Figure 4.28, A). After the first four tentacles have been formed, four more are added, then eight more. The gut becomes partially divided by four vertical ridges, and from the oral surface of the scyphistoma an epidermal pocket sinks deep into each ridge. The scyphistoma feeds mostly on small crustaceans and grows to a height of about 6 or 7 mm. It may produce new individuals like itself by budding (Figure 4.28, B), but eventually it divides transversely into a number of prospective medusae (Figure 4.28, C). The original tentacles disappear and superficial constrictions in the column deepen until the scyphistoma, now called a **strobila,** looks like a stack of saucers with scalloped edges. Only the basal portion persists as a scyphistoma. Late in the complex process of strobilation it sprouts tentacles that it will need when it starts to feed again.

By the time the several young medusae, called **ephyrae,** separate, they have eight prominent lappets, each with a notch in which a rhopalium and other components of a marginal sense organ develop (Figure

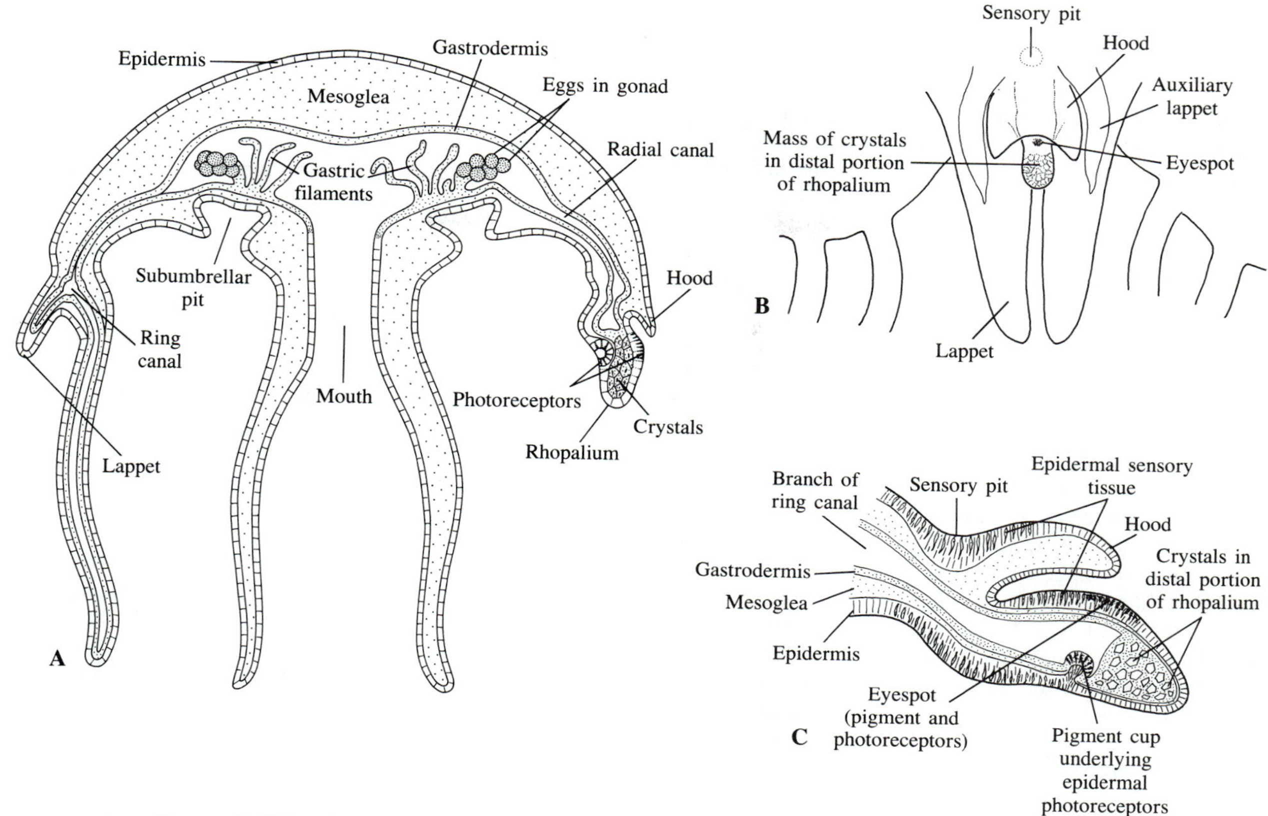

FIGURE 4.27 Order Semaeostomeae. A. Cross-section through a medusa; diagrammatic. B. Marginal sensory complex of *Aurelia,* viewed from the aboral side; diagrammatic. C. Rhopalium and hood of *Aurelia,* sectioned lengthwise; diagrammatic.

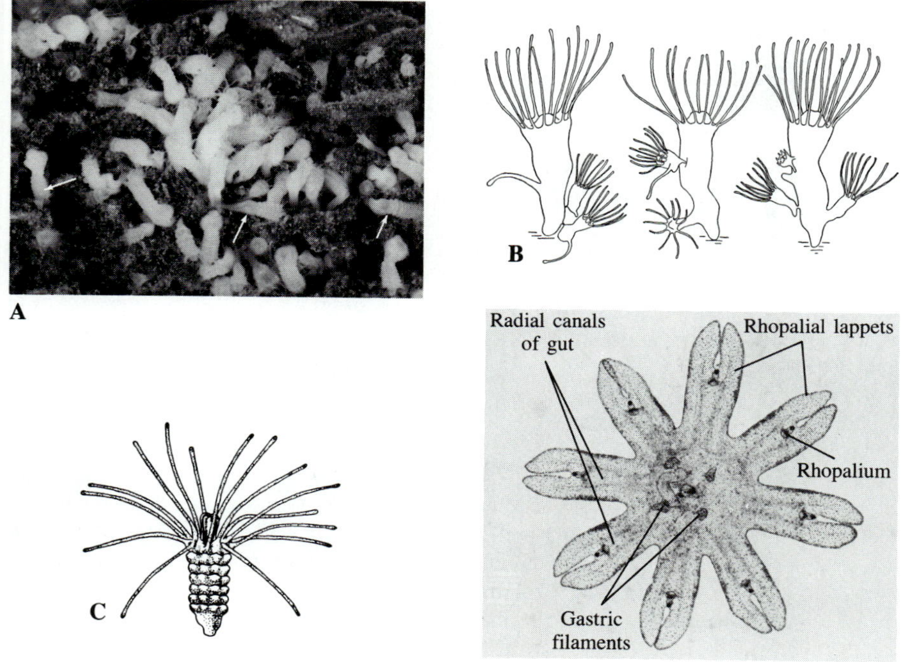

FIGURE 4.28 Order Semaeostomeae. A. Scyphistomae of *Aurelia*. Several are in the process of dividing transversely (arrows) to form ephyrae. B. Budding scyphistoma. (Gilchrist, Biological Bulletin, *72*.) C. Scyphistoma in the process of forming ephyrae. (Uchida and Nagao, Annotationes Zoologicae Japonenses, *36*.) D. Ephyra; photomicrograph.

4.28, D). The eight broad lobes of the adult medusa are not derived from the lappets. They are formed by extensive growth between the lappets, so that the lappets end up between the lobes.

Order Semaeostomeae

Aurelia and its close relatives belong to the order Semaeostomeae. The cardinal features of this group are well-developed, frilly oral lobes; hollow tentacles, situated at the margin or on the underside of the bell near the margin; and radial canals extending from the stomach toward the margin. Many species reach a large size. *Cyanea capillata,* one of the more abundant scyphozoans in waters of the Northern Hemisphere, commonly attains a diameter of 50 cm, and specimens 2 m across have been reported. *Cyanea,* by the way, is one of several genera of semaeostomes in which the marginal tentacles, unlike those of *Aurelia,* reach a great length. The tentacles of a large individual may trail for more than 10 m when they are fully extended. The toxin in the nematocysts of the tentacles and oral lobes is potent and causes a severe skin reaction in some humans. Another important genus is *Chrysaora* (Figure 4.29).

The life history of some semaeostomes, including *Cyanea,* follows the general pattern described for *Aurelia.* In others, however, there is no scyphistoma stage. In *Pelagia,* for instance, the planula develops directly into an ephyra, which then gradually assumes the form of the adult jellyfish.

Order Coronatae

Coronate medusae (Figure 4.30, A) are distinctive because of a circular furrow on the aboral surface of the bell. This furrow, which marks the location of the ring of muscle that effects swimming movements, divides the bell into two distinct regions. The tentacles are solid. Coronate medusae are found only in deep water. Some species are taken at depths down to 5000 m or more, and it is unusual to find any of these animals within 500 m of the surface.

Polyps are known only for one genus. These may be solitary or colonial, but they are distinctive in that they secrete hard tubes with a texture similar to that of the perisarc of hydroids.

Order Rhizostomeae

The medusae of rhizostomes (Figure 4.30, B and C) are substantially limited to tropical and subtropical seas. They lack marginal tentacles, and although they have frilly structures homologous to the four oral lobes of semaeostomes, each of these is divided lengthwise into half-lobes. A still more striking characteristic of rhizostomes is the absence of a mouth. Instead, the oral lobes have hundreds or thousands of small openings that suck in the juices of prey organisms, which include fishes. The many openings on each half-lobe may be concentrated in one, two, or three more or less separate groups. They lead into canals that join a main canal traversing the half-lobe and joining the stomach. As in

FIGURE 4.29 Order Semaeostomeae. A. *Chrysaora helvola.* (Naumov, Keys to the Fauna of the USSR, *75.*) B. *Chrysaora hyoscella.* (Schensky, Natur und Volk, *61.*)

semaeostomes, radial canals extend outward from the stomach and distribute food to other parts of the animal. Although rhizostomes have no marginal tentacles, they usually have one or many tentaclelike appendages on each oral half-lobe.

The polyps of rhizostomes produce one medusa at a time, either laterally or from the oral end. The young medusa has a true mouth, but this becomes occluded by fusion of the bases of the oral lobes. The system of suctorial openings and canals derives from branching grooves that originally run from the lobes to the mouth.

Species of *Cassiopea,* which inhabit shallow water, are interesting in that their mouth lobes are populated by symbiotic dinoflagellates (zooxanthellae). Although these jellyfishes do pulsate, they lie upside down on a sandy bottom. During the daytime, therefore, the symbionts receive the light they need for photosynthesis.

Order Stauromedusae

Stauromedusae are attached by an aboral stalk to seaweeds or eelgrass. Although aberrant in several respects, they are unmistakably similar to more generalized scyphozoans. *Haliclystus* (Figure 4.31, A and B) and *Thaumatoscyphus* (Figure 4.31, C) are representative of the group. The body as a whole is nearly funnel-shaped. Around the margins of the funnel are eight lobes bearing several to many tentacles tipped with concentrations of nematocysts. In each indentation between lobes is a structure called an **anchor,** or rhopalioid. This consists of a knob of nematocysts supported by a stalk that emerges from the center of a sticky cushion. Below the cushion is a concentration of dark pigment—an eyespot. The anchors, and also the tentacles, enable *Haliclystus* to cling to the substratum even when the aboral stalk is temporarily detached.

The mouth is located on a slight elevation in the center of the oral surface. Its corners are drawn out only slightly. Also on the oral surface are four depressions that sink down into the partitions dividing the gut into four gastric pockets. These pockets run nearly to the tip of the aboral stalk and have filaments bearing nematocysts. The gonads form eight bands, one pair being closely associated with each pocket. The margins of the funnel, the partitions between the gastric pockets, and the aboral stalk have prominent muscles. The end of the stalk has adhesive glands.

Haliclystus and its relatives feed mostly on small crustaceans, especially caprellid amphipods, which are usually common in the same habitats. The eggs or sperm, depending on the sex, are discharged through the mouth. The fertilized eggs develop into planulae that have no cilia, but that can creep along by subtle

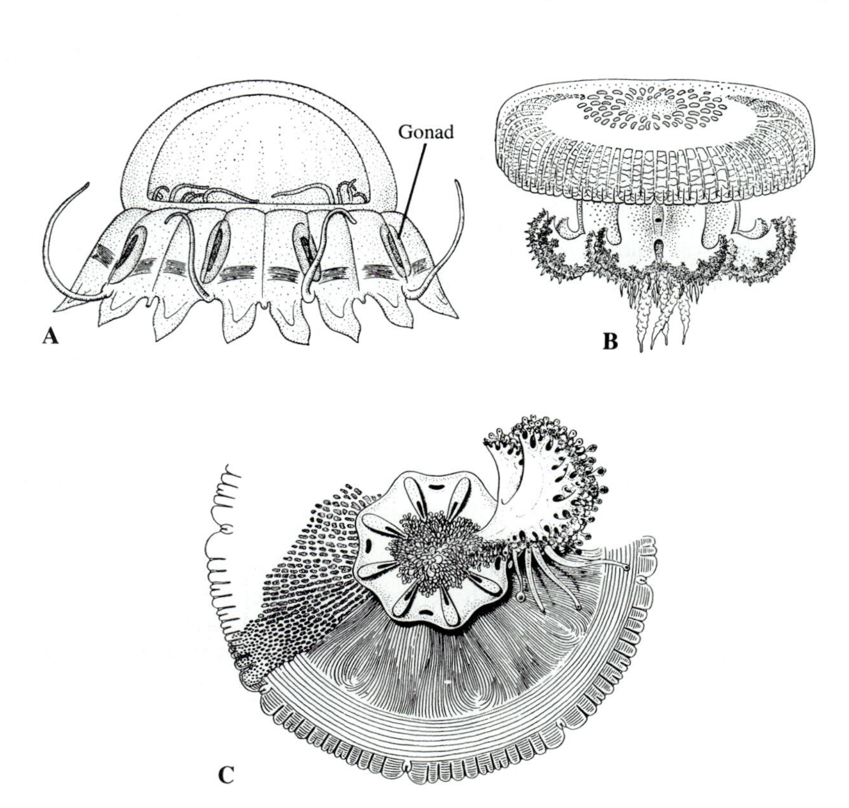

FIGURE 4.30 A. *Nausithoe* (order Coronatae). (Komai, Memoirs of the College of Science, Kyoto Imperial University, B, *10*.) B. *Cephea octostyla* (order Rhizostomeae). C. *Cotylorhiza tuberculata* (order Rhizostomeae). All but one of the four mouth lobes have been removed; some of the muscle layer has also been removed on the left side in order to show the gut canals. (B and C, Mayer, Carnegie Institution of Washington Publication no. 109.)

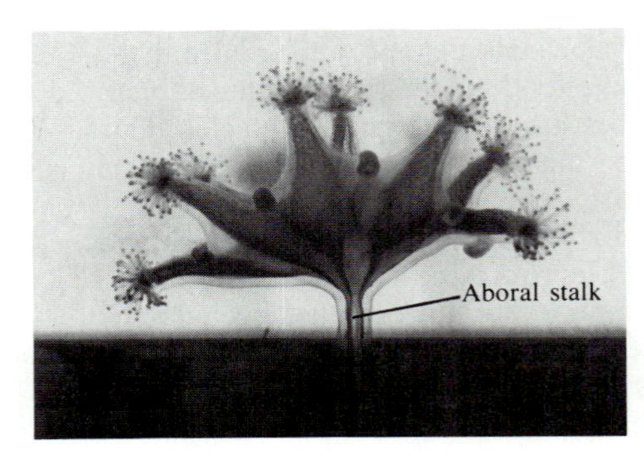

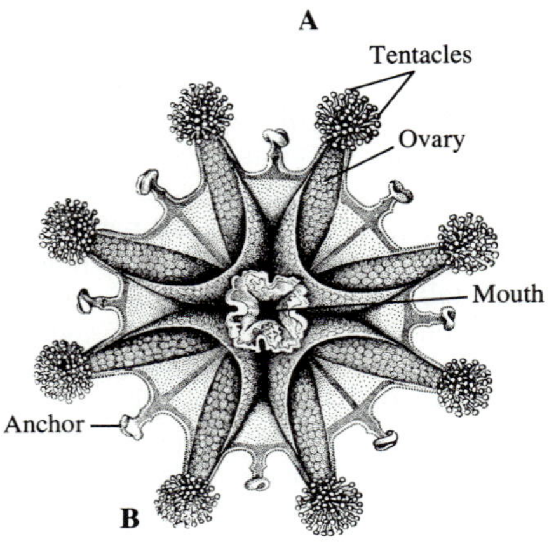

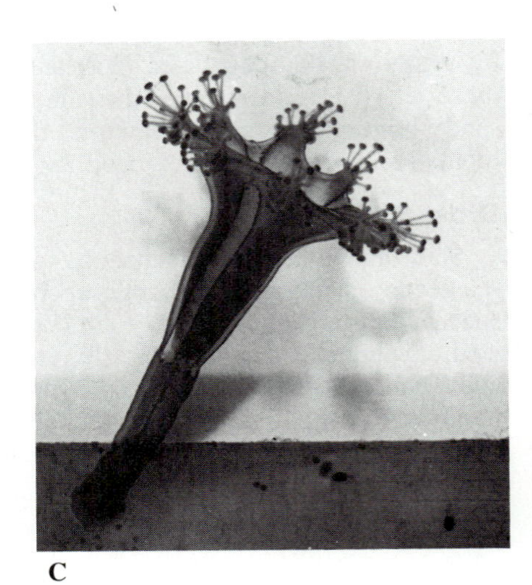

FIGURE 4.31 Order Stauromedusae. A. *Haliclystus stejnegeri*. B. *Haliclystus salpinx (Octomanus monstrosus)*. (Naumov, Keys to the Fauna of the USSR, *75*.) C. *Thaumatoscyphus hexaradiatus*.

changes in shape. The classical study of the life cycle of one species of *Haliclystus* showed that the planulae congregate, settle, develop a mouth, and start feeding. Then they produce short stolons that break off and crawl away as daughter planulae. The daughter planulae have been reported to become sexually mature individuals, but this needs to be confirmed.

References on Scyphozoa

Calder, D. R., 1973. Laboratory observations on the life history of *Rhopilema verrilli* (Scyphozoa: Rhizostomeae). Marine Biology, *21*:109–114.

Gladfelter, W. G., 1973. A comparative analysis of the locomotory system of medusoid Cnidaria. Helgoländer Wissenschaftliche Meeresuntersuchungen, *25*:228–272.

Ralph, P. M., 1960. *Tetraplatia,* a coronate scyphomedusan. Proceedings of the Royal Society of London, B, *152*:263–281.

Russell, F. S., 1970. The Medusae of the British Isles. Volume 2. Cambridge University Press, Cambridge.

Widersten, B., 1965. Genital organs and fertilization in some Scyphozoa. Zoologiska Bidrag från Uppsala, *37*:139–182.

CLASS CUBOZOA

The cubozoan medusae—they are usually called cubomedusae—have long been placed under the Scyphozoa, but they probably deserve a class of their own. The jellyfishes that belong to this small group are characterized by a tall bell that is squarish in outline when viewed from above or below (Figure 4.32, A). The tentacles are solid, and there are either four of them or four groups of them. All species have rather complex eyes, in which the photoreceptors are organized into a retinal layer beneath a lens and cornea. These jellyfishes are efficient in orienting themselves with respect to light coming from one direction.

Cubomedusae are primarily tropical and subtropical in distribution, and fishes are their usual prey. Some species have potent nematocyst toxins and are dangerous to humans. *Chironex fleckeri,* the box jellyfish or sea wasp, whose range extends from the Indian Ocean to the Coral Sea, has caused more than 60 fatalities, mostly in the Australian region. Certain Australian beaches, in fact, are closed to bathers when *Chironex* is prevalent. This jellyfish is an agile swimmer, and even though it is large, it is difficult to see, especially

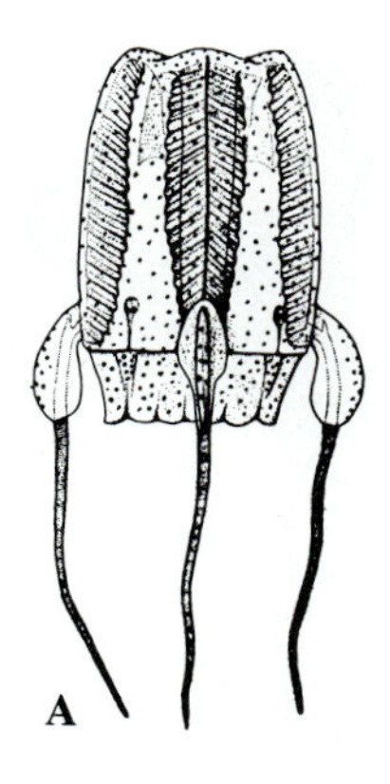

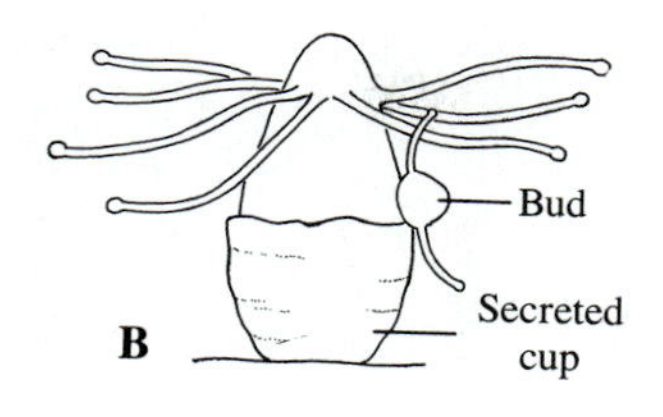

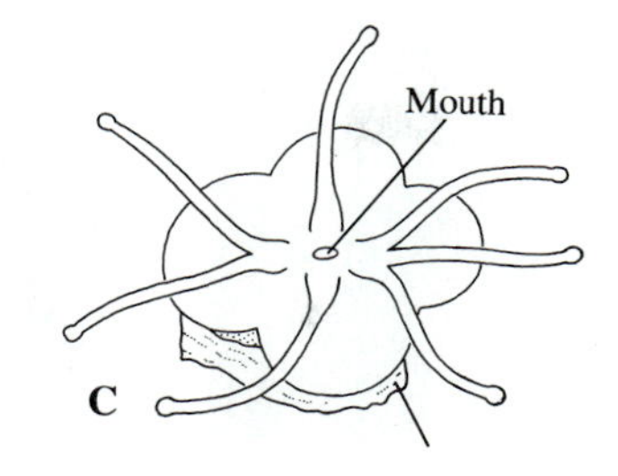

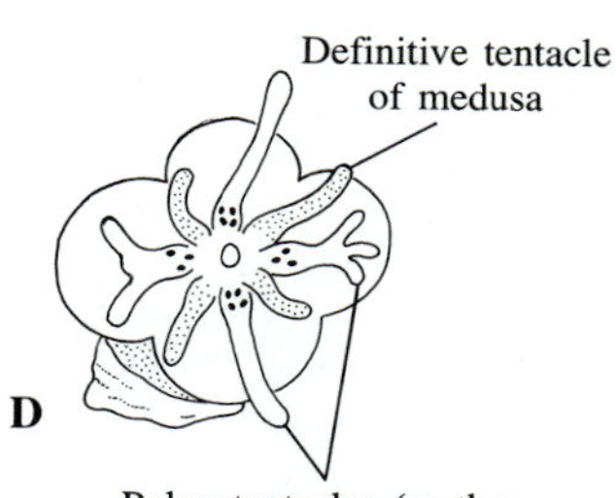

FIGURE 4.32 Class Cubozoa. A. *Carybdea*. B. *Tripedalia cystophora,* polyp, producing a bud that will become another polyp. C, D. *Tripedalia,* stages in the metamorphosis of a polyp into a medusa. As the eight tentacles of the polyp become shorter, four rhopalia develop. The first four definitive tentacles of the medusa develop between the tentacles of the polyp; four more are added later. (After Werner, Cutress, and Studebaker, Nature, *232*.)

against a light background. The polyps, attached to rocks in brackish creeks fringed by mangroves, produce medusae that are washed out to sea during periods of heavy rainfall in December. The medusae grow rapidly and persist until April or May. The bell diameter may reach 25 cm, and the tentacles, when fully extended, may be more than 18 m long. The tentacles are fragile as well as adhesive, so that when they contact human flesh they break off and must be peeled away. The various toxins delivered by the nematocysts cause intense pain, and may also cause shock, heart failure, cessation of breathing, rupture of blood cells, and other complications. Death sometimes comes within a few minutes. Direct contact with as much as 5 m of tentacle length is likely to be fatal to an adult. Survivors retain ugly scars. *Chironex* is now considered to be the most venomous animal known. It is nevertheless nibbled at by the hawksbill turtle and at least three kinds of fishes. They are probably either immune to the toxins or have some structural defenses against the action of the nematocysts.

The best known cubomedusan life cycle is that of *Tripedalia cystophora*. This species has a small, solitary polyp with several club-shaped tentacles (Figure 4.32, B); there are no vertical partitions such as are characteristic of the gut of scyphozoan polyps. The polyp is capable of encystment, and can also bud off small polyps that separate from the parent. A medusa is produced by metamorphosis of an entire polyp (Figure 4.32, C and D).

References on Cubozoa

Calder, D. R., and Peters, E. C., 1975. Nematocysts of *Chiropsalmus quadrumans* with comments on the systematic status of the Cubomedusae. Helgoländer Wissenschaftliche Meeresuntersuchungen, *27*:364–369.

Cropp, B., 1985. Australia's invisible stinger. Oceans, *18*:3–7.

Pearse, J. S., and Pearse, V. B., 1978. Vision in cubomedusan jellyfishes. Science, *199*:458.

Werner, B., 1975. Bau und Lebensgeschichte des Polypen von *Tripedalia cystophora* (Cubozoa, class nov., Carybdeidae) und seine Bedeutung für die Evolution der Cnidaria. Helgoländer Wissenschaftliche Meeresuntersuchungen, *27*:461–504.

CLASS ANTHOZOA

In the class Anthozoa, there is no medusa stage in the life history, and the gonads derive from the gastrodermis. Fertilized eggs develop into planulae, and these become new polyps, usually by way of a transitional

FIGURE 4.33 Class Anthozoa, subclass Zoantharia, order Actiniaria. (See also Color Plate 2.) A. *Haliplanella lineata.* B. *Anthopleura xanthogrammica,* contracted at low tide, showing bits of gravel stuck to the column. C. *Epiactis prolifera,* on eelgrass, with young attached to the base of the column. D. *Lebrunia danae,* a Caribbean species that has two sets of tentacles. Those shown here are pseudotentacles peripheral to the true tentacles. They are characterized by cushions of nematocysts (acrorhagi), and they also contain zooxanthellae. They are extended during the daytime. The true tentacles, which are unbranched, are extended only at night and are used for capturing small organisms. (Photograph by Thomas Suchanek.)

actinula stage. A distinctive feature of anthozoans is the pharynx. This derives from a stomodeal invagination of ectoderm that meets the endoderm from which the lining of the rest of the gut is formed. The endodermal portion of the gut is divided, moreover, by radially arranged vertical partitions called **septa.** In some anthozoans there are eight septa, all of which reach inward as far as the pharynx and unite with its wall. In others there are ten or more septa that unite with the pharynx, and generally several to many others that do not. Below the pharynx, however, all of the compartments are completely open to the central part of the gut. The tentacles of anthozoans are hollow, and their cavities are continuous with the gut compartments around the pharynx.

In most of the major groups of anthozoans, the pharynx has one or two **siphonoglyphs.** These are lengthwise grooves whose especially active cilia drive water into the gut, thus keeping the polyp inflated as long as the animal is in a relaxed state. The wall of the pharynx is otherwise collapsed, so water moving down the siphonoglyphs does not escape too freely from the mouth. A slight pressure is thus built up within the animal. The movements of an anthozoan depend largely on the way muscles interact with the hydrostatic skeleton formed by the water that fills the gut.

SUBCLASS ZOANTHARIA (HEXACORALLIA)

More than half of the living species of anthozoans belong to the Zoantharia. The members of this subclass have at least ten septa that join the pharynx, and usually

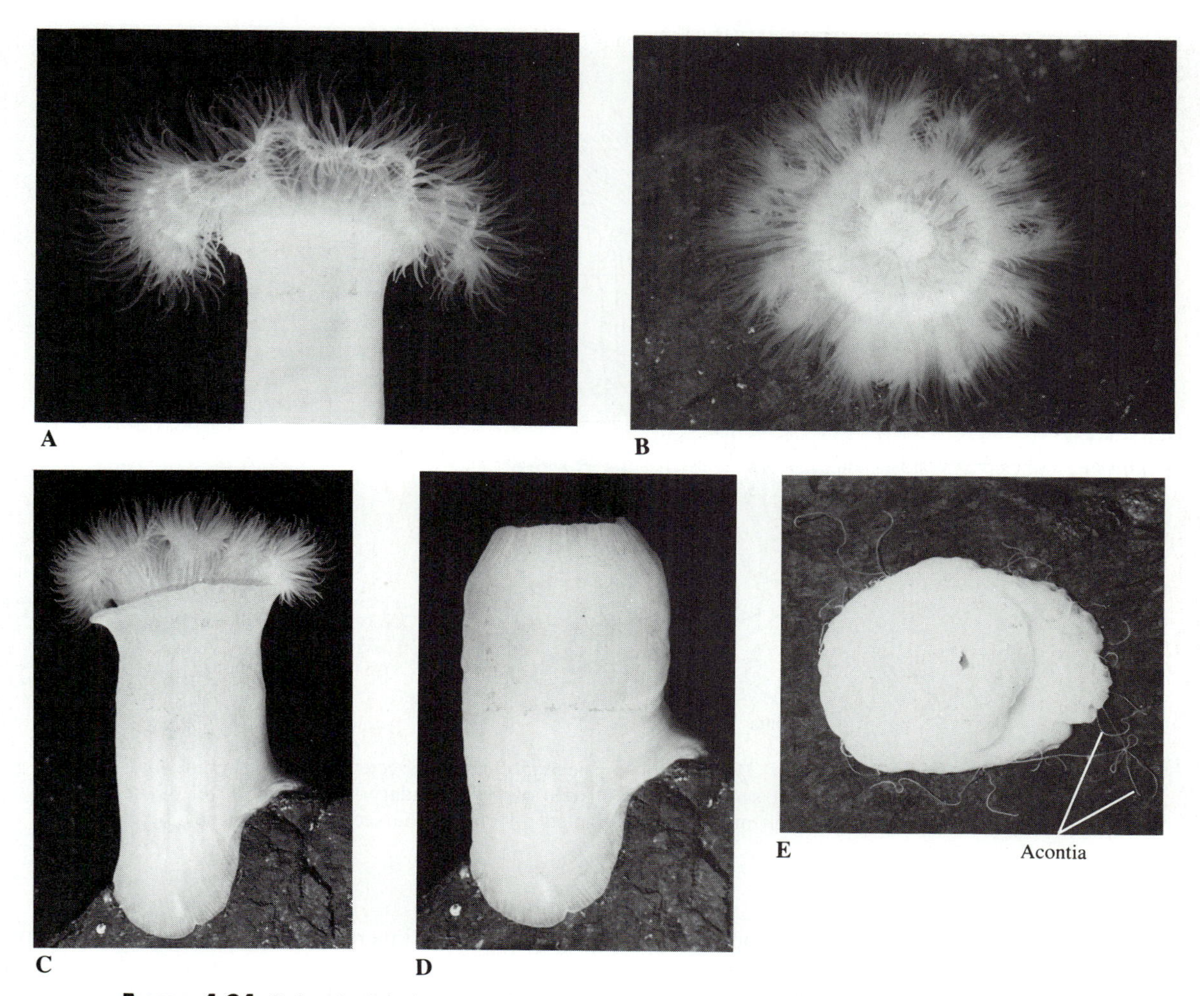

FIGURE 4.34 Order Actiniaria. *Metridium senile*. A. Expanded oral disk and upper portion of column. B. Oral disk. C. Entire animal, the sphincter muscle around the disk partly contracted. D. Tentacles withdrawn, sphincter muscle almost completely contracted. E. Specimen that has been poked to stimulate extrusion of acontia.

there are ten or more tentacles. As a rule, the tentacles are not branched. When branching does occur, it is not in the regular, featherlike pattern that is typical only of the subclass Alcyonaria.

Order Actiniaria

The actiniarians are the true sea anemones. In temperate regions, they are the most abundant anthozoans found at the shore. A typical anemone (Figures 4.33, 4.34; Color Plate 2) consists of three main portions: the **oral disk,** fringed by tentacles and with a mouth in the center; the **column,** which is capable of being lengthened and shortened; and the **pedal disk,** which secretes sticky material and by which the animal clings to the substratum.

Most species are attached to rocks, shells, wood, seaweeds, and other firm objects. They tend to remain in one place, but many of them can glide slowly by muscular activity of the pedal disk and thus relocate themselves. Some species burrow in mud or sand—displacing the substratum by alternately distending and narrowing the basal portion of the column—and often become attached to a buried pebble or shell.

When an anemone is in its normal feeding posture, the column is elongated, the oral disk is broad, and the tentacles are extended. If the animal is touched, it contracts quickly. The column becomes shorter and the oral disk and tentacles disappear beneath a circular fold that closes like the mouth of a bag pulled shut by a drawstring.

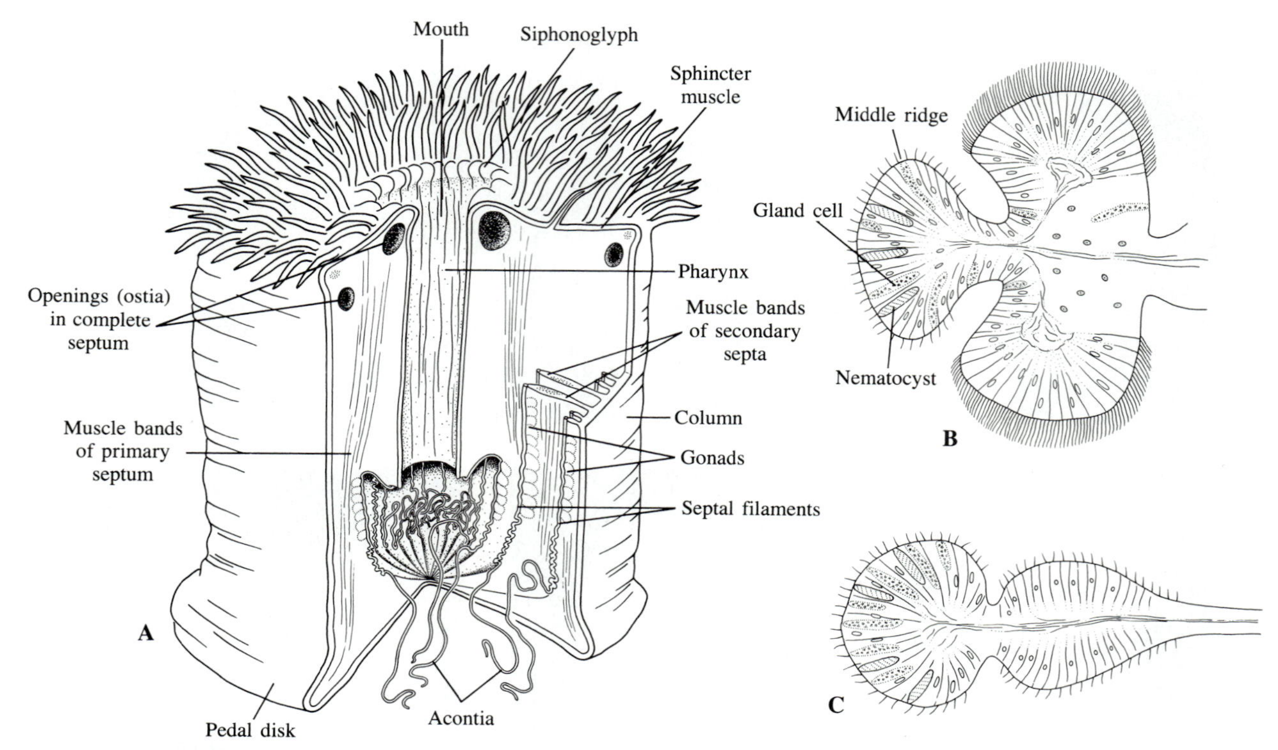

FIGURE 4.35 Order Actiniaria. *Metridium*. A. Specimen partly dissected to show internal anatomy. B. Transverse section of septal filament of one of the secondary septa at the level of the pharynx. C. Transverse section of the septal filament of one of the secondary septa at a level below the pharynx.

As a rule, the pharynx has two siphonoglyphs, located on opposite sides. The rest of the gut, below and around the pharynx, is subdivided by septa that run inward from the body wall (Figure 4.35, A). In most anemones, twelve or more septa reach all the way to the pharynx. These are called **primary,** or complete, **septa.** Usually, there are also secondary and tertiary septa, none of which extend to the pharynx. Below the pharynx, all compartments are completely open. Water and food can therefore circulate freely, and may enter the tentacles, whose hollow cores are essentially part of the gut. Circulation in the upper part of the column is generally facilitated by large openings **(ostia)** in the primary septa. Many species have a pore at the tip of each tentacle, and pores may also be present in the body wall of the column. Pores of both types help the animal squeeze out water when stimulated to contract.

In one group of anemones, and in certain representatives of others, there are only eight primary septa. This is considered an ancestral feature: even those anemones with twelve primary septa pass through a stage in which there are only eight, with four septa added later.

The gastrodermis is largely or completely ciliated. Moreover, the innermost edges of the septa (except for the upper portions of the complete septa, which are united with the pharynx) are usually differentiated into **septal filaments** (Figure 4.35). At higher levels, a septal filament is organized into three parallel ridges (Figure 4.35, B), the middle one richly supplied with nematocysts and gland cells, the other two densely ciliated. At lower levels, the septal filament retains only the middle ridge (Figure 4.35, C). The nematocysts of a septal filament help subdue prey, and the gland cells produce enzymes that initiate digestion. The digestive process conforms to the pattern typical of the phylum. Enzymes released into the gut accomplish partial breakdown of food, but digestion is completed intracellularly.

In certain anemones, including *Metridium* (Figure 4.34; Figure 4.35, A), the middle ridges of the septal filaments, as they approach the floor of the gut, become free threads called **acontia.** These have abundant nematocysts and can protrude through the mouth as well as through pores **(cinclides)** in the body wall. Their primary function is probably protective, but they may also coil around swallowed prey. *Metridium* is largely a fil-

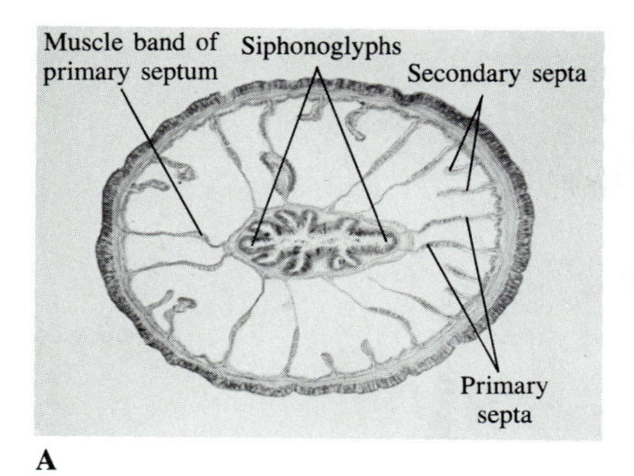

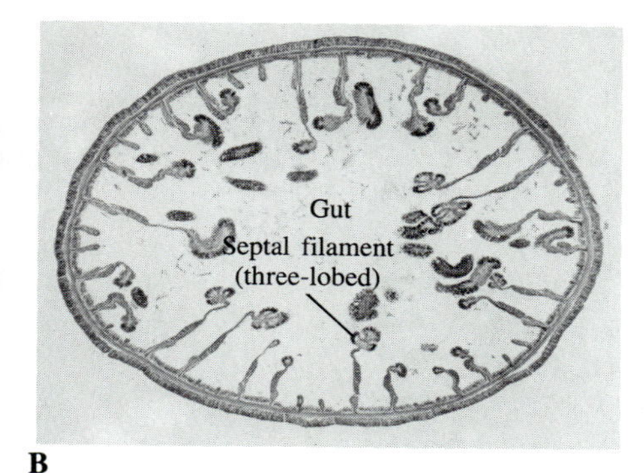

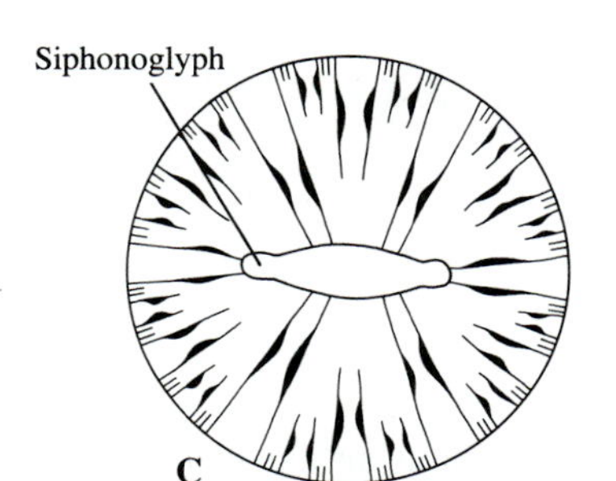

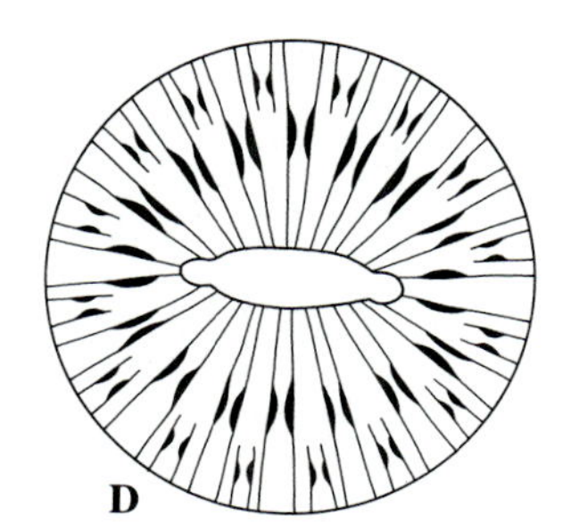

FIGURE 4.36 Order Actiniaria. A. *Epiactis,* transverse section through the column of a young specimen, showing arrangement of the septa at the level of the pharynx; photomicrograph. B. *Epiactis,* transverse section through the column below the level of the pharynx. C, D. Diagrams of typical arrangements of septa in anemones.

ter-feeder, however, using cilia on its oral disk and many small tentacles to collect small organisms and direct them to the mouth. It is not likely that the nematocysts on the acontia are needed to quiet the kind of prey that *Metridium* eats.

The septa bulge on one side, where the contractile bases of myoepithelial cells are concentrated to form longitudinal muscle bands (Figure 4.36). There are also similar muscle bands that run from the body wall toward the innermost edge of the septa, but these are not prominent. In the basal portions of the septa are well-developed, oblique, parietal muscles. When an anemone contracts, the longitudinal muscles shorten the column and the parietal muscles pull the column down to the pedal disk.

At the top of the column, just below the outermost tentacles, is a set of circular sphincter muscles. These are the muscles that draw shut the fold of tissue that closes over the oral disk and tentacles when an anemone contracts. In addition to the muscles already described, other muscles in the body wall and tentacles maintain the posture of the body and the extent to which the tentacles are extended. The muscles that run from the body wall toward the innermost margins of the septa are essentially antagonistic to the longitudinal muscles of the septa. If they contract when the longitudinal muscles relax, the column is lengthened.

The musculature of anemones, like that of other cnidarians, is epithelial in origin. Even though the enlarged contractile portions of the myoepithelial cells are beneath the level of other cells in the epidermis or gastrodermis, they are comparable to the bases of the myoepithelial cells of a hydra. The powerful muscle bands in the septa of an anemone belong to the gastrodermis, for there is no epidermal tissue in these partitions. Other muscles may be gastrodermal or epidermal, depending on their location.

The nervous system of anemones is organized in much the same way as in other cnidarians. Most of the nerve cells weave around the basal portions of the epidermal and gastrodermal cells. Receptors, even if not strikingly differentiated, are scattered all over the surface of the body, and anemones react promptly when touched. The remarkable responses of some anemones to predators and to other individuals of the same species show that coordination in these animals is efficient.

The mesoglea is moderately thick and extends into the septa. Unlike the mesoglea of medusae, it contains many collagen fibers, as well as considerable amounts of mucopolysaccharides. It often resembles fibrous connective tissue of more advanced invertebrates.

The column of some species has wartlike protuberances provided with adhesive glands and may also have nematocysts or special muscles. The protuberances may be used for holding sand grains, small pebbles, or bits of shell tightly to the column (Figure 4.33, B). In the case of certain intertidal species, this is an adaptation that provides protection against desiccation.

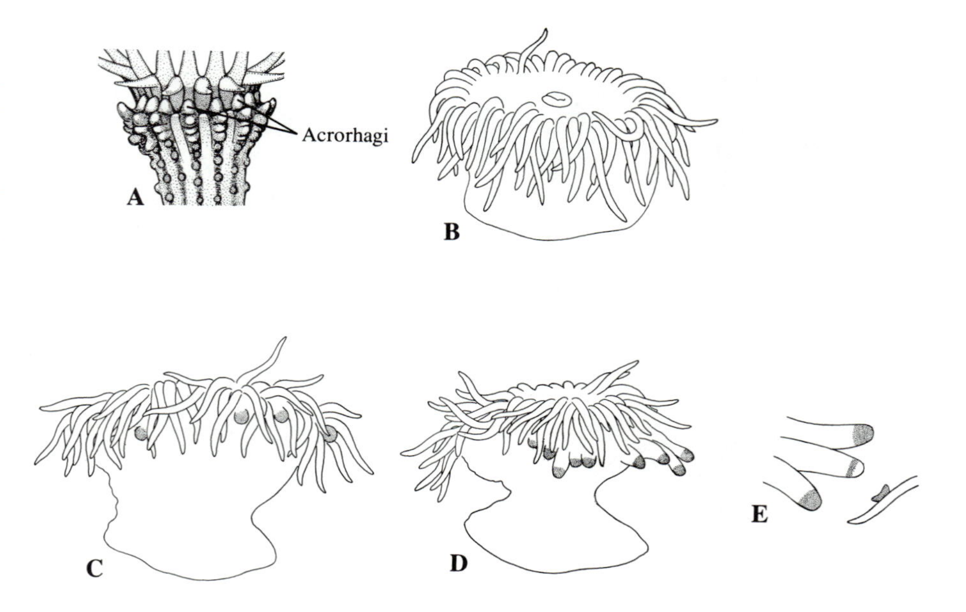

Figure 4.37 Order Actiniaria. A. *Anthopleura thallia,* upper portion, showing wartlike protuberances on the column and acrorhagi in the shallow groove between the column and tentacles. (Stephenson, The British Sea Anemones.) B–E. Antagonistic behavior of a specimen of *Anthopleura elegantissima* with respect to an individual of a different clone (nonclonemate). Before contact, the acrorhagi are inconspicuous (B), but soon after contact they become inflated (C) and directed toward the nonclonemate (D). As it touches the tentacle of the nonclonemate, an acrorhagus may leave some of its epidermis, with nematocysts, on the tentacle (E). (Francis, Biological Bulletin, *144.*)

At the top of the column, just outside the bases of the tentacles, some anemones have bulbous, hollow structures called **acrorhagi** (Figure 4.37, A). These are equipped with concentrations of nematocysts and are probably used mostly for defense against predators. Certain species, however, use their acrorhagi against members of their own kind. A striking case is that of *Anthopleura elegantissima,* a common intertidal anemone on the Pacific coast of North America. It reproduces asexually by dividing, and thus builds up extensive colonies. Each colony is a clone, for all of the individuals in it are derived from one founder. When the territory of a clone is encroached upon by anthopleuras of a genetically different strain, the members at the edge of the colony inflate their acrorhagi, direct them at the invaders, and fire nematocysts (Figure 4.37, B–E). Little harm is done to the anemones that engage in the competition for space, but their behavior does keep one clone from encroaching on another's domain.

Lebrunia, found in the West Indies, is interesting in that its acrorhagi are situated on structures that resemble tentacle clusters (Figure 4.33, D), but which are in fact entirely different elaborations of the margin. These **pseudotentacles** have zooxanthellae in their tissues and are exposed during the daytime when the symbionts are photosynthetically active. The true tentacles are extended only at night, and are used for capturing small animal organisms.

Although most anemones are attached by their basal disks, a few are modified for a pelagic life. In *Minyas,* the central portion of what would be the basal disk in most anemones is specialized as an air-filled float. The animal hangs with the oral end down.

Species of *Stomphia,* which are generally subtidal and attached to shells, can detach themselves and thrash about when threatened by a predator (Figure 4.38). If the tube feet of a sea star of the genus *Dermasterias* or *Hippasteria,* for example, touch the tentacles of a *Stomphia,* it typically responds by contracting its column and withdrawing the oral disk, then quickly extending again and producing a conical papilla on the underside of its basal disk. This forces the anemone off the shell to which it has been attached. The animal then propels itself by rapidly bending its column first in one direction, then the other. The movements of *Stomphia* and its mechanism for detaching itself are well coordinated. The muscular activity is successful in bringing about the movements because it operates against a fluid-filled gut, which serves as a hydrostatic skeleton.

Symbiotic Associations of Anemones

Some anemones live on the backs of crabs or on the snail shells occupied by hermit crabs. Among the more intensively studied associations of this type are those of *Calliactis parasitica* (Color Plate 2), found with the hermit crabs *Dardanus arrosor* and *Eupagurus bernhardus,* and *Adamsia palliata,* found with *Eupagurus prideauxi*. As a rule, there is just one *Calliactis* on the shell of *Dardanus*. When a hermit crab that has outgrown its house locates a more suitable one, it uses its second and third legs to massage the column of the anemone. Before long the anemone is relaxed and the crab

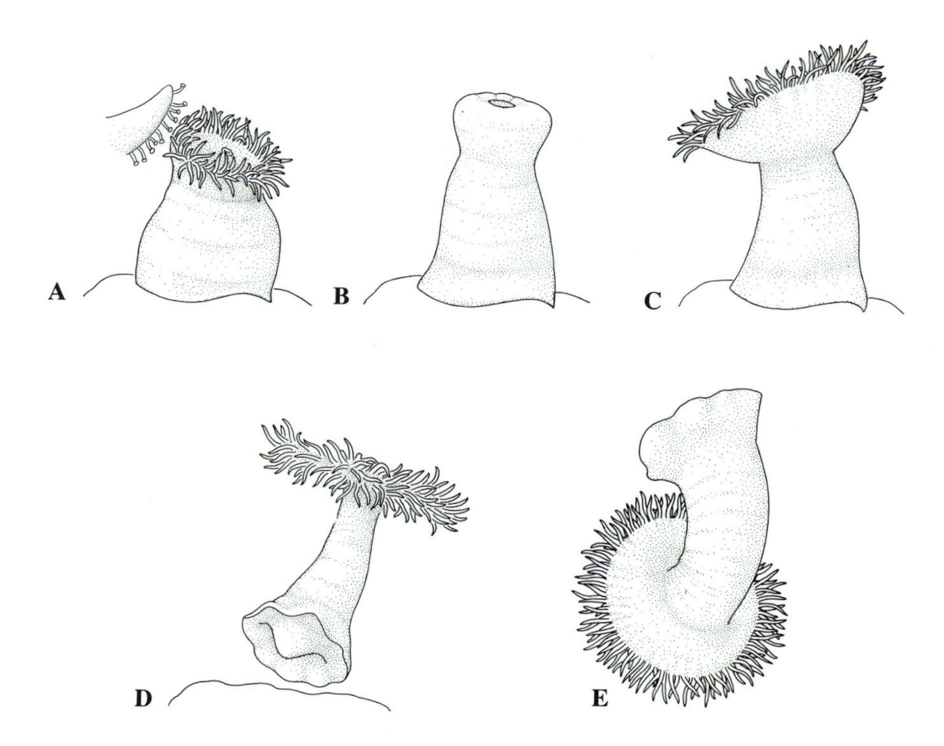

FIGURE 4.38 Order Actiniaria. Behavior of *Stomphia didemon* in response to contact with a predator, in this case a sea star, *Dermasterias*. A. Contact of the tube feet of the sea star with the tentacles of the anemone. B. Withdrawal of the tentacles. C. Expansion of the oral disk. D. Formation of a conical protuberance under the oral disk. This forces the anemone off the substratum. E. Thrashing movement of the detached anemone. (After photographs by Ross and Sutton, Science, *155*. The photographs record the behavior of *S. didemon* following its contact with a related anemone, *S. coccinea*. The response to *Dermasterias* and certain other prospective predators is, however, substantially the same.)

can pick it up by the edge of its pedal disk and move it to the new shell. As the tentacles touch the shell, some nematocysts are fired and tight contact is established. The anemone then pulls itself onto the shell and attaches itself firmly. On *Eupagurus bernhardus, Calliactis* generally occurs in groups.

Adamsia almost completely covers the snail shell inhabited by *Eupagurus prideauxi*. Its pedal disk applies a secretion to the edges of the aperture of the shell. As this material hardens, the shell is enlarged, so the hermit crab does not outgrow it and have to search for a new one.

An anemone living with a crab or hermit crab probably profits from its host's mobility, which brings it into situations where food is available. Although hermit crabs feed primarily on detritus, they also scavenge to some extent on animal tissue, as do many true crabs, and perhaps some of this type of food is used by the anemones. The crab or hermit crab is partly or almost wholly camouflaged, and it has been shown that the nematocysts of these symbiotic anemones repel octopuses that might otherwise attack the host. The associations between hermit crabs and anemones thus appear to be of mutual benefit.

Although sea anemones are carnivores, many of them have zooxanthellae living within the cells of the gastrodermis. Zooxanthellae are symbiotic dinoflagellates (see Figure 2.5, A) whose pigments generally confer an olive color on the tissue in which they reside. Some species retain their flagella during their intracellular existence, but the ones in anemones and corals do not. The zooxanthellae presumably enter the gastrodermal cells of young polyps, or perhaps during the planula stage, but they are not transmitted with eggs. In anemones that reproduce asexually, the problem of transmission is simplified. There are usually only one or two symbionts in each cell, and the total number in proportion to the size of the polyp remains more or less constant; excess zooxanthellae are extruded.

The zooxanthellae in anemones are photosynthetically active. Much of the carbon they fix is made available to the host in the form of glycerol, glucose, and other organic compounds. Relatively little ammonia is excreted by anemones that are populated by zooxanthellae; most of it is used by the zooxanthellae.

Some anemones have zoochlorellae—unicellular green algae—in their gastrodermal cells, and there may even be a mixture of zooxanthellae and zoochlorellae. This is the case in *Anthopleura elegantissima,* whose intraspecific aggressive behavior was discussed earlier. In a specimen that is green, microscopic examination reveals that the symbionts are predominantly or exclusively zoochlorellae; in olive-colored specimens, zooxanthellae predominate, or may be the only symbionts.

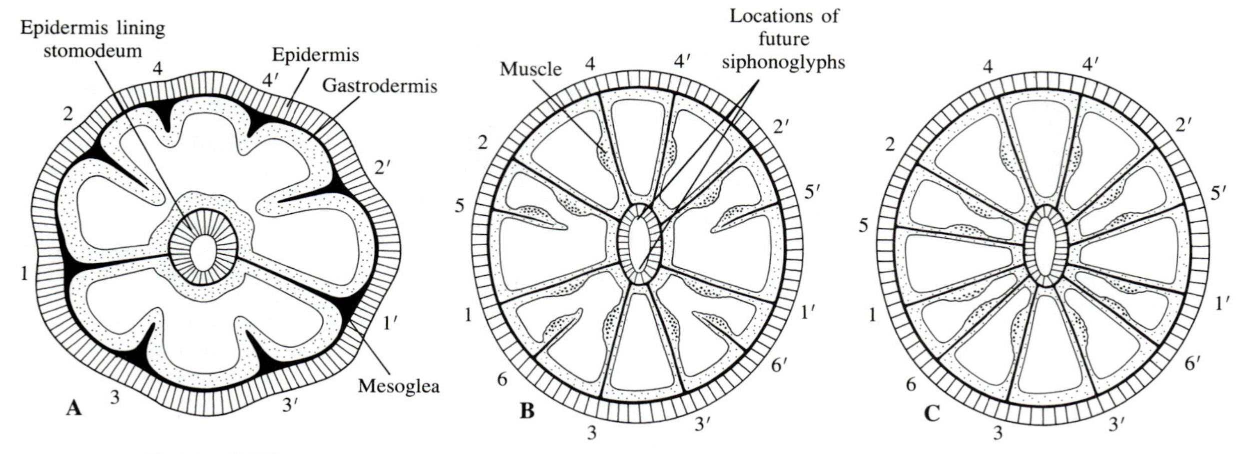

Figure 4.39 Order Actiniaria. A, B, C. Successive stages in the development of six pairs of primary septa in *Actinia;* diagrammatic. 1–1′, 2–2′, 3–3′, 4–4′, 5–5′, and 6–6′ are couples, in order of their formation; 3–3′, 4–4′, 1–6, 1′–6′, 2–5, and 2′–5′ are pairs.

Although zoochlorellae that live in hydras and some other invertebrates are known to make some of the organic compounds they synthesize available to their hosts, those in sea anemones have not been shown to do this.

Sexual Reproduction and Development of Anemones

The sexes of sea anemones are almost always separate. The simple gonads, derived from gastrodermal cells, lie in the mesoglea of certain septa. Sperm break out of the septa and escape from the gut by way of the mouth, through pores at the tips of the tentacles, or through pores in the body wall. Females of some species also liberate their eggs, whereas others retain them in the gut until fertilization has taken place and development has proceeded to the planula stage or even to a small anemone.

After a planula becomes attached by its anterior end, an ectodermal depression develops at the opposite end. This is the **stomodeum,** or future pharynx. It joins the gut cavity that appears among the endodermal cells that fill the inside of the planula.

The septa begin to develop after the stomodeum and endodermal gut have united. The first eight primary septa originate two at a time (Figure 4.39, A), one on each side of the plane that runs through the larger diameter of the pharynx and through both siphonoglyphs, or locations of future siphonoglyphs. (This is the plane that divides the animal into equal halves.) Each two septa that originate successively in this way form a **couple,** but of the six pairs of primary septa found in most anemones, only two—those closest to the siphonoglyphs—are derived strictly from couples (3–3′ and 4–4′ in Figure 4.39, B and C). The other four pairs (1–6, 1′–6′, 2–5, and 2′–5′) consist of a member of one of the first four couples and a member of a couple that develops later. The incomplete secondary and tertiary septa are formed subsequently.

The tentacles originate as outpocketings of the body wall at the edge of the future oral disk. The sequence of their formation is a little different from that of the septa. The first cycle, nearest the mouth, usually consists of six tentacles. Succeeding cycles are added in sixes or multiples of six. A common pattern is 6, 6, 12, 24, and 48.

Some anemones brood their young to the actinula stage. When they escape from the mother—through the mouth or through pores at the tips of the tentacles—they already have the rudiments of tentacles and are ready to settle down. *Epiactis prolifera,* a common species on the Pacific coast of North America and also on the shores of Japan, broods its young externally at the base of the column. The planulae escape through the mouth, settle on the column, and there develop into small anemones that look as if they had been produced by budding (Figure 4.33, C). They eventually move off and begin a life of their own.

In the genus *Peachia,* the young are transported by hydrozoan medusae. When a planula of *Peachia* is eaten by an appropriate medusa, it moves into one of the radial canals, where it develops into an actinula. The actinula eats its way out of the canal or escapes through the mouth, then anchors itself to the underside of the bell and starts to consume the gonads, stomach, radial canals, and tentacles. In the laboratory, a growing *Peachia* has been observed to transfer itself from one medusa to another. After its tentacles and septa are fairly well developed, the anemone falls off its host and settles on a sandy bottom. Mature animals live in burrows, with their oral disks and tentacles just above the surface of the sand.

Asexual Reproduction of Anemones

Some anemones reproduce asexually by a process of laceration, in which pieces of tissue separate from the pedal disk and differentiate into complete individuals. Slow division of an anemone into two more or less equal animals is fairly common, too. This latter kind of reproduction is characteristic of *Anthopleura elegantissima*.

Order Corallimorpharia

The name Corallimorpharia may suggest that the animals in this group have a calcareous skeleton, but this is not the case. They look much like anemones, although the musculature in the column, septa, and tentacles is poorly developed. The tentacles, which cannot be completely retracted, generally have knoblike tips and are often arranged in indistinct rows radiating outward on the oral disk. There is no "drawstring" sphincter muscle comparable to that just below the edge of the oral disk of an anemone.

Siphonoglyphs are barely recognizable. There are many irregularly distributed septa, some of which are incomplete. The gonads are on the innermost edges of the septa—the portions comparable to the septal filaments of anemones—and there are no ciliated tracts here.

Some corallimorpharians are solitary; others, including *Corynactis* (Color Plate 2) form colonies by dividing. The polyps may retain some tissue connections.

Order Ptychodactiaria

Ptychodactiarians constitute a small group restricted to Arctic regions. Like corallimorpharians, they lack ciliated tracts along the innermost edges of the septa, and their gonads are close to these edges. The gonads are unusual, however, in that they protrude from the septa as stalked structures, instead of lying in the mesoglea beneath the gastrodermis from which they originate.

Order Zoanthidea

Zoanthids, found in temperate and colder regions as well as in tropical seas, resemble certain sea anemones. They do not, however, have a distinct pedal disk. In solitary species, the area of attachment to the substratum is generally no broader than the column. In colonial types, the bases of the polyps remain connected to form a communal holdfast (Figure 4.40, A).

Most zoanthids incorporate sand grains, diatoms, bits of shell, and other debris into the wall of the column. The material first accumulates on the outside, but much of it is eventually overgrown by the epidermis and ends up in the mesoglea. The mesoglea is unusual in another respect: it is traversed by a system of canals that are independent of the gut. The cells that line the canals are thought to be of epidermal origin.

A

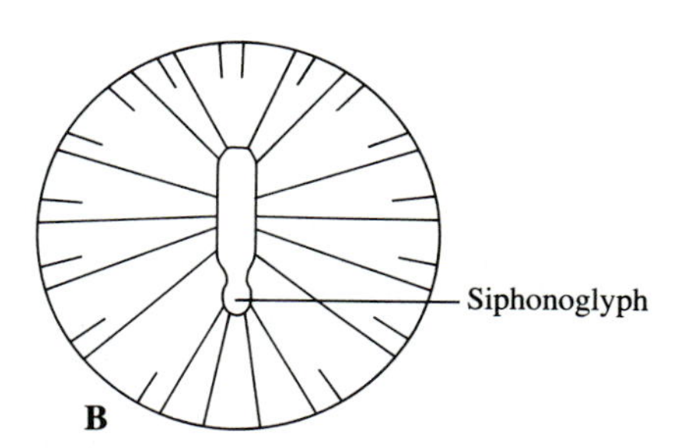

Figure 4.40 Order Zoanthidea. A. *Epizoanthus scotinus*. B. Transverse section through the column of a zoanthid, at the level of the pharynx, showing the arrangement of septa; diagrammatic.

The numerous tentacles form a neat ring at the margin of the oral disk. Both nematocysts and mucus are involved in the capture of small organisms, which are moved from the tentacles to the mouth by ciliary activity. The pharynx has a single siphonoglyph. The many septa of mature polyps lack pronounced muscle bands. They are symmetrically arranged, those on one side of the pharynx matching those on the other side (Figure 4.40, B). In most zoanthids, primary septa, which extend all the way to the pharynx, alternate regularly with secondary septa that do not reach the pharynx. As a rule, gametes are produced only on the primary septa.

Order Scleractinia (Madreporaria)

Scleractinians—the cup corals and stony corals—have a calcareous skeleton and lack siphonoglyphs. In other respects they are similar to sea anemones. Some members of the group are solitary, but most are colonial. Many of the nearly 1000 species are important components of tropical coral reefs.

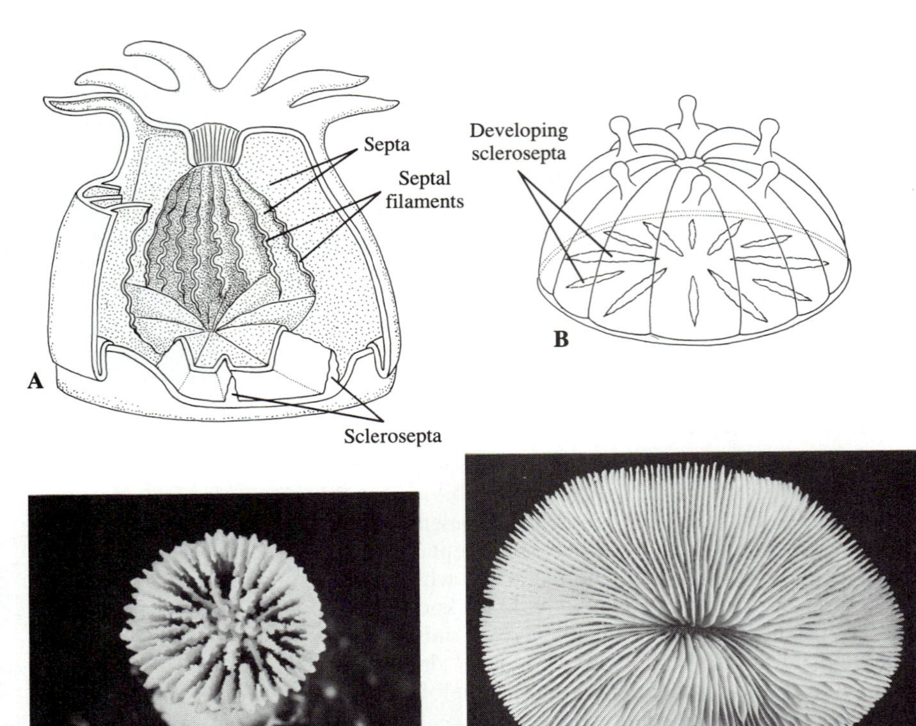

FIGURE 4.41 Order Scleractinia. A. Organization of the polyp of a solitary coral; diagrammatic. B. Young polyp of *Siderastrea radians,* showing disposition of developing sclerosepta. (After Duerden, Carnegie Institution of Washington Publication no. 20.) C. *Balanophyllia elegans,* skeleton. (A living specimen is shown in Color Plate 2.) D. *Fungia,* skeleton.

The skeleton is secreted by the epidermis and is essentially external (Figure 4.41, A). This is true even of the calcareous partitions within the cup that a polyp occupies. A young polyp (Figure 4.41, B) lays down a basal plate with a cycle of six radial ridges, then adds six more ridges. As the fold of epidermis that covers each ridge produces more calcareous material, the ridge becomes higher and joined to the wall of the cup secreted by the epidermis of the column. The calcareous partitions formed in this way are called **sclerosepta.** If there are extensions of the sclerosepta outside the wall, these are called **costae** (Figure 4.42, A and B). Between the sclerosepta, soft septa comparable to those of anemones appear. They grow inward into the gut of the polyp; those that unite with the pharynx form the primary septa.

The wall typically becomes wider above, to keep pace with the growing polyp. The lower portion, where the polyp started out, is gradually left behind by the living tissue. A second calcareous wall, the **epitheca,** may be deposited around the original wall, and be connected to it by costae. Another skeletal component that may be formed is a pillarlike upgrowth from the base of the wall, called the **columella.** The lower portions of some or all of the sclerosepta may be joined to it.

The skeleton provides protection as well as support for the polyp. When the polyp contracts, the tentacles and other soft tissues are pulled down into the cup.

Good examples of solitary scleractinian corals are *Balanophyllia* (Figure 4.41, C; Color Plate 2), found in temperate and colder regions of the Northern Hemisphere, and *Fungia* (Figure 4.41, D), from tropical waters. The latter may be large—up to 15 cm in diameter—and the lobed, often irregular appearance of the skeleton may suggest that it is colonial. It is, nevertheless, one large polyp with hundreds of sclerosepta. The polyps of *Fungia* are unusual in that they are not attached to a hard substratum; they simply lie on sand in shallow water. Each one is formed, however, at the top of an attached upright growth, and the unattached phase does not begin until the polyp is broken off by a disturbance.

In colonial scleractinians, which predominate in coral reefs, new polyps are produced by budding. This may be initiated on the column or on the sheet of epidermal and gastrodermal tissue that connects the pol-

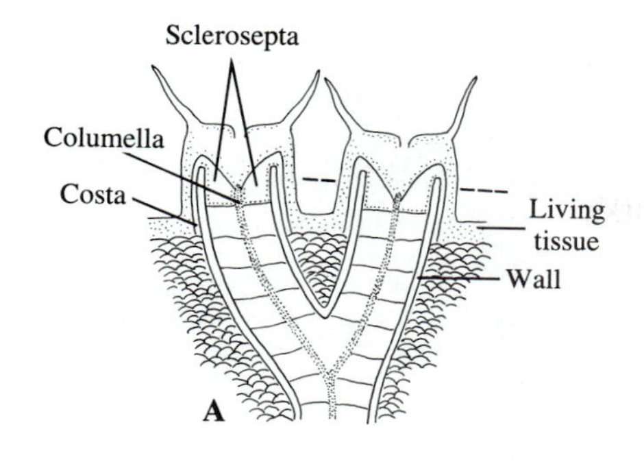

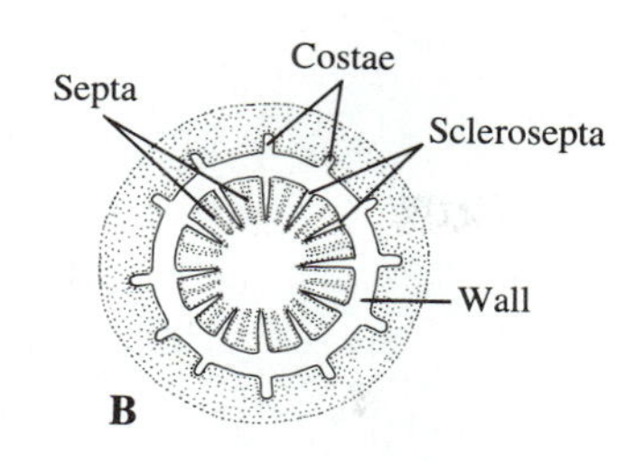

FIGURE 4.42 Order Scleractinia. A. *Galaxea,* portion of a colony that has been sectioned to show the relationship of functional polyps to skeletal material no longer occupied by living tissue. In this genus, the sclerosepta extend outside the wall of the cup, forming vertical ridges called costae. B. Section through a polyp at the level indicated in A by a dashed line. C. *Eusimilia,* skeleton. D. *Montastrea cavernosa,* skeleton.

yps. It may also begin on the oral disk of a polyp, in which case the oral disk becomes elongated and a second mouth appears. Some of the original tentacles remain with the old mouth, others become organized around the new one. Additional tentacles are then added to both sets. The arrangement of septa and sclerosepta tends to be very irregular in species that initiate division on the oral disk.

The end result of repeated budding is a colony with hundreds or thousands of members whose skeletons and soft tissues are united (Figures 4.42 and 4.43). As a colony enlarges, the deposit of skeletal material becomes thicker. The living tissue, however, is restricted to the surface portion (Figure 4.42, A and B). At intervals, the polyps and the tissue that connects them secrete calcareous material that separates the active part of the colony from portions of the skeleton that are no longer occupied. As the living part of the colony moves outward and new polyps appear, the diameter of the colony increases. In a piece of an old colony, the progressively abandoned portions of the skeletal cups form tubelike configurations that radiate to some extent because the colony has been continually enlarged by budding. Here and there one can see where a polyp has produced a bud.

The brain corals, such as *Meandrina* (Figure 4.44, A–C) and *Agaricia* (Figure 4.44, D) exhibit an odd type of oral-disk budding that leads to formation of long, winding series of polyps. The individual polyps do not have their own sets of tentacles. They share the same set, which is lengthened and augmented so that it encloses new polyps as these differentiate. The formation of a new polyp is announced by the appearance of a mouth. If two mouths appear at about the same place, the series splits.

Coral Reefs

Scleractinians are the principal constituents of most coral reefs. Corals of the hydrozoan order Milleporina are also likely to be present, however, and in certain areas they produce massive growths. Calcareous algae contribute to reef structure, too.

All reef-building corals have zooxanthellae (symbiotic dinoflagellates) in their gastrodermal cells. Much of the organic matter produced photosynthetically by the symbionts diffuses into the tissues of the hosts. This explains why luxurious growths of corals are often found even in tropical seas in which zooplankton is sparse. Most corals consume zooplankton and some depend considerably on it, but other corals rely almost completely on organic nutrients supplied by their zooxanthellae.

As a rule, corals with zooxanthellae are contracted during the day. Feeding on zooplankton is limited to

FIGURE 4.43 Order Scleractinia. A. Coral reef, consisting mostly of colonies of *Acropora,* at low tide. (Photograph by Leonard Muscatine.) B. *Acropora,* skeleton of a small colony. C. *Acropora,* portion of a colony, tentacles extended. (Photograph by Thomas Suchanek.)

hours of darkness, when the tentacles are extended (Figure 4.43, C; Figure 4.44, B).

The uptake of carbon dioxide by zooxanthellae probably favors the deposition of calcareous skeletal material. Where Ca^{++} and HCO_3^- ions are plentiful, they combine to form $Ca(HCO_3)_2$. This is unstable, and if carbon dioxide resulting from its deposition is utilized by zooxanthellae, then calcium carbonate and water should be the other products:

$$Ca(HCO_3)_2 \longrightarrow CO_2 + H_2O + CaCO_3$$

The calcium carbonate is believed to be adsorbed onto mucopolysaccharides secreted by epidermal cells of corals. Much remains to be learned about calcification in these cnidarians, and the generalizations just put forth may be refuted or modified in coming years.

The association formed by various corals, coralline algae, and other algae is a productive one in terms of the output of organic matter. An active reef may produce 20 grams of organic matter per square meter of surface per day. This is about equal to the maximum productivity of a very efficient terrestrial crop, such as sugar cane, growing under optimum conditions. A wheat field, or an assemblage of living organisms in a temperate marine environment, generates about one-tenth of the dry weight of organic material produced each day by a coral reef of equal area. The productivity of marine plankton in open water in the tropics is no better than that of organisms in a desert—about 0.2 gram per square meter per day.

The distribution of coral reefs is not random. They are mostly concentrated in warmer waters of the Indo-Pacific basin and western Atlantic Ocean, between the Tropics of Cancer and Capricorn. A temperature of 20°C is about the minimum at which corals can grow

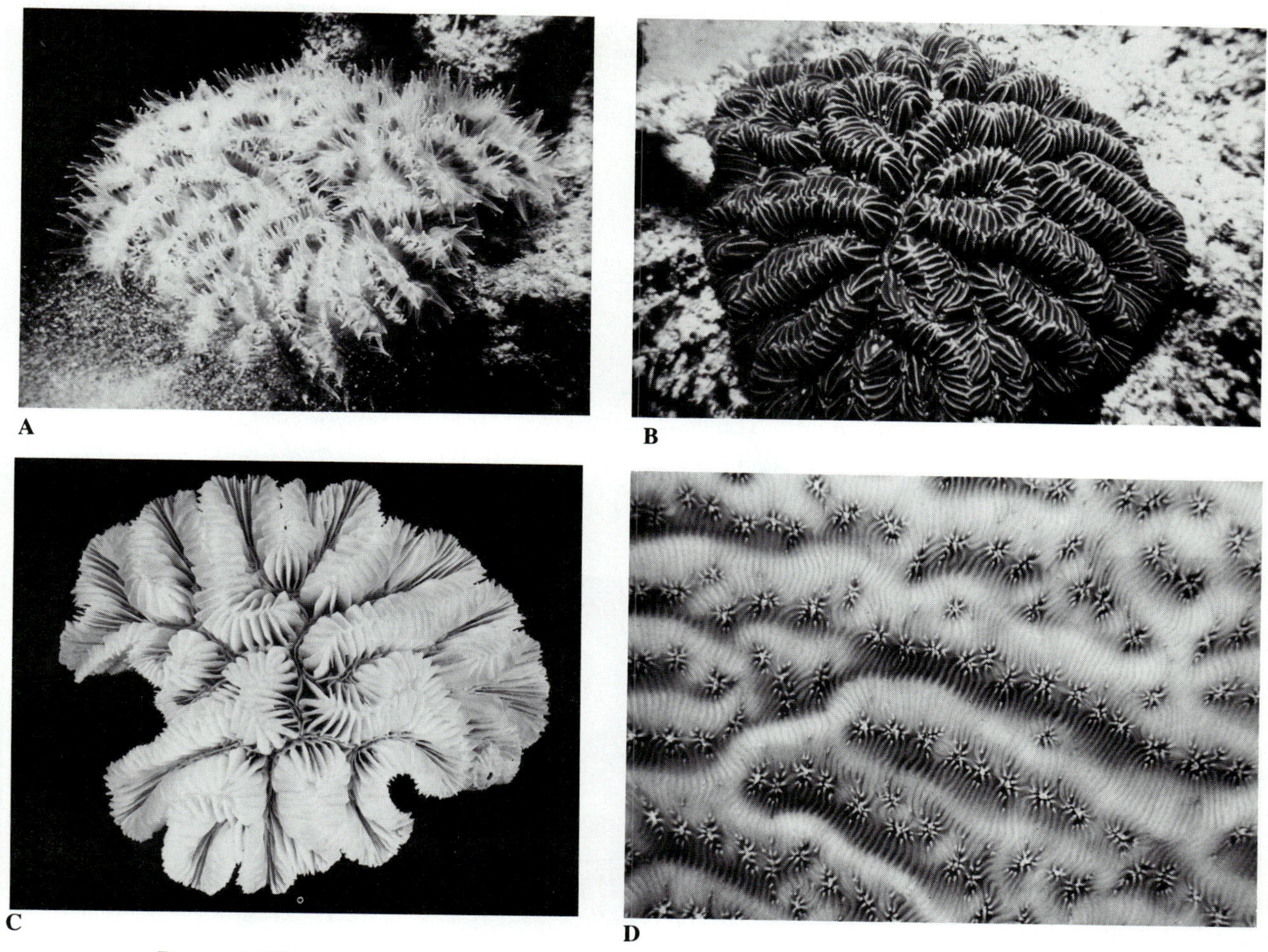

FIGURE 4.44 Order Scleractinia. A, B. *Meandrina meandrites,* a brain coral, at night (tentacles extended) and in the daytime (tentacles withdrawn). (Photographs by Thomas Suchanek.) C. *Meandrina,* skeleton. D. *Agaricia,* a brain coral, showing numerous mouths; the tentacles have been withdrawn. (Photograph by Thomas Suchanek.)

luxuriantly. Where conditions are favorable, corals occupy much of the hard substratum available to them, and almost all reefs consist of at least several species. Next to corals, the most important components of reefs are coralline red algae, which deposit calcium carbonate in their tissues. Both corals and coralline algae require good illumination, so they are limited to relatively shallow water. The majority of them are found at depths of less than 30 m, but a few may extend to depths of approximately 70 m.

There are three main types of coral reefs: fringing reefs, barrier reefs, and atolls. **Fringing reefs,** which include all the reefs found in the Hawaiian region, extend as platforms away from the edges of volcanic islands. As new growth of coral and coralline algae covers older growth, the reef expands in a seaward direction. The Great Barrier Reef of Australia and similar **barrier reefs** around other islands in the Indo-Pacific region follow the shoreline but are separated from it by a lagoon. Thus a barrier is built up between the island and the open sea. At intervals, the reef is interrupted by channels. An **atoll** is a reef that encircles a lagoon.

More than a century ago, Charles Darwin hypothesized that barrier reefs and atolls began as fringing reefs. It appeared to him that as an island sank little by little, active coral growth at the outer edge of what had been a fringing reef would form a barrier reef (Figure 4.45, A). Because conditions in the lagoon between the reef and the island are less favorable for coral growth,

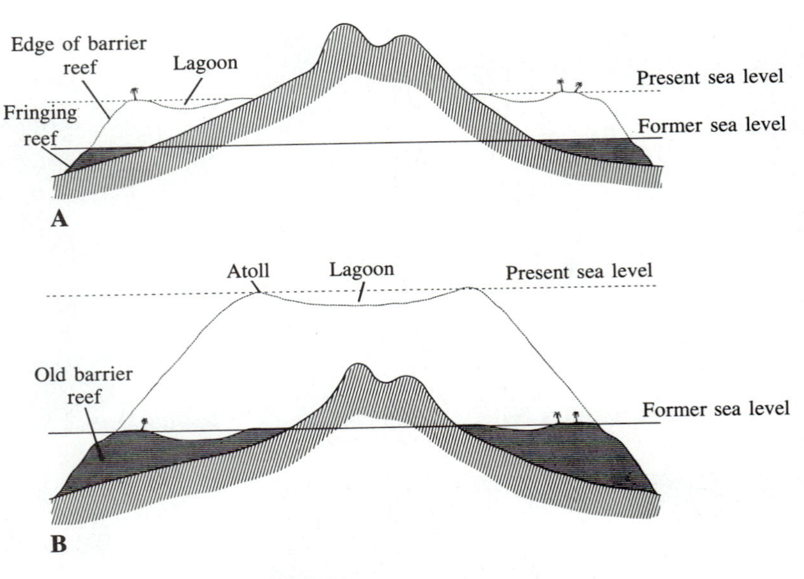

Figure 4.45 A. Darwin's explanation of the origin of a barrier reef from a fringing reef. As the island sinks, upward growth of the fringing reef leads to the formation of a barrier reef, whose outer portions are separated from the island by a lagoon. Small islands that support terrestrial vegetation, including coconut palms, are formed of coral rock and sand cast up by the sea. B. Darwin's explanation of the origin of an atoll from a barrier reef. As the island sinks completely below the sea, upward growth of the barrier reef leads to the formation of an atoll. C. Gardner Atoll, in the southwestern Pacific Ocean, photographed from an altitude of 2000 m. (Mertens, Natur und Volk, *88*.)

it would remain a lagoon. An atoll, Darwin explained, resulted when an island sank completely below the surface, so that only newer portions of the barrier reef that originally encircled the island remained at the surface (Figure 4.45, B and C).

Darwin's explanations of coral-reef formation have stood the test of time. Evidence confirming them has been obtained by drilling into atolls. On Enewetak, preparatory to the atomic explosions created there between 1948 and 1958, the United States Navy had to drill through more than 1000 m of coral deposits before striking basalt, the volcanic bedrock. The deepest fossils were about 50 million years old, but they were types characteristic of shallow-water reefs, substantiating Darwin's theory of gradual subsidence.

SUBCLASS CERIANTIPATHARIA

The two orders within this subclass—Ceriantharia and Antipatharia—are very different. They do, however, share certain characteristics. Their septa (numerous in Ceriantharia, generally ten in Antipatharia) are all joined to the pharynx and usually lack prominent muscle bands.

Order Ceriantharia

In external appearance, ceriantharians (Figure 4.46, A and B) resemble sea anemones but have no basal disk. New septa are formed in couples on the side of the column opposite the siphonoglyph, and differential growth of the column keeps the many septa more or less evenly spaced and symmetrically arranged. There

FIGURE 4.46 Subclass Ceriantipatharia. A. *Pachycerianthus fimbriatus* (order Ceriantharia). (Photograph by Ronald Shimek.) B. *Pachycerianthus* (order Ceriantharia), showing the circle of tentacles around the mouth as well as those at the margin of the oral disk. C. *Antipathes* (order Antipatharia). (Photograph by Thomas Suchanek.) D. *Antipathes,* a piece of a colony and a greatly enlarged portion of one branch, showing an expanded polyp, a contracted polyp, and the axial skeleton, which has many protuberances.

are two sets of tentacles: one set borders the mouth, the other is at the edge of the oral disk.

Most ceriantharians are found in temperate regions. They live in mud, forming vertical burrows lined by a slippery mucus into which exploded nematocysts have been incorporated. The nematocyst threads are so numerous that they form a fabric. Some species are large—50 cm or even taller when fully extended. In the normal feeding posture, the oral end is raised above the surface of the mud.

Order Antipatharia

Colonies of antipatharians (Figure 4.46, C and D) are supported by a blackish axial skeleton that usually has spinelike projections. The skeletal material is a hardened protein called antipatharin. It resembles gorgonin, found in gorgonians (discussed below), but has little or no collagen associated with the protein.

Because of the form and color of the skeleton, these animals are called black corals and thorny corals. The polyps stick out of the softer tissue that covers the skeleton. Typically, they have six tentacles and six primary septa, plus four or six secondary septa that are complete. The tentacles are so irregularly branched that antipatharians—even the few that have eight tentacles—are not likely to be confused with alcyonarians, whose tentacles are branched in a featherlike pattern.

Antipatharians are mostly limited to tropical and subtropical regions, and they live in comparatively deep water. The skeletons of elaborately branched types are sometimes used for decoration or as charms to repel sickness. (The genus name *Antipathes,* on which the name of the order is based, means "against disease.") Pieces of the skeleton of species that have a single main stem can be bent, with the aid of heat, into bracelets and other kinds of ornaments.

SUBCLASS ALCYONARIA (OCTOCORALLIA)

Alcyonarians include many kinds of reef-building corals, as well as soft corals, sea pens, and gorgonians. The feeding polyps of these anthozoans have eight tentacles, pinnately branched in symmetrical fashion. There are eight septa, all of them joined to the wall of the pharynx. The bulges on the septa, marking the location of the principal muscle bands, are on the sides that face the single siphonoglyph of the pharynx. This is decidedly different from the situation in Zoantharia and Ceriantipatharia, in which some muscle bands face the septa and others do not. Below the level of the pharynx, the edges of the septa are free, so water and food can circulate from the central part of the gut into the radial compartments.

Almost all alcyonarians are colonial. Some have two kinds of polyps. The eight-tentacled gastrozooids, already described above, are concerned with feeding and usually also with production of gametes. **Siphonozooids,** without tentacles (or with rudimentary ones), have an entirely different function, that of allowing water to freely enter and leave the colony. Continual movement of water through a system of gastrodermal tubes keeps the colony inflated, brings oxygen to deeper tissues, and carries away wastes. In a few cases, siphonozooids produce gametes.

Most alcyonarians deposit calcareous spicules in the mesoglea. The cells that do this are of epidermal origin, having wandered into the mesoglea to carry out their function.

The classification of alcyonarians is complicated and is not dealt with in detail here. Emphasis is given instead to the groups that have many species or that show especially interesting features.

Order Alcyonacea

The Alcyonacea consists of a large and extremely varied assemblage of cnidarians, often distributed in four separate orders: Stolonifera, Telestacea, Alcyonacea, and Gorgonacea. The polyps are similar in all of them, but there is great diversity in skeletal structure and colony organization. The most simple and most complex types are linked by a continuous series of intermediate types, and this argues against breaking up the group into separate orders. It also makes it difficult to establish distinct boundaries for suborders. The treatment here is a practical one: the use of English names for a few of the better-known major categories. These can easily be reconciled with order names in other classifications.

Stoloniferans and Telestaceans

Simpler stoloniferans, such as *Clavularia* (Figure 4.47, A), form colonies in which the member polyps, all gastrozooids, are connected by branching horizontal runners. The epidermis of the stolons secretes a hardened covering comparable to the perisarc of many hydroids, and the mesoglea generally contains calcareous spicules. In slightly more complex types, the colony is a mat, with the polyps interconnected by a system of branching gastrodermal tubes that run through the mesoglea.

Telestaceans are limited to the tropical portions of the Atlantic Ocean. Their colonies consist of creeping stolons and slender, upright stems (Figure 4.47, B). There is a thin, perisarclike covering on the outside, and calcareous spicules and other hard material may be deposited around the gastrodermal tubes, but the stems are usually rather flexible. Each stem arises as a single polyp, and as it lengthens additional polyps sprout along its sides. Each new polyp hooks up with the sev-

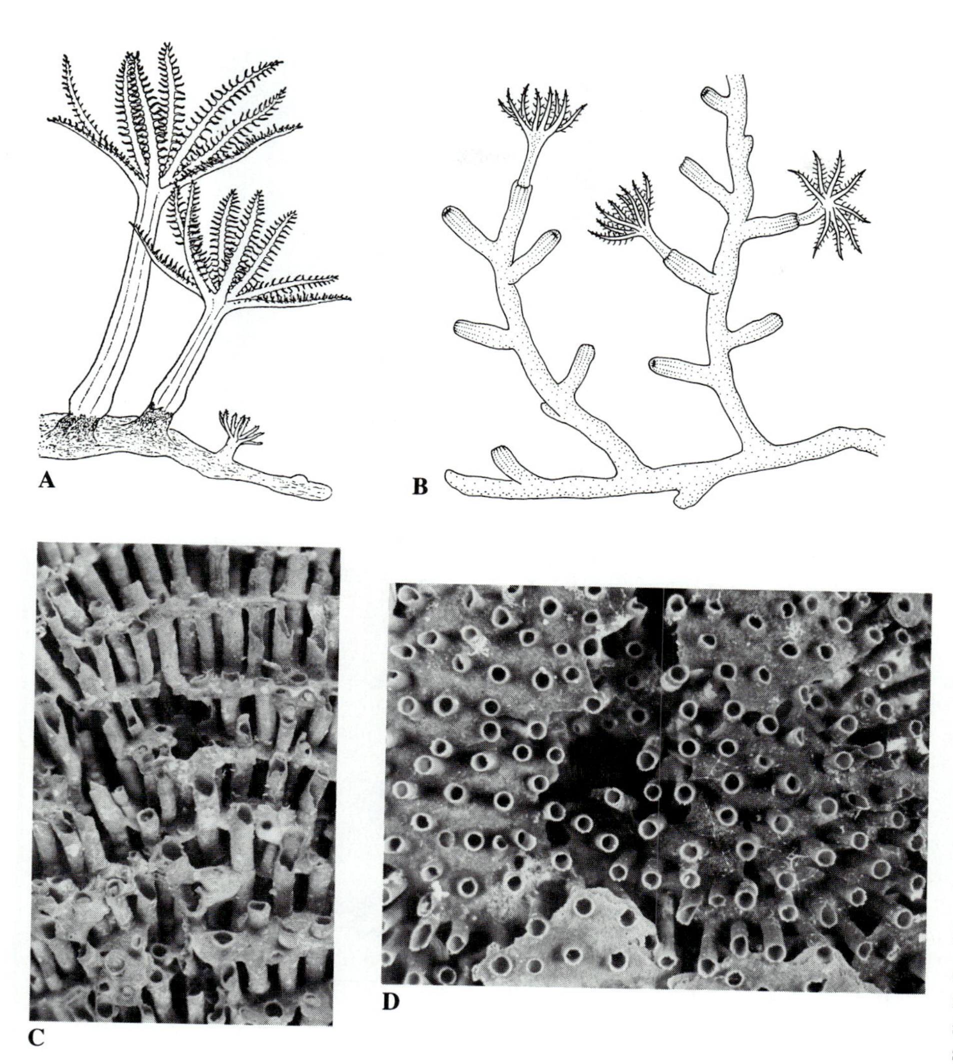

FIGURE 4.47 Subclass Alcyonaria, order Alcyonacea. A. *Clavularia,* a stoloniferan. The numerous polyps originate from stolons that spread over the substratum. (Gohar, Publications of the Marine Biological Station, Ghardaqa, Egypt, *6.*) B. *Telesto riisei,* a telestacean. C, D. Skeletons of *Tubipora,* the organ-pipe coral of the Indo-Pacific region. The tubes and platforms are produced by union of skeletal elements in the mesoglea. Living tissue is restricted to the more superficial portions of the colony.

eral gastrodermal tubes running through the stem at that level. In some species the stems branch extensively.

Organ-Pipe Coral

In the organ-pipe coral, *Tubipora* (Figure 4.47, C and D), a still higher degree of complexity is attained. Throughout most of the colony, the mesoglea is stiffened by fused spicules of calcium carbonate. After the polyps of a young colony have grown out from the base for some distance, they become joined by platformlike connections through which gastrodermal tubes extend. As the colony enlarges, additional platforms develop progressively farther away from the original basal encrustation, and older parts of the colony die, even though the mesogleal skeleton persists. *Tubipora* is an important contributor to certain tropical coral reefs.

Soft Corals

Soft corals, such as *Alcyonium* (Figure 4.48, A and B), have calcareous spicules in the mesoglea, but these are not fused. The colonies of these animals are generally lumps, irregularly lobed masses, or stalked structures with a number of blunt branches. The polyps may all be gastrozooids, but in some genera there are both gastrozooids and siphonozooids. A network of gastrodermal tubes pervades the mesoglea, connecting the polyps. If there is an appreciable stalk, this portion of the colony is not likely to have any polyps, although it will have gastrodermal tubes.

Because of the absence of a solid skeleton, soft corals are capable of contracting into masses much smaller than those of expanded colonies. Some normally expand and contract twice in each 24-hour period. These

FIGURE 4.48 Order Alcyonacea. A. *Alcyonium,* a soft coral of northern waters. B. *Alcyonium,* transverse section through a portion of a colony. The polyp at the left has been cut at the level of the pharynx; the polyp at the right has been cut below the level of the pharynx. The lining of the pharynx, derived from the ectoderm of a stomodeal invagination, is continuous with the epidermis that covers the colony as a whole.

animals are found in colder waters as well as in tropical and temperate regions; they are among the few alcyonarians other than pennatulaceans and certain stoloniferans to be found outside tropical seas.

Gorgonians

Gorgonians—the sea whips and sea fans (Figure 4.49, A and B; Color Plate 2)—form extensively branched colonies. Sometimes all branches are in one plane, in which case they may also anastomose irregularly to form a network. Such fanlike gorgonians are oriented perpendicular to prevailing currents; this increases efficiency of food-capture. The stiffness of the colonies is due to an axial skeleton. In some species this consists almost entirely of a protein–collagen mixture called **gorgonin;** in others there are calcareous spicules as well as gorgonin. Gorgonin or spicules, or both, may also be in the softer tissue external to the axial skeleton. The mesoglea of this outer layer is traversed by gastrodermal tubes that are continuous with the gut cavities of the member polyps, which are generally all gastrozooids. Several gastrodermal tubes may run to the gut of a single polyp. In some species, in fact, new branches of a colony have eight tubes, one going to each of the eight compartments of the gut of the first polyp.

According to some accounts, the axial skeleton is deposited within a tube of epidermis. The tube can be traced back to when the founder polyp, before it formed tentacles, was something like a thimble, with a cone of epidermis projecting up into it. As growth proceeded, the inside of the tube became filled with skeletal material, and when the main stem branched, the tube within it did likewise.

A

B

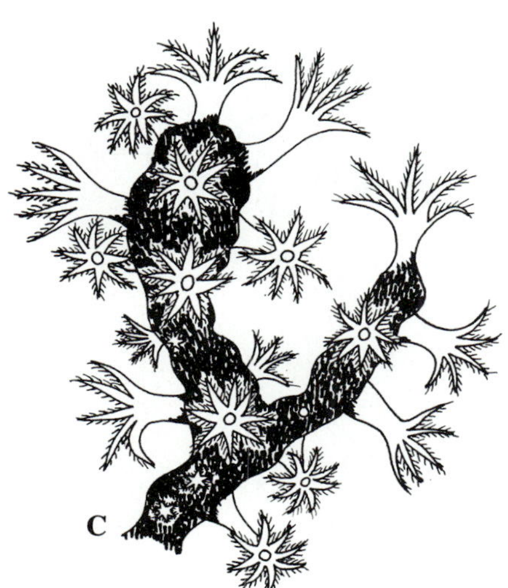

C

FIGURE 4.49 Order Alcyonacea. Gorgonians. A. *Eunicea,* branches of a colony. (Photograph by Thomas Suchanek.) (See also *Paragorgia pacifica,* Color Plate 2.) B. *Gorgonia ventalina,* whose anastomosing branches are arranged in one plane. (Bayer, Studies on the Fauna of Curaçao and Other Caribbean Islands, *12*.) C. *Corallium rubrum,* the precious coral.

Corallium rubrum (Figure 4.49, C), the precious coral much used in the manufacture of jewelry, especially in the Mediterranean region, is unlike other gorgonians in that it lacks gorgonin. The orange-red axial skeleton is hard because the calcareous spicules that make up most of it become cemented together into a solid mass. Colonies of *Corallium* branch freely and sometimes reach a height of 60 cm, and the larger stems may be 2 or 3 cm in diameter. The mesoglea external to the skeleton contains spicules, but these remain separate. The gastrodermal tubes in this portion of the colony are branches of a number of closely spaced, parallel tubes that lie next to the axial skeleton. Both gastrozooids and siphonozooids are present. The latter are simply pores with muscular walls, but they move water into and out of the colony.

Order Pennatulacea

Pennatulaceans form colonies in which both gastrozooids and siphonozooids are regularly present. In some species, all siphonozooids may be exhalant as well as inhalant; in others, some take in water while others allow water to leave the colony. Pennatulaceans have a stalk buried in mud or sand and an expanded portion on which the polyps are borne. Both are stiffened by an axial rod that consists partly of hard organic material

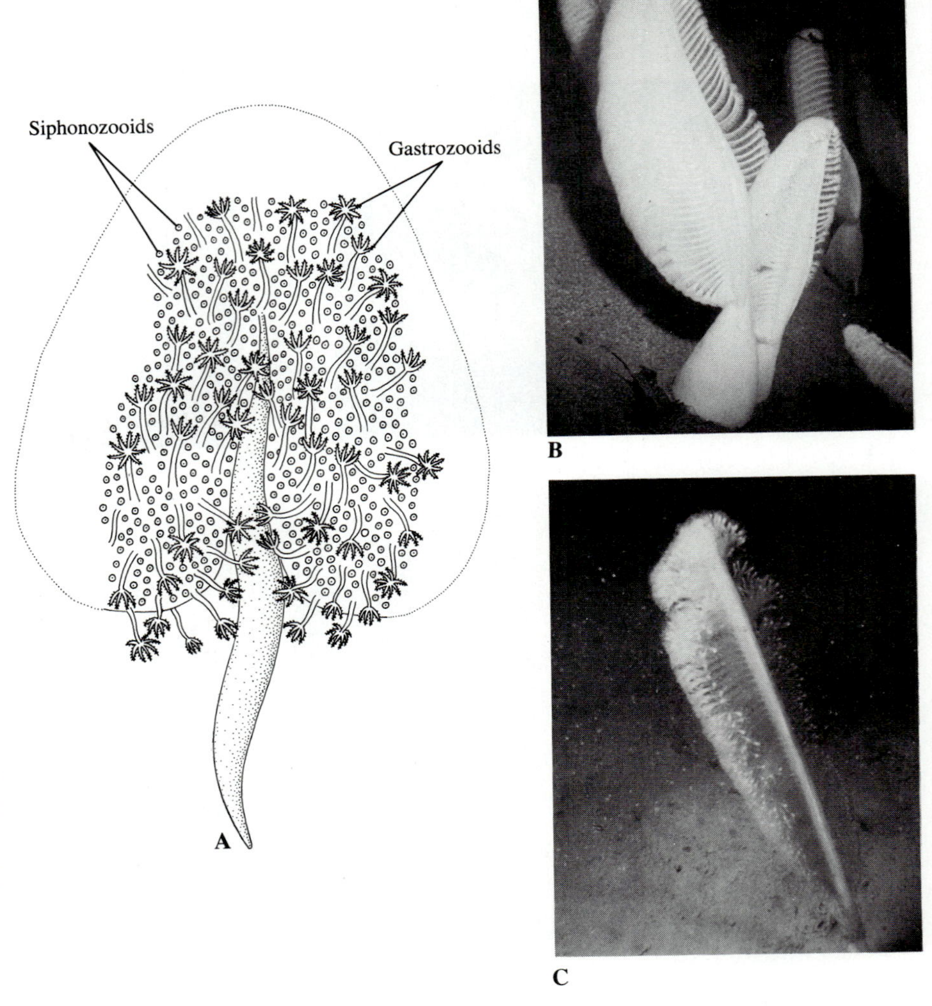

FIGURE 4.50 Order Pennatulacea. A. *Renilla,* the sea pansy. The gastrozooids and siphonozooids are limited to the upper surface of the colony; those in the peripheral portions have been omitted. B. *Ptilosarcus gurneyi,* a sea pen. (Photograph by Charles Birkeland.) C. *Virgularia,* a sea pen. (Photograph by Ronald Shimek.) (See also *Veretillum cynomorium,* Color Plate 1.)

and partly of calcium carbonate. In sea pansies (Figure 4.50, A), the expanded portion is flattened out into what looks like a succulent leaf; this rests on the muddy bottom, and the gastrozooids and siphonozooids are restricted to the free upper surface. In some close allies of sea pansies, the expanded portion resembles a cucumber and has gastrozooids and siphonozooids over most of it.

In sea pens (Figure 4.50, B and C; Color Plate 1) the colony as a whole, even if fleshy, resembles a feather. This is because the gastrozooids are arranged in a row along both sides of the stiffened rachis, or because there are several to many series of gastrozooids on both sides. The series on one side typically alternate with those on the other. In sea pens, gastrozooids of each series are fused for most of their length, resulting in numerous "leaves" that partly overlap one another. Whether or not there are such leaflike outgrowths from the central rachis, the siphonozooids of the colony are limited to the rachis.

The founder polyp of a pennatulacean colony survives as the stalk and as the axial part of the exposed portion. Its gut divides into two parallel canals, and the stiffening material is laid down in the mesoglea between these. As the colony enlarges, new polyps arise by asexual proliferation, and the mesoglea common to all members increases in volume. Branches from the gut of the founder polyp form the gastrodermal canals that run to all of the gastrozooids and siphonozooids. All gastrozooids in a particular colony are either male or female.

Order Helioporacea (Coenothecalia)

The Indo-Pacific blue coral, *Heliopora coerulea,* is one of the few species of the Helioporacea. The skeleton consists of fibers of calcium carbonate that have become fused into dense plates and it owes its blue color to certain salts of iron. The living tissue of *Heliopora* is restricted to a layer a few millimeters deep, covered above and below by epidermis. There are both gastrozooids and siphonozooids at the surface, and the polyps are connected by tubes of gastrodermal tissue that run through the common mesoglea. New skeletal material is deposited between closely spaced, fingerlike projections. The more prominent of these are the deeper portions of the polyps themselves; the rest, more slender, are downgrowths from the sheet of tissue between the polyps. The skeleton is strictly external to the basal epidermis. Growth of the colony is necessarily accompanied by reorganization of the living portion. Tissue that is left behind eventually dies, but as long as it survives it may continue to lay down some skeletal material.

References on Anthozoa and Coral Reefs

Adey, W. H., 1978. Coral reef morphogenesis: a multidimensional model. Science, *202:*831–837.

Bayer, F. M., 1961. The shallow-water Octocorallia of the West Indian Region. Studies on the Fauna of Curaçao and Other Caribbean Islands, no. 55.

Bayer, F. M., 1981. Key to the genera of Octocorallia exclusive of Pennatulacea (Coelenterata: Anthozoa), with diagnoses of new taxa. Proceedings of the Biological Society of Washington, *94:*902–947.

Bayer, F. M., Grasshoff, M., and Verseveldt, J., 1983. Illustrated Trilingual Glossary of Morphological and Anatomical Terms Applied to Octocorallia. E. J. Brill/Dr. W. Backhuys, Leiden.

Carlgren, D., 1949. A survey of the Ptychodactiaria, Corallimorpharia, and Actiniaria. Kunglica Svenska Vetenskapsakademiens Handlingar, series 4, *1:*1–121.

Ditlev, H., 1980. A Field-Guide to the Reef-Building Corals of the Indo-Pacific. Dr. W. Backhuys, Rotterdam; Scandinavian Science Press, Klampenborg.

Francis, L., 1973. Intraspecific aggression and its effect on the distribution of *Anthopleura elegantissima* and some related sea anemones. Biological Bulletin, *144:*73–92.

Goldberg, W. M., 1976. Comparative study of the chemistry and structure of gorgonian and antipatharian coral skeletons. Marine Biology, *35:*253–267.

Gomez, E. D., Birkeland, C. E., Buddemeier, R. W., Johannes, R. E., March, J. A., Jr., and Tsuda, R. T. (editors), 1981. Proceedings of the Fourth International Coral Reef Symposium. 2 volumes. Marine Science Center, University of the Philippines, Quezon City.

Goreau, T. F., Goreau, N. I., and Goreau, T. J., 1979. Corals and coral reefs. Scientific American, *241* (2):124–136.

Goreau, T. F., Goreau, N. I., and Yonge, C. M., 1971. Reef corals: autotrophs or heterotrophs? Biological Bulletin, *141:*240–260.

Jones, O. A., and Endean, R., 1973–1977. Biology and Geology of Coral Reefs. 4 volumes. Academic Press, New York and London.

Kaplan, E. H., 1982. A Field Guide to Coral Reefs of the Caribbean and Florida. Houghton Mifflin Co., Boston.

Kastendiek, J., 1976. Behavior of the sea pansy, *Renilla kollikeri* Pfeffer (Coelenterata: Pennatulacea) and its influence on the distribution and biological interactions of the species. Biological Bulletin, *151:*518–537.

Kuhlmann, D., 1985. Living Coral Reefs of the World. Arco Publishing, New York.

Lang, J., 1973. Interspecific aggression by scleractinian corals: 2. Why the race is not only to the swift. Bulletin of Marine Science, *23:*260–279.

Lewis, J. B., 1977. Processes of organic production in coral reefs. Biological Reviews, *52:*305–347.

Manuwal, R. L., 1981. British Anthozoa. Synopses of the British Fauna, no. 18. Academic Press, London and New York.

Newell, N. D., 1972. The evolution of reefs. Scientific American, *226* (6):54–65.

Porter, J. W., 1976. Autotrophy, heterotrophy, and resource partitioning in Caribbean reef-building corals. American Naturalist, *110:*731–742.

Robin, B., Pétron, C., and Rives, C., 1980. Living Corals: New Caledonia, Tahiti, Reunion, Caribbean. Les Editions du Pacifique, Papeete, Tahiti.

Ross, D. M., 1967. Behavioral and ecological relationships between sea anemones and other invertebrates. Oceanography and Marine Biology, An Annual Review, *5:*291–316.

Smith, F. G. W., 1971. Atlantic Reef Corals. University of Miami Press, Coral Gables, Florida.

Smith, S. V., and Kinsey, D. W., 1976. Calcium carbonate production, coral reef growth, and sea level change. Science, *194:*937–939.

Stephenson, T. A., 1928–35. The British Sea Anemones. 2 volumes. Ray Society, London.

Stoddart, D. R., 1969. Ecology and morphology of recent coral reefs. Biological Reviews, *44:*433–498.

Taylor, L. E., 1981. A coral reef gives up some secrets—as scientists build a reef of their own. Smithsonian, *11*(10): 34–43.

Trench, R. K., 1971. The physiology and biochemistry of zooxanthellae symbiotic with marine coelenterates. I. Assimilation of photosynthetic products of zooxanthellae by two marine coelenterates. Proceedings of the Royal Society of London, B, *177:*225–235.

Veron, J. E. N., 1987. Corals of Australia and the Indo-Pacific. Agnus & Robertson, North Ryde, South Wales, Australia, and London.

Veron, J. E. N., Pichon, M., Wallace, C. C., and Wijsman-Best, M., 1976–1984. Scleractinia of Eastern Australia, parts 1–5. Australian Institute of Marine Science, Monograph Series, nos. 1, 3, 4–6.

Yonge, C. M., 1963. The biology of coral reefs. Advances in Marine Biology, *1*:209–260.

Summary

1. The Cnidaria, with about 10,000 living species, is one of the larger phyla of invertebrates. Most of its members are marine, but some are found in fresh water.

2. The phylum is characterized by radial or biradial symmetry, and by a gut that does not have an anus. The mouth serves not only for intake of food, but also for elimination of undigested residues. The gut may, however, have branch canals, or it may be subdivided by vertical partitions into chambers; these arrangements greatly enlarge its surface area.

3. Cnidarians have definite tissues, and these originate from two embryonic germ layers, endoderm and ectoderm. The former gives rise to the gastrodermis, which is the lining of the gut; the latter becomes the epidermis that covers the outside of the body, and in some cnidarians it also differentiates into the lining of the pharynx. Between the epidermis and gastrodermis there is a secreted layer called the mesoglea. This is not a tissue, although cells may wander into it from the epidermis or gastrodermis.

4. Changes in shape, such as are involved in swimming movements, food capture, and other activities, depend on contractile cells. These are part of the epidermal and gastrodermal layers, although the portions of the cells in which the contractile filaments are located may be sunk below the level of the other cells. Even when the musculature is highly developed, as it is in sea anemones and some jellyfishes, the contractile cells nevertheless belong to the epithelial layers.

5. The mesoglea, especially when it is extensive, serves as a kind of skeleton. In jellyfishes, it makes up most of the body mass and is important in locomotion. A jellyfish swims by jet propulsion, forcing water out from under the body by muscular contractions. During a contraction, the shape of the mesoglea is deformed, but as soon as the muscles relax, the mesoglea springs back to its original shape and the animal is ready for another pulsation.

6. Many cnidarians have hardened skeletal elements. These are sometimes external to the epidermis, sometimes within the mesoglea. The hardened material may be a sclerified protein, or it may be calcareous, which is the case in corals. Both types are present in some cnidarians.

7. There is a nervous system. The neurons are closely associated with the epidermal and gastrodermal layers, and generally form a network. Most of the synapses are not polarized; in many neurons, therefore, transmission of an impulse may proceed in either direction. Sensory structures of cnidarians include photoreceptors, touch receptors, and statocysts concerned with maintaining a proper orientation.

8. A distinctive feature of the phylum is the presence of nematocysts. Each nematocyst is secreted within an epithelial cell called a nematocyte and consists of a capsule that contains an inverted and coiled tubular thread. Touch, or an appropriate chemical stimulus, causes a nematocyst to "fire"; the inverted thread unwinds and turns inside-out. Nematocysts are used primarily for capture of prey and for defense. The threads of some types are open at the tip; as they penetrate the tissue of a prey or predator, they inject a toxin. Other types of nematocysts act as lassos, or deliver a sticky secretion.

9. Most cnidarians have tentacles. These are usually close to the mouth or at the edge of a defined oral area. In jellyfishes, the tentacles are typically at or near the edge of the bell-shaped or saucer-shaped body. The mouth itself, however, is commonly bordered by frilly elaborations concerned with handling food. These oral structures, like the tentacles, are provided with nematocysts for capturing and subduing prey.

10. There are two main body forms in the Cnidaria. One of them, the polyp, is nearly always attached to a firm substratum or buried in mud or sand. The other, called the medusa, or jellyfish, is generally free-swimming. Some cnidarians have both a polyp and a medusa stage in the life cycle. When this is the case, the polyp stage produces medusae asexually, and the medusae reproduce sexually, giving rise to polyps. There are many variations on this theme, however, and sometimes the medusae are not released by the polyps that give rise to them; they produce their gametes while they are attached to the polyps. A further complication is that some cnidarians form polymorphic colonies, in which some polyps are specialized for one function, such as feeding, and others are specialized for different functions, such as defense or production of medusae.

11. Of the four classes of Cnidaria, three are mentioned in this summary. The class Hydrozoa is characterized by epidermal origin of gametes and nematocysts. There is usually a polyp stage, often forming a polymorphic colony. The polyps typically produce medusae, and although these may never become detached and swim, they function as sexual individuals. Some hydrozoans are strictly polyps or strictly medusae.

12. In the class Scyphozoa, gametes are derived from the gastrodermis, and nematocysts may be produced by either the epidermis or the gastrodermis. There is generally, but not always, a polyp stage that gives rise to medusae asexually. Most large jellyfishes belong to this class.

13. In the Anthozoa, there is no medusa stage. The

polyps, or certain of them in the case of some colonial types, are sexual individuals. Gametes are derived from the gastrodermis. Unlike hydrozoans and scyphozoans, anthozoans have a pharynx that originates as a stomodeal invagination; its lining is therefore ectodermal. The gut is partially divided into chambers by vertical septa. Anthozoans are often histologically complex, and the mesoglea may contain many cells as well as secreted components; the cells are nevertheless of epidermal or gastrodermal origin. This is the largest class of cnidarians, and includes anemones, sea pens, and most of the corals, which have calcareous skeletons.

5

PHYLUM CTENOPHORA

COMB JELLIES AND SEA GOOSEBERRIES

Phylum Ctenophora
- (Tentaculata)
  - Order Cydippida
  - Order Lobata
  - Order Ganeshida
  - Order Thalassocalycida
  - Order Cestida
  - Order Platyctenida
- (Nuda)
  - Order Beroida

Introduction

The ctenophores (''comb-bearers'') constitute a small phylum of strictly marine animals. Typically they are planktonic and resemble jellyfishes in being transparent and having ample mesoglea. Their symmetry is biradial rather than radial, however, and, instead of swimming by jet propulsion, they depend on the activity of large cilia that cohere in aggregates called **combs,** or **ctenes.** The mesoglea provides form and buoyancy, but its role as an elastic skeleton is minimal except in a few species that undulate or that can flap a pair of broad lobes at the oral end of the body. The combs are arranged in eight meridional rows, originating close to the aboral pole. The beating of the cilia is usually almost continuous, so that a ctenophore swims steadily, instead of pulsating jerkily the way a jellyfish does.

Unlike most cnidarians, ctenophores do not have radially arranged tentacles. When tentacles are present, there are only two of them, and they have branches on one side. The epidermis of the tentacles has many highly differentiated cells called **colloblasts** (''glue cells'') used in the capture of food. There are no nematocysts, except in species of *Haeckelia,* which acquire them by eating jellyfishes.

Although ctenophores are traditionally placed next to cnidarians, the relationship of the two phyla may not be especially close. The method of locomotion of ctenophores as well as their lack of nematocysts are important distinctions, but two other points of difference are perhaps even more basic. In cnidarians, wherever cilia are present, there is only one to each cell; in ctenophores, a ciliated cell usually has several to many cilia, a characteristic of flatworms, nemerteans, and most other phyla. Cnidarians, moreover, are diploblastic animals in which the contractile cells are of the myoepithelial type and belong to either the epidermis or gastrodermis. Ctenophores have muscle cells that are separate from the epidermis and gastrodermis. The cells that form muscle are segregated early in embryonic development and are here interpreted to represent mesoderm. Thus ctenophores, along with flatworms and all of the more complex phyla, are considered triploblastic.

General Structure

The genus most likely to be studied in the laboratory is *Pleurobrachia,* the sea gooseberry (Figures 5.1, A and B; Figure 5.2). It demonstrates the important characteristics of the phylum. The description of *Pleurobrachia* is applicable also to certain other genera, especially *Mertensia, Hormiphora,* and *Euplokamis* (Figure 5.1, C), all of which are widely distributed.

Most species of *Pleurobrachia* attain a diameter of 1 to 1.5 cm and are nearly spherical. The consistency of the body is firm. If a specimen is viewed from either the oral or aboral end, the biradial symmetry characteristic of ctenophores will be obvious. It is particularly evident in the arrangement of comb rows, tentacles, branches of the gut, and aboral sensory complex.

Comb Rows

The comb rows of *Pleurobrachia* do not quite reach the oral pole. The cilia of each comb are large and cohere for most of their length, so they form what looks like a glassy paddle. Ordinarily, their effective beat is directed aborally, so the oral end of the animal moves forward. The sequence of comb-row activity starts nearest the aboral end, however. The beat of a particular comb will be in phase with the corresponding comb in the other row of the same quadrant. A ctenophore may reverse its beat so that the effective stroke is in the direction of the mouth, and in that case it swims with its aboral end forward.

Tentacles and Colloblasts

There are two tentacles, which originate within deep sheaths on opposite sides of the body. The openings of the sheaths are closer to the aboral end than the oral end. Each tentacle has many branches, all of which originate on one side of the tentacle. When *Pleurobrachia* assumes its fishing posture, it extends its tentacles to a length of several centimeters. As the tentacles trail, the branches look something like the threads of a cobweb. The main stems and branches have cores of smooth muscle. When these muscles contract fully, the tentacles withdraw completely into their sheaths. The musculature in the basal portion of each tentacle, where it is inserted into the deepest part of its sheath, is particularly well developed.

For capturing food, which consists mostly of copepods and other small crustaceans, *Pleurobrachia* relies on a simple principle: stickiness. Each fully functional colloblast (Figure 5.3) in the epidermis of a tentacle has a protruding cap of adhesive granules. When a prey organism bumps into the tentacle, the colloblasts in that region ''fire.'' Each one has a coiled filament that is believed to act in much the same way as the spring in a jack-in-the-box. This spring is derived from cytoplasm, and for much of its length it winds around the cell. It is, however, rooted in the basal portion of the cell and terminates within the cap, where it is partly covered by the spheroidal body, from which fibers extend to the adhesive granules. During differentiation of a colloblast, the nucleus becomes converted into a rodlike axial structure.

The basal portion of each colloblast synapses with a nerve cell. Thus the stimulation of colloblasts by contact with the prey, or the firing of the colloblasts, engages the nervous system. The tentacle to which the

FIGURE 5.1 Order Cydippida. A. *Pleurobrachia bachei*. Somewhat diagrammatic drawing of a living specimen. The meridional canals that lie beneath the comb rows, and the canals by which these are joined to the stomach, are not shown (see Figure 5.2). The animal generally swims with the oral end leading and the tentacles trailing. B. *Pleurobrachia bachei*. (Photograph courtesy of Claudia Mills.) C. *Euplokamis dunlapae*. Most of the branch tentacles are coiled up, but one (arrow) is undergoing rhythmic coiling and uncoiling. See notes concerning *Euplokamis* in text, under order Cydippida. (Photograph courtesy of Claudia Mills.)

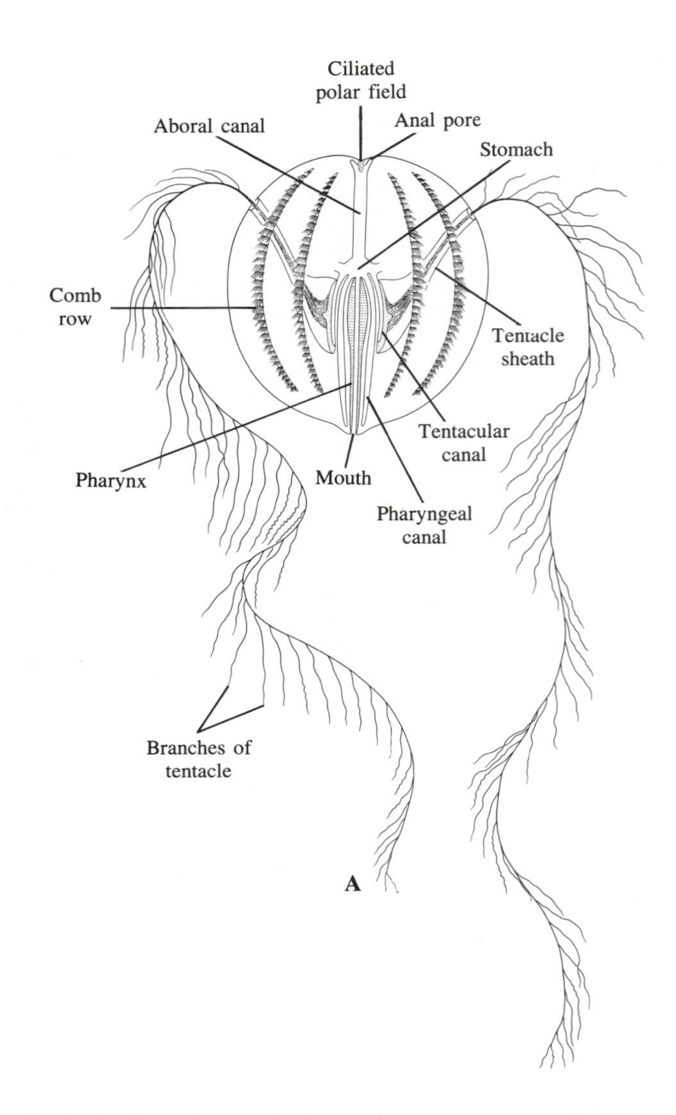

B

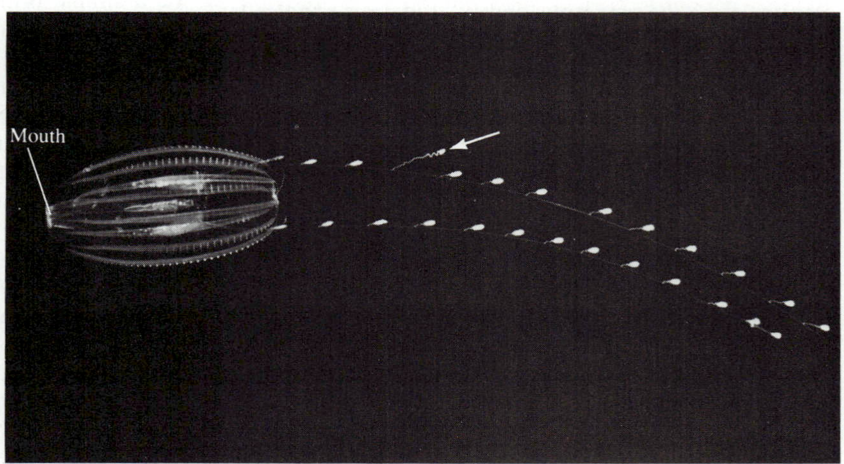

C

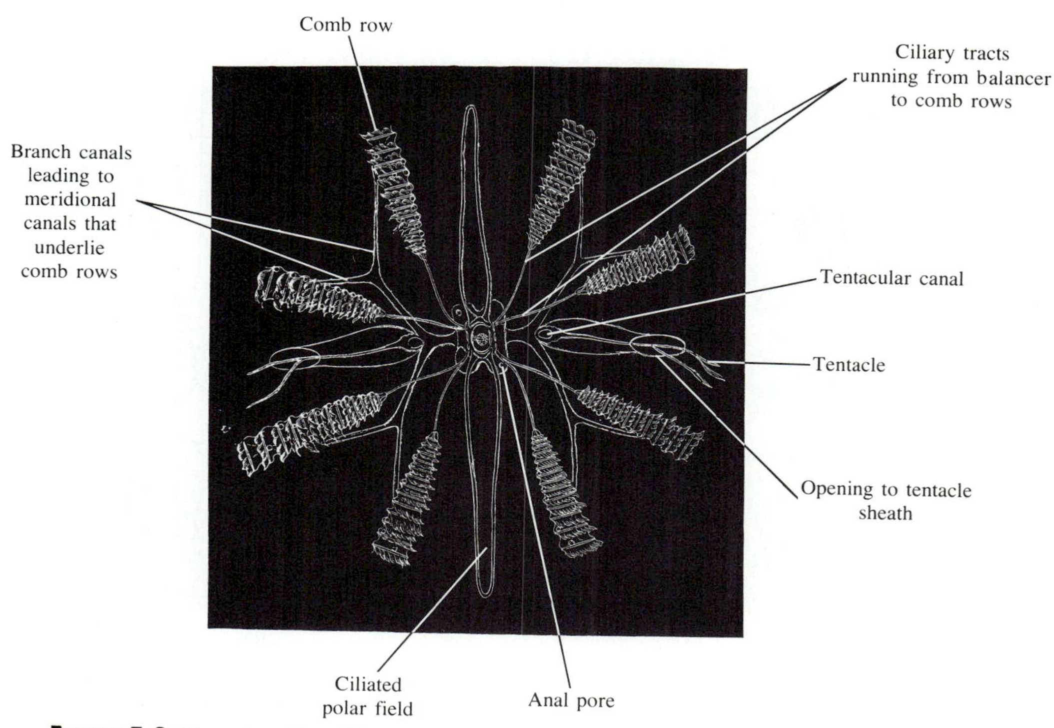

FIGURE 5.2 *Pleurobrachia pileus*. View of the aboral end, showing branches of the gut leading to the meridional canals beneath the comb rows, as well as other details. (Mayer, Carnegie Institution of Washington, Publication 162.)

prey is stuck is then shortened and bent so that the prey is brought to the mouth.

GUT

The mouth of *Pleurobrachia* opens into a pharynx that is derived embryologically from a stomodeal invagination. The lining of the pharynx is therefore of ectodermal origin. The inner part of the pharynx is thick-walled, and it is in this region that most of the extracellular digestive activity takes place. The stomach and the system of canals joined to it are concerned with distribution of solubilized food materials and with intracellular digestion. One canal extends straight toward the aboral pole, where it divides into four lobes (Figure 5.2). Two of these end blindly, and two open to the outside by what are conventionally called **anal pores.** In terms of their development, these pores are not true anuses, but fluid and indigestible residues do leave the body through them. In *Pleurobrachia* they also serve as openings through which sperm (and perhaps also eggs) are released. The remaining canals are arranged in a strongly biradial pattern. Each of the two main canals extending from opposite sides of the stomach sends out one branch that runs alongside the pharynx and another that lies next to the base of a tentacle. A little farther out, each main canal divides twice, producing the four branches that lead to the **meridional canals** situated beneath the comb rows. In *Hormiphora* (Figure 5.4), there is a different arrangement: there are four main canals, instead of two, going out laterally from the stomach, and the pharyngeal and tentacular canals arise directly from the stomach.

Fluid circulates through the canals by ciliary activity. In *Pleurobrachia,* the fluid in a particular canal is driven in one direction on one side and in the opposite direction on the other side. At various points along each digestive canal are minute pores, each surrounded by a circle of ciliated cells. These circles, called **rosettes,** perhaps have something to do with excretion.

MESOGLEA

Although the mesoglea is almost as clear as water, a feature that may make ctenophores nearly invisible to certain predators, it has amoeboid cells, some muscle, and some collagen fibers. In *Pleurobrachia* and its close relatives, the musculature is especially well developed in the tentacles, where it forms a central core. Nevertheless, there are many muscle strands beneath

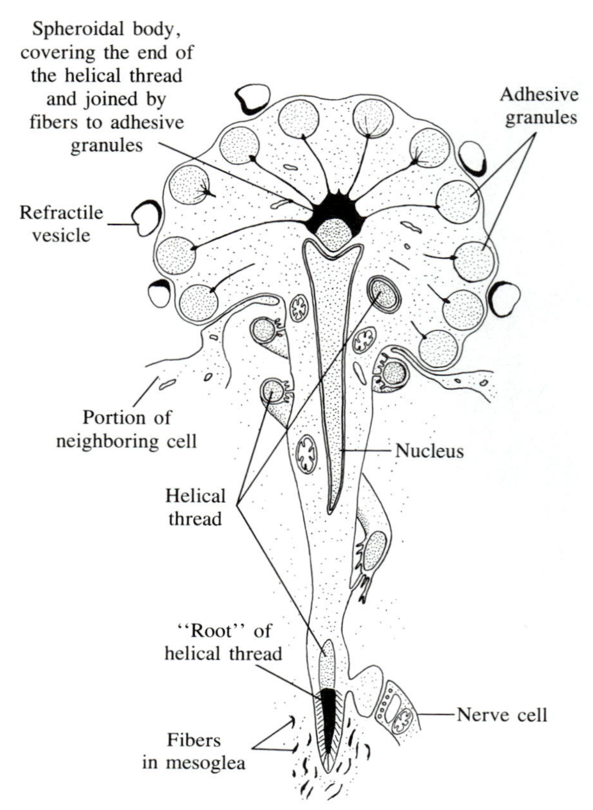

FIGURE 5.3 Diagram of the structure of a colloblast, based on electron micrographs. Note that the "external" portion of the helical thread is surrounded by a thin layer of cytoplasm, which is joined by one of its several ridges to the main body of the cell. (After Franc, Biological Bulletin, *155*.)

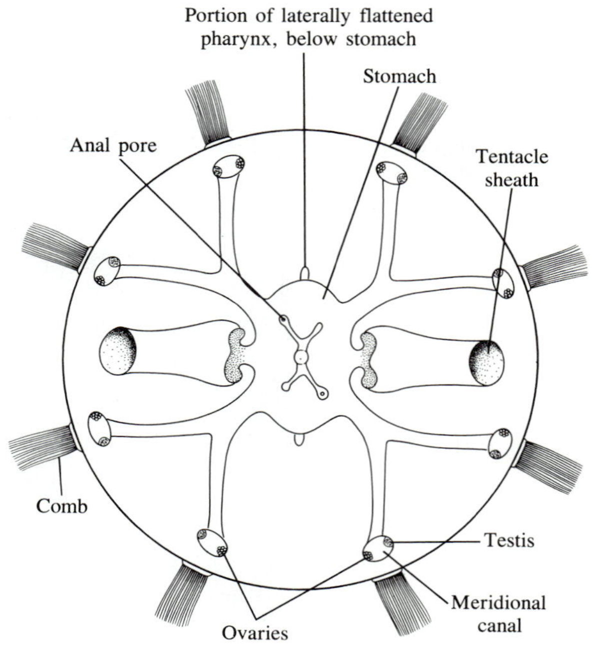

FIGURE 5.4 Order Cydippida. *Hormiphora;* diagrammatic representation of the branches of the gut, gonads within the meridional canals, and anal pores at the aboral end. (After Delage and Hérouard, Traité de Zoologie Concrète, *2*.)

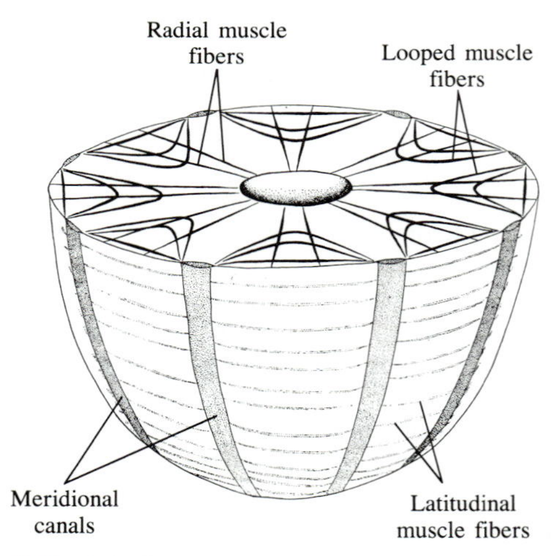

FIGURE 5.5 *Pleurobrachia,* arrangement of some of the muscle fibers. (Chapman, Biological Reviews, *33*.)

the epidermis of the main body mass (Figure 5.5), and the contractions of these effect slight changes of shape. In certain genera, particularly those that swim in part by undulating movements, the body musculature is considerably more prominent than it is in *Pleurobrachia*.

NERVOUS SYSTEM AND SENSORY STRUCTURES

The nervous system of ctenophores, like that of cnidarians, is netlike. The net is widely distributed through the body, but is most intensively developed beneath the comb rows and beneath the ciliary tracts that lead to the comb rows from the apical sensory complex. Although much of the nerve net is closely associated with the epidermis, some parts of it underlie the gastrodermis and the lining of the pharynx, and there are neurons that lead to muscle cells. There are no nerve trunks or ganglionic masses in the system.

The aboral sensory complex (Figure 5.6) is extremely interesting. It has two elongated areas called **polar fields,** which are densely ciliated. In the center of the aboral pole is an area roofed over by a glassy dome consisting of large, coalesced cilia. Arising from sensory cells on the floor of this domed structure are four tufts of long cilia **(balancers),** the tips of which are in contact with a *statolith,* a cluster of calcareous bodies held together by a gelatinous matrix. Except

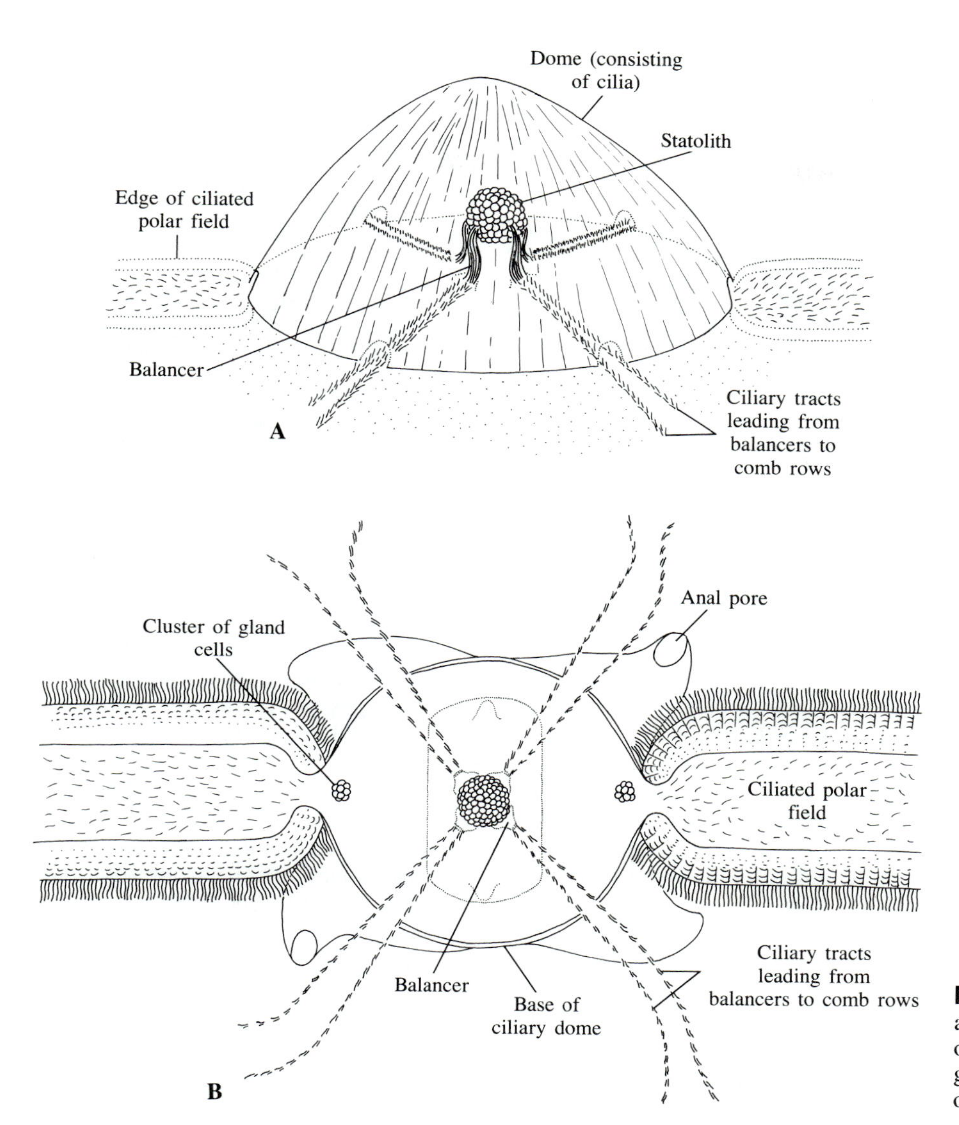

Figure 5.6 A. Portion of the aboral region of a cydippid ctenophore, as seen in side view; diagrammatic. B. The aboral region of *Hormiphora;* diagrammatic.

when the oral–aboral axis is perfectly perpendicular to the water surface, the calcareous mass is bound to weigh more heavily, or pull more strongly, on certain of the four tufts than on others. Such changes in tension stimulate the sensory cells to initiate impulses that lead to a change in the orientation of the ctenophore.

At the base of each tuft there originates a tract of short cilia that passes out through an opening in the glassy dome, then bifurcates, one branch going to each of the two comb rows of a particular quadrant. Impulses originating in the sensory cells that bear the balancers reach the comb rows through the tracts of ciliated cells. There are two general ways by which the activity of the combs in a particular row is coordinated (Figure 5.7). In *Pleurobrachia* and other members of the order Cydippida, and also in *Beroe,* coordination depends on mechanical forces generated by the combs as they beat: when one comb beats, the deformation in the cell surface where that comb is situated generates an impulse that stimulates the succeeding comb. In *Mnemiopsis* and *Bolinopis,* which belong to the order Lobata, the ciliary tracts that extend from the aboral balancers to the comb rows continue their course in the oral direction, and thus connect successive combs. In these types, coordination of the beat within a row depends at least in part on neuroid transmission; that is, on impulses that pass through the ciliated cells themselves.

The area enclosed by the glassy aboral dome may contain four inconspicuous crescentic bodies believed to function as photoreceptors. They consist partly of

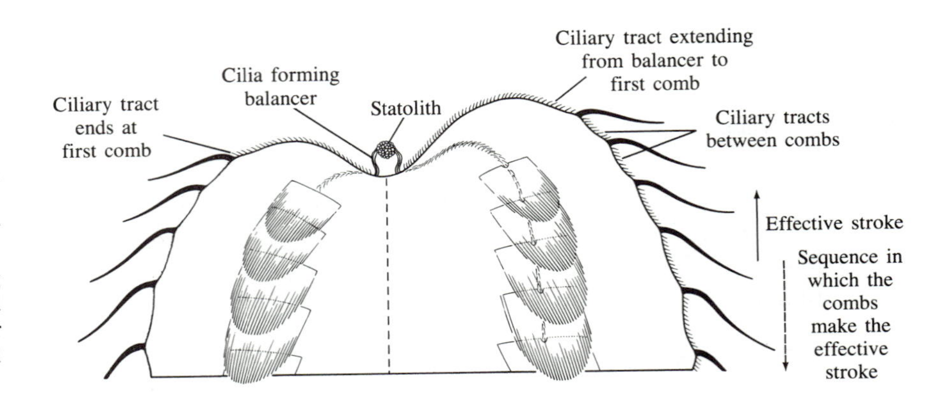

FIGURE 5.7 Relationship of comb rows to ciliary tracts originating at the balancers. On the left is the arrangement characteristic of ctenophores of the orders Cydippida *(Pleurobrachia, Hormiphora)* and Beroida *(Beroe),* in which the ciliary tract stops at the aboralmost ctene; on the right is the arrangement characteristic of the order Lobata *(Bolinopsis, Mnemiopsis),* in which the ciliary tract reaching the aboralmost ctene is continued between successive combs. (After Tamm, Journal of Experimental Biology, *59*.)

highly modified, nonmotile cilia similar to those found in photoreceptors of certain hydrozoan medusae. Clusters of glandlike cells may also be present on the floor of the dome.

BIOLUMINESCENCE

Most ctenophores, including *Pleurobrachia,* are bioluminescent. Their light-emitting cells are part of the gastrodermis of the meridional canals and can be traced back to certain cells of embryonic endoderm, which are already bioluminescent. The production of light is intracellular, and in differentiated animals is under nervous control.

REPRODUCTION AND DEVELOPMENT

Pleurobrachia and nearly all other ctenophores are simultaneous hermaphrodites. In some genera, however, a particular individual is at first a male, then a female. This type of sequential hermaphroditism is called **protandry.** So far, only one species has been reported to have separate sexes. The oogonia and spermatogonia originate in the gastrodermis of the meridional canals that underlie the comb rows. It is usual for each of these canals to have a simple testis on one side and an ovary on the other. The enlarging oocytes are of two types: those that will eventually become ova, and those that are filled with yolk and serve as nutritive cells. Sperm and oocytes ready to be fertilized may be released through breaks in the body wall or through the anal pores at the aboral end. In certain species the eggs are not set free until they have been fertilized. Self-fertilization can and does take place in some ctenophores.

Cleavage in ctenophores differs from that of cnidarians in being regular and exact, and development is determinate. Structures of the adult are mapped out early. If the products of the first cleavage are separated, each will develop into half a ctenophore with only four comb rows and four meridional canals. In time, however, the young animal may fill out its missing parts.

The following account of early development, as illustrated in Figure 5.8, is based on observations on *Beroe*. The first two cleavages are meridional and equal, but the third is slightly oblique (A). At this stage the biradial symmetry of the future adult is already established. The fourth cleavage (B) segregates eight micromeres above eight macromeres. The micromeres multiply more rapidly than the macromeres, so that by the time there are 16 macromeres there are already about 48 micromeres. They form a sheet over the macromeres, a process of gastrulation called **epiboly** (C). Before the 16 macromeres have been overgrown at the lower pole of the embryo, however, each produces another generation of micromeres. Then the macromeres rotate in such a way that the new micromeres are pushed up toward the top of the embryo (D). About the same time, the external layer derived from the original micromeres invaginates, forming a stomodeum (E), which becomes the lining of the future pharynx. This invagination pushes the macromeres deeper into the embryonic mass, and segregates them as endoderm.

The micromeres that were produced at the lower pole and pushed upward then differentiate into the muscles of the tentacles and of other parts of the body. They are here considered to represent mesoderm, but it should be noted that this idea is not universally accepted: these micromeres bud off before definite endoderm has been segregated and they thus may be more closely related to ectoderm than to endoderm, with which the formation of mesoderm is usually associated.

Most ctenophores, other than those of the order Beroida, go through a stage that resembles an adult *Pleurobrachia*. The larva is called a **cydippid larva,** after the order to which *Pleurobrachia* and its close allies

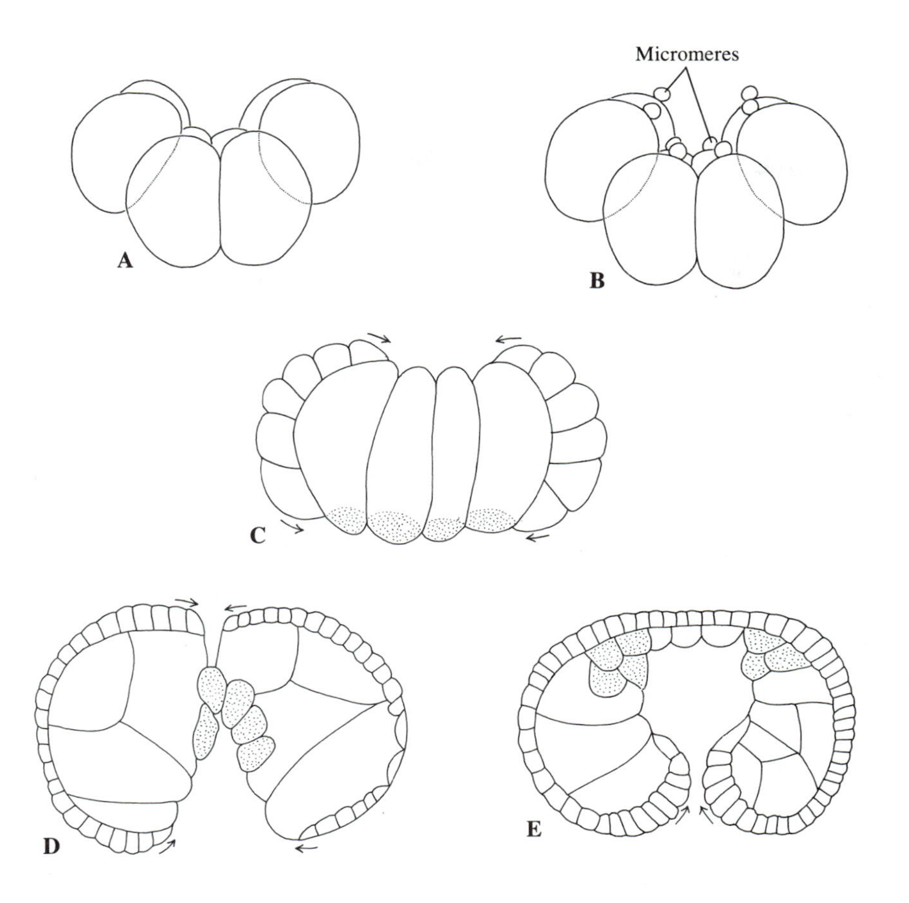

FIGURE 5.8 Early development of *Beroe* (cell nuclei omitted). A. Eight-cell stage. B. Sixteen-cell stage. C. Section of an embryo undergoing epiboly. The micromeres are multiplying and forming a covering over the macromeres. The sixteen macromeres (only four are shown in the section) are about to bud off micromeres from their lower poles, as indicated by stippling. D. Rotation of the macromeres pushes the new micromeres (mesoderm cells) upward. E. Stomodeum being formed by invagination of micromeres.

belong. It may have tentacles that are proportionately longer than those of the adult. This is the case, for instance, in *Bolinopsis* and *Mnemiopsis*. A still more remarkable feature of some cydippid larvae, such as those of *Bolinopsis*, is their ability to produce functional gametes. The eggs, after fertilization, develop in the usual way. The gonads of such sexually precocious cydippids disappear and are not replaced until the animals reach full maturity.

The metamorphosis of the cydippid is most drastic in ctenophores of the order Platyctenida, whose members are creeping or sessile as adults. Some of them scarcely resemble ctenophores. Most platyctenids, however, have a more or less typical cydippid stage, even in the case of some that brood their young in diverticula of the gut canals.

CLASSIFICATION

The phylum Ctenophora is often divided into two classes—Tentaculata and Nuda—on the basis of the presence or absence of tentacles. This system is now considered artificial, and most recent outlines of ctenophore classification proceed directly from phylum to order. Because it may be helpful to the student to see at a glance which groups have tentacles and which do not, a compromise plan is followed here. Tentaculata and Nuda are used, but only as informal headings, not as names of classes.

TENTACULATA

These ctenophores have two tentacles, each branched on one side. The tentacles are not always prominent, however.

Order Cydippida

The body of cydippids is usually spherical or ovoid. The tentacles are long and originate within deep sheaths into which they can be retracted. The main canals of the gut end blindly. *Pleurobrachia, Hormiphora,* and *Euplokamis* (Figure 5.1, C) are typical representatives of the order. *Euplokamis* is remarkable in that the branches of its tentacles have striated muscle as well as smooth muscle. The striated muscle is arranged in such a way with respect to collagen, smooth muscle, mesoglea, and fluid-filled compartments that its contraction results in high-speed *extension* of a branch tentacle

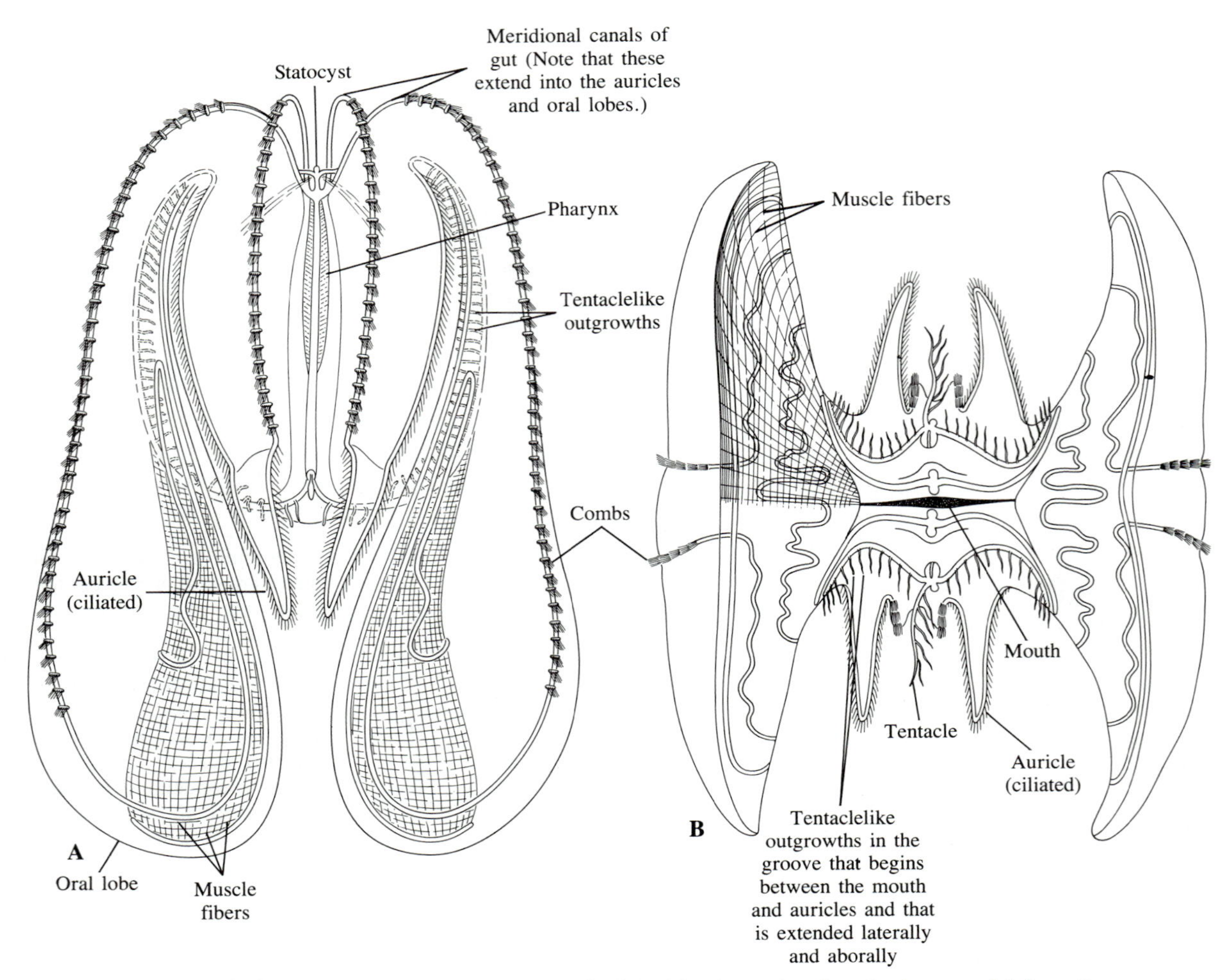

Figure 5.9 Order Lobata. A. *Mnemiopis leidyi,* side view, showing the large oral lobes. B. *Mnemiopsis mccradyi,* view of the oral side, showing the true tentacles, which have branches, as well as the tentaclelike outgrowths in the groove between the mouth and auricles. Muscle fibers are indicated only in a portion of one of the oral lobes. (After Mayer, Carnegie Institution of Washington, Publication 162.)

from its previously coiled state. This sudden extension is triggered by contact with a prey organism and is effective in making the colloblasts stick to the prey. Return to the coiled state is probably effected by contraction of one type of smooth muscle. Another type of smooth muscle operates to bring about slow wriggling movements of a branch tentacle, perhaps making it function as a lure. The branch tentacles of *Euplokamis* are histologically so complex that they may be viewed as organs.

Haeckelia (Euchlora) resembles the genera mentioned above, but is unusual in that it lack colloblasts. It eats small trachyline medusae and stores the undischarged nematocysts of these jellyfishes in its own tentacles. *Haeckelia* does not seem to use the nematocysts for capture of prey, but perhaps they serve a protective function.

Order Lobata

The body is slightly flattened laterally, and on the oral side it is drawn out into two large lobes that have considerable musculature and that can be used for locomotion. The two tentacles are small and have no sheaths into which they can withdraw. Four of the comb rows have ciliated extensions called **auricles.** The gut canals that follow the comb rows are united in pairs at the oral end. *Mnemiopsis* (Figure 5.9) and *Bolinopsis* are the best-known genera.

Order Ganeshida

Ganesha, the only genus in this order, is similar to cydippid ctenophores in having two long tentacles that are branched on one side. The sheaths of these tentacles, however, open close to the oral end, and the mouth is wider than it is in cydippids. The eight gut canals that run beneath the comb rows join a ring canal that encircles the mouth region; in this respect *Ganesha* resembles members of the orders Lobata and Beroida more than it resembles cydippids.

Order Thalassocalycida

At present, the only species in this group is *Thalassocalyce inconstans,* a tropical and subtropical ctenophore. Its shape usually resembles that of a shallow-belled medusa, but it may be compressed laterally or constricted into two lobes. The mouth is at the center of the oral end, and beside it are two tentacles that lack sheaths. The comb rows are short, but the gut canals extend far beyond them and wind around through the body. *Thalassocalyce* catches small crustaceans with the help of mucus secreted on the oral side of the bell, and the action of cilia moves the food to the mouth.

Order Cestida

The body is much flattened laterally and drawn out into a bandlike form. Four of the comb rows are long and four are short. All eight gut canals, however, are well developed. The two tentacles that are homologous with those of cydippid ctenophores are reduced, but they do have sheaths. Feeding is facilitated by short supplementary tentacles; there is a row of these on both sides of each of the two long grooves that extend toward the mouth. (Some zoologists have interpreted these small tentacles as branches of the primary tentacles.) Cestids swim by graceful undulations.

Cestum veneris—Venus' girdle—is perhaps the most famous of all ctenophores. Its size, as measured from the tip of one lateral lobe to the tip of the other, may reach 2 m. It is largely limited to the Mediterranean Sea and tropical areas of the Atlantic Ocean. *Velamen* (Figure 5.10) is a similar genus. Its species, some of which reach a breadth of about 15 cm, are also warm-water types.

Order Platyctenida

This group consists of a variety of odd ctenophores, most of which have no comb rows in the adult stage. In spite of their highly specialized structure, platyctenids have a pair of tentacles that originate within sheaths and have branches on one side. The young, moreover, are of the cydippid type. A decidedly unusual feature of certain species is their ability to reproduce asexually by budding off small pieces that differentiate into complete animals.

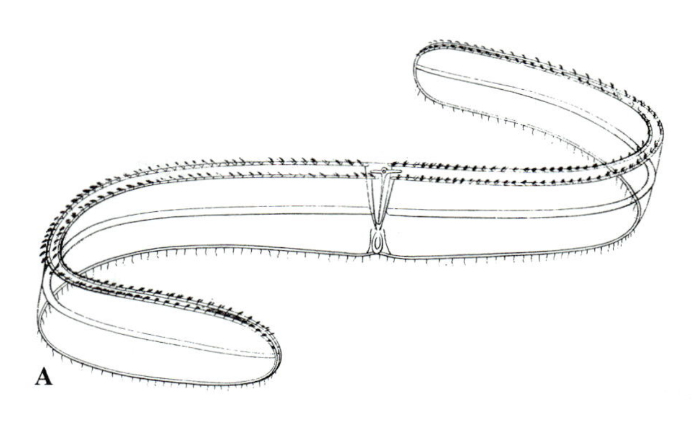

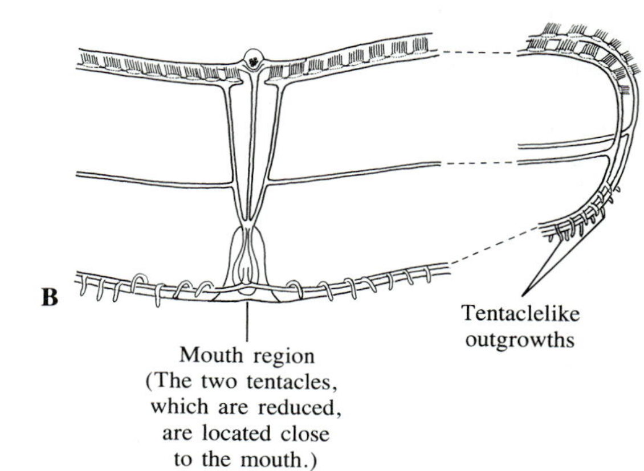

FIGURE 5.10 Order Cestida. A. *Velamen*. B. Some details of the structure of *Velamen*. In the central portion, the canals, tentaclelike outgrowths, and ctenes on the side farthest from the observer are not shown. (After Mayer, Carnegie Institution of Washington, Publication 162.)

Ctenoplana (Figure 5.11) is the only genus in which comb rows persist in the adult. The animal may swim in the plankton, crawl on the bottom, or glide in contact with the surface film. In its crawling or gliding phase, *Ctenoplana* becomes flattened and withdraws its comb rows into little pockets, and it distends its mouth so much that most of the ciliated pharynx is everted to form a creeping "sole."

In *Coeloplana,* which also may crawl or float, flattening of the body and eversion of the pharynx are permanent. Papillae on the aboral surface are used for brooding the young. Most of the known species are symbiotically associated with anthozoans of the subclass Octocorallia. For a time after the genus *Coeloplana* was discovered, and before its young were observed to have comb rows, it was thought by some zoologists to be intermediate between cnidarians and flatworms.

Lyrocteis is an odd genus in which the two tentacle sheaths are raised up on "chimneys." The few known species are sedentary.

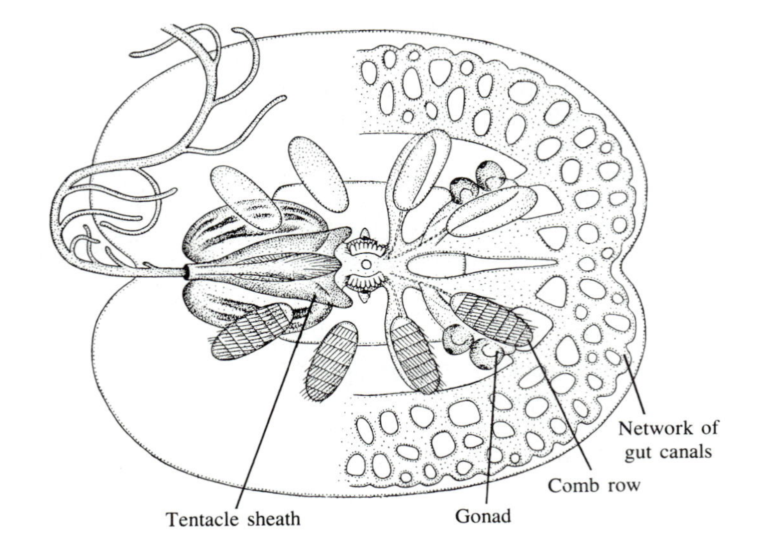

Figure 5.11 Order Platyctenida. *Ctenoplana,* aboral view. The specimen is drawn diagrammatically to simplify showing one of the tentacles (left) and portions of the gut (right). (Komai, Memoirs of the College of Science, Kyoto Imperial University, Series B, 9.)

Nuda

Order Beroida

Beroe and *Neis* are the only genera of nontentaculate ctenophores. The body is slightly flattened but nevertheless saclike because of the large mouth and pharynx. The meridional gut canals that underlie the comb rows are branched, and in certain species some of the branches unite to form a network.

Species of *Beroe* (Figure 5.12), a few of which grow to a length of more than 30 cm, feed on ctenophores of the orders Cydippida and Lobata. The prey is caught on the lips of the mouth, where there are glands that produce sticky mucus and a toxic substance, then swallowed with the help of pharyngeal muscles and macrocilia. A **macrocilium** is a bundle of fused cilia, and has several hundred sets of microtubules instead of the 9 pairs-plus-2 arrangement typical of simple cilia. The whole unit is enclosed within a common membrane. The macrocilia beat toward the pharynx, and they can also be used much like teeth. Another interesting feature of *Beroe,* perhaps found also in other ctenophores, is the presence of especially large mitochondria in the cells from which the cilia of the combs originate. Energy consumption in these cells is probably high.

Summary

1. Ctenophores are marine and typically planktonic animals. They are usually ovoid or cucumber-shaped, and their symmetry is biradial. Much of the body consists of nearly transparent mesoglea.

2. Ctenophores swim by the activity of large cilia. These cohere to form comblike structures arranged in eight meridional rows. As a rule, the oral end of the body is directed forward.

3. Most genera have two tentacles that are branched on one side and that can be retracted into sheaths. Many of the epidermal cells on the tentacles are of a type called colloblasts, used in capture of prey. In a fully differentiated colloblast, the exposed cap contains gran-

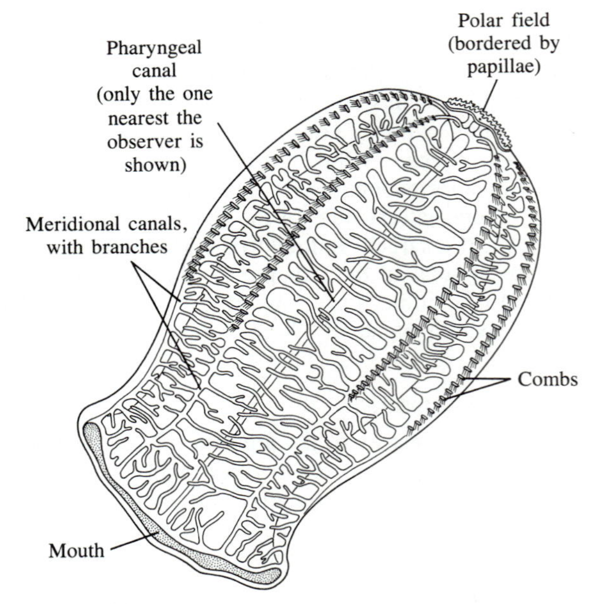

Figure 5.12 Order Beroida. *Beroe*. (After Mayer, Carnegie Institution of Washington, Publication 162.)

ules of a sticky substance, and there is a coiled intracytoplasmic filament that operates in much the same way as the spring in a jack-in-the-box. When a prey organism, such as a small crustacean, comes into contact with a tentacle, colloblasts are discharged; the springs within them lengthen and the cells rupture, but the sticky caps adhere to the prey and immobilize it. The basal portions of colloblasts form synapses with neurons, and this arrangement probably facilitates effective action of the tentacles in bringing the prey to the mouth.

4. Nearly all species are hermaphroditic. Development is determinate, and the biradial body plan is established early. Certain cells segregated during embryonic development are considered to represent mesoderm; they differentiate into muscles, which in ctenophores are separate from the gastrodermis and epidermis. Most members of the phylum, even specialized creeping or sedentary types, go through a stage that resembles adult ctenophores of the order Cydippida, to which the common genus *Pleurobrachia* belongs.

5. Although ctenophores resemble jellyfishes, mostly because of their ample mesoglea and transparency, their relationship to cnidarians may not be especially close. The development of muscles from mesoderm is one important way in which ctenophores differ from cnidarians. Furthermore, ciliated cells of ctenophores bear several to many cilia, a feature characteristic of most phyla, but not of cnidarians.

REFERENCES ON CTENOPHORA

Franc, J.-M., 1978. Organization and function of ctenophore colloblasts: an ultrastructural study. Biological Bulletin, *155*:527–541.

Freeman, G., and Reynolds, G. T., 1973. The development of bioluminescence in the ctenophore *Mnemiopsis leidyi*. Developmental Biology, *31*:61–100.

Harbison, G. R., Madin, L. P., and Swanberg, N. R., 1978. On the natural history and distribution of oceanic ctenophores. Deep-Sea Research, *25*:233–256.

Horridge, G. A., 1965. Relations between nerves and cilia in ctenophores. American Zoologist, *5*:357–375.

Horridge, G. A., 1974. Recent studies on the Ctenophora. *In* Muscatine, L., and Lenhoff, H. M. (editors), Coelenterate Biology. Academic Press, New York and London.

Mackie, G. O., Mills, C. E., and Singla, C. L., 1987. Structure and function of the prehensile tentilla of *Euplokamis* (Ctenophora, Cydippida). Zoomorphology, *107*:319–337.

Madin, L. P., and Harbison, G. R., 1978. *Thalassocalyce inconstans,* new genus and species, an enigmatic ctenophore representing a new family and order. Bulletin of Marine Science, *28*:680–687.

Mayer, A. G., 1911. Ctenophores of the Atlantic coast of North America. Carnegie Institution of Washington Publication No. 162.

Mills, C. E., 1987. Revised classification of the genus *Euplokamis* Chun, 1880 (Ctenophora: Cydippida: Euplokamidae n. fam.) with a description of the new species *Euplokamis dunlapae*. Canadian Journal of Zoology, *65*:2661–2668.

Mills, C. E., and Miller, R. L., 1984. Ingestion of a medusa *(Aegina citrea)* by the nematocyst-containing ctenophore *Haeckelia rubra* (formerly *Euchlora rubra*); phylogenetic implications. Marine Biology, *78*:215–221.

Rankin, J. J., 1956. The structure and biology of *Vallicula multiformis* gen. et sp. nov., a platyctenid ctenophore. Journal of the Linnean Society, *43*:55–71.

Siewing, R., 1977. Mesoderm bei Ctenophoren. Zeitschrift für Zoologische Systematik und Evolutionsforschung, *15*:1–8.

Tamm, S. L., 1973. Mechanisms of ciliary co-ordination in ctenophores. Journal of Experimental Biology, *59*:231–245.

6

Phyla Platyhelminthes and Gnathostomulida

Phylum Platyhelminthes
- Class Turbellaria
  - Order Acoela
  - Order Catenulida
  - Order Macrostomida
  - Order Polycladida
  - Order Neorhabdocoela
  - Order Temnocephalida
  - Order Protricladida
  - Order Tricladida
- Class Digenea
- Class Monogenea
- Class Aspidogastrea (Aspidobothria)
- Class Cestoda
- Class Cestodaria

Phylum Gnathostomulida

This chapter is concerned mostly with the animals called flatworms, which make up the large phylum Platyhelminthes. Although flatworms have retained a saclike gut, they have in other respects reached a much higher level of organization than cnidarians and ctenophores. The advances that probably appeared for the first time in flatworms, or in their immediate ancestors, were basic to the evolution of more complex invertebrates, such as annelids, molluscs, and arthropods. Thus an appreciation of the structure, development, and other attributes of the Platyhelminthes will enable one better to understand the material presented in succeeding chapters.

The microscopic gnathostomulids, of which there are only about 100 described species, may not be so closely related to flatworms as has been thought. They are now generally given phylum rank, but it is convenient to deal with them in the same chapter as flatworms.

Phylum Platyhelminthes

Introduction

Flatworms are soft-bodied and depend on a wet environment, but they are nevertheless ecologically diversified. The phylum is well represented in the sea and in fresh water, and certain types of flatworms are found in rain forests and other relatively moist terrestrial habitats. About three-fourths of the approximately 11,000 described species are internal or external parasites, and the majority of these have life cycles that involve two or more hosts. The evolution of parasitic flatworms and their life cycles is closely tied to the evolution of the animals in which, or on which, they live.

Flatworms are distinctly different from cnidarians and ctenophores in several respects: (1) they have bilateral symmetry, and one end of the body is differentiated as a head region; (2) the nervous system has a brainlike concentration of ganglia and definite longitudinal nerve trunks; (3) the mesoderm gives rise to a wider variety of tissues—and more complex tissues—than it does in ctenophores, and there are, moreover, true organs; (4) cleavage is of the spiral type; (5) sexual reproduction involves copulation, and the reproductive system includes a penis or comparable structure for transferring sperm; and (6) many flatworms have protonephridia that function in osmoregulation. Each of these features is discussed in turn below.

Cephalization and Bilateral Symmetry

For an animal that moves actively, hunts for food, or searches for a host, it is advantageous for the end that leads to be equipped with sense organs and with a brain capable of integrating information and regulating responses. The differentiation of a distinct head region is called **cephalization.** Although cephalization is less well developed in flatworms than it is in many other phyla, it is nevertheless apparent, particularly in certain free-living types.

In animals that are cephalized, the upper and lower surfaces are usually different, and the right and left halves of the body are mirror images of one another. This is the essence of bilateral symmetry, which is characteristic not only of most phyla of invertebrates, but also of vertebrates. The only way to cut a flatworm or any other bilaterally symmetrical animal into equal halves is to slice it lengthwise along the midline. Some internal structures may be distributed asymmetrically, but this does not contradict the basically bilateral plan.

Nervous System and Sensory Structures

The nervous system of flatworms is more complex and centralized than that of cnidarians and ctenophores. Most members of the phylum have a recognizable brain close to the anterior end. Many of the nerves extending from the brain run directly to sensory structures, which are generally more concentrated in the head region than elsewhere. Others form longitudinal main trunks whose branches supply nearly all parts of the body. Most flatworms have two main trunks located ventrolaterally just beneath the epidermis.

Any or all of the following types of sensory structures may be present in a flatworm: a simple **statocyst** close to the brain; **chemosensory pits** in the epidermis; bristles consisting of cilia (these are probably **touch receptors**); and **light receptors** associated with **eyespots.** The eyespots themselves are just masses of dark pigment that function as shields for the actual photoreceptors. If there were no shields, all receptors would be stimulated by light coming from any direction. The shields ensure that certain receptors will be stimulated by light coming from one direction, others by light coming from other directions. The nervous system integrates the information and the animal then makes an appropriate response.

It is important to stress the fundamental difference between the photoreceptors of flatworms and those found in many cnidarians. The photoreceptors associated with eyespots of cnidarians have modified cilia, and in this respect they are similar to retinal cells in the eyes of vertebrates. As a rule, photoreceptor cells of flatworms lack cilia; instead, an extensive part of the surface of each one is differentiated into crowded **microvilli.** However, a few flatworm photoreceptors do have modified cilia rather than microvilli.

Triploblastic Structure and the Organization of Tissues into Organs

The tissues of flatworms are derived from three embryonic germ layers: ectoderm, endoderm, and mesoderm. Thus flatworms, like ctenophores, are **triploblastic,** rather than diploblastic, as is the case in Cnidaria. Because muscle, bone, blood, and some other tissues of vertebrates are of mesodermal origin, the appearance of this third germ layer was an important step in the evolution of higher animals.

In cnidarians, contractile cells are part of the epidermis or the gastrodermis. Ctenophores are different in that their muscles originate from cells that are interpreted to represent mesoderm. In flatworms, however, musculature is much more extensively developed. Usually there is a layer of **circular muscle** just beneath the epidermis, then a layer of **longitudinal muscle.** Muscle fibers of mesodermal origin have various other orientations within the body, and they are also present in portions of the reproductive system. The musculature of the pharyngeal portion of the gut is exceptional in that it is derived from cells that are more closely related to those that form the epidermis than to mesoderm.

Another derivative of mesoderm in flatworms is a cellular "packing material" called **parenchyma.** This takes the place of the mesoglea found in cnidarians and ctenophores. In the strict sense, parenchyma consists of relatively unspecialized cells that occupy space not filled by other tissues. However, certain specialized cells, such as those of muscle and nerve, may have cell bodies that resemble parenchyma cells and may be mixed with these and perhaps even make up much of what may appear to be parenchyma. Regardless of its exact composition, the packing tissue can be reshaped to some extent when a flatworm contracts or relaxes, or when it fills its gut with food, copulates, or performs other functions that require muscular activity. Some structures that are not tightly bound to the tissues around them slip back and forth within the parenchyma.

The fact that tissues are rather richly diversified in flatworms is linked to another characteristic of these animals: the presence of organs. An **organ** is a structure that consists of more than one kind of tissue. A good example is a muscular pharynx, which may be constructed in such a way that it can be protruded. This is an efficient device for rapid ingestion of food, whether it be a whole small organism or tissue sucked from a larger organism. Some components of the reproductive system of flatworms also qualify as organs.

Spiral Cleavage and Some Other Aspects of Development

In animals that have it, spiral cleavage usually becomes evident at the third division. The first four blastomeres divide obliquely, so that the smaller products, called micromeres, come to lie on the furrows between the larger cells, or macromeres, rather than directly above them. The pattern is perpetuated in several successive divisions. This type of cleavage is clearly evident in eggs of polyclad turbellarians (see below). As in annelids and molluscs, the mesodermal components of the body—most of the musculature, the reproductive system, and certain glands—can be traced back to progeny of the 4d cell which is a member of the quartet produced by the fourth division of the macromeres. In polyclads, however, the 4d cell also produces all of the endoderm, whereas in annelids and molluscs the endoderm is derived partly from progeny of the 4d cell and partly from the macromeres. In polyclads, these cells degenerate, liberating nutritive yolk.

The micromeres of the upper quartets multiply and form ectoderm. They overgrow the larger cells beneath them, a process called **epiboly.** The result is a more or less solid gastrula, or **stereogastrula.** Eventually, an ectodermal invagination, or **stomodeum,** produces the pharyngeal portion of the gut; this becomes joined to the endodermal part. Derivatives of some micromeres that have sunk into the inner mass differentiate into the musculature of the pharynx.

In most flatworms, each egg is accompanied by many yolk cells. Cleavage therefore deviates from the clearly spiral pattern typical of polyclads and a few other groups within the phylum.

Whether development is direct or indirect is independent of whether or not the eggs are accompanied by separate yolk cells. Some polyclads, for example, have direct development, looking very much like adults when they hatch; others have a distinctive larval stage and arrive at their mature form by way of a gradual metamorphosis. Similarly, among flatworms that have yolk cells, development may be direct or indirect. It is in some of the parasitic groups, including tapeworms and digenetic flukes, that the life history becomes especially complicated. Not only may there be more than one larval stage, but there may also be asexual reproduction by larvae or by other immature stages. This phenomenon is described in more detail below in connection with the groups of which it is characteristic.

Reproductive System

In cnidarians and ctenophores, reproductive systems are relatively simple. There are no hints of the complexities found in flatworms, which have **gonoducts,** specialized structures for transferring and receiving sperm, and an assortment of accessory glands. As a rule, flatworms are hermaphroditic, but insemination requires copulation.

Because the classification of flatworms is based to a large extent on the organization of the reproductive system, it will be best to describe the morphology and

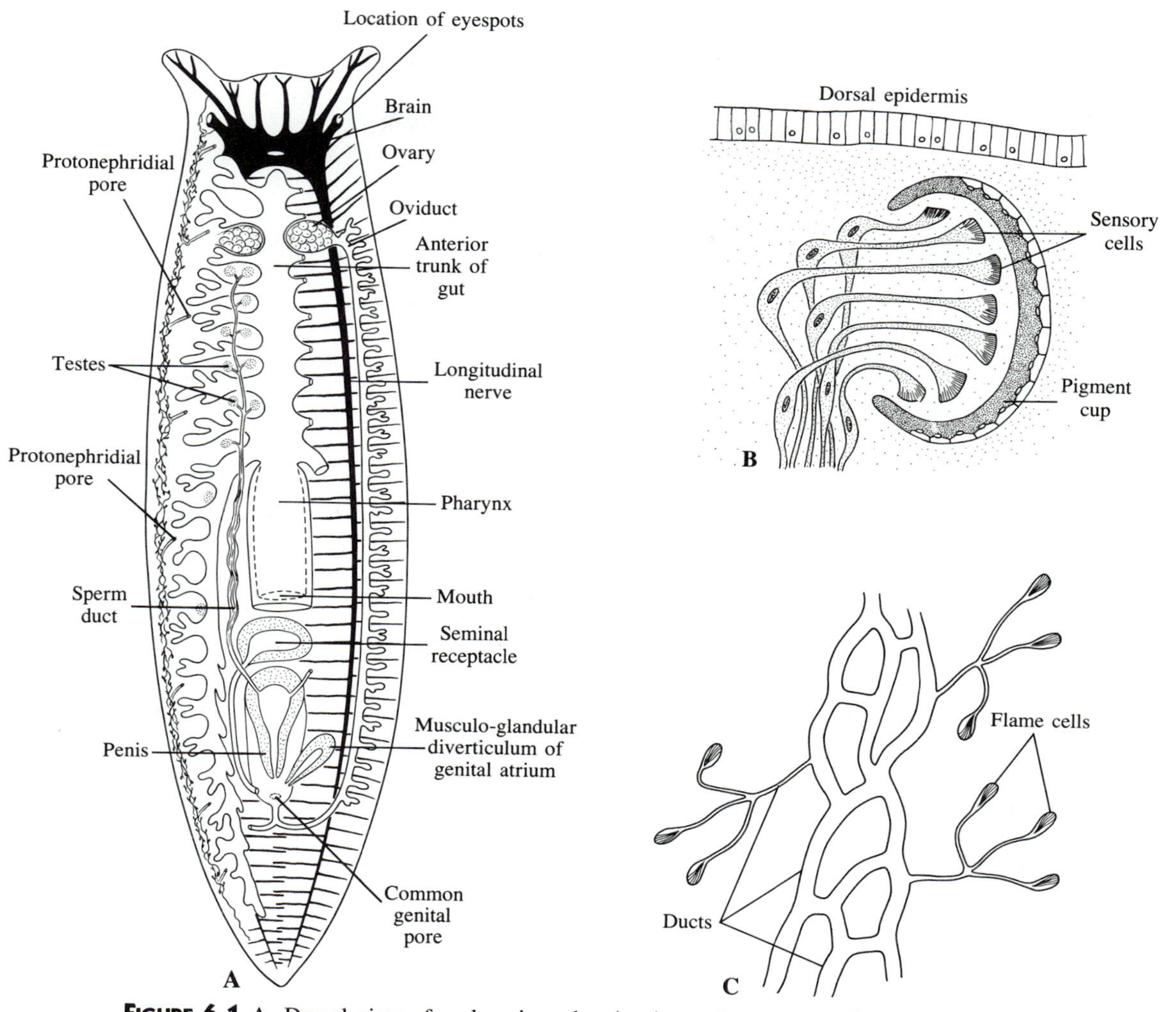

FIGURE 6.1 A. Dorsal view of a planarian, showing internal structures; diagrammatic, with certain symmetrically placed structures omitted on one side or the other. B. Section through an eyespot; diagrammatic. C. Portion of a protonephridium, showing anastomosing ducts.

function of the component structures under the major taxonomic assemblages within the phylum.

PROTONEPHRIDIA

Among the parenchyma cells of many flatworms, including most of the parasitic species, are structures called **protonephridia,** concerned primarily with osmoregulation. In flatworms that live in fresh water or in wet places on land, the tissues are more saline than the environment, so water diffuses into the body more rapidly than it diffuses out. To maintain an appropriate water–salt balance, water must be eliminated, and this is the job of the protonephridia. Free-living marine flatworms usually lack protonephridia, but they are isosmotic with the medium in which they live and do not have to face the problem of osmoregulation.

A typical protonephridium (Figure 6.1, A and C) consists of a system of ducts whose ultimate branches terminate in **flame cells.** Each flame cell is more or less cup-shaped, although the cup may be deep, and it has two or more cilia arising from its innermost portion. The cilia, being long, extend into the cavity of the succeeding cell, which is tubular and which joins a collecting duct. The duct cells generally have cilia, too, and they often have microvilli or elaborate folds that increase their surface area.

In flame cells that have been studied with the electron microscope, the walls of the cups have slits (Figure 6.2). These are places where the flow of water from the

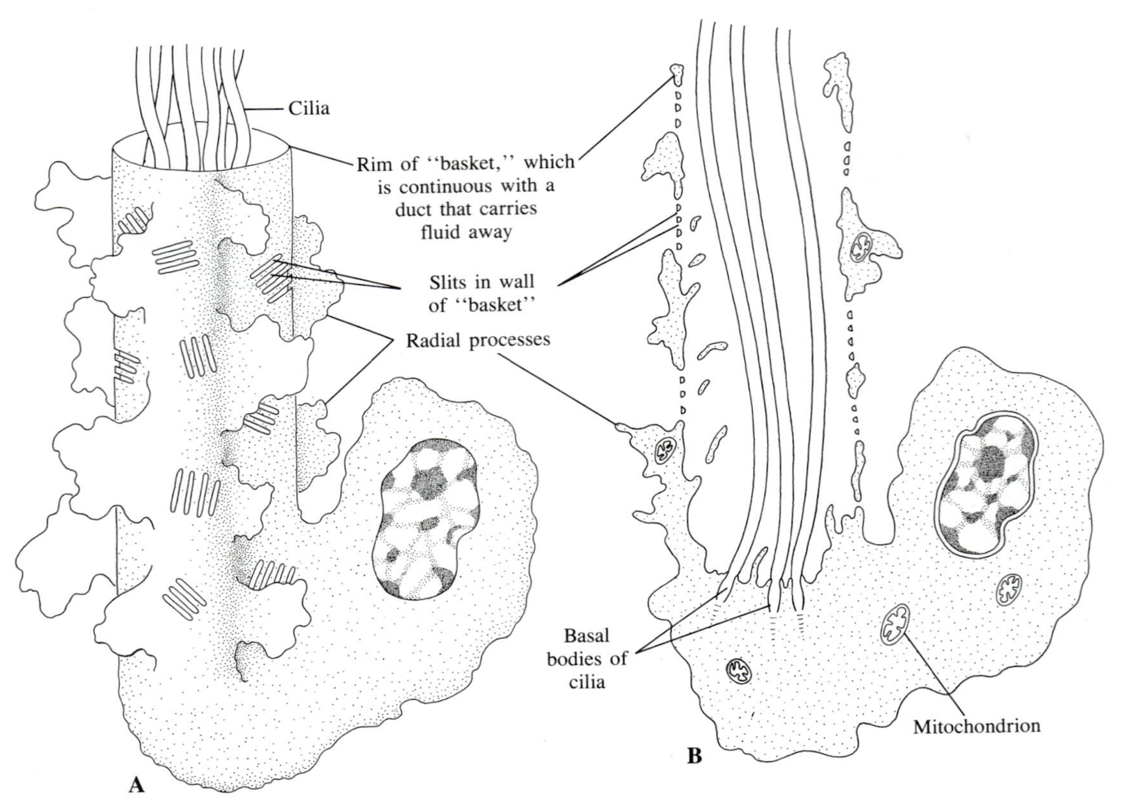

FIGURE 6.2 Flame cell of a planarian; diagrams based on electron micrographs. A. Entire cell. B. Lengthwise section of the cell. (After McKanna, Zeitschrift für Zellforschung und Mikroskopische Anatomie, *92*.)

outside of the cell into the hollow cavity is most rapid. Pumping out excess water is cellular work and requires an expenditure of energy. It is thought that the beating of the long cilia within the hollow cavity of a flame cell and its adjacent tubular cell drives fluid in the direction of the external pore, thereby creating a negative pressure in the cavity and drawing water continuously into the hollows of the flame cells. On the basis of what has been learned about the protonephridia of a rotifer, *Asplanchna,* it appears that the flow is indeed due to such a differential in pressure. It is likely, moreover, that some reabsorption takes place in the ducts leading away from the flame cells.

Protonephridia could conceivably function in the excretion of nitrogenous wastes, and perhaps of other substances, as well as in osmoregulation. Probably, however, elimination of nitrogenous substances—including ammonia, which is almost certainly produced in large amounts by carnivorous flatworms, and which, because of its toxicity, cannot be allowed to accumulate—takes place over much of the body surface. The epithelium that lines the gut is also believed to function in excretion.

The number and arrangement of protonephridia varies. Sometimes there is just one major collecting duct serving all of the flame cells; sometimes there are right and left systems, with the two major collecting ducts leading to separate pores or to a single pore. In planarians and certain other flatworms, there are several pores on both sides (Figure 6.1, A), and the ducts are interconnected in such a way that they form networks.

From the position of the flame cells within the parenchyma, one might conclude that protonephridia are of mesodermal origin. However, flame cells, tubular cells, and the cells that line the collecting ducts are derived from ectoderm cells that turn inward.

CLASSIFICATION OF FLATWORMS

It has been a tradition to divide the phylum Platyhelminthes into three classes: Turbellaria, mostly free-living flatworms with external ciliation and various gut arrangements; Trematoda, parasitic, without cilia as adults, and with a bilobed gut; and Cestoda, parasitic, without cilia, and without a gut. This is a convenient system and it is easily understood. Unfortunately, the

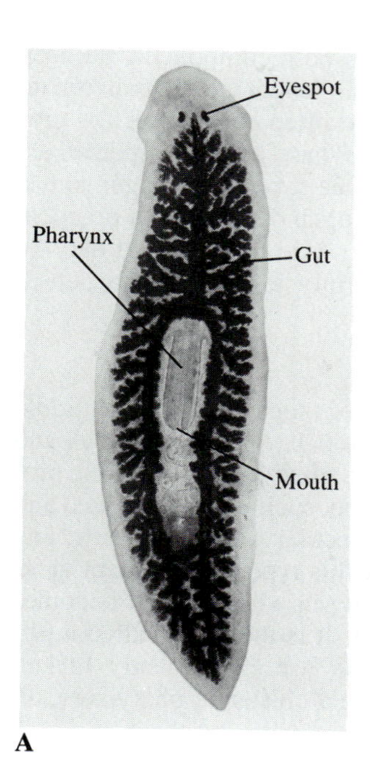

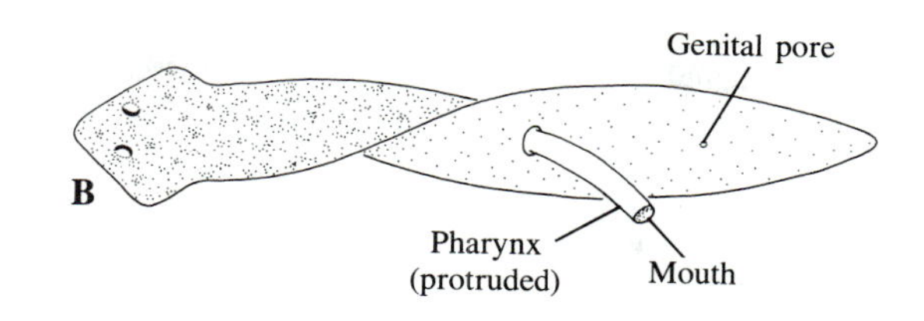

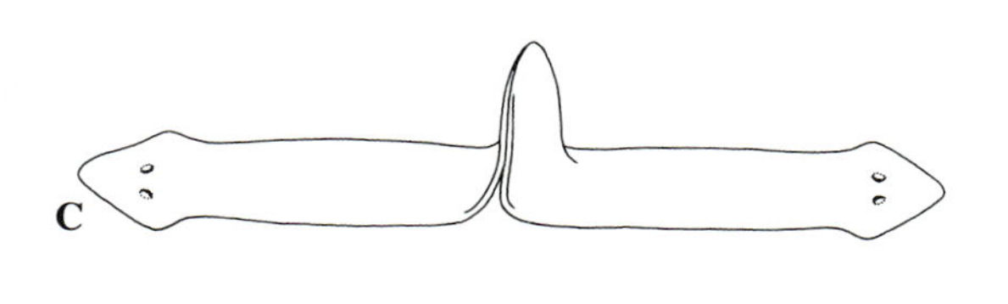

FIGURE 6.3 Planarians. A. Photomicrograph of a specimen that has eaten food containing particles of carmine. (Courtesy of Carolina Biological Supply Co., Inc.) B. Worm with its pharynx protruded. C. Pair of worms mating.

Trematoda seems to consist of three different assemblages, and the Cestoda has its nonconformists, too. The scheme of classification presented here has six classes: Turbellaria, Digenea, Monogenea, Aspidogastrea, Cestoda, and Cestodaria. The Digenea, Monogenea, and Aspidogastrea, taken together, constitute the old Trematoda; the Cestoda plus the Cestodaria equal the old Cestoda.

CLASS TURBELLARIA

The turbellarians, unlike the classes of flatworms that follow, are mostly free-living. Their body surfaces are typically ciliated, at least on the ventral side, although some parasitic types lack external cilia altogether. Planarians, found almost routinely in lakes, ponds, and streams, are good examples of the class. Some common genera are *Planaria, Dugesia, Polycelis,* and *Phagocata*. These worms, which belong to the order Tricladida, are by no means the simplest turbellarians, but they illustrate the principal features of the class and of the phylum Platyhelminthes as a whole.

Planarians as Examples of Turbellaria

External Features, Epidermal Structures, and Locomotion

A planarian (Figure 6.3) is distinctly flattened and clings tightly to whatever it is crawling on. The ventral surface is ciliated (Figure 6.4), but the steady creeping movements of the worm depend primarily on subtle peristaltic waves that run from the anterior end to the posterior end. These waves are caused by the contraction and relaxation of the muscles that lie just beneath

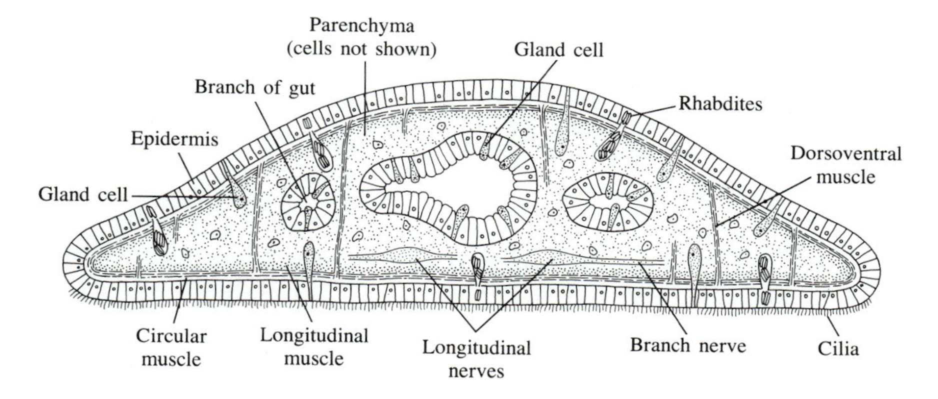

FIGURE 6.4 Transverse section of a planarian, anterior to the pharynx; somewhat diagrammatic, and with protonephridia and reproductive structures omitted.

the epidermis, and they are effective so long as the animal is on a track of mucus. Most of the mucus is produced by glands that have sunk from the epidermis into the parenchyma; the secretion reaches the surface by way of slender canals. Other subepidermal glands produce sticky material that aids in adhesion, or that breaks the bond formed by the viscid material. (This duogland adhesive system is discussed in greater detail under the next heading.)

Even so, in many turbellarians forward progression depends largely or completely on ciliary activity rather than on peristalsis. This is especially true of small species, some of which are more likely to swim than to creep. Swimming by larger turbellarians, however, depends on muscular undulations.

The rodlike bodies found singly or in bundles within the epidermal cells, or between them, are called **rhabdites** (Figure 6.4). They consist mostly of protein. When extruded, they convert into a slime that facilitates gliding and that perhaps also forms a protective coating over the animal. Various groups of turbellarians have rhabdites of one sort or another, and they are usually secreted by epidermal cells. In planarians, however, some are produced by cells beneath the epidermis; the slender necks of these subepidermal cells extend to the surface.

Duogland Adhesive System

In planarians and in most other groups of turbellarians, the cells that secrete sticky material operate in conjunction with cells that secrete a substance that breaks the adhesive bond. This arrangement is called a **duogland adhesive system.** The location of the two types of cells varies, but in general both components lie deep in the parenchyma and have slender necks that run to the surface. The openings are often on a nipplelike projection (papilla). The necks pass through specialized cells called **anchor cells.** One type of arrangement is shown in Figure 6.5. The duogland system is a practical one for organisms whose life style requires that they be able to attach themselves firmly to objects and detach themselves quickly.

Digestive System

The mouth of a planarian is located near the middle of the ventral surface (Figure 6.1, A; Figure 6.3, A and B). It leads to the muscular **pharynx** that can be protruded through the mouth to suck in the tissues and juices of animals such as insect larvae, crustaceans, and oligochaetes. A pharynx of this type, free (except at its base) of the wall of the cavity in which it lies, becomes folded when it is withdrawn. It is therefore called a **plicate pharynx.** When not in use, it is neatly tucked away. A few planarians have a cluster of pharynxes, all of which may be functional.

Continuous with the lumen (cavity) of the pharynx are the three main branches of the gut, one of which runs anteriorly, the others posteriorly (Figure 6.1, A; Figure 6.3, A). These are branched further, so that nearly all parts of the body are close to one or more of the small branches. In certain species some of the small branches unite into networks. The first stages of digestion, and apparently also the penetration of the pharynx into the prey, are initiated by a protein-splitting enzyme secreted by glands in the pharynx. Digestion is continued in the three lobes of the gut, which produce a sim-

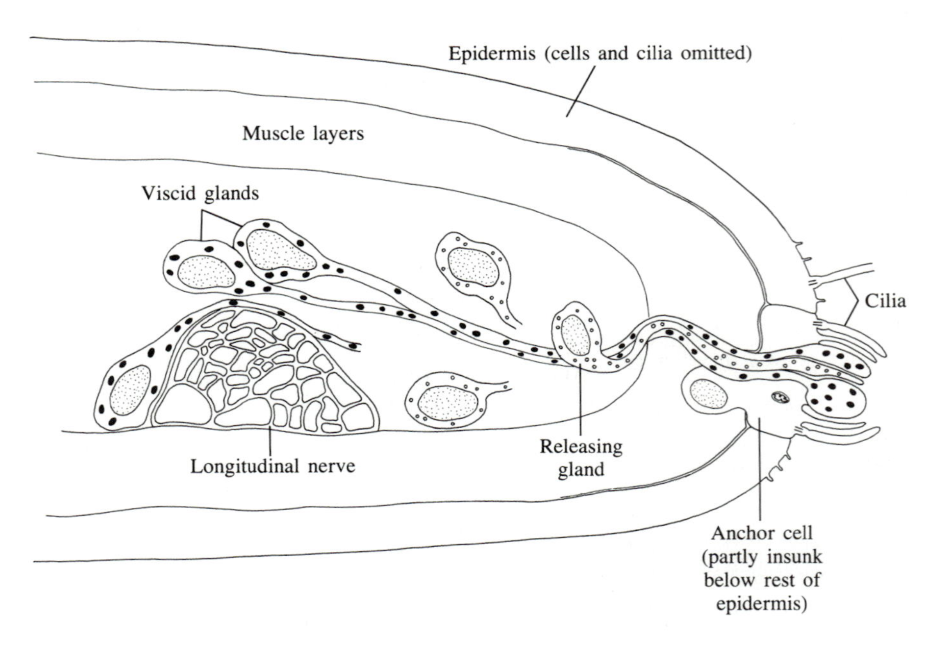

Figure 6.5 A lateral adhesive organ of *Myozona* (Class Turbellaria, order Macrostomida). There are two viscid glands and one releasing gland in each duogland complex; the necks of the glands run through an anchor cell. (After Tyler, Zoomorphologie, *84*.)

ilar or identical enzyme. Little by little the food mass is reduced to a ''soup'' of particles small enough to be engulfed by phagocytic cells of the gut epithelium; thus digestion is completed intracellularly. Any undigested residues are returned to the gut lumen. To get rid of them, the planarian first takes in water through the mouth, thus irrigating its gut lobes; then it contracts its body wall muscles to flush the water out. The gut functions not only in digestion but also in distributing the products of digestion, and in this respect it is similar to the gut of cnidarians and ctenophores. At least some excretion of nitrogenous wastes probably takes place across the epithelium of the gut, too.

Planarians typically have many flame cells and several protonephridial pores on both sides of the body (Figure 6.1, A). The ducts that drain the flame cells sometimes form complicated networks (Figure 6.1, C).

Nervous System and Sense Organs

The brain of a planarian, located close to the anterior end, is bilobed (Figure 6.1, A). Many of the nerves originating from it fan out directly into the sense organs of the head region. Much of the body, however, is innervated by branches of the two large **longitudinal nerves** that run ventrolaterally through the parenchyma. There are usually some smaller longitudinal nerves also.

Most of the sensory structures are concentrated in the head region. The sensitivity of a planarian to tactile stimuli can be demonstrated by touching its anterior end with a needle. The worm will respond by pulling back and then moving in another direction. Evidence for chemoreception can be observed when a piece of fish or crayfish is placed into a stream. The planarians, sensing the direction from which the juices are coming, move upstream and locate the source. Little is known about the chemoreceptors, but they seem to be clusters of specialized cells on the dorsal surface of the head region.

The position of photoreceptors is revealed by the eyespots (Figure 6.1, A and B; Figure 6.3, A). In certain genera there are two cuplike masses of dark pigment; in others, there are several to many small, scattered eyespots. As explained earlier, the pigment masses shield the photoreceptor cells so these will be stimulated only by light coming from certain directions.

Sexual Reproduction and Development

Like almost all flatworms, planarians are **hermaphroditic.** The male and female reproductive systems (Figure 6.1, A) share a **common genital pore** and **genital atrium.** The basic elements of the male system are numerous **testes** arranged in two sets, one on each side of the anterior part of the body. The **sperm ducts** lead to a muscular **penis** that is protruded from the atrium at the time of copulation. The female system includes a pair of compact **ovaries** situated rather far forward. The **oviducts,** on their way to the genital atrium, are entered by ducts from **yolk glands (vitellaria)** scattered through the lateral portions of the body.

During copulation (Figure 6.3, C) sperm introduced by the penis of one worm into the genital atrium of its partner are stored in a sac called the **seminal bursa,** which is essentially a diverticulum of the atrium. To reach the eggs, the sperm must leave the bursa, reenter the atrium, and swim up the oviducts to small dilations called **seminal receptacles.** These are adjacent to the ovaries, and the eggs are fertilized as they emerge from the ovaries. Then the zygotes move toward the atrium, and yolk cells and other material produced by the yolk glands are added to them. The yolk provides nutrition for the developing embryos; the other products are converted into hard egg capsules, generally called **cocoons,** that enclose groups of eggs. The assembling of eggs and formation of cocoons takes place in the genital atrium. It is here also that a sticky secretion is added so that the cocoons will adhere to a solid surface when they are laid. Development of the eggs is direct and within a few weeks the juveniles emerge as worms already recognizable as planarians.

As the foregoing description is the first of many that will deal with the functional anatomy of reproductive systems of invertebrate animals, it is important to know precisely what the various terms mean. *Genital atrium, ovary, oviduct, testis, sperm duct,* and *penis* should present no problems. The potentially confusing ones are *seminal bursa, seminal receptacle,* and *seminal vesicle.* The last term was not mentioned in connection with planarians, for it is absent in these worms. Most flatworms have one, however. It is a chamber close to the penis or other copulatory device, and it stores sperm before copulation. It is thus part of the male system. The seminal bursa and seminal receptacle belong to the female system. When there is only one sac to receive the sperm and store it until eggs are ready to be fertilized, the term *seminal receptacle* is enough. But when, as in a planarian, there is a sac that initially receives the sperm and another to which sperm go before they are actually used, then the first is called a *seminal bursa,* the second a *seminal receptacle.*

Not all flatworms have a penis that can simply be protruded. In many of them, the male copulatory organ is of a type called a **cirrus,** which functions by turning inside out. Throughout this chapter, the terms *penis* and *cirrus* are used in accordance with this distinction.

Asexual Reproduction and Regeneration

Many planarians can reproduce asexually by **fission.** The posterior half of the worm holds tightly to the substrate while the anterior half moves slowly forward until the two halves break apart. The halves then regenerate whatever parts are needed to make them into whole worms. In some species, or in certain "races," sexual reproduction is either never observed or is extremely rare. The reproductive system does not develop fully and fission is the only way such planarians can multiply. In others, seasonal sexuality, correlated with an increase in water temperature, may alternate with periods in which reproduction is strictly by fission.

Most planarians have a high capacity for regeneration, as might be expected from their ability to reproduce by fission. If a planarian is cut transversely into halves or even quarters, each piece will probably develop into a whole worm. Regeneration typically begins with the epidermis' closing in to cover the cut surface. Undifferentiated cells then proliferate and their products form the tissues required to reconstitute the worm. The source of the undifferentiated cells is not certain. They could already be present in an uncut worm, being thus somewhat comparable to interstitial cells of a hydra; or they could be derived by dedifferentiation of certain specialized cells.

The extent to which small pieces will regenerate varies with the species and the distance from the head. In general, the farther a piece is from the head, the less likely it is to regenerate a complete head or other parts. For instance, the tail half of a sexually mature planarian will lose its genital organs and never regenerate them, although it develops a new head, pharynx, and other necessary structures. Because the head region exerts such strong control over regeneration, a planarian may be said to have **axial polarity.**

Many biologists have pursued planarian regeneration and the literature on the subject is enormous. A review by Rose (1970) summarizes the conclusions of experts.

Classification of Turbellaria

The classification of turbellarians is complicated and unstable. Some schemes in current use lay much stress on the presence or absence of special yolk glands in the female reproductive system. The acoels, polyclads, catenulids, and macrostomids, plus a couple of other small groups, do not have yolk glands. They would be put into the subclass Archoophora—the "old-style egg-bearers." Whatever nutritive material is necessary for the development of the embryos is formed within the eggs themselves, and cleavage is typically spiral. The subclass Neoophora—the "new-style egg-bearers"—characterized by well-developed yolk glands, would include all other groups of turbellarians. The eggs of these, surrounded by cells filled with yolk granules, do not often show distinctly spiral cleavage.

The system of classification used here subdivides the Turbellaria directly into orders, the first four of which lack yolk glands. The characterization of the orders is based primarily on the type of pharynx, if this structure is present, and on the organization of the reproductive system. A few small groups are omitted.

Order Acoela

Acoels (Figure 6.6) are the simplest turbellarians, and certain of their features may have been characteristic of the earliest flatworms. They have no pharynx and no real gut. The mouth is a simple opening on the ventral surface, usually near the middle of the body (Figure 6.6, B). Food is pushed through it by a convulsive movement of the anterior part of the body. Then it enters a **syncytium**—a multinucleate mass of protoplasm—and becomes incorporated into vacuoles within which digestion proceeds. Undigested residues, still in vacuoles, are moved toward the mouth and expelled. Most acoels, especially smaller species, feed chiefly on diatoms; others consume small invertebrates, particularly crustaceans, or a mixture of diatoms and invertebrates. Some carnivorous types have zoochlorellae or zooxanthellae in their tissues, and in at least one species the algal symbionts are diatoms that have multiplied after escaping from their siliceous cell walls.

Acoels generally have a statocyst close to the brain (Figure 6.6, A), and in a few species there are also pigment masses that form distinct eyespots. Other visible sensory structures are aggregates of stiff cilia scattered over the surface of the body, especially around the margins. These are thought to be touch receptors. The epidermis secretes structures that resemble rhabdites, but they consist of mucopolysaccharide rather than protein. It is best to call them **rhabdoids.**

The reproductive system of acoels is simple compared with that of most other turbellarians. The gonads are not really distinct. Sperm develop in packets in lateral areas of the body and migrate by poorly defined ducts to a seminal vesicle. Associated with the seminal vesicle is a muscularized penis or cirrus that can be extended through the male genital pore. Sometimes the male and female portions of the reproductive system share a common genital pore (Figure 6.6, D); sometimes the male and female pores are widely separated (Figure 6.6, E). The **vagina** leads to a seminal receptacle in which activated sperm accumulate. As oocytes that proliferate in lateral regions of the body (or along the midline in some acoels) enlarge, they move posteriorly until they are close to the seminal receptacle. Sperm reach the eggs through one or more sclerotized nozzles that project from the receptacle, and fertilized eggs are laid through temporary openings in the epidermis of the ventral surface. A few acoels have neither a female pore nor a vagina. It is believed that sperm are injected by one partner into the tissues of the other.

Development is direct; when a young acoel breaks out of the egg membrane, it resembles the adult, although it has no reproductive system. An interesting

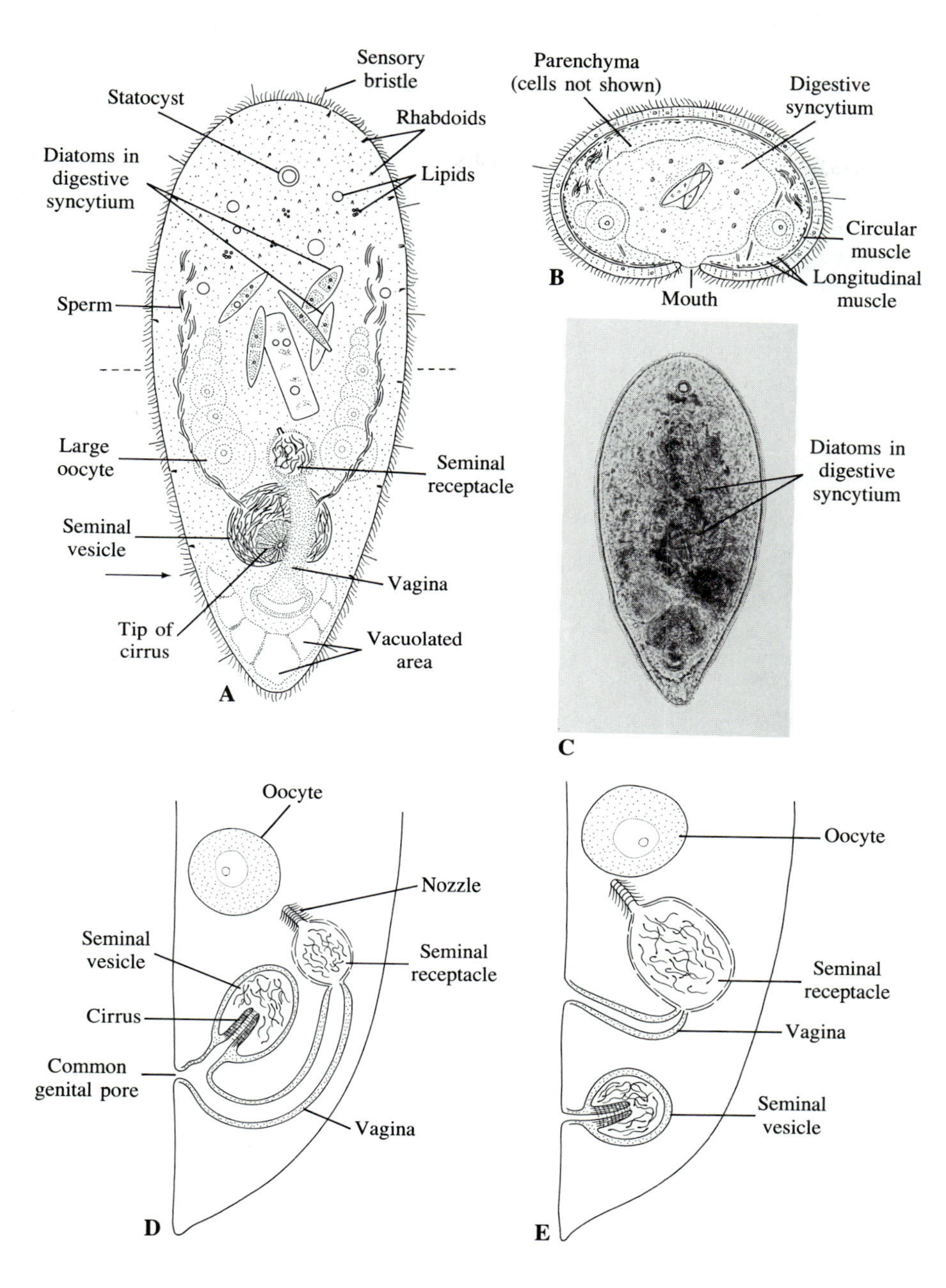

FIGURE 6.6 Order Acoela. A. *Otocelis luteola,* dorsal view of a living specimen. The genital pore, on the ventral surface, is at the level indicated by an arrow. B. *Otocelis,* transverse section at the level indicated in A by a dashed line; rhabdites omitted. C. *Otocelis,* photomicrograph of a living specimen. D. Diagram of the copulatory components of the reproductive system of an acoel similar to *Otocelis*. E. Diagram of the copulatory components of the reproductive system of an acoel with separate male and female genital pores.

feature of the embryology of acoels is that spiral cleavage begins at the second division instead of the third division (Figure 6.7). Thus at the 4-cell stage there are two micromeres on the furrow that separates the two macromeres. The same pattern is followed for several successive divisions, so the micromeres form duets instead of quartets.

With one known exception, acoels are marine. The majority of them are no more than 1 mm long, although a few reach 5 mm. They can almost always be found in diatom-rich sediments that cover mud, sand, stones, and seaweeds.

Order Catenulida

Catenulids are found in fresh water and also in marine and brackish habitats. They reproduce by transverse fission, and many species form chains of individuals that are in various stages of differentiation (Figure 6.8, D). As a rule, differentiation of an individual is completed before it breaks away, or before a group of

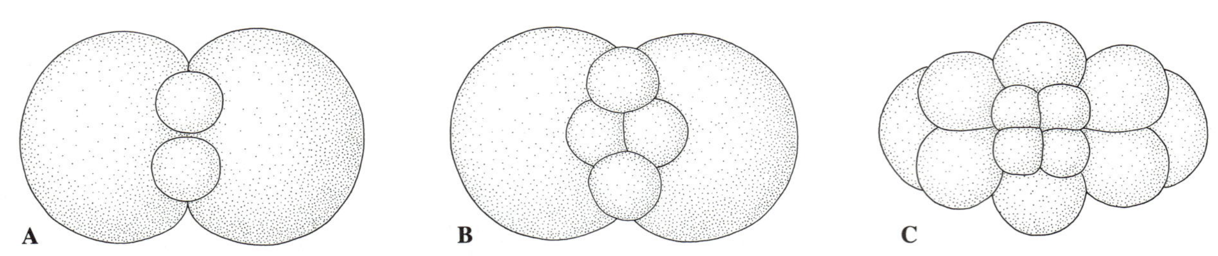

FIGURE 6.7 Order Acoela. Spiral cleavage in *Polychoerus caudatus;* embryos viewed from above. A. Second cleavage completed; two micromeres on the furrow separating the two macromeres. B. Four micromeres, two macromeres. C. Sixteen-cell stage (not all cells visible). (After Gardiner, Journal of Morphology, *11.*)

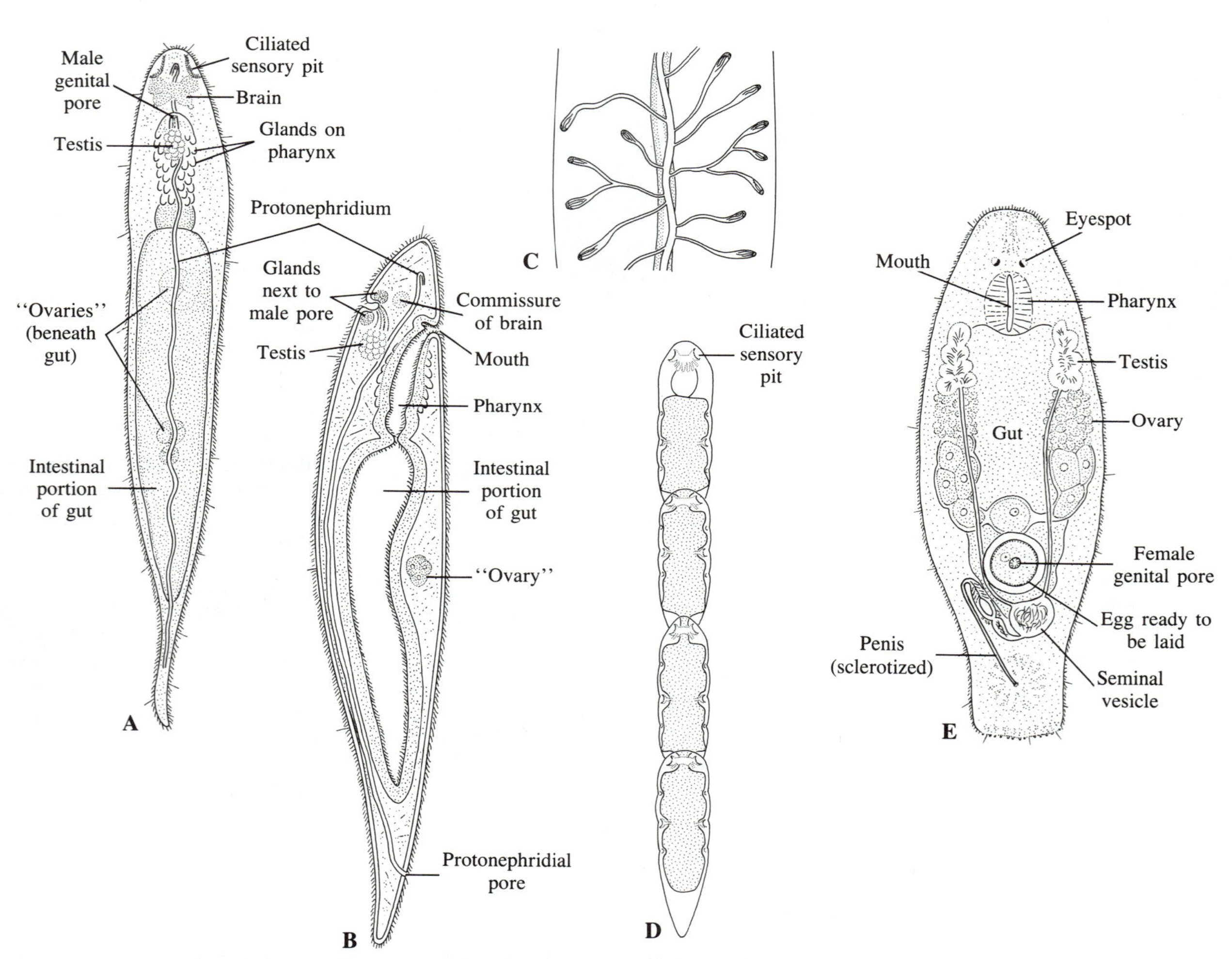

FIGURE 6.8 A. *Stenostomum oesophagium* (order Catenulida), dorsal view. Only the thicker portion of the duct of the protonephridium, which leads to the pore, is visible. B. *Stenostomum oesophagium,* sagittal section; diagrammatic. (A and B, after Kepner, Carter, and Hess, Biological Bulletin, *64.*) C. *Stenostomum hemisphericum,* portion of the protonephridium. The many flame bulbs join the slender duct, which runs forward for nearly the whole length of the body. Near the anterior end, the duct becomes thicker and the thickened portion extends backward to the pore. D. *Stenostomum grande,* a chain of 16 members in various stages of differentiation. (C and D, after Marcus, Boletin da Faculdade de Filosofia, Ciências e Letras, Universidade de São Paulo, *10.*) E. *Macrostomum grande* (order Macrostomida). (After Marcus, Boletin da Faculdade de Filosofia, Ciências e Letras, Universidade de São Paulo, *11.*)

which it is the leading member breaks away. Each of the ciliated epidermal cells of a catenulid has only a few cilia, instead of many, as is typical for flatworms in general.

A common genus in fresh water is *Stenostomum* (Figure 6.8, A–D). Because of their small size and cylindrical form, stenostomums are easily confused with ciliates. Neither eyespots nor a statocyst are present, but there is a pair of light-refracting granules, or clusters of granules, associated with the brain. The function of these has not been determined. Four pairs of longitudinal nerves run backward from the brain, but only those of the ventralmost pair are prominent. The pharynx is simple but its epithelium is ciliated, and these worms feed on small animals, especially protozoans. There is a single protonephridium. The many flame bulbs join a slender, anteriorly directed duct; in front of the brain this becomes thicker, and runs backward above the gut, finally reaching the pore on the ventral surface close to the posterior end.

All species of *Stenostomum* multiply by fission, and chain formation is commonly observed. The reproductive system is often totally absent, although it may appear and become functional under certain circumstances. Gonads, when present, are unpaired. The ovary consists of one or more clusters of cells beneath the gut; eggs are released singly through breaks in the ventral epidermis. The testis and male genital pore, equipped with an eversible cirrus, are located dorsally at about the level of the pharynx. Insemination takes place by injection of sperm through the epidermis of the partner.

Order Macrostomida

Macrostomids, like catenulids, have a relatively simple pharynx and they lack a statocyst. Most species, however, have eyespots. The gonads are paired, and the male and female genital pores are on the ventral side of the body near the posterior end.

The genus *Macrostomum* (Figure 6.8, E) is represented in fresh water (including hot springs) and in marine and brackish habitats. The hind portion of its slightly flattened body is spatulate and adhesive.

There is considerable diversity in this small order, and some members of the group are decidedly different from *Macrostomum*. Certain of the marine representatives are like catenulids in that they divide transversely to form chains.

Order Polycladida

Polyclads (Figure 6.9, A and B) are not at all similar to acoels, catenulids, or macrostomids, even though they share with these worms the negative character of the absence of yolk glands. They are relatively large when compared to the members of the three orders that have been discussed previously; most of them are in the range of 1 to 4 cm, but some are even larger. The gut is extensively ramified and the branches reach to almost all parts of the body. There is a muscular pharynx, often ruffled, that can be protruded, and many species use this to envelop their prey. All polyclads are carnivores; they feed on various molluscs, crustaceans, and other invertebrates. Some suck out the zooids in colonies of compound ascidians.

The most obvious sensory structures observed in almost all polyclads are eyespots. There are usually many of them, concentrated in clusters near the brain, at the bases of a pair of tentacles, or along the margins of the body.

The reproductive system of a polyclad (Figure 6.9, A and C) is rather complex, even though there are no yolk glands and the gonads are not well developed. The ovaries and testes are simple follicles scattered between branches of the gut. An odd feature of polyclads is that the right and left major sperm ducts, into which all of the smaller sperm ducts empty, are joined posteriorly. They terminate at a seminal vesicle–penis complex that may have various specializations, including a sclerotized penis stylet. In the female system, the oviducts unite before reaching the genital pore, and there is generally a seminal receptacle (sometimes called *Lang's vesicle*) associated with the common duct. Glands near the female genital pore are believed to produce the material that enables a polyclad to stick clusters of eggs to rocks or to other firm surfaces.

The eggs of polyclads provide the best examples of spiral cleavage found in flatworms (Figure 6.10, A–D). Subsequent development toward maturity is usually direct, but sometimes indirect. In species that have indirect development, the classical type of swimming larva is called *Müller's larva* (Figure 6.10, E). Its body is drawn out into four pairs of lobes, and there is a tuft of specialized sensory cilia at the anterior end. There are three eyespots closely associated with the brain. The two that are symmetrically placed on both sides of the midline have typical flatworm photoreceptor cells whose surfaces are folded into microvilli. The third eye has modified cilia, such as are found in photoreceptors of cnidarians, echinoderms, and vertebrates. The larva is gradually transformed into a typical young polyclad, which adopts a benthic existence. *Götte's larva* is similar to Müller's larva, but has only two pairs of lobes.

All polyclads are marine and are most likely to be found on the undersides of rocks, to which they can cling tightly because of sticky secretions. Some species, when dislodged, are good swimmers, moving by graceful undulations. The swimming movements are especially attractive if the margins of the body are frilly, as they are in polyclads of the type called ''skirt dancers.''

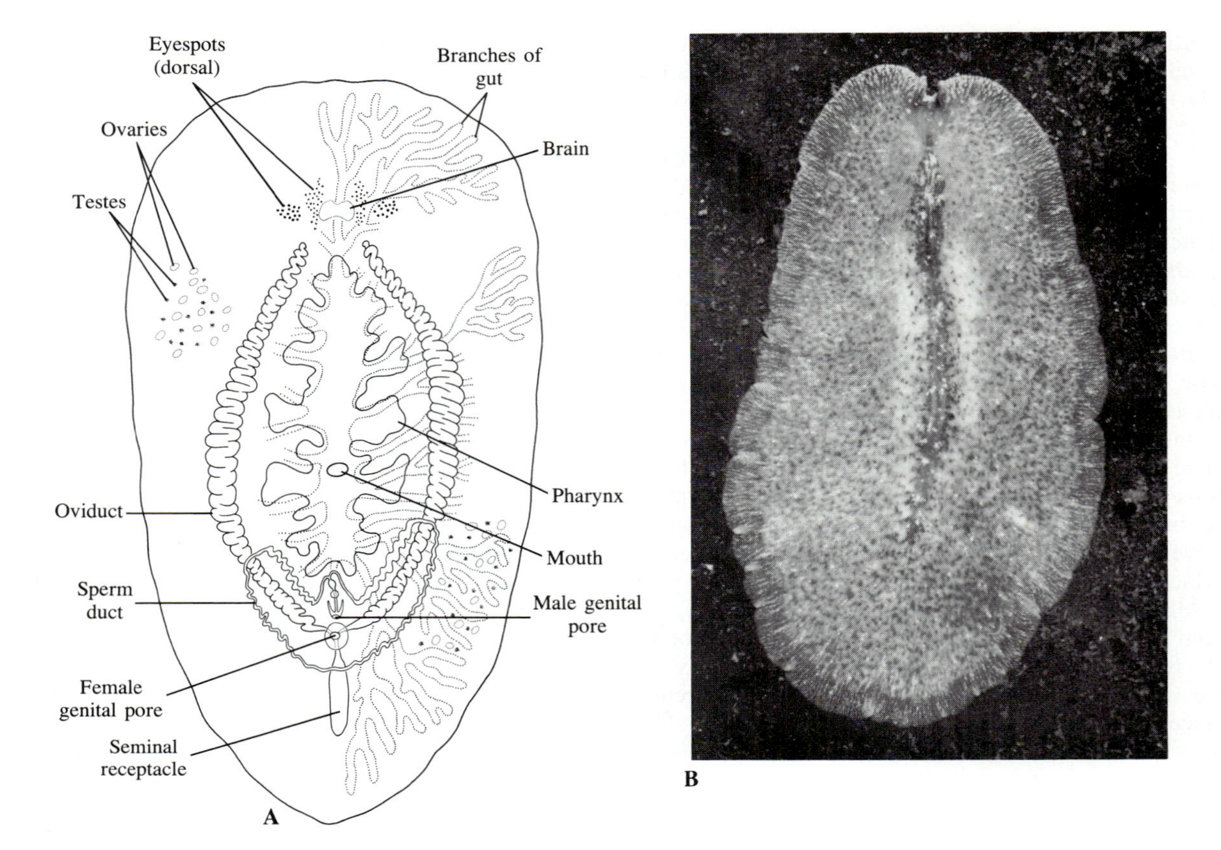

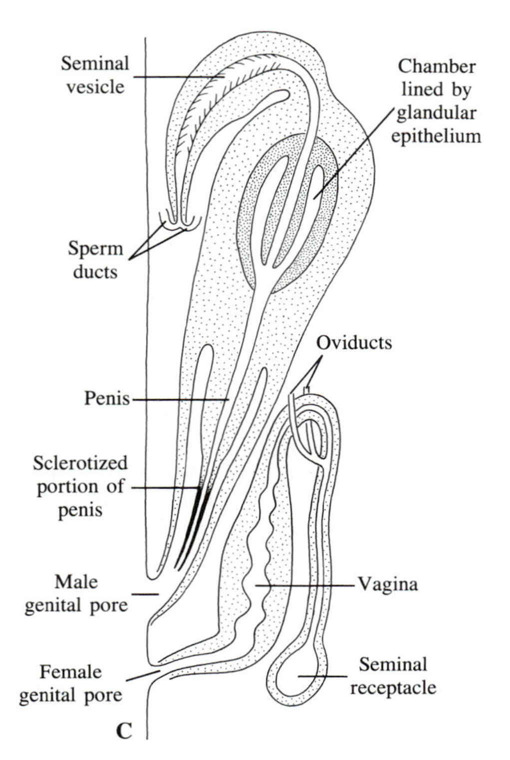

FIGURE 6.9 Order Polycladida. A. Diagram of a typical polyclad, ventral view. B. *Pseudoceros canadensis*. C. Copulatory portions of the reproductive system of *Notoplana*, as seen in sagittal sections; somewhat diagrammatic.

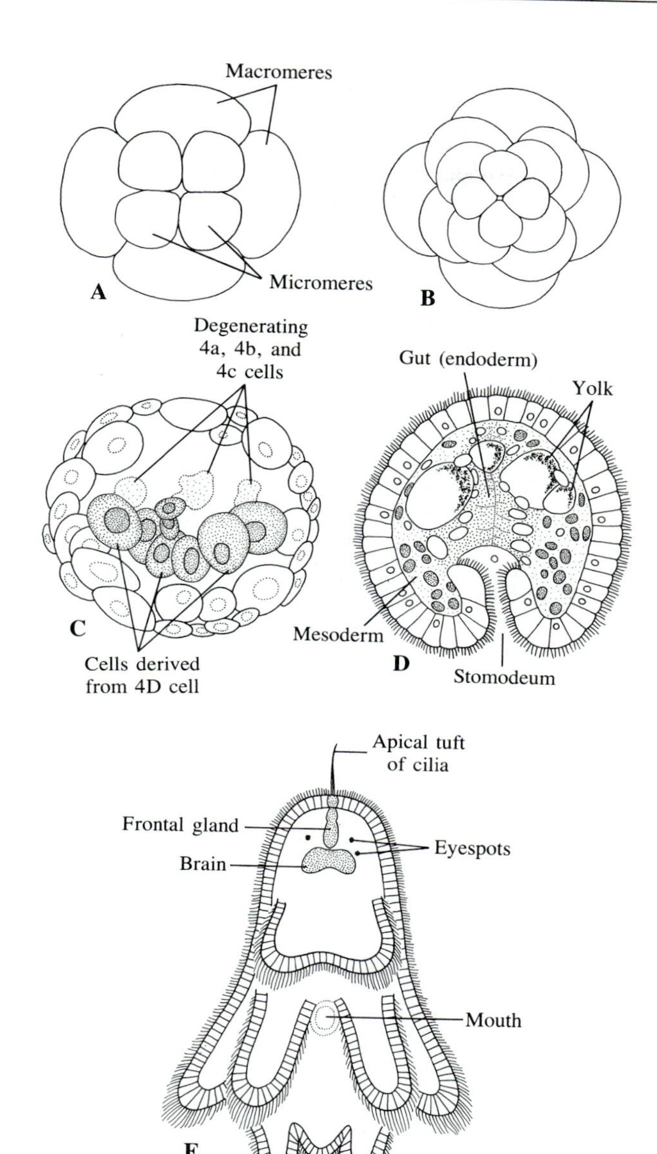

FIGURE 6.10 Order Polycladida. A. Embryo of *Hoploplana inquilina* after the third cleavage (four micromeres on the furrows between macromeres). B. Embryo after the fourth cleavage; the micromeres and macromeres shown in A have divided again. C. Section through an early gastrula, showing progeny of the 4d cell, which form mesoderm and endoderm. D. Section through a slightly later stage, after the endodermal portion of the gut has begun to differentiate and the stomodeum has started to form as an invagination of ectoderm. (After Surface, Proceedings of the Academy of Natural Sciences of Philadelphia, *59*.) E. Müller's larva of a polyclad, dorsal view. The three eyespots are more closely associated with the dorsal side of the brain than the drawing suggests. (After Kato, Japanese Journal of Zoology, *8*.)

Order Neorhabdocoela

This is the largest single group of turbellarians, and it is well represented in both marine and freshwater habitats. It is essentially comparable to the old order Rhabdocoela, minus some nonconformists that have been transferred to other groups. The term *rhabdocoel* is here understood to mean a member of the Neorhabdocoela. Except for a few aberrant types, rhabdocoels have a gut, and this is unbranched, although it may have shallow side pockets. A muscular, **bulbous pharynx** (Figure 6.11, A–C; Figure 6.12, A) is typical of the order, and is used for swallowing small invertebrates, especially crustaceans, or for sucking in tissues and juices of prey.

The reproductive system of rhabdocoels is generally rather condensed. There are one or two definite testes and one or two ovaries, and yolk glands concerned with production of cells that contain nutritive material as well as material that will be converted into thick-walled **egg capsules.** Each capsule contains one to several fertilized eggs.

An excellent introduction to the functional morphology of the reproductive system of rhabdocoels is provided by *Syndisyrinx* (Figure 6.11, B and C) or its close relative, *Syndesmis*. Worms of both genera are found in the intestine of sea urchins, where they feed on gut epithelium and symbiotic ciliates. In living specimens and in stained whole mounts, much of the reproductive system is laid out with almost diagrammatic clarity. Certain details, however, must be worked out with the help of serial sections.

At the posterior end of the body is the genital atrium. This has a single opening to the outside, but it serves both the male and female systems. In the male system, the ducts that carry sperm from the two testes unite to form a common sperm duct, or **vas deferens.** This leads to a slight dilation, the seminal vesicle, where sperm are stored. From the seminal vesicle a long, muscularized ejaculatory duct extends posteriorly toward the genital atrium. This duct, which in some species is coiled back on itself at least once, is succeeded by a sclerotized penis stylet, which serves as the inseminating device. The bore of the stylet is extremely narrow, but large enough to allow sperm to pass through it. Muscles controlling the position of the ejaculatory duct with respect to other tissues keep the stylet withdrawn or allow it to protrude through the male pore and finally from the common genital atrium. Just how it finds its way into the vagina of the female system of another worm is something no one has explained.

The genital atrium is a good place to begin a description of the female system. Note that there are two female openings, one leading into the vagina, the other into the **uterus.** Sperm forced into the vagina first accumulate in the seminal bursa and then go on to the

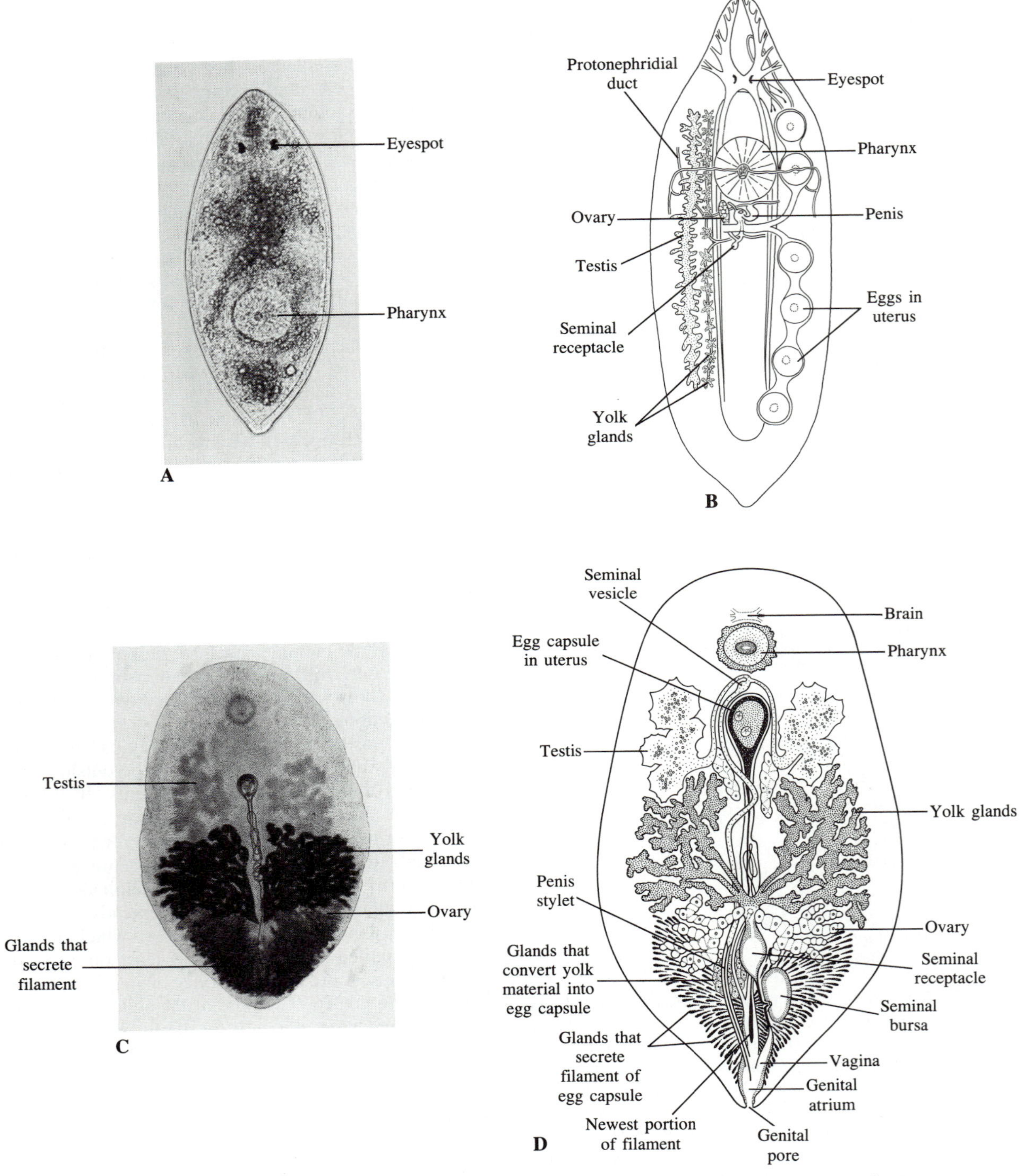

FIGURE 6.11. Order Neorhabdocoela. A. Typical small, free-living rhabdocoel; photomicrograph. B. *Mesostoma ehrenbergi;* somewhat diagrammatic, with certain of the symmetrically placed structures shown on only one side or the other. (After Ruebush, Science, *91*.) C. *Syndisyrinx franciscanus,* stained specimen; photomicrograph. D. *Syndisyrinx franciscanus,* from the gut of a sea urchin, dorsal view. The bulbous pharynx is shown, but the rest of the gut is omitted.

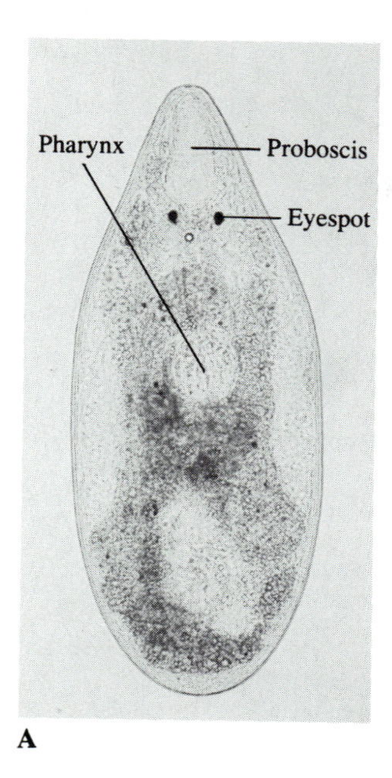

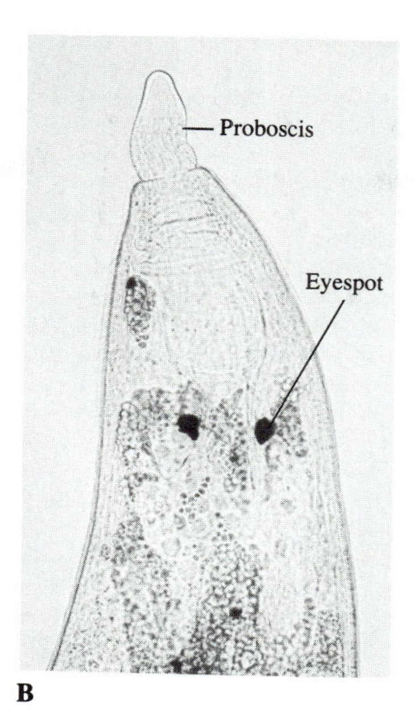

FIGURE 6.12 Order Neorhabdocoela. A. Kalyptorhynch. B. Anterior end of the kalyptorhynch, proboscis partly protruded.

seminal receptacle. (In most flatworms, there is no bursa; the seminal receptacle is the only sperm-storing sac in the female system.) The eggs are produced by two ovaries and move into a chamber adjacent to the seminal receptacle, where sperm from the seminal receptacle can reach them and where fertilization takes place. Several zygotes, accompanied by numerous nutritive yolk cells produced by the yolk glands, enter a duct in which nonnutritive secretions of the yolk glands are converted into an egg capsule. After the capsule reaches the uterus, the crowded glands that surround this structure produce a secretion that hardens to form a filament attached to the capsule. As the filament becomes longer, the capsule is pushed progressively farther forward and the uterus enlarges to accommodate it. Eventually, the capsule and filament are expelled from the uterus and pass out of the genital atrium. The filament presumably serves as a lasso that gets tangled up with a piece of seaweed or something else that may eventually be eaten by an appropriate species of sea urchin.

In about two months, the young worms within the egg capsule reach the hatching stage. If taken into the gut of an urchin, the capsule breaks up into many pieces. It is likely that the fracturing of the capsule depends on enzymes produced by the juveniles themselves, rather than by the host; nevertheless, hatching requires the presence of host tissue.

The remarkably transparent rhabdocoels of the genus *Mesostoma* (Figure 6.11, D), found in freshwater lakes and ponds, are also good subjects for study. They are fairly large—about 2 to 5 mm long—and have an uncluttered anatomy. The bulbous pharynx, which can be protruded, and the rest of the gut are easily seen. The reproductive system shows most of the components described for *Syndisyrinx,* although the proportions of the various parts are different. Protonephridia, lacking in *Syndisyrinx,* are well developed; their ducts open to the outside on both sides of the mouth. There are prominent eyespots associated with the large brain.

Some species of mesostomas produce two kinds of eggs. The so-called summer eggs have thin capsules and little yolk, and they hatch in two or three weeks. The resting eggs have thick capsules and considerable yolk, and they remain dormant from autumn to spring. The summer eggs are produced by self-inseminating worms, some of which have hatched from resting eggs that have overwintered. The reproductive systems of these worms are not fully developed, hence the thin egg capsules and slight amount of yolk accompanying each egg. Nevertheless, there may be more than one generation of worms that produce eggs of this type. Eventually, as autumn approaches, the reproductive system of some or most worms becomes fully equipped with the glandular components needed for forming resting eggs, and cross-insemination replaces self-insemination.

Kalyptorhynch rhabdocoels (Figure 6.12), common in the surface sediment of sandy and muddy marine habitats, are unusual in that they have a short **proboscis** at the anterior end. This has nothing directly to do with the gut, but it can be protruded from its pocket and functions in the capture of prey. Sticky secretions, and sometimes also a pair of stout hooks, assist in grasping the victim. The proboscis can be bent downward and

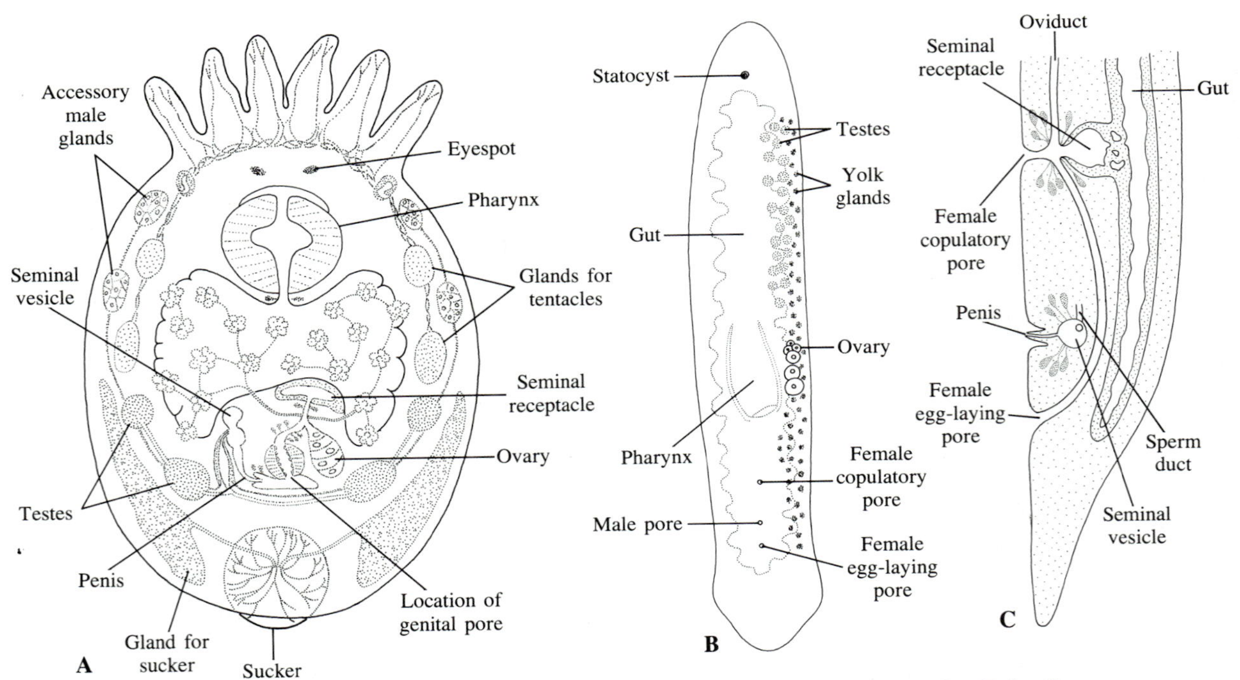

FIGURE 6.13 A. *Temnocephala novae-zealandiae* (order Temnocephalida). (After Fyfe, Transactions of the Royal Society of New Zealand, 72.) B. *Monocelis* (order Protricladida).

backward, thus bringing the prey within reach of the pharynx, which is also protrusible.

Order Temnocephalida

Temnocephalans are found only on freshwater crayfishes in South America, Australia, New Zealand, islands of Polynesia, and Madagascar (the best-known species are certain of those occurring in New Zealand). Most worms of this interesting group have several fingerlike outgrowths at the anterior end (Figure 6.13, A), but in some there are only two of these, and in a few others there are outgrowths along the sides of the body as well as at the anterior end. As a rule, the external surfaces are not ciliated. This is one reason why temnocephalans are sometimes advanced to the rank of class.

Temnocephalans are primarily commensals that feed on diatoms and other small organisms that live on the surfaces of their crayfish hosts. They stick their stalked egg capsules to the exoskeleton. When the young hatch, they resemble the adults.

Order Protricladida

Most of the protriclads used to be placed in an order called Alloeocoela. This turned out to be a rather unnatural assemblage and many of its members have been transferred to other orders, including some that have so few representatives that they are not dealt with here. Protriclads are largely marine and are common in sandy and muddy beaches where there are plenty of small crustaceans for them to prey upon. They are usually rather elongate worms, and those that reach a length of 2 or 3 mm can be seen with the naked eye if they are crawling in a glass dish set on a dark background. A typical protriclad (Figure 6.13, B) will have a **protrusible pharynx** much like that of a planarian. The gut, which runs nearly the length of the body, may have many shallow lateral diverticula, but there are no main branches. There is almost always a statocyst near the anterior end of the body. The posterior end is often thinned out into a spatulate adhesive structure.

The testes, ovaries, and yolk glands take up most of the space not occupied by the gut. Some protriclads have three genital pores: one for the penis, one for the vagina, and another for the duct through which fertilized and encapsulated eggs reach the outside (Figure 6.13, B and C). In other protriclads, however, there is a single pore serving all three of these components of the reproductive system.

Order Tricladida

Planarians (Figures 6.1, 6.3, 6.4), already discussed because they show the important general features of the Turbellaria, belong to this order. The name *Tricladida*

FIGURE 6.14 Order Tricladida. *Orthodemus terrestris,* a terrestrial turbellarian.

refers to the three branches of the gut, one of which extends anteriorly, the other two posteriorly. The mouth, usually located near the middle of the ventral surface, is at the tip of a protrusible pharynx. (In some planarians, there are several functional pharynxes.) Although the reproductive system may deviate in one respect or another from that of a planarian, the layout typically includes a pair of compact ovaries and scattered yolk glands and testes (Figure 6.1, A). There are many variations on the basic plan, however, as should be expected in a group as large as this.

The order Tricladida is ecologically diversified; it has representatives in terrestrial habitats as well as in fresh water and the sea. A few of the marine species are commensal with other invertebrates. *Bdelloura* and *Syncoelidium,* which live on the gill lamellae of the horseshoe crab, *Limulus,* are perhaps the best-known commensal genera. They are odd in that they have two female genital pores, each associated with a seminal receptacle. These complexes are lateral to the ventral midline, where the male genital pore and penis are located. *Bdelloura* has a rather elaborate adhesive disk at its posterior end.

Terrestrial triclads, which feed on small invertebrates, are primarily found in humid tropical or subtropical regions, but a few occur naturally in temperate areas. A number of them are established in greenhouses or outdoors in places remote from their original homes. The native habitat of some domesticated species is not even known. *Bipalium kewense,* for instance, was described from specimens found in the Royal Botanical Gardens at Kew, just outside of London, to which it was undoubtedly introduced with plant material. It now has a wide distribution in greenhouses and gardens. A large specimen of *B. kewense* may be more than 20 cm long, and it looks more like a nemertean than a flatworm. Another common garden type in parts of America and Europe is *Orthodemus terrestris* (Figure 6.14), which looks like a little black slug. Some tropical triclads are large, heavy-bodied, and strikingly marked.

References on Turbellaria

Apelt, G., 1969. Fortpflanzungsbiologie, Entwicklungszyklen, und vergleichende Frühenentwicklung acöler Turbellarien. Marine Biology, *4*:267–325.

Ax, P., and Borkett, H., 1968. Organisation und Fortpflanzung von *Macrostomum romanicum* (Turbellaria, Macrostomida). Zoologischer Anzeiger, Supplement, *32*:344–347.

Bowen, I. D., and Ryder, T. A., 1974. The fine structure of the planarian *Polycelis tenuis* (Ijima). III. The epidermis and external features. Protoplasma, *80*:381–392.

Boyer, B. C., 1971. Regulative development in a spiralian embryo as shown by cell deletion experiments in the acoel *Childia*. Journal of Experimental Zoology, *176*:97–105.

Douglas, A. E., 1987. Experimental studies on symbiotic *Chlorella* in the neorhabdocoel Turbellaria *Dalyellia viridis* and *Typhloplana viridata*. British Phycological Journal, *22*:157–161.

Jennings, J. B., 1957. Studies on feeding, digestion, and food storage in free-living flatworms. Biological Bulletin, *112*:63–80.

Kenk, R., 1972. Freshwater planarians (Turbellaria) of North America. Biota of Freshwater Ecosystems, Identification Manual No. 1, Environmental Protection Agency, Washington.

Kozloff, E. N., 1972. Selection of food, feeding, and physical aspects of digestion in the acoel turbellarian *Otocelis luteola*. Transactions of the American Microscopical Society, *91*:556–565.

McKanna, J. A., 1968. Fine structure of the protonephridial system in planaria. I. Flame cells. Zeitschrift für Zellforschung und Mikroskopische Anatomie, *92*:509–523.

Martin, G., 1978. A new function of rhabdites: mucous production for ciliary gliding. Zoomorphologie, *91*:235–248.

Martin, G., 1978. Ciliary gliding in lower invertebrates. Zoomorphologie, *91*:249–261.

Moraczewski, J., 1977. Asexual reproduction and regeneration of *Catenula*. Zoomorphologie, *88*:65–80.

Nentwig, M. R., 1968. Comparative morphological studies after decapitation and after fission in the planarian *Dugesia dorotocephala*. Transactions of the American Microscopical Society, *97*:297–310.

Prudhoe, S., 1982. British Polyclad Turbellarians. Synopses of the British Fauna, no. 26. E. J. Brill/Dr. W. Backhuys, London and Leiden.

Prudhoe, S., 1985. A Monograph on Polyclad Turbellaria. British Museum (Natural History) and Oxford University Press, London.

Prusch, R. D., 1976. Osmotic and ionic relationships in the freshwater flatworm *Dugesia dorotocephala*. Comparative Physiology and Biochemistry, *54A*:287–290.

Reisinger, E., and Kelbetz, S., 1964. Feinbau und Entladungs-mechanismus der Rhabditiden. Zeitschrift für Wissenschaftliche Mikroskopie und Mikroskopische Technik, *65*:477–508.

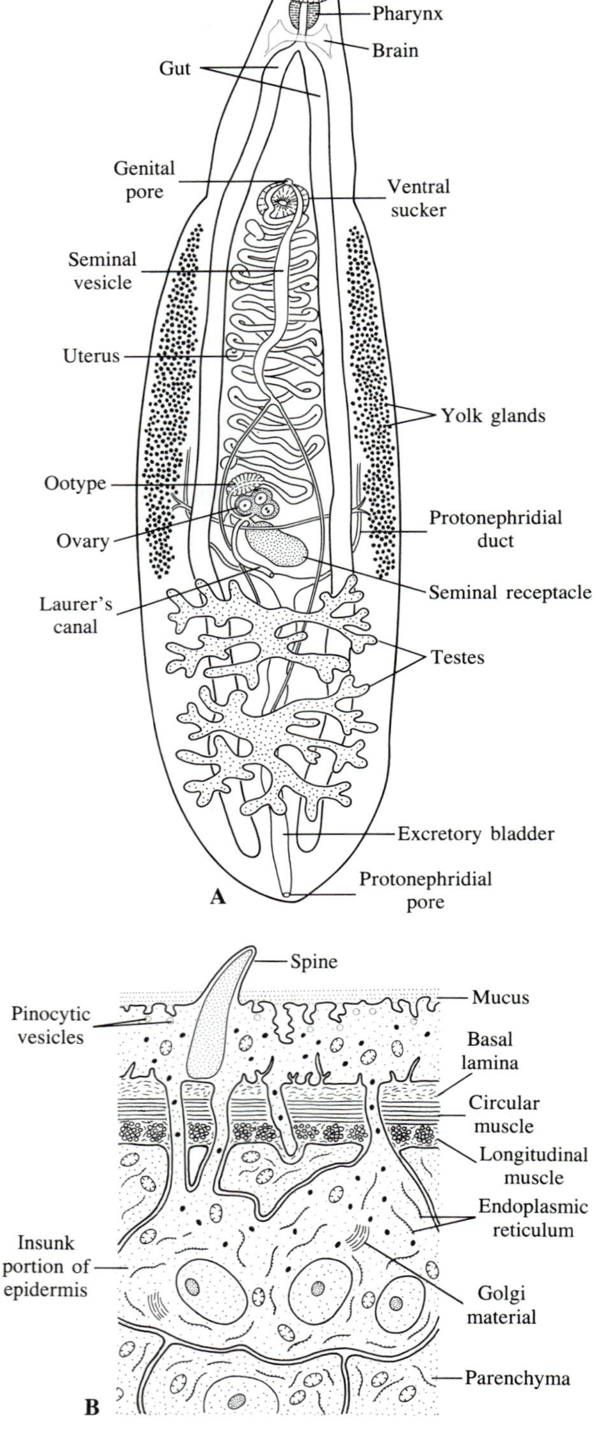

FIGURE 6.15 Class Digenea. A. *Clonorchis sinensis*, dorsal view. B. Histology of the epidermis and underlying tissue of a digenean; diagram based on electron micrographs.

Reynoldson, T. B., and Sefton, A. D., 1976. The food of *Planaria torva*, a laboratory and field study. Freshwater Biology, *6*:383–393.

Riser, N. W., and Morse, M. P. (editors), 1974. Biology of the Turbellaria. McGraw-Hill Book Co., New York.

Rose, S. M., 1970. Regeneration. Appleton-Century-Crofts, New York.

Smith, J., III, Tyler, S., Thomas, M. B., and Rieger, R., 1982. The morphology of turbellarian rhabdites: phylogenetic implications. Transactions of the American Microscopical Society, *101*:209–228.

Tyler, S., 1976. Comparative ultrastructure of adhesive systems in the Turbellaria. Zoomorphologie, *84*:1–76.

Wilson, R. A., and Webster, L. A., 1974. Protonephridia. Biological Reviews, *49*:127–160.

CLASS DIGENEA

Most flatworms called *flukes* are digeneans, so named because their life cycles involve two separate periods of reproduction. Adults, which parasitize various organs of vertebrates—especially the gut, liver, lungs, bladder, and blood vessels—reproduce sexually. Earlier stages, which usually live in molluscs, sometimes annelids, reproduce asexually. To complicate matters, the ultimate progeny resulting from asexual reproduction generally encyst in an animal that is likely to be eaten by the vertebrate host. A vertebrate harboring the adult stage of a digenean is called a primary or **definitive host.** A mollusc or annelid in which asexual reproduction proceeds is termed the first **intermediate host,** and if another animal is used for encystment, this is called the second intermediate host.

Integument of Digeneans

The epidermis of adult digeneans, unlike that of most turbellarians, is not ciliated. It forms two main strata (Figure 6.15, B). The outer layer is a sheet of cytoplasm without cell boundaries; it is connected by processes with the inner layer, which does appear to be cellular and which has the nuclei of the epidermis. The fact that the "cells" of the inner layer are joined to the continuous outer layer indicates that they are really part of the same syncytium. The circular and longitudinal muscle fibers run between the two strata. Sinking-in of epidermal components occurs in many turbellarians, but not often to the same extent as it does in digeneans. The epidermis of turbellarians is, moreover, generally cellular. (The old notion that the epidermis of acoels and some other groups is syncytial has been disproved.)

The complex surface invaginations of the outer layer of a digenean's epidermis suggest that the epidermis is active in absorption of nutrients and probably also in excretion. If stiff spines are present on the epidermis, as is the case in certain species, these secreted structures project from the outer layer.

The two-layered type of epidermis is found in other classes of parasitic flatworms. It will not be discussed again except in connection with tapeworms, whose epidermis shows additional specializations.

Structure of a Typical Digenean, *Clonorchis sinensis*

Before describing the life histories of a few representative digeneans, we will examine the structure of an adult fluke. *Clonorchis sinensis,* the Chinese liver fluke (Figure 6.15, A), is a good example. It is sold by many supply houses dealing in biological materials. *Clonorchis* is a parasite of humans, cats, dogs, and some other carnivorous mammals in China, Korea, Japan, and other Asiatic countries. It inhabits biliary passages of the liver and is responsible for some serious symptoms, including emaciation and localized swellings.

Clonorchis has two suckers. The **ventral sucker,** or **acetabulum,** is used for clinging to the wall of a biliary passage. The **oral sucker** assists in feeding; it is cupped tightly to the wall while the muscular pharynx sucks in blood and tissue juices and pumps these into the short esophagus. Once through the esophagus, the food enters the two long intestinal branches of the gut, where digestion and absorption take place.

Most of the other structures that are conspicuous in whole specimens are elements of the reproductive system. A bilobed brain may be recognizable, however, as a clear area just above the esophagus. An **excretory bladder** and its pore will probably be visible in the posterior part of the animal. The bladder receives the main collecting ducts of the right and left protonephridial complexes, whose flame bulbs are scattered through much of the body.

The organization of the reproductive system in *Clonorchis* is basically similar to that in almost all digeneans. There are two testes, each with a duct that leads to a common sperm duct, or vas deferens. This soon expands into a seminal vesicle where mature sperm accumulate. The duct from the seminal vesicle runs through a muscular structure that can be protruded. Although it is often called a cirrus, it is in fact a proper penis, for it does not need to be turned inside-out to function. Except when it is protruded, the penis lies in a sac.

The female part of the reproductive system is more complicated than the male part. The female pore, which shares a small depression with the male pore, leads into the uterus, a much-folded tube that is generally conspicuous because it is filled with hundreds of eggs. The uterus also serves as the insemination canal. Sperm forced into it during copulation travel up to the seminal receptacle, where they are stored until they are needed. *Laurer's canal,* found in *Clonorchis* and many other flukes, is believed to be a vestigial insemination duct.

The seminal receptacle, the ovary, the ducts from the yolk glands, and a chamber called the **ootype** are all close together. As eggs go down the oviduct toward the ootype, they are fertilized, and within the ootype they become surrounded by nutritive yolk cells. Certain secretions of the yolk glands convert into the shells that enclose the eggs. **Mehlis' glands,** which surround the ootype, probably have something to do with getting the production of an eggshell started, perhaps by secreting an initial membrane. There are, however, two kinds of gland cells in the complex around the ootype, and their exact roles are not yet understood. The eggs undergo embryonic development while in the uterus. By the time an egg is laid, it has developed into a ciliated larva.

Life Cycle of *Clonorchis*

Eggs laid by *Clonorchis* (Figure 6.16, A) reach the small intestine by way of the bile duct and eventually pass out of the host with feces. If an egg gets into fresh water, the ciliated larva that develops within the protective covering will have a chance to get into an appropriate first intermediate host, which is a snail. This larva, called a **miracidium** (Figure 6.16, B), contains some cells destined to produce another generation of larvae. In most digeneans, the miracidium escapes from the egg and searches for a host, but this is not the case in *Clonorchis*. An egg must be eaten by the right kind of snail before the miracidium comes out. From the intestine of the snail, where it hatches, the miracidium bores its way into other tissues. It then loses its ciliated epidermis and develops into a **sporocyst.** (This is not a satisfactory term, but it is entrenched.) A sporocyst (Figure 6.16, C) is like a sac in which germinal cells that were already recognizable within the miracidium multiply and form embryos of a second generation, called the **redia** stage.

The redia (Figure 6.16, D), unlike the sporocyst, has a short gut. It reproduces asexually to form many **cercariae.** A cercaria (Figure 6.16, E; see also Figure 6.17, for a variety of cercarial types) is essentially a juvenile fluke, with a mouth, pharynx, and two-branched intestine characteristic of the adult. It has a tail that will eventually be shed, but for the time being the tail is needed for locomotion. After a cercaria escapes from a birth pore in the redia, it works its way out of the snail. It uses its tail to propel itself through the water. If it contacts the right kind of fish, it penetrates the skin, dropping the tail in the process, and pushes itself into deeper tissues, where it encysts. It is now called a **metacercaria** and it may live for a long time while it waits for a chance to get into a human or some other suitable definitive host. If the fish is eaten raw, or after being incompletely pickled, digestive juices in the stomach or small intestine will liberate the

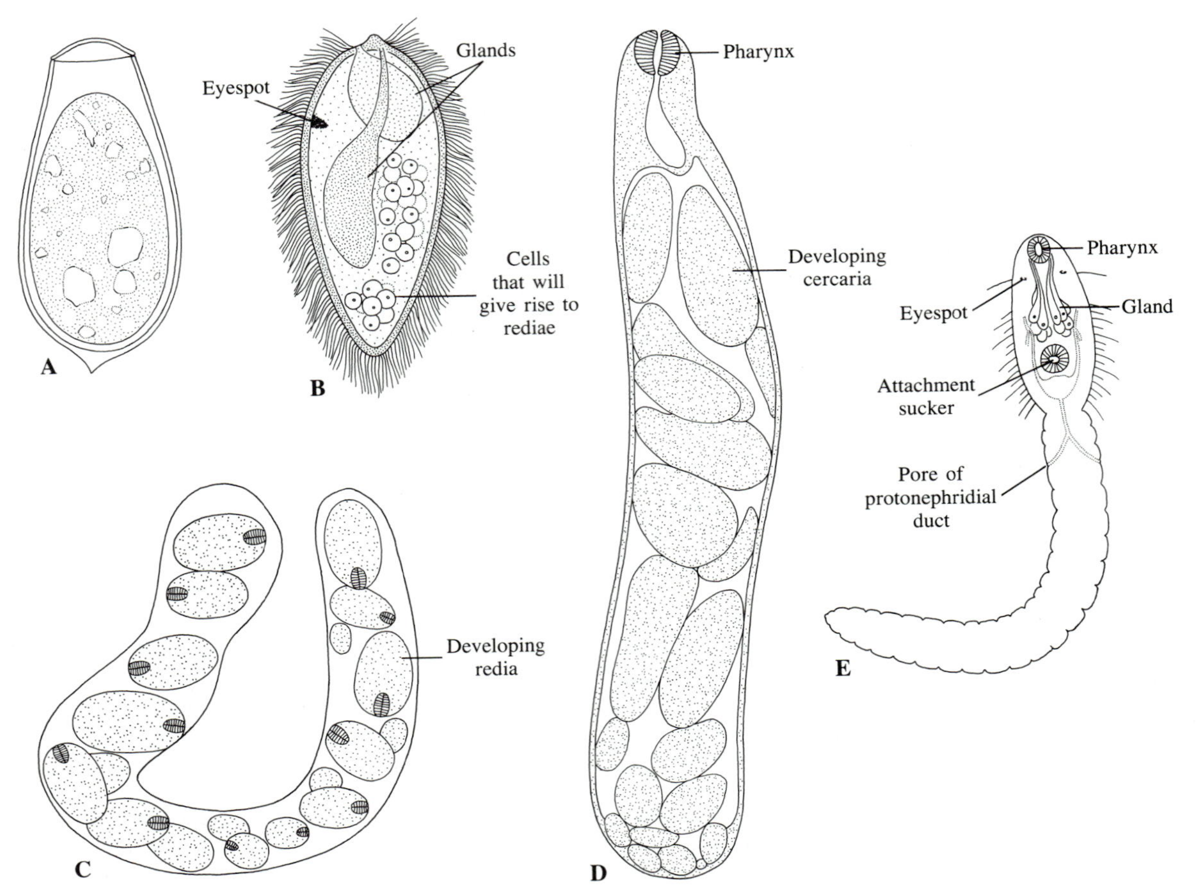

Figure 6.16 Class Digenea. Stages in the life cycle of *Clonorchis*. A. Egg. B. Miracidium. C. Sporocyst producing rediae. D. Redia producing cercariae. E. Cercaria.

metacercaria from its cyst. The young fluke then finds its way up the bile duct to the liver.

The chances of *Clonorchis* completing its life cycle may seem remote, but they are enhanced by the tremendous reproductive potential of this parasite—not only of the egg-laying adult, but also of the sporocyst and redia stages. It is estimated that at least 15 million humans are hosts to *Clonorchis*. Many millions of cats and dogs also have this worm, so it is a decidedly successful parasite.

Life Cycles of Other Representative Digeneans

The life cycle of *Clonorchis* is specialized for getting this fluke into a carnivorous vertebrate. There are, however, many digeneans in herbivorous vertebrates, too. These hosts generally become parasitized by eating a metacercaria whose cyst is attached to vegetation or embedded in a small invertebrate that is accidentally eaten with vegetation. Two interesting examples, both fairly common, will be described. Schistosomes will also be discussed, for they are not only important parasites of humans, but also representative of a group of digeneans whose cercariae do not encyst; they enter directly into the definitive host.

Fasciola hepatica

Fasciola hepatica, the sheep-liver fluke, is a relatively large species—2 or 3 cm long—that lives in the bile ducts of sheep, cattle, and some other ruminants. Eggs laid by *Fasciola* pass down the bile duct to the intestine, and leave the body with fecal material. Embryonation then begins, and if an egg is washed into water or happens to be in a wet meadow, the miracidium may hatch within a few weeks. It escapes by a trapdoor that opens up at one end of the egg. The miracidium swims actively in search of a suitable host, a

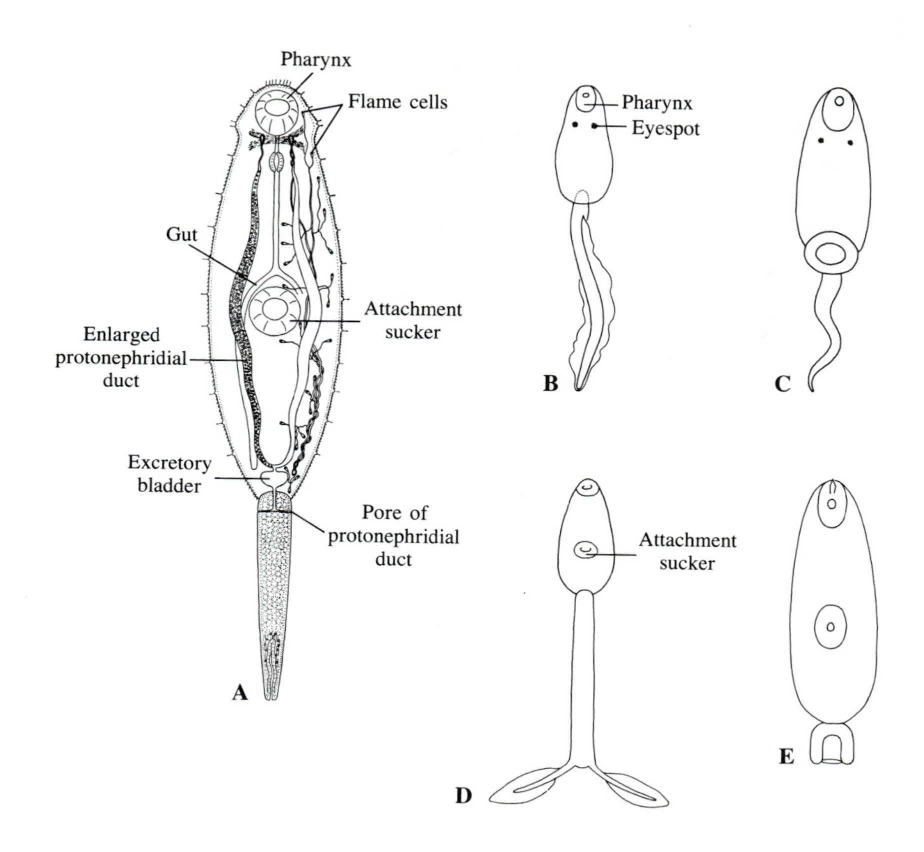

FIGURE 6.17 Class Digenea, representative cercariae. A. Ventral view of a typical cercaria. Some of the symmetrically placed structures are shown only on one side or the other. (Holliman, Tulane Studies in Zoology, 9.) B. Cercaria that has no ventral sucker and whose tail has a finlike fold. C. Cercaria whose ventral sucker is displaced to the posterior portion of the definitive body. D. Fork-tailed cercaria, such as is characteristic of schistosomes. E. Cercaria with a short tail.

pond snail of the genus *Lymnaea*. If it is successful, it literally bores its way into the tissue of the snail and sheds its ciliated epithelium in the process. Within the snail—usually in the mantle tissue—it develops into a sporocyst. As in *Clonorchis,* the progeny of the sporocyst are rediae. They escape from the parent sporocyst and migrate to a new site within the snail, generally the digestive gland. The reproductive cells of each redia may proliferate into a second generation of rediae or directly into cercariae. A cercaria liberated through the birth pore of a redia works its way out of the tissues of the snail. Reaching the stem or leaf of a plant growing at the water's edge, it sheds its tail and encysts, becoming a metacercaria. When eaten by a grazing ruminant, the cyst is digested and the metacercaria penetrates the wall of the intestine. Once loose in the visceral cavity it finds the liver. After penetrating this organ, it locates a bile duct, where it grows to sexual maturity at the expense of the tissue cells and fluids of its host.

Dicrocoelium dendriticum

Dicrocoelium dendriticum is a small fluke that lives in the gall bladder and bile ducts of sheep, goats, cattle, and deer, and sometimes parasitizes small herbivorous mammals such as rabbits. Its life cycle is normally completed entirely on land. An egg must be eaten by a suitable snail (*Cochlicopa lubrica* is the usual first intermediate host in North America and Europe) before the miracidium within it can escape and develop into a sporocyst. The sporocyst produces another generation of sporocysts, and the progeny of these are cercariae. There is no redia stage. As the cercariae leave the snail, they are incorporated into little balls of slime that certain ants collect and eat. The ants serve as second intermediate hosts. The cercariae penetrate the wall of the crop and encyst as metacercariae in the subesophageal ganglia. Parasitized ants convulsively open and close their mandibles, and eventually lock themselves to the tops of shoots of grass. There they may be eaten by a grazing herbivore, thus getting the metacercariae into a definitive host in which they can develop to sexual maturity.

The life cycles of *Clonorchis, Fasciola,* and *Dicrocoelium* have evolved to fit the life styles of the definitive and intermediate hosts—*Dicrocoelium* even modifies the behavior of its second intermediate host to such an extent that this host is likely to be eaten. Some digenean life cycles are more complicated than any of the three chosen as illustrations, but in all of them the main idea is the same: get the miracidia into the right mollusc

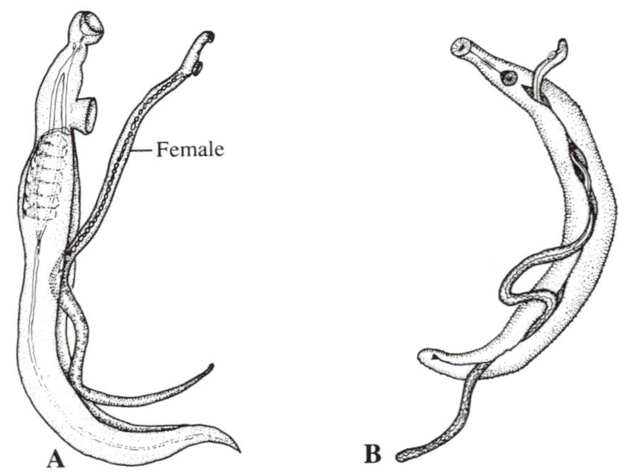

FIGURE 6.18 Class Digenea, schistosomes. The adults live in pairs. A. *Schistosoma japonicum*. B. *Schistosoma haematobium*. (Cheng, General Parasitology.)

or annelid, and get the cercariae into a situation where they can encyst and be eaten by the right vertebrate host.

Schistosomes

In one group of flukes, called schistosomes, the life history is somewhat simpler. The cercaria, characterized by a forked tail (Figure 6.17, D), penetrates the skin of the definitive host. There is no encysted metacercaria stage. Adult schistosomes (Figure 6.18) live in blood vessels of various birds and mammals. Humans are hosts to three species of one genus, *Schistosoma,* and are exposed to the cercariae while wading in water, as when working in a rice field. After penetrating the skin and losing their tails, the cercariae migrate to the blood vessels, where they mature and mate.

Schistosomes are among the few flatworms that have separate sexes. The male has a groove in which the female resides. The eggs have spines that help them get pushed through blood vessels until they reach the bladder or gut, so that they can be voided with urine or feces. The disease caused by schistosomes is called *schistosomiasis;* it is debilitating and dangerous, and responsible for much suffering in Africa, the Orient, and tropical America. The symptoms are due not only to the activities of the worms, but also to inflammation and abnormal thickening of the tissues caused by the presence of eggs, including those that are carried by the blood to the liver or other organs. An affliction of the skin called *swimmer's itch* is due to the penetration of the skin by freshwater or marine cercariae of schistosomes that are searching for the feet of ducks and other waterfowl. They do not develop further in humans, fortunately, but they cause an allergic reaction wherever they penetrate the skin and die.

Other Digeneans

The classification of digeneans, based on morphology of adults and larval stages, is rather complicated and outside the scope of this book. The subject is dealt with in textbooks of parasitology, in works concerned specifically with digeneans, and in encyclopedic references on invertebrates.

Because adult digeneans are found in fishes, amphibians, reptiles, birds, and mammals, and in herbivores as well as carnivores, it should be obvious that their life histories are extremely varied. Although molluscs generally serve as first intermediate hosts, many groups of invertebrates function as second intermediate hosts. In some species, moreover, there is a third intermediate host. Even if substantially no development takes place in the third host, this host may be essential to getting the young parasite into an appropriate definitive host.

References on Digenea

Crompton, D. W. T., and Joyner, S. M., 1980. Parasitic Worms. Wykeham Publications, London.

Dawes, B., 1946. The Trematoda, with Special Reference to British and Other European Forms. Cambridge University Press, Cambridge.

Erasmus, D. A., 1972. The Biology of Trematodes. Edward Arnold, London; Crane, Russak and Co., New York.

Schell, S. C., 1970. How to Know the Trematodes. Wm. C. Brown Co., Dubuque, Iowa.

Smyth, J. D., and Halton, D. W., 1984. The Physiology of Trematodes. 2nd ed. Cambridge University Press, London.

Yamaguti, S., 1958. Systema Helminthum, volume 1 (parts 1, 2). Interscience Publishers, New York.

CLASS MONOGENEA

Monogeneans do not require intermediate hosts and they do not undergo asexual multiplication. They are mostly parasites of the skin or gills of aquatic vertebrates, although some live in the body cavity, bladder, and other internal organs. A few occur on cephalopod molluscs. In general, these worms are well adapted for holding tightly to their hosts. At the posterior end of the body is at least one muscular sucker, sometimes several. The suckers are augmented by hooks, and may also have internal skeletal supports. There are muscles and tendonlike structures that pull back the hooks or allow them to protrude, enabling the worm to hold its position or to detach and move about in search of a better place to suck blood or tissue juices. The species illustrated (Figure 6.19) show just two of the various kinds and arrangements of suckers and hooks in monogeneans.

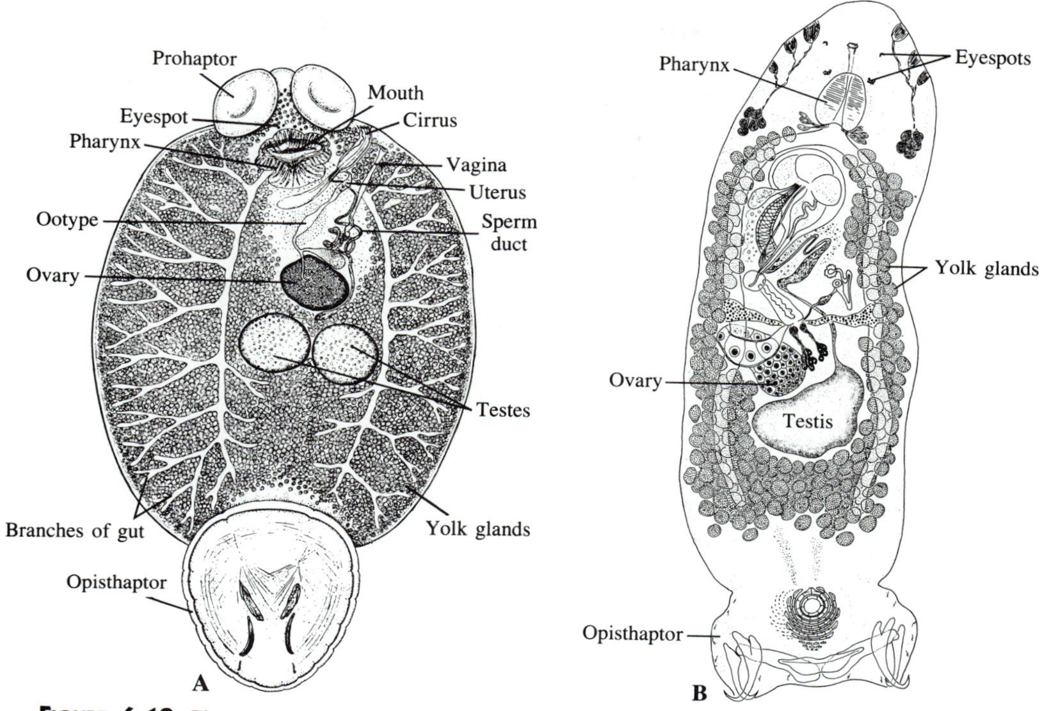

FIGURE 6.19 Class Monogenea. A. *Epibdella seriolae*. Not all structures of the reproductive system are labeled. (Yamaguti, Japanese Journal of Zoology, *5*.) B. *Cycloplectanum americanum*. (Oliver, Vie et Milieu, *19*.)

Besides suckers and hooks, monogeneans have glands concerned with secretion of sticky material that helps them attach. Some of these glands are associated with the suckers at the posterior end. There may also be adhesive glands at the anterior end, on both sides of the mouth. The mouth itself may be relatively simple, but generally it is situated within a sucker that operates not only for feeding but for temporary attachment while the worm is prospecting for a place to settle down. The mouth sucker or the area around it may have various specializations, such as small protuberances and secondary suckers, which help to maintain tight contact with the host.

Like most flatworms, monogeneans are hermaphroditic. The male reproductive system usually has a single testis, but in some species there are several. The vas deferens, as it approaches the genital pore, becomes specialized as a seminal vesicle for storage of sperm, then as an eversible cirrus, the organ of copulation. The cirrus is forced out of the genital pore when the muscular walls of the cirrus sac contract. The exact morphology of the cirrus sac varies and is sometimes complex.

In the female system, there is an ovary, with or without a perceptible oviduct, then a seminal receptacle where sperm are stored after insemination, and finally an ootype where the egg is "assembled." Ducts from the yolk glands enter the female system near the ootype, so that yolk is applied to the fertilized egg before the shell is secreted. The glands around the ootype correspond to Mehlis' glands of digenetic trematodes and presumably are concerned with secreting the shell or converting certain products of the yolk glands into shell material. After an egg has been assembled it is moved into the uterus.

The eggs of monogeneans tend to be large and to be produced in small numbers; there may be only one or two in the uterus at a particular time. The eggs usually remain in the uterus until a juvenile, similar to the adult except for having a ciliated epidermis, is nearly ready to hatch. The uterine pore, through which eggs leave the parent, is separate from the pore of the vagina, which leads to the seminal receptacle and through which insemination is achieved. Some species have two vaginas; in others, the vagina fails to reach a pore and insemination is probably achieved by injection of sperm through the epidermis. Another oddity found in the female reproductive system of certain monogeneans is a duct that joins a branch of the gut. A similar duct is present in some rhabdocoel and protriclad turbellarians.

CLASS ASPIDOGASTREA (ASPIDOBOTHRIA)

Aspidogastreans constitute a small group of parasites with a peculiar host distribution. Some live in the gut, gall bladder, or bile ducts of fishes, and a few are found in the gut of freshwater turtles. Several species occur in various clams and snails, and one is known to pass part of its life in lobsters and other crustaceans. It is generally thought that all or most of the species that live in fishes or turtles go through a period of development in an invertebrate. Nevertheless, some of the aspidogastreans parasitizing molluscs mature in these hosts. It has been suggested that aspidogastreans were associated with molluscs and perhaps other invertebrates before they became adapted to living in aquatic vertebrates. Some zoologists think, however, that the capacity of these worms to mature in molluscs is a secondary development.

Only certain aspects of the structure of aspidogastreans are discussed here. Additional details should be sought in specialized references, including textbooks of parasitology. The gut, unlike that of a digenean or monogenean, is not divided into two branches. The most distinctive feature is the bizarre sucker by which an aspidogastrean clings to its host. This sucker generally covers much of the ventral surface, and is divided by longitudinal and transverse partitions into many separate units. In some species, it looks like a waffle. Spaced out along the margins of the sucker of most species are little pores. These lead to clusters of gland cells.

One of the better-known species is *Aspidogaster conchicola* (Figure 6.20), which inhabits the pericardial cavity of freshwater unionid clams, and which is found sometimes in fishes. Eggs laid by *Aspidogaster* reach the mantle cavity of the clam by way of the renal organs, and are swept out with the exhalant water current. Newly hatched individuals, like the adults of this species, have no external cilia. The sucker is at first relatively small and simple. How the young worms get into the pericardial cavity of a prospective host is a mystery.

Some other aspidogastreans whose development has been studied do have cilia during a free-swimming larval stage. The cilia are concentrated in a few tufts, and they eventually disappear.

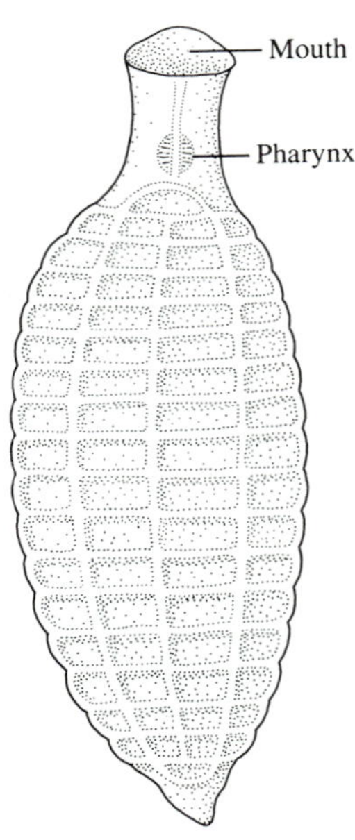

Figure 6.20 Class Aspidogastrea. *Aspidogaster conchicola,* showing the complex sucker that covers much of the ventral surface. Internal organs, except for the muscular pharynx, omitted.

References on Monogenea and Aspidogastrea

Bychowsky, B. E., 1961. Monogenetic Trematodes. Their Systematics and Phylogeny. Translated from the Russian by P. C. Oustinoff, edited by W. J. Hargis, Jr. American Institute of Biological Sciences, Washington.

Crane, J. W., 1972. Systematics and new species of marine Monogenea from California. Wasmann Journal of Biology, *30:*109–166.

Dawes, B., 1946. The Trematoda, with Special Reference to British and Other European Forms. Cambridge University Press, Cambridge.

Dollfus, R. P., 1958. Trématodes. Sous-classe Aspidogastrea. Annales de Parasitologie, *33:*305–395.

Ip, H. S., Desser, S., and Weller, I., 1982. *Cotylogaster occidentalis* (Trematoda: Aspidogastrea): scanning electron microscopic observations of sense organs and associated surface structures. Transactions of the American Microscopical Society, *101:*253–261.

Rohde, K., 1972. The Aspidogastrea, especially *Multicotyle purvisi* Dawes, 1941. Advances in Parasitology (Academic Press, Inc.), *10:*77–151.

Sproston, N. G., 1946. A synopsis of the monogenetic trematodes. Transactions of the Zoological Society of London, *25:*185–600.

Yamaguti, S., 1963. Systema Helminthum, volume 4. Interscience Publishers, New York.

CLASS CESTODA

Tapeworms, as adults, are parasites that live in the intestine and bile ducts of vertebrates. They have no gut of their own, and they must absorb amino acids, sugars, and other nutrients from the digestive tract of their hosts. The body generally consists of three rather distinct regions (Figure 6.21, A). The first of these, called the **scolex,** serves as an organ of attachment. Behind the scolex is the **neck,** a zone of active cell proliferation and a place where serially arranged units, called **pro-**

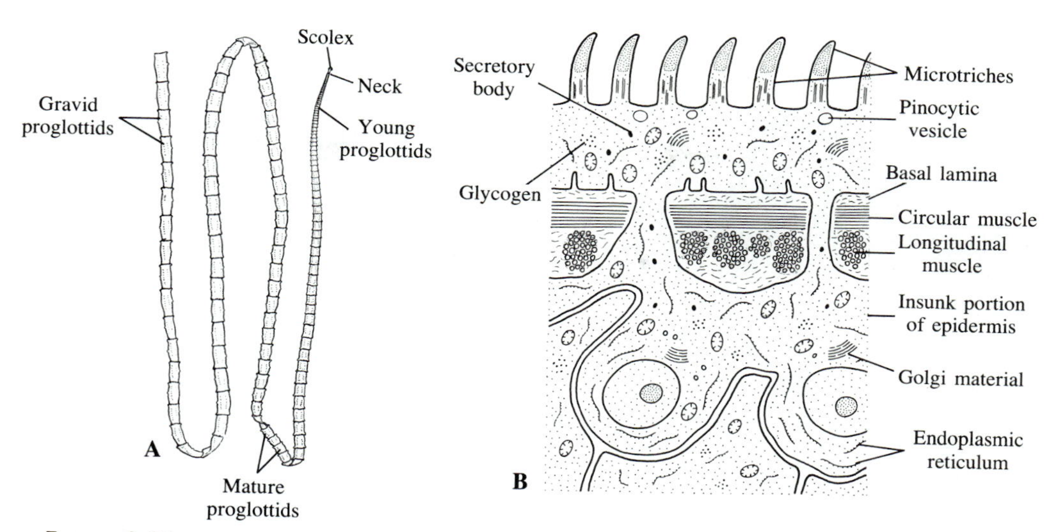

FIGURE 6.21 Class Cestoda. A. *Taenia pisiformis*, entire worm. (Schoening *et al.*, United States Yearbook of Agriculture, 1956.) B. Histology of the epidermis and underlying tissue of a tapeworm; diagram based on electron micrographs.

glottids, begin to develop. The third region, termed the **strobila,** consists of proglottids arranged in the order in which they were formed, the youngest being those nearest the neck. Typically, mature proglottids are hermaphroditic. In some tapeworms, in fact, there are two sets of male and female organs in each proglottid.

Some zoologists regard tapeworms as segmented animals. Proglottids, however, are somewhat comparable to the medusae that form when a polyp of a scyphozoan jellyfish divides transversely into several units. The medusae eventually become sexual individuals. In the case of tapeworms, the proglottids reach sexual maturity while they are still attached to one another, although in at least some species a few do become detached and wander by themselves. There are additional arguments for and against the idea that proglottids are segments, but the view taken here is that a proglottid is more nearly comparable to a sexual member of a colony than to a segment of an earthworm, centipede, or insect. The tendency to form reproductive units that may or may not become separate individuals is widespread among invertebrates. It has already been mentioned, for instance, in connection with turbellarians of the order Catenulida.

Epidermis

The epidermis of a tapeworm (Figure 6.21, B) is somewhat similar to that of a digenean. Electron microscopy reveals that the epidermis has two main components: a more or less continuous layer of cytoplasm, in which cell boundaries are scarce or absent, and processes that extend inward between bundles of circular and longitudinal muscle. These processes, which contain the nuclei of the epidermis, reach well below the musculature. The outer layer is elaborated into closely spaced projections called **microtriches** (singular, **microthrix**). These are somewhat comparable to the microvilli present on many kinds of cells, especially those concerned with absorption. The structure of microtriches, however, includes microtubules and other complexities not found routinely in microvilli. Calculations indicate that microtriches increase the surface available for absorption by several times, but not to the extent that typical microvilli do. The double membrane of a microthrix is organized as a mosaic, and the spaces between the almost hexagonal units of the mosaic probably permit a rapid movement of soluble materials. Thus, in spite of the fact that microtriches would seem to provide less surface area than microvilli, they may be just as efficient, if not more so. What appear to be pinocytic vesicles in the outer layer of epidermal cytoplasm are probably visible signs of absorption.

In tapeworms that embed the scolex deep into the tissue of the intestinal wall, this structure probably functions in absorption as well as for attachment. This is suggested by the fact that such scolices have abundant microtriches.

Representative Cestodes

Taenia pisiformis

With few exceptions, tapeworms pass through at least one intermediate host, which may be an invertebrate or a vertebrate, before reaching the definitive host. *Taenia pisiformis* (Figure 6.21, A), a parasite of domestic dogs and cats, as well as of some of their wild

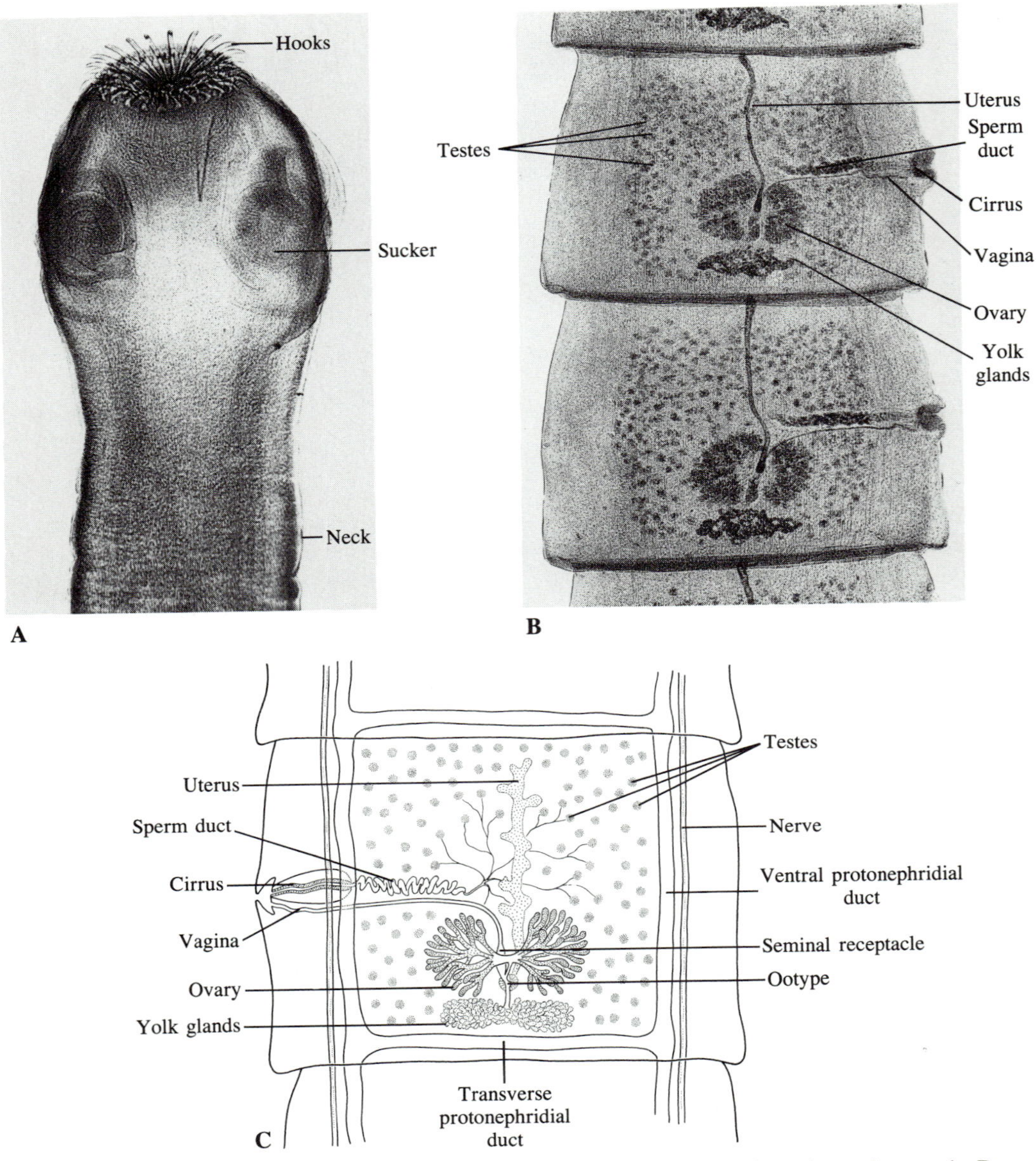

FIGURE 6.22 Class Cestoda. *Taenia pisiformis*. A. Scolex and neck region; photomicrograph. B. Mature proglottid; photomicrograph. (A and B, courtesy of Carolina Biological Supply Co., Inc.) C. Mature proglottid; somewhat diagrammatic.

relatives, illustrates the general pattern of the life cycle. This tapeworm, whose length may reach 1 m, is readily available from biological supply houses and is the species most likely to be studied in the laboratory.

The scolex (Figure 6.22, A) has an apical complex of hooks, and behind this are four suckers. The neck region is short, and there is no sharp line between it and the strobila. The youngest proglottids show only the bare rudiments of the reproductive system, but farther back one sees a sequence of many proglottids in which the sex organs are fully developed (Figure 6.22, B and C). Although the proglottids are flattened, the usual terms of orientation applied to them—anterior, posterior, dorsal, ventral, right, left—are somewhat arbitrary. The ovary is considered to be near the ventral surface and close to the posterior edge.

There are flame cells in the parenchyma, and these are joined by small tubules to major collecting ducts. In *Taenia* and other genera that belong to the same general group of tapeworms, there are four continuous longitudinal protonephridial ducts running through the strobila near the lateral edges. Two of these are ventral, two are dorsal. The ventral ducts are joined in the hind part of each proglottid by a transverse duct (Figure 6.22, C). In the region of the scolex, the ventral and dorsal duct on each side unite. The protonephridial ducts are often referred to as excretory canals. Fluid from the ducts has been analyzed for one species and was found to contain ammonia and urea, as well as glucose, lactic acid, and some soluble proteins. Thus, while the protonephridial ducts are probably excretory, they may also function in distribution of nutrients, and perhaps in osmoregulation. In *Taenia,* they simply open at the posterior edge of the terminal proglottid, but in some tapeworms they enter a distinct bladder that has a pore to the outside, or are joined by lateral ducts to many small pores on the surface.

The arrangement of the male and female reproductive systems in each mature proglottid is similar to that in a digenean. The male and female genital pores are close together, opening into a shallow cup on either the right or left side. The numerous testes are scattered through the dorsal part of the proglottid and fine ducts connect them to the coiled vas deferens. The lumen of the vas deferens is continuous with the duct that runs through the cirrus. This eversible copulatory structure lies within a muscular sac; when the walls of the sac contract, the cirrus is turned inside out and can be pressed into the female opening of the same or another proglottid. So-called **prostate glands,** whose secretion probably activates the sperm, open into the duct of the cirrus.

The female part of the reproductive system is a little more complicated, at least as far as ducts are concerned. The conspicuous, bilobed ovary is joined to the oviduct, a portion of which is specialized as the ootype. The latter is connected to the vagina–seminal receptacle complex that leads inward from the female genital pore, and also to the uterus, located in front of the ovary. The seminal receptacle is really just a dilation of the part of the vagina that is closest to the ootype, but it is where sperm are stored after copulation. As oocytes leave the ovary and enter the oviduct, they are penetrated by sperm and undergo meiosis, casting off polar bodies. Union of sperm and egg nuclei then takes place. Before a zygote enters the ootype, one or a few yolk cells, produced by the yolk glands behind the ootype, become associated with it. The yolk cells supply nutritive material, as well as substances that will be converted into an eggshell. It is believed that Mehlis' glands, next to the ootype, have something to do with getting shell formation started and perhaps with physically altering components of yolk that will eventually harden. In any case, after their shells have been formed, the zygotes are shunted into the uterus, where they are stored and

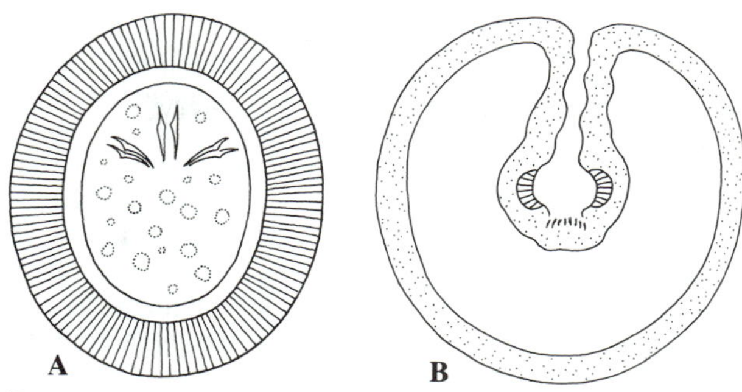

FIGURE 6.23 Class Cestoda. *Taenia pisiformis*. A. Six-hooked oncosphere within eggshell. B. Cysticercus, with invaginated scolex.

where they undergo some development. In older proglottids, the uterus is large and extensively branched, and there is little else left of either the male or female reproductive system.

The ripe proglottids break way from the strobila and leave the host with fecal material. When a proglottid disintegrates, its hundreds of eggs are set free. Each one contains a six-hooked larva called an **oncosphere** (Figure 6.23, A), which is ready to develop further in an intermediate host—a rabbit, hare, or suitable rodent. After an egg is eaten by the intermediate host, the shell is broken down by pepsin, acting in the stomach, and by trypsin, a pancreatic enzyme that acts in the small intestine. The hooks at one end of the oncosphere help it work its way through the wall of the intestine and into a vein that leads to the liver. Some oncospheres end up elsewhere, but these die. Those that do make it to the liver soon lose their hooks and grow into slender worms about 1 cm long. The posterior part of each oncosphere detaches and disintegrates, but the anterior part persists and shows the primordium of the scolex of the prospective adult.

About a month after the oncosphere enters the liver, it leaves this organ and moves about the visceral cavity, finally working its way under the peritoneum. Much of the tissue within it disintegrates and it becomes swollen with fluid, so that it is now bladderlike. At this stage it is called a **cysticercus** (Figure 6.23, B). The scolex develops further, and its hooks and suckers can be seen facing the cavity of an invagination that is open to the outside.

No further development takes place unless the rabbit or other intermediate host is eaten by dog, cat, or related carnivore. In the intestine of a suitable definitive host, enzymes in the pancreatic juice digest away the bladder, releasing the scolex. This turns inside-out, so that its hooks and suckers can grip the intestinal epithelium. In 2 or 3 weeks, proglottids begin to differentiate; within 2 months, some of them are sexually mature.

Taenia solium and *Taenia saginata*

Taenia solium and *T. saginata* (sometimes called *Taeniarhynchus saginatus*) are large tapeworms that parasitize humans. Both may reach a length of more than 4 m. Their life cycles, however, conform to the pattern described for *T. pisiformis*. The usual intermediate hosts for *T. solium* are pigs, whereas cattle serve as intermediate hosts for *T. saginata*. Meat containing the cysticerci of these tapeworms is not likely to slip past careful inspectors, but if it does, thorough cooking will make it safe.

Persons who have the tapeworm stage of *T. solium* in the intestine are sources of further danger to themselves and to others. The oncospheres in eggs voided with fecal matter cannot, of course, develop directly into adult tapeworms, but they can develop into cysticerci, just as they would in a pig or some other intermediate host. The presence of a cysticercus in the brain, an eye, or certain other organs may have serious consequences.

Echinococcus granulosus

Echinococcus granulosus, which lives in the intestine of dogs and related carnivores, is only about 5 mm long, and it usually has only two large proglottids at a time (Figure 6.24, A). The normal intermediate hosts are sheep, pigs, cattle, and some other ungulate mammals. If eggs voided with the feces of a carnivore are swallowed by an intermediate host, the oncospheres within them may develop into bladderlike cysts called **hydatids.** The inside of the wall of a hydatid (Figure 6.24, B) produces thousands of scolices, as well as secondary bladders in which scolices develop. A large hydatid may be more than 10 cm in diameter, and sometimes the cysts are clustered. In any case, hydatids located in vital organs are extremely dangerous to the animal that has them. Unfortunately, hydatid disease may develop in humans who swallow eggs that have been deposited with dog feces. If a parasitized dog licks itself, then licks a person's face, eggs in the dog's saliva may get into the person's mouth.

The life cycle of *Echinococcus granularis* is perpetuated in areas where dogs have access to slaughterhouse waste containing hydatids, and where sheep or other intermediate hosts are likely to swallow eggs voided with dog feces. A variety of other hosts keep the cycle going in the wild; for instance, wolves and moose, wolves and reindeer, and the Australian dingo and wallaby. Two other species of *Echinococcus* also occasionally cause hydatid disease in humans.

Hymenolepis nana (Vampirolepis nana)

This species, which reaches a length of about 4 cm, is the most common tapeworm in humans. It is unusual in that it requires no intermediate host; the eggs voided with a person's feces contain the infective stage. Thus a person who is not careful about sanitation can be reinfected repeatedly. Hatching takes place in the duodenum, and when the six-hooked oncosphere is released, it penetrates the epithelium and deeper tissues and settles in a lymph channel, where it develops into a stage called the **cysticercoid** (Figure 6.24, C). This is somewhat comparable to the cysticercus of a *Taenia,* but its scolex is not inverted, and it has a "tail," where the hooks of the oncosphere persist. In a few days, the cysticercoid works its way back to the lumen. The bladder and tail disappear, and after the scolex becomes at-

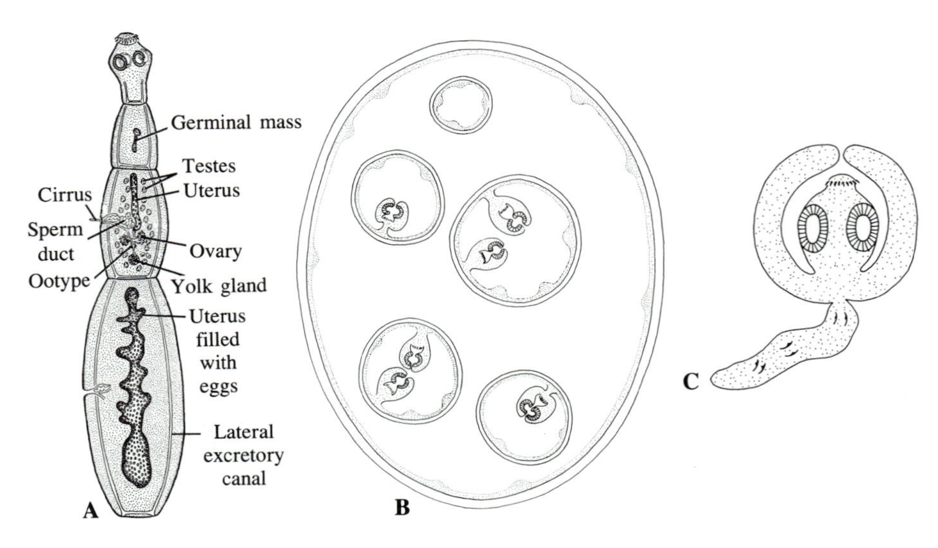

FIGURE 6.24 Class Cestoda. A. *Echinococcus granulosus,* entire worm. (Brown, Selected Invertebrate Types.) B. *Echinococcus,* hydatid cyst, with daughter cysts containing developing scolices. C. *Hymenolepis nana,* cysticercoid stage.

tached to the wall of the small intestine, proglottids begin to differentiate from the neck region.

Diphyllobothrium latum

A rather different life history is exhibited by *Diphyllobothrium latum*. Its definitive hosts include humans, bears, mink, and a variety of other mammals that eat raw fish. A fertilized egg of *Diphyllobothrium* develops into a ciliated larva called a **coracidium** (Figure 6.25, A). If an egg voided with fecal material reaches water and hatches, the coracidium swims freely and thus has a chance of being eaten by the right kind of copepod. Once in the gut of its first benefactor, it loses its cilia and works its way into the hemocoel, where it grows and elongates into the **procercoid** stage (Figure 6.25, B). This retains, at its posterior end, the hooks of the coracidium. If the copepod is eaten by a fish, the procercoid migrates into the tissues and enlarges into the **plerocercoid** stage (Figure 6.25, C). This loses the larval hooks, but at its anterior end it acquires the two grooves that will eventually form the deep grooves *(bothria)* on the scolex of the adult tapeworm. If the plerocercoid starts out in a small fish that is eaten by a larger fish, it will probably adapt to its new host, and its prospects for getting into a suitable definitive host may be improved. When a fish carrying a plerocercoid is eaten by the right kind of mammal, the immature worm is set free in the intestine and matures there.

Diphyllobothrium is found only where appropriate ecological circumstances permit the cycle to repeat itself indefinitely: fresh water, the right copepods, the right fishes, and one or more of the right vertebrate hosts. Most of the human cases are concentrated in eastern and northern Europe, where it is the custom to eat fish that is raw or not thoroughly pickled.

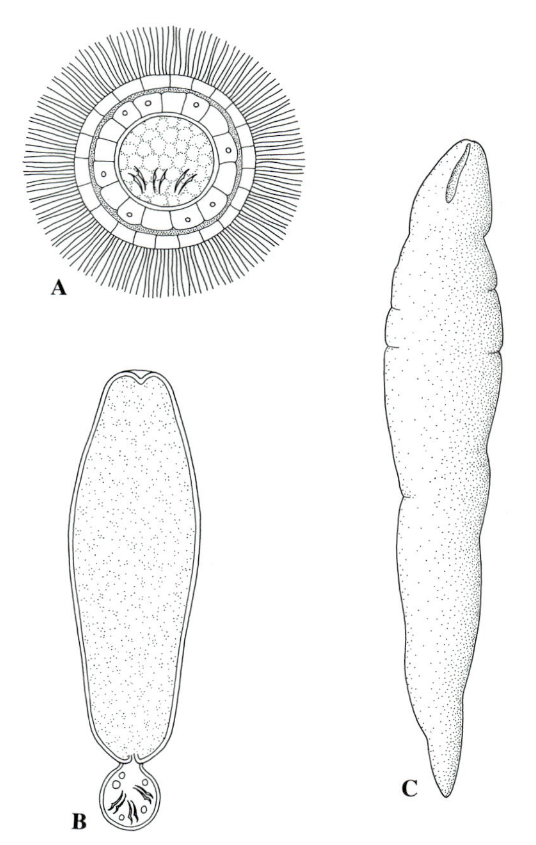

FIGURE 6.25 Class Cestoda. Stages in the development of *Diphyllobothrium latum*. A. Coracidium. B. Procercoid, from the hemocoel of a copepod. The hooks of the coracidium have been displaced to the posterior end. C. Plerocercoid (infective stage), from the flesh of a freshwater fish.

Other Cestodes

Adult tapeworms occur in most classes of vertebrates, from elasmobranch fishes to mammals. They show considerable morphological diversity, not only in the arrangement of the reproductive organs, but also in the structure of the scolex. Two interesting types of scolices, one with bothria, are shown in Figure 6.26.

No attempt is made here to describe the several orders, but it should be mentioned that of the tapeworms discussed in preceding paragraphs, *Taenia, Echinococcus,* and *Hymenolepis* belong to the order Cyclophyllida, and *Diphyllobothrium* belongs to the Pseudophyllida.

A wide variety of intermediate hosts are utilized by tapeworms. Most important are insects, crustaceans, molluscs, and vertebrates. As in the digeneans, the life

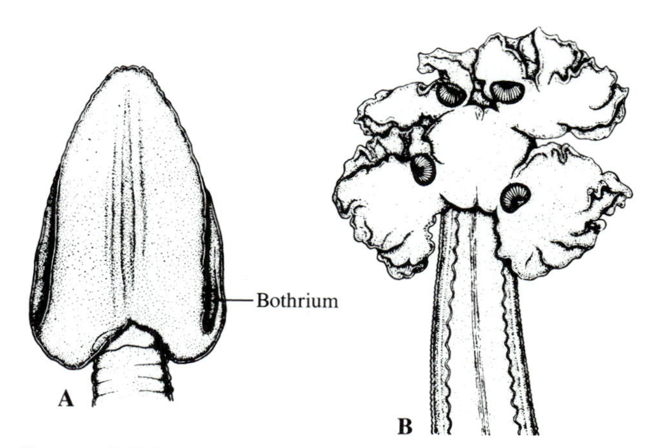

FIGURE 6.26 Class Cestoda, scolices. A. *Ptychobothrium* (order Pseudophyllida). B. *Phyllobothrium* (order Tetraphyllida). (Yamaguti, Japanese Journal of Zoology, *6*.)

cycles are beautifully adapted to fit the food habits of the definitive hosts.

CLASS CESTODARIA

The relatively few known species of cestodarians have no gut, and this is one reason they were long considered to be close relatives of tapeworms. Another feature that has linked them with tapeworms is the presence of hooks at the anterior end of the larva; in cestodarians, however, there are ten of these hooks instead of six. In any case, cestodarians have no scolex of the type found in tapeworms, and the body is not divided into proglottids.

The two subdivisions of the Cestodaria are composed of worms called amphilinids and gyrocotylids. Amphilinids (Figure 6.27, A) occur in the body cavity of certain fishes, including sturgeons, and in the lungs of an Australian freshwater turtle. They are flattened worms that may reach a length of several centimeters. At the anterior end there is a muscular organ that serves as a kind of proboscis; it is used for boring. The male and female pores involved in copulation are at the posterior end, whereas the uterine pore, through which embryonating eggs are laid, is next to the proboscis. The ovary is compact and the yolk glands are concentrated near the lateral margins of the body. The numerous testes are scattered.

According to the results of limited experimental work on the life history of *Amphilina foliacea,* which lives in sturgeons of the Old World, it appears that an egg will release the ciliated larva inside it only after it has been eaten by a suitable amphipod crustacean. The larva penetrates the gut, losing its cilia in the process, and enters the hemocoel. Here it grows to a length of about 2 to 4 mm, retaining its ten larval hooks. When the intermediate host is eaten by a sturgeon and digested, the larva is set free and burrows into the body cavity. Sturgeons have abdominal pores that permit eggs laid by *Amphilina* to escape from their body cavities.

Gyrocotylids inhabit the spiral valve of chimaerid fishes. They are striking worms, with a ruffled body margin and a posterior "rosette" (Figure 6.27, B). The

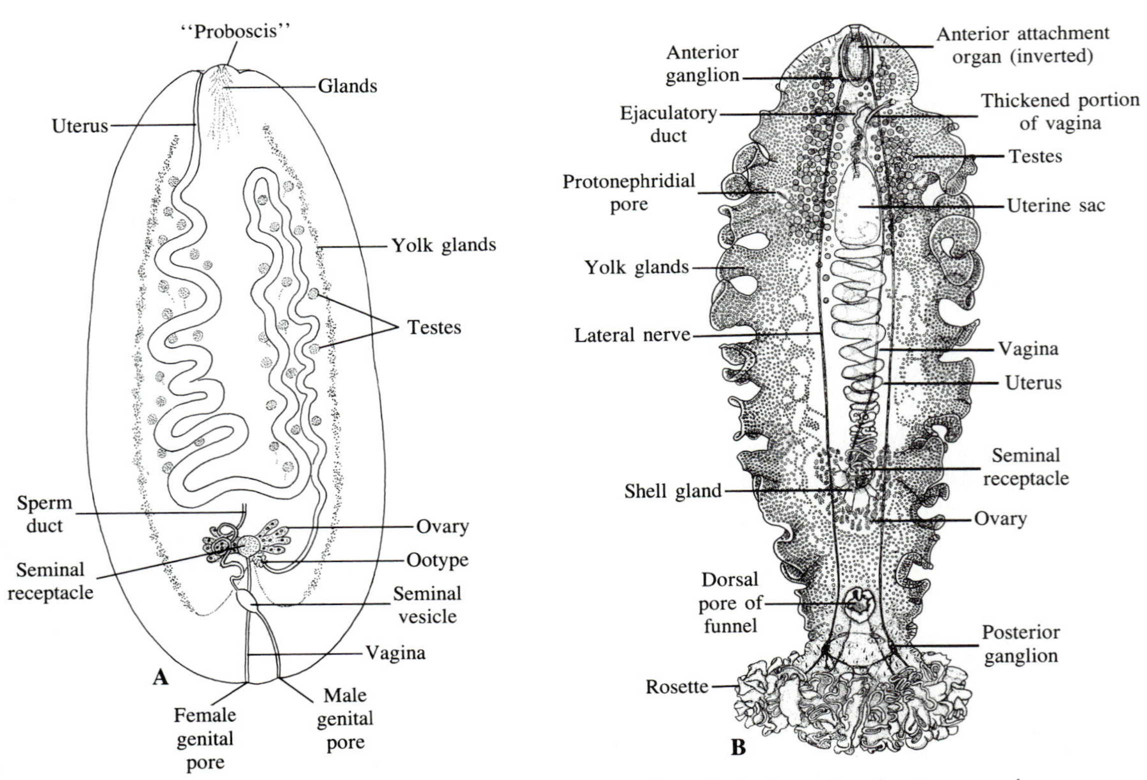

FIGURE 6.27 Class Cestodaria. A. *Amphilina foliacea,* from the body cavity of a sturgeon; simplified. B. *Gyrocotyle fimbriata,* from the spiral valve of the ratfish, *Hydrolagus colliei.* (Lynch, Journal of Parasitology, *31.*)

rosette, used for attachment, is actually a funnel, and its narrow end leads to a pore on the dorsal surface. At the anterior end there is an eversible structure, cup-shaped when withdrawn; this also functions as an attachment organ. In the region around it, and sometimes elsewhere, there are spines that originate deep in the subepithelial tissue and that can be moved to some extent by muscles.

The male genital pore is located on a small protuberance on the ventral surface near the anterior end of the body, and although there are some muscular complications and a seminal vesicle in the lower part of the sperm duct, there appears to be no eversible cirrus involved in copulation. The testes are scattered through much of the anterior part of the body. The ovary is a rather diffuse structure, with a number of separate small ducts leading to a main oviduct. After it is joined by ducts from the yolk glands, which are spread through much of the body, the oviduct runs to the ootype, where shells are formed around the eggs. Close to the ootype is a seminal receptacle, in which sperm acquired during copulation are stored. The insemination canal, or vagina, that leads to the seminal receptacle–ootype complex opens on the ventral surface not far from the male pore. Eggs put together in the ootype move into a spacious, coiled uterus that is emptied by a pore on the dorsal surface.

The life cycle of gyrocotylids has not been fully worked out. If an egg laid by one of these worms is watched carefully, the embryo can be seen to develop into a ten-hooked, ciliated larva. Eventually the larva hatches, but what happens next is not clear. Some zoologists think that an intermediate host is required; others believe that none is necessary.

References on Cestoda and Cestodaria

Aral, H. P. (editor), 1981. Biology of the Tapeworm *Hymenolepis diminuta*. Academic Press, New York.

Arme, C., and Pappas, P. W., 1983. Biology of the Eucestoda. 2 volumes. Academic Press, New York and London.

Bandoni, S. M., and Brooks, D. R., 1987. Revision and phylogenetic analysis of the Amphilinidea Poche, 1922 (Platyhelminthes: Cercomeria: Cercomeromorpha). Canadian Journal of Zoology, *65:*1110–1128.

Freeman, R. S., 1973. Ontogeny of cestodes and its bearing on their phylogeny and systematics. Advances in Parasitology, *11:*481–557.

Rohde, K., and Georgi, M., 1983. Structure and development of *Australamphilina elongata* Johnston, 1931 (Cestodaria: Amphilinida). International Journal of Parasitology, *13:*273–287.

Schmidt, C. D., 1970. How to Know the Tapeworms. Wm. C. Brown Company, Dubuque, Iowa.

Slais, J., 1973. Functional morphology of cestode larvae. Advances in Parasitology, *11:*395–480.

Smyth, J. D., 1969. The Physiology of Cestodes. Oliver and Boyd, Edinburgh and London.

Wardle, R. A., and McLeod, J. A., 1952. The Zoology of Tapeworms. University of Minnesota Press, Minneapolis.

Wardle, R. A., McLeod, J. A., and Radinovsky, S., 1974. Advances in the Zoology of Tapeworms. University of Minnesota Press, Minneapolis.

Xylander, W. E. R., 1984. A presumptive ciliary photoreceptor in larval *Gyrocotyle urna* Grube and Wagener (Cestoda). Zoomorphology, *104:*21–25.

Yamaguti, S., 1959. Systema Helminthum. Volume 2. Interscience Publishers, New York.

SUMMARY

1. Flatworms are bilaterally symmetrical and cephalized, with a distinct brain near the anterior end of the body and with prominent longitudinal nerve trunks.

2. During embryonic development, mesoderm is formed in addition to ectoderm and endoderm, so these animals are triploblastic. As in ctenophores, the mesoderm gives rise to muscle, but it also produces various other tissues. Moreover, flatworms have organs that consist of more than one kind of tissue.

3. Although the gut lacks an anus, it is histologically more complex than the gut of cnidarians and ctenophores. A portion of it is usually differentiated into a muscular pharynx, an efficient device for feeding on whole organisms or on tissues or fluids of organisms. In many flatworms, the pharynx can be protruded.

4. Another important characteristic of flatworms is the protonephridial system. This consists of hollowed-out cells joined to ducts that lead to one or more openings on the body surface. Protonephridia are concerned primarily with elimination of excess water, thereby maintaining a constant salt-to-water ratio in the tissues.

5. The reproductive system is highly developed—more so than in some phyla that are more advanced than flatworms in other respects. Besides gonads and ducts, there are male structures used in copulation and female structures for receiving sperm, producing yolk for nourishment of the embryo, and acting as centers where eggshells are formed. Nearly all flatworms are hermaphroditic, but cross-insemination is the rule.

6. Cleavage of fertilized eggs follows the spiral pattern, as is typical for molluscs, annelids, and certain other higher phyla. The spirality usually becomes apparent at the third cleavage, when the orientation of the mitotic spindles is such that the four smaller cells produced by this division lie on the furrows between the four larger cells, rather than directly above the larger cells, as is the case in radial cleavage.

7. The four main classes of flatworms are the Turbellaria, Digenea, Monogenea, and Cestoda. Turbellarians are mostly free-living, and include herbivores as well as carnivores. They inhabit the sea and fresh water, and a few are found in moist terrestrial habitats. They are usually ciliated externally. The other three main classes have external cilia only during certain larval stages, if at all.

8. The Digenea, as adults, are intestinal parasites of vertebrates, but they undergo a period of development, including asexual reproduction, in a mollusc or annelid. The products of the final phase of asexual multiplication, called cercariae, must either penetrate the definitive vertebrate host, encyst in a second intermediate host that will be eaten by the vertebrate host, or encyst on vegetation that will be eaten.

9. Monogenea, which are generally external parasites of fishes and amphibians, have a direct life cycle; there is no period of development or asexual reproduction in an intermediate host.

10. The Cestoda, or tapeworms, are parasites in the gut of vertebrates and are unusual in that they have no digestive tract; they absorb nutrients through the epidermis. They are unusual also in that they form chains of sexually reproducing units called proglottids. Most tapeworms undergo a period of development, and sometimes also asexual proliferation, in an intermediate host.

PHYLUM GNATHOSTOMULIDA

An appreciation of the gnathostomulids as a distinctive group worthy of phylum rank has come only recently. The few species that had been known before 1960 were usually placed in the Turbellaria. Intensive study of the small animals living between sand grains on ocean beaches and in bays has brought to light almost 100 species of gnathostomulids. Some of them inhabit relatively anaerobic situations, and perhaps this is one reason why they had been overlooked by biologists working with interstitial organisms. In general, they are delicate little worms—mostly less than 1 mm long—that are better studied while alive than after preservation.

In most gnathostomulids the body is slender and is cylindrical or slightly flattened (Figure 6.28). The epidermis is ciliated, but as a rule there is only one cilium on each cell. This, and the fact that the ciliary beat can be reversed, are among the unusual features of gnathostomulids. The glassy, bristlelike sensory cilia that pro-

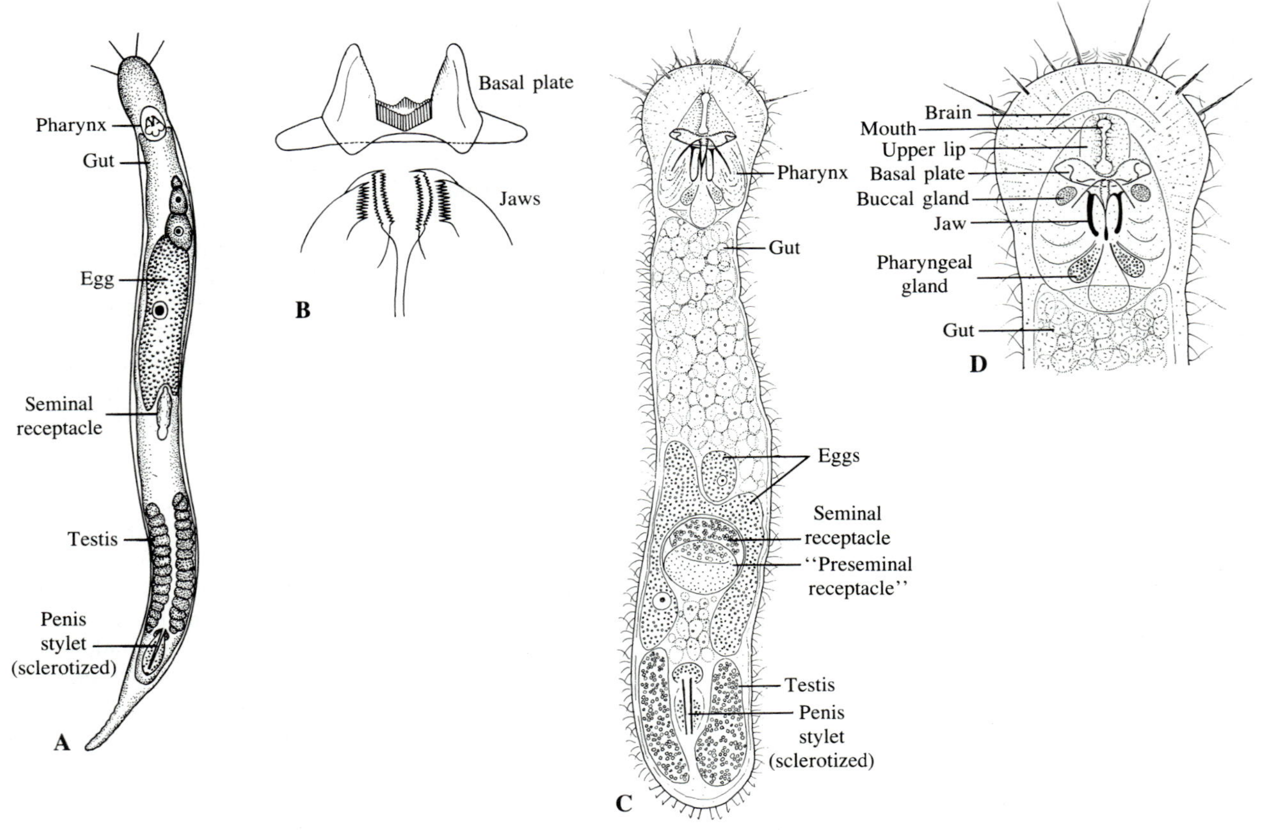

FIGURE 6.28. Phylum Gnathostomulida. A. *Gnathostomula paradoxa.* (Ax, Veröffentlichen des Instituts für Meeresforschung in Bremerhaven, *2.*) B. *Gnathostomula,* basal plate and jaws. (Ax, Zeitschrift für Zoologische Systematik und Evolutionsforschung, *3.*) C. *Problognathia minima,* dorsal view. D. *Problognathia,* head region. (C and D, Sterrer and Farris, Transactions of the American Microscopical Society, *94:*263.)

ject from the epidermis are much like those found on many turbellarians as well as on some annelids, rotifers, gastrotrichs, and various other invertebrates.

The muscles are cross-striated. Beneath the sheath of circular fibers, just beneath the epidermis, are three pairs of longitudinal muscle bands. These effect twisting movements and abrupt shortening of the body. Locomotion is accomplished largely by ciliary activity, however.

Another remarkable feature of gnathostomulids, when they are compared with turbellarians, is the set of hard parts associated with the mouth and pharynx. There is a toothed plate located on the ventral lip of the mouth, and this is used to scrape bacterial coatings from sand grains and for gathering up other small organisms. The two jaws, operated by muscles, masticate the food and push it backward into the gut, which runs for most of the length of the body. An anal pore has been found in some genera, but in most gnathostomulids the gut ends blindly.

Protonephridia whose function is presumed to be excretory are located in various parts of the body. Whether occurring singly or in groups of two or three, those on one side of the midline are paired with those on the other side. They are similar to the solenocytic protonephridia of gastrotrichs (Chapter 10). With some exceptions, each consists of three cells: a cup-shaped terminal cell in whose cavity there is a cilium surrounded by eight rodlike structures; a tubular canal cell, into whose lumen the cilium and rods of the terminal cell extend; and an outlet cell, also tubular, that leads to a pore at the surface. In some genera, there are only the terminal cell and outlet; another variation is the absence of rods around the cilium of the terminal cell.

Three pairs of longitudinal nerves, closely associated with the epidermis, extend posteriorly from the brain. A pair of short nerves connects the brain with a prominent ganglion in the pharyngeal region.

There is little parenchyma in gnathostomulids, for most of the space not taken up by structures already mentioned is occupied by the reproductive system. Hermaphroditism is the rule, but there are exceptions. The male portion of the reproductive system generally includes a pair of testes in the hind part of the body, a penislike inseminating device that may have a cuticularized stylet, and a dilation that serves as a seminal vesicle. The male gonopore is located posteriorly. The female portion of the reproductive system usually has a single ovary and a seminal receptacle for storing sperm received from another animal. In species that have a vagina, it is likely that true copulation occurs, but hypodermic impregnation, accomplished with the piercing stylet of the male system, is known to take place in some gnathostomulids.

After a large oocyte is entered by a sperm from the seminal receptacle, it leaves the body by way of a break in the epidermis. Cleavage is of the spiral type, and development is direct.

SUMMARY

1. Gnathostomulids, of which there are about 100 described species, live between sand grains in marine sediments. Most of them are less than 1 mm long.

2. They resemble small turbellarian flatworms in general appearance and in being ciliated externally. Each ciliated epidermal cell, however, generally has only one cilium, rather than a few to many, as is the case in turbellarians.

3. Other features that distinguish gnathostomulids from turbellarians are muscles of the cross-striated type, a toothed plate in the lower lip of the mouth, and a pair of jaws in the pharynx. An anal opening has been found in some genera.

REFERENCES ON GNATHOSTOMULIDA

Ax, P., 1985. The position of the Gnathostomulida and Platyhelminthes in the phylogenetic system of the Bilateria. *In* Morris, S. C., George, J. D., Gibson, R., and Platt, H. M., The Origins and Relationships of Lower Invertebrates, pp. 168–180. Clarendon Press, Oxford.

Riedl, R. J., 1971. On the genus *Gnathostomula* (Gnathostomulida). Internationale Revue der Gesamte Hydrobiologie, *56:*385–496.

Rieger, R. M., and Mainitz, M., 1977. Comparative fine structure of the body wall in Gnathostomulida and their phylogenetic position between Platyhelminthes and Aschelminthes. Zeitschrift für Zoologische Systematik und Evolutionsforschung, *15:*9–35.

Sterrer, W., 1969. Beitrage zur Kenntniss der Gnathostomulida. I. Anatomie und Morphologie des genus *Pterognathia* Sterrer. Arkiv för Zoologi, *22:*1–125.

Sterrer, W., 1971. On the biology of the Gnathostomulida. Vie et Milieu, Supplement *22:*493–508.

Sterrer, W., 1972. Systematics and evolution within the Gnathostomulida. Systematic Zoology, *21:*151–173.

Sterrer, W., Mainitz, M., and Rieger, R. M., 1985. Gnathostomulida: enigmatic as ever. *In* Morris, S. C., George, J. D., Gibson, R., and Platt, H. M., The Origins and Relationships of Lower Invertebrates, pp. 181–199. Clarendon Press, Oxford.

7

Phylum Nemertea

Ribbon Worms

Phylum Nemertea
- Class Anopla
  - Order Palaeonemertea
  - Order Heteronemertea
- Class Enopla
  - Order Hoplonemertea
  - Order Bdellonemertea

INTRODUCTION

Nemerteans resemble turbellarian flatworms in being soft-bodied and externally ciliated, and also in the way they creep over solid surfaces. With few exceptions, however, they are considerably more elongated. "Ribbon worm" is an appropriate name for some types, but the shape of most nemerteans is much like that of a shoelace or broken rubber band (Figure 7.1; Color Plate 3). Many species reach lengths of 1 or 2 m, and a few, when fully extended, are more than 10 m long. In general, these worms are fragile, so it is not often that an especially long specimen remains intact while being extricated from its habitat. Moreover, slender nemerteans can contract to a fifth or even a tenth of their extended length, and they can shorten some portions of the body while elongating others.

Nearly all of the approximately 800 species are marine, and they occupy a variety of habitats. Some burrow in sand or mud; others live under rocks, in crevices, or among seaweeds and attached animals. A few are pelagic, and certain species are involved in symbiotic relationships with bivalve molluscs, crabs, and other invertebrates. The phylum is poorly represented in fresh water and on land; the terrestrial species are found mostly in rather humid tropical and temperate regions.

COMPARISON OF NEMERTEANS WITH FLATWORMS

Nemerteans are decidedly more advanced than flatworms in having a **complete gut,** a **circulatory system,** and an elaborate **proboscis** for capturing prey. The reproductive system of nemerteans, however, is much simpler than that of most flatworms. Although there are usually many testes or ovaries on both sides of the gut, these open directly to the outside by pores in the body wall. There are no complicated ducts such as are found in the reproductive system of flatworms, and devices for insemination are rare.

The foregut of a nemertean operates efficiently to suck in whole prey or the juices of prey, and it may even be everted, but there is nothing like the muscular, bulbous pharynx of a rhabdocoel or fluke, or the protrusible pharynx of a triclad or protriclad. In cross-section, nemerteans are usually flattened, but sometimes they are cylindrical or nearly so. Their histological organization is similar to that of flatworms, but their musculature is almost always more extensively developed.

EPIDERMIS, DERMIS, AND MUSCULATURE

As a rule, the epidermis of nemerteans is ciliated, and it generally contains numerous gland cells—single or clustered—that produce mucus and other secretions.

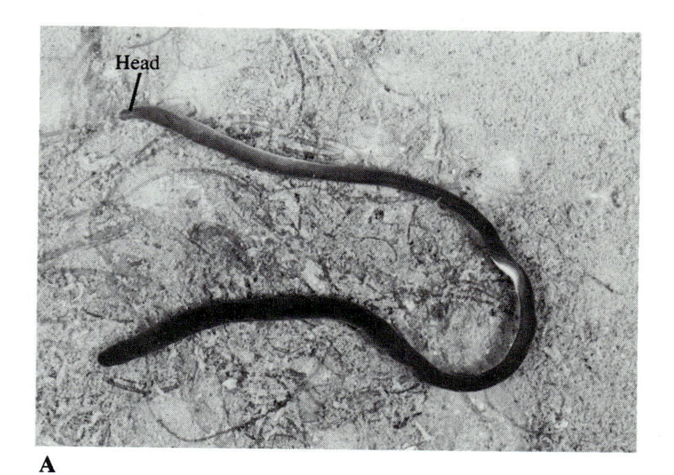

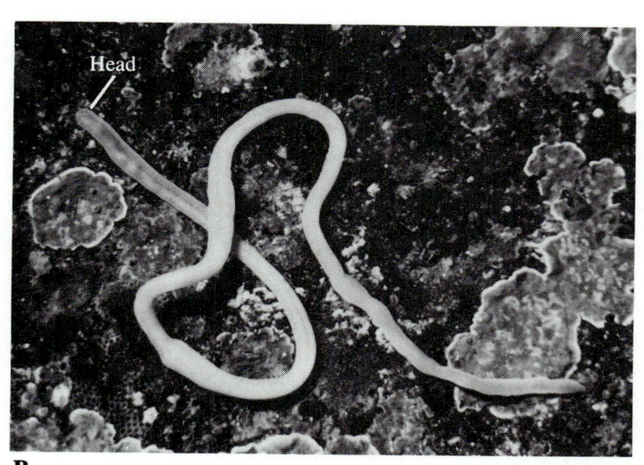

FIGURE 7.1 Living nemerteans. A. *Paranemertes peregrina*. B. *Amphiporus imparispinosus*. (Both species belong to the order Hoplonemertea and are common on the Pacific coast of North America. See also Color Plate 3).

Some of the gland cells may be beneath the epidermis, in a layer of connective tissue called the **dermis,** and in that case they have ducts that reach the surface. The extent to which the dermis is developed varies. When it is thick, it may be more or less gelatinous throughout, or it may be gelatinous nearest the epidermis and fibrous farthest from the epidermis. In many nemerteans the dermal layer is thin or absent altogether.

There are usually two or three principal layers of muscle beneath the dermis (or beneath the epidermis, when the dermis is lacking). If there are two layers (Figures 7.2, 7.3), the outer one consists of circular muscle and the inner one consists of longitudinal muscle, an arrangement that is found also in flatworms, annelids, and some other invertebrates. If there are three layers, as is typical of the order Heteronemertea, the

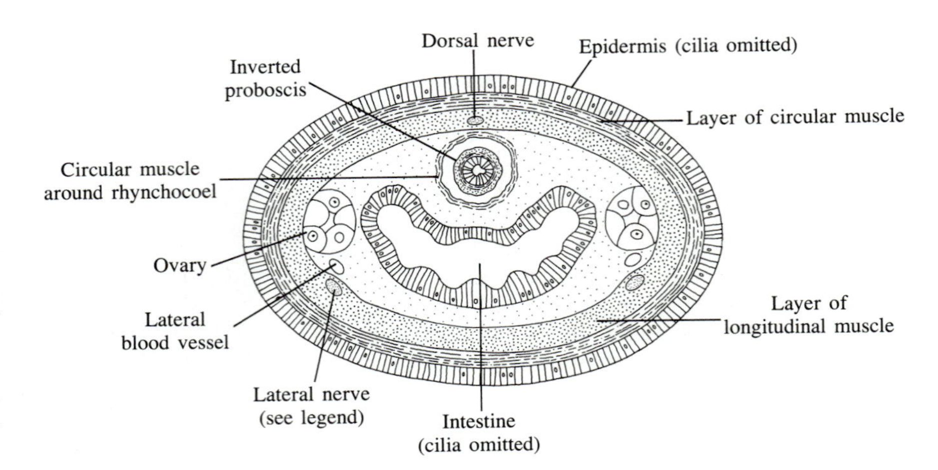

FIGURE 7.2 Transverse section showing the general histology of a nemertean of the order Palaeonemertea. The lateral nerves and dorsal nerve are in the layer of longitudinal muscle. In some members of the order, however, they are directly beneath the epidermis, especially at more anterior levels of the body.

outer and inner layers consist of longitudinal fibers and the middle layer consists of circular fibers. The outer layer is often incorporated into the dermis (Figure 7.4). In all nemerteans there are muscles that do not fit into the main strata just described. Some are diagonal, others radial or dorsoventral. The muscles concerned with operation of the proboscis will be described later.

The amount of parenchyma varies according to how thick the dermis is, and on the extent to which the musculature is developed. Some nemerteans have relatively little parenchyma.

LOCOMOTION

The locomotion of nemerteans is accomplished by muscular or ciliary activity, or both. It is usually aided by secretion of mucus, which forms a track beneath the animal or a tube around it. Musculature plays a more important role than cilia in the locomotion of some types. This is especially obvious in the case of certain burrowing species and those that can swim energetically by undulations of the body, though many creeping nemerteans also rely mostly on muscular activity. Cilia are

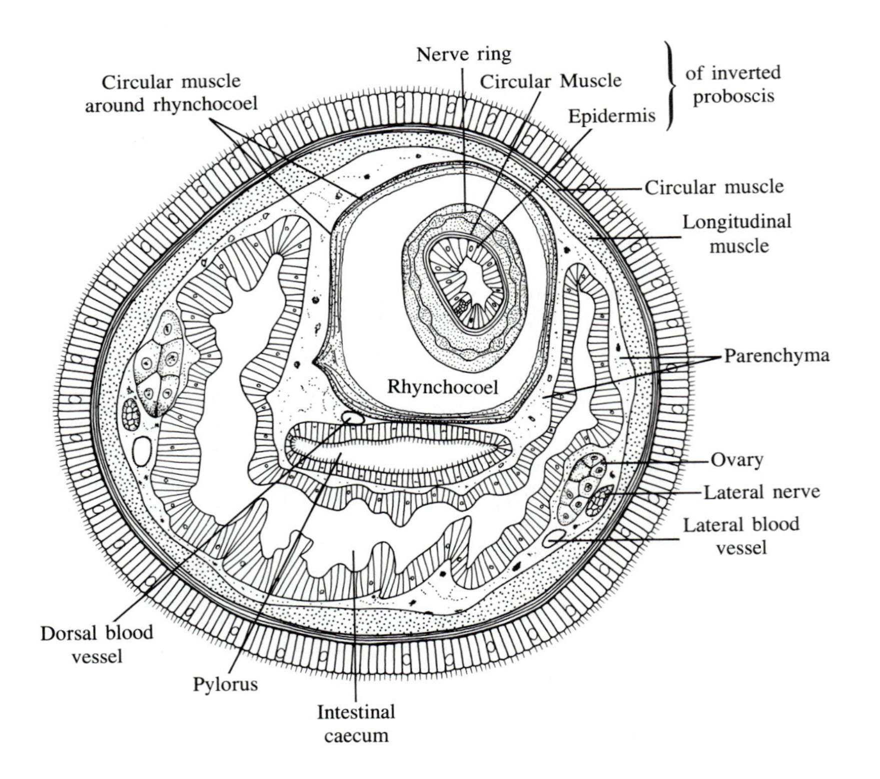

FIGURE 7.3 Transverse section of an *Amphiporus* (order Hoplonemertea), at the level of the pylorus and intestinal caecum (see Figure 7.6). (After Ivanov *et al.*, Major Practicum of Invertebrate Zoology.)

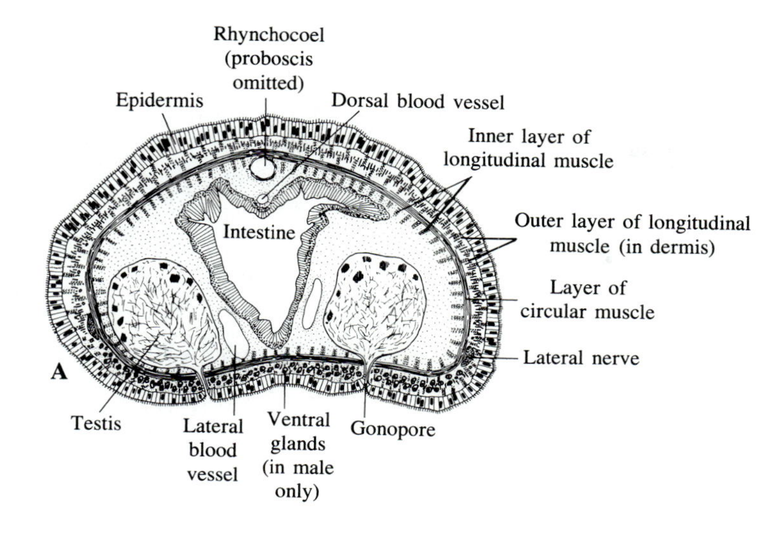

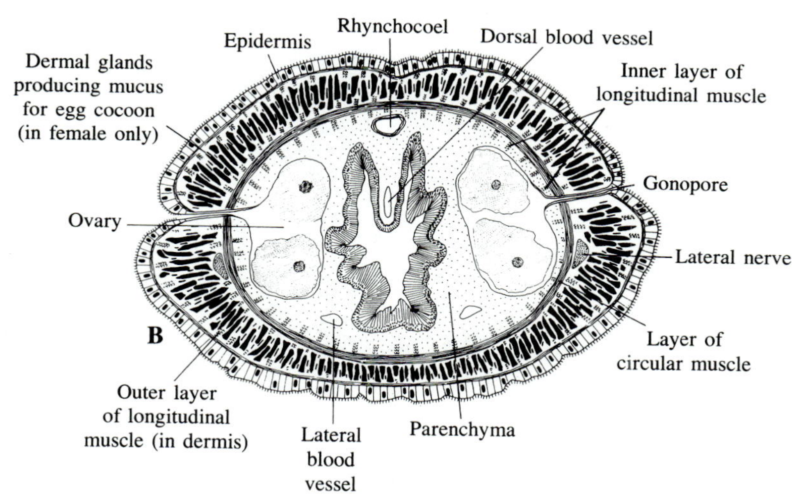

FIGURE 7.4 Transverse sections of *Lineus ruber* (order Heteronemertea). A. Male. B. Female. In both sexes, the dermis contains gland cells that secrete mucus needed for locomotion. The only glands labeled are those in the ventral part of the male, which probably produce a substance that causes the sperm to remain somewhat concentrated, and those throughout the dermis of the female, which produce a cocoon around the eggs. (Bierne, Annales des Sciences Naturelles, Zoologie, series 12, *12*.)

apt to be effective in the locomotion of small species and recently hatched juveniles. Use of the proboscis in locomotion and burrowing is dealt with below.

DIGESTIVE TRACT AND NUTRITION

The **mouth** of a nemertean is located at or near the anterior end, and the **anus** is at the posterior end (Figure 7.5). The **foregut,** which is derived from an ectodermal invagination (the embryonic **stomodeum**), is generally differentiated into a **buccal cavity, esophagus,** and **stomach.** Most of the digestive tract consists of the **intestine,** or midgut, whose lining is derived from endoderm. The intestine almost always has many lateral diverticula. Usually there is a short and simple **hindgut;** this, like the foregut, is of ectodermal origin, and the embryonic invagination that gives rise to it is called the **proctodeum.**

In one order of nemerteans, the Hoplonemertea, the stomach is typically joined to the intestine by a tube called the **pylorus.** From the level where this enters the intestine, the latter sends off an anteriorly directed **caecum** that has lateral diverticula like those of the rest of the intestine. In addition to this intestinal caecum, there may be a short caecum extending forward from the stomach and another one extending backward from the esophagus. Figure 7.6 shows the spatial relationships of the caeca and other components of the gut in a species of *Amphiporus*. In most nemerteans, the pylorus is relatively short and the stomach and esophageal caeca are less well developed than shown in the illustration.

Although the intestine has diverticula, and although there may also be caeca, the digestive tract of a nemertean is a "one-way street." As food moves through the digestive tract, it is subjected to a succession of enzymes and different degrees of acidity. A complete gut

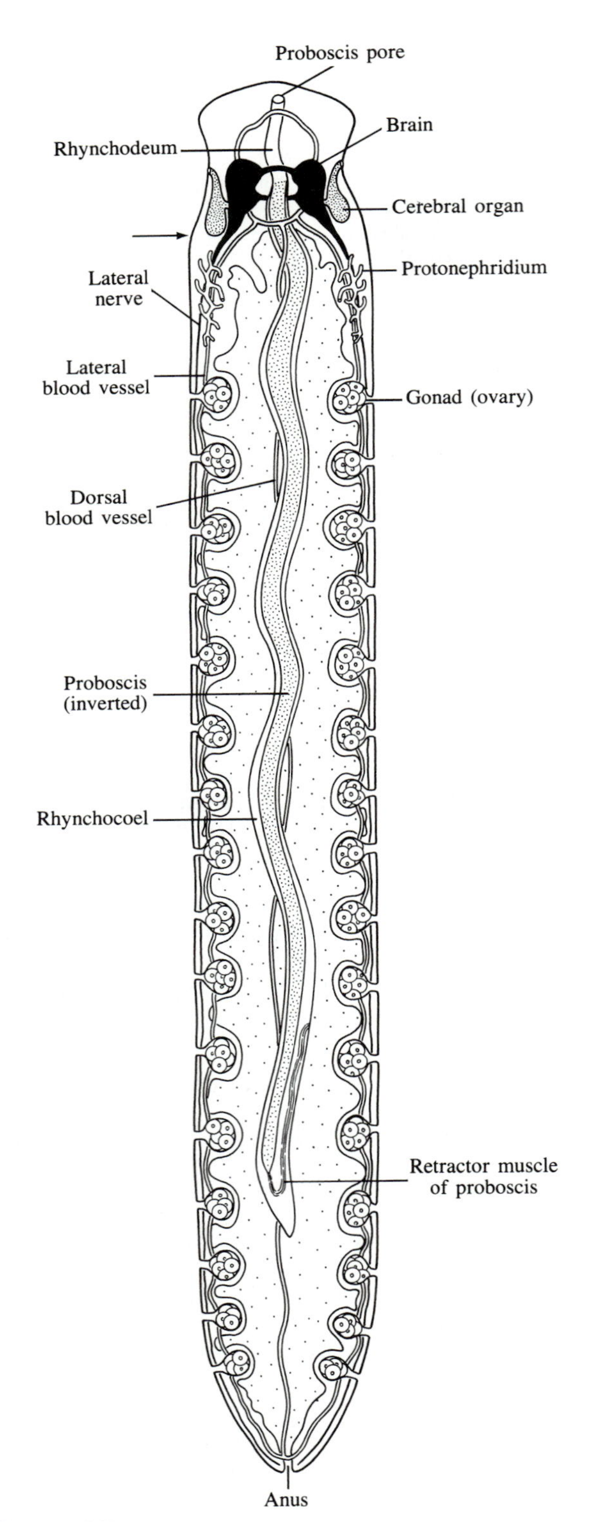

FIGURE 7.5 Nemertean structure (heteronemertean type), dorsal view. (The mouth, being ventral, is not shown; it is located at the level indicated by an arrow.)

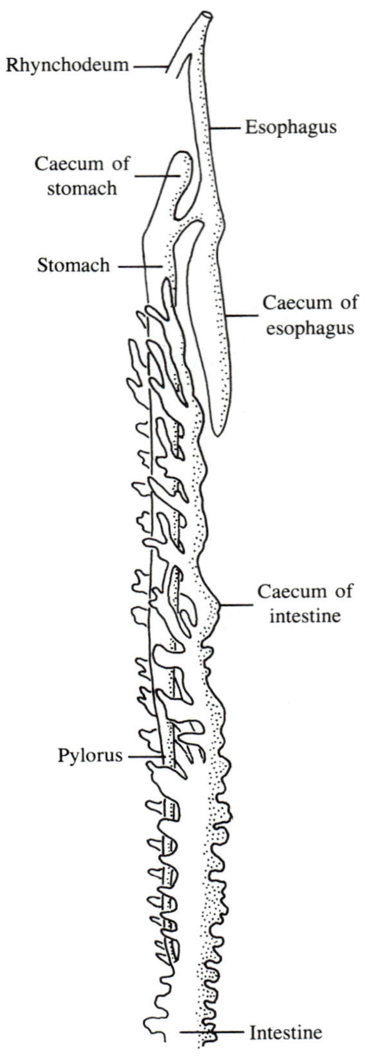

FIGURE 7.6 Anterior portion of the gut of *Amphiporus occidentalis* (order Hoplonemertea). The caeca of the esophagus and stomach are more prominent in this species than they are in most hoplonemerteans, and the pylorus is unusually long. (After Coe, Bulletin of the Museum of Comparative Zoology, Harvard College, *49*.)

such as this is more efficient than a saclike gut that has only one opening, not only because it permits considerable regional specialization, but also because feeding, digestion, absorption, and elimination of residues do not interfere with one another.

In general, the preintestinal portions of the gut are concerned with killing the prey (if it is not dead already) and with adding secretions such as carbonic anhydrase, an enzyme that increases acidity. In the intestine, extracellular digestion begins under the influence of enzymes that break down proteins. Particles of food

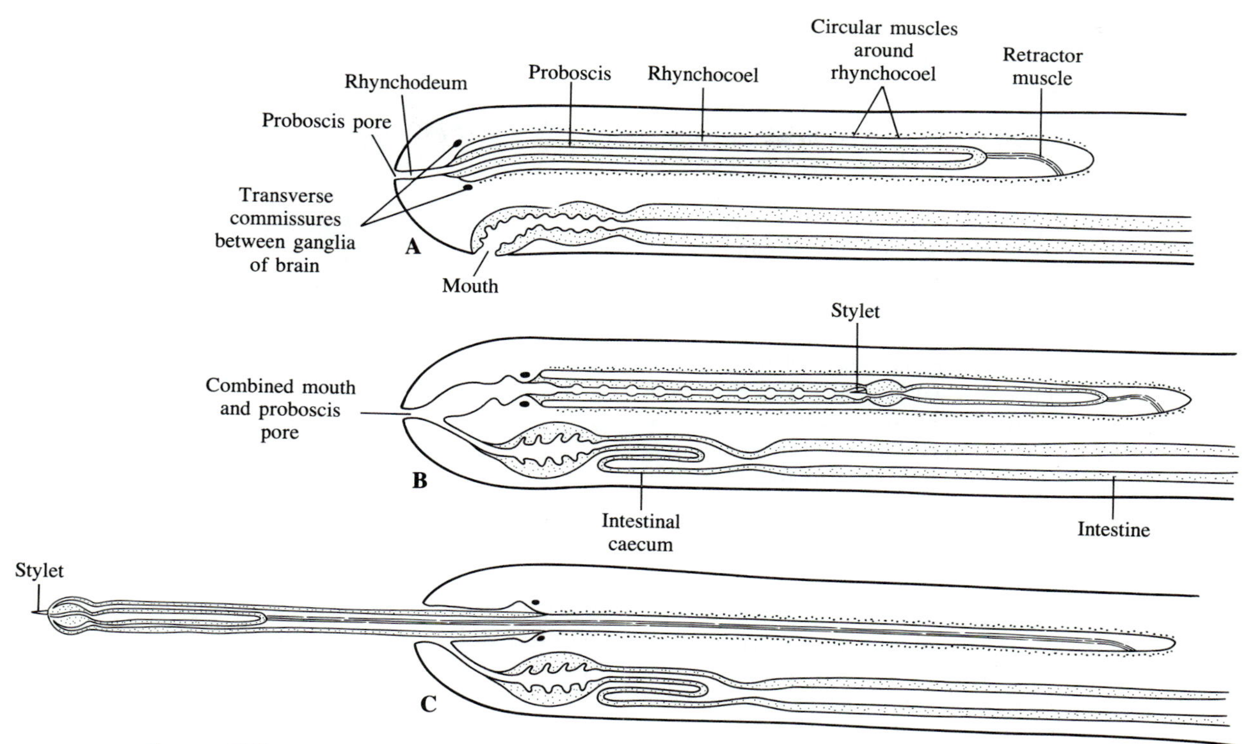

FIGURE 7.7 Longitudinal sections of the anterior portion of the body of nemerteans, diagrammatic. A. Heteronemertean type, proboscis inverted. B. Hoplonemertean type, proboscis inverted. C. Hoplonemertean type, proboscis everted.

are then taken into the gastrodermal cells of the intestine, where intracellular digestion of fats, proteins, and carbohydrates proceeds under alkaline conditions. Undigested residues are eliminated through the anus. The gut, except for the buccal portion, which is concerned with swallowing prey or sucking juices, has little or no musculature of its own. It is, however, ciliated from mouth to anus, and peristaltic activity of the general body musculature also helps keep the food going in the right direction.

UPTAKE OF NUTRIENTS THROUGH THE EPIDERMIS

Certain nemerteans are among the many soft-bodied invertebrates that absorb amino acids and other nutrients through the epidermis. In *Lineus ruber,* the epidermal cells have crowded microvilli that form a brush border of the sort commonly found on epithelia involved in active transport. The extent to which *L. ruber* depends on dissolved nutrients is not known. In the case of *Carcinonemertes errans,* which lives on egg masses of crabs, it is almost certain that young worms must rely on amino acids, and perhaps on other substances, taken up through the epidermis. Their nutritive reserves have been depleted during development, and they are not quite ready to suck nourishment from the crab eggs.

PROBOSCIS AND CAPTURE OF PREY

With only a few exceptions, nemerteans are carnivores that feed on other invertebrates, especially live ones. For capturing and immobilizing their prey, they use the remarkable proboscis. The accuracy with which this structure sometimes strikes is what gives the phylum its name (*nemertea* means ''unerring''). When not in use, the proboscis is inverted, and the opening through which it is shot out is at the anterior end of the body. The opening may be separate from the mouth (Figure 7.5; Figure 7.7, A) or it may be joined to the mouth (Figure 7.7, B and C). Either way, however, the proboscis is not part of the digestive system and is in no way comparable to the protrusible pharynx found in some turbellarians. If its pore is joined to the mouth, it is because the stomodeal invagination that became the foregut was united during development with the invagination in which the proboscis differentiated.

An inverted proboscis is surrounded by a closed, fluid-filled space called the **rhynchocoel** (Figure 7.7, A and B). This space develops within tissue of mesodermal origin and it is lined by an epithelium; in both of these respects it is comparable to a coelom. Surrounding the rhynchocoel is a sheath of muscles, most of which have a circular orientation. The posterior end of the proboscis is typically joined to the wall of the rhynchocoel by a **retractor muscle,** which is a continuation of longitudinal muscles within the proboscis itself. (Certain nemerteans, including species of *Cerebratulus* and a few other genera, lack this retractor muscle.) Between the proboscis pore and the place where the proboscis actually begins is a cavity called the **rhynchodeum;** this is continuous with the cavity enclosed by the inverted proboscis, and both are lined by epidermis.

When a nemertean is excited by a suitable prey animal, it constricts the muscular sheath around the rhynchocoel. This increases the hydrostatic pressure within the rhynchocoel and the proboscis is forced to turn inside-out (Figure 7.7, C). Thus it emerges from the proboscis pore. If the retractor muscle contracts at the same time as the muscles around the rhynchocoel relax, the proboscis is withdrawn. (It is not known how withdrawal is effected in nemerteans that lack a retractor muscle.) When the proboscis–rhynchocoel complex is nearly as long as the body itself, the retractor muscle must obviously undergo an astonishing change in length every time the proboscis is everted or withdrawn.

When "fired," the proboscis may operate in either of two general ways. One way is simply to wrap itself around the prey, clinging with the aid of sticky epidermal secretions and sometimes exuding a venom that helps to quiet the victim. But in the majority of nemerteans—those belonging to the order Hoplonemertea—the proboscis is armed with a **stylet** (Figure 7.7, B and C; Figures 7.8, 7.9). Eversion of the proboscis brings the stylet to its tip, so the stylet can be used to make one or more punctures in the skin of the prey; the punctures allow venom to penetrate the tissue.

Stylets consist mostly of calcium phosphate and are secreted within giant cells called **styletocytes** (Figure 7.8; Figure 7.9, B). There are usually two or more styletocytes, located not far from the **basis,** a mass of secreted material in which the stylet is embedded. If this stylet is torn off or damaged, it will soon be replaced by a reserve stylet, which somehow is moved to the basis from the styletocyte in which it was secreted.

Posterior to the stylet of a hoplonemertean, the inner cavity of the proboscis has glands that have been thought to produce the venoms characteristic of these worms. It has been assumed that the muscles of what is called the **stylet bulb** squeeze the secretion out of a small "ejaculatory duct" that opens beside the basis of the stylet, so that the venom may enter punctures made

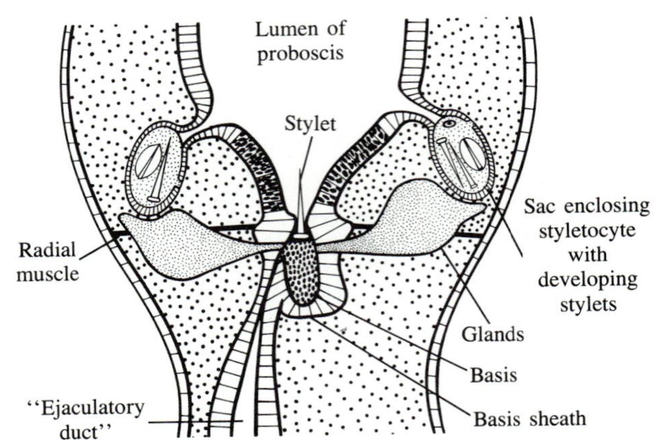

FIGURE 7.8 Principal components of the stylet apparatus of *Paranemertes peregrina*. The "ejaculatory duct" has in the past been thought to deliver venom, but this is unlikely. (Courtesy of Stephen Stricker.)

by the stylet. There is no experimental evidence to support this widely held view. In *Paranemertes peregrina,* the species in which the venom (a compound called anabaseine) has been most thoroughly studied, only small quantities of it are found in the posterior part of the proboscis. Most of the venom is localized in the anterior portion of the proboscis, the part that is everted ahead of the stylet. There is need for more work on this subject.

The venoms of nemerteans are not all the same. Their effectiveness varies according to the species that produce them, the type of prey attacked, and, if a stylet is used, the number and location of punctures. One nemertean is known to immobilize amphipod crustaceans in less than a minute.

Although the proboscis may pull food toward the mouth, actual feeding begins only after the mouth is distended and gets a good hold on the prey. Peristaltic waves of muscular activity help work the victim deeper and deeper into the gut. Digestion begins as soon as acid secretions of the preintestinal portions of the gut are added to the food. Nemerteans are very distensible, and some are capable of swallowing prey with a diameter and length equal to their own, or even larger. Not all species swallow whole prey, however. Some suck in the juices of animals that they have killed or that were already dead when they found them. Certain genera, such as *Tetrastemma* and *Amphiporus,* evert much of the foregut in order to feed on juices of their prey.

By quickly extending the proboscis, attaching its tip to a firm substratum, then shortening the whole body, some terrestrial nemerteans are able to make one or two rapid advances. This type of activity perhaps enables them to avoid or startle a predator. Members of the ge-

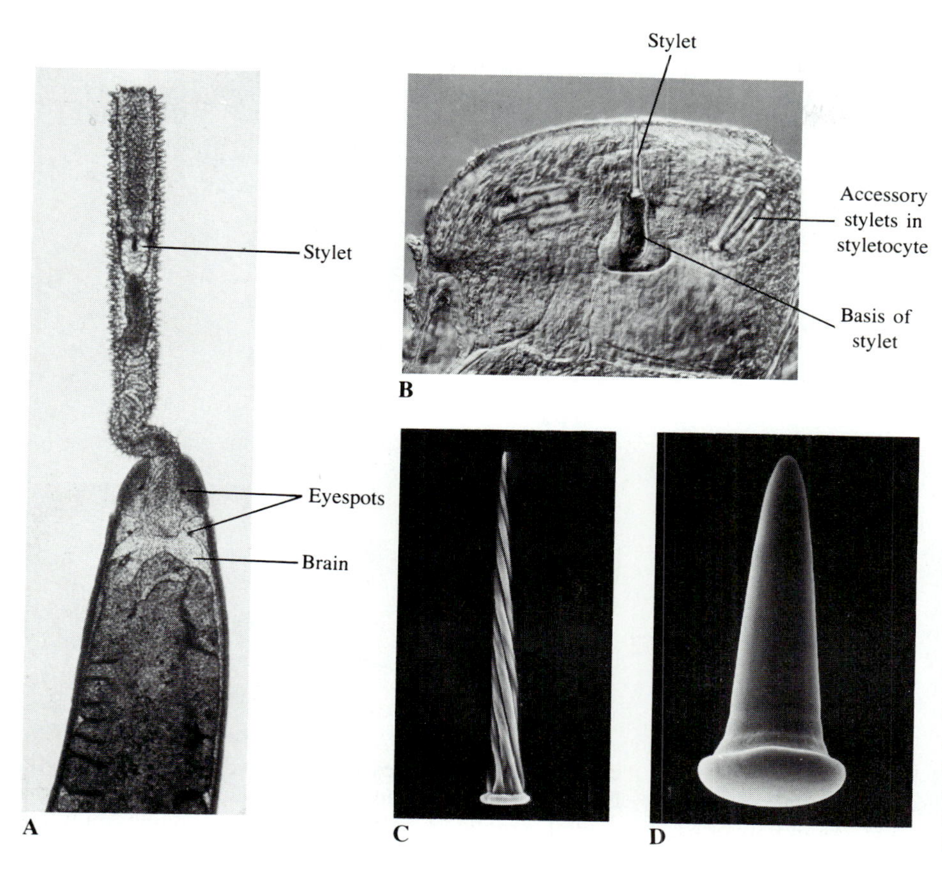

FIGURE 7.9 A. Anterior portion of a small hoplonemertean of the genus *Tetrastemma*, in the process of everting its proboscis. B. Tip of the almost completely everted proboscis of *Tetrastemma*. C. Stylet of *Paranemertes peregrina;* scanning electron micrograph. D. Stylet of *Zygonemertes virescens;* scanning electron micrograph. (C and D courtesy of Stephen Stricker.)

nus *Cerebratulus,* and probably other nemerteans that live in mud or sand, use the proboscis for burrowing.

CIRCULATION, EXCRETION, AND OSMOREGULATION

The circulatory system is closed, the blood being confined to definite vessels. These vessels have a cellular lining, called an **endothelium,** and in this respect they are different from blood vessels of annelids and various other invertebrates, whose blood vessels are channels within an extracellular matrix. Because the vessels of nemerteans are lined by cells from the time they originate within mesoderm, they could, like the rhynchocoel, be interpreted as coelomic spaces. In the vertebrates, the endothelium lining the blood vessels derives from cells that invade the embryonic vessels. A few groups of invertebrates—for instance, octopuses and squids, of the phylum Mollusca, and sea cucumbers, of the phylum Echinodermata—have vessels that are lined by cells, but this is probably a secondary development, for it is not typical of other members of these phyla.

It should be pointed out that some digenean flukes have fluid-filled channels, lined by a syncytial endothelium, on both sides of the midline. The channels contain cells, but show no color, and they are usually referred to as lymphatic vessels.

The arrangement of major blood vessels of nemerteans, some portions of which may be dilated into conspicuous sinuses, varies. Commonly there is a large vessel on both sides of the body, and there may also be an important dorsal vessel (Figures 7.3–7.5). Various loops connect the major vessels anteriorly, posteriorly, and elsewhere. Thick-walled vessels have some intrinsic musculature and are contractile; thin-walled vessels lack muscle. Muscular movements of the body as a whole undoubtedly help to keep the blood circulating. When the blood of a nemertean is reddish, orange, yellow, or green, the pigment is generally concentrated in cells. The reddish pigment is stated to be hemoglobin.

Most nemerteans, other than the highly modified pelagic types, have flame-bulb **protonephridia.** In the usual protonephridial apparatus (Figure 7.5; Figure

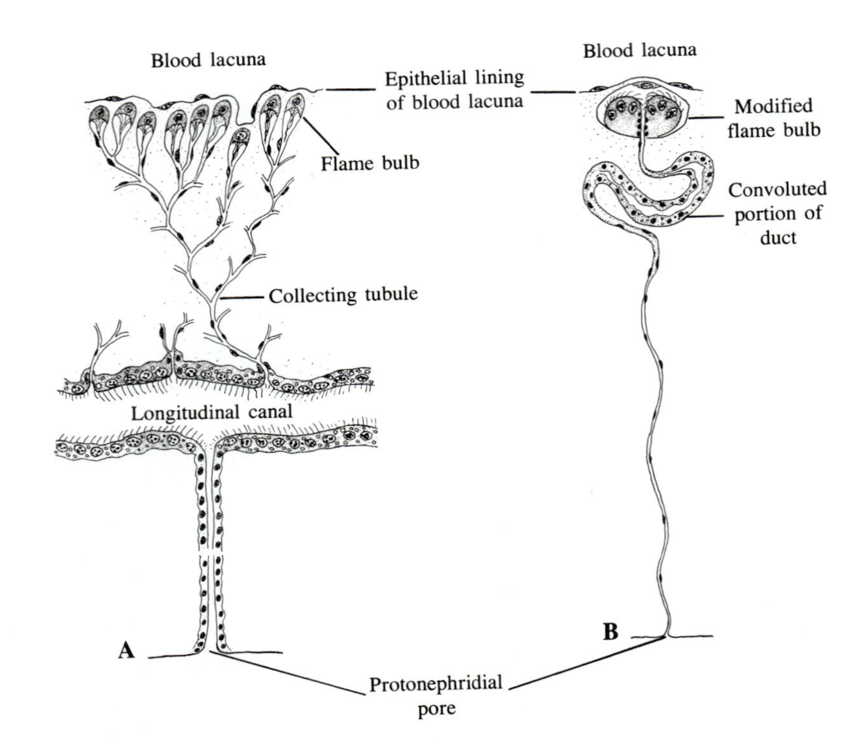

FIGURE 7.10 A. Portion of a protonephridium of a type widespread in nemerteans. B. Simpler type of protonephridium, characteristic of *Cephalothrix*. In both types, the flame bulbs are intimately associated with blood vessels. (Coe, Transactions of the Connecticut Academy of Arts and Sciences, *35*.)

7.10, A), there are two lateral longitudinal canals fed by branched flame-bulb complexes. Each lateral canal has a short duct leading to the outside. As in flatworms, however, the extent and complexity of the protonephridial system vary considerably. For instance, there may be numerous large, multinucleate flame bulbs, each with its own duct to the outside (Figure 7.10, B). An interesting development that shows up in certain groups of nemerteans is the close association of flame bulbs with blood vessels (Figure 7.10, A and B). The bulbs may even protrude into the vessels. That the protonephridia are concerned with osmoregulation seems certain in the case of some terrestrial nemerteans in which activity of the flame bulbs increases when the worms are flooded with fresh water. When marine nemerteans have well-developed protonephridia, often with convoluted collecting ducts, this suggests that they are involved in excretion of nitrogenous wastes or perhaps of excess salts, rather than of water only. Nemertean excretion and osmoregulation is a wide-open field for research.

NERVOUS SYSTEM AND SENSE ORGANS

The nervous system of nemerteans (Figure 7.5) is somewhat similar to that of more advanced turbellarians such as triclads. Of the four major ganglia in the brain, the dorsal ones are larger than the ventral ones. The commissure that unites the right and left dorsal ganglia lies above the rhynchocoel, and the one that unites the ventral ganglia is below it, but both commissures are above the gut. The brain sends out a pair of lateral nerves that run backward for most of the length of the body. These nerves are close to the epidermis in some types; in others, they are embedded in the musculature or in the parenchyma internal to the musculature (Figures 7.2–7.4). Sometimes there is an important middorsal nerve (Figure 7.2), and there may also be a midventral nerve. By way of small nerves that originate from the trunks or from the ganglia, all parts of the body are innervated. Impulses are said to travel slowly in nemerteans, and coordination of some activities may not be impressive.

The sense organs of nemerteans include photoreceptors associated with eyespots, tactile cells (usually with stiff bristles of ciliary origin), statocysts, and ciliated slits and furrows. The latter are presumably the centers of chemoreception that enable nemerteans to recognize live prey or carcasses that are worth exploring. Not all nemerteans can detect food from a distance. Some, such as *Paranemertes,* which feeds mostly on nereid polychaetes, have to bump into their prey before they respond positively to it and fire the proboscis.

A unique feature of many nemerteans is a pair of **cerebral organs** (Figure 7.5). These are sometimes distinct from the ganglia, sometimes more or less insepar-

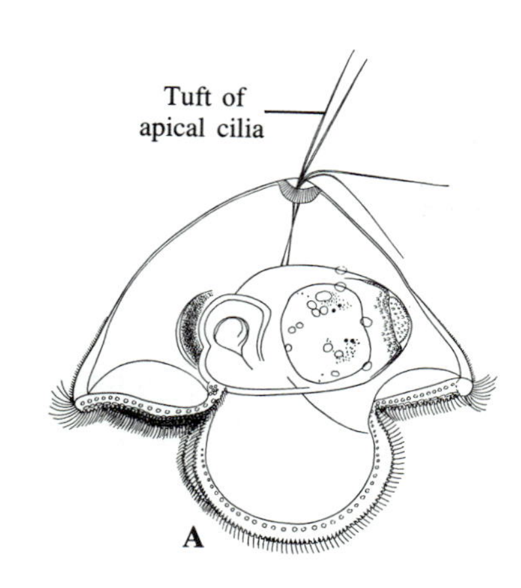

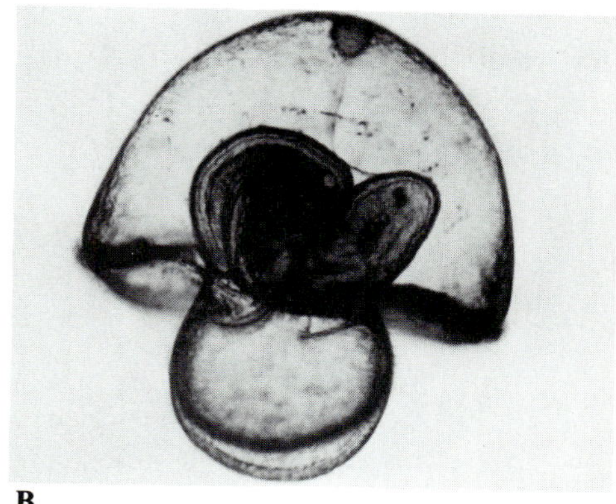

FIGURE 7.11 Pilidium larvae, with young nemerteans inside. A. Drawing showing the arrangement of cilia. (Thorson, Meddelelser fra Kommissionen for Danmarks Fiskeri- og Havundersøgelser, Plankton, *4*.) B. Photomicrograph. (The cilia are not distinct.) (Courtesy of Russel Zimmer.)

ably fused with them. Basically, however, the cerebral organs are deep pits that run from the epidermis to the large dorsal ganglia. They are generally thought to secrete neurohormones or to have a chemosensory function, or both.

REPRODUCTION AND DEVELOPMENT

Nearly all marine nemerteans have separate sexes, but hermaphroditism is common among freshwater and terrestrial species. The gonads (Figures 7.2–7.5) are typically strung out between the intestinal diverticula for much of the length of the body, opening by pores or ruptures on the lateral, dorsolateral, or ventrolateral surfaces. The gametes are squeezed out by contractions of the general body musculature.

While some species spawn without any close contact between the sexes, others mate. The male may simply release sperm over the body of the female, or the partners may secrete a tube of mucus around themselves and discharge their gametes into this. Males of certain of the pelagic nemerteans that live in deep water have fingerlike outgrowths of the body wall, presumably used for inseminating females.

Females often hold their eggs together in masses, depositing them in their burrows or attaching them to firm objects. This is the case, for instance, in some species of *Lineus;* prominent glands in the dermis of the female (Figure 7.4, B) secrete mucus to form a cocoon around the eggs. Viviparity is typical of freshwater and terrestrial nemerteans, and of some marine species; the eggs develop into juvenile worms within the gonads.

Cleavage is spiral and the fate of the blastomeres is determinate. Further development may be either direct or indirect. In either case, however, the foregut and hindgut originate as ectodermal invaginations; they later become joined to the endodermal midgut. The epithelia that line the rhynchodeum and the lumen of the inverted proboscis also develop from an ectodermal invagination. This may secondarily become united with the foregut, or the foregut may lose its opening to the outside and enter the rhynchodeum. The rhynchocoel and blood vessels develop within tissue of mesodermal origin. The protonephridia, as in flatworms, derive from ectoderm.

Indirect development involves a free-swimming larva and is characteristic of only some members of the order Heteronemertea. The **pilidium** (Figures 7.11, 7.12) is a beautiful larva that resembles a helmet because of the flaplike lobes that grow on either side of the mouth. It can feed on microscopic organisms, although its gut lacks an anal opening. In time, a pilidium typically develops three lateral pockets on each side and one median posterodorsal pocket (Figure 7.12, A). These pockets become deeper and eventually are separated from the surface, becoming **amniotic cavities.** The thickened ectoderm in the deepest part of each pocket gives rise to the epidermis of the young nemertean, as well as to some other structures, such as the lining of the rhynchodeum and inverted proboscis. The worm continues its development until it is ready to break out of the pilidium. The outer part of the pilidium, after being cast off, is not necessarily wasted. In some species it provides the young worm with its first meal.

The **Desor larva** (named for the zoologist who discovered it) also occurs in certain genera of the order Heteronemertea. It does not look like a pilidium, for it lacks the two larger lobes and some other structures characteristic of the latter. The young worm is isolated within it by ectodermal invaginations, just as in the pilidium.

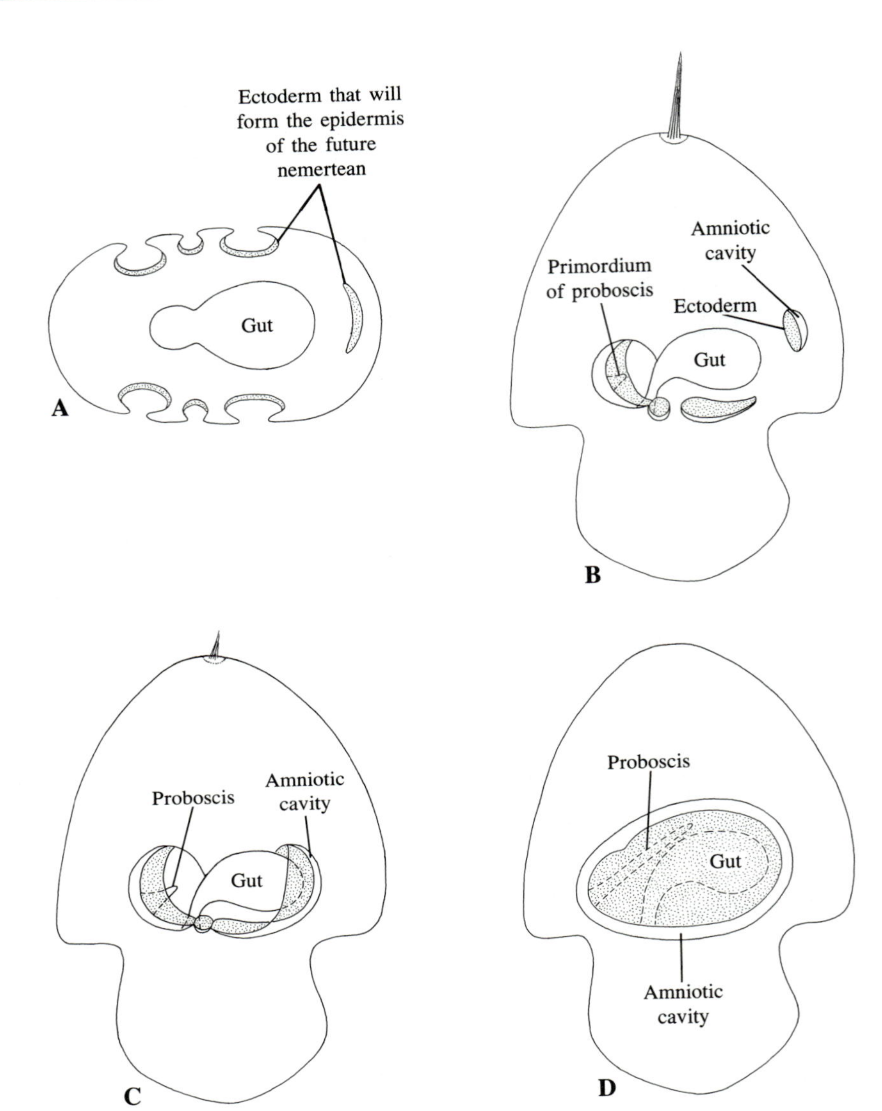

FIGURE 7.12 Diagrams showing young nemertean forming within a pilidium. A. Optical section of a pilidium viewed from above. The six paired pockets and one posterodorsal pocket will become deeper and will close. B. The pockets have closed, forming amniotic cavities. The disks of ectoderm indicated by stippling will enlarge and eventually enclose the gut of the pilidium. The disks of the two anteriormost pockets have already become joined and have given rise to the primordium of the proboscis. C, D. Stages in the envelopment of the gut by ectoderm, and continued development of the proboscis. The single amniotic cavity around the young nemertean in D has been formed by fusion of spaces that were originally separate.

REGENERATION

If the proboscis is torn off, which sometimes happens, it is regenerated. In many species, moreover, worms that have broken apart or that have been cut up will regenerate all missing parts. The capacity for this kind of regeneration varies greatly. Within the genus *Lineus,* for instance, some species can regenerate from small pieces, others cannot. If a fragment is to regenerate and form a complete worm, the presence of at least some nerve tissue is obligatory. Pieces from the anterior part of a worm, however, are less likely to regenerate successfully than those taken from posterior parts. Natural fragmentation, followed by regeneration, occurs in certain nemerteans, and may be viewed as a kind of asexual reproduction.

PHYLOGENY

Nemerteans almost certainly evolved from worms that were ancestral also to flatworms. They presumably had reached a rather high level of histological differentiation but had a simple reproductive system. In having a complete gut, blood vessels, a proboscis, and certain other specializations—such as the close association of flame bulbs with blood vessels—nemerteans show distinct advances over the flatworm body plan. The rhynchocoel that surrounds the proboscis, moreover, is essentially a type of coelomic cavity; it originates within mesoderm and is lined by an epithelium. Although some zoologists have suggested that certain other animals, even lampreys, may be derived from nemerteans, the general consensus today is that these worms form

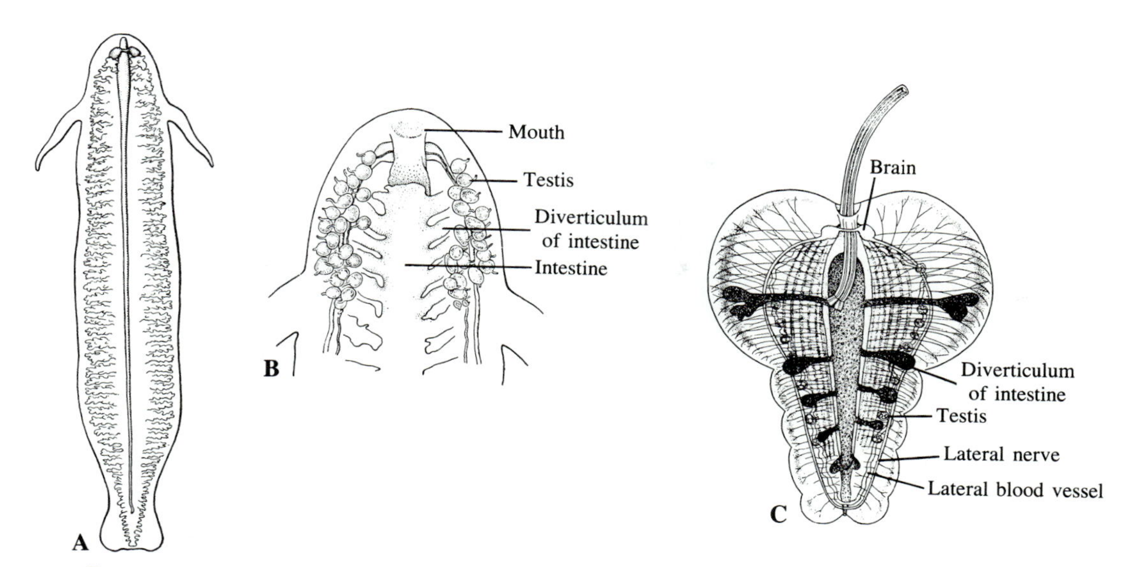

FIGURE 7.13 Two pelagic nemerteans. A. *Nectonemertes mirabilis,* male, showing intestinal diverticula. B. Anterior portion of a male *Nectonemertes,* showing testes. C. *Pelagonemertes moseleyi,* female, much flattened. (Coe, Bulletin of the Scripps Institution of Oceanography, *6.*)

an offshoot that has not given rise to any other surviving groups.

CLASSIFICATION

The phylum Nemertea is traditionally divided into two classes, each further divided into two orders. The outstanding features of these classes and orders are characterized briefly below.

CLASS ANOPLA

In anoplans, the proboscis has no stylet (*anopla* means "without a weapon") and the mouth and proboscis pore are separate. The mouth is far enough from the anterior tip of the body that it lies behind the brain. The lateral nerves, as seen in transverse sections, are either external to the musculature or buried in it, but are never internal to the musculature.

Order Palaeonemertea

Eyespots are lacking, and cerebral organs are found only in certain genera. There is no middorsal blood vessel. The lateral nerves may be just beneath the epidermis or buried in one of the strata of muscles of the body wall (Figure 7.2). (In a particular species, in fact, the lateral nerves may be close to the epidermis in the anterior part of the body, but within the musculature at more posterior levels.) *Tubulanus* (Color Plate 3) is a particularly widespread genus.

Order Heteronemertea

Eyespots and cerebral organs are generally present, and there is a middorsal blood vessel (Figure 7.4). The lateral nerves are buried in the musculature and are thus fairly remote from the epidermis. Common genera are *Cerebratulus* (which consists mostly of burrowers in sand or mud), *Lineus,* and *Micrura* (Color Plate 3).

CLASS ENOPLA

Enoplans are considered to be more advanced than anoplans. The proboscis is typically armed with a stylet (*enopla* means "with a weapon"). The mouth and proboscis pore may be one and the same or they may be separate; in either case, however, the mouth is close to or at the anterior tip of the body, so it is in front of the brain. The lateral nerves are in the parenchyma internal to the musculature.

Order Hoplonemertea

The proboscis is armed with a stylet. Some important genera are *Amphiporus* (Figure 7.1, B; Figure 7.3; Color Plate 3), *Tetrastemma* (Figure 7.9, A and B), *Paranemertes* (Figure 7.1, A), and *Oerstedia.* A few odd types deserve special mention. Among these are the short-bodied, highly specialized bathypelagic nemerteans, such as *Nectonemertes* (Figure 7.13, A and B) and *Pelagonemertes* (Figure 7.13, C). Some of them are weakly muscularized and seem merely to drift around in deep water; others are good swimmers.

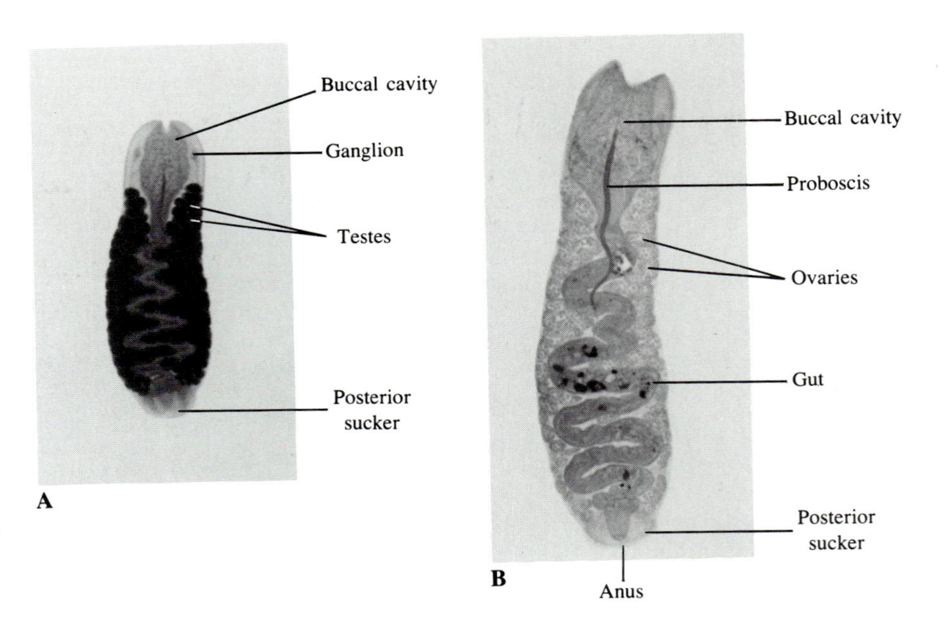

FIGURE 7.14 *Malacobdella*, fixed and stained specimens. A. Male, slightly flattened. B. Female, much flattened.

Species of *Carcinonemertes* suck juices from the eggs being brooded by female crabs; they do not reach maturity on male or nonbrooding female crabs. The small freshwater nemerteans of the genus *Prostoma* are hoplonemerteans, too. *Geonemertes* is a terrestrial genus found mostly in tropical and warm-temperate regions; one species, native to New Zealand, is sometimes seen in greenhouses far from its native land, and another occurs under stones and debris at the edges of salt marshes on the Pacific coast.

Order Bdellonemertea

The only genus of bdellonemerteans is *Malacobdella* (Figure 7.14), a leechlike worm that inhabits the mantle cavity of various marine clams. *Malacobdella* has a prominent posterior sucker for attaching itself to the mantle, and also has a capacious buccal cavity lined by ciliated, fingerlike projections. These worms feed on copepods, other small crustaceans, diatoms, and detritus that the hosts bring into the mantle cavity through the inhalant siphon. The proboscis of *Malacobdella* is well developed but lacks a stylet and is rarely everted.

SUMMARY

1. Nemerteans are generally elongated animals, but they are like turbellarian flatworms in being bilaterally symmetrical and externally ciliated. Their general histological organization is also similar to that of flatworms, although they usually have proportionately more musculature. They differ from flatworms in having a complete gut, a unique type of proboscis for capturing or immobilizing prey, and a circulatory system. The blood vessels are almost unique among those of invertebrates in that they are lined by cells.

2. The proboscis is an inverted tubular structure whose origin is independent of the gut. It lies in a fluid-filled cavity called the rhynchocoel, which is surrounded by circular muscles. When these muscles contract, the hydrostatic pressure in the rhynchocoel is raised, causing the proboscis to turn inside-out and be everted. The proboscis may emerge through its own pore or, in the case of nemerteans in which the proboscis pore and foregut have become united secondarily, through the mouth. A sticky secretion and venom on the proboscis help quiet the prey, and in many nemerteans the proboscis has a stylet for making punctures. Withdrawal of the proboscis is usually effected by a retractor muscle.

3. Most nemerteans are marine, but a few are found in fresh water and in moist terrestrial situations. They feed on other invertebrates, including polychaete annelids, crustaceans, and molluscs. The prey, after being paralyzed or immobilized, may be swallowed whole, or its juices may be sucked out. Certain species rely to some extent on soluble nutrients they take up through the epidermis.

4. Although nemerteans are in some respects more advanced than flatworms, they have a simpler reproductive system, and sexes are usually separate. In most species, there are numerous gonads on both sides of the body. These release their gametes directly to the outside. Development may be direct or there may be a larval stage that is distinctly different from the adult.

REFERENCES ON NEMERTEA

Bierne, J., and Rué, G., 1979. Endocrine control of reproduction in two rhynchocoelan worms. International Journal of Invertebrate Reproduction, *1*:109–120.

Cantell, C. E., 1969. Morphology, development, and biology of the pilidium larvae from the Swedish west coast. Zoologiska Bidrag från Uppsala, *38*:61–111.

Coe, W. R., 1943. Biology of the nemerteans of the Atlantic coast of North America. Transactions of the Connecticut Academy of Arts and Sciences, *35*:129–328.

Fisher, F. M., Jr., and Oaks, J. A., 1978. Evidence for a nonintestinal nutritional mechanism in the rhynchocoelan, *Lineus ruber*. Biological Bulletin, *154*:213–225.

Gibson, R., 1972. Nemerteans. Hutchinson University Library, London.

Gibson, R., 1982. British Nemerteans. Synopses of the British Fauna, no. 32. Linnean Society of London and E. J. Brill/Dr. W. Backhuys, London and Leiden.

Gibson, R., and Moore, T., 1976. Freshwater nemerteans. Zoological Journal of the Linnean Society, *58*:177–218.

Jennings, J. B., and Gibson, R., 1969. Observations on the nutrition of seven species of rhynchocoelan worms. Biological Bulletin, *136*:405–443.

Kem, W. R., 1973. Biochemistry of nemertine toxins. *In* Martin, D. F., and Padilla, G. M. (editors), Marine Pharmacognasy. Action of Marine Biotoxins at the Cellular Level. Academic Press, New York and London.

McDermott, J. J., 1976. Observations on the food and feeding behavior of estuarine nemertean worms belonging to the order Hoplonemertea. Biological Bulletin, *150*:57–68.

Roe, P., 1970. The nutrition of *Paranemertes peregrina*. I. Studies on food and feeding behavior. Biological Bulletin, *139*:80–91.

Roe, P., 1976. Life history and predator–prey interactions of the nemertean *Paranemertes peregrina* Coe. Biological Bulletin, *150*:80–106.

Roe, P., Crowe, J. H., Crowe, L. M., and Wickham, D. E., 1981. Uptake of amino acids by juveniles of *Carcinonemertes errans* (Nemertea). Comparative Biochemistry and Physiology, *69A*:423–427.

Stricker, S. A., and Cloney, R. A., 1981. The stylet apparatus in the nemertean *Paranemertes peregrina:* its ultrastructure and role in prey capture. Zoomorphology, *97*:205–223.

8

PHYLA PLACOZOA, DICYEMIDA, AND ORTHONECTIDA

The three groups of organisms described in this chapter are probably not closely related. It is convenient to deal with them together because they are at about the same level of structural complexity, which is somewhat comparable to that of a planula larva of a cnidarian. This is not to say that the simplicity of placozoans, dicyemids, and orthonectids is ancestral; in the case of the last two groups, at least, it is almost certainly derived.

The dicyemids and orthonectids, which are parasitic, are often united into one phylum, Mesozoa. Certain stages of dicyemids superficially resemble those of orthonectids, but in other respects these organisms are distinctly different. It is likely that in the course of their evolution, perhaps from flatworms of some sort, both have ended up at about the same level of structure. In any case, the formal name Mesozoa, originally applied to dicyemids and then extended to orthonectids, is undesirable because of its intended implication: that these organisms are intermediate between protozoans and metazoans. When used in the vernacular sense, however, *mesozoans* is a handy collective term for referring to the two groups.

PHYLUM PLACOZOA

In the late nineteenth century, the zoologist F. E. Schulze studied some thin coatings found on the glass of marine aquaria and found them to consist of two layers of cells, most of which had cilia. He decided that the growths were animals and invented the name *Trichoplax adhaerens* for them. Subsequently, *Trichoplax* was interpreted as the planula stage of a cnidarian, and for all practical purposes it was forgotten until after 1960, when reports of its occurrence, not all of them published, began to turn up in various parts of the world. The organism has recently been studied in considerable detail, and the idea that it is an odd type of planula has been discarded.

Specimens of *Trichoplax* (Figure 8.1), which are commonly 1 or 2 mm in diameter, slightly resemble large amoebae, mostly because they have an irregular shape and because they move slowly. The movement, however, is due to ciliary activity. There is no polarity; a *Trichoplax* can glide in any direction. Cellular organization is evident, and the nuclei of individual cells are recognizable. Large, light-reflecting lipid globules are conspicuous; they are localized within degenerating cells of the upper epithelium. In cross-sections (Figure 8.2), it can be seen that the normal cells of this layer are relatively thin, except where each has a bulge that accommodates the nucleus, and that they have single cilia. The tall cells of the lower epithelium are of two types. Some have a cilium as well as surface projections; the rest have neither of these structures but are filled with secretory products. There is foamy extracellular material between many of the cells of the lower epithelium. Between the upper and lower epithelia, which are continuous at the edges of the organism, there are contractile cells. These are joined to one another and also to some of the epithelial cells. It is not known how they are coordinated. They contain symbiotic bacteria and vacuoles filled with small concretions, and their mitochondria are clustered, often in such a way that they appear stacked.

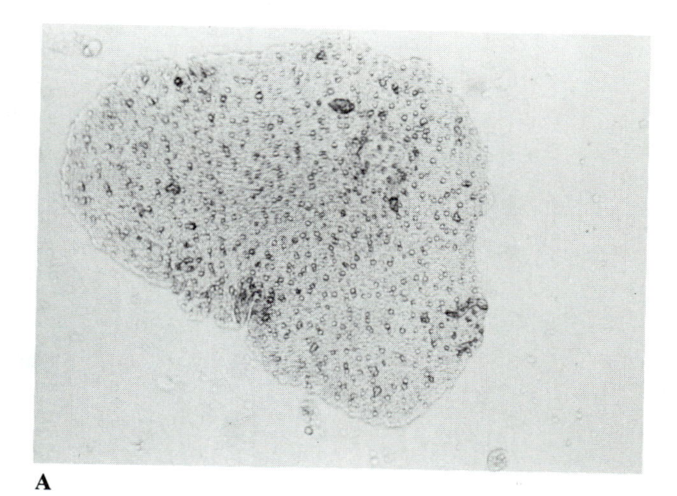

A

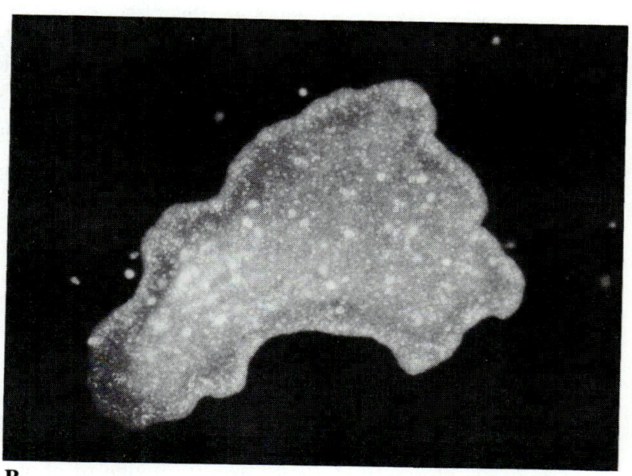

B

FIGURE 8.1 *Trichoplax adhaerens;* photomicrographs of living specimens. A. Bright-field illumination. B. Dark-field illumination. (Courtesy of Richard Miller.)

The cells of the lower epithelium have been observed to take up and digest small algal cells. It is thought that these cells also produce digestive enzymes that act extracellularly, digesting flagellates, algal cells, and bacteria that *Trichoplax* traps between itself and the

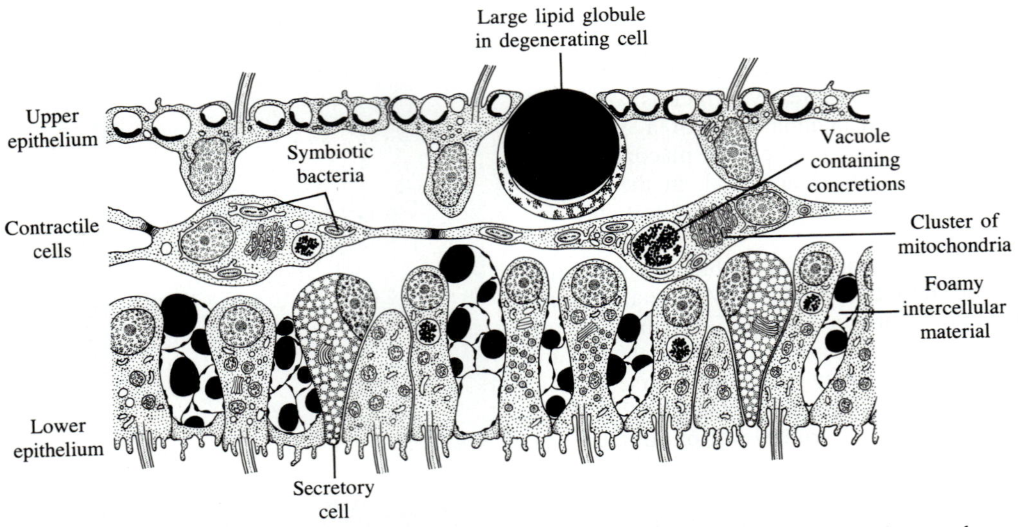

FIGURE 8.2 *Trichoplax adhaerens;* cross-section, diagrammatic, based on electron micrographs. (Courtesy of Karl Grell.)

substratum. Sometimes, when the animal humps up, a shallow, gutlike cavity is formed.

Trichoplax reproduces asexually by dividing into two more or less equal halves, and also by budding off portions of the body. If a specimen is cut, the pieces may regenerate into normal individuals. Dissociated cells have been observed to reassemble, in much the same way as dissociated cells of some sponges will do.

Cells that are considered to be eggs are derived from certain cells of the lower epithelium. They sink into the space between the epithelia and grow to a size of about 100 μm. Production of eggs is not initiated unless *Trichoplax* goes into a kind of degenerative phase, which is occasioned by overcrowding. Only those eggs that secrete what looks like a fertilization membrane undergo cleavage. Certain amoeboid cells have been interpreted as sperm, but the evidence is not convincing. Cleavage is radial and equal. Embryos have not been reared to maturity under laboratory conditions, however.

It is only natural that *Trichoplax* has figured in phylogenetic speculations. It is, after all, much like a blastula that has been flattened out, and some biologists who have speculated on the origins of metazoans have viewed an organism of this general sort, with certain cells specialized for feeding, as a progenitor of more advanced types in which an invagination of the body wall formed a gut. When *Trichoplax* humps up, the concavity beneath it remotely resembles an archenteron.

The term **placula** has been applied to the type of body organization characteristic of *Trichoplax,* and the phylum name, Placozoa, is based on it. Perhaps *Trichoplax* is truly primitive in the sense that it is related to long-extinct early metazoans. There is a good chance, however, that it is a secondarily simplified derivative of some group of lower invertebrates.

REFERENCES ON PLACOZOA

Grell, K. G., 1971. *Trichoplax adhaerens* F. E. Schulze und die Entstehung der Metazoen. Naturwissenschaftliche Rundschau, *24:*160–161.

Grell, K. G., 1972. Eibildung und Furchung von *Trichoplax adhaerens* F. E. Schulze (Plakozoa). Zeitschrift für Morphologie der Tiere, *73:*297–314.

Grell, K. G., 1981. *Trichoplax adhaerens* and the origin of Metazoa. Atti dei Convegni Lincei (Accademia Nazionale dei Lincei), *49:*107–121.

Grell, K. G., and Benwitz, G., 1971. Die Ultrastruktur von *Trichoplax adhaerens* F. E. Schulze. Cytobiologie, *4:* 216–240.

PHYLUM DICYEMIDA

Dicyemids are found in the kidneys and pericardial cavity of certain cephalopod molluscs, especially benthic types. Almost all near-shore octopuses, and some deep-water species as well, are parasitized. Cuttlefishes and their allies regularly have dicyemids, too; squids, with a few exceptions, do not. Altogether about 75 species of dicyemids have been described.

Dicyemids were discovered before the middle of the nineteenth century, but it took a long time for zoolo-

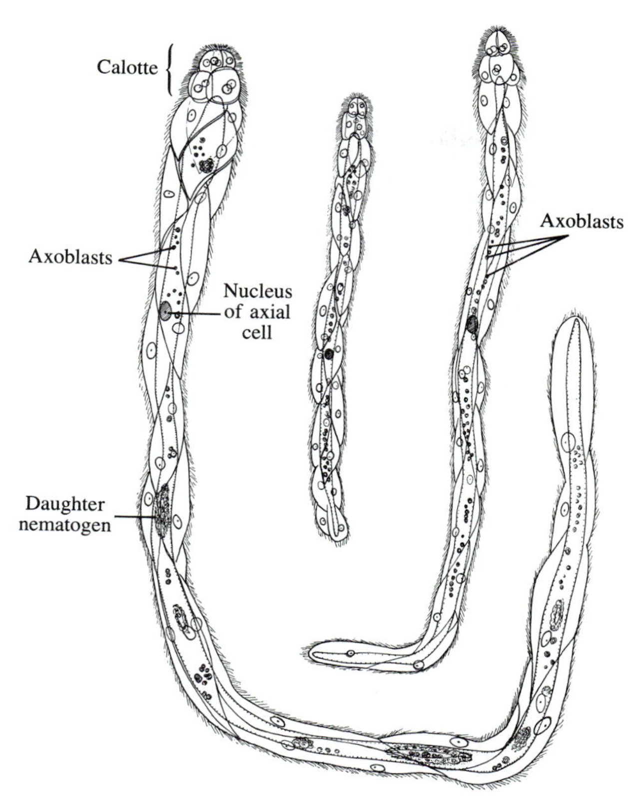

FIGURE 8.3 *Dicyema,* nematogen stage. The largest specimen contains daughter nematogens. (McConnaughey, Journal of Parasitology, *35*.)

gists to interpret the life cycle with reasonable accuracy—and important details still need to be worked out. The conventional terminology used in describing the various stages is complicated and bears almost no resemblance to that used in connection with any other lower invertebrates. Some of the terms were introduced before the relationships between successive stages were adequately understood. In this account, based largely on what is known about the genera *Dicyema* and *Dicyemennea,* jargon is reduced to a minimum. Not all information given here is necessarily applicable to all dicyemids, and some details are subject to different interpretations.

In a smear preparation of a cephalopod kidney that is parasitized by dicyemids, one sees hundreds of slender, delicate worms of the sort shown in Figure 8.3. Their usual length is about 1 or 2 mm, but they can be considerably longer or shorter, depending on the species. The dense population of parasites has been built up by asexual reproduction that began with the first individuals that took up residence in the kidney. The slender worms are called **nematogens** because they give rise to threadlike individuals like themselves.

No one is absolutely sure how the first nematogens become established in the kidney. The earliest stages observed already have a jacket of ciliated epidermal cells enclosing two or three cells that form an inner mass. One of these becomes the **axial cell;** the other one, or one of the other two, is taken up by the axial cell and is now called an **axoblast.** It multiplies, forming many daughter axoblasts (Figure 8.3; Figure 8.4, A). Meanwhile, the axial cell increases in size to keep pace with the growth of the nematogen as a whole, and its nucleus becomes greatly enlarged.

In time, some of the axoblasts develop into daughter nematogens (Figure 8.4, A; Figure 8.5). After an axoblast has undergone several cleavages, there is a large cell surrounded by small cells (Figure 8.5, A–C). The latter differentiate into the ciliated epidermal cells. The large cell divides unequally once or twice. One of the smaller products survives and is taken up by the large cell, and thus the axial cell–axoblast complex of the daughter nematogen is formed (Figure 8.5, D–F). After the nematogen has become ciliated (Figure 8.4, A; Figure 8.5, G), it escapes from the parent by working its way out between the epidermal cells, and it becomes a parasite in the same kidney. Before long, it produces daughter nematogens and its progeny repeat the process. Many of the nematogens, including the oldest ones, continue to reproduce indefinitely, and nearly all of the space in the kidney may be occupied by nematogens of one generation or another.

The number of epidermal cells, usually about 40 or 50, is almost constant for nematogens of a particular species. Two tiers of cells at the anterior end form what is called the **calotte** (Figure 8.3). The cilia on these cells are short and interdigitate with the microvillar projections of the brush border of the cephalopod's kidney epithelium, so that the parasite adheres tightly. Some pathological changes have been observed in the brush border, but dicyemids are relatively benign parasites that seem to subsist on soluble organic materials in the kidney fluid. The surfaces of most of the epidermal cells have convoluted ridges (Figure 8.6) that greatly increase the absorptive area, and there are also pinocytic vesicles.

After production of daughter nematogens slows down, the axoblasts assume a new role. They develop into **infusorigens** (Figure 8.7), which may be viewed either as much-reduced, hermaphroditic sexual individuals or as gonads containing sperm and oocytes. The first of these interpretations is perhaps the more logical one, for the ball of sperm and oocytes is, in terms of its origin from an axoblast, homologous to a daughter nematogen.

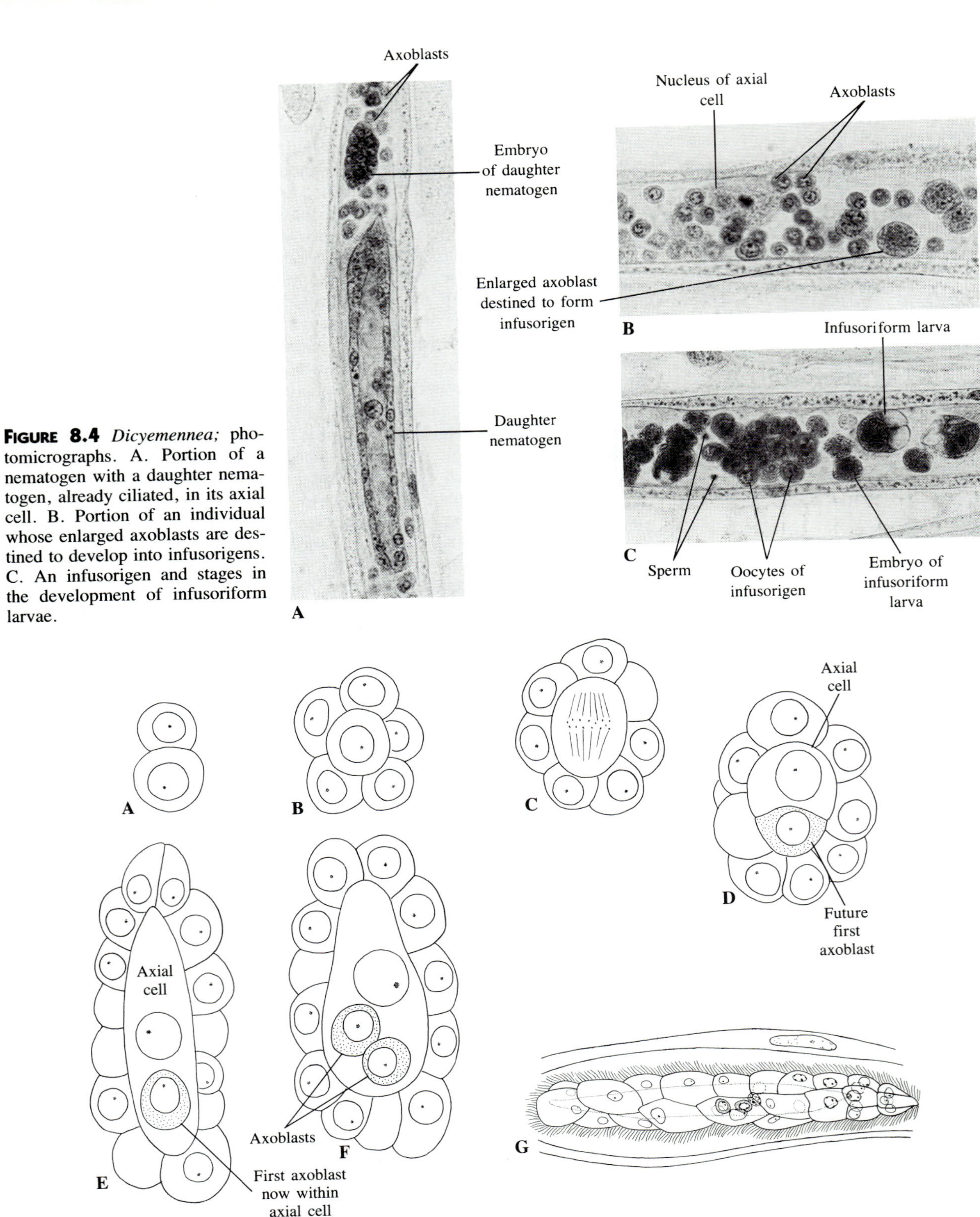

FIGURE 8.4 *Dicyemennea;* photomicrographs. A. Portion of a nematogen with a daughter nematogen, already ciliated, in its axial cell. B. Portion of an individual whose enlarged axoblasts are destined to develop into infusorigens. C. An infusorigen and stages in the development of infusoriform larvae.

FIGURE 8.5 *Dicyema,* stages in the development of a daughter nematogen. A. Products of first division (unequal) of an axoblast. B. Future epidermal cells surrounding a central cell. C, D. Division of the central cell, forming an axial cell and future first axoblast. E. First axoblast has been taken up by the axial cell. F. First axoblast has divided. G. Young nematogen, already ciliated. (After McConnaughey, University of California Publications in Zoology, *55.*)

The first division of an axoblast destined to form a sexual individual is unequal, and the smaller product perishes (Figure 8.7, A–C). Second and third divisions produce two more cells (Figure 8.7, B and C). One of these divides several times, covering the parent cell with oocytes. The other enters the parent cell (Figure 8.7, D) and divides mitotically (Figure 8.7, E) to produce a few cells which undergo meiotic divisions that produce sperm. The sperm lack flagellar tails. After an oocyte has been penetrated by a sperm—which can happen before or after the oocyte has become detached from the cluster—it divides meiotically and casts off two polar bodies. Fertilization follows.

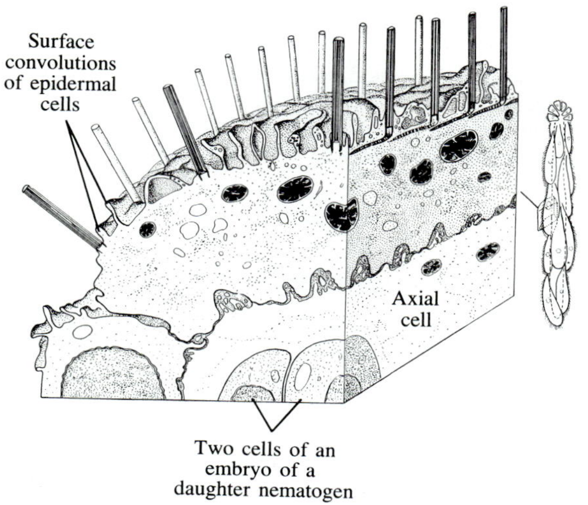

FIGURE 8.6 *Dicyema,* portion of a nematogen; diagram based on electron micrographs. (Bresciani and Fenchel, Videnskabelige Meddelelser fra Dansk Naturhistorisk Forening, *128.*)

The zygote develops into a stage that is decidedly different from a nematogen. To earlier zoologists it looked something like an infusorian—that is, a ciliated protozoan—and it became known as the **infusoriform larva.** This stage (Figure 8.4, C; Figure 8.8), though generally only about 40 or 50 μm long, is complicated. Its anterior half is covered by two unciliated cells that form a thin-walled cup. The several cells covering the posterior half are ciliated. Within the anterior cup are two proportionately large cells that usually contain light-reflecting masses of a substance that has recently been identified as magnesium inositol hexaphosphate. (It was previously thought to be uric acid or guanine.) A number of other cells are arranged in a precise fashion around an inner cavity called the **urn.** Certain of these cells are ciliated on the side facing the cavity. The urn is nearly filled by four cells, each of which contains

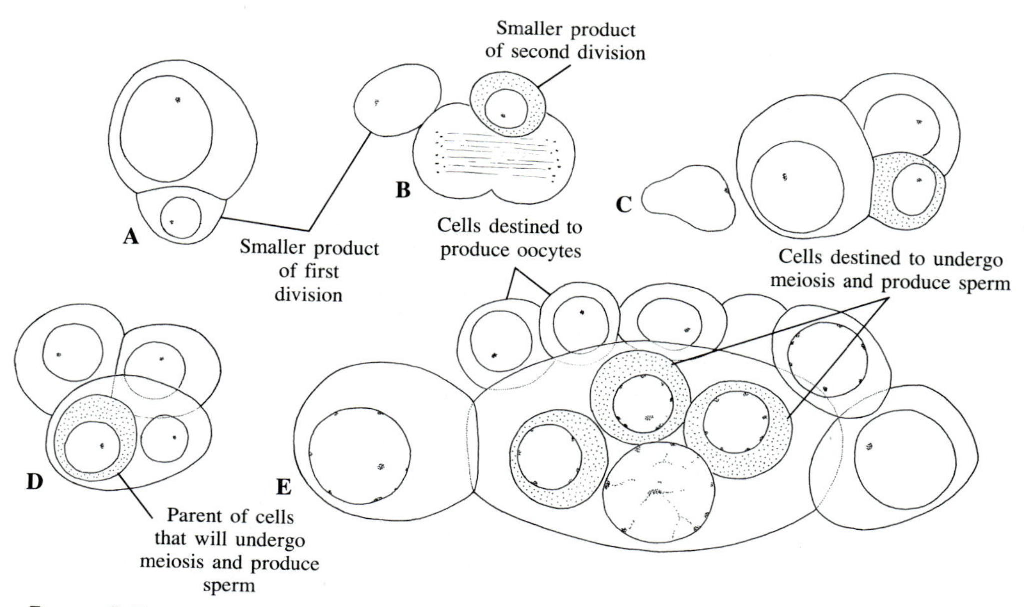

FIGURE 8.7 *Dicyema,* stages in the development of an infusorigen. A. Products of the first division of an enlarged axoblast. The smaller product will perish. B. Third division in progress. C. Third division completed. D. The smaller product of the second division has entered the parent cell; the other cell has divided. E. Young infusorigen with cells destined to produce oocytes on the outside and cells destined to produce sperm within the cytoplasm of the largest cell. (After McConnaughey, University of California Publications in Zoology, *55.*)

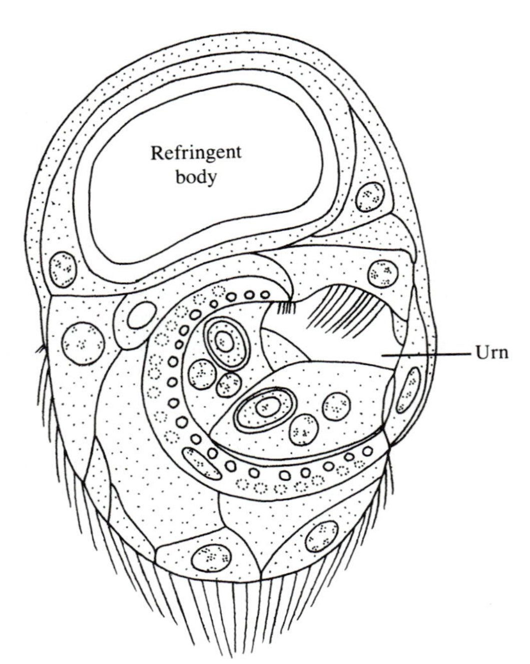

FIGURE 8.8 *Dicyemennea,* infusoriform larva; sagittal section, close to the midline; based on electron micrographs. (After Matsubara and Dudley, Journal of Parasitology, *62*.)

one or two nuclei as well as another complete cell. In the infusoriform larvae of *Dicyemennea,* and perhaps in those of other genera as well, the two cells that cover the anterior half separate at one point on both the dorsal surface and ventral surface. Tufts of cilia originating on certain of the internal cells stick out through the openings.

The fully developed infusoriform larvae occupy cavities within the axial cell of the nematogen, and their cilia beat slowly. Enlargement of a cavity causes the epidermis of the nematogen to bulge, and then to rupture temporarily, at which time the larva escapes and leaves the cephalopod host with urine. What happens to the infusoriform larva is not known. It is almost certainly the infective stage, however. Perhaps it somehow becomes transformed into a nematogen, or perhaps each of its urn cells can develop into a nematogen.

REFERENCES ON DICYEMIDA

Hochberg, F. G., 1982. The ''kidneys'' of cephalopods: a unique habitat for parasites. Malacologia, *23*:121–134.

Lapan, E. A., and Morowitz, H., 1972. The Mesozoa. Scientific American, *277*(6):94–101.

McConnaughey, B. H., 1951. The life-cycle of the dicyemid Mesozoa. University of California Publications in Zoology, *55*:295–336.

McConnaughey, B. H., 1963. The Mesozoa. *In* Dougherty, E. C., *et al.* (editors), The Lower Metazoa, pages 151–168. University of California Press, Berkeley and Los Angeles.

Matsubara, J. A., and Dudley, P. L., 1976. Fine structural studies of the dicyemid mesozoan, *Dicyemennea californica* McConnaughey. I. Adult stages. Journal of Parasitology, *62*:377–389.

Matsubara, J. A., and Dudley, P. L., 1976. Fine structural studies of the dicyemid mesozoan, *Dicyemennea californica* McConnaughey. II. The young vermiform stage and the infusoriform larva. Journal of Parasitology, *62*:390–409.

PHYLUM ORTHONECTIDA

Unlike dicyemids, whose hosts are always cephalopod molluscs, orthonectids live in a wide variety of marine invertebrates. The approximately 20 described species, and a number that have been reported but not named, are parasites of animals as diverse as turbellarians, nemerteans, polychaete annelids, gastropod and bivalve molluscs, brittle stars, and ascidians. Alfred Giard, the first zoologist to deal intensively with orthonectids soon after their discovery in the late nineteenth century, astutely concluded that they were worms ''degraded'' by parasitism to the structural level shown by the planula larva of cnidarians. Though certain stages in the life cycles of orthonectids and dicyemids are superficially similar, it is almost certain that the two groups have evolved along separate lines.

When orthonectids are found in a parasitized host, the stages that one recognizes right away are developing or mature males and females (or, in the case of one genus, hermaphroditic individuals). They are thus hardly comparable to the nematogens of dicyemids, although they may resemble these. The epidermis of sexual individuals consists of ciliated and unciliated cells arranged in rings encircling the body (Figures 8.9, 8.10, 8.12). In most species there is little variation in the number of rings, and the sequence of ciliated and unciliated rings is astonishingly precise. In female orthonectids the cellular mass enclosed by the epidermis consists mostly of oocytes, but peripheral musculature is also present. In males there is a testis, musculature, some apparently undifferentiated cells, and sometimes also what are thought to be elastic cells. The name Orthonectida means ''straight swimming.'' It was applied by Giard in allusion to the fact that when seemingly mature males and females are set free by dissection of their host, they swim in a straight line. When orthonectids are really mature and escape naturally from their host, however, they generally swim by a spiralling motion.

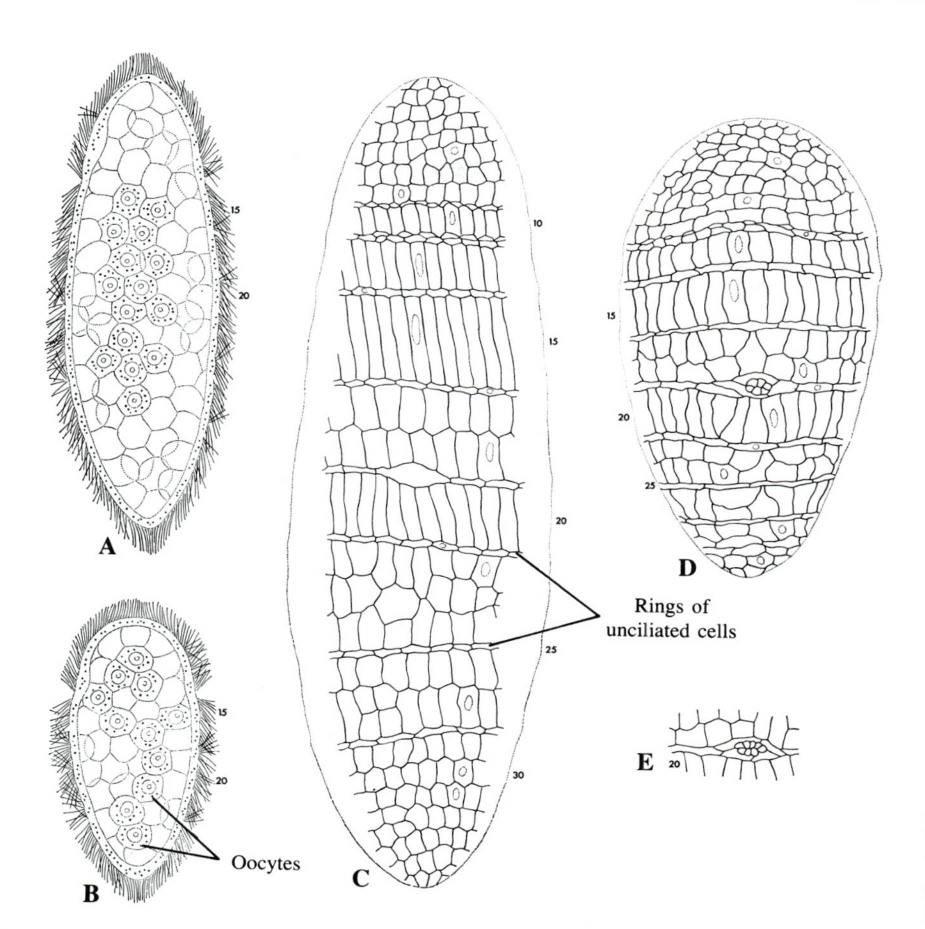

FIGURE 8.9 *Rhopalura ophiocomae,* female. A, B. Elongated and ovoid types, in optical section, drawn from life. There are muscle cells with a longitudinal, circular, and oblique orientation between the oocytes and epidermal cells. C, D. Elongated and ovoid types, impregnated with silver nitrate to show boundaries of epidermal cells. E. Small cells surrounding the genital pore.

The best known orthonectid is *Rhopalura ophiocomae,* a parasite of a widely distributed brittle star, *Amphipholis squamata*. For some unknown reason, the females (Figure 8.9; Figure 8.11, A) are of two shapes and sizes, each with a characteristic number and sequence of ciliated and unciliated epidermal cells. The longer females are about 250 μm long; the shorter, ovoid ones are about 150 μm long. The oocytes are packed into the central mass, and muscle cells with longitudinal, circular, and oblique orientations lie between the oocytes and the epidermis. The genital pore, surrounded by several small cells, is located within a group of unciliated rings (Figure 8.9, D and E).

The mature male of *Rhopalura* (Figure 8.10; Figure 8.11, B) is very different from the female, and also much smaller—only about 130 μm long. The body has three general regions. The middle region differs from the first and third regions in that none of its epidermal cells is completely ciliated, though some have cilia near their anterior or posterior margins. The cells of five of the epidermal rings contain glassy inclusions whose composition and function are not known. The genital pore (Figure 8.10, D) is located within a group of unciliated rings in the third region, although the testis extends part way into the middle region. Each sperm has a short tail (Figure 8.10, E). Behind the testis are eight longitudinal muscle cells and a core of four cells containing structures that resemble collagen fibers, but that probably consist of something else, perhaps paramyosin. (True collagen is extracellular.) The core cells terminate soon after they reach the testis, but the peripheral muscles extend forward around the testis and most of them run through the middle region of the body into the anterior region. These muscle cells account for bending movements, and the core cells probably provide support and elasticity. The testis is surrounded by its own sheath of muscle cells, which no doubt squeeze out ripe sperm at the time of copulation.

Male and female *Rhopalura* originate within a so-called **plasmodium,** supposedly a mass of protoplasm containing embryonic cells. In a parasitized brittle star, the plasmodia are closely associated with the gonads,

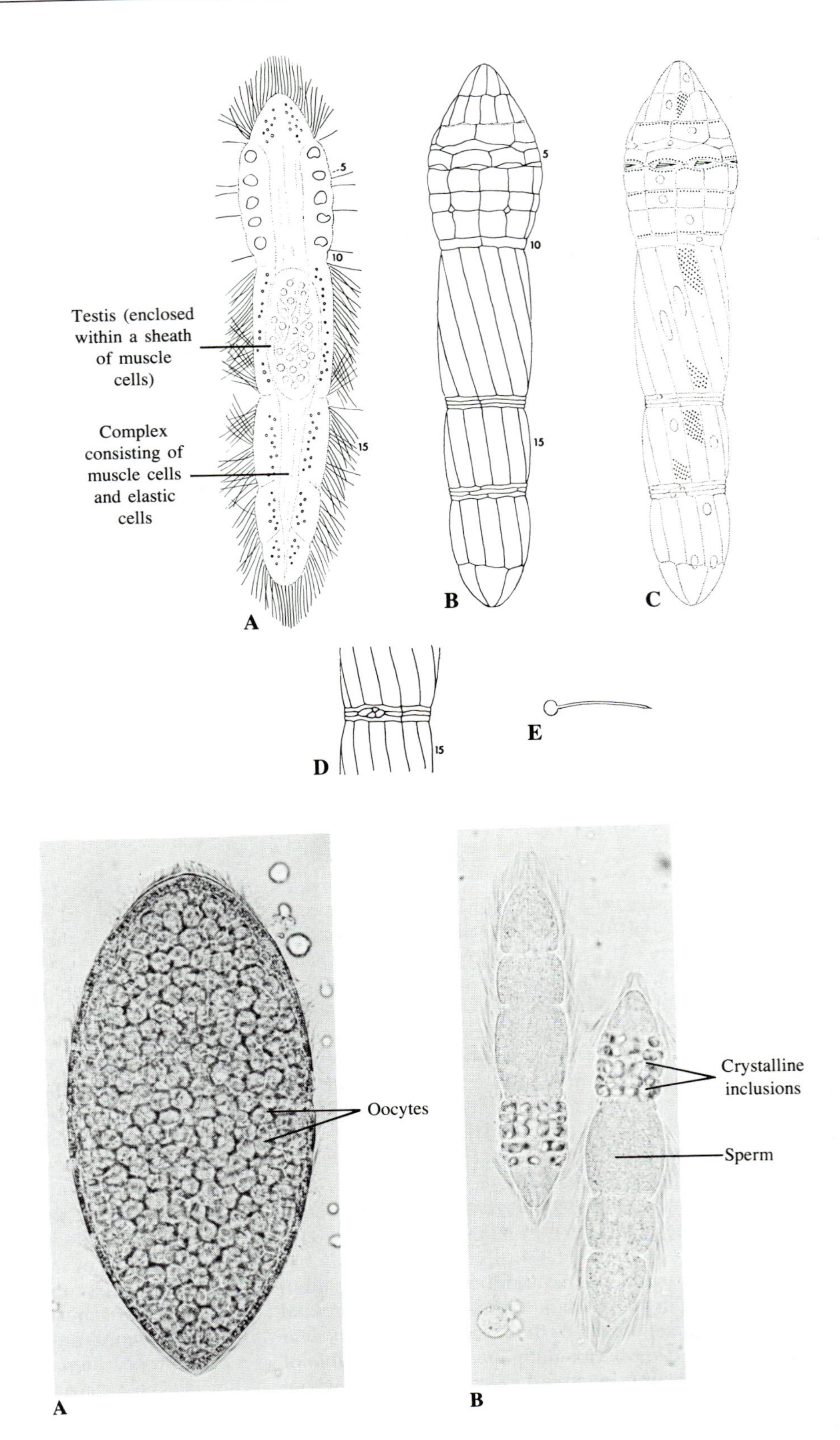

FIGURE 8.10 *Rhopalura ophiocomae,* male. A. Living specimen, in optical section, drawn from life. B. Specimen impregnated with silver nitrate to show boundaries of epidermal cells. C. Composite drawing based on specimens impregnated with silver nitrate to show boundaries of epidermal cells and on specimens impregnated with silver albumose to show the distribution of kinetosomes of cilia of representative epidermal cells. Cells of rings 1, 2, 11, 15, 19, and 20 are completely ciliated; the other rings are either unciliated or have single transverse rows. The nature of the structures shown in the cells of ring 6 is not known. D. Small cells surrounding the genital pore. E. Sperm.

FIGURE 8.11 *Rhopalura ophiocomae;* photomicrographs of living specimens. A. Female, slightly flattened. B. Males.

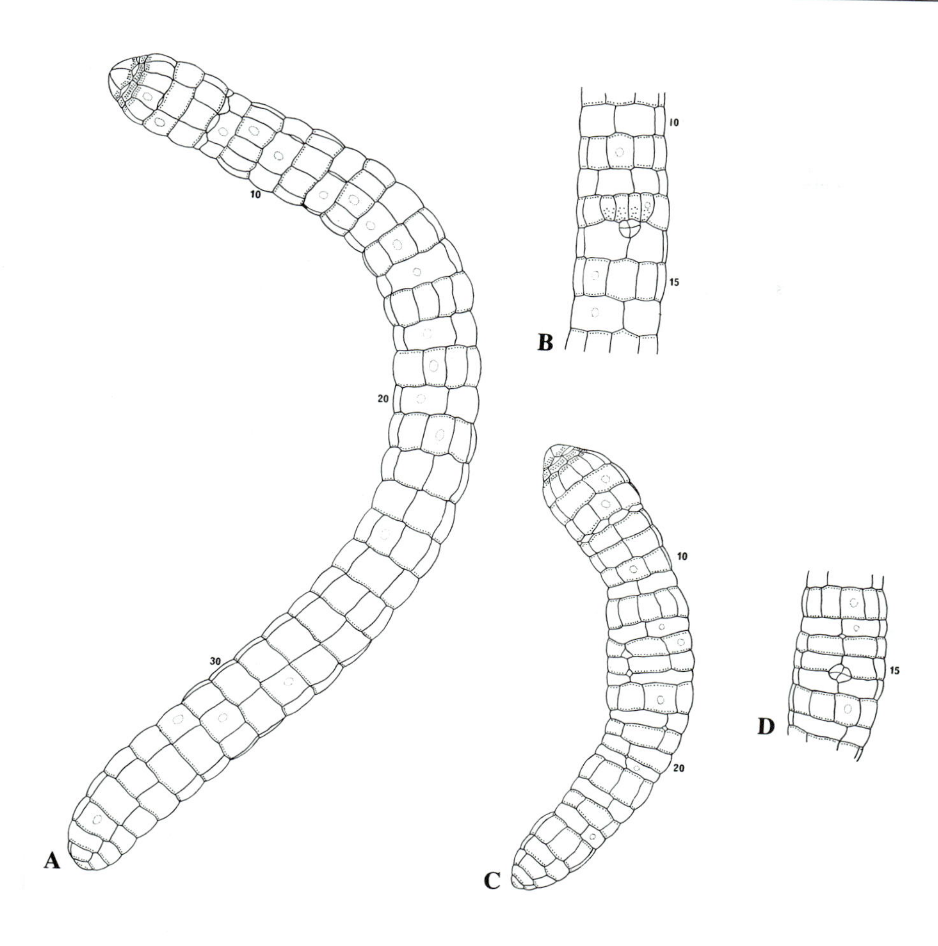

FIGURE 8.12 *Ciliocincta sabellariae*. Composite drawings based on specimens impregnated with silver nitrate to show boundaries of epidermal cells and on specimens impregnated with silver albumose to show the distribution of kinetosomes. A. Female. B. Region of genital pore of female. C. Male. D. Region of genital pore of male.

and the latter may become nonfunctional. A particular plasmodium may contain developing males or females, or both. When fully mature, the orthonectids emerge from the plasmodium and leave the body of the host through the bursal pockets. (The bursal pockets of brittle stars permit the escape of gametes, but in *Amphipholis* they also serve as brood pouches for developing young, which are released as fully formed juveniles.)

Once free in the sea water, the male and female *Rhopalura* mate, somehow bringing their genital pores together long enough for sperm transfer to be effected. The males then die, but the females live a few days longer, allowing the fertilized eggs to develop into ciliated larvae. These are simple and consist of only a few cells. In classical accounts of the life cycle of *Rhopalura,* the larvae are said to emerge from females and to enter the bursal pockets of *Amphipholis*. This seems logical enough, but it needs to be confirmed. It is possible that they are eaten, or that whole females containing broods of larvae are eaten. Although it has been assumed that the ciliated larvae somehow become plasmodia, the transformation has not been followed. Conceivably one larva, several larvae, or perhaps only a single cell of a larva, might give rise to a plasmodium. It is not really certain, however, that the plasmodium belongs to the parasite. It could be some host tissue that has become disorganized and within which cells derived from one or more larvae multiply and eventually develop into adult orthonectids.

There are a few other species of *Rhopalura,* all associated with the gonads of various bivalve and gastropod molluscs. The remaining genera of orthonectids—*Intoshia, Ciliocincta,* and *Stoecharthrum*—are, with a few exceptions, parasites of hosts other than those used by *Rhopalura,* and they tend to be localized in the parenchyma or tissues of the body wall, rather than the gonads. In *Intoshia* the ciliated rings of the epidermis are mostly covered with cilia; in *Ciliocincta* (Figure 8.12) and *Stoecharthrum* the cilia are largely restricted to the anterior or posterior margins, or both, of certain rings. The oocytes in females of all three genera form a single row of cells within the axial mass. Although muscle cells are found in both sexes, males do not have the elastic cells found in the posterior half of the body of males of *Rhopalura*. *Stoecharthrum* is odd in that it is hermaphroditic; it is otherwise like a *Ciliocincta* that has been greatly lengthened by the addition of rings.

REFERENCES ON ORTHONECTIDA

Kozloff, E. N., 1969. Morphology of the orthonectid *Rhopalura ophiocomae*. Journal of Parasitology, *55*:171–197.

Kozloff, E. N., 1971. Morphology of the orthonectid *Ciliocincta sabellariae*. Journal of Parasitology, *57*:585–597.

SUMMARY

1. The phyla Placozoa, Dicyemida, and Orthonectida consist of microscopic and relatively simple multicellular organisms that have neither a gut nor a nervous system. All of them are marine and all of them have external flagella or cilia. In other respects, however, the three groups are very different from one another.

2. The only known member of the Placozoa is *Trichoplax adhaerens*. It is up to about 2 mm in diameter, flattened, without symmetry or orientation, and capable of changing its shape as it creeps along. The cells of the thin upper epithelium and most of those of the thicker lower epithelium have single cilia. The cells in the lower epithelium ingest small unicellular organisms and may also produce enzymes that digest such organisms externally. Between the two epithelial layers are contractile cells that are joined to one another and also to some of the epithelial cells.

3. *Trichoplax* can reproduce asexually by dividing and by budding off pieces. There is also sexual reproduction. Eggs, and also cells that may be sperm, are found between the epithelial layers. Although eggs undergo cleavage, development under laboratory conditions stops short of reaching the adult stage. *Trichoplax* is certainly one of the simplest multicellular animals, and some zoologists view it as an organism that represents an early, blastulalike stage of metazoan evolution.

4. The Dicyemida are parasites that live in the kidneys and pericardial cavity of cephalopod molluscs. The prevailing stage, called the nematogen, is wormlike. It has an outer layer of ciliated epithelial cells and a large axial cell that encloses smaller cells that divide to form new nematogens. The young escape, grow up, and repeat the pattern of asexual multiplication. Eventually sexual individuals develop within the axial cell. These consist of oocytes and a cell in which nonflagellated sperm are produced. After the sperm enter the oocytes, these latter cast off polar bodies and undergo cleavage, developing into ciliated larvae. The larvae escape from the dicyemid within which they were produced and leave the host with its urine. The fate of the larvae is not known, but they probably represent the infective stage.

5. The Orthonectida are tissue parasites of invertebrates belonging to several phyla. The stage that is most commonly observed consists of males and females (or, in one genus, of hermaphroditic individuals). The outer layer of cells, most of which have cilia, encloses the gametes as well as contractile cells and some cells that have other functions. The sexual individuals leave the host and mate, and the fertilized eggs develop into ciliated larvae. These emerge from the female parent and enter new hosts. They are claimed to give rise to multinucleated masses of protoplasm within which cells differentiate. These cells undergo cleavage and give rise to a new generation of sexual individuals.

9

Some Pseudocoelomate Phyla

Nematoda, Nematomorpha, and Acanthocephala

Phylum Nematoda
- Class Secernentea
- Class Adenophorea

Phylum Nematomorpha

Phylum Acanthocephala
- Class Archiacanthocephala
- Class Palaeacanthocephala
- Class Eoacanthocephala

In certain invertebrates, the structures enclosed by the body wall lie in a fluid-filled cavity called the **pseudocoel.** This cavity sometimes derives from the embryonic blastocoel, and sometimes from a space or spaces that develop within a solid mass of mesoderm. Unlike a true coelom, a pseudocoel is not lined by a continuous peritoneum, and the organs within it are not supported by mesenteries. It should be noted, moreover, that some of the pseudocoelomate invertebrates, such as gastrotrichs and many nematodes, are so compactly organized when they are mature that there is little or no unoccupied space between the gut and body wall.

Other features found in most of the pseudocoelomate groups, but not in all of them, are a complete gut and an external cuticle. Many groups also exhibit **cell constancy,** in which the body as a whole, or certain parts of it, such as the epidermis, nervous system, or gut, consist of a fixed number of cells or, if the tissue is syncytial, of a fixed number of nuclei. The number is established early in development, and in cases of strict cell constancy it does not increase. Growth, in other words, is achieved through enlargement of cells or of the syncytium; there is no further multiplication of cells or nuclei.

The name *Aschelminthes,* which means "worms with a cavity," has often been used to bring all pseudocoelomates together in one phylum or superphylum. It will be best, however, to keep the several groups of pseudocoelomates separate, for although some of them may be closely related, others almost certainly are not. No significance should be attached to the order in which the eight phyla are discussed in this chapter and the next.

Phylum Nematoda: Roundworms

"If all the matter in the universe except the nematodes were swept away, our world would still be dimly recognizable . . . . we would find its mountains, hills, valleys, rivers, lakes, and oceans represented by a film of nematodes."

Nathan Augustus Cobb, 1914

Introduction

The nematodes are decidedly successful animals, present almost everywhere there is moisture. Many of them live in situations that would discourage nearly all organisms except certain bacteria. There are nematodes in hot springs, in vinegar, in paperhanger's paste that has turned sour, and in other odd places. Free-living species feed mostly on bacteria, other microorganisms, or small invertebrates, so they are regularly abundant in soil, mud, slimy coatings, and anywhere else that organic material is being metabolized. A fallen apple that has started to decay will almost certainly have thousands of nematodes crawling through it.

Nematodes are also important as parasites. Humans are hosts to over 30 species, and household and barnyard animals support many others. It is probable that nearly every vertebrate is parasitized by one or more kinds; invertebrates, especially arthropods and molluscs, are also hosts to nematodes. Other nematodes attack plants, directly destroying tissue or inducing abnormalities that interfere with normal functions. Some cause substantial damage to ornamental and food plants throughout the world.

No one knows how many kinds of nematodes there are. The 12,000 species that have been described probably constitute only a small fraction of the number that exist. In spite of their success and extensive speciation, however, nematodes form a rather homogeneous group.

In the late nineteenth century, two great discoveries were made by biologists working with nematodes. In 1877 Patrick Manson, a Scottish physician working in China, showed that the juveniles of *Wuchereria bancrofti,* the worm that causes filariasis, are carried by mosquitoes. This was the first time that a bloodsucking insect had been implicated in the transmission of disease. It set the stage for many other discoveries concerning the role of arthropods as carriers of pathogenic organisms.

About the same time that Manson was busy with his research in tropical medicine, several European zoologists working with *Parascaris equorum,* a parasite of horses, were laying the foundation for modern genetics. Bütschli noted that after an egg of *Parascaris* has been entered by a sperm, it divides twice, forming two polar bodies. The diploid chromosome number in *Parascaris* is only four, so this worm provided Bütschli with especially favorable material for the observation that the nuclear divisions involved in the production of polar bodies were meiotic and reduced the chromosome number in the egg to two. A related discovery of immense significance was made by Van Beneden. He saw that when fertilization takes place, the sperm and egg pronuclei each contribute two chromosomes, thus restoring the diploid number. So by the turn of the century, when Mendel's laws were rediscovered, biologists were ready to understand the chromosomal basis of inheritance. Genetics of nematodes themselves was not actively pursued until recently, but it is now an active field of biological research, concerned mostly with biochemical and morphological mutants in some free-living species for which good culture techniques have been developed.

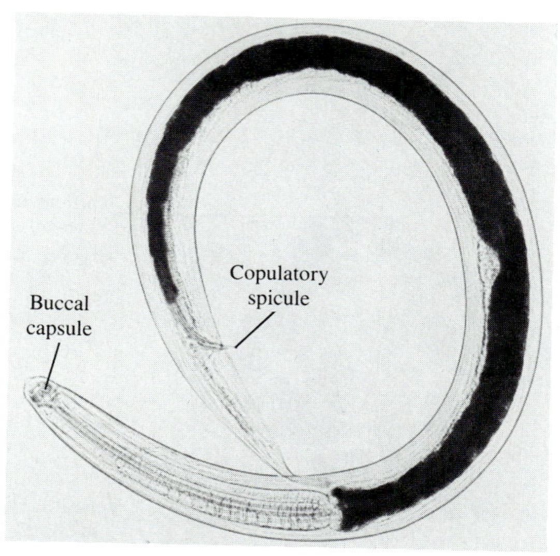

FIGURE 9.1 A free-living marine nematode; photomicrograph. The reproductive organs and gut (the intestinal portion is dark) are tightly packed into the space enclosed by the body wall. The specimen is a male, and shows a copulatory spicule protruding from the anus, which also serves as the genital pore.

GENERAL STRUCTURE

The body of a nematode (Figures 9.1, 9.6) is typically slender and has a glassy or shiny appearance, due to the cuticle (discussed below). Free-living types, and those that attack plants, are generally less than 5 mm long. Nematodes that parasitize animals have a much wider size range. While many of them are only a few millimeters long, others reach a length of 4 or 5 cm, and some are considerably larger. *Ascaris lumbricoides,* which inhabits the human intestine, is up to 40 cm long; *Dracunculus medinensis,* which lives in subcutaneous tissues, may grow to a length of about 1 m.

BODY WALL AND PSEUDOCOEL

The epidermis—it is often called the "hypodermis," but there is no need for this term—secretes a tough but flexible **cuticle** (Figures 9.2, 9.3, 9.4). This is not a simple, homogeneous layer, though it may appear to be when a whole worm or a section of it is examined with the light microscope. Electron microscopy reveals that the cuticle consists of three main layers, each with characteristic fibers, rods, plaques, or other structures (Figure 9.4). The innermost layer is subdivided into sheets of fibers, the fibers of each sheet being oriented at a different angle with respect to the long axis of the body. The strength of the cuticle is due to this arrangement of fibers, as well as to chemical composition. Parts of the cuticle contain enzymes, RNA, ATP, carbohydrates, lipids, and other substances that indicate that it is physiologically active.

The epidermis itself is said to be cellular in some nematodes, syncytial in others. It usually shows four longitudinal thickenings called **chords** (Figures 9.2, 9.3, 9.5, B). The dorsal and ventral chords contain the principal dorsal and ventral longitudinal nerves; the lateral chords also have nerves, as well as the longitudinal excretory canals, when these are present.

The body-wall musculature, beneath the epidermis, consists of large, longitudinally oriented cells. The epidermal chords divide the musculature into four longitudinal sectors, and the two sectors in the upper half of the body oppose the two in the lower half. The undulatory movements of a nematode, during which it bends into the shape of a C, S, or some more complex wave pattern, are therefore up-and-down, rather than side-to-side. When a nematode bends, the muscles nearest the inside of each curve are contracted, those on the opposite side are relaxed (Figure 9.5, A). The length and diameter of the worm remain nearly constant. The muscles are antagonistic to the pseudocoel, which is a hydrostatic skeleton, and to the strong cuticle, which can bend but which does not stretch. Because there are no circular muscles in the body wall, a nematode does not exhibit localized dilations and constrictions.

The major part of each muscle cell, which may be spindle-shaped or somewhat flattened, contains the contractile myofilaments (Figure 9.5, B). The nucleus is usually in a bulge that faces the pseudocoel. From this bulge, part of the muscle cell extends as a process to a nerve. This situation is not common among animals, although it does occur in several other phyla; as a rule, processes of nerve cells go out to the muscles.

The amount of free space in the pseudocoel varies. In some nematodes—*Ascaris* and *Parascaris,* for instance—the cavity is an extensive one (Figures 9.2, 9.3). In many members of the group, however, the reproductive organs occupy all or nearly all of the space between the gut and the body wall.

DIGESTIVE TRACT

The mouth opens into a **buccal capsule** that is cuticularized and that may have complexities such as teeth (Figure 9.15), ridges, or a hollow, piercing **stylet** (Figure 9.6). The **pharynx,** which succeeds the buccal capsule, is also cuticularized, but it differs from the buccal capsule in having a fairly thick wall and usually one or two prominent swellings called **pharyngeal bulbs.** The wall of the pharynx is provided with radially arranged

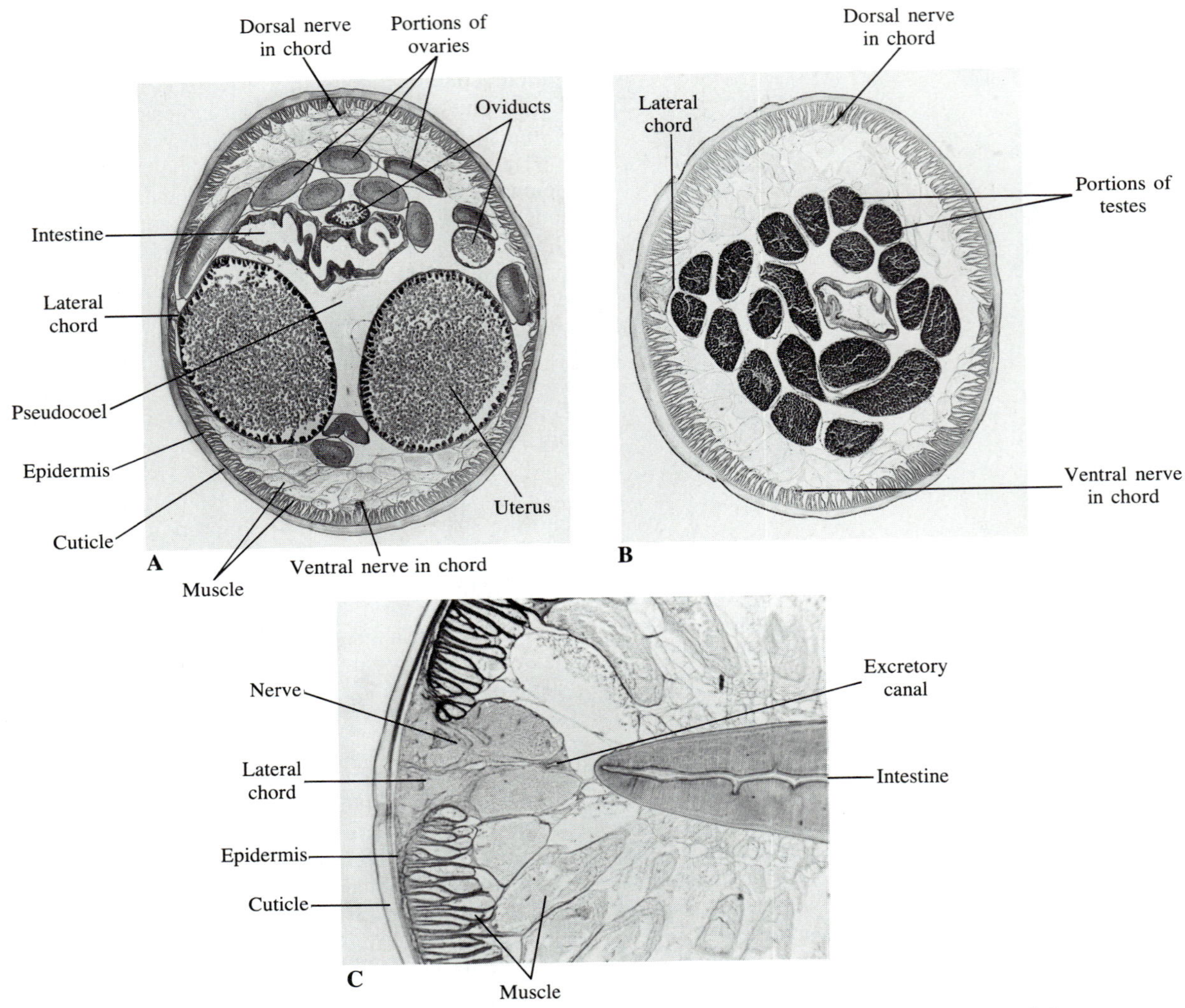

FIGURE 9.2 *Ascaris suum;* photomicrographs of transverse sections taken near the middle of the body. A. Female. B. Male. (A and B, courtesy of Carolina Biological Supply Co., Inc.) C. Small portion of the body wall, showing one of the lateral chords.

muscles that dilate the lumen (cavity) when they contract. When they relax, the lumen collapses, becoming triradiate in transverse section. The pharyngeal wall, including the musculature, is an epithelial structure derived from an invagination of ectoderm. The radial musculature, in other words, is part of the epithelium itself; it is not derived from mesoderm, as is the body-wall musculature.

The **intestine** consists of a layer of epithelial cells. Facing the lumen, these cells have crowded microvilli that greatly enlarge their surface area, thus facilitating digestion and absorption. The microvilli are sometimes just barely visible in sections examined with the light microscope. The **hindgut,** leading to the **anus,** is like the buccal capsule and pharynx in being of ectodermal origin and in being cuticularized. In males, this part of the gut is a **cloaca,** for it receives the ejaculatory portion of the sperm duct and has various copulatory devices associated with it (Figure 9.11, A). The anus is not quite terminal, so there is a short **postanal tail.**

EXCRETION AND OSMOREGULATION

Excretion in nematodes is poorly understood. No one is sure that the so-called excretory system is really concerned with elimination of nitrogenous wastes.

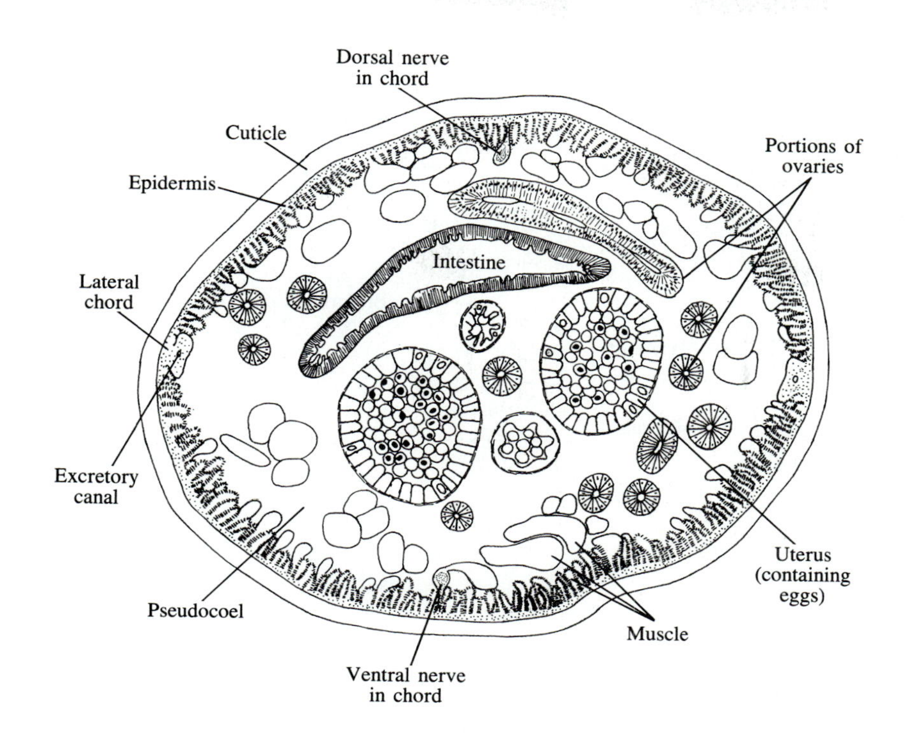

Figure 9.3 *Ascaris suum*. Transverse section of a female, taken near the middle of the body; semidiagrammatic.

Some nematodes do not even have this system. There is evidence, however, that it functions in osmoregulation, at least in certain species that have been studied. In its simplest form, it consists of a large cell, the **renette** (Figure 9.7, A), which lies beneath the pharynx. The duct of the renette runs to an opening on the ventral surface and usually has a dilation close to the pore. The renette may consist of two cells that share the same duct.

In many nematodes the situation is much more complex. Extensions of the renette cells go up into the lateral chords of the epidermis, where they form canals (Figure 9.7, B and D). The midventral outlet persists, and the duct may be cuticularized for part or all of its

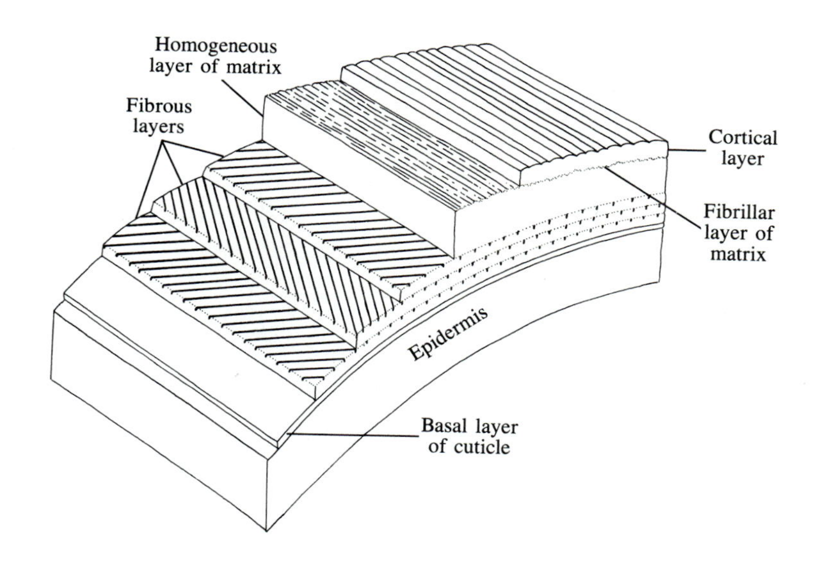

Figure 9.4 Cuticle of a nematode, diagrammatic and generalized; based on electron micrographs.

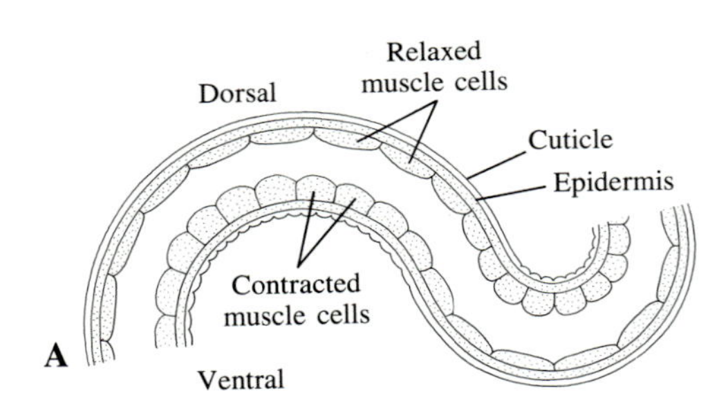

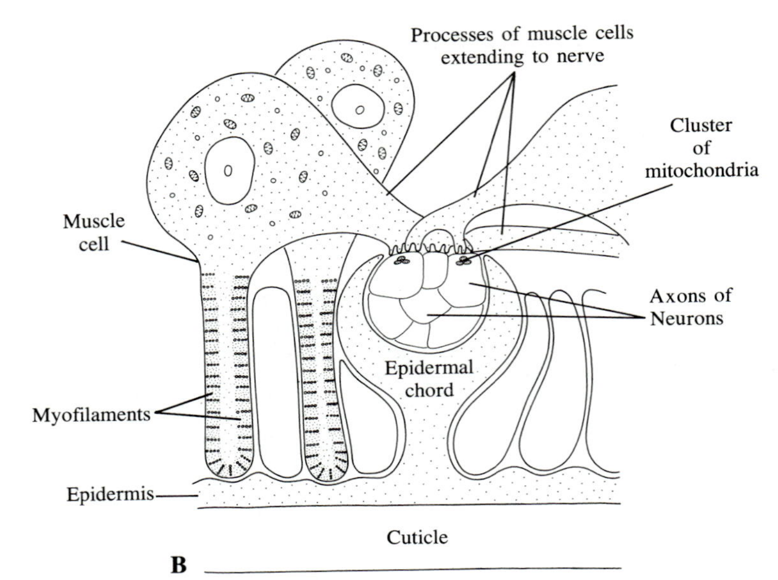

FIGURE 9.5 A. Relationship of contraction and relaxation of muscle cells to sinuous bending movements in a nematode. (After Crofton, Nematodes.) B. Diagram of muscle cells of *Ascaris*, based on electron micrographs of transverse sections. (After Rosenbluth, Journal of Cell Biology, *26*.)

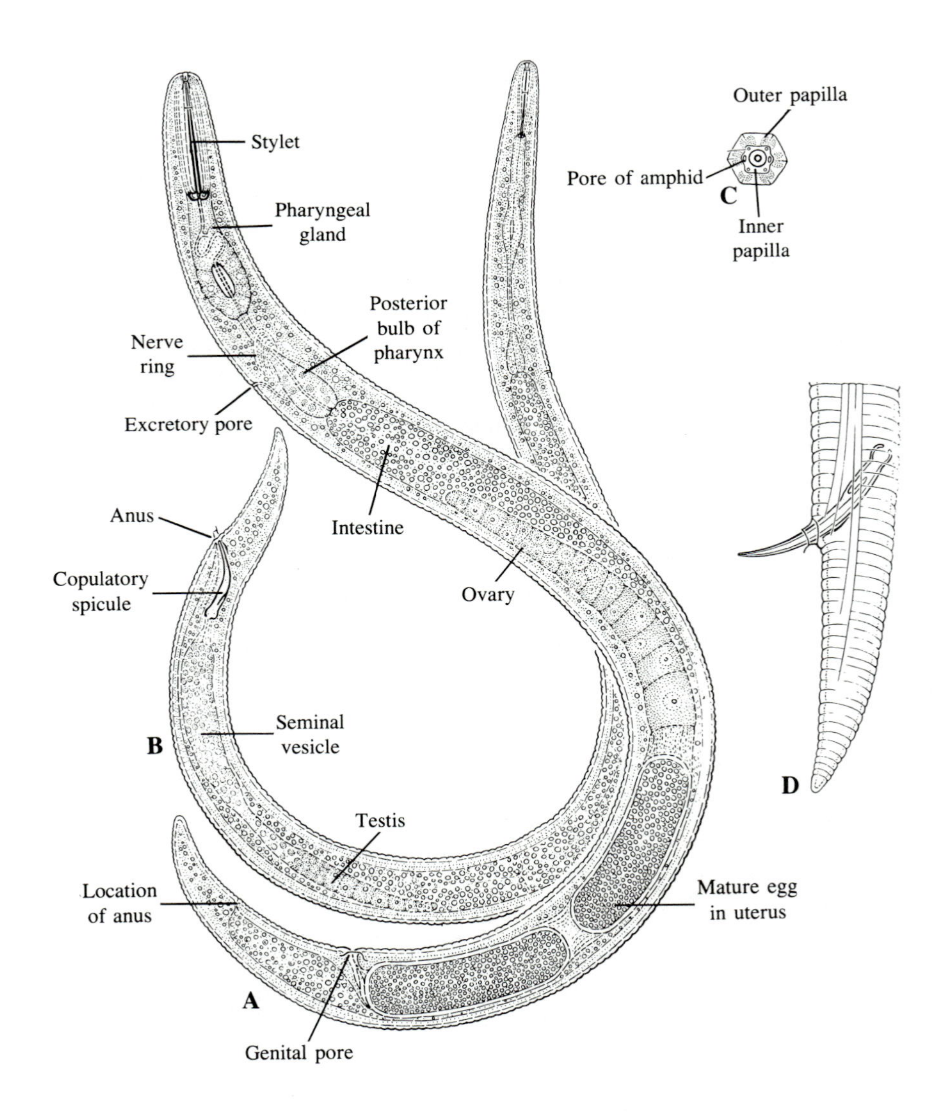

FIGURE 9.6 *Paratylenchus hamatus*, a nematode that parasitizes roots of various plants. A. Female. B. Male. C. Apical view of anterior end. D. Posterior portion of male, showing extruded copulatory spicule. (Thorne and Allen, Proceedings of the Helminthological Society of Washington, *17*.)

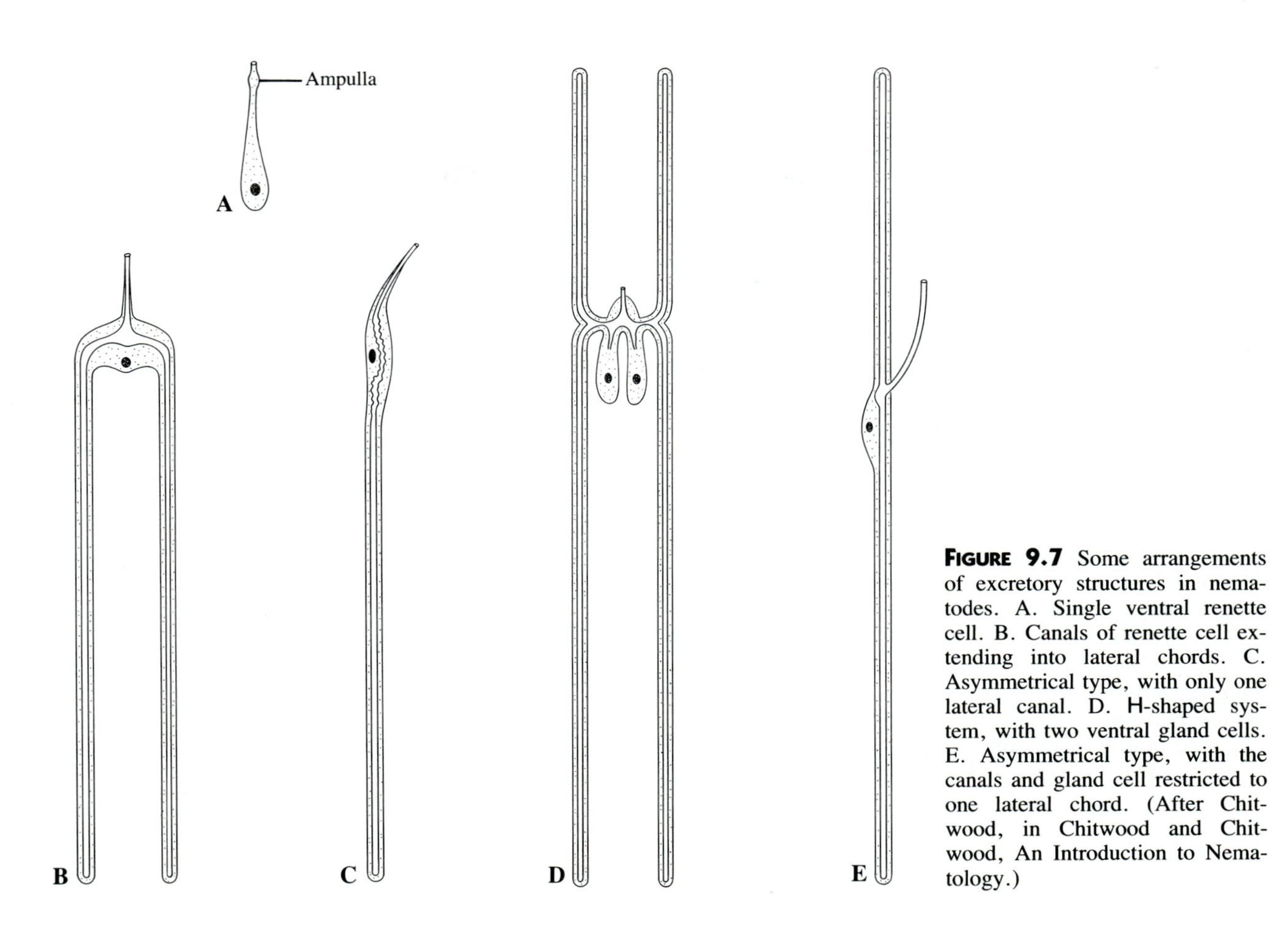

Figure 9.7 Some arrangements of excretory structures in nematodes. A. Single ventral renette cell. B. Canals of renette cell extending into lateral chords. C. Asymmetrical type, with only one lateral canal. D. H-shaped system, with two ventral gland cells. E. Asymmetrical type, with the canals and gland cell restricted to one lateral chord. (After Chitwood, in Chitwood and Chitwood, An Introduction to Nematology.)

length. Further modifications include the loss of the renette cell on one side or the other of the body (Figure 9.7, C and E), and the addition of gland cells to the system. The canals in the lateral chords have been viewed by some zoologists as being originally protonephridial elements—solenocytes or flame bulbs—that have become linked with renette cells. Studies on the development of the canals in the lateral chords indicate, however, that they derive from the renette cells.

The excretory canals and the dilated portion of the duct that leads to the external pore have been observed to contract. In some nematodes, moreover, contraction becomes more frequent as the salinity of the external medium decreases. Studies of the canals with the electron microscope reveal fine ducts running into them, as well as what are probably contractile filaments around them. All of this suggests that the system is osmoregulatory, and perhaps excretory as well. Nevertheless, in *Ascaris suum,* which lives in the intestine of pigs and which has been intensively studied, excretion of ammonia, urea, and some other compounds of nitrogen is achieved primarily through the wall of the gut. This may be the usual situation in nematodes.

The ability to control uptake or loss of water varies. The survival of nematodes that live in soil, moss, and some other terrestrial and freshwater habitats depends on their capacity to osmoregulate. Parasitic species, however, generally must have more or less constant external conditions, although they can adapt to some extent. For example, *Ascaris* can maintain its internal tonicity at a fairly constant level, even if the concentration of certain ions in the intestinal fluid varies, and it can change the proportion of urea to ammonia that it excretes, depending on environmental conditions. In a standard mammalian saline solution, *Ascaris* excretes about ten times as much ammonia as urea. As the salinity increases, the worm excretes more urea in proportion to ammonia, until finally it is producing nearly twice as much urea as ammonia. This ability to adjust its output of nitrogenous wastes according to environmental conditions is adaptive, for ammonia is toxic and cannot be allowed to accumulate. Thus elimination of ammonia requires steady loss of considerable water, something a worm cannot afford to do when it is in a medium that is more saline than its own body fluids.

The glands associated with the renette system are thought to produce enzymes concerned with molting of the cuticle by juveniles. Substances that can break down host tissue have also been claimed to come from these glands.

NERVOUS SYSTEM AND SENSE ORGANS

Associated with a **nerve ring** that encircles the posterior part of the pharynx are several ganglia (Figure 9.8). From these ganglia, or from the places where they are joined to the ring, originate nerves that run anteriorly and posteriorly. Most prominent of the longitudinal nerves are the **dorsal** and **ventral nerves,** which lie in the epidermal chords. They are readily recognizable in transverse sections of large nematodes such as *Ascaris*. The ventral nerve usually does not originate at the nerve ring but begins at the confluence of two smaller nerves that do start at the ring. Some of the nerves have ganglionic swellings at intervals, and there is a complex of ganglia and connectives around the hindgut. The finer nerves that supply sense organs and other structures originate mostly from such ganglia.

The cuticular bristles and papillae found in the head region, tail region, and elsewhere are thought to be touch receptors. Electron microscopy demonstrates that protuberances around the mouth have modified cilia. The **amphids**—two little pouches that open by pores beside the mouth (Figure 9.9, A)—also contain groups of modified cilia. Amphids are generally assumed to be chemosensory, and perhaps that is their main role, but other functions have also been ascribed to them. In hookworms, which feed to a considerable extent on blood, they contain a substance that inhibits clotting. In any case, the only structures in a nematode that have cilia are the amphids and some of the papillae. In the head region of certain species are simple eyespots or more elaborate light receptors—these sometimes have a cuticular "lens." Nematodes that do not have any obvious light receptors may nevertheless respond to light by increasing their activity or by moving toward or away from the source, depending on their habitat preference.

FIGURE 9.8 Anterior portion of the nervous system of a nematode, ventral view; diagrammatic and generalized. Not all nerves are shown. The basic plan has many variations.

In certain nematodes there is a pair of structures called **phasmids** (Figure 9.9, B). The external pores of

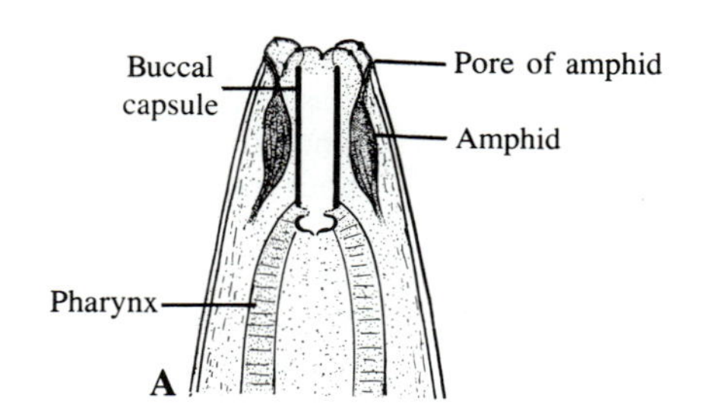

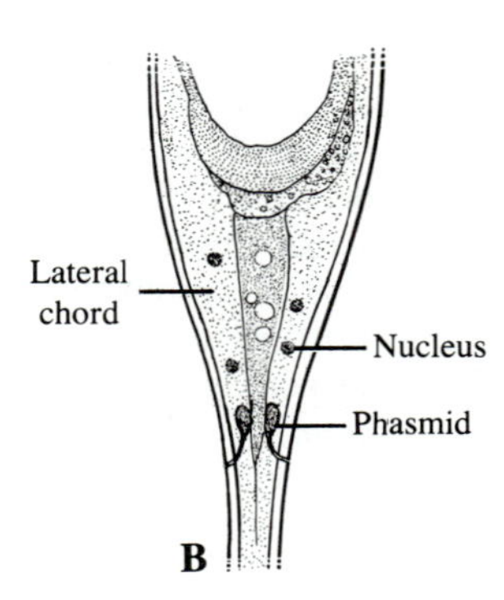

FIGURE 9.9 A. *Rhabditis elongata*. Anterior portion, showing amphids. B. Posterior portion, showing phasmids. (Chitwood, Journal of Morphology and Physiology, *49*.)

these are close to the anus and are reached by fine, cuticle-lined ducts. Some phasmids may be chemoreceptors; others may have an excretory function. It has also been suggested that these structures may help produce pheromones that mark scent trails for other members of the species to follow. Phasmids, amphids, and sensory papillae of nematodes provide opportunities for much interesting research.

REPRODUCTION AND DEVELOPMENT

The sexes of nematodes are easily distinguished. In males, the reproductive system joins the hindgut to form a cloaca. In females, there is no connection between the reproductive system and the gut; the genital pore is separate and usually some distance anterior to the anus.

The female reproductive system usually has two tubular **ovaries,** each continuous with an **oviduct** and **uterus** (Figure 9.10, A). The tubes are long and packed like spaghetti into the pseudocoel. The uteri are the broadest portions, but where an oviduct becomes a uterus, there is sometimes a separate swelling that functions as a seminal receptacle. The two uteri unite not far from the genital pore; the vagina, which has a cuticular lining, is therefore short. In certain nematodes there is just one functional ovary–oviduct–uterus complex (Figure 9.6, A); the other one is vestigial or absent altogether.

The principal components of the male system (Figure 9.6, B; Figure 9.10, B)—a single **testis,** a **seminal vesicle** for storage of sperm, and the **sperm duct (vas deferens)**—are also tubular. As the sperm duct approaches the cloaca, it becomes appreciably muscularized to form an **ejaculatory duct.** At least some glands opening into the ejaculatory duct produce a cement that plugs the female genital pore after copulation, presumably to prevent escape of sperm. There are usually one or two **spicules** for widening the genital pore of the female. These are cuticular structures that lie in pockets

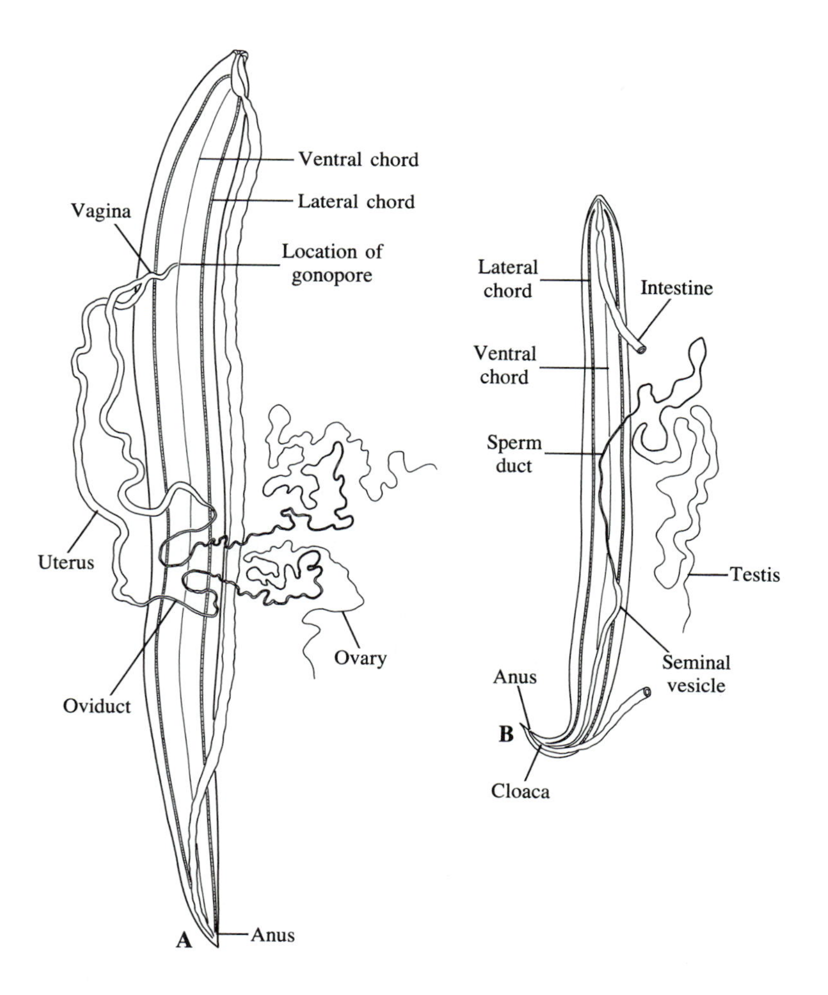

FIGURE 9.10 *Ascaris suum,* reproductive system as seen in dissections. A. Female. B. Male.

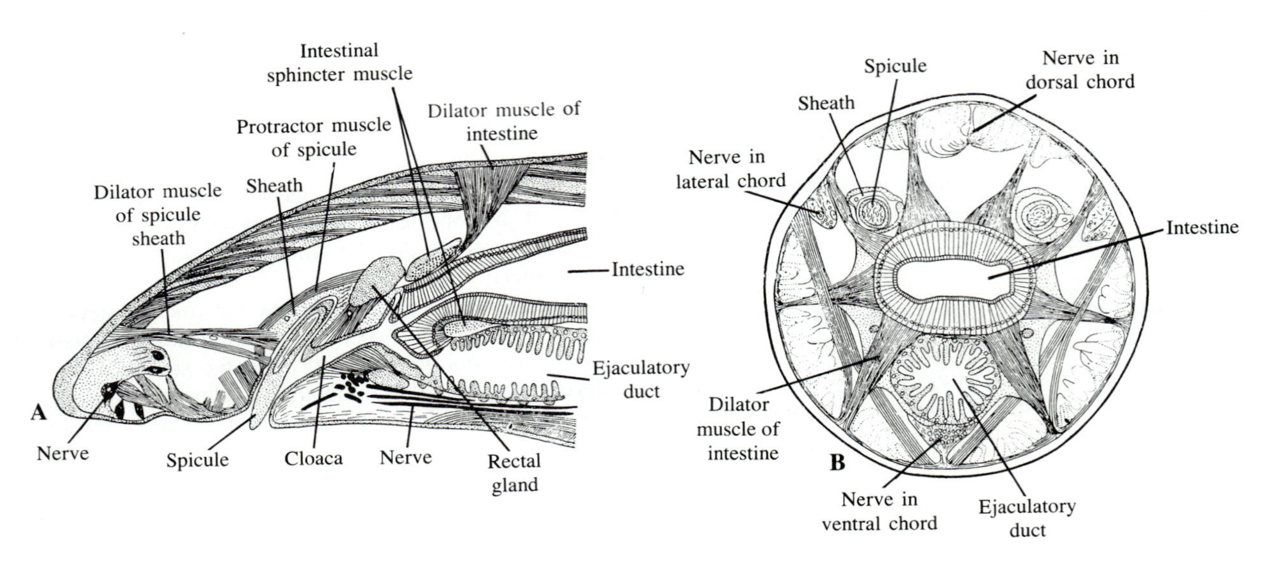

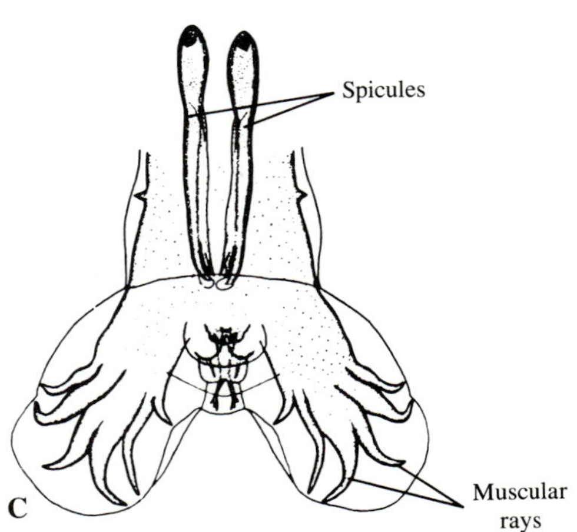

FIGURE 9.11 A. *Parascaris equorum,* male, longitudinal section of posterior portion of body, showing one of the two copulatory spicules. B. *Parascaris equorum,* transverse section at the level of the copulatory spicules. (A and B, Ivanov *et al.*, Major Practicum of Invertebrate Zoology, after Voltzenlogel, Zoologische Jahrbücher, Abteilung für Morphologie und Ontogenie der Tiere, *16*.) C. *Ostertagia leptospiralis,* male, showing the complex copulatory bursa characteristic of strongylid nematodes. (Andrews, Transactions of the Royal Society of New Zealand, Zoology, *5*.)

on the dorsal side of the cloaca (Figure 9.11, A and B), and they are protracted and retracted by special muscles. There may also be cuticular elaborations of the cloacal wall that help guide the spicules. In hookworms and other nematodes that belong to the order Strongylida, the male has a structure called the copulatory bursa at the posterior end of the body (Figure 9.11, C). The flaplike and mostly cuticular lobes of this are operated by muscular rays and clasp the female tightly during copulation. In certain members of the phylum, the cloacal wall is modified as an eversible organ that functions as a penis.

The large sperm of nematodes lack flagella, although they sometimes have a tail that might be mistaken for a flagellum. Though the motility of nematode sperm has been claimed by some to resemble the movement of an amoeba, the way they move seems more like gliding than like amoeboid motion.

When eggs leave the ovary of a female nematode, they are primary oocytes. Sperm penetrate them as they

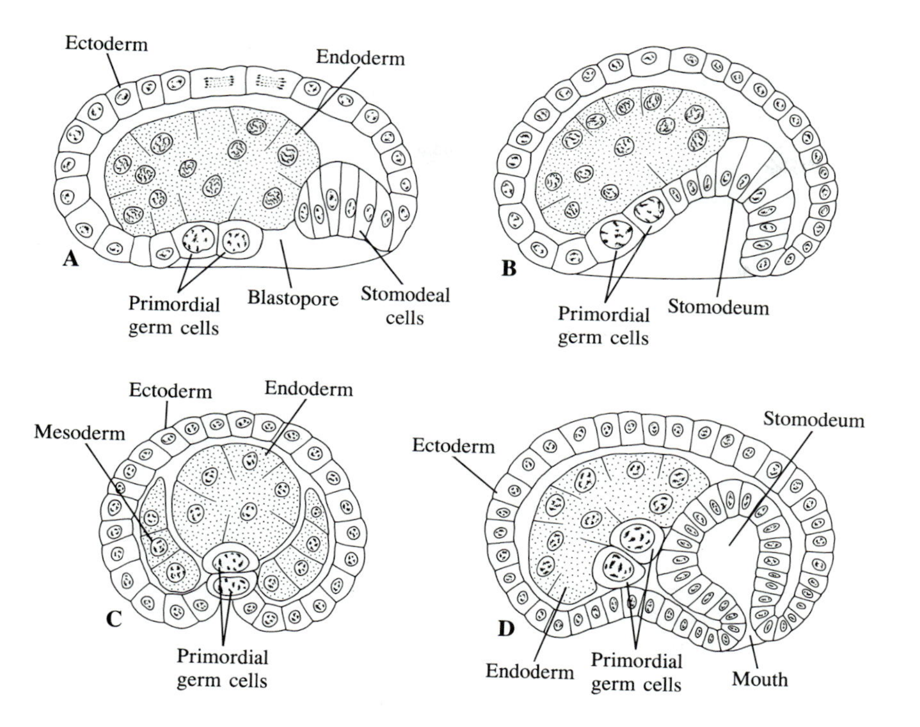

FIGURE 9.12 *Parascaris equorum,* early stages in development. A. Shortly before closure of the blastopore, longitudinal optical section. B. Stomodeal invagination forming. C. Same stage as B, but in transverse optical section, showing mesoderm lateral to endoderm. D. Stomodeal invagination completed. The primordial germ cells have sunk into the blastocoel. (After Boveri, Anatomischer Anzeiger, *2*.)

pass through the seminal receptacles (when these are present) or while they are in the upper portions of the uteri. Formation of the egg shell begins soon after fertilization. The zygote usually secretes three layers, the middle one containing considerable chitin; one or two additional layers may be deposited by the uterus.

Movement of eggs down the uteri is accomplished by muscular activity in the walls of these organs. Muscles around the vagina force the eggs from the genital pore. At the time a female nematode lays her eggs, these may contain still-uncleaved zygotes, embryos, or juveniles that are ready to hatch. In many species, moreover, the females are ovoviviparous, releasing worms that have already hatched.

Development is highly determinate. Nematodes provide, in fact, some of the best examples of determinate development, appreciated ever since Boveri published his magnificent work on the embryology of *Parascaris*. At the very first cleavage, a cell destined to produce nothing but ectoderm is segregated from one destined to give rise to additional ectoderm as well as endoderm, mesoderm, and germ cells. Subsequent cleavages are not synchronous. At the 4-cell stage the embryo is T-shaped, with the two cells forming the "crossbar" being those whose fate is to produce ectoderm. The cell right behind these will produce endoderm, mesoderm, and the stomodeum, the future pharynx. The cell at the "stem" of the T gives rise to germ cells and additional ectoderm. The cell destined to produce nothing but germ cells is segregated at the fourth cleavage.

Only the germ cells retain all of their chromosomes; they are the only genetically complete cells in a nematode's body. In *Parascaris,* the zygote has four chromosomes, two coming from the egg and two from the sperm. As the early cleavages take place, parts of these chromosomes break down and are resorbed, except in cells whose progeny will become germ cells. This phenomenon, called *chromatin diminution,* has counterparts in various other organisms. The remaining portions of the four chromosomes fragment into smaller chromosomes. The chromosomes observed during gametogenesis, fertilization, and early cleavage are therefore really compound chromosomes.

Gastrulation is sometimes achieved by epiboly—the overgrowth, by multiplying ectoderm cells, of the germ cells, mesoderm, and endoderm. More often, however, it is accomplished partly by epiboly, partly by invagination. Either way, the endoderm, mesoderm, and primordial germ cells end up in the blastocoel. An invagination of ectoderm forms a stomodeum, which becomes the buccal cavity and pharynx (Figure 9.12, A–C). Later, a similar invagination produces the hindgut. The nervous system and renette cells are also of ectodermal origin. The first few mesoderm cells, initially in two groups at the sides of the endodermal mass (Figure 9.12, C), multiply, and most of the products differentiate into the longitudinal muscles of the body wall. The pseudocoel originates as spaces that develop within mesoderm.

Nematodes, like other pseudocoelomate animals whose development has been studied in detail, exhibit cell constancy. After a certain number of cleavages have been completed, there is little further division except in the germ cells. Once the somatic cells have started to become irreversibly specialized as components of nerve, muscle, and other tissues, the growth of the animal as a whole depends on an increase in the size of these cells rather than on an increase in their number.

In the life cycle of a nematode, the cuticle is shed four times. Before each molt, the old cuticle is softened by an enzyme in the fluid that accumulates between it and the new cuticle. The cuticle is sometimes shed in one piece, sometimes in two pieces (a small anterior cap separates from the rest of it).

The changes leading up to a molt are believed to be controlled by hormones produced by neurosecretory cells, such as are present in the main nerve ring. In the case of some parasitic nematodes, molting appears to be influenced by physiological circumstances in the host. In many parasitic species whose juvenile stages are free-living, the cuticle shed during the second molt persists as a sheath around the worm. An ensheathed juvenile is unable to feed but is likely to be resistant to certain unfavorable environmental conditions, and it may survive for several weeks or months until it is swallowed by a suitable host or penetrates the skin of its host, as hookworms do.

Some Nematode Parasites of Humans

Humans are hosts to about 30 species of nematodes, ranging from small worms that cannot be seen except with a microscope to giants such as *Ascaris*. A few of the more important of these parasites are discussed here, not only because they are troublesome, but because their reproductive strategies and life cycles illustrate some of the ways in which parasitic nematodes ensure their success and survival.

Ascaris lumbricoides

Ascaris lumbricoides, which lives in the small intestine, is morphologically almost identical with *A. suum* of pigs. There is a possibility that humans originally got *Ascaris* from pigs that had been domesticated, or perhaps humans had it in the first place and then gave it to pigs. Nevertheless, the two species exhibit strong host-specificity. Both are large; the length of females sometimes exceeds 40 cm and that of males may attain 30 cm. *Ascaris* feeds mostly on material in the lumen of the intestine, but it may occasionally suck blood and tissue juices. If the infestation is heavy, serious malnutrition may result, especially in children. There are other unpleasant symptoms, too: abdominal pains, allergic reactions, insomnia, and general malaise. Sometimes the intestine is blocked by a tangled mass of *Ascaris*. The worms may wander into the appendix and may even crawl up the esophagus, from which they can move on to the trachea, Eustachian tubes, middle ears, and nose. In every way, *Ascaris* is a thoroughly undesirable associate of humankind.

Unlike the eggs of *Parascaris,* those of *Ascaris* do not undergo cleavage until they have left the uterus of the female. Cleavage may not begin, in fact, until the eggs have left the body of the host with fecal material. It takes them a few days to develop into first-stage juveniles (Figure 9.13). The first molt is accomplished while the worms are still within the eggshells, and the resulting second-stage juveniles are infective. If an egg containing one of these juveniles is swallowed with food or water, or with dirt on one's fingers, it hatches in the duodenum. The worm crawls out of a small opening that appears in the eggshell and penetrates the duodenal wall. After entering a blood or lymphatic vessel, it is carried to the heart, then to a lung. Here it breaks out of a capillary and works its way out of the tissue into an air space. After completing two more molts, it crawls up to the trachea and eventually reaches the pharynx and is swallowed again. In the small intestine it molts for the last time and grows to maturity. If many young *Ascaris* are traveling through the tissues of the duodenal wall and lungs they may leave considerable wreckage behind them. The physical damage they do is bad enough, but juveniles that have lost their way and died may induce severe reactions.

Ascaris eggs are tough and can survive for years in soil. Even in a dry environment, they may remain viable for a long time. Vegetables and berries produced on soil fertilized with human feces are ideal vehicles for infection. Children often contaminate themselves with eggs that they pick up while playing in soil, or that happen to be on toys they put into their mouths.

Enterobius vermicularis

The pinworm, *Enterobius vermicularis,* is common in humans in temperate regions of the world, and is primarily a parasite of children. It is a small species—the female is about 1 cm long, the male a bit shorter—that lives in the caecum, appendix, and adjacent portions of the large intestine. When the females are ready to lay their eggs, they move to the anus, and their activity in this part of the body may cause severe itching. Within a few hours after the eggs have been laid, or after they have burst out of a female that has dried out,

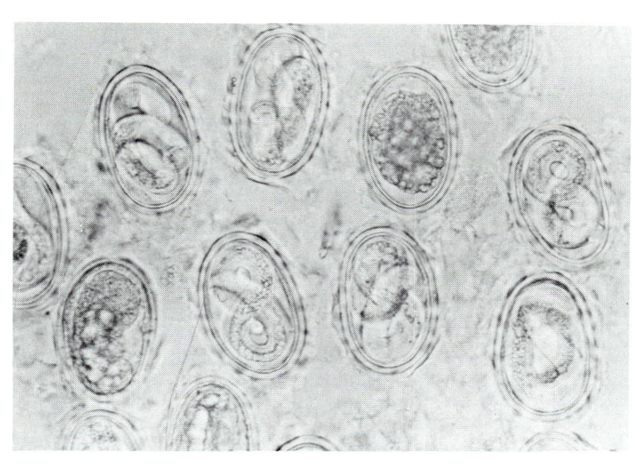

FIGURE 9.13 Eggs of *Ascaris suum,* with embryos and first-stage juveniles. (Courtesy of Carolina Biological Supply Co., Inc.)

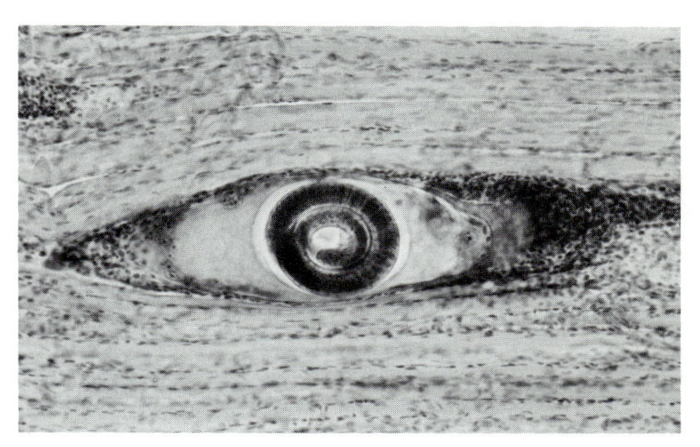

FIGURE 9.14 *Trichinella spiralis,* encysted juvenile in pork.

the young worms enclosed by their eggshells reach the infective stage. Some of the eggs inevitably end up on door knobs, bedsheets, and anything else that can be contaminated by unwashed hands or an unwashed behind. Eggs may also be blown through the air. If eaten, the eggs hatch and the juveniles settle in the mucosa of the caecum for a while, then they move to the cavity of the caecum and other parts of the large intestine. There are simple and effective treatments for pinworm infections, which are more often found in white children than in blacks or Asians.

TRICHINELLA SPIRALIS

Adult trichina worms, *Trichinella spiralis,* are acquired by eating pork, and also meat from some game animals, that has not been cooked enough to kill the juveniles that are encapsulated in the flesh. The worms mature in the small intestine. The males die soon after mating, but the females, which are about half the size of pinworms, burrow into the intestinal epithelium. They cause some damage to the tissue and produce toxic secretions. This is only the beginning of a parasitized person's troubles. Female trichinas bear their young alive, for the eggs hatch in the uterus. The juveniles burrow into the intestinal wall and are distributed by blood and lymph to all parts of the body. Those that get into fibers of striated muscle, including that of the diaphragm, grow to a length of about 1 mm, roll up into little spirals, and become encapsulated (Figure 9.14). Their presence in the muscle leads to inflammatory reactions and deterioration of the muscle. The disease is called trichinosis. In time, the capsules become calcified. Each adult female living in the intestine produces at least a thousand young in her lifetime, which lasts 2 or 3 months. Thus a person who eats a piece of meat containing a few hundred encysted juveniles may end up with many thousands of juveniles of the next generation in the muscles. The cycle of *Trichinella* is perpetuated not only in pigs that have access to garbage containing pork, but also in rats, wild boars, bears, raccoons, seals, opossums, and some other game animals whose life styles include carnivory or cannibalism, or both. To avoid *Trichinella,* cook pork, pork sausage, and other possible sources of infection thoroughly.

ANCYLOSTOMA DUODENALE AND *NECATOR AMERICANUS*

The two important hookworms of humans—*Ancylostoma duodenale* and *Necator americanus*—are about 1 cm long. They live in the small intestine, where they feed by sucking in blood and tissue. The buccal cavity of *Ancylostoma* is equipped with sharp teeth (Figure 9.15); *Necator* has cutting plates. The teeth or plates are used for biting off plugs of tissue as the worms embed their anterior portions in the intestinal wall.

The eggs that female hookworms produce leave the body of the host with feces and hatch into juveniles of the first stage. These feed on bacteria and perhaps other organic material, including that contributed to the soil with excrement. They molt twice in about a week, but after the second molt they retain the old cuticle as a sheath. They are now in the infective stage, though they are unable to feed. In soil that is moist, well aerated, and reasonably warm, they may persist for several weeks. On contact with delicate skin, such as that between the toes, the infective juveniles work their way into the body and migrate through blood and lymph to the lungs. Here they break out of the capillaries, causing some damage, and get into the air spaces of the lungs. If coughed up and swallowed they can reach the small intestine, where they undergo their third and

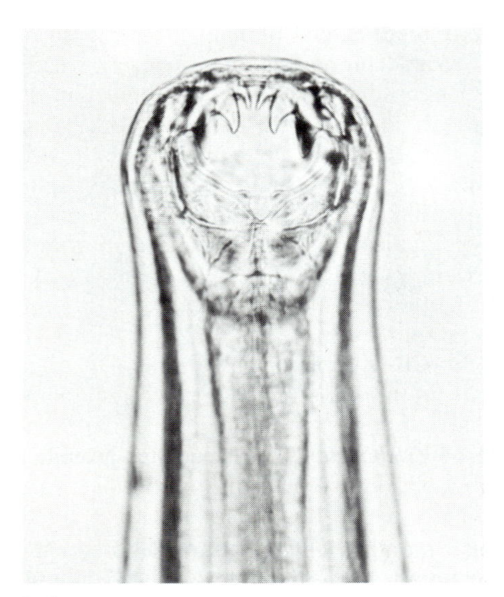

Figure 9.15 Anterior portion of a hookworm, *Ancylostoma duodenale,* showing the cutting teeth of the buccal cavity. (Courtesy of Carolina Biological Supply Co., Inc.)

fourth molts and mature. The adults may live for several years. Walking barefoot where hookworms are prevalent is obviously hazardous, but even dirt kicked up on the delicate skin around the ankles could result in an infection. Handling soiled laundry that has been damp for a few days is also not recommended.

Both *Necator americanus* and *Ancylostoma duodenale* are widespread, and their ranges overlap. The former, in spite of its Latin name (''American murderer''), was almost certainly brought to the New World from Africa. It is basically tropical in its distribution, but in the southern part of the United States, where hookworm disease continues to be a problem, most infestations consist of *Necator*. *Ancylostoma* is the more common hookworm, affecting humans in southern Europe, North Africa, and Asia.

WUCHERERIA BANCROFTI

A more complex life history is illustrated by the nematodes called filarial worms. Humans are parasitized by several species, all transmitted by blood-sucking insects, including mosquitoes. *Wuchereria bancrofti* is the most widely distributed of the filarial worms affecting humans. It is found in the South Pacific and in coastal sections of much of Asia, Africa, and South America. The adult worms—up to a meter long but less than half a millimeter wide—live in the glands and ducts of the lymphatic system, causing inflammation and also blocking the flow of lymph. Localized accumulation of lymph may lead to grotesque enlargement of a leg, the scrotum, or some other part of the body. The skin of the swollen member eventually dries out, due to deterioration of the sweat glands. *Elephantiasis* is the name given to the symptoms seen in aggravated cases of filariasis.

The worms may live for several years. Females of filarial worms, like those of *Trichinella,* bear their young alive. The first-stage juveniles are called microfilariae. In *Wuchereria* these are about 0.2 mm long, and usually appear in the peripheral circulation at night—or in the daytime in persons who work at night. Apparently the daily activity cycle of a parasitized person influences when the microfilariae circulate through the capillaries of the skin, where they must be if they are to be sucked up by the intermediate host, a female mosquito. From the gut of the mosquito, the microfilariae work their way to the flight muscles. Here, over a period of several days, they grow to a length of about 2 mm and molt twice. Then they move to the proboscis; while the mosquito is biting an animal, the microfilariae crawl out of the side of the proboscis. Those that find the puncture in the skin and enter it have a chance to reach the lymphatic tissues and mature.

DRACUNCULUS MEDINENSIS

The guinea worm, *Dracunculus medinensis,* has an equally complicated life cycle. Its intermediate host is a copepod crustacean, which must be eaten if the life cycle is to be completed. *Dracunculus* (the ''little dragon'') is prevalent in parts of India, the Near East, Africa, and the East Indies. The adult worms live deep in the subcutaneous tissue. Females may be over a meter in length, but males are only about 3 mm long. The young are released as juveniles, and when a female is ready to liberate a few thousand, she moves to the superficial layers of the skin, particularly in parts of the body that are apt to be immersed in water. The head end of the worm makes a burrow that almost reaches the surface, and a substance she releases causes a blister to form in the skin. After the blister breaks, the end of the burrow is exposed. If the skin of this region is put into water, a milky fluid comes out of the little hole. This fluid contains juvenile worms about 0.5 mm long, released from a portion of the uterus that protrudes when the body wall of the anterior part of the worm ruptures. If the skin is dried off, then put back into water after an hour or two, the liberation of juveniles will be repeated. After numerous contacts with water have induced the release of almost all of the little worms, the female dies and is somehow absorbed.

If the juvenile worms are eaten by copepods of the genus *Cyclops,* they have a chance of getting into other humans. After being swallowed by a *Cyclops,* a juvenile quickly works its way from the gut into the hemocoel (the blood-filled body cavity) and there it stays for a few weeks, growing a little and molting three times.

If the *Cyclops* is swallowed by a person taking a drink of water, the young worm breaks out of the copepod (or is digested out of it) and migrates to the subcutaneous tissues, where maturation and mating take place.

In many areas where *Dracunculus* is prevalent, much of the water available for bathing, washing clothes, or drinking is localized in small ponds or wells. These small water sources, if they contain *Cyclops*, are perfect for perpetuation of *Dracunculus*. As one puts an arm or leg into the water, or becomes immersed completely, the juveniles are given an opportunity to escape from the uterus of the female. The same water, if drunk, will deliver a *Cyclops* carrying a young *Dracunculus* to a human recipient.

When *Dracunculus* causes a blister to form on the skin, there may be some unpleasant side reactions. These range from a rash to asthma and fainting. These reactions are probably allergic responses of the body to substances produced by the worm. In areas where the guinea worm is prevalent, the natives know that the appearance of a blister signals the arrival of the worm to a point near the surface of the skin. By repeatedly putting water on the ulcer left after the blister has broken, or by applying suction, a competent medicine man can get enough of the worm out to start winding it onto a slender stick. The operation cannot be hurried, and it may take two weeks to remove the worm in one piece. If a worm breaks during the process of being pulled out, there is a good chance that the retreating portion will carry pathogenic bacteria from the surface into deeper tissues, or will induce a severe allergic reaction.

Classification

Although the phylum Nematoda is large in number of species, it is structurally homogeneous. This has forced specialists concerned with classification of these worms to deal with some rather minute details. Certain of the structures on which the two classes and their several orders are based are difficult to see at all. For instance, the feature that separates the classes is the presence or absence of a pair of phasmids, which are unicellular glands that open by pores on the ventral surface near the anus. Even if phasmids are present, they may be reduced to small papillae and thus confused with other structures. In view of the complexities of nematode classification, it seems best to deal with the subject only briefly here. Students who need a more comprehensive account will find one in specialized treatises, and also in some textbooks of parasitology.

Class Secernentea

This group, also called Phasmidea, is characterized by the presence of phasmids. Pinworms, *Ascaris*, hookworms, *Dracunculus*, and *Wuchereria*—the life cycles of all of these have already been described—belong to various orders that are parasitic in animals. Certain orders of this class, however, include free-living nematodes or species that attack plants. Those that live in plants, like some nematodes parasitic in animals, have a stylet—a specialization of the pharynx—for making punctures preparatory to sucking up juices.

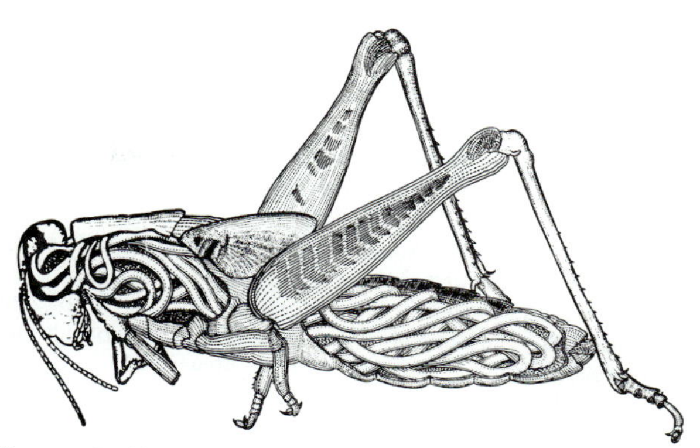

Figure 9.16 *Agamermis decaudata*, a mermithid nematode in the hemocoel of a grasshopper nymph. (Christie, Journal of Agricultural Research, *52*.)

Class Adenophorea

In Adenophorea (or Aphasmidea) there are no phasmids. This group, like the Secernentea, has some orders whose members are exclusively parasitic in animals. *Trichinella* belongs to one of these. Two other orders consist primarily of free-living species, either predators or juice-suckers that use a hollow stylet to draw nourishment from plants and animals. One order includes the mermithids (Figure 9.16), which are parasitic in insects until they are nearly mature. They then escape from their hosts and live in soil or water, eventually mating and laying eggs on vegetation.

References on Nematoda

Andrassy, I., 1976. Evolution as a Basis for the Systematization of Nematodes. Pitman, London.

Bird, A. F., 1971. The Structure of Nematodes. Academic Press, London and New York.

Chitwood, B. G., and Chitwood, M. B. (editors), 1950. Introduction to Nematology. 2nd edition. Monumental Printing Company, Baltimore. (Reprinted 1974, by University Park Press, Baltimore.)

Crofton, H. D., 1966. Nematodes. Hutchinson University Library, London.

Croll, N. A. (editor), 1976. The Organization of Nematodes. Academic Press, New York and London.

Croll, N. A., and Matthews, B. E., 1977. Biology of Nematodes. Blackie, Glasgow; John Wiley and Sons, New York.

Deutsch, A., 1978. Gut ultrastructure and digestive physiology of two marine nematodes, *Chromadorina germanica* (Bütschli, 1874) and *Diplolaimella* sp. Biological Bulletin, *155*:317–355.
Filipjev, I. N., and Schuurmans-Stekhoven, J. H., 1941. A Manual of Agricultural Helminthology. E. J. Brill, Leiden.
Goodey, T., 1963. Soil and Freshwater Nematodes. 2nd edition. John Wiley and Sons, New York.
Lee, D. L., and Atkinson, H. J., 1976. The Physiology of Nematodes. 2nd edition. Oliver and Boyd, Edinburgh and London; Columbia University Press, New York.
Levine, N. D., 1968. Nematode Parasites of Domestic Animals and of Man. Burgess Publishing Company, Minneapolis.
Nicholas, W. L., 1975. The Biology of Free-Living Nematodes. Clarendon Press, Oxford.
Platt, H. M., and Warwick, R. M., 1983. Free-living Marine Nematodes. Part I. British Enoplids. Synopses of the British Fauna, no. 28. Linnean Society of London and Cambridge University Press, London, Cambridge, and New York.
Platt, H. M., and Warwick, R. M., 1988. Free-living Marine Nematodes. Part 2, British Chromatids. Synopses of the British Fauna, no. 38. Linnean Society of London and E. J. Brill/Dr. W. Backhuys, Leiden and New York.
Somers, J. A., Shorey, H. H., and Gastor, L. K., 1977. Sex pheromone communication in the nematode, *Rhabditis pellio*. Journal of Chemical Ecology, *3*:467–474.
Stone, A. R., Platt, H. M., and Khalil, L. F. (editors), 1983. Concepts in Nematode Systematics. (Systematics Association Special Volume, no. 22.) Academic Press, London and New York.
Thorne, C., 1961. Principles of Nematology. McGraw-Hill Book Company, New York.
Wharton, D. A., 1986. A Functional Biology of Nematodes. Johns Hopkins University Press, Baltimore.
Yamaguti, S., 1961. Systema Helminthum, volume 3 (parts 1, 2). Interscience Publishers, New York.
Zuckerman, B. M., Mai, W. F., and Rohde, R. A. (editors), 1971. Plant Parasitic Nematodes. 2 volumes. Academic Press, New York and London.

Phylum Nematomorpha

No one had ever seen it happen, but everyone knew that if you put long horsehairs into a puddle of water and let them stay, they would turn into hairsnakes.

From *A Boy's Town*,
by William Dean Howells (1890)

Introduction

Typical nematomorphs, generally called horsehair worms, hairsnakes, or Gordian worms, are sluggish, wirelike animals found in a variety of freshwater habi-

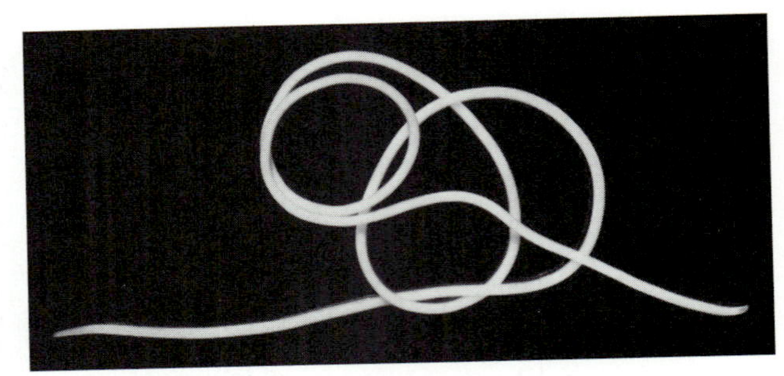

Figure 9.17 A living adult nematomorph.

tats, including watering troughs. The belief that they develop from horsehairs goes back at least to the fourteenth century. The fact that they are often tangled led Linnaeus to name the first genus *Gordius,* commemorating a mythical king who is said to have tied the celebrated Gordian Knot. The elaborate knot, the king said, could be undone only by the one destined to conquer the East. Alexander the Great, the story goes on to explain, undid it by cutting through it with his sword.

General Structure

Nematomorphs are almost perfectly cylindrical (Figure 9.17) and so slender that a worm 1 m long may be only 1 or 2 mm thick. The majority of them are about 10 to 30 cm long. The color is generally some shade of brown or blackish brown, although the anterior tip of the body is often white.

The cuticle consists of three main layers. The outermost layer, when viewed with the light microscope, appears to be relatively homogeneous in texture, but electron microscopy reveals that it is differentiated into three or four main strata (Figure 9.18, A). The next layer is very thin, except where a variety of bodies it contains push up into the outermost layer. The warty appearance of the cuticle of some species (Figure 9.18, B) is probably related to the presence of these bodies, but further studies on the ultrastructure of nematomorphs will have to be made before the significance of the external elaborations of the cuticle, including spinelike outgrowths (Figure 9.18, C), can be understood. The innermost layer of the cuticle is the thickest and consists to a large extent of fibers arranged in a number of strata; the fibers of successive strata have different orientations. The epidermis that secretes the cuticle appears to be a syncytium. Ducts run into the cuticle from the surface of the epidermis.

The musculature of the body wall lies just beneath the epidermis and consists of a single layer of ribbonlike cells (Figure 9.19). These are oriented longitudinally. As a rule, the pseudocoel is largely obliterated

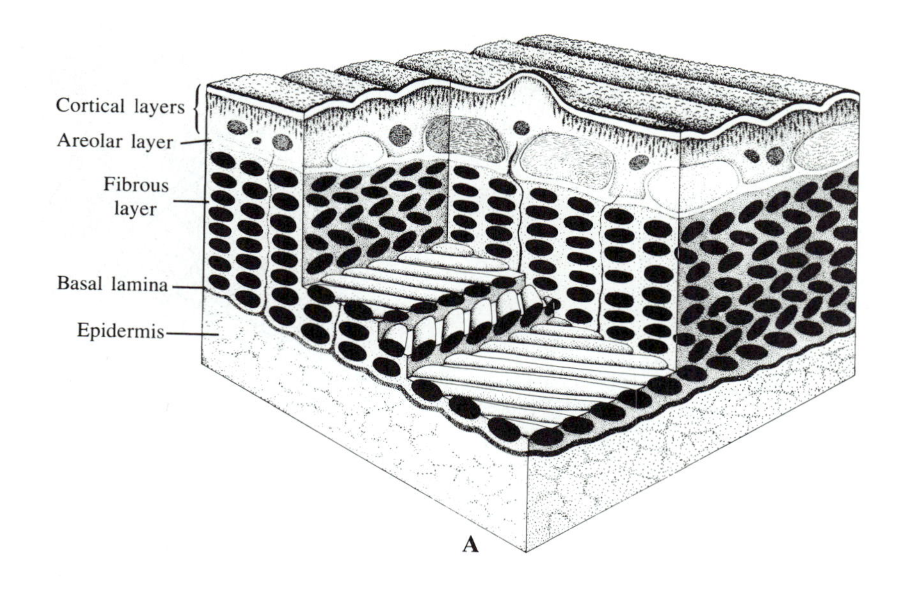

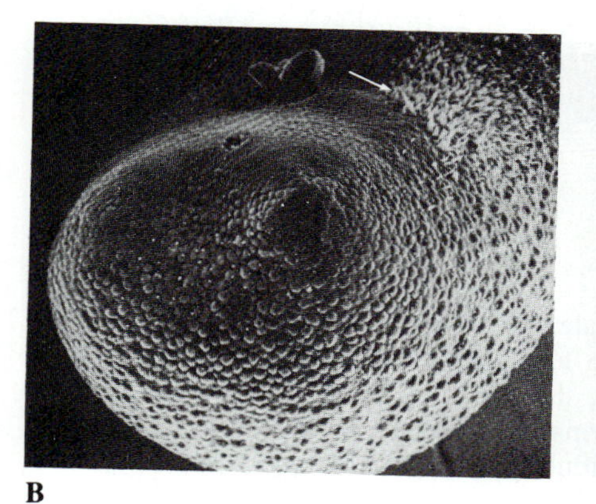

Figure 9.18 A. The cuticle of *Paragordius varius;* diagram based on electron micrographs. (Zapotosky, Proceedings of the Helminthological Society of Washington, *38*.) B. Posterior end of a male nematomorph (probably a species of *Chordodes*), as seen with the scanning electron microscope; the micrograph shows the prominent "areolae" that characterize the cuticle over much of the body surface of both sexes, and an extensive patch of spinelike outgrowths characteristic of the male. C. The spinelike cuticular outgrowths of the male. (B and C, Martin, Proceedings of the Helminthological Society of Washington, *40*.)

by proliferation of parenchymal cells, amongst which is a secreted material said to resemble collagen. There are usually remnants of the pseudocoel, however, especially around the gut.

The gut is greatly modified. The mouth, situated at the tip of the body (or on the ventral surface close to the tip), is nonfunctional and may not even be recognizable. The anterior part of the gut is sometimes tubular, but it may be reduced to a strand of cells without a definite lumen. Farther back, the gut is usually a slender tube (Figure 9.19), and its hindmost portion, lined by cuticle, is joined by the genital ducts. This last part, leading to the anus, is therefore a cloaca.

The **ventral nerve** originates in the epidermis, and although it sinks into the parenchyma, it remains connected to the epidermis by a sheet of tissue called the nerve lamella (Figure 9.19). Near the anterior end of the body is a brainlike concentration of nerve cells, and this has connections with the ventral nerve. In the cloacal region, the nerve may be thickened into a **cloacal ganglion,** although this does not consist of particularly large or specialized cells.

Reproduction, Development, and Life History

In certain genera, the sexes are not obviously different in external appearance. In some, however, the posterior tip of the body is bilobed in the male (Figure 9.20, A and B) and blunt in the female (Figure 9.20, C). There are also genera in which the posterior end of the body has two lobes in the male and three lobes in the female.

Both sexes have a pair of gonads in the pseudocoel, surrounded by parenchymal cells that have filled up much of that cavity (Figure 9.19). The testes are tubular

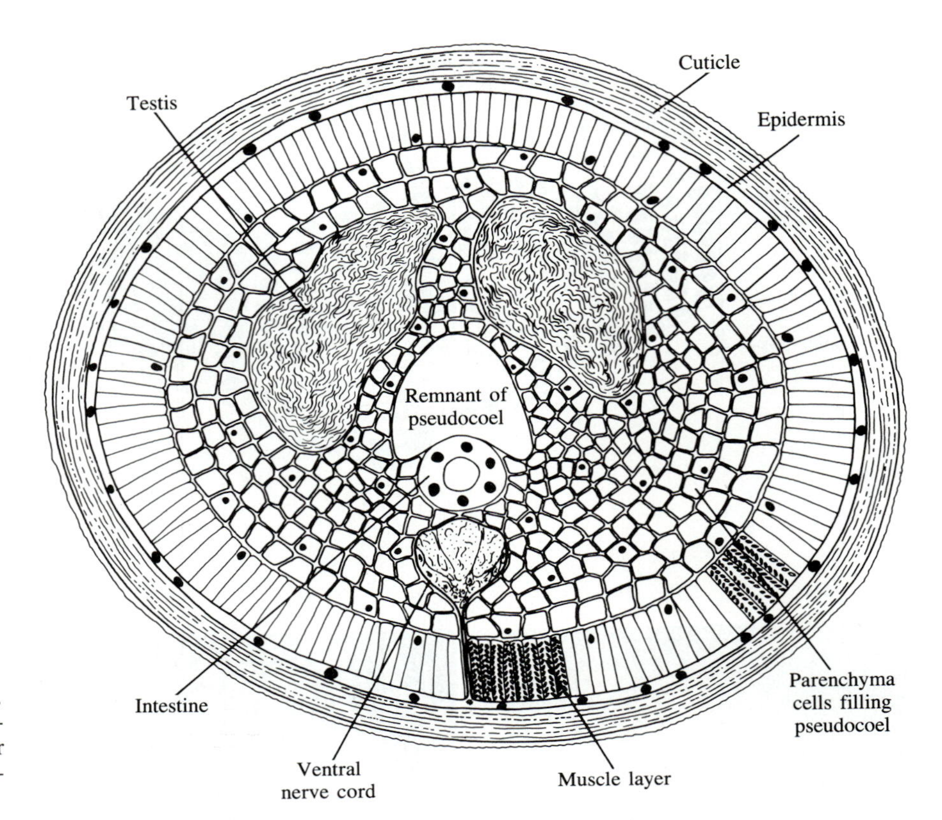

Figure 9.19 *Gordius aquaticus*, male, transverse section; diagrammatic. (After Ivanov *et al.*, Major Practicum of Invertebrate Zoology.)

organs that run for much of the length of the body. The sperm ducts widen out into seminal vesicles, then enter the cloaca separately or after becoming joined into what has been interpreted as a penis. The cloaca is surrounded by radially arranged muscles that act as dilators. Bristles, spines, and sticky warts in the tail region of the male of some species are probably concerned with copulation itself or with sensory functions related to mating, or both.

In the female, the ovaries start out as tubular organs similar to the testes of the male. As they prepare to produce eggs, they develop many diverticula. In some horsehair worms, the ovaries retain a close association with their oviducts; in others the eggs apparently must wander through spaces in the parenchyma before locating the oviducts. The oviducts lead to the anterior part of the cloaca, which is glandular instead of being cuticularized. From this part extends a blind sac that serves

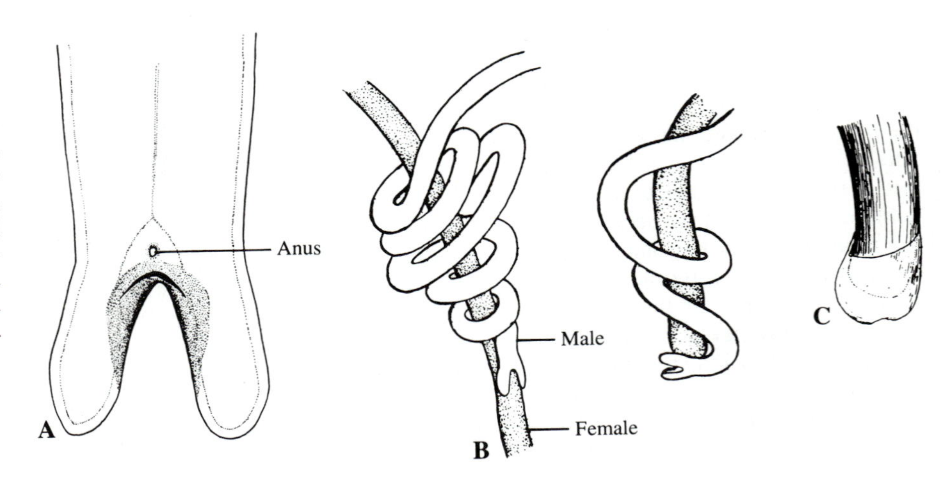

Figure 9.20 *Gordius albopunctatus*, posterior end of male. (Muus, Videnskabelige Meddelelser fra Dansk Naturhistorisk Forening i København, *118*.) B. Two stages in the mating of *Gordius aquaticus*. C. Posterior end of a female *Gordius aquaticus* capped by a spermatophore deposited by the male. (B and C, Dorier, Travaux du Laboratoire d'Hydrobiologie et de la Pisciculture de l'Université de Grenoble, *22*.)

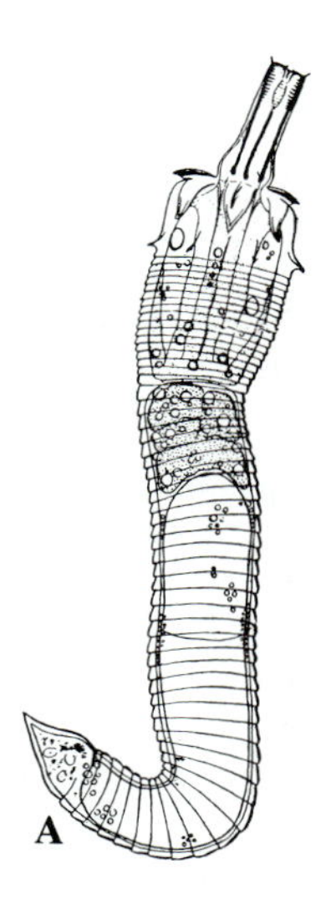

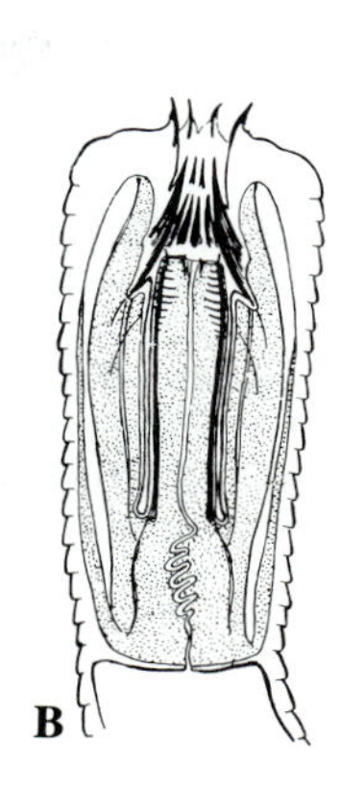

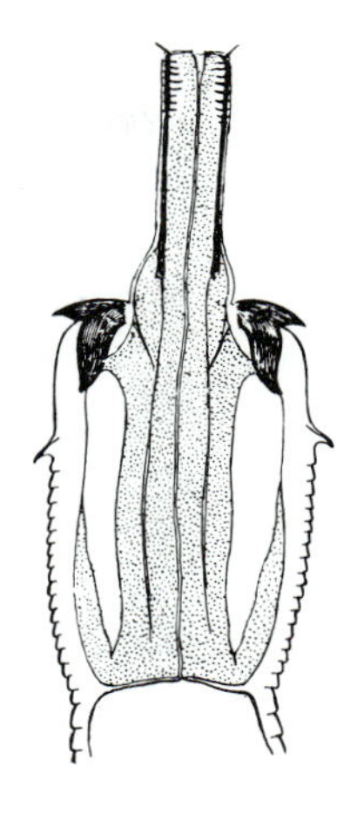

FIGURE 9.21 A. Larva of *Gordius aquaticus*. B. Perforating apparatus of the larva, retracted (left) and protracted (right). (Dorier, Travaux du Laboratoire d'Hydrobiologie et de la Pisciculture de l'Université de Grenoble, *22*.)

as a seminal receptacle, storing sperm injected by the male.

Mating of horsehair worms generally takes place in late summer or autumn. In at least certain species the males are able to recognize females that are still virgins. During copulation, the posterior ends of the participants are brought into tight contact (Figure 9.20, B) and the male deposits a sticky spermatophore over the cloacal opening of the female (Figure 9.20, C). The sperm evidently migrate to the seminal receptacle where they are stored until egg-laying begins. If the female loses the spermatophore, she may be sought out for copulation again.

Eggs are laid in long strings. The number of eggs produced by a single female may reach several million. The glandular anterior part of the cloaca is believed to secrete the material that holds the strings of eggs together. After reproducing, the males and females die; the postcopulatory life of males may be especially short.

The fertilized egg of a horsehair worm develops into a **larva** that is very different from the adult (Figure 9.21). The mouth is at the tip of a spiny, two-part head that can be pulled back into a sheath. The rest of the body is ringed externally. The anus is an inconspicuous opening near the posterior end.

After the larva hatches, it must enter an appropriate insect, centipede, or millipede. The worm will spend almost all of its life as a parasite, emerging only when it is ready to undergo final maturation in fresh water and reproduce. In most cases the larva is probably eaten by a prospective host when the arthropod visits the edge of a body of water. It has been demonstrated, however, that larvae introduced experimentally into the hemocoel of an insect will develop normally. In at least some horsehair worms, the larva encysts soon after it hatches and is probably eaten while in the encysted state; the material forming the cyst is thought to derive from brownish cells that constitute the forepart of the intestine. After the larva gets into the host, it burrows through the wall of the gut and enters the hemocoel. There it molts, losing the spiny elaborations of the head region. The host survives the ordeal of being parasitized, but its fat-storing bodies usually disappear and its gonads may never develop, so that it is effectively castrated. For a time the young worm is soft and whitish, but eventually the cuticle hardens. The light or dark brown color characteristic of adult horsehair worms does not appear until they bore their way out of their hosts and enter water. Obviously, the life cycle cannot be completed unless the arthropod host falls into water or dies in a situation from which the escaping worm can reach water.

Some nematomorphs can survive for a time in hosts that are inappropriate for their further development. These hosts are not necessarily arthropods; they may be snails, leeches, or even tadpoles or fishes. After entering such a host, the larva encysts. In many cases of this type, the life cycle is never completed. The first host could, however, be eaten by an arthropod in which normal development can proceed. In some instances, at least, the first host may function as a transfer host, bringing the larva indirectly to an arthropod that otherwise would not be accessible to the larva.

NECTONEMA, A MARINE NEMATOMORPH

The preceding account has dealt with what may be called typical horsehair worms—those that parasitize insects and other terrestrial arthropods and reproduce in fresh water. One marine genus, *Nectonema,* parasitizes various crustaceans—shrimps, crabs, and hermit crabs.

After maturing and emerging from their hosts, the males and females become active swimmers. *Nectonema* is characterized by long bristles near the midline on both the dorsal and ventral surfaces, and the internal anatomy differs considerably from that of horsehair worms. Little is known of reproduction and development of *Nectonema*.

References on Nematomorpha

Eakin, R. M., and Brandenburger, J. C., 1974. Ultrastructural features of a Gordian worm (Nematomorpha). Journal of Ultrastructure Research, *46:*351–374.

Martin, W. E., 1973. Electron scan of a nematomorph cuticle. Proceedings of the Helminthological Society of Washington, *40:*173–177.

May, H. G., 1919. Contributions to the life histories of *Gordius robustus* Leidy and *Paragordius varius* (Leidy). Illinois Biological Monographs, *5:*127–338.

Nielsen, S.-O., 1969. *Nectonema munidae* Brinkmann (Nematomorpha) parasitizing *Munida tenuimana* G. O. Sars (Crust. Dec.) with notes on host–parasite relations and new host species. Sarsia, *38:*91–118.

Zapotosky, J. E., 1971. The cuticular ultrastructure of *Paragordius varius* (Leidy, 1951) (Gordioidea: Chordodidae). Proceedings of the Helminthological Society of Washington, *38:*228–236.

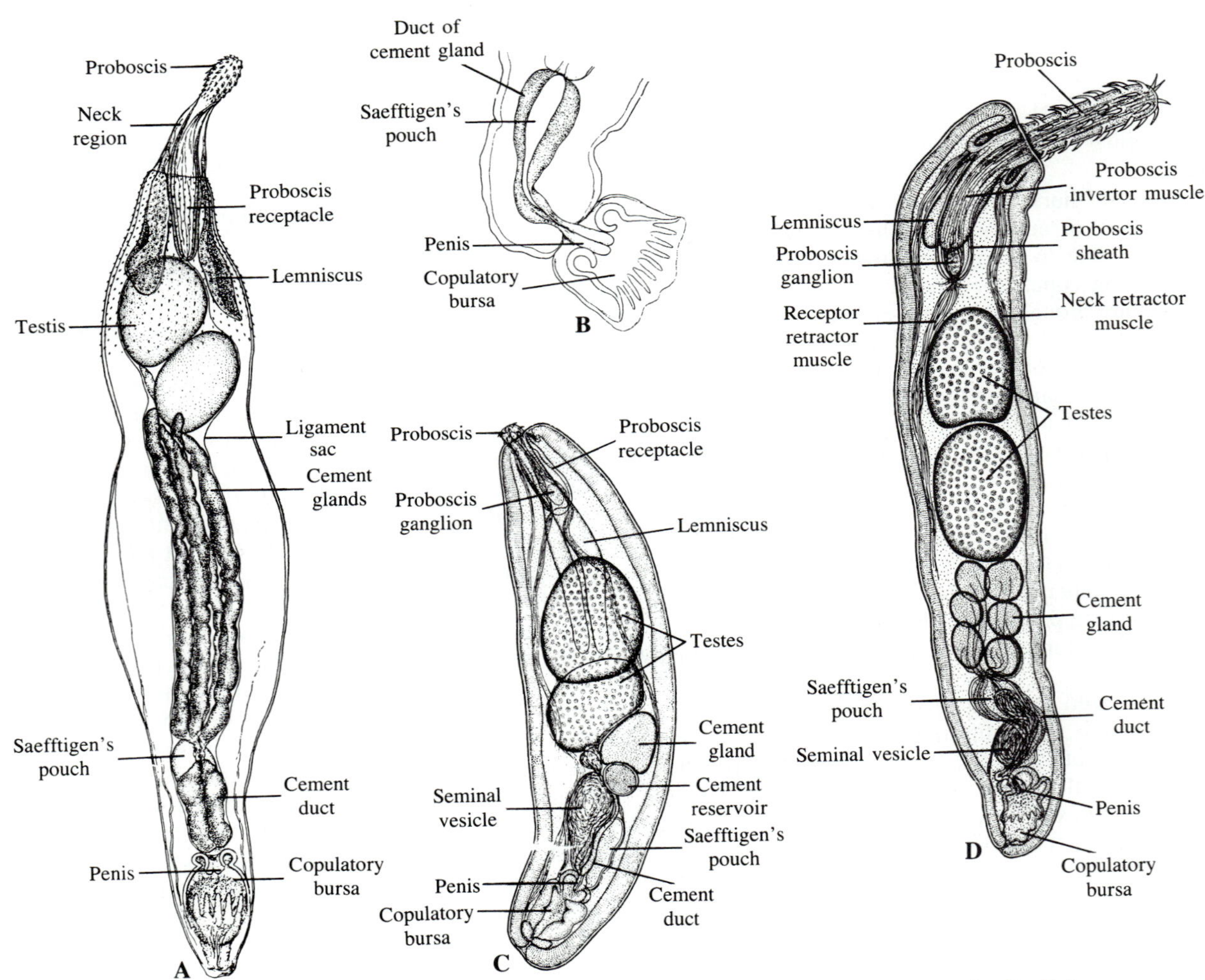

Figure 9.22 Acanthocephalans. A. *Polymorphus magnus,* a parasite of birds, male. B. *Polymorphus magnus,* posterior end of male, with copulatory bursa everted. (A and B, Petrochenko, Acanthocephalans of Domestic and Wild Animals.) C. *Neoechinorhynchus zacconis,* from a freshwater fish, male. D. *Acanthocephalus opsalichthydis,* from freshwater fishes, male. (C and D, Yamaguti, Japanese Journal of Zoology, *6.*)

PHYLUM ACANTHOCEPHALA: THORNY-HEADED WORMS

INTRODUCTION

Acanthocephalans are the only pseudocoelomates that show no trace of a gut. As adults they live in the digestive tract of vertebrates, especially teleost fishes, turtles, birds, and mammals. Early stages generally develop in crustaceans, insects, or myriapods. At least 600 species have been named, but there must be many more awaiting discovery and description.

Acanthocephalans are usually elongated and rather firm-bodied worms. Most of them are cylindrical or become cylindrical soon after they are put into a saline solution or a preservative. Some are more or less permanently ringed, so that they may appear to be segmented; others become ringed when they contract. The length of mature specimens ranges from about 1 mm to over 1 m; most species are between 5 mm and 2 cm long. One of the giants is *Macracanthorhynchus hirudinaceus,* common in pigs and occasionally found in humans; females of this species regularly reach a length of 15 to 30 cm.

GENERAL STRUCTURE

PROBOSCIS

The anterior end of an acanthocephalan is provided with a thorny attachment organ called the **proboscis.** When this is embedded in the host's intestinal wall, the hooks point backward and the parasite's hold is not easily broken. The proboscis can be inverted into a sac called the **proboscis receptacle** (Figure 9.22, A, C, and D; Figure 9.23). The muscle that accomplishes the inversion runs from the inside of the proboscis to the wall of the receptacle. When it contracts, the proboscis is invaginated, so that the hooks now face a temporary cavity, with their tips pointing forward. When muscles in the wall of the receptacle contract, the hydrostatic pressure within the receptacle increases and the proboscis is thrust out again. There are also retractor muscles that pull the proboscis receptacle farther back into the body, and muscles that infold the short "neck" region that lies between the hooked portion of the proboscis and the rest of the body (Figure 9.22, C and D). These muscles are attached to the body wall some distance behind the receptacle.

BODY WALL AND PSEUDOCOEL

The body wall of an acanthocephalan shows some interesting complexities, but these cannot be fully appreciated without the aid of electron microscopy. The

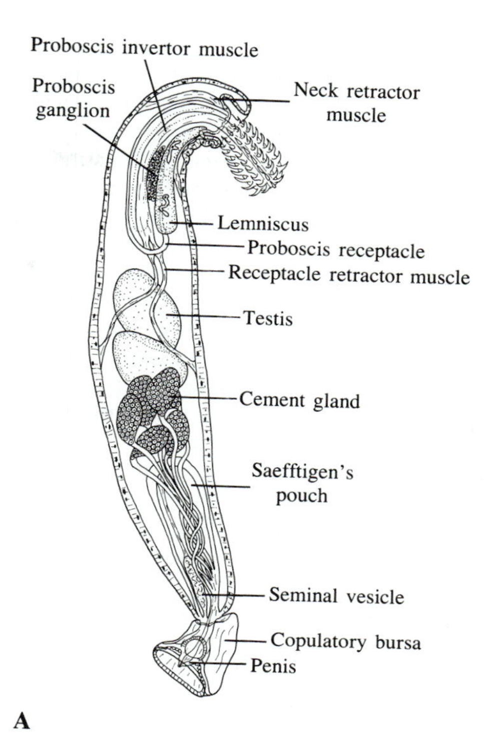

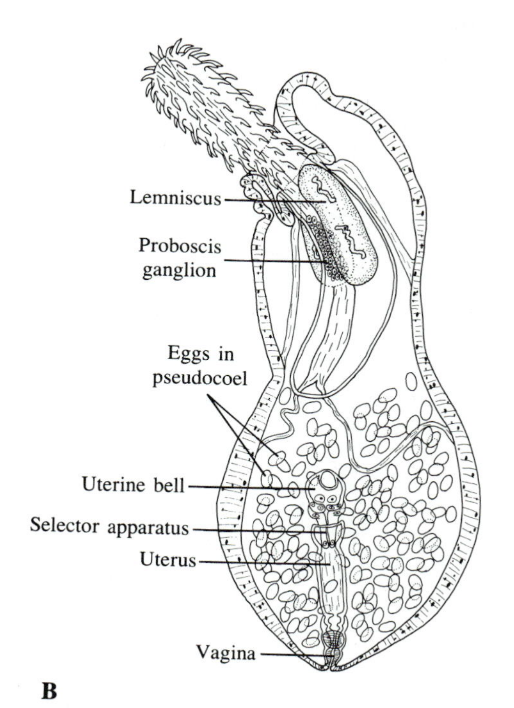

FIGURE 9.23 *Echinorhynchus lageniformis,* a parasite of marine fishes. A. Male. B. Young female. (Pratt and Olsen, Journal of Parasitology, *57.*)

cuticle is coated by a mucopolysaccharide that presumably reaches the surface by way of ducts that originate in the epidermis (Figure 9.24). The inner part of the cuticle and outer part of the epidermis contain the narrow necks of the ducts and together constitute what is called the *striped layer* of the body wall. The rest of the epidermis is differentiated into a *felt layer,* which contains numerous fibers as well as the basal portions of the ducts, and a *radial layer,* which has relatively few fibers but is extensively honeycombed by channels called **lacunae.** There are usually one or two main channels that run the length of the trunk. The lacunae are simply spaces, for they have no epithelial linings; they almost certainly function in distribution of nutrients and other dissolved substances.

The nuclei of the epidermis are in the radial layer. When there are only a few of them, they are generally large and extensively lobed. When there are many of them, they are products of fragmentation of the relatively few nuclei originally present. The apparent absence of cell boundaries in the epidermis indicates that it is a syncytium. The inner surface of the epidermis, in contact with the basal lamina, is much folded and shows some vesicles that perhaps function in pinocytosis. Nutrients seem to be taken up all over the body surface, however. Connective tissue separates a layer of circular muscle from the basal lamina and from the longitudinal muscle, which is next to the pseudocoel.

The pseudocoel, usually spacious, contains a pair of structures called **lemnisci** (Figure 9.22, A, C, and D; Figure 9.23). These originate from the epidermis in the neck region, but are otherwise free of the body wall. The lemnisci are pervaded by lacunar vessels continuous with those of the epidermis. The lemnisci probably serve as reservoirs for lacunar fluid that is displaced when the proboscis and neck are pulled back. Various other functions have also been ascribed to these structures, however. Whatever they do, they are decidedly characteristic of acanthocephalans and are not found in any other phylum.

EXCRETORY SYSTEM

Most acanthocephalans have no recognizable excretory system. Nitrogenous wastes are perhaps eliminated to at least some extent by way of the ducts that extend from the epidermis to pores in the cuticle. In some members of the class Archiacanthocephala, there are two clusters of flame-bulb **protonephridia** in the pseudocoel (Figure 9.25), which in males join the sperm duct and in females run to the uterus. Sometimes there is a bladder associated with each main duct of the protonephridium, close to where it leaves the cluster of flame bulbs.

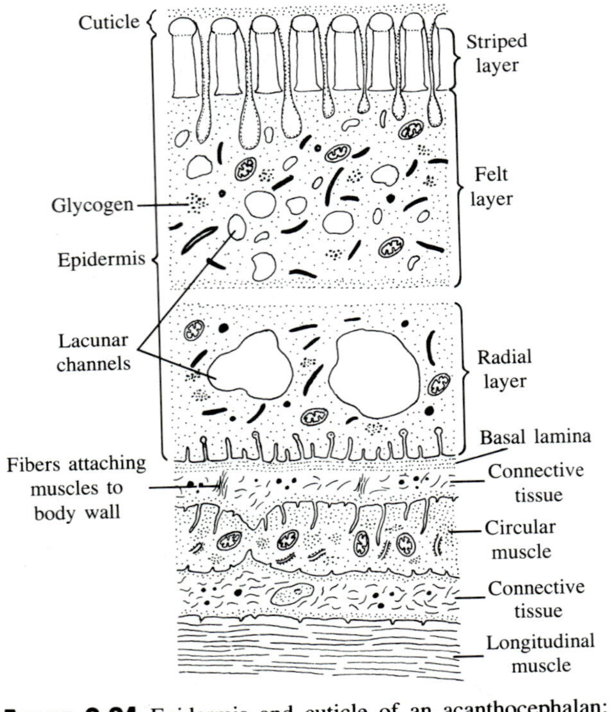

FIGURE 9.24 Epidermis and cuticle of an acanthocephalan; diagram based on electron micrographs.

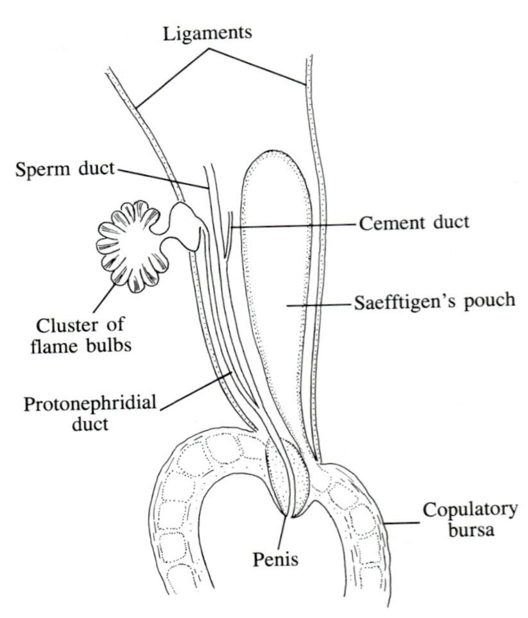

FIGURE 9.25 Relationship of one of the two flame-bulb protonephridia of a male archiacanthocephalan to ducts of the reproductive system; diagrammatic and generalized. In females, the protonephridial ducts enter the uterus.

Nervous System

A brainlike structure, usually called the **proboscis ganglion,** lies in the ventral wall of the proboscis receptacle. It consists partly of large ganglionic cells, partly of nerve fibers. The nerve trunks leaving it are difficult to recognize, but tracts go to the neck and up into the proboscis. The small pits sometimes observed on the neck or proboscis, or both, are probably sensory structures. The trunk is innervated by a pair of **lateral nerves.** In males these lateral nerves have connections with ganglia closely associated with the copulatory apparatus.

Reproduction, Development, and Life Cycle

Sexes are separate, and males are almost always appreciably smaller than females of the same species. The reproductive organs are basically confined to a tube of connective tissue called the **ligament sac,** although in many acanthocephalans this breaks down as the worms approach maturity. Males have two testes, one behind the other (Figure 9.22, A, C, and D; Figure 9.23, A). Their ducts unite to form a common sperm duct, or vas deferens, which generally has a prominent dilation that functions as a seminal vesicle; the separate ducts may also have dilations, but these are not conspicuous. The penis opens into a pocket called the **bursa.** This can be forced out to form a cup over the posterior end of the female. The wall of the bursa contains a system of spaces continuous with the cavity of a muscular sac called **Saefftigen's pouch,** and eversion of the bursa (Figure 9.22, B) is thought to depend on injection of fluid into the system when the pouch contracts. Muscles in the bursa enable this structure to clasp the tip of the female's body. One or several **cement glands** lying just behind the testes produce a secretion that helps tighten the bond between the male and female. After sperm have been deposited in the vagina, the cement hardens to form a cap; this prevents backflow. As a rule, there are several cement glands, each with a duct that delivers its secretion to the sperm duct. In some species there is only one cement gland; this is joined to a sac in which the secretion accumulates before a duct carries it to the sperm duct.

In females, there is a single ovary, which does not survive long as a discrete structure. It generally disintegrates by the time the worm is mature, so that there are hundreds of eggs floating in the ligament sac (when this persists) or in the pseudocoel (Figure 9.23, B). If the ligament sac is divided into two portions by a partition, as is the case in females of certain groups, the eggs will be in the dorsal portion. The sperm introduced into the female orifice fertilize the eggs as these become dissociated from the egg masses. Embryonation and formation of the shell then begin. Of the three layers that constitute the egg shell in most acanthocephalans, the outermost can be traced back to the membrane covering the oocyte before the sperm entered it. The innermost layer is a derivative of the fertilization membrane. A thicker shell membrane is laid down between these soon after the embryo starts to develop.

To get out of the female, the embryonating eggs must first enter an open-ended tube called the **uterine bell** (Figure 9.23, B; Figure 9.26). This is a muscular structure, and its posterior portion functions as a sorting apparatus. Eggs ready to go into the uterus are admitted

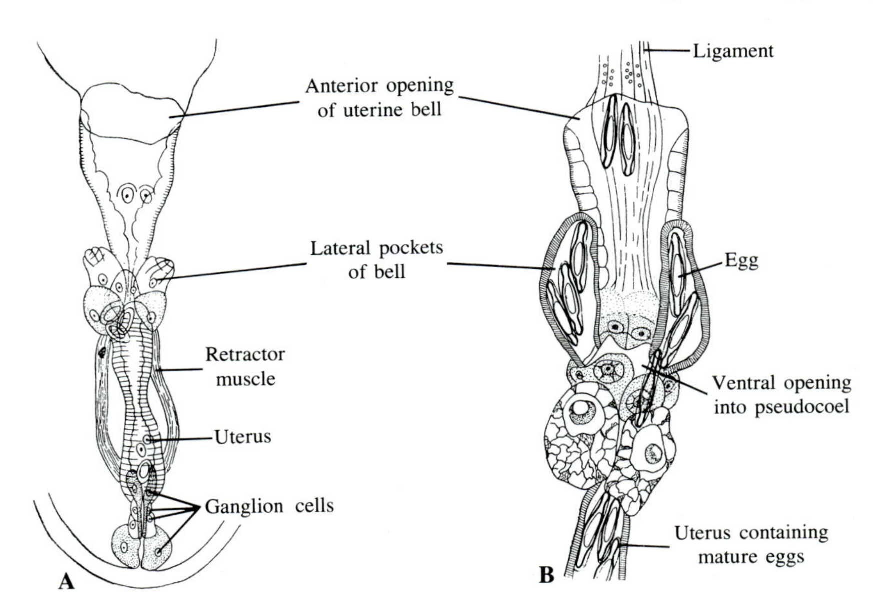

Figure 9.26 Uterine bells and associated reproductive structures in female acanthocephalans. A. *Neoechinorhynchus rutili*. B. *Pseudechinorhynchus clavula*. (Petrochenko, Acanthocephalans of Domestic and Wild Animals.)

to it; eggs that are insufficiently mature are rejected, and they slip back into the ligament sac (or the ventral part of it, if it is divided) for further development. Eventually these eggs will have to stand trial again. In at least one acanthocephalan in which the uterine bell has been studied, it appears that the selection apparatus recognizes sufficiently mature eggs by their greater length.

The uterus is muscular, and the eggs that have accumulated in it are forced out through the vagina and genital pore. They leave the definitive host with fecal material. At this stage, each egg contains a larva called the **acanthor,** characterized by six or eight hooks at its anterior end (Figure 9.27, A). If the egg is swallowed by an appropriate intermediate host, the acanthor develops a bit further, then hatches (Figure 9.27, B). With the help of its hooks and body movements, it slowly works its way through the gut wall (Figure 9.27, C) and reaches the hemocoel. During the next stage, called the **acanthella,** the worm loses the hooks that were characteristic of the acanthor and gradually develops structures typical of the adult (Figure 9.27, D). The pseudocoel originates early by separation of an outer layer of mesoderm from an inner mass. By the time the juvenile has reached the infective stage (Figure 9.27, E), it has a fully formed proboscis, proboscis receptacle, and lemnisci, and portions of the reproductive system have begun to differentiate. If the intermediate host is eaten by a suitable definitive host, the juvenile is freed, becomes attached to the intestinal wall, and grows to sexual maturity.

For acanthocephalans that parasitize terrestrial vertebrates, the intermediate hosts are usually insects. Species that live in fishes and other aquatic vertebrates generally use amphipods or ostracodes, but some rely on insects that are likely to be eaten by the definitive hosts.

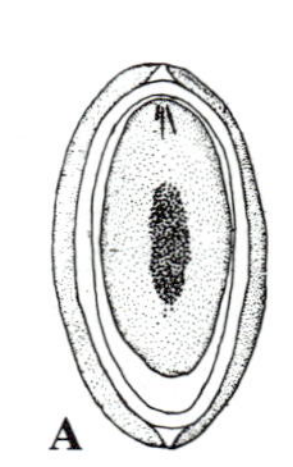

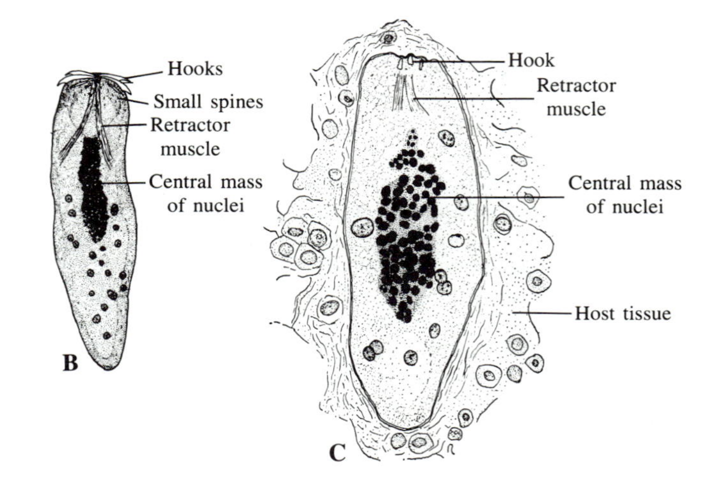

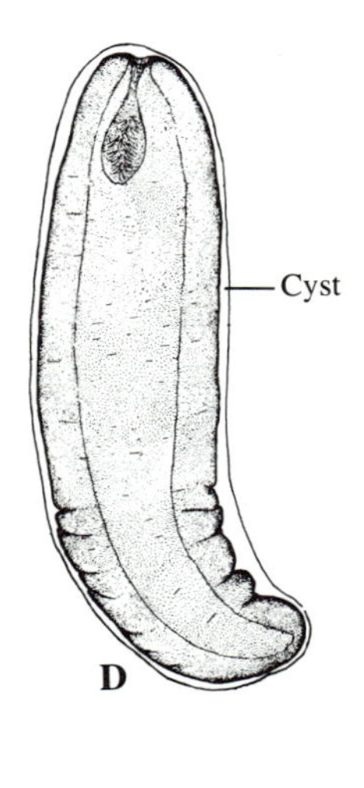

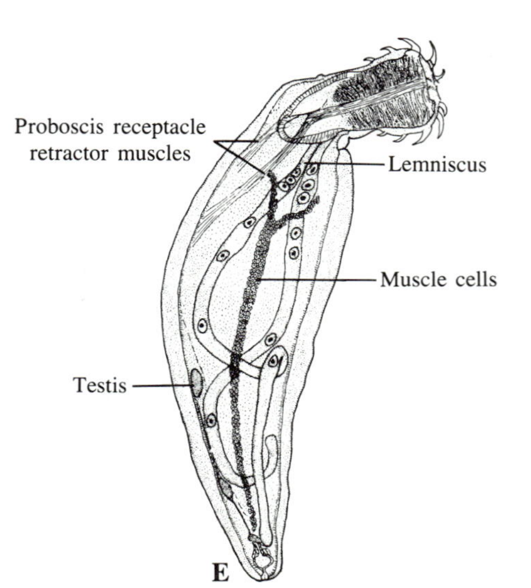

Figure 9.27 Stages in the life history of *Macracanthorhynchus ingens*. (Adult worms live in raccoons; beetles serve as intermediate hosts.) A. Egg containing an acanthor. B. Acanthor after hatching. C. Sagittal section of an acanthor in the gut wall of a beetle. D. Late acanthella encysted in the hemocoel of a beetle. E. Infective juvenile male. (Moore, Journal of Parasitology, *32*.)

An interesting aspect of the biology of certain parasitic worms is the way in which they modify the behavior of their intermediate hosts, making these more vulnerable to being eaten by suitable definitive hosts. Metacercariae of the fluke *Dicrocoelium dendriticum,* for instance, affect ants in such a way that these are more likely than unparasitized ants to be eaten by sheep and related herbivores. Among acanthocephalans, *Polymorphus paradoxus* affects its intermediate host, the amphipod crustacean *Gammarus lacustris,* in a somewhat similar way. Unparasitized amphipods tend to remain hidden; parasitized individuals swim close to the surface of the water or cling to vegetation near the surface, so they are especially prone to being eaten by dabbling ducks or by muskrats that consume vegetation.

In some species the life cycle is complicated by involving a second intermediate host in which development may proceed no further, but which is more likely than the first intermediate host to be eaten by the definitive host. The second intermediate host, in other words, is a **transfer host;** it bridges the ecological gap between the first intermediate host and the definitive host. For example, a frog that eats insects or crustaceans would be an admirable transfer host for an acanthocephalan destined to parasitize a frog-eating bird or mammal.

Parasitism of humans by acanthocephalans is rare. Two of the several species that have been reported from humans are *Moniliformis moniliformis (M. dubia),* which is fairly common in rats, and *Macracanthorhynchus hirudinaceus,* normally a parasite of pigs. The intermediate hosts of the former are cockroaches; beetles and their larvae serve as intermediate hosts for the latter.

CLASSIFICATION

Specialists concerned with the systematics of acanthocephalans recognize three classes.

CLASS ARCHIACANTHOCEPHALA

In archiacanthocephalans, there is either a single dorsal lacunar canal or both a dorsal and a ventral canal, and there are only a few large nuclei in the epidermis. In females, the ligament sac is divided into two chambers. Males have several cement glands. The only acanthocephalans that have protonephridia belong to this group. The adult worms parasitize birds and mammals; the intermediate hosts are insects and myriapods.

CLASS PALAEACANTHOCEPHALA

In palaeacanthocephalans, the main lacunar canals are lateral, and the nuclei of the epidermis break up, so there are many small ones. The ligament sac in females is undivided and does not persist to maturity. Males have several cement glands. The adult worms are parasites of amphibians, fishes, and reptiles.

CLASS EOACANTHOCEPHALA

In eoacanthocephalans, the main lacunar canals are dorsal and ventral, but are often obscure. There are only a few epidermal nuclei, and these are large. In females, the ligament sacs are divided and persistent. Males have a single cement gland. The adult worms are parasites of fishes, amphibians, and reptiles.

REFERENCES ON ACANTHOCEPHALA

Crompton, D. W. T., 1977. An Ecological Approach to Acanthocephalan Physiology. Cambridge University Press, Cambridge and New York.

Crompton, D. W. T., and Nickol, B. B. (editors), 1985. Biology of the Acanthocephala. Cambridge University Press, Cambridge and New York.

Hammond, R. A., 1966. The proboscis mechanism of *Acanthocephalus ranae*. Journal of Experimental Biology, *45*:203–213.

Holmes, J. C., and Bethel, W. M., 1972. Modification of intermediate host behaviour by parasites. *In* Canning, E. U., and Wright, C. A. (editors), Behavioural Aspects of Parasite Transmission. Academic Press, London and New York.

Hopp, W. B., 1954. Studies on the morphology and life cycle of *Neoechinorhynchus emydis* (Leidy), an acanthocephalan parasite of the map turtle, *Graptemys geographica* (LeSueur). Journal of Parasitology, *40*:284–299.

Lumsden, R. D., 1975. Surface ultrastructure and cytochemistry of parasitic helminths. Experimental Parasitology, *37*:267–339.

Nicholas, W. O., 1967, 1973. The biology of Acanthocephala. Advances in Parasitology, *5*:205–246, *11*:671–706.

Whitfield, P. J., 1970. The egg sorting function of the uterine bell of *Polymorphus minutus* (Acanthocephala). Parasitology, *61*:111–126.

Yamaguti, S., 1963. Systema Helminthum, volume 5. Interscience Publishers, New York.

SUMMARY

1. A pseudocoel is a body cavity that is not lined by a peritoneum. Most of the invertebrates dealt with in Chapters 9 and 10 are characterized by this type of body cavity, or have other features that indicate affinities with pseudocoelomate animals. Not all of the phyla in this assemblage are necessarily closely related, however.

2. The largest of the pseudocoelomate phyla is the Nematoda. Most nematodes have a slender, cylindrical body and a strong, flexible cuticle. All of the muscles of the body wall are oriented longitudinally, so nematodes can bend and undulate, but they are unable to constrict and dilate.

3. The gut of nematodes is complete, and the pharynx, equipped with radial muscles, is used as a sucking or pumping organ. Cilia are absent, except for some highly modified ones in certain sense organs.

4. Sexes of nematodes are separate. The young hatch as juveniles that resemble the adults, and go through four molts of the cuticle before reaching the stage that grows to sexual maturity.

5. Nematodes live in almost all habitats where moisture is present. Many of them are parasites of other animals and of plants.

6. The phylum Nematomorpha consists of the slender-bodied horsehair worms. Most of them are parasites of insects and myriapods up to the time they reach adulthood. Then they escape from their hosts, enter fresh water, and reproduce. Sexes are separate. The eggs laid by females develop into distinctive larvae. If a larva gets into a suitable host, it settles in the hemocoel and grows to maturity. There is a complete gut in the larva, but during the later parasitic phase the mouth becomes scarcely recognizable and the anterior part of the gut loses the lumen. Adults do not feed. The pseudocoel becomes largely filled by cells and material that resembles collagen.

7. Members of the phylum Acanthocephala are called thorny-headed worms. The adults are parasites in the gut of vertebrates, attaching themselves to the tissue of the gut by a retractable, spiny proboscis. There is no digestive tract at any time in the life history. Sexes are separate. The eggs leave the body of the host with fecal matter. If an egg is eaten by an appropriate myriapod, insect, or crustacean, the larva within it hatches, works its way into the hemocoel, and develops into the juvenile infective stage. This is similar to the adult but does not have a fully differentiated reproductive system.

8. Five other pseudocoelomate phyla—Rotifera, Gastrotricha, Kinorhyncha, Loricifera, and Priapula—are covered in Chapter 10.

10

More Pseudocoelomate Phyla

Rotifera, Gastrotricha, Kinorhyncha, Loricifera, and Priapula

- Phylum Rotifera
  - Class Monogononta
  - Class Digononta
- Phylum Gastrotricha
  - Class Chaetonota
  - Class Macrodasya
- Phylum Kinorhyncha
- Phylum Loricifera
- Phylum Priapula

This chapter deals with five additional phyla of pseudocoelomate animals. No significance should be attached to the fact that they are grouped together, or to the order in which they are presented. All of these phyla, like those covered in Chapter 9, are distinctive and may constitute evolutionarily isolated groups. Even when certain characters seem to suggest a kinship between two or more of the phyla, we must consider the possibility that the similarities are due instead to convergent or parallel evolution.

Phylum Rotifera: Rotifers

Introduction

With about 2000 described species, the rotifers form the second largest assemblage of pseudocoelomate invertebrates. Although their number does not approach that of nematodes, and although they have not become involved in parasitism to any great extent, it is probably fair to say that their morphological diversity is at least equal to that of nematodes.

Most rotifers are between 100 and 500 μm long and are thus roughly comparable in size to ciliates such as *Paramecium* and *Stentor*. The giants of the phylum reach lengths of about 3 mm. Even the smaller species, however, are rather complicated, and on studying a living rotifer for the first time, one will probably be surprised by how many structures are packed into it.

The name *Rotifera* means "bearing a wheel," and refers to an illusion created by the activity of special ciliary tracts at the anterior end of the body. Collectively, these cilia form a complex known as the **corona,** or wheel organ. One part of the corona is a field of cilia closely associated with the mouth; the other part is organized into a band that encircles the body, or into specialized configurations derived from this band. The corona is important in collection of food as well as in locomotion.

Certain tissues of rotifers, including the epidermis, are syncytial. The epidermis is unusual in that its outer portion contains a dense layer of fibrous material. In the past, this layer has been confused with a cuticle. A true external cuticle, consisting of a secreted mucopolysaccharide, does exist in some rotifers, but the intraepidermal fibrous component is always present. In many species it is thickened to form a **lorica:** a "house" into which parts of the body that must be extended for locomotion, feeding, or attachment can be withdrawn (Figure 10.4, A). In certain genera, projections of the body wall form movable spines or paddles; these provide protection against predators. In rotifers that bend freely and lengthen or shorten by telescoping, the fibrous layer is thinned out at intervals, permitting the body wall to fold back on itself.

The majority of rotifers are freshwater animals, found in habitats as different as lakes and ponds, seepage areas, bird baths, soil, and moss. There are, however, numerous species in the sea and in brackish-water habitats. Rotifers show adaptations to a wide variety of life styles. Most of them creep freely over vegetation or over sessile animals, or through sediment and slimy organic coatings. Many are strictly planktonic; some are permanently attached. There are also commensal species that profit from living in close association with other animals, but that do not harm their hosts. Relatively few are truly parasitic.

More than three-fourths of all species of rotifers belong to the order Ploima, which is represented in fresh water as well as in the sea. The following account of rotifer morphology applies to females of several common genera of this group, such as *Epiphanes* and *Proales*. Males, when they occur, are small and aberrant; they will be discussed under the section on reproduction below.

General Structure

The **head** region of a generalized member of the order Ploima is slightly demarcated from the **trunk,** which makes up most of the body and which tapers posteriorly to a slender **foot** (Figure 10.1; Figure 10.2). Within the foot are the **pedal glands** whose ducts deliver a sticky secretion to a pair of **toes.** *Epiphanes* and other rotifers that have this equipment can attach themselves, pivot, and stretch while they extend the corona. In some species, the toes are decidedly unequal. The foot is usually divided superficially into two or more "segments," and the trunk may have similar annulations.

Corona and Digestive Tract

The mouth typically is located ventrally at the anterior end of the body, and is partly or wholly surrounded by a ciliated area called the **buccal field.** This is essentially part of the corona, and in certain genera there is little or no coronal ciliation that is distinct from the buccal field. What is believed to be the ancestral plan of coronal ciliation is shown in Figure 10.3. Note that the buccal field is continuous with a **circumapical band** that encircles the anteriormost part of the head region. The unciliated area enclosed by the band is called the **apical field.**

The ancestral plan is characteristic of several genera of ploimate rotifers. Usually, however, the circumapical band and buccal field are more or less separate, and one or both may show various specializations. The circumapical band, for instance, is often differentiated into two circles of cilia (Figure 10.4, A). The more nearly

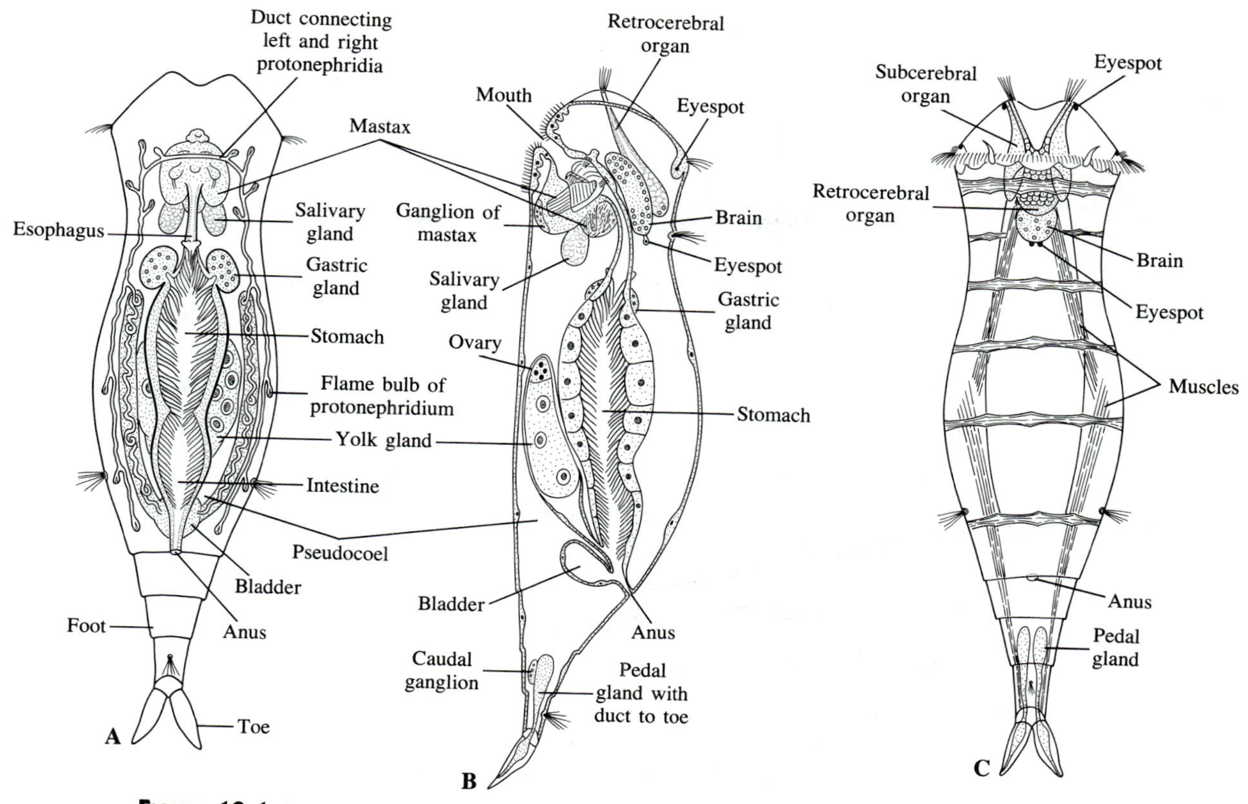

FIGURE 10.1 General morphology of a rotifer of the class Monogononta, order Ploima; diagrammatic. A. Ventral view, showing digestive system and protonephridia. B. View from left side, showing the digestive system and other structures located near the midline. C. Dorsal view, showing nervous system, sense organs, and some important muscles. (After Remane, in Bronn, Klassen und Ordnungen des Tierreichs, *4*:2:1.)

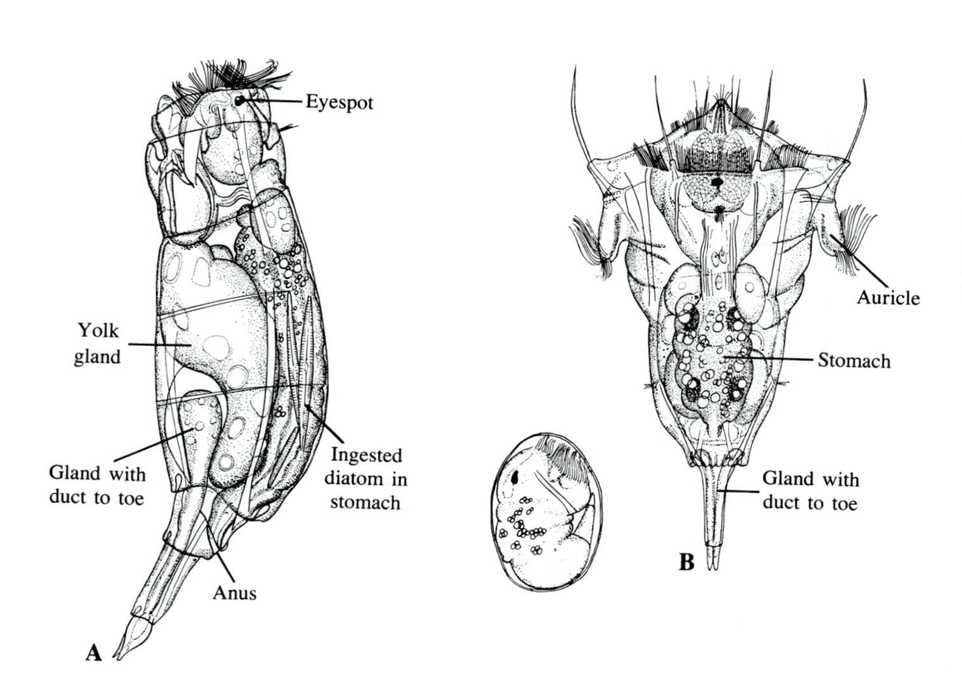

FIGURE 10.2 Class Monogononta, order Ploima. Two marine species. A. *Proales reinhardti*, lateral view from the left side. B. *Synchaeta vorax*, dorsal view, and an egg containing a juvenile. (Hollowday, Journal of the Marine Biological Association of the United Kingdom, *28*.)

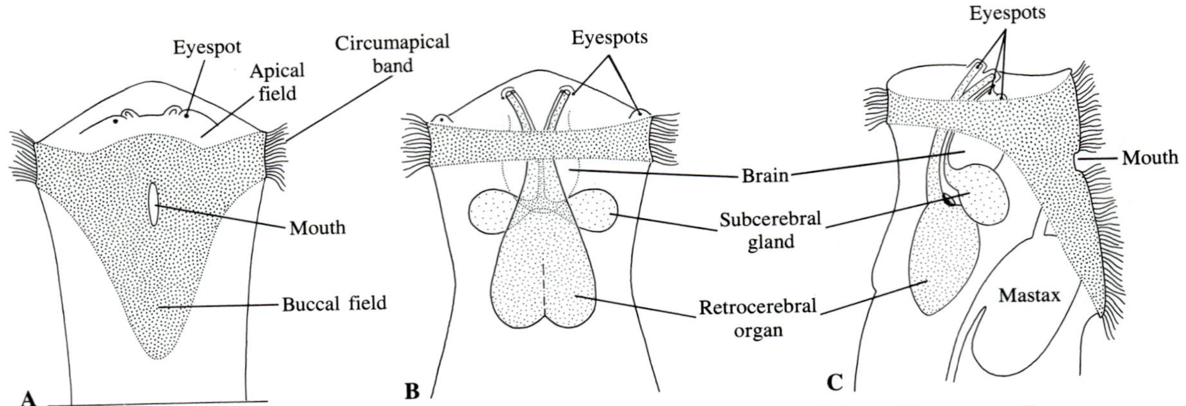

Figure 10.3 Diagram illustrating the presumed ancestral condition of the rotifer corona. Some genera of the order Ploima exhibit a plan that is similar to this, although certain tracts of cilia are distinct from others. A. Ventral view. B. Dorsal view. C. Lateral view from right side. (After de Beauchamp, Archives de Zoologie Expérimentale et Générale, *6*.)

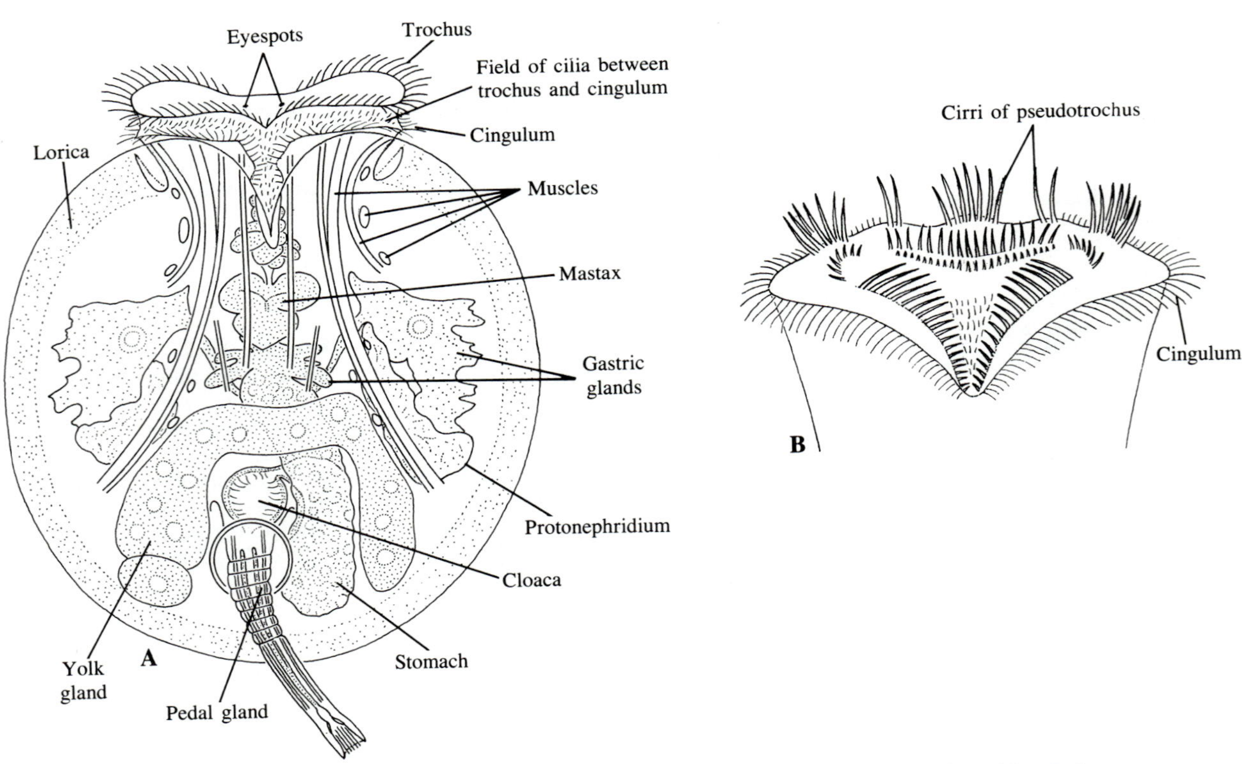

Figure 10.4 Class Monogononta, order Ploima. A. *Testudinella patina,* a rotifer with a lorica into which the corona and foot can be withdrawn. Only the more condensed portions of the two protonephridia are shown. These portions have flame bulbs, but there are also flame bulbs in the anterior part of the body. (After Seehaus, Zeitschrift für Wissenschaftliche Zoologie, *137*.) B. *Epiphanes senta,* ventral view of the coronal region. (After de Beauchamp, Archives de Zoologie Expérimentale at Générale, *6*.)

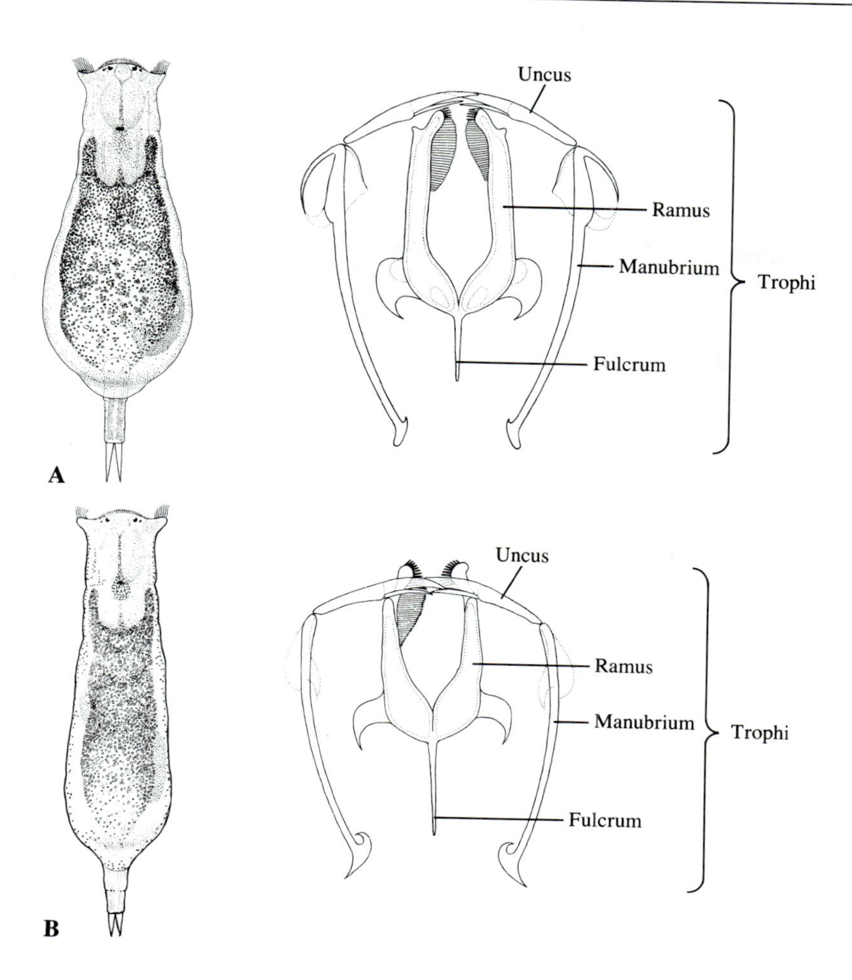

FIGURE 10.5 Class Monogononta, order Ploima. A. *Itura chamadis* and the sclerotized components of its mastax. B. *Itura viridis*. (Harring and Myers, Transactions of the Wisconsin Academy, *23*.)

anterior circle is the **trochus** and the more nearly posterior one is the **cingulum.** Another feature noted in certain genera is the concentration of some of the circumapical cilia into a pair of lateral tufts, or "auricles" (Figure 10.2, B). There are also ploimate rotifers in which the trochus is greatly reduced or absent. Among other variations noted in the buccal field is the aggregation of many of the cilia into stout, bristlelike structures called **cirri** (Figure 10.4, B). Collectively, the cirri, which are believed to be sensory, form a complex called the **pseudotrochus,** not exactly homologous to a trochus.

The mouth leads into the **mastax.** This is a pharynx derived from a stomodeal invagination. It has radial musculature and a cuticular lining, as does the pharynx of most pseudocoelomate animals. The muscles are not part of the epithelium, however, and in this respect the rotiferan mastax is decidedly different from the pharynx of nematodes and gastrotrichs. In addition, the mastax has sclerotized, jawlike pieces, called **trophi** (Figure 10.5). There are basically seven of these, but they may be supplemented by additional pieces. Six of the trophi (unci, rami, and manubria) are arranged as pairs; the seventh (fulcrum) is unpaired. The trophi are operated by their own specialized muscles, which are striated.

The mastax is followed by a short esophagus and a prominent stomach that consists of rather large cells (Figure 10.1, B). Both the esophagus and stomach are ciliated. The two **gastric glands** that open into the anterior portion of the stomach presumably produce digestive enzymes or mucus, or both. The intestine, separated from the stomach by a constriction, becomes narrower as it approaches the anus, which is situated dorsally at the place where the trunk is succeeded by the foot (Figure 10.1, A–C).

PSEUDOCOEL AND PROTONEPHRIDIA

Most rotifers have a moderately spacious pseudocoel between the gut and body wall. The two **protonephridia** and the large ovary–yolk gland complex lie in this cavity (Figure 10.1, A and B). In the protonephridia, groups of **flame bulbs** and their tubules are parts of large, multinucleated cells. (The nuclei lie in the walls of the tubules.) Each bulb, from which several cilia originate, is joined to its tubule by one or two major cytoplasmic connections, as well as by numerous smaller connections that form a "palisade" around some cytoplasmic "pillars" that are peripheral to the cilia (Figure 10.6, B). As the cilia beat, they presumably create a negative pressure within the cavity in which they lie, so fluid enters this cavity through the spaces between the many small cytoplasmic channels. The main collecting duct runs to a **bladder** that opens into the hindmost part of the duct. This portion of the duct is therefore a **cloaca,** for it is used by two systems. The bladder is usually emptied at frequent intervals. The urine is hypotonic to the fluid in the pseudocoel, so it is certain that the protonephridial system is concerned with osmoregulation. It could conceivably eliminate nitrogenous wastes, too. In some rotifers the right and left protonephridia are joined by a cross-connection.

MUSCULATURE

The musculature falls into three main categories. In the first category are muscles that are part of the internal organs, including the pharynx, stomach, intestine, bladder, oviduct, and other visceral structures. The muscles of the pharynx, already mentioned above, are especially well developed. The most obvious muscles of rotifers are those of the second category (Figure 10.1, C). They form discrete bands closely associated with the epidermis, and they may be called body-wall muscles. They are striated and are organized basically into longitudinal and circular components. In certain rotifers the circular musculature consists to a large extent of complete or nearly complete rings; in others it consists of bands that run transversely or dorsoventrally. The body-wall muscles, whatever their arrangement, are the ones that effect bending movements and retraction of structures such as the toes and corona. They also operate to increase hydrostatic pressure within the animal, thus bringing about extension of retractable structures. The muscles of the third category run from the body wall to various internal organs, including parts of the gut.

NERVOUS SYSTEM AND SENSE ORGANS

The principal ganglion, large enough to be called a **brain,** is dorsal to the pharynx (Figure 10.1, B and C). It is joined by connectives to a prominent ganglion beneath the pharynx, and to a ganglion located in the foot. There may be some smaller swellings along the connectives as well as on nerves arising from the connectives or from the major ganglia.

The more obvious sense organs of rotifers are eyespots (Figures 10.1, 10.2, 10.3; Figure 10.8, B) and

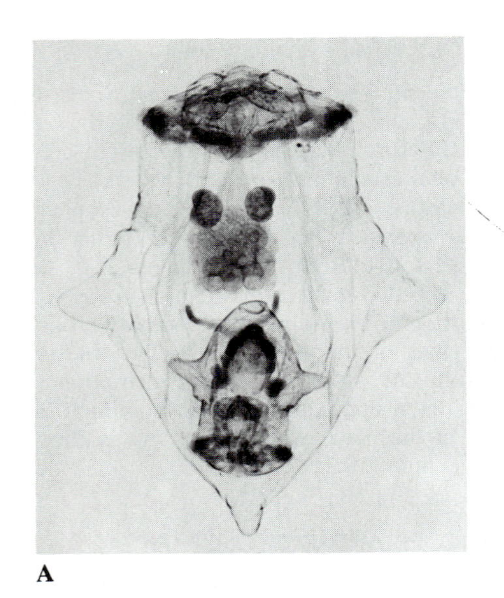

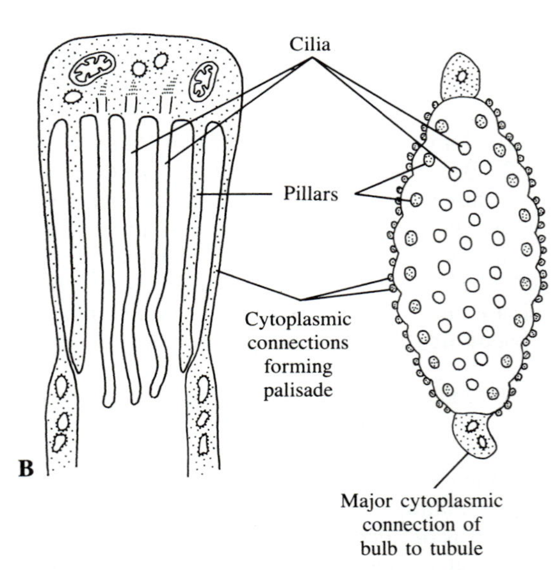

Figure 10.6 Class Monogononta, order Ploima. *Asplanchna*. A. Stained specimen; photomicrograph. B. Single flame bulb, sectioned lengthwise (left) and transversely (right). (After Warner, from Wilson and Webster, Biological Reviews, *49*.)

aggregates of nonvibratile cilia. The number of eyespots varies. There is almost always one, or a pair, closely associated with the brain, sometimes embedded in it. The pigment is generally reddish. Some rotifers have two smaller eyespots on the corona. Sense organs consisting of cilia likewise vary in position and prominence, but when they are present, there is usually one dorsal to the brain (Figure 10.1, B) and one on each side of the body lateral to the stomach (Figure 10.1, A and C). Similar structures may be located among the vibratile cilia of the corona. In certain rotifers, sensory cilia are situated on prominent stalks called antennae.

The **retrocerebral organ,** an internal structure usually located behind the brain, sometimes above it, opens onto the corona by a pair of ducts (Figure 10.1, B and C; Figure 10.3, B and C). Of the various functions that have been ascribed to it, the most logical is secretion of mucus. In one ploimate rotifer it has in fact been shown to have mucus-secreting components. The **subcerebral organ,** a single or double structure lying beneath the brain of many rotifers, also has ducts running to the coronal area; the ducts sometimes closely parallel those of the retrocerebral organ. The function of the subcerebral organ is unknown.

Feeding

Most ploimate rotifers, including *Epiphanes* and *Proales,* feed on bacteria, unicellular algae, and other small particles that are in suspension. Some genera, however, are partly or wholly carnivorous. *Asplanchna* and *Synchaeta,* for instance, consume ciliates, rotifers, and various other animal organisms. The carnivorous types generally use the two larger sclerotized pieces **(rami)** of the mastax to grasp their prey. The tips of the rami are extended from the mouth and brought together like the tips of a pair of forceps. Then the prey is pulled back into the pharynx and mangled. The larger indigestible residues are usually eliminated through the mouth.

In certain types, the tips of the rami are used to make a hole in the prey, so that the juices can be sucked out without drawing the prey into the pharynx. This is the case in *Chromogaster,* which feeds on certain dinoflagellates. Some ploimate rotifers, such as *Notommata* and *Cephalodella,* use the rami to bite into the cells of filamentous green algae. They then withdraw cell fluid by suction.

Reproduction and Development

In the order Ploima, females commonly reproduce by **parthenogenesis;** their diploid eggs develop, without fertilization, into females. The yolk-producing portion of the ovary–yolk gland complex is generally larger than the ovary proper. Eggs moving from the ovary and past the yolk gland grow considerably as they pick up nutritive material. Full-sized eggs reach the cloaca by way of a short oviduct, and are then laid. Development is direct. Some genera are **ovoviviparous;** their eggs develop into little rotifers while they are still in the oviduct.

In most ploimate rotifers, there is also sexual reproduction. The onset of this is related to environmental conditions, particularly to the type of food available. Females slightly different from the usual type appear, and although these are diploid, they produce haploid eggs which, unless they are fertilized, develop into haploid males. The males are small, short-lived, and unable to feed, for they have only vestiges of the gut. The testis occupies much of the body. The sperm duct, sometimes provided with accessory glands, leads to a pore near the posterior end or to a distinct penis that projects from the posterior part of the body.

When males mate with females, they either deposit sperm in the cloaca or somehow inject them through the body wall and into the pseudocoel. The zygotes resulting from fertilization of haploid eggs are of course diploid, and they develop into strictly parthenogenetic females. The zygotes are usually thick-shelled and are at least moderately resistant to unfavorable conditions. They are often called *resting eggs* because they remain dormant for an extended period. Diploid eggs that develop parthenogenetically, and also the haploid eggs that develop into males, have a thin shell.

In many accounts of rotifer reproduction, the terms *mictic* and *amictic* are used in connection with the two different types of eggs or the females that produce them. *Mictic* alludes to ''mixing'' of genes, and refers to females that can mate, or to the haploid eggs they produce. *Amictic* means the opposite, and is applied to parthenogenetic females and also to their eggs.

Classification

The classification of rotifers is not dealt with in detail here. The main reason for describing a few of the fundamental features of each class and order is to illustrate some of the diversity within the phylum.

Class Monogononta

Monogonont rotifers have just one ovary–yolk gland complex. The group includes the large order Ploima and contains about nine-tenths of all species in the phylum. The basic features of generalized members of the Ploima have been described already, so only a few rather different types are mentioned here. The order includes many genera with a well-developed lorica, and this often has spinelike outgrowths, ridges, and other elaborations. The head, foot, and toes stick out of the lorica when the animal is active, but they can be quickly withdrawn. The general structure and ornamentation of the lorica are important in classification.

Asplanchna (Figure 10.6, A) is a rather common planktonic genus that has neither foot, toes, nor lorica. It is one of the genera whose gut is blind, there being no intestine or anus. The various species of *Asplanchna* feed on small invertebrates, including other rotifers, and also on one-celled algae that are swimming or in suspension. Females produce fully formed young; the large eggs develop in a dilated portion of the oviduct. (Many other genera are also ovoviviparous, and the habit is not limited to the Ploima or even to the Monogononta.) Because it is transparent, *Asplanchna* is an excellent subject for laboratory study, either alive or in stained preparations.

There are two other orders of monogonont rotifers. The Flosculariacea (Figure 10.7, A and B) have no toes. They generally secrete a tube or a soft, gelatinous covering, and they are usually attached by a sticky disk. In the corona, which is sometimes circular, sometimes lobed, the more prominent cilia are typically organized into a trochus and cingulum (Figure 10.7, B). Between these components is a groovelike field of shorter cilia, and there are cilia on most of the rest of the buccal field.

The Collothecacea, of which *Collotheca* (Figure 10.7, C) and *Stephanoceros* (Figure 10.7, D) are representative genera, are attached to a firm substratum and enclosed by a gelatinous tube. The corona is unusual in that it is drawn out into five (rarely seven) lobes. These are provided with long, bristlelike cilia, and they also have typical short cilia on their inner faces. Below the bases of the lobes, the corona resembles a funnel, with the mouth at the bottom. Protozoans and other small animals are trapped when the bristly lobes turn inward, forming a cage. The prey is then taken into the pharynx. In a few genera, including *Acyclus* and *Cupelopagis*, there are no bristles or other cilia on the corona, and the lobes are reduced. The corona is nevertheless used for capturing prey.

CLASS DIGONONTA

In the Digononta, females have a pair of ovaries. Except for that, the two orders in this class are very different. The order Seisonacea consists of a few species belonging to a single genus, *Seison* (Figure 10.8,

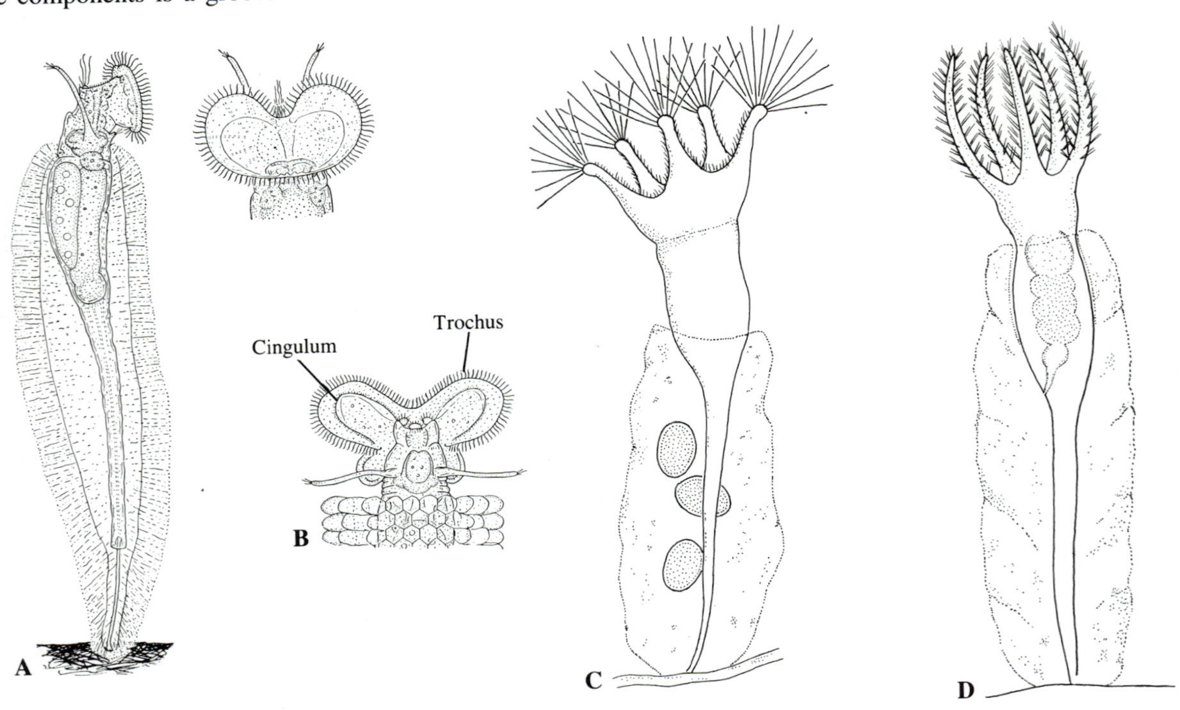

FIGURE 10.7 Class Monogononta. A. *Ptygura tacita* (order Flosculariacea), lateral view (left) and dorsal view of the coronal region (right). B. *Floscularia decora*, ventral view of the coronal region. The tube consists of pellets of gelatinous material. (A and B, Edmondson, Transactions of the American Microscopical Society, 59.) C. *Collotheca* (order Collothecacea); details of internal anatomy omitted. D. *Stephanoceros* (order Collothecacea); most details of internal anatomy omitted.

A). They are found on marine crustaceans of the order Leptostraca. These rotifers lack toes, but they can attach themselves tightly to their hosts. The body is slender and highly contractile, and the corona is small. There is neither a lorica nor a secreted tube. Males and females occur together, and reproduction is strictly sexual. In females, there are no prominent yolk glands associated with the ovaries; the ovaries themselves produce the nutritive material that becomes incorporated into the enlarging oocytes.

All other digonont rotifers belong to the order Bdelloidea. These are common in fresh water and in moss. Representative genera are *Philodina* (Figure 10.8, B) and *Habrotrocha*. Although bdelloid rotifers can swim, they are mostly creeping types that move in a leechlike fashion. The fibrous layer of the epidermis is thinned out at intervals, so the body is superficially divided into a number of "segments." Bdelloids are thus capable of bending freely, as well as of telescoping to a certain extent.

The corona is typically well developed, and is interesting because the trochal cilia of the circumapical band are organized into separate right and left components. The two circles may be complete or they may be interrupted medially, and they are often situated on disklike structures that can be retracted. When the cilia of the two circles are actively beating, a bdelloid rotifer may indeed appear to have two wheels turning at its anterior end.

The trochal cilia, which are generally large, produce a backward-directed current. The cilia of both the trochus and the cingulum drive food particles down into a groove that lies between these two components of the corona, and small cilia within the groove carry the particles to the mouth (Figure 10.8, C). The mastax not only grinds up the food that enters the pharynx but also operates as a sort of pump that draws particles into the pharynx.

The foot may have two, three, or four retractile toes. There may also be a couple of nonretractile accessory

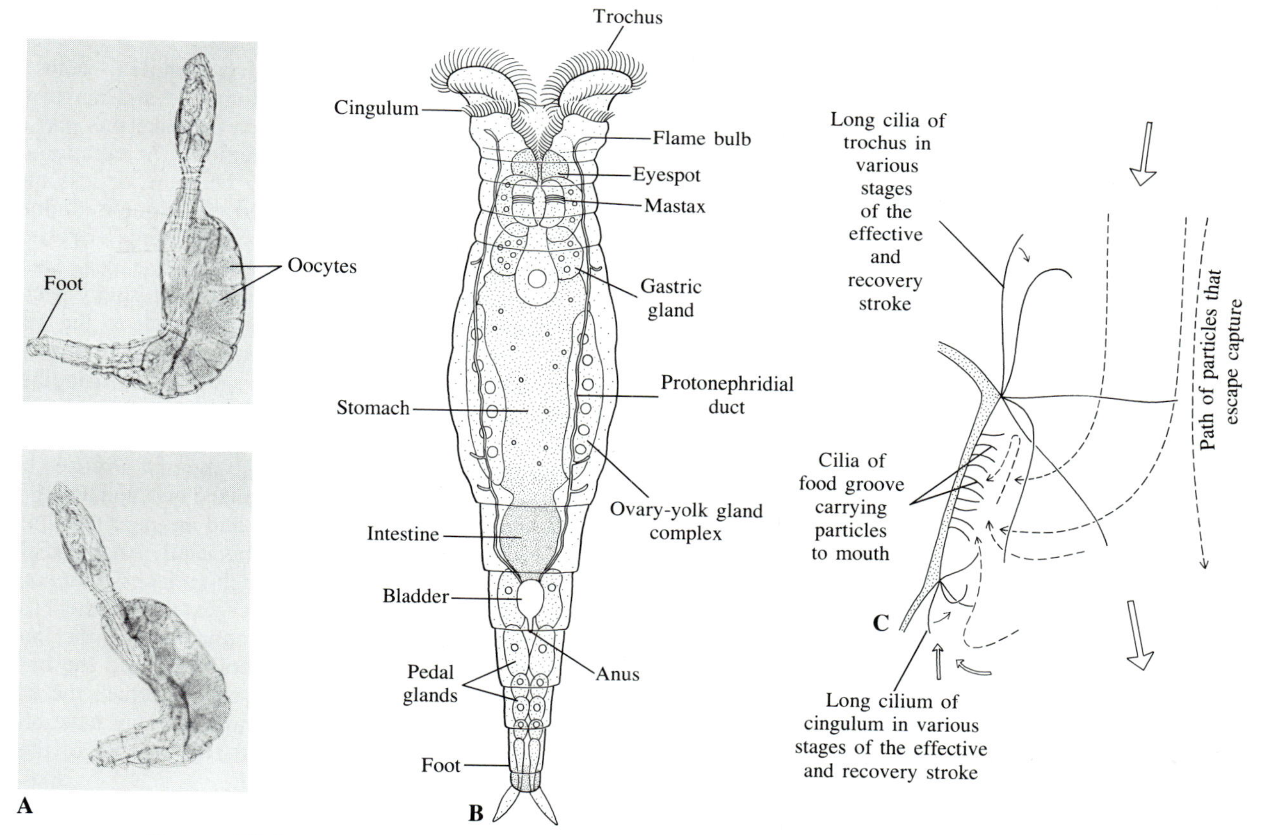

FIGURE 10.8 Class Digononta. A. *Seison* (order Seisonacea), female, almost fully extended (above) and contracted (below); photomicrographs. B. *Philodina* (order Bdelloidea), ventral view. C. Corona of a bdelloid rotifer, optical section, showing major currents (larger arrows of two sizes) and pathways followed by prospective food particles (broken arrows). (C, after Strathmann, Jahn, and Fonseca, Biological Bulletin, *142*.)

"spurs" on the foot, anterior to the toes. Males are totally absent in this group, so females must always reproduce parthenogenetically. Each of the two ovaries has a somewhat separate yolk gland associated with it. An interesting aspect of many bdelloid rotifers is their capacity to withstand desiccation. They become totally inactive until moisture becomes available again. Their ability to go into a state of dormancy compensates for their inability to produce resting eggs, and enables them to survive in moss that dries out periodically.

References on Rotifera

Aloia, R. C., and Moretti, R. L., 1973. Mating behavior and ultrastructural aspects of copulation in the rotifer *Asplanchna brightwelli*. Transactions of the American Microscopical Society, *92:*371–380.

Birky, C. W., and Gilbert, J. J., 1971. Parthenogenesis in rotifers: the control of sexual reproduction. American Zoologist, *11:*245–266.

Clément, P., and Wurdak, E., 1984. Photoreceptors and photoreception in rotifers. *In* Ali, M. A. (editor), Photoreception and Vision in Invertebrates. Plenum Press, New York.

Donner, J., 1966. Rotifers. Frederick Warne and Company, London.

Dumont, H. J., and Green, J. (editors), 1980. Rotatoria: Proceedings of the 2nd International Rotifer Symposium. Hydrobiologia, *73:*1–263.

Gilbert, J. J., 1974. Dormancy in rotifers. Transactions of the American Microscopical Society, *93:*490–513.

Gilbert, J. J., 1980. Developmental polymorphism in the rotifer *Asplanchna sieboldi*. American Scientist, *68:*636–646.

King, C. E. (editor), 1977. Proceedings of the First International Rotifer Symposium. Ergebnisse der Limnologie, in Archiv für Hydrobiologie, Beiheft 8.

Salt, G. W., Sabbadini, G. F., and Commins, M. L., 1978. Trophi morphology relative to food habits in six species of rotifers (Asplanchnidae). Transactions of the American Microscopical Society, *97:*469–485.

Strathmann, R. R., Jahn, T. L., and Fonseca, J. R. C., 1972. Suspension feeding by marine invertebrate larvae: clearance of particles by ciliated bands of a rotifer, pluteus, and trochophore. Biological Bulletin, *142:*505–519.

Thane-Fenchel, A., 1968. A simple key to the genera of marine and brackish-water rotifers. Ophelia, *5:*299–311.

Wilson, R. A., and Webster, L. A., 1974. Protonephridia. Biological Reviews, *49:*127–160.

Phylum Gastrotricha

Introduction

Gastrotrichs, whose name ("hairy belly") refers to the ciliated ventral surface, are found both in marine and freshwater habitats. About 500 species have been described. Freshwater gastrotrichs are generally common in sediment and in slimy coatings on submerged vegetation, sticks, and stones. Some of them are easily cultured, and live specimens can be purchased for classroom study. Marine species are most likely to be found in sediment or in washings of beach sand.

In spite of the fact that they are multicellular and have some fairly complicated internal structures, gastrotrichs tend to be small and compact. The freshwater species, which belong to the class Chaetonota, do not often exceed a length of 200 μm. The majority of them are about 150 μm long, the size of a large paramecium. Marine gastrotrichs, which belong partly to the Chaetonota but mostly to the class Macrodasya, have a wider size range, and the largest are approximately 1.5 mm long.

The following account of gastrotrich morphology and biology deals largely with freshwater chaetonotans, because these are the ones most likely to be studied in the laboratory of formal courses.

Class Chaetonota

The shape of a generalized chaetonotan, such as *Chaetonotus* (Figure 10.9), is elongated and narrowed sufficiently behind the head region to make this part of the body distinct from the trunk region. The ciliation on the flattened ventral surface may be more or less uniform, or it may be concentrated in two longitudinal bands or in several to many transverse bands or rows.

At the posterior end of the body are two toes. These are provided not only with glands whose sticky secretion enables the gastrotrich to attach itself to the substratum, but also with glands that secrete a "releaser" that breaks the sticky bond. Thus there is a **duogland system** comparable to that of flatworms (see Chapter 6).

The true **cuticle** that covers all external surfaces has two main portions: an outer lamellar layer consisting of successive membranous sheaths, and an inner fibrous or granular layer next to the epidermis itself. An especially interesting feature of the gastrotrich cuticle is that some of the sheaths of the outer layer extend to the cilia, wrapping around them for their entire length; the inner layer may also form a kind of collar around the basal portions of the cilia. Depending on the genus, the dorsal and lateral surfaces of a chaetonotan may have bristles, spines, or scales (Figure 10.10), and all of these elaborations are part of the cuticle.

The mouth, which is terminal or turned slightly downward, is encircled by cuticular bristles. Preceding the pharynx, there may be a short, cuticle-lined buccal cavity; when present, this generally has ridges or teeth, and it may be eversible. The pharynx proper, which is

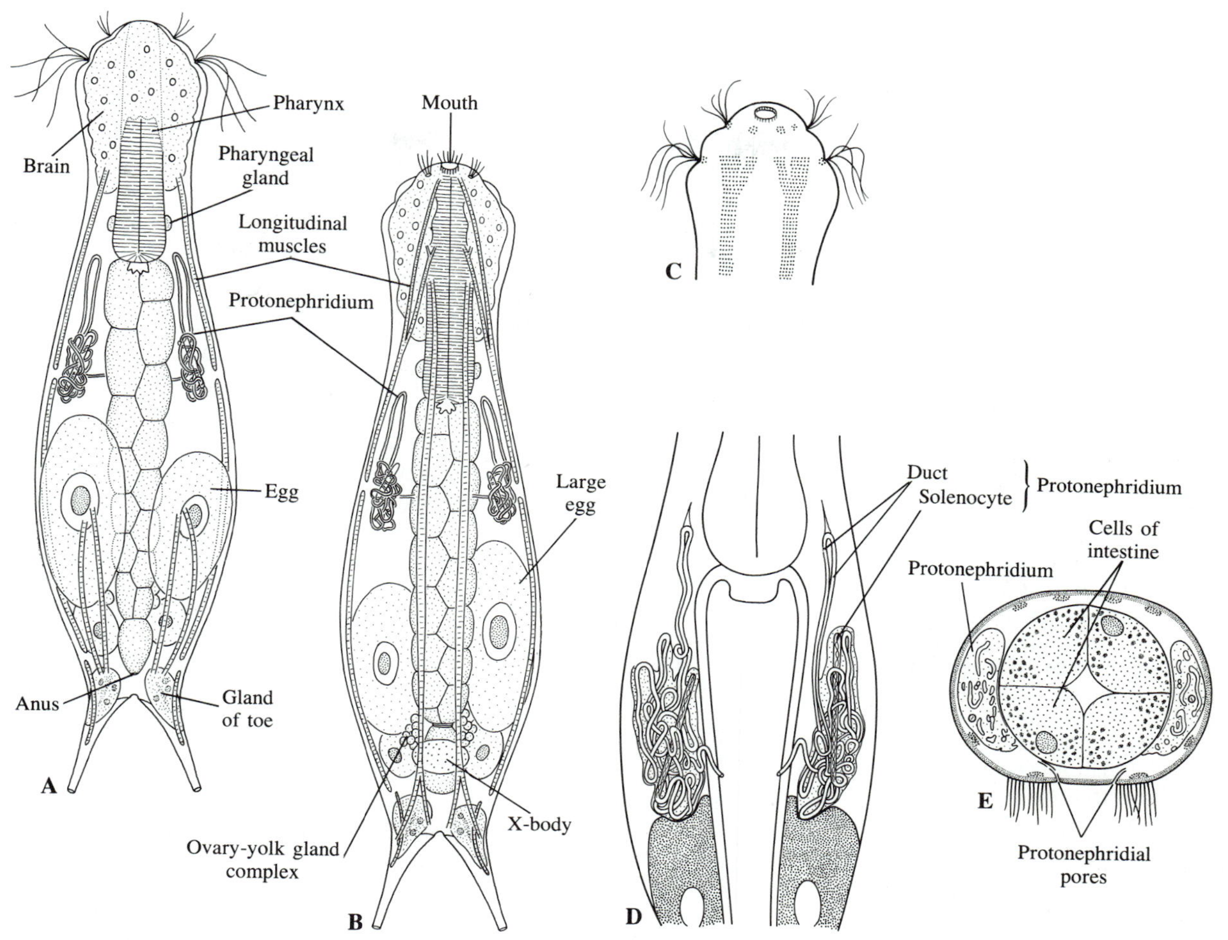

FIGURE 10.9 Class Chaetonota. *Chaetonotus*. A. Dorsal view. B. Ventral view. C. Anterior portion, showing distribution of cilia on ventral surface. D. Protonephridia. (See also Figure 10.11.) E. Transverse section at the level of the protonephridial pores. (After Remane, in Bronn, Klassen und Ordnungen des Tierreichs, *4*:2:1.)

derived from a stomodeal invagination of ectoderm, is also lined by a cuticle. The histology of the pharynx is much like that of nematodes in that it has well-developed radial muscles that control the shape of the lumen. These muscles, as in nematodes, originate from the epithelium itself. When the muscles contract, the lumen is dilated; when they relax, the lumen closes, usually becoming triradiate in cross-section. Two pairs of salivary glands are applied to the outside of the wall of the pharynx and have fine ducts leading to the lumen.

The rest of the gut consists of a thin-walled intestine and a short hind portion, usually called the rectum. The anus is on the ventral surface, nearly but not quite terminal. The food of freshwater chaetonotan gastrotrichs is made up of bacteria, small protozoans and algae, and fine particles of organic material.

The longitudinal muscles, two protonephridia, and the reproductive system are crammed into the space between the gut and body wall. Thus, although there is an obvious blastocoel in the embryo, this cavity does not persist as a pseudocoel. The muscles are in discrete bands. There are several pairs of them, and they function in shortening and bending the body. In addition, there are small muscles that operate the toes, bristles, and some other structures.

Both of the protonephridia consist of two closely associated **solenocytes** (Figure 10.11). Osmoregulatory cells of this type, found in several phyla, differ from

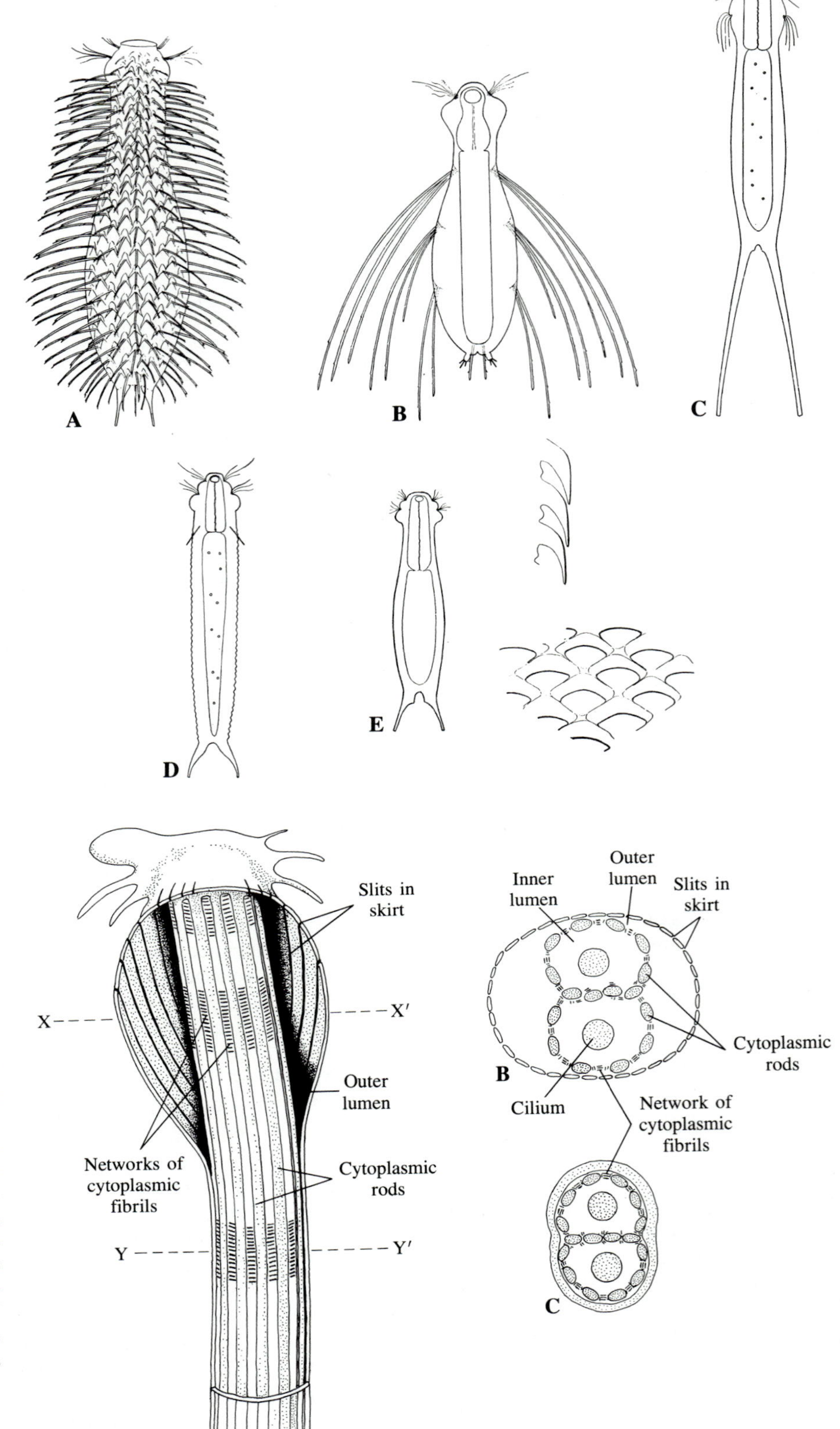

FIGURE 10.10 Class Chaetonota. External features of representative species. A. *Chaetonotus murrayi*. (Remane, in Bronn, Klassen und Ordnungen des Tierreichs, *4*:2:1.) B. *Stylochaeta scirteticus*. C. *Polymerurus callosus*. D. *Ichthyidium sulcatum*. E. *Lepidodermella squamata,* and lateral (top) and surface (bottom) views of the scales that cover much of the body. (B–E, Brunson, Transactions of the American Microscopical Society, *69*.)

FIGURE 10.11 Class Chaetonota. *Chaetonotus,* protonephridium; diagrams based on electron micrographs. A. Portion of a protonephridium, some of the skirt cut away, showing the palisade of cytoplasmic rods originating from one of the two solenocytes. Between each two rods is a network of cytoplasmic fibrils. B. Transverse section through the proximal portion of the protonephridium, at the level indicated by the dashed line X–X′ in A. Where the cytoplasmic rods of the two sets interdigitate, they are pressed so closely together that the lumina around the two cilia are essentially separated. C. Transverse section at a more distal level, indicated by the dashed line Y–Y′ in A. (After Brandenburg, Zoologische Beiträge, new series, *12*.)

flame cells in that they have only one cilium. In *Chaetonotus,* the cilium of each solenocyte is surrounded by a palisade of cytoplasmic rods, and neighboring rods are connected by a network of delicate cytoplasmic fibrils. Where the rods of one palisade interdigitate with those of the other, they form a barrier between the lumina in which the cilia beat. Together, the two solenocytes form a skirt around the cilium–palisade complexes. In the bulbous proximal portion of the protonephridium, the skirt consists of strips of cytoplasm separated by narrow slits. It is believed that fluid enters a solenocyte through the slits, going first into the outer lumen. It then presumably passes through the networks between rods, thus reaching the inner lumina. Below the bulbous portion of a protonephridium, the skirt has no slits.

The nervous system is somewhat similar to that of many flatworms. There is a proportionately large, bilobed brain above the anterior part of the pharynx (Figure 10.9, A). From the brain, two lateral nerves extend backward in close association with the epidermis. Small nerves that originate from the brain and lateral nerves run to all parts of the body.

On the head region are tufts of cilia, as well as bristles that consist of modified cilia. These structures are believed to be sensory. There may also be eyespots. Among various other structures that are presumed to have a sensory function are pits that are partly eversible; when present, these are on the sides of the head.

REPRODUCTION AND DEVELOPMENT

Freshwater chaetonotans are parthenogenetic females. The main elements of the reproductive system are a combined ovary–yolk gland (sometimes divided into right and left lobes) and a structure called the **X-body.** The X-body accumulates secretions produced by some cells that are part of the gonad, but its function is not known. Most chaetonotans produce only a few eggs during their lifetime. When an egg is ready to be laid, it contains considerable yolk and may already be about half the size of the female. There seems to be no permanent genital pore; the egg is squeezed out through a temporary opening in the posteroventral part of the animal. As soon as the egg leaves the body, the coating around it hardens into an eggshell.

Two types of eggs are produced. A particular individual may, in fact, produce both types, though not at the same time. The eggs of one type are thin-walled and not resistant to extreme cold or desiccation; they begin to develop immediately. Thick-walled eggs can withstand freezing and drying and require a period of dormancy before they start to develop. The shells of these eggs often have prominent spines, warts, or hooks; such elaborations are not typical of thin-walled eggs.

Cleavage is more or less spiral at first, but the embryo soon shows bilateral symmetry. Gastrulation is accomplished by the sinking in of two cells, which presumably multiply to give rise to the digestive tract and mesodermal components of the body. Development is direct, and newly hatched individuals have most of the structures found in adults; only the reproductive system needs to develop further. Young that hatch from especially large eggs do not have to grow much to reach full size.

Some freshwater chaetonotans have clusters of sperm, but the significance of these male gametes is not understood. In *Lepidodermella squamata,* sperm appear only in certain individuals, and not until the period of parthenogenetic egg-laying has been completed. They are located on one or both sides of the body, just anterior to the female gonad. A few eggs may be laid after the sperm appear, but they either do not hatch or the juveniles that do hatch fail to develop normally. Another interesting feature of the reproductive biology of *L. squamata* is that enlargement of the X-body coincides with the appearance of sperm.

Marine chaetonotans usually have testes and are functional hermaphrodites. Their eggs are of the thin-walled type.

CLASS MACRODASYA

The class Macrodasya consists entirely of marine species. These are characterized by several to many **adhesive tubes** on both sides of the body (Figures 10.12, 10.13). Another interesting feature of macrodasyans is the presence of a pair of pores that connect the lumen of the pharynx with the outside. The pores presumably permit much of the water that is sucked in with food to leave the gut.

In chaetonotan gastrotrichs, and also in some macrodasyans, each ciliated epidermal cell has several cilia. In this respect, these animals are like flatworms and members of most other phyla that have cilia. There are, however, macrodasyans whose epidermal cells bear single cilia. Some zoologists view the monociliated condition as an ancestral trait, and as evidence that gastrotrichs may have close kinship with early bilateral animals. This argument is based to a considerable extent on the fact that sponge choanocytes and the epidermal cells of cnidarian planulae have single cilia.

The protonephridia of macrodasyans vary in their exact organization, but they are of the solenocyte type. In *Turbanella cornuta,* which has been extensively studied, each protonephridium has four solenocytes (Figure 10.14). Within the hollow of each solenocyte is a palisade of cytoplasmic rods, between which are networks of cytoplasmic fibrils. Thus, as in chaetonotans, the palisade separates an inner lumen, within which a

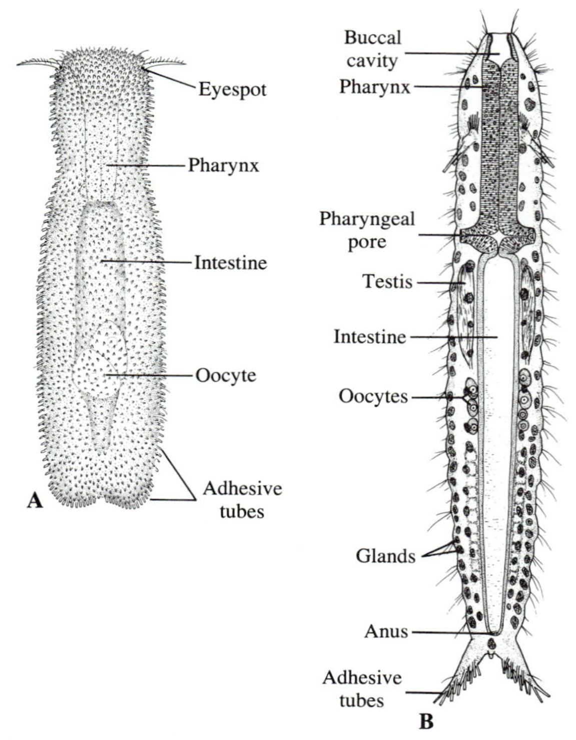

Figure 10.12 Class Macrodasya. A. *Platydasys tentaculatus*. (Swedmark, Archives de Zoologie Expérimentale et Générale, *93*.) B. *Paraturbanella pallida*. (Schmidt, Mikrofauna des Meeresbodens, *26*.)

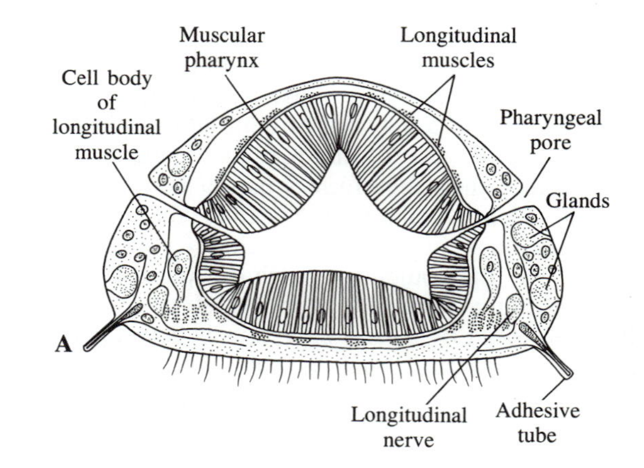

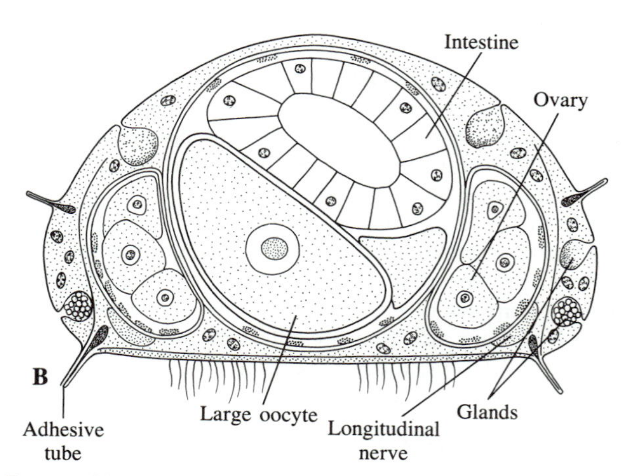

Figure 10.13 Class Macrodasya. A. Transverse section of a macrodasyan at the level of the pharyngeal pores; diagrammatic. B. Transverse section at the level of the intestine; diagrammatic. (After Remane, in Bronn, Klassen und Ordnungen des Tierreichs, *4*:2:1.)

cilium beats, from an outer lumen. The skirt of the solenocyte is perforated by pores and slits, and it is through these openings that fluid is believed to enter the outer lumen. The network of the palisade probably constitutes a second filter.

The four solenocytes of a protonephridium do not lead directly to a collecting tubule. Instead, their lumina are continued into chimneylike elevations of a large collecting cell. This is joined to an exit cell whose lumen carries fluid toward a pore. The cilium and two rodlike cytoplasmic structures within the lumen of the exit cell originate from the collecting cell. There are no continuous channels running through the collecting cell. It is thought that water and other substances in the process of being eliminated are moved through the collecting cell and into the lumen of the exit cell by active transport. Much of the cytoplasm of the collecting cell is occupied by small vacuoles, and it is likely that these are way stations in the movement of fluid through the cell. There is evidence that pinocytosis is involved in the take-up and elimination of fluid by the collecting cell.

Macrodasyans reproduce sexually and are essentially hermaphroditic, even if some of them function first as males, then as females. Their reproductive systems are rather complicated. In *Macrodasys* and certain other genera that have been intensively studied, there are two testes, one on each side of an unpaired median ovary. During mating, sperm are not transferred directly from the two male genital pores to the female pore. Instead, each of the partners bends in such a way that it takes up its own sperm into a so-called **caudal organ,** whose pore is just in front of the anus (Figure 10.15). Then the caudal organ is everted and used to pump sperm into the female genital pore of the partner. The sperm are stored in a seminal receptacle. After an egg has reached full size and has been fertilized, it is released through a rupture in the body wall.

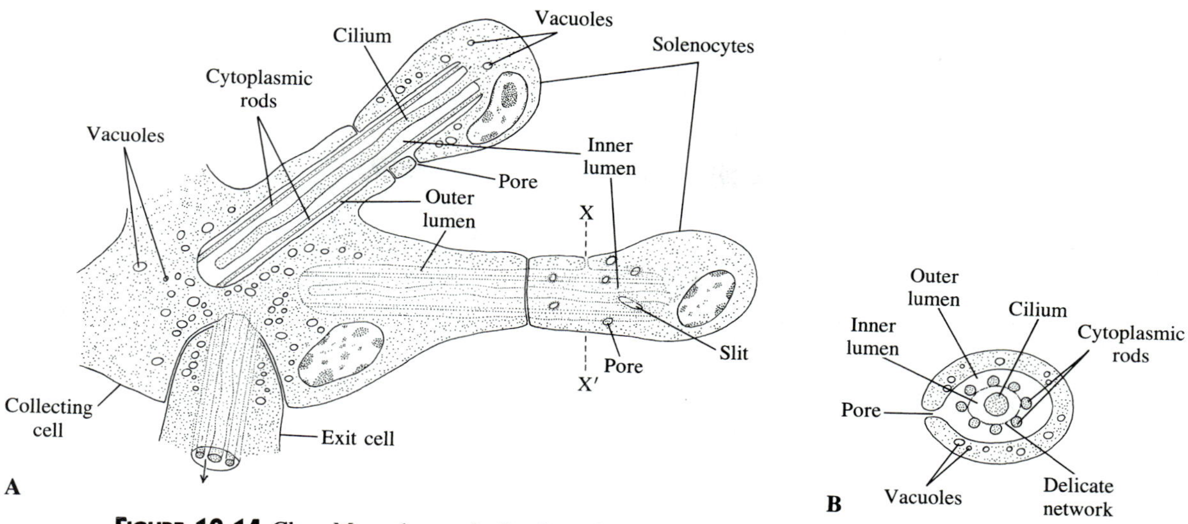

Figure 10.14 Class Macrodasya. A. Portion of a protonephridium of *Turbanella cornuta;* diagram based on electron micrographs. Only two of the four solenocytes associated with the collecting cell are shown; one is sectioned lengthwise. (Detailed explanation in text.) B. Transverse section of a solenocyte at the level indicated by the dashed line X–X′ in A. (After Teuchert, Zeitschrift für Zellforschung and Mikroskopische Anatomie, *136,* but simplified.)

Some macrodasyan gastrotrichs stick **spermatophores**—little packets of sperm—on the body of the partner. Sperm emerge from the packets and somehow locate the female pore. Other macrodasyans inseminate their partners directly, without intervention of a caudal organ or spermatophore.

References on Gastrotricha

Amato, A. J., and Weiss, M. J., 1982. Flexibility in the cuticular pattern of a cell-constant organism, *Lepidodermella squammata*. Transactions of the American Microscopical Society, *101*:229–240.

Brandenburg, J., 1966. Die Reusenformen der Cyrtocyten. Zoologische Beiträge, new series, *12*:345–417.

Brunson, R. B., 1950. An introduction to the taxonomy of the Gastrotricha with a study of eighteen species from Michigan. Transactions of the American Microscopical Society, *69*:325–352.

d'Hondt, J. L., 1971. Gastrotricha. Oceanography and Marine Biology, An Annual Review, *9*:141–192.

Hummon, M. R., 1983. Reproduction and sexual development in a freshwater gastrotrich. 1. Oogenesis of parthenogenetic eggs (Gastrotricha). Zoomorphology, *104*:33–41.

Hummon, W. D., 1966. Morphology, life history, and significance of the marine gastrotrich, *Chaetonotus testiculophorus* n. sp. Transactions of the American Microscopical Society, *85*:450–457.

Rieger, G. E., and Rieger, R. M., 1977. Comparative fine structure of the gastrotrich cuticle and aspects of cuticle evolution within the Aschelminthes. Zeitschrift für Zoologische Systematik und Evolutionsforschung, *15*:81–124.

Rieger, R. M., Ruppert, E. E., Rieger, G. A., and Schoepfer-Sterrer, C., 1974. On the fine structure of gastrotrichs with description of *Chordodasys antennatus* sp. n. Zoologica Scripta, *3*:219–237.

Ruppert, E. E., 1978. The reproductive system of gastrotrichs. II. Insemination in *Macrodasys:* A unique mode of sperm transfer in Metazoa. Zoomorphologie, *89*:207–228.

Ruppert, E. E., 1978. The reproductive system of gastrotrichs. III. Genital organs of Thaumastodermatinae nov. subfam. and Diplodasyinae nov. subfam. with discussion of reproduction in the family. Zoologica Scripta, *7*:93–114.

Ruppert, E. E., and Shaw, K., 1977. The reproductive system of gastrotrichs. I. Introduction with morphological data for two new *Dolichodasys* species. Zoologica Scripta, *6*:185–195.

Schmidt, P., 1974. Interstitielle Fauna von Galapagos. IV. Gastrotricha. Mikrofauna des Meeresbodens, number 26.

Schrom, H., 1972. Nordadriatische Gastrotrichen. Helgoländer Wissenschaftliche Meeresuntersuchungen, *23*:286–351.

Teuchert, G., 1968. Zur Fortpflanzung und Entwicklung der Macrodasyoidea (Gastrotricha). Zeitschrift für Morphologie der Tiere, *63*:343–418.

Teuchert, G., 1973. Die Feinstruktur des Protonephridialsystems von *Turbanella cornuta* Remane, einem marinen Gastrotrich der Ordnung Macrodasyoidea. Zeitschrift für Zellforschung und Mikroskopische Anatomie, *136*:277–289.

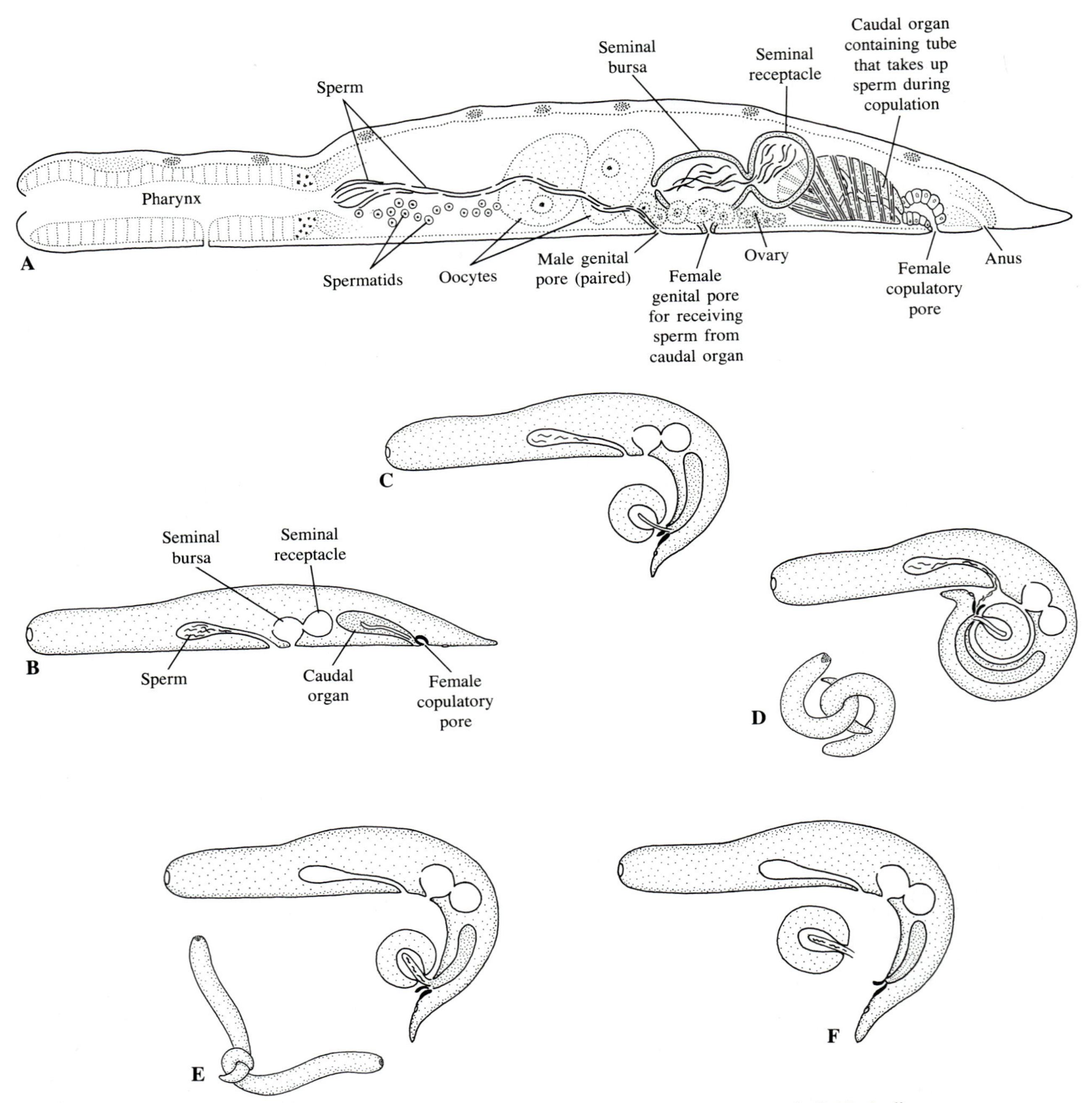

Figure 10.15 Class Macrodasya. A. Sagittal section of a sexually mature individual; diagrammatic. B–F. Stages in copulation and transfer of sperm from the caudal organ of one individual to the seminal bursa–seminal receptacle complex of the partner. Although the illustrations can only show insemination of one individual by another; the process is reciprocal. B. Precopulatory individual. C. Attachment of two individuals; tube of caudal organ everted and inserted into seminal bursa of partner. D. One individual of the pair taking its own sperm into the caudal organ. E. Transfer of sperm to the tube of the caudal organ. F. Separation of individuals; tube of caudal organ, containing sperm, detached and left in seminal bursa of the partner. (After Ruppert, Zoomorphologie, *89,* but modified.)

Tyler, S., and Rieger, G. E., 1980. Adhesive organs of the Gastrotricha. I. Duo-gland organs. Zoomorphologie, *95*:1–15.

Weiss, M. J., 1988. Stabilization in the name of the freshwater gastrotrich *Lepidodermella squamata,* with nomenclatorial corrections for congeneric species. Transactions of the American Microscopical Society, *107*:369–379.

Weiss, M. J., and Levy, D. P., 1979. Sperm in "parthenogenetic" freshwater gastrotrichs. Science, *205*:302–303.

Phylum Kinorhyncha

Introduction

The kinorhynchs constitute a small but distinctive group of strictly marine animals. There are nearly 100 described species, most of which inhabit the superficial layer of sediment that accumulates on muddy and sandy bottoms. A few live in relatively clean sand or other habitats. The genus most likely to be collected intertidally is *Echinoderes*. It is often found in protected bays where the substratum has a cohesive, diatom-rich coating that also favors many other small invertebrates, including turbellarians, nematodes, copepods, and ostracodes. The following description applies to any species of *Echinoderes* and at the same time illustrates the important features of the phylum.

General Structure

The body of a mature *Echinoderes* (Figures 10.16, 10.17)—about 250 to 350 μm long, depending on the species—is nearly cylindrical, though slightly flattened on the ventral side. It is divided into 13 units. These are now generally called **segments,** although in the past they were called **zonites** to avoid the connotation that they are exactly comparable to segments of coelomate animals such as annelids. The segmentation is nevertheless more than skin deep. Much of the musculature is segmental, or extends from one segment to the next, and the ventral nerve cord has a swelling in nearly every segment.

The first two segments make up what is arbitrarily designated the head; the other 11 segments form the trunk. Segment 1 consists of a bulbous basal portion that has several sets of complex spines called **scalids,** and a terminal cone on which the mouth is located; the mouth is encircled by a ring of short spines. In spite of its proportionately large size, segment 1 can be withdrawn into the trunk; when this happens, the scalids turn forward like the ribs of an umbrella that has been blown inside out. They can do this because their bases have socketlike articulations. Segment 1 is normally busy all the time, being alternately thrust out and pulled back. Each time the head is everted and its scalids are spread, *Echinoderes* pulls itself forward through the sediment. Its movements are slow and somewhat jerky. Segment 2 consists of 16 plates called **placids.** When segment 1 is withdrawn, the placids fit together to form a tight closure in front of it.

In segments 3 and 4—the first two segments of the trunk—the cuticle is differentiated into a dorsal plate, the **tergum** (or tergite), and a ventral plate, the **sternum** (or sternite). All succeeding segments have two sterna, separated along the midline. The trunk is flexible, because the cuticle is thinner at the sutures between segments (each trunk segment slightly overlaps the one behind it), and also where the tergum and sternum articulate. Bending movements are due to the action of muscles that run from segment to segment (Figure 10.18, B). There is usually one such muscle on each side of the midline, both dorsally and ventrally. Oblique muscles, extending from the tergum of one segment to the sterna of the next segment behind it, are present in at least certain portions of the body. The dorsoventral muscles, confined to individual segments (Figure 10.19, A), operate to pull the terga and sterna closer together, increasing pressure in the pseudocoel and thus assisting in eversion of the head.

Retraction of the head involves a complex of muscles, arranged in two circles around the gut; some of these extend backward for several segments (Figure 10.18, A). The muscles that pull the placids of segment 2 tightly together collectively form a ring. It would be out of place here to discuss in detail the various small muscles that control movable spines and other structures. It should be pointed out, however, that the muscles of kinorhynchs are, with few exceptions, of the **striated** type.

The epidermis has few cell boundaries and is apparently a syncytium. It secretes the cuticle, which contains considerable chitin and which has a variety of spinelike and hairlike differentiations. In general the spines are movable, at least to some extent, but the hairs are only flexible. Most trunk segments have a ventrolateral spine on each side and also a middorsal spine. Segment 13 has a pair of especially large spines, as well as smaller spines that are described below in connection with sexual dimorphism. The posterior margins of all trunk segments have complex fringes of fine hairs. On the ventral side of segment 4 is a pair of tubes from which sticky material exudes, and some of the other structures that have been assumed to be ventrolateral spines are also adhesive tubes. The function of the sticky secretion is not known.

The gut runs the full length of the body. The buccal cavity, into which the mouth opens, leads to a pharynx that has a cuticular lining, an epithelium, and a thick

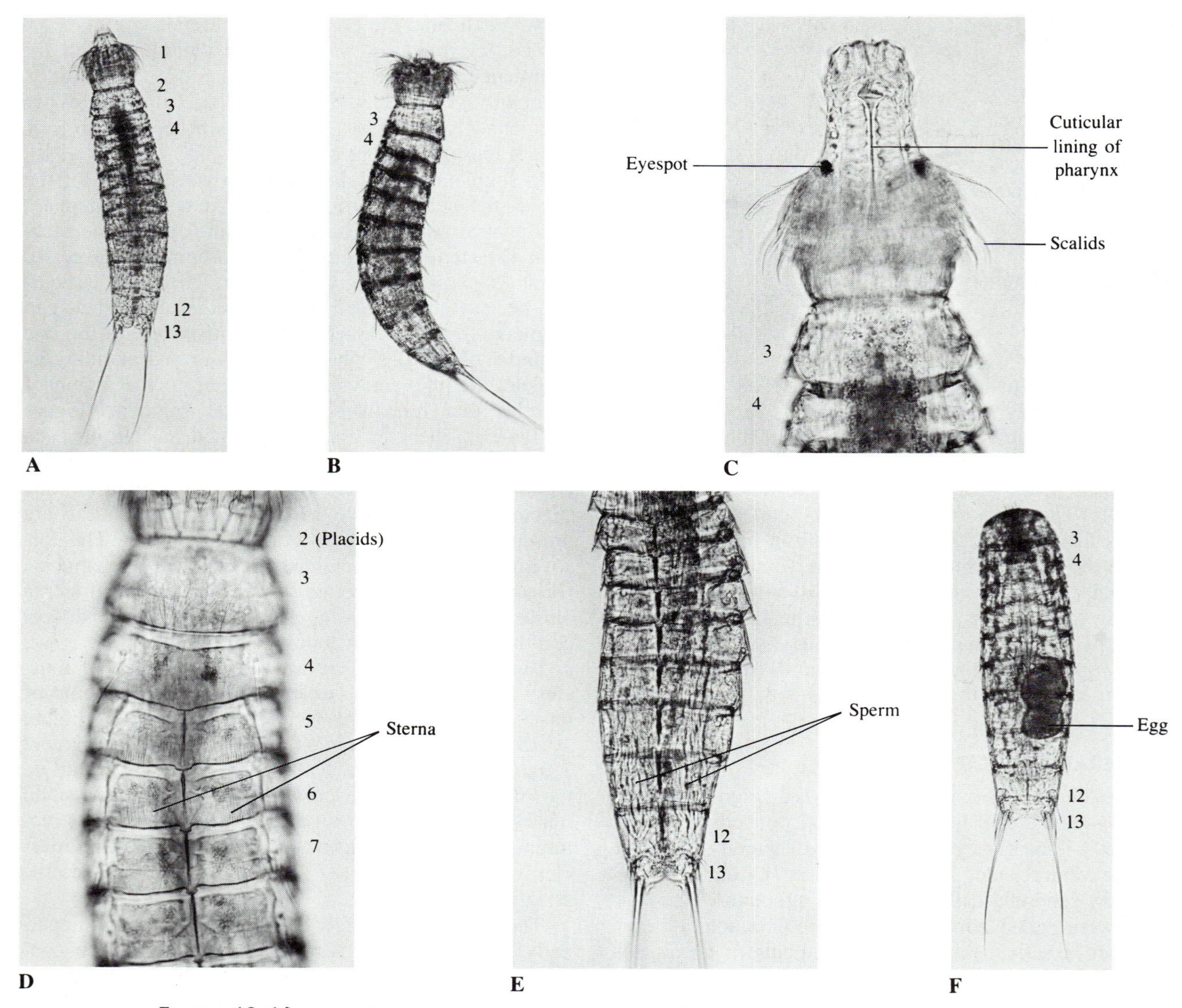

Figure 10.16 *Echinoderes kozloffi;* photomicrographs. (Numbers identify segments.) A. Male, dorsal view. B. Male, lateral view from right side. C. Anterior portion of body, optical section. D. Ventral view of middle portion of body. E. Male, posterior portion of body. F. Female, head withdrawn.

layer of radial muscles. Contraction of these muscles dilates the lumen. Although a short esophagus may be recognized, almost all of the rest of the gut consists of what may be called the **intestine** (or stomach–intestine). Its wall is made up of relatively large cells, bounded externally by a network of delicate muscle fibers. Between the intestine and the anus is a short hindgut. Intertidal species of *Echinoderes* feed almost exclusively on diatoms and other fine material that is swallowed along with diatoms. Kinorhynchs that live in deeper water, where living diatoms are not abundant, probably subsist mostly on detritus.

On both sides of the gut, there is a protonephridium of the solenocyte type (Figure 10.19, B). Studies on one species of *Echinoderes* have shown that each protonephridium consists of five cells (Figure 10.19, C). The pairs of cilia that arise from three of the cells, one of which is attached to the cuticle of the dorsal part of the tergum of segment 10, extend into a lumen that begins in the distal portion of the third cell, runs through

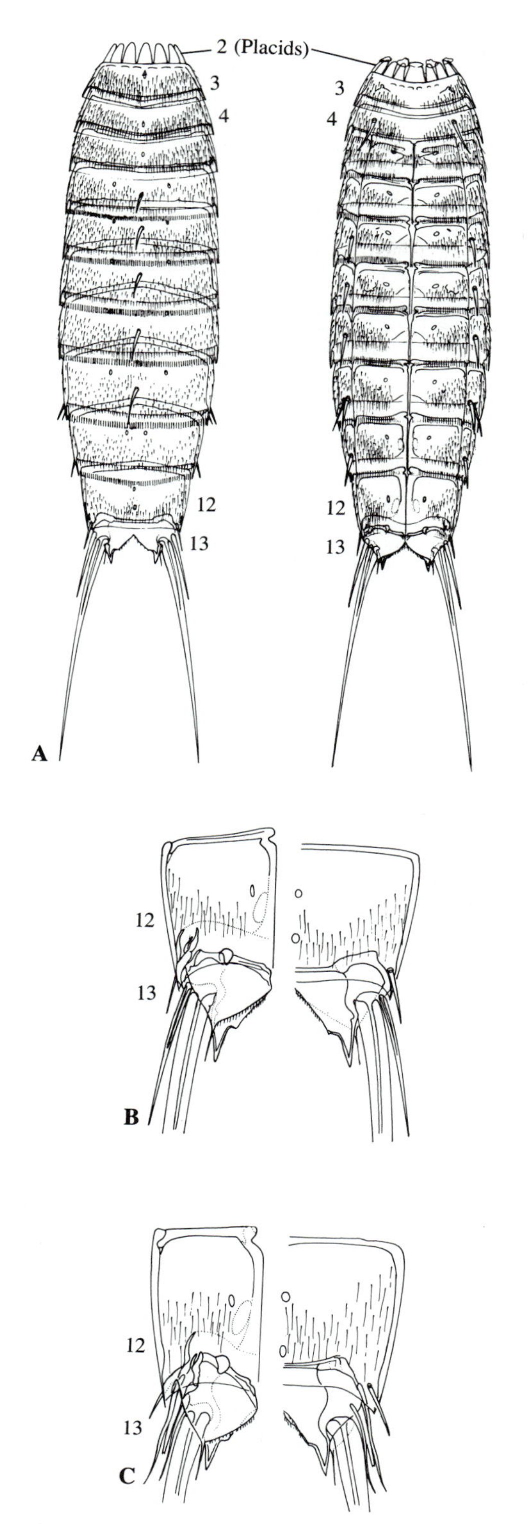

FIGURE 10.17 *Echinoderes dujardinii*. A. Female, ventral view (left) and dorsal view (right). B. Female, segments 12 and 13, ventral view (left) and dorsal view (right). C. Male, segments 12 and 13, ventral view (left) and dorsal view (right). (Higgins, Smithsonian Contributions to Zoology, no. 248.)

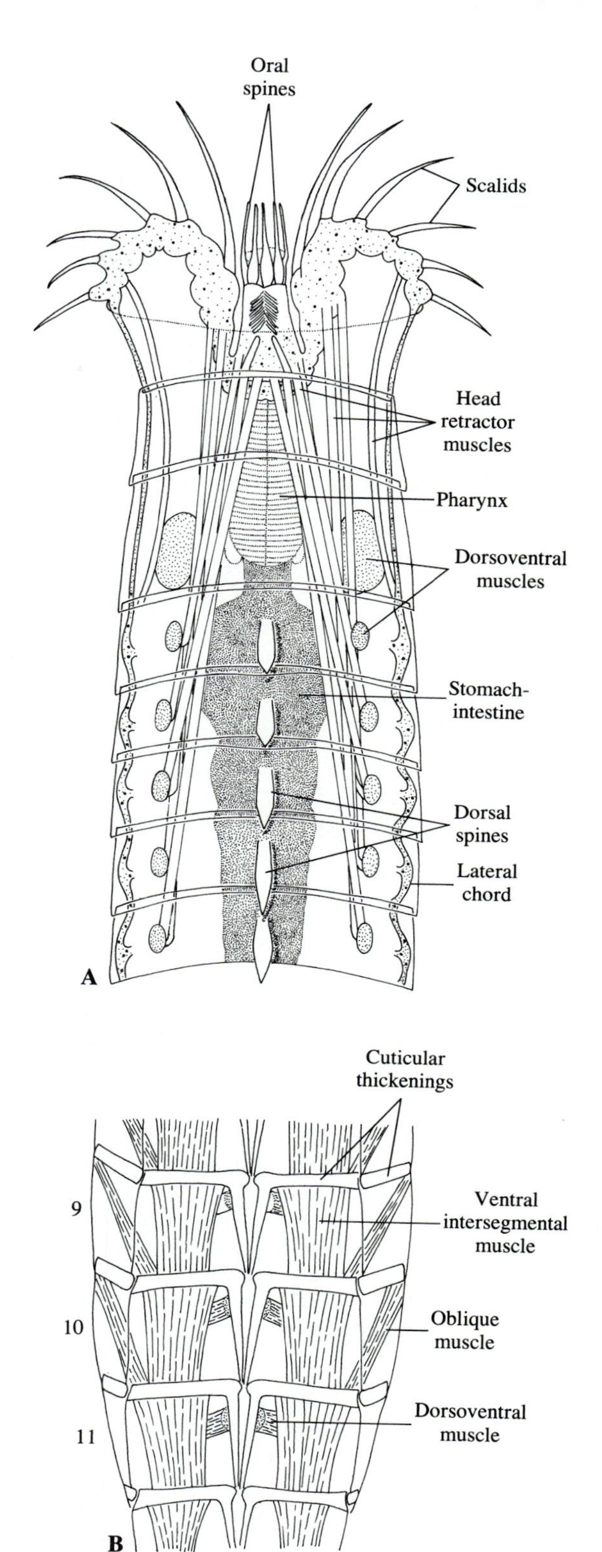

FIGURE 10.18 A. *Pycnophyes communis* (order Homalorhagida), anterior part of body, showing muscles that retract the head. B. *Echinoderes dujardinii,* musculature of several segments near the posterior end of the body. (After Zelinka, Monographie der Echinodera.)

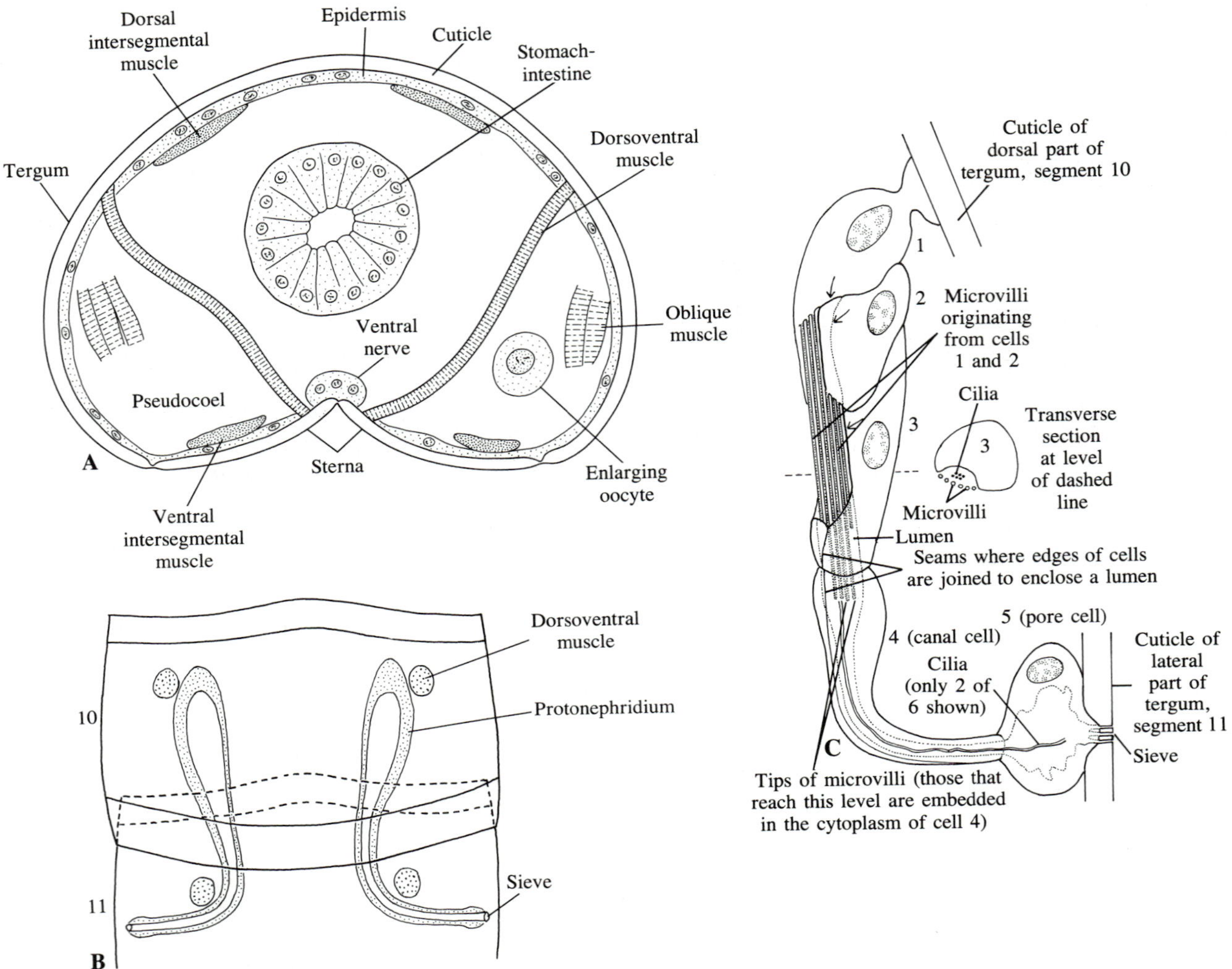

Figure 10.19 A. *Echinoderes dujardinii,* transverse section through one of the segments near the middle of the body. B. Protonephridia; diagrammatic, details now shown. (A and B, after Zelinka, Monographie der Echinodera.) C. Left protonephridium of *E. aquilonius;* diagram based on electron micrographs. To avoid confusion, the proximal portions of the cilia are not shown, but arrows indicate where the pairs originate. Intracellular supporting rods within cells 1–3, as well as some other details, are also omitted. (After several figures by Kristensen and Hay-Schmidt, Acta Zoologica, *70.*)

the next two cells, and ends at a cuticular sieve on the side of the tergum of segment 11. Long microvilli originating from the first two cells form a partial closure around the space through which the cilia pass before they reach the lumen. Presumably, flow into the protonephridium is most active here. In a species of *Pycnophyes,* each protonephridium consists of 25 cells, 22 of which are binucleate. One of them has a thin perforated wall.

The more conspicuous elements of the nervous system are a nerve ring around the pharynx and a midventral cord in the epidermis (Figure 10.19, A). Both have ganglionlike swellings; in the ventral cord the swellings correspond to the segments. The smaller nerves running from the nerve ring and ventral nerve cord are difficult to distinguish. At least many of the hairs and bristles on the surface of the body are sensory, as they are associated with nerve endings. It is almost certain that the scalids of the head are sensory, for in the genus *Kinorhynchus (Trachydemus)* these have been shown to have cilia of a type often found in receptors. Many species of *Echinoderes* have a pair of conspicuous brick-red eyespots in the head, and a dorsal thickening of the epidermis also may have accumulations of orange pigment that could mark the position of sensory cells. The pigment for the eyespots and dorsal accumulations consists of carotenoids obtained from diatoms eaten by *Echinoderes*.

REPRODUCTION AND DEVELOPMENT

Sexes are separate in kinorhynchs and sperm or prospective ova can usually be seen within mature individuals. In some genera, males and females may also be distinguished by the arrangement of spines and bristles at the posterior end of the body. In *Echinoderes,* females have a pair of substantial spines lateral to the long terminal spines on segment 13 (Figure 10.17, B), whereas males have three rather delicate spines—one of them truncate at the tip—in approximately the same position (Figure 10.17, C).

It is unusual to find more than one full-sized oocyte on each side of the gut. Smaller oogonia and oocytes are found in front of and behind the larger ones, and the ovaries run through several segments. The nature of the ovary and origin of the gametes need to be investigated in detail, but classical accounts indicate that there is an apical cell that buds off oogonia and nutritive cells, as well as an epithelium that encloses these and thus delimits the ovary. The genital pores are located ventrolaterally, on the suture between segments 12 and 13, and there are vague oviducts running to these. At least in certain genera, such as *Pycnophyes* and *Kinorhynchus,* a diverticulum observed on each oviduct probably functions as a seminal receptacle; in *Pycnophyes,* it has been reported to contain mature sperm. How the sperm are delivered to it by the male is not known. Egg-laying has not been observed, either, but presumably the fertilized egg is squeezed through the genital pore.

In male kinorhynchs, spermatogonia are said to originate from apical cells comparable to those that produce oogonia in females. Mature sperm are large—sometimes more than 80 μm long—and they are also fairly bulky. In *Echinoderes* the sperm are packed like spaghetti into the testes, which occupy much of the space between the gut and body wall of several segments. The sperm ducts lead to genital pores located ventrolaterally between segments 12 and 13. Copulation has not been observed, but it is likely that the two sets of spines that originate on segment 13, lateral to the large terminal spines (Figure 10.17, C), are involved in transfer of sperm to the genital pores of the female.

The embryology of kinorhynchs remains unknown, and a straight story of later development within the egg membrane has been told for only one species of *Echinoderes*. Many years ago Carl Zelinka, a careful student of these animals, concluded that kinorhynchs must hatch at the 11-segment stage, for although he had never seen an egg, he also had not seen young kinorhynchs with fewer than 11 segments. So far as *Echinoderes* is concerned, Zelinka was absolutely right. The published report of a free-living larval stage that becomes a juvenile by adding segments and a definitive head is fiction. There is certainly no larval stage, and development of a juvenile with 11 segments takes place entirely within the egg membrane (Figure 10.20, A). The head, as a matter of fact, is fairly well differentiated before segmentation of the body becomes conspicuous. When the young kinorhynch is ready to hatch, about 11 days after the egg is laid, it straightens out and extends its spiny head, ripping open the membrane (Figure 10.20, B). It begins right away to feed on diatoms (Figure 10.20, C). The fact that the eggs are almost always coated by detritus probably explains why they had not previously been seen.

The transition from the newly hatched 11-segment stage to the adult involves several molts. In each molt, the cuticle on the outside of the body is shed, along with the cuticle lining the buccal cavity, pharynx, and hindgut (Figure 10.20, D). The unpaired caudal spine characteristic of the 11-segment stage is lost at the third molt, and after the fifth molt the animal has 13 segments and is nearly mature.

CLASSIFICATION

The phylum is conventionally divided into two orders (they could just as well be called classes): Cyclorhagida and Homalorhagida. In the former, the first segment of the trunk either forms an unbroken ring (as in *Echinoderes*) or is organized into two lateral valves that can close like a clam shell to cover up the entire head, including the placids (as in *Semnoderes;* Figure 10.21, A). In two genera of cyclorhage kinorhynchs *(Campyloderes, Centroderes),* the head has no placids, but the first segment of the trunk is like that in *Echinoderes*. A median caudal spine is present only in some of the earlier juvenile stages of *Echinoderes,* but it is characteristic of adults of other genera of Cyclorhagida.

In the order Homalorhagida, the first segment of the trunk has four pieces: a dorsal tergum, a midventral sternum, and a pair of ventrolateral sterna. *Kinorhynchus* (Figure 10.21, B and C) and *Pycnophyes,* which include some of the more extensively studied kinorhynchs, are the only genera in this order.

REFERENCES ON KINORHYNCHA

Brown, R., 1983. Spermatophore transfer and subsequent sperm development in homalorhagid kinorhynchs. Zoologica Scripta, *12:*257–266.

Higgins, R. P., 1968. Taxonomy and postembryonic development of the Cryptorhagae, a new suborder for the mesopsammic kinorhynch genus *Cateria*. Transactions of the American Microscopical Society, *87:*21–39.

Higgins, R. P., 1971. A historical overview of kinorhynch

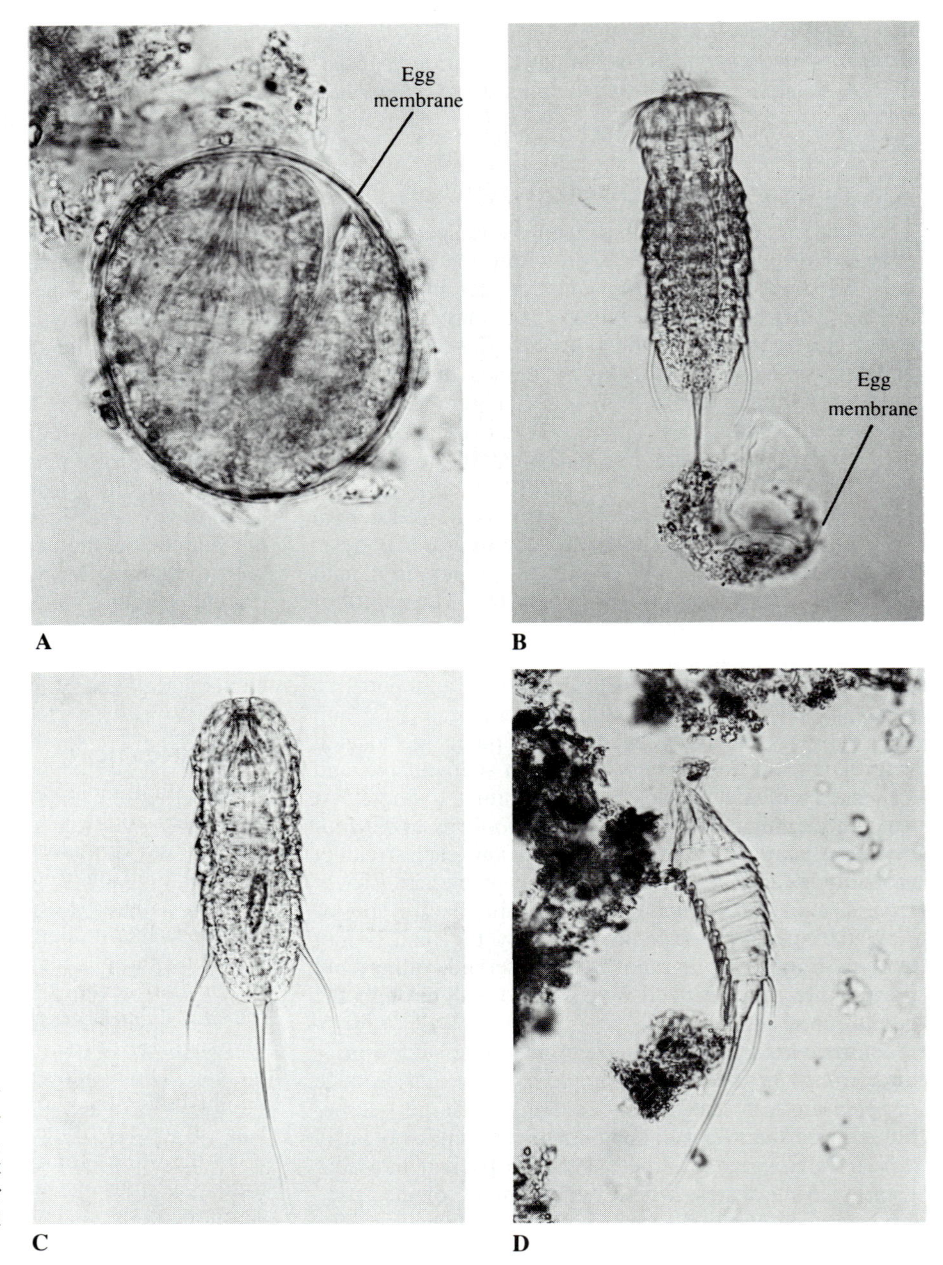

FIGURE 10.20 *Echinoderes kozloffi;* photomicrographs. A. Juvenile (11 segments), head inverted, ready to hatch. B. Juvenile just after hatching. C. Juvenile after eating its first diatom. D. Molted cuticle of juvenile.

research. Smithsonian Contributions to Zoology, no. 76, pages 25–31.

Higgins, R. P., 1977. Redescription of *Echinoderes dujardinii* (Kinorhyncha) with descriptions of closely related species. Smithsonian Contributions to Zoology, no. 248.

Higgins, R. P., 1983. The Atlantic barrier reef ecosystem at Carrie Bow Cay, Belize, II: Kinorhyncha. Smithsonian Contributions to the Marine Sciences, no. 18.
(Includes keys to most species of several important genera.)

Kozloff, E. N., 1972. Some aspects of development in *Echinoderes* (Kinorhyncha). Transactions of the American Microscopical Society, *91*:119–130.

Neuhaus, B., 1988. Ultrastructure of the protonephridia in *Pycnophyes kielensis* (Kinorhyncha Homalorhagida). Zoomorphology, *108*:245–253.

Nyholm, K.-G., 1976. Ultrastructure of the spermatozoa in Homalorhaga Kinorhyncha. Zoon, *4*:11–18.

Zelinka, C., 1928. Monographie der Echinodera. Wilhelm Engelmann, Leipzig.

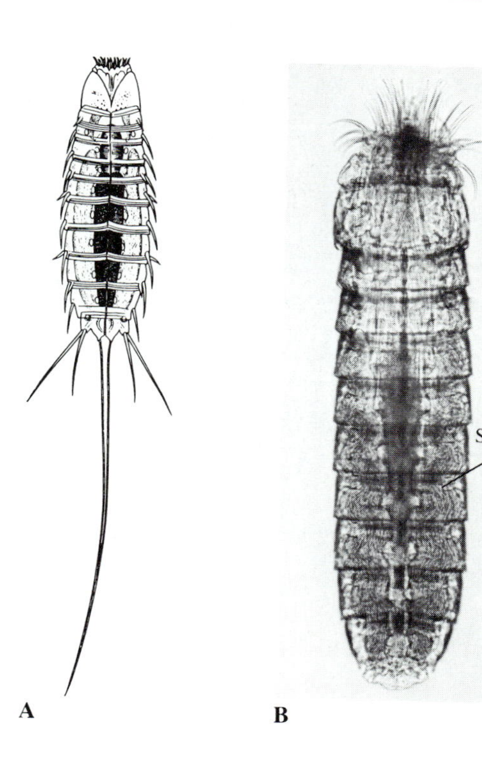

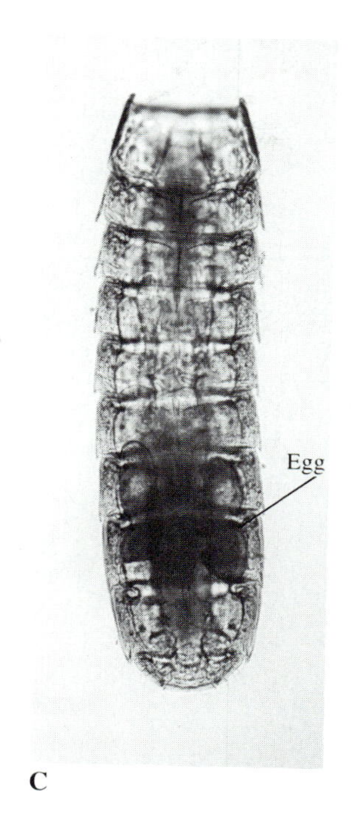

FIGURE 10.21 A. *Semnoderes armiger* (order Cyclorhagida). (Remane, in Bronn, Klassen und Ordnungen des Tierreichs, *4*:2:1.) B. *Kinorhynchus ilyocryptus* (order Homalorhagida), male, head extended; photomicrograph. C. *K. ilyocryptus,* female, head withdrawn.

PHYLUM LORICIFERA

The few known species of the marine phylum Loricifera have been found at depths ranging from about 15 m to over 8000 m. These animals, which inhabit shelly gravels and sand, had been overlooked until recently. Because of their small size, they are difficult to see in samples of coarse sediment. They are believed to cling tightly to particles, and perhaps to other organisms. Most of the specimens that have been collected were obtained by placing the sediment in fresh water to dislodge the meiofauna, then straining the water through a fine sieve.

The first loriciferan to be described was *Nanaloricus mysticus*. It was originally found off the Atlantic coast of France, but has since been collected near the Azores and off Florida. The following account deals primarily with this species, although mention is made of species belonging to two other genera, *Pliciloricus* and *Rugiloricus*. These have been taken off North and South Carolina.

The adults of *Nanaloricus* (Figures 10.22 and 10.23), not quite 250 μm long, could at first be mistaken for kinorhynchs or larvae of certain priapulans. The resemblance to a young priapulan is especially striking because the posterior portion of the body, called the **abdomen,** is enclosed by a cuticular **lorica** into which the head, neck, and thorax can be withdrawn. The lorica consists of six longitudinal plates, each terminating anteriorly in a few stout spines. Ducts of epidermal glands extend to pores at the tips of these spines, as well as to pores scattered over the surfaces of the plates.

In the genera *Rugiloricus* (Figure 10.24, A) and *Pliciloricus* (Figure 10.24, B), the anterior edges of the lorica plates are not drawn out into spines. The plates have lengthwise folds, however.

The anterior part of the head is conical. A short stalk that bears the mouth can be extended from the tip of the cone. Also at the tip are openings through which stylets located within the cone can be protruded. The rest of the head region is characterized by rings of spines, some of which are jointed. Those of the first ring, called **clavoscalids,** are the stoutest. The **spinoscalids,** which constitute the next several rings, become progressively more slender, and those of the last ring have small, platelike bases rather than bulbous bases.

The neck region has an irregular ring of spines called **trichoscalids,** each of which arises from a basal plate. Certain of these plates, lateral in position, have a small auxiliary spine, and the midventral plate has a prominent pore. Anterior to each plate that bears a trichoscalid, and lined up with it, are one or two similar but smaller plates.

In *Rugiloricus* and *Pliciloricus,* the trichoscalids are double structures, and neck plates anterior to those from which trichoscalids originate bear a small spine or a sharp projection. The spinoscalids of the last row,

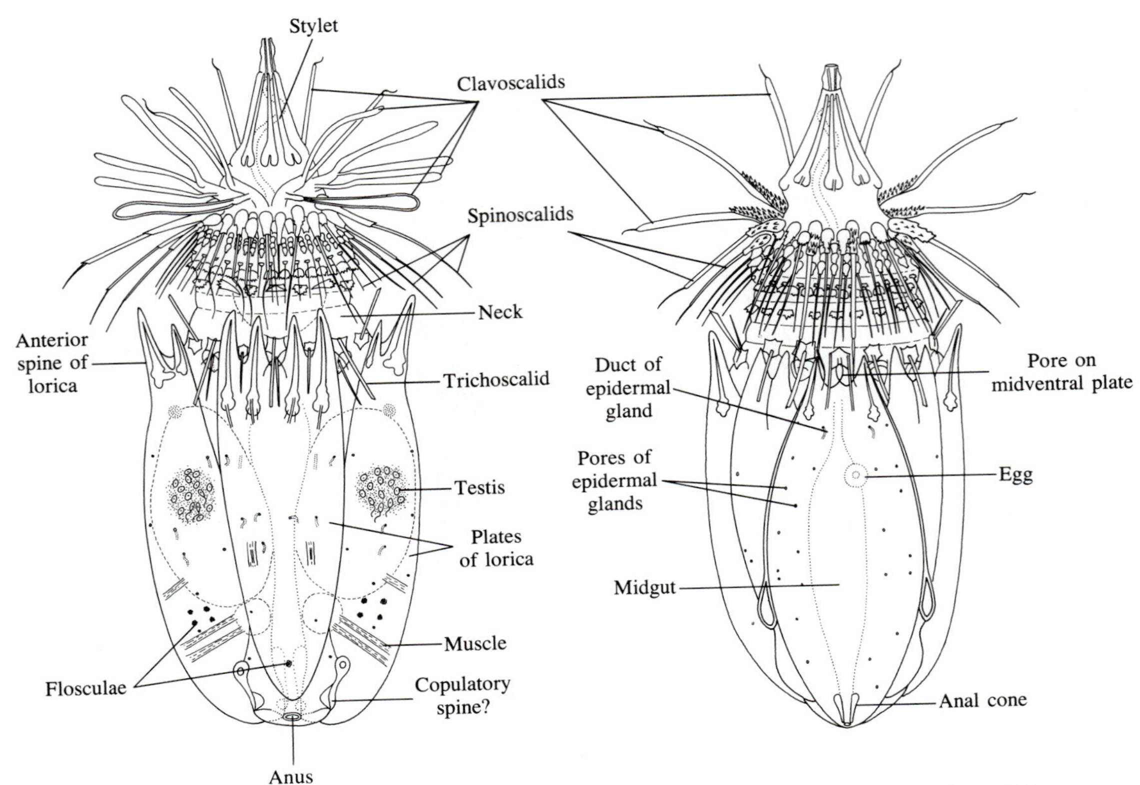

Figure 10.22 *Nanaloricus mysticus*. A. Male, dorsal view. B. Female, ventral view. (After Kristensen, Zeitschrift für Zoologische Systematik und Evolutionsforschung, *21*, p. 169, 167.)

moreover, are clawlike, and the two ventralmost spinoscalids of the first row are joined, at least to some extent, forming what is called a double organ.

The thoracic region lacks scalids or spines, although it may be subdivided transversely by lines where folding takes place. In some species, including *Nanaloricus mysticus,* it is so short as to be scarcely noticeable between the neck and the lorica-encased abdomen.

The gut begins as a slender, cuticularized buccal tube (Figure 10.23). Associated with this tube are two salivary glands and two stylets that probably can be projected into the lumen. Following the buccal tube is a bulbous pharynx equipped with radial musculature and lined with cuticle, some portions of which are conspicuously thickened. When the mouth stalk is protruded from the tip of the head cone, the anterior part of the pharynx is close to the base of the cone; otherwise it is a little farther back in the body.

Posterior to the pharyngeal bulb is an esophagus, then a midgut, a rectum, and a short anal cone, whose lining has a thick cuticle. The diet of loriciferans has not yet been described.

The nervous system includes a large dorsal brain that fills much of the head region. Nerves from the brain innervate the clavoscalids and spinoscalids; the clavoscalids also receive nerves from ganglia that encircle the buccal tube just behind the brain. The trichoscalids of the neck region are supplied by nerves from a ventral ganglion in that part of the body. There are also ganglia at more posterior levels. One of them, located ventrally close to the posterior end, innervates sensory structures called **flosculae.** There are four of these on each ventrolateral plate of the lorica and one on the midventral plate. Each floscula consists of a cluster of microvilli.

The large retractor muscles of the head are striated, and perhaps all of the muscles will prove similar in this respect. The muscular system as a whole, however, has not been worked out in detail. The body cavity between the gut and body wall is probably a pseudocoel, but this also needs study. Two pores on the ventral surface near the posterior end are possibly openings of protonephridia, but the protonephridia themselves have not been described.

The female has two large ovaries lateral to the midgut, believed to produce one functional egg at a time. The male has two testes in a comparable position, and

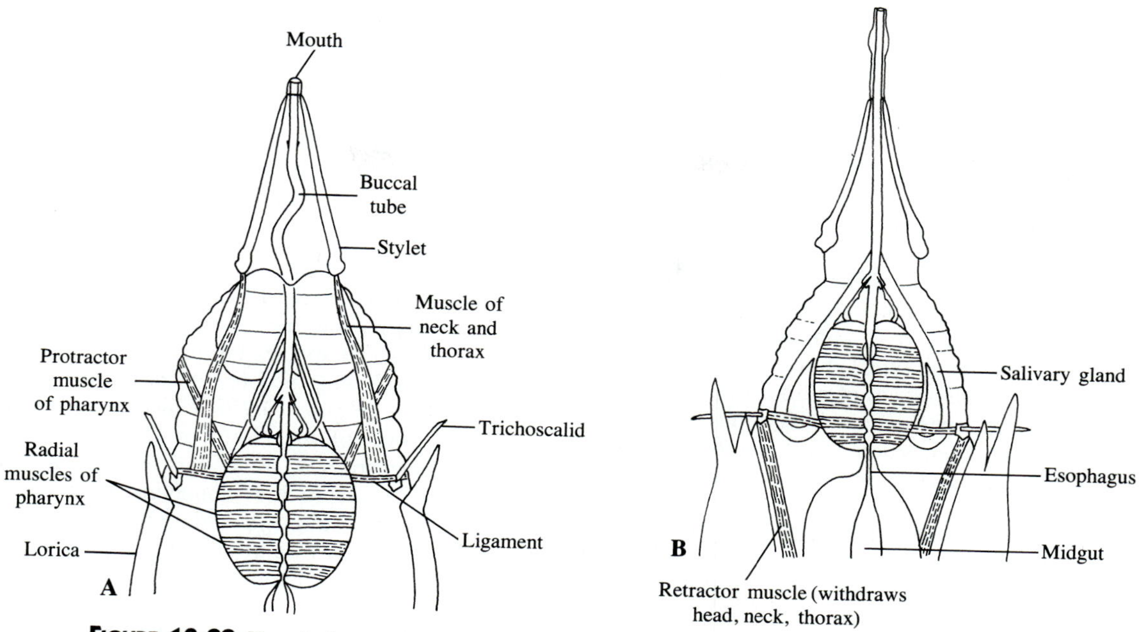

FIGURE 10.23 *Nanaloricus mysticus,* anterior portion, in optical section, cycloscalids and spinoscalids omitted. A. Female, buccal tube withdrawn, ventral view. B. Male, buccal tube extended, dorsal view. (After Kristensen, Zeitschrift für Zoologische Systematik und Evolutionsforschung, *21,* p. 170.)

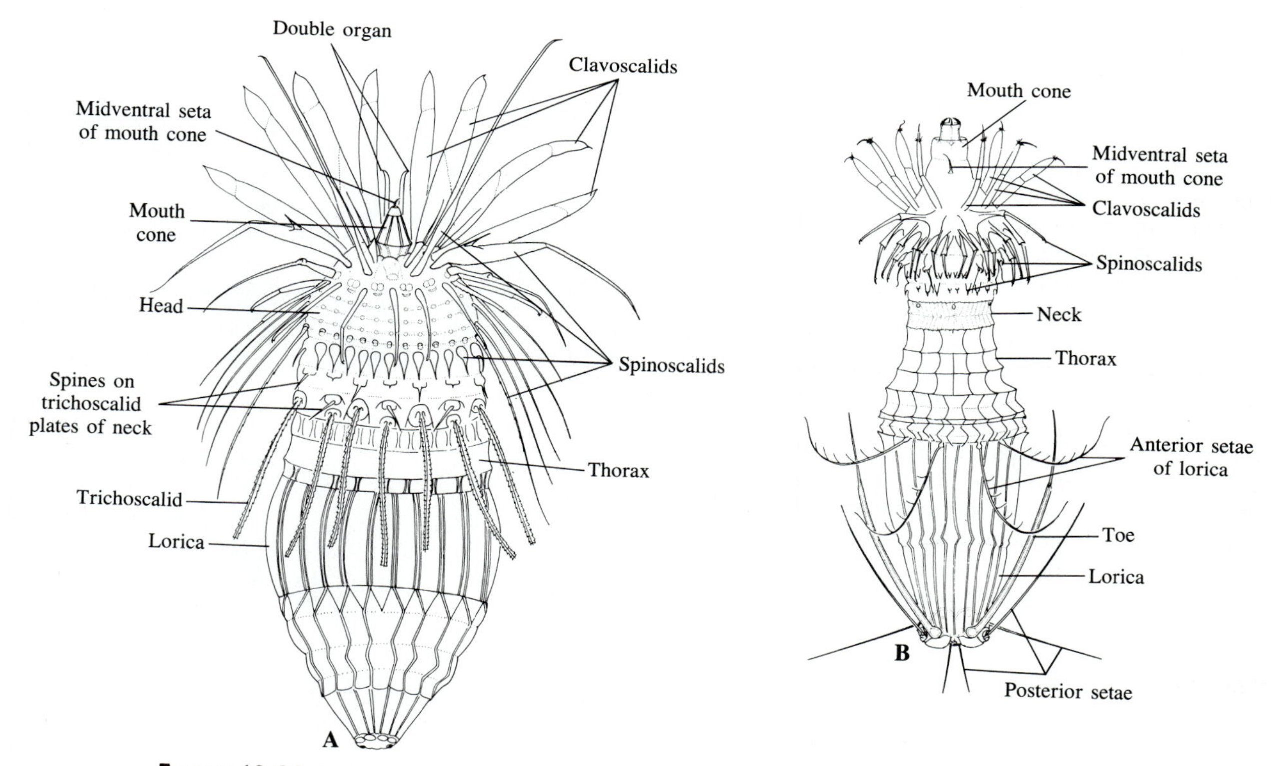

FIGURE 10.24 A. *Rugiloricus ornatus,* female, dorsal view. B. *Pliciloricus profundus,* juvenile, ventral view. (Higgins and Kristensen, Smithsonian Contributions to Zoology, no. 438.)

they contain small sperm that have tails. Other parts of the reproductive system are poorly understood. The female has a pair of structures that may possibly be seminal receptacles; in the male, two spines on the ventral side of the body near the posterior end perhaps have something to do with copulation. A pair of small pores observed close to the ventral midline of both sexes may be genital openings.

The male of *Nanaloricus* differs from the female in having eight clavoscalids, instead of six. Moreover, six of the clavoscalids of the male—the middorsal, dorsolateral, and ventrolateral ones—have three branches. The significance of this aspect of sexual dimorphism is not known. In other species in which adult specimens of both sexes have been studied, the number and appearance of the clavoscalids are about the same in the male and female.

The loriciferan material that has been studied includes juveniles ("Higgins-larvae"), molted cuticles, and adults to which the cuticle of the last juvenile stage is still attached. One species of *Pliciloricus,* in fact, was described on the basis of a juvenile; adults have not yet been found. Juveniles (Figure 10.24, B) resemble adults in having clavoscalids and spinoscalids on the head, but the form and arrangement of these are different than on adults. What seems to be homologous to the neck region is a collarlike ring that can be folded; it probably serves as a closing apparatus when the head is withdrawn. The thorax is proportionately longer than in adults, having five or six circles of cuticular plates. The lorica has lengthwise folds, and there are some setae (slender cuticular bristles), perhaps sensory or locomotory, at its anterior margin and also at the posterior end. The most striking feature about the posterior end of a juvenile is a pair of tapering toes, each with small scalelike plates along both margins. These toes are believed to function in locomotion.

The study of Loricifera has just begun. We may hope that further work on these remarkable little animals will enable us to understand more fully their structure and habits. The resemblance of loriciferans to kinorhynchs and to priapulan larvae has already been mentioned. They are also similar in certain respects to larvae of nematomorphs, and even to tardigrades and rotifers. Most of the similarities are probably due to convergent or parallel evolution, but perhaps some of them indicate close relationships.

References on Loricifera

Higgins, R. P., and Kristensen, R. M., 1986. New Loricifera from southeastern United States coastal waters. Smithsonian Contributions to Zoology, no. 438.

Kristensen, R. M., 1983. Loricifera, a new phylum with aschelminth characters from the meiobenthos. Zeitschrift für Zoologische Systematik und Evolutionsforschung, *21:*163–180.

Kristensen, R. M., and Shirayama, Y., 1988. *Pliciloricus hadalis.* (Pliciloricidae), a new loriciferan species collected from the Izu-Ogasawara Trench, western Pacific. Zoological Science, *5:*875–881.

Phylum Priapula

There are only about 15 described species of priapulans, all marine and mostly subtidal. Some burrow in mud, but the small species are usually in sand or coral rubble. The following account deals mostly with *Priapulus caudatus* (Figure 10.25), which is widely distributed in the seas of the Northern Hemisphere. It is a relatively large species, sometimes reaching a length of 8 cm. In areas where it is abundant, it is among the invertebrates that stingrays and skates stir out of the mud and swallow. *Priapulopsis* and *Acanthopriapulus,* each with one or two known species, are similar.

The generic name *Priapulus,* given by Lamarck in 1816, indicates that the animal looks like a little penis. The body consists of three regions: head, trunk, and **caudal appendage.** These share a continuous body cavity. The mouth, located at the tip of the head, is encir-

A

B

Figure 10.25 *Priapulus caudatus.* A. Head inverted. B. Head everted.

cled by teeth that are hardened elevations of the cuticle. Similar structures, generally called spines, are arranged in neat longitudinal rows on the head. The entire head can be inverted by striated retractor muscles that extend to it from various points on the body wall of the trunk (Figure 10.26; Figure 10.27, A). Eversion of the head is accomplished by contraction of striated muscles of the body wall of the trunk, which increases the pressure on the fluid in the body cavity. The circular trunk muscles are directly beneath the epidermis and are concentrated in numerous rings that make this part of the body appear segmented; the longitudinal muscles form a nearly complete sheath. *Priapulus* burrows by everting its head, thereby anchoring itself in the mud, then pulling up its trunk and withdrawing the head before thrusting it out again.

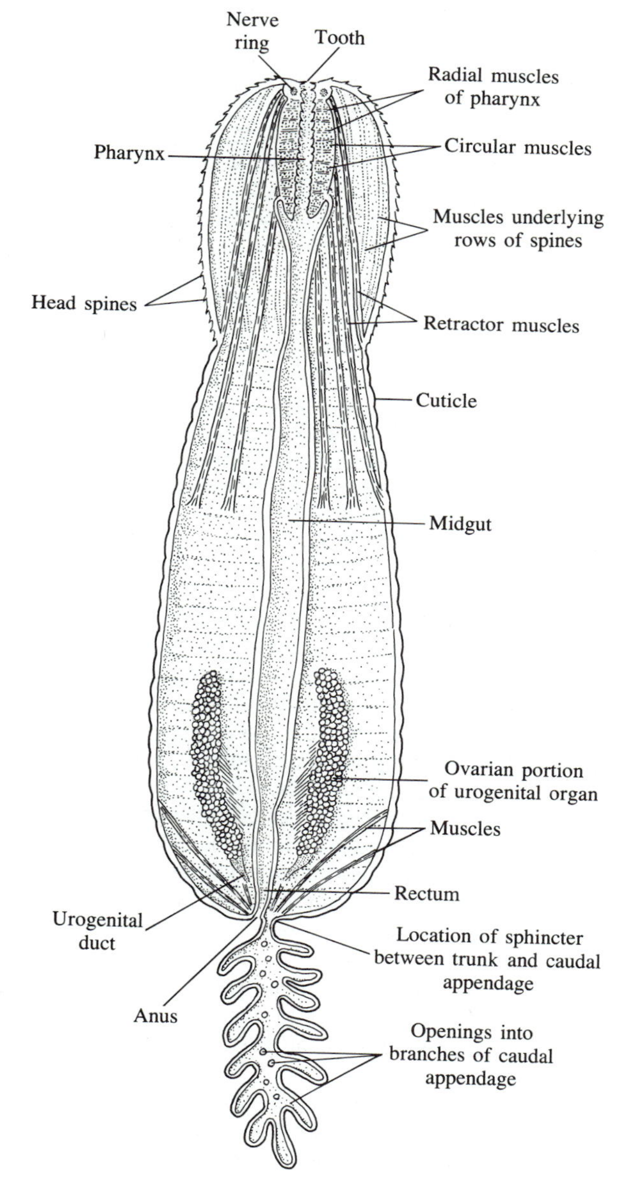

FIGURE 10.26 *Priapulus caudatus,* shown as if cut in half along the frontal plane, with the dorsal portion removed; diagrammatic. The anus has been shifted to the left.

The caudal appendage, which is thin-walled and lightly muscularized, is a tubular outgrowth of the trunk. A sphincter of muscles at the junction of the two body regions controls the size of the opening. Because the caudal appendage branches into many fingerlike lobes, it resembles a gill, but if it is removed the animal will continue to live for months and perhaps indefinitely. *Priapulopsis bicaudatus* has two caudal appendages. Some genera have none.

The body cavity, according to recent studies, does not have an epithelial lining. The inner surface of the body wall and the outer surface of the gut do, however, have many flattened amoebocytes, and these could be misinterpreted as cells of a continuous peritoneum. The origin of the body cavity is uncertain, for there is still no complete account of development in priapulans. The fluid in the cavity of *Priapulus* contains numerous disklike cells containing hemerythrin, a respiratory pigment found in only a few other invertebrates: sipunculans, polychaetes of the family Magelonidae, and certain inarticulate brachiopods.

The gut follows a straight course, and the anus is located at the posterior end of the trunk. The pharynx (within the head) and the hindgut are lined by a cuticle, so it is likely that both are derived from secondary ectodermal invaginations. The pharyngeal cuticle is thick and has a rasplike surface, with many small, hard teeth. Circular and radial musculature enable the pharynx to crush and swallow polychaetes and other prey organisms. The teeth around the mouth are used for grasping.

The more obvious elements of the nervous system are a ring around the pharynx, close to the mouth (Figure 10.26) and a ventral nerve cord (Figure 10.27, A). The latter is closely associated with the epidermis and is prominent enough to be visible externally. It shows a slight enlargement in each of the superficial rings of the trunk. At least many of the small skin papillae on the trunk of *Priapulus* have sensory components, including bristles that contain modified cilia. These animals definitely react to touch, and they also shun strong light, although special light receptors have not been identified.

Occupying a substantial part of the body cavity is a pair of **urogenital organs.** For a time during development, these are strictly excretory, consisting of a duct and clusters of tightly interdigitated solenocytes (Figure

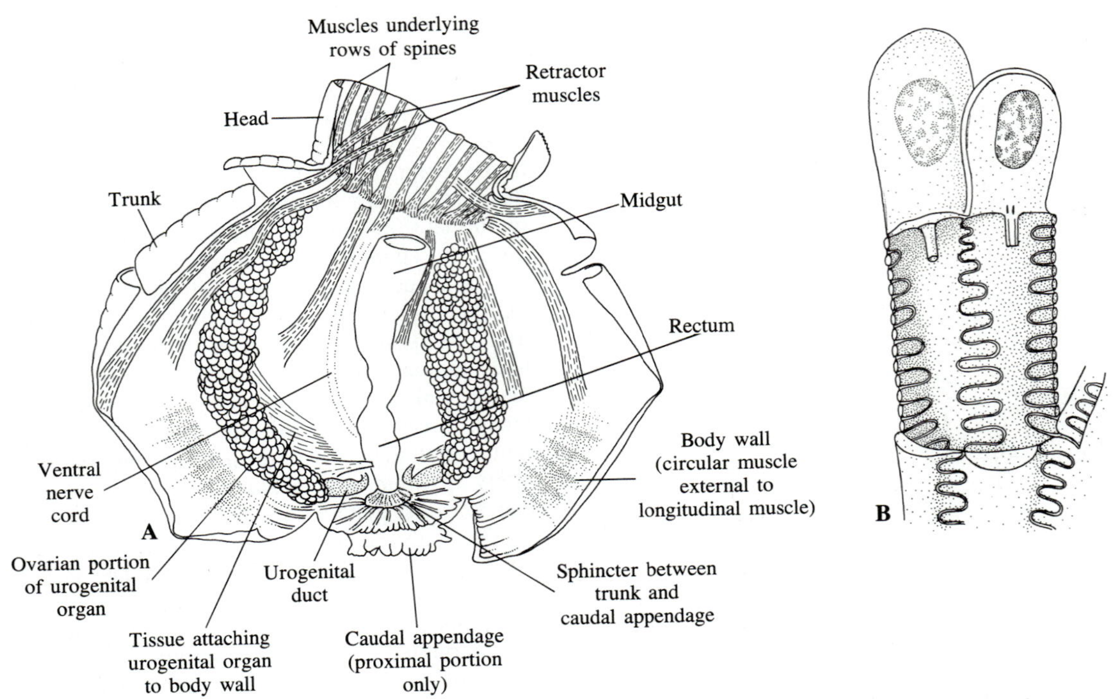

Figure 10.27 A. *Acanthopriapulus horridus* Dissection of trunk and posterior part of head. (After Théel, Kungliga Svenska Vetenskapsakademiens Handlingar, *47*.) B. Two solenocytes from one of the clusters of solenocytes of the excretory portion of a urogenital organ of *Priapulus caudatus;* reconstruction based on electron micrographs. Only the basal portions of the long cilia are shown. (After Kümmel, Zeitschrift für Zellforschung und Mikroskopische Anatomie, *62*.)

10.27, B). Eventually a gonad—a testis or ovary, depending on the sex of the animal—develops from the anterior part of each duct and becomes partly incorporated into the sheet of tissue that connects the urogenital organ to the body wall. The ducts of the right and left urogenital organs open to the outside a short distance in front of the anus.

Under laboratory conditions, spawning by male *Priapulus* has been observed to trigger spawning by females. Fertilization takes place externally. (This is probably the rule for priapulans, although in *Tubiluchus corallicola,* found in Bermuda and the Barbados, there is a possibility that internal fertilization takes place.) Little is known about what happens after the embryo has passed through the blastula and gastrula stages. The gastrula is succeeded, however, by a larva that resembles a cnidarian planula. The ectodermal layer, which is not ciliated, encloses a syncytial mass. A cuticle soon develops, but the development of the larva immediately thereafter is not known.

In older larvae of *Priapulus* (Figure 10.28, A and B), the body consists of a head, neck, and abdomen. The thick cuticle of the abdomen is differentiated into several large plates and two small anterior plates; it also has a few sensory bristles. Each time the larva molts, the lorica is shed along with the rest of the cuticle. At metamorphosis into the adult form, the neck and abdomen become the trunk, and the caudal appendage grows out from this. Molting continues to take place at intervals as long as the animal lives.

Some priapulan larvae, especially that of *Tubiluchus* (Figure 10.28, C), show a striking resemblance to loriciferans, and also to kinorhynchs. The idea that there is a close relationship between the three phyla has indeed been proposed, but it will be best to keep an open mind on this subject.

The loricate larvae of priapulans feed on detritus. Carnivory begins after metamorphosis, except in *Tubiluchus,* which continues to eat detritus even after reaching the adult stage.

Although there are only a few known species of priapulans, the phylum is greatly diversified. The types that live between sand grains or in coral rubble, or that secrete tubes, differ markedly from one another and from mud-inhabiting species of *Priapulus, Acanthopriapulus,* and *Halicryptus*. One of the sand-inhabiting genera,

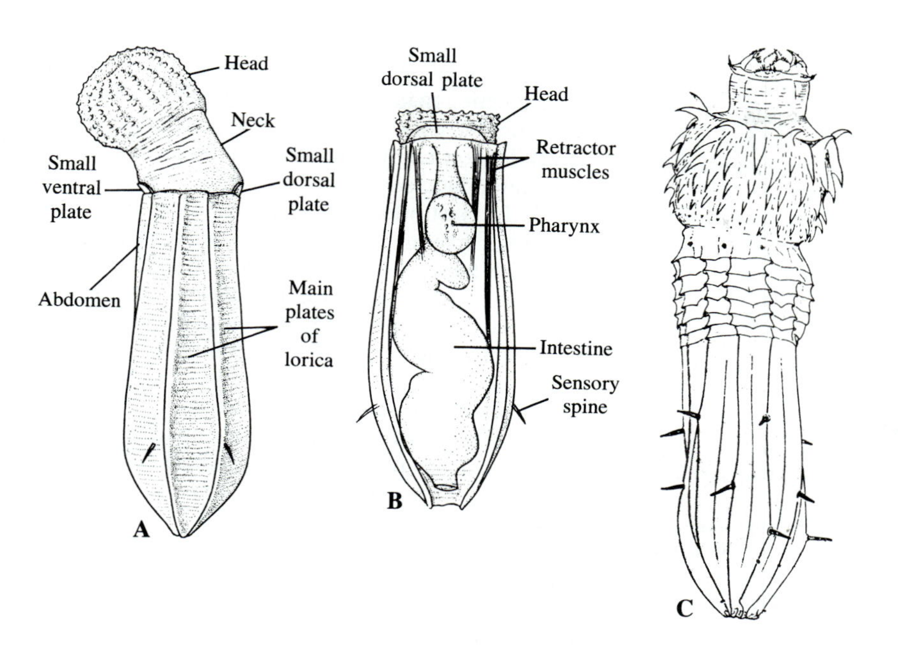

FIGURE 10.28 A. *Priapulus caudatus,* loricate larva, entire. B. *Priapulus caudatus,* loricate larva, dissected. (Lang, Arkiv för Zoologi, *41A*.) C. *Tubiluchus corallicola,* loricate larve. (Van der Land, Zoologische Verhandelingen, no. 112.)

Meiopriapulus, is illustrated (Figure 10.29, A). Two especially remarkable genera are *Maccabeus* (Figure 10.29, B) and *Chaetostephanus*. These live in tubes in the superficial layer of sediment. Hooks at the posterior end anchor the worms in their tubes. At the anterior end of the head are two circles of bristly tentacles that are probably homologous to spines. Those of the inner circle are believed to be sensory. After one of them has been touched by a small prey organism, the longer tentacles of the outer circle bend inward to form a trap.

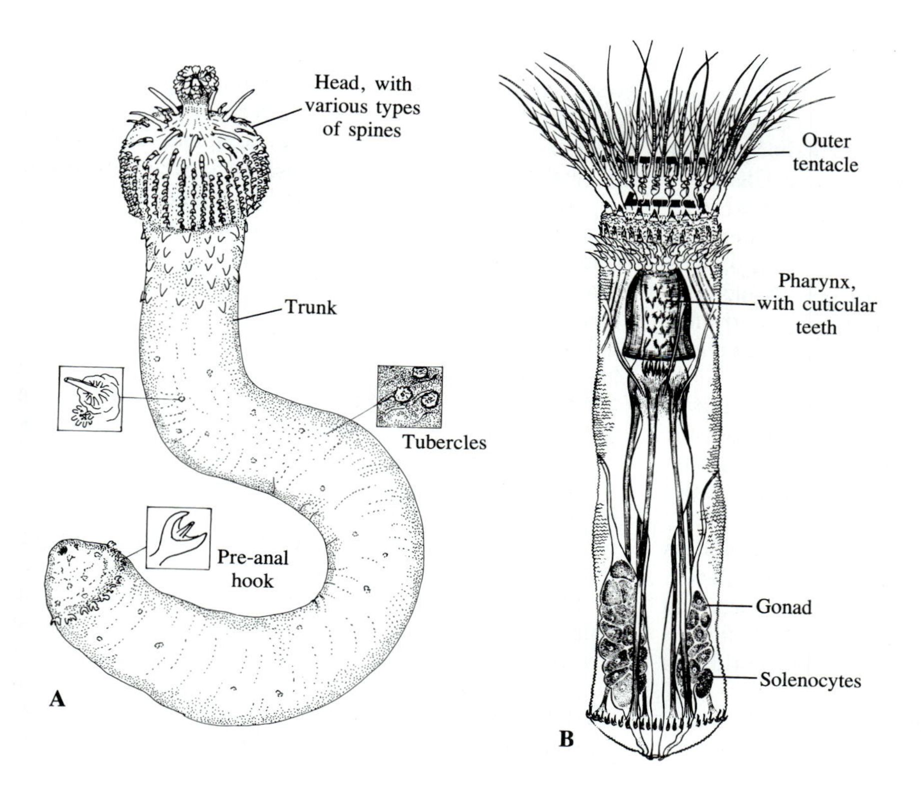

FIGURE 10.29 A. *Meiopriapulus fijiensis,* a small (3 mm) priapulan that lives in sandy beaches in Fiji. Some of the structures visible at the surface are enlarged. (Morse, Transactions of the American Microsocopical Society, *100*.) B. *Maccabeus tentaculatus,* a small, tube-inhabiting priapulan from the Mediterranean Sea. (Por and Bromley, Journal of Zoology, *173*.)

References on Priapula

Calloway, C. B., 1975. Morphology of the introvert and associated structures of the priapulid *Tubiluchus corallicola* from Bermuda. Marine Biology, *31:*161–171.

Hammond, R. A., 1970. The burrowing of *Priapulus caudatus*. Journal of Zoology (London), *162:*469–480.

Hammond, R. A., 1970. The surface of *Priapulus caudatus* (Lamarck, 1816). Zeitschrift für Morphologie der Tiere, *68:*255–268.

Kirsteuer, E., 1976. Notes on adult morphology and larval development of *Tubiluchus corallicola* (Priapulida), based on *in vivo* and scanning electron microscopic examinations of specimens from Bermuda. Zoologica Scripta, *5:*239–255.

Morse, M. P., 1981. *Meiopriapulus fijiensis* n. gen., n. sp.: an interstitial priapulid from coarse sand in Fiji. Transactions of the American Microscopical Society, *100:*239–252.

Por, F. D., and Bromley, H. J., 1974. Morphology and anatomy of *Maccabeus tentaculatus* (Priapulida: Seticoronaria). Journal of Zoology (London), *173:*173–197.

Salvini-Plawen, L. von, 1974. Zur Morphologie und Systematik der Priapulida: *Chaetostephanus praeposteriens,* der Vertreter einer neuen Ordnung Seticoronaria. Zeitschrift für Zoologische Systematik und Evolutionsforschung, *12:*31–54.

Van der Land, J., 1970. Systematics, zoogeography, and ecology of the Priapulida. Zoologische Verhandelingen (Leiden), no. 112.

Van der Land, J., 1982. A new species of *Tubiluchus* (Priapulida) from the Red Sea. Netherlands Journal of Zoology, *32:*324–335.

Summary

1. Of the eight pseudocoelomate phyla discussed in Chapters 9 and 10, only two—Rotifera and Gastrotricha—have motile external cilia. In Rotifera, these cilia are concentrated at the anterior end, where most of them are organized into special tracts that form the corona, or "wheel organ," so named because the activity of the cilia gives the illusion of a wheel that is turning. The coronal cilia are important in both locomotion and feeding.

2. A distinctive feature of the pharynx of rotifers is the mastax, a complex of cuticular jaw pieces used for grinding or mashing food, and sometimes for grasping prey. The gut, portions of which are ciliated, is usually complete, but in some species there is no anus.

3. At the posterior end of the body, many rotifers have one or two "toes" provided with adhesive glands. An external cuticle is present in certain species, but what generally looks like a cuticle is really a fibrous layer within the epidermis. There are protonephridia of the flame-bulb type.

4. Rotifers are abundant in fresh and salt water, and they also occur in moss, wet soil, and other situations where there is moisture, even if only temporarily. Many members of the phylum are exclusively parthenogenetic; males are unknown. Others have sexual reproduction as well as parthenogenetic reproduction, depending on environmental conditions, and a few always reproduce sexually.

5. In Gastrotricha, which are found in marine and freshwater habitats, the motile external cilia, concerned only with locomotion, are on the ventral surface. There is a true cuticle and this is unusual in that the outer of its two layers forms a sheath around the cilia. In certain gastrotrichs, there is only one cilium per cell. This is a feature found in several other phyla, such as Porifera, Gnathostomulida, and Echinodermata.

6. The gut of gastrotrichs is complete. There is no mastax, but the pharynx is specialized for sucking and pumping. The two protonephridia are of a modified solenocytic type.

7. The two classes of gastrotrichs, Chaetonota and Macrodasya, are rather different in their morphology as well as in their patterns of reproduction. Chaetonotans, which include all of the freshwater species and some marine species, are, as a rule, parthenogenetic. Macrodasyans, which are strictly marine, are hermaphroditic, and sperm are transferred from one individual to another during the mating process.

8. All members of the small phylum Kinorhyncha are marine, and most of them live in sediment. They are covered by a rather firm cuticle, and the body is organized into units that are conventionally called segments. Although the pseudocoel is not partitioned transversely, the arrangement of the muscles and ganglionic swellings in the ventral nerve cord are decidedly segmental tendencies. The gut is complete, and its pharyngeal portion, with radial muscles, is well developed.

9. Adult kinorhynchs have 13 segments. The first of these, which bears the mouth, has many movable spines (scalids) and can be completely withdrawn. In most species, segment 2 is divided into numerous plates that turn inward to cover segment 1 after it has been pulled back. The remaining 11 segments form the trunk, and the cuticle of these is differentiated into a dorsal tergum and either one ventral sternum or two sterna that meet along the midline. There are two solenocytic protonephridia.

10. Sexes are separate in kinorhynchs. Development is direct. The juveniles hatch when they have 11 segments and add more segments in the course of several molts.

11. The phylum Loricifera consists of small marine animals that inhabit shelly gravel. At present, only a few species are known. Behind the head, which is characterized by large, movable spines similar to those on the head of a kinorhynch, are a short neck and thorax; all three of these body regions can be withdrawn into the abdominal portion of the body, which is enclosed by a cuticular lorica. The gut is somewhat similar to that of a kinorhynch. Juveniles, which molt periodically, resemble adults, although they have a pair of large posterior spines that are lacking in adults. Sexes are separate, but the reproductive biology of these recently discovered invertebrates has not yet been studied.

12. The phylum Priapula is represented by wormlike animals that live in sediments, especially mud. The head region, studded with nonmovable spines, is inverted when it is withdrawn into the trunk. Attached to the posterior end of the trunk of some species is a caudal appendage. The body cavity, considered to be a pseudocoel because it lacks a continuous peritoneum, is peculiar in that it is lined by flattened amoebocytes. Most priapulans are carnivores, but at least one feeds on detritus. Sexes are separate. Each of the two gonads is associated with a mass of solenocytes. There is a larval stage in which the trunk region is enclosed by a lorica.

11

Phylum Annelida

EARTHWORMS, LEECHES, AND THEIR RELATIVES

Phylum Annelida
- Class Polychaeta
- Class Oligochaeta
- Class Hirudinoidea
  - Subclass Hirudinea
    - Order Rhynchobdellida
    - Order Arhynchobdellida
  - Subclass Acanthobdellida
  - Subclass Branchiobdellida

INTRODUCTION

Annelids are the most complex of all worms. In addition to having a **complete gut** and in being **bilaterally symmetrical, cephalized,** and **triploblastic,** they have certain features that distinguish them sharply from any of the phyla dealt with previously.

One obvious characteristic of an annelid is the division of its body into a linear series of units (Figure 11.1). These are called **segments,** or *metameres,* and they are demarcated externally by constrictions of the body wall, which give the worm a ringed appearance. In the course of evolution of annelids, the segmented body plan has been a highly modifiable one, and the great diversity seen in the phylum is to a large extent due to the ways in which individual segments or groups of segments have become specialized.

Another important feature of annelids is the **coelom.** This type of fluid-filled cavity, like a pseudocoel, separates the body wall from the gut. It differs from a pseudocoel, however, in that it is lined by an epithelium, called the **peritoneum.** Nearly all of the phyla discussed in the remaining chapters have a true coelom, or at least have structures whose formation is related to the appearance of a coelom during development.

In annelids, coelomic spaces appear within mesoderm as the segments differentiate. Thus each segment initially has its own coelomic compartment, separated from those of preceding and succeeding segments by partitions called **septa.** Some or most of the septa may eventually be disrupted or disappear, thus allowing coelomic spaces to run together. Nevertheless, the septa are conserved in the majority of annelids. The gut runs through these partitions on its course from the mouth to the anus, and certain other structures—longitudinal blood vessels and the nerve cord, for instance—also pass through them.

There is considerably more to segmentation than the division of the body into somewhat independent units, with or without compartmentalization of the coelom. Various internal and external structures, such as ganglia and nerves, branches of the longitudinal blood vessels, and excretory ducts are arranged according to a segmental plan.

A **circulatory system** is well developed in most annelids. Certain vessels are muscularized and function as blood-pumping organs. There may also be one or more distinct hearts, which are enlarged and muscularized portions of vessels. The vessels of annelids, like those of nearly all invertebrates, originate within an extracellular matrix and do not have an internal cellular lining. In this respect they differ from the vessels of nemerteans, which are from their inception lined by cells of mesodermal origin and which are thus somewhat comparable to coelomic spaces (see Chapter 7). Some zoologists prefer to use the term **blood vascular system** for the type of circulatory system found in annelids and most other invertebrates, to distinguish it from the **coelomic circulatory system** of nemerteans. (See Ruppert and Carle, 1983, for a concise discussion of this topic.)

Reduction or disappearance of the circulatory system has taken place in many annelids. When only a few

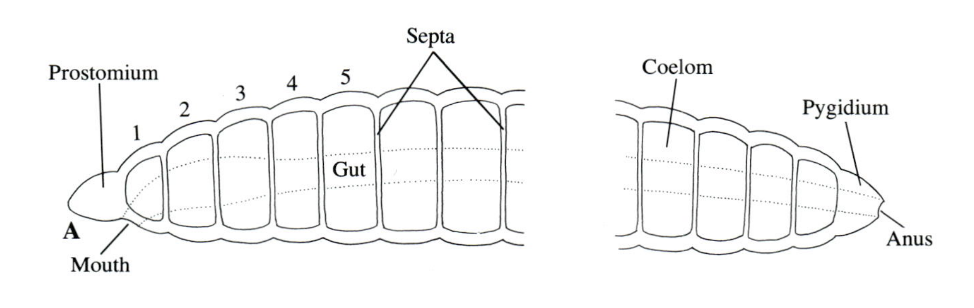

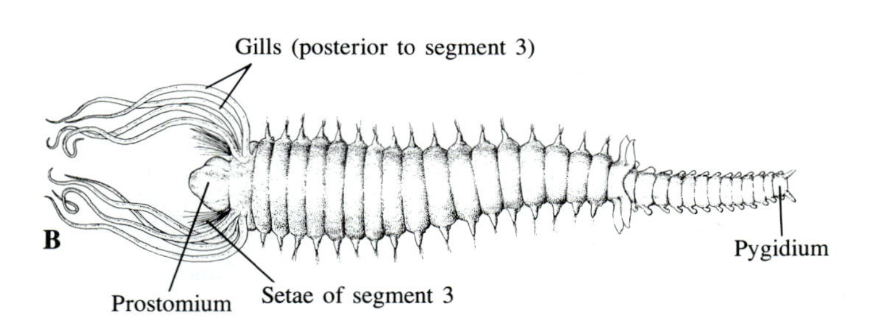

FIGURE 11.1 A. Diagram of a longitudinal section of an annelid, showing the relationship of true segments to the prostomium and pygidium. The section is taken to the left of the midline. Most systems are omitted. B. Specialized annelid (a polychaete of the family Ampharetidae), showing modifications of segments in different regions of the body. The peristomium, which bears tentacles used in feeding, and two succeeding segments are not visible in this dorsal view. (Hartman, Occasional Papers of the Allan Hancock Foundation, no. 28.)

blood vessels, or none, remain, the coelom assumes the function of transporting oxygen, nutrients, and metabolic wastes. Leeches and some other types of annelids in which reduction of vessels has occurred are discussed later in this chapter.

Cephalization is more advanced in annelids than it is in any of the lower phyla. The head region is often conspicuously differentiated and provided with antennae, eyes, and other sense organs. The brain is rather complex and is joined to a main longitudinal nerve cord, which is located in the body wall ventral to the gut. The cord appears to have been derived from two originally separate nerve cords that have become united; in many annelids, in fact, it is a distinctly double structure. In most members of the phylum, nearly all segments have a ganglion from which the principal lateral nerves extend.

These cardinal features of the phylum, and other important characteristics of annelids, are dealt with in more detail in the following sections of this chapter.

SEGMENTATION

The segments of an annelid lie between an anterior lobe of the body, called the **prostomium,** and a posterior lobe, called the **pygidium** (Figure 11.1). The prostomium is preoral; in other words, the mouth is located just behind it, on the first true segment. The pygidium bears the anus, unless this opening has been displaced forward during development.

The prostomium and pygidium are not regarded as segments because they are defined before any true segments are formed. The process of segmentation begins when serially arranged coelomic cavities start to appear in the mesoderm between the prostomium and pygidium (Figure 11.2). The youngest segment in the sequence is the one closest to the pygidium. The prostomium and pygidium also differ from segments in that they never develop bristles **(setae)** or certain other external or internal structures that are characteristic of segments.

The first true segment is called the **peristomium.** Frequently, however, one or more additional segments are united with it, so the posterior limit of the first segment is not clear. In such cases, the term *peristomium* is applied to what appears to be the first segment, even if it is produced by fusion of segments.

BODY WALL

The epidermal cells that line the outside of the body wall are typically columnar and secrete a fibrous cuticle. This consists of mucopolysaccharides and fills in the spaces between the microvillar outgrowths of the cells (Figure 11.3). The iridescence of many annelids is due to fine striations in the surface of the cuticle. These striations act as prisms, breaking the light that falls on the cuticle into spectral components.

The epidermis generally has scattered gland cells that secrete mucus, and in certain annelids there are also glands that produce luminescent material (Figure 11.33, C). The significance of the luminescence is not known, but the phenomenon occurs in many members of the phylum, including some earthworms.

Ciliation of the epidermis is characteristic of many annelids, but it is usually restricted to gills and other surfaces where respiration is effected, and to structures concerned with collecting microscopic food particles. In a few small types, however, cilia cover much of the body surface and are important in locomotion.

In most annelids, a distinct dermis lies beneath the epidermis. It is made up mostly of connective tissue, but is traversed by nerves and blood vessels.

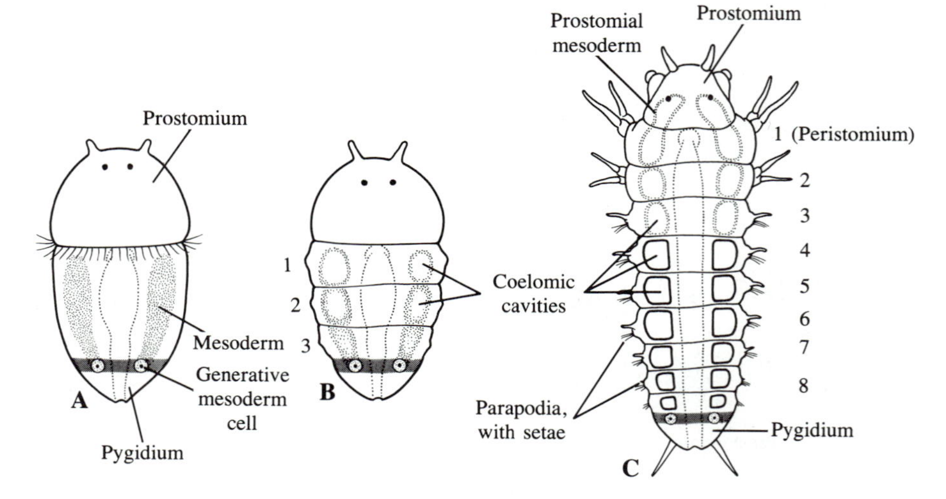

FIGURE 11.2 Diagrams showing the process of segmentation in an annelid of the class Polychaeta. A. Larva before segmentation has begun. B. Coelomic spaces in the mesoderm, and three segments rather distinct. C. Additional segments formed and head structures, parapodia, and setae differentiated. Note that the youngest segment is just anterior to the pygidium. (After Snodgrass, Smithsonian Miscellaneous Collections, 97.)

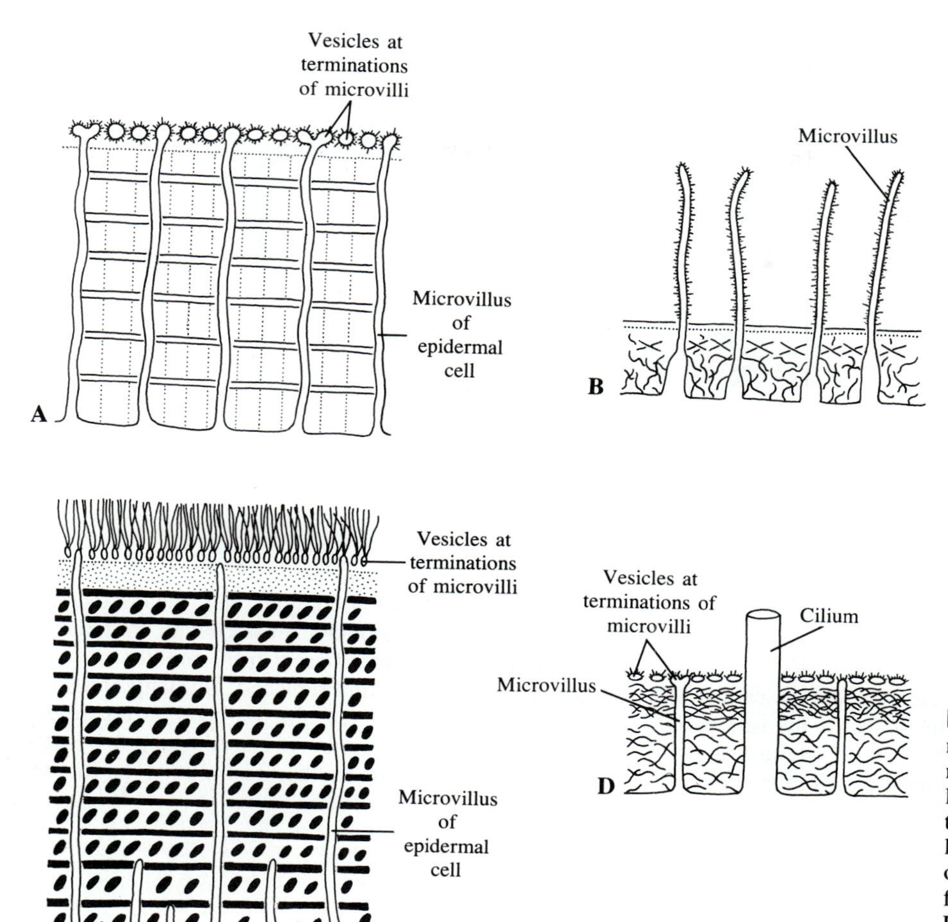

FIGURE 11.3 Cuticles of four annelids; diagrams based on electron micrographs. A. *Scoloplos,* a polychaete. B. *Polygordius,* one of the polychaetes called archiannelids. C. *Lumbricus,* a terrestrial oligochaete. D. *Aelosoma,* a freshwater oligochaete. (After Rieger and Rieger, Acta Zoologica, *57.*)

The musculature of the body wall is organized into two main layers (Figure 11.4). The outer layer consists of fibers whose orientation is circular, and which are restricted to individual segments. The inner layer consists of longitudinal fibers, and each fiber typically spans several segments and overlaps other fibers. The continuity of the longitudinal muscle is especially important to annelids that swim or crawl by undulations of the body, and to sedentary types that must be able to withdraw instantly into a tube, burrow, or crevice when threatened.

In addition to circular and longitudinal muscles, many annelids have oblique muscle bands that originate near the ventral midline and run laterally into the body wall. These are especially characteristic of certain polychaetes. Leeches have a layer of diagonal muscles between the circular and longitudinal muscles; they also have numerous dorsoventral muscle bands.

The innermost cellular layer of the body wall is the peritoneum. Along the midline, it forms a double partition that divides each segmental coelomic compartment into right and left halves. This structure serves as a **mesentery** in which the gut and certain major blood vessels are suspended.

The septa between segments consist partly of connective tissue and muscle fibers. They also have blood vessels and nerves, however, and both sides of each septum are lined by a peritoneum continuous with that of the body wall and mesentery.

SETAE

In most annelids, other than leeches, nearly all true segments have bristles called setae.[1] Each seta is composed of chitin and protein, and is secreted within a

[1]These structures are also called *chaetae* (singular *chaeta*). This term is, in fact, preferable, because it reduces the risk of confusing annelid bristles with the various kinds of setae found in arthropods. However, perhaps because *seta* is shorter and easier to spell, it is now entrenched in the literature dealing with annelids.

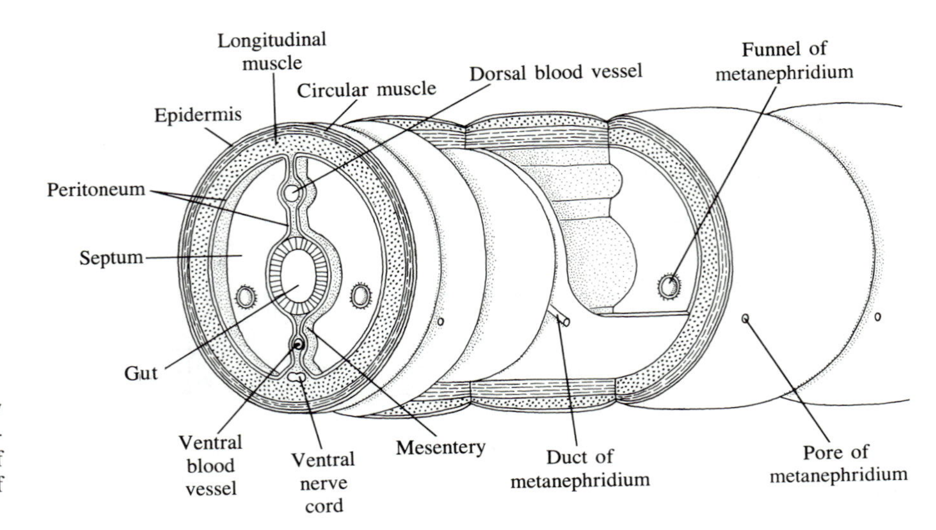

FIGURE 11.4 Diagram of a few segments of an annelid, generalized, showing the musculature of the body wall and arrangement of certain internal organs.

deep epidermal pit (Figure 11.5). A large cell at the bottom of the pit polymerizes the secretions of the other cells. The size, number, and arrangement of microvilli on the apical surface of this cell, which is called a **chaetoblast,** determines both the shape of the seta and its internal structure. Many setae have serrations or other elaborations, and some consist of two articulating pieces. A seta differentiates and lengthens only at the base; the production of a complex seta means that the microvilli on the chaetoblast must change in order to make one part of a seta different from another.

A worn seta is shed and replaced. In many annelids, moreover, changes in body size or body form are accompanied by the appearance of larger setae, or by new types of setae. This is another testimonial to the remarkable properties of chaetoblasts.

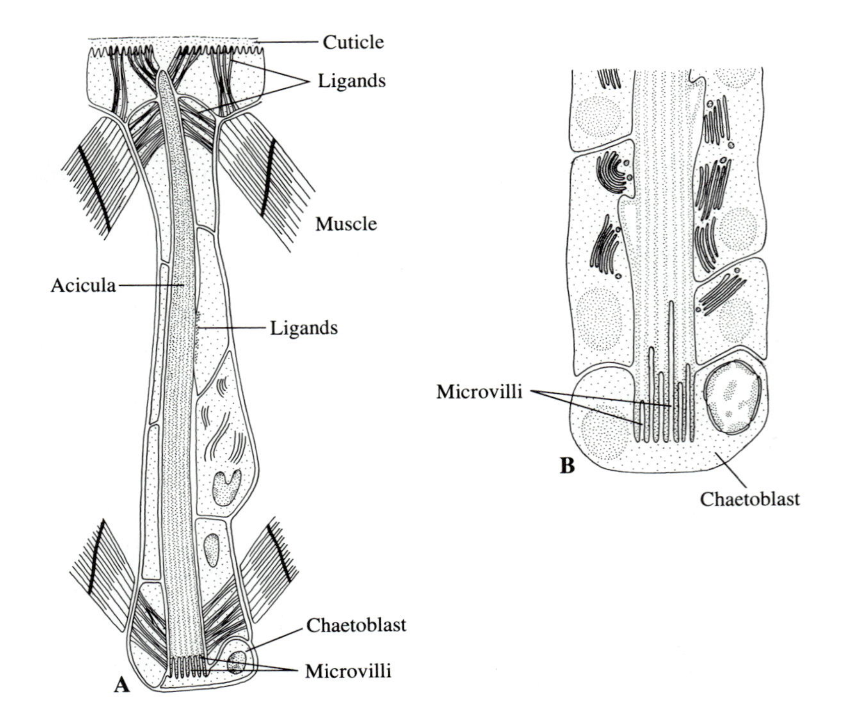

FIGURE 11.5 A seta (acicula) of a young polychaete, in the epidermal pit in which it has been formed. (After a drawing, based on electron micrographs, by Rita O'Clair.) B. A chaetoblast and adjacent epidermal cells concerned with formation of a seta. (After O'Clair and Cloney, Cell and Tissue Research, *151*.)

The protraction and retraction of a seta, and changes in the angle at which the bristle projects from the body, are controlled by muscles. These muscles are not attached directly to the seta but to the epidermal cells that line the pit in which the seta was secreted. Bundles of fibers, called **ligands,** run through the epidermal cells, thereby linking the muscles with the seta at points where the cells are tightly bonded to it.

COELOM

The coelom, being filled with fluid and enclosed by a muscular body wall, is a hydrostatic skeleton. As certain muscles contract and others relax, the animal changes shape, though its volume remains the same because the fluid in the coelom cannot be compressed. Locomotion and other activities of an annelid depend upon changes in shape. The coelom also functions in distribution of nutrients, oxygen, and metabolic wastes, and as a cavity in which gametes mature. If there are ducts that carry out gametes, or ducts concerned with excretion and osmoregulation, these connect the coelom with the exterior.

Segmentation, together with an elaborate musculature and an efficient nervous system, permits accurately controlled movements and allows for considerable versatility. The septa between segments more or less separate the coelomic cavity of each segment from that of preceding and succeeding segments. Suppose that the septa are rather "tight," so that coelomic fluid does not move freely from one segment to another. When circular muscles in the body wall of a particular segment contract and the longitudinal muscles relax, this segment becomes longer and more slender (Figure 11.6, A). When the circular muscles relax and the longitudinal muscles contract, the same segment becomes shorter and fatter (Figure 11.6, B). It is also possible for muscles of the body wall to contract or relax in such a way that a segment becomes longer on one side and shorter on the other, so that a curvature is produced. Such changes in the shape of individual segments, or series of segments, are important in nearly all activities of annelids. The forward progression of an earthworm, for example, depends on elongation of some segments as others hold fast to their position by means of setae that engage the substratum. An annelid whose swimming or crawling movements involve undulations of the body must be able to bend some segments in one direction while it bends others in the opposite direction.

In many annelids, septa are absent between some or all segments, and even when septa are present, they may be incomplete. However, as long as the musculature and other tissues of the body wall are strong enough to prevent uncontrolled bulging, the coelom can

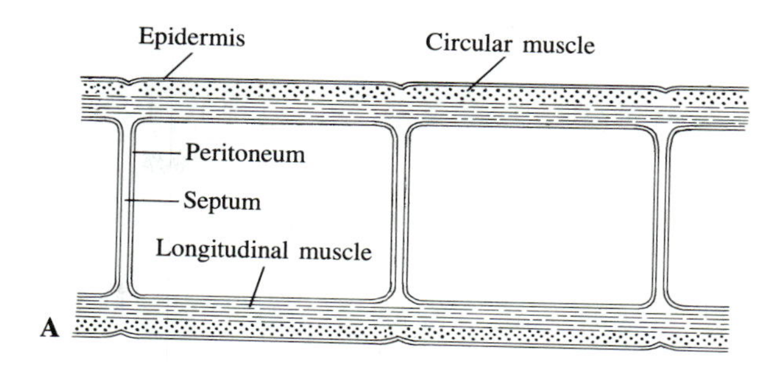

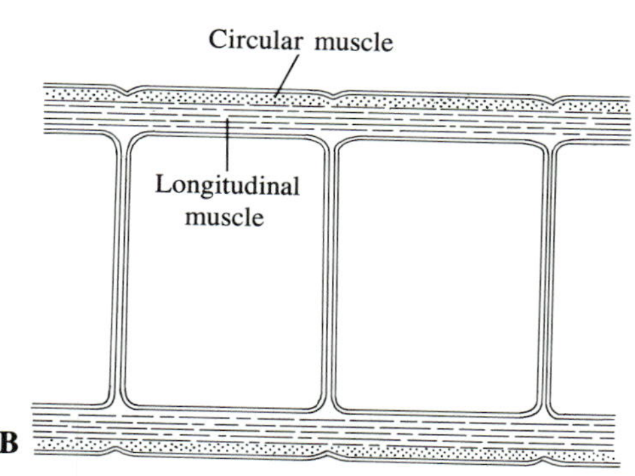

FIGURE 11.6 Diagram showing the relationship of body wall muscles to the shape of segments. A. Longitudinal muscles relaxed, circular muscles contracted, so that the segments are elongated. B. Longitudinal muscles contracted, circular muscles relaxed, so that the segments are shortened.

function as a hydrostatic skeleton in much the same way as it does in annelids that have "tight" septa.

PROTONEPHRIDIA, METANEPHRIDIA, AND COELOMODUCTS

Flatworms, nemerteans, and some other lower phyla that were dealt with in earlier chapters have protonephridia characterized by flame cells or solenocytes. These structures, of ectodermal origin, are believed to be concerned mostly with osmoregulation, though they probably sometimes also serve in excretion. In annelids, **solenocytic protonephridia** (Figure 11.7; Figure 11.8, A) occur in a few families of the class Polychaeta. They are paired and arranged segmentally, and each protonephridium has a duct to the outside. **Flame-cell protonephridia** are less common than those of the solenocytic type; they have been found in some polychaetes, and in the larva of at least one polychaete.

Most annelids, including the majority of polychaetes, have a different kind of paired excretory–osmoregulatory organ called the **metanephridium.** This, like a protonephridium, is of ectodermal origin. It consists, typically, of a funnel open to the coelom and a

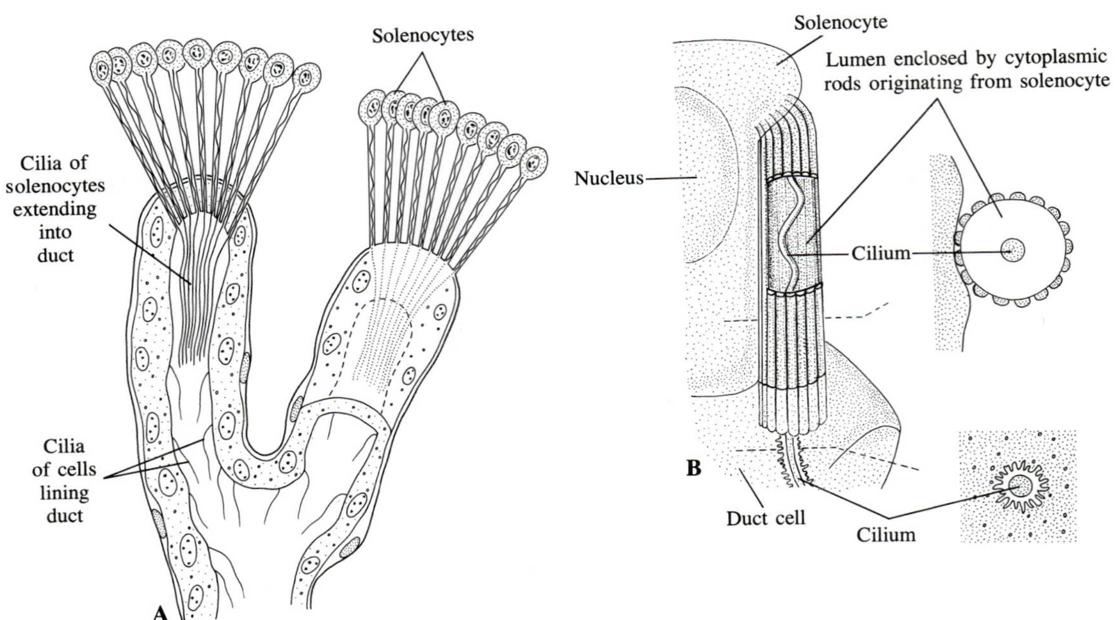

FIGURE 11.7 A. Portion of a protonephridium of *Phyllodoce paretti* (family Phyllodocidae), as seen with the light microscope, showing two clusters of solenocytes. (After Goodrich, Quarterly Journal of Microscopical Science, *86*.) B. Portions of one solenocyte and its associated duct cell from a protonephridium of *Glycera unicornis* (family Glyceridae); diagrams based on electron micrographs. The 17 cytoplasmic rods that originate from the solenocyte form a palisade around the cilium. They terminate in a cuplike cavity of the duct cell. The cilium, however, exends into the narrow duct that begins at the bottom of the cavity. The membrane lining the lumen of this duct is much folded. The drawings at the right are based on micrographs of sections taken at the levels indicated by the dashed lines. (After Brandenburg, Zoologische Beiträge, new series, *12*.)

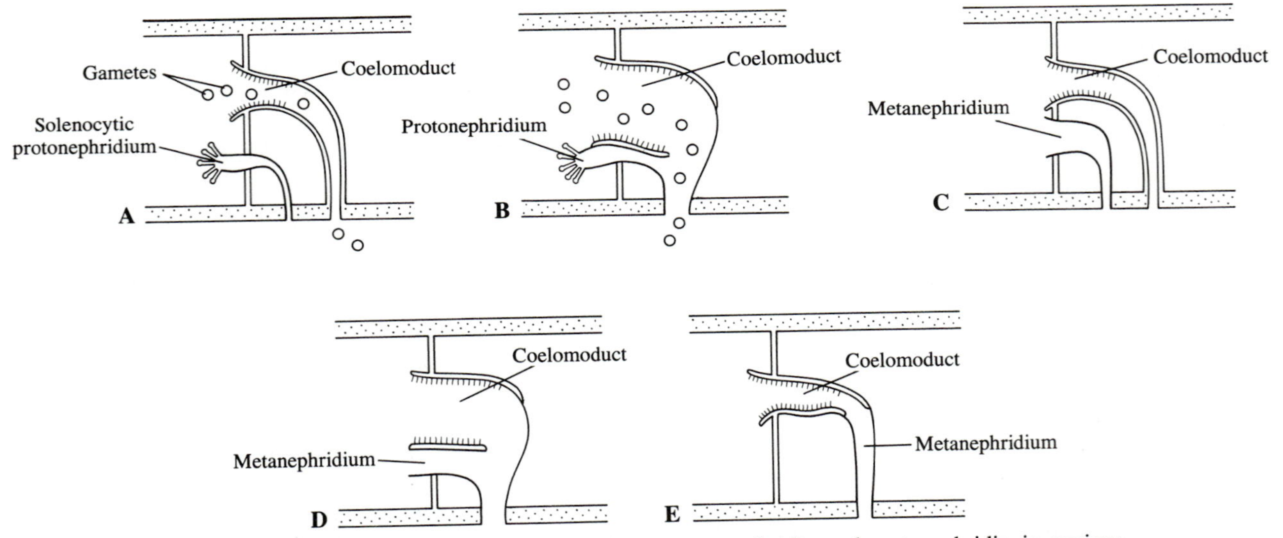

FIGURE 11.8 Relationships of coelomoducts to protonephridia and metanephridia in various polychaetes. A. Protonephridium (solenocytic type) and coelomoduct completely separate. B. Protonephridium and coelomoduct united, forming a protonephromixium. C. Metanephridum and coelomoduct completely separate. D. Metanephridium and coelomoduct united to form a metanephromixium. The duct of the metanephridium serves both funnels. E. Metanephridium and coelomoduct united in such a way that they form a mixonephridium. The funnel is derived from the coelomoduct, but most of the rest of the compound structure is derived from the metanephridium. (After Goodrich, Quarterly Journal of Microscopical Science, *86*.)

duct leading to a pore on the outside of the body wall (Figure 11.8, C; Figure 11.19, A; Figure 11.47). Unless the septa between segments have disappeared, the funnel of a metanephridium lies in one segment and the duct passes through the septum into the succeeding segment. In some annelids, such as the common garden earthworm, the number of pairs of metanephridia nearly equals the number of segments. In other words, almost every segment has the funnels of one pair of metanephridia. But this is not always the case; there may be only a few pairs of metanephridia serving a body with perhaps 50 or 100 segments.

Another kind of funnel organ, the **coelomoduct,** may also be present (Figure 11.8, A and C). Unlike metanephridia, coelomoducts are of mesodermal origin. Their function is to carry gametes out of the coelomic cavity. Coelomoducts are often reduced, however, to the point that they cannot serve as gonoducts. In such cases, the release of gametes from the coelom usually depends on rupture of the body wall.

Funnel organs are often compound structures, consisting partly of a coelomoduct and partly of a metanephridium (Figure 11.8, D and E). The product of such a fusion usually serves not only in excretion and osmoregulation, but also as a genital duct. Solenocytic protonephridia may be united with coelomoducts, too (Figure 11.8, B). The varying degrees of fusion of protonephridia and metanephridia with coelomoducts are discussed in greater detail below in connection with the class Polychaeta.

Classification

The approximately 10,000 species of annelids are assigned to three classes. The largest and most varied of these is the Polychaeta, which is almost exclusively marine, although it has a few representatives in fresh water. The other two classes, Oligochaeta and Hirudinoidea, have many species in fresh water but are also found in the sea and in moist places on land. In fact, the Oligochaeta, which includes earthworms, is one of the more successful groups of invertebrates occupying terrestrial habitats.

Among the features that distinguish the classes are the presence or absence of setae, the arrangement of setae when these are present, and the organization of the reproductive system. There are, however, other important considerations, and these are dealt with as the classes are characterized in succeeding sections of this chapter.

General References on Annelida

Anderson, D. T., 1973. Embryology and Phylogeny in Annelids and Arthropods. Pergamon Press, Oxford and New York.

Brinkhurst, R. O., 1982. Evolution in the Annelida. Canadian Journal of Zoology, *60:*1043–1059.

Dales, R. P., 1967. Annelids. 2nd edition. Hutchinson University Library, London.

Mangum, C., 1976. Respiratory physiology in annelids. American Scientist, *58:*641–647.

Mill, P. J., 1978. Physiology of Annelids. Chapman and Hall, London.

Nicol, J. A. C., 1948. The giant axons of annelids. Quarterly Review of Biology, *23:*291–323.

Ruppert, E. E., and Carle, K. J., 1983. Morphology of metazoan circulatory systems. Zoomorphology, *103:*193–408.

Class Polychaeta

The Polychaeta, with about 6000 species, is by far the largest class of annelids. Worms of this group generally have many setae (or chaetae) on each segment, and this is why they are called polychaetes. Another important feature is the presence of segmental structures called **parapodia.** These are paired lateral outgrowths in which the setal sacs are located.

A typical parapodium (Figure 11.9, A) has two main lobes: an upper one, the **notopodium** (*noto-* refers to the dorsal side of the animal), and a lower one, the **neuropodium** (*neuro-* here refers to the ventral side, where the nerve cord is located). As a rule, both lobes have setae, one of which is especially stout and serves as a supporting rod. It is called the **acicula** (plural *aciculae*)[2], and it usually projects only slightly, if at all, whereas the other setae of each lobe may be extruded for nearly their full length.

The notopodium often has a fingerlike dorsal outgrowth, and the neuropodium may have a similar ventral outgrowth. These structures are called **cirri** and they are primarily sensory. A portion of the notopodium or neuropodium may also be specialized as a **gill,** usually richly supplied with blood vessels. Other parapodial structures characteristic of many polychaetes are presetal and postsetal lobes, located on the anterior and posterior faces, respectively, of the notopodium or neuropodium.

Polychaetes have many different life styles. Accordingly, their parapodia are specialized for various functions. For instance, they may be broadened so that they serve effectively as paddles for swimming or for creating water currents within a tube. Among other modifications is the loss of the notopodium, or reduction of both lobes, even though the setae that belong to the lobes are present and well developed.

The setae of polychaetes also show much diversity, according to their function. Hooklike setae, for instance, maintain tight contact with the wall of a burrow or tube, and setae that resemble oars are typical of some

[2]An alternative spelling is *aciculum* (plural *acicula*).

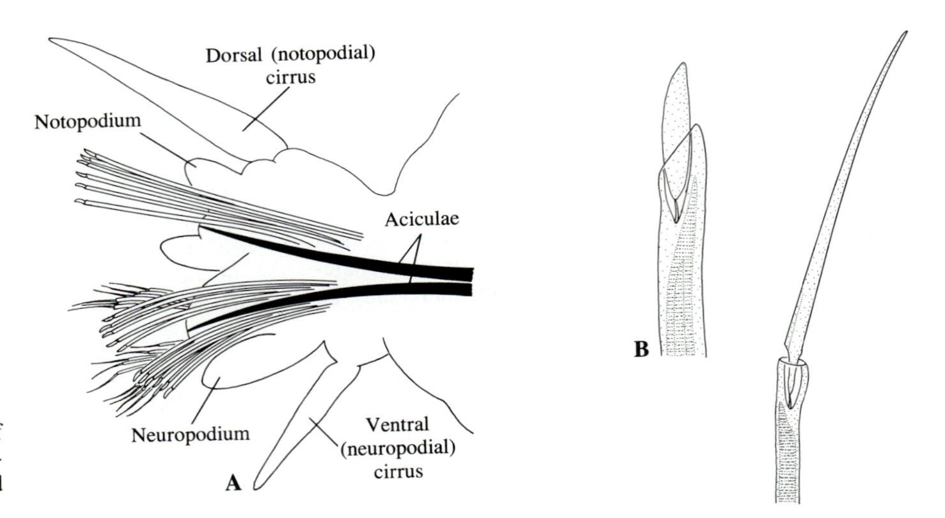

FIGURE 11.9 A. Parapodium of *Nereis pelagica* (family Nereidae). B. Two types of compound setae from the parapodium.

swimming polychaetes. More than one type of seta may be present on a single notopodium or neuropodium, and the two lobes may be decidedly different in terms of the setae they bear. Setae that consist of a single piece are said to be *simple* (Figure 11.10, A); those that are divided into a long basal portion and a terminal unit are said to be *compound* (Figure 11.9, A; Figure 11.10, B).

Head Region

Specializations of the head are even more varied than those of the parapodia. The **head** is here defined as the portion of the body that consists of the prostomium and peristomium, which, taken together, generally form a distinctive region. The latter, as explained previously, is sometimes the first true segment, sometimes the product of fusion of the first segment with one or more succeeding segments.

In some polychaetes, especially active crawlers and swimmers, the head has a variety of sensory structures (Figure 11.11). The prostomium may have one or more pairs of **eyes,** one to several **antennae,** and a pair of fleshy sensory structures called **palps.** Antennae and eyes are basically dorsal, whereas palps arise from the ventral side of the prostomium, unless they have been displaced laterally. The peristomium often bears **tentacular cirri,** which are derivatives of the notopodial portions of parapodia. When there is more than one pair of these, this is a sign that the peristomium has been produced by more than one segment.

Contrasting with heads that have eyes, antennae, and

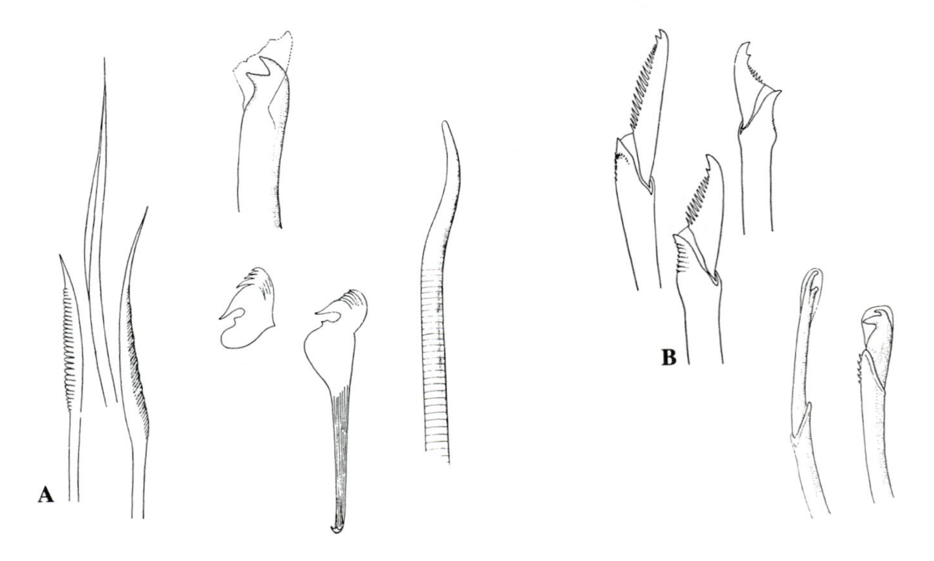

FIGURE 11.10 Polychaete setae. A. Several types of simple setae. B. Several types of compound setae.

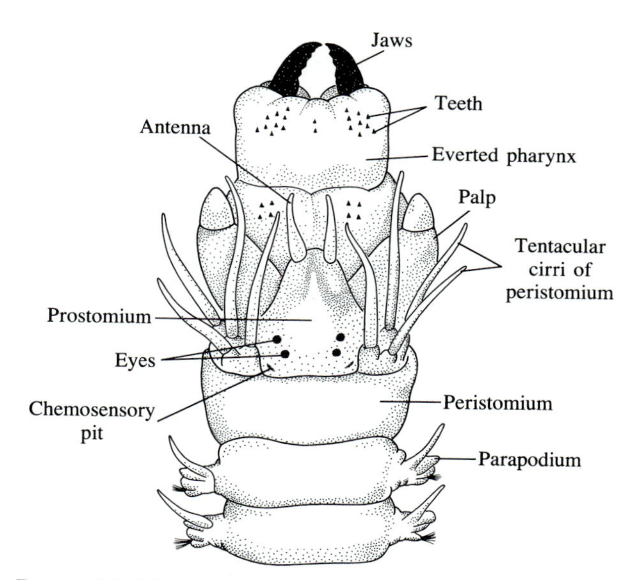

FIGURE 11.11 Head Region of a nereid polychaete, pharynx everted.

other outgrowths are those that have only eyes, or that have no obvious sensory structures at all. One must not assume, however, that a polychaete without visible cephalic sense organs has no sensory structures in the head region; these are certainly present, even though they are not easily recognized.

In several groups of polychaetes, particularly tube-dwellers that feed by collecting microscopic particles from the water or from sediment, the head itself is much reduced but gives rise to several to many extensile tentacles or featherlike outgrowths. These structures secrete mucus that traps food particles, and they also have tracts of cilia, often concentrated in grooves, for moving the mixture of mucus and food toward the mouth. Some polychaetes that feed in this way, as well as polychaetes that have other specializations of the head region, are discussed below in connection with the life styles of selected families.

Digestive Tract

In most polychaetes, the gut is a straight tube, but it is differentiated into regions. As a rule, a pharynx, esophagus, stomach, and intestine are recognizable, but there are deviations from this plan. In a few worms of this diversified class, the gut is coiled or folded back on itself and is therefore considerably longer than the body.

In some groups, the pharynx can be protruded as a grasping organ (Figure 11.11). Such a pharynx is usually equipped with jaws or hardened teeth, or both, and it may have glands that secrete a venom for quieting prey (Figure 11.23, C). A protrusible pharynx is not necessarily restricted to carnivorous polychaetes, however; many herbivorous species use jaws and teeth for grasping and tearing pieces of seaweed, or for defense.

Digestive enzymes, appropriate to the kind of food consumed, are produced either by glands that are part of the lining of the gut or that are concentrated in special diverticula. *Nereis,* for instance, has a pair of diverticula arising from the anteriormost portion of the esophagus, but there are also digestive glands in the epithelium of the long intestinal region.

Protonephridia, Metanephridia, and Coelomoducts

The paired excretory–osmoregulatory organs characteristic of annelids are basically either protonephridia or metanephridia. These structures are of ectodermal origin, but they are often united with coelomoducts, which are of mesodermal origin. The primary function of coelomoducts is to carry gametes out of the coelom. When a coelomoduct is joined to a protonephridium or metanephridium, the compound structure that is formed functions both as a gonoduct and as an excretory–osmoregulatory organ.

Coelomoducts may unite with protonephridia or metanephridia in various ways. Figure 11.8 shows some of the arrangements; others differ only in minor details. The term **protonephromixium** is applied to a compound structure in which the funnel portion of a coelomoduct is joined to the duct of a protonephridium (B). This is the situation in certain genera of several families of freely mobile (''errant'') polychaetes. The hypothetical ancestral condition in which protonephridium and coelomoduct are completely separate (A) has not been found and perhaps does not exist. The assumption that fusion of the two components has taken place is based on the fact that in one genus of polychaetes (*Vanadis,* in the family Alciopidae) there are paired segmental pockets in which gametes accumulate. These pockets do not open to the outside, and are therefore not typical coelomoducts, but they do eventually become joined to the protonephridial ducts as shown in the illustration.

Separate metanephridia and coelomoducts (C) are known only in the family Capitellidae, and only a few families have compound structures called **metanephromixia,** in which the funnels of a coelomoduct and a metanephridium are recognizable, although the duct to the outside is single (D). In most polychaetes that have funnel organs, these are of a type called **mixonephridia** (E; see also Figure 11.19, B). The funnel end consists of a coelomoduct, whereas most of the rest of the compound structure is an ectodermal metanephridium. This is the arrangement in many errant polychaetes and in almost all polychaetes that secrete tubes or that live in more or less permanent burrows.

Coelomoducts may become specialized for some function other than gamete release, or they may disappear almost entirely. In the family Nephtyidae, for instance, they are rather closely associated with protonephridia, but they are reduced to small structures called **ciliophagocytic organs,** concerned with accumulation and processing of wastes. In polychaetes of the family Nereidae, which have metanephridia rather than protonephridia, there are ciliophagocytic organs in the dorsolateral portions of the coelom of most segments. When the coelomoducts no longer function as gonoducts, the gametes are generally released through breaks in the body wall.

Whether an adult polychaete has metanephridia or protonephridia, it is likely to have simple protonephridia during its development. This is especially true of polychaetes whose development includes a free-swimming larval stage.

Respiration

In some polychaetes, there are no obvious respiratory structures; most of the body surface probably functions in exchange of gases. Many members of the class, however, have distinct gills. Some gills, as mentioned previously, are outgrowths of notopodia or neuropodia. These usually have a blood supply. Others originate from the dorsal body wall and are basically separate from the notopodia, even if they are close to them. Gills of this type may be simple filaments or they may be branched in a comblike, featherlike, or bushy pattern. Sometimes they are found on only one or a few segments, sometimes on numerous segments. They may have coelomic channels extending into them. This is the case in the families Glyceridae and Capitellidae, in which the circulatory system is reduced.

Circulation of water over respiratory surfaces, whether or not there are differentiated gills, is generally promoted by activity of epidermal cilia. Undulations of the body and waving of the gills themselves may also be important. In certain polychaetes that live in tubes or burrows, peristaltic movements of the body or rhythmic beating of parapodial lobes creates respiratory currents.

Some tube-dwelling polychaetes have filamentous or branched structures that originate from the head region. The primary function of these is to trap microscopic food particles, but because they have a large surface area and are pervaded by blood vessels and coelomic channels, they are also effective respiratory organs.

Circulatory System

Nearly all polychaetes have a circulatory system. It is unusual to find capillary networks exactly comparable to those of vertebrates, but there are nevertheless many small vessels between larger vessels, and all vessels, except those that are strongly muscularized, have thin walls.

There is much variation in the organization of the circulatory system. Generally, however, there are two especially prominent longitudinal vessels that run for most of the length of the body. One of them, the **dorsal vessel,** is above the gut; the other, the **ventral vessel,** is beneath the gut (Figure 11.12). The dorsal vessel carries blood mostly in the anterior direction. Branches of it lead directly or indirectly to the ventral vessel, which carries blood primarily in the posterior direction. There are, in addition, many vessels that supply blood to the tissues or that return blood to the major vessels.

Most of the larger blood vessels, especially the dorsal vessel, are contractile and show peristaltic activity. There may also be one or more distinct hearts. These are enlarged portions of certain vessels and function as "booster pumps." In some polychaetes, the complexes of smaller blood vessels in certain organs or tissues have contractile branches that end blindly. These also help to keep blood circulating, though often in a more nearly back-and-forth movement than in one direction. In the family Sabellidae, contractions of blind vessels are responsible for much of the circulation of blood.

A portion of the gut may have a plexus of enlarged blood vessels, so that there is almost a pool of blood in the tissues. This is the case, for instance, in *Arenicola* (Figure 11.12, B), in which the plexus develops gradually from a network of small vessels seen in the gut wall of young specimens.

The blood contains amoebocytes and generally has a respiratory pigment. This is usually **hemoglobin,** but in certain families of polychaetes it is a greenish pigment called **chlorocruorin.** This is similar to hemoglobin in that it is a porphyrin compound, its iron being concentrated in heme units. When found in the blood, these pigments are usually dissolved in the plasma and are in molecules much larger than those of hemoglobin in red corpuscles of vertebrates. Whereas the corpuscle-bound hemoglobin of a mammal has a molecular weight of about 60,000, the plasma hemoglobin of some polychaetes has a molecular weight of about 3 million.

Polychaetes whose hemoglobin is concentrated in corpuscles usually have a reduced circulatory system. The corpuscles are in the coelomic fluid, which functions as blood, transporting oxygen, nutrients, and wastes. The families Glyceridae and Capitellidae, mentioned above as groups in which the gills are pervaded by coelomic channels, are among the polychaetes whose circulatory system is reduced. Their corpuscular hemoglobin is in comparatively small molecules. Localization of hemoglobin in corpuscles may be a useful adaptation in that it probably minimizes loss of coelomic respiratory pigment through funnel organs that function as excretory–osmoregulatory structures or as gonoducts.

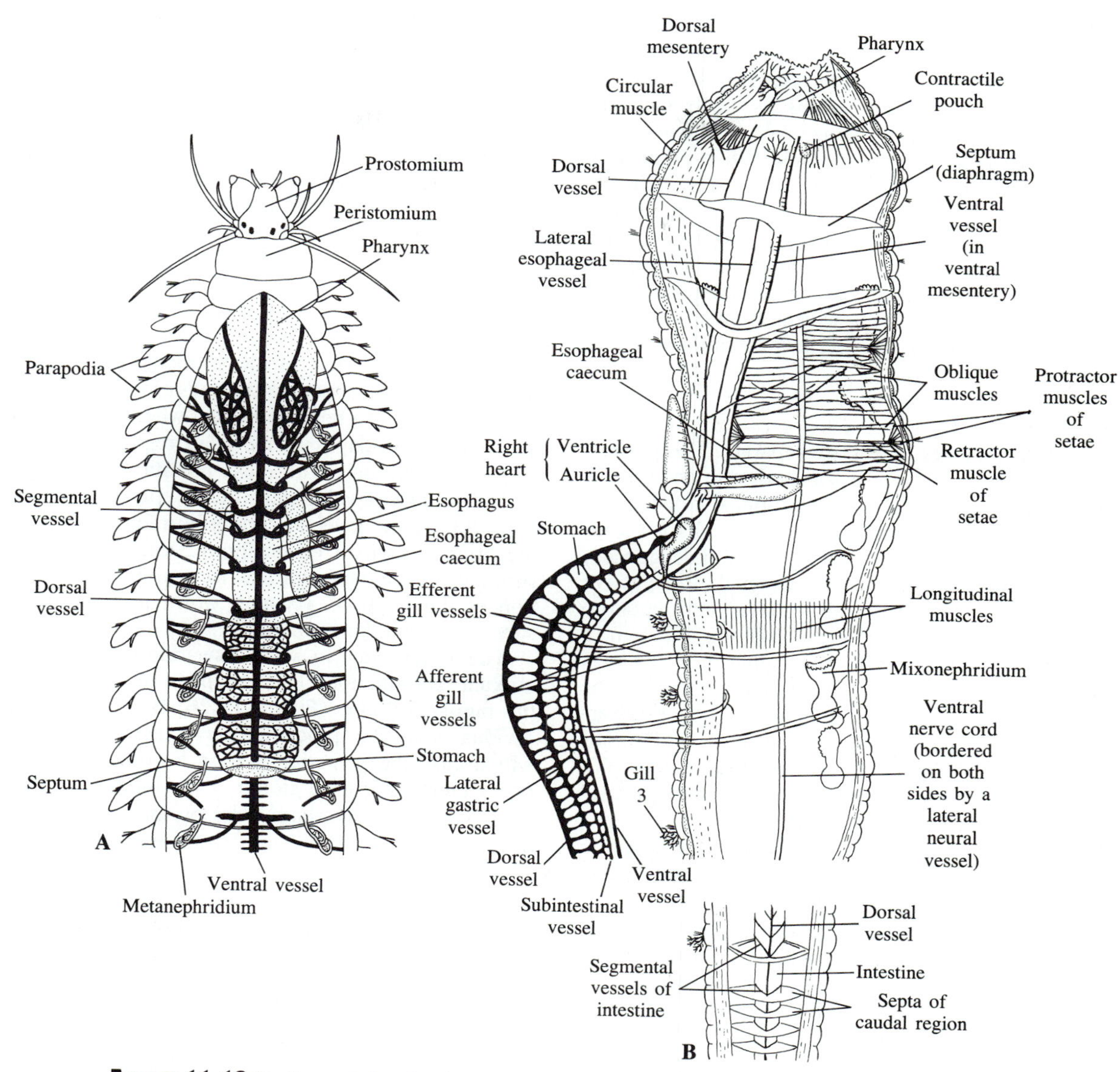

FIGURE 11.12 Portions of the circulatory system of two polychaetes, as seen in dissections from the dorsal side. A. *Nereis*. B. *Arenicola*. The gut, in the anterior part of the body, has been pushed to the left, and much of the posterior part of the body has been omitted. Note the plexus of blood vessels around the stomach. All six of the mixonephridia on the right side of the body are shown, but those on the left side are omitted. Longitudinal muscles and oblique muscles are shown only in restricted areas. Many small blood vessels are not illustrated. Some structures have been shifted slightly from their normal position to avoid congestion of lines. (After Ashworth, Liverpool Marine Biology Committee Memoirs, *11*.)

Worms of the family Serpulidae have both hemoglobin and chlorocruorin in the blood, and there are some polychaetes in which hemoglobin is both bound in corpuscles in the coelom and dissolved in the plasma of the blood. Members of the family Magelonidae have a totally different pigment, **hemerythrin,** concentrated in blood corpuscles. Though it is red and contains iron, it is not a porphyrin compound. Instead of having heme units that bind oxygen, hemerythrin carries oxygen between each of two iron atoms.

In most polychaetes, respiratory pigments function in much the same way as hemoglobin does in vertebrates. When the oxygen tension in the external environment is low, oxygen held by the respiratory pigments splits off and becomes available for cellular respiration. In some polychaetes, the oxygen tension in the environment must drop to an extremely low level before oxygen dissociates from the blood pigment. This is especially true of certain polychaetes that inhabit mudflats. At low tide, when worms living beneath the surface are unable to irrigate their tubes or burrows, the demand for oxygen by bacteria and other organisms continues to be high. The supply of oxygen held by hemoglobin cannot last long once dissociation begins, so most worms in a situation of low oxygen tension must sooner or later enter a phase of anaerobic respiration or pick up oxygen from the air that may temporarily pervade their burrows. Lugworms (family Arenicolidae) are known to engage in aerial respiration to some extent.

Sexual Reproduction

In most polychaetes, sexes are separate. The primordial germ cells are closely associated with the peritoneum, but they are believed to originate outside the coelom. Sometimes the patches of germ cells are conspicuous enough to be recognizable as simple gonads, but as a rule they are not easily detected. Prospective gametes break away from the patches and undergo much of their development in the coelom. Thus, in a polychaete that is ready to spawn, some or many segments are nearly filled with sperm or large oocytes. The number of segments involved in gamete production varies according to the type of polychaete. In families characterized by division of the body into a stout thoracic region and a slender abdominal region, gametes are usually formed in the abdomen.

There are three main ways in which ripe gametes may escape from the coelom. In some polychaetes, such as those belonging to the burrowing family Capitellidae, the coelomoducts are separate from the metanephridia and serve purely as gonoducts. In the majority of polychaetes, however, the funnel organs are compound structures; they are either mixonephridia or metanephromixia, derived partly from coelomoducts and partly from metanephridia, or they are protonephromixia, produced by the union of coelomoducts and protonephridia (Figure 11.8). These compound structures serve as excretory–osmoregulatory organs and also as gonoducts. Good examples of mixonephridia can be seen in lugworms. There are several pairs of them in the anterior part of the body. The elaborate funnel of each mixonephridium is a coelomoduct, but most of the remaining portion is a metanephridium (Figure 11.19, B).

The third main type of gamete release is found in the families Nereidae, Eunicidae, and a few others. Eggs or sperm are discharged, often in clouds, through breaks in the body wall. The ruptures develop after enzymes have softened the tissues, a phenomenon known as **sarcolysis.** This method of spawning usually takes place while the worms are swarming near the surface. As a rule, the worms die after they have released their gametes.

In most polychaetes that are strictly benthic or strictly pelagic, individuals ready for sexual reproduction are not especially different from other full-grown worms. In some species that are primarily benthic, however, spawning takes place during a terminal stage in which the worms swarm at the surface, and these swarmers are often very different from the benthic individuals.

One way in which a swarmer develops is by transformation of an entire worm. This is common in the family Nereidae. Most of the segments behind the first few develop large parapodia and broad-bladed setae. The eyes may also become larger, whereas the antennae and other appendages of the head region may become smaller. Individuals with these modifications are called **epitokes.** The development of an epitoke is activated by a drop in the level of certain hormones produced by the brain. Removal of the brain from a young worm stimulates premature epitoke formation, and implantation of the brain of a young worm into an older one suppresses epitoke formation.

The other way in which swarmers develop is by asexual reproduction. This is typical of many species in the family Syllidae (Figure 11.26, B–F). A sexual individual may be produced by transverse fission, in which case the posterior portion becomes the epitoke, or by budding from a stolon at the hind end. The epitokes generally have large eyes on the prostomium and differ in various other ways from asexual individuals. In some species the male epitokes are different from the females. All of this has made the taxonomy of syllids difficult, because in the past separate names may have been applied to male and female epitokes and to the nonepitokous phase of the same species.

In polychaetes of certain families, a hind portion filled with gametes breaks away from the rest of the worm. This is not a true epitoke, for it is not a complete worm. A particularly famous case is that of the palolo worm, *Palola viridis* (Figure 11.22, H), found in the Samoan region. When this polychaete, a member of the family Eunicidae, prepares for reproduction, the posterior two-thirds of the body becomes ripe with gametes. On one or two nights in late October or early November, when the moon is in its last quarter, the hind portions of the worms break off and start swimming. So many spawning pieces clutter the water that the natives of the region collect them for food, and other animals also take advantage of the situation. The anterior portions of the worms remain in the reefs and regenerate their hind parts.

As a rule, female polychaetes release their eggs, either singly or in clusters held together by mucus. The

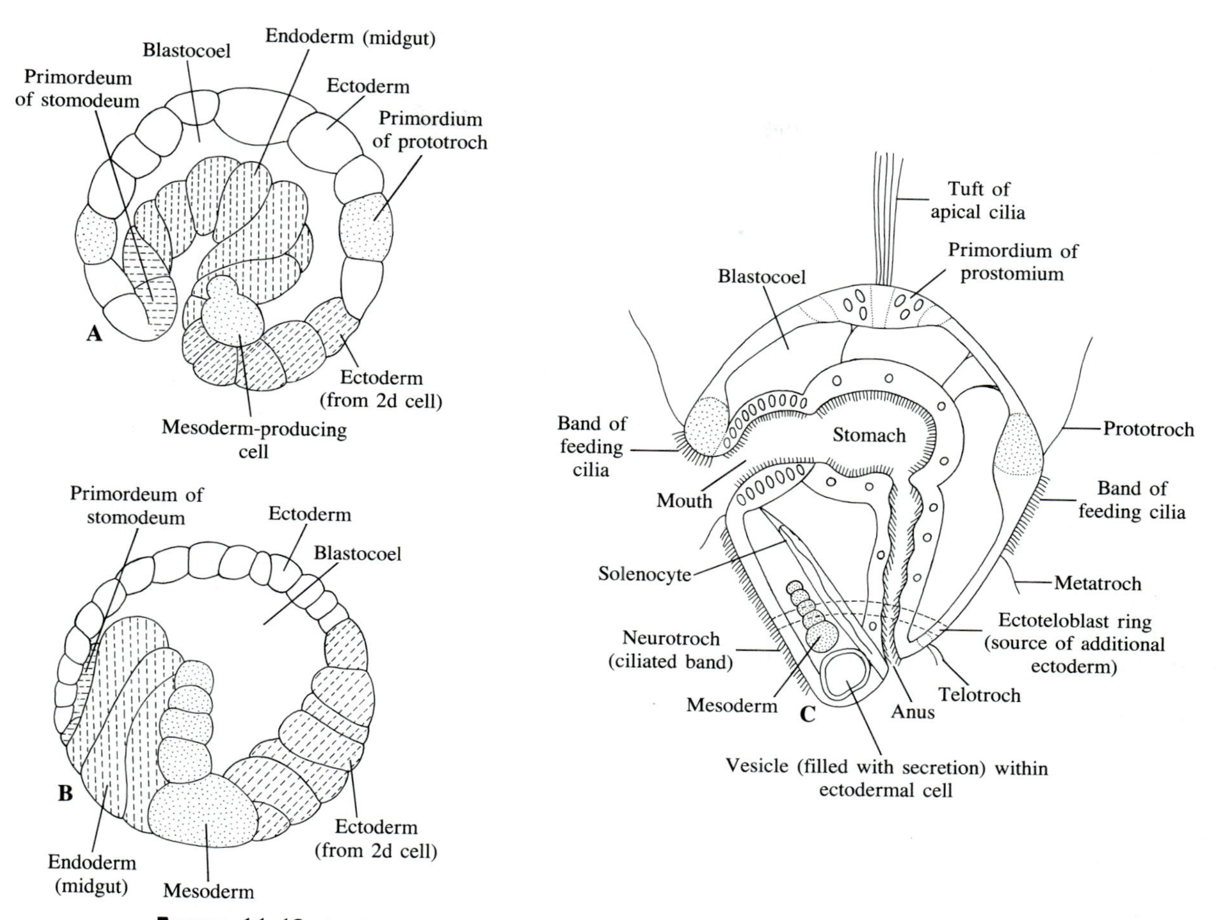

FIGURE 11.13 A. Gastrula of *Hydroides uncinatus* (family Serpulidae), a species whose egg contains little yolk. B. Gastrula of *Scoloplos armiger* (family Orbiniidae), a species whose egg contains a moderate amount of yolk. C. Trochophore of *Galeolaria caespitosa* (family Serpulidae), in optical section.(A–C, after Anderson, Acta Zoologica, *47*.)

eggs may, however, be stuck to sand grains, to the tube in which the worm is living, or to other objects. Brooding is characteristic of several families. In scaleworms, for instance, the young may be brooded under scalelike structures, called elytra, on the dorsal surface, and some serpulids care for their young within the tube or in the inverted cone of the operculum (Figure 11.31, C and D). An interesting type of brooding is found in certain syllids, in which the large eggs are stuck to the sides of the female's body, between the parapodia, and there develop into little worms (Figure 11.26, A).

Viviparity is rare, but it does occur. An especially striking case is that of *Nereis limnicola* on the Pacific coast of North America. This species is found in brackish water and even in fresh water, and it broods its young within the coelom. To complicate matters further, it is a self-fertilizing hermaphrodite. The viviparity is probably a good adaptation for a nereid that lives in conditions of low or varying salinity, for the young can develop under more nearly constant conditions than they could if they were swimming free.

Development

The following account of polychaete development is generalized, based on what happens in species that have a larval stage called the **trochophore.** The eggs of such polychaetes generally have a small to moderate amount of yolk, and cleavage is spiral. Sometimes an obvious blastocoel is present, sometimes not. Gastrulation (Figure 11.13, A and B) may be achieved by invagination, by epiboly (the overgrowth of prospective endoderm cells by ectoderm cells), or by a mixture of the two methods. An archenteron is formed, and soon the embryo reaches the trochophore stage (Figure 11.13, C).

The trochophore typically has a complete gut, the mouth being derived from part of the blastopore, and the anus having been formed secondarily. As the blastopore elongates, the anterior part of it remains open, and a stomodeal invagination pushes the original mouth deeper. Then the rest of the blastopore closes. Later, a second opening appears in what had been the posterior portion of the blastopore, and this opening, on joining the gut, becomes the anus.

The trochophore usually has an apical tuft of sensory cilia at the anterior end. Around its equator is a band of long cilia called the **prototroch,** and a band of shorter cilia concerned with driving food particles toward the mouth. Long cilia encircling the posterior part of the larva form the **telotroch,** and if there is a similar band between the feeding cilia and the telotroch, this is called the **metatroch.** A longitudinal band of short cilia, located ventrally between the mouth and anus, is called the **neurotroch.**

Part of the mesoderm originates from the 4d cell of the embryo, as it does in other invertebrates characterized by spiral cleavage. The 4d cell sinks inward during gastrulation and the two cells produced by its division multiply to form some endodermal cells for the intestinal region and two masses of mesoderm that flank the gut. It is within this mesoderm, called **endomesoderm,** that the coelomic spaces will appear. Another kind of mesoderm appears in the blastocoel, too. This is called **ectomesoderm,** because it comes from certain embryonic cells that also contribute to the ectoderm. The ectomesoderm differentiates mostly into various muscle bands of the larva, including those that run from the outer covering to the gut. Nerves and an anterior ganglionic mass beneath the sensory plate are of strictly ectodermal origin, as are also the two protonephridia that lie on both sides of the gut. Each protonephridium is a single solenocyte, or a single flame cell.

Although some trochophores, as they swim, feed on diatoms and other small organisms, others are not capable of feeding. Those that develop from eggs that have a lot of yolk usually rely entirely on this yolk for their nutrition, and their free-swimming existence is often short. Whether or not it feeds, a trochophore gradually becomes transformed into a small polychaete. The region between the mouth and the telotroch lengthens and the first traces of segmentation appear internally as well as externally, for the endomesodermal bands flanking the gut begin to develop a succession of coelomic cavities (Figure 11.2). The region anterior to the mouth becomes the prostomium; although the endomesoderm that goes up into it may develop one or more pairs of coelomic sacs, this unit of the body is not counted as a true segment. The posteriormost part of the trochophore, where the anus is located, becomes the pygidium, and the newest segment always forms directly in front of it.

The development of the earliest parapodia and setae may precede the appearance of the first segments, but it is probably more common for them to differentiate about the same time as the segments (Figures 11.14, 11.15). The first setae do not necessarily survive; in some polychaetes they are shed and then replaced by definitive setae.

As the trochophore transforms into a segmented worm, a midventral thickening becomes noticeable. This marks the location of the nerve cord. The ciliated bands, protonephridia, and ectomesodermal muscles of the trochophore disappear. New body-wall muscles develop from the bands of endomesoderm, which also give rise to the peritoneum that lines the coelomic sacs.

The number of segments that form before the young worm settles into its usual habitat depends on the species. As a rule, metamorphosis is not completed until the worm does settle, unless the species is permanently planktonic. Some polychaetes have very specific requirements so far as the substratum is concerned. Larvae of *Ophelia bicornis* (family Opheliidae), *Owenia fusiformis* (family Oweniidae), and *Notomastus lateralis* (family Capitellidae) are among those that will settle and metamorphose only if they come into contact with soft substrata in which the sand grains are of appropriate size and are coated with appropriate bacteria and other organic matter.

Not all polychaetes have a trochophore larva. Direct development is characteristic of many species, especially those whose eggs are large and full of yolk. Even when there is a trochophore, further development does not necessarily follow the pattern described above. Some young polychaetes, as they transform into adults, go through distinctive temporary stages.

Asexual Reproduction

The formation of swimming epitokes by syllids has already been mentioned above. This is an asexual process, even though it produces sexual individuals. Various other polychaetes divide into two pieces or break up into several pieces, each of which regenerates the parts that it lacks. Asexual reproduction of this sort has been observed in several families, including the Chaetopteridae, the Cirratulidae, the Spionidae, and the Sabellidae.

Some cirratulids break up into individual segments, and each segment does better than regenerate a new worm: it produces two worms, and after these become detached, it may produce two more (Figure 11.38, E–H). This type of asexual reproduction can quickly lead to the formation of dense colonies of a single species.

Regeneration

Most polychaetes that have lost structures such as antennae, palps, and parapodia can regenerate these. The capacity for regenerating substantial portions of the

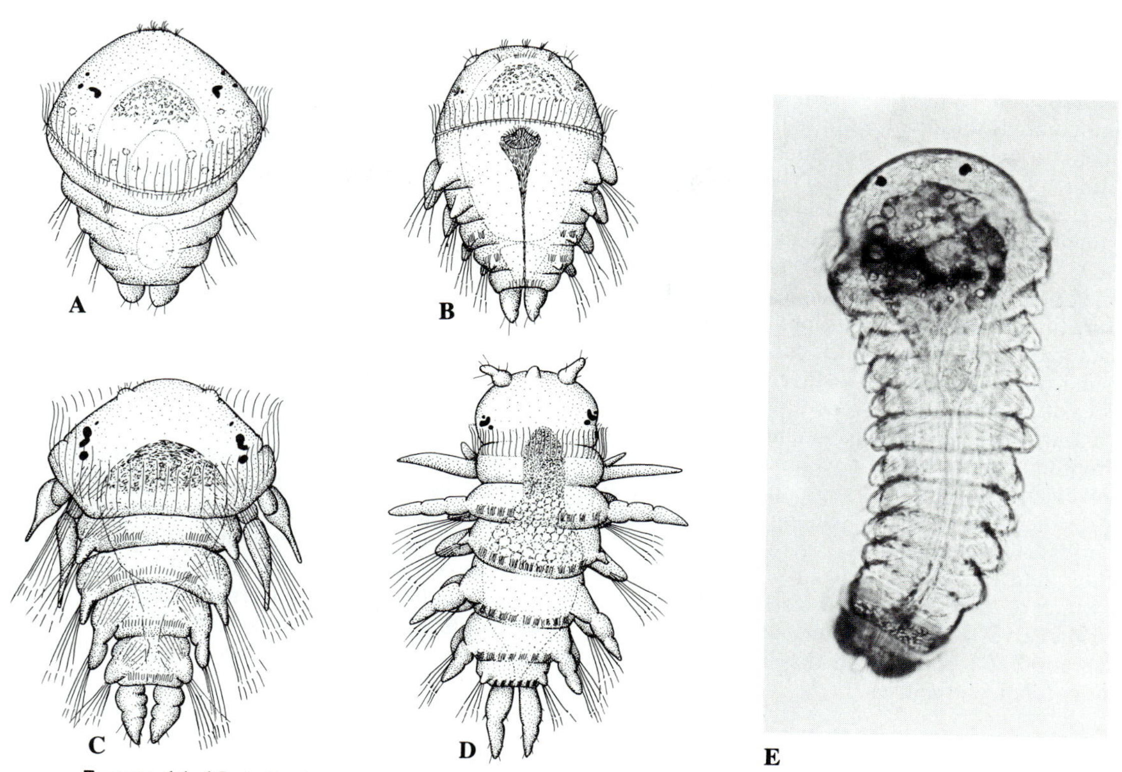

FIGURE 11.14 A–D. Successive stages in the development of *Ophiodromus pugettensis*. (Blake, Ophelia, *14*.) E. Segmented young polychaete found in plankton; photomicrograph.

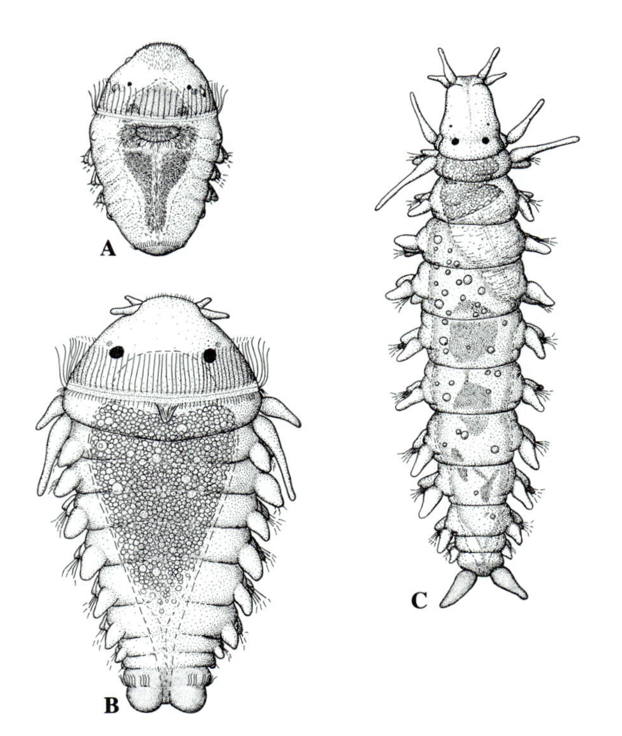

FIGURE 11.15 A–C. Successive stages in the development of *Eteone dilatae* (family Phyllodocidae). (Blake, Ophelia, *14*.)

body varies from group to group. In general, if several segments are removed from the posterior end, these will readily be replaced. Some polychaetes, such as lumbrinerids and eunicids that live in sandy beaches subjected to considerable wave action, are often observed to be regenerating posterior segments. These long-bodied worms are, in fact, known to drop off pieces of the body if they are injured or even if they are especially excited.

Regeneration of the head region is sometimes possible, sometimes not. It depends both on the type of polychaete and on how many segments have been lost. Nereids and phyllodocids, for instance, do not replace heads as easily as do syllids, some cirratulids, or members of a few other families in which the capacity for asexual reproduction is highly developed. Nevertheless, some polychaetes that do not reproduce asexually are efficient regenerators of heads. Among these are certain chaetopterids and sabellids.

The literature on regeneration in polychaetes is extensive. Much of the experimental work is related to the influence of nerve tissue and neurosecretory structures on regeneration. Information on the subject should be sought in specialized treatises.

Nereis as a Representative Polychaete

Polychaetes are so diversified that no one kind is a perfect representative of the class, but *Nereis* and related genera, which belong to the family Nereidae, are particularly suitable for beginning a survey of the group. After studying a nereid, one will be prepared to understand the specializations of other polychaetes.

A large nereid (Figure 11.16, A) may have more than a hundred segments, most of which are similar externally. Each parapodium (Figure 11.9, A) has a well-developed notopodium and neuropodium. The stout acicula within each of these lobes provides support and also serves as a firm basis to which muscles that manipulate the other setae are linked. Except for the aciculae, the setae are mostly of the compound type, in which the basal shaft is tipped by a curved blade (Figure 11.9, B). The form of the setae, and especially the blades, varies from one part of the worm to another, and the setae on the notopodium of a particular segment may be different from those on the neuropodium.

Nereids are efficient crawlers, and many of them can also swim or burrow, or both. In simple crawling, the worm achieves forward progression by touching some of its parapodia to the substratum and bending them backward. These parapodia are then lifted, shortened, and bent forward while others are being touched down for the propulsive backward stroke. During the backward stroke, the setae are extended; during the forward stroke, they are pulled back. The neuropodial lobe, being lowermost, works harder than the notopodial lobe so far as lifting itself and pulling back its setae are concerned. The operation of the many parapodia on each side of the body follows a wavelike pattern (Figure 11.16, B), but the parapodia on opposite sides of a segment are "out of step" with one another. Crawling is speeded up when parapodial movements of the type just described are combined with undulations produced by activity of longitudinal muscles in the body wall. Swimming movements are also effected by undulations coupled with parapodial activity.

Polychaetes of several other families use their parapodia in much the same way as nereids. Some combine parapodial "stepping" with undulations; others, however, lack the musculature necessary for undulating and rely entirely on their parapodia.

The head is illustrated in Figure 11.11. On the dorsal side of the prostomium are a pair of antennae, two pairs of eyes, and a pair of olfactory pits lined by ciliated cells. On its ventral side is a pair of fleshy palps; each palp is constricted transversely into two units. The antennae and palps are concerned primarily with reception of tactile stimuli. The olfactory pits are chemosensory, and if they are destroyed, the worm is unable to locate food.

The eyes (Figure 11.20, C) are moderately complex. Beneath the surface cuticle and modified epidermis is a gelatinous lens enclosed by a cuplike retina. The cells that make up the retina are of two types: those that contain dark pigment and those that function as photoreceptors. The latter are extended into the cup as tightly packed rods. Prolongations of the photoreceptor cells emerge from the cup and become incorporated into the optic nerves. Most polychaetes, whether or not they have obvious eyes, shun bright light. Some species, however, are attracted to light of low or medium intensity.

The peristomium is provided with two or four pairs of long tentacular cirri. These are two-piece structures and represent specializations of parapodia, but there are no setae associated with them. They are comparable to the prostomial antennae and palps in that they have concentrations of nerve endings for receiving tactile stimuli. When a nereid sticks its head out of its tube, burrow, or other hiding place, it usually keeps its tentacular cirri spread, thereby monitoring as large a space as possible. Touch receptors are not limited to the head structures that have been mentioned. In most polychaetes, including nereids, they are present in the cirri that are outgrowths of the parapodia, as well as on most other areas of the body surface.

Body Wall and Setae

The structure of the body wall conforms, in general, to the basic annelidan plan. The musculature is, however, specialized to the extent that it requires explana-

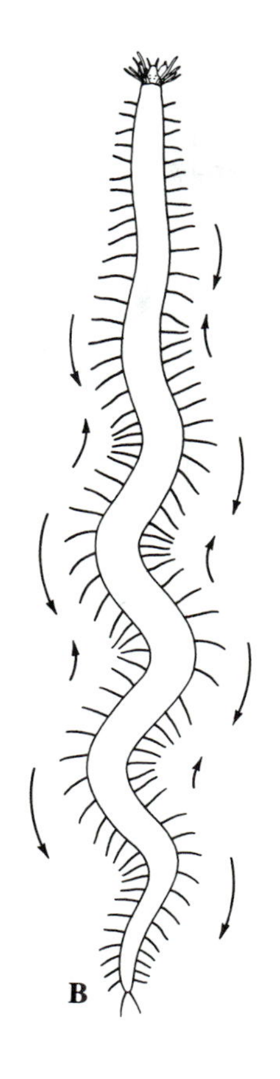

FIGURE 11.16 A. *Nereis vexillosa,* a common polychaete on the Pacific coast of North America. B. Locomotion of a nereid polychaete.

tion. The longitudinal muscles, beneath the sheath of circular muscles, are concentrated in four prominent bands (Figure 11.17, A). Originating near the ventral midline and extending to the lateral portions of the body wall are bands of oblique muscles. There may be several of these on both sides of a single segment. This is the case not only in nereids but in many other polychaetes (see, for instance, *Arenicola,* Figure 11.12, B). In some types, moreover, there are muscle bands that run diagonally from near the ventral midline of one segment to the lateral portion of the body wall of the succeeding segment. These are called *pedal muscles,* and when they are present they lie between the oblique muscles and the longitudinal muscles.

The parapodia have their own musculature that effects the movements of the parapodia and also the protraction, orientation, and retraction of the setae.

Digestive Tract

The mouth leads into the pharynx. This has a cuticular lining, parts of which are thickened to form a pair of large jaws and several to many small teeth (Figures 11.11 and 11.18). The pharynx can be turned inside-out and protruded through the mouth, thereby bringing the jaws and teeth to the exterior. The eversion is accomplished partly by protractor muscles that pull the pharynx forward, but it also depends on contraction of muscles in the body wall, which increases hydrostatic pressure in the coelom. (One can produce the same result by squeezing a nereid firmly just behind the head.)

Hydrostatic eversion of the pharynx is not only important in feeding and defense, which are activities that require exposure of the jaws. It also provides the forward thrust that burrowing nereids require to make their excavations in sand, mud, or clay. After the pharynx has been everted, it is pulled back into the body by retractor muscles.

The formidable jaw apparatus may suggest that nereids are carnivores. However, most members of the group subsist largely on seaweed or detritus. The jaws are used for tearing food into pieces, and the small teeth probably have an abrasive effect or help hold food in the pharynx after this organ has been withdrawn. Some species, such as *Platynereis bicanaliculata* on the Pacific coast of North America, not only eat seaweed but also make a tube by cementing fragments of seaweed together with mucus. During the winter, when *Ulva* and other algae preferred by *P. bicanaliculata* are scarce, the worm feeds mostly on diatoms and other microscopic material. A few nereids are genuine carnivores,

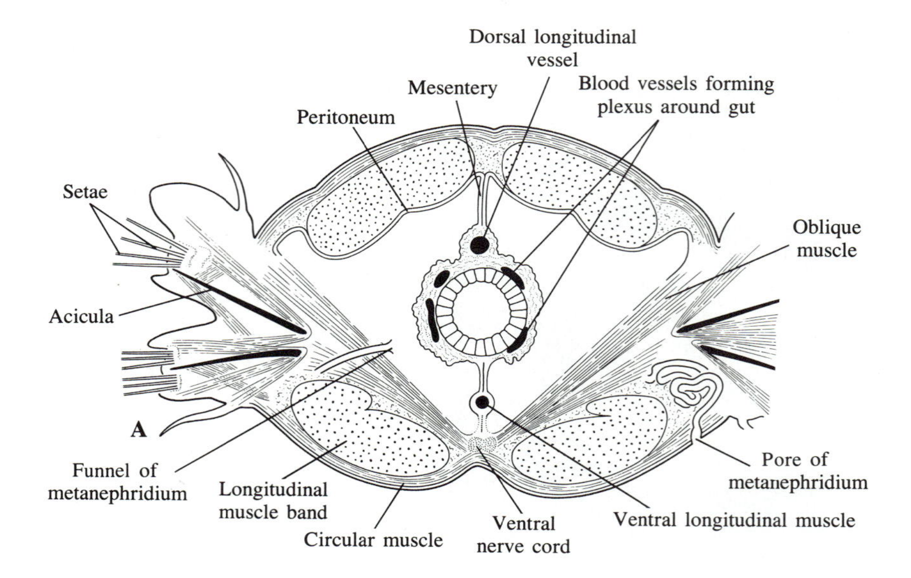

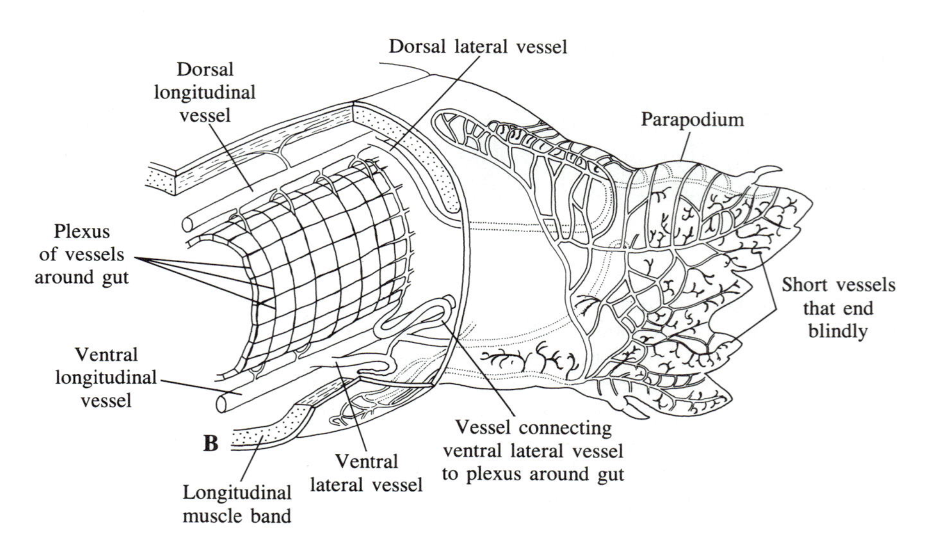

Figure 11.17 A. Transverse section through the body of a nereid polychaete. B. Blood supply to the parapodium of a nereid. (After Nicoll, Biological Bulletin, *106*.)

or scavengers on animal matter. Among these are *Nereis fucata* and *Cheilonereis cyclurus,* which live in snail shells occupied by hermit crabs. They stick their heads out of the shell to feed, often competing with their hosts.

Behind the pharynx is a pair of glandular caeca (Figure 11.12, A), and it is at this level that the short esophagus begins. The remainder of the gut, which is by far the longest portion, is the intestine. It is rather thin-walled but has some musculature and exhibits peristalsis. Gland cells that produce digestive enzymes are incorporated into the lining of the gut, and the esophageal caeca are probably sites of enzyme production, too. Absorption of soluble products of digestion is believed to take place over much of the length of the intestine.

Metanephridia and Derivatives of Coelomoducts

Almost every true segment, other than the peristomium, has a pair of metanephridia (Figure 11.12, A; Figure 11.19, A). Each of these funnel organs occupies two segments. The ciliated funnel, called the *nephrostome,* faces forward from the septum on which it is located, and the long coiled duct runs through a sac of vascularized tissue in the succeeding segment. Most of the nitrogenous waste material excreted by nereids and other polychaetes is ammonia. Some of it enters the tubule from the coelom, some from the blood supply

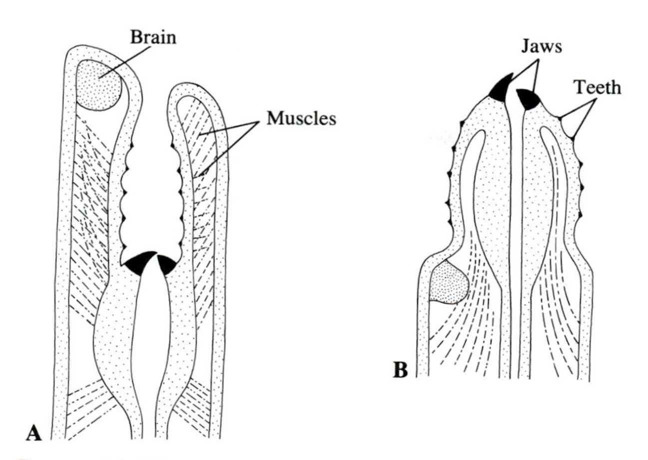

FIGURE 11.18 Diagrams of longitudinal sections through the pharynx of a nereid, showing the organ in its inverted (A) and everted (B) positions.

around the duct. The *nephridiopores* (openings of the metanephridia) are just ventral and medial to the neuropodia.

There are only remnants of coelomoducts. They are small, ciliated outgrowths of the coelomic epithelium and are located near the lateral edges of the dorsal muscle bands.

Circulatory System

As is typical for polychaetes in general, the circulatory system (Figure 11.12, A) is closed. The two main longitudinal vessels, one above the gut and one below it, have some muscle in their walls and pump blood peristaltically. In the dorsal vessel, blood is moved mostly forward; in the ventral vessel, it is moved mostly backward. In nearly all segments, branches of the dorsal vessel run to capillary networks in the wall of the gut. The blood then moves to the ventral vessel, and branches of this go out to the nerve cord, parapodia, and other parts of the body wall. The blood eventually returns to the dorsal vessel. The entire body surface can function to at least some extent in exchange of oxygen and carbon dioxide, but the parapodia, especially their notopodial gills, have a well-developed blood supply (Figure 11.17, B) and are believed to be the most important sites for diffusion of gases. The blood of nereids is red; its oxygen-carrying pigment, dissolved in the plasma, is hemoglobin.

Nervous System

The brain, located in the prostomium, is a substantial one (Figure 11.20, A and B). It sends nerves to the sensory structures of the prostomium—the antennae, palps, eyes, and olfactory pits. Also originating in the brain are the two connectives that run around the phar-

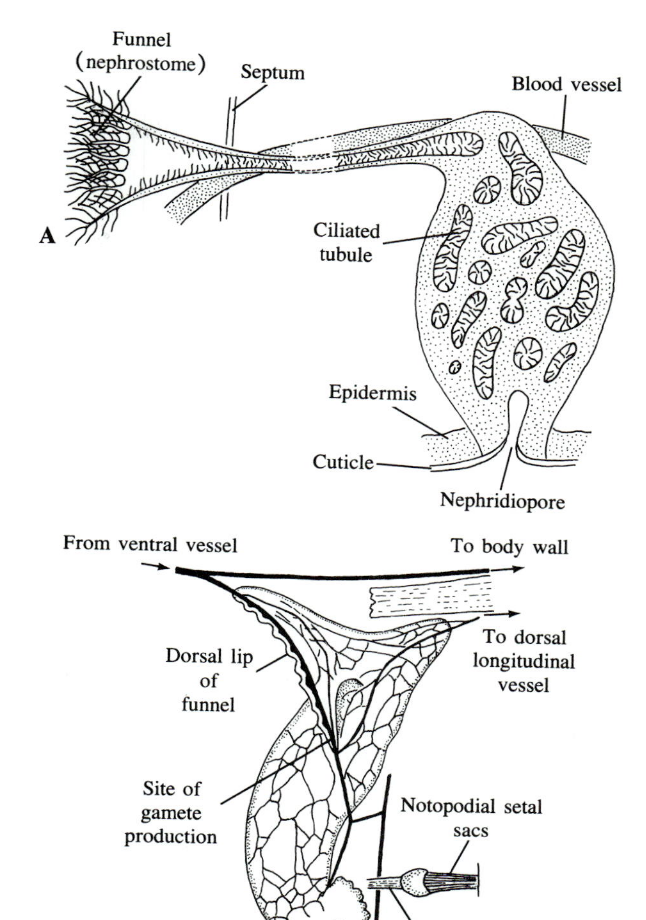

FIGURE 11.19 A. Diagram of a metanephridium of *Nereis vexillosa*. The cells forming the wall of the tubule and the tissue around the tubule are not shown. A blood vessel passes close to the metanephridium but apparently has no branches that go into the tissue of the organ. (After Jones, Biological Bulletin, *113*.) B. Mixonephridium of *Arenicola marina*. The funnel is derived from a coelomoduct, whereas much of the tubule, leading to the pore, is a metanephridium. (After Ashworth, Liverpool Marine Biology Committee Memoirs, *11*.)

ynx to the ventral nerve cord. Nerves from small ganglia associated with each connective near the brain extend to the dorsal side of the pharynx and to the tentacular cirri of the peristomium.

The double structure of the nerve cord, which is embedded in the body wall, is a feature found in many annelids and is believed to be an ancestral character. The longitudinal tracts of the cord include both small neurons and giant fibers (Figure 11.21). The latter—there are several of them at any given level of the cord

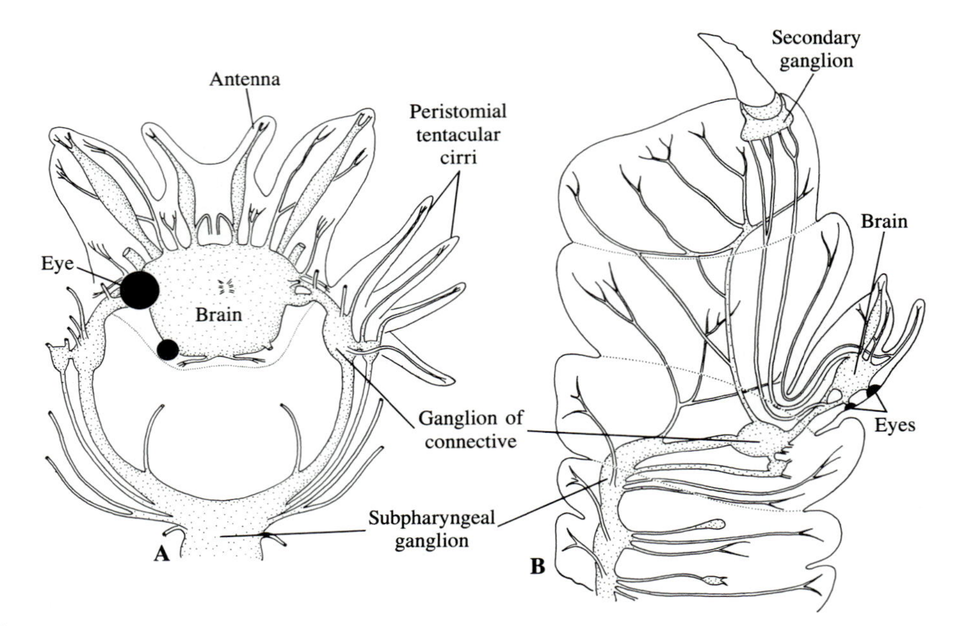

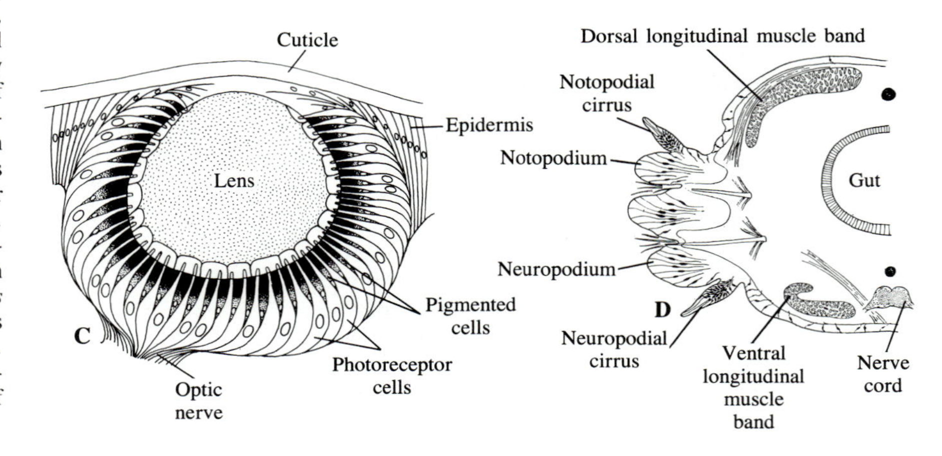

Figure 11.20 A. Nervous system in the head region of *Nereis,* dorsal view, showing the brain, connectives leading to the ventral nerve cord, and nerves that supply various structures. B. Diagram of the nervous system in the head region of *Nereis,* lateral view from the left side. The pharynx is in its everted position. (A and B, after Henry, Microentomology, *12.*) C. Eye of *Nereis,* section. D. Diagram of a transverse section through a segment of *Platynereis dumerilii,* showing sensory cells in various parts of the body wall. (Smith, Philosophical Transactions of the Royal Society of London, B. *240.*)

in *Nereis*—are concerned with fast transmission of impulses over long distances. This capacity for rapid conduction is important to polychaetes whose life style requires that they be able to react instantly to certain stimuli, such as the approach of a predator, and to contract almost all segments simultaneously. Giant fibers, whose axons extend for comparatively long distances, are prominent components of the nerve cord in some types of polychaetes that inhabit tubes; at the slightest stimulus—even a shadow—the worms quickly retract into the tube and away from danger.

The ganglia of the nerve cord of nereids send out paired lateral nerves that extend into the body wall, septa, and other internal structures. Some segments near the anterior end have two pairs of principal lateral nerves, but most segments have three pairs.

Some of the sensory structures—eyes, olfactory pits, antennae, palps, and tentacular cirri—were discussed above in connection with the head region. All surfaces of the body are supplied with tactile receptors. Figure 11.20, D shows the general distribution of these in a section through a single segment of a nereid.

Diverse Life Styles Among the Polychaetes

Herbivores and Omnivores

Most members of the family Nereidae are herbivores that consume algae, which they tear into pieces with

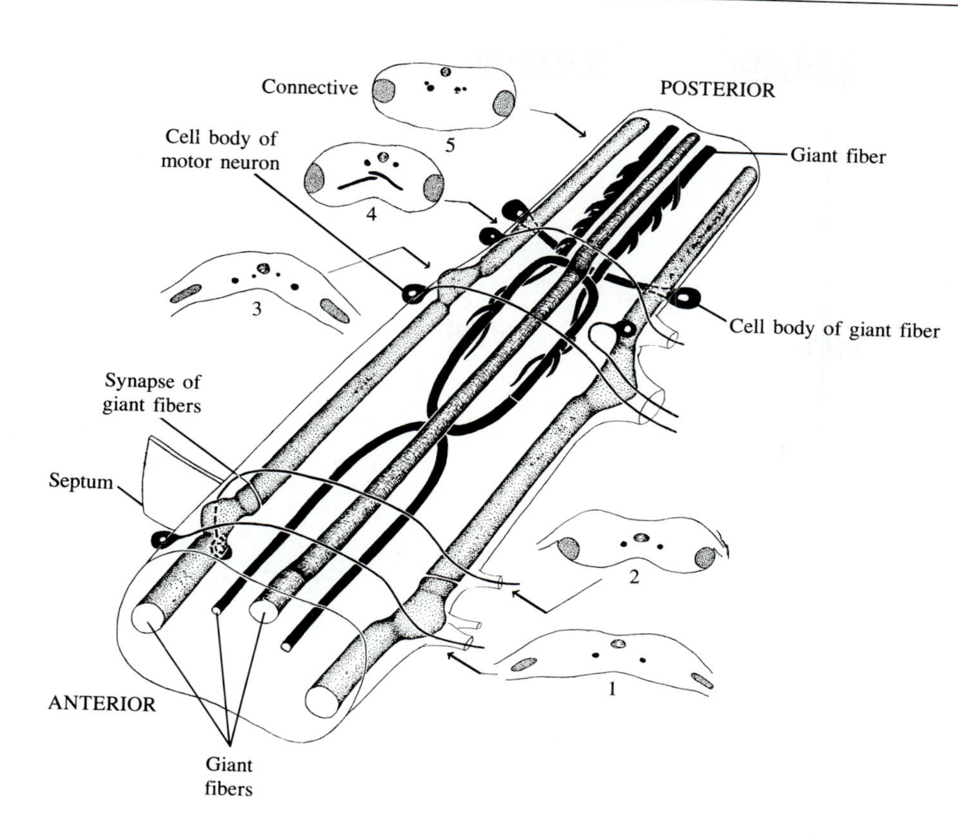

FIGURE 11.21 Stereodiagram of a ganglion and connective in the ventral nerve cord of *Nereis diversicolor*. In nereids, the ganglia are arranged partly in one segment and partly in another. The small drawings are of transverse sections through four levels of the ganglion (1–4) and through the connective (5) that extends to the succeeding ganglion. Except for the giant fibers, the only neurons shown are a few motor neurons. All of the segmental nerves, however, contain both motor and sensory neurons. (Smith, Philosophical Transactions of the Royal Society of London, B. *240*.)

their strong jaws. Some, however, feed to a large extent on detritus, at least during periods when the algae they prefer to eat are not abundant. A few, including those that live in snail shells occupied by hermit crabs, are regularly or opportunistically carnivores or scavengers on animal matter.

Another family that is made up primarily of herbivores is the Lumbrineridae. A lumbrinerid (Figure 11.22, A–C) looks like a slender earthworm because of its general shape and simple prostomium. The septa between segments are "tight"—that is, there is almost no leakage from one segment to another—and the musculature of the body wall is well developed. Both of these features adapt lumbrinerids for active burrowing in marine sand and mud. The pharynx, provided with a complicated jaw apparatus (Figure 11.20, D and E), is everted in order to grasp and tear algae that accumulate on the surface of the substratum. Some species are, to a greater or lesser extent, carnivores, and some utilize organic detritus.

Most members of the family Onuphidae (Figure 11.22, F and G) live in tubes made from pieces of foreign material. The tubes are usually embedded in mud or sand, but a few species are mobile, and certain onuphids leave their tubes temporarily. As a group, they are omnivorous scavengers, taking pieces of algae, dead animal matter, or detritus. Some, however, specialize to a considerable extent, thus being more nearly herbivores than carnivores, or vice versa. The eversible pharynx, armed with several jaws, is effective in grasping and biting, and the sensory equipment of the head region, which includes seven prostomial antennae, enables these worms to recognize prospective food and to act accordingly.

Eunicids, with one to five prostomial antennae, are closely related to onuphids and have a similar life style. Most of them occupy tubes, but there are exceptions. *Palola viridis* (Figure 11.22, H), for instance, burrows in corals or in coralline algae associated with corals. The swarming of detached posterior portions of one species of *Palola* was described earlier.

Carnivores

Several families of polychaetes are carnivores that use an eversible pharynx for capturing their prey. In many worms that fit into this category, the pharynx has strong jaws or teeth, and it may also have glands for producing a venom. As a rule, the radial musculature of the pharynx is well developed, and the organ is used to suck in tissues or whole organisms.

Especially good examples of carnivorous polychaetes are those of the family Glyceridae (Figure

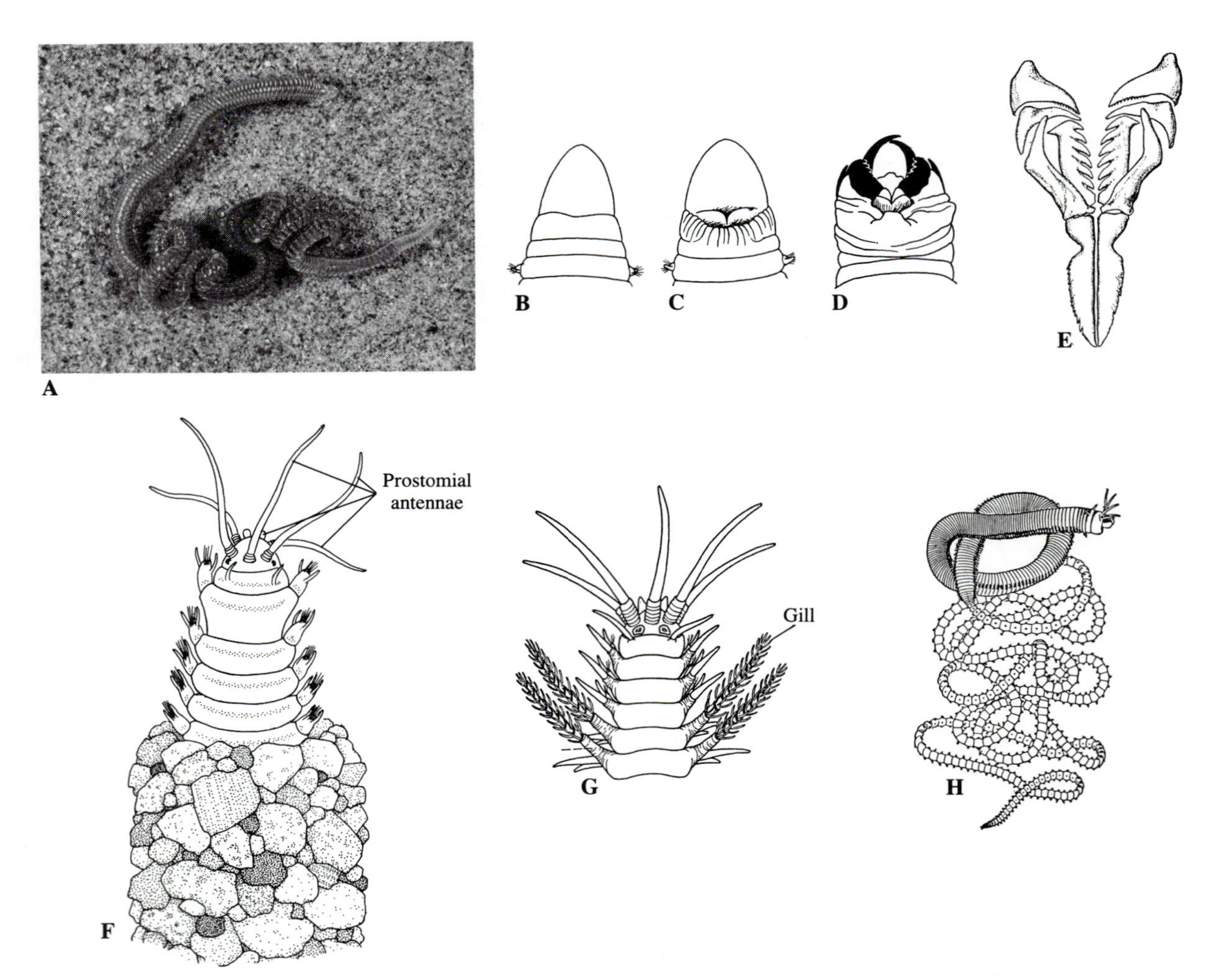

FIGURE 11.22 A. *Lumbrineris zonata* (family Lumbrineridae). B–D, *Lumbrineris fragilis,* head region, in dorsal view, ventral view, and ventral view with jaws everted. (Pettibone, Bulletin of the United States National Museum, *227.*) E. *Ninoe gayheadia* (family Lumbrineridae), jaw apparatus. (Hartman, Occasional Papers of the Allan Hancock Foundation, no. 28.) F. *Onuphis conchylega* (family Onuphidae), anterior end of the worm sticking out of its tube, made of sand grains and bits of shell. G. *Diopatra cuprea* (family Onuphidae), anterior region. (Pettibone, Bulletin of the United States National Museum, *227.*) H. *Palola viridis* (family Eunicidae). The slender portion, when ripe with gametes, becomes detached and swims. (Woodworth, Bulletin of the Museum of Comparative Zoology, Harvard College, *51.*)

11.23, A–C), which live in mud and sand. They excavate burrows that have several openings at the surface. The prostomium, tipped by four antennae, has both mechanoreceptors and chemoreceptors. The former detect vibrations produced by an approaching crustacean or some other prospective prey organism; the chemoreceptors enable the worm to target the prey as it moves closer.

A glycerid strikes by everting its pharynx and grasping the prey with four hooklike jaws, each of which has its own venom gland and a duct by which the venom reaches the tip (Figure 11.23, C). There are no septa in the anterior part of the body, so the pharynx is more or less free in the coelom and is everted quickly and completely when body wall muscles contract to raise the hydrostatic pressure in the coelom. The esophagus is S-shaped while the pharynx is in the inverted position, and is merely straightened when the pharynx is thrust out.

Polychaetes of the family Nephtyidae (Figure 11.23, D–G) slightly resemble nereids, though there are four small antennae on the prostomium and palps are absent.

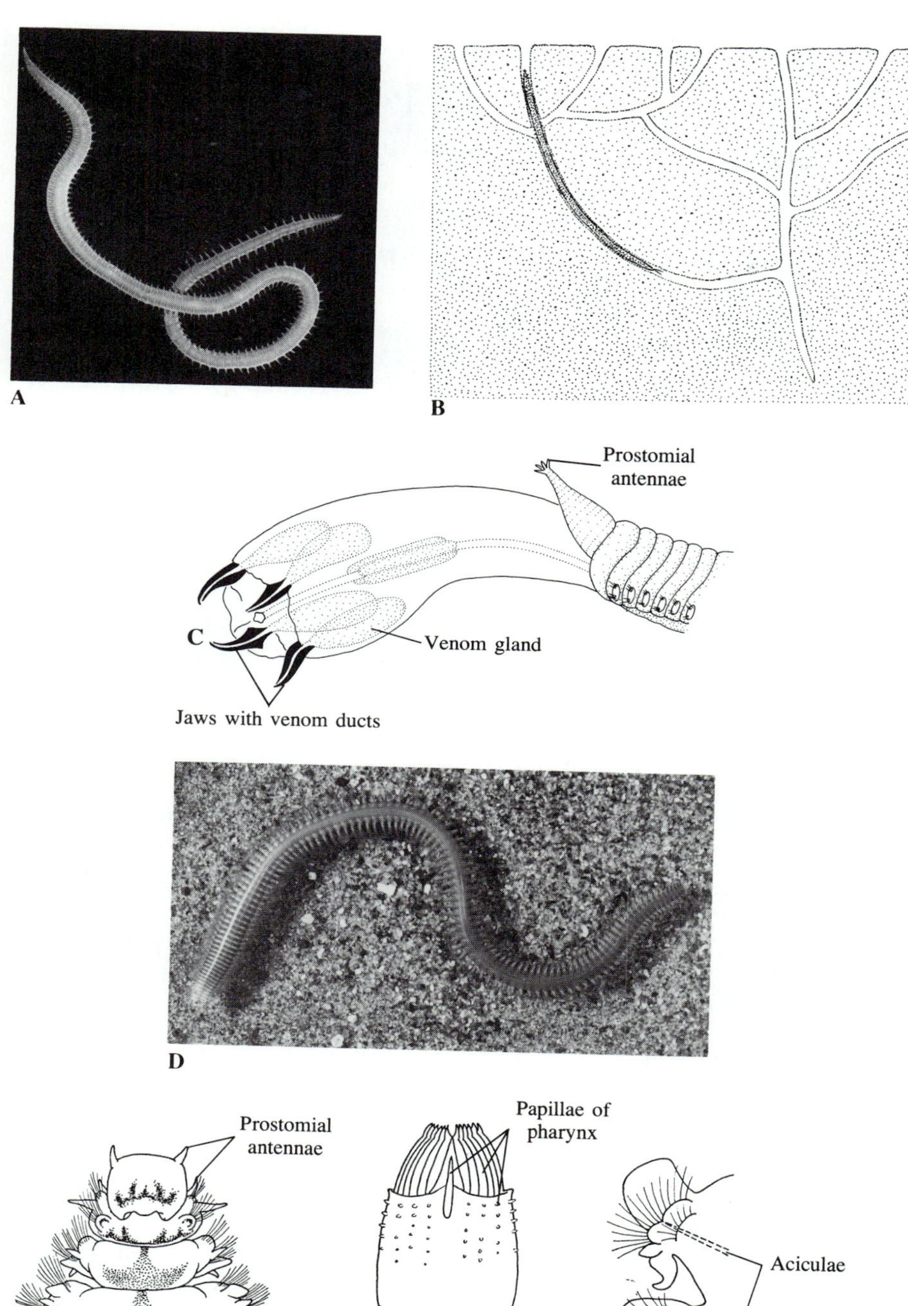

FIGURE 11.23 A. *Hemipodus borealis* (family Glyceridae). B. Burrow system of a glycerid. C. Everted pharynx of *Glycera,* showing jaws and venom glands. (After Michel, Cahiers de Biologie Marine, *11.*) D. *Nephtys* (family Nephtyidae). E. *Nephtys picta,* head region, dorsal view. There are four small antennae on the prostomium. F. *N. incisa,* head region, pharynx everted. G. *N. discors,* parapodium. (E–G, Pettibone, Bulletin of the United States National Museum, *227.*)

On the pygidium, moreover, there is a single cirrus, instead of two cirri. The large pharynx has a pair of jaws, but these are not quite exposed when the pharynx is everted. The numerous papillae on this organ are nevertheless distinctive. Nephtyids burrow actively through sand and mud, and most of them feed on small invertebrates, including other polychaetes. One species, however, is said to feed on detritus, and the gut of another has been found to contain sand, as well as remains of invertebrates. Perhaps this nephtyid obtains part of its nourishment from microorganisms that coat the sand grains.

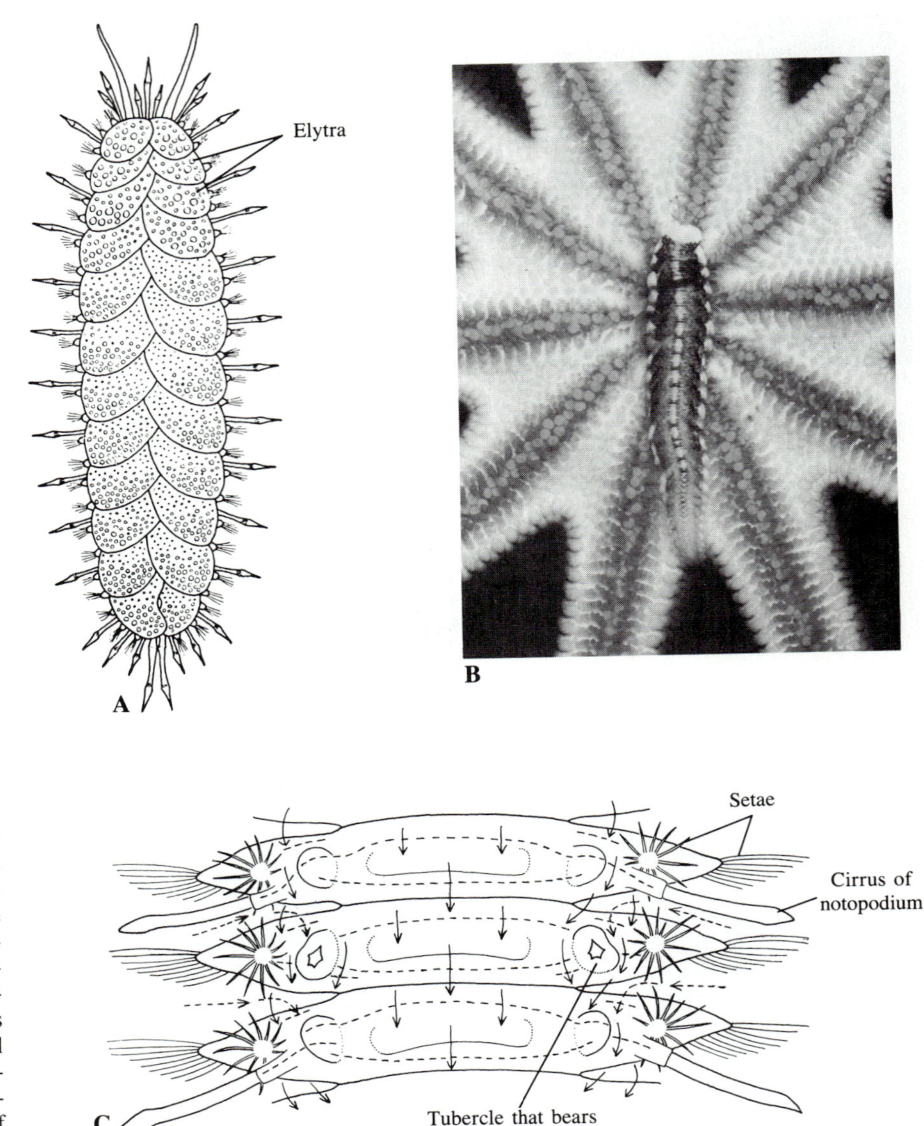

FIGURE 11.24 Family Polynoidae. A. *Lepidonotus squamatus,* dorsal view. (Pettibone, Bulletin of the United States National Museum, *227.*) B. *Arctonoe vittata,* on the oral surface of the sea star *Crossaster papposus.* C. Portion of the dorsal surface from near the middle of the body of *Harmothoe imbricata,* elytra removed, showing directions taken by ciliary currents. Dashed lines indicate the location of cilia, which are in discontinuous bands. The main currents are indicated by solid arrows, the compensatory currents (which bring water into the elytral area from the sides) by dashed arrows. (After Segrove, Proceedings of the Zoological Society of London, *108.*)

Scaleworms, which belong to the Polynoidae (Figure 11.24) and related families, are often associated with other invertebrates. They are commonly found, for instance, with molluscs and echinoderms, with polychaetes that live in tubes, and with crustaceans that live in burrows. They are characterized by several to many pairs of **elytra** (singular *elytron*) on the dorsal surface. Elytra are scales secreted by specialized portions of the notopodia. Currents of water, produced by ciliary activity, flow between the notopodia and combine to form a respiratory current that passes backward between the elytra and the dorsal body wall (Figure 11.24, C). Some scaleworms brood their young beneath the elytra.

The eversible pharynx of scaleworms is equipped with two pairs of jaws, and is an effective biting device. On the whole, these polychaetes seem to be carnivores. One widely distributed species, *Harmothoe imbricata,* uses its palps to detect vibrations produced by prospective prey organisms, and perhaps many other scaleworms do the same thing.

In the case of commensal species, there is usually only one scaleworm associated with each host. Among those whose feeding has been studied is *Arctonoe vittata* (Figure 11.24, B), which is closely associated with

FIGURE 11.25 *Aphrodita japonica* (family Aphroditidae). A. Dorsal view. B. Ventral view.

various species of invertebrates on the Pacific coast of North America. It bites at sedentary polychaetes that it encounters while its host is moving about. *Hesperonoe adventor,* which lives in the burrow of the echiuran *Urechis caupo* (Chapter 12), eats various small invertebrates that get into the burrow, and may take material from the net of mucus that its host uses to trap food.

Some of the commensal scaleworms show positive responses to hosts to which they have become conditioned. For instance, if an *Arctonoe vittata* that has been living in the mantle cavity of a keyhole limpet is given a choice between a limpet of that species and a chiton with which *A. vittata* is also sometimes associated, it will generally go to the former. Presumably, a young worm that has chosen its host will tend to remain with that host—or at least with a host of the same species—for the rest of its life.

Sea mice, which belong to the family Aphroditidae, are scaleworms of a sort. Their scales do not show because they are obscured by a felt of hairlike notopodial setae. *Aphrodita* (Figure 11.25), found subtidally in mud, is the best-known genus. Like scaleworms, sea mice create a posteriorly directed respiratory current between the elytra and dorsal body wall. They feed on a variety of small invertebrates. Digestion begins in the tubular portion of the intestine and is completed in branched diverticula of this part of the gut. Absorption takes place in the diverticula.

In the family Syllidae (Figure 11.26), the eversible pharynx generally has either a single tooth or a circle of teeth, and is often backed up by a muscular bulb that sucks tissues or juices of the prey into the gut. These worms have three antennae, a pair of palps, and usually four eyes on the prostomium. They feed on various small invertebrates. Many species have an epitokous phase, and some reproduce by budding. The females of certain species, following sexual reproduction, carry their young for a time (Figure 11.26, A).

Members of the family Phyllodocidae (Figure 11.27, A) may resemble syllids, but they have four or five antennae, usually two eyes, and no palps. They are primarily mobile hunters that search for small invertebrates, especially other polychaetes. A few, however, are believed to be scavengers on animal matter. The pharynx, though muscular and efficient as a grasping organ, has no jaws.

The Amphinomidae, which are mostly tropical, are also jawless, but the mouth has a muscular lip that is extended and used for rasping. These polychaetes feed largely on coral polyps and sea anemones, but one species swallows small goose barnacles and defecates their calcareous plates. Some are scavengers on animal matter. Amphinomids are called fireworms because their fragile notopodial setae, when broken, release a powerful irritant. It is not prudent, therefore, to touch them.

The worms of two other carnivorous families—Tomopteridae and Alciopidae—are strictly planktonic. They have neither jaws nor teeth, but the pharynx is

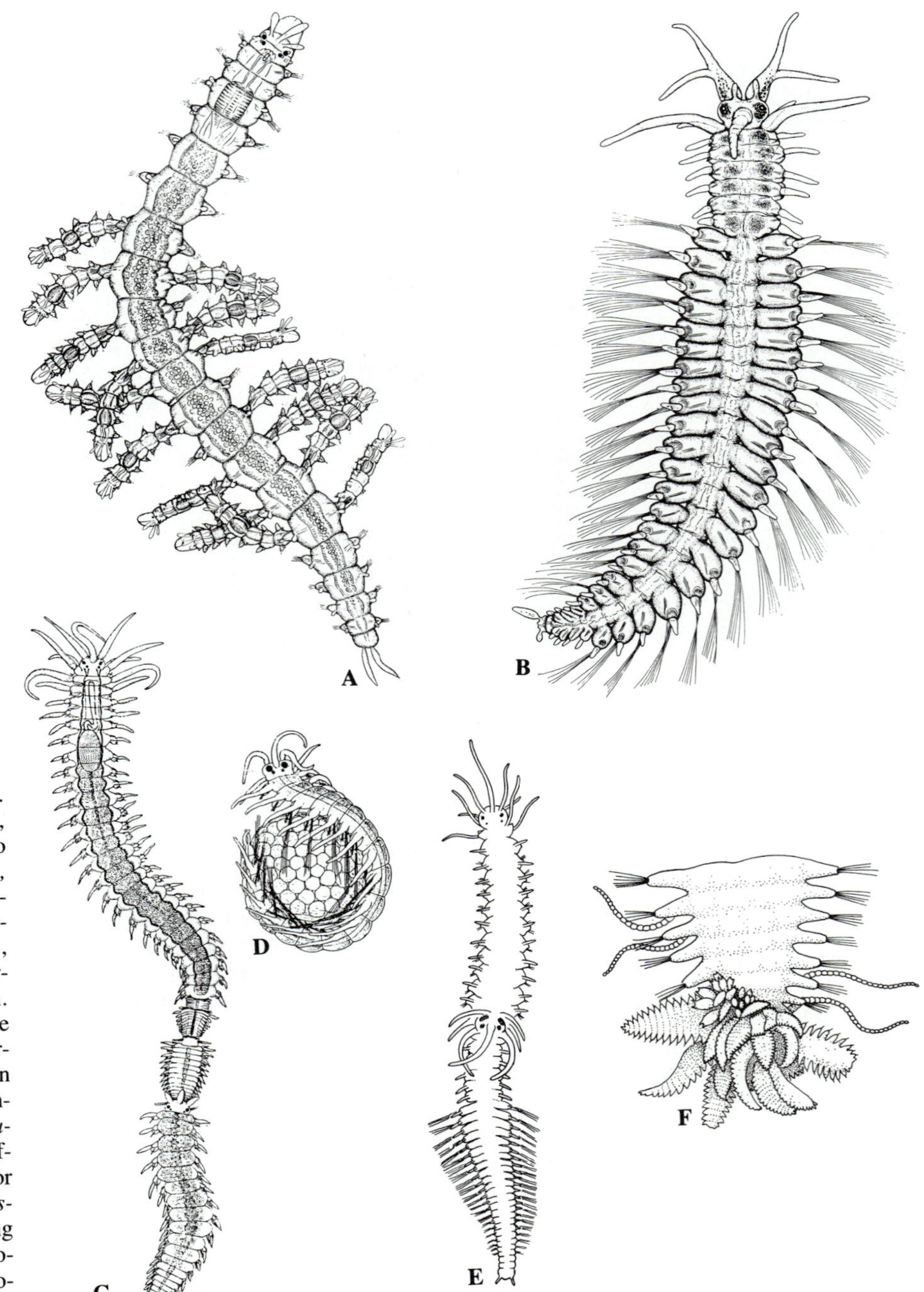

Figure 11.26 Family Syllidae. A. *Exogone gemmifera,* female, with numerous young attached to her body. B. *Proceraea cornuta,* epitokous male. (A and B, Rasmussen, Ophelia, *11.*) C. *Autolytus prolifere,* a benthic individual, with four female epitokes differentiating from its posterior region. D. *A. prolifer,* female epitoke with its egg sac. (C and D, Thorson, Meddelelser fra Komissionen for Danmarks Fiskeri- og Havundersøgelser, Plankton, *4.*) E. *Autolytus,* with a male epitoke differentiating from its posterior region. F. *Trypanosyllis crosslandi,* posterior portion, budding off male epitokes. (Potts, Proceedings of the Cambridge Philosophical Society, *17.*)

muscular and eversible. Tomopterids (Figure 11.27, B), which are almost as transparent as clear water, have no setae, except for the aciculae in one pair of enlarged anterior parapodia. Most of the parapodia have membranous extensions, and they function beautifully in swimming. Alciopids (Figure 11.27, C), which do possess setae and which are generally long and slender, are remarkable because of their large and rather complex eyes.

Ciliary–Mucous Feeders

Polychaetes of several families use featherlike structures or slender tentacles for trapping small food particles. The particles stick to mucus, which is moved to-

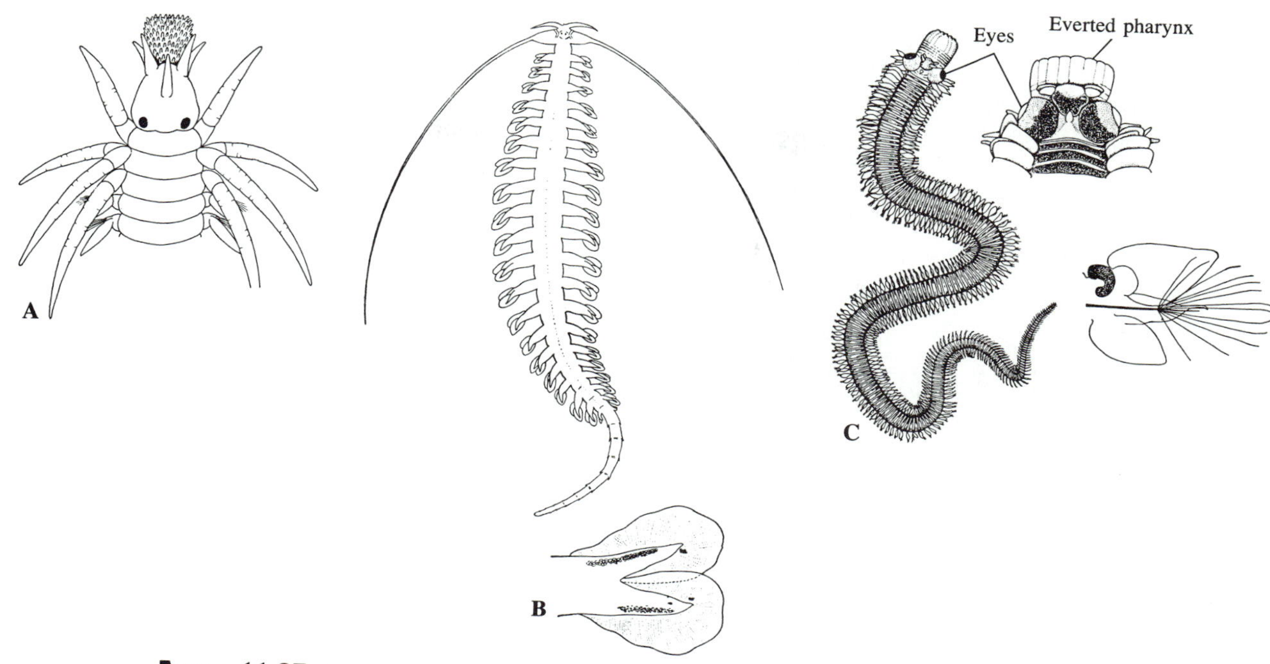

Figure 11.27 A. *Eulalia viridis* (family Phyllodocidae), head region, pharynx everted. (Pettibone, Bulletin of the United States National Museum, *227*.) B. *Tomopteris helgolandica* (family Tomopteridae) and one of its parapodia. C. *Rhynchonerella angelini* (family Alciopidae), its head region, and one of its parapodia. (B and C, Ushakov, Fauna of the USSR, new series, *102*.)

ward the mouth by ciliary action. Some ciliary–mucous feeders, such as sabellids and serpulids, collect particles that circulate freely in the water. Others collect particles from sediment, and are thus somewhat similar to certain detritus feeders.

Sabellids (Figures 11.28, 11.29, and 11.30) live in tubes. The anterior portion of the body looks like a feather duster, because the peristomium gives rise to many pinnately branched structures called **radioles.** The bases of these are arranged in two semicircles or spirals. Each radiole is stiffened by an axial rod that consists of vacuolated cells. The radioles have short side branches, called **pinnules,** whose axial supports consist of a single row of vacuolated cells. Pinnules have three tracts of cilia (Figure 11.29, C). The abfrontal cilia, on the undersides of the pinnules, circulate water through the crown of radioles, from below upwards. The long laterofrontal cilia also circulate water, and

A

B

C

Figure 11.28 Family Sabellidae. A. *Schizobranchia insignis*. B. *Eudistylia polymorpha*. (The tube is overgrown by a compound ascidian, *Botryllus*.) C. *Myxicola infundibulum*. (See also Color Plate 3.)

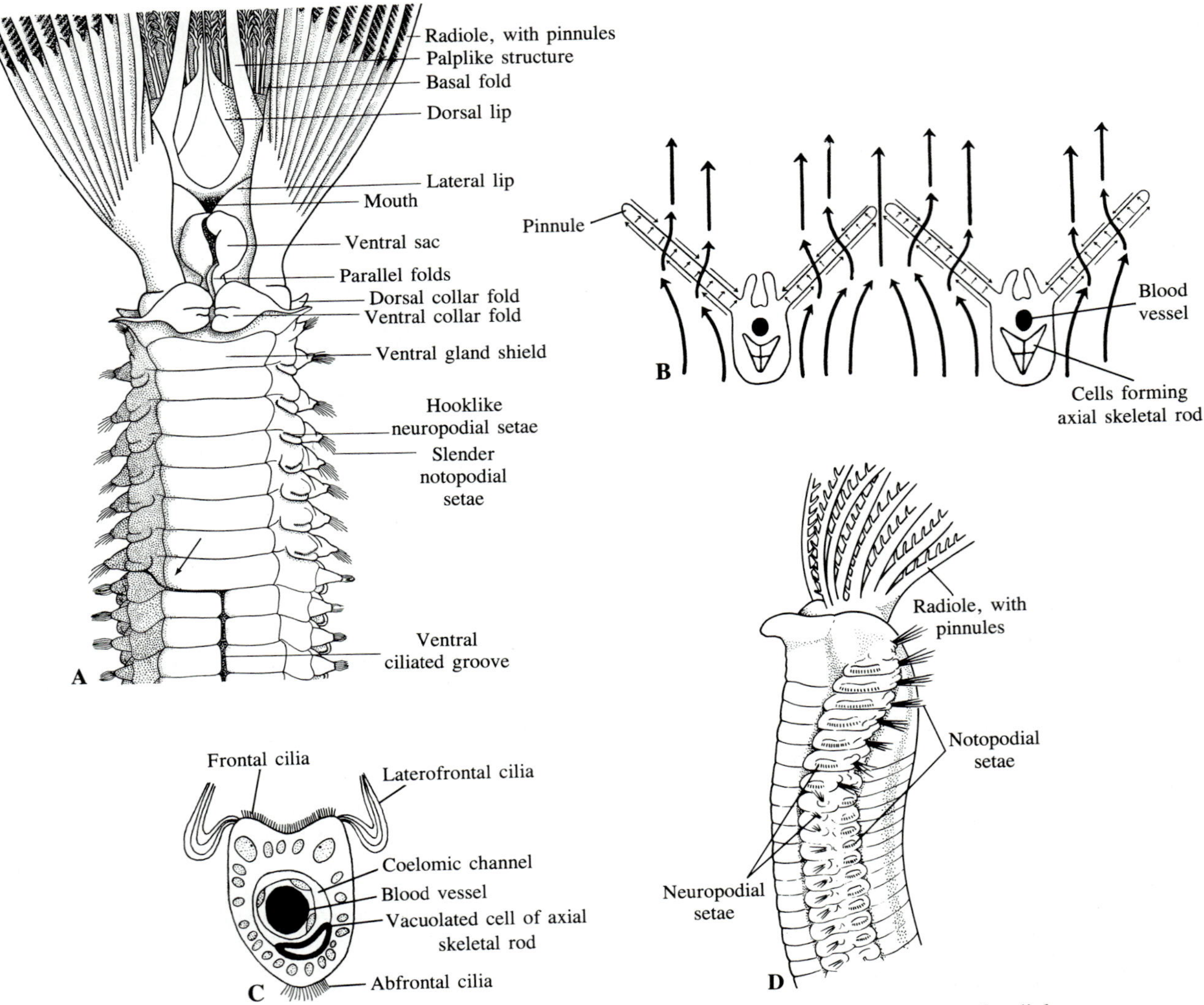

FIGURE 11.29 Family Sabellidae. A. *Sabella pavonina,* ventral view of the crown of radioles and several segments. Note that the ventral ciliated groove turns to the right (arrow); after reaching the dorsal midline, it continues anteriorly. B. *S. pavonina,* transverse section of two radioles, showing the direction of the beat of cilia (small arrows) and major currents (large arrows). C. *S. pavonina,* transverse section of a pinnule, showing arrangement of the ciliary tracts. The activity of the abfrontal cilia draws water into the crown of radioles, whereas the activity of the laterofrontal cilia throws suspended particles onto the field of frontal cilia; the latter move the particles, trapped in mucus, toward the main stem of the radiole. (A–C, Nicol, Transactions of the Royal Society of Edinburgh, *56.*) D. *S. fabricii,* anterior portion, showing how the location of hooklike setae and long, slender setae is reversed after the first few segments. (After Ivanov *et al.*, Major Practicum of Invertebrate Zoology.)

they knock prospective food particles onto the broad tracts of short frontal cilia located on the upper faces of the pinnules. The particles become trapped in mucus secreted by epidermal glands on the sides of the pinnules. The frontal cilia move the mixture of particles and mucus to the main stem of the radiole, where there is a groove lined by cilia. This leads to a basal collecting groove that winds toward the mouth.

The groove on each radiole, and also the basal groove, operate as sorting devices (Figure 11.30, A and B). Only the smallest particles—the ones destined to reach the mouth—can sink to the deepest portions of

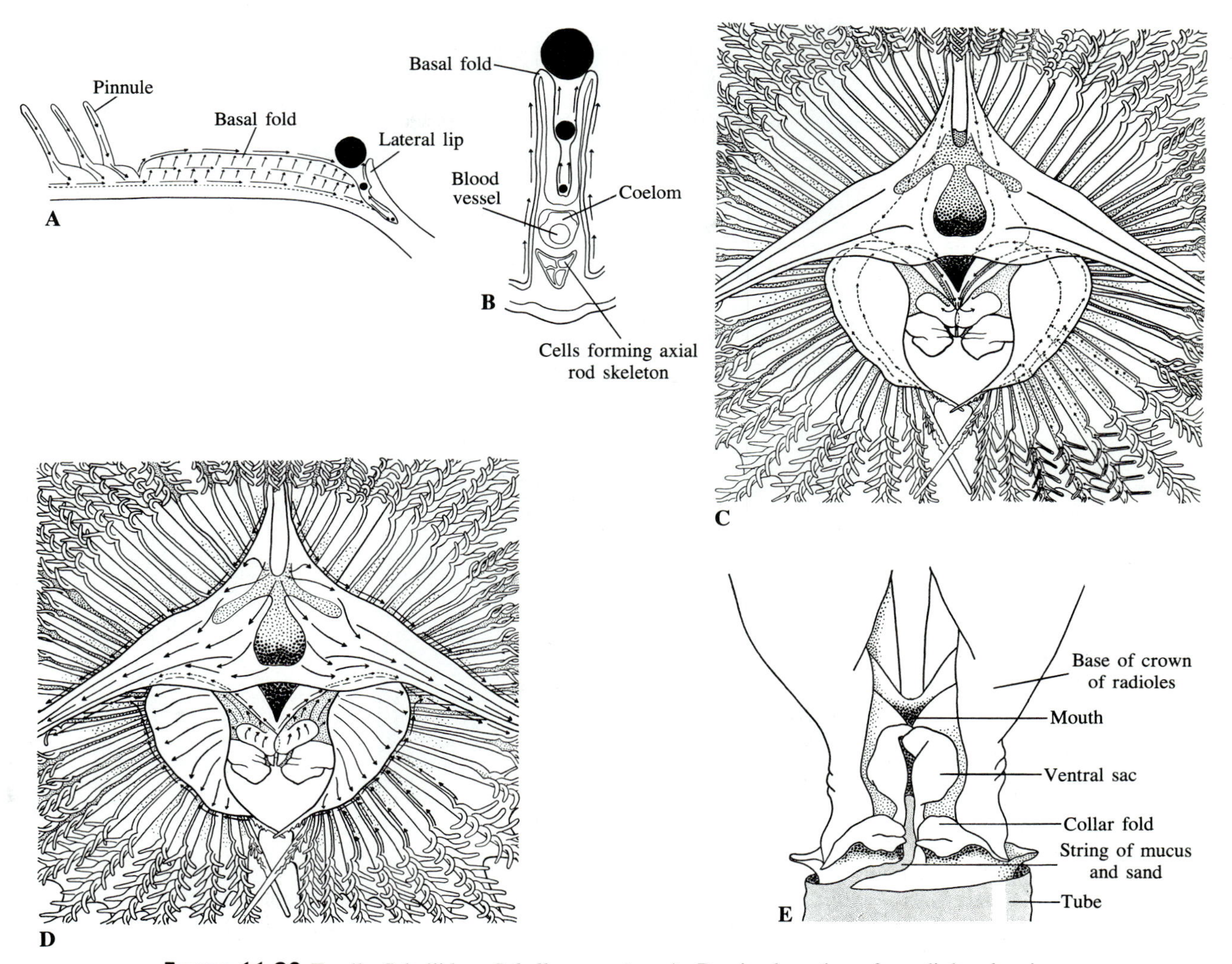

FIGURE 11.30 Family Sabellidae. *Sabella pavonina*. A. Proximal portion of a radiole, showing the lateral lip and one basal fold (the other fold, nearer the observer, has been removed). The direction of beat of cilia is indicated by arrows, and particles of three sizes are indicated by black dots. B. Transverse section through the proximal portion of a radiole, showing the direction of beat of cilia (arrows) and particles of three sizes (black dots). C. Terminal view of the crown of radioles, showing ciliary tracts that carry particles toward the mouth and ventral sac. D. Terminal view of the crown of radioles, showing tracts along which particles that will be rejected are carried. E. Ventral view of the anterior region, showing formation of a string of mucus and sand, and application of the string to the lip of the tube. (Nicol, Transactions of the Royal Society of Edinburgh, *56*.)

the grooves, which are narrow. Particles of medium size sink part way down, and they end up in two sacs on the ventral side of the peristomium, where they become mixed with mucus that will be used in tube construction (Figure 11.30, E). The largest particles, which remain for a time in the broader upper portion of the groove, are swept away by strong ciliary currents before or after they reach the base of a pinnule.

Part of the peristomium of a sabellid is differentiated into a collarlike structure. This is used to mold new tube material. The mixture of mucus and particles, many of which are small sand grains, emerges from the ventral sacs in strings. The worm, meanwhile, rotates in its tube, thereby ensuring even distribution of the mucus, and the liplike portions of the collar operate much like hands in applying and firming the secretion.

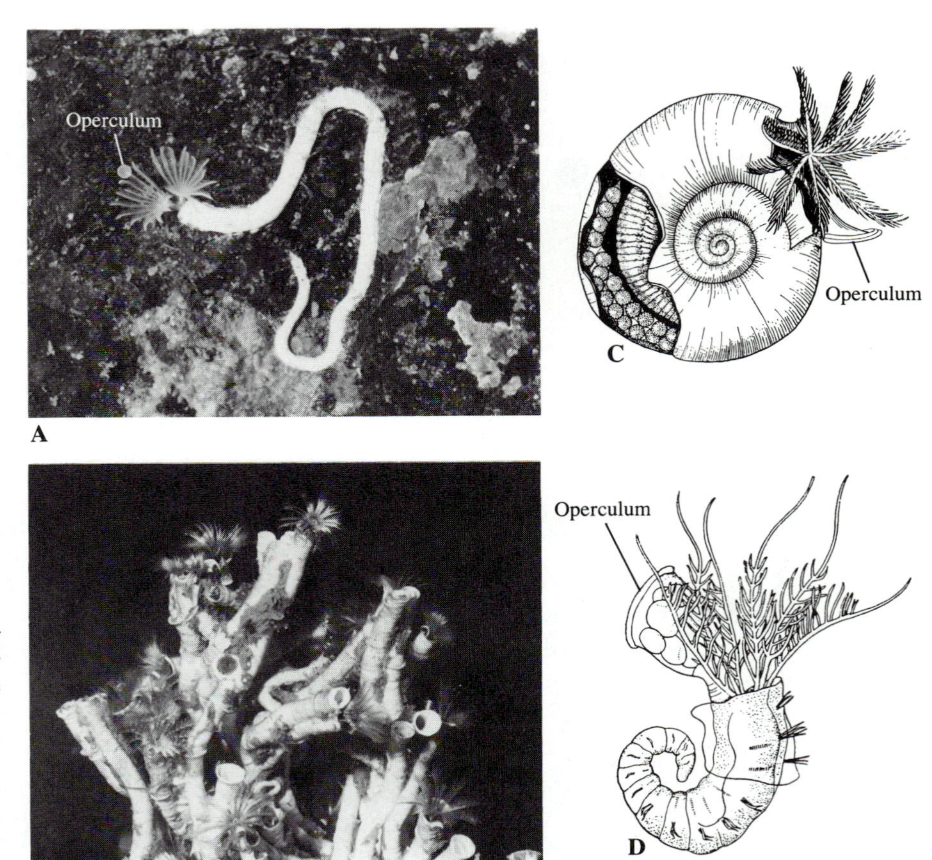

Figure 11.31 A. *Pseudochitinopoma occidentalis* (family Serpulidae). B. *Mercierella enigmatica* (Serpulidae), a widespread brackish-water speices. (See also Color Plate 3.) C. *Spirorbis borealis* (family Spirorbidae), female brooding young in its tube. (Thorson, Meddelelser fra Komissionen for Danmarks Fiskeri- og Havundersøgelser, Plankton, *4*.) D. *Spirorbis nipponicus* (Spirorbidae), female brooding young in the operculum. (Ushakov, Keys to the Fauna of the USSR, *56*.)

For thickening the tube along its entire length, the animal has mucus glands on the ventral side of all segments. In general, the tubes of sabellids are both leathery and gritty. Some genera, including *Myxicola* (Figure 11.28, C), however, secrete a slimy tube that has no hard particles other than those that accidentally become stuck in it.

Other interesting adaptations of sabellids to a tube-dwelling existence are seen in the organization of their funnel organs, the method they use to bring gametes and fecal wastes out of the tube, and their rapid withdrawal reactions when disturbed. The funnel organs are mixonephridia. Two of them, located in the anterior part of the body, are concerned with excretion. They are large, occupying several segments, and their ducts unite so that there is a single middorsal pore from which wastes are swept away by ciliary currents. At more posterior levels of the body are several pairs of small, segmentally arranged mixonephridia that function as gonoducts.

When gametes are shed, they are carried forward by a ciliated groove that runs along the ventral surface (Figure 11.29, A). As this groove nears the anterior end, it comes around the right side of the worm and then completes its course on the dorsal surface. Once the gametes reach the base of the crown of radioles, they are carried away by ciliary currents. The groove just described is important also in bringing fecal wastes to the anterior end, so that they can be dispersed by ciliary currents.

When the midventral ciliated groove switches to a dorsal position, the two major types of setae on the parapodia change places (Figure 11.29, D). Over most of the length of a sabellid, there are bristlelike setae in the neuropodial lobes and hooklike setae in the notopodia. Anterior to the place where the groove becomes dorsal, however, the bristlelike setae are in the notopodial lobes and the hooklike setae are in the neuropodial lobes. The hooklike setae are used for gripping the inside of the tube.

Sabellids withdraw with astonishing speed when touched or when a shadow passes over them. In some species the location of light-sensitive receptors is revealed by small eyespots or swellings where dark pigment is concentrated. The quick responses of sabellids to stimulation is accomplished by contraction of well-

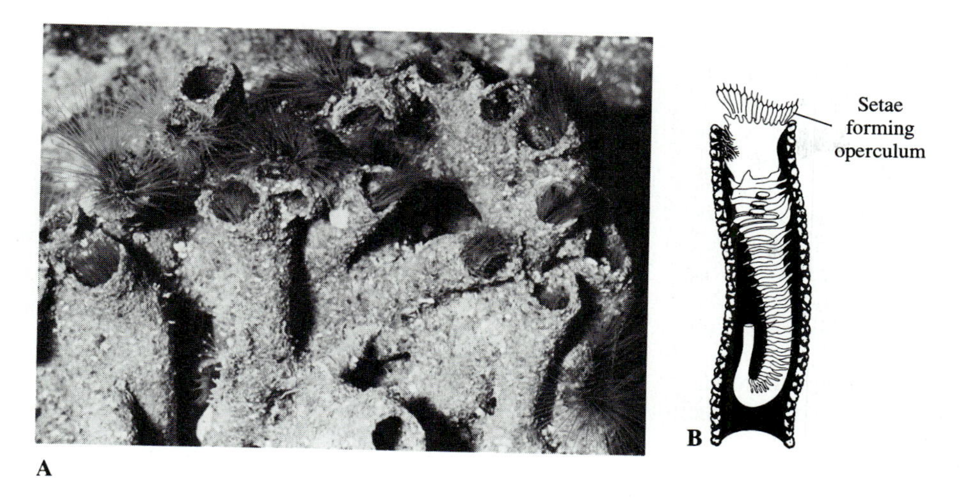

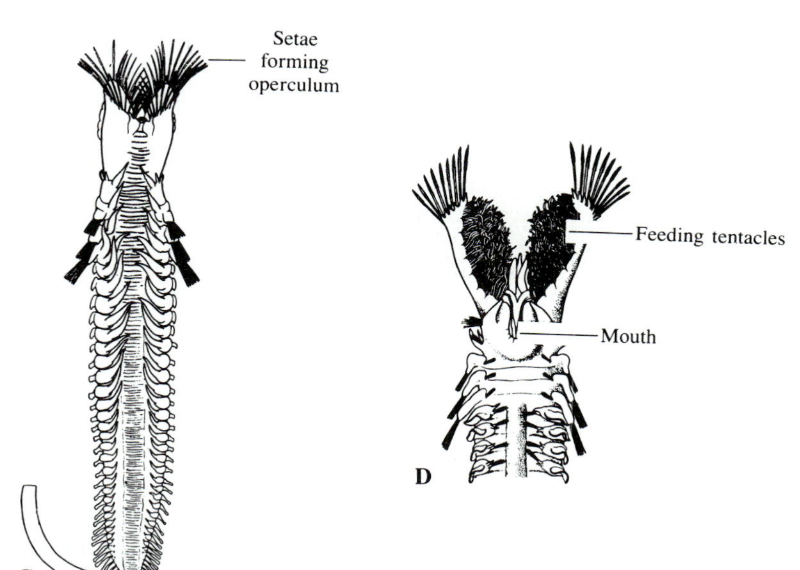

FIGURE 11.32 Family Sabellariidae. A. *Sabellaria cementarium,* feeding tentacles extended. B. *S. spinulosa* in its tube (Schäfer, Natur und Volk, *79.*) C. *Idanthyrsus armatus,* dorsal view. D. *I. armatus,* ventral view of anterior part of body. (C and D, Okuda, Journal of the Faculty of Science, Hokkaido University, Zoology, *6.*)

developed bands of longitudinal muscle, but it also involves rapid transmission of nerve impulses through giant fibers located in the ventral nerve cord.

Polychaetes of the families Serpulidae (Figure 11.31, A and B; Color Plate 3) and Spirorbidae (Figure 11.31, C and D) are similar to sabellids in the way they feed, but their tubes consist of calcium carbonate. New tube material is produced by glands below the mouth and is worked into place by the peristomial collar, in much the same way as in sabellids. Ventral glands all along the body add material to the inside of the wall of the tube. A serpulid can be distinguished from a sabellid not only because it has a calcareous tube, but also because it has a device for closing the tube when it pulls in its crown of radioles. The closing device, called the **operculum,** is usually cup-shaped or spherical, and generally has a short stalk. The whole structure is a modified radiole from the dorsalmost part of the right or left half-crown of radioles. Sometimes two opercula are formed—one being derived from each half-crown—but this is the result of a developmental aberration.

Another interesting family of polychaetes that feed by the ciliary–mucous method is the Sabellariidae. *Sabellaria* (Figure 11.32, A and B) and *Idanthyrsus* (Figure 11.32, C and D) are representative genera. These worms live in tubes whose consistency is much like that of concrete. The prostomium is a relatively small lobe anterodorsal to the mouth. It has a pair of sensory palps, but is otherwise undistinguished. The peristomial region, consisting of two fused segments, is unusual in that it grows forward over the prostomium and is crowned by two or three circles of stiff setae that block the opening of the tube when the worm withdraws. Ventrally, on both sides of the midline, the peristomium bears many tentacles that collect food particles. Once stuck in mucus, the particles are carried to the mouth along ciliated tracts. The tentacles can also pick up sand grains and deliver them to the fleshy lips that border the mouth. These lips work the grains into mucus produced by a horseshoe-shaped gland that partly surrounds the mouth. Some species, such as *Sabellaria alveolata* of European waters, build extensive reefs, because the larvae tend to settle on the tubes of adult worms.

The posterior portion of the body of a sabellariid has no setae, and it is so slender and thin-walled that the hindgut within it seems to be naked. The animal doubles back this part, so the anus faces the open end of the tube. This arrangement helps get fecal material going in the right direction.

Mucous-Bag Feeders

Among the more remarkable polychaetes are members of the family Chaetopteridae (Figure 11.33). They secrete parchmentlike tubes, often with considerable sand embedded in them. In some species, including the most famous one, *Chaetopterus variopedatus* (Figure 11.33, A–C), the tube is U-shaped, so that it has two openings at the surface of the sand or mud. In others, the tubes are oriented almost vertically, or simply form tangled mats that are mired down on the sea bottom or that attach to firm substrata. Although the following account describes feeding in *Chaetopterus,* it applies in principle to other genera in the family.

The body of *C. variopedatus,* about 25 cm long, is almost grotesque, because of the way certain segments abruptly differ from adjacent ones. The notopodial lobes of the parapodia of segment 12 are expanded into "wings," and their epithelium is both ciliated and equipped with glands that produce a lot of mucus. In segments 14 to 16 the notopodia are not only enlarged, but those on the left are united across the back of the animal with those on the right, so that they form semicircular fans. The fans flutter about 60 times a minute, and this activity brings water into the tube, drives it past the worm, and forces it out of the other opening. The current produced by the fans causes the sheet of mucus secreted by the wings on segment 12 to hang back as a bag, and when this touches the food cup, located about half way between the fans, it becomes anchored. The bag of mucus acts as a strainer that collects diatoms, bacteria, and other small particles of food.

The food cup not only serves as an anchoring device, but also rolls up the mucus as fast as more is produced by the wings. Eventually the bag breaks loose and all of it ends up in the ball being handled by the food cup. The cup then leans forward and drops the ball onto a ciliated middorsal groove. This groove carries the ball forward to the mouth so that it can be swallowed.

When larger particles, unsuitable as food, get into the tube, the wings let them pass by so that they do not clutter up the bag of mucus.

In certain chaetopterids, water is driven through the tube by ciliary activity rather than by notopodial fans. These worms generally have a series of segmentally arranged food cups and an equal number of bag-producing pairs of notopodia. *Spiochaetopterus costarum* (Figure 11.33, D), whose vertical tubes are buried in mud or sand, may have as many as 13 bags of mucus functioning at one time. It also collects particles on its mucus-coated palps, which it extends beyond the opening of the tube. The ciliated groove that runs the length of each palp carries the food to the mouth.

Phyllochaetopterus prolifica (Figure 11.33, E), whose tubes are only about 1 or 1.5 mm in diameter, is often extremely abundant on pilings and floating docks, as well as on bottom sediments. Like *Spiochaetopterus,* it uses its palps for feeding. It provides one of the best examples of asexual reproduction among polychaetes. The pieces into which it divides regenerate the parts they lack. Temporarily, there may be several worms in a single tube, but as a rule the young eventually form their own tubes, often as branches of the parent's tube. It is primarily through asexual reproduction that *Phyllochaetopterus* builds up its characteristically dense colonies. Some populations have no sexual individuals at all.

Deposit Feeders

Deposit-feeding polychaetes mop up and swallow sediment. Some are relatively unselective, taking in just about all of the loose material in front of them; others are choosy and reasonably successful in picking out particles that are likely to have food value.

Good examples of unselective deposit feeders are the so-called red worms (Color Plate 3) found in the sand of wave-swept beaches. They belong to the family Opheliidae. Where they occur, red worms generally form large concentrations, often with several thousand individuals per square meter of surface area. Their distribution on the beach, however, is usually limited to a narrow band at about the midtide level. The worms dig down to a depth of several centimeters and eat sand. As the sand passes through the gut, bacteria, diatoms, and other organic material clinging to the grains are digested.

Lugworms—*Arenicola, Abarenicola,* and related genera—belong to the family Arenicolidae (Figure 11.34). They live in nearly J-shaped burrows in mud or sand, with the tail directed toward the opening of the

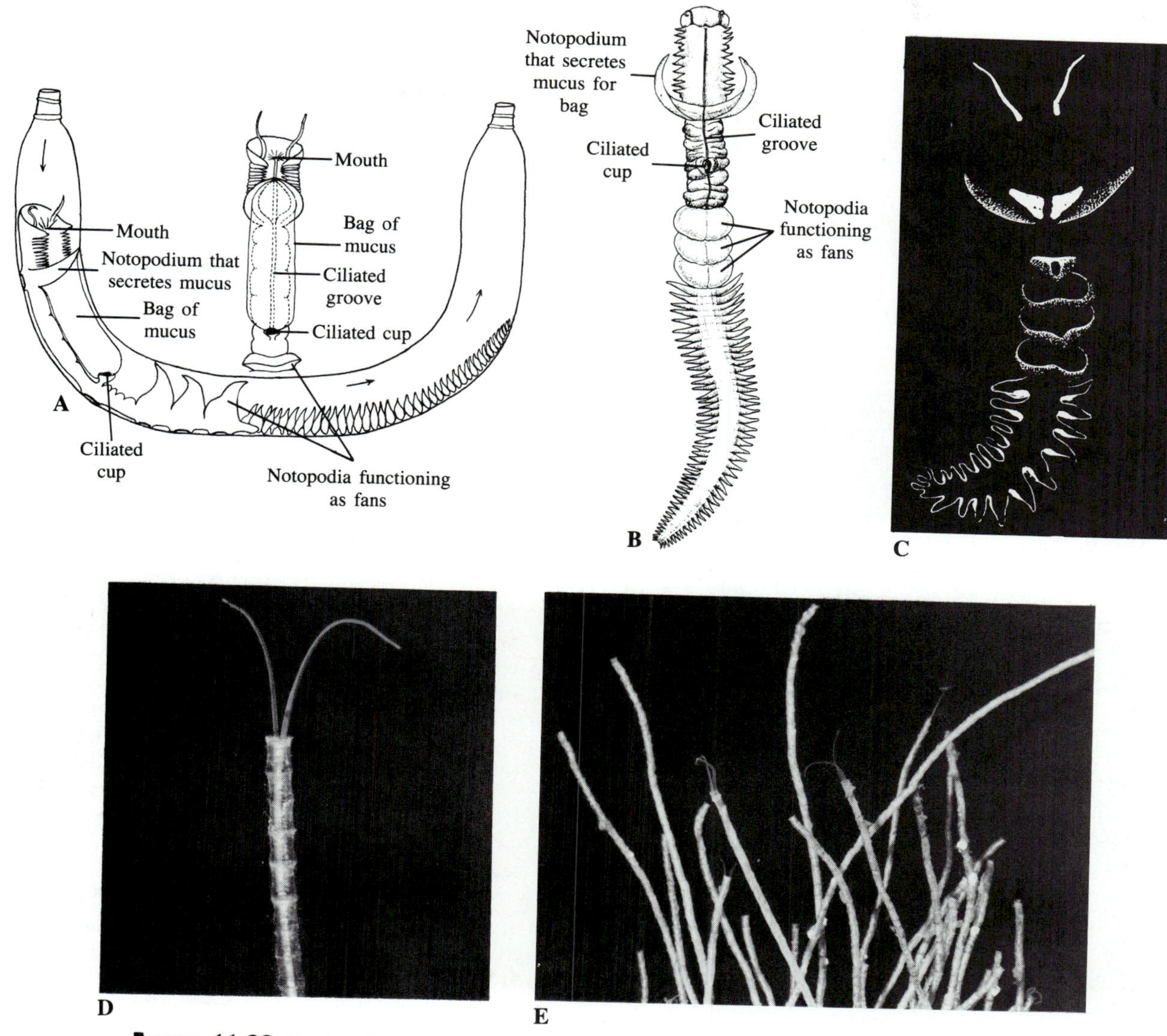

FIGURE 11.33 Family Chaetopteridae. A. *Chaetopterus variopedatus* in its tube, and also a dorsal view of the anterior end of the worm. (MacGinitie, Biological Bulletin, *77*.) B. *Chaetopterus,* dorsal view. (Okuda, Journal of the Faculty of Science, Hokkaido University, Zoology, *6*.) C. *Chaetopterus,* luminescence. (Panceri, Atti Accademia delle Scienze Fisiche e Matematiche, Napoli, *7*.) D. *Spiochaetopterus costarum,* its palps extended from the tube. E. *Phyllochaetopterus prolifica,* palps extended from some tubes.

burrow and the head directed toward the blind end (Figure 11.34, B). The presence of a bed of lugworms is revealed by coiled fecal castings that look as if they had been squeezed out of a cake-decorating device (Figure 11.34, C). In feeding, a lugworm thrusts out its short pharynx to mop up sand and organic material and then pulls it back in. Mud or sand often caves in above the place where the head is working, so useful food sifts down to the worm.

Like other worms that are confined to burrows or tubes, a lugworm must solve certain problems related to respiration and elimination of wastes. To ensure efficient exchange of oxygen and carbon dioxide, and to wash away ammonia, the animal makes peristaltic movements that cause water to be drawn into the burrow and then to be flushed out. Certain species of *Arenicola* exhibit a precise rhythm in their daily activities. Following a period of headward irrigation, the lugworm

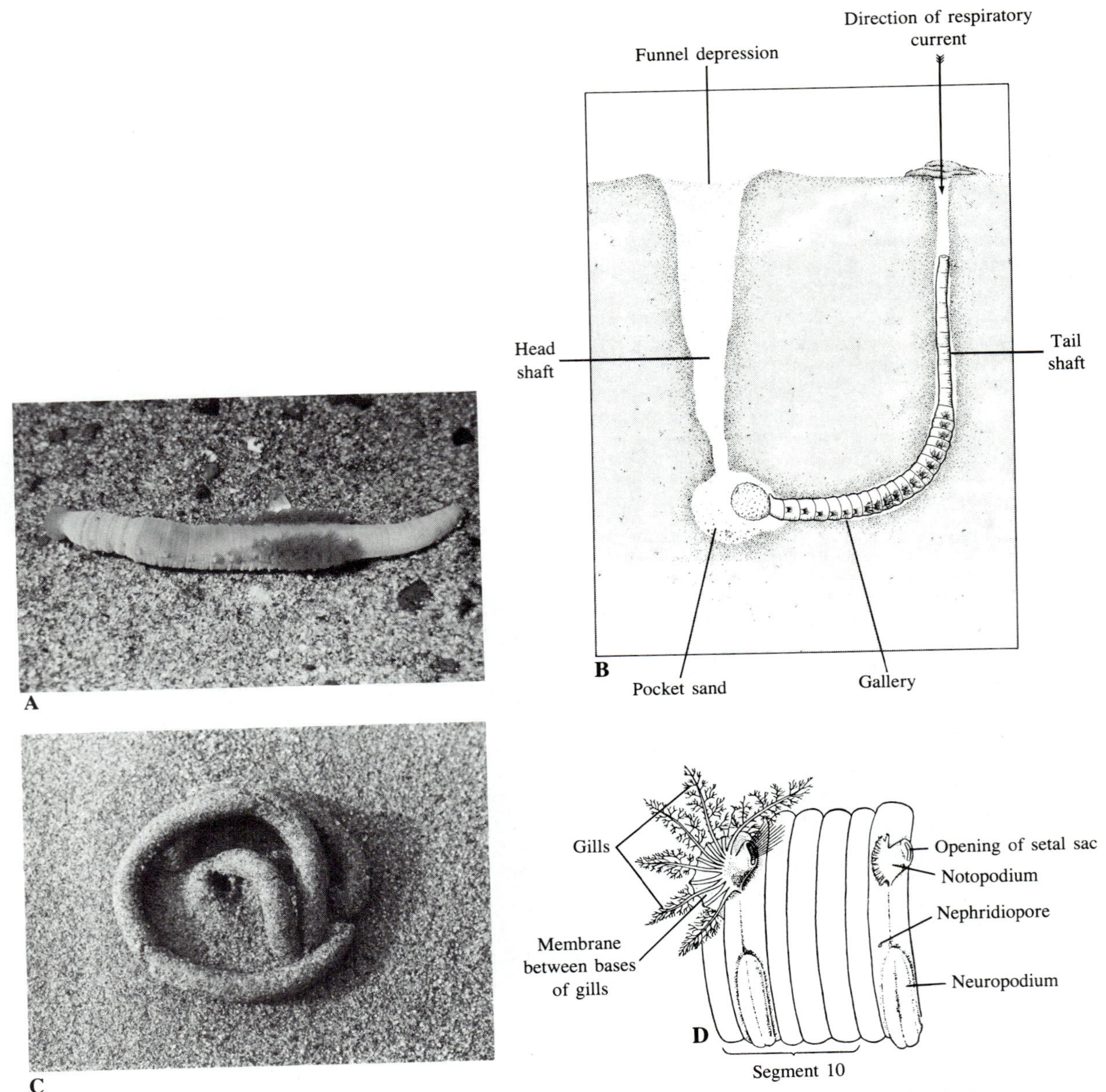

FIGURE 11.34 Family Arenicolidae. A. *Abarenicola pacifica*. B. *Abarenicola pacifica* in its burrow. (Hylleberg, Ophelia, *14*.) C. *Abarenicola claparedi*, fecal castings on the surface of sand. D. *Arenicola marina*, segment 10 and a portion of segment 9 (gills removed on the latter). (After Ashworth, Liverpool Marine Biology Committee Memoirs, *11*.)

feeds. Next there is a period of tailward irrigation followed by defecation at the surface. Thus fecal material does not accumulate in the burrow. As long as a lugworm is immersed in water, the cycle is repeated continuously, each cycle taking about an hour.

Polychaetes of the family Maldanidae (Figure 11.35, A), some of which are called "bamboo worms" because of their elongated segments, feed in much the same way as lugworms. They are usually oriented head down in vertical tubes that consist of sand grains held together by mucus, and they use an eversible pharynx for mopping up food. Periodically they eliminate fecal matter, composed mostly of small sand grains, from the upper end of the tube.

Capitellids (Figure 11.35, B) are also sediment eaters. They are slender worms, generally less than 1 mm

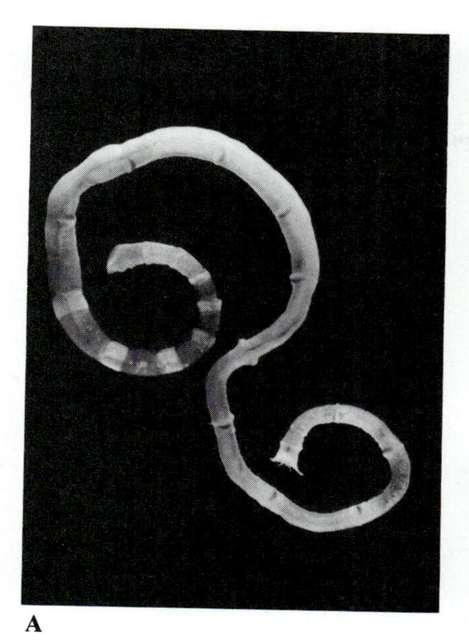

FIGURE 11.35 A. Family Maldanidae. *Axiothella rubrocincta*. B. Family Capitellidae. *Notomastus tenuis*. C. Family Sternaspidae. *Sternaspis scutata*.

in diameter, though perhaps as long as 10 cm or more, and sometimes they occur in enormous concentrations, especially in muddy bays. They eat as they burrow, alternately everting and withdrawing a bulbous pharynx–esophagus complex. Many capitellids are able to survive in habitats where decomposition of organic matter by bacteria uses up most of the free oxygen.

Sternaspis (Figure 11.35, C), the only genus in the family Sternaspidae, consists of short-bodied, grublike worms that live in subtidal muds. They feed on the sediment as they plow through it. Several of the anterior segments can be inverted, and the gut, which is considerably longer than the body, folds back on itself. Two other distinctive features of *Sternaspis* are a pair of hardened platelike structures and a tuft of gill filaments at the posterior end of the body.

Owenia (Figure 11.36), which belongs to the family Oweniidae, makes a tube of sand grains mixed with secreted material. The tube is oriented vertically in the sand. Most of the segments of *Owenia* are elongated like those of a maldanid, but the peristomium is drawn out into branched lobes. The worm bends down and touches the lobes to the sandy substratum, thereby mopping up particles that are carried to the mouth along

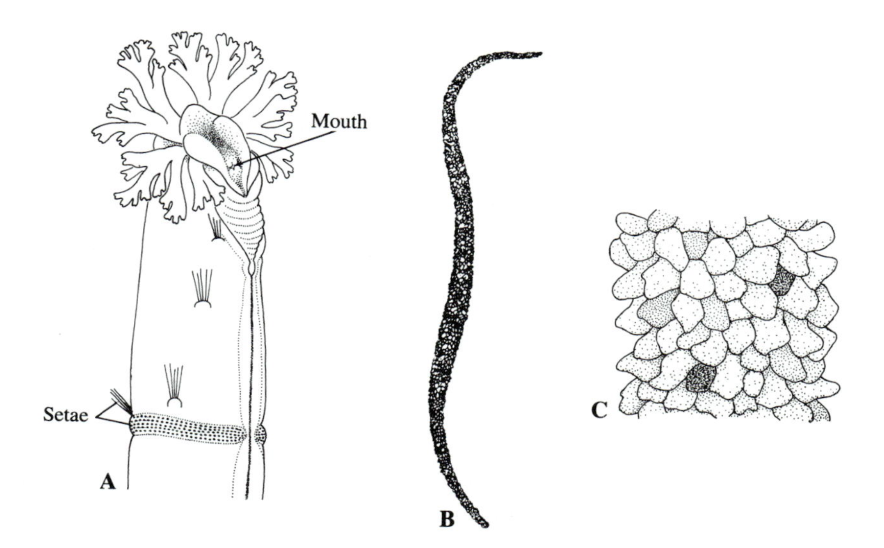

FIGURE 11.36 Family Oweniidae. *Owenia fusiformis*. A. Anterior portion of worm. B. Tube. C. Small portion of tube, greatly enlarged.

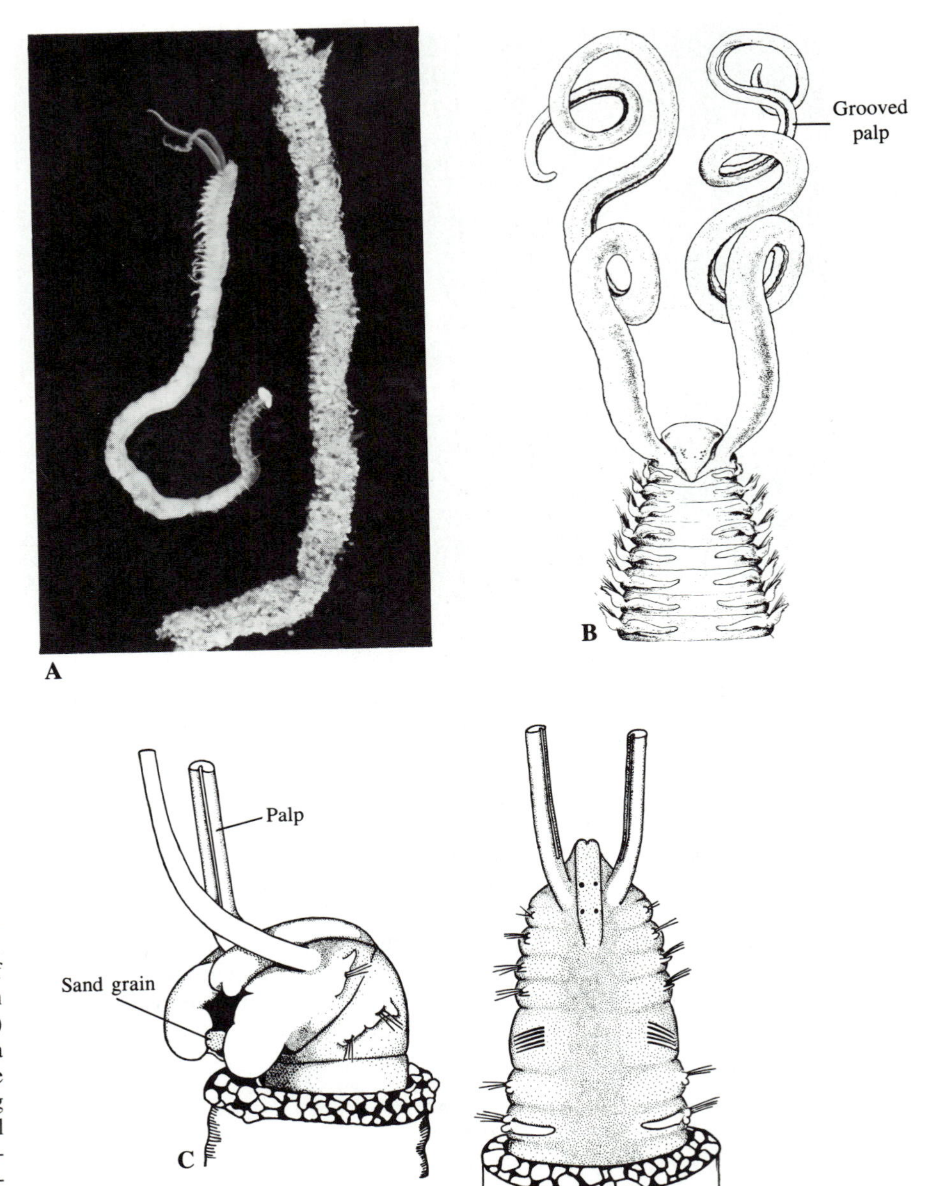

FIGURE 11.37 Family Spionidae. A. A typical spionid and its tube. B. Anterior portion of the body of *Spio punctata*. (Hartman, Allan Hancock Pacific Expeditions, *25*.) C. *Polydora ciliata*, placing a sand grain on the rim of its tube (left) and in the tubelining posture (right). (Dorsett, Journal of the Marine Biological Association of the United Kingdom, *41*.)

ciliary tracts. The lobes may also be used to collect particles that are in suspension.

Polychaetes of the family Spionidae (Figure 11.37) are characterized by a pair of long, grooved palps arising from the dorsal side of the prostomium or peristomium. These structures are thought to be homologous to the ventral prostomial palps of various other polychaetes; their displacement to the dorsal surface, and sometimes to the peristomial region, is secondary. In some species, the palps are used for collecting food particles from surface deposits; in others, they are used for capturing small organisms, such as diatoms, that are in suspension. At least one spionid secretes a mucous net for trapping particles within its tube, although it also employs its palps for feeding by both methods just described. Spionids generally make their tubes by mixing mud or sand with mucus (Figure 11.37, C). They sometimes form dense populations, especially on soft substrata.

Polychaetes of the family Cirratulidae differ considerably among themselves. Some, such as those of the genus *Dodecaceria* (Figure 11.38, C and D), have a

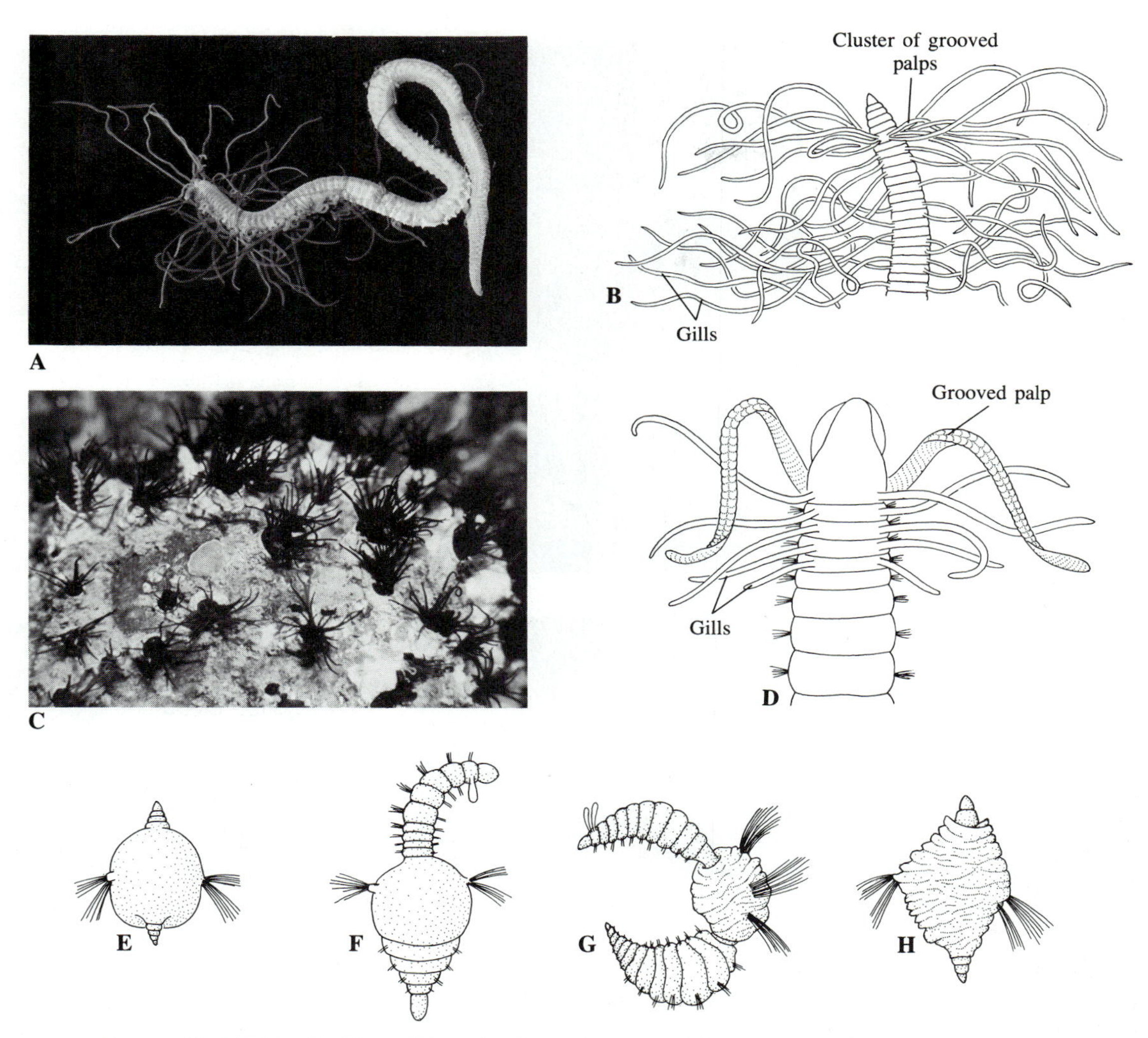

Figure 11.38 Family Cirratulidae. A. *Cirratulus spectabilis*. B. *Cirratulus cirratus,* anterior portion. C. *Dodecaceria fewkesi*. This species secretes calcareous tubes, and, by asexual reproduction, it forms dense colonies. Of the tentaclelike structures extended from each tube, two are grooved palps, the rest gills, D. *Dodecaceria concharum,* anterior portion. E–H. Asexual reproduction in *Dodecaceria caulleryi*. A single segment, isolated when a worm fragments, produces two worms. After these become detached, the same segment can repeat the process. (After Dehorne, Bulletin Biologique de la France et de la Belgique, *77*.)

pair of grooved palps, similar to those of spionids, on the peristomial region; others, including the species of *Cirratulus* (Figure 11.38, A and B), have two clusters of grooved filaments on the dorsal side of the body behind the peristomium. In addition to these structures, which collect food particles and transfer them to the mouth, there are generally several to many pairs of filamentous gills. Cirratulids are sedentary worms, and most of them live in mud, muddy sand, or gravel, often in crevices or in places where the substratum is stabilized by roots of plants such as eelgrass. A few, including *Dodecaceria,* secrete calcareous tubes.

The species of *Dodecaceria* are capable of asexual reproduction. Some divide into two more or less equal halves, which regenerate missing parts to form complete worms. Others have more prodigious powers of procreation. They break up into individual segments that bud off two complete individuals and that sometimes have enough tissue left to repeat the process (Figure 11.38, E–H).

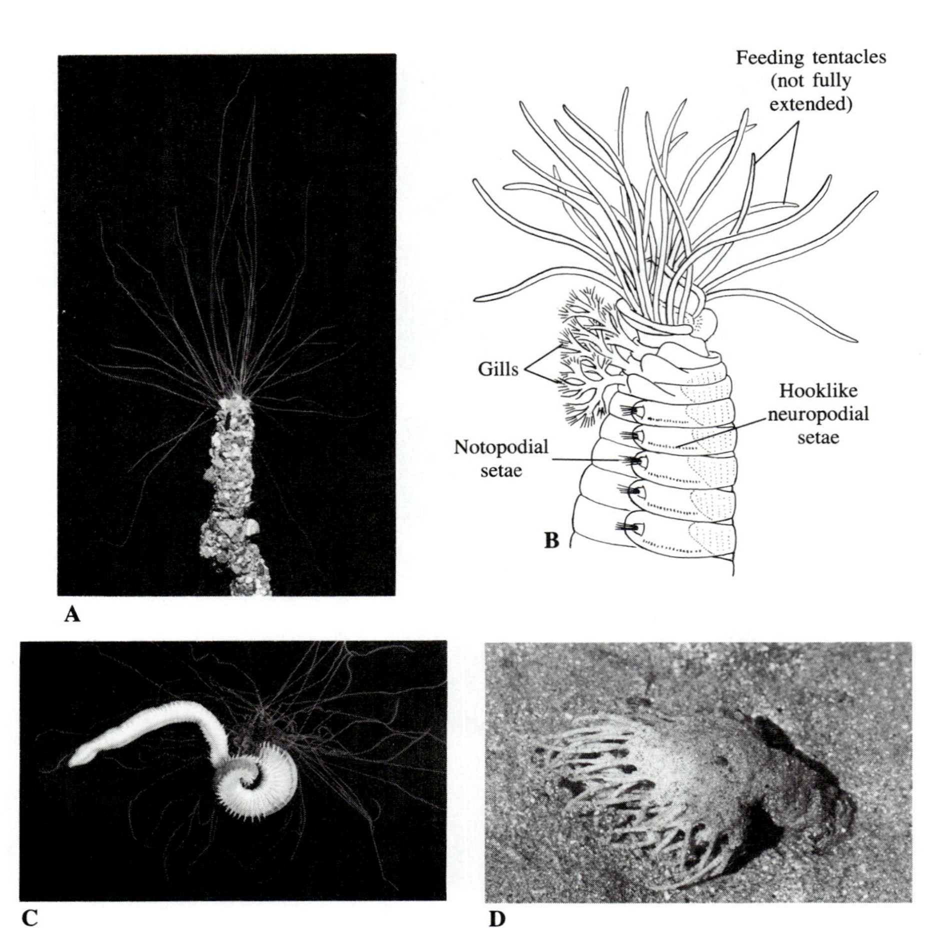

FIGURE 11.39 Family Terebellidae. A. *Eupolymnia heterobranchia,* tentacles extended from its tube. B. *Eupolymnia,* anterior portion. C. *Thelepus crispus,* removed from its tube. D. *Pista pacifica,* flared portion of tube projecting above the sandy substratum.

In several families of polychaetes, feeding is accomplished by slender tentacles that are spread out over the substratum. The tentacles arise from the peristomium, and each one has a ciliated groove along which small particles, such as diatoms, are conveyed toward the mouth. Members of the family Terebellidae (Figure 11.39) exhibit this type of feeding especially well. These worms inhabit tubes made of mud or of parchmentlike material. In some species, the tubes are attached to rocks; in others, they are buried in soft sediments. There are many tentacles, but the ciliated grooves of these do not lead directly to the mouth. Instead, the tentacles take turns wiping their basal portions against a lip above the mouth, and cilia on this lip drive suitable food particles into the mouth. Unsuitable particles are rejected or are incorporated into the tube. The cementing substance is mucus, produced by glands on the ventral side of the anterior part of the body.

In the Ampharetidae (Figure 11.40, A), closely related to the Terebellidae, there are fewer tentacles, and they are used in a slightly different way. After the tentacles have accumulated considerable food, they are simply pulled back into the mouth. Terebellids cannot do this; their tentacles are contractile, but not withdrawable.

Pectinariids (or amphictenids) (Figure 11.40, B and C) are also deposit feeders that sort particles with their tentacles. They make short, conical tubes out of sand grains. These tubes are open at both ends, and the worms bury themselves head-down in mud or sand. The chimney of the tube projects above the substratum. Pectinariids are rather mobile and able to relocate themselves.

The Strange Myzostomes

The family Myzostomidae consists of peculiar little worms that live in or on echinoderms, especially crinoids. The flattened, sometimes almost circular body of a myzostome (Figure 11.41, A) is not segmented externally. On its ventral side, however, there are five pairs of fleshy projections, each with a few stout, hooklike setae. The projections are believed to be modified parapodia. Dorsally, near the edges of the body, are slender outgrowths (usually 10 pairs) that resemble the cirri of polychaete parapodia.

The gut, equipped with a protrusible pharynx, is

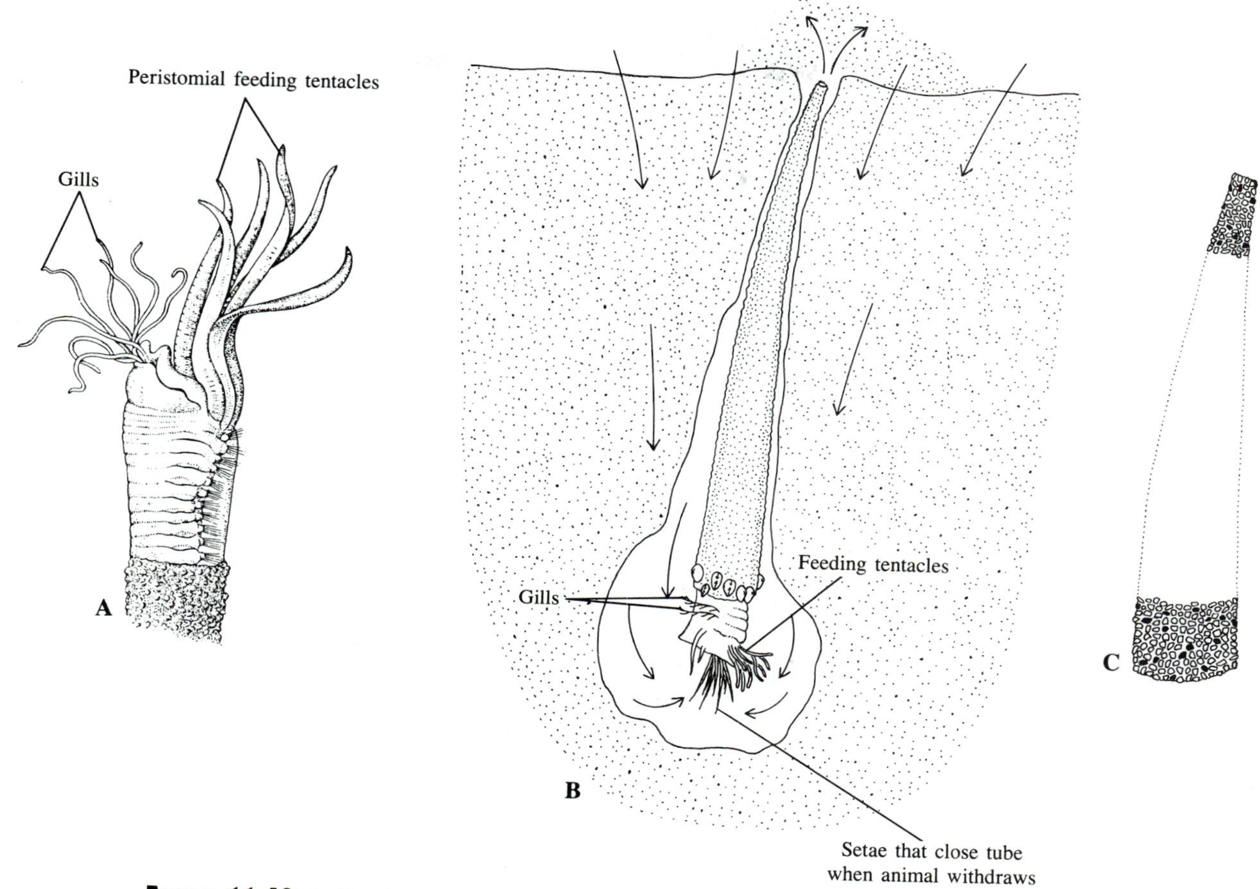

FIGURE 11.40 A. Family Ampharetidae. *Samytha bathycola*. (Ushakov, Keys to the Fauna of the USSR, *56*.) B. Family Pectinariidae. *Pectinaria australis* in its burrow. Water currents are indicated by arrows. The anterior portion of the tube has several small commensal clams (*Aphroditica bifurca*) attached to it. (After Wear, Biological Bulletin, *130*.) C. Tube of a pectinariid, constructed of sand grains and mucus.

complete. The middle portion of the gut sends out branched side pockets that take up much of the space within the body. Most of the myzostomes that live on crinoids use the protrusible pharynx to feed on the material that the host has trapped and is passing along ciliated grooves toward its mouth. Some species grip the tissue of the crinoid so tightly that they leave scars. A few even sink into the tissue and may engender the formation of galls. Certain myzostomes are internal parasites of crinoids, and there is one that feeds on the gonadal tissue of basket stars.

The coelom, though lobed, shows no signs of segmentation. It is mostly occupied by gonadal tissue. The portion along the midline, above the gut, is drained by a pair of ciliated ducts that are probably metanephridia. These run to the gut rather than directly to the outside, and they have been claimed to function in removal of excess sperm and useless by-products of spermatogenesis. There is no circulatory system in these worms.

The brain, lying above the pharyngeal portion of the gut, is small. The several ganglia of the ventral nerve cord are consolidated into a compact mass (Figure 11.41, B), but the paired arrangement of lateral nerves is nevertheless suggestive of segmentation.

Myzostomes are essentially hermaphroditic, but at least some of them function first as males, then as females. The male genital pores, on small papillae located at the bases of the third parapodia, are linked with seminal vesicles (Figure 11.41, A). The latter receive sperm from the pair of testes, whose follicles are scattered through the coelom. The ovaries on both sides of the body are likewise coelomic (Figure 11.41, B and C). The eggs pass into a single duct that leads to the hind gut.

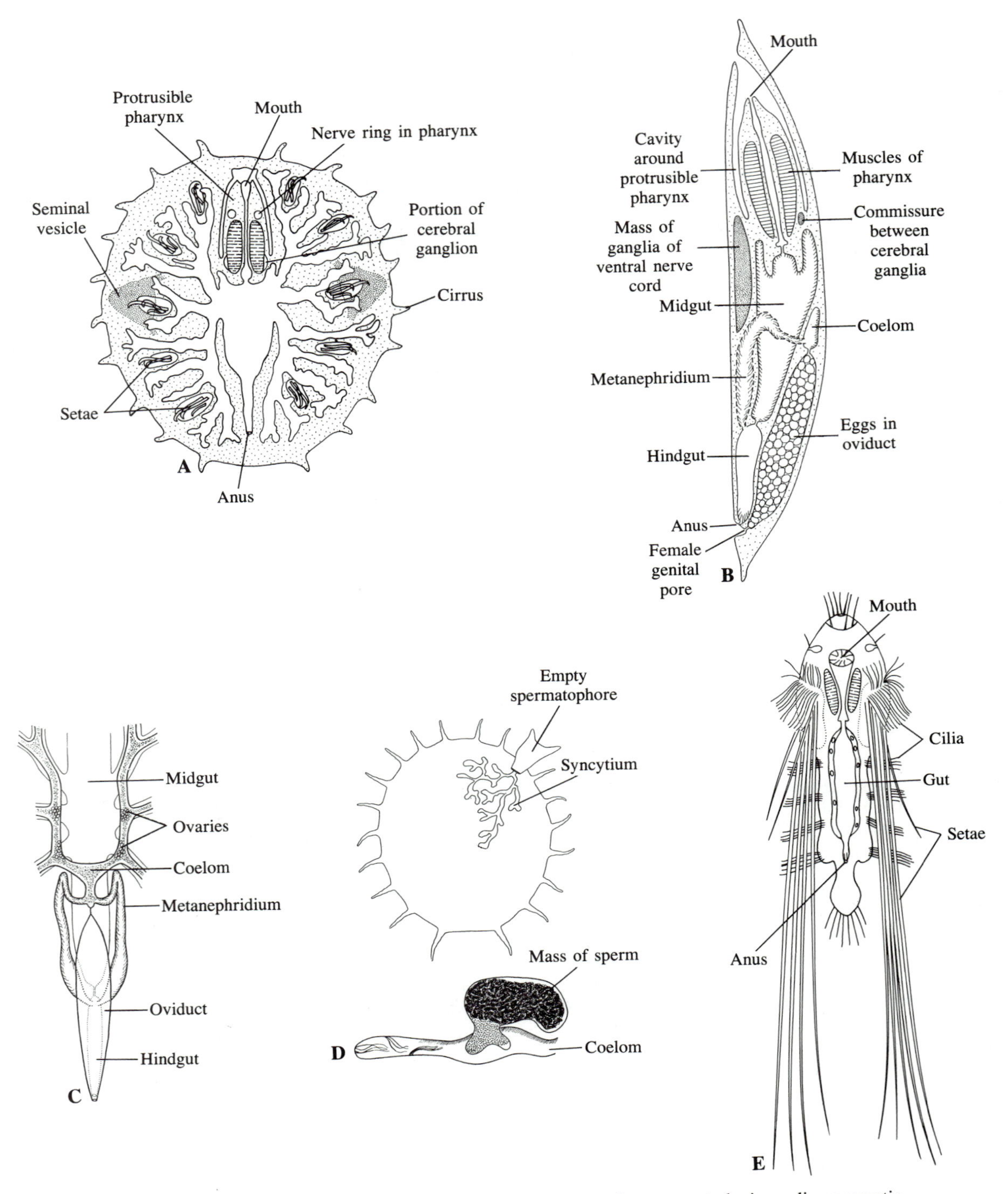

FIGURE 11.41 Family Myzostomidae. A *Myzostomum cirriferum,* ventral view; diagrammatic. (After Platel, Cahiers de Biologie Marine, *3.*) B. *M. cirriferum,* female, median sagittal section. C. *M. cirriferum,* female reproductive system, ventral view. D. *M. cirriferum,* sperm-bearing syncytium penetrating tissues of a female (above) and a portion of a syncytium that has liberated some of its sperm into the coelom of the female (below). (B–D, Jägersten, Zoologiska Bidrag från Uppsala, *18.*) E. *M. parasiticum,* 6-day-old larva. (After Jägersten, Arkiv för Zoologi, *31A.*)

Sperm are transferred from one partner to the other in a spermatophore, secreted within a seminal vesicle. When the spermatophore is forced out by a contraction of the wall of the vesicle, it sticks to the epidermis of the partner. Some of the cells inside it migrate into the tissue (Figure 11.41, D), fusing to form a syncytium that continues to move inward. The syncytium draws masses of sperm along with it till they reach the oocytes within the coelom.

Myzostomes whose development has been studied have spiral cleavage and a trochophore larva. The trochophore transforms into an adult by way of a stage that has two sets of long setae, as well as circles of cilia (Figure 11.41, E). These setae and cilia are lost before the definitive setae and parapodia appear. Only one pair of coelomic sacs appears during development. Because of this, and also because the only evidence for segmentation is seen in the arrangement of setae and the nervous system, some zoologists feel that the placement of myzostomes in Annelida is doubtful.

Archiannelids

At one time, the name *Archiannelida* was applied to an assemblage of relatively simple worms thought to be ancestral to other groups of annelids. However, this simplicity is probably a secondary feature. Moreover, the five families conventionally called archiannelids are not necessarily closely related. All of them are now considered to belong to the Polychaeta.

Archiannelids are marine, and they are most likely to be found near the surface of sandy, gravelly, and muddy substrata. Most of them feed on diatoms, bacterial coatings, and fine detritus.

Members of the family Dinophilidae (Figure 11.42, A and B) are mostly under 2 mm long, and have only a few segments. There are no head appendages, and parapodia and setae are present only in one genus, *Parapodrilus*. There is extensive ciliation on the ventral surface, and in most of the genera the segments are encircled by rings of cilia. The majority of dinophilids have an eversible, muscular pharynx, used for licking food from sand grains. The protonephridia consist of flame cells.

The Nerillidae (Figure 11.42, C) also have only a few segments, but they have one, two, or three prostomial antennae, as well as a pair of prostomial palps. There are parapodia, each with a bundle of simple or compound setae. Nerillids use their eversible pharynx for rubbing diatoms and other particulate matter from sand grains. In some of the genera, the pharynx is equipped with jaws.

In *Saccocirrus* (Figure 11.42, D), the only genus of Saccocirridae, the prostomium has a pair of long antennae and also a pair of eyes. These worms may reach a length of about 2 cm, and have numerous segments. There are small parapodia, each with a few simple setae. The food consists of diatoms and other organic particles found in sediments through which the worms burrow.

The Protodrilidae also have a pair of antennae and eyes. The segments, however, are mostly indistinct, and there are no parapodia. There are two genera, one with setae, one without. The lower lip of the mouth region is used for scraping food from sand grains.

Polygordius (Figure 11.42, E) and *Chaetogordius* are the only genera of the Polygordiidae. The former, which is more commonly encountered, lacks setae, whereas *Chaetogordius*, as its name implies, has them. Both genera are characterized by a slender body and numerous segments, but these segments are poorly defined. There are two long antennae, but no eyes. Like most archiannelids, polygordiids get their food by scraping sand grains. The pharynx is eversible in some species but not in others.

References on Polychaeta

General

Anderson, D. T., 1966. The comparative embryology of the Polychaeta. Acta Zoologica, *47*:1–42.

Barnes, R. D., 1965. Tube-building and feeding in chaetopterid polychaetes. Biological Bulletin, *129*:217–233.

Baskin, D. G., 1973. Neurosecretion and endocrinology of nereid polychaetes. American Zoologist, *16*:107–124.

Clark, R. B., and Olive, P. J. W., 1973. Recent advances in polychaete endocrinology and reproductive biology. Oceanography and Marine Biology, An Annual Review, *11*:175–222.

Dales, R. P., and Peter, G., 1972. A synopsis of the pelagic Polychaeta. Journal of Natural History, *6*:55–92.

Daly, J. M., 1973. The ability to locate a source of vibrations as a prey-capture mechanism in *Harmothoe imbricata*. Marine Behaviour and Physiology, *1*:305–322.

Eakin, R. M., Martin, G. G., and Reed, C. T., 1977. Evolutionary significance of fine structure of archiannelid eyes. Zoomorphologie, *88*:1–18.

Fauchauld, K., 1975. Polychaete phylogeny: a problem in protostome evolution. Systematic Zoology, *23*:493–506.

Fauchauld, K., 1977. The polychaete worms. Definitions and keys to the orders, families and genera. Natural History Museum of Los Angeles Science Series, *28*:1–190.

Fauchauld, K., and Jumars, P., 1979. The diet of worms: a study of polychaete feeding guilds. Oceanography and Marine Biology, An Annual Review, *17*:193–284.

Gray, J., 1939. Studies in animal locomotion, VIII. The kinetics of locomotion in *Nereis diversicolor*. Journal of Experimental Biology, *16*:9–17.

Gustus, R. M., and Cloney, R. A., 1973. Ultrastructure of the larval compound setae of the polychaete *Nereis vexillosa* Grube. Journal of Morphology, *140*:355–366.

Hermans, C. O., 1969. The systematic position of the Archiannelida. Systematic Zoology, *18*:85–102.

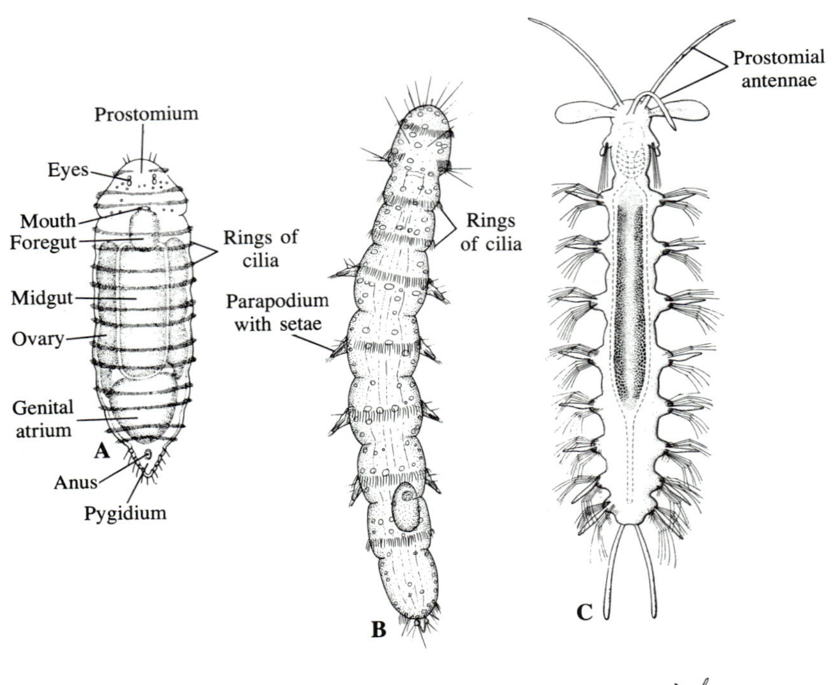

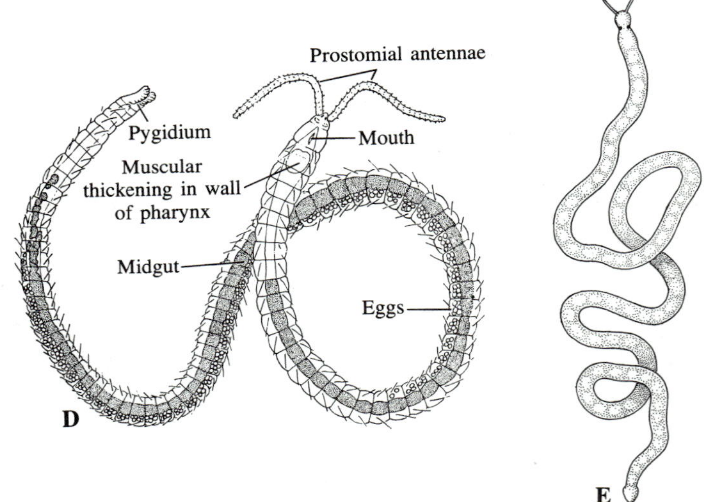

FIGURE 11.42 Archiannelids. A. Family Dinophilidae. *Dinophilus vorticoides,* female. (Ivanov *et al.,* Major Practicum in Invertebrate Zoology.) B. Family Dinophilidae. *Parapodrilus psammophilus.* (Westheide, Helgoländer Wissenschaftliche Meeresuntersuchungen, *12,* 1965.) C. Family Nerillidae. *Mesonerilla fagei.* (Swedmark, Archives de Zoologie Expérimentale et Générale, *98.*) D. Family Saccocirridae. *Saccocirrus sonomacus.* Martin, Transactions of the American Microscopical Society, *96.*) E. Family Polygordiidae. *Polygordius neapolitanus.* (After Fraipont, Fauna und Flora des Golfes von Neapel, *14.*)

Jacobsen, V. H., 1967. The feeding of the lugworm, *Arenicola marina.* Ophelia, *4:*91–109.

Kudenov, J. D., 1977. The functional morphology of feeding in three species of maldanid polychaetes. Zoological Journal of the Linnean Society, *60:*95–109.

Mangum, C. P., 1976. The oxygenation of hemoglobin in lugworms. Physiological Zoology, *49:*85–99.

Nicol, E. A. T., 1931. The feeding mechanism, formation of the tube, and physiology of digestion in *Sabella pavonina.* Transactions of the Royal Society of Edinburgh, *56:*537–598.

O'Clair, R. M., and Cloney, R. A., 1974. Patterns of morphogenesis mediated by dynamic microvilli: chaetogenesis in *Nereis vexillosa.* Cell and Tissue Research, *151:*141–157.

Regional Faunas

Temperate Pacific Region

Banse, K., and Hobson, K. D., 1974. Benthic errantiate polychaetes from British Columbia and Washington. Bulletin 185, Fisheries Research Board of Canada.

Dales, R. P., 1957. Pelagic polychaetes of the Pacific Ocean. Bulletin of the Scripps Institution of Oceanography, University of California, *7:*99–168.

Hartman, O., 1968. Atlas of Errantiate Polychaetous Annelids from California. Allan Hancock Foundation, University of Southern California, Los Angeles.

Hartman, O., 1969. Atlas of Sedentariate Polychaetous Annelids from California. Allan Hancock Foundation, University of Southern California, Los Angeles.

Hobson, K. D., and Banse, K., 1981. Sedentariate and archiannelid polychaetes of British Columbia and Washington. Bulletin 209, Canadian Bulletin of Fisheries and Aquatic Sciences.

Temperate Atlantic Region

Day, J. H., 1973. New Polychaeta from Beaufort, with a key to all species recorded from North Carolina. NOAA Technical Report NMFS Circular 375.

Fauvel, P., 1923. Polychètes errantes. Faune de France, *5:* 1–488.

Fauvel, P., 1927. Polychètes sédentaires. Addenda aux errantes, archiannélides, myzostomaires. Faune de France, *16:*1–494.

Gardiner, S. L., 1975. Errant polychaetes from North Carolina. Journal of the Elisha Mitchell Scientific Society, *91:*77–220.

Pettibone, M. H., 1963. Marine polychaete worms of the New England region. I. Aphroditidae through Trochochaetidae. Bulletin of the United States National Museum, *227:*1–356.

Uebelacker, J. M., and Johnson, P. G. (editors), 1984. Taxonomic guide to the polychaetes of the northern Gulf of Mexico. 7 vols. Barry A. Vittor & Associates, Mobile, Alabama.

South African Region

Day, J. H., 1967. A Monograph on the Polychaeta of Southern Africa. 2 vols. British Museum (Natural History), London, Publication no. 656.

CLASS OLIGOCHAETA

The name Oligochaeta refers to the fact that the annelids of this class, unlike those of the Polychaeta, typically have only a few setae on each segment. The setae of oligochaetes, moreover, are not as varied as those of polychaetes. There are, however, other important differences between the two groups.

Whereas many polychaetes have sensory outgrowths arising from the prostomium and peristomium, or have peristomial tentacles or radioles for gathering food, such structures are not found in oligochaetes. Thus the head region of oligochaetes appears to be comparatively simple. Oligochaetes also lack parapodia, although some aquatic types have fleshy outgrowths that function as gills.

Nearly all oligochaetes are hermaphroditic, and their reproductive systems are, in general, more complex than those of polychaetes. Instead of simply releasing gametes as most polychaetes do, oligochaetes copulate and exchange sperm. Following copulation, the sperm are stored in seminal receptacles until the eggs are ready to be fertilized. In many oligochaetes, a few segments of the body have epidermal glands for secreting a cocoon within which one or more eggs develop into young worms. Whether or not there is a cocoon, development is direct, there being no larval stage comparable to the trochophore of polychaetes.

The class Oligochaeta consists of about 3000 species, and the group as a whole is varied, though much less diversified than the Polychaeta. Most oligochaetes are found in freshwater habitats, especially mucky places where vegetation is decaying. In terrestrial situations, there are species that burrow in soil or humus, and some of the larger soil-inhabiting species are the earthworms familiar to almost everyone. There are also many oligochaetes, mostly small ones, living in marine muds and under organic debris cast up on beaches.

The Earthworm as an Example of the Oligochaeta

The European earthworm, *Lumbricus terrestris,* is now well established in North America and other continents where it has been introduced. Because it is large and easily dissected, it is widely used for illustrating the main features of the Oligochaeta, and often of annelids in general.

External Features

Between the prostomium and pygidium, both of which are distinct, there are about 100 to 150 true segments. The mouth is situated on the ventral surface, at the anterior edge of the peristomium, which is conventionally numbered segment 1 (Figure 11.43). The anus is at the tip of the pygidium. All of the true segments, except the peristomium, have eight setae, arranged in four pairs; two pairs are ventral and two are ventrolateral.

Several sets of pores open at the surface of the body, but some of them may be barely visible. The most difficult to find are those along the dorsal midline. They are located in the grooves between segments, beginning with segments 11 and 12. Coelomic fluid forced out of these pores helps keep the body surface moist. More easily seen are the pores of the metanephridia, located ventrolaterally on almost every segment. Four pairs of genital pores open just lateral to the ventral setae. One pair, on segment 15, consists of male pores, from which sperm issue. At the posterior edges of segments 9 and 10 are the two pairs of female pores that receive sperm during copulation, and the pair of female pores through which eggs are laid are on segment 14.

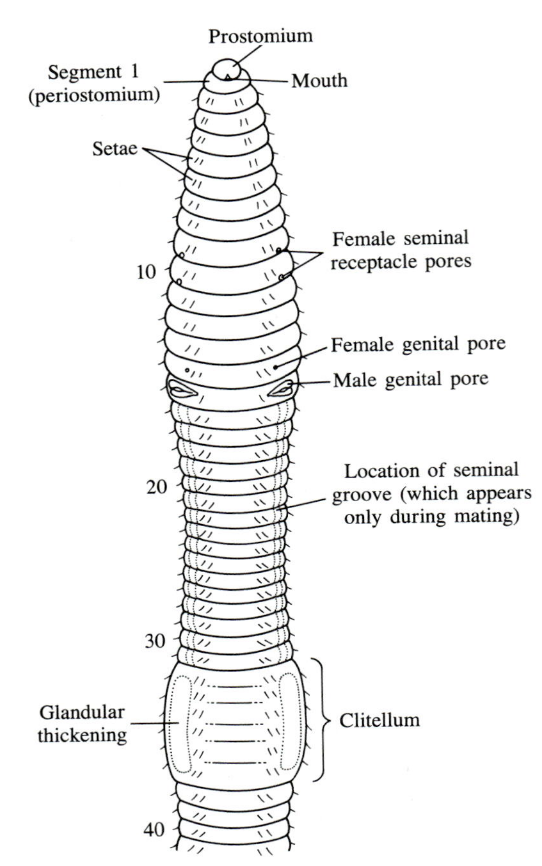

FIGURE 11.43 *Lumbricus terrestris,* anterior end, ventral view.

Body Wall and Coelom

The organization of the body wall is shown in Figures 11.44 and 11.45. It need not be discussed in detail, because it fits the plan described near the beginning of this chapter for annelids in general. One should appreciate, however, that whereas the circular muscles form rather distinct segmental units, each longitudinal muscle fiber spans several segments. Note also that the setal sacs are provided with muscles that control the extent to which the setae are protruded or withdrawn and the angle at which they are extended (Figure 11.45).

The coelom, though occupied by the gut, metanephridia, and reproductive organs, is rather spacious. Successive segmental compartments of the coelom are almost completely separate, but the septa between them have openings through which the ventral nerve cord and major blood vessels just above and below the cord pass. The openings are encircled by muscles that form sphincters, minimizing leakage of coelomic fluid from one segment to another. There is musculature elsewhere in the septa, too, but the surfaces of these partitions are covered by peritoneum.

Although the shape of a segment may be changed by activity of muscles in the body wall, the volume of fluid in each coelomic compartment remains about the same. Contraction of longitudinal muscles shortens a segment or group of segments, whereas contraction of circular muscles causes the same segment or group of segments to lengthen. In an earthworm, as a rule, when a segment is shortened, its setae are extended, so that this segment ''holds its ground.'' Forward progression of the animal involves the shortening of some segments, whose setae anchor those segments, while more anterior segments, their setae withdrawn, are being lengthened.

The burrowing activities of an earthworm also depend on the animal's body-wall muscles and its ''tight'' coelomic compartments. An earthworm has amazing endurance, and can continue to burrow for hours. Some of the burrowing polychaetes, such as lugworms, are relatively ''leaky,'' especially at the nephridiopores. Due to loss of fluid during intense muscular activity, they arc unable to sustain their thrusting movements for as long as an earthworm can.

Digestive System

Anteriorly, the gut is differentiated into a short buccal cavity, then a muscularized pharynx (Figure 11.46). Many of the muscles of the pharynx are arranged radially and connected to the body wall; they dilate the pharynx and also can pull it backward, and are important in the process of swallowing plant material and soil. The esophagus is provided with a pair of lateral pouches and two pairs of glands that convert excess calcium and carbonate ions in the blood into crystals of calcium carbonate. These crystals enter the gut and are eventually eliminated with fecal matter. Behind the esophagus are a crop, where food is stored temporarily, and a gizzard, which has a thick, muscular wall for mashing up the food.

The long intestinal region of the gut functions both in digestion and absorption. Its surface area is increased by a longitudinal fold, called the **typhlosole,** which sinks down into it from the dorsal side (Figure 11.44; Figure 11.47, A). The connective tissue in the wall of the intestine extends into the typhlosole, and so does the **chloragogen tissue,** which consists of yellowish cells derived from the peritoneum. Chloragogen tissue is involved in synthesis of lipids and glycogen, and is thought to function also in the conversion of nitrogenous wastes into ammonia and urea.

Metanephridia

Nearly all of the true segments, except the first few, have a pair of metanephridia (Figure 11.47, A). From the nephrostome (funnel) of each metanephridium a duct passes backward through the septum into the succeeding segment (Figure 11.47, B). Here it becomes a

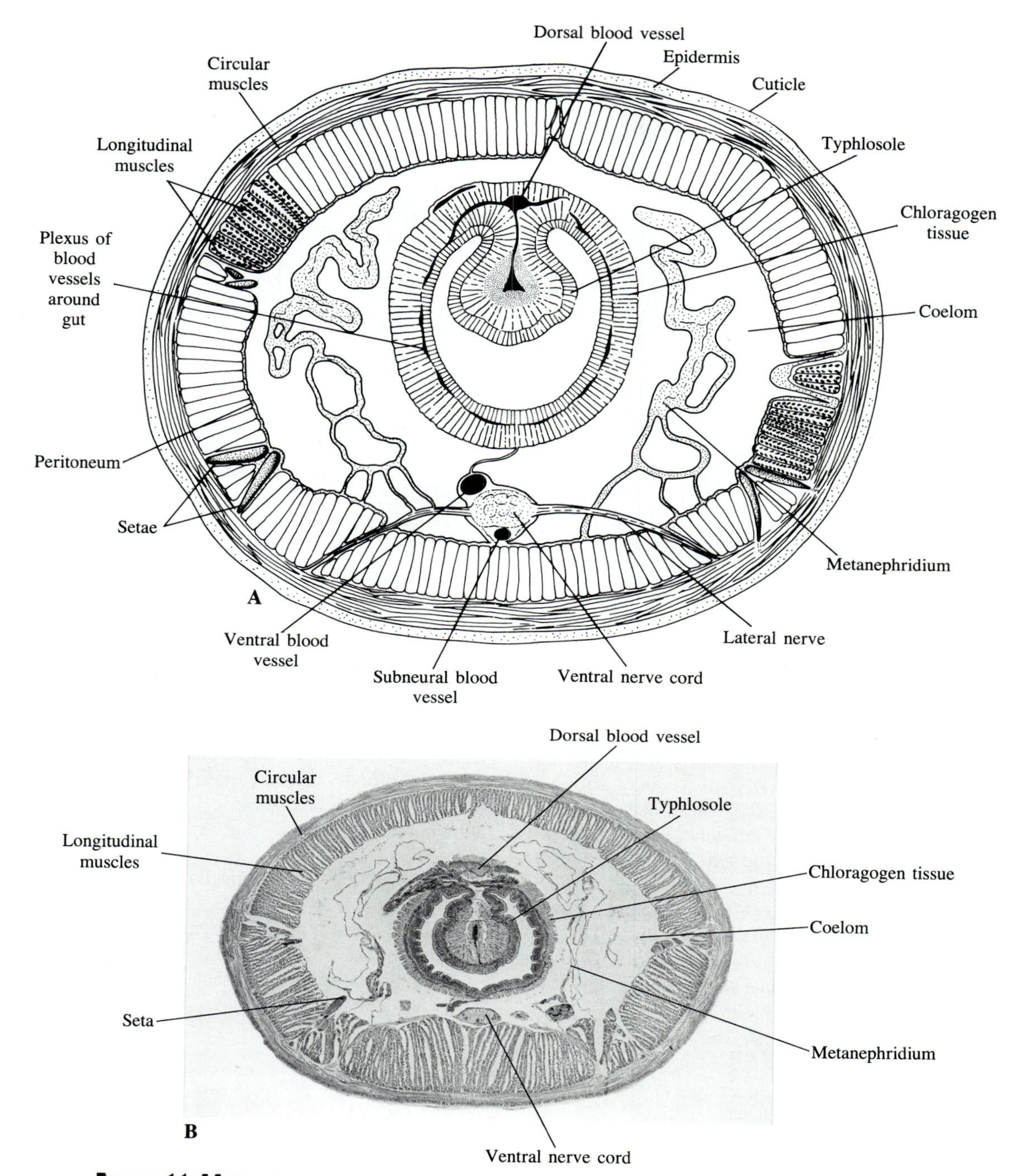

FIGURE 11.44 *Lumbricus,* transverse sections through segments in the midregion of the body. A. Diagram, cellular details omitted. B. Photomicrograph. (Courtesy of Carolina Biological Supply Co., Inc.)

complex structure, with three loops through which the duct runs on its way to the nephridiopore. Note, however, that the duct doubles back on itself more than once. The part nearest the nephrostome has a narrow lumen and is ciliated internally. It enters the first loop, then the second, then doubles back and returns to the first loop. Now, with a slightly wider lumen, still ciliated, it reenters the second loop. Becoming even

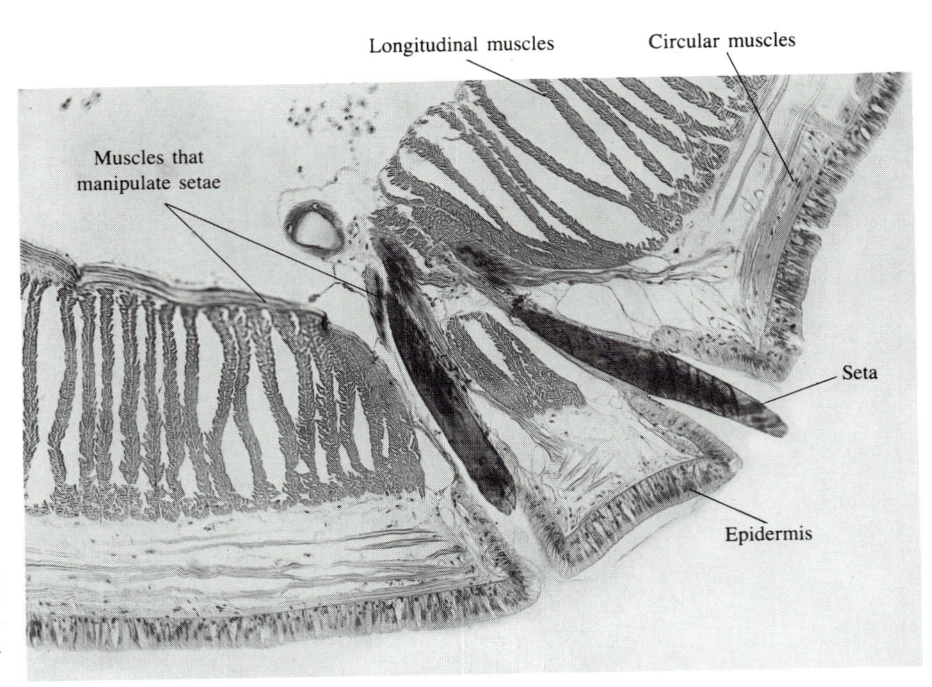

FIGURE 11.45 *Lumbricus*. A. Portion of the body wall, showing musculature and setae; photomicrograph. (Courtesy of Carolina Biological Supply Co., Inc.)

wider, and also unciliated, the duct passes once more through the first loop and finally enters the third loop. Here the lumen becomes especially wide, and the wall of the duct becomes muscular; this portion of the metanephridium functions as a bladder that stores urine.

The metanephridia—especially in the second loop—are liberally supplied with small blood vessels. The blood is delivered to these by branches of the longitudinal vessel that runs beneath the gut. The small vessels reabsorb water, thus minimizing loss of water through the pores of the metanephridia. The principal nitrogenous wastes eliminated are ammonia and urea. The former is produced most abundantly by worms that are well fed, whereas the latter may not appear in appreciable quantities until the worms have been starved for a time; however, sometimes ammonia formation increases in animals that are being starved. Although nitrogenous wastes enter the metanephridia with coelomic fluid, they are also believed to diffuse into the ducts from the blood vessels. Figure 11.47, C shows some of the components of coelomic fluid and blood that enter a metanephridium and that are then either excreted or reabsorbed.

Although the metanephridia are important in excretion of ammonia and urea, secretion of mucus also eliminates considerable nitrogen, mostly in the form of protein.

It should be mentioned that not all oligochaete meta-

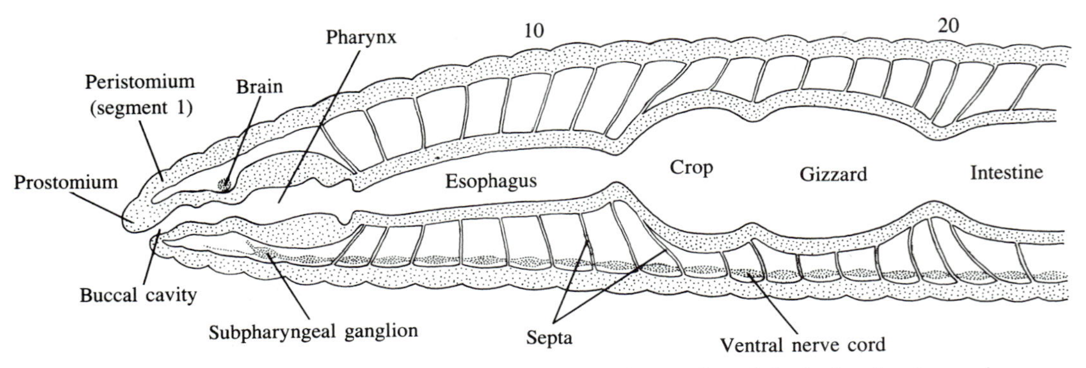

FIGURE 11.46 *Lumbricus*. Longitudinal section of anterior portion of the body, showing regions of the gut. Most other organs omitted.

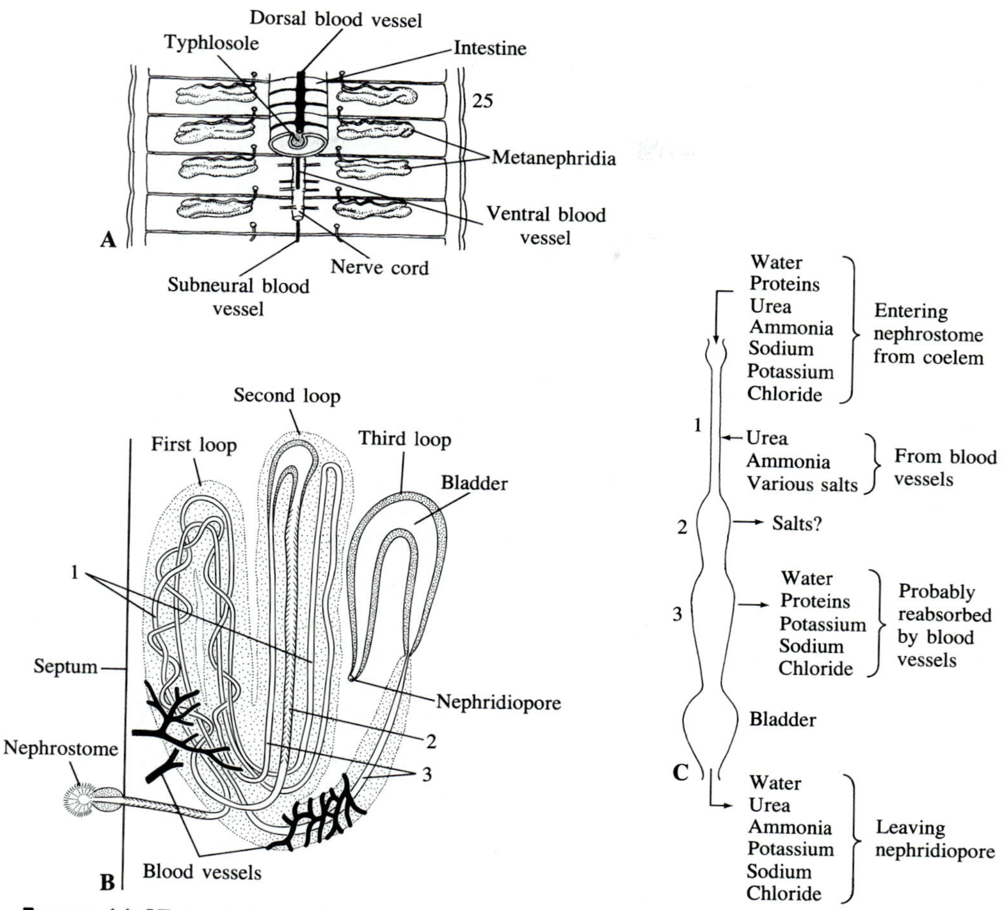

FIGURE 11.47 *Lumbricus*. A. Several segments, as seen in a dissection from the dorsal side, showing arrangement of metanephridia and some other structures. B. Metanephridium; diagrammatic. C. Diagram of a metanephridium, shortened but showing the sections numbered in B, and also showing some of the components of coelomic fluid and blood that enter a metanephridium and that are then either reabsorbed or excreted. (After Laverack, The Physiology of Earthworms, but simplified.)

nephridia fit the pattern described for *Lumbricus,* either in structure or in arrangement. Sometimes the nephrostomes are reduced or even closed. In certain groups of oligochaetes, moreover, there are metanephridia whose ducts open into the gut, usually at the level of the pharynx and intestine. The connection of the ducts with the gut may be direct, or it may be indirect, by way of longitudinal collecting ducts that run through at least several segments. Other variations are multiple metanephridia in some or many segments, and metanephridia that have numerous nephrostomes.

Circulatory System

Only the larger vessels of the circulatory system (Figure 11.48) can be seen clearly in a dissection. All tissues of the body are pervaded by small vessels, and the pattern of circulation is complicated. The large dorsal vessel running through the chloragogen tissue above the gut functions as a heart; its wall has enough musculature to drive blood forward by peristaltic activity. The swollen segmental vessels in segments 7 to 11 also pulsate, and appear to be involved in regulation of blood pressure. Blood moves through them to the noncontractile ventral vessel that lies in the mesentery just beneath the gut. This acts as a main artery, carrying blood mostly in the posterior direction.

Segmental branches of the ventral vessel deliver blood to the body wall; other branches lead to the metanephridia and gut, or to the lateral neural vessels that lie on both sides of the ventral nerve cord and that supply not only the cord but the tissue around it. From this region the blood goes into the subneural vessel beneath the cord. This vessel is essentially a vein. The

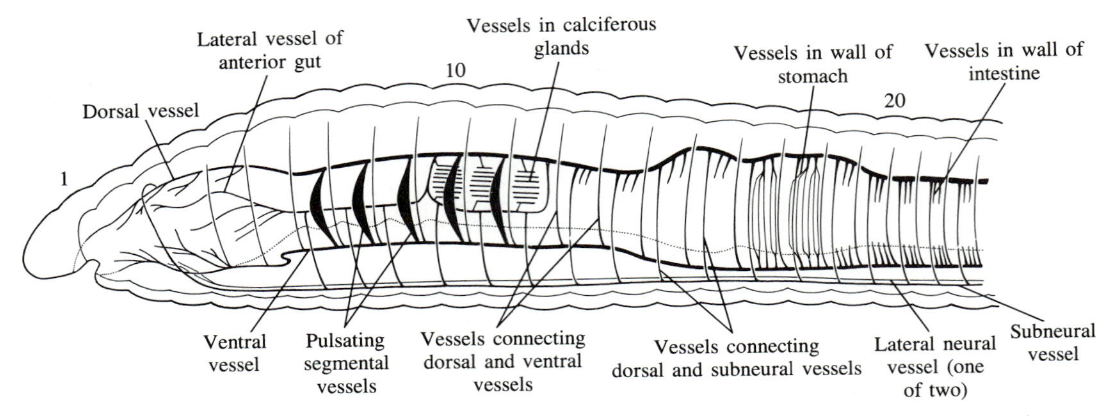

Figure 11.48 *Lumbricus*. Circulatory system; diagrammatic. (After Ivanov *et al.*, Major Practicum of Invertebrate Zoology.)

blood that runs through it is returned to the dorsal vessel by the parietal vessels. These also pick up blood that has passed through networks in the body wall, intestine, metanephridia, and other structures. There are, in addition, some small veins that run directly from the intestine to the dorsal vessel.

The red color of an earthworm's blood is due to hemoglobin. This is dissolved in the plasma.

Nervous System and Sense Organs

The more obvious elements of the nervous system include the two lobes of the brain, located above the pharynx in segment 3, and the connectives that run from the brain to a pair of subpharyngeal ganglia in segment 4 (Figure 11.49). Neither the lobes of the brain nor the subpharyngeal ganglia are single segmental ganglia; the fact that nerves coming from them supply two or more segments indicates that they are complexes of ganglia. Figure 11.49 shows that the prostomium is innervated by nerves from the brain, and that segments 1 and 2 receive nerves chiefly from the brain and circumpharyngeal connectives. Segments 3 and 4 are innervated by nerves from the subpharyngeal ganglia. The diagram is much simplified, however, and does not show how astonishingly complex the nervous system of an earthworm really is.

The ventral nerve cord runs from the subpharyngeal ganglia to the posterior end of the body. It is a double structure, as it is in *Nereis,* and swells out into a pair of ganglia in each segment. The nerve cells are largely external to the fibers that run up and down the two halves of the cord. In the middle of the upper part of

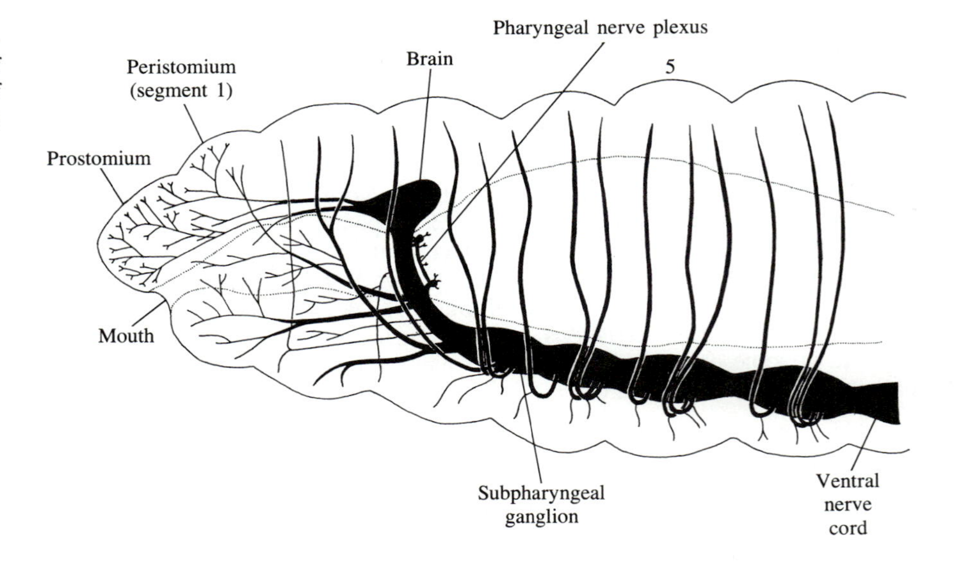

Figure 11.49 *Lumbricus*. Nervous system of anterior part of body. (After Hess, Journal of Morphology and Physiology, *40*.)

the cord are three giant fibers. These begin in the subpharyngeal ganglia and are periodically interrupted by slanting membranous partitions, which are probably homologous to synapses. Even with these interruptions, the system of giant fibers transmits impulses rapidly from one end of the animal to the other. In almost all segments, the ganglia send out three pairs of lateral nerves that supply the body wall.

Small cells that function as photoreceptors are buried in the epidermis of the prostomium, segment 1, last segment, and certain other segments. They are linked with the plexus of nerve fibers that lies just beneath the epidermis. In general, earthworms respond positively to weak light and negatively to strong light. Epidermal cells that bear minute bristles are especially abundant in the anterior part of the body. Some of them are probably concerned with recognition of food, others with detection of vibrations and touch stimuli.

Reproduction

Earthworms, like almost all oligochaetes, are hermaphrodites, and fertilization is internal. The male and female components of the reproductive system are concentrated in a few segments near the anterior end of the body (Figure 11.50). The gonads are closely associated with the peritoneum, and the ducts that carry sperm and eggs to the outside are coelomoducts.

In *Lumbricus terrestris,* there are two pairs of testes, located in segments 10 and 11. These produce spermatogonia that accumulate in three pairs of sacs that are conventionally called seminal vesicles, although these are simply pockets of the coelom that arise from the septa that lie between segments 9 and 12. If the contents of a seminal vesicle are examined with a microscope, various stages in the development of sperm will be seen, as well as gregarines (parasitic protozoans) that destroy sperm. To reach the ciliated funnels of the sperm ducts, the mature sperm must move out of the seminal vesicles and into the coelom of segments 10 and 11. In segment 12 the two ducts on each side of the body unite to form a common sperm duct that runs to a ventrolateral pore on segment 15.

The female part of the reproductive system has a single pair of ovaries, located in segment 13. The oogonia set free by the ovaries enter a pair of egg sacs that lie in segment 14 but that have openings into segment 13. The oogonia become oocytes, and these reach the outside by way of the oviducts that run from segment 13 to ventrolateral pores on segment 14. The rest of the female reproductive system consists of two pairs of seminal receptacles in segments 9 and 10. Their external pores may not be easy to see, for they are located in the intersegmental grooves. The function of the seminal receptacles is to store sperm received by one worm from another.

When two earthworms mate (Figure 11.51, A), they become joined by their ventral surfaces and with the anterior end of one partner facing the posterior end of the other. The worms are held tightly together by secretions of the **clitellum** (a thickened portion of the body, consisting of several glandular segments), by mucus produced on other parts of the body, and by some setae

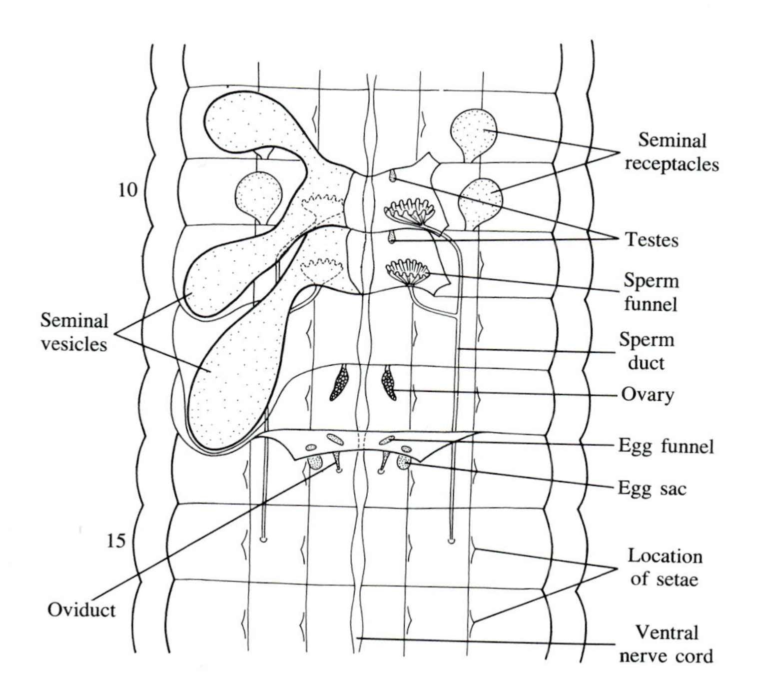

Figure 11.50 *Lumbricus*. Reproductive system; diagrammatic. (After Ivanov *et al.*, Major Practicum of Invertebrate Zoology.)

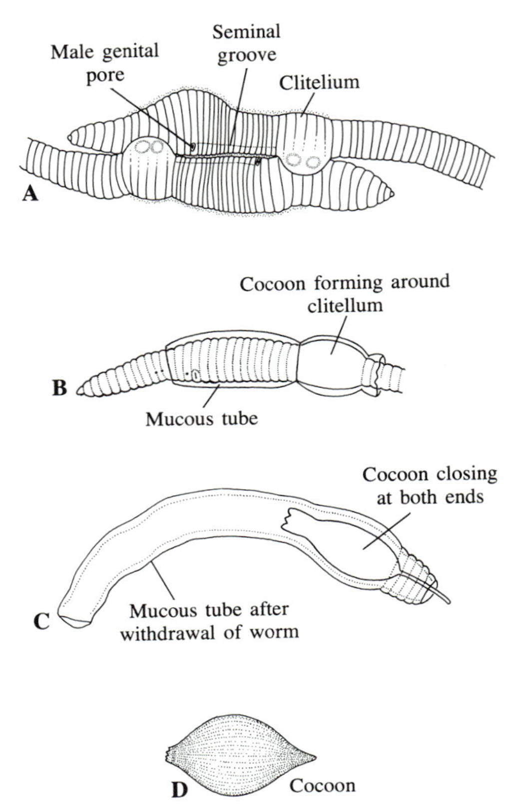

FIGURE 11.51 A. *Lumbricus*. Mating of two worms. B–D. Stages in the formation of a cocoon by an earthworm. (Foot and Strobell, Biological Bulletin, *3*.)

that are specialized for being jabbed into tissue; the modified setae are on segments 10, 11, and 26. The male gonopores are some distance away from the openings of the seminal receptacles. The sperm of one worm reach the seminal receptacles of the partner by way of longitudinal seminal grooves that appear on both sides of the midline. These grooves are maintained by the contraction of certain small muscles in the body wall. Muscular activity is also responsible for propelling sperm toward the seminal receptacles of the partner. Because the worms are in tight contact with one another, the grooves are closed over and are therefore more like tubes. The mating process usually takes about 2 or 3 hours.

In some earthworms other than *Lumbricus,* and in oligochaetes of various other groups, the male genital pores of one worm are brought directly into contact with the seminal receptacle pores of the partner, and there may even be a penislike elaboration that assists in sperm transfer. One must not assume, therefore, that the reproductive system of *Lumbricus* is characteristic of all oligochaetes. It does, however, have most of the structures found in oligochaetes in general.

In *Lumbricus,* fertilization of eggs takes place within a few days after the worms have mated. The process of fertilization is linked with the production, by glands on the clitellum, of a band of tough material that forms a ring around the worm (Figure 11.51, B). This is called a **cocoon,** which fills to the point of bulging with albuminous material that is also secreted by the clitellum. The eggs are thought to get into the cocoon by way of temporary grooves that connect the clitellar region with the external openings of the oviducts, or while the clitellum is being moved forward by muscular contractions of the body. Sperm stored in the seminal receptacles are delivered to the eggs as the cocoon passes over segments 9 and 10. Eventually the cocoon slips off and its open ends constrict (Figure 11.51, C and D). Now the cocoon resembles a seed, and is about the size of a grain of wheat.

The eggs themselves are moderately yolky, and cleavage is not as easy to follow as it is in polychaetes whose eggs have little yolk. It is nevertheless spiral, and mesoderm is formed in the same general way as in polychaetes. Development is direct, and after about a week one small earthworm is ready to emerge from the cocoon. As a rule, the several other eggs that had been deposited in the cocoon do not develop normally. The production of cocoons may continue for a few months even if the worm does not mate again, because sperm stored in the seminal receptacles remain viable for a long time.

Regeneration

Earthworms readily repair damage to the body wall, and most species can regenerate substantial pieces if these have been cut off. The extent to which regeneration is possible varies according to the species. *Lumbricus terrestris* is able to regenerate a new head region provided that not more than about 15 anterior segments have been removed. Even when it does produce a new head, however, it usually does not regenerate more than four segments, even if several others had also been removed. Posterior segments are replaced more freely.

Some earthworms rather closely related to *Lumbricus* have a much higher capacity for regeneration. A piece consisting of just a few segments cut from near the middle of the body may produce both a head region and a posterior portion. The ability to regenerate is also strong in several groups of oligochaetes other than earthworms. Some of these—the families Tubificidae, Naididae, and Aeolosomatidae—are discussed briefly below. They often reproduce asexually, so it is not surprising that they are efficient in regenerating lost parts.

Earthworms as Soil Conditioners

Lumbricus and other earthworms are important as biological conditioners of soil. Not all species do exactly the same things, but most of them burrow, pass soil and

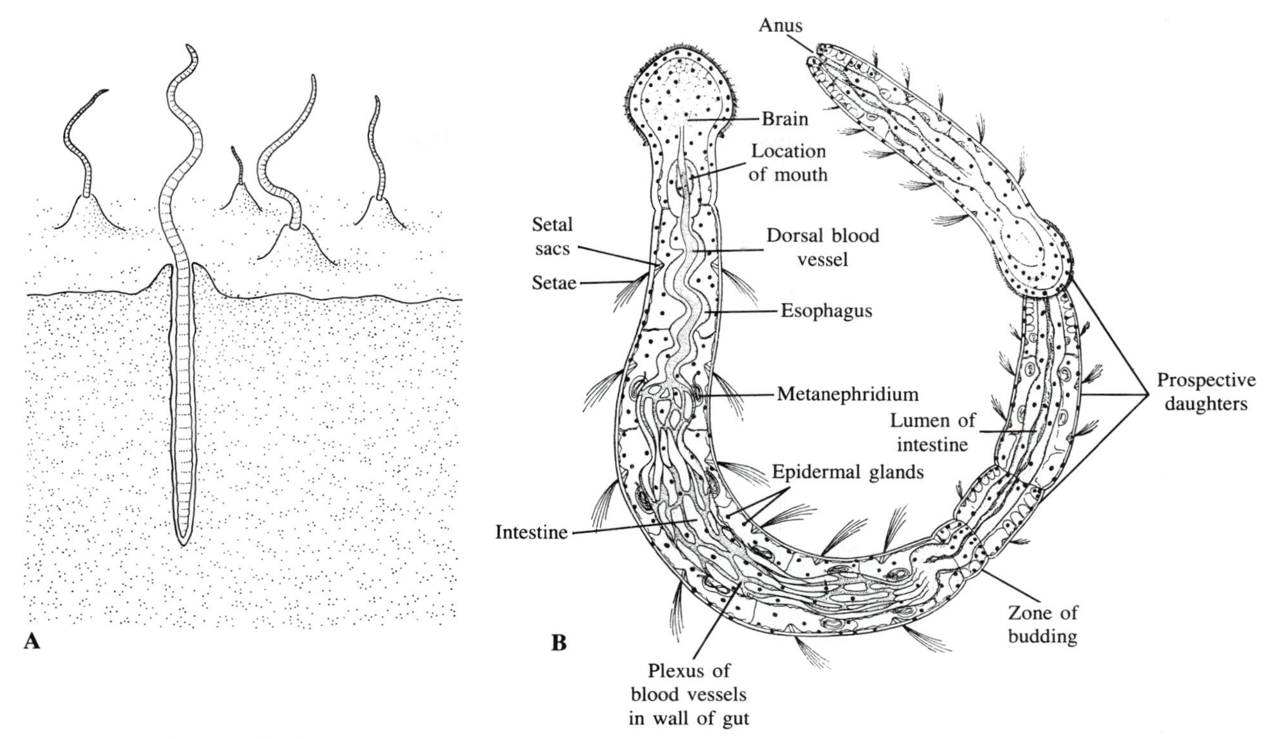

FIGURE 11.52 A. *Tubifex,* showing its orientation (head down) in mud. B. *Aeolosoma hemprichii,* with three prospective daughter individuals in a chain. Only a few structural details are labeled. (Ivanov *et al.*, Major Practicum of Invertebrate Zoology.)

loose organic matter through their guts, and deposit fecal matter at the surface. Besides cultivating and aerating the soil, they influence the availability of nutrients needed by plants. The fecal material of earthworms contains proportionately more nitrogen, usable phosphorus, and certain other mineral components than soil that has not been processed.

The amount of soil turned over by a population of earthworms is astonishing. It has been estimated that the worms in an acre of rich pasture soil in England pass from 4 to 36 tons of earth through their guts in one year. The activity of earthworms in turning over soil and organic matter fascinated Charles Darwin long enough for him to write an entire book on the subject: *The Formation of Vegetable Mould Through the Action of Worms*.

Classification

The classification of oligochaetes is based mostly on the arrangement of reproductive organs, the structure of metanephridia, and the number of setae in each segment. So far as setae are concerned, earthworms and their close relatives have eight per segment, generally in four pairs, as is the case in *Lumbricus*. In some other groups of oligochaetes, however, there are several setae in each of four bundles, only four setae altogether, or no setae at all.

The various orders, suborders, and families of Oligochaeta are not described here—this information, for those who need it, is found in comprehensive treatises listed in Chapter 1 and in specialized works listed at the end of this section. Instead, only a few distinctive oligochaetes that are decidedly unlike earthworms are mentioned briefly.

Family Tubificidae

Members of the family Tubificidae are aquatic and are found in both freshwater and marine habitats. The best known genus is *Tubifex* (Figure 11.52, A), which is common in sluggish streams, including those into which sewage and other organic wastes are dumped. The presence of *Tubifex,* in fact, is often an indication that the water is polluted and that the amount of dissolved oxygen is low. The worms, which are about 2 cm long, may form large aggregations, and they construct mud tubes. Their heads are directed downward and their slender posterior portions wave in the water, thereby accelerating uptake of available oxygen. The red color of *Tubifex,* which makes it easy to see the worms in the field, is due to hemoglobin, which is concentrated in corpuscles.

Family Enchytraeidae

Enchytraeids are abundant in humus-rich soils, and many species are found under decaying vegetation on marine beaches and at the margins of freshwater habitats. In general, these worms are under 3 cm long, and most of them are whitish. Like tubificids, they usually have four bundles of setae per segment, but there are exceptions. The so-called whiteworms cultivated as food for aquarium fishes are enchytraeids; they multiply rapidly in moist sand on which a piece of bread is allowed to decay.

Family Naididae

Naidids—*Nais, Dero, Chaetogaster,* and related genera—have only a few segments, and they are not often more than 1 cm long. Some have ciliated gills around the anus. Many species reproduce by transverse fission, during which the tail end of the anterior daughter and the head end of the posterior daughter differentiate from one particular segment of the parent.

The species of *Chaetogaster* are commonly associated with freshwater snails, especially pulmonates such as *Lymnaea* and *Physa*. They are primarily carnivores and have been observed to eat miracidia of digeneans as these settle on snails (see Chapter 6).

Family Aeolosomatidae

Aeolosoma (Figure 11.52, B) and its relatives, all of which live in fresh water, are unusual in several respects. They have a ciliated prostomium, used in locomotion, and the mouth and pharynx are specialized for sucking in food. The septa between segments are incomplete. The gonads are rather indistinct, being simple patches of germinal cells much like those of many polychaetes. Sperm reach the exterior by way of metanephridia, and eggs are discharged through breaks in the body wall. Some experts believe that aeolosomatids should be excluded from the Oligochaeta.

References on Oligochaeta

Anderson, D. T., 1966. The comparative early embryology of the Oligochaeta, Hirudinea and Onychophora. Proceedings of the Linnean Society of New South Wales, *91:*10–43.

Brinkhurst, R. O., 1982. British and Other Marine and Estuarine Oligochaetes. Synopses of the British Fauna, no. 21. Linnean Society of London and Cambridge University Press, Cambridge and New York.

Brinkhurst, R. O., and Jamieson, B. G. M., 1971. Aquatic Oligochaeta of the World. Oliver and Boyd, Edinburgh and London.

Edwards, J. A., and Lofty, J. R., 1977. Biology of Earthworms. 2nd edition. Chapman and Hall, London.

Giere, O., and Pfannkuche, O., 1982. Biology and ecology of marine Oligochaeta, a review. Oceanography and Marine Biology, An Annual Review, *20:*173–308.

Jamieson, B. G. M., 1981. The Ultrastructure of the Oligochaeta. Academic Press, New York.

Laverack, M. S., 1963. The Physiology of Earthworms. Pergamon Press, Oxford and New York.

Lee, K. E., 1985. Earthworms. Their Ecology and Relationships with Soils and Land Use. Academic Press, Orlando, Florida and London.

Satchell, J. E. (editor), 1983. Earthworm Ecology. Chapman and Hall, New York.

Seymour, M. K., 1969. Locomotion and coelomic pressure in *Lumbricus*. Journal of Experimental Biology, *51:*47–58.

Singer, R., 1978. Suction feeding in *Aeolosoma*. Transactions of the American Microscopical Society, *97:*105–111.

Stephenson, J., 1930. The Oligochaeta. Clarendon Press, Oxford; Oxford University Press, New York.

Wallwork, J. A., 1983. Earthworm Biology. Edward Arnold, London.

CLASS HIRUDINOIDEA

The Hirudinoidea consists mostly of leeches, but it includes two other small groups of annelids. The worms of this class appear to be more closely related to oligochaetes than to polychaetes. They have no parapodia, and there are no antennae or palps on the head region. The reproductive system bears some resemblance to the oligochaete plan, and with a few exceptions these animals have a clitellum whose secretions are involved in the mating process and in producing a cocoon within which eggs develop into fully formed small worms. The class includes predators, parasites, and commensals.

Distinctive external features of the Hirudinoidea are the presence of a posterior sucker (and sometimes also an anterior sucker); reduction or complete absence of setae; and the superficial division of segments into two or more rings, which gives the impression that there are more segments than there really are. The coelom, moreover, is reduced, and its internal segmentation is obscured.

SUBCLASS HIRUDINEA: LEECHES

Of the approximately 500 species of leeches, the majority are found in fresh water. Most of the others are marine, but there are a few that live on land, especially in moist tropical areas.

External Features

As a rule, leeches (Figure 11.53) are dorsoventrally flattened to some extent, and they are highly muscular, firm-bodied animals. There are no setae. The number of segments is almost always the same: 33, in addition to the prostomium. A few segments at the anterior end,

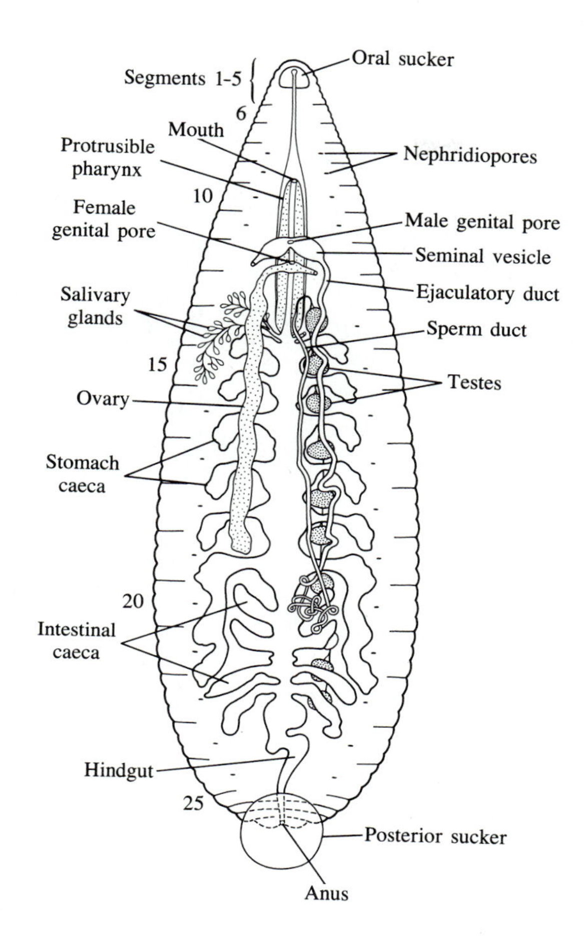

FIGURE 11.53 *Glossiphonia complanata,* a widely distributed freshwater leech, ventral view. Only the left half of the male reproductive system and the right half of the female reproductive system are shown. (After Harding and Moore, Fauna of British India, *27*.)

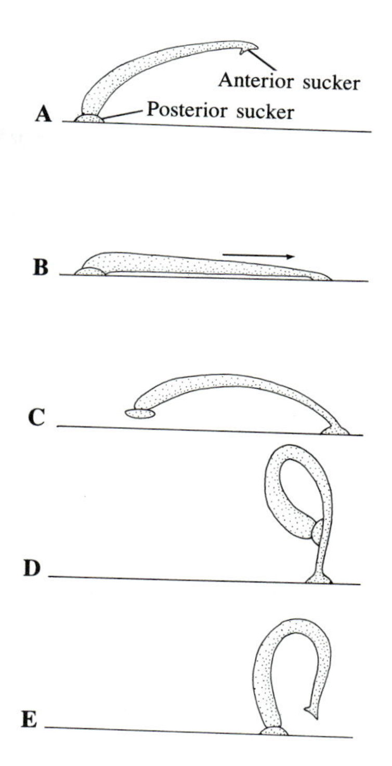

FIGURE 11.54 Successive stages in the locomotion of a leech, *Piscicola geometra*. Note, in D, that this species, after detaching its posterior sucker (C), touches the sucker to itself before attaching it to the substratum again (E). (After Herter, Biologie der Tiere Deutschlands, *35*.)

and those incorporated into the sucker at the posterior end, are indistinct, so the exact number can only be determined by studying the arrangement of segmental ganglia in the central nervous system. The fact that most of the segments are superficially subdivided into several rings compounds the difficulty of counting segments. When one ring of each segment differs from the others in having a concentration of sensory papillae (Figure 11.57, A), this provides a helpful marker.

The posterior sucker is almost always derived from seven or eight segments. Its musculature consists of several components, some of which operate to increase or decrease the diameter of the sucker, others to accentuate the curvature of its concave inner surface. Secretions of epidermal glands contribute to the sucker's adhesive properties. In nearly all leeches there is also a simple oral sucker, essentially just a shallow pit within which the mouth is located. Thus a leech can inch along by coordinating the use of both suckers with muscular activities that alternately lengthen and shorten the body (Figure 11.54). In the usual pattern of locomotion, a leech attached by its posterior sucker stretches, attaches itself by its anterior sucker, then detaches its posterior sucker. After shortening its body, it becomes reattached by its posterior sucker. Some leeches can swim gracefully and efficiently by up-and-down undulation of the body.

Many species have considerable black epidermal pigment, generally concentrated in spots or stripes. Pigment of other colors—especially brown, green, or olive, and sometimes red or near-red—may also be present. The pigment is often located within chromatophores that can be expanded and contracted; leeches with this arrangement are able to change their color patterns.

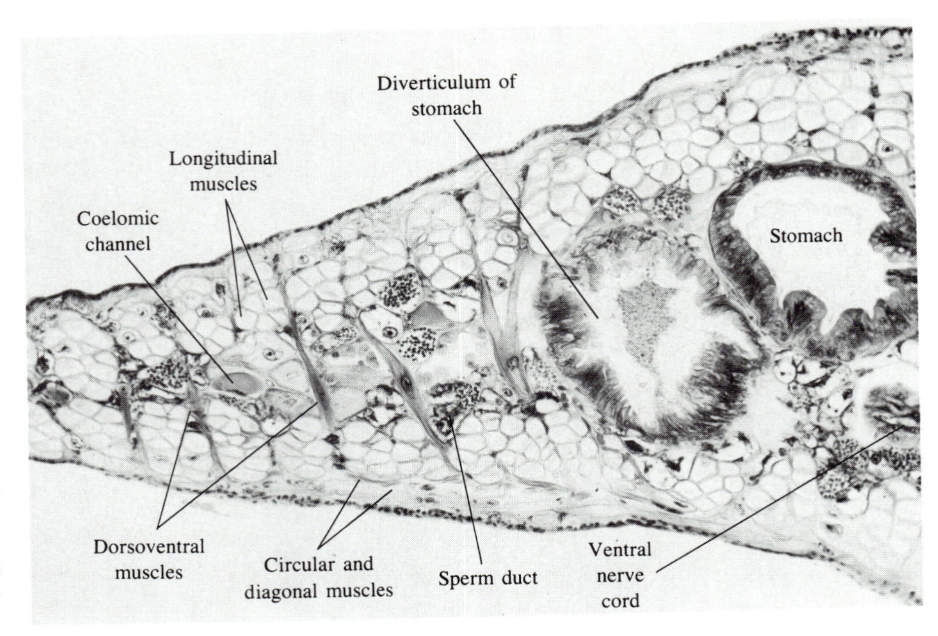

FIGURE 11.55 Portion of a transverse section of *Helobdella stagnalis,* a widely distributed freshwater leech, showing some structures of the body wall; photomicrograph.

Body Wall, Coelom, and Blood Vessels

The thick body wall, beneath the epidermis, consists largely of muscle tissue arranged in an outer circular layer, middle oblique layer, and inner longitudinal layer (Figure 11.55). The solid, almost rubbery texture of a leech is due in part to its muscularity and in part to the fact that the coelom is nearly filled by tissue, mostly fibrous connective tissue. Serially arranged coelomic cavities do appear during early stages of development, and at least a few septa are recognizable in some adult leeches. In most species, however, the only clear internal evidences of segmentation are those shown by the arrangements of ganglia and metanephridia.

Portions of the coelom that have not been filled up by tissue are organized into channels that function as a circulatory system (Figure 11.56, A). This must not be confused with a true circulatory system such as is found in most oligochaetes and polychaetes. In leeches that have a true circulatory system, the system is usually reduced to a few major vessels, and it remains distinct from the coelomic channels. When the coelomic system assumes the entire burden of circulation, certain of its channels—particularly the lateral longitudinal ones—are muscularized and can propel the fluid within them. The channels may send "capillaries" out into the body wall. In some leeches, the fluid within the channels contains hemoglobin, which facilitates transport of oxygen. Leeches in which there is a functional true circulatory system as well as a system of coelomic channels do not have hemoglobin.

In a few parasitic leeches that live on fishes, the body wall has outgrowths that function as gills (Figure 11.59, B); these are pervaded by extensions of the coelomic channels. Most leeches, however, have no special structures for respiration. The rhythmic undulations that many species make, even when not moving, circulate water around the body and thus promote exchange of oxygen and carbon dioxide.

Digestive System and Feeding

Some leeches are predators that swallow whole small invertebrates, such as oligochaetes, or that suck tissues and juices from invertebrates, usually killing them in the process. Others fit into the category of parasites, for they suck blood from fishes, amphibians, turtles, and other aquatic vertebrates. A few of the leeches that live in moist terrestrial habitats are bloodsuckers, too.

The organization of the gut shows various specializations related to feeding habits. The pharynx is usually adapted either for biting and sucking, or for penetration and sucking. In either case its lining is of ectodermal origin and much or all of it is cuticularized. The biting–sucking type of pharynx is found in the medicinal leech and its relatives, all of which feed on blood. A leech of this type attaches by its anterior sucker to a suitable host, then uses the three sharp jaws at the front of the pharynx to make an incision that starts the blood flowing. Salivary glands secrete a substance called hirudin, which inhibits coagulation of the blood; an anesthetic secretion is also applied to the wound. The lumen of the pharynx, triradiate when collapsed, is dilated by radial muscles, but circular and longitudinal muscles are also involved in the pumping action of the pharynx.

In predaceous leeches, and in some bloodsucking species, the pharyngeal region is organized on the tube-within-a-tube plan (Figure 11.57, B and C). The inner

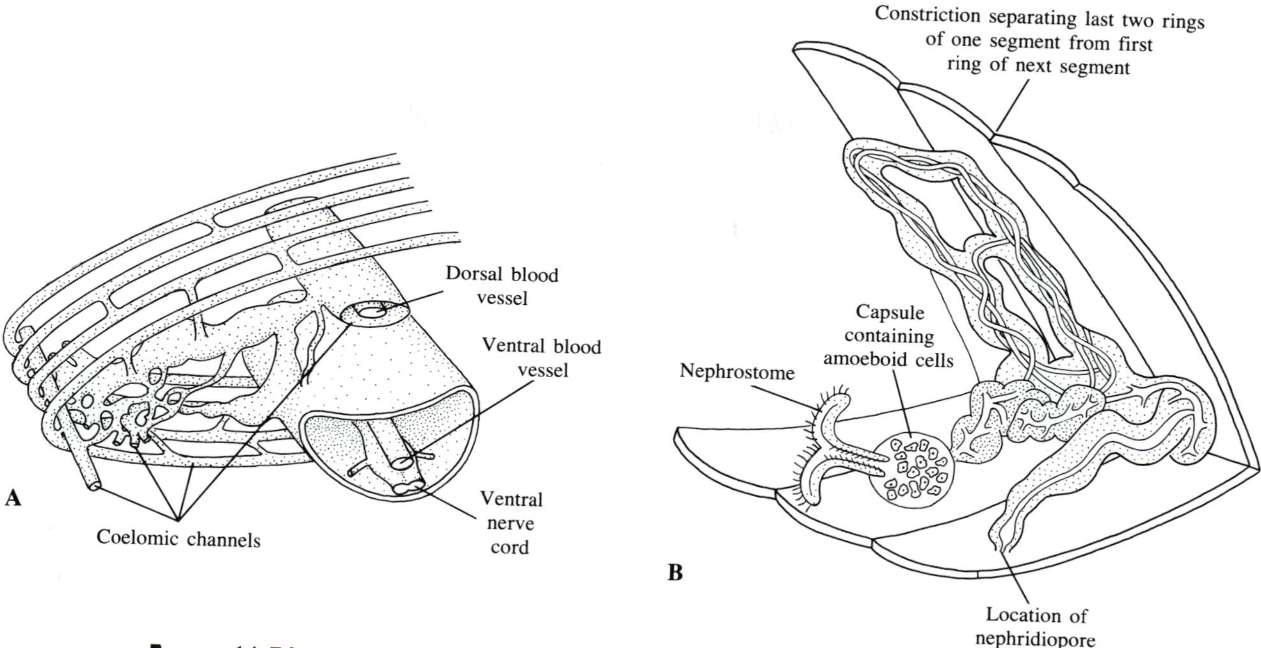

FIGURE 11.56 A. *Glossiphonia complanata,* stereodiagram of a portion of the body, showing blood vessels and coelomic channels. B. Metanephridium of a glossiphoniid leech. (After Oka, Annotationes Zoologicae Japonenses, *4*.)

tube, which is the pharynx proper (it is often called a proboscis), originates as a deep circumferential fold in the wall of the anteriormost part of the pharynx. It is lined by cuticle both internally and externally. Once a leech with this type of pharynx attaches its oral sucker to a host or prey animal, it forces the inner tube, which is fairly rigid, into the tissue. Enzymatic activity perhaps facilitates penetration. Once the tube is embedded in the tissue, the leech can start sucking blood or other juices.

The pharynx of a leech is sometimes succeeded by a short esophagus, but it generally leads almost directly into a stomach, or crop. This typically has several pairs of lateral caeca (Figure 11.53; Figure 11.57, A). The caeca enable the animal to feed heartily when it has a chance to do so. Some bloodsucking leeches take several weeks or even months to finish digesting a large meal. They do, however, quickly eliminate most of the water from the blood they take in. The material that remains in the caeca consists mostly of corpuscles.

Leech digestion is interesting in that it is concerned almost entirely with proteins. The protein-splitting enzymes are exopeptidases, which slowly attack the terminal portions of the complex molecules instead of quickly breaking the molecules up into amino acids. Digestion in many leeches is aided by bacteria that live in the stomach.

Behind the stomach, and often separated from it by a muscular sphincter, is an intestine that may also have a few lateral caeca. Finally, there is the terminal portion of the digestive tract, called the rectum or hindgut. Because the last few segments are specialized to form the posterior sucker, the anus of a leech is not terminal as it is in oligochaetes and polychaetes; it is located on the dorsal surface just anterior to the sucker.

Metanephridia

The paired excretory organs of leeches, which are usually limited to segments 10 to 17, are true metanephridia; that is, they are basically of ectodermal origin. The nephridiopores are generally located ventrolaterally in the same region of the body. In certain terrestrial leeches, however, two pairs of pores are situated on the posterior sucker; this specialization provides the sucker with fluid it needs to function properly.

Leech metanephridia are almost always peculiar in one way or another. Except in some species thought to exhibit the ancestral condition, the ciliated nephrostomes, which open into coelomic channels, are not joined to the ducts; each one simply ends in a small capsule (Figure 11.56, B). The ducts themselves are odd in that they are much-branched canals that run intracellularly through a cord of large cells. In the medicinal leech and some of its relatives, each metanephridium has several nephrostomes, all of which lie within a single capsule.

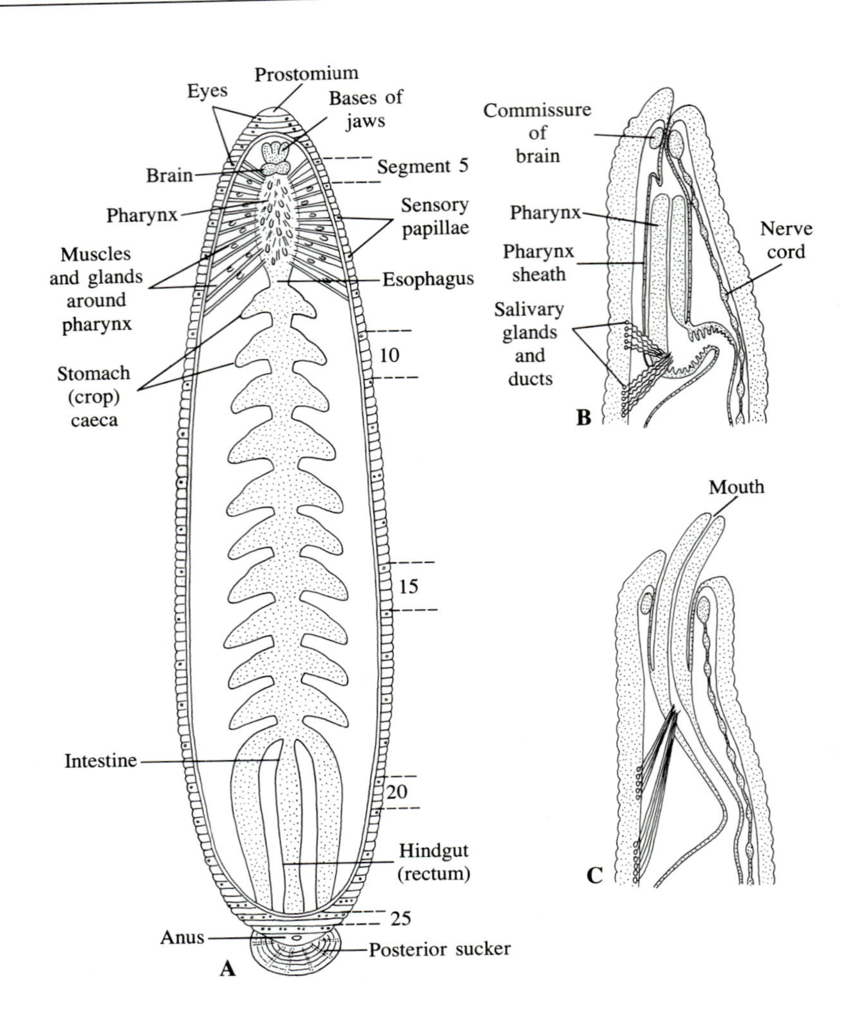

Figure 11.57 A. *Hirudo medicinalis,* dissected from the dorsal side, showing the arrangement of the gut (most other organs omitted). B, C. Diagram of the pharyngeal region of a rhynchobdellid leech, showing the pharynx (B) withdrawn and (C) partly protruded. (After Herter, Biologie der Tiere Deutschlands, *35*.)

It appears likely that the work of excreting ammonia and urea, the principal nitrogenous waste products, is accomplished primarily by the ducts, which are intimately associated with the coelomic channels. What the funnels and capsules do is not perfectly clear, but it is thought that amoeboid cells produced within the capsules are carried into the coelomic channels by the cilia of the nephrostomes. These cells can ingest particulate waste material. Some of the tissue that fills up the coelom or that lines coelomic channels also consists of phagocytic cells. How the particulate wastes are unloaded is still a mystery.

Nervous System and Sense Organs

In most leeches the brain is rather small, and its position in segment 5 is a little surprising. It includes the ganglia of the prostomium and segments 1 to 4, all of which have moved backward during early stages of development. Two connectives curve around the pharynx and join the subpharyngeal ganglion in segment 6. From this, a double ventral nerve cord extends posteriorly. Each distinct ganglion of the cord, as far back as the region of the sucker, consists of six groups of cell bodies, called **follicles;** generally the follicles are arranged in two sets of three, one set forming a transverse row in front of the other. There are follicles of the same sort in the brain and subpharyngeal ganglion; the number of them is helpful in determining the number of segments in the anterior part of the body.

The sense organs of leeches enable them to respond to light, touch, vibrations, temperature changes, currents, and chemical stimuli, including secretions and excretory products of prospective hosts or prey organisms. The location of some light receptors is indicated by eyespots, in which the several to many sensory cells are concentrated within cups of black pigment; when eyespots are present, they are usually arranged in pairs on several of the more anterior segments. Simpler light

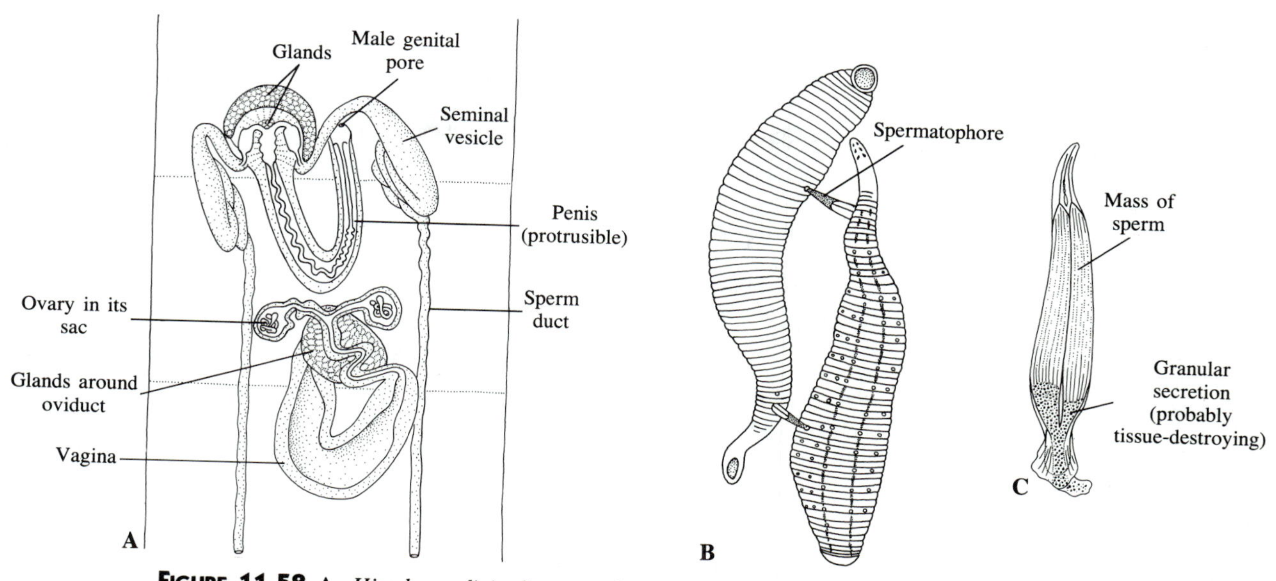

FIGURE 11.58 A. *Hirudo medicinalis,* reproductive system, dorsal view. The testes (there are usually 10 pairs) are located more posteriorly and are not shown. The female genital pore, which opens into the vagina, is on the ventral midline beneath the glands around the oviduct. (After Leuckart and Brandes, Die Parasiten des Menschen.) B. *Glossiphonia complanata,* two individuals attaching spermatophores to one another. (After Brumpt, Mémoires de la Société Zoologique de France, *13.*) C. Spermatophore of *Haementeria parasitica.* (After Whitman, Journal of Morphology and Physiology, *4.*)

receptors without pigment cups are scattered over most of the body.

Sensitivity to light and darkness is exploited in various ways. In general, leeches shy away from light, but some parasitic species respond positively to illumination under certain conditions, such as when they are hungry. The ability to detect passing shadows is important to leeches seeking a host. Leeches whose epidermal pigmentation is localized in chromatophores that expand or contract can modify their color patterns according to the color of the background and intensity of illumination. The chromatophores respond to secretions produced in the ventral nerve cord or in the connectives that encircle the pharynx, but it is the light receptors that inform the nervous system about the background and illumination.

Leeches respond quickly to being touched, and certain parts of the body, including the anterior and posterior suckers, are especially sensitive. Sensory cells that have small bristles sticking out of them are almost certainly touch receptors. These cells are often set into small papillae concentrated in one of the several rings that are part of most segments (Figure 11.57, A). Responses to vibrations and currents perhaps also depend on the bristle-bearing receptors. The receptors that detect various other stimuli, including temperature changes (important to leeches that seek warm-blooded vertebrates) and secretions or excretory products of hosts or prey animals, have yet to be identified. Leeches can provide biologists with many fascinating problems for research in sensory physiology.

Reproduction

Although the reproductive system of leeches generally fits a basic pattern, the details of the reproductive process are extremely variable. Hermaphroditism is the rule, but most leeches function first as males, then as females.

There is usually a pair of testes in each of several successive segments (Figure 11.53). The two sperm ducts, as they approach the single male genital pore on the ventral side of segment 10, become dilated or coiled into seminal vesicles (Figure 11.53 and Figure 11.58, A); these store sperm before copulation. The seminal vesicles open into a common chamber that is usually both glandular and muscular. In some species, this part of the male system functions as a penis; it can be protruded and inserted into the female genital pore of the partner. In others, it secretes a spermatophore that is deposited on the skin of the partner.

There are generally two ovaries (Figure 11.53 and Figure 11.58, A), which are coelomic sacs filled with egg-producing tissue. The oviducts unite to form a vagina that runs to the genital pore on segment 11. In leeches that copulate, a portion of the vagina may have a dilation or diverticulum that functions as a seminal receptacle; from this, the sperm can migrate to the ovarian sacs and penetrate the oocytes.

Leeches that secrete spermatophores usually do not copulate. They mate, however, by bringing their ventral surfaces together and depositing spermatophores on one another (Figure 11.58, A and B). The sperm emerge from the spermatophores and migrate through the tissues until they reach the ovarian sacs. This method of insemination, typical of many parasitic and predaceous leeches, is especially well documented for *Placobdella parasitica,* which lives on freshwater turtles in North America. As two worms mate, one everts its male genital chamber and produces a spermatophore. Into this it pumps sperm and a tissue-destroying secretion. The spermatophore hardens quickly, and it is pressed so firmly against the skin of the partner that the skin is broken. The spermatophore then shrinks, forcing its contents into the tissues. The tissue-destroying fluid induces the formation of channels through which some of the sperm reach the ovarian sacs and are thus able to penetrate the oocytes. In certain leeches the spermatophore must be applied to a particular portion of the ventral surface; just internal to this area is a mass of "vector tissue" that is organized in such a way that sperm can move rapidly to the ovarian sacs.

Most leeches have several segments specialized to form a glandular clitellum. This functions in much the same way as the clitellum of earthworms, facilitating mating and also secreting a cocoon within which fertilized eggs develop into small leeches. Leech eggs are not especially yolky, and sometimes special glands on the clitellar region secrete proteinaceous material that is incorporated into the cocoons and that helps nourish the developing young. The cocoons are usually stuck to a firm object or deposited in a protected place. However, many leeches, including some of the common freshwater species, hold their cocoons on the ventral surface (Figure 11.59, A) and also carry their young for a time after they have hatched.

Classification

Order Rhynchobdellida

Rhynchobdellid leeches are characterized by a protrusible pharyngeal tube. This structure, often called a proboscis, is used to pierce the skin of a prey organism, such as a mollusc, or the skin of a fish or other vertebrate from which blood can be sucked. The blood of rhynchobdellids lacks hemoglobin and is confined to vessels.

This order includes a number of common freshwater genera, such as *Glossiphonia* (Figure 11.53; Figure 11.59, A) and *Helobdella,* which are mostly predators on invertebrates. Some of the leeches in this group retain their cocoons on the ventral surface and carry the young after they hatch. A somewhat different assemblage consists of *Piscicola* and related genera, found in both marine and freshwater habitats. Most of them are bloodsuckers, but some are predators on invertebrates. A third group within the Rhynchobdellida is represented by *Ozobranchus* (Figure 11.59, B), characterized by branching gills. *Ozobranchus* and its close relatives are

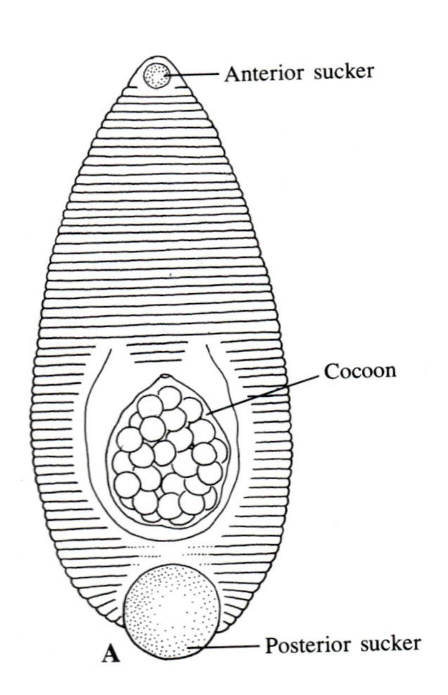

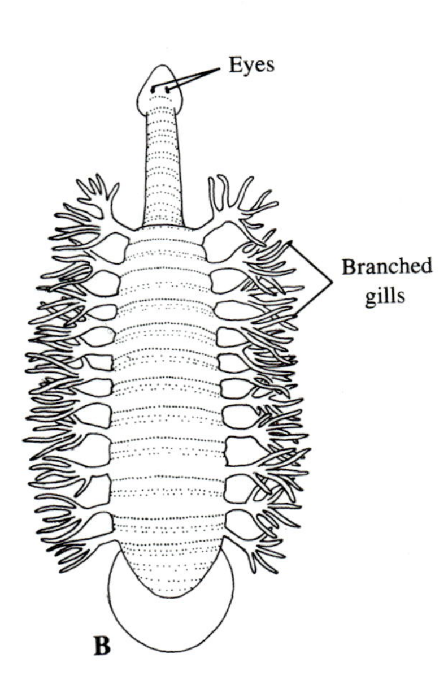

FIGURE 11.59 A. *Glossiphonia lata,* ventral view, showing a cocoon attached to the body. The cocoon is produced at a more anterior level, but is moved to this position for brooding. (After Nagao, Japanese Journal of Zoology, *12*.) B. *Ozobranchus jantzeanus.* (After Oka, Records of the Indian Museum, *24*.)

bloodsuckers and are found in fresh water as well as in the sea.

Order Arhynchobdellida

Arhynchobdellids do not have a pharyngeal tube, but they may have jaws, teeth, or hard pharyngeal ridges with stylets. The circulatory system is of the open type; although there are some major vessels, the blood circulates freely through coelomic spaces. Moreover, the blood contains hemoglobin.

The members of this order are restricted to fresh water and moist terrestrial habitats. Some of the freshwater types, such as the famous medicinal leech, *Hirudo medicinalis* (Figure 11.57, A), are three-jawed bloodsuckers. The order also includes some freshwater and amphibious leeches that swallow their prey. These lack jaws, or have only poorly developed jaws. A representative genus is *Haemopis*.

The terrestrial leeches are mostly bloodsuckers and are primarily tropical, but a few are found in temperate regions. Certain genera, such as *Xerobdella,* have three pairs of jaws. Others, such as *Haemadipsa,* have just two or three jaws. The larger terrestrial leeches tend to parasitize warm-blooded vertebrates, and some are a nuisance to humans. They may climb into trees or bushes and wait for a suitable host to pass by.

The use of the medicinal leech for drawing blood goes back to ancient times. Over 2000 years ago this style of blood-letting was recommended as a treatment for bites and stings of venomous animals. Mostly, however, leeching was resorted to because many diseases were thought to be due to ''corrupt blood.'' The practice became especially popular in the nineteenth century, and one writer calculated that leeches were drawing more than 300,000 liters of blood each year in France alone. The estimate was almost surely much too high, but there is no doubt that blood-letting was popular. Images of leeches even appeared on fancy dresses. Intensive collection of the medicinal leech threatened the survival of the species in some parts of Europe.

The use of leeches has come back into medical science, for removing blood that accumulates in an ear, lip, or tip of a finger that has been surgically reattached. While it may not be difficult to join arteries, connecting veins is much less likely to be successful. Leeches will take up the blood that otherwise cannot move away until the venous circulation becomes functional again.

In half an hour, a medicinal leech may draw as much as ten times its own weight in blood. Its efficiency is due to secretion not only of an anticlotting substance, but also of a substance that causes veins to dilate and of another that causes cells to separate and thus allow the other secretions to penetrate more rapidly into the tissues.

The medicinal leech is now being reared for use in surgical procedures. Several other species, moreover, are being produced for work in other areas of biological research, especially neurophysiology.

References on Hirudinea

Conniff, R., 1987. The little suckers have made a comeback. Discover, *8*(8):85–94.

Cross, W. H., 1976. A study of predation rates of leeches on tubificid worms under laboratory conditions. Ohio Journal of Science, *76:*164–166.

Dickinson, M. H., and Lent, C. M., 1984. Feeding behavior of the medicinal leech, *Hirudo medicinalis* L. Journal of Comparative Physiology, A, *154:*449–456.

Fernández, J., and Stent, G. S., 1980. Embryonic development of the glossiphoniid leech *Theromyzon rude:* structure and development of the germinal bands. Developmental Biology, *78:*407–434.

Gray, J., Lissmann, H. W., and Pumphrey, R. J., 1938. The mechanism of locomotion in the leech. Journal of Experimental Biology, *15:*408–430.

Haupt, J., 1974. Function and ultrastructure of the nephridium of *Hirudo medicinalis* L. II. Fine structure of the central canal and urinary bladder. Cell and Tissue Research, *152:*385–401.

Klemm, D. J., 1972. Freshwater leeches of North America. Biota of Freshwater Ecosystems, Identification Manual No. 1. Government Printing Office, Washington, D. C.

Mann, K. H. (editor), 1962. Leeches (Hirudinea)—Their Structure, Physiology, Ecology, and Embryology. Pergamon Press, Oxford and New York.

Nicholls, J. G., and Van Essen, D., 1974. The nervous system of the leech. Scientific American, *230*(1):38–48.

Richardson, L. R., 1969. A contribution to the systematics of the hirudinid leeches. Acta Zoologica Hungarica, *15:*97–149.

Sandig, M., and Dohle, W., 1988. The cleavage pattern in the leech *Theromyzon tessalatum* (Hirudinea, glossiphoniidae). Journal of Morphology, *196:*217–252.

Sawyer, R. T., 1972. North American freshwater leeches, exclusive of the Piscicolidae, with a key to all species. Illinois Biological Monographs, *46:*1–154.

Sawyer, R. T., 1984. Leech Biology and Behavior. 3 vols. Oxford University Press, New York; Clarendon Press, Oxford.

Soos, A., 1965–69. Identification key to the leech genera of the world. Acta Zoologica Hungarica, *11:*417–464; *12:*147–156; *12:*371–407; *13:*417–432; *15:*397–454.

Wilde, V., 1975. Investigation of the symbiotic relationship between *Hirudo officinalis* and bacteria. Zoologischer Anzeiger, *195:*289–306.

SUBCLASS ACANTHOBDELLIDA

The only species in this subclass is *Acanthobdella peledina* (Figure 11.60), a parasite of freshwater fishes in Russia, Scandinavia, and Alaska. It has generally been classified as a leech, but it differs from true leeches in having two pairs of setae on each of five of

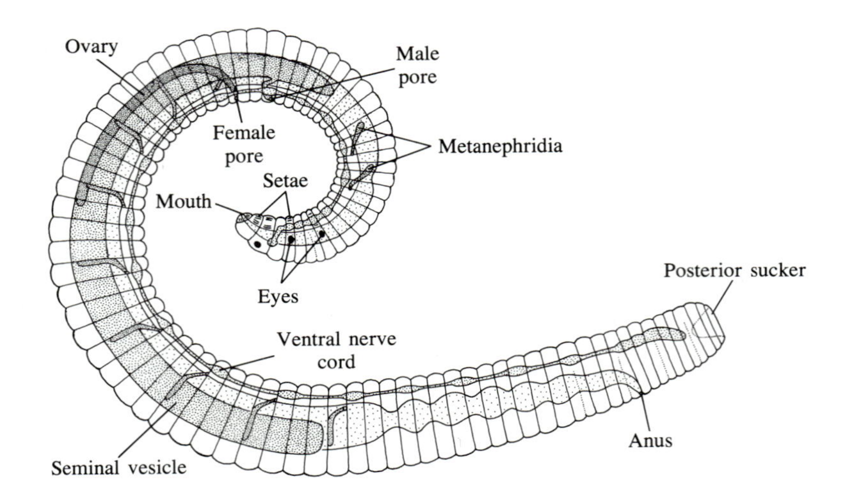

FIGURE 11.60 *Acanthobdella peledina* (subclass Acanthobdellida), somewhat diagrammatic. (After Livanov, Zoologische Jahrbücher, Abteilung für Anatomie und Ontogenie, *22*.)

the more anterior segments, and also in having only 30 segments altogether. Four segments form the posterior sucker. Most of the other segments are divided into four rings.

There is no distinct anterior sucker. The gut has no long caeca comparable to those of leeches, but it does have some shallow lateral pouches. The anus is located dorsally just anterior to the posterior sucker.

The coelom is reduced, but there are distinct septa between most of the segments. The paired ovaries and testes are long and tubular. They are lateral to the gut and run through several segments. The circulatory system is closed.

Acanthobdella reaches a length of up to 3 cm and is a transitory parasite. It lives on fishes for about 4 months out of each year; the rest of the time it is free-living. During its parasitic phase it sucks up blood and skin tissue. This worm probably represents an early stage in the evolution of leeches.

Reference on Acanthobdellida

Holmquist, C., 1974. A fish-leech of the genus *Acanthobdella* found in North America. Hydrobiologia, *44:*241–245.

SUBCLASS BRANCHIOBDELLIDA

Branchiobdellids (Figure 11.61), which live on the exoskeletons of freshwater crayfishes, were originally thought to be leeches, mostly because they lack setae and have both an anterior and a posterior sucker, as well as jaws. Then, for a long time, they were placed in the Oligochaeta. These worms do not fit well among either oligochaetes or true leeches, and it makes sense to give them the status of a subclass within the Hirudinoidea.

Branchiobdellids have 15 segments, the first three or four of which are united to form the head. The anterior sucker, in which the mouth is located, is sometimes bordered by tentaclelike outgrowths. Of the remaining eleven segments, the last one or two form the posterior sucker. Each distinct segment between the head and posterior sucker is divided into two rings.

When there is a single pair of testes, these are usually in segment 9; when there are two pairs, they are in segments 9 and 10. A penislike male copulatory organ is located in a pouch on segment 10. A pair of ovaries and the openings through which eggs are laid are on segment 11, but the pores that lead to the seminal receptacles of the female system are on segment 9. Segments 10 and 11 are modified to form a clitellum, which secretes the cocoons in which the worm encloses its eggs and which it sticks to the exoskeleton of its host.

Most branchiobdellids feed on diatoms and other organic food, using a pair of jaws, on the dorsal and ven-

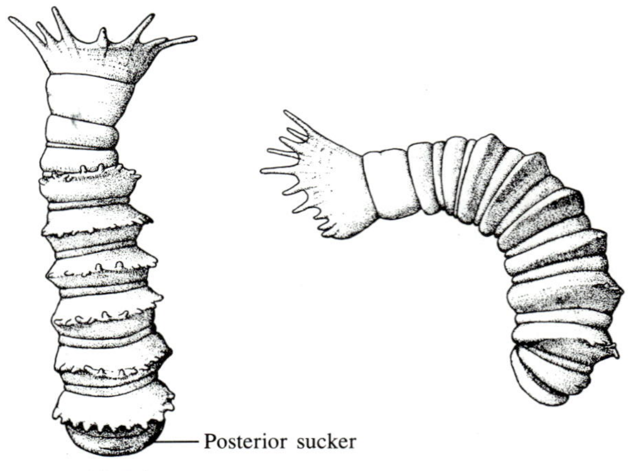

FIGURE 11.61 *Stephanodrilus uchidai* (subclass Branchiobdellida). (Yamaguti, Journal of the Faculty of Science, Hokkaido University, Zoology, *3*.)

tral sides of the buccal cavity, for scraping this material from the surface of the host. It appears likely, however, that at least a few species draw blood from the gills. The gut has no caeca comparable to those of leeches. The anus is located middorsally above the posterior sucker.

Most of the coelom is distinct, although there is reduction of this space in the anterior and posterior portions of the body. There are two pairs of metanephridia.

Branchiobdellids tend to be concentrated on the chelipeds (large, pincer-bearing legs) or on the sides of the carapace of their crayfish hosts. They cling tightly with the posterior sucker, but they can use the anterior sucker for temporary attachment, and their style of locomotion is similar to that of leeches.

References on Branchiobdellida

Holt, P. C., 1965. The systematic position of the Branchiobdellida. Systematic Zoology, *14:*25–32.

Holt, P. C., 1968. The Branchiobdellida: epizootic annelids. The Biologist, *L*(3–4):79–94.

Jennings, J. B., and Gelder, S. R., 1979. Gut structure, feeding and digestion in the branchiobdellid oligochaete *Cambarincola macrodonta* Ellis, 1912, an ectosymbiote of the freshwater crayfish *Procambarus clarkii*. Biological Bulletin, *156:*300–314.

SUMMARY

1. The worms that make up the phylum Annelida exhibit several features not seen in lower phyla. They usually have an extensive true coelom between the gut and body wall. This cavity, lined by a peritoneum, originates as a series of spaces within embryonic mesoderm.

2. The body is divided into a linear series of segments, which are evident not only externally but also internally. As a rule, each segment has its own coelomic cavity, separated from that of the preceding and succeeding segments by partitions called septa.

3. The gut is complete and extends through all of the segments. It is supported by a vertical mesentery that divides the coelomic cavities into right and left portions.

4. Most annelids have a circulatory system. The two largest vessels run longitudinally above and below the gut. Blood is usually circulated by the pumping action of the dorsal vessel, and sometimes also by contractions of hearts, which are enlarged and strongly muscularized portions of certain vessels. The usual respiratory pigments are hemoglobin and chlorocruorin.

5. The nerve cord is located ventrally and appears to have been derived from two parallel cords that have united. The ganglia and branch nerves are arranged according to a segmental pattern. The brain lies above the gut and is joined to the ventral cord by connectives that extend around the pharynx.

6. Excretory–osmoregulatory organs are mostly either protonephridia of the solenocytic type, or metanephridia. The latter, like protonephridia, are of ectodermal origin, but they consist of a funnel that is open to the coelom and a duct that leads to an external pore. Both protonephridia and metanephridia are paired segmental structures.

7. Funnel organs of another type, called coelomoducts, are of mesodermal origin. They function primarily as gonoducts, but they may become joined to protonephridia or metanephridia to form compound structures that function both as gonoducts and as organs of excretion and osmoregulation.

8. The body wall generally has layers of circular and longitudinal muscle, and there may also be muscles with other orientations. Bristles called setae are present in nearly all annelids other than leeches and have various functions, the more important of which are maintaining contact with the substratum during locomotion or burrowing, and maintaining contact with the wall of a tube or burrow.

9. There are many diverse life styles among annelids, and many modes of sexual and asexual reproduction. Cleavage, when it can be followed clearly, is of the spiral type.

10. The three classes of Annelida are the Polychaeta, Oligochaeta, and Hirudinoidea. Polychaetes are generally characterized by lateral, segmental outgrowths called parapodia, in which the setae are located. There are usually many setae on each segment. As a rule, the head region has antennae, palps, or other specialized sensory structures, but this is not necessarily the case. Feeding methods are extremely diversified. The class includes omnivores, herbivores, carnivores, detritus feeders, and ciliary–mucous feeders. Sexes are nearly always separate, and the gonads are inconspicuous structures that liberate prospective gametes into the coelom. From the coelom, gametes escape through coelomoducts, through compound funnel organs formed by union of coelomoducts and protonephridia or metanephridia, or through breaks in the body wall. There is often a ciliated larval stage called the trochophore. Polychaetes are entirely aquatic, and mostly marine.

11. Annelids of the class Oligochaeta lack parapodia and they generally have only a few setae on each segment. The head region does not have antennae, palps, or tentacles. Oligochaetes generally feed on detritus, humus, or microscopic organisms. They are hermaphroditic and mate for reciprocal transfer of sperm. The reproductive system is often complex. Development is direct. Oligochaetes live in the sea, in fresh water, and in moist habitats on land. Among the terrestrial types are the common garden earthworms.

12. The Hirudinoidea includes leeches, as well as two other small groups. Leeches do not have setae, and most of their segments are superficially constricted into two or more secondary rings. These worms are very muscular, firm-bodied animals. There is a posterior sucker for attachment, and usually also an anterior sucker, within which the mouth is located. The coelom is reduced, being largely filled with tissue. A circulatory system comparable to that of other annelids is often absent, but definite channels in the tissue that fills the coelom serve the same functions. Hermaphroditism is the rule, and the male system includes either a penis for copulation or a device that functions in transfer of a spermatophore to the partner. Development is direct. Most leeches are freshwater animals, but there are many in the sea, and some are terrestrial. The group includes predators that feed on invertebrates, as well as transitory or more or less permanent parasites that suck blood or other tissues from vertebrates or invertebrates.

12

Phyla Echiura, Sipuncula, Pogonophora, and Vestimentifera

- Phylum Echiura
- Phylum Sipuncula
- Phylum Pogonophora
    - Class Athecanephria
    - Class Thecanephria
- Phylum Vestimentifera
    - Class Axonobranchia
        - Order Riftiida
    - Class Basibranchia
        - Order Lamellibrachiida
        - Order Tevniida

The four phyla of worms described in this chapter are relatively small groups. Each is so distinctive that it cannot be assigned to any other major group of invertebrates. All of them, however, probably have a rather close relationship to annelids.

Phylum Echiura

Introduction

All members of the phylum Echiura are marine. Most of them live in burrows in mud or sand, but some nestle in crevices. Though there are only about 100 species, the group is rather richly diversified. The following account of echiuran structure and function deals primarily with generalized types such as *Echiurus* (Figure 12.1, A), which is buried in mud and probes the surface sediment for food. It will be supplemented by a description of certain aspects of the biology of some species that have decidedly different life styles.

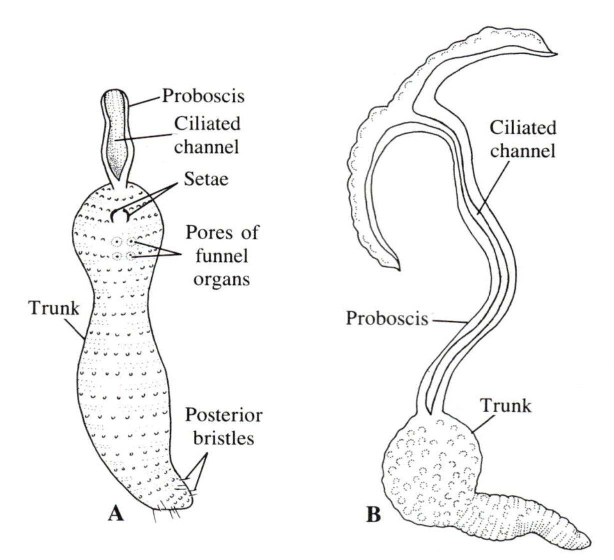

Figure 12.1 A. *Echiurus echiurus*. B. *Bonellia viridis*.

General Structure

Echiurans are unsegmented, but they are like annelids in having a paired **ventral nerve cord,** at least one pair of **setae,** and a **proboscis** that corresponds to a prostomium. The proboscis is generally spatulate or tonguelike (Figure 12.1, A), but it may be long and slender, and sometimes it is divided at the tip (Figure 12.1, B). It is muscular and can be shortened and lengthened, but it cannot be withdrawn or invaginated. In most species it is used for collecting detritus. Echiurans that extend a fleshy proboscis into the surface sediment run the risk of losing it to a hungry fish, but the structure can be regenerated.

The portion of the body behind the proboscis is called the **trunk.** The **mouth** is situated on its ventral side, next to the base of the proboscis, and the **anus** is at the posterior end. Not far behind the mouth there is usually a pair of stout setae whose bases are embedded in deep sacs, and which can be manipulated by a system of muscles. These ventral setae are absent in some echiurans, and in one species there are many setae concentrated in two groups. The slender bristles found in one or two circles close to the posterior tip of the trunk of certain genera look much like setae, but whether they should be regarded as true setae is doubtful.

The proboscis is a nearly solid and rather muscular structure. Being a collector of food, it has a ciliated channel on its ventral side. Elsewhere its epidermis is richly supplied with mucus-secreting glands and sensory cells. The body wall of the trunk is usually fairly thick. Although the epidermis lays down a thin cuticle, it also has sensory cells and the necks of glands whose secretory portions are located in a more or less gelatinous layer beneath the epidermis. This same layer also may have a heavy concentration of pigment-producing cells. Most of the rest of the trunk wall, up to the coelomic peritoneum, is composed of circular, longitudinal, and oblique muscles.

The **coelom** of the trunk is spacious and undivided, although it is traversed by strands of tissue, particularly muscles, that connect the body wall and the gut. The proboscis does not have a coelomic cavity. In some echiurans, however, it contains remnants of the embryonic blastocoel. Most echiurans burrow by extending the proboscis into the mud or sand in which they live. Peristaltic movements of the body as a whole are also important, especially in species in which the proboscis is not well developed.

Digestive Tract

The gut (Figure 12.2) has several divisions and subdivisions. The mouth is typically located at the bottom of an appreciable depression, and the pharynx, esophagus, and stomach (if one is recognizable) are all derived from a stomodeal invagination, so they have a lining of ectodermal origin. The long intestine, extensively looped, is paralleled for most of its length by a bypass **siphon,** which probably helps reduce the amount of water going through the intestine proper and thus minimizes the extent to which digestive enzymes are diluted. The intestine characteristically has a ciliated gutter running for its entire length. The hindgut, or rectum, is usually short, although it is enlarged in a few echiurans that pump water in and out of it for respiratory purposes.

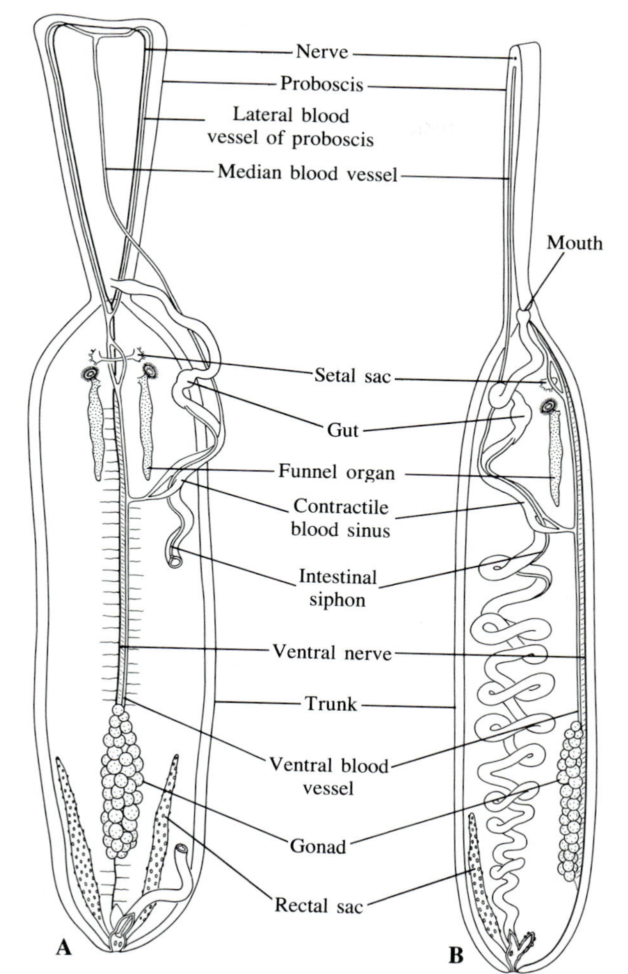

FIGURE 12.2 General structure of an echiuran; diagrammatic. A. Animal dissected from the dorsal side. B. Animal dissected from the right side. (After Delage and Hérouard, Traité de Zoologie Concrète.)

FUNNEL ORGANS AND RECTAL SACS

The **funnel organs** (Figure 12.2) open to the outside on the ventral surface of the trunk, behind the setae. There may be only one, two, or three pairs of these structures, but in some species there are many. They are probably mixonephridia; the actual funnels, sometimes rather elaborate, seem to be coelomoducts that have been joined to basically ectodermal metanephridia. When the funnel is close to the body wall, which is generally the case, the rest of the organ is a blind sac projecting into the trunk coelom. The funnel organs definitely serve as gonoducts, and perhaps have other functions as well.

Opening into the hindgut are two prominent structures called **rectal sacs** (also called anal sacs and urinary sacs). They are thought to be coelomoducts, and they typically have many ciliated funnels. The funnels are sometimes on tubercles scattered over the length of each sac, sometimes on slender outgrowths of the sac; there may also be other arrangements. The activity of cilia in the funnels causes a filtrate of coelomic fluid to enter the rectal sacs and to move on to the hindgut, but the filtrate is evidently not modified during its passage, so the sacs seem not to function in osmoregulation or regulation of ionic balance. Nevertheless, the epithelium and lumen of the sacs contain brown or reddish brown granules, and these are believed to be excretory products. The sacs have some musculature and are therefore contractile.

CIRCULATORY SYSTEM

The circulatory system (Figure 12.2) is fairly well developed in most echiurans. There is typically a ventral vessel running the length of the trunk, and this sends off two branches, one extending into each side of the proboscis. From the point where these branches become joined anteriorly, an unpaired vessel runs backward to the dorsal side of the gut; above the esophagus it swells into a heartlike structure which is then connected to the ventral vessel. The blood contains cells similar to those in the coelomic fluid, and is usually colorless; some echiurans, however, have hemoglobin.

NERVOUS SYSTEM

The more obvious components of the nervous system are a ring around the esophagus and the ventral nerve cord of the trunk; both are generally visible in dissected animals. A loop that runs through solid tissue of the proboscis is not likely to be seen except in sections. Nowhere in the nervous system are there prominent ganglia, and the branch nerves arising from the ventral nerve cord are not arranged symmetrically. The ventral cord is, however, basically double, as in annelids, although this duality is usually not evident except in histological sections. Sense organs are poorly developed in echiurans. They seem mostly to be specialized cells in the skin, sometimes concentrated in papillae and often bearing specialized cilia.

REPRODUCTION AND DEVELOPMENT

Sexes are separate. When there is an obvious gonad (Figure 12.2), this is usually a single structure situated midventrally in the posterior part of the body. It is almost certainly derived from the peritoneum. If there is no discrete gonad, the precursors of gametes also probably originate from the peritoneum. Gametes set free in the coelom leave the body by way of the funnel organs and fertilization usually takes place in the sea. Spiral cleavage and gastrulation are followed by formation of a free-swimming trochophore larva similar to that of polychaete annelids. This generally remains in the plankton for several weeks before settling and undergoing metamorphosis. In species whose eggs are particularly yolky, the trochophores subsist entirely on the yolk.

SOME UNUSUAL ECHIURANS

Two echiurans are used here to illustrate striking deviations from the body plan and life style of generalized types. *Urechis caupo* (Figure 12.3), the "weenie worm," is abundant on some tide flats in California, and it has relatives in other parts of the world. *Urechis* is a large worm, occasionally reaching a length of 30 cm, but it has a proportionately small proboscis, which it uses to dig a U-shaped burrow. Glands at the base of its proboscis secrete mucus that is plastered to the wall of the burrow. Then, as the animal continues to secrete mucus, it moves backward, so a funnel-like net is formed. The pores in the net have a diameter of about 40 Å, so they trap bacteria and particles much smaller than bacteria. Peristalsis of the trunk moves water through the net, past the worm, and out of the rear opening of the burrow. After the net has become fairly well saturated with food, the worm pulls it loose from the burrow and swallows it. Thus *Urechis* gets back its net for recycling, as well as the food that the net has trapped.

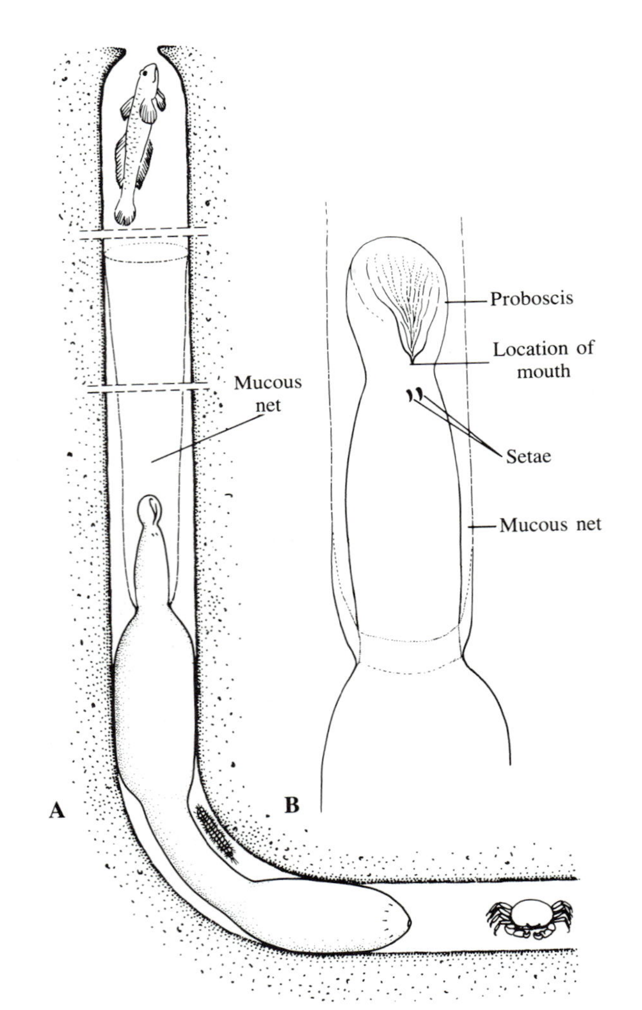

FIGURE 12.3 *Urechis caupo.* A. Specimen in its U-shaped tube, with some of its commensals. B. Anterior portion of the body. (After Fisher and MacGinitie, Annals and Magazine of Natural History, series 10, *1*.)

The gut of *Urechis* is long and extensively looped, and for much of its length it is paralleled by a siphon comparable to that in various generalized echiurans. The hindgut, a cloaca as in other echiurans, is enlarged and muscular, and by peristaltic action it can draw in water through the anus, then force it out again. There is no trace of a circulatory system. The coelomic cavity is spacious and its peritoneal lining is ciliated; it undoubtedly does the work of a circulatory system, distributing oxygen, nutritive material, and wastes. The coelomocytes of *Urechis* contain hemoglobin and function in transport of oxygen. The peristaltic pumping action of the hindgut may be helpful in respiration.

Prospective gametes, presumably derived from cells proliferated by some part of the peritoneum (a gonad has not been found), enlarge as they float free in the coelom. When mature, they are trapped by **collecting organs.** Each of the six funnel organs of *Urechis* has a few of these. They are slender, spiral outgrowths of the funnels and have a ciliated groove that will only accept and conduct gametes of the right size and shape. The ripe eggs of *Urechis,* unlike immature eggs, are not spherical; they have an indentation that is perhaps somehow detected by the lips of the groove. Eggs or sperm accumulate in the saclike funnel organs and this fact makes *Urechis* useful for work on fertilization and development. If a fine pipette is carefully inserted into the pore of a funnel organ, gametes can be sucked out without damaging the worm.

Urechis, like some other marine invertebrates that maintain permanent burrows, runs a kind of hotel for other animals. The California species is sometimes called the "innkeeper," and its associates are a little fish, a clam whose siphons open into the burrow, pea crabs (related to those that live in the mantle cavity of certain clams), a shrimp, and a scaleworm, which is a polychaete annelid belonging to the family Polynoidae. By living with *Urechis,* these animals are protected from predators, yet they can feed on material that falls into the burrow, or that is brought into it by the peristaltic pumping of their host.

Bonellia viridis (Figure 12.1, B), primarily a European species, exhibits one of the most remarkable cases of sexual dimorphism in the entire animal kingdom. Females nestle in crevices, and a specimen with a trunk length of 6 or 7 cm may extend its nearly T-shaped proboscis for 1 m. Males are only 2 or 3 mm long and live in a dilated portion of the single funnel organ of the female, so they are in a position to fertilize the eggs before these leave the body. There are usually about 10 or 15 males in the funnel organ of each female.

Sex determination in *B. viridis* has been the subject of numerous investigations. In about 17% of the larvae, sexuality is determined genetically. Most larvae, however, are **indifferent:** Whether they become males or females depends on where they settle. Larvae that settle on the proboscis of a female become males. They remain on the proboscis for up to about a month, then move into the gut and somehow reach the funnel organs. Here they live as parasites, without a functional gut and without a proboscis or circulatory system. Sperm maturing in the coelom enter a sac that opens by a pore at the anterior end. Males retain external ciliation throughout their lives.

Indifferent larvae that are not influenced by close contact with a female become females. If larvae pair off as they settle, which frequently happens, one partner develops into a male, the other into a female. Extracts of proboscis tissue from females induce masculinization of indifferent larvae, but the substance or substances that are responsible for this change have not been identified.

References on Echiura

Chapman, G., 1968. The hydraulic system of *Urechis caupo* Fisher & MacGinitie. Journal of Experimental Biology, *49:*757–767.

Fisher, W. K., 1946. Echiuroid worms of the North Pacific Ocean. Proceedings of the United States National Museum, *96:*215–292.

Fisher, W. K., and MacGinitie, G. E., 1928. The natural history of an echiuroid worm. Annals and Magazine of Natural History, series 10, *1:*204–213.

Harris, R. R., and Jaccarini, V., 1981. Structure and function of the anal sacs of *Bonellia viridis* (Echiura: Bonelliidae). Journal of the Marine Biological Association of the United Kingdom, *61:*413–430.

Jaccarini, V., Agius, L., Schembri, P. J., and Rizzo, M., 1983. Sex determination and larval sexual interaction in *Bonellia viridis* (Echiura: Bonelliidae). Journal of Experimental Marine Biology and Ecology, *66:*25–40.

Newby, W. W., 1940. The embryology of the echiuroid worm *Urechis caupo*. Memoirs of the American Philosophical Society, no. 16.

Rice, M. E., and Todorović, M. (editors), 1975–1976. Proceedings of the International Symposium on the Biology of Sipuncula and Echiura. 2 volumes. Naučno Delo Press, Belgrade.

Stephen, A. C., and Edmonds, S. J., 1972. The Phyla Sipuncula and Echiura. British Museum (Natural History), London.

Phylum Sipuncula: Peanut Worms

Introduction

All of the approximately 300 species of sipunculans are marine. They are sendentary animals that live buried in mud or firm sand, in holes bored in rock by other animals, under or between rocks, and in empty shells of snails; one tiny species *(Golfingia minuta)* occupies tests of foraminiferans. They like tight situations into which their bodies fit almost perfectly.

General Structure

Sipunculans (Figure 12.4; Figure 12.5) are bilaterally symmetrical and have an extensive coelom, but they show no evidence of segmentation, internally or externally. A substantial part of the body—sometimes nearly half of it—forms what is called the **introvert.** This can be completely withdrawn into the trunk. Powerful retractor muscles (Figure 12.5, A and C) run from

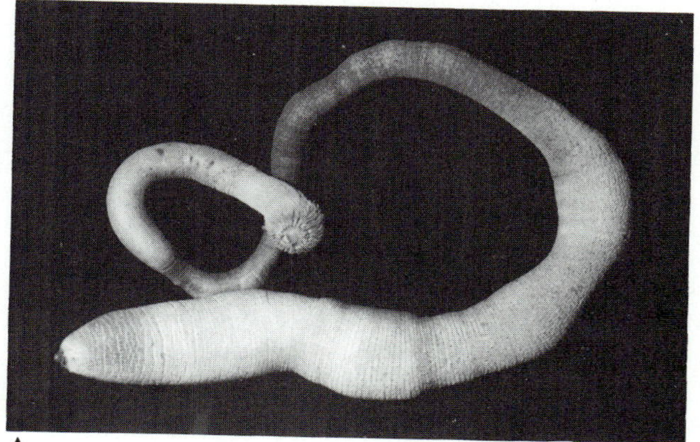

A

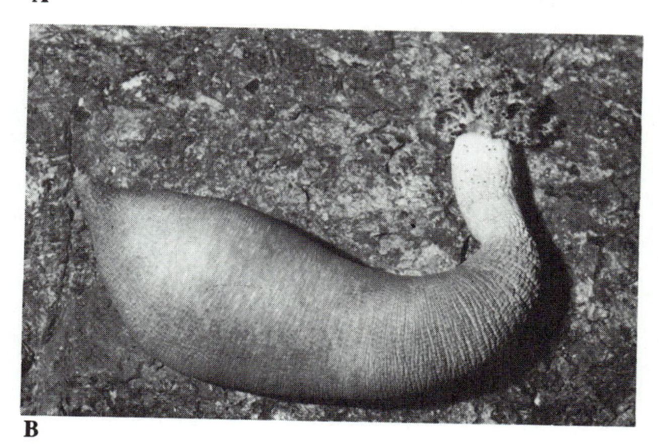

B

Figure 12.4 A. *Golfingia vulgaris,* a species with unbranched tentacles. B. *Themiste pyroides,* a species with branched tentacles. In both figures, the arrow indicates the proximal portion of the introvert.

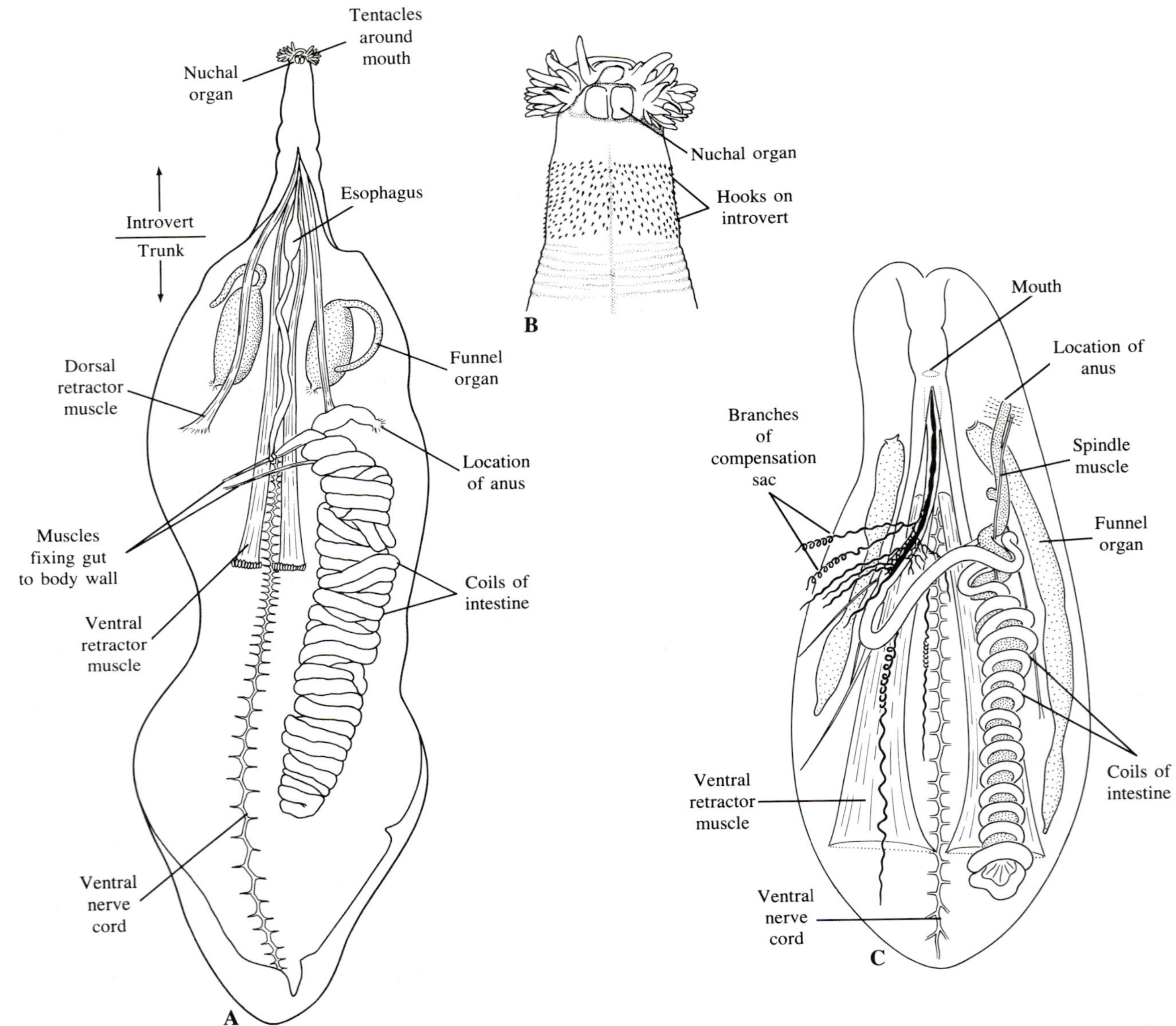

Figure 12.5 A. *Golfingia vulgaris,* dissected from the dorsal side. The two compensation sacs, one dorsal and one ventral to the esophagus, have been omitted. B. Anterior portion. (After Théel, Kunglica Svenska Vetenskapsakademiens Handlingar, *39*.) C. *Themiste pyroides,* introvert invaginated, dissected from the dorsal side. (After Fisher, Proceedings of the United States National Museum, *102*.)

the body wall of the trunk nearly to the tip of the introvert, and when these contract the introvert invaginates. When the retractor muscles relax, the high hydrostatic pressure created inside the coelom by sheaths of muscle in the unyielding body wall forces the introvert to become extended again. The introvert is an effective device for excavation, and is used by some sipunculans for burrowing, even into stiff clay.

The mouth is situated at the anterior end of the introvert and is usually partly or completely encircled by **tentacles** (Figure 12.4; Figure 12.5, A and B) concerned with collection of detritus and other food. The

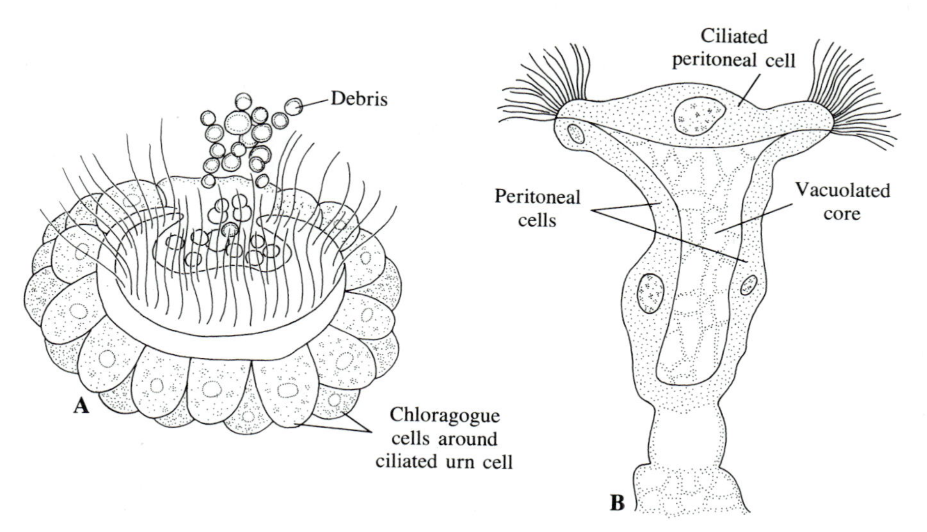

FIGURE 12.6 A. Fixed urn of *Sipunculus nudus*. B. Free urn of a species of *Phascolosoma*. (After Selensky, Zeitschrift für Wissenschaftliche Zoologie, *90*.)

tentacles may be simple or elaborately branched, and cilia on their inner faces move food to the mouth. One rather widespread genus, *Phascolosoma,* is odd in that its tentacles form a crescent-shaped group on the dorsal side of the mouth; they are ciliated on their outer faces.

The body wall, though not especially thick, is so tough that a sipunculan is not likely to burst or even bulge when it is squeezed hard between thumb and forefinger. Beneath the epidermis, which is covered by a cuticle, there is a layer of connective tissue, then muscle, and finally the peritoneum that lines the coelom. The muscular portion of the body wall generally consists of an outer layer of circular muscle and an inner layer of longitudinal muscle, and there is sometimes a sheet of oblique muscle fibers between these.

The cuticle and underlying epidermis may be elaborated into various kinds of projections, especially papillae and hard hooks or denticles. These are sometimes most prominent on the introvert (Figure 12.5, B), sometimes on the trunk region. A few genera have a hard shield on the trunk close to the base of the introvert. Sipunculans that live in shells generally have prominent cuticular thickenings that help them grip the inside wall of the spire.

The coelomic cavity is spacious. The fluid within it bathes most of the tissues of the body, transporting nutrients, oxygen, and metabolic wastes. It contains a rich variety of cellular components and some noncellular material as well. Some cells contain **hemerythrin,** a respiratory pigment that functions in transport of oxygen; these red corpuscles constitute a rather high proportion of the total volume of coelomic fluid. There are also colorless amoebocytes of various types, some of which are phagocytic. Most unusual are the **urns,** though these are not present in all sipunculans. When they do occur, they are either large single cells with a crown of cilia around one end, or they consist of a number of cells (Figure 12.6, A), one of these being comparable to a single-celled urn, the others forming a cuplike covering around it. The urns found in the coelomic fluid originate on the peritoneum as **fixed urns** (Figure 12.6, B). If a fixed urn has **chloragogue cells** associated with it and the whole ensemble breaks free, it forms the multicellular type of urn. Chloragogue cells are part of the peritoneum and are specialized for accumulation of wastes. They usually have a yellowish color and are most abundant around the gut and compensation sacs, discussed below.

Urns swim with the ciliated crown at the rear. They collect particulate material, including amoebocytes, and the masses of accumulated junk seem to become the noncellular brown bodies that float in the coelomic fluid until they are eliminated through the funnel organs. Considering how intensively urns have been studied, we still know relatively little about them and their role in the physiology of sipunculans. That they accumulate debris and foreign material injected experimentally into the coelom is unquestioned. Just how they do it is something else. It has been claimed that because urns are positively charged, they do not pick up positively charged red corpuscles, but do collect negatively charged coelomocytes and other material.

The tentacles are extended by a closed hydrostatic system separate from that of the coelom. Each tentacle is pervaded by vessels that are joined to a ringlike vessel, often in the form of a network, at the base of the tentacular crown. A prominent vessel above the esophagus is also connected to the ring. This vessel serves as a **compensation sac** accepting fluid that is forced out of the tentacles when these are withdrawn. In one genus there are two compensation sacs, one dorsal and one ventral to the esophagus. The system of vessels just described strongly resembles the complex consisting of the ring canal, tentacular canals, and Polian vesicle of sea cucumbers (see Chapter 21). In some sipunculans the tentacular system is bright red because it contains a high concentration of hemerythrin-carrying corpuscles, and the compensation sac has many fine branches that look like blood vessels (Figure 12.5, B). As the fluid is driven through the system by pulsations of the compensation sac, the whole ensemble resembles a true circulatory system, and it undoubtedly is important in carrying oxygen to some tissues and perhaps to the coelomic fluid. The hemerythrin within the tentacular system has a different molecular weight than that within the coelom.

DIGESTIVE TRACT

The gut of a sipunculan is internally ciliated (Figure 12.5, A and C). It runs nearly to the posterior tip of the body, then loops back on itself and terminates at a mid-dorsal anus usually situated on the trunk near the base of the introvert. (The anus is sometimes located on the introvert itself.) Generally the gut is divided into three main regions: esophagus, intestine, and rectum. But the intestinal portion is long and it is not histologically homogeneous from one end to the other. Its posteriorly and anteriorly directed loops are intertwined in a double spiral. The rectal portion of the gut often has a little blind sac or glandular outgrowths. The gut is to a large extent free of the body wall, but there are some mesenteries, and the intestinal portion is organized around a so-called spindle muscle that originates on the body wall close to the anus. The posterior end of this muscle is sometimes attached to the body wall, too.

The prominent funnel organs of sipunculans, easily recognizable in dissected animals (Figure 12.5, A and C), are thought to be mixonephridia. There are usually two of them, rarely just one, and they open to the outside by pores on the ventral surface, a little anterior to the level of the anus. Each has a rather small ciliated funnel (''vibratile pavillion'') which represents a coelomoduct; the rest of each funnel organ is a sac homologous to a metanephridium.

NERVOUS SYSTEM AND SENSE ORGANS

The general layout of the nervous system resembles that of an annelid, but it shows some unusual features. The ventral nerve cord, readily visible in a dissected animal (Figure 12.5, A and C), is neither double nor ganglionated, and the nerves that branch off it do not arise symmetrically on the right and left sides. The **brain** (Figure 12.7), sometimes obviously bilobed, is dorsal to the anteriormost part of the esophagus and is joined to the ventral cord by a pair of connectives that encircle the gut. Nerves originating at the brain and connectives innervate the tentacles, retractor muscles of the introvert, and other structures. Because of its location close to the tip of the introvert, the brain must pivot about 180° when the introvert turns inward.

Closely associated with the brain are several structures, some of which are enigmatic. Not all of them show up equally well in all sipunculans, and only three are briefly mentioned here. The **digitiform organ** consists of a number of fingerlike projections that extend forward from the anterior side of the brain. The function of these is not known. The **nuchal organ** (Figure 12.5, A and B) is a ciliated cushion (usually bilobed) that interrupts the circle of tentacles. It receives nerves from the brain and is probably an olfactory structure. When a pair of pigmented eyespots survives beyond the larval period, these are located at the bottom of **ocular tubes.** The tubes are mostly filled by coagulated material and the openings become blocked by cuticle. Experiments show, however, that the receptors next to the pigment mass are sensitive to light.

Reproduction and Development

Sexes are separate. There is a pair of small gonads located ventrally on the body wall where the retractor muscles of the introvert are attached. Spermatogonia

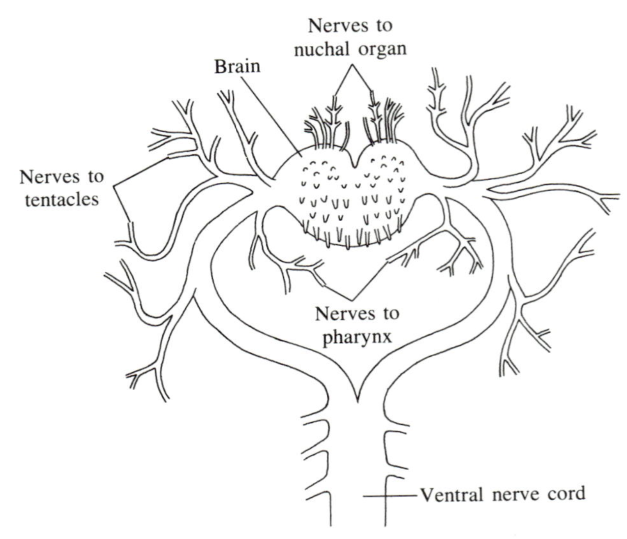

Figure 12.7 Brain and anteriormost portion of ventral nerve of *Golfingia gouldi*. (After Andrews, Studies from the Biological Laboratory, Johns Hopkins University, *4*.)

and oogonia are set free in the coelom, where gametogenesis is completed. When ripe, the gametes enter the funnel organs and may be stored for a time in these. Fertilization takes place after the gametes are discharged by way of the nephridiopores.

In most respects, the early development of sipunculans is similar to that of many polychaetes and molluscs. Cleavage is spiral and unequal, and development is determinate. Gastrulation may be by invagination or epiboly. The blastopore closes, but the mouth is later formed in the same region, so sipunculans are **protostomatous.** The formation of the definitive mouth is accompanied by a stomodeal invagination. Mesoderm is derived from a teloblast cell, and the coelom is formed by **schizocoely.**

There are four main patterns of later development. Some sipunculans develop directly into the adult form; there are no larval stages. In another pattern, a **trochophore** remains enclosed by the egg membrane (Figure 12.8, A), so it cannot feed; it subsists on yolk. It develops into another type of larva called the **pelagosphaera** (Figure 12.8, B), which swims for a short time but continues to depend on yolk until metamorphosis. In a third pattern, the nonfeeding trochophore is succeeded by a pelagosphaera that does feed and that may remain in the plankton for a long period. Finally, there are species in which the nonfeeding trochophore metamorphoses directly into the adult form without an intervening pelagosphaera.

Asexual reproduction has been noted in two species, *Sipunculus robustus* and *Aspidosiphon brocki*. A worm constricts to produce one or a few prospective new individuals from its posterior half. The formation of lateral buds from the anterior half has also been observed.

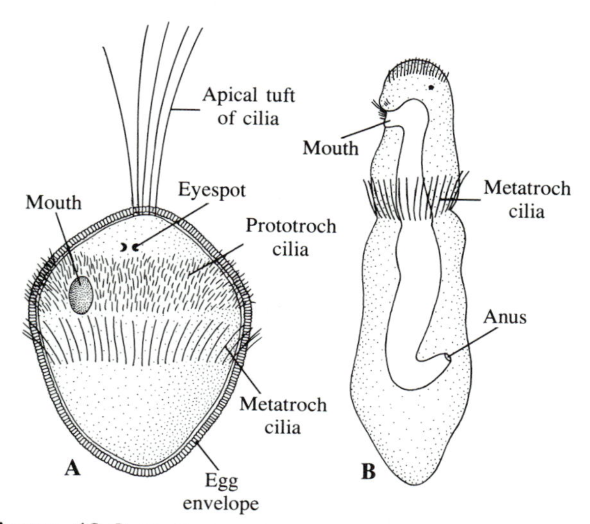

FIGURE 12.8 *Golfingia vulgaris.* A. Trochophore. B. Pelagosphaera larva. (After Gerould, Zoologische Jahrbücher, Abteilung für Anatomie und Ontogenie der Thiere, *23.*)

REFERENCES ON SIPUNCULA

Cutler, E. B., 1973. Sipuncula of the western North Atlantic. Bulletin of the American Museum of Natural History, *152:*103–204.

Gibbs, P. E., 1977. British Sipunculans. Synopses of the British Fauna, no. 12. Linnean Society of London and Academic Press, London and New York.

Manwell, C., 1960. Histological specificity of respiratory pigments. II. Oxygen transfer systems involving hemerythrin in sipunculid worms of different ecologies. Comparative Biochemistry and Physiology, *1:*277–285.

Rice, M. E., 1967. A comparative study of the development of *Phascolosoma agassizii, Golfingia pugettensis,* and *Themiste pyroides,* with a discussion of developmental patterns in the Sipuncula. Ophelia, *4:*143–171.

Rice, M. E., 1969. Possible boring structures of sipunculids. American Zoologist, *9:*803–812.

Rice, M. E., 1970. Asexual reproduction in a sipunculan worm. Science, *167:*1618–1620.

Rice, M. E., 1973. Morphology, behavior, and histogenesis of the pelagosphaera larva of *Phascolosoma agassizii* (Sipuncula). Smithsonian Contributions to Zoology, no. 132.

Rice, M. E., 1976. Larval development and metamorphosis in Sipuncula. American Zoologist, *16:*563–571.

Rice, M. E., 1981. Larvae adrift: patterns and problems in life histories of sipunculans. American Zoologist, *21:*605–619.

Rice, M. E., and Todorović, M. (editors), 1975–1976. Proceedings of the International Symposium on the Biology of the Sipuncula and Echiura. 2 volumes. Naučno Delo Press, Belgrade.

Ruppert, E. E., and Rice, M. E., 1983. Structure, ultrastructure, and function of the terminal organ of a pelagosphaera larva (Sipuncula). Zoomorphology, *102:*143–163.

Stephen, A. C., and Edmonds, S. J., 1972. The Phyla Sipuncula and Echiura. British Museum (Natural History), London.

Walter, M. D., 1973. Feeding and studies on the gut content in sipunculids. Helgoländer Wissenschaftliche Meeresuntersuchungen, *25:*486–494.

Williams, J. A., and Margolis, S. U., 1974. Sipunculid burrows in coral reefs: evidence for chemical and mechanical excavation. Pacific Science, *28:*357–359.

PHYLUM POGONOPHORA

INTRODUCTION

Pogonophorans—the "beard-bearers"—are tube-dwelling marine worms that live at depths ranging from about 25 m to more than 6000 m. They pack a lot of complexity into their extremely slender bodies, which are not often more than 1 mm wide. Even the tubes of some species are so fine that they resemble hairs. In

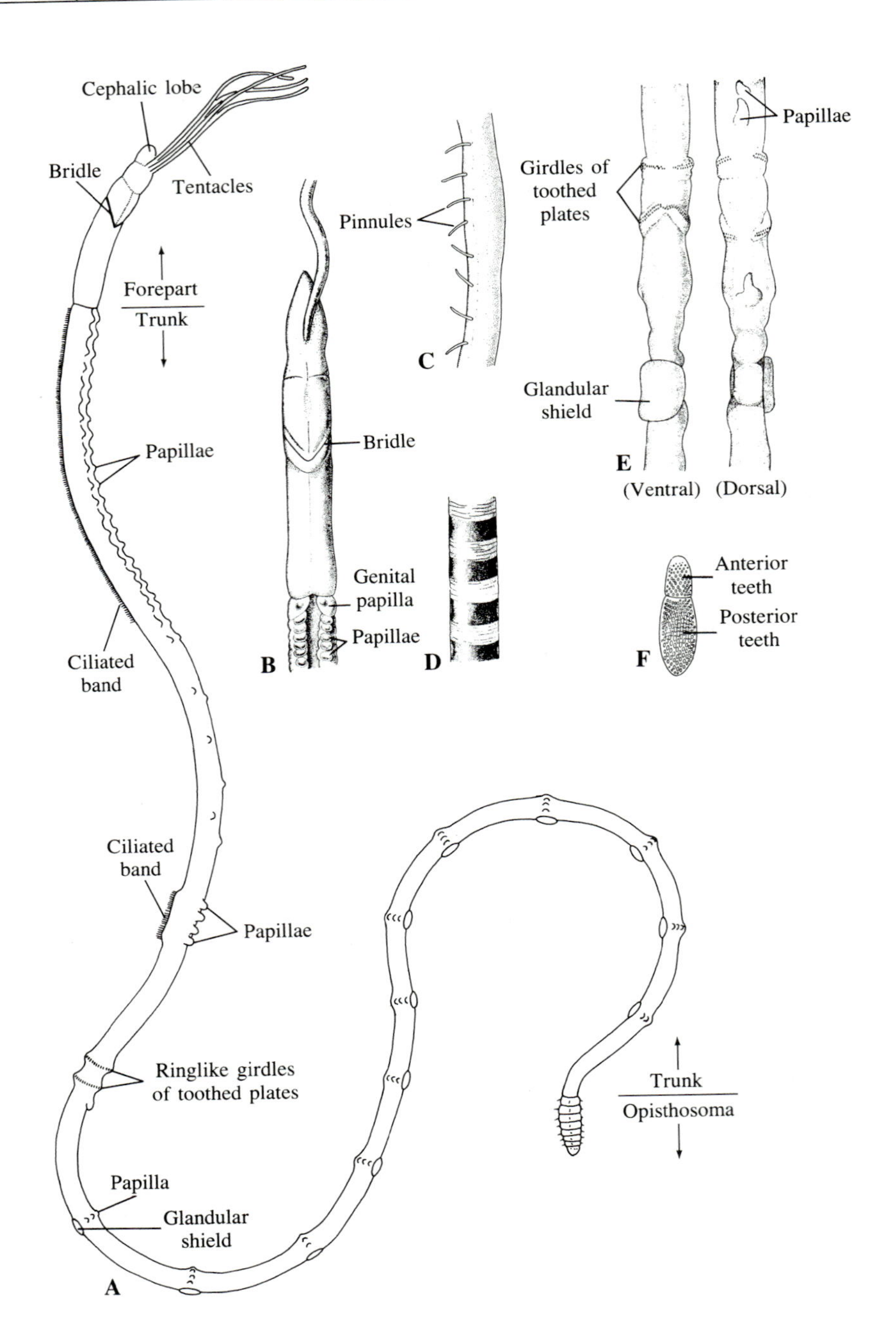

Figure 12.9 A. Diagram of a pogonophoran. The body has been shortened. (After Southward, in Giese and Pearse, Reproduction of Marine Invertebrates, 2.) B. *Siboglinum caulleryi,* forepart and anteriormost portion of the trunk, dorsal view. C. Portion of the tentacle. D. Portion of the tube. E. Portion of the trunk, including the region of the girdles of toothed plates. F. Toothed plate. (B–F, Ivanov, Pogonophorans.)

sorting dredged material, it would be easy to overlook the tubes or to dismiss them as fibers such as would be worn off hemp ropes.

The first specimens to be studied seriously were collected off Indonesia by the Dutch *Siboga* Expedition of 1899–1900. They were turned over to Maurice Caullery, a superb French zoologist, and formed the basis for his description of *Siboglinum weberi*. Caullery realized that the worms were very unusual. The fact that they had no gut bothered him, and this was the main reason he suspected that the specimens he studied were not complete. The absence of a gut, however, is characteristic of pogonophorans.

Our understanding of pogonophorans has progressed rapidly since 1949, when A. V. Ivanov started publishing on them in Russia. Over 100 species have now been

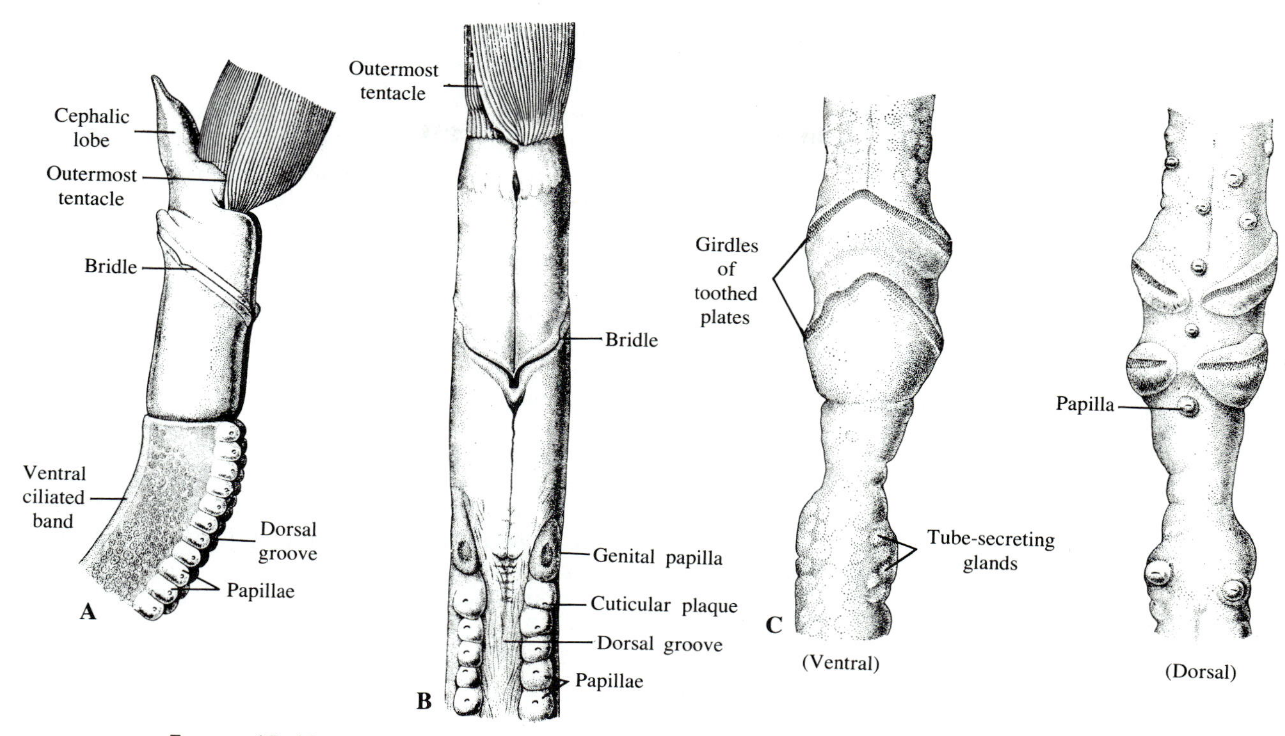

FIGURE 12.10 A. *Spirobrachia grandis,* forepart and anteriormost part of trunk, lateral view from left side. B. Dorsal view. C. *Lamellisabella zachsi,* portion of trunk, including the region of the girdles of toothed plates. (Ivanov, Pogonophorans.)

described, and some aspects of the anatomy and development of these worms have been thoroughly studied. Nevertheless, until 1964 the posterior portion was missing from all specimens that had been examined carefully. When it was found, it turned out to be segmented and to have setae resembling those of an annelid. Because of the way the body cavities are arranged, most zoologists working with pogonophorans had thought of them as fitting into the deuterostome line, along with echinoderms and hemichordates (Chapters 21 and 22). But the structure of the hind portion, as well as some other recently discovered features, strongly suggest that these animals are more closely related to annelids and other protostomes.

The terminology applied to the various divisions of the body has not been stable. This is partly because the posterior portion was unknown for a long time and partly because newer findings with respect to the embryology of pogonophorans have led to a reevaluation of what formerly were thought to be three distinct regions: protosome, mesosome, and metasome. Following the terminology in favor now, the terms **forepart** (protosome plus mesosome), **trunk** (metasome), and **opisthosoma** (segmented portion, with setae) are used here.

GENERAL STRUCTURE

A typical pogonophoran is shown in Figure 12.9, A. Note that the **cephalic lobe,** at the anterior end of the forepart, bears tentacles. The number of these ranges from 1 to more than 200. The base from which the tentacles arise is sometimes circular or horseshoe-shaped, sometimes spiral (Figure 12.10, A and B; Figure 12.11). In most species, each tentacle has two rows of delicate branches called **pinnules** (Figure 12.9, C). These are portions of single epidermal cells, but they are pervaded by minute capillaries that connect the afferent and efferent blood vessels within the tentacle (Figure 12.12, B). The remainder of the forepart, behind the cephalic lobe, is divided into two sections by a hard, cuticular ridge, called the **bridle,** or *frenulum* (Figure 12.9, B; Figure 12.10, A and B).

The trunk, distinctly set off from the forepart, is long in proportion to its width and has several specializations. Among these are a longitudinal ciliated band

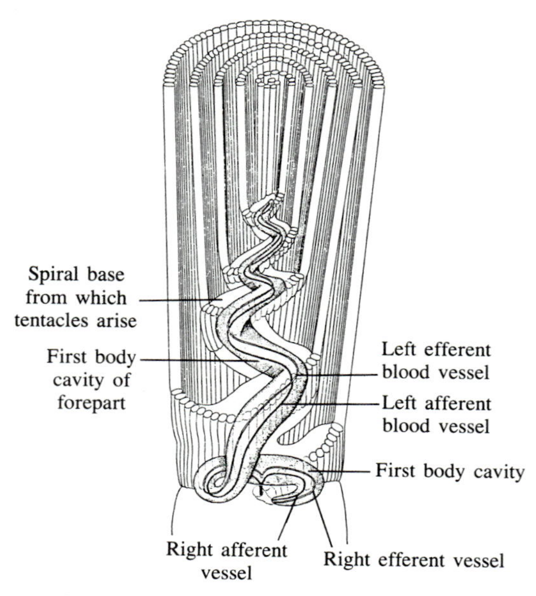

FIGURE 12.11 *Spirobrachia grandis,* body cavity and blood vessels in the spiralled base from which the tentacles arise, dorsal view; diagrammatic. (Ivanov, Pogonophorans.)

in the anterior portion, a similar band farther back, papillae, glandular shields, and two ringlike girdles that consist of hard, toothed plates (Figure 12.9, E and F; Figure 12.10, C). The plates contain considerable chitin and are secreted by single cells. In these respects they are similar to setae of annelids. They will be called plates here, however, so that they will not be confused with the segmentally arranged setae of the opisthosoma.

Underneath the epidermis of the trunk is a layer of circular muscle, and internal to this is a layer of longitudinal muscle. The same general arrangement of muscles is characteristic of the body wall elsewhere, although in some places there are also muscles with a dorsoventral or oblique orientation.

The reproductive organs, discussed below, are located in the anterior part of the trunk region. In the posterior half or third of the trunk, much of the space may be occupied by the trophosome, a structure dealt with below in connection with nutrition of pogonophorans.

The opisthosoma (Figure 12.9, A; Figure 12.13) has from a few to about 25 segments, each of which bears several setae. These probably help the animal to anchor

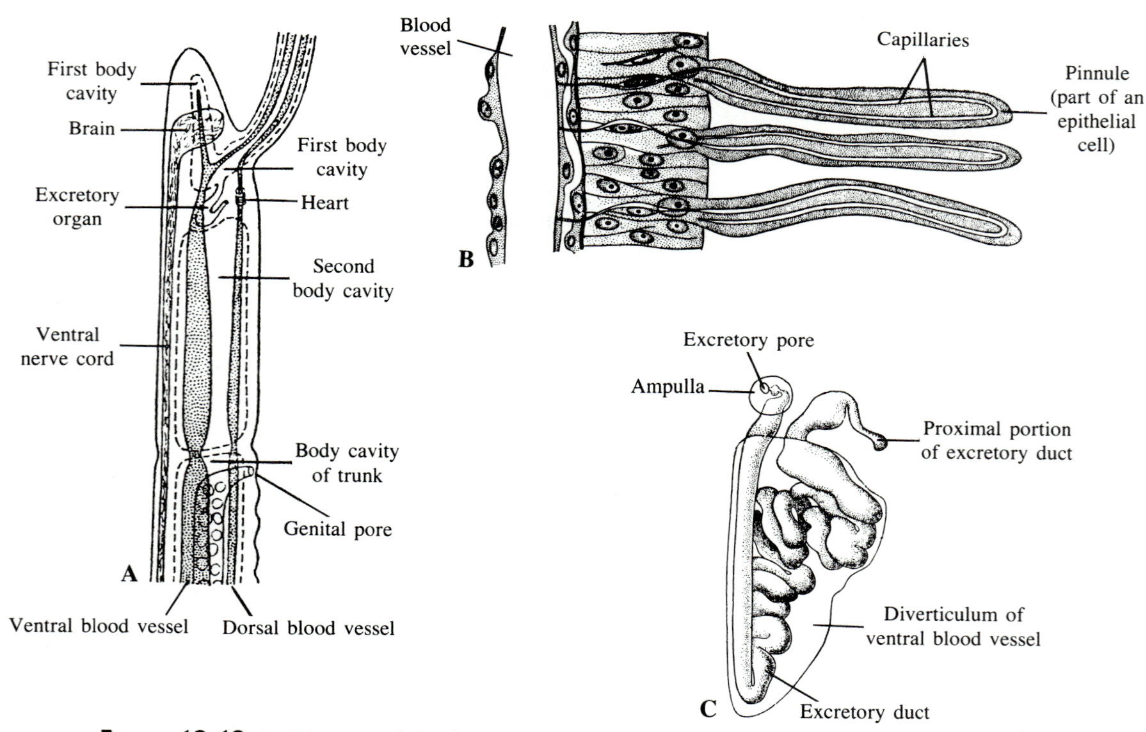

FIGURE 12.12 A. Diagram of the forepart and anterior portion of the trunk of a pogonophoran, as seen in sagittal section. B. *Lamellisabella zachsi,* portion of a longitudinal section of a tentacle, showing the blood vessel in the pinnules. C. *L. zachsi,* excretory organ. In this species, much of each excretory organ lies within a diverticulum of the dorsal blood vessel. (B and C, Ivanov, Pogonophorans.)

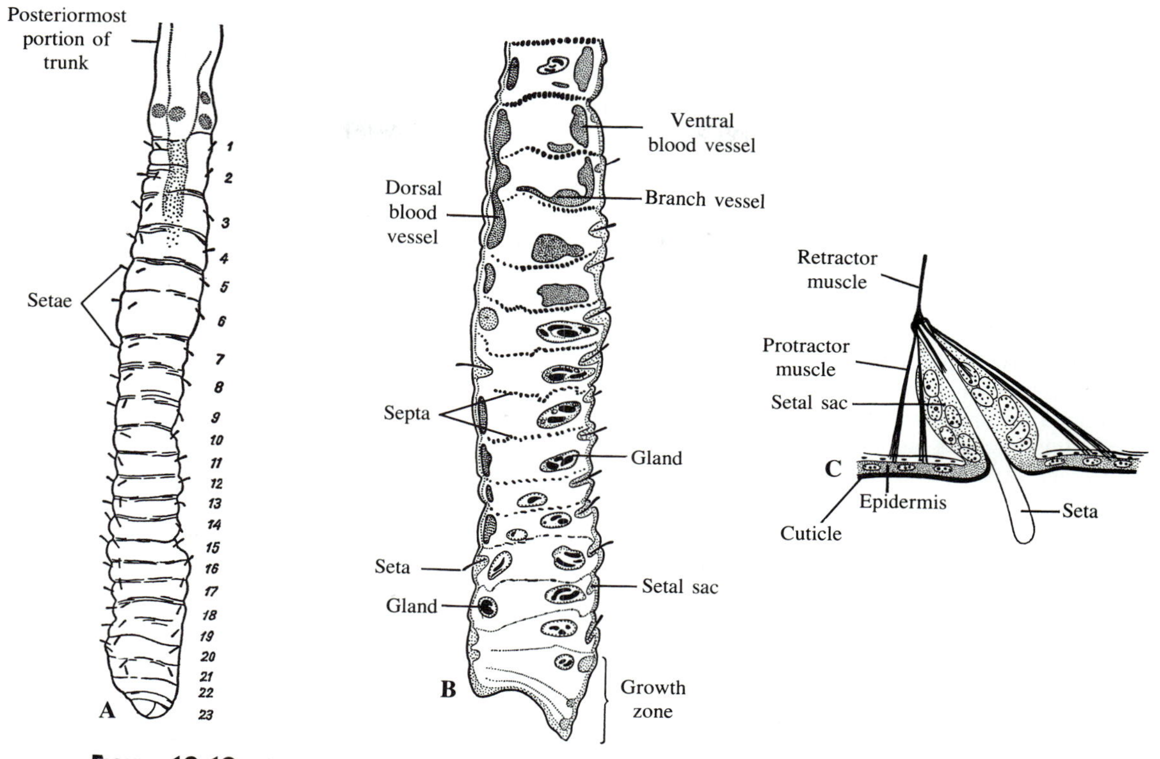

Figure 12.13 *Siboglinum caulleryi.* A. Opisthosoma, dorsal view. In this species, each segment has six nearly evenly spaced setae. B. Opisthosoma, nearly median sagittal section. C. Setal sac, seta, and muscles that protract and retract the seta. (Ivanov, Cahiers de Biologie Marine, *6*.)

itself in the lower portion of the tube. Internally, the segments are separated by septa.

The tubes of pogonophorans (Figure 12.9, D), secreted by epidermal glands, consist largely of hardened proteins but also contain some chitin. As a rule, they are set vertically in bottom mud, with a small portion exposed. The animals can move up and down in their tubes to some extent. The toothed plates of the trunk normally cling to the wall of the tube and allow the anterior and posterior halves of the body to elongate or contract freely. The bridle can probably be used in a similar way.

BODY CAVITIES

Pogonophorans have several sets of body cavities. These originate within mesoderm and are considered to be coeloms, although only certain portions of them have an epithelial lining. The most anterior one, in the cephalic lobe of the forepart, extends into the tentacles (Figure 12.12, A). Farther back in the forepart is a cavity divided along the midline by a dorsoventral mesentery. In some species, there is a small space beneath the heart, which lies between the first and second cavities just described.

The body cavity of the long trunk is also divided by a dorsoventral mesentery, but the separation into right and left halves is not complete. These spaces, moreover, may have connections with the paired spaces of the forepart. At most transverse levels of the trunk, the spaces are nearly filled by organs of the reproductive system, trophosome, and other structures.

The dorsoventral mesentery does not continue into the opisthosoma, so the cavity in each segment is not divided. The septa between segments (Figure 12.13, B) are similar to those of some annelids in that they are composed to a large extent of muscle.

Although there appear to be true epithelial linings in some of the cavities—for instance, on the anterior faces of the septa of the opisthosoma—this is not characteristic of all cavities. What looks like a peritoneal epithelium in pogonophorans generally consists of nucleated, noncontractile portions of muscle cells. This kind of lining is associated with some of the body cavities of not only pogonophorans, but also of vestimentiferans, hemichordates, and annelids. Electron microscopy may reveal that it is a rather general phenomenon in many invertebrates whose coelomic linings have been thought to be epithelial. The old idea that a body cavity is not a true coelom unless it has an epithelial lining will probably be revised.

CIRCULATORY SYSTEM

The circulatory system is closed. The two principal vessels—dorsal and ventral (Figure 12.12, A)—run longitudinally for nearly the entire length of the body. The **dorsal vessel** carries blood anteriorly, and a strongly muscularized portion of it, located in the fore-part just behind the cephalic lobe, functions as a heart. Blood pumped by the heart goes into the afferent vessels of the tentacles, and on returning from the tentacles by the efferent vessels, it enters the **ventral vessel.** In the cephalic lobe, the ventral vessel connects with two large lateral vessels, but in general its blood flow is in the posterior direction. The reproductive organs, trophosome, and other internal structures and tissues are supplied by blood that moves through small vessels and networks that connect the ventral vessel with the dorsal vessel.

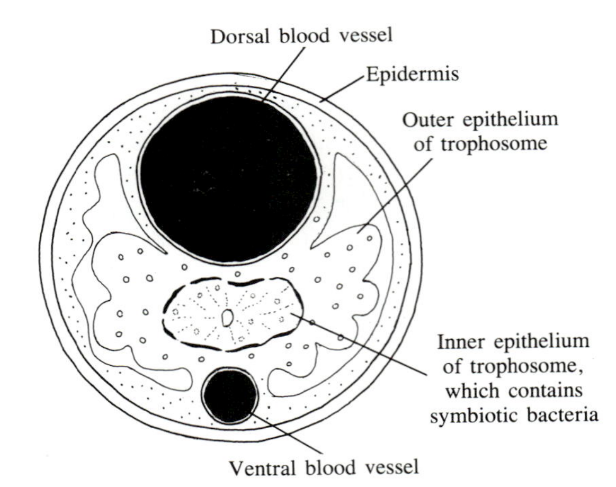

FIGURE 12.14 *Siboglinum fiordicum,* transverse section through the posterior part of the trunk, showing the trophosome; diagrammatic. In some species, the trophosome does not have a cavity. (After Southward, Journal of the Marine Biological Association of the United Kingdom, *62*.)

EXCRETORY ORGANS

There is a pair of excretory organs, which are probably coelomoducts. These start their course within, or close to, the body cavity of the cephalic lobe, and terminate at lateral or dorsolateral pores (Figure 12.12, A). In the class Athecanephria, a substantial part of each excretory organ lies alongside the posterior portion of a lateral blood vessel. The right and left excretory organs may also have a transverse connection across the midline. In the class Thecanephria, the arrangement of the excretory organs is more interesting. Much of each one passes through a lateral diverticulum of the ventral vessel (Figure 12.12, C). The diverticula, called **renal sacs,** are essentially the proximal portions of the lateral vessels of the cephalic lobe.

NUTRITION

Because pogonophorans lack a digestive tract, their nutrition has been the subject of considerable speculation. The idea that these animals secrete enzymes to digest food outside their bodies now seems unlikely, but an ability to take up dissolved organic matter, including protein molecules, has been demonstrated. Some small species, with a diameter of about 0.2 mm, will live for an extended period if they are provided with soluble nutrients.

After symbiotic bacteria were discovered in the trophosomal tissue of Vestimentifera, several species of Pogonophora were examined for similar symbionts. All of them were found to have **intracellular bacteria** in the trophosome. The **trophosome,** which lies in the trunk between the dorsal and ventral blood vessels (Figure 12.14), and which may also partly envelop these vessels, is a cylinder or solid core of tissue, yellowish or brownish, that may often be seen through the body wall of intact specimens. It consists of two layers, interpreted to be epithelia, separated by a complex of blood vessels. The cells of the inner layer are populated by bacteria. It is conjectured that the bacteria are chemoautotrophic, using energy obtained by oxidizing inorganic compounds to synthesize organic compounds, some of which are made available to the host pogonophorans.

The embryonic origin of the trophosome is not yet understood. The possibility that it is an endodermal derivative, and therefore a vestigial gut, has been suggested.

NERVOUS SYSTEM

The principal components of the nervous system (Figure 12.15) are part of the epidermis. Much of the brain consists of a thick swelling on the ventral side of the cephalic lobe. Lateral extensions of the swelling turn dorsally and meet on the midline, so a ringlike structure is formed. Small nerves from both sides of the ring go out into the tentacles. The ventral portion of the brain becomes narrowed into a nerve cord that extends backward for the length of the body. In the opisthosoma, however, there are three ventral nerve tracts, one on the midline and two lateral to it. In each segment,

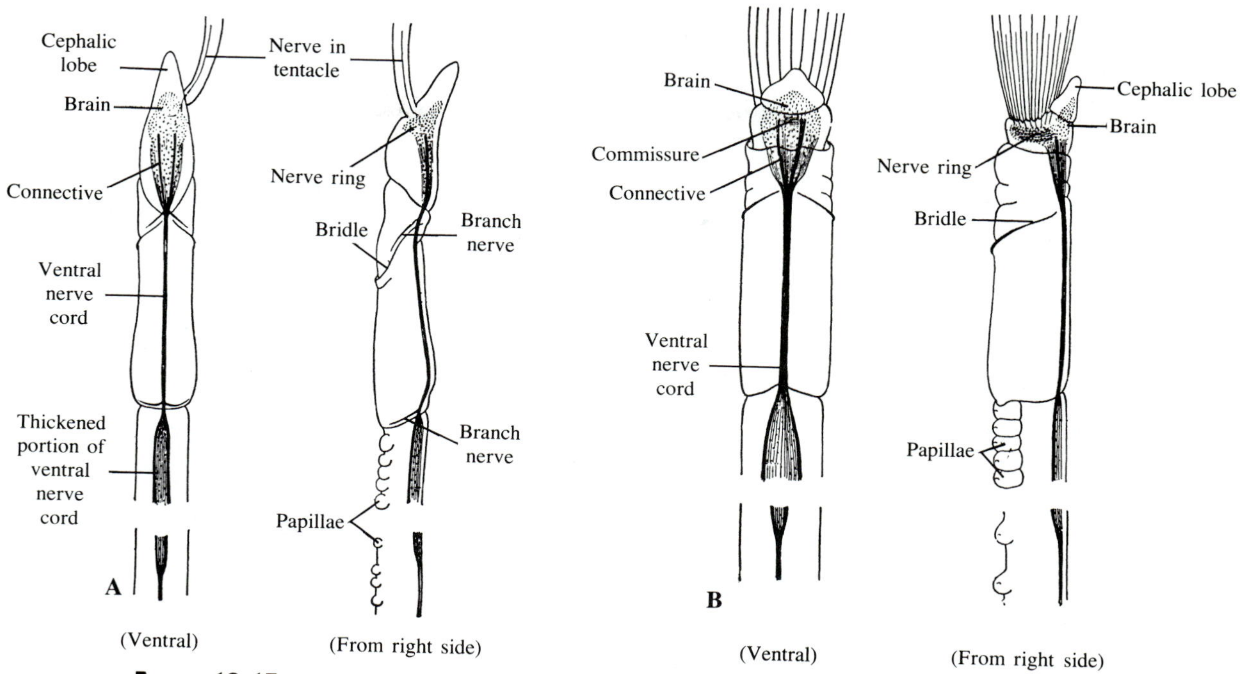

FIGURE 12.15 Major components of the nervous system in the forepart and trunk of pogonophorans. A. *Siboglinum caulleryi*. B. *Polybrachia annulata*. (Ivanov, Pogonophorans.)

between the median and lateral tracts, there is a pair of ganglionlike swellings.

It should be noted that until the segmented opisthosoma of pogonophorans was discovered, these animals were generally thought to be deuterostomes, and to be especially close to hemichordates, in which the principal nerve cord is dorsal. Thus in much of the earlier literature concerning pogonophorans, the side on which the nerve cord is located was considered to be dorsal.

Little is known about sense organs in pogonophorans. There appear to be scattered sensory cells, especially on the cephalic lobe. The ventral ciliated tracts on the trunk are probably sensory structures, for they are extensively innervated.

REPRODUCTION AND DEVELOPMENT

With only one known exception, sexes are separate. The reproductive organs are located in the anterior part of the trunk, and the gonoducts are believed to be coelomoducts. Males have a pair of elongated **testes,** and the **sperm ducts** extend forward to genital pores located laterally near the anterior edge of the trunk (Figure 12.16, A). While the mature sperm are in the ducts, they are packaged into **spermatophores** (Figure 12.16, B).

Females usually have paired **ovaries,** unless one ovary does not develop, a situation typical of certain genera. The **oviducts** at first run backward, then turn forward, ending at genital pores located just in front of the series of large trunk papillae (Figure 12.16, C).

Spermatophores released by male pogonophorans are perhaps caught by the tentacles of females, or perhaps they become tangled with the tubes of females. They have been observed to break down within a few hours, liberating the sperm locked inside them. Eggs are thought to be fertilized after they have been extruded into the tube.

Embryos are incubated in the tube of the female. Much of what is known about the development of pogonophorans has been learned from species of *Siboglinum*. The zygote is oblong and surrounded by a membrane that persists through the early cleavages. The first cleavage produces equal blastomeres, but the second division is unequal (Figure 12.17, A), so at the 4-cell stage there are two smaller cells and two larger cells (Figure 12.17, B). In general, the pattern of cleavage comes closer to being of the radial type than of the spiral type, but from the third cleavage onward the divisions of the blastomeres are not synchronous. The large macromeres, very distinct from the small micromeres (Figure 12.17, C–F), have almost all of the yolk.

When there are about 150 cells, the embryo is a solid gastrula, formed by the overgrowth of certain cells

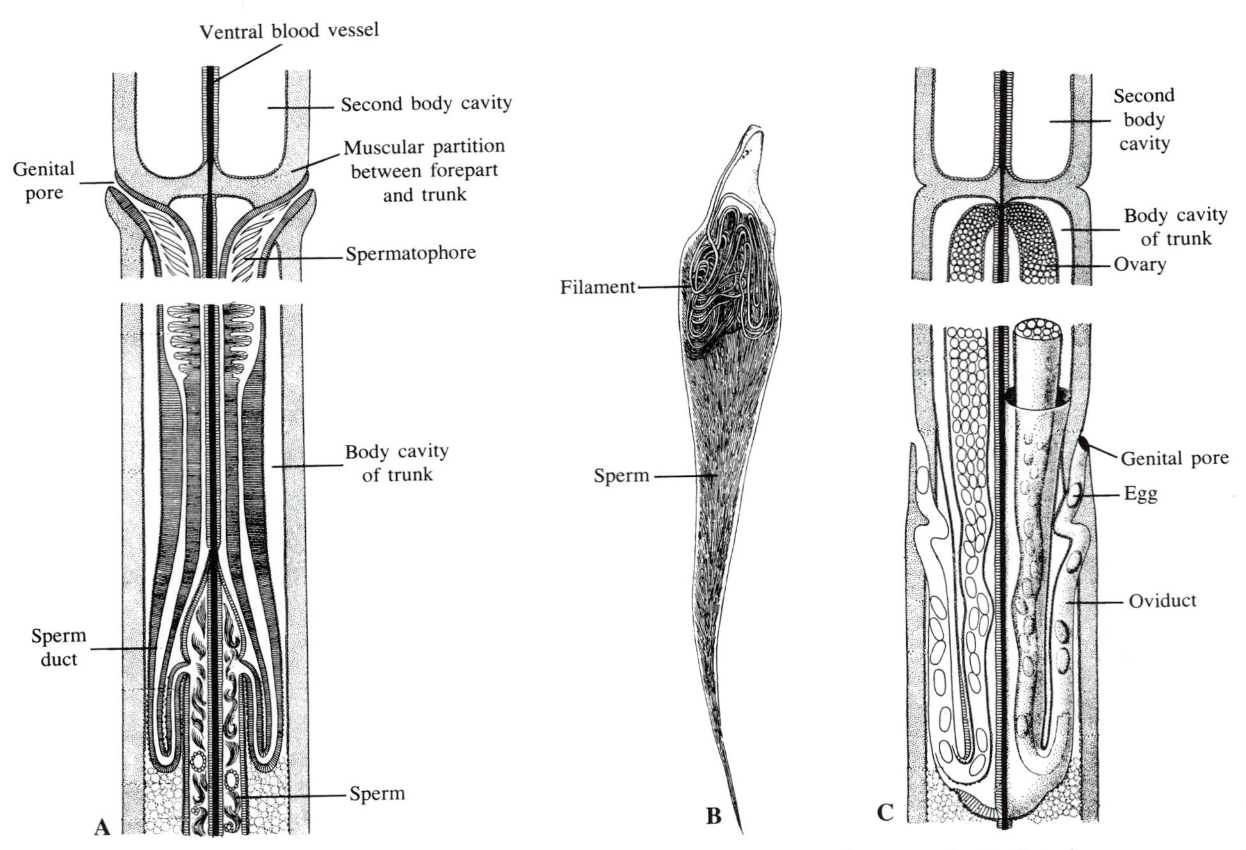

FIGURE 12.16 A. *Lamellisabella zachsi,* male reproductive system; diagrammatic. B. *Spirobrachia grandis,* spermatophore. The filament is extruded after the spermatophore has been released. C. Female reproductive system of a pogonophoran; diagrammatic. (Ivanov, Pogonophorans.)

by others. At this stage the ectodermal cells of the posterior half are, on the average, decidedly larger than those of the anterior half, but there are two large cells on one side near the anterior end. The inner cell mass of the gastrula consists mostly of three larger cells that represent endoderm and two groups of small cells that represent mesoderm.

Two rings of epidermal cells—one near the anterior end and one near the posterior end—sprout cilia, and a ciliated band also develops on one side of the body between the rings (Figure 12.17, G). The body cavities arise by schizocoely. These eventually become organized into the unpaired and paired cavities of the forepart, the paired cavities of the trunk, and the segmental cavities of the opisthosoma. A constriction that develops near the posterior end separates the future opisthosomal portion from the trunk (Figure 12.17, H and I). Setae appear rather early in this portion; they develop within ectodermal sacs. Toothed plates of the girdles begin to appear near the middle of the body, and a tentacle eventually starts to grow out from a primordium just behind the anterior ciliated ring. (*Siboglinum,* unlike most pogonophorans, has only one tentacle when it is mature.) The large endodermal cells are said to break down. Presumably they provide at least some material that can be put to use by other cells of the embryo. The possibility that certain of them give rise to trophosomal tissue needs to be investigated. (In a species of *Oligobrachia,* one of the early stages has a short gut, the ventral mouth being interpreted as the blastopore. The gut cells soon deteriorate, however.)

Blood vessels develop within the mesoderm, and it is assumed that this germ layer is the source of all tissues except the epidermis, nerves, and those tissues in other structures that are clearly derived from ectoderm, such as setal sacs and epidermal glands.

After it settles on an appropriate substratum, the wormlike swimming stage of *Siboglinum* excavates a depression, using ciliary activity and muscular movements. It then burrows deeper, anterior end first (Figure 12.18). Later, it turns around in its burrow, so that its anterior end is uppermost, and tube secretion begins.

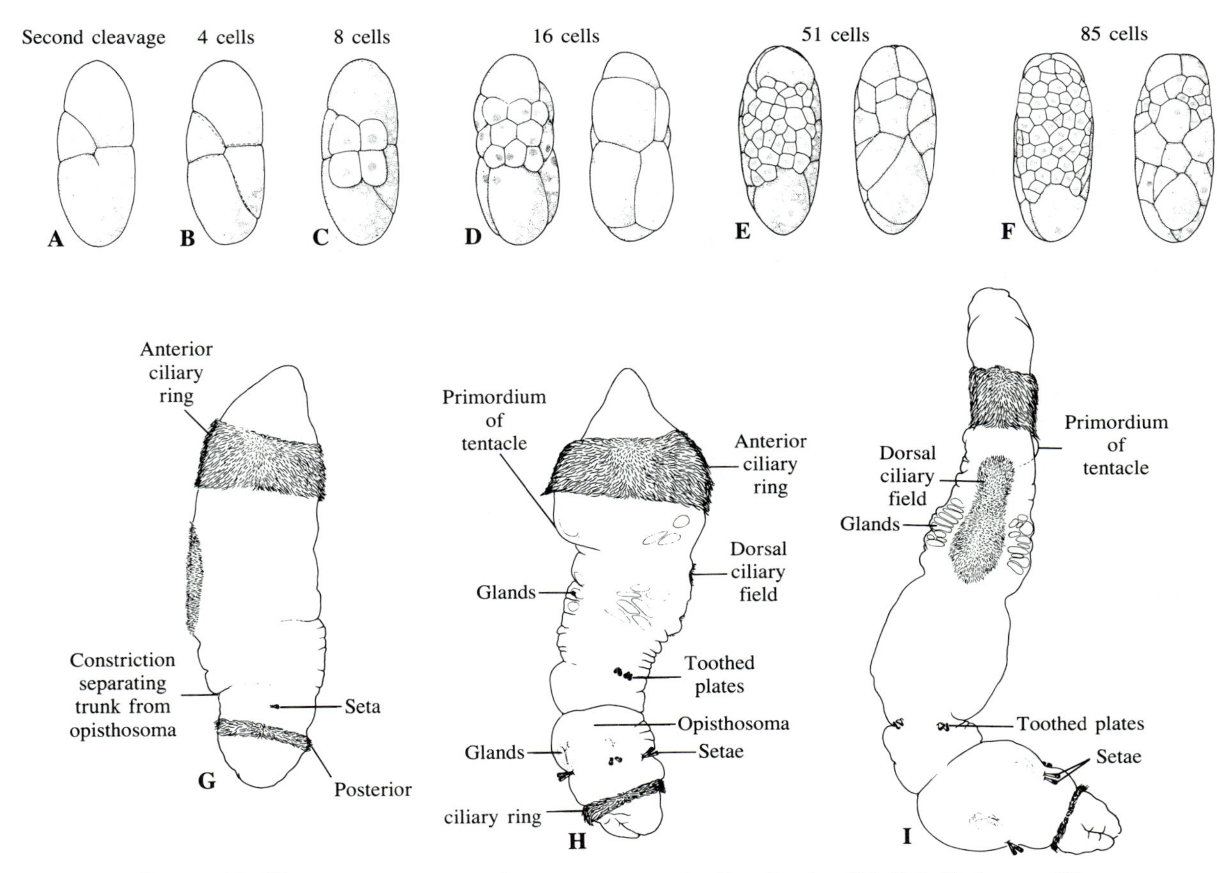

FIGURE 12.17 *Siboglinum fiordicum*. A–F, cleavage. (Bakke, Sarsia, *60*.) G–I. Early wormlike stages. (Bakke, Sarsia, *56*.)

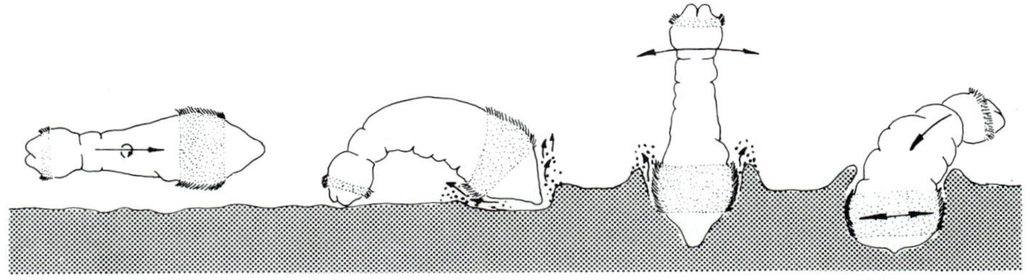

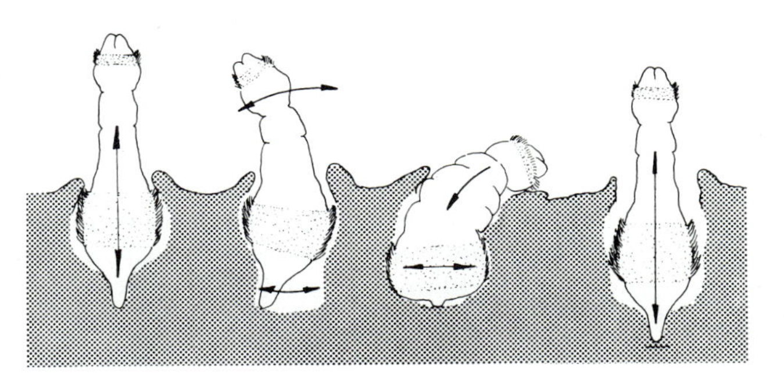

FIGURE 12.18 *Siboglinum fiordicum,* settling and burrowing of early wormlike stage. The cilia of the anterior ring sweep away some sediment, and the pointed anterior end is pressed into the depression. Gradually the worm works its way deeper into the sediment. Eventually it turns around, so that the anterior end is uppermost. (Bakke, Sarsia, *56*.)

CLASSIFICATION

CLASS ATHECANEPHRIA

In the Athecanephria, which includes *Siboglinum* (the largest genus, characterized by a single tentacle) and *Oligobrachia,* the excretory organs lie close to the lateral vessels of the cephalic lobe, but are not enclosed by them. The cephalic lobe, moreover, is distinctly separate from the rest of the forepart. The tentacles, if there are more than one, are not fused together.

CLASS THECANEPHRIA

In this group, which includes the genera *Heptabrachia, Polybrachia, Lamellisabella,* and *Spirobrachia,* a large part of each excretory organ is enclosed by a saclike diverticulum of the ventral blood vessel. The cephalic lobe is not always distinctly separated from the rest of the forepart, and the tentacles are sometimes fused together by cuticular material. The body cavity of the cephalic lobe is horseshoe-shaped, but in *Spirobrachia,* in which the bases of the tentacles form a spiral, the left arm of the cavity is greatly elongated and also spiralled (Figure 12.10, A and B; Figure 12.11).

REFERENCES ON POGONOPHORA

Bakke, T., 1977. Development of *Siboglinum fiordicum* Webb (Pogonophora) after metamorphosis. Sarsia, *63:*65–73.

George, J. D., and Southward, E. C., 1973. A comparative study of the setae of Pogonophora and polychaetous Annelida. Journal of the Marine Biological Association of the United Kingdom, *53:*403–424.

Ivanov, A. V., 1960. [Pogonophorans.] Fauna SSSR, n.s., no. 75 (in Russian). (English translation, 1963, Academic Press, London.)

Ivanov, A. V., 1965. Structure de la région postérieure sétigère du corps des Pogonophores. Cahiers de Biologie Marine, *6:*311–323.

Ivanov, A. V., 1975. Embryonalentwicklung der Pogonophora und ihre systematische Stellung. Zeitschrift für Zoologische Systematik und Evolutionsforschung, Sonderheft: 10–44.

Nørrevang, A., 1970. On the embryology of *Siboglinum* and its implication for the systematic position of the Pogonophora. Sarsia, *42:*7–16.

Nørrevang, A. (editor), 1975. The phylogeny and systematic position of the Pogonophora. Zeitschrift für Zoologische Systematik und Evolutionsforschung, Sonderheft. 143 pp.

Southward, E. C., 1969. Growth of a pogonophore. A study of *Polybrachia canadensis* with a discussion of the development of taxonomic characters. Journal of Zoology, London, *157:*449–467.

Southward, E. C., 1971. Recent researches on the Pogonophora. Oceanography and Marine Biology, An Annual Review, *9:*193–220.

Southward, E. C., 1972. On some Pogonophora from the Caribbean and Gulf of Mexico. Bulletin of Marine Science, *22:*739–776.

Southward, E. C., 1975. A study of the structure of the opisthosoma of *Siboglinum fiordicum.* Zeitschrift für Zoologische Systematik und Evolutionsforschung, Sonderheft: 64–76.

PHYLUM VESTIMENTIFERA

INTRODUCTION

There are only a few known species of Vestimentifera, and some of them are among the animals characteristically found in the vicinity of deep-sea geothermal vents. The first species, described in 1969 as *Lamellibrachia barhami,* was collected off southern California at a depth of about 1100 m. It was assigned to a new order, Vestimentifera, of the Pogonophora. But *Lamellibrachia* did not fit well among pogonophorans, and after other species of vestimentiferans were studied, the group was raised to the rank of phylum.

Compared to pogonophorans, vestimentiferans are large worms. *Lamellibrachia barhami,* for instance, reaches a length of more than 1 m and a diameter of nearly 1 cm. The most spectacular vestimentiferan is *Riftia pachyptila* (Figures 12.19 and 12.20), which occurs around geothermal vents. It was first observed and collected by biologists in the submersible *Alvin,* lowered to a depth of nearly 2500 m in the region of the Galapagos Rift. It is now known to occur as far north as Mexico. *Riftia* reaches a length of about 1.5 m and a diameter of about 3 cm.

Morphological studies on vestimentiferans proceed slowly. Preparing sections from many levels of the long body and then reconstructing the complex internal anatomy from these sections requires much painstaking work. The investigator must, moreover, be cautious in interpreting features that may have been damaged or otherwise modified before preservation. As work on early development of vestimentiferans continues, the derivation of body cavities, excretory organs, and other structures will be better understood, and specialists will be able to decide which components are homologous to those of pogonophorans, annelids, and other phyla.

Before going on to a more detailed account of the morphology of vestimentiferans, it will be best to summarize their main characteristics. Like pogonophorans, they live in secreted tubes that contain chitin, and they do not have a gut. Their bodies are organized into four regions (Figure 12.21): the **obturacular region,** which has many slender tentacles and also the large, bilobed **obturaculum** that serves as a device for closing the tube; the **vestimental region,** expanded laterally into two winglike structures; the **trunk,** which makes up

A B

Figure 12.19 *Riftia pachyptila*. A. Large mass of individuals near a hydrothermal vent in the region of the Galapagos Rift, depth about 2500 m. B. Closer view of a few individuals. Both sides of the obturacular region, extended from the tube, are covered by many thousands of fine tentacles arranged in lamellae. (Photographs courtesy of Robert R. Hessler.)

most of the length of the body; and the **opisthosoma,** a segmented posterior region, some segments of which have numerous setae. Vestimentiferans have a variety of internal spaces, but only certain of these have an epithelial lining that is typical of a true coelom. The circulatory system is well developed, and there is a pair of excretory organs in the obturacular region. The nervous system, which includes a brain in the obturacular region, is essentially part of the epidermis. Sexes are separate, and the gonads are in the trunk region, but the genital pores are in the vestimental region. The trunk contains a mass of tissue called the trophosome; symbiotic bacteria within this are believed to synthesize organic material upon which vestimentiferans depend for their energy and growth requirements.

Figure 12.20 Habitat of *Riftia pachyptila* near a hydrothermal vent. (Arp, Childress, and Fisher, Bulletin 6, Biological Society of Washington.)

General Structure

The most complete anatomical and histological description of a vestimentiferan is that for *Lamellibrachia luymesi*. It is based on studies of a single specimen, about 1 m long, collected at a depth of 500 m off Guyana. The account that follows deals to a considerable extent with this species, but it includes observations on *L. barhami* and *Riftia pachyptila*. The latter, since its discovery in 1979, has been intensively studied with respect to structure and physiology, and much of what has been learned about it is probably applicable to other vestimentiferans.

In *Lamellibrachia* (Figure 12.21), the two stiff lobes of the obturaculum are almost completely surrounded by thousands of slender tentacles that originate at its base. The tentacles are arranged in concentric semicircular series, and the proximal portions of those in each

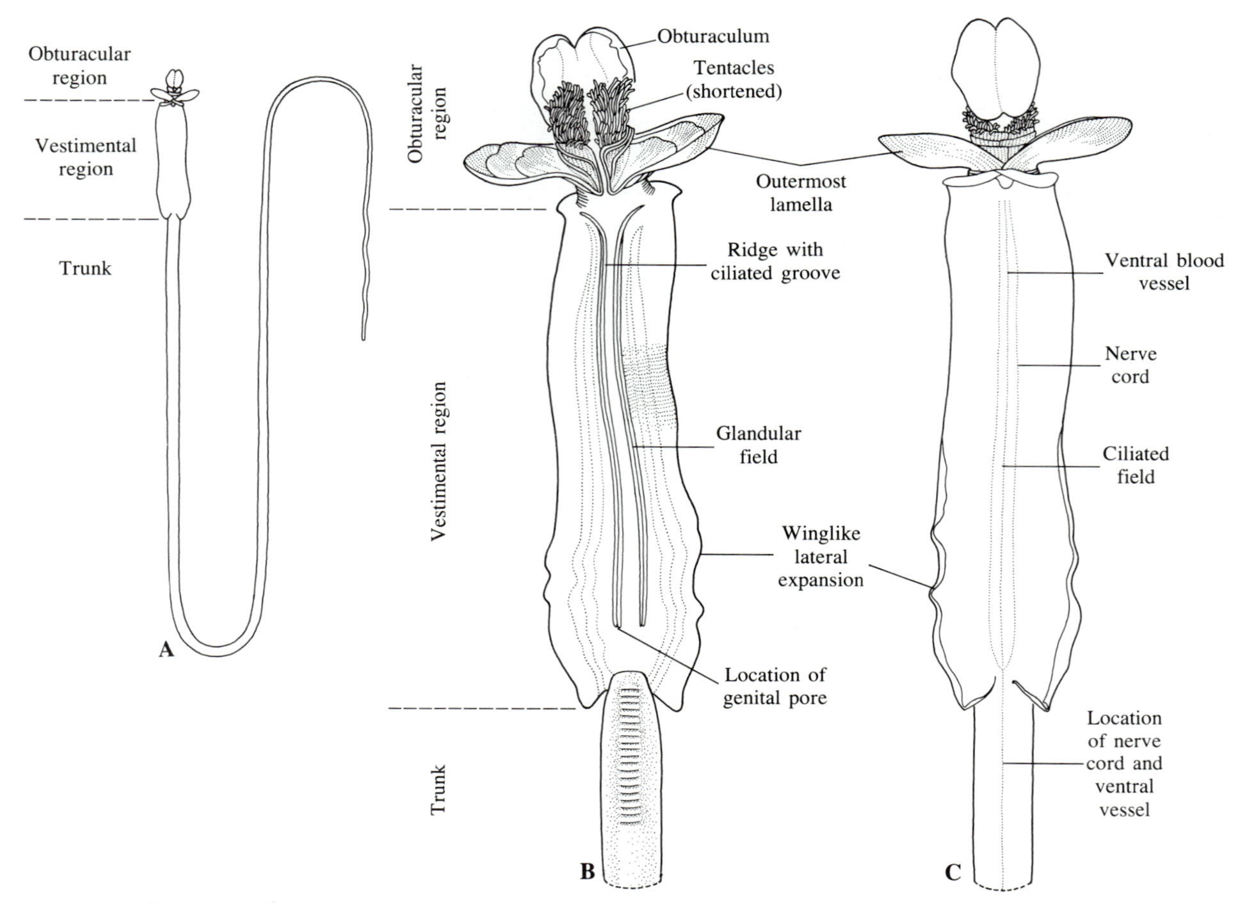

FIGURE 12.21 *Lamellibrachia luymesi*. A. Regions of the body, ventral view; opisthosoma missing (unknown). B. Anterior portion, dorsal view. C. Anterior portion, ventral view. (After van der Land and Nørrevang, Biologiske Skrifter, Kongelige Danske Videnskabernes Selskab, *21*.)

series are united by a common cuticular matrix. The free portions have side pinnules and also lengthwise tracts of cilia. External to the series of tentacles just described are several **lamellae** consisting of tentacles that lack pinnules and cilia and that are united for their entire length by cuticular material. The outermost lamellae on both sides are the largest.

In *Riftia,* the tentacles arise from the sides of the obturaculum, and their bases form nearly vertical rows. The tentacles of each row are united by cuticular material for about half or three-quarters of their length, but the distal portions are free and have flattened pinnules. Among these respiratory tentacles are a few larger tentacles that are believed to have a sensory function.

The vestimental region of *Lamellibrachia* is considerably longer than the obturacular region. When the animal is in the tube, the lateral winglike expansions are folded over the middorsal field. The wings consist mostly of connective tissue, but the upper epithelium is glandular and the lower epithelium, which faces the wall of the tube when the wings are folded, has many deep glands as well as papillae with cuticular plaques (Figure 12.22, A). Much of the tissue medial to the wings is made up of muscle, primarily longitudinal muscle. There is a longitudinal ciliated field on the ventral midline of the vestimental region, and males have two ciliated grooves on the dorsal surface of this part of the body.

On the ventral surfaces of the vestimental wings, and in many other places on the body, there are patches of epidermal cells that are characterized by microvilli (Figure 12.22, B). Between the microvilli is fibrillar material that forms a cuticlelike layer. The microvilli presumably have an absorptive function.

The long trunk has many papillae, some covered by cuticular plaques, others bearing the openings of deep

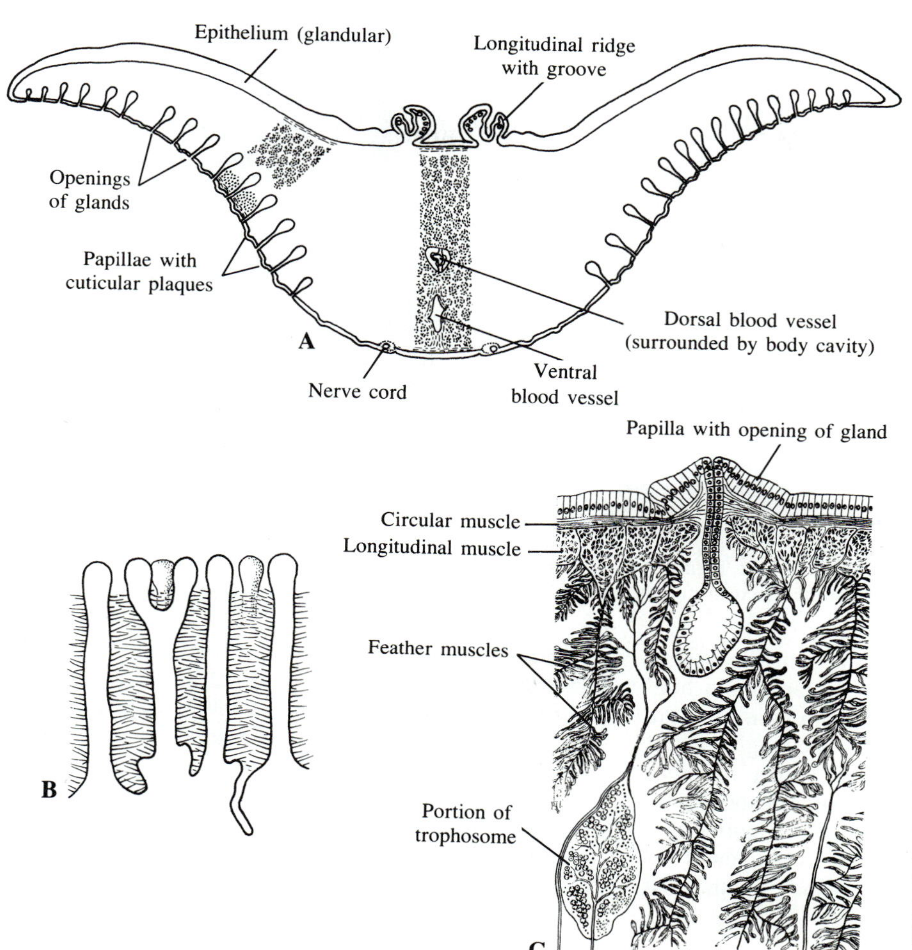

FIGURE 12.22 *Lamellibrachia luymesi*. A. Transverse section through vestimental region; diagrammatic, not all tissues shown. B. Microvilli and cuticlelike deposit characteristic of certain epidermal cells. C. Small portion of body wall of trunk. (After van der Land and Nørrevang, Biologiske Skrifter, Kongelige Danske Videnskabernes Selskab, *21*.)

glands. The epithelium is underlain by circular muscle, internal to which is a compact layer of longitudinal muscle, and then a thick layer of somewhat loosely organized longitudinal muscles. The latter are believed to be the principal retractors of the trunk. They are attached to thin, branching sheets of connective tissue that hang centripetally into the body cavity. Because of their appearance in transverse sections, they are called **feather muscles** (Figure 12.22, C).

The trunk contains the gonads and genital ducts, but much of the space within it is taken up by the trophosome, consisting of cells populated by symbiotic bacteria. A dorsal and a ventral mesentery extend from the body wall into the trophosomal mass (Figure 12.23, A).

The opisthosoma is missing in specimens of *Lamellibrachia luymesi* and *L. barhami*. In *Riftia pachyptila* it consists of about 100 short segments. Those in the anterior half are encircled by bands of closely spaced setae, each band being several setae wide (Figure 12.23, B). The exposed ends of the setae are characterized by numerous small teeth, similar to those on the plates that form two rings on the trunk of pogonophorans. In *Tevnia jerichonana, Oasisia alvinae, Ridgeia piscesae,* and *R. phaeophiale,* there are fewer opisthosomal segments than in *Riftia,* and the setae of the anterior segments are arranged in single rings. The setae themselves, however, are similar to those of *Riftia*.

NERVOUS SYSTEM

The more conspicuous components of the nervous system (Figure 12.24, A) are ventral and essentially part of the epidermis. From the brain, which is located in the obturacular region, two large nerves extend into the lobes of the obturaculum, and many small nerves supply the tentacles. Posteriorly, the brain becomes slightly narrowed into a thick nerve cord that soon divides into the two rather widely spaced nerves of the

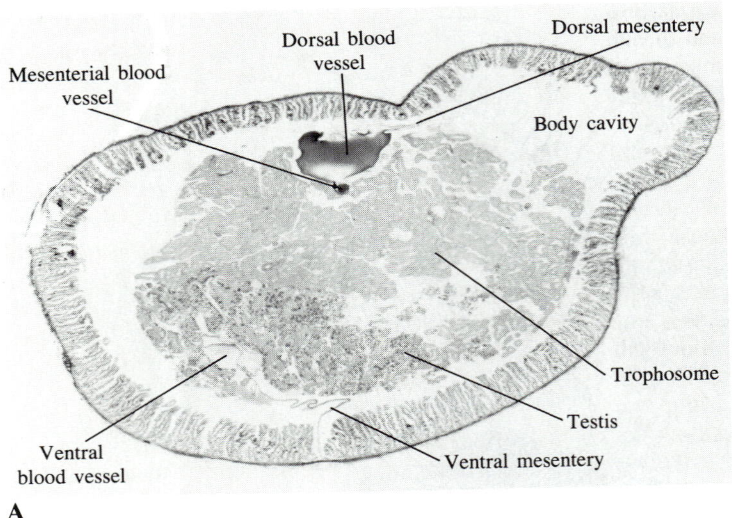

FIGURE 12.23 *Riftia pachyptila*. A. Transverse section through the trunk; photomicrograph. B. Setae of opisthosoma; scanning electron micrograph. (Courtesy of Meredith L. Jones.)

vestimental region. These are tubular and the spaces within them can be traced forward as far as the hind part of the brain. Where the vestimental region joins the trunk, the nerves of the former unite. The single ventral nerve thus formed is at first tubular, then becomes solid. It runs the length of the trunk and opisthosoma.

The brain is pierced by the two body cavities that surround the large blood vessels of the obturacular region. There is also a tunnel through which a bundle of longitudinal muscles passes.

EXCRETORY ORGANS

The two **excretory organs** (Figure 12.24, B), above the brain in the obturacular region, are perhaps coelomoducts, but how they originate is not known. Each consists of many fine tubules and a long duct. The tubules of the right and left excretory organs intermingle, however, so that they form a single mass. The body cavities that pass through the brain also go through the mass of excretory tubules, and it is possible that the free ends of the tubules open into these spaces. In *Lamellibrachia,* the ducts of the excretory organs unite to form a single short duct whose pore is on the dorsal side of the obturacular region; in *Riftia,* the ducts remain separate, so there are two pores (Figure 12.26).

CIRCULATORY SYSTEM

The circulatory system is complex, and there are some discrepancies between accounts of its organization in different species. The description here is therefore concerned only with the basic features.

In the opisthosoma, trunk, and vestimental region, a large dorsal blood vessel (Figure 12.25, A) is encircled over at least much of its length by longitudinal muscles, and this musculature almost certainly serves to propel blood forward. In the anteriormost part of the vestimental region of *L. luymesi,* the dorsal vessel is said to pass through a heartlike structure, the musculature of which is part of the wall of the body cavity that encloses the vessel itself. Such a heart has not been observed in other species. In the posterior part of the obturacular region, the dorsal vessel divides. Two of the large branches are the ones that go through the brain into the lobes of the obturaculum (Figure 12.25, A); the others supply blood to the tentacles (Figure 12.25, A–C). Within each tentacle is an afferent vessel and an efferent vessel, and there are two types of cross-connections between these. Those of one type go through the epidermis of the cylindrical part of the tentacle; those of the other type form networks within the epidermis of the pinnules. It is estimated that *Riftia* has over 200,000 tentacles, and considering that portions of these have pinnules, the surface area for exchange of gases between the environment and blood is enormous.

Blood returning from the tentacles enters vessels that lead to the ventral vessel. This runs continuously through the vestimental region, trunk, and opisthosoma. Branches of it distribute blood to many tissues. There is also a large vessel in each wing of the vestimental region. These vessels join the ones that carry blood from the obturaculum and the tentacles to the ventral vessel. The structure called the *sinus valvatus* by describers of *L. luymesi* probably functions to keep blood in the ventral vessel from flowing anteriorly.

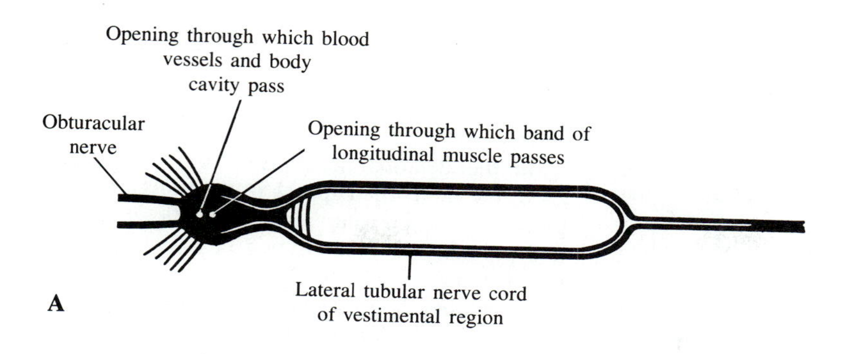

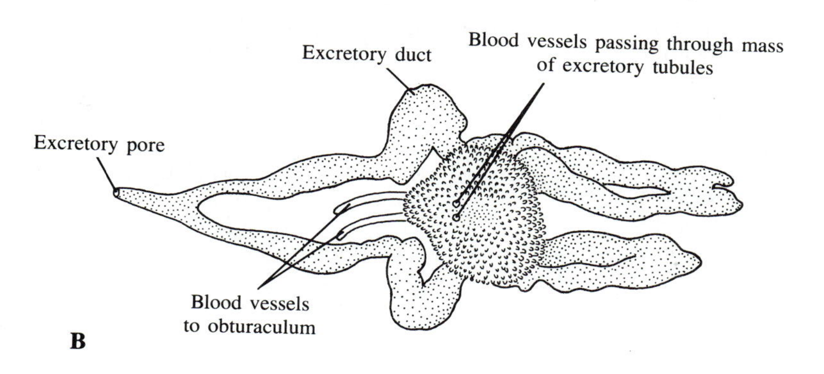

FIGURE 12.24 *Lamellibrachia luymesi*. A. Main components of the anterior part of the nervous system; diagrammatic. (Van der Land and Nørrevang, Biologiske Skrifter, Kongelige Danske Videnskabernes Selskab, *21*.) B. Excretory organs, ventral view. (After van der Land and Nørrevang.)

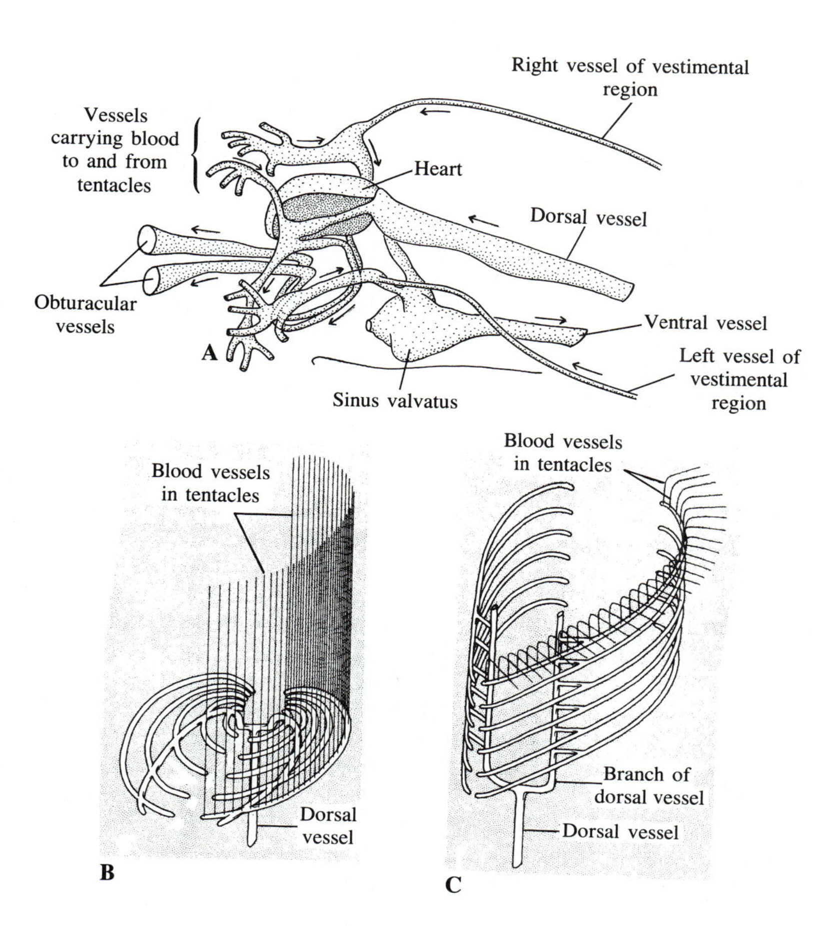

FIGURE 12.25 A. *Lamellibrachia luymesi*, blood vessels in anterior part of vestimental region and posterior part of obturacular region; diagrammatic. Arrows show the probable direction of blood flow. (After van der Land and Nørrevang.) B. *Lamellibrachia barhami*, afferent blood supply to tentacles; diagrammatic. C. *Riftia pachyptila*, afferent blood supply to tentacles; diagrammatic. In *Riftia*, the bases of the lamellae in which the tentacles are arranged on both sides of the obturaculum are more or less vertical with respect to the long axis of this structure, whereas in *Lamellibrachia* the lamellae originate behind the obturaculum. (B and C, Jones, Bulletin 6, Biological Society of Washington.)

Among the branches of the ventral vessel are those that supply the trophosome. This structure is thoroughly pervaded by small vessels. Some of the blood that passes through the trophosome goes to the dorsal vessel; some of it goes first to the mesenteric vessel, which lies just beneath the dorsal vessel. Either way, however, it generally passes through networks of small vessels in the tissue between the dorsal vessel and mesenteric vessel. Within the trunk portion of the dorsal vessel, alongside the openings of the vessels that enter it from below, is a longitudinal fold of tissue that probably functions as a valve to prevent backflow when the dorsal vessel contracts to pump blood forward.

The two large vessels in the obturaculum appear not to branch. While they may be important in supplying oxygen and nutrients to the tissue, another likely function is to serve as a hydrostatic skeleton for expansion of the obturaculum. This part of the body has muscles that can contract it, but no muscles that can cause it to expand. Conceivably, a vestimentiferan could spread its obturacular lobes by forcing blood into the large vessels. The fact that there is an extensive sphincter around each of these vessels, near the place where they emerge from the brain, lends some support to this idea.

REPRODUCTIVE SYSTEM

Some aspects of the organization of the reproductive system are still poorly understood and controversial. In the male of *Riftia*, the tubules within which spermatogenesis takes place form a rather large and compact mass beneath the trophosome. The two sperm ducts running through the mass have connections with the tubules. The genital pores are at the posterior ends of the two ciliated grooves located on the dorsal surface of the vestimental region.

In *Lamellibrachia luymesi* and *L. barhami*, the arrangement of the sperm ducts and the genital pores is similar to that in *Riftia*. The relationship of sperm-producing tissue to the sperm ducts, however, is different. In *L. barhami*, the germinal epithelium from which the sperm are derived is said to be located in the posterior portions of the two sperm ducts; in *L. luymesi*, numerous scattered testes in the body cavity of the trunk region are said to produce sperm that first are free in the body cavity, then enter the sperm ducts through pores. Descriptions of both species state that there are many sperm-storing diverticula along the sperm ducts.

In females, the vestimental region has no ciliated grooves leading to the two genital pores. The ovaries and oviducts lie beside one another, and the ovaries are thought to be anteriorly directed continuations of the oviducts. In *Lamellibrachia barhami*, the right ovary–oviduct complex is reported to be less well developed than the left one, and is perhaps nonfunctional, but this asymmetrical arrangement needs to be confirmed.

BODY CAVITIES

In the detailed description of *Lamellibrachia luymesi*, the fluid-filled spaces in the trunk, vestimental region, and obturacular region are shown to be part of one probably continuous body cavity. The question of whether any or all of the spaces may be regarded as true coelomic spaces has not, however, been answered definitively. In *Riftia*, the body cavities are said to be arranged in a different manner, and to form five separate divisions (Figure 12.26). One division consists of the cavities in the segments of the opisthosoma. These are separated from one another by septa, and are also partly subdivided by rings of circular muscle. The septa

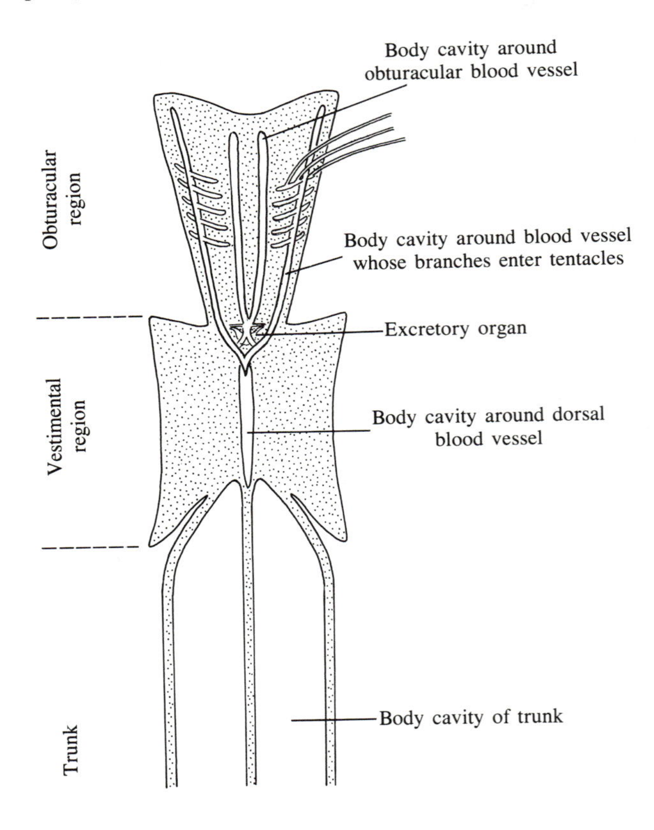

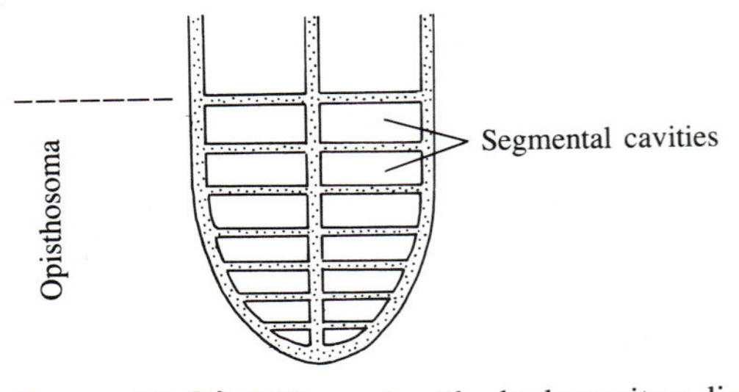

Figure 12.26 *Riftia pachyptila*, body cavites; diagrammatic. (After a drawing lent by Meredith L. Jones.)

themselves have muscles on both their anterior and posterior faces. No peritoneal lining has been seen in the opisthosoma, but perhaps it is thin and thus difficult to identify.

The body cavity of the trunk region has some epithelial linings that resemble a peritoneum. These are present, for instance, around the trophosome and sperm-producing tubules. But the feather muscles and some other structures are not lined.

Another cavity in *Riftia* is restricted to the posterior half of the vestimental region, where it encloses the dorsal vessel. The peripheral portion of this space has an epithelial lining, but there is no lining around the vessel. A fourth cavity, which surrounds the dorsal vessel in the anterior part of the vestimental region, is probably continuous with the spaces that extend into the tentacles. In the tentacles, at least, the spaces are lined by a thin layer of muscle cells, rather than by an epithelium.

The remaining body-cavity complex of *Riftia* consists of the spaces that enclose the two large obturacular blood vessels. These do appear to have continuous epithelial linings.

It should be noted that vestimentiferans are not the only animals in which certain of the spaces considered to be part of a coelom are lined by an epithelium and others are not. The same somewhat ambiguous situation exists in pogonophorans and some annelids.

So far as is known, the fluid in the body cavities of vestimentiferans is the same as the blood.

Nutrition of Vestimentifera

It is believed that the organisms associated with deep-sea vents are supported primarily by activities of chemosynthetic bacteria. An abundant component of the hot water that is spewed from the vents is hydrogen sulfide (Figure 12.20). By oxidizing this and perhaps other inorganic compounds, the bacteria obtain energy to reduce carbon dioxide and synthesize organic compounds. Thus, in darkness, these organisms accomplish what photosynthetic organisms do with light energy.

The animal life around the vents includes not only vestimentiferans, but also molluscs, polychaete annelids, crustaceans, enteropneusts, and fishes. As primary producers, the bacteria would provide food for some of the invertebrates, which would then be consumed by carnivores and scavengers or decomposed by heterotrophic microorganisms.

But how do vestimentiferans, which have no digestive tract, obtain the organic compounds they need? It is almost certain that the intracellular bacterial symbionts in the trophosomal tissue of the trunk are primary producers, much like free-living chemosynthetic bacteria. The blood of *Riftia* has been shown to have a strong affinity for oxygen and can carry considerable sulfide. The bacteria, supplied with oxygen and sulfide by way of the intricate vascular system of the trophosome, have the reactants they need for producing energy and thus for synthesizing organic compounds. The presence of elemental sulfur in the trophosome is an indication that oxidation of sulfide does indeed take place. We may expect that further research will explain how appropriate forms of nitrogen and other elements become available to the bacteria so that they can synthesize amino acids and proteins.

All vestimentiferans whose trophosomes have been examined for symbiotic bacteria have them; in other words, the bacteria are not limited to those vestimentiferans that live in the vicinity of geothermal vents. Among other invertebrates that are found close to vents, at least two—the bivalves *Bathymodiolus thermophilus* and *Calyptogena magnifica*—have chemosynthetic bacterial symbionts; the bacteria live in the tissues of the ctenidia. In cold-water habitats, pogonophorans, some bivalves, and some oligochaetes have comparable symbionts.

In connection with pogonophorans, it was mentioned that the trophosome may be an endodermal derivative and therefore a vestigial gut. An intriguing recent discovery with respect to vestimentiferans is that very young specimens, taken from small tubes attached to tubes of mature worms, have a complete and functional gut. The external ciliation of individuals less than 0.5 mm long suggests, in fact, that these have developed from trochophore larvae. The gut persists for a time after the trophosome becomes prominent and its cells have become populated by endosymbiotic bacteria. The way segments appear on a young worm strongly suggests that vestimentiferans and pogonophorans are closely related. Tentacles quickly become numerous, and the vestimentum and obturaculum become well established by the time a juvenile is about 1 mm long. Interesting changes in the arrangement of setae take place during early development. At first there are only hairlike bristles and four-toothed hooks; many-toothed hooks appear later.

Classification

The known species of Vestimentifera are presently assigned to five genera. These have been distributed among higher taxa as follows.

CLASS AXONOBRANCHIA

The tentacles arise from the lateral surfaces of the obturaculum.

Order Riftiida

Riftia pachyptila is the only described representative.

Class Basibranchia

The tentacles originate at the base of the obturacular region.

Order Lamellibrachiida

Lamellibrachia and *Escarpia,* both characterized by a single excretory pore, are placed in this order.

Order Tevniida

This order includes *Tevnia* and *Ridgeia,* which have two excretory pores.

References on Vestimentifera

Arp, A. J., Childress, J. J., and Fisher, C. R., Jr., 1985. Blood gas transport in *Riftia pachyptila*. Biological Society of Washington, Bulletin 6:289–300.

Cavanaugh, C. M., 1985. Symbiosis of chemoautotrophic bacteria and marine invertebrates from hydrothermal vents and reducing sediments. Biological Society of Washington, Bulletin 6:373–388.

Grassle, J. F., 1982. The biology of hydrothermal vents: a short summary of recent findings. Marine Technical Society Journal, *16:*33–38.

Jones, M. L., 1981. *Riftia pachyptila,* new genus, new species, the vestimentiferan worm from the Galapagos Rift geothermal vents (Pogonophora). Proceedings of the Biological Society of Washington, *93:*1295–1313.

Jones, M. L., 1981. *Riftia pachyptila* Jones: observations on the vestimentiferan worm from the Galapagos Rift. Science, *213:*333–336.

Jones, M. L. (editor), 1985. Hydrothermal vents of the eastern Pacific: an overview. Biological Society of Washington, Bulletin 6. vii + 547 pp.
(Contains several articles dealing with morphology, systematics, and physiology of vestimentiferans, especially *Riftia pachyptila;* three of the articles are listed separately in this bibliography.)

Jones, M. L., 1985. On the Vestimentifera, new phylum: six new species and other taxa, from hydrothermal vents and elsewhere. Biological Society of Washington, Bulletin 6:117–158.

Jones, M. L., 1985. Vestimentiferan pogonophores: their biology and affinities. *In* Morris, S. C., George, J. D., Gibson, R., and Platt, H. M., The Origins and Relationships of Lower Invertebrates, pp. 327–342. (Systematics Association Special Volume no. 28.) Oxford: Clarendon Press.

Jones, M. L., 1989. The Vestimentifera, their biology and systematic and evolutionary patterns. Oceanologica Acta, *12*.

Jones, M. L., and Gardiner, S. L., 1985. Light and scanning electron microscopic studies of spermatogenesis in the vestimentiferan tube worm *Riftia pachyptila* (Pogonophora: Obturata). Transactions of the American Microscopical Society, *104:*1–18.

Jones, M. L., and Gardiner, S. L., 1988. Evidence for a transient digestive tract in Vestimentifera. Proceedings of the Biological Society of Washington, *101:*423–433.

Southward, E. C., 1988. Development of the gut and segmentation of newly settled stages of *Ridgeia* (Vestimentifera): implications for relationship between Vestimentifera and Pogonophora. Journal of the Marine Biological Association of the United Kingdom, *68:*465–487.

Van der Land, J., and Nørrevang, A., 1977. Structure and relationships of *Lamellibrachia* (Annelida, Vestimentifera). Biologiske Skrifter, Kongelige Danske Videnskabernes Selskab, København, *21*(3):1–102.

Webb, M., 1969. *Lamellibrachia barhami* gen. nov., sp. nov. (Pogonophora), from the northeast Pacific. Bulletin of Marine Science, *19:*18–47.

Webb, M., 1977. Studies on *Lamellibrachia barhami* (Pogonophora). II. The reproductive organs. Zoologische Jahrbücher, Abteilung für Anatomie und Ontogenie der Tiere, *97:*455–481.

Summary

1. Sipunculans and echiurans, which are strictly marine, are unsegmented and mostly stout-bodied worms. The life cycle usually includes a trochophore larva similar to that of polychaete annelids.

2. Members of both phyla are primarily nestlers or burrowers in soft substrata, and they have a complete gut. They are deposit-feeders, but they obtain food differently. In an echiuran, a portion of the body anterior to the mouth is drawn out into a proboscis. Detritus trapped in mucus secreted by the proboscis is directed to the mouth by activity of cilia. In sipunculans, numerous small tentacles surround the terminal mouth and are used in ciliary feeding.

3. The anterior portion of a sipunculan, called the introvert, can be invaginated into the rest of the body. The gut, moreover, is nearly U-shaped, so that the anus is situated rather far forward, instead of at the posterior end, as it is in an echiuran.

4. Pogonophorans and vestimentiferans are greatly elongated worms that live in secreted tubes. All of them are marine and most are found at considerable depths.

5. Pogonophorans are slender, generally less than 1 mm in diameter, and are divided into a forebody that has tentacles, a long trunk, and a short opisthosoma. The latter is segmented and has setae, so it resembles part of an annelid. Vestimentiferans are large, often exceeding a thickness of 2 cm, and are divided into an obturacular region with thousands of tentacles, a vestimental region with winglike lateral expansions, a long trunk, and an opisthosoma similar to that of pogonophorans.

6. There is no digestive tract in pogonophorans and vestimentiferans. Within the tissue of a structure called the trophosome, which occupies a considerable part of the trunk region, there are symbiotic bacteria that are believed to be chemoautotrophic, capable of making organic nutrients for themselves and for their hosts. Uptake of organic nutrients has also been demonstrated in pogonophorans. Several of the few known species of vestimentiferans are among the animals characteristically found around geothermal vents.

13

Phylum Mollusca

SNAILS, CLAMS, OCTOPUSES, AND THEIR RELATIVES

- Phylum Mollusca
  - Class Monoplacophora
  - Class Polyplacophora
  - Class Aplacophora
    - Subclass Solenogastres
    - Subclass Caudofoveata
  - Class Gastropoda
    - Subclass Prosobranchia
      - Order Archaeogastropoda
      - Order Patellogastropoda
      - Order Mesogastropoda
      - Order Neogastropoda
    - Subclass Opisthobranchia
      - Order Nudibranchia
      - Order Cephalaspidea
      - Order Thecosomata
      - Order Gymnosomata
      - Order Notaspidea
      - Order Sacoglossa
      - Order Anaspidea
      - Order Pyramidellacea
      - Order Acochlidiacea
    - Subclass Gymnomorpha
      - Order Onchidiacea
    - Subclass Pulmonata
      - Order Basommatophora
      - Order Archaeopulmonata
      - Order Stylommatophora
      - Order Systellommatophora
  - Class Bivalvia
    - Subclass Palaeotaxodonta
    - Subclass Cryptodonta
    - Subclass Pteriomorpha
    - Subclass Heterodonta
    - Subclass Anomalodesmata
  - Class Scaphopoda
  - Class Cephalopoda
    - Subclass Nautiloidea
    - Subclass Ammonoidea
    - Subclass Coleoidea

Introduction

With more than 100,000 living species, molluscs make up the second largest phylum of animals (arthropods are by far the largest phylum). Their extensive evolutionary diversification has led molluscs to remarkable success on land as well as in the sea and fresh water. As herbivores, carnivores, scavengers, and filter feeders, molluscs consume much organic matter, and they are in turn eaten by other animals. Thus they are important constituents of food chains, and in some habitats they form a large part of the biomass. Many species provide food for humans, and we have learned how to promote the production of some of them.

Shells of molluscs have been used as money and in the manufacture of buttons and jewelry, and they have brought pleasure to collectors fascinated by their forms and colors. Shell collecting, in fact, stimulated scientific study of molluscs long before most other phyla of invertebrates were given serious attention.

On the other hand, certain land snails and slugs are serious pests on farms and in gardens, and there are small marine bivalves that destroy pilings and other wooden structures by burrowing into them. Numerous land and freshwater species, mostly snails, are intermediate hosts for flukes that parasitize humans and domestic animals.

A Generalized Model Mollusc

A basic body plan can be detected in all of the seven surviving classes of molluscs. For explaining the plan, it is convenient to use a generalized model (Figure 13.1). Diagrams similar to this have been used to portray a hypothetical ''ancestral mollusc.'' Although the generalized model does illustrate the characteristics common to most groups of living molluscs, it is best not to assume that it represents the prototype.

Among the molluscan features shown by the model are bilateral symmetry, lack of segmentation, and the organization of the body into four main regions: the **head, foot, visceral mass,** and **mantle.** The development of each of these regions varies, and in certain specialized molluscs one or more regions may be greatly modified. Nevertheless, all four are typical of the molluscan body plan, and an appreciation of their spatial relationships is essential preparation for understanding diversity within the phylum.

The head has the **mouth** and accessory structures concerned with feeding. Most notable of these is the **radula,** a cuticular ribbon studded with hard teeth, which is found in the foregut of nearly all molluscs other than those of the class Bivalvia. The head also has important ganglia of the nervous system, as well as sense organs, although eyes or obvious tentacles are not necessarily present.

The foot is generally muscular and used for crawling, burrowing, or clinging. In some bivalves, such as sea mussels and scallops, an important function of the foot is secretion of material by which the animal maintains an attachment to rock, shell, wood, or some other firm substratum.

The visceral mass is almost always a proportionately large part of the body. Much of it is filled up by the gut and digestive gland, the latter consisting of diverticula of the stomach. It also contains the excretory organs, heart, pericardial cavity, gonads, and other components of the reproductive system.

The mantle covers the visceral mass. Its epidermis, which secretes the **shell,** is continuous with that of the

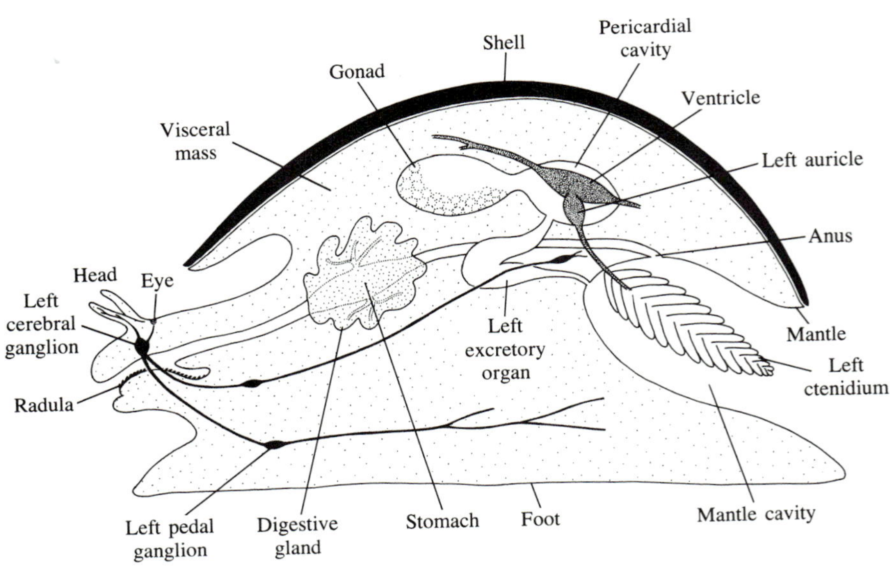

Figure 13.1 Diagram of a generalized model mollusc, viewed from the the left side.

foot and head region. Typically, a portion of the mantle hangs down as a skirtlike fold, so there is at least a narrow space between the fold and the rest of the body. This space, called the **mantle cavity,** may extend all around the lateral margins of the body, as it does in the model, or it may be somewhat localized.

Note the structures located in the posterior part of the mantle cavity of the model: the **anus,** a pair of **excretory organs,** and a pair of gills of a type called **ctenidia.** This arrangement was probably characteristic of some early molluscs, and a similar plan is found in a few living species. The diversity of molluscs, however, is to a considerable extent evident in modifications of the mantle cavity and structures within it. These modifications are discussed below in connection with the groups of which they are typical. For the time being, it is important to understand what the mantle cavity is, and that the ctenidia, anus, and openings of the excretory organs are in it.

Another probably ancient feature of molluscs is shown in the model. This is the continuity of the three derivatives of the embryonic coelom: the **pericardial cavity,** in which the heart lies, the gonads, and the excretory organs. The latter, usually called **renal organs** or **kidneys,** are **coelomoducts,** and it is likely that they originally served not only for eliminating nitrogenous wastes, but also as gonoducts. They do, in fact, serve this dual role in some living molluscs.

The relationship of the ctenidia to the **heart** deserves comment, for this too is part of the molluscan body plan. Blood that has been oxygenated in each ctenidium is carried by an efferent vessel to one of the two **auricles.** It then enters the **ventricle,** which pumps the blood out to the tissues by one or two main arteries.

In the course of molluscan evolution, there have been many changes in the arrangement of the nervous system. The model nevertheless shows certain features that are characteristic of living molluscs and that probably derive from early stages in the evolution of the phylum. In the head, two **cerebral ganglia** are joined together by a transverse commissure. From each of them, a nerve cord leads down into the foot where it enlarges into a **pedal ganglion;** another cord, ganglionated at intervals, extends from each cerebral ganglion into the visceral mass.

In some molluscs, particularly chitons, the organization of the body with respect to the longitudinal axis is obvious. The mouth is near one end and the anus is near the other end. There is no doubt about what is anterior, posterior, dorsal, ventral, right, or left. This is not the case in snails, octopuses and squids, and certain bivalves. One reason for this is that in the classes Gastropoda, Cephalopoda, and Bivalvia the body plan is to a considerable extent organized around a secondary axis, which is dorsoventral. The tendency for visceral organs to be piled up around the dorsoventral axis is most pronounced in octopuses and squids. The true anterior and posterior ends of these animals are close together, whereas the ventral surface and dorsalmost part of the visceral mass are far apart. In snails, much of the digestive system, excretory system, and reproductive system is concentrated in an elevated visceral mass, which occupies most of the space in the typically coiled shell of a snail. A further complication in the orientation of snails is **torsion,** which is the shift, during development, of the most prominent portion of the mantle cavity (which has the anus, openings of the excretory organs, and ctenidia) from its originally posterior position to a location above the head. Torsion and coiling of the visceral mass impose changes in orientation, but they do not alter the basic bilateral symmetry of a snail.

ADDITIONAL GENERAL FACTS ABOUT MOLLUSCS

The model mollusc is helpful in explaining the spatial relationships of the main body regions, the mantle cavity, ctenidia, coelomic derivatives, and some other internal organs. However, to appreciate more fully how certain features have become modified in the seven classes of the phylum, those features should be discussed in more detail.

SHELL

The shell is generally the most obvious feature of a mollusc, although it has disappeared or become greatly reduced in some members of the phylum. It is secreted by the mantle and usually has three distinct layers (Figure 13.2). The outermost layer, the **periostracum,** is composed of a durable organic material called **conchiolin.** The amount of periostracum varies according to the kind of mollusc, but ranges from an almost imperceptible transparent coating to a fibrous sheath that may be elaborated into hairs or even into flexible, spinelike outgrowths.

Beneath the periostracum is the **prismatic layer.** This consists almost entirely of polygonal prisms of calcium carbonate whose long axes are at right angles to the periostracum. Some conchiolin may dip down between the crystals. The extent to which a shell is thickened depends largely on the activity of the mantle edge in depositing the prismatic layer. Underlying the prismatic layer is the **nacreous layer,** laid down in thin sheets by all parts of the mantle epithelium that are in contact with the shell. In abalones, pearl mussels, and other molluscs in which the nacreous layer is iridescent, the color is due to microscopic irregularities that break up light into its spectral components. In most species,

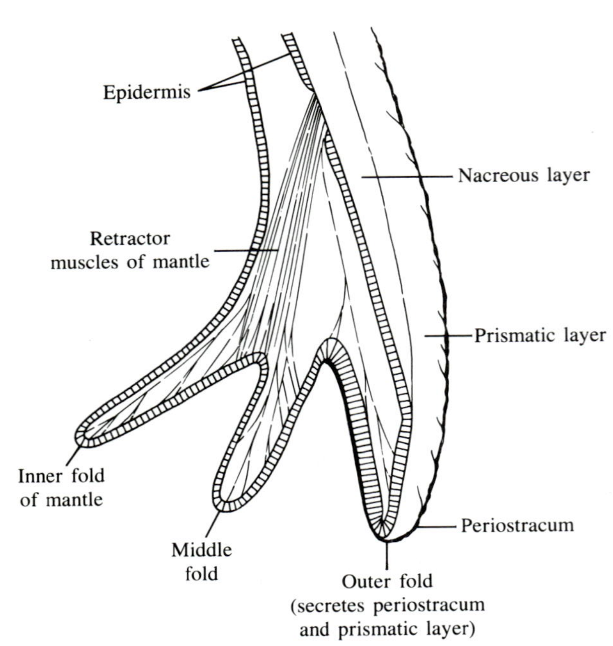

FIGURE 13.2 Margin of the mantle and shell of a mollusc (in this case, a bivalve).

the nacreous layer is so smooth that it is no prettier than the inside of a kitchen sink. When some portion of a shell other than the edge is damaged, the mantle epithelium that would normally produce the nacreous layer deposits enough additional calcareous material to effectively patch up the hole or crack.

MANTLE CAVITY AND CTENIDIA

The extent of the mantle cavity varies greatly. In sea slugs it has been so reduced that it no longer exists, and in air-breathing snails and slugs much of it has been modified to form a lung. Most molluscs, however, have a recognizable mantle cavity and at least one ctenidium. Although ctenidia are respiratory organs, it is important not to confuse them with various other kinds of gills in molluscs that are not comparable to ctenidia in structure or origin.

A generalized ctenidium consists of two rows of ciliated leaflets; the leaflets of one row alternate with those of the other. The cilia move water across the leaflets, facilitating the exchange of oxygen and carbon dioxide. When water passes between the leaflets from below and blood is delivered first into their upper portions, the principle of counterflow operates and the rate of gas exchange is maximized. The layout of the mantle cavity and the ctenidia within it is such that well-defined inhalant and exhalant currents are created; the inhalant current brings in clean water and the exhalant current sweeps away wastes. The evolution and life styles of molluscs are closely linked to modifications in the structure and arrangement of ctenidia, including the total loss of ctenidia in certain groups.

DIGESTIVE SYSTEM

The feeding habits of molluscs are extremely varied, and there are many diverse specializations in the structures concerned with procuring and processing food. These specializations are found not only in the head region, pharynx, and succeeding portions of the gut, but also in the mantle, ctenidia, and other organs.

Radula and Jaws

The **buccal cavity** (or **pharynx**) of a mollusc is a foregut derived from a stomodeal invagination. Its epithelial lining is therefore of ectodermal origin. The radula (Figure 13.3), a strip of cuticle to which teeth are attached, is part of a complex structure called the **buccal mass,** or *odontophore*. This is a tongue-shaped protuberance of the floor of the buccal cavity. It can be pulled down into a sac, and it includes, besides the radula, connective tissue, some cartilagelike supports called **bolsters,** and many muscles. Certain of the muscles that connect the bolsters to the body wall protract or retract the buccal mass, turn it up and down, and swing it from side to side during the feeding process. Others protract or retract the radula itself, so that it slips back and forth over the rest of the buccal mass. There are also muscles that spread apart the bolsters and flatten the radula, as well as muscles whose effect is just the opposite (Figure 13.4, A–C). The teeth of the radula, like the cuticle to which they are attached, are chitinous, and they are often hardened by metallic impregnation.

The number of teeth and their arrangement differ from one group of molluscs to another. Usually, however, there are many teeth arranged in longitudinal and transverse rows (Figure 13.3, B). In a scraping or rasping type of radula, which is characteristic of most molluscs that have this structure, the teeth curve backward. Thus they do their work while the radula is being retracted, rather than while it is being protracted (Figure 13.3, A). As the radula is pulled back, material caught on the teeth is drawn into the pharynx. The surfaces of rocks and the glass walls of aquaria often show markings where the radula teeth of a mollusc have scraped off food (Figure 13.3, C).

There is a price a mollusc must pay if it is to use its radula for scraping or rasping: it must replace the teeth that wear away. Wear is greatest in the anteriormost part of the radula. After a tooth has deteriorated to the point that it is nonfunctional, it separates from the strip of cuticle and generally passes through the gut, even-

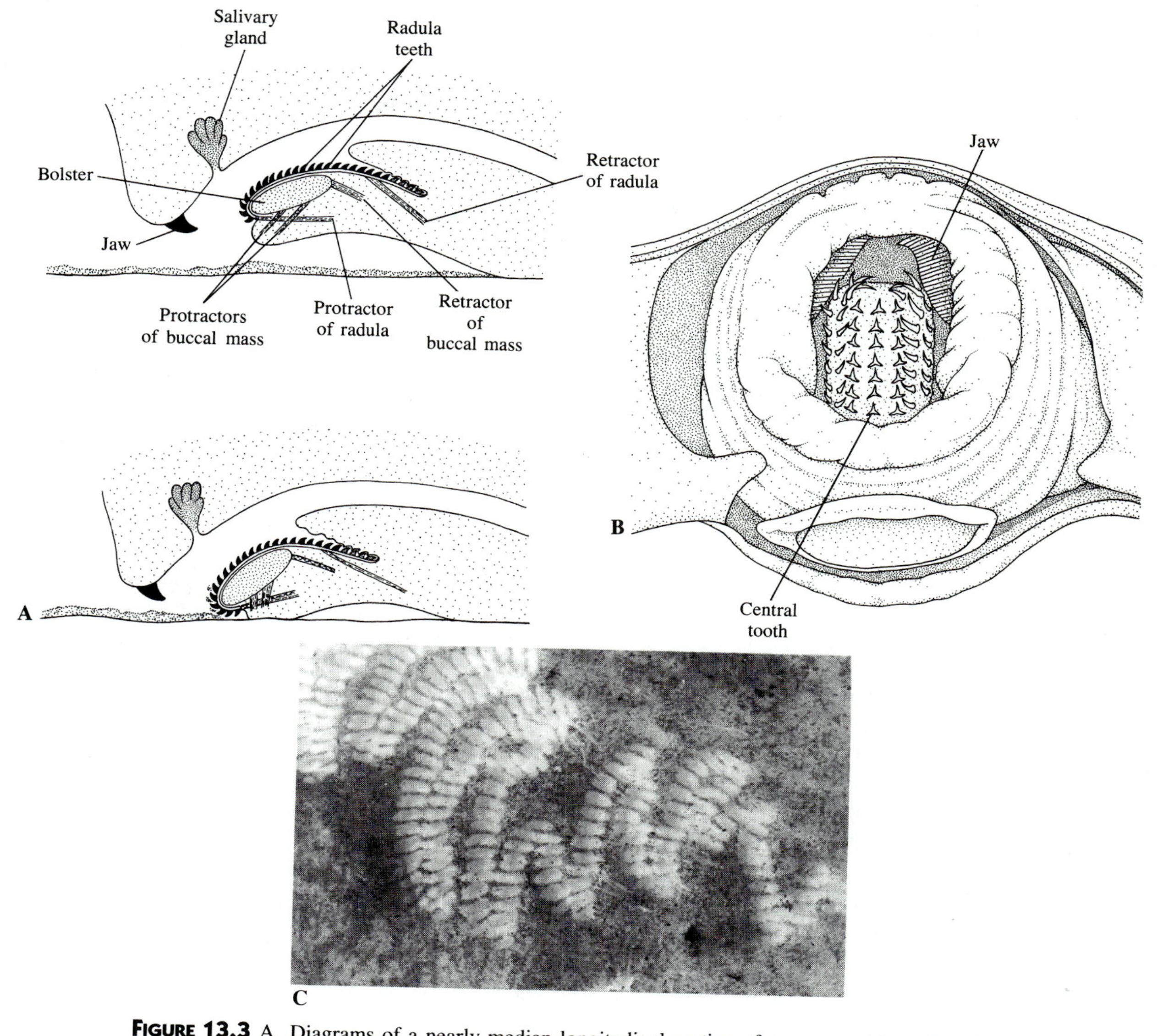

FIGURE 13.3 A. Diagrams of a nearly median longitudinal section of a gastropod buccal cavity, showing the relationship of the radula to one of the bolsters and to the muscles concerned with retraction and protraction. In the upper figure, the radula is retracted; in the lower figure, retraction is underway and the teeth are scraping off food. B. Radula (taenioglossan type) and jaws of *Natica millepunctata,* a mesogastropod, viewed through the open mouth. (After Ankel, Biologisches Zentralblatt, *57.*) C. Radula scrape marks left by an abalone that has been feeding on diatoms coating the wall of an aquarium. Each light area represents the surface scraped during a single retraction of the radula.

tually being voided with fecal material. New teeth are continually secreted and shaped in the deepest part of the radula sac (Figure 13.4, D), and they become affixed to the new cuticle that is added to the strip that overlies the buccal mass. Slow migration of the strip of cuticle brings the teeth forward until they are in a functional position.

Some molluscs that have a radula also have **jaws.** These are sclerotized structures just inside the mouth. They may be used for biting, crushing, or scraping food from a radula as it is retracted. In octopuses and squids, the biting jaws are especially powerful and there is often a **venom gland** opening into the buccal cavity close to them.

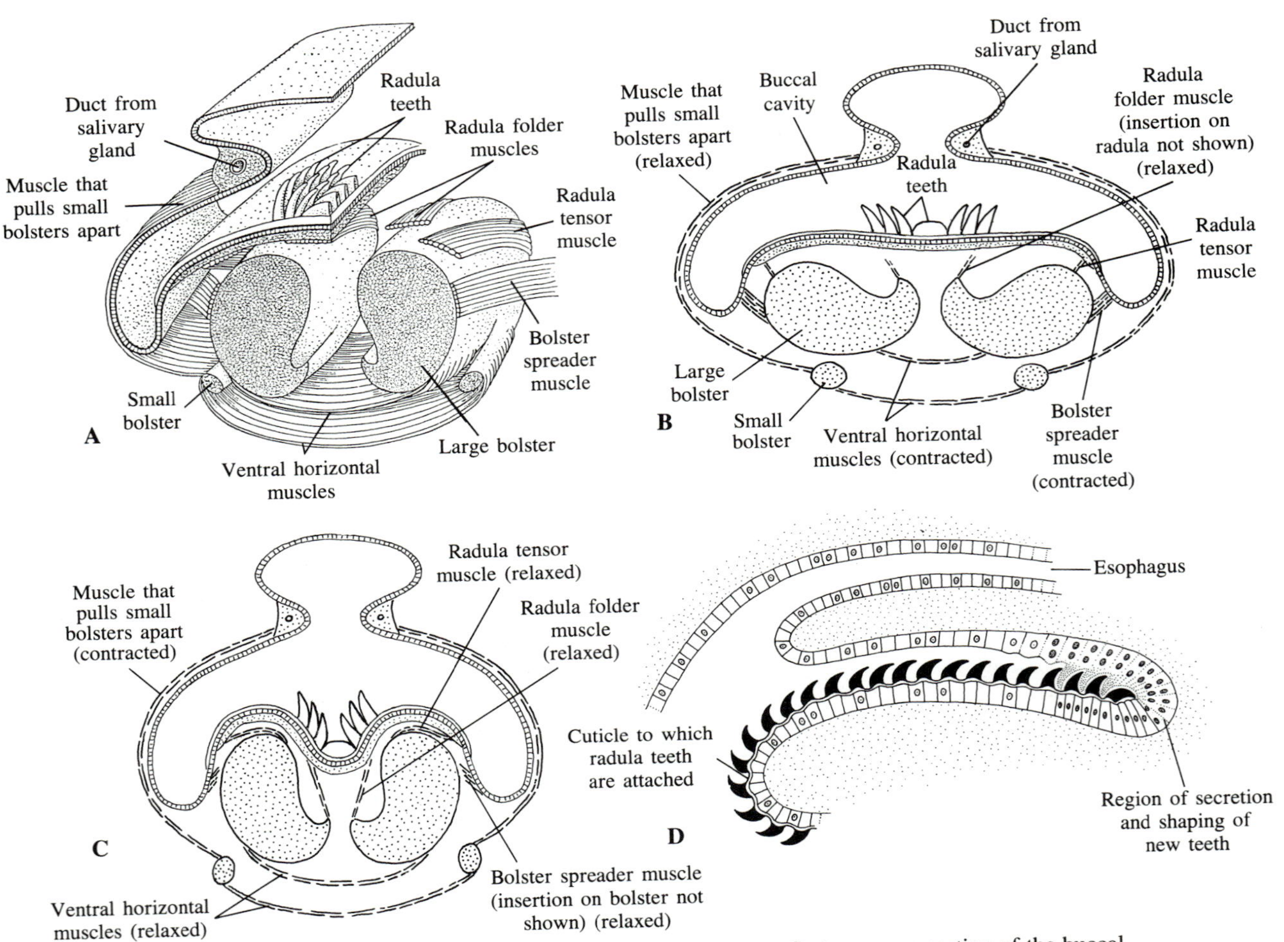

Figure 13.4 A. *Viviparus viviparus* (freshwater mesogastropod), transverse section of the buccal region, showing the radula, bolsters, and some major muscles that operate these structures. B. *Viviparus,* transverse section through the buccal region, showing the radula flattened as a result of contraction of the radula tensor muscles connecting it to the large bolsters, which have been spread simultaneously by contraction of their spreader and ventral horizontal muscles. C. The same buccal cavity, after relaxation of the radula tensor muscles, bolster spreader muscles, and ventral horizontal muscles, and contraction of the radula folder muscles. It is not possible, in a simple diagram, to show contracted and relaxed muscles in perfect proportion. Insertions of some muscles are not drawn in because the total lengths of these muscles would not accurately reflect the contracted or relaxed condition. The arrangements of bolsters and muscles associated with the radula vary greatly among molluscs, even among gastropods. Furthermore, the terminology used in describing them is not consistent. (A–C, after Starmühlner, Österreichische Zoologische Zeitschrift, *3*.) D. Sagittal section (diagrammatic) of a radula, showing the region where new teeth are formed and shaped.

Gut and Digestive Gland

The part of the gut behind the buccal cavity usually consists of an esophagus, stomach, digestive gland, intestine, and rectum. But in bivalves, which have neither a radula nor a buccal cavity, food that enters the mouth goes to the stomach by way of a short esophagus.

The digestive gland occupies much of the visceral mass. Although it often appears to be solid, it is in reality a mass of branching, tubular diverticula of the stomach. It is in the diverticula that much of the process of digestion takes place, either intracellularly or extracellularly. In some molluscs, however, certain enzymes produced by the digestive gland act in the stomach. Uptake, by the blood, of soluble products of digestion takes place mostly around the diverticula.

The stomach itself may have as many as three different functions in processing food: sorting, grinding, and digesting. In general, small particles sorted out on special ciliary–mucous tracts move directly into the diverticula of the digestive gland. Particles too large to be carried into the digestive gland may be subjected to some crushing, grinding, or enzymatic activity in the stomach, and the fine particles resulting from these activities can then be processed in the digestive gland.

In most bivalves and some gastropods, and also in monoplacophorans, the gut is provided with a **crystalline style,** a rodlike structure whose texture is similar to that of firm gelatin. It is sometimes secreted in the anterior part of the intestine, sometimes in a side-pocket of the stomach. Either way, it projects into the stomach. Among the substances incorporated into the style, and released as style material dissolves in the stomach, are enzymes that break down complex carbohydrates, such as starches, into absorbable simple sugars. Because most of the information published about the style is based on studies of bivalves, a more substantial discussion of the functions of this structure is given below in connection with molluscs of that group.

Particles that consist of inert material, or of material that is not affected by the activities of the stomach–digestive gland complex, enter the intestine and are eliminated. The intestine is apparently not often important in either digestion or absorption. It serves mainly to compact fecal material and move it by ciliary action to the rectum.

COELOM

The coelom of molluscs, like that of annelids, can be traced back to spaces that appear within the mesoderm of the embryo. It is thus a **schizocoelom.** Its derivatives are the gonads, pericardial cavity, and coelomoducts. One or both coelomoducts serve as organs of excretion, so they may be called kidneys or renal organs, or even nephridia. It is a mistake, however, to call them metanephridia, for this term is properly applied only to funnel organs derived from ectoderm, rather than from mesoderm. Metanephridia are not found in molluscs.

The coelomoducts of early molluscs were almost certainly gonoducts as well as organs of excretion. In some groups of living molluscs, they are still used for both purposes. The spatial and functional relationships of the gonads, pericardial cavity, and coelomoducts are important in considerations of classification and phylogeny, especially among the Gastropoda.

CIRCULATORY SYSTEM, HEMOCOEL, AND BLOOD

The circulatory system includes a heart that usually consists of a ventricle and one or two auricles. Blood is pumped by the ventricle into arteries, but only in cephalopods do these lead to capillary networks. The arteries of most molluscs simply terminate and allow blood to enter spaces in the tissues. Thus the circulatory system is said to be of the open type. The blood spaces are interconnected and collectively form the **hemocoel.** Veins collect the blood from the hemocoel and carry it to the auricles, which then deliver it to the ventricle.

It is important to understand that the hemocoel is not part of the true coelom. It is merely a system of blood-filled spaces that bathe the tissues. In a functional sense, however, the hemocoel of a mollusc is similar to the annelid coelom. For instance, it serves as a hydrostatic skeleton against which muscles operate. This is of special significance to molluscs whose burrowing depends on displacement of hemocoelic blood from one part of the body to another.

The usual respiratory pigment in the blood of molluscs is **hemocyanin,** which is pale blue when oxidized, colorless when reduced. The oxygen-binding component contains copper, rather than iron, as in hemoglobin. Nevertheless, in many molluscs hemoglobin is the carrier of oxygen.

REPRODUCTION

In most molluscs, sexes are separate. Hermaphroditism is the rule, however, in certain groups of gastropods, and it is also characteristic of some members of other classes. The primitive pattern of reproduction is one in which males and females shed sperm and eggs into the water, but females often retain their eggs in the mantle cavity until they have been fertilized, or even until they have developed into swimming larvae, into some other dispersal stage, or into juveniles that already resemble the adults.

The most advanced patterns of reproduction, found in many gastropods, involve copulation. Males and hermaphroditic individuals have a penis for depositing sperm in the female reproductive tract, so fertilization is internal. The female tract has glands for adding nutritive material to the eggs and for secreting capsules within which the eggs are enclosed when they are laid. In an interesting method of insemination used by cephalopods, the male transfers a packet of sperm by means of the tip of a modified tentacle. The packet may simply be deposited in the mantle cavity of the female, or the female may have a special structure for receiving it.

Cleavage is typically spiral, and gastrulation is by epiboly or invagination, or by a mixture of the two processes. After the blastopore has elongated and closed, an ectodermal invagination that develops at or close to the upper end of the blastoporal region is the primordium of the buccal cavity. A similar invagination at the other end will become the rectum. Both invaginations eventually join the endoderm, segregated in the interior of the embryo by gastrulation, and the formation of the gut is completed. Because the mouth originates in the blastoporal region, molluscs, like annelids, are said to be **protostomatous.**

Mesoderm, and sometimes additional endoderm, derives from the 4d cell, which sinks into the embryo from the lip of the blastopore. Progeny of this cell soon occupy much of the space between the gut and ectoderm, and it is within the mass of mesoderm cells that the coelom develops by schizocoely. Some mesoderm forms from certain ectodermal cells but contributes only to certain muscles and a few other structures that do not survive beyond the larval stage.

The **trochophore** larva, characteristic of many polychaete annelids (see Chapter 11), is also found in the life history of molluscs of the classes Polyplacophora (chitons) and Scaphopoda (tusk shells), which are strictly marine, and of certain marine representatives of the classes Gastropoda and Bivalvia. Except in the Polyplacophora, it is succeeded by a type of larva called the **veliger,** which undergoes metamorphosis to become the adult. The veliger stage is often arrived at without an intermediate trochophore, and in some gastropods it develops from a trochophore that does not emerge from the egg capsule. Direct development, with neither a trochophore nor a veliger, is characteristic of many gastropods and bivalves, and is typically the pathway of development in freshwater and terrestrial types.

None of the foregoing description of development should be applied to cephalopods. These have comparatively large and yolky eggs, whose early cleavage is superficial and results in the formation of a **blastodisk,** a cap of cells whose arrangement is bilaterally symmetrical. Further development of cephalopods is also very different from that of other molluscs. It is, moreover, always direct.

General References on Mollusca

Abbott, R. T., 1974. American Seashells. 2nd edition. Van Nostrand Reinhold Company, New York.

Cheng, T. C., 1967. Marine molluscs as hosts for symbioses. Advances in Marine Biology, *5:*1–424.

Fretter, V. (editor), 1968. Studies in the Structure, Physiology and Ecology of Molluscs. (Symposia of the Zoological Society of London and Malacological Society of London, 22.) Academic Press, London and New York.

Graham, A., 1949. The molluscan stomach. Transactions of the Royal Society of Edinburgh, part 3, *61:*737–778.

Graham, A., 1955. Molluscan diets. Proceedings of the Malacological Society of London, *31:*144–159.

Jørgenson, C. B., 1966. Biology of Suspension Feeding. Pergamon Press, Oxford and New York.

Morton, J. E., 1979. Molluscs. 5th edition. Hutchinson University Library, London.

Potts, W. T. W., 1967. Excretion in molluscs. Biological Reviews, *42:*1–41.

Purchon, R. D., 1977. The Biology of the Mollusca. 2nd edition. Pergamon Press, Oxford and New York.

Raven, C., 1958. Morphogenesis: The Analysis of Molluscan Development. Pergamon Press, Oxford and New York.

Runnegar, B., and Pojeta, J., Jr., 1974. Molluscan phylogeny: the paleontological viewpoint. Science, *186:*311–317.

Solem, A., 1974. The Shell Makers: Introducing Mollusks. John Wiley and Sons, New York.

Vagvolgyi, J., 1967. On the origin of molluscs, the coelom, and coelomic segmentation. Systematic Zoology, *16:*153–168.

von Salvini-Plawen, L., 1972. Zur Morphologie und Phylogenie der Mollusken. Zeitschrift für Wissenschaftliche Zoologie, *184:*205–394.

Wilbur, K. M. (editor), 1983–1988. The Mollusca. 12 volumes. Academic Press, New York and London.

Wilbur, K. M., and Yonge, C. M., 1964–1966. Physiology of Mollusca. 2 volumes. Academic Press, New York and London.

Yochelson, E. L., 1978. An alternative approach to the interpretation of the phylogeny of ancient mollusks. Malacologia, *17:*165–191.

Yonge, C. M., and Thompson, T. E., 1976. Living Marine Molluscs. Collins, London.

CLASS MONOPLACOPHORA

The limpetlike shells of monoplacophorans (Figure 13.5, A and B), characterized by five to eight pairs of internal muscle scars, have long been known from certain fossil deposits. The animals that produced these shells flourished in the Silurian period but died out near the end of the Devonian period, about 350 million years ago. In 1952, the Danish Galathea Expedition, dredging off Costa Rica at a depth of approximately 3600 m, collected 10 specimens of a living mollusc that is thought to be related to the fossil types. It was named *Neopilina galatheae* and its anatomy was described in detail. Since then, additional species have been found; some of these are assigned to *Neopilina,* others to *Vema* and *Monoplacophorus*. The account here deals with *N. galatheae*.

Neopilina has a bilaterally symmetrical shell, up to about 3 cm long, with an apex that points anteriorly (Figure 13.5, C). Although the animal has paired mus-

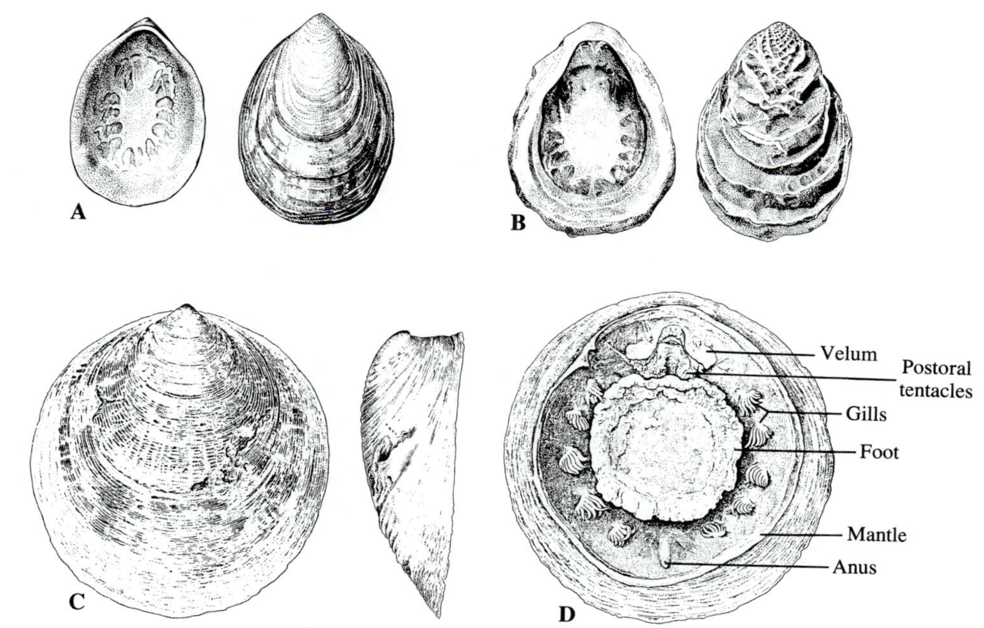

FIGURE 13.5 Monoplacophora. A. Shell of *Pilina unguis* (fossil). B. Shell of *Tryblidium reticulatum* (fossil). C. Shell of *Neopilina galatheae*. D. *Neopilina galatheae*, ventral view. (All from Lemche and Wingstrand, Galathea Report, *3*.)

cles, including eight pairs of foot retractor muscles, there are no conspicuous muscle scars on the shell, and in this respect *Neopilina* differs from its fossil relatives. The foot, used for creeping, is circular (Figure 13.5, D). The head has a pair of indistinct tentacles in front of it, a pair of tentacular tufts behind it, and a lateral flap **(velum)** on both sides of it. There are five pairs of gills in the shallow groove between the foot and mantle. Whether these should be regarded as true ctenidia is uncertain. The anus is on the midline, well behind the foot.

The radula, which lies within a long, coiled sac, has some comblike teeth similar to those of chitons. It is believed to be used for scraping growths of encrusting xenophyophorids, which are multinucleate amoeboid organisms that make an external skeleton from frustules of diatoms and tests of radiolarians and foraminiferans. Certain xenophyophorids taken from habitats where *Neopilina* occurs appear to have been scraped by the radula of this mollusc; the gut contents of *Neopilina,* moreover, indicate that the animal eats xenophyophorids.

The pharynx has a pair of glands as well as a pair of large diverticula that extend dorsally. (These diverticula were previously thought to be coelomic cavities.) The stomach has a caecum within which a crystalline style is secreted, and a single slitlike opening into the digestive gland. The intestine is coiled, with six turns.

The coelom is represented by the pericardial cavity and by two pairs of ovaries or testes (Figure 13.6, A and B). There are six renal organs on each side of the body, and all of them lead to separate pores in the mantle cavity. It is likely that they originate as coelomoducts, although none of them has been shown to have a clear connection with the pericardial cavity, and only two pairs of them, which also serve as gonoducts, are joined to the gonads.

The pericardial cavity, in the posterior part of the body, encloses a pair of ventricles and two pairs of auricles (Figure 13.6, C). One pair of auricles collects blood from the first four pairs of gills, and the other pair collects blood from the last pair of gills. After the blood is delivered to the ventricles, it is pumped forward into aortas that soon become united into a single vessel. As in most other molluscs, much of the circulatory system consists of spaces that form a hemocoel.

The principal components of the nervous system are a thick nerve ring around the mouth and two pairs of nerve cords. The cords of one pair extend into the foot, and those of the other pair run through the mantle. Both pairs form loops, and there are also numerous cross-connections between the cords (Figure 13.6, A). The only obvious sense organs, other than the tentacles already mentioned, are a pair of statocysts behind the mouth, but almost certainly there are receptors in the velar flaps and elsewhere.

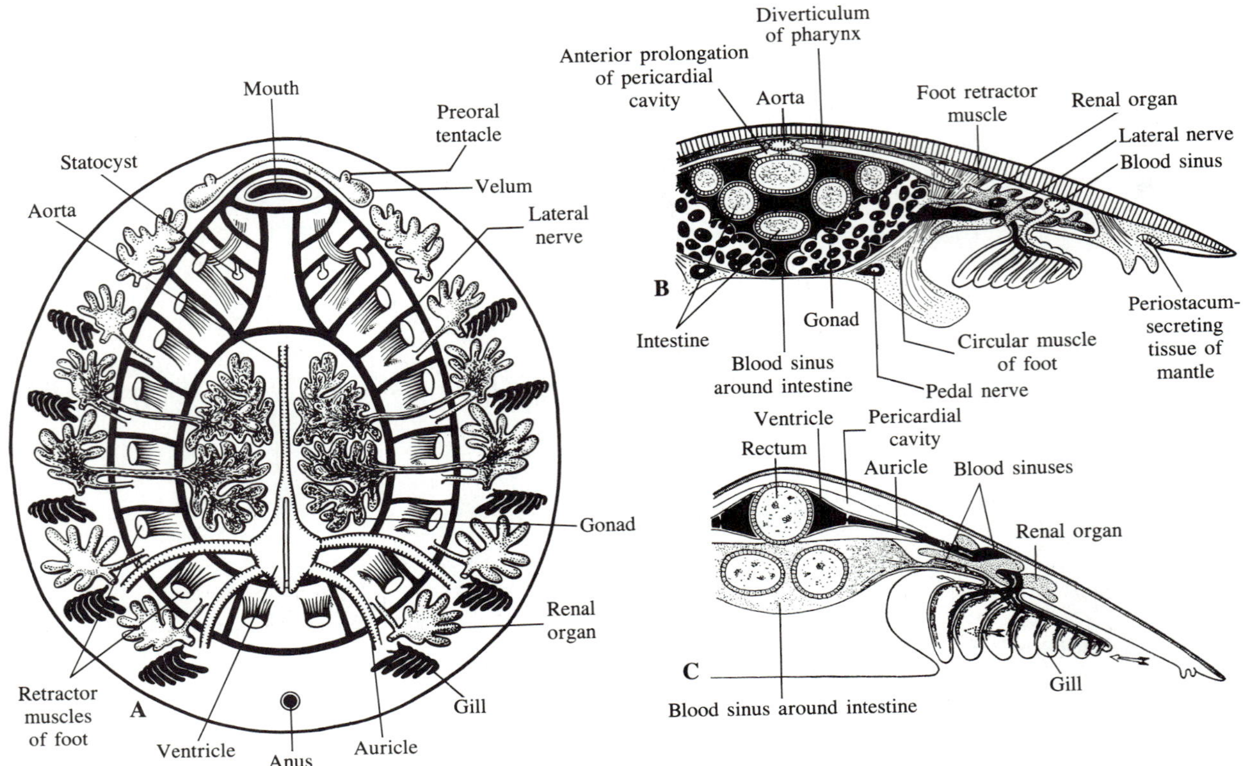

Figure 13.6 Monoplacophora, *Neopilina galatheae*. A. Diagram of general anatomy. The heart and aorta are dorsal to other structures, and the animal is shown as if it were transparent. B. Transverse section through the body anterior to the heart. C. Transverse section at the level of the heart. The arrows indicate the presumed pattern of water flow in the mantle cavity. (Lemche and Wingstrand, Galathea Report, *3*.)

When *Neopilina* was discovered, the serial repetition evident in the arrangement of its foot retractor muscles, gills, coelomoducts, gonads, and auricles led to speculations that molluscs, as a group, evolved first as segmented animals. A few zoologists have even viewed molluscs as derived from annelids. Whether the serial repetition of some organs of monoplacophorans should be considered a form of segmentation or as evidence that annelids gave rise to molluscs is doubtful. It is nevertheless likely that molluscs and annelids had an ancient common ancestry. This is suggested by similarities in embryonic development and by the fact that many marine representatives of both phyla have a trochophore larva.

References on Monoplacophora

Lauterbach, K. E., 1983. Gedanken zur Entstehung der mehrfach paarigen Excretionsorgane von *Neopilina* (Mollusca, Conchifera). Zeitschrift für Zoologische Systematik und Evolutionsforschung, *21*:38–52.

Lemche, K., and Wingstrand, K. G., 1959. The anatomy of *Neopilina galatheae* Lemche, 1957 (Mollusca Tryblidiacea). Galathea Report, *3*:9–71.

McLean, J. H., 1979. A new monoplacophoran limpet from the continental shelf off southern California. Los Angeles County Museum of Natural History, Contributions to Science, no. 317.

Moskalev, L. I., Starobogatov, IA. I., and Filatova, Z. A., 1983. [New data on the abyssal Monoplacophora from the Pacific and Atlantic oceans.] Zoologicheskii Zhurnal, *62*:981–996 (in Russian).

Tendal, O. S., 1985. Xenophyophores (Protozoa, Sarcodina) in the diet of *Neopilina galatheae* (Mollusca, Monoplacophora). Galathea Report, *16*:95–98.

Wingstrand, K. G., 1985. On the anatomy and relationships of recent Monoplacophora. Galathea Report, *16*:7–94.

Class Polyplacophora

The polyplacophorans, commonly called chitons, make up a relatively small group of marine molluscs, there being only about 600 species. They are, however, important constituents of the fauna in rocky habitats and

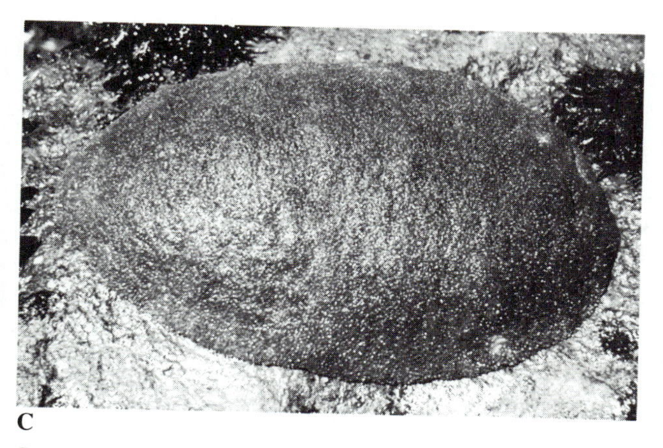

FIGURE 13.7 Polyplacopnora. Representative chitons. A. *Placiphorella velata*. B. *Mopalia muscosa*. C. *Cryptochiton stelleri*. (See also Color Plate 3).

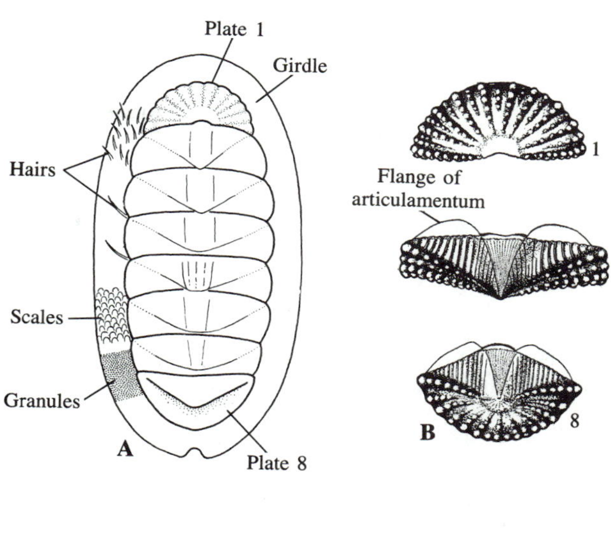

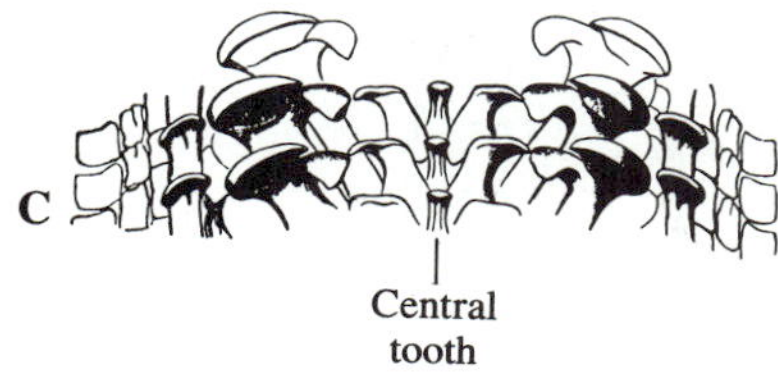

FIGURE 13.8 Polyplacophora. A. Diagram (dorsal view) showing the eight shell plates of a chiton, and also some of the ornaments found on the girdle of various species. B. Plates 1 and 8, and one from near the middle of the body, of *Chiton komaianus*. (Taki and Taki, Venus, *1*.) C. Portion of a radula of *Acanthopleura spinosa*. (Cooke, in Cambridge Natural History, *3*.)

sometimes also on shelly bottoms. Chitons are flattened and usually elongated, and typically are protected dorsally by a shell that consists of eight plates (Figure 13.7, A and B; Figure 13.8, A and B; Color Plate 3). The foot is broad and powerful, well adapted for clinging tightly to a hard surface. The low profile of a chiton, its shell, and its strong foot are adaptive for an animal that lives where there is considerable wave shock and that risks water loss during periods of low tide. If a chiton is dislodged, it can roll into a ball, with the plates providing protection almost all around.

Beneath the thin periostracum, each shell plate has two main layers (Figure 13.9, C). The outer layer, usually heavily pigmented, is called the **tegmentum.** It consists of a mixture of conchiolin (the same material that forms the periostracum) and calcium carbonate. The tegmentum is traversed by epidermal structures whose ultimate branches reach the periostracum. More will be said about these later on. The inner layer of a shell plate, called the **articulamentum,** is composed almost wholly of calcium carbonate. It is not often pigmented, although it may be delicately tinted. In each plate except the first, the articulamentum projects forward as a flange (Figure 13.8, B) that is overlapped by the posterior part of the plate in front of it.

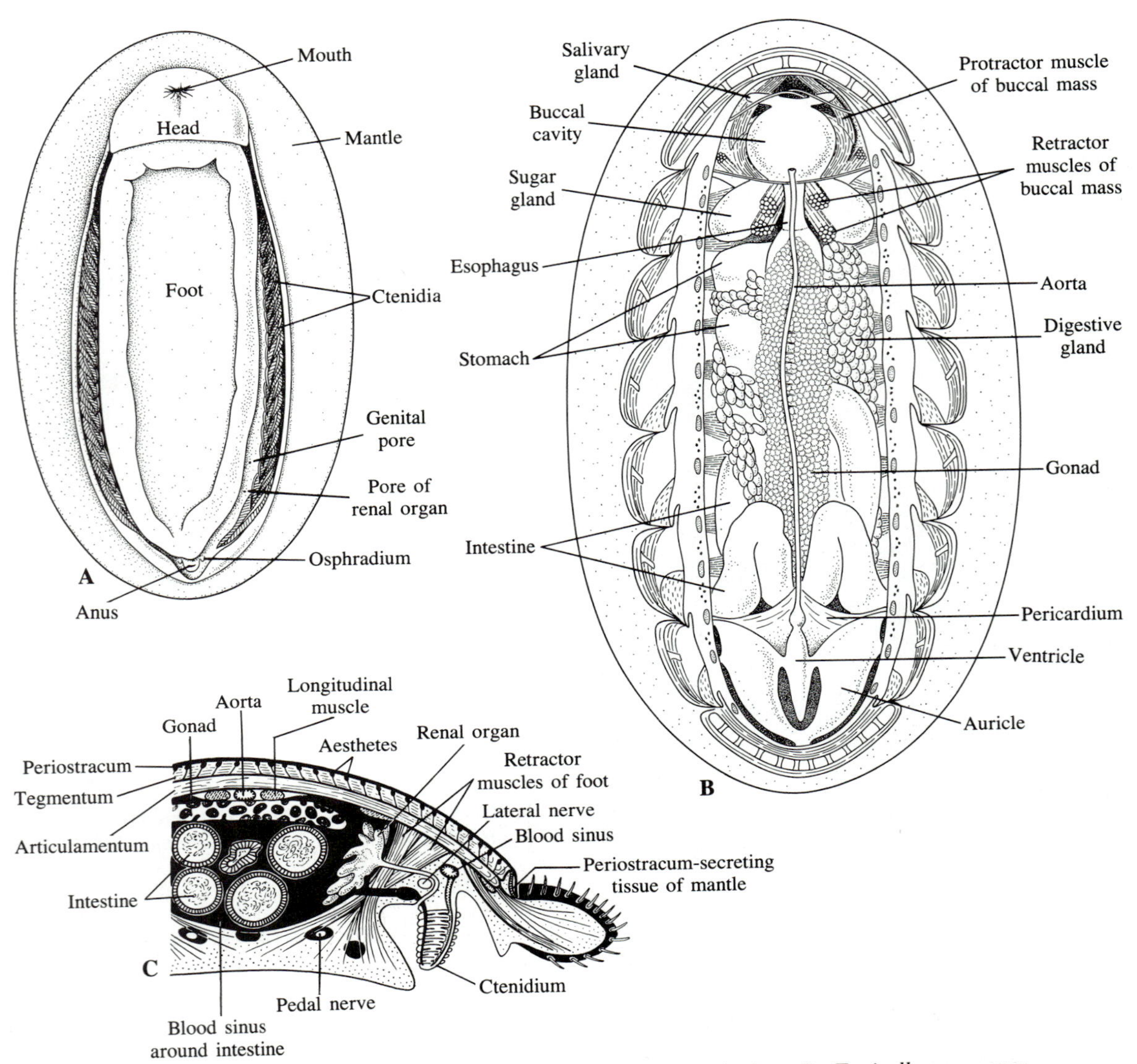

Figure 13.9 Polyplacophora. A. *Tonicella marmorea,* ventral view. B. *Tonicella marmorea,* dissected from the dorsal side. (After Ivanov et al., Major Practicum of Invertebrate Zoology.) C. Diagram of a transverse section of a chiton, just anterior to the heart. (Lemche and Wingstrand, Galathea Report, *3*.)

The leathery marginal portion of the mantle, which extends up onto the dorsal surface to partly cover the shell plates, is called the **girdle.** It may be practically smooth, or it may have small granulations, calcareous scales or spicules, or hairlike or bristlelike outgrowths (Figure 13.8, A). Whatever its appearance, it is overlain by a chitinous cuticle. In the world's largest species, *Cryptochiton stelleri* (Figure 13.7, C) of the Pacific coast of North America, the shell plates are completely covered by the tough girdle. There are various other species in which only a small portion of each plate is visible. When a chiton pulls itself down by its strong foot muscles, it presses the edge of the girdle tightly to the rock. By raising the roof of the mantle cavity a little, or by drawing the underside of the girdle farther away from the foot, it can create a slight negative pressure in the mantle cavity. The girdle will then act in much the same way as a suction cup, augmenting the grip the animal's foot has on the substratum.

The head of a chiton (Figure 13.9, A; see also *Placiphorella,* Color Plate 3), anterior to the foot, is recognizable but not highly developed. There are no eyes or tentacles, though there are sensory **lappets.** On each side of the body, in the mantle groove that separates the head and foot from the mantle, are several to many ctenidia. The action of cilia on the ctenidia moves water through the mantle cavity in a posterior direction. There is often a pronounced median cleft in the girdle at the posterior end, where the stream of water emerges. In at least many chitons, most of the ctenidia hanging down into the mantle cavity curve toward the foot and actually touch it. Successive ctenidia are held together by cilia, so that the mantle cavity tends to be divided into an upper and lower chamber. As long as the girdle is lifted at least slightly, water will enter the lower chamber and be driven by action of cilia to the upper chamber, then backwards and out.

The radula, used to scrape food from the substratum, fits into a sac that is about a third the length of the animal as a whole. If removed by dissection and stretched out, the radula is more than half the chiton's length. There are usually 17 teeth in each of the many transverse rows (Figure 13.8, C). An interesting feature of these teeth is that they are rich in a magnetic form of iron oxide; the whole radula can therefore be picked up with a magnet. Most chitons eat algae and diatoms; some, however, such as those of the genus *Mopalia,* consume considerable animal matter in the form of hydroids, bryozoans, and sponges. A few species have decidedly different approaches to getting food. *Placiphorella velata* (Figure 13.7, A; Color Plate 3), found on the Pacific coast of North America, is a sedentary type whose favorite habitat is the underside of a rock ledge. The girdle of the anterior part of the animal forms a broad veil that can be lifted until a small crustacean or worm wanders under it. The veil is then clamped down to the rock and the prey is eaten. *Placiphorella* can also scrape off encrusting sponges and algae.

The subradular organ, opening into the buccal cavity just in front of the radula sac, is believed to function as a taster. Like the radula, it can be extruded through the mouth.

Once food is ingested and mixed with mucus secreted by the salivary glands, it moves into the esophagus. This part of the gut is provided with a pair of large **sugar glands** (Figure 13.9, B; Figure 13.10, C) whose main function is to secrete an enzyme or enzymes that convert starches and other polysaccharides into soluble sugars. There may also be a pair of smaller glands on the esophagus. The part of the stomach into which the esophagus opens is a cuticle-lined sac. This is mostly ventral to the rest of the stomach, which is entered by ducts from the large digestive gland, which secretes enzymes concerned with digestion of proteins and also of starches. Digestion is believed to be primarily extracellular.

The long and coiled intestine is divided into two portions separated by a sphincter. Undigested residues pass into the hind portion where they are worked into fecal pellets. As soon as the pellets are discharged from the anus, they are carried away by the current of water leaving the posterior part of the mantle cavity.

The heart of a chiton lies dorsally near the posterior end of the body (Figure 13.9, B). The ventricle, enclosed by the pericardial sac, receives blood from the right and left auricles. It pumps blood into an anteriorly directed aorta, and branches of this open into hemocoelic spaces. Eventually blood moves to the ctenidia, and from these it returns to the auricles. Hemocyanin is the respiratory pigment in most chitons.

The paired renal organs (Figure 13.10, A and B) are coelomoducts. They drain the pericardial cavity, in which at least some nitrogenous wastes accumulate. The renal organs are extensive, however, and have numerous diverticula that are bathed by blood moving through the hemocoel. Each is directed forward, then backward, finally opening into the mantle cavity not far from the posterior end.

The nervous system (Figure 13.11, A), rather diffuse and decentralized, is somewhat similar to that of the monoplacophoran *Neopilina.* From a ring around the buccal tube, two nerves run through the foot for much of its length; two others, which are joined anteriorly and posteriorly to form a continuous loop, are located in the mantle. There are numerous cross-connections. The only obvious ganglia in the system are some swellings associated with the highly muscularized buccal apparatus.

The uncomplicated head of a chiton suggests that this part of the body is poorly supplied with sense organs. It does, however, have the subradular organ, which can be extruded through the mouth and which is thought to be concerned with chemoreception. Patches of sensory tissue in the posterior part of the right and left mantle grooves are also probably chemosensory, and they may be homologous to osphradia of gastropods (see below).

There are receptors located in the tubes that run to the exposed surfaces of the shell plates. In certain chitons, some of these epidermal outgrowths terminate in simple eyes that have a "cornea" and "lens." All chitons, at least when they are young, have tegmental structures called **aesthetes** (Figure 13.11, B). These are of two types. The larger ones (megalaesthetes) contain

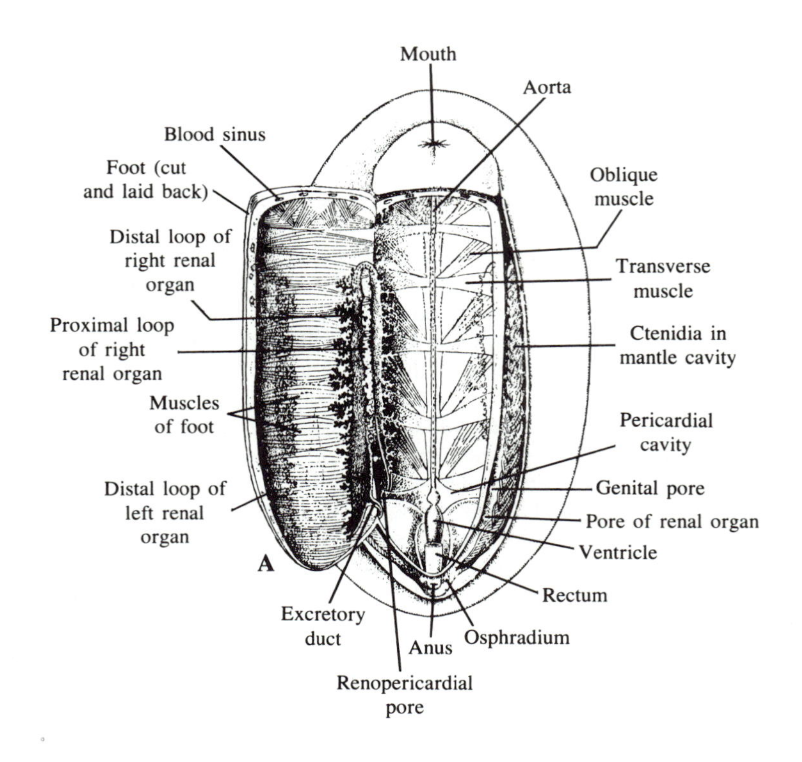

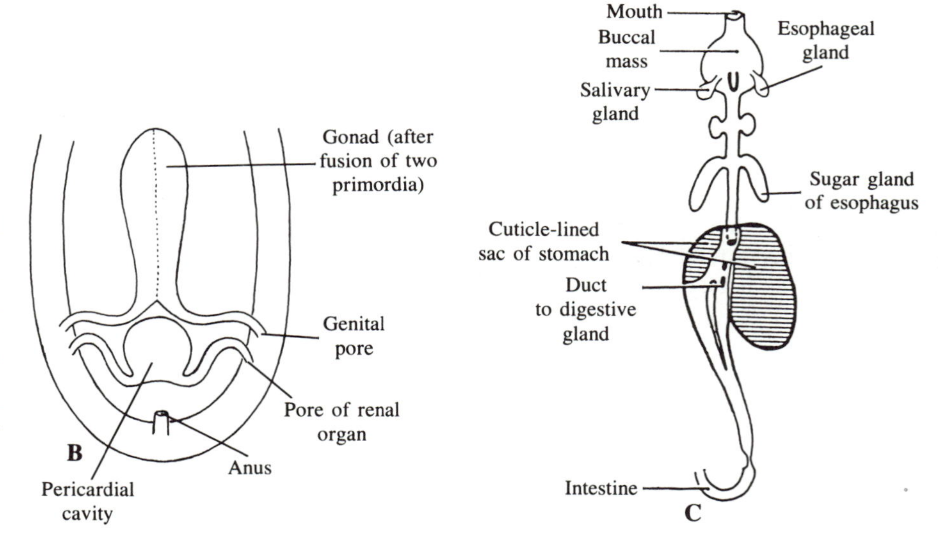

FIGURE 13.10 Polyplacophora. A. *Tonicella marmorea,* dissected from the ventral side, showing the extensive renal organs and some other structures. (Ivanov et al., Major Practicum of Invertebrate Zoology.) B. Diagram showing the relationship of the reproductive and excretory systems to the pericardial cavity. C. Diagram of the digestive system of a chiton. (Owen, in Wilbur and Yonge, Physiology of Mollusca, *2*.)

cells thought to be photoreceptors, and also some cells that have a secretory function. The smaller ones (micraesthetes) are single cells whose basal portions lie within the megalaesthetes. Thus a single megalaesthete is accompanied by several micraesthetes. Each aesthete has a cap of organic material in which a perforated calcareous plate is embedded, so the sensory cells do not actually reach the surface. In many chitons the tegmentum of the shell plates becomes badly eroded and the aesthetes disappear. It may be that the receptors in these structures are more important to young animals than to adults.

All epidermal surfaces of chitons seem to be sensitive to touch. Moreover, the partly chitinous hairs on the upper surface of the girdle of many species (Figure 13.7, A and B; Figure 13.11, D) contain the dendrites

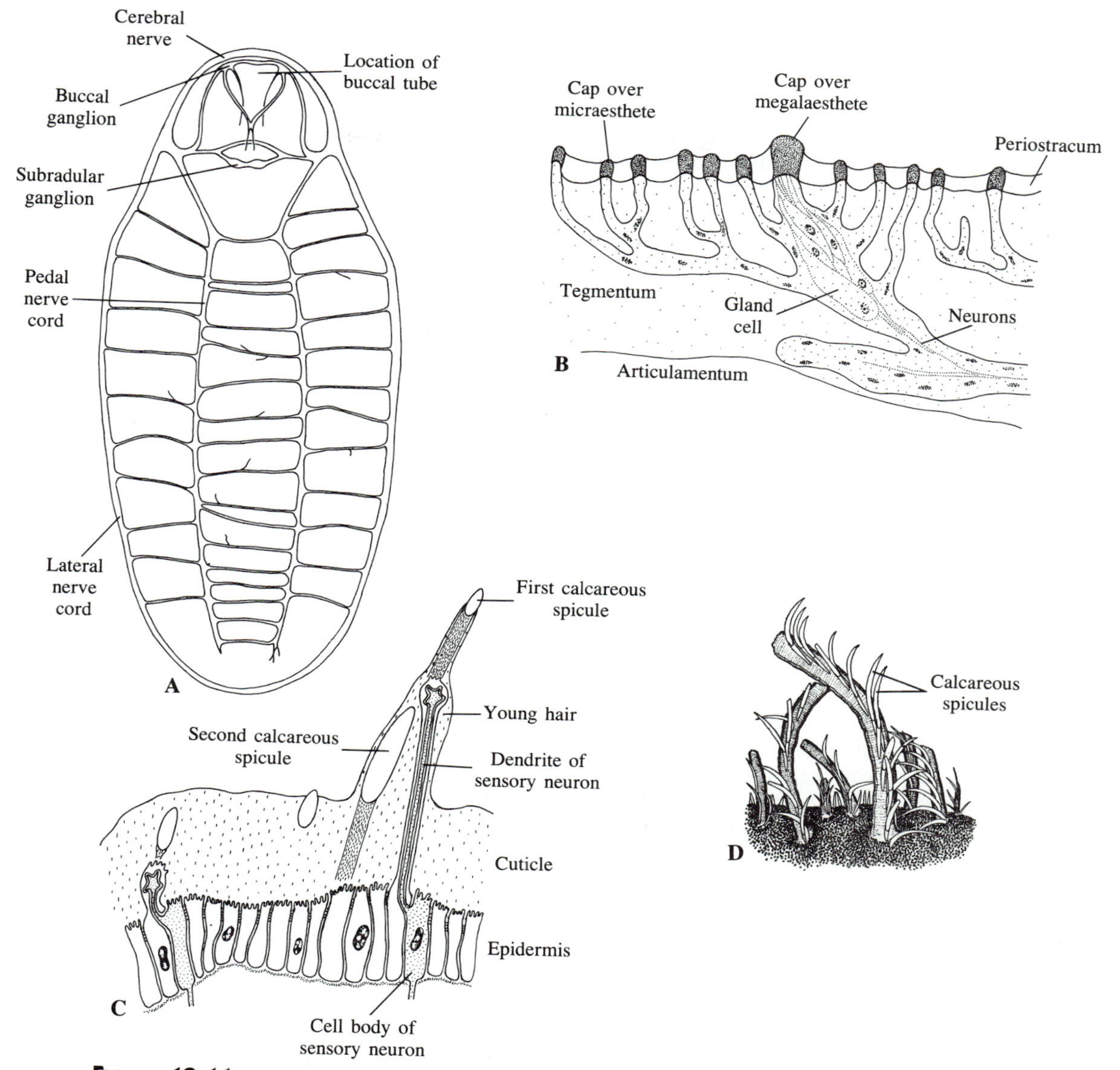

FIGURE 13.11 Polyplacophora. A. Diagram of the nervous system of a chiton. B. *Tonicella marmorea,* section through a shell plate of a young individual, showing aesthetes. (After Ivanov et al., Major Practicum of Invertebrate Zoology.) C. *Mopalia muscosa,* portion of the girdle of a young individual, showing a developing hair; based on electron micrographs. (After Leise, Journal of Morphology, *189*.) D. *Mopalia retifera,* girdle hairs. (Taki and Taki, Venus, *1*.)

of sensory neurons. The cell bodies of these neurons are in the epidermis that secretes the hairs, as well as the continuous cuticle that covers the girdle (Figure 13.11, C). The formation of a hair, at least in some species, is preceded by the deposition of a calcareous spicule. As a hair lengthens, the neuron supplying it also lengthens, and the spicule, at the tip of the hair, is pushed farther away from the place where it was secreted. With the addition of other spicules and other neurons, a hair becomes an increasingly more complex structure.

Chitons have separate sexes. The gonads, of coelomic origin, are at first paired structures, but the two primordia become fused to form a single mass in front of the heart (Figure 13.9, B; Figure 13.10, B). The gonoducts lead to the mantle cavity, and their openings are close to the pores of the renal organs (Figure 13.9, A; Figure 13.10, A and B). The eggs, individually coated with jelly, may be fertilized while still in the mantle cavity or after they have been set free.

The trochophore larva feeds on diatoms and other planktonic organisms until it is ready to settle and transform into a small chiton. The eight shell plates develop on the dorsal side of the trochophore. Not all chitons have a free-swimming larva, however. Some that produce yolky eggs brood their young in the mantle cavity, releasing them only after the process of metamorphosis is well under way.

Some chitons, like various limpets, exhibit a homing behavior; that is, they have permanent residences to which they return after foraging. Certain species, in fact, occupy shallow but nevertheless shape-fitting depressions in rock. The depressions develop slowly, and a particular individual may be occupying a home that was left to it by a succession of predecessors.

The foraging activities of chitons are often correlated with tides and the day–night cycle. *Mopalia muscosa,* a common intertidal species on the Pacific coast of North America, forages mostly at night and when it is either under water or being splashed by waves. It moves at speeds up to about 25 cm per hour and is a rather unselective consumer of small algae. It is one of the chitons that has a homing behavior. *Cyanoplax hartwegii,* which lives at higher levels than *Mopalia,* feeds primarily on one brown alga, *Pelvetia,* and does so mostly at night and when it is out of water or just being splashed. If it were underwater in an intertidal area with considerable wave action, it would have trouble staying on *Pelvetia*. *Cyanoplax* does not exhibit a strong homing tendency, but it does usually rest in places where there are others of the same species, or where others of the same species have been.

References on Polyplacophora

Boyle, P. R., 1972. The aesthetes of chitons. I. Role in the light response of whole animals. Marine Behavioral Physiology, *1*:171–184.

Boyle, P. R., 1974. The aesthetes of chitons. II. Fine structure in *Lepidochitona cinereus*. Cell and Tissue Research, *153*:383–398.

Boyle, P. R., 1976. The aesthetes of chitons. III. Shell surface observations. Cell and Tissue Research, *172*:379–388.

Boyle, P. R., 1977. The physiology and behavior of chitons. Oceanography and Marine Biology, An Annual Review, *15*:461–509.

Christensen, M. E., 1954. The life history of *Lepidopleurus asellus* (Spengler) (Placophora). Nytt Magasin for Zoologi, *2*:52–72.

Fischer, F. P., 1988. The ultrastructure of the aesthetes in *Lepidopleurus cajitanus* (Polyplacophora: Lepidopleurina). American Malacological Bulletin, *6*:153–159.

Fischer, F. P., Eisensamer, B., Miltz, C., and Singer, I., 1988. Sense organs in the girdle of *Chiton olivaceus*. American Malacological Bulletin, *6*:131–139.

Jones, A. M., and Baxter, J. M., 1987. Molluscs: Caudofoveata, Solenogastres, Polyplacophora and Scaphopoda. Synopses of the British Fauna, no. 37. E. J. Brill/Dr. W. Backhuys, London and Leiden.

Leise, E. M., 1984. Chiton integument: metamorphic changes in *Mopalia muscosa* (Mollusca, Polyplacophora). Zoomorphology, *104*:337–343.

Leise, E. M., 1986. Chiton integument: development of sensory organs in juvenile *Mopalia muscosa*. Journal of Morphology, *189*:71–87.

Leise, E. M., 1988. Sensory organs in the hairy girdle of some mopaliid chitons. American Malacological Bulletin, *6*:141–151.

Leise, E. M., and Cloney, R. A., 1982. Chiton integument: ultrastructure of the sensory hairs of *Mopalia muscosa* (Mollusca: Polyplacophora). Cell and Tissue Research, *223*:43–59.

Russell-Hunter, W. D., 1988. The gills of chitons (Polyplacophora) and their significance in molluscan phylogeny. American Malacological Bulletin, *6*:69–79.

Scheltema, A. H., 1988. Ancestors and descendents: relationships of the Aplacophora and Polyplacophora. American Malacological Bulletin, *6*:57–68.

Smith, S. Y., 1975. Temporal and spatial activity patterns of the intertidal chiton *Mopalia muscosa*. Veliger, *18* (Supplement):57–62.

CLASS APLACOPHORA

The aplacophorans, of which there are about 150 species, are wormlike marine molluscs that live at moderate to considerable depths. Most of them are small—less than 5 cm long—and unless one is dredging in an area where these animals are plentiful, the occasional specimen that gets mashed against the sorting screen may not be noticed. Aplacophorans do not have a shell; instead, they secrete a flexible cuticle in which calcareous spicules are embedded. In some types, the protruding spicules form a feltlike covering (Figure 13.12, A).

The nervous system (Figure 13.12, B), with a nerve ring in the anterior part of the body and two pairs of major longitudinal nerves, resembles that of chitons. Because of this similarity, aplacophorans and chitons were for a long time put into the same class, Amphineura.

SUBCLASS SOLENOGASTRES

In this group of aplacophorans, represented by *Neomenia* and a few other genera, a ventral ciliated ridge begins close to the mouth and extends nearly the full length of the animal. The ridge is thought to be the homologue of the foot of other molluscs. The grooves on both sides of it are probably mantle grooves. They are continuous with a ventral depression at the posterior

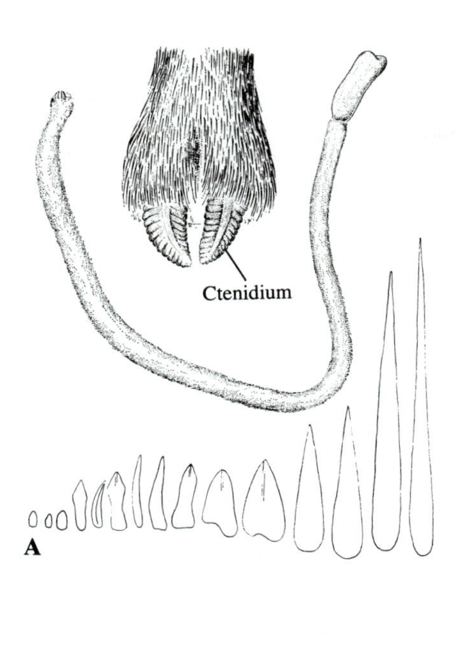

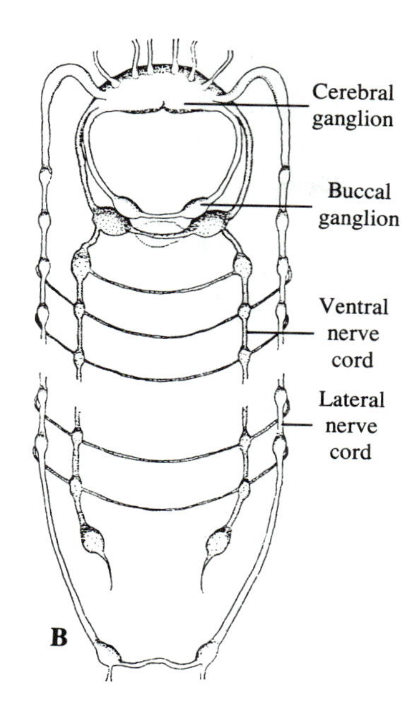

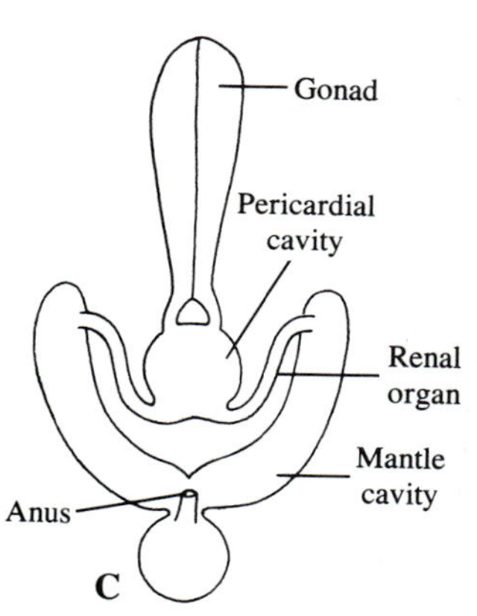

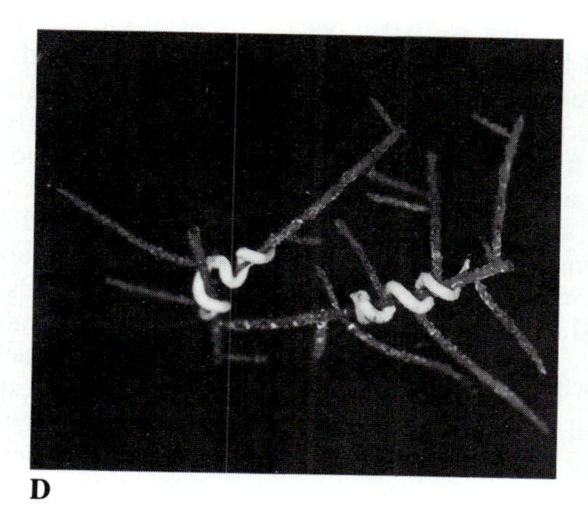

FIGURE 13.12 Aplacophora. A. *Chaetoderma nitidulum,* entire animal, posterior portion (further enlarged), and spicules of the integument. (Muus, Danmarks Fauna, *65.*) B. *Genitoconia atriolonga,* nervous system. (von Salvini-Plawen, Sarsia, *27.*) C. Diagram of the posterior part of the body of an aplacophoran, showing the relationship of the reproductive and excretory systems to the pericardial cavity. D. A species of the subclass Solenogastres, wrapped around a hydroid.

end, thought to be a reduced mantle cavity. The anus and openings of the coelomoducts are located in the depression.

Neomenia and its relatives are hermaphroditic. The gonads release their gametes into the pericardial cavity, and the coelomoducts, which are renal organs as well as gonoducts, carry them to the exterior (Figure 13.12, C). There are no true ctenidia, but the mantle grooves have folds that probably function as gills.

Some of the aplacophorans of this subclass live in mud or sand; others crawl around on hydroids (Figure 13.12, D), alcyonarians, and seaweeds. A few that are associated with cnidarians have been observed to evert the foregut so that the radula can tear off pieces of tissue.

SUBCLASS CAUDOFOVEATA

The genus *Chaetoderma* (Figure 13.12, A) is representative of this subclass. A ventral ridge and ventral grooves are lacking, but there is a deep mantle cavity at the posterior end. It contains a pair of ctenidia, as well as the anus and pores of the coelomoducts. *Chaetoderma* and its relatives live in soft sediments and feed on foraminiferans, diatoms, and other small organisms; they seem not to swallow much of the mud itself.

References on Aplacophora

Deimel, K., 1982. Zur Ableitung des Radulaapparates des Chaetodermatidae (Mollusca: Caudofoveata). Zeitschrift für Zoologische Systematik und Evolutionsforschung, *20:*177–187.

Jones, A. M., and Baxter, J. M., 1987. Molluscs: Caudofoveata, Solenogastres, Polyplacophora and Scaphopoda. Synopses of the British Fauna, no. 37. E. J. Brill/Dr. W. Backhuys, London and Leiden.
Morse, M. P., 1979. *Meiomenia swedmarki* gen. et sp. n., a new interstitial solenogaster from Washington, U.S.A. Zoologica Scripta, *8*:249–253.
Scheltema, A. H., 1978. Position of the Aplacophora in the phylum Mollusca. Malacologia, *17*:99–109.
Scheltema, A. H., 1983. On some aplacophoran homologies and diets. Malacologia, *23*:427–428.
Scheltema, A. H., 1988. Ancestors and descendents: relationships of the Aplacophora and Polyplacophora. American Malacological Bulletin, *6*:57–68.
Thompson, T. E., 1960. The development of *Neomenia carinata* Tullberg (Mollusca, Aplacophora). Proceedings of the Royal Society of London, B, *153*:263–278.
von Salvini-Plawen, L., 1978. Antarktische und subantarktische Solenogastres. Eine Monographie: 1898–1974. Zoologica (Stuttgart), no. 128:1–315.

CLASS GASTROPODA

The gastropods—snails, slugs, and an assortment of oddities—make up the largest and most diversified class of molluscs. Over 80,000 living species have been given names, and the number of fossil species is nearly 20,000. Gastropods have been evolving for about 550 million years, since the early part of the Cambrian period, and have become adapted to many different habitats in the seas, in fresh water, and on land. The food preferences of some gastropods are rather specific, but as a group these molluscs eat almost every kind of animal and plant matter: seaweeds, fungi, dung, plankton, fishes, various invertebrates, and even each other.

The Shell and Some of Its Modifications

In gastropods that have a shell, this is usually coiled, but it is sometimes cap-shaped or conical, and in a very few species it is a bivalved structure, similar to that of a clam. Unless it is thin and easily cracked, the shell provides protection from predators and mechanical stress, including wave-shock. For snails that live on land or in portions of the intertidal zone that are exposed for long periods at low tide, a shell helps to reduce the amount of water lost by evaporation. There are many gastropods without shells, or with only rudimentary ones, and these have found other solutions to the problems associated with crawling around naked.

The oldest part of a gastropod shell, whether this is coiled or not, is at the apex. In snails whose life history includes a veliger stage, the delicate **protoconch** (Figure 13.14) represents the start of shell formation. If the apex of the shell has been eroded, which is almost always the case, the protoconch will no longer be present.

As a snail grows, it secretes new shell material at the edge of the aperture. Both the organic periostracum and the prismatic layer are produced only by the edge of the mantle (Figure 13.2). The nacreous layer, sometimes pearly but generally more like porcelain, is secreted by the epidermis that is in tight contact with the inner surface of the shell. Thickening of the shell depends on additions to the nacreous layer. In a coiled shell, each successive whorl is cemented to the one before it. The axial portion of the shell, a sort of pillar around which the spire coils, is called the **columella** (Figure 13.13).

Coiling of the visceral mass is a consequence of faster growth on one side than on the other. It is an advantageous adaptation for an animal for whom indefinite growth along a straight axis is impractical. Given the orientation of the foot in relation to the viscera, the visceral mass of a snail enlarges dorsally. If it were not for coiling, the mass would sprawl or tumble, becoming not only vulnerable but a hindrance to locomotion.

Some extinct gastropods had truly **planispiral** shells (Figure 13.16), in which all of the coils were in the same plane. These snails probably carried their shells vertically along the dorsal midline. This design would not be as efficient, in either compactness or balance, as that of the **conispiral** type, characteristic of the majority of gastropods. In a conispiral shell the coils are organized around an elongated axis, and the apex is directed posteriorly at an oblique angle. With the center of its newest and largest whorl a little to one side of the midline and the apex of the spire leaning to the opposite side, the snail distributes the weight of its visceral mass almost evenly above the foot. Even the relatively few living gastropods whose shells are coiled in one plane—such as freshwater snails of the genus *Planorbis*—adhere to this principle of weight distribution.

Cap-shaped shells, such as are typical of limpets (Figure 13.23), are basically similar to the conical shells of extinct ancestors of modern gastropods. A shell of this type is not, however, a sure sign that the animal inside it is primitive. Gastropods with cap-shaped shells belong to several different groups, and the limpetlike form of most of them has been arrived at secondarily. A cap-shaped shell can be clamped down tightly to a rock or some other firm, smooth substratum. It permits the visceral mass to be balanced in a different way than in snails with a conispiral shell.

Some conispiral shells have proportionately high, slender spires; others have low, fat spires. The height of the spire as a whole depends to a large extent on the height-to-diameter ratio of the body whorl. The angle at which the spire is carried also varies. For instance, in *Gibbula* and *Calliostoma* (Color Plate 4), the spire is carried at a relatively high angle. In most snails, however, the spire is carried at a low angle. Low-angle shells, characteristic of more advanced members of the subclass Prosobranchia, are the only ones likely to have a **siphonal canal** (Figure 13.17, A; Figure 13.31). This

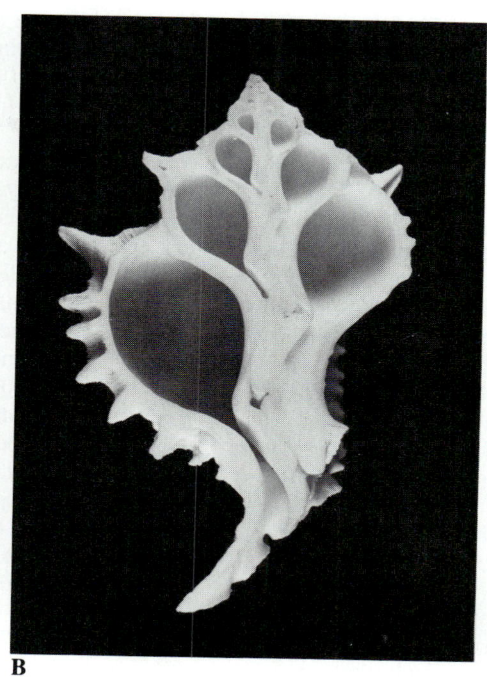

FIGURE 13.13 Gastropoda. Shells of two neogastropods sectioned to show the relationship of the whorls to the columella.

is an extension of the anterior edge of the aperture, and arises close to the axis of coiling. It is basically a trough, but sometimes its edges become fused to form a complete tube. The canal protects the **siphon,** which is a rolled-up projection of the mantle (Figure 13.17; Figure 13.33, B) that serves to bring water into the mantle cavity. The siphon is a useful adaptation, for it permits a snail to cling tightly to a rock or to plow through mud or sand and at the same time to draw in water from which oxygen can be absorbed. Because the siphon has a narrow bore, particles that might clutter the mantle cavity are excluded. The siphon functions also as part of a system used by a snail in finding food and in getting its bearings. Water brought into the mantle cavity from in front of the animal is monitored by sensory structures.

Most coiled gastropod shells are **dextral** (right-handed). In other words, when the shell is viewed with the aperture lowermost and facing the observer, the aperture is on the right. There are, however, **sinistral** (left-handed) shells in various divisions of the Gastropoda (Figure 13.31, D). Within a single species there may be rare individuals in which coiling is opposite the typical direction. Moreover, some snails that are sinistral when they are very young later become dextral.

In general, gastropods that live on exposed shores have shorter spires than do related species inhabiting protected situations. Their apertures are likely to be proportionately larger and more nearly circular, presumably to enable the foot to cling tightly to a larger area. This variation in spire height may be found in a single species. In *Nucella lapillus,* common on the rocky shores of Great Britain and Europe, specimens living in protected areas have a higher spire and smaller aperture than those in exposed areas. If high-spired specimens are transplanted to an exposed area, they are more readily dislodged by waves than the natives.

Among the more obvious forms of sculpturing on shells are **spiral ridges** and **axial ribs** (Figure 13.17, A; Figure 13.31; Figure 13.33, C, E, and F). The former, which follow the whorls, develop because the mantle does not secrete new shell material evenly at the edge of the aperture; thus there are ridges alternating with furrows. Axial ribs—these are sometimes stout, sometimes like thin flanges—cross the whorls and appear when periods in which the shell is greatly thickened alternate with periods in which relatively little new material is deposited. To complicate matters further, if the mantle edge does not deposit a continuous axial rib, this will come out as a series of bumps, beads, or spines. Sharp spines (Figure 13.13, C) can be useful in discouraging fishes and other animals that swallow whole snails, and stout spines may reduce the vulnerability of snails to predators that crush shells. Thick shells offer protection against crabs and fishes whose *modus operandi* is to crack shells or chip them away.

Snails such as *Olivella* (Figure 13.33, A and B), which plow actively through mud or sand in search of prey, generally have little or no external sculpturing. Even among species whose shells are sculptured, there may be considerable variation, much of it related to conditions in the habitat. In *Nucella lamellosa* (Figure 13.33, C), which inhabits rocky shores along the Pacific coast, specimens with conspicuous axial flanges are likely to be found only in quiet or deep water; relatively smooth shells are typical of specimens that live where there is strong wave action.

Most aquatic snails, after being dislodged and falling through the water, land with the aperture facing up. This is potentially dangerous, because in order to right themselves, the snails must expose the foot and hence run the risk of having it nipped off by a fish. Some fishes do, in fact, dislodge larger snails and then attack them. An interesting adaptation that reduces vulnerability to predatory fishes is seen in *Ceratostoma foliatum* (Figure 13.31, B). The shell of this species has three winglike axial ribs. When the snail falls through the water, it rocks back and forth about its long axis. It lands with its aperture down about half the time, in which case it does not need to extend its foot so much to reattach itself.

In some gastropods there is sculpturing within the aperture. This usually consists of a ridge or a series of toothlike elevations. A prominent tooth that projects from the lip is characteristic of many predatory snails. *Acanthina* (Figure 13.33, F) inserts its tooth into the aperture of a barnacle or another snail, and thus remains firmly positioned during the slow process of drilling through the shell of its prey. *Ceratostoma* (Figure 13.31, B) uses its tooth to grasp the edges of a hole made in soft rock by a boring bivalve mollusc, then sticks its long pharyngeal tube down into the burrow to get at the tissue of the animal. It can also grip the shell of a barnacle. *Opeatostoma,* found off the Pacific coast of Central America, is thought to use its exceptionally long tooth (more than a third the length of the shell) for anchoring itself in sand or gravel as it plunges its pharyngeal tube into the burrows of polychaetes.

Torsion

During the early development of most gastropods, the mantle cavity, at first situated posteriorly, shifts in a counterclockwise direction until it is above the head and facing forward (Figure 13.14). This remarkable phenomenon, called **torsion,** must not be confused with coiling of the visceral mass, a separate matter. As a rule, torsion is accomplished in two stages. In the first stage, contraction of certain muscles brings about a rotation of about 90°; this step may be completed in a few minutes, or it may require a few hours. The second stage, which depends on differential growth and generally takes at least a day, shifts the mantle cavity another 90°. In some gastropods, torsion may be effected almost completely by one or the other of these processes. No matter how torsion is achieved, however, the relationships of several structures of the adult are drastically altered by it (Figure 13.15, A and B). The gut becomes looped because the anus is carried along with the mantle cavity to a forward position. The ctenidia and renal organs are likewise brought anteriorly. The connectives that join the pleural, parietal, and visceral ganglia become noticeably twisted. The posterior portion of the left-hand nerve complex is displaced to the right side and goes under the gut; the posterior portion of the right-hand complex is carried to the left, above the gut.

The significance of torsion is a much-debated subject. It has been considered an adaptation that permits the head of the larval gastropod to withdraw quickly to a position of relative safety. The less vulnerable foot

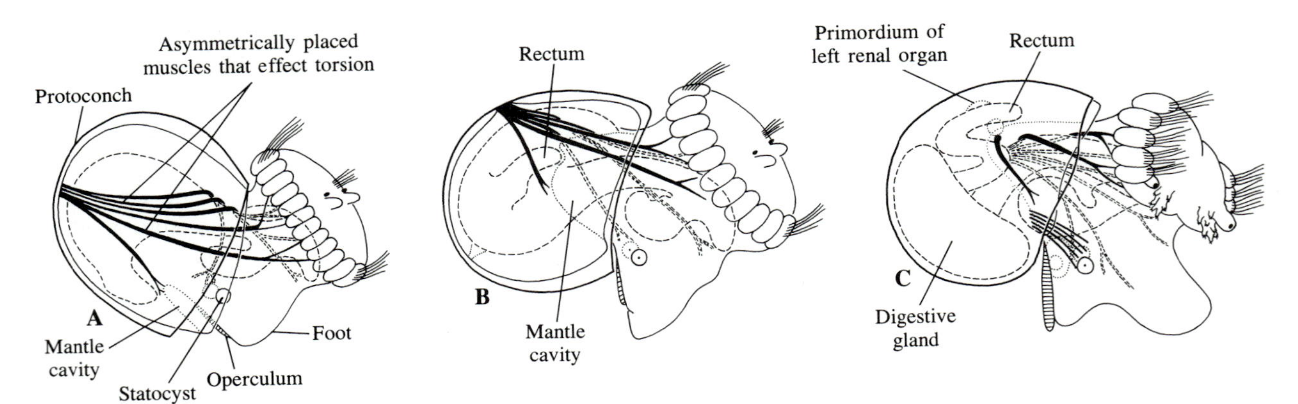

FIGURE 13.14 *Patella vulgata* (Patellogastropoda), stages in torsion, viewed from the right side. A. Veliger, 70 hours old, immediately preceding torsion. B. Veliger, 76 hours old, 90° torsion completed. C. Veliger, 96 hours old, 180° torsion completed. (After Crofts, Proceedings of the Zoological Society of London, *125*.)

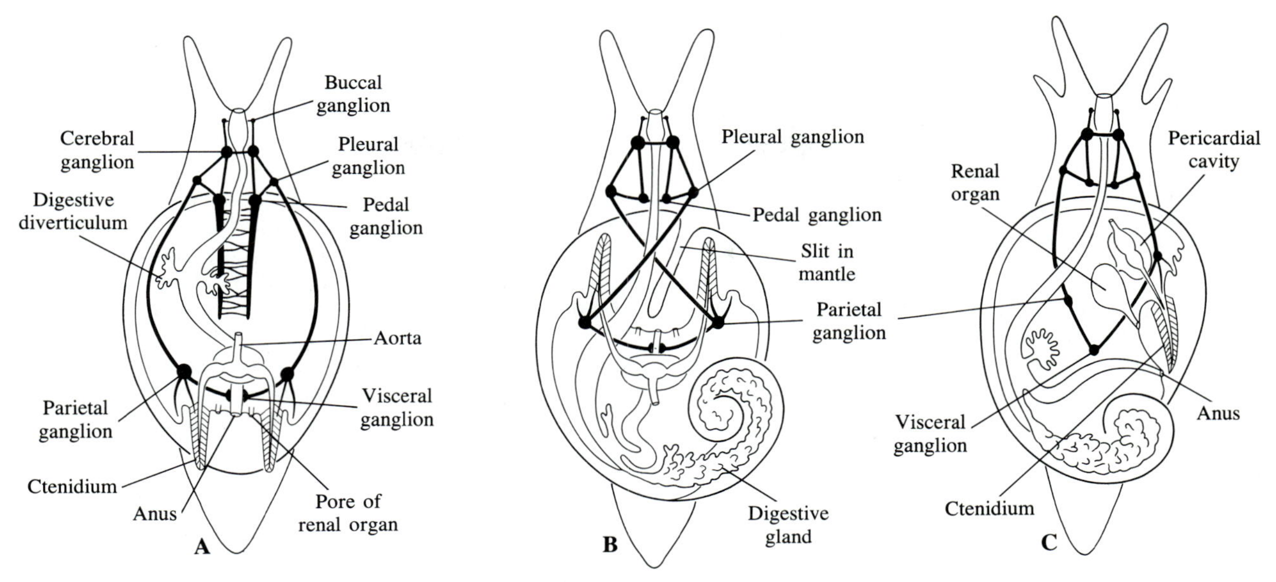

FIGURE 13.15 A. Diagram showing probable ancestral arrangement of ganglia, ctenidia, and other structures in gastropods. B. Archaeogastropod similar to *Haliotis,* showing 180° torsion. C. Opisthobranch, partly detorted.

would follow. This idea, which stresses the value of torsion to the larva, may not seem especially persuasive in light of the fact that most predators of larval and juvenile gastropods probably swallow them whole.

A different concept of the advantages of torsion stresses the importance of a forward-directed mantle cavity to the adult, even if the larva profits from torsion in some other way. An advantage of having the ctenidia face forward is that the water passing over them will be relatively uncontaminated by material that may be stirred up by movement of the snail. Any sensory structures in the mantle cavity that are concerned with monitoring water quality or detecting food will be able to function better if they are closer to the anterior end than the posterior end. However, though torsion may facilitate both respiratory and sensory functions, it does create a problem. The anus and openings of the renal organs now discharge wastes close to the head and into the incoming current of water, and these wastes could foul the mantle cavity. Gastropods have solved the problem of sanitation in a variety of ways, and the classification of gastropods is to a large extent based on the morphological differences connected with these solutions.

Torsion may also make it easier for a gastropod to carry its shell. Among early Cambrian fossils are planispiral mollusc shells that were apparently carried with the aperture posteriormost (Figure 13.16, A). As the visceral mass and shell of the mollusc enlarged, the center of gravity would have shifted progressively for-

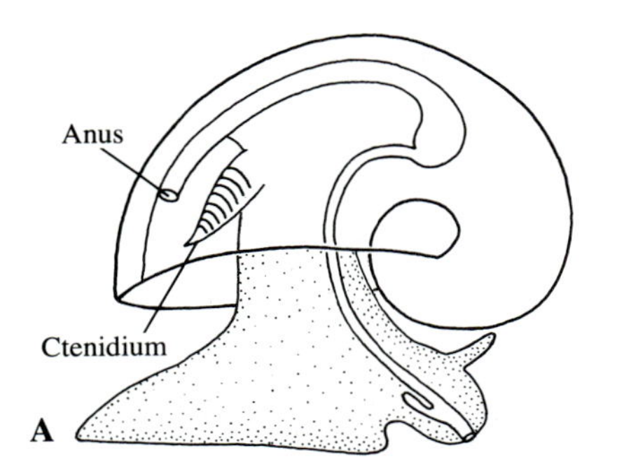

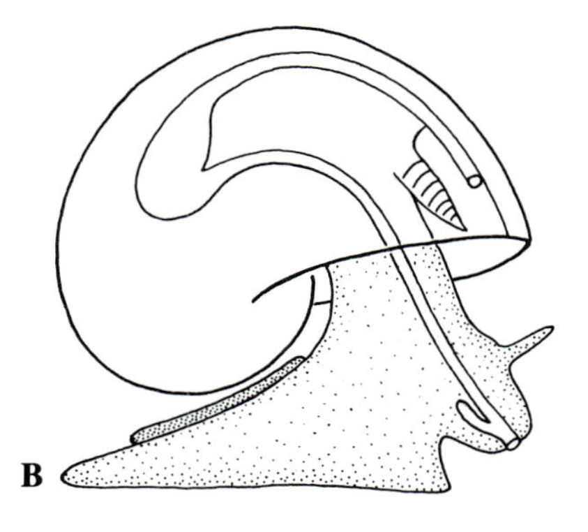

FIGURE 13.16 A. Reconstruction of an early Cambrian mollusc, showing shell carried with the aperture posteriormost. B. Similar mollusc, as it would be after torsion. (After Ghiselin, Evolution, *20.*)

ward, perhaps to the point that the animal would have been pushing its shell rather than pulling it. Torsion would have reversed the orientation of the shell (Figure 13.16, B) and thus made locomotion more efficient. That this idea is a logical one is supported to some extent by the fact that many gastropods that have a lightweight shell, or no shell at all, have undergone detorsion.

The Head, Foot, and Locomotion

The head is almost always distinct, even if there is no constriction between it and the foot (Figure 13.17). Besides bearing the mouth and other structures concerned directly with feeding, the head generally has recognizable sense organs in the form of eyes and tentacles. When tentacles are present, the eyes may be situated at their tips or at their bases.

The foot, as a rule, is proportionately large, and is used for creeping over firm substrata or for plowing through mud or sand. It consists mostly of muscle and connective tissue, but it has blood sinuses that contribute to a hydrostatic skeleton. In gastropods that are protected by a shell, the foot is pulled back into the shell by a powerful retractor muscle. One end of this fans out into the foot; the other end is attached to tissue bound to the columella, or, in the case of limpets, to the apical region of the shell. A few modern gastropods, including abalones, have two retractor muscles, but the one on the left is small.

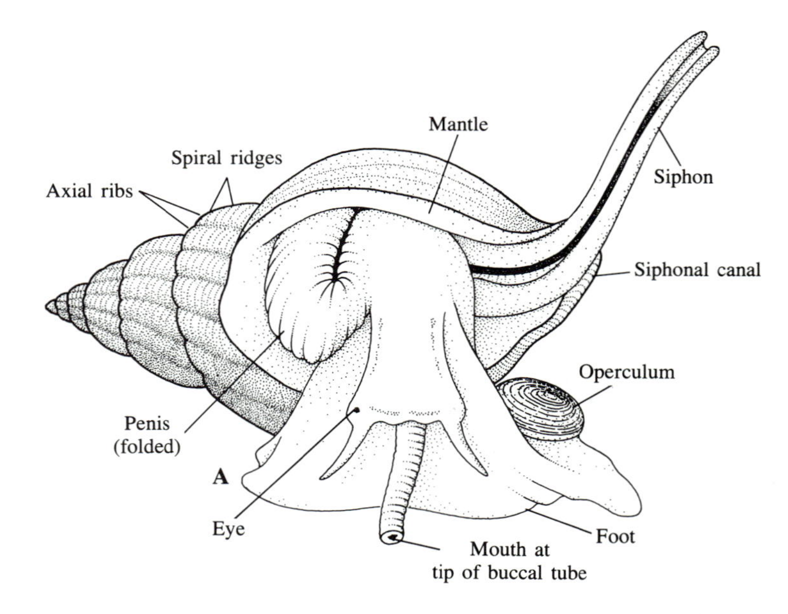

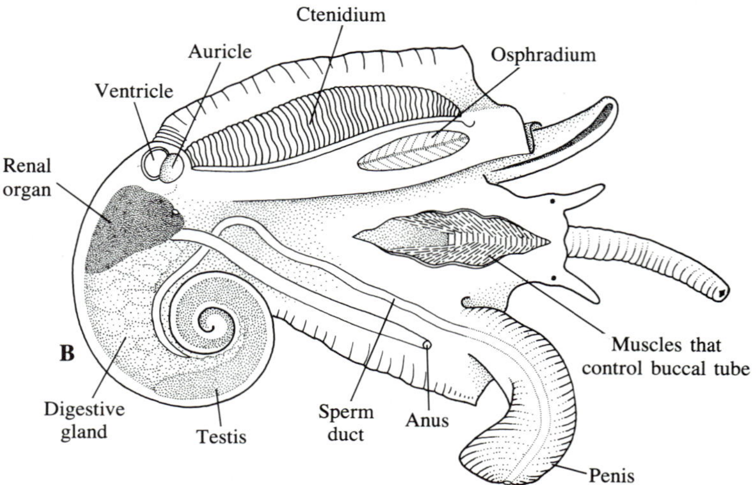

FIGURE 13.17 *Buccinum undatum* (Neogastropoda). A. Intact animal; right anterolateral view. B. Mantle cavity opened and part of buccal tube exposed by dissection; dorsal view. (After Cox in Moore, Treatise on Invertebrate Paleontology, I, Mollusca *1*.)

In a snail that has a coiled shell, the posterior part of the foot is the last to be pulled back into the shell when the animal is dislodged or threatened. The back side of this portion of the foot often has a "trapdoor" that closes the aperture. This is called the **operculum** (Figure 13.17, A), and it usually consists of horny material similar to the periostracum. In some gastropods, however, the operculum is heavily calcified.

The foot is ciliated and provided with glands that secrete mucus. The pedal gland, located in the anterior part of the foot, is an especially prominent mucus-secreting gland noted in many gastropods. In snails and slugs that glide over firm surfaces, locomotion is effected mostly by waves of muscular contraction that sweep forward or backward across the sole of the foot (Figure 13.18). The waves may be transverse or diagonal to the long axis of the foot. There are generally several successive waves in progress at a given moment. In some gastropods, there are two sets of waves; those on one side may be either in phase or out of phase with those on the other side.

Some snails extend the slightly raised anterior part of the foot, then establish firm contact with the substratum and with one powerful contraction pull the rest of the foot forward. A rapid, "pole-vaulting" type of locomotion is used by certain conchs that live on sandy bottoms. In these snails the operculum on the back of the foot is specialized as a clawlike structure that can be pushed into the sand. Contraction of the columellar muscle tips the animal forward, whereupon the anchor is lifted and then pushed into the sand again.

Moon snails (*Polinices*, Figure 13.28, A and B) and many others that plow through soft sediments, or glide over them, move by the action of cilia, aided by the lubrication provided by mucus. In general, the waves of ciliary activity on the sole and sides of the foot sweep from front to back.

Many gastropods, mostly types that filter microscopic food from water currents, move only when they are forced to relocate themselves. Some of the filter-feeders, in fact, do not move at all. They have cap-shaped shells whose margins, in the course of growth, become distorted so that they fit the contour of a rock or shell perfectly (Figure 13.28, C–E). If one of these gastropods is detached, it is not likely to be able to reestablish itself except in the same place. There are also gastropods that are permanently cemented to rocks or shells. These animals, called vermetids, have shells that are uncoiled or loosely coiled. They trap microscopic food in strands or nets of mucus.

In limpets and certain other snails, the foot can be used in much the same way as a suction cup. The animal holds itself so tightly to a rock or other firm surface that it is not easily dislodged. The fact that some intertidal limpets seem always to be in the same place does not mean that they do not move around. They probably wander freely at night or when submerged by tides. Certain species exhibit a homing behavior and return, after foraging, to where they have been living for an indefinite period. They recognize their homes by the physical texture of the surface or by the organic coating, or both. Some of them, over several years, erode the rock to the extent that the edges of their shells fit perfectly into slight depressions.

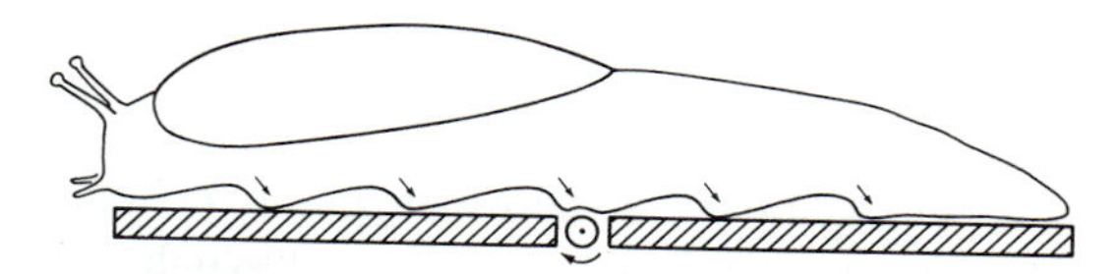

FIGURE 13.18 Diagram showing the waves of muscular activity in the foot of a slug. The roller is turned by the waves as they impinge on it. (After Bonse, Zoologische Jahrbücher, Abteilung für Physiologie, *54*.)

Snails that swim by muscular movements include the heteropods and so-called pteropods. In heteropods, which are often much compressed laterally, the foot is modified to serve as a fin; it is held uppermost. The shell, which is often reduced, hangs down. These odd gastropods are specialized for capturing and swallowing invertebrates and small fishes. In pteropods, some of which are filter-feeders, others carnivores, the foot is modified to form two flaps; these operate in much the same way as oars.

There are a few marine gastropods that float passively. The most famous of these is *Janthina* (Figure 13.29, C), a prosobranch that feeds on *Velella*, the by-the-wind-sailor, from which it gets its purple color. The foot of *Janthina* secretes gas bubbles and mucus, thereby forming a floating raft.

Digestive Tract, Feeding, and Digestion

The food habits of gastropods are extremely diverse, so there are many specializations in the organs concerned with feeding and digestion. Some of the specializations are evident in the radula. These are of considerable importance in classification and are discussed below in connection with the subclasses and certain orders of gastropods.

As a rule, there are one or two pairs of salivary glands whose ducts open into the buccal cavity. These glands are concerned primarily with secretion of slimy mucus that lubricates food and keeps particles together so that these may be processed efficiently. In some carnivorous gastropods, the salivary glands produce proteolytic enzymes.

The esophagus, which follows the buccal cavity, does not always have glandular elaborations. When they are present, they may be small diverticula or may be concentrated in a distinct gland. In gastropods that use a venom to subdue their prey, the secretion is produced by such an esophageal gland. In many opisthobranchs and pulmonates, the esophagus is enlarged to function as a crop into which a large amount of food can be stuffed, and in some opisthobranchs this part of the gut is muscular and lined with hard plates that crush and grind food.

Like almost everything else in gastropods, the stomach–digestive gland complex has many variations. In primitive types, such as abalones and trochids (Figure 13.20, C), the stomach has complex ciliated tracts for sorting particles. Acceptable particles are conveyed to the tubules of the digestive gland, where they are digested intracellularly. In more advanced groups, the stomach is less concerned with sorting particles. It is more likely to be somewhat muscularized and it often has hardened plates. Nevertheless, intracellular digestion of at least some components of the food is characteristic of many advanced gastropods. Furthermore, when digestive enzymes act in the stomach, as is the case in the large group of carnivorous snails called neogastropods, these enzymes are produced in the digestive gland.

It is impossible to do justice, in a few words, to the stomach–digestive gland complex. More will be said about it as the major groups of gastropods are dealt with.

The intestine is typically long and concerned mostly with consolidation of fecal matter. The terminal portion is called the **rectum,** although it may be histologically similar to the rest of the intestine. The functions of glands that open into the rectum of many gastropods are not known, but lubrication of fecal matter and excretion of nitrogenous wastes have been proposed. An excretory function is suggested by the relatively complex structure of the rectal glands of neogastropods.

Circulatory System

Two auricles are characteristic of the heart of the more primitive gastropods, whether they have two ctenidia (as abalones and keyhole limpets do) or only one. In more advanced gastropods, which have one ctenidium or none, there is a single auricle. The ventricle is more powerfully muscularized than the auricles. When the ventricle contracts, backflow is prevented by valves between the chambers. Blood is pumped into an aorta that soon divides into two main trunks, one supplying the head region and foot, the other supplying the visceral mass. From smaller arterial branches, blood enters the sinuses that collectively make up the hemocoel. In general, blood bathes the renal organs before going to the ctenidia, mantle wall, or lung for oxygenation. Many of the channels through which blood flows back toward the heart are sufficiently well defined that they can be called veins. Circulation is promoted by muscular activities involved in locomotion, feeding, and other body functions.

Hemocyanin is the usual respiratory pigment, but a few gastropods, such as the freshwater pulmonates of the genus *Planorbis,* have hemoglobin. This hemoglobin has a weak affinity for oxygen when compared with the hemoglobin of vertebrates.

Excretory System

The primitive arrangement of renal organs, in which a pair of them is joined to the pericardial cavity (Figure 13.21, C–F), is characteristic of abalones, keyhole limpets, and true limpets and their close relatives. The members of the pair are not equal—the one on the right is larger—but both operate as excretory organs, processing a filtrate that gets into the pericardial cavity through the wall of the heart, especially the auricles. The two renal organs may have appreciably different functions. Studies on abalones have shown that one is concerned to a greater extent than the other with reabsorption of water and certain ions. The right renal organ also serves as a gonoduct.

In more advanced gastropods, the arrangement is different. The left renal organ persists as an excretory structure (Figure 13.21, H). Part of the one on the right is retained and functions as a gonoduct. There may no longer be a reno–pericardial connection, and in that case blood spaces around the renal organ, rather than around the heart, are the source of the filtrate. These spaces are also concerned with reabsorption of water and ions.

The principal nitrogenous waste excreted by aquatic gastropods is ammonia. Urea is not often produced, but amino acids and purines are eliminated in substantial quantities by some species. In pulmonates and certain prosobranchs that are adapted for living out of water all or much of the time, the principal nitrogenous waste is uric acid, which is excreted in crystalline form. Because of its low solubility, crystalline uric acid is relatively nontoxic and is an appropriate waste product for animals that cannot afford to lose much water. Many freshwater pulmonates, derived from groups that once were terrestrial, continue to excrete considerable amounts of uric acid.

Nervous System and Sense Organs

The layout of the nervous system is influenced by torsion and detorsion, as well as by coiling and other manifestations of asymmetric growth. Figure 13.15, A shows what the nervous system of presumably symmetrical early gastropods may have looked like, whereas Figure 13.15, B and C show the effects of torsion and detorsion, respectively. Figure 13.45, A shows **condensation,** the shortening of certain connectives to the point that some of the ganglia are run together.

The **cerebral ganglia,** above the esophagus, are almost always prominent. They innervate the eyes, tentacles, lips, skin of the head region, and statocysts. The **buccal ganglia,** applied to the wall of the buccal cavity, send nerves to the sheath of the radula, salivary glands, and some other structures in that general region of the body. The **pedal ganglia,** typically just below the cerebral ganglia, innervate the foot. The **pleural ganglia,** more or less lateral to the cerebral ganglia and usually slightly posterior to them, send nerves to the mantle edge and to certain of the muscles that retract the foot.

The **parietal ganglia** innervate the ctenidia, osphradia, esophagus, renal organs, gonads, and a variety of other structures. They are often deep in the visceral mass, but in some gastropods they are fairly far forward. The **visceral ganglia** are mostly concerned with innervation of the rectal portion of the gut. The parietal and visceral ganglia are the ones most likely to be affected by torsion, detorsion, and asymmetrical growth. Note in the diagram of a prosobranch gastropod (Figure 13.15, B) how torsion has brought about twisting of the pleuro–parietal connectives. The connective that would have been on the right side if there had been no torsion is looped over the gut, and the one that would have been on the left side is looped under the gut. In other words, the positions of the right and left parietal ganglia, as well as of the corresponding visceral ganglia, have been reversed. The original right parietal ganglion is often called the **supraintestinal ganglion,** especially when it lies above the gut; its counterpart, below the gut, is the **infraintestinal ganglion.**

The effects of detorsion can be seen in the nervous systems of most opisthobranch and pulmonate gastropods. In both groups, as a rule, detorsion is accompanied by moderate to drastic condensation. It is best not to generalize too much, however, for there is much variation among both opisthobranchs and pulmonates. In some opisthobranchs, in fact, the nervous system is torted just as much as it is in prosobranchs.

Much of the exposed body surface of almost any gastropod is sensitive to touch. Microscopic examination of the surface will in fact reveal many little bristles, or clusters of bristles, linked with receptors in the epidermis. The tentacles on the head, as well as various papillae and fleshy outgrowths of the mantle and foot, are certainly well provided with touch receptors. Olfactory receptors are present on the head tentacles of at least some types of gastropods.

When eyes are present, they are usually located at the tips of the tentacles. They are at the bases of the tentacles, however, in freshwater pulmonates and some pulmonates that live at the fringes of the sea. In a few unusual opisthobranchs that have a lung similar to that of pulmonates, there are several to many eyes scattered over the dorsal surface of the sluglike body. The simplest eyes, such as are found in certain limpets, are pits lined by light-sensitive retinal cells, among which are interspersed pigmented cells. In more complex eyes, the eye is nearly or completely covered by a cornea and has a lens. It is probable that the eyes of most gastropods are concerned mostly with detection of light rather than with forming an image. The behavior of some snails, however, indicates that they use their eyes to locate and recognize prey, or to recognize predators in time to escape. This is almost certainly the case in heteropods, which are carnivores that swim in the marine plankton. The large and complicated eyes of these molluscs are equipped with muscles, so they can be rotated.

Statocysts are more or less regularly present, except in vermetids, whose shells are cemented to a hard surface, and in some other gastropods that are more or less sedentary. The statocysts are paired and embedded in the tissue of the foot, often close to the pedal ganglia. The nerves that supply them come, however, from the cerebral ganglia. The heteropods seem to have the most complex statocysts. This is to be expected, for these molluscs are rather agile swimmers to whom organs of equilibrium are important.

The **osphradium** is a sense organ in the mantle cavity of many gastropods, especially those that continually circulate water through the cavity. It is often especially well developed in carnivores that search for live or dead animals, and it is believed to have an olfactory function. For some gastropods, however, it may serve mostly as a ''silt meter,'' keeping the animal informed about the amount of suspended sediment in the water taken into the mantle cavity. Osphradia are located close to the ctenidia and are sometimes so extensively folded that they could be mistaken for ctenidia. In keyhole limpets, abalones, and other primitive prosobranchs that have two ctenidia, there are also two osphradia; prosobranchs with a single ctenidium have only one osphradium.

Some opisthobranchs and aquatic pulmonates also have an osphradium. In opisthobranchs, it is located in the mantle cavity, unless this has disappeared, in which case the osphradium is just in front of the anus. In aquatic pulmonate snails, it is beside the opening into the lung cavity.

Reproduction and Development

Gastropods have a single gonad. In limpets, keyhole limpets, abalones, and other relatively primitive types that belong to the order Archaeogastropoda of the subclass Prosobranchia, two coelomoducts function as renal organs. The gonad is linked with one of them. This coelomoduct therefore serves as both an excretory structure and a gonoduct. The extent to which it can be specialized for reproduction is limited, so lower gastropods are generally free-spawners. In the higher gastropods—the majority of the prosobranchs and all members of the subclasses Opisthobranchia and Pulmonata—it is the rule that one coelomoduct is a renal organ; the other contributes to a gonoduct that has various elaborations concerned with copulation, internal fertilization, and production of capsules or gelatinous material for enclosing the eggs.

In prosobranchs, including the more primitive, free-spawning archaeogastropods, the gonad produces only sperm or only eggs, so the sexes are separate. In the opisthobranchs and pulmonates, however, hermaphroditism prevails. The gonad of these gastropods is called an **ovotestis** because it produces both eggs and sperm, although not necessarily simultaneously. A free-swimming trochophore larva, somewhat similar to that of annelids, is characteristic of the archaeogastropods. The trochophore is succeeded by a veliger (Figure 13.14), which has a large, bilobed flap called the velum (veil), a rather distinct head (with tentacles and eyes), the beginnings of a foot, and a larval shell. The velum is essentially an expansion of the prototroch of the trochophore stage. The veliger uses its long velar cilia for swimming and for bringing diatoms and other small particles into the range of the cilia in a food groove. Within the groove, mucus secretion and ciliary activity move the food to the mouth.

In more advanced gastropods, whose eggs are inclined to be large, yolky, and enclosed within a firm mass or within capsules, there is no recognizable trochophore stage. A veliger stage is likely, however, except in land snails and their freshwater relatives, and in some other types in which development is direct. The veliger does not always escape from the egg mass or egg capsule within which it has developed. In many gastropods, the veliger transforms into a small snail before it emerges to begin a life on its own. In such cases, although the veliger is nourished by yolk, it may eat up its siblings in the same capsule. It often happens that not all of the fertilized eggs in a capsule develop normally. Some of the abnormal embryos may be swallowed whole by the veligers, or break down into cells that the veligers can eat. Even among the species of a single genus, there may be considerable variability in the number of veligers that develop normally and the number of embryos that provide food.

It is during the veliger stage that torsion takes place (Figure 13.14). The outcome and consequences of this phenomenon were discussed above, but the mechanics of the process are described briefly here. The right and left mesodermal bands of the early veliger are asymmetrically developed. The right band gives rise to several muscles that serve as retractors of the larval foot. Posteriorly, these muscles converge upon a rather restricted area of the shell, and anteriorly they bend to the left in front of the gut as they spread out into the tissues of the foot. At the appropriate time, a strong contraction of the muscles straightens them, pulling the hindmost part of the veliger's shell around to the right, and sometimes as far as it can go in a forward direction. In the latter case, muscular contraction accounts for all or almost all of the 180° rotation. In many gastropods, however, it takes care of only part of the process of torsion; the rest is achieved by differential growth.

Torsion may stop far short of the maximum, as is the case in some members of the subclass Opisthobranchia. A further complication is that asymmetry may be achieved without any obvious torsion. In most opisthobranchs, for instance, the genital ducts lead to gonopores on the right side just behind the head; their definitive position is established without any obvious rotation. Some other structures of opisthobranchs may, in fact, be situated in such a way that they are in a pretorsional position, or at least not far from a pretorsional position. Thus opisthobranchs as a group are said to have undergone **detorsion** in their evolution. This does not mean, however, that one can observe first a torsion then a reversal of torsion in the development of individual opisthobranchs. The adult form is achieved more or less gradually, even though there may be a distinct veliger stage.

CLASSIFICATION

SUBCLASS PROSOBRANCHIA

The prosobranchs make up the largest and most diversified subclass of gastropods. The name given to the group is based on the fact that the ctenidia of these molluscs, when present, are in front of the heart. This forward position of the ctenidia is due to torsion, which brings the mantle cavity and the structures associated with it from the rear to the front of the animal, and which also twists the nervous system.

Sexes are separate. In this respect prosobranchs are decidedly different from most members of the other two subclasses, which are hermaphroditic.

Order Archaeogastropoda

As the name Archaeogastropoda suggests, the molluscs in this group are considered to be the most primitive of the living gastropods. In each of the many transverse rows of teeth in the radula is a single large central tooth, and on both sides of this are a few substantial lateral teeth and numerous slender marginal teeth (Figure 13.20, A). This type of radula, said to be of the *rhipidoglossan* (''fan-tongue'') type, is effective in scraping off diatoms, algal crusts, sponges, and other low-growing organisms.

There may be two ctenidia—the ancestral condition (Figure 13.20, B)—or only one, but the ctenidia are mostly, if not completely, *bipectinate;* that is, they have two rows of filaments. Typically, but not always, the filaments have internal skeletal rods.

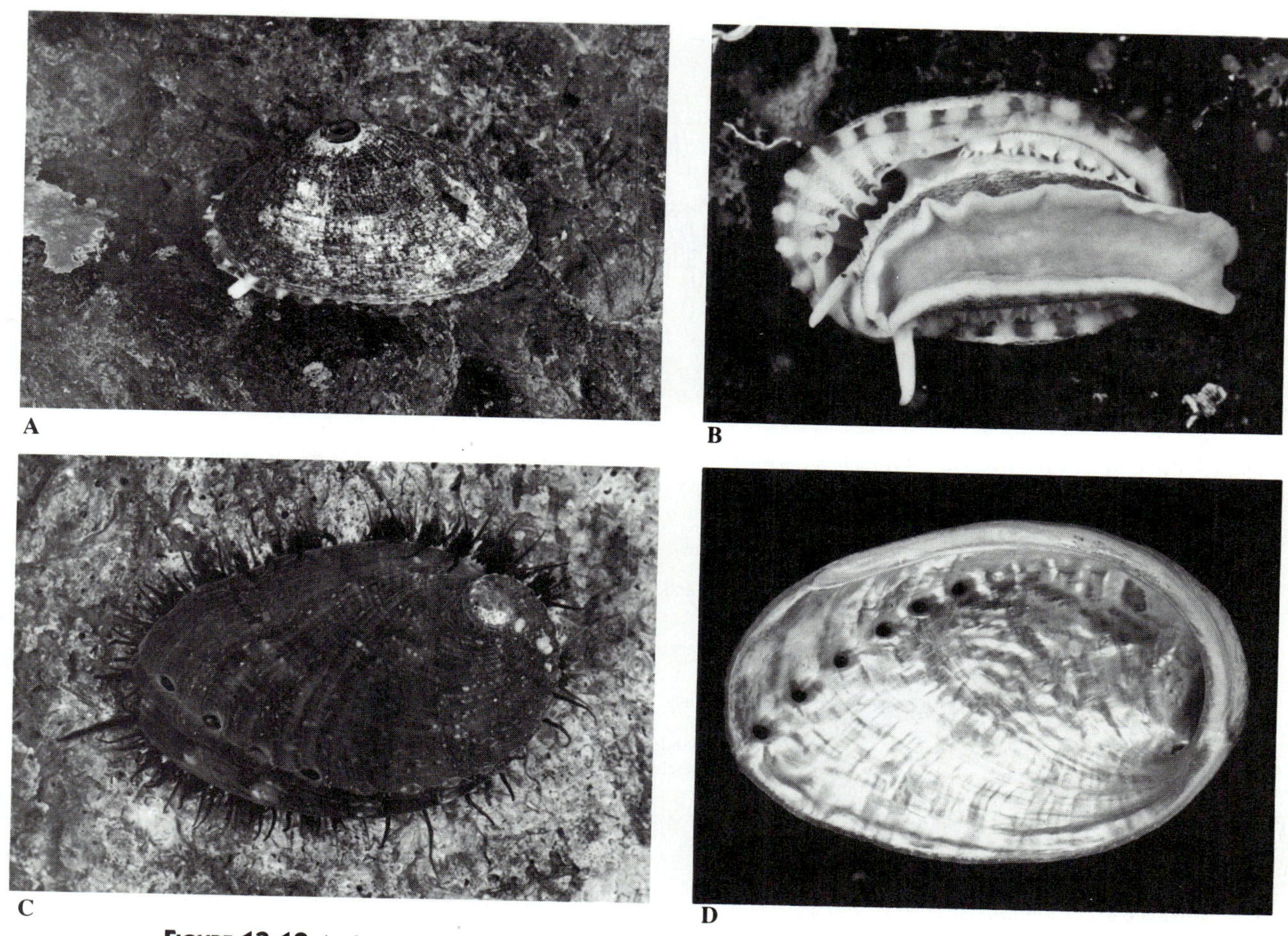

FIGURE 13.19 Archaeogastropoda. A, B. *Diodora aspera*. In the ventral view (B), a portion of a commensal polychaete annelid (*Arctonoe vittata*, family Polynoidae) is visible. C. *Haliotis kamtschatkana*. D. *Haliotis walallensis*, interior of shell. (See also *Fissurellidea*, *Gibbula*, and *Calliostoma*, Color Plate 4.)

In the archaeogastropods that belong to the group called keyhole limpets—*Diodora* (Figure 13.19, A and B), *Puncturella* (Figure 13.20, B), and *Fissurellidea* (Color Plate 4) are examples—the shell has a hole at or close to the apex. The hole originates as a marginal slit, but it soon becomes isolated from the edge of the shell. Water brought into the mantle cavity by the activity of cilia on the two ctenidia (Figure 13.22, A) exits by way of the opening. The anus and pores of the renal organs are close to the opening, so that wastes enter the current just before it leaves the animal's mantle cavity. This minimizes the chance that there will be fouling of the ctenidia and other mantle structures.

In abalones (*Haliotis*, Figure 13.19, C and D), which also have paired ctenidia, there is a similar pattern of water circulation. The shell is coiled, but it has one or more holes that serve the same function as the single opening in the symmetrical shell of a keyhole limpet. Each new hole is formed at the margin, and at any particular time several holes may be operational. The older ones eventually become sealed; they are of no use after the mantle cavity has moved, as a consequence of growth of the animal as a whole, from the part of the shell where the first few openings were formed.

Some snails related to keyhole limpets and abalones have a coiled shell with a marginal slit that becomes deeper as the animal adds shell material to the edge of the aperture. The slit serves in the same way as one or more holes.

In archaeogastropods with paired ctenidia, each auricle of the heart receives blood from one of them. There are two renal organs (Figure 13.21, D and E); at least one of them is joined to the pericardial cavity. The

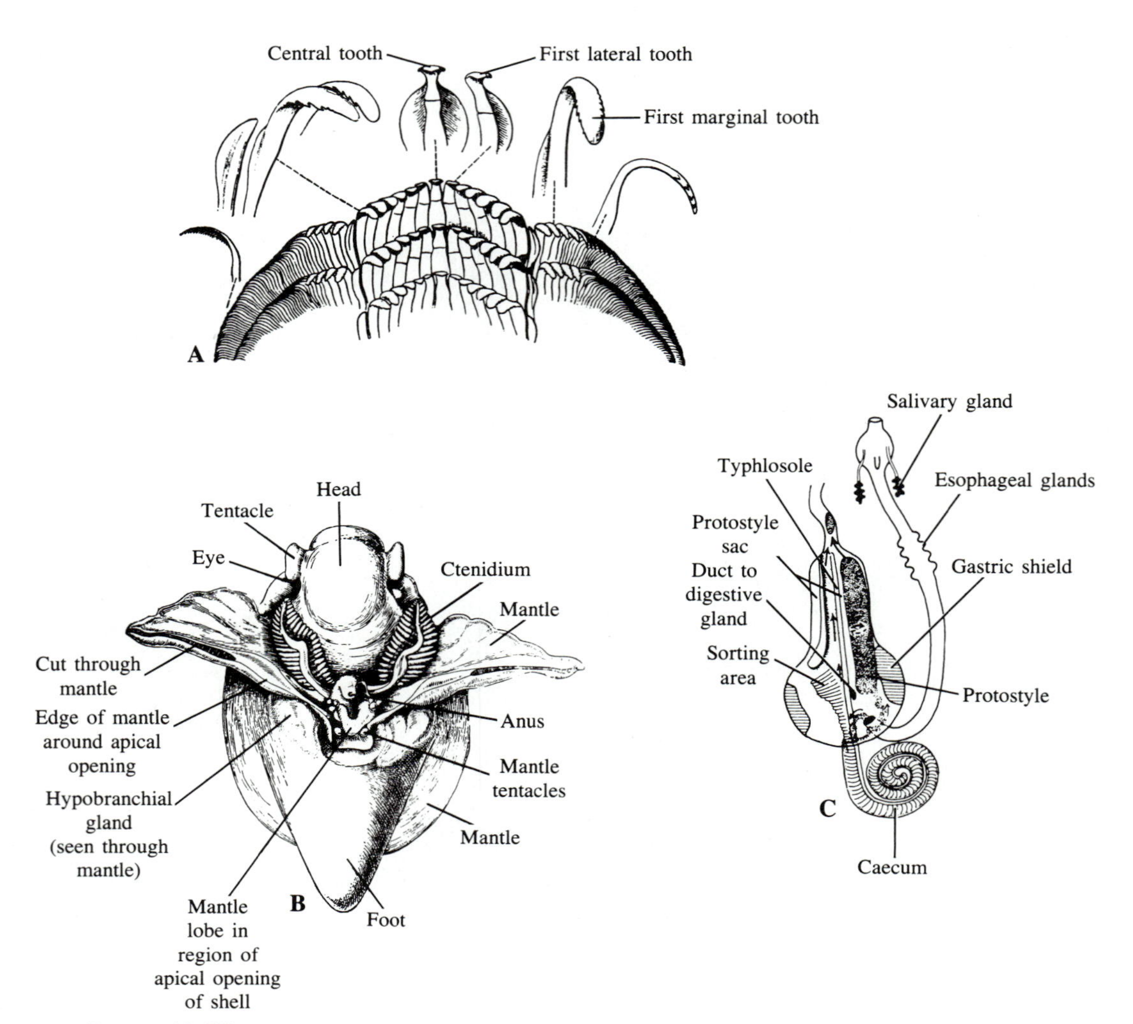

FIGURE 13.20 Archaeogastropoda. A. Portion of the radula of a species of *Margarites*. (Cooke, in Cambridge Natural History, *3*.) B. *Puncturella noachina,* shell removed and mantle cavity opened by a cut. (Ivanov et al., Major Practicum of Invertebrate Zoology.) C. Diagram of the anterior part of the digestive tract of a trochid, such as *Trochus, Gibbula,* or *Calliostoma*. (Owen, in Wilbur and Yonge, Physiology of Mollusca, *2*.)

right renal organ is the larger and receives the duct of the gonad; it is therefore concerned not only with excretion but also with spawning, for it carries eggs or sperm to the mantle cavity.

Not all archaeogastropods have paired ctenidia. In top shells, such as *Gibbula* and *Calliostoma* (Color Plate 4), and also in turban shells and some other members of the order, there is just one ctenidium, corresponding to the left ctenidium of keyhole limpets and abalones. It is provided with two rows of filaments and is free of the mantle wall for much of its length. The left auricle, as would be expected, receives blood from the ctenidium. Of the two renal organs, the right one is the larger and serves also as the gonoduct (Figure 13.21, C). Water coming into the mantle cavity from the front and left side passes over the ctenidium (Figure 13.22, B), then over the anus and excretory pores, which are displaced toward the right, where the water exits. In this group, then, fouling is minimized because water tends to move across the mantle cavity in one direction.

There are some archaeogastropods whose morphology does not conform to that of examples mentioned above. In *Nerita* and several other genera, for instance,

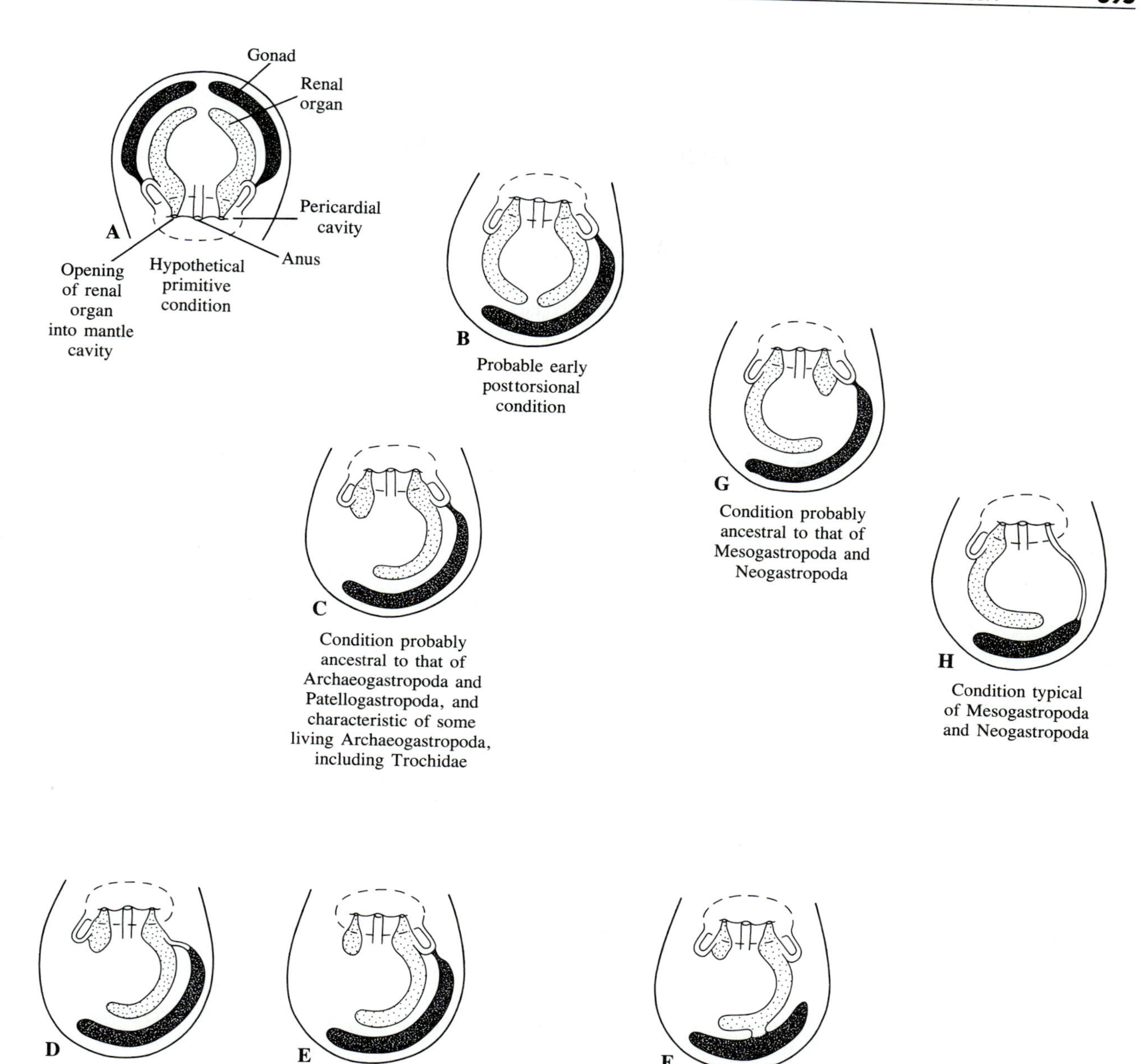

FIGURE 13.21 Relationship of the gonad and renal organ to the pericardial cavity in various prosobranch gastropods. (After Yonge, Philosophical Transactions of the Royal Society of London, B, *232*.)

the ctenidial filaments lack skeletal rods, and the right coelomoduct is modified for strictly reproductive functions; the left coelomoduct is the only renal organ.

In its most elaborate and presumably most primitive form (Figure 13.20, C), the archaeogastropod stomach has an extensive sorting area consisting of ciliated ridges and furrows, a blind sac (caecum) into which the sorting area extends, and a hardened shield. There is also a broad channel, bordered by two ridges called *typhlosoles,* that continues into the intestine, where it forms a trough somewhat separate from the rest of the lumen. The trough secretes a string of mucus, the *protostyle,* which is rotated by ciliary activity. The protostyle projects into the stomach, and it is believed to

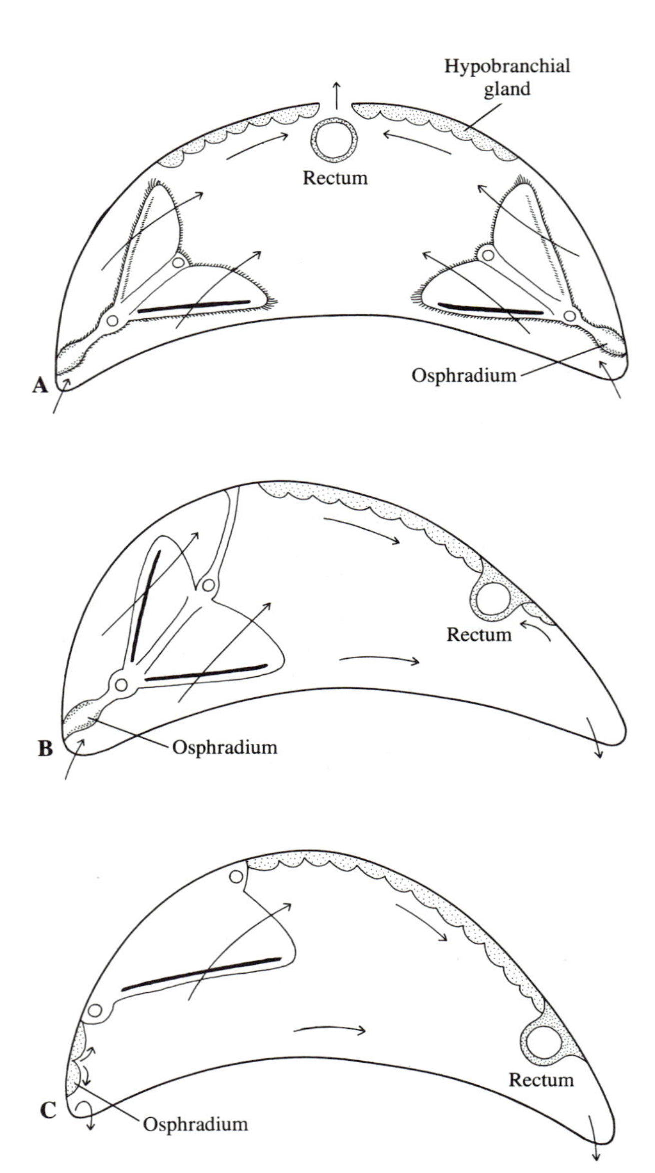

FIGURE 13.22 Diagrams of the mantle cavity of prosobranch gastropods, as seen in transverse sections viewed from the posterior side. Arrows show the direction of water currents. A. Archaeogastropod with two bipectinate ctenidia, as in *Haliotis, Diodora,* and *Puncturella*. B. Archaeogastropod with a single, mostly bipectinate, ctenidium, as in *Gibbula, Calliostoma, Tegula,* and *Margarites*. C. Mesogastropod or neogastropod with a single monopectinate ctenidium. (After Yonge, Philosophical Transactions of the Royal Society of London, B, *232*.)

function to some extent as a winch that winds in the mixture of food and mucus entering the stomach from the esophagus. It could also help to abrade the mixture against the shield, thus setting food particles free so that they can be subjected to the sorting process in the stomach. Acceptable particles are driven by the action of cilia into the tubules of the digestive gland, where they are taken up by cells. Rejected particles are shunted into the intestine and become incorporated into fecal matter.

In keyhole limpets, which graze on sponges rather than on algae or diatoms, the caecum is reduced and the sorting area does not extend into it. There are other deviations from the primitive plan among various archaeogastropods, but in general the stomach of most members of the order is similar to the arrangement described above.

Order Patellogastropoda

This order includes the marine limpets, such as those of the genera *Lottia, Macclintockia, Acmaea,* and *Tectura* (Figure 13.23). The cap-shaped shell is different from that of keyhole limpets in that it has no opening near the apex. The radula, moreover, is relatively narrow (Figure 13.24, A). In each transverse row, the central tooth is either absent or much reduced. There are two to several lateral teeth on both sides of the midline and usually one to several marginal teeth, but in some genera marginals are lacking. This kind of radula, used mostly for rasping algae and scraping off coatings of diatoms, is said to be of the *docoglossan* (''speartongue'') type.

Limpets usually have a gill on the left side of the forward part of the mantle cavity (Figure 13.24, B), but it is thought not to be a true ctenidium. Its filaments lack supporting skeletal rods such as are typical of most archaeogastropod ctenidia, and there are also differences in the ciliation and other features. Secondary gills of another type are sometimes present in the lateral mantle groove; this is the case in *Patella,* which does not have a gill in the forward part of the mantle cavity.

The pattern of water flow in the forward part of the mantle cavity is from left to right. Thus clean water passes over the gill, if one is present, then goes on to pick up fecal wastes from the anus and excretory wastes from the right renal organ before leaving the mantle cavity. There is a left renal organ (Figure 13.21, F), but it is usually small or modified into a brood chamber.

The heart has a single auricle; when there is a gill in the forward part of the mantle cavity, blood that has been oxygenated in its filaments goes into the auricle. This arrangement is similar to that in archaeogastropods that have a single true ctenidium.

Advanced Prosobranchs

The two orders of advanced prosobranchs are the Mesogastropoda and Neogastropoda. It is not easy to draw a sharp line between them. In general, however,

Figure 13.23 Patellogastropoda. A. *Lottia ochracea*. B. *Macclintockia scabra*. C. *Acmaea mitra*. This species feeds on coralline red algae, and its shell becomes encrusted by algae of this type. B. *Tectura insessa,* which erodes the stipes of a kelp, *Egregia*.

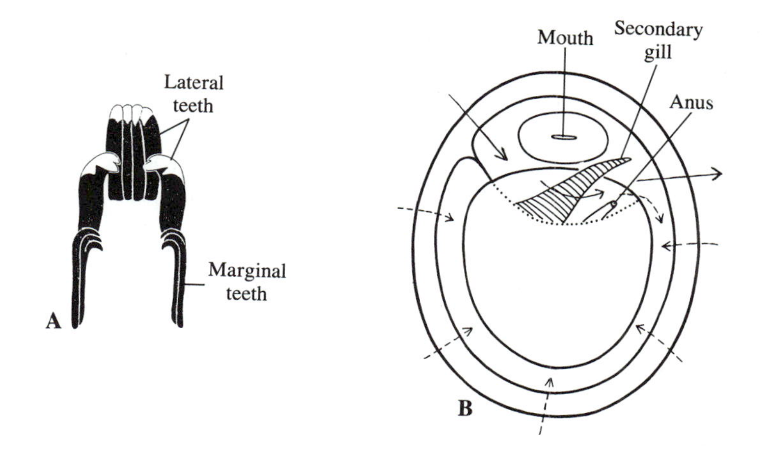

Figure 13.24 Patellogastropoda. A. Portion of the radula of *Patella vulgata*. (Richter, Natur und Museum, *9*.) B. Respiratory and cleansing currents in several species of limpets of the genera *Acmaea, Lottia,* and *Tectura*. (The animal is viewed from the dorsal side and drawn as if transparent.) Similar but not identical patterns occur in other species of these genera. (After Yonge, Veliger, *4*.)

the following features can be relied on with considerable (if not absolute) confidence. Most neogastropods have an extensile buccal tube (Figure 13.17), and some of them—the toxoglossans—have a proboscis that is not properly part of the gut. Mesogastropods never have the latter, and generally lack the former.

A siphon (a rolled-up edge of the mantle) and siphonal canal (a troughlike or tubular prolongation of the edge of the shell aperture, which protects the siphon) are prominent in neogastropods (Figures 13.13, 13.17, 13.31, and 13.33; also *Amphissa,* Color Plate 4), but are rarely prominent, and often absent altogether, in mesogastropods.

The differences in the organization of the radula also merit a detailed explanation. In most mesogastropods, the radula is of the *taenioglossan* ("ribbon-tongue") type. It is fairly broad and usually has seven teeth in each transverse row (Figure 13.3, C; Figure 13.4, A–C). The central tooth is generally well developed. The three sharp-cusped teeth on both sides of it are sometimes interpreted as laterals, sometimes as a lateral and two marginals.

The radula is narrower in neogastropods than in mesogastropods. The maximum number of teeth in each transverse row is five, and in some groups each row has only two teeth. The diversity of radular types among neogastropods is discussed below in a brief review of this order.

In some mesogastropods that graze on diatoms or other algal coatings, or that consume detritus with a high content of polysaccharides, the stomach has a crystalline style similar to that of most bivalves. This is a gelatinous rod that projects into the stomach. As it dissolves, it releases, among other things, enzymes concerned with digestion of polysaccharides. It is rare to find a crystalline style in neogastropods, although it does occur in a few species. A well-known case is that of *Ilyanassa obsoleta* (Figure 13.33, D), a mudflat snail which is native to the Atlantic coast of North America and which has been introduced to the Pacific coast. Unlike nearly all other neogastropods, which are carnivores, *Ilyanassa* feeds mostly on detritus. More is known about the crystalline style of bivalves than about that of gastropods, so this structure is discussed in greater detail below.

In various groups of mesogastropods, especially herbivores, there is a pair of jaws on the dorsal side of the buccal cavity, close to the mouth (Figure 13.3), or a single jaw that is formed by the fusion of two plates. In general, the jaws are probably used in conjunction with the radula for cutting off pieces of algae. Neogastropods have, at most, only vestiges of jaws.

In both mesogastropods and neogastropods, there is a single ctenidium, on the left side of the forward part of the mantle cavity (Figure 13.17, B; Figure 13.22, C). It is bound to the mantle wall for its entire length and has a single row of filaments. It is therefore said to be *monopectinate* rather than bipectinate. Only the left auricle of the heart has survived, and the left coelomoduct is the functional renal organ. It has no connection with the gonad and is purely excretory (Figure 13.21, H). Remnants of the right coelomoduct, still hooked up to the gonad, serve as a gonoduct and accessory reproductive structures. Since the gonoduct is no longer involved with a renal organ, it is free to become specialized in various ways related to reproduction. Thus there may be complex copulatory devices and glands (Figure 13.25) whose development would hardly be possible if the right coelomoduct were concerned with excretion.

The foot of female neogastropods has a gland that secretes hard, proteinaceous egg capsules (Figure 13.25, C); egg capsules of mesogastropods are usually somewhat gelatinous (Figure 13.26, B) and are secreted by glands along the oviduct.

Order Mesogastropoda

The more important features of mesogastropods have been mentioned in the preceding section, in which these molluscs were compared with neogastropods. The further discussion of the group here is therefore concerned mostly with examples of their lifestyles, especially with respect to feeding.

Mesogastropods include many herbivores, such as the intertidal periwinkles of the genera *Littorina* and *Lacuna* (Figure 13.26, A and B). These snails rasp tissue from algae or scrape coatings of diatoms and other microscopic plants. Nearly all of the freshwater prosobranchs are mesogastropods, and with a few exceptions they are also herbivores. Some of them, including *Goniobasis* and *Fluminicola,* widely distributed in North America, have a crystalline style, a feature not found in many gastropods.

Ciliary–mucous feeding is characteristic of several diverse groups, including the slipper-shells of the genus *Crepidula* (Figure 13.26, C–E). In these snails, as water is moved from left to right in the mantle cavity, diatoms and other small food particles are trapped in mucus that coats the ctenidial filaments. Tracts of cilia carry the mucus to the tips of the filaments, where it reaches a ciliated groove that runs forward along the right side of the body. The radula breaks off pieces of the string of mucus as this reaches the mouth. A network of mucous strands prevents large particles from entering the mantle cavity. This mucus is steadily moved to a pocket near the mouth, where it is compacted into pellets that may be either eaten or allowed to drift away in the current leaving the mantle cavity.

Somewhat similar techniques of ciliary–mucous

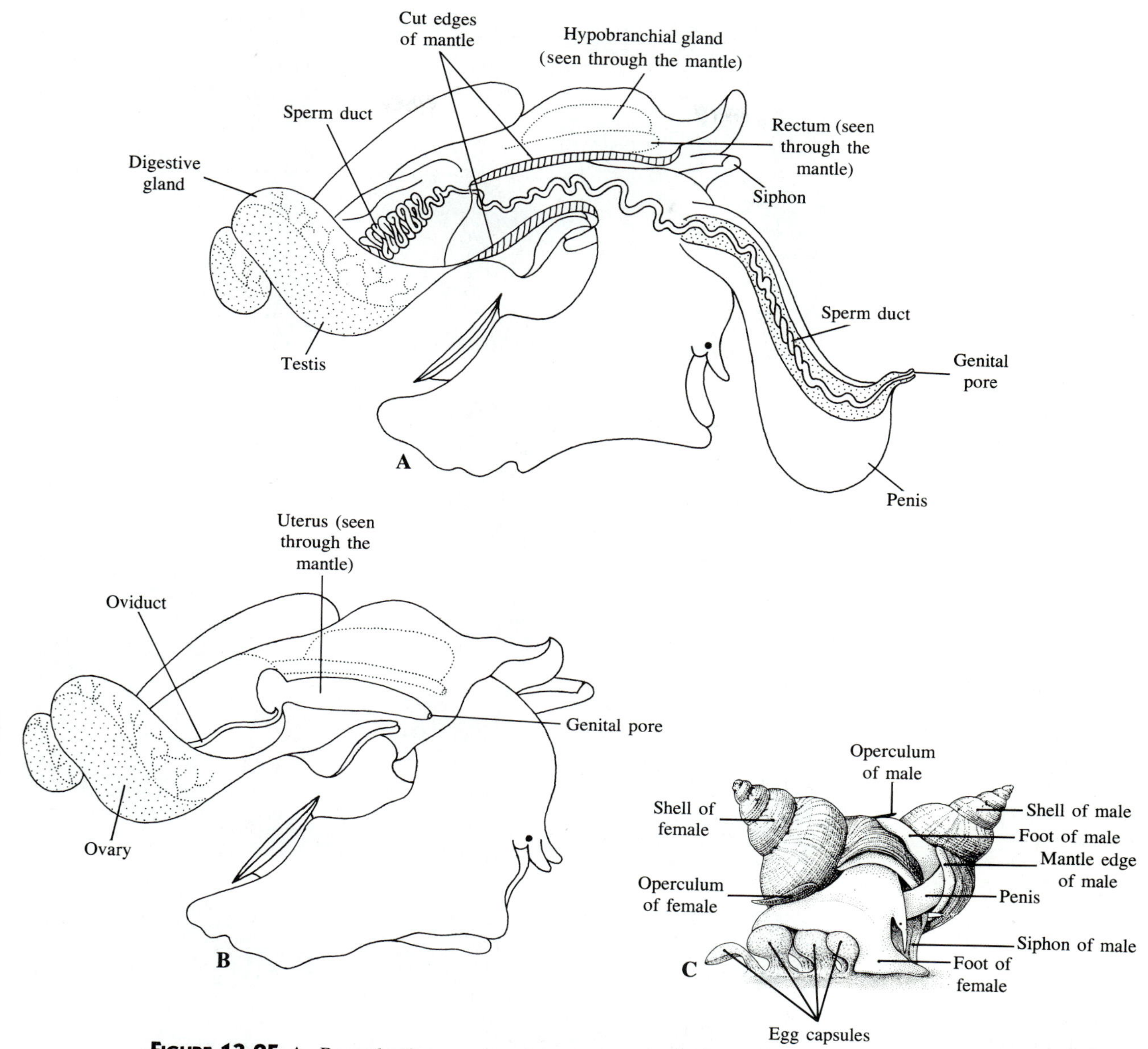

FIGURE 13.25 A. Reproductive system of a male *Buccinum undatum* (Neogastropoda), viewed from the right side. B. Female. (After Ivanov et al., Major Practicum of Invertebrate Zoology.) C. *Neptunea antiqua* (Neogastropoda), simultaneously copulating and depositing egg capsules. (Pearce and Thorson, Ophelia, *4*.)

feeding are used by other mesogastropods, including *Trichotropis* (Figure 13.27, A–C), and by the strange vermetid snails, whose uncoiled shells, resembling worm tubes (Figure 13.27, D), are attached to rocks or pilings. Certain vermetids, however, employ a decidedly different method. They secrete long strands of mucus from a gland on the foot. After the strands have accumulated food for a while, they are pulled back and swallowed. In *Serpulorbis squamigerus* of the Pacific coast, several individuals contribute to a sheet of mucus; after one begins to pull in its share of food, the others do likewise.

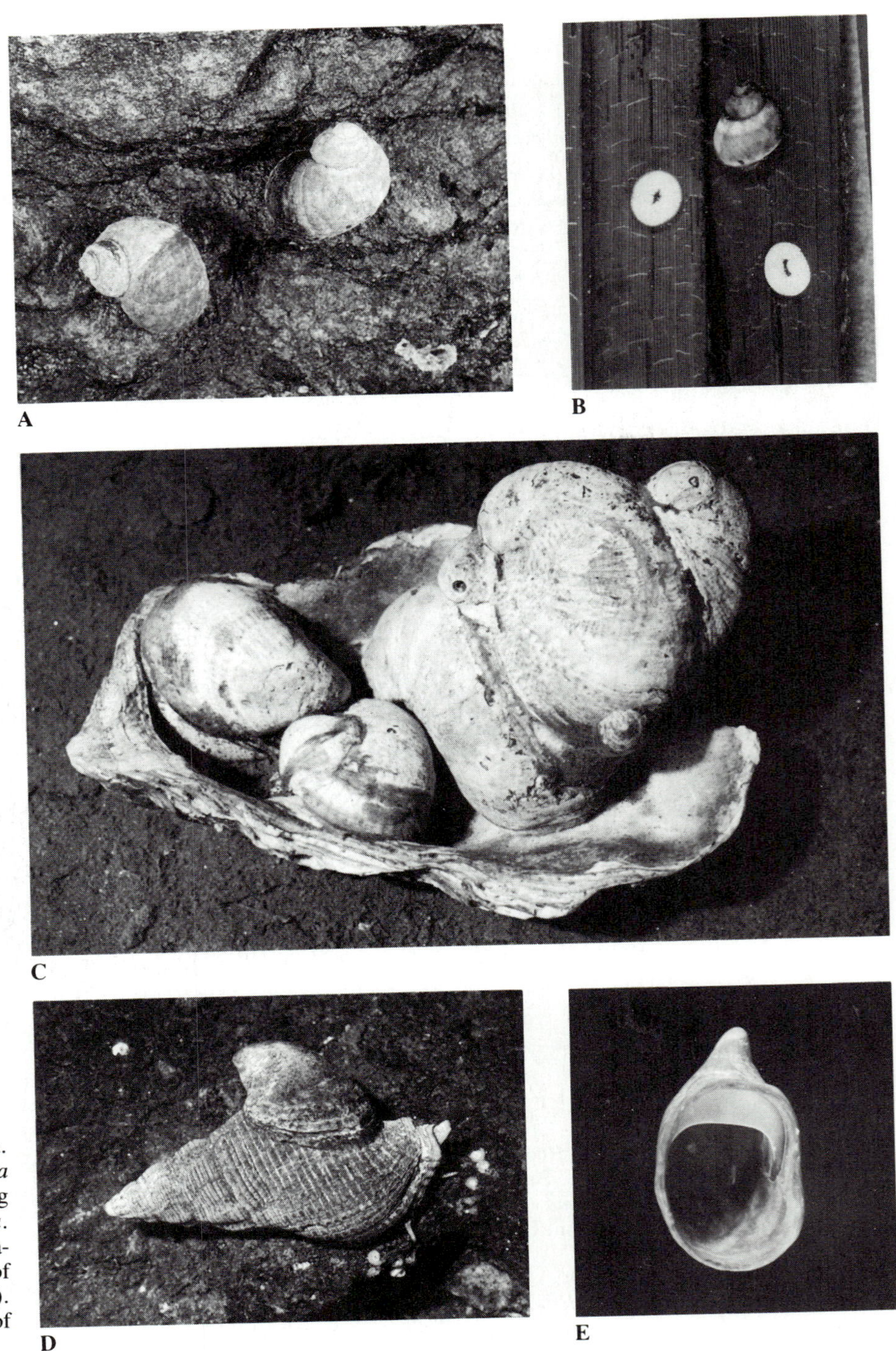

FIGURE 13.26 Mesogastropoda. A. *Littorina sitkana*. B. *Lacuna variegata* and two of its egg masses. C. *Crepidula fornicata*. D. *Crepidula adunca*, permanently attached to the shell of *Searlesia dira* (Neogastropoda). E. *Crepidula adunca*, interior of shell.

Among the carnivorous mesogastropods are some whose method of feeding is much like that of many neogastropods. *Polinices* (Figure 13.28, A and B) and its relatives use the radula to drill through the shells of clams to get at the soft parts. The holes bored by these snails are perfectly bevelled (Figure 13.28, C). *Fusitri-*

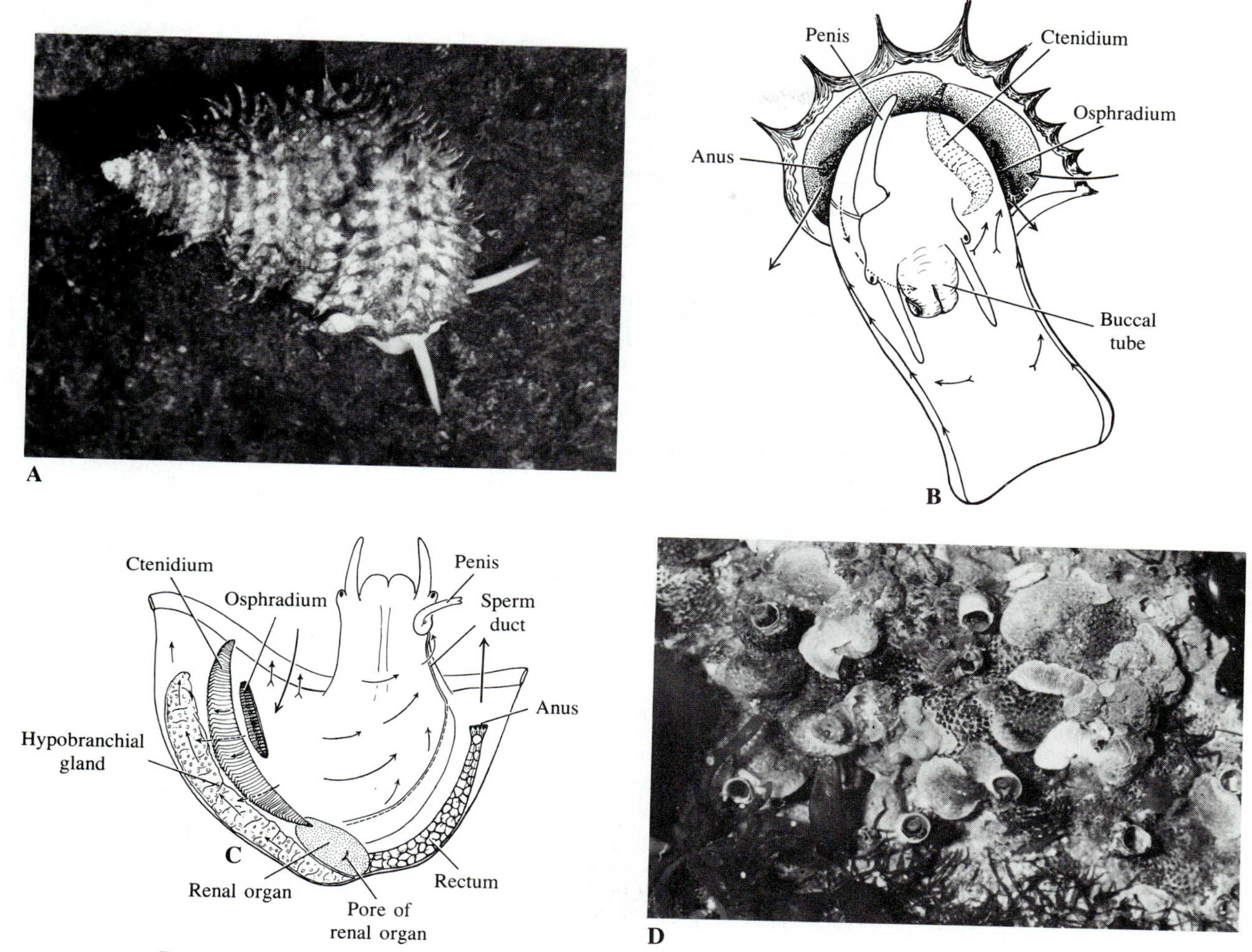

FIGURE 13.27 Mesogastropoda. A. *Trichotropis cancellata.* The small snail near the lip of the aperture is *Odostomia columbiana,* a pyramidellid opisthobranch that parasitizes *Trichotropis.* B. *Trichotropis* extended out of its shell. Plain arrows indicate respiratory and feeding currents; broken arrows indicate the current carrying food to the mouth; feathered arrows indicate rejection currents. C. *Trichotropis,* anterior portion of body, with mantle cavity opened on the right side. Plain arrows indicate feeding currents; feathered arrows indicate rejection currents. (B and C, Yonge, Biological Bulletin, *122.*) D. *Petaloconchus compactus,* a vermetid gastropod.

ton (Figure 13.28, D) uses its radula to rasp holes in a variety of animals, including sea urchins and ascidians. A few mesogastropods suck zooids from compound ascidians, juices from sea anemones (Figure 13.29, A and B), and so on. These may legitimately be considered parasites, for they are more or less permanently associated with their hosts. The most remarkable of parasitic snails, which are wormlike parasites living in the body cavities of sea cucumbers, are not easily classified (Figure 13.30). They are now generally considered to be mesogastropods. They have neither shells nor radulas, and would scarcely resemble gastropods if they did not have veligers of the gastropod type.

It was mentioned earlier that the radulas of nearly all mesogastropods have seven teeth in each transverse row. Among the exceptions to this rule is the oceanic *Janthina* (Figure 13.29, C), which secretes a float that keeps it at the surface. Its radula has no central tooth,

FIGURE 13.28 Mesogastropoda. A. *Polinices lewisii*. B. *Polinices* plowing through sand. C. Hole bored by *Polinices* in the bivalve *Protothaca staminea*. D. *Fusitriton oregonensis*. Because of its long siphonal canal, this species resembles a neogastropod.

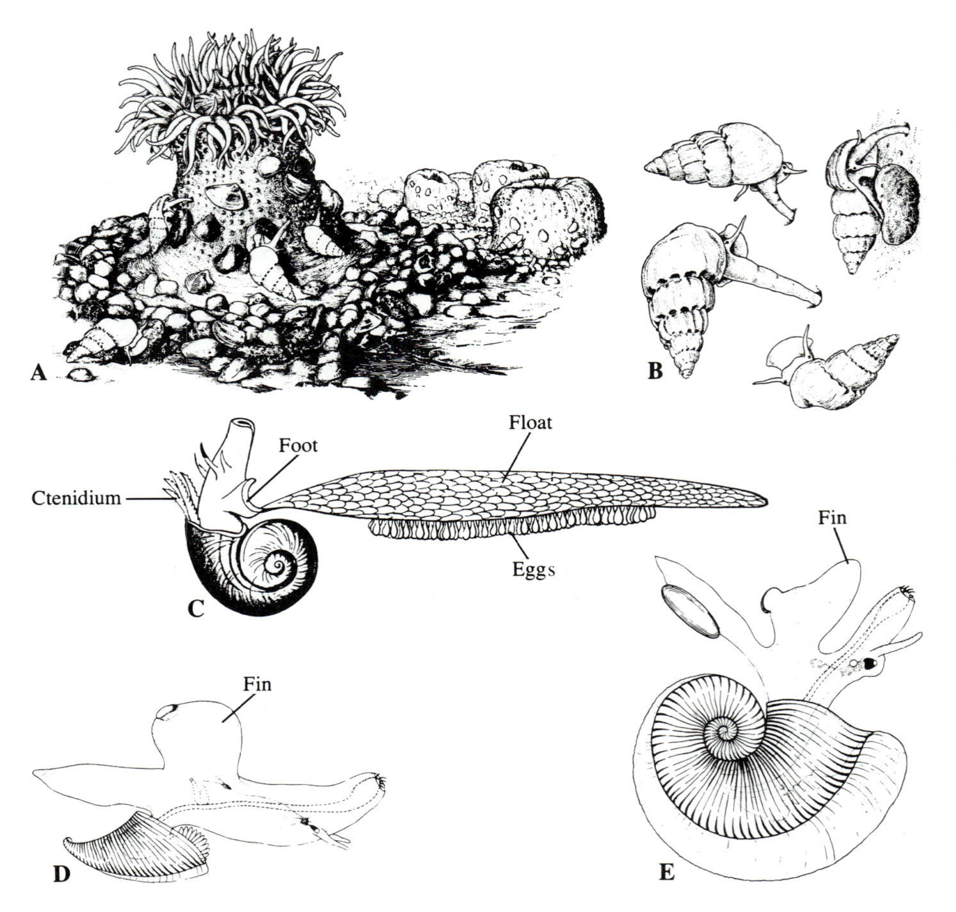

FIGURE 13.29 Mesogastropoda. A. *Opalia crenimarginata* sucking tissue from the column of the sea anemone *Anthopleura xanthogrammica*. B. Several specimens in various postures. (Thorson, Videnskabelige Meddelelser, Naturhistorisk Forening i København, *119*.) C. *Janthina fragilis*. (Cooke, in Cambridge Natural History, *3*.) D. *Atlanta*. E. *Carinaria* (D and E, Ankel, Natur und Museum, *92*.)

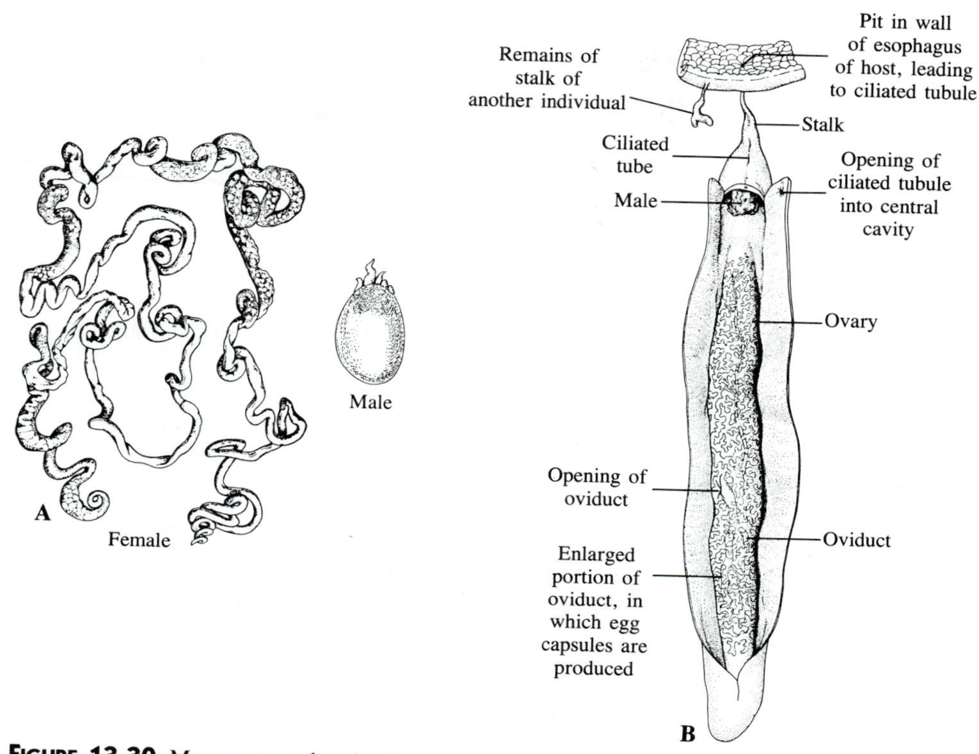

FIGURE 13.30 Mesogastropoda. A. *Thyonicola dogieli,* female and dwarf male; the male is greatly enlarged in proportion to the female. B. *Enteroxenos oestergreni,* female, cut to show some internal structures and an implanted male. The body has been considerably shortened. (Lützen, Ophelia, *18.*)

but its several teeth on both sides of the midline are long and specialized for grasping *Velella,* its prey.

The heteropods are also oceanic. They live close to the surface, where they feed on small fishes and planktonic invertebrates. They are oriented upside-down, and their slow swimming movements are effected by a fin-like modification of part of the foot. Some heteropods have substantial shells (Figure 13.29, E); in others, the shell is reduced (Figure 13.29, D).

Order Neogastropoda

Neogastropods, whose shells typically have a pronounced siphonal canal (Figure 13.31), form the largest order of prosobranchs. There are, in fact, more kinds of neogastropods than all other kinds of molluscs combined. The major distinctions between neogastropods and mesogastropods have already been explained and are not repeated here. However, to discuss the lifestyles of neogastropods, it is essential to characterize the several radular types found in this huge assemblage.

In one large group within the order, the radula is of the *stenoglossan* (''narrow-tongue'') type (Figure 13.32). It has three teeth in each transverse row—a central tooth and two laterals. This type of radula is well adapted for rasping flesh from other animals, and is often used also for drilling through calcareous shells. *Urosalpinx cinerea,* which bores into oysters, has received special attention because of the economic losses caused by its predatory activities. The process of making a neat hole depends in part on the abrasive action of the radula and in part on a chelating agent secreted by a gland on the anterior part of the sole of the foot. After *Urosalpinx* has fixed itself on an oyster's shell, it scrapes for about a minute with its radula, then turns the gland inside-out and presses it to the abraded area for about a half hour. By repeatedly alternating the abrasive phase with episodes of chemical softening, the snail makes a neat hole large enough for its buccal tube to pass through. *Urosalpinx* is able to drill through a shell 2 mm thick in about 8 hours. In many neogastropods, drilling is followed by injection of venom, so that the prey is paralyzed. The snail can then apply its buccal tube and radula to almost any soft tissues that are exposed. This is the technique used by species of *Nucella* (Figure 13.33, C), which feed mostly on mussels and barnacles.

The most remarkable neogastropod radulas are those of the *toxoglossan* (''archer-tongue'') type, found in cone shells (Conidae) and certain members of the large family Turridae. There are only two teeth, both marginals, in each transverse row. They are long and harpoon-like, and rolled up in such a way that they enclose an extensive space in which a paralytic venom can be stored (Figure 13.34). By applying pressure on its radula sac, a toxoglossan snail can displace one of its teeth, which is moved to the tip of the proboscis and held there until it is stabbed into a prey animal. The venom is produced by an esophageal gland and is probably applied to the tooth after it has been detached from the radula.

The proboscis (Figure 13.35, A and B) must not be confused with the buccal tube. It is not part of the gut. It lies within a deep invagination, the **rhynchodeum,** and when it is not in use it is folded up around the

FIGURE 13.31 Neogastropoda. A. A species of *Murex,* with a long siphonal canal and long spines. B. *Ceratostoma foliatum.* The tooth at the outer lip of the aperture is used for clinging to the shell of a prey animal, such as a barnacle, or to the edge of a burrow in rock that is occupied by a bivalve. C. *Trophonopsis lasius.* D. *Antiplanes perversa.* The shell is coiled sinistrally and shows a prominent anal notch that is characteristic of members of the family Turridae.

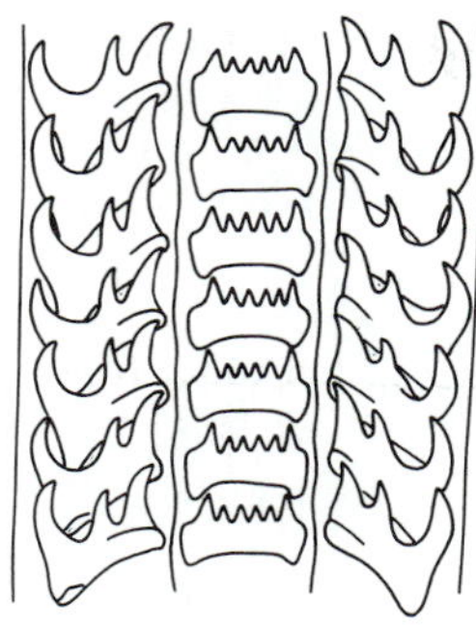

FIGURE 13.32 Neogastropoda, *Buccinum undatum*, portion of the radula. (After Ivanov et al., Major Practicum of Invertebrate Zoology.)

buccal tube. An increase in the hydrostatic pressure within hemocoelic spaces of the proboscis is believed to cause its extension. The proboscis is used not only for grasping a radula tooth and plunging it into the prey, but also for sucking in the prey.

Radulas of some turrids have five teeth (a central tooth, two laterals, and two marginals) in each transverse row; others have four teeth (two laterals, two marginals) or three teeth (a central and two marginals) in each row. The teeth are used for rasping, slicing, or stabbing, or some combination of these functions. A venom may be produced, but the teeth operate while they are still attached to the radular ribbon. It is likely that the various radula types found among turrids, as

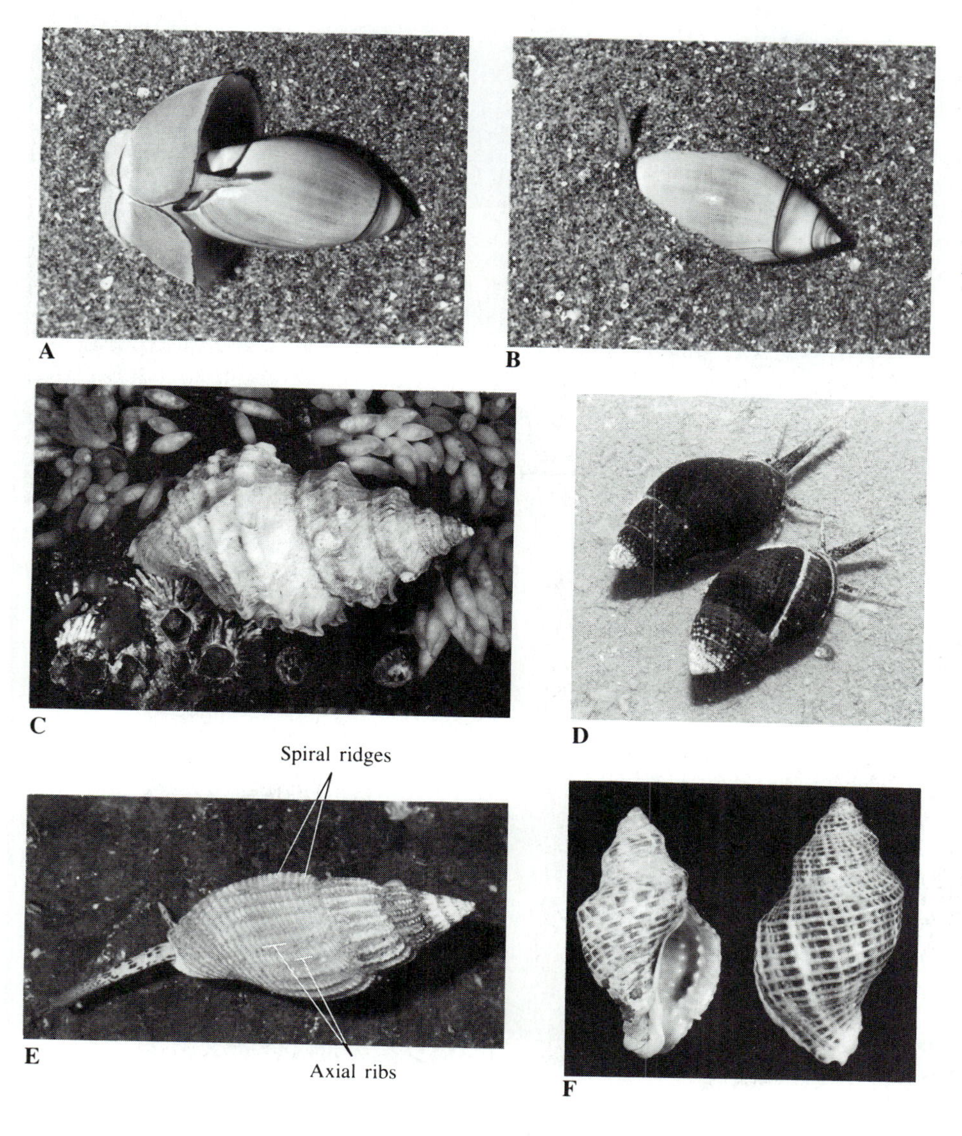

FIGURE 13.33 Neogastropoda. A. *Olivella biplicata*, exposed. B. *Olivella*, partly buried, with its siphon extended above the surface of the sand. C. *Nucella lamellosa* and its egg cases. D. *Ilyanassa obsoleta*. E. *Ocenebra lurida*. F. *Acanthina spirata*. (See also *Amphissa columbiana*, Color Plate 4.)

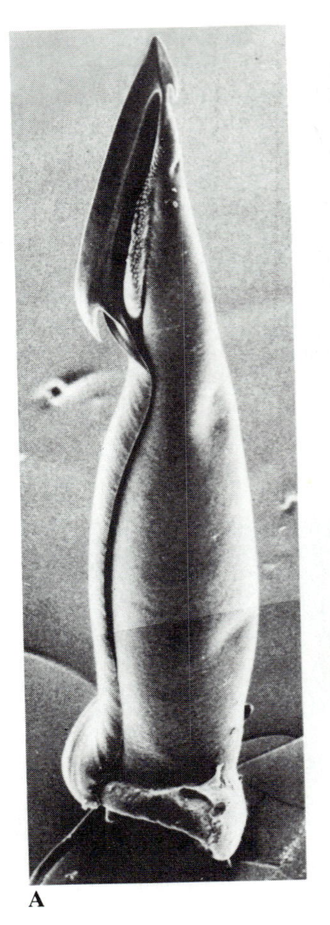

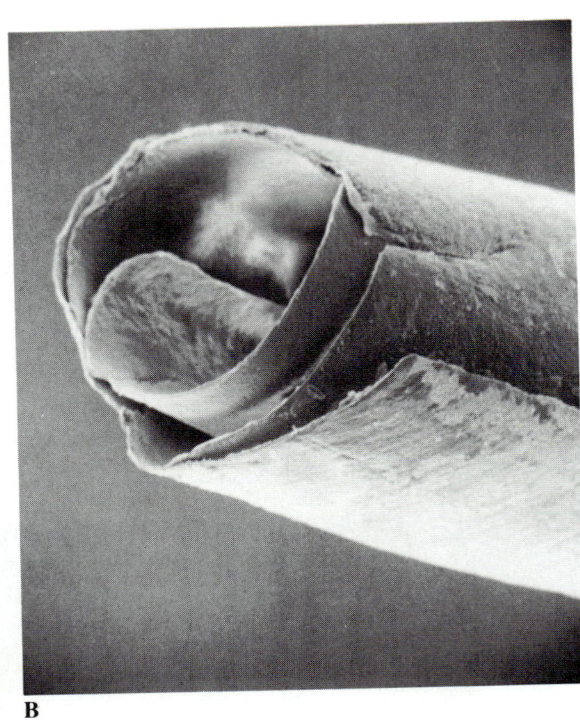

FIGURE 13.34 Neogastropoda, *Conus imperialis*. A. Entire radula tooth. Note the barbs at both ends of the harpoonlike tip. B. Broken shaft of a radula tooth, showing the extensive space for accumulating venom. (Scanning electron micrographs courtesy of Alan Kohn.)

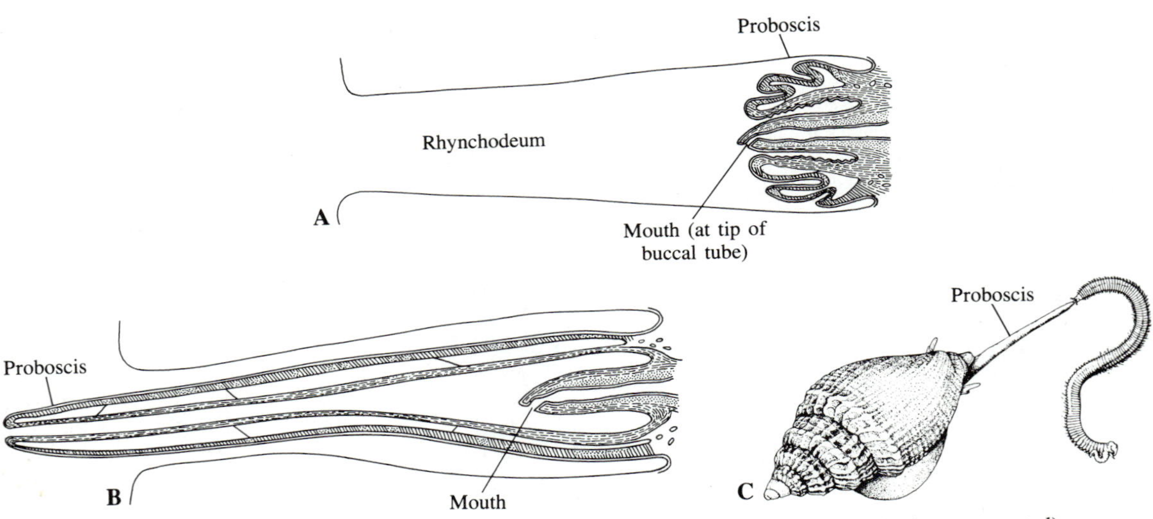

FIGURE 13.35 Neogastropods, family Turridae. A. Buccal tube and probosis (the latter retracted) of *Oenopota levidensis*. B. Proboscis extended. (After drawings lent by Ronald Shimek.) C. *Lora trevelliana,* feeding on a polychaete that it has impaled with a radula tooth. (Pearce, Ophelia, *3*.)

well as those of cone shells, are derived from radulas of the taenioglossan type, such as are characteristic of most mesogastropods, rather than from stenoglossan radulas, such as are typical of the majority of neogastropods.

Cone shells and turrids feed on a variety of invertebrates, especially polychaete annelids (Figure 13.35, C). Some cone shells also eat small fishes. The venoms of a few species of cone shells are dangerous to humans, and several fatalities have been recorded.

References on Prosobranchia

Byrne, M., 1985. The life history of the gastropod Thyonicola americana Tikasingh, endoparasitic in a seasonally eviscerating holothurian host. Ophelia, *24*:91–101.

Fretter, V., and Graham, A., 1962. British Prosobranch Molluscs. Their Functional Anatomy and Ecology. Ray Society, London.

Gainey, L. F., 1976. Locomotion in the Gastropoda: functional morphology of the foot in *Neritina reclivata* and *Thais rustica*. Malacologia, *15*:411–431.

Ghiselin, M., 1966. The adaptive significance of gastropod torsion. Evolution, *20*:337–348.

Graham, A., 1988. Molluscs: Prosobranch and Pyramidellid Gastropods. Synopses of the British Fauna, no. 2 (second edition). Linnean Society of London and E. J. Brill/Dr. W. Backhuys, London and Leiden.

Haszprunar, G., 1988. On the origin and evolution of major gastropod groups, with special reference to the Streptoneura. Journal of Molluscan Studies, *54*:367–441.

Hughes, R. N., 1986. A Functional Biology of Marine Gastropods. Johns Hopkins University Press, Baltimore.

Linsley, R. M., 1978. Shell form and evolution of gastropods. American Scientist, *66*:432–441.

Miller, S. L., 1974. Adaptive design of locomotion and foot form in prosobranch gastropods. Journal of Experimental Marine Biology and Ecology, *14*:99–156.

Miller, S. L., 1974. The classification, taxonomic distribution, and evolution of locomotive types among prosobranch gastropods. Proceedings of the Malacological Society of London, *41*:233–272.

Nielsen, C., 1975. Observations on *Buccinum undatum* L. attacking bivalves, and on prey responses, with a short review on attack methods of other prosobranchs. Ophelia, *13*:87–108.

Palmer, A. R., 1977. Function of shell sculpture in marine gastropods: hydrodynamic destabilization in *Ceratostoma foliatum*. Science, *197*:1293–1295.

Pender, W. F., 1973. The origin and evolution of the Neogastropoda. Malacologia, *12*:295–338.

Shimek, R. L., 1975. The morphology of the buccal apparatus of *Oenopota levidensis* (Gastropoda, Turridae). Zeitschrift für Morphologie der Tiere, *80*:59–96.

Shimek, R. L., and Kohn, A. J., 1981. Functional morphology and evolution of the toxoglossan radula. Malacologia, *20*:423–438.

Smith, C. R., 1977. Chemical recognition of prey by the gastropod *Epitonium tinctum* (Carpenter, 1864). Veliger, *19*:331–340.

Yonge, C. M., 1947. The pallial organs in the aspidobranch Gastropoda and their evolution throughout the Mollusca. Philosophical Transactions of the Royal Society of London, B, *232*:443–518.

SUBCLASS OPISTHOBRANCHIA

With about 1500 species, all of which are marine, the Opisthobranchia is a comparatively small subclass of gastropods. It is nevertheless richly diversified and some of its members have decidedly unusual lifestyles.

The name Opisthobranchia (''hind gill'') alludes to the fact that when a ctenidium is present, it is posterior to the heart (Figure 13.15, C). This is the result of detorsion. In the veliger stage of many opisthobranchs, the mantle cavity is extensive and rather far forward, its position with respect to the head being partly dorsal and partly dorsolateral on the right side. As further development takes place, the anus usually moves backward, although it generally remains to the right of the midline.

In opisthobranchs in which the shell of the veliger stage is shed, the mantle cavity usually disappears. When the shell is retained, the mantle cavity, even if it is not extensive, tends to persist on the right side. There are a few shelled opisthobranchs, in fact, in which the mantle cavity remains just about where it was in the veliger. The torsion of these animals, in other words, is almost fully comparable to that of prosobranchs.

Some opisthobranchs, notably nudibranchs of the suborder Doridacea, have very yolky eggs, and the veliger is not at all typical. Development to the adult form is nearly direct. Detorsion is almost complete, and the anus is on the midline close to the posterior end.

The layout of the nervous system of opisthobranchs nearly always shows the effects of detorsion (Figure 13.15, C) or shortening of the connectives between ganglia, or both. Only in some members of the order Cephalaspidea does the nervous system retain a decidedly torted appearance comparable to that of prosobranchs. In these cephalaspideans, the mantle cavity remains far forward.

Most opisthobranchs, but not all, have a pair of olfactory tentacles, called **rhinophores,** on the dorsal part of the head region (Figures 13.36, 13.37, and 13.41). These sometimes are rolled up to form cylinders, and sometimes are ''leafy'' or elaborate in other ways. They may even consist of two portions, one forming a sheath into which the other can be retracted. In addition to rhinophores, there may be a pair of so-called *oral tentacles,* and sometimes the anterolateral corners of the foot are drawn out into *pedal tentacles*. These last two types of tentacles are characteristic of certain nudibranchs. Eyes, when present in opisthobranchs, are usually situated at the bases of the rhinophores, or in the same general location when rhinophores are lacking.

A

B

C

FIGURE 13.36 Nudibranchia. A. *Triopha catalinae* (Doridacea). B. *Onchidoris muricata* (Doridacea), on its prey, the bryozoan *Membranipora membranacea*. (See also *Archidoris*, Color Plate 4.) C. *Aeolidea papillosa* (Aeolidacea). (See also *Hermissenda*, Color Plate 4.)

There are many radular types among opisthobranchs. One extreme is shown by members of the order Sacoglossa, which suck out the contents of cells of seaweeds. The radula of these molluscs has a single lengthwise row of central teeth (Figure 13.41, C). At the other extreme are opisthobranchs whose radulas have several hundred teeth in each of many transverse rows, so that there are perhaps 75,000 teeth altogether. In some species that suck tissues of animals or that swallow their prey whole, the radula is either absent or reduced to a vestigial structure.

In opisthobranchs that swallow whole snails or other small animals and then crush them, the crushing action is carried out in the esophagus, which may be equipped with calcareous plates. Most opisthobranchs also have a pair of jaws, one on each side of the midline in the dorsal part of the buccal cavity. The stomach is usually small and fairly simple, being primarily a passageway to the digestive diverticula. There are, however, exceptions. In some opisthobranchs the stomach has a sorting area, typhlosoles, and other specializations similar to those of archaeogastropods. In a few—the thecosomatous pteropods—the stomach even has a crystalline style.

The digestive gland is almost always proportionately large and elaborate. In certain groups, some of which are herbivorous, others carnivorous, portions of the digestive gland extend into fingerlike or leaflike projections of the dorsal surface (Figure 13.36, C; Figure 13.37, A–C; Figure 13.41, A and B). In general, digestion is both intracellular and extracellular; the enzymes for extracellular digestion are secreted by the digestive gland and also by the salivary glands.

In the food consumed by some opisthobranchs, rel-

FIGURE 13.37 Nudibranchia. A. *Tochuina tetraquetra* (Dendronotacea). B. *Melibe leonina* (Dendronotacea), ventral view. (Photograph by Anne Hurst.) (See also *Tritonia* and *Dendronotus*, Color Plate 4.) C. *Dirona albolineata* (Arminacea). D. *Armina californica* (Arminacea).

atively little material cannot be digested, so these animals produce hardly any fecal material. A few short-lived species that do not have to face indefinitely the problem of what to do with undigested residues just store these in cells of the digestive gland.

Opisthobranchs are hermaphroditic and have a single gonad divided into egg- and sperm-forming follicles. The gonad is therefore an ovotestis, although male and female gametes are not necessarily produced simultaneously. Details of the organization of the reproductive system vary, but in general the ovotestis is joined to a single duct that soon splits into an oviduct and a sperm duct (Figure 13.38). The external openings of these, on the right side of the body and usually near the head, are close together, and they may share a common vestibule.

Associated with the terminal portion of the sperm duct is a muscular penis, and the duct may also have a seminal vesicle for storing sperm. Linked with the oviduct is a seminal receptacle that receives sperm during copulation, a small chamber in which fertilization takes place, a mucus-secreting gland, and a gland that secretes an albuminous string, ribbon, or some other kind of mass in which the eggs are laid.

In some opisthobranchs, the sperm duct, oviduct, and a pathway leading to the seminal receptacle are functionally separate but not completely partitioned off from one another. The penis, moreover, is considerably farther forward than the genital openings. Sperm reach it by way of an external ciliated groove.

Order Nudibranchia

Nudibranchs, the animals generally called sea slugs, make up the largest order of opisthobranchs. They are usually abundant on rocky shores, in beds of eelgrass or growths of seaweeds, and on pilings and floating docks, as well as in shallow subtidal habitats. They are carnivores, and many of them live in more or less permanent association with the sponges, hydroids, or

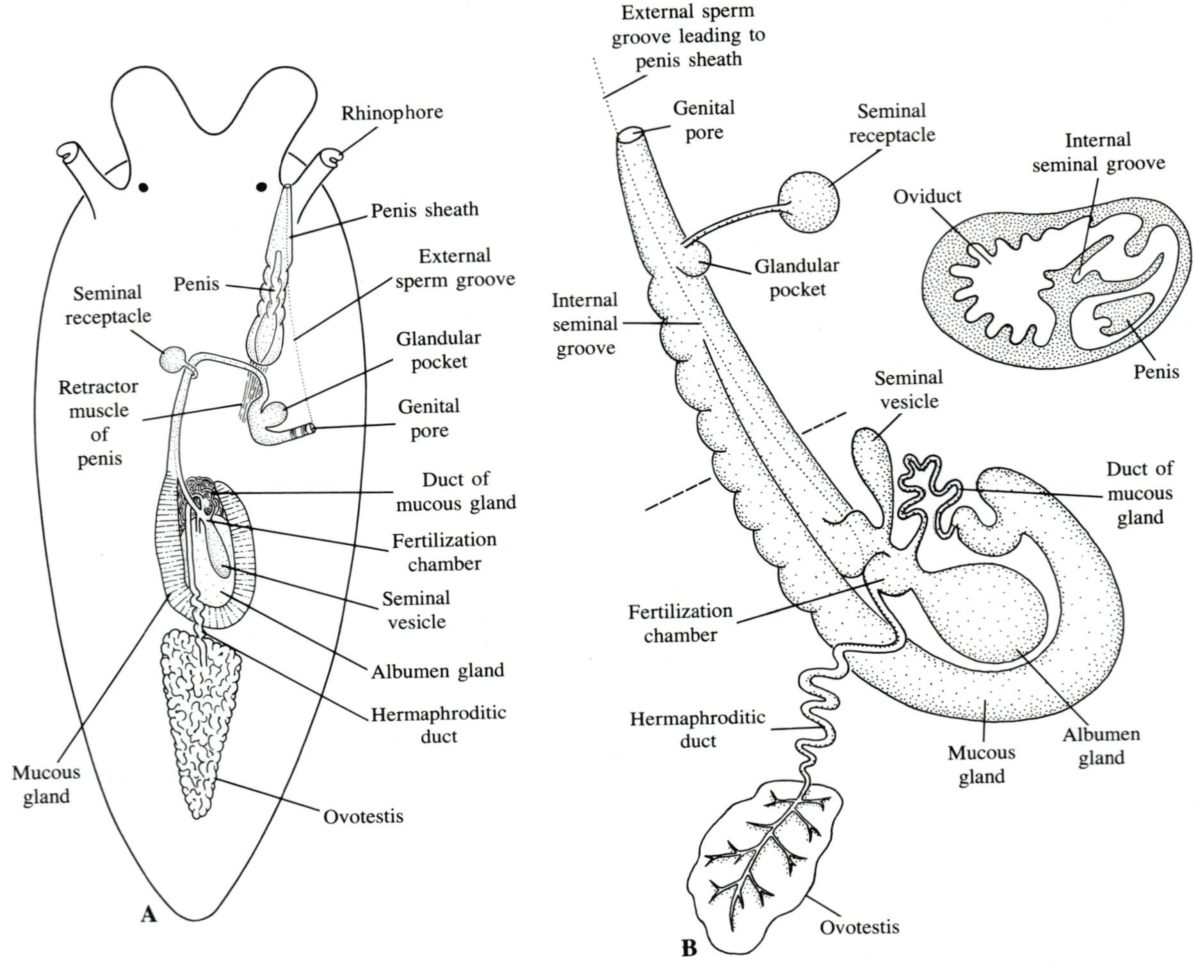

FIGURE 13.38 Opisthobrancia, Anaspidea, reproductive systems. A. *Phyllaplysia taylori*. (After McCauley, Proceedings of the California Academy of Sciences, series 4, *24*.) B. *Aplysia punctata*. The inset shows a transverse section taken at the level indicated by dashed lines. (After Eales, Liverpool Marine Biology Committee Memoir 24.)

bryozoans that they eat. Certain kinds are typical of muddy or sandy bottoms, and these are mostly predators on sea pens and sea whips.

Sea slugs have no shell, no distinct mantle cavity, and no ctenidium. They may, however, have various kinds of outgrowths that function as gills. The graceful form and spectacular coloration of many nudibranchs makes them irresistible to students collecting at the shore. The color patterns are not for decoration, however. They are almost certainly warnings to prospective predators that a mouthful of nudibranch is a mouthful of misery. The fact that these molluscs are rarely attacked or eaten by other animals appears to be due in part to the way they taste, and in part to the fact that certain species concentrate nematocysts that they acquire when they eat cnidarians.

The suborders of Nudibranchia are described briefly here to illustrate the diversity within this group. Members of the Doridacea (Figure 13.36, A and B; also *Rostanga*, Color Plate 1, and *Archidoris*, Color Plate 4) are the only nudibranchs in which the anus is on the midline and either encircled or flanked by plumose gills. Most species in this group feed on sponges or bryozoans, and some of them have the same color or markings as the colonies over which they graze. A few dorids eat barnacles and other invertebrates.

Nudibranchs of the suborder Aeolidacea (Figure 13.36, C; also *Hermissenda*, Color Plate 4) are primarily predators on hydroids, sea anemones, and other cnidarians, although some species eat a wide variety of animal material, including ascidians and dead fishes. An interesting aspect of the biology of aeolids is that the fleshy outgrowths that cover much of the dorsal surface contain branches of the digestive gland. These outgrowths are called **cerata** (*ceras* is the singular form of the word). Some intracellular digestion takes place within them, and there is also production of enzymes that act in the stomach. Each ceras has a little sac at its tip, and the cells that line this take up and store nematocysts that have not discharged their threads. The sac is continuous with the digestive diverticulum that goes up into the ceras, and it is open to the outside, so presumably the nematocysts that once belonged to a cnidarian could be used by an aeolid for its own defense, though nudibranchs in general also seem to be exempt from predation because of ill-tasting substances they secrete. Experiments with aeolids show that pinching and poking will provoke these animals to release nematocysts from their cerata, and that the nematocysts may then ''fire.'' Some swimming aeolids of the Australian region are in fact dangerous to humans, for they store up nematocysts obtained by eating *Physalia,* the Portuguese man-of-war. What has not been conclusively demonstrated in aeolids is that a particular predator will engender nematocyst release and that the predator will be repulsed because of the nematocysts.

Sea slugs of the suborder Dendronotacea (Figure 13.37, A and B; also *Dendronotus* and *Tritonia,* Color Plate 4) are typically characterized by branched cerata containing diverticula of the digestive gland. These end blindly; there are no sacs for storing nematocysts. Most dendronotaceans do, however, feed on cnidarians, ranging from hydroids to sea anemones, sea whips, and sea pens, and scyphistomas of scyphozoan jellyfishes. Some of them are large-bodied and may reach a length of 30 cm. Certain species, especially *Tritonia diomedea* (Color Plate 5), have been used extensively in research on the brain, partly because the cells that make up the brain are large enough to be stimulated individually with an electrode. By stimulating one brain cell or another, *Tritonia* can be induced to swim or go through other motions similar to those performed by an intact animal that has received a natural stimulus, such as contact with a sea star that might eat it. (Certain dorid nudibranchs are also used in studies on the brain.)

One unusual genus of dendronotaceans deserves special mention. This is *Melibe* (Figure 13.37, B), which has paddle-shaped cerata and a large oral hood bordered by slender tentacles. The hood is used for capturing small crustaceans, especially amphipods.

The suborder Arminacea includes some nudibranchs that have cerata and some that do not. When cerata are present (Figure 13.37, C), they are like those of aeolids in being unbranched, but they do not concentrate nematocysts; the anus is on a distinct papilla among the cerata on the right half of the dorsal surface. Most arminaceans of this type feed on small snails, whose shells they can crack. In another type, of which *Armina* (Figure 13.37, D) is a good example, there are no cerata and the anus is lateral on the right side. On both sides, there is a sequence of folds that function as gills. *Armina* lives on quiet sandy or muddy bottoms and feeds on sea pens.

Other Orders of Opisthobranchs

The second largest order of Opisthobranchia is the Cephalaspidea, whose name refers to the fact that the head is covered by a broad sheet of soft tissue. There are no tentacles or rhinophores. The sides of the foot are generally expanded and turned up along the sides of the body. The shell may be strictly internal, as it is in *Aglaja* and *Chelidonura* (Figure 13.39, A and B), mostly internal, as it is in *Haminoea* (Figure 13.39, C), or completely external, as it is in *Acteon* (Figure 13.39, D) and *Rictaxis*.

Haminoea and its close relatives are vegetarians that feed on eelgrass and seaweeds. The majority of cephalaspideans, including *Aglaja* and *Chelidonura,* eat worms and molluscs. Some species of *Chelidonura* are among the few animals that will consume nudibranchs that are avoided by other predators.

Several genera of cephalaspideans, including *Acteon* and *Rictaxis,* have a well-developed forward mantle cavity with a ctenidium. The nervous system of these animals shows the torted condition typical of prosobranchs.

The orders Thecosomata and Gymnosomata have, in the past, been united into one group called Pteropoda. These planktonic molluscs swim with the aid of oarlike lateral expansions of the foot. In thecosomes, moreover, the oars have ciliary tracts that bring microscopic food to the mouth. Some thecosomes have a calcified external shell. This is spiral in certain genera, such as *Limacina* (Figure 13.40, A), conical or nearly bivalve in others (Figure 13.40, B). A very different type of shell, called a pseudoconch, is present in *Cymbulia* (Figure 13.40, C) and related types. It is internal, and cartilaginous rather than calcified.

The gymnosomes, of which *Clione* (Figure 13.40, D) is representative, have no shell. They also lack a mantle cavity. They are carnivores that have hard buccal structures and other specializations for grasping prey, which include thecosomes.

In members of the order Notaspidea, the shell is cap-shaped. It is sometimes internal and reduced, but when it is external the animal resembles a limpet (Figure

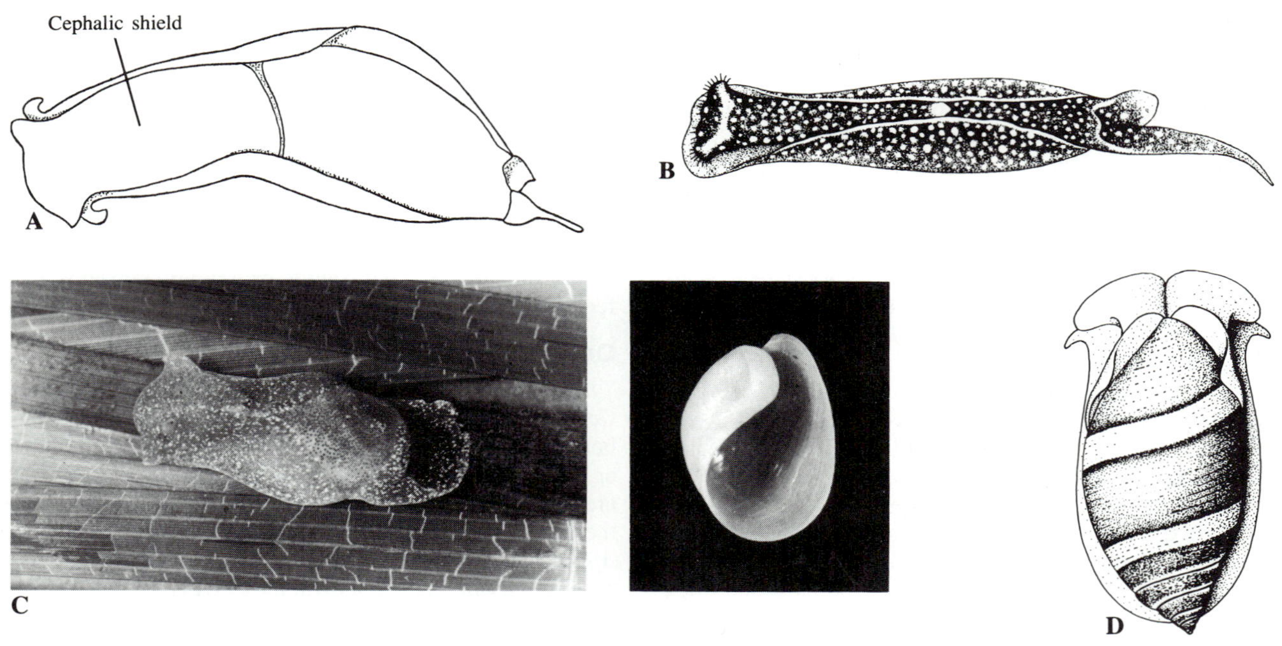

FIGURE 13.39 Cephalaspidea. A. *Aglaja ocelligera* (color pattern not shown). The shell is entirely internal. B. *Chelidonura flavipunctata*. (Baba, Journal of the Department of Agriculture, Kyūsyū Imperial University, *6*.) C. *Haminoea vesicula* and its shell. D. *Acteon tornatilis*. (Thompson and Brown, British Opisthobranch Molluscs.)

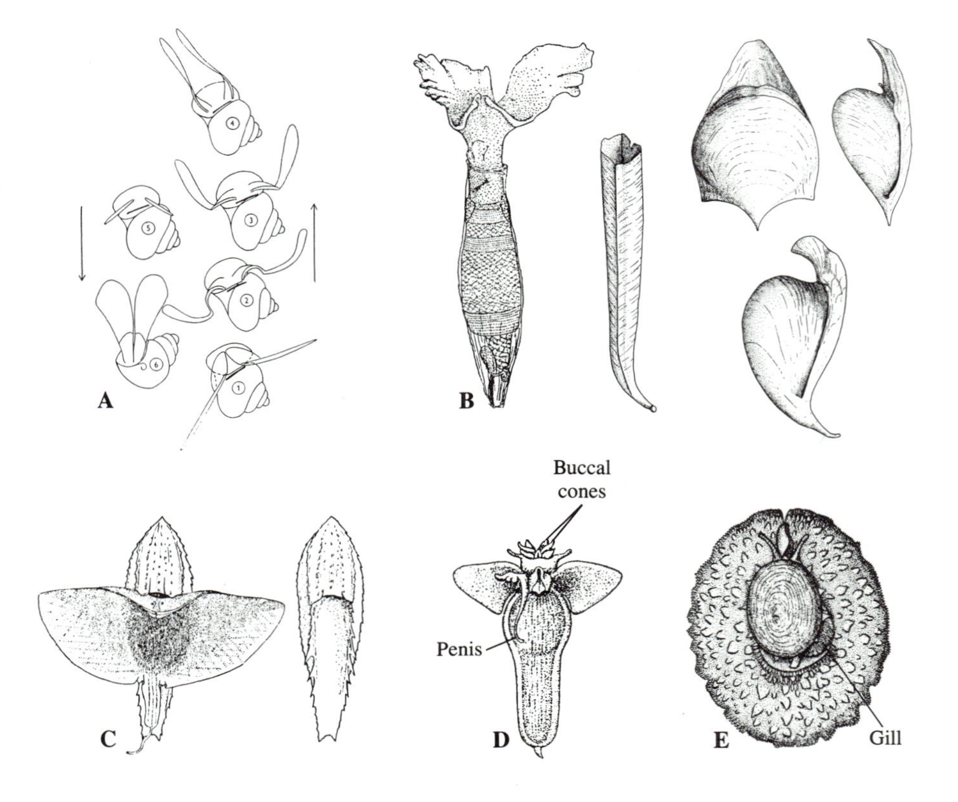

FIGURE 13.40 A. *Limacina helicina* (Thecosomata), successive stages in sinking and swimming upward. (Morton, Journal of the Marine Biological Association of the United Kingdom, *37*.) B. *Cuvierina columnella* and shells of related thecosomes. C. *Cymbulia peroni* and its internal pseudoconch. (B and C, van der Spoel, Conseil International pour Exploration de la Mer, Zooplankton, Sheet 140–142.) D. *Clione limacina* (Gymnosomata). E. *Berthella plumula* (Notaspidea). (Hoffman, in Bronn, Klassen und Ordnungen des Tierreichs, *3*:2:3.)

13.40, E). The mantle cavity, however, is on the right side, not in front, and the gill (probably a secondary gill) is relatively large and much folded. Some notaspideans feed on sponges, others suck out tissues of ascidians, and certain species are scavengers on dead animals.

The order Sacoglossa (or Ascoglossa) consists of opisthobranchs that suck out the contents of algal cells. The radula of these molluscs has a single lengthwise series of sharp teeth (Figure 13.41, C), the forwardmost one being used for slicing into the cells. After it has worn out, it is relegated to a little sac, and the next tooth in the series becomes functional. Many sacoglossans have cerata into which diverticula of the digestive gland extend (Figure 13.41, A). In a group of genera that includes *Elysia* (Figure 13.41, B; Color Plate 5), the sides of the foot are expanded into winglike structures that can be turned up to nearly cover the rest of the animal. *Elysia* is remarkable in that the chloroplasts it gets from green algae continue to function for some time after they have been taken up by cells of the digestive diverticula that extend into the expanded portions of the foot.

Another interesting group of sacoglossans consists of a few genera that have a bivalved shell (Figure 13.41, D), complete with an adductor muscle, hinge, and ligament. They resemble little clams, and some of the fossil species, discovered before living specimens were found, were described as bivalves. One valve, however, shows a spiral protoconch, and the internal anatomy of these animals indicates that they are sacoglossans.

In the Anaspidea, which eat algae, the shell, if present at all, is internal. There is a substantial mantle cavity on the right side of the body, and this usually has a ctenidium. *Phyllaplysia* (Figure 13.42, C) and the sea hares, *Aplysia* (Figure 13.42, A and B) and *Tethys,* are representative genera. In sea hares, some of which reach weights of more than 10 kg, the sides of the foot are expanded into lobes that are turned up around the body. Several species are widely used in research on nerve physiology.

The order Pyramidellacea consists of relatively small snails that have coiled shells and that live in close contact with other snails, or with bivalves, echinoderms, sea anemones, annelids, and other invertebrates. They use an extensile buccal tube for pumping in tissues and juices of their hosts. These gastropods may be thought of as parasites. Two species are illustrated in Figures 13.27, A and 13.42, E.

In the order Acochlidiacea, there is no shell, and the visceral mass is prominent and distinctly separated from the foot (Figure 13.42, D). These small molluscs are interstitial in coarse sand. They appear to be primarily if not entirely carnivorous.

SUBCLASS GYMNOMORPHA

Several small groups of marine gastropods without shells have been placed in the Gymnomorpha, but their relationship to one another is not necessarily close. The principal assemblage within the subclass is the order Onchidiacea, represented by *Onchidella* (Figure 13.42,

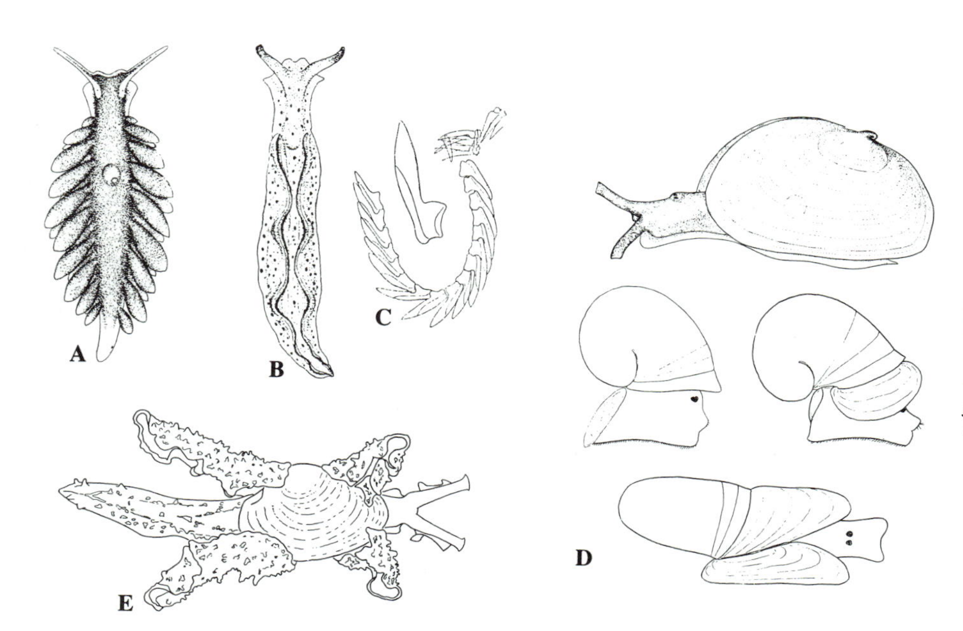

FIGURE 13.41 Sacoglossa. A. *Stiliger boodleae*. (Baba, Journal of the Department of Agriculture, Kyūsyū Imperial University, *6*.) B. *Elysia marginata*. (Baba, Publications of the Seto Marine Biological Laboratory, *6*.) C. *Elysia atroviridis,* radula, one tooth enlarged. D. *Tamanovalva limax,* and stages in the formation of its bivalved shell. (Kawaguti, Proceedings of the Japanese Academy, *35*.) E. *Lobiger serradifalci*. (Gonor, Vie et Milieu, *12*.) (See also *Elysia,* Color Plate 5.)

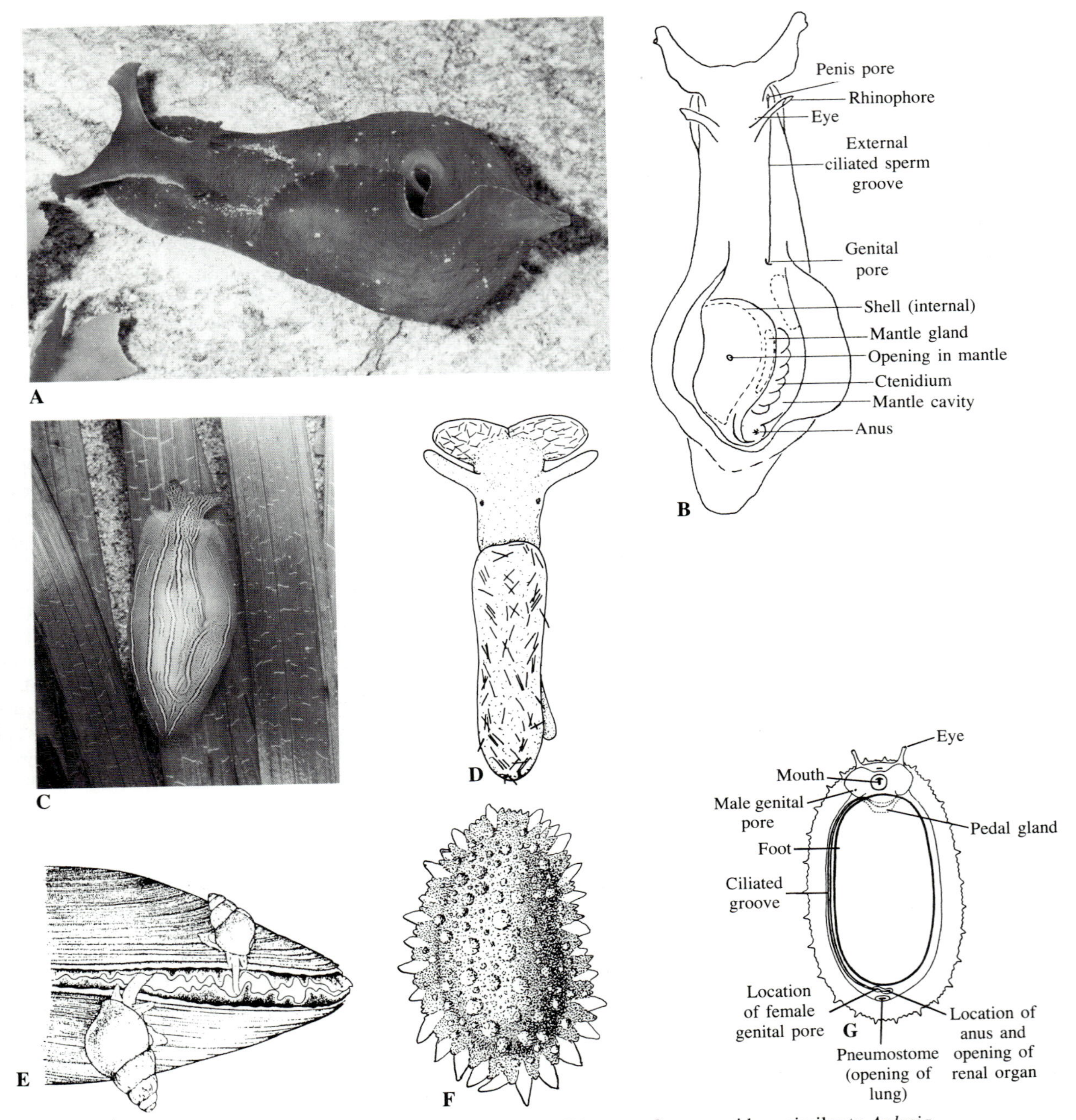

FIGURE 13.42 A. *Aplysia punctata* (Anaspidea). B. Diagram of an anaspidean similar to *Aplysia*. (Beeman, Veliger, *3,* Supplement.) (See also Figure 13.38, B.) C. *Phyllaplysia taylori* (Anaspidea). (See also Figure 13.38, A.) D. *Hedylopsis spiculifera* (Acochlidiacea). (Monniot, Vie et Milieu, *13,*) E. *Brachiostoma rissoides* (Pyramidellacea), parasitic on *Mytilus edulis*. (Rasmussen, Videnskabelige Meddelelser fra Dansk Naturhistorisk Forening i København, *107.*) F. *Onchidella indolens* (Gymnomorpha: Onchidiacea). (Marcus, Kieler Meeresforschung, *12.*) G. *Onchidella celtica,* ventral view. The pedal gland, joined to both the head and foot, is indicated by a dotted outline. The female genital opening, anus, and opening of the renal organ lie in a small posterior depression, which is perhaps a reduced mantle cavity. (Fretter, Journal of the Marine Biological Association of the United Kingdom, *25;* modified.)

F and G), *Oncidium,* and some other sluglike types that live in the middle or upper reaches of the intertidal zone. These molluscs, which feed on algae and diatoms, have most often been assigned to the subclass Pulmonata, because they have a lung that is used for aerial respiration when the animals are exposed at low tide. The lung opening is at the posterior end, however, not on the right side as it is in pulmonates. It is likely that the lungs of the two groups have evolved independently. The organization of the reproductive system seems closest to that of certain opisthobranchs, but placement of the Onchidiacea in the subclass Opisthobranchia has not been received with enthusiasm by all specialists.

A typical feature of onchidiaceans is the ciliated groove on the right side of the body, between the foot and mantle. This carries fertilized eggs from the female genital pore, located at the posterior end of the body, to the mucus-secreting pedal gland, sandwiched between the head and foot, where egg masses are shaped. In onchidiaceans whose life history has been studied, there is a veliger stage that undergoes torsion, then detorsion. In some species, the veliger is free-swimming; in others it is not released from the egg mass, and hatching is delayed until the sluglike form of the adult is attained.

References on Opisthobranchia and Gymnomorpha

Awati, P. R., and Karandikar, K. R., 1940. Structure and bionomics of *Oncidium verruculatum,* Cuv. Journal of the University of Bombay, *8:*3–57.

Awati, P. R., and Karandikar, K. R., 1948. *Oncidium*. Zoological Memoirs, University of Bombay, *1:*1–53.

Britton, K. M., 1984. The Onchidiacea (Gastropoda, Pulmonata) of Hong Kong with a worldwide review of the genera. Journal of Molluscan Studies, *50:*179–191.

Conklin, E. J., and Mariscal, R. N., 1977. Feeding behavior, ceras structure, and nematocyst storage in the aeolid nudibranch, *Spurilla neapolitana*. Bulletin of Marine Science, *27:*658–667.

Day, R. M., and Harris, L. G., 1978. Selection and turnover of coelenterate nematocysts in some aeolid nudibranchs. Veliger, *21:*104–109.

Fretter, V., 1943. Studies in the functional morphology and embryology of *Onchidella celtica* (Forbes and Hanley), and their bearing on its relationships. Journal of the Marine Biological Association of the United Kingdom, *25:*685–720.

Ghiselin, M. T., 1965. Reproductive function and the phylogeny of opisthobranch gastropods. Malacologia, *3:*327–378.

Gosliner, T. M., 1981. Origins and relationships of primitive members of the Opisthobranchia (Mollusca: Gastropoda). Biological Journal of the Linnean Society, *16:*197–225.

Kaelker, H., and Schmekel, L., 1976. Structure and function of the cnidosac of the Aeolidoidea. Zoomorphologie, *86:*41–60.

MacFarland, F. M., 1966. Studies of opisthobranchiate mollusks of the Pacific coast of North America. Memoirs of the California Academy of Sciences, *6*.

Schmekel, L., and Portmann, A., 1982. Opisthobranchia des Mittelmeeres. Nudibranchia und Saccoglossa. Springer-Verlag, Berlin, Heidelberg, and New York.

Thompson, T. E., 1976. Biology of Opisthobranch Molluscs. Volume 1. Ray Society, London.

Thompson, T. E., 1988. Benthic Opistobranchs. Synopses of the British Fauna, no. 8 (second edition). Linnean Society of London and E. J. Brill/Dr. W. Backhuys, London and Leiden.

Thompson, T. E., and Brown, G. H., 1984. Biology of Opisthobranch Molluscs, Volume 2. Ray Society, London.

Willows, A. O. D., 1971. Giant brain cells in mollusks. Scientific American, *224* (2):69–76.

SUBCLASS PULMONATA

Although some prosobranchs of the order Mesogastropoda live in moist terrestrial situations, the pulmonates are the only molluscs that have become successful in relatively dry habitats. Among the features that fit them for life on land is the lung, which permits them to breathe air. This structure, derived from part of the mantle cavity, has spongy walls and a rich blood supply. Its opening, called the **pneumostome,** is on the right side of the mantle (Figure 13.43, A and B). In most terrestrial pulmonates, air is drawn into the lung when the pneumostome is opened and the floor of the cavity is pulled down enough to create a slight negative pressure. The pneumostome is then closed, and the floor of the cavity is allowed to rise, so that there is now a slight positive pressure. This favors diffusion of oxygen into the blood that bathes the lung. When the pneumostome opens again, the cycle repeats.

The ample secretion of slime helps to protect terrestrial pulmonates from drying out and also makes their locomotion easier. Another adaptation to life on land is direct development. This is by no means limited to pulmonates, but it is certainly important in their survival, because it eliminates free-swimming larval stages.

Many pulmonates, including the pond snails of the genera *Lymnaea* (Figure 13.43, D) and *Physa,* are aquatic. These are survivors of a group that returned to the water during the Mesozoic era. Some of the aquatic types still use the lung for breathing air; others keep it filled with water, so that it functions as a gill. In a few species the lung has disappeared, but other features indicate that these snails are nonetheless pulmonates.

The shells of pulmonates tend to be thin, except in the case of certain terrestrial species that live where the soil is decidedly calcareous, and where the calcium content of plants that the snails eat is therefore also high. When the shell is obvious, it is usually coiled, but some freshwater pulmonates have cap-shaped shells (Figure 13.43, D) similar to those of marine limpets.

FIGURE 13.43 Pulmonata. A. *Ariolimax columbianus,* native to the Pacific regions of the United States and Canada. The pneumostome, on the right side of the mantle, is open. B. *Arion ater,* a European species that is now widely distributed in other parts of the world; it is very destructive in gardens. C. *Lymnaea* and *Ferrissia* (with a limpetlike shell), freshwater pulmonates. D. *Phytia myosotis* (below) and two specimens of *Assiminea californica* (Mesogastropoda), which is often associated with it in salt marshes on the Pacific coast of North America. (See also *Helix* and *Theba,* Color Plate 5.)

The land-inhabiting types called slugs rarely have conspicuous shells. When a shell is present in these molluscs, it is generally reduced to a thin plate, and it is more often strictly internal than external. The reduction or loss of the shell has occurred independently in various groups of pulmonates, so not all slugs are closely related.

Most freshwater pulmonates feed on algal coatings and other plant material. The diet of terrestrial species is more varied. It includes green plants, fungi, decaying fruits, dung, dead animals, and live animals, such as earthworms and other snails. Some of the pulmonates that feed on green plants are pests in gardens and on farms.

Nearly all pulmonates have a jaw—a hard plate (Figure 13.44, A)—on the dorsal side of the buccal cavity, close to the mouth. It may be used to scrape against the radula to dislodge food that is caught on the teeth, or it may be used to cut up food pressed against it by the radula.

Arrangement of the radular teeth is correlated with food habits. Carnivorous types generally have a few specialized teeth in each transverse row. In the majority of pulmonates, however, each transverse row has a cen-

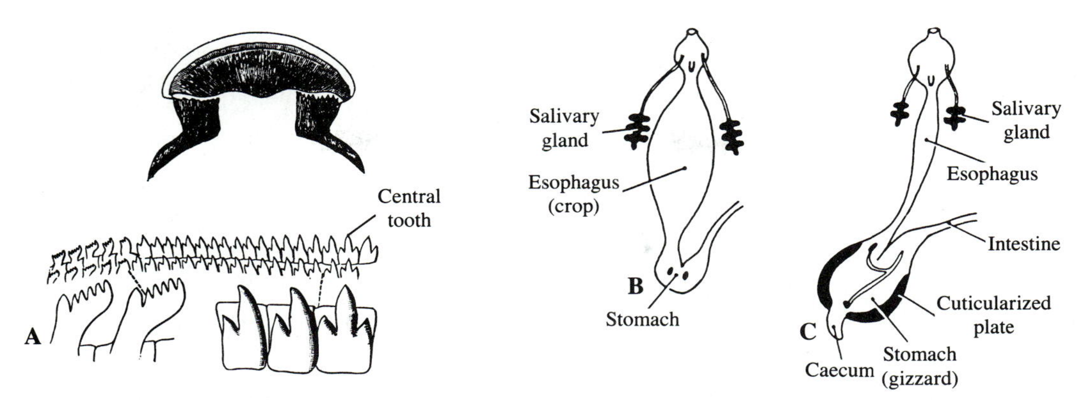

FIGURE 13.44 Pulmonata. A. The jaw and a portion of the radula of *Lymnaea stagnalis*. (Cooke, in Cambridge Natural History, *3*.) B. Digestive tract of a stylommatophoran. C. Digestive tract of a basommatophoran. (B and C, Owen, in Wilbur and Yonge, Physiology of Mollusca, *2*.)

tral tooth, several to many lateral teeth, and many marginal teeth (Figure 13.44, A). The number of teeth in each row is commonly about 100 to 200.

It was mentioned earlier that the esophagus of some pulmonates is enlarged to form a crop (Figure 13.44, B); this enables them to consume a large meal. The stomach is often muscular and may be provided with hard plates for mashing or crushing (Figure 13.44, C). The digestive gland is nearly always large in proportion to the animal, and at least some of the digestion is intracellular. Extracellular digestion is promoted by enzymes produced by the digestive gland and by the salivary glands. Bacteria in the esophageal crop and intestine probably produce most of the enzymes that break down cellulose, a considerable part of the diet of many pulmonates.

The intestine is long, especially in terrestrial species that feed voraciously on plant material. This includes the snails and slugs that do so much damage in gardens. The intestine compacts undigested residues into fecal strings or pellets, and in terrestrial pulmonates it is probably also concerned with uptake of water. In the case of pulmonates whose intestinal bacterial flora accounts for digestion of cellulose, the products of this digestive activity must be absorbed in the intestine.

The principal nitrogenous waste of land pulmonates is uric acid. The solubility of the crystalline form of this compound is so low that it will not harm the animal. The crystals of uric acid are formed in the renal organ and voided through its external opening, which is usually close to the pneumostome. Many freshwater pulmonates also produce considerable amounts of uric acid, but these snails are derived from groups that were once adapted to a terrestrial existence.

Although pulmonates have direct development, they nevertheless exhibit torsion. The mantle cavity, for instance, is more nearly anterior than posterior. The nervous system, however, does not show the twisted arrangement characteristic of prosobranchs. This is because some of the connectives have become greatly shortened, so that several of the major ganglia are concentrated in a mass under the esophagus (Figure 13.45, A).

Pulmonates are hermaphroditic, but their reproduction involves mating and exchange of sperm. As in opisthobranchs, there is a single gonad, the ovotestis, but this does not necessarily produce eggs and sperm simultaneously. Usually, after the ovotestis has produced sperm for a while, mating takes place and the recipient stores the sperm in its seminal receptacle, which is part of the female reproductive system. Then the eggs develop to the point that they are ready to be fertilized.

The basic plan of the reproductive system (Figure 13.45, B) is similar in most pulmonates, although there are many minor variations. The ovotestis is connected to a hermaphroditic duct that splits into a sperm duct and an oviduct. The sperm duct leads to the muscular penis, which, except during copulation, is held in its sheath by a retractor muscle. Sperm transfer is sometimes reciprocal, sometimes a one-sided affair. The sperm pumped into the vagina travel up to the seminal receptacle, where they are stored and nourished.

Eventually, at least some sperm must go back down to the vagina, then ascend the oviduct nearly to the point where it merges with the sperm duct. Here the sperm are stored in a little pocket until eggs come down the hermaphroditic duct to be fertilized. Following fertilization, the albumen gland secretes some material around each egg, and the eggs continue their course down the oviduct, which has other glands. When the

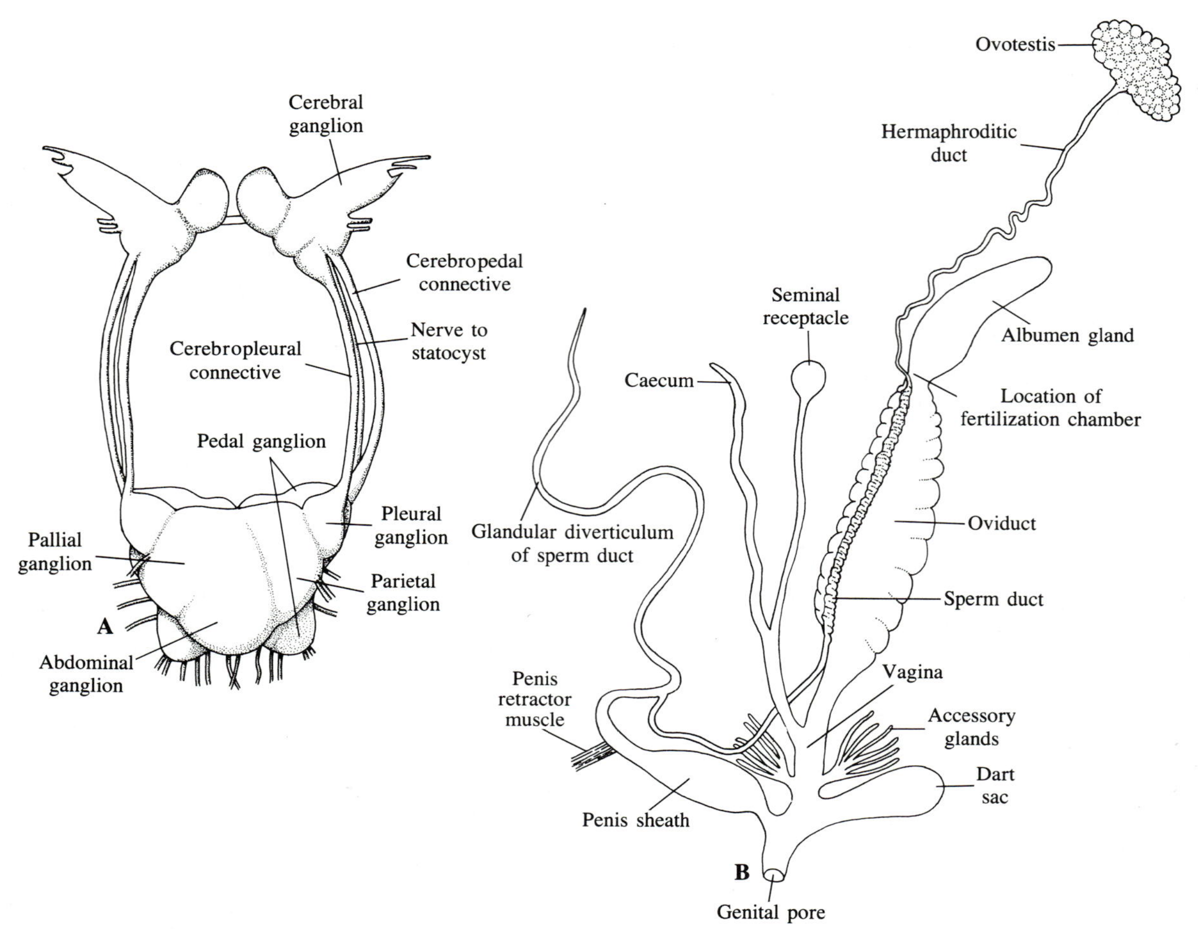

FIGURE 13.45 Pulmonata. A. Principal ganglia of *Helix pomatia*. Of the seven ganglia in the cluster beneath the esophagus, only the small pleural ganglia and the large pedal ganglia are paired. The abdominal ganglion is the product of fusion of a visceral ganglion with a subintestinal ganglion. (After Schmalz, Zeitschrift für Wissenschaftliche Zoologie, *111*.) B. Reproductive system of *Helix aspersa;* diagrammatic.

eggs are finally extruded from the genital pore, they are surrounded by nutritive material and a protective covering. In most land snails and slugs, the eggs are laid in clusters, and each one has a thin, limy coat. Freshwater pulmonates generally lay their eggs in gelatinous masses.

Structures associated with the essential reproductive organs of some land pulmonates include a *dart sac,* from which small calcareous arrows are shot out by muscular contractions. The arrows penetrate the tissues of the prospective partner, and perhaps serve as "attention-getters" preceding copulation. The glands associated with the female system are mostly concerned with producing material that will surround the eggs, and there is much variation in this glandular equipment from group to group. In many species, sperm are packaged in hardened *spermatophores,* so some part of the male system, generally the penis, must be concerned with secreting these.

Many terrestrial pulmonates, and some freshwater types that live at the margins of ponds in which the water level fluctuates, estivate during the dry season. They secrete a membrane of mucus across the aperture; this protects them from desiccation and often enables the animals to adhere to vegetation throughout the dry season. Certain snails of Europe, North Africa, and the Near East form conspicuous assemblages (Color Plate 5) on shrubs and herbaceous plants, not only in their native habitats, but in areas to which they have been introduced. The European *Helix aspersa* (Color Plate 5), for instance, is commonly seen estivating on vegetation in drier parts of California.

A different sort of protective device secreted across the aperture of some terrestrial snails is the **epiphragm.** This is calcareous and is most commonly encountered in species that hibernate. Sometimes additional epiphragms are laid down beneath the first one.

Classification

Order Basommatophora

The name Basommatophora indicates that the eyes of these gastropods are at the bases of the tentacles. They are mostly freshwater animals derived from pulmonates that had been terrestrial but that returned to aquatic environments during the Mesozoic era. Familiar examples are *Lymnaea* (with a dextrally coiled shell; Figure 13.43, C), *Physa* (with a sinistrally coiled shell), *Planorbis* and *Menetus* (with nearly planispiral shells), and *Ferrisia* (with a shell like that of a limpet; Figure 13.43, C). Most of them have retained the lung, and this is generally used for taking in air. In some, however, it is kept filled with water and functions as a gill. *Ferrisia* and its close relatives are lungless, but they have a secondary gill on the left side of the body.

Siphonaria, a limpetlike type that lives in rocky intertidal areas and is truly marine, is remarkable in that it has a veliger larva. This suggests that it may not be far removed from the group of prosobranchs that presumably gave rise to pulmonates. *Siphonaria* has both a lung and a gill; the gill is located in a part of the mantle cavity separate from the lung, and a portion of it may be a remnant of a ctenidium. Some relatives of *Siphonaria* lack the gill.

Order Archaeopulmonata

The eyes of archaeopulmonates, like those of basommatophorans, are at the bases of the tentacles. The archaeopulmonates have, in fact, usually been included in the Basommatophora, and distinctions between the two orders cannot be explained without going into internal anatomy of several examples of each group. In general, however, it may be said that archaeopulmonates have coiled shells in which the aperture is narrowed by folds or teeth on the columella. Most members of this order live in salt marshes and other habitats at the fringes of the sea. In other words, they are almost terrestrial, being immersed only at the highest tides. *Phytia* (Figure 13.43, D) and *Melampus* are examples of salt marsh types. *Carychium* is a genus found in moist inland habitats.

Order Stylommatophora

This group includes the familiar land snails and slugs (Figure 13.43, A and B; also *Helix* and *Theba,* Color Plate 5), as well as some amphibious snails, such as *Succinea* (Color Plate 5). There are two pairs of invaginable tentacles; the eyes are at the tips of the second pair. The fact that over 50 families are recognized in the Stylommatophora shows how greatly the order is diversified. Many species are relatively large. Slugs of the genus *Ariolimax* (Figure 13.43, A), found in the Pacific coast states and British Columbia, often exceed a length of 15 cm when outstretched. The giant land snails of the African genus *Achatina,* some of which have become serious pests in Hawaii and other tropical areas into which they have been introduced, may have shells over 20 cm long.

Many tropical and subtropical stylommatophoran snails are permanently arboreal; that is, they spend their entire lives in trees or large shrubs. Some of these snails have extremely colorful shells, and the color patterns exhibited by a single species may vary extensively within a single population, and even more between populations that are isolated from one another by physical or ecological barriers, such as absence of appropriate vegetation. In Florida, numerous populations of tree snails of the genus *Liguus* (Color Plate 5) have disappeared because their wooded habitats, often small, have been destroyed to make room for homes and other developments. Shell collectors have also taken a heavy toll. Similarly, in Hawaii and other parts of the world, some species or varieties that have a limited geographic distribution are endangered by human encroachment.

Order Systellommatophora

Systellommatophorans are slugs native to tropical America, although some species have become established in other parts of the world. Like stylommatophorans, they have two pairs of tentacles, but these are not invaginable. The foot is narrow, and the body hangs down on both sides of it. There is no lung and no mantle cavity. The anus, and usually also the opening of the renal organ, are posterior instead of on the right side. *Veronicella* is a representative genus.

References on Pulmonata

Burch, J. B., 1962. How to Know the Eastern Land Snails. W. C. Brown Company, Dubuque, Iowa.

Cameron, R. A. D., and Redfern, M., 1976. British Land Snails. Synopses of the British Fauna, no. 6. Linnean Society of London and Academic Press, London and New York.

Fretter, V., and Peake, J. (editors), 1975–1979. Pulmonates. 3 volumes. Academic Press, London and New York.

Mackenstedt, U., and Märkel, K. 1987. Experimental and comparative morphology of radula renewal in pulmonates (Mollusca: Gastropoda). Zoomorphology, *107*:209–239.

Pilsbry, H. A., 1939–1946. Land mollusca of North America. 2 volumes, in 4 parts. Academy of Natural Sciences of Philadelphia, Monograph 3.

Runham, N. W., and Hunter, P. J., 1970. Terrestrial Slugs. Hutchinson University Library, London.

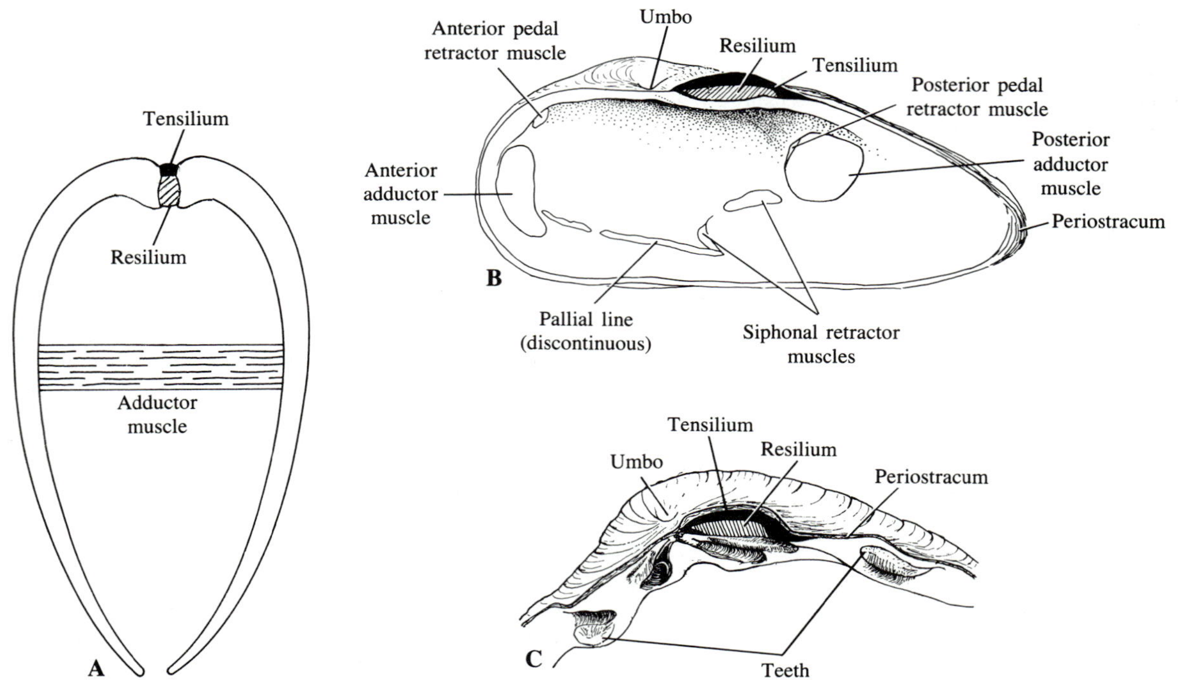

FIGURE 13.46 A. Diagram of a transverse section of a bivalve shell, showing the tensilium and resilium of the hinge ligament. B. Right valve of *Hiatella,* showing hinge ligament and scars of muscles. (Yonge, Malacologia, *11.*) C. Hinge region of the right valve of *Glans carpenteri.* (Yonge, Proceedings of the Malacological Society of London, *38.*)

CLASS BIVALVIA

The bivalves—clams, oysters, mussels, and their relatives—constitute the second largest class of molluscs. They are strictly aquatic and are especially successful in the sea, where the majority of the approximately 15,000 living species are found. In most parts of the world, however, at least a few kinds may be expected in freshwater habitats. As a source of food for humans, bivalves are more important than any other group of molluscs. Certain species are sources of pearls or shell material that can be used to make buttons and jewelry. A few bivalves that burrow into wood are of economic importance because of the damage they do to pilings and other timbers immersed in salt water.

Shell, Hinge, and Adductor Muscles

The *two-piece shell* of a bivalve is generally large enough to accommodate most if not all of the animal's soft parts. To pump water into and out of its mantle cavity by ciliary activity, a bivalve must allow its shell to gape at least slightly. If the foot is to be extended, as it must be if it is to be used for burrowing or attachment, this also requires that the valves be parted.

Bivalves probably evolved from limpetlike ancestors in which the shell consisted mostly of organic material. Calcification of a large portion of the shell on both the right and left side could have started the process of evolution in the direction of producing a clam. If some of the muscles originally used to pull the shell down more tightly were reoriented to become adductors that closed the valves, that would be a further advance. The fossil record has molluscs that are believed to show these steps. Bivalves, by the way, are not the only molluscs that have a two-piece shell: a few opisthobranch gastropods (order Sacoglossa) have independently arrived at the same condition.

The valves are held together dorsally by a two-part **ligament** (Figure 13.46; Figure 13.48, C; Figure 13.56) of tough, elastic material called *abductin.* The outer part, or **tensilium,** is often visible externally (Figure 13.47, A); the inner part, the **resilium,** is hidden except in a few species in which the tensilium does not cover it completely. The size and proportions of the tensilium and resilium with respect to one another and with respect to the shell vary considerably among bivalves, and the two components operate in different ways. The tensilium is like a springy C-shaped clamp that opens

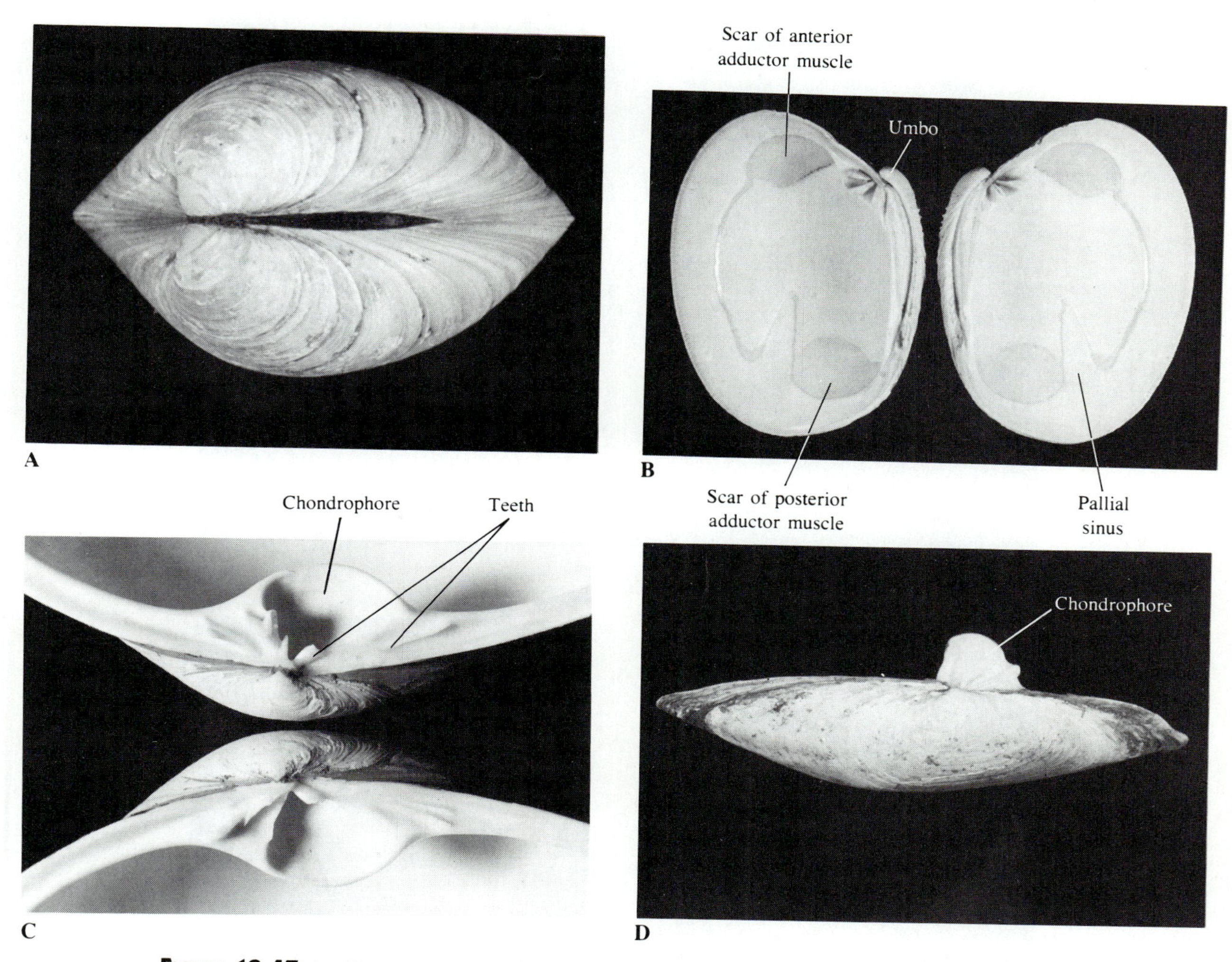

FIGURE 13.47 A. *Compsomyax subdiaphana,* dorsal view, showing the tensilium of the hinge ligament. B. *Protothaca staminea,* valves separated, showing the hinge teeth, scars of adductor muscles, pallial line, and pallial sinus. C. *Tresus capax,* valves separated, showing hinge teeth and the large chondrophores in which the resilium of the hinge ligament lies. D. *Mya arenaria,* left valve, dorsal view, showing the shelflike chondrophore to which one side of the resilium is attached.

wider when pressure on it is released. Unless the valves of a clam shell are pulled tightly together by the adductor muscles, the tensilium will cause the shell to gape at its anterior, ventral, and posterior margins. The resilium, on the other hand, behaves in much the same way as a piece of rubber. When the adductor muscles contract, the resilium is compressed; when the muscles relax, the resilium expands, thus reinforcing the valve-opening action of the tensilium. The part of the hinge region where the resilium is attached to each valve is called the **chondrophore.** When the resilium is large, the chondrophores may be deep cups (Figure 13.47, C), or the chondrophore in one valve may be in the form of a projecting shelflike or spoon-shaped structure (Figure 13.47, D).

In the majority of bivalves there are *teeth* close to the upper margin of each valve, just below the hinge ligaments. These teeth interlock so that the valves do not slip out of joint as they are acted upon by the adductor muscles and ligaments. Primitive bivalves have numerous teeth, a condition referred to as **taxodont dentition** (Figure 13.50, A; Figure 13.51, A). This type of dentition is not often encountered in more advanced groups. Generally there are just a few teeth

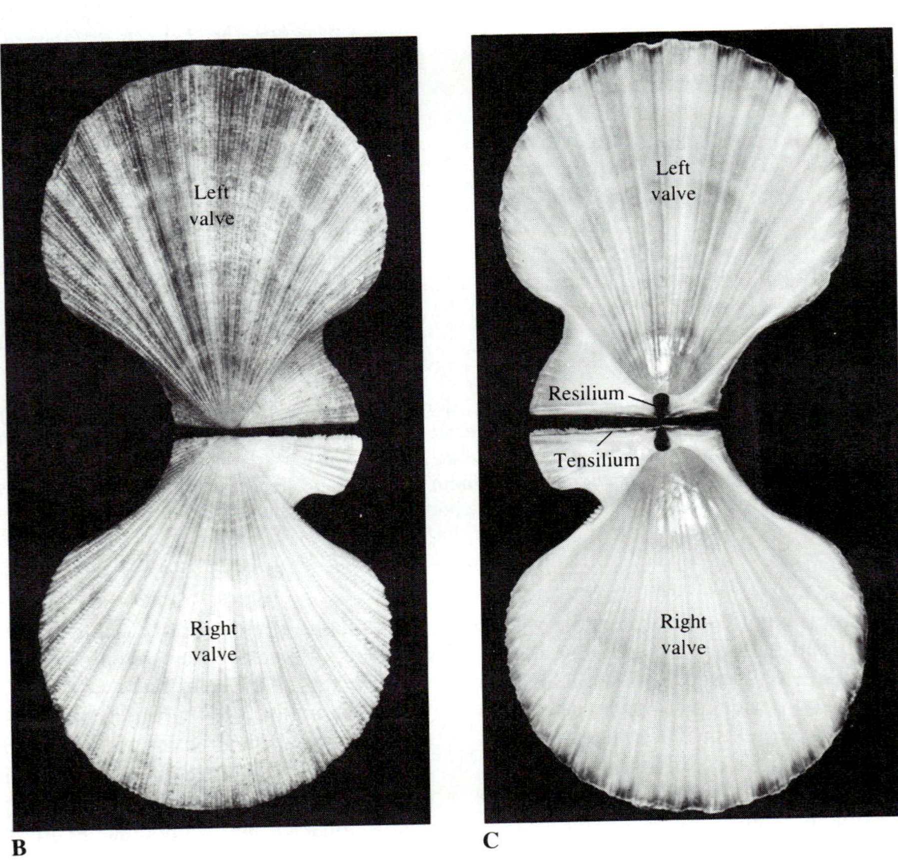

FIGURE 13.48 *Chlamys hastata,* a scallop. A. Intact shell viewed from the right side. B. Exterior of separated shell valves. C. Interior of separated shell valves.

(Figure 13.46, C; Figure 13.47, B and C), and in some bivalves—sea mussels, for instance—there are none.

Enlargement of the valves takes place along their margins. The outer lobe of the mantle lays down the periostracum and prismatic layer (Figure 13.2). Thickening is accomplished over all parts of the mantle surface, and involves addition of calcareous material to the nacreous layer of the shell. The oldest part of each valve, called the **umbo** (plural *umbones*) is close to the hinge; unless it has been worn away, it is likely to form a prominent beak or bump (Figure 13.46; Figure 13.47, B). The pattern of concentric growth lines readily visible on the valves of most bivalves begins at the umbo.

The shells of many species have a nacreous layer that is pretty enough to make them useful in the manufacture of jewelry and other decorations. Before synthetic materials came into wide use, most shirt buttons were punched out of the shells of freshwater clams. The best pearls are produced by certain sea mussels, especially of the genus *Pinctada*. A pearl forms when a particle of foreign material lodges between the mantle and the shell. If the mussel secretes many layers of nacreous material around the particle, a pearl of some value may result. In the Japanese culture-pearl industry, bits of shells of freshwater clams are often used to induce pearl formation.

Most bivalves have two **adductor muscles,** one near the anterior end and one near the posterior end. In an empty shell one can see the scars where these muscles were attached (Figure 13.47, B). In certain bivalves, the anterior adductor muscle is small, so the posterior one bears most of the responsibility for closing the shell. This is the case in sea mussels. In scallops (Figure 13.48; Figure 13.56; Figure 13.63, D) and oysters (Figure 13.58, A; Figure 13.64, F), the anterior adductor muscle has disappeared entirely, and the powerful posterior adductor muscle occupies a nearly central position. A scallop can use its adductor muscle for swimming by jet propulsion. It alternately opens and closes the shell, drawing water into the mantle cavity and then forcing it out again in a localized jet. A scallop's movements are erratic and jerky, but they may enable the animal to escape from a sea star or some other predator. The large single adductor muscle of an oyster, scallop, or related bivalve consists of two well-defined portions. One part is made up of fast-acting striated muscle fibers, and the other part is made up of smooth fibers that contract more slowly, but that can maintain prolonged tight closure of the valves.

Foot

The foot is generally well developed and specialized for digging a burrow or for plowing through mud or sand. In bivalves such as sea mussels, scallops, and oysters, which do not burrow, the foot is reduced, though a part of it produces material that enables the animal to adhere to rock, wood, or shell.

The movements of the foot in digging depend largely on the interaction of its muscles with the hydrostatic skeleton provided by the hemocoel. The adductor muscles that close the shell, and the muscles that protract and retract the foot, also play important roles in the burrowing process (Figure 13.49). The following sequence of events is more or less typical.

While the valves gape, the foot can be extended slightly by the action of its **protractor muscles.** These are attached to the anterior portions of the valves and are oriented so that when they contract, the foot is pulled forward and protrudes. If the valves are then drawn closer together by the action of the adductor muscles, this may squirt out enough water to help loosen up the sand or mud. What is more important, however, is that it puts pressure on the water remaining in the mantle cavity, and the pressure on the fluid in the hemocoel is therefore also increased. Hemocoelic fluid is thus forced into the foot, which is extended further into the substratum and dilated. The dilation anchors the foot. Then the two pairs (anterior and posterior) of foot **retractor muscles** pull the rest of the animal downward. If the two sets of retractors operate one after the other, this produces a rocking motion that may help speed up the burrowing. After the adductor muscles have relaxed, the sequence of events just described can be repeated until the clam has buried itself at the appropriate depth. Minor changes in orientation can be made by poking with the foot, which has the effect of lifting the animal, or by burrowing laterally or upward if necessary.

In certain bivalves, including *Acila* (Figure 13.50, A) and *Yoldia* (Figure 13.51, B), the foot has a flattened sole. When the foot is protracted, the edges of the sole are folded together, so the foot is momentarily almost bladelike. After the foot has been pushed into the mud, the sole flattens out and serves as an anchor as the shelled part of the animal is pulled down.

Mantle, Ctenidia, Labial Palps, and Siphons

The mantle not only secretes the shell but also encloses a spacious cavity that surrounds the foot, visceral mass, and other soft parts. In many bivalves, a portion of the mantle is modified to form siphons by which water moves into the mantle cavity and then out again. Hanging down into the mantle cavity are two ctenidia, one on each side of the foot–visceral mass complex. Cilia on the ctenidia circulate water through the mantle cavity, thus facilitating exchange of oxygen and carbon dioxide across the surfaces of the ctenidia as well as across the epithelia of the mantle, foot, and other structures. In most bivalves, the ctenidia are also concerned with collecting microscopic food.

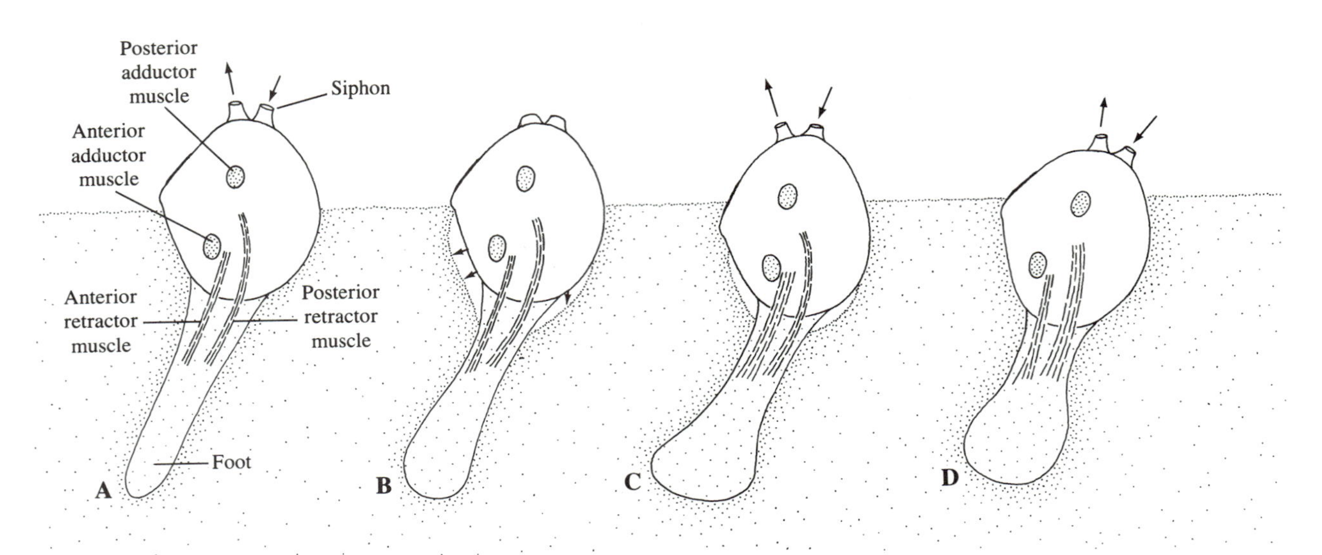

FIGURE 13.49 Sequence of stages in the burrowing activity of a bivalve. A. Shell valves allowed to separate slightly by relaxation of adductor muscles, thereby anchoring the clam; foot lengthened by contraction of circular and transverse muscles; foot retractor muscles relaxed. B. Tip of foot dilated and becoming an anchor as a result of blood forced into it; siphons close and adductor muscles contract, forcing water out of mantle cavity and thus loosening up the sediment. C. Anterior pedal retractor muscle contracts, pulling dorsal side of shell down toward the anchor made by the tip of the foot. D. Posterior pedal retractor muscle contracts, pulling the animal deeper into the substratum. The cycle is then repeated. (Mostly after Trueman, Symposia of the Zoological Society of London, *22*.)

Bivalves whose ctenidia have only a respiratory and water-circulating function are called **protobranchs.** Among the genera in this group are *Acila* and *Yoldia* (Figures 13.50 and 13.51). Their ctenidia are relatively small and are located in the posterior part of the mantle cavity. Each one has two rows of leaflike branches whose cilia drive water past them (Figure 13.52, A and B). Blood supplied to a ctenidium by one vessel returns to an auricle of the heart by another vessel. In protobranchs, the collection and sorting of food is the function of the **labial palps,** which are closely associated with the mouth (Figure 13.50). The distal portion of each palp is an extensile structure called a **palp proboscis.** As the proboscides probe the substratum, their mucous coatings trap food particles, which are moved by ciliary activity to the proximal portions of the palps, where they are sorted. Suitable particles are directed to the mouth; particles that are too large or unacceptable for some other reason are rejected (Figure 13.50, B).

In bivalves whose ctenidia are used for collecting food, these organs are larger in proportion to body size than they are in protobranchs. The lateral branches are filamentous rather than leaflike. The filaments are usually directed ventrally, then turned up dorsally. In a transverse section, therefore, the ctenidium on each side of the body has the configuration of a W in which the four lines have been pushed close together (Figure 13.52, D and E; Figure 13.53). In sea mussels, ark shells, scallops, and some of their relatives, the descending and ascending portions of individual filaments are bound together by tissue bridges (Figure 13.52, D; Figure 13.53, A), but there are few if any bridges between successive filaments. With a fine needle one can separate the filaments without damaging them. The filaments nevertheless cohere to some extent, thanks to tufts of thigmotactic cilia (Figure 13.54). These cilia also keep the filaments evenly spaced so that water can move freely from the inhalant portion of the mantle cavity into the upper exhalant portion. Ctenidia of this sort are said to be of the **filibranch** ("thread-gill") type.

In most bivalves, including freshwater mussels, successive filaments, as well as the descending and ascending portions of individual filaments, are united by tissue bridges. The term **eulamellibranch** ("true sheetlike gill") is applied to ctenidia of this type (Figure 13.55, A). Between the tissue bridges are openings and channels through which water that has been drawn into the mantle cavity can reach the exhalant stream.

Among the variations on the two main plans, one in particular should be mentioned. This is the situation that exists both in oysters (Figure 13.55, B) and in fan

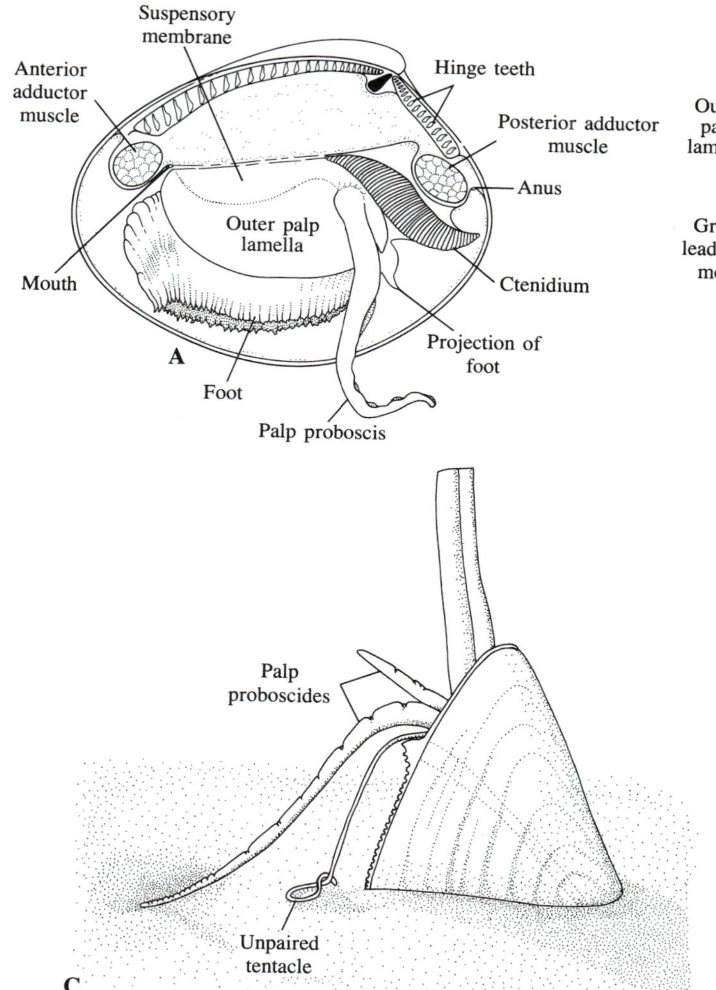

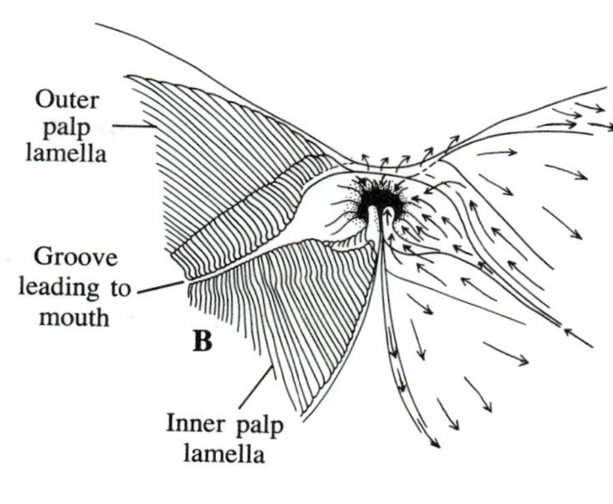

FIGURE 13.50 Protobranchia. A. *Acila castrensis,* as seen from the left side, and with the left valve removed. B. *Acila castrensis,* relationship of palp lamellae to the mouth. The lamellae have been spread apart; folds are shown on the right lamellae, and directions taken by currents are indicated by arrows on the left lamellae. (After Stasek, Proceedings of the Zoological Society of London, *137.*) C. *Yoldia limatula,* one of its palp proboscides collecting food from the substratum. The unpaired tentacle, believed to be sensory, originates on the right side of the body, rather far posteriorly. (Drew, Memoirs of the Biological Laboratory, Johns Hopkins University, *4.*)

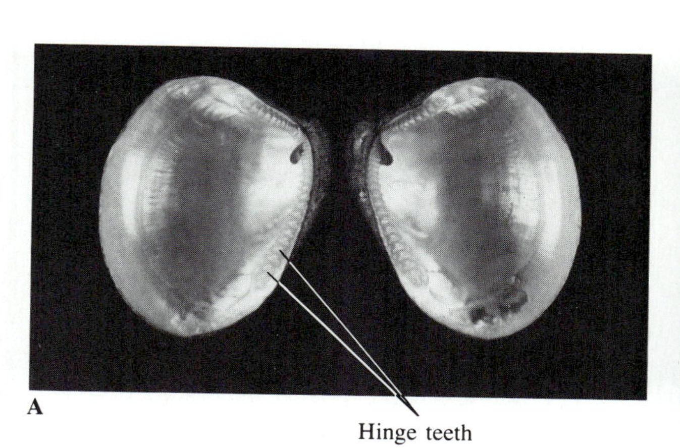

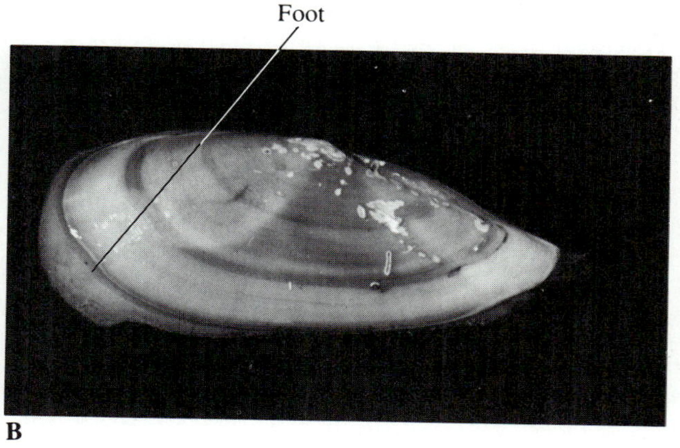

FIGURE 13.51 Protobranchia. A. *Acila castrensis,* valves separated, showing taxodont dentition of hinge. B. *Yoldia amygdalea,* the foot partly extended.

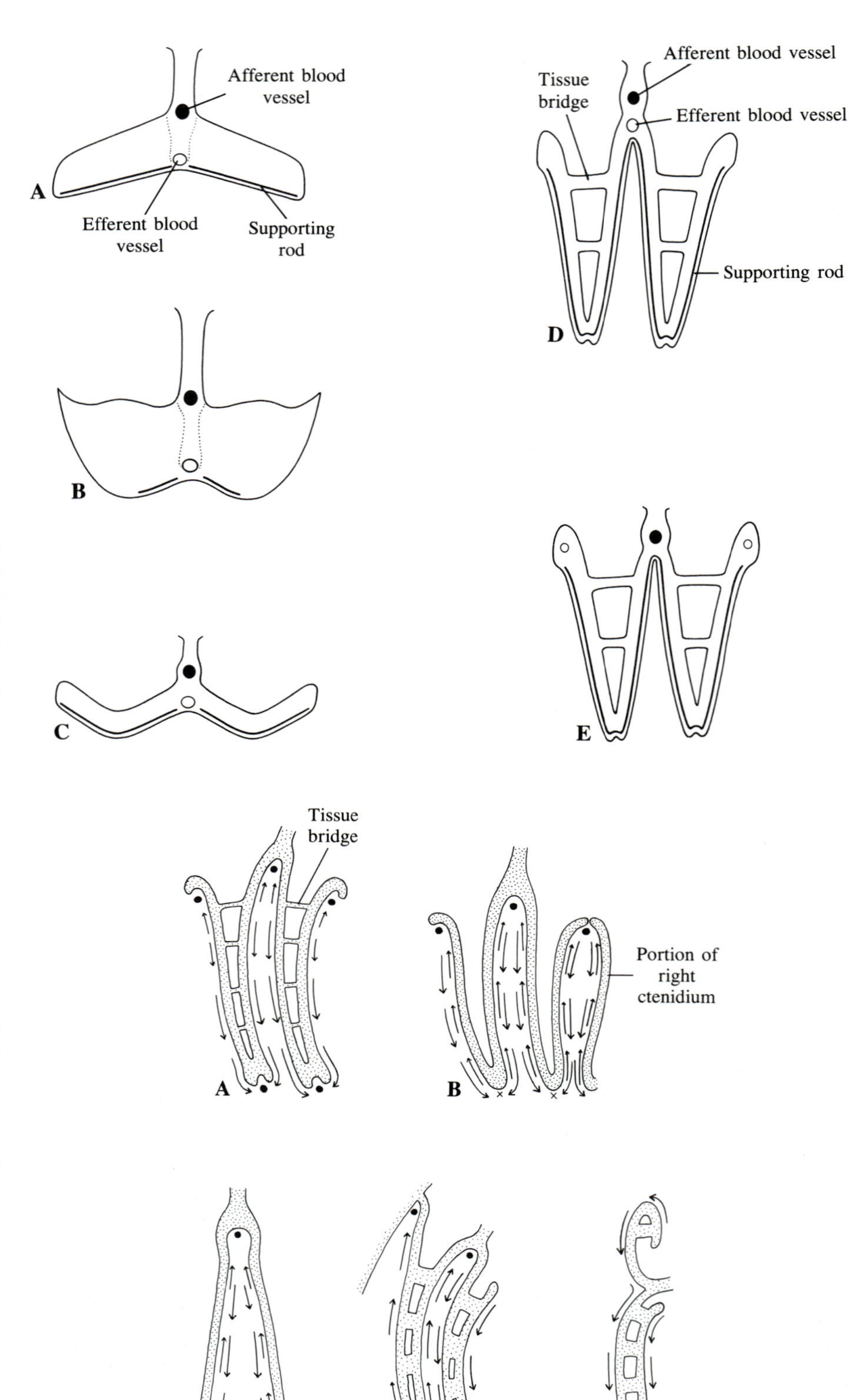

FIGURE 13.52 Some probable trends in the evolution of bivalve ctenidia. A. *Nucula* and related genera (Protobranchia). B. *Malletia* (Protobranchia). C. Hypothetical condition intermediate between that of protobranchs and filibranchs. D. Filibranch type. E. Eulamellibranch type.

FIGURE 13.53 Transverse sections of the left ctenidium of various bivalves of the filibranch and eulamellibranch type. In all cases, the ctenidium is viewed from the posterior side. The channels along which food is passed toward the labial palps are indicated by black dots; currents moving posteriorly are indicated by an x. A. *Mytilus* (filibranch type). B. *Pododesmus* (filibranch type). C. *Heteranomia* (filibranch type). D. *Unio* (eulamellibranch type). E. *Macoma* (eulamellibranch type). (Mostly after Atkins, Quarterly Journal of Microscopical Science, *79*.)

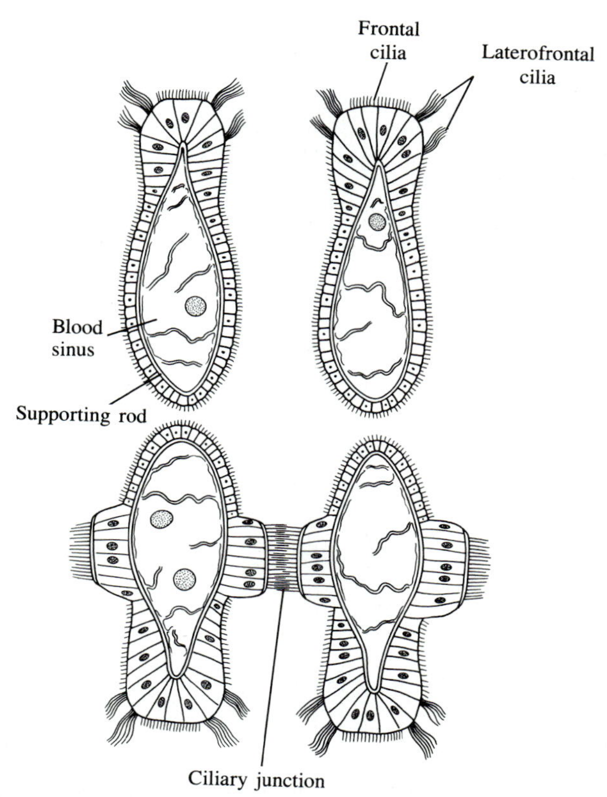

FIGURE 13.54 Section through the ascending and descending portion of two ctenidial filaments of *Mytilus edulis* (filibranch type). Tissue bridges between the ascending and descending portions are not shown (see Figure 13.53). Ciliary junctions between successive filaments are present on both portions, but are not continuous, and are here shown only on one portion. (Pelseneer, in Lankester, A Treatise on Zoology.)

mussels (Figure 13.55, C and D), which are related to typical sea mussels and scallops but have more or less solidly united filaments. Most of the filaments are organized into groups that resemble pleats. A similar arrangement prevails in cockles, razor clams, and many other bivalves that belong to the eulamellibranch line. The operation of pleated ctenidia is discussed below.

The food of filibranch and eulamellibranch bivalves consists of bacteria, diatoms, other unicellular algae, and organic detritus. The diet varies, of course, with the type of clam and the ecological circumstances under which it lives. The intense activity of long cilia on the ctenidial filaments brings prospective food particles into contact with mucus that coats the filaments. Other cilia then move the particles upward or downward until they reach longitudinal tracts that carry them forward. Considerable sorting takes place on the ctenidia themselves, but it is up to the labial palps to decide finally which particles will be directed to the mouth. The size of acceptable particles may be as small as 1 μm. Some bivalves, in fact, subsist to a large extent on bacteria that are in the range of about 1 to 3 μm.

On each ctenidium, there may be as many as five food-carrying tracts leading to the palps, as shown in Figure 13.53, A. Note that one of the tracts is in the upper, axial part of the ctenidium and is bordered by the bases of the diverging filaments; two run along the lowermost edges of the filaments, where these bend back on themselves; the other two are near the free tips of the filaments. The arrangement just described is characteristic of sea mussels and many other filibranch bivalves. Of course, if the filaments are not bound together by tissue bridges, then only the upper, axial tract is really continuous. As long as the filaments are very close together, however, the other four tracts also function efficiently in the transport of food. In oysters, which are basically filibranchs even though their ctenidial filaments are to a considerable extent united by tissue bridges, all five tracts are indeed continuous. In the majority of eulamellibranch bivalves, the number of food-carrying tracts on each ctenidium is reduced to three or even one (Figure 13.53, D and E).

In pleated ctenidia, the pleats, each consisting of several filaments, form broad ridges. At the bottom of each groove between successive pleats is a single *principal filament*. In oysters (Figure 13.55, B), the cilia on the principal filaments drive particles trapped in mucus dorsally, toward the three upper tracts that lead to the labial palps. (The arrangement is similar to that in a sea mussel, shown in Figure 13.54, A.) Cilia on the side of the pleats may also drive small particles dorsally, but the pleats are mostly concerned with moving larger particles ventrally, to the tracts located along the lower edge of each half of a ctenidium, where the filaments bend back on themselves. Though the larger particles are trapped in mucus, many fall out of these tracts and thus never reach the palps. Those that do go on to the palps are subjected to rigorous sorting, and most are rejected.

In fan mussels *(Pinna),* all particles are carried toward the ventral tracts. As in oysters, small particles are moved primarily along the principal filaments. After they enter the ventral tracts, the small particles are concealed beneath the tips of the pleats (Figure 13.55, C). Larger particles, moved along the pleats themselves, are transported toward the palps along more superficial tracts, including those formed by cilia on the tips of the pleats.

Pleated ctenidia are subject to a certain amount of contraction and expansion. Contraction narrows the pleats and also pulls them closer together. Furthermore, it reduces the size of the tracts that carry particles toward the palps. Fine particles may continue to be transported efficiently by the principal filaments and along

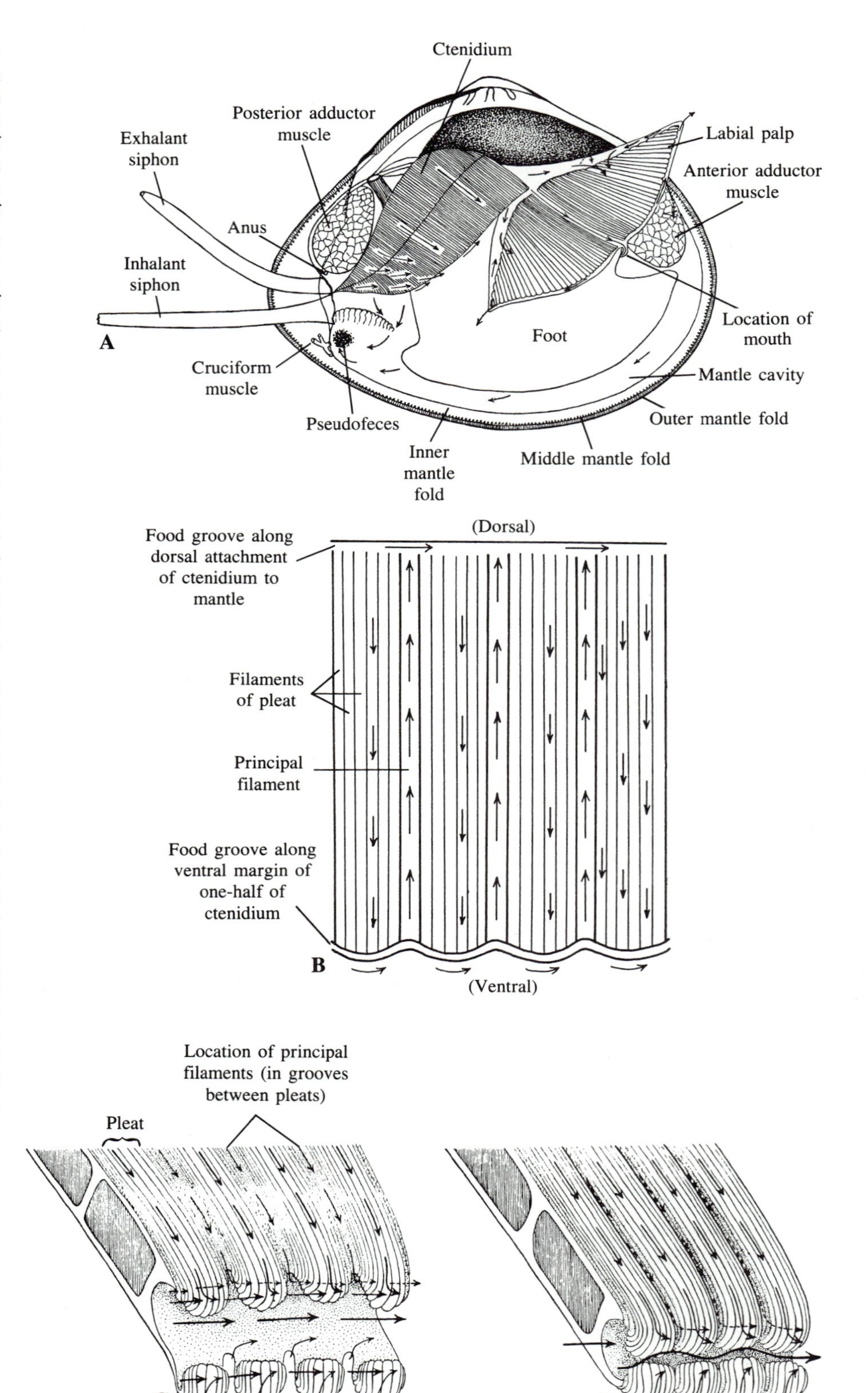

FIGURE 13.55 A. *Macoma balthica,* from the right side, right valve removed. The fibers of the cruciform muscle (not discussed in the text) traverse the suture of the right and left inner mantle folds just below the inhalant siphon. Arrows show direction of ciliary currents on the ctenidia, palps, and surface of the mantle. (Yonge, Philosophical Transactions of the Royal Society of London, B, *234*.) B. Portion of a pleated ctenidium of an oyster, *Crassostrea virginica;* diagrammatic. Arrows indicate the direction taken by food-collecting ciliary currents. Note that particles are moved dorsally along the principal filaments and ventrally along the filaments of the pleats. (After Galtsoff, Fishery Bulletin, *64*.) C. Ventral portion of half a ctenidium of *Pinna fragilis*. Note that the food-collecting currents produced by the principal filaments, as well as those of the pleats, move particles ventrally. In general, the fine particles reaching the ventral tract remain under the tips of the pleats as they are transported toward the labial palps (broken arrows); larger particles, carried downward mostly along the pleats, are moved toward the palps along more superficial tracts. D. The same ctenidium after muscular contractions have pulled the pleats closer together and narrowed the ventral tract. The movement of small particles between the pleats and under the tips of the pleats continues (broken arrows), but the arrangement is less favorable for transport of larger particles. (C and D, Atkins, Quarterly Journal of Microscopical Science, *79*.)

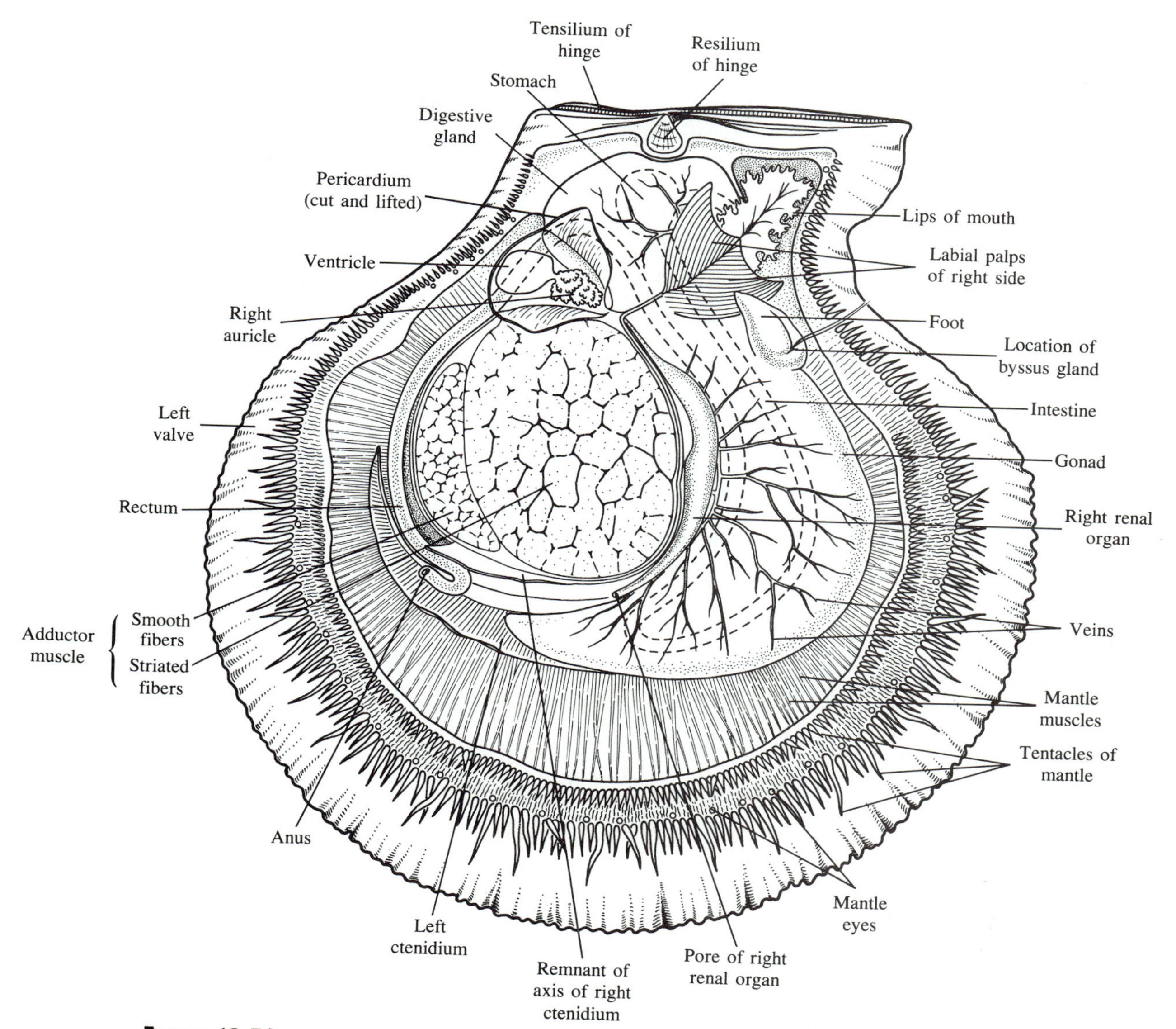

FIGURE 13.56 *Pecten islandicus,* after removal of the right valve and right ctenidium. (After Ivanov et al., Major Practicum of Invertebrate Zoology.)

the ventral tracts hidden by the tips of the pleats (Figure 13.55, D), but the situation for larger particles is more precarious. They are less likely to reach the palps than when the ventral tracts are wide open.

Contraction of pleated ctenidia is usually a response to the presence of excessively high concentrations of particles in the water that enters the mantle cavity. It minimizes the risk that the ctenidia will become clogged.

An interesting arrangement is characteristic of some filibranchs, including the ark shell *(Arca)* and jingle shells *(Anomia, Heteranomia,* and *Pododesmus).* The three upper tracts transport prospective food toward the mouth, whereas the two lower tracts carry heavier particles in a posterior direction and eventually drop them into the mantle cavity (Figure 13.53, B and C).

The labial palps—typically two on each side of the mouth—complete the job of sorting particles directed to them by the food-carrying tracts on the ctenidia. The corrugated appearance of the palps (Figure 13.55, A; Figure 13.56) is due to the fact that ciliated ridges alternate with grooves. As particles move across them, the coarser and heavier ones tend to settle in the grooves; these particles are carried ventrally until they drop off. The smaller and lighter particles are kept moving toward the mouth, but considerable sorting goes on in tracts that run upward or downward along the sides of the ridges.

The term *pseudofeces* is applied to the debris rejected by the ctenidia and palps. This material settles on the mantle and on the sides of the foot, whereupon it may simply fall out through the gap between the valves or be driven out by ciliary action. Most bivalves, however, periodically flush out pseudofeces by briskly pulling their valves together.

Not all bivalves have true siphons. In mussels, oysters, and scallops (Figure 13.56; Figure 13.58, A; Figure 13.63, D), for instance, the edges of the right and left mantle folds remain free. Water enters the mantle cavity almost everywhere that the shell gapes, except at the posterior end, where the exhalant current emerges. At one point, the mantle folds touch, and this keeps the inhalant and exhalant currents separate. Siphons, when present, are produced by a fusion of the right and left mantle edges and by a "puckering-out" of this tissue. In bivalves such as *Macoma* (Figure 13.55, A; Figure 13.63, A), *Scrobicularia* (Figure 13.57, C), and *Tellina,* which have separate inhalant and exhalant siphons, the inhalant siphon can be extended beyond the surface of the sand or mud and is used to suck up detritus. It is thus something like a hose attachment on a vacuum cleaner. Bivalves in which the two siphons are fused together generally feed on fine material that is suspended in the water. In a littleneck clam (*Protothaca,* Figure 13.47, B), cockle, or jackknife clam (Figure 13.57, A; Figure 13.63, B), the fused siphons are so short that the animal's posterior end must be close to the surface of the substratum. In other types, however, the siphons are relatively long and the clam can be buried deep in the substratum (Figure 13.57, B). Some clams, such as the geoduck of the Pacific coast, may be nearly a meter below the surface of the mud by the time they are full-grown.

Whether the siphons are separate or united, they can often be completely withdrawn into the mantle cavity. In such cases the *pallial line*—the line that marks the

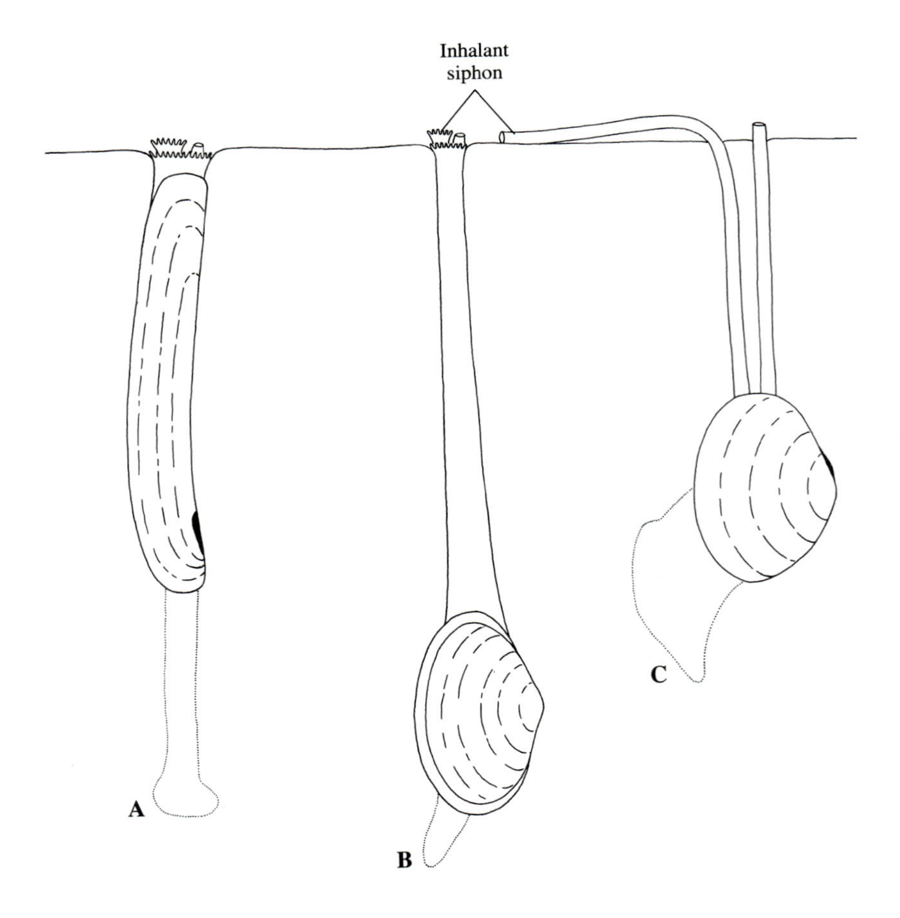

FIGURE 13.57 Orientation, in mud or sand, of three different bivalves. The extent of the foot, when fully protruded, is indicated by a dotted outline. A. Jackknife clam (family Solenidae), a rapid burrower. B. *Mya arenaria,* a slow, deep burrower. C. *Scrobicularia plana,* with completely separate siphons, the inhalant one being used to draw in particles from the surface of the substratum. (After Yonge, Philosophical Transactions of the Royal Society of London, B, *234*.)

edge of the area where the mantle is tightly bound to the shell—shows a deep indentation (Figure 13.47, B). In this region, the mantle operates something like a hinge, permitting the siphons to be extended or pulled back as needed. In many clams with united siphons, the siphonal mass is so fleshy that little of it can be pulled into the mantle cavity (Figure 13.63, C). The mass can be shortened by contraction of muscles within it, but that is about all.

Digestive System

The three main divisions of the gut (Figures 13.56 and 13.58) are the esophagus, stomach, and intestine. The intestine, however, is long and usually differentiated into three regions, referred to here as the *anterior portion, middle portion,* and *rectum*. The many tubules that make up the digestive gland are branches of a few diverticula of the stomach. Except in protobranchs, the stomach is provided with a crystalline style, secreted either in the anterior portion of the intestine, in which case it lies in a trough separated from the rest of the lumen by one or two ridges called *typhlosoles,* or in a deep pocket that opens only into the stomach (Figure 13.58, D and E). Much of the style consists of a somewhat hardened mucoprotein, so its texture resembles that of firm gelatin. Incorporated into the style are enzymes concerned with the digestion of starch and glycogen, and perhaps of cellulose. In addition, there are components that stabilize pH in the stomach and emulsify fats. The style may also release something that dissociates bacteria and other prey microorganisms from inert particles to which they are attached.

The crystalline style has been studied rather intensively, but the way in which it releases its enzymes and other ingredients is still controversial. Nearly all accounts state that the style is rotated, continuously or intermittently, by cilia on the epithelium that lines the sac or trough in which the style is secreted, and that the portion projecting into the stomach is worn away or dissolved as new material is added to the basal portion. The fact that the stomach of many bivalves has a hardened cuticular shield against which the tip of the style may press as it rotates seems to support this view. The rubbing action would wear away style material and perhaps have a grinding effect on some components of the food. While rotating, the style could serve as a stirring rod and also as a winch that draws strings of mucus mixed with food from the esophagus into the stomach.

There is probably much truth in what has just been said. It must be noted, however, that rotation of the style has seldom been seen. The free end of the style often accumulates considerable detritus, and this does not fit well with the idea that the tip must be worn away if enzymes are to be released. Moreover, material in the central core of the style has been observed to emerge from the tip.

Although the enzymes released from the crystalline style function in the stomach, much of the digestive activity in a bivalve is intracellular, taking place in the tubules of the digestive gland. Ciliary tracts on the wall of the stomach and major diverticula direct fine particles into the tubules, where they are ingested by certain of the epithelial cells and converted into nutrients that can be absorbed by the blood that bathes the digestive gland. It is believed that most of the soluble nutrients resulting from enzymatic activity in the stomach are also carried up into the tubules for absorption by the blood. Undigestible particles that happen to be taken up by epithelial cells in the tubules are packaged into masses that are discharged and sent back to the stomach, from which they are shunted, along with other unsuitable material, into the intestine.

To a considerable extent, the stomach and anterior part of the intestine form a single functional complex. It has already been pointed out that the style sac is often just a side channel of the intestinal lumen. One of the ciliated typhlosoles of the intestine may extend into the stomach as far as where the diverticula of the digestive gland originate.

Some absorption of nutrients may take place in the intestine, but this region of the gut, especially its middle portion, is concerned mostly with absorption of water and other processes that lead to the formation of fecal pellets. By concentrating fecal material into compact masses, a bivalve lessens the chance that it will foul its own ciliary–mucous feeding mechanism. From the rectum, fecal pellets drop into the posterior part of the mantle cavity, from which ciliary currents carry them away.

In protobranch bivalves, which lack a true crystalline style, a trough in the anterior portion of the intestine secretes a string of mucus called the **protostyle** (Figure 13.58, C). This rotates in much the same way as a crystalline style is said to do. It is also reported to have an amylase that converts starches to simpler carbohydrates, but on the whole its enzyme content is believed to be low. As new protostyle material is secreted, the older part of the mucous string moves backward into the middle portion of the intestine and becomes incorporated into fecal pellets.

Little is known about digestion in protobranchs, but the results of recent studies on the tubules of the digestive gland indicate that there is intracellular digestion comparable to that which occurs in other bivalves. Protobranch stomachs, in general, are more muscular and have a more nearly complete cuticular lining than those of most filibranchs and eulamellibranchs. Their ciliated sorting tracts, moreover, are comparatively simple. It is

Figure 13.58 A. Digestive tract of an oyster, *Crassostrea virginica*. A portion of the intestine, beginning at the stomach and running alongside the crystalline style sac, then bending back sharply, has been omitted. (See B.) B. Stomach, crystalline style sac, and anterior portion of the intestine, as seen in a latex cast, viewed from the right side. Other elaborations of the stomach, including a caecum in which sorting takes place, would be seen only from the left side. (A and B, Galtsoff, Fishery Bulletin, *64;* modified.) C. Stomach of *Nucula* (Protobranchia). D. Stomach typical of many eulamellibranchs. (C and D, Owen, in Wilbur and Yonge, Physiology of Mollusca, *2*.) E. Stomach of *Glossus humanus* (eulamellibranch), showing circulation of particles. The bold arrows indicate currents that carry coarse particles; small unbarred arrows indicate currents that carry small particles, and arrows with a crossbar mark rejection currents. (After Owen, Journal of the Marine Biological Association of the United Kingdom, *32*.)

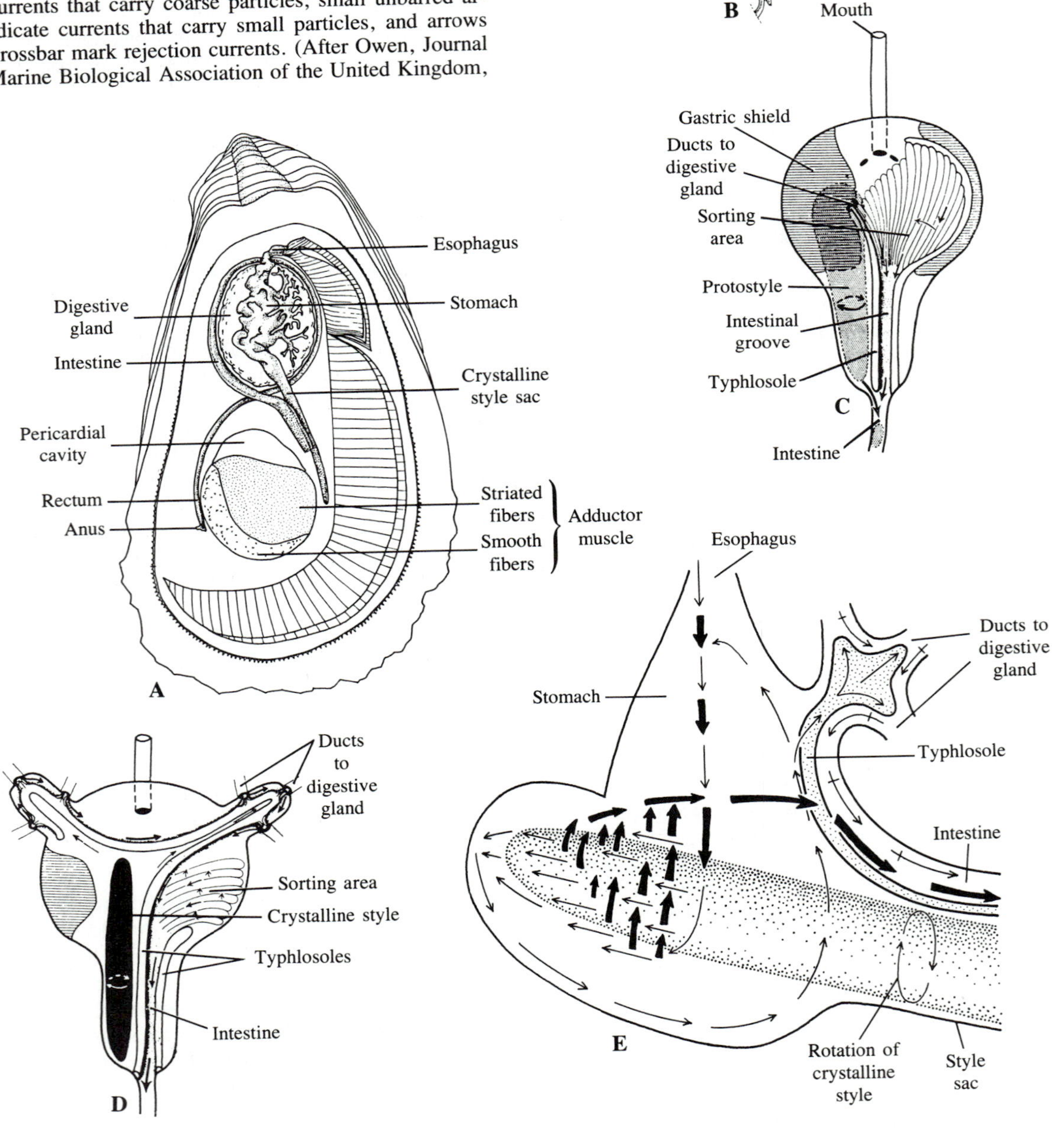

probable that the stomachs of protobranchs are specialized for processing comparatively large particles, or large aggregates of particles.

In many bivalves, feeding is cyclical. While this is inevitable in intertidal species that may be exposed for several hours at a time, it has also been noted in marine and freshwater species that are continuously submerged. In certain intertidal bivalves, the crystalline style is partly or completely dissolved when feeding ceases, then is reformed when feeding begins again.

Circulatory System

The heart, and the pericardial cavity that encloses it, are located dorsally. In most bivalves, the ventricle is pierced by the rectum (Figure 13.59). This is because the heart originates as two primordia, one on each side of the gut. The pockets fuse, so that the median, ventricular portion of the heart surrounds the rectum. There are generally two aortas leaving the ventricle—one directed anteriorly, the other posteriorly—but in certain bivalves, including sea mussels, only the anterior aorta is present.

Branches of the aortas deliver blood to the spaces that make up the hemocoel. The few larger spaces where blood collects are called *sinuses*. Exchange of oxygen and carbon dioxide takes place across the epithelia of the mantle, ctenidia, and foot, and removal of nitrogenous wastes is effected by the renal organs. The blood eventually returns to the auricles of the heart by way of major veins that drain the ctenidia and mantle, but there are many small venous vessels throughout the body. The exact pattern of blood flow varies from group to group, but a generalized plan is shown in Figure 13.60, A. It should be noted that a large part—about 50%—of the volume of soft tissue of a bivalve consists of blood.

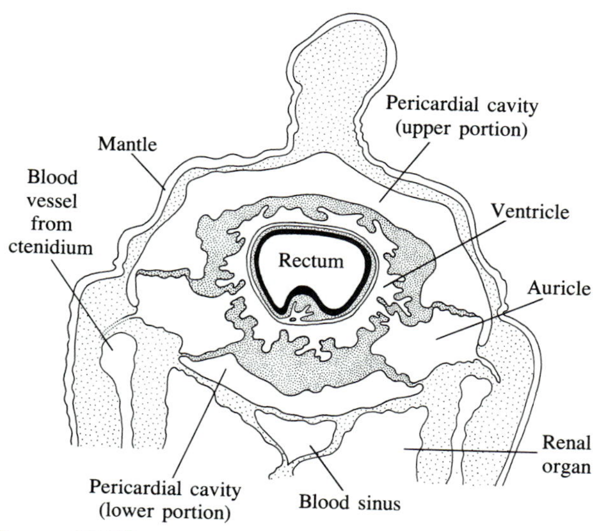

FIGURE 13.59 Transverse section through the dorsal part of the body of *Anodonta cellensis,* showing the relationship of the heart to the pericardial cavity, rectum, and renal organs. (Krug, Zeitschrift für Wissenschaftliche Zoologie, *119*.)

Excretory System

The renal organs, generally brownish, are located on both sides of the ventral portion of the pericardium (Figures 13.56, 13.59, and 13.61). They are coelomoducts that connect the pericardial cavity with the mantle cavity. The blood around the renal organs is mostly venous blood that goes to the ctenidia before it returns to the auricles of the heart. In some of the primitive bivalves, the gonads are joined to the renal organs, so these structures serve as gonoducts as well as excretory structures. In more advanced bivalves, however, the gonads have their own ducts to the mantle cavity.

In nearly all bivalves, there is a pair of brownish or blackish glands on the pericardium or on the auricles of the heart. These are derived from the coelomic epithelium of the pericardial sac. When situated on the auricles, as in sea mussels of the genus *Mytilus,* they give these structures a spongy appearance. A variety of names, including "Keber's organs," have been applied to them, but it seems best to call them *pericardial glands*. They are basically tubular, but their walls are much folded, and they have one or more openings into the pericardial cavity. Not much is known about their function, but there is evidence indicating that cells of the pericardial glands take up wastes from the blood, alter and concentrate them within lysosomes, and then discharge them by pinching off little blobs of cytoplasm. Upon reaching the pericardial cavity, the wastes can be eliminated by the renal organs.

Pericardial glands are not the exclusive property of bivalves. They occur also in gastropods, especially certain opisthobranchs.

Nervous System and Sense Organs

The nervous system (Figure 13.60, B) is generally symmetrical. The two large cerebral ganglia alongside the esophagus are joined together by a commissure. Connectives run from them to paired pedal ganglia in the foot and to visceral ganglia that are usually situated rather far posteriorly. The functions of the ganglia are interrelated, and the extent to which they are independent varies. The pedal ganglia, for instance, have the motor centers for control of the foot, but the sensory nerves in the foot go to the cerebral ganglia, so the influence of the pedal ganglia on foot movements must depend on input from the cerebral ganglia. The visceral ganglia are involved in control of the siphons and the opening and closing reflexes of the valves, but they,

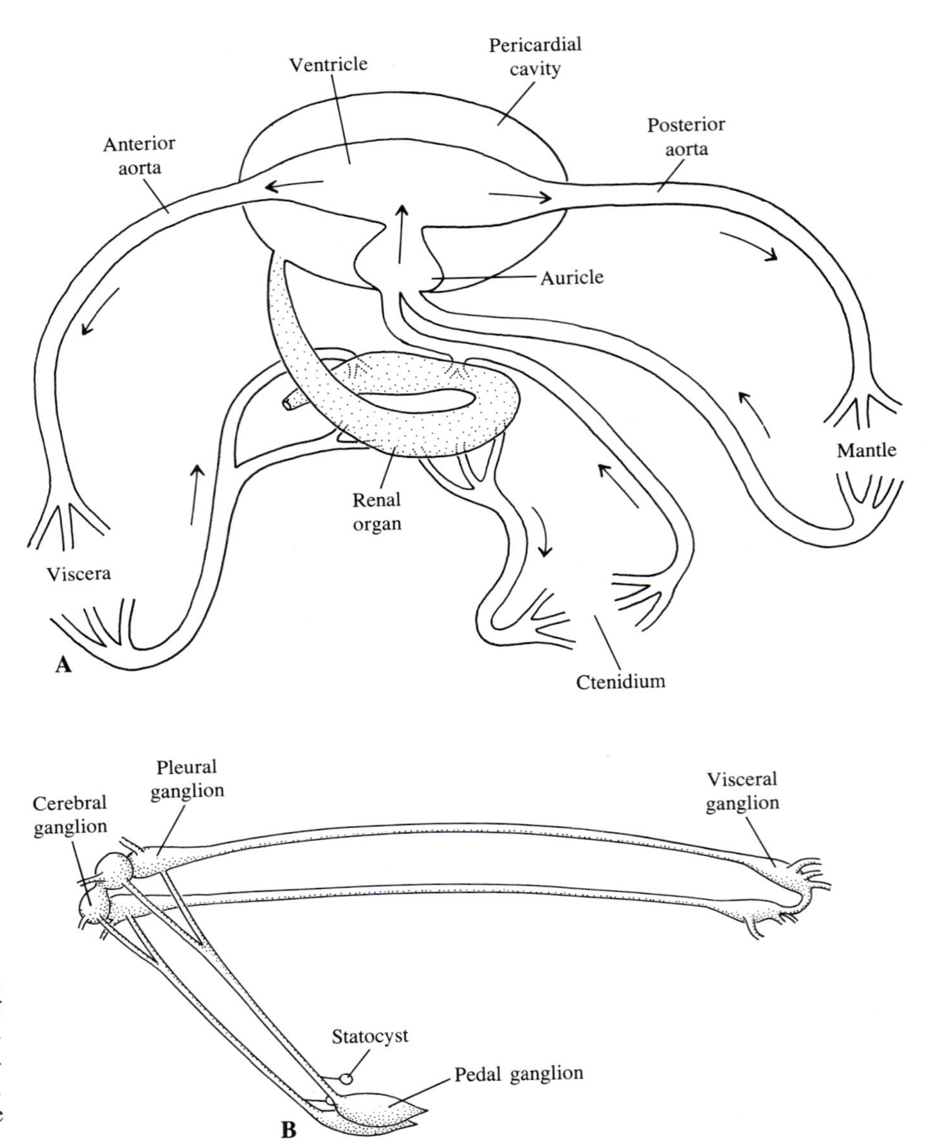

FIGURE 13.60 A. Diagram of the general pattern of circulation of blood in a bivalve. B. Major ganglia and connectives of the nervous system of *Nucula nucleus*. (After Pelseneer, Archives de Biologie, *11*.)

too, may depend on the cerebral ganglia for information.

There is usually a pair of **statocysts** buried in the foot near the pedal ganglia. These are basically sunken epidermal pits; only in a few primitive bivalves do they retain a connection with the outside. As might be expected, statocysts are reduced in oysters and some other bivalves that are permanently attached. In scallops, which usually lie with the right valve down, the left statocyst is better developed than the right one.

Touch receptors seem to be everywhere in the epidermis, but when recognizable sensory structures are present, these are generally concentrated at the edge of the mantle, especially on the middle fold, or near the tips of the siphons. It is on the middle fold of the mantle that the many eyes and tentacles of a scallop (Figure 13.56; Figure 13.63, D) are located. A scallop's eyes are rather elaborate, with a cornea, lens, and retina. They are efficient in detecting changes in light intensity. In many bivalves, if light receptors are present at all, they are very simple eyespots.

Bivalves often have a pair of structures that resemble the osphradia of chitons and prosobranch gastropods. When obvious, they are on the part of the mantle that is ventral to the posterior adductor muscle (which is the only adductor muscle in some bivalves). They sometimes are drawn out into ridges beside the posterior portions of the ctenidia. While it may be that they can de-

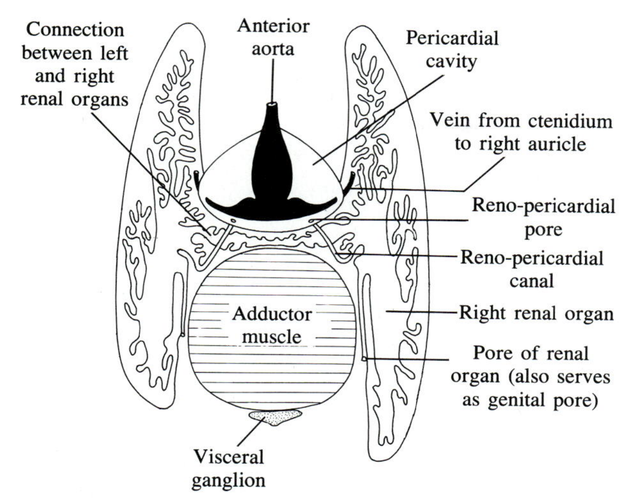

FIGURE 13.61 Relationship of renal organs to the pericardial cavity and mantle cavity of *Crassostrea virginica;* diagram based on a series of sections. (Galtsoff, Fishery Bulletin, *64*.)

tect particles of silt that enter the mantle cavity, their function is still uncertain. Whatever they do, it is not likely that they are homologous to true osphradia.

Reproduction and Development

In most bivalves, the sexes are separate. Some exceptions are mentioned below. There is a pair of gonads, which are usually extensive, sometimes spreading from the visceral mass into the mantle lobes. As a rule, the gonoducts have their own pores opening into the mantle cavity, but in protobranchs and certain lamellibranchs the gametes reach the mantle cavity by way of the renal organs; the latter arrangement is the more primitive one. Bivalves are generally free-spawners that liberate their gametes into the surrounding water, though the females of some species brood their young in the mantle cavity. (A few examples of brooding are given later in this section.)

Simultaneous hermaphroditism is rather rare in bivalves, but it does occur in some genera, including certain scallops. Alternation of maleness and femaleness is a common phenomenon, and has been intensively studied in oysters. Some oysters start out as males, become females, then repeat the sequence indefinitely. Others start out as females then become males and continue this pattern. *Crassostrea gigas* (Figure 13.64, F), an oyster native to the Orient but now widely cultivated along the Pacific coast of North America and elsewhere, is not locked into a sequence. As its spawning season approaches, it may become either a male or a female.

Many marine bivalves have a free-swimming trochophore stage, but this is soon transformed into a veliger (Figure 13.62). Secretion of the shell begins in a plate-like area on the dorsal side, and before long the two valves are distinct. If the veliger developed from a large egg in which there was a considerable store of nutritive material, it is not likely to feed, and it may not even be a good swimmer. Veligers that contain relatively little yolk generally do feed on microscopic components of the plankton, and they usually swim in the plankton for a few weeks.

At metamorphosis, the velum is sloughed off. By this time the layout of the ctenidia, gut, adductor muscles, and other structures is basically the same as in an adult clam. It is important to appreciate that metamorphosis and further development are not likely to take place unless the larva locates a suitable place in which to settle. Veligers whose settlement has been studied are generally able to recognize the right kind of sand, rock, or other type of substratum. Some bivalves are known to be able to delay metamorphosis for a time, and a few may go into a special waiting stage. As a rule, however, a veliger must settle soon after it is ready to do so; otherwise it will die.

Direct development and brooding are characteristic of some marine clams, especially small eulamellibranchs. The young are set free from the mantle cavity of the parent after they are fully formed little clams. Sometimes stages more or less comparable to the trochophore and veliger can be recognized, but they are not typical.

Among freshwater clams, a free veliger is unusual, and there is never a free trochophore. In nearly all species, the young are brooded. A special type of juvenile, called the *glochidium,* occurs in the life history of most freshwater "mussels." (These are eulamellibranchs, and must not be confused with sea mussels, which are filibranchs.) Glochidia are generally no more than 0.5 mm in length or height. When set free by the parent, they leave the mantle cavity by way of the gape between valves or are carried out by the exhalant current. They have no future unless they become attached to an appropriate fish. If the fish swims close enough to a glochidium, the little clam may be drawn up into the gills with the respiratory current, or it may clamp its valves together at just the right time and thus attach itself to a fin or some other place on the body. In the genus *Anodonta,* each valve of the glochidium's shell has a hook that grasps the skin.

After a glochidium is properly situated, it becomes overgrown by the host's tissue, and phagocytic cells of the mantle gather nourishment from this tissue. During the parasitic phase, which may last for a month or more, the young clam becomes reorganized. Some structures, including the adductor muscle, are destroyed, then replaced. In time, the parasite begins to

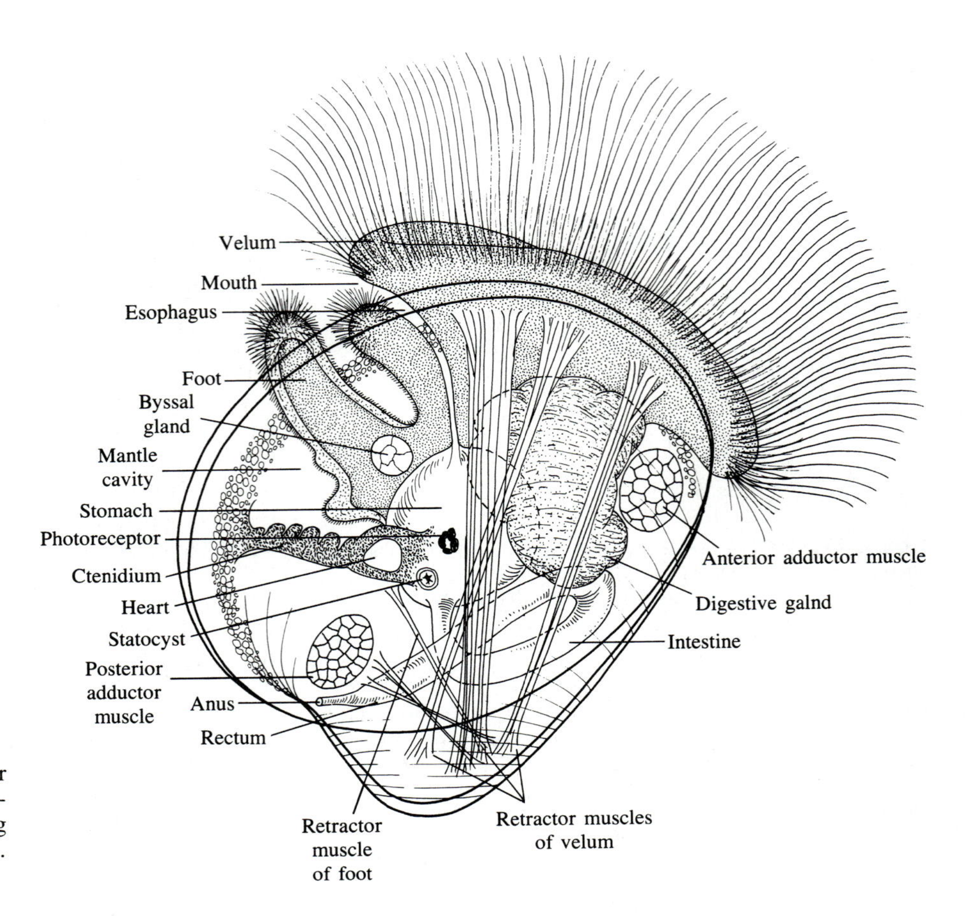

Figure 13.62 Late veliger (length about 0.25 mm) of *Crassostrea virginica,* in swimming position, viewed from the side. (Galtsoff, Fishery Bulletin, *64*.)

look more like a clam. It escapes from its prison in the skin tissue of the fish, sinks to the bottom, and starts a new kind of life as a filter-feeder. The fish serves not only as a host for the clam's parasitic phase, but also as an agent of dispersal.

Some Lifestyles of Bivalves

Most bivalves live in mud, sand, or gravel. Cockles, freshwater mussels, and others that have short siphons, or no true siphons at all, are shallowly buried, so that the posterior end of the shell is at or near the surface of the substratum. If dislodged, they use the foot and movements of the valves to dig themselves back into the substratum until they are properly positioned to draw in and expel water. Razor clams and jackknife clams (Figure 13.57, A; Figure 13.63, B) fit into this general category, although they are proficient burrowers that can quickly dig deep into the sand if threatened.

Clams with extensile siphons generally have a burrow that is deeper than the shell is long (Figure 13.57, B and C). Some species, if dug up, can work their way back into the substratum and continue their lives as if nothing had happened. Others are unable to excavate a new burrow; the body and shell have become too large in proportion to the foot. This is the case with horse clams *(Tresus)* and the geoduck *(Panope)* of the Pacific coast. Such clams start burrowing when they are small, and by the time they are mature they may be deeper than 50 cm. The linings of their burrows are slicked and firmed by secretions of mucus.

Certain species that have more or less permanent burrows can move up and down within them. They use the foot for positioning themselves. Most of the bivalves in this category, which includes jackknife clams, have fairly short siphons, and they must be close to the surface to pump water.

Sea mussels—*Mytilus* (Figure 13.64, A and B), *Modiolus,* and allied genera—attach themselves to rocks, shells, wood, and other firm objects by means of **byssal threads.** These are organic structures, tough but flexible, that originate in a gland at the base of the reduced, fingerlike foot. The secretion of the byssus gland is a fluid that is delivered along a ciliated groove to the tip of the foot, which is pressed to the substratum. As soon as the material has hardened up in the groove, which acts as a kind of mold for shaping the thread, the foot

FIGURE 13.63 A. *Macoma inquinata,* showing separate inhalant (lower) and exhalant (upper) siphons. B. *Solen sicarius,* a jackknife clam. C. *Panomya ampla.* The siphonal mass is extremely large in proportion to the rest of the animal. D. *Chlamys hastata,* a scallop, gaping. The numerous eyes and tentacles on the edges of the mantle lobes are visible.

is disengaged from it. By repeating the process of secretion and attachment, the mussel produces a number of threads that collectively form a strong bond between it and the substratum. Certain mussels, such as the widespread *Mytilus edulis,* are rather mobile. They can use the foot for crawling around when they have to, although they are attached most of the time. If a mussel is ripped loose in some way, it is extremely vulnerable unless it lands where it can soon reestablish itself.

Mussels are not the only bivalves that produce byssal threads. The habit is widespread in this class of molluscs. Attachment may be intermittent, however, as is the case in many scallops, which can jerk themselves free and swim away by clapping the valves of the shell together. This is likely to happen, for instance, when a scallop is approached by a predator.

The peapod borers of the genus *Adula* (Figure 13.64, C) are in the same family as sea mussels. Byssus threads are used for attachment to the wall of the burrow. The valves are only slightly roughened, but they are protected by a tough periostracum. It is likely that the boring action is achieved by rubbing the valves against the rock. Hard material caught between the periostracum and the rock probably functions as an abrasive.

Members of the genus *Lithophaga* (Figure 13.64, D) are mussels that burrow into limestone, coral, or aggregates of calcareous material. The mucus secreted by the margin of the mantle has a high acidity. The action of the acid, coupled with abrasive movements of the valves, erodes the substratum. The periostracum protects the shell from being eaten away.

Among the bivalves that are tightly cemented to hard surfaces are oysters, rock jingles, and certain scallops. Rock jingles have a hole in one valve, and through this runs a byssal secretion, often calcified, that attaches the animal tightly to rock or shell (Figure 13.64, E). In oysters (Figure 13.64, F), the situation is different; almost all of one valve is cemented down. The reason for this is that after the veliger larva of an oyster has settled

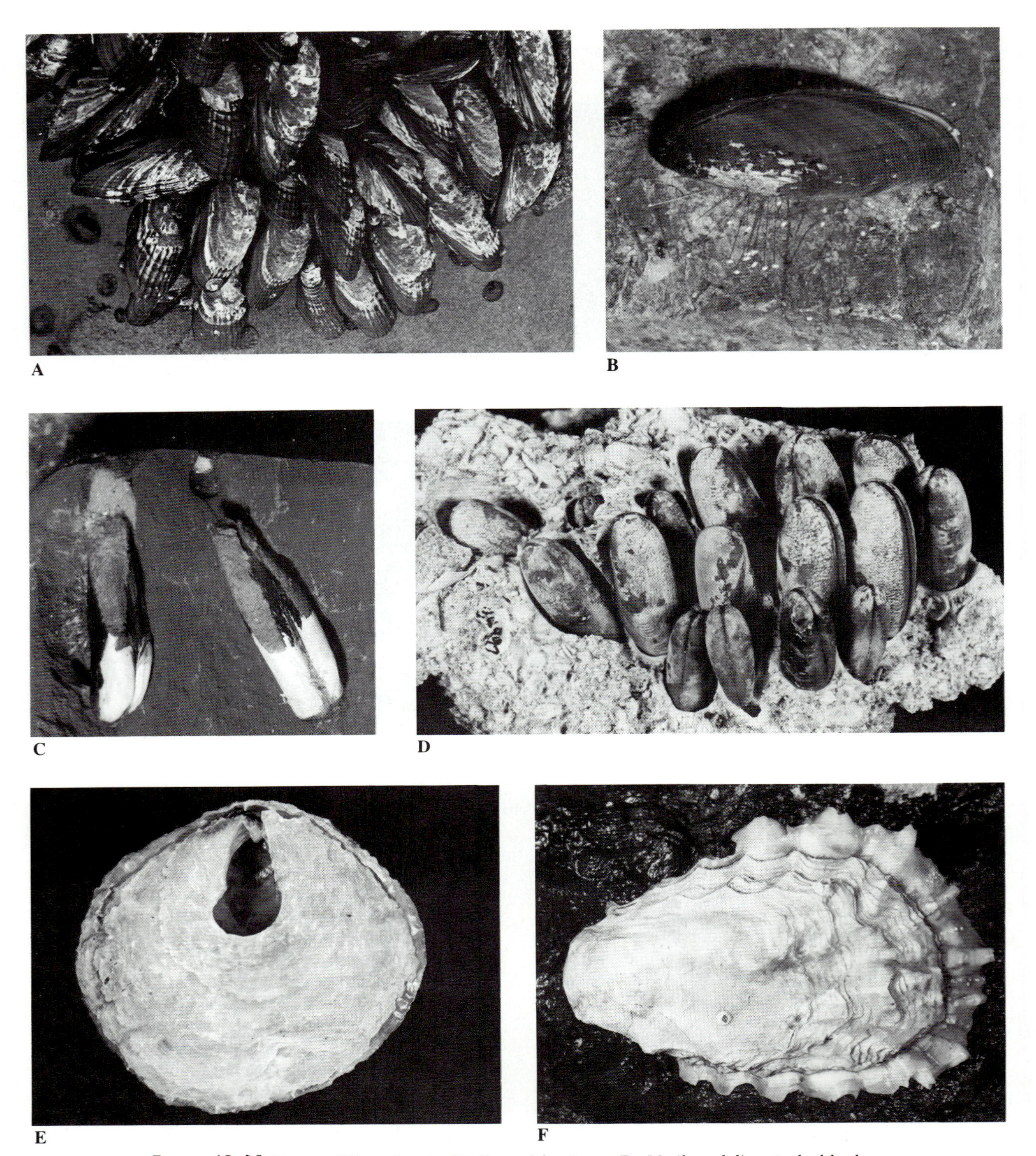

FIGURE 13.64 Diverse filibranchs. A. *Mytilus californianus*. B. *Mytilus edulis,* attached by byssus threads to the substratum. C. *Adula californica,* a rock-boring mussel. D. *Lithophaga plumula,* a mussel that burrows in calcareous substrata, including limestone. E. *Pododesmus cepio,* a rock jingle. The byssus cement that holds the animal to a hard substratum emerges through the hole in one valve. F. *Crassostrea gigas,* an oyster. One valve is permanently cemented to a hard substratum.

and secreted enough byssal material to attach itself, the edge of the mantle takes over the function of producing cement. While the animal is growing, the mantle keeps on depositing cement around the edge of the attached valve. The free valve is usually at least slightly smaller than the one that is attached. In the genus *Chama* and its allies, which are closely related to oysters, the free valve is like a hinged lid for the boxlike attached valve.

Of the scallops that are permanently attached, *Hinnites* is probably the best-known genus. The process by which one valve is cemented down is about the same as in oysters and chamas. A young *Hinnites* looks like a typical scallop, but after the animal has grown to a large size—10 cm or more—the characteristic outline of a scallop is likely to be recognizable only in the older part of the free valve.

The habit of boring into firm clay, soft rock, coral, shell, conglomerates of calcareous fragments, and wood has evolved independently in several groups of bivalves. A variety of techniques are used for boring, and the process of penetration begins as soon as a young bivalve settles on a suitable substratum. In one group, called piddocks (family Pholadidae), a part of the foot is specialized to function as a sucker, and the surface of the anterior portion of the shell is roughened like a rasp (Figure 13.65, A). By attaching itself firmly to the substratum, opening and closing the valves, and rocking the shell as a whole, the animal digs its burrow. There is no true hinge in a piddock; the valves are held together by the adductor muscles, and their movements are not so restricted as they would be if there were a hinge. One or more accessory shell plates are usually secreted to protect the adductors, which would otherwise be exposed. Some species of piddocks, after their period of active boring has ended, secrete calcareous material to close the gape where the foot had previously emerged (Figure 13.65, B).

As a piddock grows, its burrow becomes progressively wider. There is often a slight spiral twist to the burrow, due to the fact that the clam rotates a little as it bores. In any case, the clam is permanently imprisoned in its burrow; its only contact with the outside world is by way of the siphons, which are extended to the opening. Some piddocks, including *Penitella* (Figure 13.65, B), burrow into soft rock (especially sandstone) or into thick shells, as those of abalones. Others, such as *Zirfaea* (Figure 13.65, A), dig into hard clay. A few (*Xylophaga* and *Martesia*) bore into wood.

The so-called shipworms of the family Teredinidae cause much damage to pilings, piers, and other wooden structures immersed in sea water (Figure 13.65, D). The roughened valves of a shipworm's shell are small in proportion to the animal as a whole (Figure 13.65, E). Attachment is by a suckerlike foot, and abrasion is achieved by alternately spreading the valves and drawing them together. Some of the fine "sawdust" that a shipworm produces is taken into the inhalant siphon, along with plankton. The stomach has a side pocket in which the particles can be accumulated while they are waiting to be shunted into the digestive gland. At least some of the cellulose is utilized, but the digestion of it is thought to be the result of bacterial activity, and in most shipworms plankton is the principal food source.

In timbers colonized by shipworms, the burrows generally follow the grain and remain separate from one another. The limy coating of the burrows is secreted by the mantle. Another elaboration produced by the mantle is a pair of featherlike or leaflike *pallets*. These operate as closing devices, blocking the opening of the burrow when the siphons are retracted.

A number of bivalves remain unattached. Certain scallops and the file shells of the genus *Lima* fit into this category. These are good swimmers that escape predators, such as sea stars, octopuses, and fishes, by clapping their valves together vigorously enough to create strong jets of water. Although the anterior adductor muscle has disappeared in these bivalves, the posterior adductor is large and in a central position. It has two main components. One consists of striated muscle that is concerned with the rapid contractions essential to swimming by jet propulsion; the other consists of smooth muscle that locks the valves together for a prolonged period if necessary.

The growths of sponges on some scallops (see Figure 3.17) probably provide considerable protection against predators. Sponges are disagreeable to most animals other than dorid nudibranchs, for whom they are a staple food. Sponges may also make it more difficult for a predator to pin down or grasp the shell. As a rule, however, only one valve is thoroughly coated by a sponge.

Numerous bivalves, mostly rather small species, live in empty shells, empty burrows made by rock-boring bivalves, holdfasts of kelps, and similar situations where they are protected yet still able to feed. *Kellia* (Figure 13.66) is an interesting example because its inhalant siphon projects anteriorly rather than posteriorly. The middle lobe of its mantle, moreover, extends outside the shell and spreads over the valves.

Bivalves of various groups live in close association with other animals. Only a few examples will be mentioned. The siphons of *Cryptomya californica,* found on the Pacific coast of North America, open into the burrows of ghost shrimps (*Callianassa*) and mud shrimps (*Upogebia*), and also into those of the echiuran worm *Urechis* (see Figure 12.2). The clam feeds on microscopic material in the current of water that its host moves through the burrow. The situation is a good one

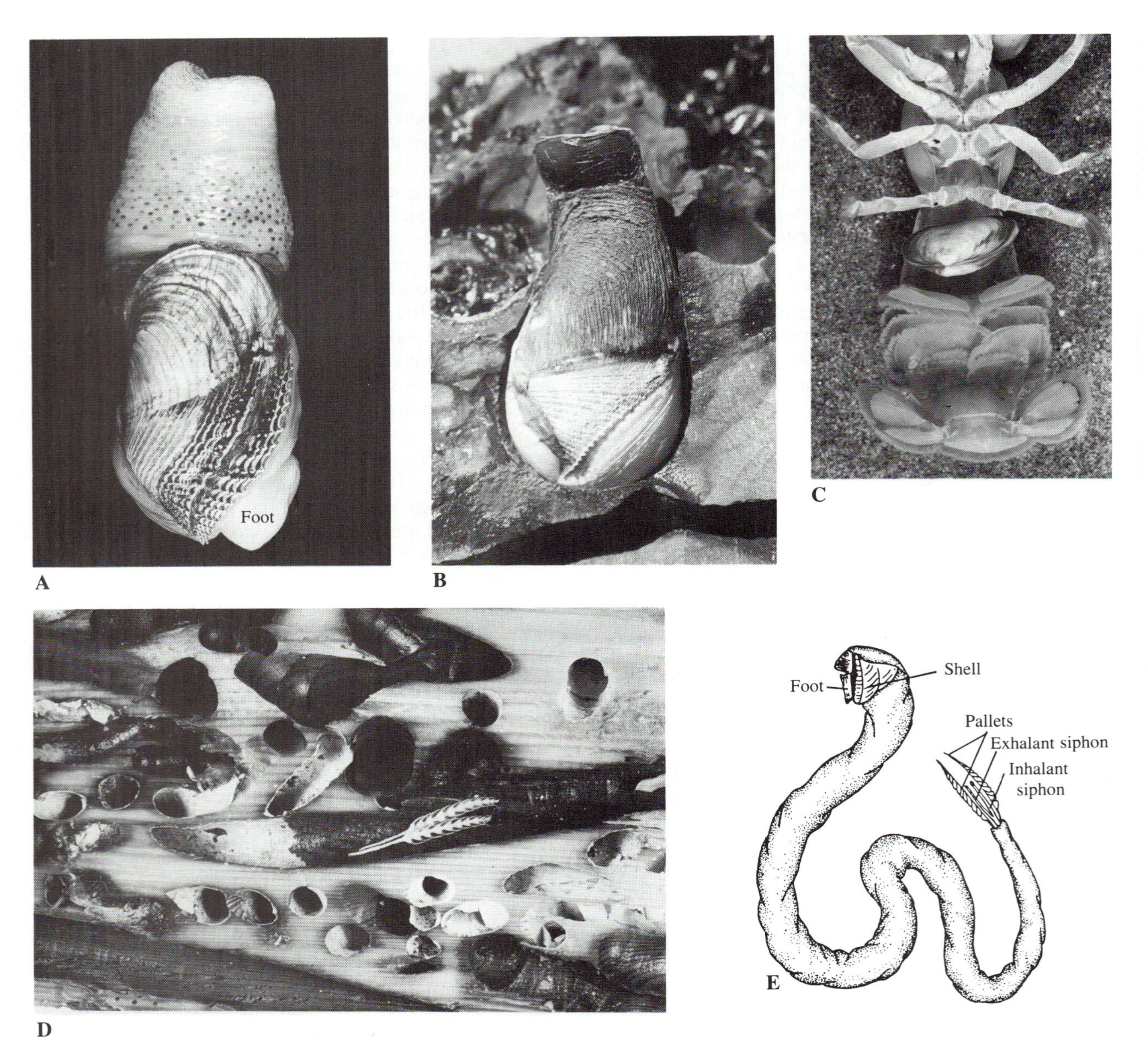

Figure 13.65 A. *Zirfaea pilsbryi,* which burrows into stiff clay. B. *Penitella penita,* in its burrow in shale. After reaching maximum size, the animal seals the gape between shell valves, so the foot no longer protrudes. C. *Pseudopythina rugifera,* attached by a byssus to the abdomen of a mud shrimp, *Upogebia pugettensis.* D. Burrows of the shipworm, *Bankia setacea,* in an old piling. The pallets of a deceased *Bankia* are shown in one of the burrows. E. *Bankia setacea,* removed from its burrow. (Turner, A Survey and Illustrated Catalogue of the Teredinidae.)

for *Cryptomya,* because it provides protection from the usual predators that attack clams whose siphons open at the surface. *Pseudopythina rugifera* (Figure 13.65, C) lives on the abdomen of *Upogebia,* where it is attached by byssus threads. It, too, can extract microscopic food from the current that its host drives through the burrow. *Pseudopythina* also inhabits the ventral surface of some larger species of *Aphrodita,* which are polychaetes that

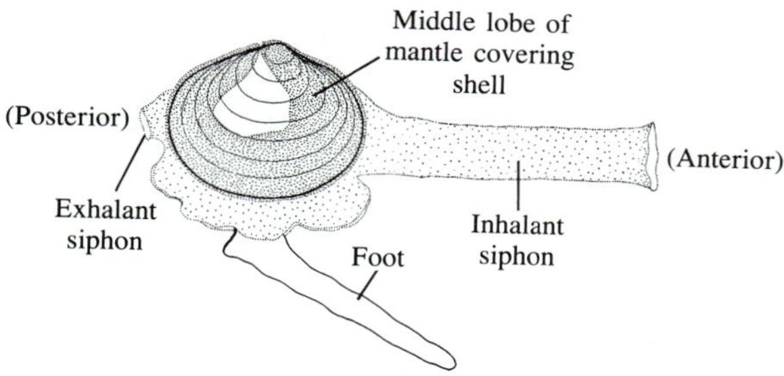

FIGURE 13.66. *Kellia suborbicularis,* a eulamellibranch that nestles in holes, empty shells, and similar situations. The inhalant and exhalant siphons, formed by the inner lobe of the mantle, are widely separated, the inhalant one being displaced to an anterior position. The middle lobe of the mantle extends out of the shell and covers much of both valves. (Drawing from Yonge, University of California Publications in Zoology, *55.*)

plow through soft mud. There are a number of other bivalves whose lifestyles are similar to those of *Cryptomya* and *Pseudopythina.*

Mytilimeria nuttallii of the Pacific coast is a rather fragile, thin-shelled clam that is always covered by a compound ascidian. Only its siphons are exposed. There are also bivalves that inhabit sponge masses.

In the giant clams of the family Tridacnidae (Figure 13.67), found in the Indo-Pacific region, the siphonal portions of the mantle are hypertrophied and populated by large numbers of zooxanthellae, which are symbiotic dinoflagellates. The clams live in shallow water, hinge side down, so that the ventral gape of the shell is uppermost and the siphonal tissues are exposed to light. Pigments in the tissues reduce the light intensity to a level favorable for photosynthesis by the zooxanthellae. Soluble products of the photosynthesis, as well as products released from zooxanthellae that are digested, contribute to the nutrition of the clams. The zooxanthellae are actually in amoeboid cells of the blood, so they may be transported to other parts of the body before they are digested. They may also be eliminated by way of the digestive gland and stomach. Tridacnid clams do not rely entirely on their symbionts; they also use their ctenidia and labial palps for ciliary–mucous feeding. The largest species, *Tridacna gigas,* reaches a length of more than 1 m and a weight of more than 200 kg.

Certain deep-water clams that live in the vicinity of geothermal vents, and some that occur in more conventional marine habitats, have autotrophic bacterial symbionts in their ctenidial tissues. The bacteria oxidize inorganic compounds, especially hydrogen sulfide, thereby obtaining energy that enables them to synthesize organic compounds from carbon dioxide. The clams depend to a large extent on nutrients supplied to them by the bacteria. Those of the genus *Solemya,* which have no gut, or have only a much-reduced gut, probably get nearly all of their nourishment from bacteria that live in the ctenidia. The habitats of some species of *Solemya* are characterized by considerable or-

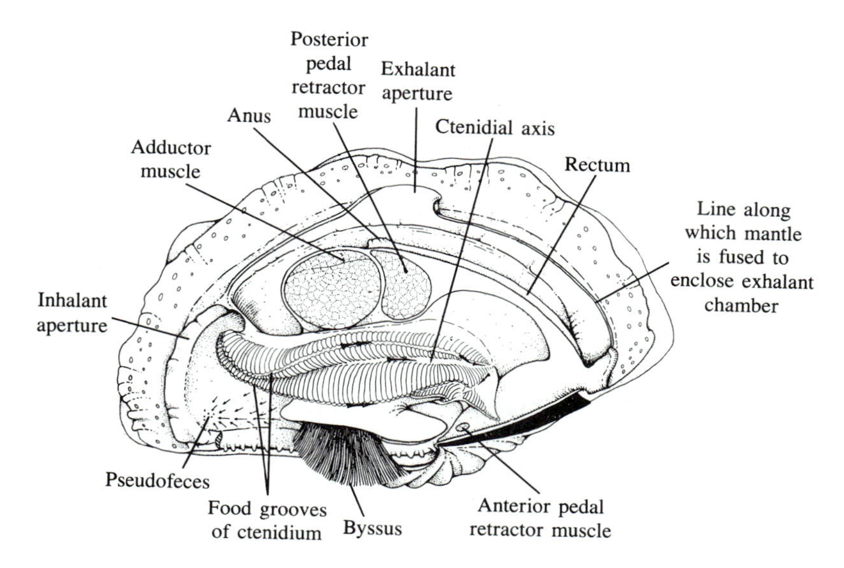

FIGURE 13.67 *Tridacna derasa,* in its usual orientation, attached by a byssus; right valve removed. The only adductor muscle is homologous to the posterior adductor muscle of other bivalves. (Stasek, Archives de Zoologie Expérimentale et Générale, *101.*)

ganic matter, however, so perhaps these bivalves also take up dissolved nutrients. A more complete discussion of symbiosis between invertebrates and bacteria that oxidize hydrogen sulfide is given in connection with worms of the phylum Vestimentifera (see Chapter 12).

Among the more remarkable bivalves are those that belong to the order Septibranchia (Figures 13.68 and 13.69). These clams are carnivores that feed mostly on copepods, ostracodes, and other small crustaceans. They are strictly marine, living subtidally in soft sediments. The inhalant siphon, extended just slightly above the surface of the mud, has some tentacles in which prey-detecting receptors are located. When the prey gets close enough, it is sucked into the mantle cavity. The sucking action is effected by a rapid upward movement of horizontal diaphragms that run between the foot and mantle on both sides of the body. The diaphragms are thought to be derived from ctenidia. In certain genera they slightly resemble ctenidia, because they have openings crossed by ciliated filaments. In others, each diaphragm has a row of simple pores. If the openings are closed as the diaphragms are quickly elevated, a negative pressure is created in the lower half of the mantle cavity. The diaphragms are then lowered slowly, while their pores are open (Figure 13.69, B).

Once a prey organism has been drawn into the mantle cavity, it is grasped by muscular labial palps and worked into the mouth. The stomach is rather muscular and lined with a cuticle, and can probably be used to crush certain types of prey. There is a small style whose function is not really understood. It perhaps serves primarily as a source of mucus. Digestion is thought to take place in the stomach, the digestive glands delivering their enzymes to this part of the gut.

Most septibranchs are only about 1 cm long, so they often escape notice, especially when samples of mud are processed through coarse sieves. A feature that helps one to recognize a septibranch is the way the posterior portion of its shell is drawn out into a beak (Figure 13.69, A). The siphons are extended only a short distance beyond the tip of this. The mantle edges are fused except in one small area. It is through this opening that the foot protrudes when the animal plows through the mud.

A

B

Figure 13.68 Septibranchia, *Cardiomya californica*. A. Exposed, viewed from the left side. B. Partly buried, siphons extended.

CLASSIFICATION

Although bivalves are not as diversified as gastropods, they are certainly varied. Unfortunately for the systematist, there are many parallel trends in the evolution of these molluscs. This makes classification difficult. Bivalves that may resemble one another and have the same general lifestyle may have evolved along entirely separate lines. In the discussion of the functional morphology of bivalves, considerable stress was laid on the structure and operation of the ctenidia and labial palps, the arrangement of the adductor muscles, and the organization of the hinge. These features are of primary importance in classification.

SUBCLASS PALAEOTAXODONTA

The palaeotaxodonts include most of the more primitive bivalves—the ones called protobranchs. The ctenidia of these are located in the posterior part of the mantle cavity and are much like those of chitons and

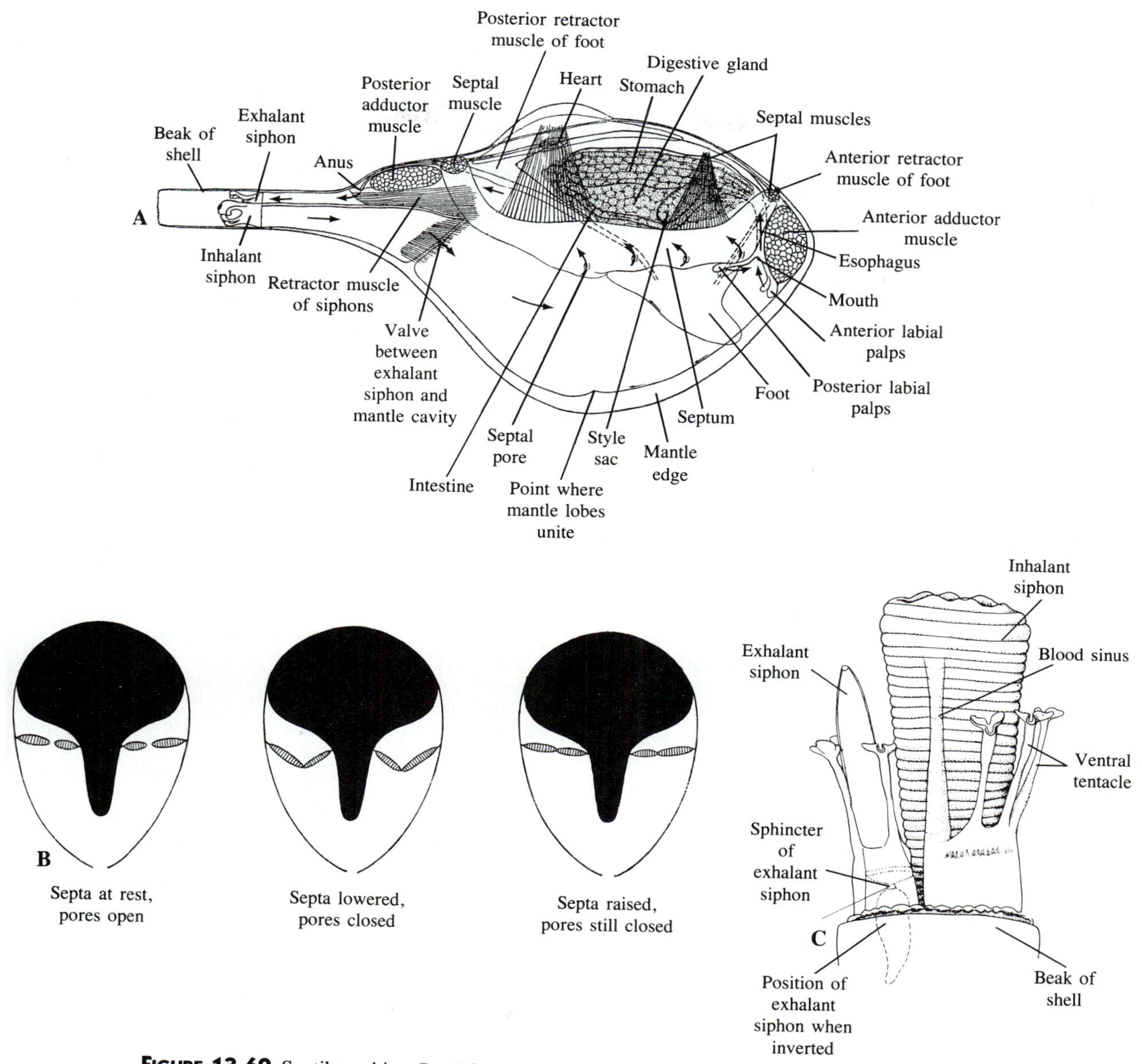

FIGURE 13.69 Septibranchia, *Cuspidaria rostrata*. A. Right valve removed, showing most organs. B. Diagram of the movements of the septa. (Yonge, Philosophical Transactions of the Royal Society of London, B, *216*.) C. Siphonal region. (Reid and Reid, Sarsia, *56*.)

many archaeogastropods in having relatively broad, leaflike branches. They are concerned with circulation of water and respiration, not with feeding, which is a function of the labial palps and their extensions. The hinge of the shell is characterized by taxodont dentition. There are always two adductor muscles. Siphons may be well developed, as in *Yoldia* (Figure 13.50, C), or absent, as in *Acila* (Figure 13.50, A).

SUBCLASS CRYPTODONTA

The only living cryptodonts belong to a single genus, *Solemya*. Their ctenidia are of the protobranch type, but their shells are toothless, or nearly so. Symbiotic bacteria in the tissue of the ctenidia provide nutrients for the clams. This group has many extinct representatives.

"LAMELLIBRANCHS"

In lamellibranchs, the ctenidia are used in ciliary–mucous collection of food. The individual filaments, homologous to the leaflike branches of the ctenidia of protobranchs, are bent back on themselves. In the past, most systems of bivalve classification included a subclass Lamellibranchia, further subdivided into Filibranchia (with consecutive ctenidial filaments free of one another) and Eulamellibranchia (with consecutive filaments fused). These are useful pedagogical distinctions, because they stress differences in the functional morphology of the ctenidia. In modern systems, however, the lamellibranchs are broken down into several subclasses. Those with living representatives are discussed briefly here.

SUBCLASS PTERIOMORPHA

In spite of much diversity, the pteriomorphs form a rather well unified assemblage. Nearly all of them have filibranch ctenidia (a few are eulamellibranch), and siphons are absent or rudimentary.

One group of pteriomorphs, represented by *Arca* and *Glycymeris,* is characterized by a shell with taxodont dentition and by a well-developed foot used for digging. The two adductor muscles are about equal, unless one is absent, which is the case in *Philobrya*.

In sea mussels, such as *Mytilus* (Figure 13.64, A and B) and *Modiolus,* the hinge generally lacks substantial teeth and the adductor muscles are decidedly unequal (the anterior one is small). The foot is reduced; it can be used for crawling, but not for digging. A byssus gland that is part of the foot secretes organic threads for attaching the animal to a firm substratum. Several genera, including *Adula* (Figure 13.64, C) and *Lithophaga* (Figure 13.64, D) burrow into rock.

Scallops, oysters, and rock jingles also lack hinge teeth, and only the posterior adductor muscle is present. The foot, as in sea mussels, is reduced. In rock jingles (Figure 13.64, E) and oysters (Figure 13.64, F), one valve is permanently cemented to the substratum; this is also true of certain scallops, such as *Hinnites*. Most scallops, however, are free or attach themselves temporarily by byssus threads.

SUBCLASS HETERODONTA

Most bivalves, including freshwater mussels, fit into the Heterodonta. The ctenidia are always of the eulamellibranch type, the foot is usually well developed, and both adductor muscles are prominent. Siphons—long or short, fused or separate—are typically present. The hinge of the shell has a few interlocking teeth.

SUBCLASS ANOMALODESMATA

In Anomalodesmata, hinge teeth are small or absent, but these bivalves are not likely to be confused with pteriomorphs because both of their adductor muscles are well developed, and there are siphons. The ctenidia are either of the eulamellibranch type, as in *Entodesma* and *Mytilimeria,* or are specialized to the extent that they no longer resemble ctenidia, as is the case in the order Septibranchia.

References on Bivalvia

Ansell, A. D., and Nair, N. B., 1969. A comparison of bivalve boring mechanisms by mechanical means. American Zoologist, *9:*857–868.

Bayne, B. L., 1976. Marine Mussels: Their Ecology and Physiology. Cambridge University Press, Cambridge and New York.

Ellis, A. E., 1978. British Freshwater Bivalve Mollusks. Synopses of the British Fauna, no. 11. Linnean Society of London and Academic Press, London and New York.

Felbeck, H., 1983. Sulfide oxidation and carbon fixation by the gutless clam *Solemya reidi:* an animal–bacteria symbiosis. Journal of Comparative Physiology, B, *152:*3–11.

Galtsoff, P. S., 1964. The American Oyster. Bulletin of the United States Bureau of Fisheries, *64:*1–480.

Hogarth, M. A., and Gaunt, A. S., 1988. Mechanisms of glochidial attachment (Mollusca: Bivalvia: Unionidae). Journal of Morphology, *198:*71–81.

Jorgenson, C. B., 1974. On gill function in the mussel *Mytilus edulis*. Ophelia, *13:*187–232.

Kristensen, J. H., 1972. Structure and function of crystalline styles in bivalves. Ophelia, *10:*91–108.

Lane, C. E., 1961. The teredo. Scientific American, *204*(2):132–140.

Moore, M. N., Bubel, A., and Lowe, D. M., 1980. Cytology and cytochemistry of the pericardial gland cells of *Mytilus edulis* and their lysosomal responses to injected horseradish peroxidase and anthracene. Journal of the Marine Biological Association of the United Kingdom, *60:*135–149.

Morton, B., 1978. Feeding and digestion in shipworms. Annual Review of Oceanography and Marine Biology, *16:*107–144.

Nair, N. B., and Saraswathy, M., 1971. The biology of wood-boring teredinid molluscs. Advances in Marine Biology, *9:*336–509.

Pirie, B. J. A., and George, S. G., 1979. Ultrastructure of the heart and excretory system of *Mytilus edulis* (L.). Journal of the Marine Biological Association of the United Kingdom, *59:*819–829.

Reid, R. G. B., and Reid, A. M., 1974. The carnivorous habit of members of the septibranch genus *Cuspidaria*. Sarsia, *56:*47–54.

Stanley, S. M., 1970. Relation of shell form to the life habits of the Bivalvia (Mollusca). Geological Society of America, Memoir 125.

Stanley, S. M., 1972. Functional morphology and evolution of byssally attached bivalve mollusks. Journal of Paleontology, *46:*165–212.

Stasek, C. R., 1961. The ciliation and function of the labial palps of *Acila castrensis* (Protobranchia, Nuculidae) and an evaluation of the role of protobranch organs of feeding in the evolution of Bivalvia. Proceedings of the Zoological Society of London, *137*:511–528.

Stasek, C. R., 1965. Feeding and particle-sorting in *Yoldia ensifera* (Bivalvia, Protobranchia), with notes on other nuculanids. Malacologia, *2*:349–366.

White, K. M., 1942. The pericardial cavity and the pericardial gland of the Lamellibranchia. Proceedings of the Malacological Society of London, *25*:37–88.

Wood, E. M., 1974. Some mechanisms involved in host recognition and attachment of the glochidium larva of *Anodonta cygnea*. Journal of Zoology (London), *173*:15–30.

Yonge, C. M., 1941. The protobranchiate Mollusca: a functional interpretation of their structure and evolution. Philosophical Transactions of the Royal Society of London, B, *230*:79–147.

Yonge, C. M., 1971. On functional morphology and adaptive radiation in the bivalve superfamily Saxicavacea. Malacologia, *11*:1–44.

Yonge, C. M., 1975. Giant clams. Scientific American, *232*(4):96–105.

CLASS SCAPHOPODA

There are only about 200 living species of scaphopods, all marine. Most of them are subtidal, and their habitat is usually muddy sand. The tusk-shaped shell of a scaphopod (Figures 13.70 and 13.71), open at both ends, may be completely buried as the animal works its way through the sediment, or it may be only partly buried. The foot and the tentaclelike structures called **captacula,** which are concerned with feeding, are extended through the broad end. Water enters and leaves the

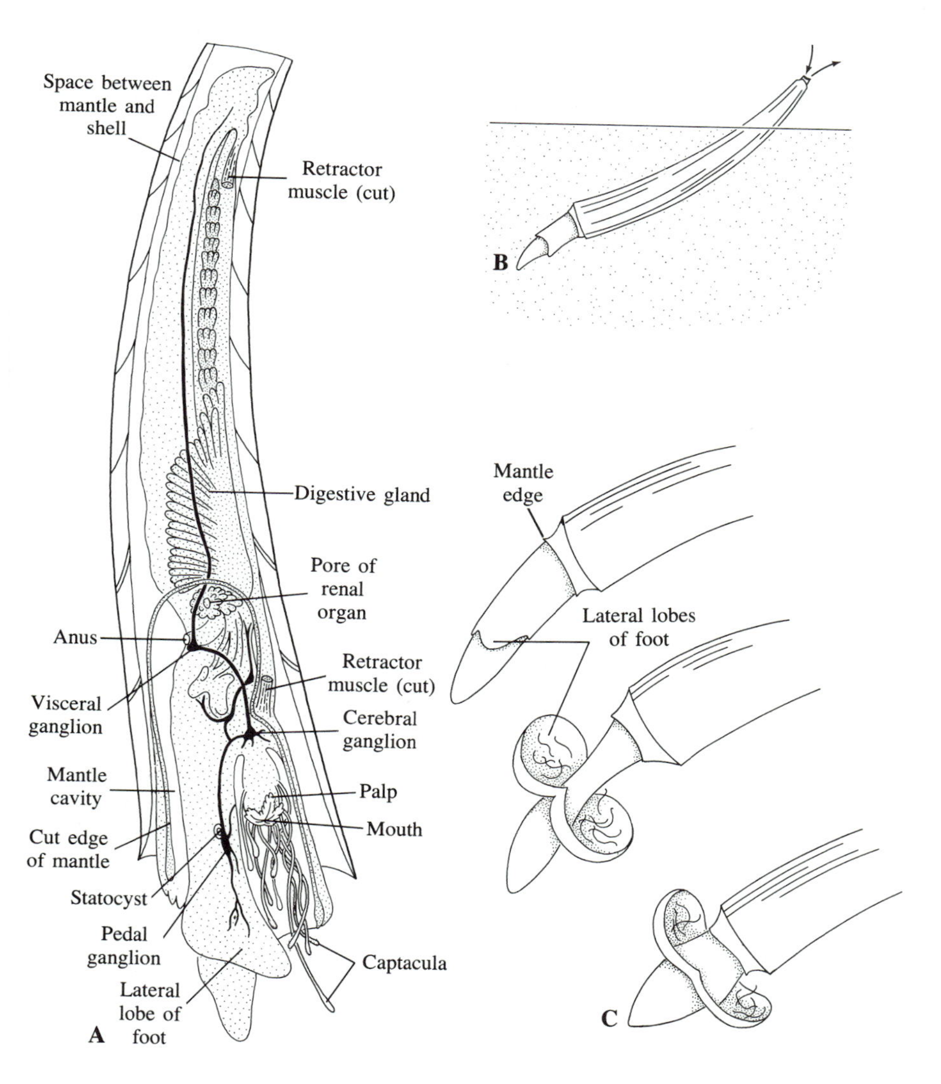

FIGURE 13.70 Scaphopoda, *Dentalium entale*. A. General anatomy; semidiagrammatic. B. The animal in its usual posture in sand. C. Successive stages in the action of the foot in burrowing. (B and C, after Morton, Journal of the Marine Biological Association of the United Kingdom, *38*.)

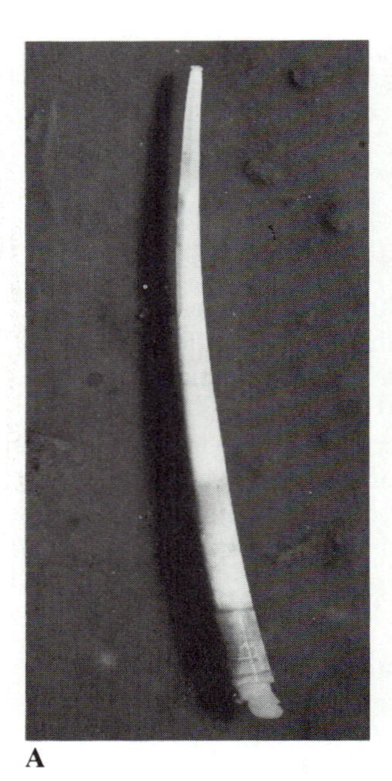
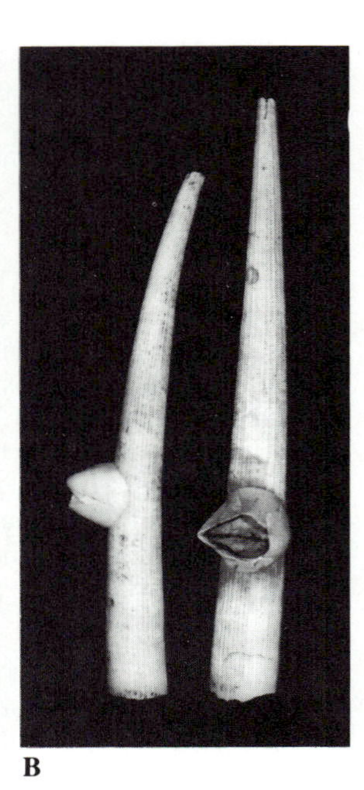

A B

Figure 13.71 Scaphopoda. A. *Dentalium pretiosum*. The foot is partly extended. B. A species that often has a barnacle growing on it.

mantle cavity by way of the small opening at the tip of the shell. Except for a few giants that reach a length of about 10 cm, the usual size range is from 1 to 5 cm. The shells of some fossilized species, however, are 60 cm long.

The ventral edges of the right and left mantle lobes are united to form a tube that encloses the rest of the body. There are no ctenidia, but the ventral portion of the mantle has ciliated ridges that may have a respiratory function. In general, slow movement of water into and out of the mantle cavity depends on ciliary activity, whereas rapid expulsion of water is caused by muscular contraction of the foot.

The foot is discoidal, conical, or tongue-shaped. When thrust into the sand, then dilated, it anchors the animal. The rest of the body is then pulled along. Repeated action of this type accomplishes forward progression and burrowing. In a few species, the foot has a pair of side lobes that can be filled with blood (Figure 13.70, C), and these assist the foot in maintaining its position after each successive advance.

The head is greatly modified and has an anterior prolongation, but this does not extend beyond the edge of the shell. Food is collected by the captacula. There are several hundred of these, originating in two groups lateral to the base of the prolongation of the head. The animal excavates a feeding cavity with its foot. The captacula, drawn out of the mantle cavity and pulled along by the action of cilia on their bulbous tips, then explore the cavity or probe the loose sediment. The tips have a depression in which there are glands that appear to function as a duogland adhesive system for capturing food. Contraction of muscles within the captacula brings food to the mouth. Some scaphopods feed mostly on foraminiferans; others consume sediment or various small animal organisms, including kinorhynchs and bivalves. Species whose captacula have a continuous ciliary tract, or a series of tufts of cilia, are able to bring fine sediment into the mantle cavity by the action of the cilia, but placing the food collected in this manner into the mouth is the function of the tips of the captacula.

The radula is well developed and its teeth are highly mineralized. In some species it functions as a grinding organ, but in others it seems to operate as a ratchet that pushes food into the esophagus. The stomach has a secreted gastric shield as well as a small caecum. The digestive glands attached to the stomach are thought to secrete digestive enzymes and to absorb the soluble products of digestion. The intestine makes one or two loops backward and forward before it reaches the anus, which opens into the ventral part of the mantle cavity.

Blood is circulated through the hemocoelic spaces by muscular movements of the body. Near the anus of some scaphopods, however, there is a contractile structure that pumps blood just as a heart does. There is no pericardial cavity surrounding it, and whether it is homologous to the heart of other molluscs is uncertain.

The nervous system is prominent, and in terms of the relative positions of its major ganglia—cerebral, pedal, pleural, and visceral—is typically molluscan.

There are two coelomoducts that function as renal organs. They open by separate pores into the mantle cavity, close to the anus. A single testis or ovary, in the posterior part of the body, joins the right renal organ, so this structure also serves as a gonoduct. The trochophore larva that develops from a fertilized egg becomes a veliger. The right and left downgrowths of the mantle, if they are separate to begin with (they usually are), become fused to enclose a mantle cavity that is open only at its anterior and posterior ends.

References on Scaphopoda

Bilyard, C. R., 1974. The feeding habits and ecology of *Dentalium entale stimpsoni* Henderson (Mollusca: Scaphopoda). Veliger, *17:*126–138.

Emerson, W. K., 1962. A classification of the scaphopod mollusks. Journal of Paleontology, *36:*461–482.

Gainey, L. F., Jr., 1972. The use of the foot and captacula in the feeding of *Dentalium* (Mollusca: Scaphopoda). Veliger, *15:*29–34.

Morton, J. E., 1959. The habits and feeding organs of *Dentalium entalis*. Journal of the Marine Biological Association of the United Kingdom, *38:*225–238.

Poon, P., 1987. The diet and feeding behavior of *Cadulus tolmiei* Dall, 1897 (Scaphopoda, Siphonodentalioida). Nautilus, *101*:88–92.

Shimek, R. L., 1988. The functional morphology of scaphopod captacula. Veliger, *30*:213–221.

Trueman, E. R., 1968. The burrowing process of *Dentalium*. Journal of Zoology (London), *154*:19–27.

CLASS CEPHALOPODA

The cephalopods—octopuses, squids, and their kin—are the most complex molluscs. In general, they are larger and more active than other molluscs, and they are the most intelligent invertebrates. The majority are specialized for swimming, but some, including octopuses, are adapted primarily for a benthic existence.

Cephalopods are restricted to the sea. The period of their greatest prominence spanned the Paleozoic and Mesozoic eras—the Age of Fishes and the Age of Reptiles—and fossils of at least 10,000 species are known, though this does not necessarily mean that there were more than a couple of thousand species flourishing at any one time. Cephalopods are relatively large carnivores and are close to the top of the trophic pyramid, so we cannot expect them to be especially numerous in terms of either species or individuals. About 700 species are living today, but many of these will not be seen except by biologists who have special gear for collecting actively swimming animals from relatively deep water.

The name cephalopod means ''head-foot.'' The head does in fact merge with the foot. The mouth and eyes, of course, belong to the head; the arms, as well as specialized tentacles that may be present, and also the **funnel** through which water is expelled from the mantle cavity, are derivatives of the foot (Figure 13.72). Not only are the head and foot run together, but the body is enlarged along its dorsoventral axis. Thus what may appear to be the anterior and posterior ends of a torpedo-shaped squid or a bulbous-bodied octopus are more nearly dorsal and ventral. For practical purposes, however, the terms *anterior* and *posterior* are often used for the opposite ends. The lower side, where the mantle cavity is located, is really posterior, but is ventral in the artificial terminology.

For swimming, cephalopods depend mostly on jet propulsion. Contraction of circular muscles in the body wall surrounding the mantle cavity expels water from the cavity. If the free edge of the mantle is simulta-

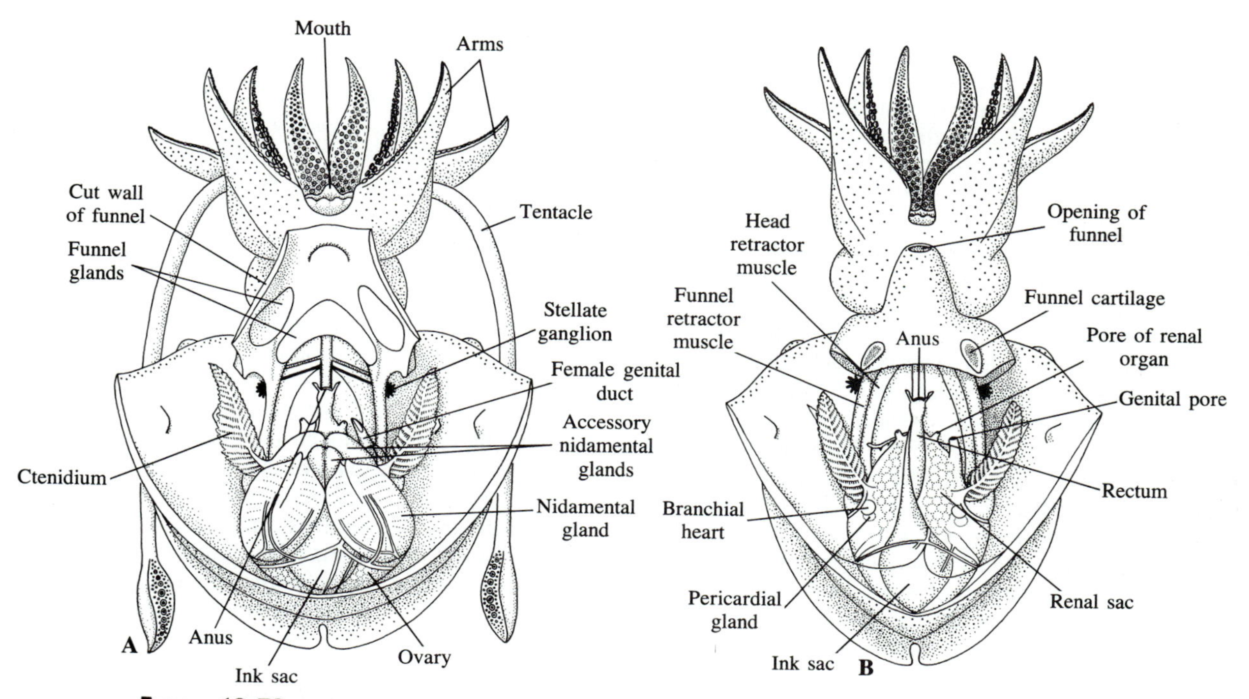

FIGURE 13.72 Cephalopoda, *Sepia officinalis*. A. Female. B. Male. In both, the mantle cavity has been opened by a median cut through the body wall of the lower side of the body. Some internal structures are visible through the nearly transparent wall of the visceral mass. The tentacles of the male have been withdrawn into their pockets, so are not shown. Nerves (black) and most blood vessels are not labeled. (After Tompsett, Liverpool Marine Biology Committee Memoir 32.)

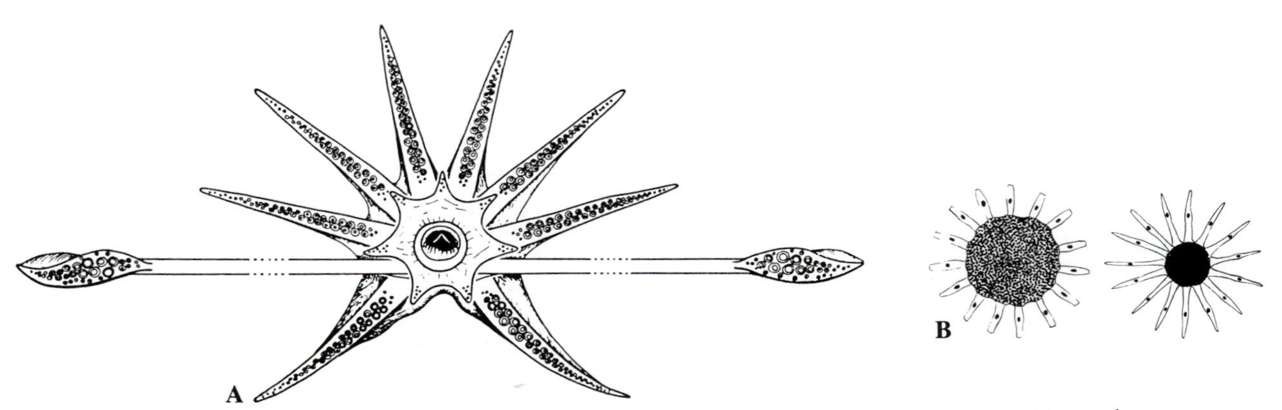

FIGURE 13.73 A. *Loligo vulgaris,* arms and tentacles spread, oral view. (Bidder, in Wilbur and Yonge, Physiology of Mollusca, *2.*) B. Chromatophores of a cephalopod, the pigment mass expanded (left) and contracted (right). (Muus, Danmarks Fauna, *65.*)

neously pulled tightly against the head by muscles that operate as a "drawstring," then the only way for the water to escape is through the narrow funnel. Dilation of the mantle cavity, which draws water back in, is achieved by contraction of radial muscles. By manipulating its funnel, the animal can to a considerable extent control the direction in which it swims. Rapid movement, however, is usually in a backward direction.

Locomotion is not limited to jet propulsion. An octopus, for instance, moves over the bottom by using its muscular and remarkably flexible arms, which are equipped with suction cups. In cuttlefishes, the graceful undulations of the lateral fins provide the impetus for a slow, steady kind of movement. In general, however, fins of cephalopods are more important as stabilizers and steering devices than they are in locomotion.

The skin of most cephalopods contains chromatophores in which black pigment or pigment of some other color is concentrated. Muscles that extend radially away from a chromatophore can pull it out into a nearly flat disk, so the pigment is spread more or less evenly and shows clearly (Figure 13.73, B). When the radial muscles relax and some circular muscles contract, the diameter of the cell is made smaller. Then the pigment is concentrated into a tiny dot and is less obvious. The complicated color pattern exhibited by the body as a whole at a particular moment is the result of the dilation of some chromatophores and the contraction of others.

When a cephalopod turns almost white all over, that is due to contraction of chromatophores of all colors. The astonishing changes in color pattern—related to the color of the background, to the approach of a predator or competitor, or to events in courtship—are primarily under the control of the nervous system. The eyes see, the brain and nerves operate to organize and transmit impulses, and the chromatophores adapt. There is also some hormonal control of chromatophores.

Digestive System

Cephalopods are carnivores that use their arms and their specialized tentacles for capturing prey. Their food consists mostly of crustaceans, molluscs, and fishes. The front part of the buccal cavity is equipped with a pair of hard jaws that form what looks like an upside-down parrot's beak (Figure 13.75). The upper jaw is the smaller of the two, and is partly enclosed by the lower one when the jaws are pulled together. Deeper in the buccal cavity is a radula (Figure 13.74, A; Figure 13.75) that acts as a rasp and that also pushes food backward. There are usually two pairs of salivary glands. The first pair is primarily concerned with secretion of mucus for lubricating the food. The second pair, sometimes fused into a single gland, produces some mucus, but it also secretes a mixture of paralytic venoms, such as 5-hydroxy-tryptamine. The bite of almost any octopus is likely to be painful, but a few species have potent venoms that are dangerous to humans.

The esophagus, sometimes with a dilation (crop) in which fairly large organisms can be accommodated temporarily, leads to the muscular stomach, which mashes the food. Connected to the stomach is a thin-wall caecum (Figure 13.76). Two ducts from the large digestive gland, in which most if not all digestive enzymes are produced, unite before reaching the caecum. Digestion is extracellular. It may begin in the stomach, but most of it takes place in the caecum, or in both the caecum and digestive gland. Absorption of soluble nutrients also occurs mostly in these two parts of the digestive system. Ciliary activity in the caecum keeps particles stirred up and directs resi-

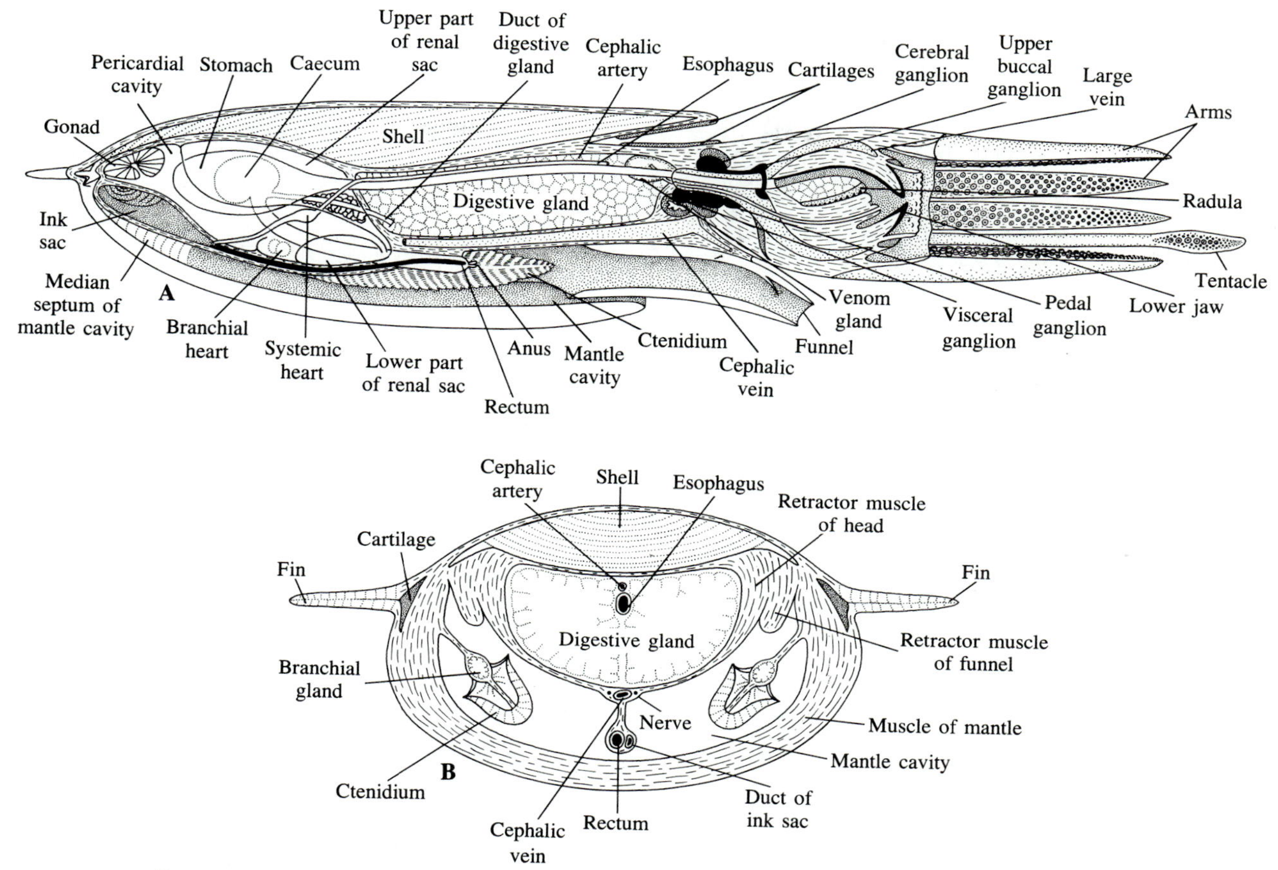

FIGURE 13.74 *Sepia*. A. Longitudinal optical section, diagrammatic, showing most internal organs. Not all structures are labelled, however. B. Transverse section through the body at the level of the rectum. (After Tompsett, Liverpool Marine Biology Committee Memoir 32.)

dues, mixed with mucus, into the intestine, whose course begins where the caecum joins the stomach. Some residues go directly from the stomach to the intestine.

The intestine is usually short. The terminal portion of the gut, called the rectum, has a diverticulum that is differentiated into an ink-secreting gland and an ink-storage sac (Figure 13.72; Figure 13.74, A; Figure 13.76). The ink, squeezed out of the sac on command, consists of a suspension of granules of melanin. Once out of the anus, it is forced from the mantle cavity through the funnel. Cohering enough to form a cloud, it might serve as a smokescreen or as a "dummy" that could fool a predator. The ink of at least certain cephalopods seems to be disagreeable to other animals, and in some cases it appears to impair the operation of olfactory organs. A few deep-water squids discharge a luminescent substance from the ink sac.

One other point about the digestive system of cephalopods needs to be stressed: the operation of the gut depends to a large extent on muscular activity. In the majority of other molluscs, the stomach and intestine have little if any musculature; movement of food and the formation of feces are accomplished largely by the action of cilia, aided by secretion of mucus.

Ctenidia

Most cephalopods have a pair of ctenidia within the mantle cavity (Figure 13.72). *Nautilus*, the only surviving member of its subclass, has two pairs. The ctenidia are not ciliated. Animals as large and active as cephalopods need a more efficient system than ciliation for bringing water into the mantle cavity and moving it over the ctenidia. Rhythmic contractions and relaxations of muscles in the mantle wall accomplish these

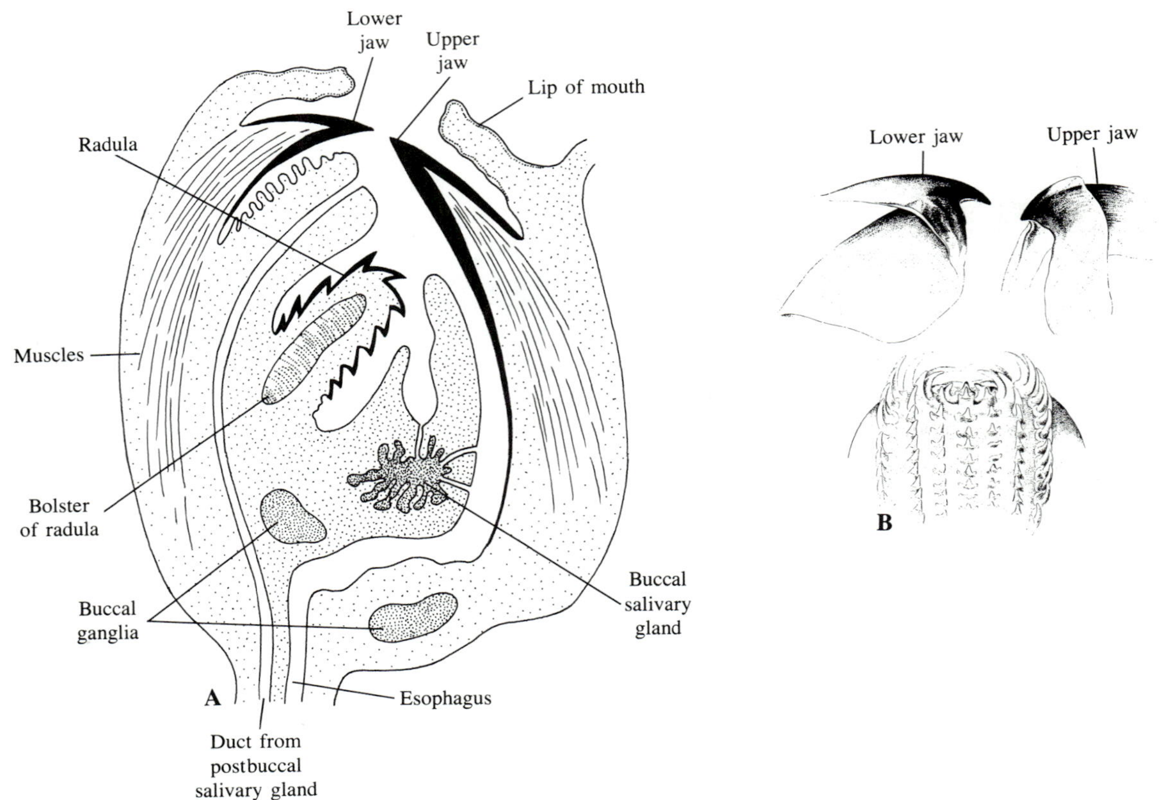

FIGURE 13.75 A. *Sepia,* buccal bulb, as seen in sagittal section; diagrammatic. (After Tompsett, Liverpool Marine Biology Committee Memoir 32.) B. *Loligo forbesi,* jaws and portion of the radula. (Muus, Danmarks Fauna, *65.*)

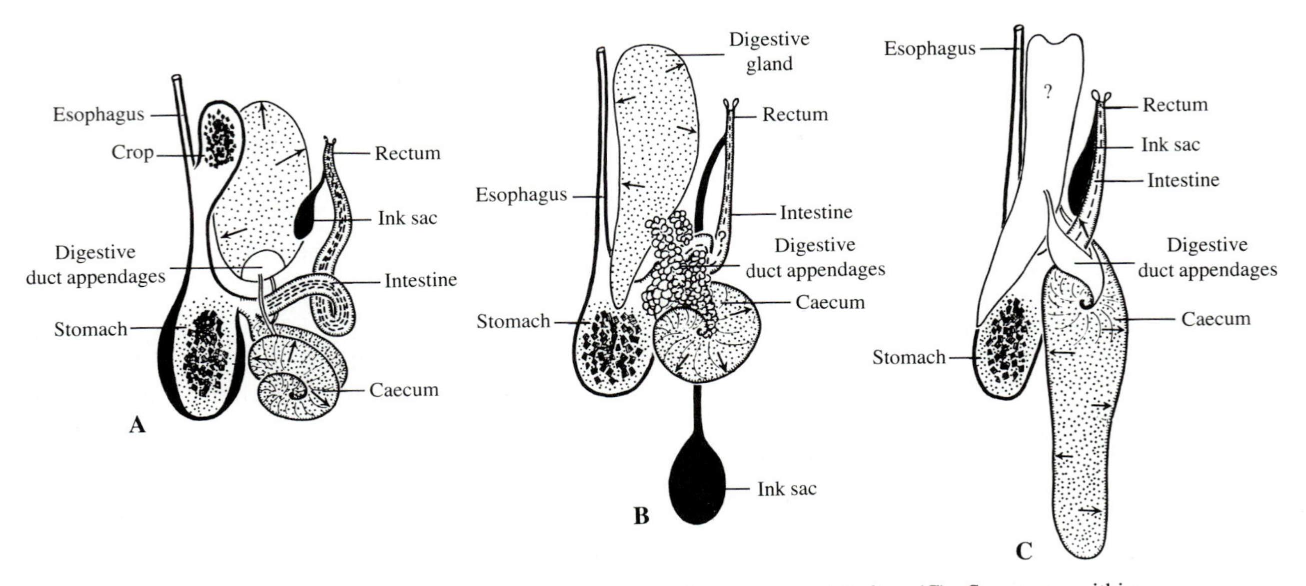

FIGURE 13.76 Digestive tracts of *Octopus* (A), *Sepia* (B), and *Loligo* (C). Structures within which digestion is known to take place are stippled; sites of absorption are indicated by arrows. Broken lines in the intestine indicate strings of mucus, with residues, coming from the caecum. The digestive duct appendages, whose tubules are joined to the ducts of the digestive gland, may be concerned with absorption of nutrients and also with osmoregulation and urine formation. In *Sepia* and *Loligo* they lie within the sacs that enclose the renal organs. (Bidder, in Wilbur and Yonge, Physiology of Mollusca, *2.*)

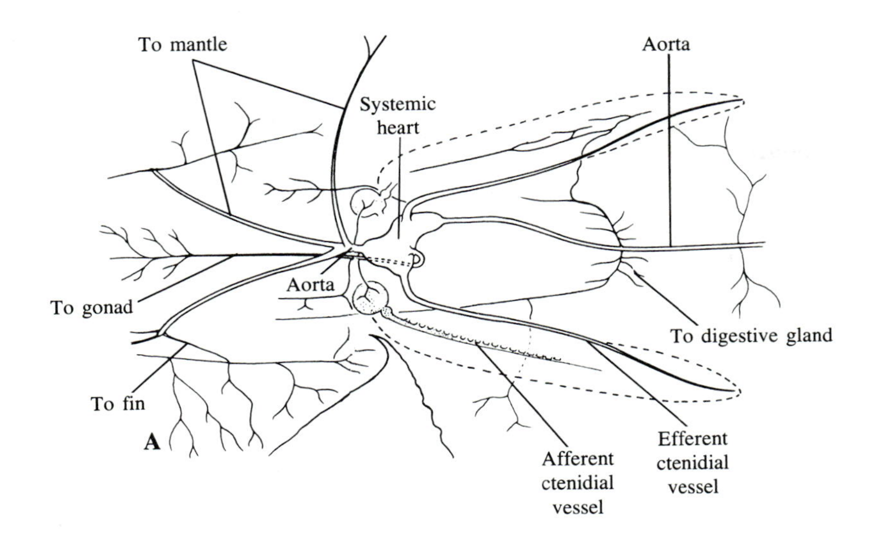

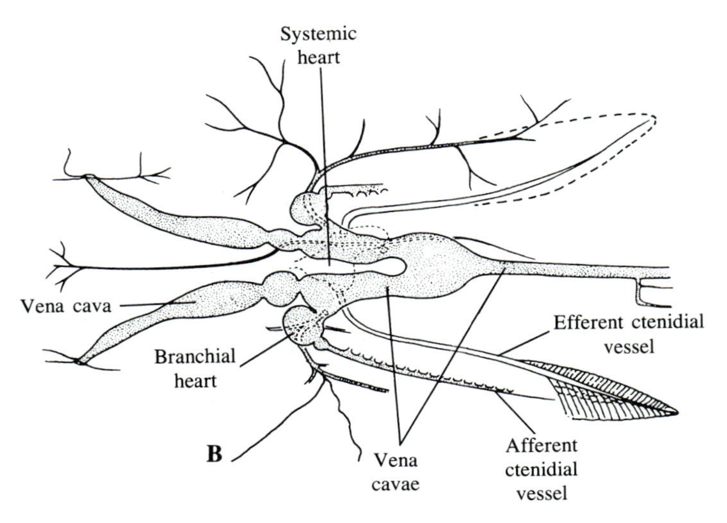

FIGURE 13.77 *Loligo pealii.* A. Arterial system, with some portions of the venous system. The view is from the underside of the animal. The usual terms of orientation ("dorsal aorta" and so on) are avoided because they do not agree with the true orientation of cephalopods. B. Venous system, with some portions of the arterial system. (Wells, in Wilbur, The Mollusca, *5*.)

functions. The efficiency with which blood flows through the ctenidia is increased by so-called **branchial hearts,** which are "booster pumps" located at the bases of the ctenidia (Figure 13.72, A; Figure 13.77, B; Figure 13.78).

Circulatory System

The circulatory system is of the closed type, with extensive capillary networks in tissues where exchange of gases, uptake of nutrients, and elimination of nitrogenous wastes take place. This arrangement is more suitable than an open system for animals that are as active as cephalopods are. The general pattern of circulation (Figures 13.77 and 13.78) is as follows. The ventricle of the heart forces blood into a large artery that runs forward and one that runs backward. These vessels, called *aortas,* branch to supply most parts of the body. Blood that has circulated through capillaries in the tissues is returned by large veins called *venae cavae,* but these do not lead directly to the heart. The single anterior vena cava divides, and its two main branches, which may be joined by other veins, run through the right and left renal sacs and then go on to the branchial hearts. These booster pumps drive blood through the capillary networks of the ctenidia, where uptake of oxygen and elimination of carbon dioxide take place. The two posterior venae cavae also connect with the branchial hearts. From the ctenidia, oxygenated blood is carried to the two auricles, which force it into the ventricle, and the cycle of distribution starts all over again.

This general plan just described has many modifications. In octopuses, for instance, only one aorta is really prominent, and there are no posterior venae

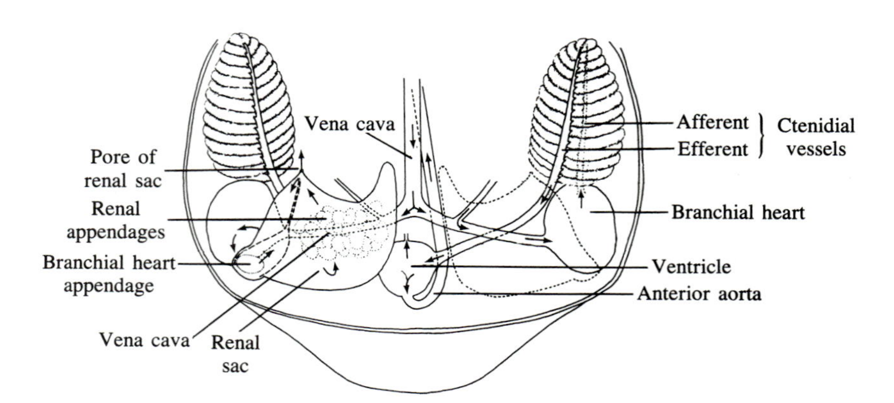

FIGURE 13.78 *Octopus,* relationship of renal organs to components of the circulatory system; diagrammatic. (Martin, in Wilbur, The Mollusca, *5*.)

cavae. In *Nautilus,* in which there are four ctenidia, the number of branchial hearts, renal organs, and auricles is also doubled, so the layout of the major blood vessels is necessarily different.

Excretory System

Excretion has been most intensively studied in certain large octopuses, because of the relative ease with which substantial amounts of urine and blood can be withdrawn from these animals. The renal organs of an octopus (Figure 13.78) are thin-walled sacs through which the two branches of the vena cava pass on their way to the branchial hearts. These vessels, as well as the veins coming from the digestive tract and joining them, are covered with *renal appendages,* which look like pieces of a cauliflower. Each appendage has a branching diverticulum of the vein from which it is suspended, and it has muscles for squeezing blood back into the vein. The appendages contract about 10 times a minute, nearly keeping beat with the branchial hearts.

The renal organs of molluscs, like the pericardial cavity and gonads, are parts of the coelom. In octopuses, there is no pericardial cavity around the systemic heart. Next to each branchial heart, however, is a space believed to be part of the pericardial cavity. The space encloses a *branchial heart appendage,* whose much-branched lumen is continuous with that of the branchial heart. Contractions of a branchial heart not only drive blood into the ctenidium with which it is associated, but also promote filtration through the wall of the branchial heart appendage. The formation of urine, therefore, starts here. From the pericardial cavity, urine moves through the *renopericardial canal* to the renal organ. The renopericardial canal is in intimate association with blood vessels, and it is here that reabsorption of glucose takes place. The urine in the renal organ bathes the renal appendages, and across their walls ammonia and other nitrogenous wastes are transferred from the blood to the urine. The urine leaves the renal organs by way of pores that open into the mantle cavity. The renal organs do not have a monopoly on excretion. Considerable ammonia is eliminated across the surfaces of the ctenidia.

Nervous System and Sense Organs

That cephalopods have a highly developed nervous system is indicated by their agility in locomotion and capture of prey, their courtship behavior, and the capacity of some species to learn and remember. No other invertebrates have a brain so highly developed as that of cephalopods. To begin with, the brain is large, and is believed to contain not only the cerebral ganglia, but also the homologues of the pleural, pedal, and visceral ganglia of other molluscs. A complex set of terms is used for designating the various lobes and areas of the brain and also for the principal nerves but these are not dealt with here. However, attention is called to the impressive optic lobes (or optic ganglia) from which nerves run to the eyes, and the brachial lobe (or lobes) from which nerves go to the arms.

In addition to the ganglia incorporated into the brain, there are substantial ganglionic centers elsewhere. Most prominent among these, and readily exposed by dissection, are the *stellate ganglia* in the mantle on both sides of the body. These are involved in control of muscular activity in the mantle wall, and are linked with the brain by giant neurons. This arrangement permits efficient activation of the mechanism of jet propulsion that a cephalopod uses for swimming rapidly.

Certain lobes of the upper part of the brain are centers for learning and memory. Most of the experimental work on learning by cephalopods has been done with *Octopus vulgaris,* a common species in European waters. It can discriminate between objects of different shape, and will learn to go to an object associated with a reward and to avoid one that is consistently associated with punishment. Offer an octopus a hermit crab, for example, and it will attack it and eat it. If, however,

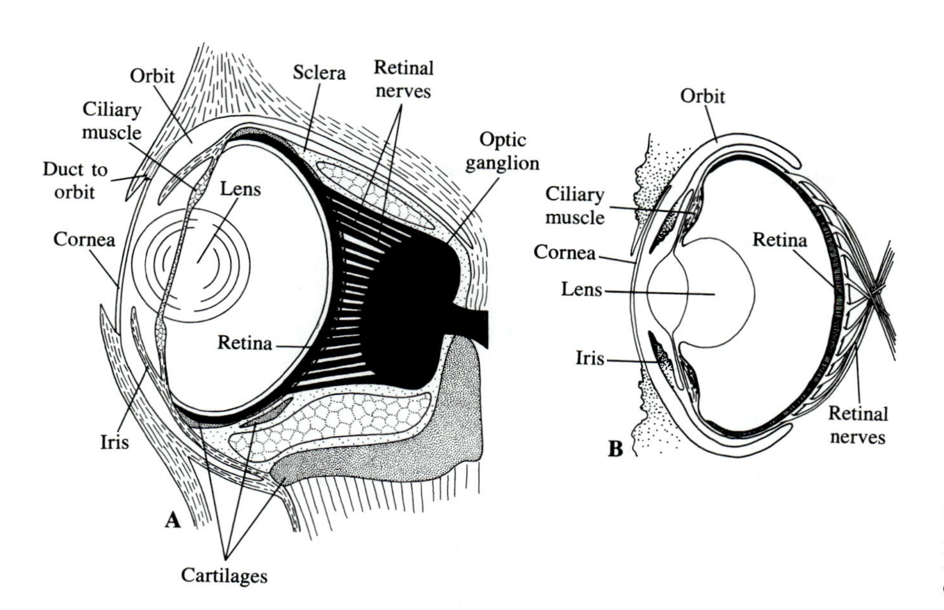

FIGURE 13.79 Cephalopod eyes, in sagittal section; diagrammatic. A. *Sepia*. (After Tompsett, Liverpool Marine Biology Committee Memoir 32.) B. *Octopus*. (Wells, In Wilbur and Yonge, Physiology of Mollusca, 2.)

the shell in which the hermit crab is living has a sea anemone on it, the octopus will, after one or two unpleasant contacts with the nematocysts of the anemone, know enough to leave the crab alone.

The eyes (Figure 13.79) of most cephalopods, having a cornea, iris, lens, and retina, are similar to the eyes of vertebrates. They can form an image, and at least some species can distinguish colors. The shape of the lens cannot be changed, however, so focusing is accomplished by muscles that move the lens closer to or farther from the retina. The aperture (pupil) of the iris is generally a slit, but its shape and size can be manipulated to some extent. Visual acuity is moderately good. *Nautilus* has an eyecup that is almost completely closed over by a circular flange, but there is no lens. This primitive type of eye, whose structure is somewhat comparable to that of a pinhole camera, is also found in some other cephalopods. In *Nautilus,* it is probably concerned primarily with detection of light.

It is important to understand that there is no evolutionary relationship between the eyes of cephalopods and those of vertebrates. The similarity is a case of convergence. In cephalopods, the retina derives from the epidermis that lines the eyecup; in vertebrates, it originates from an extension of the brain that goes out to meet the eyecup.

Beneath the brain is a pair of rather complicated balance organs that register changes in orientation. These are of great importance to animals that are powerful swimmers and that must maneuver quickly. A statocyst is incorporated into each one, but there are other movable components in them. The brain, balance organs, and eyes are protected by an internal skeleton consisting of material that resembles cartilage.

Reproduction

Sexes are separate. The gonads, following the rule for molluscs, are coelomic derivatives. In males, the vas deferens, generally much coiled, carries sperm from the testis to a glandular structure within which they are packaged into spermatophores. The spermatophores are then shunted into a sac where they accumulate and from which they reach the left side of the mantle cavity by way of the genital pore (Figure 13.72, B). The pore may be on a prominent tube, sometimes called the "penis." This is a misnomer, for it is not used in copulation.

In females, there is generally one oviduct (Figure 13.72, A). Octopuses have two; so does *Nautilus,* but only one is functional. Glands in the wall of the terminal portion of the oviduct usually secrete the capsules that cover the eggs, but sometimes, as in *Nautilus,* this is achieved by glands in the mantle cavity. The lining of the mantle may also have glands for secreting material to hold the eggs together in a cluster and make it possible for the animal to stick them to a firm object.

Sperm transfer is effected by the tip of a particular arm. The specialized portion is modified in one way or another to carry a load of spermatophores. In octopuses, for instance, it has an obvious depression; in some other cephalopods there is a deep sac. After a period of courtship, which establishes rapport between the sexes, the male sticks the tip of the spermatophore-transferring arm into his mantle cavity and picks up

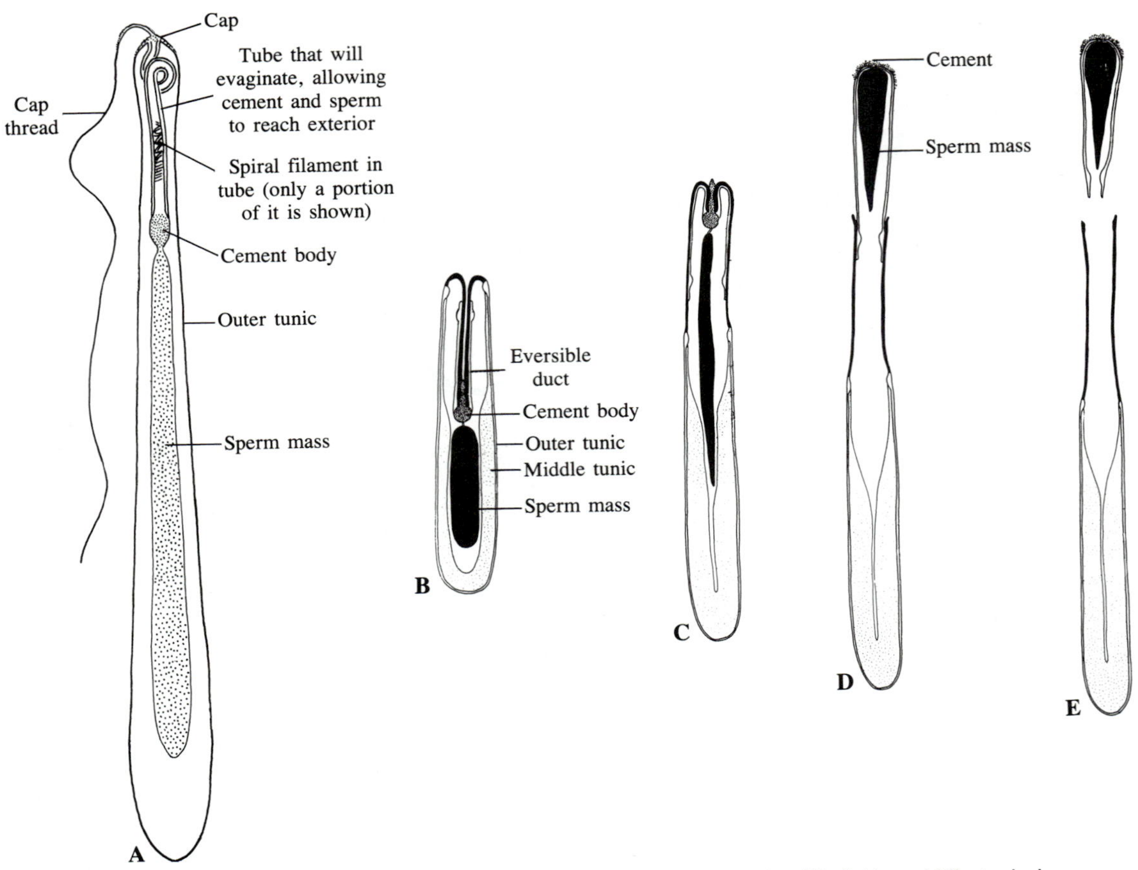

FIGURE 13.80 A. *Loligo pealei,* fully developed spermatophore; simplified (the middle tunic is not shown). B–E. Stages in the discharge of a spermatophore; diagrammatic, and not showing the cap, cap thread, or spiral filament within the eversible duct. After the middle tunic of the mature spermatophore (B) has taken up water, it exerts pressure on the structures internal to it, forcing the eversible duct to turn inside-out (C). This brings the sticky cement to the surface (D) and also leads to detachment of the portion containing the sperm mass (E). (Drew, Journal of Morphology, *32*.)

some spermatophores. He then inserts it into the female's mantle cavity, placing the spermatophores into or close to a genital pore, or on a specialized area that serves as a kind of seminal receptacle. Every species has a characteristic pattern of spermatophore transfer and acceptance. In a few, as *Argonauta,* the specialized arm tip is pinched off and left in the female's mantle cavity. Many years ago a genus of worms, *Hectocotylus,* was described on the basis of the mistaken notion that a six-suckered arm tip found in the mantle cavity was a parasite. Ever since, the term "hectocotylus arm" has been used to designate the arm that a male cephalopod uses for spermatophore transfer.

The spermatophores secreted by most cephalopods are complicated and fairly large. The ones produced by squids of the genus *Loligo* are representative. They are about 1 cm long and have the shape of a slender club (Figure 13.80, A). At one end is a detachable cap that covers the opening of an eversible duct. When the cap comes off, presumably because of mechanical disturbance of the thread attached to it, water enters the spermatophore and is taken up osmotically by the middle tunic, creating a high internal pressure. This pressure, together with that applied by the elastic outer tunic, forces the duct to evert. Thus the sticky material within the cement gland is brought to the surface, enabling the spermatophore to adhere to wherever the male has placed it. The portion of the discharged spermatophore that contains the sperm mass separates from the rest of it, and the sperm emerge over a period of several hours or days. Some stages in the sequence of eversion are shown in Figure 13.80, B–E. The function of the spiral filament that occupies the duct before eversion is not understood. The filament is a brittle structure, and may have nothing directly to do with the process. Perhaps it only serves to keep the duct from collapsing before the spermatophore "fires."

Nautilus, a Special Case

The several species of *Nautilus,* limited to waters of the Indo-Pacific region, are the only cephalopods that have a true external shell. This is partitioned into chambers (Figures 13.81 and 13.82). The animal occupies the newest and largest of the spaces, which is still open, but it retains a connection with the others by a cord of vascularized tissue called the **siphuncle.** The older chambers contain only gases, with nitrogen and carbon dioxide predominating. The most recently vacated chamber is at first full of fluid, and one or two others will also contain some fluid (Figure 13.81, B). The proportions of inorganic solutes in the chamber fluid are about the same as they are in the blood of the animal, but the total salinity is lower. The basal portions of the epithelial cells of the siphuncle are extensively folded, so each cell has a complex system of canals within it. It is believed that the epithelial cells take up solutes by active absorption, a process that involves enzymes and requires an expenditure of energy. The resulting increased concentration of solutes in the canals would lead to osmotic uptake of water from the chambers, and thus to gradual emptying of the chambers. Gases coming out of solution in the blood move in the opposite direction, but do so very slowly, and the maximum pressure in old chambers does not quite attain one atmosphere.

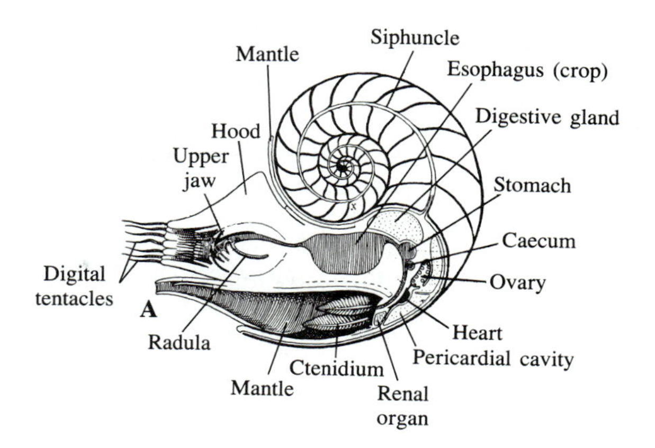

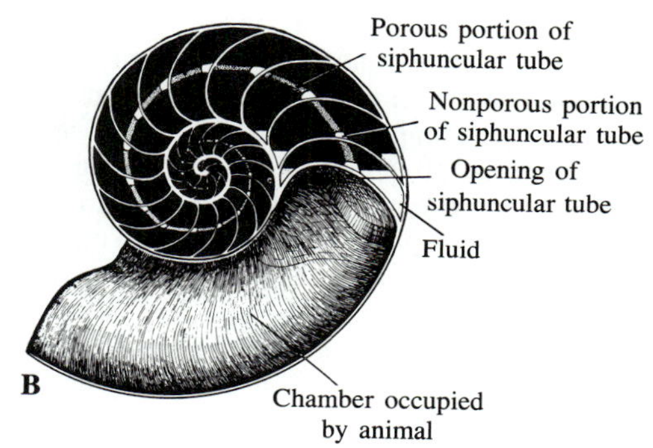

FIGURE 13.81 *Nautilus macromphalus*. A. General anatomy (simplified). (Naef, in Fauna e Flora del Golfo di Napoli, Monograph 35.) B. Shell, sectioned to show the chambers, siphuncle, and fluid remaining in recently vacated chambers. (Denton and Gilpin-Brown, Advances in Marine Biology, *11*.) (See also Color Plate 5.)

A

B

FIGURE 13.82 *Nautilus*. A. Intact shell. B. Shell sectioned to show chambers and siphuncle.

The siphuncle is surrounded by a deposit of conchiolin, which is in turn coated by a thin layer of calcareous material. Although these coverings protect the siphuncle from bursting because the fluid pressure in this structure is greater than the pressure in the chambers that are being emptied, they must be porous in order to allow for the uptake of water and salts.

The old idea that the siphuncle constantly controls the fluid-to-gas ratio in the chambers has been discarded. It is possible, however, that in extinct cephalopods the siphuncle did function both in removing fluid from the chambers and in adding fluid to them.

Most of the species of *Nautilus* live fairly close to shore, in water about 50 to 500 m deep. They are inclined to stay on or near the bottom in the daytime and to swim closer to the surface at night.

The many tentacles concerned with food capture are organized into two complexes around the mouth. The ones in the outer complex are called *digital tentacles,* and most of them fall into two rows. They are retractable into basal sheaths that are united to form a continuous *cephalic sheath,* the uppermost part of which is further differentiated into a *hood*. The hood and two specialized tentacles associated with it form a closing device functionally comparable to a snail's operculum. The digital tentacles do not have suckers, but they are adhesive and operate in recognition and capture of prey, mostly fishes and crustaceans. Not all of them act in the same way. Some are concerned with chemoreception or with making initial contact with the prey, others with grasping the prey and passing it to the *buccal tentacles,* which form the inner complex.

The buccal tentacles hold the prey so that it can be bitten by the jaws and swallowed. The arrangement of these is different in males and females. Both sexes have a set on the right and left side of the mouth, and females also have a set below the mouth. This group is interrupted medially by a structure called *Owen's organ,* thought to be olfactory and to be involved somehow in mating behavior. Beneath Owen's organ, on the inside of the cephalic sheath, is *Valencienne's organ,* on which the male deposits spermatophores.

Males have buccal tentacles on both sides of the mouth, but none beneath it. Several of the more nearly ventral tentacles in each of the two lateral sets are organized into unequally developed spermatophore-transferring organs (the "spadix" and "antispadix"). A median structure, in the same position as Valencienne's organ of the female, is called *van der Hoeven's organ,* and almost certainly has something to do with reproduction, though no one knows precisely what.

The eyes of *Nautilus* are primitive. In front of and behind each eye is a small chemoreceptor tentacle. These four tentacles, like the digital and buccal tentacles, can be retracted into sheaths.

There are two pairs each of ctenidia, auricles, renal organs, and structures thought to be comparable to osphradia of gastropods. Chromatophores are absent, and there is no ink sac. The mantle is bound to the wall of the newest chamber of the shell, so it is of no use in providing the impetus necessary for jet propulsion. This function is taken over by the funnel. The ventral margins of the funnel are free. They overlap when the funnel contracts, and draw apart when the funnel is dilated so that water can enter the mantle cavity again.

The eggs of *Nautilus,* covered by complicated capsules, are fairly large and contain considerable yolk. It is believed that when the young hatch, they are small replicas of the adults. Much remains to be learned about reproduction and development in these animals.

Evolution of Cephalopods

It is likely that cephalopods evolved from a group of ancient gastropods. Among the features that accounted for their early success was the chambered shell. As each successive chamber became wholly or partly filled with gas, the animal could continue to grow and enlarge its protective shell without losing buoyancy. When cephalopods finally emancipated themselves from shells, they became more agile and therefore more successful as competitors of bony fishes. The shelled types, with the exception of the few species of *Nautilus,* became extinct.

The study of shells of fossil cephalopods is almost a science in itself. Shells provide substantially all of the material that paleontologists have to work with when they describe new species or higher taxa, and when they speculate on the phylogeny of cephalopods. Any comprehensive work on invertebrate paleontology will have more information on the shell structure than can be presented here. An attempt is made here, however, to hit the high spots.

The shell of the earliest known cephalopod, *Plectronoceras,* found in a Cambrian (early Paleozoic) deposit, is about 1 cm long, slightly curved, and divided into many small chambers. It is placed, without much hesitation, into the same general group as that to which *Nautilus* belongs. If its older chambers were gas-filled—and this is believed to have been the case—*Plectronoceras* would almost certainly have floated vertically. There were many subsequent developments in the nautiloid shell, some of which made possible horizontal swimming, and therefore greater maneuverability. Among these developments was the addition of enough mineral material to one side of the shell to weight it down. Some of the ancient nautiloids reached a length of 5 m. Coiling—in a loose spiral in earlier species, in a tight spiral in later ones—led to fusion of one coil to the next, and thus there evolved shells of the type that *Nautilus* has. Coiling turned out to be a good solution to the problem of maintaining a more or less horizontal orientation of the animal without having

A

B

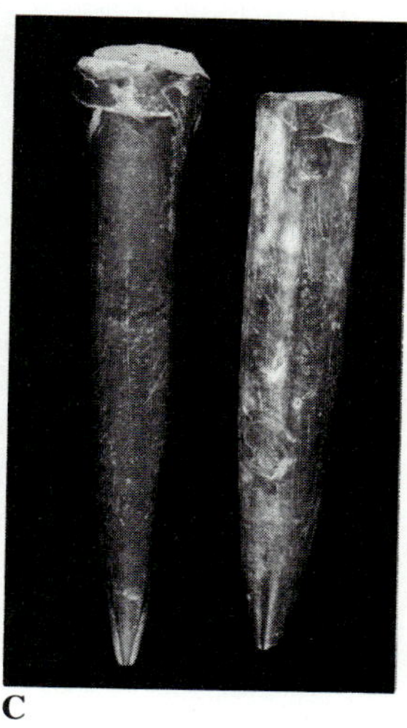

C

FIGURE 13.83 A, B. Fossils of ammonoid shells. C. Fossils of belemnoid shells.

to add ballast. In some of the extinct nautiloids, as in the living *Nautilus,* the shell was coiled so that the unoccupied, gas-filled chambers were mostly above the animal; in other words, the upper side of the animal faced the preceding coil. In others, however, the direction of coiling was reversed, so the newest part of the shell, and the animal itself, would have been above the older coils. A siphuncle was regularly present in extinct nautiloids, just as in *Nautilus*.

The ammonoids (Figure 13.83, A and B) originated later in the Paleozoic era and flourished through much of the Mesozoic era, which lasted for about 100 million years. They also had external, chambered shells, and these were generally coiled. The siphuncle, instead of running straight through the empty chambers as an unsupported strand, was attached to the wall of the shell. The sutures, visible externally at the points where the partitions between chambers were joined to the shell, were often elaborate. Some species, moreover, had a hard plate, or pair of plates, for closing the shell.

With a few possible exceptions, ammonoids were swimming and floating animals. Among those with coiled shells, the largest were about 1 m in diameter. What they ate is not known, but they were more successful than nautiloids, which declined as ammonoids became more and more important. Ammonoids were extinct, however, by the close of the Mesozoic. Until then, though, they provided food for certain carnivorous marine lizards, called mosasaurs.

The belemnoids (Figure 13.83, C), as well as the antecedents of squids, cuttlefishes, octopuses, and other present-day cephalopods, also appeared in the Mesozoic. Belemnoids more or less coincided with ammonoids. Their shells were internal and consisted of three main portions. The *phragmocone,* a conical, chambered part, formed the core of the shell right behind the animal's viscera; it had a siphuncle and was almost certainly concerned with maintaining proper buoyancy. The *rostrum,* making up most of the rest of the rather solid shell, would have protected the phragmocone and also added enough weight to keep the animal horizontal. An anteriorly directed dorsal extension of the rostrum, the *pro-ostracum,* would have supported the mantle, provided an extensive surface for muscle attachments, and shielded the visceral organs.

Modern squids and cuttlefishes seem to be rather closely related to belemnoids, but this is not to say that the former are derived from the latter. The ancestors of the modern groups may have been around before belemnoids got started. The shells of cuttlefishes are nevertheless somewhat similar to those of belemnoids. The cuttlebone, sold in pet shops so that canaries and other cage birds will have a hone for their beaks, represents the phragmocone. The rostrum and pro-ostracum are much reduced. There is no distinct siphuncle, but the tissue bordering the phragmocone has a rich blood supply, so that the animal can control the volume and the osmolality of the fluid within the chambers of its shell. In general, the volume of fluid in the shell, and the density of the shell as a whole, are greater when the animal is adjusted to staying on the bottom than when it is adjusted to swimming near the surface. The fluid content of the shell as a whole ranges from about 10 to 30%, and the density ranges from about 0.5 to 0.7.

An interesting internal shell is found in *Spirula* (Figure 13.84, A and B), which is a close relative of cuttlefishes. It is loosely coiled and consists entirely of a chambered phragmocone.

In most squids, such as *Loligo,* the shell consists of organic material. There is no phragmocone and no

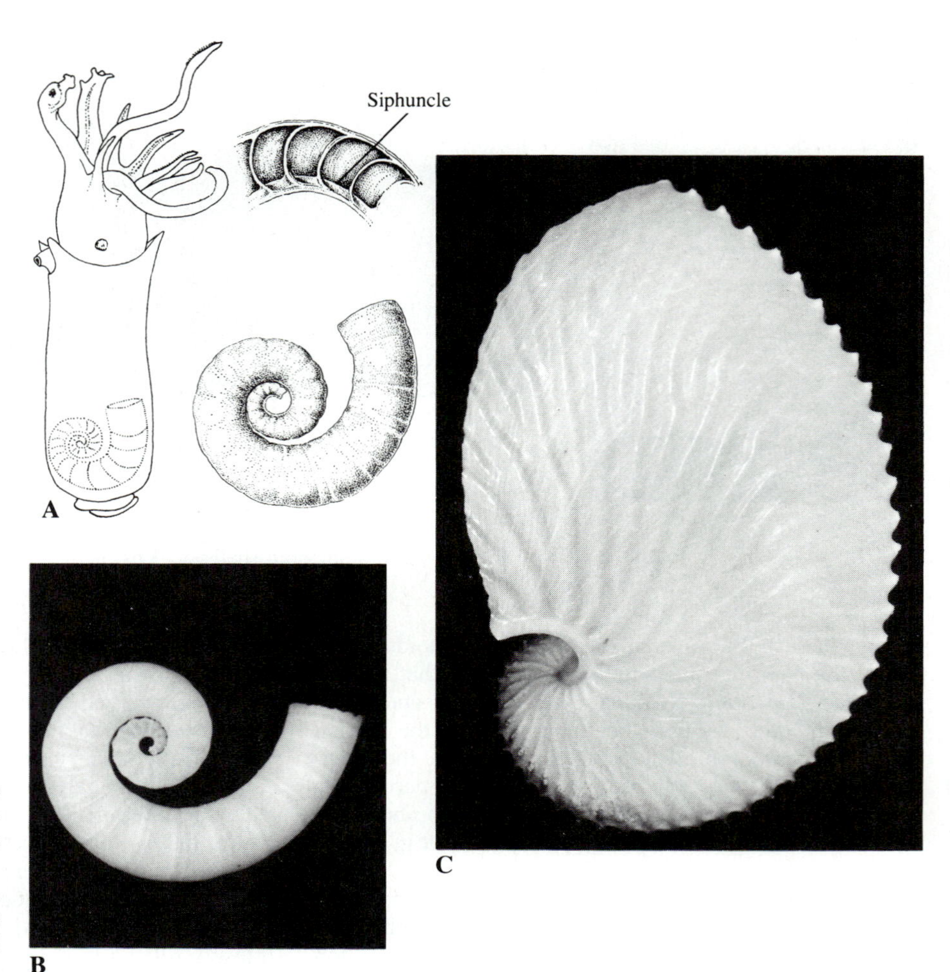

Figure 13.84 A. *Spirula spirula* (length about 6 cm) and its internal shell. (Muss, Danmarks Fauna, *65*.) B. *Spirula*, shell, C. *Argonauta* (the paper nautilus), egg case secreted by female. (This is not homologous to the true shell of other cephalopods.)

rostrum; only the pro-ostracum has survived. Octopuses and their allies do not have a shell, although females of a few genera, notably *Argonauta,* the paper nautilus, make what looks like one (Figure 13.84, C). It is a hardened secretion produced by membranes on two of the eight arms, and is used for flotation and as a chamber in which eggs and young are brooded. The animal must hold on to it, for the shell is not joined to the visceral mass. The shell of *Argonauta* is clearly not homologous to a true shell secreted by the mantle.

CLASSIFICATION

Classification and phylogeny of cephalopods cannot really be understood without going into the illustrious fossil record left by these animals. Among the earliest cephalopods were relatives of the genus *Nautilus*. They prospered during the Paleozoic era, but during the Mesozoic era they were largely replaced by ammonoids and belemnoids. These, in turn, were gone by the time the Mesozoic ended. Most of what we know about the wholly extinct groups is based on shell structure, but it nevertheless influences the way we view existing cephalopods and their relationships.

SUBCLASS NAUTILOIDEA

The only living genus of nautiloids is described elsewhere in this chapter, so it will suffice to recapitulate just a few of the features that define the subclass. The external shell is divided into many serial compartments through which a siphuncle runs. The shell, coiled in *Nautilus* and many of the fossil genera, was straight or simply curved in others. On the basis of what is known about *Nautilus,* it is likely that extinct nautiloids shared at least some of the following characters: numerous retractile tentacles, without suckers, on the head; two pairs each of ctenidia, auricles, renal organs, and osphradia; no ink gland; and eyes in the form of open cups, with neither cornea nor lens.

SUBCLASS AMMONOIDEA

Ammonoids were plentiful during the Mesozoic but died out before the start of the Paleozoic. That they did well for a long time is indicated by the fact that paleontologists have described about 600 genera and more than 130 families. These taxa are based almost entirely on features shown by the shells. These were external, generally coiled, and divided into chambers. There was a continuous siphuncle, but this was attached to the inside of the wall of the shell. In Figure 13.83, A, note the zigzag lines that mark the complex sutures between the chambers of ammonoids. Next to nothing is known about the soft parts, but it is assumed that these animals had numerous tentacles.

SUBCLASS COLEOIDEA

All surviving cephalopods, other than those of the genus *Nautilus,* belong to the Coleoidea. Of the five orders in the subclass, the Sepioidea and Teuthoidea are characterized by 10 arms, 2 of which are differentiated as slender, retractile tentacles. All 10 have suckers. The shell is internal, or at least mostly internal, and the extent to which it is developed varies; this topic is discussed elsewhere, in connection with general features of cephalopods. Representative genera of the Sepioidea are *Sepia,* the cuttlefish (Figures 13.72, 13.85, and 13.86, A), *Sepietta* (Figure 13.86, B), *Rossia* (Figure 13.86, C), and *Spirula* (Figure 13.84, A and B). The order Teuthoidea includes *Loligo* and other squids (Figure 13.87).

FIGURE 13.85 *Sepia officinalis,* the cuttlefish.

Coleoids of the order Belemnoidea, extinct since the close of the Mesozoic era, were the first to have an internal shell (Figure 13.83, C). The shell was distinctive in that the part directly behind the animal's viscera was largely occupied by a phragmocone. This was partitioned into chambers through which a siphuncle passed, and it is believed to have been concerned with maintaining proper buoyancy. The bulk of the shell, called the rostrum, was rather solid and would have protected the phragmocone. An upper projection, which

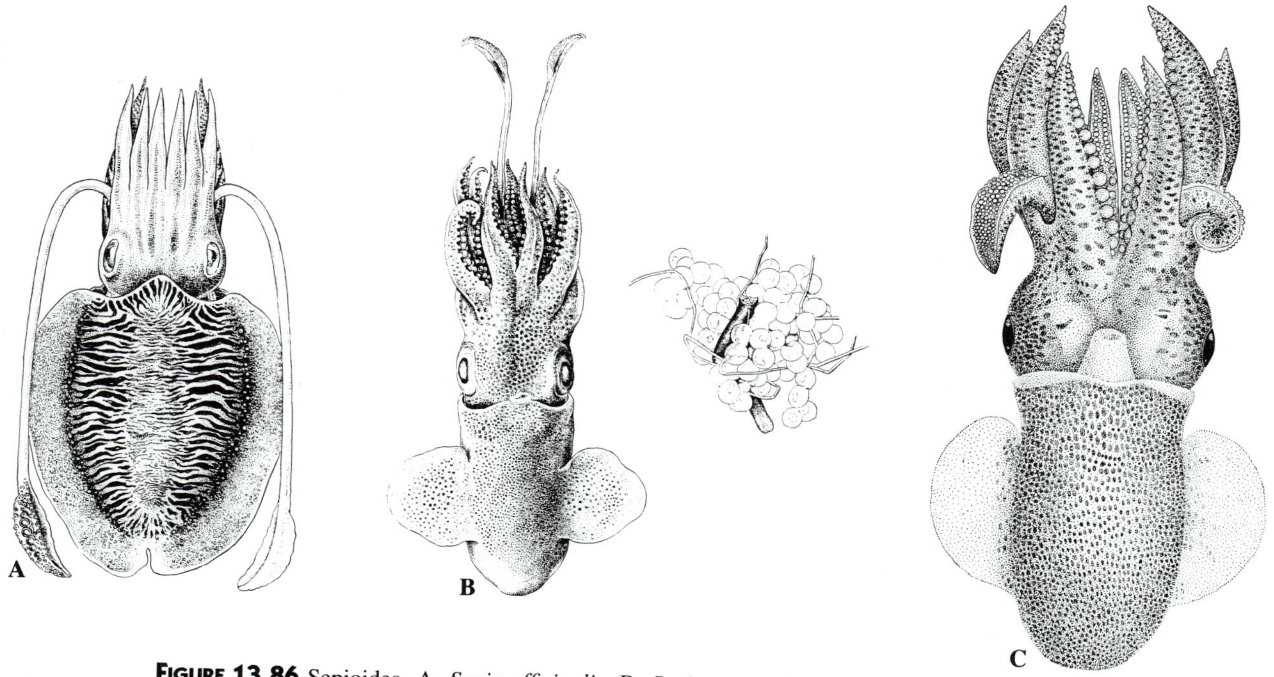

FIGURE 13.86 Sepioidea. A. *Sepia officinalis*. B. *Sepietta oweniana* and its egg mass. (A and B, Muus, Danmarks Fauna, *65*.) C. *Rossia pacifica*. (Berry, Bulletin of the United States Bureau of Fisheries, 1910.)

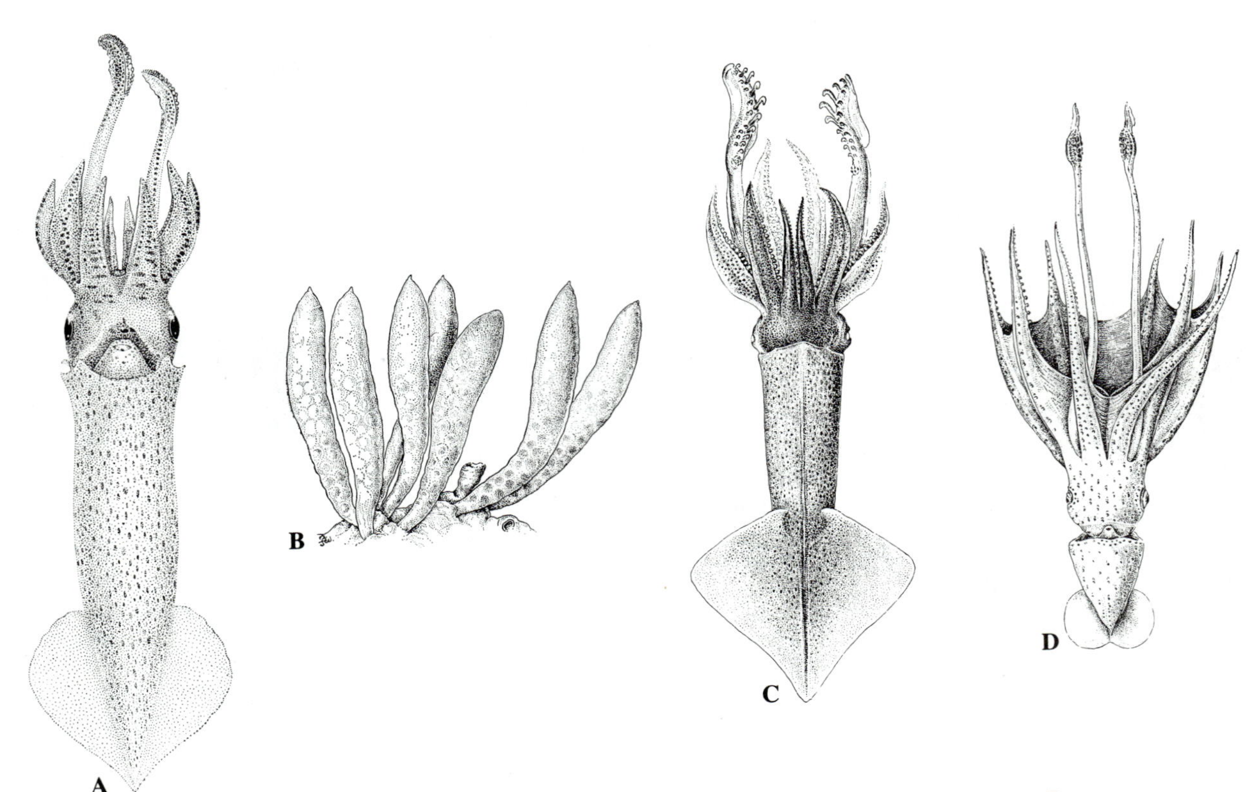

FIGURE 13.87 Teuthoidea. A. *Loligo opalescens*. (Berry, Bulletin of the United States Bureau of Fisheries, 1910.) B. *Loligo forbesi*, egg masses. (Muus, Danmarks Fauna, *65*.) C. *Onychoteuthis banksi*. The tentacles have large hooks. D. *Histioteuthis bonelliana*. (C and D, Muus, Conseil International pour l'Exploration de la Mer, Zooplankton, Sheet 97.)

extended forward like an awning, probably supported the mantle, shielded visceral organs, and provided a solid structure to which muscles could be attached. The number of arms in belemnoids is hard to determine. Some are thought to have had 6, others 8 or 10.

The remaining orders are the Octopoda and Vampyromorpha. In neither of these is there a true shell. The Octopoda includes, besides octopuses (Figure 13.88), the remarkable *Argonauta* and a few related genera in which the female secretes what looks like a shell (Figure 13.84, C) but is not. All cephalopods of this order are characterized by eight more or less equal arms that are distinct nearly or quite to their bases. The body has no fins. The largest species is *Octopus dofleini*, which has a wide distribution in the northern Pacific Ocean. Its arm spread reaches 3 m, and its weight may be as much as 45 kg.

In the order Vampyromorpha, found only in deep water, the eight arms are connected by conspicuous webs that resemble the ones between the toes of a duck's foot, and there is a horizontal fin on both sides of the hind part of the body. Another interesting feature of these strange and rarely seen animals are two slender, almost filamentous tentacles that can be withdrawn into pockets within the circle formed by the arms. These, like the tentacles of squids and cuttlefishes, are thought to be specialized arms.

References on Cephalopoda

Arnold, J. M., 1965. Normal embryonic stages of the squid, *Loligo pealii* (Lesueur). Biological Bulletin, *128*:23–32.

Bidder, A. M., 1950. Digestive mechanisms of European squids. Quarterly Journal of Microscopical Science, new series, *91*:1–43.

Boycott, B. B., 1965. Learning in the octopus. Scientific American, *212*(3):42–50.

Boyle, P. R., 1983–1987. Cephalopod Life Cycles. 2 volumes. Academic Press, London, New York, and Orlando, Florida.

Clarke, M. R., 1966. A review of the systematics and ecology of oceanic squids. Advances in Marine Biology, *4*:91–300.

Denton, E. J., 1974. On the buoyancy and the lives of modern and fossil cephalopods. Proceedings of the Royal Society of London, B, *185*:273–299.

Denton, E. J., and Gilpin-Brown, J. B., 1973. Flotation mechanisms in modern and fossil cephalopods. Advances in Marine Biology, *11*:197–264.

FIGURE 13.88 Octopoda. *Octopus rubescens*. In the lower photograph, the animal is just beginning to propel itself backward with a jet of water forced out of the siphon (just below the eye).

Donovan, D. T., 1964. Cephalopod phylogeny and classification. Biological Reviews, *39:*259–287.

Gilpin-Brown, J. B., 1972. Buoyancy mechanisms of cephalopods in relation to pressure. Symposia of the Society for Experimental Biology, *26:*251–259.

Haven, N., 1972. The ecology and behavior of *Nautilus pompilius* in the Philippines. Veliger, *15:*75–80.

Holme, N. A., 1974. The biology of *Loligo forbesi* Steenstrup in the Plymouth area. Journal of the Marine Biological Association of the United Kingdom, *54:*481–503.

House, M. R., and Senior, J. R. (editors), 1981. The Ammonoidea: The Evolution, Classification, Mode of Life and Geological Usefulness of a Major Fossil Group. (Systematics Association, Special Volume, 18.) Academic Press, New York and London.

Kier, W. M., 1982. The functional morphology of the musculature of squid (Loliginidae) arms and tentacles. Journal of Morphology, *172:*179–192.

Lehmann, U., 1981. The Ammonites: Their Life and Their World. Cambridge University Press, London and New York.

Nesis, K. N., 1987. Cephalopods of the World. T. H. F. Publications, Ascot, England and Neptune City, New Jersey.

Nixon, M., and Messenger, J. B. (editors), 1977. The Biology of Cephalopods. Symposia of the Zoological Society of London, no. 38. Academic Press, London and New York.

Packard, A., 1972. Cephalopods and fish: the limits of convergence. Biological Reviews, *47:*241–307.

Roper, C. F. E., and Boss, K. J., 1982. The giant squid. Scientific American, *246*(5):96–104.

Saunders, W. B., and Landman, N. H. (editors), 1987. Nautilus. The Biology and Paleobiology of a Living Fossil. Plenum Press, New York and London.

Ward, P., 1982. Nautilus: have shell, will float. Natural History, *91*(10):64–69.

Ward, P., 1983. The extinction of the ammonites. Scientific American, *249*(4):136–146.

Ward, P., 1987. The Natural History of Nautilus. Allen and Unwin, Boston and London.

Ward, P., 1988. In Search of *Nautilus*. Simon and Schuster, New York and London.

Ward, P., Greenwald, L., and Greenwald, O. E., 1980. The buoyancy of the chambered nautilus. Scientific American, *243*(4):190–203.

Wells, M. J., 1978. Octopus: Physiology and Behaviour of an Advanced Invertebrate. Chapman and Hall, London; John Wiley and Sons, New York.

Yarnell, J. L., 1969. Aspects of the behavior of *Octopus cyanea* Gray. Animal Behavior, *17:*747–754.

SUMMARY

1. Molluscs form the second largest phylum of animals. Five of the seven living classes are strictly marine, but two—Gastropoda and Bivalvia—are abundantly represented in fresh water as well as in the sea. Many gastropods, moreover, are adapted for life in terrestrial habitats.

2. As a group, molluscs are characterized by organization of the bilaterally symmetrical, unsegmented body into four main regions: head, foot, visceral mass, and mantle. The mantle secretes the shell, when this is present, and usually encloses a cavity within which one or more peculiarly molluscan gills, called ctenidia, are located. The structural diversity of molluscs is to a considerable extent expressed in the way the various structures just mentioned are specialized, simplified, or lost altogether.

3. Ctenidia, in their least modified form, are paired, and each consists of a supporting axis from which two rows of leaflets originate. The leaflets on one side of the axis alternate with those on the other side.

4. A feature of all classes except Bivalvia is the radula. Unless greatly modified, it is a ribbonlike structure, studded with teeth, and is formed in a sac on the ventral side of the buccal cavity. It can, however, be protracted beyond the mouth, then retracted, so that it functions as a scraper that loosens food and draws it back into the buccal cavity.

5. The gut is complete and food is generally moved through it by ciliary activity (by muscular activity in Cephalopoda). There is typically a large digestive gland, which consists of diverticula of the stomach. Much of the digestion and absorption takes place in the diverticula.

6. The derivatives of the embryonic coelom are the pericardial cavity (which surrounds the heart), the renal organs, and gonads and gonoducts. Primitively, these are joined, but in most molluscs they are partly or completely separate.

7. Except in cephalopods, in which blood is confined to vessels, the circulatory system is of the open type. The ventricle of the heart pumps blood into arteries, and from these the blood enters spaces that are part of an extensive hemocoel. Eventually it is collected by veins that carry it to the one or two auricles of the heart.

8. Reproductive systems range from simple types, in which gametes are released by spawning, to complex types with devices for copulation, storage of sperm after copulation, and production of egg capsules. Like annelids, molluscs have spiral cleavage and are protostomatous. Another feature that links them with annelids is the presence, in some marine groups, of a trochophore larva.

9. Only the four main classes of molluscs—Polyplacophora, Gastropoda, Bivalvia, and Cephalopoda—are characterized briefly here. The Polyplacophora, called chitons, have a shell that consists of a series of eight overlapping dorsal plates and several to many pairs of ctenidia. The head is relatively simple; the foot is large and effective in clinging and slow locomotion. Chitons use the radula for scraping.

10. The Gastropoda, represented by snails and slugs, is the largest class and has the greatest diversity. The head is usually well developed. The shell, when present, may be coiled, conical, or cap-shaped, or it may be reduced to a small external or internal plate, or lost altogether. Reduction and loss of the shell have occurred independently in various groups of gastropods.

11. A characteristic of most members of the Gastropoda is torsion. During development, a combination of muscle contraction and differential growth brings the posterior portion of the mantle cavity forward, above the head. Thus the anus, openings of the renal organs, and ctenidia (if present) are anterior instead of posterior. The problem of how to prevent fouling of the respiratory current with fecal and renal wastes has been solved in several different ways. In most members of the largest group of gastropods, called prosobranchs, the respiratory current of clean water passes over the one or two ctenidia before wastes are discharged into it; it leaves the mantle cavity without being recirculated over the ctenidia.

12. In one group of marine gastropods, the opisthobranchs, the mantle cavity is reduced and there is rarely a ctenidium. There is also detorsion, so if there is a mantle cavity, it is rather far back on the right side. The general body surface and various kinds of secondary gills function in respiration. In pulmonate gastropods, which are primarily terrestrial and freshwater types, part of the mantle cavity is specialized to form a lung for aerial respiration. (In some freshwater pulmonates, however, it is kept filled with water and functions as a gill.)

13. The more primitive gastropods, all of which are marine, are mostly free-spawners, whereas more advanced types have complex reproductive systems. They copulate and lay eggs enclosed by secreted capsules.

14. Bivalves—clams, mussels, and their relatives—have a two-piece shell. The valves, hinged together dorsally, can be pulled together by one or two adductor muscles. Although most bivalves use the foot for digging in mud or sand, some lie on the substratum or attach themselves to it by cement or by threads of organic material. A few burrow in rock, shell, or wood by abrasive action of the valves, sometimes aided by an acid secretion.

15. In most members of the Bivalvia, the ctenidia are greatly enlarged and serve not only in respiration, but also in ciliary–mucous collection of food particles. The particles are carried to the labial palps, located beside the mouth. The palps sort the particles, rejecting some, directing others into the mouth. In primitive bivalves, whose ctenidia are strictly for respiration, the palps are extended out of the mantle cavity and are themselves used for collecting microscopic food.

16. The Cephalopoda, which includes squids and octopuses, is the most advanced class of molluscs. Cephalopods are specialized for capturing animal prey and for rapid swimming by jet propulsion. The propulsion is effected by forcing water from the mantle cavity through a narrow tube, called the funnel. This structure, together with the arms and tentacles that surround the mouth, are collectively homologous to the foot of other molluscs. In octopuses, which are primarily benthic, the arms are used effectively for locomotion over the ocean bottom and for clinging.

17. Compared to other molluscs, cephalopods have a more highly developed brain and a more complex behavior. The circulatory system is of the closed type, in which arteries and veins are connected by small vessels comparable to capillaries. The shell is usually reduced and internal, or absent. (*Nautilus,* with a large, chambered shell, is the major exception.)

14

The Lophophorate Phyla and Entoprocta

They have crowns, but no heads to set them on.

—Paul Illg, in a lecture on lophophorates to students of invertebrate zoology

- Phylum Phoronida
- Phylum Brachiopoda
  - Class Articulata
    - Order Terebratellida
    - Order Rhynchonellida
  - Class Inarticulata
    - Order Lingulida
    - Order Acrotretida
- Phylum Bryozoa (Ectoprocta)
  - Class Gymnolaemata
    - Order Ctenostomata
    - Order Cheilostomata
  - Class Stenolaemata
  - Class Phylactolaemata
- Phylum Entoprocta

Three phyla—Phoronida, Brachiopoda, and Bryozoa—have ciliated tentacles organized into a functional unit called a **lophophore.** The term *lophophore* means "a structure that bears a tuft," but "tuft" is not quite the right word, for the tentacles are not in a simple cluster. In phoronids and brachiopods, they are organized into right and left coils that meet at the mouth; in bryozoans, they form a circle or a horseshoe pattern around the mouth.

Intense ciliary activity on the tentacles creates currents that circulate through and around the lophophore. The currents bring in microscopic particles of food and a fresh supply of oxygen, and they carry away metabolic wastes, fecal material, and gametes. In phoronids and brachiopods, the process of feeding involves secretion of mucus, whose function is to trap particles that touch the tentacles. The mixture of food and mucus is then delivered to the mouth along special ciliary tracts. In bryozoans, however, particles are driven to the mouth without the aid of mucus.

Lophophorates are not the only invertebrates that use ciliated tentacles for collecting microscopic food. Sabellid and serpulid polychaete annelids, for instance, feed in much the same way as phoronids and brachiopods, but their body organization is decidedly different from that of the animals dealt with in this chapter, and their tentacular crowns are not referred to as lophophores.

In general, lophophorates are poorly cephalized—so much so that they do not really have a head. The nervous system as a whole is relatively simple when compared with that of other advanced invertebrates such as annelids and molluscs. The gut, except in those brachiopods in which the intestine ends blindly, is U-shaped. Although the mouth and anus are fairly close together, fecal wastes are swept away before they can foul the feeding currents.

The lophophorate phyla are similar in having radial cleavage, but in other respects their embryology and later development vary considerably. In brachiopods and phoronids, gastrulation is by invagination, and the mouth arises in the area where the blastopore had previously appeared and then closed. In bryozoans, however, gastrulation is usually achieved by *delamination;* that is, some cells produced by division of those making up the wall of the blastula sink inward and form the endoderm. Even if a functional gut develops at an early stage, neither the mouth nor the anus is related to a blastopore. In many bryozoans, in fact, appearance of a gut is delayed until after the swimming larval stage has settled and begun its metamorphosis.

Phoronids and brachiopods have a coelom that usually arises by outpocketing or partitioning of the archenteron, or within a mass of mesoderm budded from the archenteron. The body cavity of bryozoans is generally called a coelom, but it is not always completely lined by a peritoneum. In at least some species, the cavity originates when the larval gut degenerates, leaving a space. This suggests that it is not a true coelom. The origin of the mesodermal cells that line the cavity and that contribute to other structures has not been explained. The similarities of the three lophophorate phyla are perhaps the result of convergent evolution. Brachiopods and bryozoans have left long and detailed fossil records, but these do not go back far enough to explain the origin of lophophorates and the paths by which they diverged, if indeed they all had common ancestors.

The phylum Entoprocta is also discussed in this chapter. Entoprocts are similar to most bryozoans in having a complete circle of tentacles around the mouth. For a long time, in fact, they were considered to belong in the Bryozoa, and then it became the fashion to place them next to the pseudocoelomate phyla. Although the body cavity of adult entoprocts does resemble a pseudocoel, these animals are certainly different from nematodes, kinorhynchs, rotifers, and other pseudocoelomate phyla discussed in Chapters 9 and 10. There are, moreover, striking similarities in the structure and metamorphosis of the larvae of entoprocts and bryozoans, and some specialists are now convinced that these two groups are closely related.

PHYLUM PHORONIDA

INTRODUCTION

The phoronids, of which there are about 20 species, are marine animals that live in secreted tubes. A few reach a length of about 15 cm when fully extended, but most are under 5 cm. They occur intertidally as well as in the shallow subtidal zone. Some inhabit muddy sand or gravel, and may form dense populations that cover large areas. Others are attached to rocks, shells, or wood, usually forming tangled masses that accumulate considerable sediment. A few phoronids burrow into shells or into limestone, but how they do this is not known. One species is embedded in the tube of a ceriantharian, which is constructed of mucus and discharged nematocysts; another is characteristically found in the mixture of mucus and clay particles with which the burrows of mud shrimps are lined.

There are only two genera, and they are not especially different. *Phoronopsis,* which includes the larger species found in muddy sand and gravel, is characterized by a deep groove that encircles the body just below the lophophore. The groove is absent or barely noticeable in *Phoronis*. The genus name *Phoronis* commemorates Io, a mythological character whose surname was Phoronis. She was changed into a white heifer and wandered far and wide before being restored to her original form. The name is apt, because most phoronids have a planktonic larval stage that is decidedly unlike the adult.

GENERAL STRUCTURE

EXTERNAL FEATURES AND TUBE

When a phoronid is feeding, it keeps its lophophore raised well above the lip of the tube (Figure 14.1, A and B). The number of tentacles may exceed 500, but in some of the smaller species there are fewer than 100. The base of the lophophore is essentially horseshoe-shaped (Figure 14.2, C), but in most phoronids both arms of the horseshoe are coiled (Figure 14.2, A). The two rows of tentacles, separated by a shallow groove, meet at the ends of the arms.

The mouth is located along the midline in the lophophoral groove. The anus, also on the midline but outside the lophophore, is situated on a rather prominent elevation. Flanking it on the right and left are the openings of the **funnel organs,** which permit gametes to escape from the coelom and which probably also have an excretory function. In certain phoronids, the mouth is bordered, on the side nearest the anus, by a little ridge or flap called the **epistome.** Other structures sometimes visible at the lophophoral end of the animal are glands concerned with reproduction; these are discussed below.

The elongated **trunk,** below the lophophore, is often superficially ringed, and the lowermost portion may be swollen. The tube consists to a large extent of chitin, but there are usually sand grains, bits of shell, and other foreign material embedded in it. As a rule, the tube is oriented vertically with respect to the substratum, and much or nearly all of it is buried. It is not necessarily straight, however. In some species, in fact, it is almost always kinked or contorted, and the tubes of neighboring individuals may be intertwined.

BODY WALL AND COELOMS

The epidermis of the trunk region has cells specialized for secretion of tube material and mucus. Beneath the epidermis is a thin sheet of **circular muscle,** and internal to this is an extensive layer of **longitudinal muscle,** organized into distinct bundles (Figure 14.2, D). The coelom of the trunk region is divided into four unequal sectors by mesenteries of peritoneal tissue. These mesenteries also segregate the longitudinal muscle bundles into four groups.

The lophophore has its own coelomic cavity. A fingerlike extension of this goes into each tentacle.

A

B

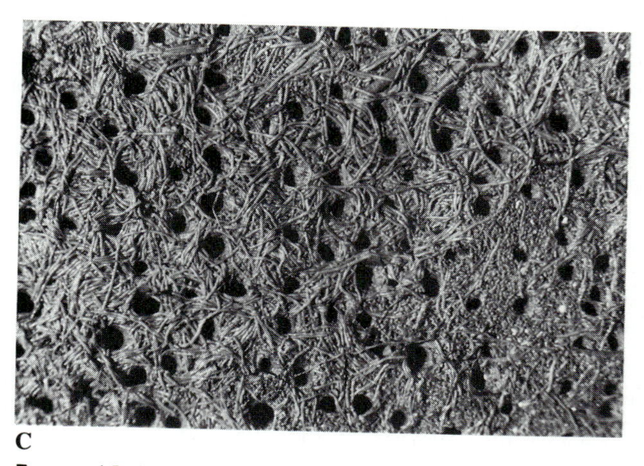

C

FIGURE 14.1 A. *Phoronis vancouverensis,* portion of a dense aggregation. B. *Phoronopsis harmeri,* its lophophore extended well above the aperture of the tube, which is hidden by algal debris. (Photograph by Ronald Shimek.) C. *Phoronopsis viridis,* fecal castings among the apertures of tubes in a dense aggregation (lophophores withdrawn at low tide).

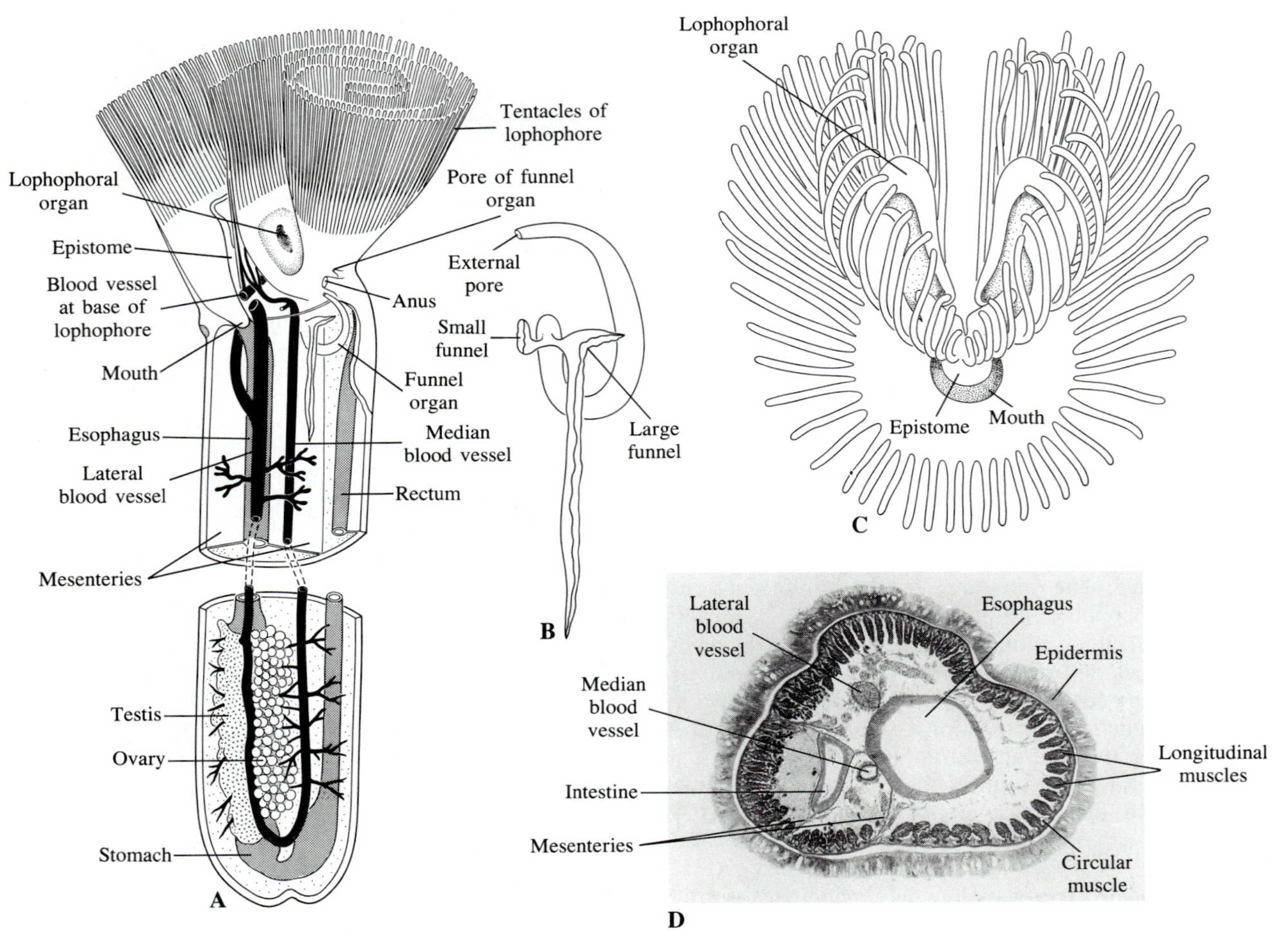

FIGURE 14.2 A. *Phoronis australis,* sectioned lengthwise; diagrammatic. Much of the trunk has been omitted. B. *Phoronis australis,* funnel organ. (A and B, after Benham, Quarterly Journal of Microscopical Science, *30.*) C. *Phoronis architecta,* lophophore. (After Brooks and Cowles, Memoirs of the National Academy of Sciences, Washington, *10.*) D. *Phoronopsis harmeri,* transverse section through the trunk; photomicrograph.

DIGESTIVE SYSTEM AND FEEDING

As the activity of certain long cilia circulates water through the tentacular crown of the lophophore, food particles, including diatoms, become trapped in mucus. They are carried down ciliary tracts to the collecting groove that runs between the two rows of tentacles. This groove, which is also ciliated, delivers the particles to the mouth.

The long esophagus extends down to a stomach near the lower end of the organism. From the stomach, the intestine runs upward to the anus. The gut is only slightly muscularized, and the passage of food through it is accomplished primarily by the action of cilia. The food is rotated as it is moved along, so its passage is slowed and a better mixing of the food with enzymes is assured. As fecal wastes emerge from the anus they are swept away by a current of water leaving the lophophore. In quiet bays where the phoronid population is dense, the surface of the substratum may be covered with strings of fecal matter (Figure 14.1, C).

CIRCULATORY SYSTEM AND FUNNEL ORGANS

The blood vessels consist of myoepithelial cells derived from the peritoneum. As is typical of blood vessels of most invertebrates, the basal lamina of the cells is within the tube, rather than external to it. The vessels move blood along by peristaltic action. Two major vessels extend for the full length of the trunk (Figure 14.2, A): the *median vessel* carries blood from the network around the stomach to the lophophore, and the *lateral vessel* collects blood from the lophophore and distributes it through the trunk. Most of the secondary vessels (except those leading to the network around the stomach) end blindly, so that blood leaving a major vessel must return to it by the same route. The flow of blood back and forth in the blind vessels of the tentacles is easy to see in a live animal, not only because the tentacles are transparent but because the hemoglobin of phoronids is concentrated in corpuscles.

The trunk coelom contains a pair of funnel organs that are generally assumed to be metanephridia, but which are probably compound structures in which a coelomoduct of mesodermal origin is joined to part of a larval protonephridium. It seems best just to call them funnel organs. They serve as gonoducts, and the fact that they sometimes contain what appears to be a brownish waste material suggests that they also have an excretory function. The funnel organs open by pores on elevations beside the one on which the anus is located (Figure 14.2, A). In some phoronids each funnel organ has two funnels, one larger than the other (Figure 14.2, B). These open into separate chambers of the trunk coelom, so that all four chambers are served. When each funnel organ has just one funnel, only the two larger chambers are served.

NERVOUS SYSTEM

The principal nerves are closely associated with the epidermis. At the base of the lophophore is a ring of nerve tissue that is the closest thing a phoronid has to a brain. Nerves extend from this ring into the tentacles. Also originating from the ring are one or two prominent giant neurons. These run down the trunk and innervate the longitudinal muscle bands responsible for instantly shortening the animal when its extended lophophore is threatened.

SEXUAL REPRODUCTION AND DEVELOPMENT

In phoronids that have separate sexes, there is a single gonad; hermaphroditic species have a testis and an ovary close together (Figure 14.2, A). The gonads originate from portions of the peritoneum that are intimately associated with blind blood vessels in the lower part of the trunk. In addition to gamete-producing cells, they contain what appears to be nutritive tissue. The gametes are eventually set free in the trunk coelom, and reach the outside by way of the funnel organs.

The sperm are packaged into spermatophores by glands called *accessory spermatophoral organs*. There is one of these near the opening of each funnel organ, and it is part of a larger glandular complex called a **lophophoral organ** (Figure 14.2, A and C). In at least some species of *Phoronopsis,* the spermatophores have leaflike or spiral sails (Figure 14.3, A), but sails are lacking in spermatophores of the genus *Phoronis*. After a spermatophore reaches the lophophore of a female or hermaphroditic individual, substances released by it induce lysis of the tissue of a tentacle. The sperm get into the coelomic canal of the tentacle and then enter the coelom of the trunk, where they fertilize the eggs.

After fertilization, the eggs are released through the funnel organs. In some species, including *Phoronopsis harmeri* of the northern Pacific coast, they are simply carried away by currents. Most female and hermaphroditic phoronids, however, brood their cleaving eggs and embryos, holding them at the base of the lophophore with the aid of secretions produced by *nidamental glands*. These, like the accessory spermatophoral organs, are portions of the lophophoral organs. In hermaphroditic species, in fact, each nidamental gland is adjacent to an accessory spermatophoral organ. Neither of the two components is likely to be observed except during the reproductive period.

Whether hermaphroditic phoronids are self-fertilizing is not known. In *Phoronis vancouverensis,* which has been studied extensively, the eggs are definitely fertilized before they come out through the funnel organs. Their cleavage is blocked at the first metaphase stage until they emerge, about 25 or 30 each day during the breeding season. In this species, and generally in others that brood, parental care continues until the young reach a beautiful swimming stage called the **actinotroch**. The structure and development of the actinotroch are discussed below.

Cleavage is decidedly radial, but sometimes division of one of the first two blastomeres is delayed and there is a temporary 3-cell stage. In addition, when the fertilization membrane is tightly applied to the egg, the pattern of radial cleavage may be distorted so that it superficially resembles spiral cleavage. This often happens in sea stars and other invertebrates whose cleavage is indisputably radial.

Gastrulation is accomplished by invagination. The blastopore closes early in development, but the mouth later appears in the same general region, and an anus opens up some distance away. Thus phoronids are considered to be protostomes. Some of the mesoderm is derived from the archenteron, some from ectoderm.

The typical pattern of development includes the actinotroch stage. Among the distinctive external features of this type of larva (Figure 14.3, B), unique to phoronids, are an apical tuft of cilia at the anterior end, a ciliated preoral hood, a set of tentacles forming an incomplete collar behind the mouth, and a ring of cilia around the anus. The apical tuft is usually uppermost as the larva swims.

The actinotroch has three coelomic cavities. The most anterior of these, the **protocoel,** occupies part of the preoral region and is lined by ectomesoderm. The

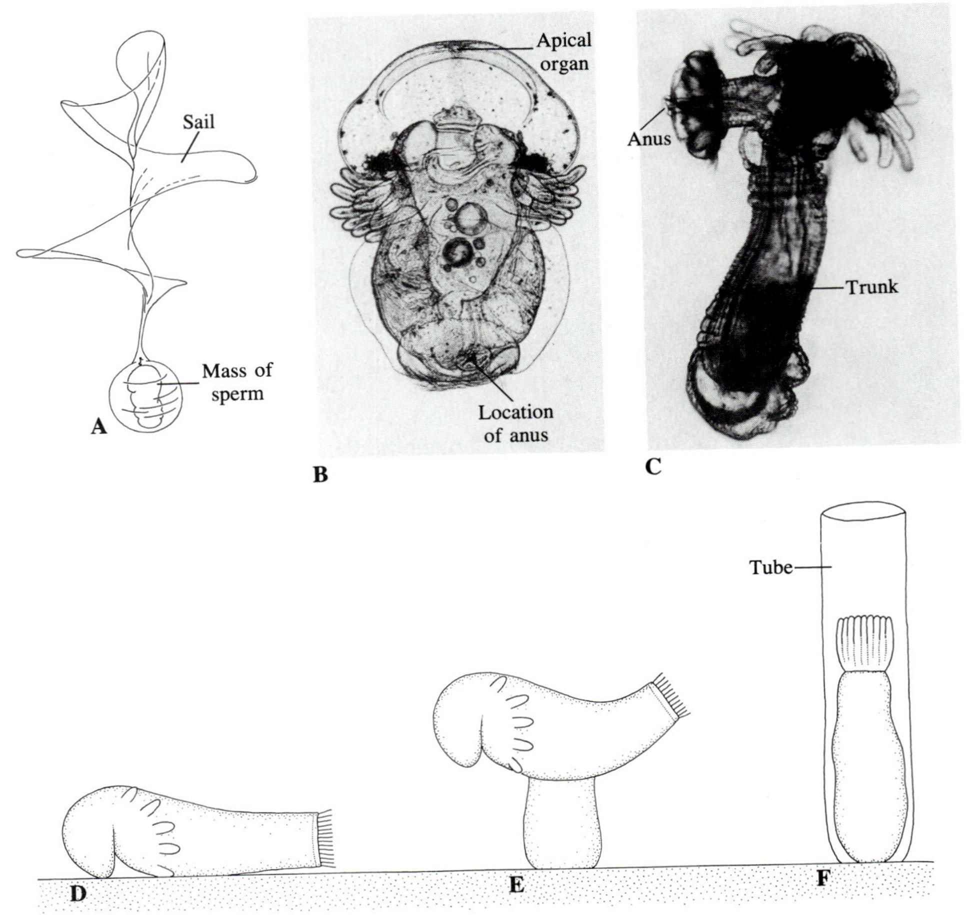

FIGURE 14.3 A. *Phoronopsis harmeri,* spermatophore. (After Zimmer, Journal of Morphology, *121.*) B. *Phoronopsis harmeri,* young actinotroch; photomicrograph. C. An early stage in the transformation of an actinotroch into an adult; the trunk is already rather well developed. (Photomicrograph by Russel Zimmer.) D–F. *Phoronis hippocrepia,* successive stages in settlement and metamorphosis of an actinotroch. In time, the animal will erode the calcareous substratum, forming a burrow. (After Silén, Acta Zoologica, *35.*)

cavity of the collar, called the **mesocoel,** can be traced back to the blastocoel, but it becomes lined by mesodermal cells budded off the endoderm. The **metacoel,** or future trunk coelom, originates as a space within a large ventral mass of mesoderm. The excretory organs of the actinotroch are solenocytic protonephridia, and they are derived from ectoderm.

The orientation of a phoronid cannot be understood without reference to the metamorphosis of the actinotroch. For this reason, terms such as anterior and posterior, dorsal and ventral, were avoided in discussing the morphology of the adult. After an actinotroch finds a suitable substratum on which to settle, it resorbs its oral hood, so the protocoel is obliterated. (Some zoologists think, however, that the epistome, whether or not it has a small cavity inside, is derived from the oral hood.) The long trunk region of a phoronid, whose coelom is the larval metacoel, originates as the result of a remarkable process by which the swimming actinotroch is converted into a wormlike benthic organism. On the ventral side of the actinotroch is an inverted pouch, called the *metasomal sac*. This is wrapped around the gut. Eventually it turns inside out, pulling a loop of the gut with it, and develops into the trunk of the prospective adult (Figure 14.3, C). This explains why the gut of a mature phoronid is bent back on itself like a hairpin. The entire trunk is thus an elaboration of the ventral side of the actinotroch, and the lophophoral end of an adult, which has both the mouth and anus, is most nearly comparable to the dorsal side of the actinotroch.

During metamorphosis, the tentacles of the collar behind the mouth are mostly resorbed, and the tentacles of the adult lophophore grow out from buds between them. The coelomic space in the basal portion of the lophophore and in the tentacles corresponds to the mesocoel of the larva. The protonephridia disappear, although the pores remain and portions of the ducts contribute to the funnel organs of the adult.

Like the larvae of many other types of invertebrates, an actinotroch must find a suitable substratum before it will undergo metamorphosis. Figure 14.3, D–F shows successive stages in settlement and metamorphosis of *Phoronis hippocrepia,* which requires a calcareous substratum.

Direct development occurs in one species, *Phoronis ovalis* (Figure 14.4). This phoronid, like *P. hippocrepia,* burrows into calcareous substrata, and is most commonly found in empty shells of bivalve molluscs. It is hermaphroditic, and is believed to release its eggs through openings formed when it sheds its lophophore, rather than through the funnel organs. The eggs, which have considerable yolk, undergo early development in the upper part of the tube of the parent. Within a few days the gastrulae, which are disk-shaped (A), swim out. A marginal flange, perhaps comparable to the preoral lobe of an actinotroch, appears along the anterior side (B), and then a blunt outgrowth develops behind the mouth (C, D). The anus opens up on this outgrowth, which is possibly homologous to the metasomal sac of an actinotroch. The larva becomes more elongated (E) and its gliding activities resemble those of a small flatworm. About a week after the gastrula has left the tube, the larva settles, loses its external cilia, and becomes hemispherical (F). The transformation of the hemispherical stage into the adult has not been described.

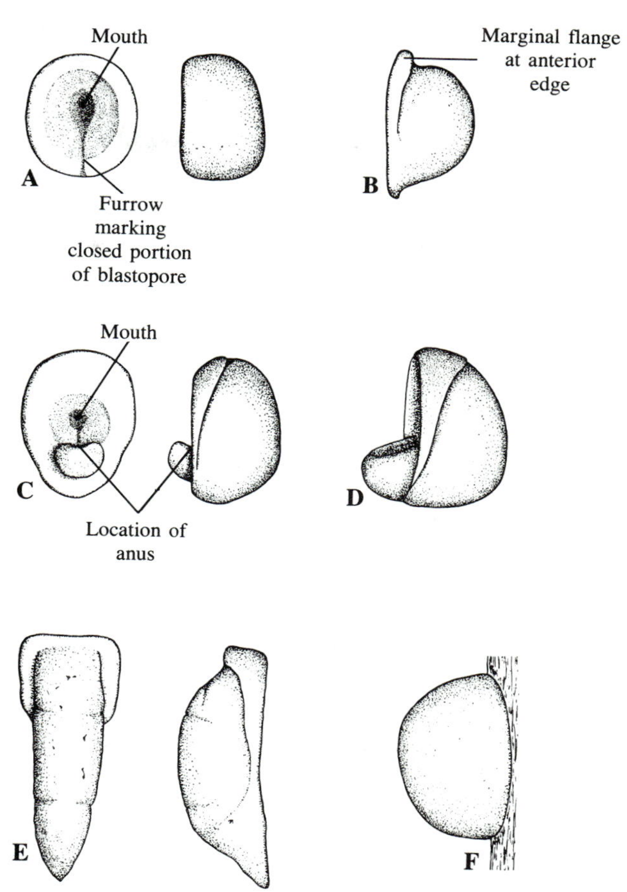

FIGURE 14.4 Stages in direct development of *Phoronis ovalis.* A. Gastrula recently released from the tube of the parent, ventral view (left) and lateral view (right). B. Slightly older stage, lateral view. C. Two days after release, ventral view (left) and lateral view (right). D. Blunt outgrowth appears, and the anus (not visible in the drawing) is located on the anterior side of this. E. Five days after release, dorsal view (left) and lateral view (right). F. Attached individual, seven days after release (Silén, Acta Zoologica, *35.*)

ASEXUAL REPRODUCTION AND REGENERATION

Some phoronids reproduce asexually as well as sexually. *Phoronis vancouverensis,* for instance, can divide into two approximately equal pieces. The half with the lophophore must regenerate the lower part of the trunk; the other half must regenerate the upper part of the trunk as well as a new lophophore. This type of reproduction is conducive to formation of dense colonies of intertwined individuals.

In several species whose capacity to regenerate has been studied experimentally, short pieces of the trunk—especially those cut out just above the basal portion—will develop into complete worms. Thus the capacity for regeneration is strong.

AUTOTOMY

Some phoronids, especially those that live in sandy mud, pinch off their lophophores after they have been dug up. This form of **autotomy** is achieved by contraction of circular muscles in the trunk below the base of the lophophore. If the trunk remains in good condition, the lophophore will be regenerated. The only known predators on phoronids are certain nudibranchs, but the capacity for autotomy could nevertheless be an adaptation that spares the trunk portion if the animal is attacked. In at least some populations of certain species, shedding of the lophophore is a seasonal phenomenon, and in this case it may be a way of eliminating a structure that is of little use and perhaps even a liability at times when planktonic food is scarce.

References on Phoronida

Emig, C. C., 1974. The systematics and evolution of the phylum Phoronida. Zeitschrift für Zoologische Systematik und Evolutionsforschung, *12:*128–151.

Emig, C. C., 1977. Embryology of Phoronida. American Zoologist, *17:*21–37.

Emig, C. C., 1979. British and Other Phoronids. Synopses of the British Fauna, no. 13. Linnean Society of London, Estuarine and Brackish-Water Sciences Association, and Academic Press, London and New York.

Silén, L., 1954. Developmental biology of Phoronidea of the Gullmar Fiord Area (west coast of Sweden). Acta Zoologica, *35:*215–257.

Zimmer, R. L., 1967. The morphology and function of accessory reproductive glands in the lophophores of *Phoronis vancouverensis* and *Phoronopsis harmeri*. Journal of Morphology, *121:*159–178.

Zimmer, R. L., 1972. Structure and transfer of spermatozoa in *Phoronopsis viridis*. *In* Arcenaux, C. J. (editor), 30th Annual Proceedings of the Electron Microscopy Society of America, 108–109.

Phylum Brachiopoda

Introduction

The fewer than 300 species of brachiopods living today are remnants of a marine group that has left about 1000 genera and 30,000 species in the fossil record. The two classes of the phylum can be traced back to the Cambrian period, the earliest part of the Paleozoic era. Both classes flourished in succeeding periods of the Paleozoic, and they were still going strong in the Mesozoic era—the Age of Reptiles. After the Mesozoic era, the brachiopods declined sharply. Some of the species that have survived, however, are similar to the most ancient prototypes.

The name Brachiopoda is derived from Greek words meaning "arm" and "foot." For a long time, the lophophore of a brachiopod was thought to be the homologue of the molluscan foot. Thomas Huxley's refutation of this idea, in the middle of the nineteenth century, took a while to sink in. In spite of the fact that brachiopods resemble clams, their shell valves are dorsal and ventral (rather than right and left) with respect to the body. In other words, the shells of brachiopods and clams are not homologous.

Some brachiopods are called lamp shells because they resemble certain ancient oil lamps molded from clay. The surviving species are not often more than 3 cm across, although a few reach 7 cm. Among the fossils there are types with shells measuring about 30 cm in diameter.

The two classes of living brachiopods, Articulata and Inarticulata, differ in several important ways. The most tangible difference is the presence of a rather complex hinge in the shell of the Articulata and the absence of a hinge in the Inarticulata.

Class Articulata

General Structure

Pedicle, Shell, and Muscles

Articulate brachiopods, whose shell valves are unequal, are permanently attached to rocks or shells, including shells of other brachiopods. There is usually a short stalk called the **pedicle** (Figure 14.5, A; Figure 14.7). It is composed mostly of tough connective tissue and it emerges through a hole or notch in the posterior part of the larger valve—the one that is ordinarily uppermost (Figure 14.6, A; Figure 14.7). In spite of its usual position, this valve is really the ventral valve. To avoid confusion, it is sometimes called the *pedicle valve,* and in that case the dorsal valve is called the *brachial valve,* because the internal calcareous support of the lophophore ("brachium") is joined to it. Here, however, the terms *dorsal* and *ventral* are used, for they apply to all brachiopods, including those that do not have a pedicle and those in which the lophophore lacks an internal support.

The pedicle itself is not capable of movement, but **adjustor muscles** that originate on both valves are inserted into it (Figure 14.7, B). These muscles make it possible for an articulate brachiopod to change its orientation to some extent.

The two valves are hinged together by a tooth-and-socket arrangement. This permits one valve to rock on the other and at the same time prevents a sideways slippage. The ventral valve has a tooth on each side, not far from the hole or notch where the pedicle emerges; the dorsal valve has sockets in the corresponding positions. The transverse axis of the hinge is thus the fulcrum for the rocking motion of the valves.

Two opener muscles, called *diductors* (Figure 14.7), originate on the ventral valve and run backward to the cardinal process, a platelike extension of the dorsal valve. This process is behind the hinge axis. When the diductor muscles contract and pull the two valves closer together posteriorly, the valves separate anteriorly. The gape is always slight, however. The adductor muscles, which close the shell, are in front of the hinge and are almost vertical. There are two of them originating on the ventral valve, but generally each divides into two portions before reaching the dorsal valve, so there are

FIGURE 14.5 Class Articulata. A. *Terebratalia transversa,* lateral view. B. *Terebratalia transversa,* valves separated in order to show some internal structures. C. *Laqueus californicus.* D. *Terebratulina unguicula.* The valves of one specimen are gaping, so that the lophophore can be seen.

four points of attachment on this half of the shell (Figure 14.6, B).

Only about a third of the space between the valves is occupied by the tissues of the animal. The rest is an open, water-filled cavity, comparable to the mantle cavity of a clam. It is in fact called the mantle cavity (Figure 14.7), and the tissue that lines the shell is called the mantle. As in molluscs, the mantle edge secretes both an organic periostracum and the primary layer of the shell, composed of calcium carbonate. A secondary layer, consisting of slender calcium carbonate crystals separated by strips of protein, is produced almost everywhere over the outer surface of the mantle. Additions to this layer account for the continued thickening of the shell as the animal grows.

The free edges of the mantle lobes are fringed by organic bristles. These are called setae and, like the setae of annelids, are secreted within epidermal pockets. They keep coarse particles from entering the mantle cavity.

The shell valves of many articulate brachiopods have perforations called *punctae* (Figure 14.7, B). In a live animal, each puncta is occupied by a fingerlike extension of the mantle, called a *caecum*. Secretory cells at the tip of a caecum send threads of cytoplasm toward the surface of the valve, but these stop just beneath the periostracum. The function of the caeca is not understood. Perhaps their secretions inhibit the activities of boring organisms.

Lophophore

When the two valves of an articulate brachiopod are pulled apart, the most obvious structure exposed is the lophophore (Figure 14.5, D; Figure 14.7). Its two arms

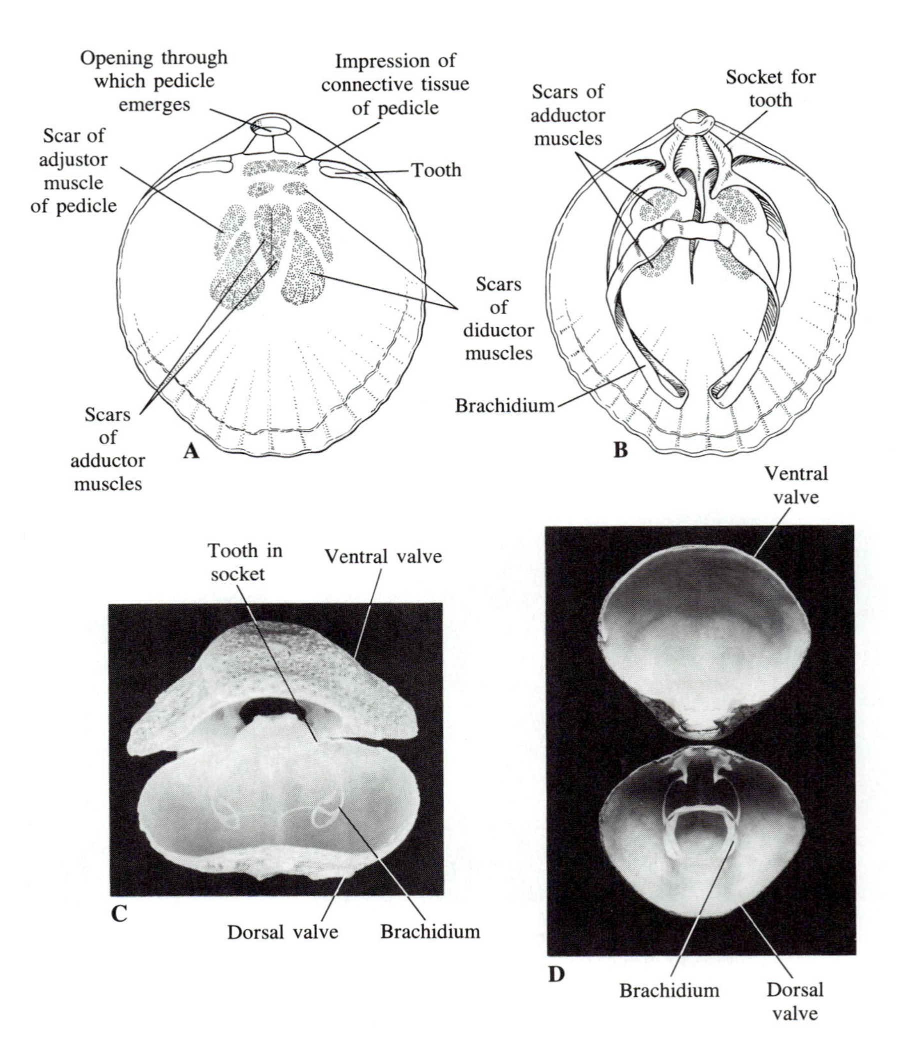

FIGURE 14.6 Class Articulata. A. *Magellania,* ventral valve. B. *Magellania,* dorsal valve. (A and B, After Davidson, Transactions of the Linnean Society of London, Zoology, series 2, *4*.) C. *Terebratalia transversa,* anterior view of an empty shell, valves gaping. D. *Terebratalia transversa,* empty shell, valves separated.

are coiled in a complex fashion, and each has hundreds of tentacles. The activity of cilia on the tentacles creates the currents that bring water, oxygen, and microscopic food into the mantle cavity and that also carry away wastes. The way in which the lophophore tentacles function in feeding is discussed below.

In most genera, the base of the lophophore is supported by the **brachidium,** a delicate ribbon of calcium carbonate that is joined to the dorsal valve (Figure 14.6, B–D). It is almost impossible to tease away the tissues of the lophophore without damaging the brachidium, but if the dorsal valve is soaked in a strong solution of sodium hypochlorite or sodium hydroxide for a day or two, the tissues will be destroyed to the extent that they can be washed away. The brachidium is sometimes still intact in shells that are empty when collected.

Coeloms

As in phoronids, there are two coeloms: the general body cavity and a cavity that is restricted to the lophophore (Figure 14.7). The former, which is fairly spacious, is inevitably exposed when the valves are pulled apart. This cavity contains the digestive tract, funnel organs, and gonads, although the latter are in extensions that pervade the mantle lobes.

The lophophoral coelom is not large enough to be visible except in sections. It extends, however, into each tentacle (Figure 14.8). Certain cells of the coelomic epithelium within the tentacles are contractile and curl the tentacles when these are touched. A layer of connective tissue just beneath the epidermis is antagonistic to the contractile cells, so when these cells relax

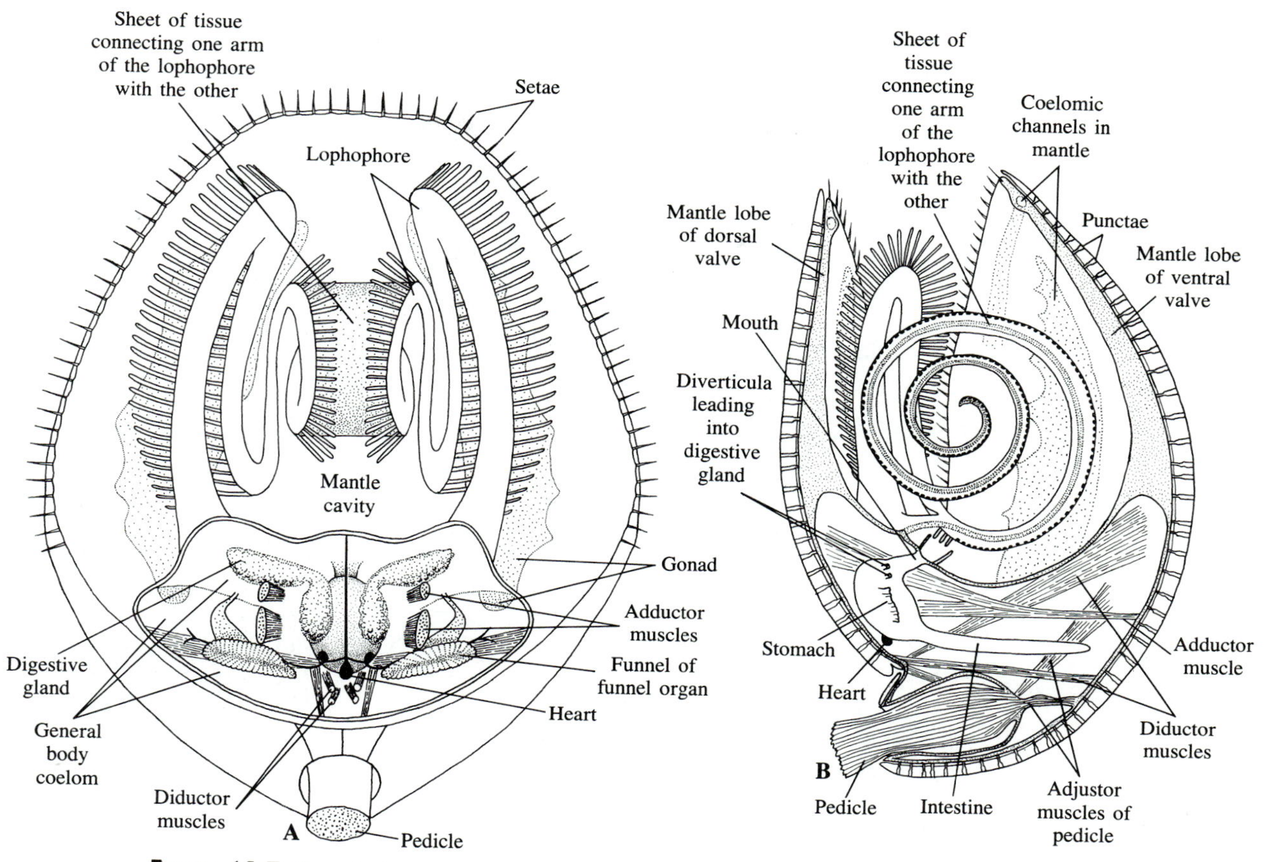

FIGURE 14.7 Class Articulata. *Magellania*. A. Diagram of structures visible after removal of the dorsal valve and portions of the body wall adhering to it. B. Diagram of structures visible in an animal that has been sectioned lengthwise. (After Delage and Hérouard, Traité de Zoologie Concrète.)

the tentacles straighten out again as a result of hydrostatic pressure within the lophophoral coelom.

Digestive System and Feeding

As the activity of cilia circulates water through the lophophore, food particles, including diatoms, are trapped in mucus on the tentacles. The particles are then moved down special ciliary tracts to the basal collecting groove (Figure 14.8, A) that leads to the mouth. It has been claimed that at least some brachiopods depend as much if not more on suspended colloidal material than on diatoms and other planktonic organisms of similar size.

The gut (Figure 14.7) typically consists of a short esophagus, a stomach with digestive diverticula, and an intestine that ends blindly. Considerable intracellular digestion takes place in the diverticula. In general, food and undigested residues are moved through the gut mostly by ciliary activity, assisted by the action of muscles. As the intestine has no anus, fecal pellets must be carried back to the mouth. They drop into the mantle cavity and are flushed out by ciliary currents or when the animal abruptly pulls its valves together and squirts out water.

Circulatory System and Funnel Organs

There is a contractile heart in the mesentery on the posterodorsal side of the stomach (Figure 14.7). From the heart some major blood vessels can be traced anteriorly and posteriorly. Histological sections show that the blood vessels are similar to those of phoronids in that the cells forming their walls are myoepithelial and derive from the peritoneum. The vessels pervade the lophophore and its tentacles, walls of the funnel organs,

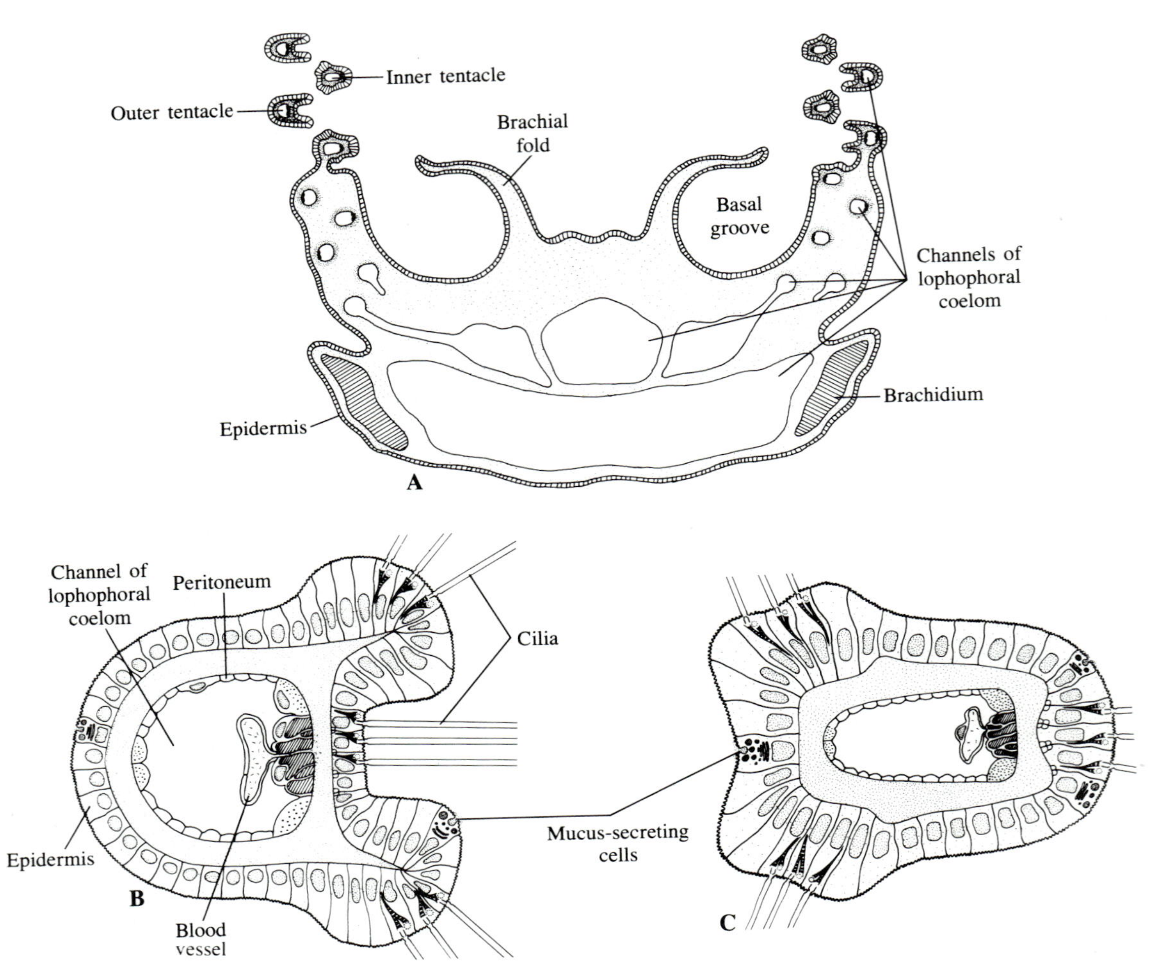

FIGURE 14.8 Class Articulata. *Terebratalia transversa*. A. One arm of the lophophore, transverse section, showing the two types of tentacles that arise from it, and also the coelomic channels that pervade it. B. An outer tentacle, transverse section. Note that the blood channel is formed by involution of the peritoneum. C. An inner tentacle, transverse section. (Drawings courtesy of Christopher G. Reed.)

walls of the mantle canals in which gametes develop, and probably most other parts of the body.

There are usually two funnel organs, but certain articulate brachiopods have two pairs. These structures are found on both sides of the gut and are closely associated with the body wall that separates the general body coelom from the mantle cavity (Figure 14.7, A). The funnels themselves are frilly, and the ducts open by pores into the mantle cavity. The funnel organs collect ripe gametes and carry them out of the coelom. They are probably concerned also with excretion. The funnel organs are generally called metanephridia, but it is likely that they derive in part from coelomoducts.

Nervous System

The nervous system is rudimentary and difficult to follow in dissections. These animals, after all, are not strongly cephalized. The most prominent part of the nervous system is a ganglionic ring around the esophagus, from which relatively small nerves run to all parts of the body. There are no obvious sense organs, but when touched or brightly illuminated, an articulate brachiopod generally responds by closing its shell.

Reproduction

Sexes are separate. The gonads lie in branches of the general body coelom that pervade the mantle tissue lining the valves of the shell (Figure 14.5, B; Figure 14.7, A). Gametes move to the main portions of the body coelom, where the funnel organs are located. The funnel organs deliver the gametes to the mantle cavity, from which they are discharged into the sea, except in the case of a few species that brood their young for a time.

The development of articulate brachiopods is discussed below in connection with that of inarticulate types, in order to stress the striking differences between the two groups.

Classification

Order Terebratellida

The lophophore of terebratellids is generally fully supported by a brachidium, and the shell usually has punctae. There are from two to four main trunks of the body coelom in each mantle lobe. There is a single pair of funnel organs. Representative genera are *Magellania* (Figure 14.6, A and B; Figure 14.7), *Terebratalia* (Figure 14.5, A and B; Figure 14.6, C and D), *Terebratulina* (Figure 14.5, D), and *Laqueus* (Figure 14.5, C).

Order Rhynchonellida

The lophophore is only partly supported by calcareous structures (there is no brachidium), and the shell usually lacks punctae. Two main trunks of the body coelom extend into each mantle lobe. There are typically two pairs of funnel organs. *Hemithiris* and *Frileia* are representative genera.

CLASS INARTICULATA

General Structure

The shells of inarticulate brachiopods do not have tooth-and-socket hinges. The operation of oblique muscles ensures that the valves line up properly when they are drawn tightly together by the two pairs of adductor muscles. The valves gape when the visceral portion of the body is pulled back, thereby increasing coelomic hydrostatic pressure against the valves.

There is generally a stalk, and although it is called a pedicle, it is not homologous to the pedicle of articulate brachiopods. It is an outgrowth of the ventral portion of the mantle. It is muscular and usually contains a cavity that is part of the general body coelom. As in the articulate brachiopods, the ventral valve is the one from which the pedicle emerges. This half of the shell, however, is usually about the same size as the dorsal valve, or smaller. Moreover, in species that are attached to hard substrata, the ventral valve, whether or not there is a pedicle, is lowermost.

In the majority of inarticulate brachiopods, the principal inorganic component of the shell is calcium phosphate rather than calcium carbonate. In addition, there is considerable chitin and protein. Shells that consist mostly of calcium carbonate are found only in a few living genera.

The lophophore is elaborate, just as it is in the articulate brachiopods, but there is no brachidium to support it. Finally, the gut of inarticulate brachiopods is complete. The intestine is fairly long and leads to an anus that opens either into the right side of the mantle cavity or directly to the outside, between the posterior ends of the valves.

The inarticulate brachiopods most often studied in courses in invertebrate zoology are *Lingula* (Figure 14.9, A) and *Glottidia* (Figure 14.9, C; Figure 14.10). They live in vertical burrows in muddy sand, with the free end of the long pedicle anchored in the substratum. The shell can be used for digging, and its two valves are nearly identical. When the pedicle is extended, the anterior end of the shell is about flush with the surface. Long setae on the edges of the mantle form three siphonlike channels that emerge through holes in the sand (Figure 14.9, B). The lateral channels allow water to enter the mantle cavity from both sides of the shell, and the median channel, leading away from the anterior end, is exhalant. When the animal is disturbed, the pedicle contracts and the shelled portion is pulled deeper into the burrow.

The visceral mass of *Lingula* and *Glottidia*—and of most inarticulates—takes up proportionately more space than it does in most articulate brachiopods. It contains not only the gut, the large digestive glands, and the funnel organs, but also the gonads (Figure 14.10). Remember that the gonads of articulate brachiopods are situated in channels of the general body coelom that branch out into the mantle lobes. Inarticulate brachiopods have comparable channels, but there are no gonads within them. Instead, the gonads form two substantial masses behind the digestive glands. The gametes, after being set free in the general body coelom, reach the mantle cavity by way of the funnel organs.

A heart, located in the mesentery dorsal to the stomach, pumps blood into anterior and posterior vessels. Branches of these open into spaces within the tissue. The amoebocytes of *Lingula* and *Glottidia* contain hemerythrin, a respiratory pigment that has a limited distribution in the animal kingdom, being found otherwise only in sipunculans and some annelids and priapulans.

Much of the nervous system of inarticulate brachiopods is closely associated with the epidermis. The greatest concentration of cell bodies is located underneath the esophagus, where there is a *subenteric ganglion* (or sometimes a pair of ganglia). Connectives from the subenteric ganglion extend dorsally to encircle the gut, and nerves branching off the connectives supply the lopho-

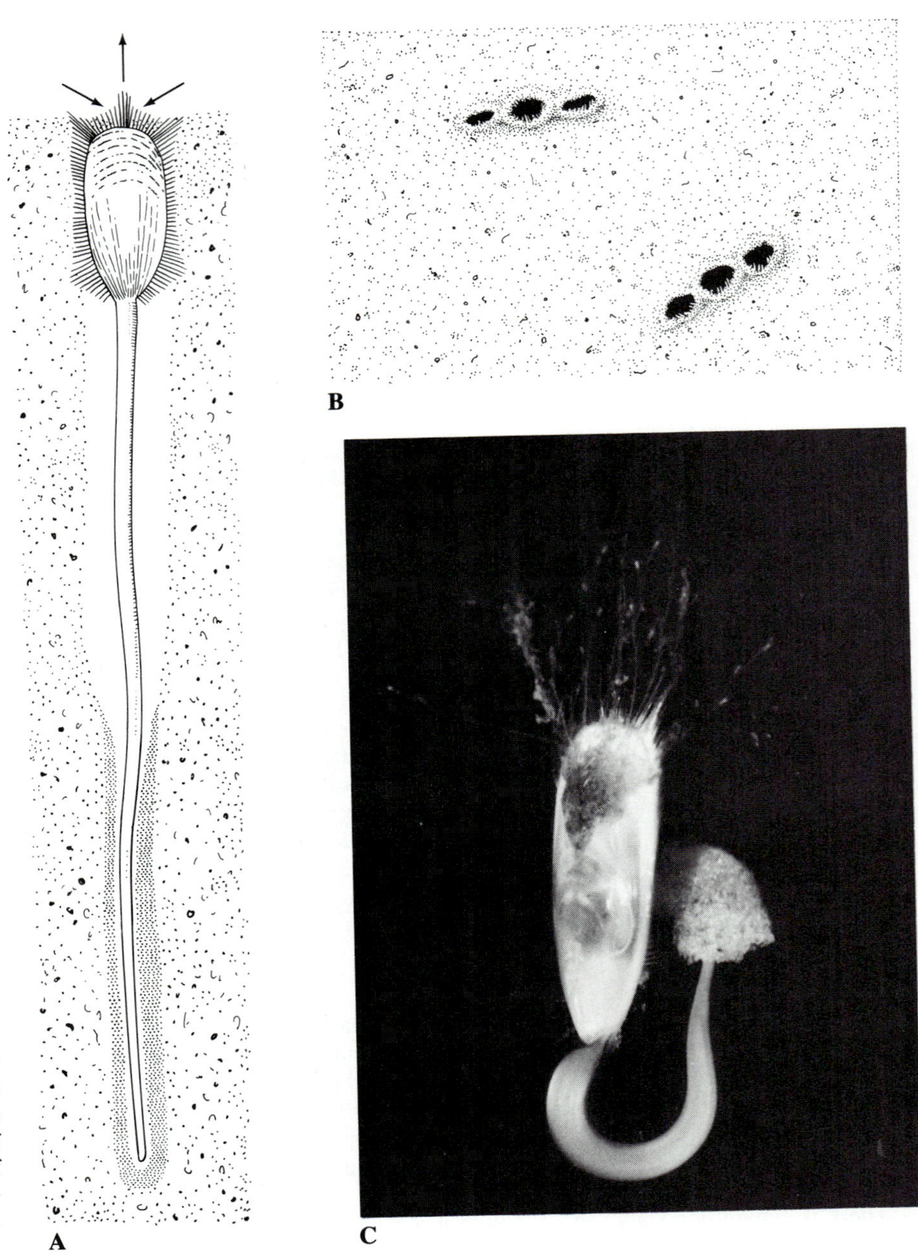

Figure 14.9 Class Inarticulata. A. *Lingula* in its burrow. (After François, Archives de Zoologie Expérimentale et Générale, series 2, 9.) B. *Lingula,* openings of two burrows at the surface of muddy sand. C. *Glottidia pyramidata.* (Photograph by Robert T. Paine.)

phore and its tentacles. Nerves originating from the subenteric ganglion itself reach nearly all other parts of the body.

Classification

Order Lingulida

The only genera in this order are *Lingula* and *Glottidia,* both of which live in vertical burrows in soft substrata. The shell is of the chitinophosphatic type, in which there is considerable chitin and protein, and in which the principal inorganic constituent is calcium phosphate. The two valves of the shell are similar and the long pedicle emerges between them.

Order Acrotretida

The valves of the shell of acrotretids are usually decidedly unequal, the dorsal valve being larger and more convex than the ventral valve. Brachiopods of this order

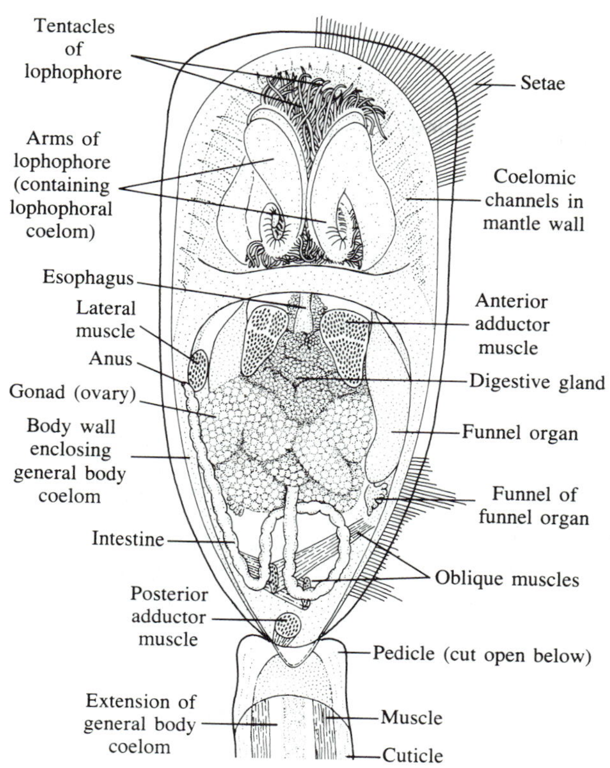

FIGURE 14.10 Class Inarticulata. *Glottidia pyramidata*. The dorsal valve has been removed and the general body cavity opened to show most of the internal organs. Only a small portion of the pedicle is shown, and this has been sliced open. Most of the setae at the edge of the mantle have been omitted.

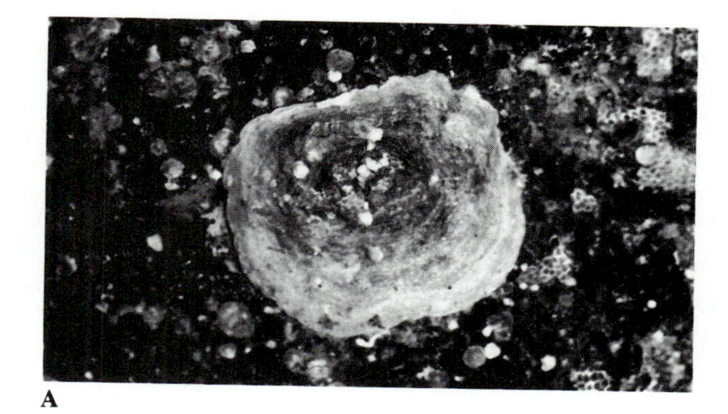

FIGURE 14.11 Class Inarticulata. *Crania californica*. A. Dorsal view. The ventral valve is firmly cemented to the rock. B. Two specimens, anterior view. The shells are gaping and the lophophores are fully extended. Lophophores of some bryozoans can be seen on the shell of the specimen at the right, and also on the substratum in front of it. (Photograph by Michael LaBarbera.)

are attached to hard substrata, with the ventral valve lowermost. In certain genera, including *Discina* and *Discinisca,* the shell is chitinophosphatic, and there is a short pedicle emerging from a notch in the ventral valve. In some other genera, such as *Crania* (Figure 14.11), the shell consists mostly of calcium carbonate and there is no pedicle; the ventral valve is cemented to the substratum, apparently by periostracal material.

REPRODUCTION AND DEVELOPMENT OF BRACHIOPODS

It has already been explained that the gonads of inarticulate brachiopods lie in the main part of the general body coelom, and that the gonads of articulate brachiopods are located in coelomic channels that extend into the two lobes of the mantle. The gametes reach the mantle cavity by way of the funnel organs. Fertilization may take place in the mantle cavity or after the eggs have been shed into the sea. A few articulate brachiopods brood their embryos among the tentacles of the lophophore, or in a pair of special pouches that are specializations of the mantle of the ventral valve, or elsewhere in the mantle cavity. In one genus, *Argyrotheca,* the funnel organs serve as brood chambers, so fertilization is believed to take place in the general body coelom. Brooding species release their young after they have reached a swimming larval stage.

Cleavage is radial, and the blastula (Figure 14.12, A) has a substantial blastocoel. Gastrulation is usually by invagination (Figure 14.12, B). The blastopore closes as it becomes elongated into a groove. In most species whose development has been studied, the mouth and stomodeal ingrowth appear later in the anterior part of the groove. These brachiopods are therefore protostomes. The anus of inarticulates generally opens up afterward and is unrelated to the blastopore. In *Crania,* however, it appears where the blastopore has closed,

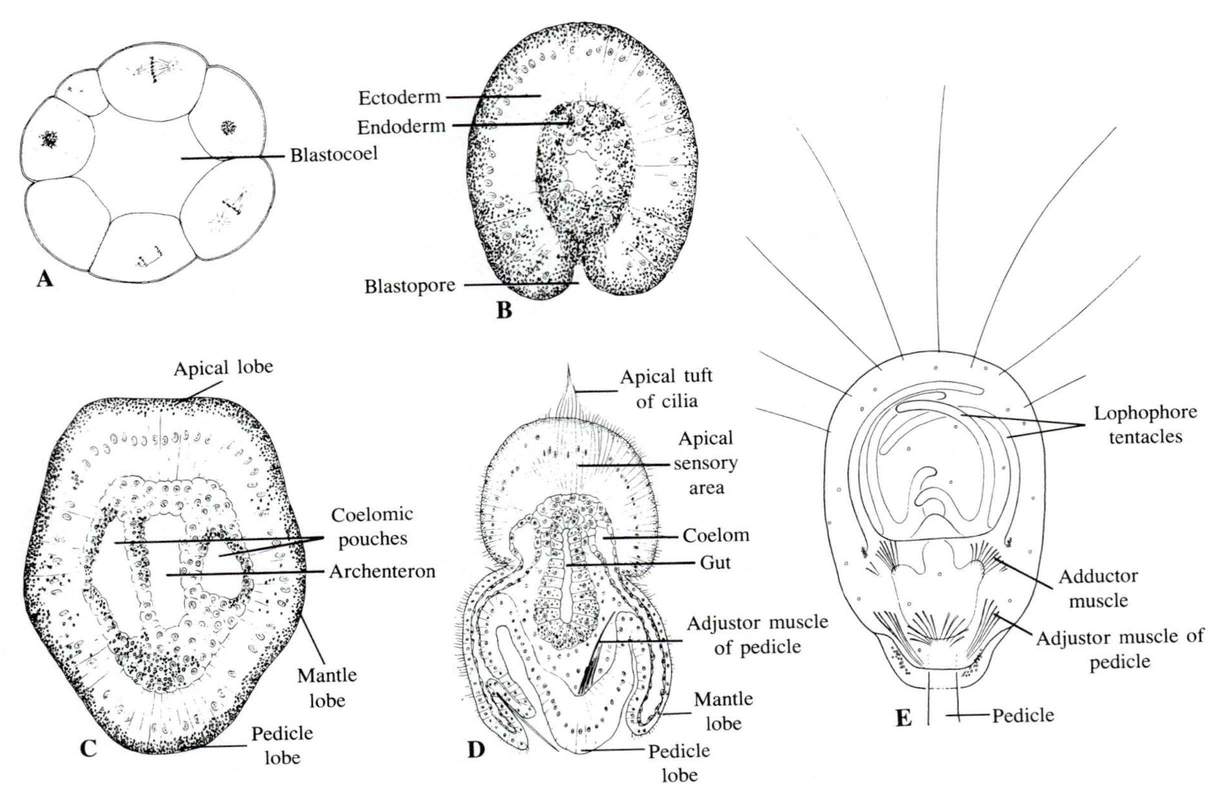

Figure 14.12 Class Articulata. Some stages in development of *Terebratulina*. A. Early blastula, section. B. Gastrula, section. C. Frontal section of a slightly later stage, showing formation of coelomic pockets by vertical partitions that subdivide the archenteron. D. Swimming larva, frontal section. (See also Figure 14.13.) E. Recently metamorphosed individual. (Franzén, Zoologiska Bidrag från Uppsala, *38*.)

and the mouth arises separately. Thus *Crania* may be viewed as a deuterostome. Some other aspects of its development are also unique.

Mesoderm arises from the archenteron. In the inarticulate brachiopod *Lingula,* two lateral masses of cells bud off from it; these subsequently become hollowed out, forming right and left coelomic spaces. Among articulates, the same general thing happens in at least one species of *Terebratulina,* but a rather different method of mesoderm and coelom formation has been reported for another species of this genus, as well as for representatives of certain other genera, including *Terebratalia* and *Hemithiris*. Two sheets of tissue grow down into the archenteron and subdivide it lengthwise into three cavities (Figure 14.12, C). This happens about the time the blastopore closes. The middle cavity becomes part of the definitive gut, whereas the two lateral cavities, lined by mesoderm, are embryonic coeloms. The anteriormost portion of the archenteron is also cut off by a fold of tissue that is set at right angles to those just described. This creates another coelomic space on the anterior side of the embryo, and this space becomes continuous with the lateral pockets. Some other modes of mesoderm and coelom formation have been described for various articulates, but they need to be reexamined.

In both classes, there is a planktonic larval stage. The larva of articulate brachiopods (Figure 14.12, D; Figure 14.13) has no mouth and cannot feed; it subsists on yolk. It consists of three parts. The anterior portion, called the *apical lobe,* is ciliated, and it may have eyespots. Some of the cilia at the tip of it are longer than the others and form a sensory tuft. The middle portion of the larva, called the *mantle lobe,* may be generally ciliated for a time, but eventually the ciliation becomes restricted to a midventral tract (Figure 14.13). The mantle lobe grows backward, becoming somewhat cup-shaped, and sometimes nearly covers the last part of the body, termed the *pedicle lobe*. This is the portion by which the larva becomes attached when it settles, and it develops into the pedicle of the adult (Figure 14.12, E).

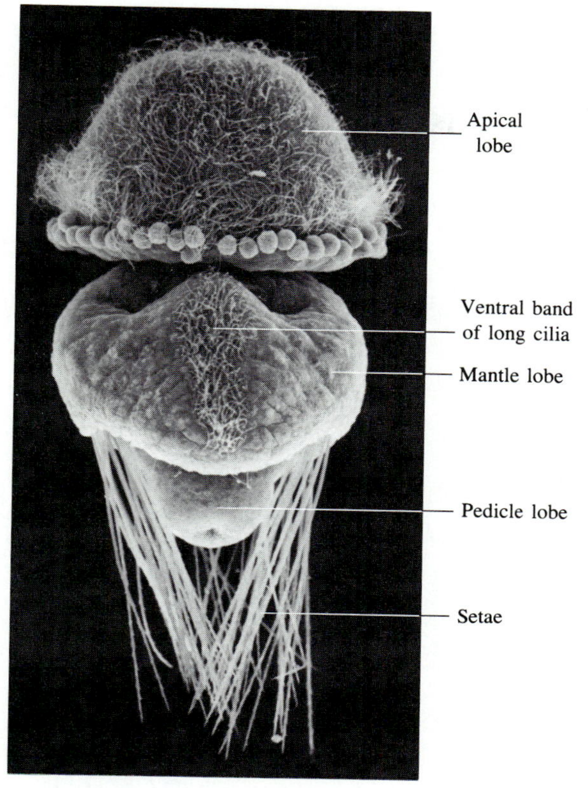

FIGURE 14.13 Class Articulata. *Terebratalia transversa,* swimming larva, ventral view. (Scanning electron micrograph by Christopher G. Reed.)

During metamorphosis, the mantle lobe of the larva divides into a dorsal and a ventral part. These turn forward to envelop the anterior portion of the young brachiopod, and they now represent the mantle lobes of the adult, which secrete the valves of the shell. The apical lobe, now covered, differentiates into the lophophore and much of the visceral part of the body. The animal becomes reoriented so that the dorsal valve is closer than the ventral valve to the substratum.

In the inarticulate *Lingula,* whose development has been extensively studied, the embryo divides into two portions. The anterior portion, as in articulates, differentiates into the lophophore and much of the visceral part of the body. The posterior portion produces dorsal and ventral mantle lobes, and these grow forward to enclose the anterior portion. They also start to secrete the valves of the shell while the young animal is still in the plankton. The pedicle originates from the ventral lobe and lengthens appreciably before the odd-looking, shelled juvenile settles on an appropriate sandy substratum. Because the planktonic juvenile of *Lingula* already shows the general body form of the adult, there is no striking metamorphosis comparable to that undergone by the larva of an articulate brachiopod. The development of *Glottidia* (Figure 14.14) is similar to that of *Lingula*.

The two coelomic cavities of adult brachiopods, though often called mesocoel and metacoel, should not be forced into homology with the mesocoel and metacoel of phoronids. In articulate brachiopods, the origin of the coelomic pockets, and the fates of various portions of these pockets, vary so much that no consistent pattern emerges. It is probably fair to say, however, that after the right and left embryonic coeloms become joined, there is really only one cavity that serves the lophophore, visceral mass, mantle, and pedicle. The terms *mesocoel* and *metacoel* may be helpful in describing the internal anatomy of brachiopods, but they must not be taken too seriously.

REFERENCES ON BRACHIOPODA

Brunton, C. H. C., and Curry, G. B., 1979. British Brachiopods. Synopses of the British Fauna, no. 17. Linnean Society of London, Estuarine and Brackish-Water Sciences Association, and Academic Press, London and New York.

Cooper, G. A., 1977. Brachiopods from the Caribbean Sea and Adjacent Waters. Studies in Tropical Oceanography, no. 14. University of Miami Press, Coral Gables, Florida.

Franzén, Å., 1969. On larval development and metamorphosis in *Terebratulina,* Brachiopoda. Zoologiska Bidrag från Uppsala, *38*:155–178.

Gutmann, W. F., Vogel, K., and Zorn, H., 1978. Brachiopods: biomechanical interdependencies governing their origin and phylogeny. Science, *199*:890–893.

LaBarbera, M., 1977. Brachiopod orientation to water movement. 1. Theory, laboratory behavior, and field orientations. Paleobiology, *3*:270–287.

LaBarbera, M., 1978. Brachiopod orientation to water movement: functional morphology. Lethaia, *11*:67–89.

Paine, R. T., 1963. Ecology of the brachiopod *Glottidia pyramidata*. Ecological Monographs, *33*:255–280.

Reed, C. G., and Cloney, R. A., 1977. Brachiopod tentacles: ultrastructure and functional significance of the connective tissue and myoepithelial cells in *Terebratalia*. Cell and Tissue Research, *185*:17–42.

Richardson, J. R., 1979. Pedicle structure of articulate brachiopods. Journal of the Royal Society of New Zealand, *9*:415–436.

Rudwick, M. J. S., 1970. Living and Fossil Brachiopods. Hutchinson University Library, London.

Steele-Petrovic, H. M., 1976. Brachiopod food and feeding processes. Paleontology (London), *19*:417–436.

Stricker, S. A., and Reed, C. G., 1985. The ontogeny of shell secretion in *Terebratalia transversa* (Brachiopoda, Articulata) I. Development of the mantle. Journal of Morphology, *183*:233–250.

Stricker, S. A., and Reed, C. G., 1985. The ontogeny of shell secretion in *Terebratalia transversa* (Brachiopoda, Articulata) II. Formation of the protegulum and juvenile shell. Journal of Morphology, *183*:251–271.

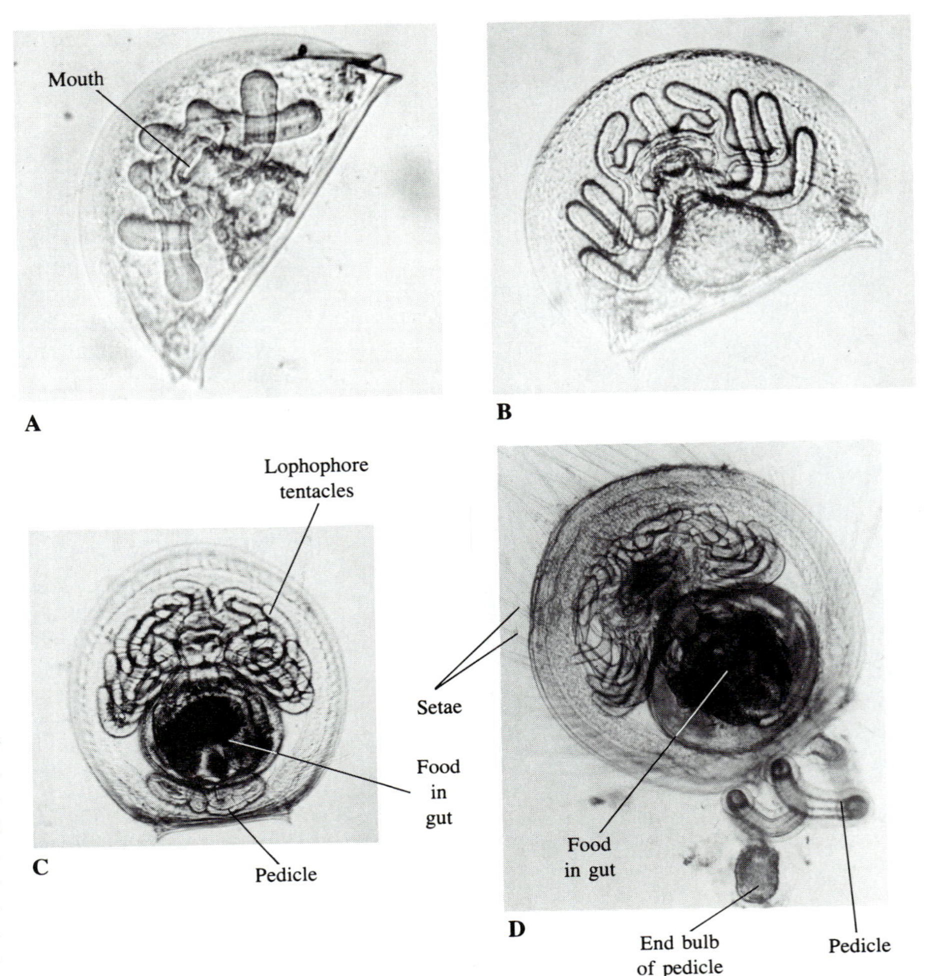

FIGURE 14.14 Class Inarticulata. Stages in the development of *Glottidia pyramidata*. A. Planktonic larva about 7 days after fertilization of the egg. B. Larva about 12 days after fertilization. C. Larva just prior to settlement, about 20 days after fertilization. D. Recently settled individual. (Photomicrographs by Robert T. Paine.)

Stricker, S. A., and Reed, C. G., 1985. Development of the pedicle in the articulate brachiopod *Terebratalia transversa* (Brachiopoda, Terebratulida). Zoomorphology, *105:* 253–264.

Suchanek, T. H., and Levinton, J., 1974. Articulate brachiopod food. Journal of Paleontology, *48:*1–5.

Thayer, C. W., and Steele-Petrovic, H. M., 1975. Burrowing of the lingulid brachiopod *Glottidia pyramidata:* its ecologic and paleoecologic significance. Lethaia, *8:*209–221.

PHYLUM BRYOZOA (ECTOPROCTA)

INTRODUCTION

The Bryozoa, with about 4000 living species, is by far the largest and most successful of the three phyla of lophophorate animals. The group almost certainly dates back to the Cambrian period, or even earlier, and its numerical strength was established by the Ordovician period, about 500 million years ago. Bryozoans have left an impressive fossil record; approximately 15,000 species have been described by paleontologists.

Nearly all bryozoans are colonial, and the phylum name alludes to the mosslike appearance of certain types. Another name for the group is Ectoprocta, which refers to the fact that the anus is outside the circular or horseshoe-shaped lophophore. It is preferred by many zoologists, mostly because Bryozoa formerly included some animals that are now placed in a separate phylum, Entoprocta, in which the anus is within the circle of tentacles. It seems better, however, to conserve the old name Bryozoa than to risk confusing Entoprocta and Ectoprocta because of their similar spellings.

Bryozoans, along with sponges, hydroids, barnacles, ascidians, and some other marine invertebrates, are important fouling organisms on ship bottoms, pilings, buoys, and floating docks. Fouling organisms may not by themselves damage the structures they colonize, but they contribute to a biological coating that prepares the substratum for attack by boring organisms, such as the isopod *Limnoria* or the bivalve molluscs called shipworms.

General Structure

Unlike brachiopods and most phoronids, members of a bryozoan colony are microscopic, or nearly so. The colonies are built up by asexual reproduction. They may form thin crusts or lumpy growths that resemble those of some corals, or they may branch from creeping stolons or upright stems so that they look like certain hydroids. Each member of a colony is called a **zooid.** A zooid secretes a covering around itself, and the texture of the colony depends largely on the nature of the secreted material, which is generally a protein associated with one or more mucopolysaccharides. Among the mucopolysaccharides produced by bryozoans is a substance closely related to chitin. Some bryozoan colonies are soft and jellylike, other are almost leathery; in most types, however, the coverings secreted by the zooids have the texture of celluloid or clear plastic, or they are calcified to the extent that they are partly or completely opaque.

In some species there are only a few zooids in a colony, but in others asexual reproduction continues until there are thousands. Not all zooids are necessarily involved in feeding. Many bryozoans have zooids specialized for other functions, and these zooids must get their nourishment from the feeding zooids. Avicularia are zooids specialized for pinching, and are believed to discourage unwanted settlers and certain types of predators. Vibracula have long vibrating bristles that are known, in some species, to keep the colony free of sediment and to enable a colony to unbury itself. Another type of zooid, the ooecium, is concerned with brooding of embryos. In a few bryozoans the stolons by which the colony is anchored to a solid object consist of modified zooids. Small zooids whose function is unknown have also been observed in the colonies of certain species.

At this point it is necessary to introduce some rather complicated terminology and to insist on its being used correctly. Each zooid's body wall, often boxlike or tubular, consists of a layer of peritoneum, epidermis, and the secreted covering. Collectively, these layers form what is called a **cystid** (Figure 14.15). The term **zooecium** ("zooid's house") is often applied to it, but its use should be restricted to the secreted covering.

Most of the rest of the animal is called a **polypide.** This term covers all "movable parts": the lophophore

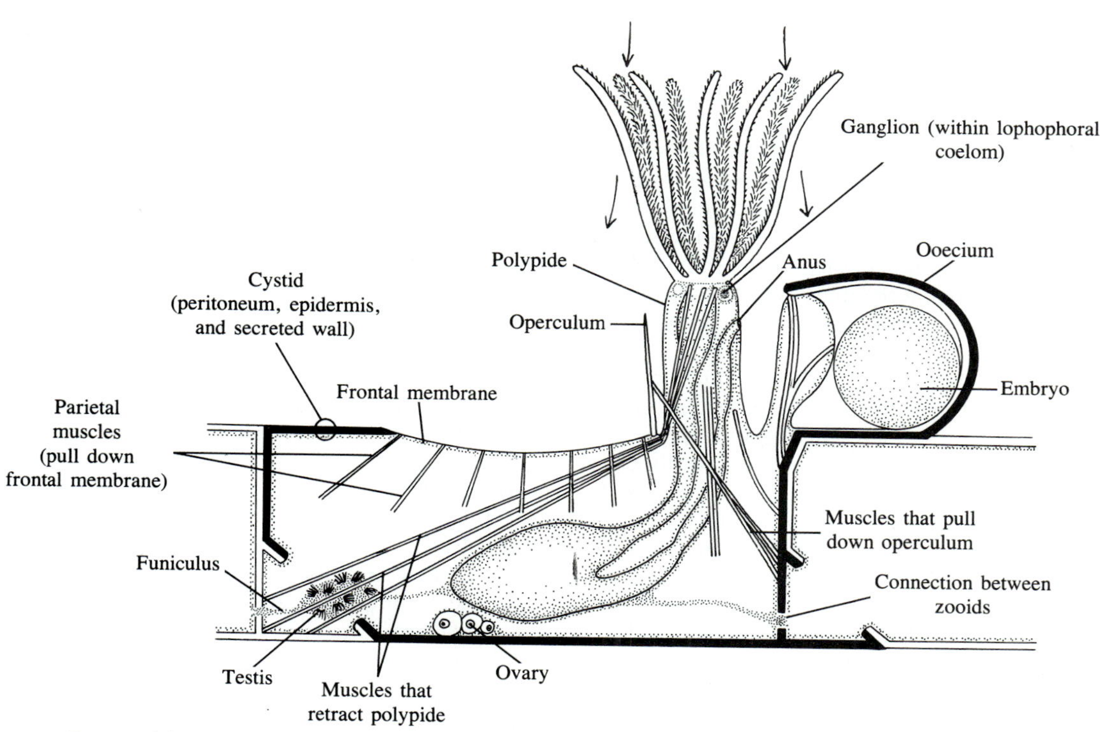

Figure 14.15 Order Cheilostomata. Diagram of a feeding zooid, portion of the zooid distal to it, and an ooecium produced by the latter. (After Ryland, Bryozoans, but slightly modified.)

and the "neck" region, the gut and tissues around the gut, and the muscles that retract the lophophore. Excluded from the polypide are the **funiculus,** which is a tube of peritoneal cells that connects the polypide to the body wall, and the muscles that run from one part of the body wall to another. Most of the muscles, whether considered part of the polypide or not, are lined by peritoneum. Gametes originate from the peritoneum that lines the inside of the cyst, that covers the polypide, or that forms the funiculus. In some bryozoans, pores in the cystids allow for tissue continuity and also permit coelomic fluid to circulate between zooids. Furthermore, branches of the funiculus of a zooid may be joined to those of adjacent zooids. Such connections may be traced back to the time when one zooid was budding off another.

Feeding in bryozoans depends primarily on ciliary activity, which creates water currents that pass downwards through the lophophore, and which also move particles of food toward the mouth. In some species, however, flicking motions of the tentacles are important in driving food to the mouth. *Bugula neritina,* a common fouling organism in harbors, is able to trap small animal organisms of the plankton by bending the tips of its tentacles inward to form a cage. Perhaps other distinctive methods of feeding will be discovered in this large phylum of invertebrates.

The gut has muscle, but ciliary activity accounts for much of the movement of food. In most bryozoans, the gut has four general regions: pharynx, esophagus, stomach, and intestine. Digestion, some of which is intracellular, and absorption take place in the stomach. The middle portion of the stomach usually forms a deep caecum, and it is here that most of the intracellular digestion proceeds. In a few genera, including *Bowerbankia,* the part of the stomach just before the caecum is specialized as a **gizzard,** which is strongly muscularized and has a sclerotized and toothed inner lining. Fecal wastes are consolidated in the intestine. As they emerge from the anus, they are carried away by currents leaving the lophophore.

Bryozoans have two body cavities, which are true coeloms in that they are lined by a peritoneum. One of them forms a ring at the base of the lophophore and extends into the tentacles. The other, much larger, is the principal body cavity. The two coeloms are rarely, if ever, completely separate. The terms *mesocoel* and *metacoel* are often applied to the body cavities of bryozoans, as they are to the two coelomic cavities of brachiopods, but they are misleading because they imply homology of these cavities with the mesocoel and metacoel of phoronids. The cavities are at best only analogous, as is explained below in connection with development of gymnolaemate bryozoans.

The surface area of the lophophore tentacles is extensive in proportion to the size of the zooid, and the cilia on the epidermis create currents that promote exchange of oxygen and carbon dioxide as well as elimination of soluble nitrogenous wastes. The coelomic fluid, sloshed around whenever a zooid is withdrawn or extended, is believed to distribute oxygen and nutrients through the body. Although bryozoans do not have a circulatory system that pervades all tissues, the results of recent studies of the funiculus suggest that this structure is comparable to a blood vessel. Its wall consists of peritoneal cells, and when a distinct basal lamina can be seen, this is on the inner faces of the cells, an arrangement typical of blood vessels not only of phoronids and brachiopods but also of most other invertebrates. In phylactolaemate and cyclostomatous bryozoans, muscle fibers within the funicular tube are tightly joined to the basal lamina. It appears probable that the funiculus, which extends to the intestinal region of the gut, distributes nutrients to other structures, such as the testis.

There are no excretory organs in any marine bryozoans, although these animals have ways of packaging up waste materials. Some of the phylactolaemate bryozoans, which are found in fresh water, have structures that have been interpreted as funnel organs, discussed below in connection with the morphology of phylactolaemates.

The exact arrangement of the nervous system has been studied in relatively few bryozoans. Its organization varies from group to group but remains basically the same; in the interest of keeping things simple, it seems best to deal with it in general terms. There is a ganglion at the base of the lophophore, between the mouth and anus. The ganglion may actually lie within the lophophoral coelom, in which case it is enclosed by a sac of peritoneal tissue. In most bryozoans the ganglion is simply a prominent swelling on a ring of nerve tissue from which branches run into the tentacles and other parts of the polypide and body wall. In phylactolaemates, however, which have a horseshoe-shaped lophophore, the ganglion sends off two main trunks, one going into each half of the horseshoe. There are other important nerves in phylactolaemates, including a pair that form a ring around the pharynx.

In certain bryozoans, and possibly in most of them, nerves pass through pores in the cystids, thereby connecting the nervous system of one zooid with that of its neighbors. This has been demonstrated by histological studies and also by experiments. If a mechanical or electrical stimulus is applied to the surface of a zooid of *Membranipora* or *Electra,* the withdrawal of the lophophore of that zooid is followed immediately by withdrawal of lophophores of other zooids in the same general area of the colony. The response cannot be made to spread throughout the colony, however, and application of a stimulus to the tentacles of a lophophore results in withdrawal of only that lophophore. Nevertheless, the withdrawal responses of numerous zooids after a predator has contacted just a few of them would be important in survival of the colony. The speed of conduction through the colonial nervous system is about 1 m/second.

CLASSIFICATION

The three classes of bryozoans are so different that each must be discussed in some detail to avoid a homogenized and misleading view of the phylum. The class names are descriptive; each stresses a structural feature that may be helpful in remembering the derivations. Phylactolaemata means "guarded throat," and refers to the presence, next to the mouth, of a flap of tissue, the epistome. In the Gymnolaemata ("naked throat") an epistome is absent. Members of the class Stenolaemata ("narrow throat") also lack an epistome, but their name refers not to the mouth region but to the slender neck of the zooid, below the lophophore.

CLASS GYMNOLAEMATA

About three-fourths of all living bryozoans belong to the Gymnolaemata. The group is more diversified than either of the other classes and shows some remarkable specializations. The two orders are sufficiently different that they are discussed separately.

Order Ctenostomata

Ctenostomata means "toothed mouth," but it actually refers to a pleated collar around the neck of the everted portion of the polypide. When the lophophore and neck region are withdrawn, the collar is also pulled down into the cystid and folds to form an effective closing device.

Bowerbankia

Bowerbankia (Figure 14.16, A and B), a common marine and brackish-water genus, illustrates well the main features of the order. Its branching stolons, consisting of specialized, nonfeeding zooids, are attached to worm tubes, hydroids, shells, ascidians, and other firm substrata. These zooids not only produce more of their own kind, but also produce buds that develop into feeding zooids. The cystid of a feeding zooid is shaped something like a slender cucumber. If the lophophore and neck of the polypide are fully extended, the pleated collar will be readily visible. The mouth, surrounded by the bases of about 10 ciliated tentacles, leads to a slightly dilated pharynx. This is succeeded by a slender esophagus and a large stomach, most of which consists of the caecum, but part of which is differentiated into a gizzard. The intestine leads to the anus, located on one side of the neck region.

When the retractor muscles that extend from the body wall to the base of the lophophore contract, the lophophore and neck region are pulled down into the cystid. The collar is drawn in by the longitudinal parietal muscles, and contraction of the sphincter muscle causes the pleats of the collar to fold and form a tight closure for the cavity that now encloses the lophophore and neck.

Withdrawal of the lophophore and associated structures increases the pressure in the body cavity, and the cystid wall is not elastic enough to bulge more than slightly, if at all. The pressure probably remains high, so that if the various retractor muscles merely relax, the withdrawn parts will start to emerge. Complete extension of the lophophore, neck, and inverted portion of the body wall depends on the two sets of transverse parietal muscles. They run from one side of the body wall to the other and their contraction, by diminishing the diameter of the cylinder, provides the necessary increase in hydrostatic pressure.

The funiculus extends right through the body wall and is continuous with the funiculus of the modified zooids that form the creeping stolon. In other words, each time a new zooid is formed, whether its role is to feed or to be part of the stolon, it remains connected to its parent.

Bowerbankia, like most bryozoans, is hermaphroditic. The ovary and testis originate from the peritoneum of the body wall. An egg, when ready for fertilization, is large, and escapes through a pore between two tentacles. It is brooded in the cavity formed by inversion of the upper part of the body wall, a normal occurrence when the lophophore is withdrawn, as has been explained previously. The lophophore, pleated collar, and much of the rest of the polypide degenerate, increasing the space available for brooding. The swimming larva into which the embryo develops cannot feed and must soon settle and metamorphose into the founder zooid of another colony. The pattern of brooding shown by *Bowerbankia* is found also in other ctenostomes, but different arrangements—including brooding in the coelom itself—have been reported.

Other Ctenostomes

Bowerbankia is only one of approximately 40 genera of ctenostomes, and not everything that has just been said about it applies to all other members of the order. Some ctenostomes, for instance, do not even have the pleated collar that gives the group its name. Several genera form fleshy or cartilaginous growths in which the cystids are united, as in *Flustrellidra* (Figure 14.17, A) and *Alcyonidium* (Figure 14.17, B). These bryozoans are sometimes mistaken for seaweeds.

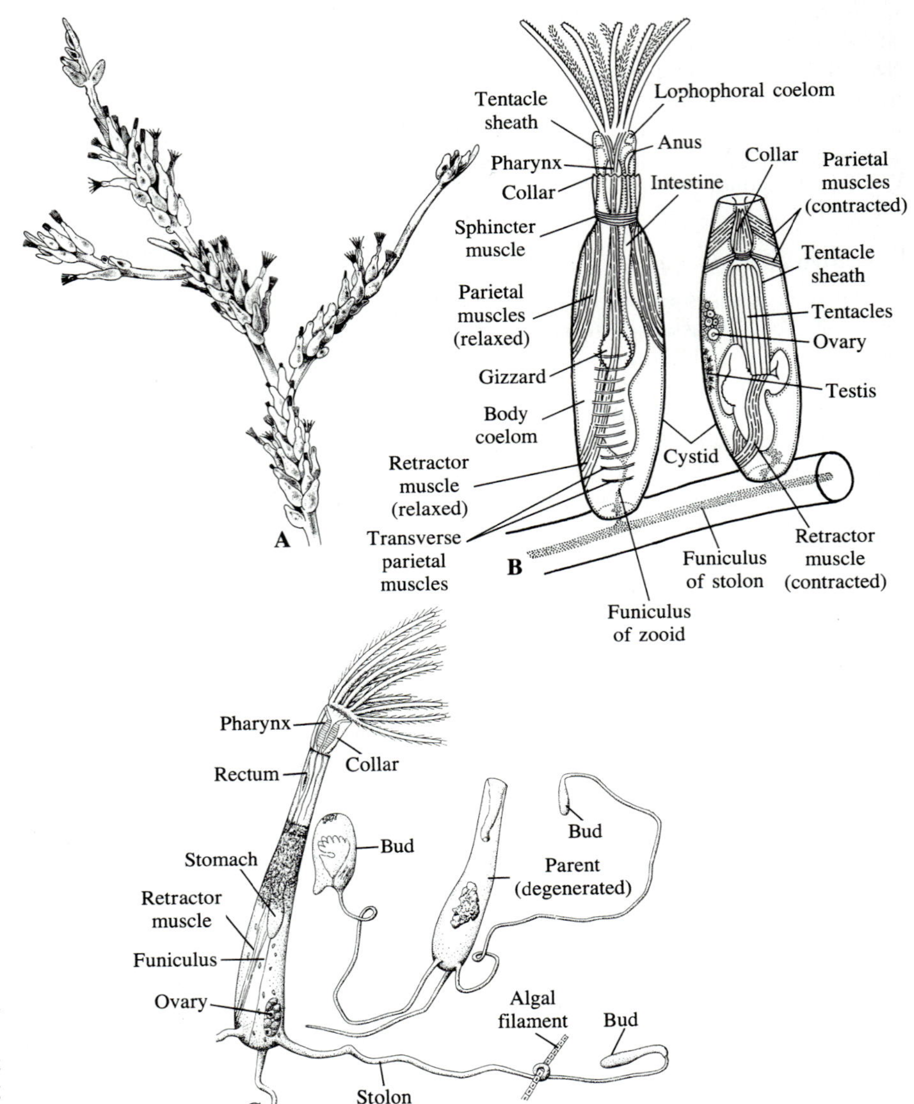

FIGURE 14.16 Order Ctenostomata. A. *Bowerbankia imbricata,* portion of a colony. (Prenant and Bobin, Faune de France, *60*.) B. *Bowerbankia*. Two zooids, one with its lophophore extended, the other with its lophophore withdrawn. Transverse parietal muscles are shown only in the former. (Mostly after Ryland, Bryozoans.) C. *Monobryozoon limicola*. After the buds produced on the stolons reach a certain stage, the parent dies. (Franzén, Zoologiska Bidrag från Uppsala, *35*.)

An especially remarkable ctenostome is *Monobryozoon* (Figure 14.16, C), which adheres to sand grains. Its colony usually has only two zooids, one for feeding and one for attachment.

Order Cheilostomata

In cheilostomatous bryozoans, the opening through which the lophophore is extended or withdrawn is provided with a little trapdoor, called the **operculum** (Figure 14.15). This is the basis for the name of the group, which means "lipped mouth," even if the opening referred to is not the real mouth. The operculum is pulled shut by muscles, but the way it is hinged to the edge of the cystid enables it to swing open when pressure in the body cavity is increased and the lophophore is extended. In some cheilostomes small muscles are also involved in the opening process.

There are over 600 genera of cheilostomes, and the group is so diversified that description of just one or two types will not provide a proper introduction to it. The several genera briefly discussed here illustrate a wide variety of specializations in the structure and function of zooids, and also some of the patterns of reproduction and colony formation.

Membranipora and *Bugula*

Membranipora (Figure 14.18) forms circular or irregular encrustations, especially on blades of large kelps. Asexual reproduction by the two founder zooids

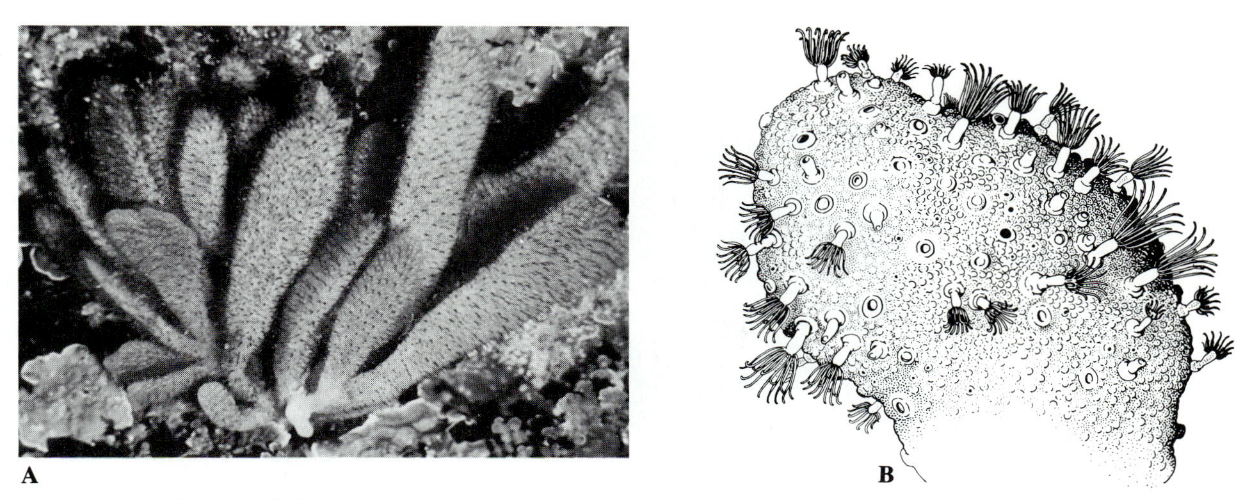

FIGURE 14.17 Order Ctenostomata. A. *Flustrellidra corniculata*. B. *Alcyonidium polyoum*. (Prenant and Bobin, Faune de France, *60*.)

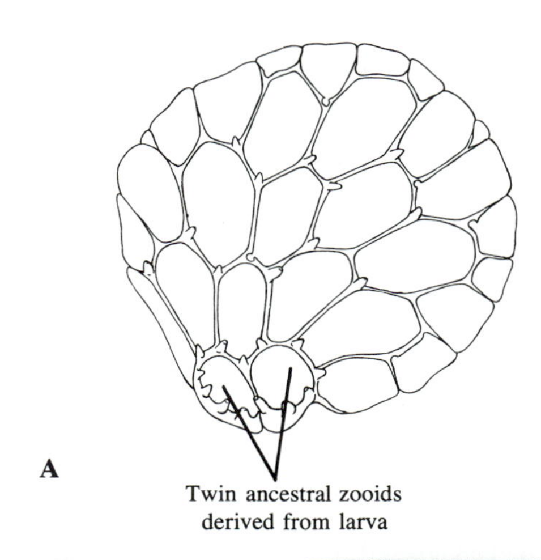

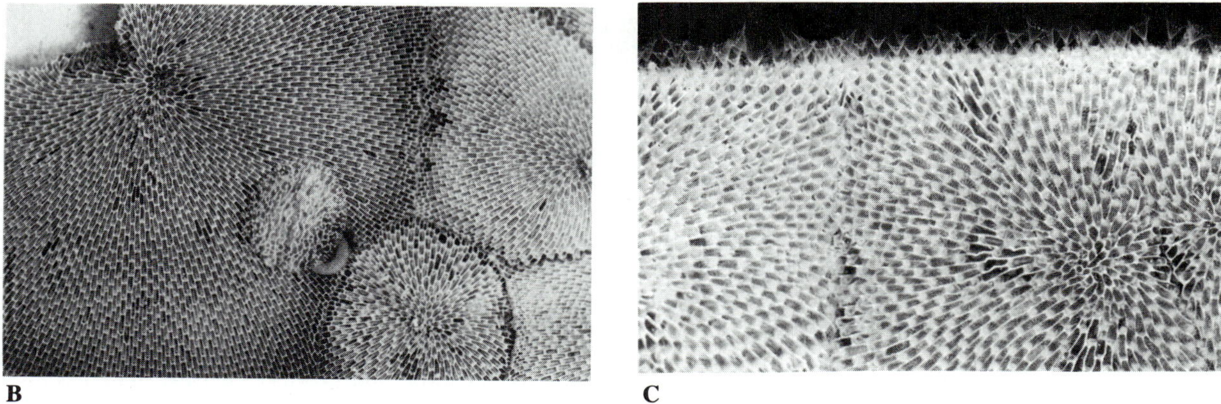

FIGURE 14.18 Order Cheilostomata. *Membranipora membrancea*. A. Several contiguous colonies. The organism in the center is a nudibranch gastropod, *Doridella steinbergae,* which nips off the lophophores. Its pattern of light lines resembles that produced by the calcified walls of the bryozoan. The crescent-shaped structure at the right of *Doridella* is its egg mass. B. Closer view of a colony. The lophophores of the zooids at the edge of the colony are easily seen. C. Young colony. The zooids can be traced back to the two ancestral zooids that develop from the swimming larva. (Lutaud, Annales de la Société Royale Zoologique de Belgique, *90*.)

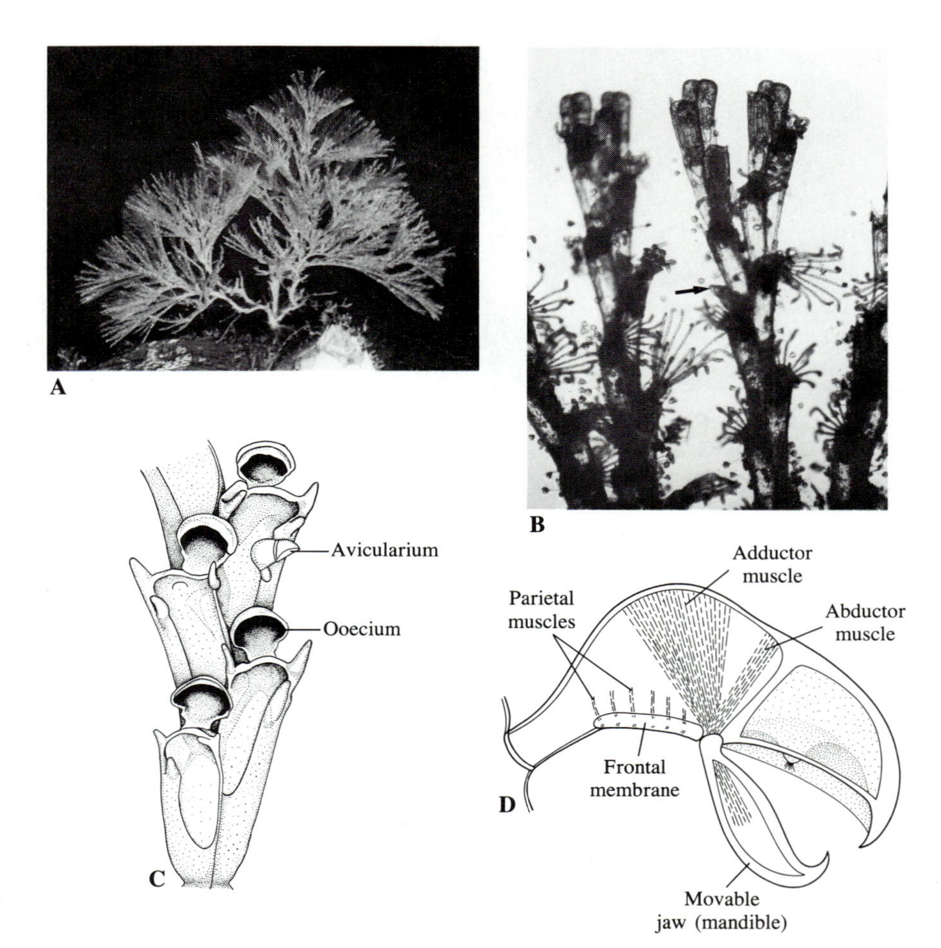

Figure 14.19 Order Cheilostomata. *Bugula*. A. Well developed colony. B. Portion of a colony, showing some lophophores extended and one avicularium (arrow); photomicrograph. C. Portion of a colony, lophophores withdrawn, showing an avicularium and four ooecia. (Rogick and Croasdale, Biological Bulletin, *96*.) D. An avicularium, showing musculature and other structures.

(Figure 14.18, C) is repeated until the radiating lines of zooids have branched many times. A large colony may be several centimeters across. The zooids are all of one type and are concerned with both feeding and reproduction. Each occupies a boxlike cystid whose four side walls are whitened by calcification, so that when the colony is examined under low magnification it resembles lacework. The bottom of each box is in tight contact with the surface of the kelp. The top is mostly covered by a flexible membrane, called the **frontal membrane,** but at one end there is an opening through which the lophophore of the zooid may be extended. When a lophophore is withdrawn, its opening is closed by the operculum.

The lophophore and neck region of the zooids pop in when disturbed, but soon pop out again to resume their feeding posture. The withdrawal is effected by the retractor muscles that pull the extended part of the animal back into its house. These muscles, which run from near the base of the lophophore to the body wall, shorten at a speed of over 20 times their length per second. This is not equaled by any other muscles—vertebrate or invertebrate—whose speed of contraction has been measured. When the lophophore is suddenly withdrawn, the pressure in the body cavity is raised. To compensate for this, the frontal membrane that covers the top of the cystid bulges slightly, and the pressure inside and outside the body cavity is thereby equalized. When the membrane is pulled down by contraction of muscles that run between it and other parts of the body wall (Figure 14.20, A), pressure within the body cavity is again raised and the lophophore is thrust out.

The expanded lophophores of *Membranipora* and most other encrusting bryozoans are nearly contiguous and form an efficient filtering mechanism for collecting diatoms, bacteria, and other small food particles. Individual lophophores may bend and swivel to test the water and orient themselves to make the most of an oncoming current.

Bugula (Figure 14.19) is usually abundant on pilings, floating docks, rocks, and seaweeds. Its bushy growths are slightly spiralled. As a colony grows, many of the older zooids die, leaving empty zooecia. An individual feeding zooid is much like that of *Membranipora,* except that it is not so nearly rectangular and is not stuck down to the substratum, and there is no operculum associated with the opening through which the lophophore is extended and withdrawn. Much of the cystid is calcified, but a portion that is not hardened can bulge, as it does in *Membranipora,* when pressure within the cystid is increased.

Here and there among the feeding zooids of *Bugula* are structures that look like bird beaks on short stalks. These are the **avicularia** (Figure 14.19, B–D), zooids specialized for pinching and thus capturing or discouraging settlers. The movable jaw of an avicularium, called the *mandible,* is a modified operculum; the jaw that cannot move is what is left of the rest of the cystid. A well-developed adductor muscle pulls the mandible against the immovable jaw, and an abductor muscle pulls it away. There is even a small uncalcified membrane that can bulge to compensate for small changes in pressure within the body cavity of the avicularium. A tuft of sensory cells with long cilia is the only external remnant of the zooid, which has become so specialized that it cannot feed. An avicularium is totally dependent on the feeding zooid that produced it, and is connected to the latter by a strand of tissue.

Another type of specialized zooid found in *Bugula* is the **ooecium** (also called *ovicell*), within which an embryo develops. It may appear to be merely a cuplike elaboration of a feeding zooid, but it is in fact a separate entity. Its origin and the way it functions are discussed below in the section on sexual reproduction and development of cheilostomes.

Bugula exhibits a phenomenon observed in various gymnostomes: the formation of **brown bodies.** A brown body is produced when a polypide degenerates, leaving a ball of tissue that consists mostly of remains of the stomach. Because the remains form a compact mass, the dark color of the inclusions in the cells of the stomach is intensified. The formation of a brown body does not mean that the zooid must die. A completely new polypide may develop from living tissue on the wall of the cystid, and one or more brown bodies may be stored in the body cavity of a functioning zooid. In some bryozoans the brown body becomes incorporated into the lumen of the gut of a new polypide and is then eliminated through the anus.

Although brown bodies were not mentioned in connection with *Bowerbankia* and other ctenostomes, they do occur in that group. In *Bowerbankia,* as a matter of fact, a feeding zooid that has produced a brown body often dies and disappears, and a new zooid may be budded off the stolon at about the same place. In stenolaemates, discussed in the next section of this chapter, the polypide may also undergo degeneration and replacement of zooids, with or without the formation of obvious brown bodies. When phylactolaemate polypides deteriorate, however, they do the job so thoroughly that nothing like a brown body is left; moreover, these polypides are not replaced.

Reversible deterioration of a polypide does undoubtedly enable a zooid to get rid of accumulated wastes and still survive. The phenomenon is probably of wide significance to the individual zooid and to the colony as a whole, but this aspect of the biology of bryozoans has received relatively little attention.

Some Other Genera and Their Specializations

The diversity of cheilostomes is especially evident in structural plans that permit these bryozoans to compensate for pressure changes within the cystid. Recall that in *Membranipora* and *Bugula* a flexible frontal membrane covers all or much of the upper surface of the cystid; this membrane bulges slightly when the polypide is withdrawn and pressure within the cystid is increased.

In most cheilostomes, the cystid is more completely calcified than in *Membranipora* and *Bugula.* In several genera, including *Micropora,* there is a calcified shelf beneath the frontal membrane (Figure 14.20, B). This shelf is interrupted, however, so the frontal membrane responds freely to changes in pressure, just as it does in *Membranipora* and *Bugula.* In another large assemblage of cheilostomes, the flexible membrane is overlain by a porous calcareous covering (Figure 14.20, C). The pores permit water to move in and out, and the functioning of the membrane is not interfered with. The pores such as are seen in *Cribrilina* (Figure 14.21, C) are spaces left when spines that develop along the edges of the upper surface of the cystid become fused. In some species of *Callopora* (Figure 14.21, B), the individual spines are still recognizable, and in others (Figure 14.21, A) they merely overhang the frontal membrane.

In *Porella, Microporella* (Figure 14.21, E), *Schizoporella* (Figure 14.21, D), and their relatives, the calcareous covering over the frontal membrane has only one pore. This **ascopore** is located close to the edge of the orifice of the cystid, or just inside the lip of the aperture. The ascopore leads into a membrane-lined sac, the **ascus.** Muscles homologous to those that insert on the frontal membrane of other cheilostomes insert on the floor of the ascus. When they contract, the space within the ascus is enlarged, so water moves into it through the ascopore. Consequently, pressure on the coelomic fluid is increased, and the polypide is extended. When the polypide is withdrawn, thus raising pressure in the coelom, the muscles inserting on the ascus relax, and water is forced out of this chamber. The ascus therefore functions as a compensation sac.

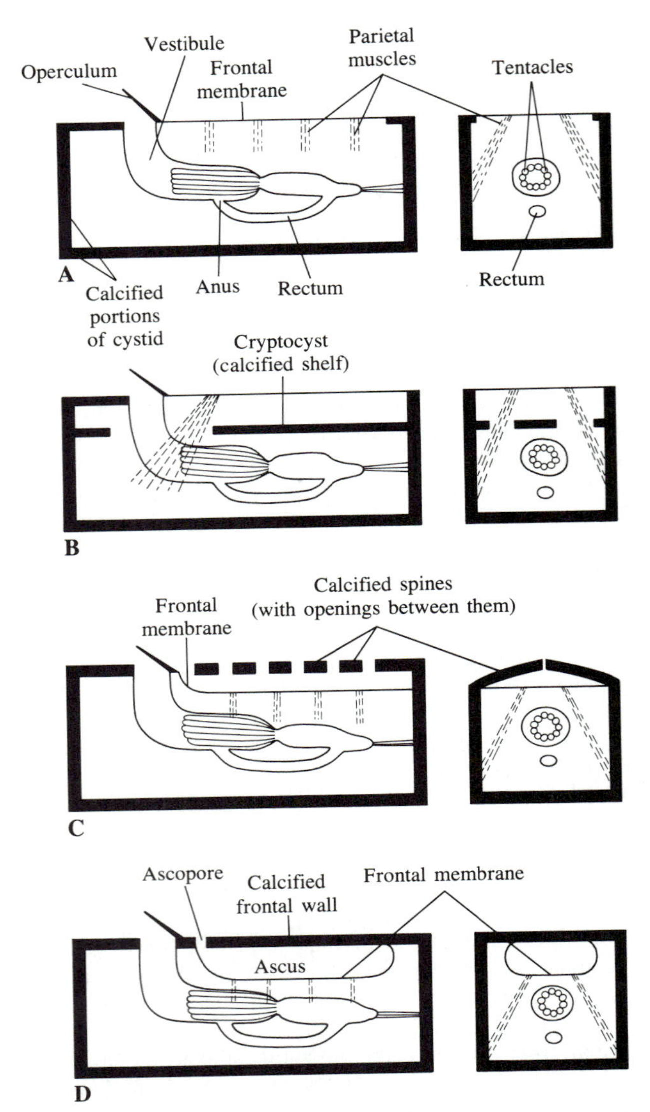

Figure 14.20 Order Cheilostomata: four main arrangements of the frontal membrane with respect to the calcified portion of the cystid. Illustrations at right show the zooids in transverse section. A. Simplest type, such as found in *Membranipora*. The frontal membrane is exposed. When it is pulled down by parietal muscles, pressure in the body cavity increases and the polypide extends. B. Arrangement found in *Micropora* and other genera, with an incomplete calcified shelf (cryptocyst) beneath the frontal membrane. The muscles that pull down the membrane are concentrated in the region where the shelf is interrupted. C. Frontal membrane overlain by calcified spines. These are generally fused so that there narrow slits between them (as in *Callopora*) or only pores (as in *Cribrilina*). D. Ascophoran type, typical of many genera, including *Porella*, *Microporella*, *Schizoporella*, *Mastigophora*, and *Phidolopora* (see Figure 14.21, D–G). The frontal wall is calcified, but an ascopore allows water to enter the sac (ascus) beneath it. The parietal muscles insert on the floor of the ascus.

Many cheilostomes that have an ascus–ascopore complex appear to have other pores as well. These are not really openings, however, for they are covered by a cuticle that is external to the calcified layer, and they are also generally plugged by tissue. The term applied to them is *pseudopore*.

Avicularia are widely distributed among the genera of cheilostomatous bryozoans, although they are not necessarily so conspicuous or elaborate as those of *Bugula*. In many genera they occur singly or in pairs on the feeding zooids, usually not far from the aperture. Frequently they are unstalked, and the movable mandible is flush with the calcified frontal wall of the feeding zooid (Figure 14.21, A, B, D, and E). The function of avicularia of this "trapdoor" type needs to be investigated. In certain species they have been observed to close down on small carnivorous polychaetes that could be predators, but more needs to be learned about their role in maintaining colony health.

A **vibraculum** is a modified zooid with a movable bristle homologous to the operculum of a feeding zooid and to the mandible of an avicularium. As the bristles of the vibracula wave back and forth, they probably discourage settlers and help keep the colony free of sediment. Their cleansing activity has been documented in the case of certain species that form mobile colonies that are likely to become buried in sediment. They also enable a colony to regain the surface after being buried, and even to move over the surface of the sediment.

Vibracula are much less common than avicularia. *Mastigophora* (Figure 14.21, F), *Microporella*, *Scrupocellaria*, and *Cupuladria* are among the genera that have them. Species of *Cupuladria* are unusual in that they lie unattached on sandy bottoms. The zooids are organized into a dome-shaped structure about 2 cm in diameter, their lophophores and vibracula facing outward. Periodically, the zooids cleanse themselves by shedding their sclerotized frontal coverings, as well as the coverings on the inner surface of the dome.

Sexual Reproduction and Development

Origin of gametes, sexual reproduction, and development have been studied in a variety of cheilostome bryozoans, but there are many gaps in our knowledge. As a rule, cheilostomes are simultaneous hermaphrodites; there are, however, exceptions. The gonads originate on the peritoneum, the ovary being usually on the body wall, and the testis being on either the body wall or the funiculus. Spermatogenesis is odd in that a single spermatogonial cell produces 64 sperm. The six synchronous nuclear divisions (the last two of which are meiotic) are not accompanied by cytoplasmic division, and it is not until transformation of spermatids into sperm is well under way that these are pinched off.

As many species produce sperm and ripe oocytes simultaneously, internal self-fertilization is possible and

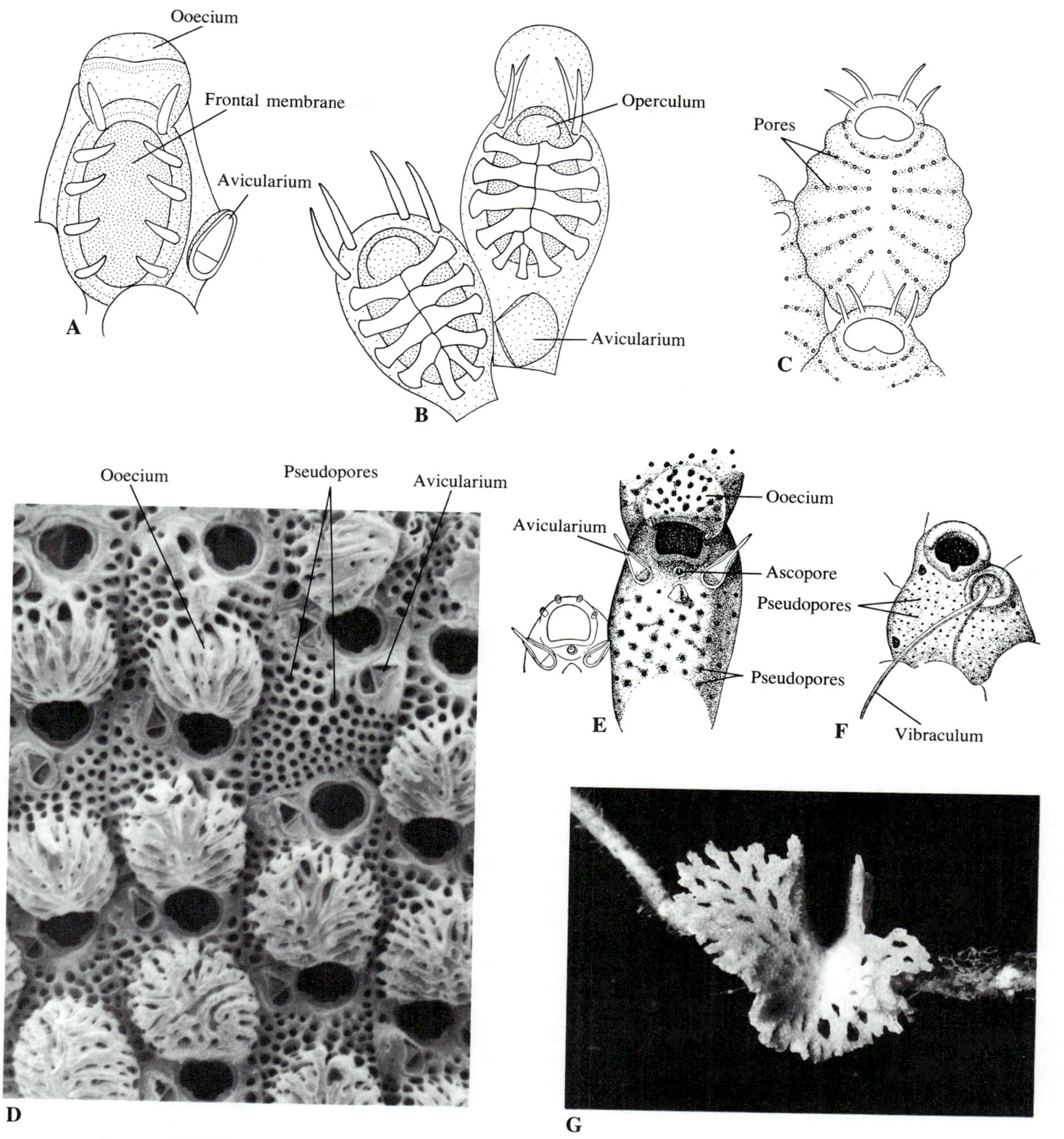

FIGURE 14.21 Order Cheilostomata. Representative genera. A. *Callopora lineata*. B. Another species of *Callopora*. C. *Cribrilina intermedia*. D. *Schizoporella unicornis*. (Scanning electron micrograph by David Denning.) E. *Microporella californica*. F. *Mastigophora porosa*, with a long vibraculum. (E and F, Osburn, Allan Hancock Foundation Pacific Expeditions, *14*.) G. *Phidolopora labiata*. The branches of the colony anastomose, forming a lacy network.

perhaps does take place. This aspect of reproduction needs to be investigated more fully. External cross-fertilization is definitely known to take place in several genera. In *Electra* and *Membranipora*, for example, sperm emerge from pores at the tips of certain tentacles and are swept away by currents leaving the lophophore (Figure 14.22, A). If they reach the lophophore of another individual they may become attached to portions

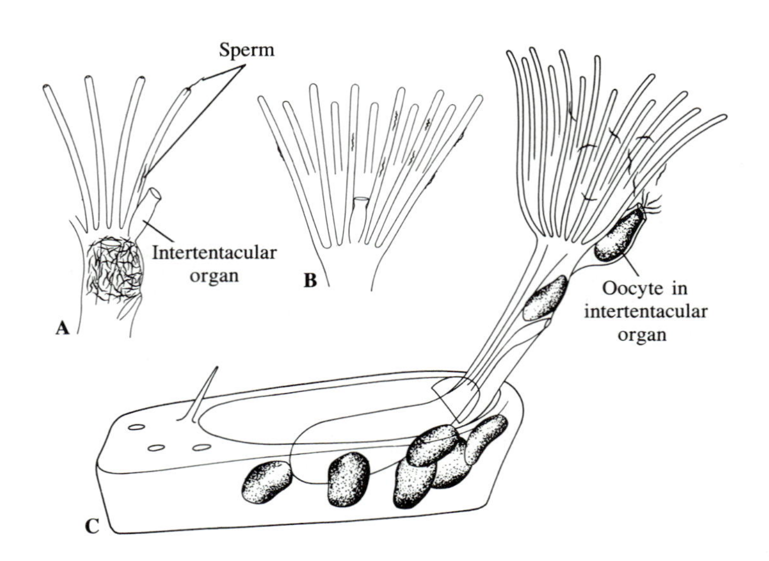

Figure 14.22 Order Cheilostomata. *Electra posidoniae*. A. Portion of the lophophore in lateral view, showing sperm being released from one of the two dorsomedial tentacles. B. Sperm adhering to tentacles of a recipient zooid. C. Lateral view of a zooid, showing several oocytes, one of which is in the intertentacular organ and is attracting sperm. (Silén, Ophelia, *3*.)

of the tentacles where cilia are sparse or absent (Figure 14.22, B). Then, as an oocyte emerges from a pore at the tip of the *intertentacular organ,* a tube between two tentacles, sperm move to it (Figure 14.22, C). The oocyte probably exudes a substance that attracts sperm.

Following entry of the sperm, the oocyte undergoes two successive meiotic divisions, so two polar bodies are produced. The cleavages that follow are radial. By the time the fifth cleavage has been completed, the embryo consists of two to four tiers of cells, but these tiers are not necessarily equal in the number or size of the cells in them. The blastula typically has only a small cavity. Gastrulation is generally accomplished by invagination in cheilostomes such as *Electra* and *Membranipora,* whose eggs are small and whose swimming larvae feed while they are in the plankton. Gastrulation by epiboly and ingression is the rule in species that have large, yolky eggs and swimming larvae that do not feed.

The swimming larva characteristic of *Electra* and *Membranipora* is called the **cyphonautes** (Figure 14.23, A). It is markedly flattened from side-to-side and has a transparent, two-valved shell; it feeds mostly on small planktonic diatoms. The tracts of cilia concerned with locomotion and feeding are shown in the illustration. Some other types of swimming larvae found in cheilostomes are not at all like the cyphonautes. That of *Bugula* (Figure 14.23, B), for instance, is almost completely ciliated, has no shell, and does not feed.

Larvae of bryozoans do not have coelomic sacs comparable to those found in the larvae of phoronids and brachiopods. After a larva settles, it goes through a drastic metamorphosis and extensive cellular reorganization. The origin of mesoderm and the coeloms is obscure. Thus it is best not to homologize the lophophoral coelom and general body coelom of bryozoans with the coeloms of either phoronids or brachiopods.

Ooecia and Brooding

Specialized zooids called ooecia were mentioned in connection with colony structure in *Bugula.* They are, however, characteristic of many genera of cheilostomes.

The function of an ooecium is to brood a developing embryo until the larva is ready to be released. Sometimes, as in *Bugula,* the egg taken up by an ooecium is small and is nourished by tissue that forms a structure somewhat comparable to a placenta. In most cheilostomes that brood, the eggs are comparatively large and the embryos subsist on their own yolk while they are in the ooecia. As a rule, the larvae released from ooecia do not feed. They may, nevertheless, be rather complex, with glands for attachment to the substratum and also with primordia of some tissues and organs of adult zooids.

An ooecium (Figure 14.24) consists of two parts, and these have different origins. The outer part, the *ooecial fold,* is a highly modified zooid produced by the zooid located just distal to the one whose eggs it will receive. The inner part, the *ooecial vesicle,* which encloses the brood chamber, is an outgrowth of the egg-producing zooid. In *Bugula,* the brood chamber is connected by a pore with the space through which the lophophore of the egg-producing zooid is extended. The pore can be enlarged by contraction of muscles in the ooecial vesicle to allow an egg to enter the brood chamber, and again later to allow the swimming larva to escape. The epithelial tissue facing the brood chamber is glandular, and during the period of brooding the embryo is in tight contact with it. The fact that the embryo will not develop further if it is removed from the ooe-

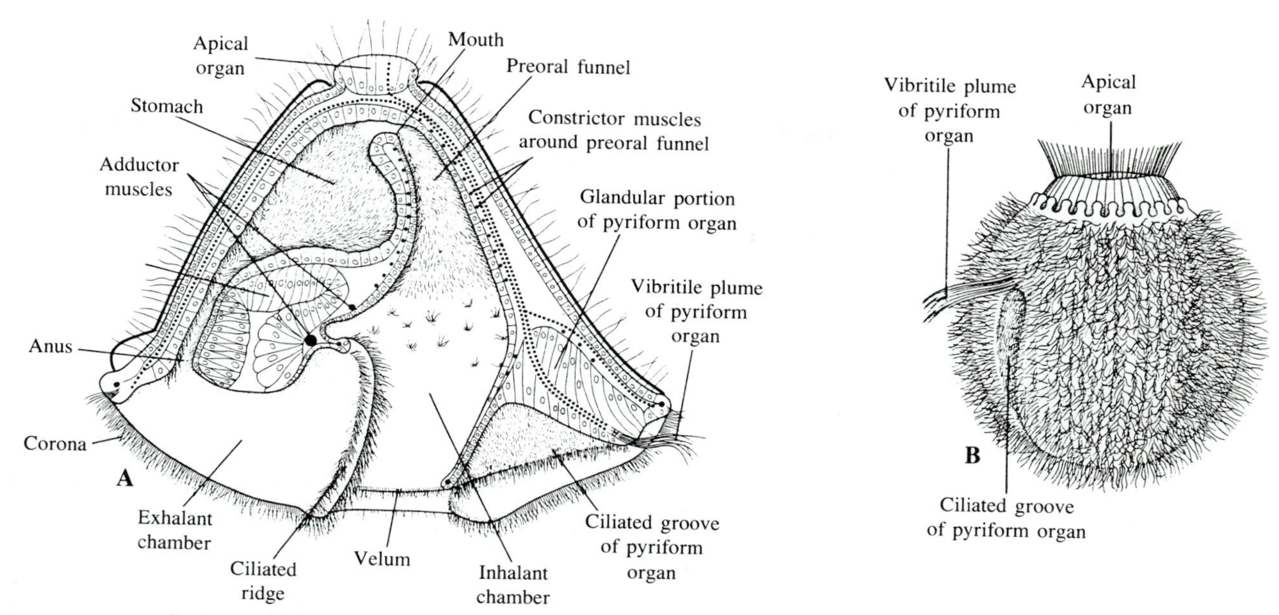

FIGURE 14.23 Order Cheilostomata. A. Cyphonautes larva of *Electra pilosa,* seen from the right side. Not all muscles are shown. When the larva settles, the adhesive sac is everted and affixed to the substratum. B. Larva of *Bugula neritina*. The adhesive sac (not visible in the drawing) is at the lower end and is everted at the time of settling. (A and B, Nielsen, Ophelia, *9*.)

cium suggests that it is dependent on the glandular tissue.

Hippodiplosia insculpta is interesting in that some of the zooids making up its colonies are strictly female, whereas the rest either function as males or have the same general form, even if they do not produce sperm. The females have larger apertures than males (Figure 14.25, A and B). When a female zooid produces its first oocyte, it ceases to feed and it becomes dependent on its neighbors. It also gives rise to a small, nonfeeding zooid whose function is not known (Figure 14.25, C). Several male zooids abut each female zooid. The

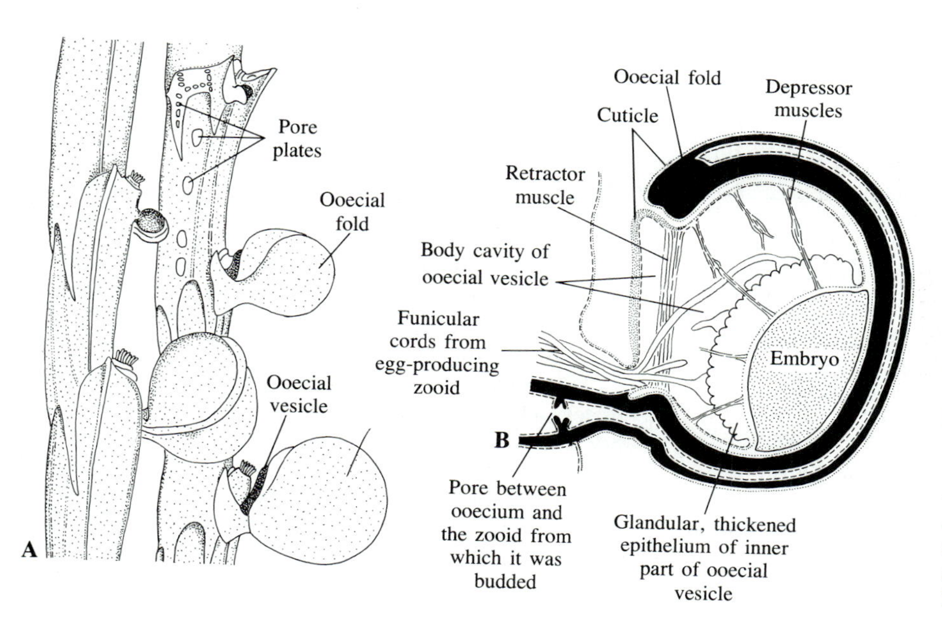

FIGURE 14.24 Order Cheilostomata. *Bugula neritina*. A. Stages in the formation of ooecia. Two contiguous rows of zooids are shown, as seen from the left. The rows have been pulled apart, making it possible to see some of the pores through which tissue connections pass. The ooecial fold is a specialized zooid derived from the zooid distal to the one whose eggs are to be brooded; the ooecial vesicle is, however, an outgrowth of the egg-producing zooid. B. Median section of an ooecium. (After Woollacott and Zimmer, Marine Biology, *16*.)

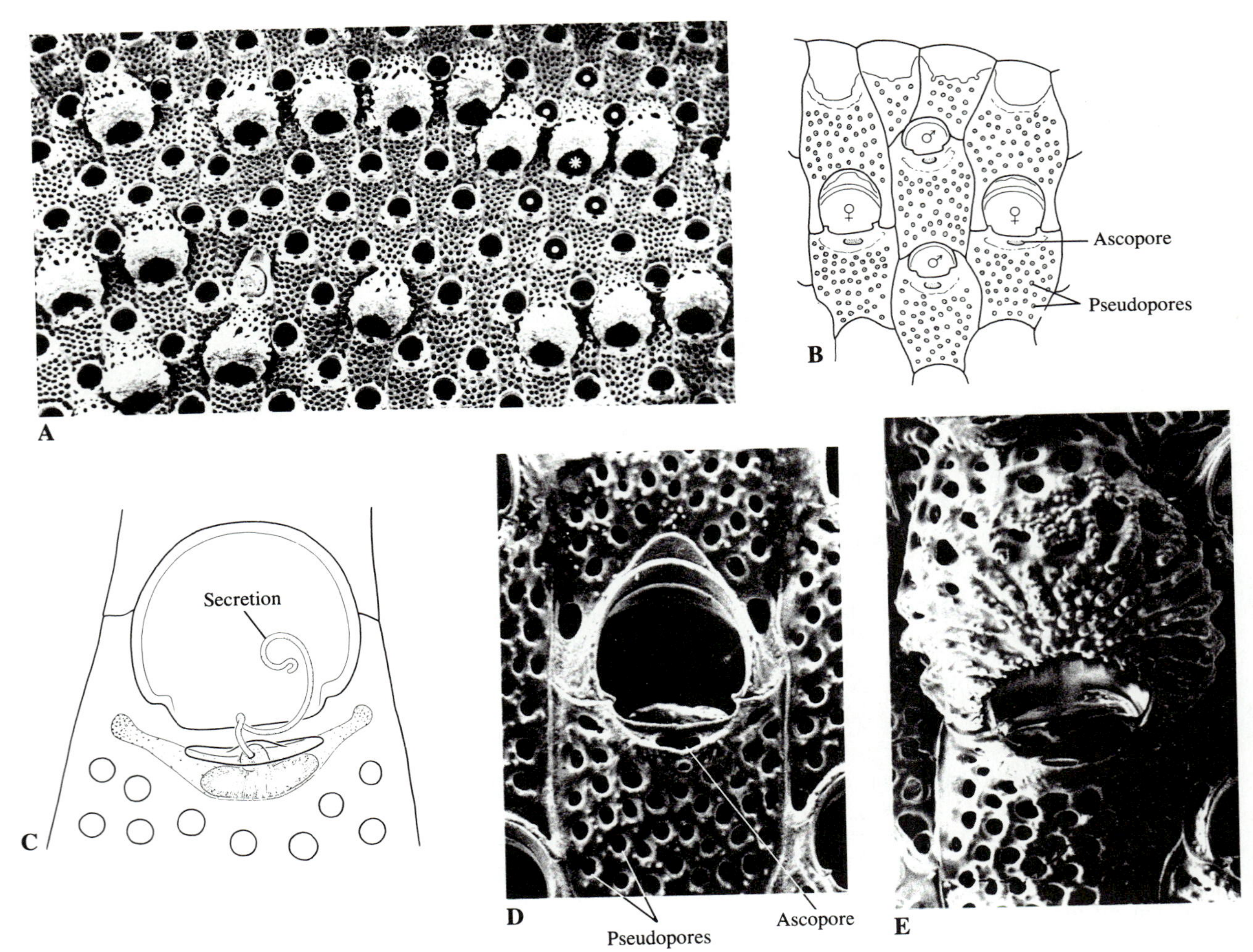

Figure 14.25 Order Cheilostomata. *Hippodiplosia insculpta*. A. Portion of a colony (organic material removed) showing male zooids with small apertures (some marked by white dots) and female zooids with large apertures (one marked with a white asterisk), as well as ooecia. B. Drawing of a small portion of a colony, showing male and female zooids. C. Small, specialized zooid that is produced by a female zooid when the latter begins to produce oocytes. The secretion, of unknown function, is extruded when the zooid is preserved. D. An early stage in the formation of an ooecium just distal to the aperture of a female zooid; scanning electron micrograph. E. Fully developed ooecium. (Nielsen, Ophelia, *20*.)

one just distal to the aperture of the female is the one that produces the ooecial fold of the ooecium in which the female's eggs will be brooded (Figure 14.25, D and E).

CLASS STENOLAEMATA

Three of the four orders of stenolaemates are known only as fossils. These will not be discussed, although *Archimedes,* an interesting genus of the order Cryptostomata, is illustrated in Figure 14.27, A. The following account applies to the surviving order, Cyclostomata, which has provided us with all the information we have on the organization of the soft parts of the class as a whole.

The tentacles of cyclostomes, as those of gymnolaemates, are arranged in a circle around the mouth. These bryozoans differ from gymnolaemates, however, with respect to the system used for extending the polypide. Remember that gymnolaemates have muscles that insert on a flexible frontal membrane. Contraction of the muscles draws the membrane down far enough to increase

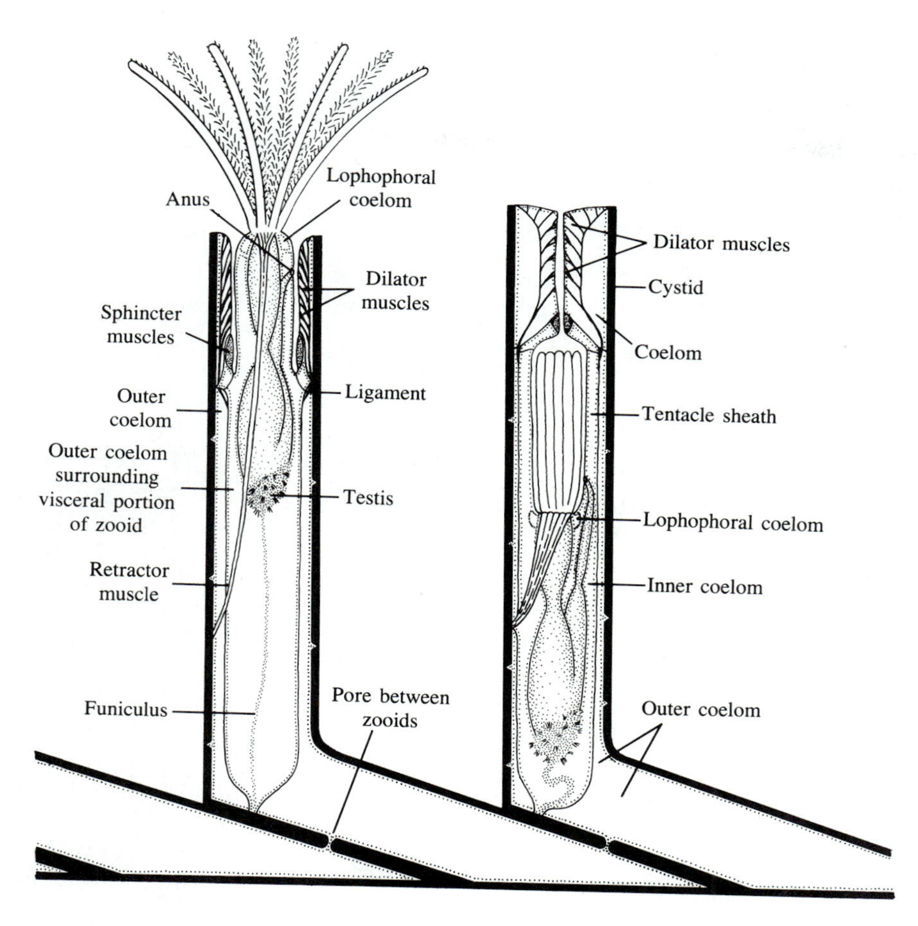

FIGURE 14.26 Order Cyclostomata. Diagram of two zooids of a type in which the cystids are long tubes, as they are in *Tubulipora*. (After Ryland, Bryozoans.)

pressure in the coelom. If the muscles concerned with withdrawal of the polypide relax at the same time, the polypide will be thrust out.

In cyclostomes (Figure 14.26), the body wall is tubelike, and its outer layer is thoroughly calcified except at the aperture. When the polypide is withdrawn, the aperture is almost completely blocked. The membrane that stretches across the aperture is continuous with a membrane that extends downward to form a narrow canal. Between this membrane and the body wall is a coelomic cavity traversed by dilator muscles. When these muscles contract, the canal is enlarged and pressure on the coelomic space is increased. This in turn puts pressure on an inner coelomic sac that surrounds the visceral portion of the zooid. This causes the polypide to be extended. If there were no sac separating the coelom into inner and outer chambers, the system would presumably operate just as efficiently, so it is not easy to find a reason for the existence of the sac. The membrane that blocks the aperture and lines the canal near the aperture is somewhat comparable to the frontal membrane of gymnolaemate bryozoans.

The colonies of cyclostomes may be low and encrusting or they may be erect and branched. In erect types, the zooids are often arranged in bundles or fanlike patterns. Some genera, such as *Heteropora* (Figure 14.27, B), resemble certain corals because their branches are stout and solidly calcified.

As a rule, all or most feeding zooids in a cyclostome colony are capable of producing sperm, usually on the funiculus. Egg production, however, is limited to a proportionately few enlarged zooids that function as females (Figure 14.27, C and D). These typically lack a lophophore, at least by the time they reach the reproductive stage. A single egg is produced in each, and after it is fertilized, the specialized zooid serves as a brood chamber for it. The embryo pinches off secondary embryos, and these may in turn produce still more embryos. The general term for this phenomenon, which is not common in the animal kingdom, is *polyembryony*. (Among the celebrated cases is that of armadillos, which produce several young from one fertilized egg.) The larvae that are released from the brood chamber are ciliated but unable to feed. Those that find a suitable substratum on which to settle become the founders of new colonies.

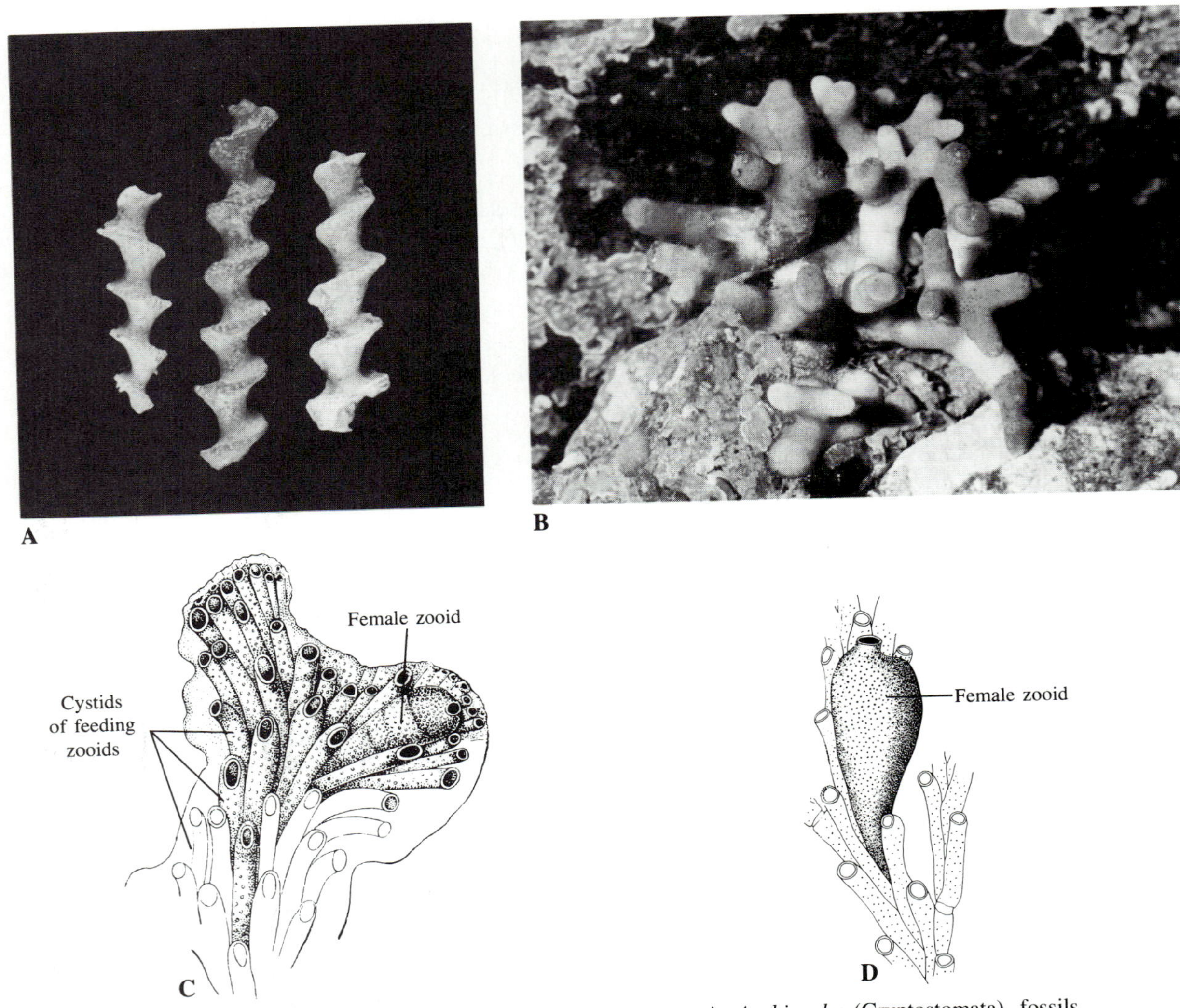

FIGURE 14.27 Orders Cryptostomata and Cyclostomata. A. *Archimedes* (Cryptostomata), fossils (Mississippian period) from Illinois. The openings of the cystids are arranged in a spiral around the axis, so the colony is shaped like a screw. B. *Heteropora pacifica* (Cyclostomata). The colony, with stout branches, resembles that of some corals. C. *Tubulipora dilatans* (Cyclostomata). (Kliuge, Keys to the Fauna of the USSR, *76*.) D. *Crisia cribraria* (Cyclostomata). Osburn, Allan Hancock Pacific Expeditions, *14*.)

Except for the zooids specialized for egg production and brooding, there is limited polymorphism among cyclostomes. Nothing comparable to avicularia or vibracula occurs in this group. In some genera, strange zooids with just one tentacle have been shown to be males. Anchoring zooids whose function is to grip the substratum are found in a few cyclostomes.

CLASS PHYLACTOLAEMATA

Some of the facts presented in connection with the gymnolaemates and stenolaemates apply to the freshwater phylactolaemates. The system for feeding by ciliary activity is in general the same, and the organization of the gut is similar. The lophophore, however, is bent into the shape of a horseshoe (Figure 14.28), and it has many more tentacles than would be found in the lophophore of almost any gymnolaemate or stenolaemate. The mouth, situated on the midline between the two rows of tentacles, is bordered on one side by an epistome.

The colonies of most freshwater bryozoans are fairly soft, and in a number of genera the secreted outer portions of the cystids are thick and gelatinous. As a rule,

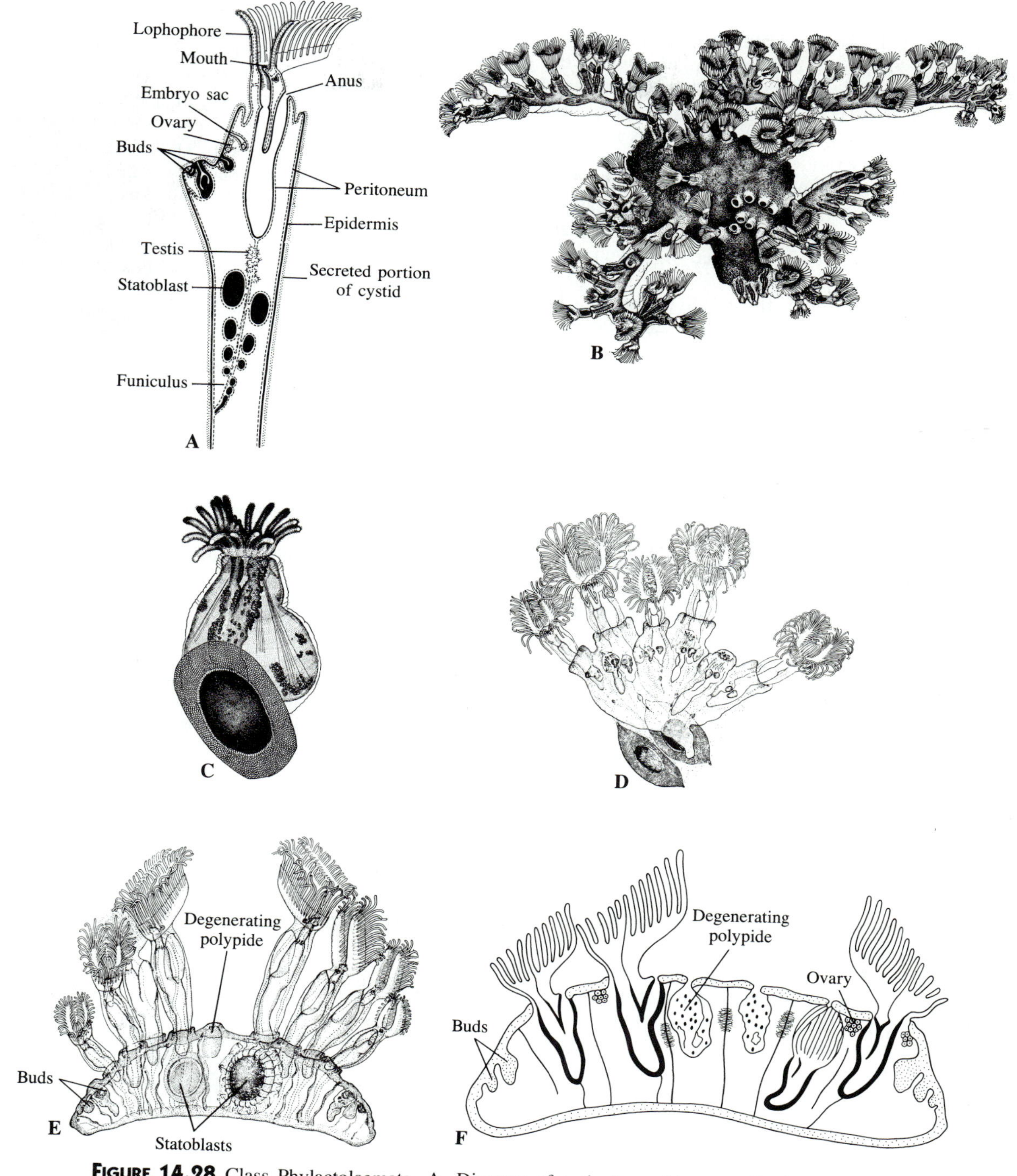

FIGURE 14.28 Class Phylactolaemata. A. Diagram of a single zooid, generalized. (Brien, Annales de la Société Royale Zoologique de Belgique, *74*.) B. *Hyalinella punctata,* a small colony. C. *Hyalinella punctata,* the first zooid that has developed after germination of a statoblast. (B and C, Rogick, Ohio Journal of Science, *45*.) D. *Lophopus crystallinus,* a young colony. The two valves of the statocyst separated during germination. E. *Cristatella mucedo,* transverse section through a colony. (D and E, Brien, Annales de la Société Royale Zoologique de Belgique, *74*.) F. *Cristatella mucedo,* diagram of a transverse section through a colony.

the body wall and even the coelom of adjacent zooids are continuous. Cilia on the peritoneum keep fluid moving through the socialized coelomic system, which extends into the tentacles. Between the epidermis and peritoneum are circular and longitudinal muscles that can increase pressure on the coelom and thus force out the lophophore. Retraction of the lophophore is effected by a set of retractor muscles. *Cristatella* (Figure 14.28, E and F), whose colony has a flattened "sole," can use its body wall muscles to creep slowly over aquatic plants and submerged wood. Its speed of movement is about 2 or 3 cm per day.

The coelom consists of three compartments, but these are not completely separate. The smallest compartment is restricted to the epistome and is often referred to as the protocoel, but there is no good evidence for its homology with the embryonic protocoel of a phoronid larva or that of any other invertebrate with three distinct embryonic coeloms. The lophophoral and general body coeloms correspond to those of other classes. The coeloms contain amoeboid cells whose function remains unknown.

Elimination of what appear to be coelomic waste materials has been observed to take place through temporary pores at the tips of the tentacles. A few species have a structure that resembles a pair of partly united funnel organs. This is linked to the coelom of the lophophore as well as to the principal body cavity. It is thought not to open by a permanent pore, but elimination of material from it has been seen to take place by a rupture that then closes. Portions of the gut are also believed to function in elimination of various nitrogenous wastes, including uric acid. Soluble wastes, such as ammonia, can probably be eliminated by diffusion across the exposed surfaces of the body, especially the tentacles.

During the growing period, a phylactolaemate continually produces buds, thus enlarging its colony. A bud develops as a localized inpocketing of epidermis and peritoneum (Figure 14.28, A, E, and F). Within this a new zooid differentiates and grows outward. The process of differentiation is complex and not all accounts agree in detail, but the entire new zooid arises from epidermis and peritoneum. The pattern of budding and the nature of the zooecial material determine the appearance and texture of the colony. Colonies of *Cristatella* may divide into two or more pieces.

Phylactolaemates are hermaphroditic, and the simple gonads are derived from peritoneal cells. The ovary develops as a localized bulge on the body wall, and the testis generally covers part or nearly all of the funiculus. The eggs are fertilized while they are in the parental coelom and then enter a pocket of the body wall, called an *embryo sac*. Cleavage leads to production of a blastula, and the wall of this becomes two-layered. The outer layer is ectoderm, the inner layer peritoneum; the central cavity is the future coelom. After the embryo has differentiated an internal zooid by a process rather similar to the production of a bud by a mature zooid, it develops a ciliated locomotor organ and escapes by rupture of the cystid. This stage, in spite of its precocious budding, is generally called a larva, and it is a larva in the sense that it will change drastically and lose its locomotor organ after it settles. It has already produced a zooid, however, and this in turn may have started to bud off additional zooids before settling takes place.

A distinctive feature of phylactolaemates is their ability to produce *statoblasts* (Figure 14.28, A, C–E), which are thick-walled over-wintering bodies that can survive low temperatures and drying. A statoblast originates on the funiculus, and at first it consists of a mass of yolky peritoneal cells covered by two layers of epidermal cells. The outer epidermal layer secretes a hard shell between it and the inner layer; the inner layer survives as the epidermis of the prospective new individual. The shell contains chitin and consists of two valves that separate when the statoblast germinates. Various distinctive types of statoblasts have been given names, depending on their shape, on whether they contain air and can float, or on whether they have hooked spines or other elaborations.

Depending on the genus, each zooid may produce a single statoblast, or it may produce a number of them sequentially. Statoblasts are sometimes released by their parent zooids while these are still alive, but it is more common for them to stay in the zooids until these die and disintegrate in late autumn or early winter. Although statoblasts are resistant to adverse conditions, germination does not depend on their having been exposed to such conditions. When germination does take place, a young zooid differentiates and starts to build a colony by asexual reproduction.

References on Bryozoa

Cancino, J. M., and Hughes, R. N., 1987. The effect of water flow on growth and reproduction of *Celleporella hyalina* (L.) (Bryozoa: Cheilostomata). Journal of Experimental Marine Biology and Ecology, *112*:109–130.

Carle, K. J., and Ruppert, E. E., 1983. Comparative ultrastructure of the bryozoan funiculus: a blood vessel homologue. Zeitschrift für Zoologische Systematik und Evolutionsforschung, *21*:181–193.

Cook, P. L., 1977. Colony water currents in living Bryozoa. Cahiers de Biologie Marine, *18*:31–47.

Gordon, D. P., 1975. Ultrastructure and function of the gut of a marine bryozoan. Cahiers de Biologie Marine, *16*:367–382.

Harmelin, J.-G., 1976. Le sous-ordre des Tubuliporina (bryozoaires cyclostomes) en Méditerranée. Ecologie et systématique. Mémoires de l'Institut Océanographique, Monaco, *10*:1–326.

Harvell, C. D., 1986. The ecology and evolution of inducible defenses in a marine bryozoan: cues, costs, and consequences. American Naturalist, *128*:810–823.

Hayward, P. J., 1985. Ctenostome Bryozoans. Synopses of the British Fauna, no. 33. Linnean Society of London, Estuarine and Brackish-Water Sciences Association, and E. J. Brill/Dr. W. Backhuys, London and Leiden.

Hayward, P. J., and Ryland, J. S., 1979. British Ascophoran Bryozoans. Synopses of the British Fauna, no. 14. Linnean Society of London, Estuarine and Brackish-Water Sciences Association, and Academic Press, London and New York.

Hayward, P. J., and Ryland, J. S., 1985. Cyclostome Bryozoans. Synopses of the British Fauna, no. 34. Linnean Society of London, Estuarine and Brackish-Water Sciences Association, and E. J. Brill/Dr. W. Backhuys, London and Leiden.

Jebram, D., 1986. Arguments concerning the basal evolution of the Bryozoa. Zeitschrift für Zoologische Systematik und Evolutionsforschung, *24*:266–290.

Kaufmann, K. W., 1973. The form and functions of the avicularia of *Bugula*. Postilla, *151*:1–26.

Lacourt, A. W., 1968. Monograph of the freshwater Bryozoa Phylactolaemata. Zoologische Verhandelingen, Leiden, no. 93.

Larwood, G. P. (editor), 1973. Living and Fossil Bryozoa. Academic Press, London and New York.

Larwood, G. P., and Abbott, M. B. (editors), 1979. Advances in Bryozoology. Academic Press, London and New York.

McEdward, L. R., and Strathmann, R. R., 1987. The body plan of the cyphonautes larva of bryozoans prevents high clearance rates: comparison with the pluteus and a growth model. Biological Bulletin, *172*:30–45.

Markham, J. B., and Ryland, J. S., 1987. Function of the gizzard in Bryozoa. Journal of Experimental Marine Biology and Ecology, *107*:21–37.

Nielsen, C., 1970. On metamorphosis and ancestrula formation in cyclostome bryozoans. Ophelia, *8*:217–256.

Nielsen, C., 1981. On morphology and reproduction of *'Hippodiplosia' insculpta* and *Fenestrulina malusii* (Bryozoa Cheilostomata). Ophelia, *20*:91–125.

Nielsen, C., and Pedersen, K. J., 1979. Cystid structure and protrusion of the polypide in *Crisia* (Bryozoa, Cyclostomata). Acta Zoologica, *60*:65–88.

Osburn, R. C., 1950–53. Bryozoa of the Pacific coast of America. Allan Hancock Pacific Expeditions, *14*:1–841.

Reed, C. G., and Cloney, R. A., 1982. The larval morphology of the marine bryozoan *Bowerbankia gracilis* (Ctenostomata: Vesicularioidea). Zoomorphology, *100*:23–54.

Reed, C. G., and Cloney, R. A., 1982. The settlement and metamorphosis of the marine bryozoan *Bowerbankia gracilis* (Ctenostomata: Vesicularioidea). Zoomorphology, *101*:335–348.

Reed, C. G., Ninos, J. M., and Woollacott, R. M., 1988. Bryozoan larvae as mosaics of multifunctional ciliary fields: ultrastructure of sensory organs of *Bugula stolonifera* (Cheilostomata: Cellularioidea). Journal of Morphology, *197*:127–145.

Rider, J., and Cowen, R., 1977. Adaptive architectural trends in encrusting ectoprocts. Lethaia, *10*:29–41.

Ross, J. R. P. (editor), 1987. Bryozoa: Present and Past. Western Washington University, Bellingham.

Ryland, J. S., 1970. Bryozoans. Hutchinson University Library, London.

Ryland, J. S., 1976. Physiology and ecology of marine bryozoans. Advances in Marine Biology, *14*:285–443.

Ryland, J. S., 1979. Structural and physiological aspects of coloniality in Bryozoa. *In* Larwood, G. P., and Rosen, B. R. (editors), Biology and Systematics of Colonial Organisms. Academic Press, New York.

Ryland, J. S., and Hayward, P. J., 1977. British Anascan Bryozoans. (Synopses of the British Fauna, no. 10.) Linnean Society of London and Academic Press, London and New York.

Silén, L., 1966. On the fertilization problem in the gymnolaematous Bryozoa. Ophelia, *3*:113–140.

Silén, L., 1972. Fertilization in the Bryozoa. Ophelia, *10*:27–34.

Stricker, S. A., 1988. Metamorphosis of the marine bryozoan *Membranipora membranacea:* an ultrastructural study of rapid morphogenetic movements. Journal of Morphology, *196*:53–72.

Stricker, S. A., Reed, C. G., and Zimmer, R. L., 1988. The cyphonautes larva of the marine bryozoan *Membranipora membranacea*. I. General morphology, body wall, and gut. Canadian Journal of Zoology, *66*:368–383.

Stricker, S. A., Reed, C. G., and Zimmer, R. L., 1988. The cyphonautes larva of the marine bryozoan *Membranipora membranacea*. II. Internal sac, musculature, and pyriform organ. Canadian Journal of Zoology, *66*:384–398.

Thorpe, J. P., Shelton, G. A., and Laverack, M. S., 1975. Colonial nervous control of lophophore retraction in cheilostome Bryozoa. Science, *189*:60–61.

Winston, J. E., 1978. Polypide morphology and feeding behavior in marine ectoprocts. Bulletin of Marine Science, *28*:1–31.

Winston, J. E., 1984. Why bryozoans have avicularia—a review of the evidence. American Museum Novitates, no. 2789.

Winston, J. E., 1986. Victims of avicularia. Marine Ecology, *7*:193–199.

Woollacott, R. M., and Zimmer, R. L. (editors), 1977. Biology of Bryozoans. Academic Press, New York and London.

PHYLUM ENTOPROCTA

INTRODUCTION

There are fewer than 100 species of entoprocts, almost all limited to the sea. Their fuzzy growths resemble those of certain bryozoans and also those of some hydroids. Colonial types, in which numerous members originate from a creeping stolon, are found on seaweeds, shells, rocks, wood, and sometimes on ascidians, mussels, or other animals. The only freshwater entoproct, *Urnatella gracilis,* which occurs in North America and Europe, produces new members by budding from an upright stalk rather than from a stolon. The species that live on the skins of polychaete annelids, sipunculans, and some other invertebrates, as well as on the tubes of polychaetes, are solitary. They bud off new individuals, however, and dense aggregations may form if the young settle close to their parents.

GENERAL STRUCTURE

The basic body plan of entoprocts is fairly constant. The following description applies, in general, to solitary species as well as to zooids of colonial types.

The body consists of two main parts: a stalk and a calyx bordered by tentacles (Figures 14.29 and 14.30). The mouth is located off-center in the area encircled by tentacles, and much of this area may form a shallow depression called the *vestibule*. The anus is situated on a pronounced elevation opposite the mouth. The fact that the anus is within the tentacular crown is one of the features that distinguishes entoprocts ("inside anus") from the Bryozoa, or Ectoprocta, in which the anus is outside the crown. With the mouth on one side of the vestibule and the anus on the other, the gut is more or less U-shaped. It is completely ciliated and consists of regions conventionally called esophagus, stomach, intestine, and rectum.

The space between the gut and body wall, and the spaces in the tentacles and stalk, are fluid-filled, but they contain many cells, especially amoebocytes. The spaces do not have a peritoneal lining, and they are usually regarded as components of a pseudocoel.

Entoprocts have protonephridia similar to those of flatworms, nemerteans, and several other groups of invertebrates. There are two flame-bulb complexes in the calyx, close to the esophagus. The ducts leading away from them unite to form a single duct that reaches the exterior in the vestibule between the mouth and anus (Figure 14.29).

The nervous system includes a fairly prominent bilobed ganglion close to the stomach (Figure 14.29). Most of the nerves extending away from this ganglion run to small subsidiary ganglia, and the ultimate peripheral nerves arise from these. Scattered bristles, and bristles concentrated in tufts on the sides of the calyx, are sensory structures that consist of modified cilia.

Except for ciliated tracts on the tentacles and within the vestibule, the body surface has a secreted cuticular covering. Beneath the epidermis are muscles that effect

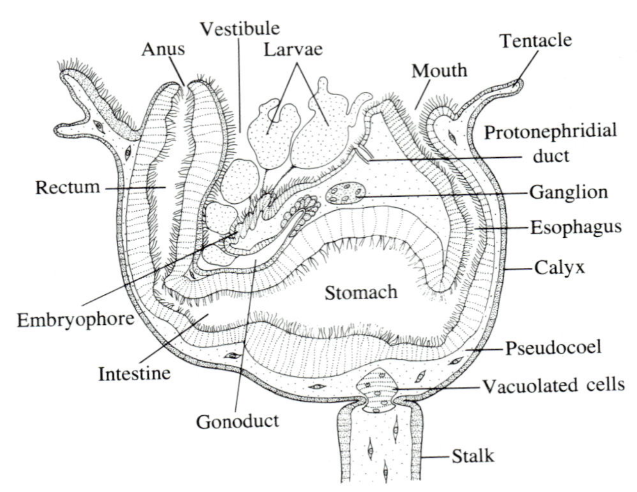

FIGURE 14.29 *Pedicellina cernua,* sectioned lengthwise; somewhat diagrammatic. (After Cori, in Bronn, Klassen und Ordnungen des Tierreichs, *4*:2:4.)

bending movements and changes in shape. Some of the cells of the epidermis of the tentacles are themselves contractile. These are called myoepithelial cells, and their contractile fibers have a striated appearance. The tentacles cannot be shortened to any great extent, but they can be turned inward. A sphincter muscle running through a flange at the edge of the calyx (Figure 14.30, B) acts much like a drawstring, pulling the flange into a tighter circle over the tentacles. A large, stellate muscle cell found at the base of the calyx of at least some entoprocts can operate to circulate fluid in the pseudocoel.

The stalks of colonial entoprocts generally have an elaborate musculature. In *Myosoma,* for instance, the stalk has a well-developed diagonal musculature that is visible externally. In *Barentsia,* the stalk of each member has one or more muscular enlargements that are like knees (Figure 14.33, B); bending takes place at these points. In most solitary entoprocts the stalk is cemented to the skin of the host. In *Loxosoma,* however, there is a muscular disk at the base of the stalk. If the disk releases its hold on the worm or tube to which it is attached, *Loxosoma* can inch its way to another location. The tentacles temporarily grip the surface while the animal is changing its position.

The current that brings food to an entoproct and also carries away the animal's wastes is generated by the activity of long cilia on the sides of the tentacles (Figure 14.30, C). The current moves upward through the tentacular crown (Figure 14.30, B). When a diatom or other food particle of suitable size and texture lands on the upper surface of a tentacle, it is caught in mucus

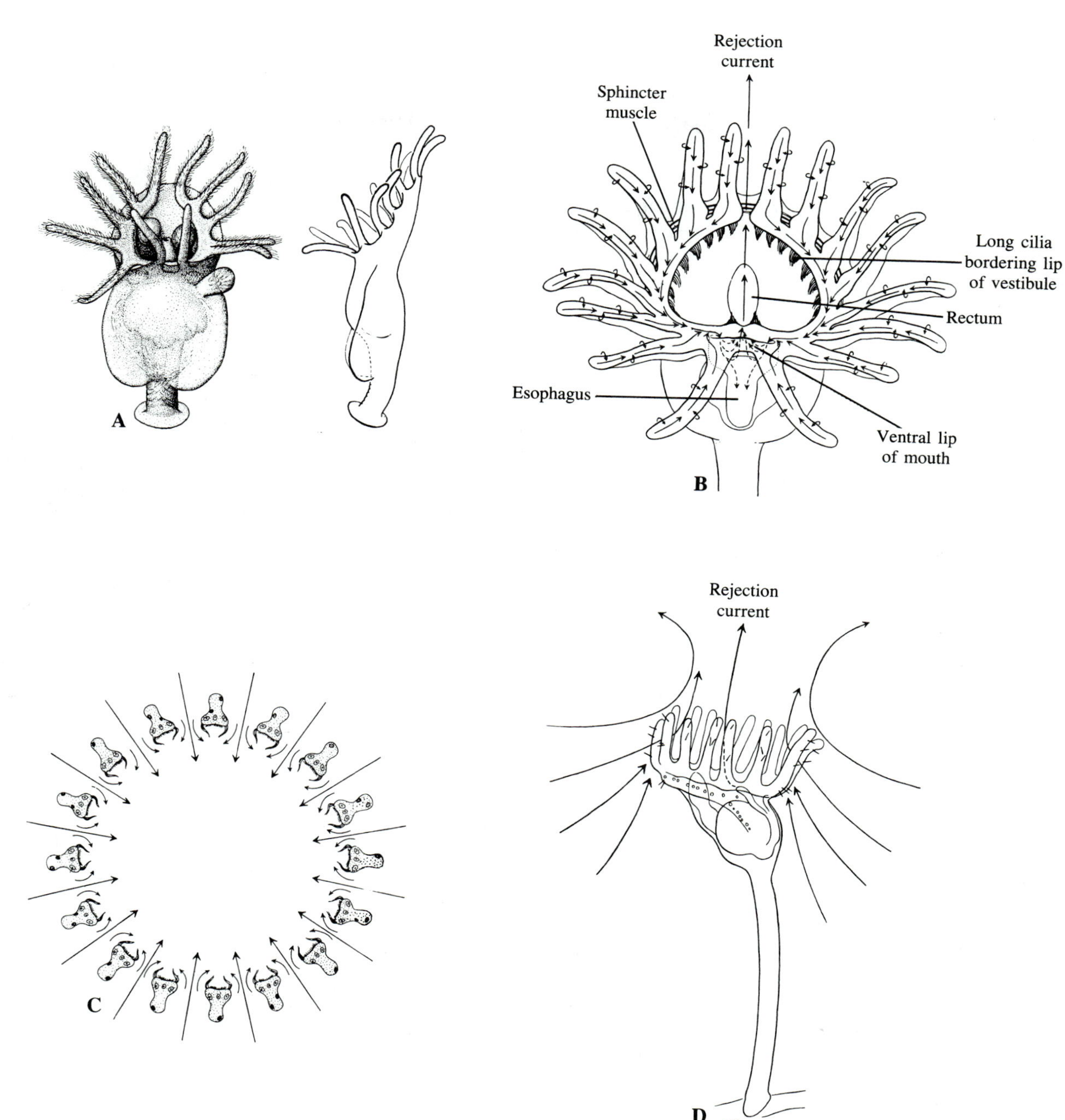

FIGURE 14.30 A. *Loxosoma rhodinicola,* frontal view (left) and lateral view (right). This species lives on the body surface of *Rhodine loveni,* a polychaete of the family Maldanidae. (Franzén, Zoologiska Bidrag från Uppsala, *33.*) B. *Loxosoma crassicauda,* frontal view of the tentacular crown, showing directions of currents produced by cilia. C. *Loxosoma crassicauda,* section through the tentacular crown, showing direction of beat of lateral cilia (small arrows) and direction of water currents (large arrows). D. *Loxosoma crassicauda,* lateral view, showing direction of water currents produced by lateral cilia of the tentacles and the rejection current produced by cilia on the ridge above the mouth. (B–D, Atkins, Quarterly Journal of Microscopical Science, *79.*)

and driven down a tract of short cilia to the base of the tentacle, where it enters a semicircular collecting groove that leads to the mouth. Digestion and at least some absorption are thought to take place in the stomach. As fecal wastes leave the anus, they are wafted away by the currents coming up through the tentacular crown (Figure 14.30, B and D).

The solitary commensal entoprocts, which belong mostly to the genera *Loxosoma* and *Loxosomella*, are usually oriented to take advantage of currents produced by their hosts. Two examples are shown in Figure 14.31.

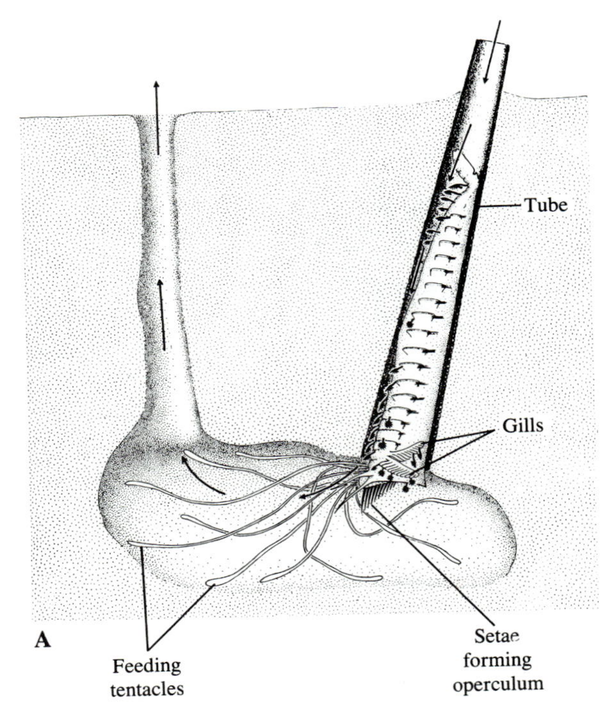

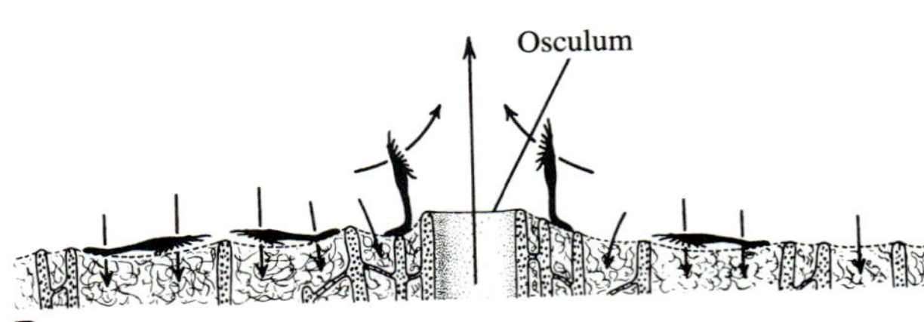

FIGURE 14.31 A. *Loxosoma pectinaricola* on the polychaete *Pectinaria belgica* (family Pectinariidae). The near side of the tube has been removed to show the worm, the currents produced by the worm, and the entoproct attached to various parts of the body surface. (Nielsen, Ophelia, *1*.) B. *Loxosomella vivipara* on the surface of the sponge *Ircinia fasciculata* (class Demospongiae). The arrows indicate water currents, some produced by the sponge, others by cilia on the tentacles of the entoproct. (Nielsen, Ophelia, *3*.)

REPRODUCTION

BUDDING

Colonial entoprocts generally add new members by budding from their stolons. In some species of *Barentsia*, the upright stalks also produce buds, and this accounts for the fact that the stalks as well as the stolons are branched. In *Urnatella*, as has been mentioned previously, all new members are produced by branching of the stalk.

Budding is also characteristic of solitary entoprocts. The buds originate on the sides of the calyx (Figure 14.32). The progeny fall off when they are ready to become independent.

SEXUAL REPRODUCTION AND DEVELOPMENT

In most entoprocts the sexes are separate. There is a pair of gonads, but the ducts of these unite before reaching the genital pore, which is located between the protonephridial opening and the anus. Hermaphroditic species have a pair of testes and a pair of ovaries. The ducts from all four gonads lead to a single pore.

Reproduction and development have been studied in several species, but *Pedicellina cernua* (Figure 14.29; Figure 14.33, A)—a colonial type—has been the most intensively investigated. By a strong contraction of its calyx, a male *Pedicellina* forces a cloud of sperm from the sperm duct. The eggs, which are small but have a moderate amount of yolk, are probably fertilized in the common oviduct of the female, although it is possible that they are fertilized in the separate oviducts or perhaps even while they are still in the ovary.

After a zygote has been covered by an envelope secreted by glands in the wall of the common oviduct, it is moved to a part of the vestibule that serves as a brood pouch. Here it becomes attached to an area called the *embryophore*. Cleavage is spiral and leads to the formation of a hollow blastula. Gastrulation is achieved by invagination, and the blastocoel is nearly obliterated as the macromeres and some of the micromeres push into it. The blastopore closes, but a stomodeal invagination appears in the same general area and forms the mouth and esophagus. The original archenteron becomes the stomach and intestine, and a proctodeal invagination produces the rectum and anus.

One of the macromeres that turns inward during gastrulation is the cell from which much of the mesoderm derives. It seems to be closely comparable to the mesoderm-forming 4d cell of molluscs, annelids, and other

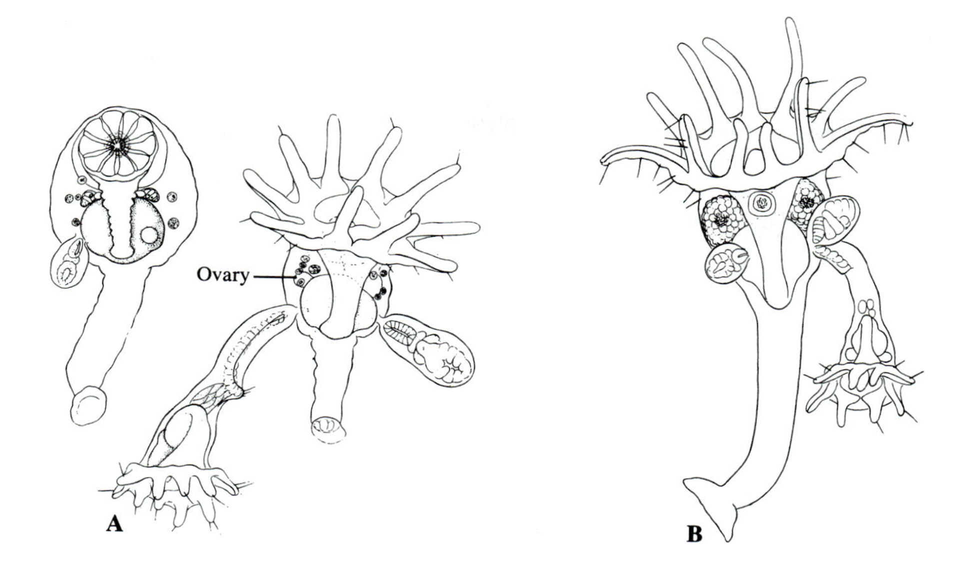

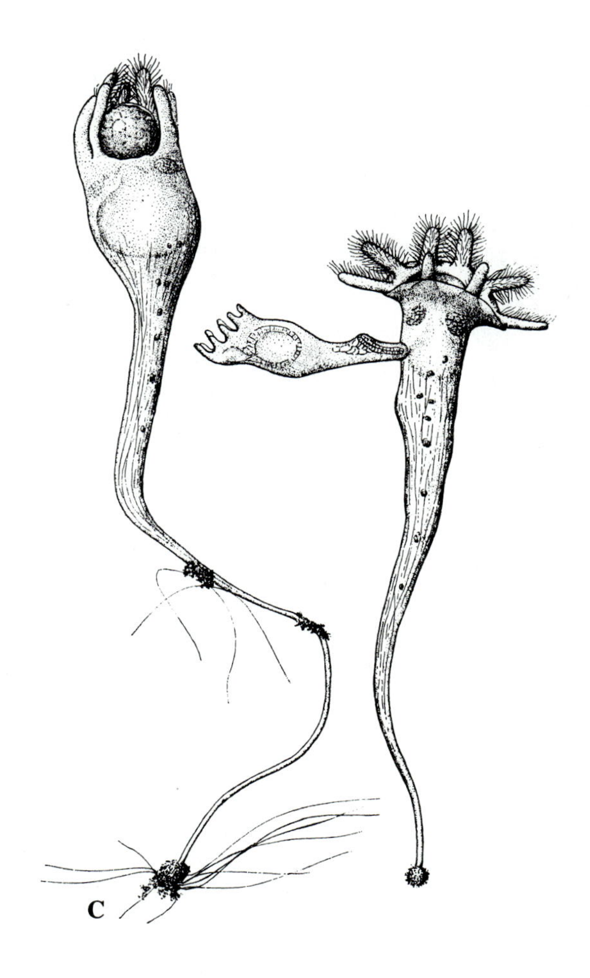

FIGURE 14.32 A. *Loxosoma singulare*. One female, with tentacles extended, has two buds; the other, with tentacles withdrawn, has one bud. This species lives on the body surface of polychaetes of the genus *Notomastus* (family Capitellidae). B. *Loxosomella claviformis,* male with tentacles extended. There are three buds. This species lives on the body surface of polychaetes of the genus *Laetmonice* (family Aphroditidae). (A and B, Atkins, Quarterly Journal of Microscopical Science, *75*.) C. *Loxosomella glandulifera,* a female with an embryo (left) and one with a bud. This species lives on polychaetes of the genus *Panthalis* (family Polyodontidae). (Franzén, Zoologiska Bidrag från Uppsala, *33*.)

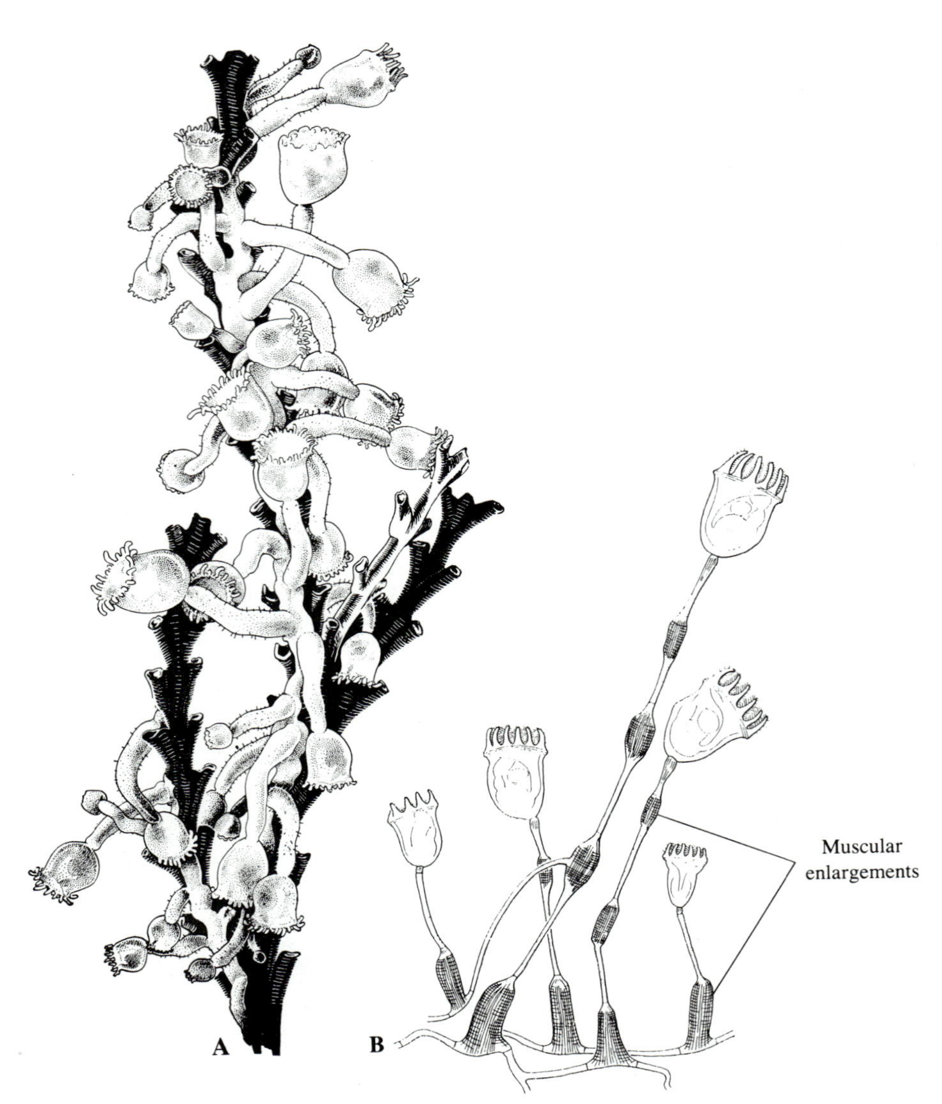

FIGURE 14.33 A. *Pedicellina cernua,* portion of a colony growing on a hydroid. (Prenant and Bobin, Faune de France, *60.*) B. *Barentsia gracilis.* (Nielsen, Ophelia, *1.*)

invertebrates that exhibit spiral cleavage. Some cells budded off the stomodeum also contribute to the formation of mesodermal structures, including muscle and connective tissue. The pseudocoel is apparently a remnant of the blastocoel. The amoebocytes and other loosely organized cells in it are probably mesodermal elements. The protonephridia, however, are thought to derive from ectoderm, and they are almost certainly homologous to protonephridia of other invertebrates.

The larvae are freed when their connections with the embryophore (Figure 14.29) are broken as a result of muscular contractions of the parent's calyx. The larvae may have begun to filter-feed while still in the vestibule of the parent, and in one species of *Barentsia* they have been observed to take food from the ciliated tract that leads to the parent's mouth. Feeding by larvae on nutritive cells in the brood pouch has also been reported.

The more important external characteristics of the larva (Figure 14.34, B–F) are a circular band of locomotor cilia, a ciliated pit forming a dorsal apical organ, and a pair of pits that constitute an anterior preoral organ. Both mouth and anus are on the ventral side, which is usually differentiated into a broad "foot." The larva can swim freely or it can glide while its foot is in contact with the substratum. The larvae of solitary entoprocts usually have an anterior pair of eyespots. The apical organ and preoral organ are frequently everted, and it is likely that they are sensory structures that help the larva find a suitable place to settle.

There is considerable variation in what happens

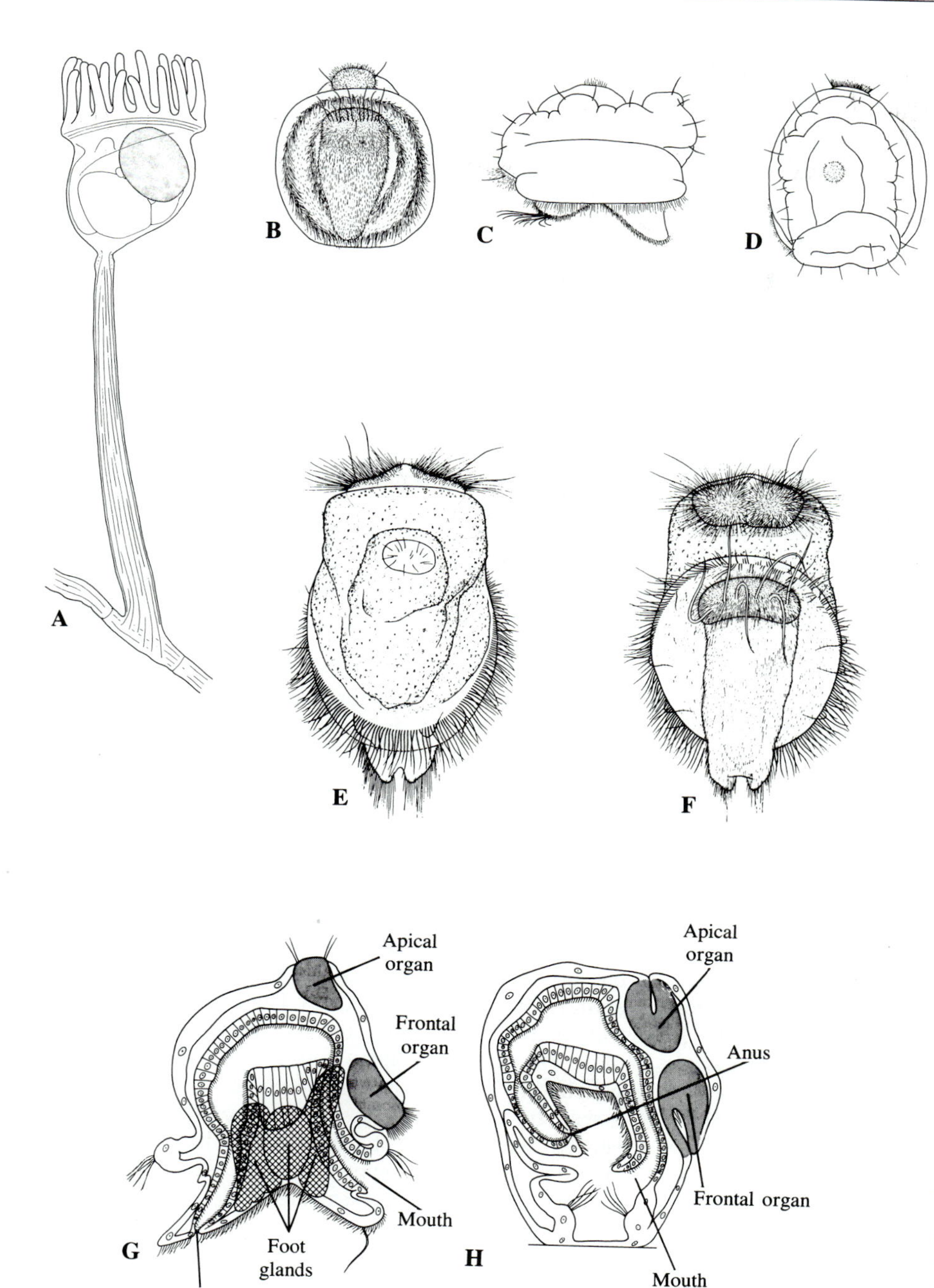

FIGURE 14.34 A. *Pedicellina nutans,* a specimen with an embryo in the vestibule. B–D. *Pedicellina nutans,* a creeping larva, ventral view, view from the left side, and dorsal view. E, F. *Loxosomella varians,* a larva in dorsal and ventral views. G, H. *Pedicellina nutans,* diagrams of a larva just before and just after settling. The apical organ and frontal organ are sensory structures; the foot glands secrete adhesive material. (Nielsen, Ophelia, *9*.)

when the larva settles and undergoes metamorphosis. In some colonial types, a stalk grows out from an attachment plate formed by fusion of the edges of what had been the circular ciliated band of the larva. The larval mouth and anus close and the gut rotates about 180°; new stomodeal and proctodeal invaginations form on the upper side, and the gut becomes functional again. The tentacles originate from the edges of the developing calyx. The stolons, from which new members are produced by budding, grow out from the base of the founder individual.

In at least one of the solitary entoprocts, *Loxosomella murmanica,* metamorphosis of the larva slightly resembles what has just been described for colonial types. The stalk grows out from the region of the preoral organ, and the gut is salvaged to form the gut of the adult. In other solitary species that have been studied, however, the larva forms buds that develop into mature individuals; the rest of the larva simply deteriorates. A still more astonishing thing happens in a few species. The larva, whether or not it produces external buds, gives rise to one or two internal buds. When these are released by rupture of the larva, they may already be sexually mature dwarf males. So although the phylum Entoprocta is small, its members show considerable diversity in the ways their larvae metamorphose.

REFERENCES ON ENTOPROCTA

Mariscal, R. N., 1965. The adult and larval morphology and life history of the entoproct *Barentsia gracilis* (M. Sars, 1835). Journal of Morphology, *116*:311–338.
Deals with *B. benedeni*.

Nielsen, C., 1964. Studies on Danish Entoprocta. Ophelia, *1*:1–76.

Nielsen, C., 1966. On the life-cycle of some Loxosomatidae (Entoprocta). Ophelia, *3*:221–247.

Nielsen, C., 1966. Some Loxosomatidae (Entoprocta) from the Atlantic coast of the United States. Ophelia, *3*:249–275.

Nielsen, C., 1971. Entoproct life-cycles and the entoproct/ectoproct relationship. Ophelia, *9*:209–341.

Nielsen, C., and Rostgaard, J., 1976. Structure and function of an entoproct tentacle with a discussion of ciliary feeding types. Ophelia, *15*:115–140.

SUMMARY

1. Three phyla of invertebrates—Phoronida, Brachiopoda, and Bryozoa—have food-collecting tentacles organized into a structure called the lophophore. Each tentacle contains a diverticulum of a coelomic cavity located at the base of the lophophore. This coelom is partly or fully separate from the general body coelom in which the visceral organs lie.

2. The phylum Phoronida consists of only about 20 species, all marine. They are wormlike animals that secrete a tube, from the open end of which the elaborate lophophore is extended for ciliary–mucous feeding. Both the mouth and anus are located at the base of the lophophore; the hairpin-shaped gut extends nearly to the base of the elongated trunk.

3. Vertical partitions divide the trunk coelom of a phoronid into four chambers; they also segregate the longitudinal muscles of the body wall into four groups. There are two funnel organs, which are gonoducts and probably function in excretion. The corpuscles within the blood vessels contain hemoglobin.

4. Members of the strictly marine phylum Brachiopoda, which has about 260 living species and several thousand fossil species, have a bivalved shell resembling that of clams. The valves are dorsal and ventral, however, rather than left and right. The cavity enclosed by the shell contains a large and complex lophophore used in ciliary–mucous feeding.

5. The general body coelom of a brachiopod encloses the gut, digestive gland, and two or four funnel organs, which serve as gonoducts and as organs of excretion. Extensions of this coelom go out into the mantle lobes that line the valves of the shell. The gonads may be in the extensions, or in the part of the coelom that contains the other viscera. Although the circulatory system is not prominent, it includes a definite heart.

6. The two classes of Brachiopoda are very different. In the Articulata, the valves are locked together by a tooth-and-socket arrangement. The gut ends blindly, there being no anus. The pedicle, a stalk consisting mostly of connective tissue, emerges from a hole or notch in the ventral valve and is firmly attached to a hard substratum. The lophophore is usually bound to the dorsal valve by a calcareous supporting structure called the brachidium.

7. In the class Inarticulata, there is no tooth-and-socket hinge. The gut, moreover, is complete. The pedicle, when present, is muscular and has an extension of the coelom within it. In two genera that live in vertical burrows in muddy sand, the pedicle is capable of considerable extension and contraction. In genera that are attached to hard surfaces, the pedicle is short or absent, in which case the ventral valve is cemented to the substratum. The lophophore of inarticulates is not supported by a calcareous brachidium.

8. The large phylum Bryozoa, primarily marine but also well represented in fresh water, consists almost entirely of colonial organisms. The member zooids that form a colony are small, but they are compact and complex. The zooids occupy "houses" called cystids, from which they extend their lophophores. Feeding is primarily by ciliary activity, mucus not being involved, and the feeding currents run through the crown of tentacles from above downwards. The mouth is located within the circle or horseshoe-shaped group of tentacles, but the anus is outside (hence the name Ectoprocta, meaning "outside anus," often used in place of the name Bryozoa).

9. Each feeding zooid in a bryozoan colony has a spacious coelom within its cystid. Only a few freshwater species have what may be homologues of funnel organs, however. The gametes are released through pores at the tips of the tentacles or through openings between the tentacles.

10. An interesting feature of the colonies of many bryozoans is polymorphism—the modification of some zooids for functions other than feeding. The specialized zooids include those used for attachment or forming a stolon, as well as avicularia and vibracula, which discourage competitors from settling on the same substratum. Female zooids may have a different appearance than male zooids. There are also structures called ooecia, within which embryos are brooded. The outer part of an ooecium is a modified zooid, but its inner portion, where the embryos develop, is an outgrowth of the zooid that produces the eggs.

11. The small phylum Entoprocta, almost entirely marine, consists of sessile animals, either solitary or colonial, that resemble bryozoans. Both the mouth and anus, however, are within the circle formed by the tentacles. Moreover, the feeding currents of entoprocts are the reverse of those of bryozoans; that is, they pass through the crown of tentacles from below upwards. Mucus is used in the collection of food, and this is another way in which entoprocts differ from bryozoans. The body cavity is more like a pseudocoel than a true coelom, for it is not lined by a peritoneum, but it contains amoebocytes and other cells. There is a pair of flame-bulb protonephridia.

15

Phylum Arthropoda
General Characteristics

Introduction

No one knows how many species of animals are living today, but it is estimated that a million and a quarter have been described. About a million of these are arthropods. The diversity of arthropods exceeds that of any other phylum, so it is not possible to master this enormous assemblage in a few weeks of study. It is best first to concentrate on learning the important features that characterize arthropods, then to tackle the major subdivisions in order to understand how they differ from one another.

Do arthropods constitute a single phylum? This has been hotly debated and the controversy may never be resolved to everyone's satisfaction. Some zoologists who have given much thought to the problem are convinced that the living arthropods—and at least some extinct groups—are descended from a single ancestral stock of segmented worms. Others insist that arthropods are polyphyletic, that they are descended from three or four separate ancestral stocks. Arthropods were already rather well diversified in the earliest part of the Paleozoic era, more than 500 million years ago, and the oldest fossils do not provide unequivocal support for one view or the other. We will return to the problem of arthropod phylogeny at the end of this chapter.

The arthropods with which almost everyone is familiar are insects, ticks, spiders, scorpions, centipedes, "sowbugs," lobsters, crabs, and shrimps. There are, however, many other distinctive groups in this huge assemblage. The major categories of arthropods are dealt with in the four succeeding chapters, but it is essential first to explain the characteristics that are common to all or most of these animals.

Structure

TAGMATA AND APPENDAGES

Arthropods, like annelids, are segmented animals, but their segments are assembled into body divisions called **tagmata** (the singular form of the word is **tagma**). While it is true that many polychaete annelids have rather distinct body regions, these are not so pronounced as the head, thorax, and abdomen of an insect, the cephalothorax and abdomen of a crayfish, or the prosoma and opisthosoma of a spider (Figure 15.1). The individual segments within a tagma are not always distinct, but a study of embryonic development, the layout of ganglia, and the arrangement of appendages are helpful in establishing how many segments belong to that part of the body.

The appendages of arthropods are not only **paired,** but also **jointed** (Figures 15.1, 15.2, 15.7; Figure 15.8, B). This is the most striking feature of the group, whose name means "jointed legs." In some arthropods, every segment has a pair of appendages. In others, appendages develop only on certain segments.

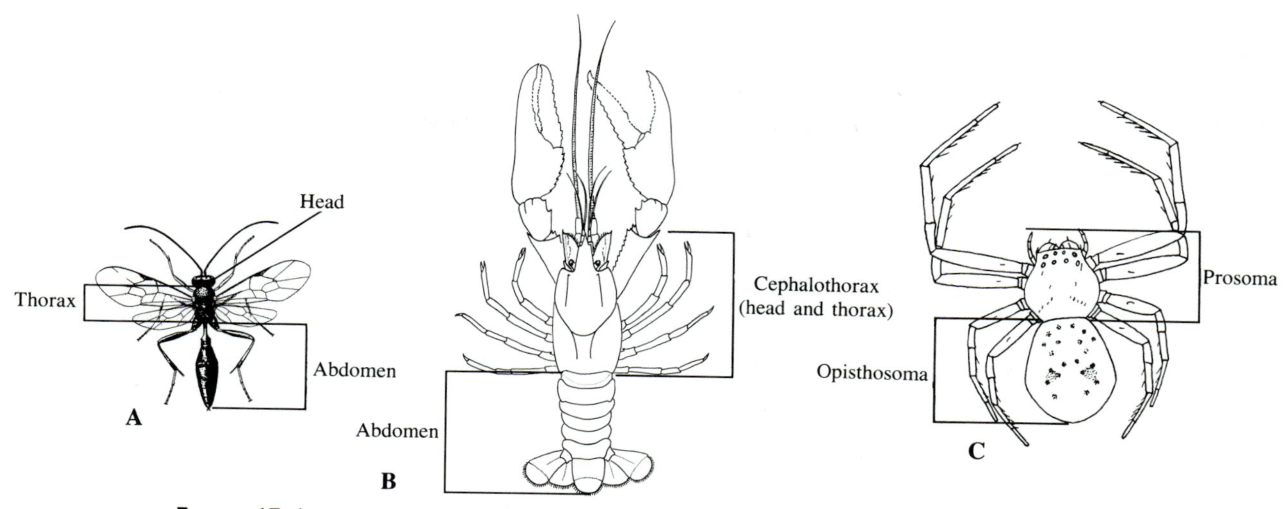

FIGURE 15.1 Tagmata in three diverse types of arthropods. A. Insect. B. Crayfish. C. Spider.

Most appendages, whether they are legs, mouthparts, or other structures, consist of several tubular pieces that are properly called *articles*. To call these pieces *segments* is to risk confusing them with body segments. The term *joint* is often used in place of *article,* but this also is unsatisfactory, because it is more suitably applied to a place where two articles meet and where bending takes place.

In the now-extinct arthropods called trilobites, all of the appendages except the antennae had two branches. The branches were decidedly different, but both originated from the short first article, called the **precoxa** (Figure 15.2, A). Among living arthropods, only the Crustacea have appendages of this type, which are said to be **biramous.** In Crustacea, however, the two branches—the inner **endopodite** and outer **exopodite**—originate from the second article, or **basis,** rather than from the first article, which is called the **coxa** (Figure 15.2, B). The exopodite, moreover, is often absent, so that some or most of the appendages are **uniramous.**

Crustacean appendages may have coxal outgrowths called **epipodites** (Figure 15.2, B), which function as gills. They are soft and not divided into articles, and should not be confused with exopodites or endopodites.

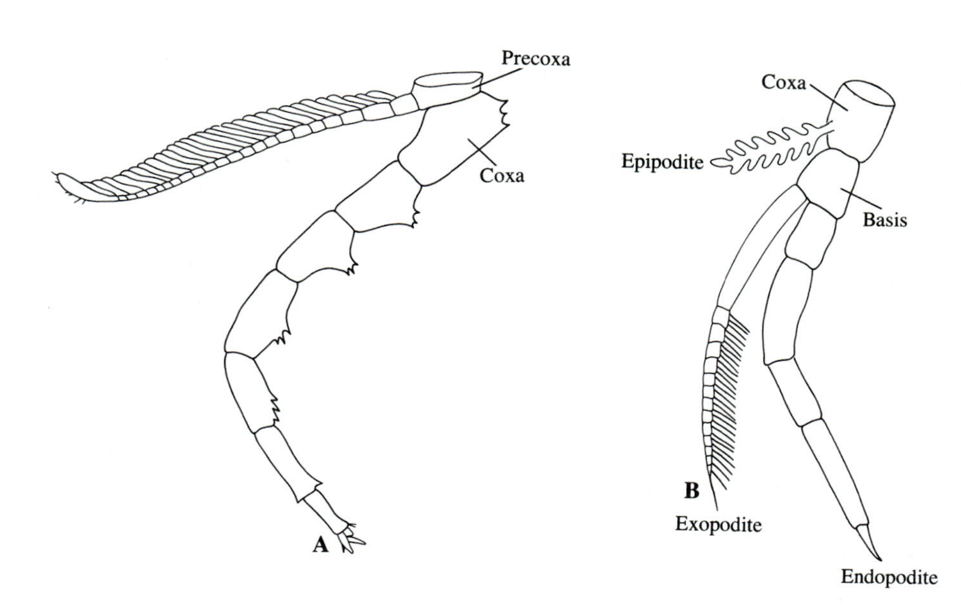

FIGURE 15.2 Biramous appendages; diagrammatic and generalized. A. Trilobite. B. Crustacean.

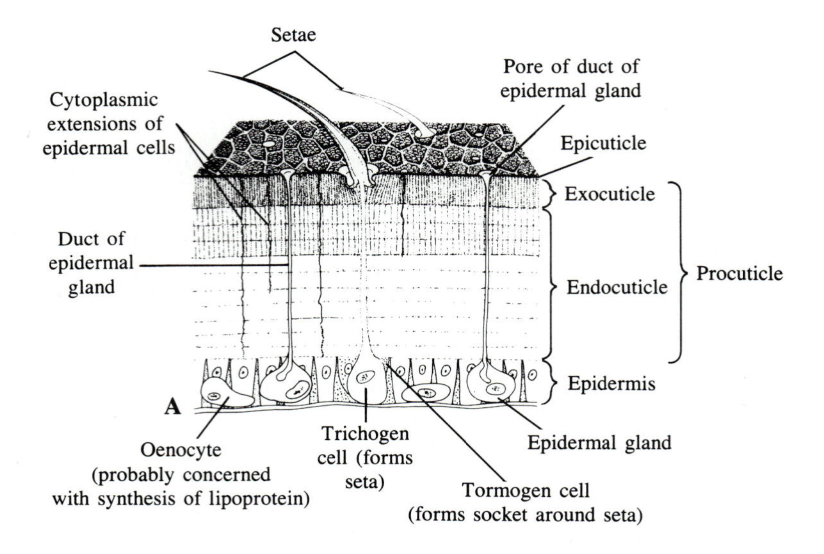

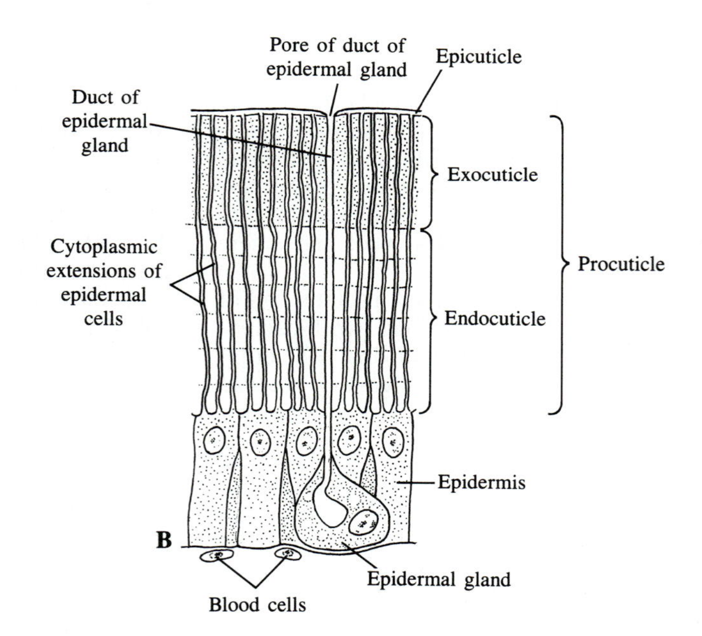

FIGURE 15.3 A. Diagrammatic representation of the cuticle and epidermis of an arthropod. (Hackman, in Florkin and Scheer, Chemical Zoology, *6*.) B. Diagram showing the cytoplasmic extensions of the epidermal cells in the cuticle of an insect.

These and various other specializations of crustacean appendages are dealt with in Chapter 17.

CUTICLE

Another cardinal feature of arthropods is the external skeleton, or **exoskeleton.** This is a **cuticle** secreted by the **epidermis.** It typically consists of two main layers (Figure 15.3). The outer layer, the **epicuticle,** is usually thin; its principal chemical constituents are proteins, lipids, and waxes. The much thicker inner layer, or **procuticle,** is subdivided into two principal strata, both containing **chitin** in addition to proteins and other substances. Chitin is similar to a polysaccharide, although it has a nitrogenous fraction, and it is linked with proteins. It is moderately permeable to water and to a variety of dissolved substances, and it is flexible.

The two strata of the procuticle are the **exocuticle** and **endocuticle.** The exocuticle is **sclerotized,** or "tanned." Sclerotization is a process in which protein chains link together to form a tough layer.

The endocuticle is the stratum that becomes calcified in crabs, lobsters, and some other hard-shelled arthropods. As a rule, however, the calcification does not ex-

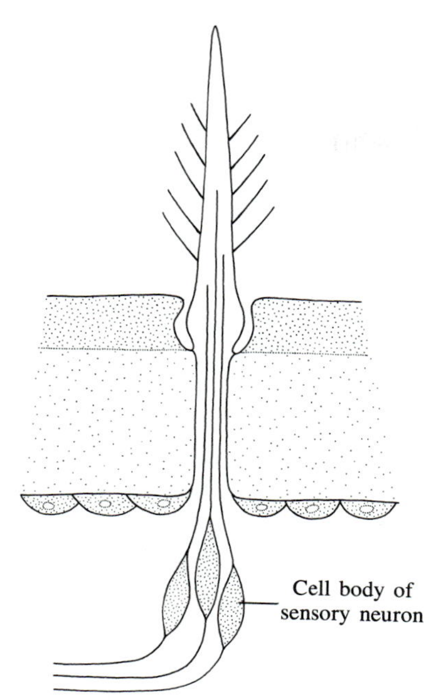

FIGURE 15.4 Sensory seta of the crab *Pinnixa chaetopterana*. (After Hanström, Vergleichende Anatomie des Nervensystems der Wirbellosen Tiere.)

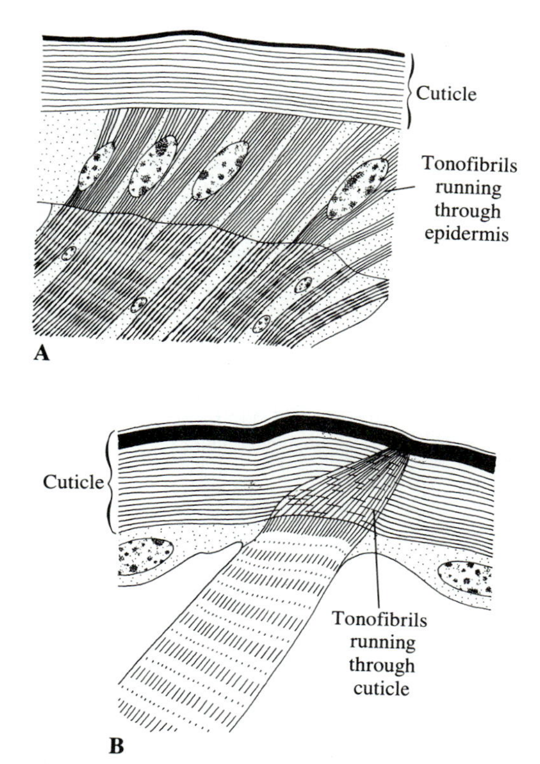

FIGURE 15.5 Diagrams showing tonofibrils by which the cells of an insect muscle are attached to the cuticle. A. In a beetle, *Dytiscus*. (After Casper, Zeitschrift für Wissenschaftliche Zoologie, *107*.) B. In a fly. (After Perez, Archives de Zoologie Expérimentale et Générale, Series 5, *4*.)

tend to a thin layer of endocuticle that is next to the epidermis. Extending outward through both strata of the procuticle, and also through the epicuticle, are ducts by which secretions of epidermal gland cells may reach the surface. Cytoplasmic extensions of the epidermal cells also pervade the cuticle, so the cuticle is, in part, a living layer.

The cuticle of an arthropod has several functions. It provides a protective covering and is rigid enough to serve as an exoskeleton for the body, mouthparts, legs, and other appendages. The crushing or chewing jaws of various arthropods and the powerful pincers of crabs and lobsters are examples of appendages whose thick, hard cuticle fits them for the work they do. At least many of the bristlelike outgrowths of the cuticle, called **setae,** are hooked up with nerve endings and operate as sensory structures (Figure 15.4). In insects and other arthropods that have air passages from which oxygen is taken up by the tissues or by the blood, these passages are lined by a delicate cuticle. Finally, both the foregut and hindgut, whose epithelia are of ectodermal origin, have a thin cuticular lining. In some arthropods, the foregut is an elaborate grinding mill or has other modifications that are thickenings of the cuticle.

If the cuticle is to serve as an exoskeleton, it must be organized in such a way that muscles can be attached to it, and it must be flexible at all points where bending is to take place. Muscles are joined to the procuticle by delicate fibers called **tonofibrils** (Figure 15.5), which are produced within the intervening epidermal cells. Where muscles are attached to the cuticle, the cuticle may not be noticeably modified, but often it is thickened on its inner surface, or infolded to form thin sheets called **apodemes** (Figure 15.6). Anyone who has feasted on crab or lobster will probably remember these structures, which provide extensive surfaces for muscle attachment. The tendons (Figure 15.7) that enable muscles to operate structures that they do not reach directly are apodemes, too.

Even if most of the cuticle is hard or thick, or both, it is thin and pliable at points of articulation. It is also folded at these joints (Figure 15.8, A), so that when bending takes place, it unfolds instead of being stretched on one side. Frequently, however, there are additional specializations of the cuticle where two articles or segments meet. The more common of these are condyle-socket joints (Figure 15.8, B) that slightly resemble condyle-socket joints of vertebrates. They permit more varied movements than would be possible with simpler, hingelike joints.

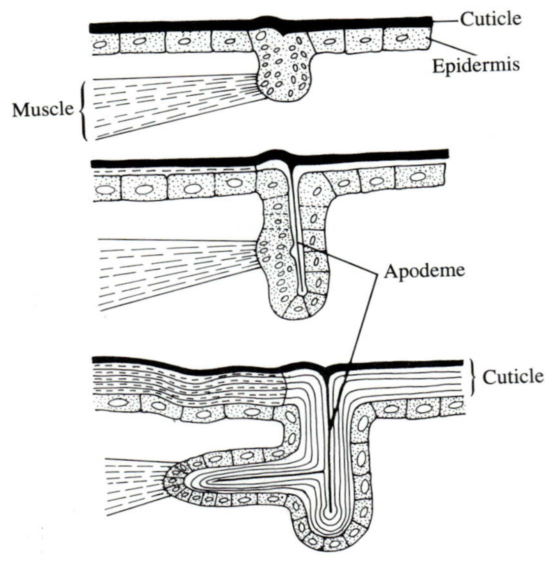

FIGURE 15.6 Formation of an apodeme between two segments of an arthropod; diagrammatic. (After Janet, Anatomie du corselet et histolyse des muscles vibrateurs, après le vol nuptial, chez la reine de la fourmi.)

In the body proper, portions of the cuticle that are especially thickened or stiffened are called **sclerites.** In general, each segment has a dorsal sclerite, or **tergum,** and a ventral sclerite, or **sternum.** Between them is a lateral region called the **pleural membrane,** where the cuticle is thinner and softer. The pleural membrane allows for some "give," and is functionally comparable to the thinner cuticle between successive segments. A relatively straightforward arrangement of these components can be seen in most segments of the abdomen of an insect (Figure 15.9, A).

In the appendage-bearing region of an arthropod (Figure 15.9, B), the pleural area is often sclerotized to a considerable extent, so that a hardened plate called the **pleuron** is formed. A further complication is that the tergal, sternal, and pleural plates are often subdivided, partially or completely, into two or more pieces, each of which has a functional importance. For instance, the wing of an insect, which is an outgrowth of a tergum, articulates with a process of one of the pleural pieces of the same segment. This process in fact provides the fulcrum for the wing's movement. The sclerites of the head region, which collectively form a hard capsule, are difficult to homologize with those of the thorax and abdomen, and a special nomenclature has been devised for them.

It is impossible to discuss the many different arrangements of sclerites found in the various body regions of the many diverse groups of arthropods. Some of them are mentioned below in connection with the groups themselves. It should be pointed out, however, that the tergum generally is the most expansive sclerite. Sometimes it nearly encircles the segment. This is the case in certain millipedes, in which the pleural and sternal portions of a segment are jammed together to form a small ventral plate to which the legs are attached.

COLORATION

Sclerotized proteins usually are colorless or amber-colored, but there may be pigments, such as melanin, in either the sclerotized or nonsclerotized portion of the cuticle. Moreover, coloration in arthropods is not always due to pigments deposited in the cuticle itself. If

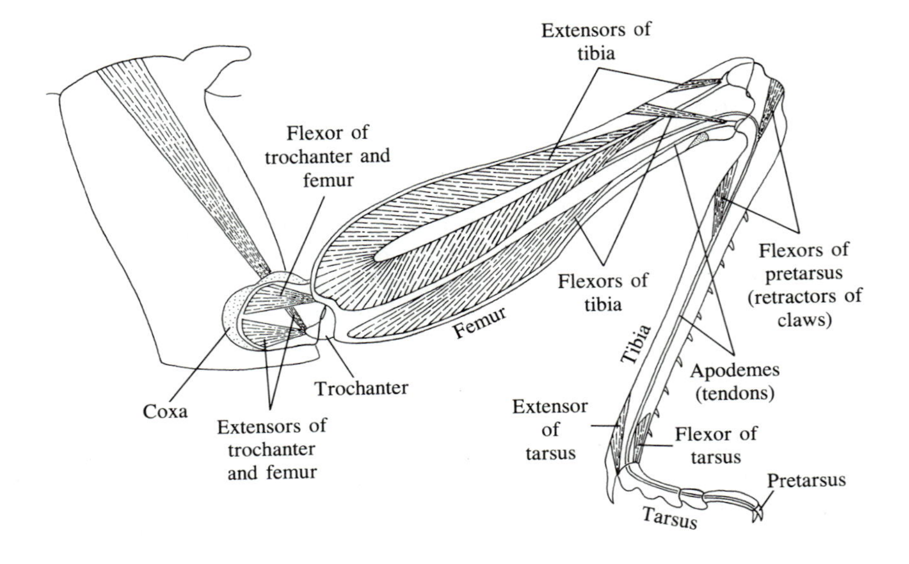

FIGURE 15.7 Muscles and tendons (apodemes) in the leg of a grasshopper. (Some smaller muscles have been omitted.) (After Snodgrass, Principles of Insect Morphology.)

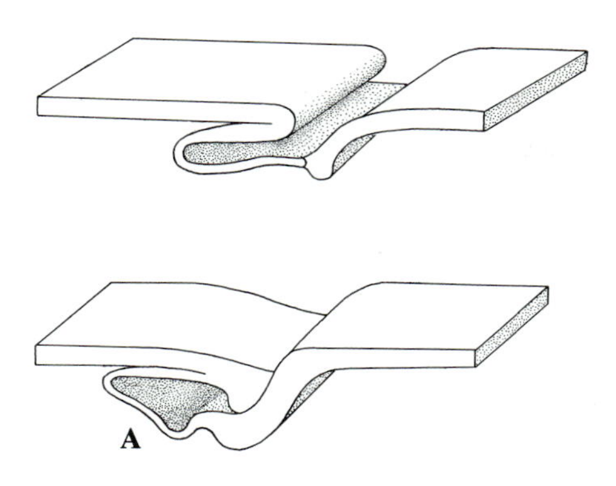

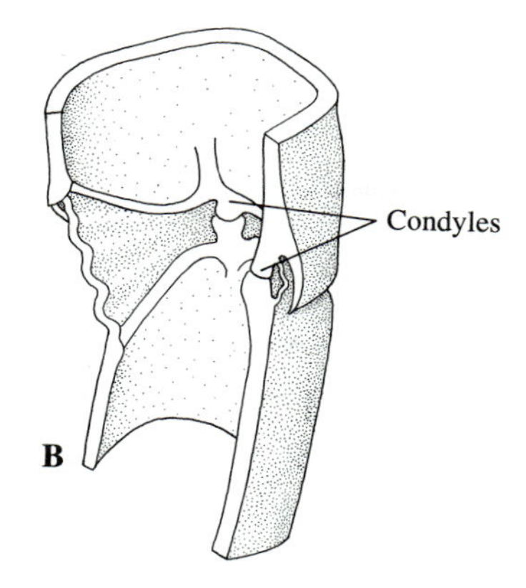

FIGURE 15.8 Some types of articulations in arthropods. A. Two types of articulations between segments. B. An articulation between articles of an appendage, showing two condyles. (After Weber, Lehrbuch der Entomologie.)

the cuticle is thin and transparent, carotenoids and other pigments in the blood or other underlying tissues may show through. There may also be specialized pigment-bearing cells, called *chromatophores,* under the epidermis; these are discussed in connection with Crustacea, in Chapter 17.

Another type of coloration in arthropods is **structural color,** which is due to the refraction of light from small scales, delicate striations in the cuticle, or other irregularities of the surface. The situation is comparable to that of a blue jay, which is blue because the feather barbs are constructed in such a way that they reflect blue light. The metallic or iridescent gold, green, blue, and purple colors observed in many arthropods, especially beetles, butterflies, and moths, have a similar structural basis. The orange, yellow, and bright white seen in butterfly wings are due to reflective deposits of crystallized nitrogenous wastes related to uric acid. During the pupal stage, the wastes accumulated during the larval stages are redistributed, and most of them end up in the wings.

MOLTING

Because the cuticle of an arthropod fits the animal tightly, and because it is usually not especially stretchable, it must be shed periodically to allow for growth. In other words, linear growth takes place in spurts that follow each of several to many molts. Although the old cuticle is shed in a relatively short time, there is much

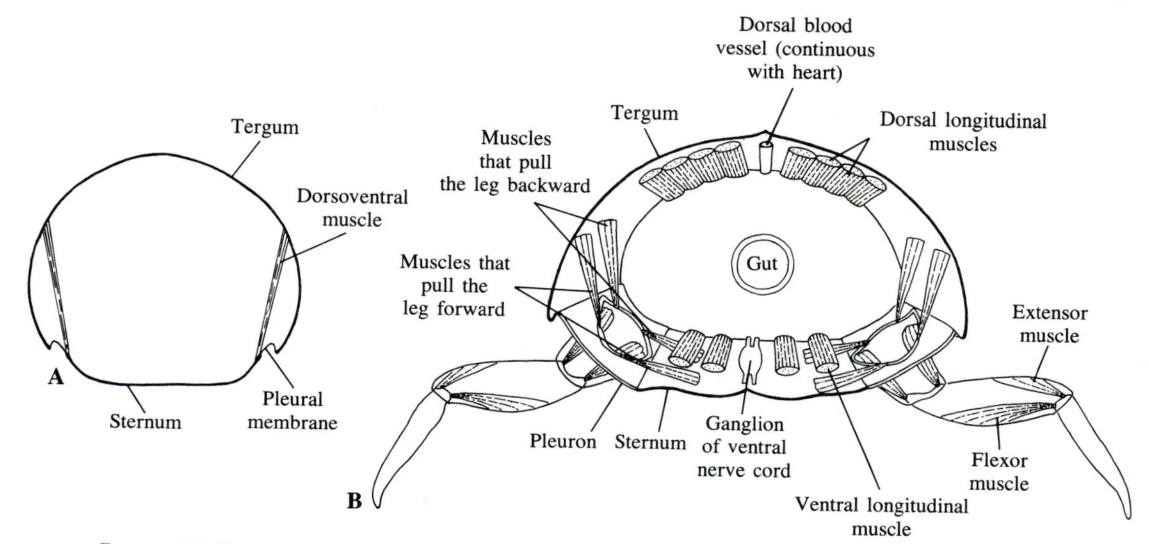

FIGURE 15.9 A. Transverse section through the abdomen of an insect; diagrammatic. The only muscles shown are those that have a primarily dorsoventral orientation. These compress the abdomen by pulling the tergum closer to the sternum. B. Transverse section through an appendage-bearing segment of an arthropod, seen in perspective; generalized and diagrammatic.

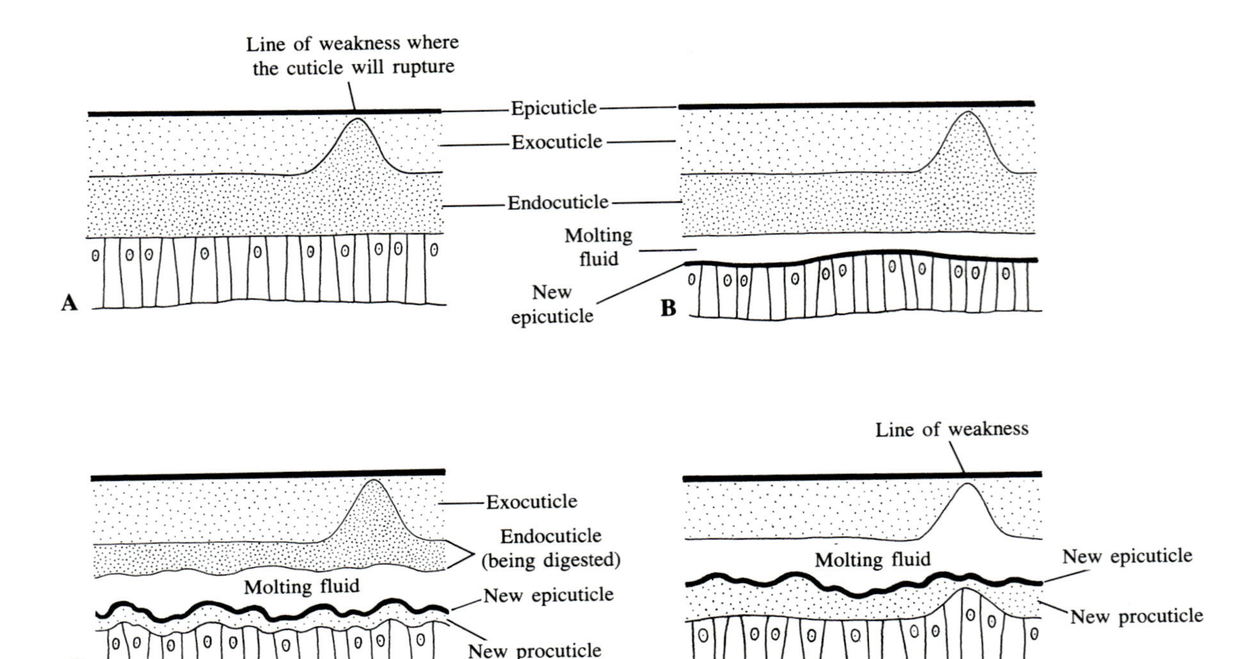

FIGURE 15.10 Stages in the molting process of an arthropod. A. The cuticle and epidermis between molts. B. Accumulation of molting fluid beneath the old cuticle, and formation of a new epicuticle. C. Breakdown of the old endocuticle and formation of a procuticle beneath the new epicuticle. D. Secretion of new procuticle nearing completion; old procuticle now reduced to exocuticle, ready to be shed.

advance preparation for this event. During the period that it is growing, an arthropod is almost continually involved in processes associated with molting, and before it slips out of its old cuticle it must already have secreted much of its new cuticle. Some arthropods cease to grow and molt after they reach the adult stage. This is the case, for instance, in many insects, including beetles and butterflies. Others, such as crabs and lobsters, molt at intervals indefinitely.

The molting process begins with production of a fluid that accumulates between the epidermis and the cuticle, thus separating these two components (Figure 15.10, A and B). The epidermis secretes a new epicuticle (Figure 15.10, B) and then a new procuticle (Figure 15.10, C and D). The epidermis produces chitinase and proteases that pass through pores in the new cuticle and digest much of the old procuticle, of which only the sclerotized exocuticle resists destruction. The end result of all these proceedings is the formation of a substantial part of a new cuticle and the readiness of the old cuticle to be sloughed off. The old cuticle splits along predetermined lines (Figure 15.11), and the animal crawls out of it. The animal then takes up considerable water or air into the gut, and because the incompletely formed new cuticle is soft and stretchable to some extent, the body becomes larger. This, of course, is not true growth, because it does not involve addition of tissue; but so long as the cuticle has been stretched before it hardens, the animal can take its time in adding new tissue and eliminating the excess water or air.

More procuticle—sometimes most of it—is added after the molt, and in crabs and other crustaceans whose procuticle is calcified, much of the calcium carbonate and calcium phosphate are deposited within a few days. During the intermolt period of some terrestrial and freshwater crustaceans, including crayfishes, calcium salts are stored as *gastroliths* ("stomach stones"). These are formed in outpocketings of the stomach and serve as reserves on which the animal can draw when it needs calcium. For marine crustaceans, calcium is not likely to be in short supply.

MUSCLES

Muscles involved in locomotion are **striated** and are concentrated in discrete bundles. Only the heart and a few other visceral structures are enveloped by smooth muscles that are organized into sheets or layers. The

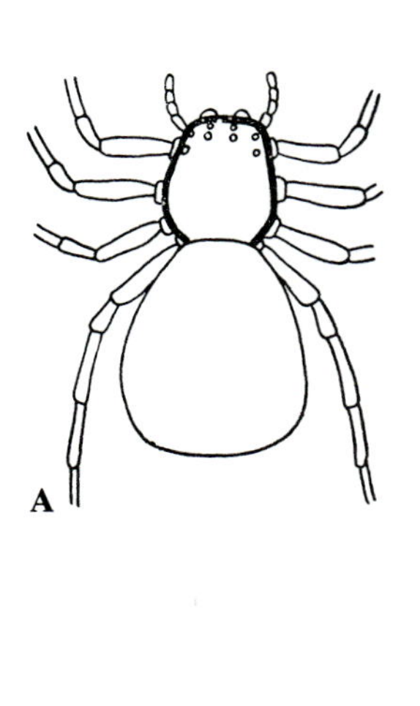

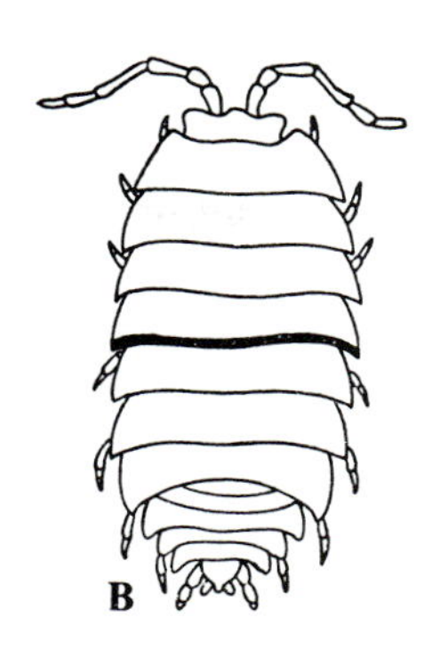

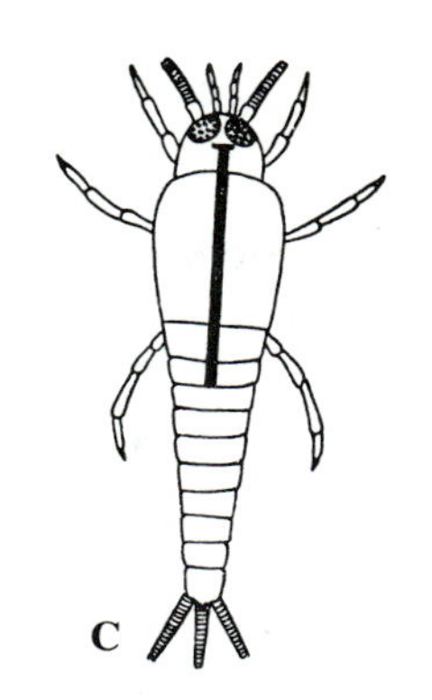

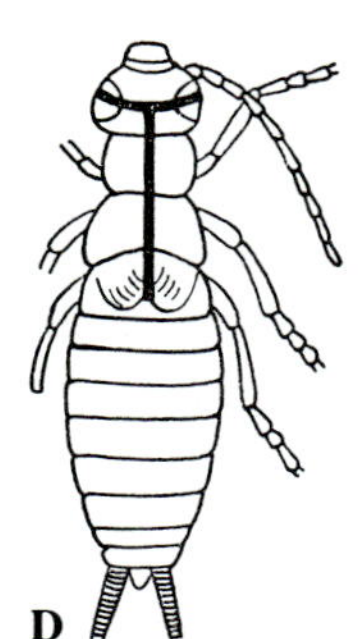

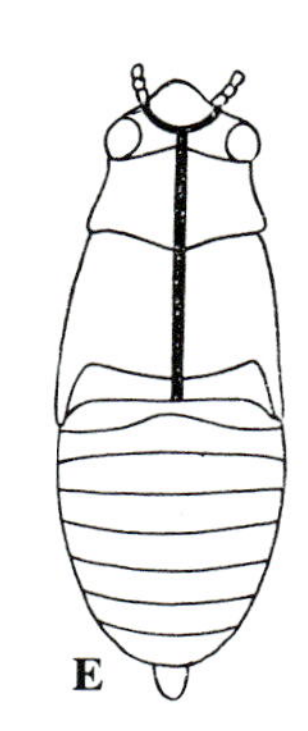

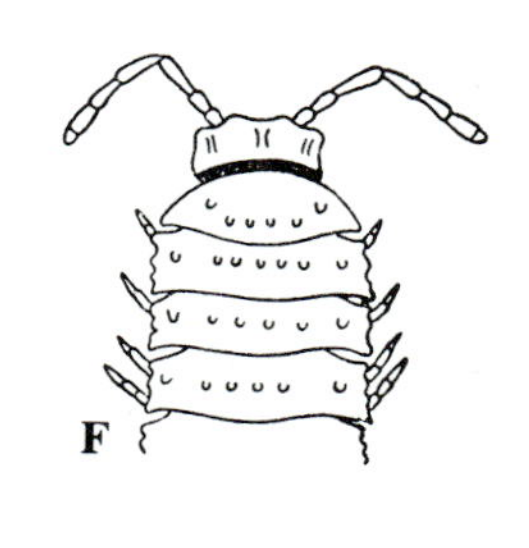

FIGURE 15.11 Molting fissures (heavy black lines) in various arthropods. A. Spider. B. Isopod crustacean. C. Silverfish (insect). D. Earwig (insect). E. Cicada (insect). F. Millipede. (Lawrence, Biology of the Cryptic Fauna of Forests.)

bundled-up arrangement of striated muscle cells is advantageous for operating jointed appendages and movable body segments. Each muscle, with its origin in one article or segment and its insertion in another, has a specific and sharply focused effect when it contracts (Figure 15.7).

Arthropod muscle bundles consist of fewer cells than would be present in skeletal muscles of most vertebrates. Among the neurons supplying them, moreover, some are inhibitory. The action of these is directly opposite that of neurons which stimulate a particular muscle cell to contract. A vertebrate skeletal muscle is divided up into *motor units,* innervated strictly by excitatory neurons. The graded responses of the muscle depend on how many of the units are stimulated. In arthropods, graded responses depend on how many cells are stimulated and how many are inhibited. Innervation of skeletal muscles by inhibitory neurons is unique to arthropods.

The contractile cells of arthropod muscles are of two types: phasic and tonic. *Phasic cells,* or "fast" cells, are those that exhibit a quick response, such as would be necessary for an insect to fly. *Tonic cells,* or "slow" cells, respond slowly and are important in maintaining posture or in carrying out slow movements. A particular muscle may consist entirely of phasic or tonic cells, or it may contain cells of both types. Tonic muscle cells are innervated only by tonic neurons. Phasic cells may be innervated entirely by phasic neurons or by both phasic and tonic neurons, in which case they can operate either as phasic or as tonic cells (Figure 15.12). Thus the nature of a muscle's response depends not only on the type of contractile cells that are in it, but also on the nervous input these cells receive.

THE HEAD AND ITS APPENDAGES

Whether the head region is distinct or is part of a more comprehensive tagma, such as a cephalothorax, arthropods are highly cephalized invertebrates. To understand the head region, it is necessary to consider not only its general organization and appendages, but also its embryonic development and the components of the nervous system that are part of it. Early in the development of an arthropod embryo, an anterior **cephalic lobe** differentiates (Figure 15.13). An important external feature derived from this in practically all arthropods is the *labrum,* the "upper lip" that overhangs the

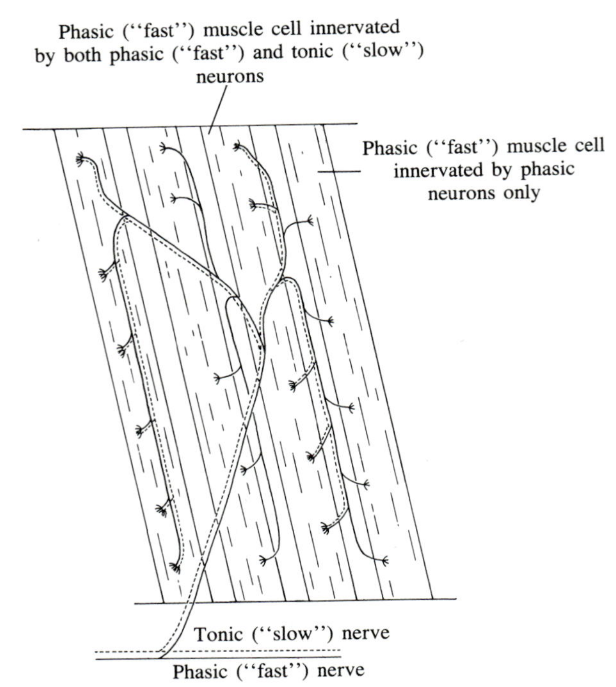

FIGURE 15.12 Diagram showing double innervation of insect muscle. In the example shown, some muscle fibers are capable of phasic ("fast") action or tonic ("slow") action; others are capable only of phasic action. (After Hoyle, in Scheer, Recent Advances in Invertebrate Physiology.)

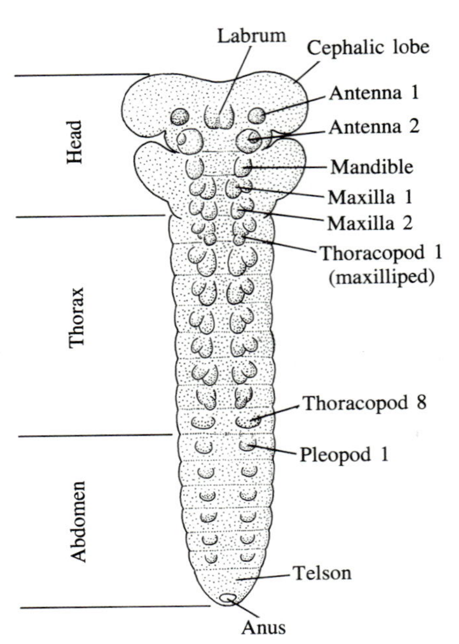

FIGURE 15.13 Diagram of the embryo of an isopod crustacean, showing the first antennae and labrum developing on the cephalic lobe, and the remaining appendages developing on the true segments of the head, thorax, and abdomen.

mouth. The labrum develops, in fact, from the anteriormost part of the cephalic lobe. If eyes are present, these also appear on the cephalic lobe. The same is true of the antennae of insects and the first antennae of crustaceans. The cephalic lobe of the embryo is therefore similar to the prostomium of an annelid, especially one like *Nereis,* which has eyes and sensory outgrowths.

The antennae that develop on the cephalic lobe of arthropods may be homologous with prostomial palps of polychaetes; however, a cephalic lobe and a prostomium may not be exactly comparable. A cephalic lobe sometimes shows one or more embryonic somites, although these do not differentiate into segments. The prostomium of an annelid does not form somites.

Crustaceans have a second pair of antennae. Although these are in front of the mouth, they originate on the somite behind the embryonic mouth. During development, in other words, they shift to a preoral position. Insects and myriapods—millipedes, centipedes, and their relatives—lack second antennae, but slight traces of these appendages may be present on the first postoral somite of the embryo.

In certain crustaceans, such as fairy shrimps (order Anostraca), an anterior portion of the body bearing the first and second antennae is rather distinctly set off from the rest of the animal. In that case it may be said that there is a *primary head* consisting of structures derived from the cephalic lobe and first postoral somite. In most crustaceans, however, this primary head is tightly joined to three additional postoral segments that bear the mouthparts. The mouthparts of crustaceans are called, in order, *mandibles, first maxillae,* and *second maxillae*. Insects and myriapods have the same arrangement. It is important not to assume that the mandibles and succeeding mouthparts of crustaceans are exact homologues of those bearing the same names in insects and myriapods. Some of the more hotly debated controversies of arthropod morphology and phylogeny are concerned with this problem of homology.

Spiders, scorpions, and their allies have no first antennae, and no appendages quite comparable to second antennae, mandibles, or first or second maxillae. Their only preoral appendages, called *chelicerae,* originate on the first postoral somite. The developmental history of chelicerae is therefore similar to that of the second antennae of crustaceans, and, in fact, these two sets of appendages may be homologous. The first postoral appendages in spiders and other chelicerate arthropods are called *pedipalps*. They almost always have a sensory function, but they may also be used for locomotion and other purposes. The pedipalps are succeeded by walk-

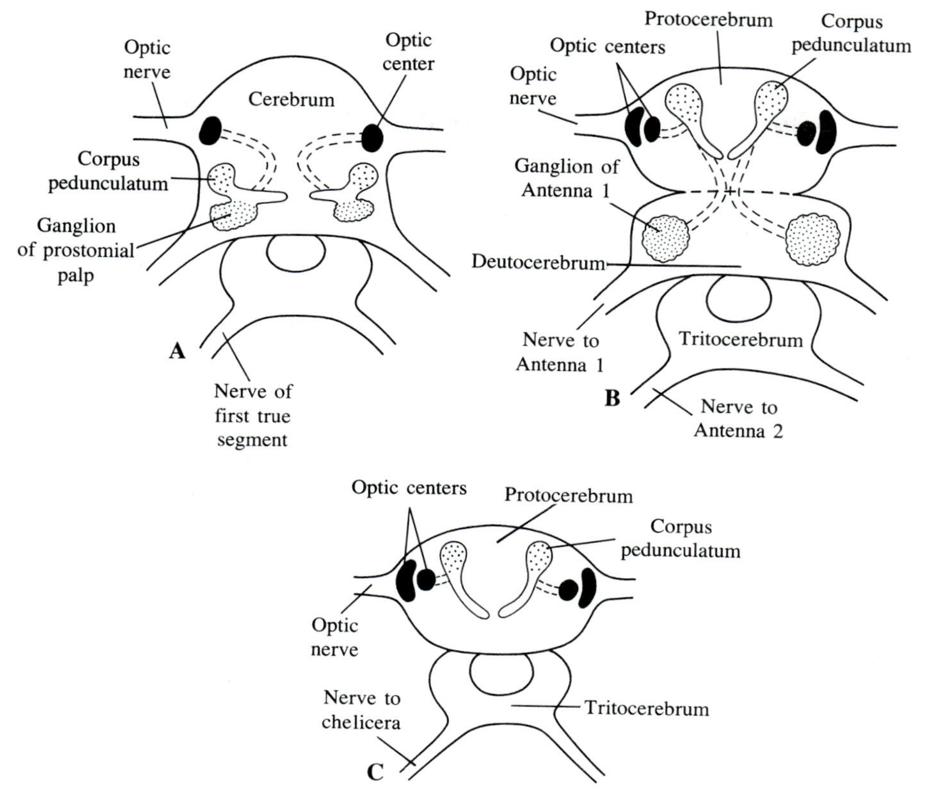

FIGURE 15.14 Comparison of the brain of a polychaete annelid with brains of two major groups of arthropods. A. Polychaete annelid, such as *Nereis*. B. Crustacean or insect (second antennae are absent in insects.) C. Chelicerate arthropod. The protocerebral structures called corpora pedunculata have had various functions ascribed to them. In general, they probably receive sensory input from visual and olfactory pathways and act as coordinating centers, thereby influencing motor responses. They are almost always well developed in social insects, and perhaps are centers for complex behavior patterns influenced by sensory clues. Much, however, remains to be learned about them. (After Hanström, Vergleichende Anatomie des Nervensystems der Wirbellosen Tiere.)

ing legs, and the usual number of these is four pairs. The term *prosoma* is applied to the portion of the body derived from the cephalic lobe, the embryonic somite that produces the chelicerae, and the succeeding five somites, which give rise to the pedipalps and walking legs. The question of homologies will not go away, however. Are the pedipalps of a spider homologous to mandibles of crustaceans, insects, and myriapods? Are the first walking legs homologous to first maxillae, and the second legs to second maxillae? These are difficult questions, and experts do not agree on the answers. Anyone who has a special interest in comparative morphology, embryology, and phylogeny of arthropods should consult the treatises listed at the end of this chapter.

BRAIN AND VENTRAL NERVE CORD

In arthropods other than chelicerates, the brain has three main parts (Figure 15.14, B). The first of these is called the **protocerebrum.** It innervates the eyes. The next portion, the **deutocerebrum,** innervates the antennae of insects and myriapods and the first antennae of crustaceans. It is generally agreed that the last part of the brain, called the **tritocerebrum,** is a postoral ganglion that has moved to a preoral position. In crustaceans it innervates the second antennae, which are lacking in insects and myriapods.

In chelicerate arthropods (Figure 15.14, C), the protocerebrum innervates the eyes. The deutocerebrum is not recognizable, and fitting in with this fact is the absence of appendages comparable to the first antennae of crustaceans or the antennae of insects and myriapods. The tritocerebrum innervates the chelicerae.

In annelids (Figure 15.14, A), the brain is lodged in the prostomium and innervates the eyes, palps, and antennae, when these are present, as they are in many polychaetes, such as *Nereis*. The connectives that run ventrally around the gut are joined to the first ganglion of the ventral nerve cord; this ganglion is posterior to the mouth. There is a possibility that the protocerebrum and deutocerebrum of the arthropod brain form the equivalent of the *Nereis* brain, and that the tritocerebrum is comparable to the first postoral ganglion of *Nereis*. Some would say, however, that the deutocerebrum of arthropods is also a postoral ganglion.

The ventral nerve cord of arthropods is similar to that of annelids in that it is a double structure in which successive pairs of segmental ganglia are joined by longitudinal connectives. There are commissures between the members of a pair of ganglia, but these are not recognizable as such when the ganglia are so tightly fused that their duality is obscured. In some groups, such as scorpions, lower crustaceans, and lower insects, many of the successive ganglia are distinct. Condensation of ganglia, however, is characteristic of other groups. In spiders, for instance, all of the ventral ganglia are run together into a compact mass, and the brain rests directly on the anterior part of the mass.

SENSE ORGANS

Arthropods have a wide variety of sense organs. This is to be expected in a group that is so diversified. There are hairs sensitive to touch, chemoreceptors, structures capable of picking up sounds, proprioceptors, statocysts, and eyes.

Touch Receptors, Chemoreceptors, and Auditory Organs

When one looks at a fuzzy insect, spider, or crustacean, one must realize that although many of the cuticular hairs and bristles may have a protective function, at least some are provided with touch receptors or chemoreceptors. Touch receptors are likely to be in the form of nerve endings located at the bases of the bristles (Figure 15.4). A slight change in the orientation of the bristle, as when it is touched, stimulates the nerve ending. Chemoreceptors are not always located in bristles, but when they are, the nerve processes run right to the open tips of the bristles. Chemoreceptors may also be clustered in stout elevations of the cuticle, or may be concentrated in pits or other depressions. Touch receptors and chemoreceptors are generally more abundant on the antennae, mouthparts, and other appendages than elsewhere.

Auditory organs, capable of picking up sound vibrations, generally operate on the "drumhead principle." A thin cuticular membrane stretched across a depression vibrates and stimulates nerve endings that are indirectly joined to it.

Proprioceptors

Proprioceptors are sense organs that record changes in tension in tissues. Such changes take place when an animal moves, and even when it is at rest, because it has weight and because external mechanical forces impinge upon it. The proprioceptors of vertebrates are chiefly located in muscles and tendons. Presumably there are proprioceptors of one sort or another in nearly all groups of invertebrates, but certain of those in arthropods have been studied rather intensively.

Just two types of arthropod proprioceptors will be mentioned: those located at joints and at wing bases, and those associated with muscles. In proprioceptors of the first type, dendrites of sensory neurons are associated with soft cuticle or with strands of cells and collagen that extend from one article of an appendage to another. Any change in tension that takes place when the cuticle is deformed by bending, or when a strand of the type just described is stretched or bent, will be registered by the dendrites.

Proprioceptors associated with muscles generally consist of a modified muscle cell and dendrites of a sensory neuron that innervate it. As the modified cell is shortened or lengthened in response to the contraction or relaxation of the muscle itself, the dendrites are stimulated.

Statocysts

Statocysts are almost always found in shrimps, crabs, and related crustaceans. They are usually situated at the bases of the first antennae and operate on the same general principle as statocysts of jellyfishes, turbellarians, and various other invertebrates. Each contains a hard mass, consisting of one or more sand grains and perhaps some secreted material. This mass rests on sensory hairs of the cuticular lining. As the animal's orientation changes, the mass moves and its weight is shifted. The central nervous system is informed of the changes by impulses that originate in the receptors at the bases of the hairs. A crustacean statocyst is formed as an epidermal invagination, and it is lined by a cuticle that is shed at each molt. After the molt, new sand grains must either be deliberately inserted or get into the sac when the animal buries itself. The degree to which the sac remains open varies; sometimes it becomes closed as the animal's exoskeleton thickens and hardens. In any case, the receptors in a statocyst are basically touch receptors, but they are specialized in such a way that the structure as a whole is an organ of balance.

Some crabs and other crustaceans have complicated statocysts with side pockets. Besides containing one or more statoliths, these pockets may have hairs that respond to flow of fluid within the sac or its ramifications. Such statocysts operate in much the same way as the semicircular canals in the ears of vertebrates. They are especially important to crustaceans that swim actively and that must dodge predators, because they efficiently and continuously provide information on an animal's constantly changing orientation.

Eyes

The eyes of arthropods are of two main types: **simple eyes** (often called *ocelli*) and **compound eyes.** In both types, the cornea is part of the continuous layer of cuticle that covers the body. Thus, even if the cornea is

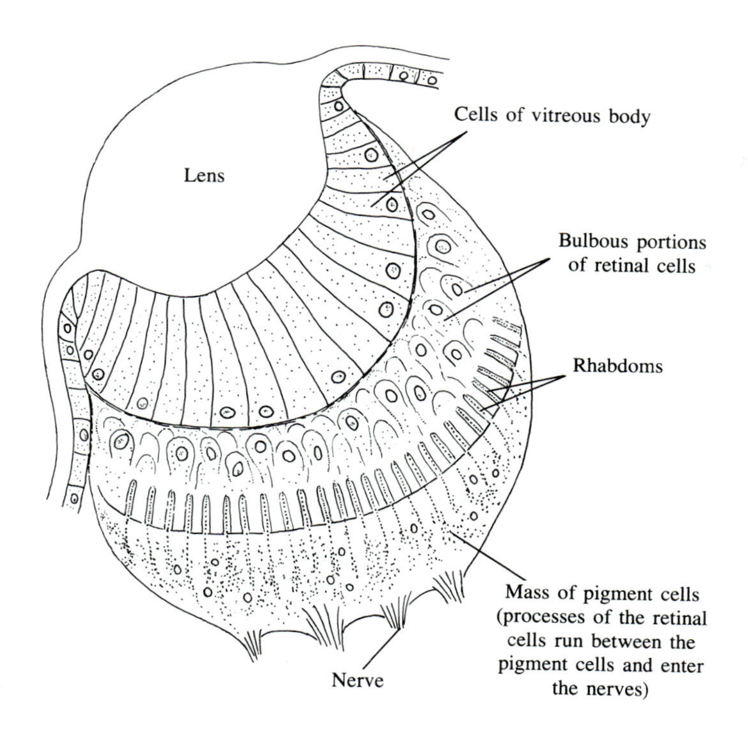

FIGURE 15.15 One of the simple eyes of a wolf spider, *Arctosa varians*. (The diagram is based on electron micrographs of Baccetti and Bedini, Archives Italiennes de Biologie, *102*.)

shaped to function as a lens, there is no way to change the focus. Simple eyes, recognizable as small, glassy dots on the heads of various arthropods, including spiders and insects, have a single lens and several to many photoreceptors (Figure 15.15). Their structure varies considerably, however, and not all of the simple eyes of one animal are necessarily alike. In general, simple eyes do not form images, except in jumping spiders, but they provide information about changing light conditions and therefore about the proximity of prey, predators and other enemies, and prospective mates.

A compound eye, as its name suggests, consists of many units, all of which are more or less equal in terms of their structure and function (Figure 15.16). The units are called **ommatidia,** and their exposed corneal portions, which function as lenses, are generally hexagonal or shaped in some other way that makes them fit tightly together. The crystalline cone beneath each cornea is also a lens, and is secreted by a group of epidermal cells distinct from those that lay down the cornea. Deeper in the ommatidium is a complex called the **retinula,** which consists of several cells arranged in a circle (Figure 15.17). These retinular cells are the actual photoreceptors. They are essentially nerve cells, for they send out axons that lead to ganglia within the brain, or within an eyestalk, in the case of crabs, shrimps, and some other crustaceans. All or most of the retinular cells have, on their inner faces, thousands of crowded microvillar projections. The aggregate of microvilli belonging to one retinular cell (Figure 15.18, A), or to two retinular cells, when there is no clear line of demarcation between those of one cell and those of another (Figure 15.18, C), is called a **rhabdomere.** The central core of an ommatidium, formed by the rhabdomeres and any free space that may be left between them (sometimes there is no extra space), constitute a translucent mass termed the **rhabdom** (or *rhabdome*). Dark pigment shields the sides of each ommatidium. This pigment may be located in the retinular cells, or in cells that are peripheral to them, or in both. The pigment may migrate according to changes in the intensity of the light, but in general its function is to prevent the light striking one ommatidium from leaking into others.

The focus of the individual ommatidia is fixed. Each ommatidium, moreover, is oriented to accept light coming from certain directions but not others. A single ommatidium, in other words, can form only a partial image of a scene. The brain must assemble the information coming from the many ommatidia into an *apposition* image that presumably looks something like a mosaic. When the surface of a compound eye is markedly convex, as it often is, the eye covers a wide field. The system as a whole should function well in detection of movement, as that of a prey or predator, because a very slight shift in any part of the scene will simultaneously affect several ommatidia.

So long as the ommatidia are screened off from one another by dark pigment, there will be little leakage of light from one unit to another. Each ommatidium will

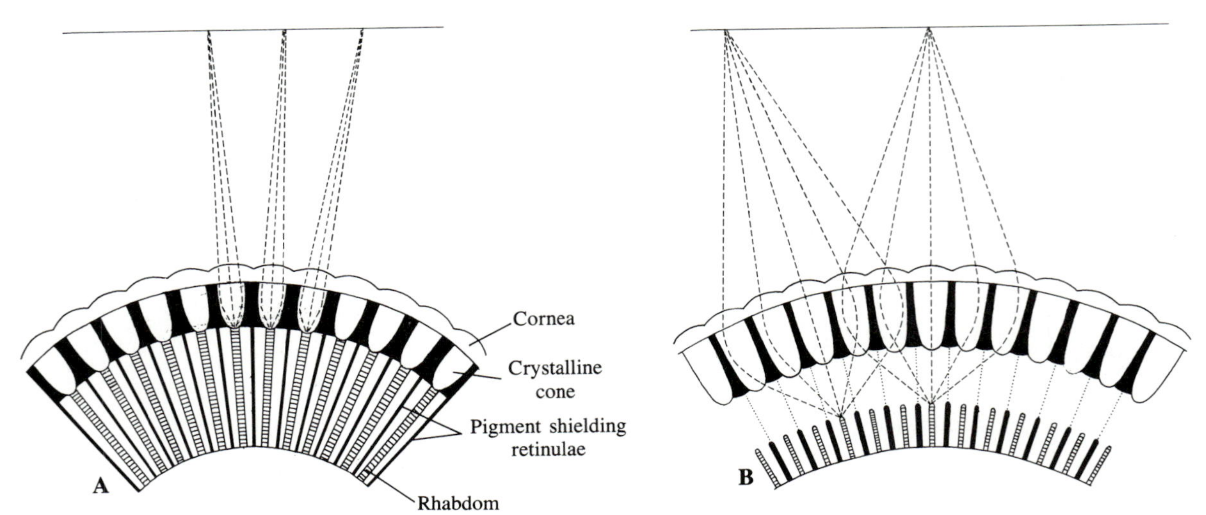

FIGURE 15.16 Diagrams of sections through compound eyes of two main types. A. "Apposition" type. All retinulae are shielded from one another by dark pigment, so that the rhabdom of one ommatidium is affected only by light coming through the cornea and crystalline cone of that ommatidium. B. "Superposition" type. The pigment shields around the retinulae are not complete, so that one rhabdom may be affected by light that has entered several ommatidia. The diagrams show extremes. It is important to appreciate that the eyes of many arthropods are versatile. The pigment in the shields can be dispersed, making the eye function like that illustrated in A, or it can be concentrated near one end of the shields, so that light may pass from one ommatidium to another (see Figures 15.17 and 15.19).

therefore have a unique and narrow field of view. The formation of an apposition image does, however, require reasonably bright light. Nocturnal arthropods, and those that regularly inhabit dark places, tend to have eyes in which there is less screening pigment between ommatidia, so that light rays passing through the cornea of one ommatidium can reach the retinular cells of others. In eyes of this type, called *superposition* eyes, the ommatidia and their cones are adapted for gathering weak light. Visual acuity is likely to be poor. A certain

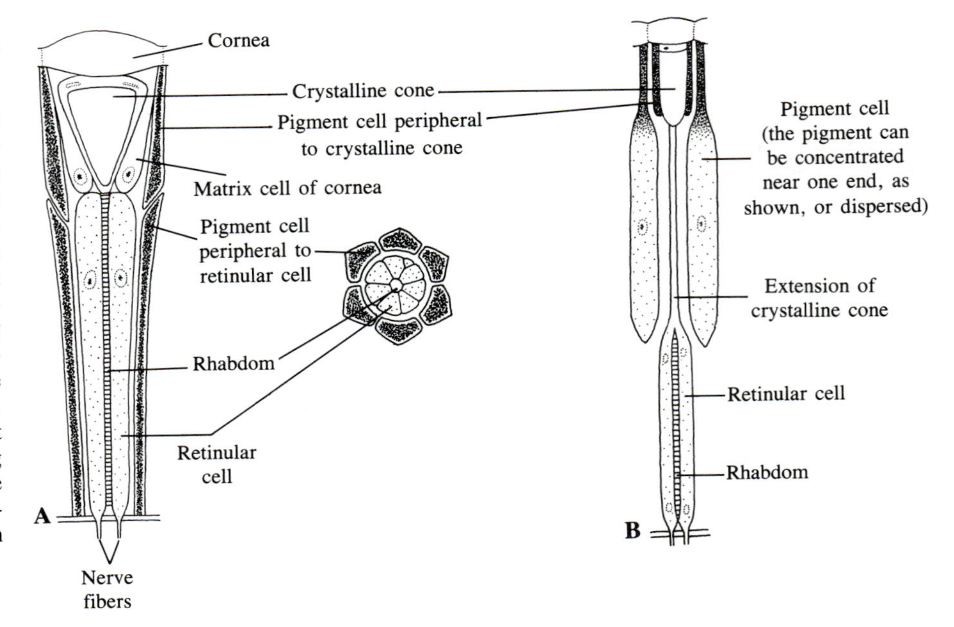

FIGURE 15.17 A. Ommatidium from an eye of the "apposition" type. The pigment in the cells external to the retinular cells is evenly distributed, so that the ommatidium is shielded from light that enters neighboring ommatidia. (After Snodgrass, Principles of Insect Morphology). B. An ommatidium from an eye that can function by "apposition" or "superposition." When light intensity is low, the pigment is concentrated at the distal portions of the peripheral cells, as shown here. The rhabdoms are therefore not shielded from light entering neighboring ommatidia, and the eye may function by "superposition." (After Weber, Lehrbuch der Entomologie).

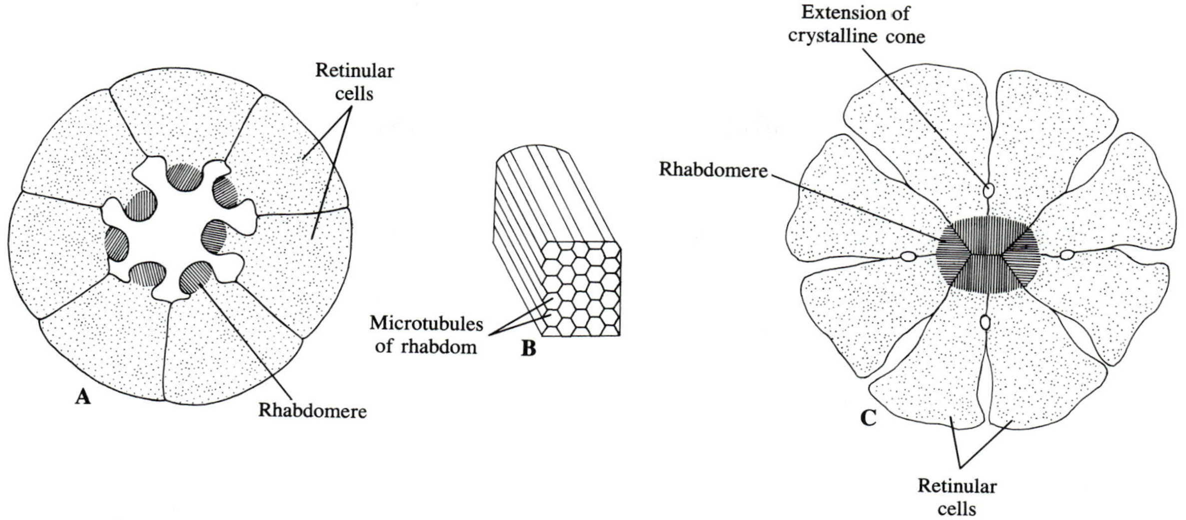

FIGURE 15.18 A. Ommatidium of the fruit fly, *Drosophila*; diagrammatic. The rhabdom consists of seven rhabdomeres and the free space between them. Note that the microvilli of the rhabdomeres have differing orientations. The pattern is repeated in other ommatidia, over a wide area in the eye. (After electron micrographs in Wolken, Invertebrate Photoreceptors, and Waddington and Perry, Proceedings of the Royal Society of London, B, *153*.) B. Arrangement of tightly packed microvilli in a portion of a rhabdom. (After Wolken, Invertebrate Photoreceptors.) C. Ommatidium of the hornet, *Vespa*; diagrammatic. The microvillar portions of the eight retinular cells form four pairs of rhabdomeres, the members of each pair not being distinguishable. Note the four extensions of the crystalline cone, which run proximally between the retinular cells. (Based on an electron micrograph in Wolken, Invertebrate Photoreceptors.)

amount of image formation is probably possible, however, and the animal will certainly be able to recognize changes in light intensity and movements in its dim surroundings. Many arthropods, by controlling the distribution of dark pigment around their ommatidia (Figure 15.19), can use their eyes in bright light or dim light, but it is doubtful that they can form an apposition image except when plenty of light is available and the ommatidia are insulated by their pigment shields.

The number of ommatidia in proportion to the size of an eye has much to do with visual acuity. The situation is somewhat comparable to that of photographic films: the finer the particles of silver in a photographic emulsion, the better the resolution will be when the film is developed. In general, however, an apposition image produced by a compound eye will be much less refined than the image produced by the eye of a higher vertebrate. In the vertebrate eye, of course, only one lens is involved in focusing an image on the retina, and the sensory receptors of the vertebrate retina are more numerous, finer, and more closely spaced than those in compound eyes, so visual acuity is much greater.

Some arthropods can distinguish colors, or at least have receptors that are sensitive to certain wavelengths of light. The eyes of cockroaches, for instance, can discriminate between green and other colors, and they can also pick up ultraviolet light. Color discrimination is important to crustaceans that change color to match their surroundings, as shrimps of the genus *Crangon* do. These shrimps can adjust the distribution of various pigments in their chromatophores so that they are relatively unobtrusive on backgrounds of certain colors. Similar color-discriminating capacities have been found in hermit crabs and other crustaceans, and are presumably helpful in locating shelter and food, and in avoiding predators.

The evolution of compound eyes has had much to do with the success of arthropods. The fossil record shows that they appeared early in the history of these animals, though their distribution is not continuous among living arthropods. Of the chelicerates, for example, only horseshoe crabs have them, but spiders, scorpions, ticks, harvestmen, and all their other relatives do not. Compound eyes are, however, characteristic of nearly all insects, the majority of centipedes and millipedes, and most crustaceans.

ABSENCE OF FLAGELLA AND CILIA

In nearly all phyla of invertebrates, one or more processes depend on activity of flagella or cilia. Among arthropods, however, flagella and cilia are practically nonexistent; the tails of sperm in some crustaceans and a few other groups have a flagellar basis, but that is all.

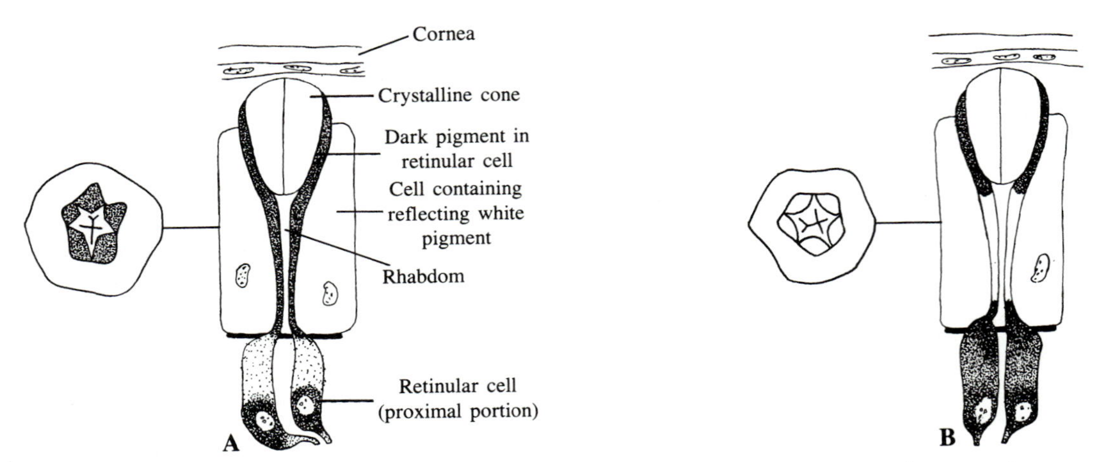

FIGURE 15.19 Ommatidia from the eyes of *Gammarus*, an amphipod crustacean; diagrammatic. The transverse sections are taken at the levels indicated by dashed lines. A. Light-adapted specimen; rhabdom shielded by pigment. B. Dark-adapted specimen; pigment has been moved to the proximal and distal portions of the retinular cells. (Kleinholz, in Waterman, The Physiology of Crustacea, *2*.)

Obviously, the functions that might be served by flagella or cilia must be taken care of in other ways, as they are in nematodes.

Locomotion, whether achieved by crawling, running, jumping, wriggling, swimming, or flying, depends on movements of certain appendages, body segments, and wings. These structures are operated by muscles, as are also the appendages used in feeding, creating respiratory currents, reproduction, and other nonlocomotory activities. Passage of food and fecal material through the gut is promoted to a considerable extent by muscles, too. The operation of some structures is effected by localized changes in blood pressure, but these changes depend, in turn, on muscles.

DIGESTIVE TRACT AND DIGESTION

The digestive tract is complete and consists of three main portions: **foregut, midgut,** and **hindgut.** The first and last, being derived, respectively, from a stomodeal and a proctodeal invagination, are lined by epithelia of ectodermal origin. The epithelia secrete a cuticle that is replaced during each molt, just as the cuticle on the outside of the body is replaced. The cuticle of the foregut may have various elaborations that are used for grinding or straining food, and this region may also have a dilated part in which food can be stored temporarily.

Digestion and absorption generally take place in the midgut, which is lined by an epithelium of endodermal origin and which may have diverticula or glands of varying complexity and function. Sometimes the enzymes produced by the midgut are delivered to the foregut, so that digestion may begin there. The hindgut is concerned to a large extent with organizing fecal material, but the microorganisms upon which termites and some other insects rely for digestion of cellulose are concentrated in this part of the digestive tract. Reabsorption of water, especially important to terrestrial arthropods, is generally a function of the hindgut, too.

CIRCULATORY SYSTEM, HEMOCOEL, AND RESPIRATION

There is a circulatory system in nearly all arthropods. It is said to be of the *open* type, because the blood is not confined to vessels. From the arteries leaving the heart, blood escapes into spaces that bathe the tissues. These spaces collectively form the *hemocoel*. No matter how extensive it may be, the hemocoel must not be confused with a true coelom. Coelomic spaces do develop within the mesoderm of the embryo, but they do not become organized into a large cavity. Their fate is discussed presently.

The heart is characteristically a tubular structure lying above the gut. It is surrounded, at least partly, by a hemocoelic space, the *pericardial sinus* (Figure 15.20). (This is not the same as the pericardial cavity of molluscs because it is not a coelomic derivative.) From this space, blood reenters the heart through one or more pairs of openings called *ostia*. These close when the heart contracts, so backflow is prevented. In general, contraction of the heart depends on muscles in its thin wall. Dilation of the heart, which must coincide with the opening of the ostia, depends on elastic suspensor ligaments that are external to the heart. The lig-

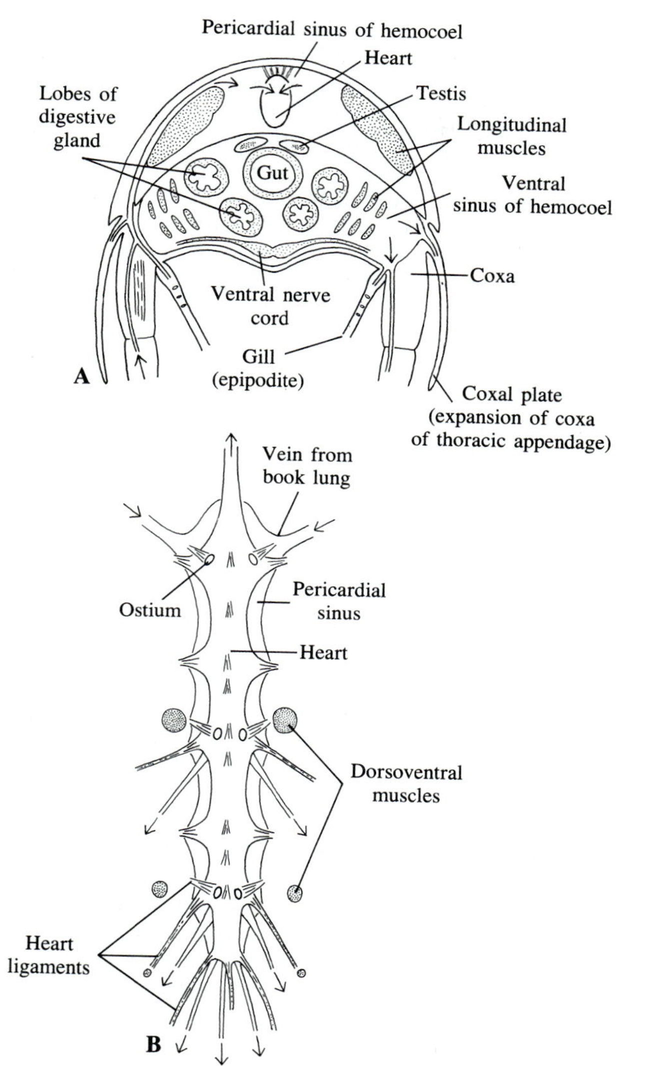

FIGURE 15.20 A. Transverse section through the thoracic region of an amphipod crustacean, showing the heart, pericardial sinus, and ventral sinus. The two sinuses make up most of the hemocoel. Blood leaving the heart by way of anterior and posterior aortas collects in the ventral sinus, which bathes most of the visceral organs. From the ventral sinus it goes to the appendages and gills (coxal epipodites), and then to the pericardial sinus. It reenters the heart by way of paired ostia. Most of the pathways described are indicated by arrows. (After Cussans, in Liverpool Marine Biology Committee Memoirs, *12*.) B. Heart and pericardial sinus of a spider, as seen in a dissection, dorsal view; direction of blood flow indicated by arrows. Blood leaves the heart by way of several arteries. After circulating through much of the body, it is oxygenated in respiratory structures called book lungs (see Chapter 16), and then returned to the pericardial sinus by a pair of large veins. It reenters the heart through the three pairs of ostia. The heart ligaments distend the heart, acting antagonistically to the muscles in the heart wall. (After Ivanov et al., Major Practicum in Invertebrate Zoology.)

aments are stretched when muscles in the heart contract, and they pull on the heart as soon as the muscles relax in preparation for the next contraction.

Hemocyanin is the common respiratory pigment that binds oxygen, but hemoglobin occurs in some representatives of nearly every major group. The distribution of the pigments, in other words, does not follow taxonomic lines. Moreover, while the blood functions in oxygen transport in most groups of arthropods, it does not do so to any great extent in insects. The tracheal respiratory system of these animals consists of tubes whose ultimate branches pervade all tissues and deliver atmospheric oxygen directly to nearly every cell. The subject of respiratory organs and their relationship to the circulatory system is dealt with more thoroughly in following chapters in connection with the major subdivisions of the Arthropoda.

The blood contains a variety of free cells, including phagocytes. The capacity of the blood to clot quickly is an attribute of the blood of many arthropods. This is especially important in certain crustacean and arachnids that will readily shed a limb to escape the grasp of a predator.

COELOM, COELOMIC DERIVATIVES, AND EXCRETORY ORGANS

Of the coelomic spaces that appear within the mesoderm of embryonic somites, some do not persist. A few become involved in formation of the gonads, genital ducts, and various structures concerned with excretion or osmoregulation, or with secretion of saliva or silk. Among the known or presumed excretory–osmoregulatory derivatives of the coelom are the coxal glands of spiders, scorpions, and other arachnids, and the antennal and maxillary glands of crustaceans. The names given to them refer to the location of their pores. Coxal glands open on the proximal articles (coxae) of one or more pairs of legs; antennal and maxillary glands open at the bases of the antennae or maxillae.

An antennal, maxillary, or coxal gland consists basically of a sac and a duct; at least part of the duct is of ectodermal origin. The sac is bathed by blood that circulates through the hemocoel. Most of what is known about these structures has been learned from certain crustaceans, and it seems best not to generalize too much about what they do in other arthropods. Their primary function in crustaceans is the selective elimination of certain ions. They do, however, excrete some soluble nitrogenous wastes, including ammonia, urea, and amino nitrogen.

The silk glands associated with the spinnerets of spiders are thought to be homologues of coxal glands. In insects, there are glands that open near the labium, the lower lip formed by fusion of the basal portions of the second maxillae. The secretory portions of these glands are located in the thorax, but their embryonic origin and the fact that their pores are located in the labial region of the head suggest that they are comparable to maxillary glands of crustaceans. Their usual function in most insects is secretion of saliva or silk, but they may serve for excretion or osmoregulation in the primitive, wingless groups. Centipedes, millipedes, and other myriapods have maxillary glands, too. They may be linked with the first and only maxillae, with the first of two pairs, or with both pairs.

In insects, spiders, and some other groups of primarily terrestrial arthropods, the major excretory organs are developmentally unrelated to the coelomic derivatives that have just been discussed. They are called *Malpighian tubules,* after Marcello Malpighi, an Italian microanatomist of the late seventeenth century who made detailed studies of insects. The tubules, which are outgrowths of the posteriormost part of the midgut, lie free in the hemocoel. Their main function is to convert uric acid, a nitrogenous waste produced in the tissues, into a nearly insoluble form. The uric acid precipitates as a result of the high acidity that prevails in the tubules near the point where they join the midgut. The more or less biscuit-shaped crystals of uric acid enter the hindgut and are voided with fecal matter. This method of disposing of nitrogenous wastes is advantageous for terrestrial animals because it uses less water than would be required for eliminating soluble and toxic wastes, such as ammonia or unprecipitated uric acid.

Reproduction and Development

Arthropods generally have separate sexes. Hermaphroditism is characteristic of most barnacles and some parasitic isopods, but it is not often found in other groups. As a rule, reproduction requires mating, or at least a close association of the male and female. Even aquatic arthropods do not simply shed their gametes the way some other invertebrates do.

The fact that there is mating, or at least courtship behavior, does not mean that arthropods necessarily copulate and that males always have a structure that functions as a penis. Insects and certain crustaceans do indeed copulate, and the male copulatory devices, whether or not they are appendages, bear the genital pores. In most crustaceans, however, the sperm-transferring devices do not have the genital pores—they are not even on the same segment as the pores. They fulfill their function by picking up loads of sperm, or packages of sperm called **spermatophores,** and putting these on or near the genital pores of the female.

In spiders, sperm transfer is even less direct. The male deposits a drop of seminal fluid on a web or thread of silk and then draws the fluid up into reservoirs in his pedipalps, which are on the anterior part of the body, remote from the genital pores. During the mating ritual, the pedipalps deliver the sperm to the seminal receptacles of the female. In some chelicerates other than spiders, and also in certain groups of myriapods, sperm transfer is usually effected by way of a spermatophore that the male deposits on the substratum, usually after courtship has proceeded to the right stage. The female is then led to the spermatophore or finds it some other way.

The eggs of most arthropods contain so much yolk that early cleavages are **superficial** (Figure 17.6, A; Figure 18.11, B–E). These lead, however, to the formation of a definite layer of cells in the peripheral region of the egg. A portion of this layer, called the *germinal band* or *germinal disk,* produces the embryo, which enlarges as it uses up the yolk and which may eventually occupy all of the space within the egg membrane.

A few members of this vast group of invertebrates exhibit distinctly spiral cleavage of the sort generally characteristic of annelids, molluscs, and other protostomes. Additional evidence for the kinship of arthropods with protostomes is found in the fact that the blastopore is sometimes the precursor of the mouth. The fate of the blastopore varies considerably, however. It may close and form nothing, or it may lengthen and then partly close, forming a tube with a mouth at one end and an anus at the other. In any case, it does not simply become the anus, as in deuterostomes.

Later development is discussed in appropriate following chapters in connection with the major subdivisions of arthropods. It will suffice to state here that there are two main patterns: direct and indirect. In *direct development,* the young, on hatching, resemble adults, even though they may lack certain appendages. Maturity is reached gradually; following each molt the animal looks more and more like the adult. Direct development is characteristic of many insects, such as grasshoppers and cockroaches, as well as of spiders, centipedes, some crustaceans, and a number of other groups.

Indirect development involves a series of larval stages and then a drastic metamorphosis to the adult form. Butterflies illustrate this type of development. The caterpillar that hatches from the egg grows and molts a number of times, but it shows little resemblance to the adult. It then enters into a nonfeeding pupal stage, during which it is transformed into a butterfly. A pupal stage is characteristic only of certain groups of insects. In crustaceans and other arthropods that have indirect development, there is nevertheless a series of larval stages that look little if at all like the adult.

THE MAJOR GROUPS OF ARTHROPODS

We return now to a question that was asked in the introduction to this chapter: ''Do the arthropods constitute a single phylum?'' The one-phylum view implies belief in the idea that all arthropods are descendants of a single ancestral stock of segmented animals. The possibility that they have evolved along more than one line has often been raised, but the phylum Arthropoda remained a permanent fixture until recently.

In 1977, Manton summarized the evidence in favor of dismantling the Arthropoda and recognizing three phyla with living representatives: Chelicerata, Crustacea, and Uniramia. The first two of these have traditionally been major subdivisions of the Arthropoda. The Uniramia, in Manton's view, would include not only the insects and myriapods, generally considered to be closely related, but also the Onychophora.

The main reasons for Manton's scheme are summarized in Table 15.1, which also lists the features that argue in support of the idea that arthropods are monophyletic. Boudreaux (1979) is one of the recent authors who has championed the classical concept.

Note that the features listed in Table 15.1 are not contrasting. No one will argue with the general characteristics of arthropods on the right. Manton's criteria deal with more specific concerns.

In this book, Chelicerata, Crustacea, and Uniramia are accepted as the three major groups of living arthropods. Each is accorded the rank of subphylum under Arthropoda. This arrangement should not offend anyone, for the distinctions between the three subphyla are emphasized sufficiently that the possibility of their having had separate evolutionary origins is never precluded. Onychophorans are removed from the Uniramia and given phylum rank, a solution that is now generally accepted in spite of Manton's arguments. The extinct trilobites are also treated as a subphylum of Arthropoda.

SUMMARY

1. The Arthropoda, which includes insects, centipedes, crustaceans, spiders, and several related groups, is by far the largest assemblage of animals. Some zoologists view it as a single, unified phylum; others believe that its major subdivisions have evolved along three or four separate lines of descent. It is generally agreed, however, that the ancestors of arthropods were annelids, or animals similar to annelids. This idea is strongly supported by the fact that arthropods are segmented and have a ventral nerve cord that consists basically of paired segmental ganglia and paired longitudinal connectives.

2. Three of the distinctive features of arthropods are visible externally. These are the jointed appendages that originate in pairs from some or all segments, the organization of segments into two or three body divisions called tagmata, and the cuticle. The latter, secreted by the epidermis, covers the body and usually forms a stiff exoskeleton, except in portions of the body where there must be some flexibility. The cuticle is molted periodically to allow for growth of the animal.

3. The articles that constitute jointed appendages, and body segments that are not tightly fused together, can be moved with respect to one another by muscles. These muscles are in discrete bundles and their action, especially in the case of appendages, is often effected

TABLE 15.1 ARGUMENTS FOR THE POLYPHYLETIC OR MONOPHYLETIC ORIGIN OF ARTHROPODS

Polyphyletic origin (for example, Manton, 1977)	Monophyletic origin (for example, Boudreaux, 1979)
1. Differences in structure of the cuticle	1. Cuticle containing sclerified proteins and chitin
2. Differences in structure and development of mandibles	2. Presence of paired, jointed appendages
3. Differences in the composition of the tagmata	3. Presence of tagmata and cephalization
4. Differences in the organization of appendages (uniramous, biramous, etc.)	4. Circulatory system of the open type, and characterized by an extensive hemocoel and a heart perforated by ostia
5. Presence of compound eyes in some, not in others	5. Reduction of the coelom
6. Differences in embryonic development	6. Superficial early cleavage

by way of tendons. Thus a muscle need not be joined directly to the structure it moves. Tendons, as well as most other firm internal supports to which muscles are attached, are part of the cuticle.

4. Appendages of arthropods include antennae, such as are found in insects and crustaceans, mouthparts concerned with handling and chewing food, legs for walking or clinging, and various kinds of structures used for swimming, creating respiratory currents, collecting microscopic food, holding eggs, and other functions. Wings of insects are not true appendages. They are outgrowths of the cuticle of the thorax. Their movements, however, are brought about by muscles. These, like muscles concerned with other locomotor activities, are striated.

5. There are no active cilia or flagella in arthropods, except for the flagella in the sperm tails of a few groups.

6. Compound eyes are characteristic of most insects and many crustaceans. Eyes of this type, which consist of numerous separate optical units called ommatidia, are uniquely arthropodan.

7. Arthropods typically have a circulatory system of the open type. Blood pumped by the heart moves through interconnected spaces that collectively form a hemocoel, and returns to the heart through one or more pairs of openings called ostia. The hemocoel must not be confused with a true coelom.

8. The true coelomic spaces that appear during development contribute to the gonads, certain types of excretory and osmoregulatory structures, glands that produce silk, and some other structures, but not to a major body cavity comparable to the coelom of annelids.

9. With few exceptions, sexes are separate. Transfer of sperm may be achieved by true copulation or by some less direct method that brings sperm to the female genital opening. Eggs of most arthropods are so yolky that early cleavages are superficial. Eventually, however, a layer of cells is formed in the peripheral region of the egg. A part of this layer, called the germinal band, produces the definitive embryo.

General References on Arthropoda

Anderson, D. T., 1973. Embryology and Phylogeny in Annelids and Arthropods. Pergamon Press, Oxford and New York.

Boudreaux, H. B., 1979. Arthropod Phylogeny with Special Reference to Insects. John Wiley and Sons, New York.

Clarke, K. V., 1973. The Biology of the Arthropoda. Edward Arnold, London; American Elsevier, New York.

Gupta, A. P. (editor), 1979. Arthropod Phylogeny. Van Nostrand Reinhold Company, New York.

Gupta, A. P., 1987. Arthropod Brain, Its Evolution, Development, Structure, and Functions. Wiley-Interscience, New York.

Horridge, G. A., 1977. The compound eye of insects. Scientific American, *237*(1):108–120.

Manton, S. M., 1973. Arthropod phylogeny—a modern synthesis. Journal of Zoology (London), *171*:111–130.

Manton, S. M., 1977. The Arthropoda: Habits, Functional Morphology, and Evolution. Clarendon Press, Oxford.

Neville, A. C., 1975. Biology of the Arthropod Cuticle. Springer-Verlag, Berlin, Heidelberg, New York.

Sharov, A. C., 1966. Basic Arthropodan Stock: With Special Reference to Insects. Pergamon Press, Oxford.

Snodgrass, R. E., 1951. Comparative Studies on the Head of Mandibulate Arthropods. Comstock Publishing Company, Ithaca, New York.

Snodgrass, R. E., 1952. A Textbook of Arthropod Anatomy. Cornell University Press, Ithaca, New York. (Reprinted 1971, by Hafner Publishing Company, New York.)

Tiegs, O. W., and Manton, S. M., 1958. The evolution of the Arthropoda. Biological Reviews, *33*:255–337.

16

Subphyla Trilobitomorpha and Chelicerata

- Subphylum Trilobitomorpha
  - Class Trilobita
- Subphylum Chelicerata
  - Class Merostomata
    - Subclass Eurypterida
    - Subclass Xiphosura
  - Class Arachnida
    - Order Scorpionida (Scorpiones)
    - Order Pseudoscorpionida
    - Order Araneida (Araneae)
    - Order Phalangiida (Opiliones)
    - Order Thelyphonida (Uropygi)
    - Order Schizomida
    - Order Phrynichida (Amblypygi)
    - Order Palpigradida (Palpigradi)
    - Order Ricinuleida (Ricinulei)
    - Order Solpugida (Solifugae, Solifugida)
    - Order Acari (Acarina)
  - Class Pycnogonida

Introduction

A survey of arthropods may be begun with any of the major assemblages, or even with a single type, such as a crayfish or honeybee. It is important, however, to develop the discussion in such a way as to present a clear picture of the diversity and functional morphology of the group as a whole. Starting off with trilobites has some disadvantages because these animals were extinct before the end of the Paleozoic era. We cannot come to grips with them in the same way that we can with living arthropods. Some of the ideas we have about them are tentative, but it would be a mistake not to deal with trilobites at all. Why not put them first and take advantage of their apparently ancestral arthropodan features? Chief among these is the presence of numerous segments, each with a pair of limbs that resemble the biramous appendages of modern crustaceans.

No great significance must be attached to the fact that the brief discussion of trilobites is followed by a review of chelicerates. The arrangement is a practical one. Chelicerates are the only living arthropods that do not have jaws. In this respect they are similar to trilobites and different from crustaceans, insects, and myriapods. This must not be construed to mean that chelicerates are necessarily closer to trilobites than the crustaceans are, or that all of the arthropods with jaws have evolved from a single ancestral stock. Some zoologists and paleontologists do believe that jaws of crustaceans, insects, and myriapods are homologous structures. Most of those who disagree think that insects and myriapods are derived from a stock that also gave rise to onychophorans and perhaps even to tardigrades, and that the crustaceans and chelicerates have diverged from other ancestral groups. The evidence that supports both views is drawn not only from comparative anatomy and the fossil record, but also from studies on embryonic development.

Subphylum Trilobitomorpha

Trilobites were important constituents of the fauna of Paleozoic seas. Before they became extinct, during the Permian period, they left a superb fossil record. Other early arthropods whose fossils resemble those of trilobites have been found in ancient marine deposits, especially the Burgess shale formation of British Columbia. A few that have been discovered in coal-bearing strata are almost certainly remains of freshwater or terrestrial organisms. Paleontologists use Trilobitomorpha as a kind of umbrella to cover all of the groups that seem to have characteristics linking them with trilobites. Only the true trilobites are discussed here.

Class Trilobita

In general, trilobites (Figures 16.1, 16.2) were dorsoventrally flattened and had a more or less oval or elliptical outline. The largest were nearly a meter long, but most of the approximately 4000 described species were under 10 cm. The **cephalon,** or head region, was covered by a shield. The raised median portion of this, called the **glabella,** was separated from the lateral portion on each side by a longitudinal furrow, so that the cephalon had a distinctly three-lobed appearance. The transverse grooves on the glabella of many trilobites are probably indications of head segments. The lateral portions of the shield were divided into a fixed cheek, joined to the glabella, and a movable cheek, on which a **compound eye** was located.

The **trunk region** consisted of a variable number of segments, each characteristically with a median eminence somewhat similar to the glabella. Many or all of the numerous trunk segments—always the more anterior segments, at least—were separated by flexible sutures, permitting a trilobite to roll up into a ball. As a rule, some of the posterior segments were fused to one another, as well as to the terminal **telson,** on which the anus was located. In certain genera, however, all of the trunk segments and telson were distinct. (In most treatises, the term *pygidium* is used for the complex that consists of the telson and whatever trunk segments are fused to it. *Pygidium,* as used in connection with annelids, is essentially synonymous with *telson,* so its application to a series of true segments plus the telson is not appropriate.)

Specimens in which the appendages are well preserved are not common. The most fossilizable part of the body was the dorsal side, where the cuticle was calcified. It is nevertheless clear that trilobites had a pair of **antennae** in front of the mouth (Figure 16.1, A). These may have been homologous to the antennae of insects and myriapods and to the first antennae of crustaceans. On each true segment between the mouth and telson was a pair of **biramous appendages.** Four pairs were on the cephalon, the rest on the trunk. All of these biramous appendages were similar, so there would not have been much division of labor. The absence of appendages specialized to serve as mouthparts was a unique feature of trilobites.

One must be careful in attempting to reconcile the two branches of a trilobite limb with those of biramous appendages of crustaceans. The outer branch may have resembled the exopodite of certain crustacean limbs, but it was probably more nearly comparable to an epipodite, because it arose from the first article, not from the second article (Figure 16.1, E; see also Figure 15.2, A). Some paleontologists believe, in fact, that the large

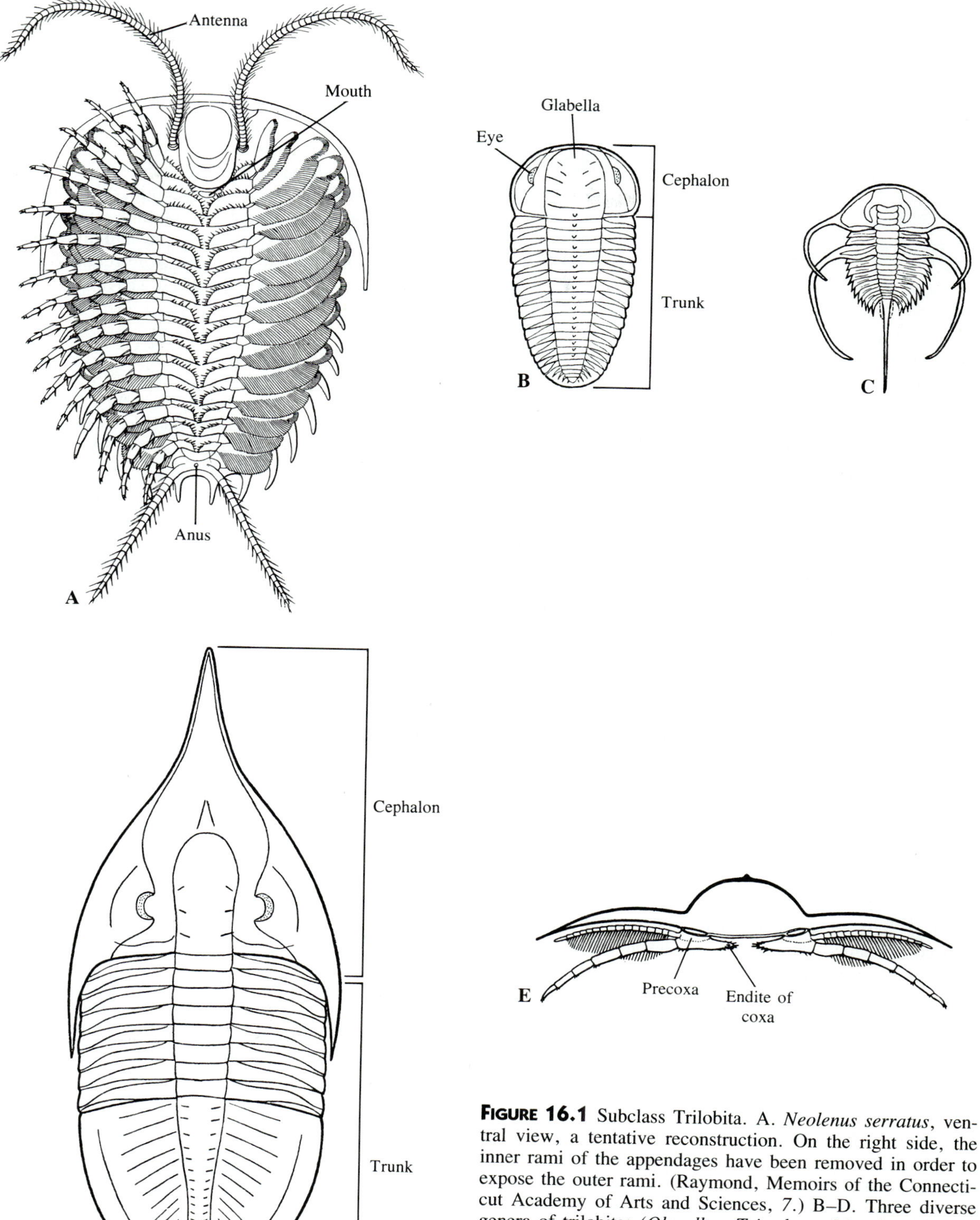

FIGURE 16.1 Subclass Trilobita. A. *Neolenus serratus*, ventral view, a tentative reconstruction. On the right side, the inner rami of the appendages have been removed in order to expose the outer rami. (Raymond, Memoirs of the Connecticut Academy of Arts and Sciences, *7*.) B–D. Three diverse genera of trilobites (*Olenellus*, *Triarthrus*, *Megalaspis*), dorsal view. In *Megalaspis*, many of the trunk segments are fused to one another, as well as to the telson. (Størmer, Skrifter utgitt av Det Norske Videnskaps-Akademi i Oslo, Matematisk-Naturvidenskapelig Klasse, 1944.) E. Transverse section of a trilobite; diagrammatic. (Snodgrass, Smithsonian Miscellaneous Collections, *97*.) (See also Figure 15.2, A).

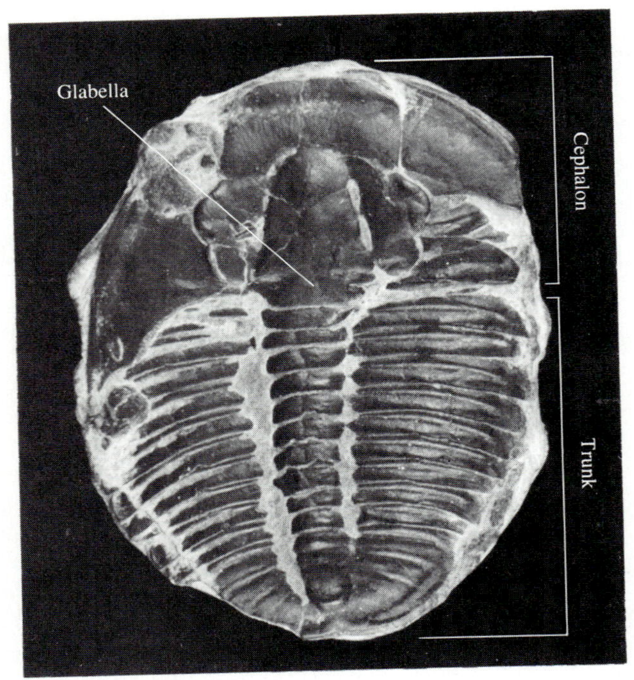

FIGURE 16.2 Photograph of a trilobite fossil, dorsal view.

first article, or coxa, really consisted of two articles: the coxa proper and a small precoxa, to which the outer branch was attached.

The outer branch of each appendage was fringed on one side by many closely spaced filaments, and it therefore looked like a gill. Apparently, however, the filaments were stiff. They may have been more useful for straining out food, or perhaps even for swimming or digging, than for gas exchange. The medial portion of each coxa formed a toothlike projection, or endite, that must have been used for handling food and passing it to the mouth. In many species, at least, the proximal portions of the appendages bordered a distinct midventral channel.

There are fossils of young trilobites in which the body consisted of the cephalon plus a short posterior unit that was probably mostly telson. Later stages show segments differentiating in the posterior region, the newest segment being closest to the telson. This pattern of development more or less fits in with what we know about the formation of segments in other arthropods, as well as in annelids.

Using X-ray techniques on trilobite fossils that have become impregnated with iron sulfide, paleontologists have been able to work out some previously unknown or misunderstood details of external and internal anatomy, including the arrangement of muscles, organization of the gut, and even the contents of the gut. Continued studies along these lines will no doubt clarify many points that have been vague or disputed.

It is thought that most trilobites lived in the bottom sediments of ancient seas, but there is good evidence that some were permanently pelagic. Studies on iron-sulfide-impregnated fossils of certain species indicate that the food consisted of finely particulate organic material. It is likely that in a group as diversified as the Trilobita there were at least a few scavengers or carnivores.

SUMMARY

1. Trilobites were important components of the fauna of the seas during the Paleozoic era, and especially during the Cambrian and Ordovician periods. Most of them were benthic organisms, but some were pelagic. These animals left a detailed fossil record and are of interest because they formed a large and diversified group of early arthropods.

2. The body of a trilobite was divided into a cephalon, or head region, and a trunk. The dorsal surface of the cephalon was covered by a shield and had a pair of compound eyes. On its ventral side, the cephalon bore a pair of preoral antennae and four pairs of leglike appendages. The trunk, which was sometimes segmented throughout but which usually had some of its posterior segments fused to one another as well as to the terminal telson, was provided with numerous pairs of leglike appendages. These, as well as the four pairs on the cephalon, had an articulated outer branch, fringed on one side by stiff filaments, that may have served as a gill or food-straining device, or even as an organ for swimming or digging. The branch arose from the proximal article of the appendage, so it was perhaps more nearly comparable to an epipodite of a crustacean appendage than to an exopodite, which arises from the second article.

3. The name Trilobita (''three lobes'') refers to the fact that the body of these animals had a raised median portion, especially prominent in the cephalon, separated by furrows from two lateral portions. The lateral portions projected out over the appendages.

REFERENCES ON TRILOBITA

Bergström, J. B., 1973. Organization, life, and systematics of trilobites. Fossils and Strata, No. 2. Universitets-Forlaget, Oslo.

Cisne, J. L., 1974. Trilobites and the origin of arthropods. Science, *186*:13–18.

Levi-Setti, R., 1975. Trilobites: A Photographic Atlas. University of Chicago Press, Chicago.

Martinsson, A. (editor), 1975. Evolution and Morphology of the Trilobita, Trilobitoidea, and Merostomata. Fossils and Strata, no. 4. Universitetsforlaget, Oslo.

Stürmer, W., and Bergström, J., 1973. New discoveries on trilobites by X-rays. Palaeontologische Zeitschrift, *47:*104–141.
Withington, H. C., 1957. The ontogeny of trilobites. Biological Reviews, *32:*421–469.

SUBPHYLUM CHELICERATA

In chelicerate arthropods—spiders, scorpions, ticks, and their allies—the body consists of two main portions, although these may not be clearly separated. The anterior part, called the **prosoma** (''fore body''), has six pairs of appendages. The posterior part, called the **opisthosoma** (''hind body''), although derived from a sequence of embryonic somites, may have no appendages at all. In certain groups, however, there is at least one pair. In spiders, for instance, the spinnerets from which silk is exuded are much-modified appendages. Only in the bizarre horseshoe crabs are there several pairs of serially arranged opisthosomal appendages. These animals are commonly placed in the Chelicerata, but their relationship to other groups within the subphylum is probably not so close as it is generally assumed to be.

Of the six pairs of appendages on the prosoma, the first are called **chelicerae.** The term *chelicera,* upon which the name of the subphylum is based, means ''clawed horn.'' It alludes to the fact that the terminal article of an appendage of this type is a movable claw. The claw may be used for piercing, or it may form part of a pincerlike structure. The important thing to remember for the moment is that the chelicerae are distinctly different from the antennae of insects and myriapods, and from the first and second antennae of crustaceans. There is a possibility that chelicerae are homologous to the second antennae of crustaceans, for they originate postorally and move to a preoral position. There are other interpretations of their homologies, however.

The second appendages of chelicerates are called **pedipalps,** although they are not necessarily palplike. The remaining four pairs are **walking legs.** A few deviations from this arrangement must be expected, for not all chelicerates fit perfectly into a generalized plan. Another complication is that the prosoma and opisthosoma are not always really distinct. In ticks and mites, for instance, they are run together so that the body appears to consist of just one piece. All of the appendages of ticks and mites nevertheless belong to the prosomal portion.

CLASS MEROSTOMATA

The few species of horseshoe crabs, which belong to the subclass Xiphosura, are the only surviving members of the Merostomata. Comparison of these bizarre animals with some fossil remains found in Silurian deposits shows that xiphosurans have not changed much in nearly 400 million years. The subclass Eurypterida, which had representatives that were nearly 3 m long, overlapped the Xiphosura in the geological time scale, but became extinct in the Permian period.

SUBCLASS EURYPTERIDA

Eurypterids will not be dealt with in detail, but a few of their unusual features will be pointed out. The prosomal appendages consisted of a pair of chelicerae, a pair of pedipalps, and four pairs of legs (Figures 16.3 and 16.4). The degree to which the various prosomal appendages were differentiated varied. When the last legs were paddlelike, as they were in most genera, they were almost certainly used for swimming. It is likely that eurypterids swam with the ventral side of the body uppermost. There was a pair of compound eyes and a pair of simple eyes on the dorsal side of the prosoma.

The opisthosoma of eurypterids consisted of 12 segments, plus a telson, which was usually long and which may have served as a sting, as it does in scorpions. In complete fossils, all 12 segments are visible dorsally, but there appear to be only 11 ventrally. This is because the ventral portion of the first opisthosomal segment was displaced forward, forming a platelike structure between the coxae of the last pair of prosomal legs. This structure, the *metastoma,* is interpreted to be a product of fusion of a pair of appendages belonging to the first opisthosomal segment. The appendages of the second through sixth segments were paired platelike structures similar to those on the opisthosoma of horseshoe crabs. There were gills on all of them, but the structure of these is not fully understood.

The genital pore was on the ventral midline in the second opisthosomal segment. It was covered by the genital operculum, believed to represent the pair of platelike appendages of that segment. An articulated median piece, prominent in males, reduced in females, has been interpreted to have originated by fusion, along the midline, of the epipodites of the appendages that form the genital operculum. We can be sure that the last word has not been written on the subject of sexual dimorphism and possible clasping devices in these remarkable arthropods.

Eurypterids seem to have been largely marine and brackish-water animals, and it is likely that most of them inhabited large lagoons, rather than open seas. They are thought to have preyed on other invertebrates, and possibly also on vertebrates.

SUBCLASS XIPHOSURA

Of the few species of horseshoe crabs, the one that is best known is *Limulus polyphemus,* which occurs from Nova Scotia to Yucatan and the West Indies. The

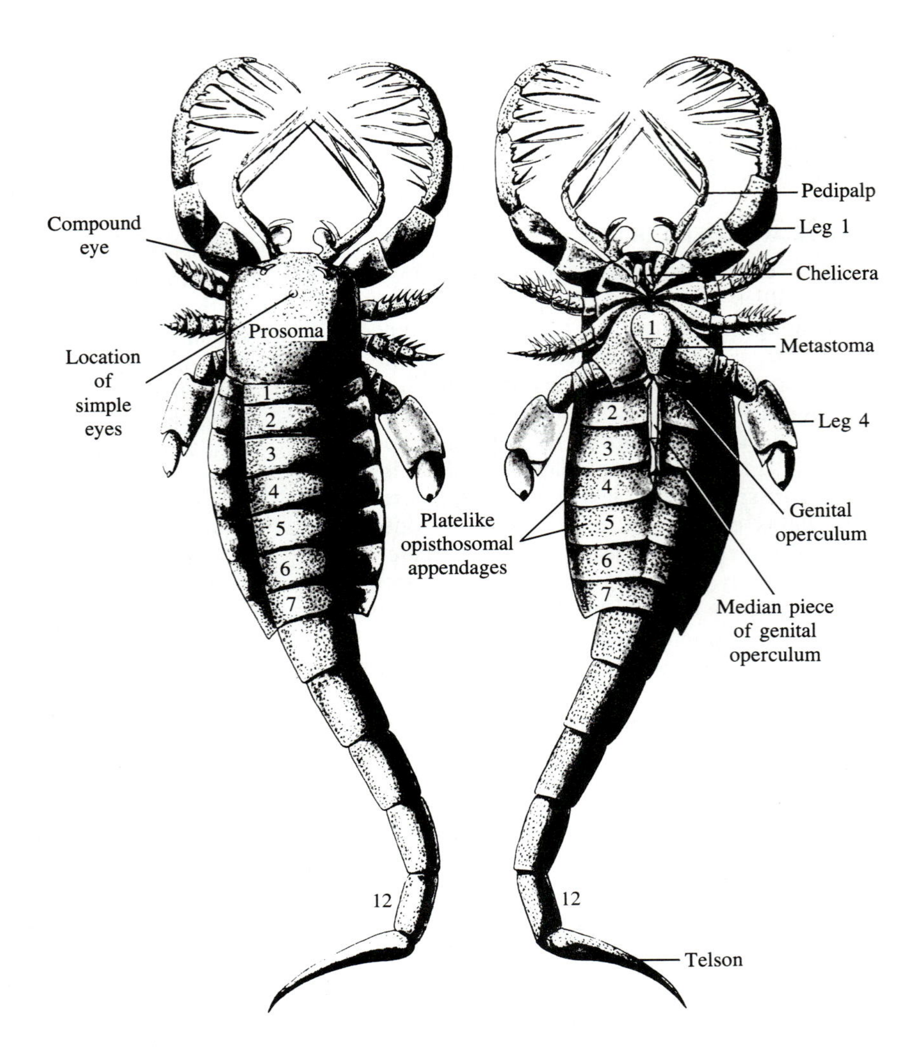

FIGURE 16.3 Subclass Eurypterida. *Mixopterus kiaeri* (Silurian period), male, dorsal and ventral views, reconstruction. (The fossils are about 60 cm long.) (Størmer, Skrifter utgitt av Det Norske Videnskaps-Akademi i Oslo, Matematisk-Naturvidenskapelig Klasse, 1933.)

others, belonging to the genera *Tachypleus* and *Carcinoscorpius,* are Asiatic. The description given here is based on *Limulus,* but applies reasonably well to the group as a whole.

Tagmata and Appendages

The body of *Limulus,* whose length may exceed 25 cm, consists of three main portions: a broad prosoma, an opisthosoma, and a long spine that is sometimes called a telson (Figures 16.5 and 16.6). All of these body divisions are covered with an exoskeleton that is sclerified but not calcified. The prosoma has a pair of widely spaced compound eyes on the dorsal surface, about half-way back, and a pair of simple eyes near the midline at a more anterior level. These eyes are sensitive to ultraviolet light as well as to light of the visible spectrum. Horseshoe crabs are the only living chelicerates that have compound eyes.

On its underside, the prosoma has six pairs of appendages. The chelicerae, concerned with manipulation of food, are just in front of the mouth. The second appendages, flanking the mouth, are homologues of the pedipalps of spiders and other arachnids. They are chelate in females, but not in males, which use the simple claws as clasping devices during mating. The pedipalps and the next three pairs of appendages serve as walking legs. Their enlarged, spiny coxal articles are specialized to function as jaws for crushing prey, and are therefore called **gnathobases.** The sixth appendages are also legs, but they are longer and have neither gnathobases nor chelae. They can be used for pushing the animal forward and for digging, and are peculiar in that they terminate in four bladelike processes, one of which is distinctly longer than the others and is tipped with a couple of spines. The coxae of these last legs also have a little epipodite, called the *flabellum,* used for cleaning the gills on the opisthosomal appendages.

The opisthosoma, clearly set off from the prosoma when the animal is viewed with its dorsal side up, is bordered on the right and left by a series of substantial

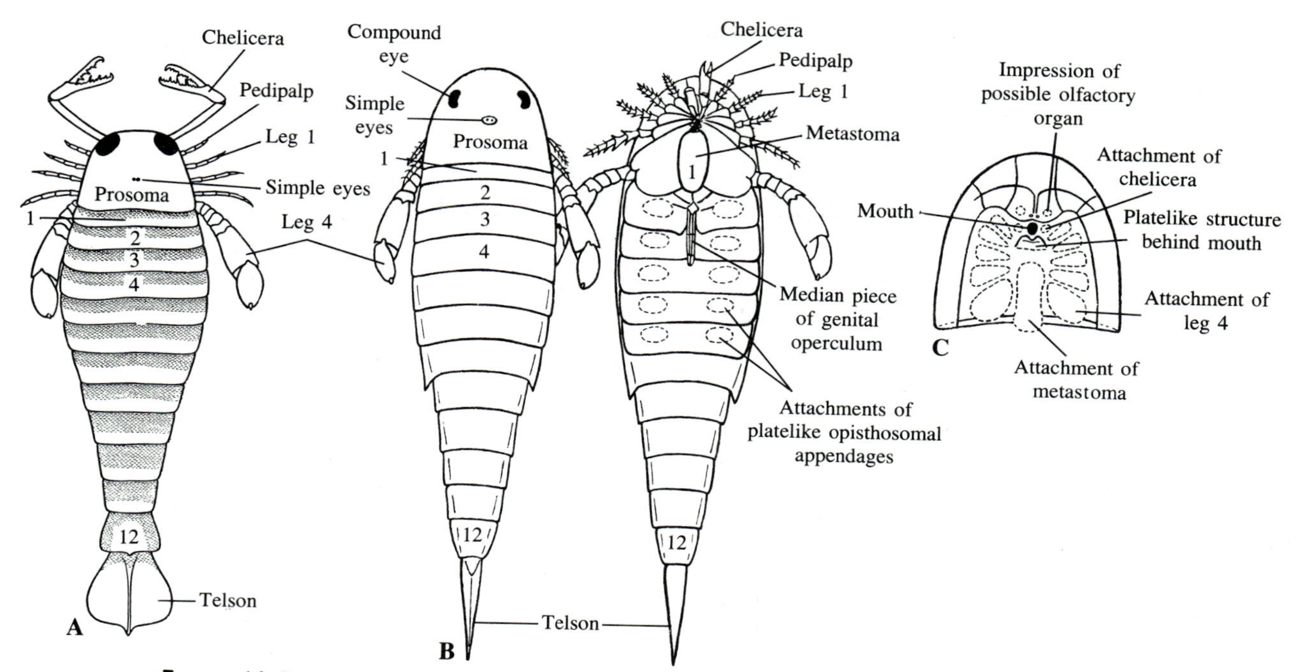

FIGURE 16.4 Subclass Eurypterida. A. *Pterygotus rhenaniae* (Devonian period). B. *Hughmilleria norvegica,* dorsal and ventral views (Silurian period). C. Prosoma of *Hughmilleria.* (Størmer, Skrifter utgitt av Det Norske Videnskaps-Akademi i Oslo, Matematisk-Naturvidenskapelig Klasse, 1944.)

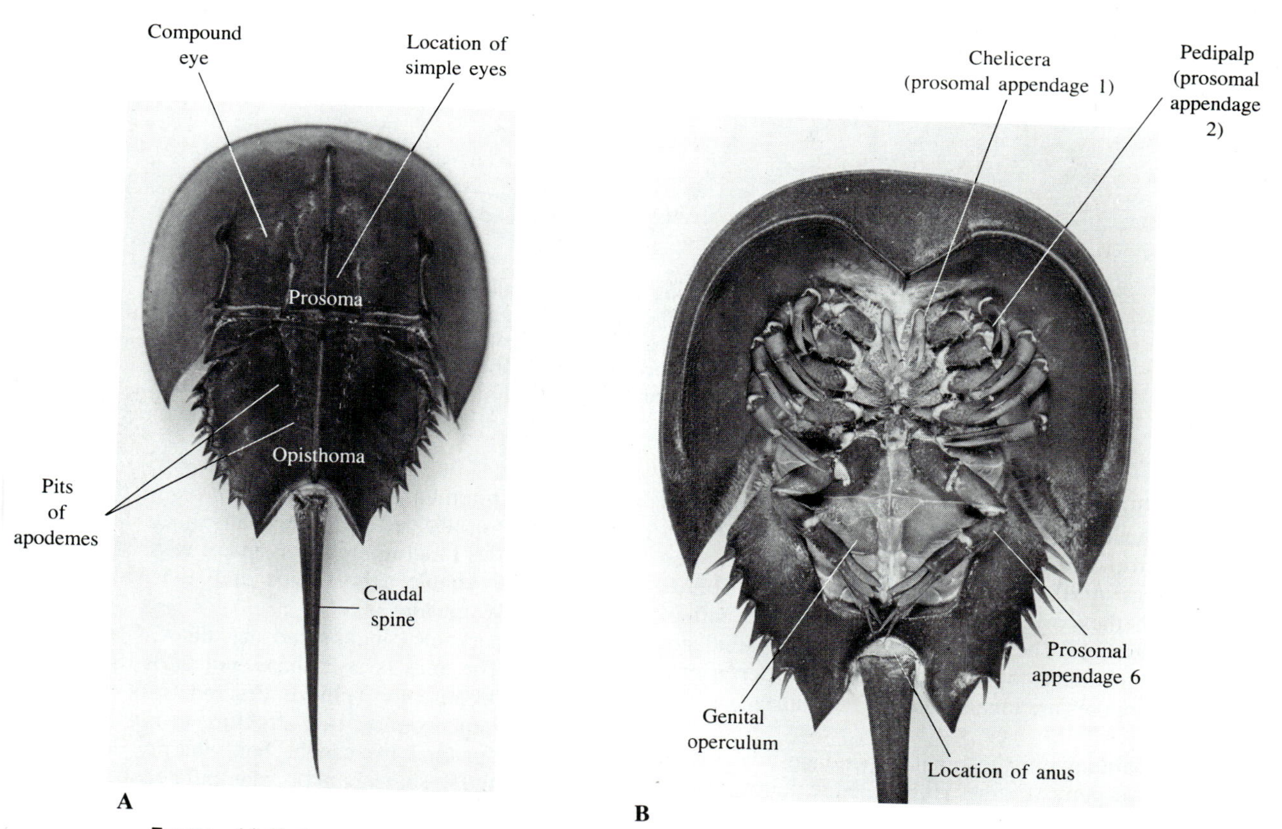

FIGURE 16.5 Subclass Xiphosura. *Limulus polyphemus*, female. A. Dorsal view. B. Ventral view. (See Figure 16.6 for more detailed labeling of appendages.)

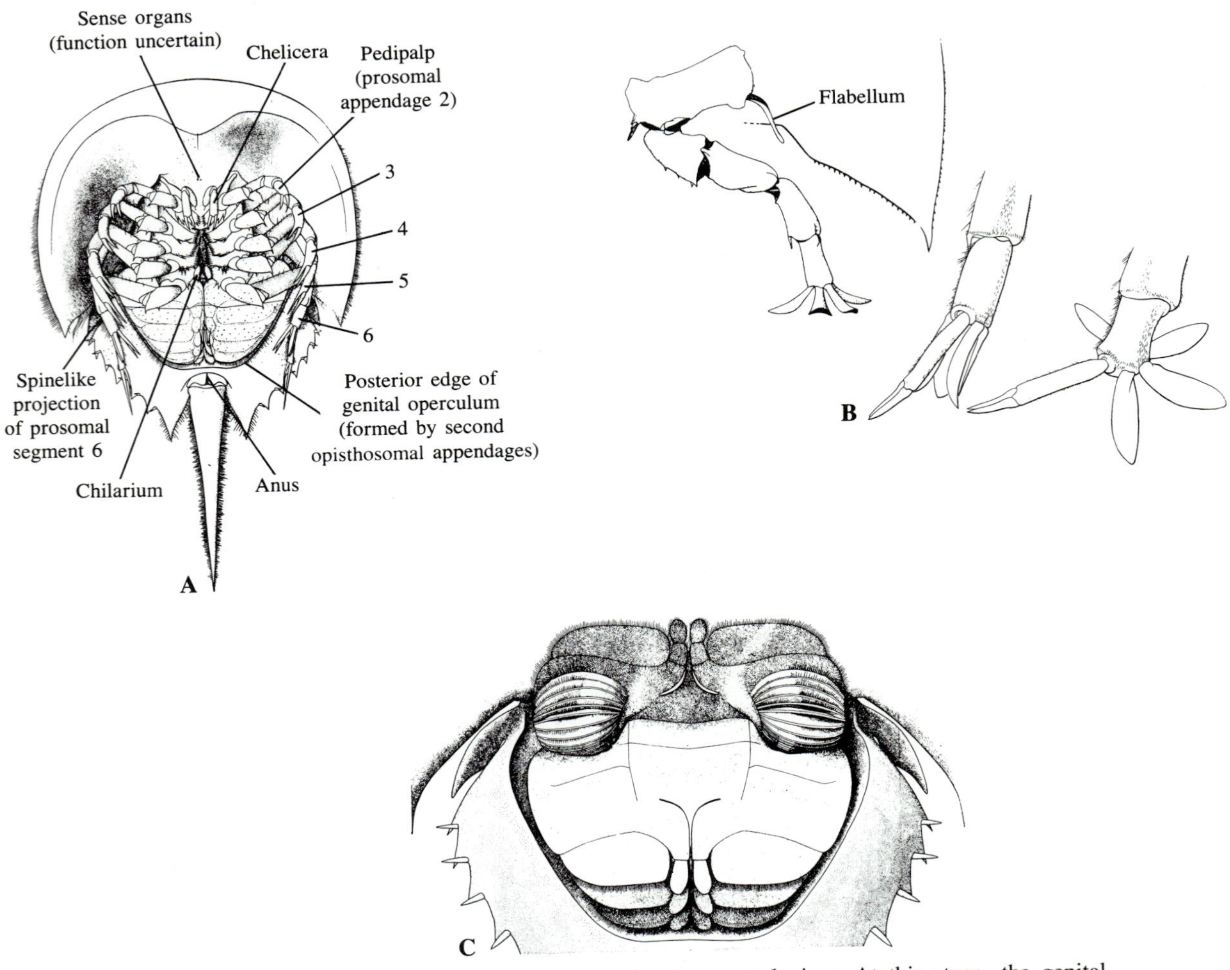

FIGURE 16.6 *Limulus polyphemus*. A. Young female, ventral view. At this stage, the genital operculum covers the appendages that bear the book gills, but is translucent enough to permit them to show through. B. Sixth prosomal appendage. C. Ventral view of the opisthosoma of a young individual. The genital operculum has been removed, and one pair of appendages bearing book gills has been lifted with a needle so as to expose its folds. (A–C, Ankel, Natur und Volk, *88*.)

spines. Its first appendages are two small plates called **chilaria,** which are believed to push fragments of food forward. Although the chilaria are located between the last legs of the prosoma, the embryonic somite from which they originate is considered to be part of the opisthosoma. The next opisthosomal appendages are united medially to form the **genital operculum;** the two gonopores are located on the inner surface of this structure.

The remaining opisthosomal appendages, of which there are five pairs (Figure 16.6, C), are flattened like those that form the operculum, but the right and left members of each pair are not united medially. They are obscurely biramous, and the inner faces of their broad outer lobes are folded into many thin sheets called **book gills.** Besides functioning as organs of respiration, these appendages enable a horseshoe crab to swim with its ventral side upward.

The remarkable caudal spine articulates with the opisthosoma. The anus is at its base, not on it. Because of this, and because the spine is derived from more than one embryonic somite, this structure is not a true telson. The opisthosoma can be bent sharply downward, and the caudal spine can be manipulated by muscles that insert upon its base. The movements of the opisthosoma and spine make it possible for the animal to bur-

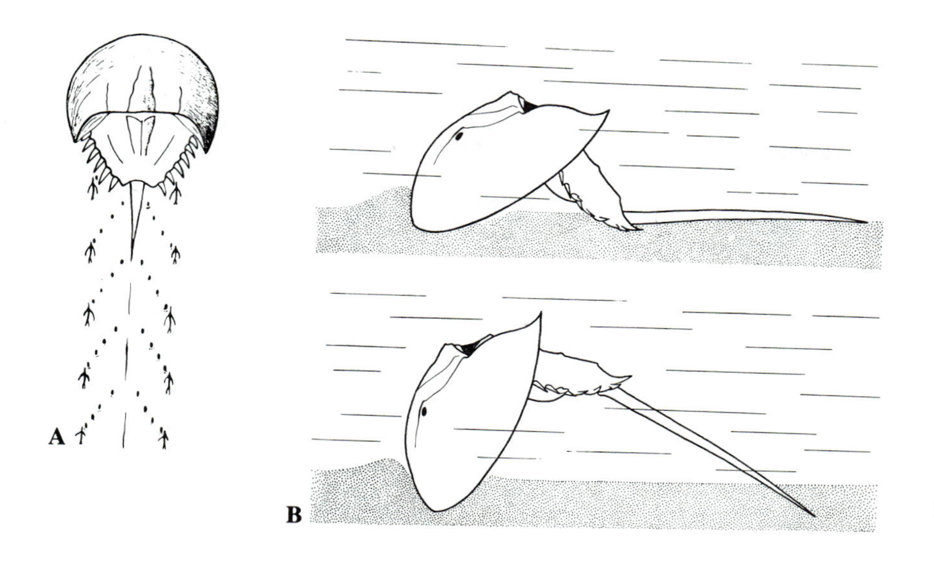

FIGURE 16.7 *Limulus polyphemus*. A. Tracks left in the substratum by a moving animal. (Malz, Natur und Museum, *94*.) B. Two ways in which the animal can push itself into a soft substratum. (Ankel, Natur und Volk, *88*.)

row (Figure 16.7, B), turn over, and work itself out of various predicaments.

Digestive Tract

The **foregut,** derived from a stomodeal invagination, consists of two distinct portions, both lined by a cuticle that has lengthwise ridges. The first part is usually called the *esophagus*. It runs dorsally and forward to the second part, which is a muscular *gizzard* (Figure 16.8). The cuticular ridges of the gizzard are studded with teeth that grind up food. Unfavorable material can be directed back to the mouth and regurgitated. Two pairs of extensively branched **digestive glands,** which fill up much of the prosoma and some of the opisthosoma, open into the broad anterior portion of the **midgut.** Considerable intracellular digestion takes place in these glands. The **hindgut** leads to the anus, located at the base of the caudal spine. Like the foregut, the hindgut is an ectodermal derivative and has a cuticular lining.

Excretory and Circulatory Systems

The two **coxal glands,** which lie next to the gizzard but have ducts leading to the bases of the last prosomal legs, are assumed to be excretory organs. Each has four lobes. These lobes, the sac into which they open, and part of the duct can be traced back to four of the six coelomic pockets that appear on both sides of the prosoma during early stages of development.

The circulatory system (Figure 16.8, A) is moderately complex. The **heart,** with eight pairs of **ostia,** is situated partly in the opisthosoma, partly in the prosoma. Three of the major arteries are anterior prolongations of the heart. Others are directed laterally, and some of them join the important collateral arteries that lie on both sides of the heart. These vessels and their branches deliver blood to many parts of the body. Not all of the arteries live up to the customary definition, for some blood may return to the heart by way of these vessels. Most of the blood, however, after entering hemocoelic spaces, moves into a large ventral sinus, from which veins carry it to the book gills. Other vessels then carry it to the pericardial sinus, from which it enters the heart through the ostia. Muscles connecting the pericardial membrane to the walls of the ventral sinus are thought to promote flow of venous blood. The respiratory pigment in the blood of xiphosurans is **hemocyanin.**

Nervous System

The central nervous system is condensed. The protocerebrum sends nerves to the eyes and to a sense organ, believed to have an olfactory function, in front of the mouth. On both sides of the protocerebrum is a little projection that has been interpreted as a vestigial deutocerebrum. The tritocerebrum, from which nerves run to the chelicerae, is bound up with the nerve ring that runs around the esophagus. This in turn is hard to separate from the five ventral prosomal ganglia and even from two of the opisthosomal ganglia. Behind this complex, however, the double ventral nerve cord becomes more distinct and most of the remaining opisthosomal ganglia are well separated.

Endoskeleton

There is a formidable array of muscles concerned with operating the appendages and caudal spine, flexing the body, opening the mouth and anus, dilating the heart and portions of the foregut, and performing similar functions. The muscles of *Limulus* are not discussed

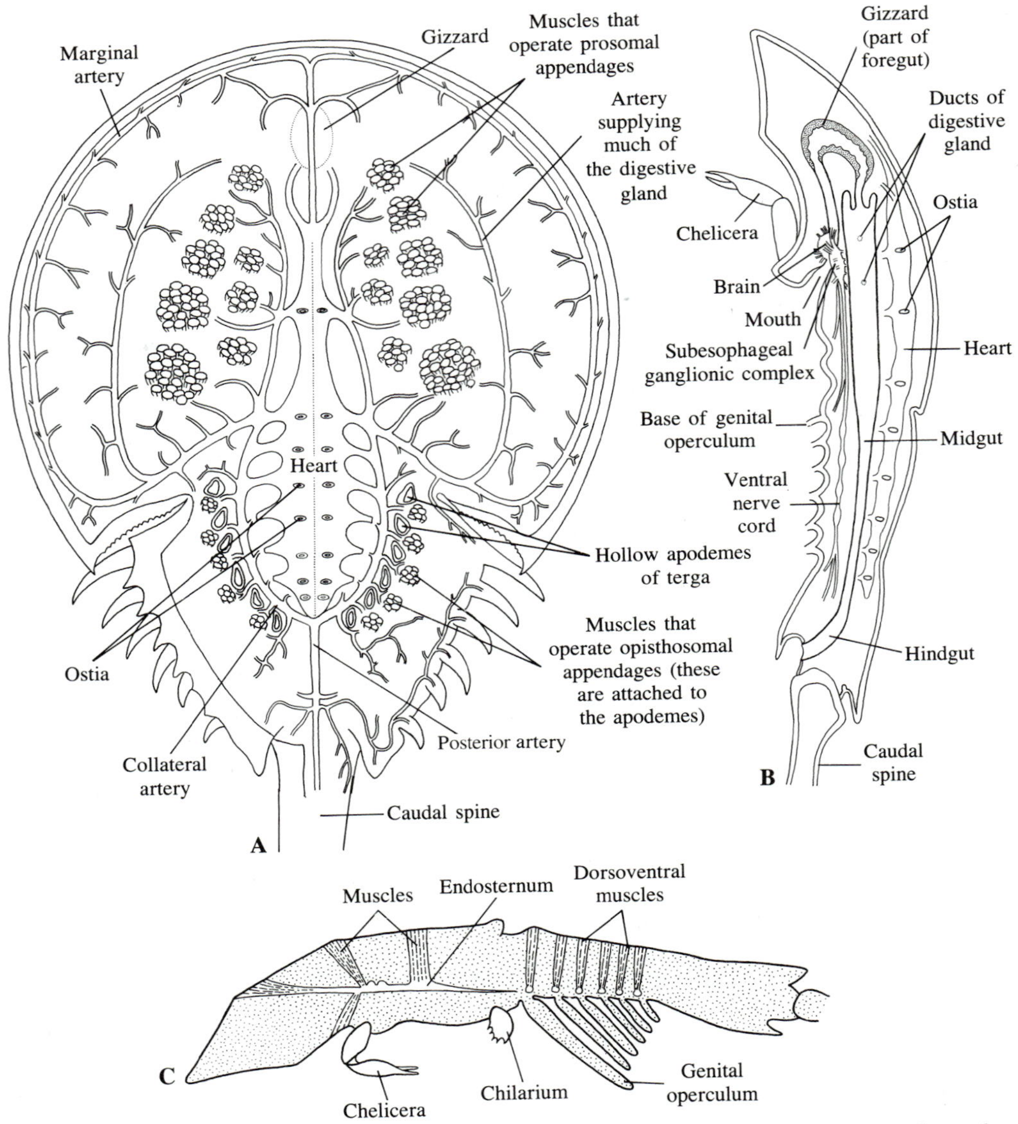

Figure 16.8 *Limulus polyphemus*. A. Specimen dissected to show the heart, major blood vessels, and some of the muscles. The covering of the pericardial sinus has been removed. Much of the space in the prosoma and anterior part of the opisthosoma is occupied by branches of the two pairs of digestive glands. B. Sagittal section, on the midline; diagrammatic and greatly simplified. Most of the prosomal appendages have been omitted, and only the bases of the opisthosomal appendages are shown. C. Sagittal section, to one side of the midline, showing the muscles that run from the cuticle to the endosternum of the prosoma and to the transversely oriented endosternal bars of the opisthosoma. Muscles that operate the prosomal appendages originate on the endosternum as well as on the cuticle; those that operate the opisthosomal appendages originate mostly on the apodeme pits shown in A. (C, partly after Firstman, Journal of Arachnology, *1*.)

in detail here, but it should be explained that certain of the structures to which muscles are attached are part of a true **endoskeleton.** The most obvious component of this endoskeleton is the *endosternum,* an elongated and rather elaborate plate that lies in the prosoma. It is above the nerve cord and ventral to the midgut; more anteriorly, where the foregut bends back on itself, it is between the esophagus and gizzard. Many of the powerful muscles attached to the cuticular exoskeleton of the prosoma insert on it, and the endosternum also provides sites for attachment of muscles that operate all of the prosomal appendages except the chelicerae. The prosomal appendages are also supplied, however, by muscles that attach to the cuticle.

In the opisthosoma, beneath the ventral nerve cord, there is a series of transverse endoskeletal bars. Muscles coming from the cuticular exoskeleton are attached to these in much the same way as they are attached to the endosternum, but the bars are not sites for attachment of muscles that operate the opisthosomal appendages. These muscles originate on the apodemes that sink down as deep pits from the cuticle of the dorsal side of the opisthosoma, as well as from other portions of the cuticle. In addition to the endosternum and transverse bars in the opisthosoma, there are a few minor endoskeletal elements.

The endoskeletal structures of *Limulus* resemble certain types of vertebrate cartilage, and they are of mesodermal origin. Somewhat comparable endoskeletal elements are found in certain spiders and other arachnids. These structures are among the few exceptions to the rule stated in Chapter 15: muscle bundles that operate appendages, that move body segments or tagmata with respect to one another, or that accomplish similar functions are attached to the cuticle or to apodemes of the cuticle.

Reproduction and Development

The gonad consists of branching and anastomosing tubes located dorsally in the prosoma and opisthosoma. The genital pores are situated on the operculum, formed by a pair of appendages belonging to the second opisthosomal segment. The eggs are large, almost 3 mm in diameter. The sperm have a distinct head and a flagellar tail, and they are motile.

Mating takes place when the animals congregate in shallow water. Males clasp the females from above, using their pedipalps, and spray sperm on the eggs as these emerge. The eggs are laid in a trough made by the female in muddy sand near the high-tide line.

It takes several weeks for an egg to hatch as a *trilobite larva,* so called because of its slight resemblance to a trilobite. The larva, not quite 1 cm long, has all of the prosomal appendages characteristic of the adult, but only two of the five pairs of book gills. It is nevertheless capable of swimming and of burying itself in sand. In the course of further development, punctuated by molts, the other book gills are added, the caudal spine enlarges, and the young animal assumes its adult form. Horseshoe crabs have long lives—perhaps as long as 19 years—and they do not reach sexual maturity until they are at least 9 years old.

References on Xiphosura

Barthel, K. W., 1974. *Limulus:* a living fossil. Horseshoe crabs and interpretation of an Upper Jurassic environment (Solenhofen). Die Naturwissenschaften, *61:*428–533.

Bolton, M. L., and Loveland, R. E., 1987. Orientation of the horseshoe crab, *Limulus polyphemus,* on a sandy beach. Biological Bulletin, *173:*289–298.

Bonaventura, J., Bonaventura, C., and Tesh, S. (editors), 1982. Physiology and Biology of Horseshoe Crabs: Studies on Normal and Environmentally Stressed Animals. Alan R. Liss, New York.

Cohen, J. A., and Brockmann, H. J., 1983. Breeding activity and mate selection in the horseshoe crab, *Limulus polyphemus.* Bulletin of Marine Science, *33:*275–281.

Lochhead, J. H., 1950. Xiphosura. *In* Brown, F. A. (editor), Selected Invertebrate Types. John Wiley and Sons, New York.

Patten, W., and Redenbaugh, W. A., 1900. Studies on *Limulus,* II. The nervous system of *Limulus polyphemus,* with observations on the general anatomy. Journal of Morphology, *16:*91–200.

Rudloe, A., 1981. Aspects of the biology of juvenile horseshoe crabs, *Limulus polyphemus.* Bulletin of Marine Science, *31:*125–133.

Rudloe, A. E., and Rudloe, J., 1981. Horseshoe crabs: living fossils. National Geographic, *159:*562–572.

CLASS ARACHNIDA

There are about 70,000 known living species of arachnids. Most of these are spiders and mites, but the class also includes ticks, scorpions, and several other distinctive groups. Although some scorpions, dating back to the Silurian period, were aquatic, arachnids now form a basically terrestrial assemblage. The several hundred species of mites living in fresh water and in the sea are secondarily adapted to an aquatic existence. There are also a few spiders that can live under water, but these must maintain a coating of air around themselves.

General Structure

Arachnids are diversified, but in general they conform to the following body plan. The **prosoma** has a pair of **chelicerae,** a pair of **pedipalps,** and four pairs of **walking legs.** The six pairs of prosomal appendages provide external evidence that there are at least six segments in this part of the body, even if the limits of the segments themselves are obscure. In the **opisthosoma,** which may have up to 12 segments, the appendages do not help much. The only definite opisthosomal appendages are the spinnerets of spiders and a pair of peculiar comblike structures found in scorpions.

Even when the number of opisthosomal segments does appear to be perfectly clear, one must be wary. In scorpions, for instance, there are 12 segments, plus the telson that forms the sting. During embryonic development, however, there are 13 opisthosomal somites. The first of these does not develop into a recognizable segment. Similar elimination of from one to three somites takes place in various other groups of arachnids.

The usual organs of respiration are **book lungs** or **tracheae,** or both, although some small arachnids have neither. When book lungs are present, they are on the ventral side of the opisthosoma and are arranged as one to four pairs. They are essentially pockets in which the body wall is thinned and folded into several sheets. These sheets have blood flowing through them, and although they have a cuticular covering, they are permeable to oxygen and carbon dioxide and provide considerable surface for gas exchange. Flow of air into and out of a pocket may be accelerated by action of muscles attached to the wall.

Tracheae (Figure 16.9) are air tubes that extend into the tissues from small openings, called *spiracles,* at the surface. In insects, centipedes, millipedes, and onychophorans, they deliver oxygen directly to the cells that need it; but in arachnids the air tubes give up oxygen to the blood and the blood distributes it to the tissues. More is said about tracheal systems of arachnids in the section below on spiders.

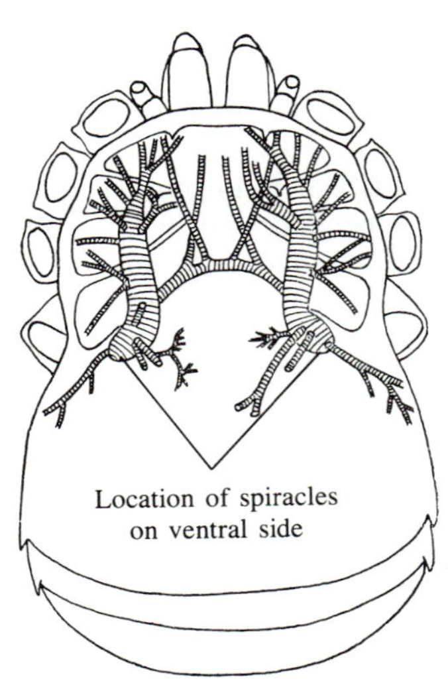

FIGURE 16.9 Tracheal system of *Opilio parietinus* (Order Phalangiida). (Lawrence, Biology of the Cryptic Fauna of Forests, based on Kaestner, in Kükenthal, Handbuch der Zoologie, *3*.)

Digestive System and Feeding

Arachnids, with the exception of some mites and daddy longlegs, feed on other animals. They may be specialized, as ticks are, for drawing up blood and tissue fluids. Many of them, however, are predators that capture prey and kill it, sometimes with the help of a venom. In general, predatory arachnids do not swallow large pieces. They discharge, through the mouth, digestive fluids produced in the midgut. The enzymes begin to act outside the body, breaking down proteins and other constituents of the prey's tissues. Scorpions, solpugids, whipscorpions, and large spiders may hasten liquefaction by chewing or kneading the prey while it is being basted with digestive juices. The fluid or semifluid food is then drawn in by a pumping action of the pharynx, and sometimes of the esophagus; both are parts of the foregut. The pumping is effected mostly by activity of radial muscles that alternately contract to dilate the foregut and then relax to let it collapse. As in other arthropods, the foregut is derived from an ectodermal invagination and is lined by cuticle.

It is in the midgut, located partly in the prosoma but mostly in the opisthosoma, that digestion is completed and absorption takes place. In many arachnids, including spiders and scorpions, the midgut has branching diverticula that greatly increase the space and surface for digestion and absorption. The diverticula may also make it possible for the animal to take in a large meal, an important matter if another meal is not likely to come soon. Behind the midgut is the hindgut. This, like the foregut, is derived from an ectodermal invagination. In certain ticks, the hindgut has no lumen, and in many mites it is absent altogether.

Circulation

In arachnids that have a heart—and most of them do—this structure is located in the opisthosoma. It is typically arthropodan in that it is dorsal and has one or more pairs of ostia by which blood can enter it from the pericardial sinus, which is part of the hemocoel (Figure 15.20, B). The heart of arachnids usually shows more distinct signs of segmentation than it does in other arthropods. Up to seven segmental chambers can be seen in the heart of some scorpions, but usually there are fewer, and often the heart seems not to be segmented at all. In general, the number of pairs of ostia conforms to the number of recognizable chambers, but this is not necessarily the case.

The wall of the heart is lightly muscularized. Contraction of the muscles coincides with closure of the ostia, so that blood is forced into the major arteries, which are anterior and posterior continuations of the heart, and also into segmentally arranged lateral arteries. After circulating through the hemocoelic spaces, the blood reaches the pericardial sinus, from which it reenters the heart when the heart is dilated and its ostia are opened. The dilation is accomplished by the pull of suspensor ligaments that are stretched when the heart contracts. In scorpions, most spiders, and some other groups, the blood contains hemocyanin, which functions in oxygen transport.

Excretion

The principal nitrogenous wastes of arachnids are guanine, uric acid, and related compounds. Crystals of these substances, which are of such low solubility that they are harmless, accumulate in cells of various parts of the body, including the epidermis and digestive glands. Most arachnids, however, have excretory organs in the form of **Malpighian tubules.** As in insects, millipedes, and centipedes, these tubules are delicate diverticula arising from the posteriormost part of the midgut. Bathed by blood in the hemocoel, they convert soluble, toxic nitrogenous substances into crystals of guanine or uric acid. The crystals are delivered to the midgut and voided with fecal material. Production of nearly insoluble nitrogenous wastes is an important adaptation to terrestrial life, for it enables an animal to eliminate these wastes without giving up much water.

The function of the coxal glands found in arachnids needs much study. Like antennal and maxillary glands of crustaceans, and maxillary glands of myriapods and some insects, they originate in part from coelomic pockets. Their openings are at the bases of one or more pairs of walking legs. The essential features of a coxal gland are a thin-walled epithelial sac, in direct contact with the blood contained in hemocoelic sinuses, and a duct. The duct may be complicated and sometimes it has a bladderlike dilation. The coxal glands do not always have openings, however. On the basis of what is known about antennal and maxillary glands of crustaceans, it seems likely that coxal glands that do reach the outside are concerned primarily with selective elimination of certain ions rather than with excretion of nitrogenous wastes.

Nervous System and Sense Organs

The brain lies above the foregut, generally above the esophageal portion. The protocerebrum innervates the eyes; the tritocerebrum innervates the chelicerae. (There is no deutocerebrum in arachnids.)

In scorpions, which seem to be the most primitive arachnids in certain respects (especially their segmented opisthosoma), several of the ganglia of the opisthosomal part of the ventral nerve cord are distinct and widely separated. The more anterior ones, however, are fused with a mass of prosomal ganglia located beneath and alongside the esophagus. In most arachnids, all or nearly all of the opisthosomal ganglia end up in the prosoma, where they become incorporated into a ganglionic mass that envelops the esophagus (Figure 16.24). The nerves that supply opisthosomal structures therefore originate in ganglia located in the prosoma.

Sensory Setae, Trichobothria, and Slit Sense Organs

As in other arthropods, touch and vibration receptors and chemoreceptors are associated with many of the cuticular setae. In a chemoreceptor seta, a process of a nerve cell usually extends to the tip of the seta, where there is a pore. In a touch or vibration receptor, the process usually ends at the base of the seta.

A special type of setal sense organ, extremely responsive to air currents and vibrations, is found in various groups of arachnids, including spiders and scorpions. This is the **trichobothrium.** The hair is very long, and it emerges from a membrane-capped cylinder that is enclosed by a deep cup, or *bothrium* (Figure 16.10, D). The cylinder is filled with fluid. The slightest air current causes the hair to bend and displace the basal portion, within the cylinder, enough to stimulate the processes of some sensory cells. These terminate under a helmet-shaped cap that is attached to the proximal end of the hair.

Slit sense organs—and **lyriform organs,** which are groups of slit sense organs—are also characteristic of arachnids. They are located on appendages (Figure 16.10, A) as well as on the body proper, and are concerned with picking up vibrations and also with proprioception. Slit sense organs can be recognized as depressions where the cuticle is thinned out (Figure 16.10, B and C). Processes of sensory cells running to the cuticle are sensitive to changes in the pressure of fluid around them. The changes may be caused by external vibrations or by slight deformation of the cuticle as the animal moves. Chemosensory elements of the nervous system are thought to be present in some slit sense organs.

Eyes

Although some fossil scorpions had compound eyes, and although the retinal cells in eyes of modern scorpions form groups similar to retinulae of compound eyes, the eyes of arachnids are considered to be of the simple type. The cornea consists of modified cuticle and serves also as a lens. In spiders, the epidermis that secretes the cornea later breaks down and converts into a vitreous body that separates the lens from the layer of retinal cells. The light-sensitive elements of the retinal cells may face the lens or they may face away from it; that is, the cell bodies of the retinal cells may be behind the rhabdoms or they may be between the lens and the rhabdoms (Figure 15.14). In scorpions, a vitreous body does not develop, and the light-sensitive elements of the retinal cells always face the lens. The degree of success with which arachnid eyes can form an image probably depends to a large extent on the number of retinal receptors; the more numerous and more crowded they are, the better the chance that a fairly good image will be formed. The eyes of jumping spiders, which ambush

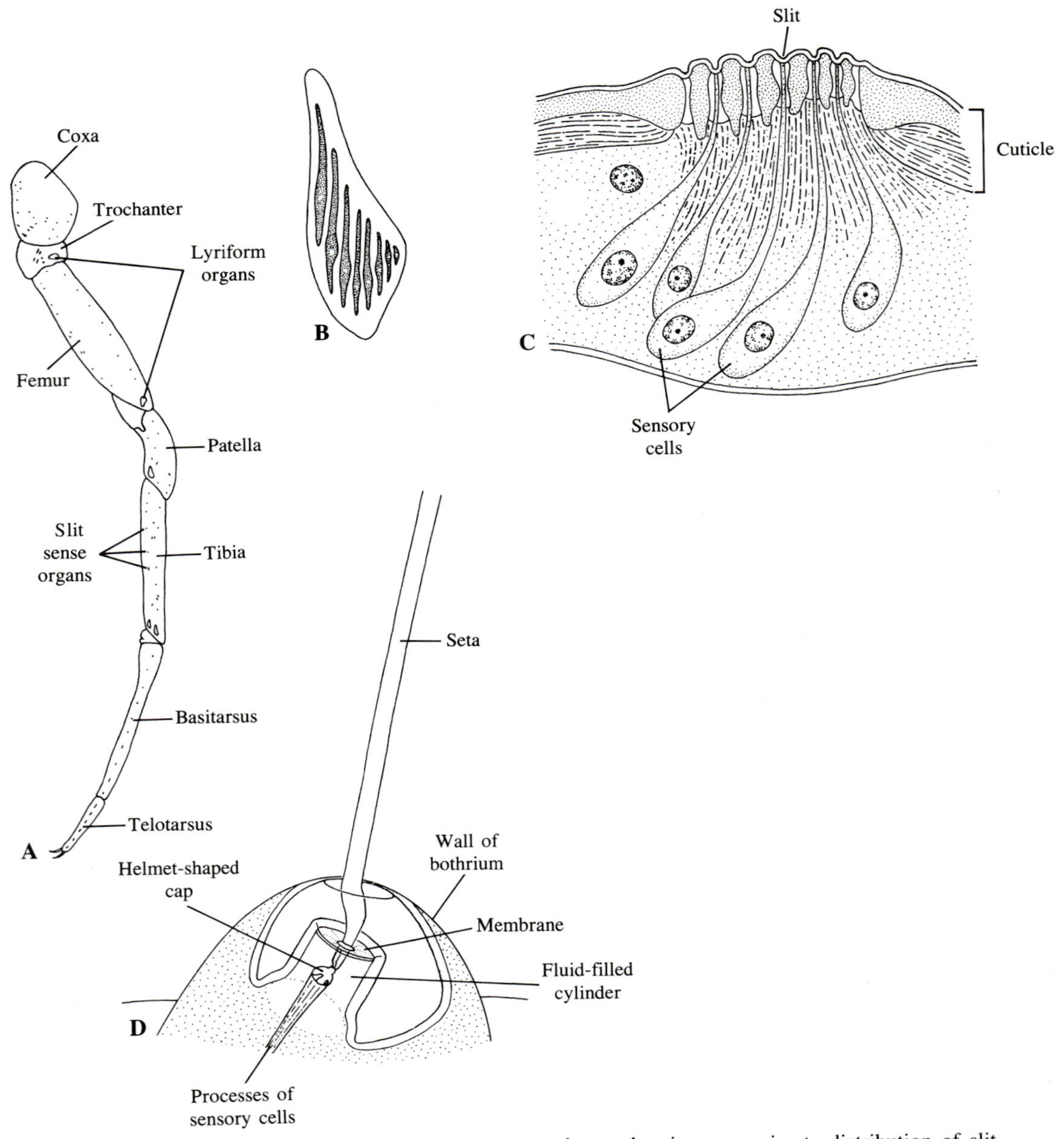

FIGURE 16.10 A. Leg of the spider *Araneus undatus*, showing approximate distribution of slit sense organs and lyriform organs. B. Lyriform organ, surface view. (A and B, after Vogel, Jenaische Zeitschrift für Naturwissenschaft, *59*.) C. Lyriform organ, cross-section. D. Trichobothrium of the spider *Tegenaria durhami*. (After Görner, Cold Spring Harbor Symposia in Quantitative Biology, *30*.)

their prey or run it down, seem to be efficient for recognizing shapes as well as for detecting movements.

Not all of the pairs of eyes are necessarily alike. One or more pairs may have sharper acuity than others, or be structured to permit better close vision or distance vision than others. Bright reflective material in the eyes of some spiders helps them see in dim light by causing the light that has just impinged on a retinal receptor to bounce back through the receptor, thus giving a double dose of stimulation. Spiders whose eyes have an abundance of such reflective material, forming what is called a *tapetum,* can often be recognized at night because their eyes shine in the beam of a flashlight.

Reproduction and Development

In male and female arachnids of all groups, there is a single ventral gonopore. This develops on the second embryonic somite of the opisthosoma, which becomes the first definitive segment of this part of the body. (Opisthosomal somite 1 remains rudimentary.) The gonads are also in the opisthosoma; they are sometimes single, sometimes paired. Females generally have a seminal receptacle (or two of them, when there are two oviducts) for storing sperm received from males. There is typically a complicated mating ritual that leads to transfer of sperm. In certain groups, the male packages his sperm in a spermatophore. He may place this directly on the female's gonopore or deposit it on the ground and draw the female over it. In spiders, the male puts a drop of spermatic fluid on a silken thread, which may be part of the web. He then picks up the drop with his pedipalps. These appendages have a bulblike cavity for storing the spermatic fluid until the right moment in the mating ritual. At that time, the tips of the pedipalps are inserted into the gonopore of the female and insemination is accomplished.

Embryonic development of arachnids is not discussed in detail here: the information presented in Chapter 15 concerning early development of arthropods applies, in general, to arachnids. The eggs are usually so yolky that cleavage is at first superficial. Total cleavage is, however, characteristic of some types. As in most arthropods, a germinal band is formed from the part of the cell layer that surrounds the original egg, and this band becomes the definitive embryo. The band shows distinct division into somites, even though external evidence for segmentation may be lacking, as it is in the opisthosoma of nearly all spiders. Some information concerning later stages of development is given below in connection with individual groups of arachnids.

General References on Arachnida

Barth, F. G. (editor), 1985. Neurobiology of Arachnids. Springer-Verlag, New York and Heidelberg.

Barth, F. G., and Pickelmann, P., 1975. Lyriform slit sense organs. Journal of Comparative Physiology, *103:*39–54.

Cloudsley-Thompson, J. E., 1968. Spiders, Scorpions, Centipedes, and Mites. Pergamon Press, Oxford and New York.

Savory, T., 1978. Arachnida. 2nd edition. Academic Press, London and New York.

Slansky, F., Jr., and Rodriguez, J. G. (editors), 1987. Nutritional Ecology of Insects, Mites, Spiders, and Related Invertebrates. John Wiley and Sons, New York.

Snodgrass, R. E., 1948. The feeding organs of Arachnida, including mites and ticks. Smithsonian Miscellaneous Collections, *110*(10):1–93.

Snow, K. R., 1970. The Arachnids: An Introduction. Columbia University Press, New York.

Weygoldt, P., and Paulus, H. F., 1979. Untersuchungen zur Morphologie, Taxonomie und Phylogenie der Chelicerata. I. Morphologische Untersuchungen. Zeitschrift für Zoologische Systematik und Evolutionsforschung, *17:*85–116.

Weygoldt, P., and Paulus, H. F., 1979. Untersuchungen zur Morphologie, Taxonomie und Phylogenie der Chelicerata. II. Cladogramme und die Entfaltung der Chelicerata. Zeitschrift für Zoologische Systematik and Evolutionsforschung, *17:*177–200.

Order Scorpionida (Scorpiones)

Because of their form and their sting, scorpions have always fascinated humans. Images of these animals appear on coins and household objects of ancient civilizations. Perhaps no other invertebrates, with the possible exception of spiders, have been so extensively exploited in stories of adventure and murder.

While most scorpions are found in deserts where the climate is warm and dry, some inhabit regions that are humid or that become extremely cold in winter. In general, they are nocturnal, hiding during the day under rocks, logs, loose bark, and other objects. Those that enter houses are apt to hide in shoes and under piles of clothing and papers. In areas where scorpions do become household pests, it is advisable to shake shoes before putting them on, and not to walk barefoot at night. The largest scorpions may exceed a length of 15 cm, but the majority are less than 10 cm long.

General Structure

The oldest arachnids in the fossil record are scorpions that are believed to have been aquatic. Certain attributes of these early fossil scorpions, as well as differences between scorpions and other arachnids, have led some zoologists to place scorpions in the Merostomata, the group that includes eurypterids and horseshoe crabs. Nevertheless, in most respects scorpions clearly show characteristics of arachnids. The prosoma bears six pairs of appendages: chelicerae, pedipalps, and four pairs of walking legs (Figure 16.11, A and C). The pedipalps are large and chelate, and they resemble the chelipeds (fourth thoracic appendages) of lobsters, crayfishes, and crabs. There are typically a pair of eyes on the middorsal part of the prosoma and two pairs (sometimes more) of anterolateral eyes. Although these eyes have simple lenses, their retinal cells are arranged in groups, and in this respect they slightly resemble compound eyes.

The opisthosoma consists of 12 segments, plus the terminal telson. There are 13 embryonic somites to begin with, but the first loses its identity. The definitive segment 1 has the genital pore, covered by a bilobed

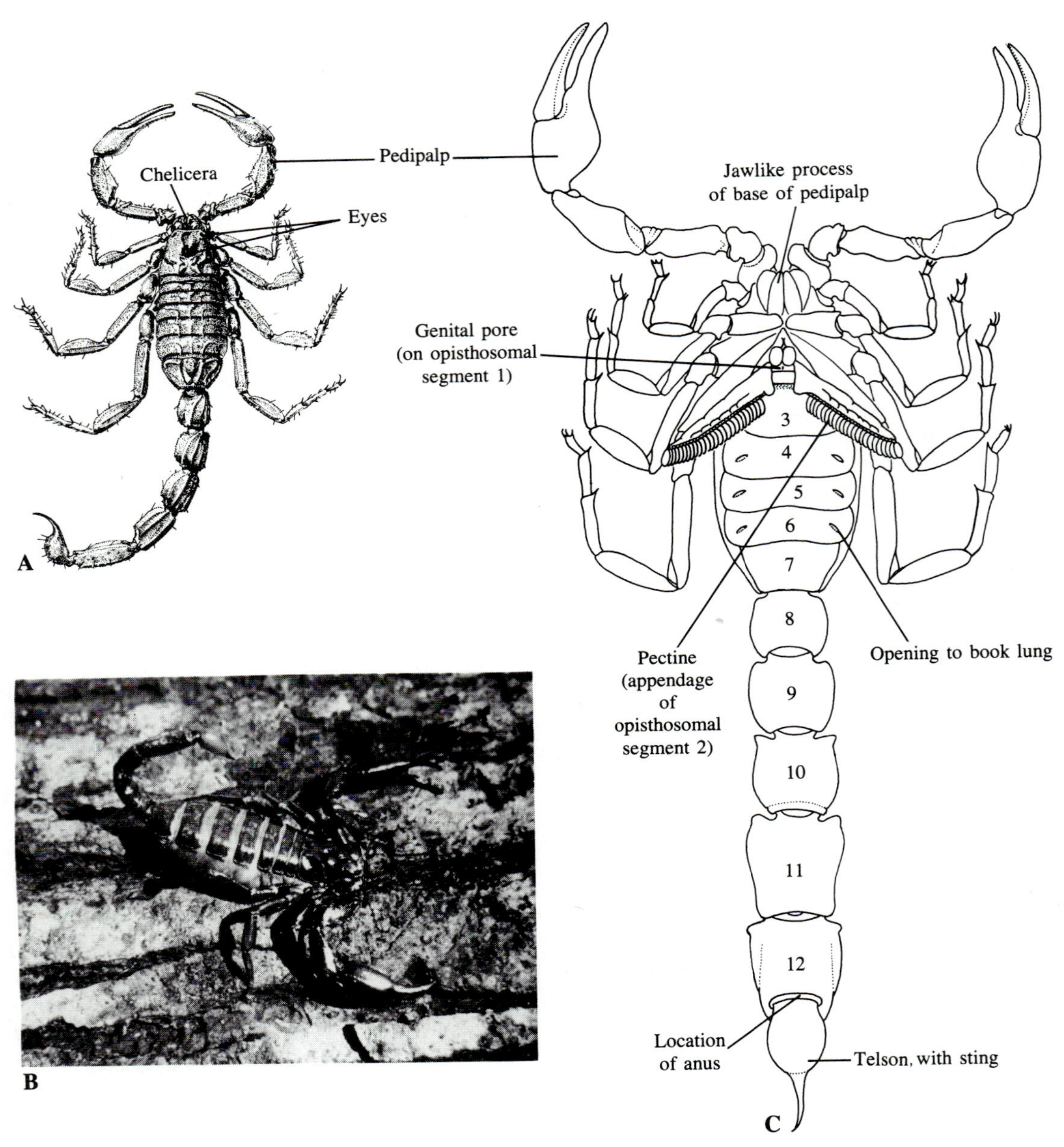

Figure 16.11 Order Scorpionida. A. *Mesobuthus macmahoni*, female, dorsal view. (Vachon, Videnskabelige Meddelelser fra Dansk Naturhistorisk Forening i København, 120.) B. Living scorpion. C. Ventral view of a scorpion.

genital operculum. Segment 2 has a pair of complicated comblike appendages called **pectines** (Figure 16.11, C), which have sensory functions. There are no other appendages on the opisthosoma, but segments 3 to 6 have the four pairs of book lungs. The walls of these ventrolateral pockets are much folded, so that the surface area for exchange of oxygen and carbon dioxide is greatly increased. Blood circulates through the folds, which resemble the pages of a book. In segment 7, the opisthosoma becomes abruptly narrowed, and the succeeding articulation is where a scorpion flexes its tail.

The telson functions as the stinger. Venom is produced by glands that occupy much of the telson, and the duct from these has its opening beside the sharp tip. Scorpions will not attack humans, but they may be provoked to jab with the stinger if stepped on or sat on.

Androctonus australis of North Africa, certain species of *Centrurus* in Mexico and Arizona, and a few others are dangerous, particularly to children. The venom contains a neurotoxic fraction that is paralytic and a hemorrhagic fraction that causes localized necrosis. The stings of scorpions seem to fall into two general categories: those that cause only a little pain and those that are very painful and that may be accompanied by complications such as partial paralysis, fever, and even death.

Digestive System and Feeding

Scorpions feed on a wide variety of insects, spiders, millipedes, and other terrestrial arthropods, and some will attack newborn mice. They use their large pedipalps to capture and hold the prey, and also to crush and knead it. By flexing the tail upward and bringing the tip forward, a scorpion can press in its sting. The chelicerae generally pick at the prey, sometimes dismembering it. The mouth lies within a ventral depression bordered by the bases of the chelicerae, pedipalps, and first two pairs of walking legs. The pharynx and esophagus form a short, cuticle-lined foregut in the prosoma. Radial muscles dilate the pharynx and circular muscles reduce its lumen; the two sets enable the pharynx to act as a sucking pump. The midgut, with several diverticula on both sides of its anterior portion, begins in the posterior part of the prosoma and runs to the last segment, where a cuticularized hindgut succeeds it. The anus is located on the ventral side of the opisthosoma, at the base of the telson.

Digestion is initiated externally when a mixture of intestinal enzymes is regurgitated onto the tissues of the prey. Glands at the bases of the first and second walking legs contribute something comparable to a salivary secretion. As the tissue is liquefied into a kind of soup, it is sucked up by the pumping action of the pharynx. The rest of the digestive process, as well as absorption, is achieved in the diverticula of the anterior part of the intestine.

Circulatory and Excretory Systems

The rather large heart, surrounded by the pericardial sinus, lies in the broad anterior half of the opisthosoma. It beats rapidly, forcing blood into large anterior and posterior arteries, as well as into arteries that are directed laterally. Valves in the proximal portions of the arteries prevent backflow. The blood enters sinuses of the hemocoel and eventually goes into two large ventral sinuses, portions of which are intimately associated with the book lungs. From these sinuses, the blood reaches the pericardial sinus by several pairs of veinlike vessels, and finally returns to the heart through the ostia. The blood contains hemocyanin.

Besides the Malpighian tubules (one or two pairs), there are cells that store nitrogenous wastes. The coxal glands open on segment 5 of the opisthosoma. They are rather complicated, partly because the saclike portion of each one is pervaded by branches of one of the arteries. The vessels do not, of course, open into the sac, but the surface area available for exchange of wastes or excess ions is extensive.

Nervous System

The nervous system (Figure 16.12) conforms more closely than that of other arachnids to what presumably was the ancestral plan. Several of the ganglia belonging to the opisthosomal segments are distinct and separated by long connectives. The location of the ganglia does not necessarily conform, however, to the segments in which they have originated. For instance, the ganglion that innervates structures in the fifth opisthosomal segment may be as far forward as the third segment. The ganglia of the first four opisthosomal segments are

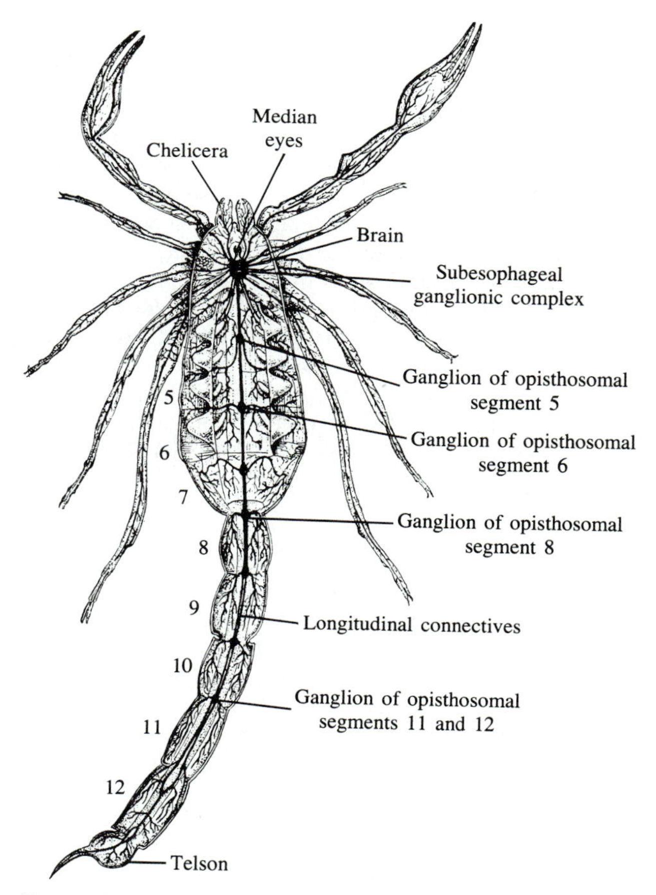

FIGURE 16.12 Order Scorpionida. Nervous system of a scorpion. (Ivanov et al., Major Practicum in Invertebrate Zoology.)

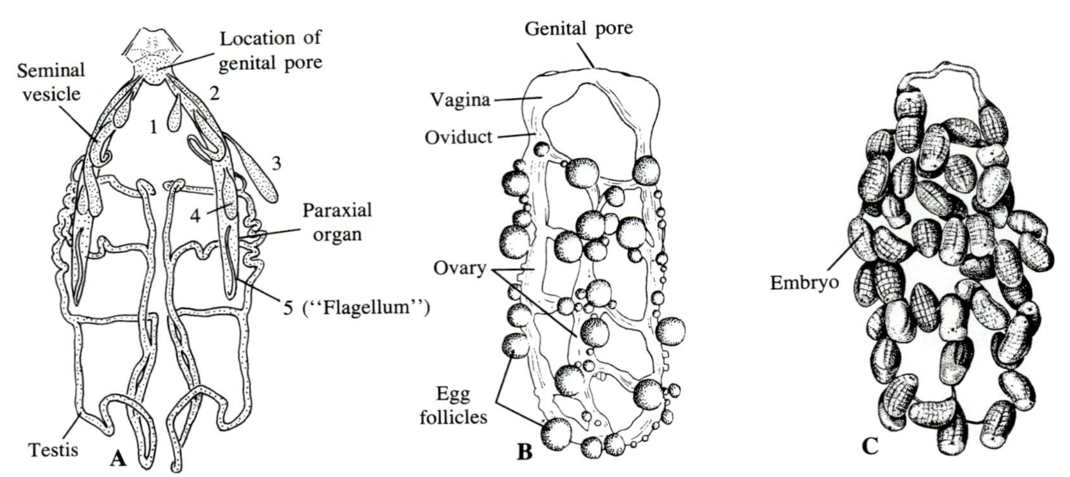

FIGURE 16.13 Order Scorpionida. Reproductive system of *Buthus*. A. Male. The various diverticula and glands associated with the paraxial organ are numbered instead of being given names, for their specific functions are uncertain; on the right side, some of them are separated from the paraxial organ, to which they are normally closely appressed. B. Female. C. Gravid female. (B and C, Ivanov et al., Major Practicum in Invertebrate Zoology.)

united with those of the prosomal segments to form a massive subesophageal ganglion. This is joined by connectives to the brain located above the esophagus.

Scorpions have abundant sensory setae, including trichobothria, and there are also slit sense organs. When eyes are present on the prosoma, as is the case in most species, two are close to the dorsal midline and two to five others lie on both sides of the anterior part of the prosoma. The comblike pectines, mentioned earlier as the only appendages on the opisthosoma, are waved over the substratum as a scorpion walks. Males use them, in connection with the mating ritual, for locating a place suitable for deposition of a spermatophore, but the fact that they are present in both sexes indicates that they have other functions.

Reproduction and Development

The gonads of both sexes are unusual in that each member of the pair consists of two trunks joined by cross-connections (Figure 16.13). Thus the gonads resemble rope ladders. In females, the oviducts unite just before they reach the genital pore, forming a chamber to receive sperm, or even an entire spermatophore. In some species, the median portions of the ovaries are joined along the midline. In males, the lower portion of each sperm duct broadens out into a seminal vesicle, and this enters one arm of a complicated, more or less V-shaped structure called the *paraxial organ*. Each arm of this produces half of the spermatophore, which is eventually extruded through the genital pore. The various glands and diverticula associated with the paraxial organ have not been studied sufficiently, but it is assumed that most of them are concerned with production of the spermatophore.

Mating (Figure 16.14) involves considerable courtship. There is no true copulation, but fertilization is nevertheless internal. The male deposits a spermatophore on the ground and, while holding the female's pedipalps with his own, pulls her over it. As her genital pore presses down on the spermatophore, or as her genital operculum catches on it, some of the contents of the spermatophore are released. After the eggs have been fertilized, pockets on the main trunks and cross-connections of the ovaries enlarge and serve as incubatory pouches for the developing young, which are born as little scorpions. The pouches often have elaborations that function as placentas; through these the young receive nourishment in addition to that provided by yolk. In a few species development takes place outside the body of the mother.

References on Scorpionida

Alexander, A. J., 1959. Courtship and mating in buthid scorpions. Proceedings of the Zoological Society of London, *133*:145–169.

Carthy, J. D., 1968. The pectines of scorpions. *In* Carthy, J. D., and Newell, G. E. (editors), Invertebrate Receptors. Symposia of the Zoological Society of London, no. 23. Academic Press, London and New York.

Hadley, N. F., 1973. Adaptational biology of desert scorpions. Journal of Arachnology, *2*:11–23.

Vachon, M., 1953. The biology of scorpions. Endeavour, *12* (46):80–89.

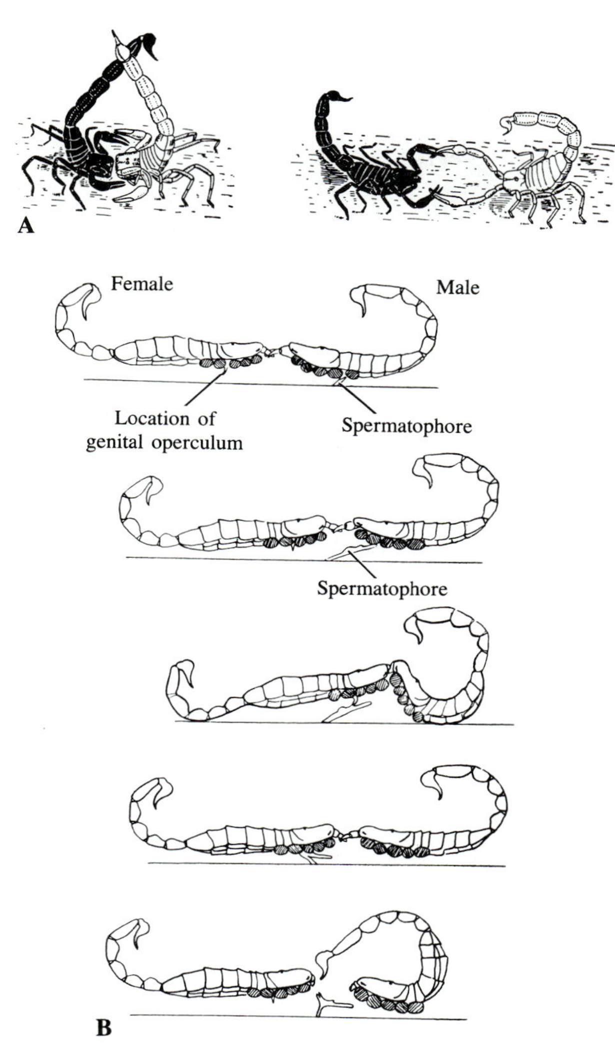

FIGURE 16.14 Order Scorpionida. A. Courtship of *Buthus*. (Lawrence, Biology of the Cryptic Fauna of Forests, after Fabre.) B. Deposition of a spermatophore and insemination of the female of *Opisthophthalmus latimanus*. The male deposits a spermatophore, then moves backward, pulling the female forward. He then jerks the female upward and closer to him, whereupon her genital operculum catches on some hooklike elaborations of the spermatophore. In the process, the spermatophore is bent, and portions of it are everted and enter the female's genital pore. The pair stays in this position for a few minutes. Finally, the male, with sting raised, chases the female away. (Alexander, Proceedings of the Zoological Society of London, *128*.)

Order Pseudoscorpionida

Pseudoscorpions (Figure 16.15) resemble true scorpions because their pedipalps are proportionately large and chelate. The opisthosoma of a pseudoscorpion does not, however, become narrowed down to form a tail, and there is no terminal sting. Each chelicera has only two articles, one of which is the movable finger. This is beneath the fixed finger, a prolongation of the basal article. (In true scorpions, each chelicera has three articles, and the movable finger is lateral to the fixed finger.) The inner edges of the fingers that form the chelae have comblike elaborations; these are used for cleaning some of the other appendages. The chelicerae have silk glands whose ducts open to the outside near the tips of the movable fingers; these glands are almost certainly homologous to the venom glands of a spider. When venom glands are present in pseudoscorpions, they open on the fingers of the large chelae of the pedipalps. As is typical for arachnids, there are four pairs of walking legs. There are usually two pairs of lateral eyes on the prosoma. The opisthosoma has 12 segments, none of which has appendages. There is no recognizable telson.

Pseudoscorpions are only a few millimeters long, and they live mostly under bark, rocks, peeling wallpaper, and similar situations. They feed mostly on small insects. Associated with the mouth is a complex of structures, derived partly from the lips in front of and behind the mouth, and partly from the basal portions of the pedipalps and chelicerae. The elements of this complex are not dealt with here; it will suffice to say that some of them are movable and create a sucking action. The pharynx is a less efficient pump than its counterpart in true scorpions.

There are no Malpighian tubules; the only discrete structures that perhaps function in excretion are the coxal glands, whose ducts lead to the basal portions of the third legs. Within various tissues, however, there are cells that accumulate relatively insoluble nitrogenous wastes. Respiration is by a tracheal system, derived from book lungs. There are only two pairs of main tracheal trunks, however. Their openings are on the ventral side of the third and fourth segments of the opisthosoma.

In both males and females, the genital opening is on the third segment of the opisthosoma. The gonoducts coming from the right and left halves of the unpaired gonad unite before they reach the opening. In the female, the common duct (''uterus'') has diverticula that function as seminal receptacles. The reproductive system of the male is provided with glands for producing spermatophores. As in scorpions, there is a mating ritual, and the male pulls the female toward him until her genital pore touches the spermatophore he has deposited (Figure 16.15, C). The eggs are fertilized internally, but develop externally in a mass of jelly attached to the underside of the opisthosoma. After the young hatch, but before they are ready to be released by the mother,

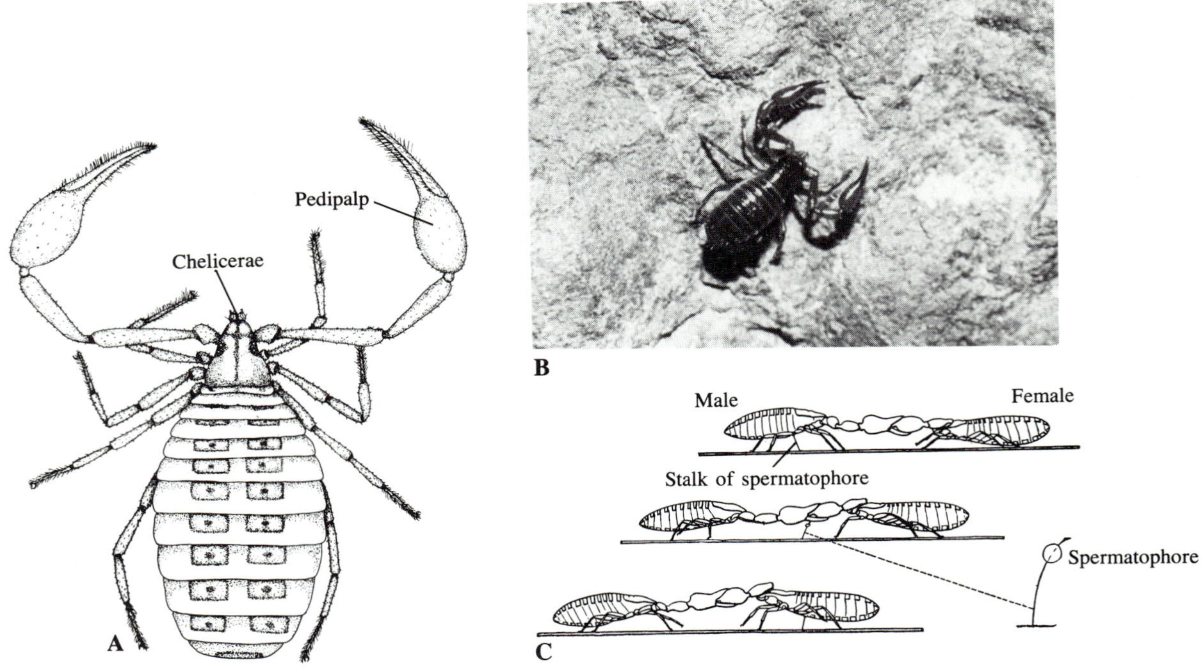

FIGURE 16.15 Order Pseudoscorpionida. A. *Garypus californicus*, a species found close to the seashore along the Pacific coast of North America. (Moore, Journal of Entomology and Zoology, Pomona College, *9*.) B. *Halobisium occidentale*, found under rocks and debris near bays and saltmarshes on the Pacific coast. C. Courtship and spermatophore transfer in *Chermes pygmaeus*. After the male has deposited a stalked spermatophore and confronted the female, he withdraws, and the female advances until the spermatophore contacts her genital pore. (Chamberlin, Stanford University Publications, Biological Sciences, *7*.)

they feed on a deposit of nutritive material produced by the ovaries. While brooding their eggs and young, female pseudoscorpions remain hidden and they may weave nests of silk around themselves.

References on Pseudoscorpionida

Chamberlin, J., 1931. The arachnid order Chelonethida. Stanford University Publications, Biological Sciences, *7*:1–284.

Gilbert, O., 1951. Observations on the feeding of some British false scorpions. Proceedings of the Zoological Society of London, *121*:335–358.

Goddard, S. J., 1976. Population dynamics, distribution patterns and life cycles of *Neobisium muscorum* and *Chthonius orthodactylus*. Journal of Zoology, *178*:295–304.

Hoff, C. C., 1949. The pseudoscorpions of Illinois. Illinois Natural History Survey Bulletin, *24*:413–498.

Weygoldt, P., 1969. The Biology of Pseudoscorpions. Harvard University Press, Cambridge.

Order Araneida (Araneae)

On the basis of all that is known about their structure and behavior, spiders constitute the most diversified order of arachnids. There are more than 30,000 species. The largest, a South American tarantula, reaches a body length of 9 cm and a leg spread of 28 cm. The usual body size is from a few millimeters to about 2 cm. The tiniest species are only about 1 mm long. Spiders differ from their relatives in having the anteriormost part of the opisthosoma narrowed to a slender stalk (Figure 16.18, A, C, and D; Figure 16.19), often barely large enough to accommodate the gut, aorta, nerve cord, and some muscles. The stalk may not be readily apparent, especially when a spider is viewed from the dorsal side (Figure 16.16; Figure 16.17, A–C), because it is concealed by the broad portion of the opisthosoma that follows it. Except in one relatively small group of spiders, the opisthosoma is not segmented in the adult stage. Studies on the development of this tagma have shown, however, that it is derived from 12 embryonic somites.

Appendages

The layout of the prosomal appendages (Figure 16.18, A, C, and D; Figure 16.19) conforms to the usual arachnid plan. The chelicerae, consisting of two articles, are *subchelate* rather than chelate. In other words, the terminal article, which forms the movable claw, does not have a fixed finger against which it can close; instead, it bends back to touch one edge of the basal article. The claw, with a tiny pore near its tip, serves as a fang that delivers a venom secreted by a gland located partly in the chelicera itself, partly in the prosoma. In some spiders the fangs strike downward; in others they are directed toward one another.

The pedipalps are important sensory appendages, and in some spiders they are used for walking. They consist of six articles. The terminal one, the *tarsus,* is used by mature males for storing sperm and transferring these to the seminal receptacle pores of the female. The legs, of which there are four pairs, have seven articles (Figure 16.10, A), with the *telotarsus* (the more distal of two tarsal articles) bearing two or three claws and sometimes other minute specializations.

The only appendages of the opisthosoma are **spinnerets** (Figure 16.18, C and D; Figure 16.19), from which silk is exuded. More is said about these structures below, but their origin should be explained here. A spider may have up to eight spinnerets, but only four of them are true appendages. They arise in pairs on segments 4 and 5 of the opisthosoma and gradually become displaced to a more posterior position. Just medial to the primordia of these four spinnerets, outgrowths of the body wall may appear and develop into what may be called *secondary spinnerets*. In the course of the

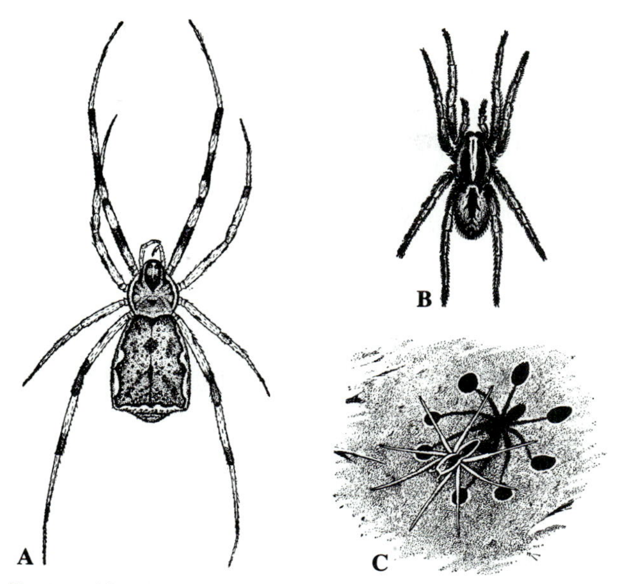

FIGURE 16.16 Order Araneida. A variety of spiders. A. *Epsinus angulatus*, female. (Locket and Millidge, British Spiders.) B. *Lycosa phipsoni*, female. (Pocock, in Fauna of British India.) C. *Dolomedes fimbriatus*, which walks on water. (Zhadin and Gerd, Rivers, Lakes, and Reservoirs of the USSR, Their Fauna and Flora.)

A

B

C

FIGURE 16.17 Order Araneida. Photographs of spiders. A. Tarantula. (Courtesy of Carolina Biological Supply Company, Inc.) B. Member of the family Amaurobiidae. C. *Araneus diadematus*, an orb weaver, dorsal view (left) and ventral view (right).

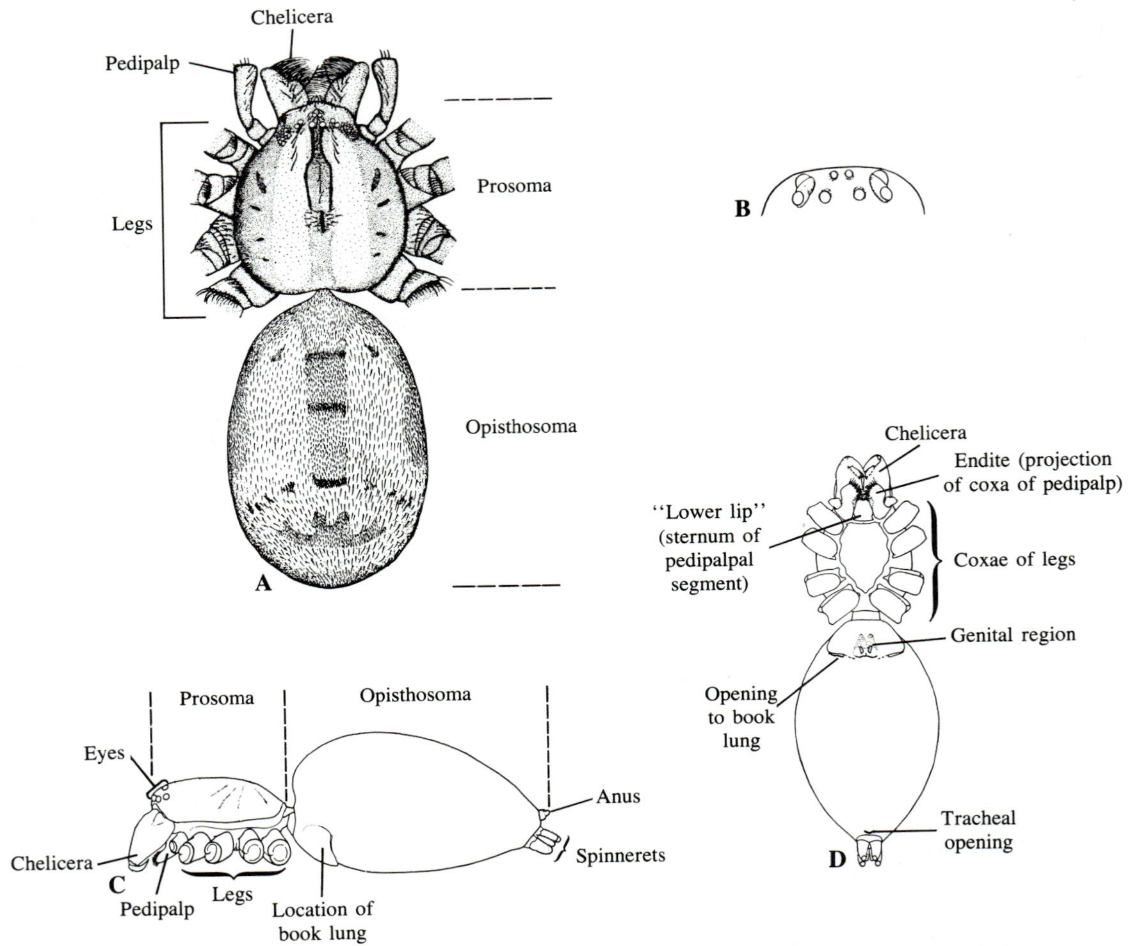

FIGURE 16.18 Some external features of spiders. A. *Gradungula sorenseni*, female, dorsal view. B. Eyes of *Gradungula*. The two lateral eyes on each side are located on a common tubercle. (A and B, Forster, Pacific Science, *9*.) C. Diagram of a spider, lateral view from the left side. D. Diagram of a spider, ventral view. (C and D, Kaston, Connecticut State Geological and Natural History Survey, *70*.)

evolution of spiders, much has happened to the spinnerets. The medial ones of the anterior group of four are often absent, or are fused into a single structure. The product of the fusion, called a **cribellum,** produces a special kind of silk that is combed out into a sheet by bristles on the hind legs. Various modifications have taken place in other spinnerets as well.

Digestive Tract

The mouth of a spider is situated in a slight depression bordered anteriorly by the labrum, laterally by the coxae of the pedipalps, and posteriorly by the sternum of the pedipalpal segment. The chelicerae tear at the prey, reducing it to small pieces, and glands in the labrum and coxae of the pedipalps secrete saliva. Juice from the midgut is pumped out of the mouth and its enzymes initiate digestion, then the "soup" of partly digested material is sucked into the digestive tract. Stiff hairs on the medial portions of the coxae of the pedipalps prevent large particles from entering the mouth. Both the pharynx and so-called stomach are provided with muscles that enable these portions of the gut to act as pumps. The pharynx and stomach, as well as the slender esophagus between them, are ectodermal derivatives. The midgut, of endodermal origin, is partly in the prosoma, partly in the opisthosoma, and generally has a number of diverticula that function as digestive glands. Some of the diverticula may reach into the

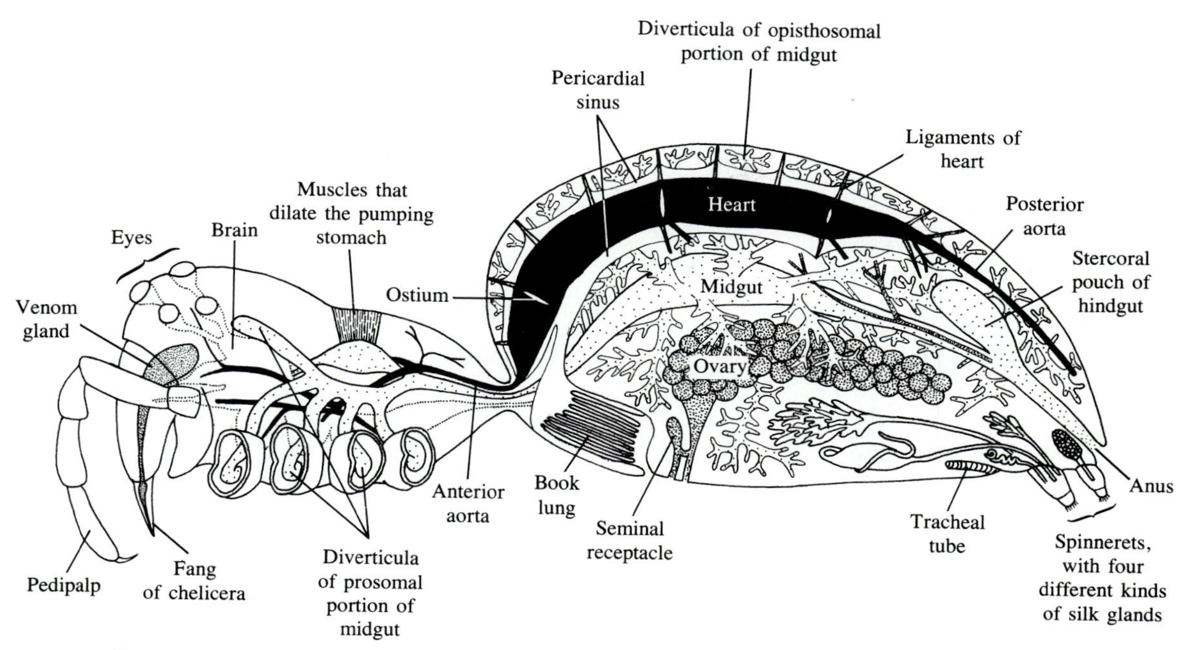

FIGURE 16.19 Internal anatomy of a spider. (The type on which the drawing is based has two pairs of spinnerets.)

coxae of the walking legs. The hindgut, lined by tissue of ectodermal origin, typically has a dorsal sac *(stercoral pouch)* that is used for storing fecal material and nitrogenous wastes that have been delivered to the gut by the Malpighian tubules. The anus is situated on a little papilla.

Venom

Presumably all spiders have venom that can be used to subdue prey. Only a few species are dangerous to humans, however. The black widows, of the genus *Latrodectus,* are some of the ones that humans are likely to contact, for they are often common in attics, garages, outhouses, and similar habitats. Most of the unpleasant experiences with black widows begin when a person disturbs a web, puts on clothing in which a spider is hiding, or picks up wood from a woodpile. The bite itself may scarcely be noticed, but the venom affects the nervous system and may cause muscle spasms, respiratory paralysis, pain in various parts of the body, and other disagreeable symptoms. Bites of black widows are especially dangerous to small children and to persons with heart disease. The venom of certain species is more potent than that of others.

Spiders of the genus *Loxosceles* are among those whose venom affects blood. The bite of *L. laeta,* found in Argentina, Peru, and Chile, may cause severe symptoms in humans. *Loxosceles reclusa,* the brown recluse, which occurs in parts of the United States, is less dangerous, but its bite may become surrounded by a tender, red area that sometimes turns into a necrotic ulcer; pain and fever accompany this. The brown recluse lives in houses and therefore may get into towels, clothing, or other objects. It has long been known in Texas, Oklahoma, Kansas, Missouri and a few states farther east, and its range seems to be spreading northward. An African species of *Loxosceles* has become established in several university libraries. It is thought to have been introduced with books or packing material.

Respiration

In spiders that are considered primitive, there are two pairs of book lungs (Figure 16.19). They develop in somites 2 and 3 of the opisthosoma, so their openings, located ventrally, are rather far forward on this tagma. There are deviations from the primitive plan, however. Some spiders have just the first pair of book lungs. Others have the first pair plus two sets of tracheal tubes, one member of each set being a derivative of a second book lung, the other being a derivative of a hollowed-out apodeme. There are also spiders in which both pairs of book lungs have been replaced by tracheae, or in which those of one pair have disappeared and those of the other pair have been replaced by tracheae. The tracheal tubes may be branched or unbranched. If branched, some of the ramifications enter

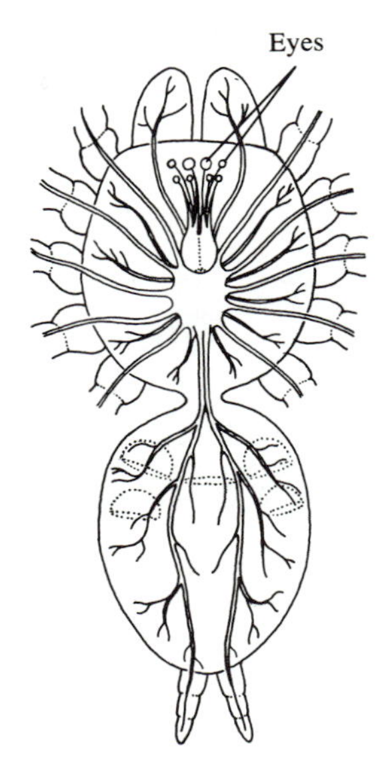

FIGURE 16.20 Nervous system of a spider. (Pocock, Proceedings of the Zoological Society of London, *2*.)

the prosoma. Neither the branched nor the unbranched type makes direct contact with tissue cells, such as is typical of tracheal systems of insects, millipedes, centipedes, and onychophorans; rather, they exchange gases with the blood.

The few spiders that can live under water for extended periods belong to two genera of one family. *Argyroneta,* found in fresh water, makes a dome of silk that serves as a kind of diving bell for storing air. *Desis,* which inhabits holes in seashore rock or coral, makes a silk tube for the purpose of trapping air during periods of low tide. Neither these genera nor the spiders that walk on the surface of water (Figure 16.16, C) are truly aquatic in the sense that they are immersed in water.

Circulation and Excretion

The tubular heart (Figure 15.20, B) is in the dorsal part of the opisthosoma. The pattern of circulation is similar to that of scorpions. Blood that has circulated through the "pages" of the book lungs returns to the pericardial sinus (Figure 16.19), then enters the heart through two or more pairs of ostia.

A pair of Malpighian tubules, typically branched, joins the posteriormost part of the midgut. As in other arachnids, as well as in insects and myriapods, these convert soluble nitrogenous wastes into relatively insoluble compounds, such as guanine and uric acid. The stercoral pouch dorsal to the hindgut is often filled with crystalline nitrogenous material that is to be voided with feces. Nitrogenous wastes also accumulate in cells of the epidermis, caeca of the midgut, and other tissues.

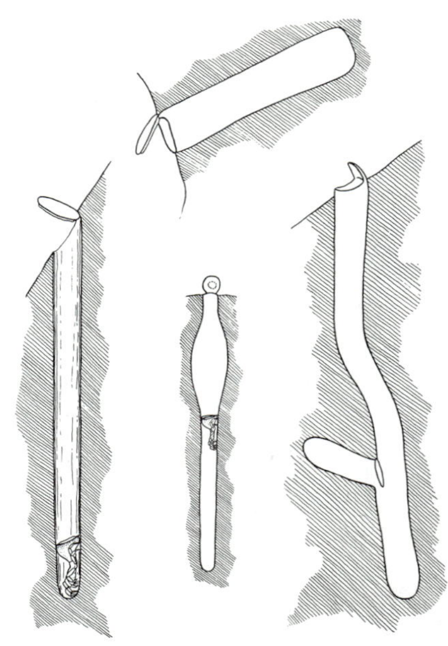

FIGURE 16.21 Burrows of females of four species of trapdoor spiders (Gertsch and Wallace, American Museum Novitates, no. 884.)

The brilliant white color of some spiders is due to high concentrations of guanine or similar compounds in the epidermis. Coxal glands that are assumed to be concerned with excretion are generally present in spiders, but little is known about what they do. When there are two pairs, they open on the first and third legs; if there is just one pair, the pores are on the first legs. The venom glands, linked with the chelicerae and the silk-producing glands associated with the spinnerets, are thought to be modified coxal glands.

Nervous System

The nervous system (Figure 16.20) is condensed. The nerve mass beneath the esophagus, joined to the cerebral ganglia above this part of the gut, consists of 4 ganglia belonging to the prosoma as well as up to 12 that belong to the opisthosoma. In other words, most of the ganglia, during the course of development, become dissociated from the embryonic somites in which they originate. The several simple eyes on the prosoma are innervated by nerves coming from the cerebral ganglia, but most other organs are supplied by nerves from the ganglia of the subesophageal mass.

Silks and Webs

The silks produced by spiders, like those made by insects and some other arthropods, are proteins. Most species secrete at least four or five kinds. The proteins are in soluble form when they come out of the several

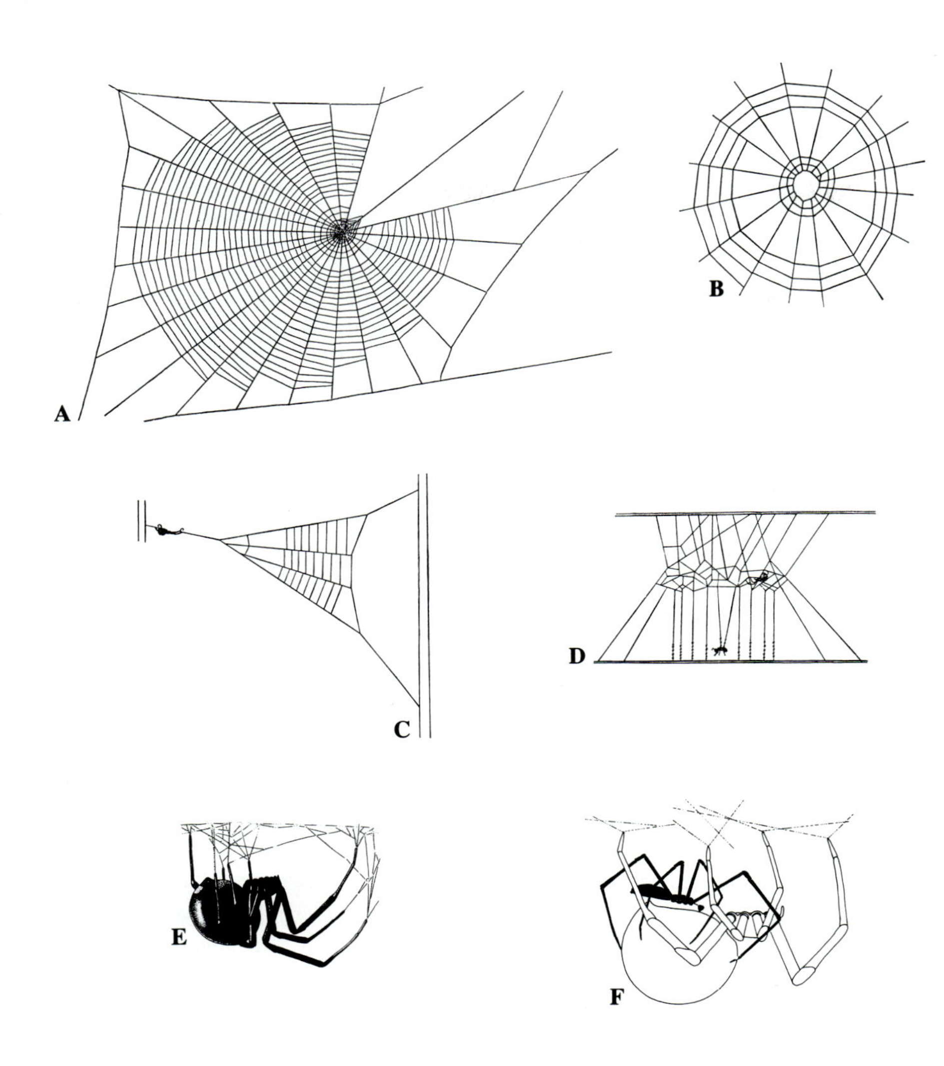

FIGURE 16.22 Webs produced by various spiders. A. Orb web of *Zygiella*. (Locket and Millidge, British Spiders.) B. Orb web (central portion) of *Tetragnatha montana*. There is an open "hub" space. C. The spring-trap of *Hyptiotes paradoxus*. D. Cobweb of a therediid spider. The basal portions of the vertical threads are sticky. An insect that touches them is lifted up and suspended. (B–D, Bristowe, A Comity of Spiders.) E. Female black widow spider, *Latrodectus mactans*, in the usual position in her web. F. Black widow spiders in mating position (male black). (Kaston, Transactions of the San Diego Society of Natural History, *16*.)

little nozzles of each spinneret, but they immediately become insoluble. The conversion from the soluble to the insoluble condition is believed to be due to a change in the orientation of the molecules. Coincidentally, the molecular weight of the protein changes. Thus, if the soluble protein has a molecular weight of 30,000, the solidified silk has a molecular weight of about 200,000 to 300,000.

Spider silks have a higher tensile strength than any other natural fiber that has been tested, and a higher tensile strength than any metal. They are therefore ideal for constructing webs and draglines, and also for some other purposes. A dragline enables a spider to maintain contact with a firm object while it wanders, or to suspend itself from a branch or from the ceiling. It may also serve as a trail marker, or lead a male to a female, as it does in the case of some wolf spiders. Nearly all spiders use draglines for one purpose or another. A capacity to withstand stretching would seem to be an important characteristic of most spider draglines: whereas a nylon thread, for example, can be stretched about 15% before it breaks, the dragline of *Araneus*, the garden spider, can be stretched twice as much.

Among the simpler devices that spiders use to detect approaching prey are threads similar to draglines. For example, some trapdoor spiders (Figure 16.21) extend threads away from the openings of the silk-lined burrows in which they live. When a prey animal touches one of the threads, the spider feels the change in tension and responds accordingly.

True webs show much variety. Some types are characteristic of particular families. Orb webs (Figure 16.22, A and B), which are suspended and used mostly to trap flying insects, are familiar to nearly everyone. They are typical of the family Araneidae, which includes many common garden spiders. The ability to make orb webs has evolved independently in another family, however. The foundation for an araneid's orb web is laid down in the form of a few lines, some or all of which may be reinforced by application of additional threads. After a system of radiating spokes has been produced, the spider spins a spiral thread from near the center outward. This thread, like the spokes, is dry in the sense that it is not sticky. It is a temporary addition, and is picked up by the spider while a sticky spiral thread is spun out to replace it. In spinning this sticky thread, the spider begins on the outside and works inward. It uses the middle claws of its legs for pulling the sticky thread into tight contact with the spokes, but otherwise avoids touching the thread.

A funnel web differs from an orb web in that it has a tubelike depression into which the spider can withdraw. The web consists of tightly woven threads, some of which may be raised slightly above the rest of the sheet. These are extra hurdles that trip insects as they attempt to cross the web. With or without the extra threads, however, the operation of a funnel web depends largely on the mechanical obstructions it provides, rather than on the stickiness that characterizes the spiral threads of an orb web.

Webs of the type called cobwebs are made by black widows and other members of the family Theridiidae. Black widows construct rather chaotic tangles of threads (Figure 16.22, E), but some cobwebs are neat, three-dimensional networks supported by sticky vertical lines (Figure 16.22, D). Cobweb spiders have a row of spines on the tarsi of their hindmost legs. The spines are used for combing out a band of silk with which to wrap and immobilize the prey.

The spring-trap web (Figure 16.22, C) is an interesting type. It is triangular, and its broad end is attached to a twig. The opposite end is narrowed down to a solitary "signal line," held by claws on the spider's first leg. The spider sticks itself to another twig with its spinnerets. When an insect hits the web, the spider instantly releases silk from the spinnerets, so that the tension on the signal line and the rest of the web is slackened. The insect is thus more likely to become trapped by the threads of the collapsing web than if the web were to remain taut.

In building its web, a spider follows instinct; in other words, a newly hatched spider knows exactly what to do to make the web characteristic of the species. A spider can, however, respond to emergencies that affect its web, and web-building and repairing are influenced by the animal's physiological state as well as by environmental circumstances. If the feat of constructing a web seems remarkable, one should be even more impressed by the fact that a complicated web must be replaced frequently, for the stickiness of the silk does not last long. Some spiders replace their webs daily. Many species eat their webs, digest the silk, and rapidly incorporate the amino acids into new silk.

Spiders that construct webs for trapping prey generally have poorly developed eyes. The prey is located on the basis of vibrations transmitted along the threads and perceived by the trichobothria, the specialized sensory hairs on the appendages. The various groups of hunting spiders, which stalk their prey or attack it from an ambush, have good eyesight. Their legs also tend to be stouter than those of web-spinning spiders. The ability of some hunting spiders to jump depends on rapid extension of the legs caused by an abrupt increase in blood pressure, rather than on the action of leg muscles.

Young spiders of many species use silk in a technique of dispersal called *ballooning,* although *kiting* would seem a more appropriate term. They spin out delicate threads to be lifted by a breeze and carried for perhaps hundreds of kilometers. Some full-grown spiders of small species use this method, too. There is no guarantee that the spider will reach a favorable situation before it starves. It may end up in the middle of an ocean or high in the sky.

All spiders employ silk in connection with reproduction, even if they do not use it for any other purpose. Females enclose their eggs in a silken sac. Males, which transfer sperm to females with the tarsi of their pedipalps, spin a special web or line on which to deposit a drop of spermatic fluid, then draw the fluid up into the tarsi. This is discussed further in the following section.

Reproduction and Mating Behavior

As a rule, females are larger than males of the same species, but mature males have conspicuously enlarged pedipalps. In both sexes, the reproductive system is in the opisthosoma, mostly beneath the midgut and digestive glands. The genital pore is behind the opening of the first pair of book lungs, or their tracheal replacements. In the female (Figure 16.23, A), there is a pair of ovaries, and the oviducts from these unite to form a single vagina. The two seminal receptacles that receive and store sperm have their own openings to the outside, but are linked internally with the vagina, so eggs are fertilized just before they are laid. The male (Figure 16.23, B) has a pair of testes and sperm ducts; the short common duct formed by the union of these has a dilation that serves as a seminal vesicle.

The male uses the tarsal portions of his pedipalps for transferring sperm. The tarsi are not markedly specialized except in mature individuals. Each pedipalp (Figure 16.24) has a sac for storing sperm, an elastic-walled blood sinus next to it, and a slender style (or *embolus*). When the sinus fills abruptly with blood, it applies pressure on the sperm-storing sac, forcing spermatic fluid out through the style. The style, being extended at the same time, is inserted into a seminal receptacle of the female.

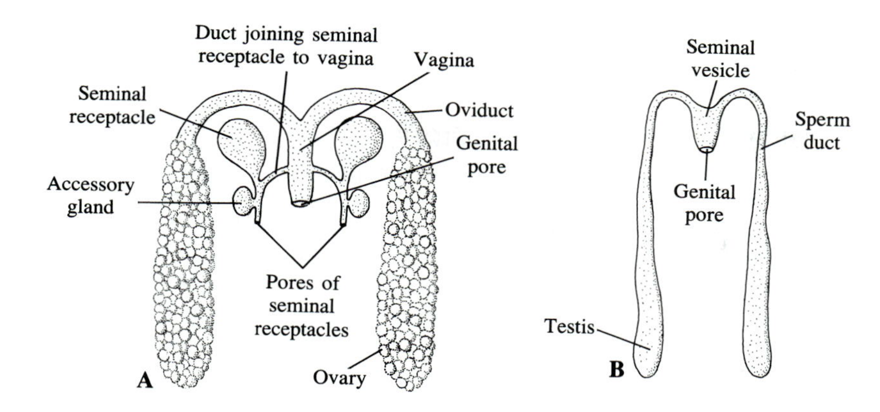

FIGURE 16.23 Reproductive systems of spiders. A. Female. B. Male.

The relatively simple arrangement of the tarsal components just described is characteristic only of certain primitive spiders. In most groups, the tarsus has many complicated elaborations, with special names for each of them. Collectively, they function to orient the style properly in the opening of the seminal receptacle and to deliver the sperm. The cuticular elements associated with the openings to the seminal receptacles of the female are organized in such a way that the male's pedipalpal styles fit the openings as a key fits a lock. In some spiders a style cannot be withdrawn after it has been inserted: it breaks off and is left behind.

Preparation for insemination begins when the male spins a special little web or thread, then deposits a drop of spermatic fluid on it (Figure 16.25, A). The habit of spinning out silk for this purpose persists in spiders that do not make webs for trapping prey. When the tarsus of a pedipalp is applied to the drop (Figure 16.25, A and B), the fluid is drawn up into the storage sac, perhaps by capillarity or as a result of lowered pressure in the sac due to resorption of some of its contents.

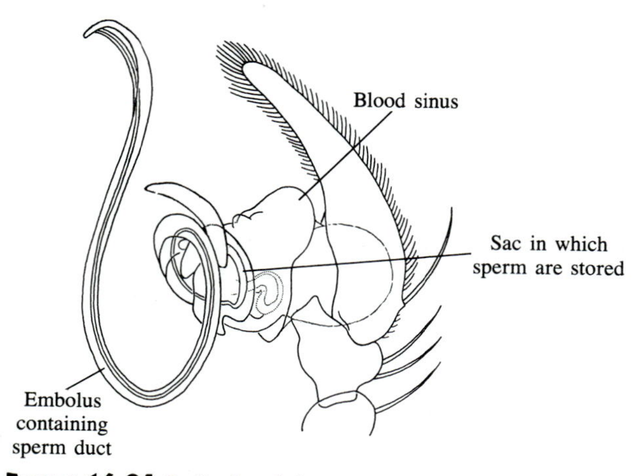

FIGURE 16.24 Pedipalp of the male of *Agelena naevia*, with the embolus extended. (Explanation in text.) (After Petrunkewitsch, Journal of Morphology, *40*.)

The encounter between a male and a female generally involves considerable courtship before the male gets a chance to insert the tips of his pedipalps into the openings of her seminal receptacles. Courtship and mating of spiders has engaged the attention of many amateur and professional arachnologists. There is enough information on the subject to fill a large volume, but only a few generalities can be discussed here. Spiders, being predators, must be able to differentiate between prospective mates and possible prey, and they use a variety of chemical, tactile, auditory, and visual stimuli to do this. The chemical cues include pheromones, which serve not only as attractants but as inducers of courtship behavior. Experiments have demonstrated that males of some species will go through a courtship routine if they are presented with exudates that have been washed off a female and then concentrated by evaporation. A single leg from a female may initiate courtship behavior, and in many wolf spiders a female's silk provides sufficient stimulation.

Vibrations transmitted by way of web lines enable some spiders to recognize one another. A male, for instance, may elicit the right response from a female by plucking on a line of her web with a certain frequency and intensity. In jumping spiders and a few others that have good eyesight, "dancing," special postures, and movements of appendages may be used by males to bring out appropriate responses in females. Color as well as form is important in the visual stimuli transmitted by many species. The striking color patterns of certain jumping spiders are presumably involved in bringing males and females together.

The males of wolf spiders (Lycosidae), which are fast-moving, mostly nocturnal types that live in holes in the ground, use acoustic stimuli to announce their presence to prospective partners. At the joint between the

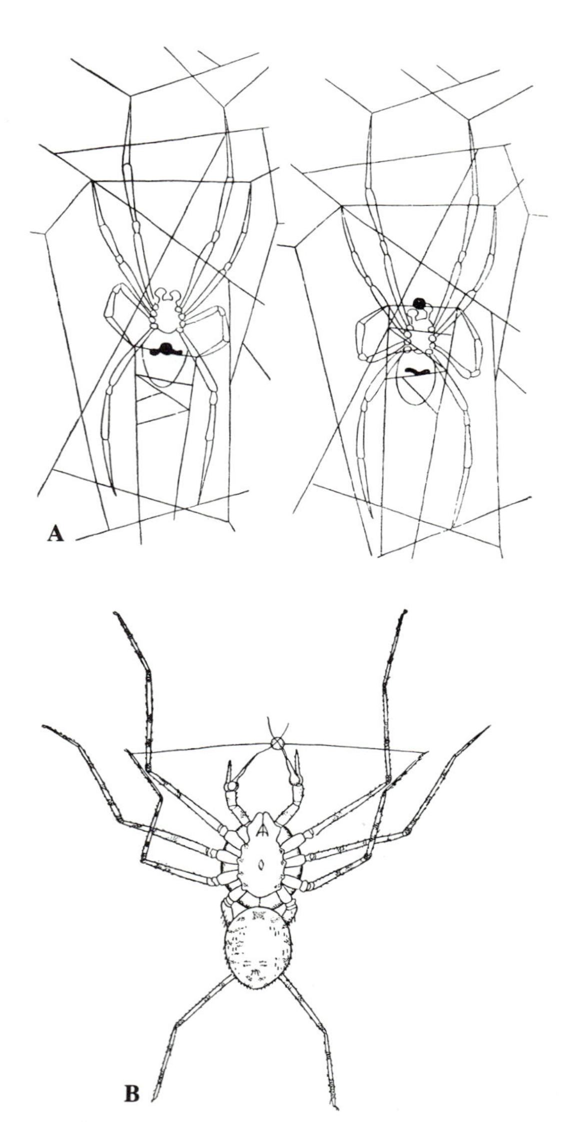

FIGURE 16.25 A. Male of *Latrodectus tridecemguttatus* depositing a drop of sperm on a silken thread, then pulling the thread so as to bring the drop to his pedipalps. B. Male of *Scytodes thoracica* loading his pedipalps with sperm he has deposited on a silken thread. (Ivanov, Spiders.)

last and next-to-last articles of each pedipalp are a little "file" and "scraper." By flexing his pedipalps at this joint, the spider makes sounds that are transmitted through the ground, and apparently also through the air.

There are various mating postures that enable the male to insert the tips of his pedipalps into the female's seminal receptacle pores (Figure 16.22, F; Figure 16.26). If the male is on top of the female—a position used by most so-called running spiders—the male must reach around the female on one side, then the other side, in order to fill both receptacles. In most of the other positions, the male is beneath the female, but not necessarily with his ventral surface facing hers.

Some spiders mate only once. This is to be expected in spiders in which the male's tarsal styles break off during the process of mating. In females of certain species, plugs that form after mating permanently block the openings to the seminal receptacles. There are, however, many spiders that mate several times; the females may produce a set of eggs after each mating.

It is popularly believed that female black widows suck the juices out of their mates after mating. This does happen, but less commonly in black widows than in spiders of some other groups. The practice is biologically advantageous to the species, even if it is hard on the individual males. The female needs nourishment in order to produce yolk-rich eggs; by consuming the juices of the male, she makes use of nutritive material that would otherwise eventually be disposed of by organisms of decay. Newly hatched young of some spiders eat up their mother after she dies, and this is another way of keeping energy "in the family."

The eggs, fertilized as they pass down the oviducts to the vagina and genital pore, are generally laid on a small web. By applying more silk, the female makes an egg sac characteristic of the species. The egg sac may be formed in contact with rock, wood, or bark, or it may be suspended from a web or hidden away in a den. Females of many species guard their egg sacs, and some carry a sac with them, holding it with their chelicerae or keeping it attached to their spinnerets.

Development

A single female may produce a number of egg sacs, with anywhere from a few eggs to several hundred in each one, depending on the species and the nutritional state of the spider. When the little spiders hatch, they are similar to adults. The mother may have to liberate them from the egg case, but usually they manage to get out by themselves. By the time they emerge, they may already have undergone at least one molt. The young of some species cling to the mother for a while.

The number of molts varies according to the species and the sex. The amount and kind of food available may affect the molting pattern. For instance, black widows molt more frequently if they are well fed than if they are poorly fed. But some other spiders have a fixed number of molts; they simply grow faster when they have plenty to eat. In many species, males molt less often than females. There are, in fact, a few spiders in which the male is fully mature, even if tiny, when he hatches; the female must grow and molt a number of times. In general, most spiders molt about 5 to 10 times. Molting after reaching sexual maturity is rare, but it happens in females of certain species.

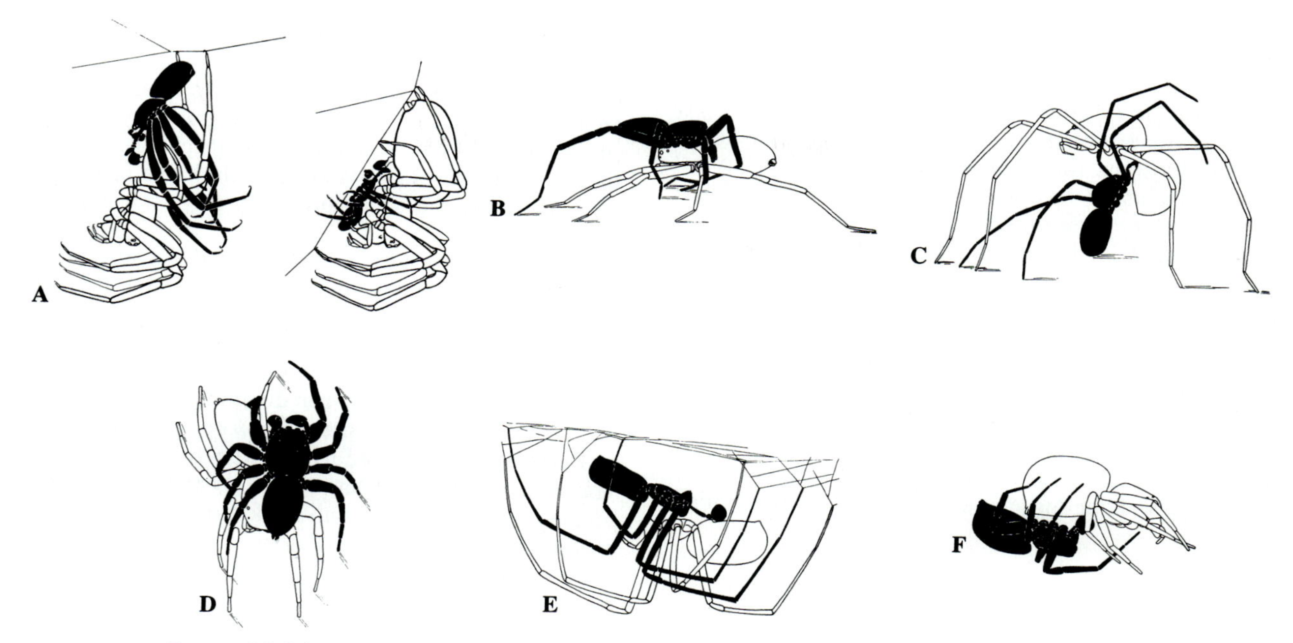

FIGURE 16.26 Mating postures of various spiders (males black). A. *Aranea diadematus* (left) and *A. pallidus* (right). (Grasshoff, Natur und Museum, *94*.) B. *Lycosa rabida*, a wolf spider. C. *Scytodes thoracica*. D. *Phidippus clarus*, a jumping spider. E. *Linyphia marginata*. F. *Distichus triguttatus*. (B–F, Kaston, Connecticut State Geological and Natural History Survey, *70*.)

References on Araneida

Abalos, J. W., and Baez, E. C., 1963. Spermatic transmission in spiders. Psyche, *70*:197–207.

Bristowe, W. S., 1939–1941. The Comity of Spiders. 2 volumes. Ray Society, London.

Bristowe, W. S., 1958. The World of Spiders. Collins, London.

Buchli, H. H. R., 1969. Hunting behavior in the Ctenizidae. American Zoologist, *9*:175–193.

Comstock, J. H., 1948. The Spider Book. 2nd edition, revised and edited by W. J. Gertsch. Cornell University Press, Ithaca.

Foelix, R. F., 1982. Biology of Spiders. Harvard University Press, Cambridge.

Gertsch, W. J., 1979. American Spiders. 2nd edition. Van Nostrand Reinhold Company, New York.

Kaston, B. J., 1964. The evolution of spider webs. American Zoologist, *4*:191–207.

Kaston, B. J., 1970. Comparative biology of the American black widow spiders. Transactions of the San Diego Society of Natural History, *16* (3):33–82.

Kaston, B. J., 1978. How to Know the Spiders. Wm. C. Brown, Dubuque, Iowa.

Kaston, B. J., 1981. Spiders of Connecticut. 2nd edition. Connecticut State Geological and Natural History Survey Bulletin no. 70 (revised).

Levi, H. W., and Levi, L. R., 1968. A Guide to Spiders and Their Kin. Golden Press, New York.

Locket, C. H., and Millidge, H. F., 1951–1953. British Spiders. 2 volumes. Ray Society, London.

McCreve, J. D., 1969. Spider venoms: biochemical aspects. American Zoologist, *9*:153–156.

Nentwig, N. (editor), 1987. Ecophysiology of Spiders. Springer-Verlag, New York.

Parry, D. A., 1965. The signal generated by an insect in a spider's web. Journal of Experimental Biology, *43*:185–192.

Preston-Mafham, R., and Preston-Mafham, K., 1984. Spiders of the World. Facts on File Publications, New York.

Schulta, J. W., 1987. The origin of the spinning apparatus in spiders. Biological Reviews, *62*:89–113.

Shear, W. A., 1986. Spiders: Webs, Behavior, and Evolution. Stanford University Press, Stanford, California.

Turnbull, A. L., 1966. Population of spiders and their potential prey. Canadian Journal of Zoology, *44*:557–583.

Turnbull, A. L., 1973. Ecology of true spiders. Annual Review of Entomology, *18*:306–848.

Walcott, C., and Kloot, W. G. van der, 1959. Physiology of the spider vibration receptor. Journal of Experimental Zoology, *141*:191–244.

Witt, P. N. (editor), 1969. Web-building in spiders. (Proceedings of a symposium, not strictly limited to web-building.) American Zoologist, *9*:70–238.

Witt, P. N., Reed, C. F., and Peakall, D. B., 1968. A Spider's Web. Springer-Verlag, New York.

Witt, P. N., and Rovner, J. S. (editors), 1982. Spider Communication: Mechanisms and Ecological Significance. Princeton University Press, Princeton, New Jersey.

Order Phalangiida (Opiliones)

The large order Phalangiida includes the spiderlike animals called daddy longlegs and harvestmen (Figure 16.27). The prosoma and the more anterior segments of the opisthosoma form one unified piece; only the last few segments of the opisthosoma are distinct. There are, however, 10 segments in the opisthosoma; another one, the first in the series, drops out during development. There is usually a pair of simple eyes located on a little bump on the anterodorsal part of the prosoma. The chelicerae have three articles, counting the movable finger of the pincer. In daddy longlegs, which are the phalangiids most likely to be seen in North America, the pedipalps are slender and resemble the walking legs. Stout, spiny pedipalps are characteristic of cave-inhabiting species and of many that live in the tropics. When the walking legs are extremely long, the terminal articles are subdivided, giving the legs much flexibility for clinging to grass stems and similar objects. A leg will be autotomized if it is grabbed by a bird or other predator, and the missing limb will gradually regenerate; autotomy is thus a protective mechanism. Many species have prosomal glands that discharge bad-smelling secretions; the secretions discourage some predators.

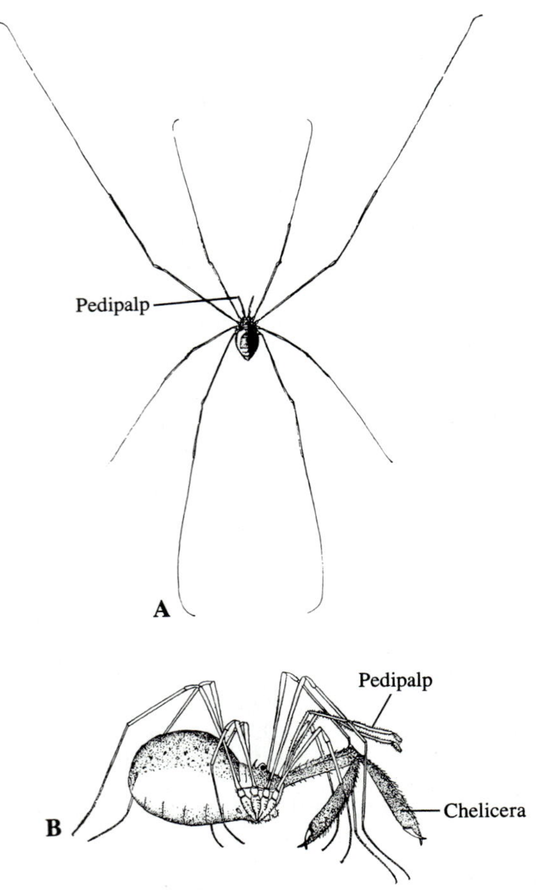

FIGURE 16.27 Order Phalangiida. A. *Leiobunum vittatum*. (Weed, Transactions of the American Entomological Society, *19*.) B. *Taracus pallipes*. (Banks, Journal of Entomology and Zoology, Pomona College, *3*.)

Most phalangiids are carnivores and scavengers that consume insects, mites, and other terrestrial arthropods; some eat land snails and slugs. Species of the genus *Phalangium* have been observed feeding on fallen fruit. The chelicerae are used for grasping and tearing up food. The midgut usually has diverticula, and it is said that these are primarily sites of absorption.

Respiration is effected by a system of tracheal tubes. The openings to these are on the second opisthosomal segment, but homology with book-lung tracheae has not been established. Some phalangiids have secondary tracheal openings on each walking leg. The heart has only two pairs of ostia, and the only distinct blood vessels are the anterior and posterior aortas. There are no Malpighian tubules, but there are coxal glands with openings just behind the third legs. There are also cells that accumulate nitrogenous wastes.

The paired gonads are united to form a U-shaped structure, and are connected to the genital pore by oviducts or sperm ducts that unite before reaching the genital pore. In both sexes the genital pore is located at the tip of a prominent ventral tube. The mating position is head-to-head, so the genital pores are rather remote. The male's tube can, however, be bent and extended forward far enough to place a spermatophore in the genital pore of the female. Note that this is true copulation, not often found in other arachnids. Females of many species use their genital tubes as ovipositors for placing fertilized eggs in holes in the ground or in empty snail shells. Some phalangiids stick their eggs to stones or other firm objects.

References on Phalangiida

Edgar, A. L., 1971. Studies in the biology and ecology of Michigan Phalangida. Miscellaneous Papers of the Museum of Zoology, University of Michigan, *144*:1–64.

Sankey, J. P. H., and Savory, T. H., 1974. British Harvestmen. Synopsis of the British Fauna, no. 4. Linnean Society of London and Academic Press, London and New York.

Order Thelyphonida (Uropygi)

The whipscorpions (Figure 16.28), of which there are fewer than 100 species, are characterized by vicious-looking pedipalps, slender, antennalike first legs, and a "tail." The tail, though divided into a number of

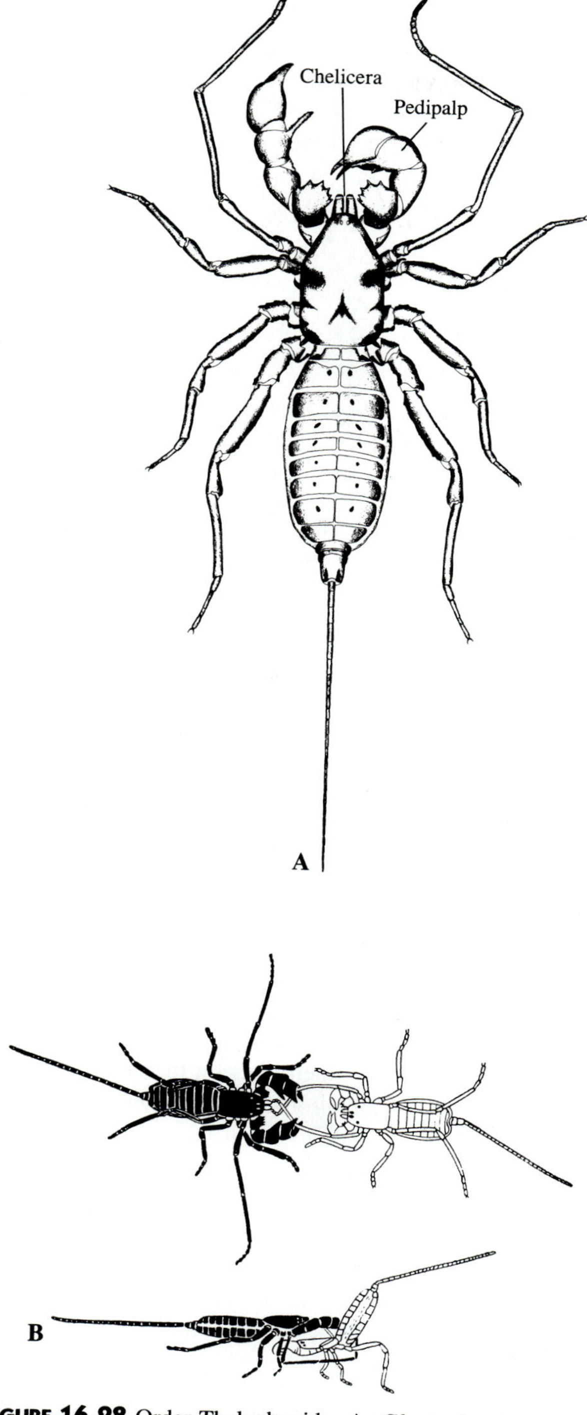

FIGURE 16.28 Order Thelyphonida. A. *Glyptogluteus augustus*. (Rowland and Cooke, Journal of Arachnology, *1*.) B. *Thelyphonus*, in courtship. (Lawrence, The Biology of the Cryptic Fauna of Forests.)

units, is a modified telson. The chelicerae are small. There are two pairs of book lungs on the opisthosoma.

These animals are restricted to relatively warm areas, and they generally remain hidden during daylight hours. Some dig their own burrows; others live under rocks, under bark, and in similar habitats. They come out at night to feed on insects and other small animals, using their pedipalps to crush and tear the prey. The chelicerae serve to hold and manipulate pieces of tissue that will be partly digested, before being swallowed, in a preoral cavity similar to that of true scorpions. The coxae of the pedipalps, being fused, contribute to the posterior wall of the preoral cavity.

Glands at the anus secrete substances that a whipscorpion can spray out to protect itself, or to start softening the hard cuticle of a beetle or some other prey insect it has caught. The name *vinegaroon* is given to certain species because the secretion they release consists mostly of acetic acid. It also contains caprylic acid, however. This is a "spreader" that reduces surface tension and that probably facilitates penetration of the acetic acid into the cuticle of the prey or the skin of an attacker. Vinegaroons can defend themselves by pinching with their pedipalps. The largest known species, *Mastigoproctus giganteus,* is fairly common in parts of the southern United States and Mexico; it reaches a length of about 7 cm. It is harmless except for its pinch and its "vinegar," which irritates the skin and mucous membranes.

Courtship and mating have been studied in a few species. There may be much embracing (Figure 16.28, B) and other activity before the male can deposit a spermatophore in the female's genital pore. Females remain hidden while carrying their eggs in a sac on the abdomen. Even after the young hatch, they may cling to their mother for a time. Whipscorpions are sometimes placed, along with the schizomids (below), into a single order, Uropygi.

References on Thelyphonida

Eisner, T., 1962. Survival by acid defense. Natural History, *71*:10–19.

Eisner, T., Meinwald, J., Monro, A., and Ghent, R., 1961. Defence mechanisms of arthropods. I. The whipscorpion, *Mastigoproctus giganteus* (Lucas) (Arachnida, Pedipalpida). Journal of Insect Physiology, *6*:272–298.

Order Schizomida

If they had prominent tails, schizomids (Figure 16.29, A) would look much like whipscorpions. The telson is short, however, and consists of no more than three units. The pedipalps are not so stout as those of whipscorpions, but they are nevertheless effective implements for capture of prey, and their coxae are fused to help form the preoral cavity. There are anal glands

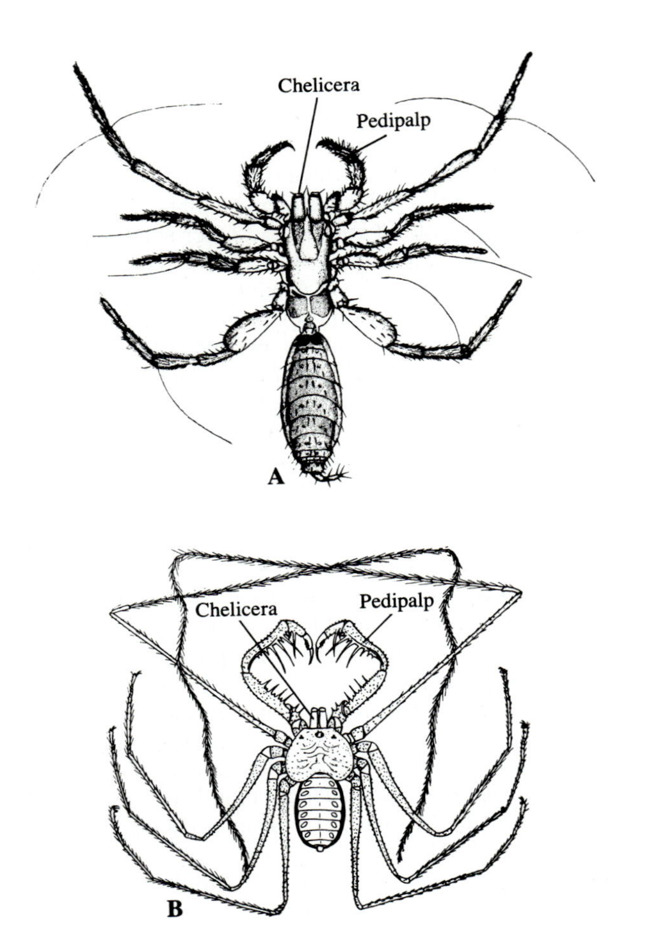

Figure 16.29 Orders Schizomida and Phrynichida. A. *Trithyreus pentapeltis*, a schizomid. (Moles, Journal of Entomology and Zoology, Pomona College, *9*.) B. *Stygophrynus dammermani*, a phrynichid. (Roewer, in Bronn, Klassen und Ordnungen des Tierreichs, *5*:4:4.)

for secreting a repellent spray, and a single pair of book lungs.

These animals are small—rarely more than 5 mm long—and are mostly restricted to warmer regions. They are generally found under large rocks. They probably feed on insects, centipedes, and perhaps other terrestrial arthropods, but little is known about this aspect of their biology. As in scorpions, whipscorpions, and many other arachnids, transfer of a spermatophore to the genital pore of the female is preceded by much courtship, at least in the few species whose reproductive biology is known. The female remains hidden while carrying her eggs in a sac on the abdomen.

Order Phrynichida (Amblypygi)

The "scorpion spiders," or "tailless whipscorpions" (Figure 16.29, B), have no recognizable telson, but they resemble whipscorpions and schizomids in having well-developed pedipalps and slender first legs. The first legs are much longer than the others, and are sometimes so slender that they are almost hairlike; they are used only as "feelers." These animals are mostly tropical, and a few species reach a length of over 3 cm. They grab and kill insects and other arachnids with their pedipalps, after having recognized the prey with their delicate first legs. Lacking anal glands, their only means of defense seems to be pinching with the pedipalps. They can run fast, however, and usually manage to disappear into a crevice.

An odd feature of the mating of at least some species is that the male deposits, on the ground, a "spermatophore" that contains no sperm. The sperm are added later, and the female then places her genital pore on the mixture. Females carry their eggs on the abdomen until they hatch, and the young may cling to their mother for a while after they emerge.

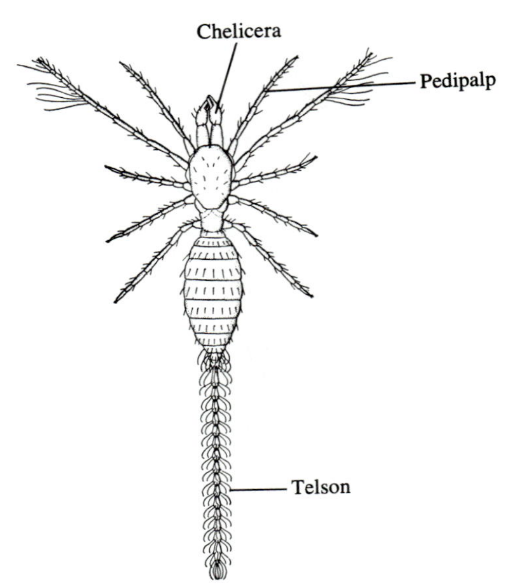

Figure 16.30 Order Palpigradida. *Koenenia mirabilis.* (Wheeler, American Naturalist, *34*.)

Order Palpigradida (Palpigradi)

Palpigradids (Figure 16.30), of which there are about 50 species, resemble whipscorpions in having a slender, multiarticulated telson. They are generally less than 2 mm long, so the name *microwhipscorpions* fits them rather well. The chelicerae are proportionately larger than in whipscorpions, and the pedipalps, instead of being specialized for seizing and mangling prey, are much like the walking legs. In whipscorpions and schizomids, the coxal articles of the pedipalps are fused and help to delimit the preoral cavity; in microwhipscorpions, the coxae are decidely separate and have nothing

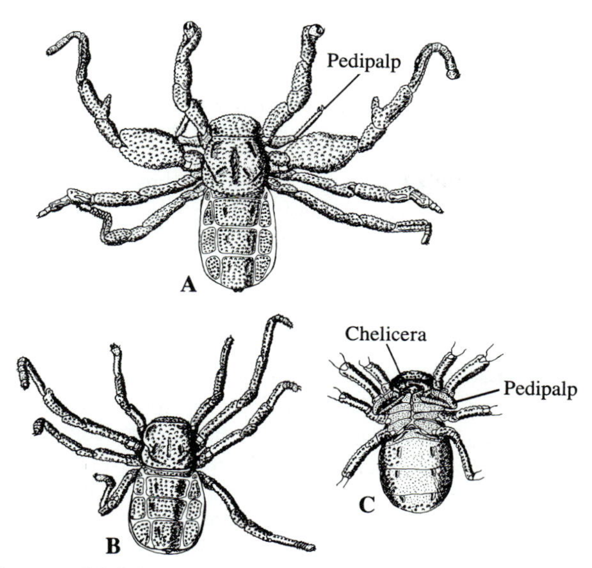

FIGURE 16.31 Order Ricinuleida. *Ricinoides afzelii.* A. Male, dorsal view. B. Female, dorsal view. C. Female, ventral view. (Tuxen, Journal of Arachnology, *1*.)

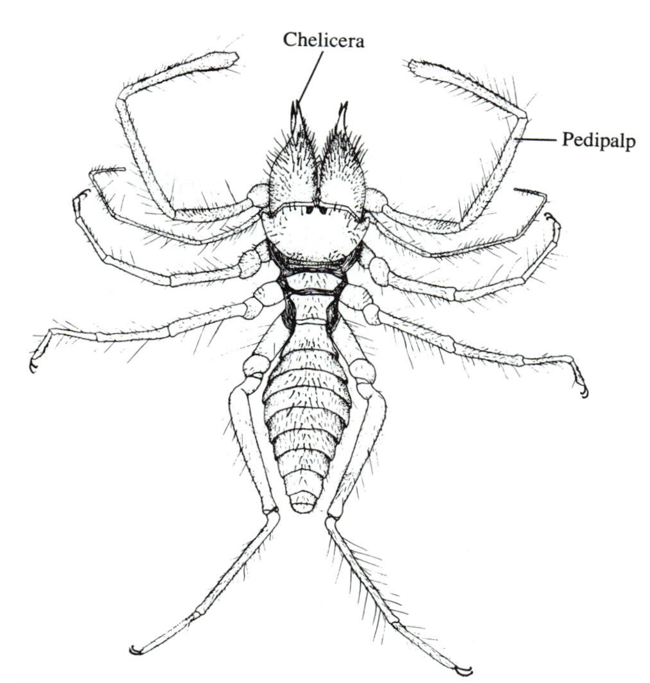

FIGURE 16.32 Order Solpugida. *Eremobates formicaria*, male. (Nesbet, Journal of Entomology and Zoology, Pomona College, *9*.)

to do with the preoral cavity. There are neither book lungs nor tracheae, but some ventral sacs on the opisthosoma may function in respiration.

Microwhipscorpions are found mostly in warmer areas, as in Mediterranean regions, California, and tropical Africa. A few, however, have been reported from caves in central Europe. The majority of the known species have been collected under stones and in similar habitats, where the risk of desiccation is minimal. They shun the light and are not likely to be encountered in the open. Relatively little is known about their biology.

Order Ricinuleida (Ricinulei)

Ricinuleids (Figure 16.31) are quite uncommon: there are only about 25 species, most of which are in tropical America and Africa. One American species occurs as far north as Texas. These animals live under logs, in leaf litter, and in caves, and range in size from about 4 mm to 1 cm. They feed on insects and spiders. They are unique among arachnids in having a hinged flap, the *cucullus,* hanging down from the front edge of the carapace. The chelicerae and the mouth region are hidden beneath this flap. The pedipalps, like the chelicerae, are pincerlike. As in whipscorpions and some other orders, the coxal articles of the pedipalps are united and help delimit the preoral cavity. The body as a whole is firm, much like that of some ticks, because of the relatively hard cuticle. The tracheal system is derived from book lungs.

In males, the tips of the third legs are specialized for transferring sperm. The second legs are used for stroking and tapping the female. After establishing rapport with her, the male climbs on her back and soon produces a mass of sperm that becomes firm enough to be picked up and put into her gonopore. Females of one species have been found carrying a single egg under the flap at the anterior end of the body. The young hatch in a six-legged stage. They reach adulthood by way of a series of distinct stages, following the pattern typical for arachnids. A particular stage may last for years, however, so ricinuleids are slow in attaining maturity.

References on Ricinuleida

Cooke, J. A. L., 1967. The biology of Ricinulei. Journal of Zoology (London), *151*:31–42.

Pollock, J., 1966. Life of the ricinulid. Animals, *8*(15):402–405.

Order Solpugida (Solifugae, Solifugida)

Solpugids (Figure 16.32) are called wind scorpions because of their speed in running. They look a little like spiders, but are distinctly different in several respects. The chelicerae, consisting of only two articles, are stout and generally directed straight ahead. The movable claw is ventral to the basal article. The latter, because of the way it articulates with the prosoma, can pivot.

The pedipalps are long, and the terminal article has an eversible disk that a solpugid can use for climbing on smooth surfaces.

The walking legs are slender. The coxal articles of the first pair are partly united to the coxae of the pedipalps. On the proximal portion of each fourth leg are five little structures called *racquet organs*. The stalks of these become broadened distally into flat, more or less triangular blades. The racquet organs are apparently unique to solpugids, and they are probably sensory. There is a pair of eyes on the anteromedian part of the prosoma, and sometimes there are vestiges of lateral eyes.

The tracheal tubes may have evolved independently of book lungs. There is typically a pair of openings on both the third and fourth segments of the opisthosoma, and a single midventral opening on the fifth segment, as well as openings on the second walking leg of the prosoma. There are Malpighian tubules on the gut and also a pair of coxal glands whose openings are on the coxae of the pedipalps.

Solpugids are primarily animals of desert regions. They dig burrows, and most species do not emerge from their hiding places except at night or just before dark. They may look dangerous, and some can nip a person's skin with their chelicerae, but they have no venom. Nevertheless, they are aggressive predators, feeding largely on insects and earthworms. The chelicerae are used for tearing and decapitating the prey, and the pharynx functions as a pump for sucking in the juices. It has been reported that when some solpugids feed, they discharge a secretion from the coxal glands. Perhaps this functions as a kind of saliva. A few species reach lengths of 6 or 7 cm, and these are powerful hunters.

Both sexes have paired gonads but only one genital pore. The male uses his chelicerae to pick up a drop of spermatic fluid and transfer it to the female's genital pore. Some of the mating behaviors of solpugids are extremely interesting. In a European species of *Galeodes,* for instance, the male pounces on the female, who is larger than he is. The male uses his pedipalps and walking legs to embrace the female's prosoma. This quiets her, and the male may carry her to another place before putting her on her back, widening her genital pore with his chelicerae, then carrying out the transfer of sperm. After pinching together the edges of the female's genital pore, he runs away. By the time the female of this species lays her eggs, in a deep pit that she has dug, the embryos within them have reached a rather advanced stage.

References on Solpugida

Junqua, C. M., 1966. Recherches biologiques et histophysiologiques sur un solifuge saharien. Mémoires du Muséum National d'Histoire Naturelle, A, *43:*1–124.

Muma, M. H., 1951. The arachnid order Solpugida in the United States. Bulletin of the American Museum of Natural History, *97:*35–141.

Muma, M. H., 1962. The arachnid order Solpugida in the United States. Supplement 1. American Museum Novitates, *2092:*1–44.

Muma, M. H., 1967. Behavior of North American Solpugida. Florida Entomologist, *50:*115–123.

Muma, M. H., 1970. A synoptic review of North American, Central American, and West Indian Solpugida. Arthropods of Florida, *5*. Florida Department of Agriculture and Consumer Services, Bureau of Entomology, Contribution 154.

Order Acari (Acarina)

In mites and ticks, which make up the order Acari, the prosoma and opisthosoma are not clearly separated, and individual segments are obscure. The usual prosomal appendages, however, are present: chelicerae, pedipalps, and four pairs of legs (except in some mites that have only one or two pairs). Ticks constitute a small minority within the Acari, but they are closely related to mites, so it is a mistake to think of these two groups as separate. In general, ticks are larger than mites—females of a few species reach a length of more than 2 cm after they have fed—and they are bloodsuckers whose meals are provided by higher vertebrates. Most mites are between 0.5 and 2 mm long, and relatively few of them suck blood. They are considerably more diversified than ticks.

Mites

Mites (Figure 16.33) are found just about everywhere in terrestrial habitats, and they are also represented in fresh water and in the sea. It is difficult to estimate how many species exist, but at least 20,000 have been described, and it is certain that many thousands more await discovery and description. Some acarologists are willing to bet that there are as many kinds of mites as there are insects. Many insects, in fact, have specific mites associated with them.

In spite of their small size, mites have many structural details that must be considered by specialists concerned with defining species, genera, and higher taxa, as well as with understanding the biology of these animals. At the anterior end of the body (Figure 16.34) there is characteristically a beaklike forward projection called the **capitulum** (or **gnathosoma**). The first three prosomal segments are incorporated into this. Its upper portion, the **tectum,** is like an awning that overhangs the rest of it. The lower portion consists of the **labrum,** the **epistome** (to which the labrum is attached), and the **hypostome,** which is a trough formed by the fusion of the medial edges of the coxal articles of the pedipalps. The chelicerae are inserted beneath the tectum, but they may be mounted in a collarlike structure that permits them to be extended and withdrawn. The main point is

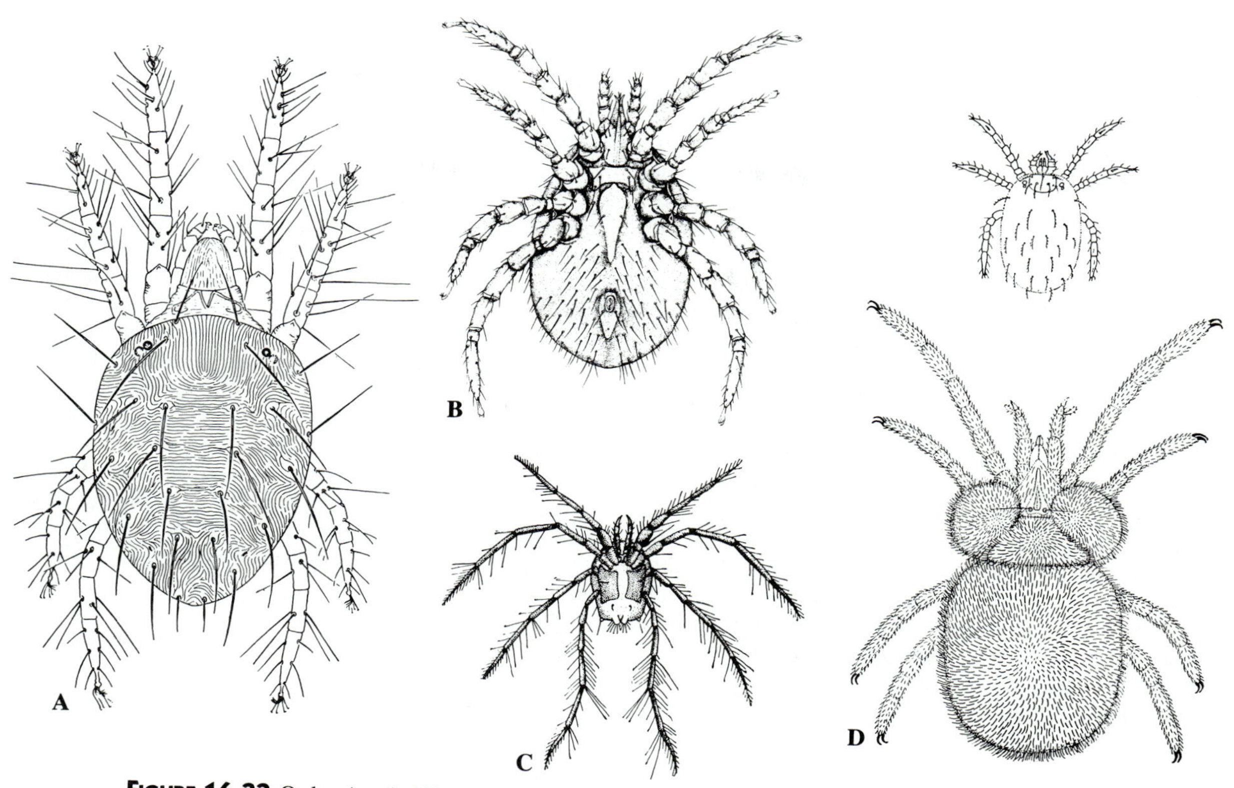

FIGURE 16.33 Order Acari. Mites. A. *Tetranychus schoeni*, female, dorsal view. This species is a pest of certain fruit trees. (Pritchard and Baker, Hilgardia, *21*. B. *Liponyssus bacoti*, a tropical rat mite, female, ventral view. (Dove and Shelmire, Journal of Parasitology, *18*.) C. *Unionicola,* found in fresh water, often in the mantle cavity of clams. (Zhadin and Gerd, Rivers, Lakes, and Reservoirs of the USSR, Their Fauna and Flora.) D. *Eutrombicula alfreddugesii*, a recently hatched individual (above) and an adult (below). (Lawrence, The Biology of the Cryptic Fauna of Forests.)

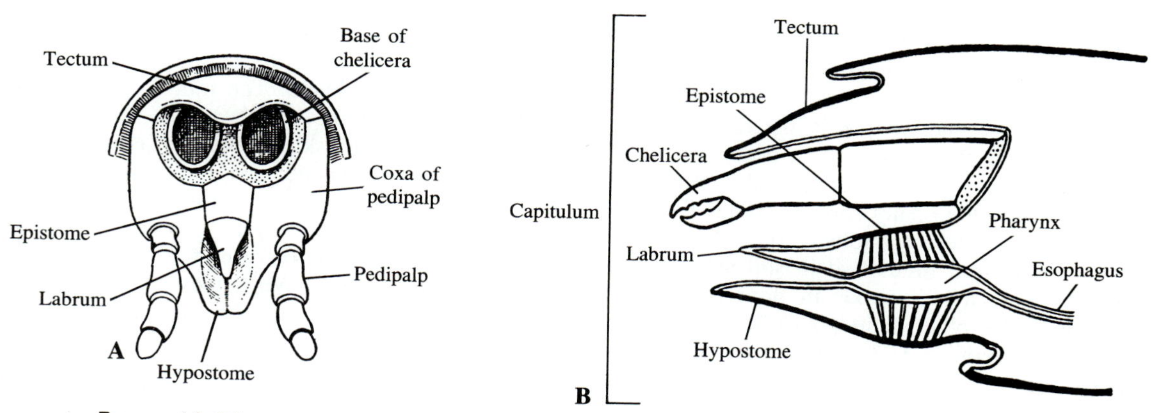

FIGURE 16.34 Order Acari. Mites. A. Front of the prosoma of a mite; diagrammatic. The chelicerae have been removed. B. Sagittal section of the anterior portion of the prosoma of a mite; diagrammatic. The chelicera shown is to the right of the midline. (Snodgrass, Smithsonian Miscellaneous Collections, *110*.)

that the trough and the lip above it operate effectively in piercing and in drawing up liquid food.

The pharynx, which makes up most of the foregut, is often X-shaped in cross-section when it is collapsed. It usually functions as a pump, and dilation of the lumen is effected by muscles that pull on the wall of the pharynx. The midgut generally has several pairs of caeca. Some mites have a hindgut and anus; others do not.

Adult mites, unless they are aquatic, generally have tracheae. The openings of these are located at the bases of the chelicerae, on the coxae of the pedipalps, on the coxae of the legs, and in various other places on the body. Most juvenile mites, and adults of some especially small species, breathe through their soft cuticle. There is a heart with only one or two pairs of ostia. Organs of excretion may include one or two pairs of Malpighian tubules that open into the hindgut, or coxal glands that open to the outside near the bases of the first legs. There are other arrangements, however. In mites that do not have a hindgut, there is sometimes an excretory sac dorsal to the midgut. A duct leads from the sac to a pore that is perhaps homologous to the anus of mites that do have a hindgut. In certain species, cells of the midgut concentrate nitrogenous wastes in a nearly insoluble crystalline form.

Techniques of mating and sperm transfer, like other aspects of mite biology, are varied. Usually, sperm are packaged in a spermatophore. This may be placed by the chelicerae on the genital pore of the female, or it may be placed on the substratum, in which case the female must be led to it and properly oriented over it. Some mites have a penislike elaboration associated with the genital pore, and this is used for true copulation. Many mites, whether they copulate, transfer spermatophores, or deposit spermatophores on the substratum, use specializations of legs or other parts of the body for embracing their partners. A few species are known to reproduce parthenogenetically as well as sexually; unfertilized eggs develop into males, fertilized eggs into females.

Eggs are typically laid in small clusters, and parasitic mites often stick their eggs to their hosts. Although most adult mites have four pairs of legs, the young, on hatching, have only three pairs (Figure 16.33, D). Maturity is attained after several molts.

Free-living terrestrial mites live in soil and humus, under debris, in decaying wood, in flour and stored grains, and in many other habitats. Collectively, they feed on almost everything: plant material and fungi, insects, insect eggs, nematodes, other mites, and much more. Some freshwater species are predators that suck out the juices of small crustaceans; others, at least when they are young, are intimately associated with certain hosts, such as clams. Truly marine mites, most of which belong to one family, are primarily predators that feed on crustaceans and worms. Most mites found in the spray zone or at upper levels of sandy beaches are members of groups that are mostly terrestrial; some are parasites of amphipod crustaceans or other invertebrates.

Many species are found on or in the skin of vertebrates, especially reptiles, birds, and mammals, and certain kinds inhabit the lungs or air sacs. Humans and other domestic animals are troubled by numerous species. Some, such as chiggers, which are immature mites, cling to the skin and feed mostly on dead cells, but their salivary secretions are irritating. Mites that cause mange and scabies tunnel into the skin to feed on cells and lymph. There are also some bizarre, long-bodied types that penetrate into hair follicles and sebaceous glands. In humans, these do not always engender symptoms, but they cause a severe form of mange in dogs and some other domestic mammals. Various mites, including chiggers, transmit scrub typhus and other diseases.

Plants are not exempt from the predations of mites, which work their way into the tissues or selectively destroy individual cells. Many species do great damage to crops. The damage may be direct, or it may be caused by viruses or other pathogens transmitted by the mites.

Ticks

Ticks (Figure 16.35) suck blood from reptiles, birds, and mammals, and they have interesting specializations for penetrating the skin, maintaining tight contact with their benefactors, and drawing up blood. The structure of the capitulum (Figure 16.35, B and C) is similar to that of many mites, but has some distinctive features. The hypostome is shaped like a trough, and its ventral surface and edges have rows of backward-pointing spines. Above the hypostome are the chelicerae, each enclosed for most of its length by a tubular sheath. Within the snoutlike complex formed by the hypostome and cheliceral sheaths is the labrum, drawn out anteriorly into a slender blade. The mouth lies between the labrum and hypostome.

Penetration of the skin of the host is initiated by the chelicerae. These appendages, equipped with an elaborate claw for making an incision in the skin (Figure 16.35, D), are extended by an increase in blood pressure caused by contraction of certain muscles in the anterior part of the body. While they are being extended, they are supported laterally by the pedipalps, each of which has a groove on its medial side. As soon as the skin has been cut, the feeding apparatus—the hypostome, cheliceral sheaths, and structures enclosed by these—can be pushed into the wound. The spines on the hypostome anchor the feeding apparatus while the tick takes its time sucking up blood by pumping action

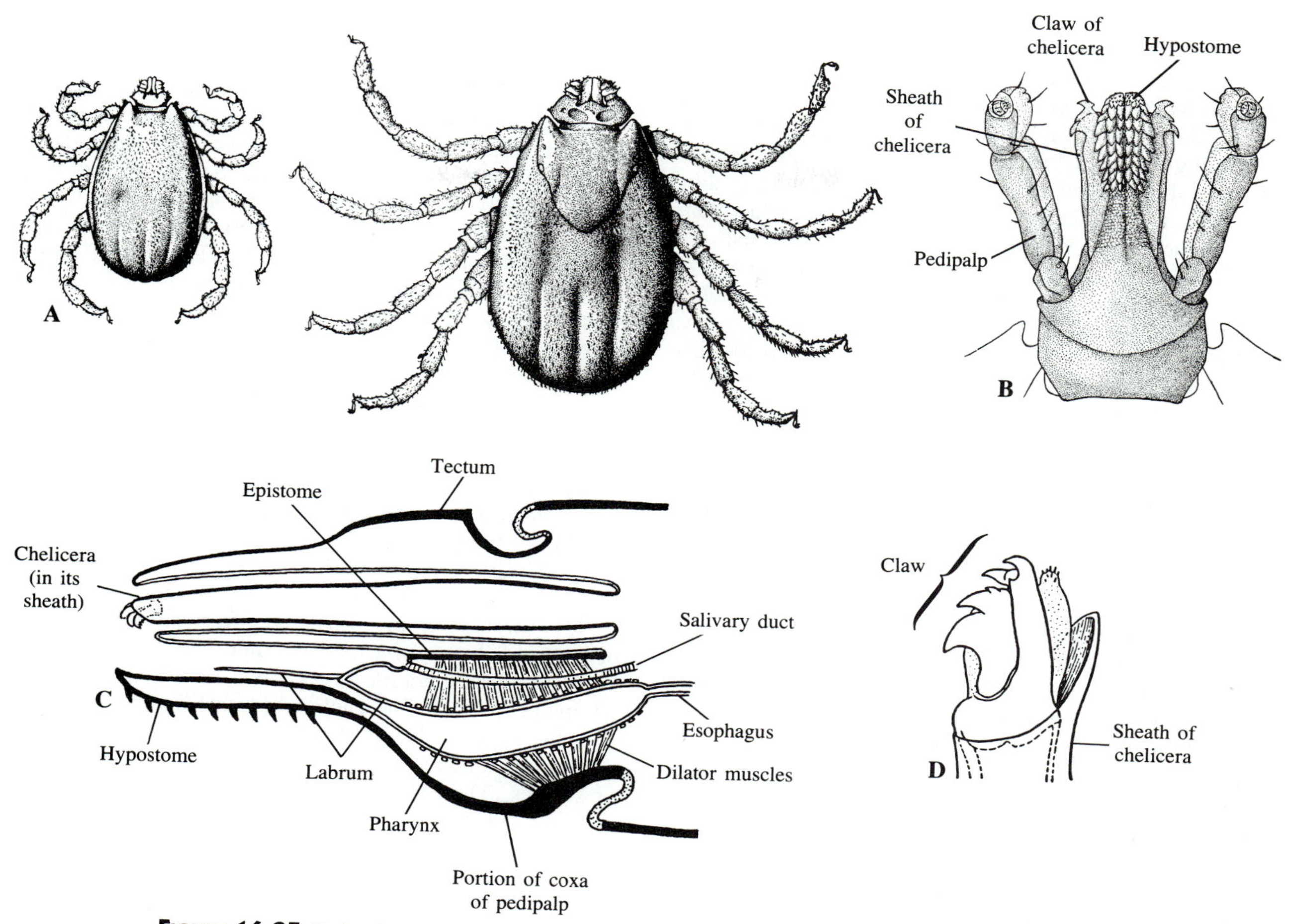

FIGURE 16.35 Order Acari. Ticks. A. *Boophilus calcaratus*, male (left) and female (right). (Serdiukova, Ixodid Ticks of the Fauna of the USSR.) B. *Amblyomma cajennense*, mouth parts, ventral view. (Hunter and Hooker, United States Bureau of Entomology, Bulletin 72.) C. Capitulum of a tick, as seen in a longitudinal section; diagrammatic. D. Right chelicera of *Dermacentor andersoni*, ventral view. The complex claw consists of three parts, two of which have sharp teeth for cutting the skin. (C and D, Snodgrass, Smithsonian Miscellaneous Collections, *110*; slightly modified.)

of the pharynx. A secretion of the salivary glands, whose ducts open between the labrum and epistome, inhibits blood clotting.

The flexible cuticle of the opisthosoma of a starved tick is usually folded to some extent, so that when the animal fills its gut with blood, the rear part of the body can swell up. Some ticks may double or triple their size by the time they finish a hearty meal. Feeding is generally followed by a molt, achieved after the tick has dropped off its host. The hindgut of certain species does not have a lumen; the anus therefore cannot function in elimination of residues.

Some ticks—those called argasids—live in nests and lairs of their hosts, so they are never far from a meal. As juveniles and as adults, they feed and molt repeatedly, and females lay eggs intermittently, in batches that number in only the hundreds. The ixodid ticks, on the other hand, have a very different sort of life style. They usually do not live continually near their hosts. Moreover, females do not lay their eggs until they have fed and molted for the last time, and they generally produce at least several thousand eggs.

The ixodid life cycle begins when an egg hatches into a juvenile (often called a seed tick, or larva) with three pairs of legs. This must wait patiently for a suitable host to come along, which may not happen for several months or even a year. Once on a host, it feeds for a few days, then falls off and molts, emerging as a second-stage juvenile (or nymph) with four pairs of legs. If lucky enough to find a new host, it will feed for a few days, drop off, and molt for the second time, becoming an adult. Some ixodid females can mate as soon as they are adult; others feed for a while if they can find a host, then mate, molt again, and lay their eggs. There are variations on the patterns just described. Males usually die soon after they mate, and in some species they do not feed after reaching the adult stage.

Many ticks, especially ixodids, are vectors of diseases of humans, domestic animals, and wild animals. The rickettsial organism that causes Rocky Mountain spotted fever in the western and midwestern portions of the United States is transmitted by *Dermacentor andersoni*. More widespread geographically is Lyme disease, which has been reported from at least 19 countries. This malady of humans is named for Old Lyme, Connecticut, where many persons were found to have an arthritic condition now known to be caused by a spirochaete transmitted by a tick. The disease usually begins with symptoms that resemble those of influenza. Symptoms that appear weeks or months later include, besides arthritis, neurological disorders and inflammation of the heart. In eastern North America, *Ixodes dammini* is the usual vector. Both of its juvenile stages prefer to take blood from a white-footed mouse, whereas adult females prefer deer. It is the second juvenile stage, carrying spirochaetes that the first juvenile stage picked up from a mouse, that is most likely to infect a human.

Ixodes dammini also transmits *Babesia microti*, a protozoan blood-cell parasite of voles. In humans, this organism causes symptoms similar to those of malaria. Some patients in the United States have been found to be suffering simultaneously from a *Babesia* infection and Lyme disease. Tick fever of cattle is another disease caused by a *Babesia* and transmitted by ticks.

Carrying pathogenic microorganisms is not the only kind of problem that ticks create for humans and other animals. Their salivary secretions cause inflammatory allergic reactions in the skin and may also induce more serious complications, such as paralysis.

References on Acari

Arthur, D. R., 1962. Ticks and Disease. Pergamon Press, London.

Baker, E. W., and Wharton, G. W., 1952. An Introduction to Acarology. Macmillan, New York.

Evans, G. O., Sheals, J. G., and MacFarlane, D., 1961. The Terrestrial Acari of the British Isles. Volume 1. British Museum (Natural History), London.

Krantz, G. W., 1970. A Manual of Acarology. Oregon State University Bookstores, Inc., Corvallis.

McDaniel, B., 1979. How to Know the Mites and Ticks. Wm. C. Brown Co., Dubuque, Iowa.

CLASS PYCNOGONIDA

Pycnogonids, or sea spiders (Figures 16.36, 16.37), are found in almost all seas, but they are not necessarily common in all marine habitats. The species encountered intertidally, on floating docks, and in shallow water are generally associated with hydroids and sea anemones, whose juices and tissues they suck. Sea slugs, sea cucumbers, jellyfishes, tunicates, and various other invertebrates sometimes have pycnogonids on them, and pycnogonids have been found on the ctenidia of mussels. Of the approximately 600 known species, a few have been collected at considerable depths. Little is known about the lifestyles of these.

The fact that the first appendages of pycnogonids are chelicerate suggests that these animals are related to arachnids, horseshoe crabs, and their extinct allies. Pycnogonids do certainly resemble spiders, but on close examination they show some decidedly unusual features. With a few exceptions, the body consists almost entirely of the prosoma. The opisthosoma, on which the anus is located, is generally reduced to an inconspicuous posterior projection.

The forepart of the prosoma, often called the *head* or *cephalon* by specialists concerned with pycnogonids, bears the chelicerae (also known as *chelifores*) and pedipalps *(palps)*. The chelicerae or pedipalps, or both, are absent in adults of many species. In some pycnogonids, the chelicerae are present but reduced to knobs or stumps. The mouth is located on a prominent suctorial tube *(proboscis)* that articulates with the prosoma.

The portion of the prosoma from which the legs originate is segmented, but the segments may not be distinctly separated from one another. Much of each segment consists of the **lateral processes** to which the legs are attached. In males, there are typically five pairs of legs. Those of the first pair are smaller than the others and are specialized for carrying clusters of eggs (Figure 16.36, B and C, Figure 16.37, C). For this reason, they are called **ovigerous legs,** or simply *ovigers*. Females, which are not involved in caring for the eggs they lay, generally lack these appendages.

The other four pairs of legs, used for walking and clinging, are odd in several respects. They have eight articles in addition to a terminal claw; sometimes small auxiliary claws are also present. The conventional terminology for the articles is shown in Figure 16.36, B. A branch of the gonad and a diverticulum of the midgut extend into each leg. This is a decidedly unusual feature of pycnogonids. (In spiders, gut diverticula may extend into the coxae, but no farther.)

In certain genera, there are species with one or two extra prosomal segments, and therefore five or six pairs of walking legs (Figure 16.36, D) instead of four. This sort of variation in segment and leg number is almost never encountered in arthropods other than pycnogonids.

Phylum Porifera

Class Demospongiae

An urn-shaped tropical species. (Photograph by Ronald Shimek.)

Ophlitaspongia pennata, an orange-red encrusting species, with *Rostanga pulchra*, a nudibranch gastropod that eats this sponge. The coiled structures are strings of eggs laid by the nudibranch.

Microciona prolifera, whose growths form fingerlike projections.

A freshwater sponge. The green color is due to zoochlorellae in the amoebocytes.

Phylum Cnidaria

Class Hydrozoa

Tubularia crocea (order Hydroida, suborder Athecata).

Class Anthozoa

Veretillum cynomorium (subclass Alcyonaria, order Pennatulacea), a club-shaped sea pen found in the Atlantic and Mediterranean.

Phylum Cnidaria
Class Anthozoa

Paragorgia pacifica (subclass Alcyonaria), a gorgonian.

Anemonia sulcata (subclass Zoantharia, order Actiniaria), a common European sea anemone.

Calliactis parasitica (subclass Zoantharia, order Actiniaria), a European anemone that lives on snail shells occupied by the hermit crab *Eupagurus bernhardus*. The acontia are bright pink.

Urticina lofotensis (subclass Zoantharia, order Actiniaria).

Corynactis californica (subclass Zoantharia, order Corallimorpharia).

Balanophyllia elegans (subclass Zoantharia, order Scleractinia), a cup coral.

PHYLUM NEMERTEA

Class Anopla

Tubulanus sexlineatus.

Micrura verrilli.

Class Enopla

Amphiporus bimaculatus.

PHYLUM ANNELIDA

Class Polychaeta

Serpula vermicularis (family Serpulidae).

Euzonus mucronatus (family Opheliidae), a ''red worm'' of sandy beaches.

PHYLUM MOLLUSCA

Class Polyplacophora

Tonicella lineata.

Placiphorella velata, ventral view, showing the head, foot, mantle cavity, and enlarged anterior portion of the girdle, under which prey are trapped.

Phylum Mollusca

Class Gastropoda

Fissurellidea bimaculata (subclass Prosobranchia, order Archaeogastropoda), a keyhole limpet.

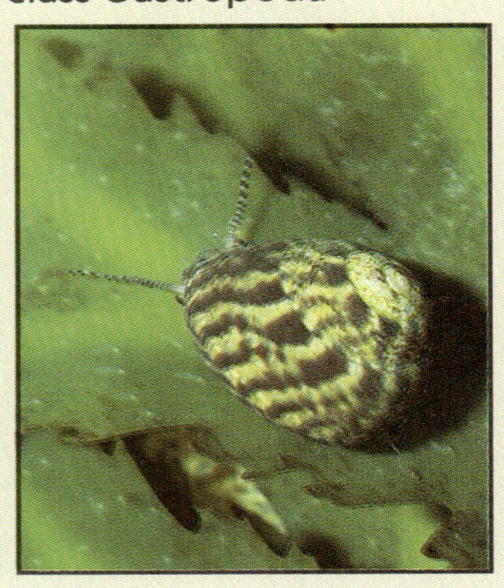

Gibbula umbilicalis (subclass Prosobranchia, order Archaeogastropoda).

Calliostoma annulatum (subclass Prosobranchia, order Archaeogastropoda).

Amphissa columbiana (subclass Prosobranchia, order Neogastropoda).

Dendronotus diversicolor (subclass Opisthobranchia, order Nudibranchia).

Tritonia diomedea (subclass Opisthobranchia, order Nudibranchia).

Hermissenda crassicornis (subclass Opisthobranchia, order Nudibranchia).

Archidoris montereyensis (subclass Opisthobranchia, order Nudibranchia).

Phylum Mollusca

Class Gastropoda

Elysia hedgpethi (subclass Opisthobranchia, order Sacoglossa) on *Codium*, the green alga it eats.

Liguus fasciatus (subclass Pulmonata, order Stylommatophora), a Florida tree snail.

Helix aspersa (subclass Pulmonata, order Stylommatophora), a European snail that has been introduced into other parts of the world.

Theba pisana (subclass Pulmonata, order Stylommatophora), attached to vegetation during a dry season in Tunisia.

Succinea (subclass Pulmonata, order Stylommatophora), an amphibious snail found on vegetation in freshwater marshes.

Class Cephalopoda

Nautilus macromphalus (subclass Nautiloidea). (Photograph by Ronald Shimek.)

Phylum Arthropoda

Subphylum Chelicerata

Class Pycnogonida

Achelia nudiuscula, male carrying eggs.

Subphylum Crustacea

Class Malacostraca

Megalorchestia californiana (subclass Peracarida, order Amphipoda), a beach hopper.

Idotea wosnesenskii (subclass Peracarida, order Isopoda).

Pagurus ochotensis (subclass Eucarida, order Decapoda), a hermit crab. Much of the snail shell it occupies is covered with barnacles.

Cryptolithodes sitchensis (subclass Eucarida, order Decapoda). The broad carapace covers all appendages.

Lopholithodes mandtii (subclass Eucarida, order Decapoda), a box crab.

Hapalogaster mertensii (subclass Eucarida, order Decapoda).

Phylum Arthropoda

Subphylum Crustacea

Class Malacostraca

Callianassa californiensis (subclass Eucarida, order Decapoda).

Pachygrapsus crassipes (subclass Eucarida, order Decapoda).

Subphylum Uniramia

Class Insecta

A tropical butterfly (order Lepidoptera).

Class Diplopoda

Harpaphe haydeniana, a millipede.

Phylum Echinodermata

Class Crinoidea

Cenometra bella. (Photograph by Ronald Shimek.)

Class Asteroidea

Pisaster ochraceus.

Henricia leviuscula.

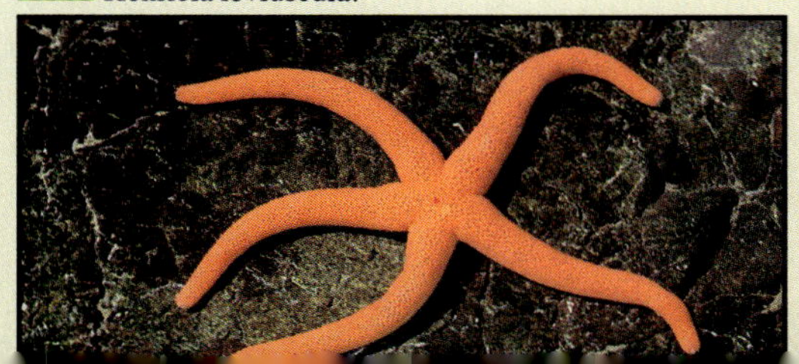

Phylum Echinodermata

Class Asteroidea

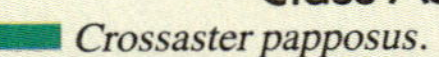 *Crossaster papposus.*

Class Echinoidea

Echinus esculentus, a species of the European Atlantic region.

Class Holothuroidea

Strongylocentrotus franciscanus, a large species of the Pacific coast.

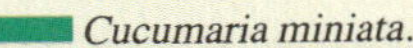 *Cucumaria miniata.*

Phylum Urochordata

Class Ascidiacea

Cnemidocarpa finmarchiensis, a solitary species.

Perophora annectens, a social ascidian, with specimens of the cup coral *Balanophyllia elegans*. (Photograph by Ronald Shimek.)

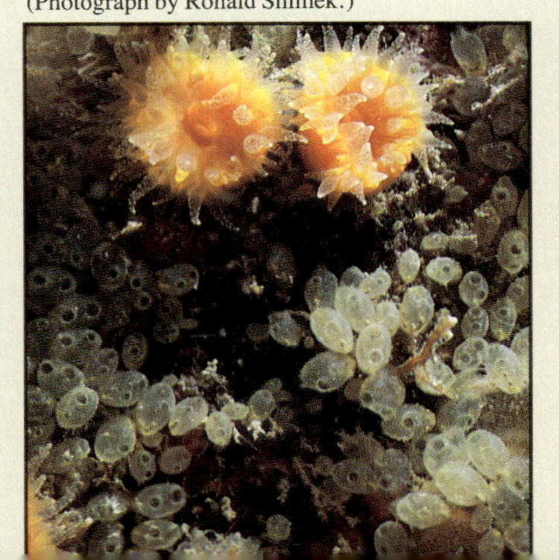

Metandrocarpa dura, a social species.

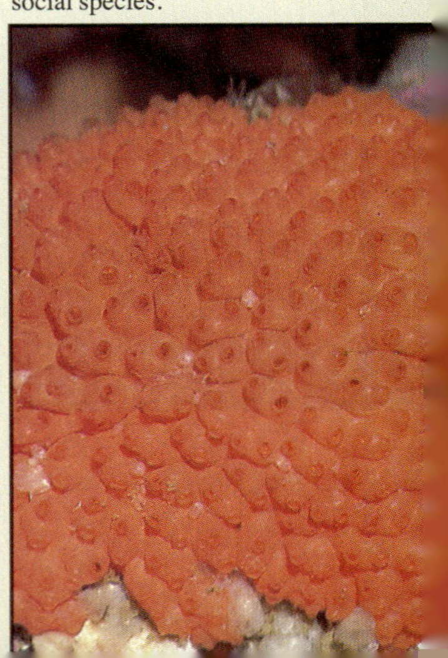

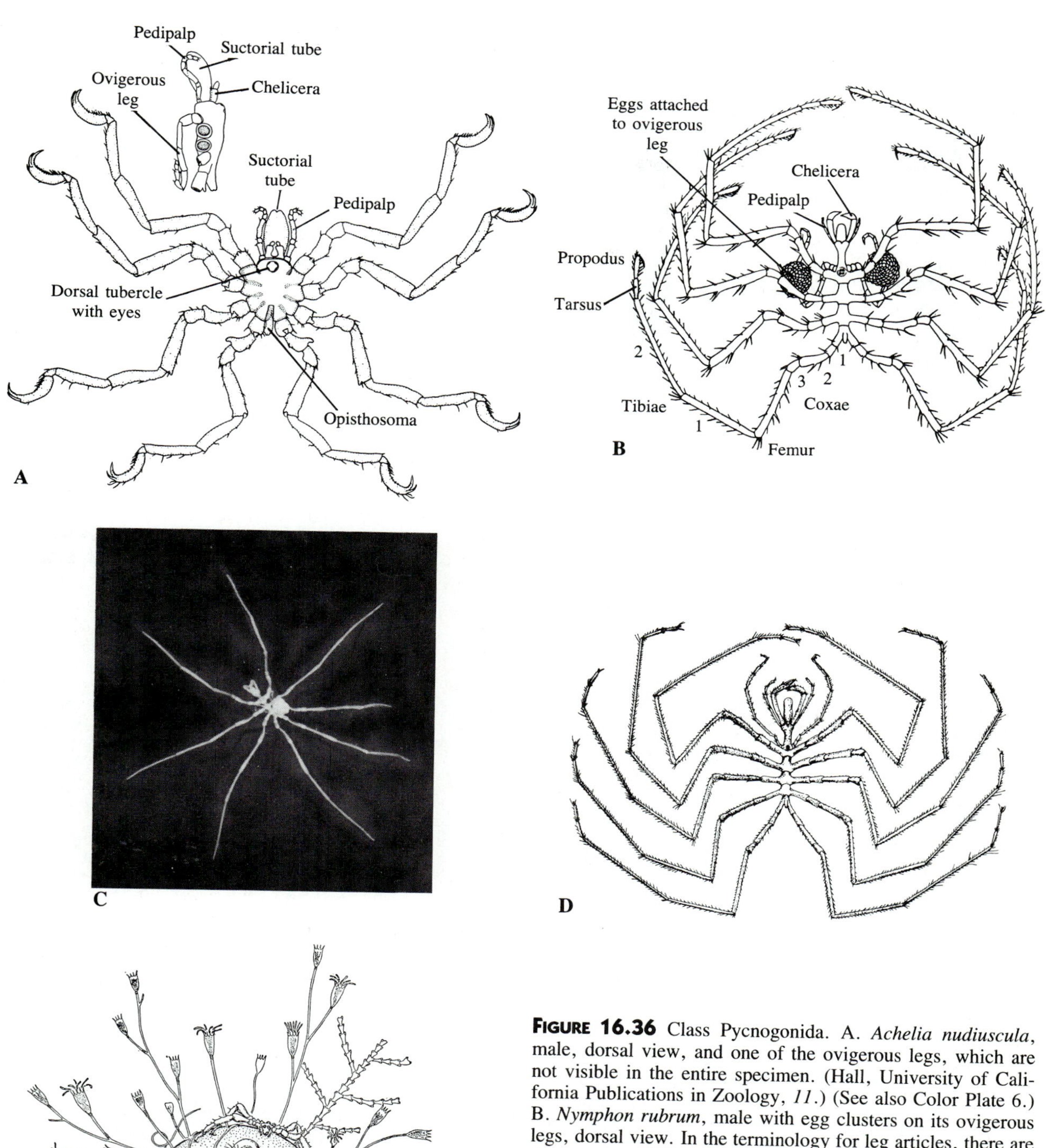

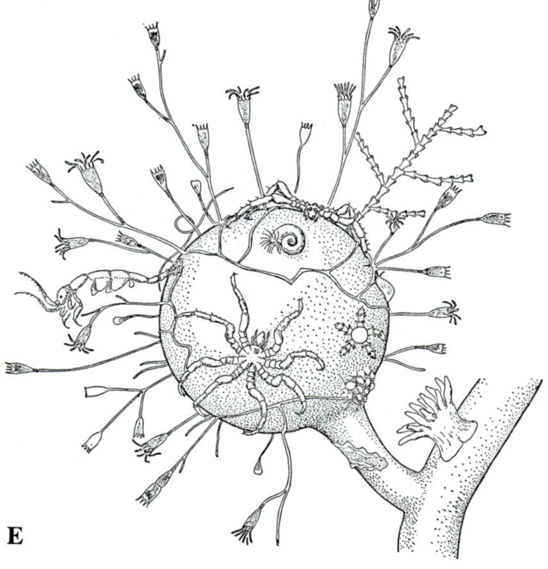

FIGURE 16.36 Class Pycnogonida. A. *Achelia nudiuscula*, male, dorsal view, and one of the ovigerous legs, which are not visible in the entire specimen. (Hall, University of California Publications in Zoology, *11*.) (See also Color Plate 6.) B. *Nymphon rubrum*, male with egg clusters on its ovigerous legs, dorsal view. In the terminology for leg articles, there are three coxae and two tibiae. (Tiegs and Manton, Biological Reviews, *33*.) C. *Nymphon grossipes*, male, dorsal view. The right ovigerous leg of this specimen is carrying an egg cluster. D. *Pentanymphon antarcticum*, male, dorsal view. There are five pairs of legs in addition to the ovigerous legs. (Helfer and Schlottke, in Bronn, Klassen und Ordnungen des Tierrechs, *5*:4:2.) E. *Anoplodactylus petiolatus*, on a bladder of the seaweed *Sargassum*. The assemblage of animals shown includes a caprellid amphipod (left), a serpulid polychaete, a young brittle star, a polyclad tubellarian, two species of hydroids, and a sea anemone. The hydroids and anemone are the most likely sources of food for the pycnogonid. (Hedgpeth, Proceedings of the United States National Museum, *97*.)

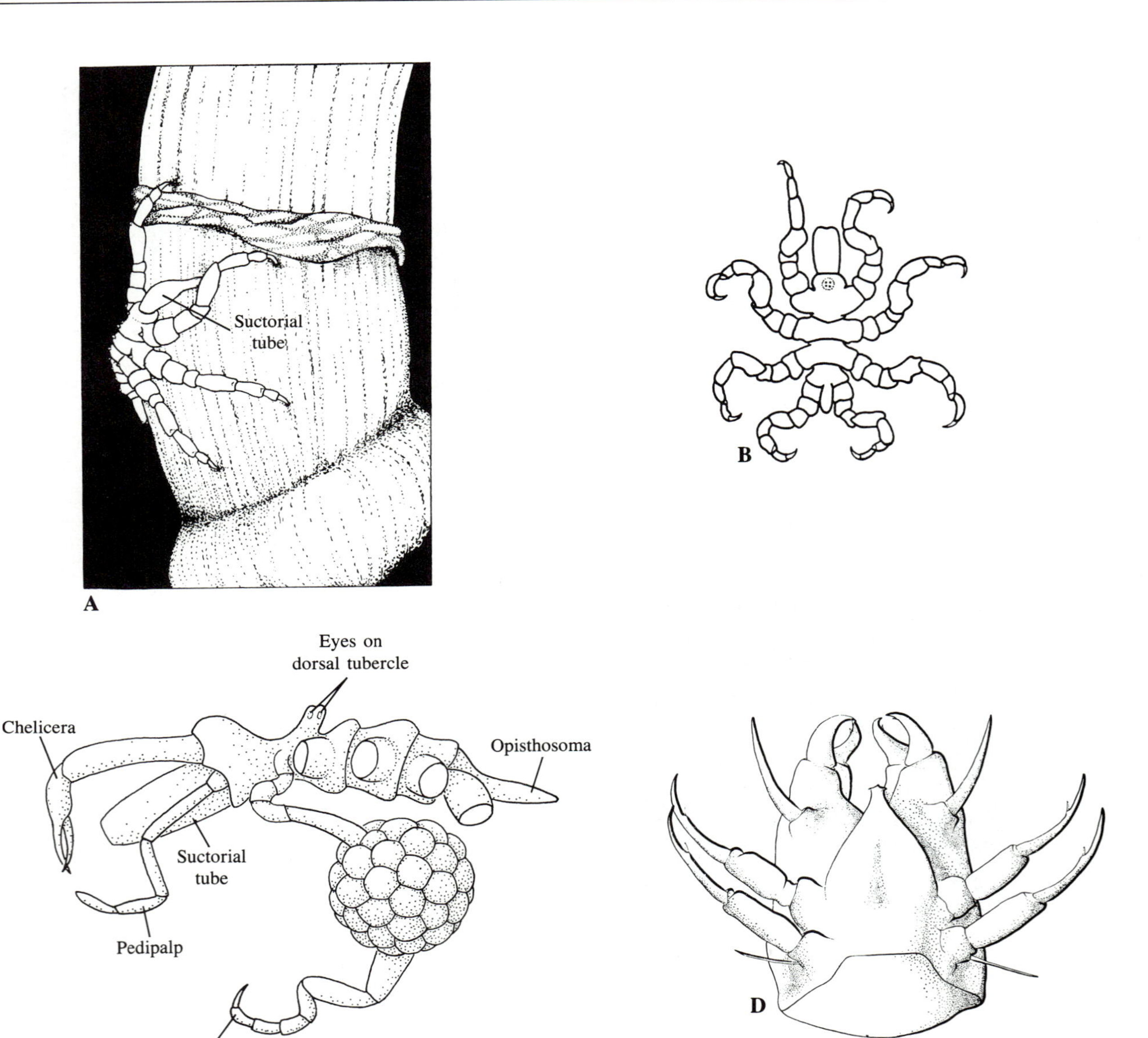

FIGURE 16.37 Class Pycnogonida. A. *Pycnogonum littorale*, with its suctorial tube embedded in the column of a sea anemone, *Metridium senile*. The ring of slime produced by the anemone does not restrict the movements of the pycnogonid. (Schloemer, Natur und Volk, *78*.) B. *Pycnogonum stearnsi*, which lives on anemones on the Pacific coast of North America; dorsal view. (Courtesy of Joel Hedgpeth.) C. *Chaetonymphon spinosum*, male, showing an ovigerous leg to which a cluster of eggs is cemented. (Most other appendages and all setae omitted. (After Helfer and Schlottke, in Bronn, Klassen und Ordnungen des Tierreichs, *5*:4:2.) D. Protonymphon larva of a species of *Achelia*, ventral view. (Sanchez, Archives de Zoologie Expérimentale et Générale, *98*.)

So far as general body form is concerned, pycnogonids fall into two categories: long-legged and short-legged types. *Achelia* (Figure 16.36, A; Color Plate 6), *Nymphon* (Figure 16.36, B and C), *Pentanymphon* (Figure 16.36, D) *Anoplodactylus* (Figure 16.36, E) are examples of the former; *Pycnogonum* (Figure 16.37, A and B) is an example of the latter. The largest of the long-legged pycnogonids, found in the Antarctic region, has a body length of about 6 cm and a leg spread of about 50 cm. Most species have a leg spread of between 5 mm and 3 or 4 cm.

On the dorsal side of the prosoma, at about the level

of the ovigerous legs (or where they would be if they were present), there is usually a cluster of four simple eyes set on a low tubercle. The eyes may be fused in such a way that there appear to be only two, and some pycnogonids have no eyes at all. The brain, located beneath the eye tubercle, consists of the protocerebrum and a ganglion that has moved forward, in the course of development, from an originally postoral position. Nerves from this ganglion, interpreted to be a tritocerebrum, supply the chelicerae. Connectives from the brain run around the esophagus to a ganglion that innervates the pedipalps and ovigerous legs. The remaining ventral ganglia of the prosoma are not always distinctly separated by connectives, but in pycnogonids with four pairs of legs there are basically four ganglia. In species that have one or two extra pairs of legs, the number of ganglia is correspondingly increased. Ganglia belonging to the reduced opisthosoma are fused to the last prosomal ganglion.

The suctorial tube has a fairly complicated internal structure. It has musculature for dilating the pharynx, thus enabling this part of the foregut to operate as a sucking pump. The pharynx is derived from a stomodeal invagination, so its epithelium is of ectodermal origin and coated by a cuticle. It has three thickened plaques studded with small teeth, and these are used for grinding up food. The esophagus, located in the prosoma proper, is also a stomodeal derivative, but its cuticular lining has no special elaborations. The midgut, or intestine, is separated from the esophagus by a valve. As has been explained previously, the midgut has diverticula going out into the legs; there are diverticula in the suctorial tube, too. Behind a valve at the posterior end of the midgut is a short hindgut.

Most pycnogonids suck in juices and tissues from the hydroids, sea anemones, or other animals with which they are associated. What deep-sea types feed on is a mystery. Little is known about the actual process of digestion. It has been claimed that epithelial cells lining the midgut pick up food particles, become detached, and then reattach somewhere else, thus making nourishment available to other cells. This is a subject that needs to be investigated.

The circulatory system of a pycnogonid is simple. The hemocoel, in the legs as well as the body proper, is divided by a horizontal partition into dorsal and ventral chambers, but these chambers are connected by pores. The heart lies in the dorsal chamber and pumps blood in a forward direction. The blood gradually moves into the ventral chamber, passes backward, then goes into the dorsal chamber. From this it reenters the heart by way of one or more pairs of ostia.

The integument of pycnogonids may be fairly thick, and it is often hairy, spiny, or warty. The muscles in the appendages are organized into sets of antagonists. In the legs, for instance, almost all articles are provided with both flexors and extensors, and the suctorial tube has both flexor and retractor muscles. In pycnogonids in which the prosomal segments are distinct, these can be moved to some extent by bundles of longitudinal muscles.

The gonad is situated in the dorsal part of the leg-bearing part of the prosoma. It is approximately U-shaped; its two arms, turned anteriorly, send out branches into the legs. In males, as a rule, the last three pairs of legs have gonopores on their second coxae; in females, the gonopores are located on the second coxae of the last one, two, or three pairs of legs. Egg-laying is preceded by coupling, so the eggs are fertilized as they emerge from the gonopores. The male picks them up right away, and with the help of a secretion of some glands on the femoral articles of his legs, he sticks clusters of eggs to his ovigerous legs.

The eggs of pycnogonids hatch into a strange larva called a **protonymphon** (Figure 16.37, D). This has the suctorial tube and a pair of chelicerae. It also has two additional pairs of appendages that are probably homologues of the pedipalps and ovigerous legs. These are usually lost during development, however, and later replaced by the adult structures. The protonymphon has only rudiments of the first pair of walking legs, so all of the walking legs are added in the course of further development.

The protonymphon attaches to a suitable host animal, usually a cnidarian; it may even invade the tissues. In the course of several molts, the young pycnogonid passes through several juvenile stages in which the more posterior segments and their respective walking legs are added sequentially. There is much reorganization in the anterior part of the prosoma, with the chelicerae moving to a more forward position. In some pycnogonids, the protonymphon and juvenile stages are associated with hosts other than those they feed upon when they are adult. For instance, a species that spends part of its life on the ctenidia of a mussel, *Mytilus californianus,* eventually feeds on hydroids.

Where do pycnogonids fit into the scheme of classification of arthropods? The chelate first appendages link them with arachnids and their allies. At no stage in development, however, does the body have a substantial opisthosoma. Studies on the nervous system, often helpful in revealing homologies between indistinctly separate segments of one type of arthropod with those of another, seem to have brought up more questions than they have answered. Some zoologists who have worked extensively with pycnogonids would raise them to the rank of subphylum, thus putting them on a par with chelicerates, mandibulates, and the now-extinct trilobites. It will be best to consider pycnogonids an anomalous group whose status is currently uncertain.

References on Pycnogonida

Fry, W. G. (editor), 1978. Sea Spiders (Pycnogonida). Proceedings of a meeting in honor of Joel W. Hedgpeth held on 7 October 1976 in the rooms of the Linnean Society of London. Zoological Journal of the Linnean Society, *63:*i–x, 1–238.

Fry, W. G., and Stock, J. H., 1978. A pycnogonid bibliography. Zoological Journal of the Linnean Society, *63:*197–238.

Hedgpeth, J. W., 1949. The Pycnogonida of the western North Atlantic and the Caribbean. Proceedings of the United States National Museum, *97:*157–342.

Hedgpeth, J. W., 1954. Phylogeny of the Pycnogonida. Acta Zoologica, *35:*193–213.

King, P. E., 1973. British Seaspiders. Synopses of the British Fauna, no. 5. Linnean Society of London and Academic Press, London and New York.

King, P. E., 1973. Pycnogonids. St. Martin's Press, New York; Macmillan, London.

Manton, S. M., 1978. Habits, functional morphology and the evolution of pycnogonids. Zoological Journal of the Linnean Society, *63:*1–21.

Nakamura, K., 1981. Postembryonic development of a pycnogonid, *Propallene longiceps*. Journal of Natural History, *15:*49–62.

SUMMARY

1. The Chelicerata is a primarily terrestrial assemblage of arthropods. It includes spiders, scorpions, mites, ticks, and several related groups.

2. Except in mites and ticks, in which the tagmata are not distinct, the body is divided into an anterior prosoma and a posterior opisthosoma.

3. There are typically six pairs of appendages on the prosoma: chelicerae, pedipalps, and four pairs of legs. Neither the chelicerae nor the pedipalps are comparable to the antennae, mandibles, and maxillae characteristic of the head region of most other arthropods.

4. The appendages of the opisthosoma do not fit a consistent pattern. In most chelicerates, there are either no appendages at all on this tagma or just one or two pairs of highly modified appendages. In spiders, for instance, the only opisthosomal appendages are the spinnerets, from which silk is exuded. In scorpions, there is a single pair of comblike sensory appendages, called pectines, on one of the more anterior opisthosomal segments. The so-called horseshoe crabs, which belong to the marine subclass Xiphosura, have six pairs of platelike opisthosomal appendages, and all but the first pair are equipped with respiratory organs called book gills. Similar appendages were present in the long-extinct subclass Eurypterida.

5. With the exception of harvestmen and certain mites, chelicerates are carnivores that consume other animals or suck out their juices.

6. Respiratory organs, when present in terrestrial types, are either book lungs or tracheae. Book lungs, located on the ventral side of certain opisthosomal segments, are deep pits with sheetlike elaborations that resemble the pages of a book. The tracheal systems of some chelicerates (including spiders) are derived from book lungs; those of other chelicerates have probably evolved independently.

7. A dorsal heart, with ostia, is usually present. As in other arthropods, it lies in a pericardial sinus of the hemocoel.

8. Excretory organs include Malpighian tubules, which open into the posterior part of the midgut, and coxal glands, which have their pores on the coxae of certain appendages. Coxal glands are derived from coelomic cavities and coelomoducts.

9. Transfer of sperm may be accomplished by true copulation, in which a penislike structure is used by the male to inseminate the female. This is the case in harvestmen, daddy longlegs, and some mites. In most chelicerates, however, transfer of sperm is effected less directly. Male spiders, for instance, deposit a drop of sperm on a silken web or on a strand of silk, then draw the fluid into reservoirs in the pedipalps. During the mating ritual, the tips of the pedipalps are inserted into the genital pores of the female. In scorpions and several other groups, sperm transfer involves deposition, by the male, of a spermatophore. The female must find this and place her genital pores on it, or she must be drawn to it by the male.

10. The pycnogonids, or sea spiders, which generally feed by sucking juices from cnidarians, bryozoans, or other invertebrates, deviate from the usual body plan of chelicerates. The opisthosoma is reduced to the point that it is barely recognizable. Males, moreover, have five pairs of legs, those of the first pair being specialized for carrying eggs received from the female.

17

Subphylum Crustacea

Subphylum Crustacea
- Class Cephalocarida
- Class Branchiopoda
  - Order Anostraca
  - Order Notostraca
  - Order Diplostraca
    - Suborder Conchostraca
    - Suborder Cladocera
- Class Remipedia
- Class Ostracoda
- Class Mystacocarida
- Class Copepoda
  - Order Calanoida
  - Order Harpacticoida
  - Order Monstrilloida
  - Order Cyclopoida
  - Order Siphonostomatoida
  - Order Poecilostomatoida
- Class Branchiura
- Class Cirripedia
  - Order Thoracica
  - Order Acrothoracica
  - Order Ascothoracica
  - Order Rhizocephala
- Class Tantulocarida
- Class Malacostraca
  - Subclass Phyllocarida
  - Subclass Hoplocarida
    - Order Stomatopoda
  - Subclass Eumalacostraca
    - Superorder Syncarida
      - Order Anaspidacea
      - Order Bathynellacea
    - Superorder Pancarida
    - Superorder Peracarida
      - Order Mysidacea
      - Order Cumacea
      - Order Amphipoda
        - Suborder Gammaridea
        - Suborder Hyperiidea
        - Suborder Caprellidea
      - Order Isopoda
        - Suborder Flabellifera
        - Suborder Asellota
        - Suborder Valvifera
        - Suborder Oniscoidea
        - Suborder Epicaridea
      - Order Tanaidacea
    - Superorder Eucarida
      - Order Euphausiacea
      - Order Decapoda
        - Suborder Dendrobranchiata
        - Suborder Pleocyemata
          - Infraorder Caridea
          - Infraorder Stenopodidea
          - Infraorder Astacidea
          - Infraorder Palinura
          - Infraorder Anomura
          - Infraorder Thalassinidea
          - Infraorder Brachyura

INTRODUCTION

Of the approximately 30,000 described species of Crustacea, most live in the sea. The subphylum is well represented in fresh water, however, and a few groups are successful in damp terrestrial habitats. There is a marvelous diversity in the Crustacea, and certain types are among the most plentiful organisms on the face of the earth. A cubic meter of sea water may contain several thousand copepods, and freshwater lakes and ponds often have high concentrations of *Daphnia* and related cladocerans. The small planktonic types just mentioned feed on diatoms and other unicellular algae, but they are consumed by various animals, including fishes, so they constitute a nutritional link between photosynthetic primary producers and carnivores.

BODY REGIONS, TAGMATA, AND APPENDAGES

In most groups of Crustacea, it is possible to recognize three body regions: **head, thorax,** and **abdomen.** These do not necessarily coincide with the obvious tagmata. For instance, what appears to be the head often has one or more segments of the thorax incorporated into it. In certain Crustacea, moreover, most of the segments and appendages behind the head are similar, so the entire postcephalic region is called the **trunk.**

The arrangement of tagmata, segments, and appendages of crustaceans forms the basis for classification of these animals. The many different plans are covered in later sections of this chapter, in which the classes, subclasses, orders, and other major taxa are characterized. This section deals primarily with terminology applicable to appendages of nearly all Crustacea. Mastery of this material is a prerequisite for understanding diversity in this large assemblage of invertebrates.

HEAD APPENDAGES

Most crustaceans, except for some much-modified types, are readily placed in the subphylum because they have two pairs of antennae on the anterior part of the head. In this respect, they differ from members of the subphylum Uniramia—insects, millipedes, centipedes, and related groups, covered in Chapters 18 and 19—which have only one pair of antennae. Spiders, scorpions, and their kin, which belong to the subphylum Chelicerata, lack antennae. Their first appendages, as explained in Chapter 16, are typically clawed or pincer-like.

The first and second head appendages of crustaceans are here called **first** and **second antennae,** and in illustrations they are labelled antenna 1 and antenna 2. (An alternative nomenclature, used in some treatises, is *antennules* and *antennae*.) Behind the antennae, the head typically has three pairs of mouthparts. These, in order, are the **mandibles** and **first maxillae** and **second maxillae** (maxilla 1, maxilla 2). The mandibles are usually stout jaws that crush food and shove it into the mouth; the maxillae are generally concerned with sorting food and delivering it to the mandibles. Altogether, then, the head has five pairs of appendages (Figure 17.1).

THORACIC APPENDAGES

All of the appendages of the thorax may be called **thoracopods.** Usually, however, some of them are specialized for one function and some for another, so they are given more specific names. For example, if certain of the more anterior thoracopods are specialized as food-handling or food-sorting appendages somewhat similar to maxillae, they are called **maxillipeds** (''maxilla-like legs''). Those used primarily for locomotion—walking or swimming—are called **pereopods.** The term **pereon** is applied to the assemblage of thoracic segments from which the pereopods originate. In an isopod (Figure 17.2), for instance, the thorax consists of eight segments, the first of which is fused to the head and bears maxillipeds; the remaining seven segments, therefore, constitute the pereon and bear the pereopods or walking legs. A crayfish (Figures 17.8, 17.9) or lobster also has eight thoracic segments, but these are covered dorsally and laterally by a **carapace** continuous with that which covers the head. The head and thorax thus form a tagma called the **cephalothorax.** Nevertheless, there are eight pairs of appendages on the thoracic region. The first three pairs are maxillipeds; the remaining five pairs are pereopods, or walking legs, although the first of them, with large pincers, are specialized for defense and for grasping and crushing prey.

ABDOMINAL APPENDAGES

Abdominal appendages are called **pleopods,** because the term **pleon** is essentially synonymous with *abdomen*. Like the thoracic appendages, the abdominal pleopods show various morphological and functional specializations, and in certain groups of Crustacea, those on the last segment, or last three segments, are so distinctive that they are designated **uropods.** Depending on the crustacean and the particular abdominal segment, a pair of pleopods may be used for swimming, jumping, respiration, brooding eggs, copulation, or some other function. The same pair may, in fact, have two or more rather different functions.

The abdomen terminates in a **telson,** on which the anus is usually located. The telson is perhaps the homologue of the pygydium of annelids. It is not a true segment and has no appendages. In a few crustaceans, however, it has two projections called **caudal rami;**

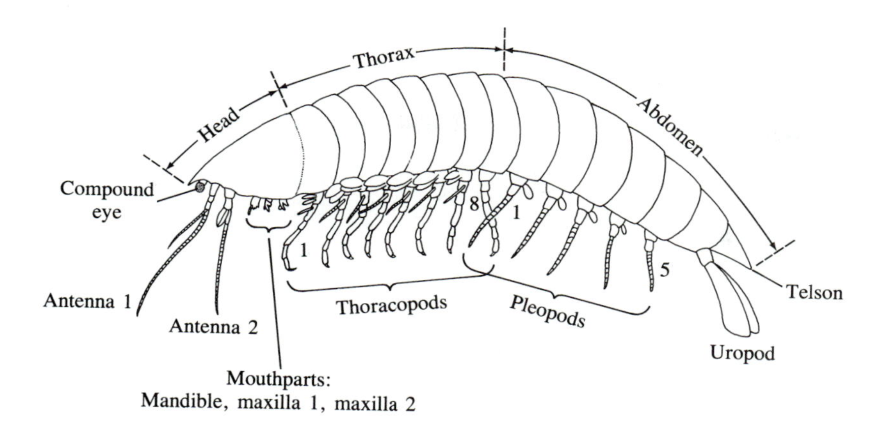

FIGURE 17.1 The appendages of a crustacean in which the three basic tagmata—head, thorax, and abdomen—are distinct. The diagram is based on *Anaspides,* a member of the class Malacostraca, superorder Syncarida. The first thoracic segment is fused to the head, but its anterior limit is marked by a prominent groove. (After Siewing, Verhandlungen der Deutschen Zoologischen Gesellschaft, *18*.)

when these are jointed, they may superficially resemble appendages, especially uropods.

BIRAMOUS APPENDAGES

Biramous appendages are those that have two branches. In crustaceans, the branches usually arise from the second article, the **basis** (Figure 17.3, A). The branch nearer to the midline of the animal is the **endopodite;** the lateral branch is the **exopodite.** In certain groups within the subphylum, only a few of the appendages are biramous, the rest are uniramous.

Although the first antennae of some higher crustaceans, including crayfishes, have two branches (sometimes even three), these do not correspond to the exopodite and endopodite of other appendages, because they do not arise from the same article. First antennae, moreover, are preoral outgrowths of the cephalic lobe of the embryo, which is more nearly comparable to the prostomium of an annelid than to a true segment. They

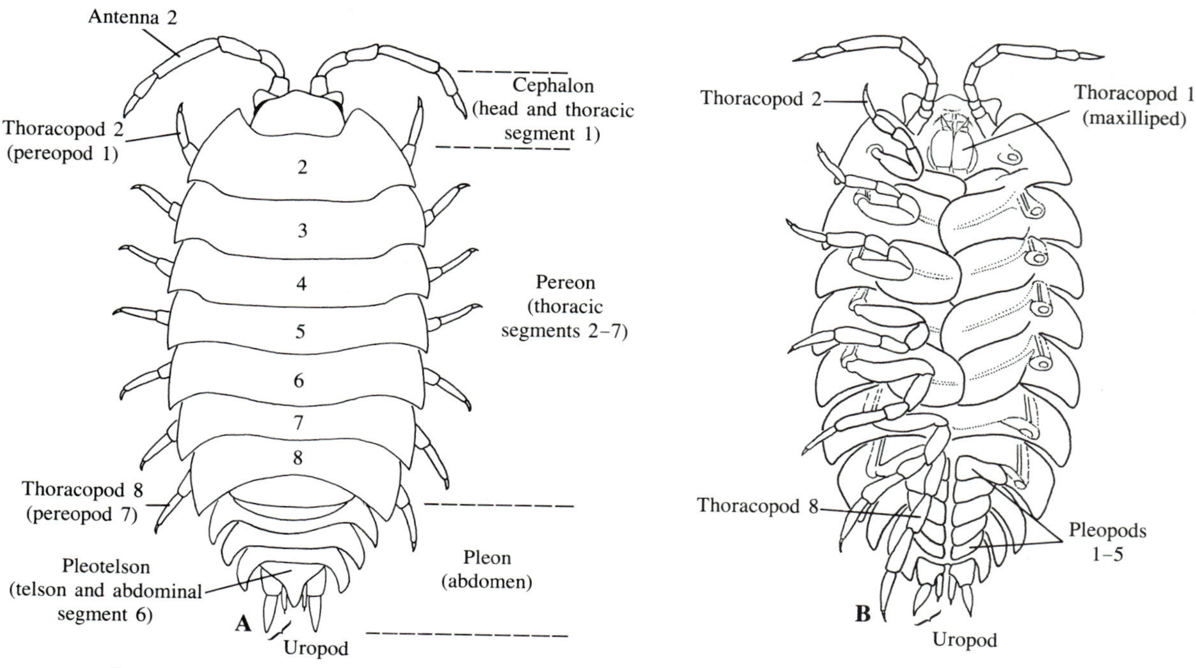

FIGURE 17.2 A sowbug, *Porcellio* (class Malacostraca, superorder Peracarida). A. Dorsal view. The first antennae, which are minute, are not visible. B. Ventral view. The mandibles and first and second maxillae are largely hidden by the platelike maxillipeds, which are the first thoracic appendages.

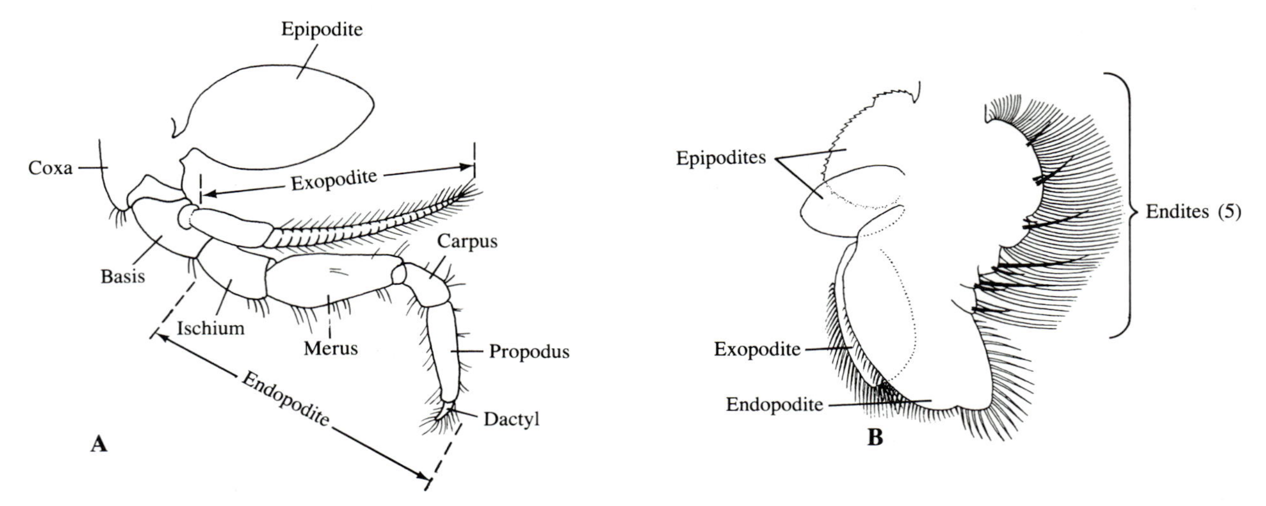

FIGURE 17.3 The two major types of appendages in crustaceans. A. Stenopodial appendage, generalized. B. A phyllopodial appendage.

are perhaps homologous to the prostomial palps of certain polychaetes, such as *Nereis*. All other appendages originate from postoral segments, although the second antennae end up in a preoral position.

PHYLLOPODIA AND STENOPODIA

There are two general types of appendages in Crustacea. **Phyllopodia** (Figure 17.3, B), which are flattened and somewhat leaflike, are effective in swimming, creating water currents, respiration, collecting and transferring food, and sometimes in other ways. In **stenopodia** (Figure 17.3, A), the proximal portions and the rami may be stout or slender, but they are not leaflike. Stenopodia are generally fitted for walking, grasping, or clinging. They may, however, have setae that strain out microscopic food or that increase surface area and make the limbs efficient devices for swimming. Both phyllopodial and stenopodial appendages may be modified for carrying eggs while these develop, and both may be specialized for sperm transfer. The superficial appearance of a limb does not always reveal all its functions; one must look carefully for elaborations that may not be obvious. For instance, a phyllopodial appendage used for swimming, creating feeding currents, and respiration may have a jawlike basal process for chewing or for passing food forward to the mouthparts.

The layout of a biramous appendage is best approached with the help of a generalized stenopodial limb (Figure 17.3, A), because the rami and articles are more sharply demarcated in a stenopodial than in a phyllopodial appendage. The first article, joined to the body proper, is called the **coxa.** The endopodite and exopodite originate from the next article, the basis. The two rami may be similar, or they may be specialized for different functions, or one ramus may be reduced or absent. As a rule, if one ramus is missing, it is the exopodite, but there are exceptions.

In certain of the higher crustaceans, including crayfishes, crabs, shrimps, amphipods, and isopods, the endopodite of the thoracic appendages has a small and fixed number of articles. These are called the **ischium, merus, carpus, propodus,** and **dactyl** (claw) (Figure 17.3, A). When there is no exopodite, which is usually the case, the appendage as a whole, including the coxa and basis, consists of seven articles.

If the coxa of a stenopodial appendage has one or more outgrowths that serve as gills, these are called **epipodites.** An epipodite is not divided into articles and must not be confused with either of the rami that arise from the basis.

In a typical phyllopodial appendage (Figure 17.3, B), the exopodite can be differentiated from the endopodite easily enough, but it appears to consist of a single large article; the endopodite, moreover, is neither distinctly articulated nor sharply set apart from the basis. The medial side of the coxa and of the basis of a phyllopodial appendage is often elaborated into a series of projections called **endites.** When the basal endite is specialized as a strong jaw, it is called a *gnathobase*. As in some stenopodial appendages, the coxal portion may have one or more epipodites that function as gills. These are usually, but not always, on the lateral edge of the appendage.

CUTICLE AND MOLTING

It was pointed out in the introduction to arthropods (Chapter 15) that the cuticular exoskeleton restricts growth. If there is to be an increase in size, the cuticle

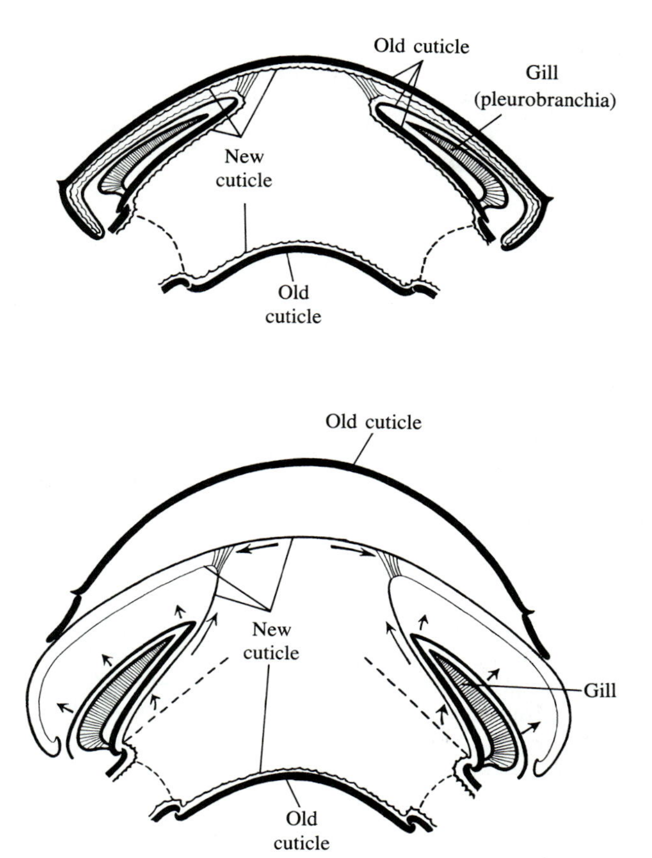

FIGURE 17.4 Transverse sections through the cephalothorax of the crab *Maia squinado,* showing the relationship of the old cuticle to the new cuticle before the molt (above), and the expansion of the new cuticle after the old cuticle has been shed (below). The arrows indicate the directions of the expansion. (Drach, Annales d'Institut Océánographique, Monaco, *19.*)

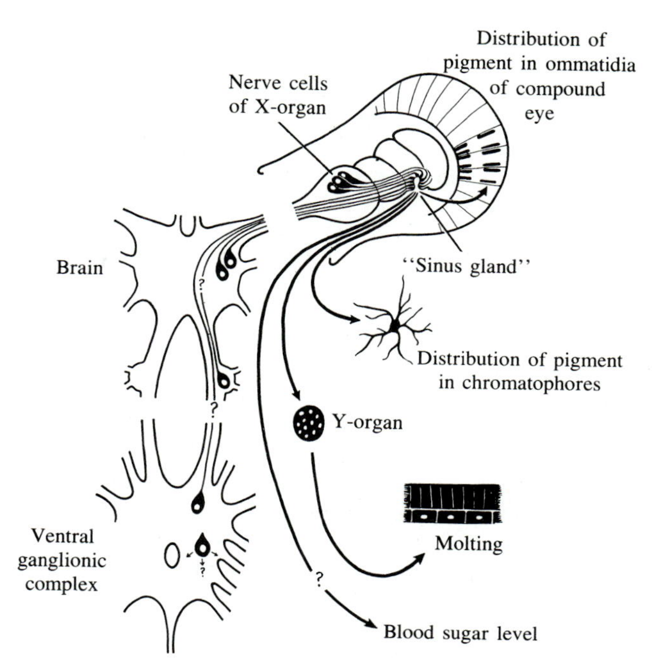

FIGURE 17.5 Some of the structures and processes affected by hormones produced by neurosecretory cells of the X-organ (located in the eyestalk) and of the central nervous system of a crab. The so-called sinus gland is where the neurosecretory cells terminate and where release of their hormones is concentrated. Some of their output is released at other points, however. Question marks indicate uncertainties or the possibility that other tissues may be involved as intermediaries. (Welsh, in Waterman, The Physiology of Crustacea, *2.*)

must be molted periodically. Before the new cuticle becomes too hard, the size of the body increases rapidly, mostly due to uptake of water or air, depending on whether the arthropod is aquatic or terrestrial. After this, there is true growth; that is, tissue increases in volume and dry weight to fill up the space available for it. Crustacean molting can be explained here on the basis of what happens in a crab (Figure 17.4). Not all crustaceans have such heavily calcified exoskeletons as some crabs do, but the basic process of molting is similar in most groups that have been studied.

As a crab prepares to molt, the outer layers of a new cuticle—not yet stiffened by calcium salts—are secreted beneath the old cuticle. Some of the soluble calcium derived from dissolution of salts in the old cuticle may be stored in the blood, but in most marine crustaceans the calcium needed for hardening the new cuticle comes from seawater. After the old cuticle splits along specific lines of weakness, the crab emerges. While its new cuticle is soft, the animal remains nearly inactive, but it takes up considerable water and becomes larger than it had been. Then the calcium drawn from seawater or blood is deposited in the cuticle, which reaches the "paper-shell" stage. Feeding is now resumed and much of the water in the body is replaced by tissue. The crab may stay in the intermolt condition for several weeks or longer, feeding actively and building up its organic reserves. Then it prepares for another molt.

The molting cycle is controlled by hormones (Figure 17.5). In crabs and other decapod crustaceans, the Y-organs, located in the head, produce a hormone that promotes the production of new cuticle and separation of the old cuticle from the new. The X-organs, situated in the eyestalks, secrete a hormone that inhibits molting. In each eyestalk there is also a "sinus gland." This is where the axons of neurosecretory cells of the X-organ, brain, and some other parts of the nervous system deliver their hormones and where the hormones may be stored. As long as the sinus gland releases enough of the secretion of the X-organ to inhibit the activity of the Y-organs, molting will be suppressed. The molt cycle is closely linked with the storage of organic reserves and sometimes with attainment of sexual maturity, and these aspects of the biology of crustaceans are influenced by secretions of a number of other glands or neurosecretory structures.

Coloration

Coloration in crustaceans is not always due to pigments deposited in the cuticle itself. If the cuticle is thin and transparent, carotenoids and other pigments within the blood and tissues may show through. There may also be specialized pigment-bearing cells, called *chromatophores,* beneath the cuticle and epidermis. Chromatophores are especially characteristic of certain shrimps. They are large cells with radiating processes, and are capable of rearranging their pigment granules in such a way that the animal changes color. When the pigment in a particular chromatophore is dispersed, it is more conspicuous than when it is clumped together. By dispersing its black pigment, therefore, a shrimp can match a dark background; by concentrating this same pigment into tiny specks, it can match a light background. Granules of a reflective white pigment may be dispersed or concentrated in the same way.

If there are several kinds of chromatophores, each kind with pigment of a different color, the responses of the crustacean with respect to lightness, darkness, and color of the background may be rather remarkable. If a chromatophore contains pigments of two colors, granules of one color may be dispersed while those of the other color are concentrated. Some shrimps, in fact, have chromatophores with pigments of three or four colors. If an animal is kept on the same background and given the same amount of light for a long period, certain pigments, or even the chromatophores that contain these pigments, may largely disappear. This is a morphological color change, distinct from a physiological change, which can be reversed rapidly.

Color changes that involve chromatophores are under the control of hormones. It is well known that in shrimps, crabs, and other decapod crustaceans, certain cell bodies of the optic ganglia in each eyestalk secrete neurohormones in response to what the eyes see. The hormones are carried along the axons of the nerve cells to the sinus glands, from which they are released into the blood and tissue fluids. Removal of the eyestalks has different effects on different decapod crustaceans, and there are many possibilities owing to the fact that some species have more pigments or more kinds of chromatophores than others do. Sometimes it causes the general body color to be lighter, sometimes darker. The effect of removal of eyestalks can be reversed by injection of sinus-gland extract. The sinus-gland extract obtained from one species may work on another species, even if they are not closely related. Sinus-gland hormones affect not only body color but also the distribution of pigments that shield the many separate functional units in compound eyes.

Reproduction and Development

Reproductive System and Sperm Transfer

The gonads and gonoducts of crustaceans, like those of arthropods in general, are coelomic derivatives. Basically, the gonads are paired, but they may become fused along the midline. Even then, the right and left gonoducts usually remain separate. As a rule, the gonads are located in the thorax and the genital pores are on certain of the more posterior thoracic segments, but there are exceptions.

The sperm have flagellar tails and exhibit flagellar motility in only a few groups: Branchiopoda, Cirripedia, Mystacocarida, and Ostracoda. In most crustaceans the sperm are nonmotile, and even if they have tails, these are not of flagellar origin. Tailless sperm are more or less routinely characteristic of the Decapoda, which includes crabs, lobsters, and shrimps. In this group the sperm are often starlike, with a number of processes pervaded by slender extensions of the nucleus.

In certain crustaceans, such as barnacles, there is an outgrowth of the body wall that functions as a penis, and in Ostracoda there are two ''hemipenes'' for copulation. It is more common, however, for sperm transfer to be effected by specialized appendages, usually on the anterior part of the abdomen. Males of many crustaceans package their sperm in **spermatophores** and deposit these on the genital pores of females. When spermatophores are involved, fertilization is likely to be external. This is the case, for instance, in freshwater crayfishes.

Embryology

The embryology of Crustacea is so complicated and so varied that the account given here can hit only the high spots. Cleavage and subsequent development are greatly influenced by the amount of yolk within the egg. When there is little yolk, early cleavage may be total, and sometimes a hollow blastula is formed. In general, however, the eggs of crustaceans are yolky, and the nutritive material occupies much of the interior, especially the central portion. When this is the case, cleavage is usually superficial and results in the formation of a jacket of cells around the yolk mass (Figure 17.6, A and B). Within any one group of crustaceans there may be some whose cleavage is total, some whose

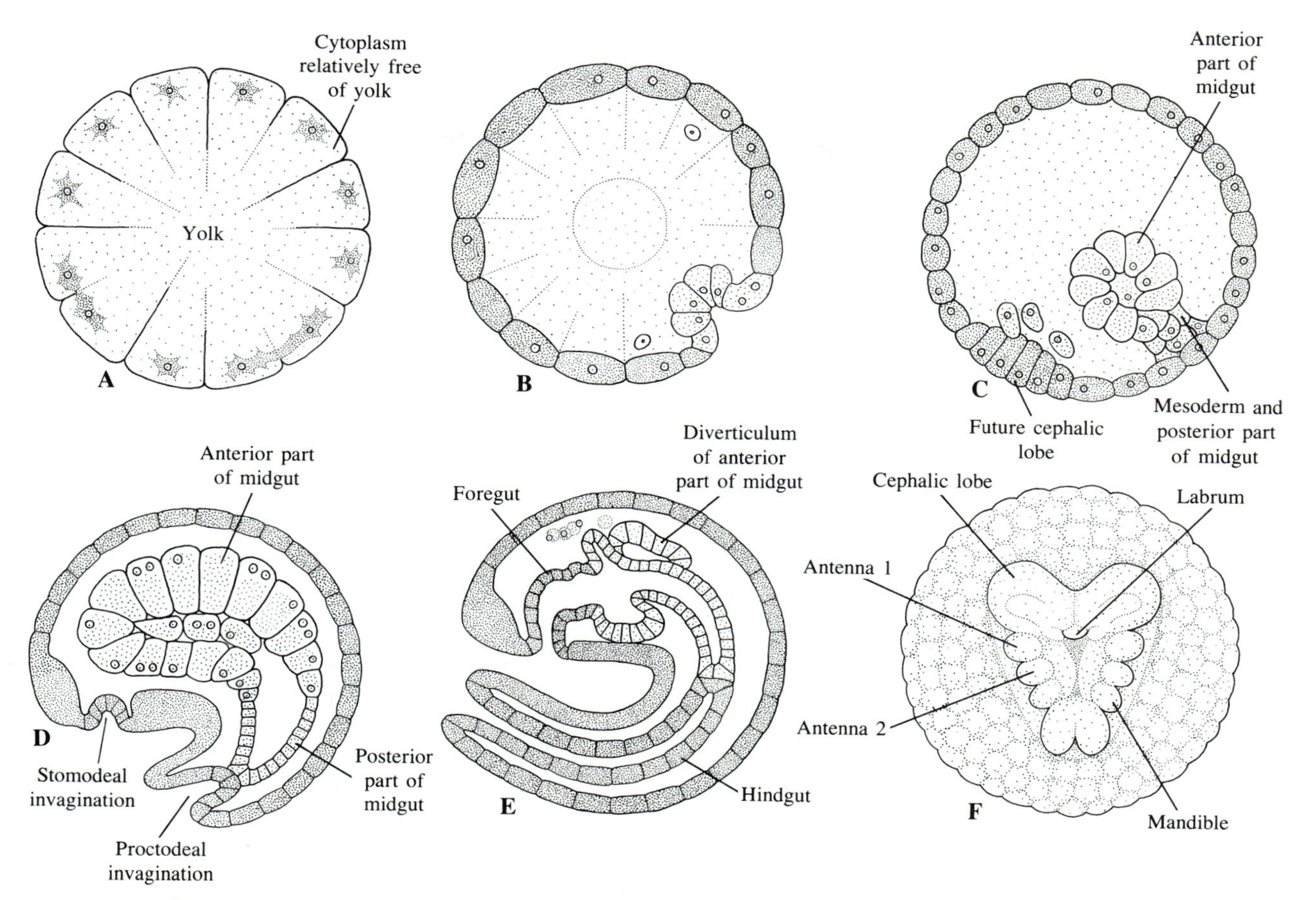

FIGURE 17.6 Selected stages in the development of certain Crustacea. Not all Crustacea adhere exactly to this pattern; some, for instance, have total cleavage. In general, however, the cells from which the endoderm and mesoderm develop are segregated from ectoderm during gastrulation, and the foregut and hindgut are formed by the stomodeal and proctodeal invaginations of ectoderm. A–E. *Galathea* (class Malacostraca, order Decapoda, infraorder Anomura). (Fioroni, Zeitschrift für Morphologie der Tiere, *67*.) A. Early cleavage (superficial because of the large amount of yolk in the egg). B. Early stage of gastrulation. C. Later stage of gastrulation. D. Stomodeal and proctodeal invaginations beginning to form. E. Formation of gut essentially complete; diverticula of anterior part of midgut develop into the digestive gland. The embryo shows the marked flexion characteristic of many Crustacea. Mesoderm is not shown; at first ventral to the gut, it spreads laterally and upward. F. The germinal band of the crab *Macropodia* (class Malacostraca, order Decapoda, infraorder Brachyura), showing the primordia of the labrum and the first three pairs of appendages. Although crabs do not release a nauplius larva (Figure 17.7), the appendages evident in the stage shown correspond to those present in a nauplius. (Lang and Fioroni, Zoologische Jahrbücher, Abteilung für Anatomie und Ontogenie, *88*.)

cleavage is superficial, and others in which cleavage is total at first, then superficial.

Even when cleavage is total, it is not distinctly spiral, as in annelids and molluscs; the quartets of cells that one sees in the early embryos of many annelids and molluscs do not materialize. The segregation of endoderm from ectoderm may be achieved by invagination (Figure 17.6, B and C) or epiboly. Mesoderm originates from cells in the region of the blastopore, or at least from cells that push inward or that are overgrown during epiboly, and in this respect crustaceans do resemble annelids and molluscs.

One important point about crustacean development needs to be stressed: the definitive embryo is generally derived from a restricted area on one side of the egg called the *germinal band* (Figure 17.6, F). The ventral

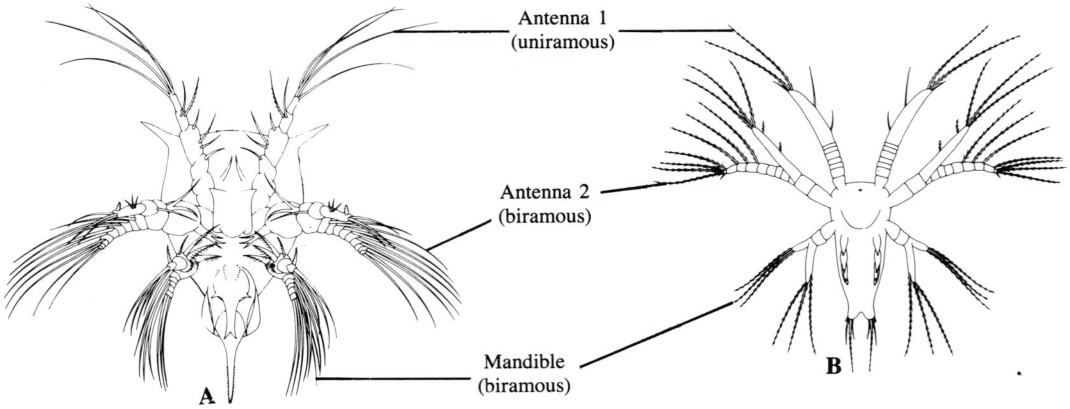

FIGURE 17.7 Nauplius larvae. A. Barnacle. B. Penaeid shrimp. (Sanders, Memoirs of the Connecticut Academy of Arts and Sciences, *15*.)

surface of the embryo faces outward. The cylindrical form of the body is not established until mesoderm grows upward on both sides and meets dorsally. The endodermal cells gradually incorporate the yolk that is around them, and the gut is completed when stomodeal and proctodeal invaginations of ectoderm, which become the foregut and hindgut, join the endoderm that forms the midgut (Figure 17.6, D and E). The nervous system and lining of the statocysts, if these are present, are also of ectodermal origin. Mesoderm forms the musculature, connective tissue, heart, and blood vessels. It also gives rise to the gonads, portions of the gonoducts and their diverticula, and the coelomoducts that serve as organs of excretion or osmoregulation, or both. The coelomoducts are the only clear remnants of coelomic spaces, although the gonads and gonoducts are certainly coelomic derivatives.

DIRECT AND INDIRECT DEVELOPMENT

Relatively few crustaceans release their eggs as they lay them. Most incubate the eggs, at least for a time, in a brood pouch or on certain appendages. The stage at which hatching takes place varies considerably. In some crustaceans, the young do not hatch until they resemble adults. This kind of gradual, **direct development** is to be expected among terrestrial or semiterrestrial species, and also among those that live in swift streams, where swimming larvae might be swept away. Nevertheless, direct development is characteristic of several groups of crustaceans that are strictly marine, or that are at least better represented in the sea than anywhere else, including amphipods, isopods, and their close allies.

If there is a series of larval stages, development is said to be **indirect.** The youngest free-larval stage found in crustaceans is the **nauplius** (Figure 17.7), typical of the life histories of ostracodes, barnacles, copepods, penaeid shrimps, and certain other groups. It is not segmented, but it has appendages that correspond to the first antennae, second antennae, and mandibles of later stages. The first antennae are uniramous; the second antennae and mandibles are biramous. There is usually a median eye that consists of a few pigmented cups containing photoreceptors, and a pair of coelomoducts that open at the bases of the second antennae. All three pairs of appendages are used for swimming or crawling, but if the gut is functional, the second antennae and mandibles, generally provided with long setae, are also used for collecting microscopic food. The **labrum,** a lip in front of the mouth, helps the appendages direct food into the opening. In nauplii that contain much yolk, the onset of feeding is delayed until a later stage.

After the nauplius molts, there follows a series of **metanauplius** stages during which the last two segments of the head, then the segments of the thorax and abdomen, are added in front of the telson. As a rule, the appearance of new segments is from anterior to posterior. Any appendages that materialize on the new segments remain rudimentary for a while. In general, the metanaupliar sequence ends when the first and second maxillae differentiate. In some crustaceans, however, the metanaupliar sequence gives way to a series of stages that increasingly resemble adults. This is the case, for instance, in certain of the branchiopods, including fairy shrimps and tadpole shrimps.

Not all crustaceans that have free-swimming larvae go through nauplius and metanauplius stages. In many groups, including crabs, hermit crabs, and most shrimps, the larva that hatches from an egg already has a carapace, three pairs of functional mouthparts in ad-

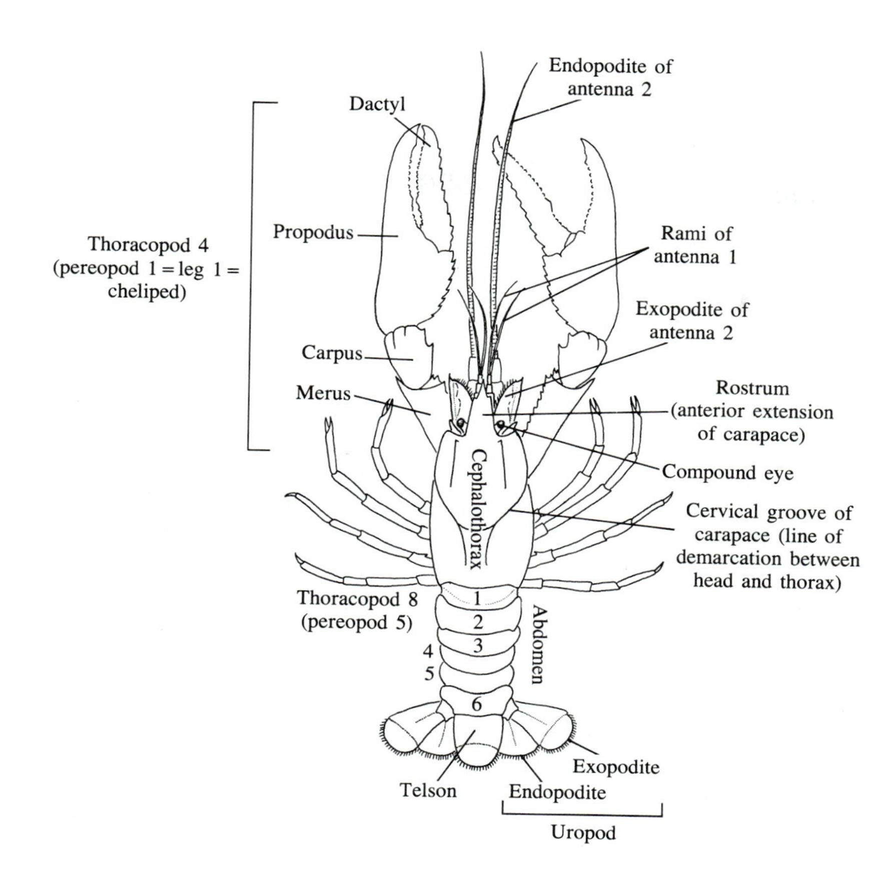

FIGURE 17.8 Crayfish (class Malacostraca, order Decapoda, infraorder Astacidea), dorsal view.

dition to first and second antennae, a few thoracic segments provided with appendages for locomotion, and a well-developed abdomen. Even the compound eyes are in an advanced stage of differentiation.

Regardless of whether or not there are naupliar and metanaupliar stages, there are many patterns of later development. It seems best to deal with these in connection with the groups of which they are characteristic.

THE CRAYFISH AS A REPRESENTATIVE CRUSTACEAN

Freshwater crayfishes are commonly used for explaining the general structure of crustaceans. They are large enough to make it easy to study the tagmata, appendages, and various other external and internal features. Much of what can be learned from a crayfish can be applied to the study of other crustaceans, especially those of the class Malacostraca, which includes, besides crayfishes, crabs, shrimps, isopods, and amphipods. It is important to understand, however, that crayfishes are advanced crustaceans, and that they show specializations not found in most other groups.

EXTERNAL ANATOMY AND SPECIALIZATIONS OF APPENDAGES

Cephalothorax

The body of a crayfish is divided into two tagmata: a **cephalothorax** and an **abdomen** (Figures 17.8, 17.9). The cephalothorax consists of the head and eight thoracic segments; the limits of some of the thoracic segments are visible on the ventral side. Dorsally and laterally, however, the cephalothorax is covered by a continuous **carapace.** The part of the carapace behind the *cervical groove,* which separates the head from the thorax, hangs down as a free fold, enclosing a cavity occupied by the gills. The arrangement of these is discussed presently. A pointed, anteriorly directed projection of the carapace, the *rostrum,* forms an overhang above the compound eyes.

Over much of the body, including the appendages, the cuticle is noticeably thickened and hardened by deposits of calcium salts. The thickening and hardening is most apparent in the carapace, terga of the abdomen, and legs. There is more to the exoskeleton than what can be seen on the outside. In the ventral part of the thoracic region, there is a complex assemblage of

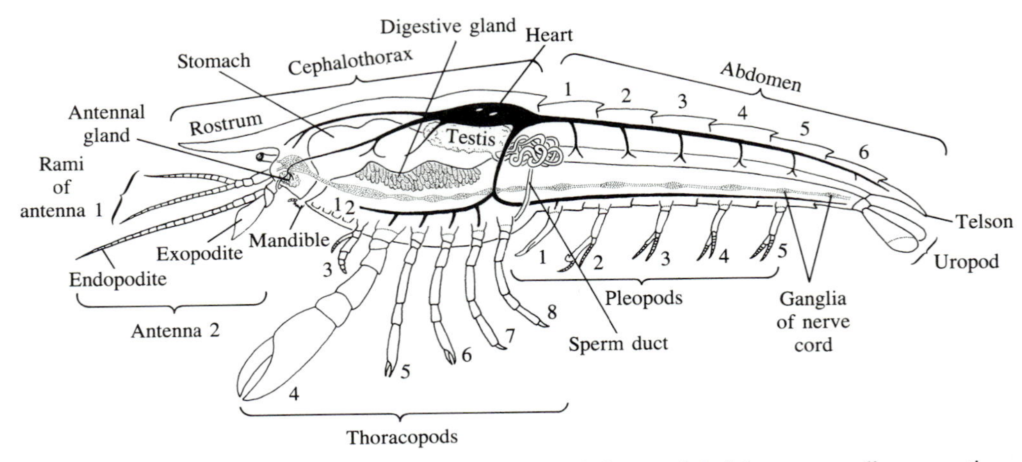

FIGURE 17.9 Longitudinal section of a crustacean of the crayfish–lobster type; diagrammatic. Maxillae 1 and 2, whose bases are shown behind the mandibles, are not labeled. Of the thoracopods, numbered 1–8, the first three pairs are maxillipeds (only the bases of 1 and 2 are shown); the rest are pereopods.

apodemes to which muscles that operate the limbs are attached. Although these ingrowths may go deep into the thorax and are fused at certain points to form what looks like an internal skeleton, they are in fact folds of the exoskeleton.

The **first antennae** have two multiarticulate branches, but these originate from the third article, rather than from the second article, so they are not exactly comparable to the **exopodite** and **endopodite** of other biramous appendages. The first antennae of crustaceans, originating from the cephalic lobe of the embryo, are not part of the series of appendages that arise from true segments behind the mouth. Although located in a preoral position, the **second antennae** do belong to the series. They have slender endopodites, but their exopodites are specialized as broad, scalelike structures. Both of the foregoing appendages have touch and olfactory receptors; the first antennae are also concerned with equilibrium, for each has a statocyst opening to the outside of the first article. (The way a statocyst functions is discussed in Chapter 15.)

The coxal portions of the **mandibles** (Figure 17.10, A) are heavily calcified jaws. The slender palp attached to each jaw consists of the basis plus two articles of the endopodite; there is no exopodite. The small **first maxillae** (Figure 17.10, B) also lack exopodites. The three flattened, setose portions of these appendages are the basis and two articles of the endopodite. Each **second maxilla** (Figure 17.10, C) has an exopodite that is fused with a coxal epipodite to form a leaflike structure called the **scaphognathite,** or **gill bailer.** The rapid beating of the bailers drives water forward out of the respiratory chambers beneath the carapace on both sides of the body, bringing a continual fresh supply of water into the chambers from below. The coxa and basis of the second maxillae are not clearly separated, but each is flattened and drawn out into two medially directed endites. The endopodites of the second maxillae are not divided into articles.

The first three of the eight thoracopods are **maxillipeds,** and are concerned with food handling. The third maxilliped (Figure 17.10, F) is perhaps the most instructive single appendage on a crayfish, because it shows clearly all of the major parts of a typical stenopodial limb. Both endopodite and exopodite are well developed, and there is an **epipodite** on the coxa. A part of this epipodite is modified to form a gill. The second maxilliped (Figure 17.10, E) is basically similar to the third, but smaller. The first maxilliped (Figure 17.10, D) is much like a maxilla because of its flattened coxa and basis. It has a short endopodite, an exopodite, and a broad epipodite; the latter, however, has no gill.

Thoracopods 4 to 8 are the **pereopods,** or legs. The first of these are stout **chelipeds,** whose pincers (chelae) are used for defense, capture of prey, and tearing and manipulating food. The movable finger, or dactyl, closes against a nonmovable finger, which is an extension of the propodus. The remaining legs are more slender, but the second and third pairs (Figure 17.10, G) have small pincers. In crayfishes, the legs have no exopodites, but, like the maxillipeds, they do have epipodites. Portions of the epipodites are specialized to serve as gills.

If a crayfish leg is seized roughly, the animal may sever the appendage from the body along a preordained line, or "breaking joint," at the base of the ischium. The autotomy is accomplished by contraction of a certain muscle and is a useful adaptation for helping a

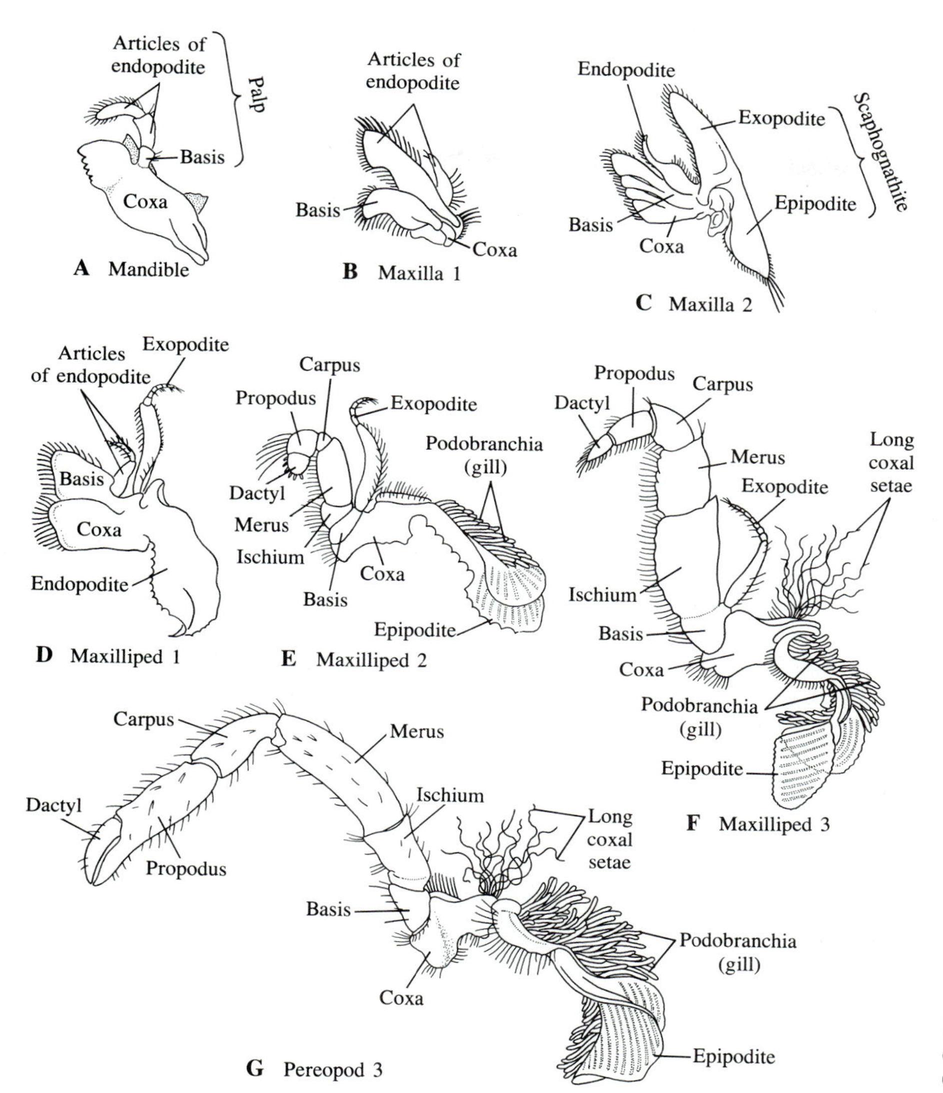

FIGURE 17.10 Some appendages of a crayfish. (After Huxley, The Crayfish.)

crayfish escape from a predator that has gotten hold of a leg. The blood clots on the exposed surface, and a small fold of tissue closes up the wound until healing and regeneration get underway. In certain anomuran crabs, notably those of the group called porcellanids (Figure 17.66, B), the capacity for autotomy is highly developed. If a porcellanid's leg is immobilized, the animal will often snap it off by reflex action and walk away. Some brachyuran crabs are also able to autotomize limbs when they have to.

As a rule, decapod legs—and other appendages, too—regenerate after they have been autotomized or removed in some other way. The capacity for regeneration even extends to the compound eyes. If the entire stalk of an eye is removed, however, what grows back may look more like an antenna than an eye.

Abdomen

The six segments of the abdomen are distinct (Figures 17.8, 17.9), and the exoskeleton of each is differentiated into a hard dorsal tergum and a somewhat softer ventral sternum (Figure 17.11). The abdomen ends in a **telson,** on which the **anus** is situated.

The extent to which the abdominal appendages are developed varies according to the genus and sex. In females, the second to fifth pairs are biramous **pleopods** used to some extent for respiration and for holding eggs while these are being incubated. The first pair is vestig-

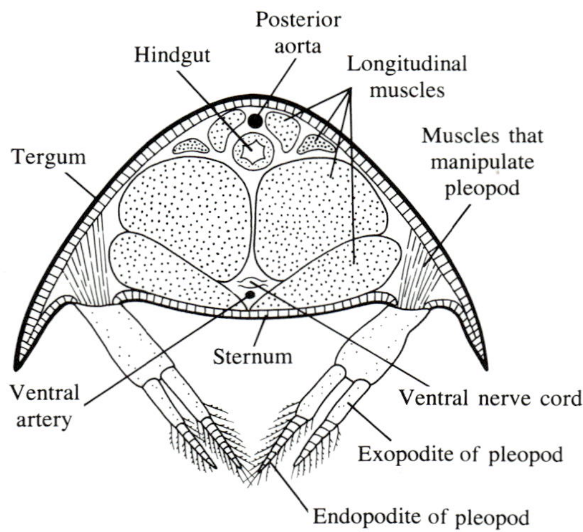

FIGURE 17.11 Transverse section through the abdomen of a crayfish, diagrammatic and simplified.

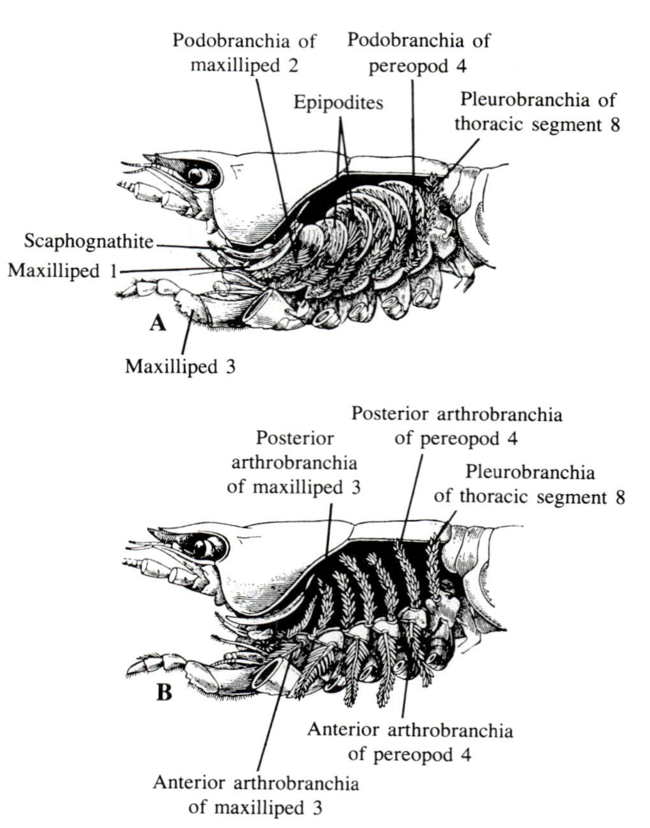

FIGURE 17.12 Gills of a crayfish (*Astacus astacus*). A. All gills intact. B. Podobranchiae removed so that the pleurobranchiae and arthrobranchiae can be seen more clearly. (Huxley, The Crayfish.)

ial or absent. In males (Figure 17.9), the first two pairs (or just the second pair, when the first pair is missing) are modified to function as spermatophore-transferring devices, discussed further below. In both sexes, the last pair of abdominal appendages are called **uropods.** They have flattened endopodites and exopodites, and together with the telson they form the tail fan that a crayfish can flex ventrally to drive itself forcefully backward.

GILLS AND RESPIRATION

The gills (Figure 17.12) are located within the lateral respiratory chambers enclosed by the carapace. They are thin outpocketings of the body wall and have a delicate cuticle. Blood is carried out to the gill by way of one channel in the main stem, then back to the body by a separate channel. In terms of their origin, the gills are of three types: **podobranchiae,** which grow out of the epipodites of the coxae of the thoracic appendages; **arthrobranchiae,** which arise from the membranes joining the coxae to the trunk; and **pleurobranchiae,** which arise from the wall of the thorax itself. On certain segments, there are two arthrobranchiae, one behind the other, on both sides. This suggests that the basic plan for Crustacea is to have four pairs of gills on each thoracic segment. In no crayfishes, however, are all four pairs fully developed on any one segment; some segments, in fact, have only one pair, or none at all.

DIGESTIVE SYSTEM

The **mouth** leads into a short **esophagus,** which is followed by the two chambers of the so-called **stomach** (Figure 17.13, A). The esophagus and stomach together make up the **foregut,** which is lined by a cuticle that is shed each time a crayfish molts. The cuticle of the first portion of the stomach, called the ''cardiac'' chamber, has special elaborations for physically breaking down large pieces of food. There are two broad ventrolateral thickenings, each with setae and a large tooth, that can be brought together to squash bits of food. In addition, there are three unpaired cuticular plates (''ossicles'') arching over the dorsal part of the cardiac chamber. On both sides of the cardiac chamber, the plate given the number 2 in Figure 17.13 has a row of cusps. The small plate given the number 3 has a large median tooth. The teeth collectively form the **gastric mill.** When certain muscles on the wall of the stomach contract, the cardiac chamber becomes longer and narrower. This brings the lateral cusps together and swings the dorsal tooth forward to interact with them to grind up food. Juices from the digestive glands, which are diverticula of the short midgut, are drawn forward to the cardiac portion of the stomach when this chamber is dilated by certain mus-

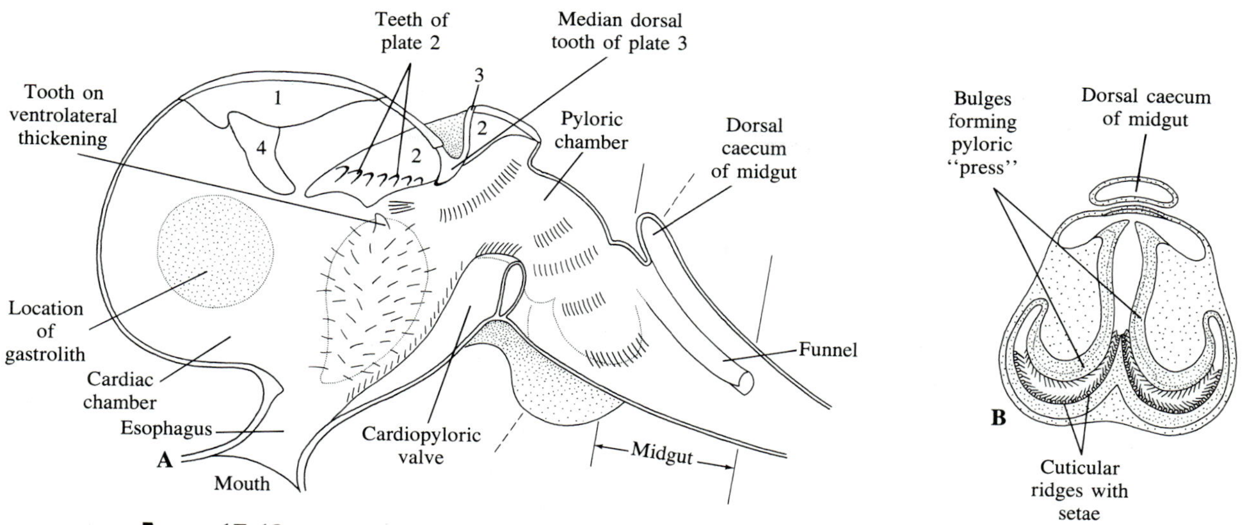

FIGURE 17.13 A. The foregut and midgut of a crayfish, as seen in a midsagittal section; diagrammatic and simplified. The cardiopyloric valve separates the more anterior cardiac chamber of the stomach from the pyloric chamber. The hardened cuticular plates are numbered; plates 1–3 arch over the dorsal midline; plate 4 is a member of a pair. B. Transverse section through the gut at the level indicated in A by a dashed line.

cles. Thus some enzymatic digestion is already in progress during the phase of physical breakdown.

Separating the cardiac and pyloric chambers is the cardiopyloric valve. It allows the "mash" to go through, but restrains coarse residues, such as bits of shell. These must be regurgitated. The organization of the pyloric chamber is complicated, and not all of its intricacies are described here. This part of the foregut is primarily concerned with sorting the mash. Its lateral walls bulge inward, forming a so-called *press* (Figure 17.13, B). The two sides of the press are studded with setae; their function is to squeeze down on the mash, pushing indigestible residues backward into the midgut and forcing fluid and finely divided material into the elaborate filters on the floor of the pyloric chamber. The filters consist of cuticular ridges that run more or less lengthwise. Setae originating on the ridges lie nearly crosswise over the furrows between the ridges, making a kind of grillwork. Only liquids and fine particles can pass the filters.

The ducts of the right and left **digestive glands** open into the anteriormost part of the **midgut,** which begins just behind the pyloric filter. The many fine tubules of each digestive gland are surrounded by networks of contractile fibers, so fluid is alternately drawn into them and expelled from them. As cells that line the tubules are shed and break down, releasing their enzymes, they are replaced from a stock of immature cells. The digestive glands are concerned not only with digestion but also with absorption and storage of nutrient material.

A couple of other details deserve mention. One is the funnel, or pyloric valve; the other is the dorsal caecum of the midgut. The funnel is a lightly cuticularized, upside-down trough that projects backward from the pyloric chamber and runs right through the midgut to the hindgut. By way of this trough, residues destined to become fecal material can be delivered directly to the hindgut. The dorsal caecum is immediately above the valve. It is glandular, but the function of its secretion is uncertain. Perhaps it assists in consolidating fecal material within the hindgut.

The **hindgut,** like the foregut, is an ectodermal derivative and is lined by a thin cuticle. It is in this part of the gut that relatively fine undigested residues are compacted into feces. The anus is situated on the telson.

EXCRETION, OSMOREGULATION, AND CIRCULATION

Ammonia is the most important of the nitrogenous wastes, but some urea and amines are also produced. The wastes diffuse away from the animal across the surfaces of the gills and other parts of the body where the cuticle is thin, and perhaps across the walls of the gut. The **antennal glands,** or **green glands,** which open by

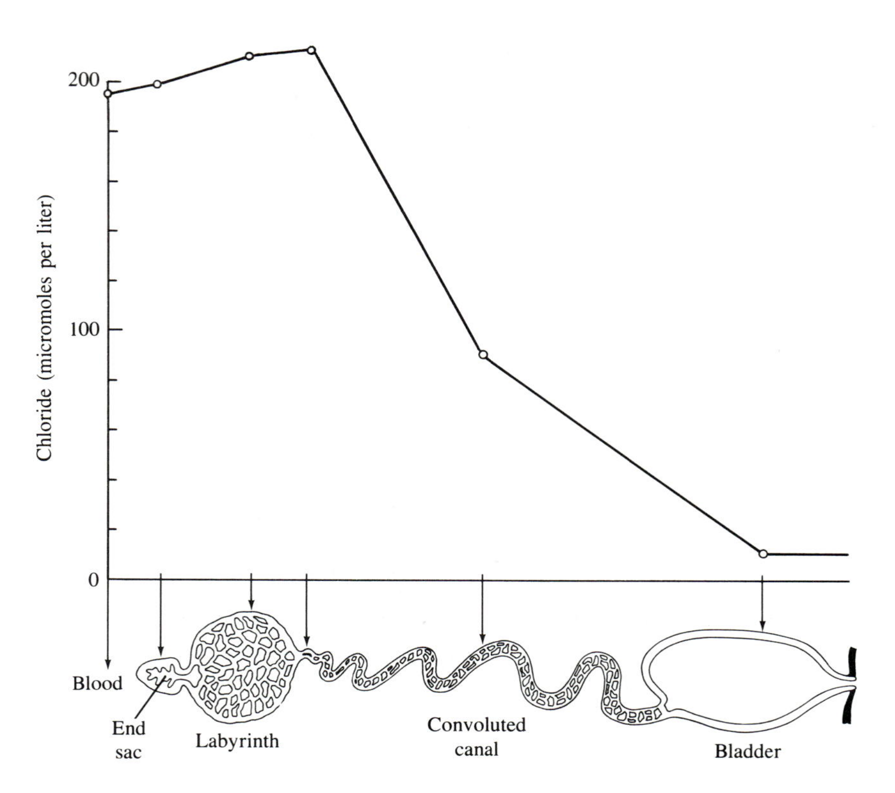

FIGURE 17.14 Diagram of a green gland of a crayfish, with a graph of the chloride content of the excretory fluid. In *Astacus astacus,* used in the physiological studies, the amount of urine produced in 24 hours is just under 4% of the live weight of the animal. The chloride content in the end sac is about equal to that in the blood. It goes up slightly in the labyrinth, where there is probably some secretion of chloride, then drops markedly in the convoluted canal, where resorption of salts takes place. (After Parry, in Waterman, The Physiology of Crustacea, *1*.)

pores on the coxal articles of the second antennae (Figure 17.9), are concerned primarily with osmoregulation. They serve to eliminate water, thus preventing dilution of the blood, which is slightly more saline than the freshwater environment. Each antennal gland (Figure 17.14) consists of several distinct portions: an end sac that is divided up by internal partitions; a labyrinth whose walls are so much folded that it is almost spongy; a convoluted canal; and a bladder drained by a short duct to the outside. The end sac is a remnant of an embryonic coelom; the rest of the gland is a coelomoduct.

The **heart** (Figure 17.9), located in the **pericardial sinus,** a hemocoelic space in the dorsal part of the thorax, has a moderate amount of muscle. When the heart is dilated, blood enters it by three pairs of valved openings called *ostia*. The valves prevent backflow when the heart contracts to pump blood through the anterior and posterior aortas and other arteries. These vessels also have valves. The fine ultimate branches of the arteries are open, allowing blood to enter hemocoelic spaces that bathe the tissues. From a particularly prominent sinus in the ventral part of the thorax, blood moves into the gills, where much of the exchange of oxygen and carbon dioxide takes place. From the gills, blood returns to sinuses in the lateral regions of the thorax. It then moves to the pericardial sinus and finally into the heart itself. The respiratory pigment is **hemocyanin.**

NERVOUS SYSTEM AND SENSE ORGANS

Brain and Ventral Nerve Cord

Four pairs of ganglia are incorporated into the brain. Two pairs, derived from the presegmental portion of the head and from the first segment, form the **protocerebrum,** which innervates the eyes. The **deutocerebrum** consists of the ganglia of the second segment and innervates the first antennae; the **tritocerebrum** consists of the ganglia of the third segment and innervates the second antennae. Running around the esophagus are connectives that join the brain to the ventral cord (Figure 17.15), the first part of which is formed by a complex of ganglia that innervate the mouthparts (these ganglia properly belong to the head) and the maxillipeds (which are thoracic appendages).

The **ventral nerve cord** is a double structure in the sense that its ganglia and connectives are paired. Over much of the length of the cord, however, these components are fused to the extent that their duality is not apparent. On each side of the cord is a giant fiber that runs without interruption from the circumesophageal connective to the posterior part of the abdomen. The

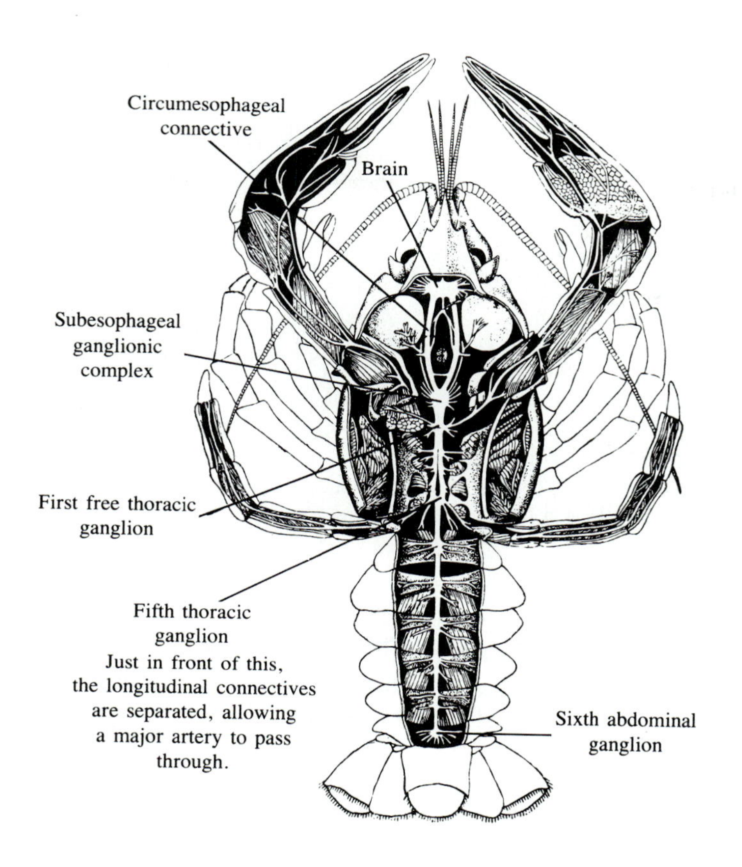

FIGURE 17.15 Nervous system of a crayfish. (Ivanov *et al.*, Major Practicum of Invertebrate Zoology.)

giant fibers transmit impulses efficiently (about 15 m per second) and are important in bringing about the powerful ventral flexion of the abdomen that enables a crayfish to swim rapidly backward to escape danger.

Sense Organs

The **compound eyes,** partly hidden by the rostrum, are stalked and movable to some extent. The cornea that covers each eye consists of transparent cuticle. It is divided into two or three thousand nearly square facets, each of which is a lens for one of the visual units, or ommatidia, of the eye. (The general structure of arthropod ommatidia is discussed in Chapter 15.)

Opening on the upper side of the coxa of each antennule is a **statocyst.** The cuticle that lines this is elaborated into many fine sensory setae to which sand grains are attached by mucus. If the crayfish is tilted or turned over, the sand grains respond to gravity and weigh on the setae differently than if the animal has its ventral side down. Impulses traveling along neurons that run from the statocyst hairs to the brain inform the brain of a change in orientation, and the animal rights itself. At each molt, the lining of the statocyst, and also the sand grains, are sloughed off. If a crayfish that has just molted is placed in filtered water so that it has no access to sand grains, its equilibrium will not be normal. If fine particles of iron are introduced into the aquarium, some of these will get into the statocysts. The animal's orientation can then be manipulated by a magnet: if a magnet is held above the crayfish, the filings will be drawn closer to the dorsal side of the statocyst, as they would be if the animal were lying on its back. The crayfish responds by turning over, so that its ventral surface is nearest the magnet.

Many of the cuticular setae on the external surfaces of a crayfish are tactile; others are chemosensory. The chemosensory setae are especially concentrated on the first antennae and mouthparts, and on the tips of the second antennae and chelae.

REPRODUCTION AND DEVELOPMENT

The gonads of the male and female (Figure 17.16), lying in the thorax between the gut and the hemocoelic space around the heart, are basically paired structures, but they are fused along the midline for much of their length, so that it appears that there is a single testis or ovary. There are two sperm ducts or oviducts. In the male, the genital pores are on the coxae of the fifth (last) legs; in the female, they are on the coxae of the

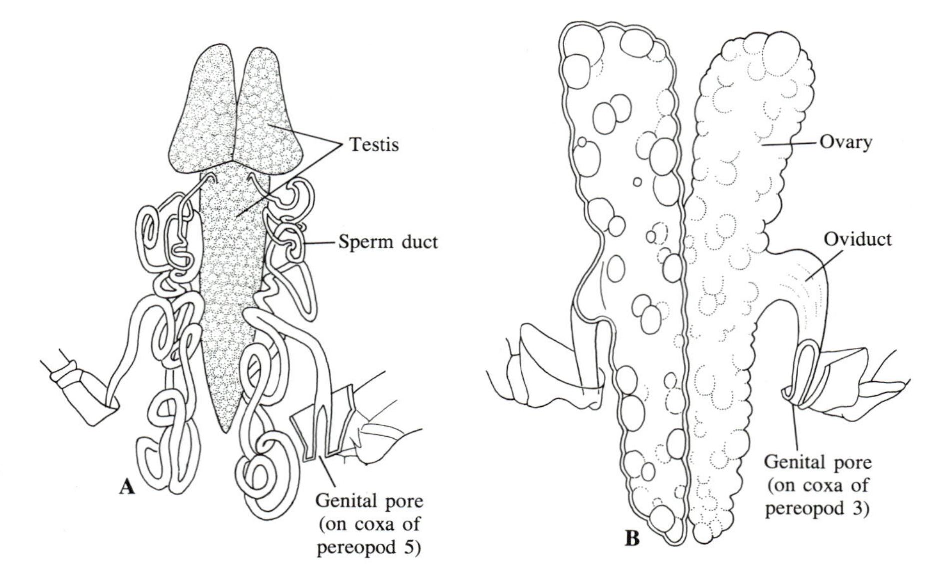

FIGURE 17.16 Reproductive systems of a crayfish. A. Male. B. Female. In both sexes, the right and left gonads are separate anteriorly, but posteriorly they are fused.

third legs. The sperm, without flagellar tails, have many radiating processes and are packaged into **spermatophores** while they are in the sperm ducts.

In males, the first two pairs of pleopods (or just the second pair, in genera that lack first pleopods) are specialized for transferring spermatophores. In females, the first pleopods are vestigial or absent, and the second pleopods look much like the others. In the mating ritual, the male, after loading his specialized pleopods with spermatophores coming from the genital pores on the coxae of his last thoracic legs, turns the female over, grasps some of her legs with his chelae, and holds her abdomen, which is turned upward, with his telson. The female's genital pores are on the coxae of her third legs, and not far behind each pore is a pit that serves as a seminal receptacle. The male applies his specialized pleopods to the pits and leaves the spermatophores, which are mixed with mucus that enables them to adhere.

Within a few days or weeks, the female uses her legs to clean her pleopods in preparation for egg-laying. She then lies on her back, discharges a sticky secretion from glands on the pleopods, and lays her eggs, which are about 2 mm in diameter and very yolky. They are fertilized by sperm that escape from the spermatophores, and become cemented to the pleopods. The female then rights herself, and she may remain hidden for much of the time that the eggs are developing on her abdomen. Slow movements of the pleopods circulate water over the embryos. The young hatch several weeks later as juvenile crayfishes. They grip the pleopods or empty egg shells until they have molted once, then drop off to start life on their own.

GENERAL REFERENCES ON CRUSTACEA

Bliss, D. E. (editor), 1982–1985. The Biology of Crustacea. 10 volumes. Academic Press, New York and London.

Lockwood, A. P. M., 1968. Aspects of the Physiology of Crustacea. W. H. Freeman, San Francisco; Oliver and Boyd, Edinburgh and London.

McLaughlin, P. A., 1980. Comparative Morphology of Recent Crustacea. W. H. Freeman, San Francisco.

Rebach, S., and Dunham, D. W. (editors), 1983. Studies in Adaptation. The Behavior of Higher Crustacea. John Wiley & Sons, New York.

Schmitt, W. L., 1965. Crustaceans. University of Michigan Press, Ann Arbor.

Schram, F. R. (editor), 1983. Crustacean Phylogeny. A. A. Balkema, Rotterdam.

Schram, F. R., 1986. Crustacea. Oxford University Press, Oxford and New York.

Starobogatov, YA. I., 1988. Systematics of Crustacea. Journal of Crustacean Biology, *8*:300–311.

Whittington, H. B., and Rolfe, E. D. I. (editors), 1963. Phylogeny and Evolution of Crustacea. Museum of Comparative Zoology, Harvard University, Special Publication.

CLASSIFICATION OF CRUSTACEA

In classifying crustaceans, much emphasis is given to characteristics that are thought to be primitive and those that are believed to be advanced. Within many of the classes, orders, and subsidiary taxa, however, there are mixtures of primitive and advanced traits, as well as some members that are on the whole more primitive or more advanced than others. This situation, together with the great diversity and adaptive radiation of the Crustacea, makes classification difficult and unstable.

The scheme in this book conforms closely to a system that is in general use. It has many thoughtful and well-informed critics who revise the sequence of groups and raise or lower the ranks of certain taxa, hoping to arrive at a system of classification that is as closely as possible based on the phylogeny of crustaceans. Nevertheless, if you understand the scheme given here, and learn the outstanding features of the major groups, you should have little trouble adjusting to other systems and understanding how they differ.

CLASS CEPHALOCARIDA

Cephalocaridans were not described until 1955, and only a few species are known. They are marine crustaceans, all less than 4 mm long, that live in the surface layer of sandy and muddy deposits. Some are found in shallow water, others at considerable depths. The group is widely distributed, and certain species are abundant in favorable habitats.

The body of a cephalocaridan (Figure 17.17) is rather slender, and is divided into head, thorax, and abdomen. There is no carapace. The head has the usual series of crustacean appendages: first and second antennae, mandibles, and first and second maxillae. There are compound eyes on both sides of the head, near the base of the labrum, but they are not evident externally and may not be functional. Their presence was noted only recently.

There are eight thoracic segments, and the first seven have appendages that resemble the second maxillae. The proximal portion of each of these appendages is not clearly divided into a coxa and basis, and its medial side is studded with several toothlike endites. On the part that would seem to correspond to the basis, there are three outgrowths: an articulated endopodite and exopodite, and a nonarticulated epipoditelike structure ("pseudepipodite"). Thus the first seven thoracic appendages and second maxillae may be said to be triramous. The eighth pair of thoracic appendages may be lost or reduced to the point that it has only the exopodite.

The abdomen, with 11 segments and a telson, has no appendages except for reduced egg-carrying structures on the first segment. The telson has a pair of caudal rami, each protracted by one or two long setae.

Cephalocaridans move by action of their head appendages (other than the mandibles) and thoracic appendages. The backward beat is sequential, beginning with the posteriormost pair. If cephalocaridans were oriented dorsal side up, the activity of their limbs would resemble that of fairy shrimps. As detritus is stirred up by the appendages, it is drawn into a midventral channel. The movement of water past the appendages accumulates diatoms and other food on the featherlike setae of the rami. The material is combed from one appendage by the setae of another, and is pushed forward in the midventral channel by action of the endites until it reaches the mouth. To clean its appendages, a cephalocaridan bends its abdomen under its head and thorax, then pulls backward until it is more or less straightened. During the pulling-back process, setae on the last segment of the abdomen unclog the setae on the legs, and the long setae on the caudal rami clean out the midventral channel.

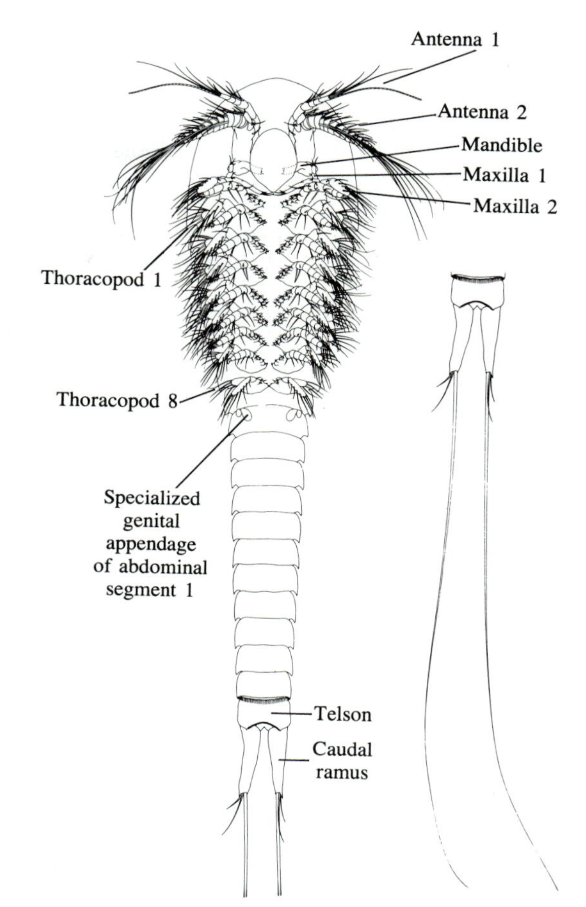

FIGURE 17.17 Class Cephalocarida. *Hutchinsoniella macracantha,* female. (Sanders, Memoirs of the Connecticut Academy of Arts and Sciences, *15.*)

Cephalocaridans are unusual among crustaceans (except for barnacles, certain parasitic isopods, and a few others) in being hermaphroditic. The gonoducts lead to a pair of common genital pores on the sixth thoracic segment. After two eggs are laid, they are held by the reduced appendages on the first segment of the abdomen. Hatching is at the metanaupliar stage. There are numerous molts leading to gradual attainment of the adult form.

The absence of a carapace, the triramous condition of some of the limbs, the way the endites of appendages move food forward along the midventral channel, and the very gradual development of cephalocaridans have been cited as primitive characters. Hermaphroditism, brooding, and hatching at the metanaupliar stage would seem, however, to be advanced characters.

References on Cephalocarida

Burnett, B., 1981. Compound eyes in the cephalocarid crustacean *Hutchinsoniella macracantha*. Journal of Crustacean Biology, *1*:11–15.

Hessler, R. R., 1964. Cephalocarida: comparative skeletomusculature. Memoirs of the Connecticut Academy of Arts and Sciences, *16*:1–97.

McLaughlin, P. A., 1976. A new species of *Lightiella* (Crustacea: Cephalocarida) from the west coast of Florida. Bulletin of Marine Science, *26*:593–599.

Sanders, H., 1963. The Cephalocarida. Functional morphology, larval development, comparative external anatomy. Memoirs of the Connecticut Academy of Arts and Sciences, *15*:1–80.

CLASS BRANCHIOPODA

Branchiopoda means "gill-legged," and it refers to the phyllopodial appendages characteristic of the trunk of some groups within this class. The term *trunk* is generally substituted for *thorax* and *abdomen,* because in branchiopods that have numerous segments behind the head, these are all much alike and their appendages are mostly of the same type. In anostracans and conchostracans, all of the many trunk appendages are phyllopodial; in notostracans, all but the first pair are phyllopodial. The situation is different in cladocerans. They have a much-shortened trunk and only a few pairs of appendages, and sometimes all of these are stenopodial.

Order Anostraca

The order Anostraca (Figure 17.18) includes the crustaceans called fairy shrimps and brine shrimps. These animals are permanently planktonic and swim upside-down. They have no carapace and their bodies are relatively slender. They are thought to be primitive because the segments and limbs of the trunk are all much alike, and because the arrangement of ganglia in the trunk follows a strictly segmental pattern.

Most anostracans are found in freshwater ponds that dry out during the summer. Their eggs are resistant to desiccation and hatch after the ponds fill again in winter. These crustaceans generally mature and reproduce in late winter and early spring.

The head of an anostracan is distinctly demarcated from the trunk. A median naupliar eye persists, but is inconspicuous compared to the large compound eyes (Figure 17.18, A and D). The first antennae are short and slender, and often lack distinct articles; the second antennae are usually reduced in females, but they are large, stout, and divided into two articles in males (Figure 17.18, B, C, and D), which use these appendages for clasping females. In males of some genera, the proximal portions of the right and left second antennae are fused together. The second antennae may also be branched, like staghorns. Mandibles and first maxillae are well developed, but the second maxillae are small, a feature characteristic of branchiopods in general.

The number of trunk segments varies, but there are at least 19, of which the first 11 or more bear phyllopodia. (All North American anostracans bear 20 trunk segments, 11 with appendages and 9 without them.) These appendages have epipodites and setose endites, but are not divided into articles. The only distinct joints, in fact, are those separating the exopodite and endopodite from the basis. The cuticle that covers the appendages is thin, and it is the pressure of blood within hemocoelic spaces that keeps the appendages stiff. The telson, at the hind end of the trunk, has a pair of caudal rami, and the anus lies between them.

The phyllopodia (Figure 17.18, A, B, C, and E) are concerned not only with locomotion, but with collecting food. They are therefore good examples of generalized appendages. Those of each segment act synchronously. The backward stroke is propulsive, and while certain appendages are in this phase of movement, others are making the forward recovery stroke. Every trunk appendage goes through the complete cycle more than 100 times a minute. As a phyllopodium makes its forward stroke, the space between it and the succeeding appendage widens, so a slight suction is created. Between the appendages of the right and left sides is a channel in which water with suspended particles can flow, and the suction produced by the temporary separation of two successive appendages draws in water from the sides. Soon, of course, the appendages come closer together again and the water is forced out, but the constant circulation of water through the setal filters on the endites of the appendages concentrates particles in the midventral groove. A current moving forward in the groove carries the particles to the mouth region. Just how this current is generated is not perfectly clear. Mucus produced by glands on the labrum helps the particles cohere; setae on the first maxillae move the particles directly to the mouth or to the mandibles. Unsuitable material is rejected by movements of the mouthparts or endites of some of the more anterior trunk appendages, or by lifting up the labrum.

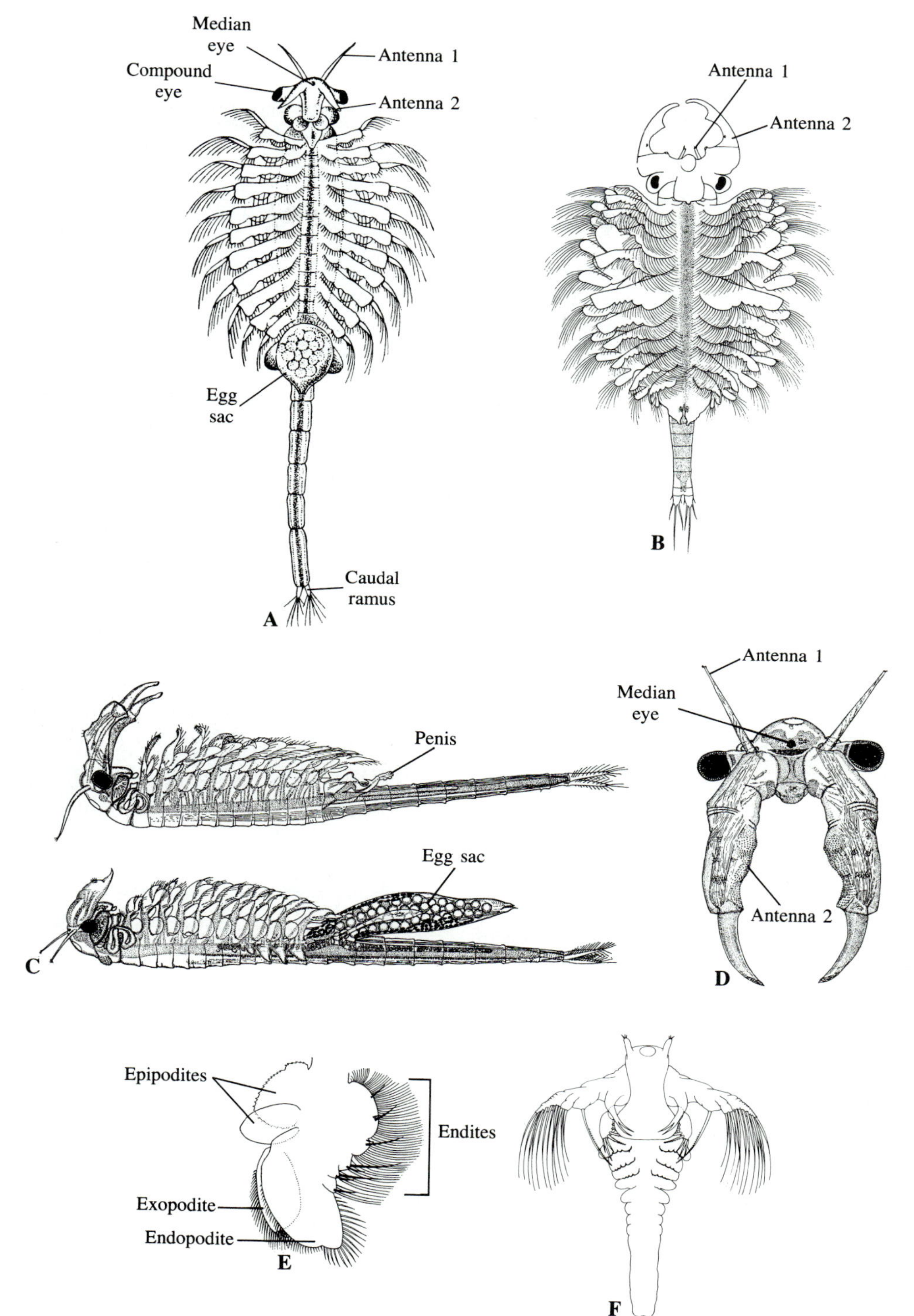

FIGURE 17.18 Class Branchiopoda, order Anostraca. A. *Artemia salina,* female, ventral view. (Ivanov *et al.,* Major Practicum of Invertebrate Zoology.) B. *Artemiopsis stefanssoni,* male, ventral view. (Linder, Zoologiska Bidrag från Uppsala, *20.*) C. *Branchinecta cornigera,* male (above) and female (below). D. *Branchinecta cornigera,* male, anterior aspect of head. (C and D, Lynch, Proceedings of the United States National Museum, *108.*) E. *Branchinecta paludosa,* eighth trunk appendage. (Linder, Zoologiska Bidrag från Uppsala, *20.*) F. *Artemia salina,* nauplius. (Sanders, Memoirs of the Connecticut Academy of Arts and Sciences, *15.*)

Not all anostracans are totally filter-feeders. Some, like the large *Branchinecta gigas,* which may reach a length of 10 cm, will eat copepods and even other anostracans. Prey animals are captured by setae on the endites of the trunk appendages and moved forward so that the mandibles can crush them.

Anostracans have paired gonads, but in females the oviducts unite to form a single vagina. This lies within a swelling on the ventral side of the first two limbless segments, whose sterna are fused together. In gravid females, the swelling and the vagina enlarge to form a brood pouch (Figure 17.18, A and C). In males, the genital pores are separate and each is situated on a prominent penis. When anostracans mate, the male swims beneath the female and grasps her with his stout and often elaborate antennae. He then twists his body so that his penes are close to her genital pore. Reproduction does not always depend on mating, however; some brine shrimps are parthenogenetic.

Fertilized eggs—or eggs that will develop parthenogenetically—undergo at least some development while they are still in the brood pouch. The nauplii (Figure 17.18, F) hatch after the eggs have been dropped. In the case of anostracans that produce thick-walled, resistant eggs, hatching is likely to be delayed until the next rainy season, when temporary pools that have dried out are filled up again. Both thick-walled and thin-walled eggs may be produced by females of a single species; the thin-walled eggs are not resistant to desiccation and normally hatch in a few days. In any case, the nauplius is succeeded by a series of metanaupliar stages and the adult form is attained gradually.

The best-known anostracans are the brine shrimps of the genus *Artemia* (Figure 17.18, A). They have generally been referred to one species, *A. salina,* but these widely distributed animals differ with respect to their chromosomal numbers, genetics, and reproductive patterns. It may be best to say that there are several sibling species. Brine shrimp live in highly saline inland lakes, such as the Great Salt Lake of Utah, but they are also found in coastal ponds from which water is allowed to evaporate in order to concentrate salt. Their osmoregulation is reminiscent of that of marine fishes, which excrete salts in order to keep the salt concentration of body fluids below that of the surrounding medium. A brine shrimp in water containing 4% salt (slightly more concentrated than sea water) will keep the salt concentration of its blood at about 1.1%. In brine with 30% salt, it holds the salt concentration of its blood at about 2.7%. In adults, the salt is excreted through the epipodites of the first 10 trunk appendages; larvae eliminate salt from a dorsally located salt gland.

Brine shrimps produce desiccation-resistant eggs that will hatch in salt water, providing nauplii and later stages for classroom study or for feeding tropical fish and some other aquarium animals. They can be raised to maturity if fed *Dunaliella,* a small green flagellate commonly found with brine shrimps.

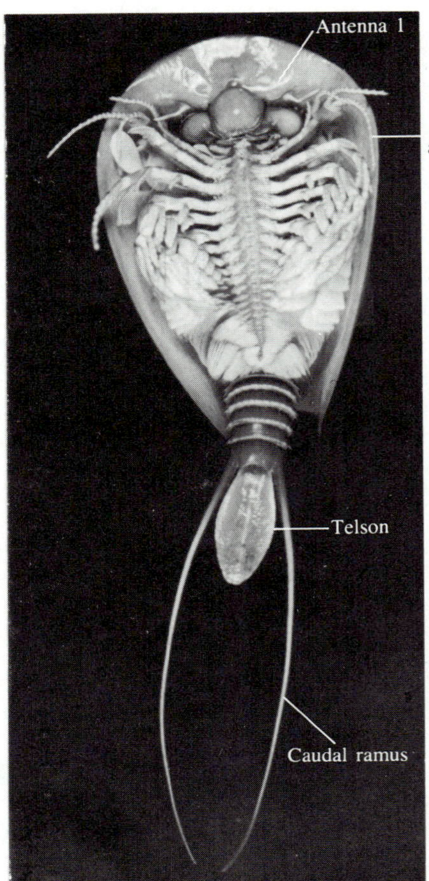

Figure 17.19 Class Branchiopoda, order Notostraca. *Lepidurus,* ventral view.

Order Notostraca

In notostracans (Figures 17.19, 17.20), of which there are only a few species, there is a domelike carapace that covers about half of the body. The carapace is fused to the head—it is an outgrowth, in fact, of the last head segment—but not to any of the trunk segments. The sessile compound eyes, on the upper surface of the carapace, are conspicuous, and just behind them may be four simple eyes left over from the naupliar stages. The animal must be turned over to see the small first antennae (sometimes not obviously articulated) and still tinier second antennae, which are succeeded by the mandibles, first maxillae, and second maxillae.

The flexible trunk has many serial units. Each of the

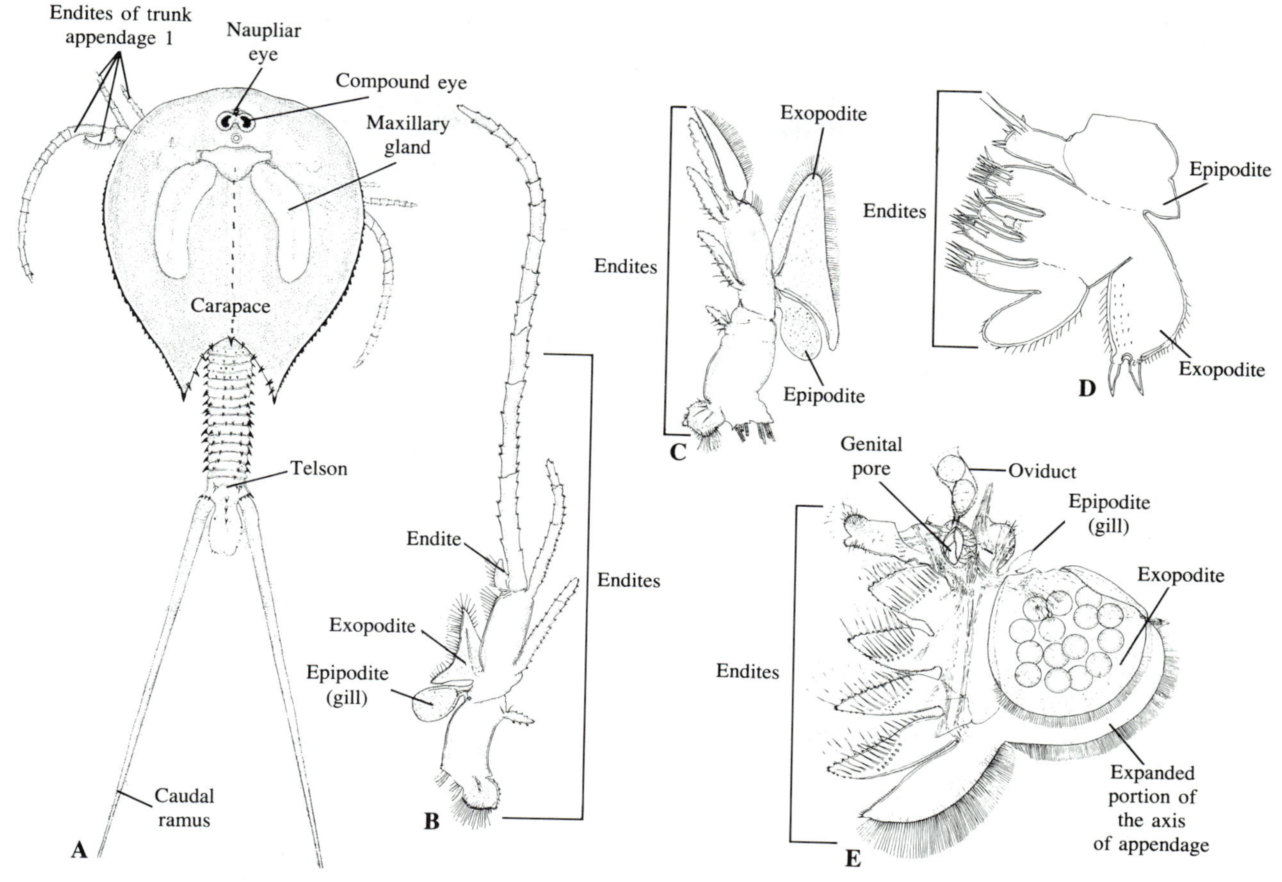

FIGURE 17.20 Class Branchiopoda, order Notostraca. *Lepidurus lemmoni*. A. Male, dorsal view. B, C, D. First, second, and last trunk appendages of female. (These appendages are similar in the male.) E. Right eleventh trunk appendage of female. The exopodite covers the eggs, which are held on a broadened portion of the axis of the appendages. (Lynch, Transactions of the American Microscopical Society, *85*.)

first 11 or 12 units has one pair of appendages, but each unit in the region that follows may have up to six pairs. Thus what looks like a single segment is produced by fusion of several embryonic somites; the term *body ring* would seem more suitable than *segment*. The posterior part of the trunk has no appendages, but the telson is provided with a pair of long, articulated caudal rami. The first trunk appendages are distinctive because of their long endites, which have chemoreceptors in them; the rest are phyllopodial and are all much alike, and they are graduated in size, the smallest being the last in the series.

Most notostracans are about 2 cm long, but a few are larger than this. *Lepidurus lynchi*, of the western United States, may be 10 cm long. These animals are strictly limited to freshwater ponds, and most are found only in ponds that dry out in summer. They feed on detritus and various small organisms, including diatoms and other algae, and to some extent on dead or moribund animals—other crustaceans, tadpoles, and so on. They can use the endopodites and endites of their anterior trunk appendages for rasping off tissue. Coarse food is passed toward the mouthparts by action of the endites, which border a midventral food groove. Fine particles collected from sediment that has been stirred up are also moved forward after being concentrated in this groove.

Notostracans usually skim over the surface of the sediment or plow through the superficial layer of it. The more anterior trunk appendages are most effective in propulsion. These animals can also swim and sometimes do this upside-down.

Parthenogenesis, as well as sexual reproduction, is known to occur in this group. Some species, in fact, are strictly parthenogenetic, even though a few males may be present in the population. Mature females have a kind of brood pouch formed by specialization of certain portions of the eleventh trunk appendages. The eggs are held only for a short time, then set free. In the case of species that live in temporary ponds, the eggs must be able to survive desiccation during the dry months. Eggs of some notostracans will hatch even after being dried for several years. The nauplius is succeeded by metanaupliar stages, and the adult form is arrived at gradually.

Order Diplostraca

The two remaining major groups of branchiopods, Conchostraca and Cladocera, are typically characterized by a carapace that largely or completely encloses the body. Both groups also have biramous second antennae, and these are used for swimming. The eyes are sessile, and the caudal rami, which are not divided into articles, are clawlike. Because they share the combination of features just mentioned, the Conchostraca and Cladocera are often treated as suborders of the Diplostraca. In other systems of classification they are raised to the rank of order.

Suborder Conchostraca

In a conchostracan (Figure 17.21), the body is laterally flattened and enclosed within a carapace that is hinged along the dorsal midline. The carapace is therefore bivalved and resembles a clamshell. It is fused only with the posterior part of the head. The two valves can be pulled closer together by an adductor muscle in the next-to-last segment of the head; this is the segment that bears the first maxillae. As in notostracans, the eyes are sessile. The first antennae are short, sometimes divided into two articles, sometimes not divided. The biramous second antennae are the most conspicuous appendages and are used in locomotion. The number of pairs of trunk appendages ranges from 10 to more than 30. These appendages are phyllopodial, but they are not equally developed. The large and more elaborate ones are those in the anterior part of the series. The abdomen

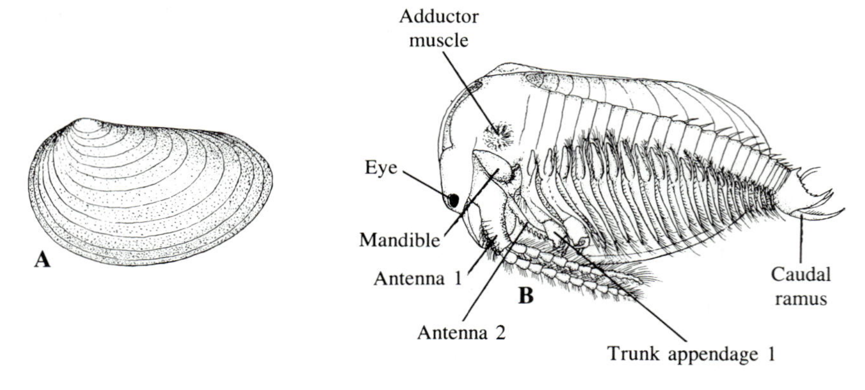

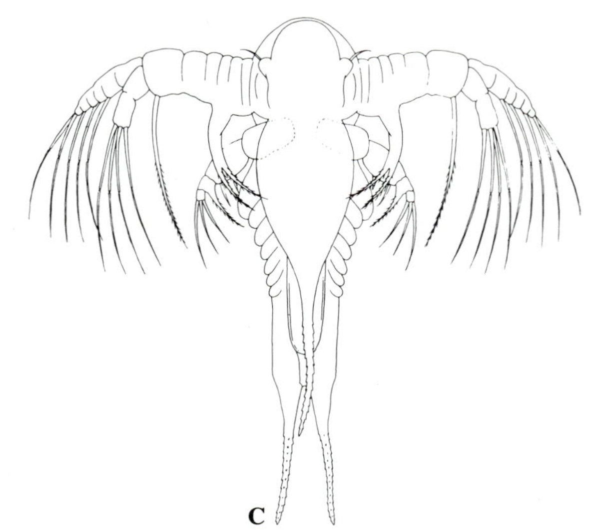

FIGURE 17.21 Class Branchiopoda, suborder Conchostraca. A. *Estheria obliqua,* female, carapace, as seen from the left side. B. *Estheria obliqua,* male, as seen from the left side after removal of the left valve of the carapace. (A and B, Calman, in Lankester, A Treatise on Zoo-ogy.) C. *Limnadia lenticulans,* nauplius. (Sanders, Memoirs of the Connecticut Academy of Arts and Sciences, *15.*)

is reduced and bent downward, ending in a pair of caudal rami. The rami are used for keeping some of the free space enclosed by the carapace clean, so that water circulation is not impeded.

Conchostracans are usually about 5 mm to 1.5 cm long and live in the mud of freshwater ponds, including some that dry out during the summer. Most species plow through the mud or skim over it, feeding on detritus and microscopic organisms. Some, however, remain more or less stationary in mats of moss or other vegetation, or lie on their sides on the bottom. One genus, *Lynceus,* is specialized for swimming upside-down; it collects food while moving, after the fashion of anostracans.

Most species of conchostracans have functional males, but some are parthenogenetic. In mating, the male holds onto the edges of the valves of the female's carapace and sticks part of his trunk between the valves. The eggs, after being laid, are kept for a while in the space between the trunk and carapace, being held here by specialized exopodites of some of the appendages. The eggs are set free at the next molt, unless the female continues to carry them until they have hatched and the larvae have molted several times. This happens in one genus. Hatching, in any case, is at the naupliar stage (Figure 17.21, C). The carapace shows up after the larva has reached the metanaupliar stage. The eggs of conchostracans that live in temporary ponds must be capable of surviving seasonal drought.

Suborder Cladocera

Cladocerans—called "water fleas" because of their jerky swimming movements—are very different from the preceding groups of branchiopod crustaceans. They have only a few trunk segments and not more than six pairs of trunk appendages; these appendages are more specialized than those of anostracans and notostracans, and exhibit considerable division of labor. There is a carapace, and although this originates from the posterior edge of the head, it does not cover the head. In most cladocerans, including the common *Daphnia* (Figures 17.22, 17.23, A), it grows down laterally on both sides, covering the trunk and its appendages. Dorsally, it is fused to two or more trunk segments. It is not hinged like that of a conchostracan and is therefore not truly bivalved. In some types, such as *Apages* (Figure 17.23, C) and *Leptodora* (Figure 17.23, D), the carapace is reduced.

The compound eyes are fused into one median eye, which is movable to some extent and generally enclosed within a cavity. There may also be a simple eye surviving from early stages of development, but this is separate from the compound eye. The first antennae are usually short and slender, especially in females, but the second antennae are prominent. With both rami well developed, these appendages are used for swimming. The trunk appendages are relatively small, and in most species they function in filtering out microscopic algae from the plankton. Not all cladocerans are filter-feeders; some, including *Leptodora,* are carnivores, and their mouthparts, especially the mandibles, are specialized for dealing with copepods and other small animals caught by the trunk appendages. The trunk appendages of the carnivores, by the way, have cylindrical endopodites and exopodites; in the filter-feeders, however, at least some of the appendages in the set of five or six pairs are phyllopodial. The posterior part of the trunk, without appendages, is tipped by a pair of caudal rami,

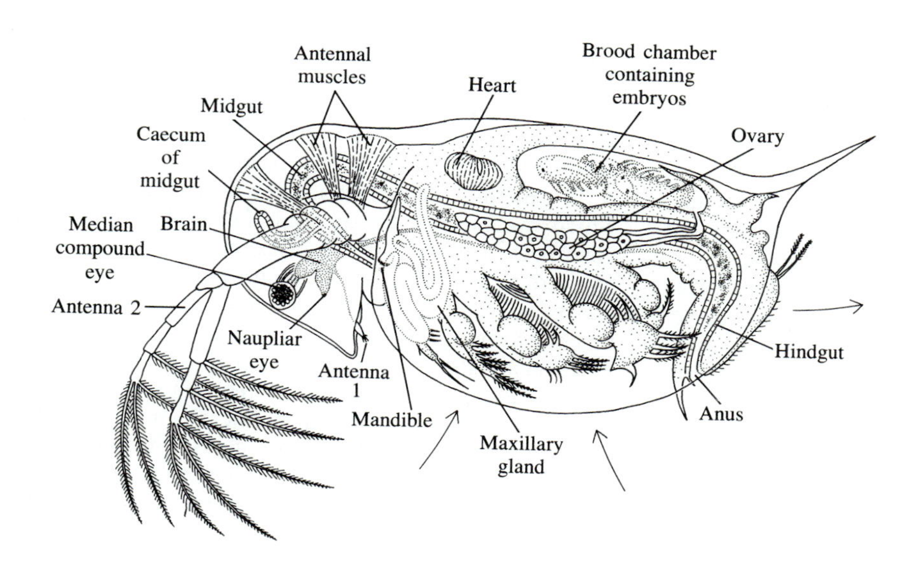

Figure 17.22 Class Branchiopoda, suborder Cladocera. *Daphnia,* female, showing major structures, most of which are visible in living specimens. Arrows indicate direction of water currents.

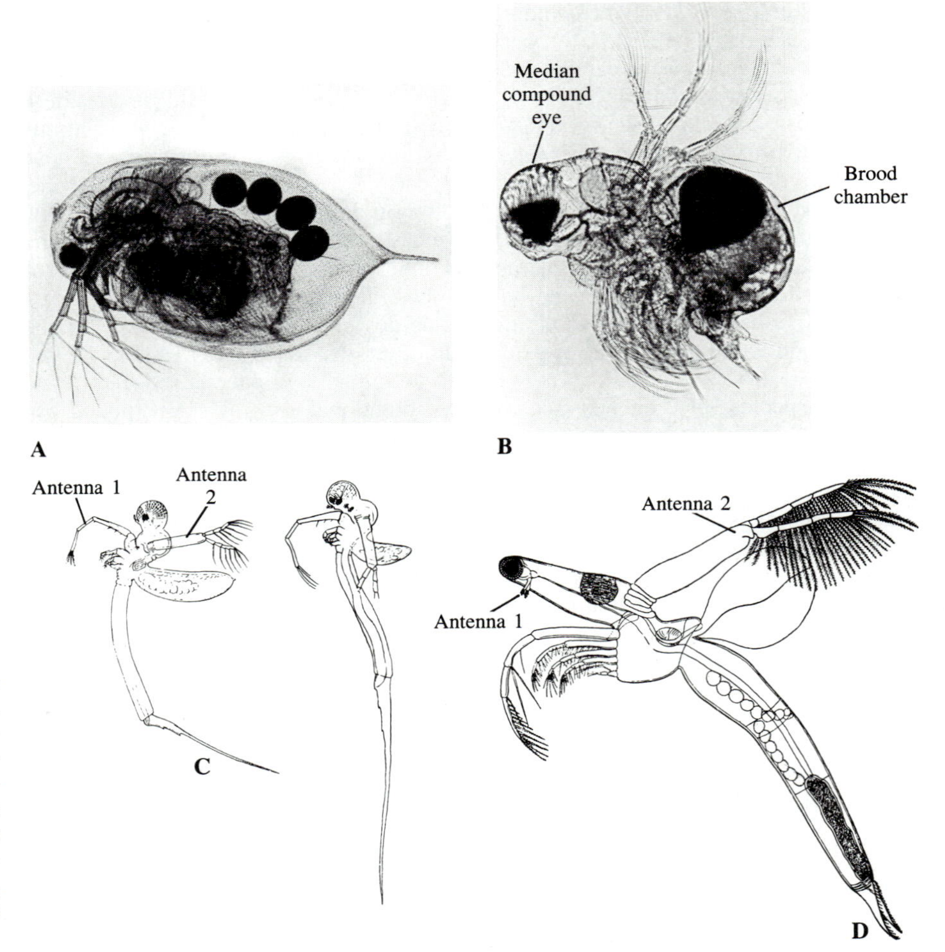

FIGURE 17.23 Class Branchiopoda, suborder Cladocera. A. *Daphnia,* female, photomicrograph. B. *Podon,* photomicrograph. C. *Apages cylindrata,* female, two forms. (Birshtein, in Zenkevitch, Atlas of Invertebrates of the Caspian Sea.) D. *Leptodora.* (Zhadin and Gerd, Rivers, Lakes, and Reservoirs of the USSR, Their Fauna and Flora.)

which may be partly fused into a single clawlike piece. In *Daphnia* and similar types, the hindmost portion, often called the "postabdomen," is bent downward.

The effective stroke of the second antennae usually drives a cladoceran upward; the animal then sinks while holding its antennae upraised before another stroke propels it again. The trunk appendages are concerned mostly for creating an anterior-to-posterior current and for collecting phytoplankton and other small food organisms, including bacteria. Not all of the appendages are alike; only certain of them have well-developed setal combs for filtering. As in other branchiopods, food becomes concentrated in a midventral groove between the trunk appendages and is moved in the direction of the mouthparts.

Some relatives of *Daphnia* can cling to vegetation, pull themselves along, or walk with the help of their first two pairs of trunk appendages. In certain species, these appendages can be used for agitating mud or for scraping up food material. *Anchistropus minor* is remarkable in that it can attach itself to hydras and use the claws on its second and third trunk limbs to tear off cells, which are carried to the mouthparts by a feeding current. Cladocerans exhibit much diversity in locomotion and feeding; only the surface of the subject has been scratched in this short account.

In general, female cladocerans of the *Daphnia*-type, which live in fresh water, reproduce both by parthenogenesis as well as with the participation of males. The eggs are retained in a brood chamber under the carapace above the trunk. Parthenogenetic eggs are diploid. Eggs that are to be fertilized are haploid, and the diploid zygotes generally develop into parthenogenetic females. The situation is complicated by the fact that parthenogenetic eggs produced by a female are not all genetically alike, for although they are diploid, they have undergone a first meiotic division. Parthenogenesis, which produces both males and females, generally alternates with sexual reproduction, which produces only females. In some species, a single female may lay eggs of one kind, then eggs of the other kind. The fertilized eggs are relatively thick-shelled and often remain dormant through the winter, but this is not always the case. There is much variation, even within a single species, depending on ecological circumstances. In lakes in which conditions are very stable, parthenogenetic reproduction may continue indefinitely. An interesting sidelight of the alternation of parthenogenetic and sexual phases is a phenomenon called *cyclomorphosis*. The term refers to the fact that the size and appearance of individuals in one parthenogenetic generation may be different from those in a preceding or succeeding generation. Populations of a particular species living in one lake may show a different pattern of cyclomorphosis than those in another body of water not far away. The individuality is due to various environmental factors—temperature, light, turbulence, available food, and so on.

Development is usually direct in cladocerans, but there are exceptions. The odd *Leptodora,* for instance, hatches as a metanauplius, a larva that resembles a nauplius but that has additional segments. The young molt many times, and molting continues even after the adult stage has been reached.

The brood chamber in which fertilized eggs are retained may become enclosed by a capsule that looks something like a saddle draped over the trunk. This is called the **ephippium** (from *epi,* upon, and *hippium,* horse; the word should be pronounced as "ep-hippium"). The capsule is dropped at the next molt, or remains with the carapace when this is shed. In time, an ephippium will rise to the surface, and it may even be picked up on a feather of a bird. The eggs, with thick shells that prevent desiccation, may then be carried to another body of water.

With nearly 500 species, cladocerans constitute the largest group of branchiopods. Most of what has been said here about them is based on *Daphnia* and similar genera. *Apages* and *Leptodora* have been mentioned as distinctly different types. There are other oddities, too. The marine genera *Podon* (Figure 17.23, B) and *Evadne,* often common in plankton, have four pairs of trunk appendages that are unprotected by the carapace. The rami of these appendages are cylindrical. The carapace is modified to serve as a brood chamber and little else. The median compound eye of *Podon* and *Evadne* is large in proportion to the rest of the body.

References on Branchiopoda

Anderson, D. T., 1967. Larval development and segment formation in the branchiopod crustaceans *Limnadia stanleyana* and *Artemia salina*. Australian Journal of Zoology, *15*:47–91.

Belk, D., 1984. Antennal appendages and reproductive success in the Anostraca. Journal of Crustacean Biology, *4*:66–71.

Benesch, R., 1969. Zur Ontogenie und Morphologie von *Artemia salina* L. Zoologische Jahrbücher, Abteilung für Anatomie und Ontogenie der Tiere, *86*:307–458.

Botnaruic, N., 1948. Développement des phyllopodes conchostracés. Bulletin Biologique de la France et de la Belgique, *82*:31–36.

Brooks, J. L., 1946. Cyclomorphosis in *Daphnia*. Ecological Monographs, *16*:409–447.

Brooks, J. L., 1957. Systematics of North American *Daphnia*. Memoirs of the Connecticut Academy of Arts and Sciences, *13*:1–180.

Browne, R. A., and MacDonald, G. A., 1982. Biogeography of the brine shrimp, *Artemia:* distribution of parthenogenetic and sexual populations. Journal of Biogeography, *9*:331–338.

Cannon, H. G., and Leak, F. M. C., 1933. The feeding mechanisms of the Branchiopoda. Philosophical Transactions of the Royal Society of London, B, *222*:267–352.

Fryer, G., 1966. *Branchinecta gigas* Lynch, a non-filter-feeding raptatory anostracan, with notes on the feeding habits of other anostracans. Proceedings of the Linnean Society of London, *177*:19–34.

Fryer, G., 1968. Evolution and adaptive radiation in the Chyodoridae (Crustacea: Cladocera): a study in comparative functional morphology and ecology. Philosophical Transactions of the Royal Society of London, B, *254*:221–385.

Fryer, G., 1974. Evolution and adaptive radiation in the Macrothricidae (Crustacea: Cladocera): a study in comparative functional morphology and ecology. Philosophical Transactions of the Royal Society of London, B, *269*:137–274.

Fryer, G., 1983. Functional ontogenetic changes in *Branchinecta ferox*. Philosophical Transactions of the Royal Society of London, B, *303*:229–343.

Linder, F., 1941. Morphology and taxonomy of the Branchiopoda Anostraca. Zoologiska Bidrag från Uppsala, *20*:101–302.

Linder, F., 1952. The morphology and taxonomy of the Branchiopoda Notostraca, with special reference to the North American species. Proceedings of the United States National Museum, *102*:1–69.

Longhurst, A. R., 1955. A review of the Notostraca. Bulletin of the British Museum of Natural History, *3*:1–57.

Lynch, M., 1980. The evolution of cladoceran life histories. Quarterly Review of Biology, *55*:23–42.

Martin, J. W., and Belk, D., 1988. Review of the clam shrimp family Lynceidae Stebbing, 1902 (Branchiopoda: Conchostraca), in the Americas. Journal of Crustacean Biology, *8*:451–482.

Mathias, P., 1937. Biologie des crustacés phyllopodes. Actualités Scientifiques et Industrielles, *447*:1–107.

Persoone, G., Sorgeloos, P., Roel, O., and Jaspers, E. (editors), 1980. The Brine Shrimp *Artemia*. 3 volumes. Universa Press, Wetteren, Belgium.

CLASS REMIPEDIA

The first species of Remipedia, *Speleonectes lucayensis* (Figure 17.24), was described in 1981. It was found on one of the islands of the Bahamas, in a cave that is about 1 km from the sea but that is joined to the sea by a tidal creek. Fresh water also enters the cave

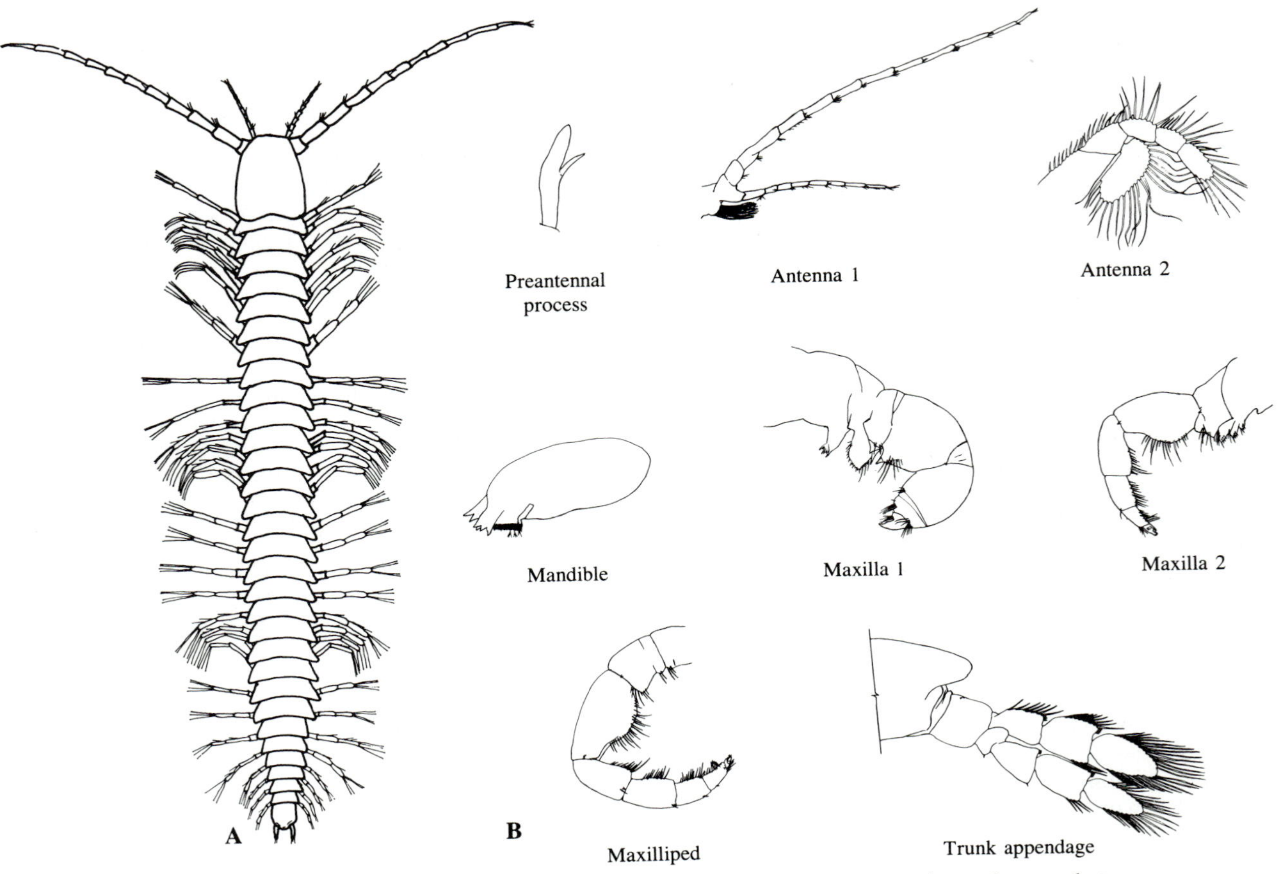

FIGURE 17.24 Class Remipedia. *Speleonectes lucayensis*. A. Dorsal view, drawn from a photograph of an active specimen. (Photograph and drawing by Dennis Williams, courtesy of Jill Yager.) B. Preantennal process, head appendages, and one trunk appendage. (Yager, Journal of Crustacean Biology, *1*.)

and tends to form a layer over the water of moderate or high salinity in which *Speleonectes* lives. Additional species and new genera have been discovered in similar situations on other islands of the Bahamas, the Turks and Caicos Islands, and the Canary Islands.

Speleonectes, whose body form slightly resembles that of a centipede, is about 2 cm long, and it swims rapidly, with the ventral side up. The head region is unusual in that it has a pair of bilobed projections, called *frontal processes,* anterior to the bases of the first antennae. One of the lobes is smaller than the other, and both articulate with a basal piece. The first antennae branch into two rami; the rami originate on the second article, rather than farther distally, as is characteristic of Malacostraca, the only other class of crustaceans in which the first antennae are branched. In the somewhat flattened second antennae, the endopodite consists of three articles, and the exopodite consists of a single article. Eyes are lacking.

The mouthparts are uniramous. The mandible is stout, and has a flattened molar process, as well as three incisorlike cusps and three smaller projections. The first and second maxillae appear to be fitted for grasping.

The first trunk segment is joined to the head; its appendages are a pair of maxillipeds that resemble the maxillae. The remaining trunk segments, of which there are 31 or 32, are distinct, and each has a pair of setose biramous swimming appendages. Although these are flattened, they are more nearly stenopodial than phyllopodial. Their size decreases posteriorly, much as it does in phyllopodial appendages of most groups of the class Branchiopoda. The telson has a pair of caudal rami.

The anterior part of the foregut is muscular and can be dilated, a fact that fits in with a few observations that remipedes are carnivores. *Speleonectes* has been observed to grasp a small shrimp, and some time later release its empty exoskeleton. The midgut, which begins with the first trunk segment, has paired segmental diverticula. The short hindgut is restricted to the telson.

The brain is fairly large. It is joined, by connectives that curve around the foregut, to a ventral nerve cord in which successive pairs of segmental ganglia are rather well separated.

Remipedes are believed to be hermaphroditic. There is a pair of ovaries in the posterior part of the head region, and these are connected to the gonopores, located at the bases of the appendages on trunk segment 14, by oviducts that are mostly dorsal to the midgut. What has been interpreted as a paired male system, extending from the head region to the same pores, is ventral to the gut. The head region has paired maxillary glands, presumably concerned with excretion, and a heartlike dilation of the main middorsal blood vessel.

In having numerous trunk segments, each with a pair of biramous appendages, remipedes are similar to certain of the Branchiopoda. The caudal rami are comparable to those of branchiopods, cephalocaridans, copepods, and a few other groups. The paired processes in front of the first antennae are truly unusual, however. Although no musculature has been found to be associated with them, they could conceivably be appendages. No known crustaceans have three pairs of appendages anterior to the mandibles.

References on Remipedia

Schram, F. R., 1983. Remipedial crustacean phylogeny. Crustacean Issues, *1*:23–28.

Schram, F. R., Yager, J., and Emerson, M. J., 1986. Remipedia. Part I. Systematics. Memoir 15, San Diego Society of Natural History.

Yager, J., 1981. Remipedia, a new class of Crustacea from a marine cave in the Bahamas. Journal of Crustacean Biology, *1*:328–333.

Yager, J., 1987. *Cryptocorynetes haptodiscus* and *Speleonectes benjamini,* from anchialine caves in the Bahamas, with remarks on distribution and ecology. Proceedings of the Biological Society of Washington, *100*:302–320.

Yager, J., and Schram, F. R., 1986. *Lasionectes entrichoma* (Crustacea: Remipedia) from anchialine caves in the Turks and Caicos, West Indies. Proceedings of the Biological Society of Washington, *99*:65–70.

CLASS OSTRACODA

Ostracodes (for etymological reasons, this spelling is preferable to "ostracods") constitute a thoroughly successful group of Crustacea, well represented in fresh water as well as in the sea. In general, they are small animals less than 3 mm long, but there are giants that reach 2 cm. Most of the approximately 2000 species crawl around in sediments and in deposits that accumulate on aquatic animals and plants. There are some planktonic types, however, and some that live as commensals with other invertebrates, including freshwater crayfishes. A few species live in humus in damp forests of Africa and New Zealand. Ostracodes have left an illustrious fossil record and have received much attention from paleontologists concerned with stratigraphy.

The most obvious feature of an ostracode is its bi-

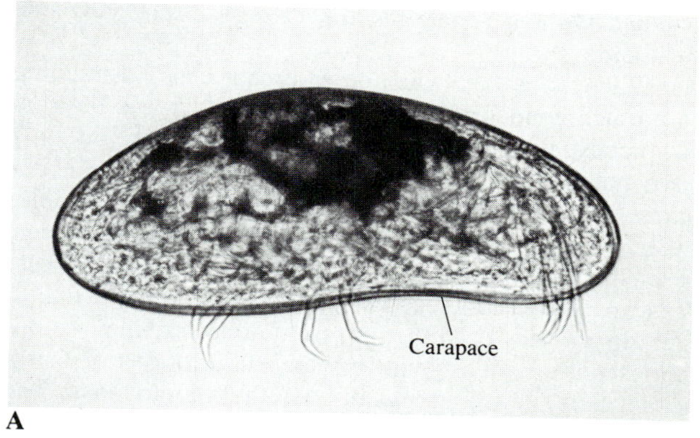

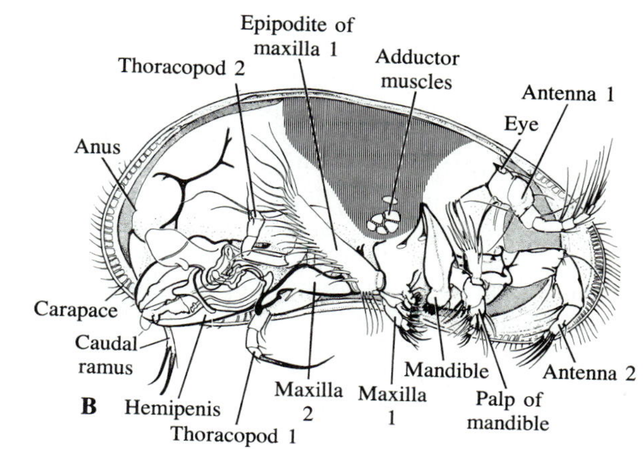

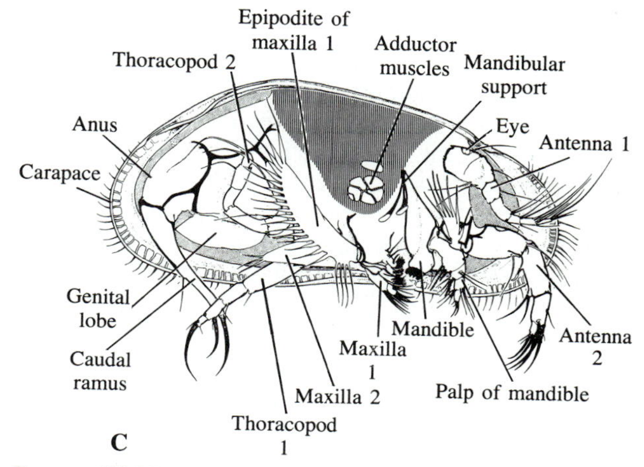

FIGURE 17.25 Class Ostracoda. A. A small, nearly transparent marine ostracode, photomicrograph. B. *Candona suburbana,* male, right valve removed. C. *Candona suburbana*, female. (B and C, McGregor and Kesling, Contributions from the Museum of Paleontology, University of Michigan, *22*.)

valved carapace (Figure 17.25). The valves are pulled together by adductor muscles. When the appendages are withdrawn, the animal looks like a little clam, or seed, and if the carapace is calcified or opaque because of its pigmentation, the organization of the body cannot be studied without separating the valves or removing one of them.

When compared with other crustaceans, ostracodes are much condensed, and their segmentation is blurred. Typically, there are seven pairs of appendages, but only two pairs belong to the thorax, and none to the abdomen, which is so short as to be scarcely recognizable. All five pairs of crustacean head appendages, however, are present: first antennae, second antennae, mandibles, first maxillae, and second maxillae. The second maxillae do not always function as mouthparts; they sometimes serve as legs, and in many descriptions of ostracodes they are called legs or even first thoracopods. The broad epipodites, or "branchial plates," characteristic of the first maxillae (or of both maxillae) of certain ostracodes (Figure 17.25, B and C), drive a respiratory current from anterior to posterior.

Among the ostracodes are many filter-feeding types that use their mouthparts for collecting diatoms and other fine particles. The first and second maxillae, and sometimes also the mandibles, are intricately fringed with setae for straining out food. The feeding current may be created by certain of the mouthparts or by the antennae. Some ostracodes hang upside down from the surface film and process the thin layer of scum. Many are relatively nonselective consumers of detritus, such as is found in sediment through which the animals plow. Others are herbivores that eat algae, scavengers that use all sorts of animal and plant material, or even carnivores that capture copepods, chaetognaths, fish larvae, and other small animals. Such a remarkable diversity in food habits is matched by varied specializations of the mouthparts and other appendages.

In ostracodes that swim, the second antennae usually serve as the organs of locomotion. Creeping types generally use their second antennae or their legs, or both. Legs may also be used for clinging, as in the case of the species that live among the setae on the ventral side of the cephalothorax of crayfishes. There are many ostracodes in which neither of the two pairs of thoracic appendages are really leglike, and both pairs may be missing altogether. In certain groups, the second thoracic appendages, directed upward and backward, function as cleaning devices.

Most ostracodes have a single median eye between the bases of the first antennae. This can be traced back to a naupliuslike early stage of development. In some members of one major group, the order Myodocopida, there is a pair of lateral compound eyes; in other members of this order, no eyes of any kind are present. Myodocopids are unique in another respect: they are the only ostracodes that have a heart.

Sexual dimorphism is often marked, especially when the first antennae, second antennae, or first "legs" (second maxillae) are specialized as devices for clasping females. There are other sex differences, however, in certain groups. In the order Myodocopida, for instance, males generally have larger compound eyes than females. The gonads are usually paired, and the gonoducts reach the exterior posteroventrally. The reproductive systems (Figure 17.26) are compact, but they are also rather complicated. Females have seminal receptacles for storing sperm received during copulation, and the openings to these pockets are separate from those through which eggs are laid. In effect, therefore, a female has four genital pores. The sperm ducts of males may be provided with seminal vesicles for storing sperm or with elaborations whose function is probably to pump out sperm during copulation. The two sperm-transferring structures, called *hemipenes* (singular *hemipenis*), are cuticularized outgrowths of the body, and are sometimes complex. The sperm tend to be large. The largest sperm known, in fact, is that of a species of *Platycypris* found in Australia. It reaches a length of about 1 cm, although the animal itself is only about 2 mm long.

Copulation is usually effected while the male, with valves spread, is above the female and clinging to the edges of her valves with his first antennae, second antennae, or first legs. While in this position, the male can extend his hemipenes until they reach the openings to the seminal receptacles of the female. In some ostracodes, copulation is achieved while the partners are in contact with one another posteroventrally. Males are unknown in many freshwater species, or are absent in at least certain geographic areas or under certain conditions. In such cases, females reproduce parthenogenetically.

In species that simply release their eggs, or cement them to aquatic plants or other firm objects, hatching generally takes place when development has reached a stage that resembles a nauplius. This has three pairs of appendages that will become the first antennae, second antennae, and mandibles of the adult, but it is not a typical nauplius because none of the appendages is biramous. A bivalved shell, moreover, is already present. The young ostracode molts several times on its way to becoming a mature individual with the usual five pairs of head appendages and two pairs of thoracic appendages. It has already been explained, however, that one or both pairs of thoracic appendages may be lacking. Some marine ostracodes brood their eggs between the valves of the carapace, and the young may not be released until they have reached a relatively advanced stage.

References on Ostracoda

Benson, R. H., 1966. Recent marine podocopid ostracods. Oceanography and Marine Biology, An Annual Review, *4*:213–232.

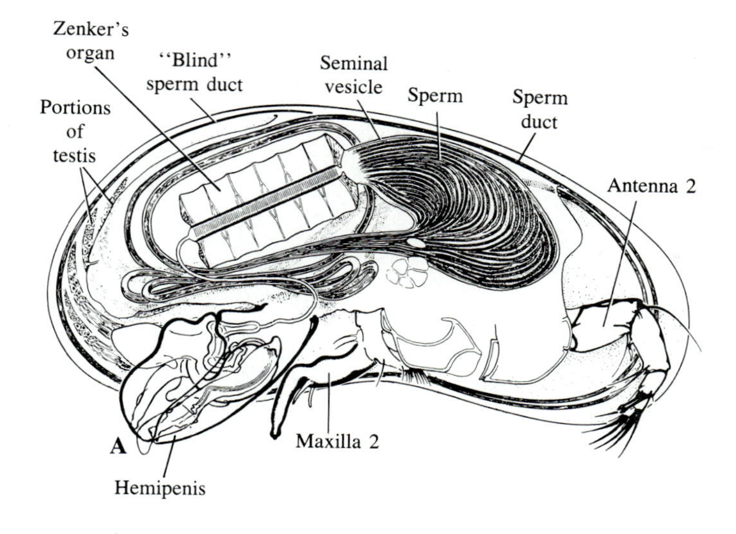

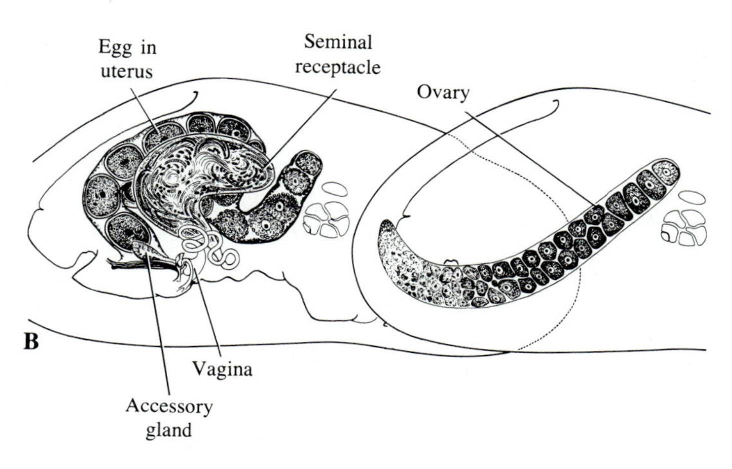

Figure 17.26 Class Ostracoda. *Candona suburbana,* reproductive system. A. Male. B. Female. (McGregor and Kesling, Contributions from the Museum of Paleontology, University of Michigan, *22*.)

Cannon, H. G., 1933. Feeding mechanism of certain marine ostracods. Transactions of the Royal Society of Edinburgh, *57*:739–764.

Cannon, H. G., 1940. Anatomy of *Gigantocypris muelleri*. Discovery Reports, *19*:185–244.

Hoff, C. C., 1942. The ostracods of Illinois. Illinois Biological Monographs, *19*:1–186.

Kesling, R. V., 1951. The morphology of ostracod molt stages. Illinois Biological Monographs, *21*:i–viii, 1–324.

Lochhead, J. H., 1968. The feeding and swimming of *Conchoecia*. Biological Bulletin, *134*:456–464.

Neale, J. W. (editor), 1969. The Taxonomy, Morphology, and Ecology of Recent Ostracoda. Oliver and Boyd, Edinburgh.

Class Mystacocarida

Mystacocaridans, of which there are only a few known species, are marine crustaceans that live between sand grains, subtidally as well as on beaches. They are small, mostly between 0.5 and 1.0 mm long, which is perhaps one reason they were not discovered—or at least not recognized—until 1943.

The body of a mystacocaridan (Figure 17.27, A) is slender. The head is odd in that the anterior portion, which has long first antennae, is separated by a constriction from the part that bears the second antennae, mandibles, and first and second maxillae. The mandibles and second antennae are the only biramous appendages. Compound eyes are lacking, but there are simple eyes, called ocelli (usually four).

The 10 segments behind the head are collectively referred to as the *trunk*. The appendages on the first segment function as mouthparts, so they are called *maxillipeds;* the appendages on the next four segments consist of a single article. The telson at the tip of the trunk has prominent caudal rami that look like jaws.

Mystacocaridans feed on detritus and bacteria that their bristly second maxillae and maxillipeds scrape

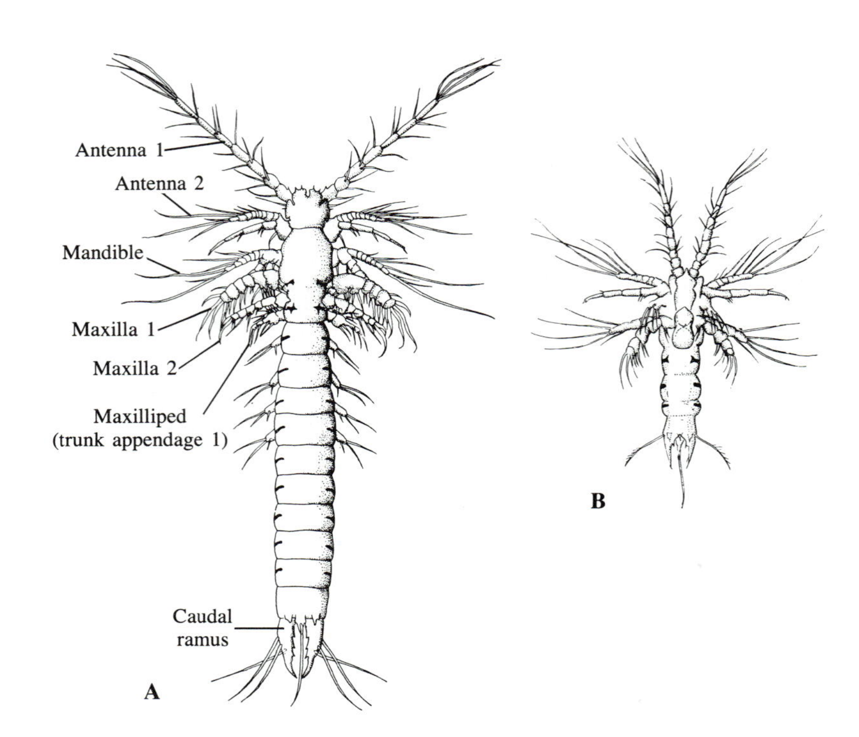

FIGURE 17.27 Class Mystacocarida. *Derocheilocaris remanei*. A. Adult. B. Metanauplius. (Delamare-Deboutteville, Vie et Milieu, *4*.)

from sand grains and push to the mouth, or first to the mandibles, whose coxal portions function as jaws. The second antennae, mandibles, and first maxillae are used for locomotion. By flexing, straightening, and shortening, mystacocaridans can push themselves around, so long as the posterior end can be brought into contact with a stable sand grain.

In both sexes the single genital pore is on the fourth trunk segment, but in the male the appendages of the fifth segment are apparently specialized for sperm transfer. Nauplii and metanauplii (Figure 17.27, B) are among the 10 stages of development that have been recorded for the best-known species, *Derocheilocaris remanei,* which has a wide distribution on the Atlantic and Mediterranean coasts of Europe and Africa and on the Indian Ocean side of South Africa.

The fact that the first three pairs of appendages of adult mystacocaridans resemble those of a nauplius suggests that mystacocaridans constitute a primitive group. Another feature pointing to their primitiveness is the nature of the ventral nerve cord. The ganglia touch, but the connectives that run from segment to segment are decidedly separate.

References on Mystacocarida

Delamare, C., and Chappuis, P. A., 1954. Morphologie des Mystacocarides. Archives de Zoologie Expérimentale et Générale, *91:*7–24.

Hessler, R. R., and Sanders, H. L., 1966. *Derocheilocaris typicus* revisited. Crustaceana, *11:*141–155.

Jansson, B. A., 1966. The ecology of *Derocheilocaris remanei*. Vie et Milieu, *17:*143–186.

Zinn, D. J., Found, B. W., and Kraus, M. G., 1982. A bibliography of the Mystacocarida. Crustaceana, *42:*270–274.

CLASS COPEPODA

Copepods form a large and successful assemblage, represented in marine, estuarine, and freshwater habitats. Some are planktonic; others crawl around in sediments or between sand grains; many are parasites or commensals. In situations where free-living copepods thrive, they are likely to be the most numerous of nonmicroscopic animals. They are important consumers of detritus, diatoms, and other algae, and they are in turn food for carnivores, especially fishes and chaetognaths.

In most copepods, other than some bizarre parasitic types, the body is organized into three main regions, called the *cephalosome, metasome,* and *urosome* (Figure 17.28). These are terms of convenience, because the number of segments incorporated into each region varies from group to group. The cephalosome includes the head, with its first antennae, second antennae, mandibles, first maxillae, and second maxillae, as well as at least the first of seven thoracic segments, which provides a pair of maxillipeds. Sometimes the second thoracic segment is part of the cephalosome, too. The metasome, then, begins with either the second or the third thoracic segment and extends to where the body usually abruptly narrows. This point marks the start of the uro-

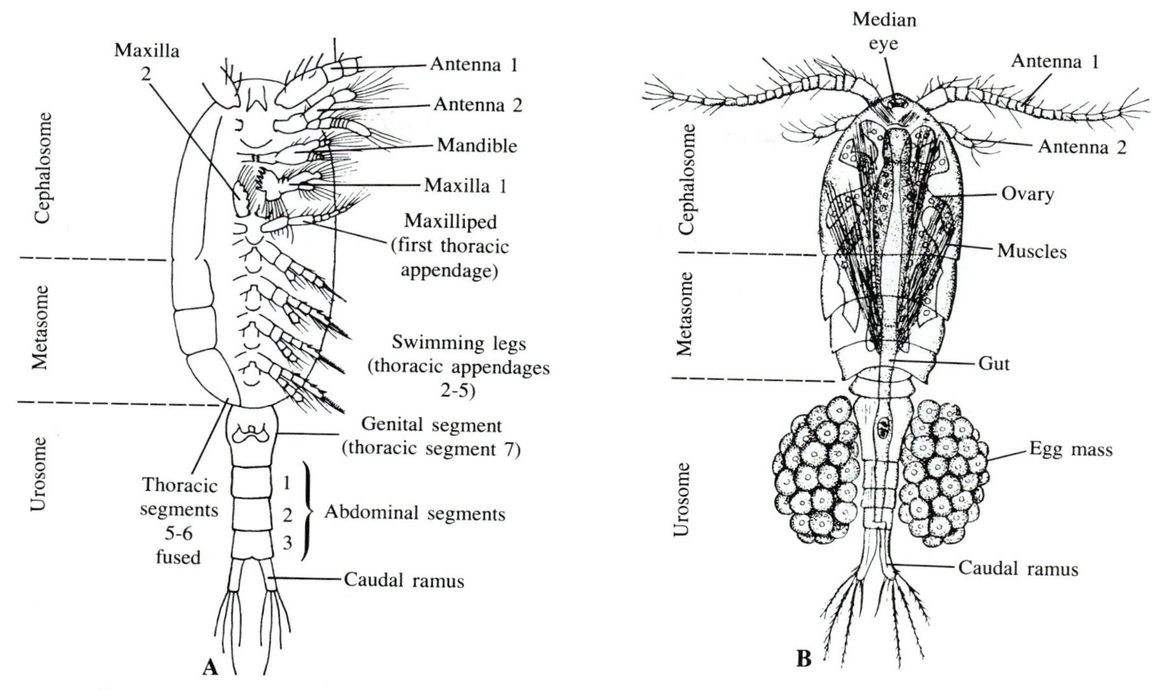

FIGURE 17.28 Class Copepoda. A. Diagram of a female *Pseudocalanus* (order Calanoida), ventral view. (Corkett and McLaren, in Russell and Yonge, Advances in Marine Biology, *15*.) B. Female *Cyclops strenuus* (order Cyclopoida). (Ivanov *et al.*, Major Practicum of Invertebrate Zoology.)

some, which always includes the last thoracic segment—the one on which the genital pore or pores are located—as well as the three abdominal segments and the telson. Except in the order Calanoida, one or two pregenital thoracic segments are also incorporated into the urosome.

The swimming legs, borne on four or five of the postmaxillipedal thoracic segments, are interesting in that their coxae are united. This is the basis for the name Copepoda: they are the "oar-footed" crustaceans. Males generally have five functional pairs; in females the fifth pair is usually absent or much reduced. In males of certain groups, valvelike structures beside the gonopores have what appear to be small limbs associated with them. It is rare that even vestiges of such limbs can be recognized in females. There are no appendages on the abdomen of copepods, but the telson has a pair of caudal rami.

There is often a median eye on the cephalosome. This has only a few ocelli, usually three, and it can be traced back to the nauplius larva. In a few planktonic copepods, two of the ocelli are greatly enlarged and have prominent cuticular lenses. Other sense organs in copepods include a variety of setae and related structures.

The extent to which the circulatory system is developed varies. Some copepods have no heart or aorta, and circulation in the hemocoel depends on pulsations of the gut. The excretory organs of adult copepods are called *maxillary glands* because they open at the bases of the second maxillae; in terms of their origin they are coelomoducts. In naupliar stages, similar structures open at the bases of the second antennae, so they are called *antennal glands*.

Sexes are almost always separate; only a few parasitic copepods are claimed to be hermaphroditic. Males are usually smaller than females. It has already been explained that the genital pore or pores are located on the last thoracic segment, which is part of the urosome. The gonads are anterior to this, so they are mostly in the dorsal part of the metasome. Males have a single testis and either one or two sperm ducts. If there are two ducts, there are two genital pores. Females usually have one ovary, but there are two in some parasitic species that produce large numbers of eggs. Either way, there are two oviducts. These may remain separate, so that there are two ventral or ventrolateral pores on the genital segments. This is the situation in the order Cyclopoida and a few other groups. In the Harpacticoida, the pores are separate but close together within a deep

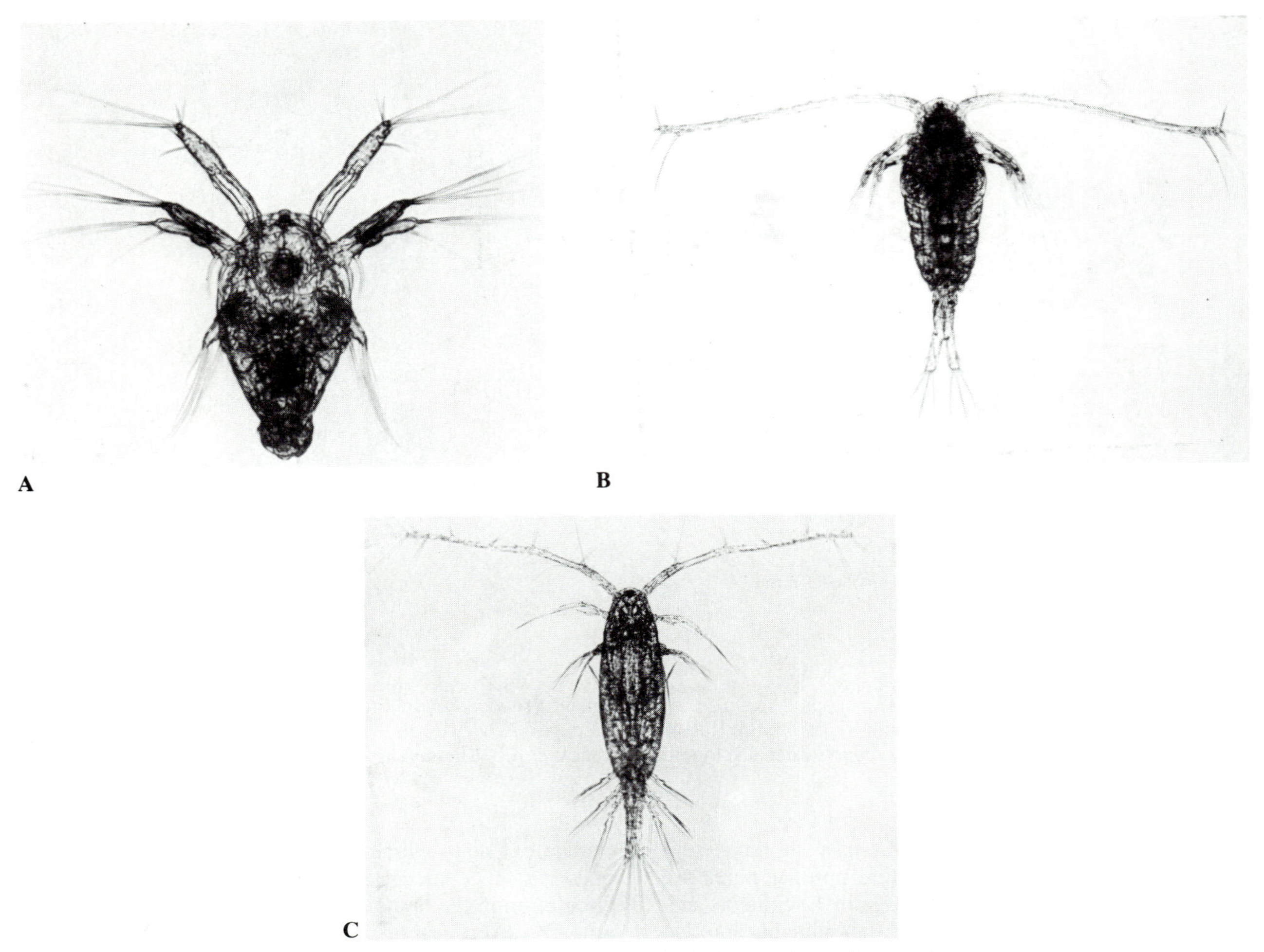

FIGURE 17.29 Class Copepoda, order Calanoida. *Calanus;* photomicrographs. A. Nauplius. B. Late copepodid stage. C. Adult.

pit, called the *atrium*. In the Calanoida, the oviducts unite before reaching the atrium. The arrangement of the one or two seminal receptacles with respect to the genital pores is similarly varied. Cyclopoids have a single seminal receptacle, with a median insemination pore distinct from that of the openings of the oviducts, but it has connections with both oviducts so that the eggs are fertilized just before they are laid. Harpacticoids have two seminal receptacles; their openings into the atrium are close to, but separate from, the openings of the oviducts. Calanoids likewise have two seminal receptacles, one on each side of the single oviduct pore.

Mating is achieved in a variety of ways. The first antennae, maxillipeds, and fifth legs of the male are the appendages most likely to be used for clasping the female. One of the fifth legs may be specially modified to serve as a spermatophore-transferring device. Some female copepods set their eggs free, but it is more common for them to hold them in sacs attached to the genital segment of the urosome. The cementing material is produced by glands on the oviducts. When there is a single genital pore, or two pores very close together, just one egg sac is formed; when there are two widely spaced pores, two sacs develop.

The nauplius (Figure 17.29, A), on hatching from the egg, has the usual three pairs of appendages: first antennae, second antennae, and mandibles. The gut may already be functional, in which case the larva begins to collect microscopic food from the plankton or from sediment, depending on whether it is a swimming or crawling type. In many copepods, however, the nauplius and several succeeding metanaupliar stages subsist

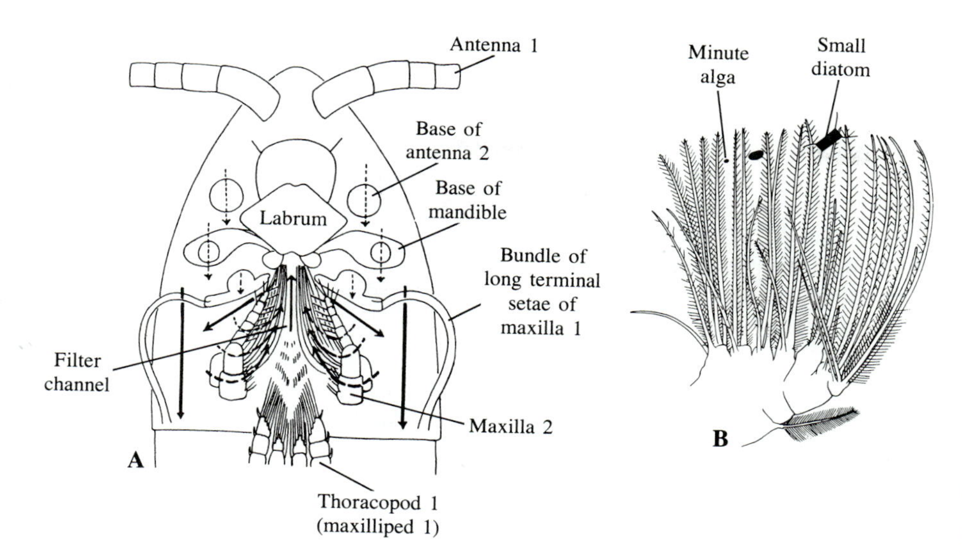

FIGURE 17.30 Class Copepoda, order Calanoida. A. *Diaptomus,* anterior portion, ventral view, showing the filter comb that concentrates food particles and directs them to the mandibles. Arrows indicate major currents. (After Storch and Pfisterer, Zeitschrift für Vergleichende Physiologie, *3.*) B. *Calanus finmarchicus,* second maxilla, showing its setae and some prospective food particles, ranging from an extremely minute alga (*Nannochloris,* about 5 μm in diameter, to a diatom about 50 μm in diameter. The second maxillae do not necessarily touch the particles, however.) (Marshall and Orr, Journal of the Marine Biological Association of the United Kingdom, *35.*)

on yolk; feeding does not begin until the first copepodid stage is reached (Figure 17.29, B). This stage resembles the adult (Figure 17.29, C) in that it has a distinct abdomen and some thoracic appendages; the rest of the thoracic appendages are added during succeeding copepodid stages. There are many interesting specializations in developmental patterns, especially among parasitic types.

In terms of their morphology and lifestyles, copepods are remarkably diversified. Certain of the orders, in fact, consist wholly of highly specialized parasitic types. Some of the parasites, as adults, bear almost no resemblance to the more nearly generalized copepods on which most definitions of the subclass are based. The following account of the several orders is brief, and many details with which specialists are concerned are omitted. It will, however, be of use to the student who needs an outline of copepod classification.

Order Calanoida

Calanoids (Figure 17.28, A; Figures 17.29, 17.30, 17.31), found in fresh and salt water, are almost entirely planktonic. They are characterized by large first antennae, usually with many long setae, which assist in flotation. In nearly all species, one of the first antennae of the mature male is specialized for grasping the female. Calanoids are the only copepods in which the separation of the metasome and urosome is just in front of the genital segment of the thorax. In all other copepods, at least one additional thoracic segment is part of the urosome.

The slow, steady forward progression of a calanoid is brought about by activity of the second antennae and mouthparts. The exopodite and endopodite of the second antennae beat alternately, more than 2000 times a minute. Rapid, jerky movements are effected by thrusts of the thoracic appendages, at which time the first antennae, usually held out at right angles to the body, are pulled close against the body. As soon as the thoracic appendages cease their activity, the first antennae are spread out again so as to retard the animal's sinking.

Calanoids feed while they are moving slowly and steadily. The activity of the second antennae, mandibular palps, first maxillae, and maxillipeds drives water backward, and it also creates lateral swirls (Figure 17.30, A). The second maxillae (Figure 17.30, B) are concerned with trapping the food particles, but they do this without necessarily touching them. As the second maxillae separate, water and suspended particles are sucked into the space between them. Then, as the second maxillae are drawn together again, the particles are pushed forward to the endites of the first maxillae, which in turn push them into the mouth. In other words, the setae on the second maxillae do not act as strainers, as has been thought. They simply increase the surface area of the second maxillae, making these appendages more effective as paddles.

There are exceptions to the general method of filter-feeding just described. Members of the genus *Acartia,* for example, use the second maxillae as scoops or strainers. A few of the larger calanoids are carnivores whose mouthparts can restrain small invertebrates and fish larvae. Certain species are capable of both filter-feeding and predation.

Females of some calanoids hold their eggs in a ventral sac until the nauplii hatch. In other species, the eggs are allowed to drop as they emerge from the genital pore.

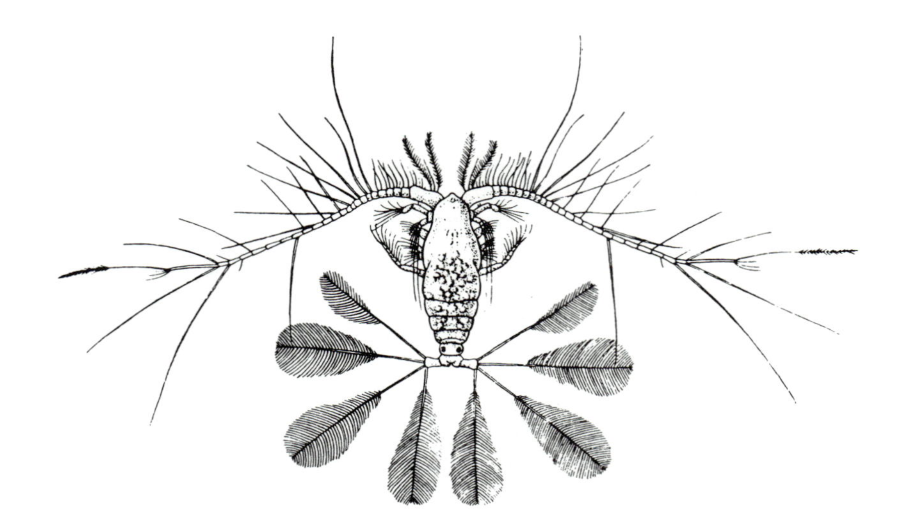

FIGURE 17.31 Class Copepoda, order Calanoida. *Calocalanus bavo*. (Calman, in Lankester, A Treatise on Zoology.)

Order Harpacticoida

Most harpacticoids (Figure 17.32) crawl around in sediment, including that which coats plants and sessile animals. Some—mostly slender-bodied types (Figure 17.32, C)—are specialized for living in beach sand. Many harpacticoids swim intermittently, but few qualify as truly planktonic organisms. In general, the free-living species are "grubbers" or "sand lickers" that feed on diatoms, bacteria, and detritus. Certain members of the group burrow into red algae. *Porcellidium,* which looks like a tiny isopod, clings to the leaves of eelgrass, but consumes only diatoms and other adherent

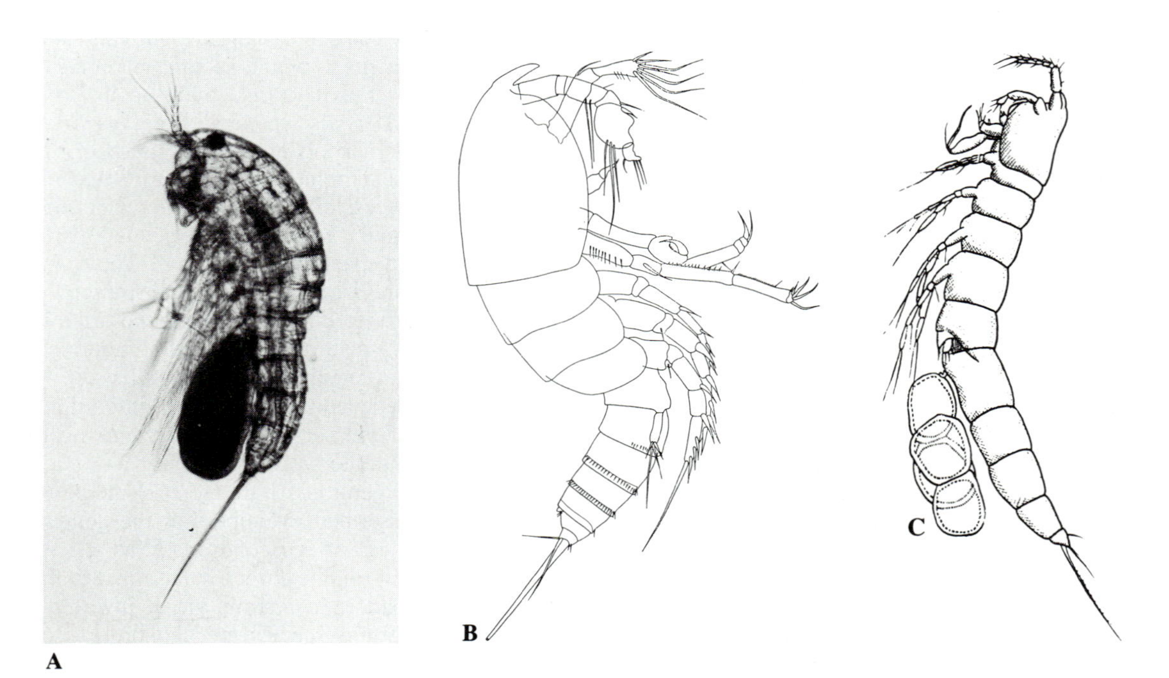

FIGURE 17.32 Class Copepoda, order Harpacticoida. Various marine species. A. Photomicrograph of a living female. B. *Harpacticus clausi,* male. (Verwoort, United States National Museum, Bulletin 236.) C. *Paraleptastacus spinicauda,* female. (Delamare-Deboutteville, Vie et Milieu, *4*.)

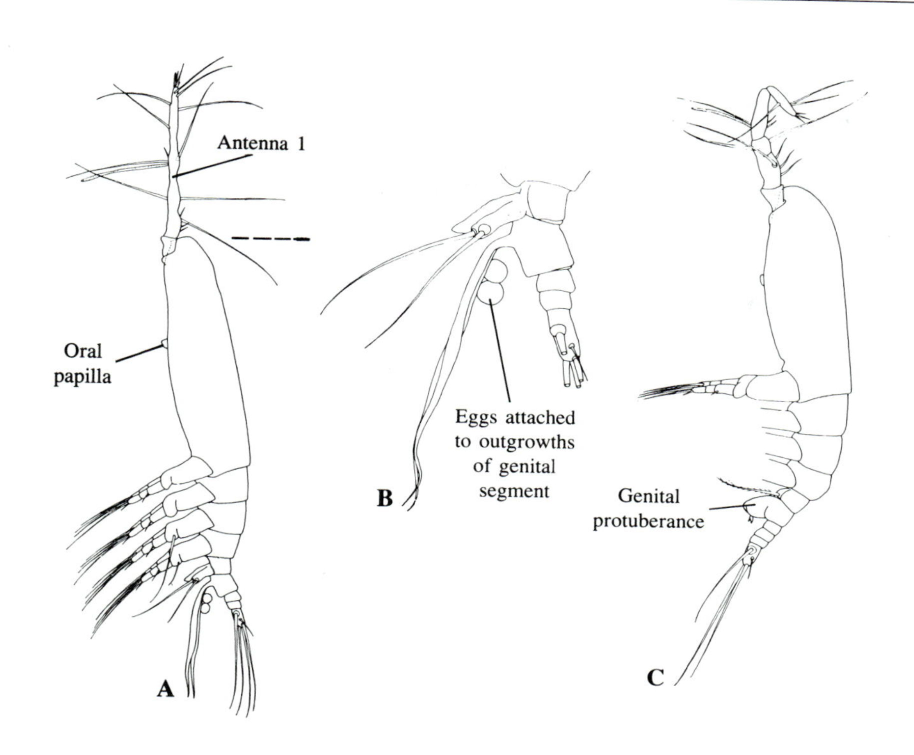

Figure 17.33 Class Copepoda, order Monstrilloida. *Monstrilla wandelii*. A. Female, lateral view. B. Female, posterior portion, lateral view. C. Male, lateral view. The male has three distinct segments behind the genital segment, whereas the female has only two. (Park, Transactions of the American Microscopical Society, *86*.)

material. A few species are commensals or parasites of other invertebrates and fishes.

The first antennae of harpacticoids are usually short, there being not more than 10 articles; the second antennae, which are biramous but have a reduced exopodite, are also short. The separation of the urosome from the metasome is often indistinct, but when there is an obvious separation, it is in front of the segment that precedes the genital segment; thus the urosome consists of the abdomen plus two thoracic segments. Females typically carry their eggs in a single ventral sac, but paired egg sacs are characteristic of some species.

Order Monstrilloida

Monstrilloids (Figure 17.33), all of which are marine, are odd in that although they are free-living and planktonic as adults, their younger stages are parasitic in annelids, molluscs, and brittle stars. The gut is not functional at any time. The immature stages subsist on nutrients they absorb from the blood of their host, and they also store up reserves that will be needed later by the adult. Whatever mouthparts are present in younger stages disappear; adults have first antennae but no second antennae. The line of separation between the metasome and urosome is in front of the last or next-to-last pregenital segment. Females may carry numerous eggs in two compact sacs, or just a few eggs that are stuck to a pair of outgrowths of the body wall of the genital segment (Figure 17.33, B). Soon after the nauplii hatch they must enter a suitable host. The adults escape into the sea by breaking out through the skin of the animal in which they have reached maturity.

Order Cyclopoida

The more nearly generalized, free-living cyclopoids (Figure 17.28, B; Figure 17.34, A) are usually found in the plankton of freshwater or marine habitats, but some live in sediment. They are like harpacticoids in that the line separating the metasome from the urosome is in front of the pregenital segment. The first antennae, however, are inclined to be longer than those of harpacticoids, and in males they may be specialized for grasping females in preparation for sperm transfer. The second antennae are uniramous, and the palp of the mandibles is reduced or absent. Mature females carry a pair of egg sacs; these are held in a more nearly lateral than ventral position.

Of the cyclopods that live on fishes and various invertebrates, most are parasites. A few, however, scavenge on sloughed-off epithelial cells and similar material. Certain of the cyclopoids associated with other animals are, in terms of their morphology, not far removed from free-living species (Figure 17.34, B and C). At the other end of the spectrum are types that would hardly be recognized as copepods if it were not for their telltale larvae. Among the many remarkable parasitic cyclopoids are those that inhabit the pharynx, blood vessels, and other internal organs of ascidians

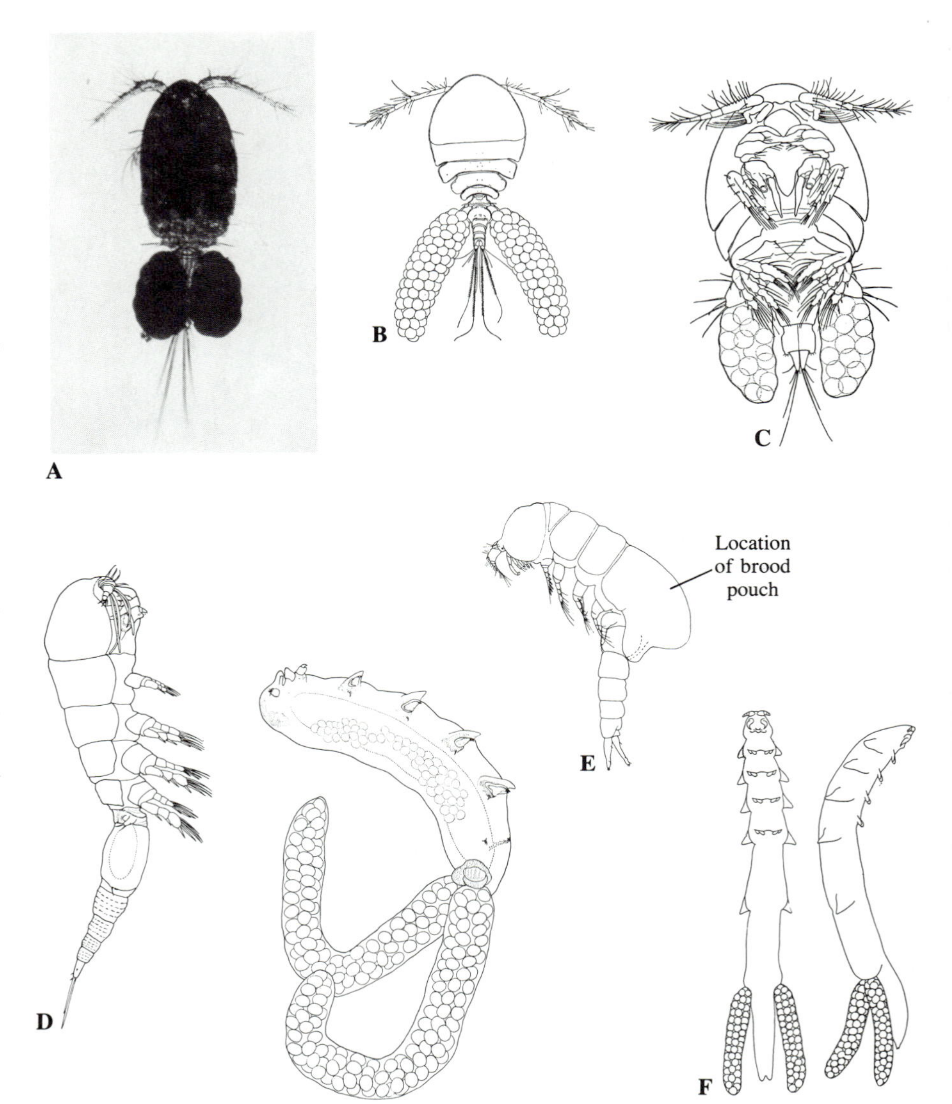

FIGURE 17.34 Class Copepoda, order Cyclopoida. (See also Figure 17.28, B.) A. Female of a marine, sediment-inhabiting species; photomicrograph. B. *Lichomolgus hetaericus,* female, dorsal view. This species is associated with certain octocorals in Madagascar. (Humes, Proceedings of the Biological Society of Washington, *81.*) C. *Clausidium vancouverense,* female, ventral view. This species lives on the body surface of ghost shrimps on the Pacific coast of North America. (Light and Hartman, University of California Publications in Zoology, *41.*) D. *Haplostomella reducta,* male (left) and female (right), a parasite of the compound ascidian *Distaplia occidentalis.* (Ooishi and Illg, Special Publications of the Seto Marine Biological Laboratory, series 5.) E. *Dorotypus pulex,* female, a parasite of various solitary ascidians. (Illg and Dudley, Pubblicazioni della Stazione Zoologica di Napoli, *34.*) F. *Mytilicola porrecta,* female, ventral (left) and lateral (right) views. This species parasitizes various marine bivalve molluscs. (Humes, Journal of Parasitology, *40.*)

(Figure 17.34, D and E). In females of the family Notodelphyidae (Figure 17.34, E), the body is swollen because there is an internal brood pouch within the thoracic region.

Another interesting group of parasitic cyclopoids is represented by *Lernaea,* common on freshwater fishes. The head of the female, with fleshy outgrowths, anchors the copepod in the skin of the host. All appendages are greatly reduced; those belonging to the thorax are widely spaced because the segments in this part of the body are elongated. Males are less strongly modified.

Species of *Mytilicola* (Figure 17.34, F) live in the gut of bivalve molluscs, including sea mussels and oysters. Females, generally not quite 1 cm long, are about twice as large as the males of the same species. The body is slender and the usual copepodan tagma as well as the segments are indistinct. The second antennae are modified for clinging. There are no mandibles, and the maxillae and legs are reduced. In some areas of Europe where the edible mussel *(Mytilus edulis)* is cultivated, reduced productivity is attributed to heavy parasitism by *Mytilicola intestinalis.*

Order Siphonostomatoida

For a long time, a group of copepods parasitizing fishes and certain other vertebrates, including whales, were referred to an order called Caligoida. An important common feature of these copepods is a mouth cone used for sucking in blood and tissue juices. This tube is formed by partial or complete fusion of the labrum, a liplike flap in front of the mouth, with a similar struc-

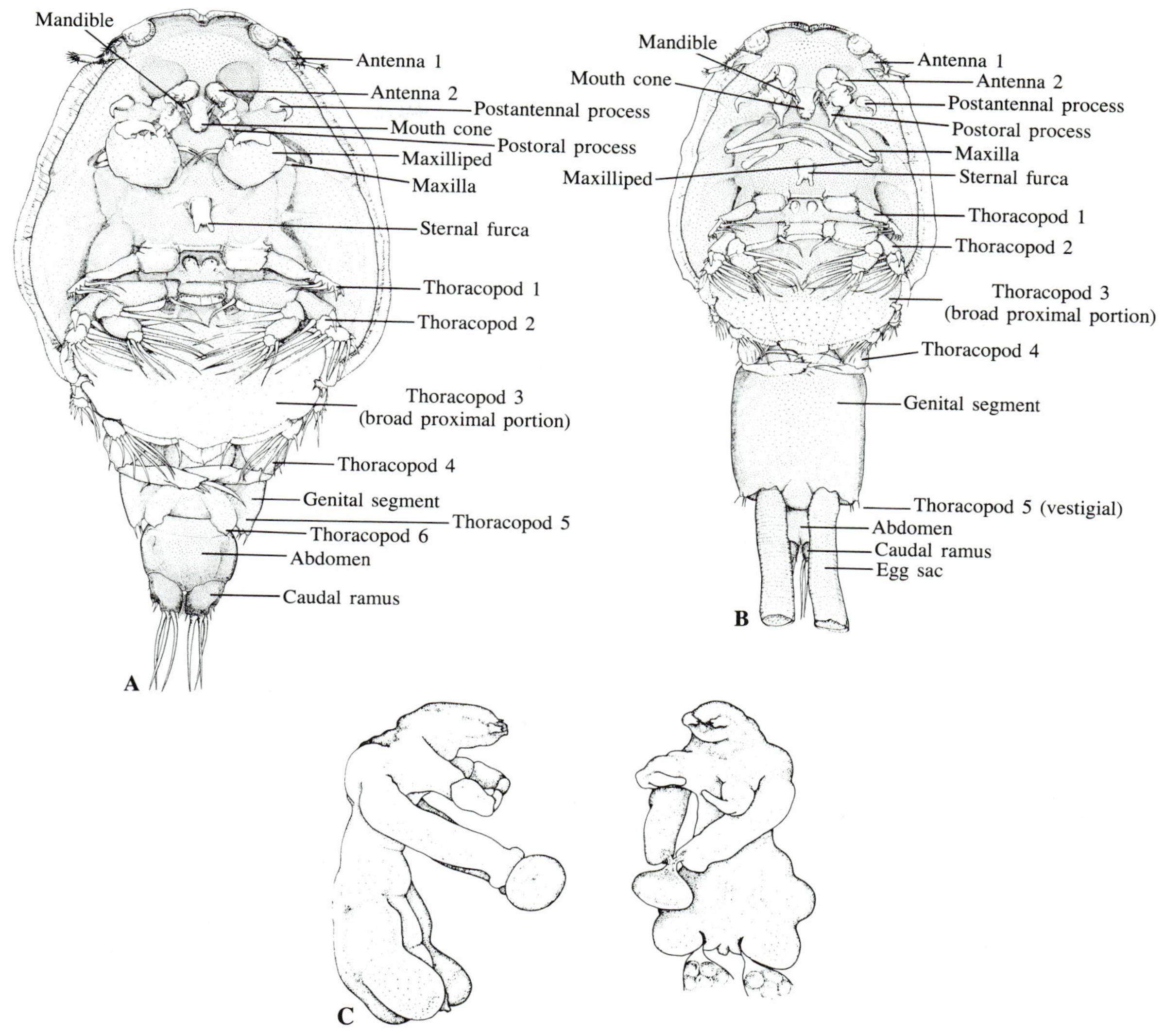

FIGURE 17.35 Class Copepoda, order Siphonostomatoida. A. *Caligus curtus,* male, ventral view. B. *Caligus curtus,* female, ventral view. (A and B, Parker, Kabata, Margolis, and Dean, Journal of the Fisheries Research Board of Canada, *25.*) C. *Salmincola nordmanni,* female, lateral view (left) and ventral view (right). Kabata, Journal of the Fisheries Research Board of Canada, *26.*) Both *Caligus* and *Salmincola* are parasites of fishes.

ture behind the mouth. Some copepods previously thought to fit best in the order Cyclopoida have a similar mouth cone, so the order Siphonostomatoida has been proposed to take in these as well as the ones formerly called caligoids. The more important families are the Caligidae and Lernaeopodidae, which are parasites of fishes.

In adult caligids (Figure 17.35, A and B), which are usually much flattened, the metasome and urosome are distinct, and the separation between these regions of the body is in front of the next-to-last pregenital segment. The first antennae are reduced, and the second antennae, first maxillae, and second maxillae are used for clinging to the host. The small mandibles are inside the mouth cone, near its tip. The paired egg sacs carried by females are large and conspicuous. In some species, the first copepodid stage, on contacting a suitable host, shoots out a filament from a little pocket on its head, and this serves as a temporary anchor.

In the lernaeopodids, such as *Salmincola* (Figure 17.35, C), there is no distinct separation of metasome and urosome, and segmentation in general is obscure.

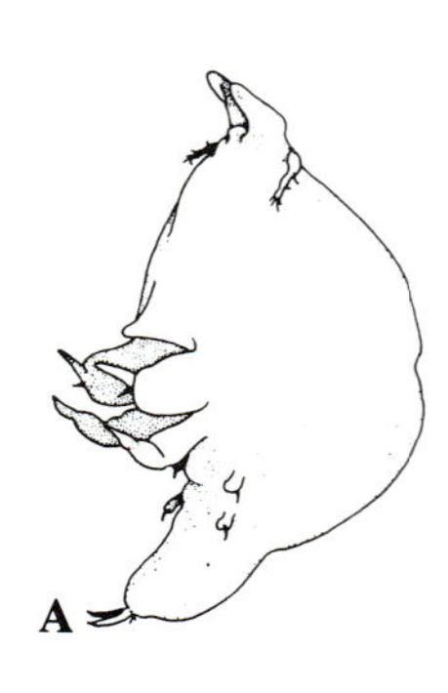

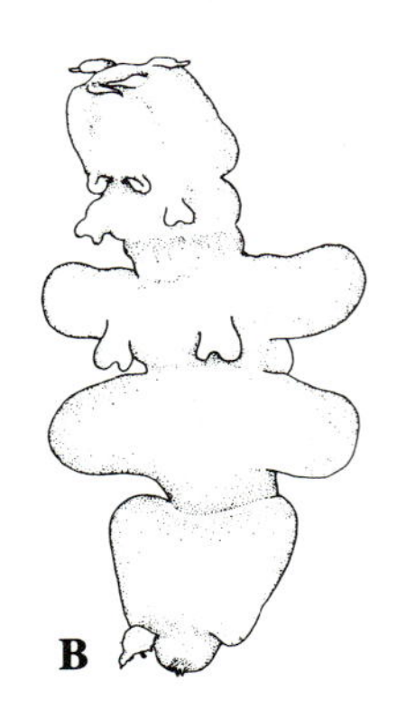

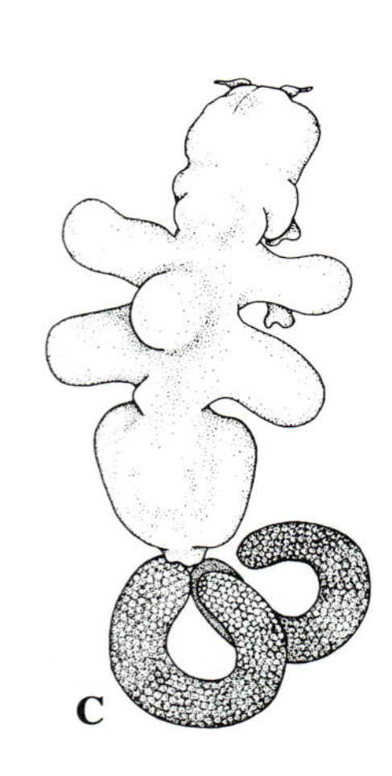

FIGURE 17.36 Class Copepoda, order Poecilostomatoida. *Chondrocanthus narium,* parasitic on marine fishes. A. Male, lateral view. B. Female, ventral view. C. Female with egg masses, dorsal view. The male, not quite 1 mm long, is much smaller than the female. (Kabata, Journal of the Fisheries Research Board of Canada, *26*.)

The large second maxillae are specialized for "hugging" the gill filaments of fishes, or they may be fused at their tips to form a device that grips the tissue. Another odd feature of lernaeopodids is that the bases of the maxillipeds, during development, move up ahead of the second maxillae. In some species the males are dwarf and permanently attached to females. The copepodid larvae establish contact with the host by a filament extruded from the head.

Order Poecilostomatoida

The Poecilostomatoida, associated with various invertebrates and fishes, have until recently been treated as a subgroup of Cyclopoida. Some are extremely modified for a parasitic existence; others resemble generalized cyclopoids. The mandibles are distinctive in being sickle-shaped. There are other differences between the mouthparts of poecilostomes and those of cyclopoids, but these need not be discussed in detail here. *Chondracanthus* (Figure 17.36), *Ergasilus,* and *Bomolochus* are among the more important genera; they live on marine fishes.

References on Copepoda

Free-Living Copepoda

Boxhall, G. A., and Schminke, H. K., 1988. Biology of Copepods. Kluver Academic Publishers, Dordrecht, Boston, and London.

Cannon, H. G., 1929. Feeding mechanism of the copepods *Calanus* and *Diaptomus*. Journal of Experimental Zoology, *6*:131–144.

Corner, E. D. S., 1961. The feeding of the marine copepod *Calanus helgolandicus*. Journal of the Marine Biological Association of the United Kingdom, *41*:5–66.

Fahrenbach, E. W., 1962. The biology of a harpacticoid copepod. La Cellule, *62*:301–376.

Fox, B. W., 1977. Feeding behavior of *Calanus pacificus* in mixtures of food particles. Limnology and Oceanography, *22*:472–491.

Fryer, C., 1957. The feeding mechanism of some freshwater cyclopoid copepods. Proceedings of the Zoological Society of London, *129*:1–25.

Gurney, R., 1931–1933. British Freshwater Copepoda. 3 volumes. Ray Society, London.

Hardy, R. C., 1956. The Open Sea. Collins, London.

Hopkins, C., 1982. The breeding biology of *Euchaeta norvegica* in Loch Etive, Scotland. Journal of Experimental Marine Biology and Ecology, *60*:71–102.

Koehl, M. A. R., and Strickler, J. R., 1981. Copepod feeding currents: food capture at low Reynolds number. Limnology and Oceanography, *26*:1062–1073.

Lang, K. A., 1948. Monographie der Harpacticiden. Nordiska Bokhandeln, Stockholm.

Marcotte, B. M., 1977. An introduction to the architecture and kinematics of harpacticoid feeding: *Tisbe furcata*. Mikrofauna des Meeresbodens, *61*:183–196.

Marshall, S. M., 1973. Respiration and feeding in copepods. Advances in Marine Biology, *11*:57–120.

Marshall, S. M., and Orr, A. P., 1955. The Biology of a Marine Copepod. Oliver and Boyd, Edinburgh. (Reprinted 1972, with additions, by Springer-Verlag, New York, Heidelberg, and Berlin.)

Paffenhöfer, G. A., Strickler, J. R., and Alcaraz, M., 1982. Suspension feeding by herbivorous calanoid copepods: a cinematographic study. Marine Biology, *67*:193–199.

Parasitic Copepoda

Dudley, P. L., 1966. Development and Systematics of Some Pacific Marine Symbiotic Copepods. The Biology of the Notodelphyidae. University of Washington Press, Seattle.

Friend, C., 1957. Life-history and ecology of the salmon gill-maggot *Salmincola salmonea*. Transactions of the Royal Society of Edinburgh, *60*:503–541.

Gotto, R. V., 1979. The association of copepods with marine invertebrates. Advances in Marine Biology, *16*:1–109.

Grabda, J., 1963. Life cycle and morphogenesis of *Lernaea cyprinacea*. Acta Parasitologica Polonica, *11*:169–198.

Graininger, J. N. R., 1951. The biology of the copepod *Mytilicola intestinalis*. Parasitology, *41*:135–142.

Humes, A. C., and Stock, J. H., 1973. A revision of the family Lichomolgidae Kossmann, 1877, cyclopoid cope-

pods associated mainly with marine invertebrates. Smithsonian Contributions to Zoology, no. 127.
Kabata, Z., 1979. Parasitic Copepoda of British Fishes. Ray Society, London.
Ooishi, S., and Illg, P. L., 1977. Haplostominae (Copepoda, Cyclopoida) associated with compound ascidians from the San Juan Archipelago and vicinity. Special Publications from the Seto Marine Laboratory, Series 5.
Yamaguti, S., 1963. Parasitic Copepoda and Branchiura of Fishes. Interscience Publishers, New York.

CLASS BRANCHIURA

Branchiurans (Figure 17.37), found in fresh water as well as in the sea, are flattened crustaceans that cling to fishes and sometimes to amphibians. They feed on blood, tissue fluids, and mucus, and most of them are intermittent parasites that spend at least some time swimming around. In general, it appears that they are not attracted to fishes and do not recognize fishes by their shadows. If touched by a fish, however, a branchiuran will usually respond positively and may be successful in establishing tight contact with it.

The genus *Argulus,* with about 100 species, is representative, and shows some interesting specializations for parasitism. The body is much flattened, and the basal portions of the first maxillae are modified to form suckers for attachment. These appendages do not function as mouthparts. Neither do the second maxillae, although these at least look like limbs. The sharp mandibles are situated near the tip of a tube called the *proboscis* and can be turned out so that they can be used to cut through the skin of a fish and initiate the uptake of liquid food. Even the second antennae of *Argulus,* being small and hooklike, are specialized to serve as aids to parasitism. A peculiar feature found only in the genus *Argulus* is the poison spine, located just anterior to the proboscis (Figure 17.37, B). This definitely produces a venom of some sort. Heavily infested fishes may be seriously affected by a combination of venom, mechanical damage, and secondary infection of wounds by fungi or bacteria.

On the dorsal surface of the head region are a median eye, which can be traced back to the nauplius stage, and a pair of compound eyes. The head is united to the first of the four thoracic segments. The lateral outgrowths of the thoracic segments, together with the similar lateral expansions of the head, form a sort of carapace. It is this carapace, in fact, that gives an *Argulus* its much-flattened appearance. All of the thoracic segments have biramous swimming appendages. The abdomen is not visibly segmented. The anus is at the base of a deep median notch, and close to it are two little lobes that are interpreted as caudal rami.

The gut of *Argulus* is adapted for taking up a large meal while it can be had. A lateral diverticulum on each side of the midgut branches into a system of caeca that go out into the carapace. There is a heart, with a single pair of ostia, in the fourth thoracic segment.

In females, there is a single ovary but two oviducts, only one of which functions at a particular time. The

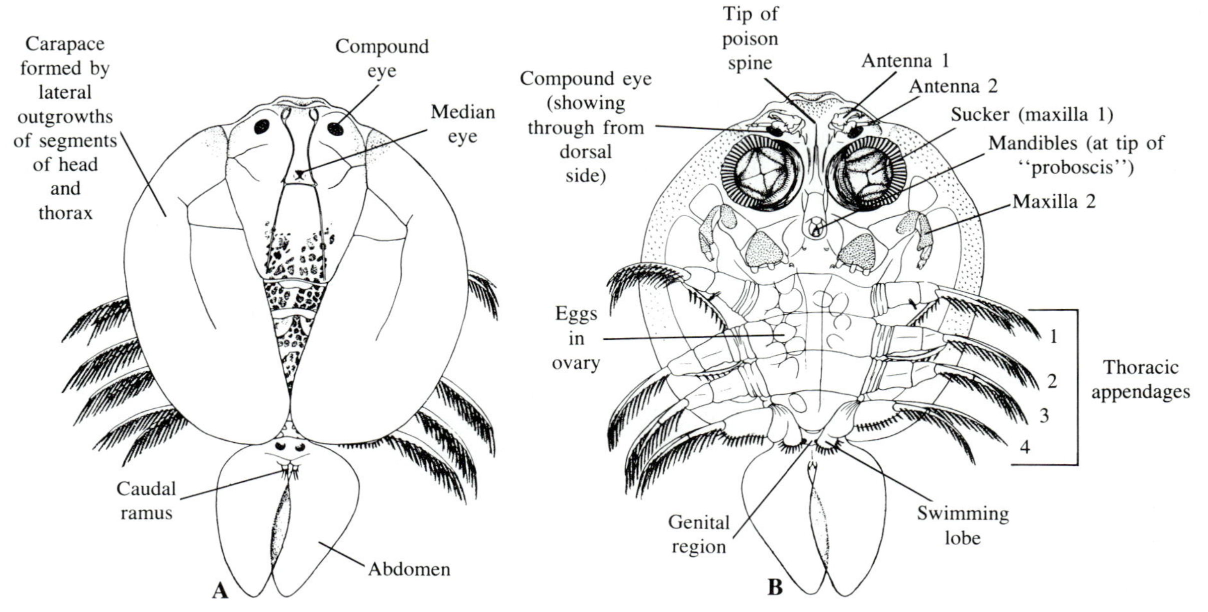

FIGURE 17.37 Class Branchiura. *Argulus mongolianus,* female. A. Dorsal view. B. Ventral view. (Tokioka, Annotationes Zoologicae Japonenses, *18.*)

genital pores are situated in a pocket on the fourth thoracic segment, just behind the limbs. Also opening into this pocket are seminal receptacles. The testes of males are paired. They lie in the abdomen, but their ducts run to the thorax, where they unite to form a seminal vesicle. From this, two ducts extend backward and join just before reaching a genital pore located between the limbs of the fourth segment. Mating takes place while a female is attached to a fish. The male settles on her back, twists to get his free thoracic segments under her, and delivers sperm to the seminal receptacles.

When a female *Argulus* is ready to lay eggs, she detaches herself from the host. With the help of a secretion of the functional oviduct, she sticks the eggs to an aquatic plant, submerged twig, or rock. The number of eggs laid at one time varies, but it may be more than 400 in some species. By going back temporarily to a life of parasitism, a female can prepare herself for another episode of egg-laying.

In the South American genus *Dolops,* the male places a spermatophore on a pair of spines on whose tips the seminal receptacle pores are located. The sperm enter the seminal receptacles, but later, as the eggs are being laid, the spines puncture the eggs and inject sperm into them.

The nauplius and metanauplius stages are undergone within the egg. When an *Argulus* hatches, it resembles the adult, but its first antennae, second antennae, and mouthparts are not yet fully specialized for the functions they will have to perform. The thorax already has four pairs of thoracic appendages, and in certain species all four can be used right away for swimming. As a rule, however, only the first thoracic appendages are well developed. The abdomen lacks the prominent notch it will acquire later, but the caudal rami are distinct. Parasitism begins immediately, with hooks on the second antennae and first maxillae serving as anchors. Maturity is not attained until after several molts.

References on Branchiura

Cressey, R. F., 1972. The genus *Argulus* (Crustacea; Branchiura) of the United States. Biota of Freshwater Ecosystems, No. 2. United States Government Printing Office, Washington, D. C.

Shimura, S., 1981. The larval development of *Argulus coregoni*. Journal of Natural History, *15*:331–348.

Stammer, J., 1959. Morphologie, Biologie, und Bekämpfung der Karpfenläuse. Zeitschrift für Parasitenkunde, *19*:135–208.

Yamaguti, S., 1963. Parasitic Copepoda and Branchiura of Fishes. Interscience Publishers, New York.

CLASS CIRRIPEDIA

As a group, the barnacles and their relatives are the most highly specialized crustaceans. The majority of them, in the adult stage, are attached to hard surfaces or burrow in calcareous substrata, including shells of molluscs and skeletons of corals. A few are parasites of other invertebrates, especially decapod crustaceans, echinoderms, and cnidarians. The parasitic types may be so greatly modified that they do not appear to be crustaceans, let alone cirripedes, but the early stages in their life histories disclose their affinities.

The anatomy of acorn barnacles, such as *Balanus* (Figure 17.38, A–C; Figure 17.39, A–C) or *Chthamalus* (Figure 17.38, D), will provide a good introduction to the group. The appendage-bearing part of the body is enclosed by a modified carapace that is fleshy on the inside but covered externally by a circle of six calcareous plates: the *carina, rostrum,* and two *lateral plates* on each side (Figure 17.39, C). There is also a valvelike closing apparatus that consists of four additional plates: two *terga* and two *scuta*.

The head region of an adult barnacle lacks first and second antennae, but mandibles and first and second maxillae are present. The six pairs of thoracic appendages, called *cirri,* are biramous, and both the endopodites and exopodites are covered with setae that filter small food particles from the water. There is no distinct abdomen, and the anus lies at the end of the thorax, where a long penis also originates.

The two scuta of the closing apparatus of a barnacle are connected by a strong adductor muscle, and both terga and scuta can be pulled down tightly by depressor muscles. When a barnacle is closed up, the pressure within it is fairly high. When the adductor and depressor muscles relax, blood flows out into the cirri. These appendages are thus extended and emerge from the opening between the plates. In acorn barnacles, the cirri rhythmically comb the water for food, which consists of diatoms and other minute particles.

The cirri may all be much alike in size and form, but one or more pairs of anterior cirri may be differentiated to function as food-handling devices. The maxillae and mandibles are concerned with final sorting of the food before it is swallowed. The foregut is both muscularized and cuticularized, and leads into a U-shaped midgut provided with a pair of caeca. The hindgut, like the foregut, is cuticularized.

There is no heart in the circulatory system. Movement of blood through hemocoelic spaces usually depends on muscular activities going on in the body. Respiration probably takes place over much of the body surface, especially that of the cirri and the inner part of the mantle, which may be extensively folded. When a barnacle's cirri are withdrawn and its terga and scuta are pulled down, oxygen uptake is reduced and metabolism is lowered.

In goose barnacles, such as *Lepas* (Figure 17.40, A and B) and *Pollicipes* (Figure 17.40, C and D), the

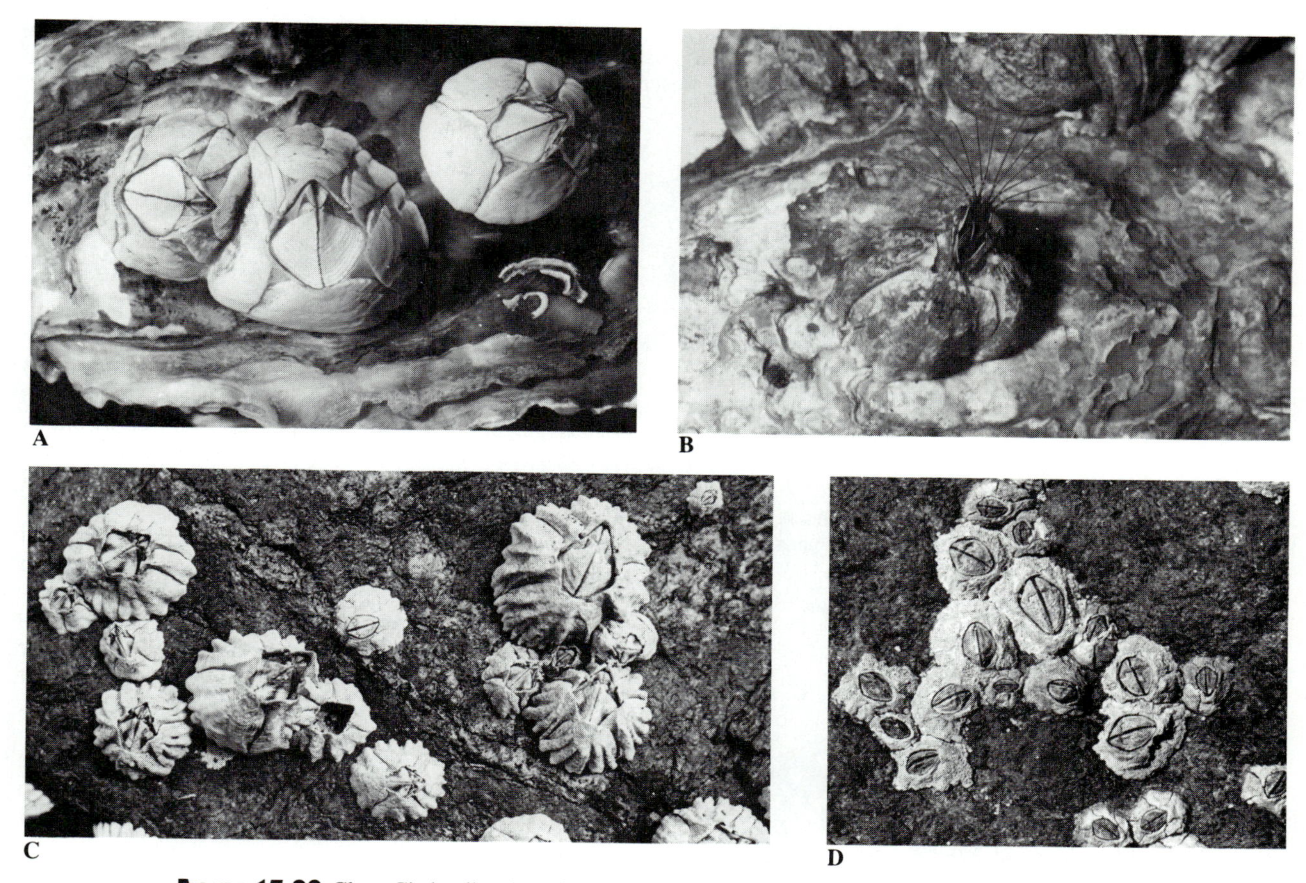

FIGURE 17.38 Class Cirripedia. A variety of acorn barnacles. A. *Balanus crenatus*. B. *Balanus crenatus,* with its feeding cirri extended. C. *Balanus glandula*. D. *Chthamalus dalli*.

shelly portion is at the end of a fleshy stalk. In *Lepas,* the two terga, two scuta, carina, rostrum, and lateral plates correspond to those of an acorn barnacle. The carina, rostrum, and laterals are not tightly joined, however, and the mantle is partly exposed. In *Pollicipes,* many small lateral plates develop.

Some goose barnacles have a heartlike structure called the *blood pump*. This lies near the esophagus and drives blood into the stalk; there is a valve in the main vessel, so that backflow is minimized. Another feature of certain types is the presence of slender outgrowths, probably respiratory, arising from the body near the bases of the thoracic appendages.

Most barnacles are hermaphroditic. The female genital openings are on the first thoracic segment, a feature decidedly unique among Crustacea. In goose barnacles, the ovary is in the stalk (Figure 17.40, D), but in acorn barnacles it is in the thoracic region. The testis is always in the thorax, and the vas deferens leads to an extensible penis at the hind end of the thorax. Although some barnacles shed their male gametes into the sea, most engage in copulation, using the penis to reach over to a neighbor and deposit spermatophores in the mantle cavity.

The eggs of barnacles, fertilized in the mantle cavity, develop into nauplii (Figure 17.41, A and B). In having a median naupliar eye and three pairs of appendages, homologous to first and second antennae and mandibles, these nauplii resemble those of copepods and some other crustaceans. They are distinctive in appearance, however, because of the hornlike anterolateral projections. The nauplii are released after they have molted a few times. Following the last molt, they transform into a completely different type of larva called the *cypris* (Figure 17.41, C). This resembles an ostracode, for it has a bivalved carapace. The first antennae and the mandibles of the nauplius are passed on to the cypris, but the second antennae are lost. The remaining

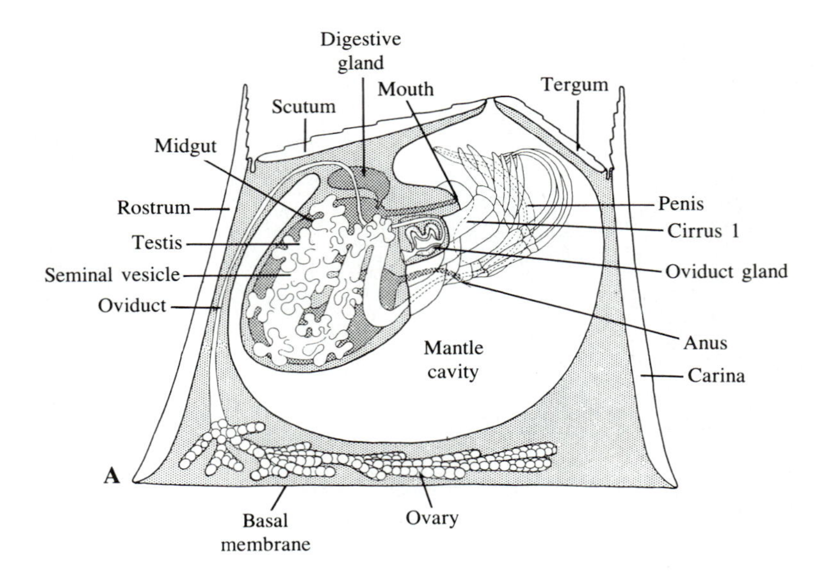

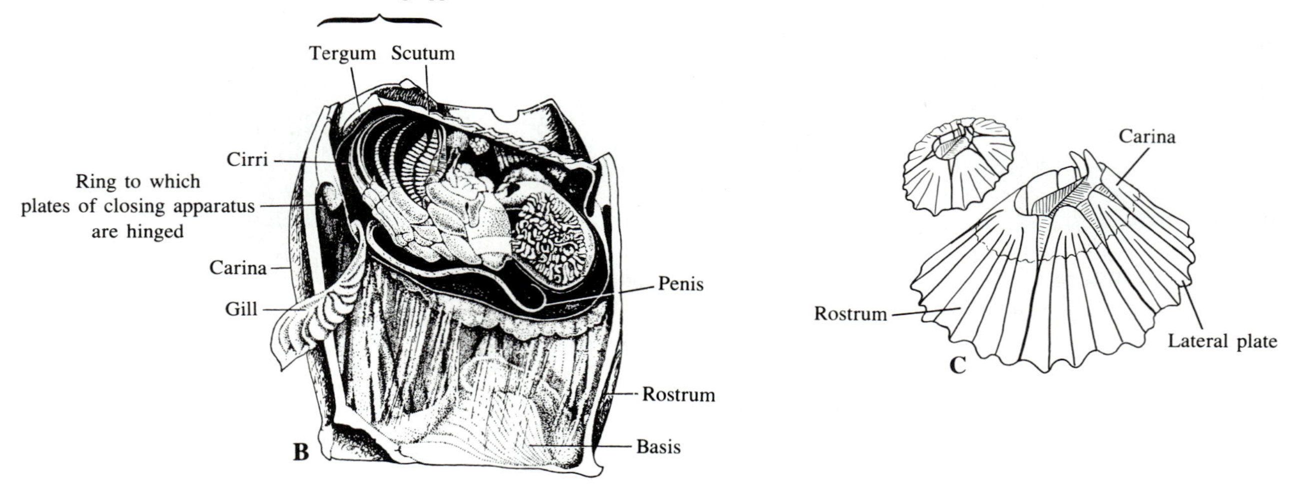

FIGURE 17.39 Class Cirripedia. Internal anatomy of acorn barnacles. A. *Balanus,* sectioned sagittally; diagrammatic. (Walley, Journal of the Marine Biological Association of the United Kingdom, *45*.) B. *Balanus,* sectioned sagittally, showing some features not illustrated by A. (Costlow and Bookhout, Biological Bulletin, *113*.) C. *Balanus balanus,* views of the shell at two growth stages. Note that the oldest part of the shell remains around the aperture, as new shell material is deposited beneath it. Shell material is also added at the edges of the shell plates. (Gutman, Natur und Volk, *91*.)

appendages of the cypris are the two pairs of maxillae and six pairs of biramous cirri. Besides the naupliar eye, there is a compound eye on each side of the anterior part of the body.

The cypris larva is unable to feed, but it uses its thoracic appendages for swimming. If it lands on a suitable substratum it will transform into a barnacle. On settling, it walks on its first antennae as if these were stilts, searching for the right place. Finally it becomes permanently attached by its first antennae. The compound eyes disappear, and the body becomes drastically reorganized. Beneath the bivalved shell, which is soon cast off, the plates that form the wall and the closing apparatus of the adult start to develop in the newly differentiated mantle tissue. The carina of the wall and the two terga of the closing apparatus originate on the dor-

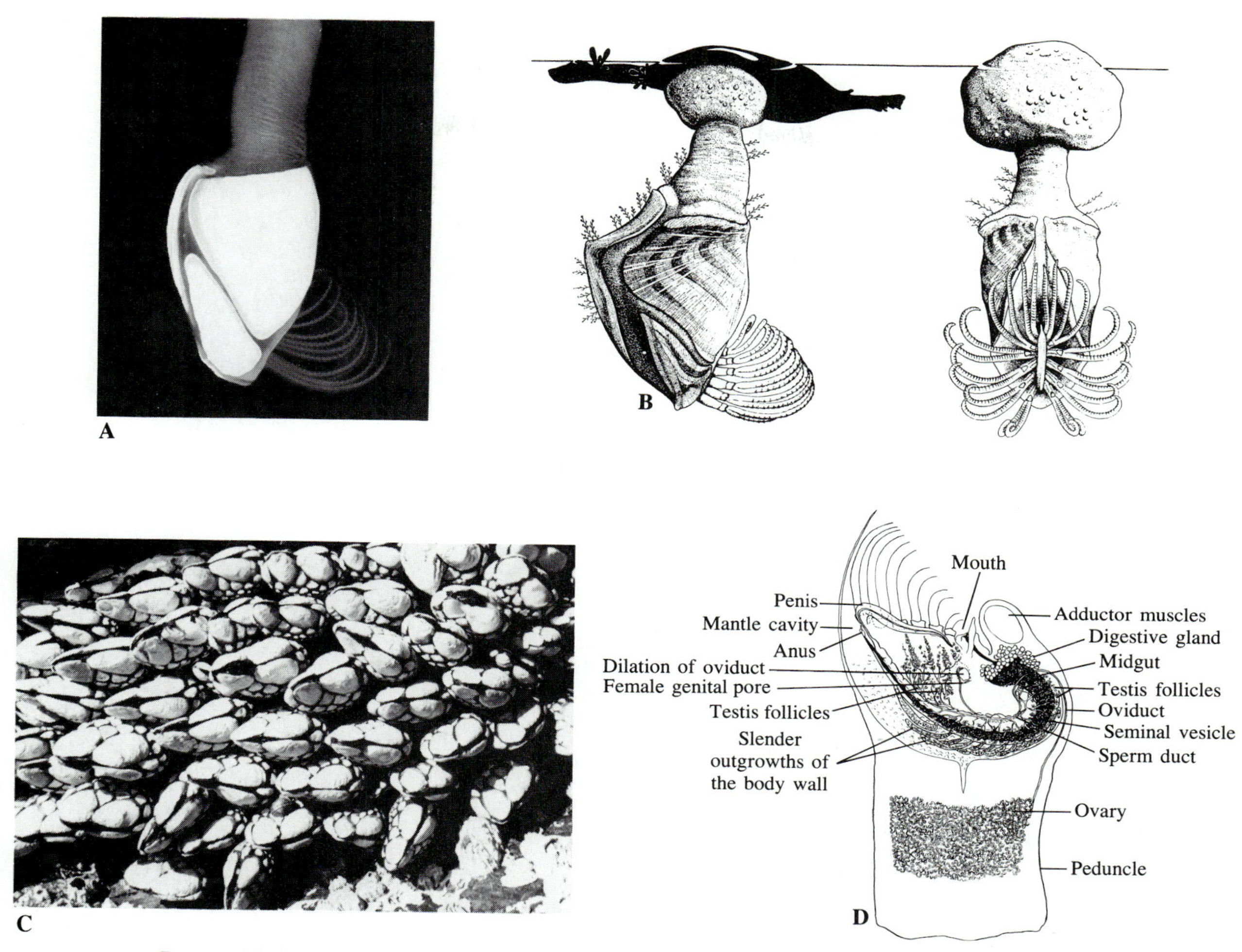

FIGURE 17.40 Class Cirripedia. A. *Lepas anatifera,* a pelagic goose barnacle, which lives attached to timbers and other floating objects. The feeding cirri are extended. B. *Lepas fascicularis,* a young individual in the process of secreting its own float (left) and after the float has become functional (right). (Ankel, Natur und Volk, *80*.) C. An aggregation of *Pollicipes polymerus,* which is found in rocky intertidal habitats on the Pacific coast of North America. D. Some aspects of the internal anatomy of *Pollicipes*. (Howard, Biological Bulletin, *119*.)

sal side of the cypris, whereas the rostrum of the wall and the two scuta of the closing apparatus originate on its ventral side. Two lateral plates develop in the wall on the right and left. In some acorn barnacles, the ones nearest the rostrum fuse with it; in others, the two on each side fuse. Various other arrangements of the plates may materialize, and certain of the plates may become secondarily divided.

The naupliar eye, usually reddish, may survive in the adult, but it is buried rather deep in the head region, remote from the brain. It continues to function as a photoreceptor.

In goose barnacles, a somewhat similar metamorphosis takes place, although the final form is decidedly different from that of acorn barnacles. The fleshy stalk is essentially a hypertrophied head region, for the first antennae were originally at the point where the barnacle became attached to the substratum.

Order Thoracica

The acorn and goose barnacles, already described in some detail, belong to the Thoracica. The group is unified by two main features: the mantlelike carapace, covered by calcareous plates, and six pairs of cirri. Thora-

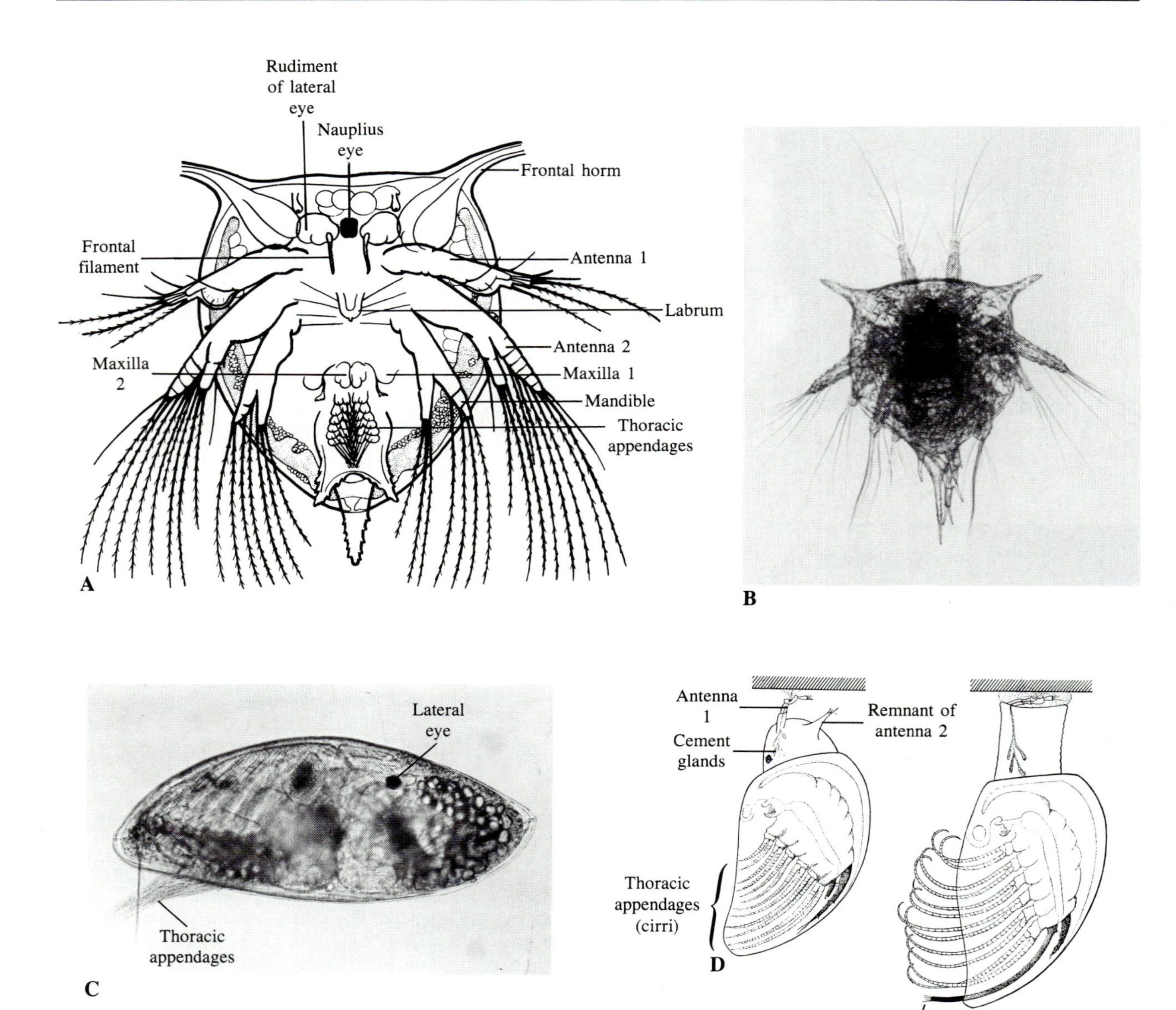

FIGURE 17.41 A. Nauplius larva of a goose barnacle, *Pollicipes spinosus*. (Batham, Transactions of the Royal Society of New Zealand, *75*.) B. Nauplius of a barnacle; photomicrograph. C. Cypris of a barnacle, photomicrograph. D. Attachment and metamorphosis of a cypris larva into a goose barnacle. (Ankel, Natur und Volk, *80*.)

cicans live attached to rocks, wood, shells, turtles, the tough skins of whales, and other firm substrata. Most of them are hermaphroditic. In some, however, sexes are separate. The males are small and unable to feed, and they do not look much like barnacles; they are permanently affixed to the females. Small complemental males are also known to occur in certain hermaphroditic species.

Order Acrothoracica

Acrothoracicans (Figure 17.42, A), most of which are only a few millimeters long, burrow into shells (especially snail shells inhabited by hermit crabs), skeletons of corals, and plates of barnacles. In at least certain genera, penetration of calcareous structures is promoted by secretion of carbonic anhydrase. Some mechanical abrasion, however, is effected by firm cuti-

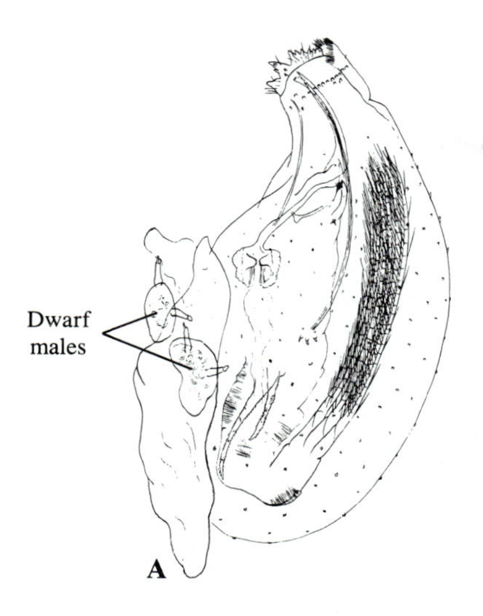

FIGURE 17.42 Class Cirripedia. A. *Cryptophialus heterodontus* (order Acrothoracica), which burrows into shells of molluscs. Two dwarf males are associated with the female. (Tomlinson, United States National Museum, Bulletin 296.) B. Female of a species of *Dendrogaster*) (order Ascothoracica), a parasite in the body cavity of sea stars. (Photograph courtesy of George Shinn.)

cular spines on the mantle. There are no calcareous plates. The first cirri are specialized for transferring food collected by the other cirri to the mouthparts. The total number of pairs may be six, as in thoracicans, but some genera have fewer. The gut is complete in certain types, blind in others. Sexes are separate. The males are dwarfs whose development does not progress very far beyond that of the cypris stage. They are attached to females by their first antennae, which are their only appendages. Acrothoracican nauplii and cyprids resemble those of other barnacles.

Order Ascothoracica

Ascothoracicans form a small group of parasites that live on or in cnidarians of the class Anthozoa, or within the perivisceral coelom of echinoderms. They would be difficult to recognize as relatives of barnacles were it not for the fact that females release cypris larvae. Even the nauplii, retained in a brood chamber, are not typical of the Cirripedia, for they lack anterolateral "horns."

Adult ascothoracicans usually have a bivalved carapace basically like that of the cypris. In females, however, the lateral portions of the body may become so enlarged that they nearly hide the other parts. Males live attached to females, or within the mantle cavity of females. They are small, but much less bizarre than females. In most species, both sexes have a distinct abdomen, a feature that other cirripedes have lost. The much-branched female of *Dendrogaster* (Figure 17.42, B), however, has neither abdomen nor valves.

There are first antennae, but no second antennae after the nauplius stage. The mouthparts may be specialized, along with the conical labrum, for piercing and sucking, or they may be reduced in types that absorb nutrients, as *Dendrogaster* is presumed to do. The six pairs of thoracic appendages are generally obvious in males, but they may be vestigial or lost altogether in females.

It should be noted that not all specialists assign the Ascothoracica to the Cirripedia. Some consider them to belong to a separate group close to the Cirripedia.

Order Rhizocephala

Rhizocephalans—the "root-headed" cirripedes—are parasites of other crustaceans, especially shrimps, crabs, hermit crabs, and other members of the Decapoda. Only during their larval stages can their affinity with barnacles be recognized. Approximately 400 species have been described. Some, including those belonging to the genera *Loxothylacus* (Figure 17.43, A) and *Sylon* (Figure 17.43, C), occur singly; others, such as *Sacculina gregaria* (Figure 17.43, B) and *Peltogasterella gracilis* (Figure 17.43, D), typically occur in groups of several to many individuals.

The portion of a rhizocephalan visible outside its host is called the **externa.** It is part of a female parasite and consists mostly of a gonad and a so-called *mantle cavity* in which eggs are fertilized and in which they are usually brooded to the nauplius stage. The externa does not appear until the parasite has been in the host for at

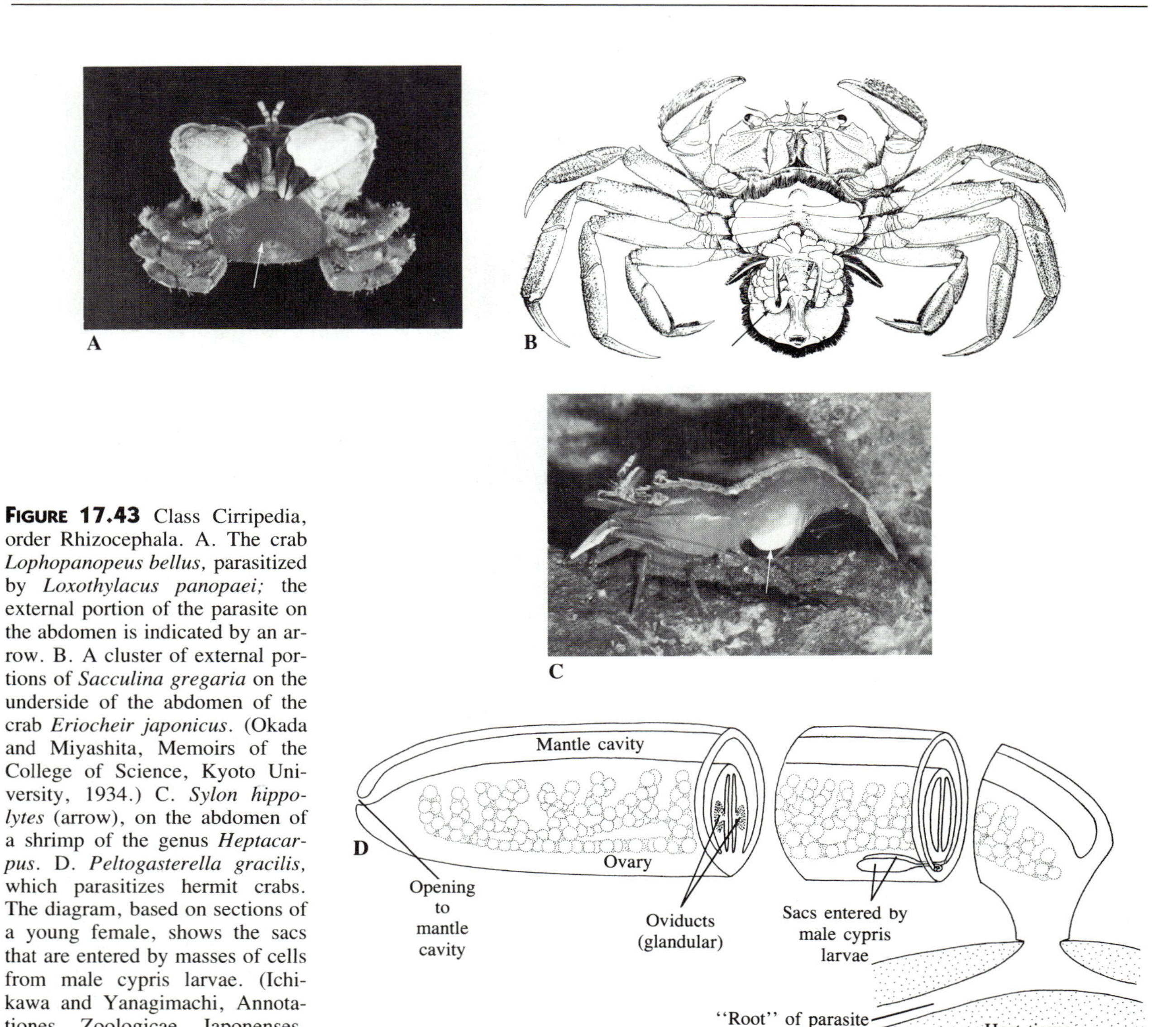

FIGURE 17.43 Class Cirripedia, order Rhizocephala. A. The crab *Lophopanopeus bellus,* parasitized by *Loxothylacus panopaei;* the external portion of the parasite on the abdomen is indicated by an arrow. B. A cluster of external portions of *Sacculina gregaria* on the underside of the abdomen of the crab *Eriocheir japonicus*. (Okada and Miyashita, Memoirs of the College of Science, Kyoto University, 1934.) C. *Sylon hippolytes* (arrow), on the abdomen of a shrimp of the genus *Heptacarpus*. D. *Peltogasterella gracilis,* which parasitizes hermit crabs. The diagram, based on sections of a young female, shows the sacs that are entered by masses of cells from male cypris larvae. (Ichikawa and Yanagimachi, Annotationes Zoologicae Japonenses, *31*.)

least a few months. During the internal phase, the rhizocephalan forms a system of rootlike tubes within the tissues of the host; sometimes these tubes penetrate nearly all parts of the body. Eventually, a sac that will become the externa is formed, and this emerges when a patch of cuticle on the host's abdomen disintegrates.

Figure 17.44 shows some of the stages in the life cycle of *Sacculina carcini,* which parasitizes the European crab *Carcinus maenas*. The larvae released from the mantle cavity of an externa become cyprids. A cyprid destined to develop into a female becomes attached by its first antennae to a young crab, usually at the base of an appendage or at the base of a seta, where the cuticle is thin. The cypris now undergoes a remarkable transformation. It sheds its bivalved carapace, eyes, and all appendages except the first antennae. The part that remains is much smaller, and is called a **kentrogon.** Between the first antennae, a hollow stylet appears, and this is pushed through the cuticle of the crab. It acts as a hypodermic needle through which several cells of the kentrogon enter the hemocoel. (In some rhizocephalans, only one cell is injected.)

It is likely that the injected cell or cells are carried in the blood for a time, but soon a young internal parasite becomes evident. In the case of most rhizocephalans that have been studied, this is on the midgut. As it grows, it sends out the rootlike processes that pervade the tissues. Parasitized hosts are usually rendered repro-

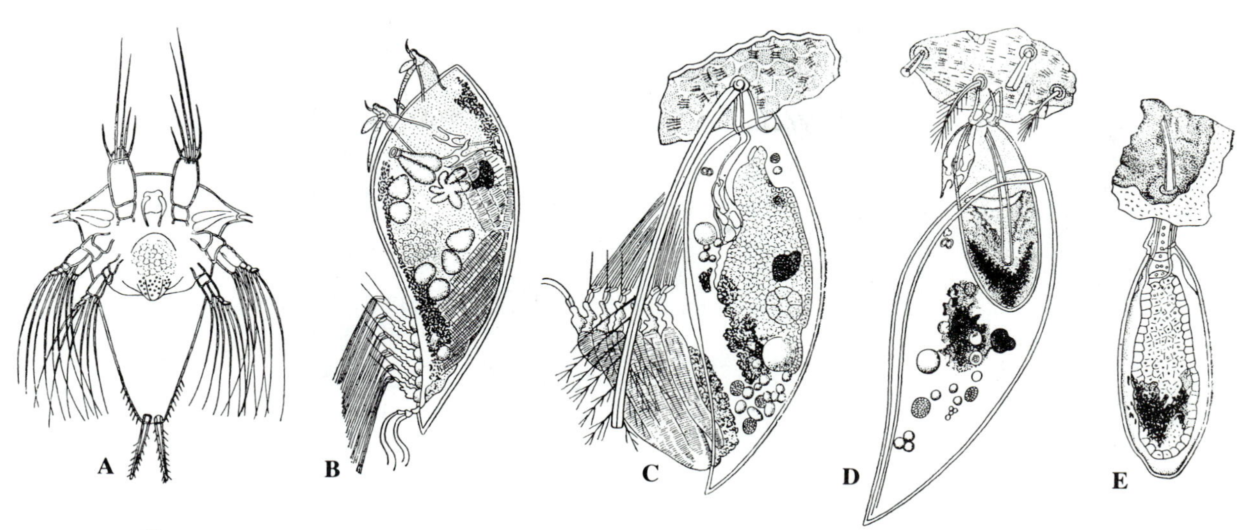

Figure 17.44 Early stages in the life cycle of *Sacculina carcini,* which parasitizes a European crab, *Carcinus maenas*. A. Nauplius. B. Cypris. C. Attachment of the cypris, by its antennae, to the cuticle of the crab, and the sloughing off of its locomotor appendages. D. The kentrogon, a sac of surviving tissue, some cells of which will enter the crab. E. Entry of cells of the kentrogon. (Dogiel, General Parasitology, after Delage, Archives de Zoologie Expérimentale et Générale, Series 2, *2*.)

ductively sterile, and, except for some hermit crabs, do not molt after the externa appears. If the host is a male, it generally develops female secondary sex characters and female grooming behavior; thus it keeps the parasite's protruding externa clean, just as a female would keep its own eggs attached to its abdomen clean.

In several species of rhizocephalans, the nauplii and cyprids (and even the eggs from which the nauplii develop) are of two sizes: the smaller ones are female, the larger ones males. A single adult female may produce numerous broods of nauplii, there being from a few hundred to 15,000 nauplii in a brood. A particular female may produce purely female nauplii for a while, then shift to the production of males or mixtures of males and females.

How are the eggs in the mantle cavity of the externa of a female parasite fertilized? This is one of the most remarkable phenomena known to occur in crustaceans. After an opening into the mantle cavity has appeared on an externa, male cyprids are attracted to it. Within each cypris, a mass of cells covered by a cuticle, sometimes spiny, has already become differentiated. Depending on whether there are one or two receptacles in the externa, one or two cyprids discharge their sacs, which travel to the receptacles and shed their cuticular coverings. These function as plugs that prevent sacs released by other male cyprids from entering. The cells contributed by a male cypris are spermatogonia, and it is from them that mature sperm derive. These fertilize the eggs discharged into the mantle cavity by the female, and the eggs are held together, while being brooded to the nauplius stage, by a secretion of the colleteric ("glue-producing") glands.

In several genera of rhizocephalans, the externa of the female does not have a mantle cavity or mantle aperture until late in its development. It is likely that spermatogonial cells are injected into the externa by one or more male kentrogons, or that they are injected into the host and then enter the female.

Some rhizocephalans hatch as cyprids rather than as nauplii. The cyprids may lack appendages. They work their way into the tissues of the host without benefit of a kentrogon stage. The externa lacks receptacles, and spermatogenesis takes place in islets of male cells located in the mantle cavity. The origin of the male cells has not yet been explained.

References on Cirripedia

Free-Living Barnacles

Anderson, D. T., 1981. Cirral activity and feeding in the barnacle *Balanus perforatus,* with comments on the evolution of feeding mechanisms in thoracican cirripedes. Philosophical Transactions of the Royal Society of London, B, *291*:411–449.

Barnes, H., 1970. A review of some factors affecting settlement and adhesion in the cyprids of some common barnacles. *In* Manley, R. S. (editor), Adhesion in Biological Systems, pp. 89–111. Academic Press, New York.

Barnes, H., and Barnes, M., 1954. The biology of *Balanus balanus*. Oikos, *5*:63–76.

Barnes, H., and Reese, E. S., 1959. Feeding in the pedunculate cirripede *Pollicipes polymerus*. Proceedings of the Zoological Society of London, *132*:569–585.

Bernard, F. J., and Lane, C. E., 1962. Early settlement and metamorphosis of the barnacle *Balanus amphritrite*. Journal of Morphology, *110*:19–39.

Bourget, E., and Crisp, D. J., 1975. Factors affecting deposition of the shell in *Balanus balanoides*. Journal of the Marine Biological Association of the United Kingdom, *55*:231–249.

Gutmann, W. F., 1960. Funktionelle Morphologie von *Balanus balanus*. Abhandlungen der Senckenbergische Naturforschende Gesellschaft, *500*:1–43.

Newman, W. A., and Ross, A., 1976. Revision of the balanomorph barnacles, including a catalogue of the species. Memoirs of the San Diego Society of Natural History, *9*:1–108.

Southward, A. J., 1955. Feeding of barnacles. Nature, *175*:1124–1125.

Southward, A. J. (editor), 1987. Barnacle Biology. (Crustacean Issues, volume 5.) A. A. Balkema, Rotterdam.

Starbird, E. A., 1973. Friendless squatters of the sea. National Geographic, *144*:623–633.

Burrowing and Parasitic Cirripedia

Foxon, C. E. H., 1940. Life history of *Sacculina carcini*. Journal of the Marine Biological Association of the United Kingdom, *24*:253–264.

Grygier, M. J., 1983. *Ascothorax*, a review with descriptions of nine new species and remarks on larval development, biogeography, and ecology. Sarsia, *68*:103–126.

Grygier, M. J., 1983. Ascothoracica and the unity of the Maxillopoda. Crustacean Issues, *1*:73–104.

Høeg, J., 1987. The relationship between cypris ultrastructure and metamorphosis in male and female *Sacculina carcini* (Crustacea, Cirripedia). Zoomorphology, *107*:299–311.

Høeg, J., and Lutzen, J., 1985. Crustacea Rhizocephala. Marine Invertebrates of Scandinavia, no. 6. Norwegian University Press, Oslo.

Ichikawa, A., and Yanagimachi, R., 1958. The nature of the testes in *Peltogasterella*. Annotationes Zoologicae Japonenses, *31*:82–96.

Ichikawa, A., and Yanagimachi, R., 1960. Studies on the sexual organization of the Rhizocephala. II. The reproductive function of the larval (cypris) males of *Peltogaster* and *Sacculina*. Annotationes Zoologicae Japonenses, *33*:42–56.

Lützen, J., 1981. Observations on the rhizocephalan barnacle *Sylon hippolytes* M. Sars parasitic on the prawn *Spirontocaris lilljeborgi* (Danielssen). Journal of Experimental Marine Biology and Ecology, *50*:231–254.

Ritchie, L. E., and Høeg, J. T., 1981. The life history of *Lernaeodiscus porcellanae* and co-evolution with its porcellanid host. Journal of Crustacean Biology, *1*:334–347.

Tomlinson, J. T., 1969. The burrowing barnacles (Cirripedia: Order Acrothoracica). United States National Museum, Bulletin 269.

Turquier, Y., 1968. Le mécanisme de perforation du substrat par *Trypetesa nassarioides*. Archives de Zoologie Expérimentale et Générale, *109*:113–122.

Turquier, Y., 1972. Contribution à la connaissance des Cirripédes Acrothoraciques. Archives de Zoologie Expérimentale et Générale, *113*:499–551.

Walker, G., 1985. The cypris larva of *Sacculina carcini* Thompson (Crustacea: Cirripedia: Rhizocephala). Journal of Experimental Marine Biology and Ecology, *93*:131–145.

Walker, G., 1987. Further studies concerning the sex ratio of the larvae of the parasitic barnacle, *Sacculina carcini* Thompson. Journal of Experimental Marine Biology and Ecology, *106*:151–163.

Yanagimachi, R., 1961. The mode of sex determination in *Peltogasterella*. Biological Bulletin, *120*:272–283.

CLASS TANTULOCARIDA

The class Tantulocarida consists of nearly microscopic external parasites of other marine crustaceans: ostracodes, copepods, isopods, and tanaids. There are only a few known genera. The name of the group is based on the larval stage, called the *tantulus* (Figure 17.45, A and C), characterized by a prominent shield that covers the head region (which has no appendages) and sometimes the first thoracic segment as well. The mouth is located in the center of an anteroventral attachment disk, and associated with the gut in the head region are a piercing stylet and a cylindrical structure that can be protruded through the mouth and embedded in the tissue of the host. Each of the six segments of the thorax has two pairs of appendages; the first five pairs are uniramous in some species (Figure 17.45, B), biramous in others (Figure 17.45, D). The sixth pair is always uniramous. In one genus, the abdomen consists of five segments plus the telson; in the other genera, there is only one distinct segment plus a terminal unit that is formed from the telson and some embryonic somites. In either case, the telson has a pair of caudal rami.

The tantalus is believed to be infective from the time it hatches. Its attachment disk becomes firmly cemented to the cuticle of an appendage or body proper of the host. A puncture is made, and the parasite begins to draw fluids into its gut.

The development of males and females follows different pathways. In a prospective female, a sac appears dorsally just behind the head shield, and the thorax and abdomen are deflected ventrally (Figure 17.45, E) and soon shed. Most of the sac becomes a brood pouch (Figure 17.45, F) within which fertilized eggs develop into the next generation of tantulus larvae. In the formation of a male, the posterior part of the thorax of the attached stage enlarges, and the whole body undergoes a radical transformation (Figure 17.45, G). After a time, the sexually mature male (Figure 17.45, H) emerges from the molted exoskeleton. While it is developing, and until it is released, the male obtains nourishment by a tubular stalk connected to the mouth of

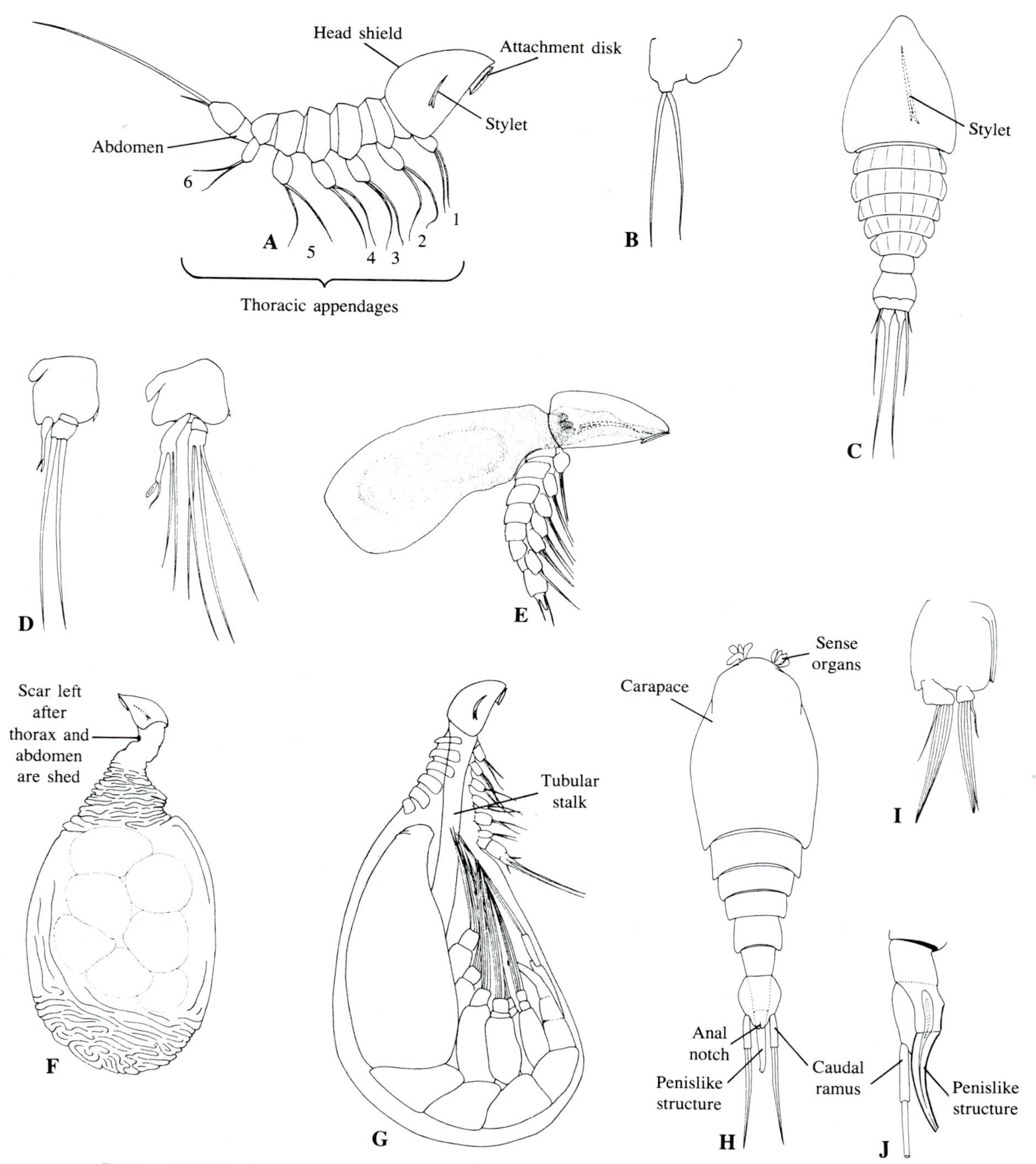

Figure 17.45 Class Tantulocarida. A. *Microdajus langi,* tantulus larva, lateral view. B. *Microdajus langi,* tantulus, first thoracic appendage. C. *Onceroxenus birdi,* tantulus. D. *Onceroxenus birdi,* tantulus, first and second thoracic appendages. E. *Microdajus gaelicus,* early-stage female, showing sac formed behind head shield and deflected thorax and abdomen. F. *Onceroxenus birdi,* mature female, with developing eggs in sac. G. *Microdajus langi,* male within sac formed by tantulus larva. H. *Microdajus langi,* male removed from sac. I. *Microdajus langi*, male, biramous thoracopod. J. *Microdajus langi,* male, abdomen with penislike outgrowth, lateral view. (All after Boxhall and Lincoln, Philosophical Transactions of the Royal Society of London, B, *315*.)

what had been the tantulus. Once it becomes free-living, however, the male no longer feeds. It has a carapace that covers the head region and two thoracic segments; the other four thoracic segments are free. There are six pairs of thoracopods, the first five of which are biramous (Figure 17.45, I), and these function effectively in swimming. The abdomen of the male in some species is unsegmented, but in others it has one distinct segment plus a terminal unit. The posteriormost part of the abdomen may be considered to be the telson; it has a pair of caudal rami or a pair of long setae. A median ventral outgrowth of the abdomen (Figure 17.45, J) is believed to function as a penis.

Tantulocaridans are probably more closely related to barnacles than to any other living groups of crustaceans. This is suggested by the fact that there are six thoracic segments, each with a pair of appendages. Furthermore, the ventral outgrowth of the abdomen of males may be the homologue of the penis of barnacles.

References on Tantulocarida

Boxhall, G. A., and Lincoln, R. J., 1983. Tantulocarida, a new class of Crustacea ectoparasitic on other crustaceans. Journal of Crustacean Biology, *3*:1–16.

Boxhall, G. A., and Lincoln, R. J., 1987. The life cycle of the Tantulocarida (Crustacea). Philosophical Transactions of the Royal Society of London, B, *315*:267–303.

Huys, R., and Boxhall, G. A., 1988. A new genus and species of tantulocaridan (Crustacea: Tantulocarida) parasitic on a harpacticoid copepod from the Skagerak. Sarsia, *73*:205–211.

CLASS MALACOSTRACA

About three-fourths of the known crustaceans belong to the Malacostraca, so it is a large group. With almost no exceptions, these animals have 14 segments behind the head. Eight of these segments belong to the thorax, six to the abdomen. There is, in addition, a distinct telson, on which the anus is located. Because the head bears five pairs of appendages, the total number of appendage pairs possible on most malacostracans is 19. Not all of these necessarily materialize; when any are missing, they are generally abdominal. The diversity of this important group of Crustacea is to a large extent expressed by the layout of the tagmata, and by the types of appendages and their specializations.

Figure 17.1, which is a somewhat generalized drawing of *Anaspides,* a relatively primitive type, is useful for introducing the malacostracan body plan. The five pairs of head appendages—first and second antennae, mandibles, and first and second maxillae—are succeeded by eight pairs of thoracopods. In *Anaspides* the first thoracopods are called *maxillipeds,* because their activities in manipulating food are integrated with those of the true mouthparts. The remaining thoracopods are concerned with locomotion or grasping aquatic vegetation. The first thoracic segment is united with the head, but a groove indicates the line along which fusion has taken place.

The first five segments of the abdomen have biramous appendages called *pleopods.* All of these are more or less equally developed in females, but in males the first two pairs are specialized for transferring sperm during mating. The pleopods are important for swimming. The exopodites of *Anaspides* are articulated and rather long, but the endopodites are short and have only one or two articles. The last abdominal appendages are well-developed uropods. Together with the telson, they form a tail fan. By vigorously flexing its abdomen downward, *Anaspides* can propel itself backward to escape danger, just as crayfishes and many other malacostracans can.

Figure 17.9 illustrates a malacostracan of the advanced crayfish–lobster type in which the number of appendages is exactly the same as in *Anaspides.* Three pairs of thoracopods, however, are specialized as maxillipeds, used in food-handling, leaving five pairs to function as legs for walking, grasping, and crushing. There is, moreover, a carapace. This is fused to all eight thoracic segments, obliterating their limits dorsally and laterally, and uniting the head and thorax into a single tagma called the *cephalothorax.* The first five segments of the abdomen have pleopods, and the sixth segment has uropods. In the male, the more anterior two pairs of pleopods are specialized for transfer of sperm. In both sexes, the more nearly generalized and clearly biramous pleopods are used for swimming, and females use theirs for carrying eggs and young.

Although appendages of malacostracans, like those of other crustaceans, are primitively biramous, either the exopodite or endopodite may be reduced or absent; generally, it is the exopodite that is missing. The mandible of a malacostracan, for instance, has no exopodite. The large piece that functions as a jaw is probably the coxa; the rest of the appendage, the palp, consists of the basis plus what is left of the endopodite. Even if this interpretation of which article forms the large jawpiece is not quite correct, the main idea is not changed: a mandible has no exopodite and its endopodite is small and simplified. Similarly, epipodites may be present on certain appendages, not on others. It has already been explained that the classification of Malacostraca depends to a large extent on specializations of the appendages. Thus characterization of the appendage complements of individual groups can be deferred until these assemblages are discussed separately.

In various malacostracans, the endopodite of one or more pairs of thoracopods behind the maxillipeds may be chelate or subchelate. The term *chelate* indicates an appendage in which the next-to-last article *(propodus)*

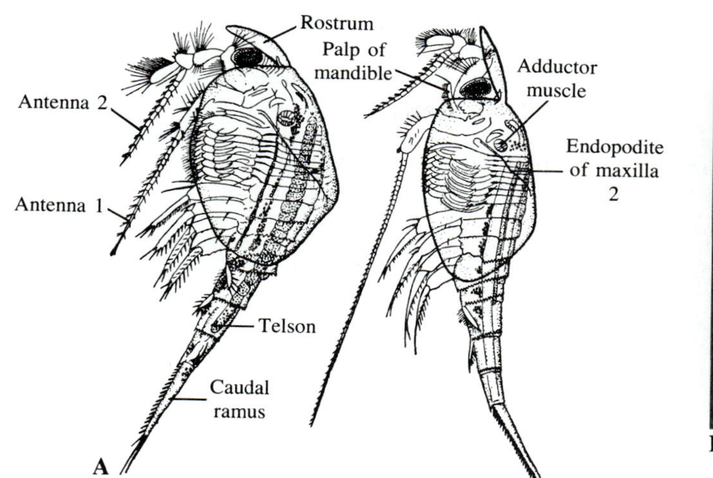

FIGURE 17.46 A. Subclass Phyllocarida, order Leptostraca. *Nebalia bipes,* female (left) and male (right). (Iashnov, in Gaevskaia, Guide to the Fauna and Flora of the Northern Seas of the USSR.) B. *Nebalia pugettensis,* a living specimen.

is drawn out into a projection against which the terminal article (*dactylus,* or claw) can close. This is the situation seen in the stout chelipeds, or pincers, of a lobster, crayfish (Figures 17.8, 17.9), or crab (Figures 17.69, 17.70, 17.71). To be chelate, an appendage need not be especially large, however. The rather slender second and third legs of a crayfish, for instance, are chelate, and in shrimps, which rarely have any stout legs, two or three pairs are chelate (Figure 17.61, A and B).

A subchelate appendage is one in which the claw must bend backward in order to close against the propodus (Figures 17.47, D; 17.53, A; and 17.55, A). Appendages of this sort are widespread among various groups of malacostracans, including certain shrimps. They are especially characteristic of the Amphipoda, which typically have two pairs of them (thoracopods 2 and 3, called *gnathopods*), and the Hoplocarida, in which the first five pairs of thoracopods are subchelate.

SUBCLASS PHYLLOCARIDA

The phyllocaridans have provided paleontologists with a rather important series of fossils, but only about 10 species are living today. All of these are marine and all fit nicely into a single order, Leptostraca. The best-known genus is *Nebalia* (Figure 17.46), which inhabits sediments in situations where algae and other organic materials are decaying and where anoxia prevails. Nebalians are strongly hydrophobic once they are exposed to air, so if a pool in which they are living is stirred vigorously enough to bring some of them to the surface, they will float and can be skimmed off. Most species are about 1 or 1.5 cm long.

These animals show an interesting assortment of characters. They have a large carapace that is not fused to any of the thoracic segments but that encloses all of the thorax and part of the abdomen. The carapace is soft and its right and left halves can be pulled closer together by an adductor muscle located at the level of the maxilla. Hinged to the carapace is a movable rostrum, which can be pulled downward to cover the gap between the valves and thus protect the stalked eyes. These usually project for a short distance, but they can be pulled back. The first and second antennae are prominent (the latter are especially long in males) and uniramous. They are succeeded by mandibles, first maxillae, and second maxillae; the long endopodites of the first maxillae are directed backward and function as "sweepers" of the cavity enclosed by the carapace.

The thorax has eight pairs of similar phyllopodia. The endopodites, consisting of a few articles, are setose and are used for collecting fine detritus, which is passed along a midventral groove to the mouthparts; the exopodites are not divided into articles and are usually flattened. The epipodites, also flattened, function as gills. In females, the phyllopodia hold the eggs and young until these are ready to be set free as fully formed juveniles. The abdomen is unusual for malacostracans in having seven segments. The first four pairs of pleopods are biramous and are used in swimming, though they are more nearly stenopodial than phyllopodial; the next two pairs are reduced and uniramous. There are no appendages on the "superfluous" seventh segment. The telson is odd in that it has two terminal rami that articulate with the main portion. These, as explained previously, are not true appendages.

References on Phyllocarida

Cannon, H. G., 1927. The feeding mechanism of *Nebalia*. Transactions of the Royal Society of Edinburgh, *55*:355–370.

Manton, S. M., 1934. The embryology of the crustacean *Nebalia*. Philosophical Transactions of the Royal Society of London, B, *223*:163–238.

Rowett, H. G. Q., 1946. The feeding mechanisms of *Calma glaucoides* and *Nebaliopsis typica*. Journal of the Marine Biological Association of the United Kingdom, *26*:352–357.

SUBCLASS HOPLOCARIDA

Order Stomatopoda

The only living hoplocaridans are the mantis shrimps (Figure 17.47), which form the order Stomatopoda. Most of the 300 species of this distinctive marine group are found in warmer waters, and many of them are inhabitants of coral reefs. They occupy crevices or burrows from which they can strike at prey.

The head of a stomatopod is peculiar in that the first antennae and stalks of the compound eyes appear to originate from successive segments separate from the rest of the head. These do not correspond exactly to true segments. Above these "pseudosegments" is a movable rostrum that articulates with the rest of the carapace (Figure 17.47, A). The carapace covers only the first four thoracic segments and is fused to the first three. The compound eyes, whose orientation can be changed by activity of their stalks, have crowded ommatidia and are efficient in recognizing prospective prey as well as determining how far away the prey is. Many species do more than just lurk in their crevices or burrows; they move toward the prey after they have identified it. Chemoreceptors on the antennae help target the prey.

The first to fifth thoracic appendages are subchelate, but those of the second pair are larger and more powerful than the others. The propodus of the second thoracic appendages is grooved so that the sharp-tipped dactyl, sometimes toothed along the cutting or crushing margin, fits partly into the propodus the way a blade fits into a pocket knife. Depending on its structure, the fast-closing dactyl is used for snatching and slicing soft prey such as worms, shrimps, and small fishes, or crushing prey such as crabs, snails, and bivalve molluscs. Some stomatopods are more than 30 cm long, so the death-dealing action of the second thoracic appendages must not be underestimated. The dactyl of many species can inflict severe damage on a human finger, and the tip of a dactyl of even a small specimen can penetrate the skin.

The first thoracic appendages, generally provided with many setae, are used for grooming the antennae, mouthparts, eyes, and other parts of the body, including the pleopods and tail fan. The third to fifth pairs are used for grasping and manipulating food. All of the first five pairs are sometimes called maxillipeds, but this term is not quite appropriate for them, partly because they are subchelate, and partly because their functions are not strictly limited to selecting and handling food in close collaboration with the mouthparts. The last three pairs of thoracic appendages are walking legs. They are biramous, with slender exopodites. (The first five thoracic appendages lack exopodites, although certain of them have epipodites that function as gills.)

The abdomen is proportionately long and terminates in a broad telson. This is flanked by a pair of large uropods, so there is a well-developed tail fan that can be flexed ventrally for rapid backward propulsion. The five pairs of pleopods anterior to the uropods have leaflike rami. The endopodites are hooked together by small outgrowths, so the two pleopods on each segment operate as a unit. The pleopods are used for swimming and respiration, and also for displacing sand during construction of a burrow.

The gonopores of males are on the coxae of the eighth thoracic appendages. The first pleopods are specialized for transfer of sperm. In females, there is a single gonopore on the ventral side of the sixth thoracic segment. Courtship and mating usually involve complex behavior. Males have been observed to probe burrows until they locate a female that will take them in. After a series of aggressive maneuvers, rapport is established and mating ensues. The mating process itself, which leads to deposition of spermatophores on the underside of the female, may involve brushing of the thorax of the female by the first thoracic appendages of the male and similar activities. Courtship and mating may go on for several days, but then aggressive activity is resumed and the male is obliged to leave the burrow.

When the eggs are laid, secretions from glands on the last three thoracic segments cause them to stick together. In most species, the mass of eggs is held by the third to fifth thoracic appendages until hatching takes place, but there may be other arrangements. In general, females do not eat during the brooding period. A newly hatched stomatopod has all eight of its thoracic segments and a large carapace. The set of abdominal segments is completed by way of a series of molts. The appearance of stomatopod larvae is so distinctive (Figure 17.47, F) that special names have been invented for them. Basically, however, they are similar to the zoea larvae of crabs, hermit crabs, and some other members of the order Decapoda. The larvae are planktonic and

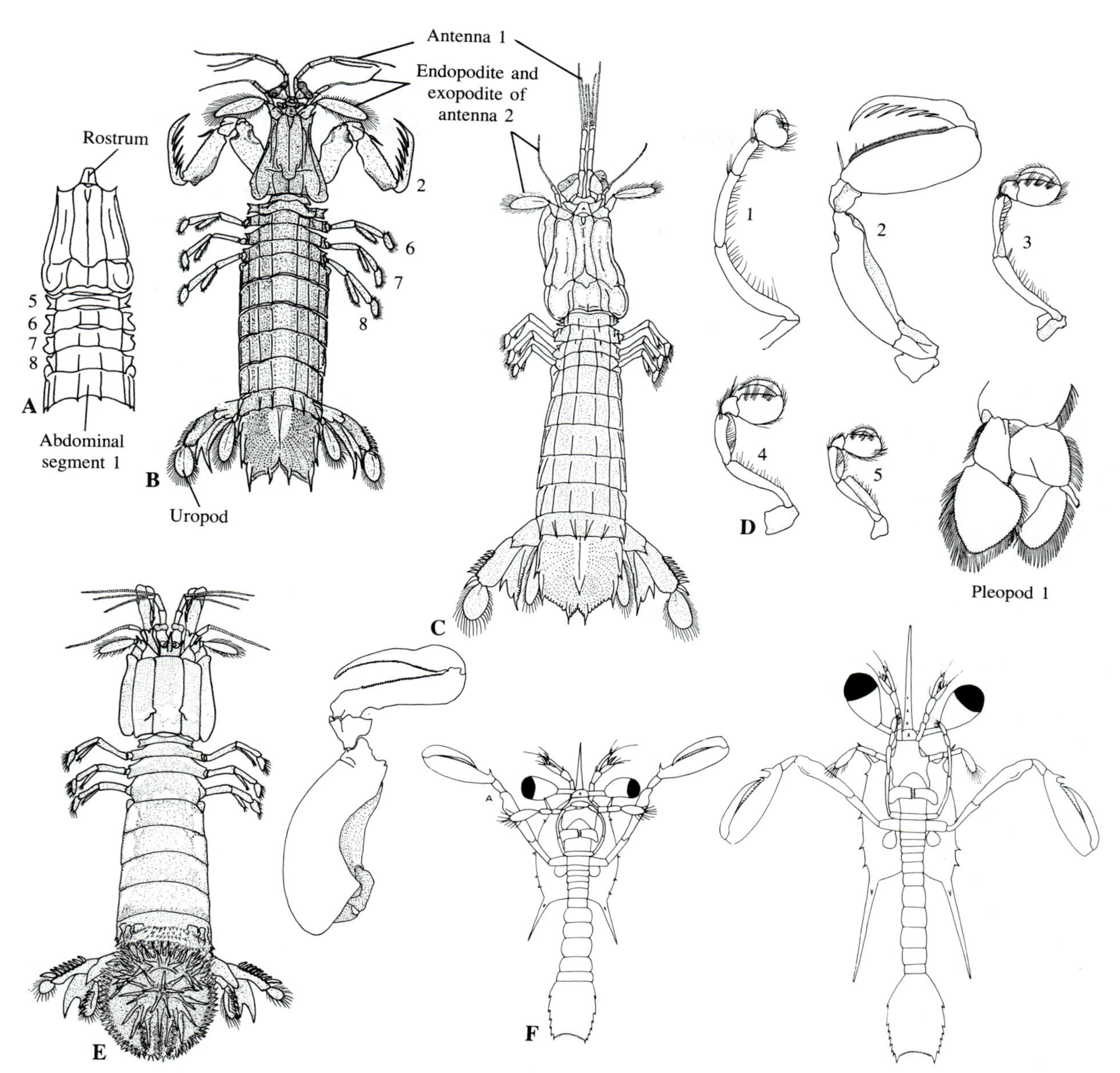

FIGURE 17.47 Subclass Hoplocarida, order Stomatopoda. A. Anterior portion of the body of a stomatopod, dorsal view, showing the hinged rostrum and free thoracic segments, diagrammatic. B. *Squilla oratoria,* male, dorsal view. C. *Squilla calumnia,* female, dorsal view. D. *Squilla calumnia,* female, thoracopods 1–5 and pleopod 1 (not all drawn to the same scale). Except for thoracopod 5, none would be visible in a dorsal view of an entire animal. E. *Gonodactylus guerini,* female, and its thoracopod 2. Note that the telson, used to close the opening of the burrow, resembles a small sea urchin. (A–E, Townsley, Pacific Science, *7*.) F. First and second in the series of larval stages of *Squilla empusa*. (Morgan and Provenzano, Fishery Bulletin, *77*.)

may not settle down for several weeks. In some species, they migrate to the surface just after sunset and just before dawn, times when fishes likely to feed on them are less active.

Mantis shrimps are extremely aggressive, not only toward other species but also toward members of the same species. Many of them have brightly colored markings on the exopodites of the second antennae, uropods, dactyls, and other structures. These serve as warnings to prospective enemies. The larger and more powerful species generally have the brightest markings, and this is believed to discourage aggression by subordinate species, which are likely to lose fights and which would be only a source of annoyance to the dominant species. Nevertheless, some fights proceed to the point that one individual is injured or killed. During aggressive encounters, the opponents frequently roll up into a ball and use the tail fan as a shield.

When holed up in its lair, a stomatopod may use its second thoracic appendages or its tail fan to block the opening so that intruders, even of the same species, cannot enter. In at least one species, *Gonodactylus guerini,* the telson is so spiny that it looks like a little sea urchin (Figure 17.47, E) against the mass of coral in which this stomatopod lives. Many mantis shrimps, especially just after molting, when they are soft and vulnerable, use pebbles and other debris to conceal the openings to their hiding places.

Stomatopods are certainly interesting animals and it is a pity that they are not often found except in tropical and subtropical regions. Some of the larger species are gathered for food.

References on Hoplocarida

Burrows, M., 1969. The mechanics and neural control of the prey capture and strike in the mantid shrimps *Squilla* and *Hemisquilla.* Zeitschrift für Vergleichende Physiologie, *62:*361–381.

Caldwell, R. L., and Dingle, H., 1976. Stomatopods. Scientific American, *234*(1):80–89.

Caldwell, R. L., and Dingle, H., 1978. Ecology and morphology of feeding and agonistic behavior in mudflat stomatopods. Biological Bulletin, *155:*134–149.

Kunze, J. C., 1983. Stomatopoda and the evolution of the Hoplocarida. Crustacean Issues, *1:*165–188.

Manning, R. B., 1969. Stomatopod Crustacea of the Western Atlantic. University of Miami Press, Coral Gables, Florida.

Mauchline, J., 1984. Euphausiid, Stomatopod and Leptostracan Crustaceans. (Synopses of the British Fauna, no. 30.) Linnean Society of London, Estuarine and Brackish-water Sciences Association, and E. J. Brill/Dr. W. Backhuys, London and Leiden.

Morgan, S. G., and Goy, J. W., 1987. Reproduction and larval development of the mantis shrimp *Gonodactylus bredini* (Crustacea: Stomatopoda) maintained in the laboratory. Journal of the Crustacean Biology, *7:*595–618.

Reaka, M. L., 1981. The hole shrimp story. Natural History, *90*(7):37–43.

Reaka, M. L., and Manning, R. B., 1981. The behavior of stomatopod Crustacea and its relationship to rates of evolution. Journal of Crustacean Biology, *1:*309–327.

Serène, R., 1954. Observations biologiques sur les stomatopodes. Annales de l'Institut Océanographique, Monaco, *29:*1–94.

SUBCLASS EUMALACOSTRACA

Eumalacostracans differ from phyllocaridans in lacking a seventh abdominal segment; the two halves of the carapace, moreover, are not joined by an adductor muscle that pulls them together. They are unlike hoplocaridans in that not all of the first five pairs of thoracic appendages are subchelate, and the anterior part of the head is not divided into two segmentlike units.

Superorder Syncarida

Syncaridans are unified by the following complex of characters: they have no carapace; all eight of the thoracic segments are similar, unless one is fused to the head; the second to sixth thoracic appendages, and sometimes others, are biramous. The first two of the features just mentioned are certainly not limited to syncaridans, but there is no other group that has all three, a combination that is probably primitive for the Malacostraca.

The syncaridans are divided into two orders. The known species—there are about 75 altogether—live in out-of-the-way places or in odd habitats.

Order Anaspidacea

There are only a few living species of anaspidaceans, but the order is of considerable zoological interest because it shows some features that are almost certainly primitive in malacostracans. Nearly all of the appendages, for instance, are biramous. The following account of anaspidacean structure is based largely on *Anaspides tasmaniae* (Figures 17.1; 17.48, A–D). This species, which may reach a length of about 5 cm, is representative of a suborder that is restricted to Tasmania and southeast Australia.

There is no carapace, but the first thoracic segment is fused to the head, even though the boundary between it and the head is marked by a dorsal groove. (The fusion of one thoracic segment to the head is seen in some other groups of malacostracans, too.) The thoracic segments are all much alike and most of the thoracic appendages not only are biramous, with small but well-developed exopodites, but also bear two fairly large, flattened epipodites that function as gills. The first thoracopods are concerned mostly with handling food, and they are sufficiently different from the other thoracopods to merit being called maxillipeds. A curious feature of the five pairs of pleopods of *Anaspides* is that only their exopodites are prominent; the endopodites are small, conical, and reduced to one or two articles. (In *Koonunga* and *Micraspides,* only the first two pleopods of males have endopodites; these are specialized for sperm transfer.)

The basal article of each first antenna contains a sta-

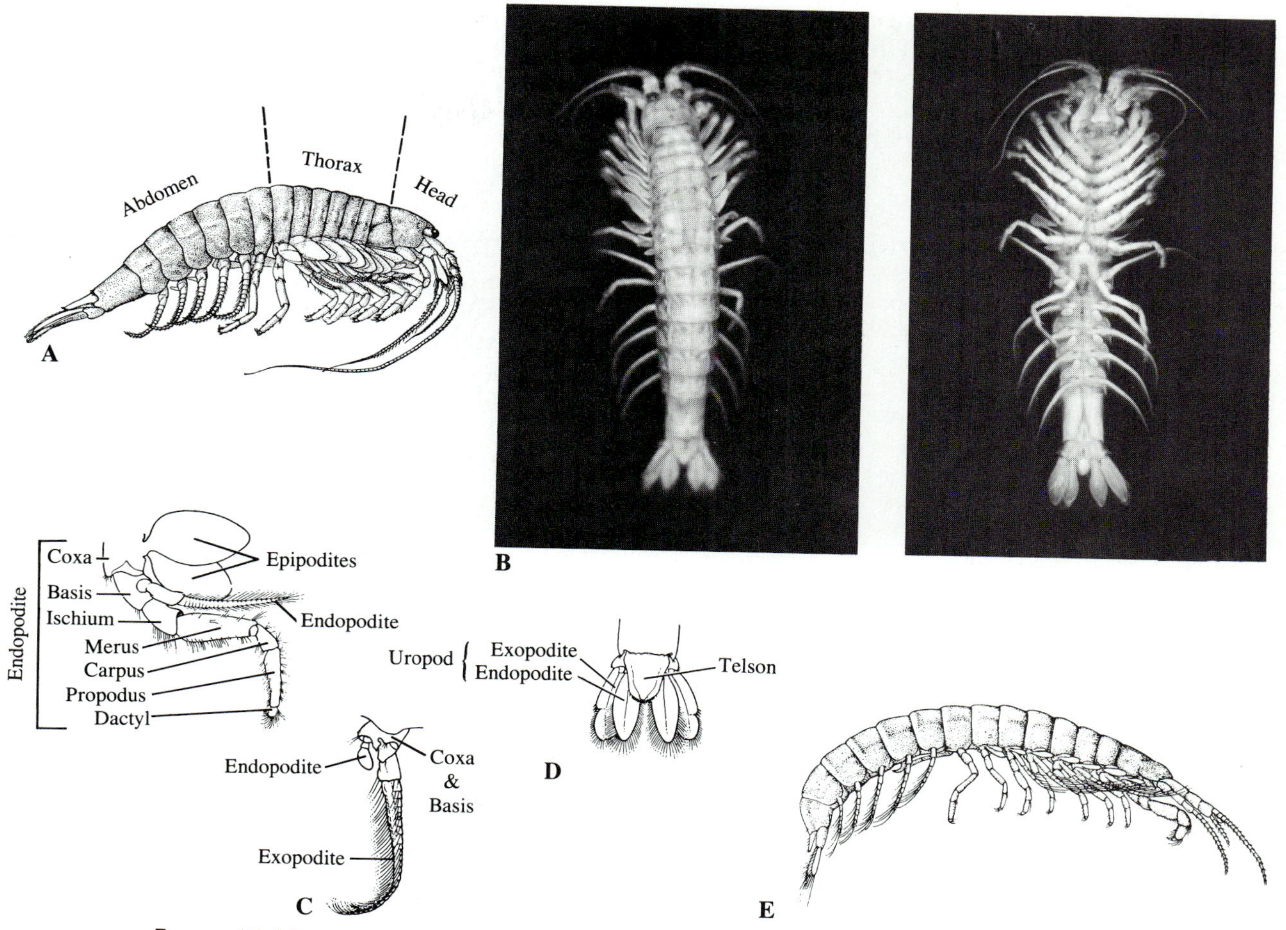

FIGURE 17.48 Superorder Syncarida, order Anaspidacea. A. *Anaspides tasmaniae,* female, lateral view from right side. (Calman, in Lankester, A Treatise on Zoology.) B. *Anaspides tasmaniae,* female, dorsal view (left) and ventral view (right). C. *Anaspides tasmaniae,* a thoracopod (left) and pleopod (right). D. *Anaspides tasmaniae,* uropods and telson. (C and D, Waterman and Chace, in Waterman, The Physiology of Crustacea, *1*.) E. *Koonunga cursor* (Calman, in Lankester, A Treatise on Zoology.) See also Figure 17.1 for a generalized interpretation of *Anaspides*.

tocyst. The compound eyes are stalked in *Anaspides* and *Paranaspides,* sessile in *Koonunga* and *Micraspides*. There is no major subesophageal ganglionic complex (at least in *Anaspides*) and the right and left members of each pair of ganglia in the ventral nerve cord are rather distinctly separated, although they are connected by commissures. The chain of ganglion pairs is strongly segmental.

Anaspides, which lives in streams and lakes at moderate elevations in the mountains, walks on the endopodites of its legs or swims with the exopodites of its pleopods. The tail fan, formed of telson and uropods, is pressed into service for rapid backward swimming. The first legs bring algae and detritus to the mouthparts, and sometimes are used for catching worms or other small animals. Thus *Anaspides* is more or less omnivorous. *Paranaspides* occupies habitats similar to those of *Anaspides,* but seems to feed exclusively on fine detritus. Its first legs are nevertheless used for getting prospective food material to the mouthparts. The survival of *Anaspides* and *Paranaspides* is threatened by the introduction of trout into habitats where there were formerly no crustacean-eating predators. *Koonunga* (Figure 17.48, E), which lives in temporary ponds, is probably a detritus eater. It can burrow into the mud with its first legs and survive dry periods. Little is known about *Micraspides,* which lives in moisture held by mats of *Sphagnum* moss.

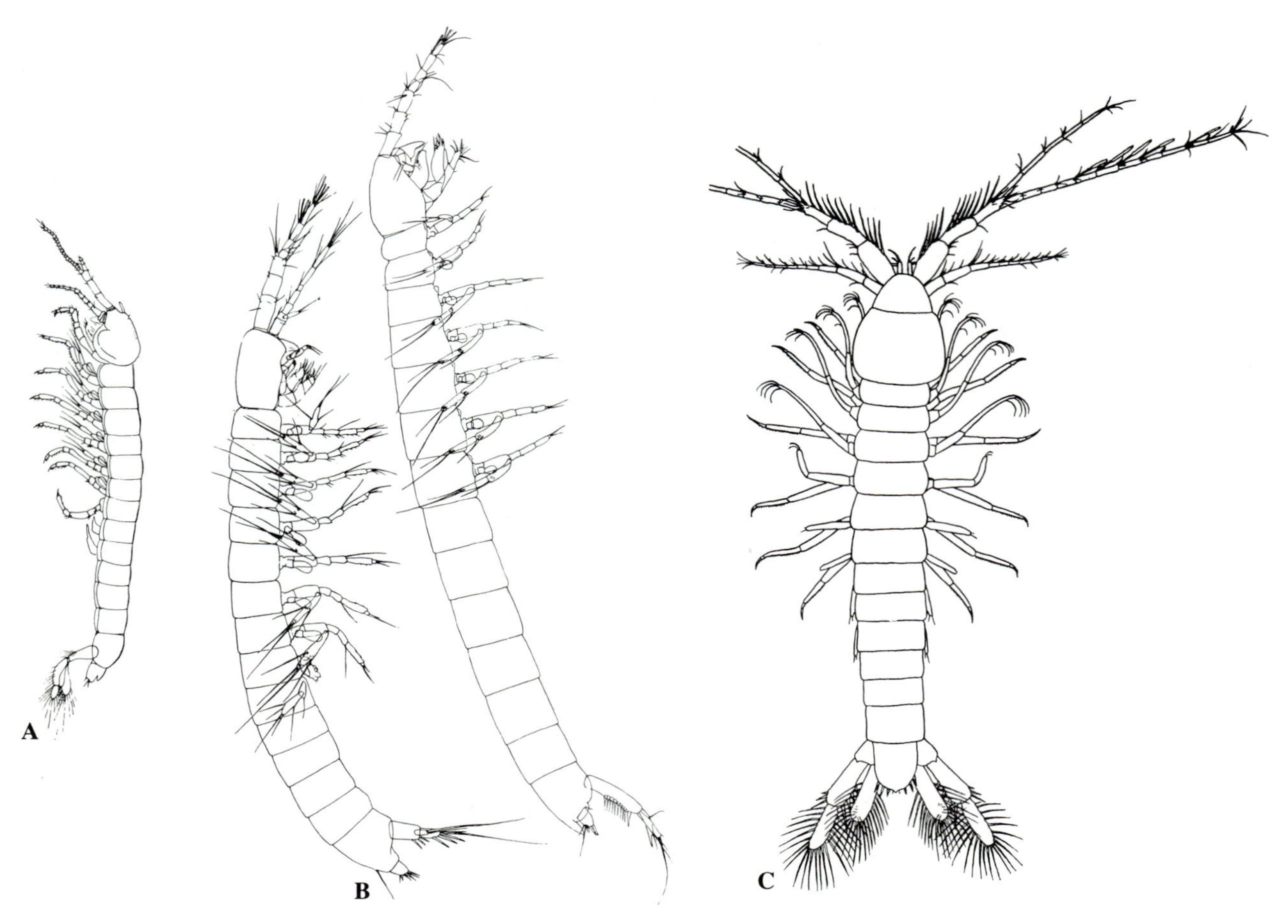

FIGURE 17.49 A. *Parastygocaris,* superorder Syncarida, order Anaspidacea. B. *Bathynella cautinensis* (left) and *Parabathynella neotropica* (right) (superorder Syncarida, order Bathynellacea) (Noodt, Gewässer und Abwässer, *37–38.*) C. *Monodella argentarii* (superorder Pancarida, order Thermosbaenacea). (Fryer, Transactions of the Royal Society of Edinburgh, *66.*)

Female *Anaspides* produce only a few eggs at a time, and stick them to firm objects. The eggs are fairly large—about 1 mm in diameter—and development within them proceeds directly to a juvenile that looks much like an adult. It has, however, a median eye resembling that of a nauplius, as well as some other features that change drastically in the course of further growth and successive molts.

A second suborder includes the stygocarids (Figure 17.49, A), which so far have been collected only in Chile, Argentina, Australia, and New Zealand. They are not more than 4 mm long and inhabit situations where subterranean water percolates through a bed of sand. They can move with agility between sand grains, but are helpless when removed from the sand. Eyes are lacking, and the first thoracic segment is fused with the head. The first thoracic appendages, moreover, are modified to serve as maxillipeds. There are uropods, but pleopods are present only in males, which have two pairs specialized for transferring sperm. Development, so far as is known, is direct.

Order Bathynellacea

Bathynellaceans (Figure 17.49, B), most of which are about 1 or 2 mm long, are like stygocarids in being eyeless and in living in the same kind of habitat—sand through which fresh water percolates. Of the approximately 60 species, some have been found in water from wells and springs, even hot springs. Others are known from sandy beaches of freshwater lakes or situations close to marine lagoons. Two species occur in Lake Baikal, in Siberia, at depths of nearly 1500 m.

Bathynellaceans differ from stygocarids in that the first thoracic segment is free of the head. The last thoracic segment lacks appendages, except in males of certain species; the next-to-last pair may also be missing. The telson is united to the sixth abdominal segment, and has a pair of small but distinct caudal rami. (In stygocarids the little projections that might be interpreted as caudal rami are at best vestigial.) There are uropods on the sixth segment. Pleopods may be present on the first or first and second abdominal segments, but in some genera there are none at all.

Bathynella natans, found in wells throughout much of Europe, is perhaps the best-known species. It has been observed to swim a little by bending movements but is essentially adapted for holding on to sand grains against a current. It feeds on detritus and protozoans, which it dispatches with its sharp-toothed mandibles. *Bathynella* lays single eggs, and when the young hatch they have all of the head appendages, one pair of thoracic appendages, and a few additional thoracic segments without appendages. It attains the full complement of segments and appendages through a series of molts.

References on Syncarida

Brooks, H. K., 1962. On the fossil Anaspidacea, with a revision of the classification of the Syncarida. Crustaceana, *4:*229–242.

Jakobi, H., 1954. Biologie, Entwicklungsgeschichte und Systematik von *Bathynella natans*. Zoologische Jahrbücher, Abteilung für Systematik, *83:*1–62.

Noodt, W., 1956. Naturliches System und Biogeographie der Syncarida. Gewässer und Abwässer, *37–38:*77–186.

Schminke, H. K., 1978. Die phylogenetische Stellung der Stygocarididae (Crustacea, Syncarida)—unter besondere Berücksichtigung morphologischer Ähnlichkeiten mit Larvenformen der Eucarida. Zeitschrift für Zoologische Systematik und Evolutionsforschung, *16:*225–239.

Serban, E., 1972. *Bathynella* (Podophallocarida Bathynellacea). Travaux de l'Institut de Spéologie 'Emile Racovitza.' *11:*11–224.

Ueno, M., 1954. The Bathynellidae of Japan. Archiv für Hydrobiologie, *49:*519–538.

Superorder Pancarida

Pancaridans (Figure 17.49, C), all placed in a single order (Thermosbaenacea), are extremely rare. The first species to be described, *Thermosbaena mirabilis,* was found shortly after 1920, in a thermal spring in Tunisia. The spring, with a salinity about the same as that of sea water and a temperature of about 45° C, had been exploited by the Romans as a supply of warm water for a public bath, and it is still being used. *Thermosbaena* is a small crustacean—about 3.5 mm long—and lives among the blue-green algae, diatoms, bacteria, and detritus that coat the walls of the spring. It seems to feed on nearly all of these components. Other species have been reported from warm springs, pools in caves, and certain brackish situations.

Some of these animals do not have eyes; others have reduced eyes that may or may not be visible externally. The carapace covers three thoracic segments, but is fused only to the first segment. In females, part of the carapace serves as a brood pouch; this is unusual for malacostracans. The thoracic appendages, used in swimming or scurrying, are biramous, and those of the first pair are modified to function as maxillipeds. Some of the remaining pairs may be absent. In *Thermosbaena,* for instance, the seventh and eighth pairs are missing. Uniramous pleopods are present on segments 1 and 2 of the abdomen and there is a pair of biramous uropods on segment 6, which is united to the telson, as it is in isopods.

References on Pancarida

Barker, D., 1962. A study of *Thermosbaena mirabilis* (Malacostraca, Peracarida) and its reproduction. Quarterly Journal of Microscopical Science, *103:*261–286.

Maguire, B., Jr., 1964. *Monodella texana* n. sp., an extension of the range of the crustacean order Thermosbaenacea to the western hemisphere. Crustaceana, *9:*149–154.

Stock, J. H., and Langley, G., 1981. The generic status and distribution of *Monodella texana,* the only known North American thermosbaenacean. Proceedings of the Biological Society of Washington, *94:*569–578.

Superorder Peracarida

The two largest groups of malacostracans are the Peracarida and Eucarida. In eucaridans (discussed further in their own section below), a carapace is always present and is fused to all thoracic segments. Thus the head and thorax are united into a cephalothorax. In peracaridans that have a carapace—and not all of them do—this is not fused to more than four thoracic segments, even if it covers all eight. Another distinctive feature of peracaridans is the brood pouch, found in mature females, consisting of shieldlike outgrowths of the coxae of certain of the thoracic appendages. These outgrowths are called *oostegites*. The compound eyes are usually sessile; only mysidaceans have stalked eyes. The young, on hatching, generally resemble the adults.

Order Mysidacea

In mysidaceans (Figure 17.50), the eyes are conspicuously stalked. There is a flexible carapace that covers much of the thorax, but it is fused to only the first three thoracic segments. The inner surface of the carapace functions in respiration. The first pair of thoracic appendages (sometimes the first two or three pairs) are maxillipeds. The remaining pereopods are stenopodial, biramous, and setose, and are used for swimming. When well-developed pleopods are present on the first five segments of the abdomen, these also are biramous and are concerned with swimming. The sixth abdominal appendages are modified as uropods that form, together

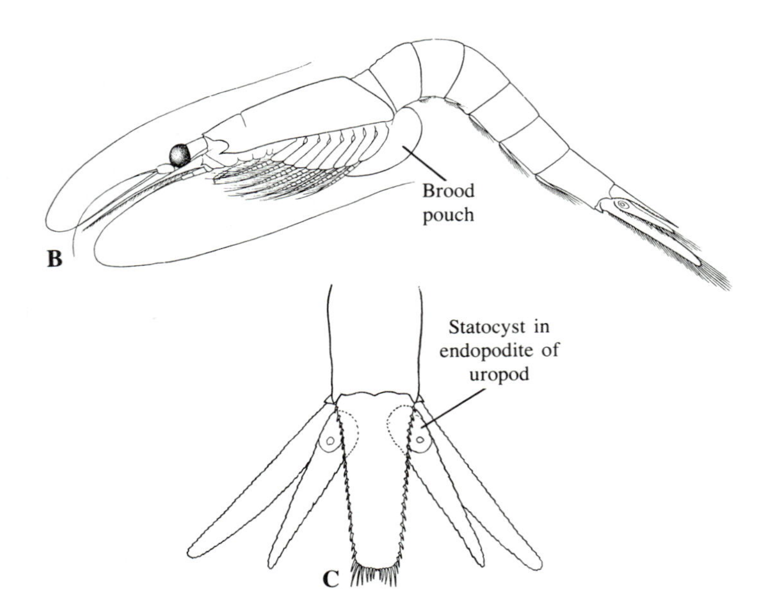

A

FIGURE 17.50 Superorder Peracarida, order Mysidacea. A. A species of *Archaeomysis*. B. *Neomysis nakazawai,* female. (Ii, Japanese Journal of Zoology, 6.) C. *Paramblyops brevirostris,* female, uropods and telson. (Tattersall, Discovery Reports, *28*.)

with the telson, a tail fan. In most genera—those belonging to the suborder Mysina—the base of the endopodite of each uropod has a statocyst (Figure 17.50, C); the glassy body within the statocyst is composed of calcium fluoride.

The brood pouches of females, formed of coxal oostegites, often bulge prominently. Because of this, mysidaceans are sometimes called "opossum shrimps." Development is gradual, and the young look like the adults when they are released from the brood pouch.

Most mysidaceans are basically filter-feeders that collect diatoms and other microscopic food on the setae of the endopodites of the thoracic appendages and mouthparts. The second maxillae and exopodites of the thoracic appendages are instrumental in setting up water currents. Filter-feeding may be supplemented by capture of copepods and other small animals, or by scavenging. A few species are regularly carnivores, with certain thoracic appendages specialized for grasping prey and equipped with cutting mouthparts.

With a few exceptions, mysidaceans are marine. Some are planktonic; others live buried in sand, mud, or silt. Those that inhabit sandy beaches can usually be collected with a dipnet run through shallow water, or in holes dug in sand close to the water's edge. A few are found in freshwater lakes. These are thought to be relicts of populations trapped in small bodies of water when glaciers separated these from the ocean.

The two suborders of Mysidacea will not be compared in detail. It will suffice to say that most species—including the freshwater types—belong to the Mysina. In this group, females lack substantial pleopods, and rely on their thoracic exopodites for swimming. There are usually only two or three pairs of oostegites forming the brood pouch.

In the other suborder, Lophogastrina, both sexes have well-developed pleopods. There are oostegites on seven pairs of thoracic appendages; only those of the first pair, which function as maxillipeds, lack them. Another distinctive feature is the presence of branched gills, which are epipodites, on most of the thoracic appendages. An especially remarkable genus is *Gnathophausia,* one species of which reaches a length of 35 cm. A gland that opens at the base of the exopodite of each second maxilla produces a secretion that glows. Most members of the Lophogastrina, including the several species of *Gnathophausia,* are limited to deep water.

References on Mysidacea

Cannon, H. G., and Manton, S. M., 1927. The feeding mechanism of a mysid crustacean, *Hemimysis*. Transactions of the Royal Society of Edinburgh, *55*:219–253.

Ii, N., 1964. Mysidae (Crustacea). *In* Fauna Japonica. Biogeographical Society of Japan, National Science Museum, Tokyo.

Mauchline, J., 1980. The biology of mysids. Advances in Marine Biology, *18*:1–369.

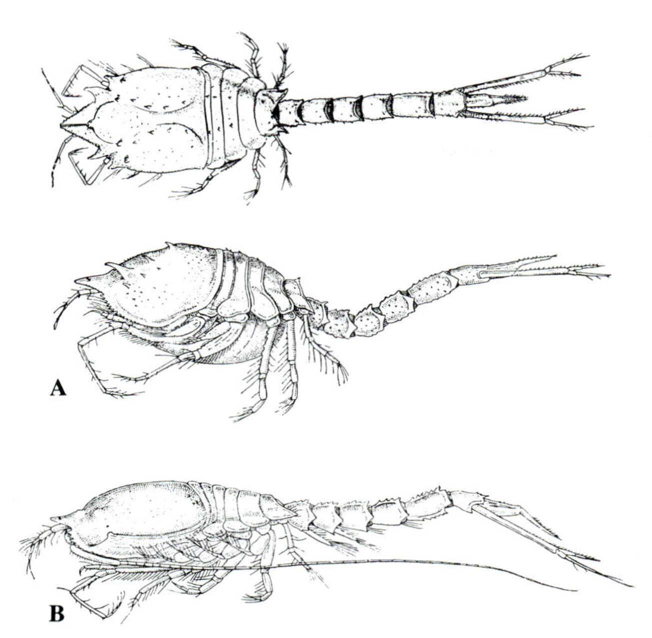

Figure 17.51 Order Cumacea. A. *Diastylis cornuta,* female, dorsal view (above) and lateral view (below). (Fage, Faune de France, *54*.) B. *Diastylis cornuta,* male, lateral view. (Lomakina, keys to the Fauna of the USSR, *66*.)

Morgan, M. D. (editor), 1982. Ecology of Mysidacea. Dr. W. Junk, The Hague, Boston, and London.
Tattersall, W. M., and Tattersall, O. S., 1951. British Mysidacea. Ray Society, London.

Order Cumacea

Cumaceans (Figures 17.51, 17.52), of which there are about 800 species, are marine and brackish-water organisms. Most of them are under 1 cm long, and their usual habitats are mud or sand. They live scarcely buried and show some interesting adaptations for collecting food. Cumaceans have a prominent carapace which covers all segments of the thorax, but which is generally fused with only three or four of them. There are exceptions, however; sometimes as many as six segments are involved. The first three pairs of thoracic appendages are modified as maxillipeds, leaving the other five pairs to function as legs. The third maxillipeds and some of the legs may be biramous; the rest consist of endopodites only.

Both sexes have slender, biramous uropods on the sixth abdominal segment, which is sometimes fused to the narrow telson. Males may have pleopods on from one to five of the other segments of the abdomen, but in some species these appendages are lacking; females never have pleopods. The abdomen is so flexible that it can be bent under the rest of the body; the uropods can then be used for cleaning the cephalothorax. Both males and females use their legs for swimming, and males can also use their pleopods, if they have any.

To burrow, a cumacean turns up its abdomen and uses its last three pairs of thoracic legs as shovels to kick up the sand or mud. By flexing the abdomen, the animal makes its hole larger, but much of the sediment that has been kicked up falls back onto the animal. In time only the anteroventral part of the body, from the first legs forward, is exposed (Figure 17.52). In several species whose feeding has been studied, a sand grain passed by the first legs to the maxillipeds is held by these appendages while the mandibles and first maxillae brush off diatoms and other particles. The sand grain is then discarded and another is brought into position. The size of the grains that can be manipulated by cumaceans varies. Some species prefer to work with grains that are only a fraction of a millimeter in diameter; others tackle grains nearly as large as their cephalothorax. Filter-feeding cumaceans use setae on their mouthparts to

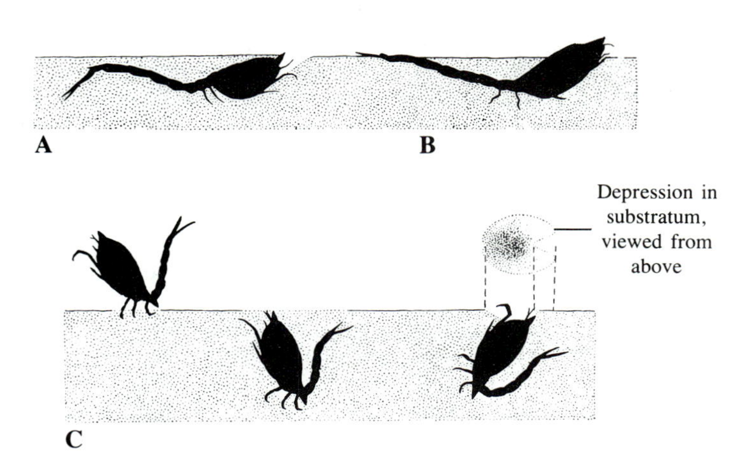

Figure 17.52 Order Cumacea. Position of various cumaceans in the substratum. A. *Iphinoe trispinosa*. B. *Leucon nasica*. C. *Diastylis rathkei,* before and after burrowing into the substratum. The first pereopods excavate a shallow depression, and the third maxillipeds are the primary collectors of detritus processed by the other maxillipeds and mouthparts. (Lomakina, Keys to the Fauna of the USSR, *66,* but A based on Zimmer, and B and C based on Forsman.)

comb out small organic particles. Muddy water is circulated through a complicated funnel in which the mouth parts operate. The funnel is formed in part by the ventral edges of the carapace, which are extended toward the midline, and in part by the maxillipeds. As the second maxillae beat backward, water is drawn into the chamber, and the setae of the first maxillae collect most of the food. The forward beat of the second maxillae forces water out laterally through narrow spaces between the first and second maxillae. (The above explanation is somewhat simplified.)

The first maxillipeds have long epipodites that function as gills, supplementing the respiratory surface provided by the inner lining of the carapace. The epipodites also circulate water, moving it forward through the carapace from the level of the second and third maxillipeds. Setae on the third maxillipeds block the entry of large particles. The respiratory current leaves the carapace in the region of the pseudorostrum, a siphonlike structure whose dorsal and lateral walls are formed by the carapace and whose ventral walls are formed by the two flattened epipodites of the first maxillipeds.

It has already been mentioned that males may have pleopods, whereas females do not. There are other differences between mature individuals of the two sexes, and the most notable of these is seen in the second antennae. In males, the second antennae are long and directed backward; in females, they are short. Males are also likely to have better-developed eyes. Mating often takes place as the animals swim. In some species the males are regularly more active than females in this respect, forming what may be called bachelor swarms. In species in which the males have pleopods, the first or second pairs of these are used for clasping. The females hold their eggs in the brood pouch formed by the oostegites of the third maxillipeds and first to third legs. The young are released when they look like cumaceans, although they still lack the last pair of legs.

References on Cumacea

Jones, N. J., 1976. British Cumaceans. (Synopses of the British Fauna, no. 7.) Linnean Society of London and Academic Press, London and New York.

Valentin, C., and Anger, K., 1977. *In situ* studies on the life cycle of *Diastylis rathkei*. Marine Biology, *39*:71–76.

Order Amphipoda

Amphipods, of which there are about 5000 species, are among the more abundant Crustacea found in marine and freshwater habitats. The majority are aquatic, but wherever there are sandy beaches or salt marshes, there are likely to be semiterrestrial species that require moisture but that are rarely submerged. Truly terrestrial types are found in wet tropical forests. Amphipods feed on a variety of animal and plant matter and are in turn consumed by fishes, birds, and other carnivores. They are thus important in the food web.

Mature amphipods generally fall into a size range of about 5 mm to 4 cm. They can be smaller, however, and a few giants collected in deepwater traps in the Antarctic region are more than 20 cm long. With some exceptions, they tend to be slightly compressed laterally, so they are higher than they are wide. There is no carapace, but the first segment of the thorax, whose appendages are maxillipeds, is always fused to the head, forming a tagma called the *cephalon*. The seven remaining segments, all of which are usually free of the head, bear the pereopods, or legs (Figure 17.53, A). The first two pairs of legs are typically subchelate and function as prehensile organs called *gnathopods*. The last three legs are characteristically long and bent backward. There are no exopodites on the legs, but at least some of these appendages have soft, saclike epipodites that serve as gills (Figure 17.53, C). In females, oostegites form the usual peracaridan type of brood pouch.

The appendages of the abdomen, or pleon, are generally biramous, but there are exceptions. The arrangement is unlike that in any other crustaceans. The first three pairs, called pleopods (Figure 17.53, A) are similar, and are used for swimming and respiration. The last three pairs, called uropods (Figure 17.53, B), are of a different type, and they usually become smaller posteriorly. Only the last pair, of course, is homologous to the uropods of other malacostracans. The uropods are strong and can be used for jumping, especially in the semiterrestrial amphipods called beach hoppers. The telson is small and may be fused, partly or completely, to the last segment of the abdomen.

Although both rami of the pleopods are narrow, they are fringed with setae and are effective paddles for swimming forward. The setae are spread when a pleopod makes its effective backstroke, but they drag behind on the recovery stroke, reducing the surface area. Some species swim with the dorsal side up, others with the ventral side up. By repeatedly flexing its abdomen ventrally, then straightening it, an amphipod can superimpose jerky motions on graceful pleopodal swimming.

When an amphipod walks with the ventral side down, the uropods may provide propulsive force, and the beating of the pleopods may assist. In general, however, the third and fourth legs are used for pulling the body forward, and the sixth and seventh legs push. (In a beach hopper, the sixth and seventh legs also brace the animal so that it does not fall over.) The fifth legs are used when the animal must turn or back up. Many amphipods, including species of *Gammarus* common in fresh and brackish water, crawl on their sides. The fifth to seventh legs of the side of the body that is in contact with the substratum are used for pushing. The other

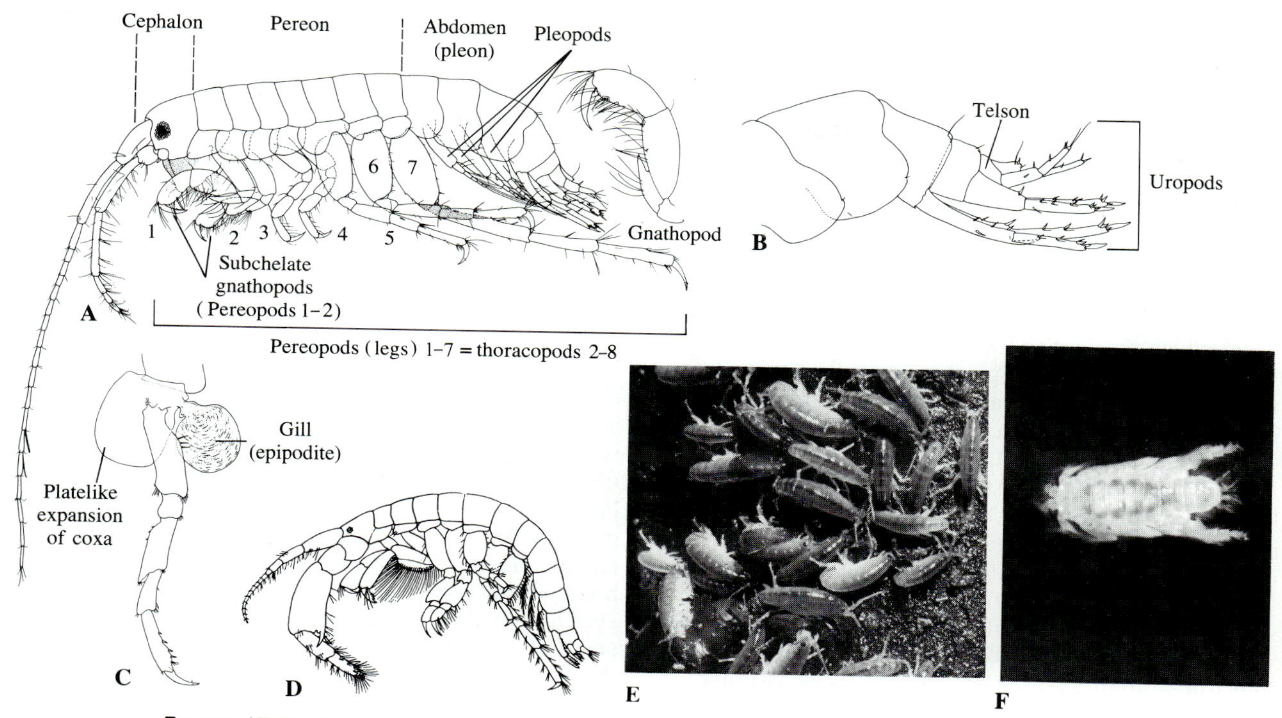

FIGURE 17.53 Order Amphipoda, suborder Gammaridea. A. *Aoroides,* male and first gnathopod of male enlarged. Both gnathopods are examples of subchelate appendages. In the terminology of some specialists, the term *pereopod* is applied only to the last five pairs of thoracopods. B. *Konatopus,* posterior portion, showing telson and left uropods (all three pairs of uropods are biramous). (A and B, Barnard, Smithsonian Contributions to Zoology, number 34.) C. Pereopod 3 of *Parahyale,* showing the gill (coxal epipodite). (Iwasa, Journal of the Faculty of Science, Hokkaido Imperial University, series 6, *3*.) D. *Corophium,* characterized by especially stout second antennae. E. *Traskorchestia traskiana,* which lives at higher levels (drift zone) on beaches of the Pacific coast of North America. (See also *Megalorchestia,* Color Plate 6.) F. *Eohaustorius washingtonianus,* specialized for living in sand of ocean beaches.

legs are held close to the ventral surface. The movement is aided by activity of the abdomen and its uropods.

The paired gonads lie in the thoracic region. The male gonopores are situated between the seventh legs, which are on the last thoracic segment. The female gonopores are located between the fifth legs. The oostegites that form the brood pouch are produced by the coxae of the second to fifth legs. In most families, males are larger than females; in some, however, the size relationship is reversed. Males are apt to have proportionately larger second antennae and stouter subchelae on the gnathopods than females. Association of males and females begins a few days before the females are ready to molt preparatory to egg-laying. This is why males are so often seen riding on the backs of females. Males use their gnathopods to grip the anterior edge of one thoracic segment and the posterior edge of another. After a female completes the crucial molt, the members of a mating pair become oriented so that their ventral surfaces are in contact, and the male uses his pleopods to direct strings of sperm toward the female's gonopores. Soon afterward, eggs are deposited in the brood pouch and fertilized there. When the young hatch, they already resemble adults, but they remain in the brood pouch for a while longer. In at least some species, females can produce several successive broods of young, which implies that they mate several times.

Suborder Gammaridea

About three-fourths of the amphipods belong to the suborder Gammaridea. The preceding discussion of morphology and biology of amphipods is based largely on this group, which are what most zoologists think of as "standard" amphipods. In gammarideans, only the first segment of the thorax is fused to the head. The eyes are not so large that they cover much of the head. The coxal plates of the legs, especially the first four pairs, are large, and the abdomen is well developed. These amphipods are well represented in the sea and at the fringes of the sea, and they are the only amphipods likely to be found in fresh water and in tropical forests.

Gammaridean amphipods exhibit considerable diversity in their techniques for getting food. Many species are scavengers of decaying plants. Beach hoppers fit into this category. They live under wrack stranded at the high-tide line on sandy beaches, or dig burrows from which they emerge at night in order to feed. They use their mouthparts and gnathopods to tear off and manipulate bits of seaweed and other plant material. Some amphipods will be drawn to dead or half-dead fish, and some are omnivores that will eat just about anything, from algae to small crustaceans and annelids. Certain amphipods that eat seaweeds build temporary shelters by rolling up part of a blade to form a tube.

A number of groups are specialized for scraping diatoms and thin coatings of organic material from sand grains, and others are filter-feeders that process sediment. Setae on the second antennae, gnathopods, or other legs may be used for accumulating diatoms or other suitable particles, which are then transferred to the mouthparts. Even the uropods may be involved in collecting food. Species of *Corophium* (Figure 17.53, D), which are almost always present in bays and estuaries, construct tubes of mud mixed with sticky secretions. Sediment is dragged into the tube by the antennae. Setae on the gnathopods trap food as it is moved backward with the water current driven by the pleopods.

Some species are efficient burrowers. Beach hoppers (Figure 17.53, E; Color Plate 6), which belong to the family Talitridae, dig rapidly into sand that is barely moist. They use their gnathopods and more anterior walking legs for pulling themselves into the sand, head first. The more posterior legs, which support the body and push it deeper, are spread to allow sand grains to pass between them. After a little sand has accumulated on the dorsal side of the uropods, the abdomen, which has been flexed ventrally, straightens out and throws the sand clear of the burrow. Some beach hoppers dig down to a depth of several centimeters and stay buried during daylight hours. After dark they emerge to scavenge on decaying plant material.

Species that live in watery sand, and that remain buried unless dislodged by waves, generally do not maintain a permanent burrow. Most of them, including those of the widespread family Haustoriidae (Figure 17.53, F), move freely through loose sand. Haustoriids use almost all of their appendages, even the second antennae, to pull, push, or steer themselves through sand. The flattened third and fourth legs are admirably constructed to function as shovels. Once buried, the animals almost swim through the sand, using the pleopods as propelling devices as well as for moving sand backward. The broad coxal plates of some of the more anterior legs, and certain of the articles of the fifth to seventh legs, keep the sand from filling up the space beneath the animal, where the current of water driven by the pleopods is moving.

Suborder Hyperiidea

Members of the Hyperiidea (Figure 17.54) are similar to gammarideans in having the head fused to just one thoracic segment and in having a well-developed abdomen. The eyes, however, cover much of the head, and the coxal plates of the thoracic legs are small. This is a strictly marine and planktonic group, and many of the known species ride on the jellyfishes and other gelatinous zooplankton organisms on which they feed. In general, they seem to be carnivores. One of the more remarkable hyperiids is *Phronima* (Figure 17.54, B). Its body is skeletonlike, and the fifth legs are chelate. The eyes are divided, so that half of each one faces laterally and half faces upward. A female *Phronima* cleans out the tunic of a salp (a type of planktonic urochordate;

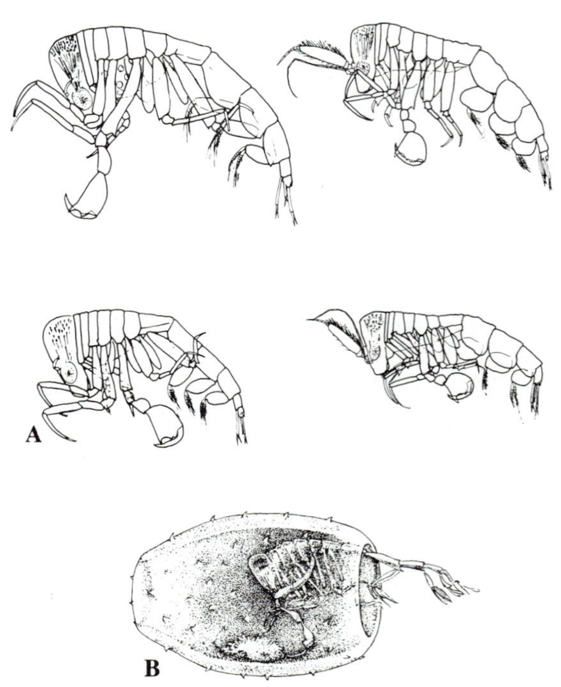

Figure 17.54 Order Amphipoda, suborder Hyperiidea. A. *Phronima stebbingi,* females (left) and males (right) of two forms of the species. (Shih, Dana Report, number 74.) B. *Phronima sedentaria* in the test of a salp. (Lützen, in Danmarks Fauna, 75.)

Chapter 24) or the bell of some siphonophore until this looks like a barrel that is open at both ends. It occupies the cavity and keeps the "house" moving slowly through the water by beating its pleopods.

Suborder Caprellidea

In the suborder Caprellidea, two segments of the thorax are fused to the head. The eyes are small, more nearly like those of gammarideans than of hyperiideans. Some of the thoracic appendages may be absent or vestigial, and even the well-developed legs do not have prominent coxal plates. The abdomen is short (sometimes almost totally absent) and there are, at most, only one or two pairs of reduced pleopods. One family, Caprellidae, is made up of "skeleton shrimps," well named because of their slender bodies (Figure 17.55, A–C). Except for one or two primitive genera, caprellids have substantially no abdomen. Their third and fourth legs are vestigial, but the oostegites that form the brood pouch of the female are nevertheless outgrowths of the coxal portions of these appendages; the legs also usually give rise to epipodites that serve as gills. The fifth legs are reduced in certain genera.

Caprellids are usually associated with hydroids and bushy bryozoans, but they are often found on sea stars and other animals, as well as on algae and eelgrass. Most seem to be detritus feeders, but some are carnivores that nip off hydroid polyps or capture small animals. When motionless, a caprellid resembles a preying mantis. It clings with the dactyls of its posterior legs to

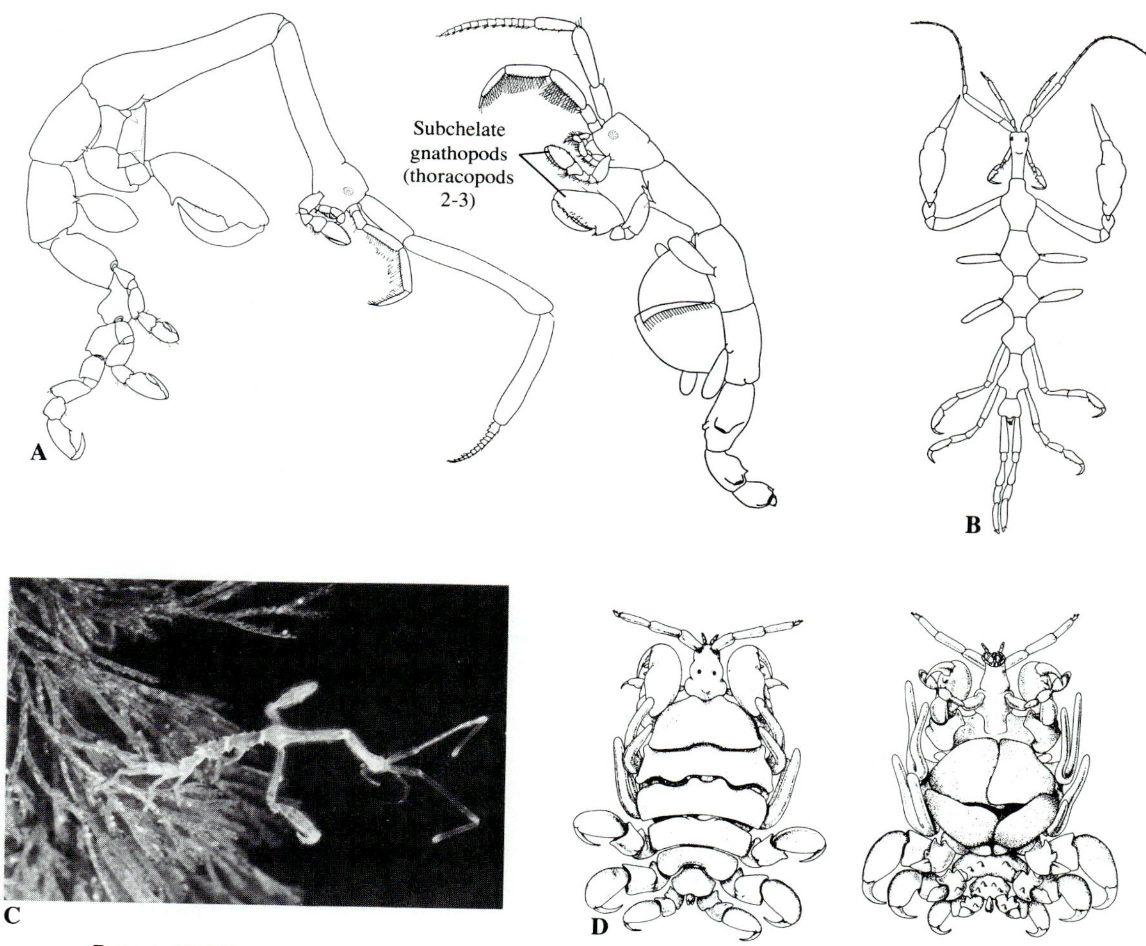

Figure 17.55 Order Amphipoda, suborder Caprellidea. A. *Caprella equilibra,* male (left) and female (right). (The three posterior thoracopods of the female are omitted.) B. *Hemiaegina minuta,* male. (A and B, McCain, United States National Museum, Bulletin 278.) C. *Caprella californica,* on a colony of the bryozoan *Bugula neritina. D. Cyamus ovalis,* male, dorsal view (left) and female, ventral view (right). (Iwasa, Journal of the Faculty of Science, Hokkaido Imperial University, series 6, *3*.)

whatever it happens to be on, and the front part of its body is raised with gnathopods ready. If detached, it moves after the fashion of an inchworm, alternately attaching itself by its gnathopods and then by its posterior legs.

The family Cyamidae consists of caprellideans that have relatively short bodies (Figure 17.55, D). As in the Caprellidae, the third and fourth legs are vestigial, but they have remarkably long epipodial gills. The abdomen is almost unrecognizable. Cyamids live on the skin of whales, and what and how they eat remains a mystery.

References on Amphipoda

Barnard, J. L., 1969. The families and genera of marine gammaridean Amphipoda (Crustacea). United States National Museum, Bulletin 271.

Bousfield, E. L., 1958. Fresh water amphipod crustaceans of glaciated North America. Canadian Field Naturalist, *72*:55–113.

Bousfield, E. L., 1973. Shallow-water Gammaridean Amphipoda of New England. Cornell University Press, Ithaca, New York.

Bowman, T. E., 1973. Pelagic amphipods of the genus *Hyperia* and closely related genera (Hyperiidea: Hyperiidae). Smithsonian Contributions to Zoology, number 136.

Caine, E. A., 1978. Habitat adaptations of North American caprellid Amphipoda. Biological Bulletin, *155*:288–296.

Leung, Y. M., 1967. An illustrated key to the species of whale-lice (Amphipoda: Cyamidae), ectoparasites of Cetacea, with a guide to the literature. Crustaceana, *12*:279–291.

Lincoln, R. J., 1979. British Marine Amphipoda: Gammaridea. British Museum (Natural History), London.

McCain, J. C., 1968. The Caprellidae of the western North Atlantic. Bulletin of the United States National Museum, *278*:1–147.

Madin, L. P., and Harbison, G. R., 1977. The associations of Amphipoda Hyperiidea with gelatinous zooplankton. I. Associations with Salpidae. Deep-Sea Research, *24*:449–463.

Shih, C.-T., 1969. The systematics and biology of the family Phronimidae (Crustacea: Amphipoda). Dana Reports, *74*:1–100.

Order Isopoda

Almost everyone is familiar with sowbugs, which are garden representatives of the large order Isopoda. When all major groups of crustaceans are considered, the isopods are the most successful in terms of colonization of terrestrial habitats. There are many species in fresh water, too, but by far the most impressive assemblage of these animals is found in the sea.

The cardinal features of the Isopoda may be summarized briefly as follows: the body is usually at least slightly flattened dorsoventrally, and the first thoracic segment, whose appendages are maxillipeds, is fused to the head (Figures 17.2, 17.56) to form a tagma called the *cephalon*. There is no carapace. The seven pairs of legs, used for walking or clinging, lack exopodites and epipodites. All pairs are generally much alike (Figure 17.57, A), although sometimes the first one, two, or three pairs are somewhat different from the rest. They are not chelate, and any tendency toward a subchelate condition is minimal. The oostegites that form a brood pouch in the female are, as is usual in peracaridans, outgrowths of the coxal portions of the legs. The abdomen is well developed, but not all of its segments are distinct; at least one is fused to the telson, forming what is called a *pleotelson*. The first five pairs of abdominal appendages are biramous pleopods, generally much flattened and functioning as gills. The uropods on the last segment are also biramous, but their form varies

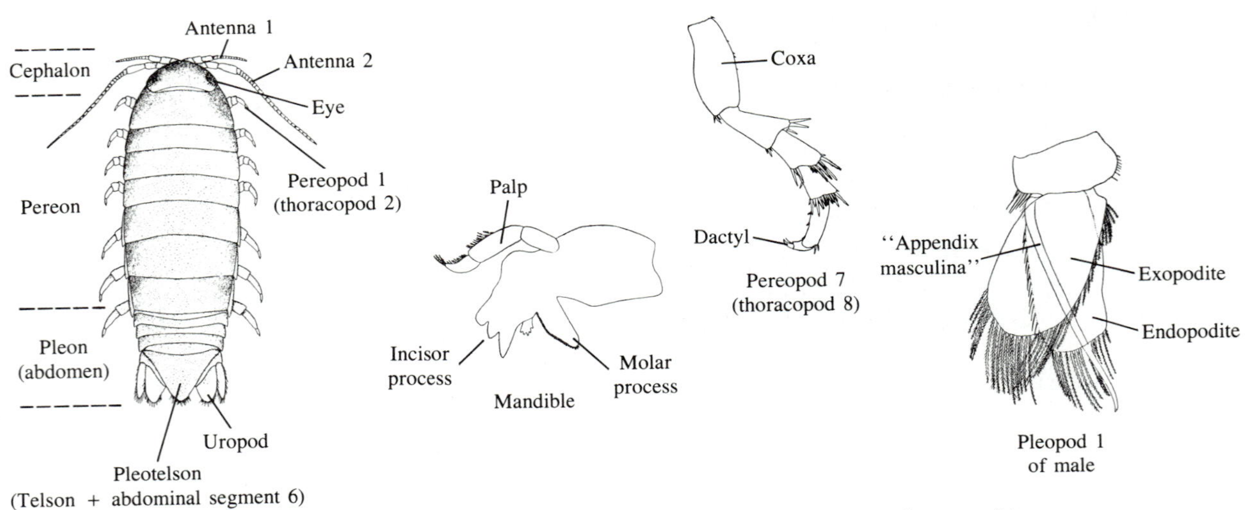

FIGURE 17.56 Order Isopoda. *Cirolana harfordi* (suborder Flabellifera) and some of its appendages.

FIGURE 17.57 Order Isopoda. A. *Ligia pallasi* (suborder Oniscoidea), dorsal view of a mature female (left) and ventral view of a male (right). B. *Armadillidium vulgare* (suborder Oniscoidea), the terrestrial pillbug, which can roll up into a ball. C. *Limnoria lignorum* (suborder Flabellifera), the gribble and the tunnels it has made in a timber immersed in seawater. D. *Gnorimosphaeroma oregonense* (suborder Flabellifera). E. *Cirolana harfordi* (suborder Flabellifera). See also Figure 17.56, which shows details of some of the appendages of *Cirolana*.

greatly, and in one large group of isopods they are turned under the abdomen and cover the pleopods.

The oostegites of the female (Figure 17.2) usually appear before she is fully mature, but their form is not perfected until the molt that just precedes egg laying. The number of oostegites varies according to the group. They are usually borne on legs 1 to 5 or 1 to 4, but in some species all seven pairs of legs have them; in others they are limited to just one pair. The oviducts coming from the pair of ovaries lead to gonopores on the sixth thoracic segment.

The gonopores of the male are situated on the last (eighth) thoracic segment. Transfer of sperm to the gonopores of the female is effected by the second pleopods or sometimes by both the first and second pairs. The mating posture varies. In garden sowbugs of the genus *Porcellio,* the male climbs crosswise onto the back of the female at about the level of the gonopores. He can then bend his abdomen under her thorax so that the two pairs of sperm-transferring pleopods can accomplish their mission. Some other isopods mate in much the same way, but this is by no means a standard procedure. Moreover, pairing may begin several days before the female has undergone her crucial molt. In any case, it is usual for eggs to be laid into the brood pouch a few hours after sperm have been deposited on the female's gonopores. After the eggs hatch, the young normally remain in the brood pouch until they have all but the last pair of legs. They undergo several molts before they are ready to leave. It is important that there be some circulation of water in the brood pouch. The oostegites can be raised and lowered to some extent, and in some species the coxae of the maxillipeds have outgrowths that create a current within the brood pouch. Terrestrial isopods use various techniques to collect moisture from raindrops or dew and transfer some of the water to their brood pouches.

Of the several suborders of isopods, five are numerically important. These are characterized briefly here to illustrate some of the diversity within this large group of Crustacea.

Suborder Flabellifera

In the Flabellifera, all of which are marine or estuarine, the uropods are directed posteriorly, and sometimes they are substantial enough to contribute to a tail fan. The mouthparts are generally specialized for chewing or piercing. The gribbles, belonging to the genus *Limnoria* (Figure 17.57, C), do an enormous amount of damage to pilings and other wooden structures in the sea. They use their mandibles to cut off tiny bits of wood and thus burrow as they feed. Glands opening into the midgut secrete cellulase, but gribbles cannot subsist entirely on cellulose and related compounds; they need sources of nitrogen, and therefore must depend on bacteria, fungi, and other microorganisms in the wood. *Sphaeroma, Gnorimosphaeroma* (Figure 17.57, D), and their relatives, which can roll up into a ball, are mainly scavengers on plant material. *Cirolana* (Figures 17.56; 17.57, E) and some of its allies scavenge on dead fishes and other animal matter. *Lironeca* and *Aega,* with piercing mouthparts, attach to fishes, from which they suck blood and tissue. These parasitic isopods are among the few malacostracans that are protandric hermaphrodites; that is, they function first as males, then as females.

Suborder Asellota

Members of the Asellota are found in both marine and freshwater habitats. All of the abdominal segments except the first, or first and second, are joined to the telson, so the pleotelson is a proportionately large structure. The uropods usually stick straight backward beyond it. *Asellus* (Figure 17.58, A) is a common freshwater genus. *Munna* (Figure 17.58, B) and numerous related genera are marine; some of them are long-legged, spidery types.

Suborder Valvifera

In the Valvifera, all of which are marine, the uropods are interesting in that they are platelike and bent under the abdomen, thus covering the pleopods. The pleotelson is substantial and has at least three abdominal segments incorporated into it. The legs are generally strong and adapted for grasping seaweeds or eelgrass. In most genera, including *Idotea* (Color Plate 6), the mouthparts are used for feeding on plant material, but some valviferans are predators on other animals or scavengers on animal matter.

Suborder Oniscoidea

The Oniscoidea (Oniscidea) is an almost strictly terrestrial group, and includes the familiar garden sowbugs and pillbugs (Figure 17.2; Figure 17.57, B). A few, such as *Ligia* (Figure 17.57, A), are found in rocky habitats at the fringes of the sea, where they get considerable moisture but are not often really drenched. As a rule, the members of this group scavenge on plant material. The first five abdominal segments are distinct; only the sixth is incorporated into the relatively small pleotelson. The uropods generally project well beyond this structure. The first antennae are much reduced. Pillbugs and sowbugs have tracheal systems on some or all of the five pairs of abdominal pleopods. These exchange oxygen and carbon dioxide with the blood, rather than directly with tissue cells, so they are more nearly like those of spiders than like those of insects. The presence of tracheae in this group of crustaceans is a good example of convergent evolution.

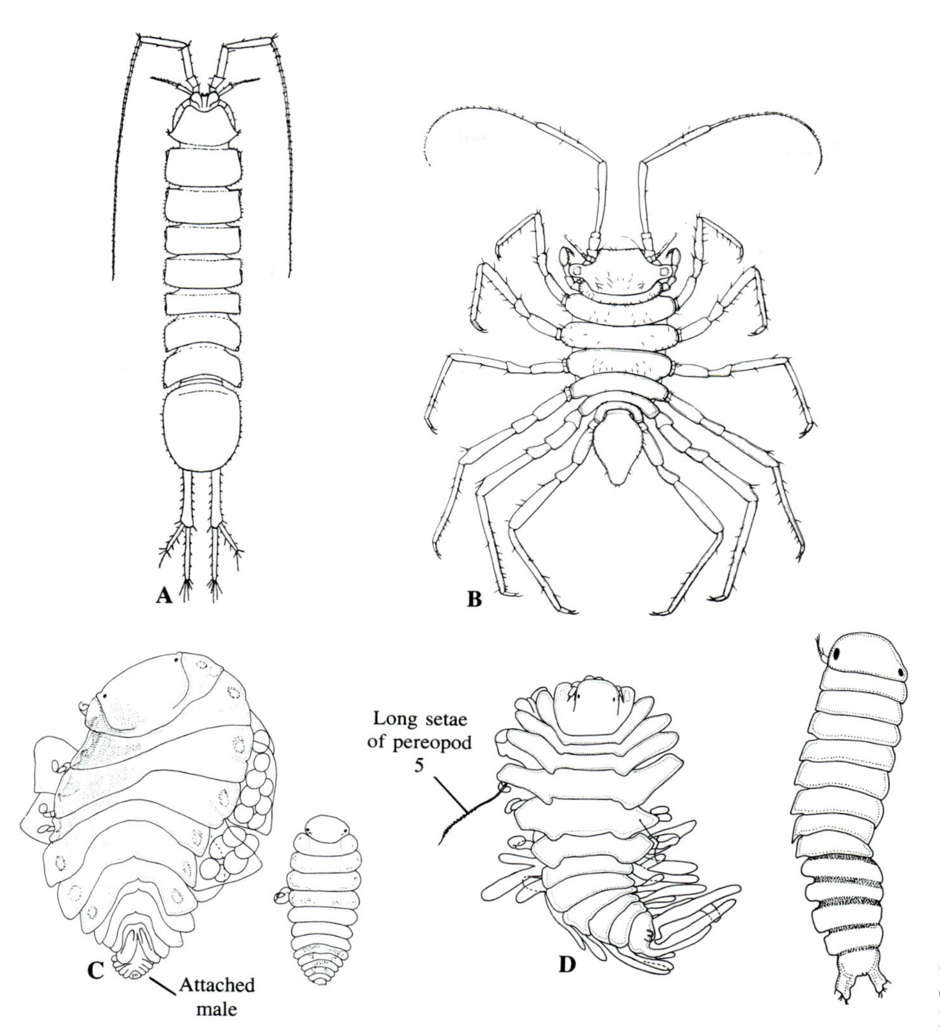

FIGURE 17.58 Order Isopoda. A. *Asellus californicus* (suborder Asellota); the pereopods have been omitted. (Miller, University of California Publications in Zoology, *39.*) B. *Munna armoricana* (suborder *Asellota*). (Carton, Bulletin de la Societé Linnéene de Normandie, series 10, *2.*) C. *Pseudione* (suborder Epicaridea), female (left) and male (right). D. *Stegias clibanarii* (suborder Epicaridea), female, with male attached (left) and detached male (right). (C and D, Menzies and Glynn, Studies on the Fauna of Curaçao and other Caribbean Islands, *27.*)

Suborder Epicaridea

The Epicaridea are a group of blood-sucking parasites that live on or within various crustaceans, including shrimps, crabs, and barnacles. Females are usually asymmetrical (Figure 17.58, C and D) and often vaguely segmented. The first and second antennae are vestigial, if they are recognizable at all, and the appendages in general are greatly modified. Thoracic legs, if present, are used for tightly gripping the host; sometimes only those on one side of the body are well developed. Oostegites, however, are often more prominent than they are in other isopods. The almost needlelike mandibles form a device for piercing tissue and drawing up fluid.

Males are small in comparison with females, but look a little more like ''normal'' isopods; there is generally one living in close association with each female. The first stage of development, called the *epicaridium,* subsists on yolk. The succeeding stage, the *microniscus,* may be parasitic on a copepod, but the third stage, the *cryptoniscus,* is free-living.

The Epicaridea, like some of the parasitic Flabellifera, includes some protandric hermaphrodites. Certain species cause the gonads of their hosts to deteriorate, or to fail to develop normally.

References on Isopoda

Edney, E. B., 1954. Woodlice and the land habitat. Biological Reviews, *29:*185–217.

Naylor, E., 1972. British Marine Isopods. (Synopses of the British Fauna, no. 3.) Linnean Society of London and Academic Press, London and New York.

Schultz, G. A., 1969. How to Know the Marine Isopod Crustaceans. Wm. C. Brown Co., Dubuque, Iowa.

Sutton, S. L., 1972. Woodlice. Ginn and Co., London.
Van Name, W. G., 1936. American land and fresh-water isopod Crustacea. Bulletin of the American Museum of Natural History, *71*:1–535.

Order Tanaidacea

Tanaids (Figure 17.59) are engaging little crustaceans, almost strictly limited to the sea and estuarine situations. Of the approximately 400 species, only a few occur in fresh water. The typical habitat of tanaids is sediment, including that which accumulates in colonies of clump-forming algae. Most species are no more than 5 mm long. The general body plan of tanaids is similar to that of isopods in that the appendages of the first thoracic segment are modified as maxillipeds, leaving seven pairs of legs. In addition, the last segment of the abdomen is fused to the telson, forming a pleotelson. Tanaids are easily distinguished from isopods, however, because the first of their seven pairs of thoracic legs are chelate, and the pincers are sometimes large in proportion to the rest of the body. Another unusual feature of tanaids when compared with isopods is that they have a carapace. This is small and involves only the first two thoracic segments—the ones from which the maxillipeds and chelate first legs originate—but it encloses a respiratory cavity. The uropods are generally slender and do not contribute to a tail fan of the type found in some isopods.

Although tanaids do not constitute a large group, they are diversified and show some interesting specializations. One genus, *Pagurapseudes,* lives in tiny snail shells and resembles a hermit crab. A number of species are burrowers; others build tubes out of the sediment in which they live. A common circumboreal species, *Leptochelia dubia,* can almost be said to hold some mudflats together. It often occurs in enormous concentrations and incorporates diatoms and much of the loose sediment into its tubes. The cementing secretion, produced by thoracic glands, issues from the coxae of certain legs.

Tanaids feed mostly on detritus or diatoms, using the chelae to bring food to the maxillipeds. Some species collect at least some of their food by filtration. Rhythmic movements of the maxillipeds draw a current of water past the mouthparts, and particles are caught on the setae.

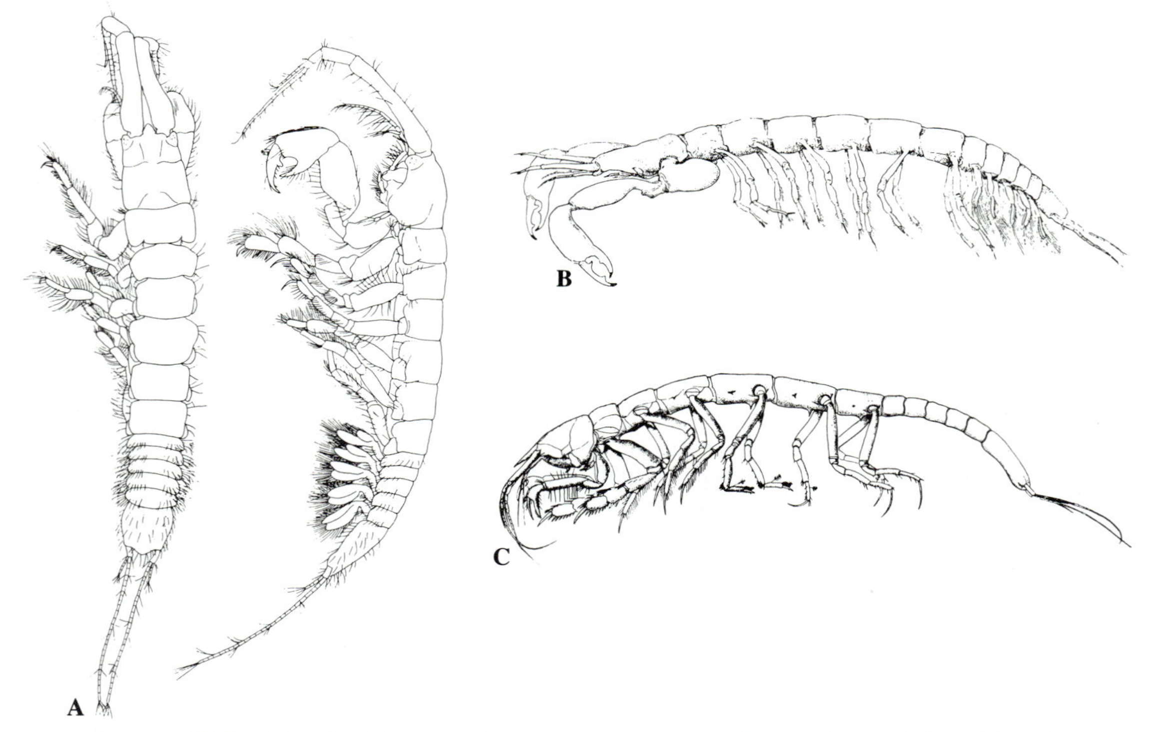

FIGURE 17.59 Order Tanaidacea. A. *Apseudes mussauensis,* male, dorsal view (left) and lateral view (right). (Shiino, Videnskabelige Meddelelser fra Dansk Naturhistorisk Forening, *128*.) B. *Apseudes galatheae,* female, lateral view. (Specimen taken in the Kermadec Trench at a depth of 6770 m.) C. *Neotanais serratispinosus,* male. (Specimen taken in the Kermadec Trench at a depth of 8210 m.) (B and C, Wolff, Galathea Reports, *2*.)

Sexual dimorphism is marked because females have oostegites. These marsupial plates are generally associated with the second to fifth legs (thoracic segments 3 to 6), sometimes with the fifth legs only. Females may lack some or all pleopods, but there is usually a complete set in males. The chelae and eyes are often much better developed in males than in females.

Complex courtship and mating procedures have been reported for certain species. The end result is the deposition, by the male, of sperm on the ventral side of the thorax of the female. Then the oostegites are pulled down. As the eggs are laid they are stuffed into the brood pouch and fertilized. Development is more or less direct, but when the young hatch they may not look much like adults, and the articles of the appendages may not yet be distinct. As a rule, when the young leave the brood pouch, they still do not have the last pair of thoracic legs.

References on Tanaidacea

Bückle-Ramirez, L. F., 1965. Untersuchungen über die Biologie von *Heterotanais oerstedi*. Zeitschrift für Morphologie und Ökologie der Tiere, *55*:714–782.

Holdich, D. M., and Jones, J. A., 1983. Tanaids. Synopses of the British Fauna, no. 27. Linnean Society of London, Brackish-water Sciences Association, and Cambridge University Press, Cambridge, London, and New York.

Lang, K., 1967. Taxonomische und phylogenetische Untersuchungen über die Tanaidacean. Arkiv för Zoologi, series 2, *19*:343–368.

Superorder Eucarida

Most of the larger malacostracans—crabs, lobsters, shrimps, hermit crabs, and their allies—belong to the Eucarida. There are about 10,000 species in the group. Fusion of the carapace to all eight thoracic segments unites the head and thorax into a cephalothorax. There is never a thoracic brood pouch consisting of oostegites, as is the case in mature females in the Peracarida. The compound eyes are stalked, and in this respect they differ from the eyes of all peracaridans except mysids. As a rule, eucaridans have a series of larval stages that differ markedly from the adult, but development is sometimes direct.

Order Euphausiacea

Euphausiaceans (Figure 17.60), best known by their Norwegian name "krill," are important constituents of the marine plankton, especially in colder seas. A few fall into a size range of 5 to 8 cm, but most are only 2 or 3 cm long. Their eyes are often large in proportion to the body as a whole. In some predaceous species each eye is divided into a part that faces laterally and a part that is directed upward.

The carapace is typically eucaridan in that it is fused to all of the thoracic segments. It looks "unfinished," however, because it does not reach down far enough ventrally to cover the gills, which are epipodites of the thoracic appendages. There are no maxillipeds, so all eight pairs of thoracic appendages may be called legs, even though the last two pairs are small. The endopodites of the legs are larger than the exopodites, but both rami are setose and are held beneath the body in such a way that they form a net for collection of diatoms and other small components of phytoplankton and zooplankton. In a few genera, the endopodites of the second and third pairs of legs are chelate or otherwise modified for capture of copepods, chaetognaths, and other animal prey of about the same size. Of the six pairs of abdominal appendages, those of the last pair are uropods that flank the telson to form a tail fan, which can be flexed for rapid backward swimming. The other abdominal appendages are pleopods that effect steady forward progression. The pleopods are surprisingly narrow, but their endopodites and exopodites, which consist of single articles, are provided with many setae. Operation of the strainer formed by the thoracic appendages appears to depend to some extent on forward progression of the animal, but there is evidence that exopodites also produce a feeding current. Just how the particles collected by the thoracic appendages are delivered to the mouthparts has not been fully described.

Euphausiaceans generally have a soft cuticle and are almost transparent. They do, however, have chromatophores and other patches of pigment, usually red, purple, or brown. Most of them also have light-emitting organs called photophores (Figure 17.60, B). These are directed ventrally, and are situated on the eyestalks, the coxae of most of the thoracic appendages, and along the midline on the first four abdominal segments. A photophore slightly resembles an eye. It has a lens, a mass of rodlike structures concerned with production of light, and a reflecting layer internal to this. The photophore is covered by some modified cuticle that resembles a cornea. The capacity of euphausiaceans to produce light on the ventral side may be an adaptation that makes them more difficult to see against the surface when this is brightly illuminated.

Euphausiaceans are easily confused with mysids, which are also well represented in plankton. In mysids, however, the carapace covers proportionately more of the thorax, even though it is fused with only a few of the more anterior segments. The statocysts characteristic of the uropods of mysids are lacking in euphausiaceans, and, of course, female euphausiaceans never have a brood pouch consisting of oostegites.

Some euphausiaceans drop their eggs singly; others carry them in clusters on the ischium of the sixth thoracic limb. Characteristically, two naupliar stages are

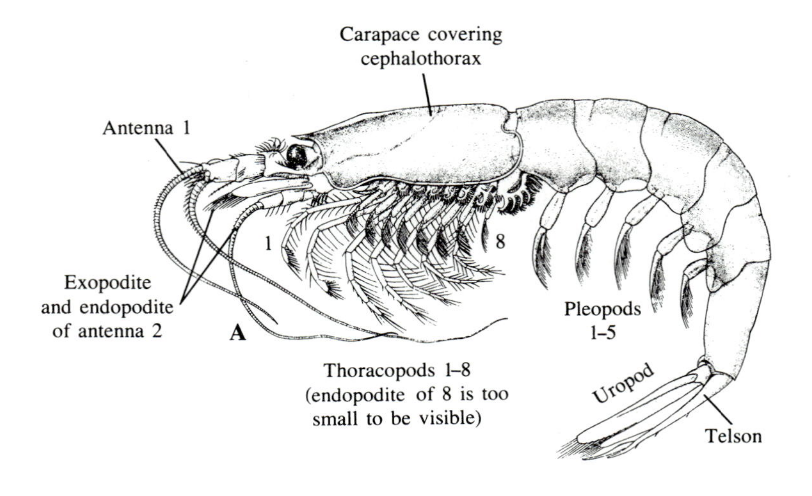

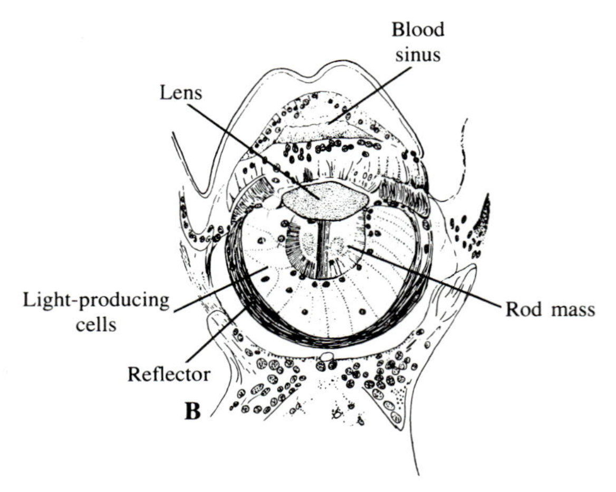

Figure 17.60 Superorder Eucarida, order Euphausiacea. A. *Thysanopoda acutifrons,* a young specimen. (Einarsson, Dana-Report, number 27.) B. A photophore of *Meganyctiphanes norvegica.* (Harvey, in Waterman, The Physiology of Crustacea, *2*.)

succeeded by a metanauplius; all three of these stages live on yolk that had been stored in the egg, and use their second antennae for swimming. The metanauplius is succeeded by about ten stages peculiar to euphausiaceans, and these are capable of feeding. They fall into three main types: the *calyptopis,* in which the eyes are hidden by the carapace; the *furcilia,* during which the second to eighth thoracic appendages and the pleopods form; and the *cyrtopia,* in which swimming is effected by pleopods and in which the antennae resemble those of adults. In the case of *Euphausia superba,* common in Antarctic seas, the nauplii and metanauplius are found only at considerable depths in comparatively warm water, and later stages move progressively closer to the surface, where the water is colder. In species that carry their eggs, development is generally accelerated, and the series of stages is shortened.

Euphausiaceans are generally more abundant in the open sea than near shore or in bays. The depth at which they swim varies with the species, the conditions of temperature and illumination, the stage of development, the availability of food, and other considerations. Some species make daily vertical migrations, coming to the surface at night and then sinking deeper during daylight hours. Heavy concentrations of these crustaceans often constitute much of the "deep scattering layer" recorded by depth-sounding instruments.

Euphausiaceans are important "middlemen" in the marine food web. Depending on the species, they feed on various components of the phytoplankton and zooplankton, and are in turn consumed by many larger carnivores, including fishes, petrels, penguins, seals, and whales. A large blue whale may consume more than 2 tons of krill in a day.

References on Euphausiacea

Boden, B. P., Johnson, M. W., and Brinton, E., 1955. The Euphausiacea of the North Pacific. Bulletin of the Scripps Institution of Oceanography, *6*:287–400.

Fisher, L. R., and Goldie, E. G., 1959. Food of *Meganycti-*

phanes norvegica. Journal of the Marine Biological Association of the United Kingdom, *38*:291–312.

Mauchline, J., 1958. The development of the Euphausiacea, especially that of *Meganyctiphanes*. Proceedings of the Zoological Society of London, *132*:627–639.

Mauchline, J. R., 1967. Feeding appendages of the Euphausiacea. Journal of Zoology (London), *153*:1–43.

Mauchline, J., 1984. Euphausiid, Stomatopod and Leptostracan Crustaceans. Synopses of the British Fauna, no. 30. Linnean Society of London, Estuarine and Brackish-water Sciences Association, and E. J. Brill/Dr. W. Backhuys, London and Leiden.

Price, H. J., Boyd, K. R., and Boyd, C. M., 1988. Omnivorous feeding behavior of the Atlantic krill *Euphausia superba*. Marine Biology, *97*:67–77.

Order Decapoda

The decapods ("ten-legged" crustaceans) include most of the familiar larger crustaceans: shrimps, lobsters, crayfishes, crabs, and hermit crabs. They differ from euphausiaceans in several important respects. Of the eight pairs of thoracic appendages, the first three are modified to serve as maxillipeds; the other five—the legs—are concerned with locomotion, grasping, crushing prey, and some other functions. The carapace grows down far enough on both sides of the body to cover the gills. As in euphausiaceans, these are basically epipodites of the coxae of the thoracic appendages, but often there are three for each appendage, and in that case one has been displaced to the joint between the coxa and the thorax, and another to the thorax itself. The respiratory current is produced by paddlelike projections (*scaphognathites*) of the maxillae. The five pairs of abdominal pleopods serve various functions, including swimming and circulating water through a burrow. In females they are generally modified for carrying eggs and brooding the young until these are released as advanced larvae or as juveniles. In males the one or two pairs of pleopods nearest the thorax are specialized for transferring spermatophores to the genital openings of the female. The last pair of abdominal appendages, the uropods, may form a tail fan with the telson. This is the case in shrimps, lobsters, crayfishes, and some other decapods, but it is not true of all members of the group.

There is much diversity in feeding techniques among decapods. Most true crabs, lobsters, and crayfishes are scavengers or predators, or both. They use their chelae for cracking shells, tearing up food, and passing pieces to the maxillipeds and mouthparts. Some are primarily consumers of animal matter, but freshwater crayfishes are more or less omnivorous, and many crabs subsist largely on seaweed. Detritus is the principal food of shrimps, hermit crabs, porcellanid crabs, and some true crabs. Most detritus feeders, however, are opportunists that will eat just about anything they can get, especially if it is animal matter. The so-called mud shrimps and ghost shrimps, *Upogebia* and *Callianassa,* filter microscopic particles from a current of water that they drive through the burrows they inhabit. The currents are created by the fanning motions of the pleopods. Fringes of setae on the anterior legs trap food, which is then combed from them by the third maxillipeds and passed forward. A few other specialized techniques of feeding are mentioned below in connection with particular groups of decapods.

With about 9000 species, the Decapoda is a large order. It is so diversified that no one group within it provides a neat stereotype for describing the others. In the evolution of the decapods, there has been much convergence as well as divergence. Thus the animals we call crabs, for instance, do not form a cohesive group. They are similar in that they have short abdomens, usually bent back under the thorax, but this condition has been arrived at more than once and is not by itself proof of close relationship. Some decapods show such messy mixtures of traits that their systematic position within the order is not agreed upon.

The classification of decapods given here is perhaps artificial to some extent. It is, however, widely used and it works well, even if certain of the major groupings consist of decapods whose relationship is doubtful.

The distinction between the two suborders is based on the structure of the gills. In certain shrimps, the thoracic gills, whether they originate from the coxa, the articulation between the coxa and body wall, or from the body wall above the coxa, have two series of branches, each member of a series being subdivided in such a way that it appears bushy. These shrimps form the suborder Dendrobranchiata. In all other decapods, assigned to the suborder Pleocyemata, the gills consist either of two broad lobes running the length of the central axis (the more common arrangement) or of numerous tiers of filaments (characteristic of only a few shrimps).

One alternative system of classification relies on the fact that shrimps are primarily swimmers (suborder Natantia), whereas other decapods are mostly crawlers (suborder Reptantia). This system is easy to comprehend, but is probably more artificial than the one that separates the Dendrobranchiata from Pleocyemata on the basis of gill structure.

Suborder Dendrobranchiata

There are about 2000 species of shrimps, but only 350 belong to the Dendrobranchiata; the rest are in the Pleocyemata. In general, however, shrimps are slightly compressed laterally, have a well-developed abdomen, and use their biramous pleopods for swimming. By spreading its tail fan and vigorously flexing its abdomen ventrally, a shrimp can drive itself rapidly backward.

The endopodites of the thoracic appendages serve as legs for walking and grasping; exopodites are not often present. The carapace is usually (not always) prolonged anteriorly as a pointed rostrum. The second antennae are distinctive in that their exopodites are specialized as scalelike structures.

The dendrobranchiate shrimps include those of the family Penaeidae, some of which are large and of great commercial importance. The first to third legs of penaeids are chelate (Figure 17.61, A), a feature not often seen in pleocyemate shrimps. As a rule, the eggs are not carried by the female; they fall away as they are laid. The young hatch as nauplii, which is unusual for Decapoda. *Penaeus setiferus,* found in the Gulf of Mexico and up the Atlantic coast as far as North Carolina, is the most extensively exploited species in North America. It reaches a length of nearly 20 cm. The annual catch of this and related penaeids in the United States is about 100,000 tons.

The only other family of Dendrobranchiata is the Sergestidae. In sergestids, the second and third legs are chelate, but the first legs are not. The fourth and fifth pairs of legs are reduced. As in penaeids, the life history includes a nauplius stage.

Suborder Pleocyemata

It has already been explained that the Pleocyemata, to which most shrimps and all nonshrimp decapods belong, differ from the Dendrobranchiata on the basis of gill structure. More than 95% of the decapods are pleocyemates, and the group is enormously diversified, for it includes, besides shrimps, the crayfishes, lobsters, crabs, hermit crabs, and many other distinctive types. The following sections discuss each of the major infraorders.

Infraorder Caridea

Most shrimps belong to the Caridea. The members of this group have one or two pairs of chelate or subchelate legs (Figure 17.61, B). Females attach their eggs to pleopods and the young do not hatch until they reach the zoea stage (Figure 17.61, C); in some species, in fact, they do not hatch until they are fully formed shrimps. Some of the several families have a "brokenback" look, because the abdomen is bent sharply downward at the third segment.

Four families of carideans are especially important in marine habitats. The Pandalidae are characterized by first legs that lack chelae or have only small chelae. The second legs are distinctive in that the carpus is subdivided into many small articles, which make it very flexible. Pandalids are protandric hermaphrodites. Members of the family Alpheidae generally live in pairs, and their first legs, often unequal, have large chelae. These are the snapping shrimps, or pistol shrimps, which can snap the dactyl of a chela against the fixed finger with such force as to stun or kill prey by percussion. In the Hippolytidae, the chelae of the first legs are small, about the same size as those of the second legs. The Crangonidae have subchelate first legs; these are larger than the second legs, which have small chelae. The rostrum of a crangonid is scarcely noticeable.

Two of the larger families have many freshwater and brackish-water representatives. These are the Atyidae and Palaemonidae. In atyids, the fingers of the chelae of the first and second legs do not make good contact, and are tipped with tufts of long setae. In palaemonids, the first and second legs have only small chelae. Of the few remaining families, some consist mostly or entirely of marine species that inhabit considerable depths.

Infraorder Stenopodidea

There are not many species of Stenopodidea—about 30. Unlike most shrimps, they are not flattened laterally. The first three legs, like those of penaeids, are chelate, but the third legs are much larger than the others. A few species of this group inhabit cavities within hexactinellid sponges (Chapter 3). Some others, associated with coral reefs, have been observed to use their chelate legs to clean fishes. "Cleaners" are also found, however, in the Hippolytidae, one of the families of Caridea mentioned in the preceding section.

Infraorder Astacidea

The Astacidea includes freshwater crayfishes (Figure 17.8) and marine lobsters. In these decapods, the first to third legs are chelate, but the first legs are especially large and their chelae are powerful. The abdomen is about as long as the cephalothorax and has a well-developed tail fan. The exopodites of the uropods are divided transversely into two articles. Female crayfishes brood their eggs until the young hatch as juveniles with the form of the adult; lobsters brood until the young hatch in the mysis stage.

Some crayfishes burrow in marshy habitats and in pastures where the ground is wet but where there is no standing water. Others spend most of their lives in lakes and ponds, but dig burrows along the margins and find refuge in these when the water is low or cold. In one genus of Australian crayfishes, sometimes placed in a separate infraorder, Austroastacidea, there are species that are decidedly terrestrial, living in holes on wooded hillsides. They will not live long if immersed in water.

Infraorder Palinura

Palinurans, of which there are about 130 species, differ from astacideans in that their first legs are not specialized as powerful chelipeds. (Sometimes one or more pairs of legs are chelate, but they are not espe-

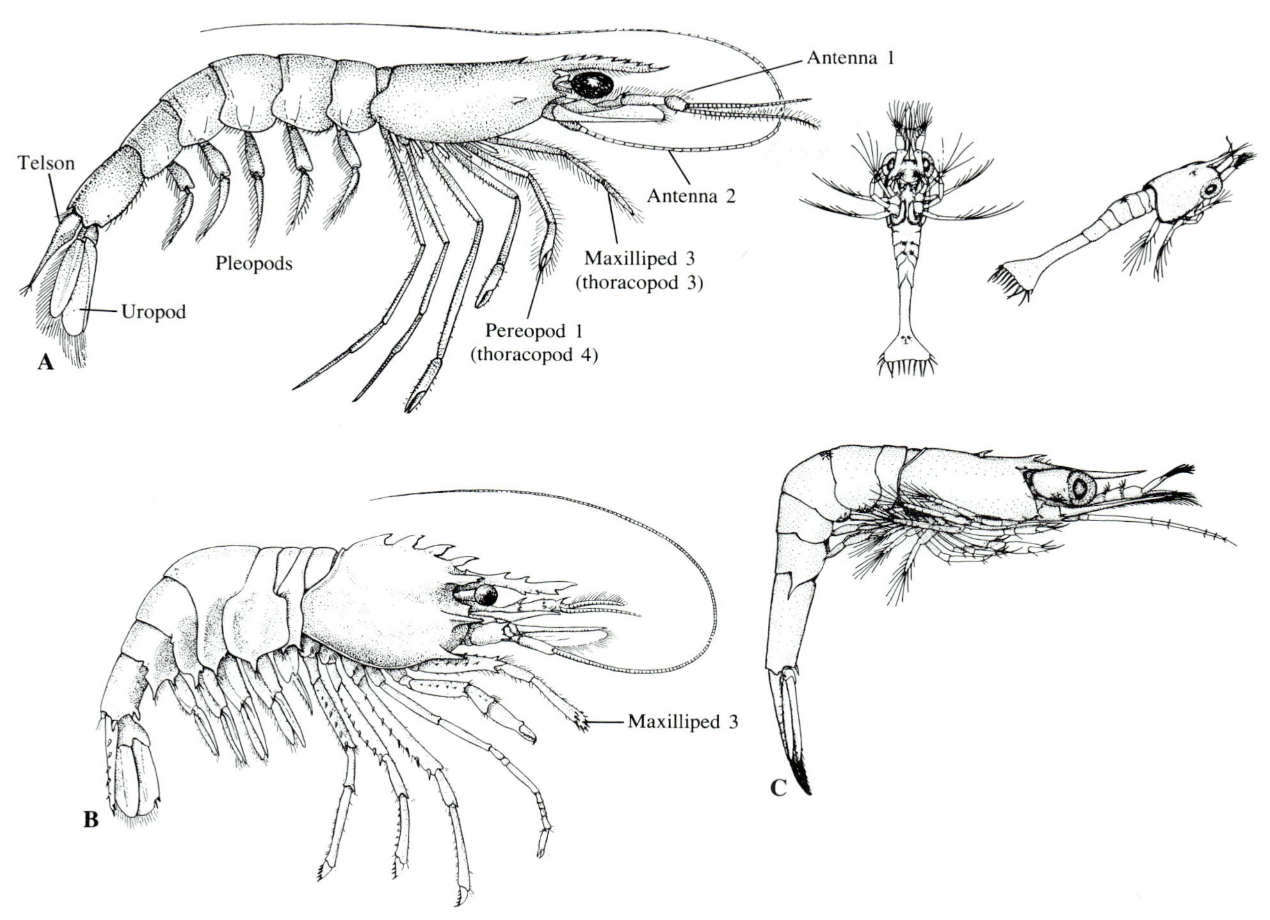

FIGURE 17.61 Order Decapoda. A. *Penaeopsis kishinouyei* (suborder Dendrobranchiata). (Boone, Bulletin of the American Museum of Natural History, *63*.) B. *Lebbeus groenlandicus* (suborder Pleocyemata, infraorder Caridea). (Boone, Bulletin of the Vanderbilt Marine Museum, *3*.) C. *Palaemonetes pugio* (infraorder Caridea), an early zoea stage (above) and a stage that is nearing maturity (below). (Broad, Biological Bulletin, *112*.)

cially large.) The exopodites of the uropods, moreover, are not divided transversely into two articles, but otherwise the tail fan of a palinuran is similar to that of an astacidean. To define the Palinura fully, it would be necessary to discuss many details that are found only in certain genera.

Palinurans are mostly, but not entirely, warm-water decapods. Among the types likely to be encountered in a fish market or in the frozen-food department of a grocery store are the spiny lobsters (Palinuridae; Figure 17.62, A) and shovelnose lobsters (Scyllaridae).

A remarkable type of larva, called the phyllosoma (Figure 17.62, B), is characteristic of the life history of many palinurans. It follows a naupliuslike stage that typically hatches from eggs carried by the female. The phyllosoma stage, which is sometimes benthic and which may also be carnivorous, generally lasts for several months and goes through several molts before the definitive number of appendages and adult form are attained.

Infraorder Anomura

The name Anomura (''abnormal tail'') refers to the fact that in many members of this assemblage the abdomen is asymmetrical or otherwise peculiar. Anomurans form the most varied major group of decapod Crustacea, but nearly all of them share two obvious characteristics: the second antennae are lateral to the eyes, rather than between the eyes, and the legs of the fifth pair are small and sometimes tucked away in the gill chamber beneath the carapace, where they function as cleaners.

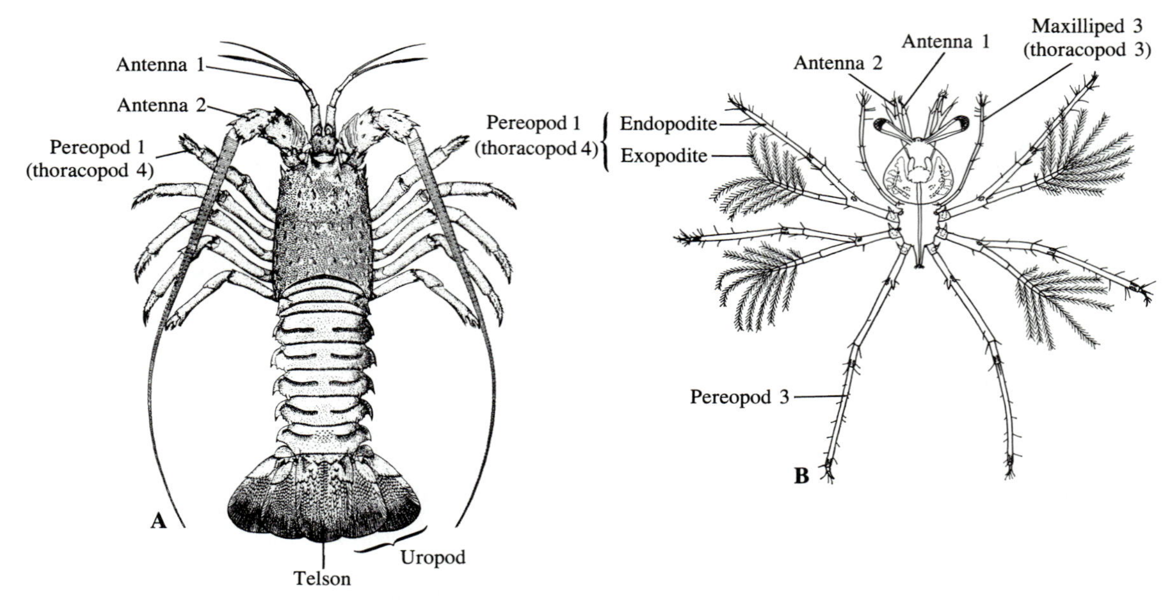

FIGURE 17.62 Infraorder Palinura. A. *Panulirus interruptus,* the California spiny lobster. (Rathbun, Crustaceans, in Fisheries and Fisheries Industries of the United States. B. *Janus edwardsii,* phyllosoma larva. (Batham, Transactions of the Royal Society of New Zealand, Zoology, *9*.)

The Anomura includes hermit crabs, a variety of types that resemble true crabs, and a few that look much like crayfishes or lobsters. The following discussion of the subdivisions of Anomura is much simplified, but it brings out some of the diversity in this extremely interesting assortment of decapods.

The superfamily Paguroidea includes hermit crabs and some crablike types thought to be derived from them. Marine hermit crabs (Color Plate 6) generally live in snail shells, and sometimes in limpet shells, single valves of clam shells, hard worm tubes, or other hollow structures; a few do not protect themselves this way. Species that use snail shells have a coiled abdomen (Figure 17.63, C). Pleopods used for creating a respiratory current and for brooding eggs are generally developed only on the left side. The left uropod is larger than the right, but both are used to wedge the tip of the abdomen in the upper part of the spire of the shell. The right chela is often decidedly larger than the left, and may be used to block the aperture of the shell. Most species are detritus-feeders and scavengers, but some are to a large extent carnivores, at least under certain conditions.

As a shell-inhabiting hermit grows larger, it must find successively larger shells. The size of the house may be a crucial problem for a female about to start carrying eggs. Most hermit crabs whose behavior has been studied go through a ritual of testing an empty shell for suitability, inserting a cheliped or making other movements to establish that the shell is unoccupied and has the right characteristics. It may abandon the old shell to try on the new prospect for size. In some situations, there is competition for shells, so a hermit may have to fight for the new shell it has picked out for itself.

Some hermit crabs live in tough, almost leathery sponges of the genus *Suberites* (Figure 3.17, A). Each one starts out in a shell, but after this has been overgrown by a sponge, the shell gradually dissolves and soon the crab occupies a spiral cavity in the sponge itself. An interesting species found on the Pacific coast of North America is *Discorsopagurus schmitti* (Figure 17.63, E) which inhabits the empty tubes of polychaetes of the families Sabellariidae and Serpulidae. It has a symmetrical abdomen. Hermit crabs that occupy holes in rock, coral, or other hard substrata, and those that live in empty pieces of bamboo, hollow mangrove roots, and certain sponges, also have symmetrical or nearly symmetrical abdomens.

Eggs of hermit crabs hatch when the young have reached the zoea stage (Figure 17.63, F). The larvae are then released. With further growth they attain the metazoea stage, then a stage called the **glaucothoe** (Figure 17.63, F). This finds a shell, or some other abode that suits the particular species, and settles in it.

Several species of hermit crabs typically plant sea anemones on their shells (Color Plate 2). In some cases, at least, the anemones provide protection from preda-

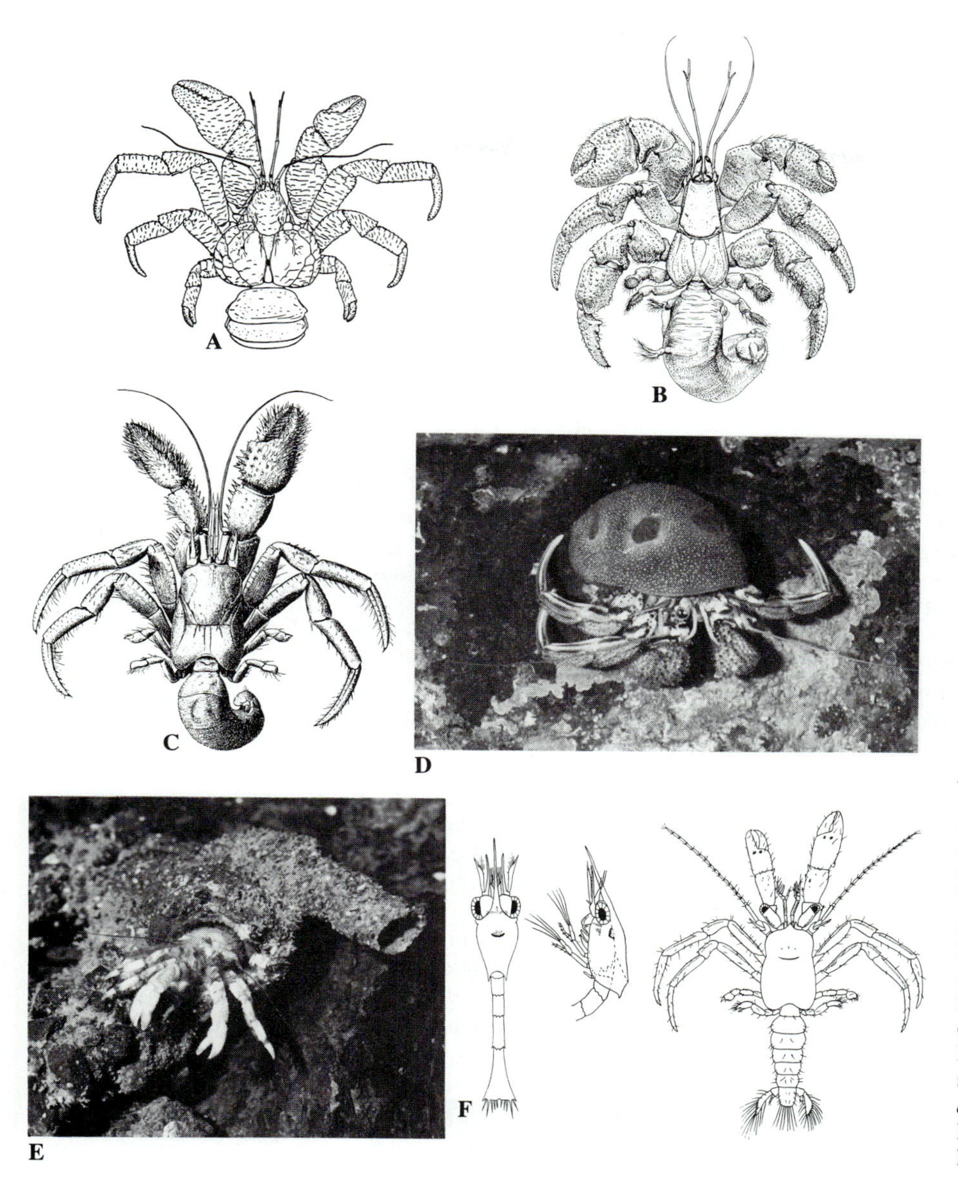

FIGURE 17.63 Infraorder Anomura, superfamily Paguroidea. A. *Birgus latro,* the coconut crab. (Calman, in Lankester, A Treatise on Zoology.) B. *Coenobita clypeatus,* removed from its shell. (Boone, Bulletin of the Vanderbilt Marine Museum, *6.*) C. *Pagurus setosus,* removed from its shell. (Benedict, Proceedings of the United States National Museum, *15.*) D. *Pagurus ochotensis.* E. *Discorsopagurus schmitti,* a straight-bodied hermit crab that inhabits tubes made by the polychaete *Sabellaria cementarium.* F. Zoea (left) and glaucothoe stage (right) of *Pagurus bernhardus.* (Macdonald, Pike, and Williamson, Proceedings of the Zoological Society of London, *128.*)

tors that are sensitive to nematocysts. An octopus, for instance, may cease to attack a hermit crab after it has been stung one or more times by an anemone's nematocysts. When a hermit crab that normally carries anemones on its shell must move to larger quarters, it gently "massages" the anemones until they relax and can be picked up and placed on the hermit's new shell. The hermit may also detach an anemone of the right species from some other substratum and stick it to its shell (Figure 17.64). Hydroids of the genus *Hydractinia* are commonly found on shells of certain hermit crabs. Some aspects of the biology of *Hydractinia* and its relationship to its host are dealt with in Chapter 4.

Terrestrial hermit crabs, of which there are numerous species in tropical and subtropical regions, must return to the sea to reproduce, for their larval stages, similar to those of marine hermit crabs, are aquatic. Mating takes place on land and the young are brooded to the zoea stage before they are released in the water. As a rule, gills are reduced or absent in terrestrial Paguroidea; the inner lining of the carapace and the body wall of the thoracic region are specialized for respiration in air, and are kept moist. Adults of some species crawl to the water's edge periodically to replenish the supply of moisture they carry with them inside their shells. Also, by digging into the sand or crawling into holes,

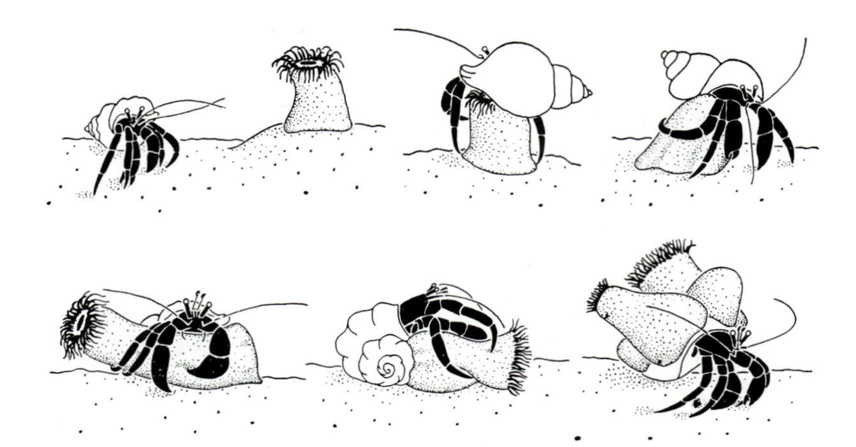

Figure 17.64 Infraorder Anomura, superfamily Paguroidea. *Dardanus arrosor,* "massaging" a sea anemone until it has become detached, then placing it on its shell. (Schöne, in Waterman, The Physiology of Crustacea, *2*.) (See also Color Plate 6.)

these animals can protect themselves from drying out. Most of them, like *Coenobita clypeatus* (Figure 17.63, B) of the Florida Keys and Caribbean islands, scavenge on plant and animal material that has been washed up on the shore.

The coconut crab, *Birgus latro* (Figure 17.63, A), is perhaps the most astonishing member of the Paguroidea. It reaches a large size—over 30 cm long—and does not carry a shell except when it is young. It walks around with its nearly symmetrical abdomen folded under the cephalothorax. It can climb coconut palms and other trees; when it crawls back down, it moves with its hind end first. Its chelae are strong and can be used to cut off fruits, including coconuts, and to extract food from them. To eat a coconut, *Birgus* must tear away the fibrous husk to locate the place where the seedling would emerge during germination. It works its chelae into this softer spot. Large crabs can apparently make holes in any part of the nut. *Birgus* also consumes dead animals, and will attack other land crabs, including sickly members of its own kind.

By the time a land hermit crab leaves the water to take up a terrestrial existence, it may already have picked up a shell; if not, it will have to find one soon. Its last aquatic stage—the glaucothoe—has a symmetrical abdomen, but this becomes asymmetrical after the animal acquires its shell, and in most species it remains asymmetrical. In the coconut crab, which gives up its shell after it reaches a length of about 1 cm, the abdomen becomes nearly symmetrical again and stays that way.

The remaining members of the Paguroidea belong to the family Lithodidae. Some of them look like true crabs because their abdomens are folded tightly against the ventral side of the cephalothorax. *Paralithodes camtschatica,* the commercially important king crab of the northern Pacific Ocean, fits into this category. Superficially it resembles some of the spider crabs, but its antennae are external to the eyes and the fifth legs are reduced. The terga of the abdomen are not like those of true crabs, either, because each consists of several separate plates instead of just one piece. *Lopholithodes* (Color Plate 6) and *Cryptolithodes* (Figure 17.65; Color Plate 6), the latter with the carapace hardened to form an oblong saucer that completely covers the legs, are other examples of this series of anomurans. Finally, there are some that look less like true crabs. The abdomen, although short, is fleshy and cannot be pressed tightly to the underside of the cephalothorax. *Hapalogaster* (Color Plate 6) and *Oedignathus,* both represented on the Pacific coast of North America, are probably closely related to certain of the marine hermit crabs.

The superfamily Galatheoidea has both lobsterlike and crablike representatives. *Munida* (Figure 17.66, A and B) and *Galathea* have a large abdomen with a well-developed tail fan that can be used for a quick getaway.

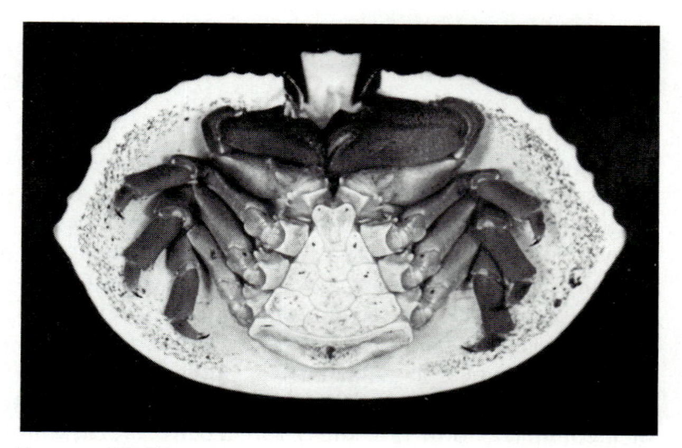

Figure 17.65 Infraorder Anomura, superfamily Paguroidea. *Cryptolithodes sitchensis,* ventral view. (See also Color Plate 6.)

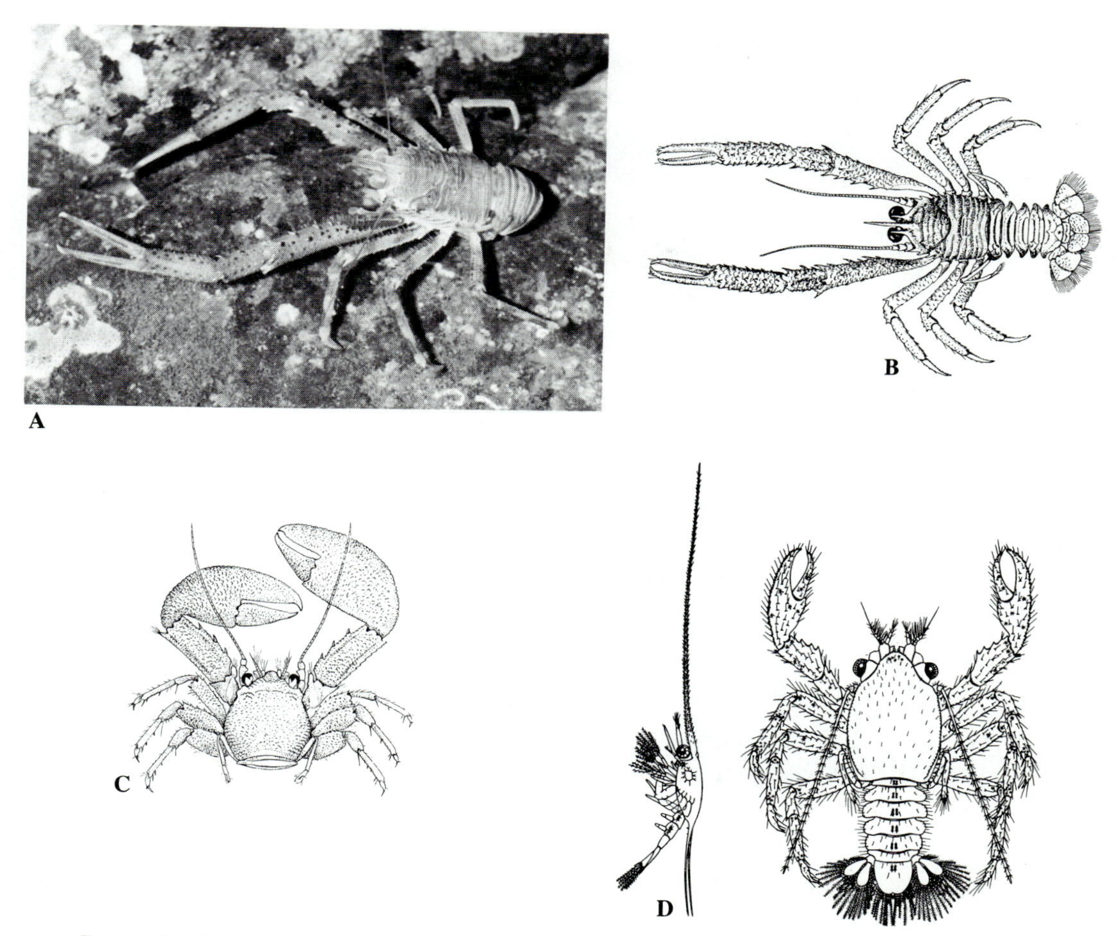

FIGURE 17.66 Infraorder Anomura, superfamily Galatheoidea. A. *Munida quadrispina*. B. *Munida quadrispina,* straightened to show the form of the abdomen. (Benedict, Proceedings of the United States National Museum, *15*.) C. *Petrolisthes armatus*. (Boone, Bulletin of the Vanderbilt Marine Museum, *3*.) D. *Petrolisthes elongatus,* zoea larva (left) and megalopa (right). (Wear, Transactions of the Royal Society of New Zealand, Zoology, *5*.)

The posterior part of the abdomen, however, is normally turned under the body. *Aegla,* a somewhat similar genus, is found in fresh water in South America; this is the only genus of Anomura that is not marine or whose young are not dependent on the marine environment. The crablike galatheoids include *Petrolisthes* (Figure 17.66, C and D) and *Pachycheles,* which have a thin abdomen held tightly to the underside of the cephalothorax. Both genera are found on the Pacific coast of North America, and there are species of *Petrolisthes* on the southern Atlantic Coast. These anomurans are filter-feeders. The zoea stage (Figure 17.66, D) is very distinctive in that it has an exceptionally long rostrum. The megalopa stage still has the abdomen directed straight back.

The superfamily Hippoidea consists of some distinctive, short-bodied anomurans that bury themselves in sand on wave-swept beaches. The so-called mole crabs of the genus *Emerita* (Figure 17.67), found in many parts of the world, including the Atlantic and Pacific coasts of North America, have been extensively studied. When they are entrenched in the sand, their anterior ends, barely exposed, face seaward. After a wave washes over them, they use their long, featherlike second antennae, on which the primary hairs have branch hairs, as a net to trap small food particles. The particles that have been collected are then transferred to the mouthparts. The shape of the body, almost like that of a plum, is well adapted for rolling around on the beach if the animal is washed out by the surf. When dislodged and temporarily disoriented, a mole crab rolls its antennae up under its maxillipeds.

The first four pairs of legs (thoracopods 4 to 7) are specialized for digging and for stirring up "liquid"

A

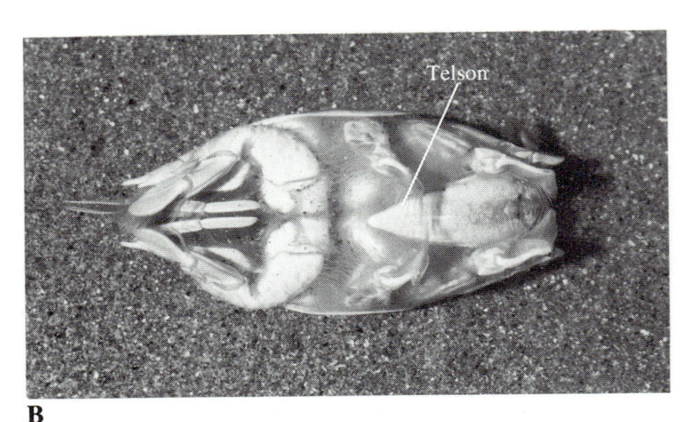

B

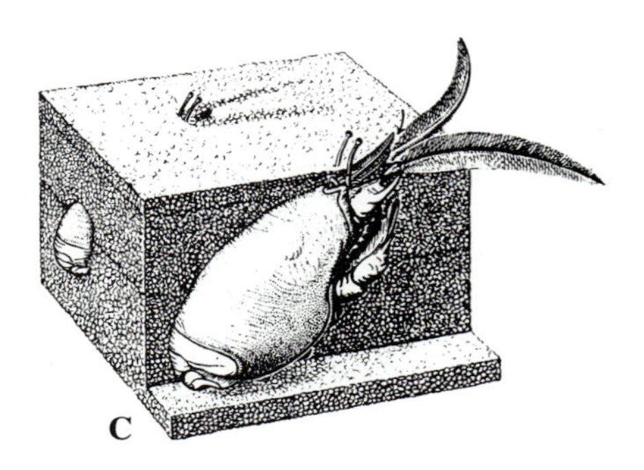

FIGURE 17.67 Infraorder Anomura, superfamily Hippoidea. A. *Emerita analoga,* the mole crab of sandy beaches of the Pacific coast of North America, exposed (left) and half-buried (right). B. *Emerita analoga,* ventral view. C. *Emerita emerita,* a European mole crab, in feeding posture. (Seilacher, Natur und Volk, *91*.)

sand so that the mole crab sinks down into it. The last pair of legs is used to remove sand and other foreign material from the gill chamber beneath the carapace. The long telson, bent under the body, is fringed by long hairs. These, together with hairs on the legs, help keep sand from entering the gill chamber. In females, the telson also serves to cover the developing eggs held by the pleopods of the short abdominal region. The eyes, on rather long stalks, are kept at the surface while the animal is in its usual posture in the sand. Although mole crabs move up and down the beach with the tide, any that may be left behind can dig in deep to avoid being dried out. When surf activity returns to that part of the beach, they work their way back to the surface of the sand and begin feeding again. Mole crabs are an important part of the diet of some shorebirds.

Infraorder Thalassinidea

Thalassinideans have a straight abdomen and their fifth legs are not reduced, but their second antennae are lateral to the eyes, and in this respect they resemble anomurans. The ones most likely to be found at the shore in temperate regions are the ghost shrimps and mud shrimps. Three or four of the pairs of pleopods of these animals have broad, flat rami that fan water through the burrows and that function in respiration; in females, the pleopods are also used for carrying eggs until they hatch at the zoea stage.

The two principal genera, *Callianassa* (ghost shrimps, Color Plate 7) and *Upogebia* (mud shrimps), will be dealt with together. There are differences in their appendages and lifestyles, but on the whole these crustaceans, some of which reach a body length of about 10 cm, are similar. Adults generally live in pairs, digging into mud and sand and making burrow systems that have several openings to the outside. In the tunnels, or where one tunnel joins another, there are enlargements wide enough to enable the shrimps to turn around, sometimes by somersaulting. *Callianassa* loosens sediment with its first, second, and third legs, and hauls the sediment to the surface in a ''scoop'' formed by the third maxillipeds. *Upogebia* uses the first and

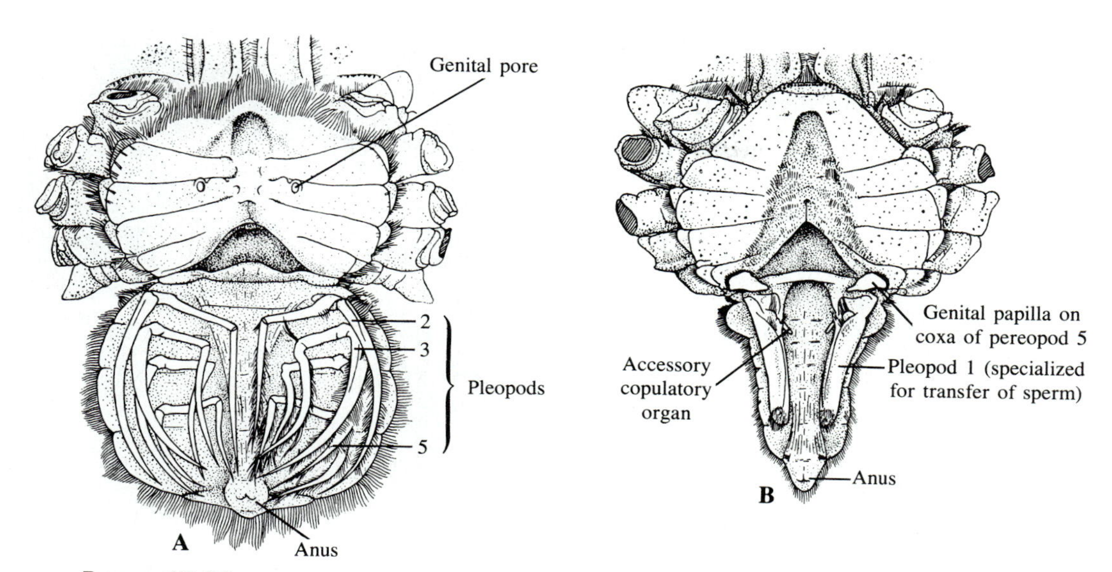

FIGURE 17.68 Infraorder Brachyura, section Brachyrhyncha. *Eriocheir japonicus,* abdomen (pulled back) and portion of cephalothorax. A. Female. B. Male. (Okada and Miyashita, Memoirs of the College of Science, Kyoto University, 1934.)

second legs for digging, and presses these legs together to hold sediment while hauling it up to the surface. *Upogebia* also mixes fine sediment with mucus and applies the mixture to the wall of the burrow to form a smooth and stable lining. *Callianassa* feeds on organic matter it obtains as it digs, as well as on particles that accumulate on the setae of certain legs as the animal circulates a current of water through its burrow. The food of *Upogebia* consists mostly of material strained out by the setae of the legs.

Species of *Callianassa* and *Upogebia* almost invariably have an interesting assortment of small animals living as commensals in or next to their burrows. The fishes, pea crabs, shrimps, polychaete annelids, and clams form a complex similar to that found with the echiuran worm, *Urechis,* on the Pacific coast (see Chapter 12). There may, however, be some more intimate associations. *Upogebia,* for instance, often has a small clam attached to the underside of its abdomen, and certain species of *Callianassa* have copepods under the carapace in the thoracic region. Both genera are frequently parasitized by isopod crustaceans of the suborder Epicaridea.

Thalassina anomala is a species common in the Malay peninsula, where its extensive burrows sometimes run through the dikes built to keep salt water from flooding rice fields. It is a nuisance not only because it allows salt water to enter the fields but also because it forms tall chimneys of mud around the openings of its burrows. The chimneys may be more than 50 cm high and so close together that they make walking difficult.

Axius, Axiopsis, and their relatives resemble crayfishes in that the exopodites of their uropods are divided into two articles. In other respects these crustaceans seem to be close to ghost shrimps and mud shrimps. Some of them live in coral reefs or in sponges; others burrow into soft bottoms. Certain species are restricted to very deep water.

Infraorder Brachyura

In brachyurans, the true crabs, the abdomen (Figure 17.68) is almost always thin, bent under the cephalothorax, and pressed tightly to it. (There are rare exceptions.) There are no uropods, and males have only the first one or two pairs of pleopods, which are specialized for transfer of sperm. The carapace usually has a marginal ridge that separates the dorsal side of the animal from the ventral side. The second antennae are between the eyes, not lateral to them, and the fifth legs are not often much smaller than the other legs. As in most decapods other than shrimps, the first legs are stout and have powerful chelae. When true crabs move quickly, they generally do so sideways, rather than backward or forward.

In most brachyurans, the larval stage between the zoea and adult is the megalopa. It has all of the adult appendages and its carapace is shaped like that of a crab, but its abdomen is proportionately long and is not yet bent tightly under the cephalothorax.

The approximately 4500 species of Brachyura are assigned to several categories, usually called sections, some of which are of limited geographic distribution.

The three sections discussed below are those that have numerous species in temperate regions.

In crabs of the section Oxyrhyncha ("sharp snout"), the carapace is usually longer than wide, and its anterior edge is prolonged into one or two sharp rostral processes. The so-called spider crabs belong to this group. Two of the many genera are illustrated: *Pyromaia* (Figure 17.69, A) and *Chionoecetes* (Figure 17.69, B). Most of the decorator crabs, which, in order to camouflage themselves, encourage growths of algae, sponges, hydroids, and other organisms on their carapaces, are in the Oxyrhyncha.

The section Brachyrhyncha consists mostly of crabs that have an oval or nearly rectangular carapace (Figure 17.70; Figure 17.71, A and B). The carapace does not have sharp rostral projections and is sometimes much wider than long. A few commercially important crabs, including *Callinectes sapidus,* the blue crab of the Atlantic coast of North and South America, belong to this group.

Many of the brachyrhynch crabs are amphibious. *Pachygrapsus crassipes* (Color Plate 7) of the Pacific coast of North America is a good example. It lives under rocks, in rock crevices, and in holes in clay banks at upper levels of the intertidal region, and spends much of its time being just wet rather than totally immersed. It is mostly a scavenger and seaweed-eater. Species of *Hemigrapsus* (Figure 17.71, A), whose food habits are similar to those of *Pachygrapsus,* live at slightly lower levels, but are nevertheless not submerged all of the time.

Some members of the Brachyrhyncha are more nearly terrestrial than the ones just mentioned. Fiddler crabs of the genus *Uca* (Figure 17.70, D) are good examples. They depend on the tide to bring food in the form of detritus, but are not often actually under water, even though they live in burrows in salt marshes and similar situations. Certain of the almost totally terrestrial crabs found in tropical regions go back to the sea only to reproduce. They are burrowing types that emerge periodically to forage for plant material and dead animals. A few of them are known to move inland for a distance of several kilometers.

Among the Brachyrhyncha are several freshwater species. Most of them belong to the family Potamidae, characterized in part by direct development of the young, which hatch as fully formed little crabs. Freshwater potamids are widely scattered but are found mostly in warmer climates. Perhaps the best-known species is *Potamon fluviatile* of the Mediterranean region. Some are terrestrial in rainy and humid parts of Africa and South America.

The family Grapsidae, which is largely marine and which includes *Pachygrapsus* and *Hemigrapsus,* also has freshwater representatives. They are restricted, however, to Jamaica. Like potamids, they do not need to return to the sea to reproduce, but they do have larval stages. One species, *Metopaulias depressus,* lives in plants of the family Bromeliaceae, the family that includes the cultivated pineapple. A characteristic of many bromeliads is that the leaf bases make little pools in which rainwater can accumulate. The Jamaican *Metopaulias* is not the only animal that makes use of these pools; almost throughout the American tropics, there are insects (including mosquitoes), small frogs, and other creatures whose entire lives or larval stages are spent in the water held by bromeliad leaf bases.

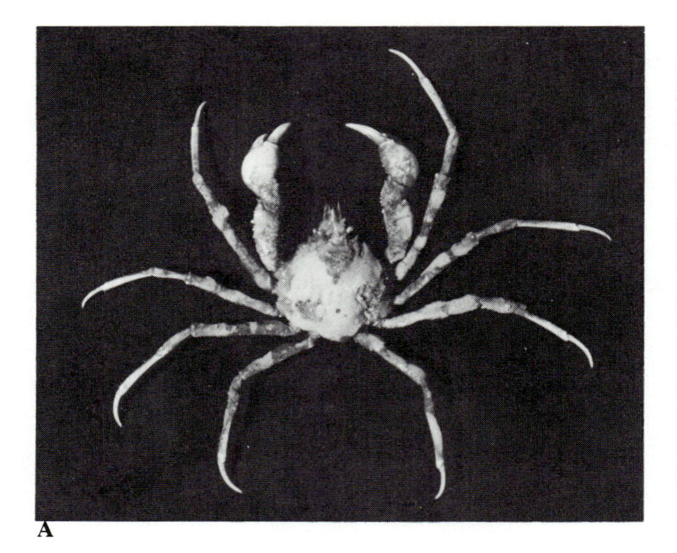

A

B

FIGURE 17.69 Infraorder Brachyura, section Oxyrhyncha. A. *Pyromaia tuberculata;* a sponge is growing on the carapace and portions of the legs. B. *Chionoecetes bairdi.*

The blue crab, *Callinectes sapidus,* a member of the family Portunidae, has been mentioned earlier as a commercially important brachyuran of the Atlantic

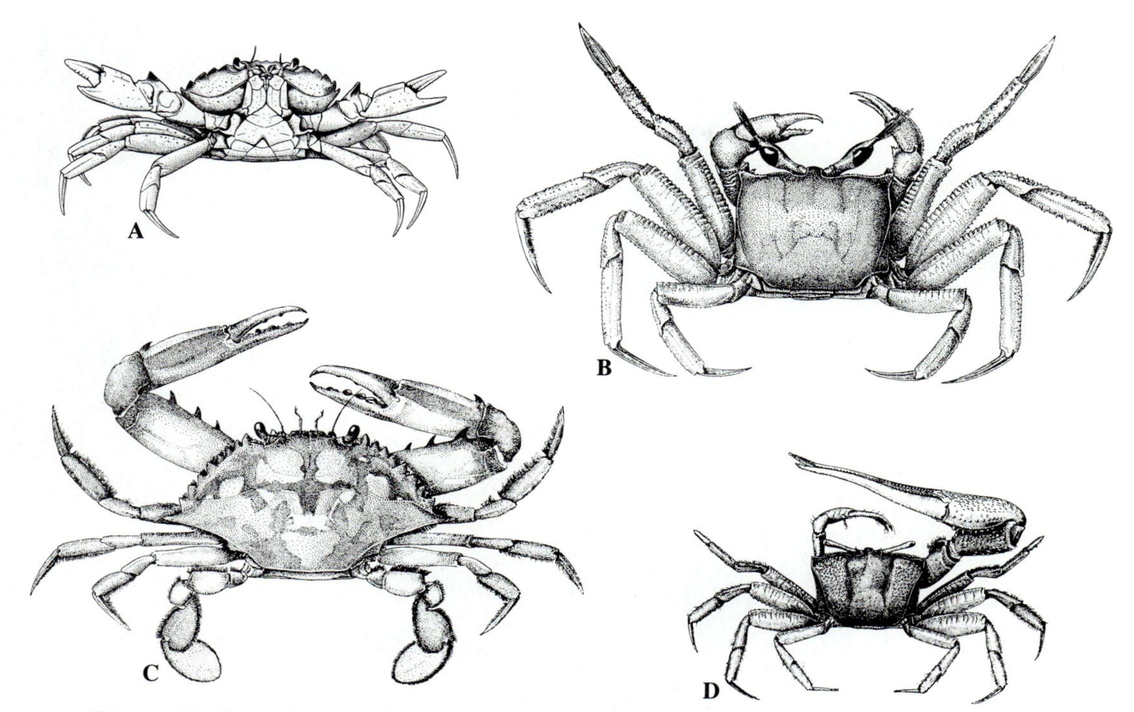

FIGURE 17.70 Infraorder Brachyura, section Brachyrhyncha. A. *Carcinus maenas,* male displaying the lateral surface of the merus of its chelipeds (agonistic behavior). (Jensen, Ophelia, *10.*) B. *Callinectes marginatus,* male. C. *Ocypode cursor,* male. D. *Uca tangeri,* male. (B–D, Capart, Expédition Océanographique Belge dans les Eaux Côtières Africaines de l'Atlantique Sud (1948–49), *3.*)

coast of the Americas. Although it mates in estuaries, where the salinity is between that of the sea and fresh water, females move to water of high salinity before laying their eggs, which adhere to the pleopods. As is usual for crabs, the young hatch at the zoea stage, pass through the megalopa stage, and finally become small adults. These migrate into estuaries and often enter strictly freshwater habitats.

The same general pattern is followed by some other brachyurans, one of the more famous of which is *Eriocheir sinensis,* the Chinese mitten crab, which belongs to the family Grapsidae. In the early part of the nineteenth century it became established in Germany, and it now is abundant in many rivers of Europe. It is a nuisance because it attacks fishes caught in nets, eats bait, and will even damage nets. As the mitten crab breeds in salt water, adults living far from the sea must make long migrations when they are ready to reproduce, and the later larval stages must do likewise. To a considerable extent, their movements are aided by tidal currents. *Eriocheir* is large enough to eat—about 7 cm across the carapace—and is much used in the Orient, where it is host for the metacercarial stage of *Paragonimus westermanni,* a lung fluke of considerable medical importance. Humans acquire the adult stage of the fluke by eating raw *Eriocheir.*

The family Pinnotheridae, all marine and characterized by a thin, soft carapace, consists of so-called pea crabs and their allies. Most of them are under 1.5 cm in greatest diameter, and nearly all live in close association with other animals. The mantle cavity of bivalve molluscs is home to many species. Others inhabit tubes of polychaete annelids and burrows of mud shrimps and ghost shrimps. A few live as commensals of ascidians. Some that inhabit polychaete tubes are much broader than long, and fit into the tubes sideways. As a rule, pinnotherids are found in pairs, or only the female is permanently associated with a particular host, and the male is transient. In the case of species that live in the mantle cavity of bivalves, some erosive damage may be done to the tissue of the ctenidia or mantle, but much of the food of these crabs is mucus and detritus. Figure 17.71, B shows one of the species associated with clams on the Pacific coast of North America.

The section Cancridea includes the genus *Cancer,* characterized by an oval carapace that is wider than

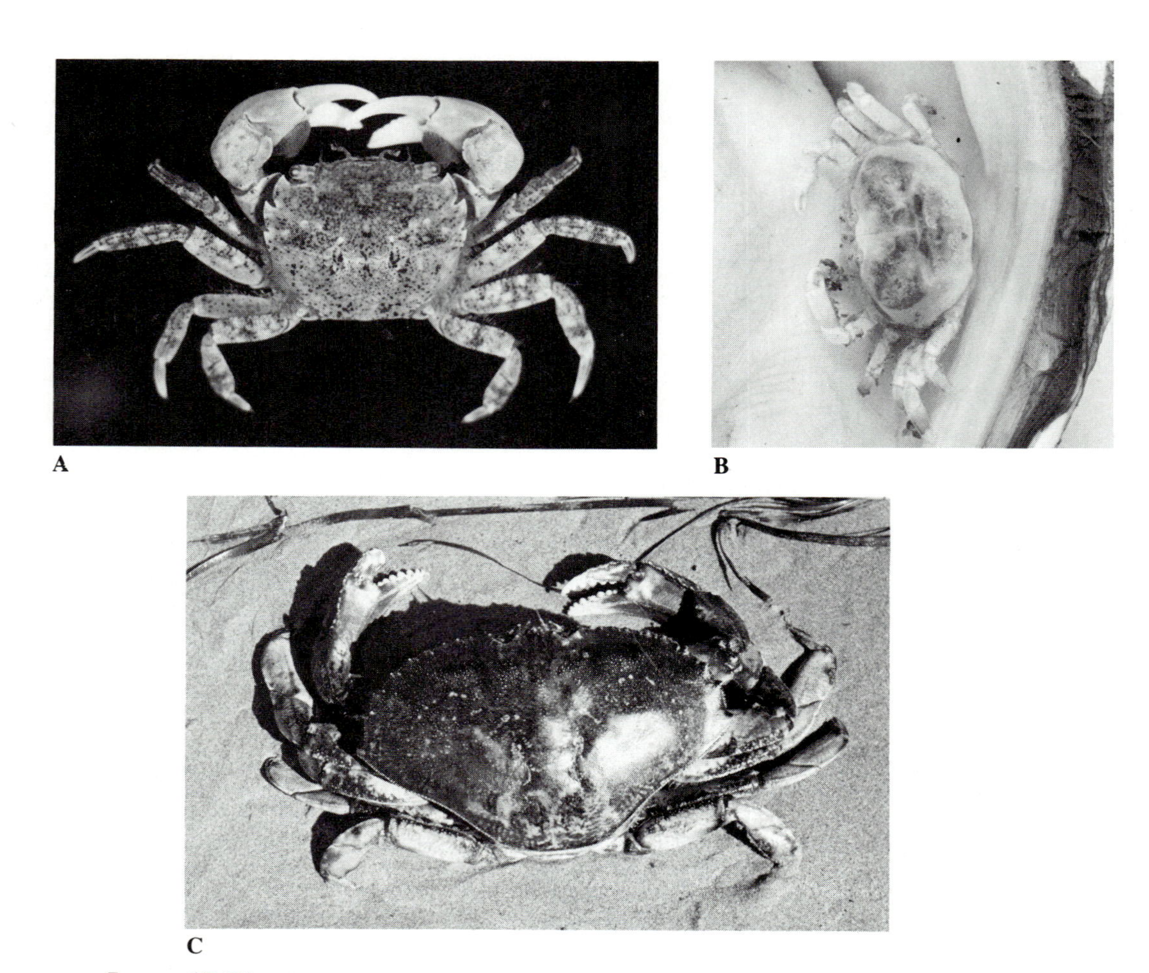

FIGURE 17.71 Infraorder Brachyura. A. *Hemigrapsus oregonensis* (section Brachyrhyncha), a grapsid crab found intertidally on the Pacific coast of North America. (See also *Pachygrapsus,* Color Plate 7.) B. *Pinnixa faba* (section Brachyrhyncha), a pinnotherid crab that lives in the mantle cavity of bivalve molluscs. C. *Cancer magister* (section Cancridea), a commercially important crab of the Pacific coast of North America.

long and that is toothed along the margins of its anterior half. The large species, such as *Cancer magister* (Figure 17.71, C) of the Pacific coast of North America, which reaches a carapace width of more than 20 cm, are highly valued seafoods.

References on Decapoda

Adelung, D., 1971. Studies on the moulting physiology of decapod crustaceans as exemplified by the shore crab *Carcinus maenas*. Helgoländer Wissenschaftliche Meeresuntersuchungen, *22:*66–119.

Adiyodi, K. G., and Adiyodi, R. G., 1970. Endocrine control of reproduction in decapod Crustacea. Biological Reviews, *45:*121–165.

Bauer, R. T., 1977. Antifouling adaptations of marine shrimps: functional morphology and adaptive significance of antennular preening by the third maxillipeds. Marine Biology, *40:*261–276.

Burggren, W. W., and McMahon, B. R. (editors), 1988. Biology of the Land Crabs. Cambridge University Press, London, Cambridge, and New York.

Cobb, J. S., and Phillips, B. F., 1980. The Biology and Management of Lobsters. 2 volumes. Academic Press, Inc., New York and London.

Crane, J., 1975. Fiddler Crabs of the World. Princeton University Press, Princeton, New Jersey.

Diaz, H., and Rodriguez, G., 1977. The branchial chamber in terrestrial crabs: a comparative study. Biological Bulletin, *153:*485–504.

Efford, I. E., 1966. Feeding in the sand crab *Emerita analoga*. Crustaceana, *10:*23–30.

Fincham, A. A., and Rainbow, P. S. (editors). Aspects of Decapod Crustacean Biology. Symposia of the Zoological Society of London, no. 59. Clarendon Press, Oxford.

Gurney, R., 1942. Larvae of Decapod Crustacea. Ray Society, London.
Hartnoll, R. C., 1969. Mating in Brachyura. Crustaceana, *16*:161–181.
Hazlett, B. A., 1966. Social behavior of the Paguridae and Diogenidae of Curacao. Studies on the Fauna of Curacao and other Caribbean Islands, *23*:1–143.
Hobbs, H. H., 1972. Crayfishes (Astacidae) of North and Middle America. Biota of Freshwater Ecosystems, Identification Manual number 9. United States Government Printing Office, Washington.
Ingle, R. W., 1983. Shallow-water Crabs. Synopses of the British Fauna, no. 25. Linnean Society of London, Estuarine and Brackish-water Sciences Association, and Cambridge University Press, Cambridge, London, and New York.
Kellogg, C. W., 1977. Coexistence in a hermit crab species ensemble. Biological Bulletin, *153*:133–144.
Omori, M., 1974. The biology of pelagic shrimps in the ocean. Advances in Marine Biology, *12*:233–324.
Provenzano, A. J., 1961. Larval development of the land hermit crab *Coenobita clypeatus*. Crustaceans, *4*:207–228.
Reese, E. S., 1963. The behavioral mechanisms underlying shell selection by hermit crabs. Behaviour, *21*:78–126.
Smaldon, G., 1979. British Coastal Shrimps and Prawns. Synopses of the British Fauna, no. 13. Linnean Society of London, Estuarine and Brackish-water Sciences Association and Academic Press, London and New York.
Spight, T. M., 1977. Availability and use of shells by intertidal hermit crabs. Biological Bulletin, *152*:120–133.
Warner, G. F., 1977. The Biology of Crabs. Elek Science, London.
Williams, A. B., 1984. Shrimps, Lobsters, and Crabs of the Atlantic Coast of the Eastern United States, Maine to Florida. Smithsonian Institution Press, Washington.

SUMMARY

1. The Crustacea, unlike other arthropods, have five pairs of head appendages: first and second antennae, mandibles (jaws), and first and second maxillae.

2. The arrangement of tagmata varies greatly within this large and predominantly aquatic assemblage. In most crustaceans, there is a thoracic region behind the head, then an abdomen, but sometimes these are not clearly separate, so there is a continuous trunk. Even when the thoracic region is well differentiated, one or more of its segments may be fused to the head. If there is a carapace, which originates as a fold from the posterior part of the head, this may be firmly united to all of the thoracic segments; in that case, the head and thorax are united into a single tagma, the cephalothorax. The classification of crustaceans is based to a large extent on the arrangement of tagmata and on the types of appendages they bear.

3. Nearly all crustaceans have some biramous appendages. The two branches—a medial endopodite and lateral exopodite—arise from the second article, which is called the basis. In certain groups, nearly all of the appendages are biramous, and this is considered to be an ancestral feature. Uniramous appendages, other than the first antennae, are viewed as appendages that have lost one of the rami. There are two main types of appendages: phyllopodia and stenopodia. Phyllopodia are characterized by flattened, almost leaflike rami, and generally function in swimming, respiration, or creating respiratory currents. Stenopodia are typically slender or more or less cylindrical, or both, and usually are concerned with walking or grasping. Mouthparts are not included in this generalization.

4. The excretory–osmoregulatory organs present in some crustaceans are coelomic derivatives. They are called coxal glands because their external pores are usually located on the coxa (proximal article) of the second antenna or second maxilla. The so-called green glands of crayfishes and lobsters are coxal glands.

5. Respiration in aquatic crustaceans often involves gills. When present, these are generally outgrowths of the coxae of certain appendages. In certain crustaceans, however, there are gills that arise from the place where coxal articles originate, or even from the sides of the body. This is the case, for example, in crayfishes. A few terrestrial crustaceans have tracheal systems somewhat similar to those of insects, myriapods, and some arachnids. When tracheae are present in crustaceans, they are not extensive. Most terrestrial crustaceans depend on portions of the body surface, including certain appendages, for gas exchange. These surfaces must be kept moist.

6. Development may be indirect, involving a series of distinctive larval stages, or direct. A type of free-swimming larva common to several diverse groups is the nauplius. It has the three pairs of appendages corresponding to the first and second antennae and mandibles of the adult. The first and second maxillae and postcephalic appendages are added later. Even in crustaceans that have direct development and that hatch as small individuals similar to the adult, there is usually an embryonic stage that has the primordia of the first three pairs of appendages and which is therefore comparable to a nauplius.

7. Compound eyes are characteristic of many crustaceans, including crabs, crayfishes, and shrimps. In some groups, however, there is a single median eye, consisting of only a few retinal units, that can be traced back to an early developmental stage, such as the nauplius.

18

Subphylum Uniramia

Class Insecta and the "Near Insects"

- Subphylum Uniramia
  - Class Insecta
    - Subclass Apterygota
      - Order Thysanura
    - Subclass Pterygota
      - Infraclass Palaeoptera
        - Order Ephemeroptera
        - Order Odonata
      - Infraclass Neoptera
        - Order Plecoptera
        - Order Orthoptera
        - Order Phasmida (Phasmatodea)
        - Order Dermaptera
        - Order Embioptera
        - Order Dictyoptera
        - Order Zoraptera
        - Order Isoptera
        - Order Psocoptera
        - Order Mallophaga
        - Order Anoplura
        - Order Hemiptera (Heteroptera)
        - Order Homoptera
        - Order Thysanoptera
        - Order Neuroptera
        - Order Mecoptera
        - Order Lepidoptera
        - Order Trichoptera
        - Order Diptera
        - Order Siphonaptera
        - Order Hymenoptera
        - Order Coleoptera
        - Order Strepsiptera
  - Class Collembola
  - Class Protura
  - Class Diplura

INTRODUCTION

In Chapter 15, it was explained that the subphylum Uniramia includes all arthropods which have a single pair of antennae and whose appendages are not divided into two branches. In contrast, members of the subphylum Crustacea have two pairs of antennae, and their appendages are biramous, unless one branch is suppressed. Both uniramians and crustaceans have mandibles, but these are different. The functional part of a uniramian mandible is its tip, whereas in a crustacean mandible it is the basal portion. The only other subphylum of living arthropods is the Chelicerata, which have neither antennae nor mandibles.

This chapter deals with the largest assemblage of uniramians, the insects and some relatives we may call "near insects." Chapter 19 covers the four groups of myriapods, which form the rest of the Uniramia.

THE INSECTS

Insects make up the vast majority of arthropods and account for about three-fourths of all described species of animals, vertebrate and invertebrate. Their immense success, especially in terrestrial habitats, tells us that they are doing something right. Three particularly important features adapt insects to life on land. Their **tracheal respiratory system** takes advantage of the abundant supply of oxygen in the air. They have a **cuticular body covering** that is flexible where it needs to be, but that is effective in preventing loss of water. Finally, most adult insects have **wings.** Thus they can fly into areas where food is available, and they can escape from adverse environments and from animals that would eat them.

The adaptations just mentioned are not the only ones that make insects so successful. They are efficient multipliers, usually with rather short life histories. There is almost no organic product that one insect or another will not relish, whether it be a ripe apple, a breakfast cereal, a dead rat, or an animal's dung. Insects are our worst competitors. They eat our crops and stored foods and the timbers in our houses, and certain diseases of humans and domestic animals are transmitted by insects. Yet we are also aided by insects, especially by those that pollinate plants, make honey, and parasitize or eat pests.

STRUCTURE AND PHYSIOLOGY

Tagmata and Appendages

Head

The body of an insect consists of three tagmata: **head, thorax,** and **abdomen** (Figure 18.1). The head has four pairs of appendages. These correspond to four of the five pairs of head appendages of crustaceans. The **antennae** of insects originate in the preoral portion of the head and are generally believed to be homologous with the first antennae of crustaceans. What is interpreted as the first of four postoral segments has no appendages comparable to crustacean second antennae or to chelicerae of spiders, scorpions, and their allies. The three pairs of mouthparts (Figures 18.1, 18.2)—**mandibles** ("jaws"), **first maxillae,** and **second maxillae**—correspond to the mandibles and first and second maxillae of a crustacean. The mandibles are one-piece structures, but both pairs of maxillae have articulated prolongations called **palps.** These, like the antennae, are sensory devices.

The bases of the second maxillae are fused to form a structure called the **labium,** or "lower lip," and the palps of the second maxillae are therefore called *labial palps*. The **labrum,** or "upper lip," which is located in front of the mouth and hinged to a part of the head called the *clypeus,* is not derived from appendages. Another structure unrelated to appendages is the **hypopharynx.** This is a projection just behind the mouth. The duct from the salivary glands, which are located in the thorax, usually opens on the hypopharynx or at its base, sometimes at the base of the labium. The glands do not necessarily secrete saliva. In larvae of moths, butterflies, and some other insects, for instance, they produce silk.

Because most of the segments of the head are fused together, the serial appendages of this tagma provide the best external evidence that it is derived from several embryonic somites. Of the various regions of a generalized insect head, as labeled in Figure 18.2, A, only one—the postocciput—corresponds to a single segment. The grooves that set apart the various other regions mark places where the cuticle is turned inward to make reinforcements and to provide shelflike apodemes for attachment of muscles. Some invaginations of the cuticle of the head are so deep, in fact, that they form what looks like an endoskeleton. The large skeletal piece called the *tentorium* is produced by invaginations that meet along the midline.

How the mouthparts function depends on their structure and on the insect's food habits. It is probable that the crushing, grinding, and chewing mouthparts of the sort found in various carnivorous, herbivorous, and omnivorous insects are fairly close to the ancestral type. In the many insects that suck nectar, plant juices, or blood, the mouthparts are modified for drawing up fluid, and in certain cases also for piercing the tissue so as to initiate flow. In most bees, wasps, and ants, the mandibles are fairly solid and toothed, but the maxillae are elongated and the labium is specialized as a grooved tongue used for licking up nectar. In mosquitoes (Figure 18.3, A and B), the labium forms a sheath around

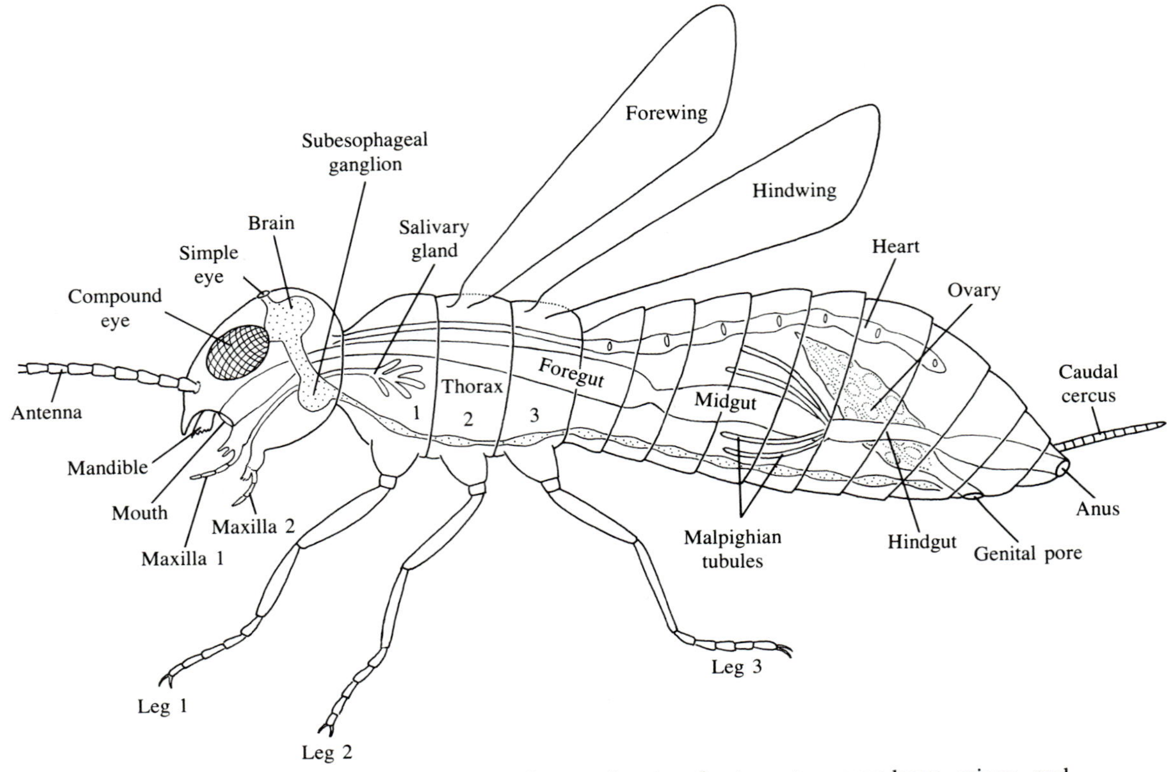

FIGURE 18.1 Diagram of a generalized insect, showing the tagmata, appendages, wings, and some internal structures.

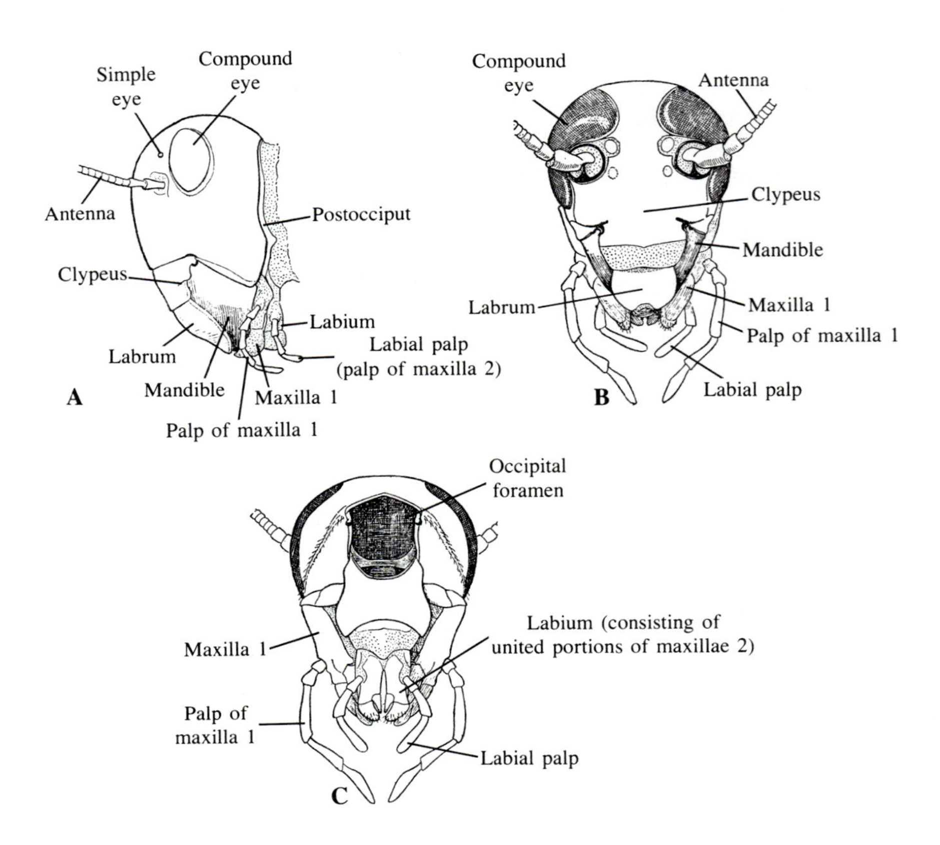

FIGURE 18.2 The head of an insect and its appendages (generalized), lateral view. B. The head of a cockroach, frontal view. C. The head of a cockroach, detached and viewed from the posterior side. (A–C, Snodgrass, Smithsonian Miscellaneous Collections, *142*.)

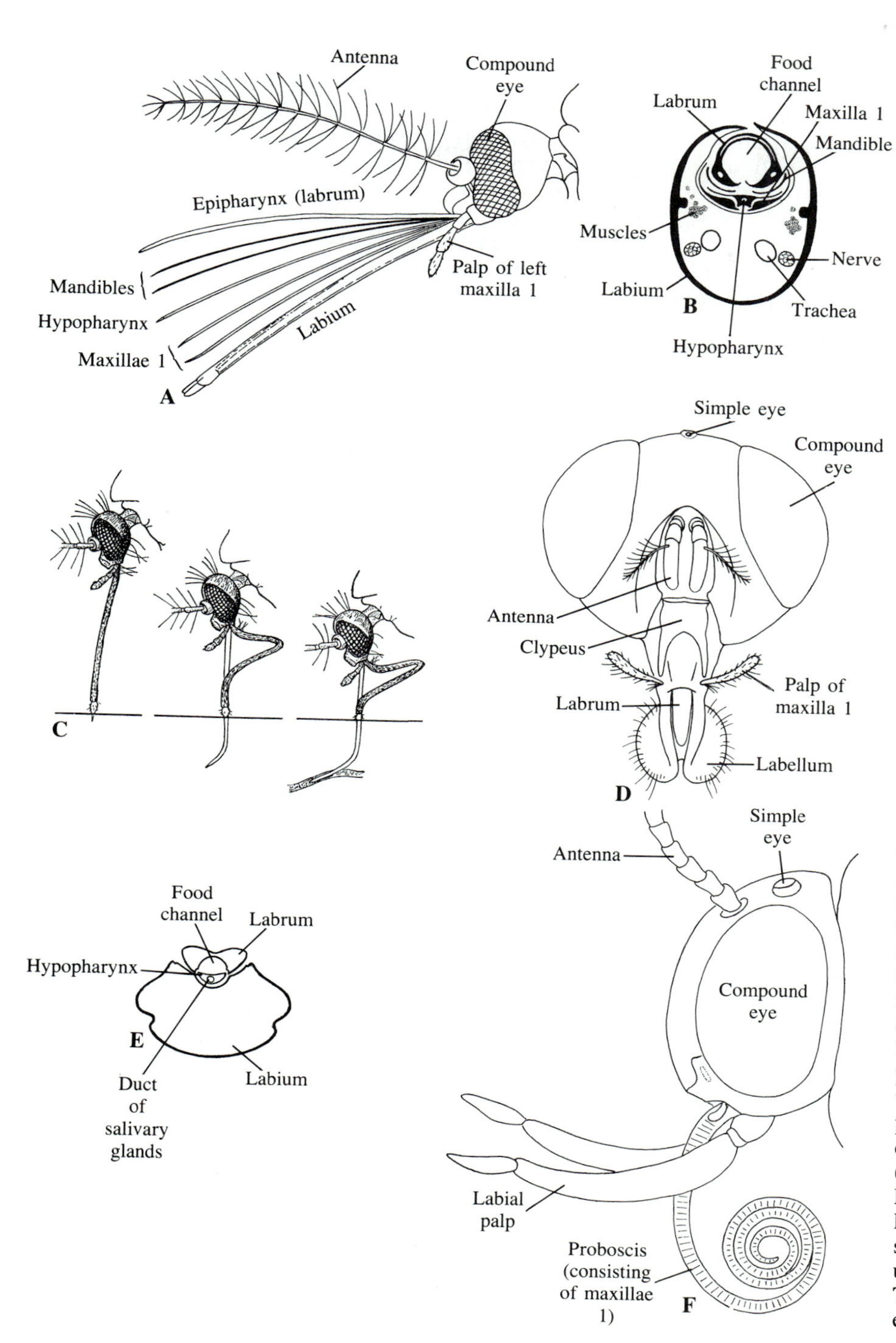

FIGURE 18.3 Some specializations of mouthparts in insects. A. Female mosquito, *Aedes aegypti* (order Diptera), with the several stylets separated from the labium. B. Transverse section through the proboscis of a female mosquito, showing the stylets enclosed by the labium. (A and B, after Snodgrass, Smithsonian Miscellaneous Collections, *104*). C. Female mosquito puncturing the skin with its stylets and finally reaching a blood capillary. (Snodgrass, Smithsonian Miscellaneous Collections, *104*.) D. Sponging mouthparts of a housefly, *Musca domestica* (order Diptera). The labellum is the expanded terminal portion of the labium. E. Transverse section through the mouthparts of a housefly at a level just distal to the palps of maxillae 1. (After Snodgrass, Smithsonian Miscellaneous Collections, *104*). F. Moth (order Lepidoptera), showing the proboscis which is unrolled when the insect feeds. The two maxillae, fused together, enclose a food channel.

the labrum and around the mandibles and maxillae, which have sharp stylets for puncturing the skin. The tongue-like hypopharynx and the epipharynx, which is an elaboration of the labrum, are inserted into the wound and form a sucking tube. The labium folds out of the way while feeding is in progress (Figure 18.3, C). True bugs have a somewhat similar arrangement, but it is the grooved maxillae that fit together in such a way that they enclose two channels, one for delivering saliva, the other for taking up food. In butterflies and moths that suck up flower nectar, the first maxillae are permanently fused to form a long tube (Figure 18.3, F). This is straightened when in use, but is otherwise rolled up. Neither the mandibles nor the labium amount to much, although the labial palps are generally well developed.

Thorax

The thorax consists of three segments (Figure 18.1). These are commonly called the **prothorax, mesothorax,** and **metathorax.** Each has a pair of legs, and if wings are present these are on the second and third segments. (Wings are not, however, part of the sequence of true segmental appendages.) The four major plates, or **sclerites,** of a thoracic segment mark areas in which the exocuticle is especially thickened. The dorsal one is designated the **tergum** (or *notum*); the ventral one is the **sternum;** the two lateral sclerites are called **pleura.** Complicating matters, however, is the fact that any or all of the four major sclerites of a particular segment may be partly or completely subdivided by sutures that run more or less at right angles to the long axis of the body. This explains why the tergal area of a thoracic segment often seems to consist of three unequal sclerites and why a pleuron may appear to be split. The pleural portions of the thoracic segments also contribute sclerites that form the coxal articles of the legs. In addition, the membranous area between the base of a wing and the corresponding pleuron generally has one or two small sclerites. It is to these sclerites that certain of the muscles concerned with operation of the wings are attached. The fulcrum for wing movement is provided by the pleural wing process, an elaboration of the upper portion of the pleuron.

Legs

The legs of insects, like those of other true arthropods, consist of several articles (Figure 18.4) separated

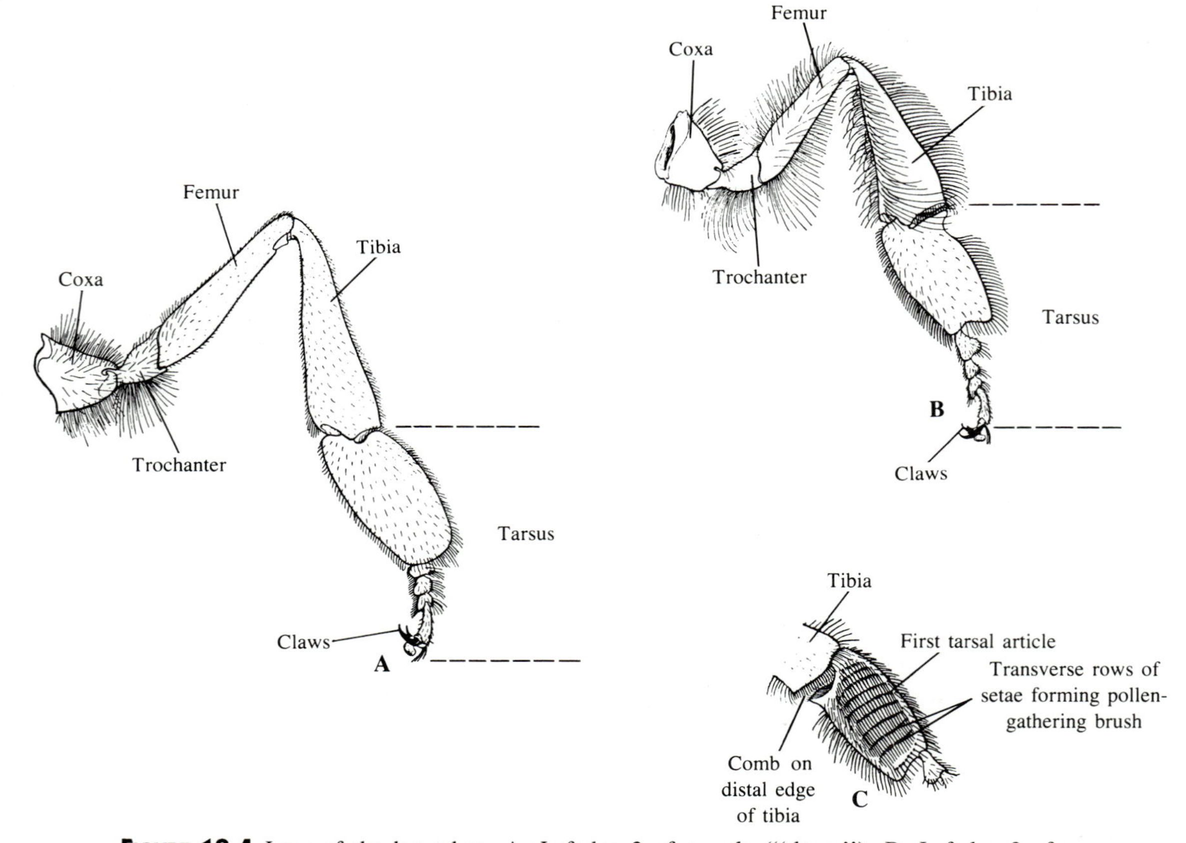

FIGURE 18.4 Legs of the honeybee. A. Left leg 3 of a male ("drone"). B. Left leg 3 of a worker. C. First and second tarsal articles and distal portion of the tibia of a worker, showing some of the specializations concerned with collection of pollen. (A–C, Snodgrass, United States Bureau of Entomology, Technical Series, Bulletin 18.)

by joints where the cuticle is thinner and more flexible. Some of the terms applied to the various articles, or groups of articles, are borrowed from the vocabulary of vertebrate anatomy. The basal article, called the **coxa,** is succeeded by the **trochanter, femur, tibia,** and **tarsus.** The relative sizes of these vary, but in general the femur and tibia are the more prominent components of insect legs. The tarsus consists of two or more articles (tarsomeres), the last of which is characteristically tipped by one or two claws. Insect legs are specialized for many different functions, and it is only reasonable to expect that not all of them conform to the plan just described. Loss of the terminal claws and some or all of the tarsal articles are fairly common modifications.

Three pairs of thoracic legs are present in the larvae of some insects, including those of the large orders Coleoptera, Lepidoptera, and Hymenoptera. When there are leglike outgrowths on certain of the abdominal segments, as is the case in lepidopterans and hymenopterans, these are called **prolegs.** They are unjointed and usually rather fleshy, and are not homologous with true legs. In lepidopterans, the prolegs have some hooked spines at their tips.

The muscles that effect movements of a leg are of two sorts. The extrinsic muscles, which originate on a tergum, pleuron, or sternum, insert on the inside of a coxa or trochanter and are concerned with moving the entire leg with respect to the thorax. The intrinsic muscles on the other hand, are entirely within a leg, and move the various articles with respect to one another. As a rule, the largest and strongest intrinsic muscles are those that originate in the femur and insert on the tibia. The jumping action of a grasshopper's hind legs, for instance, depends on such muscles.

Wings and the Mechanics of Flight

It has already been pointed out that wings are not true appendages. Each originates as an outgrowth of the edge of a tergum. The principal "veins" that strengthen the wings (Figure 18.5, A) are modified tracheae. In other words, they are not blood vessels, but belong to the system of air-filled tubes whose primary function is to deliver oxygen to the tissues and to carry carbon dioxide away. They may be paralleled, however, by channels that contain blood, and sometimes there are fine sensory nerves running alongside certain of the tracheae. The cell bodies of the neurons are in the epidermis. The veins are usually heavier and more closely spaced in the area nearest the anterior edge of the wing. When there are cross-veins between the main veins, these are merely cuticular reinforcements.

There are several small sclerites in the proximal portion of the wing. One of these articulates with the wing process of the pleuron, which serves as the fulcrum for the downstroke and upstroke (Figure 18.5, D). Two others articulate with similar wing processes of the tergum, which are situated one behind the other. In most insects, the upper part of the pleuron has one or two small sclerites with which the wing also articulates. The relationships of the wing processes, pleural sclerites, and sclerites of the wing itself are shown in Figure 18.5, A–C.

Two sets of muscles—**direct** and **indirect**—operate the wings. The indirect muscles are usually the more important in producing wing motion (Figure 18.5, D). The ones that run vertically from the tergum to the sternum pull down on the tergum when they contract, thus causing the wings to be raised. The ones that run longitudinally from apodeme to apodeme in the dorsal part of the thorax are antagonistic to the vertical muscles. If these contract when the vertical muscles relax, the tergum is elevated and the wings turn down. It is remarkable that the alternating contractions and relaxations of these muscles can produce up to several hundred wing beats per second.

The direct muscles, inserted on the small pleural sclerites below the bases of the wings (Figure 18.5, C), are generally concerned with keeping the wings properly tilted during flight, but in some of the more primitive types of winged insects, especially dragonflies, they produce most of the motion necessary for flight. In dragonflies, by the way, the downstroke of the hindwings is timed to begin when the forewings are completing the downstroke. This could be advantageous for a strong-flying insect, for the rear wings would not be operating against air that has been disturbed.

The evolution of wings perhaps got started when lateral flaps of the terga proved to be effective devices for wind-dispersal of small insects. The mechanism that permits wings to go up and down and thus enable an insect to fly probably came later. Unfortunately, however, there are more hypotheses than facts about the origin of wings. Wings are not only advantageous for dispersal but also for escape, certain kinds of courtship behavior, and thermoregulation.

The wings of most insects are studded with hairs. The wings of moths and butterflies are covered with scales whose exposed surfaces may be microscopically roughened to diffract light and thus confer iridescent colors upon the wings. Such "structural" colors are different from colors due to deposits of pigment.

In the order Thysanura, the body is primitively wingless. In various other groups, the wings have been secondarily reduced or lost. In flies, mosquitoes, and other members of the order Diptera, the hind wings are small and specialized as *halteres,* or "gyroscopic organs." They are concerned with maintaining equilibrium during flight. Their beat, which traces an arc of

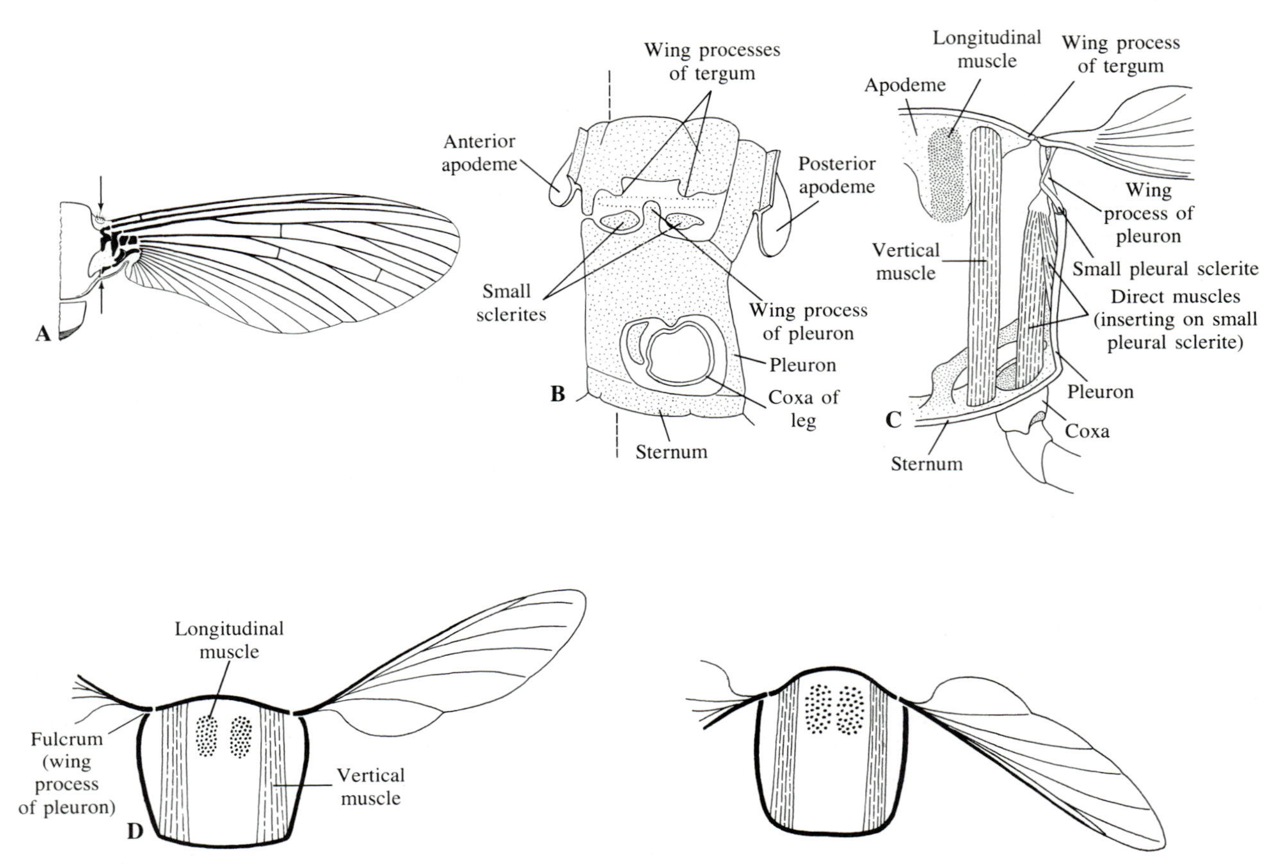

FIGURE 18.5 A. Generalized insect wing, showing points of articulation of certain of the small sclerites in its proximal portion with the wing processes of the tergum (arrows). (Modified from Snodgrass, United States Bureau of Entomology, Technical Series, Bulletin 18.) B. Lateral view (diagrammatic) of a wing-bearing thoracic segment, showing the wing processes of the tergum and pleuron, and the two sclerites of the upper part of the pleuron. C. Transverse section through a wing-bearing thoracic segment at about the level indicated in B by the dashed line. D. Diagram of wing movements. The upstroke (left) is achieved when the vertical muscles contract, depressing the tergum and pulling the wing bases downward against the wing processes of the pleura, which act as fulcra. In the downstroke (right), the longitudinal muscles contract and the vertical muscles relax, so that the tergum bulges upward and the wing bases are raised.

about 180°, is extremely rapid. If there is a change in the direction of flight, the angle at which they project is affected, and the change is monitored by sense organs at their bases. In the order Strepsiptera, the forewings perform a similar function.

Abdomen

The segmentation of the abdomen is generally obvious. In some orders, including the primitive Thysanura, there are 11 segments, and this is probably the basic number that insects have had to work with since they became a separate group. As a rule, however, one or more of the abdominal segments are much specialized and not really distinct. In dealing with Trilobita, Chelicerata, and Crustacea, the postsegmental portion of the abdomen, called the **telson,** has been referred to many times. Is there a telson in insects? Presumably it is at the tip of the last segment, although its limits are not clear. Some experts view the eleventh segment, when this is distinct, as the telson. The matter will probably be a source of controversy forever.

In general, the abdominal segments differ from the segments of the thorax in that their pleural portions are proportionately smaller and not hardened into sclerites. In each segment, in other words, the dorsal tergum and ventral sternum are more or less half-rings. The segments most likely to be absent or vestigial are the first and eleventh. In hymenopterans—bees, ants, and wasps—the first segment becomes so tightly joined to the thorax at the time of metamorphosis that it appears to be part of that tagma rather than part of the abdomen.

In female insects, the segment on which the genital pore is located is usually the eighth or ninth, and if there is an ovipositor, this consists of the two pairs of appendages belonging to these segments. In males, the genital pore is usually on an eminence that arises ventrally between the ninth and tenth segments. This structure, called the *aedeagus,* originates partly from one or two simple outgrowths and partly from a pair of embryonic appendages. It is a device used for transfer of sperm. The "claspers" by which the males of most insects hold females during copulation are also derived from the embryonic appendages that contribute to the aedeagus.

If the eleventh abdominal segment is distinct and bears appendages, these are specialized as **cerci** whose function is usually sensory. The cerci of a cricket, for instance, are concerned with detection of vibrations in air. The cerci can do other things, however. In most earwigs, they are generally stout and form a pair of forceps that can be used for pinching and grasping. In the primitive insects of the order Thysanura, there are not only cerci on the eleventh segment, but small, rudimentary limbs on some of the other ten segments.

Cuticle

The success of insects in terrestrial habitats is to a large extent due to the composition of the cuticle. The **epicuticle**—the thin outermost layer—typically includes a stratum of wax that prevents loss of water from the body. The **procuticle,** beneath the epicuticle, is generally fairly thick except at points of articulation. It consists of two main strata. The outer stratum, called the **exocuticle,** contains chitin and proteins that have become hardened as a result of cross-linking induced by phenolic substances. The inner stratum, or **endocuticle,** is composed largely of chitin and unhardened proteins. On the microscopic level, the physical organization of the cuticle may be likened to that of fiberglass.

A very important component of the cuticle, especially at the joints and wing hinges, and also of the tendonlike structures by which muscles are attached to the cuticle, is an elastic protein called **resilin.** To watch a flea jump is to see an insect use energy that has been stored up in resilin. Contraction of certain muscles, of course, is responsible for bending the leg joints and stretching the tendons. Essentially, then, when these muscles contract, much of the energy they use is transferred to the stressed resilin. The efficiency of resilin in snapping back to its relaxed condition is much greater than that of rubber or any other elastic material. One must not conclude, though, that all jumping insects rely on resilin to the same extent that a flea does. It has already been pointed out that the leap of a grasshopper depends directly on muscle contraction rather than on elastic recoil.

The epidermis secretes the cuticle and also has the responsibility of resorbing the endocuticle as an insect prepares to shed its exocuticle and epicuticle. Within the epidermis are various glands whose ducts extend to pores at the surface of the cuticle. The glands may produce secretions that are unpleasant to predators, and they also produce substances called *pheromones.* The latter serve in various ways. Some of them are sexual attractants; others regulate the activities and relationships of castes and individuals within a colony.

Digestive Tract

The gut of an insect (Figure 18.6) has three main regions: **foregut, midgut,** and **hindgut.** As in other arthropods, the first and last of these regions are lined by an epithelium that is of ectodermal origin and that secretes a cuticle; only the midgut is lined by an endodermally derived epithelium and therefore lacks a cuticle. The foregut is generally, but not always, differentiated into sections that may be called the *pharynx, esophagus, crop,* and *proventriculus.* The crop, if obvious at all, functions as a place for temporary storage of recently eaten food. In most insects that swallow bits of plant or animal tissue, the proventriculus serves as a gizzard, and its lining is studded with hard teeth or ridges that operate to break up the food. In insects whose diet consists of liquids, the proventriculus may be little more than a valve to the midgut. There are, however, many specializations of this portion of the foregut. In some bees, for instance, the proventriculus acts as a sieve that prevents pollen grains from entering the midgut along with nectar.

As explained previously, the salivary glands, located beneath the foregut, deliver their secretions to a pore at the base of the labium or hypopharynx, or on the hypopharynx itself. In many insects, the salivary juice serves as a lubricant, and it may contain enzymes that initiate digestion or that cause the cells of plant tissues to separate. In insects that suck blood, salivary glands are the sources of substances that inhibit coagulation. It is through the salivary secretions, in fact, that mosquitoes transmit malarial organisms and tsetse flies transmit the trypanosomes that cause African sleeping sickness. Silk, used by some insects for making webs or for clothing themselves when they enter the pupal stage, is another kind of salivary secretion.

It is in the midgut that most digestion and absorption of nutrients take place. There are usually some

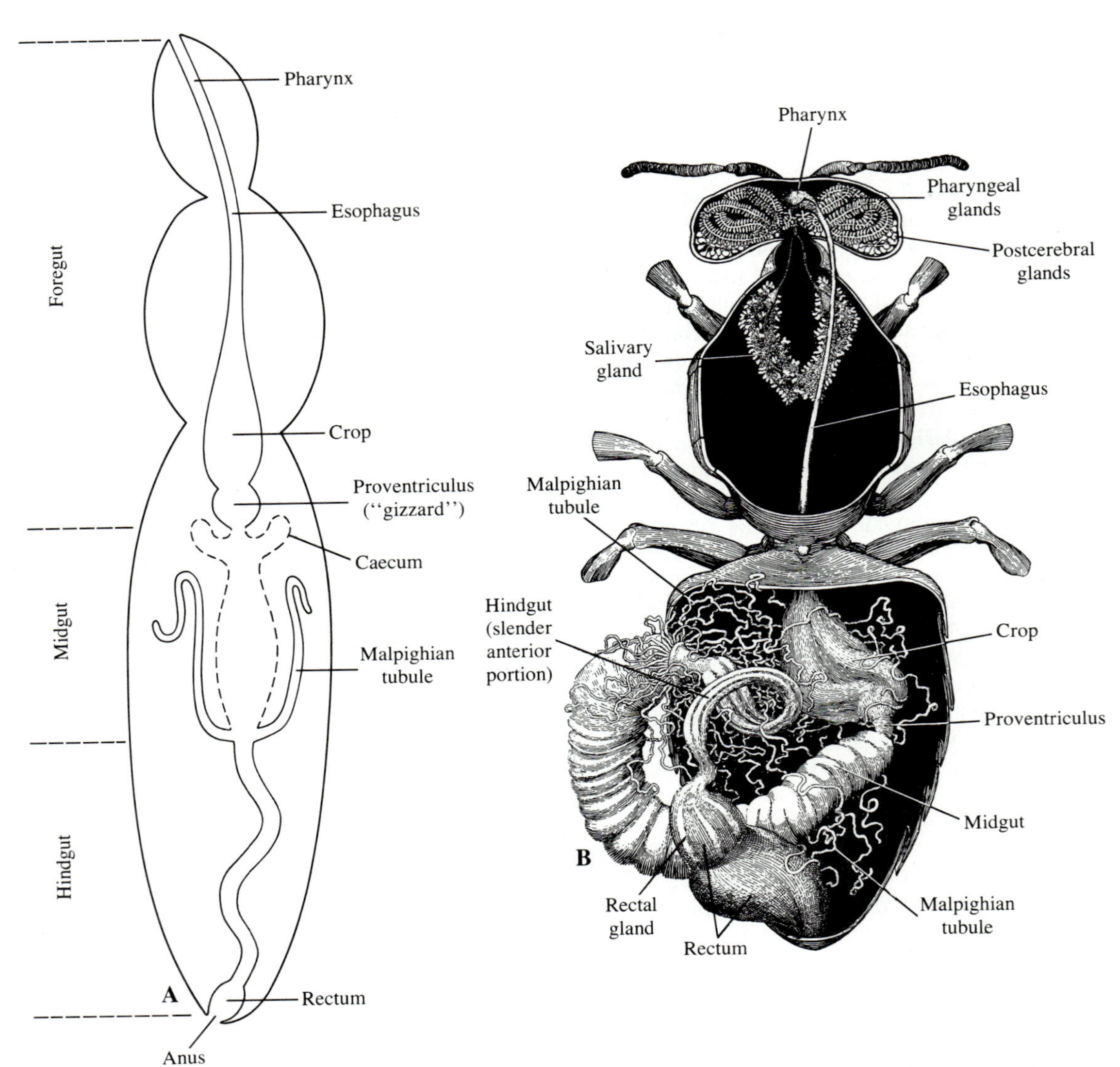

FIGURE 18.6 A. Diagram of the gut of an insect. B. Gut of a worker honeybee. The pharyngeal glands in the head produce the "royal jelly," given as food to larvae; the postcerebral glands produce a fatty substance that is presumably mixed with wax during comb-building. (Snodgrass, United States Bureau of Entomology, Technical Series, Bulletin 18.)

pouches, called *caeca,* branching off the midgut. These are sources of digestive enzymes. An interesting feature of the digestive process of many insects—those that suck up blood, nectar, and other juices are exceptions—is the secretion by the midgut of a *peritrophic membrane*. This is a delicate, netlike tube consisting of protein and chitin. It is permeable to water, enzymes, and products of digestion, but keeps any rough constituents of the food mass, and perhaps also some parasites, from direct contact with the epithelium. The peritrophic membrane, in other words, is like a thin, very permeable sausage casing, and it stays with the undigested residues as these move into the hindgut.

Resorption of Water and Excretion

In the hindgut of most insects—especially those that cannot afford to throw water away—considerable water is recovered from the fecal mass, and crystals of metabolic wastes are often added to the fecal mass. In a terrestrial insect, excretion of a toxic substance such as ammonia would require the loss of too much water to be practical. Urea, which is both soluble and relatively nontoxic, is produced by some species, but as a rule insects solve the problem of excretion by converting nitrogenous wastes into a crystalline form of uric acid and into related compounds called urates, whose solubility and toxicity are extremely low. The conversion is accomplished by the **Malpighian tubules** (Figures 18.1, 18.6).

The Malpighian tubules, of which there may be just a few or more than a hundred, are delicate tubes that branch off the posteriormost part of the midgut. (They may appear to be attached to the beginning of the hindgut, but they have been shown to originate from the midgut.) They are bathed by the fluid that circulates in the hemocoel, and in insects they are lightly muscularized. Water and other small molecules diffuse into them through the walls of their distal portions. From their proximal portions, much of the water is resorbed, and certain of the dissolved substances are also recovered. High acidity in the proximal portions of the tubules is a factor that encourages the formation of insoluble crystals of uric acid and urates, which enter the hindgut and are voided with feces. Some nonnitrogenous wastes, including calcium oxalate and calcium carbonate, may also be eliminated by way of the Malpighian tubules and hindgut.

Resorption of water is especially pronounced in the hindgut, and the feces of some insects are extremely dry. The ability to salvage water is an important adaptation for living in situations where water is in short supply.

In many insects, uric acid and other wastes may be stored up within certain tissues, especially in the cells that form the fat body (see next section). These wastes may never leave the body, but they are at least harmless and out of the way in a physiological sense. They may even be put to some use, as they are in butterflies. The opaque white, yellow, and orange pigments of butterflies consist to a large extent of nitrogenous compounds derived from wastes accumulated during larval stages.

Fat Body

The **fat body** (Figure 18.7, B), lying in the hemocoel, consists of a mass of cells of a type called *oenocytes*. They store up nutritive reserves—lipids, glycogen, and proteins—and they are also important in certain phases of metabolism. As a rule, the fat body reaches the peak of its development in larvae that are about to pupate and in nymphs that are approaching the adult stage. Its reserves are generally used up during metamorphosis. In insects that do not feed after reaching adulthood—and there are many of these—the fat body persists and serves as the principal source of nutritive material.

Tracheal System and Respiration

A tracheal system (Figure 18.7) is almost universally present in insects. It consists of a set of tubes that run from pores in the cuticle, called **spiracles,** to the tissues of the internal organs, even the gut. It is fair to say that nearly all cells in the body of an insect are close to one of the finer branches of the tracheal system. The tubes are lined by a thin cuticle continuous with that on the outside of the body. This fact fits in with what we know about the origin of the tracheal tubes: they develop from epidermal invaginations. The cuticular lining of the larger components is generally strengthened by spiral or ringlike thickenings, and is shed when an insect molts. In the finer branches, called **tracheoles,** the lining is neither supported by thickenings nor molted, but these branches are generally filled with fluid and thus are not likely to collapse.

The spiracles are found on the sides of certain segments of the thorax and abdomen. Sometimes they are simple pores, but in many insects they are situated in pits, and these may have valves and elaborate devices for filtering out dust or unwelcome guests. The walls of the tracheae are flexible, so movement of gases within them is promoted by almost all muscular activities that go on in the body. In most insects there are muscles that connect the tergal and sternal plates of the abdomen. When these muscles contract, the abdomen is flattened; when they relax, the terga and sterna rebound because of the elasticity of the cuticle and the way they are hinged. Such movements, coordinated with opening and closing of the spiracles, are thought to be especially important in drawing in and expelling air. Broad dilations of tracheal tubes, common in many insects, would be particularly subject to changes in pressure around them.

Muscular activity specifically related to gas exchange is controlled by respiratory centers in the segmental ganglia. These centers are responsive to the carbon dioxide content of the blood and tissue fluid. A diffusion gradient probably accounts for some of the movement of oxygen toward the tissues and carbon dioxide away from them. In extremely small insects it perhaps accounts for much of the gas flow.

Aquatic larvae of insects sometimes have no tracheae. In general, however, aquatic larvae have *tracheal gills:* leaflike or featherlike outgrowths that contain tracheal tubes continuous with those ramifying throughout the body. There are no spiracles, so oxygen must move into the tracheae of the gills by diffusion through the cuticle and epidermis, and carbon dioxide

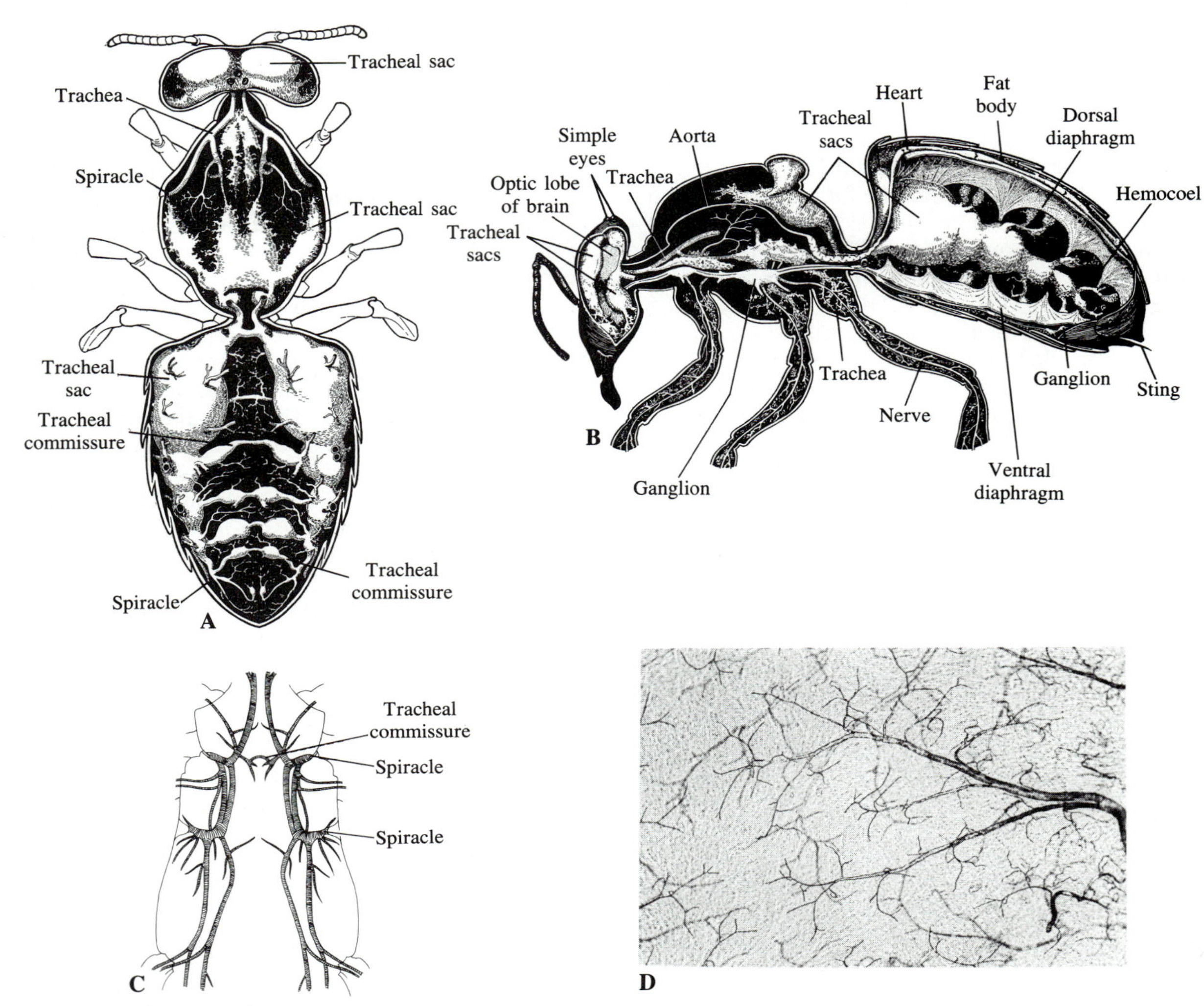

Figure 18.7 A. Tracheal system of a worker honeybee (order Hymenoptera), dorsal view. B. Portions of tracheal system, circulatory system, and nervous system of a worker honeybee, lateral view. (A and B, Snodgrass, United States Bureau of Entomology, Technical Series, Bulletin 18.) C. Tracheal system of *Esthiopterum diomedeae* (order Mallophaga). (Cope, Microentomology, 5.) D. Photomicrograph of tracheal tubes in the gut of the larva of a mealworm beetle, *Tenebrio molitor* (order Coleoptera).

must move out in the same way. Tracheal gills may be located within the hindgut, as they are in nymphs of dragonflies. A pumping action of the hindgut promotes gas exchange.

In aquatic insects whose tracheal system is of the conventional type, there is usually a mechanism for trapping a coating of air, often in many small bubbles, next to the body. Hairs can operate in this way, and the *elytra*—the hard, modified forewings of beetles—can also hold a supply of air beneath them. The aquatic larvae of mosquitoes and some other insects have breathing tubes on which the spiracles are concentrated. A breathing tube can function, of course, only if the insect is at the surface or close enough to get the tip of its tube out of water.

Circulatory System

As in other arthropods, the circulatory system (Figure 18.7, B) is of the open type. The dorsal, tubular heart pumps blood into vessels, essentially arteries, but these do not go far before they allow the blood to escape into spaces, or sinuses, that bathe the tissues. Collectively, the spaces form the **hemocoel,** which has nothing directly to do with a true coelom. The heart, as in other arthropods, has ostia that allow blood to return to it from the pericardial sinus. Blood is pumped by the heart and its anterior prolongation, the dorsal aorta, by peristaltic action. There may be blood-pumping structures besides the heart, however. The partition that segregates the pericardial sinus from the rest of the hemocoel may be contractile, and accessory pumps may be found in various other parts of the body.

The blood is concerned with transport of nutrients, nitrogenous wastes, and hormones. It also has a hydrostatic function, for it fills the extensive hemocoelic spaces. Many processes in insects depend on relatively acute, localized changes in blood pressure. Among these are the unrolling of the "tongue" of butterflies and moths, elimination of fecal material, and expansion of the wings when an insect emerges from its pupal case. Most insects, on undergoing a larval or nymphal molt, attain the size of the next stage by taking up water or swallowing air. If the gut is filled with air, there will be an increase in body size so long as the blood volume remains constant.

Most insects do not rely extensively on the blood for transporting oxygen, for oxygen reaches the cells directly through the ultimate branches of the tracheal system. In some types, however, the blood does distribute oxygen, and in the freshwater larvae of some midges, and also in some freshwater bugs, the blood contains hemoglobin.

The osmotic pressure of insect blood and tissue fluids is maintained mostly by amino acids and other organic molecules, rather than by inorganic ions. As a rule, the concentration of inorganic salts in insect blood is only about half that in vertebrate blood.

Nervous System

The brain of an insect, located above the foregut, consists of three pairs of ganglia. Those of the first pair, which form the protocerebrum, innervate the compound eyes and simple eyes, when either or both are present. Those of the second pair, belonging to the deutocerebrum, innervate the antennae. The labrum and foregut receive nerves from the third pair, which form the tritocerebrum. A pair of connectives run ventrally around the foregut, joining the tritocerebrum to the subesophageal ganglionic complex, which innervates, among other structures, the mouthparts. The ventral cord, as in arthropods generally, consists basically of paired segmental ganglia and paired connectives between successive ganglia (Figure 18.8, A). Sometimes, however, there is so much condensation of ganglia and connectives that these components run together and form a single mass (Figure 18.8, B). This is especially likely to be the case in certain small and compact insects.

Sense Organs

Insects have a wide variety of sense organs, including touch, pressure, and vibration receptors, auditory organs, chemoreceptors, receptors for perceiving changes in temperature and humidity, and compound eyes and simple eyes (ocelli). Though certain of these, common to arthropods in general, have been discussed in Chapter 15, it will be worthwhile to mention briefly some of them again here.

Touch receptors include setae whose movements excite nerve cells that end in their basal portions, as shown in Figure 15.4. There are, however, other types of cuticular structures concerned with detection of tactile stimuli, as well as with detecting vibrations and with providing information concerning an insect's own movements, as when it bends a leg or an antenna.

The sense of hearing is highly developed in insects that produce sound, and even in many insects that do not. Although relatively simple hairs, similar to those concerned with touch reception, may serve in detection of sounds, some insects have **tympanal organs,** which operate on the drumhead principle. The tympanum itself is a specialized portion of the wall of a trachea. It may be located at the surface of the body, as is typical of the auditory organs in the first abdominal segment of certain grasshoppers. Often, however, it is in a pit, which is the case in the tympanal organs on the tibia of the first legs of katydids and crickets, as well as of some grasshoppers. Figure 18.9, A shows the arrangement of a pair of pits on the tibia of a cricket, and Figure 18.9, B illustrates a transverse section through the pits. Note that the slits visible externally lead into pockets. The flattened outer walls of two large tracheae are the "drumheads." As they vibrate in response to sounds, the pressure change within the trachea affects structures called **scolopidia,** which are linked with sensory nerves. There is usually a row of scolopidia alongside the trachea of a tympanal organ. In some tympanal organs, such as those of typical grasshoppers, the scolopidia are joined to the edge of the tympanum, so the sensory nerves are stimulated directly by vibration of the tympanum.

Chemoreceptors, which include organs of taste, are usually hairs, or at least little elevations, that are open at the tip or covered only by a delicate cuticular membrane. The sensory nerve endings are at or close to the tip. Little pits supplied with nerve endings are also known to be chemoreceptors. The antennae, tarsi of the legs, mouthparts, and other parts of the body are among the places where chemoreceptors are concentrated.

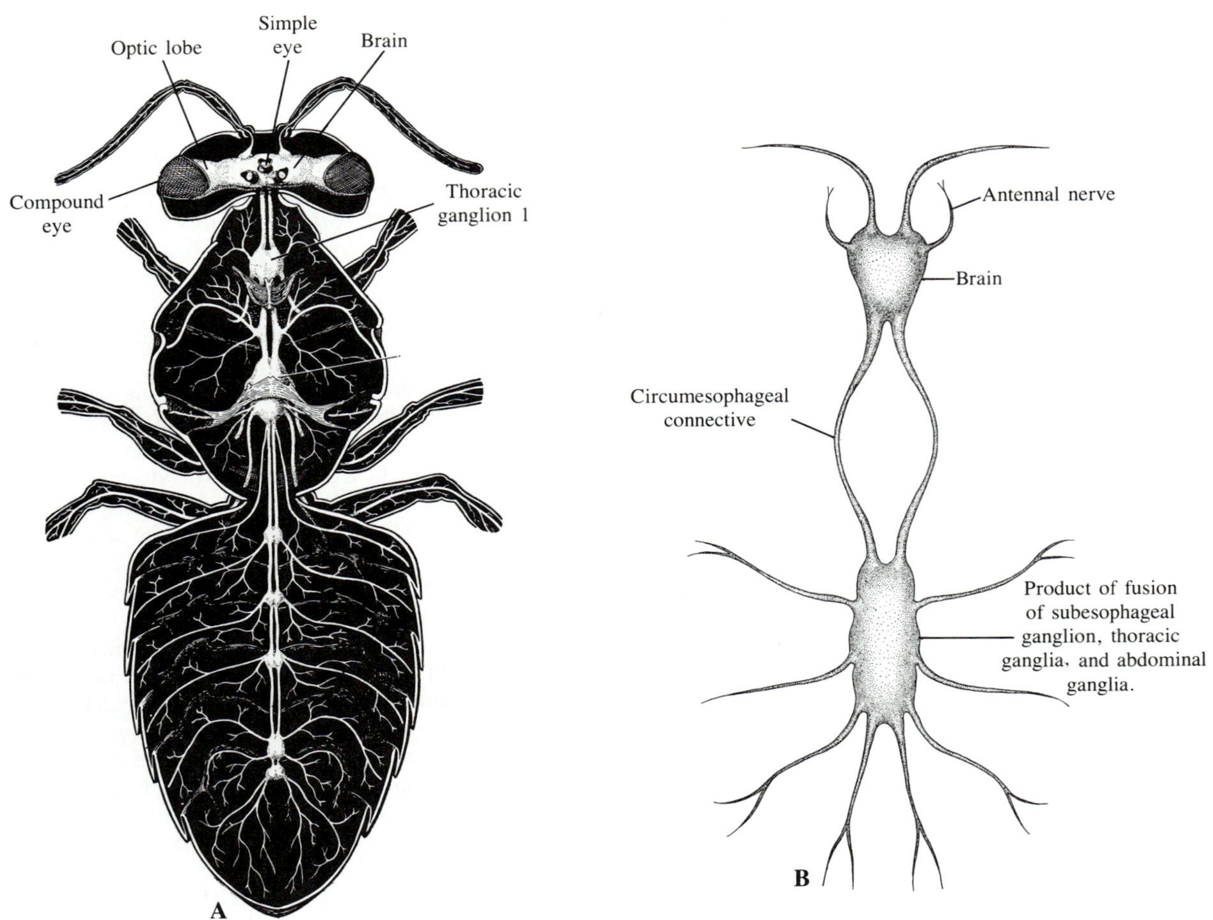

Figure 18.8 A. Nervous system of a worker honeybee (order Hymenoptera). (Snodgrass, United States Bureau of Entomology, Technical Series, Bulletin 18.) B. Condensed nervous system of a scale insect, *Aonidiella citrina* (order Homoptera). (Nel, Hilgardia, 7.)

Most adult insects and nymphs have a pair of **compound eyes,** and sometimes they also have a few simple eyes on the front of the head. When eyes are present in larvae, these are always simple, and they are usually located on the lateral portions of the head.

The structure of both simple and compound eyes has been discussed in Chapter 15 and is not dealt with again here. A few generalizations about the sensitivity of insect eyes may be offered, however. The human eye perceives light within that portion of the spectrum from about 450 mμ (4500 Å) to 700 mμ (7000 Å). Insect eyes are sensitive to wavelengths as low as about 260 mμ, which includes ultraviolet light. On the other hand, they are usually insensitive to red; sensitivity in the honeybee, for instance, drops off at about 650 mμ, which is at the border between orange and red. It has been established that many insects can distinguish colors, although they do not necessarily see these colors the way humans do.

REPRODUCTION AND DEVELOPMENT

Reproductive System

Most insects, as adults, are terrestrial, and their modes of reproduction and development show adaptations to life on land. They have true copulation and internal fertilization, and their eggs are usually enclosed by a resistant envelope. The eggs are also relatively large and yolky, and when the young hatch they are ready to start feeding.

In the male system, there are two testes. These are usually separate, but sometimes fused on the midline, and they consist of many sperm-producing follicles. The sperm ducts lead to a common ejaculatory duct that

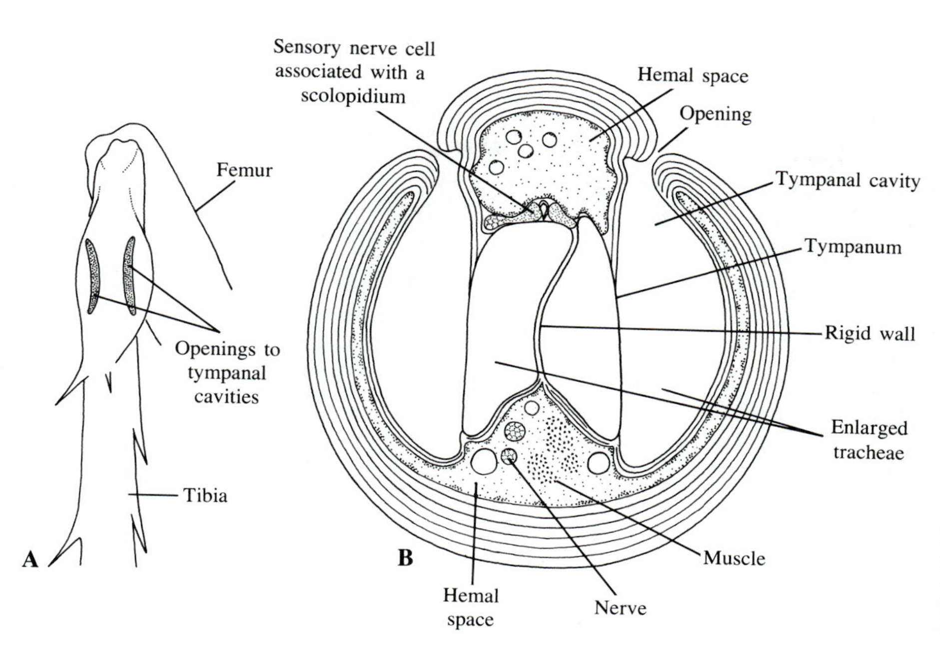

FIGURE 18.9 Auditory organs of crickets (order Orthoptera). A. Slits opening into the tympanal cavities on the tibia of leg 1. B. Section through the tibia at the level of the tympanal cavities. (Explanation in text.) (After Weber, Lehrbuch der Entomologie.)

extends to the genital pore. In some insects, there are no special elaborations on these components (Figure 18.10, A). In others, however, the sperm ducts have dilations or side-pouches that function as seminal vesicles, and the ejaculatory duct may have accessory glands (Figure 18.10, B). The accessory glands are thought to provide nourishment for the sperm and keep them healthy. The ejaculatory duct may also have a diverticulum that stores sperm. The penis, which may be concealed except during copulation, is located ventrally, usually between abdominal segments 9 and 10. It is linked with a pair of much-modified appendages that form the aedeagus.

The female reproductive system has a pair of ovaries, each of which consists of several to many tubes within which the eggs are produced (Figure 18.10, C and D). These tubes, comparable to follicles of the testes, are called *ovarioles*. The primordial germ cells are located near their tips, and as prospective eggs enlarge they move toward the oviducts and become surrounded by cells that secrete a coat called the *chorion*. They may supply some yolk, too, but most of the yolk originates within the egg itself or is contributed to it by other cells before the chorion-forming cells envelop it.

The ovarioles deliver the eggs to the oviducts, which unite to form a common oviduct, continuous with the vagina. This arises as a deep inpocketing between segments 9 and 10 of the abdomen. The vagina may have a variety of diverticula, including a seminal receptacle for storing sperm received in copulation, and accessory glands concerned with secreting an eggshell or having some other function. In insects that have *ovipositors,* which are devices for sticking eggs into soil, holes in wood, or other animals, these are usually derived, at least in part, from appendages. The ovipositor may be specialized to form a sting, as it is in bees, wasps, and some ants, and in this case the venom is produced in accessory glands linked with the vagina.

Embryology

The eggs are generally inseminated after they have reached the common oviduct. Sperm enter through one or more tiny pores in the chorion (Figure 18.11, A); the position of these corresponds to the anterior end of the future embryo. By the time sperm get into an egg, the chorion-producing cells have broken down. Polar bodies are cast off as the meiotic divisions are completed, and the female nucleus unites with the male pronucleus contributed by the sperm. In nearly all insects, other than a few primitive types, there is so much yolk that cleavage is superficial (Figure 18.11, B–E). Repeated mitotic divisions result in the production of many nuclei, which end up in the peripheral cytoplasm of the egg. After these nuclei have been separated by cell membranes, the embryo consists mostly of an inner yolky mass surrounded by a cellular layer called the **blastoderm.** A few cells do wander into the yolky mass at the posterior end, however, and these are precursors of germ cells.

Not all of the blastoderm produces the future insect. Much of it serves merely to enclose the yolky mass; some of it forms folds that eventually surround the "real" embryo in much the same way as the amnion does in higher vertebrates (Figure 18.11, G and H). The

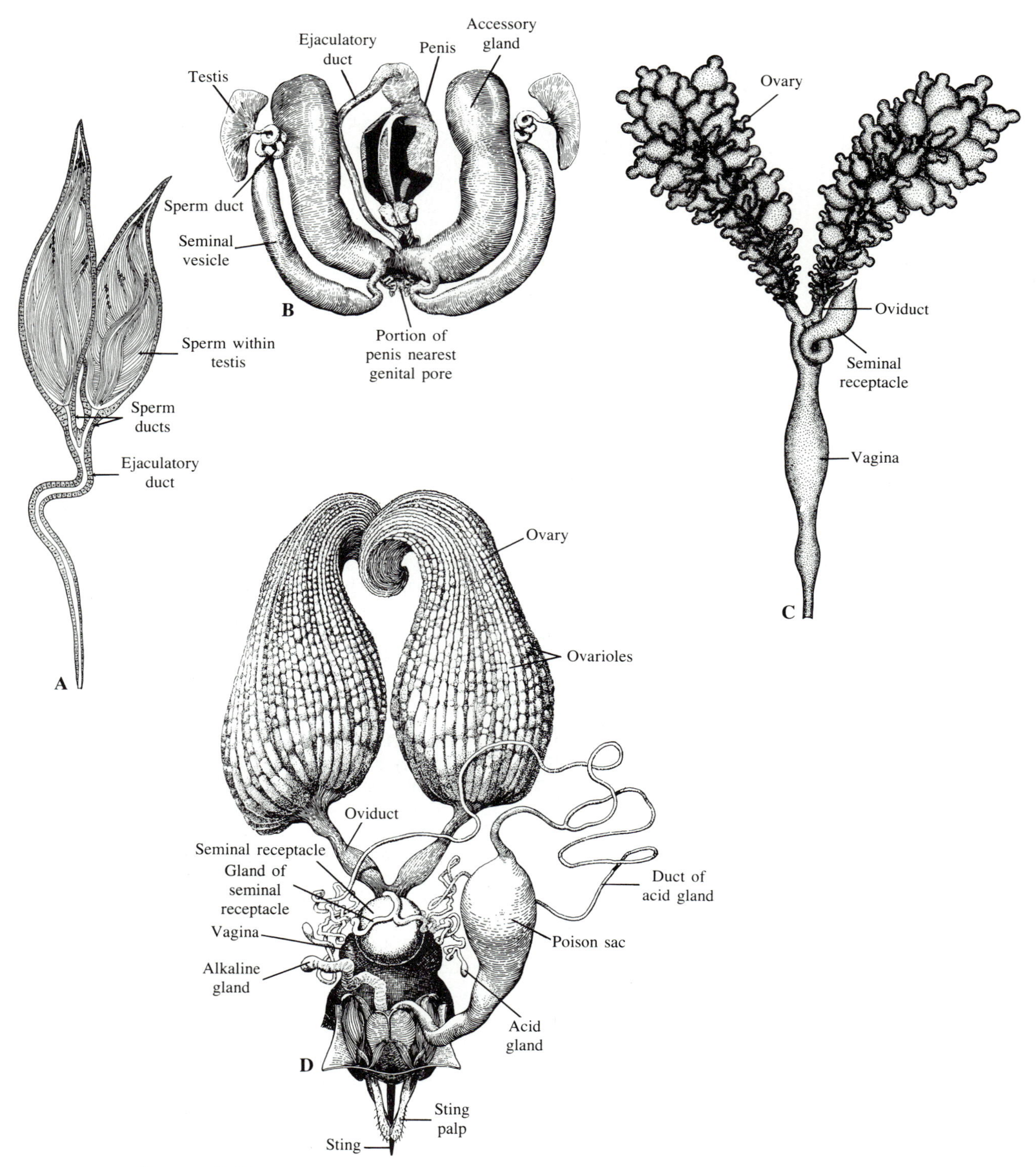

FIGURE 18.10 A. Reproductive system of a male scale insect, *Aonidiella citrina* (order Homoptera). B. Reproductive system of a male honeybee (drone) (order Hymenoptera). C. Reproductive system of a female *Aonidiella citrina*. D. Reproductive system of a queen honeybee. (A and C, Nel, Hilgardia, *7*; B and D, United States Bureau of Entomology, Technical Bulletin 18.)

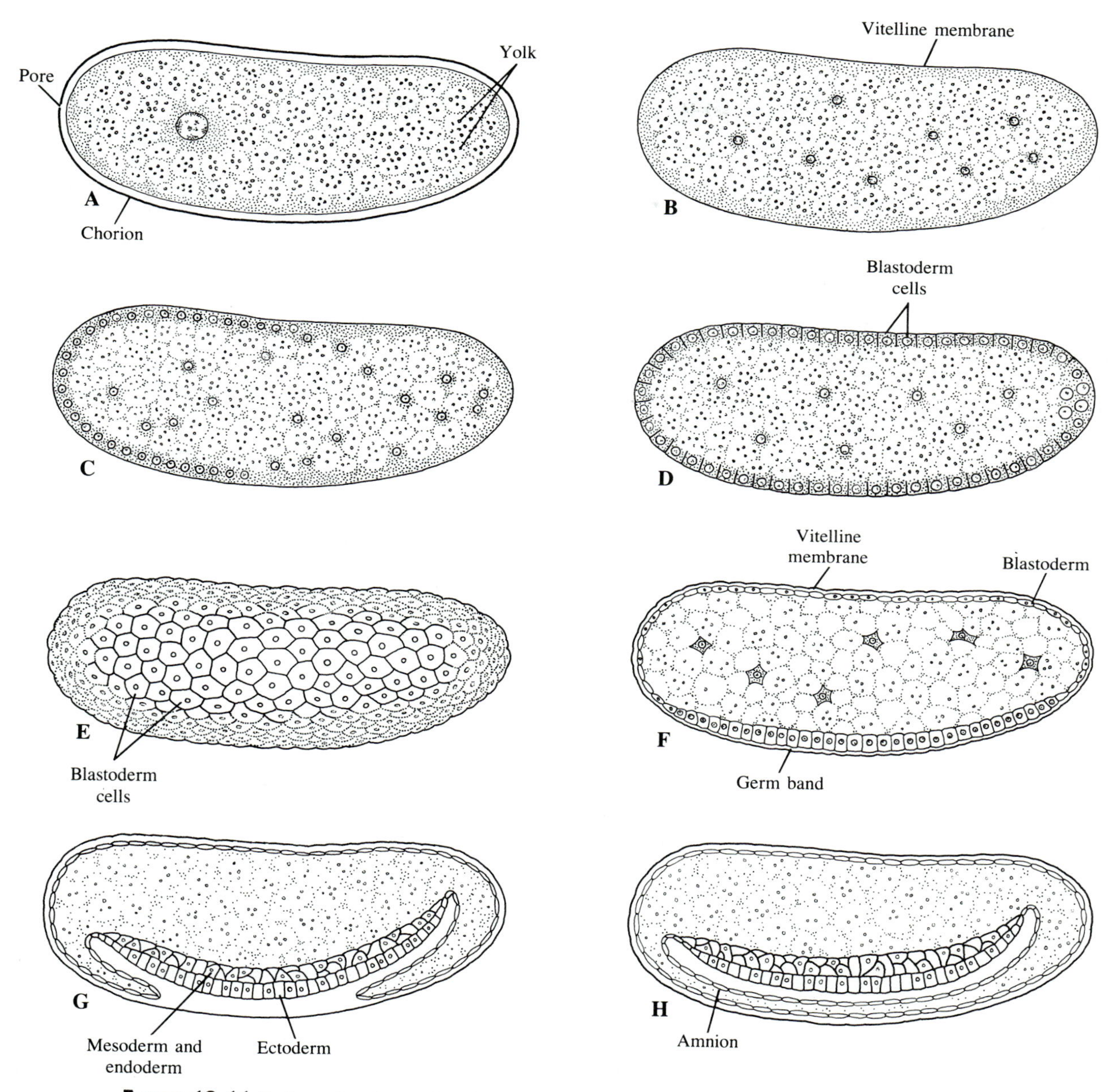

FIGURE 18.11 Early embryology of an insect, generalized. A. Yolky egg with its chorion intact, showing the pore through which sperm may enter. B–E. Superficial cleavage leading to formation of a peripheral, cellular blastoderm (chorion removed). F. Germ band differentiated on the ventral side of the egg. G, H. Formation of the amnionlike folds. Some of the cells proliferated by the ectoderm of the germ band will become mesoderm; others, especially near the opposite ends, will become endoderm and will enclose some of the yolk, forming the midgut.

process by which an insect develops from a strip of thickened blastoderm is complicated and only the bare essentials can be described here. Gastrulation begins when a lengthwise groove develops in the strip of blastoderm. The cells carried in by the deepening groove become separated from the ectodermal band that remains at the surface; most of them represent mesoderm; some, near each end, are endodermal, for they form rudiments of the midgut and enclose the yolky mass. In time, the ectodermal invaginations at the anterior and posterior ends of the embryo will produce the stomodeum and proctodeum, the future foregut and hindgut, and the digestive tube is thereby made complete. Most of the mesoderm becomes concentrated into two bands, one on each side of the midline. Traces of segmentation and tagmatization are sometimes noted about the time gastrulation is taking place, but they generally do not become really obvious until considerably later. Moreover, not all segments appear simultaneously, and the primordia of certain appendages appear before those of others. Linked with the formation of embryonic somites is the development of serially arranged coelomic cavities, which do not remain recognizable for long. In insects, the follicles of the gonads and certain portions of the gonoducts are the only conspicuous structures that can be traced back to the embryonic coelom. Most of the mesoderm differentiates into muscles, the dorsal heart, and at least part of the fat body. The nervous system originates in the ectoderm, as a pair of parallel longitudinal cords. These become segmentally ganglionated, but the double nature of the ventral nerve cord of an insect can be traced back to the way the nervous system looks when its primordia first become apparent.

Direct and Indirect Development

When an insect hatches, it may already resemble the adult, even if it lacks wings. In that case it is called a **nymph.** The body form of the adult is attained gradually, by way of a series of stages called **instars.** Successive instars are separated by molts. This type of development is called **direct development,** and it is characteristic of the more primitive insects, such as grasshoppers, cockroaches, termites, and true bugs.

The higher insects undergo a striking metamorphosis, so their development is said to be **indirect.** The young hatch as **larvae** that bear little resemblance to adults. After feeding, growing, and molting a number of times, the larva becomes a **pupa.** The pupa cannot feed and usually cannot move about, and it may be enclosed by a cocoon of silken threads, but it is during this stage that the transformation of the larva into an **adult** takes place. The adult works its way out of the pupal skin and there are no further molts. The two main types of development intergrade to some extent. In dragonflies and some other lower insects whose young are aquatic, the young resemble the adults in certain respects. In having tracheal gills and no wings, however, they are decidedly unlike adults.

The arrangement of the major subdivisions of the Insecta is based to a considerable extent on whether development is relatively direct or whether it involves a partial or complete metamorphosis. The subject will be dealt with again in connection with classification.

The remarkable changes that take place when a larva molts or undergoes metamorphosis are under the control of hormones. Periodic molting is promoted by the hormone **ecdysone,** which is produced by a gland in the prothorax. The activity of this gland is in turn dependent on a hormone produced by neurosecretory cells in the brain. Throughout larval development, the *corpora allata*—glands connected to the brain but not properly a part of it—secrete a **juvenile hormone** that suppresses metamorphosis; in other words, this hormone helps to perpetuate the larval condition. So long as there is not too much ecdysone in proportion to juvenile hormone, each time a larva molts it emerges as yet another larva. Eventually, the titre of the juvenile hormone drops off, and the last larval stage molts and becomes a pupa.

COMMUNICATION AND SOCIAL BEHAVIOR

Insects communicate with one another by sounds, chemical secretions, touch, light, and various behavioral patterns. Fireflies, of course, are famous for producing flashes of light that bring the sexes together, but various adult and larval insects produce light for reasons that are not yet understood.

Grasshoppers, crickets, katydids, cicadas, some beetles, and other insects generate sounds, and the notes may be organized into songs. The sounds are produced in many different ways and for different purposes. Songs of crickets, used as sexual attractants and in aggression, are produced by scraping one forewing against the other. In katydids, the hindlegs are rubbed against filelike elaborations of heavy veins of the forewings. Cicadas produce their buzzing sounds by vibrating membranous structures on the abdomen. The sounds draw individuals together into swarms.

Chemical communication by pheromones, which are secreted by epidermal glands, is important in maintenance of colony structure in honeybees, termites, and many other social insects, and also in bringing males and females together. The carrying power of some pheromones that serve as sexual attractants is astonishing; or, to put it another way, the capacity of some insects to recognize minute traces of these pheromones is remarkable. Certain female moths will draw males from a distance of about 4 km. This fact has made some sexual attractants useful for luring insect pests to their doom. Insects may also use pheromones as trail markers. Ants coming back from a successful foraging trip secrete substances that inform other members of the colony which way to go. Ants are also sensitive to the presence of a dead individual, and will react promptly to remove the carcass. If a perfectly healthy ant is painted with material extracted from a dead ant, the workers of the colony will be persuaded to carry out their protesting colleague.

Many insects are gregarious, at least under certain conditions. Some locusts, for instance, move in swarms that may do great damage to crops and other vegetation. Ladybird beetles congregate as they prepare for overwintering. There is no doubt that such behavior involves complex interactions between individuals, but it does not represent what may be called social life. Truly social insects—ants, termites, and some bees and wasps—are organized into castes specialized for performance of particular tasks.

The honeybee, *Apis mellifera,* is a good example of a social insect. A colony consists of three types of individuals. There is a queen, who lays eggs, and males ("drones"), who visit mating swarms and inseminate queens. The vast majority of individuals in the hive are sterile females called workers. These cannot reproduce, but they do perform maintenance activities in the hive. They secrete the wax and shape it into a honeycomb, keep the hive clean, and guard the entrance. Workers also collect all of the pollen and nectar upon which nutrition of the queen, drones, and larvae depends. Workers tend to specialize on one function, then switch to another as they grow older. The sting that each worker has at the tip of its abdomen is a modified ovipositor. It is used to discourage intruders that have designs on the honey or on the bees in the hive.

In the wild, honeybee colonies are situated in holes in tree trunks and similar places. By setting up artificial beehives, people have "domesticated" this insect and collect its honey, which consists mostly of two simple sugars, dextrose and levulose, derived from the sucrose in the nectar from which the honey is made. The levulose is very sweet, the dextrose scarcely sweet, and the "average" sweetness of the two is about equal to that of the original sucrose. In addition to sugars, honey contains about 20% water, pollen, some minerals, and fragrant substances collected from flowers. Thus honey deposited by bees whose nectar source is mostly clover will taste a little different from honey stored by bees that concentrate on orange blossoms.

Within a few days after a young queen bee leaves the hive and takes flight, she mates with one or more drones. A drone's organs of copulation are generally severed after they have been inserted into the queen's genital bursa, in which case they are removed by workers after the queen returns to the hive. The queen will not mate again, but the supply of sperm in her seminal receptacle may last for several years. Not all of the eggs she lays are fertilized, however. Larvae derived from unfertilized eggs become drones, which are haploid (16 chromosomes); larvae derived from fertilized eggs develop into diploid workers or queens. In general, unfertilized eggs are laid when there is plenty of nectar. When conditions are less favorable, most of the drones are forced to leave and they may not survive long. The life span of workers varies; some die in a few weeks. A queen may live for several years, producing about a million eggs.

Larvae that will become queens are fed differently from those that will become drones and workers. The *royal jelly,* a product of pharyngeal glands of workers, is given to prospective workers and drones for only a few days, whereas prospective queens continue to be nourished by it until they are much larger than their brothers and sisters. The larvae, regardless of how they are fed, molt three times during their growth period. Their cells are closed up with wax and they pupate, spinning silken cocoons around themselves. To emerge from its cell, a young bee breaks open the cap of its cell with its mandibles.

Egg-laying by the queen is usually accelerated when warm spring weather comes and the worker bees are especially active in bringing pollen and honey back to the hive. Eventually the hive becomes overcrowded. The queen and a few thousand workers then depart to relocate themselves in a place that scouts have determined to be suitable for a new hive. In the old hive, there remain several prospective, larval queens; the first to emerge usually stings the remaining contenders, flies off to mate, then returns to take the place of the original, departed queen.

When a bee visits a flower, pollen is trapped by the many simple or branched hairs on the body. Special combs and brushes on the legs transfer the pollen to the "pollen basket" on the tibias of the hindlegs. Each foreleg, for instance, has a brush on its tibia for cleaning the hairs on the compound eyes, and another brush on the tarsus for collecting pollen from the anterior part of the body in general. An antenna is cleaned by drawing it through a bristly notch on the tarsus; the notch is closed, after the antenna is inserted, by a flattened hair that originates on the tibia. The midlegs and hindlegs have specializations for sweeping progressively more posterior parts of the body, as well as for getting the pollen collected by the forelegs to the pollen baskets.

The pollen that bees collect serves as a source of protein for the larvae. The nectar consists to a large extent of sucrose, but within the crop the sucrose is digested to levulose and dextrose, the main components of honey. A bee uses a little of its honey for its own nutrition and regurgitates the rest into the cells of the honeycomb. Young bees whose activities are restricted to the hive take the honey into their mouths and modify it further. They also fan the air with their wings, thus promoting evaporation of water, so that the syrup becomes more concentrated.

Bees have interesting ways of informing their colleagues about nectar sources. After a worker finds a good source and fills her crop, she returns to the hive to feed the larvae or to add the nectar to the accumulation of honey. She then goes through a dance routine, repeatedly tracing two semicircles that together form a figure eight (Figure 18.12, B). The axis along which the figure eight is executed informs the members of the hive about the angle at which the nectar source is located with respect to the sun's azimuth. The frequency with which the dancer turns indicates the distance to the source. As the bee dances, she is touched by the antennae of other bees, so they become informed about the character of the nectar and pollen she has brought back.

CLASSIFICATION

CLASS INSECTA

In insect classification, the following features are particularly important: the general appearance of the body; the structure and specializations of the mouthparts, legs, and wings; the presence or absence of cerci on the last abdominal segment; and the type of development. In this book, the orders are segregated into two subclasses, according to whether or not they have wings. These subclasses are unbalanced, because there are relatively few insects that are primitively wingless. It seems best, therefore, to further subdivide the winged insects. This is done partly with the help of two infraclasses and partly by the use of some groupings that are given vernacular names.

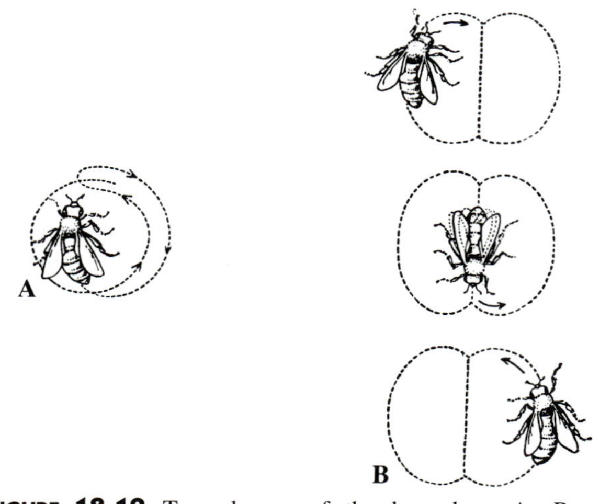

FIGURE 18.12 Two dances of the honeybee. A. Round dance. The bee makes a partial or complete circle, then reverses the path of movement. B. Three phases of the tail-wagging dance in the pattern of a figure eight. The bee traces a half-circle, then a straight line, then another half-circle. During the straight-line phase, the bee wags its abdomen. The entire pattern may be repeated several times. (von Frisch, Österreichische Zoologische Zeitschrift, *1*.)

SUBCLASS APTERYGOTA

The apterygote insects do not have wings, and on the basis of the fossil record it appears that their ancestors did not have wings, either. The only surviving order is the Thysanura. In older systems of classification, the Collembola, Protura, and Diplura were treated as apterygotes. They are no longer viewed as genuine insects, and in this book all three of these "near insect" groups are accorded the rank of class.

Order Thysanura

The Thysanura (Figure 18.13) includes the insects called "bristletails," "silverfish," and "fire-brats." Some of them, such as *Lepisma saccharina,* are com-

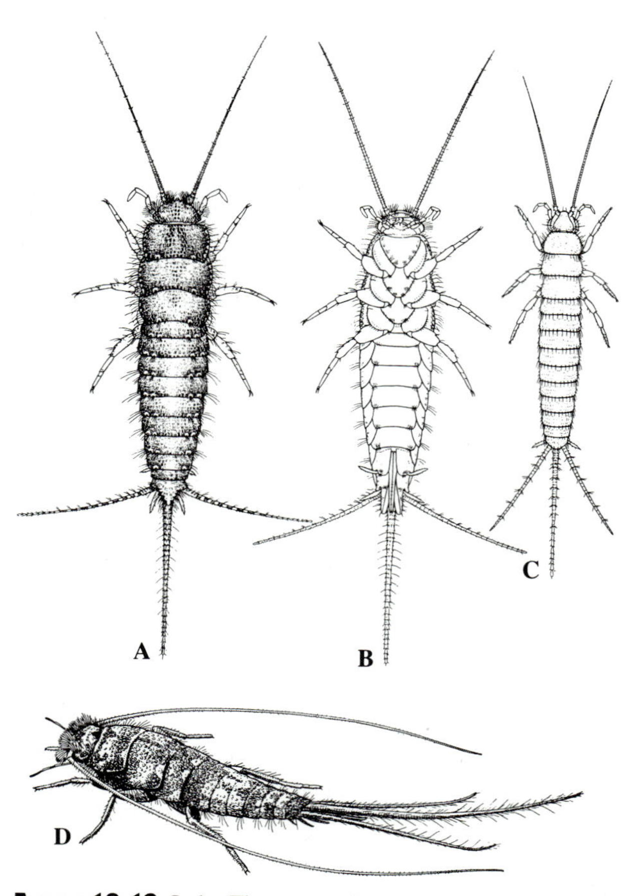

FIGURE 18.13 Order Thysanura. A. *Acrotelsella devriesiana,* male, dorsal view. B. *Acrotelsella devriesiana,* female, ventral view. C. *Trinemura excelsa,* female, dorsal view. (A–C, Watson, in The Insects of Australia.) D. *Thermobia domestica,* a common household silverfish, lateral view. (Ross, Illinois Natural History Survey, Circular 39.)

monly found in houses; others live in soil or leaf-litter, under rocks, or in association with termites and ants. Most species are about 1 or 2 cm long. The silvery appearance of thysanurans is due to the presence of small scales all over the body.

Thysanurans usually have long, slender antennae, and the mouthparts are like those of most insects in being exposed, so that they can be used for biting. Either compound eyes or simple eyes (ocelli) are present, except in certain types that live in caves, termite nests, ant nests, and other permanently dark situations. The abdomen has 11 segments from the time a young thysanuran hatches. The long median process and pair of lateral cerci on segment 11 are among the more distinctive features of the order. Some of the other abdominal segments have a pair of reduced appendages, called *styli*, and a pair of eversible sacs similar to those found in symphylans. What these sacs do has not been explained convincingly, but it has been suggested that they are concerned with respiration and with absorption of moisture. There is a well-developed tracheal system with paired spiracles on most segments of both the thorax and abdomen, and a set of Malpighian tubules for excretion.

Thysanurans are the only true insects in which insemination is achieved by indirect transfer of a spermatophore rather than by copulation. They are unusual also in continuing to molt after they have attained sexual maturity.

SUBCLASS PTERYGOTA

In terms of the way their wings are formed, pterygotes fall into two main categories. In one group, which includes grasshoppers and true bugs, the wings originate as little external pads. These insects are therefore called *exopterygotes*. Their development progresses gradually toward maturity. The young may resemble adults except for the fact that the wings and reproductive system are not fully formed. Insects of this group are said to have direct (hemimetabolous) development.

In the *endopterygote* insects, of which flies, bees, beetles, and butterflies are good examples, the wings originate within a pair of pouches produced by infolding of the body wall. The young, moreover, do not resemble adults, but undergo a drastic metamorphosis from larva to adult while in the pupal stage. The wings become external during the later phases of the pupal stage. Thus the endopterygote insects are said to have indirect (holometabolous) development.

Many pterygote insects have secondarily become wingless. Some of them fit best among the exopterygote groups; others are closer to certain endopterygotes.

In older systems of classification, the exopterygotes were usually united under a common heading, suggesting phylogenetic unity of the group as a whole. The tendency now is to arrange the orders of exopterygotes to indicate that they represent a grade of structure rather than an assemblage that has had a common evolutionary origin. In the system used in this book, the subclass Pterygota is divided into two infraclasses, Palaeoptera and Neoptera. The former, which is decidedly the more primitive, has only two orders in it. To make it easier to understand the Neoptera, the many orders assigned to this infraclass are segregated into three groups that are given vernacular names. The first two—the blattoid–orthopteroid orders and hemipteroid orders—are exopterygote and have direct development. The third group consists of the endopterygote orders, all with indirect development.

INFRACLASS PALAEOPTERA

Two groups of exopterygote insects—mayflies (order Ephemeroptera) and dragonflies and damselflies (order Odonata)—cannot fold their wings back over the abdomen when they are at rest. They are believed to be more primitive than insects that do normally fold back their wings, and they are therefore placed in the infraclass Palaeoptera. The remaining orders of exopterygote insects, as well as all endopterygote orders, are placed in the infraclass Neoptera.

Order Ephemeroptera

Adult mayflies (Figure 18.14, A)—the "ephemeral insects"—have vestigial, nonfunctional mouthparts, and generally live for only a few hours. Their wings are extremely delicate, and the forewings are much larger than the hindwings. The posterior part of the abdomen has a pair of long cerci, and often a median cercus as well. The successive nymphal stages (Figure 18.14, B) are aquatic and regularly have three cerci and feather-like tracheal gills on several of the abdominal segments. They feed largely on plant material. An especially interesting feature of mayflies is that they become winged "subadults" on completing the next-to-last molt. The subadult stage can fly, but usually it must undergo the final molt before it can mate. In certain genera, however, female subadults are reproductive; they do not make the final molt.

Order Odonata

Dragonflies (Figure 18.15, A) and damselflies have strong wings and chewing mouthparts, and are predators on other insects. The aquatic nymphs (Figure 18.15, B) are also predatory, feeding mostly on worms and various other invertebrates. Their marvelously developed compound eyes, adapted for underwater vision, become reorganized for aerial vision before the adult stage is reached. The abdomen is especially long in

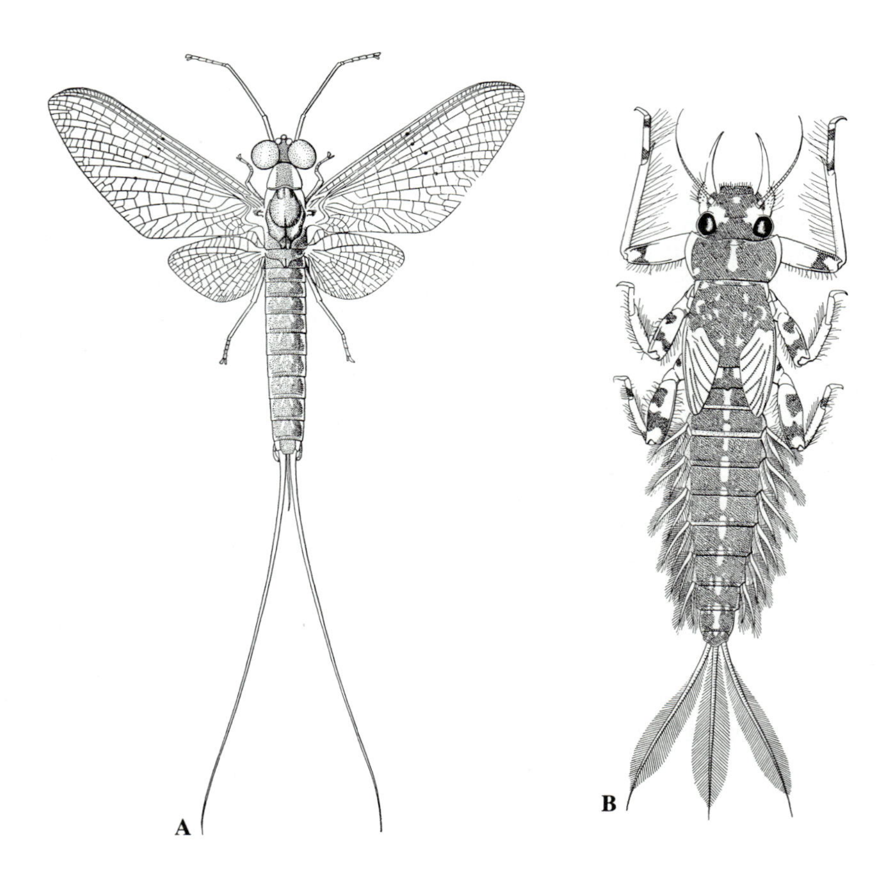

FIGURE 18.14 Order Ephemeroptera. A. *Pentagenia vittigera,* male. B. *Polymitarcys albus,* nymph. (Needham, Bulletin of the United States Bureau of Fisheries, *36*.)

adults. The forewings and hindwings of a dragonfly are decidedly unequal and are kept spread horizontally when the insect is at rest. The two pairs are almost equal in damselflies; they are not as strong as those of dragonflies, however, and when at rest they are folded back slightly, with the forewings partially covering the hindwings. Large dragonflies have a wing spread of about 10 cm, but the fossil record shows that some of their ancestors were as large as 60 cm across.

INFRACLASS NEOPTERA

In neopterans, the wings can be folded back along the body, unless some secondary modifications interfere. The abdomen, moreover, usually has a pair of posterior cerci. Some of the orders are exopterygote and have direct development. The rest are endopterygote and have indirect development.

BLATTOID–ORTHOPTEROID ORDERS

The insects belonging to the several orders grouped together here are generally characterized by mouthparts specialized for chewing, complicated wing venation, prominent abdominal cerci, and numerous Malpighian tubules. The differences between adults and nymphs of various stages is relatively slight.

Order Plecoptera

Stoneflies (Figure 18.16, A) are characterized by hindwings that are a little larger than the forewings and that are pleated in such a way that they resemble folding fans. These insects have long, slender antennae and a pair of prominent caudal cerci. The nymphs (Figure 18.16, B) live in water, usually of swift streams, and have a number of tracheal gills on the thorax and sometimes on the head region. The mandibles of the nymphs are adapted for chewing plant material. Adults may retain the strong mandibles, in which case they continue to utilize plant tissue. In many plecopterans, however, the mouthparts become weaker and the adults do not feed.

Order Orthoptera

Grasshoppers, crickets, and katydids (Figure 18.17) are characterized by long hindlegs that are specialized for leaping. Some are wingless, some short-winged, others long-winged. If wings are present, the forewings are tough and cover the hindwings when these are folded up. The mouthparts are adapted for chewing,

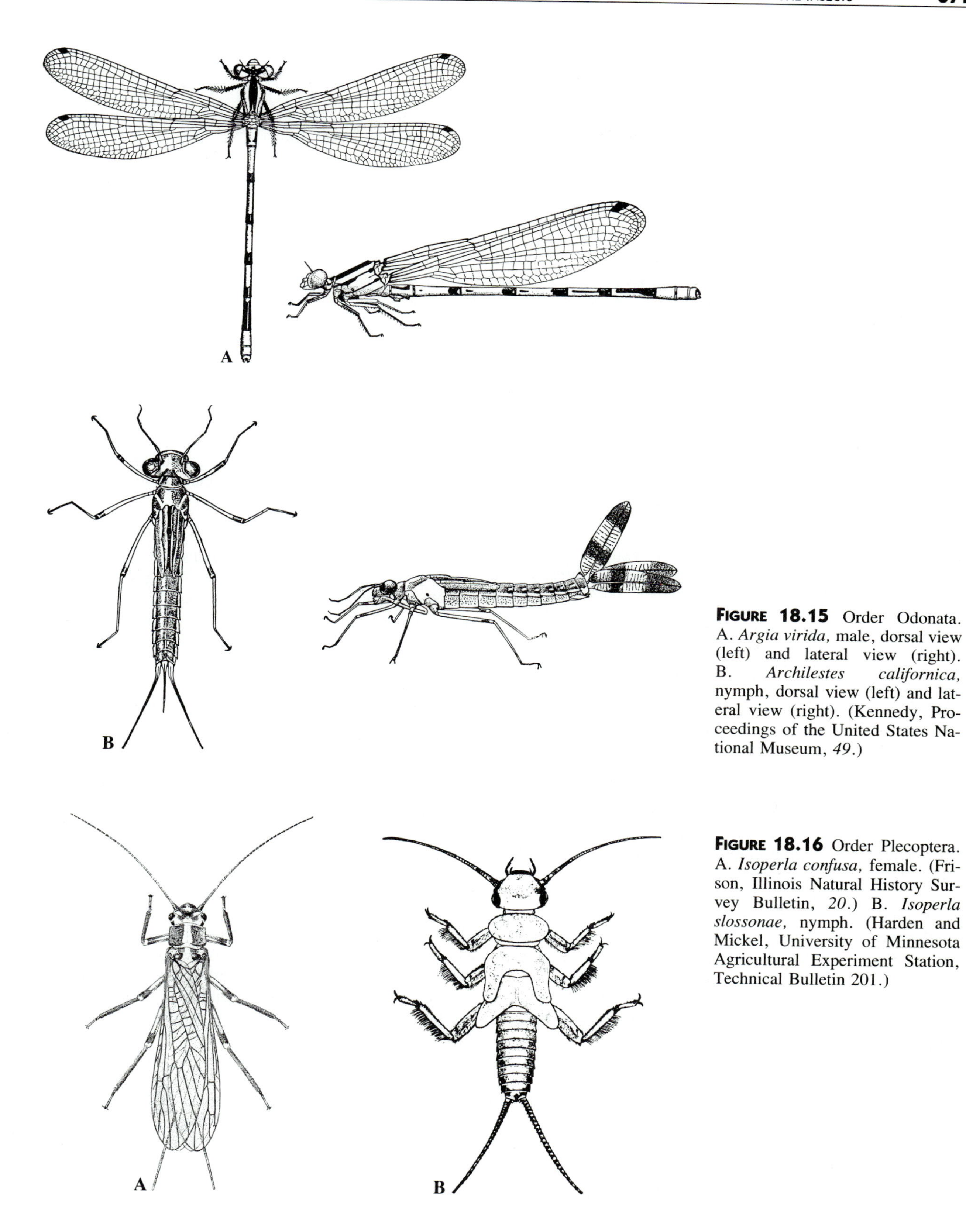

FIGURE 18.15 Order Odonata. A. *Argia virida,* male, dorsal view (left) and lateral view (right). B. *Archilestes californica,* nymph, dorsal view (left) and lateral view (right). (Kennedy, Proceedings of the United States National Museum, *49.*)

FIGURE 18.16 Order Plecoptera. A. *Isoperla confusa,* female. (Frison, Illinois Natural History Survey Bulletin, *20.*) B. *Isoperla slossonae,* nymph. (Harden and Mickel, University of Minnesota Agricultural Experiment Station, Technical Bulletin 201.)

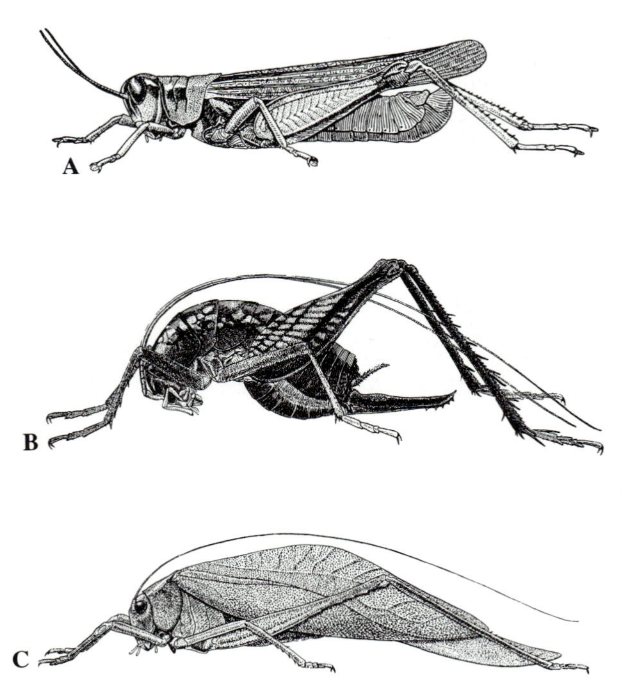

Figure 18.17 Order Orthoptera. A. *Melanoplus bilituratus*, the migratory locust. (Ross, Illinois Natural History Survey, Circular 39.) B. *Microcentrum rhombifolium*, the brush katydid. C. *Ceuthophilus maculatus*, the camel cricket. (B and C, Hebard, Illinois Natural History Survey Bulletin, *20*.)

and the food of orthopterans consists mostly of plant material. The cerci are not divided into a series of articles. Most species have organs for producing sound, and females usually have a prominent ovipositor.

Order Phasmida (Phasmatodea)

The stick insects (Figure 18.18) and leaf insects are herbivores that defy detection when they are perched on the plants with which they are typically associated. Stick insects have elongate bodies and resemble twigs; successive pairs of legs are widely spaced, and wings, if present at all, are small. Leaf insects, with flattened bodies and with wings and legs that resemble leaves, are especially remarkable mimics. The ovipositor of female phasmids is small and more or less hidden; the caudal cerci are not divided into articles. Phasmids are primarily tropical and subtropical, and use their chewing mouthparts for feeding on fresh plant material.

Order Dermaptera

The forewings of earwigs (Figure 18.19) are reduced to leathery, veinless structures that resemble the elytra of beetles. The hindwings are thin and almost semicircular, and can be folded lengthwise like a fan, so they fit under the forewings. Some dermapterans are wingless. The posterior end of the abdomen has a pair of cerci. These are usually stout enough to operate as pincers for defense as well as for seizing small prey; they are also used in mating. In general, earwigs are omnivores that live under logs or other debris, or in caves. The mouthparts are specialized for biting and chewing.

Forficula auricularia (Figure 18.19, A) is a widespread inhabitant of gardens, garages, basements, and kitchens. A few earwigs are parasites of mammals.

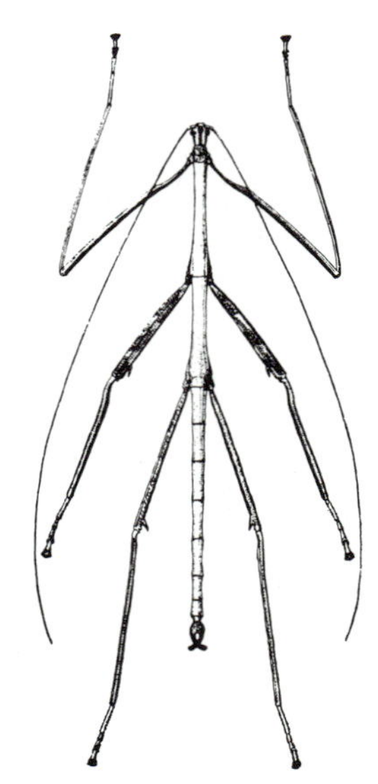

Figure 18.18 Order Phasmida. *Diapheroma femorata*, a walking stick. (Hebard, Illinois Natural History Survey Bulletin, *20*.)

Order Embioptera

The webspinners (Figure 18.20) are tropical or subtropical insects that weave tunnels of silk beneath stones, bark, and other solid objects. They may wander out at night, but by day they are generally assembled in their galleries. Males may have a pair of nearly equal wings, or they may be wingless; females are always wingless. The body is rather long and the forewings and hindwings often seem to be abnormally far apart. There is a pair of caudal cerci. These insects are mostly herbivores and have chewing mouthparts. *Embia*, the genus name on which the order name is based, refers to the fact that these insects are efficient in moving either forward or backward.

Order Dictyoptera

The Dictyoptera includes roaches and praying mantids (Figure 18.21). The hindwings are membranous and can be folded up like fans beneath the tougher forewings. The caudal cerci consist of numerous articles. In mantids, the forelegs are enlarged and specialized as

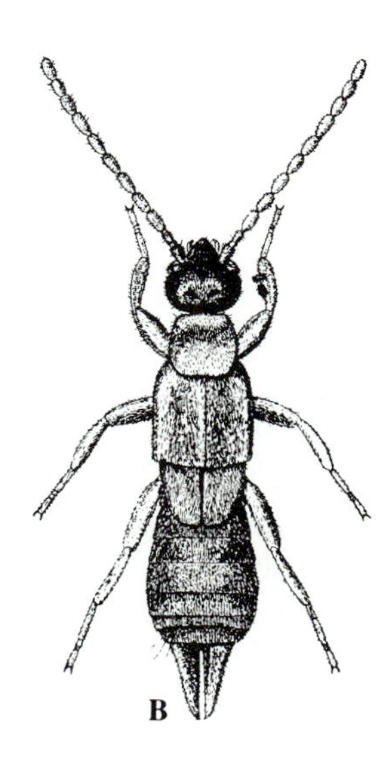

FIGURE 18.19 Order Dermaptera. A. *Forficula auricularia,* the European earwig, now common in North America. B. *Labia minor,* a small earwig native to North America. This species, which can fly rather well, is often seen on flowers or on piles of manure. (Ross, Illinois Natural History Survey, Circular 39.)

devices for seizing insects or other prey; in roaches, all three pairs of legs are substantially alike. The mouthparts are adapted for chewing. Most roaches are omnivorous, but some eat wood. These, like termites, depend on certain types of flagellate protozoans to convert cellulose into simpler compounds that they can utilize. Female roaches and mantids generally lay their eggs in a special case.

Roaches and mantids are sometimes placed in the order Orthoptera, along with grasshoppers, crickets, and their allies. Female dictyopterans do not, however, have well-developed ovipositors, and in neither sex are

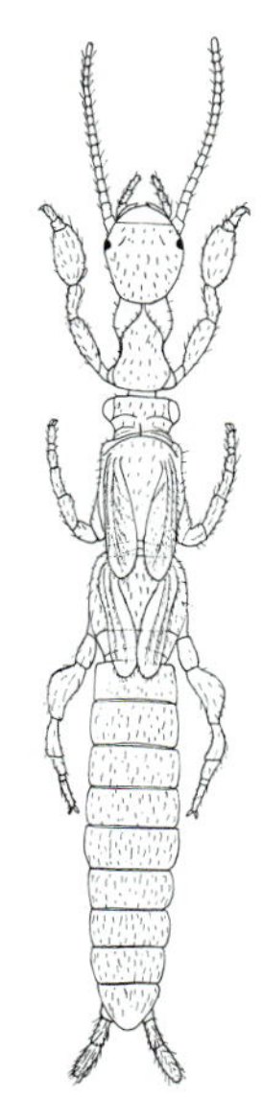

FIGURE 18.20 Order Embioptera. *Embia californica.* (Banks, Transactions of the American Entomological Society, *32.*)

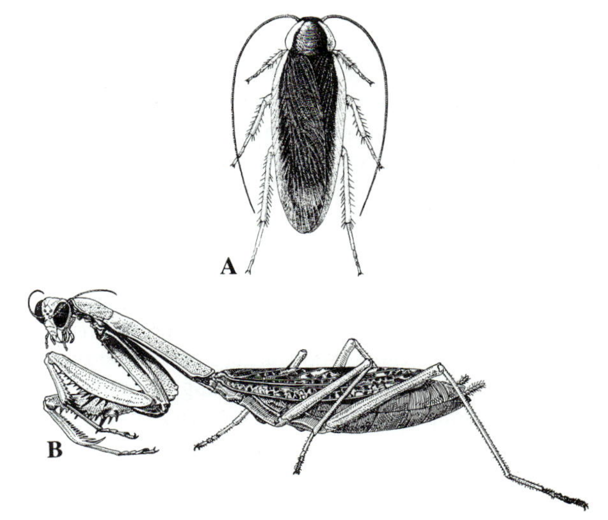

FIGURE 18.21 Order Dictyoptera. A. *Parcoblatta pennsylvanica,* a native North American cockroach. B. *Stagmomantis carolina,* a praying mantis. (Hebard, Illinois Natural History Survey Bulletin, *20.*)

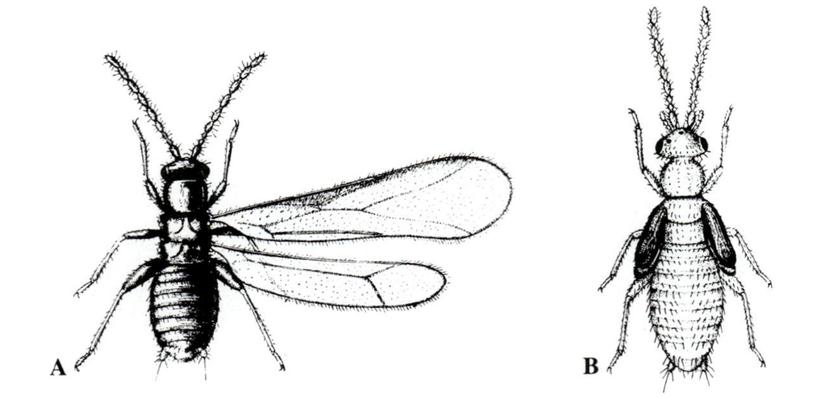

FIGURE 18.22 Order Zoraptera. *Zorotypus hubbardi*. A. Winged adult. (There are also wingless adults.) B. Nymph of winged form. (Caudell, Proceedings of the Entomological Society of Washington, *22*.)

there sound-producing organs or long hindlegs for jumping. The caudal cerci of true orthopterans, moreover, are not divided up into articles as they are in dictyopterans. Some specialists put roaches and mantids into separate orders.

Order Zoraptera

Zorapterans (Figure 18.22) represented by only a few species, are small insects restricted to humid tropical areas. They live in decaying wood, old termite nests, and comparable situations. Some are wingless, either because they do not form wings in the first place or because they drop them, as termites also do. Short caudal cerci are present on the abdomen. The mouth parts are adapted for chewing plant material.

Order Isoptera

Among the insects that have direct development, the termites ("white ants") (Figure 18.23) constitute the only group in which sociality is highly developed. The forewings and hindwings of termites are nearly equal (hence the name Isoptera), reinforced along their anterior edges, and easily shed when the flying reproductive individuals reach a place that is suitable for establishing a colony. The wings are simply broken off close to their bases, where each has a special fracture line.

In a typical termite colony, the "royal couple" occupies a special cell. The workers and soldiers are genetically males or females, but their reproductive organs are nonfunctional. The soldiers have a large head, and the mandibles, used for defending the colony, are huge. The nymphs can develop into any of the castes. In certain termites, such as those of the genus *Zootermopsis,* represented in the Pacific region of North America, there is no true worker caste. What appear to be workers are really nymphs. Some may in time become winged males or females; others cannot form wings but are nevertheless capable of becoming sexual individuals and taking the place of the founders of the colony if these should die or be destroyed.

Some termites consume only wood, in which they excavate complicated galleries that may be linked with nests in the soil below. Those that feed on living plant material build nests above ground by cementing together fecal material. The nutrition of termites depends upon certain flagellated protozoans and bacteria in the hindgut. Acetic acid is the principal product of cellulose decomposition that becomes available as a carbon source for the metabolism of termites. There are many other interesting aspects of the biology of termites, including the encouragement, by many species, of fungal growths. Presumably the fungi provide substances of nutritional value. Certain of the fungi attack the lignin in wood, thus liberating cellulose; they also alter the cellulose to some extent.

HEMIPTEROID ORDERS

Hemipteroid insects—the true bugs and their allies—lack abdominal cerci, and their wing venation is usually simple. As a rule, they have only a few Malpighian tubules. Except in the orders Psocoptera and Mallophaga, the mouthparts are specialized for sucking. Development is, as a rule, more nearly direct than indirect, but in certain orders the resemblance of nymphs to adults is slight. In the Thysanoptera, moreover, there are two or three inactive stages that are more or less comparable to pupae of endopterygote insects.

Order Psocoptera

Booklice and barklice (Figure 18.24) are small insects whose maxillae are specialized for scraping. In nature, they feed mostly on leaves or on lichens and algae that grow on bark. One species, *Trogium pulsatorium,* is a household pest that attacks the glue, and even the paper, in books. Wings, when present, are del-

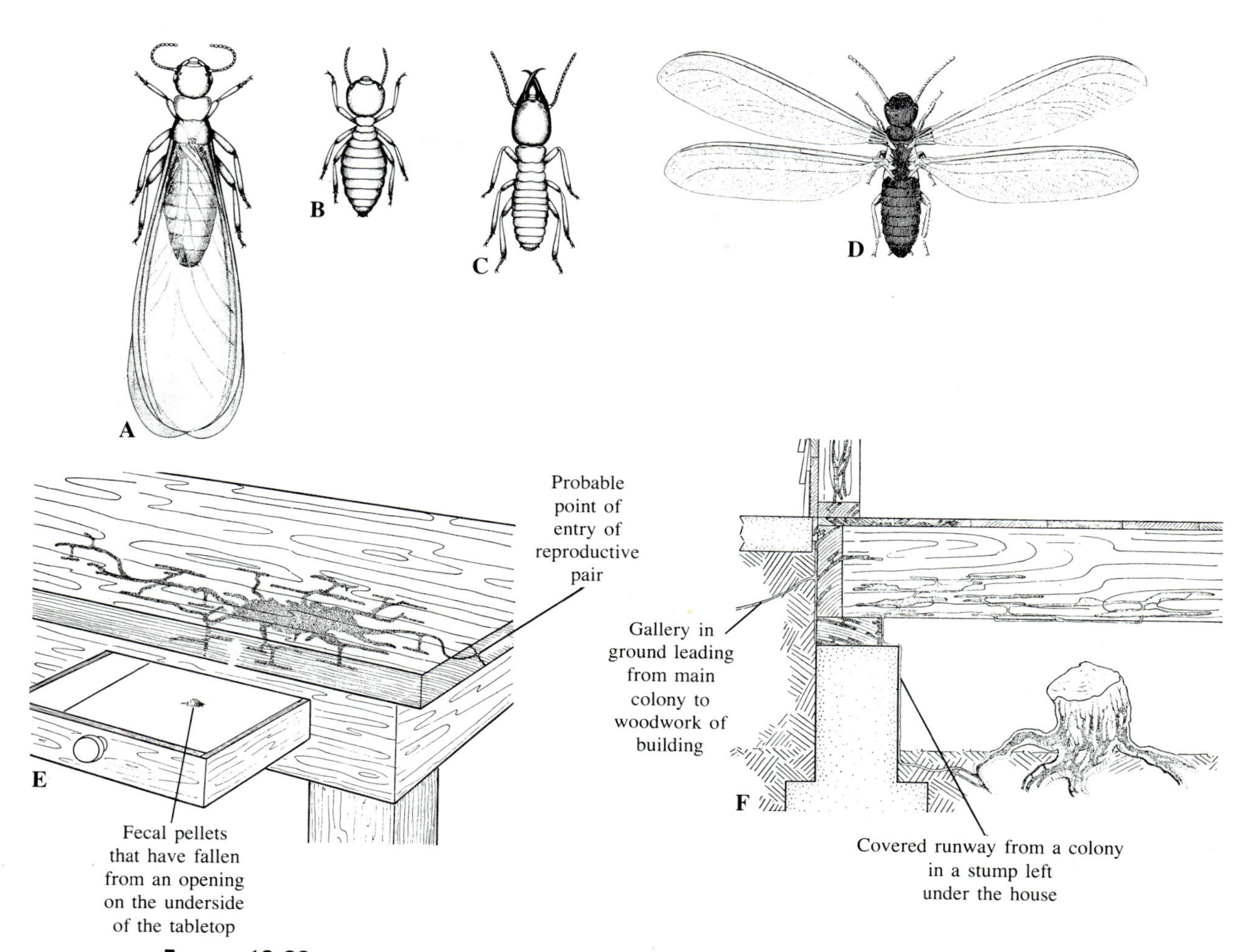

FIGURE 18.23 Order Isoptera. A–C. *Coptotermes acinaciformis,* an Australian termite. A. Winged reproductive individual. B. Worker. C. Soldier. (Gay, in The Insects of Australia.) D. Young queen of *Reticulitermes flavipes,* a common termite of eastern North America. (Ross, Illinois Natural History Survey, Circular 39.) E. Workings of the powder-post termite, *Cryptotermes brevis,* in a table top. F. Typical location of galleries and workings of the western subterranean termite, *Reticulitermes hesperus.* (E and F, Light, in Kofoid, Termites and Termite Control.)

icate and have sparse venation. At rest, they are folded straight back, forming a ''gable roof'' over the abdomen. There is a tendency for wings to be shortened or lost entirely. The clypeus—the sclerite on the lower part of the ''face,'' just above the labrum—is unusual in that it has a swollen appearance.

Order Mallophaga

Mallophagans (Figure 18.25) are called chewing lice. They are mostly found on the skin and among the feathers of birds, including domestic chickens, turkeys, and pigeons, but some live on mammals. They subsist largely on dead epidermal tissue, feathers, hair, and dried blood, but certain species will sometimes draw a little fresh blood. Their presence is often associated with skin irritations. These insects lack wings. The head is dorsoventrally flattened and as wide as, or wider than, the thorax. The antennae are rather far back on the sides of the head. The mouthparts are specialized for biting and chewing. Eyes are absent or vestigial, but sensory hairs are scattered over much of the body. The thoracic region is small in proportion to the head and abdomen. When the young hatch, they are similar to the adults.

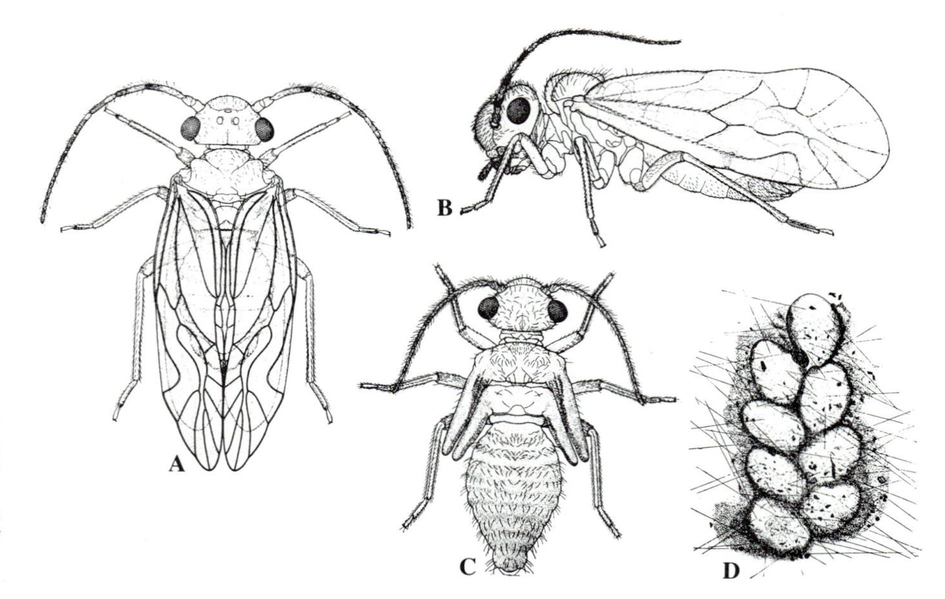

Figure 18.24 Order Psocoptera. *Anapsocus amabilis*. A. Male, dorsal view. B. Female, lateral view. C. Nymph (sixth instar). D. Eggs. (Sommermann, Annals of the Entomological Society of America, *37*.)

In some modern systems of insect classification, the Mallophaga and Anoplura (see below) are united into one order, Phthiraptera.

Order Anoplura

Anoplurans, called sucking lice, slightly resemble mallophagans, partly because of their small size and general body form, and partly because they are wingless. The heads of anoplurans are proportionately smaller and narrower, however, than those of mallophagans. The mouthparts, which can be retracted, are specialized for piercing skin and sucking up blood. These insects live on mammals. Two species are annoyances to humans. *Pediculus humanus* (Figure 18.26) is the head and body louse; *Phthirus pubis,* the crab louse, lives in pubic hair. Both species irritate the skin and both can transmit the rickettsial organism that causes typhus and also the spirochaete that causes relapsing fe-

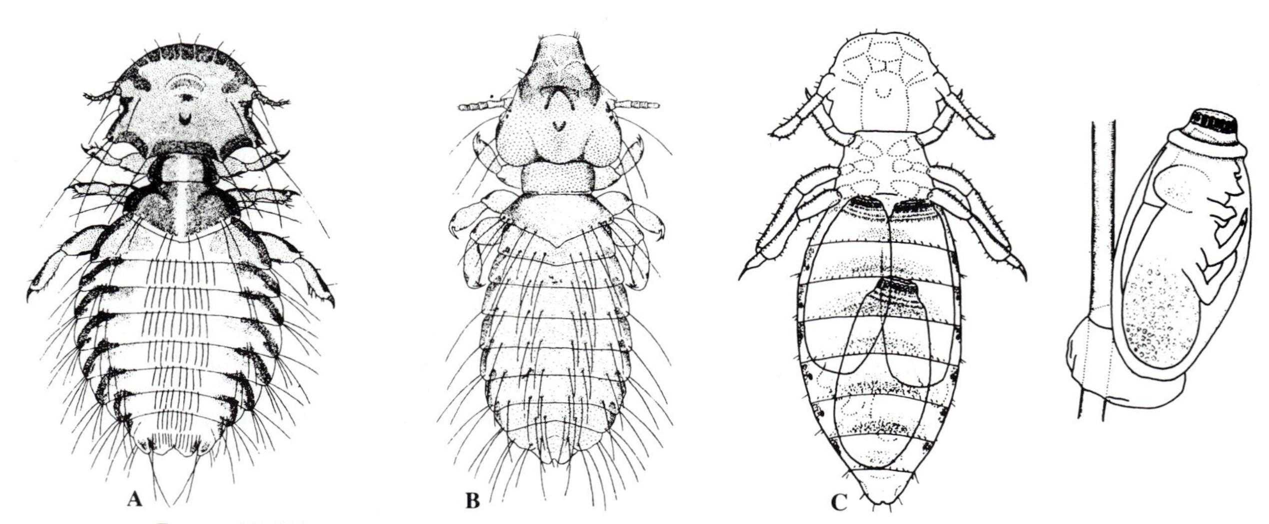

Figure 18.25 Order Mallophaga. A. *Goniodes*. B. *Docophorus*. (Kellogg, Proceedings of the California Academy of Sciences, series 2, *6*.) C. *Damalinia longicornis,* female with eggs (left) and an embryonated egg attached to a hair of a deer (right). (Andrews, Transactions of the Royal Society of New Zealand, Zoology, *5*.)

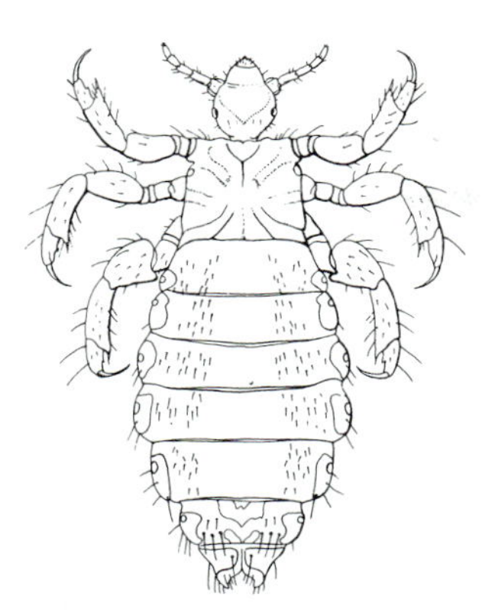

FIGURE 18.26 Order Anoplura. *Pediculus humanus humanus,* the body louse, female, dorsal view. (Seguy, Faune de France, *43*.)

ver. Other lice of this general sort affect dogs and other household or barnyard mammals. The abdomen of a louse is soft and almost like an accordion, so that it can become greatly distended while the insect consumes a hearty meal. On hatching, the young are substantially like the adults.

Order Hemiptera (Heteroptera)

Hemipterans are the true bugs (Figure 18.27). As insects go, most bugs are of moderate to large size. They are greatly diversified, but nevertheless unified by the following features: the labium is a snoutlike structure, often distinctly jointed, that partly envelops the slender mandibles and first maxillae. These mouthparts are specialized for piercing. The maxillae, moreover, have two lengthwise grooves, so when they are pressed together, two tubelike channels are formed. One delivers saliva; the other serves to draw up food. The proximal portions of the forewings are hardened, but the distal portions are membranous. When a bug is at rest, the membranous portion of one forewing covers that of the other. The hindwings are completely membranous.

Most bugs feed on plant juices. Some, however, attack invertebrates, including other insects, or eggs of invertebrates. There are also types that suck blood from vertebrates. A number of diseases of plants and animals are transmitted by bugs. Species of *Triatoma* and related genera, for instance, are the vectors of *Trypanosoma cruzi,* the flagellate protozoan that causes Chagas' disease.

In some systems of classification, the definition of Hemiptera is broadened to include the cicadas, aphids, leafhoppers, and their allies. The tendency today is to place these in a separate order, Homoptera (see below). When this is done, the name Hemiptera may be conserved for the true bugs (as is done here) or replaced by Heteroptera.

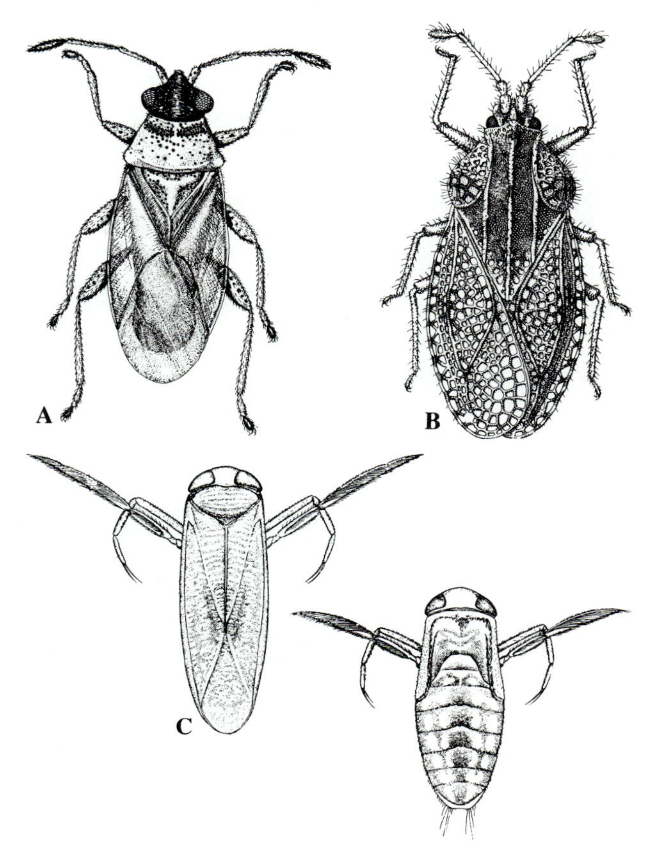

FIGURE 18.27 Order Hemiptera. A. *Oceanides*. (Ashlock, University of California Publications in Entomology, *48*.) B. *Onymochila*. (Drake, United States National Museum, Bulletin 243.) C. *Autocorixa* (fresh water), adult (left) and nymph (right). Hungerford, Kansas University Science Bulletin, *11*.)

Order Homoptera

The mouthparts of cicadas, leafhoppers, aphids, and scale insects are similar to those of bugs. These mouthparts are not as far forward on the head as they are in bugs, but the arrangement of the labium, mandibles, and first maxillae is similar. Wings, when present, are of the same general consistency throughout: firm but transparent. When a homopteran's wings are folded back, they are not crossed, but slope away from the midline like the two sides of a gable roof.

The cicadas (Figure 18.28, A) are moderately large insects whose nymphs live in soil, usually for at least several years, and feed on roots. The seventeen-year locust of North America does in fact live in the ground for 17 years before emerging and molting for the last time. Rather close to cicadas are the small spittle bugs (Figure 18.28, B), whose nymphs are protected by a frothy slime ejected from the anus. Leafhoppers,

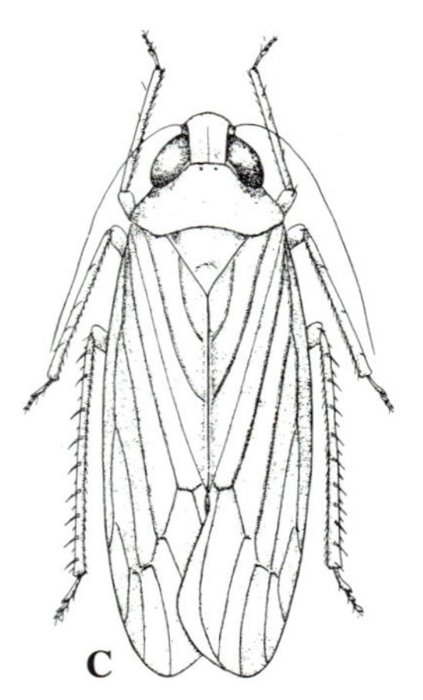

FIGURE 18.28 Order Homoptera. A. Cicada (family Cicadidae.) B. Spittlebug (family Cercopidae). C. Leafhopper (*Jassus,* family Cicadellidae). (Osborn, Bernice P. Bishop Museum, Bulletin 113.)

aphids, cochineal insects, scale insects and their allies constitute a varied assemblage of homopterans, some of which exhibit bizarre specializations. In general, the members of this group are characterized by proportionately longer antennae than are typical of cicadas and spittle bugs.

Most leafhoppers (Figure 18.28, C) can jump as well as fly, but as nymphs they remain attached to plants. Aphids (Figure 18.29, A), of great importance as pests of ornamental and food plants, have a complex biology. In general, the eggs laid in autumn by sexual individuals hatch in the spring, giving rise to females (usually wingless) that reproduce parthenogenetically and bear their young alive. The nymphs become parthenogenetic females just like their mothers, and in time many additional generations may be produced. In some species, winged females are formed sooner or later, and these can move to other plants and start new colonies. Curiously enough, the wingless individuals in the new colonies may differ in certain respects from wingless individuals developing on the original plant host. By the next autumn, some males (often winged) and sexual females (wingless) are produced. After mating, the females lay the overwintering eggs that hatch into the first generation of parthenogenetic females. The phylloxerans are similar to aphids, but lay eggs, even in the parthenogenetic female phase, and some of them have stages that live on roots. There are other complications in their life cycles, too.

Scale insects (Figure 18.29, B and C), many of which are serious pests of citrus trees and other plants, are remarkably modified. They are generally tightly affixed to the host plant, and may show considerable regression of the appendages, even the mouthparts. The nymphs cover themselves with waxy "houses" that look like fish scales or limpet shells. Wings, if present, are restricted to males, and there is only the anterior pair. In the cochineal insects, the females cover themselves with a waxy or gummy substance and lay their eggs inside the mass of secreted material. Parthenogenesis is common in scale insects and cochineal insects; males, in fact, are unknown in many species.

In spite of the damage scale insects and cochineal insects do to plants, a few species are commercially valuable. Shellac is the gummy secretion of certain cochineal insects. Some kinds produce wax. One species, living on particular cacti in Mexico, is the source of a red dye called cochineal. (The coloring principle is carminic acid.) The "manna" mentioned in the Bible was almost certainly the accumulation of a sweet secretion of a cochineal insect.

Order Thysanoptera

Thysanopterans (Figure 18.30), called thrips, are small insects whose mouthparts are specialized for

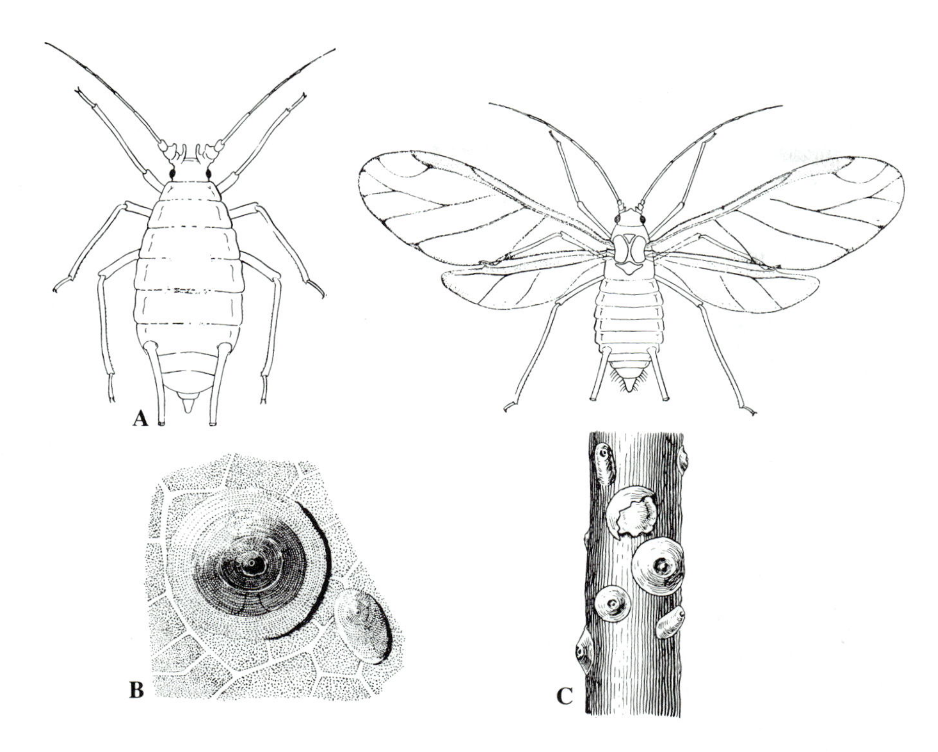

FIGURE 18.29 Order Homoptera. *Phorodon humuli,* the hop aphid (family Aphididae), wingless viviparous female (left) and winged migrant female (right). (Parker, United States Bureau of Entomology, Bulletin 111.) B. *Aonidiella aurantii,* an armored scale insect (family Diaspididae). (McKenzie, Microentomology, *3.*) C. *Quadraspidiotus perniciosus,* the San Jose scale (family Diaspididae). (Ross, Illinois Natural History Survey, Circular 39.)

piercing cells and sucking up their contents. Most of them feed on plants, and many species are pests of garden flowers and farm crops. Some thrips attack other soft-bodied insects. When well-developed wings are present, they are narrow and have a fringe of hairs anteriorly and posteriorly, so that they look like feathers. Reduction or absence of wings is characteristic of certain groups. Compound eyes are usually prominent, and there are also simple eyes. The last two nymphal stages, though characterized by external wings, are inactive and thus essentially comparable to pupae of insects that have direct development. In some thysanopterans, there is another, earlier inactive stage, although this one does not have wings.

ENDOPTERYGOTE ORDERS

All of the endopterygote orders have indirect (holometabolous) development. In other words, there is a series of larval stages, followed by a pupal stage during which metamorphosis to the adult form is completed.

Order Neuroptera

Ant lions, lacewings, alder flies, dobson flies, and their relatives generally have proportionately large wings with many cross-veins. When folded back, the wings form a ''gable roof'' over the abdomen. The antennae are long and the compound eyes are prominent. Adult neuropterans have biting and chewing mouthparts and are carnivores. Some, however, are so short-lived that they hardly have a chance to eat.

The two main subdivisions of the Neuroptera have rather different lifestyles. The larvae of alder flies (Figure 18.31, A) and dobson flies are found in fresh water and have chewing mouthparts for feeding on aquatic insects and small worms. They are also characterized by

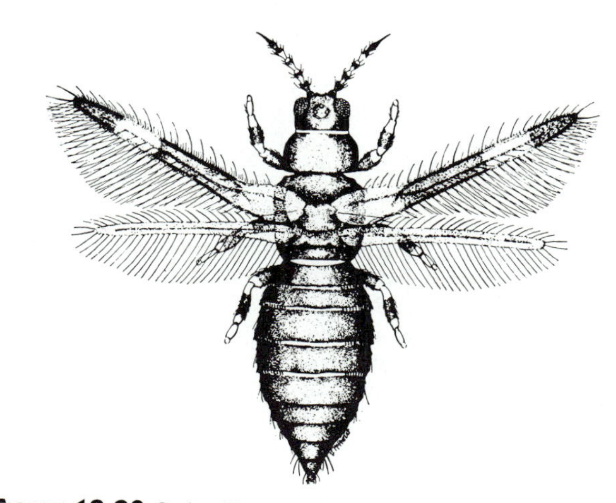

FIGURE 18.30 Order Thysanoptera. *Heliothrips fasciatus,* the bean thrips. (Essig and Hoskins, California Agricultural Extension Service, Circular 87.)

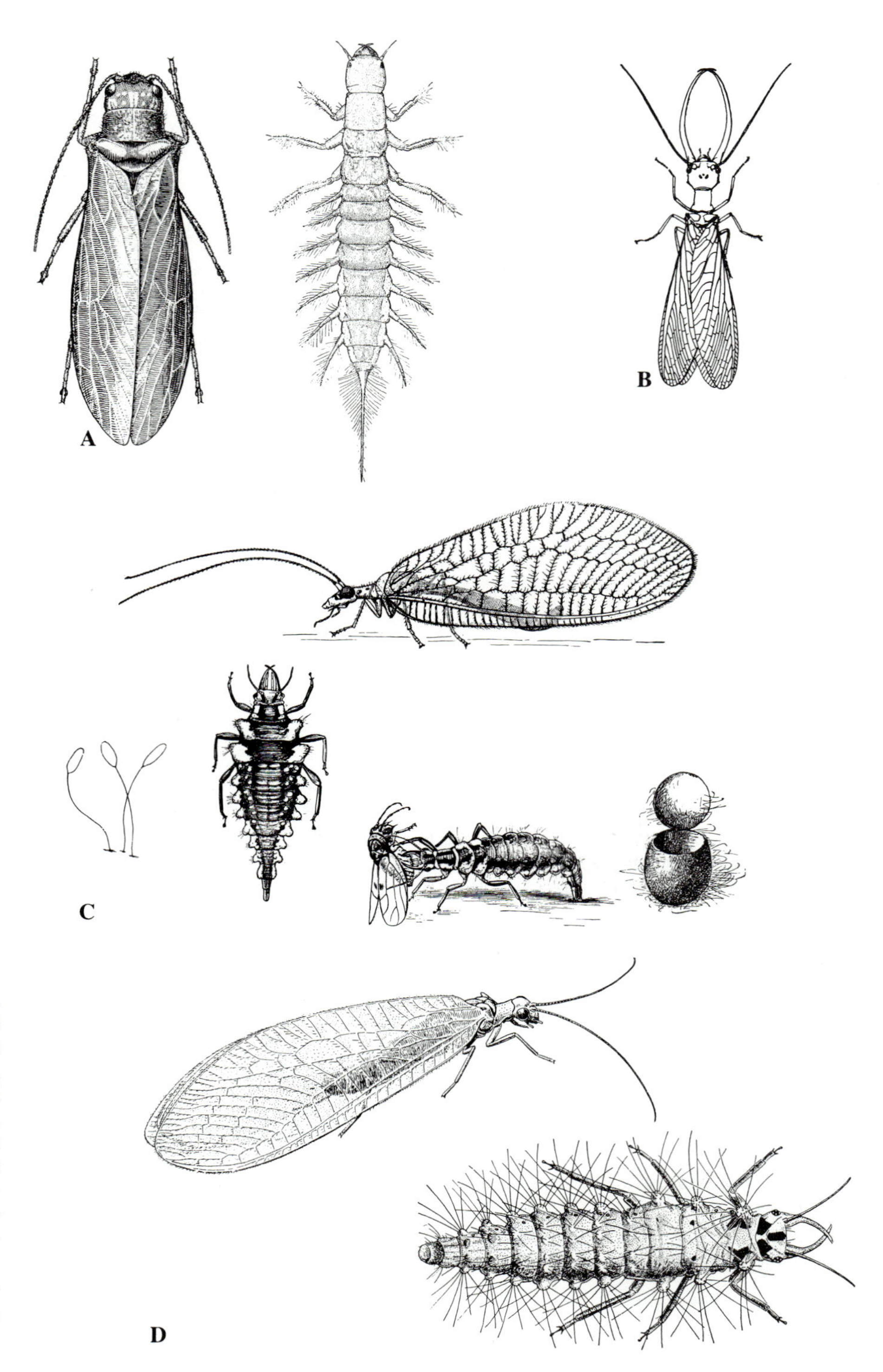

Figure 18.31 Order Neuroptera. A. *Sialis,* an alder fly, adult (left) and larva (right). (Ross, Illinois Natural History Survey Bulletin, *21*.) B. A dobson fly. (United States Yearbook of Agriculture, 1952.) C. An antlion, adult (above) and stalked eggs, larvae (one feeding on a fly it has captured), and empty pupal case (below). (Forbes, Illinois State Laboratory of Natural History, Bulletin 11.) D. *Chrysopa,* a lacewing, adult (left) and larva (right).

prominent abdominal appendages that function as tracheal gills. These larvae generally remain in the water for 2 or 3 years. Adult dobson flies (Figure 18.31, B) have a fearsome look because of their large mandibles.

The larvae of ant lions and lacewings (Figure 18.31, C and D) are terrestrial and feed primarily on insects such as aphids and ants. Some invade cocoons of spiders and consume the eggs, then use the cocoons for

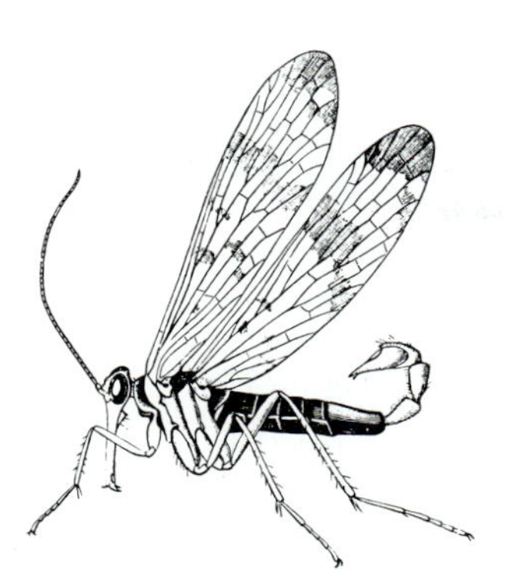

FIGURE 18.32 Order Mecoptera. *Panorpa communis,* a scorpion fly. (Sharp, in Cambridge Natural History, *5.*)

their own pupal stage. In general, the mouthparts are adapted for sucking juices, but the mandibles are often much enlarged for grabbing prey. This is the case in ant lions (''doodle bugs''). These insects, while larvae, live in sandy soil and dig craters that serve as traps into which other insects can tumble. As adults, some members of this terrestrial group of Neuroptera have marvelously developed first legs that are subchelate and resemble those of praying mantids. Lacewings pass their larval stages on plants and specialize in consuming aphids. They are thus friends of gardeners and farmers.

Order Mecoptera

Mecopterans (Figure 18.32) are called scorpion flies because of the way the males of some species hold the slender abdomen in an upcurved position. Adults have long antennae and prominent compound eyes; the mouthparts, adapted for chewing up small prey, are situated on a snoutlike extension of the head. The wings are rather large. The ''forceps'' at the tip of the abdomen of males consists of a pair of structures used in copulation. In general, scorpion flies are carnivores that feed on other insects. The larvae have well-developed thoracic legs, plus abdominal prolegs. Like adults, they consume animal matter, mostly insects. Pupation is accompanied by the secretion of a cocoon.

Order Lepidoptera

The well-developed wings of moths and butterflies (Figures 18.33 and 18.34; Color Plate 7) are characterized by microscopic scales that overlap the way roof shingles do. The body also has scales, and it is often densely hairy. The antennae are generally long and the compound eyes large. The mouthparts of adult lepidopterans are specialized for sucking up nectar and water. The first maxillae are fused together to form a tube that usually coils up when it is not in use (Figure 18.3, F); the mandibles are not functional except in a relatively few species.

The life cycles of lepidopterans provide some of the classic examples of complete metamorphosis. The eggs hatch into larvae everyone knows as caterpillars. These have three pairs of short legs in the thoracic region, and prolegs on the abdomen; the prolegs are characterized by several to many little hooks at their tips. Other specializations of the larvae are chewing mouthparts and silk glands whose openings are on the labium. The silk is used as a dragline, to make a cocoon within which the pupal stage is confined, or to attach the pupa to a twig or some other firm object.

Most butterfly and moth larvae are herbivorous, feeding on green leaves or on dried plant material, such

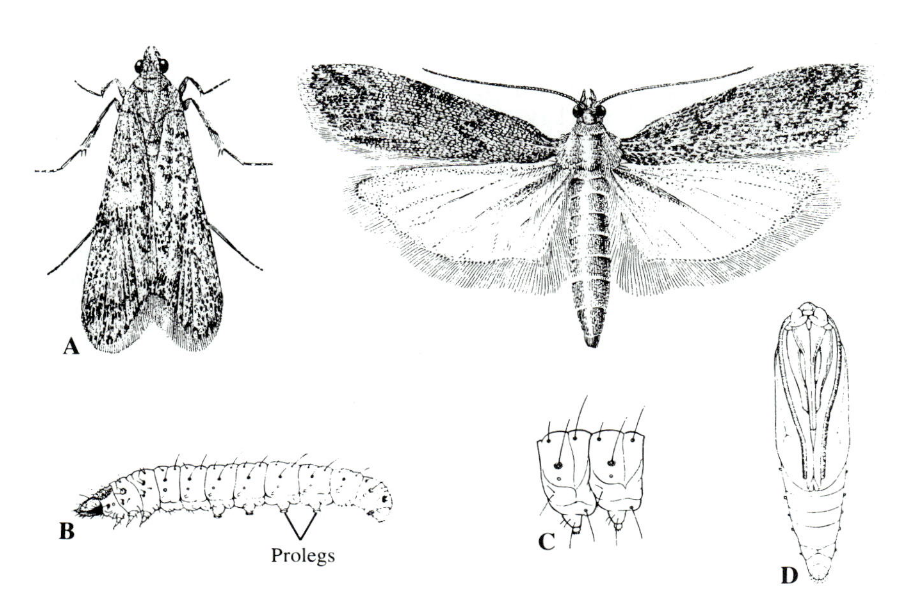

FIGURE 18.33 Order Lepidoptera. The Mediterranean flour moth, *Ephestia kuhniella.* A. Adult, wings folded (left) and spread (right). B. Larva. C. Two segments of the larva in the region of the prolegs. D. Pupa. (Essig and Hoskins, California Agricultural Extension Service, Circular 87.)

FIGURE 18.34 Order Lepidoptera. *Manduca sexta,* the tobacco hornworm. A. Adult moth. B. Eggs. C. Larva. D. Pupae. (Courtesy of Carolina Biological Supply Company, Inc.)

as grains and cereals. Many species do great damage to horticultural and agricultural crops and to stored food products. A small number are carnivores or scavengers on animal matter. The most famous lepidopteran consumer of animal products is the clothes moth, *Tinea pellionella*. Its larvae attack woolen goods, fur, and even feathers. The caterpillars of *Feniseca tarquinius,* a small butterfly of eastern North America, feed on aphids. The larvae of the bee moth, *Galleria mellonella,* eat the waxy honeycombs in beehives.

Order Trichoptera

Adult caddis flies (Figure 18.35, A) have rather large but delicate wings covered with closely spaced hairs; hence the name Trichoptera. At rest, the wings are held straight back and form a "gable roof" over the abdomen. The antennae are long and compound eyes are well developed. The maxillae are specialized for lapping up liquids; the mandibles are vestigial. These insects do not often stray far from fresh water, where the larvae develop. The larvae have chewing mouthparts, which they use in carnivory, herbivory, or scavenging. Many of them are more or less omnivorous. The larvae are especially interesting in that they usually construct mobile homes, using bits of plant material, sand, or gravel (Figure 18.35, B); the pieces are held together by silk produced by glands whose openings are close to the mouth.

Some caddis-fly larvae wander about naked, and some attach their silken shelters to sticks or stones. A few that live in swift streams make nets for catching prey. Caddis flies lay their eggs on leaves, stems, or sticks that are submerged or at least close to the water's edge. The larvae, with three pairs of legs and a pair of caudal appendages, generally have tracheal gills on all or most segments behind the head. Pupation involves

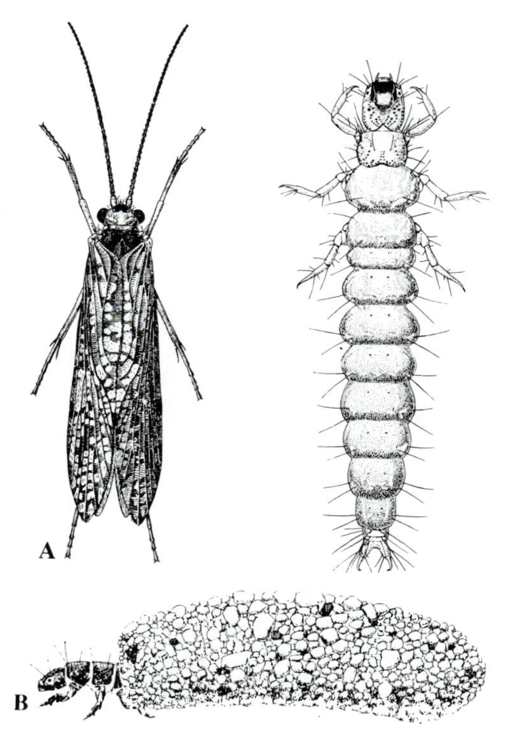

FIGURE 18.35 Order Trichoptera. A. *Rhyacophila fenestra,* adult female (left) and larva (right). (Ross, Illinois Natural History Survey Bulletin, *21*.) B. Larva and "house" of *Ochrotrichia unio*. (Ross, Illinois Natural History Survey Bulletin, *23*.)

the formation of a silken case, which may be portable or attached. Some species spin a cocoon within a mound or enclosure of gravel. The pupa, with well-developed antennae and legs, and with fairly large wing cases, already bears some resemblance to the adult. It works its way out of its case and comes out of the water before shedding the pupal skin.

Order Diptera

Flies, mosquitoes, and their allies (Figure 18.36) are generally easy to recognize because only their forewings are well developed. The hindwings are reduced to small, knobbed structures called *halteres* or "balancers"; these beat with a frequency close to that of the forewings, and they serve to stabilize flight. The mouthparts are specialized for sponging, sucking, or for both piercing and sucking. In larvae, there are no jointed legs, and the head is often rather poorly differentiated. At pupation, the cuticle of the last larval stage may be retained as a hard case called a *puparium*.

This is a huge order, with about 100,000 species. Some—including mosquitoes, crane flies, midges, gnats, and blackflies—have aquatic larvae, or larvae that inhabit very wet places. Dipteran larvae are present, however, in most terrestrial habitats where life is possible. Soil, living or dead plant material, carcasses of animals, dung, and garbage have their characteristic species. Some live in the stomachs of mammals and others are found beneath the skin.

Adult dipterans also have many lifestyles. Their food habits, for instance, range from the sipping of flower nectar to eating other insects (caught in flight) and sucking blood, as female mosquitoes do. The diversity of mouthparts found among dipterans is truly remarkable. A considerable number of species are involved in transmission of disease, either because they contaminate food with microorganisms picked up from fecal matter or because they are true biological vectors. The trypanosomes that cause African sleeping sickness are transmitted to humans and other mammals by way of the salivary secretions of tsetse flies. Malarial organisms are transmitted with the bite of female mosquitoes.

Order Siphonaptera

Fleas (Figure 18.37) are wingless insects with tough, laterally flattened bodies. Their long legs are specialized for jumping and their mouthparts are modified for piercing skin and drawing blood from mammals and birds. Eyes, if present, are simple ocelli; the antennae are short and lie in grooves. Some species lay their eggs on the vertebrate host, in which case the young generally feed on dead skin and comparable material; others lay their eggs off the host, in situations where the larvae will have access to organic matter. Flea larvae are legless and their mouthparts are adapted for chewing. The lifestyle changes drastically at metamorphosis, when the blood-sucking phase begins. The pupa is enclosed within a silken cocoon.

Fleas would be bad enough if they did nothing more than cause skin irritations. Unfortunately, some of them are vectors of serious diseases. *Xenopsylla cheopis,* a flea that commonly lives on rats and mice, transmits *Pasteurella pestis,* the bacterium that causes bubonic plague; it is also a vector for the rickettsial organism responsible for murine typhus. Several species of fleas are intermediate hosts for *Dipylidium caninum*, a tapeworm that is common in dogs and cats, and that occasionally parasitizes humans, especially children. Infection occurs when a flea containing the cysticercoid stage is swallowed.

Order Hymenoptera

Bees, wasps, ants, and their relatives (Figures 18.38, 18.39) constitute a large and diversified group of insects. Wings are not always present in adults, but when they are, the forewings and hindwings are linked during flight. The mouthparts are adapted for biting and chewing or for lapping up nectar. In many hymenopterans, the ovipositor of females is specialized for sting-

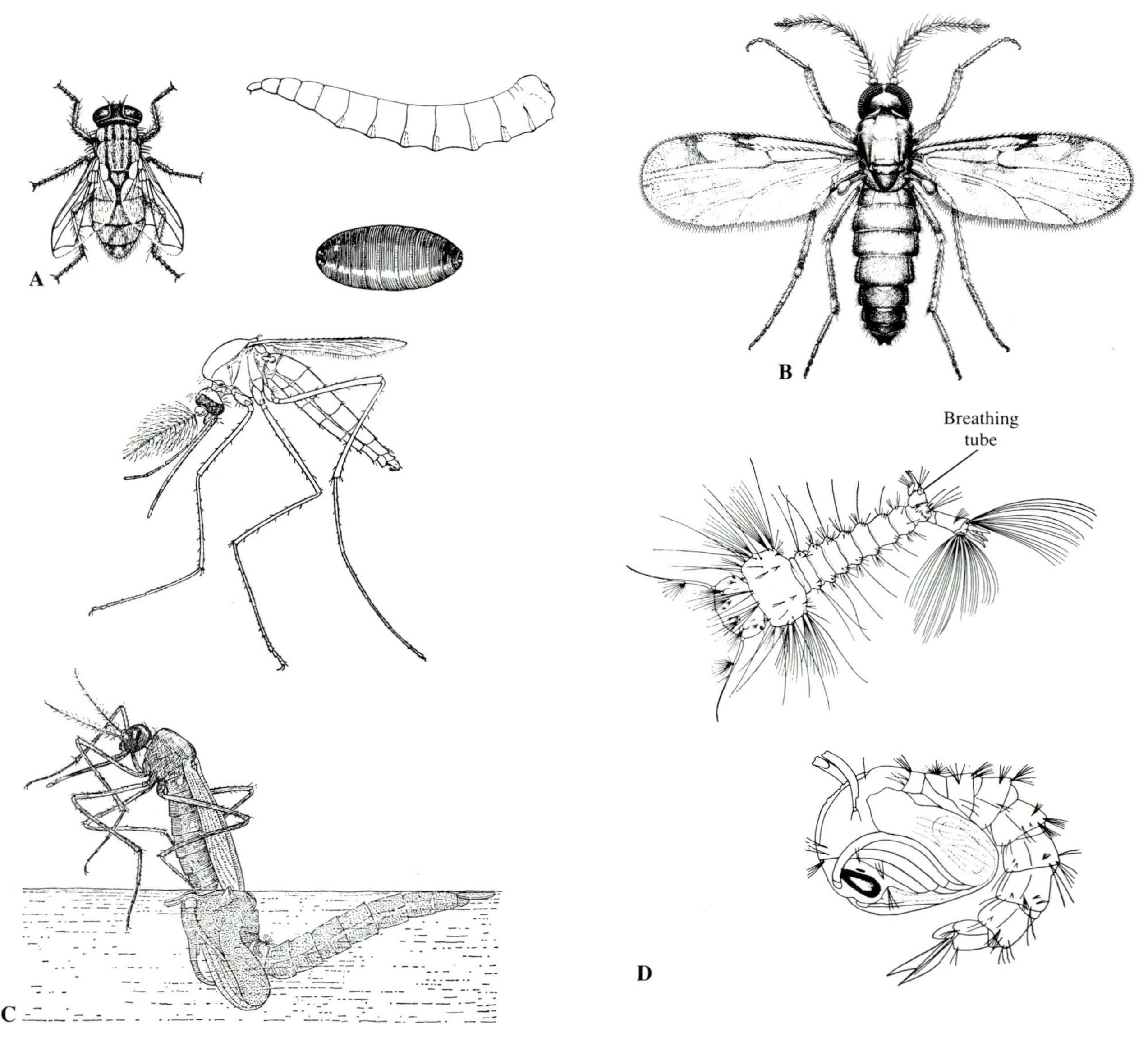

Figure 18.36 Order Diptera. A. *Musca domestica,* the house fly, adult, larva (above), and pupa (below). (Ross, Illinois Natural History Survey, Circular 39.) B. *Culicoides diabolicus,* a biting midge. C. Mosquito, *Aedes aegypti,* adult female (above) and female emerging from pupal case (below). (Snodgrass, Smithsonian Miscellaneous Collections, *139*.) D. Larva (above) and pupa (below) of a mosquito. (Ross, Illinois Natural History Survey Bulletin, *24*.)

ing, either offensively or defensively, or both. In some it is used for sawing a hole in wood or soft plant tissue, or for piercing another insect or some other animal within which eggs will be laid. The larvae of hymenopterans usually spin cocoons as they prepare to pupate.

There are two major groups within the order. In bees, ants, and true wasps, the abdomen is set off from the thorax by a sharp constriction, and the larvae do not have thoracic legs or abdominal prolegs. In wood-wasps, horntails, and sawflies, the body does not have a narrow "waist," and the larvae have both legs and prolegs. The prolegs, however, lack the little hooks that are characteristic of the prolegs of lepidopterans. Female wood-wasps and horntails (Figure 18.39, C) have long ovipositors for boring holes in live trees. The larvae feed on wood. The ovipositor of sawflies and some

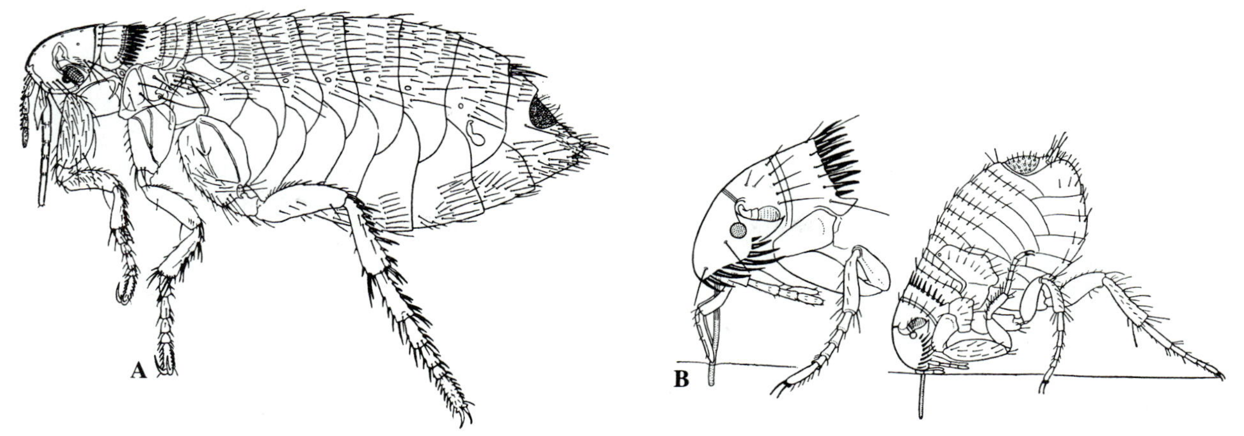

FIGURE 18.37 Order Siphonaptera. A. *Oropsylla silantiewi,* female. This species lives on marmots. (Vysotskaia, Short Guide to Fleas.) B. *Ctenocephalides felis,* the cat flea, puncturing the skin and after penetration has been completed. At the time the stylets are inserted (left), the labium is bent forward so that the stylets are in the notch between its lobes. After the stylets have fully penetrated the skin (right), the labium is bent backward and the labial palps are pressed flat against the skin. (Snodgrass, Smithsonian Miscellaneous Collections, *104*.)

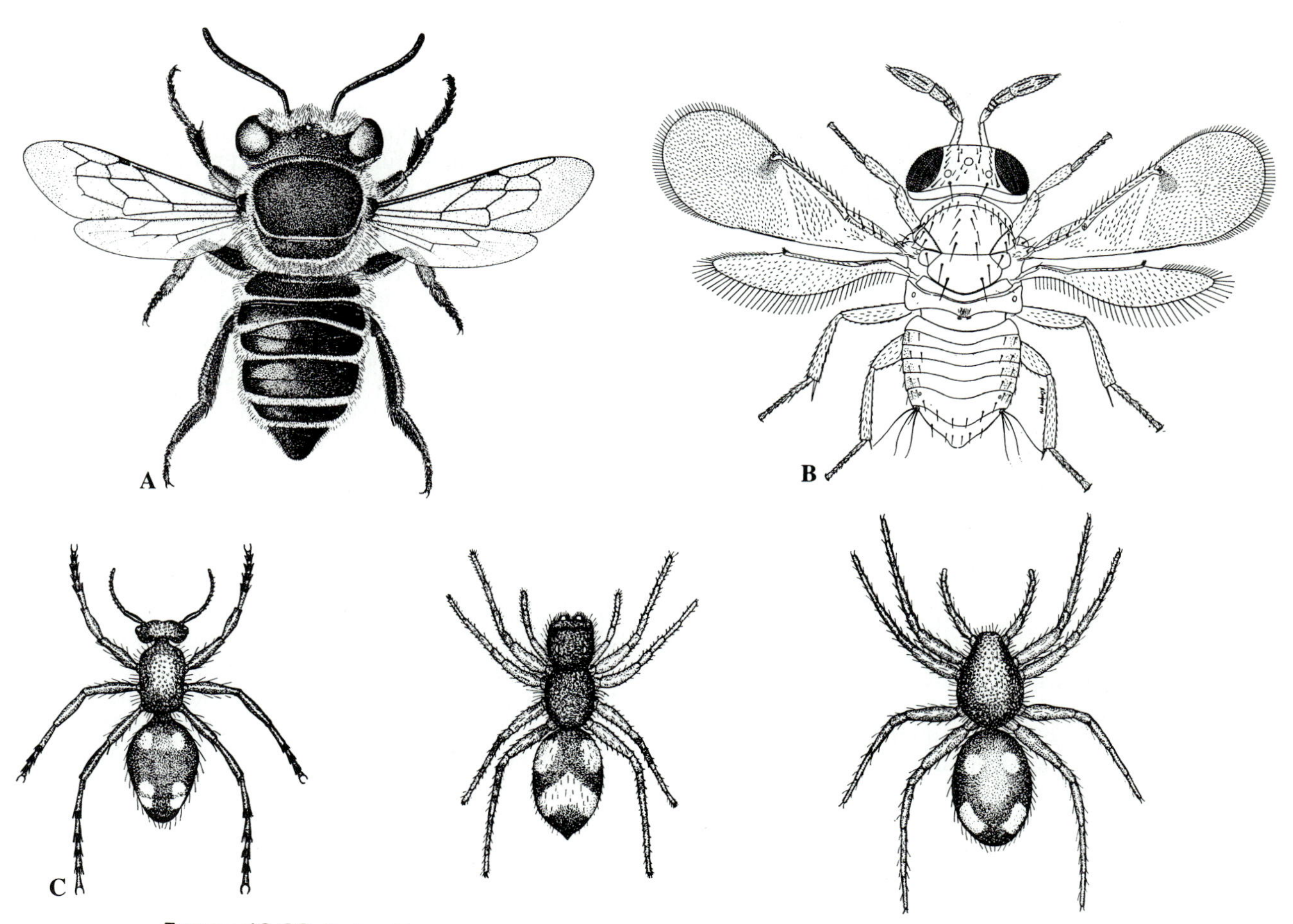

FIGURE 18.38 Order Hymenoptera. A. *Megachile manago,* a bee. (Yasumatsu, Annotationes Zoologicae Japonenses, *1*.) B. *Aphytis cercinus,* a parasite of scale insects. (Compere, University of California Publications in Entomology, *10*.) C. *Glossotilla liopyga,* a wingless African wasp (left) that is mimicked by two species of spiders (right). (Bristowe, The Comity of Spiders.)

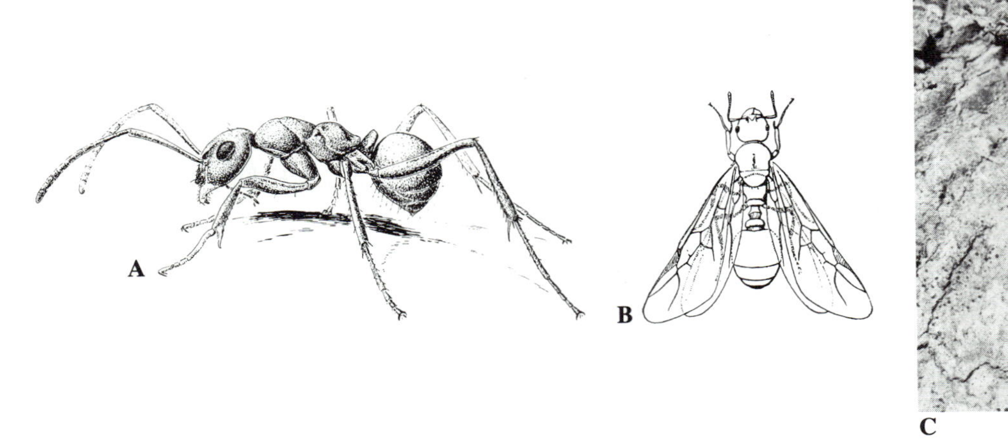

FIGURE 18.39 Order Hymenoptera. A. A worker ant (wingless) of the genus *Formica*. (Ross, Illinois Natural History Survey, Circular 39.) B. A winged ant. (Muesbeck, in the United States Yearbook of Agriculture. 1952.) C. An adult horntail (family Siricidae).

related insects operates as a kind of saw for slicing into softer plant tissue. Their larvae, known as "rose slugs" or "cherry slugs," generally burrow into the tissue.

There is a great diversity of lifestyles found among the thousands of species of bees, ants, and wasps. Many of these hymenopterans are social insects, with complicated colonies and several castes specialized to perform certain functions. Honeybees and bumblebees are among the classic examples of social insects. Some aspects of the biology of honeybees was discussed earlier in this chapter.

Ants (Figure 18.39, A and B) are also social, with queens, males, and wingless female workers; some species even have secondary castes. Most ants are scavengers that consume plant material, but some attack live plant tissues, and some are carnivorous. Among the many interesting species are those that maintain colonies of aphids, feeding on the "honeydew" that these insects produce. There are also ants that enslave workers of other species.

Hymenopterans with carnivorous larvae include many kinds of wasps, which bring insects or spiders that have been killed or paralyzed back to their nests. Some species deposit their eggs on insects and spiders; the larvae gradually consume their hosts. Certain wasps of this general type are helpful because they attack troublesome insect pests.

Order Coleoptera

There are over 300,000 species of beetles (Figures 18.40, 18.41), more than in any other order of insects—more, in fact, than in any phylum outside the Arthropoda. They occupy many terrestrial and freshwater habitats, and they are among the relatively few groups of insects that are successful at the fringes of the sea. Most beetles are firm-bodied because of a fairly hard cuticle. The forewings are modified into structures called **elytra** (singular, *elytron*). They are thick and veinless, and their function is to protect the delicate hindwings, which are used for flying. When held straight back, the elytra meet along the dorsal midline. In some beetles the elytra and hindwings are much reduced, and the hindwings may be lacking altogether. The form of the antennae is variable, but these appendages are generally prominent and sometimes they are elaborate. The mouthparts are specialized for biting and chewing.

Larvae of beetles (Figure 18.40) typically have well-developed heads and three pairs of thoracic legs; some, however, are legless. The pupae usually have a fairly soft cuticle, so they may wiggle when poked.

There is scarcely any animal or plant material that is not eaten by some kind of larval or adult beetle. Many species, including the weevils, are herbivores. Adult weevils (Figure 18.41, D) have their mouthparts on a snoutlike extension of the head and feed mostly on foliage. Their larvae, which are legless or have "false legs," burrow into tissues, eat roots, and inflict various other damage on plants. Carnivory is characteristic of several large groups. The terrestrial beetles known as cicindelids and carabids, and the freshwater dytiscids and gyrinids, are predators on various invertebrates, including other insects. The dytiscids are especially interesting because their mandibles are specialized for delivering a salivary secretion that contains not only a venom, but also proteolytic enzymes that initiate digestion outside the body. Among the distinctive features of gyrinids is the fact that they carry a coating of air around themselves when they dive from the surface.

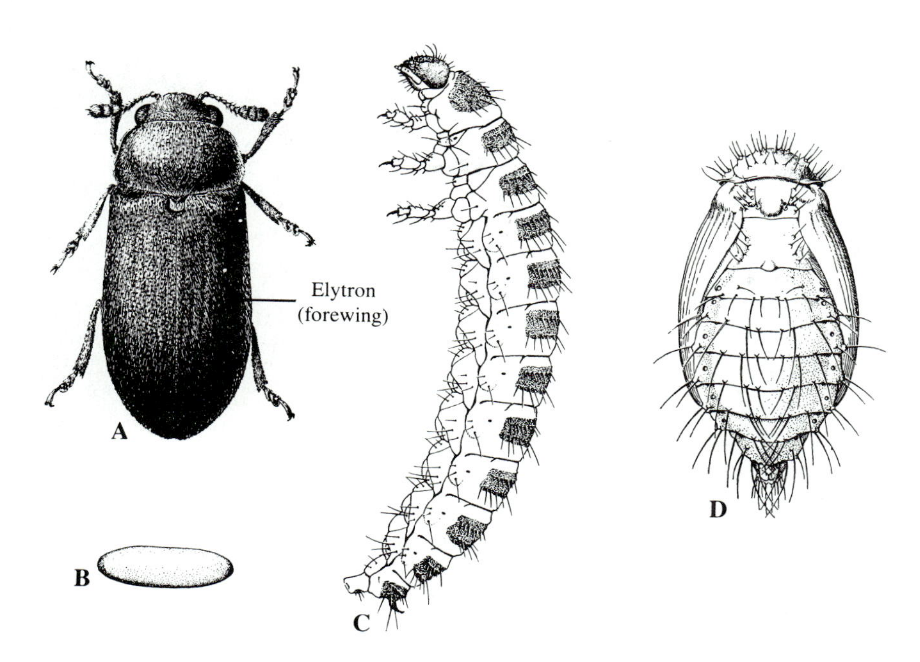

FIGURE 18.40 Order Coleoptera. *Byturus bakeri,* the western raspberry fruitworm (family Byturidae). A. Adult. B. Egg. C. Larva. D. Pupa. (Baker, Crumb, Landis, and Wilcox, Institute of Agricultural Sciences, State College of Washington, Bulletin 497.)

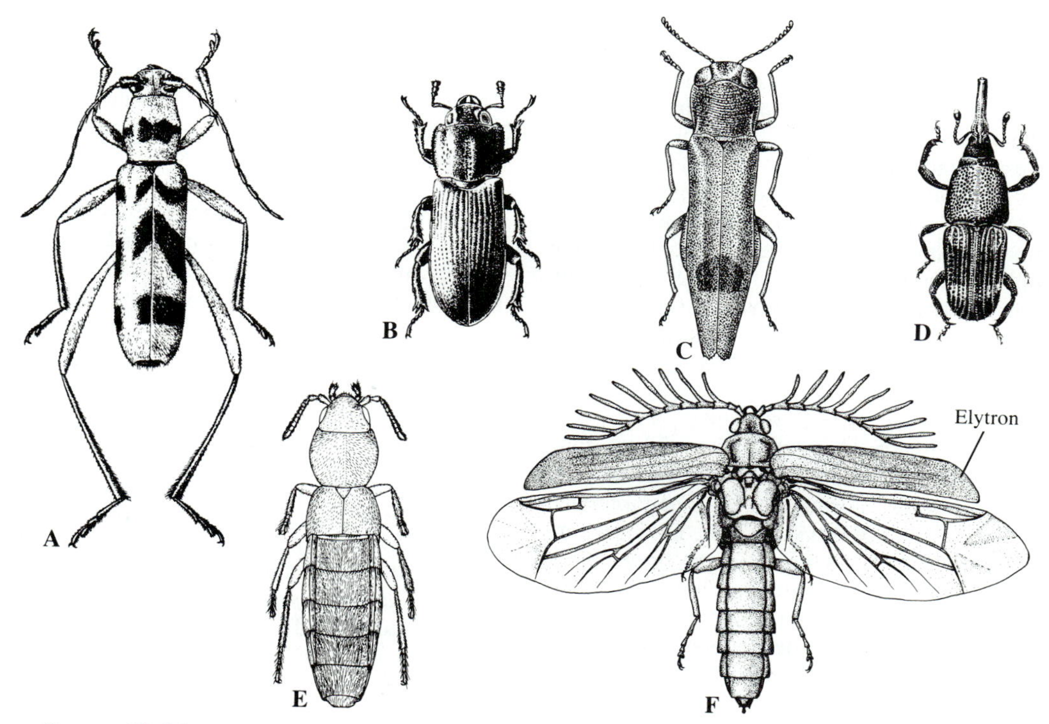

FIGURE 18.41 Order Coleoptera. A variety of beetles. A. *Demonax kotoensis* (family Cerambycidae). B. *Figulus laticollis* (family Lucanidae). C. *Agrilus yami* family Buprestidae. (A–C, Kano, Annotationes Zoologicae Japonenses, *17*.) D. *Sitophyllum oryza* (a weevil, family Curculionidae). (Essig and Hoskins, California Agricultural Extension Service, Circular 87.) E. A species of the family Staphylinidae, characterized by short wings and elytra. (Hatch, Beetles of the Pacific Northwest, 2.) F. *Pterotus obscuripennis* (family Phengodidae), wings spread. (Bohart, University of California Publications in Entomology, *7*.)

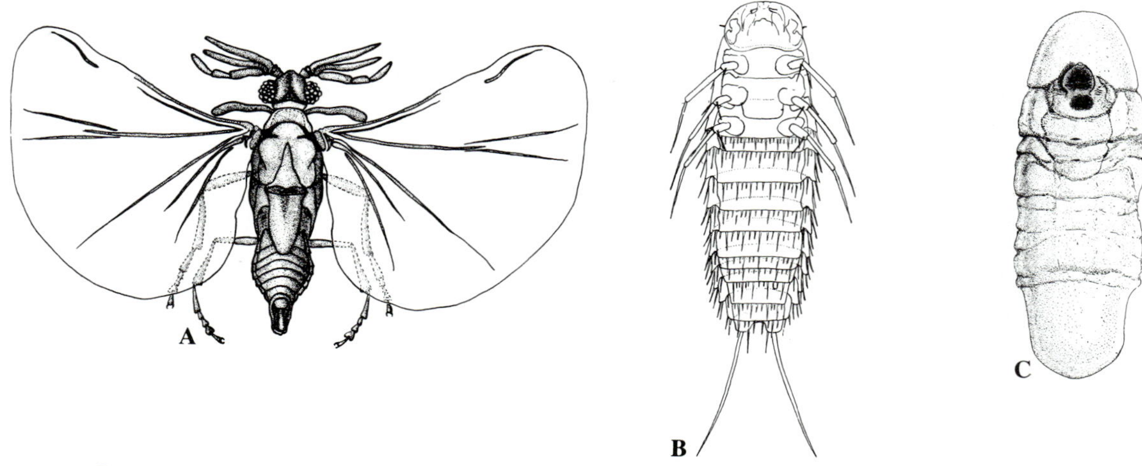

FIGURE 18.42 Order Strepsiptera. A. *Trizocera mexicana,* male. This species parasitizes certain bugs (order Hemiptera). (Bohart, University of California Publications in Entomology, *7.*) B. *Halictophagus oncometopiae,* female. Members of the genus *Halictophagus* parasitize leafhoppers and other insects of the order Homoptera. C. *Stylops,* first larva. Members of the genus *Stylops* parasitize certain bees (order Hymenoptera). (B and C, Pierce, Proceedings of the United States National Museum, *54.*)

In addition to herbivores and carnivores, there are beetles that feed on just about everything produced by plants or animals: dung, decaying carcasses, hides, and dead wood. Dermestids, which are pests that sometimes attack leather and dried biological specimens, have been put to use in museums, for they can clean the meat off skeletons without damaging delicate bones.

The reduction of wings in certain beetles has already been mentioned. Staphylinids (Figure 18.41, E) are good examples. Many of them hide under debris on sandy beaches, where winglessness would seem to be a helpful adaptation for life in windy situations. Some staphylinids are carnivores, others herbivores.

Order Strepsiptera

Strepsipterans are not often more than 2 mm long and most of them are internal parasites of other insects, especially thysanurans, orthopterans, hemipterans, homopterans, and hymenopterans. A few, however, are free-living. Males (Figure 18.42, A) have broad hindwings and can fly, but the forewings are reduced to structures that resemble the halteres, or ''balancers,'' of dipterans. The name of the order refers to the often twisted appearance of the hindwings. Females (Figure 18.42, B) are entirely wingless; in parasitic species, they have no antennae, eyes, or legs, segmentation is poorly defined, and the head and thorax are united.

In the usual life history of parasitic species, females remain in the hosts in which they have matured, but protrude from between the abdominal segments. Males emerge from their hosts and search for females with which to mate. The females do not lay eggs, but release nearly microscopic larvae (Figure 18.42, C). These have eyes and legs, and after getting into soil or onto plants, they have a chance of entering a suitable host. Those that succeed settle in the hemocoel, and after a molt they are legless. There are several more molts before the pupal stage is reached.

Some experts consider strepsipterans to be beetles. There are, in fact, beetles that resemble male strepsipterans (see Figure 18.41, F). Two families of parasitic beetles, moreover, have life histories similar to those of strepsipterans.

THE "NEAR INSECTS"

Three groups of wingless arthropods that have generally been treated as orders or subclasses of insects are here given class rank. These are the Collembola, Protura, and Diplura. They resemble insects in having three pairs of legs, but their mouthparts are hidden in a preoral cavity formed by folds that grow down from the sides of the head and unite with the labium. As a rule, Malpighian tubules are poorly developed. Insemination is achieved by indirect transfer of a spermatophore, as is typical of some groups of myriapods and chelicerates, rather than by copulation. Distinctive features of each of the three classes are brought out in the brief descriptions that follow.

CLASS COLLEMBOLA

Collembolans, or ''springtails'' (Figure 18.43, A) are easy to recognize because they jump with the aid of a prominent abdominal structure called the **furcula.**

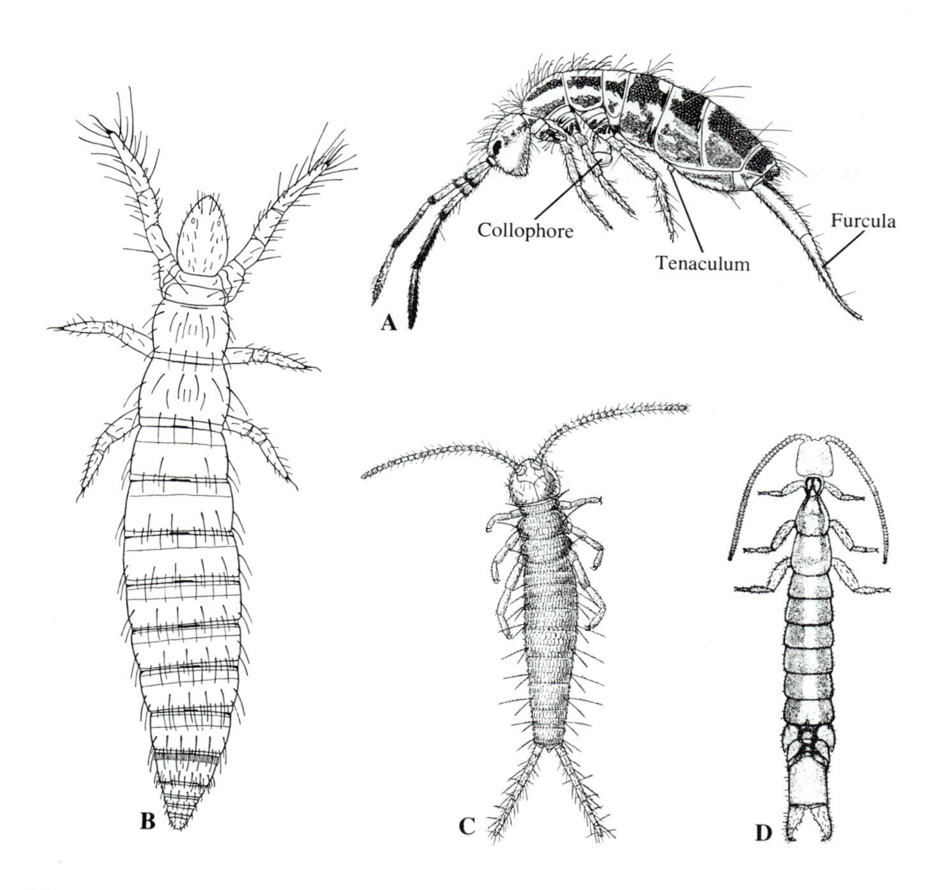

FIGURE 18.43 Classes Collembola, Protura, and Diplura. A. *Orchesilla ainsliei* (Collembola). (Folsom, American Museum of Natural History Novitates, Number 108.) B. *Acerentulus barberi* (Protura). (Ewing, Annals of the Entomological Society of America, *33*.) C. *Campodea* (Diplura). Proceedings of the United States National Museum, *72*.) D. *Japyx saussurii* (Diplura). (Swenk, Journal of the New York Entomological Society, *11*.)

They are often common in humus-rich soil, under logs, and in similar habitats. Some are regularly associated with colonies of ants or termites, and a few float on fresh water, supported by the surface film. One species, *Anurida maritima,* is a widespread scavenger that lives in decaying seaweed near the high-tide line.

Most springtails are between 2 and 5 mm long. The antennae are usually short and have only a few articles. The mouthparts are styletlike. Ocelli may be present, and the head sometimes has structures that resemble the organs of Tömösvary found in certain of the myriapod groups (see Chapter 19). When tracheae are present—not all collembolans have them—they may be with or without spiracles. If spiracles are present, there are only two, and they are located on the anteriormost part of the thorax. In collembolans that do not have tracheae, oxygen taken up through the cuticle is presumably distributed by simple diffusion or by the blood; in those with tracheae but no spiracles, oxygen is assumed to diffuse into the tracheae, which then distribute it to the cells.

The abdomen has six segments, but not all of these are distinct. Segments 1–4 and also 5–6 are often fused. There are three outgrowths, including the furcula, on the ventral side of the abdomen. They originate by fusion of paired embryonic appendages. The furcula is on segment 4, and when it is bent under the body in the "cocked" position it is gripped by a structure called the **tenaculum,** which is on segment 3. The tenaculum, in other words, functions as a retaining hook until the furcula is suddenly flexed and the animal leaps. Some species can jump for a distance of more than 10 cm.

The cylindrical tube on segment 1 of the abdomen is thought to be concerned with uptake of moisture. The name given to it long ago is *collophore*—the "glue-bearing structure"—because it was thought to produce a secretion that helped a springtail stick tightly to the substratum.

Cleavage of the egg is total. This is not true of any insects, or even of the Protura and Diplura. Some features in the development of collembolans are more like those typical of myriapods than of insects. There are no larval stages. When a springtail hatches it looks like an adult, for it has the same number of segments and appendages. It must nevertheless go through several molts before it reaches sexual maturity and adult size.

CLASS PROTURA

Proturans (Figure 18.43, B), which are less than 2 mm long, live in damp soil, leaf mold, rotting logs, and comparable habitats. They are unusual because they do not have antennae. Their first legs, however, are sensory and are kept raised much of the time, as if they were antennae. There are no eyes on the head, but there are some structures called *pseudoculi* (''false eyes''). These may be remnants of the bases of antennae or they may be homologues of the organs of Tömösvary found in myriapods.

The mouthparts of proturans are specialized for sucking up semifluid material. If spiracles are present, these are limited to the thorax. There are 12 abdominal segments; the last one, which generally has a slight prolongation, is sometimes interpreted as a telson rather than a true segment. Up to three of the anterior abdominal segments have rudimentary appendages.

On hatching, a juvenile proturan has nine abdominal segments and looks much like an adult. It undergoes three molts, adding a new segment each time.

CLASS DIPLURA

Diplurans (Figure 18.43, C and D) are decidedly different from proturans in that they have antennae, a pair of caudal cerci, and mouthparts specialized for chewing rather than sucking. There are generally some simple eyes on the head. The abdomen consists of 10 segments, and the first two to seven of these have rudimentary appendages. The cerci on the last segment may be divided into a number of articles or they may be stout, one-piece structures that can be closed together like jaws. In certain species they have the ducts of glands that secrete silk. A tracheal system is usually rather well developed, with spiracles on the thoracic segments and sometimes on the abdominal segments as well.

These animals live in much the same kinds of habitats as proturans, but they are on the whole considerably larger, ranging in length from about 2 mm to nearly 5 cm. When the young hatch, they have all of their segments. Diplurans probably represent an early offshoot from the line that also produced insects.

SUMMARY

1. Insects make up about three-fourths of all described species of animals. They have successfully colonized nearly all terrestrial habitats where life is possible, and many species occur in fresh water. Relatively few, however, are truly marine.

2. Three features are particularly important in adapting insects to life on land: a tracheal respiratory system; a cuticular body covering that is flexible but that provides protection against water loss; and wings, which enable insects to escape from predators and to fly into areas where food is available.

3. The body is divided into three tagmata: head, thorax, and abdomen.

4. The appendages of the head are the antennae, mandibles, and two pairs of maxillae; the basal portions of the second maxillae are united to form the labium, or lower lip. The mouthparts of insects are specialized for many different functions, such as cutting and chewing plant material, making punctures, and sucking blood.

5. The appendages of the thorax, which consists of three segments, are three pairs of legs. The two pairs of wings are also on the thorax, but they are not appendages; they are outgrowths of the cuticle.

6. The abdomen has 11 or fewer distinct segments. The cerci of the last segment of some insects, and certain structures involved in mating and laying eggs, are true appendages, but otherwise abdominal appendages are lacking, except in the order Thysanura.

7. The ultimate branches of the tracheal system, called tracheoles, typically allow oxygen to reach nearly every cell directly, without the intervention of blood.

8. The usual nitrogenous wastes excreted by insects are crystals of uric acid and urates, which are nontoxic because of their low solubility, and which can be eliminated without loss of much water. This is another important adaptation to a terrestrial existence. The crystals are produced in the Malpighian tubules, which are outgrowths of the posteriormost part of the midgut. After entering the gut, the crystals are voided with feces.

9. The circulatory system includes a heart and some major blood vessels, but blood circulates mostly through spaces in the tissues. Collectively, these spaces form a hemocoel, as is typical for arthropods.

10. Maturity is sometimes attained gradually, through a series of molts. This pattern is characteristic of the more primitive insects, such as cockroaches and grasshoppers. Beetles, butterflies, flies, and wasps are among the many types whose young, called larvae, are not at all like the adults. After going through several molts, a larva pupates and becomes transformed into an adult. This ''double life'' permits larvae to use food resources different from those used by adults of the same species.

11. Compound eyes, and also simple eyes, are characteristic of most adult insects, and of juveniles of species that reach adulthood gradually. In insects that have larval stages that are decidedly different from the adults, the larvae have only simple eyes.

12. Members of the order Isoptera (termites) and some members of the order Hymenoptera (ants, wasps, and bees) are social insects. They live in colonies in which there are several types of individuals, all of which contribute to the well-being of the colonies.

13. Parasitism has evolved in several orders. In certain Diptera (flies) and Hymenoptera (wasps), the larvae are parasitic; in Siphonaptera (fleas), the adults are parasitic. In Anoplura (sucking lice) and Mallophaga (biting lice), both adults and larvae are parasitic.

14. Although several orders are either primitively wingless (Thysanura) or secondarily wingless (Siphonaptera, Anoplura, Mallophaga), these are true insects in other respects. Three other wingless groups—Collembola, Protura, and Diplura—are similar to insects in having three pairs of legs, but are sufficiently different in other ways that it seems best to call them "near insects."

REFERENCES ON INSECTA AND "NEAR INSECTS"

Askew, R. R., 1971. Parasitic Insects. American Elsevier Publishing Co., New York.

Bell, W. J., and Cardé, R. T. (editors), 1984. Chemical Ecology of Insects. Sinauer Associates, Sunderland, Massachusetts.

Blum, M. S. (editor), 1985. Fundamentals of Insect Physiology. John Wiley & Sons, New York.

Borror, D. J., Triplehorn, C. A., and Johnson, N. F., 1989. An Introduction to the Study of Insects. 6th edition. Saunders College Publishing, Philadelphia.

Borror, D. J., and White, R. E., 1970. Field Guide to the Insects of America North of Mexico. Houghton Mifflin Co., Boston.

Chapman, R. F., 1982. The Insects. Structure and Function. Hodder and Stoughton, London and Sydney.

Daly, H. V., Doyen, J. T., and Ehrlich, P. R., 1978. An Introduction to Insect Biology and Diversity. McGraw-Hill Book Co., New York.

Eisner, T., and Wilson, E. O. (editors), 1977. The Insects. (Readings from Scientific American.) W. H. Freeman and Co., San Francisco.

Free, J. B., 1987. Pheromones of Social Bees. Comstock Publishing Associates, Ithaca.

Frisch, K. von, 1971. Bees: Their Vision, Chemical Senses, and Language. Cornell University Press, Ithaca.

Hermann, H. R. (editor), 1979–1982. Social Insects. 4 volumes. Academic Press, New York and London.

Horridge, G. A. (editor), 1975. The Compound Eye and Vision of Insects. Oxford University Press, London and New York.

Horridge, G. A., 1977. The compound eye of insects. Scientific American, *237*(1):108–120.

Imms, A. D., 1975. A General Textbook of Entomology. 10th edition. Methuen, London.

Jacobson, M., 1972. Insect Sex Pheromones. Academic Press, New York and London.

Jones, J. C., 1978. The feeding behavior of mosquitoes. Scientific American, *238*(6):138–148.

Jones, M. T., and Harwood, R. F., 1969. Herms' Medical Entomology. 6th edition. Macmillan, New York.

Mengel, R., and Erbes, J., 1978. Learning and memory in bees. Scientific American, *239*(1):102–110.

Michelsen, A., 1979. Insect ears as mechanical systems. American Scientist, *67*:696–706.

Peters, T. M., 1988. Insects and Human Society. Van Nostrand Reinhold, New York.

Pimentel, D. (editor), 1975. Insects, Science, and Society. Academic Press, New York and London.

Price, P. W., 1975. Insect Ecology. John Wiley & Sons, New York.

Pringle, J. W. S., 1957. Insect Flight. Cambridge University Press, Cambridge.

Pringle, J. W. S., 1975. Insect Flight. (Oxford Biology Reader.) Oxford University Press, London.

Rockstein, M. (editor), 1973–1974. The Physiology of Insects. 2nd edition. 6 volumes. Academic Press, New York and London.

Smith, K. C. V. (editor), 1973. Insects and other Arthropods of Medical Importance. British Museum (Natural History), London.

Snodgrass, R. E., 1935. Principles of Insect Morphology. McGraw-Hill Book Company, New York.

Stehr, F. W. (editor), 1987. Immature Insects. Kendall/Hunt Publishing Co., Dubuque, Iowa.

Tauber, M. J., Tauber, C. A., and Masaki, S., 1986. Seasonal Adaptations of Insects. Oxford University Press, London and New York.

Thornhill, R., and Alcock, J., 1983. The Evolution of Insect Mating Systems. Harvard University Press, Cambridge.

Wigglesworth, V. B., 1972. The Principles of Insect Physiology. 7th edition. John Wiley & Sons, New York; Chapman and Hall, London.

Wilson, E. O., 1971. The Insect Societies. Harvard University Press, Cambridge.

Winston, M. L., 1987. The Biology of the Honey Bee. Harvard University Press, Cambridge.

19

Subphylum Uniramia
MYRIAPODS

Subphylum Uniramia
- Class Chilopoda
- Class Symphyla
- Class Diplopoda
- Class Pauropoda

INTRODUCTION

In the past, the centipedes, millipedes, symphylans, and pauropods were put into one class, Myriapoda. The four assemblages do seem to be rather closely related, but they are now generally viewed as separate classes of the Uniramia. The term *myriapod* is still useful, because all of these arthropods are similar in that their bodies are organized into two tagmata: a **head,** with antennae and mouthparts, and an elongated **trunk,** with legs. Most of the segments of the trunk are more or less alike, and the number of pairs of legs on this tagma ranges from 9 in certain pauropods to about 200 in some millipedes.

In myriapods, as in insects, there is a single pair of **antennae.** The arrangement of mouthparts, however, varies among the four groups. In centipedes and symphylans there are **mandibles, first maxillae,** and **second maxillae.** Millipedes and pauropods have mandibles and first maxillae, but lack second maxillae. The head appendages and trunk appendages have distinctive specializations; it will be best, therefore, to present the more important details in connection with descriptions of the individual classes.

Myriapods are like insects and many chelicerates in having **Malpighian tubules** for excretion of nitrogenous wastes. **Tracheal respiration** is the rule, except in pauropods, which are small and soft-skinned and which can exchange gases efficiently through the body wall. Usually, the spiracles of the tracheae of myriapods cannot be closed. Because of this, and also because myriapods have little or no wax in the epicuticle, these animals are vulnerable to desiccation. Characteristically, they live in humid places and wander only at night.

Although reproduction is discussed briefly in connection with each of the four classes below, it is worth mentioning here that centipedes, symphylans, pauropods, and some millipedes transfer sperm indirectly, rather than by close contact of the male with the female. The male deposits a spermatophore, and the female either places her genital pore on it or picks it up and somehow gets the sperm to the genital pore. In most millipedes, which do mate, the male commonly uses certain specialized legs, or sometimes his mandibles, for transferring sperm to the genital pores of the female.

A

B

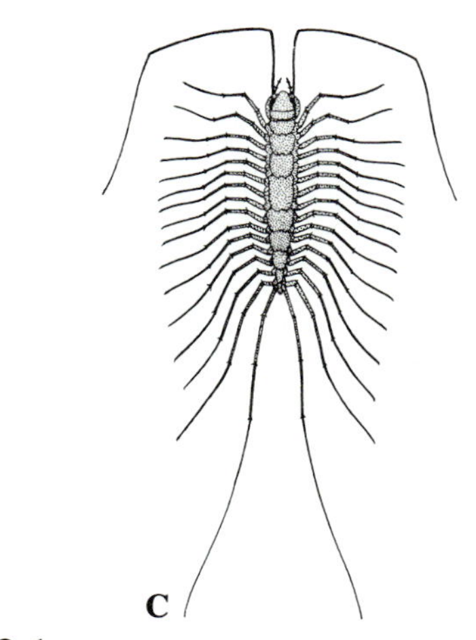
C

FIGURE 19.1 Class Chilopoda. A. *Scolopocryptops sexspinosus,* a centipede native to North America. B. *Lithobius forficatus,* a European species that has become established in many parts of the world. C. *Scutigera forceps,* the house centipede, also European in origin, but now widespread. (Ross, Illinois Natural History Survey, Circular 39.)

CLASS CHILOPODA

External Features

In centipedes (Figure 19.1), of which there are about 3000 species, the head has a pair of antennae, a pair of stout mandibles, and two pairs of maxillae (Figure 19.2, B). The bases of both sets of maxillae are usually joined, so that two structures similar to the labium of an insect are formed. The trunk may have as few as 17 segments or more than 150, and all but the last two—the genital segment and anal segment—have legs. The first legs are specialized for injecting venom; they are

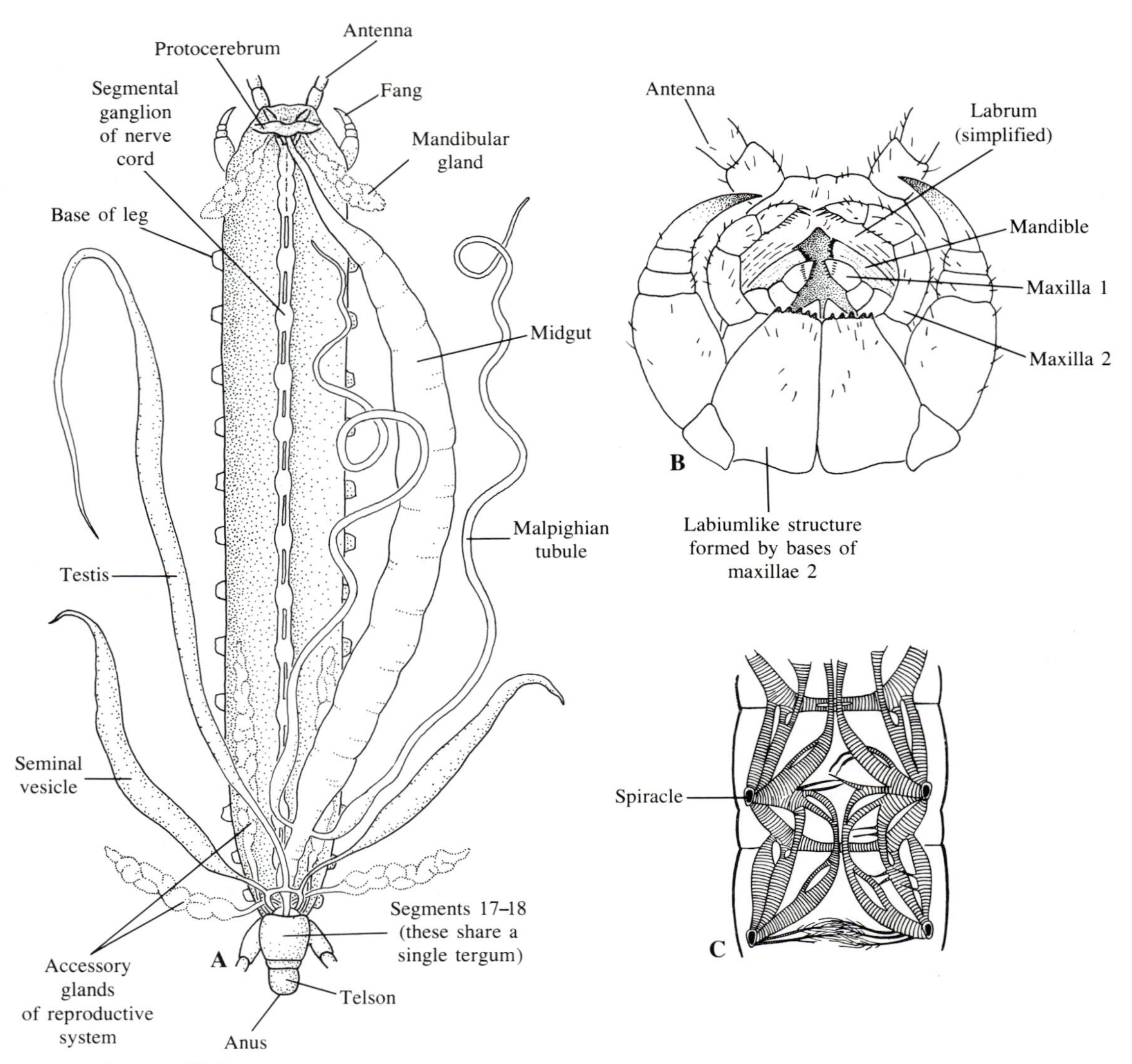

FIGURE 19.2 Class Chilopoda. A. *Lithobius forficatus,* dissection of a male. B. *Lithobius forficatus,* head region, ventral view. (After Ivanov *et al.*, Major Practicum of Invertebrate Zoology.) C. Tracheal tubes of a centipede. (Lawrence, The Biology of the Cryptic Fauna of Forests.)

often called maxillipeds, but **fangs** is a more appropriate term. The next-to-last legs, sometimes spiny, are longer than the rest and may be used for pinching, either defensively or offensively. They also operate as sensory appendages in at least some centipedes. The last legs, on the third-from-last segment, are small and specialized as genital appendages.

Although most of the leg-bearing segments are similar, their terga may not be of equal length. Longer ones may appear to be squeezing out shorter ones, and sometimes a long tergum overlaps a shorter tergum of the preceding or succeeding segment. As a result of this sort of tergal elongation, up to three pairs of legs may be associated with a single tergum. Presumably the elaboration of long terga is an adaptation for resisting the lateral pull of the legs. It gives the animal a better straight-forward thrust, instead of a snaky kind of motion that might cause some legs to tangle with others.

Centipedes are secretive animals, and most of them spend the daylight hours in soil or leaf litter, or under logs, bark, or rocks. They emerge at night to hunt for their prey. Members of one of the four orders, Geophilida, usually remain in the soil through which they burrow.

Locomotion

Most centipedes, other than types that burrow in soil, move with considerable speed and agility. Those belonging to the order Scolopendrida, of which *Scolopocryptops* (Figure 19.1, A) is representative, have 21 or 23 pairs of legs, all of about the same length. The legs make fairly long strides, and while some legs are in contact with the ground and pushing backward, others are raised, brought forward, and touched down. The leg on the left side of a particular segment is out of phase with the one on the right. Interference between successive legs is minimized not only by the pattern of movement, but also by the ability of centipedes of this group to elongate by about 20% while running. This separates the pairs of legs more widely.

Lithobiid centipedes, which form another order, have 15 pairs of legs (see *Lithobius,* Figure 19.1, B), but these are proportionately much longer than those of scolopendrids. Thus the strides they take are longer than those of scolopendrids, and the period of contact of each leg with the ground is longer. In centipedes of the order Scutigerida (see *Scutigera,* Figure 19.1, C), which also have 15 pairs of legs, the legs become progressively longer posteriorly. This makes it possible for the legs of the hind part of the body to take very long strides without getting tangled up with legs of the fore part. The efficiency of the legs in pushing is increased, moreover, by subdivision of the terminal article, the tarsus, into many units. The flexibility of the tarsus improves its contact with the substratum. Scutigerids run much faster than other centipedes.

Members of the soil-inhabiting order Geophilida work their way through soil in a manner slightly reminiscent of the burrowing movements of an earthworm. As muscle contractions narrow the width of certain segments, pressure on fluid in the hemocoel is increased, and the anterior portion of the trunk is lengthened and pushed forward. Stretching takes place at the joints between successive terga and sterna. Repeated shape changes of this sort, coupled with the anchoring effect of the legs, enable a geophilid to continue enlarging its burrow. Some centipedes of the other orders are also able to burrow and to make underground dens for themselves.

Digestive System and Feeding

The usual arthropodan arrangement of ectodermal foregut, endodermal midgut, and ectodermal hindgut is found in centipedes. While the hindgut is always short, the proportionate lengths of the foregut and hindgut vary considerably. In some centipedes, the foregut may constitute two-thirds the length of the entire gut, and its cuticular lining has anteriorly directed spinelike elaborations that presumably keep coarse particles from entering the midgut. A variety of small glands open just in front of the mouth, at the bases of the mouthparts, or on certain of the mouthparts themselves. Little is known about the functions of the glands, or about digestion in general.

Centipedes feed mostly on insects, spiders, millipedes, small snails, and earthworms. A few that live under rocks and debris just above the high-tide line at the seashore are known to consume isopod and amphipod crustaceans, barnacles, oligochaetes, snails, and some other marine or semimarine animals.

It is characteristic of centipedes to "test" their prey with their antennae, or sometimes with their legs, before attacking. As explained previously, the first legs of the trunk are specialized to serve as fangs for injecting venom. The prey may have to be immobilized, however, by several pairs of legs. The mandibles generally bite a hole in the prey or tear off pieces of tissue. The first maxillae are sometimes inserted into the wounds made by the fangs and then used to lift up the tissue so that the mandibles can nip it off. *Strigamia,* a centipede that lives at the seashore, has been observed to stick its head into the wound it has made with its fangs. While the fangs continue to cut into the tissue, it pushes its head deeper. The fact that some fluid seems to come out of the mouth of *Strigamia,* and the fact that no particulate food is present in the gut, suggest that digestive enzymes are acting externally. If this is the case, it is decidedly atypical for centipedes.

Respiration, Circulation, and Excretion

The spiracles of the tracheal tubes are located on the sides of most leg-bearing segments of the trunk. The right and left tracheal complexes in a particular segment may be connected, and the complexes of successive segments may also be joined (Figure 19.2, C). In the order Scutigerida, the typical sort of tracheal system is replaced by one whose branches exchange oxygen and carbon dioxide with the blood rather than directly with the tissues. The single opening to each pair of these tracheal complexes, which are functionally comparable to tracheal lungs of spiders and some other chelicerates, are located dorsally at the posterior edges of certain of the terga.

The dorsal heart usually has a pair of ostia for each of the several trunk segments through which it runs. As is customary in arthropods, blood accumulating in the pericardial portion of the hemocoel returns to the heart through the ostia. There are generally segmental arteries as well as an anterior and posterior artery, but the arrangement varies.

The two Malpighian tubules (Figure 19.2, A) are joined to the posteriormost part of the midgut. They eliminate uric acid and also ammonia. Centipedes that excrete substantial amounts of ammonia must have a humid environment, for ammonia cannot be gotten rid of without a concomitant loss of water. In addition to Malpighian tubules, centipedes have phagocytic cells that accumulate nitrogenous wastes of low solubility. These cells are sometimes concentrated around the Malpighian tubules, but they may be clustered around or beside the salivary glands or located in some other part of the body.

Nervous System and Sense Organs

In most centipedes, the members of each pair of ganglia in the ventral nerve cord are distinct, and the parallel connectives that join the ganglia of one segment to those of the preceding and succeeding segments are usually separate (Figure 19.3, A). The nervous system as a whole, therefore, resembles that of a generalized annelid.

Some centipedes have a few to many simple eyes on both sides of the head; compound eyes are found only in certain especially active members of the order Scutigerida. Many species, particularly those that live in soil, lack eyes altogether. Two orders, Lithobiida and Scutigerida, have a pair of complicated pits on the front of the head. These are called **organs of Tömösvary.** They occur in some members of all four classes of myriapods, but just what they do is uncertain. There is evidence that they are concerned with sound reception in centipedes, but the structure of these organs in certain millipedes suggests that they may have an olfactory function. The matter needs to be studied further.

As in arthropods generally, many of the bristles on the body and appendages of centipedes are linked with touch and vibration receptors and with chemoreceptors. There are proprioceptors and probably also vibration receptors associated with the flexible cuticle of the joints between articles and segments.

Reproduction and Development

In both sexes (Figure 19.2, A; Figure 19.3, B-D), the gonad is unpaired, and there is a single genital pore on the ventral side of the next-to-last segment. (This segment shares a tergum with the preceding segment, so its limits are not visible dorsally.) Centipedes do not actually copulate, but they do engage in courtship. In most species the male spins a primitive web and deposits a spermatophore on it. The female works the spermatophore into her genital pore with the aid of her specialized genital appendages. There is much variation, however, in mating habits, and in a few species the male has been observed to place a spermatophore directly on the female's genital opening. Techniques of egg-laying are also diverse. Most species deposit clusters of eggs in little nests. A few, including *Lithobius forficatus,* an Old World centipede now common in North America, coat their eggs individually with mud made by mixing soil with material exuded from the hindgut. The genital appendages hold each egg while it is being coated. Parental care of eggs is characteristic of many species, and in some the female sticks the eggs to her body.

The stage at which the young hatch varies from group to group. In two of the four orders (Scolopendrida and Geophilida), hatching generally takes place after all segments and appendages have been formed. In the other two orders (Lithobiida and Scutigerida), the young are ''unfinished''—they have only a few of the more anterior leg-bearing segments, and add the rest in the course of several molts.

CLASS SYMPHYLA

Symphylans are small, soil-inhabiting myriapods that resemble centipedes. The best-known species, *Scutigerella immaculata* (Figure 19.4, A–C), is about 5 mm long. A native of Europe, it is now widely distributed in temperate regions. It is often found in backyards and gardens, and should be looked for under flower pots and boards that have been in tight contact with moist soil. Like most symphylans, *Scutigerella* feeds to a large extent on decaying plant material. It also eats delicate roots, however, and because of this it is a serious pest of crops in some areas.

The head of a symphylan has antennae, mandibles, and two pairs of maxillae. The bases of the second maxillae are joined to form a lip comparable to the labium of an insect. The organs of Tömösvary are in the usual position, just behind the antennae (Figure 19.4, D). Posterior to the head are 14 trunk segments, the first 12 of which have legs. On segment 13 is a pair of structures called *cerci,* or spinnerets; silk is exuded from the tips of these. Segment 14 has two especially prominent sensory hairs. As one looks at a symphylan from above, there may not seem to be 14 units in the trunk. This is because some of the leg-bearing segments have two terga, and also because segments 13 and 14 are not distinct from one another. The terminal telson, in which the anus is located, is fused to segment 14.

At the base of almost every leg is a small coxal sac that is probably used for picking up moisture. The sacs are everted when pressure on the fluid around them is increased, and they are withdrawn by contraction of muscles. Beside the pore there is usually a bristle-tipped projection called a *style;* this is presumably a sensory structure.

Tracheae are limited to the anterior part of the body. There are only two spiracles, and these are on the head.

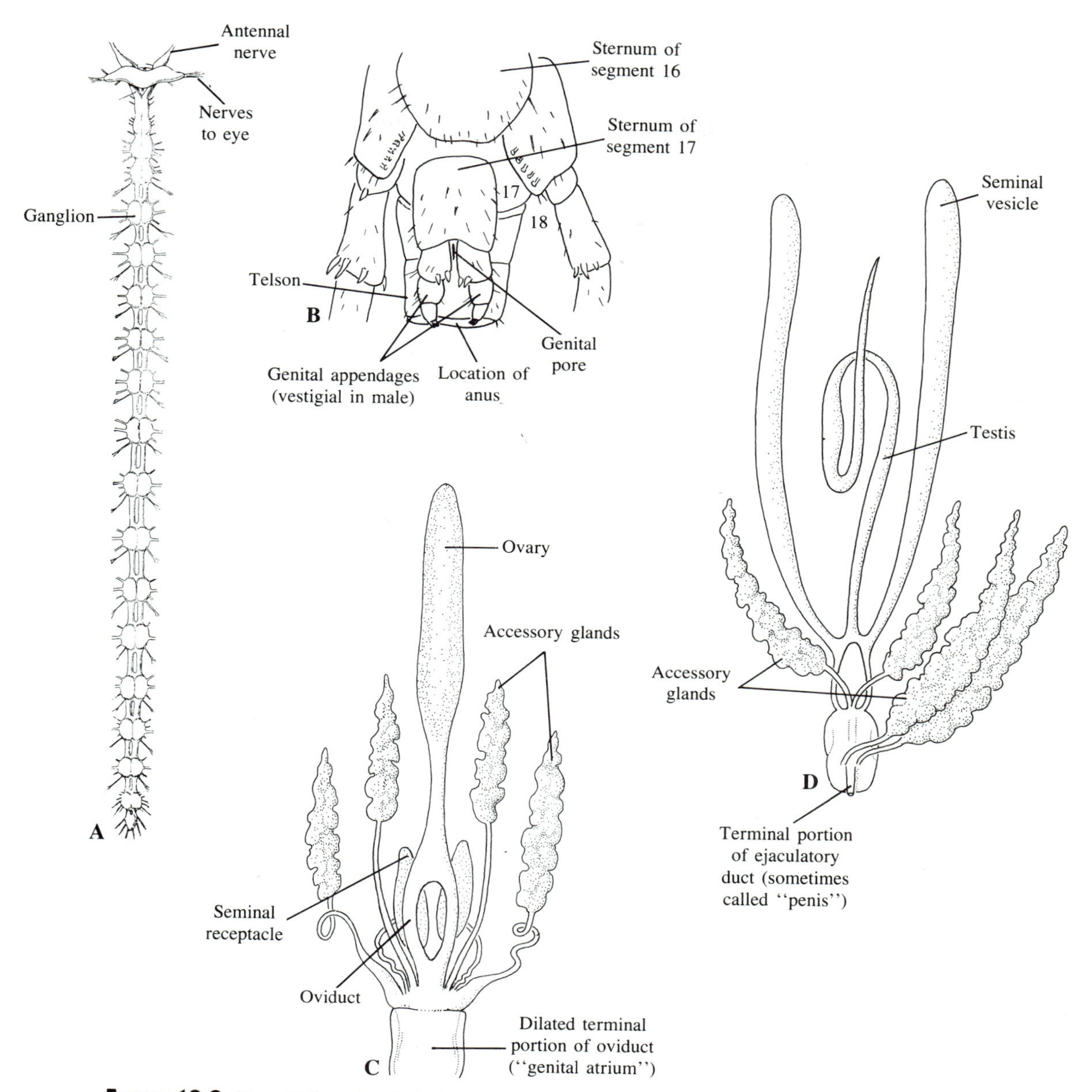

FIGURE 19.3 Class Chilopoda. *Lithobius forficatus*. A. Nervous system. B. Posterior end of the body of a female, ventral view, showing the legs modified to serve as genital appendages. C. Reproductive system of female. D. Reproductive system of male. (After Ivanov *et al.*, Major Practicum of Invertebrate Zoology.)

One pair of maxillary glands is derived from the coelomic sacs of the first maxillary somite; the origin of a second pair, observed at about the same level in *Scutigerella,* is not clear.

The gonads are paired and their ducts lead to gonopores on segment 4 of the trunk (Figure 19.4, E and F). In *Scutigerella,* the male deposits a stalked spermatophore. The female bites it off, and the sperm may then reach the two seminal receptacles located in front of the labium. While the female is laying an egg, she raises the anterior part of her body and bends down so that she can pick up the egg in her mouthparts. She then

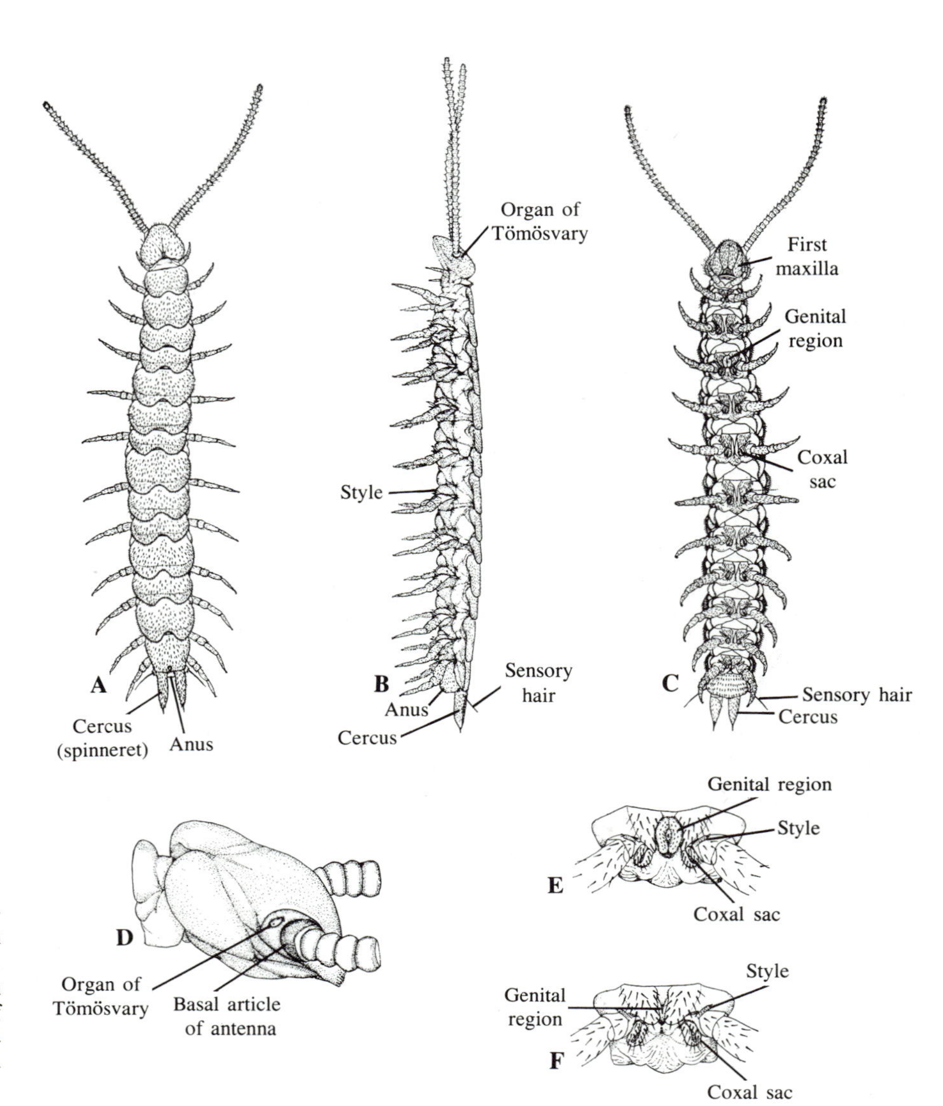

FIGURE 19.4. Class Symphyla. *Scutigerella immaculata*. A. Dorsal view. B. Lateral view from the left side. C. Ventral view. D. Head region, lateral view from the right side. E. Segment 4 of male, ventral view. F. Segment 4 of female. (Michelbacher, Hilgardia, *11*.)

attaches the egg to a sprig of moss or some other suitable object. It is not until the egg has been affixed that she applies sperm to it. When the young hatch, they have six pairs of legs and ten terga. There are several molts before sexual maturity is attained, and many more afterwards.

The styles and eversible sacs at the bases of the legs of symphylans have their counterparts in the wingless insects of the order Thysanura, and perhaps the sacs are comparable also to those located in slits at the bases of the legs of most onychophorans (Chapter 20). Two other features—the joining of the second maxillae to form a labium and the presence of 14 postcephalic segments—are characteristic of insects in general. The name Symphyla means ''union of groups,'' and alludes to the fact that this class exhibits some traits of insects as well as of myriapods.

CLASS DIPLOPODA

External Features

Millipedes, of which there are about 8000 species, do not live up to their common name, for none comes close to having a thousand legs. The name Diplopoda does suit them, however. It refers to the double-legged condition of most of the apparent ''segments.'' Because each segment derives from two embryonic somites, it

A

B

FIGURE 19.5 Class Diplopoda. A. Cylindrical millipede. B. Millipede of the flattened type. (See also Color Plate 7.)

generally has two pairs of appendages. This feature enables one to distinguish a millipede from a centipede or any other myriapod.

In some millipedes the body is nearly cylindrical (Figure 19.5, A); in others it is appreciably flattened (Figure 19.5, B; Color Plate 7). The head (Figure 19.6, A) has antennae, mandibles, and a complicated platelike structure called the **gnathochilarium,** derived from the first and only maxillae. The extent to which the various components of the maxillae are fused varies. The gnathochilarium is something like a large lower lip; it actually makes the "floor" of a cavity that leads to the mouth. A wedge-shaped segment (**collum**) that is conventionally regarded as the first trunk segment is probably homologous to the segment that bears the second maxillae in other mandibulate arthropods. The narrow sternum of this segment backs up the gnathochilarium and provides some support for it.

The number of trunk segments (including the collum) ranges from 11 to about 100. There is a single pair of legs on segments 2, 3, and 4. Each of the remaining "segments" (except for the last few, which have no legs at all) has two pairs of legs. The last segment is tipped by a three-piece telson; the preanal portion of this forms a roof over the other two, which are called *anal valves*. The trunk joints of millipedes are organized in such a way that the body has considerable flexibility but little compressibility. The long-bodied, cylindrical millipedes of the order Julida are able to coil tightly in a flat spiral, much like a watchspring, and many representatives of the order Glomerida can roll up into a ball the way pillbugs do. Deposits of calcium salts in the endocuticle confer considerable firmness on the segments of julids and some other millipedes.

Millipedes are commonly found under logs, under bark, in humus, and in similar habitats. Some wander at night, or even in the daytime, but most remain close to their hiding places. At least moderate humidity is needed by nearly all millipedes, especially by small species whose cuticle is thin and ineffective in preventing water loss.

Locomotion

The legs of millipedes generally move slowly, but collectively they exert a strong push for locomotion on the surface of the substratum or for displacing humus and loose soil. When a millipede crawls, the backward, downward stroke of one leg is followed by that of the next leg behind it, so there are waves of backward strokes along the length of the animal. The backward strokes last longer than the forward, raised recovery strokes; this is efficient because more legs are pushing at any one time than are raised.

Having double segments confers certain locomotor advantages on millipedes that burrow or that must work their way through tight spaces. Members of the order Julida are especially well adapted for pushing their way through humus and soil. Their double segments are smooth, cylindrical, and rigid, their head is rounded, and their legs are close to the midline, so they do not sprawl laterally. Millipedes that are flattened and have lateral expansions of the terga are adapted for squeezing through cracks. They can use their legs to push the dorsal surface up against the edge of a crack and perhaps enlarge the opening; segmental muscles also operate to produce a similar effect. The tergal expansions provide protection for the legs, which extend farther laterally than those of julids.

The order Chordeumida includes some millipedes that remain out in the open a considerable part of the time. They are predators for whom speed and agility are of the essence. One species has been clocked at a speed

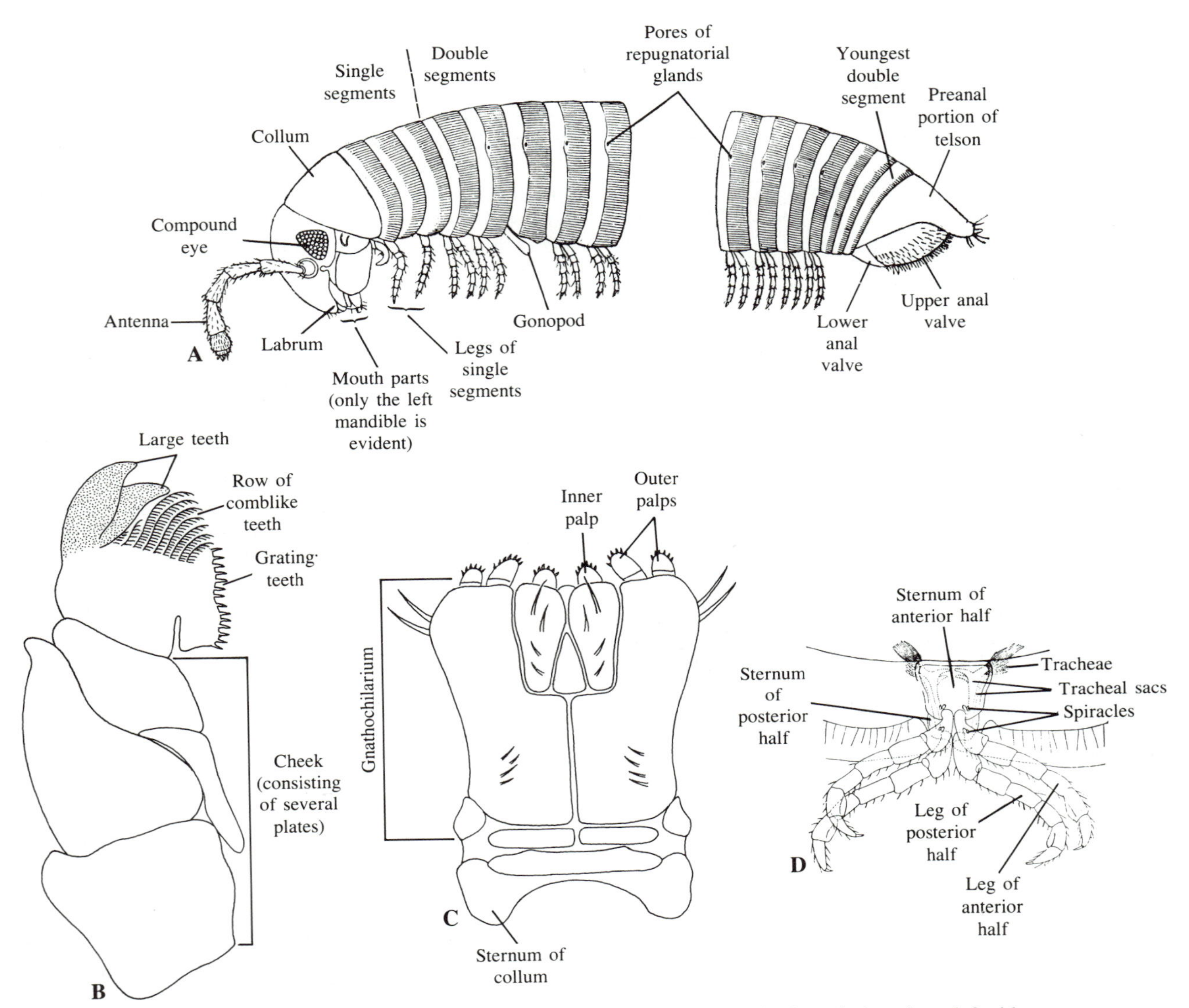

Figure 19.6 Class Diplopoda. *Schizophyllum sabulosum*. A. Male, lateral view from left side. Segments 2–4 derive from single somites; each "segment" from 5 onward derives from two somites. B. Mandible. C. Gnathochilarium, produced by union of the maxillae. D. One of the double "segments," ventral view. (After Ivanov *et al.*, Major Practicum of Invertebrate Zoology.)

of over 5 cm per second. Certain of these millipedes use their claws effectively for climbing up vertical surfaces and even for moving across the underside of horizontal surfaces.

Digestive Tract and Feeding

The midgut makes up most of the digestive tract, although both the foregut and hindgut are at least several segments long. In millipedes whose anatomy has been studied in detail, there are two unpaired glands that open just in front of the mouth. The posterior one is the larger, and much of it is located in the trunk, above the foregut. The functions of these glands has not been determined.

Most millipedes feed on decaying plant material or on soil that is rich in humus. Some eat pollen, moss, the unicellular green algae that coat bark, or other living plants, and some scavenge on dead animals or feces. Relatively few, belonging to one group of the order Chordeumida (mentioned above, under "Locomotion''), are predators that hunt for various insects, centipedes, arachnids, and earthworms. So far as is known, predatory millipedes do not inject a venom.

Respiration, Circulation, and Excretion

The spiracles of the tracheal system are segmentally arranged and are generally ventral in position. The double segments usually have two pairs of them. As in insects and most centipedes, the tracheal system supplies oxygen directly to the tissues. The heart has an anterior artery and one or more pairs of lateral arteries, but no posterior artery. There are only two Malpighian tubules. The maxillary glands, whose pores are situated on the gnathochilarium, are coelomic derivatives. They are assumed to be excretory.

Nervous System and Sense Organs

In the ventral nerve cord, the ganglia are more or less fused, so the connectives that join successive pairs are not as distinct as they are in centipedes. Each of the double segments has two pairs of ganglia.

There are no compound eyes. When simple eyes are present, these are arranged in rows or clusters on the sides of the head. Organs of Tömösvary are of general occurrence in millipedes, and the structure of those that have been studied suggests that they have an olfactory function.

Bioluminescence and Secretion of Repugnatorial Substances

A few millipedes emit light continuously from all parts of the body surface. This bioluminescence probably has a protective function.

Many species exude or spray out repugnatorial substances—including hydrogen cyanide, aldehydes, and various p-benzoquinone compounds—from segmentally arranged lateral pores. The secretions, which often smell like a mashed peach pit, vinegar, or camphor, are for defense. Some of the p-benzoquinones will irritate human skin or cause brown spots to appear as a result of their interaction with proteins, a phenomenon comparable to that of tanning of arthropod cuticle.

Reproduction and Development

In both sexes, there is a single gonad located in the ventral part of the trunk. Two gonoducts lead to separate genital pores located behind the second pair of legs (segment 3) or on their coxae, or to a single midventral opening in the same general area. In males of most groups, one or both pairs of legs on segment 7 (the third double segment) are modified to serve as sperm carriers, or gonopods, but in some millipedes legs of segment 8 are also involved. Each of the modified legs has a little pit on its basal portion, and this is filled with sperm when the male doubles over, bringing his specialized limbs into contact with the genital pores. In mating, the ventral surfaces of male and female come together, or the male winds himself around the female in such a way that he can inseminate her.

In certain species the mandibles operate as sperm carriers, and there may be even less-direct methods of sperm transfer. *Polyxenus,* for instance, deposits sperm on some silken threads spun out from glands on the legs of segment 2. He then lays down more threads from glands on the legs of segment 7. These direct the female to the sperm mass, and when she finds it she inseminates herself by everting the ventral concavity in which her genital pores are located.

Female millipedes generally lay their eggs in nests made of soil, but those that have silk glands that open at the posterior end of the body spin their nests. In a millipede that has just hatched, there are usually three pairs of legs (on segments 2, 3, and 4 of the trunk). The few double segments that follow do not have legs until after the first molt, at which time additional double segments appear. These acquire legs after the second molt, when more double segments, temporarily legless, are added. It takes several molts for a millipede to acquire all its segments and all its legs.

CLASS PAUROPODA

Pauropods, which are mostly about 1 or 2 mm long, live in soil, among decaying leaves, under logs or rocks, and in comparable habitats. The antennae of these myriapods (Figure 19.7, A and B) are peculiar in that they are biramous, with each branch bearing one or two long setae. Behind the bases of the antennae are the *pseudoculi* (''false eyes''), which are almost certainly the same as the organs of Tömösvary of other myriapods, and which are probably concerned with perception of vibrations. The mandibles are succeeded by a single pair of maxillae. What appears to be the first segment of the trunk is thought to be homologous to the collum of millipedes and to the segment that bears second maxillae in centipedes, symphylans, and insects. There is a pair of small protrusible vesicles on this segment, but they are not viewed as vestigial limbs.

The trunk proper generally consists of 10 segments plus a telson. Each of the first nine segments has a pair of legs. In one genus, *Decapauropus,* there is an extra segment; the tenth is legless but the eleventh does have legs. Because of the loss of some terga and the expansion of others, so that each covers two segments, the number of terga does not conform to the number of segments. The cuticle is thin enough and the body small enough to permit diffusion of gases and translocation of

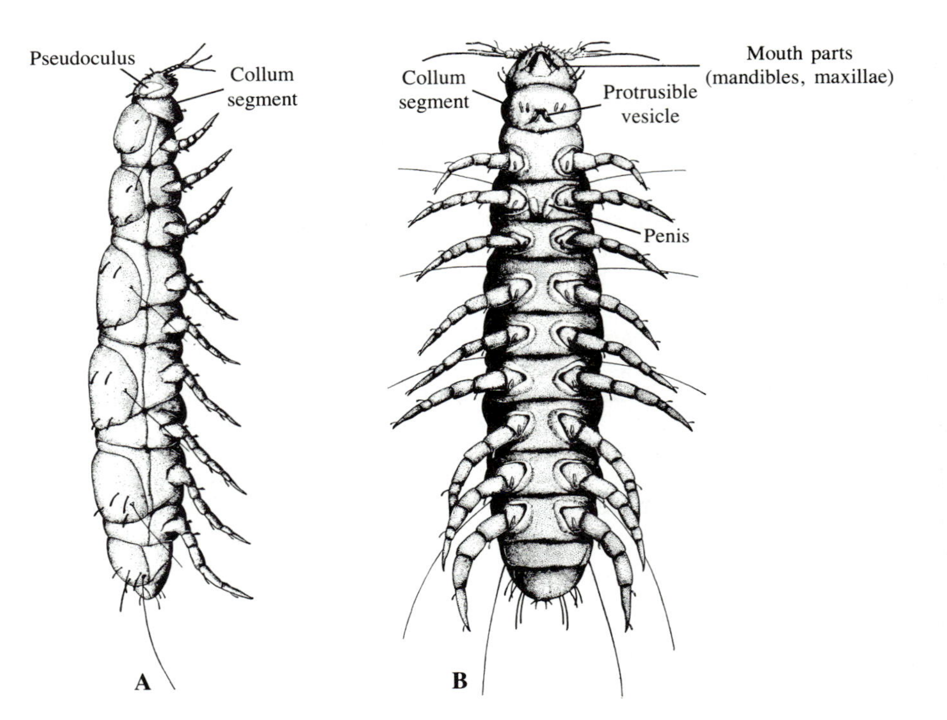

FIGURE 19.7 Class Pauropoda. *Pauropus silvaticus,* male. A. Lateral view from right side. B. Ventral view. (Tiegs, Quarterly Journal of Microscopical Science, *88.*)

nutrients to take place without benefit of a tracheal system or circulatory system. (A few African pauropods do, however, have tracheae. The openings of these are on the coxae of the legs.) For excretion, there are two short Malpighian tubules; these join the posteriormost part of the midgut.

The ovary of the female is beneath the gut. An oviduct connects it with a ventral genital pore on the midline between the second legs (segment 3 of the trunk). There is a seminal receptacle associated with the pore. The testis of a male pauropod starts to develop in the ventral part of the trunk, but it ends up above the gut. The two sperm ducts lead to separate pores behind the second legs. The pores are on little projections, generally called penises. The oviducts and sperm ducts of pauropods are unusual in that they derive from ectodermal invaginations; they are not, therefore, homologous with gonoducts in most arthropods, which are specialized coelomoducts. Whether pauropods actually mate is uncertain. In one species whose reproductive habits have been observed, the male selects a crevice in which to deposit a spermatophore, anchoring this with silken threads. The female somehow locates the spermatophore and puts her genital pore over it.

When a young pauropod hatches from the egg, it has about half of its trunk segments and only the first three pairs of legs. More segments and more legs are added as the animal grows to maturity.

The few pauropods whose feeding has been studied suck juices from the hyphal threads of fungi. They use their mandibles for biting into the threads, and effect the sucking action by peristaltic movements of the midgut. The gut contents of some pauropods consist of a mixture of fungal hyphae, fungal spores, setae of arthropods, and other material. These species seem to be scavengers as well as consumers of fungi.

SUMMARY

1. The myriapods include the centipedes (Class Chilopoda) and millipedes (class Diplopoda), as well as two small groups, the classes Symphyla and Pauropoda. In all four groups, the body is organized into two tagmata: a head and an elongated trunk.

2. Like insects, myriapods have a single pair of antennae. The mouthparts of centipedes and symphylans consist of mandibles and two pairs of maxillae; those of millipedes and pauropods consist of mandibles and a single pair of maxillae.

3. Pauropods have 9 or 10 pairs of legs on the trunk, and symphylans have 12 pairs; centipedes and millipedes usually have many pairs. The arrangement of legs in millipedes is distinctive in that most of the apparent segments of the trunk have two pairs of legs, due to the fact that each of these segments is derived from two embryonic somites.

4. Malpighian tubules function in excretion of nitrogenous wastes, and tracheae are used for respiration in all groups except the pauropods, which are soft-skinned and can exchange gases through the body wall.

5. Centipedes are carnivorous, using fangs (specialized first legs) for injecting venom. They feed mostly on insects and other arthropods, worms, and molluscs. Symphylans, millipedes, and pauropods are primarily scavengers, especially on plant material.

6. Centipedes, symphylans, pauropods, and some millipedes transfer sperm indirectly. The male deposits a spermatophore on a suitable substratum, and the female later brings this into contact with her genital opening. Most millipedes mate, the male using certain legs or the mandibles for placing sperm in the genital pores of the female.

7. Development is direct. As a rule, the appendages of the head and a few of the pairs of legs on the trunk are present at hatching; the rest of the legs are added later. Some centipedes, however, have all of their appendages by the time they hatch.

REFERENCES ON MYRIAPODS

Blower, J. G. (editor), 1974. Myriapoda. Symposia of the Zoological Society of London, no. 32. Academic Press, London and New York.

Camatini, M. (editor), 1980. Myriapod Biology. Academic Press, London and New York.

Demange, J.-M., 1981. Les Mille-Pattes (Myriapodes). Boubée, Paris.

Eason, E. H., 1964. Centipedes of the British Isles. Frederick Warne and Co., London and New York.

Kraus, O. (editor), 1979. Myriapoda. Paul Parey, Hamburg and Berlin.

Lewis, J. G. E., 1981. The Biology of Centipedes. Cambridge University Press, London.

Manton, S. M., 1977. The Arthropoda: Habits, Functional Morphology, and Evolution. Clarendon Press, Oxford.

Michelbacher, A. E., 1938. The biology of the garden centipede, *Scutigerella immaculata*. Hilgardia, *11*:55–148.

Schaller, F., 1968. Soil Animals. University of Michigan Press, Ann Arbor.

Tiegs, O. W., 1940. Embryology and affinities of Symphyla. Quarterly Journal of Microscopical Science, *82*:1–225.

Tiegs, O. W., 1947. Development and affinities of Pauropoda. Quarterly Journal of Microscopical Science, *88*: 165–267; 275–336.

20

Phyla Onychophora, Tardigrada, and Pentastomida

The groups dealt with in this chapter are not necessarily closely related. All three of them, however, exhibit some features that indicate a kinship with arthropods. They have a cuticle that is molted periodically, and the appendages they use for walking or attachment resemble limbs of certain arthropods, even if they are not jointed. The onychophorans have been of particular interest to zoologists because they show certain characteristics found in annelids, and others that are typical of arthropods. This must not be interpreted to mean that onychophorans are a link that connects annelids and modern arthropods. They are probably relicts of an early digression from one of the main lines of arthropod evolution. They are nevertheless instructive in that they give us some idea of how arthropods may have originated.

Phylum Onychophora

Onychophorans have a wide but discontinuous geographic distribution, which can probably be explained on the basis of continental drift. Some of the approximately 70 species are found in tropical areas of Asia, Africa, the West Indies, and Central and South America. The rest are restricted to subtropical and temperate regions well below the Equator: New Zealand, Australia, southern Africa, and southern South America. The tropical species are not necessarily limited to the hot lowlands; cooler places in the mountains provide suitable habitats for many of them. In general, these engaging little animals live in humid places—under logs or rocks, in stumps, among fallen leaves, or along streams. Those that inhabit areas where the seasons change tend to become inactive during cold weather or during dry periods. Their cuticle does not protect them against desiccation. Most onychophorans are from 2 to 5 cm long, but the largest species reaches a length of nearly 15 cm.

External Features, Body Wall, and Musculature

An onychophoran (Figure 20.1) slightly resembles a caterpillar. The arrangement of appendages is the only obvious external evidence that the body is segmented. Internally, however, there is serial repetition of paired **coelomic sacs** and **coelomoducts,** and sometimes there are other clearly segmental structures. The **cuticle,** which in most places is only about 1 μm thick, contains chitin and is periodically molted. It is bumpy with many crowded papillae, which are localized thickenings of the epidermal layer and other components of the body wall, and which may also contain substantial blood sinuses. Each papilla is tipped with a cuticular bristle. The papillae and bristles collectively give an onychophoran a velvety appearance.

Figure 20.1 A living onychophoran from a rain forest in Colombia.

The head (Figure 20.2, A and B) is rather well differentiated. It has a pair of simple **eyes** as well as paired **antennae, jaws,** and **oral papillae.** These structures are dealt with below.

The **legs,** of which there are from 14 to more than 40 pairs, are not divided into articles, but they may be ringed because of the way the papillae of the skin are arranged. Each leg ends in two claws (Figure 20.2, A, D, and E), which is the reason these animals are called onychophorans (''claw-bearers''). A few little pads located just short of the tip of a leg mark the portion that contacts the substratum (Figure 20.2, B, D, and H). The claws are likely to be used only if the surface is slippery. The rather slow movements of onychophorans are effected by bending the legs backward while pushing against the substratum. As certain legs make the effective stroke, others are lifted, swung forward, and touched down again. The body clears the substratum—it does not drag—and there are localized elongations and contractions somewhat comparable to those that lengthen and shorten individual segments or groups of segments in annelids.

Beneath the **epidermis,** which secretes the cuticle, there are several thin sheets of muscle. The fibers of the outermost sheet form a layer of **circular muscle,** but those of other sheets have various oblique orientations. The **longitudinal muscles** tend to be concentrated in distinct bundles. There are, in addition, muscles whose action is very specific and which operate the antennae, jaws, legs, claws, and other movable parts. All of the muscles are of the **smooth** type, and in this respect onychophorans are decidedly different from arthropods.

Digestive Tract and Feeding

The entrance to the preoral cavity, within which the mouth is situated, is bordered by a series of small lappets. Lateral to it are specialized appendages, the oral papillae (Figure 20.2, A, B, and E). On the tips of the papillae are the pores of the **slime glands.** These glands

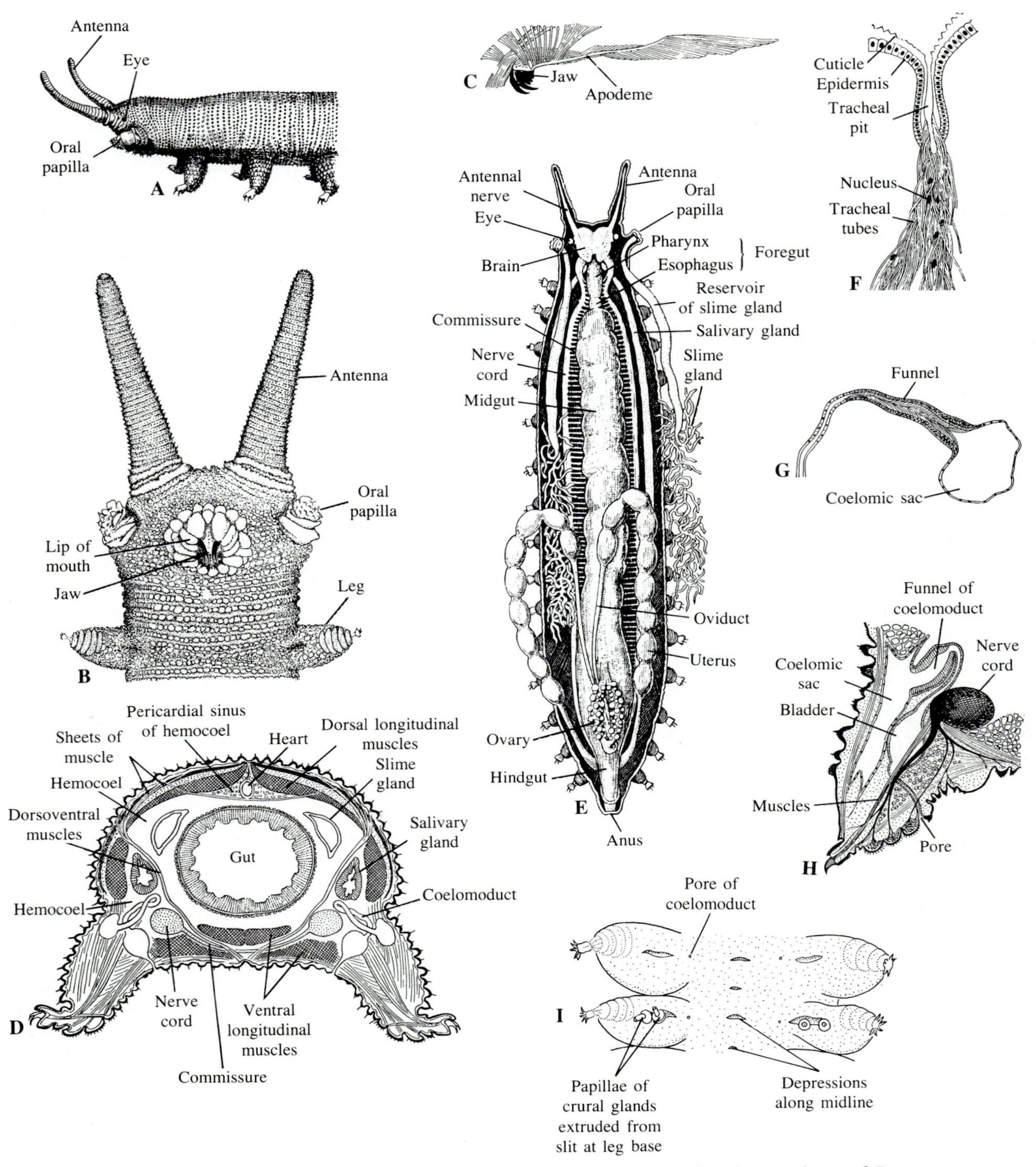

FIGURE 20.2 Morphology of onychophorans. (All drawings, unless otherwise noted, are of *Peripatoides novae-zelandiae,* and have been taken from Snodgrass, Smithsonian Miscellaneous Collections, *97.*) A. Anterior portion of the body. B. *Heteroperipatus engelhardi,* ventral view of the anterior portion of the body. (Zilch, Natur und Museum, *85.*) C. Right jaw and musculature associated with it. D. Transverse section, diagrammatic. E. Dissection of a female, dorsal view. F. Tracheal tubes originating at a tracheal pit. G. Coelomic sac and the funnel of a coelomoduct. H. *Peripatus tholloni,* section of a leg, showing the coelomoduct, nerves, and muscles associated with the appendage. I *Peripatus corradoi,* ventral view of a small portion of a male, near the posterior end, showing the slits at the bases of the legs and also the depressions along the midline. (After Bouvier, Annales des Sciences Naturelles, Zoologie, Series 9, *2.*)

are large in proportion to the animal as a whole and are among the more conspicuous structures exposed by dissection (Figure 20.2, E). Their secretion, used defensively, is squirted out in the form of threads that harden quickly and that may totally immobilize a threatening intruder. Some species can shoot out the secretion for a distance of about 50 cm.

The two jaws located in the preoral cavity itself are tipped by several sharp cusps (Figure 20.2, B and C). The preoral cavity also has the openings of a pair of glands that produce saliva, and it may have a ridge of small teeth on its upper side. The mouth, in the deepest part of the cavity, leads to the foregut, which consists of a pharynx and a short esophagus (Figure 20.2, E). The pharynx operates as a sucking organ. Its lumen is dilated by radial muscles, and there are also circular muscles that promote movement of food. The broad midgut, where digestion and absorption take place, runs for most of the length of the body. The anus, at the posterior end, is preceded by a short hindgut. Both foregut and hindgut, being ectodermal derivatives, are lined by a cuticle, as they are in arthropods.

Captive onychophorans have been observed to eat termites and various other insects, as well as sowbugs and small snails, and presumably these animals are in their natural diets.

Body Cavity, Circulation, Respiration, and Excretion

The body cavity (Figure 20.2, D) is moderately spacious, but it is a **hemocoel,** not a coelom. Muscular partitions divide it lengthwise into a median space, two lateral spaces, and a dorsal pericardial sinus that surrounds the tubular heart. These are not completely separate, so there is movement of blood from one space to another. The heart is like that of arthropods in that it receives blood from the pericardial space by way of paired, valvelike ostia. The number of pairs of these openings is generally the same as the number of pairs of legs.

For respiration, there is a system of **tracheal tubes** somewhat comparable to that of insects and certain other terrestrial arthropods. There are not, however, just a few spiracles and a few main tubes that branch extensively. Instead, onychophorans have hundreds of little pits, usually called stigmata, scattered all over the body surface. Within each pit are several pores that lead to individual tracheal tubes (Figure 20.2, F). The pores cannot be closed, and as a rule the tubes do not branch, but there are so many tubes that most tissues of the body are supplied directly with oxygen.

The excretory organs are coelomoducts. A pair of them is associated with nearly every pair of legs, and their pores are on the medial sides of the leg bases (Figure 20.2, D and H). Each coelomoduct begins with a ciliated funnel that opens into a segmental coelomic sac (Figure 20.2, G). A looped duct, also ciliated for part of its length, leads to a contractile bladder in which urine accumulates, and finally there is a short duct connecting the bladder with the pore (Figure 20.2, D and H). The urine contains a variety of nitrogenous compounds and is released in substantial drops. Nitrogenous wastes, including uric acid, are also eliminated by certain cells of the midgut, and some are stored up in blood cells within the pericardial portion of the hemocoel.

It should be noted that ciliated epithelia in onychophorans are limited to coelomoducts and to the oviducts of females. The gonoducts of both sexes, incidentally, are derived from coelomoducts, as are the salivary glands, already mentioned in connection with the preoral cavity.

Crural Glands and Other Structures that Open to the Exterior

In most onychophorans, each leg has a groove just lateral to the pore of the coelomoduct (Figure 20.2, I). Within the groove is an eversible sac, similar to the structures called *coxal sacs* in symphylans and insects of the order Thysanura. The sacs are believed to be concerned with taking up moisture. The leg grooves may also have structures called **crural glands** associated with them. They are more likely to be found in males than in females. When they are present, they lie in the lateral compartments of the hemocoel, and their openings are situated on small but rather complicated little papillae that can be extended out of the grooves in which they lie. When there are two papillae in a groove, this indicates that there are two crural glands associated with the leg. In males of certain species, the crural glands of the last pair are greatly elongated and reach forward for about half the length of the body. There is much variation in the number and arrangement of crural glands from species to species. These organs probably have something to do with mating, but their exact function is not known.

Along the ventral midline of certain species, there are small depressions at the level of each pair of legs, as well as between successive pairs (Figure 20.2, I). Their significance has not been explained.

Nervous System and Sense Organs

The more conspicuous elements of the nervous system are a rather well-developed brain and two parallel **ventral cords** (Figure 20.2, D and E; Figure 20.3, A). The **brain** has three major divisions, but the limits of these may not be distinct. The first part, with optic lobes, innervates the eyes, and the second part innervates the antennae. The third part of the brain, which is

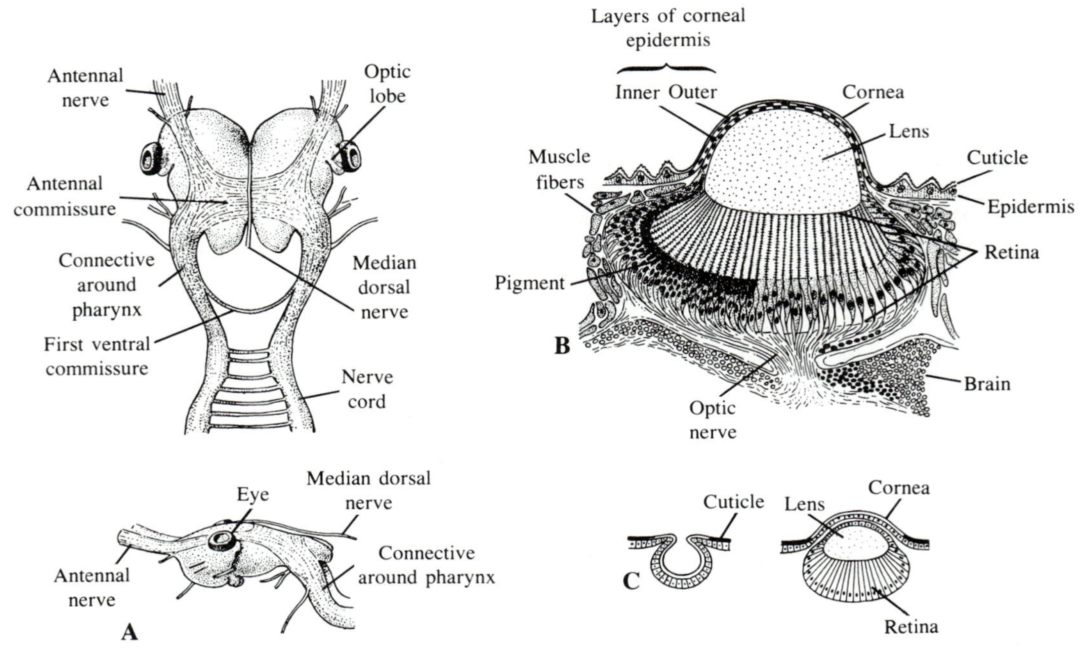

FIGURE 20.3 A. Brain of *Peripatoides novae-zelandiae,* dorsal view (above) and lateral view (below). B. Section through an eye of *Peripatoides occidentalis*. C. Two stages in the development of the eye of *Peripatoides occidentalis*. (Snodgrass, Smithsonian Miscellaneous Collections, 97.)

rather small, sends nerves to the mouth region and anterior portion of the gut, and gives rise to the connectives that run ventrally around the pharynx to the longitudinal nerve cords. It is likely that the three parts of the brain are homologues of the protocerebrum, deutocerebrum, and tritocerebrum of certain arthropods. There are many delicate transverse commissures between the cords, and the cords eventually run together near the posterior end of the body. At the level of each pair of legs the nerve cords have slight swellings, but these are not really ganglia, for they do not consist primarily of nerve-cell bodies.

The **skin papillae,** which are usually most abundant on the antennae, lateral portions of the legs, and upper parts of the body proper, are sensory structures. The base of the cuticular bristle set into each one is closely associated with nerve endings. The fact that the antennae are supplied with substantial nerves from the brain suggests that these appendages are especially important in detecting stimuli. They almost certainly have an olfactory function, but papillae specifically concerned with olfaction have not been identified. Sensory papillae located in the preoral cavity are probably also concerned with olfaction or with taste.

The eyes originate as pits in the epidermis. When fully developed (Figure 20.3, B), each has a cuplike retina, external to which is a lens covered by two layers of epidermis and a cornea. After the epidermal lining of the pit separates from the epidermis that produced it, it secretes the lens, and part of it becomes the retina. The external epidermal layer, continuous with that which covers all of the body, secretes the cornea. Two stages in the formation of the eye are shown in Figure 20.3, C.

REPRODUCTION AND DEVELOPMENT

There is a pair of gonads. These are coelomic derivatives, and at least substantial portions of the gonoducts are coelomoducts that are homologous to the segmental excretory organs. In both sexes, the single genital pore is located on the ventral surface between the legs of the next-to-last pair, or just slightly farther back.

In females (Figure 20.2, E), each oviduct becomes dilated to form what is called a uterus. Between the ovary and uterus there is usually a seminal receptacle. A short, unpaired vagina connects the uteri to the genital pore. Most species bear their young alive, or retain the eggs within the uteri until considerable development has taken place. The uteri generally have several chambers, and in the case of live-bearing types, the chambers may have structures that function in placental nourishment of the young. Some Australian onychophorans lay their eggs soon after they have been fertilized. The genital pore in these species is situated on a process that operates as an ovipositor.

The sperm ducts of males run first to seminal vesicles, then continue as coiled tubes that join to form a single ejaculatory duct leading to the genital pore. Glands along the ejaculatory duct secrete spermatophores in which the sperm are packaged. Mating may involve copulation, in which case spermatophores are deposited in the genital pore of the female. The sperm may be stored for a time in the seminal receptacles before going up into the oviducts or uteri to fertilize the eggs. The sperm are motile. In certain onychophorans—the ones that do not have seminal receptacles—the male climbs on the back of the female and shoots his spermatophores into her cuticle. The tissue at the site of penetration is broken down by amoebocytes, and clusters of sperm can get through the hemocoel to the ovaries, where fertilization takes place.

Patterns of development in the Onychophora show considerable variation. It has already been pointed out that females of most species produce fully formed young. Some, however, release partially formed young, and others lay eggs soon after these have been fertilized, so that all development takes place outside the body. Early cleavage is sometimes superficial, sometimes total. This depends to a considerable extent on the size of the egg and the amount of yolk stored in it. In certain onychophorans whose young develop internally, the embryos are joined to the walls of the uteri, and there are specialized thickenings that function as placentae.

In some species, gastrulation is by invagination and a blastopore is formed. When this is the case, the blastopore closes, but elongates to form a groove. Stomodeal and proctodeal invaginations appear at opposite ends of this groove and form the foregut and hindgut. In the South African *Peripatopsis,* whose embryology has been studied intensively, no blastopore is formed. An ingression of cells at one end of the embryo produces the endoderm. The formation of the gut is completed when the endodermal portion of the gut is joined by stomodeal and proctodeal invaginations.

Mesoderm, laid out in two lateral bands, originates from certain cells that have sunk inward with those that constitute endoderm. The bands become differentiated into serially arranged blocks, each pair of these marking an embryonic somite. Coelomic spaces open up within the mesodermal units, and those of each somite become joined dorsally. Although the spaces survive only in the vesicles of the tubular excretory organs, the lumen of the gonad, and certain portions of the gonoducts, the mesodermal tissue around them forms musculature, connective tissue, the heart, and various other structures.

The number of paired ectodermal outgrowths representing the primordia of the appendages corresponds to the number of embryonic somites. The antennae differentiate from the first pair, which are originally postoral but become preoral when the mouth migrates backward a short distance. The jaws and oral papillae differentiate from the second and third pairs of appendage primordia. The rest of the primordia become legs. *Peripatopsis* bears its young alive, but the embryos are nourished entirely by yolk; there are no placental contacts in the uteri. The gestation period of species that bear their young alive lasts several months. After being born—or after hatching, in the case of egg-laying species—a young onychophoran is on its own. As it grows, it must periodically and frequently molt its cuticle, including that which lines the foregut and hindgut. Maturity is reached within the first year, but onychophorans generally live several years and continue to grow and to molt at rather regular intervals.

REFERENCES ON ONYCHOPHORA

Anderson, D. T., 1966. The comparative early embryology of the Oligochaeta, Hirudinea and Onychophora. Proceedings of the Linnean Society of New South Wales, *91:*10–43.

Eakin, R. M., and Westfall, J. A., 1965. Fine structure of the eye of *Peripatus* (Onychophora). Zeitschrift für Zellforschung, *68:*278–300.

Manton, S. M., 1937. Feeding, digestion, and food storage in *Peripatopsis.* Philosophical Transactions of the Royal Society of London, B, *227:*411–464.

Manton, S. M., 1950. The locomotion of *Peripatus.* Journal of the Linnean Society, Zoology, *41:*529–570.

Robson, E. A., 1964. The cuticle of *Peripatopsis.* Quarterly Journal of Microscopical Science, *105:*281–299.

PHYLUM TARDIGRADA

To find tardigrades, you may not have to look any farther than the nearest clump of moss. If the moss is already wet, work it between your fingers in a shallow dish of water. This will dislodge some of the tardigrades—along with rotifers, nematodes, and other small organisms—so that they can be located with a dissecting microscope. If the only moss available is dry, soak it thoroughly for an hour before going through the routine just described. It is definitely worth the effort to hunt for these animals, for they are among the most engaging of all invertebrates.

Tardigrades are often found in lichens, too, and certain kinds occur on algae, submerged mosses, and other vegetation in fresh water. There are also marine species. Much of the following account of tardigrade morphology and biology applies to members of the class Eutardigrada, which includes some of the types commonly found on mosses and lichens in essentially terrestrial habitats, and nearly all of those that live in fresh water. The other important class, Heterotardigrada, consists partly of marine species, partly of terrestrial species. A few heterotardigrades are discussed later.

External Features, Body Wall, and Musculature

Most tardigrades are less than 0.5 mm long, but some reach 1 mm or more. As a rule, they are slow-moving, pudgy little animals (Figure 20.4, A; Figure 20.5, A and C) that remind one of some fanciful stuffed toy that has come to life. A name often applied to them is "water bears," but "moss piglets" would also be right for them. They lumber along on four pairs of legs, which are stubby, simple outgrowths of the body wall. In eutardigrades, there are generally two claws at the tip of each leg, but these may be split into two diverging points (Figure 20.5, A and C). *Milnesium* (Figure 20.4, A and B) differs from other genera in having two short, stout claws and two long, slender ones on each leg.

The body wall is fairly thin, for in most places it consists of little more than the epidermis and cuticle. The cuticle is made up of proteins; there is no chitin in it, and in this respect tardigrades differ from arthropods and onychophorans. Some eutardigrades have prominent tubercles on the body surface. The epidermis that secretes the cuticle is said to show **cell constancy;** in other words, it is derived from a fixed number of embryonic cells. This is a feature found in certain tissues of nematodes, rotifers, and other pseudocoelomate invertebrates.

In order to grow, a tardigrade is obliged to molt periodically. The claws and the lining of the foregut and hindgut are cuticular structures, so they are shed, too. A new cuticle is secreted before the animal contracts to separate itself from the old one. The rather large claw glands in the legs (Figure 20.5, A; Figure 20.7, B) are concerned with making claws to replace those lost at each molt.

The muscles involved in locomotion and changes in body shape are either individual fibers or small bundles. Thus they resemble muscles of arthropods more than they do those of onychophorans or annelids; the muscles are not striated, however.

Body Cavity, Circulation, and Excretion

The body cavity is a hemocoel. Five pairs of true coelomic pouches do appear during development, but only those of the posteriormost pair survive. They fuse and form the unpaired ovary or testis, depending on the sex of the animal. In any case, the fluid of the hemocoel serves as a circulatory system to distribute nutrients, dissolved gases, and wastes.

Exchange of oxygen and carbon dioxide must take place across the body surface, and perhaps some soluble nitrogenous wastes leave the body by simple diffusion, especially in aquatic species. In most terrestrial tardigrades, relatively insoluble nitrogenous compounds accumulate beneath the cuticle and in close association with the stylets, and are eliminated during the process of molting. Insoluble wastes are also said to become concentrated in the epidermis of some species.

In eutardigrades, there are three small glands—two lateral, one dorsal—opening into the digestive tract at the place where the midgut joins the hindgut (Figure 20.5, A). They are believed to have an excretory function, but little is known about them. In some accounts of tardigrade morphology, the lateral glands are referred to as Malpighian tubules.

Digestive Tract and Feeding

Most tardigrades suck juices out of plant cells, but some, including *Milnesium tardigradum* (Figure 20.4, A) and certain species of *Macrobiotus,* have been observed to draw nourishment from rotifers, nematodes, and other tardigrades. (*Milnesium,* in fact, is mostly carnivorous.) The suction is effected by the action of radial muscles that dilate the pharynx, which is a bulbous differentiation of the foregut (Figure 20.4, C; Figure 20.5, A and B). The other parts of the foregut are the more anterior buccal tube and the more posterior esophagus. On both sides of the former is a prominent stylet and a hard, transversely oriented bar that serves as a support (Figure 20.5, B). These cuticular structures are secreted by gland cells belonging to the epithelium of the buccal tube. While a tardigrade is preparing to molt, therefore, it must also begin to produce new stylets and supports. The stylets are operated by muscles and can be protruded out of the mouth to make a puncture that starts the cell sap flowing.

The rather spacious midgut is where digestion and absorption are assumed to take place. The narrow hindgut, cuticularized like the foregut, leads to the anus, which is usually located on the ventral surface close to the posterior end. In eutardigrades the hindgut is a cloaca, for it is joined by the oviduct or by the two sperm ducts, depending on the sex. A seminal receptacle is usually associated with it in females.

Nervous System and Sense Organs

The brain, dorsal to the buccal tube, is proportionately large (Figure 20.5, C). Its more prominent components are two conspicuous lateral lobes. The brain is joined by connectives to a ganglion beneath the buccal tube, and from this ganglion a double nerve cord runs posteriorly. The cord has four ganglionic swellings. Nerves from the ganglia supply various parts of the body, including the legs, which themselves have small ganglia. Most tardigrades have a pair of small eyespots close to the brain. These mark the location of photoreceptors.

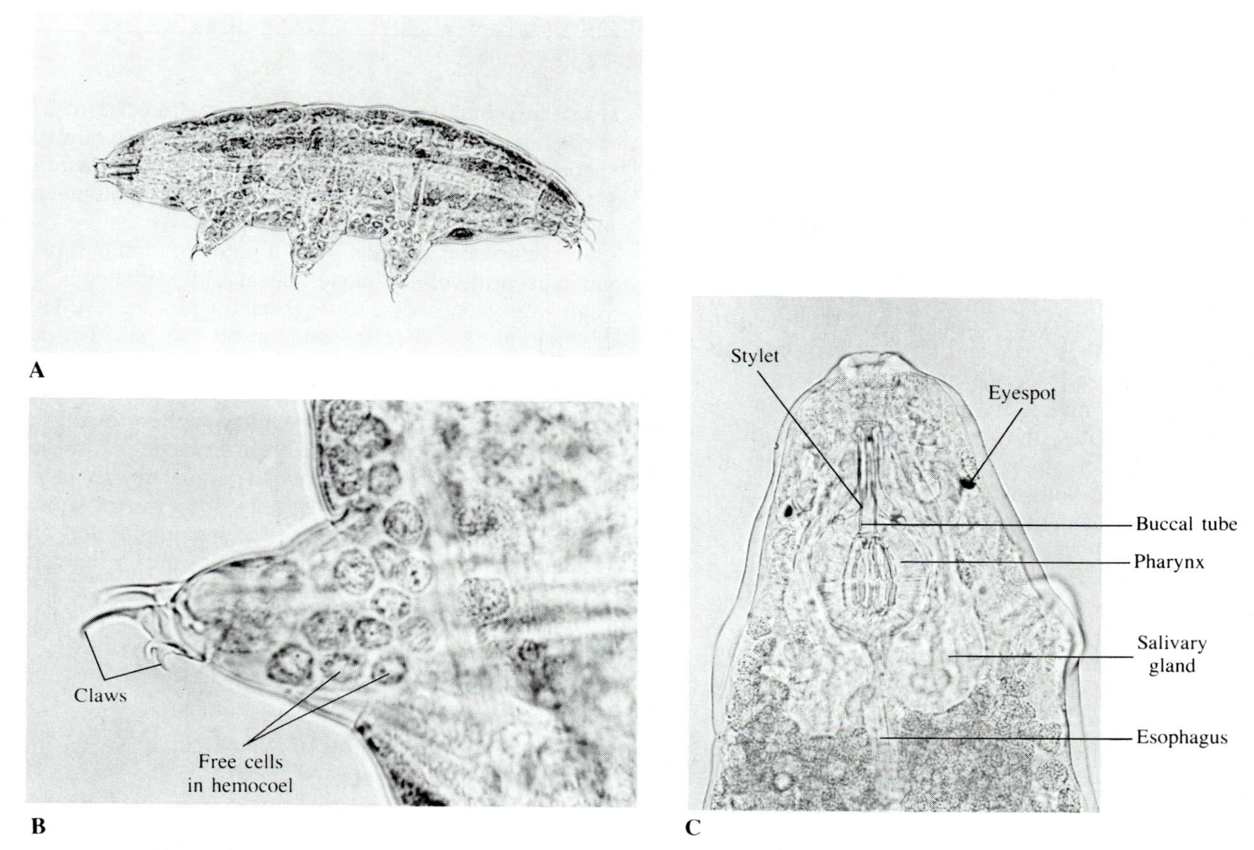

FIGURE 20.4 Living tardigrades; photomicrographs. A. *Milnesium tardigradum,* from the left side. B. *Milnesium,* one leg. C. *Macrobiotus,* head region.

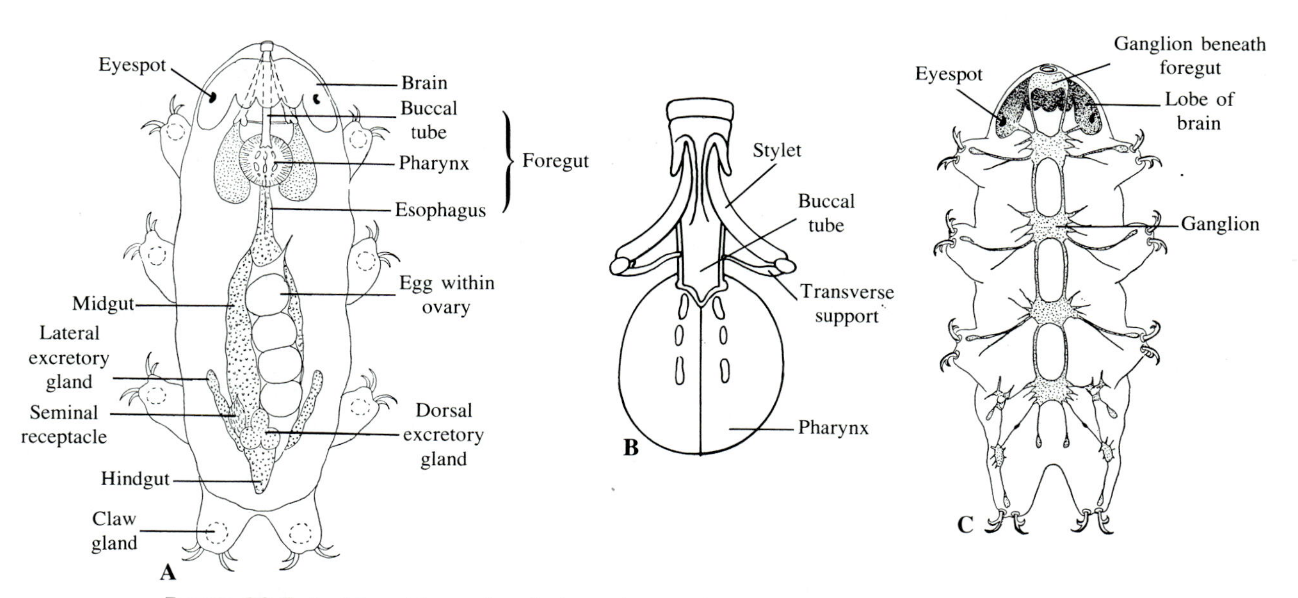

FIGURE 20.5 A. *Macrobiotus,* dorsal view, showing internal anatomy; semidiagrammatic. (Ramazotti, Memorie dell'Istituto Italiano d'Idrobiologia, *28.*) B. Buccal tube and stylet apparatus of a species of *Macrobiotus* (musculature omitted). (Puglia, Transactions of the American Microscopical Society, *83.*) C. *Macrobiotus,* nervous system, ventral view. (Ramazotti, Memorie dell'Instituto Italiano d'Idrobiologia, *28.*)

REPRODUCTION AND DEVELOPMENT

Sexes are separate. The single gonad, derived from the two posterior coelomic pouches that fuse, lies above the midgut. It has been pointed out previously that in eutardigrades the oviduct and seminal receptacle of females and the two sperm ducts of males are joined to the hindgut, so the anus serves as a genital pore.

Tardigrades are inconsistent when it comes to the ways in which they mate and fertilize their eggs. In the first place, not all species are known to have males. Those without males are parthenogenetic. In species that do have males, the sperm may be deposited in the hindgut of the female or they may somehow be injected through the thin body wall into the hemocoel. With either of these techniques of insemination, actual fertilization probably takes place in the ovary, and the eggs are then laid. A rather different style of insemination is reported for certain species. The sperm are injected through the old cuticle of a female that is getting ready to molt. The female then lays her eggs into the space between the old and the new cuticle, and it is here that they are fertilized. Development takes place in the cuticular sac left behind by the newly molted female. The fact that a tardigrade lays its eggs in a cuticular sac (Figure 20.6, A) does not necessarily mean that insemination has taken place in the manner just described.

The eggs of some tardigrades, especially terrestrial types, are thick-shelled and can resist desiccation. The shells are often sculptured (Figure 20.6, B). Certain species produce thin-shelled eggs when moisture is plentiful and thick-shelled eggs when conditions are more precarious.

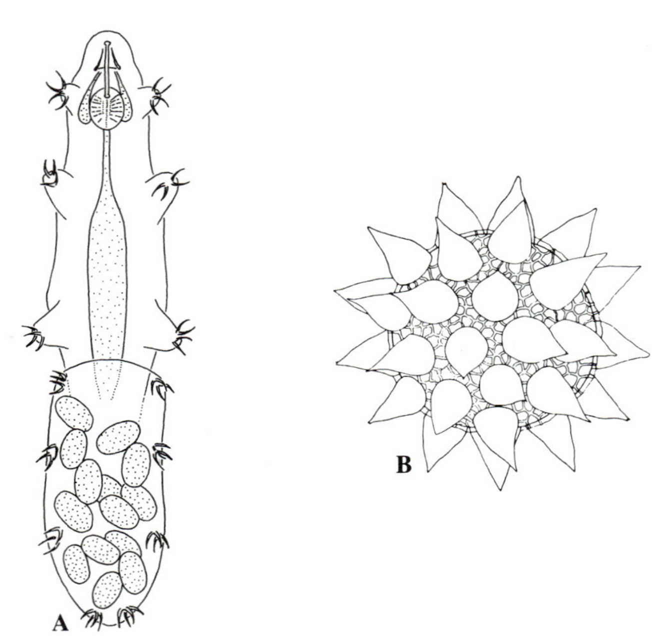

FIGURE 20.6 A. *Hypsibius nodosus,* in the process of shedding its cuticle, into which it has laid its eggs. (After Puglia, Transactions of the American Microscopical Society, *83.*) B. Egg of *Macrobiotus spectabilis,* (Ramazotti, Memorie dell'Istituto Italiano d'Idrobiologia, *28.*)

Development is direct. The foregut and hindgut originate, respectively, as a stomodeal and a proctodeal invagination. Mesoderm is segregated when the five pairs of coelomic pouches are budded off the gut. The last pair of pouches forms the ovary or testis, depending on the sex of the prospective adult. The other pouches break down, but the cells of which they are composed do not disappear; most of them differentiate into muscle cells. A young tardigrade, when it is ready to hatch, ruptures the eggshell with the help of its buccal stylets.

The serially arranged appendages, ganglia, and coelomic pouches suggest that tardigrades are in the same general evolutionary assemblage as annelids, arthropods, and their close allies. The fact that the coelomic pouches are budded from the gut is decidedly not an annelidan or arthropodan character, however.

ANABIOSIS, ENCYSTMENT, AND SIMILAR PHENOMENA

Tardigrades that live in mosses and lichens, or in other habitats that dry out periodically, are capable of going into a state of suspended animation. This condition is called **anabiosis,** or *cryptobiosis*. A tardigrade will not survive abrupt desiccation; the loss of water must be gradual. The animal contracts, withdrawing the legs, and becomes a compact, almost barrel-shaped package. The metabolic activity of a tardigrade in the anabiotic state is much reduced. Reduction of the oxygen content of the atmosphere slows metabolism even further and increases the longevity of the anabiotic phase.

Certain species can remain in the anabiotic state almost indefinitely, and some can withstand extremely unfavorable circumstances while they are in this condition. Even absolute alcohol, ether, or liquid helium (−272° C!) may not kill them. When moisture becomes available again, an anabiotic individual absorbs water and becomes active. Tardigrades taken from herbarium specimens of mosses that had been kept dry for many years revived when placed in water, although they lived for only a short time after that. In general, however, anabiotic animals recover normally, and some have been shown to be capable of going through at least several episodes of suspended animation during their lifetimes.

Some tardigrades, particularly those found in fresh water, form what may be called cysts. This happens when the animals dry out, and the resulting condition is not quite the same as anabiosis, because there is usually dedifferentiation of certain internal structures. The cyst is nothing more than the old cuticle, which separates from the new cuticle that has been secreted beneath it, but it helps protect the tardigrade from dryness.

Other forms of latency in tardigrades have been given names. *Cryobiosis* and *osmobiosis* refer, respectively, to suspended animation induced by cold and high osmotic pressure. *Anoxybiosis* is induced by lack of sufficient oxygen. This condition is fatal if prolonged for more than a few days.

HETEROTARDIGRADES

The class Heterotardigrada includes all marine genera and certain of the genera found on terrestrial mosses and lichens. The members of this group are characterized by sensory outgrowths, called cirri, at the anterior end of the body, and sometimes elsewhere. They lack the three glands that are associated with the junction of the midgut and hindgut in eutardigrades. The oviduct or sperm ducts, moreover, do not enter the hindgut, but lead to a separate genital pore on the ventral surface just in front of the anus. When claws are present, there are at least four on each leg.

Four-clawed heterotardigrades belong to *Echiniscus* (Figure 20.7, A and B) and a few closely related genera that inhabit mosses and lichens. The cuticle in these is organized into several plates separated by sutures, so that the animals appear to be armored. *Echiniscoides* (Figure 20.7, C), with just one species, which lives on delicate intertidal marine algae, may have as many as 11 claws on each leg.

Perhaps the most interesting heterotardigrades are the marine species belonging to *Batillipes* (Figure 20.7,

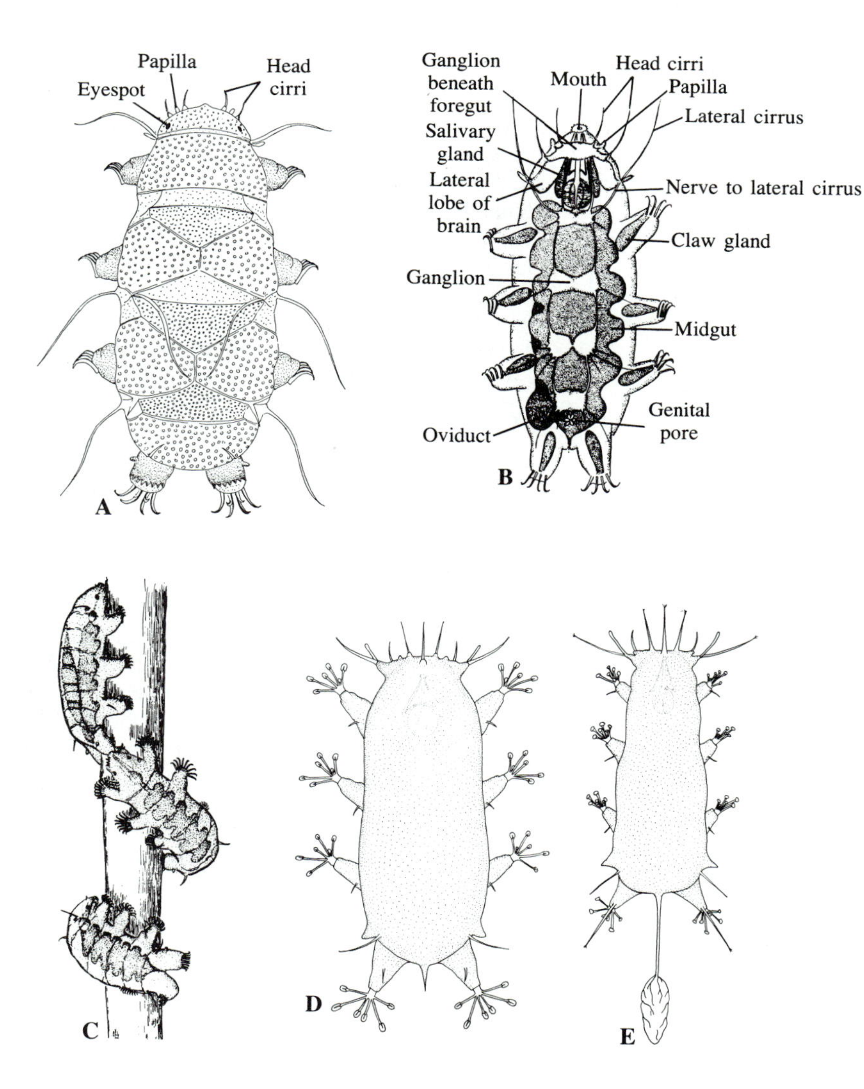

FIGURE 20.7 Heterotardigrades. A. *Echiniscus,* showing cuticular plates characteristic of this genus, as well as the anterior cirri characteristic of all heterotardigrades. (Ramazotti, Memorie dell'Istituto Italiano d'Idrobiologia, *28.*) B. *Echiniscus,* ventral view, showing some details of internal and external anatomy. C. *Echiniscoides sigismundi,* on a green alga, *Enteromorpha.* (B and C, Marcus, in Bronn, Klassen und Ordnungen des Tierreichs, *5*:4:3.) D. *Batillipes mirus,* a marine tardigrade specialized for living between sand grains. E. *Batillipes bullacaudatus.* The caudal spine is unusual in that it ends in a bulbous swelling. (D and E, McGinty and Higgins, Transactions of the American Microscopical Society, *87.*)

D and E) and allied genera. Instead of claws, these have several "toes" on each leg. The toes end in adhesive pads, which enable the animals to cling to sand grains.

A few marine species are associated with other animals. One, for instance, lives on the tentacles of sea cucumbers that inhabit sand. Another has been reported from the abdominal region of a small isopod that burrows into wood. Little is known about these symbiotic types.

References on Tardigrada

Crowe, J. H., and Cooper, A. F., Jr., 1971. Cryptobiosis. Scientific American, *225*(12):30–36.

Crowe, J. H., and Higgins, R. P., 1967. Revival of *Macrobiotus areolatus* from the cryptobiotic state. Transactions of the American Microscopical Society, *86:*286–294.

Higgins, R. P., 1959. Life history of *Macrobiotus islandicus* Richters with notes on other tardigrades from Colorado. Transactions of the American Microscopical Society, *78:*137–154.

Higgins, R. P. (editor), 1975. International Symposium on Tardigrades. Memorie dell'Istituto Italiano d'Idrobiologia, *32* (Supplement):1–469.

Morgan, C. T., and King, P. E., 1976. British Tardigrades. Synopses of the British Fauna, no. 9. Linnean Society of London and Academic Press, London and New York.

Nelson, D. R., 1982. Developmental biology of the Tardigrada. *In* Harrison, F. W., and Cowden, R. R. (editors), Developmental Biology of Freshwater Invertebrates, pp. 363–398. Alan R. Liss, New York.

Nelson, D. R. (editor), 1982. Proceedings of the Third International Symposium on Tardigrada. East Tennessee State University Press, Johnson City.

Pollock, L. W., 1976. Tardigrada. *In* Marine Flora and Fauna of the Northeastern United States. NOAA Technical Report, NMFS Circular 394.

Renaud-Mornant, I., and Pollock, L. W., 1971. A review of the systematics and ecology of the marine Tardigrada. Smithsonian Contributions to Zoology, *76:*109–117.

Riggin, C. T., Jr., 1962. Tardigrada of Southwest Virginia, with the addition of a description of a new marine species from Florida. Virginia Agricultural Experiment Station Technical Bulletin, *152:*1–145.

Schuster, R. O., and A. A. Grigarick, 1965. Tardigrades from western North America. University of California Publications in Zoology, *76:*1–67.

Phylum Pentastomida

The pentastomids constitute a small phylum of wormlike parasites found in the lungs, nostrils, nasal passages, or head sinuses of a variety of reptiles, birds, and mammals. As a rule, their life cycles involve two hosts. Thus, for instance, a fox or wolf might become parasitized after eating a rabbit that harbors the juvenile stage.

The name pentastomid alludes to the presence, in some species, of five obvious projections in the anterior portion of the body (Figure 20.8, A, B, and D). The mouth is situated on one of the projections; the others, tipped with claws for grasping the tissue of the host, are paired lateral appendages. In some genera they are reduced to the point that only the hooks persist.

Pentastomids that live in the lungs are primarily bloodsuckers; those that live in nasal passages and related cavities feed on epithelial cells, mucus, and lymph. Much remains to be learned about their habits, however.

General Structure

The body of an adult pentastomid is generally a few centimeters long. It is covered by a cuticle that contains chitin and is usually ringed externally (Figure 20.8). Beneath the epidermis, there are broad bands of circular muscle separated by intervals in which circular muscle is absent (Figure 20.9, A and B). The ringlike thickenings of the body coincide with the location of the circular bands. The longitudinal muscles are concentrated in numerous bundles, those on the ventral side being more substantial than those elsewhere. Although each longitudinal bundle runs for a long distance, some fibers diverge from it at each ring and terminate (Figure 20.9, B). Internal to the longitudinal muscles, each ring has two pairs of short, more or less oblique muscles that originate at the ventral midline (Figure 20.9, A). Most of the fibers of all three categories of muscle just described insert upon the cuticle by way of tonofibrils produced by cells of the epidermis. The muscles of pentastomids, like most muscles of arthropods, are striated.

Besides the structures of the body wall already mentioned, there are numerous small epidermal glands that open through pores in the cuticle. The function of their secretion is unknown. In most pentastomids, there are two pairs of comparatively large glands lying in the anterior part of the body cavity (Figure 20.10). The frontal glands are elongated structures, and their ducts extend to the frontal papillae on the head. The hook glands are more compact, and their ducts extend to the two pairs of clawed appendages that flank the mouth.

The body cavity is a hemocoel, not a true coelom. Some coelomic pouches appear, however, during early development.

The digestive tract is complete and runs to the posterior tip of the body (Figure 20.10). The foregut, consisting of a sucking pharynx and a slender esophagus (Figure 20.9, C), is lined by a cuticle continuous with that which covers the outside of the body. The dorsal side of the pharynx forms a pluglike papilla, and the muscles within this are attached to a shield of much-thickened cuticle that covers part of the papilla. When

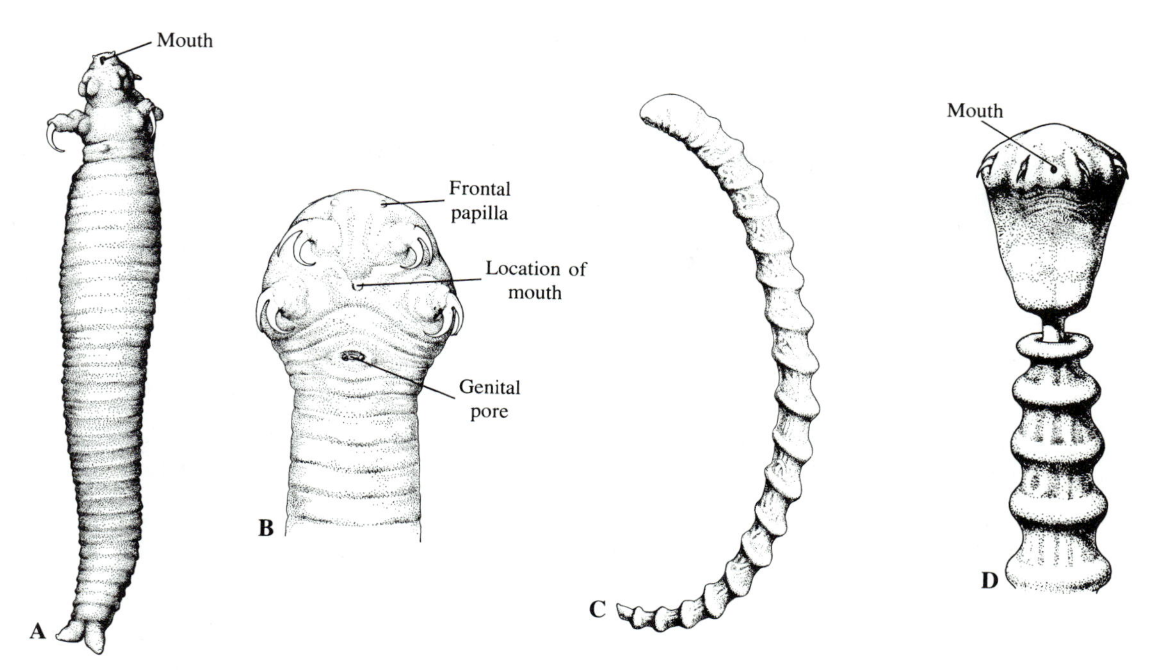

Figure 20.8 A. *Raillietiella mabuiae* (parasitic in the lungs of an African lizard). B. *Leiperia gracilis* (parasitic in the lungs of South American crocodiles), anterior portion of an immature male. C. *Armillifer armillatus* (parasitic in the lungs of African snakes), female. D. *Armillifer annulatus* (parasitic in the lungs of African snakes), anterior portion of a female. (Heymons, in Bronn, Klassen und Ordnungen des Tierreichs, *5*:4:1.)

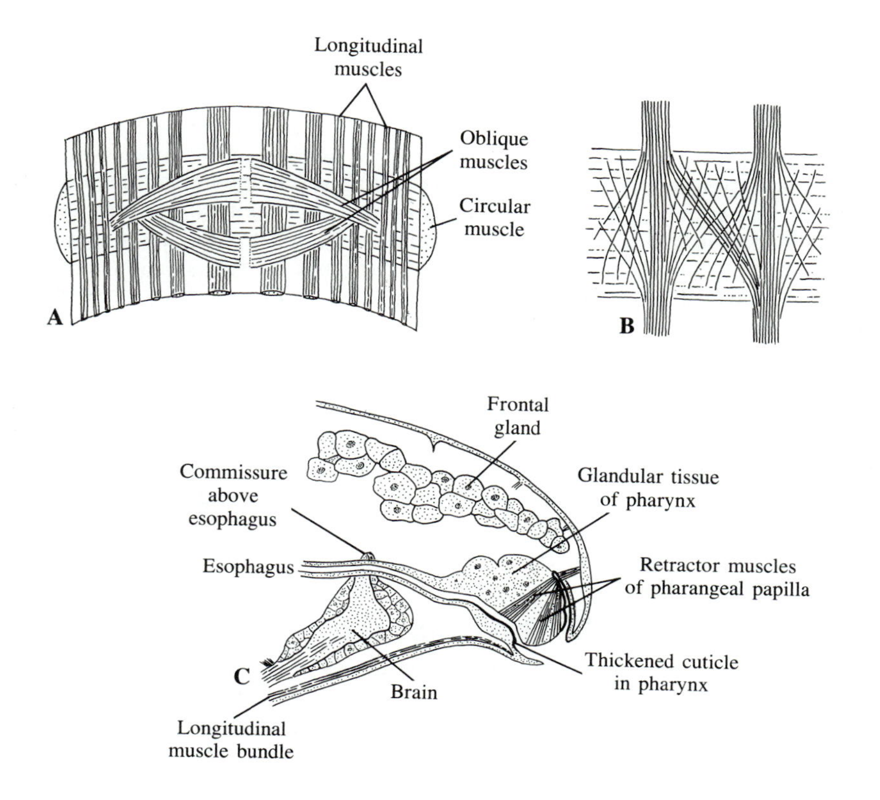

Figure 20.9 A. *Armillifer armillatus,* portion of the body wall on both sides of the midline, showing longitudinal muscles, the band of circular muscle, and the two sets of oblique muscles characteristic of each ring of the body; diagrammatic. B. *Armillifer armillatus,* portion of the body wall, showing the way some fibers of a longitudinal muscle bundle diverge from the bundle before inserting on the cuticle or joining another bundle; diagrammatic. C. *Raillietiella mediterranea,* sagittal section of the anterior portion of the body, showing specializations of the foregut. (After Heymons, in Bronn, Klassen und Ordnungen des Tierreichs, *5*:4:1.)

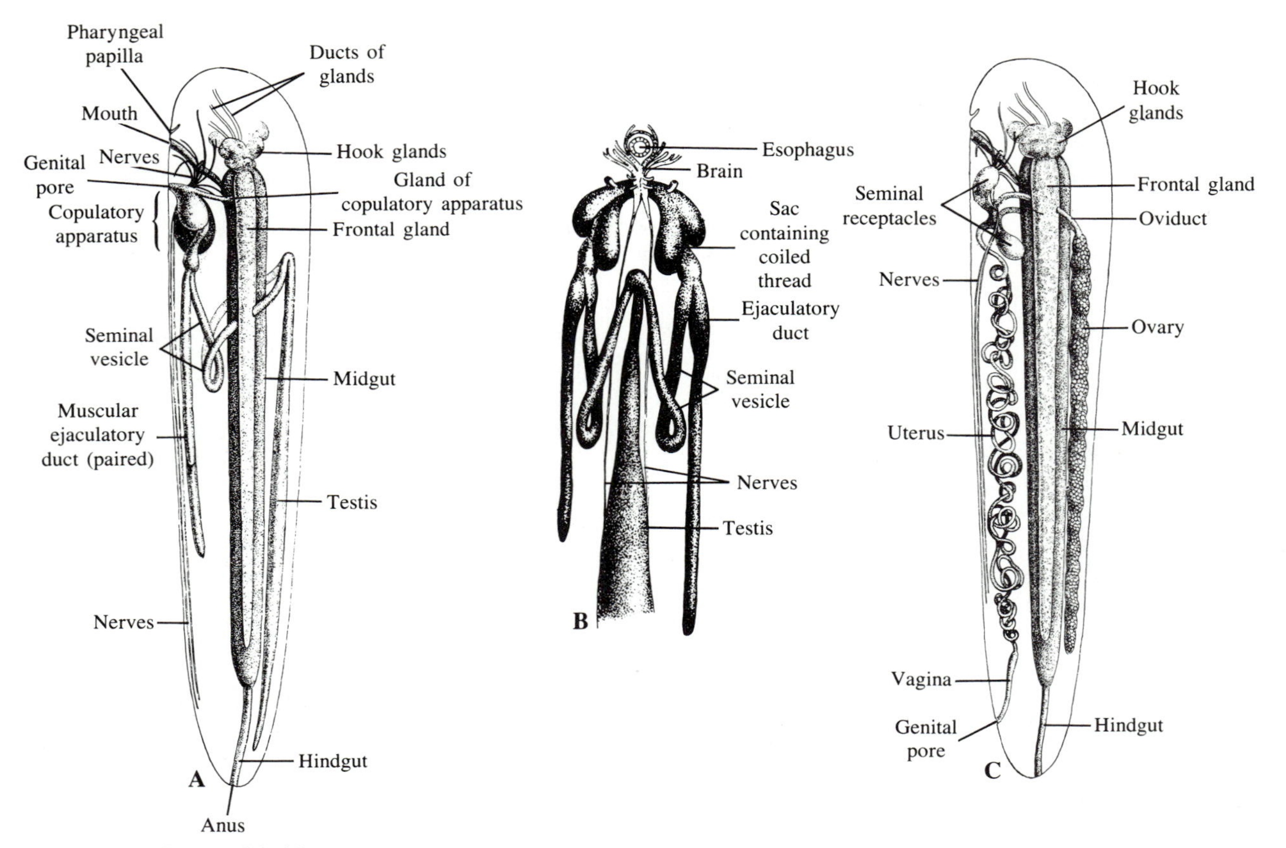

FIGURE 20.10 *Waddycephalus teretiusculus* (parasitic in the lungs of Australian snakes); all figures are diagrammatic. A. Male, from left side. B. Reproductive system of male, dorsal view. C. Female, from left side. (Heymons, in Bronn, Klassen und Ordnungen des Tierreichs, *5*:4:1.)

the muscles in the papilla relax, the lumen is blocked; when they contract, the lumen is opened. Presumably, the sucking activity of the pharynx is due at least in part to the way the papilla operates.

Digestion and absorption are assumed to take place in the midgut, which makes up most of the the digestive tract. The hindgut, like the foregut, is lined by cuticle.

There is no circulatory system as such: the hemocoel assumes its functions. Excretory organs seem to be lacking, although some specialists suppose that glands in the epidermis are concerned with elimination of nitrogenous wastes.

In some genera, the successive pairs of ganglia that make up the brain are rather distinct; in others, there is scarcely any demarcation of ganglia. In both groups, however, the right and left halves of the anteriormost part of the brain are connected by a commissure that arches over the esophagus. Among the structures innervated by nerves from the brain are the two pairs of appendages, muscles of the pharynx, and frontal papillae. Behind the brain, there may be a short chain of paired ganglia comparable to those in the ventral nerve cord of an annelid or arthropod, but the two distinctly separate ventral nerves that extend posteriorly for much of the length of the body (Figure 20.10) are viewed as long branch nerves rather than as halves of a ventral cord. In some pentastomids, in fact, these nerves come directly from the hind portion of the brain, and there is nothing comparable to a ganglionated ventral cord.

Sensory receptors are found in the skin. They are often concentrated on little papillae or distributed along a lateral line. When papillae are present, these can sometimes be everted by increasing muscular pressure on the fluid in the hemocoel; withdrawal of the papillae is effected by muscles that extend into them.

REPRODUCTION AND DEVELOPMENT

The reproductive system is well developed. In males (Figure 20.10, A and B), the single ventral gonopore is usually near the anterior end of the body. There are generally two testes, although a few pentastomids have a single testis, presumably the result of fusion of two gonadal primordia. In either case, there are two sperm ducts, each with a dilation that serves as a seminal vesicle. There is a complicated copulatory apparatus that includes a pair of muscularized ejaculatory ducts and a pair of sacs that contain chitinized threads. During copulation, the threads are extruded and enter the vagina of the female.

The position of the female genital pore varies. In one of the two major groups (order Cephalobaenida) it is at about the same level as the male gonopore; in the other group (order Porocephalida) it is in the posterior half of the body. There is always a single ovary (Figure 20.10, C). Where the paired ducts from this unite to form a long, coiled uterus–vagina complex, there are two seminal receptacles. It is thought that the eggs coming down the oviducts are fertilized when they reach the seminal receptacles. Thousands of eggs accumulate in the uterus, and this duct may become greatly extended. The eggs are small—usually less than 0.2 mm in diameter—but they have thick shells. After they are laid, they are carried with mucus to the pharynx, swallowed, and eventually eliminated with feces, or they may be expelled more directly when the host sneezes.

Embryonic development has been studied in a few pentastomids, especially *Reighardia sternae,* which lives in the air sacs of gulls and terns. It appears that after a blastula has been formed, endoderm is produced by delamination. Primordia of four pairs of appendages become recognizable. One pair is placed beside the mouth, which is formed by a stomodeal invagination; the other three pairs are postoral. At the base of the primordium of each appendage is a rudiment of a ganglion and also a coelomic pocket. On hatching, a pentastomid has only two pairs of appendages (or just two pairs of hooks, in case the appendages themselves fail to materialize). The two posterior pairs of coelomic pouches contain cells that will give rise to gametes. Many other details have been described, but they are not dealt with here.

Pentastomids are not a pivotal group in the evolution of arthropods, but anything that can be learned about their development may shed some light on where they fit in. The fact that their appendages, coelomic pockets, and ganglia are laid out metamerically in the embryo suggests a close relationship to annelids and arthropods. They could be arthropods that have become so highly specialized that they are no longer recognizable as such.

Most pentastomids seem to require an intermediate host, and a wide variety of invertebrates and vertebrates are implicated. Sometimes, by accident, they get into hosts that offer no promise for completing their life cycle. Humans, for instance, occasionally have larval pentastomids. Larvae that develop in a favorable intermediate host feed, grow, and molt one to several times. Just three examples are given here to illustrate how some representative life cycles of pentastomids are completed.

In the case of species of *Raillietiella* that live in the lungs of lizards, the intermediate hosts are usually cockroaches or other insects. After an insect swallows a larva, still enclosed by a two-layered eggshell (Figure 20.11, A), the larva hatches and uses its boring apparatus, consisting of cuticular stylets, to work its way into the hemocoel. Here it undergoes further development and loses some of the larval structures, including the boring apparatus. If the insect is eaten by a lizard, the juvenile pentastomid reaches the lungs, perhaps directly from the mouth, and matures.

In the life cycle of species of *Porocephalus,* which live in the lungs of snakes, eggs containing larvae are eaten by small mammals. The larvae (Figure 20.11, B) hatch and become encapsulated in the viscera, where they develop into juveniles. After the mammal is eaten by a snake, the young worms crawl up the esophagus to the trachea, and thus enter the lungs.

Species of *Linguatula* also have a complicated life history. That of *L. serrata* is well known. An egg eaten with vegetation by a rabbit or hare hatches in the intestine, and the larva works its way through the intestinal wall, eventually reaching one of several internal organs where it can develop and where it undergoes several molts. The more advanced juveniles, with two pairs of hooks as in the adult, are characterized by a spiny cuticle (Figure 20.11, C). After a dog, wolf, or fox eats the intermediate host, the last-stage juvenile goes from the stomach to the esophagus, then to the pharynx, and finally crawls to the nasal cavity, its permanent residence. Here it molts for the last time, and its spiny cuticle is replaced by one that is smooth. Juveniles of *L. serrata* are sometimes found in humans and other mammals that cannot be considered normal intermediate hosts.

REFERENCES ON PENTASTOMIDA

Haffner, K. von, 1973. Über die Entwicklung, vergleichende Anatomie und Evolution der Extremitäten von Pentastomiden. Zeitschrift für Zoologische Systematik und Evolutionsforschung, *11:*241–268.

Osche, G., 1959. "Arthropodencharacktere" bei einem Pentastomiden-Embryo *(Reighardia sternae).* Zoologischer Anzeiger, *163:*169–178.

Osche, G., 1963. Systematische Stellung und Phylogenie der Pentastomida. Zeitschrift für Morphologie und Ökologie der Tiere, *52:*487–596.

Riley, J., 1983. Recent advances in our understanding of pentastomid reproductive biology. Parasitology, *86:*59–83.

Riley, J., Banaja, A. A., and James, J. L., 1978. The phylogenetic relationships of the Pentastomida: the case for their inclusion within the Crustacea. International Journal of Parasitology, *8:*245–254.

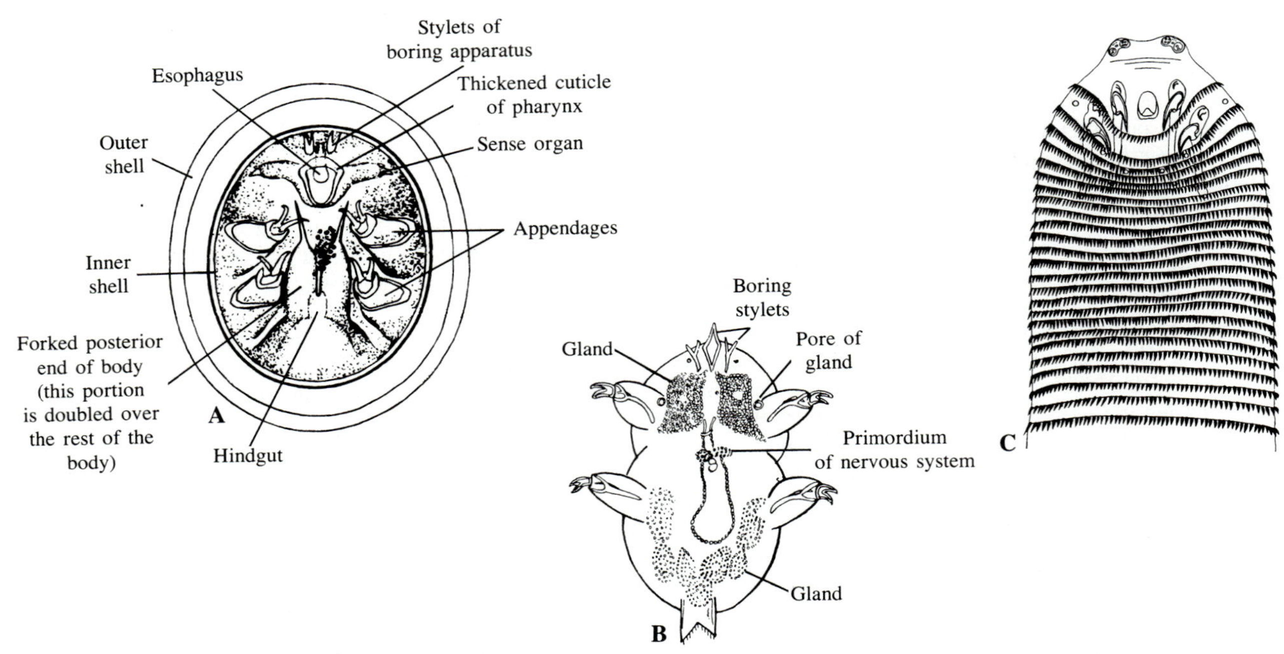

Figure 20.11 A. *Raillietiella kochi,* infective larva within the two-layered eggshell. (Adults live in the lungs of lizards.) B. Larva of *Porocephalus clavatus,* from an opossum. (Adults live in the lungs of South American snakes.) (Stiles, Zeitschrift für Wissenschaftliche Zoologie, *52.*) C. Juvenile of *Linguatula serrata,* found in the lungs of rabbits, hares, and various other mammals, including humans. (Adults are parasitic in the nasal cavity and frontal sinuses of dogs, wolves, and foxes.) The spines of the posterior margin of each ring are part of the cuticle. They disappear when the juvenile molts for the last time. (Faust, American Journal of Tropical Medicine, *7.*)

Summary

1. The Onychophora, Tardigrada, and Pentastomida are entirely separate groups, but all of them exhibit some features that link them with arthropods. They have paired appendages, and although these are not jointed, the ones used for locomotion or grasping are usually provided with claws. There is a cuticle that is molted at intervals to allow for growth, and the body cavity is a hemocoel.

2. The ventral nerve cord of onychophorans and tardigrades is a double structure, as it is in annelids and arthropods, but in pentastomids such a ventral cord extends for only a short distance behind the brain, if at all. Segmentation of the body is apparent in the arrangement of the appendages and certain internal organs, but it is not as obvious as it is in annelids and arthropods, and the segments are not grouped in such a way that they form distinct tagmata.

3. The onychophorans, which are terrestrial but depend on a humid habitat, have a pair of antennae and 14 or more pairs of legs, as well as jaws and a pair of specialized anterior appendages on which are the openings of the slime glands, used in defense. They are similar to annelids in that they have serially arranged coelomic sacs and ciliated coelomoducts, and also because their muscles are of the smooth type. The heart of onychophorans, however, is like that of most arthropods in having ostia through which blood returns to it from the hemocoel, and there is a tracheal system that supplies oxygen directly to the tissues. These animals are primarily carnivores that feed on insects and other small invertebrates. Their distribution is limited to the southern hemisphere and to tropical regions of the northern hemisphere.

4. Neither tardigrades nor pentastomids have a heart, and although coelomic spaces appear during embryonic development, these do not give rise to coelomoducts comparable to those of onychophorans.

5. Tardigrades, which are so small as to be essentially microscopic, live in moist places on land, and also in freshwater and marine habitats. They have four pairs of legs, generally tipped with claws but sometimes with adhesive pads. Most of them feed on juices drawn from cells of mosses and other plants, but some get their nourishment from small animals, such as rotifers. The foregut is equipped with piercing stylets, and a portion of it, with radial muscles for dilating the lumen, functions as a sucking organ. All of the muscles of tardigrades are smooth.

6. Pentastomids are parasites of vertebrates. They generally inhabit the lungs, nostrils, nasal passages, or head sinuses, and they suck in blood, epithelial cells, or mucus. Some species require two hosts for completion of the life cycle. Most pentastomids are wormlike animals, and their bodies are often divided into numerous ringlike thickenings, each of which has a prominent band of circular muscle. Near the mouth, close to the anterior end, there are two pairs of clawed appendages (or vestiges of these) used for grasping the tissues of the host. The muscles are striated, as in arthropods.

21

PHYLUM ECHINODERMATA

SEA STARS, SEA URCHINS, AND THEIR RELATIVES

Non coelo tantum, sed et mari suae stellae sunt.
(There are stars not only in the sky,
but also in the sea.)

J. H. Linck, De Stellis Marinis, 1733

- Phylum Echinodermata
  - Class Crinoidea
    - Order Isocrinida
    - Order Comatulida
  - Class Asteroidea
    - Order Platyasterida
    - Order Paxillosida
    - Order Valvatida
    - Order Spinulosida
    - Order Forcipulatida
  - Class Ophiuroidea
    - Order Oegophiurida
    - Order Phrynophiurida
    - Order Ophiurida
  - Class Echinoidea
    - Subclass Perischoechinoidea
    - Subclass Euechinoidea
  - Class Holothuroidea
    - Subclass Dendrochirotacea
    - Subclass Aspidochirotacea
    - Subclass Apodacea

INTRODUCTION

Echinoderms—the "spiny-skinned animals"—are strictly limited to the sea. The basic features of this phylum, which has about 6000 living species, were established early, definitely before the end of the Cambrian period. Although some distinctive types died out long ago, their fossils have helped explain the derivation and relationships of the five classes that have survived.

Among the more complex invertebrates, echinoderms are the only ones that have **radial symmetry,** a **calcareous internal skeleton,** and a unique coelomic derivative called the **water-vascular system.** The skeleton consists of many separate **ossicles** (spines). These are secreted in the dermal layer, whose tissues are derived from mesoderm. Even when the ossicles appear to be at the surface, they are covered by epidermis, except where this may have been worn away. Thus even the long, sharp spines characteristic of sea urchins are actually internal skeletal elements.

An ossicle resembles a crystal because the microcrystals of which it is composed have the same orientation. The deposition of an ossicle begins within a single cell, but in time this cell divides repeatedly and its progeny take over the job. Ossicles usually consist primarily of calcium carbonate, but they contain some magnesium and they may be colored by compounds of various metals. They are connected to muscles and to one another by collagen.

While radial symmetry is a characteristic of echinoderms, including sea stars and sea urchins, it is not obvious in some members of the phylum. In most sea cucumbers, for instance, the body is elongated and usually shows some strong tendencies toward bilateral symmetry. Even in a sea star, in which the rays diverging from the central disk are equal and evenly spaced, the disposition of certain structures indicates slight bilaterality. On the upper surface of a sea star, the **madreporite,** or *sieve plate,* lies to one side of the center. Dissection of the animal will reveal that the internal canal to which the madreporite is joined is similarly eccentric. An imaginary line that passes through the madreporite, through the center of the upper surface of the disk, and down the length of one ray divides a sea star into two halves that are mirror images of one another. In other words, it divides the sea star bilaterally.

A further complication in this matter of symmetry is that many echinoderms have a free-swimming larval stage whose symmetry is decidedly bilateral. The radial symmetry typical of the adult is not firmly established until metamorphosis. There is no connection, however, between the bilaterality of the larva and any tendency toward bilaterality in the adult. Metamorphosis is so drastic that the longitudinal axis of the larva is not comparable to the line along which an adult echinoderm may be divided into right and left halves.

Although the radial symmetry of modern echinoderms can be traced back to certain of the oldest fossils, not all early echinoderms had this type of symmetry. Many seem to have been almost perfectly bilateral, and some were asymmetrical.

The radial symmetry of echinoderms is perhaps related to the fact that the supporting structures of these animals are interconnected calcareous ossicles. In the development of most echinoderms other than sea cucumbers, the first six ossicles are platelike and form a group in which one is surrounded by the others (Figure 21.1, A). This arrangement confers strength on the complex as a whole, because the sutures between the five peripheral plates stop at the central plate and are also out of line with one another. If there were four or six plates surrounding the central plate (Figure 21.1, B and C), the sutures would line up and the system would be weaker. If there were three plates around the central plate (Figure 21.1, D), none of the sutures between the peripheral plates would be in alignment, but there would be a line of weakness extending from one suture to another around the central plate.

The remarkable water-vascular system is a closed system of canals, lined internally by a ciliated epithelium, derived from part of the embryonic coelom. The ultimate branches of certain of the canals supply fluid

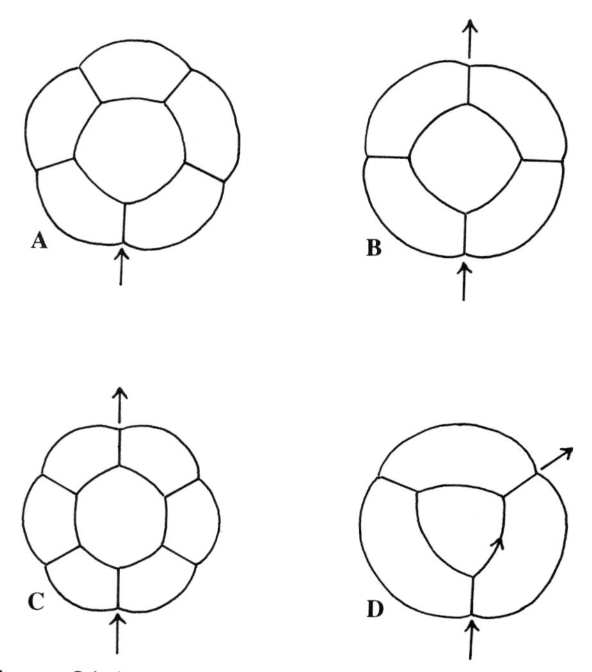

FIGURE 21.1 A. The first six skeletal plates of an echinoderm. None of the sutures in the ring of plates surrounding the central plate is in line with any of the others. B–D. Lines of weakness in other hypothetical arrangements of plates. (See discussion in text.)

to fleshy projections of the body wall, called **tube feet,** or *podia*. Each of the radially arranged portions of the body where the tube feet project is called an **ambulacrum.** The activity of the tube feet in locomotion, feeding, and other functions depends on the way that muscle and connective tissue in them interact with the fluid. The specializations of tube feet, and how various echinoderms use them, are discussed below as the major groups within the phylum are described.

The internal organs lie within a separate division of the coelom, called the **perivisceral coelom.** The fluid in this cavity, circulated by activity of cilia on the peritoneal cells, bathes many tissues and is important in translocation of oxygen and nutrients. The origin of the perivisceral coelom and its relationship to the water-vascular system are explained below in connection with the embryonic development of sea stars.

Also important in translocation, at least in some echinoderms, is the **hemal system,** which may be viewed as a kind of circulatory system. It consists of channels in connective tissue, and as a rule these channels are not lined internally by an endothelium. There may be muscle around the channels, however, and the activity of this muscle accounts for movement of fluid within them.

The ciliated cells of echinoderms, whether they are part of the epidermis, gut, or lining of the perivisceral coelom or water-vascular system, usually have a single cilium. Moreover, the basal portion of each cilium is surrounded by a circle of microvilli (Figure 21.2) that resembles the microvillar collar of a sponge choanocyte.

No definite excretory organs are present in echinoderms, although certain free amoeboid cells and fixed cells are known to accumulate waste material. The reproductive system is relatively uncomplicated, and there are no organs for copulation. Sperm are usually released into the sea; eggs may likewise be set free, but many echinoderms brood their young until they have reached a rather advanced stage. The nervous system, as might be expected in animals whose symmetry is radial, is not strongly centralized.

The development of echinoderms varies from group to group, and also according to whether little or much yolk is incorporated into the egg. In general, however, the basic pattern of embryology in this phylum shows unity on three important points: radial cleavage, survival of the blastopore as the larval anus, and origin of the coelom from outpocketings of the archenteron. These matters, as well as the metamorphosis of the larva into an adult, are discussed further in connection with sea stars and sea urchins. Early development in other groups of echinoderms proceeds along approximately the same lines.

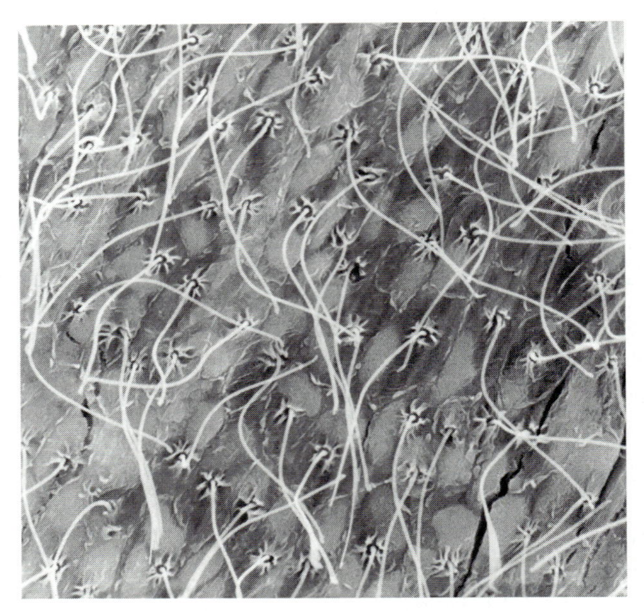

FIGURE 21.2 Cells lining the lumen of a tube foot of the sea urchin *Strongylocentrotus franciscanus*. Each cell has a cilium, the base of which is surrounded by a ring of microvilli similar to those that form the collar of a sponge choanocyte. (Scanning electron micrograph courtesy of Ernst Florey and Mary Anne Cahill.)

SEA STARS AS AN EXAMPLE OF ECHINODERMS

SYMMETRY AND ORIENTATION

The principal characteristics of the phylum Echinodermata are effectively illustrated by sea stars (Figure 21.3). Radial symmetry is one of the most impressive features of these animals. Generally there are 5 **rays** (''arms''), but some sea stars have 6, and others have as many as 20 or more. The **mouth** is in the center of the underside, which should be called the **oral surface,** for it is not comparable to the ventral surface of a bilaterally symmetrical animal. The upper side is the **aboral surface.**

SKELETON

The dermal skeleton consists of thousands of separate ossicles, some of which are visible externally as plates, spines, or other structures. Many of the ossicles, however, form an interlacing system of struts and girders that strengthens the body wall (Figure 21.4, A). The ossicles are bound together by connective tissue, and also by muscles, so that the central disk and rays are flexible to some extent. The animal can slowly

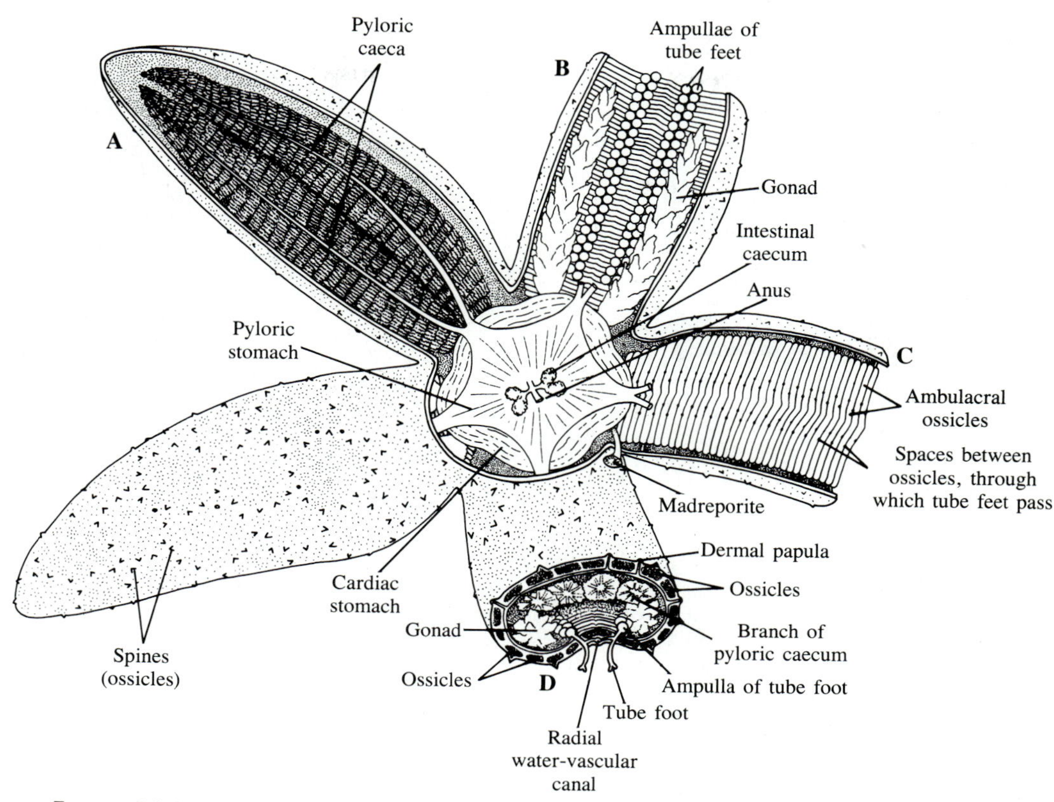

FIGURE 21.3 Anatomy of a sea star. Rays A–C show structures revealed in three successive stages of dissection; ray D has been sectioned transversely.

bend, turn itself over, and envelop its prey, but the role of the skeleton and muscles in locomotion is minimal.

PEDICELLARIAE

In sea stars of certain groups, much of the body surface has little pincers called **pedicellariae.** In their most primitive form, these are merely adjacent ossicles that can be pulled toward one another. Some simple types of pedicellariae consist of two broad ossicles that close on one another like the valves of a clamshell (Figure 21.4, D; Figure 21.21, C). There may even be a group of ossicles that operate together (Figure 21.4, B and C).

The more complicated pedicellariae (Figure 21.4, E–I) are mounted on fleshy stalks, and although they consist of three ossicles, only two form pincers; the third serves as a base to which the muscles involved in opening and closing the pincers are attached. The ossicles of the pincers may be crossed like of the two halves of a pair of pliers (Figure 21.4, G), or they may simply be pulled toward one another like the tips of a pair of forceps (Figure 21.4, E and F).

The usual function of the pedicellariae of sea stars is to keep small organisms from settling on the body, and probably also to discourage grazers that might eat away soft tissue. In certain genera, however, they are helpful in capturing food. *Stylasterias,* found on the Pacific coast of North America, has large pedicellariae (Figure 21.4, G–I) in which the jaws are like sharp-toothed combs. If enough of these pedicellariae close at the same time, they can immobilize a small fish.

TUBE FEET AND THE WATER-VASCULAR SYSTEM

The movements of a sea star depend mostly on its tube feet. On the oral side of the animal, the **ambulacral groove**—the shallow trough that runs from the tip of each ray to the mouth—is bordered by hundreds of these fleshy projections (Figure 21.21, B). In species that live in rocky, gravelly, or shelly habitats, each tube

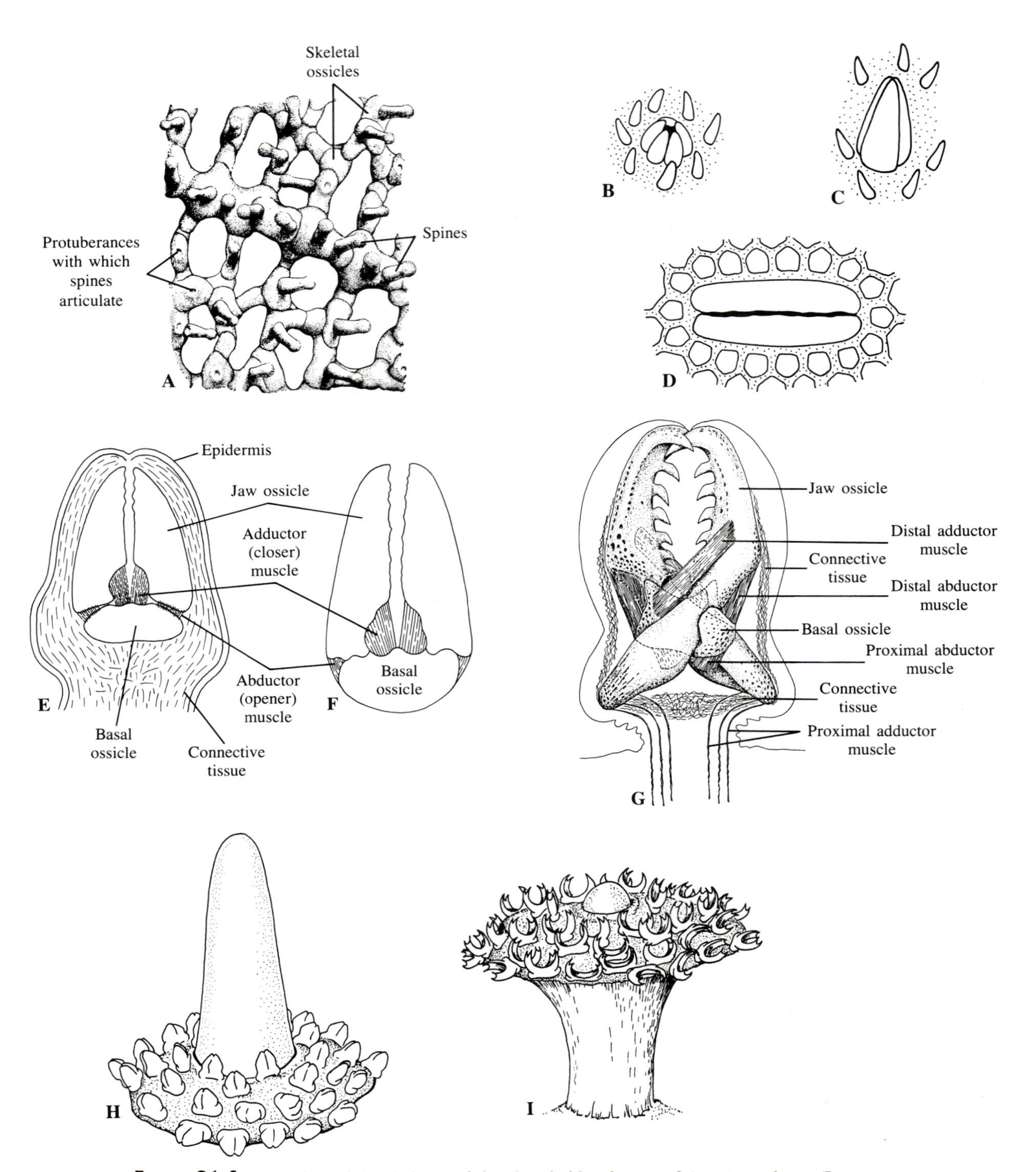

FIGURE 21.4 A. Portion of the skeleton of the aboral side of a ray of *Asterias rubens*. (Ivanov *et al.,* Major Practicum of Invertebrate Zoology.) B. Simple pedicellaria formed by several ossicles, as in *Pectinaster*. C. Simple pedicellaria consisting of three ossicles, as in *Luidia*. D. A clamshell type of pedicellaria, as in *Hippasteria* (see also Figure 21.21, C). E, F. Forceps type of stalked pedicellariae, as in *Asterias*. In F, the tissues external to the ossicles and muscles are not shown.) G. Pliers type of stalked pedicellaria of *Stylasterias forreri*. In this species, the jaw ossicles, more than 1.5 mm long, may be used to capture prey, including small fishes. H, I. Wreaths of pedicellariae around a spine of *Stylasterias*. The jaws open in response to the presence of a prospective prey organism. (G–I, Chia and Amerongen, Canadian Journal of Zoology, *53*.)

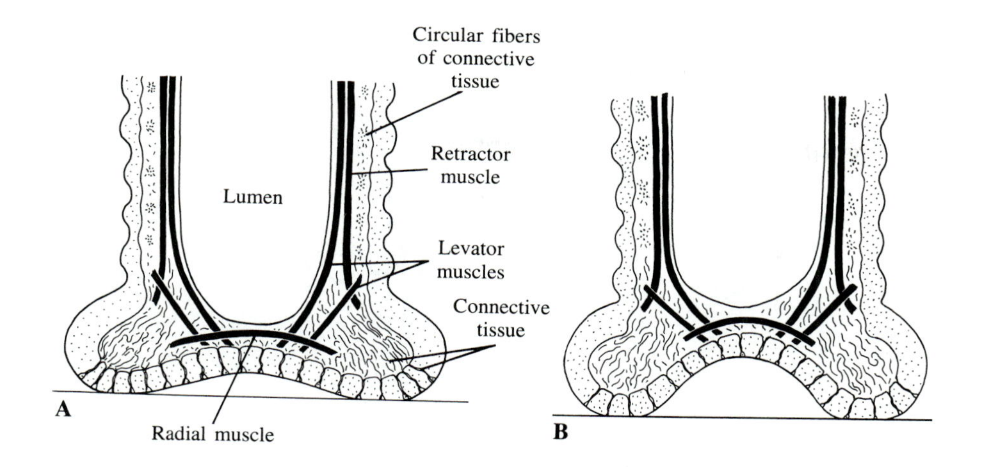

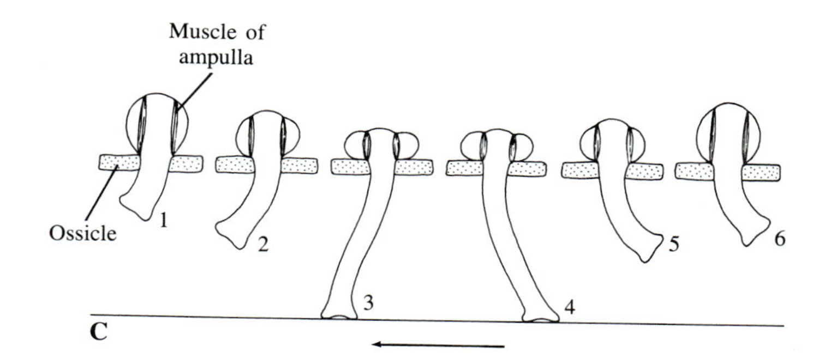

FIGURE 21.5 A, B. Successive stages in formation of a sucker by a tube foot that is in contact with a firm substratum. C. Successive stages in the stepping action of a tube foot. The tube foot swings forward, elongates, becomes attached, swings backward, and becomes detached. The arrow shows the direction of movement.

foot is tipped by a little sucker. With the help of the suckers and sticky mucus, a sea star can cling to almost any hard surface, climb vertically, or separate the shell valves of a clam.

When moving over a horizontal surface, the animal barely touches its tube feet to the substratum. The sequence of activity of a single tube foot as it steps lightly over the bottom is illustrated in Figure 21.5, C. After a tube foot makes contact with the substratum (3), it swings its upper part forward (4), becomes detached and shortens (5, 6), swings its tip forward (1, 2), and then lengthens until it touches down again (3). The activity of the hundreds of tube feet is coordinated by the nervous system, but the tube feet in any one region do not seem to operate according to a particular pattern. So long as many of them are doing the same thing at the same time, the animal advances steadily and smoothly.

When the sucker of a tube foot is used for attachment, it is pressed firmly to the substratum so that mucus can function as an adhesive (Figure 21.5, A). There is a dense mass of connective tissue in the sucker, and when the muscles that insert into this contract, the sucker becomes cupped sufficiently to form an effective suction device (Figure 21.5, B). Connective tissue in peripheral parts of the sucker prevents it from collapsing when the central portion is raised.

The tube feet are connected to the water-vascular system. It has already been pointed out that this is a division of the coelom separate from the perivisceral coelom, in which the viscera are suspended. Some parts of the water-vascular system are buried in tissue or hidden by ossicles. Among the more prominent of its components is the **ring canal** that encircles the gut just above the mouth (Figure 21.7, C). This sends a **radial canal** into each arm, and the radial canals in turn are joined to the tube feet by small **lateral canals.** Also originating from the ring canal is the **stone canal** (so-named because it is stiffened by ossicles), which leads aborally to the **madreporite** (Figure 21.6, A; Figure 21.7, A).

The lateral canals that connect the tube feet to a radial canal pass between successive ambulacral ossicles (Figure 21.3; Figure 21.6, B). Those branching off one side of a radial canal alternate with those on the other side. There is basically one row of tube feet on each side of the ambulacral groove. In many sea stars, however, the tube feet are staggered in such a way that

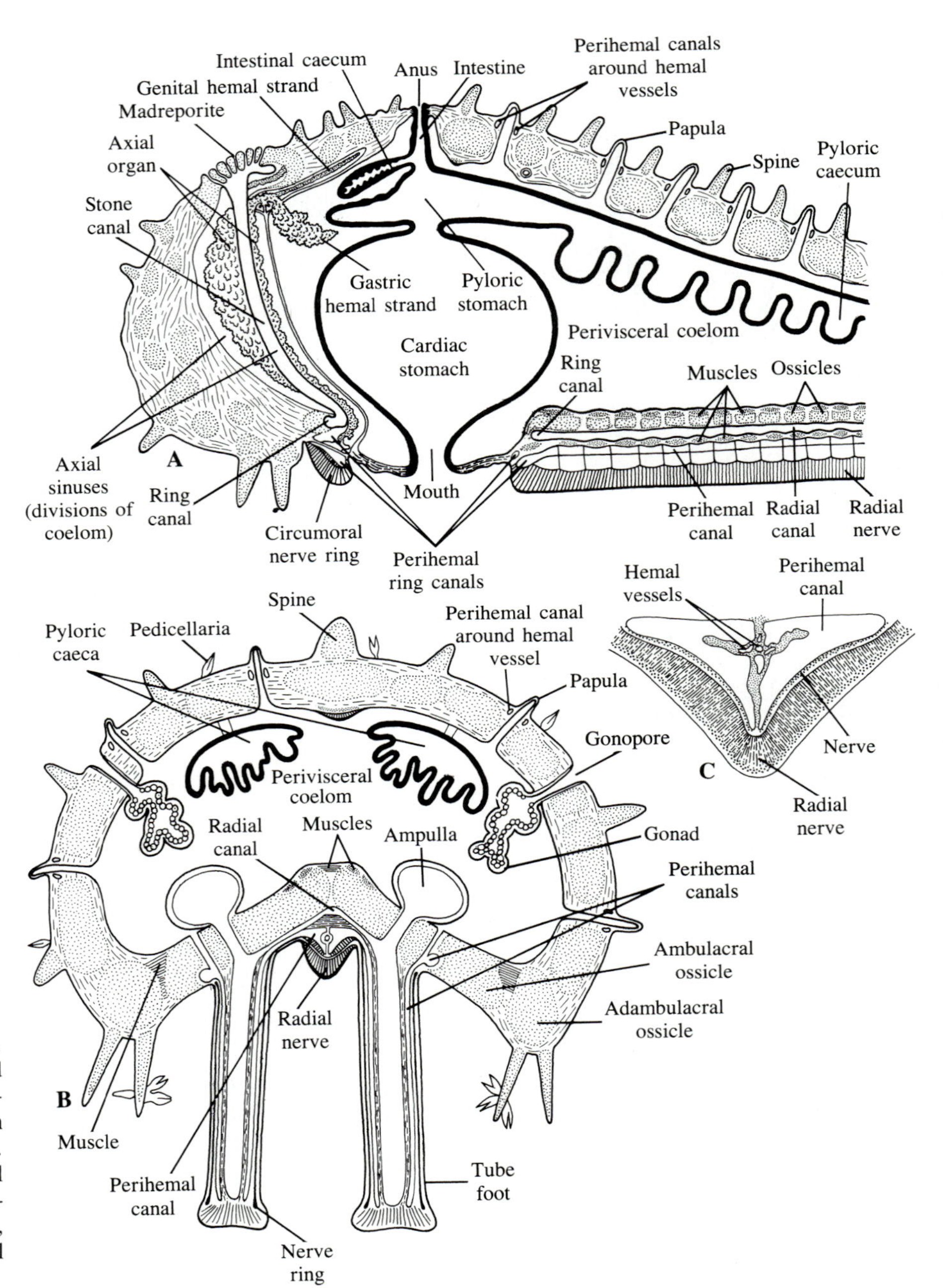

FIGURE 21.6 *Asterias rubens*. A. Section through the disk and proximal portion of a ray; diagrammatic. B. Transverse section of the proximal portion of a ray. C. Transverse section of a radial nerve and structures closely associated with it. (After Chadwick, Liverpool Marine Biological Committee Memoir 25.)

there appear to be two rows; about half the tube feet have slightly longer lateral canals than the other half, and the two types alternate regularly.

Internally, the proximal portion of each tube foot is a bulblike swelling called an **ampulla** (Figure 21.3; Figure 21.5, C; Figure 21.6, B; Figure 21.7, C). This structure is more or less free in the perivisceral coelom and serves as a reservoir for fluid. Both the ampulla and the tube foot to which it is linked have muscle fibers and considerable connective tissue. When an ampulla contracts, it forces fluid into the tube foot. Conversely, when a tube foot contracts, fluid either fills the ampulla or leaks out, or both. (It could also back up into the radial canal, although the lateral canal has valves that

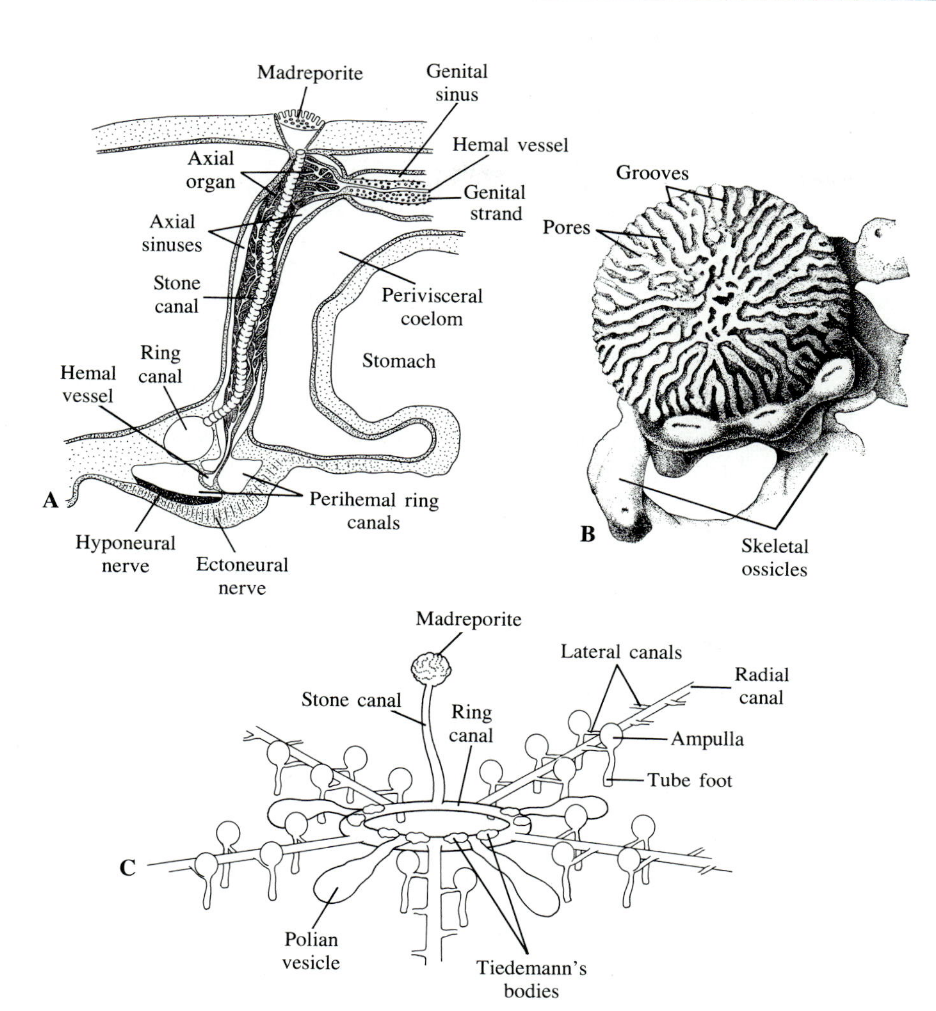

FIGURE 21.7 A. The axial complex of *Asterias rubens*. (After Ivanov *et al.*, Major Practicum of Invertebrate Zoology.) B. Madreporite of *Asterias rubens*. (Ivanov *et al.*, Major Practicum of Invertebrate Zoology.) C. Diagram showing relationships of the components the water-vascular system of a sea star.

minimize such backflow.) The lengthening, shortening, and bending movements of a tube foot are thus the result of changes in distribution of fluid brought about by muscle contractions. Connective tissue not only gives elasticity to a tube foot and its ampulla, but also prevents the tube foot from bulging when fluid is forced into it. A tube foot whose normal functioning requires that it lengthen at the right time would be of little use if it swelled like a balloon when fluid was forced into it.

Sea stars such as *Luidia* (Figure 21.19), which live on muddy bottoms, generally have tube feet with pointed tips. These can be used for displacing the substratum, so that the animal can bury itself, as well as for walking. Pointed tube feet, when present in sea stars, usually have bilobed and highly muscularized ampullae, which are effective in driving fluid into the tube feet with considerable force.

The madreporite is an ossicle, but it resembles a sieve or filter because it is perforated by minute vertical channels (Figure 21.6, A; Figure 21.7, B). Its role remains uncertain. In some sea stars that have been extensively investigated, water does not seem to pass through the madreporite to any great extent. If the madreporite is sealed, no immediate problems arise. If an intact sea star is shocked severely enough to cause all of its tube feet to contract suddenly, very little fluid will come out through the madreporite. Moreover, a ray that has been cut off and had the open end of its radial canal plugged will use its tube feet more or less normally for at least a few days. The evidence suggests that water enters the water-vascular system chiefly from the perivisceral coelom, which in turn takes up water through thin places in the body wall. Although fluids in the perivisceral coelom and water-vascular system are almost isotonic with sea water, they contain proteins, other organic substances, and free cells. They also have more potassium than sea water, and it is generally thought that, because of this, water tends to diffuse more rapidly into the coelom and water-vascular system than it diffuses out.

The ring canal has two kinds of blind sacs (Figure 21.7, C). Both types, regardless of their exact number, arise between the places where the radial canals originate. The **Polian vesicles,** thin-walled but nevertheless muscularized, become prominent when distended with fluid. They are probably compensation sacs, receiving fluid forced out of other parts of the water-vascular system, as when many of the tube feet on one or more arms contract simultaneously. The other diverticula are called **Tiedemann's bodies.** These are paired and generally wrinkled or folded. They are thought to produce the free cells that circulate in the water-vascular system. The stone canal is divided into two parallel canals. Thus fluid within it may move orally and aborally at the same time. The ring canal may also be divided into channels, sometimes several of them, but the functional significance of this division has not been explained convincingly.

DIGESTIVE SYSTEM AND FEEDING

The mouth of a sea star is controlled by muscles. In some species the opening can be dilated enough to swallow a clam that is a third or a half the diameter of the disk. There is generally a short **esophagus,** sometimes with a number of small side pouches. The **stomach** (Figure 21.3; Figure 21.6, A) is large and is usually constricted horizontally into two portions. These are confusingly named, following terminology classically used for the stomachs of mammals and other higher vertebrates. The part nearer the mouth is called the **cardiac stomach,** even though a starfish has no heart; the part nearer the aboral surface is called the **pyloric stomach.** The cardiac stomach is the larger of the two and is divided into pouches that are joined to the body wall by mesenteries and ligaments of connective tissue. This part of the stomach is eversible in some sea stars. It is turned inside out when pressure in the perivisceral coelom is raised by contraction of muscles in the body wall of the disk. Withdrawal of the cardiac stomach depends on shortening of the stretched ligaments. In species that swallow their prey whole, or that wear off bits of prey, food is taken through the mouth into the cardiac stomach. Undigested residues, such as shells and exoskeletons, are eventually discharged through the mouth.

The pyloric stomach sends out a pair of digestive diverticula into each ray (Figure 21.3; Figure 21.6). These are usually called **pyloric caeca.** As a rule, the members of a pair of caeca are joined to the stomach by a single duct, but sometimes they are separate. The caeca are supported on their aboral sides by a mesentery and fill up much of the perivisceral coelom of the rays. They branch into finer ducts that end blindly in clusters of lobules. The cells of these lobules produce digestive enzymes, absorb soluble products of digestion, and store nutritive reserves. Some intracellular digestion, of fats at least, is also accomplished within them.

Most sea stars have an **anus,** and the short **intestine** that leads to it generally has some small sacs called **intestinal caeca** (Figure 21.3; Figure 21.6, A). These secrete mucus that coats the fecal material. In certain groups, the intestine ends blindly, or is absent altogether. The anus is not used for elimination of coarser residues, which must be discharged by way of the mouth.

What sea stars eat and how they eat it varies considerably. There are some that engulf small invertebrates, especially molluscs, crustaceans, and brittle stars, eventually regurgitating undigested residues. Others can swallow relatively large clams or snails, and to do this they must distend the mouth, stomach, and disk. The enzymes of the pyloric stomach are potent: the empty shell of a clam or mussel that has been swallowed whole will usually be ejected in a few hours.

Many sea stars, such as the Atlantic *Asterias* and Pacific *Pisaster,* evert the cardiac stomach to envelop prey. Digestion by enzymes from the pyloric stomach may begin outside the body, in which case the soup resulting from breakdown of the prey's tissues is carried by ciliary activity to the pyloric stomach; or the cardiac stomach may be retracted and the prey pulled into the disk with it. An everted stomach may even slip between the closed valves of the shell of a clam or mussel to get at the tissue; eventually, the animal will gape and the sea star's job becomes easier. The tube feet play an important role in gripping the shell and pulling the valves apart enough to allow the stomach to slip in. A gape of only about 0.1 mm may suffice. In some sea stars that evert their stomachs, the digestive enzymes can attack the tissue even if the mussel is wired tightly shut.

A number of species use the everted stomach for eroding sponges or tissues of other vulnerable animals, such as sea anemones, corals, and sea pens. A few collect detritus, including material from seaweeds. Herbivory is not common among sea stars, but *Asterina miniata,* found along the Pacific coast of North America, consumes pieces of algae and has enzymes for digesting polysaccharides.

PERIVISCERAL COELOM

The perivisceral coelom extends into each ray (Figure 21.6). Thus the pyloric caeca and gonads, as well

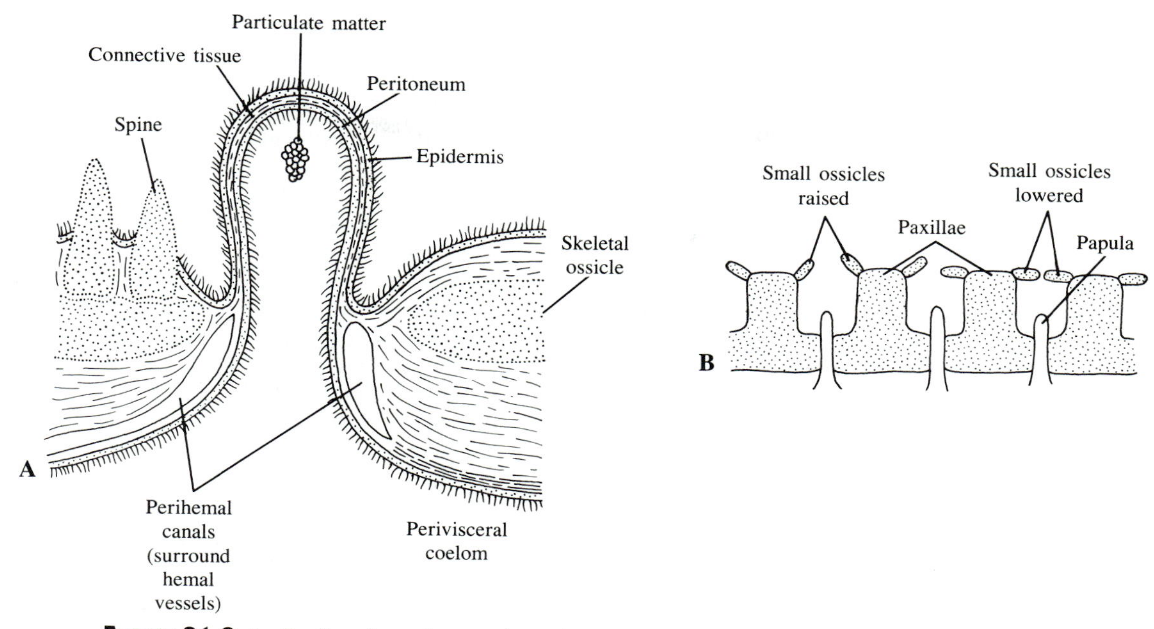

FIGURE 21.8 A. Section through a portion of the body wall, including a papula, of a sea star; diagrammatic. B. Diagram of the relationship of paxillae to papulae, characteristic of some members of the orders Paxillosida and Valvatida. In many species, there are small ossicles on the flat top of each paxilla, as well as along its margins (See Figure 21.21, A).

as the visceral organs in the disk of the animal, are bathed by the coelomic fluid. The fluid is kept in circulation by cilia on the peritoneal cells, as well as by the activity of the ampullae and slow muscular changes taking place in the body wall.

PAPULAE

On the aboral side of many sea stars are small fingerlike projections called **papulae** (Figure 21.8, A). These are thin-walled structures, consisting mostly of epidermis and peritoneum; there is very little dermal connective tissue between the two epithelial layers. Because both the epidermis and peritoneum are ciliated, and because the perivisceral coelom extends into the papulae, these outgrowths are effective devices for exchange of gases and elimination of soluble nitrogenous wastes. The papulae may not be obvious in preserved specimens, but in a living sea star, when they are distended with coelomic fluid, they are easy to see with low magnification, and sometimes without any magnification.

In sea stars that live on soft bottoms, the papulae, which are often branched, will usually be found between specialized aggregations of ossicles called **paxillae** (Figure 21.8, B). A paxilla usually consists of a blunt ossicle with which smaller ossicles articulate. The paxillae help to keep sediment from settling on the papulae, but do not interfere with circulation of water around the papulae.

HEMAL SYSTEM

The hemal system of echinoderms consists of channels within connective tissue. In sea stars, the channels generally contain a viscous fluid and crowded amoeboid cells. Very few of the channels are likely to be seen in dissected animals.

There are two main hemal rings. One encircles the mouth; the other is close to the aboral surface. The rings are joined by a network of hemal vessels located within the **axial organ,** which is adjacent to the stone canal. The axial organ has glandular components, but its functions are not understood. Connected to the oral ring are radial hemal vessels that run into each ray (Figure 21.6, B and C), and branches of these extend to the tube feet. Somewhat similar radially arranged vessels run from the aboral ring to the gonads. Various other vessels have been described but not all of them have been seen consistently.

A definitive study of the asteroid hemal system is needed. For the time being, it may be said that contractions of muscle cells surrounding some of the vessels promote flow, primarily of a back-and-forth type. The entire axial organ, and a blind pouch that branches off it aborally, have been observed to pulsate.

Closely associated with most hemal vessels, and often surrounding them completely, are canals and sinuses that are extensions of the coelom. They form the **perihemal system.** These spaces are found not only in sea stars but also in the other classes of echinoderms.

EXCRETION

The principal nitrogenous wastes of sea stars are ammonia and urea. (This is probably true for echinoderms in general.) The fact that sea stars are immersed or at least very wet most of the time makes elimination of soluble wastes through the body wall a practical solution. The papulae and tube feet are probably effective places for ammonia and urea to diffuse away. Phagocytosis, by amoeboid cells or cells of the peritoneal lining, is believed to eliminate various particulate wastes that appear in the coelomic fluid. The amoeboid cells are claimed to wander out through thin places in the body wall, carrying wastes with them. It has also been reported that a papula, after accumulating a number of amoeboid cells at its tip, will pinch off this portion.

NERVOUS SYSTEM AND SENSE ORGANS

The nervous system of a sea star is rather decentralized. This is a sensible arrangement for an animal that has radial symmetry; any part of it may be the first to receive a favorable or unfavorable stimulus and must therefore be prepared to make an appropriate response. This is not to say that a sea star is poorly coordinated. After watching a sea star move purposefully by concerted activity of its hundreds of tube feet, or turn itself over (Figure 21.22), one should be convinced that the animal's nervous system functions fairly efficiently even though there is no brain.

The more prominent elements of the nervous system remain closely associated with the epidermis of the oral surface. There is a **nerve ring** around the mouth (Figure 21.6, A) and from this a **radial nerve** extends out into each arm (Figure 21.6, B and C). The radial nerves are connected to nerves that border each side of the ambulacral groove. Branches of the major nerves already mentioned innervate the body wall, tube feet, ampullae, gut, and various other structures. Some of the branch nerves lead, however, to another division of the nervous system, which is deeper in the body wall and much of which forms a kind of nerve net close to the perivisceral coelom. This division, in turn, innervates many of the structures that also receive nerves from the more nearly superficial division.

It is an oversimplification to say that finer nerves coming from elements of the superficial nervous system are primarily sensory and that those coming from the deeper division are primarily motor, but this is probably not far from the truth. Consequently, a tube foot and its ampulla would be innervated by both divisions. Integration and reasonably rapid conduction over larger distances take place mostly in the ring and radial nerves. If the nerve ring is cut in two places, isolating a particular ray, the tube feet of that ray may continue to function but their activity will not be coordinated with that of the other rays. In some sea stars one ray is generally the "leader," the others "followers." In most species, however, all rays are capable of leading. It is thought that the places where the radial nerves join the nerve ring are more or less equal centers of dominance. Once the animal is stimulated to lead with a particular ray, the center of dominance belonging to that ray temporarily takes command.

Many sea stars have a light-sensitive **eyespot** on a modified tube foot at the tip of each ray. This eyespot consists of many pigmented cups that enclose epidermal photoreceptor cells and that are covered by little lenses derived from the cuticle of the epidermis. A sea star with eyespots of this sort generally turns up the tips of its rays to monitor the light. Certain species are sufficiently photopositive that they will follow a slow-moving beam of light. Others are decidedly photonegative, preferring darkness. Whether a sea star has eyespots or not, it is likely to react to light, and its responses to bright light may be different from those to dim light. Much of the body surface seems to be sensitive to light or darkness. In addition to light receptors, the epidermis has cells sensitive to touch and chemical stimuli.

REPRODUCTION AND DEVELOPMENT

Almost all sea stars have separate sexes, and the usual number of gonads is ten (Figure 21.3; Figure 21.6, B). The two in the proximal portion of each ray, beneath the pyloric caeca, have ducts that lead to the genital pores. These are commonly on the aboral side, close to where the ray joins the disk, but some sea stars have other arrangements. With few exceptions, fertilization of the eggs takes place as they are shed, or soon afterward. Release of sperm by males often triggers the spawning of females, a good adaptation for animals that cannot ensure fertilization by mating. As a rule, male and female sea stars release millions of sperm and eggs, but spawning is the culmination of a long period of gamete production and is not likely to be repeated in the same year. In at least some species a substance produced in the radial nerves ("radial-nerve factor") influences maturation of eggs and also stimulates spawning by inducing contraction of muscles around the ovary.

Development follows two main pathways. Species that brood their young, and some that do not, produce relatively large, yolky eggs that develop gradually and directly into little sea stars. In those that produce

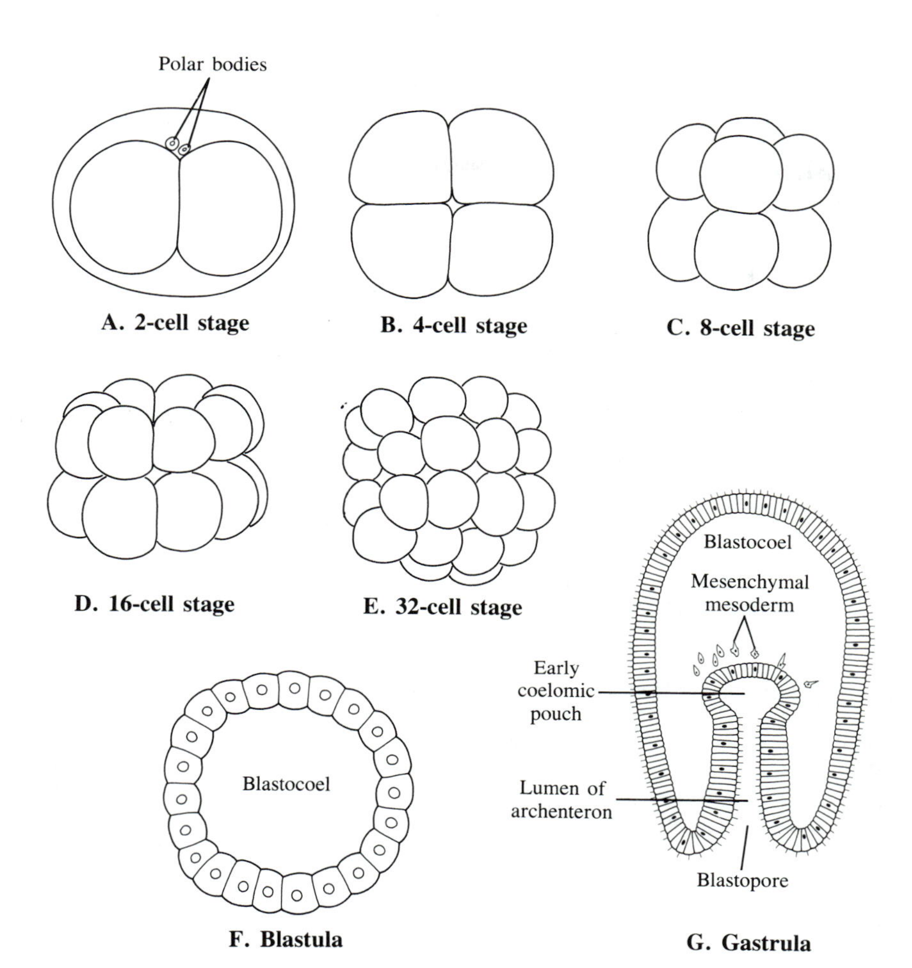

FIGURE 21.9 Radial cleavage (A–E), blastula (F), and gastrula (G) of a sea star.

smaller, less yolky eggs, development is indirect and involves a sequence of stages in which the origin of the unique water-vascular system and other structures can be observed. Not all species with indirect development conform exactly to the plan described here, but the deviations are minor.

Cleavage of the egg is **radial** and, at first, equal, or nearly so (Figure 21.9, A–E). Development is relatively **indeterminate,** so that if the first two or first four blastomeres are separated, each blastomere will continue to divide and will develop into a complete larva. By the time there are about 32 or 64 cells, the embryo is a blastula with a prominent central cavity, the blastocoel (Figure 21.9, F; Figure 21.10, A). A little later, after the number of cells has increased, the cells at the lower pole of the blastula turn inward to form the primitive gut, or **archenteron** (Figure 21.9, G; Figure 21.10, B and C). The embryo is now a gastrula, with endoderm and ectoderm. The blastopore—the opening into the archenteron—is destined to serve as the anus of the larva after it begins to feed.

The future mouth and forepart of the gut are soon formed by a second invagination. This joins the archenteron to make a complete digestive tract that the larva will use for feeding on small organisms, such as planktonic diatoms. But before this new invagination appears, the archenteron does two important things. From near its tip it buds off **mesenchyme cells** (Figure 21.9, G; Figure 21.10, C). These are for a time loosely organized within the blastocoel, but they constitute a part of the embryonic **mesoderm.** At least much of the future muscle, connective tissue, and ossicle-forming tissue of the body wall will derive from the mesenchyme. The tip of the archenteron also produces a pair of **coelomic pouches** (Figure 21.9, G; Figure 21.10, C and D). The cells that line them constitute a second type of mesoderm.

Once separate from the archenteron, the right and left coelomic pouches lengthen and become subdivided, so that there are basically three pouches on each side (Figure 21.12, A–C; Figure 21.13, A). The subdivision is not necessarily complete, however, and what happens

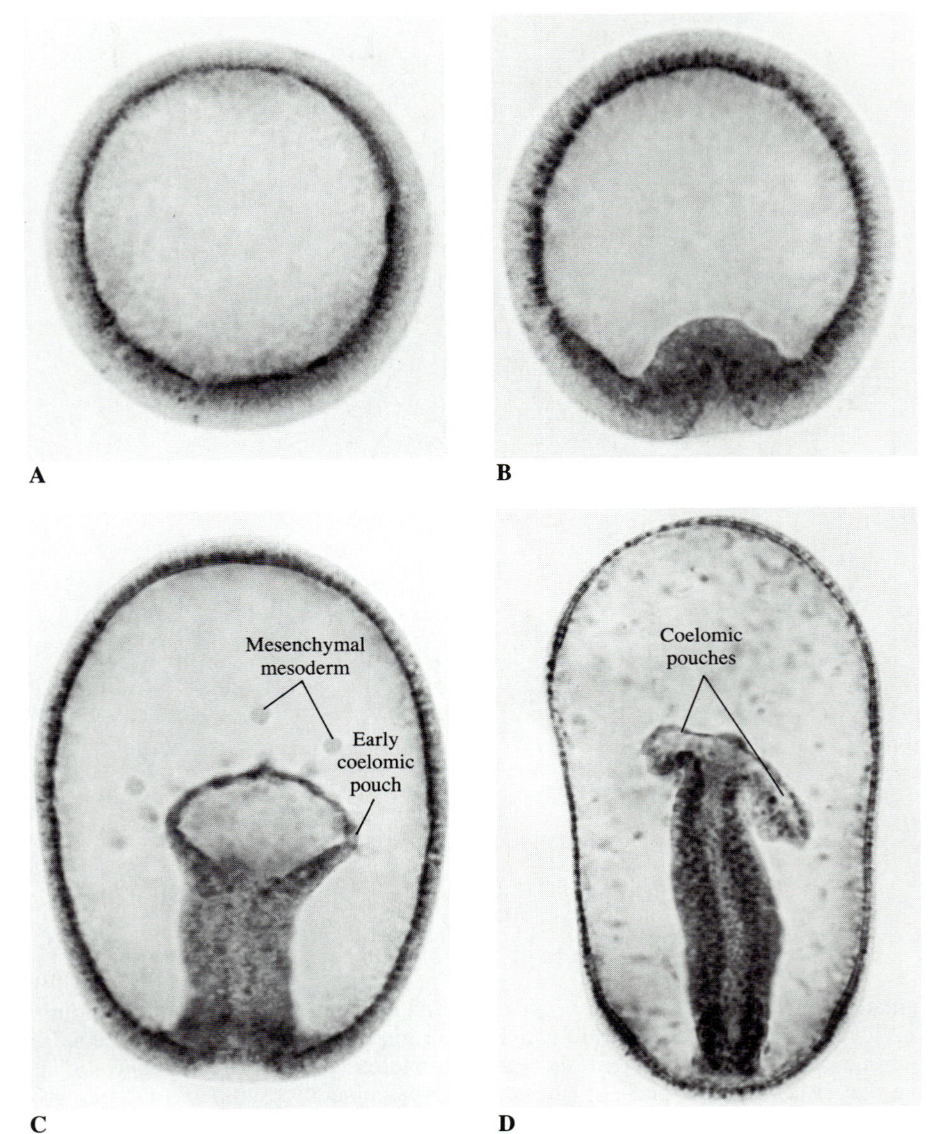

FIGURE 21.10 Stages in development of a sea star. A. Blastula. B. Early gastrula. C, D. Later gastrulas. (Photomicrographs courtesy of Carolina Biological Supply Co., Inc.)

on one side may be different from what happens on the other. On this point there is much variation, not only among sea stars but in other groups of echinoderms. Nevertheless, it is usually possible to recognize at least the suggestion of three pouches on both sides of the archenteron. This arrangement is typical of the early feeding larva, called the **bipinnaria** (Figure 21.11, A and B). Both the mouth and the anus of the bipinnaria are on the ventral side, and the gut is differentiated into an esophagus, bulbous stomach, and intestine.

The anteriormost sacs are called **axocoels,** the middle ones are called **hydrocoels,** and the posteriormost sacs are termed **somatocoels.** Their names indicate their potentialities. The somatocoels will, during metamorphosis, meet and become the perivisceral coelom. The left hydrocoel and axocoel do not become separated, so the complex formed by the two may be called the left **axohydrocoel.** From where they meet, a canal grows upward and becomes continuous with an ectodermal invagination from the dorsal surface (Figure 21.12, B and C; Figure 21.13, A). The complex thus formed is the future stone canal. In the bipinnaria, the epithelial lining of the canal, like the coelomic epithelium, is ciliated. Fluid is pumped out through the dorsal opening,

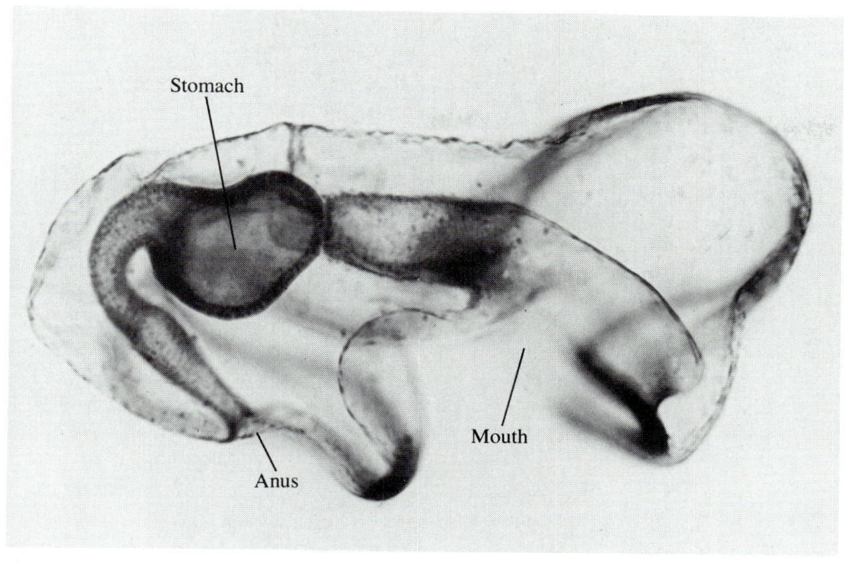

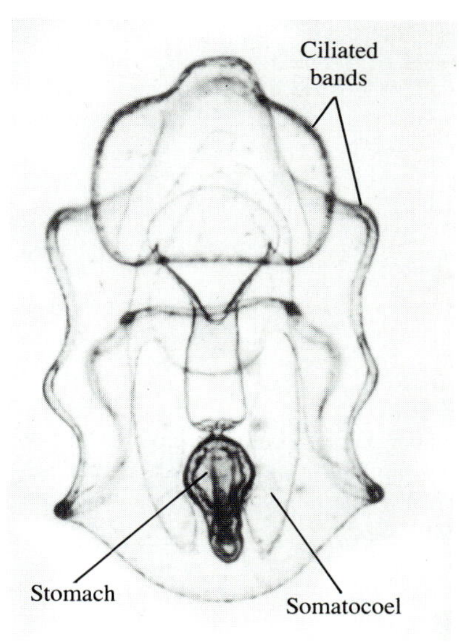

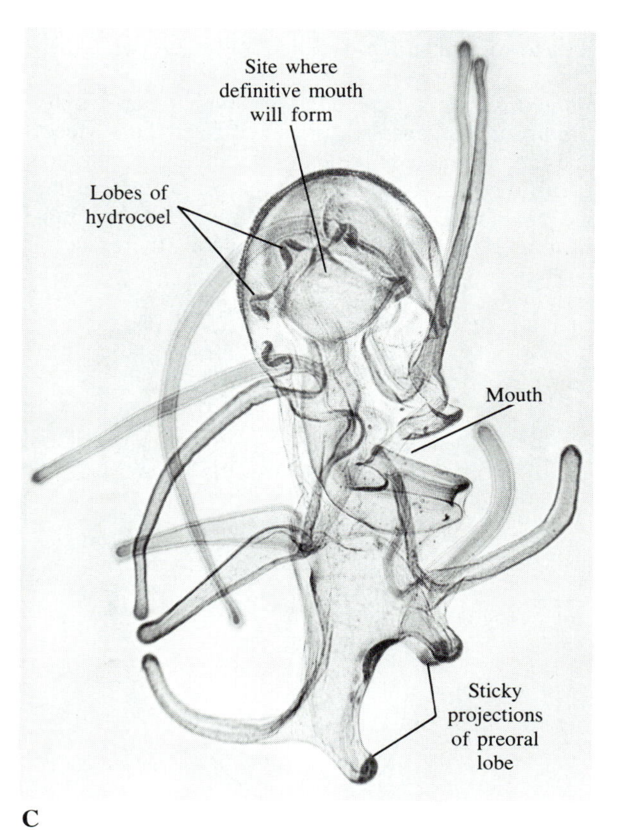

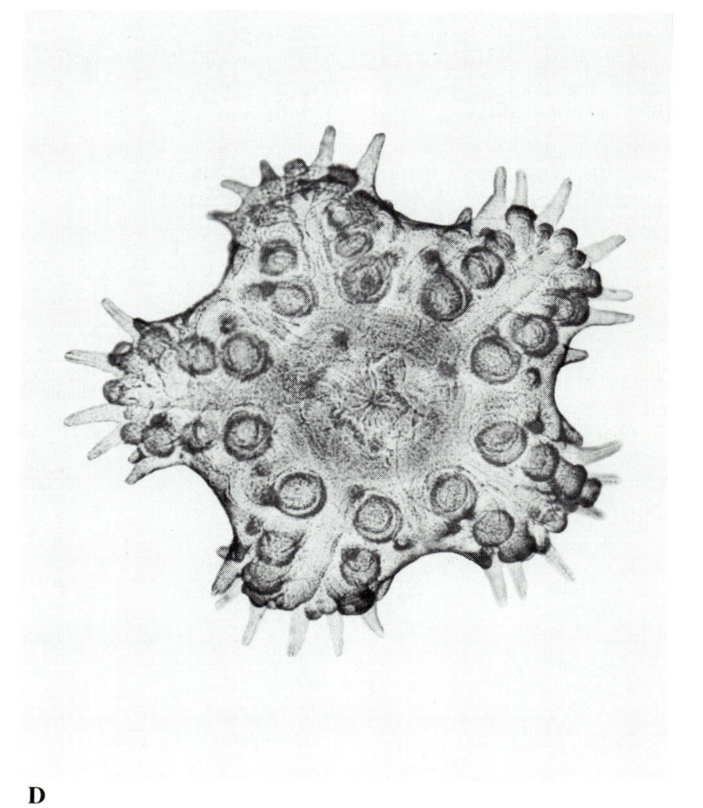

FIGURE 21.11 A. Bipinnaria larva of a sea star, lateral view from the right side. B. Bipinnaria, dorsal view. C. Brachiolaria. D. Young sea star after metamorphosis. (A, C, and D, photomicrographs courtesy of Carolina Biological Supply Co., Inc.)

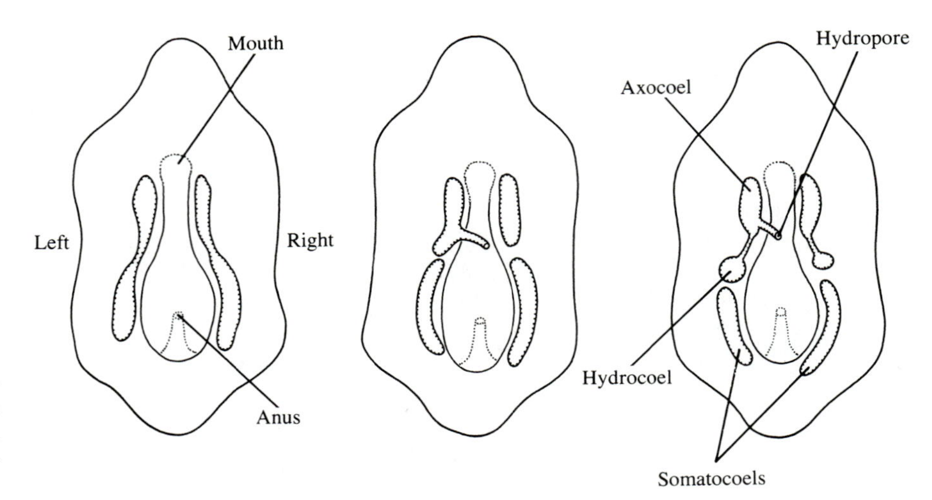

Figure 21.12 Stages in the specialization of coelomic cavities in the bipinnaria of a sea star, dorsal views; diagrammatic and generalized.

called the *hydropore,* and perhaps the canal should be viewed as an excretory organ somewhat similar to a metanephridium, or mixture of a metanephridium and coelomoduct (see Ruppert and Balser, 1986).

The left hydrocoel portion of the axohydrocoel wraps itself around the gut to form a complete circle, thus producing the ring canal of the water-vascular system. At about the same time, it becomes five-lobed, each lobe being the primordium of a radial canal (Figure 21.13, B and C). The right axocoel is transformed partly into a coelomic cavity that will surround the axial organ of the adult and partly into the primordium from which the gonads eventually arise. The axocoel on the right side generally disappears, but the right hydrocoel survives to form a structure called the madreporic vesicle.

In the bipinnaria, the ciliation, which earlier had been more or less evenly distributed over the surface, becomes concentrated into a continuous locomotor band. In time, this band follows the projections of the surface characteristic of the more advanced **brachiolaria** larva (Figure 21.11, C). In general, the brachio-

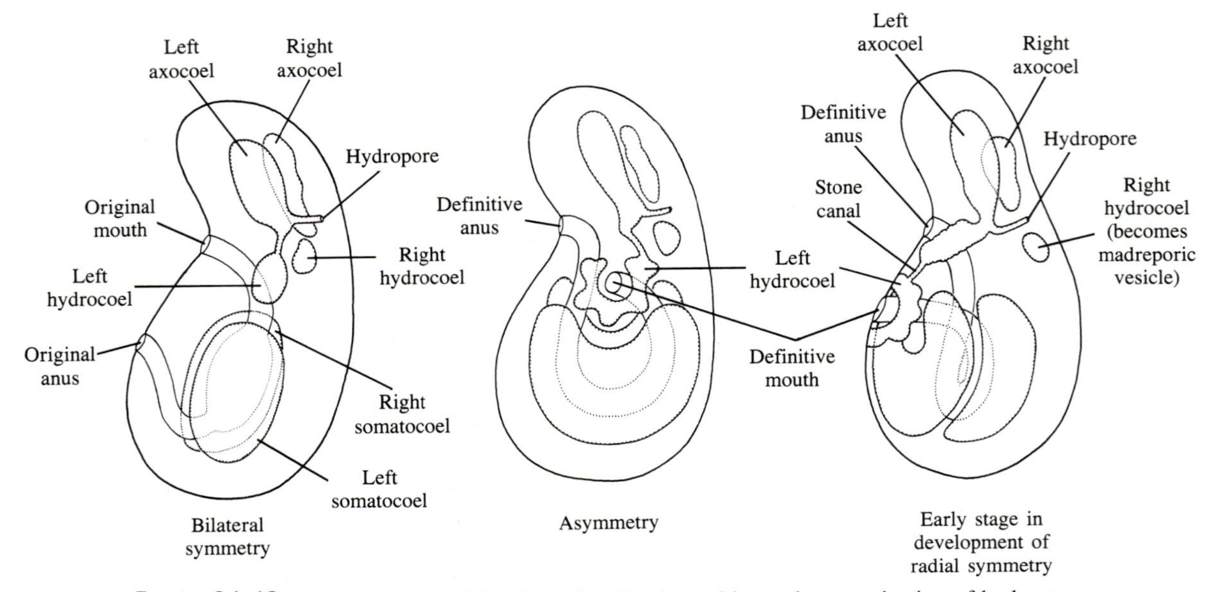

Figure 21.13 Stages in the specialization of coelomic cavities and reorganization of body structure during metamorphosis of a sea star, views from the left side; diagrammatic and generalized.

laria, like the bipinnaria, is bilaterally symmetrical, except for differences in the layout of the coelomic pouches on the right and left sides of the gut. One may expect, therefore, that metamorphosis of the brachiolaria into a radially symmetrical sea star will be drastic. Three sticky projections appear at the anterior end of the brachiolaria (Figure 21.11, C; Figure 21.14, A). These enable a brachiolaria that has settled to adhere to the substratum until the glandular area between the projections can affix the larva even more tightly. The mouth, part of the esophagus and intestine, and the anus disappear. The anterior part of the brachiolaria, the preoral lobe, where the sticky projections and adhesive disk are located, is resorbed, and the orientation of the animal is completely changed. A new mouth opens on what will be the oral surface of the adult, and an anus appears on the aboral side, so the gut is again complete (Figure 21.13, C). It has already been explained that the somatocoels become the perivisceral coelom and the left hydrocoel converts into the ring canal and radial canals of the water-vascular system. As the little sea star becomes recognizable as such, a tube foot develops on each ray. This tube foot becomes the tentaclelike structure, often with a light-sensitive eyespot, that is at the tip of each ray of the adult. Other tube feet are soon added (Figure 21.11, D). The first ossicles appear in the region around the anus.

Although embryonic development generally takes place without benefit of maternal attention, some species brood their young. The eggs of these are large and

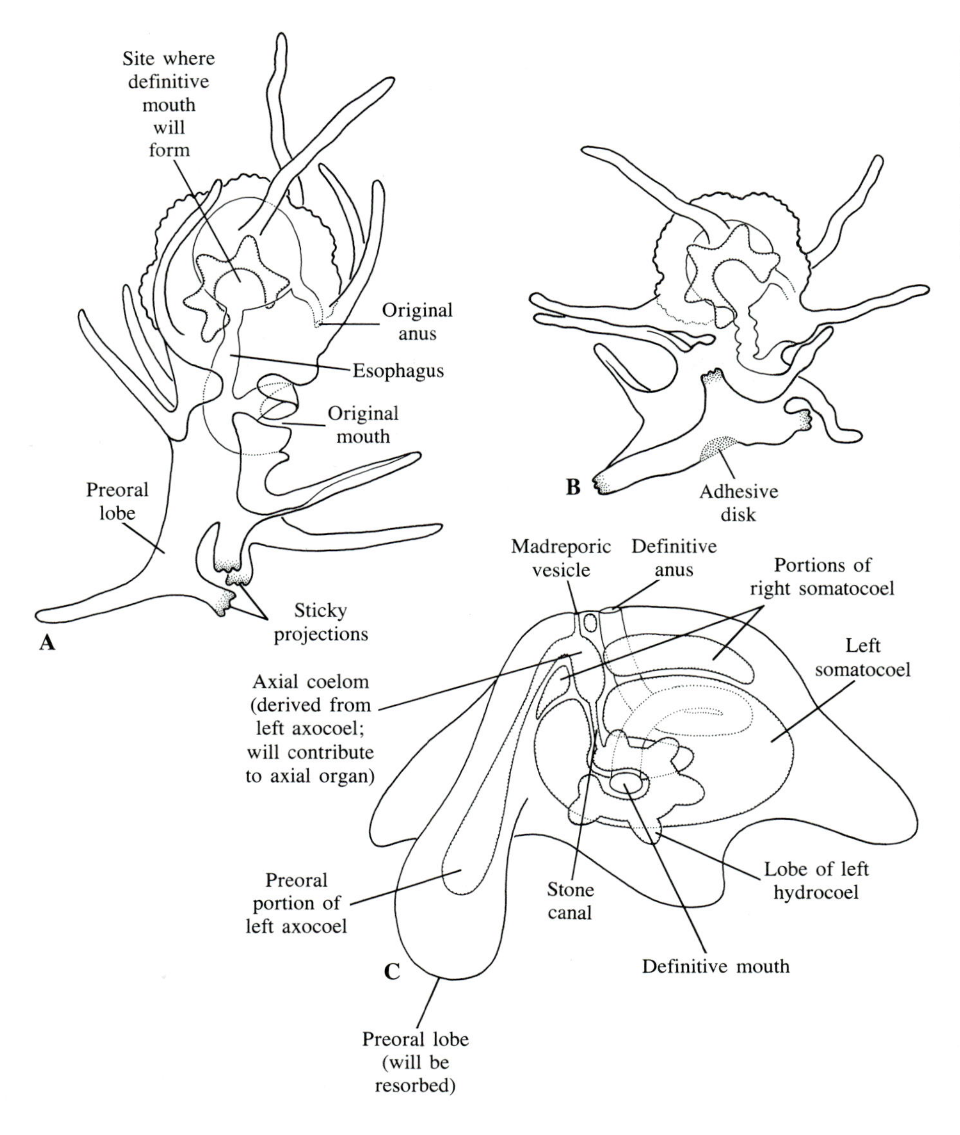

FIGURE 21.14 A, B. Two stages in the metamorphosis of the brachiolaria of *Asterias pallida*. (After Goto, Journal of the College of Science, University of Tokyo, *10*.) C. A young sea star, metamorphosis nearly completed except for resorption of the preoral lobe.

yolky, and develop directly into small sea stars. Examples of species that brood are given below in connection with the classification of Asteroidea.

REGENERATION

Most sea stars, if injured to the extent that they lose a ray or part of the disk, will slowly replace the missing portion. In some species, a complete animal will regenerate from a piece that consists of one ray and about a fifth of the disk, and even less of the disk is required if the piece includes the madreporite. A few sea stars reproduce by pulling themselves apart; each half then regenerates the parts it lacks. Species of *Linckia* carry asexual reproduction a step further; they can pinch off single rays, which reorganize into complete animals.

GENERAL REFERENCES ON ECHINODERMATA

(See also references on individual classes, especially Asteroidea.)

Binyon, G., 1972. Physiology of Echinoderms. Pergamon Press, Oxford.

Boolootian, R. A. (editor), 1966. Physiology of Echinodermata. John Wiley & Sons, New York.

Campbell, A. C., 1983. Form and function of pedicellariae. Echinoderm Studies, *1*:139–167.

Clark, A. M., 1977. Starfishes and Related Echinoderms. 3rd edition. British Museum (Natural History), London.

Coe, W. R., 1912. Echinoderms of Connecticut. Bulletin 19, Connecticut State Geology and Natural History Survey. (Reprinted 1972, as Starfishes, Serpent Stars, Sea Urchins, and Sea Cucumbers of the Northeast. Dover Publications, Inc., New York.)

Emlet, R. B., 1982. Echinoderm calcite: a mechanical analysis from larval spicules. Biological Bulletin, *163*:264–275.

Emlet, R. B., 1983. Locomotion, drag, and the rigid skeleton of larval echinoderms. Biological Bulletin, *164*:433–445.

Fell, H. B., 1962. The evolution of the echinoderms. Smithsonian Institution Annual Report, 1962:257–490.

Fell, H. B., 1966. Ancient echinoderms in modern seas. Oceanography and Marine Biology, An Annual Review, *4*:233–245.

Jangoux, M. (editor), 1980. Echinoderms Present and Past. Proceedings of the European Colloquium on Echinoderms, Brussels, September 1979. A. A. Balkema, Rotterdam.

Jangoux, M., and Lawrence, J. M. (editors), 1982. Echinoderm Nutrition. A. A. Balkema, Rotterdam.

Keegan, B. F., and O'Connor, B. D. S. (editors), 1984. Echinodermata. Proceedings of the Fifth International Echinoderm Conference, Galway, September 1984. A. A. Balkema, Rotterdam and Boston.

Lawrence, J. M., 1987. A Functional Biology of Echinoderms. Croom Helm, London and Sydney.

Millott, N. (editor), 1967. Echinoderm Biology. Symposia of the Zoological Society of London, no. 20. Academic Press, Inc., London and New York.

Nichols, D., 1962. The Echinoderms. Hutchinson University Library, London.

Pentreath, V. W., and Cobb, J. L. S., 1972. Neurobiology of the Echinodermata. Biological Reviews, *47*:363–392.

Rieger, R. M., and Lombardi, J., 1987. Ultrastructure of the coelomic lining of echinoderm podia: significance for concepts in the evolution of muscle and peritoneal cells. Zoomorphology, *107*:191–208.

Ruppert, E. E., and Balser, E. J., 1986. Nephridia in the larvae of hemichordates and echinoderms. Biological Bulletin, *171*:188–196.

Stauber, M., and Märkel, K., 1988. Comparative morphology of muscle–skeleton attachments in the Echinodermata. Zoomorphology, *108*:137–148.

Strathmann, R. R., 1971. The feeding behavior of planktotrophic echinoderm larvae: mechanisms, regulation, and rates of suspension feeding. Journal of Experimental Marine Biology and Ecology, *6*:109–160.

Strathmann, R. R., 1973. Larval feeding in echinoderms. American Zoologist, *15*:717–730.

THE MAJOR GROUPS OF ECHINODERMS

It has already been pointed out that some of the once-important assemblages of echinoderms are extinct. Specialists who strive to develop a natural system of classification for the living representatives of the phylum are nevertheless obliged to consider the extinct types. The system that is currently in favor with zoologists and paleontologists distributes echinoderms into four subphyla. One of these, Homalozoa, died out near the close of the Devonian period. It will not be dealt with here, except to say that it consisted of animals that were decidedly different from any other living or fossil echinoderms. Instead of having radial symmetry, homalozoans were asymmetrical, with tendencies toward bilaterality. Their fossil remains indicate that they had just one ambulacrum.

The three other subphyla include the five surviving groups, usually given the rank of class, as well as several groups that are extinct. The arrangement of the surviving classes under the subphyla is as follows:

Subphylum Crinozoa
  Class Crinoidea
Subphylum Asterozoa
  Class Asteroidea
  Class Ophiuroidea
Subphylum Echinozoa
  Class Echinoidea
  Class Holothuroidea

All five of the classes are distinctive. There are problems, however, with the distribution of the classes within the subphyla. The relationship of ophiuroids to asteroids and of echinoids to holothuroids may not be so close as the scheme above suggests. For this reason, the highest taxa to be dealt with here are the classes.

CLASS CRINOIDEA

The crinoids are represented by about 600 living species, largely restricted to relatively deep water. At least 5000 extinct species have been described, though there may never have been more than a few hundred kinds flourishing at any one time. Sea lilies—long-stalked crinoids whose fossils are abundant in certain Paleozoic deposits—are now relatively uncommon and only a few zoologists ever get to see them. Most of the surviving crinoids are unstalked, mobile types called feather stars (Figure 21.15; Color Plate 7). Some of these are encountered at the shore or in shallow water.

Several main groups of echinoderms related to crinoids died out before the end of the Paleozoic era. The features that link them with crinoids include a more or less globular or hemispherical body with two or more food-gathering rays. The oral surface is thought to have been held uppermost; the aboral side was attached, sometimes by a stalk such as is found in certain crinoids. Radial symmetry was characteristic of some of the extinct classes, but not all of them.

General Structure

The body of a feather star consists of two main portions: a cup-shaped **calyx** and a set of branching **rays** (Figure 21.15; Figure 21.16, A). The calyx is comparable to the disk of a starfish or brittle star. Its oral surface is covered with a skin called the **tegmen;** this is soft because the skeletal ossicles buried in it are small and rather widely separated. The aboral side of the calyx is heavily armored by large ossicles. The flexible, jointed **cirri** that originate near the base of the calyx are used to cling to a firm object (Figure 21.15) or to anchor the animal in mud or sand. The oral side of the animal thus faces away from the substratum.

FIGURE 21.15 *Florometra serratissima,* a feather star of the Pacific coast of North America. (Photograph courtesy of H. F. Dietrich.)

There are usually five rays. In most crinoids, each ray divides into two main branches, but in some genera additional branches are produced. The main branches look like feathers because they have two rows of slender **pinnules.** All portions of the rays, including the pinnules, are jointed. Each unit contains a large skeletal ossicle (Figure 21.16, C; Figure 21.17). Successive ossicles are held together by connective tissue organized into ligaments. The musculature is concentrated in distinct bands lying on the oral side of the ossicles. When these muscles contract, the ray and its pinnules curl toward the mouth. This puts the ligaments under tension. When the muscles relax, the ray and the pinnules straighten out again. A feather star can swim effectively by rapidly flexing certain of the main branches while it straightens others. This produces a graceful fluttering motion. The rays can also be used for temporarily holding onto the substratum until the cirri can be manipulated into their grasping position.

The five **ambulacral grooves** radiate from the mouth on the oral surface of the calyx into the ray branches and finally into the pinnules. On both the pinnules and main branches, the grooves are bordered by a continuous series of **lappets** that can be moved to temporarily close up the grooves (Figure 21.16, C). Associated with each lappet are three fingerlike tube feet, homologues of the tube feet of other echinoderms. The three tube feet in a particular group are usually of unequal length, but they operate collectively to knock small organisms and other particles into the ambulacral groove. They also secrete mucus that helps keep the food in the groove where it belongs. The mixture of food and mucus is moved to the mouth by ciliary activity.

Some feather stars are parasitized by strange polychaetes called myzostomes. These usually steal food from the mouth or from the main ciliated grooves that lead to the mouth. Palaemonid shrimp are often perched on the rays of tropical feather stars.

Digestive Tract

As it makes a loop through the calyx, the gut is differentiated into portions that may be called the esophagus, stomach, and intestine (Figure 21.16, B). The stomach and intestine generally have diverticula. The anus is situated on a prominent papilla near one edge of

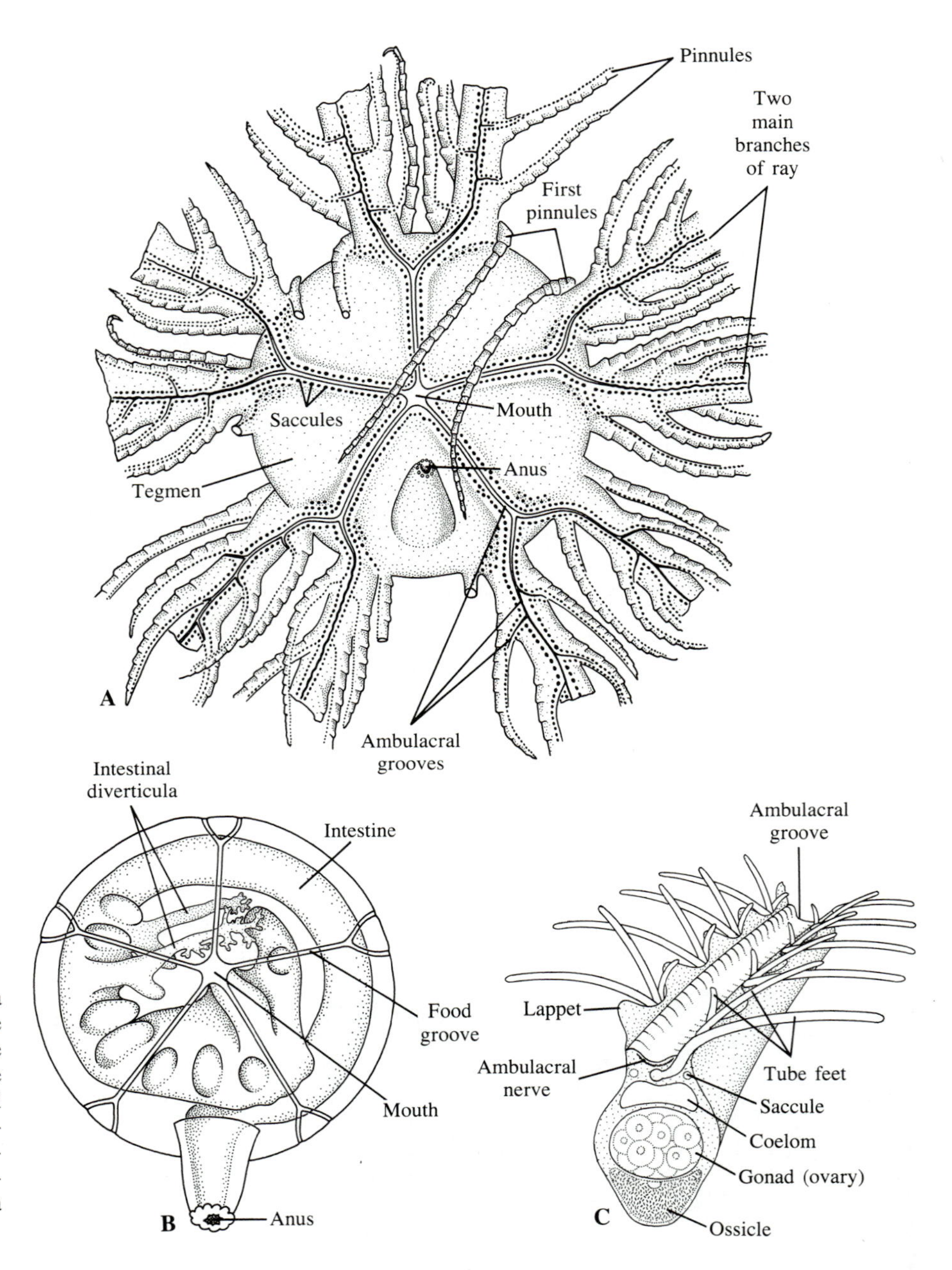

FIGURE 21.16 *Antedon bifida,* a feather star. A. Oral surface of the disk and proximal portions of the rays. B. Digestive tract, with the ambulacral grooves of the oral surface of the calyx superimposed. (After Chadwick, Liverpool Marine Biological Committee Memoir 15.) C. Portion of a pinnule.

the oral surface of the calyx. Movement of food through the gut and the compaction and elimination of fecal pellets is accomplished by ciliary activity. Little is known, however, about digestion in crinoids. Some species draw water through the anus into the hind portion of the gut, and this activity is probably concerned either with respiration or with getting rid of fecal pellets.

Perivisceral Coelom, Hemal System, and Water-Vascular System

Within the calyx, the perivisceral coelom is broken up by strands of connective tissue, so it consists of a number of indistinct spaces. A portion of the perivisceral coelom extends into each ray (Figure 21.17) and branches of it reach into all of the pinnules (Figure

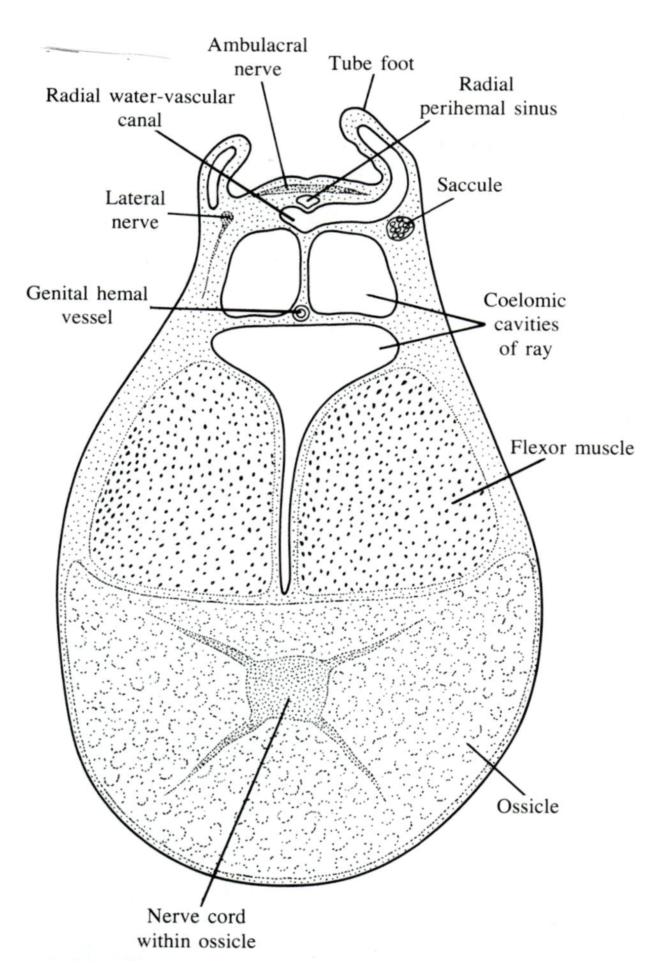

FIGURE 21.17 Transverse section of a ray of *Antedon bifida*. (After Chadwick, Liverpool Marine Biological Committee Memoir 15.)

21.16, C). Similar canals enter the cirri on the aboral side of the calyx.

The hemal system is fairly complicated. Its main components are networks of vessels around the gut and within the connective tissue that divides up the perivisceral coelom, and vessels that extend into the main branches and pinnules of the rays. Amoeboid cells are present in the perivisceral coelom and hemal vessels, and also in the water-vascular system.

The principal elements of the water-vascular system of a crinoid are the ring canal around the esophagus and five main radial canals. Branches of the radial canals go out into the branches of the rays (Figure 21.17) and reach into every pinnule (Figure 21.16, C), thereby supplying the tube feet. The tube feet have no ampullae. The contraction of the muscles of the radial canals and their larger branches increases hydrostatic pressure within portions of the water-vascular system, extending the tube feet. Bending of the tube feet, essential to their activity in feeding, depends upon their own musculature.

Crinoids generally have several stone canals joined to the ring canal, but these end in the spaces of the perivisceral coelom. The coelomic spaces, however, are in direct communication with the sea water outside the animal because the soft skin covering the oral surface of the calyx has many pores. The edges of these pores are ciliated.

Respiration and Excretion

Crinoids presumably exchange gases and eliminate nitrogenous wastes across all parts of the body surface where the skin is soft. The thousands of thin-walled tube feet are probably where much diffusion takes place. The odd little **saccules** located just beneath the surface next to the ambulacral grooves of the disk, ray branches, and pinnules (Figure 21.16, A and C; Figure 21.17) have been claimed to concentrate waste products collected by amoeboid coelomic cells. It is possible that the saccules do discharge wastes to the outside, but the mechanism by which they accomplish this needs to be explained.

Nervous System

The nervous system of a feather star is rather different from that of other echinoderms in that the greatest concentration of nerve cells is located aborally. The nerve center is a cup-shaped mass. Nerves arising from it, or from a kind of sheath joined to it, extend into the cirri. Five other important nerves run through the wall of the calyx in the direction of the oral surface, and become interconnected by a nerve ring before they continue out into the rays. A cross-section through a branch (Figure 21.17) or pinnule of a ray shows a prominent nerve running right through the ossicle. It is from this main nerve that subsidiary nerves extend to the tube feet, muscles, and sensory receptors in the skin, and to other structures.

Reproduction and Development

Sexes are separate. The gonads of most feather stars are located in the pinnules (Figure 21.16, C), generally those nearer the calyx. In certain species, however, they are in the main branches of the rays. The eggs or sperm escape through small openings that connect the gonads with the outside. Eggs may be retained for a time by the female parent, remaining stuck to the rays until they have developed into swimming larvae. Some species have special brood pouches on the ray branches or pinnules, and the young do not leave the pouches until they are fully formed little crinoids.

Regeneration

Like sea stars and brittle stars, crinoids have strong powers of regeneration. Lost cirri, pinnules, and even whole rays can be replaced. In some species, at least, the animal can survive the loss of a fifth of the calyx and will regenerate the missing portion. The aboral region of the calyx must remain intact, however, for the concentration of nerve tissue within it controls regeneration.

Classification

Only one subclass of crinoids, the Articulata, has living representatives. These belong to two orders. Several other orders of Articulata are represented in the rich fossil record of crinoids.

Order Isocrinida

The few stalked crinoids that have survived, and many of the fossil species, belong to this order. Some representatives of the group are illustrated in Figure 21.18. In most genera, the stalks have whorls of cirri, and the lowermost cirri are used for gripping the substratum. A crinoid of this type may become detached and swim or drift temporarily, and any of the cirri along the stalk may be used to grasp the substratum until the animal can reorient itself in an upright position. In one genus, there are no cirri, and the lowermost portion of the stalk is a disk by which the crinoid is permanently cemented to a hard object. Fossils of stalked crinoids show similar disks, as well as rootlike elaborations that functioned as holdfasts.

Most stalked crinoids live in rather deep water where currents are weak; about half of the known species occur at depths greater than 1000 m.

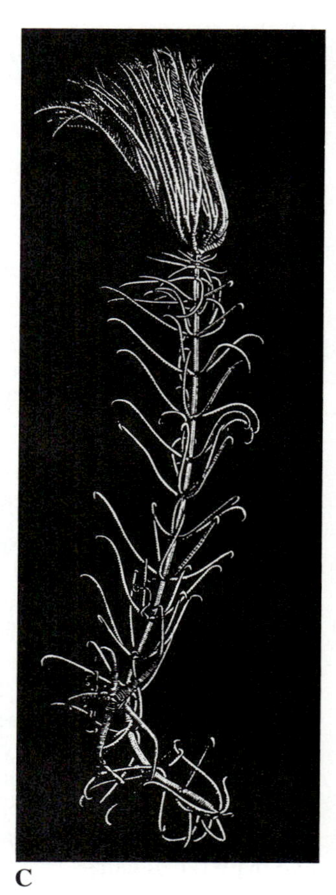

FIGURE 21.18 Stalked crinoids (order Isocrinida). A. A well-conserved fossil. B. *Rhizocrinus lofotensis*. C. *Pentacrinus asteria*. (B, C, Wyville Thomson, The Voyage of H.M.S. Challenger, 1.)

Order Comatulida

The feather stars, including *Antedon* (Figure 21.16), *Florometra* (Figure 21.15), and *Cenometra* (Color Plate 7), belong to this group. Some of them, such as a species of *Antedon* found in European waters, are occasionally encountered intertidally, but the majority of these crinoids inhabit shallow water. There are more species along tropical shores and in cold polar seas than there are in temperate regions.

Comatulids have a stalk when they are young, but the stalk is eventually discarded. In general, these animals are rather mobile, moving from place to place with the aid of their cirri and by graceful fluttering movements of the rays.

References on Crinoidea

Byrne, M., and Fontaine, A. R., 1981. The feeding behaviour of *Florometra serratissima* (Echinodermata: Crinoidea). Canadian Journal of Zoology, *59:*11–18.

Byrne, M., and Fontaine, A. R., 1983. Morphology and function of the tube feet of *Florometra serratissima* (Echinodermata: Crinoidea). Zoomorphology, *102:*175–187.

Clark, A. M., 1970. Echinodermata Crinoidea. Marine Invertebrates of Scandinavia, no. 3. Scandinavian University Books, Oslo.

Dan, J. C., and Dan, K., 1941. Early development of *Comanthus japonicus*. Japanese Journal of Zoology, *9:*565–574.

Grimmer, J. C., and Holland, N. D., 1979. Haemal and coelomic circulatory systems in the arms and pinnules of *Florometra serratissima*. Zoomorphologie, *94:*93–109.

Holland, N. D., 1971. The fine structure of the ovary of the feather star *Nemaster rubiginosa* (Echinodermata: Crinoidea). Tissue and Cell, *3:*163–177.

Holland, N. D., and Grimmer, J. C., 1981. Fine structure of the cirri and a possible mechanism for their mobility in stalkless crinoids (Echinodermata). Cell and Tissue Research, *214:*207–217.

Holland, N. D., Leonard, A. B., and Strickler, J. R., 1987. Upstream and downstream capture during suspension feeding by *Oligometra serripinna* (Echinodermata: Crinoidea) during surge conditions. Biological Bulletin, *173:*552–556.

La Haye, C. A., and Holland, N. D., 1984. Electron microscopic studies of the digestive tract and absorption from the gut lumen of a feather star, *Oligometra serripinna* (Echinodermata). Zoomorphology, *104:*252–259.

Macurda, D. B., Jr., 1978. These reef animals blossom at night. Smithsonian, 9(a):86–89.

Macurda, D. B., Jr., and Meyer, D. L., 1983. Sea lilies and feather stars. American Scientist, *71:*354–365.

Mladenov, P. V., and Chia, F. S., 1983. Development, settling behaviour, metamorphosis and pentacrinoid feeding and growth of the feather star *Florometra serratissima*. Marine Biology, *73:*319–323.

Nichols, D., 1960. The histology and activities of the tube-feet of *Antedon bifida*. Quarterly Journal of Microscopical Science, *101:*105–117.

Rutman, J., and Fishelson, L., 1969. Food composition and feeding behavior of shallow water crinoids at Eilat (Red Sea). Marine Biology, *3:*46–57.

CLASS ASTEROIDEA

Sea stars have already been discussed in some detail, in order to illustrate the more important features of the various body systems of the phylum Echinodermata. The new material presented here is therefore concerned mainly with diversity shown by representatives of the surviving orders.

Order Platyasterida

Platyasterids have no anus or intestine. Their numerous gonads, each with its own pore on the aboral surface, are arranged in two rows in each ray. The tube feet do not have suckers at their tips. The rays are bordered by a single series of large ossicles, called **marginal plates,** and each of these is studded with spines. Pedicellariae, if present, are rudimentary; some of them have three jaws.

The only commonly encountered members of this order belong to the genus *Luidia* (Figure 21.19). These sea stars are mostly subtidal and inhabit muddy and sandy substrata, in which they bury themselves. They feed on bivalve molluscs, sea cucumbers, and various other invertebrates, swallowing their prey whole.

Some fossil echinoderms called somasteroids, which are believed to have originated near the beginning of the Ordovician period, show some features that probably link them with crinoidlike types as well as surviving platyasterids. Their rays had side branches, called **metapinnules,** each consisting of a series of rod-shaped ossicles, or **virgalia.** Only the distal tips of the metapinnules were free; the proximal portions were embedded in the tissue of the rays. In later somasteroids, the metapinnules were completely buried in the rays, although the virgalia still formed neat rows (Figure 21.20). Eventually the number of virgalia in each series was reduced to three.

In *Luidia,* virgalia of the sort found in somasteroids are still recognizable and still oriented more or less at right angles to the two rows of ossicles that run the length of the ambulacral groove. The outermost virgalia have become the marginal plates. *Platasterias,* represented by a species found in deep water off the Pacific coast of Central America, shows features that link it with *Luidia* as well as with some somasteroids. The fact that it can, by erecting ossicles that border the ambulacral ossicles, form an ambulacral groove like that of a sea star suggests that it should be placed in the order Platyasterida. Some experts even put it into the genus *Luidia*.

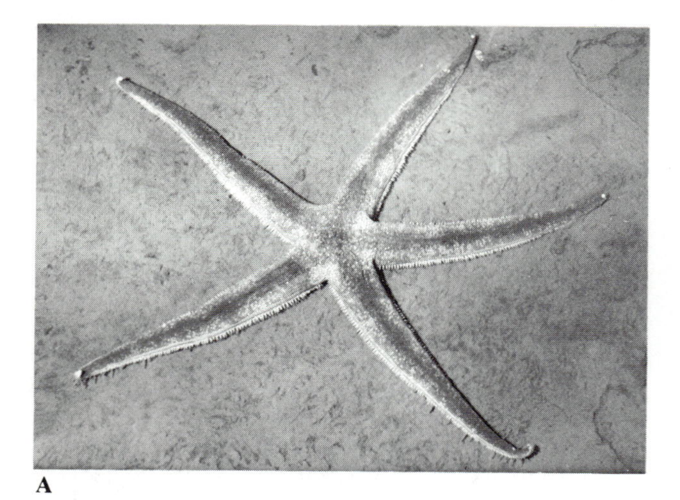

A

B

FIGURE 21.19 Order Platyasterida. A. *Luidia foliolata.* B. Tip of a ray of *Luidia.*

Order Paxillosida

Although paxillosids are similar to platyasterids in lacking an anus, as well as suckers on their tube feet, they have two series of especially prominent ossicles along the margins of the rays. These are homologous with the two outermost virgalia of platyasterids; any internal virgalia that have survived are, at best, vestigial. When pedicellariae are present, they are simple, consisting of a group of ossicles or a pair of adjacent ossicles.

Representative genera are *Astropecten, Leptychaster,* and *Ctenodiscus,* which live on soft subtidal sediments. Paxillae are present over much of the aboral surface and prevent clogging of this surface. At least some of the species feed to a considerable extent on detritus.

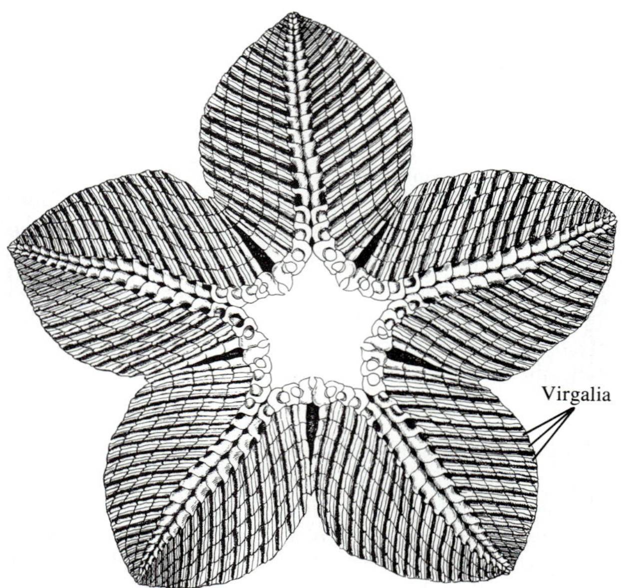

FIGURE 21.20 *Villebrunaster thorali,* a somasteroid, reconstruction of the oral surface. (Spencer, Philosophical Transactions of the Royal Society of London, B, *235.*)

Order Valvatida

As a rule, valvatids have two rows of prominent plates, derived from virgalia, along the margins of the rays. In this respect they are similar to paxillosids. They differ from these, and also from platyasterids, in having an anus and in having suckers on the tube feet. If pedicellariae are present, they are generally prominent, consisting of two rather broad ossicles, so that they resemble clamshells (Figure 21.21, C). Much of the aboral surface is covered with paxillae.

Representative genera are *Mediaster* (Figure 21.21, A and B; Figure 21.22), *Hippasteria* (Figure 21.21, C), and *Ceramaster*. Most valvatids are subtidal on pebbly or shelly bottoms, where they feed on various invertebrates, including anemones and sea pens.

Order Spinulosida

Typically, spinulosids lack platelike ossicles along the margins of the rays, and they generally do not have pedicellariae. (If pedicellariae are present, they are simply groups of small spines.) There is an anus, and the tube feet have suckers. Some species have paxillae or similar structures; others lack them.

Many of the common intertidal and subtidal genera belong to this order. Among them are *Solaster* (Figure 21.21, E), *Dermasterias, Asterina* (Figure 21.21, D), and *Henricia* (Color Plate 8). The order also includes

FIGURE 21.21 A. *Mediaster aequalis* (order Valvatida), aboral surface, showing paxillae and marginal plates. B. *Mediaster*, oral view, showing tube feet, spines bordering ambulacral groove, and marginal plates. C. *Hippasteria spinosa* (order Valvatida), portion of aboral surface of disk, showing pedicellariae of the clamshell type. D. *Asterina miniata* (order Spinulosida). E. *Solaster stimpsoni* (order Spinulosida). F. *Pteraster tesselatus* (order Spinulosida).

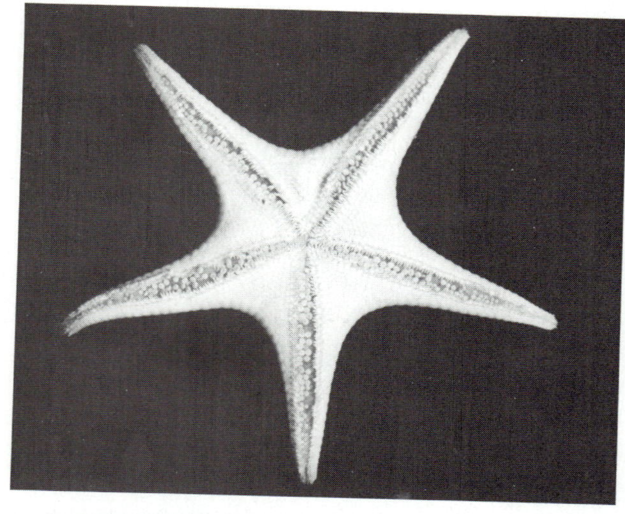

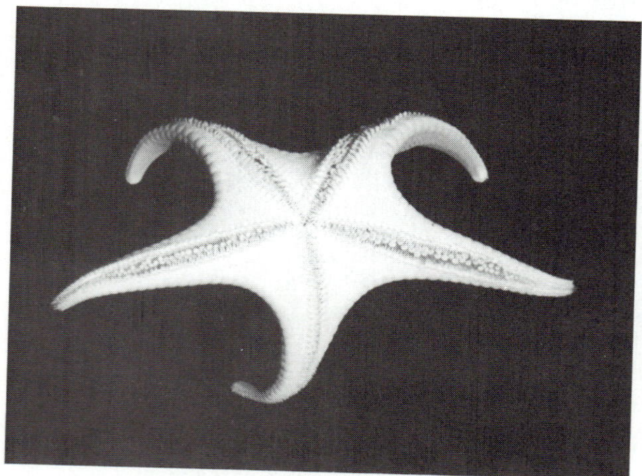

FIGURE 21.22 *Mediaster aequalis* (order Valvatida), successive stages in righting by an animal that has been turned over. The action of tube feet is coordinated with bending movements of the rays.

the remarkable genus *Pteraster* (Figure 21.21, F), characterized by a spongy chamber occupying much of the aboral side of the disk. The external opening of the cavity must not be confused with the anus. Both the anus and madreporite are on the floor of the chamber, which is the true aboral surface. The membrane that covers the cavity is like a tent in that it is supported by tall paxillae. It is porous and water can pass through it as well as through the main opening. Periodically water is forced out of the opening, after which the chamber starts to fill again. It is probably a respiratory device, but in females of a northern European species of *Pteraster* it serves as a brood chamber.

Order Forcipulatida

Forcipulatids almost always have pedicellariae. These may be of either the straight or the crossed type (see Figure 21.4). There are no platelike ossicles at the margins of the rays, and paxillae are also lacking. The gut has an anus, and the tube feet have suckers.

This group includes many familiar genera, including *Asterias, Pisaster* (Color Plate 7), *Leptasterias* (Figure 21.23, C), *Pycnopodia* (Figure 21.23, A and B), and *Crossaster* (Color Plate 8). Several species of *Leptasterias,* a genus characterized by six rays, are brooders. In the case of *Leptasterias hexactis,* which is common on the Pacific coast of North America, females hump over their large, yolky eggs until these have developed into small sea stars. *Leptasterias groenlandica* uses its cardiac stomach as a brood chamber.

References on Asteroidea

Allen, P. L., 1983. Feeding behaviour of *Asterias rubens* (L.) on soft bottom bivalves: a study in selective predation. Journal of Experimental Marine Biology and Ecology, *70:*79–90.

Anderson, J. M., 1960. Histological studies on the digestive system of a starfish, *Henricia,* with notes on Tiedemann's pouches in starfishes. Biological Bulletin, *119:*371–398.

Binyon, J., 1964. On the mode of functioning of the water vascular system of *Asterias rubens*. Journal of the Marine Biological Association of the United Kingdom, *44:*577–588.

Binyon, J., 1976. The permeability of the asteroid podial wall to water and potassium ions. Journal of the Marine Biological Association of the United Kingdom, *56:*639–647.

Blake, D. B., 1981. A reassessment of the sea-star orders Valvatida and Spinulosida. Journal of Natural History, *15:*375–394.

Chia, F. S., 1966. Brooding behavior of a six-rayed starfish, *Leptasterias hexactis*. Biological Bulletin, *130:*304–315.

Chia, F. S., and Amerongen, H., 1975. On the prey-catching pedicellariae of a starfish, *Stylasterias forreri*. Canadian Journal of Zoology, *53:*748–755.

Christensen, A. M., 1970. Feeding biology of the sea star *Astropecten irregularis*. Ophelia, *8:*1–134.

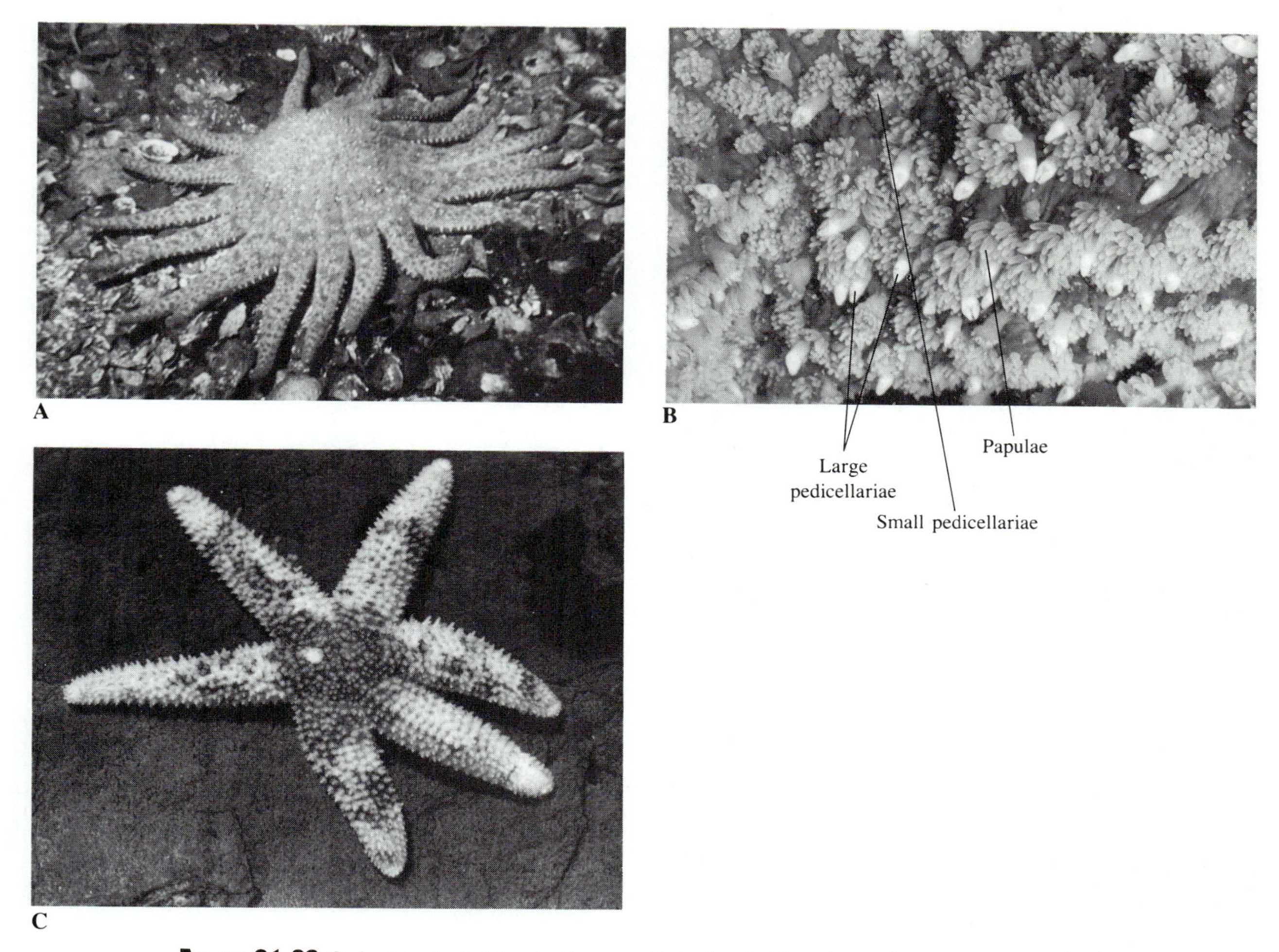

FIGURE 21.23 Order Forcipulatida. A. *Pycnopodia helianthoides,* the sunflower star. B. Pedicellariae and papulae on the aboral surface of *Pycnopodia*. C. *Leptasterias hexactis*. (See also *Crossaster papposus,* Color Plate 8.)

Feder, H. M., 1955. On the methods used by the starfish *Pisaster ochraceus* in opening three types of bivalved mollusks. Ecology, *36*:764–767.

Gale, A. S., 1987. Phylogeny and classification of the Asteroidea (Echinodermata). Zoological Journal of the Linnean Society, *89*:107–132.

Hendler, G., and Franz, D. R., 1982. The biology of a brooding seastar, *Leptasterias tenera,* in Block Island Sound. Biological Bulletin, *162*:273–289.

Laxton, J. H., 1974. Aspects of the ecology of the coral-eating starfish *Acanthaster planci*. Biological Journal of the Linnean Society, *6*:19–45.

Mauzey, K. P., 1966. Feeding behavior and reproductive cycles in *Pisaster ochraceus*. Biological Bulletin, *131*:127–144.

Mauzey, K. P., Birkeland, C., and Dayton, P. K., 1968. Feeding behavior of asteroids and escape responses of their prey in the Puget Sound region. Ecology, *49*:603–619.

Van Veldhuizen, H. D., and Oakes, V. J., 1981. Behavioral responses of seven species of asteroids to the asteroid predator, *Solaster dawsoni*. Oecologia, *48*:214–220.

Vine, P. J., 1973. Crown-of-thorns plagues: the natural causes theory. Atoll Research Bulletin, *166*:1–10.

Wood, R. L., and Cavey, M. J., 1981. Ultrastructure of the coelomic lining of the podium of starfish *Stylasterias forreri*. Cell and Tissue Research, *218*:449–473.

CLASS OPHIUROIDEA

General Structure and Skeleton

There are about 2000 species in the class Ophiuroidea, which includes the brittle stars (Figure 21.24) and basket stars (Figure 21.27). The rays of these echinoderms are slender and clearly set off from the disk.

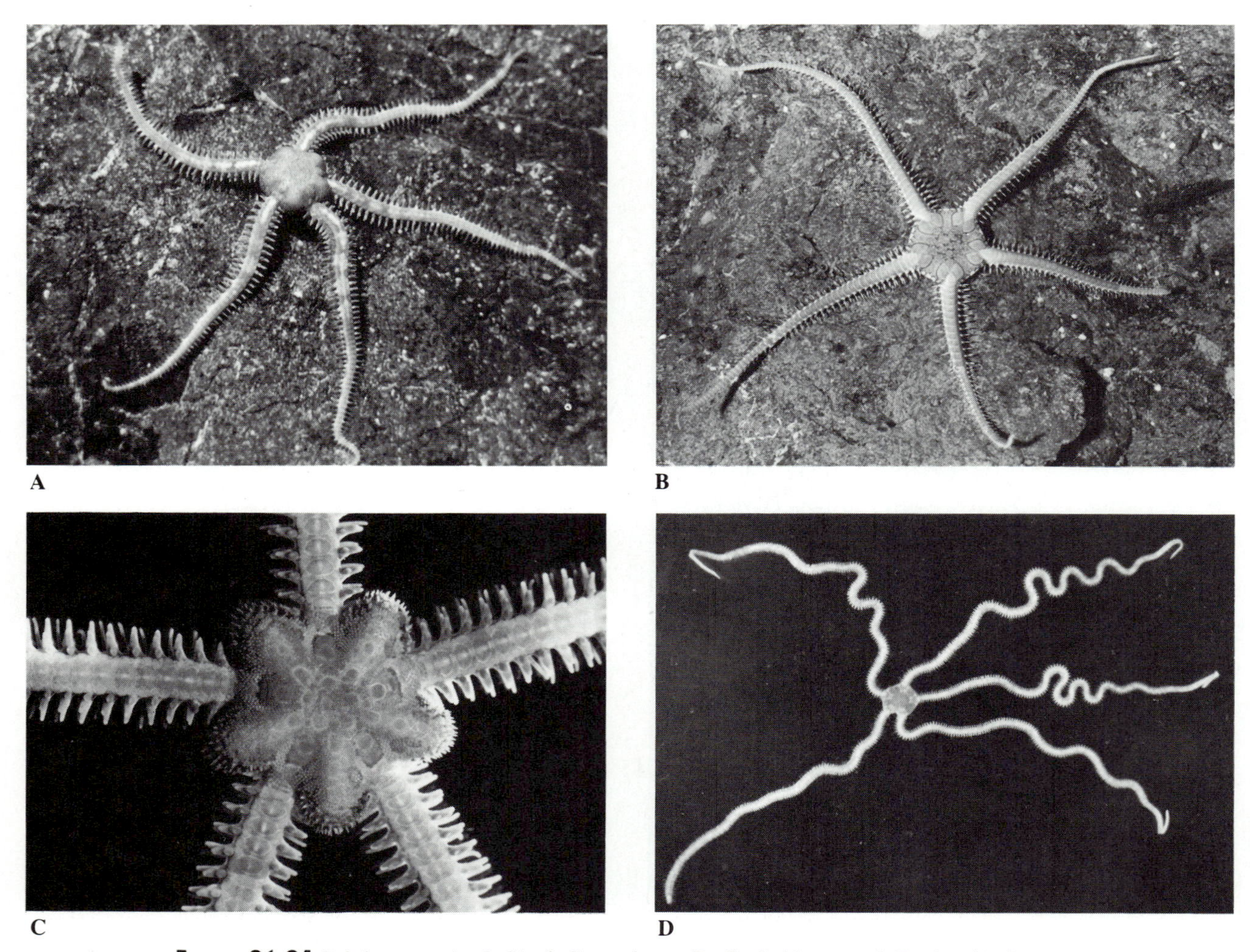

Figure 21.24 Brittle stars. A. *Ophiopholis aculeata*. B, C. *Ophiura sarsi*. D. *Amphiodia occidentalis,* a species that burrows in sand and mud.

They are also more flexible than those of sea stars, and because of the way in which their ossicles are organized, they usually are capable of snaky movements. In the first place, the rays are not cluttered by a rigid framework of relatively small ossicles; those visible externally are sufficiently separate from one another that they do not interfere with bending. The only other ossicles are large and are arranged like vertebral bones within the rays (Figure 21.25; Figure 21.26, C and D). They articulate freely and are joined by four bands of muscles.

Each vertebral ossicle is derived from two separate ossicles. These can be traced back through the fossil record to pairs of plates located on the oral surfaces of the rays of certain extinct ophiuroids. In the course of evolution the plates sank inward, taking the ambulacral groove with them. There are no ambulacral grooves in modern ophiuroids.

Nearly all ophiuroids have five equal rays, but there are a few with six or seven, and some in which one ray is decidedly longer than the others. In basket stars (Figure 21.27), the rays branch repeatedly in a dichotomous pattern and their internal ossicles fit together in much the same way as ball-and-socket joints. Thus the rays can bend, twist, and curl. Basket stars, as a group, are the largest ophiuroids. Some of them have disks 5 or 6 cm in diameter, with rays that can span 40 or 50 cm. Most brittle stars are less than 10 cm across their ray span.

Ophiuroids lack pedicellariae and papulae. As a rule, the ossicles that cover the rays are contiguous, so that the only projecting soft structures are the tube feet.

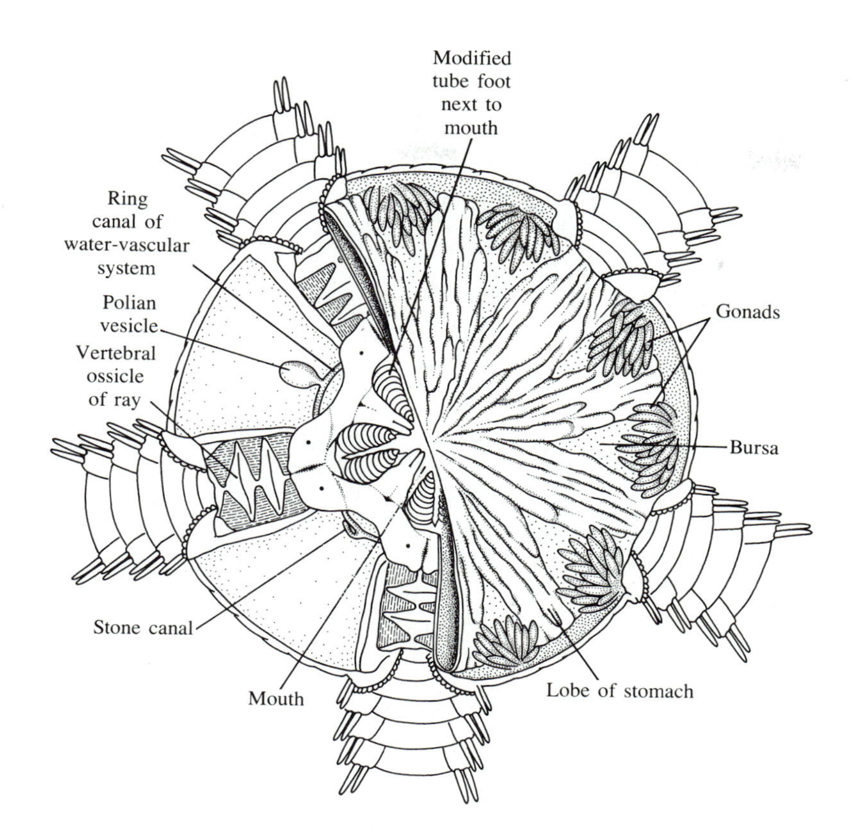

FIGURE 21.25 *Ophiura sarsi*, dissected to show the gut and gonads on one side of the disk and the structures beneath the gut on the other side. (After Ivanov *et al.*, Major Practicum of Invertebrate Zoology.)

These, together with the lining of the gut, provide considerable surface for gas exchange, but respiration also takes place in 10 thin-skinned pockets called **bursae** (Figure 21.25). The openings of the bursae are slits on the oral surface of the disk, next to the bases of the rays. By muscular activity of the disk as a whole, or of muscles that specifically control the bursae, an ophiuroid can pump water in and out of the pockets. The bursae and slits are generally ciliated, the action of the cilia helping to circulate water. In some ophiuroids the slits are subdivided, the half farthest from the mouth serving as an inhalant opening, the other half serving as an exhalant opening.

Locomotion

When brittle stars move, they generally use two rays to produce a rowing motion, with the other rays leading or trailing. Most species can turn themselves over quickly. Some wind their rays around seaweeds, hydroids, bryozoans, and other objects, and many use their rays and tube feet for burrowing into sand or mud. Brittle stars that burrow usually have especially long rays (Figure 21.24, D). After they have buried themselves they may extend one or more of these rays above the substratum to collect food and promote respiration.

Digestive System and Feeding

The gut of an ophiuroid is a blind sac; there is no anus. The mouth is preceded by five more or less triangular **jaws** (Figure 21.26, A and B), so the opening seen externally is approximately star-shaped. Each jaw originates between the bases of two rays and is a rather complex structure. It begins with a large, unpaired ossicle called the **oral shield.** Distal to this are smaller shields. The edges of the triangle are bordered by **teeth,** at the tip of which are additional teeth arranged in a vertical row. The plan of jaw structure and arrangement of teeth varies from genus to genus, however. At the base of each ray, where this reaches the cleft between the jaws, there is a pair of modified tube feet (Figure 21.25).

Some brittle stars have a recognizable esophagus, but most of the gut consists of a thin-walled, baglike stomach (Figure 21.25). The radiating lobes of this occupy the space not taken up by the gonads or other structures located within the disk. Cilia on the epithelium that lines the stomach cavity circulate small particles and the soluble products of digestion. The functions of the different portions of the epithelium need to be investigated. Some of the cells are specialized for secretion, others for the absorption and production of

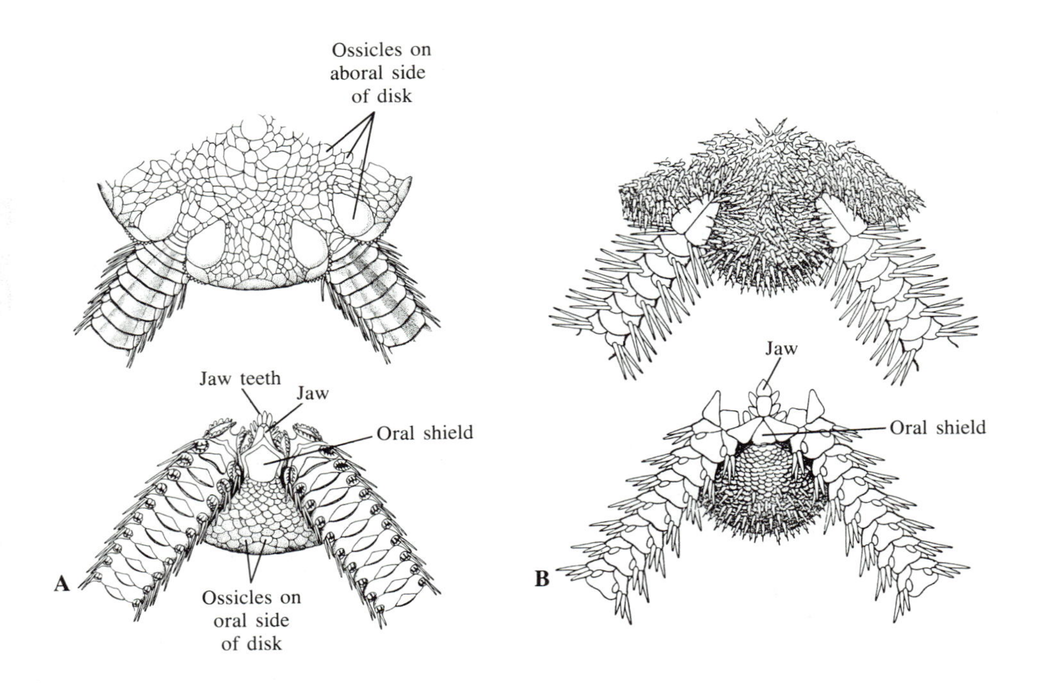

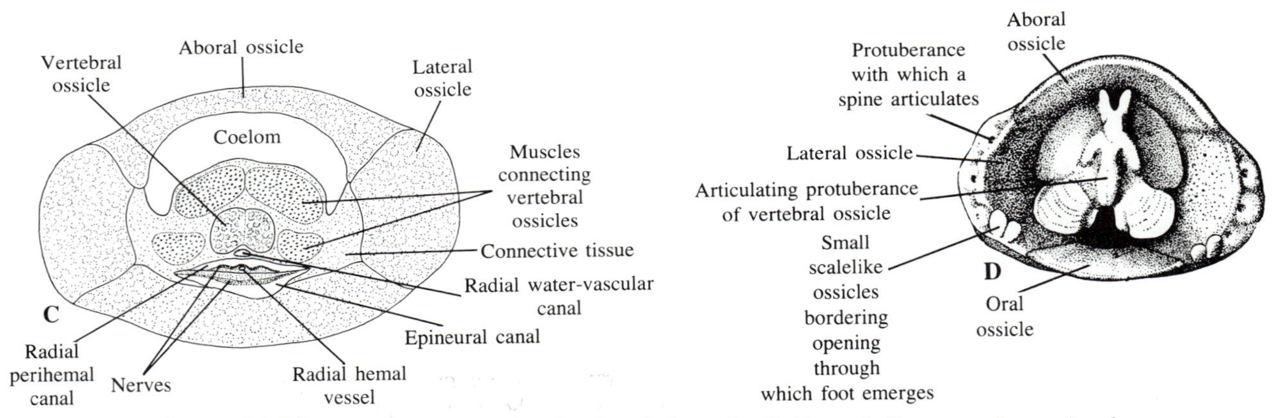

FIGURE 21.26 A. *Ophiura sarsi,* aboral and oral views. B. *Ophiacanthella acontophora,* aboral and oral views. (Diakonov, Keys to the Fauna of the USSR, *55.*) C. *Ophiura sarsi,* transverse section of a ray; the vertebra is not cut through its broadest part. The epineural canal is not comparable to a perihemal canal or any other coelomic derivative. It can be traced back to an ambulacral groove that closes over during development. (After Ivanov *et al.,* Major Practicum of Invertebrate Zoology.) D. *Ophiura sarsi,* relationship of a vertebra to other ossicles; soft tissues omitted. (Ivanov *et al.,* Major Practicum of Invertebrate Zoology.)

mucus. Amoebocytes have been observed among the epithelial cells, and some of them are thought to deliver waste material from the coelom to the gut lumen.

Brittle stars use a variety of techniques for collecting food. Some species are remarkably versatile, being able to switch from one method to another. *Ophiocomina nigra,* a North Atlantic species, exhibits nearly all of the major methods used by ophiuroids. It can use its tube feet for gathering loose particles and for picking bits of tissue from a dead animal. The food, mixed with mucus, is passed from one tube foot to another until it reaches the mouth. *Ophiocomina* has also been claimed to use its jaws to browse on algae and dead animals. Another technique employed by this brittle star is "arm-loop capture." The animal bends one of its rays into a loop, thus grasping a small crustacean, worm, or

A

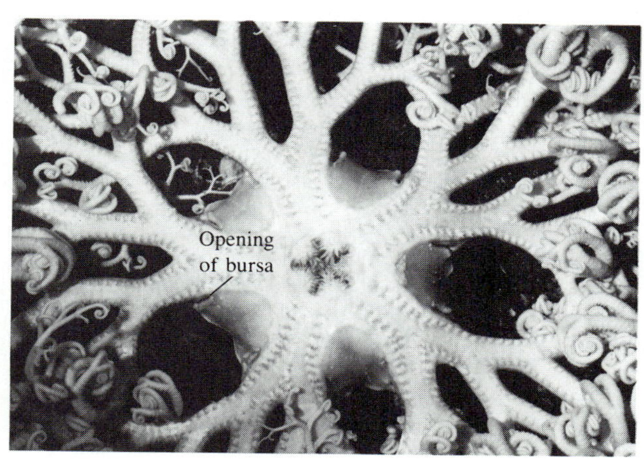

B

FIGURE 21.27 *Gorgonocephalus eucnemis,* a basket star. A. Aboral view. B. Oral view of disk and proximal portions of rays.

some other animate or inanimate object. Tube feet help hold the food in place until the loop is brought to the mouth.

For trapping phytoplankton, bacteria, and other small organisms and fine detritus, *Ophiocomina* uses mucus. The mucus hangs in strings from the rays or forms a network between the ray spines. The rays are raised and moved from side to side, their ranges of movement just barely overlapping. The activity of the tube feet and cilia concentrates the mucus into balls that are passed toward the mouth. In aquaria, *Ophiocomina* has been observed to trail some of its rays at the surface to collect the thin film of floating organic matter. Although *Ophiocomina* is subtidal and therefore not likely to use this method in its natural habitat, the technique is a practical one for intertidal species.

By using one or another of the methods described, *Ophiocomina* exploits a wide range of foods: bacteria, diatoms, various other phytoplankton, copepods, medusae, polychaetes, algae, and other material. The diet varies according to the season. Phytoplankton, for instance, is utilized most extensively in the spring, when it is especially abundant; the consumption of zooplankton picks up in late spring and summer. Few brittle stars are as versatile as *Ophiocomina,* but most of them probably use at least one of these methods. The subtidal basket stars of the genus *Gorgonocephalus* (Figure 21.27) are primarily carnivores, employing their tube feet and the flexible tips of their branched rays to capture small crustaceans.

Water-Vascular System and Tube Feet

If a recognizable madreporite is present, it is one of the five oral shields. In basket stars, the madreporite is about as prominent as it is in sea stars, but in brittle stars it generally has just one pore. In certain genera, such as *Ophiura,* there is no opening; the stone canal ends blindly (Figure 21.25).

The ring canal of the water-vascular system, located at about the level of the true mouth, sends a radial canal out into each ray. It also has some Polian vesicles and connects with a stone canal, even though this may never reach a madreporite. The radial canals run right through the vertebral ossicles of the rays, and they send out lateral canals that supply the tube feet.

The tube feet are not tipped by suckers as they are in many asteroids, and they do not have ampullae, but they are capable of considerable extension and mobility. Muscles in the walls of the various canals, as well as in the walls of the tube feet themselves, provide the pressure necessary for protraction.

Perivisceral Coelom and Hemal System

The perivisceral coelom of the disk is almost completely filled by the gut, gonads, and bursae. In the rays, which consist mostly of ossicles, muscles, and canals of the water-vascular system, the only trace of the perivisceral coelom is a small space located above the vertebral ossicles. It usually shows up in transverse sections (Figure 21.26, C).

The hemal system of ophiuroids is somewhat similar to that of sea stars. Information concerning it should be sought in specialized treatises.

Nervous System

The nervous system is not described in detail here. Its principal components—a nerve ring on the oral side, radial nerves going out into the rays, and division of peripheral components into two complexes, one more nearly external than the other—are basically similar to those of sea stars.

The epidermis of ophiuroids is sensitive to light and other stimuli. Most species are relatively quiet when they are buried in sand or are in a dark place and in close contact with rocks, shells, or seaweeds. If exposed to light, they will usually move around actively until they find a place where the light intensity is weak and where they can burrow or squeeze in between rocks or other firm objects.

Reproduction and Development

In most ophiuroids, sexes are separate. The gonads are typically in the disk, rarely in the rays, and there are usually 10 of them (Figure 21.25). They do not open directly to the outside, but into the bursae. In species that release their eggs and sperm into the sea, fertilization is followed by a pattern of development that is almost identical to that already described for sea stars. The free-swimming larva, called an **ophiopluteus,** has a number of long projections supported by calcareous skeletal elements. Metamorphosis is accomplished without benefit of attachment devices such as are typical of the brachiolaria of a sea star.

It is fairly common for ophiuroids to brood their young in the bursae. The eggs of brooding species tend to be large and yolky, and development is direct. An especially well-known brooder is the nearly cosmopolitan *Amphipholis squamata*. This small brittle star is unusual in that it is hermaphroditic; it can fertilize its eggs with its own sperm. The eggs become implanted into the lining of the bursae and are to some extent nourished by the parent. Various other reproductive nonconformists occur among the ophiuroids.

A particularly interesting life history is that of basket stars of the genus *Gorgonocephalus*. After a female has brooded her young to a certain stage, she releases them and they settle in colonies of soft corals, such as *Gersemia,* where they develop further before striking out on their own.

Classification

In one wholly extinct order of ophiuroids, the Stenurida, there were no vertebral ossicles in the rays because the homologues of these were located superficially along the open ambulacral grooves. The remaining three orders, represented by living species, do have vertebral ossicles. These orders are briefly characterized below.

Order Oegophiurida

The rays lack ossicles on their oral and aboral surfaces. The only surviving species, *Ophiocanops fugiens,* is found off the coast of Indonesia. It has stomach pouches that extend into the rays; the gonads are also located in the rays.

Order Phrynophiurida

In phrynophiurids, the aboral surfaces of the rays lack ossicles, or have only small ones. There are, however, ossicles on the oral surfaces and sides of the rays. In basket stars of the genera *Gorgonocephalus* (Figure 21.27) and *Astrophyton,* the rays are branched. The vertebral ossicles of basket stars have ball-and-socket joints, which allow the rays not only to bend, but to twist and curl.

Order Ophiurida

Most of the ophiuroids belong to this order, in which ossicles cover all surfaces of the rays. Representative genera are *Ophiura* (Figure 21.24, B and C; Figure 21.26, A), *Amphipholis, Ophiacanthella* (Figure 21.26, B), *Ophiopholis* (Figure 21.24, A), and *Amphiodia* (Figure 21.24, D).

References on Ophiuroidea

Fontaine, A. R., 1965. Feeding mechanisms of the ophiuroid *Ophiocomina nigra*. Journal of the Marine Biological Association of the United Kingdom, *45*:373–385.

Mladenov, P. V., Emson, R. H., Colpit, L. V., and Wilkie, I. C., 1983. Asexual reproduction in the West Indian brittle star *Ophiocomella ophiactoides* (H. L. Clark) (Echinodermata, Ophiuroidea). Journal of Experimental Marine Biology and Ecology, *72*:1–23.

Pentreath, R. J., 1970. Feeding mechanisms and the functional morphology of podia and spines in some New Zealand ophiuroids. Journal of Zoology, London, *161*:395–429.

Smith, J. E., 1940. The reproductive system and associated organs of the brittle-star *Ophiothrix fragilis*. Quarterly Journal of Microscopical Science, *82*:267–309.

Warner, G. F., and Woodley, J. D., 1975. Suspension feeding in the brittle star *Ophiothrix fragilis*. Journal of the Marine Biological Association of the United Kingdom, *55*:199–210.

Wilkie, I. C., 1978. Arm autotomy in brittlestars (Echinodermata: Ophiuroidea). Journal of Zoology, London, *186*:311–330.

CLASS ECHINOIDEA

General Structure and Skeleton

Echinoids—sea urchins, heart urchins, and sand dollars—do not have rays of the sort found in sea stars, brittle stars, and crinoids. They are, moreover, distinctive in that they have an internal shell, called the **test.** It consists of platelike ossicles that fit together so perfectly and so tightly that the whole structure is inflexible, except in some "leathery" deep-sea types. The **spines** that cover much of the body are ossicles, too; they articulate with the test and can be used for locomotion, burrowing, and food-gathering, as well as for protection. Another complex of movable ossicles forms "Aristotle's lantern," the elaborate jaw apparatus found in most members of the class.

To explain the structure of echinoids, it is best to begin with a typical sea urchin. The shape of an urchin, if the spines are excluded, is usually something like that of a tomato (Figure 21.28, A and D; Figure 21.30, A; Color Plate 8). The mouth is at the center of the oral

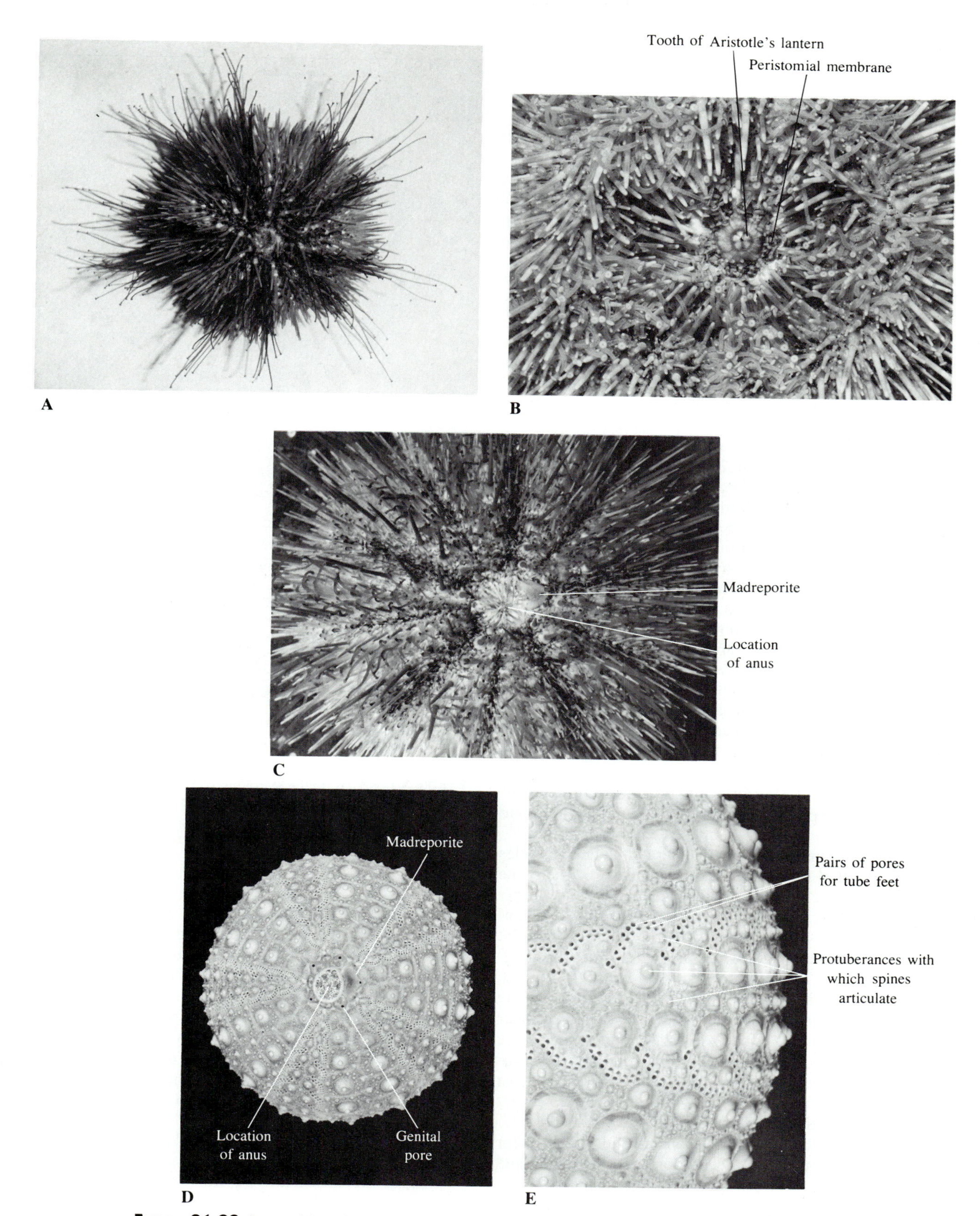

FIGURE 21.28 Sea urchins. A. *Strongylocentrotus droebachiensis,* aboral view. B. Oral region. C. Periproct region of aboral surface. D. *Strongylocentrotus franciscanus,* a cleaned test, aboral view. E. Portion of a cleaned test, showing pore pairs of an ambulacrum. (See also *Echinus,* Color Plate 8).

surface, which is normally in contact with the substratum. Surrounding the mouth is a fold of wrinkled tissue that flattens out into the **peristomial membrane** (Figure 21.28, B; Figure 21.29). This is soft, but it does have some ossicles in it; the more important of these provide support for the basal portions of 10 tube feet that arise from the peristomial membrane. There are also many pedicellariae and small spines in this region, and at the outermost edges of the peristomial membrane are 10 bushy **gills** (Figure 21.29; Figure 21.30, A). These are somewhat comparable to the papulae of a starfish, for the cavities inside them communicate directly with a part of the coelom. The anus is at the aboral pole, in the center of a flexible area called the **periproct** (Figure 21.28, C; Figure 21.30, B). The ossicles of the periproct are visible externally as small scales.

A test from which the spines and soft tissue have been removed (Figure 21.28, D and E) shows something of the functional organization of a sea urchin. Note that the test has five double rows of ossicles with pores and five double rows without pores. The ossicles with pores are the ones through which tube feet emerge, so they constitute what may be called the *ambulacral series;* the other double rows are *interambulacral*. Of the 10 ossicles that encircle the periproct (Figure 21.30, B), five belong to the ambulacral series. Each has an opening through which a tube foot, modified to function as a sensory tentacle, extends. These tube feet are, in fact, the first five that develop when a sea urchin larva undergoes metamorphosis, and are comparable to those that function as sensory tentacles at the tips of the rays of a sea star. The other five ossicles next to the periproct, which belong to the interambulacral series, have the genital pores. One of them is further modified to form the madreporite. Sometimes a second genital plate also contributes to the total area of the madreporite.

Both the ambulacral and interambulacral ossicles have tubercles on them. These are places where the spines articulate with the ossicles of the test. The larger the spine, the larger the tubercle with which it articulates. The cupped base of a spine is bound to the test by bands of connective tissue, as well by muscles (Figure 21.31, A). The connective tissue behaves in much the same way as muscle, "locking" the spine at the angle at which it has been set by the muscle. Remember that all calcareous ossicles, including spines, are formed within tissue derived from embryonic mesoderm, so they are initially covered by tissue, at least epidermis. The tissue may be worn off in places, especially near the tips of the spines.

The spines, being movable, are useful not only for protection but also for locomotion, reorientation, and other functions. If a sea urchin is turned over, it will soon right itself by efficient use of its spines and tube feet. Some species, by laying down their spines at the right angle, can work themselves into crevices or holes, and a few can even make their own holes in rock. *Strongylocentrotus purpuratus,* found along the Pacific coast of North America, is one of the most famous of these. After becoming attached in a natural depression, or perhaps in a hole bored by a bivalve mollusc, the animal slowly wears away the rock with its spines. The tips of the spines are worn away too, of course, but the animal can restore them by adding calcareous material where it has been lost. Eventually the depression in the rock may become deep enough to accommodate the whole urchin.

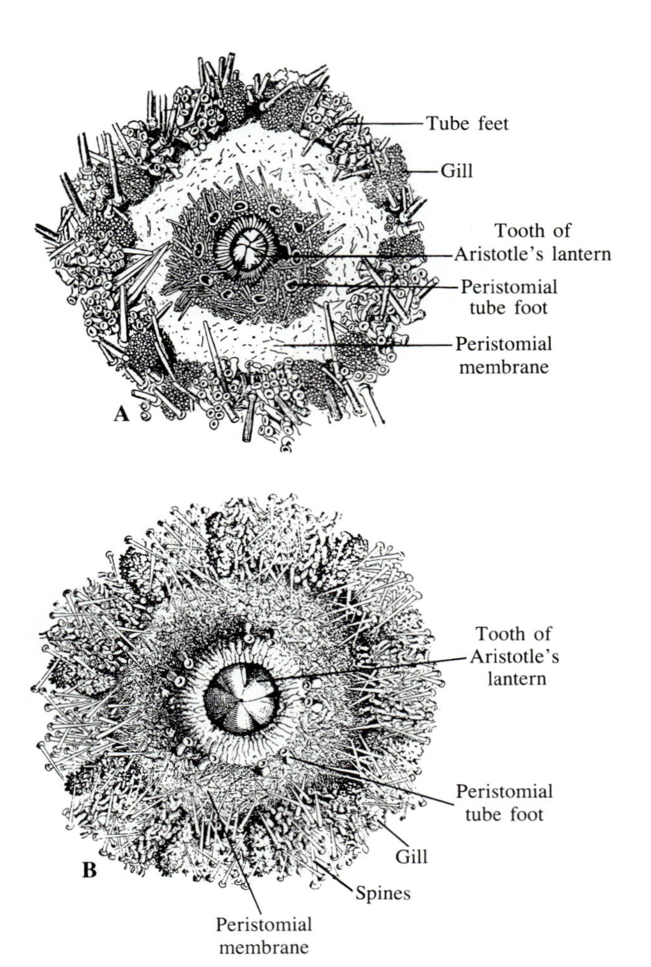

Figure 21.29 A. *Echinus esculentus,* peristomial region of oral surface. (Macbride, in Cambridge Natural History, 1.) B. *Strongylocentrotus droebachiensis,* peristomial region. (Ivanov *et al.*, Major Practicum of Invertebrate Zoology.)

Pedicellariae

Stalked **pedicellariae** are interspersed between the spines that cover the test, and are also present on the peristomial membrane. The stalk of a sea urchin pedicellaria is generally stiffened by an ossicle and does not

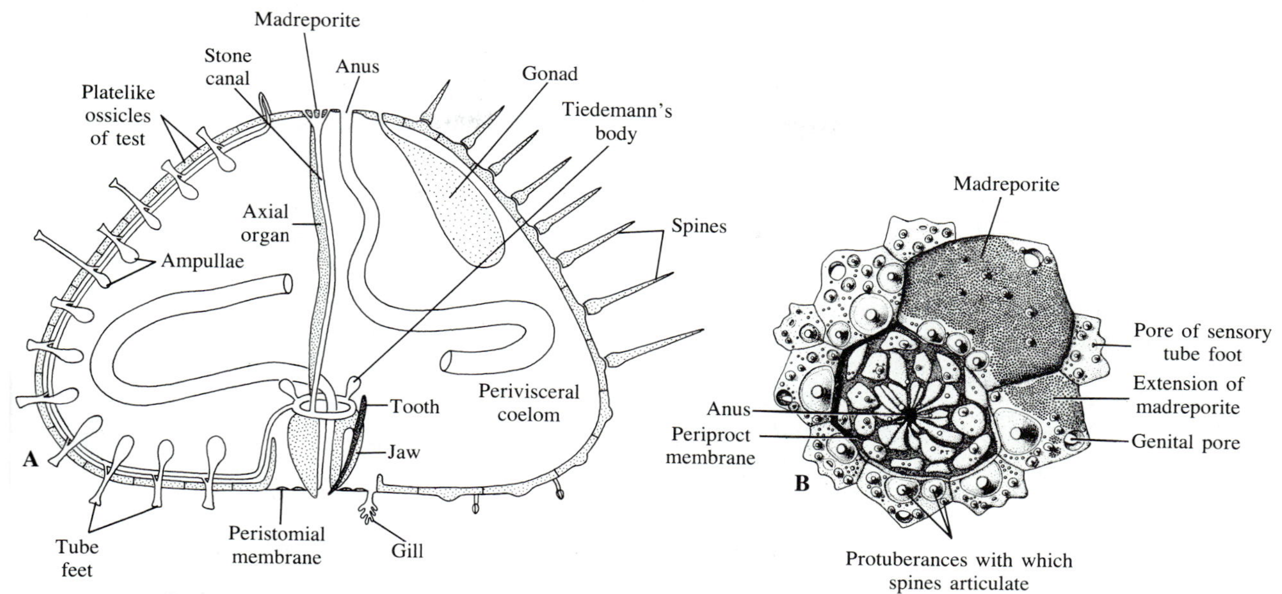

FIGURE 21.30 A. Vertical section through the central portion of the test of a sea urchin, on the left passing through one of the five ambulacra; diagrammatic and greatly simplified. B. *Evechinus chloroticus,* periproct region and associated structures. (McRae, Transactions of the Royal Society of New Zealand, *86*.)

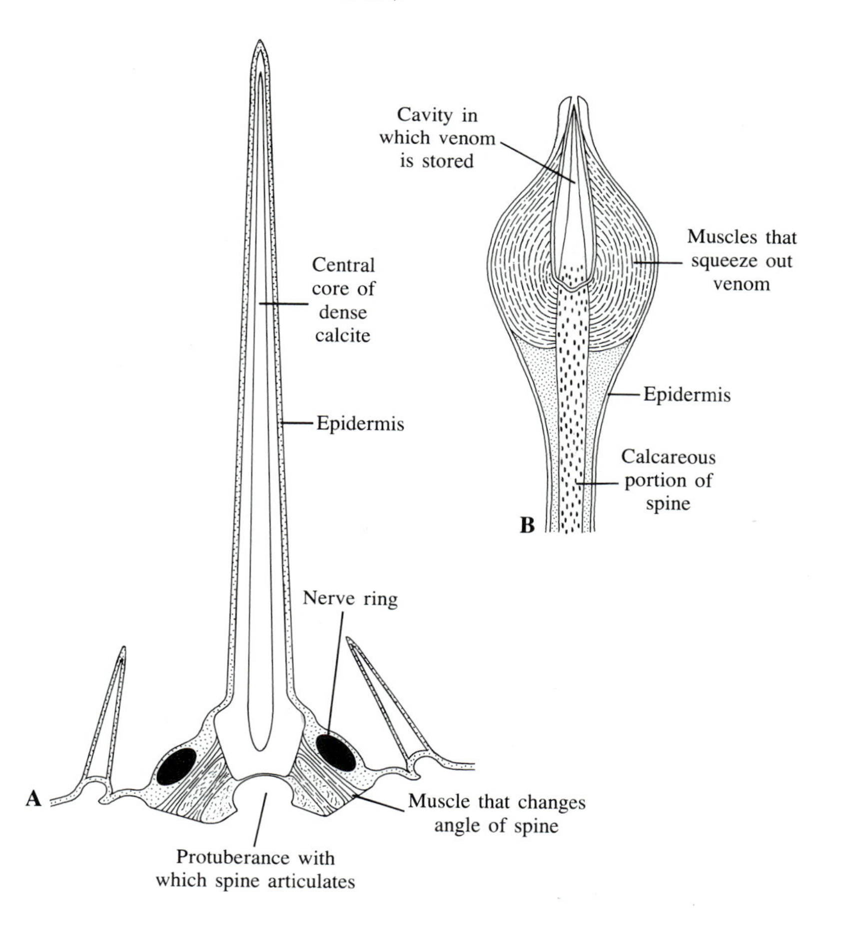

FIGURE 21.31 A. Primary spine and two secondary spines of a sea urchin; diagrammatic. B. Venomous spine of *Asthenosoma varium*. (After Sarasin and Sarasin, Ergebnisse der Naturwissenschaftlichen Forschungen auf Ceylon, *1*.)

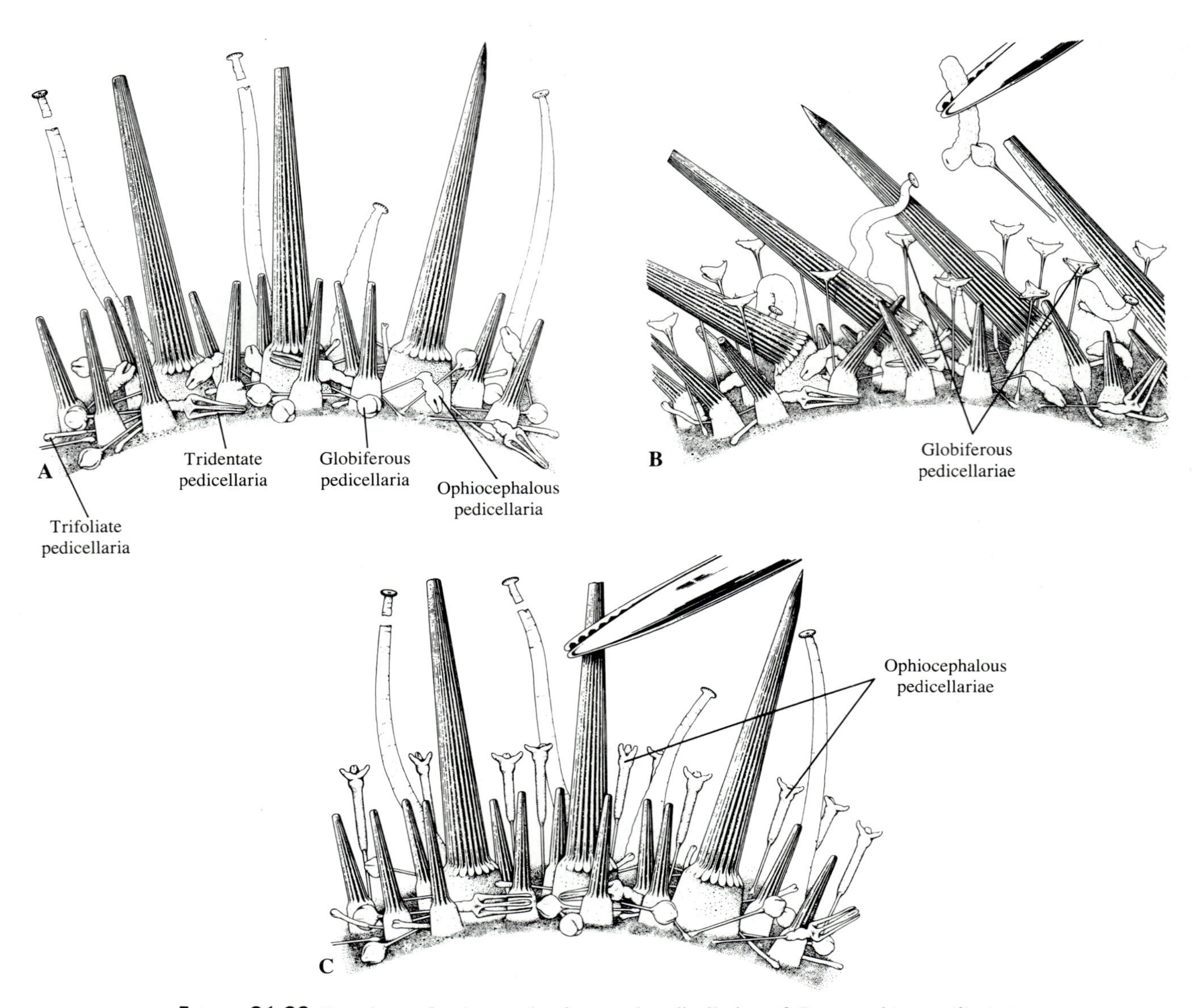

FIGURE 21.32 Reactions of spines, tube feet, and pedicellariae of *Psammechinus miliaris* to stimuli. A. Portion of surface when undisturbed. B. Portion of the surface when stimulated by the presence of a tube foot of the sea star *Marthasterias glacialis*. The globiferous pedicellariae are raised and their jaws are open, and they will close on the tissue of the tube foot if it comes close enough; the spines are directed away from the stimulus. C. Portion of the surface when disturbed mechanically. The ophiocephalous pedicellariae have lengthened and their jaws are open; the spines are directed towards the stimulus. (Jensen, Ophelia, *3*.)

lengthen, shorten, or bend freely. At its base, however, are muscles that can raise it to a standing position or pull it down so that it lies flat against the skin. Like the spines and tube feet, the pedicellariae are active and responsive to the presence of food, predators, and unwelcome settlers (Figure 21.32). Some of them, if touched on the outside, usually open, but close if touched on the inside.

Most sea urchins have at least two or three of the four main kinds of echinoid pedicellariae (Figure 21.33, A). The ophiocephalous, tridentate, and trifoliate types have only a mechanical action, but globiferous pedicellariae, with jaws that form a ball when they come together, are venomous. The venom is produced in glands in each jaw and is delivered to the tip of the jaw by a duct. Its effect varies with the species. It may be just

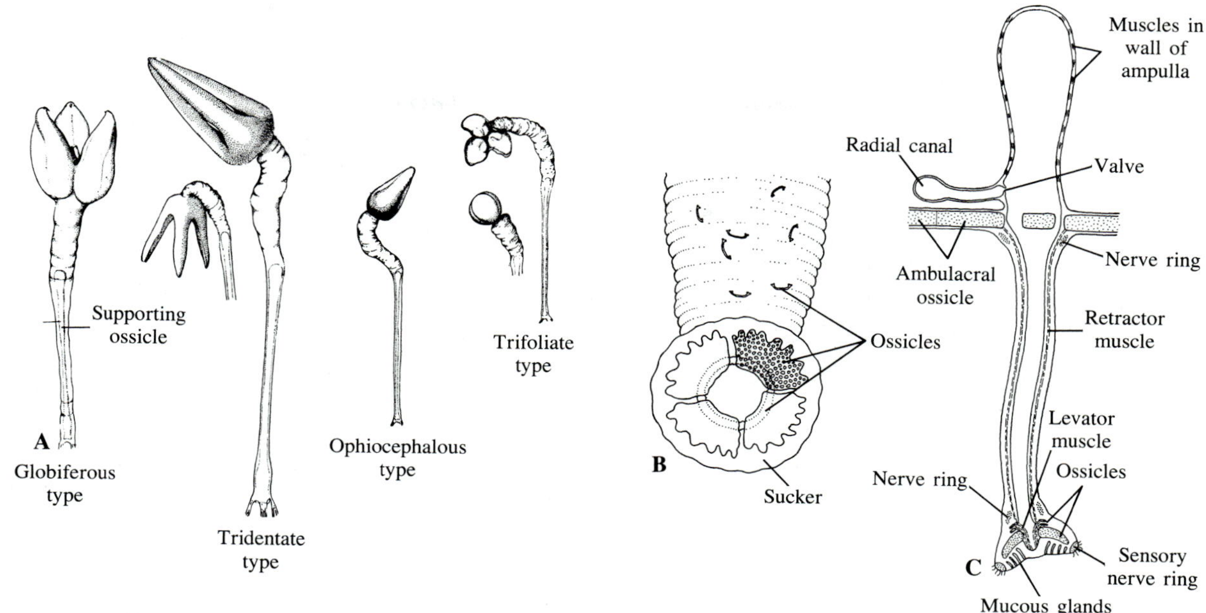

FIGURE 21.33 A. *Strongylocentrotus droebachiensis,* types of pedicellariae. (Ivanov *et al.*, Major Practicum of Invertebrate Zoology.) B. Distal portion of a tube foot of *S. droebachiensis*. (After Ivanov *et al.*, Major Practicum of Invertebrate Zoology.) C. Tube foot of a sea urchin, longitudinal section; diagrammatic.

strong enough to deter ''browsing'' or predation, but there are urchins that produce a potent, paralytic venom. Human problems with these animals are usually caused by the spines, the tips of which may break off after they have penetrated the skin. The presence of foreign tissue induces tenderness, swelling, or other disagreeable reactions. In some urchins the tip of a spine is surrounded by a cavity that stores a venom (Figure 21.31, B).

Tube Feet and Water-Vascular System

Characteristically, each tube foot goes through two pores in the ossicle with which it is associated (Figure 21.30, A). A particular ossicle may have only two or three pairs of pores, or it may have several (Figure 21.28, E); the average number varies with the species. In any case, there is only one ampulla for each tube foot, but it has a double connection with the external part of the tube foot. The wall of the external portion generally has a variety of small ossicles in it. The sucker is usually reinforced by several platelike ossicles (Figure 21.33, B).

The ampullae are joined by small canals to the radial canal that is internal to each double row of ambulacral plates (Figure 21.33, C). On the oral side of the animal, the five radial canals join a ring canal that encircles the esophagus, just above Aristotle's lantern. The ring canal has five spongy interambulacral growths thought to produce amoeboid cells for the water-vascular system. A canal running aborally from the ring canal to the madreporite is the homologue of the stone canal of sea stars.

The tube feet of sea urchins are more versatile than those of any other single group of echinoderms. They function in locomotion, attachment, collection and transport of food, and respiration; they are also sense organs and are almost certainly involved in elimination of soluble nitrogenous wastes. Although the tube feet are capable of great extension, they are also remarkably strong. Their strength depends mostly on connective tissue mixed with muscle. They are less efficient in locomotion than the tube feet of sea stars, but they can trap pieces of seaweed and other suitable food, which they pass in the direction of the mouth. Some sea urchins use a few of their feet for holding bits of shell and small rocks close to their bodies. This habit is especially characteristic of certain intertidal species, such as *Strongylocentrotus purpuratus* of the Pacific coast of North America.

Coelomic Cavities

The perivisceral coelom, occupied mostly by the gut, gonads, and ampullae of the tube feet, is a spacious cavity (Figure 21.30, A). The thin mesenteries that connect the gut to the body wall are folds of the peritoneum. The coelomic fluid contains free cells of

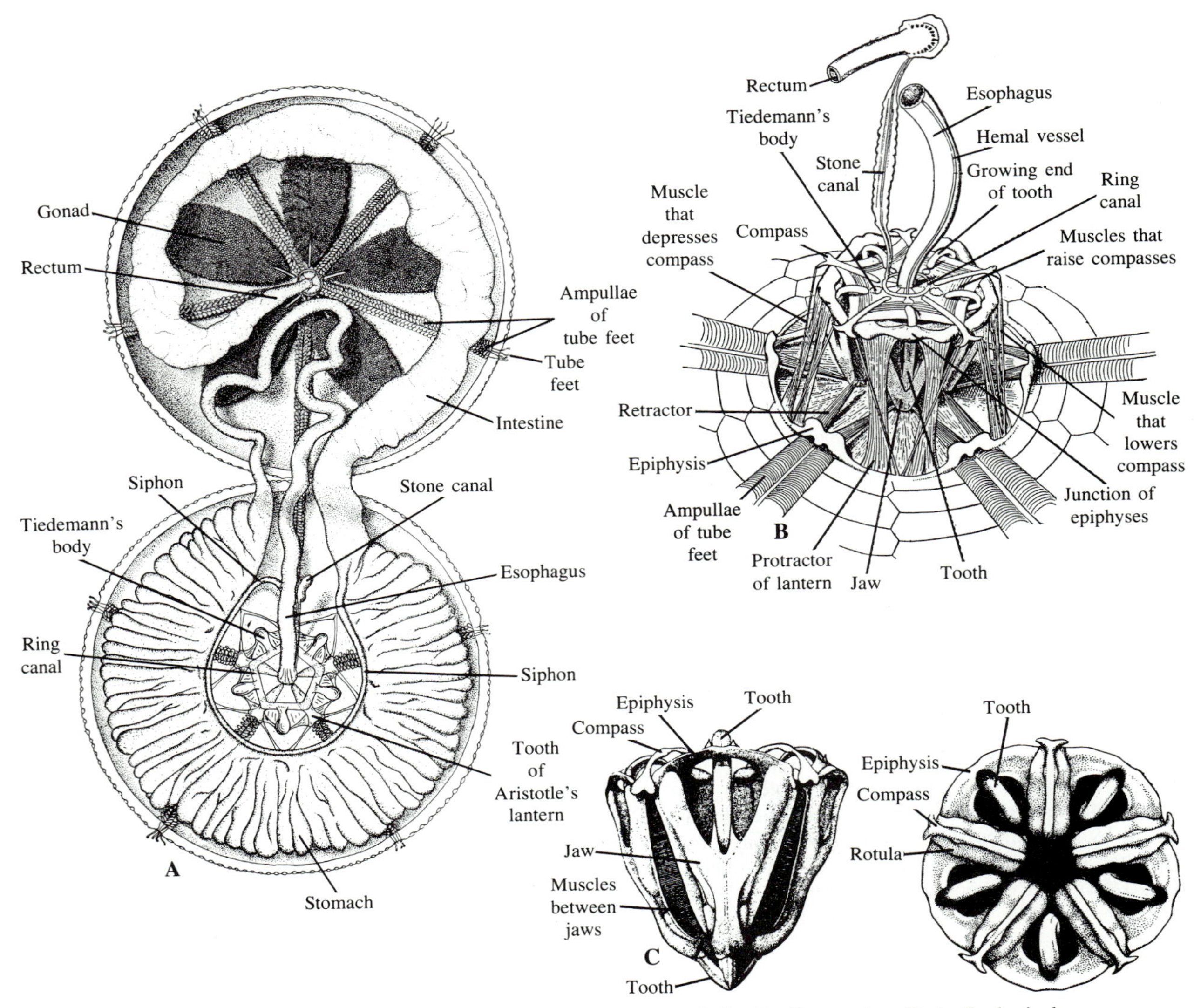

FIGURE 21.34 A. Dissection of a sea urchin. (Coe, Bulletin 19, Connecticut State Geological and Natural History Survey.) B. *Echinus esculentus,* Aristotle's lantern and associated structures. (MacBride, in Cambridge Natural History, 1.) C. *Strongylocentrotus droebachiensis,* Aristotle's lantern, lateral view (left) and aboral view (right). (Ivanov *et al.*, Major Practicum of Invertebrate Zoology.)

various types, some of which are flagellated in certain species. Aristotle's lantern is enclosed by its own coelomic compartment, and extensions of this run into the 10 gills located near the edge of the peristomial membrane. Various kinds of urchins have specialized elaborations of the lantern coelom, but these will not be considered here.

Besides the perivisceral cavity and lantern coelom, urchins have a small coelomic cavity beneath the periproct, and another one around the anus. The walls of both cavities have circular muscles that control the dilation and closure of the rectum. Another small coelomic space is found near the gonoducts.

Digestive System

Beginning with the esophagus that emerges from Aristotle's lantern, the gut makes a complete turn, then switches back and continues in the opposite direction toward the anus (Figure 21.34, A). The intestinal region, which makes up most of the digestive tract, is closely paralleled by a tube called the **siphon.** The siphon, like the gut itself, is ciliated. It is thought to function as a bypass, moving water rapidly past the intestinal region so that digestive enzymes are not greatly diluted. The food of most sea urchins consists largely of algae, but some species are opportunists that consume considerable animal material.

The five hard **teeth** of Aristotle's lantern (Figure 21.34, B and C) are used for grazing and for chopping up food. Although each tooth is a substantial structure, it is loosely embedded for much of its length in a larger calcareous structure called the **jaw** (*pyramid* and *alveolus* are other names for this). The jaw originates as two ossicles, but these become sutured together lengthwise. Another pair of ossicles is joined to the aboral end of the jaw, forming the **epiphysis.** Muscles running from the epiphyses of the jaws to the interambulacral edge of the opening in the test protract the lantern as a whole, thus making the teeth protrude from the mouth. Retraction of the jaws, and also separation of the jaws, is achieved by muscles that run from near the oral ends of the jaws to the five flanges (*auricles*) in line with the ambulacral ossicles of the test. The rocking of the jaws against one another, necessary if the teeth are to operate as scraping, chewing, and chopping devices, is the responsibility of muscles that join each jaw to its neighbors.

Two other sets of ossicles are found at the aboral end of Aristotle's lantern. Those of one set are called **rotulae.** They run from the esophagus outward, reaching the spaces between the epiphyses of adjacent jaws. Muscles connecting the rotulae to the epiphyses accomplish some of the rocking motion of the jaws, thereby supplementing the action of the muscles that connect neighboring jaws.

The other ossicles on the aboral side of the lantern form the **compasses,** or *radii,* lying aboral to the rotulae and articulating with their inner ends. Each compass actually consists of two ossicles, and the outer piece usually ends in a Y. The compasses have nothing to do with the operation of the jaws. Remember that the lantern region has its own coelomic cavity, and that this extends into the 10 external gills seen at the edges of the peristomial membrane. The compasses are connected to one another by muscles intimately associated with the "roof" of the lantern coelom. Collectively, these muscles, called **levators** ("lifters"), form a five-angled ring. When they contract, the compasses are lifted, and the roof is raised enough to lower the pressure in the lantern coelom. This causes fluid to drain out of the gills. The antagonists of the levator muscles connect the Y-shaped outer ends of the compasses with the edges of the opening in the test. Contraction of these muscles pulls the compasses down, increasing the pressure in the lantern coelom and forcing fluid back into the gills.

The five teeth are continually worn away at their exposed tips. To compensate for this, tooth material is steadily secreted within the dental sacs that cap the opposite ends of the teeth. The soft portion of a tooth, still within its sac, hardens as it is gradually pushed toward the channel which the tooth occupies in the jaw. Eventually it emerges from the oral end of the channel to perform its function until it too is worn away.

Hemal System

The hemal system is well developed. Most tissues and organs seem to be pervaded by hemal vessels. The gut has an especially rich supply, mostly branches of two main vessels, one of which is directly joined to a hemal ring that encircles the esophagus. Among the other vessels that branch off the hemal ring are the five that run in an oral direction alongside the pharynx, then turn outward to follow the ambulacra; these correspond to the radial hemal vessels of sea stars. They occupy the same position in the body wall, between the radial nerves and the radial canals of the water-vascular system. The **axial organ,** alongside the stone canal, has an extensive blood supply, connected at one end to the ring around the esophagus and at the other end to an aboral hemal ring around the rectal portion of the gut. This aboral ring sends out branches to the gonads.

A hemal system as elaborate as that of a sea urchin is almost certainly important to the animal. Unfortunately, little is known about its role in absorption and distribution of nutrients or other substances. It must be remembered that the hemal system is not the only system concerned with transport of organic and inorganic solutes or of free cells that might serve as carriers. The coelom, water-vascular system, and perihemal spaces presumably do the same general thing.

Nervous System

The principal elements of the nervous system are a nerve ring around the mouth and five radial nerves that follow the radial canals of the water-vascular system, lying between these and the test. A ring around the rectum sends branches out into the gonads. The nerves just mentioned constitute only a small part of the nervous system. The many muscles that operate Aristotle's lantern, for instance, have a correspondingly complex supply of motor nerves.

Each tube foot and pedicellaria, in order to be as reactive as it is, must have both sensory and motor nerves. Actually, the entire body surface of a sea urchin seems to be sensitive to light and other stimuli, as a result of being innervated by a plexus of fine nerves external to the test. There are, however, some specially differentiated sensory structures. Among these are the five tentaclelike tube feet that emerge through pores in the aboralmost ambulacral plates. Most sea urchins also have at least a few structures called **spheridia.** A spheridium is a tiny vesicle, sometimes stalked, that contains a calcareous ossicle. It functions as a statocyst. Spheridia are sometimes scattered throughout the ambulacral areas, but they are most likely to be in the region just peripheral to the peristomial membrane. Sometimes there is only one for each of the five ambulacra.

Reproduction and Development

The five gonads are interambulacral, and gametes are shed by way of the pores on the genital plates. Because the eggs characteristically contain little yolk, they are much used in studies on fertilization, cleavage, and early development. Eggs of sea urchins and sand dollars are, in fact, more intensively exploited for experimental studies than eggs of any other single group of invertebrates.

At the 16-cell stage, there are usually four micromeres at the lower end, four macromeres in the middle, and eight mesomeres on top. These are destined to form, respectively, the primary mesenchyme, endoderm, and ectoderm. At the hollow blastula stage, all the cells are about the same size. The descendants of the micromeres move into the blastocoel first. Then, after the descendants of the macromeres invaginate to form the archenteron, the cells at the tip of the archenteron bud off secondary mesenchyme cells that represent additional mesoderm. The gastrula becomes a type of larva called the **echinopluteus** (Figure 21.35), similar to the ophiopluteus of an ophiuroid. As in the ophiopluteus, the long projections are supported by calcareous spicules. The previously general ciliation becomes restricted to bands that extend to the projections. The gut is complete at this stage, the mouth having been formed secondarily.

The coelomic pouches are produced by the archenteron in much the same way as in sea stars, and their fates are similar. At metamorphosis the young urchin develops in an interesting way. A localized ectodermal invagination appears on the left side of the larva. This is the primordium of the future oral side of the urchin. It sinks deeper and then is separated from the surface, the space carried in with it being called the amniotic cavity. The primordium becomes five-angled and intimately associated with the left hydrocoel. Five tubercles appear where the hydrocoel insinuates into the primordium, and these are precursors of the first five tube feet, which become the tentaclelike structures close to the periproct of the adult. In time the echinopluteus becomes completely reorganized to fit the new orientation conferred upon it by the developing primordium. Most of the original ectoderm of the echinopluteus is salvaged to form the epidermis of portions not directly derived from the primordium. The original mouth and anus close and are replaced by openings that appear on the prospective oral and aboral surfaces. The amniotic cavity breaks open to the outside so that the oral face of the young urchin is exposed. The ossicles of the spines, test, and other hard structures develop from mesenchymal mesoderm.

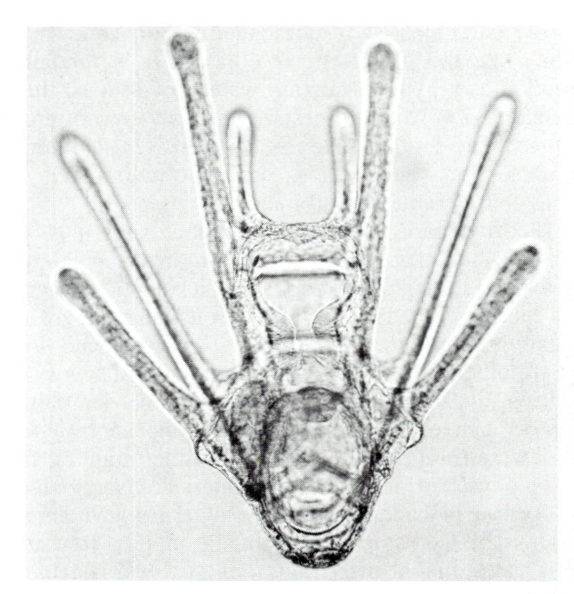

Figure 21.35 Echinopluteus larva of an echinoid; photomicrograph.

Irregular Echinoids

Because their radial symmetry is nearly perfect, sea urchins are referred to as regular echinoids. Sand dollars and heart urchins, on the other hand, are said to be irregular, because they show a strong tendency toward bilateral symmetry. Both mouth and anus are situated on the oral surface. The relative position of the two openings is such that these animals may be said to have an anterior and posterior end. When a heart urchin or sand dollar moves or burrows, it is the anterior end that leads.

A heart urchin (Figures 21.36 and 21.37) is shaped a little like a flattened egg. There is usually a prominent cleft at the anterior end; this marks the location of one of the five ambulacra. Most heart urchins plow through mud or burrow into sand, using their spines for this purpose.

The majority of the aboral tube feet are leaflike and specialized to function as gills. On the anterior ambulacrum, however, some tube feet are tipped with a scalloped disk or a set of fingerlike projections; these tube feet are capable of considerable elongation. Tube feet drop out before the ambulacra reach the equatorial region, but they reappear again on the oral side, where they are concerned primarily with collecting food and probably also with chemoreception. The most striking

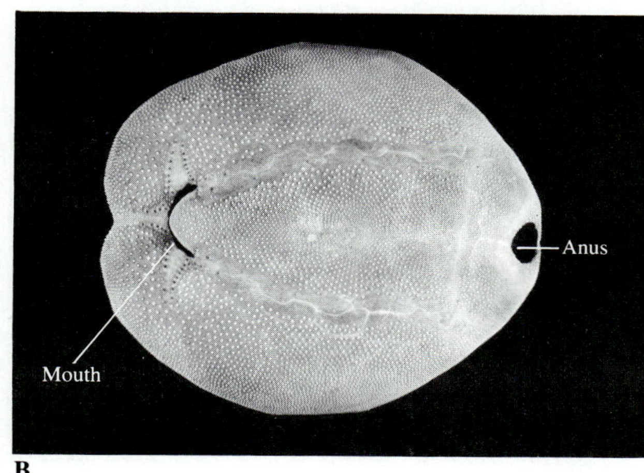

FIGURE 21.36 *Brisaster latifrons,* a heart urchin, cleaned test. A. Aboral view. B. Oral view.

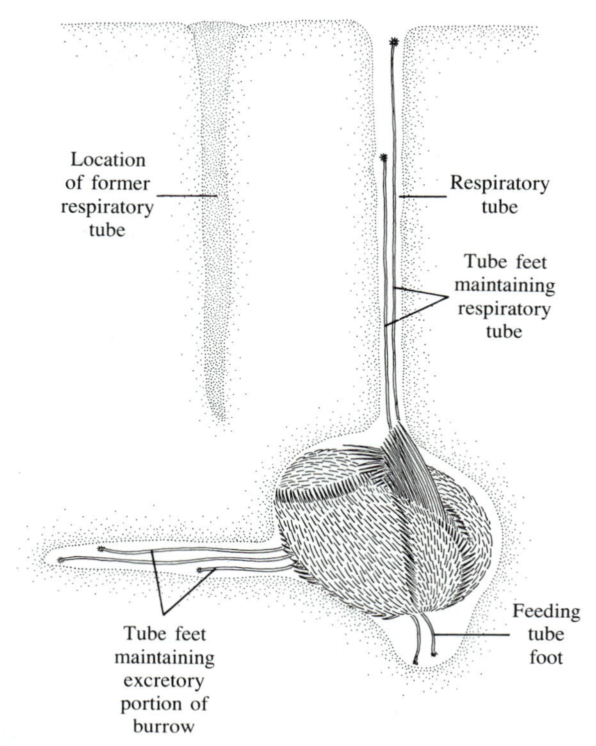

FIGURE 21.37 *Echinocardium cordatum,* a heart urchin, in its burrow in sand.

tube feet on the oral side terminate in clusters of bulb-shaped outgrowths. These tube feet can be extended upward around the body of a buried heart urchin until they reach the surface, where they can prospect for food. Material that sticks to the food-gathering tube feet is delivered to certain tracts of spines and is then directed to the mouth.

Heart urchins consume diatoms, all sorts of fine detritus, bits and pieces of animals, and sand grains coated with organic material. These animals lack an Aristotle's lantern.

Some heart urchins, such as *Echinocardium* (Figure 21.37), dig in sand to a depth of several centimeters, using certain flattened spines for this purpose. The burrows are temporary at best, but a coating of mucus helps keep the walls from caving in too freely. The epidermis that covers some small, racquet-shaped spines, restricted to special tracts on the aboral side, is said to be especially active in producing mucus. Cilia on the lower portions of these same spines are believed to be important in circulating water over the surface of the animal. The vertical tube leading to the surface is kept clear by extensile tube feet coming from the anterior-most ambulacrum on the aboral side, and the little pocket into which fecal material is directed is maintained by tube feet originating in the posterior part of the oral side.

Sand dollars (Figures 21.38 and 21.39) are strongly flattened and their ambulacra are evident only on the aboral surface. The more prominent tube feet on these ambulacra are in the form of broad sheets of soft tissue, and they serve as gills. There are also smaller, cylindrical tube feet, tipped with suckers, in the ambulacral areas and elsewhere, including the oral surface. The spines of sand dollars are so small and so crowded that they form a coating that resembles felt. They are mobile, however, and are responsible for forward progression and burrowing.

The oral surface of most sand dollars has what look like several river systems. These are food grooves in which diatoms and other particles that have become stuck in mucus are driven toward the mouth by ciliary

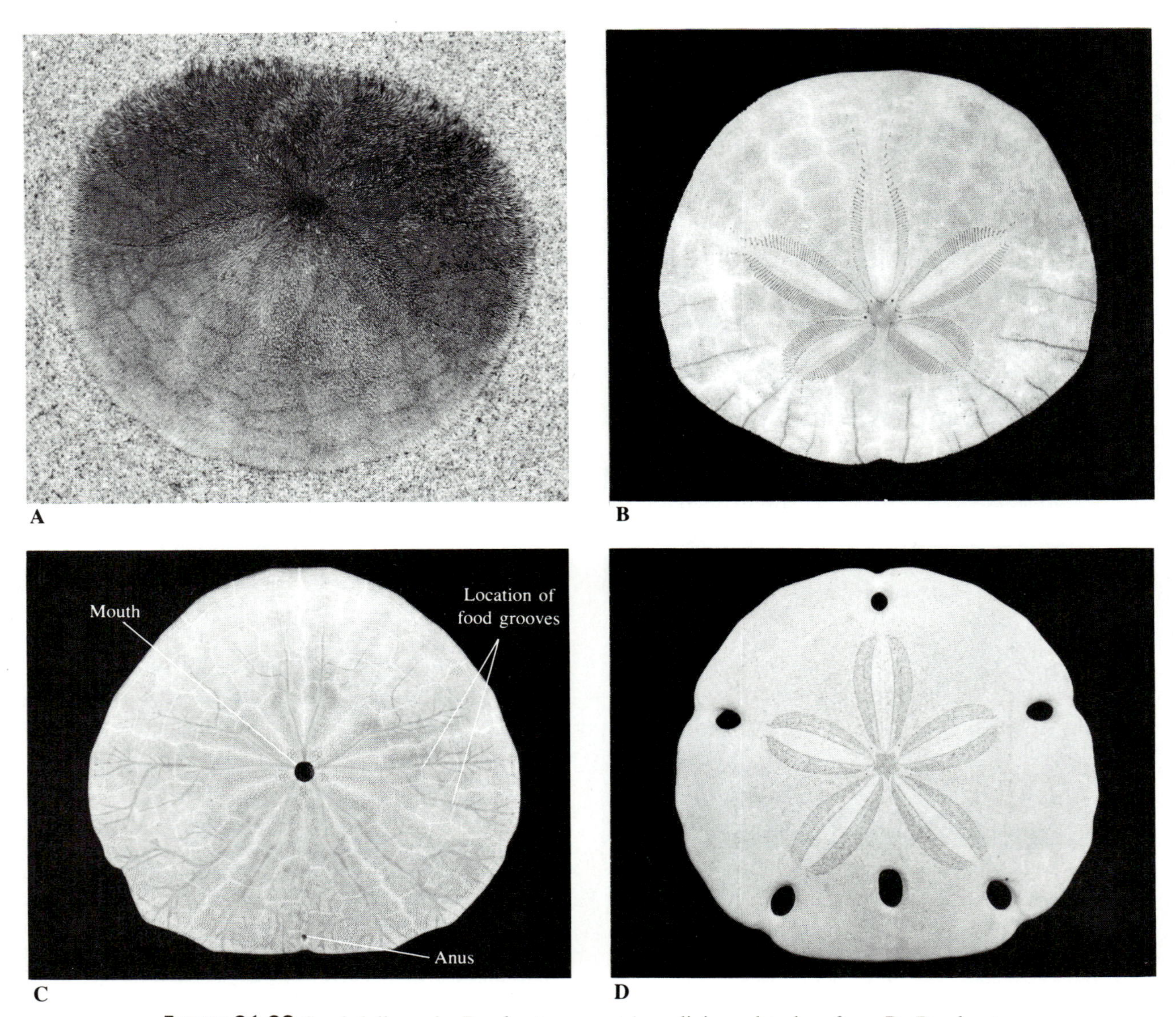

FIGURE 21.38 Sand dollars. A. *Dendraster excentricus,* living, aboral surface. B. *Dendraster,* cleaned test, aboral surface. C. *Dendraster,* cleaned test, oral surface. D. *Encope micropora,* cleaned test, aboral view.

activity, aided by some of the small tube feet. Collection of food is not limited to the oral surface; particles that have become trapped in mucus on the aboral surface are moved by cilia and tube feet toward the outer margin of the body, then to the oral surface, where they can get into a food groove. The system of feeding can be followed under a dissecting microscope if a suspension of carmine particles is applied to the animal. Sand dollars have an Aristotle's lantern, but this is small and its teeth cannot be protruded through the mouth.

Some sand dollars plow through the sand, remaining just barely covered most of the time. Others, at least under certain circumstances, become oriented with the anterior quarter or third buried, the rest sticking up almost vertically above the surface of the sand.

In certain species, the test has large holes called **lunules** (Figure 21.38, D; Figure 21.39, A). They usually start out as marginal indentations. As the diameter of the test increases, the indentations close up and are left behind. In a few species, the lunules are formed by resorption of test material and whatever internal structures happen to be in the way. Some sand dollars do not

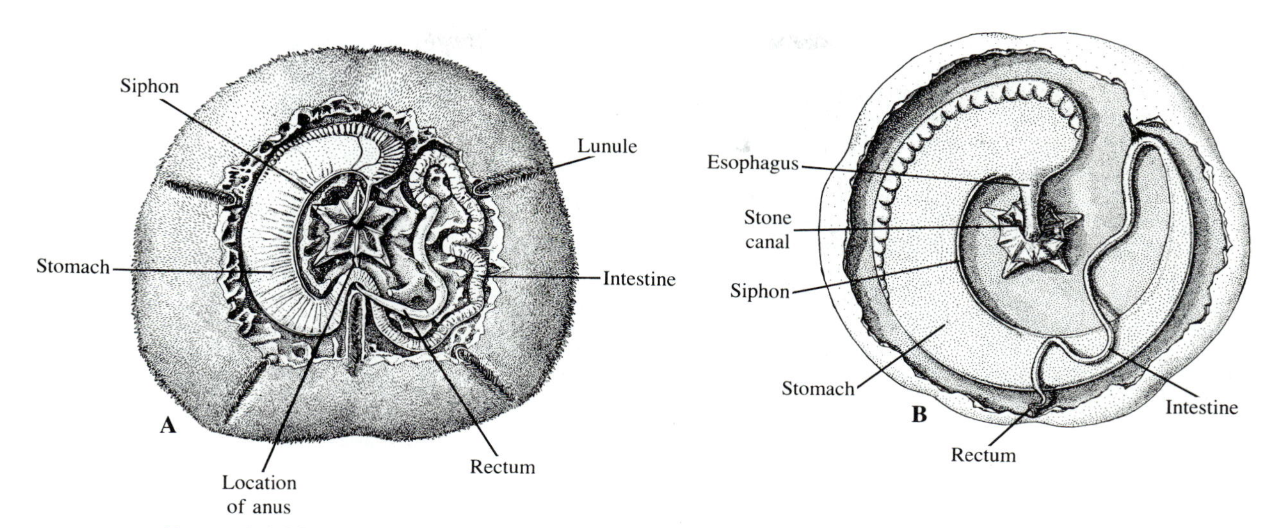

FIGURE 21.39 Sand dollars, dissected from the aboral side. A. *Mellita pentapora*. B. *Echinarachnius parma*. (Coe, Bulletin 19, Connecticut State Geological and Natural History Survey.)

close their marginal indentations, or only close some of them.

Lunules and indentations are helpful in burrowing activities, for they allow loose sand to pass through the body as well as around the margins. When there are larger spines at the edges of the lunules, these seem to be especially important in burrowing. Some sand dollars are shaped in such a way that if they are thrown onto their backs they will soon be flipped over again by the action of waves. Most species, however, must do some digging to reestablish themselves in their normal position. They gradually work one edge into the sand until they are nearly vertical and can fall over with the oral surface down. At least one of the lunuled species, if turned upside down, pushes sand through the holes until the pile under it is high enough to lift the animal to a vertical position; it then tumbles.

Pedicellariae are present in heart urchins and sand dollars, but they are not usually as conspicuous as in sea urchins. A cleaned test of a heart urchin or sand dollar will show the paired pores leading to the various kinds of tube feet, genital pores (four or five, corresponding to the number of gonads), madreporites, and tubercles with which the spines articulate.

Classification

Subclass Perischoechinoidea

Of the four orders generally assigned to this group, only one, Cidaroida, has living representatives. Although living cidaroids have a rigid test, many of the extinct members of this order, and also of two of the other orders, had flexible tests. Moreover, the number of rows of plates in the ambulacral and adambulacral regions of the test is rather unstable in the subclass as a whole.

Two features of the cidaroids are distinctive. There is a single large spine articulating with a correspondingly large tubule on each interambulacral plate, and the ambulacral plates have only one tube foot each, and therefore only a single pair of pores.

Eucidaris and *Goniocidaris* are representative genera of the Cidaroida, which are primarily, but not exclusively, tropical. Many members of this group brood their young.

Subclass Euechinoidea

In the several surviving orders of these echinoids, as well as in extinct orders, the ambulacral and interambulacral plates are arranged in double series. The ambulacral plates, however, are subdivided in certain orders. No attempt is made here to characterize the approximately 16 orders, some of which are wholly extinct. It will suffice to say that this subclass contains most of the regular echinoids, such as *Echinus* (Color Plate 8), *Arbacia, Diadema,* and *Strongylocentrotus* (Figure 21.28, A; Color Plate 8). It also includes the irregular echinoids, among which are the heart urchins (Figures 21.36 and 21.37) and sand dollars (Figures 21.38 and 21.39).

References on Echinoidea

Bachmann, S., and Goldschmid, A., 1978. Fine structure of the axial complex of *Sphaerechinus granularis* (Lam.). Cell and Tissue Research, *193*:107–123.

Bonsdorff, E., 1983. Appetite and food consumption in the sea urchin *Echinus esculentus* L. Sarsia, *68*:25–27.

Burkhardt, A., Hansmann, W., Markel, K., and Niemann, H.-J., 1983. Mechanical design in spines of diadematoid echinoids (Echinodermata, Echinoidea). Zoomorphology, *102*:189–203.

Chia, F. S., 1969. Some observations on the locomotion and feeding of the sand dollar, *Dendraster excentricus* (Eschscholtz). Journal of Experimental Marine Biology and Ecology, *3*:162–170.

Dafni, J., and Erez, J., 1987. Skeletal calcification patterns in the sea urchin *Tripneustes gratilla elatensis* (Echinoidea: Regularia). I. Basic patterns. Marine Biology, *95*:285–287.

Durham, J. W., 1966. Evolution among the Echinoidea. Biological Reviews, *41*:368–391.

Durham, J. W., and Melville, R. V., 1957. A classification of echinoids. Journal of Paleontology, *31*:242–272.

Ebert, T. A., 1982. Longevity, life history, and relative body wall size in sea urchins. Ecological Monographs, *52*:353–394.

Ebert, T. A., and Dexter, D. M., 1975. A natural history study of *Encope grandis* and *Mellita grantii*, two sand dollars in the northern Gulf of California. Marine Biology, *32*:397–407.

Fell, H. B., 1965. The early evolution of the Echinozoa. Breviora, *219*:1–19.

Fenner, D. H., 1973. The respiratory adaptations of the podia and ampullae of echinoids. Biological Bulletin, *145*:323–339.

Florey, E., and Cahill, M. A., 1977. Ultrastructure of sea urchin tube feet: evidence for connective tissue involvement in motor control. Cell and Tissue Research, *177*:195–214.

Florey, E., and Cahill, M. A., 1982. Scanning electron microscopy of echinoid podia. Cell and Tissue Research, *224*:543–551.

Jensen, M., 1981. Morphology and classification of Euechinoidea Bronn, 1860—a cladistic analysis. Videnskabelige Meddelelser Dansk Naturhistorisk Forening, *143*:7–99.

Märkel, K., and Röser, V., 1983. The spine tissues in the echinoid *Eucidaris tribuloides*. Zoomorphology, *103*:25–41.

Mooi, R., 1986. Structure and function of clypeasteroid miliary spines (Echinodermata, Echinoidea). Zoomorphology, *106*:212–223.

Okazaki, K., Dillaman, R. M., and Wilbur, K. M., 1981. Crystalline axis of the spine and test of the sea urchin *Strongylocentrotus purpuratus:* determination by crystal etching and decoration. Biological Bulletin, *161*:402–415.

Rehkämper, G., and Welsch, U., 1988. Functional morphology of the stone canal in the sea urchin *Eucidaris* (Echinodermata: Echinoidea). Zoological Journal of the Linnean Society, *94*:259–269.

Smith, A. B., and Ghiold, J., 1982. Roles for holes in sand dollars (Echinoidea): a review of lunule formation and evolution. Paleobiology, *8*:242–253.

Strathmann, R. R., 1981. The role of spines in preventing structural damage to echinoid tests. Paleobiology, *7*:400–406.

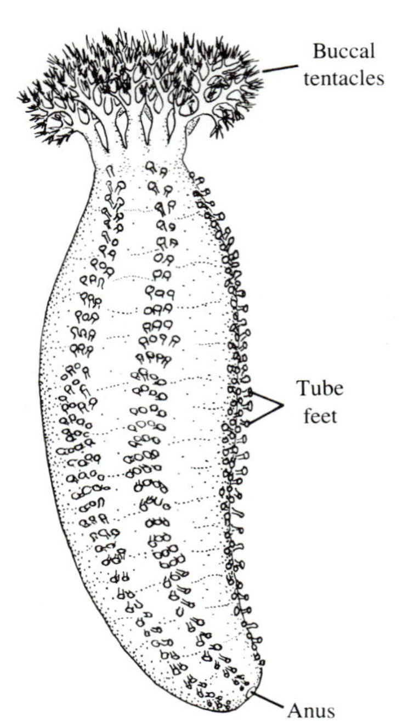

FIGURE 21.40 A typical sea cucumber, with three sets of well-developed tube feet on the surface that is in contact with the substratum; the tube feet of the two sets on the opposite side are reduced. (See also *Cucumaria,* Color Plate 8.)

Telford, M., 1983. An experimental analysis of lunule function in the sand dollar *Mellita quinquiesperforata*. Marine Biology, *76*:125–134.

Timko, P., 1976. Sand dollars as suspension feeders: a new description of feeding in *Dendraster excentricus*. Biological Bulletin, *151*:247–259.

CLASS HOLOTHUROIDEA: SEA CUCUMBERS

General Structure

In a sea cucumber (Figure 21.40; Figure 21.41, A; Figure 21.43; Figure 21.44, A and C; Color Plate 8), the body is elongated along the oral–aboral axis, so the mouth and anus are widely separated. Nevertheless, as in echinoids, the ambulacra run from the oral region nearly to the anus. In many species the tube feet, used for attachment and locomotion, are concentrated on the side of the body that is normally lowermost. Sea cucumbers of this type appear to be bilaterally symmetrical and to have a dorsal and a ventral surface. Dissection will reveal, however, that certain systems are arranged according to a radial plan. In sea cucumbers that have five equally spaced sets of functional tube feet, the radial symmetry is of course apparent externally.

Almost all sea cucumbers have obvious tentacles forming a circle around the mouth. There are usually

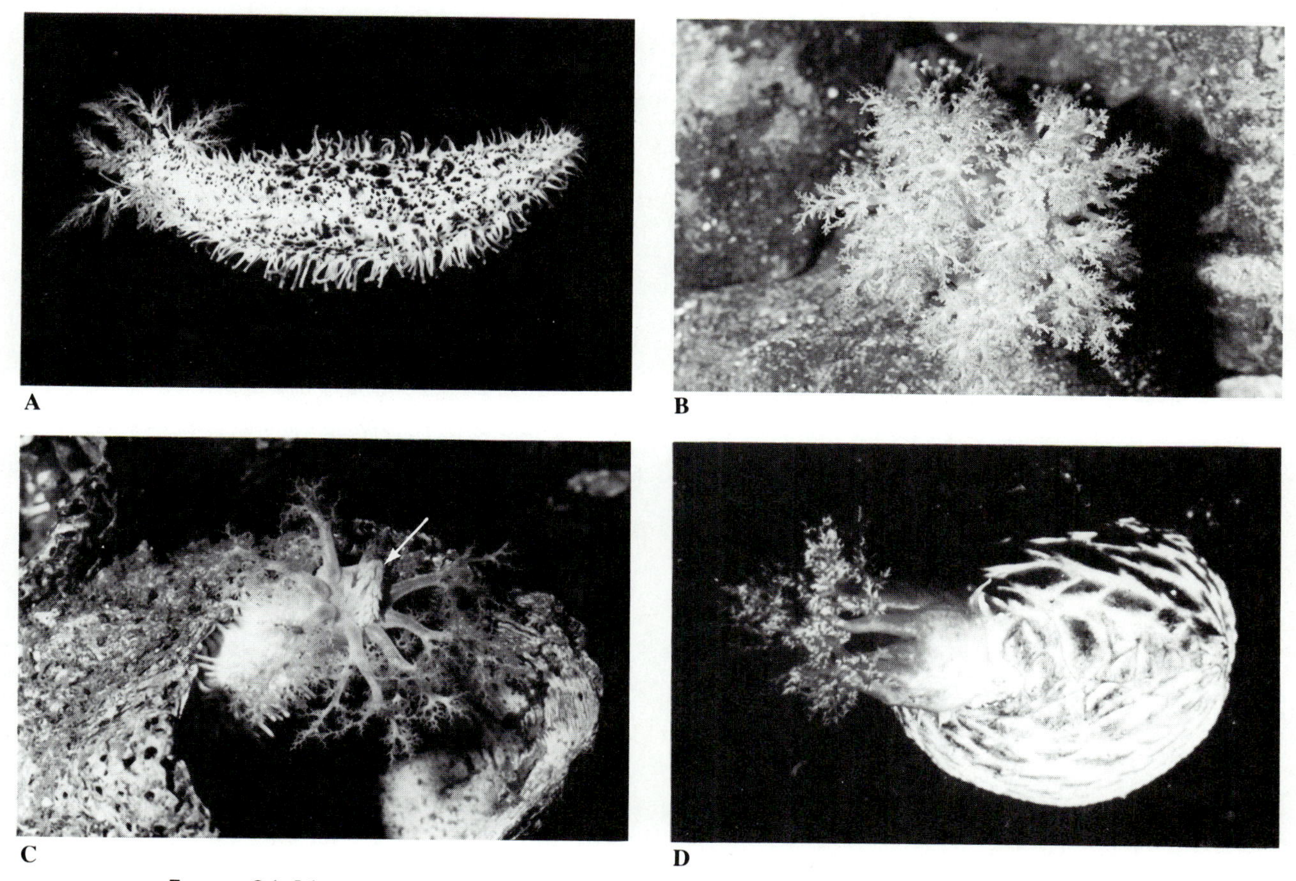

FIGURE 21.41 Subclass Dendrochirotacea. A. *Cucumaria piperata*. B. *Cucumaria miniata*, crown of extended buccal tentacles. (The rest of the animal is hidden between rocks. See also Color Plate 8.) C. *Eupentacta quinquesemita*, crown of buccal tentacles; one of the tentacles (arrow) is being pulled out of the buccal cavity after having been licked clean. D. *Psolus chitonoides*, in which the ossicles of the body wall are large and visible externally.

10 or more of these **buccal tentacles** (Figure 21.42, A), and their principal function is the collection of food. Whether they are simple, fingerlike outgrowths or extensively branched structures, they are tube feet and are connected to the water-vascular system. To be even more precise, five of the tentacles are the first five tube feet that develop, so they are comparable to those at the tip of each ray of a sea star, brittle star, or crinoid, and to the five that are closest to the periproct of a sea urchin. Although the buccal tentacles do not have much connective tissue, their musculature is fairly well developed.

The body wall of a sea cucumber, beneath the epidermis, generally has a fairly thick dermal layer (Figure 21.42, B). This consists mostly of elastic connective tissue, but it contains scattered calcareous ossicles. Between the dermis and coelomic peritoneum is a sheath of circular muscle and five single or double longitudinal muscle bands. The latter, being well defined, are obvious in an animal whose body cavity has been opened (Figure 21.43). A sea cucumber's ability to change shape depends on the muscles of the body wall, with the perivisceral coelom serving as a hydrostatic skeleton.

The calcareous ossicles of the body wall are not often obvious externally. *Psolus* (Figure 21.41, D) and its relatives are exceptions in which large, platelike ossicles protect almost all parts of the body that are not in tight contact with a hard substratum. In the case of most sea cucumbers, it is necessary to tease apart some of the connective tissue that makes up the body wall and examine the preparation with a microscope. A strong

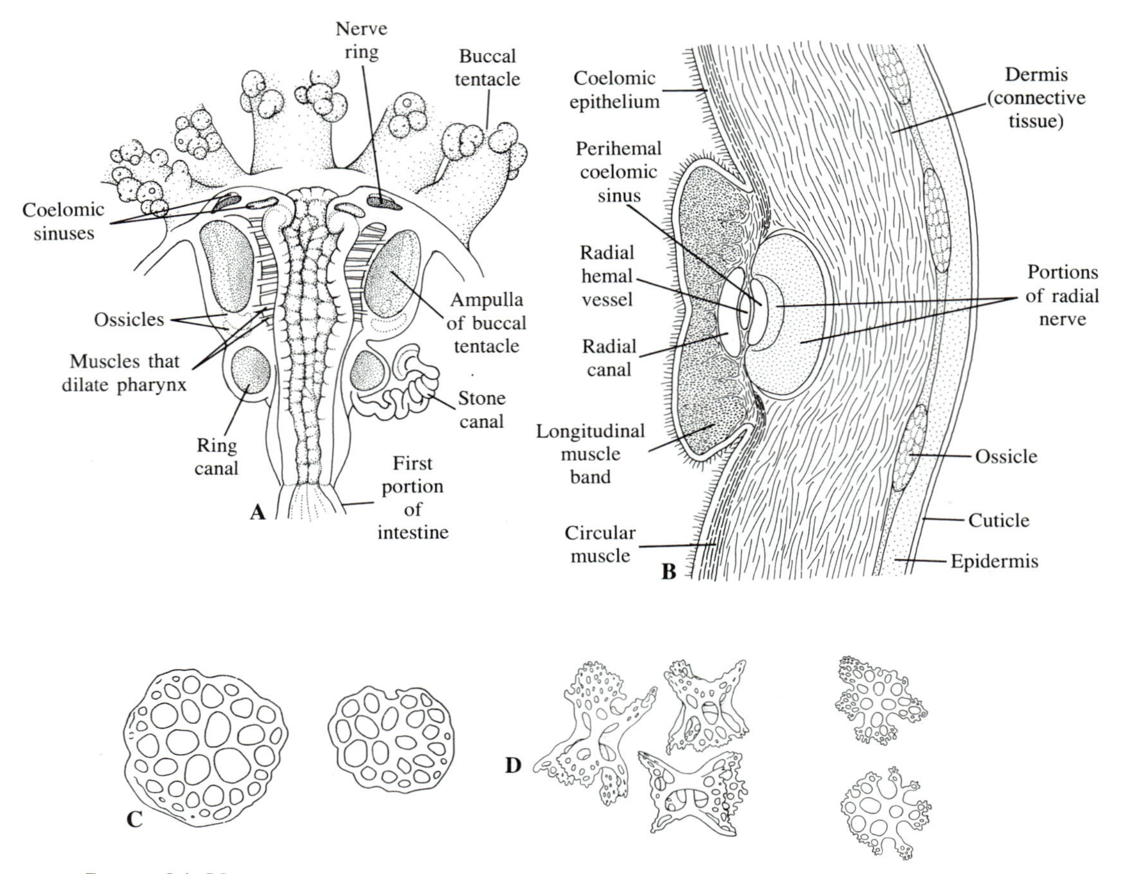

FIGURE 21.42 A. *Cucumaria curata,* buccal capsule, sectioned lengthwise. (After Brumbaugh, thesis.) B. *Cucumaria,* transverse section through a portion of the body wall, passing through a radial canal, muscle band, and associated structures. C. Ossicles from the body wall of *Stichopus japonicus*. D. Ossicles from the body wall of *Lissothuria nutriens*. (Pawson, Proceedings of the United States National Museum, no. 3592.)

solution of a laundry bleach containing sodium hypochlorite will destroy the organic material and make it easier to see the ossicles. They are generally perforated plates (Figure 21.42, C and D). Sometimes they are nearly circular and have only a few holes, so they resemble buttons.

In a group of burrowing sea cucumbers called synaptids, there are unusual two-piece ossicles (Figure 21.44, D); one portion, completely buried in the body wall, is a perforated plate, and the other portion, which protrudes, is shaped something like an anchor. When a synaptid held between the fingers or on the palm of one's hand is picked up again, it may cling slightly because the anchorlike portions engage the skin.

Digestive System

The organization of the gut varies according to the type of sea cucumber and its food habits, but in general there are four or five distinct regions. The short pharynx runs through the **buccal capsule** (Figure 21.42, A; Figure 21.43), which is supported by a circle of large ossicles. This capsule is a very important part of the animal, for it is here that the ring canal of the water-vascular system and the nerve ring are located. Each of the longitudinal muscles of the body wall, as it approaches the oral end of the animal, divides, one branch inserting in the buccal capsule. Thus retraction of the buccal portion of a sea cucumber, which brings the mouth and buccal tentacles into a temporary depression, can be effected independently of contraction of the animal as a whole.

Behind the pharynx there may be a recognizable esophagus and a dilated, muscular portion of the gut that can be called a stomach (Figure 21.43; Figure 21.45, A). More commonly, however, the pharynx seems to be succeeded directly by the long intestine.

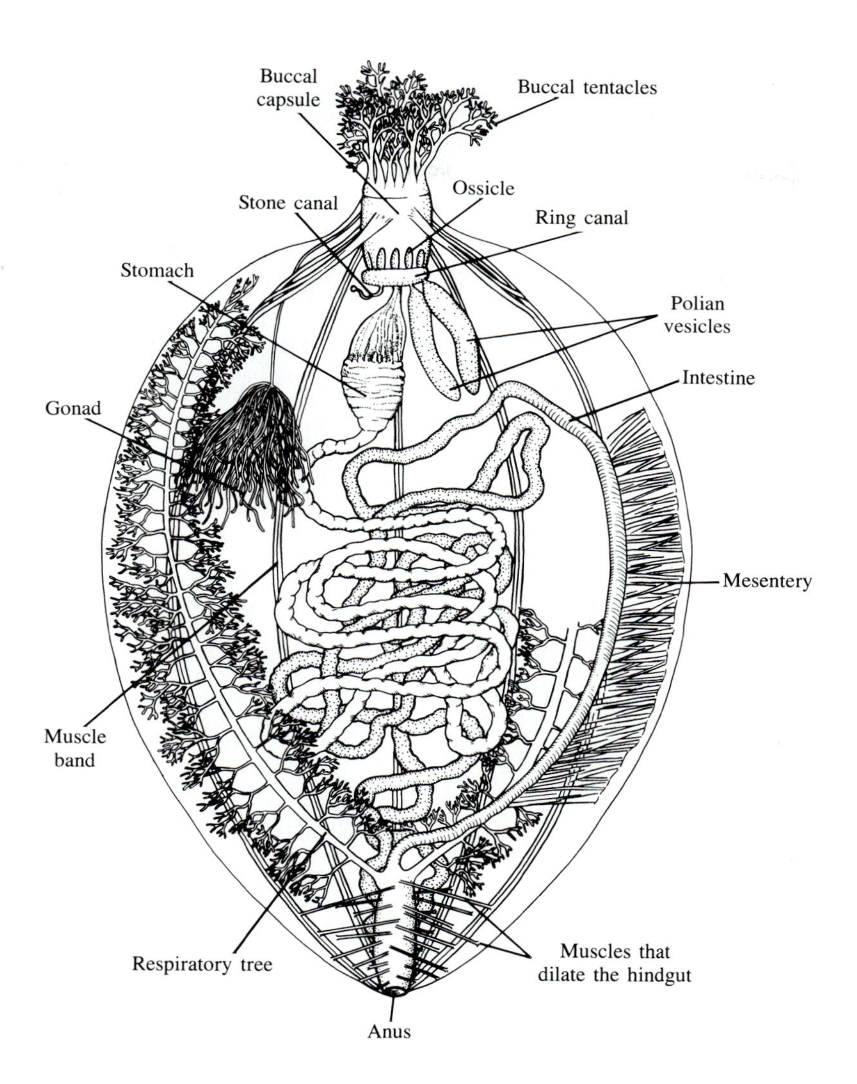

FIGURE 21.43 *Thyone briareus,* dissected. Ampullae of the tube feet are not shown. (Coe, Bulletin 19, Connecticut State Geological and Natural History Survey.)

This typically runs backward, then forward, then backward again until it reaches the anus. Each of the loops may have secondary contortions, and the intestine is not at all a homogeneous structure. It consists of several different portions, and if the intestine is slit open, one can see pronounced differences in the texture of the internal surface. The terminal portion of the alimentary tract—the hindgut—is sometimes referred to as the cloaca. This is not appropriate, because the hindgut is not joined by ducts from either the reproductive or excretory systems.

Respiratory Trees

Exchange of oxygen and carbon dioxide probably takes place to at least some extent across any thin-walled portions of the body, such as the buccal tentacles and tube feet. In some sea cucumbers the body wall is so thin everywhere that exchange of gases between the fluid in the perivisceral coelom and the outside presents no problems. In most types, however, there is a unique structural adaptation that facilitates respiration. This consists of a pair of much-branched **respiratory trees** (Figure 21.43; Figure 21.45, C), which are outgrowths of the hindgut. If a sea cucumber draws water into the hindgut by reverse peristalsis, then closes the anal sphincter and continues rhythmically to contract the hindgut, water is forced up into the many delicate branches of the respiratory trees. Exchange of oxygen and carbon dioxide can take place in the tubules. When the anal sphincter relaxes and the hindgut pumps water back out the anus, the respiratory trees "deflate." (The walls of the fine ramifications are slightly elastic, if not actually muscularized to some extent.)

Among the commensal organisms that have been found in respiratory trees of various sea cucumbers are

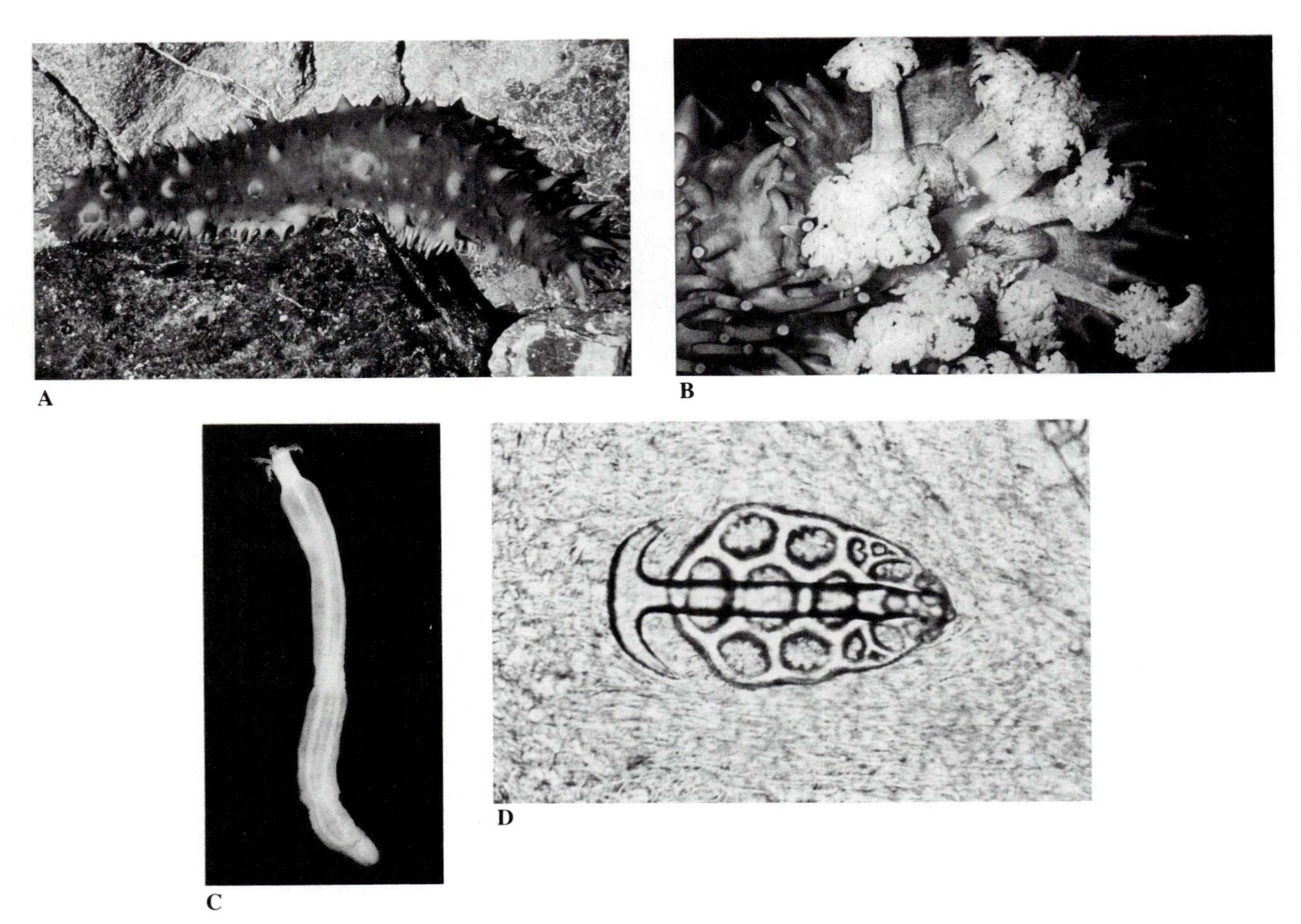

FIGURE 21.44 A. *Parastichopus californicus* (subclass Aspidochirotacea). The tube feet used for clinging to the substratum are restricted to the lower surface; the spinelike outgrowths on the upper surface are modified tube feet. B. *Parastichopus,* buccal tentacles extended. C. *Leptosynapta clarki* (subclass Apodacea). The only tube feet are the buccal tentacles. D. *Leptosynapta,* a pair of articulating ossicles. The anchor-shaped ossicle may project and engage the substratum.

ciliated protozoans, small pea crabs, and pearl fish. The survival of all of these depends on the pumping action of the hindgut. The pearl fish leave their hosts at night to forage. Although they reach a length of 15 cm, they are slender and can penetrate the anal opening when the sphincter is relaxed. They do not fit the definition of commensal very well, for they sometimes nip off bits of the respiratory trees. The trees, however, quickly regenerate missing pieces.

Tubes of Cuvier

A few sea cucumbers, such as those of the genus *Holothuria,* have slender, blind tubes arising from the main stem of one or both respiratory trees. These are called tubes of Cuvier and are used in defense. By violently contracting its body-wall muscles, the cucumber can cause the upper side of its hindgut to rupture. As a jet of coelomic fluid enters the hindgut and then leaves by way of the anus, some of the tubes are carried out with it. They instantly become sticky, like cotton candy; they also become longer and are easily broken off. A predator may become immobilized if it gets tangled up in them. In time, the cucumber repairs its hindgut and regenerates the tubes.

Sea cucumbers that do not have tubes of Cuvier may have other means of defense, especially against fishes. Some species, mostly tropical, surround themselves by an aura of a toxic or repellent substance. Extracts of sea cucumbers have been used, in fact, to paralyze fish in tide pools. Sea cucumbers are not completely exempt

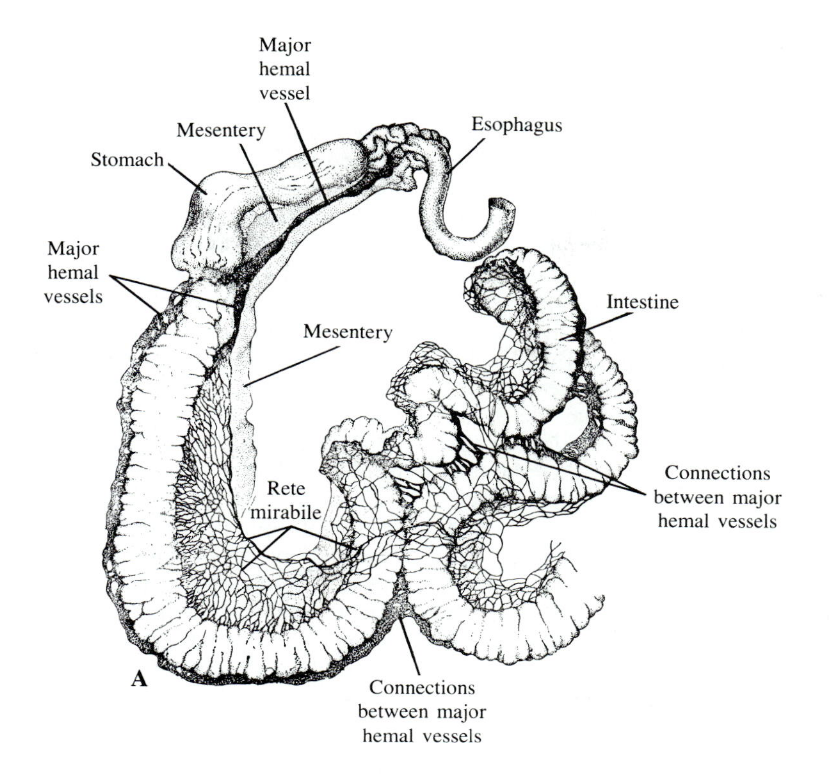

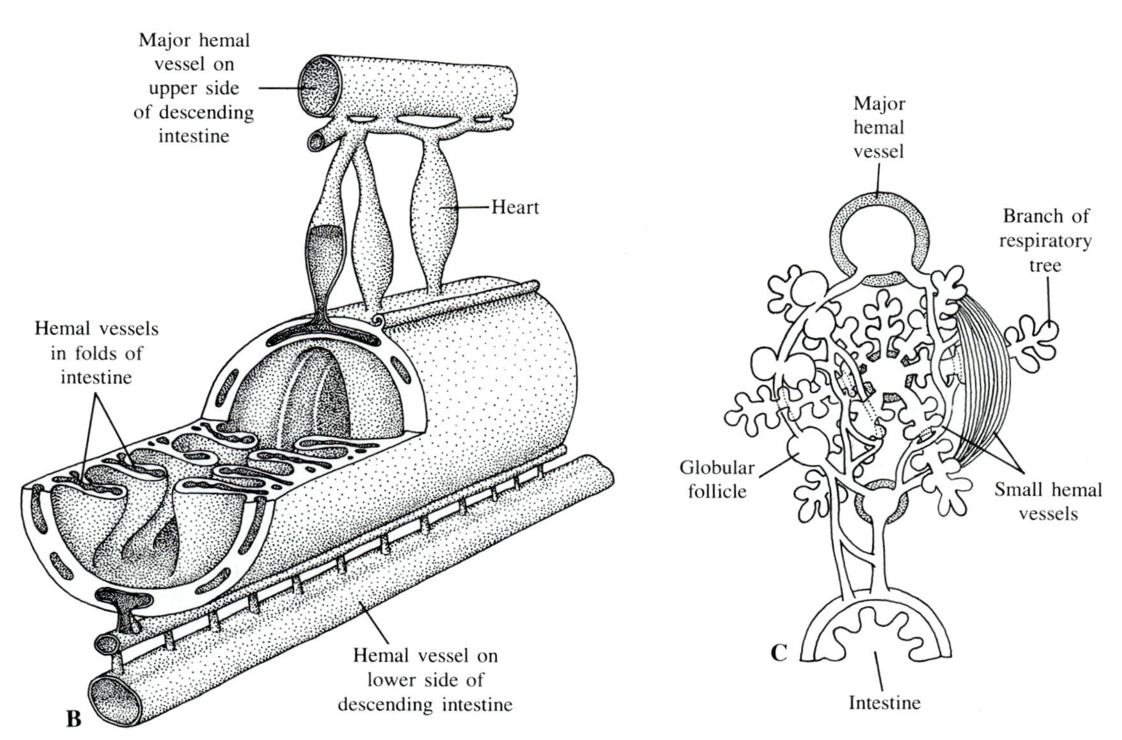

FIGURE 21.45 A. *Cucumaria frondosa,* portion of the gut, showing gross arrangement of hemal vessels associated with it. (Ivanov *et al.,* Major Practicum of Invertebrate Zoology.) B. *Isostichopus badinotus,* hemal vessels (including hearts) associated with the intestine. C. *Isostichopus,* diagram of the arrangement of hemal vessels associated with the intestine and a respiratory tree. (B, C, after Herreid, LaRussa, and DeFesi, Journal of Morphology, *150.*)

from predation, however; many of them are eaten by various animals, especially sea stars. Humans exploit certain species, particularly in parts of the Orient.

Water-Vascular System

The ring canal of the water-vascular system encircles the pharynx (Figure 21.42, A; Figure 21.43). It typically sends out five radial canals. These at first run in an oral direction beneath the large ossicles of the buccal capsule, then turn outward and aborally, eventually reaching nearly to the anus. The radial canals lie in the tissue between the longitudinal muscle bands and the perivisceral coelom (Figure 21.42, B), and the tube feet on both sides of each canal are supplied by small side branches. The tube feet are like those of sea stars and sea urchins in having well-developed ampullae. The buccal tentacles, as has been explained, are modified tube feet. They are usually connected to the radial canals at the point where these start to turn aborally, and their ampullae are often large. This is not always the case, however. In some sea cucumbers that have no tube feet except for buccal tentacles, there are no radial canals, and the buccal tentacles are joined directly to the ring canal.

If the functional tube feet are scattered over much of the body surface, instead of being restricted to five sets, there are still only five radial canals. Some of the side branches simply reach farther from the canals than others do.

In addition to the radial canals, the ring canal gives off at least one Polian vesicle and one stone canal (Figure 21.42, A; Figure 21.43). The Polian vesicle, lying free in the perivisceral coelom, is distensible and probably accepts fluid forced back into the ring canal when the buccal tentacles contract. It is, in other words, a compensation sac. The stone canal is generally stiffened by ossicles and has a recognizable madreporite at its tip. It does not, however, actually reach the surface. It simply hangs free in the perivisceral coelom or is bound to the tissue on the inside of the body wall. It is common to find more than one stone canal; some sea cucumbers have many. The number of Polian vesicles also varies.

Hemal System

Most sea cucumbers have a well-developed hemal system, and certain portions of it are especially conspicuous. The most obvious vessels are those associated with the gut, especially its first loop (Figure 21.45, A). Here, in the mesenteric tissue, is a complex of many small vessels. This is called the *rete mirabile*—the "astonishing network"—and it lies between a major hemal vessel on one side of the intestine and a similar vessel on the opposite side. These two major vessels are joined to a hemal ring that encircles the pharynx, close to the ring canal of the water-vascular system.

In some sea cucumbers, such as *Cucumaria, Eupentacta,* and other members of the subclass Dendrochirotacea, certain amoeboid cells in the hemal system, as well as the water-vascular system and perivisceral coelom, contain hemoglobin. The red color of the hemoglobin makes it easy to see many of the hemal vessels, particularly those in the region of the gut, which have little connective tissue around them. Thin-walled components of the water-vascular system, such as the Polian vesicle and ampullae of the tube feet, are also brightly colored.

The hemal vessels associated with the gut and one respiratory tree of *Isostichopus badinotus* (= *Stichopus moebii*) have been intensively studied with the aid of electron microscopy. The major vessel on the upper side of the first (descending) portion of the intestine branches into numerous hearts (Figure 21.45, B). These pump blood through small vessels lying within folds of the intestinal wall. The blood collects in the major vessel beneath the same portion of the intestine. Farther along the intestine, however, in small vessels and in the larger vessels that connect major vessels at various points, the blood flow may be in either direction. In the respiratory tree that has many small hemal vessels (Figure 21.45, C), some of the vessels have globular follicles in which deposits of iron and dead cells accumulate. The follicles also contain developing coelomocytes.

Excretion

Excretion of soluble nitrogenous wastes, mostly ammonia, probably takes place in the respiratory trees, as well as across any other thin-walled structures, including the gut, tube feet, and buccal tentacles. Particulate material can apparently be carried by amoeboid cells into the gut, respiratory trees, and body wall, from which these cells can then somehow be eliminated. Some burrowing sea cucumbers have lopsided little funnels, or urns, on the lining of the perivisceral coelom. These structures, covered with ciliated peritoneal tissue, collect particulate wastes or amoeboid cells that have accumulated wastes. According to various accounts, the wastes may then be moved into the body wall or dumped back into the coelom, where they form amorphous masses, sometimes recognizable as "brown bodies."

Nervous System

In the dermis around the mouth is a fairly prominent nerve ring from which nerves extend into the buccal tentacles and other structures in the same general region. Each radial nerve, just external to a radial canal of the water-vascular system, is differentiated into two tracts. The outer tract is connected to the nerve ring and apparently corresponds to the almost superficial radial-nerve tract on the oral side of the arm of a sea star; the inner tract, not joined directly to the nerve ring, probably corresponds to the somewhat deeper tract of a sea star. Both tracts have branches that pervade the dermis, epidermis, and muscle bands.

The epidermis of a sea cucumber is richly supplied with touch receptors, and in most species the oral and aboral ends of the body are especially sensitive. Not much is known about chemoreception in these animals. The fact, however, that some species exhibit escape responses to certain sea stars and not to others indicates that they have chemoreceptors. One of the more striking escape responses is shown by *Parastichopus californicus* (Figure 21.44, A), a large species found on the Pacific coast. After being touched by the tube feet of the sunflower star, *Pycnopodia helianthoides, Parastichopus* detaches itself from the substratum and rapidly undulates up and down, thus usually swimming to safety.

Burrowing sea cucumbers that live with the oral end down usually have statocysts to help keep them properly oriented in sand or mud. The number of statocysts and their exact position varies, but as a rule they are concentrated around the nerve ring, or situated next to the radial nerves close to where these are joined to the nerve ring. All sea cucumbers seem to be sensitive to light. Eyelike structures with dark pigment are found at the bases of the buccal tentacles of a few species, but in the majority of sea cucumbers the receptors are not obvious.

Reproduction and Development

Sexes are usually separate, but there are some hermaphroditic species. Hermaphroditism is, in fact, proportionately more common in sea cucumbers than in other echinoderms. There is a single testis or single ovary, unless both are present, and this is a distinctive feature of the class. The gonad has many slender lobes; its duct leads to a genital pore between two of the buccal tentacles. In sea cucumbers that have a strong tendency toward bilateral symmetry, the pore is between the two most nearly dorsal tentacles.

Some species simply liberate their eggs and sperms, but many brood their young in one way or another. Certain burrowing cucumbers of the family Synaptidae, for instance, brood their young in the perivisceral coelom; the young cucumbers eventually escape through a break in the body wall close to the anus. Other sea cucumbers brood their young externally in pockets of the body wall, or in some other fashion.

Brooding species and many free-spawning types have direct development, but some free-spawners have a feeding larva similar to the bipinnaria stage of sea stars. This is called the **auricularia.** It is succeeded by the **doliolaria** stage, characterized by several transverse bands of cilia derived from the originally single, continuous band of the auricularia. The doliolaria is essentially a juvenile sea cucumber, and soon develops the future buccal tentacles as well as a few other tube feet.

Evisceration and Seasonal Resorption of the Gut

An attribute of many sea cucumbers is their tendency to eviscerate. They often do this when they are handled or when they are kept in water that does not have sufficient oxygen. In nature, however, the phenomenon is most likely to be induced when a sea cucumber is threatened. Evisceration is probably a protective mechanism that either distracts a predator or that releases a repellent substance. The gut tissue of certain species is known to have a toxic effect on other animals.

There are three ways in which evisceration may take place. In some sea cucumbers, the body wall ruptures close to the oral end and the buccal capsule, buccal tentacles, gonadal tubules, and much of the gut are squeezed out and become separated from the rest of the animal. Other sea cucumbers can cut off the anteriormost part of the body by a powerful constriction of the body wall just behind the buccal capsule. A third method of evisceration is by rupture of the posteriormost part of the gut. The respiratory trees and a substantial portion of the gut are eliminated through the anus. Whichever way the evisceration is accomplished, the sea cucumber usually regenerates the missing parts.

Some sea cucumbers, even though they can and do eviscerate, are also able to slowly resorb most of the digestive system. *Parastichopus californicus, Eupentacta quinquesemita,* and some other species do this in the autumn, so that during the winter months they have no functional gut. Presumably they subsist until spring on nutrients stored in the body wall.

Rectal Pores

Two very different genera of sea cucumbers—*Parastichopus* (subclass Aspidochirotacea) and *Leptosynapta* (subclass Apodacea)—have permanent pores in the posteriormost part of the rectum. It is not known whether these pores normally allow sea water to enter the coelom. It is known, however, that in *Parastichopus californicus* the egg capsules of a turbellarian parasite that inhabits the coelom escape through the pores. Perhaps pores of this sort will be found in other sea cucumbers as well.

It has long been assumed that the larvae of the strange wormlike gastropods that parasitize *Parastichopus* and other sea cucumbers are released from the coelom only when the host eviscerates. While it is true that these molluscs are generally found in sea cucumbers that do eviscerate, there is a possibility that rectal pores are an alternative route for escape of the larvae.

Classification

Subclass Dendrochirotacea

Most dendrochirote holothuroids are characterized by buccal tentacles that branch so extensively that they are bushy. Moreover, the buccal capsule is supplied with retractor muscles, which are branches of the longitudinal muscle bands, so that the anteriormost part of the animal, tentacles included, can be pulled back into the buccal cavity. In feeding, the tentacles are bent, one at a time, stuck into the mouth, and then withdrawn while the mouth is closed tightly. Thus detritus that has accumulated in the tentacle is licked off. There are respiratory trees attached to the hindmost portion of the gut.

This subclass is diversified and includes two orders and several families. Among the more common genera are *Cucumaria* (Figure 21.41, A; Color Plate 8), *Eupentacta,* and *Psolus* (Figure 21.41, D). *Psolus* is representative of a group of dendrochirotes that are flattened on one side, so that they fit tightly against a rock or shell, and that have the rest of the body protected by platelike ossicles visible externally.

Subclass Aspidochirotacea

The buccal tentacles of aspidochirote sea cucumbers are generally branched to some extent, but are not as bushy as those of dendrochirotes. The branches are tipped by small disklike structures used for collecting food, mostly detritus. The tentacles deliver food to the mouth, but are not actually inserted into the mouth as those of dendrochirotes are. The buccal capsule is not provided with retractor muscles. The tube feet that function in locomotion are restricted to the lower side of the animal. The others are often modified as tubercles that have a sensory function.

Stichopus, Parastichopus (Figure 21.44, A and B), and *Holothuria* are well-known genera of aspidochirotes. These have respiratory trees, and *Holothuria* also has tubes of Cuvier. Several genera, mostly found only in very deep water, lack respiratory trees. One of these, *Pelagothuria,* is remarkable in that it swims, using a weblike structure between its buccal tentacles to propel itself.

Subclass Apodacea

Apodan sea cucumbers have relatively simple tentacles. These are either unbranched and fingerlike, as in *Molpadia,* or somewhat featherlike, as in *Leptosynapta* (Figure 21.44, C). There are no retractor muscles for the buccal capsule. *Leptosynapta* and its close relatives have no tube feet except for the tentacles, and they lack respiratory trees. Most of them burrow in sand or mud. *Molpadia* and its relatives, which also live in mud or sand, do have respiratory trees and they also have a few tube feet in a circle around the anus.

References on Holothuroidea

Bai, M. M., 1980. Monograph on *Holothuria* (*Metriatyla*) *scabra*. Memoirs of the Zoological Survey of India, *16:*1–75.

Byrne, M., 1985. Evisceration behavior and the seasonal incidence of evisceration in the holothurian *Eupentacta quinquesemita* (Selenka). Ophelia, *24:*75–90.

Erber, W., 1983. Der Steinkanal der Holothurien: eine morphologische Studie zum Problem der Protocoelampulle. Zeitschrift für Zoologische Systematik und Evolutionsforschung, *21:*217–234.

Fish, J. D., 1967. Biology of *Cucumaria elongata*. Journal of the Marine Biological Association of the United Kingdom, *47:*129–143.

Herreid, C. F., II, LaRussa, V. F., and De Fesi, C. R., 1976. Blood vascular system of the sea cucumber, *Stichopus moebii*. Journal of Morphology, *150:*423–452.

Jensen, H., 1975. Ultrastructure of the dorsal hemal vessel in the sea cucumber *Parastichopus tremulus*. Cell and Tissue Research, *160:*355–369.

Pawson, D. L., 1977. Marine flora and fauna of the northeastern United States. Echinodermata: Holothuroidea. National Marine Fisheries Circular no. 405.

Rustad, D., 1940. The early development of *Stichopus tremulus*. Bergens Museum Arbok, 1938, no. *8:*1–23.

SUMMARY

1. Echinoderms, all of which are marine, have a very distinctive combination of characters. Among these are radial symmetry, a calcareous endoskeleton, a water-vascular system, and a coelom that originates as outpocketings of the embryonic archenteron.

2. The water-vascular system is unique to echinoderms. It is a division of the coelom, and consists of canals that are joined to structures called tube feet. These project externally and are used for locomotion, attachment, feeding, and respiration. The other main division of the coelom is the perivisceral coelom, which surrounds the gut, gonads, and some other internal organs.

3. The endoskeleton consists of ossicles secreted by cells of mesodermal origin. Even though spines, plates, and other calcareous structures may be visible at the surface, they are formed beneath the epidermis.

4. The gut is usually complete, but an anus is lacking in the class Ophiuroidea and in some Asteroidea. A hemal (circulatory) system is present, but its vessels are obscure except in certain portions of the body of members of the classes Echinoidea and Holothuroidea. There are no definite excretory organs. The nervous system, much of which is closely associated with the epidermis, is decentralized, there being nothing comparable to a brain.

5. Sexes are generally separate, and the reproductive system is simple. The gonads release gametes by short ducts leading to the body surface, or into pockets that are open to the outside. Fertilization is typically external. Nevertheless, females of many species brood their eggs.

6. In echinoderms that have a swimming larval stage, this is bilaterally symmetrical. Radial symmetry does not become apparent until metamorphosis to the adult form.

7. There are five living classes of echinoderms. Members of the class Crinoidea, which may be stalked and attached, or unstalked and mobile, are oriented with the oral surface uppermost. The anus is also on this side of the body. The primary rays, which originate at the margins of the central disk, typically branch into two secondary rays. From these arise hundreds of side branches, called pinnules. The rays, pinnules, and aboral cirri that are used for attachment (and also the stalk, if there is one) consist to a considerable extent of ossicles that articulate in such a way that the structures are flexible. The oral surface of each pinnule and ray has a ciliated ambulacral groove bordered by tube feet. The tube feet knock small organisms and other particulate matter onto the grooves. Trapped in mucus, the food is driven to the mouth by the activity of the cilia.

8. The oral surface of a crinoid has many small pores that presumably allow water to enter the perivisceral coelom. The gonads are scattered through the rays and there are thus many gonopores. After embryonic development and metamorphosis have been completed, a young crinoid goes through a stalked phase. In the mobile feather stars, however, the stalk does not persist.

9. In the Asteroidea, or sea stars, the body consists of a central disk and five or more rays that diverge from it. The mouth is on the lower (oral) surface, and the anus, when it is present, is on the upper (aboral) surface. On the oral side of each ray there are two rows of tube feet. These usually have suckerlike tips. The body wall is strengthened by many ossicles, some of which are visible at the surface in the form of spines, plates, and other structures. The rays can bend and twist, as they must do in righting movements, adjustment to irregular surfaces, and capture of prey, but these movements are generally slow. Locomotion, burrowing, attachment, and some aspects of feeding are primarily accomplished by the tube feet. Tube feet also function in respiration, although gas exchange is accomplished partly through thin-walled extensions of the body wall called papulae. These allow the fluid in the perivisceral coelom to circulate close to the external environment.

10. Distinctive features of the body surface of many sea stars are structures called pedicellariae. These consist of pairs or groups of small ossicles that form devices for pinching, thus keeping the epidermis free of organisms that might otherwise settle on it. They may have other protective functions, and in a few sea stars they assist in capture of food.

11. As a group, sea stars are carnivores, and they feed on a variety of other animals, ranging from sponges to fishes. The group does, however, include herbivores and detritus-feeders. Some sea stars are able to evert a portion of the stomach and insert this between the valves of a clam or envelop an entire prey organism. Thus digestion may be begun outside the body. Once taken into the stomach, much of the food is distributed into diverticula that go out into the rays. These diverticula are called pyloric caeca, and they are involved in both intracellular and extracellular digestion.

12. Brittle stars, which make up most of the class Ophiuroidea, are capable of rapid movements of the rays, and their locomotion over the substratum is usually of a rowing type. The rays, unlike those of a sea star, are distinctly set off from the disk. Each ray has a series of large internal ossicles that articulate in such a way as to make the rays flexible. The tube feet are used mostly for capturing and manipulating food and for respiration, rather than locomotion.

13. The mouth of a brittle star, in the center of the oral surface, is encircled by five sets of ossicles that form jaws and teeth. The gut is saclike and restricted to the disk. There is no anus. The 10 gonads are also in the disk. They open into sacs called bursae, which in turn open to the outside by slits located at the bases of the rays. The bursae function in respiration and sometimes as pockets in which young are brooded.

14. In the class Echinoidea, there are no rays. The body is tomato-shaped in a sea urchin, disk-shaped in a sand dollar, and more or less ovoid in a heart urchin. Most of the organs lie within a shell, or test, that consists of platelike ossicles that are tightly bound together. Movable spines articulate with the test, and there are also tube feet and pedicellariae. In a sea urchin, the five ambulacral areas run from near the mouth, in the center of the oral surface, to near the anus, on the opposite side. The pores in the ossicles through which the tube feet emerge are paired because each tube foot is connected with its ampulla by two canals. In sand dollars and heart urchins, the ambulacra are proportionately less extensive than in sea urchins, and the anus is displaced away from the center of the aboral surface. This is one feature that superimposes a strong bilaterality on the radial symmetry of sand dollars and heart urchins.

15. A unique feature of the echinoids, especially sea urchins, is a complicated apparatus called Aristotle's lantern. It consists of many ossicles controlled by muscles. Five of the ossicles, set in jaws, are the teeth that sea urchins use to chop up seaweeds and other food. Sand dollars and heart urchins feed mostly on detritus, and their lanterns are reduced.

16. In the class Holothuroidea, the sea cucumbers, the body is typically elongated, with the mouth at one end and the anus at the other. Most species show at least some tendency toward bilateral symmetry. It is common to find, for instance, that of the five ambulacra, only those on the lower side of the body have well-developed tube feet that can be used for attachment and locomotion.

17. The body wall of sea cucumbers has ossicles, but they are not visible externally except in certain groups. The organs used for feeding are specialized tube feet arranged in a circle around the mouth. These are called buccal tentacles and are usually extensively branched, so that they have considerable surface area. A coating of mucus on the tentacles traps detritus and small planktonic organisms. The tentacles may simply deliver food to the mouth, or they may be inserted into the mouth so that the food can be licked off. The gut is much longer than the body and its posteriormost portion has much-branched diverticula called respiratory trees. Water taken in through the anus is forced into the respiratory trees, through the thin walls of which gas exchange with the coelomic fluid may be effected. The oral tentacles, as well as the other tube feet, are also respiratory organs. There is a single gonad; its gonopore is on the dorsal midline close to the anterior end of the body.

22

Phylum Hemichordata

Acorn Worms and Their Odd Relatives

INTRODUCTION

The two groups of surviving hemichordates, forming the classes Enteropneusta and Pterobranchia, are so different that some zoologists give both of them the rank of phylum. Nevertheless, enteropneusts and pterobranchs do share certain distinctive features. One of these is an anteriorly directed diverticulum of the buccal cavity, which for a long time was interpreted to be a homologue of the notochord of urochordates, cephalochordates, and vertebrates. Another feature common to both groups is the division of the body into three regions, each with its own coelom. Finally, the pharynx of all enteropneusts and some pterobranchs has gill slits—small perforations that open laterally to the outside. Much of the water taken into the pharynx in the course of feeding is eliminated by way of the gill slits, so that food particles are concentrated.

Enteropneusts and pterobranchs are marine. Enteropneusts are wormlike animals, most of which burrow in mud or sand. Pterobranchs inhabit tubes, although certain species may crawl out of their tubes. The fossils called graptolites, not dealt with in this book, are probably close to pterobranchs, but they are usually put into a class of their own.

CLASS ENTEROPNEUSTA

General Organization of the Body, Coelom, and Gut

An enteropneust (Figure 22.1) shows clearly the three divisions of the body characteristic of hemichordates: the **proboscis** (protosome), **collar** (mesosome), and **trunk** (metasome). The proboscis looks like a tongue or an acorn, depending on its proportions. It is richly muscularized (Figure 22.2) and admirably fitted for burrowing. Circular muscles form a layer just beneath the epidermis, and when these contract the proboscis lengthens and pushes its tip deeper into the mud or sand. Then, as circular muscles near the tip relax and the inner longitudinal muscles contract, a bulge devel-

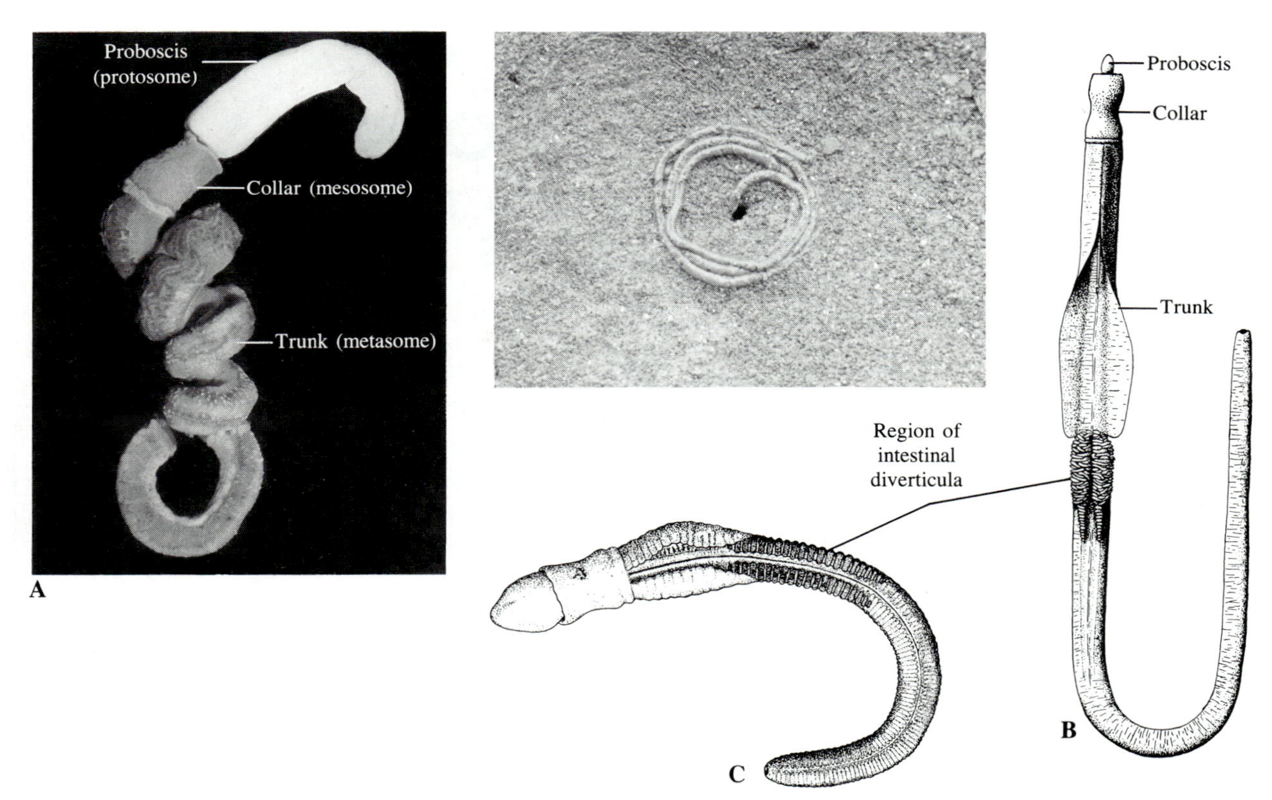

FIGURE 22.1 Class Enteropneusta. A. A species of *Saccoglossus* from the Pacific coast of North America (left) and one of its fecal castings (right). A portion of the trunk is missing. B. *Saccoglossus studiosorum,* from South Africa. (Van der Horst, Annals of the South African Museum, *32.*) C. *Ptychodera flava,* from Madagascar. (Kirsteuer, Zoologischer Anzeiger, *175.*)

ops and the worm anchors itself. The bulge moves backward along the proboscis in a peristaltic wave as the tip of the organ advances again. It should be noted, however, that some species do not burrow, even though they have a well-developed proboscis.

The coelomic cavity of the proboscis, termed the **protocoel,** is not extensive, and it may be subdivided to some extent by strands of tissue. In most enteropneusts it has an opening to the outside. This is called the **proboscis pore** and is located dorsally on the slender posterior part of the proboscis, either along the midline or just slightly to the left or right of the midline. When there are two symmetrically placed pores, as is the case in a few species, this is probably an ancestral feature.

The collar has a moderate amount of muscle and may assist the proboscis in burrowing activities. The mouth, located on the ventral side of the deep furrow that separates the collar from the proboscis, opens into a buccal cavity. It is from this that the **buccal diverticulum** (sometimes called the *stomochord*) originates and runs forward into the proboscis (Figure 22.2, A and B). The diverticulum may be flattened and may have a pair of ventrolateral pouches or other lobations. Beneath it is a firm but elastic plate, called the **proboscis skeleton.** This supports the stalk of the proboscis and also

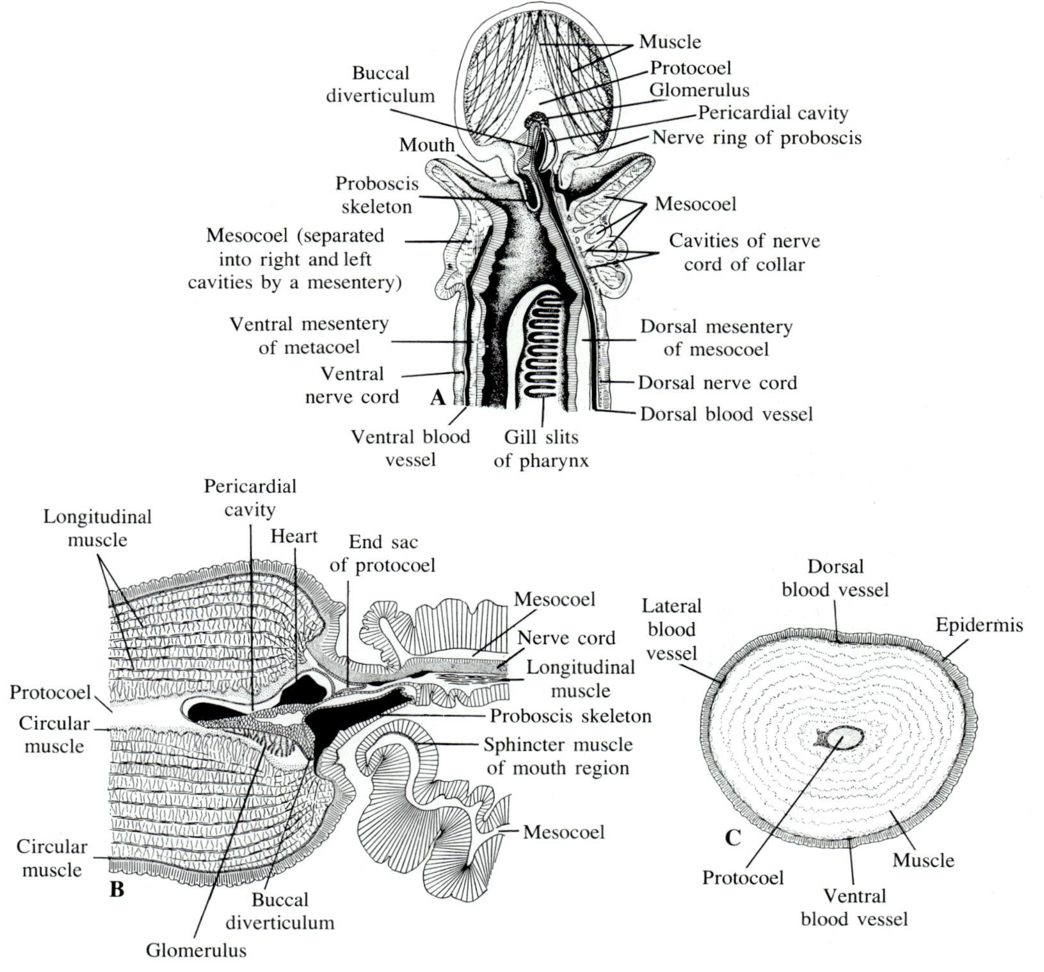

FIGURE 22.2 A. Sagittal section through the proboscis, collar, and anteriormost portion of the trunk of *Balanoglossus*. (Godeaux, in Florkin and Scheer, Chemical Zoology, *8*.) B. Sagittal section through a portion of the proboscis and collar of *Saccoglossus horsti*. C. Transverse section through the proboscis (contracted) of *Saccoglossus horsti*. (B and C, Brambell and Goodhart, Journal of the Marine Biological Association of the United Kingdom, *25*.)

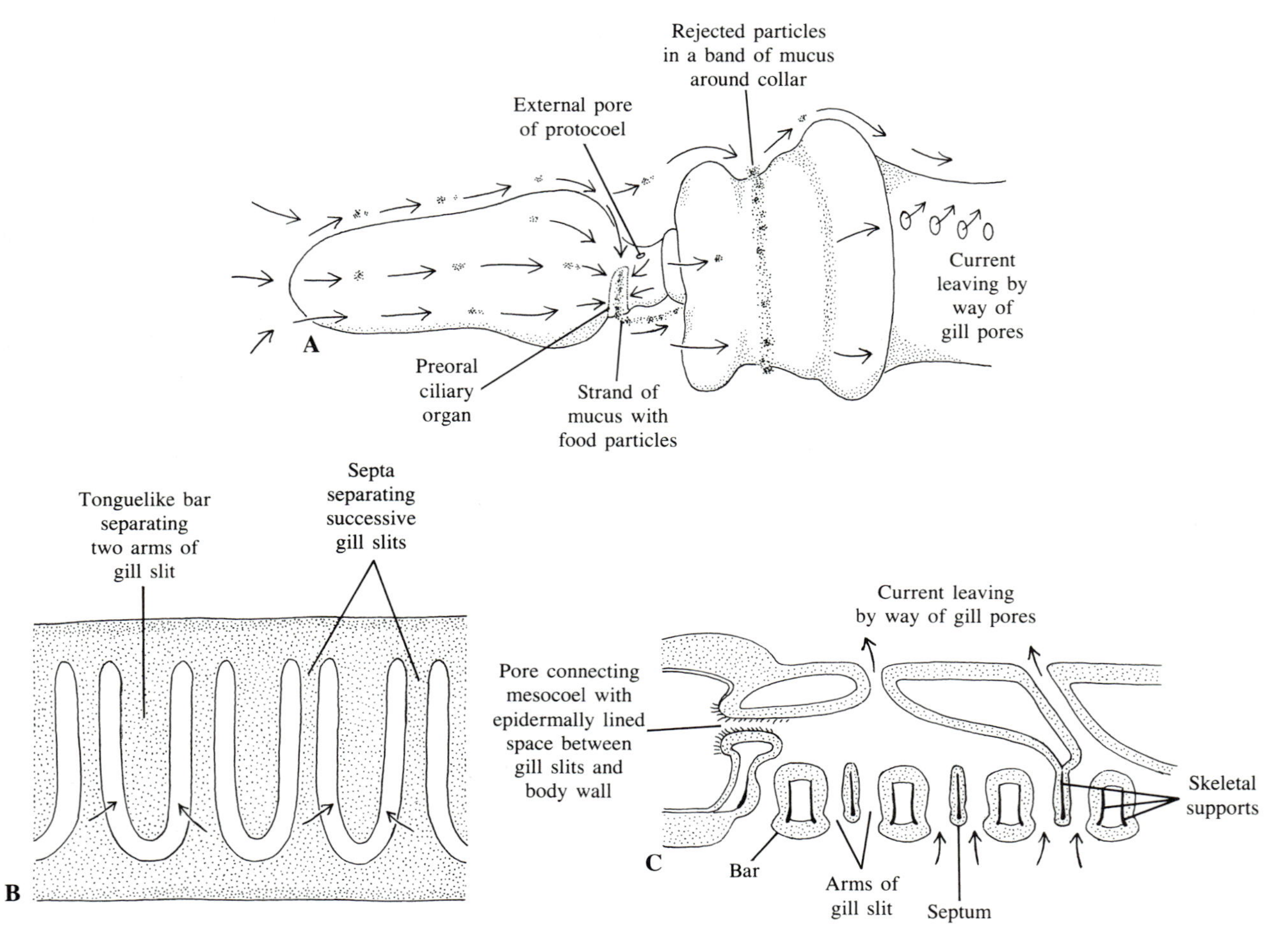

Figure 22.3 A. Movement of food and mucus in the anterior portion of *Protoglossus köhleri*. (After Burdon-Jones, Proceedings of the Zoological Society of London, *127*.) B. Gill slits on one side of the dorsal part of the pharynx of an enteropneust, as viewed from the inside of the pharynx. Water leaving by the gill slits enters cavities between the pharynx and body wall, then escapes by pores. C. Frontal section through the gill slits and body wall in the anteriormost part of the trunk of *Schizocardium brasiliense*. The direction taken by water currents moving from the pharynx to the outside is indicated by arrows. One of the two pores connecting the mesocoel with the anteriormost space between the gill slits and body wall is also shown. (After Spengel, in Fauna and Flora des Golfes von Neapel, Monograph 18.)

generally has a pair of extensions running backward along the sides of the buccal cavity. The proboscis skeleton is secreted in part by the epidermis and in part by mesodermally derived tissue in the proboscis. Lateral to the buccal diverticulum, and above it, is some stiffening tissue that resembles cartilage. This is produced by transformation of some of the tissue in the walls of the protocoel and anterior prolongations of the **mesocoel,** or collar coelom.

The mesocoel consists of two separate cavities, one on the right, the other on the left (Figure 22.2, A). Each has a posterior duct leading to the outside, as explained below, and each may be subdivided by transverse partitions.

The trunk is generally several times as long as the proboscis and collar combined. The anterior portion of it—about the first quarter—has a row of pores on both sides (Figure 22.3, A). These indicate the extent of the pharyngeal region of the gut, which is characterized by U-shaped **gill slits** (Figure 22.3, B). The two arms of each gill slit are separated by a tonguelike bar. The bars, like the septa between successive gill slits, are stiffened by skeletal supports (Figure 22.3, C).

The gill slits do not communicate directly with the externally visible pores; there is an epidermally lined space between them. In the family Ptychoderidae, which includes the genera *Ptychodera, Balanoglossus,* and *Glossobalanus,* there is an external pore and a space for each gill slit. In other families, there is a pore and a space for a series of several gill slits (Figure 22.3, C). In either case, the two anteriormost spaces receive the mesocoelic ducts referred to above.

The gill slits are usually on the dorsolateral portions of the pharynx. There may, in fact, be a pair of inwardly directed folds that partially divide the pharynx into an upper, slit-bearing division and a ventral channel in which food collects and moves backward. The efficiency of the pharynx in concentrating food and eliminating water depends on the activity of cilia.

The esophageal part of the trunk, just behind the pharyngeal region, may also have some paired openings. These are on the dorsal surface rather than on the sides, but when they are present they are linked with simple pores in the esophagus, rather than with U-shaped gill slits.

The coelom of the trunk, called the **metacoel,** is like the collar coelom in that it is separated into left and right compartments (Figure 22.4). Neither compartment, however, has an opening to the outside. The trunk coelom may be obliterated to a considerable extent, so that it does not show up at all in transverse sections taken at certain levels. An odd feature of some enteropneusts is that the trunk coelom sends anterior prolongations into the collar, which, as already explained, has its own coelomic cavities.

The long intestinal portion of the gut has various histological specializations along its course, but one part of it is distinctive in that it has many glands and more muscle than other portions. It may also have prominent diverticula whose presence is obvious externally (Figure 22.1, B and C). This part of the gut is sometimes referred to as the "hepatic" region. Both the esophagus and intestine have dorsal and ventral grooves in which ciliary activity is particularly intense. Peristaltic movements of the gut and body wall in the trunk region assist in driving the contents of the gut backward. The anus is terminal, and the elimination of fecal material requires muscular activity.

Habitat and Feeding

The majority of enteropneusts live in mud or muddy sand, intertidally or subtidally. They may form rather dense populations, with as many as 200 worms per square meter. Some of the mud-inhabiting types dig a permanent burrow that is basically U-shaped, although the portion that enables the anterior end of the animal to reach the surface may have more than one opening (Figure 22.5). The location of enteropneust burrows can generally be detected by the presence of coiled fecal castings (Figure 22.1, A) similar to those produced by the polychaetes called lugworms (Chapter 11). The castings consist mostly of sand that has been consumed along with organic detritus. Sand grains, if coated by a film of microorganisms and other organic material, are themselves a source of food.

Feeding typically involves secretion of mucus by glands on the proboscis. Particles caught up in this are moved backward by ciliary activity. Some are rejected, but some are carried to the preoral ciliary organ, located on the ventral side of the posterior portion of the proboscis. A strand of mucus and food particles formed by the preoral ciliary organ is delivered to the mouth (Figure 22.3, A).

Much remains to be learned about digestion in enteropneusts. Enzymes concerned with hydrolysis of polysaccharides and fats have been found in extracts of the gut and other parts of the body. Amylase, a starch-splitting enzyme, has been identified in mucus secreted by the proboscis, and its activity perhaps begins even before food reaches the mouth.

A number of species are not really burrowers but live in crevices in coral masses or among seaweeds. Some have even been observed draped over rocks in deep water (Figure 22.6). Most of the nonburrowing types are probably selective deposit feeders, and perhaps certain of them can trap particles suspended in the water.

Enteropneusts are not often longer than 30 or 40 cm. The one giant member of the phylum is *Balanoglossus gigas,* found in Brazil; it reaches a length of nearly 1.5 m. Many species, when handled, give off a strong odor somewhat similar to that of iodoform. Two substances known to be responsible for the odor are 2,6-dibromophenol and 3-chloroindole, which are incorporated into slime produced by epidermal glands. The slime of certain enteropneusts glows in the dark. In the case of species that burrow in stiff mud, it is generally difficult to get intact worms. The trunk region is usually so fragile that the slightest tug will cause it to break. In at least many enteropneusts, however, an extensive portion of the trunk can be regenerated.

Circulatory System

The blood, which contains some amoebocytes and a respiratory pigment, circulates partly through distinct vessels and partly through sinuses. The system is therefore considered to be of the open type. Neither the vessels nor the sinuses have their own cellular linings. Two principal vessels—one dorsal, one ventral—run longitudinally for much of the length of the body (Figure

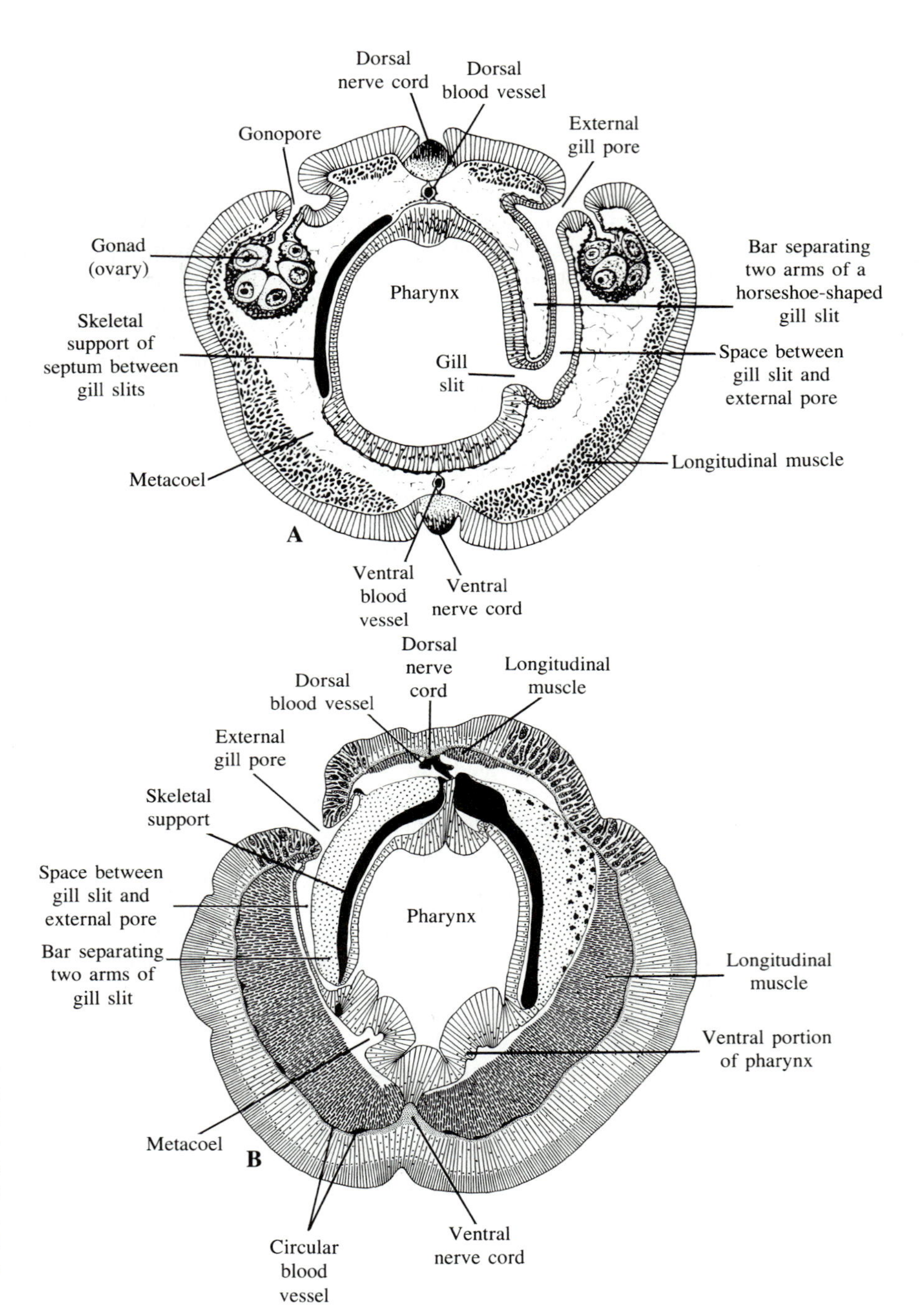

FIGURE 22.4 A. Transverse section through the trunk of a species of *Saccoglossus*. (Godeaux, in Florkin and Scheer, Chemical Zoology, *8*.) B. Transverse section through the trunk of *Saccoglossus horsti*. (Brambell and Goodhart, Journal of the Marine Biological Association of the United Kingdom, *25*.)

22.2, A; Figure 22.4, A). In the proboscis, directly above the buccal diverticulum, the dorsal vessel has a dilation conventionally called the heart. The contractile mechanism resides, however, in myoepithelial cells lining the pericardial cavity that envelops the heart. As blood is pumped forward from the heart, it passes through vessels in the **glomerulus,** a structure consisting of fingerlike projections of the epithelium of the proboscis coelom, close to the tip of the buccal diverticulum (Figure 22.2, A and B). The glomerulus is believed to have an excretory function and to operate on the pressure-filtration principle. Wastes presumably enter the proboscis coelom and leave by its external pores. Blood that has passed through the glomerulus goes ei-

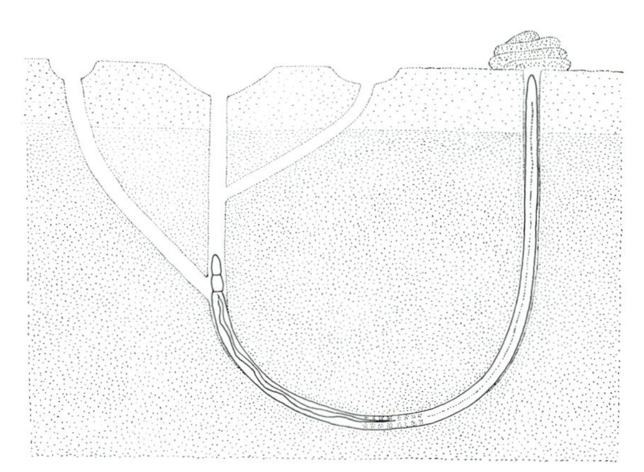

FIGURE 22.5 Burrow of *Balanoglossus clavigerus*. (After Stiasny, Zoologischer Anzeiger, *42*.)

FIGURE 22.6 Numerous specimens of an enteropneust that lives exposed on the substratum near volcanic rifts in deep water near the Galapagos Islands. Structure at left is a data-gathering probe. (Photograph courtesy of Robert Hessler.)

ther into vessels supplying the proboscis or into vessels that carry blood to the ventral vessel of the collar and trunk.

Both the ventral and the dorsal vessel are blood-pumping structures, but their peristaltic action is due to contraction of myoepithelial cells of the coelomic peritoneum, which closely borders them. The ventral vessel pumps blood posteriorly, and its many branches distribute blood to almost all parts of the collar and trunk, including the tissues next to the gill slits. After passing through small sinuses, the blood moves into vessels that lead to the dorsal vessel. This drives blood forward to the heart.

Recent studies on the arrangement of blood vessels and sinuses in the gill bars and gill septa suggest that the tissues around the gill slits are not especially important in respiration. It is likely that exchange of gases takes place over most of the body surface of an enteropneust.

Nervous System and Sensory Receptors

The nervous system of enteropneusts is comparatively primitive in the sense that most of it is closely associated with the epidermis. There are no obvious ganglia. Concentrations of nerve cells and fibers do, however, form rather distinct rings and cords (Figure 22.7). A prominent ringlike plexus in the posterior part of the proboscis is joined to something that resembles a dorsal nerve cord in the proboscis, but this consists of several somewhat separate tracts. A true dorsal cord is present in the collar and trunk, and the trunk also has a ventral cord. A ring located at the junction of the collar and trunk connects the dorsal and ventral cords and sends off many of the nerves that supply the collar region. Most of the small nerves of the trunk originate from the dorsal or ventral cords.

In the collar region, where the dorsal cord lies below the epithelium from which it is derived, the cord may have an obvious central cavity or a series of cavities. This is because it originates as a furrow, then sinks into the subepithelial tissue. If it does have a cavity or cavities, these may open to the outside at the anterior or posterior end of the collar, or at both ends. (The openings are called *neuropores*.) Because of the way the cord of the collar develops, some earlier zoologists have likened it to the hollow dorsal nerve cord of vertebrates and cephalochordates. It is not, however, markedly different in histological structure from any of the epidermal tracts; in fact, it remains connected with the epidermis by a mesenterylike strip that runs through the collar coelom lying above it. It also does not give rise to important branch nerves, but it does have giant nerve cells that permit rapid transmission of impulses. The axons of these giant cells run into the proboscis and into the ventral cord of the trunk by way of the connecting ring between collar and trunk. From the ventral cord they may extend into the anterior region of the trunk, where the gill slits and gonads are located.

Sensory receptors are found in the epidermis of all parts of the body. The more obvious of them have a slender projection, probably a modified cilium or aggregation of cilia. Most of the epidermal receptors are assumed to be touch receptors, but perhaps at least some are chemosensory.

Reproduction and Development

Sexes are separate and the numerous testes or ovaries are concentrated in the pharyngeal region of the trunk. Each gonad opens by its own pore, usually dorsolaterally (Figure 22.4, A). The arrangement is strikingly similar to that in nemerteans. In some enteropneusts the gonads are situated within a pair of prominent winglike ridges that bend toward the dorsal

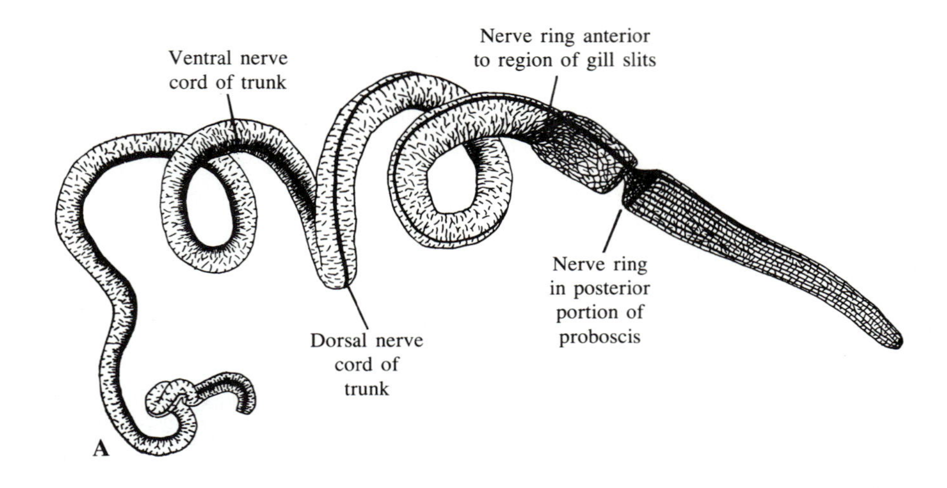

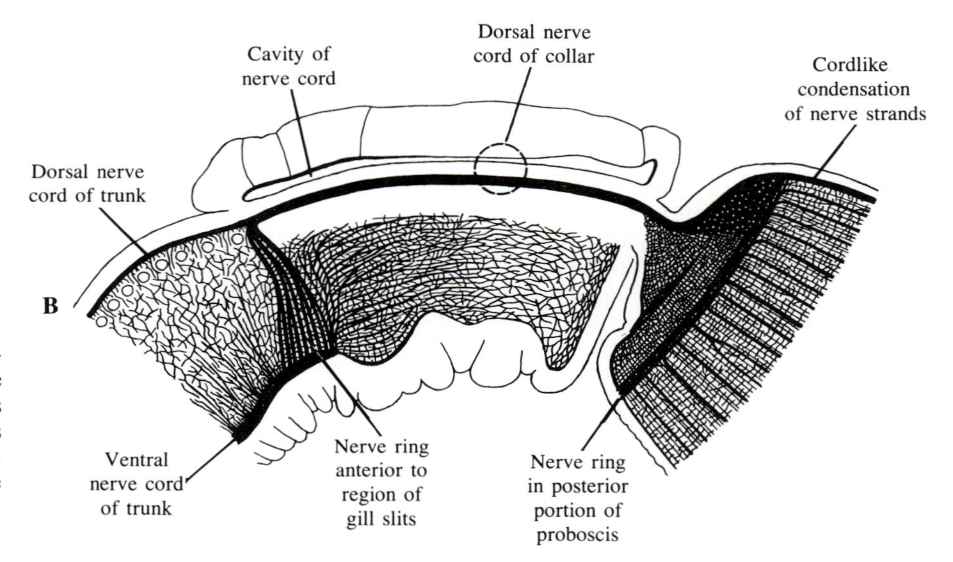

FIGURE 22.7 *Saccoglossus cambrensis*. A. Distribution of the more important nerves. B. Nerves in the collar and adjacent portions of the trunk. (Knight-Jones, Philosophical Transactions of the Royal Society of London, B, *236*.)

midline. When this is the case, the gonopores are located on the inner faces of the ridges. In *Stereobalanus*, the gonads lie within four ridges, one above and one below the pores on each side of the trunk.

Fertilization is generally external, although it must be internal in one species that is viviparous. The oocytes have a jellylike coating when they are released. On making contact with this coating, the sperm produces a tube from the tip of its acrosome. The nucleus reaches the oocyte through the tube. The oocyte undergoes two maturation divisions and then union of egg and sperm nuclei takes place.

The early cleavages are radial and almost equal; there is scarcely any difference between macromeres and micromeres (Figure 22.8, A). The side with the smallest cells is destined to be the anterior end, however. The hollow blastula (Figure 22.8, B) becomes a gastrula when the larger cells near the lower pole invaginate (Figure 22.8, C). The original blastocoel may be completely obliterated. The blastopore closes, but later on the anus appears in the same place. External ciliation develops before the anus opens up. The archenteron buds off the protocoel (Figure 22.8, D), the coelomic cavity of the future proboscis. In time the protocoel produces a canal that reaches the outside, thus forming the proboscis pore. The paired cavities of the mesocoel and metacoel (Figure 22.8, E) may form in any one of several ways: as posterior extensions of the protocoel, as new outpocketings of the archenteron, or within compact masses of mesoderm cells. The mouth is formed where the archenteron unites with ectoderm.

Ectoderm gives rise to the epidermis and to the nervous system. Endoderm is the precursor of the lining of the gut almost from mouth to anus; there is only a shallow stomodeal invagination in the mouth region. The gill slits originate as outpocketings of the gut, which

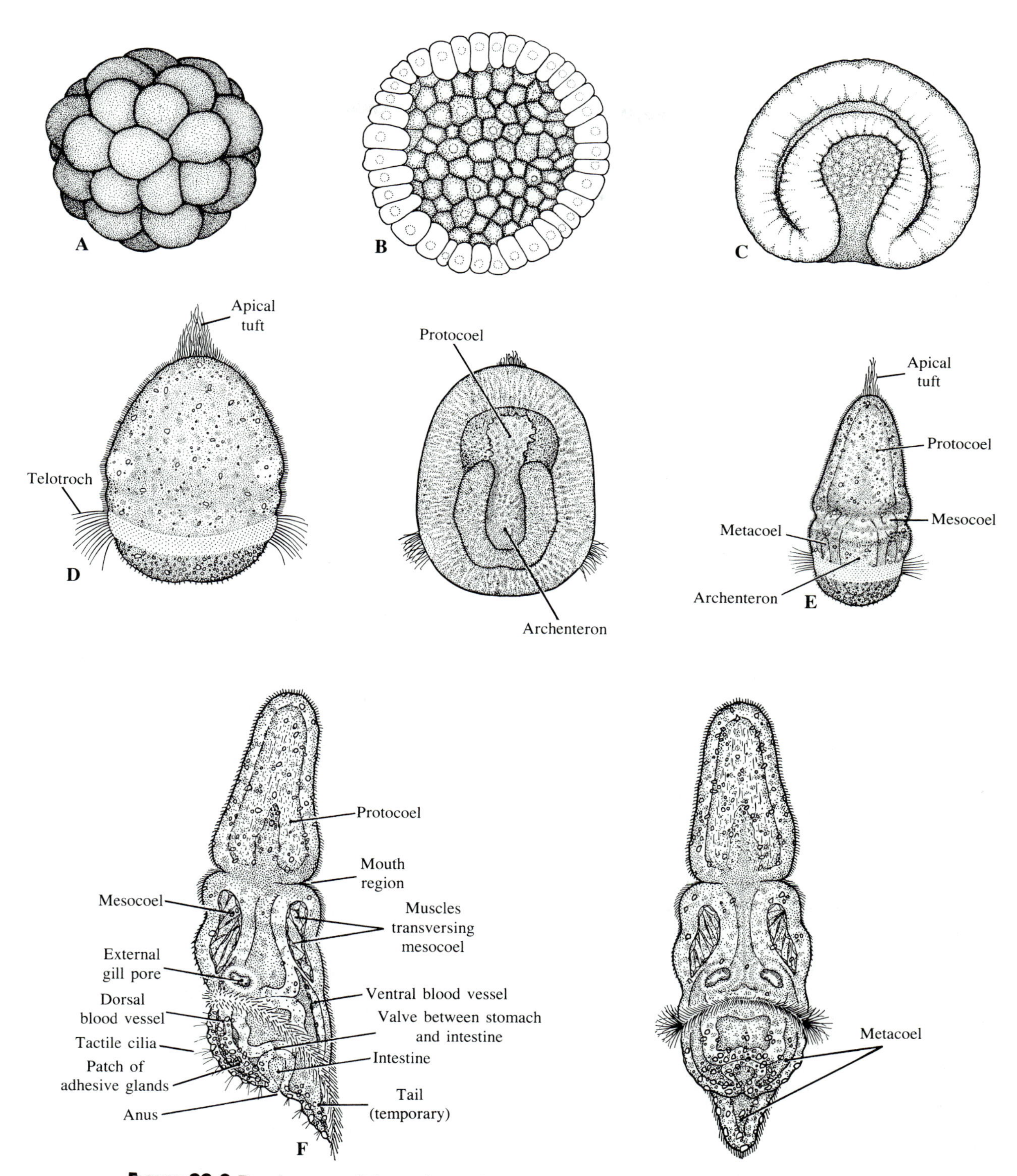

FIGURE 22.8 Development of *Saccoglossus horsti*. A. 32-cell stage. B. Blastula, in optical section. C. Gastrula, in optical section. D. Embryo at the stage in which it escapes from the egg membrane, as seen from one side (left) and in optical section (right). E. Planktonic larva. F. Creeping larva, as seen in side view (left) and dorsal view (right). (Burdon-Jones, Philosophical Transactions of the Royal Society of London, B, *236*.)

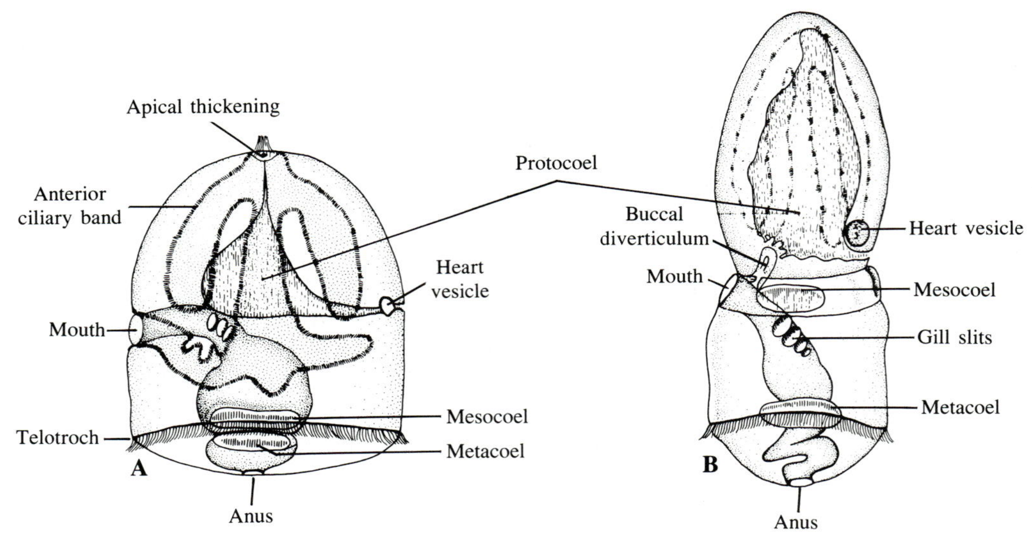

FIGURE 22.9 A. Tornaria larva. B. A young enteropneust derived from a tornaria. (Godeaux, in Florkin and Scheer, Chemical Zoology, *8*.)

eventually reach the ectoderm and break open (Figure 22.8, F). The epithelium that lines the coelomic pockets derives from mesoderm, and this germ layer also gives rise to the glomerulus, connective tissue, blood vessels, and probably the gonads. The heart vesicle may also derive from mesoderm, but an ectodermal origin has been suggested for it.

Enteropneusts with large, yolky eggs have more nearly direct development than those with small, relatively nonyolky eggs. Their young, moreover, do not begin to feed until they have settled and attained the form characteristic of the adult. In both types of development, however, there is a distinct larval stage. (True direct development takes place only in the one viviparous species.) When the larva of a yolky-egged enteropneust escapes from its egg membrane, it has an apical tuft of cilia and a posterior circle of cilia called the *telotroch* (Figure 22.8, E); it also shows the constriction that will deepen to separate the proboscis from the collar (Figure 22.8, F). The larva may swim in the plankton for a time, or it may settle on the bottom almost immediately. In either case, the apical tuft and telotroch disappear and the body assumes its elongated adult form.

In species whose development is conspicuously indirect, the embryo usually escapes from the egg membrane when it has reached the stage of a ciliated gastrula. By this time the protocoel has already been pinched off the archenteron. The mouth is formed when the archenteron bends down to touch the ventral epidermis; the anus opens up at the site of the original blastopore. The external ciliation becomes concentrated into a continuous band with a characteristic set of loops, and a telotroch develops near the posterior end. At this stage the larva is called the **tornaria** (Figure 22.9, A). It is similar in many respects to the bipinnaria larva of sea stars and the auricularia larva of sea cucumbers. In some tornariae the ciliary loops become elaborated into numerous tentaclelike outgrowths.

After the paired cavities of the mesocoel and metacoel have developed, the portion of the larva in front of the mouth takes on the form of the proboscis (Figure 22.9, B). The partly metamorphosed larva then settles, and the collar becomes set off from the trunk by a groove.

CLASS PTEROBRANCHIA

Pterobranchs are small animals, usually less than 1 cm long. Until recently, most of the specimens studied by zoologists had been collected by dredging at considerable depths, especially off Japan, off the Atlantic coast of Europe, in the Indopacific region, and in Antarctic and Subantarctic areas. These animals are now being found, however, in shallow water in Bermuda, Florida, Fiji, and elsewhere. They may prove to be common near shore in tropical areas, where their usual habitat is coral rubble.

General Structure

A pterobranch shows little resemblance to an enteropneust. The body is, nevertheless, divided into three main regions (Figure 22.10, A; Figure 22.11). The first part, perhaps homologous with the proboscis of enteropneusts, is called the **cephalic shield,** buccal shield, or oral shield. The next portion, the **collar,** has out-

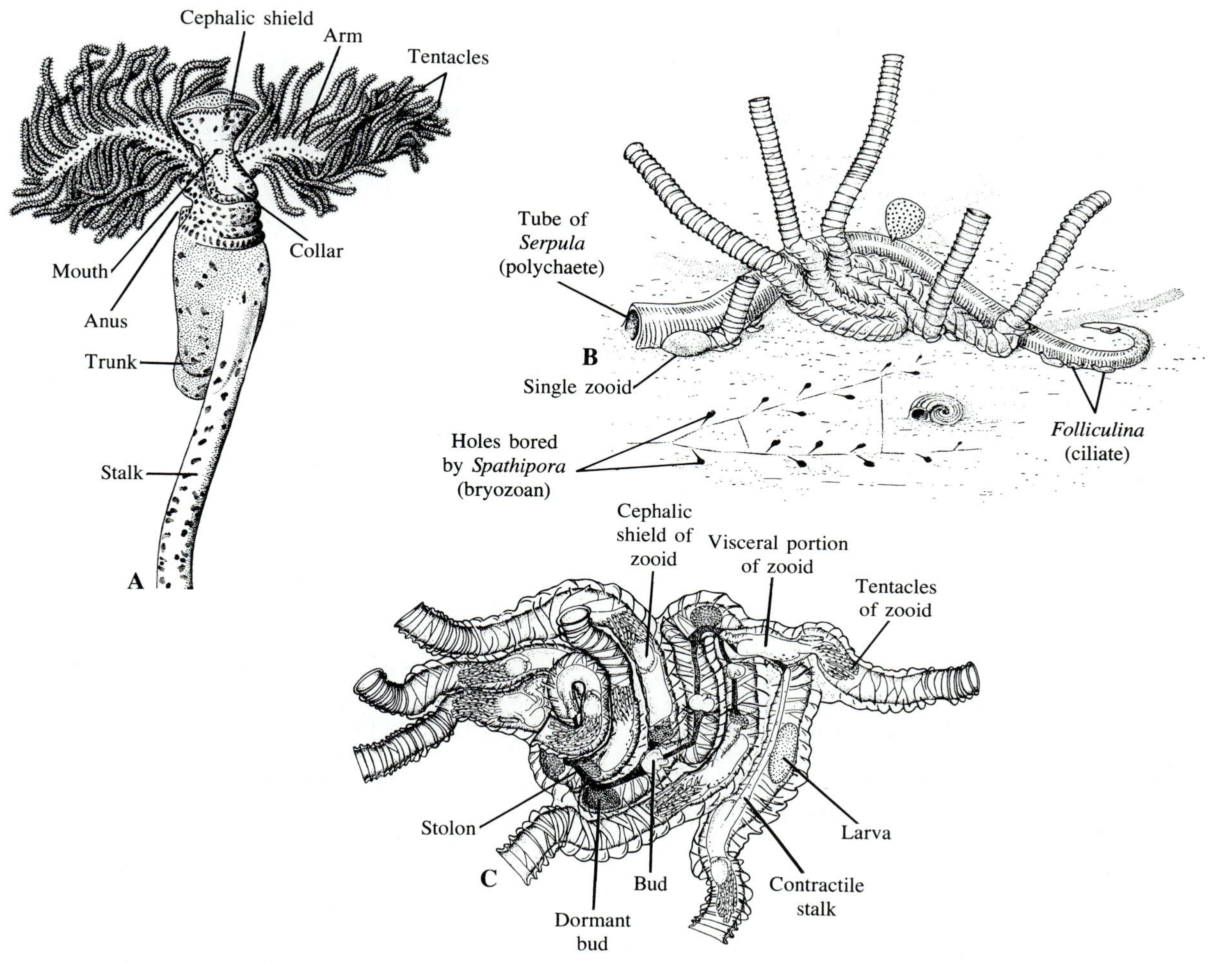

FIGURE 22.10 Class Pterobranchia. A. *Rhabdopleura normani,* a zooid removed from its tube. (Ridewood, National Antarctic Expedition 1901–1904, Natural History, *2*.) B. Tubes of a colony of *Rhabdopleura compacta*. C. Basal portions of tubes in a colony of *Rhabdopleura compacta*. The animals, which have withdrawn, can be seen within the tubes. (B and C, Stebbing, Journal of the Marine Biological Association of the United Kingdom, *50*.)

growths called **arms,** each of which is branched pinnately into a double row of **tentacles.** In the order Rhabdopleurida, represented by a single genus, *Rhabdopleura* (Figure 22.10, A), there is just one pair of arms; in the order Cephalodiscida, represented by *Cephalodiscus* (Figure 22.11), there are several pairs, and the bases of the arms form a circle. Behind the collar is the **trunk,** a part of which is thinned down to form a stalk.

The mouth is immediately behind the cephalic shield; the side on which it is located is considered to be ventral. Arising from what would seem to correspond to the buccal cavity of an enteropneust is an anteriorly directed pocket that is thought to be the homologue of the buccal diverticulum. In *Cephalodiscus* there is a pair of simple gill slits connecting the pharynx with the outside (Figure 22.12, B). In *Rhabdopleura* there are no actual openings, but there is a pair of grooves that may be vestiges of gill slits. The rest of the gut, which usually has three distinct regions—an esophagus, stomach, and intestine—is bent back on itself; the anus is located on the dorsal side of the anteriormost portion of the trunk. The nearly U-shaped form of the gut is an adaptation for life in a tube.

Feeding has been carefully observed in *Cephalodiscus gracilis*. An extended zooid uses the cephalic shield to attach itself to the mouth of the tube or to one of the spines that characterize the colony (Figure 22.11). The arms are arched in such a way as to form a basket. Lateral cilia on the tentacles drive water into the center

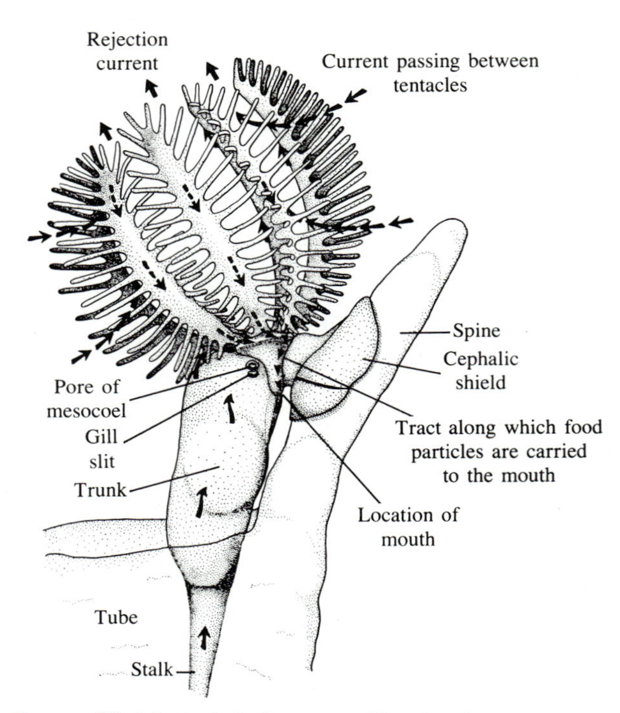

FIGURE 22.11 *Cephalodiscus gracilis,* showing currents entering the basket formed by the arms, the rejection currents, and the paths taken by food particles as they are moved down the arms (broken arrows) to the tracts that carry them to the mouth. Some of the particles moved upward by cilia on the trunk and stalk are perhaps eventually captured as food. (Lester, Marine Biology, *85*.)

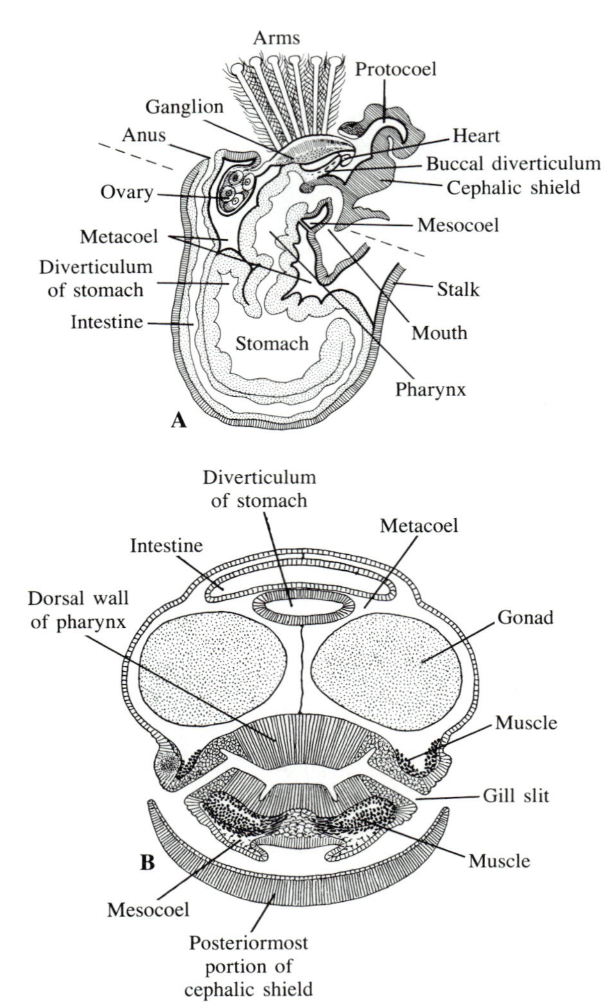

FIGURE 22.12 A. *Cephalodiscus dodecalophus,* sagittal section. (Godeaux, in Florkin and Scheer, Chemical Zoology, *8*.) B. *Cephalodiscus nigrescens,* transverse section taken at about the level indicated in A by a dashed line. The diverticulum of the stomach, however, extends farther anteriorly than it does in the specimen shown in A.

of the basket, and cilia on the inward-facing surfaces of the arms create a rejection current that drives water out of the top of the basket. Particles caught on the outward-facing surfaces of the arms, or knocked onto these surfaces by flicking action of the tentacles, are moved by cilia to the bases of the arms. Then the particles are carried to the mouth along two ciliated grooves on the collar. (These grooves border the medial edges of two structures called oral lamellae.) Collection and transport of food does not seem to involve much mucus, and is rather similar to feeding in bryozoans. Fecal pellets, and also particles rejected from the arms after being ''bounced'' upward to their tips, are carried away with the current leaving the basket.

The coelom consists of three main cavities (Figure 22.12, A). The **protocoel,** in the cephalic shield, has two openings to the outside, as in enteropneusts. The **mesocoel,** in the collar region, consists of a pair of cavities, each with an external pore. The **metacoel,** in the trunk, is also divided into right and left compartments; these, like the metacoel compartments of enteropneusts, do not have pores to the outside.

The nervous system is similar to that of enteropneusts in that it is closely linked with the epidermis from which it originates. It has some thickenings and trunks comparable to those of enteropneusts, but in general they form a less complex pattern. The circulatory system consists of unlined vessels and sinuses. One sinus, called the *heart,* is enclosed within a pericardial cavity whose lining consists of myoepithelial cells, and which is closely associated with a part of the protocoel that also has a myoepithelial lining. The origin of the pericardial cavity is closely related to that of the protocoel, and perhaps it should be viewed as part of the latter. As in enteropneusts, there is a glomerulus in the protocoel.

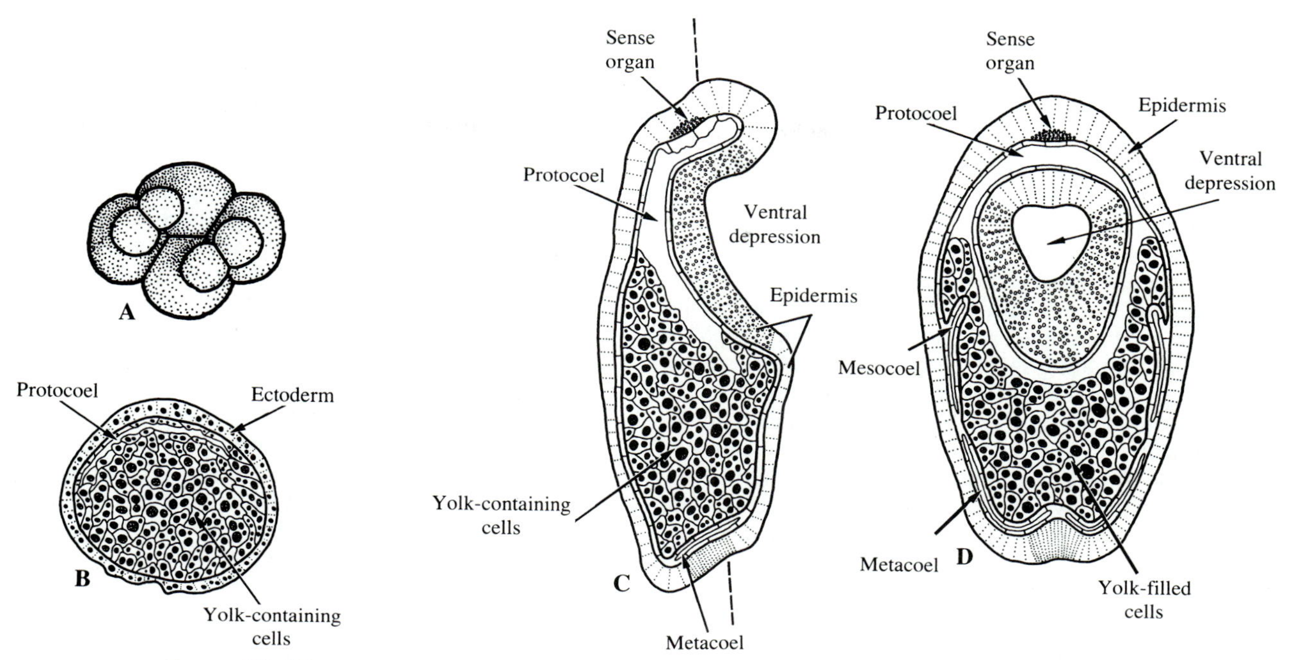

FIGURE 22.13 *Rhabdopleura normani*. A. Eight-cell stage, showing pairs of micromeres above the macromeres. B. Section of a spherical embryo in which the protocoel has appeared. C. Sagittal section of a swimming larva. D. Frontal section of a swimming larva, at about the level shown in C by a dashed line. (Lester, Acta Zoologica, *69*.)

The tubes of pterobranchs consist mostly of collagenous material secreted by epidermal glands on the cephalic shield. Some foreign material may be incorporated into the tubes. In *Rhabdopleura,* which is colonial, the members of a colony are connected by a blackish stolon that contains a core of living tissue. New members are formed by budding from the stolon. *Cephalodiscus* reproduces by budding from the sucker at the base of the stalk, but it is not truly colonial, because the young become detached from their parents. The tubes usually remain connected, however, so that a system of branching tubes is formed.

Species of *Cephalodiscus* can move about freely in their tubes, and have been observed to crawl out of them. The genus *Atubaria* ("without a tube") was proposed for some pterobranchs found clinging to hydroids in Japanese waters; they were simply specimens of a *Cephalodiscus* that had left their tubes temporarily.

Sexual Reproduction and Development

The gonads originate from the peritoneum that lines the coelom of the trunk, but they lie beneath the peritoneum. In *Cephalodiscus* (Figure 22.12, A), there are two gonads, one associated with each compartment of this coelom, and two genital pores, located on little protuberances between the anus and the groove that separates the collar region from the trunk. All individuals in an aggregation may be of one sex, or the sexes may be mixed; there may even be some hermaphrodites, in which one gonad is an ovary and the other a testis.

In *Rhabdopleura,* there is a single gonad, associated with the right compartment of the trunk coelom. Sexual members of a particular colony may be all males or all females, or they may be mixed. Furthermore, sexual members may be outnumbered by those that reproduce only asexually, and some colonies of certain species consist entirely of asexual individuals.

Reproduction and development have been most intensively studied in *Rhabdopleura normani*. The tubes of females are different from those of males in that their basal portions are coiled at least 360°. The coiled portion serves as a brood chamber where the yolky eggs develop into swimming larvae. Early cleavage is essentially radial, but the four micromeres lying above the four macromeres after the third cleavage gravitate toward the furrows, forming pairs, so the 8-cell stage (Figure 22.13, A) may appear to have been produced by spiral cleavage. After repeated cell divisions, the embryo consists of a nearly solid ball of cells that has a ciliated ectodermal layer enclosing a mass of yolk-containing cells (Figure 22.13, B). On one side there is a small cavity surrounded by a peritoneum derived from some of the yolky cells. This cavity is the future protocoel. The swimming larva (Figure 22.13, C and D) is elongated and has the rudiments of the paired mesocoels and metacoels, an epidermal sense organ at the anterior end, a ventral depression not far behind this, and an epidermal adhesive organ at the posterior end. The adhesive organ is used for temporary attachment as the larva searches for a suitable substratum on coral rubble. The larva continues to depend on yolk; there is no gut.

Within 24 hours after it has left the brood chamber, the larva settles permanently. The ventral depression is everted, and its glandular epidermis secretes a translucent covering around the entire larva. The cephalic shield, incorporating the protocoel, differentiates from the anterior region of the larva. The two arms grow out from the middorsal surface behind the cephalic shield, and the stalk grows out from the region of the adhesive organ. The arms branch into tentacles, the pharynx develops from an invagination of ectoderm, and the rest of the gut, which is late in becoming functional, develops from the yolky cells that remain. After a small opening appears in the covering, tube formation begins and the newly metamorphosed juvenile starts to feed. Other details of reproduction and development in *Rhabdopleura* may be found in papers by Lester (1988).

Summary

1. The cardinal features of the small and strictly marine phylum Hemichordata are the following: division of the body into three regions, each with its own coelom; gill slits; and an anteriorly directed diverticulum of the buccal cavity. The diverticulum has, in the past, been likened to a notochord, but it is an entirely different structure.

2. One of the two classes of living hemichordates, the Enteropneusta, consists of wormlike animals that usually live in burrows in mud or sand. The anteriormost division of the body, called the proboscis, is used in burrowing, and it also secretes much of the mucus in which an enteropneust traps food particles. The mixture of food and mucus is delivered to the mouth, located ventrally at the beginning of the collar, which is the second division of the body. The last division, called the trunk, is much longer than the proboscis and collar combined, and contains the pharynx and remaining portions of the gut. The anus is at the posterior tip of the trunk.

3. The pharynx of an enteropneust is perforated by many U-shaped gill slits. Water drawn into the mouth in the process of ciliary-mucous feeding leaves the pharynx by way of the slits, entering spaces between the pharynx and body wall. These spaces have pores to the exterior.

4. Sexes of enteropneusts are separate. The serially arranged gonads, located in the trunk region, open directly to the outside.

5. Members of the class Pterobranchia live in secreted tubes or cling to hydroids. Some of them are truly colonial, the member zooids being connected by a branching stolon. The anterior portion of the body, called the cephalic shield, is probably homologous to the proboscis of an enteropneust. The collar region has a pair of arms, or several pairs of arms that form a circle; each arm is branched pinnately into a double row of tentacles. The arms are ciliated and are used for collecting food particles, which are moved down the arms to a pair of ciliated tracts on the collar; these tracts deliver the particles to the mouth.

6. Pterobranchs have either a single pair of gill slits or lack these perforations. The gut, after passing backward into a visceral swelling of the trunk, loops forward. The anus is located just behind the collar.

7. Some pterobranchs have a pair of gonads in the trunk region; others have a single gonad.

References on Hemichordata

Barrington, E. J. W., 1940. Observations on feeding and digestion in *Glossobalanus minutus*. Quarterly Journal of Microscopical Science, *82:*227–260.

Barrington, E. J. W., 1965. The Biology of Hemichordata and Protochordata. Oliver and Boyd, Edinburgh and London; W. H. Freeman, San Francisco.

Burdon-Jones, C., 1962. Development and biology of the larva of *Saccoglossus horsti*. Philosophical Transactions of the Royal Society of London, B, *236:*553–590.

Dilly, P. N., 1972. The structure of the tentacles of *Rhabdopleura compacta* (Hemichordata), with special reference to neurociliary control. Zeitschrift für Zellforschung, *129:* 20–39.

Dilly, P. N., 1985. The habitat and behaviour of *Cephalodiscus gracilis* (Pterobranchia, Hemichordata) from Bermuda. Journal of Zoology, London, *A, 207:*223–239.

Dilly, P. N., Welsch, U., and Rehkämper, G., 1986. On the fine structure of the alimentary tract of *Cephalodiscus gracilis* (Pterobranchia, Hemichordata). Acta Zoologica, Stockholm, *67:*87–95.

Dilly, P. N., Welsch, U., and Rehkämper, G., 1986. Fine structure of heart, pericardium and glomerular vessel in *Cephalodiscus gracilis* M'Intosh, 1882 (Pterobranchia, Hemichordata). Acta Zoologica, *67:*173–179.

Duncan, P. B., 1987. Burrow structure and burrowing activity of the funnel-feeding enteropneust *Balanoglossus aurantiacus* in Bogue Sound, North Carolina. Marine Ecology, *8:*75–95.

John, C., 1932. On the development of *Cephalodiscus*. Discovery Report, *6:*191–204.

Knight-Jones, E. W., 1952. On the nervous system of *Saccoglossus cambrensis*. Philosophical Transactions of the Royal Society of London, B, *236:*315–354.

Lester, S. M., 1985. *Cephalodiscus* sp. (Hemichordata, Pterobranchia), observations of functional morphology, behavior, and occurrence in shallow water around Bermuda. Marine Biology, *85:*263–268.

Lester, S. M., 1988. Ultrastructure of adult gonads and development and structure of the larva of *Rhabdopleura normani* (Hemichordata: Pterobranchia). Acta Zoologica, *69:*95–109.

Lester, S. M., 1988. Settlement and metamorphosis of *Rhabdopleura normani* (Hemichordata: Pterobranchia). Acta Zoologica, *69:*111–120.

Pardos, F., and Benito, J., 1988. Blood vessels and related structures in the gill bars of *Glossobalanus minutus* (Enteropneusta). Acta Zoologica, *69:*87–94.

Stebbing, A. R. D., 1970. The status and ecology of *Rhabdopleura compacta* from Plymouth. Journal of the Marine Biological Association of the United Kingdom, *50:*209–221.

Stebbing, A. R. D., and Dilly, P. N., 1972. Some observations on living *Rhabdopleura compacta*. Journal of the Marine Biological Association of the United Kingdom, *52:*443–448.

Strathmann, R., and Bonar, D., 1976. Ciliary feeding of tornaria larvae of *Ptychodera flava*. Marine Biology, *34:*317–324.

Thomas, I. M., 1972. Action of the gut in *Saccoglossus otagoensis* (Hemichordata: Enteropneusta). New Zealand Journal of Marine and Freshwater Research, *6:*560–569.

23

Phylum Chaetognatha

Arrow Worms

Introduction

The chaetognaths, or arrow worms, make up a small and homogeneous phylum of marine animals. Nearly all of the approximately 70 species are planktonic, but there are a few bottom-dwelling types. Of the nonmicroscopic invertebrates found in a rich sample of plankton, chaetognaths are probably outnumbered only by copepods. They are voracious consumers of copepods and other small crustaceans, as well as of larval fishes and occasionally of one another. They are in turn eaten by fishes and some other animals.

Because they are such important constituents of the plankton, chaetognaths have been intensively investigated. Their structure, taxonomy, and distribution are moderately well understood. Much remains to be learned, however, about their physiology, reproduction, and phylogenetic relationships. The fact that chaetognaths have been linked to animals as diverse as cnidarians, nematodes, annelids, arthropods, molluscs, brachiopods, and even primitive vertebrates tells us that they form an odd and isolated group.

The more commonly encountered genera of planktonic chaetognaths are *Sagitta* and *Eukrohnia*. These are shaped something like slender torpedoes and are almost as transparent as slivers of ice. The usual length of mature animals is between 1.5 and 3 cm; a few species, however, reach a length of 10 cm. The following account of morphology is based largely on *Sagitta,* but in most respects it applies also to *Eukrohnia*.

General Structure

The body (Figure 23.1, A) has three rather distinct regions: **head, trunk,** and **tail,** each with a pair of coelomic cavities. A delicate **caudal fin** extends horizontally around the posterior portion of the tail, and there are two pairs of **lateral fins,** one pair on the trunk, the other spanning part of the trunk and part of the tail. In some species of *Sagitta,* the two lateral fins on each side are indistinctly separated, and in *Eukrohnia* there is just one long fin on each side.

Head

The head (Figure 23.1, B and C) is remarkably complicated. Part of its ventral surface is occupied by a concavity called the **vestibule,** and the slitlike mouth is located within this. The vestibule also has two pits that contain gland cells, and at its sides there are one or two rows of close-set teeth and a curved ridge or cluster of papillae. The most striking components of the head are two sets of long, chitinous **spines** that are used for grasping prey. It is upon these that the name Chaetognatha—the ''bristle-jawed animals''—is based. Muscles attached to the spines control the extent to which they are spread apart or drawn together. The spines articulate, moreover, with two hard dorsolateral supporting plates.

A fold of skin called the **hood** encircles the posterior part of the head. The base of the fold lies at a slightly

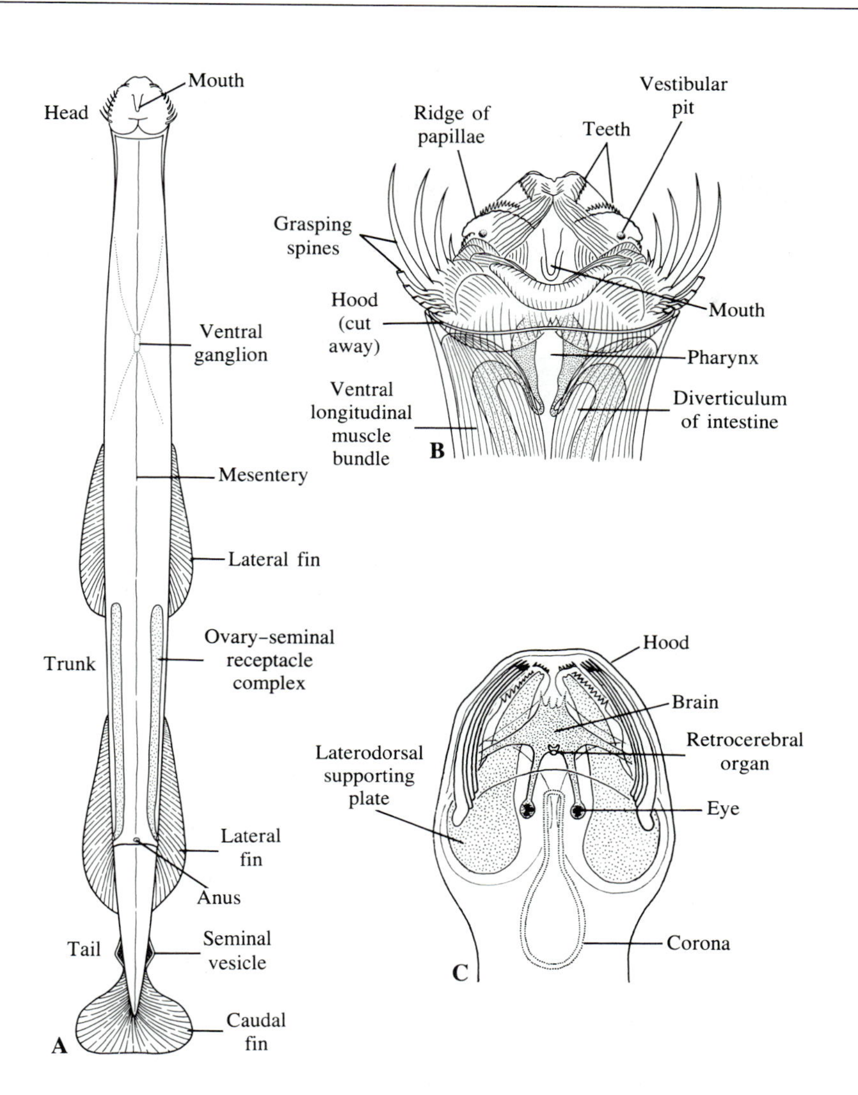

FIGURE 23.1 *Sagitta*. A. Diagram of entire animal, ventral view. B. Head and anteriormost portion of trunk, ventral view. C. Head and anteriormost portion of trunk, dorsal view. (Musculature omitted.) (B, C, After Ritter-Zahony, Denkschriften der Akademie der Wissenschaften, Wien, *84*.)

more anterior level on the dorsal side than on the ventral side. When the hood is pulled back by its retractor muscles, the dorsolateral supporting plates become more widely separated posteriorly, and the grasping spines are spread. When the retractor muscles relax and a circular ''drawstring'' muscle contracts, the hood is extended and much of the head is covered. By tucking away its spines in this manner the animal probably increases its swimming efficiency.

There are two prominent **eyes** on the dorsal side of the head. These and other sense organs of chaetognaths are dealt with below.

TRUNK

The trunk is the largest of the three divisions of the body. The intestine, which succeeds a short, muscular pharynx, sends off a pair of anteriorly directed diverticula, then runs through the trunk as a straight tube, supported above and below by a mesentery that divides the coelom into right and left halves (Figure 23.2). The bulk of the body wall of the trunk consists of **muscle bands.** Four of these—two dorsolateral, two ventrolateral—are prominent. Some additional muscles, strictly lateral in position, are less obvious and may be scarcely developed in some species or at certain transverse levels.

When a chaetognath bends sharply, then straightens out as if it were a torsion bar, the animal moves abruptly forward through the water. The bending and recovery strokes are often accompanied by a slight twisting of the trunk and tail region. The fins themselves are incapable of movement, but their broad surfaces undoubtedly help drive the animal through the

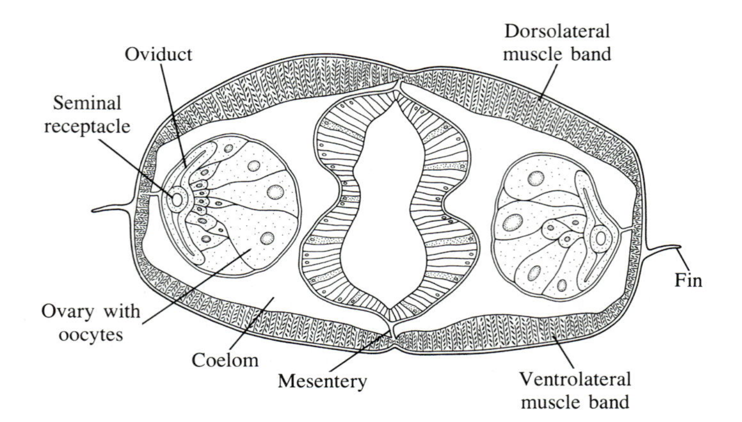

FIGURE 23.2 *Sagitta*. Transverse section through posterior portion of trunk. (After Burfield, Liverpool Marine Biology Committee Memoirs, *28*.)

water. Between leaps a chaetognath remains motionless, simply maintaining its posture with the help of its lateral and tail fins.

The glassy appearance of a chaetognath suggests that it has a secreted cuticle of the sort found in nematodes. This is not the case, except on the underside of the head. Elsewhere the epidermis, which is stratified, extends to the surface. The rodlike elements that support the fins are elongated, tonofilament-containing epidermal cells. The histology of the body wall varies from one part of the animal to another. In the trunk region, however, the more basal cells of the epidermis, packed with tonofilaments, are underlain by a thin sheath of collagen, the prominent muscle bands, and a thin but nevertheless cellular peritoneum.

The coelomic fluid is believed to serve as the medium for transport of oxygen, nutrients, and wastes. No excretory organs have been identified.

The **anus** is located ventrally at the posterior end of the trunk, and the female genital pores open laterally at about the same level (Figure 23.3, A). The prominent ovaries seen in mature animals may appear to lead di-

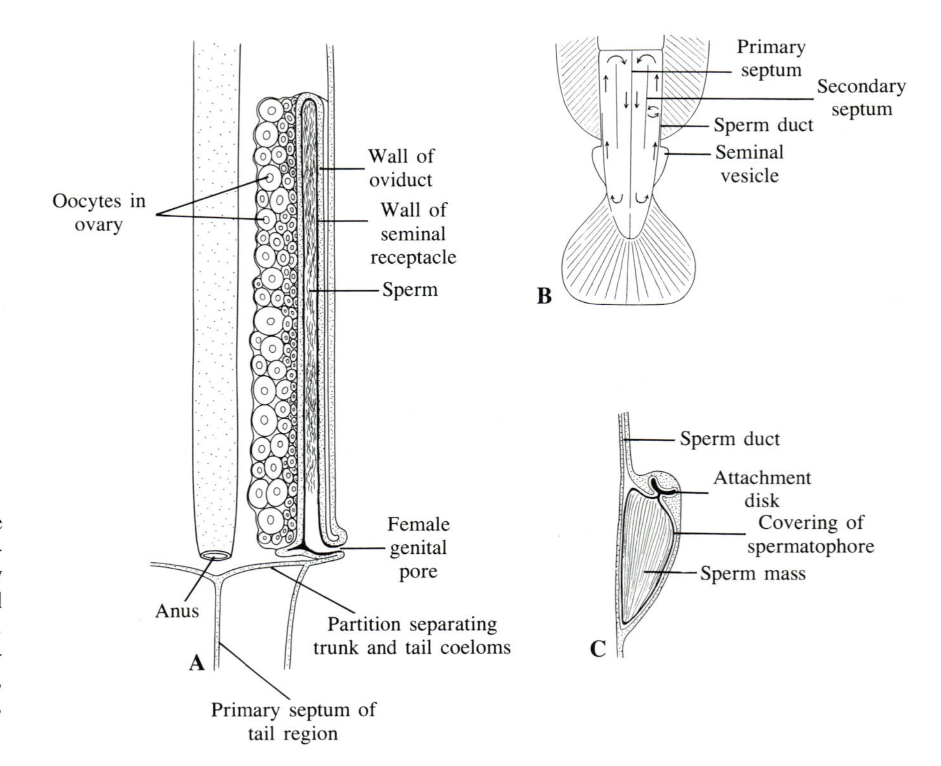

FIGURE 23.3 *Sagitta*. A. Female reproductive system. (After Burfield, Liverpool Marine Biology Committee Memoirs, *28*.) B. Tail coelom and currents within it. C. Seminal vesicle, with spermatophore enclosed. (B, C, Jägersten, Zoologiska Bidrag från Uppsala, *18*.)

rectly to the genital pores, but this is not the case. The organization of the female reproductive system is discussed further below.

TAIL

The coelom of the tail region is separated from that of the trunk by a transverse partition. It is also divided lengthwise into right and left halves by a primary septum, and each of the halves is again partially divided by a secondary septum (Figure 23.3, B).

The two **testes** are located in the anterior part of the tail coelom. They release spermatogonia, and the maturation divisions and metamorphosis of spermatids into long-tailed sperm take place as the prospective gametes circulate in the tail coelom. Movement of fluid in the coelom of the tail region—and of the trunk, too—depends on the activity of cilia, which are fairly numerous on the peritoneum. The sperm ducts, rather densely ciliated, deliver mature sperm to the two **seminal vesicles,** which are conspicuous lateral protuberances. After a mass of sperm has accumulated in a seminal vesicle, it is coated with a secretion that hardens. The package, which in some species has a sticky attachment disk at one end, is called a **spermatophore** (Figure 23.3, C). The seminal vesicle must rupture to release it.

NERVOUS SYSTEM AND SENSE ORGANS

The nervous system of chaetognaths is fairly complex. This is not surprising since these animals must make rapid and accurate responses to the stimuli perceived by their sense organs. The **brain** (Figure 23.1, C) has paired connectives running to the vestibular ganglia that are on both sides of the mouth and also to a conspicuous **ventral ganglion** located beneath the intestine in the trunk region. The vestibular ganglia give rise to a number of nerves that run alongside and beneath the pharynx. Two large nerves that run posteriorly from the ventral ganglion supply much of the trunk and tail. This ganglion also sends out several pairs of lateral nerves that innervate the lateral and dorsal portions of the trunk.

Each of the two dorsal eyes on the head (Figure 23.1, C) is divided into several pigment cups, but the pigment of these cups runs together. The sensory cells, joined to fine branches of the optic nerves that come from the brain, lie within the cups and also between the cups and the outer capsules of the eyes. The eyes are not thought to be able to form an image, although they can probably distinguish different light intensities and register the direction from which light impinges on them. Another structure found in the head region is possibly sensory, possibly neurosecretory, or both. This is the **retrocerebral organ** (Figure 23.1, C); it consists of two sacs closely associated with the brain, but it opens to the exterior by a single dorsal pore.

Other sense organs include short rows of cilia—some oriented longitudinally, some transversely—scattered all over the body surface (Figure 23.4). They are mechanoreceptors, concerned at least in part with detection of prey. There is also a ring of ciliated epithelium, called the **corona** (Figures 23.1, C; 23.7, B), on the dorsal surface of the head and anteriormost part of the trunk. The fact that this ring is supplied by rather large nerves from the brain suggests that it is a sensory structure of considerable importance, but its exact function is in doubt. It is perhaps an olfactory organ.

FEEDING

The planktonic chaetognaths whose feeding has been studied subsist mostly on copepods and larval fishes. They are believed to be able to detect a prey animal, if it is no farther than 3 mm away, by means of the ciliary receptors arranged in short rows over most parts of the body. If there is light, they can probably also locate their prey visually, seeing it as a shadow on an otherwise lighted field.

On sensing that its prey is within striking range, a chaetognath pulls back its hood and spreads its head

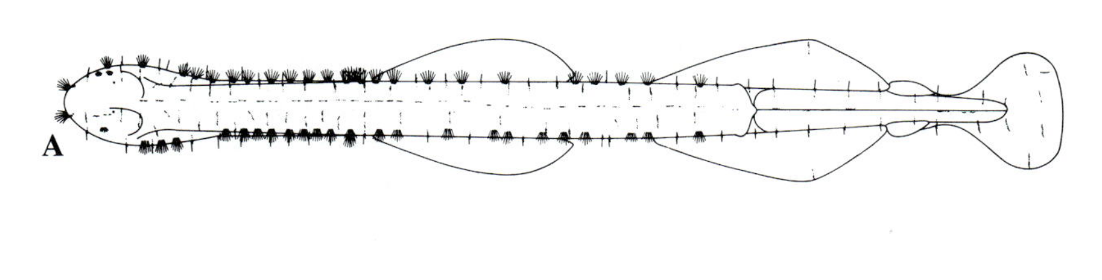

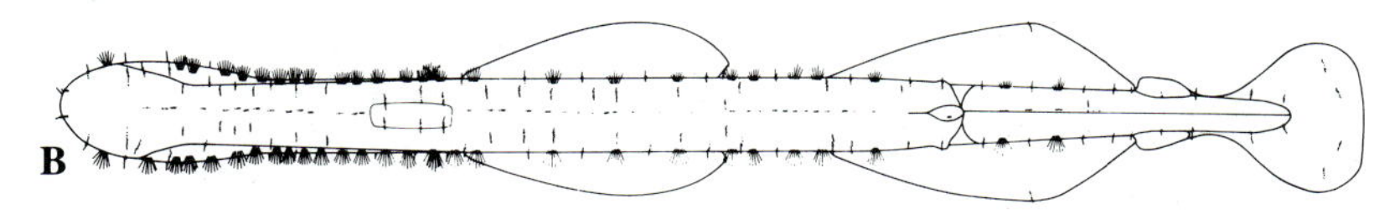

FIGURE 23.4 *Sagitta*. Distribution of ciliary mechanoreceptors. A. Dorsal view. B. Ventral view. (Feigenbaum, Canadian Journal of Zoology, *56*.)

spines, then darts forward and brings the spines together with fatal suddenness. The smaller teeth at the sides of the vestibule assist the spines in holding and crushing the prey before it is pushed into the mouth. The glandular pits in the vestibule perhaps produce a venom.

The amount of food consumed each day by one species, *Sagitta nagae,* is equal to about a third of its own body weight. Even while actively feeding, however, a chaetognath is likely to have only one or two copepods in its gut. In general, these animals seem to digest their prey rapidly and thoroughly, but some have been observed to eliminate nearly intact copepods.

Vertical Migration and Some Other Aspects of Ecology

Many planktonic chaetognaths make daily vertical migrations. *Sagitta bipunctata,* for instance, is concentrated at depths of about 20 to 40 m during much of the day and night, but around sundown and sunrise much of the population moves closer to the surface. In general, the copepods that serve as prey for *S. bipunctata* are more abundant at the surface than they are in deeper water, so it is likely that this species does most of its feeding during the early evening and early morning hours.

In parts of the Atlantic Ocean and North Sea, the abundance of one species, *Sagitta elegans,* often indicates that herring will soon appear. This chaetognath is generally found in water masses that are rich in phosphates and that have a diversified assemblage of planktonic organisms. Herring are more likely to be found with *S. elegans* than with *S. setosa,* which is characteristically associated with water masses that are poor in phosphates and that have a less varied assortment of planktonic organisms. Before 1931, *S. elegans* was abundant in the English Channel through much of the year. Then *S. setosa* became the prevailing form, and the herring fishery collapsed. Similar changes have taken place in other areas. Nevertheless, the appearance of *S. elegans* from time to time is an indication that a water mass congenial to herring may be moving in and that fishing may temporarily be profitable.

Reproduction and Development

All chaetognaths are hermaphroditic. The organization of both male and female reproductive systems is peculiar and it is no wonder that experts have disagreed on how transfer of spermatophores, insemination, fertilization, and release of fertilized eggs take place. Moreover, there seems to be considerable reproductive variation among chaetognaths.

The following account of reproduction applies primarily to *Sagitta hispida,* which has been intensively studied. Two individuals engage one another with their head spines. As they twist around, at least one of them can stick a spermatophore onto the body of the other. The spermatophore need not be deposited close to a female genital pore, because the sperm that emerge from the spermatophore can migrate for some distance over the surface of the body. In *S. hispida,* in fact, once the stream of sperm comes close to a genital pore, it divides, and about half the sperm cross over to the pore on the opposite side. Substantially the same pattern of spermatophore transfer and insemination has been observed in a benthic chaetognath, *Spadella cephaloptera* (Figure 23.5).

The two **ovaries** of mature chaetognaths are long, sausage-shaped structures. Running along the outer edge of each ovary is an **oviduct** that originates at the genital pore. Sperm swim up this and accumulate all along its length. The oviduct thus functions as a **seminal receptacle.** The sperm reach the eggs by working their way through specialized cells of the oviduct, which are called accessory fertilization cells. These form stalks to which developing oocytes within the ovary are attached. After the eggs have been fertilized, they somehow get into the oviduct and thus reach the genital pore. In *S. hispida,* a double row of eggs emerges from each pore.

Some zoologists who have studied the reproductive system of chaetognaths have described an arrangement in which the lumen of a tube corresponding to the oviduct is partly or wholly occupied by a structure that receives the sperm deposited by the partner (Figure 23.2; Figure 23.3, A). This structure, whose wall is syncytial, is considered to be the seminal receptacle. Because it has an opening to the outside, and the oviduct does not, it would seem that eggs must reach the lumen of the seminal receptacle before they can be laid, unless a new pore appears so that eggs may leave by way of the oviduct. There may indeed be much variation in the organization of the female reproductive system, and it seems best not to assume that all chaetognaths are exactly like *S. hispida.*

Although *Sagitta hispida* is planktonic, it is generally found close to shore. Its habit of attaching strings of eggs to vegetation would seem to be a good adaptation in that it minimizes the extent to which the developing young drift away from the coastal habitat. Most chaetognaths, however, simply drop their eggs, singly or in clusters. In certain species, a cluster of eggs remains attached to the parent until embryonic development has reached an advanced stage. There is obviously much variation in what happens.

Some species are said to be capable of self-fertilization. A spermatophore that has emerged from a seminal

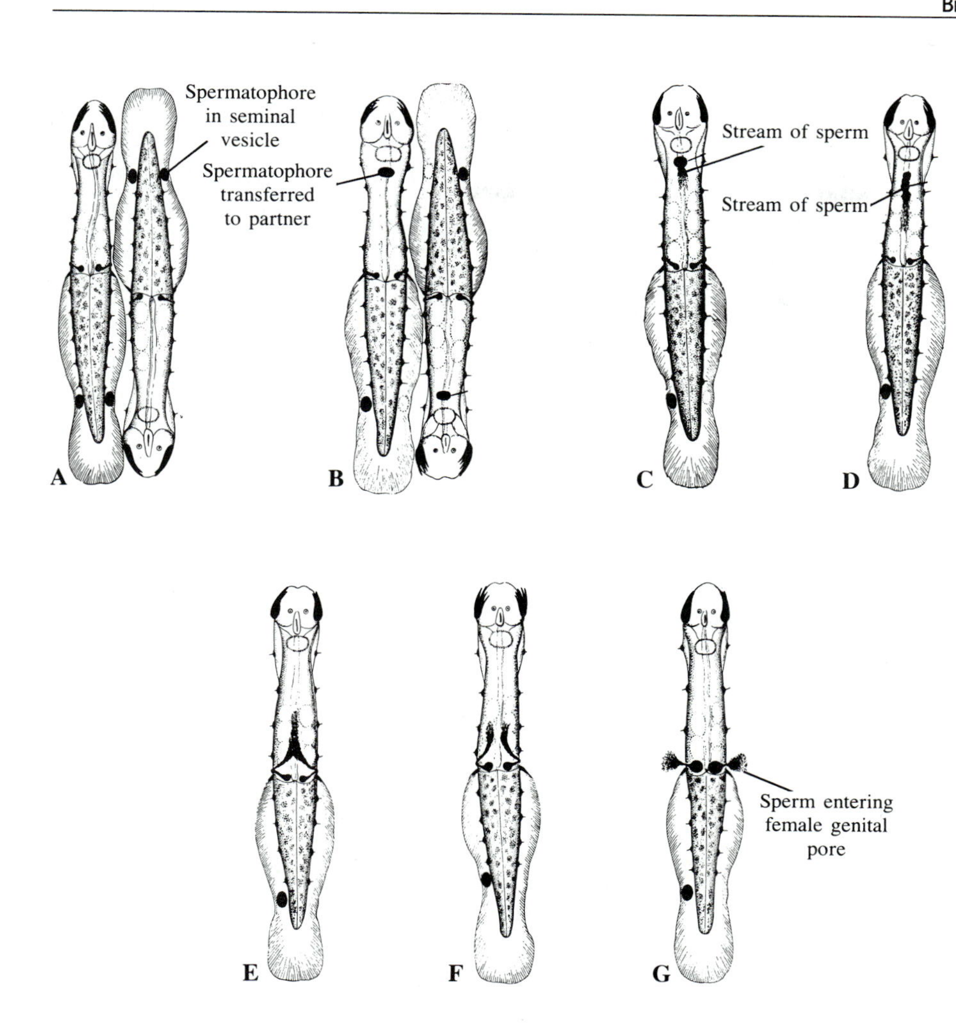

FIGURE 23.5 Spermatophore transfer and insemination in *Spadella cephaloptera*. A, B. Pairing of individuals and transfer of a spermatophore from each one to the other. C, D. Stream of sperm emerging from a spermatophore. E, F, G. Stream of sperm dividing, some sperm going to each female genital pore. (Ghirardelli, Pubblicazioni della Stazione Zoologica di Napoli, *24*.)

receptacle remains attached to the animal that produced it. After the sperm have been liberated, they locate one or both of the female genital pores.

Development (Figure 23.6) is **direct. Radial cleavage** leads to formation of a hollow blastula (A). When this invaginates to form a gastrula (B), the archenteron completely fills up what had been the blastocoel. The blastopore closes, but a stomodeal invagination appears at the opposite end of the embryo and joins the archenteron (E, F). Meanwhile, pockets of mesoderm develop on the right and left sides of the archenteron (D, E). The anteriormost part of each pocket is cut off to make the head coelom (F) and to provide mesoderm for the musculature and other internal structures of the head. The remainder of each pocket, after separating from the archenteron, forms the mesoderm and coelom of the trunk and tail. By the time the embryo has become wormlike and coiled up inside the egg membrane (G), the coelomic cavities and gut lumen have been obliterated, but in time these spaces reappear. An anus opens up after the posterior tip of the gut touches the epidermis on the ventral surface. The primordia of the gonads originate as a pair of cells budded from the archenteron at the point where the pockets of mesoderm develop (C, D). They eventually become incorporated into the trunk–tail coelom, but before this happens each primordial cell divides (E, F), one of its progeny being destined to form an ovary, the other a testis.

The foregoing account of embryology is simplified, but it should get across the idea that chaetognaths are basically deuterostomes and that their coelomic sacs and mesoderm arise from outpocketings of the archenteron.

BENTHIC CHAETOGNATHS

Although most chaetognaths are planktonic, some (such as *Spadella,* Figures 23.5 and 23.7) are bottom-dwellers. For attaching themselves to seaweeds and rocks, they have several sticky outgrowths on both

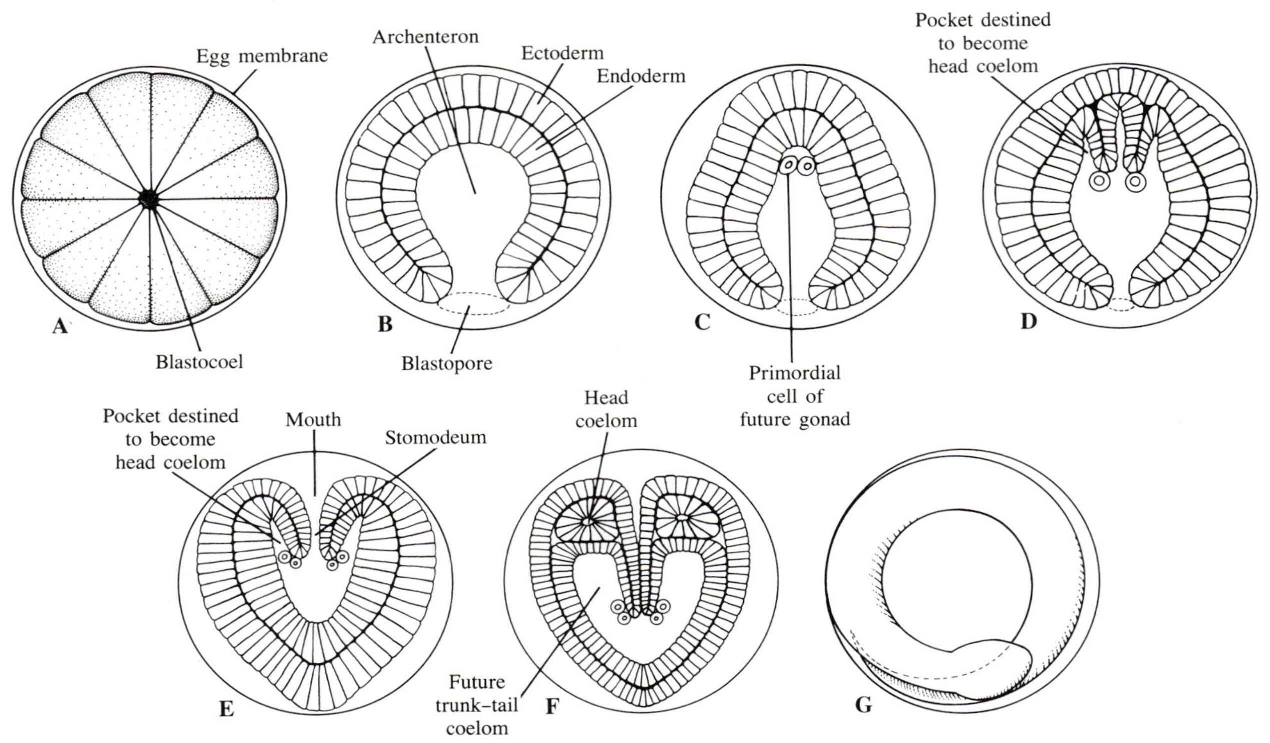

Figure 23.6 Development of *Sagitta*. A. Blastula. B. Early gastrula. C. Later gastrula, with primordial cells of future gonads. D. Formation of pockets destined to become head coelom. E. Formation of stomodeum, closure of blastopore. F. Separation of head coeloms completed. G. Young chaetognath shortly before hatching. (After Burfield, Liverpool Marine Biology Committee Memoirs, *28*.)

sides of the tail. The posture of the animals is such that the head is poised for capture and ingestion of prey that pass by. Benthic chaetognaths deposit clusters of eggs on seaweeds and other firm objects. Mating and spermatophore transfer (Figure 23.5) are achieved in much the same way as in *Sagitta*.

Summary

1. Chaetognaths form a small phylum of marine animals, most of which are planktonic. Their bodies, generally slender and nearly transparent, can arbitrarily be divided into head, trunk, and tail. Each of these regions has paired coelomic cavities. There are horizontal fins at the sides of the trunk and tail and an unpaired fin at the posterior end of the tail. The principal musculature of the body wall consists of prominent longitudinal bands. Chaetognaths make their darting movements by bending the body sharply, then letting it spring back to its straight shape.

2. Strong spines on the head are used for capture of copepods, fish larvae, and other prey. The gut is complete, the anus being located at the hind end of the trunk region.

3. Chaetognaths are hermaphroditic. The female reproductive system is in the trunk; the male system is in the tail region. Reproduction usually involves transfer of a spermatophore from one individual to another. Development is direct. Cleavage is radial, and the coelomic sacs arise as outpocketings of the archenteron.

References on Chaetognatha

Alvariño, A., 1965. Chaetognatha. Oceanography and Marine Biology, An Annual Review, *3*:115–194.

Alvariño, A., 1983. Chaetognatha. *In* Adiyodi, K. G., and Adiyodi, R. B., Reproductive Biology of Invertebrates, *1* (Oogenesis, Oviposition, and Oosorption), pages 585–610. John Wiley & Sons, New York.

Bone, Q., Brownlee, C., Bryan, G. W., Burt, G. R., Dando, P. R., Liddicoat, M. I., Pulsford, A. L., and Ryan, K.

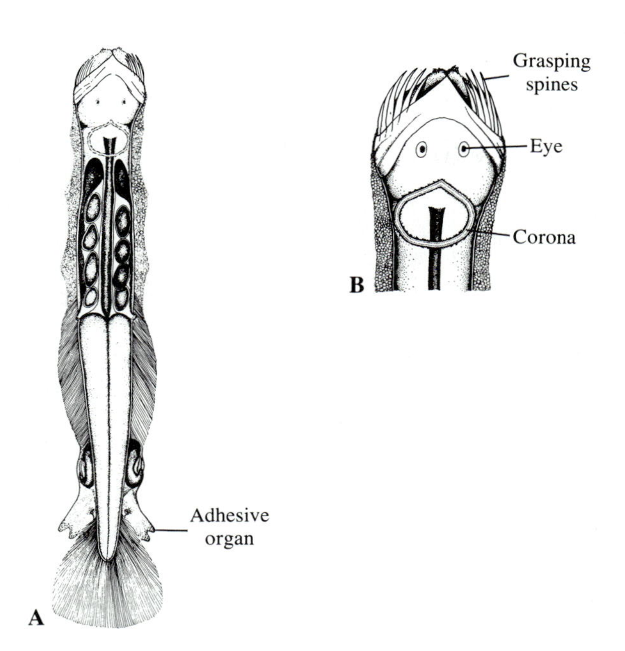

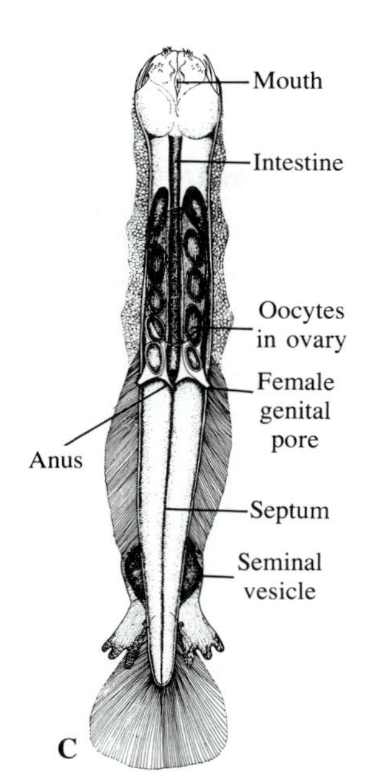

FIGURE 23.7 *Spadella hummelincki*. A. Dorsal view of entire animal. B. Head and anteriormost part of trunk, dorsal view. C. Ventral view of entire animal. (Alvariño, in Studies on the Fauna of Curaçao and other Caribbean Islands, *34*.)

P., 1987. On the differences between the two 'indicator' species of chaetognath, *Sagitta setosa* and *S. elegans*. Journal of the Marine Biological Association of the United Kingdom, *67*:545–560.

Bone, Q., and Pulsford, A., 1978. The arrangement of ciliated sensory cells in *Spadella* (Chaetognatha). Journal of the Marine Biological Association of the United Kingdom, *58*:565–570.

Duvert, M., Gros, D., and Salat, C., 1980. Ultrastructural studies on the junctional complexes in the musculature of the arrow-worm (*Sagitta setosa*; Chaetognatha). Tissue and Cell, *12*:1–12.

Duvert, M., and Salat, C., 1979. Fine structure of muscle and other components of the trunk of *Sagitta setosa* (Chaetognatha). Tissue and Cell, *11*:217–223.

Feigenbaum, D. L., 1978. Hair-fan patterns in the Chaetognatha. Canadian Journal of Zoology, *56*:536–544.

Feigenbaum, D. L., and Reeve, M. R., 1977. Prey detection in the Chaetognatha: response to a vibrating probe and experimental determination of attack distance in large aquaria. Limnology and Oceanography, *22*:1052–1058.

Ghirardelli, E., 1968. Some aspects of the biology of the chaetognaths. Advances in Marine Biology, *6*:271–375.

Goto, T., and Yoshida, M., 1981. Oriented light reactions of the arrow worm *Sagitta crassa* Tokioka. Biological Bulletin, *160*:419–430.

Nagasawa, S., and Marumo, R., 1972. Feeding of a pelagic chaetognath, *Sagitta nagae* Alvariño in Suruga Bay, central Japan. Journal of the Oceanographic Society of Japan, *28*:181–186.

Reeve, M. R., and Lester, B., 1974. The process of egg-laying in the chaetognath *Sagitta hispida*. Biological Bulletin, *147*:247–256.

Reeve, M. R., and Walter, M. A., 1972. Observations and experiments on methods of fertilization in the chaetognath *Sagitta hispida*. Biological Bulletin, *143*:207–214.

Welsch, U., and Storch, V., 1982. Fine structure of the coelomic epithelium of *Sagitta elegans*. Zoomorphology, *100*:211–222.

24

Phylum Urochordata

Die Löcher sind die Hauptsache an einem Sieb.
(The holes are the most important part of a sieve.)

Joachim Ringelnatz

- Phylum Urochordata
  - Class Ascidiacea
    - Order Aplousobranchia
    - Order Phlebobranchia
    - Order Stolidobranchia
  - Class Thaliacea
    - Order Pyrosomida
    - Order Doliolida
    - Order Salpida
  - Class Larvacea

INTRODUCTION

The urochordates, all of which are marine, are often placed, along with cephalochordates and vertebrates, into one phylum, called Chordata. The features that would seem to unify the three groups are the following: a **tubular dorsal nerve cord,** which is well separated from the epidermis; **gill slits** or **gill pouches** on both sides of the pharynx; and a firm, rodlike structure called the **notochord,** which lies beneath the nerve cord. All are characteristic of cephalochordates, and also of fishes. In other vertebrates, however, neither the notochord nor embryonic indications of gill slits persist beyond early stages of development.

Urochordates adhere to the chordate pattern in that they generally have gill slits. Moreover, when there is a nerve cord, this is dorsal. The notochord is a less consistent feature. Some urochordates never have one. The majority have it during a brief larval stage, and a few retain the structure throughout life. The fact that the notochord, when present, is restricted to the tail region is the basis for the name Urochordata.

A feature that is often overlooked in discussions of apparent affinities of the groups under consideration is the **endostyle.** This is a longitudinal groove in the floor of the pharynx. In both urochordates and cephalochordates, it secretes mucus that is used to trap microscopic food particles. Among the vertebrates, larval lampreys have a somewhat similar structure. Related to the matter of possible homologies of endostyles is the fact that the thyroid gland of vertebrates originates as a downgrowth from the floor of the pharynx. The capacity of the thyroid gland to bind iodine and to incorporate it into thyroxin is well known. The endostyle of cephalochordates also binds iodine and produces an iodine-containing secretion that is later broken down in the intestine. There is evidence for iodine-binding and synthesis of thyroxin in the endostyle of some urochordates, too. Other tissues of urochordates take up iodine, however, and iodine-binding has been reported in several phyla of invertebrates that are not part of the chordate complex. It seems best, therefore, not to lay too much stress on either the endostyle or on iodine-binding as signs of phylogenetic unity.

Unlike cephalochordates and vertebrates, urochordates do not have a coelom. The spaces in the tissues are filled with blood and constitute a hemocoel. The only cavities that could be confused with a coelom are those called **epicardia,** which originate as downgrowths from the pharynx. They are characteristic of some members of the class Ascidiacea and are discussed further below.

In this book, urochordates and cephalochordates are treated as phyla and dealt with in separate chapters. It is unlikely that either group is in the direct line of evolution of the vertebrates, although cephalochordates are probably close to the ancestral stock of vertebrates.

Urochordates are generally characterized by the presence of a secreted outer covering, or **tunic.** The name Tunicata, synonymous with Urochordata, alludes to this feature. The tunic consists to a large extent of a type of cellulose, called **tunicin.** Although cellulose is present in the cell walls of most groups of plants and in many photosynthetic unicellular organisms, it is uncommon in animals. It does occur in vertebrate connective tissue, and the quantity of it may increase with age; this happens, for instance, in humans. The tunic of urochordates contains, in addition to cellulose, a variety of substances, including mucopolysaccharides and proteins. It must not be thought of as an inert covering, for although it is external to the epidermis, it contains blood spaces and some cells, especially wandering amoebocytes.

In one class of Urochordata, the Larvacea, the tunic is lacking. These animals are interesting in that they either enclose themselves in a mucopolysaccharide "house" or secrete a bladderlike structure to which they remain attached.

CLASS ASCIDIACEA

Most of the approximately 1500 species of urochordates are ascidians. As adults, they are usually attached to one another or to rocks, shells, wood, concrete, or seaweeds. The life history generally includes a larval stage that resembles a tadpole. After a brief free-swimming existence, the larva settles and undergoes a radical metamorphosis.

The Larva

The easiest way to explain the functional morphology of an ascidian is to begin with the organization of the larva and follow its transformation into an adult. The principal features of a generalized larva are shown in Figure 24.1, A. Note that it consists of two main portions, a **trunk** and a **tail.** The notochord is limited to the tail, and for a time it consists of contiguous cells. Fluid-filled spaces develop between the cells and sometimes remain in this position. In most ascidian larvae, however, the spaces unite as the cells become wrapped around them. Thus the notochord contains a continuous central canal.

There is a substantial ganglion, called the **visceral ganglion,** located in the dorsal part of the trunk. Closely associated with it is a **sensory vesicle** (or cerebral vesicle) that contains a photoreceptor, part of which is recognizable as a darkly pigmented eyespot, and a large cell within which there is a vacuole enclosing a large melanin granule. This cell is believed to function in much the same way as a statocyst. Because it is a single cell, it is referred to here as a **statocyte.** The entire sensory vesicle develops from the right side

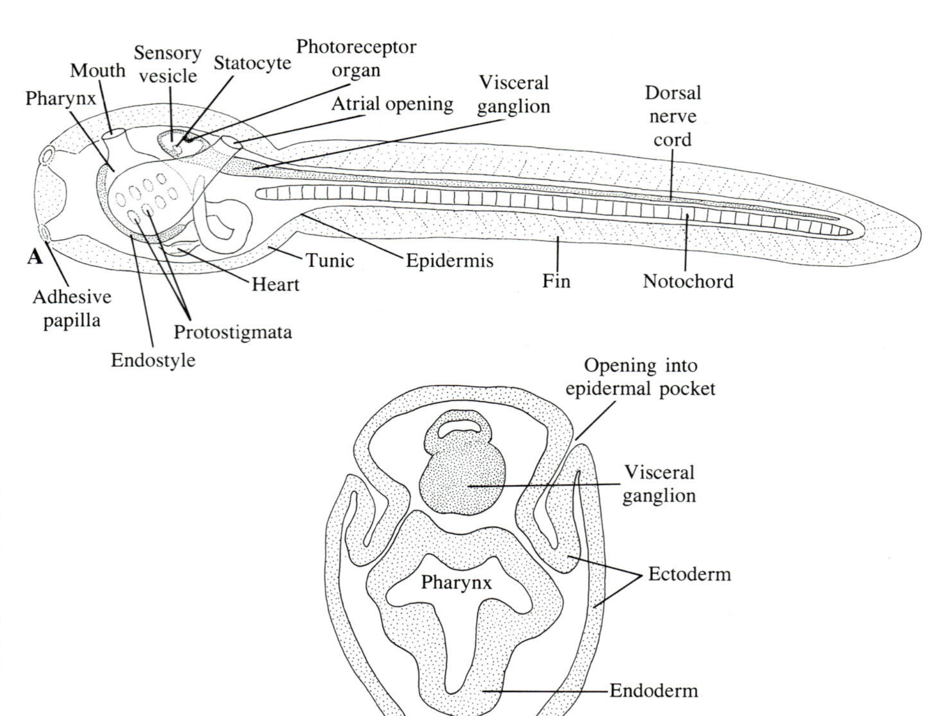

Figure 24.1 A. Larva of the type typical of many ascidians; diagrammatic. The mouth is blocked by tunic material, as are the openings of the two atrial cavities. B. Transverse section through a larva (earlier stage than that shown in A) at the level of the epidermal pockets (future atrial cavities), protostigmata have not yet developed. (After Brien, in Grassé, Traité de Zoologie.)

of a dilation that appears, during early development, in the anterior part of the neural tube.

The mouth, situated anterodorsally on the trunk, is blocked by tunic material. It leads into a gut that ends blindly, but whose pharyngeal portion has a few openings (sometimes only one or two) on each side. These openings, called **protostigmata,** are precursors of the many definitive stigmata, or gill slits, of the adult. The protostigmata connect the pharynx of the larva with a pair of epidermal pockets that have grown down from the dorsal surface on the right and left sides (Figure 24.1, B), or sometimes with a single, bilobed pocket of similar origin. For the time being, each of these pockets retains its own opening to the outside.

Other noteworthy structures of the larva are a delicate **heart** that lies beneath the pharynx, and a ciliated endostyle running the length of the floor of the pharynx. The pharynx will soon play an important role in feeding, but this does not begin until metamorphosis has been completed. Externally, at the anterior end of the trunk, there are prominent **adhesive papillae**—often three of these—that will function in fixing the larva to a firm surface. There is already considerable tunic material external to the epidermis, some of it forming the fins of the tail region, but more will soon be secreted, especially by glands that are usually concentrated in fingerlike epidermal outgrowths that encircle the trunk. These outgrowths are called **ampullae.**

Not all ascidian larvae conform to the pattern just described. In *Metandrocarpa* (Figure 24.2), for instance, there is no mouth, endostyle, or heart. In *Distaplia* (Figure 24.3, A), on the other hand, the larva, when released by the parent, is at an advanced stage of development. It even contains buds that will differentiate into new individuals soon after the larva settles.

Metamorphosis of the Larva into an Adult

The larva, its mouth obstructed by tunic material, is unable to feed. Its food reserves will not last long, perhaps only a few hours, so it must soon find a place to settle down. Its swimming movements and its attempts to drive itself against a suitable substratum depend on its tail musculature and its stiff but elastic notochord. The latter serves as a kind of torsion bar. When the tail is bent by contraction of muscle bands on one side or the other of the notochord, the notochord is put under tension. As soon as the muscles relax, the notochord straightens. These movements, performed repeatedly and rapidly, make the tail operate in a fashion somewhat similar to that of a sculling oar. If the larva succeeds in driving itself against a firm surface, it becomes

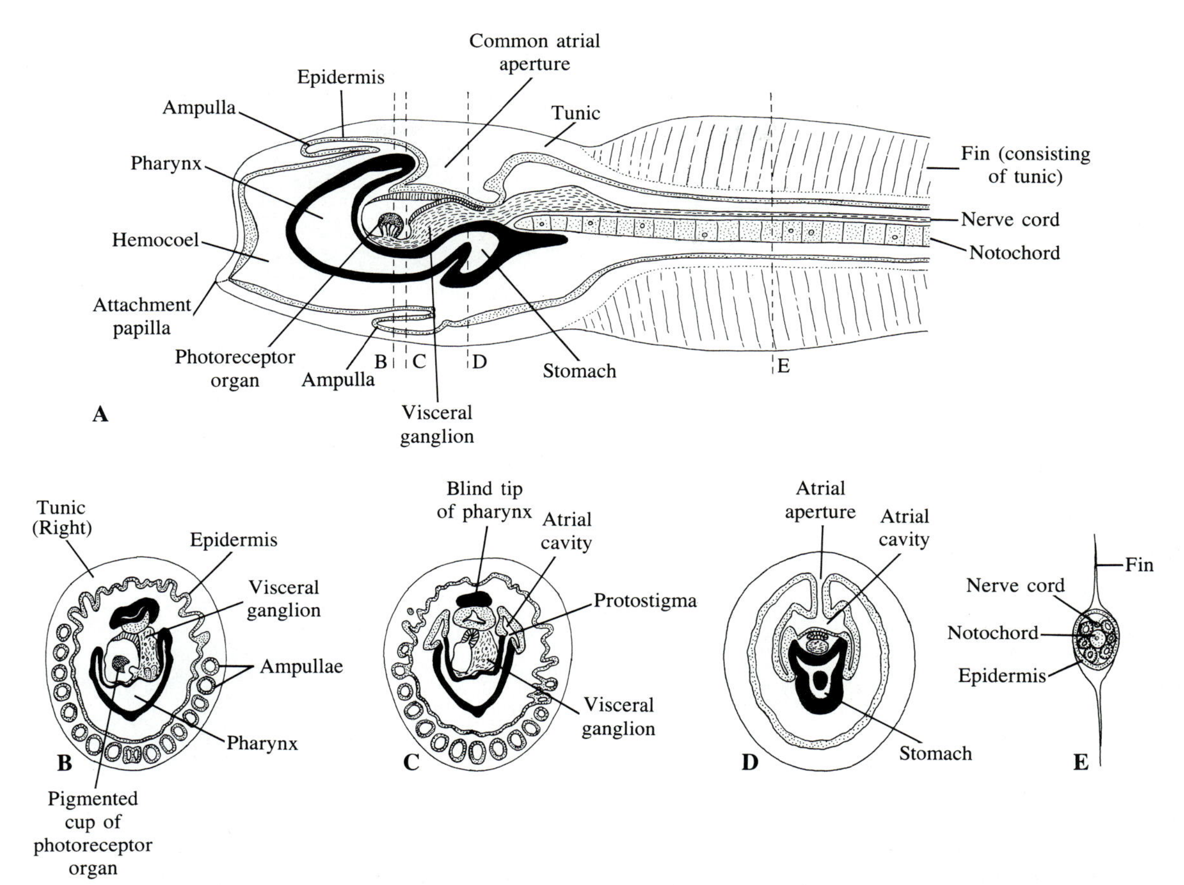

FIGURE 24.2 *Metandrocarpa taylori,* larva. This species does not release its larvae until shortly before they are ready to settle. (The free-swimming period is sometimes less than 2 hours.) Neither the mouth nor the anus opens until after metamorphosis has begun. There are only two protostigmata, and the right and left atrial cavities share a common atrial aperture. The photosensitive organ develops in the place where most ascidian larvae have a statolith, and there is no eyespot in the usual location. The notochord is not tubular as it is in the majority of species; the fluid material associated with the cells remains mostly between successive cells. There is no heart at this stage. A. Anterior half of a tadpole, median sagittal section. Two of the three attachment papillae, being lateral to the midline, are not shown. The ampullae encircle the trunk, the dorsal ones being located farther anteriorly than the ventral ones. B–E. Transverse sections taken at the level indicated on A by dashed lines and corresponding letters. (After Abbott, Journal of Morphology, *97.*)

attached by its adhesive papillae and begins its metamorphosis.

As soon as the larva is attached, the epidermal cells of the tail—or cells of the notochord itself, in some species—contract. The contractile property of the cells depends on the action of cytoplasmic microfilaments. As the tail becomes shorter, its notochord, muscles, and other internal tissues are forced into the trunk region, where they are resorbed. Initiation of tail resorption and succeeding phenomena of metamorphosis may be induced in larvae of some species by gently pinching the tail. This somehow stimulates the epidermal cells to contract just as if normal settlement had taken place.

The position of the mouth opening is gradually shifted away from the original anterior end of the larva, now firmly cemented down by its papillae as well as by tunic material. In time, the mouth becomes located on the opposite side, and is usually raised up on an eminence called the **oral siphon** (Figure 24.4, C–F). The epidermal pockets (or pocket) that border the pharynx become the **atrial cavity** of the adult. When there are two pockets, they unite and their originally separate

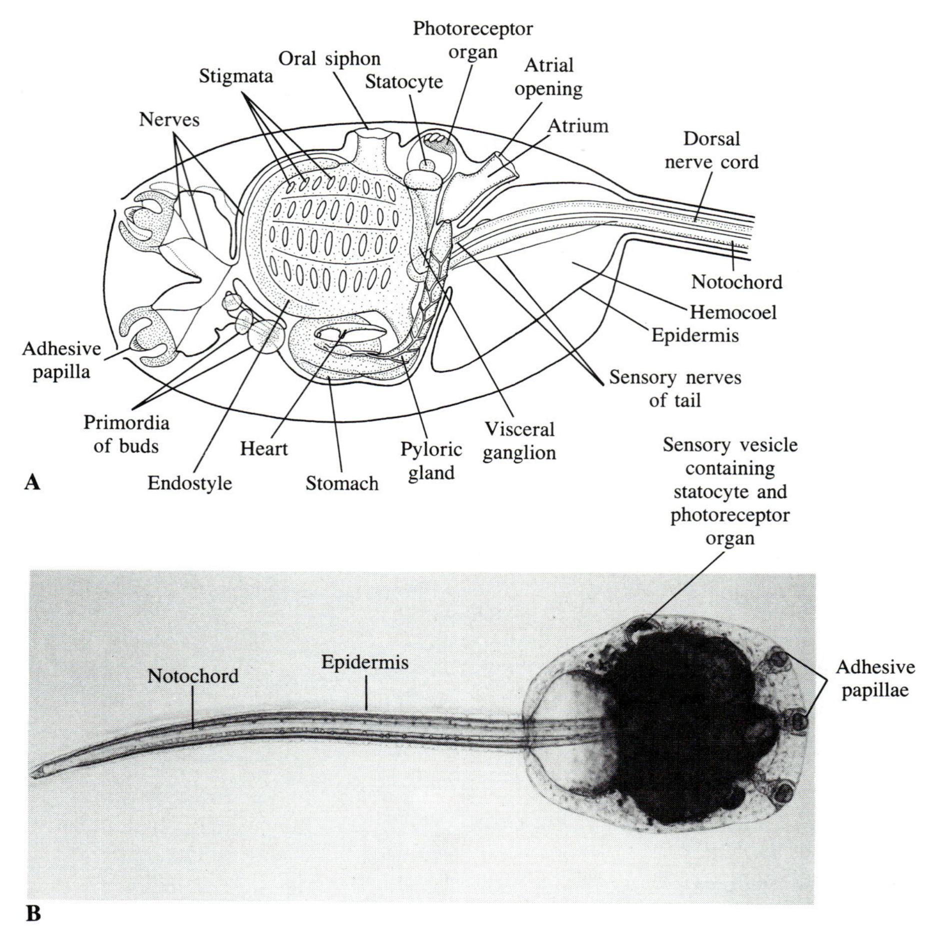

FIGURE 24.3 A. *Distaplia occidentalis,* trunk and small portion of the tail of a larva, viewed from the left side. The ampullae and left side of the atrial cavity are omitted. The perforations of the pharynx are definitive stigmata, and the primordia of asexual buds that will develop into separate zooids of a colony are already evident. In some ways, this is more nearly a juvenile than a larva. The tail has rotated 90°, so the dorsal nerve cord is on the left side of the notochord, rather than above, and the fins are horizontal, rather than vertical. (After a drawing lent by Richard A. Cloney.) B. *Diplosoma macdonaldi,* a larva seen from the right side. As in *Distaplia,* the tail has rotated 90°. All three of the adhesive papillae are in the midsagittal plane. (Photomicrograph by Richard A. Cloney.)

openings come together, so there is now a single opening located on an **atrial siphon.**

The nervous system is reorganized drastically. The photoreceptor and statocyte disappear, as does the visceral ganglion. (The latter, in some ascidians, is consumed by phagocytic cells.) The sensory vesicle is converted into a structure called the **neural gland,** which opens into the upper part of the pharynx (Figure 24.9, B). The functions of this structure are still not understood. The brain of the adult ascidian derives from the part of the nerve cord just behind the visceral ganglion. The tail portion of the nerve cord degenerates.

The few protostigmata that originally connected the pharynx of the larva with the atrial pockets become subdivided (Figure 24.4, D–G). They lengthen, often bending at the same time, and become interrupted by bridges of tissue. Eventually the pharynx is encircled by several to many sets of definitive **stigmata.** The shape of the openings varies from that of a simple pore to that of a complex spiral (Figure 24.5).

Some ascidians—those belonging to the orders Phlebobranchia and Stolidobranchia—have an interesting elaboration of the pharynx. This consists of bars of tissue, each containing a blood sinus, running lengthwise on the inner surface (Figure 24.5; Figure 24.6, A). A bar is formed by the coalescence of the tips of small papillae that originate at the places where circular and longitudinal vessels in the wall of the pharynx intersect. Each papilla contains an outgrowth of a vessel, so when the papillae unite, a continuous sinus is formed in the bar that results. A bar is free of the pharyngeal lining (Figure 24.8, B) except where it is anchored by the bases of the papillae that give rise to it. Such longitudinal bars, as well as lengthwise folds characteristic of the pharynx of stolidobranch ascidians, add considerable respiratory surface to the pharynx.

Digestive System and Feeding

Inside the oral siphon, at the mouth opening itself, is a ring of tentacles (Figure 24.6, A; Figure 24.7; Figure 24.8, A). These, like the inside of the siphon, are covered by epidermis. The function of the tentacles is

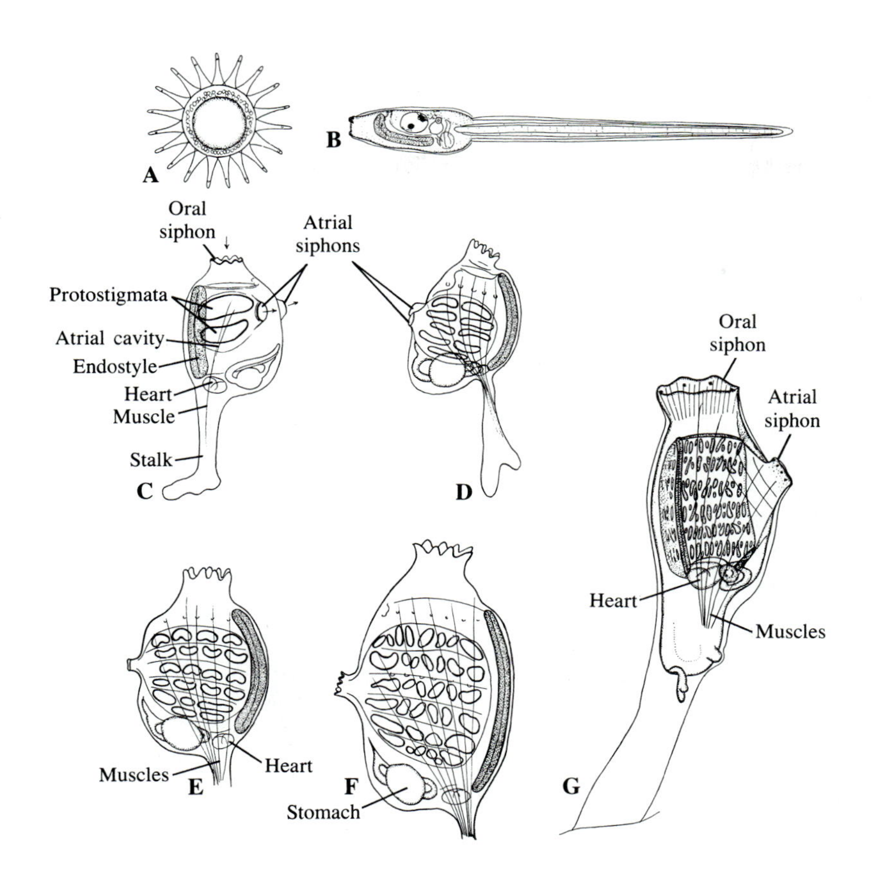

FIGURE 24.4 *Ciona intestinalis,* development. A. Egg, surrounded by follicle cells and an outer covering of cells that form a float. B. Larva. C. Early ascidian stage. The two atrial siphons are still separate. D–F. Three stages in the transition from the early ascidian stage to a stage in which there is a single median atrial siphon and six circles of definitive stigmata. (Berrill, Journal of the Marine Biological Association of the United Kingdom, *26*.) G. Later stage, with an increased number of stigmata. (Berrill, Philosophical Transactions of the Royal Society of London, B, *225*.)

to test the water before it enters the pharynx. If the water is silty or otherwise unsuitable, the oral siphon will close and ciliary activity and mucus secretion in the pharynx will slow down. If some coarse material has accumulated on the tentacles, the atrial siphon closes and the body contracts forcefully to flush out the oral siphon.

To collect food, an ascidian employs cilia and mucus. Mucus is secreted primarily by certain cells of the complex endostyle (Figure 24.6, B), and the cilia on

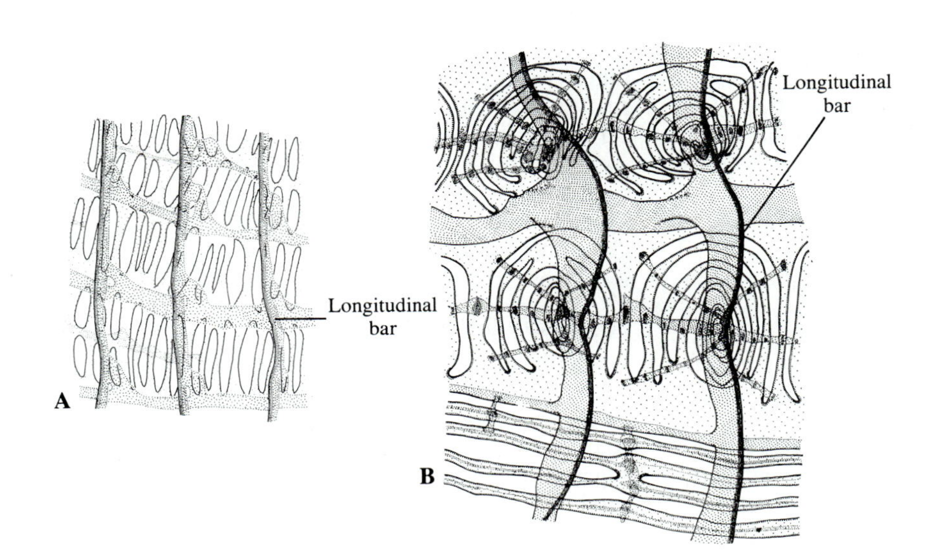

FIGURE 24.5 Stigmata. A. *Molgula phytophila*. (Monniot, Annales de l'Institut Océanographique, *47*.) B. *Heterostigma separ* (Monniot and Monniot, Sarsia, *13*.)

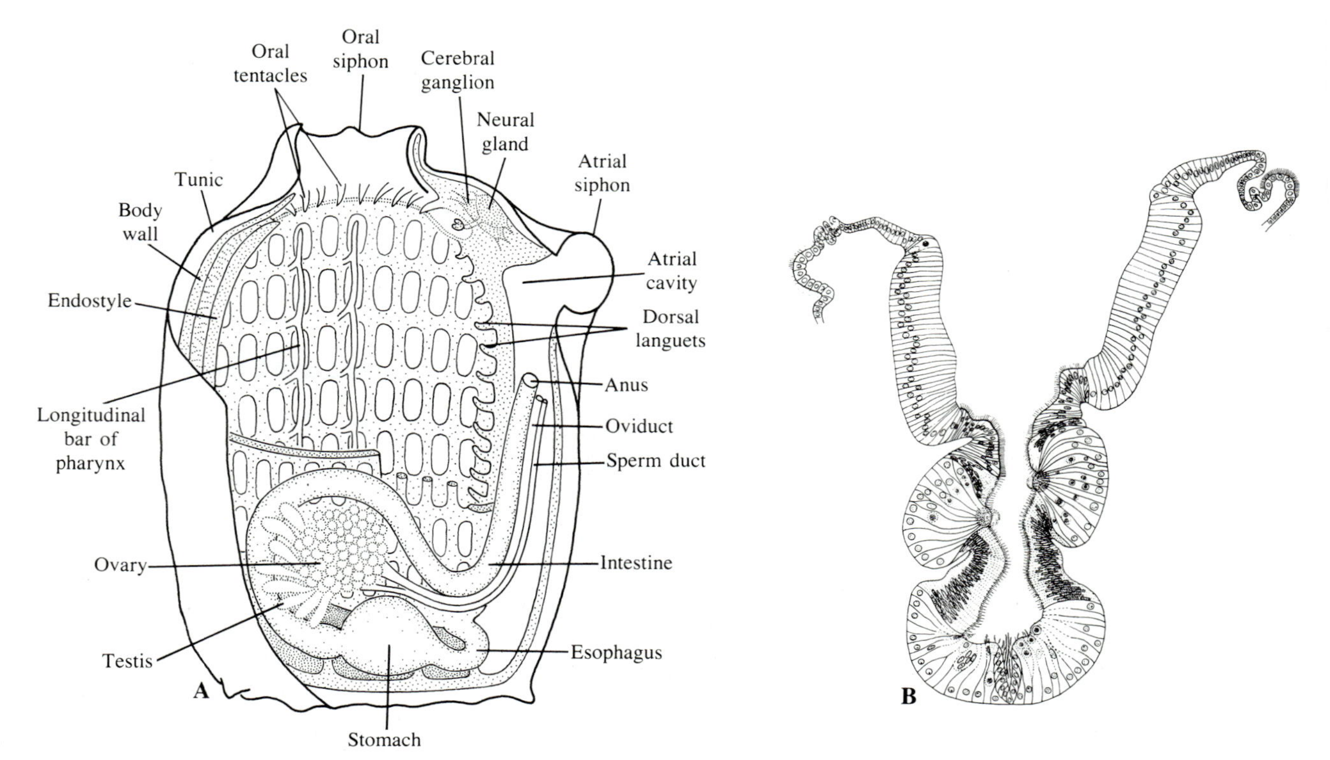

Figure 24.6 A. Diagram of a solitary ascidian of the order Stolidobranchia. The tunic, body wall, and most of the wall of the pharynx have been removed on the observer's side. Only two of the longitudinal bars of the pharynx are shown. (Mostly after Monniot and Monniot, Oceanography and Marine Biology, An Annual Review, *16.*) B. *Molgula manhattensis,* endostyle, transverse section. There are nine distict histological regions. (Godeaux and Firket, Annales des Sciences Naturelles, Zoologie, series 12, *10.*)

other endostylar cells drive it out of the groove. A sheet of mucus spreads over the entire inner surface of the pharynx and is moved by ciliary activity to the dorsal side, opposite the endostyle. Here there is a median ridge, called the **dorsal lamina,** or a series of J-shaped projections called **dorsal languets** (Figure 24.6, A; Figure 24.8, A and B). Certain tracts of cilia on the lamina or languets shape the mucus into a string; others move the string backward into the esophagus.

As the sheet of mucus moves from the endostyle to the dorsal part of the pharynx, water is driven out of the pharynx and into the atrial cavity. The cilia responsible for this are primarily those that border the stigmata. An ascidian about 3 or 4 cm high may pass 2 or 3 liters of water through the pharynx each hour. The combination of mucus secretion and ciliary activity makes a trap for capturing minute particles of food, ranging from diatoms to bacteria. Some species are such efficient filterers that they retain almost all particles 1 μm or larger. Their capacity to convert organic matter into tissue and into energy-yielding material is among the highest known in the animal kingdom.

When the food string reaches the stomach, digestion begins. Associated with the stomach is a so-called **digestive gland,** but the functions of this substantial structure are not well understood. It probably is concerned at least to some extent with digestion, but absorption and excretion are other functions that have been suggested for it. Undigested residues are carried by the tubular intestine to the anus, which opens into the atrium; they are then carried out of the animal with the stream of water leaving the atrial siphon.

Circulatory System

The heart of the adult ascidian is in about the same position as it was in the larva: beneath the posterior part of the pharynx. The heart–pericardial complex originates as a flattened tube. The tube rolls up until its edges fuse, whereupon one tube (the heart) lies within another (a noncontractile pericardium) (Figure 24.9, C).

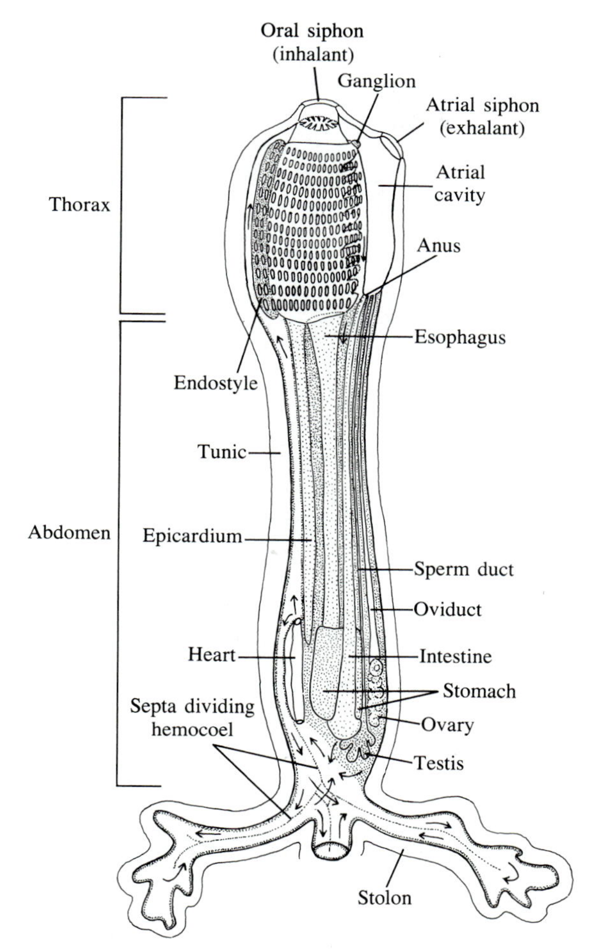

FIGURE 24.7 *Clavelina lepadiformis*. This ascidian has a long abdominal region below the pharynx, and a pair of epicardial cavities that are united for most of their length. The stolons produce buds that differentiate into new individuals. Thus a cluster of several to many individuals is formed. (After Brien, in Grassé, Traité de Zoologie.)

There are only a few definite blood vessels in ascidians; they include the major aortas at both ends of the heart. The blood circulates mostly through interconnected sinuses. Neither the heart, nor the blood vessels, nor the blood sinuses have an endothelial lining.

The two-way action of the heart, easily seen through the tunic of a transparent ascidian, is fascinating. As the heart contracts, it resembles a wet rag being wrung out by invisible hands. For several to many contractions it pumps blood in one direction; then, after a lapse, the beat of the heart is reversed, and blood is forced in the opposite direction. The best explanation for the reversal is that when the pacemaker at one end of the heart tires, the one at the other end becomes dominant. The system works well for gas exchange through the thin walls of the pharynx. Blood that has been pumped through the pharynx goes into a collecting vessel that carries it to sinuses in peripheral portions of the animal. Then, when the beat is reversed, this blood is drawn back through the pharynx again.

Ascidian blood contains several major types of cells, some of which are actively amoeboid. Each of the major types may be represented by a variety of subtypes, so that 20 or more apparently different kinds of cells may be present in the blood of the same animal. The significance of this diversity is not known. Among the blood cells are some that are richly colored by carotenoids, melaninlike compounds, and metallic substances. None of the pigments has been shown to carry oxygen. Certain blood cells, called nephrocytes, accumulate nitrogenous wastes. They are discussed under Excretion.

Ascidian blood and tissues have a high content of metals, including vanadium, niobium, molybdenum, chromium, gold, and iron. The role of most of the metals has not yet been explained. Both vanadium and iron, however, are known to serve as reducing agents in forming microscopic fibrils of tunicin. These metals are first taken up from sea water by cells of the pharynx, then picked up by certain blood cells that eventually break down in the tunic. The concentration of vanadium in sea water rarely exceeds 3 parts per million; in a few ascidians it exceeds 3000 parts per million of dry weight.

Tunic

The thickness and texture of the tunic varies from species to species. Sometimes it is soft and transparent, sometimes tough and leathery. It may have gritty foreign matter, such as sand, embedded in it, and in some genera, including *Didemnum* and *Polysyncraton,* it contains secreted calcareous spheres (Figure 24.14, B). In compound ascidians, discussed below, the tunics of the several to many zooids of the colony run together, thereby forming a single cohesive mass.

Although tunicin and some other constituents of the tunic are inert, this layer has living components. There are cells wandering through it, and in certain genera, especially solitary types, it is thoroughly pervaded by sinuses of the hemocoel.

An ascidian, whether it is solitary or a member of a colony, can usually change its shape when it is stimulated by poking or by being subjected to bright light. The siphons, which may have relatively little tunic material around them, almost always respond by closing. The modifications in shape are due to contractions of muscle in the body wall, beneath the epidermis that underlies the tunic (Figure 24.9, A).

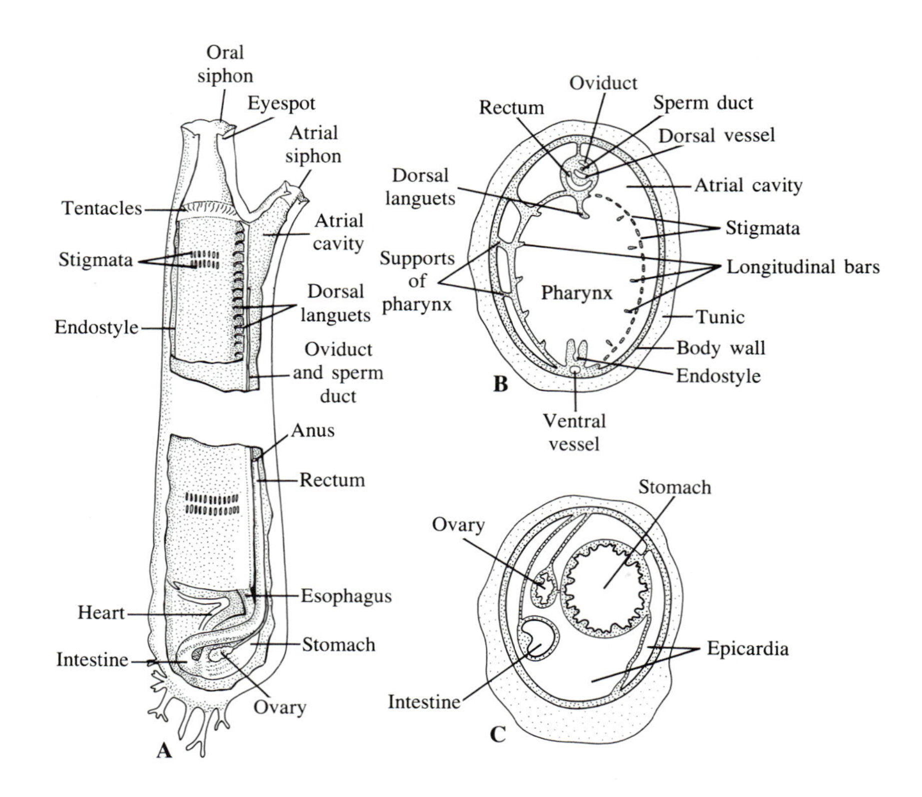

FIGURE 24.8 *Ciona intestinalis.* A. Partly dissected specimen, viewed from the left side. Much of the body wall has been removed to reveal internal organs, and in the upper third the left wall of the pharynx has also been cut away. Only portions of four of the many circles of stigmata are shown. B. Transverse section through the upper third of the body, not far below the openings of the genital ducts. On the left side, the section passes between successive circles of stigmata; on the right side, it passes through the stigmata. C. Transverse section through the stomach and other structures in the lower part of the body. (After Millar, Liverpool Marine Biology Committee Memoirs, *35*.)

In some ascidians, the tunic has a high concentration of sulfuric acid; its pH may be below 1.0. Although blood cells whose vacuoles contain both vanadium and sulfuric acid are characteristic of certain species, this is not always the case. Members of the family Didemnidae, for instance, have a very acid tunic, but they do not concentrate vanadium. Thus the idea that production of sulfuric acid is necessarily correlated with metabolic processes involving vanadium is not tenable. High acidity of the tunic protects many species from predation, and also from being overgrown by other organisms. This subject is discussed further below.

Epicardia

It was mentioned earlier in this chapter that some ascidians have one or two sacs that resemble coelomic cavities. These structures, called **epicardia** (Figure 24.7; Figure 24.8, C) because of their close association with the heart, originate as downgrowths from the floor of the pharynx. They are therefore lined by tissue that is derived from endoderm, not from mesoderm.

As a rule, after epicardia develop, they become completely separated from the pharynx, but there are exceptions. There are two of them at first. One or both may disappear, or two may become united for all or much of their length. They are most likely to be found in ascidians that have an abdomen into which the intestine, heart, and gonads, as well as the epicardia themselves, descend.

Being fluid-filled, epicardia could conceivably function in distribution of nutrients and other substances. In certain genera they serve as centers for accumulation of uric acid and urates, and are thus organs of excretion. In many ascidians they are of great importance in asexual reproduction and regeneration.

Excretion

Ascidians eliminate soluble nitrogenous wastes, mostly ammonia, by diffusion from tissue bathed by sea water. A few are known to excrete considerable urea. This has low toxicity as well as high solubility, but it is metabolically more costly to produce than ammonia.

Ascidians also dispose of nitrogenous wastes by storing them as crystals or concretions of uric acid or urates, especially sodium and potassium urate. Dissection of almost any ascidian will reveal substantial deposits of these nearly insoluble substances on the gut, mantle, wall of the atrium, or elsewhere. The cells in which the crystals or concretions have accumulated are often the ones called nephrocytes. They begin to precipitate nitrogenous compounds while they are circulating in the blood. Later they become fixed and grow into large,

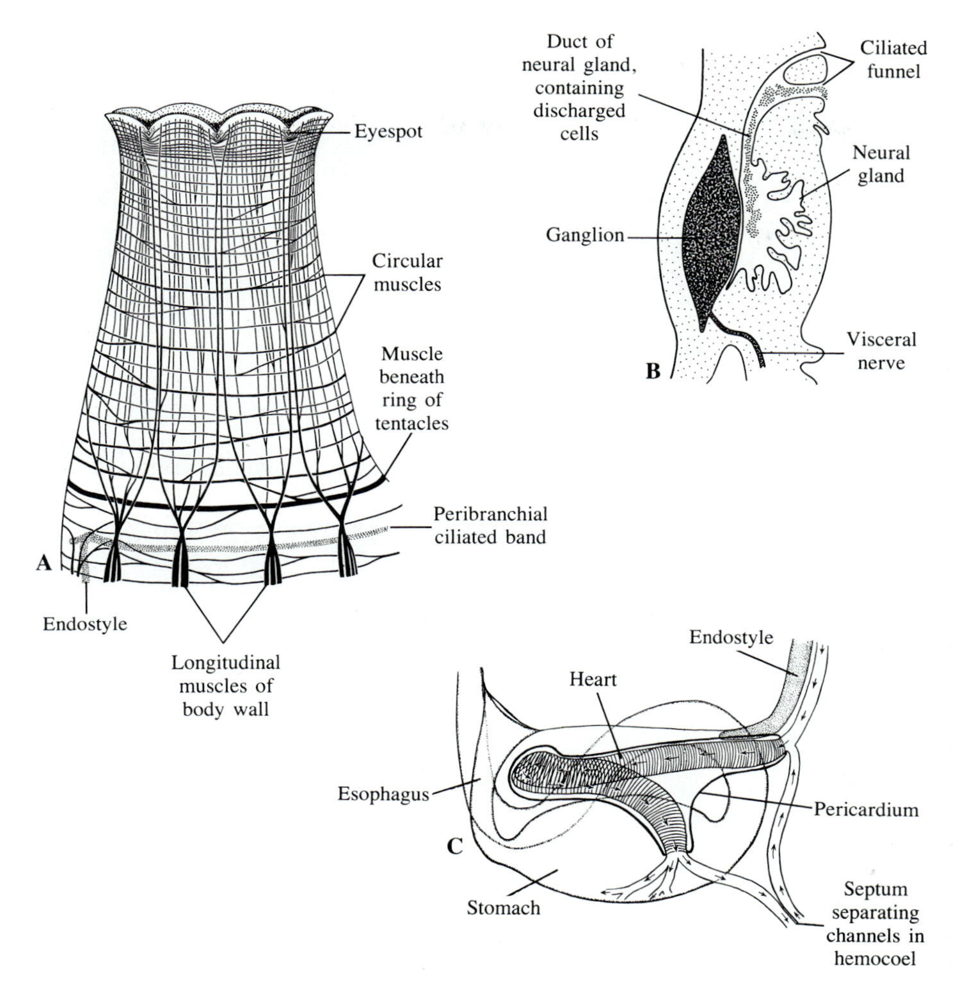

FIGURE 24.9 *Ciona intestinalis.* A. Region of the oral siphon, from the left side, showing the musculature of the body wall. B. Ganglion–neural gland complex, sectioned lengthwise. (In *Ciona,* the ciliated funnel is partly blocked by a mass of tissue attached to the wall by a mesentery-like strip, but this strip does not show in the section. (After Millar, Liverpool Marine Biology Committee Memoirs, *35.*) C. Heart and pericardium and their relationship to the gut. (Berrill, Philosophical Transactions of the Royal Society of London, B, *226.*)

multinucleate cells that enclose vacuoles within which uric acid or urates continue to collect.

Among the genera that store uric acid or urates in the epicardia are *Corella* and *Molgula*. In the latter, there is often just one large concretion.

Solitary, Social, and Compound Ascidians

An ascidian that is distinctly individual and that does not reproduce by budding is said to be **solitary.** Ascidians of this type may form large aggregations and they may be attached to one another, but all of the individuals in a cluster are completely separate and can be traced back to separate larvae. Body shape is varied. Some solitary ascidians, such as *Ciona* (Figure 24.8, A) and certain species of *Styela* (Figure 24.10, A), are tall or long-stalked, or both. Others have a squat form (Figure 24.10, B–D; Color Plate 8).

Social ascidians are those that form aggregations by repeated budding, either from stolons, as in *Clavelina* (Figure 24.7) and *Perophora* (Figure 24.11; Color Plate 8), or from basal sheets of tissue, as in *Metandrocarpa* (Color Plate 8). Sometimes the tissue bridges between the members of a colony persist indefinitely—this is the case in *Perophora*—but usually the members are joined only by tunic material, and even the tunic bridges may soon disappear.

Most social ascidians are smaller than solitary types, but they generally have the same kind of body organization, with all organs concentrated into one unit. *Clavelina* is an exception, for its gut, heart, gonads, and epicardia descend well below the pharynx into a separate part of the body called the abdomen, to distinguish it from the thoracic region in which the pharynx and atrial cavity are located.

In a **compound** ascidian, there are several to many members, called zooids, embedded in a common tunic (Figure 24.12). Unless sand becomes incorporated into it, or cells within it secrete calcareous spheres (Figure

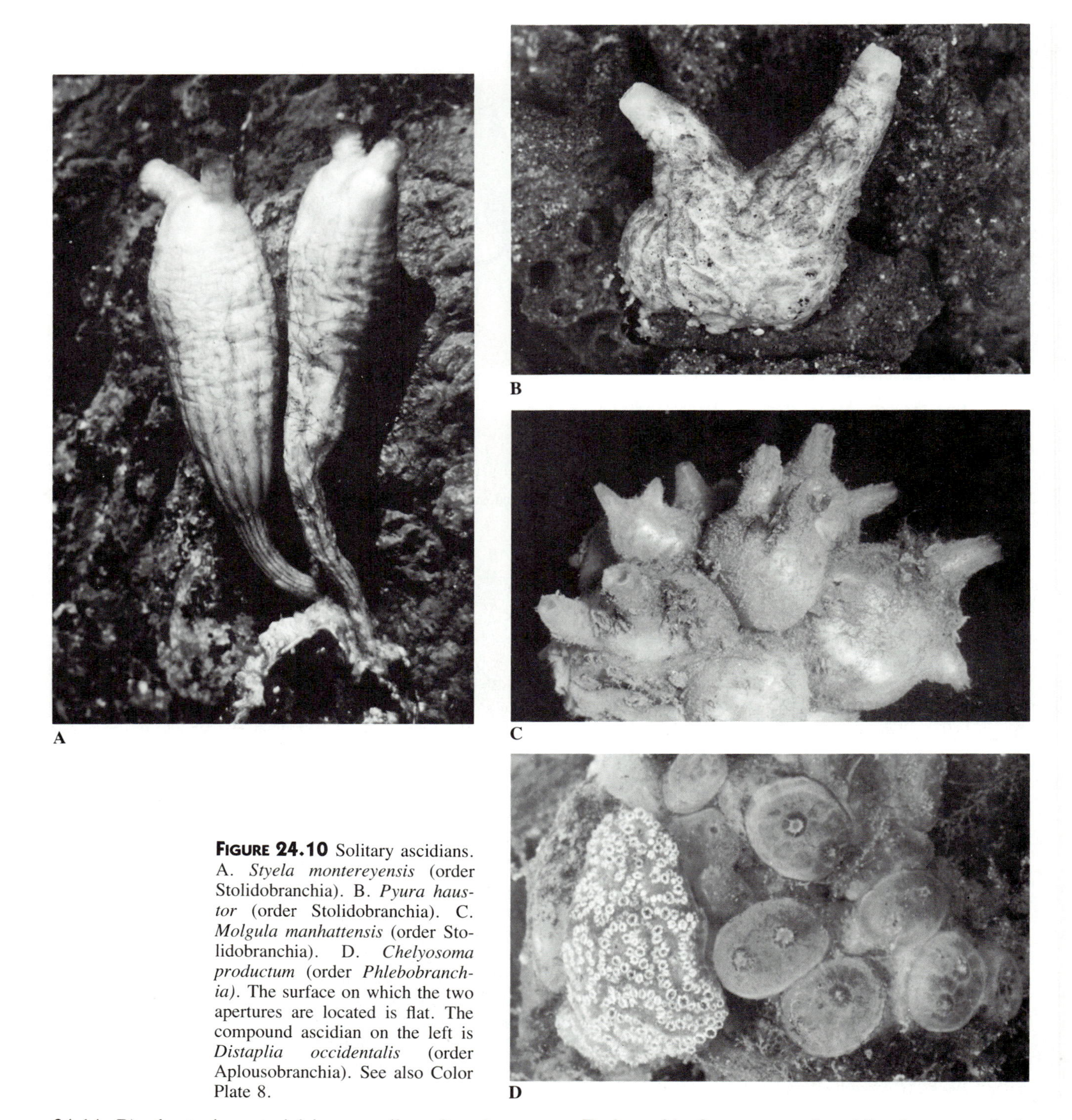

FIGURE 24.10 Solitary ascidians. A. *Styela montereyensis* (order Stolidobranchia). B. *Pyura haustor* (order Stolidobranchia). C. *Molgula manhattensis* (order Stolidobranchia). D. *Chelyosoma productum* (order *Phlebobranchia).* The surface on which the two apertures are located is flat. The compound ascidian on the left is *Distaplia occidentalis* (order Aplousobranchia). See also Color Plate 8.

24.14, B), the tunic material is generally soft and gelatinous. With few exceptions, such as *Botryllus* (Figure 24.12, D; Figure 24.13), compound ascidians have a distinct abdominal region (Figure 24.14), and in some genera there is a postabdomen, into which the heart, gonads, and epicardia descend (Figure 24.15).

Each zooid of a compound ascidian has an oral siphon that opens at the surface of the colony, and it may also have its own atrial siphon. In most genera, however, several to many zooids share the same exhalant opening. In such cases, the atrium of each zooid in the system opens into a common exhalant chamber (Figure

24.13, A; Figure 24.14, B and C; Figure 24.15). Thus the water entering the oral siphons of all members of the system leaves by way of a single opening.

There are several patterns of asexual reproduction leading to the formation and subsequent enlargement of a compound ascidian. In *Botryllus* (Figure 24.12, D; Figure 24.13), which does not have an abdominal region, buds originate on the epithelium of the atrial cavity. As a bud protrudes outward, it becomes enveloped by the epidermis and other tissues of the body wall. It differentiates into a zooid, breaks its tie with the parent, and takes its place around the common exhalant opening of the same system (Figure 24.12, D), or perhaps becomes the founder of a new system.

In some compound ascidians, the abdomen produces a series of buds, each of which is a prospective zooid

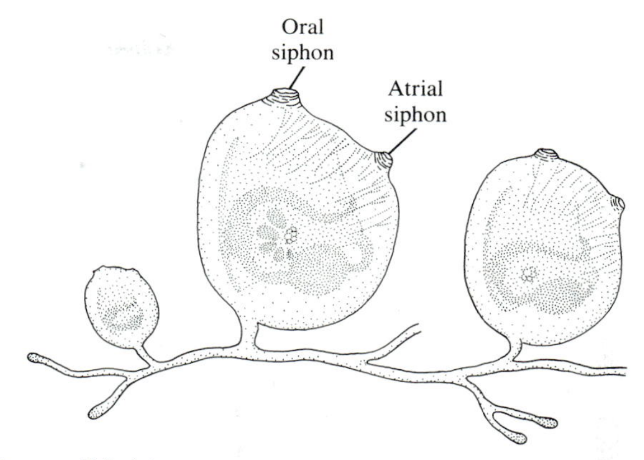

FIGURE 24.11 *Perophora,* a social ascidian that produces buds along its stolons. See also Color Plate 8.

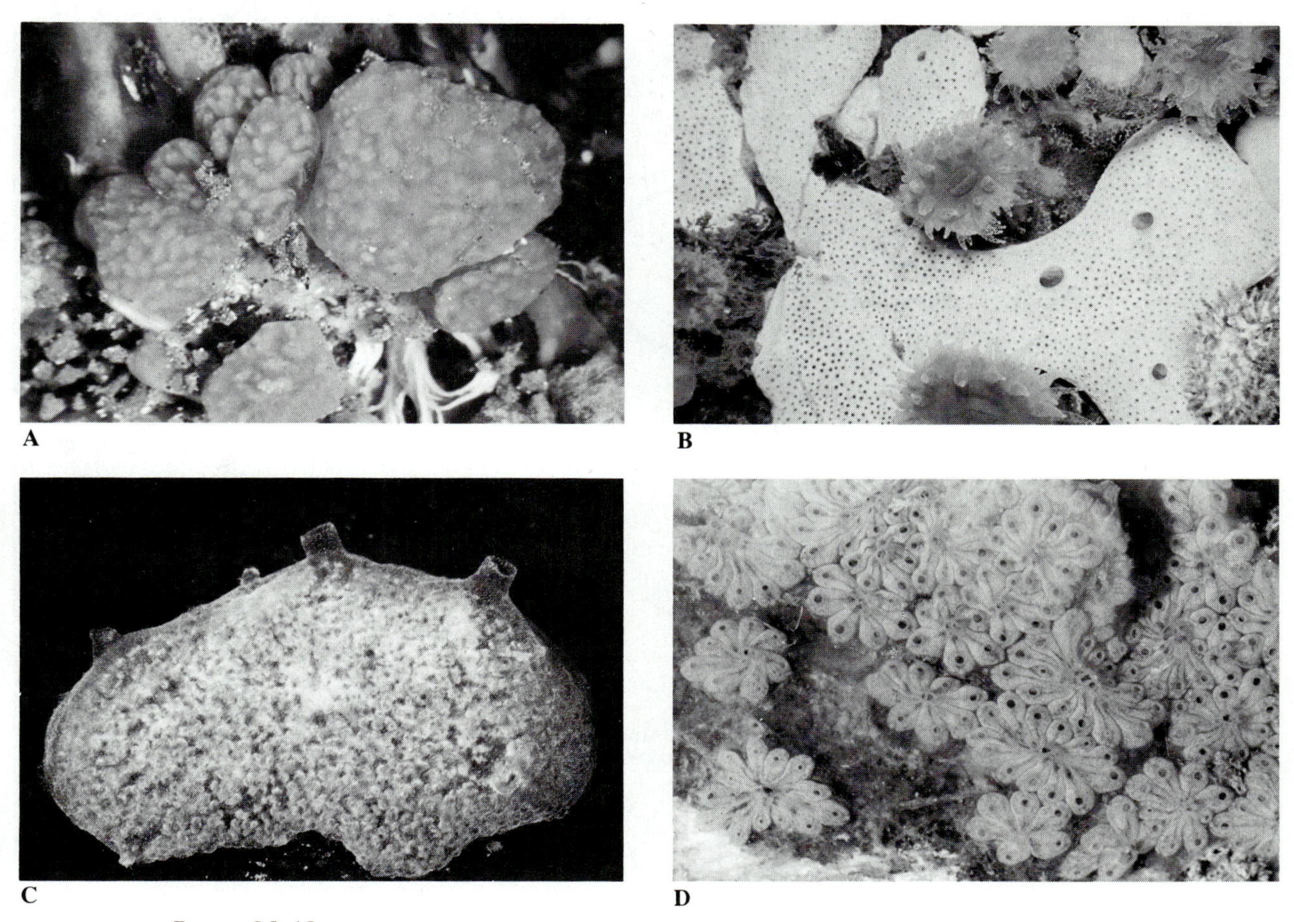

FIGURE 24.12 Compound tunicates. In all but *Ritterella,* several to many zooids share the same exhalant opening. A. *Ritterella pulchra* (order Aplousobranchia). B. *Didemnum.* (Some specimens of an anthozoan cnidarian, *Balanophyllia elegans,* are also shown.) C. *Diplosoma macdonaldi* (order Aplousobranchia). The larger exhalant openings are on chimneylike elevations. D. *Botryllus* (order Stolidobranchia). Young zooids recently produced by budding can be seen in some of the systems.

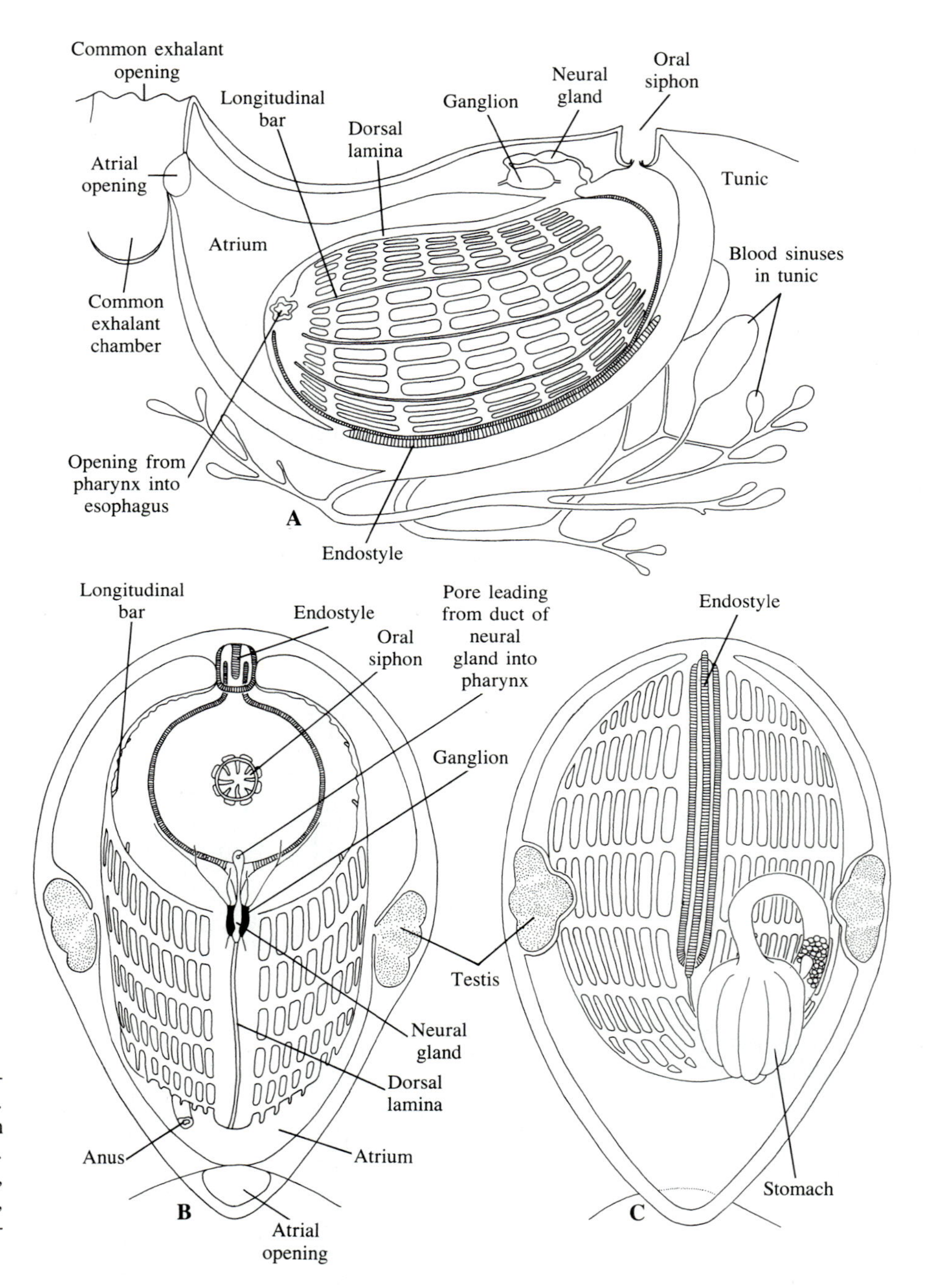

FIGURE 24.13 *Botryllus schlosseri* (order Stolidobranchia). A. Diagram of one zooid, as seen in median longitudinal section. B. Zooid, dorsal view. C. Zooid, ventral view. (After Herdman, Liverpool Marine Biology Committee Memoirs, *26*.)

(Figure 24.16, A). When there is a postabdomen, this is commonly involved in asexual reproduction, usually by dividing transversely to form several new zooids. In *Didemnum, Trididemnum,* and some related genera characterized by a short abdomen, two buds typically develop from the epidermis of the esophageal region (Figure 24.16, B). As the buds enlarge, each gets a portion of one of the epicardia of the parent. The epicardial tissue forms most of the structures that develop within the buds. One bud differentiates mostly into a pharynx–atrial cavity complex, and the other develops largely into a visceral complex. The missing portions are filled in later.

Regeneration

Many ascidians have a high capacity for regeneration. This is to be expected in a group of animals in which clusters or colonies are formed by budding or by

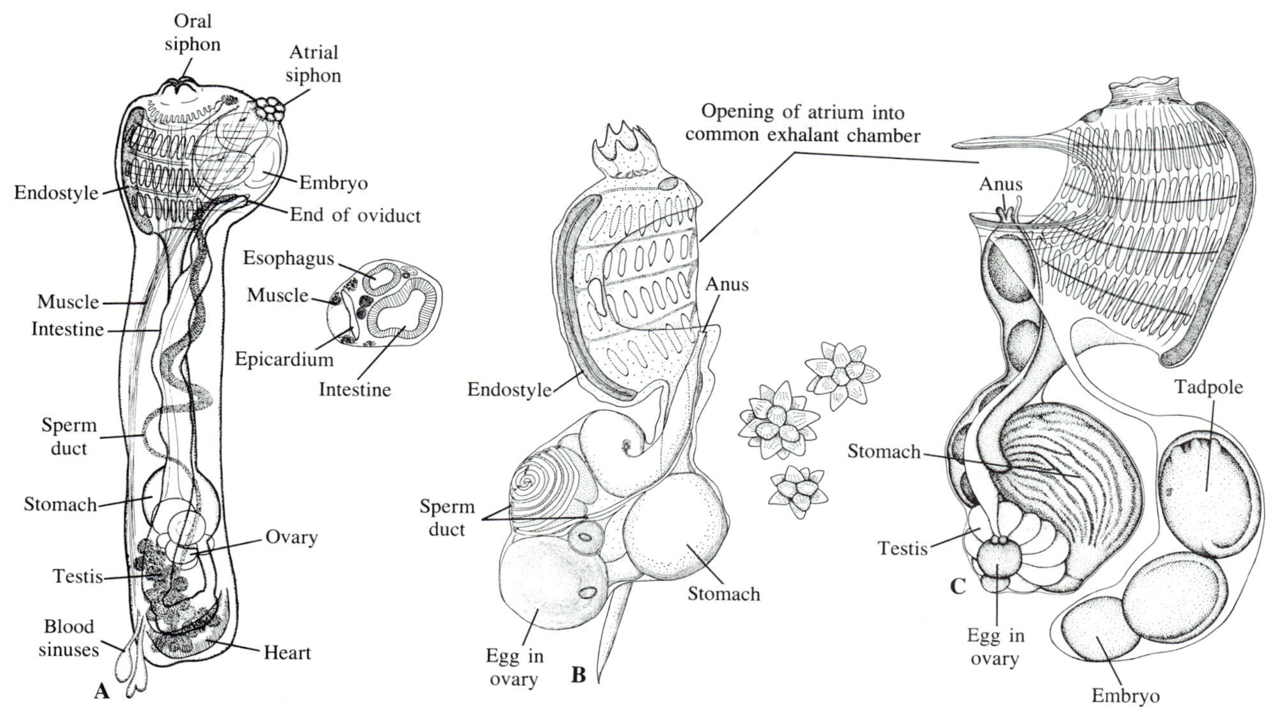

FIGURE 24.14 Zooids of compound ascidians (all of the order Aplousobranchia). A. *Archidistoma aggregatum,* an entire zooid (left) and a section (right) through the upper part of the abdomen, showing the epicardium. (Berrill, Journal of the Marine Biological Association of the United Kingdom, *27.*) B. *Polysyncraton bilobatum* (left) and some of the calcareous spheres from its tunic (right). (Medioni, Vie et Milieu, *21.*) C. *Distaplia dubia.* The atrium has a deep pocket in which embryos develop to an advanced larval stage. (Tokioka, Ascidians of Sagami Bay.)

the other methods described in the preceding section.

It is impossible to summarize here the results of the many experimental studies on regeneration of ascidians. A few words about *Clavelina,* which has been worked on by several investigators, will indicate the remarkable extent to which some members of the class are able to regenerate.

If a piece is cut from a stolon of *Clavelina,* or if a short length of stolon is effectively separated from the rest by ligation, these may differentiate into complete individuals, just as if they were buds normally produced by the stolon. If the body proper of a *Clavelina* is sectioned transversely into four pieces, all of them are likely to replace the structures they lack. In this type of regeneration—not only in *Clavelina,* but also in some other ascidians—the tissue lining the epicardia has been shown to be of primary importance. It is the source of cells that later form definitive structures.

Sexual Reproduction and Development

With few exceptions, ascidians are hermaphroditic. If there is just one ovary and one testis, these originate from what starts out as a single gonad. If there are multiple ovaries and testes, these can be traced back to several primordia. It is odd that the gonadal primordia of ascidians may arise from any of several sources. In some species, for instance, they derive from mesenchyme; in others, they come from the atrial epithelium. Where the gonads lie in mature animals also varies. They are frequently in the loop of the intestine. In compound ascidians whose zooids have a postabdomen, the gonads descend into this. When multiple gonads are present, they are typically in the body wall. The male and female gonoducts—these are single, even when there are several testes and ovaries—lead to the atrium, and the genital pores are usually close to the anus.

Some ascidians are capable of producing gametes almost continually throughout the year, as long as the water temperature is suitable. In others, the reproductive period is short. Spawning can often be induced in the laboratory by keeping the animals in the dark for several hours, then exposing them to light. Species that respond to light in this way are probably the ones most likely to release their gametes at dawn. Much work has been done on experimental control of spawning and various other patterns have been noted.

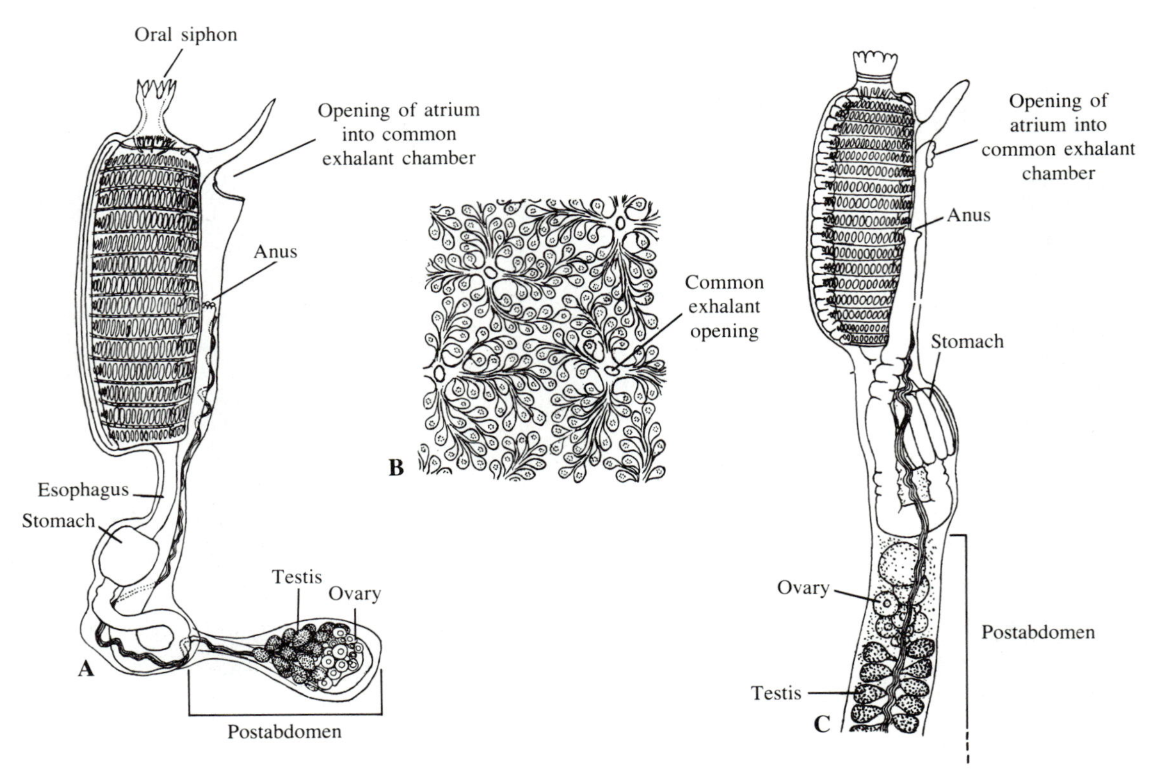

FIGURE 24.15 Zooids of compound ascidians (both of the order Aplousobranchia). A. *Polyclinum constellatum,* which has a short postabdomen. B. *Polyclinum constellatum,* arrangement of zooids around common exhalant chambers. C. *Aplidium bermudae,* which has a long postabdomen (part of the latter has been omitted). (Van Name, Scientific Survey of Porto Rico and the Virgin Islands, *10*.)

Not all ascidians shed their eggs. Many species are brooders, retaining embryos in the atrial cavity (or a specialized part of it) until they reach the tadpole stage. There are even a few compound ascidians whose tadpoles begin to metamorphose while they are still within the parent colony; this is the case in the genus *Diplosoma*. In general, brooding types produce rather large eggs, therefore large tadpoles.

Compound ascidians with small zooids often have especially large eggs, although a single zooid may produce only a few during its lifetime. In *Distaplia occidentalis,* for example, a sexually mature zooid produces only four eggs. These are brooded to the tadpole stage in the oviduct, which is all that remains of the zooid. When stimulated by light, the tadpoles swim out of the isolated oviduct. After being free for only about 3 hours, they settle and undergo metamorphosis.

Although many ascidians are capable of fertilizing their own eggs, they do not often do this. A particular animal generally releases nearly all of its sperm before it sheds any oocytes. This reduces the possibility of self-fertilization, even if only a few minutes elapse between the times when sperm and oocytes are set free. The oocytes are characteristically covered by a chorionic membrane, surrounded in turn by a layer of follicle cells. After an oocyte has been entered by a sperm, it undergoes meiosis and casts off two polar bodies. Union of the male and female pronuclei then takes place.

Early cleavage is nearly equal and follows a pattern that is basically radial, although for a time the embryo exhibits a symmetry that is more or less biradial. Cleavage is also highly determinate, so that individual cells are decidedly different in terms of what they subsequently produce.

By the time gastrulation begins, the cells destined to form mesoderm and the notochord are recognizable. The fact that blastomeres in various parts of the embryo sometimes have different colors—gray, yellow, orange, and so on—makes it easier to follow their fates. Gastrulation takes place rather early, when there are about 64 or 128 cells. It is usually achieved by invagination, although epiboly is the primary mode in species whose eggs are relatively large and yolky. Little if anything

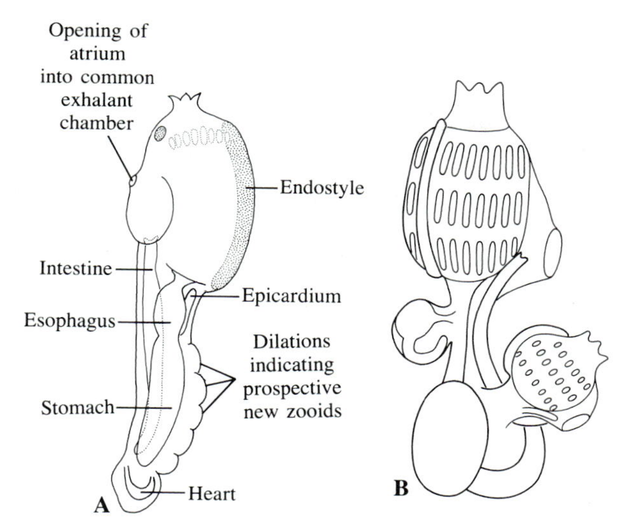

FIGURE 24.16 Asexual reproduction by zooids of compound ascidians (both of the order Aplousobranchia). A. *Aplidium zostericola* (only one of the several circles of stigmata is shown). The bulges in the abdominal region mark an early stage in the formation of new zooids. Each has a portion of the epicardium of the parent. (After Brien, in Traité de Zoologie.) B. *Trididemnum*. The zooid has produced two buds, one of which will develop into the thoracic region, the other into the abdominal region of a new zooid. (After Salfi, Archivio Zoologica Italiana, *18*.)

remains of the blastocoel. The blastopore becomes displaced from the prospective posterior end of the larva to a dorsal position. The notochord-forming cells, as they are carried into the gastrula, become arranged in a compact mass flanked on both sides by mesodermal cells. The notochordal cells divide only two or three times more, so the total number of cells in the notochord stops at about 40. The notochord lengthens as the fluid-filled spaces between the cells enlarge and finally coalesce to form a central canal. The mesodermal cells flanking the notochord differentiate into the muscle bands of the tail.

The primordium of the dorsal nerve cord, called the neural plate, is initially part of the ectoderm on the dorsal side of the embryo. The blastopore lies in the posterior part of the plate. As folds growing up from the right and left sides of the plate come together medially and unite, the neural tube is formed. For a time this is open anteriorly, and its posterior part is still linked with the blastopore. In time, however, the definitive nerve cord becomes distinct from the epidermis that overlies it and the blastopore disappears. The preceding explanation of the origin of the nerve cord is simplified, and does not deal with differentiation of the various components of the cerebral complex, which develops at the anterior end of the cord.

The mouth eventually opens near the anterior end. The atrial cavity is formed either from one invagination of ectoderm that spreads out on both sides of the pharynx, or from two invaginations that later become joined, as was already explained in connection with the structure and metamorphosis of the tadpole. The protostigmata of the tadpole develop at points of contact of the pharynx and medial walls of the atrium.

Ecology of Ascidians

Most ascidians are attached to firm substrata, such as rocks, pebbles, shells, wood, and seaweeds. Some of those that colonize seaweeds seem to require not only a certain amount of firmness but also flexibility. The abyssal types and others that live in soft sediments are generally either rooted in the mud by means of filamentous outgrowths of the tunic or are attached to small particles, such as bits of shell, on which they originally settled.

In habitats that favor their existence, ascidians are often the dominant organisms. This is evident on some rocky shores, especially in South Africa, Chile, and Australia, where a single species of ascidian may form a dense, vertical band of from 0.5 to 1 m high. Dredge hauls taken on certain gravelly and shelly bottoms sometimes bring up more ascidians, by weight, than any other kinds of organisms.

Ascidians usually form a large part of the biomass of fouling communities—the associations of organisms found on ship hulls, floating docks, pilings, and buoys. They are often abundant, moreover, in harbors and estuaries where there is considerable pollution and where the salinity is reduced owing to an influx of fresh water. Some of the estuary-inhabiting ascidians, at least as adults, are able to tolerate salinities as low as 20 parts per thousand (full-strength sea water has a salinity of about 32 parts per thousand). Certain species have traveled far, on ship hulls, in ballast, or with transplanted oysters, and have become aggressive and successful competitors for space in their new homes.

Being sessile animals, ascidians depend on the oral siphon to bring in water containing food and oxygen, and on the atrial siphon to dispose of water containing metabolic and fecal wastes. Thus it is important that they orient these apertures in such a way as to minimize fouling of water before it enters the oral siphon, and to take advantage of a prevailing current, when there is one. Many achieve this by turning the oral siphon toward an oncoming current and the atrial siphon in the opposite direction, by turning the oral siphon toward the substratum and the atrial siphon away from it, or by narrowing the atrial siphon so that water leaving the atrial cavity is shot for some distance away from the animal. In compound ascidians in which several zooids share a common exhalant chamber, the exhalant opening is often raised above the oral openings. These are just some of the solutions.

The surface of the tunic of some ascidians is colonized by members of the same or different species, or by various other kinds of organisms, such as sponges. Many ascidians, however, resist being colonized, and this is believed to be due primarily to high acidity (pH from 2.0 to 0.0) of fluids in the tunic. In at least one species, *Phallusia nigra,* a high concentration of vanadium at the surface of the tunic is also believed to discourage settlers. In most ascidians, however, large amounts of vanadium are found only in the tissues of the animal.

Vanadium is a metabolic poison and most organisms can tolerate only small amounts of it. Ascidians whose tissues or tunic have high concentrations of vanadium are almost certainly protected from most predators. High acidity of the tunic is also a deterrent to predation. Nevertheless, many species, including some that have the protection of vanadium or acidity, are eaten by fishes, sea stars, crabs, polyclad turbellarians, and gastropod molluscs. Most of the polyclads and gastropods that consume ascidians are more or less permanently associated with particular prey species. The color patterns of certain of the gastropods, in fact, so closely resemble those of their prey that the molluscs may escape detection. Most striking in this respect are some species of the mesogastropod family Marseniidae (Lamellariidae), whose color pattern almost perfectly blends in with that formed by the zooids in compound ascidians of the family Didemnidae.

A few large species of ascidians are eaten by humans, or used as bait in fishing. Skin rashes, perhaps of an allergic nature, have been attributed to contact with certain types, and a species that commonly grows on oyster shells in the Orient is believed to be the source of airborne particles that cause a respiratory ailment called "oyster shucker's disease."

Ascidians have their share of internal symbionts. Copepod crustaceans are perhaps the most commonly encountered inhabitants. Many kinds of them live in the pharynx, and these are probably mostly commensals or relatively benign parasites. The highly specialized types that live in blood sinuses, however, do not even have a gut, and are certainly parasites. The pharynx may also have amphipods, small shrimps, and pea crabs. The extent to which these affect their hosts is not known.

Possibly mutualistic symbionts, so far known only from the compound ascidians of the family Didemnidae, are photosynthetic prokaryotic organisms of the genus *Prochloron*. They are embedded in the walls of the common exhalant chambers shared by several zooids. Ranging from 6 to 25 μm in diameter, they are larger than bacteria and most blue-green algae, and they have chlorophylls *a* and *b*. In this respect, they are closer to green plants, green algae, and green flagellates than to blue-green algae, which have only chlorophyll *a,* or to photosynthetic bacteria, which have other types of chlorophylls. Among the carbohydrates they synthesize are some that are characteristic of prokaryotes and others that are produced by certain green plants. *Prochloron* is assigned to its own phylum, Chloroxybacteria (or Prochlorophyta).

Classification

Order Aplousobranchia

This order consists of ascidians in which the body is divided into two or three distinct regions: thorax and abdomen, or thorax, abdomen, and postabdomen. The pharynx does not have the longitudinal bars characteristic of the succeeding two orders. The gonads are located well below the pharynx, in the loop formed by the intestine. Representatives include *Clavelina* (Figure 24.7), a social type, and several genera of compound asicidians. Among the latter are *Distaplia* (Figure 24.14, C), *Ritterella* (Figure 24.12, A), *Didemnum* (Figure 24.12, B), *Trididemnum, Diplosoma* (Figure 24.12, C), *Polysyncraton* (Figure 24.14, B), *Polyclinum* (Figure 24.15, A and B), and *Aplidium* (Figure 24.15, C).

Order Phlebobranchia

In some genera of this group (such as *Diazona,* which is compound), the body is divided into a thorax and an abdomen. In most others—including the social *Perophora* (Figure 24.11; Color Plate 8) and the solitary *Ciona* (Figure 24.8, A), *Chelyosoma* (Figure 24.10, D), *Ascidia,* and *Corella*—there is no division. The pharynx has longitudinal bars, although these may not be well developed. The gonads are located next to the gut, usually appreciably below the pharynx.

Order Stolidobranchia

The bodies of stolidobranch ascidians are not divided into a thorax and an abdomen. The pharynx has longitudinal bars and prominent lengthwise folds. The gut and gonads are located alongside the pharynx, rather than below it. Representative genera are *Botryllus* (Figure 24.12, D; Figure 24.13) and *Botrylloides,* both compound, and *Styela* (Figure 24.10, A), *Pyura* (Figure 24.10, B), *Molgula* (Figure 24.10, C), *Cnemidocarpa* (Color Plate 8), and *Boltenia,* which are solitary. A social type that belongs to this group is *Metandrocarpa* (Color Plate 8). Also included in the Stolidobranchia are some small solitary ascidians that live in sand and fine gravel. They range in height from about 1 mm to 1 cm and are attached by rootlike specializations of the tunic. Two species are illustrated in Figure 24.17.

The ascidians called sorberaceans, which live on the bottom in deep water, probably belong to the Stolidobranchia, although some experts place them in a sepa-

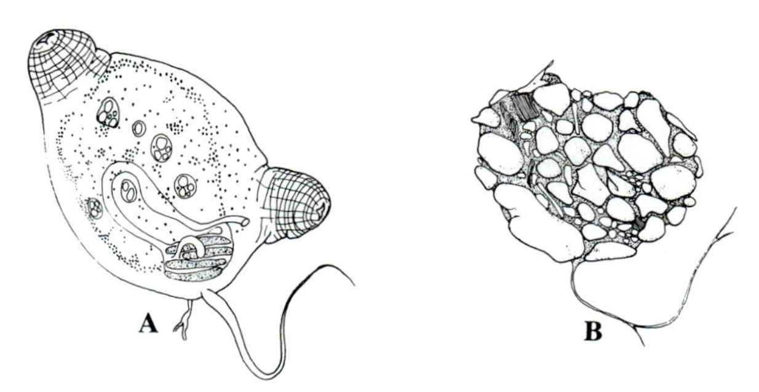

FIGURE 24.17 Two small solitary ascidians that live in sand or fine gravel. A. *Psammostyela delamarei,* about 1 mm high. B. *Cratostigma gravellophila,* about 1 cm high. Sand grains are embedded in its tunic. (Monniott and Monniot, Vie et Milieu, *12*.)

rate class. There are only three genera: *Gasterascidia* (Figure 24.18), *Hexacrobylus,* and *Sorbera.* All of them are unusual in that they are carnivorous, feeding on various small invertebrates, including nematodes, polychaetes, copepods, and ostracodes. The oral siphon has several muscular lobes for capturing prey. There are no true stigmata on the relatively small pharynx, but this organ does have two or more openings into the atrial cavity. The endostyle is reduced and there are no dorsal languets for consolidating mucus. A large sac that contains a concretion of nitrogenous wastes is much like that of *Molgula* and some other ascidians. A heart has not been found. The ganglion is large in comparison to that of ascidians. One half of it sends nerves to the oral lobes; the other gives rise to a surprisingly stout dorsal nerve that runs backward. Next to the ganglion is a neural gland that is joined to the pharynx by a long duct. Little is known about the biology of sorberaceans, and nothing has yet been published concerning their development.

CLASS THALIACEA

Thaliaceans are entirely planktonic. If found at the shore, that is because they have been washed up. The three orders of this group show considerable diversity and some amazing reproductive specializations, but the class as a whole seems to be well unified. It is impossible, however, to do justice to these animals in a brief account. Only the highlights of their biology can be presented, and the complex terminology normally used in discussing thaliaceans is simplified considerably.

Order Pyrosomida

Pyrosomes (''bodies on fire'') are famous for their luminescence. There are only a few species, and these are mostly restricted to warmer waters. They form tubular colonies with the general shape of a thimble or

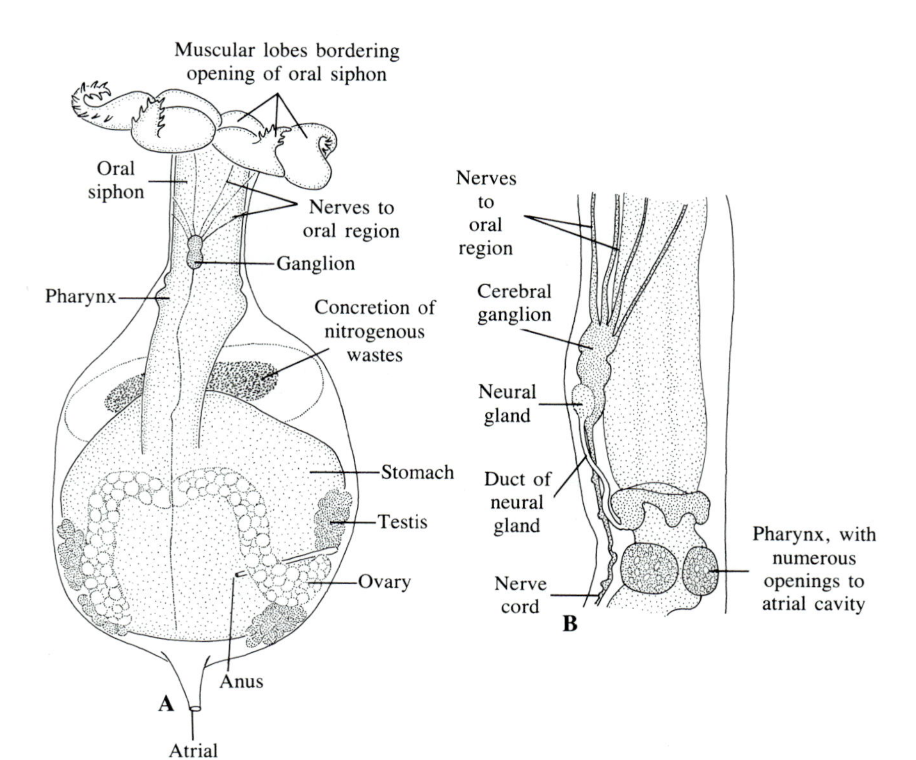

FIGURE 24.18 *Gasterascidia lyra,* a sorberacean. A. An entire specimen. B. Pharynx, cerebral ganglion, and neural gland. (After Monniot, Monniot, and Gaill, Archives de Zoologie Expérimentale et Générale, *116*.)

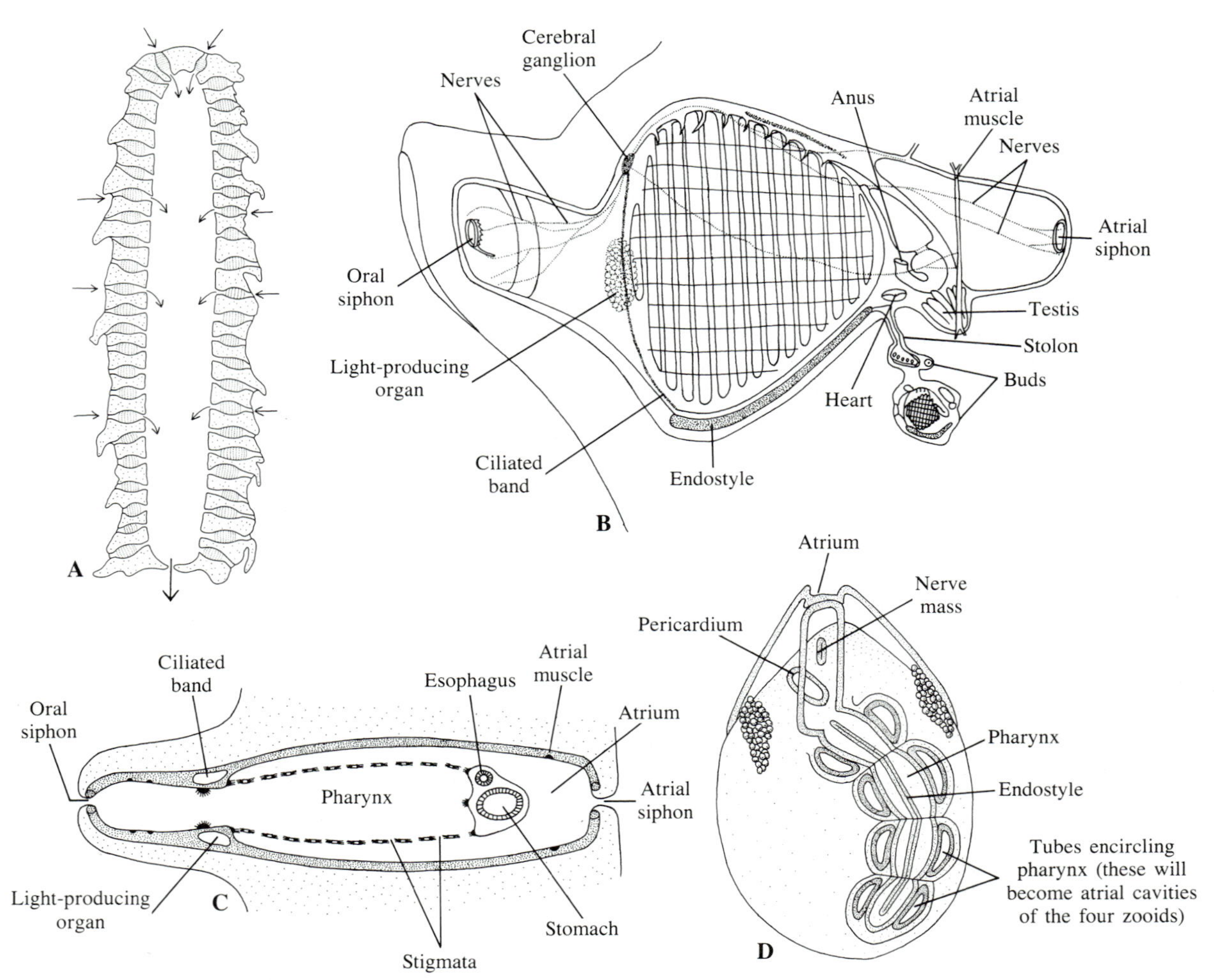

FIGURE 24.19 Class Thaliacea, order Pyrosomida. A. *Pyrosoma elegans,* a small colony sectioned lengthwise; diagrammatic. Arrows show currents passing through the zooids and leaving the opening at one end. B. *Pyrosoma triangulum,* lateral view of a single zooid. C. *Pyrosoma,* frontal section of a zooid. D. *Pyrosoma,* a cyathozooid. (After Kükenthal and Krumbach, Handbuch der Zoologie.)

elongated balloon: open at one end, closed at the other (Figure 24.19, A). The hundreds or thousands of zooids in a colony are embedded in the gelatinous wall, with their oral siphons on the outside and their atrial apertures on the inside. The water processed by the many zooids leaves the open end of the colony, which may be narrowed by a diaphragm to increase the velocity of the current and thus also increase the speed with which the colony swims.

A pyrosome colony originates from four zooids. These are derived from an individual that develops from a fertilized egg, but which itself does not survive, having given its all to the production of the four founder zooids. Each founder zooid produces buds that become part of the colony. As more buds are formed and additional matrix material is deposited, the colony increases in size. Colonies of one species reach a length of nearly 20 m, but the usual size is probably less than 50 cm.

Each mature zooid, though only about 1 cm long or less, has a proportionately large pharynx with many narrow stigmata (Figure 24.19, B and C). A portion of the atrial cavity is on the right and left sides of the pharynx; the rest of it is behind the pharynx and viscera. As water percolates through the pharynx, food is trapped in mucus produced by the endostyle and is moved to the esophagus in much the same way as in an ascidian. The anus and gonads open into the atrium. Two light-producing structures lie on each side of the anterior part of the pharynx. Buds that increase the number of individuals in the colony are produced on a stolon that arises beneath the posterior part of the endostyle. In at least certain species, amoeboid cells that separate from each bud help move the buds to the position they finally occupy.

Although pyrosomes are hermaphroditic, the ovary and testis do not usually function simultaneously, so a particular zooid is first male, then female, or *vice versa,* depending on the species. The ovary produces one large egg at a time. After the egg has been fertilized, it leaves the ovary and passes down the oviduct to the atrial cavity, where development proceeds to the point that an individual with a rudimentary pharynx and gut is formed. This individual, called a **cyathozooid,** lives long enough to produce a stolon with four buds (Figure 24.19, D), which differentiate into the zooids that become the founders of a new colony. These then leave the atrium of their grandparent.

Order Doliolida

Members of the order Doliolida (Figure 24.20) usually resemble barrels, partly because of their shape and partly because of the eight or nine hooplike muscle bands in the body wall. Some species are only a few millimeters long; the largest are about 2 cm long. The tunic is relatively thin, and doliolids are fragile. As is the case in the zooids that make up the colony of a pyrosome, the oral aperture is at one end, the atrial aperture at the other end. Locomotion is by jet propulsion, effected by closing the oral aperture and squeezing water backward out of the pharynx and atrium. As water passes through the stigmata, diatoms and other small organisms are caught in the mucus secreted by the short endostyle, which is located on the floor of the pharynx. The postpharyngeal part of the gut, consisting of a stomach and intestine, has various configurations, depending on the species; it is generally approximately U-shaped, but the intestine may be looped.

The heart is beneath the pharynx. There is a prominent ganglion dorsal to the pharynx, and delicate nerves extend from it to the muscle bands and to tactile sensory cells that are located between the lobes of the oral and atrial apertures as well as in other areas of the epidermis.

In sexual individuals, called **gonozooids** (Figure 24.20, A and B), the ovary and testis are near the gut and open into the atrium. The ovary produces only a few eggs, and cleavage does not begin until the eggs have been shed. The embryo develops into a stage that resembles an ascidian larva. It has a notochord—this structure does not appear in any other thaliaceans—and two muscle bands in its short tail. Before long, the tail disappears. The trunk region transforms into an individual that reproduces asexually. This is called an **oozooid,** because it is derived from a fertilized egg. It has nine muscle bands instead of the eight that are typical of the sexual stage. Three other distinctive features of asexual individuals are a statocyst in the body wall on the left side, a posterodorsal spurlike extension, and a **stolon** arising from the ventral surface below the endostyle.

The tip of the stolon soon starts to produce hundreds of tiny buds. Amoeboid cells somehow help these move around the right side of the body to the dorsal surface, and then backward to the spur, which by this time has enlarged to accommodate them (Figure 24.20, C). The buds become arranged in three main groups. Those in the lateral groups become **gastrozooids,** which are small feeding individuals. They are essentially simplified versions of the parent, with reduced muscle bands, a wide oral aperture, and an atrium so small that the anus opens almost directly to the outside.

The buds attached to the dorsal midline of the spur develop into stalked individuals called **phorozooids** (''carrier zooids''). These, in spite of their small size, look a little like those of the sexual phase. This is partly because they have eight muscle bands. They become detached and are replaced by other buds that migrate to the spur. Before each one separates from the parent, however, an astonishing thing happens. A migrating bud from the stolon settles on the stalk of the phorozooid and proliferates several secondary buds (Figure 24.20, D). After the phorozooid is set free and begins to feed, each secondary bud develops into a young gonozooid, which eventually becomes detached, reaches maturity, and reproduces sexually.

While asexual reproduction by the oozooid is in progress, its feeding mechanism deteriorates. The animal enters what is called the ''old nurse'' phase. The muscle bands become broader, sometimes to the point that a few of them merge. Many aspects of the biology of oozooids have not been explained. Do the gastrozooids contribute to the nourishment of the parent, whose feeding mechanism is nonfunctional? How are the phorozooids nourished until they can feed by themselves, and how do the buds developing on their stalks get nourishment?

Order Salpida

Members of the order Salpida resemble doliolids with respect to the general layout of the body, presence of muscle bands, and regular alternation of sexual and asexual phases. They usually have more tunic material than doliolids have, however, and the several muscle bands that are most nearly similar to the complete rings of doliolids are interrupted ventrally. Mature salps

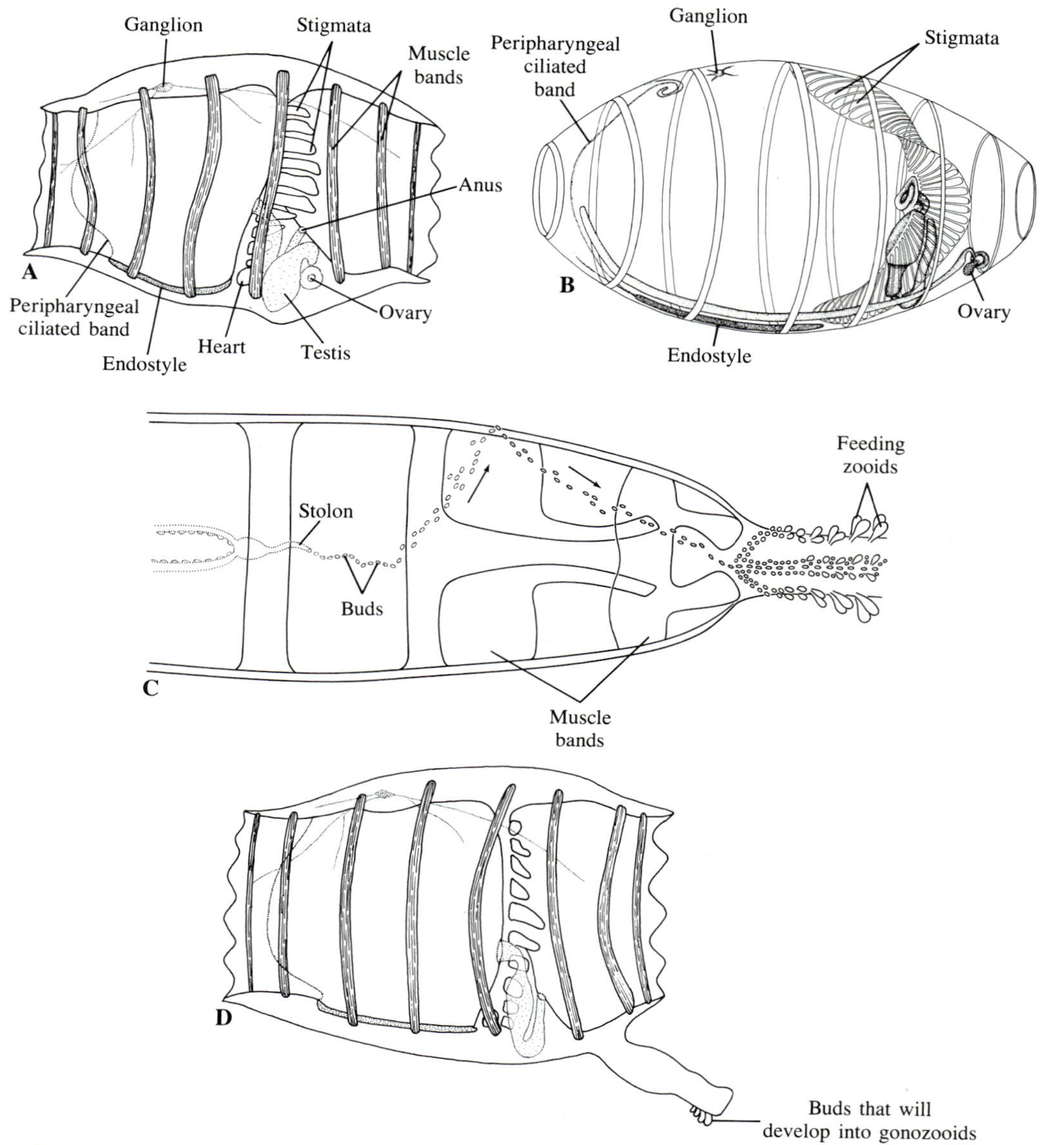

Figure 24.20 Order Doliolida. A. *Doliolum mulleri,* gonozooid. (After Kükenthal and Krumbach, Handbuch der Zoologie.) B. *Doliolum intermedia,* gonozooid. (Tokioka and Berner, Pacific Science, *12.*) C. *Doliolum gegenbauri,* an oozooid producing buds that migrate to the posterodorsal appendage, becoming gastrozooids and phorozooids. Some buds become attached to the phorozooids and give rise to a new generation of gonozooids. D. *Doliolum mulleri,* a phorozoid after its release from the posterodorsal appendage of the oozooid. The gonozooids developing on its stalk were produced by a bud that settled on it while it was attached to the oozooid. (C and D, after Kükenthal and Krumbach, Handbuch der Zoologie.)

range in length from about 1 cm to nearly 20 cm, depending on the species and stage in the life cycle.

The gonozooids (Figure 24.21, A and B), which reproduce sexually, are packed together in groups, at least when they are young. The oral and atrial apertures are, respectively, anterodorsal and posterodorsal in position. There are just two large stigmata at the rear of the pharynx, which has a well-developed endostyle. The esophagus, stomach, and intestine form a loop that ends at the anus, on the ventral side of the atrial cavity.

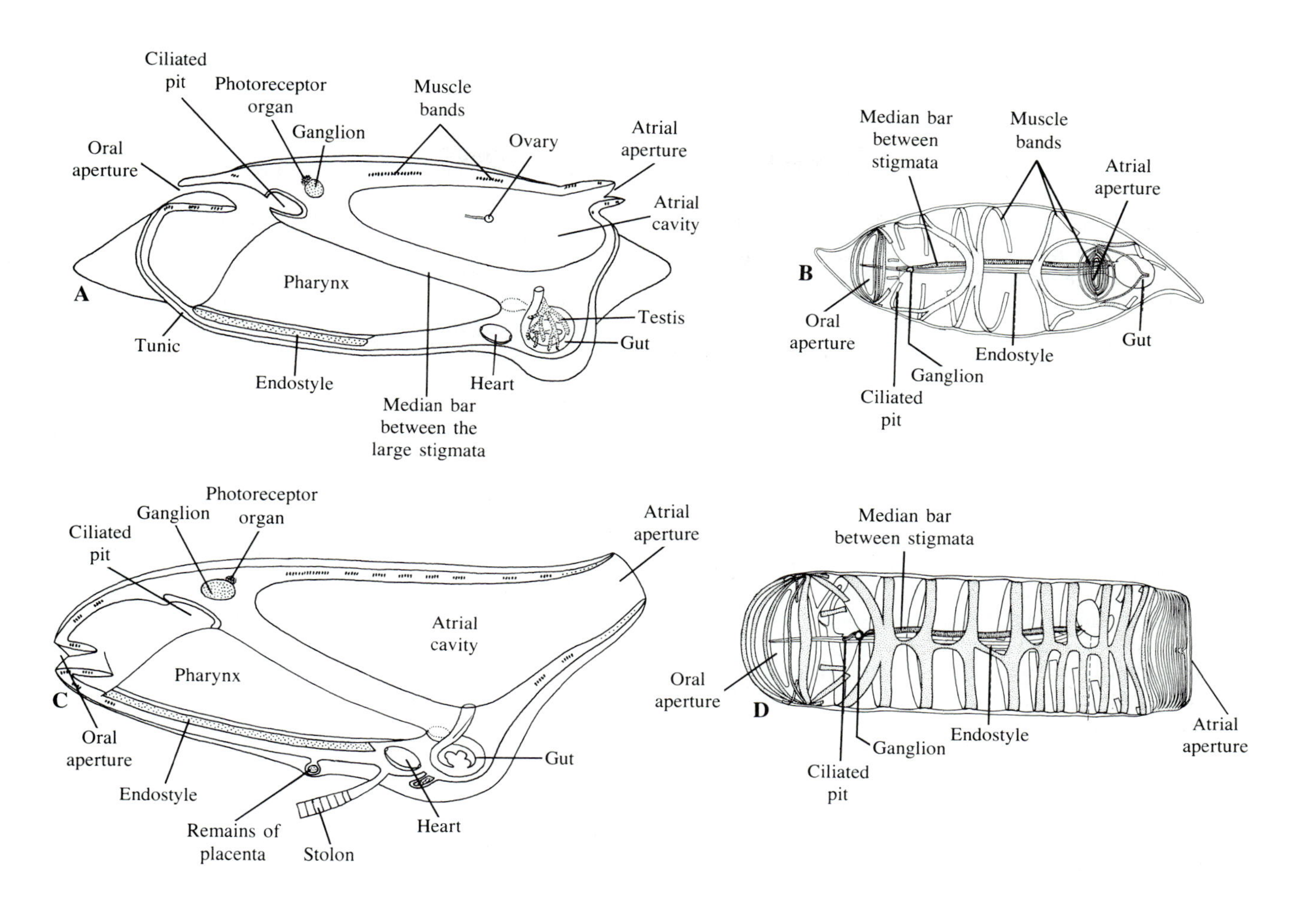

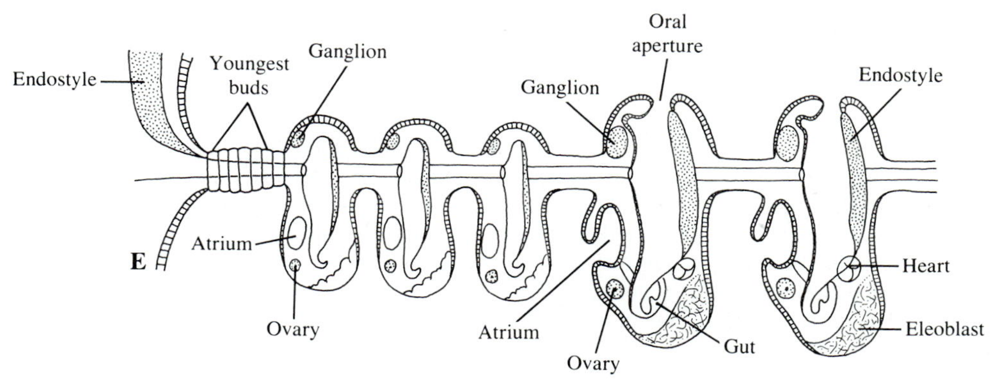

FIGURE 24.21 Order Salpida. A. Gonozooid, viewed from the left side; diagrammatic, and with most structures on the midline shown in optical section. B. *Ritteriella amboinensis,* gonozooid, dorsal view. (Meurice, Annales de la Societé Royale Zoologique de Belgique, *100.*) C. Oozooid, viewed from the left side; diagrammatic, and with most structures on the midline shown in optical section. D. *Ritteriella amboinensis,* oozooid, dorsal view. (Meurice, Annales de la Société Royale Zoologique de Belgique, *100.*) E. A chain of prospective gonozooids being produced by the stolon of an oozooid; diagrammatic. As the gonozooids enlarge, some shift to the right, some to the left of the stolon. They become tightly packed and joined by epidermal papillae. The eleoblasts are centers of blood-cell formation.

Some other structures that are generally visible are the ganglion, a ciliated pit on the dorsal side of the pharynx, a photoreceptor organ, and the heart.

Gonozooids of most salps function first as females, then as males. The ovary lies on the right side, opening by a duct into the atrial cavity. The testis is closely associated with the gut loop, and its duct runs to the ventral side of the atrium. As a rule, only one egg is produced by the ovary, and it is surrounded by a follicle of other cells. Fertilization takes place in the follicle, and the embryo develops in close association with a kind of placenta that is produced in part by the atrial epithelium and in part by the wall of the oviduct. Development into an oozooid is more or less direct, although for a time the embryo has an outgrowth that resembles a tail.

Oozooids (Figure 24.21, C and D) are rather similar to gonozooids, except that instead of having gonads, they have a stolon for asexual reproduction. The stolon is usually discernible by the time a young oozooid is released from the atrium of its mother. As the animal grows, the stolon lengthens into a spiral band on which the gonozooids differentiate. Those farthest from the body of the oozooid are the largest. The gradation in size is not necessarily continuous, however. In most species, several individuals of approximately the same size form a group (Figure 24.21, E). Although the chain of enlarging gonozooids tends to break up into shorter lengths, the salps in a particular sequence may cohere for a time. They are not only attached to the stolon, but they also adhere to one another by papillae. Successive buds become displaced to the right and left sides of the stolon, and because they are pressed together they become asymmetrical. The atrial aperture is shifted to the right or left, depending on the position of the gonozooid in the chain.

CLASS LARVACEA

Larvaceans are so named because they resemble larvae of ascidians. They are strictly planktonic animals, and most of them are less than 5 mm long. The trunk region, containing the gut, gonads, most of the nervous system and sense organs, and usually a heart, is compactly organized (Figure 24.22, A and B). The tail, with a prominent notochord, is set off from the trunk by an obvious constriction. The broader surfaces of the tail are at a right angle to the sagittal plane of the trunk, a feature found also in a few ascidian larvae.

The mouth, at the tip of the trunk, leads into the pharynx. Two ducts that originate as outpocketings of the pharynx connect this part of the gut with ventrolateral openings. Although there is no atrial cavity, the external openings are comparable to the stigmata of ascidians and thaliaceans. It is conventional, however, to call them **spiracles** (Figure 24.22, B and C). Each of the two ducts has a ring of cells provided with long and crowded cilia. The activity of these cilia draws food and water into the pharynx and drives water out through the spiracles.

As in other urochordates, the pharynx has a mucus-secreting endostyle. The mucus is molded into the shape of a food-catching cone, and as quickly as this is formed at one end, its other end is entering the esophagus as a continuous string. Digestion probably takes place mostly in the stomach, from which the intestine and rectum carry residues to the anus, located on the ventral surface.

Most larvaceans have a feature that is truly remarkable. This is a "house" that consists of a transparent mucopolysaccharide (Figure 24.23, A). It is abandoned and replaced when necessary, sometimes as frequently as every 2 or 3 hours. In a sample of plankton gathered with a net, there may be numerous larvaceans, but these will have been dislodged from their houses and will not be inclined to secrete new ones. Dipping up specimens with a jar or similar container is the only way to collect them in good condition. The house not only encloses the animal, but has much to do with the collection of food. It is secreted by epidermal cells that are arranged in a precise manner on the trunk (Figure 24.22, D). After the house separates from the trunk and expands, it has the customary shape and the appropriate openings.

A widespread genus of house-building larvaceans is *Oikopleura*. The following description of the relationship between the animal and its house applies specifically to *Oikopleura albicans*, but most other members of the genus conform to it with respect to major points. Related genera with a similar arrangement are *Megalocercus* and *Stegosoma*.

While within its house, *Oikopleura* lies in a chamber that is entered from the dorsolateral surfaces by two incurrent funnels (Figure 24.23, A). Each funnel is covered by a screen that is part of the house. As the broad tail of the animal undulates, the motion causes water to pass through the screens and funnels into the chamber. The openings in the screens are about 15 or 20 μm in diameter, so particles too large to be suitable as food are not admitted. (In other species, the size of the mesh openings ranges from about 15 μm to 50 μm.)

At the level of the posterior part of the animal's tail, the chamber has two lateral pockets. These lead upward to a **food trap** that is attached to the mouth but that is properly part of the house. The trap, spread out into two wings, is complicated, and some aspects of its structure and the way it operates are still not understood. It consists, however, of at least two membranes

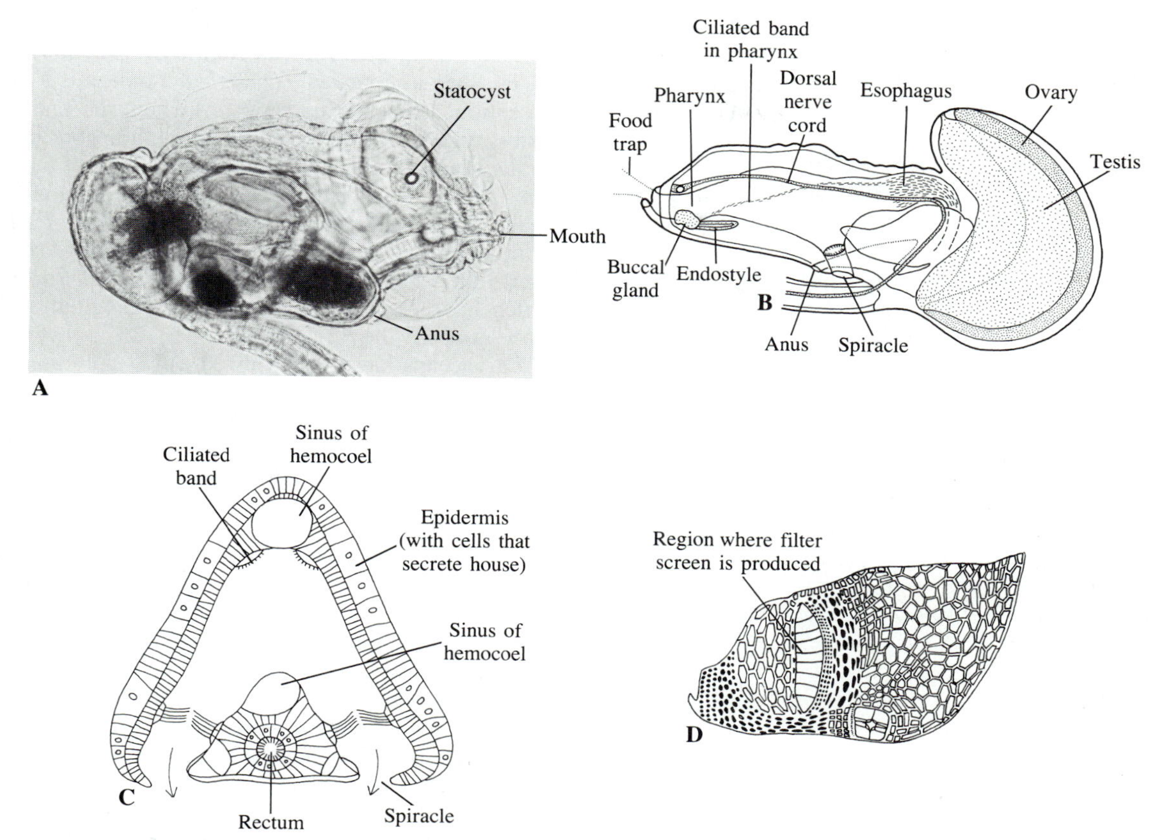

FIGURE 24.22 Order Larvacea. A. *Oikopleura dioica,* trunk and portion of the tail; photomicrograph. B. *Oikopleura dioica,* trunk region; diagrammatic. Part of the testis is overlapped by the ovary. (After Van Gansen, Annales de la Société Royale Zoologique de Belgique, *90.*) C. *Oikopleura albicans,* transverse section through the pharynx at the level of the spiracles; diagrammatic. D. *Oikopleura dioica,* pattern of epidermal oikoplasts (glands that secrete the house), from the left side. (Lohmann and Bückmann, Deutsche Südpolar-Expedition 1901–03. Zoologie, *10.*)

that enclose parallel tubes that resemble those of an air mattress. The ends of the tubes are open and allow bacteria, flagellates, and small diatoms to enter and move toward a median channel that leads to the animal's mouth. At intervals of a few seconds, the animal draws in the particles, as if it were sucking on a soda straw. The sucking action is effected largely by the beating of the long cilia located in each spiracular duct.

Most of the water that has been processed by the trap moves backward and then out of the house by way of an exit passage. The current of water leaving the house is generally strong enough to keep the house moving slowly forward. Some water, of course, is taken into the mouth with food. This water leaves the pharynx by way of the spiracles and thus reenters the chamber in which the animal lies.

By temporarily suspending tail movements and simultaneously reversing the direction of ciliary beat in the spiracular ducts, *Oikopleura* may be able to flush out its dorsolateral filter screens. Eventually, however, these become badly clogged, and the interior of the house may be fouled by fecal pellets. *Oikopleura* then abandons the house. In *O. albicans,* the anteroventral wall is thin and it has been claimed that the animal ruptures it in order to escape. In other species, powerful movements of the tail simply break up the house. In just a few minutes, a new house becomes fully functional.

Larvaceans of the genus *Fritillaria* do not construct a true house comparable to that made by *Oikopleura* and its close relatives. Instead, there is a large mucopolysaccharide bubble attached to the mouth. How this

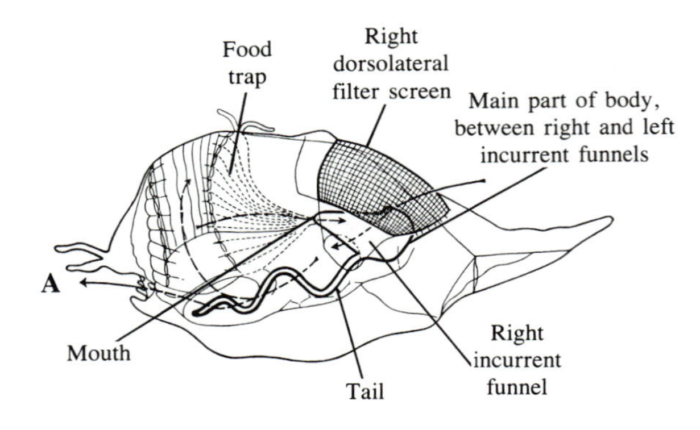

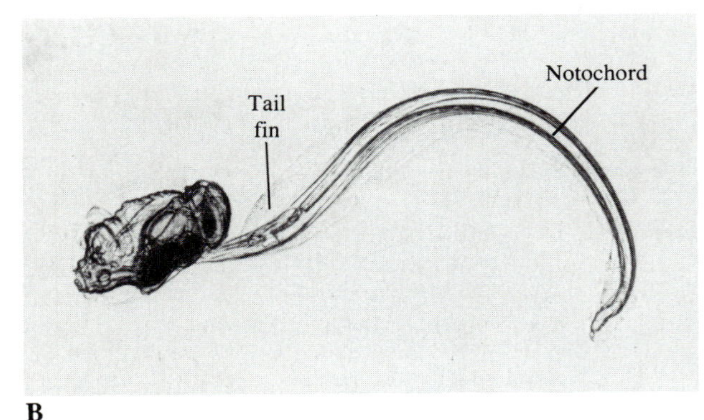

B

Figure 24.23 Order Larvacea. A. *Oikopleura albicans* in its house. Arrows show the main water currents. (Van Gansen, Annales de la Société Royale Zoologique de Belgique, *90*.) B. *Oikopleura dioica,* out of its house; photomicrograph.

is used in feeding is not known. In *Kowalevskaia,* the animal occupies the concavity of a secreted structure that slightly resembles a jellyfish. Water comes in and goes out through the broad opening, and groovelike channels direct food particles toward the mouth. Another odd feature of *Kowalevskaia* is that its jellyfish-like "umbrella" is loosely enclosed by a transparent, balloonlike membrane.

Oikopleura and other larvaceans may form a significant part of the biomass of marine zooplankton, especially in certain areas or at certain seasons. In some tropical waters, they often make up about 10% of the biomass. They provide food for various carnivores, and their abandoned houses are eaten by certain copepods.

The houses and developing house rudiments of some larvaceans contain clusters of small granules that emit light on mechanical stimulation. This type of bioluminescence, which may at times represent a considerable part of the light flashes observed in the sea at night, is one of the exceptions to the rule that bioluminescence takes place in cells.

Nearly all larvaceans are hermaphroditic, and most of them are protandric; that is, they function first as males, then as females. (*Oikopleura dioica* is a notable exception; it has separate sexes.) There are one or two testes, and sperm are released through a pore or pores on the dorsal surface near the posterior end of the trunk. The single ovary discharges its eggs through a rupture in the body wall, and the animal then dies. Early development conforms to the pattern typical of ascidians, but on the whole it is speeded up. In *Oikopleura dioica,* which has been studied extensively, gastrulation takes place at about the 32-cell stage.

Larvaceans have been viewed as sexually mature ascidian tadpoles, and also as relics of an early group of urochordates. Other interpretations of how they arrived at their present condition have been proposed, too. One of them suggests that the ancestors of larvaceans were doliolids.

Summary

1. The urochordates, which are marine animals, are characterized by a perforated pharynx that acts as a respiratory structure and also as a sieve for straining microscopic food from the water that is brought into it by ciliary activity. They have a dorsal nerve cord and usually a notochord, although the latter may be present only in the larval stage, and the former may be greatly modified when a larva undergoes metamorphosis to the adult stage.

2. Because of the features just mentioned, urochordates are often considered to be closely related to cephalochordates and vertebrates. Unlike the last two groups, however, urochordates do not have a coelom. The circulatory system, moreover, is of the open type, with relatively few definite blood vessels.

3. The largest of the classes of urochordates, the Ascidiacea, consists of animals that are sessile in the adult stage. Most of them have a notochord, but it is restricted to the tail region of the tadpolelike larval stage, and it disappears during the drastic metamorphosis into the adult. Other structures of the larva, including the dorsal nerve cord, become reduced or reoriented.

4. Ascidians usually have numerous pharyngeal perforations. These are called stigmata. As water that has entered the pharynx through the oral siphon passes through the stigmata, it enters a space called the atrium, and from this it leaves the body by an atrial siphon. Food caught in mucus that lines the interior of the pharynx is moved by ciliary activity into more posterior portions of the gut, where digestion and formation of fecal pellets take place. The rectum opens into the atrium, along with ducts from the gonads, so that fecal wastes and gametes are discharged with the stream of water leaving the atrium.

5. Adult ascidians have a secreted covering called the tunic. This is usually gelatinous or leathery, and it is often thick. It consists mostly of a polysaccharide that is chemically related to cellulose. Some ascidians reproduce asexually, and thereby form colonies in which several to many member zooids are connected by stolons or sheets of tissue and tunic material, or in which the zooids are embedded in a common tunic.

6. Representatives of the small class Larvacea are planktonic and resemble ascidian tadpoles. They are, nevertheless, sexually mature animals, and their two pharyngeal perforations open directly to the outside instead of into an atrium. Most larvaceans secrete a mucopolysaccharide "house" that not only loosely surrounds them but that is also equipped with screens for primary processing of water. Thus only fine particles that can be collected by the elaborate food-gathering mechanism can enter the channels within the house.

7. The class Thaliacea is represented by pyrosomes, doliolids, and salps. All of these are strictly planktonic. A tunic is present, but there is no notochord except in one stage of the life cycle of doliolids. Pyrosomes, by asexual reproduction, form colonies that are tubular but open only at one end. Water that passes through the pharynxes of the many members of the colony exits through the opening, so the colony moves by jet propulsion. Doliolids and salps reproduce asexually, too, and the products of this type of multiplication may adhere to one another in chainlike aggregates. The atrial cavities of the individuals in an aggregation do not, however, join a single cavity comparable to that from which water leaves the true colony of a pyrosome.

REFERENCES ON UROCHORDATA

Alldredge, A., 1976. Appendicularians. Scientific American, *235*(1):94–102.

Alldredge, A., 1977. House morphology and mechanisms of feeding in the Oikopleuridae (Tunicata: Appendicularia). Journal of Zoology (London), *181*:175–188.

Barrington, E. J. W., 1965. The Biology of Hemichordata and Protochordata. Oliver and Boyd, Edinburgh and London. W. H. Freeman, San Francisco.

Barrington, E. J. W., and Jefferies, R. P. S., 1975. Protochordates. Symposia of the Zoological Society of London, 36. Academic Press, London and New York.

Berrill, N. J., 1950. The Tunicata. With an Account of the British Species. Ray Society, London.

Braconnot, J.-C., 1970. Contribution à l'étude des stades successifs dans le cycle des tuniciers pélagiques doliolides. I. Les stades larvaire, oozoide, nourrice, et gastrozoide. Archives de Zoologie Expérimentale et Générale, *111*:629–668.

Braconnot, J.-C., 1971. Contribution à l'étude des stades successifs dans le cycle des tuniciers pélagiques doliolides. II. Les stades phorozoide et gonozoide. Archives de Zoologie Expérimentale et Générale, *112*:5–31.

Cavey, M. J., and Cloney, R. A., 1976. Ultrastructure and differentiation of ascidian muscle. I. Caudal musculature of the larva of *Diplosoma macdonaldi*. Cell and Tissue Research, *174*:289–313.

Cloney, R. A., 1977. Larval adhesive organs and metamorphosis in ascidians. I. Fine structure of the everting papillae of *Distaplia occidentalis*. Cell and Tissue Research, *183*:423–444.

Cloney, R. A., 1982. Ascidian larvae and the events of metamorphosis. American Zoologist, *22*:817–826.

Fenaux, R., 1986. The house of *Oikopleura dioica* (Tunicata, Appendicularia): structure and function. Zoomorphology, *106*:224–231.

Fraser, J. H., 1986. British Pelagic Tunicates. Synopses of the British Fauna, no. 20. Cambridge University Press, London and New York.

Galt, C. P., Grober, M. S., and Sykes, P. F., 1985. Taxonomic correlates of bioluminescence among appendicularians (Urochordata: Larvacea). Biological Bulletin, *168*:125–134.

Goodbody, I., 1974. The physiology of ascidians. Advances in Marine Biology, *12*:1–149.

Jones, J. C., 1971. On the heart of the orange tunicate, *Ecteinascidia turbinata*. Biological Bulletin, *141*:130–145.

Millar, R. H., 1970. British Ascidians. Synopses of the British Fauna, no. 1. Linnean Society of London and Academic Press, London and New York.

Millar, R. H., 1971. The biology of ascidians. Advances in Marine Biology, *9*:1–110.

Monniot, C., and Monniot, F., 1972. Clé mondiale des genres d'ascidies. Archives de Zoologie Expérimentale et Générale, *113*:311–367.

Monniot, C., and Monniot, F., 1978. Recent work on the deep-sea tunicates. Oceanography and Marine Biology, An Annual Review, *16*:181–228.

Monniot, C., Monniot, F., and Gaill, F., 1975. Les Sorberacea: une nouvelle classe de tuniciers. Archives de Zoologie Expérimentale et Générale, *116*:77–122.

Plough, H. J., 1978. Sea Squirts of the Atlantic Continental Shelf. Johns Hopkins University Press, Baltimore.

Stoecker, D., 1980. Relationship between chemical defense and ecology in benthic ascidians. Marine Ecology—Progress Series, *3*:257–265.

Young, C. M., 1988. Ascidian cannibalism correlates with larval behavior and adult distribution. Journal of Experimental Marine Biology and Ecology, *117*:9–26.

25

Phylum Cephalochordata

Lancelets

Introduction

The few species of cephalochordates, called lancelets, are marine animals that live in coarse sand. They dig in, tail first, until all but the anterior tip of the body is buried. Water is drawn in through the mouth, and as it passes through the gill slits of the pharynx, food is trapped in mucus that coats the inside of this part of the gut. Thus cephalochordates collect food by the same technique used by urochordates.

Two genera are recognized: *Branchiostoma* and *Asymmetron*. The former, perhaps better known by its synonym *Amphioxus,* is the more likely to be available for study. It has representatives in shallow water in temperate and tropical seas. The larger species of this genus reach a length of about 8 cm; one of them is collected for food in China.

External Features, Buccal Cavity, and Musculature

All species of *Branchiostoma* are elongated and pointed at both ends (Figure 25.1, A and B). There is no distinct head in these animals, but the mouth, on the ventral side close to the anterior tip of the body, is easy to recognize. It is fairly large and is bordered by slender outgrowths called **buccal cirri** (Figure 25.1, C). Separating the buccal cavity from the pharynx is a ringlike diaphragm, the **velum,** which is operated by circular muscles. Anterior to the velum are some lobes that exhibit intense ciliary activity, and pointing backward from the velum into the pharynx are about 12 small **velar cirri** similar to those around the mouth. In addition to the mouth, there are two other openings on the ventral side: the **atriopore,** through which water processed by the pharynx leaves the body, and the nearly terminal anus, displaced slightly to the left.

Running along the dorsal surface for most of the length of the body is a ridgelike **fin.** A similar elaboration is found on the ventral surface behind the atriopore. Both of the fins are more prominent in the tail region than elsewhere, and both are supported by connective tissue. Anterior to the atriopore, much of the ventral surface is flattened and bordered by two longitudinal ridges called **metapleural folds** (Figure 25.2, A and B).

The body-wall muscles of *Branchiostoma* are segmentally arranged, chevron-shaped masses. There are about 60 on each side, but those on the right and left are not directly opposite one another. The way in which they alternate increases the efficiency of the lateral bending movements by which the animal swims and burrows.

Notochord

The **notochord** (Figures 25.1, 25.2) runs for nearly the entire length of the body. It consists of many nucleated, disklike units, arranged in much the same way

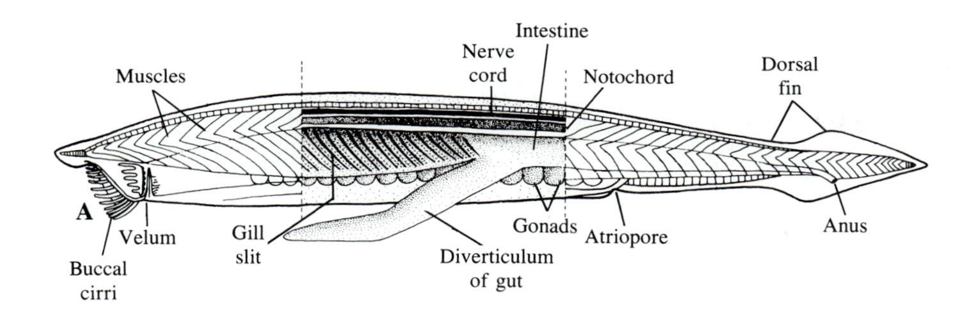

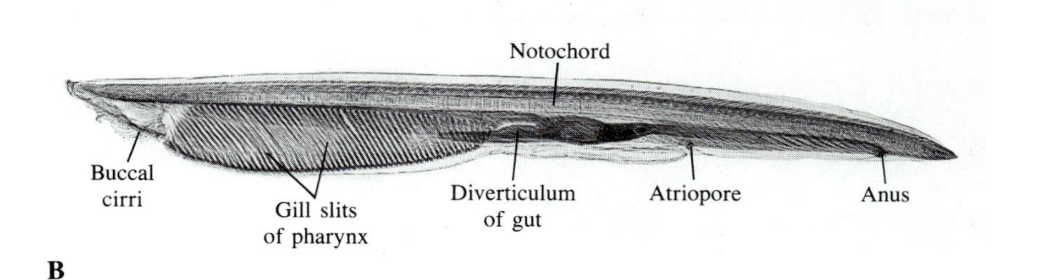

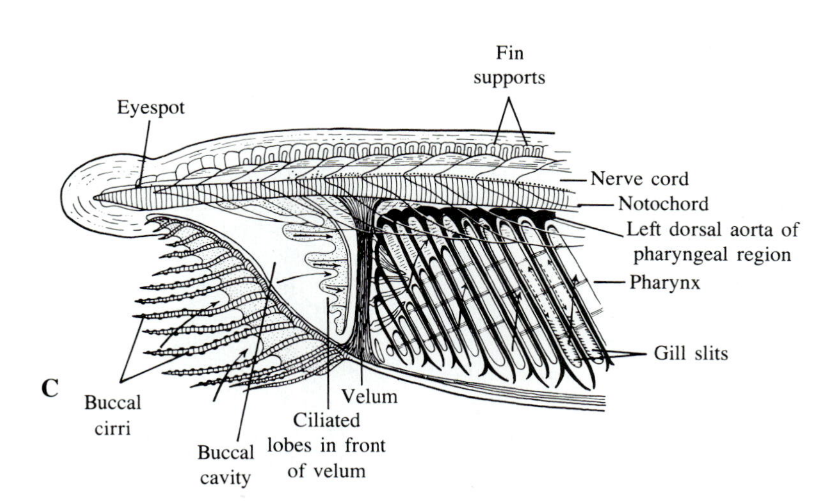

FIGURE 25.1 *Branchiostoma* (*Amphioxus*). A. Diagram showing general morphology. A portion of the body wall has been cut away and the diverticulum of the gut has been pulled out. (Godeaux, in Florkin and Scheer, Chemical Zoology, *8*.) B. Photomicrograph of a small stained specimen. (Courtesy of Carolina Biological Supply Company.) C. Anterior portion of the body. Four velar cirri are shown projecting backward from the velum. (Godeaux, in Florkin and Scheer, Chemical Zoology, *8*.)

as coins in a stack. The units are now regarded as muscle cells, for they contain cross-striated filaments of paramyosin and they contract when activated by motor nerves. There are large vacuoles within them, as well as fluid-filled spaces between them and beside them. New units are added as the animal grows; they are believed to derive from certain cells wedged between existing units.

The notochord and fluid-filled spaces associated with it collectively form a hydrostatic skeleton. This is enclosed by a tight-fitting sheath of connective tissue. When the muscle units contract, the diameter of the notochord is decreased, but the structure becomes longer and stiffer. The functional significance of the notochord's adjustability is not fully understood, but it probably influences the operation of segmental muscles as they bring about specific swimming or burrowing movements.

NERVOUS SYSTEM AND SENSE ORGANS

The **nerve cord,** dorsal to the notochord, is like that of vertebrates in being tubular. It lacks anything that might be called a brain, but the anteriormost part of the central canal is dilated to form a small vesicle. Some black pigment in the cells of the cord at the tip of the vesicle is concentrated into what resembles an eyespot

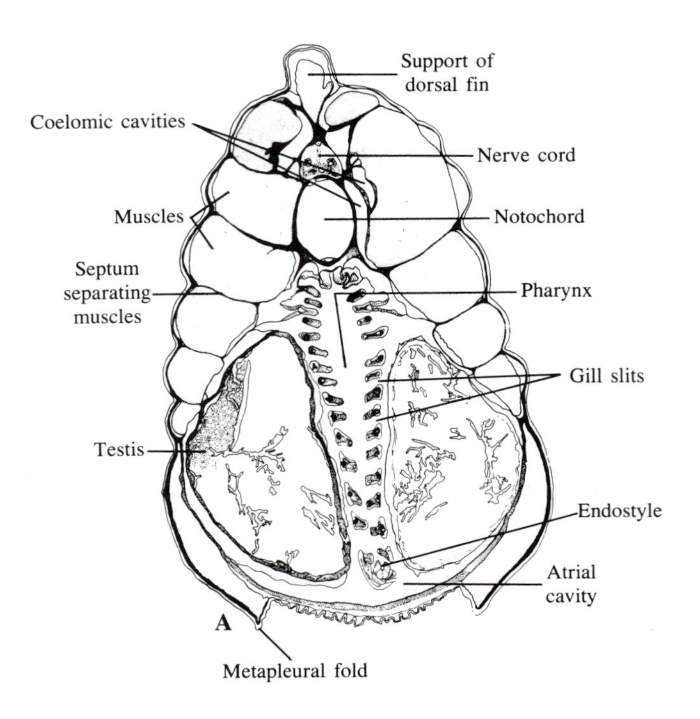

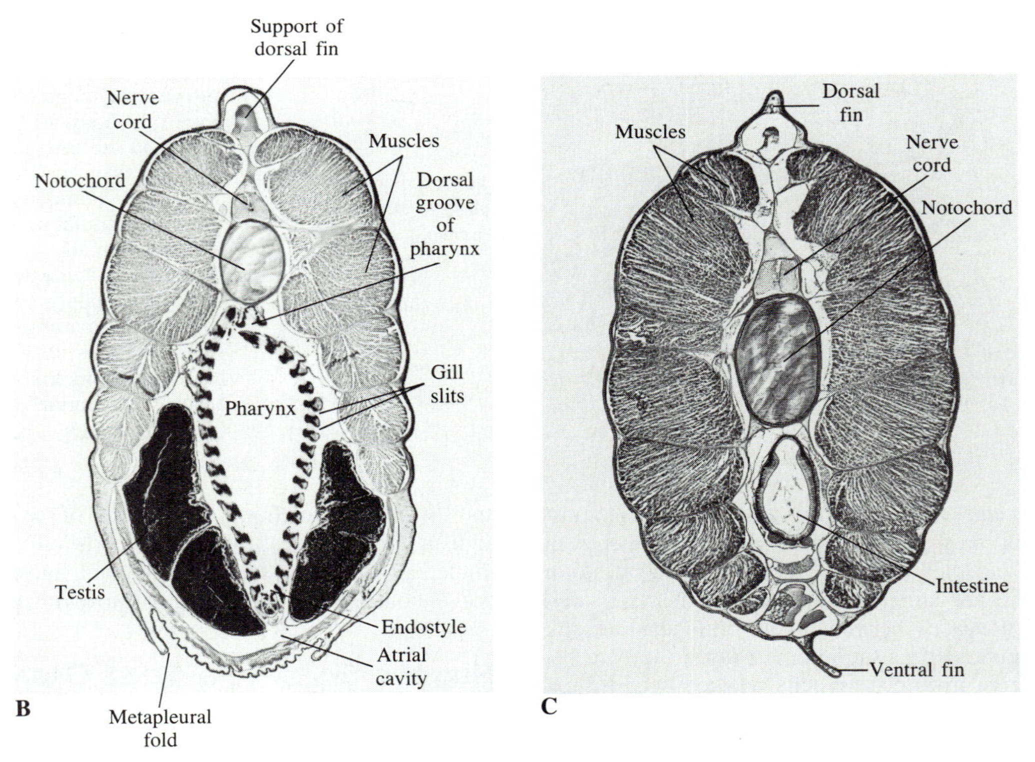

FIGURE 25.2 *Branchiostoma*. A. Transverse section through the pharyngeal region. (Godeaux, in Florkin and Scheer, Chemical Zoology, *8*.) B. Transverse section through the pharyngeal region; photomicrograph. (Courtesy of Carolina Biological Supply Company.) C. Transverse section through the intestinal region; photomicrograph. (Courtesy of Carolina Biological Supply Company.)

(Figure 25.1, C). This almost certainly is part of a photoreceptor that perceives light or shadows, but little is known about it.

Although the nerve cord is not ganglionated, it sends off branches to both sides of the body. The arrangement of these branches corresponds closely to the alternating arrangement of the muscles of the body wall. Of the two nerves that go out to the right or left at a particular transverse level, the more nearly dorsal one (which soon divides) contains both sensory and motor elements; the more nearly ventral one is a strictly motor nerve.

Touch receptors are scattered throughout the epidermis and are especially well represented in the buccal and velar cirri. The latter also seem to have chemoreceptors. A possible photoreceptor has already been mentioned in connection with the vesicle in the anteriormost part of the nerve cord. A ciliated pit in the skin above the vesicle has been viewed as an olfactory organ, but its function is not really known. The pit is evidently a remnant of the pore formed when the neural groove of the embryo closes over and becomes a tube.

Feeding and Digestion

The passage of water and food into the mouth and through the pharynx and gut depends largely on the action of cilia. The lightly developed musculature of some parts of the digestive tract probably helps keep things moving. Diatoms, protozoans, and detritus derived from breakdown of plants constitute much of the food accepted by the animal. After passing the buccal cirri, the food must go through a second filter consisting of the velar cirri. The velum encircles the original embryonic mouth, which is pushed inward when a stomodeal invagination develops and becomes the buccal cavity. The **pharynx** proper (Figure 25.2, A and B) has nearly 200 serially arranged **gill slits,** those on the left alternating with those on the right. An **endostyle,** located midventrally in the pharynx, produces mucus, and it has ciliated tracts that drive the mucus upward onto the perforated and ciliated walls of the pharynx. The sheet of mucus, with food particles trapped in it, finally reaches a ciliated groove on the dorsal side of the pharynx and is moved backward into the short esophagus, then into the intestine (Figure 25.2, C). Water leaving the pharynx through the gill slits enters the **atrium,** then exits by way of the atriopore.

The endostyle is interesting in that it not only produces mucus but also binds iodine. For this reason it is thought of as the homologue of the thyroid gland of vertebrates, which originates as a pocket on the ventral side of the pharynx.

Digestion of carbohydrates proceeds in the lumen of the intestine, but proteins and fats are digested intracellularly after they have been phagocytized. On the ventral side of the gut, at the point where the esophagus joins the intestine, is an anteriorly directed diverticulum (Figure 25.1, A and B). The glandular epithelium of this sac produces some digestive enzymes, moved by ciliary activity to the intestine. Food particles may also enter the diverticulum itself.

Most of the indigestible residues are eliminated through the anus. Some, however, are taken up by amoebocytes and carried to certain tissues where they are simply stored. Portions of the metapleural folds and the connective tissue associated with the gonads are among the places where such residues accumulate.

Respiration and Circulation

Although the pharynx is a food-collecting organ, it is also well suited to exchange of oxygen and carbon dioxide. Blood is delivered to the pharynx by a ventral vessel called the branchial artery or ventral aorta (Figure 25.2, B). The vessel that enters each primary gill bar is dilated into a little "heart." The hearts pump blood through the capillary networks in the gill bars and the oxygenated blood is then collected by two parallel dorsal aortas. Behind the pharynx these become united into a single aorta. The aortas distribute blood to various parts of the body, especially the intestinal region. The most prominent vein is the one beneath the intestine, but there are other important veins through which blood flows in the direction of the branchial artery. In general, the pattern of circulation in *Branchiostoma* is similar to that in lower vertebrates. The blood itself, however, lacks cellular components and has no pigments.

Coelom and Excretory Organs

The **coelom** is derived from numerous cavities. The more anterior of these originate as outpocketings of the archenteron; the others develop within solid masses of mesoderm adjacent to those that form the segmental muscles of the body wall. Thus both enterocoely and schizocoely are involved in the formation of the coelom. The way in which the essentially segmental spaces coalesce to form the coelomic cavities found in the adult is not described here. It will be best to say simply that the coelom consists of three principal portions: two of these are small cavities that run longitudinally between the notochord and lateral muscle masses, and the third is a moderately spacious cavity that partly envelops the intestine and atrium and also pervades the gill bars of the pharynx.

The excretory organs, situated in remnants of the coelom above the pharynx, are serially arranged. There is basically one pair of these structures for each pair of gill slits. They are ectodermal **protonephridia** that have many clusters of solenocytes, and their pores open into the atrial cavity. Closely applied to one side of each protonephridium are some blood vessels that are essentially part of the system of pharyngeal circulation.

Reproduction and Development

Sexes are separate. The gonads, limited to the pharyngeal region of the body, are located in ventrolateral divisions of the main body coelom (Figure 25.2, A and B). There are about 25 of them on both sides, and their arrangement is similar to that of the gill slits and muscle masses of the body wall in that those on the left alternate with those on the right. The gonads originate beneath the peritoneum, and when fully developed they bulge into the coelom. There are no permanent gonoducts. Gametes are released by the gonads into the coelom and reach the atrium through ruptures in the peritoneum, atrial epidermis, and thin intervening layer of connective tissue.

The evolution of these animals is probably linked to that of early vertebrates, so it is natural that their development, as well as their structure, has been intensively studied. Cleavage is of the **radial** type, as in urochordates. There is little yolk in the zygote, and divisions are equal. The blastula is hollow, and gastrulation is by invagination. The outer layer of the gastrula produces the epidermis and the nerve cord. The latter originates as a plate that roofs over to form a closed tube. The archenteron of the gastrula survives as the gut, and although the blastopore closes, the definitive anus later opens up in the same general region. After the embryonic mouth forms, a stomodeal invagination pushes it inward. The mouth of the adult is thus a secondary opening; the original mouth is at the level of the velum.

After the endodermal cells have invaginated during gastrulation, some cells at the blastoporal end of the embryo move into the blastocoel. These constitute embryonic mesoderm. It has already been explained, in connection with the coelom, that the more anterior coelomic spaces originate from the archenteron and that the others arise by schizocoely within masses of mesoderm. The mesoderm also gives rise to muscle, the circulatory system, and connective tissue. The first few gill slits open directly to the outside, but soon the metapleural folds grow downward to create the atrium, and the old and new gill slits now open into this cavity. The medial edges of the right and left metapleural folds become joined along the midline; the atriopore is the only survivor of the originally broad ventral opening.

This summary of development is greatly simplified. It does not cover many important events in the complicated embryology and organogenesis of cephalochordates. It should be sufficient, however, to show that these animals are deuterostomes and that their development is especially similar to that of urochordates and vertebrates.

Summary

1. Cephalochordates form a small phylum of marine invertebrates. They slightly resemble fishes because of their general shape and because they have a dorsal and a ventral fin; both of the fins are most prominent in the tail region, behind the anus. These animals live in coarse sand, with only the anterior tip of the body exposed. Water is brought into the pharynx through the mouth, and microscopic food that has been trapped in mucus is moved backward into the intestine. Water leaves the pharynx through many paired gill slits, which open into a cavity called the atrium. This has a single pore to the outside.

2. The notochord, beneath the dorsal tubular nerve cord, is decidedly different from that of tunicates; it consists of disklike muscle cells. The musculature of the body wall is organized into numerous segmental units.

3. Cephalochordates have a coelom, a rather well-developed circulatory system, and protonephridia consisting of solenocytes. Sexes are separate, and gametes leave the body by way of the atrial cavity. Cleavage is radial, and development, in general, follows the deuterostomatous plan because the anus develops in the vicinity of the blastopore. Certain portions of the coelom originate from the archenteron; others arise as spaces within masses of mesoderm.

References on Cephalochordata

Barrington, E. J. W., 1965. The Biology of Hemichordata and Protochordata. Oliver and Boyd, Edinburgh and London; W. H. Freeman, San Francisco.

Barrington, E. J. W., and Jefferies, R. P. S. (editors), 1975. Protochordates. Symposia of the Zoological Society of London, no. 36. Academic Press, London and New York.

Drach, P., 1948. Embranchement des Céphalochordés. *In* Grassé, P.-P. (editor), Traité de Zoologie, *11*:931–1036. Masson et Cie., Paris.

Rähr, H., 1982. Ultrastructure of the gill bars of the *Branchiostoma lanceolatum* with special reference to gill skeleton and blood vessels. Zoomorphology, *99:167–180.*

26

INVERTEBRATE PHYLOGENY

INTRODUCTION

The pronouncements of zoologists with respect to evolution of invertebrates range from expressions of hopelessness to wild fantasies. Between the extremes, however, there are many theories, ideas, and suggestions that have endured because they seem logical or because they are supported by facts, or both. Although discussions of phylogeny require considerable speculation, they are useful intellectual exercises. They help us consolidate our understanding of the phyla and other major taxa, and they make us reevaluate the criteria that are thought to indicate affinities between groups.

PHYLOGENY AND TAXONOMY

One of the goals of biology is to learn as much as possible about **phylogeny**—the history and evolutionary relationships of organisms, living and extinct. Although we may wish to trace the lineage of all organisms back to the ultimate common ancestor, which was probably a self-replicating organic molecule, it is not likely that this goal can ever be reached. This should not stand in the way of attempts to learn all that we can about the history of life on earth. We should certainly try to make sure that all taxa, from phyla down to genera, are **monophyletic,** that is, that the structure, development, and other attributes of the members of each taxon indicate that they derive from one ancestral stock, rather than from two or more separate evolutionary lines.

Highly modified organisms may be difficult to place, especially if they have lost the digestive tract, appendages, or some other cardinal features of the groups to which they belong. Nevertheless, it is often easier to put some of the most bizarre animals into appropriate taxa than it is to decide on the closeness of the relationship between certain of the phyla, such as Annelida and Mollusca, or Cnidaria and Ctenophora.

Taxonomy is the field of biology concerned with defining species, genera, families, and higher taxa, giving them names, and organizing them into systems of classification. In modern biology, taxonomy depends on criteria that imply the evolutionary unity or divergence of taxa, and it thus makes use of our understanding of phylogeny.

LIMITATIONS OF THE FOSSIL RECORD

Rocks more than 3 billion years old contain fossils of prokaryotic organisms that are similar to modern bacteria and blue-green algae. We do not know when eukaryotic organisms first appeared, but deposits 1 billion years old include fossils of protozoans and unicellular and multicellular algae, as well as of organisms that were probably similar to certain modern fungi.

By the beginning of the Cambrian era, about 600 million years ago, several diverse groups of multicellular animals had evolved. Most were soft-bodied organisms whose remains are poorly preserved. Some of the best impressions in pre-Cambrian rocks, in fact, are tracks, burrows, and borings, rather than the fossilized remains of the animals themselves. It is almost certain, however, that animals comparable to present-day cnidarians, a variety of worms, and arthropodlike invertebrates were flourishing in pre-Cambrian seas. Their remains, or other traces of their presence, are not substantial enough to enable us to assign them with confidence to modern phyla.

The pre-Cambrian fossil record is limited to those sedimentary deposits favorable to the preservation of soft-bodied invertebrates or other evidences of their existence. We are fortunate to have any early fossil record at all, because small or soft-bodied organisms are not likely to be preserved after they die, particularly in shallow seas where decomposition by microorganisms proceeds rapidly. If the seas in which living organisms began to diversify did not contain sufficient amounts of mineral salts to promote fossilization, or if the water were unsuitable for other reasons, the remains of dead organisms would not be preserved. For a long time, moreover, the invertebrates that dominated the seas probably fed on microorganisms or small plants. Until carnivory became an important mode of nutrition, natural selection would not have favored the evolution of exoskeletons, shells, and other protective coverings. After such hard parts became commonplace, more and more invertebrates became incorporated into the fossil record.

This chapter does not deal extensively with fossils. Emphasis is given instead to a discussion of relationships based on what is known about living invertebrates and their presumed unicellular ancestors.

Possible Origins of Multicellularity

What features should we expect to have been available before the simple metazoans began to evolve? The following characteristics found in various groups of unicellular organisms may be considered especially important, although some of them would not be essential:

1. The eukaryotic cell and its basic properties, including the nucleus, mitochondria, and flagellum.

2. Asexual reproduction by fission or budding.

3. Sexual union of similar gametes (isogamy) or gametes of different size and structure (anisogamy).

4. The basic modes of nutrition found in eukaryotic cells: photosynthetic autotrophy, or saprozoic or holozoic heterotrophy.

5. Locomotion, by flagella, amoeboid movement, or by some other mechanism, such as secretion of a substance or contraction effected by intracellular myofilaments.

6. Ability to secrete an inert skeletal structure, either an external coating (something like the perisarc of a hydroid) or an internal support (such as an organic fiber or inorganic spicule). Many protists, in fact, have external or internal skeletal supports.

The origin of multicellularity could have occurred in either or both of two general ways. It could have come about by cohesion of the cells resulting from fission of a single cell. This is how colonies of certain flagellates, both photosynthetic and nonphotosynthetic types, are formed, and it is exactly what happens during embryonic development of animals.

Multicellularity could also have originated by a cell's first becoming multinucleate, then becoming partitioned into separate cells. The theory of Hadži proposes that the parent type of unicellular organism was a ciliate. If a ciliate became multinucleate, and if a cellular and ciliated epidermis were to differentiate around a noncellular mass in the interior, the result would be an organism somewhat similar to an acoel turbellarian (Figure 26.1, A). In the theory of Chadefaud, which is more far-fetched than that of Hadži, the parent type would have been a flagellate characterized by a deep depression at one end, such as is typical of cryptomonads. If multinuclearity were followed by segregation of nuclei into cells, and if some of the cells became organized into an epidermal layer and others into a gastrodermal layer around the depression, an organism comparable to a gastrula would be produced (Figure 26.1, B).

While it is possible that a multinucleate organism could become multicellular, this seems unlikely. There are many known cases in which noncellular multinucleate tissues, called **syncytia,** develop from a tissue that is at first cellular. This is the case, for instance, in the digestive syncytium of acoel turbellarians, and also in skeletal muscle cells of vertebrates. But there are few if any cases in which the reverse is true.

Although the origin of multicellularity must have preceded the diversification of simple invertebrates, we do not know what the earliest multicellular organisms looked like. We can therefore only speculate on their structure, basing our speculations on evidence from embryology and on observations of certain colonial protozoans. It should be noted that multicellularity may have originated more than once, perhaps many times. This is suggested by the fact that sponges arrive at their definitive form in ways rather different from the course of early embryology in other lower invertebrates.

There is considerably more to interpreting the origin of multicellularity than merely visualizing cells sticking together to form a cohesive mass. Either all of the cells would have to feed, or certain of the cells would do this and somehow transfer nutrients to the other cells. What would be the physical form of the mass of cells? Would it be something like a pancake creeping over the surface of rock or seaweed, a ball that swims or floats freely, or perhaps a tubular organism with a mouth and gut, perhaps even an anus?

If embryology can be relied upon, we may imagine that a berrylike cluster of cells or a hollow ball of cells would precede a gastrulalike organism with a distinct gut. Among invertebrates, however, there are no existing organisms that are permanently berrylike or that resemble hollow balls. Even the flattened *Trichoplax* is not quite so simple as a blastula, for it has contractile cells and gamete-forming cells in the space enclosed by the epidermis.

It is likely that the earliest truly multicellular animal organisms derived from a group of flagellates in which coloniality had already become established. Volvocid flagellates have in the past been considered to be good prospects. In some thoughtful proposals, including that of Naef (Figure 26.2), a *Volvox*-like colony of cells (A), derived from a cluster of cells embedded in a common matrix, could have given rise to a creeping organism with phagocytic ventral cells (B). The invagination of the ventral cells to form a saclike gut (C) would have resulted in the segregation of gastrodermal cells from epidermal cells.

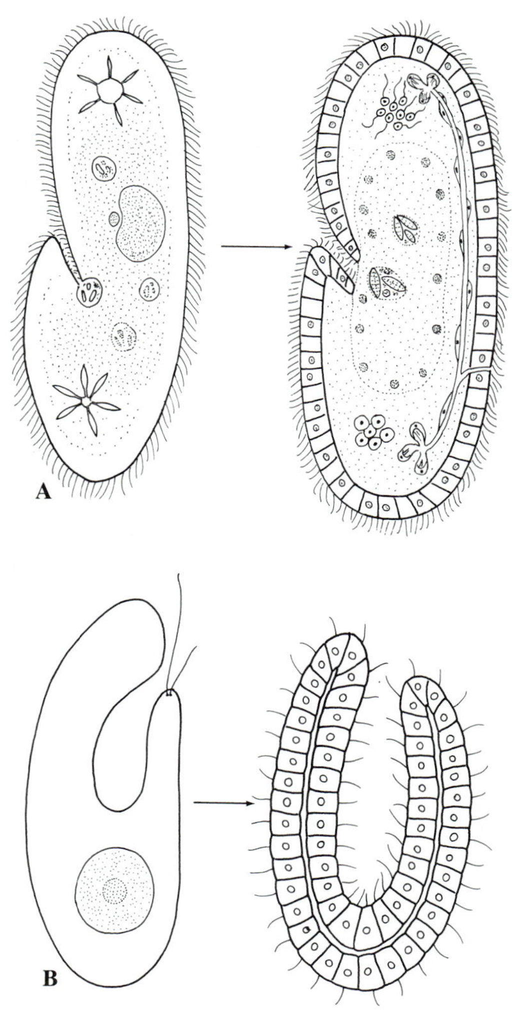

FIGURE 26.1 A. Hadži's idea of the possible origin of a multicellular animal from a ciliate. B. Chadefaud's idea of the possible origin of a multicellular organism from a flagellate characterized by a deep depression.

At the present time, choanoflagellates are most often singled out as probable ancestors of metazoans because their cells, provided with a microvillar collar around the flagellum, resemble choanocytes of sponges. There is another good reason for suspecting the choanoflagellates were involved in the evolution of metazoans: the infoldings (cristae) of their mitochondria are flattened rather than tubular, and in this respect they are like most metazoan cell types and unlike nearly all other nonphotosynthetic protozoans.

In choanoflagellate colonies, the cells are all similar, and they are held together by a matrix of jellylike material (Figure 26.3, A) or by cytoplasmic bridges that appear to be remnants of the process of fission. In metazoans, on the other hand, there is always at least some division of labor between the cells, and the cells are usually bound together by junctions of one type or another.

Cells with a microvillar collar encircling a single flagellum are not limited to choanoflagellates and sponges. The solenocytes of gastrotrichs, for example, closely resemble choanocytes and choanoflagellates; so do certain cells of epithelia that line coelomic cavities of echinoderms. It is possible, of course, that these collared cells have originated independently. Nevertheless, choanoflagellates are the best prospects, among known organisms, to be ancestors of metazoans. Sponges themselves probably were not involved in the ancestry of other surviving phyla. They have neither a gut nor anything comparable to an embryonic archenteron. They also do not secrete a basal membrane. These two features, taken together, seem to remove them from the mainstream of invertebrate evolution. The hexactinellid sponges, moreover, are to a considerable extent syncytial.

SOME POSSIBLE STEPS IN THE EVOLUTION OF EARLY METAZOANS

In the very earliest multicellular animal organisms, perhaps all cells functioned in feeding and also in locomotion. Eventually, some cells probably became specialized for feeding, others for locomotion. A layer of phagocytic cells on the underside would be expected to have been present in animals that glided over a substratum covered with microbes and organic detritus. Hypothetical organisms of this type figure in many phylogenetic schemes.

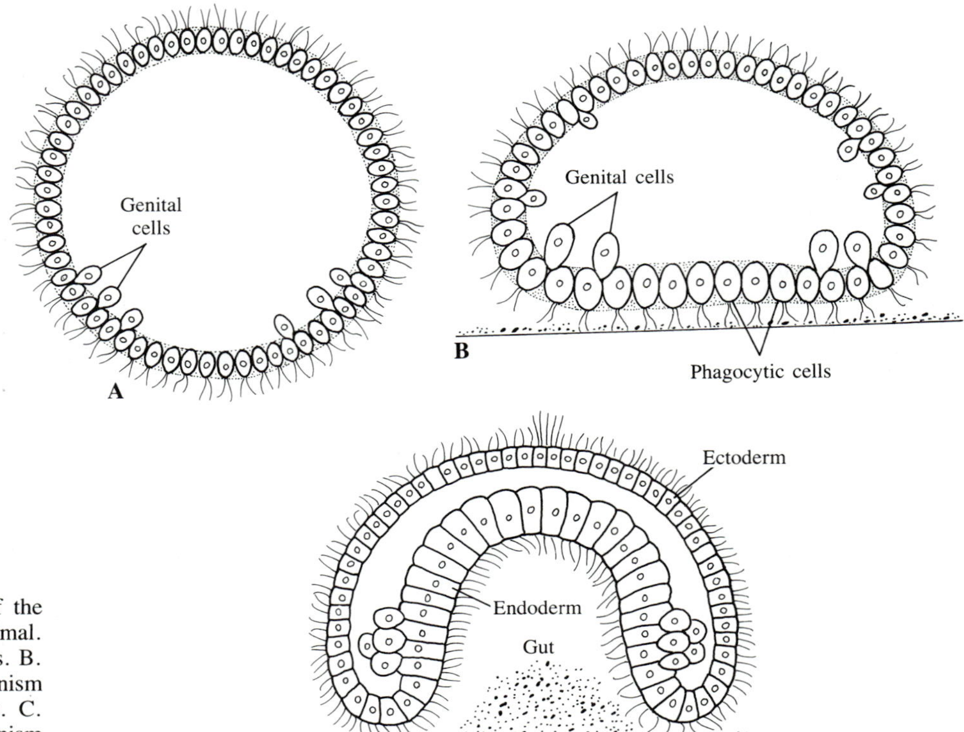

FIGURE 26.2 Naef's idea of the origin of a multicellular animal. A. *Volvox*-like colony of cells. B. Hypothetical creeping organism with phagocytic ventral cells. C. Hypothetical creeping organism with a saclike gut.

Early multicellular organisms may have reproduced by dividing after they had reached a certain size, or by setting free cells that developed into complete individuals. There may also have been cells that gave rise to gametes (Figure 26.3, B). Because all metazoans are diploid, with the exception of some that develop parthenogenetically, it is likely that formation of gametes by early multicellular animals would have involved meiosis.

A simple sponge may have originated from a phagocytic organism that had retained its choanocytes and become sessile. Inversion could have brought the choanocytes, no longer concerned with locomotion, into the interior (Figure 26.3, C). The nonflagellated cells could

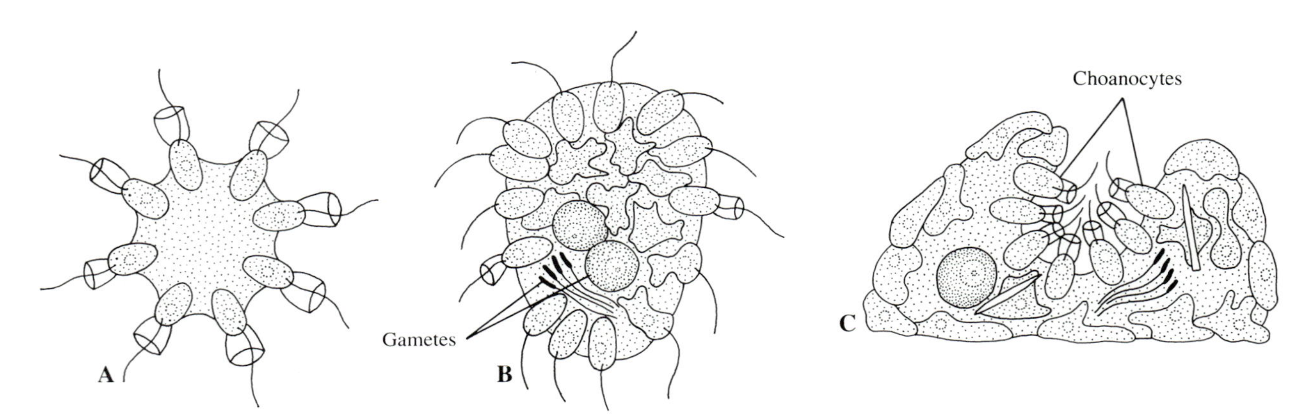

FIGURE 26.3 A. Colonial choanoflagellate. B. Hypothetica early metazoan derived from a choanoflagellate. C. Hypothetical simple sponge, in which the choanocytes line a central cavity. (After Ivanov, The Origin of Multicellular Animals.)

have become organized into a pinacocyte layer, and some cells could have become the amoebocytes that moved through the mesohyl, secreting skeletal structures, transporting food, and perhaps differentiating into gametes. In the course of further evolution, pores may have developed in the body wall, allowing water to enter the central cavity from the sides and to exit from the larger main opening, as is the case in modern asconoid sponges. It is generally accepted that sponges, whether they evolved along one or more lines, are an evolutionary dead-end: it is not likely that they gave rise to other groups of living metazoans.

Nearly all schemes that portray stages in the early evolution of animals include a **blastaea** and a **gastraea.** If these existed, and they probably did, they would have been comparable to the blastula and gastrula stages of embryonic development. The blastaea may at first have been pelagic, then benthic (Figure 26.4, A and B). Invagination of phagocytic cells to form the saclike gut of the gastraea (Figure 26.4, C and D) would increase the area available for feeding on microbes and other small particles. For maximum efficiency, a gut of this type would have to be provided with ciliated tracts that would drive particles into it.

Gut tissue would not necessarily have been produced by invagination; it could have been produced by ingression of cells from the wall of a blastula, such as is often the case in cnidarians, or by delamination of cells from those forming the wall of a blastula, which is characteristic of a few cnidarians. Cnidarians, in fact, are the only animals that come close to the hypothetical gastraea stage of evolution. They have a single opening into the gut and two cell layers, epidermis and gastrodermis. Certain of the gastrodermal cells are concerned with intracellular digestion, which was probably characteristic of early two-layered metazoans.

Some early invertebrates may have had a syncytial digestive tissue, or a mass of loosely joined phagocytic cells (Figure 26.5), toward which particles would be driven by cilia in the mouth region. Either of these arrangements could operate effectively in planktonic as well as benthic organisms, provided that the system for collecting particles and getting them to the syncytium or the phagocytic cells was efficient. A digestive syncytium certainly works well for modern acoel turbellarians. These, however, feed on energy-rich organisms much larger than bacteria, and have a muscular mechanism for swallowing food, an advance that probably

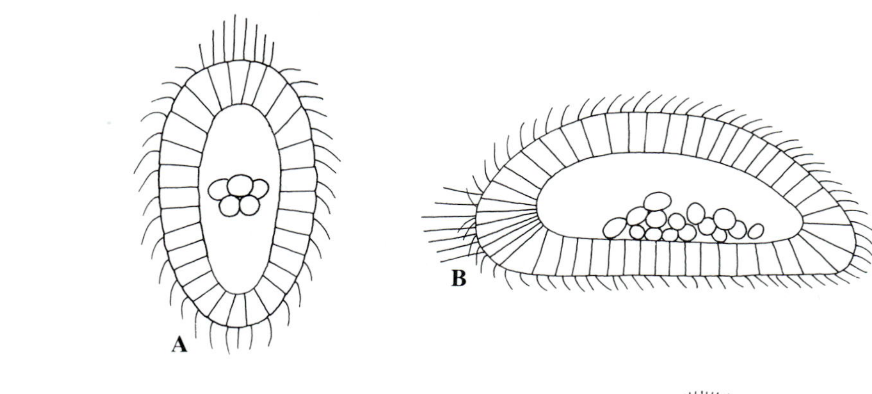

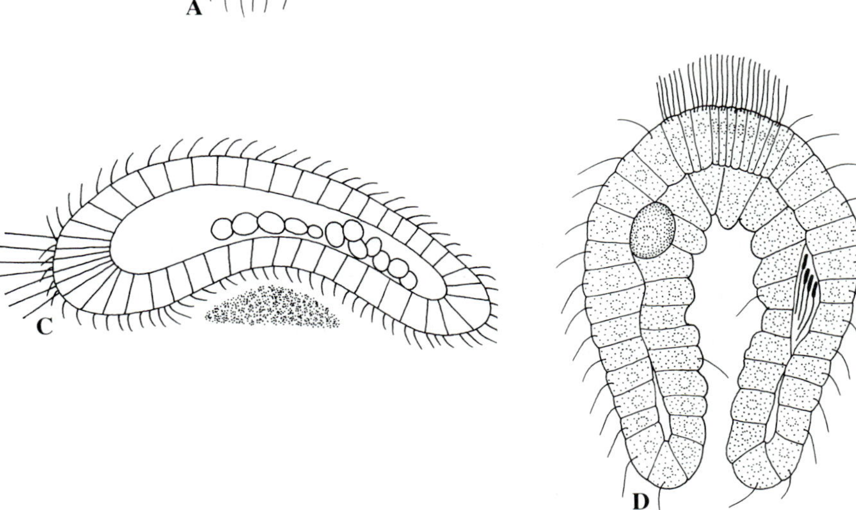

Figure 26.4 A. Hypothetical blastaea. B. Decidedly bilateral blastaea adapted for a benthic existence. C. Bilateral gastraea with a shallow gut. (A–C, after Jägersten, Zoologiska Bidrag från Uppsala, *30*.) D. A gastraea with a deep, saclike gut. (After Ivanov, The Origin of Multicellular Animals.)

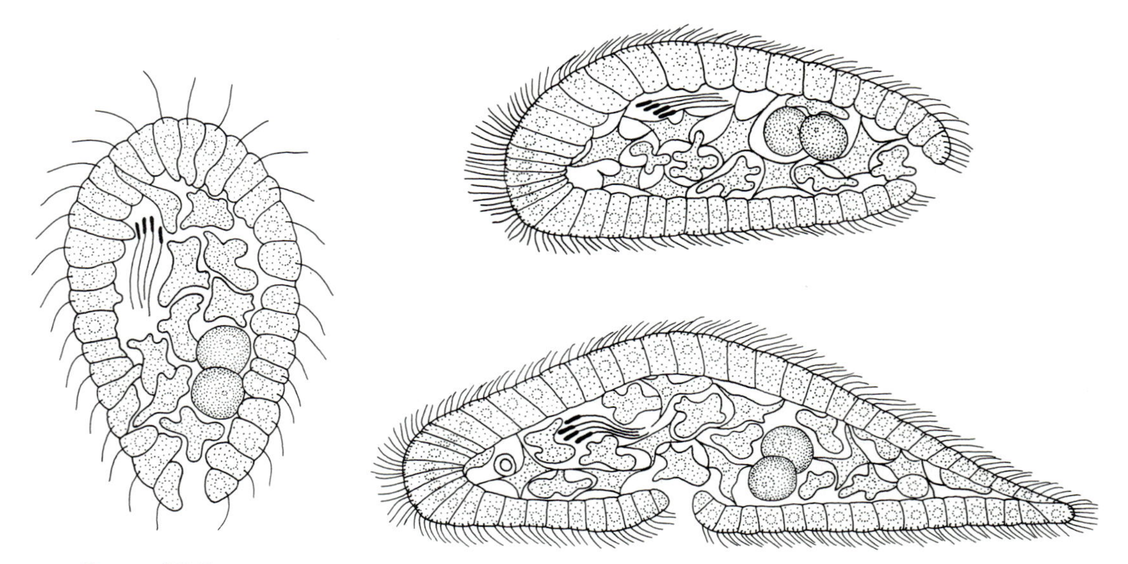

Figure 26.5 Three hypothetical metazoans characterized by an internal mass of loosely joined phagocytic cells. (After Ivanov, The Origin of Multicellular Animals.)

would not have been present in the most ancient invertebrates that had a simple gut, syncytium, or mass of digestive cells.

As soon as myoepithelial cells evolved, swallowing would have been made very efficient, as it is in present-day cnidarians. These animals, like some of the hypothetical gastrulalike organisms that probably existed long ago, are diploblastic and have a single opening into the gut. Cnidarians, however, are carnivores, and part of the sequence of digestion of the prey they capture is extracellular.

Regardless of what some of the early diploblastic organisms looked like, an important later step in evolution of animals would have been the segregation of cells other than those of the digestive tissue and epidermis. It is likely that the formation of muscle and connective tissue, for instance, was preceded by the appearance of undifferentiated cells comparable to embryonic mesoderm. This mesoderm may have been derived in one or more of the ways in which it is produced by embryos of living animals. Perhaps the distinction between radial and spiral cleavage appeared at about the same time.

The formation of a complete gut, with both mouth and anus, made possible a more efficient processing of food. In the case of animals whose diet included some undigestible components, it also would have freed the mouth from having to eliminate residues. Existing animals that have a complete gut are triploblastic, but it is conceivable that a gut of this type evolved before mesoderm and its derivatives appeared. In that case, movement of food through the gut would have been accomplished by ciliary activity or activity of myoepithelial cells of the gastrodermis, rather than by swallowing or by peristaltic movements caused by muscle around the gastrodermis.

Increased specialization of tissues, or increases in body size, or both, would require more efficient distribution of the soluble products of digestion, as well as of oxygen. A body cavity—either a remnant of the embryonic blastocoel or a cavity formed secondarily within mesoderm—would provide a space through which nutrients and oxygen could be circulated. It could also function as a hydrostatic skeleton, and would permit peristalsis of the gut, separate from muscular activity of the body wall. Other possibilities associated with the development of a body cavity, especially a true coelom, would be ducts for elimination of nitrogenous wastes, and ducts through which gametes formed in the coelom could be discharged to the external environment.

Origin of the Coelom

The formation of a coelom was a major advance in the evolution of animals, and this type of body cavity is characteristic of most phyla. Even in groups in which it is not spacious, the coelom contributes to some important structures, especially those concerned with reproduction and excretion. Three principal theories attempt to explain the origin of the coelom and these are summarized below.

Gonocoel Theory

The gonocoel theory is based on the idea that the coelom is derived from a gonad or series of gonads.

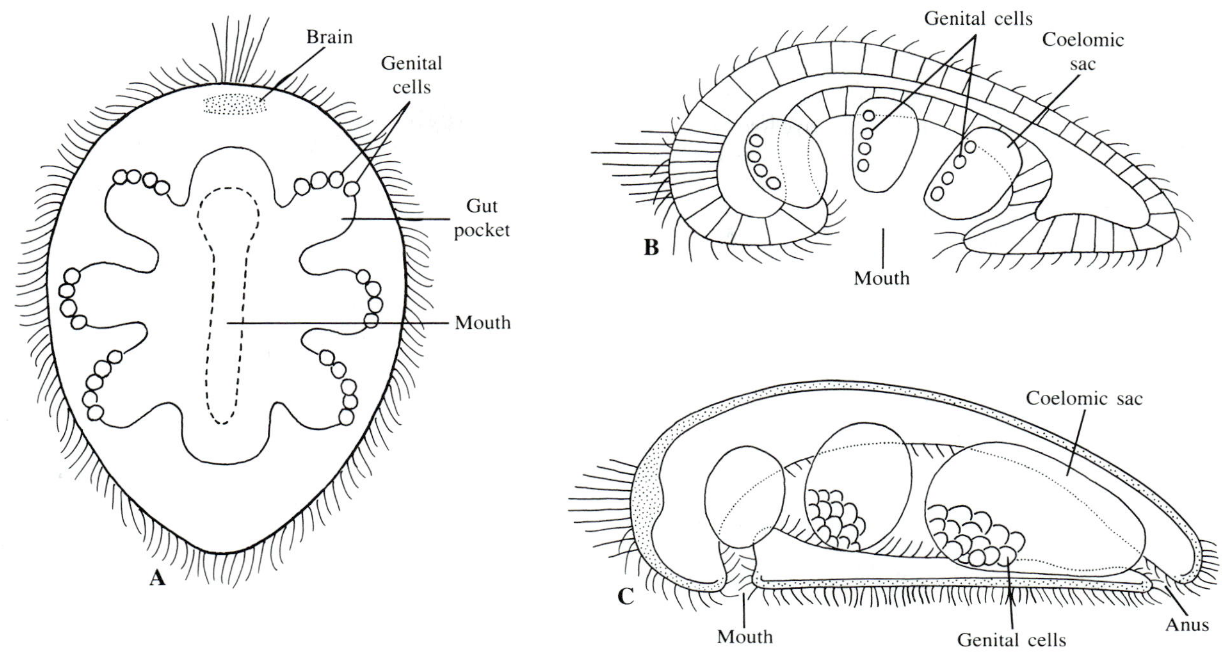

FIGURE 26.6 A. Hypothetical bilateral gastraea with gut pockets; dorsal view. B. Hypothetical bilateral gastraea in which the gut pockets have become coelomic sacs; view from the left side (right coelomic sacs omitted). C. Similar hypothetical organism, with a complete gut instead of a saclike gut. (After Jägersten, Zoologiska Bidrag från Uppsala, *30*.)

The theory usually connects the origin of the coelom with the appearance of segmentation. When it was first proposed, late in the nineteenth century, the paired and serially arranged gonads of nemerteans and some flatworms were viewed as precursors of the coelomic cavities of annelids. The simple openings of the gonads to the exterior, as seen in nemerteans, could have persisted as coelomoducts and continued to serve as gonoducts. Epidermal protonephridia, after becoming associated with the coelom, could have evolved into metanephridia.

The gonocoel theory is rather persuasive. Even if it links the origin of the coelom with segmentation, it can be adapted to explain the appearance of a coelom in invertebrates that are not segmented.

ENTEROCOEL THEORY

The enterocoel theory—another proposal that dates to the nineteenth century—states that coelomic spaces originated from side pockets of the gut (Figure 26.6, A–C). Of modern invertebrates, anthozoan cnidarians are the animals that best show gut pockets of the sort that could have separated from the central cavity and thus have converted into coelomic spaces. This theory, like the gonocoel theory, is also generally linked with the origin of segmentation, for it suggests that if several pockets separated from the gut on both sides of the body, a series of paired coelomic spaces would be formed. One may wonder, however, what advantage would be conferred on an anthozoanlike animal if its gut pockets, essential to processing of food and circulating nutrients, were to become dissociated from the gut.

Animals whose coelomic cavities develop as pouches of the archenteron, or within mesoderm budded from the archenteron, include echinoderms, phoronids, brachiopods, enteropneusts, cephalochordates, and vertebrates. The mesoderm of urochordates can also be traced back to the archenteron, but these animals lack a coelom.

SCHIZOCOEL THEORY

The schizocoel theory proposes that a coelomic space or spaces first appeared in mesodermally derived tissue between the gut and epidermis of an animal such as a flatworm. As a space appeared, the cells lining it would become organized into a peritoneum. By further enlargement, the space could envelop the gut, as it does in most annelids and in some other invertebrates, as well as in vertebrates.

This theory has much to recommend it, and is favored by many zoologists who view simple flatworms, such as acoel turbellarians, as being close to the worms from which more complex invertebrates evolved. Moreover, in annelids, molluscs, and other invertebrates that are protostomatous and that have spiral cleavage, the coelomic cavities do originate as schizocoels within solid masses of embryonic mesoderm.

The acoelomate condition seen in flatworms is generally believed to be primitive, but one could argue that it is secondary. *Lobatocerebrum,* a genus of worms recently found in marine sand, shows features of annelids and also of flatworms. Its paired ventral nerve cords are ganglionated, its protonephridia are serially arranged, and it has a complete gut. But it has no setae and no coelomic spaces, although its parenchyma is composed to a considerable extent of large, vacuolated cells. *Lobatocerebrum* is ciliated, which is characteristic of many annelids as well as of flatworms, but it has a cuticle, which is not a feature of flatworms. It must be pointed out, however, that reduction of the coelom occurs in several groups of polychaete annelids. *Lobatocerebrum* is a decidedly unusual worm, and its odd combination of features must not necessarily be accepted as an indication that flatworms or other triploblastic acoelomates are derived from annelids, which has been suggested.

In nemerteans, the rhynchocoel around the proboscis, and also the blood vessels, have cellular linings and originate as schizocoels. The fact that there is no general body cavity is one reason why the significance of the rhynchocoel and blood vessels was overlooked by early proponents of the gonocoel theory, which is based on the idea that cavities appearing within the gonads of nemerteans or similar animals were precursors of serial coelomic spaces.

It should be obvious that the main ideas on how coelomic spaces first originated are speculative. The ways coelomic spaces develop in present-day invertebrates help us to appreciate how one or more kinds of coeloms may have evolved. It is best to be cautious, however, in assuming that an enterocoelous or schizocoelous type of body cavity found in existing invertebrates is exactly comparable, in origin or arrangement, to body cavities of the earliest animals that had true coeloms.

Regardless of how it originated and how it was arranged, the coelom's contribution to the success of certain early invertebrates was probably related to the advantages it offered as a hydrostatic skeleton. It would have been important not only in locomotion but also in feeding. In modern invertebrates, operation of structures such as the pharynx of polychaetes and the proboscis of nemerteans depends on hydrostatic pressure within the body cavity or rhynchocoel. In addition, a coelom would provide a space or spaces in which fluids could circulate, thereby distributing nutrients and oxygen, and bathing internal organs more effectively than would be possible if these organs were surrounded by solid masses of cells.

Origin of Segmentation

It was pointed out in the preceding section that the gonocoel and enterocoel theories of the origin of the coelom are tied to ideas on how segmentation may have evolved. In most invertebrate groups characterized by segmentation, each segment has its own coelomic compartment, except in derived types in which septa between segments have disappeared. In cephalochordates, however, and also in the vertebrates, segmentation is of a different type. The coelom, even if it is divided into two or more cavities, is not organized into serially arranged spaces as it is in annelids. Segmentation is evident in the arrangement of muscles, ganglia, gill slits, and, in the case of vertebrates, of bones or cartilages of the vertebral column.

The essential features of the gonocoel and enterocoel theories need not be repeated here. It should be noted, however, that the enterocoel theory, based on the idea that pockets of the gut of a cnidarian or similar animal became converted into paired coelomic cavities, implies a shift from radial or biradial symmetry to bilateral symmetry. If one really favors the enterocoel theory, this may not be hard to accept. It is even possible that there were cnidarians in which biradial symmetry had become nearly or fully bilateral.

A different idea concerning the origin of segmentation is based on the fact that asexual reproduction by certain turbellarians leads to the formation of chains of several similar members. Each member is capable of continuing asexual reproduction or of becoming sexually mature. Chain-formation in tapeworms, sometimes introduced into discussions of the origin of segmentation, is a different matter. The proglottids formed in one very restricted zone behind the scolex act only as sexual individuals, and usually become separated from the chain as new proglottids differentiate. Tapeworms are specialized parasites of vertebrates, and their evolution must have been linked closely to that of their hosts, which evolved long after most invertebrate phyla had become established.

The locomotory theory is based on the idea that an animal whose movements are undulatory would be likely to complete a process of segmentation begun by repetition of certain internal structures, such as gonads, protonephridia, or metanephridia. This theory is somewhat related to a theory that credits the formation of segments to stress placed on a mesodermal mass as an embryo or growing animal elongates, and it also takes advantage of the fact that some elongated but unsegmented animals, such as nemerteans, have serially arranged structures, especially gonads. While it is true that undulatory locomotion is highly developed in annelids, and that the organization of the musculature of annelids facilitates this type of movement, many unsegmented animals swim or crawl effectively by undulating. Some of them, moreover, are long and slender. It is likely that segmentation is much more important for burrowing than it is for swimming or crawling, because it enables an animal to execute localized elongations and contractions of segments that enable it to force its way through sand, mud, or soil.

SYMMETRY

All metazoans other than sponges have some kind of symmetry. Only in the Cnidaria, however, do we find animals with perfect radial symmetry. Because cnidarians are diploblastic and histologically less complex than ctenophores, flatworms, and other phyla, it is tempting to assume that radial symmetry is more primitive than bilateral symmetry. Perhaps this is true, although it is just as likely, if not more likely, that a group of more or less bilaterally symmetrical diploblastic organisms was ancestral to invertebrates now represented by radial or biradial types, as well as to those showing bilateral symmetry.

Echinoderms are here considered part of the bilateral line. Their swimming larvae are bilaterally symmetrical. As adults, they exhibit radial symmetry, but this radial symmetry is not perfect and is achieved only after a complex metamorphosis during which the orientation of larval structures drastically changes.

SPIRAL AND RADIAL CLEAVAGE

The fact that certain phyla are characterized by spiral cleavage and others by radial cleavage has been used in nearly all phylogenetic schemes. In general, animals with spiral cleavage are also protostomatous, have mesoderm derived from the 4d cell at the lip of the blastopore, and are schizocoelous; most animals with radial cleavage are deuterostomatous, and their mesoderm is derived from the archenteron, so they are enterocoelous. These combinations of features point to an early separation of what may be called *spiralian* and *radialian* lines of animal evolution.

Certain phyla, however, do not fit either of the above patterns very well. Early development of some animals is probably influenced by features that were acquired secondarily. Among these features is storage of considerable yolk in the eggs of many invertebrates. Adult morphology may also be deceptive. In bryozoans, for instance, there are coelomic spaces that may seem to be arranged in much the same way as those of brachiopods and phoronids. These spaces, however, are formed during what may be called a second embryology, following a radical breakdown of structures formed during primary embryonic development.

THE IMPORTANCE OF LARVAL TYPES

How much phylogenetic significance should be attached to planktonic larval types? In general, it is probably true that phyla characterized by similar larvae are in some way related. It is, of course, possible that similar larvae could have evolved independently in groups that do not have a close relationship. Moreover, not all members of a unified assemblage must have the type of larva that helps define the group. Thus the trochophore larva, which, along with other features, suggests a relationship between annelids and molluscs, is found in the life histories of only certain representatives of these phyla.

There is an interesting dichotomy in the way ciliated planktonic larvae collect food. In certain types—the trochophore, for instance—the compound cilia that form a band or bands that encircle the body drive food particles toward a field of single cilia, which then transport the particles toward the mouth. Because the particles are moved in the same general direction as the water currents created by the compound cilia, the arrangement is referred to as a downstream particle-collecting system. In larvae of some decidedly different types, such as the bipinnaria and pluteus of echinoderms and the actinotroch of phoronids, the cilia that form distinct bands are single cilia. They drive water away from the ciliated area around the mouth, but particles are somehow retained in the ciliated area and are transported to the mouth. This type of arrangement is called an upstream particle-collecting system, because the direction followed by the particles is opposite that of the water currents.

The distinction between the downstream and upstream collecting systems is regarded by some zoologists as support for the idea that the spiralian–protostome line of invertebrate evolution separated early from the radialian–deuterostome line. More will be said about this in the next section.

THE TROCHAEA THEORY

The trochaea theory was formulated to explain the origin of most major metazoan assemblages from animals of the gastraea type. It is based to a large extent on ciliary bands and nervous systems of swimming larvae of marine invertebrates, and also on the fate of the blastopore. Although proposed formally in 1985 by

Nielsen and Nørrevang, and discussed intensively in another paper published by Nielsen in 1985, much of the groundwork for the theory was laid by Nielsen in 1979.

The hypothetical animal called the **trochaea** (Figure 26.7, A), with a ring of compound cilia, the **archaeotroch,** functioning as a downstream food-collecting system around the blastopore, could presumably have derived from a gastraea in which there were only monociliate cells. Such a ring is, in fact, characteristic of the swimming marine larvae of several phyla more advanced than the Cnidaria. Between the archaeotroch and the blastopore, and in the gut itself, there would probably have been monociliate cells. Particles caught and concentrated by the archaeotroch would be transported to the mouth by the cilia of the monociliate cells, and similar cilia would distribute the particles within the gut cavity.

The authors of the theory proposed that the next

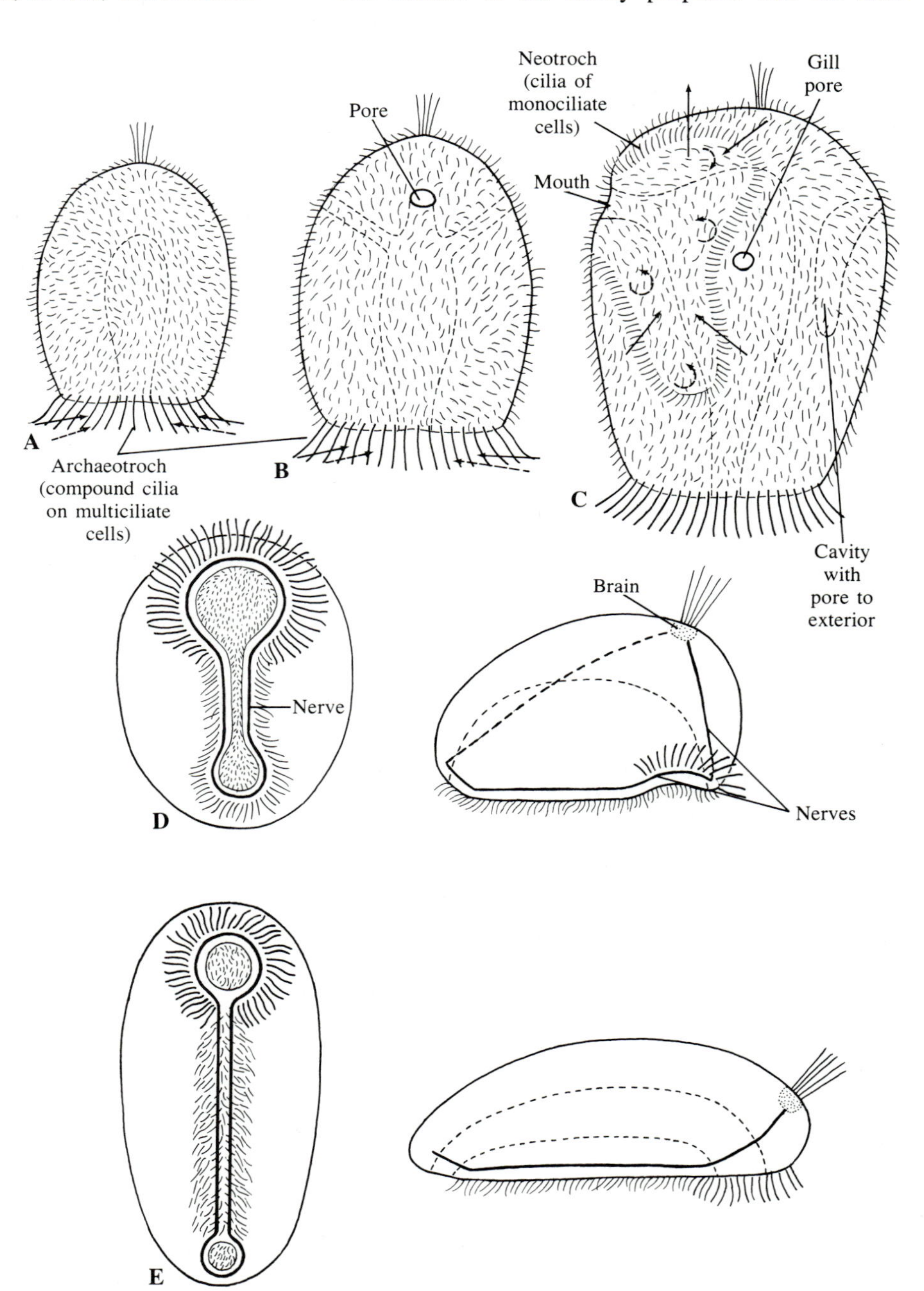

FIGURE 26.7 A. Hypothetical trochaea. B. Hypothetical protornaea. C. Hypothetical tornaea. D. Hypothetical trochaea with an elongated blastopore, adapted for a benthic existence; ventral (left) and lateral (right) views. E. The same, after union of the lateral lips of the blastopore, resulting in formation of a complete gut; ventral (left) and lateral (right) views. In A–C, solid arrows indicate the direction of water currents, and broken arrows indicate paths of particles that are being captured. (After Nielsen, Biological Journal of the Linnean Society, *68*.)

stage, the **prototornaea** (Figure 26.7, B), would have had four canals extending from the anterior part of the previously saclike gut to the surface. The arrangement would be similar to that of the four canals that arise from the aboral end of the central gut cavity of certain ctenophores. If one of the openings developed into a functional mouth, a complete gut would be formed, and passage of food through it would proceed in one direction. Cilia of monociliate cells would drive particles into the mouth and also through the gut. Residues would leave by way of the blastopore, now functioning strictly as an anus. A new band of single cilia, the **neotroch,** could have appeared around the mouth region and could have functioned as an upstream particle-collecting system. The animal would now be a **tornaea** (Figure 26.7, C).

The other three hypothetical canals that branched off the gut of the prototornaea could have evolved into ducts or pores that allowed most of the water taken in through the mouth during feeding to escape from the gut. The authors of the trochaea theory suggest that one of them could have lost its connection with the gut and thus have developed into something comparable to the water-vascular division of the embryonic coelom of echinoderms, or into some other coelomic cavity that opens to the exterior. The name tornaea, in fact, alludes to the tornaria larva of enteropneusts, which comes close to being a real-life counterpart of the tornaea.

The trochaea theory suggests that there would have been an apical brain in all pelagic organisms from the trochaea to the tornaea. The brain would be joined by several connectives to a nerve ring that would coordinate the activity of the archaeotroch cilia.

An animal of the trochaea type could conceivably have become benthic and begun to move over the substratum with its blastopore facing downward. The blastopore could have elongated, and in that case food particles could have entered on its anterior side, and rejection of unsuitable material could have occurred on its posterior side (Figure 26.7, D). This arrangement, with a more or less one-way flow, would have increased the efficiency with which food was processed. Suppose, now, that the lateral lips of the elongated blastopore became joined. A complete gut would be formed (Figure 26.7, E). Part of the blastopore would become the mouth, and part of it would become the anus. This development could have paved the way for the evolution of invertebrates of the protostome line, in which the mouth and anus develop at opposite ends of a blastopore that closes, partly or fully.

If the blastopore of the trochaea elongated and then partly closed, as proposed in the trochaea theory, a portion of the circular nerve tract that was closely associated with the archaeotroch could have become organized into a pair of ventral longitudinal cords. Hypothetical animals with such an arrangement may be designated **gastroneuralians** ("animals with a nerve on the belly side") (Figure 26.8, A). Of living phyla, annelids, echiurans, and sipunculans fit the gastroneuralian plan especially well. Several other phyla, including

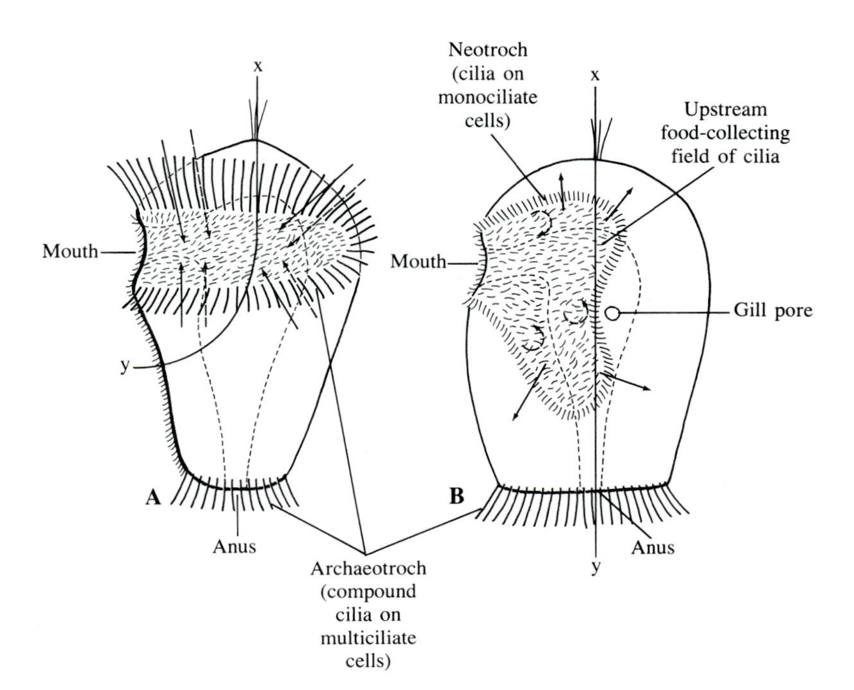

FIGURE 26.8 A. Hypothetical gastroneuralian, in which both the mouth and anus are derived from the blastopore. B. Hypothetical notoneuralian, in which the anus is derived from the blastopore and the mouth is derived from a secondary opening. The lines x–y indicate the apical–blastoporal axes, solid arrows indicate the direction of water currents, and broken arrows indicate paths of particles that are being captured. (After Nielsen, Biological Journal of the Linnean Society, *68*.)

nemerteans, flatworms, and molluscs, can be connected with a hypothetical gastroneuralian ancestor, but it takes more imagination.

Another hypothetical ancestral animal that figures in the trochaea theory is called the **notoneuralian** (''animal with a nerve on the back'') (Figure 26.8, B). The planktonic larva of the notoneuralian could have resembled a tornaea, with the anus, derived from the blastopore, being surrounded by the archaeotroch, whose function would now be strictly locomotory. The upstream food-collecting system would consist of monociliate cells around the mouth, which was formed secondarily. Loss of a concentrated ventral nerve could have accompanied adaptation of the planktonic notoneuralian to a benthic existence. Echinoderms, phoronids, and brachiopods might have derived from a notoneuralian ancestor. Conceivably, a nerve could have developed middorsally, thus paving the way for cephalochordates, vertebrates, and urochordates.

The trochaea theory is extremely speculative, but the papers of Nørrevang and Nielsen (1985) and Nielsen (1985) are worth reading for their provocative content. Another paper of Nielsen (1987) has a useful review of the structure and phylogenetic significance of ciliary bands in nearly all groups of invertebrates.

COMMENTS ON EVOLUTION AND RELATIONSHIPS OF INDIVIDUAL PHYLA

PORIFERA

It has already been explained that sponges, in which choanocytes function in feeding and creating water currents, probably derive from colonial choanoflagellates. Nevertheless, sponges have a rather wide variety of cell types, including some that secrete skeletal structures. Although a basal lamina is lacking, certain of the cells are united by tight junctions comparable to those of most metazoans.

Sponges may have evolved from several separate stocks of choanoflagellate ancestors. This is suggested by the fact that hexactinellids, being mostly syncytial rather than cellular, and lacking myocytes, are very different from sponges of other groups.

PLACOZOA

Trichoplax, whose lower epithelium is concerned with uptake of food, seems close to the hypothetical blastaea. Its cells are monociliate, and there is no basal lamina. The lack of polarity and specialized neurons are other features that suggest that the organism is primitive, even though it has myocytes in the narrow space between the upper and lower epithelium.

CNIDARIA

Because cnidarians are diploblastic and do not have an anus, their body plan is somewhat comparable to that of a gastrula. Formation of the gut by invagination is probably the primitive pattern within the phylum, although it is characteristic only of certain groups. In most cnidarians, endoderm is formed by delamination, and a lumen and mouth opening appear later, but this type of development is believed to be a consequence of production of yolky eggs, and perhaps also of brooding.

It is difficult to connect cnidarians to an early gastraea that fed on bacteria, unicellular algae, and other microorganisms. Modern cnidarians are primarily carnivores that use nematocysts for trapping prey, most of which have a higher evolutionary status than the cnidarians themselves. It is nevertheless possible that existing cnidarians are the successful survivors of a long line of gastraea-type organisms that may at first have been phagocytic, then developed the capability of carnivorous feeding.

Unlike sponges, cnidarians have a basal lamina beneath the epidermis and gastrodermis, and this forms the mesoglea, which is especially thick in jellyfishes. It may function as a deformable skeleton, and it also anchors the contractile portions of the myoepithelial cells. Another way in which cnidarians differ from sponges is in having neurons.

The fact that ciliated cells of cnidarians are monociliate may be an indication that these animals are not far removed from a flagellate ancestral stock. Even when there are thick compound cilia, the individual cilia of which they consist originate from separate cells.

CTENOPHORA

Because of their abundant mesoglea, ctenophores have usually been placed close to cnidarians, and perhaps they do have some phylogenetic connection with these animals. They are distinctly different, however, in being triploblastic, with true muscle cells derived from mesoderm rather than myoepithelial cells. Moreover, what appears to be mesoglea contains not only muscle but also neurons, so it is distinctly different from the mesoglea of cnidarians. In addition, there are multiciliate cells, a pattern of cleavage that early establishes biradial symmetry, and determinate development. Finally, the presence of colloblasts and absence of nematocysts also set cteno phores apart from cnidarians.

PLATYHELMINTHES

Flatworms, like ctenophores, have multiciliated cells and are triploblastic, but their tissue organization and nervous system are much more complex than that of ctenophores. Furthermore, although they have a saclike gut, they are protostomatous and have spiral cleavage, which ties them to a series of more advanced phyla, among them Nemertea, Annelida, Mollusca, and Arthropoda.

It is not likely that early flatworms had reproductive systems comparable to those of existing types. Even the reproductive systems of acoel turbellarians, which are the simplest members of the phylum, are more complex than those of nemerteans and several other groups whose general structure shows definite advances over the flatworm plan. One must be aware of the possibility that the ancestors of modern flatworms had reached an evolutionary status much higher than that of cnidarians and ctenophores.

NEMERTEA

Nemerteans are probably closely related to flatworms, although they may not be derived from any group of flatworms that is still living. Their general histology, flame-cell protonephridia, type of nervous system, and spiral cleavage are among the features that connect nemerteans with flatworms. The complete gut and simple reproductive system are departures from the flatworm plan. Moreover, the rhynchocoel around the proboscis, and also the blood vessels, develop in much the same way as coelomic cavities of schizocoelous higher invertebrates. Thus nemerteans have a mixture of primitive and advanced traits.

PSEUDOCOELOMATE PHYLA

Most of the pseudocoelomate phyla are so different from one another that it is not likely that they have originated from a single ancestral stock. The feature that makes it convenient to deal with them together is the pseudocoel, a type of body cavity that lacks a peritoneum. In some pseudocoelomates, this may be a remnant of the blastocoel, whereas in others, including nematodes, it arises as a space or spaces between cells.

Another feature that some zoologists consider to be useful for linking pseudocoelomate phyla is the absence of spiral cleavage. Although modified spiral cleavage has been claimed for certain rotifers, gastrotrichs, and acanthocephalans, it is different from the spiral cleavage of annelids, molluscs, and other phyla of the protostomatous–schizocoelous–spiralian division of the animal kingdom.

It should be kept in mind that in addition to the eight pseudocoelomate phyla described in Chapters 9 and 10, there is another pseudocoelomate phylum: Entoprocta. Good arguments have been made for considering entoprocts to be rather close relatives of Bryozoa, and because of this they were discussed along with bryozoans in Chapter 14.

ANNELIDA

Fossils and traces of annelids, or at least annelidlike worms, are definitely present in deposits that date to the middle of the Cambrian period, more than 500 million years ago. Of the living phyla, annelids show most clearly both an extensive coelom and segmentation, a pair of characteristics that probably was important in adapting these worms for burrowing through soft substrata. Annelids are now extremely diversified, with not only burrowing types but also crawlers, swimmers, and many sedentary species that inhabit secreted tubes.

Like flatworms, annelids exhibit spiral cleavage and protostomatous development, and their ciliated cells are multiciliate. The double ventral nerve cord is probably derived from two widely spaced ventral nerve trunks such as are characteristic of many flatworms, and also of nemerteans. Nevertheless, we do not know what sort of worms gave rise to the earliest annelids, and how many of the distinctive features of annelids, including metanephridia, evolved. We can only make guesses based on the way these features develop in living annelids.

The trochophore larva of annelids deserves mention. While the only annelids that have this type of larva are certain polychaetes, similar larvae appear during the development of sipunculans, echiurans, and some molluscs. It is likely that all of these groups are related. They have spiral cleavage and are protostomatous and coelomate, and their mesoderm is derived from the 4d cell. The coelom of molluscs is comparatively reduced, however, and neither molluscs, sipunculans, nor echiurans are segmented.

ECHIURA AND SIPUNCULA

While the Echiura and Sipuncula show no traces of segmentation, they are probably closely related to annelids. They have spiral cleavage and a trochophore larva somewhat similar to that of certain polychaetes. The formation of mesoderm from the 4d cell and the schizocoelous origin of the coelom are also typical of the annelid plan. The ventral nerve cord of echiurans and sipunculans, however, is neither double nor ganglionated, and there are no head appendages other than the food-gathering tentacles that sipunculans have around the mouth.

MOLLUSCA

Whether molluscs originated from unsegmented or segmented ancestors will probably be debated for a long time. The embryonic development of many diverse species has been carefully studied, and none shows any tendency to form a series of coelomic sacs. Nevertheless, certain structures in adult Monoplacophora—gills, gonads, coelomoducts (renal organs), and auricles of the heart—are repeated at least once. The coelom, however, is typically molluscan in that it is restricted to the pericardial cavity, the gonads, and the cavity with which all but one of the several pairs of coelomoducts communicate.

A strong case can be made for the origin of molluscs from flatworms or some now extinct group of invertebrates similar to flatworms. The arrangement of major nerve cords in monoplacophorans, polyplacophorans, aplacophorans, and even in bivalves and many gastropods certainly resembles that of some flatworms. Moreover, flatworms exhibit spiral cleavage and derivation of mesoderm from the 4d cell. The Müller's larva of certain polyclads slightly resembles the molluscan trochophore, although it lacks an anus.

The possible affinities of molluscs and annelids is also, of course, indicated by spiral cleavage, derivation of mesoderm from the 4d cell, and the trochophore larva. Although the coelom of molluscs is relatively restricted when compared to that of annelids, this is probably the result of reduction, rather than a primary feature. The nervous system is decidedly unlike that of annelids, in which there is a double ventral nerve cord, rather than widely separated main nerves typical of molluscs and flatworms.

ARTHROPODA

Annelids, or annelidlike invertebrates, are believed to have given rise to onychophorans and arthropods, both of which have a double ventral nerve cord. Onychophorans, while having legs, a hemocoelic circulatory system, a tracheal respiratory system, and a cuticle of the arthropod type, have ciliated metanephridia similar to those of annelids. The segmental coelomic spaces in which the funnels of the metanephridia lie are greatly reduced, however. In arthropods, vibratile cilia are absent except in sperm of certain groups, and the reduction of the coelom is almost complete. The segments are organized into tagmata, and the appendages, unlike those of onychophorans, are jointed, with corresponding modifications in the musculature. Onychophorans are not believed to have been the stock from which any living arthropods evolved. Fossilized remains of animals resembling onychophorans have been found in certain ancient marine deposits, but whether these may have been involved in the evolution of any genuine arthropods is not known.

In Chapter 15, it was pointed out that some specialists concerned with the evolution of arthropods believe that these animals—at least the living groups—are monophyletic. Others feel that there must have been several separate protoarthropod stocks, and that present-day chelicerates, crustaceans, and uniramians, as well as some extinct groups, evolved along separate lines. If this belief is correct, then the Arthropoda is a polyphyletic assemblage, not truly a unified phylum.

Crustaceans are the only living arthropods that have biramous appendages, second antennae, and extensive digestive glands. Their mandibles, moreover, are different from those of uniramians, but their compound eyes and hormonal control of molting are similar to those of uniramians. Among living chelicerates, well-developed compound eyes are found only in *Limulus,* although there are indications of compound eyes in scorpions, and compound eyes were definitely present in some ancient extinct scorpions. The Malpighian tubules and tracheal respiratory systems of many chelicerates would seem to link these animals with uniramians. Nevertheless, chelicerates do not have appendages comparable to the antennae and mouthparts of crustaceans and uniramians.

There are no known fossils that link insects with crustaceans, or that link chelicerates with either of these groups. Thus zoologists who believe that modern arthropods are monophyletic or polyphyletic must decide whether structures such as jointed appendages, antennae, compound eyes, mandibles, maxillae, tracheal systems, and various other structures indicate close relationships or whether they could have evolved independently in separate protoarthropod stocks.

BRYOZOA AND ENTOPROCTA

Fitting the Bryozoa into a phylogenetic scheme presents problems. The embryology of these animals is inconsistent and is frequently modified by the presence of yolk or by placental brooding. When the swimming larva has a gut, this disappears and is replaced by the gut that functions in the adult. Most of the adult structures, in fact, are produced during a second phase of embryonic development that follows the breakdown of structures present in the larva. The origin of mesoderm, including that which gives rise to the peritoneal lining of the adult coelom, is not yet understood.

Most entoprocts have spiral cleavage, which sets them apart from bryozoans, in which cleavage is radial. Their mesoderm, moreover, is derived from the 4d cell. Thus they seem to fit in with spiralians, such as molluscs and annelids. The body cavity of entoprocts is considered to be a pseudocoel, rather than a true coelom, such as is characteristic of spiralians.

In spite of their differences, entoprocts and bryozoans may be rather closely related. This is suggested by similarities between the swimming larvae of some entoprocts and those of some bryozoans, and in the way the larvae settle on the substratum before metamorphosing into adults.

ECHINODERMATA AND HEMICHORDATA

Echinoderms form a distinctive group whose origins are difficult to imagine. Perhaps they are derived from a group of animals that also gave rise to hemichordates, which are like echinoderms in that they have radial cleavage, are deuterostomatous and enterocoelous, and have a similar free-swimming larva. The form of the possible ancestral group may have been wormlike, with tentacles into which coelomic spaces extended. The tentacular system could conceivably have become separated from the rest of the coelom, if it was not already separate, and could have become the water-vascular system. With the appearance of calcareous ossicles, the evolution of the internal skeleton of echinoderms would have been initiated. The fossil record has provided no evidence to support the idea that echinoderms originated from wormlike ancestors, however. While the radial symmetry of echinoderms is an ancient trait, it is probably not primary, for among the earliest echinoderms in the fossil record there were many with bilateral symmetry, others without symmetry. The swimming larvae of all five classes of living echinoderms, moreover, are bilaterally symmetrical.

The two groups of living hemichordates—the enteropneusts and pterobranchs—are so different that they should perhaps be placed in separate phyla. Nevertheless, the division of the coelom into protocoel, mesocoel, and metacoel, and the presence of gill slits in enteropneusts and some pterobranchs, suggest a relationship. As adults, neither enteropneusts nor pterobranchs resemble echinoderms, but the tornaria larva of enteropneusts, in general form as well as in the division of the embryonic enterocoel into three compartments, is strongly reminiscent of the swimming larvae of echinoderms.

CHAETOGNATHA

The evolutionary relationships of chaetognaths are decidedly obscure. They have radial cleavage, are deuterostomous, and have a true coelom that originates from outpocketings of the archenteron. But they have a cuticlelike layer that is intracellular, like that of rotifers, and many other distinctive features that separate them from all other phyla.

UROCHORDATA AND CEPHALOCHORDATA

Urochordates and cephalochordates are like vertebrates with respect to the ways in which the dorsal nerve cord and notochord originate. The nerve cord develops from a median longitudinal trough of ectoderm that is bordered by a pair of neural folds. When these folds meet, the nerve cord is isolated from the surface. The notochord is completely absent in some urochordates, and in nearly all urochordates that do have a notochord, this structure does not survive beyond the larval stage. Nevertheless, when it is present, the notochord develops dorsal to the archenteron and lies between the gut and the nerve cord.

The probable homology of the endostyle of urochordates and cephalochordates with the thyroid gland of vertebrates is another feature that ties the three groups together. Urochordates, however, do not show any trace of the metameric features seen in the arrangement of muscles, ganglia, and some other structures of cephalochordates and vertebrates.

In most urochordates, other than larvaceans, and in cephalochordates, the pharyngeal perforations open into an atrial cavity, rather than directly to the exterior, as they do in vertebrates that have gill slits. Nevertheless, the gill slits of cephalochordates, because of the way they are arranged and because of their relationship to vessels of a closed circulatory system, may possibly be closely related to gill slits of lower vertebrates, and to gill pouches and other intimations of gill slits seen during embryonic development of higher vertebrates. The pharyngeal perforations of urochordates are arranged very differently from those of cephalochordates and vertebrates, and perhaps have a different evolutionary background.

Summary of Possible Relationships Between the Phyla

From what has already been said in this chapter, it should be obvious that diagrams showing possible sequences in evolution of the animal phyla must be viewed with caution. Although such a diagram is offered here (Figure 26.9), its only purposes are to summarize some of the important steps in evolution and to provoke thought. It shows all of the phyla dealt with in the text, but many of these phyla are accorded tentative or questionable positions with respect to what seem to be the main lines of descent. It is to be expected that highly specialized parasites, such as dicyemids, orthonectids, and pentastomids, will be difficult to place in

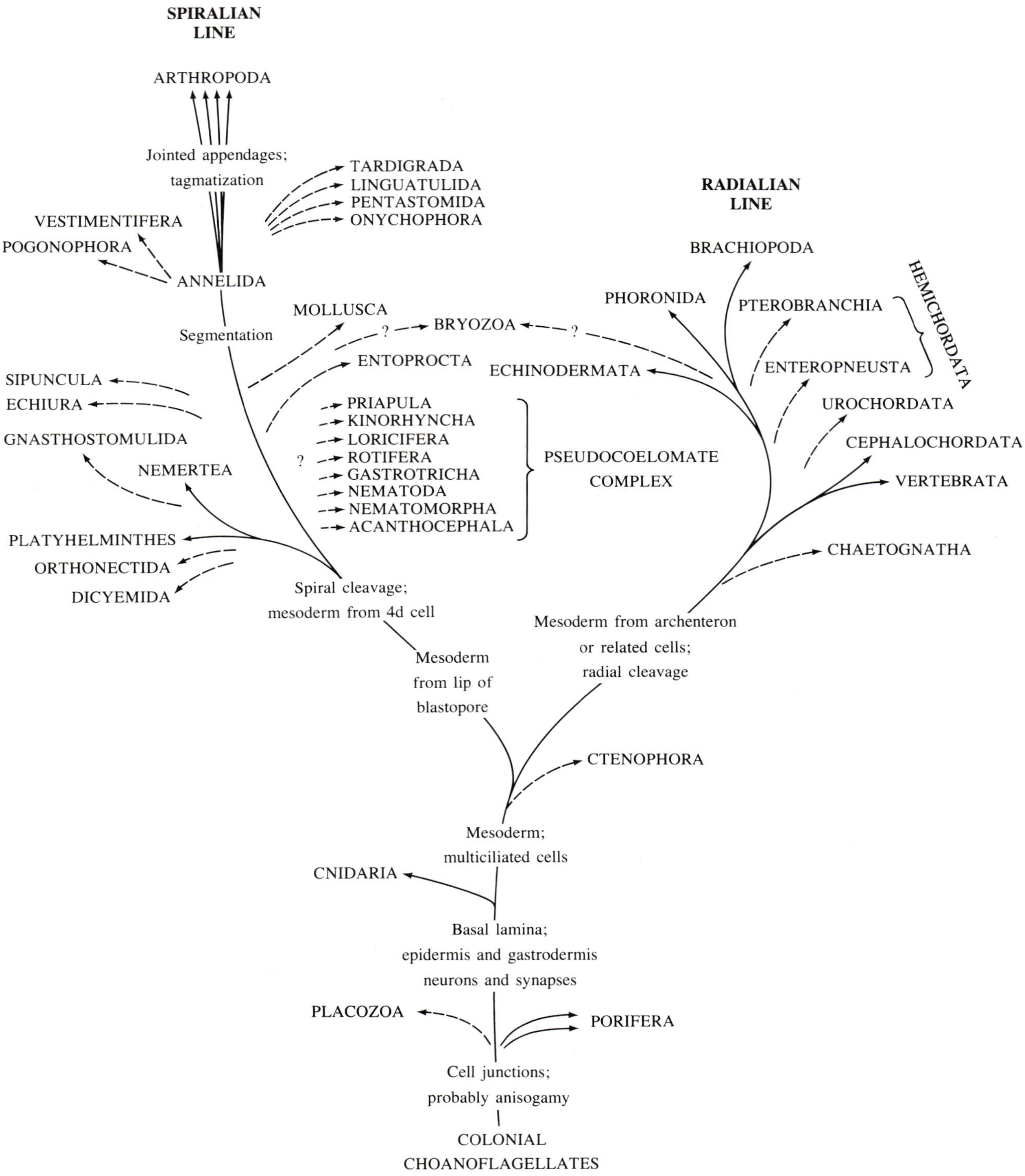

SPIRALIAN LINE

ARTHROPODA

Jointed appendages; tagmatization

TARDIGRADA
LINGUATULIDA
PENTASTOMIDA
ONYCHOPHORA

VESTIMENTIFERA
POGONOPHORA
ANNELIDA

Segmentation

MOLLUSCA
? BRYOZOA ?
ENTOPROCTA

SIPUNCULA
ECHIURA
GNASTHOSTOMULIDA
NEMERTEA

PRIAPULA
KINORHYNCHA
LORICIFERA
? ROTIFERA
GASTROTRICHA
NEMATODA
NEMATOMORPHA
ACANTHOCEPHALA

PSEUDOCOELOMATE COMPLEX

PLATYHELMINTHES
ORTHONECTIDA
DICYEMIDA

Spiral cleavage; mesoderm from 4d cell

RADIALIAN LINE

BRACHIOPODA
PHORONIDA
PTEROBRANCHIA
ENTEROPNEUSTA
HEMICHORDATA
ECHINODERMATA
UROCHORDATA
CEPHALOCHORDATA
VERTEBRATA
CHAETOGNATHA

Mesoderm from archenteron or related cells; radial cleavage

Mesoderm from lip of blastopore

CTENOPHORA

Mesoderm; multiciliated cells

CNIDARIA

Basal lamina; epidermis and gastrodermis neurons and synapses

PLACOZOA
PORIFERA

Cell junctions; probably anisogamy

COLONIAL CHOANOFLAGELLATES

FIGURE 26.9 Arrangement of metazoan phyla according to possible evolutionary sequences. Broken arrows are directed at phyla whose position with respect to main lines of descent is perhaps more doubtful than that of others. Note that the pseudocoelomate phyla (most of which are probably not closely related to one another) are not tied to a main line, although they are probably closer to certain spiralian groups than to any radialians.

any scheme. There are, however, some major assemblages that do not fit well among others. The entire pseudocoelomate complex is one of these. Another is the Bryozoa, which may have affinities with Phoronida and Brachiopoda, but which could be related to Entoprocta. The latter, unlike phoronids and brachiopods, are probably tied to the spiralian line.

REFERENCES ON INVERTEBRATE PHYLOGENY

Boardman, R. S., Cheetham, A. H., and Rowell, A. J., 1987. Fossil Invertebrates. Blackwell Scientific Publications, Oxford and Palo Alto.

Clark, R. B., 1964. Dynamics in Metazoan Evolution: The Origin of the Coelom and Segments. Clarendon Press, Oxford.

Field, K. G., Olsen, G. J., Lane, D. J., Giovannoni, S. J., Ghiselin, M. T., Raff, E. C., Pace, N. R., and Raff, R. A., 1988. Molecular phylogeny of the animal kingdom. Science, *239*:748–753.

Glaessner, M. F., 1962. Precambrian Fossils. Biological Reviews, *37*:467–494.

Glaessner, M. F., 1984. The Dawn of Animal Life. Cambridge University Press, Cambridge and New York.

Grell, K. G., 1974. Vom Einzeller zum Vielzeller, Hundert Jahre Gastraea-Theorie. Biologie in Unserer Zeit, *4*:65–71.

Hadži, J., 1963. The Evolution of the Metazoa. Pergamon Press, Oxford; Macmillan, New York.

Hanson, E. D., 1977. The Origin and Early Evolution of Animals. Wesleyan University Press, Middletown, Connecticut.

House, M. R. (editor), 1979. The Origin of Major Invertebrate Groups. Academic Press, London and New York.

Ivanov, A. V., 1968. [The Origin of Multicellular Animals.] Izdatel'stvo "Nauka," Leningrad (in Russian).

Jägersten, G., 1955. On the early phylogeny of the Metazoa. The Bilaterogastraea Theory. Zoologiska Bidrag från Uppsala, *30*:321–354.

Jägersten, G., 1959. Further remarks on the early phylogeny of the Metazoa. Zoologiska Bidrag från Uppsala, *33*:79–108.

Jägersten, G., 1972. Evolution of the Metazoan Life Cycle. Academic Press, London and New York.

Morris, S. C., George, J. D., Gibson, R., and Platt, H. M. (editors), 1985. The Origins and Relationships of Lower Invertebrates. Systematics Association Special Volume, no. 28. Clarendon Press, Oxford.

Nielsen, C., 1985. Animal phylogeny in the light of the trochaea theory. Biological Journal of the Linnean Society, *25*:243–299.

Nielsen, C., 1987. Structure and function of metazoan ciliary bands and their phylogenetic significance. Acta Zoologica, *68*:205–262.

Nielsen, C., and Nørrevang, A., 1985. The trochaea theory—an example of life cycle phylogeny. *In* Morris, S. C., George, J. D., and Platt, H. M. (editors), The Origins and Relationships of Lower Invertebrates, pp. 28–41. Clarendon Press, Oxford.

von Salvini-Plawen, L., 1978. On the origin and evolution of the lower Metazoa. Zeitschrift für Zoologische Systematik und Evolutionsforschung, *16*:40–88.

Glossary

In general, terms whose use is restricted to a single group of invertebrates are not defined here. Consult the index for references to portions of the text where they are discussed.

A

ABORAL. Referring to the side or surface opposite that on which the mouth is located; used in connection with animals such as jellyfishes, sea stars, and sea urchins, which do not have an anterior-to-posterior axis.

ACOELOMATE. Without a body cavity other than the gut.

AMOEBOCYTES. Free cells that move by pseudopodia, in much the same way as an amoeba does; found in blood, tissue fluid, and coelomic fluid, and serving a variety of functions, including ingestion of foreign particles, transport of food and waste products, and secretion of spicules.

ANAEROBIC. Referring to physiological processes that take place in the absence of free oxygen, and to organisms that live where there is no free oxygen.

ANALOGOUS. Referring to structures that perform similar functions but that have different evolutionary origins.

ANASTOMOSIS. Union of blood vessels, pseudopodia, or other slender structures, often forming a network.

ANCESTRAL (PRIMITIVE) FEATURE. A characteristic believed to have been typical of a group of organisms since early in the evolution of that group.

ANTENNA. A sensory appendage on the head of arthropods other than chelicerates. Crustaceans have two pairs of antennae; insects and myriopods, a single pair, perhaps corresponding to the first antennae of crustaceans. The term is also applied to certain sensory outgrowths on the prostomium of some polychaete annelids.

APODEME. A portion of an arthropod exoskeleton that is directed inward, and that serves as a structure to which muscles are attached.

ARCHENTERON. The primitive gut of an embryo, formed by gastrulation.

ARTICLE. One of the units that make up a jointed appendage of an arthropod.

AUTOTROPHIC NUTRITION. Manufacture of organic foods from inorganic raw materials, such as carbon dioxide, water, and mineral salts; characteristic of green plants, some protozoans, and some bacteria.

B

BASAL LAMINA (BASEMENT MEMBRANE). A sheet of secreted material, consisting of collagen and carbohydrate, that lies beneath an epithelium.

BENTHIC. Living on the bottom. The term is applied to aquatic organisms that inhabit mud or sand or that live on or in rocks, wood, shell, or other firm substrata. It is essentially the opposite of PLANKTONIC, which refers to organisms that float, drift, or swim weakly.

BILATERAL SYMMETRY. A type of symmetry in which the body has an anterior and posterior end, right and left sides, and dorsal and ventral surfaces. There is only one plane along which the body can be divided into equal halves, which are mirror images of one another. Most phyla of invertebrates and all vertebrates have bilateral symmetry.

BIOLUMINESCENCE. The production of light by an organism.

BIRAMOUS. Divided into two branches. The term is most often used in connection with certain appendages of crustaceans.

BLASTOCOEL. The central cavity of a blastula.

BLASTOMERE. One of the first few cells produced by cleavage of an egg.

BLASTOPORE. In a gastrula, the opening into the archenteron. It may persist as the anus or mouth, or it may close, after which either the anus or mouth (or both) may appear in the same area.

BLASTULA. A stage in development in which the embryo consists of a ball of cells, generally with a central cavity, the blastocoel.

C

CAECUM. A diverticulum or blind pouch.

CARAPACE. In crustaceans, an exoskeletal shield that covers the head and some or all of the thorax.

CEPHALIZATION. The condition characteristic of bilaterally symmetrical animals, in which one end of the body is differentiated as a distinct head.

CEPHALOTHORAX. In some crustaceans, the portion of the body consisting of head and thorax, which are covered by a carapace.

CHAETA. A type of bristle, produced in an epidermal pocket and consisting to a large extent of chitin, found in most annelids (other than leeches) and in certain related phyla; *see also* SETA.

CHELATE. Referring to an arthropod appendage in which the terminal article, a movable claw, closes like a finger on a prolongation of the next-to-last article, so that a pincer is formed.

CHELICERA. A member of the first pair of appendages of spiders, scorpions, and related arthropods. The terminal article is usually clawlike, and may form part of a pincer.

CHELIPED. In some crustaceans, such as lobsters, crayfishes, and crabs, an especially large and stout chelate appendage of the thoracic region.

CHEMOSENSORY. Referring to cells concerned with detection of chemical stimuli.

CHITIN. A nitrogenous polysaccharide, commonly present in the cuticle of arthropods, but found also in invertebrates of various other phyla, or in tubes they secrete.

CHLOROCRUORIN. A greenish, iron-containing respiratory pigment characteristic of many polychaete annelids.

CHOANOCYTE. A type of cell, especially characteristic of sponges and choanoflagellates, in which the proximal portion of the flagellum is encircled by microvilli.

CIRRUS. A fleshy, fingerlike outgrowth; also a copulatory structure, characteristic of many flatworms and some other invertebrates, which functions when it is turned inside out.

CLOACA. A portion of the gut, close to the anus, which receives ducts of the excretory or reproductive systems, or both.

COCOON. A protective covering, such as that secreted around a group of eggs, or woven around an insect larva when it pupates.

COELOM. A fluid-filled body cavity lined by epithelium of mesodermal origin; generally originating as a space or series of spaces within a mass of mesoderm, or as a pocket of mesoderm budded from the archenteron.

COELOMOCYTE. A free cell in the coelom.

COELOMODUCT. A duct, primarily of mesodermal origin, connecting the coelom with the exterior; generally concerned with conveying gametes to the outside or for excretion, or both. In annelids and certain other invertebrates, coelomoducts may be joined to metanephridia or protonephridia, which are ectodermal derivatives.

COLLAGEN. A fibrous protein secreted by cells. It is characteristic of various kinds of connective tissue, the basal lamina of epithelia, and the mesoglea of cnidarians and ctenophores.

COMMENSALISM. An association of two species of animals, in which one appears to obtain some benefit and the other seems to be neither harmed nor benefited. The term covers a multitude of symbiotic associations that do not fit the usual definitions of PARASITISM or MUTUALISM.

COMPLETE GUT. A gut that has both a mouth and an anus.

COMPOUND EYE. A type of eye in which there are many similar light-focusing and receptor units called ommatidia; characteristic of most insects, many crustaceans, and some other groups of arthropods.

COXA. The proximal article of an arthropod appendage.

COXAL GLAND. A coelomoduct opening on the coxa of an arthropod appendage.

CUTICLE. A noncellular coating, often structurally complex, that is secreted by the epidermis or some other epithelium of ectodermal origin (such as that which lines the foregut and hindgut of an arthropod).

D

DEFINITIVE (PRIMARY) HOST. The host in which a parasite reaches maturity and reproduces sexually.

DELAMINATION. A method of gastrulation accomplished by division of the cells of a blastula in such a way that an inner layer of endoderm is separated from an outer layer of ectoderm.

DEPOSIT FEEDING. Feeding on small particles of detritus that accumulate on the substratum.

DERIVED (ADVANCED) FEATURE. A feature not likely to have been characteristic of a group when this group first became established.

DETERMINATE DEVELOPMENT. Development in which the fate of the first few blastomeres is fixed. If experimentally separated, each blastomere will develop into only one part of a normal embryo.

DETRITUS. Decomposing organic matter in the form of fine particles, and microorganisms that are associated with the particles.

DEUTEROSTOMATOUS. Referring to animals in which the mouth does not derive from the blastopore, but from a secondary opening.

DIOECIOUS. With separate sexes. The term is a nuisance (it means "having two houses") and is not used in this book, but is defined because it is frequently encountered in other texts.

DIPLOBLASTIC. Referring to cnidarians, in which all cellular structures derive from two embryonic germ layers, ectoderm and endoderm.

DIPLOID. Referring to cells or organisms that have the "full" complement of chromosomes, these being represented by homologous pairs. The diploid number is double the HAPLOID number characteristic of gametes and, in most protozoans, of all stages except the zygote.

DUOGLAND ADHESIVE SYSTEM. A complex of glands, some producing an adhesive secretion, others producing a substance that breaks the adhesive bond; especially well known in flatworms, but probably widespread among invertebrates.

E

ENDOPODITE. The medial branch of a biramous appendage of a crustacean.

ENTEROCOELOUS. Referring to animals in which the coelom is formed when pockets of mesoderm are budded from the archenteron.

EPIBOLY. A type of gastrulation in which the cells destined to form the archenteron are overgrown by the other cells of the early embryo.

EPIPODITE. An outgrowth, often serving as a gill, of the coxa of a crustacean appendage.

EPITHELIOMUSCULAR CELL. *See* MYOEPITHELIAL CELL.

EXOPODITE. The lateral branch of a biramous appendage of an arthropod.

EYESPOT. A mass of dark pigment associated with photoreceptors. The pigment shields the photoreceptors so that they are stimulated by light coming from certain directions, but not others.

F

FILTER-FEEDING. Collecting small particles of food, such as bacteria and diatoms, by means of setose appendages, ciliary activity, a combination of ciliary activity and mucus secretion, or some other mechanism.

FLAME BULB. *See* FLAME CELL.

FLAME CELL. A hollowed-out cell within which there is a tuft of active cilia, characteristic of protonephridia of flatworms, and concerned with osmoregulation, or with both osmoregulation and excretion. The cavity of a flame cell is continuous with a duct that leads toward the exterior. The term *flame bulb* is suggested for comparable structures that are portions of syncytia, or that are multicellular or multinucleate. The distinction is not precise.

FOREGUT. The anterior portion of a complete gut, usually derived from a stomodeum (an invagination of ectoderm).

FUNNEL ORGAN. A tubular structure that has a funnellike opening into the coelom and a pore on the outside of the body wall. The term includes coelomoducts and metanephridia, and compound structures formed by unions of these.

G

GASTRODERMIS. The epithelial lining of the gut, especially the portion concerned with digestion and absorption.

GASTRULA. A stage in embryonic development in which the archenteron, consisting of embryonic endoderm, has been formed by invagination, epiboly, or some other mechanism.

GIANT FIBER. A neuron with a large and long axon (giant axon) that transmits impulses more rapidly than a sequence of neurons interrupted by synapses; characteristic of the ventral nerve cord of many annelids and wormlike invertebrates of some other phyla.

H

HAPLOID. With a set of unpaired chromosomes, representing half of the ''full'' complement characteristic of the species; typical of gametes, some adult animals derived from unfertilized eggs, and most groups of protozoans.

HEMERYTHRIN. An iron-containing respiratory pigment in the blood or body cavity of some polychaete annelids, sipunculans, brachiopods, and priapulans.

HEMOCOEL. A complex of blood-filled spaces, some of which may be large, arising within the tissues; usually part of an open circulatory system in which the blood, after leaving the heart by way of major vessels, escapes into the spaces that form the hemocoel.

HEMOGLOBIN. An iron-containing respiratory pigment in the blood of vertebrates and many invertebrates. It is bright red when oxidized.

HERMAPHRODITIC. Producing both male and female gametes, but not necessarily at the same time.

HETEROTROPHIC NUTRITION. Nutrition that depends on organic compounds produced by other organisms; characteristic of animals, nonphotosynthetic protozoans, most bacteria, and fungi.

HINDGUT. The posterior portion of a complete gut, usually derived from a proctodeum (invagination of ectoderm at the anus).

HOLOBLASTIC CLEAVAGE. A type of cleavage in which the whole egg divides; characteristic of eggs in which there is relatively little yolk.

HOMOLOGOUS. Referring to structures that have the same evolutionary origin, even if they have different functions.

HYDROSTATIC SKELETON. A fluid-filled space (such as a coelom, pseudocoel, hemocoel, or gut cavity) or a mass of tissue that serves as a skeleton. If fluid does not escape, the volume of the space or mass of tissue will remain constant, even if its shape is changed by activity of the muscles that surround it.

I

INDETERMINATE DEVELOPMENT. Development in which the fate of the first few blastomeres resulting from cleavage of the egg is not rigidly determined. If one blastomere is separated from the others, it may develop into a complete embryo.

INDIRECT DEVELOPMENT. A type of development in which an animal goes through one or more larval stages that are decidedly different from the adult.

INSTAR. One of several larval or nymphal stages in the life history; used mostly in connection with insects and some other terrestrial arthropods.

INTERMEDIATE (SECONDARY) HOST. A host in which a parasite lives (and sometimes reproduces asexually) before it enters the DEFINITIVE HOST, in which it becomes sexually mature.

K

KINETOSOME. The basal body from which a flagellum or cilium arises.

L

LABIUM. The ''lower lip'' of insects and myriapods, formed by fusion of the bases of the second maxillae.

LABRUM. The lip in front of or above the mouth of an arthropod. It is not an appendage.

LARVA. An immature animal whose form, structure, and lifestyle are decidedly different from those of the adult.

LECITHOTROPHIC. Referring to swimming larvae or other developmental stages whose nutrition depends on abundant yolk that had been incorporated into the egg.

LITTORAL. Living at the shore or in relatively shallow water near shore.

LORICA. A secreted covering, such as is characteristic of some rotifers and protozoans.

LUMEN. The cavity of the gut or other tubular or saclike organ.

M

MACROMERE. One of the larger blastomeres formed during early cleavage of an egg.

MALPIGHIAN TUBULE. An excretory tubule opening into the posteriormost part of the midgut in insects, myriapods, and some other arthropods whose nitrogenous wastes are eliminated in the form of uric acid and related compounds.

MANDIBLE. In arthropods other than chelicerates, a head appendage that functions as a jaw.

MAXILLIPED. In crustaceans, a thoracic appendage specialized to serve as a supplementary mouthpart, and usually concerned with collecting and sorting food.

MESENTERY. A sheetlike structure, consisting mostly of peritoneum, that surrounds and supports an organ in the coelom.

METAMORPHOSIS. The process by which a larva is transformed into the adult stage.

METANEPHRIDIUM. A funnel organ of ectodermal origin, primarily concerned with osmoregulation and excretion. The term is often erroneously applied to coelomoducts, which are of mesodermal origin. Metanephridia are sometimes joined, however, to coelomoducts.

MICROMERE. One of the smaller blastomeres produced during early cleavage of an egg.

MICROVILLUS. A fingerlike projection of a cell (generally not visible except with the electron microscope). Microvilli greatly increase the surface area of a cell, and are usually involved in absorption or excretion, but in some groups of animals they are components of cells specialized for reception of light and other stimuli.

MIDGUT. The endodermal portion of a complete gut.

MONOECIOUS. Hermaphroditic. The term ("having one house") is not used in this book, but is defined because it is frequently encountered in other texts.

MONOPHYLETIC. Referring to an assemblage of organisms believed to derive from one ancestral stock.

MUCOPOLYSACCHARIDE. Any one of various secreted substances that consists of a polysaccharide bonded to a protein.

MUTUALISM. A symbiotic association in which both species are benefited by the relationship.

MYOEPITHELIAL CELL. An epithelial cell that has myofilaments and is contractile; also called *epitheliomuscular cell.*

MYOGLOBIN. A kind of hemoglobin found in muscle tissue.

N

NAUPLIUS. A type of larva, with three pairs of appendages, characteristic of many crustaceans.

NEPHRIDIUM. A term applied to various kinds of structures (protonephridia, metanephridia, coelomoducts, and mixtures thereof) whose function is osmoregulation or excretion, or both. It is not used in this book.

NOTOCHORD. In cephalochordates, a firm but elastic rod of muscle cells located between the gut and dorsal nerve cord; in ascidian larvae and some other urochordates, a rod of cells enclosing a fluid-filled channel, or separated by fluid-filled spaces, located in the tail beneath the nerve cord. Notochords of other types occur in lower vertebrates and in embryos of vertebrates in general.

NYMPH. In insects and some other arthropods, a juvenile that resembles the adult. Maturity is attained through a succession of nymphal stages, rather than through a drastic metamorphosis.

O

OCELLUS. An eye, generally simple, but in any case not a compound eye. The degree of complexity of ocelli varies, but the term is not applied to eyes as elaborate as those of cephalopod molluscs.

OMMATIDIUM. One of the many similar structural units of a compound eye.

OOSTEGITE. An outgrowth of the coxa of certain thoracic appendages, contributing to a brood pouch; characteristic of isopods, amphipods, and related crustaceans.

OPERCULUM. A plate, flap, or other device used for closing the aperture of a tube, shell, or other covering that protects an animal.

OPISTHOSOMA. In chelicerates, the posterior part of the body, behind the prosoma (which bears the chelicerae, pedipalps, and legs); in pogonophorans and vestimentiferans, the posteriormost part of the wormlike body, which consists of segments that have setae.

OSMOREGULATION. The physiological processes by which the water and salt balance of the body fluids is controlled.

OSSICLE. A small calcareous structure (a "little bone"), such as a spine of a sea star or sea urchin, or one of the plates constituting the test of a sea urchin.

OVOVIVIPARITY. Development of young within the oviduct of the parent, with nourishment coming from yolk rather than by way of a placenta or comparable structure.

P

PAPILLA. A nipplelike protuberance of the skin.

PARASITISM. A symbiotic association in which one species benefits at the expense of another.

PARENCHYMA. Tissue that appears to consist of relatively undifferentiated cells, found between various internal structures and the outer cell layers, especially of flatworms. Some components of such tissue, however, are likely to be the cell bodies of specialized cells.

PARTHENOGENESIS. Development of an egg that has not been fertilized. The phenomenon is encountered in various groups of animals, including rotifers and insects (for example, the honeybee).

PEDIPALP. One of the second pair of appendages of a spider, scorpion, or other chelicerate arthropod; usually specialized for handling food, but in male spiders also used for transferring sperm.

PELAGIC. Referring to organisms that live in open water of the sea and large lakes.

PEREON. In malacostracan crustaceans, the thorax (which consists of eight segments) minus any of the more anterior segments that are fused to the head (such as is the case in isopods and amphipods), or whose appendages are specialized as supplementary mouthparts, or both.

PEREOPOD. In malacostracan crustaceans, a thoracic appendage used primarily for locomotion, grasping, or crushing, rather than as a supplementary mouthpart, or MAXILLIPED.

PERITONEUM. The epithelium that lines a coelom and contributes to the mesenteries that support organs lying in the coelom.

PHAGOCYTOSIS. Ingestion, by cells, of bacteria and other small particles; characteristic of amoebae, other organisms with pseudopodia, and amoebocytes in blood, coelomic fluid, and tissue fluid.

PHOTORECEPTOR. A cell specialized for detection of light.

PHYLLOPODIUM. In crustaceans, an appendage that is markedly flattened, generally used for swimming, respiration, and creating feeding currents.

PHYLOGENY. The origin and evolution of major groups of organisms, and also of taxa within major groups.

PLANKTONIC. Referring to aquatic organisms that drift with the current, even if they are able to swim to some extent.

PLEON. In malacostracan crustaceans, the abdomen, consisting of six (or rarely seven) segments plus the telson.

PLEOPOD. An appendage of the pleon of a malacostracan crustacean; *see also* UROPOD.

POLYPHYLETIC. Referring to an assemblage of organisms whose members are believed to derive from more than one ancestral stock.

POLYPLOID. Referring to cells that have more than two copies of a particular chromosome.

PRIMORDIUM. An embryonic rudiment of a structure, before it has differentiated.

PROBOSCIS. A portion of the body that can be extended as a snout for capture of food. Although commonly applied to a pharynx that can be everted, in this book the term is used only for structures that are not part of the gut, such as the proboscis of nemerteans.

PROCTODEUM. An invagination of ectoderm that establishes the hindgut of many animals.

PROSOMA. The anterior part of the body of a chelicerate arthropod, which bears the chelicerae, pedipalps, and legs.

PROTANDRIC. Referring to an animal that functions first as a male, then as a female.

PROTONEPHRIDIUM. An osmoregulatory–excretory complex of ectodermal origin, consisting of one or more flame cells, flame bulbs, or solenocytes, as well as their ducts.

PROTOSTOMATOUS. Referring to invertebrate phyla in which the mouth is derived from the blastopore, or appears in the part of the embryo where the blastopore had been located before it closed.

PSEUDOCOEL (PSEUDOCOELOM). A type of body cavity that is not lined by an epithelium.

PSEUDOPODIUM. A labile extension of cytoplasm used in locomotion and capture of food. Pseudopodia are characteristic of amoebae and various other protozoans, and also of phagocytic cells of metazoan animals.

PYGIDIUM. The posteriormost, nonsegmental unit of the body of an annelid.

R

RADIAL CLEAVAGE. A pattern of early cleavage in which the mitotic spindles of each successive division are at right angles to those of the preceding division. Usually, the first two cleavages are vertical, the third horizontal, so that at the eight-cell stage each blastomere of the upper tier lies directly above one of the blastomeres of the lower tier.

RADIAL SYMMETRY. A type of symmetry in which the parts are distributed evenly around an oral–aboral axis. Cuts that follow the oral–aboral axis divide the animal into equal halves, and sometimes into equal quarters or smaller pieces. Perfect radial symmetry is found only in some cnidarians.

RECEPTOR. A cell or group of cells sensitive to touch, pressure, light, or some other kind of stimulus.

S

SAPROZOIC. Feeding on dissolved organic nutrients, rather than on particulate material.

SCHIZOCOEL. A coelom produced by appearance of cavities within a solid mass of mesoderm; characteristic of annelids, molluscs, arthropods, and several other phyla.

SCLERITE. One of the hard plates into which the exoskeleton of an arthropod segment is differentiated.

SCLEROPROTEIN. A hardened protein.

SCLEROTIZED. Chemically hardened; said of proteins such as those that form the perisarc of a hydroid and much of the exoskeleton of an arthropod.

SEGMENT. One of the serially arranged units of which the bodies of annelids, arthropods, and some arthropodlike invertebrates consist.

SEMINAL RECEPTACLE. A sac that is part of the female reproductive system and that stores sperm received during copulation or insemination by other means.

SEMINAL VESICLE. In the male reproductive system, a dilation or diverticulum within which sperm are stored.

SEPTUM. A partition that separates two cavities (such as successive coelomic compartments of an annelid) or that partly subdivides a cavity (such as the gut of a sea anemone).

SESSILE. Attached to the substratum and moving little or not at all, as in the case of sponges, sea anemones, oysters, and many tube-forming invertebrates.

SETA. A cuticular bristle, usually with a cavity or tissue inside, characteristic of arthropods; the term is also commonly used in place of CHAETA, which is a bristle secreted in an epidermal pocket.

SOLENOCYTE. In the proximal portions of certain protonephridia, such as those of gastrotrichs, kinorhynchs, some annelids, and some trochophore larvae, cells that have only one or two cilia, instead of several to many, as there are in FLAME BULBS and FLAME CELLS.

SOMITE. The embryonic primordium of a segment.

SPERMATOPHORE. A packet of sperm enclosed within a covering.

SPIRACLE. An opening into the tracheal tube of an insect or other arthropod.

SPIRAL CLEAVAGE. A pattern of early cleavage in which, after the four-cell stage has been reached, the mitotic spindles are oblique with respect to those of the first two divisions. Thus at the eight-cell stage, the cells of the upper tier are not directly above the cells of the lower tier, but on the furrows between them.

STATOCYST. A type of sense organ concerned with detecting gravity or changes in orientation; usually consisting of a cavity lined by receptors that are sensitive to shifts in the position of a calcareous body or sand grain, which is free to move within the cavity.

STENOPODIUM. In crustaceans, an appendage that is not markedly flattened, and that is generally used for walking, clinging, crushing prey, transferring sperm, or brooding eggs, but sometimes for swimming or other functions.

STEREOGASTRULA. A solid gastrula, in which a mass of endodermal cells is enclosed by a jacket of ectoderm.

STERNUM (STERNITE). The ventral sclerite of the exoskeleton of an arthropod segment; also the ventral cuticular plate of a trunk segment of a kinorhynch.

STOLON. A "runner" that is in tight contact with the substratum, and that gives rise, at intervals, to new members of the colony.

STOMODEUM. An invagination of ectoderm that establishes the pharynx of anthozoan cnidarians and flatworms and the foregut of many invertebrates that have a complete digestive tract.

STYLET. A secreted, hard, piercing structure, sometimes calcareous, sometimes consisting of sclerotized protein; also the sclerotized tube characteristic of the penis of many flatworms.

SUBCHELATE. Referring to an arthropod appendage in which the terminal article closes against one edge of the next-to-last article, rather than against an immovable claw formed by a prolongation of the latter, as in a CHELATE appendage.

SUPERFICIAL CLEAVAGE. A type of cleavage characteristic of certain very yolky eggs, such as those of insects. The nuclei and cell membranes separating cells are restricted, at least for a time, to a layer close to the surface.

SYMBIOSIS. An intimate association between two species. It may be of mutual benefit to both partners

(MUTUALISM), advantageous to one at the expense of the other (PARASITISM), or advantageous to one and of no apparent benefit or harm to the other (COMMENSALISM).

SYNCYTIUM. A mass of cytoplasm that contains several to many nuclei, but that is not divided into cells.

T

TAGMA. In arthropods, a major division of the body, such as the head, thorax, or abdomen.

TELSON. The terminal piece (not a true segment) at the posterior end of an arthropod. The anus is typically located on the telson.

TERGUM (TERGITE). The dorsal sclerite of the exoskeleton of an arthropod segment; also the dorsal cuticular plate of a trunk segment of a kinorhynch.

THORACOPOD. In crustaceans, an appendage of any thoracic segment.

TRIPLOBLASTIC. Referring to animals whose tissues are derived from three embryonic germ layers: ectoderm, endoderm, and mesoderm.

TROCHOPHORE. A type of swimming larva characteristic of some marine molluscs, polychaete annelids, and certain other invertebrates. A characteristic feature of a trochophore is a circular band of cilia just anterior to the level of the mouth.

U

UNIRAMOUS. Referring to arthropod appendages, and sometimes to other outgrowths of the body wall, which are characterized by a single main stem.

UROPOD. In malacostracan crustaceans, a paired appendage of the last segment of the abdomen (pleon). It is usually decidedly different from the other abdominal appendages (pleopods). In amphipods, the appendages of the last three abdominal segments are called uropods.

V

VELIGER. The advanced larva of many molluscs, in which the mantle, foot, and shell are already differentiated.

VITELLARIA (YOLK GLANDS). Glands that produce yolk, or cells rich in yolk, as a supply of nourishment for a fertilized egg, embryo, and nonfeeding larva or juvenile.

Z

ZOOCHLORELLAE. Symbiotic unicellular green algae that live within cells of certain invertebrates, such as some freshwater sponges and some freshwater and marine flatworms.

ZOOID. A member of a colony. The term is applied to bryozoans, compound ascidians, many cnidarians, and some other groups of invertebrates.

ZOOXANTHELLAE. Symbiotic photosynthetic dinoflagellates that live within cells of various marine invertebrates, especially corals.

ZYGOTE. An egg that has been fertilized by a sperm.

ILLUSTRATION CREDITS

2.1 Sleigh, in Bittar, *Cell Biology in Medicine,* Copyright © 1973. Reprinted by permission of John Wiley and Sons, Inc. **2.4,C** Masson S. A., Paris. **2.21,C** Reproduced by kind permission of the International Council for the Exploration of the Sea. **2.35,A** Masson S. A., Paris. **2.36,B** Masson S. A., Paris. **2.42,D** Lindsay S. Olive, "Sorocarp Development in a Newly Discovered Ciliate," *Science,* Vol. 202, Nov. 3, 1978, pp. 530–532. Copyright © 1978 by the AAAS. **2.43** Masson S. A., Paris. **2.45** Society for Experimental Biology and Company of Biologists, Ltd. **4.5** Courtesy of Senckenberg-Museum. **4.14, B–C** *Ophelia,* 1967. **4.21,B** Courtesy of Senckenberg-Museum. **4.29,B** Courtesy of Senckenberg-Museum. **4.32,B–D** Reprinted by permission from *Nature,* Vol. 232, Copyright Macmillan Magazines, Ltd. **4.38** D. M. Ross, *Science,* Vol. 155, March 17, 1967, pp. 1419–1421. Copyright © 1967 by the AAAS. **4.45,C** Courtesy of Senckenberg-Museum. **5.5** C. Chapman, "The Hydrostatic Skeleton in Invertebrates," *Biological Reviews,* Vol. 33, Cambridge University Press. **5.7** Society for Experimental Biology and Company of Biologists, Ltd. **6.11,D** Ruebush, T. K., "Mesostoma ehrenbergi wardii for the Study of the Turbellarian Type," *Science,* Vol. 91, pp. 531–532, May 31, 1940. Copyright © 1940 by the AAAS. **6.24,A** Brown, *Selected Invertebrate Types*. Reprinted by permission of John Wiley and Sons, Inc. **6.28,A** Peter Ax, 1966: Die Bedeutung der interstitiellen Sandfauna für allegemeine Probleme der Systematik, Ökologie und Biologie. *Veröffentlichen des Instituts für Meeresforschung in Bremerhaven,* Suppl. 2. **7.4** Masson S. A., Paris. **9.5,A** Redrawn from the *Journal of Cell Biology,* 1965, Vol. 26, by copyright permission of the Rockefeller University Press. **10.6,B** R. A. Wilson and L. A. Webster, "Protonephridia," *Biological Reviews,* Vol. 49, Cambridge University Press. **10.28,C** Reproduced with permission from Van der Land, *Zoologische Verhandelingen,* Leiden, no. 112. **11.15** *Ophelia,* 1975. **11.26,A–B** *Ophelia,* 1973. **11.29, A–C** Reproduced with permission of the Royal Society of Edinburgh. **11.30,A–E** Reproduced with permission of the Royal Society of Edinburgh. **12.14** E. C. Southward, "Bacterial Symbionts in Pogonophora," *Journal of the Marine Biological Association of the United Kingdom,* Vol. 62, Cambridge University Press. **13.30** *Ophelia,* 1979. **13.39,D** Reproduced by kind permission of the Linnean Society of London and Estuarine & Brackish-Water Sciences Association. **14.22** *Ophelia,* 1966. **14.23** *Ophelia,* 1971. **14.25** *Ophelia,* 1981. **14.31,A** *Ophelia,* 1964. **14.31,B** *Ophelia,* 1966. **14.33,B** *Ophelia,* 1964. **14.34** *Ophelia,* 1971. **16.36,B** O. W. Tiegs and S. M. Manton, "The Evolution of the Arthropoda," *Biological Reviews,* Vol. 33, Cambridge University Press. **17.24,B** Yager, *Journal of Crustacean Biology,* Vol. 1, 1981. **17.49,C** Reproduced with permission of the Royal Society of Edinburgh. **17.70,A** *Ophelia,* 1972. **18.13,A–C** M. Quick in *The Insects of Australia* (CSIRO), Melbourne University Press. **18.23,A–C** B. Rankin in *The Insects of Australia* (CSIRO), Melbourne University Press. **21.32** *Ophelia,* 1966. **22.10,B–C** A. R. D. Stebbing, "The Status and Ecology of Rhabdopleura compacta (Hemichordata) from Plymouth," *Journal of the Marine Biological Association of the United Kingdom,* Vol. 50, Cambridge University Press. **24.1** Masson S. A., Paris. **24.6,B** Masson S. A., Paris. **24.7** Masson S. A., Paris. **24.16,A** Masson S. A., Paris. **24.19** Neumann, G., "Thaliacea, Pyrosomida, Cyclomyaria," in Kükenthal and Krumbach, *Handbuch der Zoologie,* Vol. 5, 1934–35. Walter de Gruyter & Co., Berlin. **24.20,A,B,D** Neumann, G., "Thaliacea, Pyrosomida, Cyclomyaria," in Kükenthal and Krumbach, *Handbuch der Zoologie,* Vol. 5, 1934–35. Walter de Gruyter & Co., Berlin. **26.7** Reproduced by kind permission of the Linnean Society of London and Estuarine & Brackish-Water Sciences Association. **26.8** Reproduced by kind permission of the Linnean Society of London and Estuarine & Brackish-Water Sciences Association.

INDEX

Italics refer to pages on which illustrations are located.

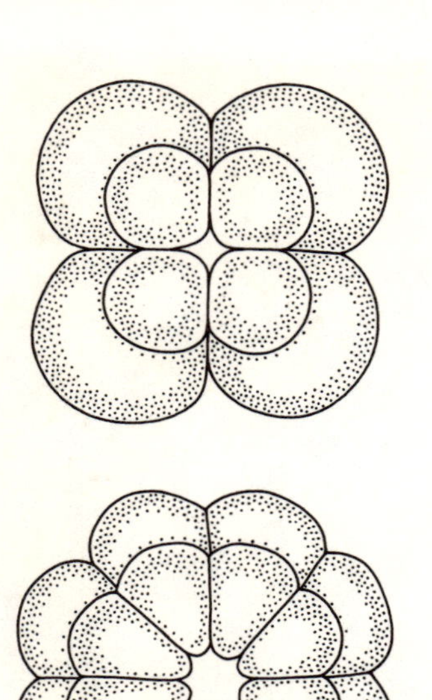

RADIAL CLEAVAGE

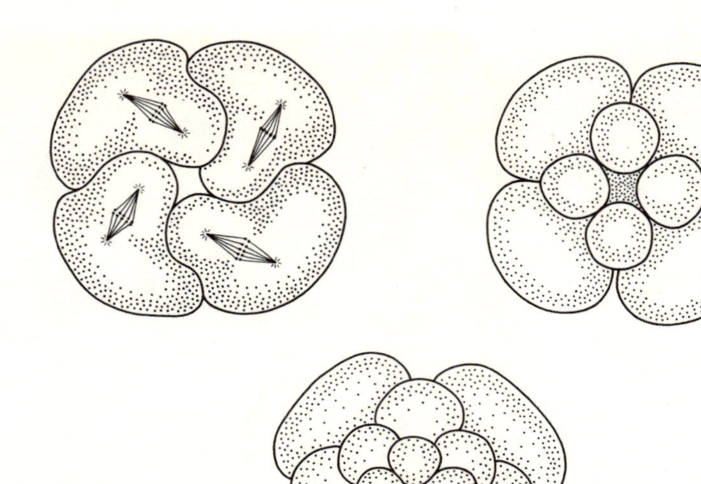

SPIRAL CLEAVAGE

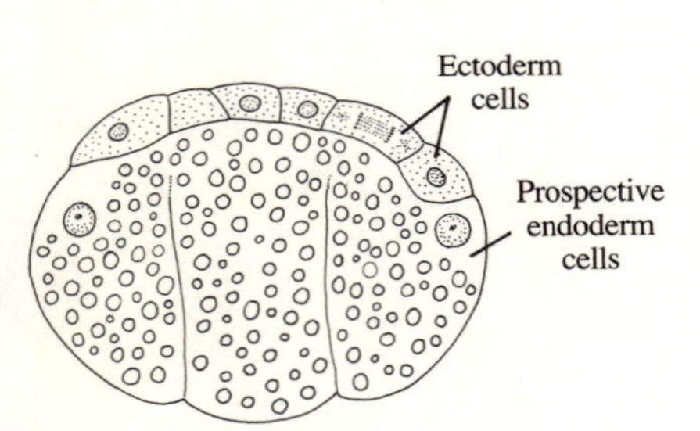

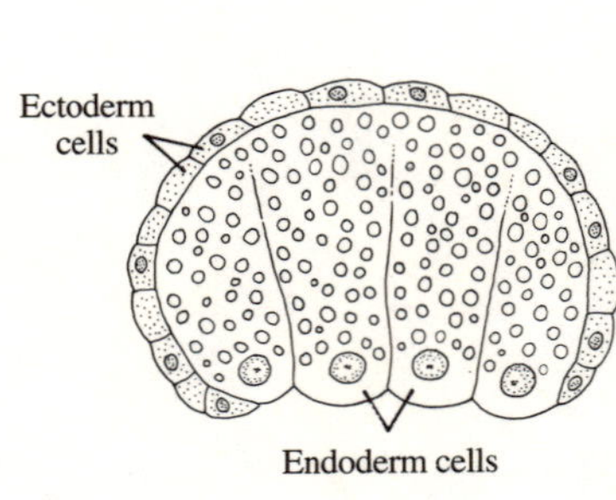

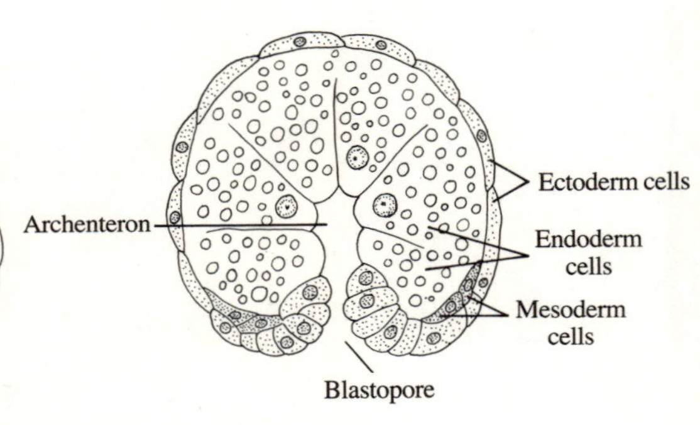

GASTRULATION BY EPIBOLY

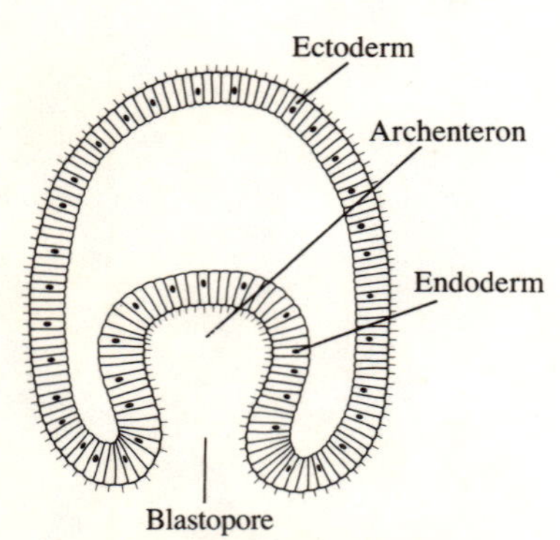

GASTRULATION BY INVAGINATION

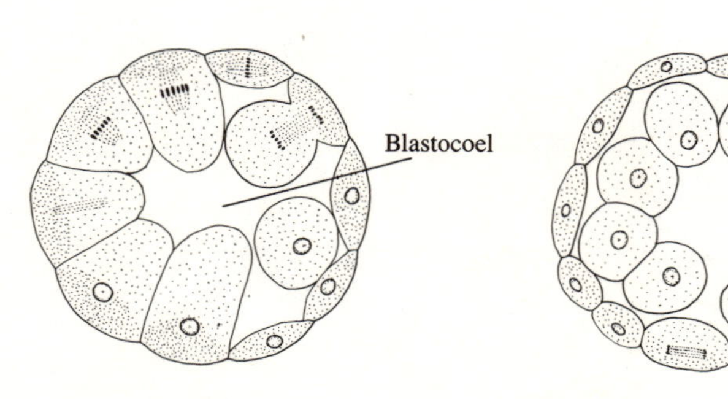

GASTRULATION BY DELAMINATION